BIOLOGICAL
SCIENCE

EXPLORING THE SCIENCE OF LIFE

BIOLOGICAL SCIENCE

PROFESSOR JON SCOTT, HIGHER EDUCATION CONSULTANT, FORMERLY UNIVERSITY OF LEICESTER

DR GUS CAMERON, UNIVERSITY OF BRISTOL

PROFESSOR ANNE GOODENOUGH, UNIVERSITY OF GLOUCESTERSHIRE

DR DAWN HAWKINS, ANGLIA RUSKIN UNIVERSITY

DR JENNY KOENIG, UNIVERSITY OF NOTTINGHAM

PROFESSOR MARTIN LUCK, FORMERLY UNIVERSITY OF NOTTINGHAM

DR DESPO PAPACHRISTODOULOU, KINGS COLLEGE LONDON

PROFESSOR ALISON SNAPE, KINGS COLLEGE LONDON

PROFESSOR KAY YEOMAN, UNIVERSITY OF EAST ANGLIA

DR MARK GOODWIN

OXFORD
UNIVERSITY PRESS

OXFORD
UNIVERSITY PRESS

Great Clarendon Street, Oxford, OX2 6DP,
United Kingdom

Oxford University Press is a department of the University of Oxford.
It furthers the University's objective of excellence in research, scholarship,
and education by publishing worldwide. Oxford is a registered trade mark of
Oxford University Press in the UK and in certain other countries

© Oxford University Press 2022

The moral rights of the authors have been asserted

Impression: 1

Published in the United States of America by Oxford University Press
198 Madison Avenue, New York, NY 10016, United States of America

British Library Cataloguing in Publication Data
Data available

Library of Congress Control Number: 2022933190

ISBN 978–0–19–878368–8

Printed in the UK by
Bell & Bain Ltd., Glasgow

This textbook is dedicated to the memory of our good friend and colleague, Dr Mark Goodwin. Mark was one of the initial members of the writing team and contributed greatly to the development of the style and approach of the book, as well as writing some of the genetics content. I had the privilege of working alongside Mark over a number of years at the University of Leicester, where he made a major contribution to the educational ethos of the School of Biological Sciences and, in particular, to the work of the GENIE Centre for Excellence in Teaching and Learning, where he led the development of the Virtual Genetics Education Centre. Mark was a true renaissance man with a remarkable range of interests and skills. Above all though, he was a committed educationalist—teaching, supporting students, and developing educational practice at Leicester, with the Open University, and also through extensive work internationally with universities in Afghanistan, Bangladesh, and Ethiopia.

Jon Scott

Brief Table of Contents

LIFE AND ITS **EXPLORATION**
Foundational Principles   2

TOPIC 1	Exploring the Science of Life	5
TOPIC 2	The Emergence of Life on Earth	15
TOPIC 3	How Do We Define Life?	30
TOPIC 4	Evolutionary Processes	34
TOPIC 5	The Diversity, Organization, and Classification of Life	42

QUANTITATIVE TOOLKITS   60

TOOLKIT 1	Understanding Data	63
TOOLKIT 2	Size and Scale	70
TOOLKIT 3	Describing Data	79
TOOLKIT 4	Ratio and Proportion	86
TOOLKIT 5	Understanding Samples	95
TOOLKIT 6	Designing Experiments	100
TOOLKIT 7	Assessing Patterns	107
TOOLKIT 8	Using Formulas and Equations	120
TOOLKIT 9	Rates of Change	135

MODULE ONE
Life at the Molecular Level   146

1	Building Blocks Molecules and Macromolecules	149
2	Energy Powering Biochemical Processes	184
3	Information Genes and Genomes	197
4	Mendelian Genetics	224
5	Reading the Genome Gene Expression and Protein Synthesis	252
6	Proteins and Proteomes	284
7	Metabolism Energy Capture and Release from Food	324
8	Molecular Tools and Techniques	343

MODULE TWO
Life at the Cellular Level   364

9	Characteristics of Prokaryotic and Eukaryotic Cells	367
10	Cell Division in Prokaryotes and Eukaryotes	405
11	Microbial Diversity	426
12	The Growth, Measurement, and Visualization of Cells	469
13	Microbes in Life Harnessing Their Power	498
14	Microbes as Agents of Infectious Disease	523
15	Viruses	555

MODULE THREE
The Human Organism Tissues, Organs, and Systems   578

16	An Overview of Physiology	581
17	Communication and Control 1 The Nervous and Endocrine Systems	594
18	Communication and Control 2 Sensory Systems	618
19	Communication and Control 3 Controlling Organ Systems	653
20	Muscle and Movement	686
21	Cardiovascular System	718
22	Respiratory System	744
23	Exercise Physiology	768
24	Renal System	785
25	Digestive System	803
26	Reproductive System	831
27	Immune System	862

MODULE FOUR
Organismal Diversity Structure, Adaptation, and Survival   900

28	The Structure of Living Organisms	903
29	Body Plans	950
30	Interaction with the External Environment	1000
31	Movement, Locomotion, and Migration	1053
32	Defence against Predation and Invasion	1102
33	Reproduction and Development	1141

MODULE FIVE
Organisms in their Environments   1196

34	Fundamental Concepts Ecology, Evolution, Species, and Speciation	1199
35	Genes Evolutionary Change in Alleles, Genotypes, and Phenotypes	1221
36	Populations Quantifying Demographics and Modelling Change	1244
37	Communities Species Interactions and Biodiversity Metrics	1270
38	Ecosystems Abiotic Interactions and Environmental Processes	1291
39	Challenges Key Threats to Ecosystems	1316
40	Solutions Managing, Conserving, and Restoring Ecosystems	1338

Full Table of Contents

List of Quantitative Tools   xiv
List of Experimental Toolkits   xiv
List of Scientific Processes   xv
List of Real World Views   xvi
List of Clinical Boxes   xvii
Meet the Authors   xviii
Acknowledgements   xxi
Editorial Advisers   xxii
A guide to learning with *Biological Science*   xxiv
For lecturers—teaching with this resource   xxvi

LIFE AND ITS EXPLORATION
Foundational Principles   2

TOPIC 1   Exploring the Science of Life   5
Introduction   5
1.1   What does it mean to take a scientific approach?   6
1.2   Ways of thinking   8
1.3   Logical reasoning   8
1.4   How do we test research hypotheses?   10
1.5   What is a null statistical hypothesis?   10
1.6   How do theories and models differ?   11
1.7   What are mathematical models?   11
1.8   Team Science   13

TOPIC 2   The Emergence of Life on Earth   15
Introduction   15
2.1   The emergence of life   16
2.2   How did Earth form?   18
2.3   Life is based on aqueous organic chemistry   22
2.4   The origins of organic molecules   22
2.5   Complex biomolecules   25
2.6   From biomolecules to life forms   28

TOPIC 3   How Do We Define Life?   30
Introduction   30
3.1   Metabolism   31
3.2   Membranes   31
3.3   Defence   32
3.4   Replication and continuity   32
3.5   Variation and selection   32
3.6   Conclusion   33

TOPIC 4   Evolutionary Processes   34
Introduction   34
4.1   What is evolution?   34
4.2   Seven reasons why the study of evolution is important   36
4.3   Conclusion   41

TOPIC 5   The Diversity, Organization, and Classification of Life   42
Introduction   42
5.1   Taxonomy   43
5.2   What is a species?   44
5.3   Higher taxa: larger groupings, families, and kingdoms   48
5.4   The domains of life: the highest taxa   49
5.5   Systematic interpretation of organismal relationships   52
5.6   Five pitfalls of taxonomic interpretation   57

QUANTITATIVE TOOLKITS   60

TOOLKIT 1   Understanding Data   63
1.1   The features of data   64
1.2   Types of variables   66

TOOLKIT 2   Size and Scale   70
2.1   Scientific notation and standard form   71
2.2   Using prefixes   72
2.3   Logarithmic scales   72
2.4   What is a logarithm?   75
2.5   What are natural logarithms?   76

TOOLKIT 3   Describing Data   79
3.1   Central tendency   80
3.2   Measures of variability   80
3.3   Frequency distributions   82

TOOLKIT 4   Ratio and Proportion   86
4.1   Expressing ratios   86
4.2   Comparing ratios   87
4.3   Relating ratios, fractions, and percentages   89
4.4   Probability   89
4.5   Proportion   90
4.6   Concentrations and dilutions of solutions   90

TOOLKIT 5   Understanding Samples   95
5.1   Populations and samples   96
5.2   Sample size   96
5.3   Sample quality   96
5.4   Related and unrelated samples   98

TOOLKIT 6   Designing Experiments   100
6.1   The importance of good experimental design   101
6.2   Problems   101
6.3   Solutions   102
6.4   Power analyses   104
6.5   Recognizing lack of independence   104
6.6   Adding variables   105
6.7   Implications for interpretation   105

TOOLKIT 7   Assessing Patterns   107
    7.1   Using graphs   108
    7.2   Implications of sample error   113
    7.3   Using null hypothesis significance tests   114
    7.4   Other options   117

TOOLKIT 8   Using Formulas and Equations   120
    8.1   Anatomy of an equation   121
    8.2   The main types of formulas in biology   122
    8.3   Carrying out calculations and keeping track of units   127
    8.4   Rearranging formulas   127

TOOLKIT 9   Rates of Change   135
    9.1   Rate is the slope (gradient) of a line or curve   136
    9.2   Quantifying the rate of change   137
    9.3   Rates involving exponential growth   139
    9.4   Rates involving exponential decay   142
    9.5   Logistic growth   143

MODULE ONE
Life at the Molecular Level   146

1   Building Blocks Molecules and Macromolecules   149
    Introduction   149
    1.1   The elements that make up biological molecules   150
    1.2   Formation and properties of biological molecules   151
    1.3   Non-covalent bonds allow interactions between
           molecules   161
    1.4   Chemical reactions involve breaking and
           making bonds   164
    1.5   Water plays a key role in supporting life   167
    1.6   Acids and bases: molecules that release or bind
           hydrogen ions   169
    1.7   Biological macromolecules: structure and function   172
    1.8   Interactions between macromolecules can be specific,
           flexible, and transient, allowing biological processes
           to occur   180

2   Energy Powering Biochemical Processes   184
    Introduction   184
    2.1   Living cells transform energy from one state to another 185
    2.2   Will a reaction happen or not?   186
    2.3   Harvesting and releasing free energy from food   188
    2.4   ATP is the universal energy intermediate in all life   189
    2.5   Electron transport during metabolism   192

3   Information Genes and Genomes   197
    Introduction   197
    3.1   The chemical structures of DNA and RNA   198
    3.2   Genome organization and packaging   206
    3.3   DNA replication   210

4   Mendelian Genetics   224
    Introduction   224
    4.1   Chromosomes and genes are carried in pairs   225
    4.2   Mendel's experiments established the laws of
           inheritance   227

    4.3   The behaviour of chromosomes as cells divide can
           explain the basis of inheritance   231
    4.4   Extensions and refinements of Mendel's laws   239
    4.5   Genetic disease   241

5   Reading the Genome Gene Expression and Protein Synthesis   252
    Introduction   252
    5.1   Gene expression: overview of the process   253
    5.2   The genetic code   254
    5.3   RNA synthesis: transcription   255
    5.4   Protein synthesis: translation   264
    5.5   Control of gene expression   271

6   Proteins and Proteomes   284
    Introduction   284
    6.1   Amino acids are the building blocks of proteins   285
    6.2   There are four levels of protein structure   291
    6.3   Protein structures reflect their function   296
    6.4   Enzymes: proteins as biological catalysts   306
    6.5   The proteome is the full set of proteins expressed in
           a cell, tissue, or organism   315

7   Metabolism Energy Capture and Release from Food   324
    Introduction   324
    7.1   The main stages of glucose oxidation   325
    7.2   Energy release from oxidation of fats and
           amino acids   328
    7.3   Glucose oxidation: a deeper look   329
    7.4   Anaerobic respiration and fermentation: production
           of ATP without the use of oxygen   337
    7.5   Photosynthesis   338

8   Molecular Tools and Techniques   343
    Introduction   343
    8.1   Basic methodologies using DNA   344
    8.2   Applications of recombinant DNA technology   353
    8.3   Genomics and bioinformatics complement
           recombinant DNA technology   359

MODULE TWO
Life at the Cellular Level   364

9   Characteristics of Prokaryotic and Eukaryotic Cells   367
    Introduction   367
    9.1   The diversity of cells   368
    9.2   The characteristics of the Bacteria   373
    9.3   The characteristics of the Archaea   383
    9.4   The characteristics of eukaryotic cells   384
    9.5   The key eukaryotic cellular structures   386
    9.6   From cells to tissues   401

10   Cell Division in Prokaryotes and Eukaryotes   405
    Introduction   405
    10.1   Bacterial reproduction   406
    10.2   Eukaryotic cell division   412
    10.3   Mitosis: a deeper look   415
    10.4   Meiosis   418

11	Microbial Diversity	426
	Introduction	426
	11.1 Classification and taxonomy	427
	11.2 Microbial habitats	429
	11.3 Bacterial diversity	432
	11.4 Archaeal diversity	439
	11.5 Fungal biodiversity and reproduction	445
	11.6 Protist diversity	458
12	The Growth, Measurement, and Visualization of Cells	469
	Introduction	469
	12.1 Maintaining a sterile environment	470
	12.2 Obtaining and using microbial culture collections	473
	12.3 Culturing cells from multicellular organisms	474
	12.4 Measuring the growth of microbial cells	479
	12.5 Visualizing cells	488
13	Microbes in Life Harnessing Their Power	498
	Introduction	498
	13.1 Microbial biotechnology	499
	13.2 The pharmaceutical industry	499
	13.3 The production of chemicals and fuels	504
	13.4 Biotechnology and food and drink production	507
	13.5 Microbial polysaccharides	513
	13.6 Biological control agents and bioremediation	514
	13.7 Biogeochemical cycles	517
14	Microbes as Agents of Infectious Disease	523
	Introduction	523
	14.1 Koch's postulates: how can we tell if something is a true pathogen?	524
	14.2 Bacterial pathogens	526
	14.3 Antibacterial drugs	531
	14.4 Fungal pathogens	534
	14.5 Protozoan diseases	539
	14.6 Plant pathology	547
15	Viruses	555
	Introduction	555
	15.1 How do we name and classify viruses?	556
	15.2 The structure of viruses	556
	15.3 How do viruses replicate?	559
	15.4 The diversity of viral replication strategies	564
	15.5 Bacteriophages	567
	15.6 Antiviral therapy	569
	15.7 Viruses beyond disease	572

MODULE **THREE**
The Human Organism *Tissues, Organs, and Systems* 578

16	An Overview of Physiology	581
	Introduction	581
	16.1 Homeostasis	582
	16.2 The cell membrane	584
	16.3 Conclusion	592
17	Communication and Control 1 The Nervous and Endocrine Systems	594
	Introduction	594
	17.1 The nervous system	595
	17.2 Signal conduction within the nervous system	595
	17.3 Resting potential	596
	17.4 Action potentials	600
	17.5 Action potential conduction	604
	17.6 Transmission between neurons	605
	17.7 Cellular signalling	614
18	Communication and Control 2 Sensory Systems	618
	Introduction	618
	18.1 General sensation	619
	18.2 Touch	620
	18.3 Pain and temperature	626
	18.4 Proprioception	629
	18.5 Special senses	632
19	Communication and Control 3 Controlling Organ Systems	653
	Introduction	653
	19.1 The autonomic nervous system	654
	19.2 Examples of autonomic nervous system function	662
	19.3 The endocrine system	665
	19.4 Conclusion	683
20	Muscle and Movement	686
	Introduction	686
	20.1 Skeletal muscle fibres and contraction	687
	20.2 Organization of skeletal muscle	694
	20.3 Reflexes and posture	701
	20.4 Overview of the motor system	705
	20.5 Conclusion	715
21	Cardiovascular System	718
	Introduction	718
	21.1 Composition and function of blood	719
	21.2 Vascular system	723
	21.3 Heart	732
	21.4 Integrated cardiovascular physiology—regulation of blood pressure	740
22	Respiratory System	744
	Introduction	745
	22.1 Anatomy of the respiratory system	745
	22.2 Ventilation of the lungs	749
	22.3 Gas exchange	757
	22.4 Control of ventilation	764
23	Exercise Physiology	768
	Introduction	768
	23.1 Cardiovascular responses to exercise	769
	23.2 Maximum oxygen uptake	775
	23.3 Pulmonary responses to exercise	777
	23.4 Exercise and acid–base balance	779
	23.5 Exercise and environmental stressors	780
	23.6 Physical activity, fitness, and health	782

24 Renal System   785
    Introduction   785
    24.1 Overview of kidney structure and function   786
    24.2 Filtration in the renal corpuscle   787
    24.3 The proximal convoluted tubule   789
    24.4 The loop of Henlé   793
    24.5 Distal convoluted tubule and collecting ducts   795
    24.6 The kidney and blood pressure   797
    24.7 The kidney and acid–base balance   798
    24.8 Micturition   799
    24.9 Endocrine functions of the kidney   800

25 Digestive System   803
    Introduction   803
    25.1 Histology and regulation of the digestive tract   804
    25.2 Anatomy of the digestive tract   805
    25.3 Gut microbiome   822
    25.4 The pancreas   823
    25.5 The liver   826

26 Reproductive System   831
    Introduction   831
    26.1 Gamete production   832
    26.2 Endocrine control   838
    26.3 Hormonal changes during the menstrual cycle   844
    26.4 Puberty   844
    26.5 The menopause   846
    26.6 Male reproductive anatomy   846
    26.7 Female reproductive anatomy   848
    26.8 Coitus and fertilization   850
    26.9 Establishment and maintenance of pregnancy   850
    26.10 Birth   857

27 Immune System   862
    Introduction   862
    27.1 Cells and organs of the immune system   863
    27.2 The innate immune response   869
    27.3 The adaptive immune response   876
    27.4 Antigen recognition by lymphocyte receptors   885
    27.5 Immunodeficiencies: diseases affecting how
          lymphocytes develop or function   895

MODULE FOUR

Organismal Diversity *Structure, Adaptation, and Survival* 900

28 The Structure of Living Organisms   903
    Introduction   904
    28.1 Cell membranes   904
    28.2 Membrane transport: talking to the outside   907
    28.3 The cytoskeleton   914
    28.4 Volumetrics, pressures, and turgor in plants   917
    28.5 Directionality and anchorage dependence   919
    28.6 Colony and organism formation   922
    28.7 Multicellular plants   927

    28.8 Multicellular animals: Metazoa   930
    28.9 Cell lineages and differentiation   933
    28.10 Communication between cells   940

29 Body Plans   950
    Introduction   950
    29.1 Form and function   951
    29.2 Plant body plans   954
    29.3 Animal body plans   964
    29.4 Hard materials   991

30 Interaction with the External Environment   1000
    Introduction   1001
    30.1 Environmental pressures and evolutionary survival   1001
    30.2 Light, geography, and the electromagnetic
          spectrum   1002
    30.3 The sound environment   1018
    30.4 The gravitational environment   1025
    30.5 The magnetic and electrical environment   1029
    30.6 The thermal environment   1030
    30.7 The gaseous environment   1038
    30.8 The aqueous environment   1044

31 Movement, Locomotion, and Migration   1053
    Introduction   1054
    31.1 Distribution of organisms on Earth   1054
    31.2 Native or alien?   1060
    31.3 Migration   1068
    31.4 Plant dispersal   1072
    31.5 Mechanisms of active locomotion   1077
    31.6 Movement without limbs: worms   1077
    31.7 Movement without limbs: snakes   1080
    31.8 Walking and running   1080
    31.9 Flying and gliding   1088
    31.10 Swimming   1093

32 Defence against Predation and Invasion   1102
    Introduction   1103
    32.1 Protection from invasion: plants   1103
    32.2 Protection from invasion: animals   1109
    32.3 The response to injury   1121
    32.4 External defences   1128
    32.5 Toxins, poisons, and venoms   1134
    32.6 Chemoreception and predator avoidance   1136

33 Reproduction and Development   1141
    Introduction   1142
    33.1 Reproduction and sex   1142
    33.2 Measuring reproduction: fecundity and fertility   1156
    33.3 Reproduction in plants   1158
    33.4 Reproduction in animals   1167
    33.5 Development in amniotes   1174
    33.6 Sperm   1176
    33.7 Annelid reproduction   1179
    33.8 Insect reproduction   1181
    33.9 Fish reproduction   1184

33.10 Amphibian reproduction   1190

33.11 Mammalian reproduction strategies   1191

MODULE FIVE
Organisms in Their Environments   1196

34   **Fundamental Concepts**   Ecology, Evolution, Species, and Speciation   1199

Introduction   1199

34.1 Ecology: a scientific approach to complexity   1200

34.2 Evolution: the lynchpin of ecology   1203

34.3 The species: a key concept in evolutionary ecology   1214

34.4 Bringing together ecology and evolution   1218

35   **Genes**   Evolutionary Change in Alleles, Genotypes, and Phenotypes   1221

Introduction   1222

35.1 Describing variation: phenotypes, genotypes, and alleles   1222

35.2 Analysing variation: phenotypes, genotypes, and alleles   1223

35.3 Modelling variation: Hardy–Weinberg modelling   1226

35.4 Genetic change and evolution   1231

36   **Populations**   Quantifying Demographics and Modelling Change   1244

Introduction   1244

36.1 Population demographics   1245

36.2 Types of population   1251

36.3 Methods of quantifying population demographics   1253

36.4 Modelling population change   1257

36.5 Modelling life-history strategies   1265

37   **Communities**   Species Interactions and Biodiversity Metrics   1270

Introduction   1270

37.1 Ecological interactions   1271

37.2 Classifying communities   1280

37.3 Species community metrics   1280

37.4 Communities over space   1284

37.5 Communities over time   1286

38   **Ecosystems**   Abiotic Interactions and Environmental Processes   1291

Introduction   1292

38.1 Defining ecosystems   1292

38.2 Abiotic factors that define and differentiate ecosystems   1293

38.3 Species–environment interactions: tolerance ranges, niches, and adaptations   1297

38.4 Ecosystem processes   1301

38.5 Keystone species and ecological engineers   1311

38.6 Ecological tipping points, ecosystem collapse, and mass extinctions   1313

39   **Challenges**   Key Threats to Ecosystems   1316

Introduction   1317

39.1 Identifying and classifying threats   1317

39.2 Human population growth   1319

39.3 Resource use and emissions   1320

39.4 Habitat loss, degradation, and fragmentation   1326

39.5 Introduction of non-native species   1331

39.6 Implications of ecosystem threats   1331

40   **Solutions**   Managing, Conserving, and Restoring Ecosystems   1338

Introduction   1339

40.1 Types of solution   1339

40.2 Protection   1339

40.3 Management of ecological problems   1343

40.4 Conservation   1345

40.5 Restoration   1354

Index   1361

List of Quantitative Tools

2.1 Calculating the relationship between the $\Delta G^{0\prime}$ value and the $E_0^{\prime}$ value in the oxidation of NADH   193

22.1 Calculation of dissolved oxygen per litre   758

22.2 Acids, bases, buffers, and pH   762

28.1 Quantifying passive diffusion   909

28.2 What is physiological saline?   913

28.3 Measuring turgor pressure in plant cells   918

29.1 Size, volume, and surface area   952

33.1 Reproductive strategies: r- and K-selection   1158

35.1 Calculating heritability   1234

36.1 Explaining population modelling notation   1260

List of Experimental Toolkits

T5.1 The classification of bacteria and their relatedness   46

T5.2 Tracking the molecular evolution of organisms   50

9.1 The 'perfect' imperfect stain and exceptions to the rule   377

11.1 Baiting for *Saprolegnia* spp.   456

12.1 Counting cells using a haemocytometer chamber   482

12.2 How do you use a light microscope?   491

15.1 Reverse transcription polymerase chain reaction   574

19.1 Measurement of hormones by immunoassay   683

34.1 An ecological approach to the scientific method   1202

List of Scientific Processes

3.1 Evidence that DNA is the genetic material   203

4.1 Randy the XX male mouse   226

5.1 Transcription factor GATA-1 is required for red blood cell development   278

7.1 Mitchell's chemiosmotic hypothesis   336

10.1 The use of conditional lethal mutants to examine the process of binary fission   407

10.2 Where's my middle? The accidental discovery of mini-cells   408

17.1 Expression cloning of a GPCR   613

18.1 Imaging touch   624

18.2 Proprioception   629

19.1 Thyroid hormone receptors   673

20.1 Research into the control of walking by the motor cortex   711

23.1 Assessing VO_{2max} in humans   775

24.1 Antidiuretic hormone stimulates the reabsorption of sodium in the thick ascending limb of the loop of Henlé   795

26.1 Research into the regulatory role of melatonin in the corpus luteum   853

30.1 Photoreceptors in Australian ants   1016

30.2 How does the blind cave fish 'see' its environment?   1022

30.3 How do goldfish survive without oxygen?   1042

32.1 Hagfish slime   1117

32.2 Don't play with fireworms!   1134

32.3 Changing shape to avoid predation   1138

33.1 Experimental mammalian parthenogenesis   1149

33.2 Discus fish mucus and the risks of parental care   1188

34.1 Artificial selection in dogs   1210

34.2 Microevolution in human dive reflex   1213

34.3 Evolution of camouflage in one species drives change in an entire insect community   1218

35.1 Selection for early laying in birds in response to climate change   1236

36.1 Metapopulation dynamics and gene flow in northern goshawks   1252

36.2 Establishing the population size of Bengal tigers in the Sundarban mangrove forest   1258

37.1 Investigating avian nest-site competition   1276

37.2 Change in similarity of fish communities in different states in the USA   1285

38.1 Becoming ectothermic: an unusual hibernation strategy in lemurs   1295

38.2 Discovering novel chemosynthesis-driven ecosystems   1302

38.3 Using stable isotope analysis to understand food webs   1307

39.1 Investigating edge effects on ancient woodland indicator plants   1330

40.1 Quantifying pied flycatcher habitat to optimize conservation interventions   1350

List of Real World Views

1.1 Ocean acidification: the other CO_2 problem — 172

2.1 Is NAD deficiency a cause of congenital malformations and miscarriage in pregnancy? — 194

3.1 Plasmids and the spread of antibiotic resistance — 206
3.2 DNA profiling — 209

6.1 Engineering enzymes for industrial applications — 310

7.1 Mitochondrial disease — 337

9.1 Mitochondrial Eve — 395

10.1 Horizontal gene transfer in nature and its use for genetic engineering — 411

11.1 What can sequencing can tell us about the biology of micro-organisms? — 428
11.2 How do the Archaea survive extreme environments? — 443
11.3 The frogs and the chytrids — 450
11.4 Cramp balls and cakes: *Daldinia concentrica* — 452
11.5 The many uses of lichen — 459
11.6 Tackling toxic algae — 464

12.1 Ethics of cell lines — 476

13.1 Mushroom yields and biological efficiency — 509

14.1 Using phalloidin as a stain for F-actin — 539
14.2 Neglected tropical diseases — 543
14.3 Closing the gap — 549

15.1 Treating HIV — 571
15.2 Myxomatosis — 573

27.1 Vaccination — 877

28.1 Micropropagation of plants — 934
28.2 Tree rings and dendrochronology — 938

29.1 Exploitation of genetic variations — 961
29.2 Ocean acidification — 993
29.3 Bone density and gravity — 995

30.1 Growing plants in space — 1027

31.1 Invasion of the topmouth gudgeon — 1062
31.2 Invasive fish species in the Mediterranean Sea — 1065
31.3 Bracken—a problem plant on the move — 1073

33.1 Thermal time and growing degree days in plant development — 1160
33.2 Pollen and archaeology — 1165

34.1 Misconceptions about evolution by natural selection — 1214

35.1 Tay–Sachs disease — 1229

37.1 National Vegetation Classification concept and method — 1280

38.1 Vegetarianism from an ecosystem and ecological footprint perspective — 1305

39.1 Oil pollution in Alaska — 1323
39.2 Devising and using the IUCN species at risk classification scheme — 1334

40.1 Fishing quotas and other methods to manage sustainable harvesting — 1342
40.2 Rotational heathland management — 1352
40.3 Ex situ conservation of ring-tailed lemurs — 1355

List of Clinical Boxes

1.1 Stereochemistry in therapeutic drug development: the tragedy of thalidomide   161

4.1 The genetics of deafness   246

5.1 β-Thalassaemia: a disease caused by aberrant RNA splicing   262

6.1 Misfolded proteins can cause disease   297

6.2 Sickle cell disease   302

7.1 The Warburg effect: metabolism in cancerous cells   338

8.1 Gene therapy: pioneers, pauses, and progress   358

10.1 Down syndrome: the most common human trisomy   424

11.1 Bacterial toxins   439

13.1 The war against antimicrobial resistance: the race to find new antibiotics   501

14.1 Determining the cause of disease: Koch's postulates and *Helicobacter pylori*   527

15.1 The Covid-19 spike protein: the target of vaccine development   559

16.1 Temperature control and fever   584

16.2 Cystic fibrosis   591

18.1 Lenses and correcting refractive errors   633

18.2 Hearing loss   646

19.1 Drugs that affect the autonomic system   664

19.2 Diabetes mellitus   680

20.1 Upper and lower motor neuron lesions   705

20.2 Parkinson's disease   713

21.1 Cardiovascular disease and plasma lipids   722

21.2 Interpreting the ECG   738

21.3 Hypertension   742

22.1 Asthma   755

22.2 Anaemia   761

22.3 Sleep apnea   766

23.1 The use of recombinant EPO by athletes   782

23.2 Physical activity or physical fitness for health   783

24.1 Kidney failure and renal dialysis   799

25.1 Coeliac disease and non-specific gluten sensitivity   820

25.2 Inflammatory bowel disease and irritable bowel syndrome   821

26.1 Contraception   851

26.2 Sexually transmitted infections   851

27.1 IgE-mediated allergic responses   881

27.2 Immunotherapy   883

27.3 Organ transplantation and graft-versus-host disease   887

27.4 Autoimmunity   891

27.5 Gene therapy for the treatment of severe combined immunodeficiency   896

Meet the Authors

The educators behind *Biological Science: Exploring the Science of Life*

The team of educators behind *Biological Science: Exploring the Science of Life* is assembled from lecturers with extensive experience of teaching biological science students, particularly in the first years of university study. They recognize the challenges faced by students as they make the transition from school to university-level study and are deeply committed to making that transition as smooth as possible.

MODULE 1 LEADS

Alison Snape is Professor of Bioscience Education at King's College London (KCL), and Senior Fellow of AdvanceHE. Alison studied genetics at university and went on to do a PhD and postdoctoral research in the field of embryonic development. However, she gradually found herself more interested in teaching than research and subsequently transferred fully to an education-focused role. Alison has more than 20 years' experience of teaching biochemistry, molecular biology, and genetics to undergraduate students on the bioscience, medical, dental, and nutrition programmes at KCL. She has a particular interest in supporting first year students in their transition to university, using methods that encourage students to think, ask questions, and support each other in their learning. She is honoured to have received three University Teaching Excellence Awards, voted for by students at KCL. She is also the co-author, with Dr Despo Papachristodoulou, of *Biochemistry and Molecular Biology*, published by Oxford University Press.

Dr Despo Papachristodoulou is Reader in Biochemistry and Medical Education at King's College London. Her interests include metabolism and nutrition, obesity, diabetes mellitus and complications, medical education, and curriculum development. She has a long and extensive experience of teaching biochemistry and nutrition to medical, dental, biomedical, and pharmacy undergraduates. She has co-authored a number of textbooks on biochemistry for undergraduates.

MODULE 2 LEAD

Professor Kay Yeoman holds a chair in Science Communication and teaches microbiology, molecular biology, and science communication in the School of Biological Sciences at the University of East Anglia (UEA). She has undertaken research in the use of agricultural waste materials in fermentation media to produce extracellular enzymes. She has also conducted research into rhizobium:legume symbiosis, investigating the uptake of iron. Kay is a keen communicator of science, leads fungal forays for her students and members of the public, and is particularly interested in fungi and their use in industry. She has been actively involved in the British Mycological Society and was a previous chair of their education committee. Kay is involved with research that explores the public understanding of science and how students can gain crucial employability skills by taking part in community engagement. She has co-authored a book on science communication, *Science Communication: a practical guide for scientists*. Kay is a Senior Fellow of AdvanceHE and is interested in the transition of pupils from school to higher education; she also investigates effective feedback, publishing work in both areas. Kay is currently Associate Pro-Vice Chancellor for Learning and Teaching Enhancement at UEA, where she is leading a programme of digital transformation in teaching and learning.

MODULE 3 LEAD

Jon Scott is a higher education consultant and Emeritus Professor of Bioscience Education at the University of Leicester where he was Pro-Vice-Chancellor for Student Experience. Jon graduated as a biologist from the University of Durham, specializing in neuroscience. His research was into sensory feedback in the control of movement. He taught physiology to medical and bioscience students and became increasingly involved in developing practice in learning and teaching, publishing on assessment and feedback, academic integrity and belonging, and retention. Among his publications, Jon is co-author of *Study and Communication Skills in the Biosciences*. Jon is a Fellow of the Royal Society of Biology, a Principal Fellow of AdvanceHE, was the UK Bioscience Teacher of the Year in 2011, and won a National Teaching Fellowship in 2012. He currently works with a range of universities in the UK and internationally.

MODULE 4 LEAD

Martin Luck is Emeritus Professor of Physiological Education at the University of Nottingham. He graduated in physiology at Nottingham in 1974 and gained MSc and PhD degrees at the University of Leeds. He also has a degree in mathematics. After research positions in the UK, Germany, and Australia, he returned to Nottingham as an academic in 1990, studying the reproductive physiology of birds and mammals but with parallel interests in student support, teaching, and undergraduate research, for which he received several awards, including a National Teaching Fellowship. He is a Principal Fellow of AdvanceHE and a Fellow of the Royal Society of Biology. Among his publications are a student's guide to research projects, a book on hormones in OUP's Very Short Introduction series, and an edited collection of lectures by the Victorian academic and biologist Arthur Milnes Marshall. Martin's other enthusiasms are scuba diving, hill walking, and his grandchildren.

MODULE 5 LEAD

Anne Goodenough is Professor of Applied Ecology at the University of Gloucestershire, where she teaches across undergraduate and postgraduate courses in wildlife biology, ecology, and environmental science. Having initially become fascinated by birds at a young age, Anne is now interested in all aspects of the natural world and enjoys educating students about the complex relationships that exist between species and the wider environment. Anne also has a particular interest in inspiring students to learn the practical ecological skills necessary to monitor, manage, and conserve wildlife in applied contexts. This often involves working with students in the field, for example, to refine plant indicator systems in ancient British woodlands, develop effective conservation for breeding birds across Europe, monitor whales in Chile, and test technological approaches for surveying mammals in the savannah grasslands of southern Africa.

QUANTITATIVE TOOLKIT LEADS

Dr Dawn Hawkins is an Associate Professor in the School of Life Sciences at Anglia Ruskin University, where she helped to design and deliver the UK's first degree programme in animal behaviour in the 1990s, and has contributed to the teaching of a range of undergraduate modules in ecology, conservation, and zoology. In order to support her students' learning in these subjects, Dawn has focused her teaching on the fundamentals of data and statistics in a biological context. This led to the publication with OUP of her textbook *Biomeasurement: A student guide to biological statistics*, now in its 4th edition. She has also been involved in a series of Higher Education Academy funded projects (NuMBerS, SUMS), designed to support the teaching of maths and statistics in higher and further education in collaboration with Dr Toby Carter, and co-founded the BioMaths Education Network with Jenny Koenig. In collaboration with colleagues and postgraduate students in the UK and Tanzania, Dawn has undertaken research looking at patterns of biodiversity in relation to natural and anthropogenic factors and has a particular interest in elephants and baboons.

Dr Jenny Koenig is an Assistant Professor in Pharmacology and a Senior Tutor at the University of Nottingham, where she contributes to undergraduate biomedical modules in pharmacology and toxicology, as well as the graduate medical course. She taught previously at the University of Cambridge, where she developed an online course 'Essential Maths for Medics and Vets', taught the mathematical aspects of pharmacology, and supported science and maths students with specific learning differences. She has also taught for the pharmaceutical industry and the British Pharmacological Society where she is a Fellow. She has published on curriculum and pedagogical approaches for maths in bioscience and co-founded with Dawn Hawkins the BioMaths Education Network. Jenny took a few years out from teaching in higher education to complete a PGCE and taught secondary science and maths. This experience gives her a good understanding of how new undergraduate students have experienced maths at school and the latest pedagogical approaches that can be used to help students understand maths and its applications in biology.

DIGITAL LEARNING LEAD

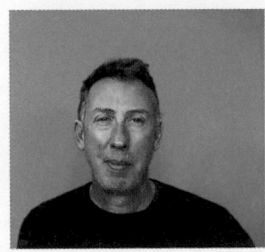

Dr Gus Cameron is Reader in Bioscience Education at the University of Bristol, where he has been since departing the United States in 2001. Gus was born in England but raised in Wales and southern Scotland, where he benefited from a progressive education heavily influenced by

A. S. Neill's Summerhill. Studying part-time for a PhD in biochemistry gave Gus the opportunity to begin teaching, but for a while he was a motorcycle racing team mechanic and says the opportunity to learn how knowledge was transmitted in this immensely practical, fiercely competitive, and deeply complex area has stayed with him. This work was concurrent with leading a very successful further education course in biochemistry, where he helped to prepare non-traditional students for university entry. These influences helped Gus, in the late 2000s, to develop eBiolabs, a dynamic laboratory manual for the biosciences. Since then, this online system has used rich multimedia to help thousands of students prepare for laboratory work and has been used by their teachers to track their progress and give timely, efficient feedback. Gus has won several teaching prizes, is a senior Fellow of the HEA, and is known mainly for leading first year biochemistry programmes.

OUR COLLABORATORS

The lead author team collaborated with the following colleagues during the writing of *Biological Science: Exploring the Science of Life*. We thank them for their invaluable input.

Prof Henderson Cleaves, Earth-Life Science Institute, Tokyo Institute of Technology and Institute for Advanced Study, Princeton (Topic 2, The Emergence of Life on Earth)

Dr Anne Cunningham, Institute of Health Sciences, Universiti Brunei Darussalam (Chapter 27)

Dr Kelly Edmunds, School of Biological Sciences, University of East Anglia (Chapters 9, 10, 11, 14)

Dr Paul Elliott, Director of Studies in Biology and Scientific Admissions Tutor, Homerton College, University of Cambridge (Topic 4, Evolutionary Processes)

Dr Steve Ennion, Associate Professor, Leicester Medical School, University of Leicester (Chapters 17, 19, 20)

Dr Gethin Evans, Deputy Head of Department, Department of Life Sciences, Manchester Metropolitan University (Chapters 23 and 25)

Dr Albert Fahrenbach, School of Chemistry, University of New South Wales, Australia (Topic 2, The Emergence of Life on Earth)

Dr Heather Felgate, Researcher, Quadram Institute, University of East Anglia (Chapters 11 and 14)

Dr Joanne Foulkes, School of Pharmacy and Biomolecular Sciences, Liverpool John Moores University (Chapter 15)

Dr Laetitia Gunton, Life and Geosciences, Australian Museum (Chapters 28–33)

Dr Helen James, School of Biological Sciences, University of East Anglia (Chapters 9, 10, 11, 12, 14)

Dr Ian Kay, Formerly Deputy Head of Department, Department of Life Sciences, Manchester Metropolitan University (Chapter 21 and 25)

Dr Laura Mongan, Senior Lecturer in Medical Education and Biomedical Sciences, University of York and The Hull York Medical School (Chapter 26)

Dr Kevin Pyke, School of Biosciences, University of Nottingham (Chapters 28–33)

Prof Katherine Sloman, School of Health and Life Sciences, University of the West of Scotland (Chapters 28–33)

Prof Jo Spencer, Head of Department, Department of Immunobiology, King's College London (Chapter 27)

Dr Claudia Stocker, Vivid Biology (Illustrator/animator, key concept videos)

Dr Paul Thomas, formerly Imaging Facility Manager at the University of East Anglia (Chapter 12)

We also extend our thanks to those colleagues who generated content to support the main topics and chapters:

Dr Martin Barker, School of Biological Sciences, University of Aberdeen (student quizzes and testbank questions, Module 4)

Dr Paul Elliott, Director of Studies in Biology and Scientific Admissions Tutor, Homerton College University of Cambridge (student quizzes and testbank questions, Module 5)

Dr Alun Hughes, Biological and Environmental Sciences, Liverpool John Moores University (student quizzes and testbank questions, Module 3)

Professor Philip Meneely, formerly Haverford College, PA, USA (student quizzes and testbank questions, Modules 1 and 2, and questions appearing in Chapter 27)

Dr Feisal Subhan, School of Biomedical Sciences, University of Plymouth (student quizzes and testbank questions, Module 3).

We also thank several of our colleagues who provided valuable input on specific elements of the project:

Dr Peter Brown, Anglia Ruskin University (Quantitative Toolkits)

Dr Toby Carter, Anglia Ruskin University (population mathematical modelling, Chapter 36, and Quantitative Toolkits)

Dr Michael Weale, Genomics Plc (Quantitative Toolkits)

On Topic 1 (Exploring the Science of Life): **Guy Norton**, **Dr Gerbrandus Green** (Anglia Ruskin University), and **Dr Danielle Green** (Anglia Ruskin University).

And for inspiration and encouragement throughout the writing of Module 2: **Professor Harriet Jones**, School of Biological Sciences, University of East Anglia.

Acknowledgements

Biological Science: Exploring the Science of Life was produced in collaboration with the following colleagues at Oxford University Press and its partner suppliers:

Editor in Chief
Jonathan Crowe

Senior Development Editors
Stephanie Southall-Paddock
Judith Lorton

Commissioning Assistants
Giulia Lipparini
Martin King
Carolin Cichy

Art Editor
Johannah Walkowicz

Text design
Claire Dickinson
Elisabeth Heissler

Cover design
Gemma Wakefield

Permissions
Maya Noronha
Devin Therien
Sophie Basilevitch

Media Editors
Chris Spannos
Karissa Venne
Lauren Thompson

Production Editor
Karen Moore

Copyeditor
Julian Thomas

Proofreaders
Jayne MacArthur
Henry MacKeith

Indexer
Susan Leech

Illustrations
Collaborate Agency
Fabian Slongo
QBS Learning

Typesetter
Straive

Editorial Advisers

The authors and publisher gratefully acknowledge the following colleagues who provided feedback on draft materials during the writing process. Their input improved the final product immeasurably; any issues or errors that remain are, of course, our sole responsibility. We also thank the further anonymous reviewers not listed here.

Dr Andy Bailey, School of Biological Sciences, University of Bristol

Dr Emma Bailey, School of Life Sciences, University of Glasgow

Dr Zita Balklava, School of Biosciences, Aston University

Dr John Barrow, School of Medicine, Medical Sciences and Nutrition, University of Aberdeen

Dr Sven Peter Batke, Senior Lecturer in Plant Sciences, Edge Hill University

Dr Louise Beard, School of Life Sciences, University of Essex

Professor Andrew Beckerman, School of Biosciences, University of Sheffield

Dr Maureen J Berg, University of Brighton

Dr Lynn Besenyei, School of Sciences, University of Wolverhampton

Luc Bussière, Department of Biology and Environmental Sciences, University of Gothenburg

Dr Thomas Butts, Sunderland Medical School, University of Sunderland

Dr Gillian Campling, Department of Biosciences, Durham University

Dr Mark A Chapman, Biological Sciences, University of Southampton

Dr Jayne Charnock, Department of Biology, Edge Hill University

Professor Christopher J Coates, Biosciences, Faculty of Science and Engineering, Swansea University

Nick Colegrave, Institute of Evolutionary Biology, University of Edinburgh

Dr Sheena Cotter, Senior Lecturer, School of Life Sciences, University of Lincoln

Dr Verena Dietrich-Bischoff, School of Biology, University of St Andrews

Dr Aysha Divan, Faculty of Biological Sciences, University of Leeds

Dr Beth Dyson, School of Biosciences, The University of Sheffield

Dr Emma Edwards, School of Science and the Environment, University of Worcester

Mark Fellowes, The School of Biological Sciences, University of Reading

Dr Neil J Gostling, School of Biological Sciences, University of Southampton, UK

Mohammad K Hajihosseini, School of Biological Sciences, University of East Anglia

Ian Hartley, Lancaster Environment Centre, Lancaster University

Tom Hartman, School of Life Sciences, University of Nottingham

Dr Meanie A Healy, Senior Lecturer in Medical Sciences, Swansea University

Dr Elaine Hemers, School of Biological and Environmental Sciences, Liverpool John Moores University

Dr Graham J Holloway, School of Biological Sciences, The University of Reading

Dr Calum Holmes, Engineering and Physical Sciences, Heriot-Watt University

Dr Rachel Hope, University of York

Dr Kris Jeremy, School of Biomedical Sciences, University of Plymouth

Professor Giles Johnson, School of Natural Sciences, University of Manchester

Dr Louise Johnson, School of Biological Sciences, University of Reading, UK

Dr Emma Jones, School of Biosciences, University of Sheffield

Dr Nawroz Kareem, School of Life Science, Keele University

Jennifer Killey, Department of Allied Health Professions, Sport and Exercise, University of Huddersfield

Dr Linda King, School of Life Sciences, Anglia Ruskin University

Tilo Kunath, School of Biological Sciences, The University of Edinburgh

Dr Beth Lawry, School of Biomedical, Nutritional and Sport Sciences, Newcastle University

Dr Philip T Leftwich, Lecturer in Biological Sciences, University of East Anglia

Dr Andy Lewis, Faculty of Health and Medicine, Lancaster University

Professor Zenobia Lewis, School of Life Sciences, University of Liverpool

Professor Frankie MacMillan, School of Physiology, Pharmacology and Neuroscience, University of Bristol

Dr Samantha McLean, Nottingham Trent University

A guide to learning with *Biological Science: Exploring the Science of Life*

We have written *Biological Science: Exploring the Science of Life* to inspire you and to make thinking like a biologist second nature. We have included a range of features to guide and support your learning, and these are described here. We wish you every success in your studies!

TO HELP YOU MASTER THE ESSENTIAL CONCEPTS

- **Life and Its Exploration: Foundational Principles module.** This scene-setting module foreshadows the chapters that follow by exploring a range of topics that underpin the way we think about and study biological systems—from the nature of scientific enquiry, the definition of life, and the fundamental process of evolution by natural selection, to the ways we can organize and classify the biodiversity that surrounds us. Read through these topics to start building your understanding of these important themes and concepts.

- Each chapter opens with a **key concept video** that summarizes the key ideas it explores. Watch this video to gauge your familiarity with the concepts the chapter discusses, and to focus your further study.

- **Interactive versions of key figures** give you the control you need to step through, and gain mastery over, the concepts these figures are illustrating.

- **Flashcard glossaries** at the end of each chapter demystify the language of biological science by helping you to recall the key terms and concepts on which further study can be built.

🌐 To access the digital resources listed here, look out for call-outs to the e-book throughout the chapters that follow.

TO HELP YOU STRENGTHEN YOUR UNDERSTANDING

- **Pause and think questions** are embedded throughout each chapter to encourage you to reflect on, engage with, and question the concepts they are exploring. Do pause for a moment to read and reflect on these questions: they will help to reinforce your understanding or unearth misconceptions.

- **Questions in multiple-choice format** at the end of each section within a chapter give you a quick, effective way of checking that you understand the key concepts covered in that section. Submit your answers to receive feedback that explains the right (and wrong) answers to take your learning further.

🌐 To access these resources, go to the relevant locations in the e-book

- Every chapter ends with **study questions and problems** that are arranged in three groups:

Concepts and definitions. These questions are written to ensure you have mastered all the important concepts and definitions introduced in the chapter.

Beyond the concepts. These questions are written to stretch you a little further than the *Concepts and definitions* questions—to encourage you to connect concepts and ideas, and to see the bigger picture of biological science as a field of study.

Apply the concepts. These questions are the most challenging and encourage you to apply concepts and ideas to novel problems. Completing questions like these is a great way to develop your critical thinking and analytical skills, and to get well-prepared for further study.

There are many different ways of answering these questions correctly, though we provide some suggested answers to help guide your learning.

TO DEVELOP YOUR QUANTITATIVE SKILLS

Quantitative tools help us to further our understanding of biological systems: they help us to test predictions, to model the processes we see happening inside and around us, and to make sense of information we have gathered. Feeling confident about the use of such tools and techniques is an important part of developing a deeper understanding of the subject.

With this in mind, our **Quantitative Toolkits** are designed to review the essential numerical and statistical concepts. They are

enhanced with short videos, in which we walk through examples step-by-step, giving you the extra support and insight you need to use quantitative skills with confidence.

 To access the videos, go to the relevant locations in the e-book

Quantitative Tool panels, which appear in various chapters, build on the Quantitative Toolkits by going further to explain the use of quantitative skills in relation to particular topics or areas of study.

TO HELP YOU SEE THE BIGGER PICTURE

Biological science pervades our lives in myriad ways. It is a dynamic, experimental science, whose frontiers are continually being advanced by researchers around the world. We include several features throughout to convey to you this bigger picture.

- Often informed by published research, the **Scientific Process** panels explore the scientific approaches that underpin our understanding of the concepts presented, and how the scientific method is used in practice.

- Various chapters feature **Experimental Toolkit** panels, which introduce you to the investigative methods used to elucidate the concepts being discussed, and which can be used more generally to further our understanding of biological systems.

- **Real World View** panels explore how the concepts under discussion in the chapter are being applied in real-life contexts, and prompt you to reflect on the important social, moral, and ethical impacts of these applications.

- One of the most exciting and important ways of using our knowledge of biological systems is to better understand human disease, and to translate that understanding into more effective ways of treating those diseases. **Clinical Boxes** explore the biological basis of disease and show how our increased understanding can lead to better, more effective treatment strategies.

For lecturers—teaching with this resource

Biological Science: Exploring the Science of Life is a modular bioscience teaching and learning resource that marries a consistent pedagogical approach to the learning of a broad range of core biological concepts. The resource spans the range of topics covered in the typical Biological Sciences (CAH03-01) level 4 degree programme in the UK, but is of sufficient depth for it to be equally appropriate for the first years of study for more specialized degree programmes. As such, the resource is appropriate as background reading to prepare your students for level 5 courses but is also accessible to students on level 3 access or foundation programmes.

HOW DO WE SUPPORT YOU AND YOUR STUDENTS?

As members of your academic community, we have drawn on our shared experiences to recognize and respond to your needs as fully as we can.

Biological science, as a discipline, spans a wide range of topics, and a deep understanding of a range of foundational concepts is important and necessary for success in later studies. We believe that such deep understanding can only be achieved if students can build a broad picture of the discipline without being overwhelmed by detail, and if they can engage in self-paced active learning. Success is also predicated upon the mastery of certain quantitative skills and a full appreciation of the nature of scientific enquiry, without which meaningful engagement with this experimental science is not possible.

Biological Science: Exploring the Science of Life responds to these needs by placing a clear central narrative, carefully structured active learning, and confidence with quantitative concepts central to its approach.

Its straightforward narrative, reinforced by key concept videos for every chapter, communicate key ideas clearly: the right information is provided at the right time and to the right depth.

Its pause and think features, self-check questions, and graded end-of-chapter questions, augmented by flashcards of key terms, directly support active learning. The combination of narrative text and learning features promote a rich, active learning experience: read, watch, and do.

Its combination of Quantitative Toolkits, Scientific Process panels, and the Life and its Exploration topics provide more insight and support than any other general biology text; they prepare students to engage with this quantitative and experimental discipline with confidence and set them on a path for success in their future studies.

SUPPORT FOR DIGITAL LEARNING

The digital edition of this resource has been carefully designed to provide your students with a broad, active, confidence-building learning experience. Each chapter section can be used by your students as a single learning session, with Pause and Think questions (with suggested answers) to stimulate reflection, and multiple-choice questions at the end of each section to reinforce learning.

In addition:

Each chapter opens with a **key concept video** that summarizes the key ideas it explores, building your students' confidence in the subject matter right from the outset.

Interactive versions of key figures give your students the control they need to step through, and gain mastery over, the concepts these figures are illustrating, proceeding at their own speed.

Flashcard glossaries at the end of each chapter demystify the language of biological science by helping your students to recall the key terms and concepts on which further study can be built.

We also provide teaching and learning resources specifically for you, as adopting lecturers:

- all the figures from the resource are available in digital format, for you to use in your teaching;

- an extensive set of questions, in multiple-choice format—and distinct from those we provide directly to students—are available for you to deliver via your institution's virtual learning environment, to support either formative or summative assessment.

Life and Its Exploration
Foundational Principles

TOPIC 1   Exploring the Science of Life

TOPIC 2   The Emergence of Life on Earth

TOPIC 3   How Do We Define Life?

TOPIC 4   Evolutionary Processes

TOPIC 5   The Diversity, Organization, and Classification of Life

LIFE AND ITS EXPLORATION

The breadth and scale of biological science is astounding and overwhelming: from bacteria that can thrive in the hottest hydrothermal vents to wood frogs that can survive being frozen; from the atoms and molecules that make up our cells to the vast ecosystems that are formed from a diverse array of organisms. At whatever scale we study the biological world, and whichever examples of biodiversity we might wish to explore, there are some foundational principles that underpin the science of life. Appreciating these will lead you to a deeper and more complete understanding of the subject as a whole.

We open this book by setting out these principles. We ask how life on earth emerged and consider what we mean by 'life'; we see how evolutionary processes have shaped the diversity of life forms that have existed and continue to exist; and we discuss how best to describe that diversity—how we might organize and classify organisms according to their similarities and differences.

But we begin by considering how to approach scientific enquiry and why thinking like a biological scientist is crucial to our progress. We will see that thoughtful, evidence-based experimentation lies at the heart of confronting and addressing the global challenges we face as a species.

Image: A Scuba diver inspects an artificial reef in the waters around Bali, Indonesia. Such reefs are being used by conservationists to enhance the growth of corals and aquatic organisms. *Source:* Helmut Corneli/Alamy Stock Photo.

Exploring the Science of Life

Introduction

There has never been a more amazing time to explore the science of life. Among the many recent achievements, the human genome has been sequenced, genes have been edited, and treatments—including vaccines for novel diseases—have been developed at an astonishing pace. We have science to thank for these advances: it gives us a clear, objective way to ask and answer questions about living organisms and enables advances in technology that have yielded new and more powerful tools to acquire and handle data.

As we will explain in this topic, science is a way of understanding the natural world using logic and evidence. Science provides *intellectual* tools, while technology provides *physical* tools. We are increasingly turning to these tools to improve our lives and to tackle the challenges facing humanity.

The greatest of these challenges are those associated with global warming (see Figure T1.1). The consequences of these changes in our physical world are being felt at every biological scale (as illustrated by the example of coral reefs in Figure T1.2). At the molecular level, mutation rates are being altered. At the level of individual organisms, physiological processes are under strain from more extreme environments. At the level of the ecosystem, habitats are shifting and biodiversity is being lost. The implications for humanity are stark: we face threats to our access to food and water, increased risk from disease, and loss of space in which to live. The work of biological scientists is central to identifying and mitigating these challenges. Exploring the science of life has become imperative, not just for understanding but also for *sustaining* life on our planet.

Contents

Introduction	5
1.1 What does it mean to take a scientific approach?	6
1.2 Ways of thinking	8
1.3 Logical reasoning	8
1.4 How do we test research hypotheses?	10
1.5 What is a null statistical hypothesis?	10
1.6 How do theories and models differ?	11
1.7 What are mathematical models?	11
1.8 Team Science	13

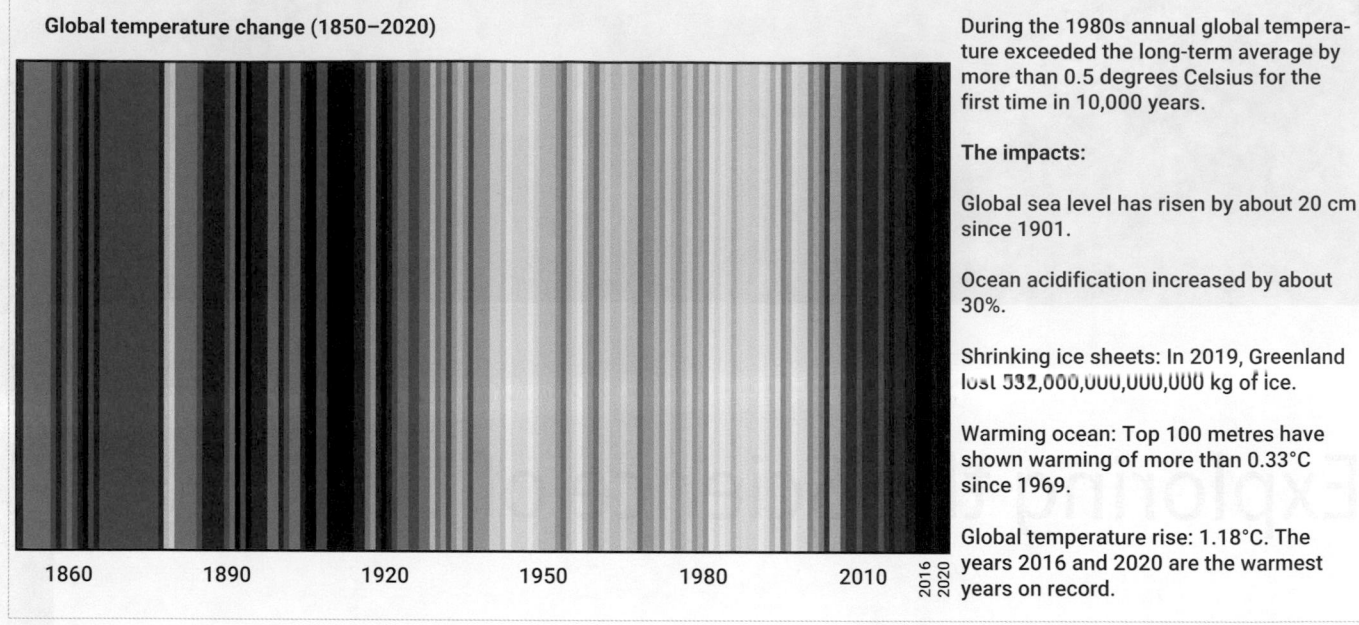

Global temperature change (1850–2020)

During the 1980s annual global temperature exceeded the long-term average by more than 0.5 degrees Celsius for the first time in 10,000 years.

The impacts:

Global sea level has risen by about 20 cm since 1901.

Ocean acidification increased by about 30%.

Shrinking ice sheets: In 2019, Greenland lost 532,000,000,000,000 kg of ice.

Warming ocean: Top 100 metres have shown warming of more than 0.33°C since 1969.

Global temperature rise: 1.18°C. The years 2016 and 2020 are the warmest years on record.

1860   1890   1920   1950   1980   2010   2016 2020

Figure T1.1 Post-industrial age climate change challenges to life. Each stripe represents the mean annual global temperature for one year. There has been a global temperature rise of just over one degree since 1850. This might not sound like much but the implications for biological systems are massive. It is not just global warming itself, but the associated changes in climate patterns, and acidification and rising level of our seas, that are the problem.
Source: Ed Hawkins, University of Reading, Creative Commons.

Throughout this book you will find examples of how technological advances are enabling scientific enquiry: we can now use high-specification microscopes in labs, and satellites in space to investigate life on Earth from the submolecular scale to the global scale. Whether it is the molecular analysis of the components of a virus, or a space rover landing on Mars, technology is enhancing our ability to measure and analyse our world and the universe beyond it.

Technology is even allowing us to overcome the barrier of time: we can assess the diets of dinosaurs by analysing stable isotopes using mass spectrometers, and we can predict how climate change will drive changes in the locations species inhabit in the future using advanced computer modelling.

As a student starting out on the path to becoming a biological scientist you are likely to be wondering: What do I need to be successful? Certainly, you need to become familiar with our current understanding of biological systems, as we will set out in the pages that follow. However, you will also need to be aware of how this information was acquired. (This really lies at the heart of the distinction between *being* a biological scientist and being *informed* about biological science.) You will need to be aware of the technological tools of your trade and, most importantly, you will need to fully appreciate the *process* of science. We take a first step towards this in the next section by considering what science involves, and what it should *not* involve.

The one thing that all good biological scientists need to get them started is curiosity—something that everyone has in abundance. Curiosity is fuelled by observing the world, as well as by reading, and listening to and talking with others: you could be on a walk with your dog, chatting with friends over coffee, following social media, or anywhere doing anything. Curiosity makes us ask questions. If you ask good questions and follow a scientific approach when answering them, you will be able to make your own contribution to the future exploration of life.

1.1 What does it mean to take a scientific approach?

Science has allowed us to achieve a huge amount, but it is surprisingly challenging to say what science *is* and how it is done. Broadly speaking, science is a way of asking and answering questions using logical reasoning to explain the natural world. The primary limitation on science is that it can only address questions that can be answered using data acquired by observation and measurement—that is, by carrying out what are called empirical studies. Science cannot answer questions such as, 'What is morally right and wrong?' This is the realm of philosophy rather than science.

What distinguishes biological science from other sciences is that the questions being posed are about living organisms and their environments.

If you were asked to think of someone doing science, you would probably imagine someone collecting data—perhaps in the lab, or maybe out in the field. Alternatively, they might be at a computer analysing data and producing graphs. Data collection and analyses are central activities in science, but scientific research is really distinguished by the mental activities involved, and particularly the use of reasoning to minimize bias and to maximize understanding.

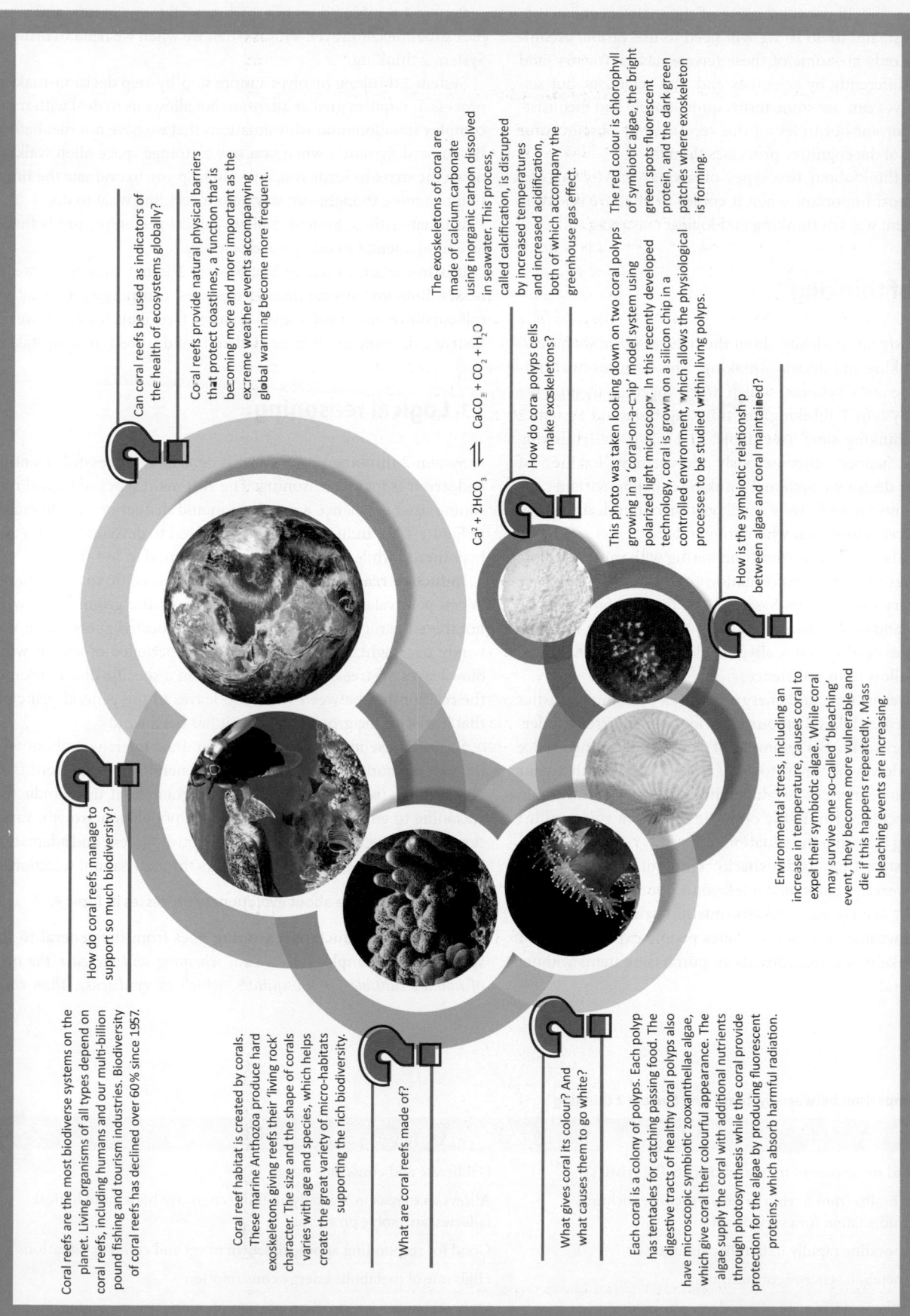

Can coral reefs be used as indicators of the health of ecosystems globally?

Coral reefs provide natural physical barriers that protect coastlines, a function that is becoming more and more important as the extreme weather events accompanying global warming become more frequent.

The exoskeletons of coral are made of calcium carbonate using inorganic carbon dissolved in seawater. This process, called calcification, is disrupted by increased temperatures and increased acidification, both of which accompany the greenhouse gas effect.

The red colour is chlorophyll of symbiotic algae, the bright green spots fluorescent protein, and the dark green patches where exoskeleton is forming.

$$Ca^{2} + 2HCO_3 \rightleftharpoons CaCO_{3} + CO_2 + H_2O$$

How do coral polyps cells make exoskeletons?

This photo was taken looking down on two coral polyps growing in a 'coral-on-a-chip' model system using polarized light microscopy. In this recently developed technology, coral is grown on a silicon chip in a controlled environment, which allows the physiological processes to be studied within living polyps.

How is the symbiotic relationship between algae and coral maintained?

How do coral reefs manage to support so much biodiversity?

Coral reefs are the most biodiverse systems on the planet. Living organisms of all types depend on coral reefs, including humans and our multi-billion pound fishing and tourism industries. Biodiversity of coral reefs has declined over 60% since 1957.

Coral reef habitat is created by corals. These marine Anthozoa produce hard exoskeletons giving reefs their 'living rock' character. The size and the shape of corals varies with age and species, which helps create the great variety of micro-habitats supporting the rich biodiversity.

What are coral reefs made of?

What gives coral its colour? And what causes them to go white?

Each coral is a colony of polyps. Each polyp has tentacles for catching passing food. The digestive tracts of healthy coral polyps also have microscopic symbiotic zooxanthellae algae, which give coral their colourful appearance. The algae supply the coral with additional nutrients through photosynthesis while the coral provide protection for the algae by producing fluorescent proteins, which absorb harmful radiation.

Environmental stress, including an increase in temperature, causes coral to expel their symbiotic algae. While coral may survive one so-called 'bleaching' event, they become more vulnerable and die if this happens repeatedly. Mass bleaching events are increasing.

Figure T1.2 The exploration of life and the impacts of climate change from the biosphere to the molecular. This figure shows different scales at which we can study biological systems and some of the impacts global warming and ocean acidification are having.

Source: Shutterstock; Shapiro, O., Kramarsky-Winter, E., Gavish, A. et al. A coral-on-a-chip microfluidic platform enabling live-imaging microscopy of reef-building corals. *Nat Commun 7*, 10860 (2016). https://doi.org/10.1038/ncomms10860

In this topic, we will consider some of the main mental activities of a scientist, and to do so we will need to use various technical terms. Not only are some of these terms, such as 'theory' and 'model', used differently by scientists and non-scientists, but scientists themselves can use some terms quite loosely and inconsistently. It is important not to let all this terminology obscure your understanding of the cognitive processes though.

Let us now think about two types of mental activity that are arguably the most important when it comes to effective scientific enquiry: different ways of thinking and logical reasoning.

1.2 Ways of thinking

In order to satisfy our curiosity about the natural world we have to do a lot of thinking and decision-making. This involves two complementary cognitive systems, which work together to promote our survival: System 1 thinking, or 'thinking fast'; and System 2 thinking, or 'thinking slow' (see Table T1.1). System 1 thinking is automatic: it happens unconsciously and accounts for the vast majority of our decisions. System 2 thinking is deliberate and conscious; it accounts for a relatively small portion of our decisions but becomes relatively more used when doing science.

System 1 enables us to carry on functioning with minimal effort despite the huge amount of information that we encounter every moment of every day. Without System 1 our brains would soon be overwhelmed, and we would not be able to carry out even the most basic of activities needed to stay alive, let alone the many additional activities that allow us to do science.

System 1 saves us time and energy. It works by using heuristics (or mental shortcuts) based on our experience and current understanding of the world. For example, if you saw a strange space alien walking down the street towards you, it is likely that your brain would use heuristics to evaluate the situation and you would immediately decide to run in the opposite direction: you wouldn't spend very long assessing the situation.

However, System 1 has its drawbacks. Think of a time when the persuasive powers of a salesperson led you to purchase something you didn't really need or want. In this situation, you will have experienced the downside of System 1. Sales people exploit System 1 to get us to make quick decisions about purchasing items without proper consideration.

System 1 thinking does not work so well for us in new and complex situations, however. This is when we when we need to bring in System 2 thinking.

System 2 thinking involves a more step-by-step decision-making process. It requires greater attention but allows us to deal with more complex situations and with situations that we have not met before. If you used System 2 when you saw a strange space alien walking down the street towards you, it would help you to evaluate the situation in a more thought-out way before deciding what to do.

Ultimately then, System 2 allows logical reasoning and is therefore fundamental to doing good science.

It is important to realize that System 2 is not without its weaknesses. Both systems can lead to mistakes, contributing to what we call cognitive biases and logical fallacies (see Table T1.2). However, System 2 thinking can help us to identify and correct these mistakes.

1.3 Logical reasoning

A System 2 thinking process that is central to successful scientific endeavour is logical reasoning. The two main types of logical reasoning used in science are induction and deduction, as illustrated in Figure T1.3. Inductive reasoning is used to develop theories and hypotheses, while deductive reasoning is used to test them.

Inductive reasoning takes us from the specific to the general. When you wake up and see leaves all over the ground that were not there the night before, you make an educated guess that it was windy overnight, based on your past experience of seeing wind blow leaves off trees. You have gone from a specific observation of the relationship between wind and leaves to the general principle that leaves on the ground indicate it has been windy.

Scientists use inductive reasoning to draw inferences about why the natural world is the way it is (the general situation) from their observations (something specific). For example, it takes inductive reasoning to go from observations about population growth, variations within species, differential reproductive success, and adaptation to the explanation that evolution occurs through natural selection.

▶ **We learn more about evolutionary processes in Topic 4.**

By contrast, **deductive reasoning** goes from the general to the specific. For example, *since global warming will increase the area of habitat suitable for mosquitoes, which carry disease, then cases*

Table T1.1 A comparison between System 1 and System 2 thinking

System 1: Thinking fast	System 2: Thinking slow
Automatic and makes use of mental shortcuts (heuristics)	Deliberate and conscious
Prevents our brains from having to fully process enormous amounts of information for every decision	Allows us to reason logically, identify cognitive biases and logical fallacies, and solve problems
Good for responding rapidly in familiar situations	Good for responding appropriately in novel and complex situations
Low rate of metabolic energy consumption	High rate of metabolic energy consumption
Accounts for the vast majority of our decision-making	Only accounts for a small proportion of our decision-making

You can find out more about System 1 and System 2 thinking in the book *Thinking, Fast and Slow* by the psychologist Daniel Kahneman.

Table T1.2 Examples of cognitive biases and logical fallacies

Availability bias	The tendency to assess information using examples that are familiar and come to mind quickly because they are recent or particularly dramatic. For example, you might decide not to go swimming in the sea after seeing a report of a shark attack on the news, even though such attacks are extremely rare. Science helps us avoid this cognitive bias through use of sampling and a consideration of sample size and quality. ▶ **We learn more about sampling in Quantitative Toolkit 5.**
Causation bias	The tendency for us to conclude that one thing causes another. For example, you might attribute the fact that someone you see in an advert looks particularly fit and healthy to the brand of trainers that they are wearing. However, if you thought longer and harder about it you would realize that the trainers are unlikely to be influencing this person's medical condition at all. Science helps us avoid this cognitive bias through careful experimental design and interpretation. ▶ **See Quantitative Toolkits 6 and 7.**
Clustering illusion bias	The tendency for us to see a pattern in a random sequence of numbers or events. For example, if you look at a scatterplot of 1000 randomly generated points, your immediate impression will be of gaps and clusters. Science helps us avoid this cognitive bias by requiring us to assess potential patterns statistically. ▶ **We learn more about patterns and their evaluation in Quantitative Toolkit 7.**
Confirmation bias	The tendency to fit new information into your existing understanding of the world. For example, human-induced global warming can lead to people attributing *all* uncommon weather events to climate change. Science helps us avoid this cognitive bias through the development of alternative research hypotheses and falsifiability (see Section 1.4 'How do we test research hypotheses?').
Appeal to emotion	The stimulation of an emotional response in place of a valid or compelling argument. For example, someone could discourage you from being vaccinated by drawing on your fear of needles. Science helps us to avoid this logical fallacy by focusing on the evidence provided by data gathering and evaluation (see How do we test research hypotheses?).

You can find an introduction to more cognitive biases and logical fallacies in *An Illustrated Book of Bad Arguments* by Ali Almossawi.

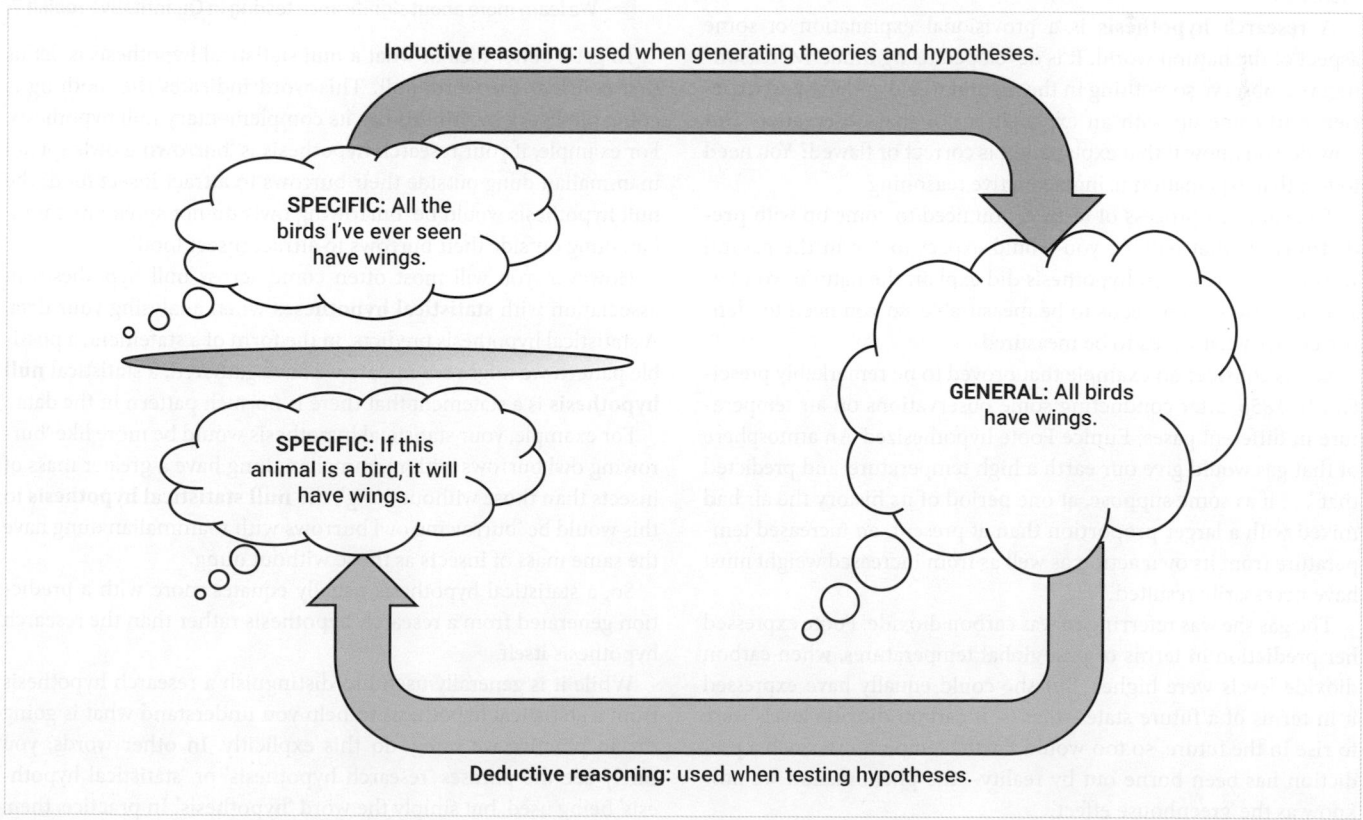

Figure T1.3 Inductive versus deductive reasoning.

of mosquito-borne disease will increase. This argument goes from general statements (or **premises**) about mosquito physiology and ecology to a specific conclusion about the number of cases of mosquito-borne diseases that are likely to be observed in the future.

However, just because an argument sounds logical in itself does not mean that it is valid. Let us consider the statement '*All birds have wings*'. The second statement '*This animal has wings and therefore it is a bird*' would be internally logical—it would be consistent with the first. But it misses the point that *not all animals with wings are birds*. Once this premise is added, the argument does not make sense. Being able to scrutinize, or **critically evaluate**, your own logical reasoning, as well as that of others, is another example of an important mental activity in science.

1.4 How do we test research hypotheses?

System 2 thinking and logical reasoning underpin why science is so effective at explaining the world. There is no better example of their application than when scientists use hypotheses. Different types of hypothesis are covered in the next section; here we focus on research hypotheses and the process of assessing them.

Research hypothesis testing is probably the single most powerful intellectual technique in science. The process of research hypothesis testing is often presented as a series of standard steps known as the **scientific method**. However, the use of 'the' here is a bit misleading: in practice, science does not operate in a single way. There is no one scientific method and science is becoming more varied as sources of data have become more diverse and the volume of data we gather increases.

A **research hypothesis** is a provisional explanation of some aspect of the natural world. It is developed using inductive reasoning: you observe something in the natural world and use past experience to come up with an explanation for that observation. But how do you know if that explanation is correct or flawed? You need to test that explanation using deductive reasoning.

To begin this process of testing, you need to come up with **predictions** of what pattern you would expect to see in the natural world if your research hypothesis did explain the natural world as stated. A prediction needs to be measurable, so you need to identify clearly what needs to be measured.

Let us consider an example that proved to be remarkably prescient. In 1856, after conducting some observations on air temperature in different gases, Eunice Foote hypothesized 'An atmosphere of that gas would give our earth a high temperature' and predicted that '. . . if as some suppose, at one period of its history the air had mixed with a larger proportion than at present, an increased temperature from its own action as well as from increased weight must have necessarily resulted'.

The gas she was referring to was carbon dioxide. Foote expressed her prediction in terms of past global temperatures, when carbon dioxide levels were higher. But she could equally have expressed it in terms of a future state—that is, if carbon dioxide levels were to rise in the future, so too would Earth's temperature. Such a prediction has been borne out by reality—the phenomenon we now know as the 'greenhouse effect'.

Once you have your predictions (and know what to measure to test your predictions), you need to make observations and

measurements to determine what is the actual pattern in the natural world. You can then compare this with the pattern you predicted. If the two match, then the hypothesis is supported: it suggests that the hypothesized explanation does describe the natural world. If the two do not match, then the hypothesis is not supported and we conclude it does *not* explain the natural world.

There is an important point of principle to note here: it must be possible for a research hypothesis to be supported, or for it *not* to be supported. (If it is not possible to *disprove* a hypothesis, it is not a valid hypothesis.) We say that the research hypothesis must be **falsifiable**. This process works even better if you consider multiple explanations (alternative hypotheses) at the same time.

The **scientific premise** of your hypotheses and predictions is the knowledge upon which they are based. You are likely to need to reference the work of others to justify your premises. You will also need to recognize, and explicitly state, any key **assumptions** you make. When critically evaluating your work, or the work of others, it is important to consider if the premises and assumptions of the logical framework being used are valid.

1.5 What is a null statistical hypothesis?

We have just seen how a research hypothesis is a provisional explanation of the natural world. Another type of hypothesis used by scientists is the **null statistical hypothesis**. This type of hypothesis forms the foundation of null hypothesis significance testing (NHST), which has been a key analytical tool in science since the middle of the 20th century.

▶ We learn more about significance testing in Quantitative Toolkit 7.

To get a better idea of what a null statistical hypothesis is, let us first consider the word 'null'. This word indicates that nothing is going on. Every hypothesis has its complementary null hypothesis. For example, if your research hypothesis is 'burrowing owls spread mammalian dung outside their burrows to attract insect food', the null hypothesis would be 'burrowing owls do not spread mammalian dung outside their burrows to attract insect food'.

However, you will most often come across null hypotheses in association with **statistical hypotheses** when analysing your data. A statistical hypothesis predicts, in the form of a statement, a possible pattern we might see in data we have gathered; a statistical **null hypothesis** is a statement that there is no such pattern in the data.

For example, your statistical hypothesis would be more like 'burrowing owl burrows with mammalian dung have a greater mass of insects than those without dung'; the **null statistical hypothesis** to this would be 'burrowing owl burrows with mammalian dung have the same mass of insects as those without dung'.

So, a statistical hypothesis usually equates more with a prediction generated from a research hypothesis rather than the research hypothesis itself.

While it is generally useful to distinguish a research hypothesis from a statistical hypothesis to help you understand what is going on, in practice we rarely do this explicitly. In other words, you rarely see the phrases 'research hypothesis' or 'statistical hypothesis' being used, but simply the word 'hypothesis'. In practice, then, you need to determine what type of hypothesis you are working with from the context in which it is being used. However, you can typically take 'hypothesis' to mean 'research hypothesis' and 'null hypothesis' to mean 'statistical null hypothesis'.

1.6 How do theories and models differ?

So far in this topic we have looked at the role of thinking and reasoning in science, the use of hypotheses, and the importance of hypothesis testing. Now let us consider two parts of the scientific intellectual landscape we use to identify and summarize our thinking: the theory and the model.

Declaring something in science to be a **theory** is a big deal! Scientific theories are comprehensive explanations of the natural world and are backed up by lots and lots of evidence. Well-known theories in biology are the **theory of evolution by natural selection**, the **germ theory of disease**, and the **cell theory of organism structure**. This use of the term 'theory' is very different (opposite, in fact) from the way it is used in everyday language—namely, to indicate a hunch or a guess. For example, 'I have a theory she won't be coming to the party' is speculative, untested, and relevant to a specific situation. So, it is more like a research hypothesis, and not like a scientific theory at all!

The paradigm shift

Theories (the scientific sort) are formulated using inductive reasoning but are assessed by testing hypotheses (the research sort), a process that uses *deductive* reasoning. The assessment of a theory can lead to it being modified or even (very rarely) to it being replaced. Making a major modification to a theory, or replacing it entirely, is an even bigger deal than saying something is a theory! When this happens, it is called a **paradigm shift**.

For example, in the late 1800s the germ theory of disease replaced the miasma theory of disease. The miasma theory put forward that disease was caused by 'bad air'. By contrast, the germ theory states that disease is caused by micro-organisms. The germ theory accounts for observed patterns of cases of all infectious diseases, whereas the miasma theory did not. Consider cholera, a disease caused by micro-organisms that are transmitted in water. Not only did germ theory explain the pattern of observed cases, but it enabled the identification of actions that could be taken to stop its spread—namely, the provision of clean drinking water.

What is a model?

The term **model** is also used in different ways, both scientifically and non-scientifically. In science, the word 'model' is used to denote a representation of something else; look at Figure T1.4 to see some examples.

A model is used to simplify and make visible something that would otherwise not be because it is too small, too large, too abstract, too distant in time, too distant in space, or too complex. By contrast, the word 'model' is more often used in a non-science context with reference to someone who might inspire you to do or want something (for example, a role model, fashion model, or life model).

Models are used in science to organize thoughts, communicate ideas, analyse data, and make predictions. To build a model, or carry out modelling, you must work out how to represent what you are thinking and recognize any assumptions that you make in the process.

Models can represent theories or, more often, specific *aspects* of a theory. As such, the two words are often used interchangeably. However, models typically equate most closely to hypotheses than they do to theories because they represent more speculative explanations for specific situations.

Take, for example, the theory of evolution by natural selection and the 'out of Africa' model of human evolution. The theory of evolution by natural selection is the scientifically accepted explanation of how biological systems change over time. By contrast, the 'out of Africa' model of human evolution—namely, that modern humans first appeared in Africa and then spread around the world, in so doing replacing all other hominid species—describes how this general theory *might* have applied to humans and is only one of several explanations under consideration. (Others include the multiregional evolution model and the assimilation model.)

1.7 What are mathematical models?

In science, we use models to represent what we see in the world around us, enabling us to make predictions. These models can take a variety of forms: physical, graphical, pictorial, mathematical, computational, or word form. Mathematical models, including statistical models and computational models, are widely used in biology.

Mathematical models use numbers and symbols to represent elements of the real world. For example, the equation d = st is a simple mathematical model that describes how the distance travelled by an object (represented by the symbol d) can be predicted by multiplying its speed (the symbol s) by the time elapsed (the symbol t). Using this model we can predict a horse walking at a speed of 5 kilometres per hour for half an hour will travel 2.5 km:

$$d = st$$
$$d = 5\,km/h \times 0.5\,h = 2.5\,km$$

This assumes that the horse is travelling at a constant speed or that the speed quoted is an average speed. We would require a more complex model if we wanted to take into account changes in speed.

A **statistical model** is a type of mathematical model that incorporates probability. It is used to describe more general situations. For example, we could use a statistical model that incorporates the general walking behaviour of horses and how individuals tend to vary from this general behaviour to model how far any one horse is likely to travel in half an hour.

▶ **We learn more about variability in Quantitative Toolkit 3.**

Statistical models are particularly used in data analyses and have especial importance when it comes to stopping us succumbing to what is called cluster illusionary bias (see Table T1.2). When computers are used to generate the output of a mathematical or statistical model, they are typically then called **computational models**, especially if they are being used to study complex systems. We see an example of the output of such a model in Figure T1.5.

Mathematical models, especially statistical ones, are particularly useful because you can **run** (calculate) them with inputs that have been chosen to reflect different scenarios and compare the outputs.

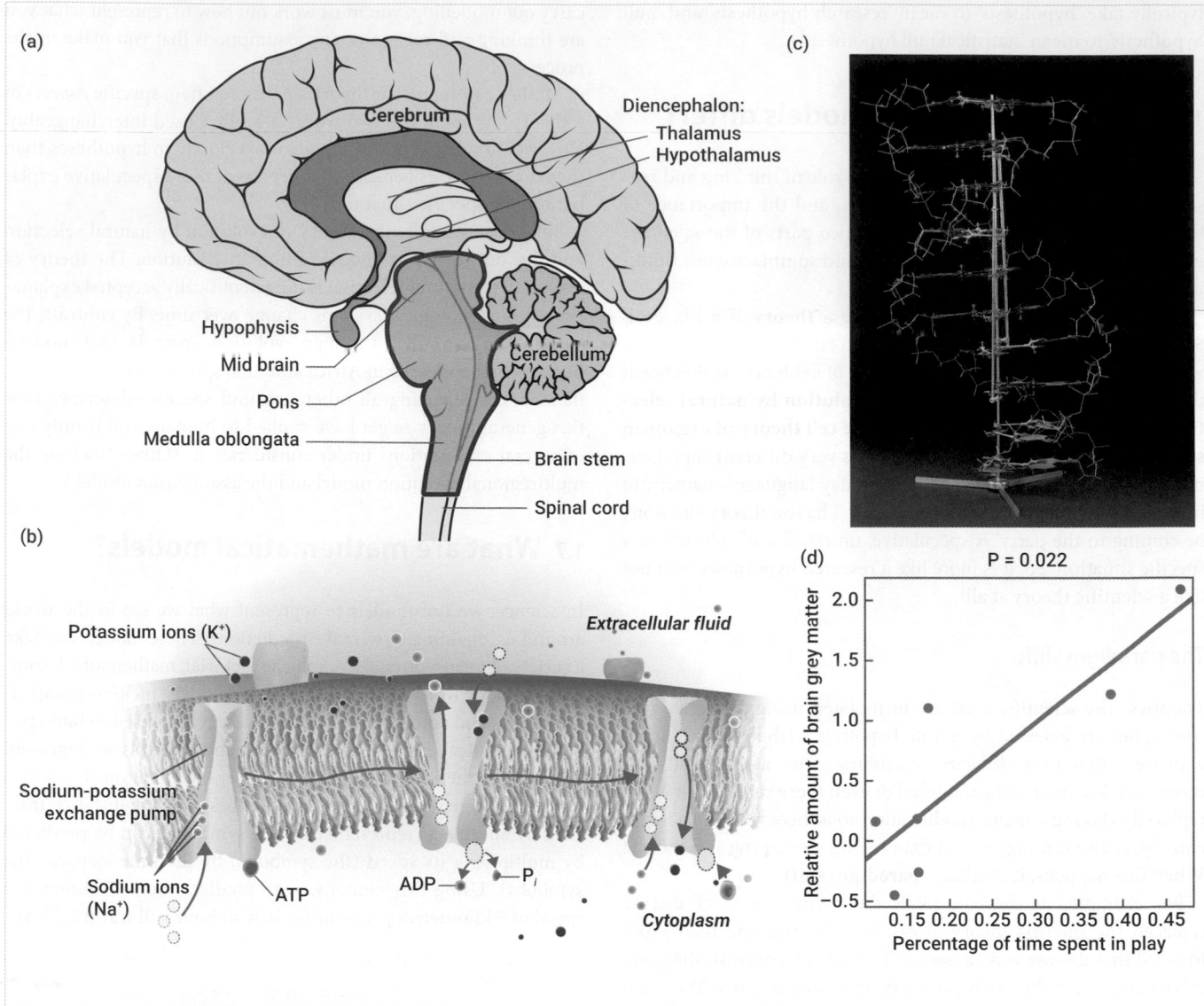

Figure T1.4 Examples of different models used to explore biological science. (a) A physical model of a human brain used in teaching brain anatomy. Labels show the main structures and much of the detail is not shown. (b) A pictorial model summarizing how active transport works. It is simplified to show only the membrane phospholipids and the pump. All the other components of the cell membrane have been omitted. (c) A physical model used to work out the double helical structure of DNA. The relative size and orientation of the components of DNA were known and piecing them together in this model showed that this arrangement is a good fit. (d) A statistical model (represented by the line on the graph) of the relationship between time spent playing and volume of brain grey matter in primates. The authors have assumed a straight-line relationship between brain grey matter volume and time spent in play.

Source: (a) Belfaqih Biology Matura book, CC BY-SA 4.0<https://creativecommons.org/licenses/by-sa/4.0>, via Wikimedia Commons; (b) Wikimedia Commons; (c) © Science Museum Group; (d) Kerney, M., Smaers, J. B., Schoenemann, P. T. et al. The coevolution of play and the cortico-cerebellar system in primates. Primates 58, 485–491 (2017). https://doi.org/10.1007/s10329-017-0615-x

(Inputs are values you put into a model and outputs are the values the model generates based on these inputs.)

The outputs of complex computational models are called **simulations**. Simulations can be used to assess the validity of models, following a similar process to hypothesis testing. However, they can also be used in an applied way to plan for different scenarios. Consider the work of a public health scientist, for example: a computational model of disease spread incorporates the number of individuals currently infected and the rate of infection as inputs, and a graph of the number of people infected over time as the simulation output. The model is then run for different intervention scenarios and the simulations compared to evaluate alternative control strategies.

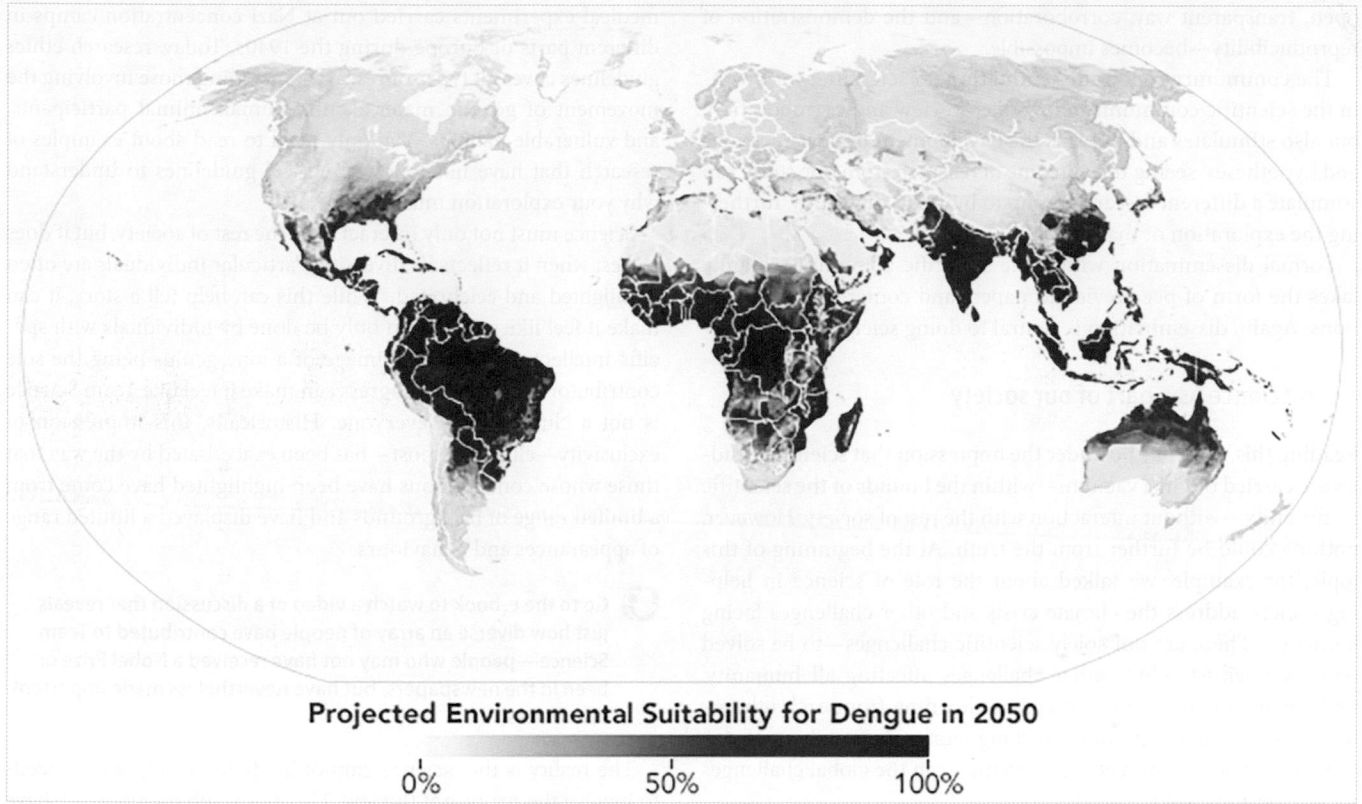

Projected Environmental Suitability for Dengue in 2050

0%   50%   100%

Figure T1.5 Models can be used to predict the occurrence of disease in the future. Dengue fever is caused by a virus carried by mosquitoes. It affects more people than any other arboviral disease. Over the last half century cases have increased 30-fold because of the increasing human population, globalization, and global warming. This map shows the output of a mathematical model designed to help predict the occurrence of dengue fever in 2050, by which time warmer temperatures will have led to a further expansion of the habitat suitable for its mosquito vector.

Source: NASA Earth Observatory map by Lauren Dauphin based on data from Janey Messina, University of Oxford.

1.8 Team Science

The key mental activities we have talked about in the preceding sections all go on inside the heads of scientists. But this does not mean that scientists operate alone. These mental activities must be accompanied by some crucial social activities. In this final section we consider some of the ways in which science involves scientists interacting with each other and with society at large. You should find this social side of science reassuring. While an awareness and appreciation of all the activities that go into science is crucial if you want to pursue a career in biological science, you do not have to be an expert in all the activities yourself: you just need to be able to make your contribution to Team Science!

Peer review involves scientists scrutinizing, or critically evaluating, the work of other scientists. Peer review is not only a standard part of the publication process of scientific papers, but it also occurs whenever scientists are considering the work of other scientists, be it over coffee within an academic or industrial organization, or at a conference.

We have already mentioned the importance of scientists critically evaluating their own reasoning and the reasoning of others. However, critical evaluation should be applied to *all* aspects of the scientific process, whether intellectual or practical. While there is no doubt that taking a scientific approach is very powerful, it does not a guarantee that any one individual study will come up with valid answers. Critical evaluation plays a crucial part in helping us to make sure that science is done as well as possible and that any limitations are made clear.

An important element of validating and evaluating someone's work is to be able to corroborate it. **Corroboration** is the process of repeating the work of others to see if you get the same results. When different researchers explore the same question, using the same method, and get the same answer, then the study is said to have **reproducibility**. Reproducibility helps to ensure that evidence is robust and it is central to good science.

Researchers are increasingly making their data publicly available online, which is massively increasing the opportunities for corroboration, and therefore for reproducibility to be demonstrated. Without the outcomes of a study being released in an open, transparent way, corroboration—and the demonstration of reproducibility—becomes impossible.

The communication, or **dissemination**, of scientific work within the scientific community allows peer review and corroboration, but also stimulates and informs the development of new questions and hypotheses: seeing the outcome of research from one study can stimulate a different research group to build on that study, furthering the exploration of that topic.

Formal dissemination within the scientific community mainly takes the form of peer-reviewed papers and conference presentations. Again, dissemination is central to doing science.

Team Science as a part of our society

Reading this, you may be under the impression that scientific studies are carried out in a vacuum—within the bounds of the scientific community—without interaction with the rest of society. However, nothing could be further from the truth. At the beginning of this topic, for example, we talked about the role of science in helping society address the climate crisis and other challenges facing humanity. These are not solely scientific challenges—to be solved only by scientists—but rather challenges affecting all humanity. Only by interacting, cooperating, and learning from each other—by our scientific community working alongside members of our economic, political, and ethical systems—can the global challenges we face be addressed.

In this vein, every scientist must understand, and conform to, **health and safety** and **research ethics** guidelines. The first international code of ethics for research on human subjects was the 1947 Nuremberg Code, something that emerged from the Nuremberg war crimes tribunal following revelations about the numerous

medical experiments carried out at Nazi concentration camps in different parts of Europe during the 1940s. Today, research ethics guidelines cover all types of research, including those involving the movement of genetic material, non-human animal participants, and vulnerable habitats. You only need to read about examples of research that have not followed ethical guidelines to understand why your exploration must do so.

Science must not only interact with the rest of society, but it does so best when it reflects its diversity. Particular individuals are often highlighted and celebrated. While this can help tell a story, it can make it feel like science can only be done by individuals with specific intellectual skills. The image of a lone genius being the sole contributor to scientific progress can make it feel like Team Science is not a club open to everyone. Historically, this impression of exclusivity—elitism, almost—has been exacerbated by the way that those whose contributions have been highlighted have come from a limited range of backgrounds and have displayed a limited range of appearances and behaviours.

 Go to the e-book to watch a video of a discussion that reveals just how diverse an array of people have contributed to Team Science—people who may not have received a Nobel Prize or been in the newspapers, but have nevertheless made important discoveries.

The reality is that science cannot be done in isolation: it needs to involve the many, not the few. The more club members we have contributing to the mental, social, practical, and other activities required to explore the science of life—both in number and diversity of members—the more likely we are to be successful in our attempts to better understand life on Earth and face up to both the huge challenges and great opportunities that lie ahead.

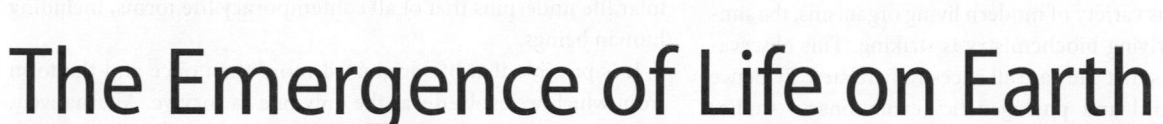

The Emergence of Life on Earth

Introduction

Scientists have been trying to understand how life started for centuries. Progress has been made and there are now many ideas, but it remains a difficult problem to explore and there is little consensus. We know of only one place in the Universe where life exists—Earth—and no one has succeeded in creating it artificially in a laboratory. Despite the continued exploration of other planets, life as we witness it on Earth may yet prove to be a rare or even unique phenomenon. There is also uncertainty about how evidence of its existence elsewhere might present itself.

As scientists try to understand the origins of biological systems, they also seek new evidence about the early history of Earth, the Solar System and the wider Universe. Such evidence can be hard to come by. Information about the origin of the Solar System lies in outer space. Earth's dynamic geology, in which rock evidence is continually re-melted and destroyed as land masses move, limits our knowledge of its ancient past.

Although we lack a lot of direct evidence about the history of life on Earth, we can explore the origin of life by focusing on where its constituents may have come from. By understanding the structure and behaviour of organic compounds, we can investigate how life's central characteristics, such as metabolism and replication, may have emerged.

This topic gives a brief introduction to what we currently know about how life emerged, and how our understanding has developed.

Contents

Introduction	15
2.1 The emergence of life	16
2.2 How did Earth form?	18
2.3 Life is based on aqueous organic chemistry	22
2.4 The origins of organic molecules	22
2.5 Complex biomolecules	25
2.6 From biomolecules to life forms	28

2.1 **The emergence of life**

All modern living organisms are cellular, as depicted in Figure T2.1. Cells are composed of discrete containers bounded by membranes. They have a high water content, creating complex solutions of metabolites, ions, and macromolecules, and their biochemical processes take place in aqueous or other fluid environments. The biochemistry of life is based on carbon.

These characteristics give us some ideas about the locations where life may have originated, and when this might have occurred.

Ancestry

Despite the enormous variety of modern living organisms, the similarity of their underlying biochemistry is striking. This observation strongly suggests that they are all ancestrally related. Evidence from the fossil record and phylogenetic (evolutionary) studies supports this notion and allows the construction of a so-called 'universal tree of life', as shown in Figure T2.2.

All modern organisms are believed to be derived from a prokaryotic (bacterium-like) organism, labelled the last universal common ancestor (LUCA). We cannot be sure what LUCA looked like or where it existed but several lines of evidence suggest that it was cellular and replicated using familiar molecular processes involving RNA and DNA.

Scientists are unsure where to place LUCA in relation to modern organisms and disagree over which, if any, modern organisms

resemble it. Some phylogenetic (ancestral) analyses suggest that the heat-loving hyperthermophilic archaeobacteria are the oldest organisms on Earth. If this is the case, it may well be that the environments these organisms inhabit today mirror the earliest environments for life and represent likely sites for the origin of life.

Equally unclear are the processes that occurred before the emergence of the LUCA as a discrete living organism. Whatever form 'life' took before that of the conceptual 'LUCA', it may have taken the LUCA several billion years to become established. The logic of ancestry (that parents have children that resemble them) suggests that life is theoretically traceable back to a single origin event (or to a cluster of closely connected events in which energy and information were exchanged), and that the emergence of cellular life underpins that of all contemporary life forms, including human beings.

It is possible that life originated more than once and the form from which we evolved was the only one to survive. Alternatively, two or more early forms may have fused. Either way, there was a single result in terms of the common biochemical and cellular processes we see around us in modern organisms.

Antiquity

Evidence for ancient life on Earth was once restricted to fossils visible to the naked eye by animals and plants. The oldest of these allowed scientists to demarcate the Cambrian period (~550 million years ago). Micropalaeontologists, who study evidence of microbial

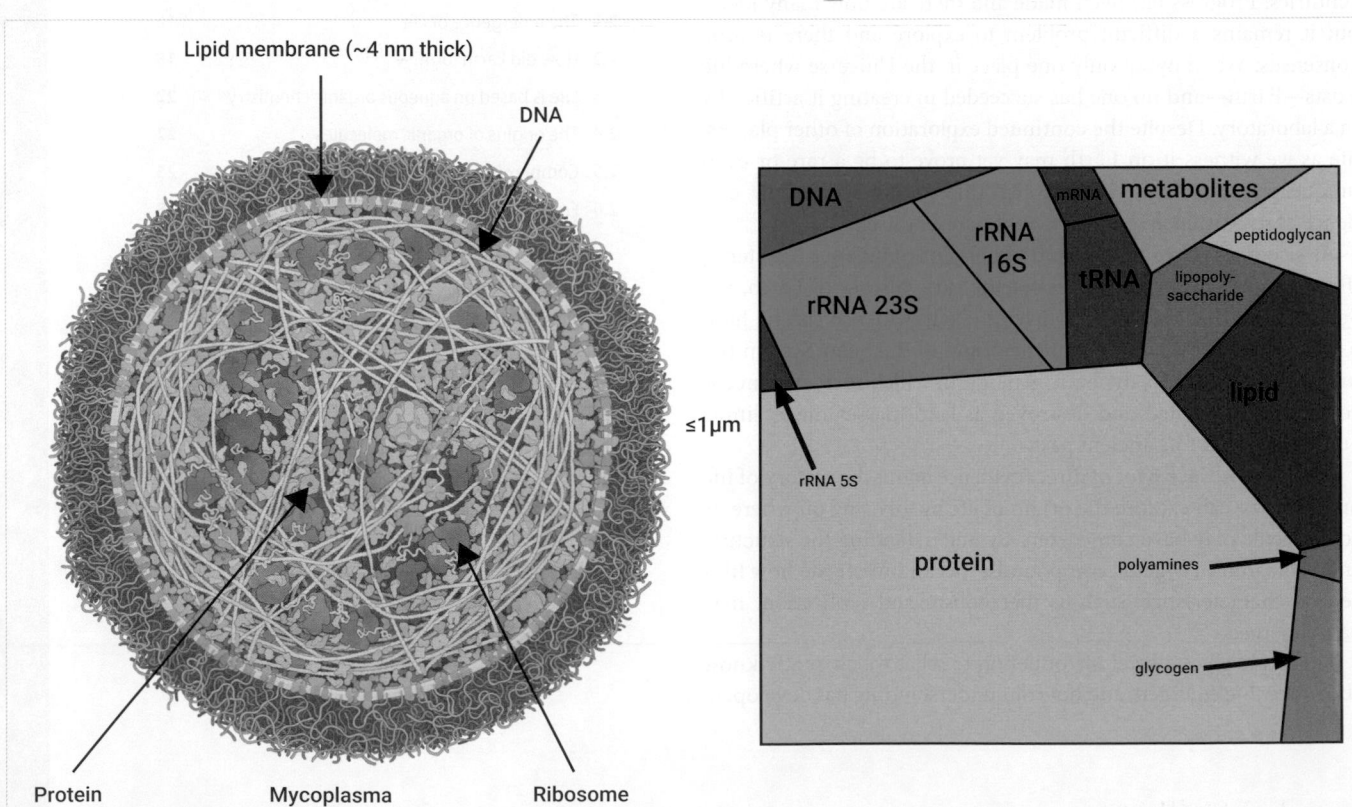

Figure T2.1 The cellular nature of living organisms. Left: The layout of a prokaryotic cell, showing the cell membrane and the relative distribution and sizes of major macromolecules in the cytosol. Right: the relative composition of the dry weight of a bacterial cell. Cells are additionally typically ~ 70% water by mass.

Sources: Left: illustration by David S. Goodsell, the Scripps Research Institute; right: data from http://bionumbers.hms.harvard.edu

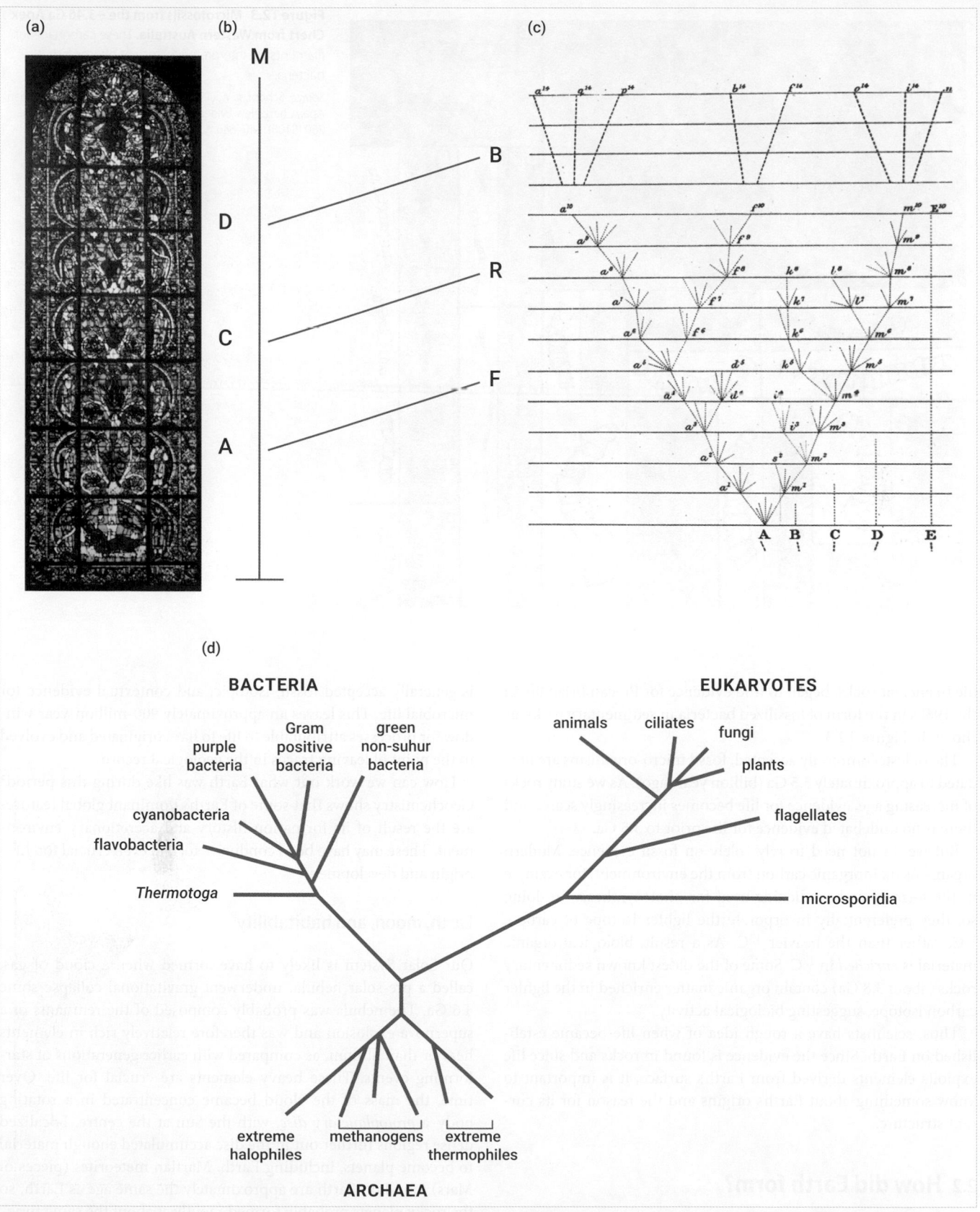

Figure T2.2 Trees of life. (a) Oldest known representation of a 'family tree', a Tree of Jesse, depicting the genealogy of Christ from the Cathedral at Chartres, 1145 AD. (b) Phylogenetic tree from Chambers' *Vestiges of the Natural History of Creation* (1844). (c) Phylogenetic tree from Darwin's *Origin of Species* (1859). (d) A modern rRNA-based phylogenetic tree.

Sources: (a) Alamy; (b) Reproduced from Chambers, *Vestiges of the Natural History of Creation*. London: John Churchill 1844; (c) Darwin, C. *On the Origin of Species* (1859); The Picture Art Collection/Alamy Stock Photo; (d) Woese, *Microbiological Reviews*, 1987, 51, 221–271 (p. 231). Licensed from PLS Clear.

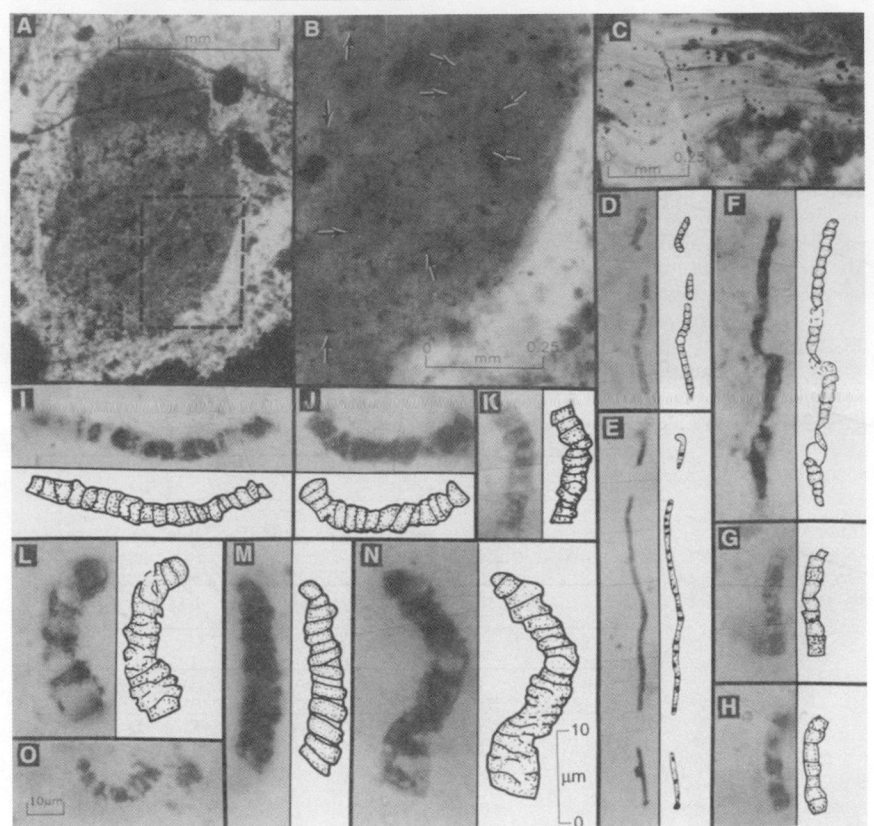

Figure T2.3 Microfossils from the ~3.46 Ga Apex Chert from Western Australia. These carbonaceous filaments are interpreted as ancient photosynthetic bacteria.

Source: Schopf, J . W. (1993). Microfossils of the Early Archean Apex Chert: New Evidence of the Antiquity of Life. Science, 260 (5108). 640–646.

life in ancient rocks, began to find evidence for Precambrian life in the 1960s in the form of fossilized bacteria in sedimentary rocks, as shown in Figure T2.3.

The oldest, commonly accepted, fossil micro-organisms are now dated to approximately 3.5 Ga (billion years ago). As we study rocks of increasing age, evidence for life becomes increasingly scarce and there is no undebated evidence for life prior to 3.5 Ga.

But we do not need to rely solely on fossil evidence. Modern organisms fix inorganic carbon from the environment, for example in the form of carbon dioxide used for photosynthesis. In doing so, they preferentially incorporate the lighter isotope of carbon, ^{12}C, rather than the heavier ^{13}C. As a result, biological organic material is *enriched* in ^{12}C. Some of the oldest known sedimentary rocks (about 3.8 Ga) contain organic matter enriched in the lighter carbon isotope, suggesting biological activity.

Thus, scientists have a rough idea of when life became established on Earth. Since the evidence is found in rocks and since life exploits elements derived from Earth's surface, it is important to know something about Earth's origins and the reason for its current structure.

2.2 How did Earth form?

Radiometric dating gives Earth's approximate age as 4.57 Ga. Life probably emerged sometime between 4.4 Ga, when the surface was cool enough to support liquid water, and 3.5 Ga, when there

is generally accepted fossil, isotopic, and contextual evidence for microbial life. This leaves an approximately 900-million-year window for processes attributable to life to have originated and evolved to the point of leaving traces in the geological record.

How can we work out what Earth was like during this period? Geochemistry shows that some of Earth's dominant global features are the result of its formation history and accretionary environment. These may have been conducive to, and even crucial for, life's origin and development.

Earth, moon, and habitability

Our Solar System is likely to have formed when a cloud of gas, called a pre-solar nebula, underwent gravitational collapse some 4.6 Ga. The nebula was probably composed of the remnants of a supernova explosion and was therefore relatively rich in elements heavier than lithium, as compared with earlier generations of star-forming events. These heavy elements are crucial for life. Over time, the mass of the cloud became concentrated in a rotating body, a *protoplanetary disc*, with the Sun at the centre. Localized dense regions further out in the disc accumulated enough material to become planets, including Earth. Martian meteorites (pieces of Mars) found on Earth are approximately the same age as Earth, so the rocky planets probably formed rapidly at about the same time.

The moon has been dated to approximately 4.51 Ga, so it seems to have formed within a few tens of millions of years of Earth. The Earth–moon system probably arose from the collision of a hypothetical roughly Mars-sized body (which scientists call Theia) with the proto-Earth. Once formed, the rotation of Earth and orbital distance of the moon became intimately linked through tidal forces. Earth's rotation has been slowing ever since, while the distance to the moon has been increasing.

An Earth day at 4.5–4 Ga would have been shorter than at present, perhaps only six hours long, and the moon would have been considerably closer, leading to massive and more frequent terrestrial oceanic tides. The moon-forming impact was also sufficiently violent to melt Earth's entire surface. This rules out a pre-lunar impact origin of life.

The moon helps to keep Earth habitable by maintaining its long-term orientation with respect to the Sun, for example by keeping it rotating roughly in the plane of Earth's orbit. Our moon is thus something like a gyroscope, keeping Earth oriented towards the Sun in a regular fashion. Earth has two other special features that have helped to make it habitable over long periods of time: an active magnetic field and active plate tectonics.

Earth's internal structure

In addition to the moon-forming impact event, early Earth sustained the impact of many protoplanetary bodies rich in heavy elements. The immense heat generated from collisions and radionuclide decay melted the solid material into a magma ocean. The molten state caused a differentiation in which the heaviest elements (e.g. nickel and iron) sank towards the centre to form the core, and lighter elements (e.g. oxygen, aluminium, and silicon) migrated towards the surface to form the mantle and crust, a process illustrated in Figure T2.4.

During this process various minor elements segregated according to their affinities for either the more metal-rich (siderophilic = 'iron-loving') deeper regions or the more silicate-rich (lithophilic = 'rock-loving') surface ones. However, there is an anomalous

concentration of highly siderophilic elements in Earth's crust: these must have arrived after the differentiation, otherwise they would have been sequestered in the mantle and core.

Earth's magnetic field probably also arose from this differentiation process. The nickel–iron core rotates at a different velocity from the outer layers, creating Earth's geodynamo. Earth's magnetic field is thought to have been operating continuously since at least 3.5 Ga. This field is essential for deflecting energetic particles from the solar wind away from Earth; these particles would otherwise slowly strip away the atmosphere.

Earth's magnetic field is notably stronger than that of any other rocky planet in the Solar System. Without it, Earth's surface would probably be an unsuitable environment for long-term biological evolution. However, the notion of a magnetic field being essential for life to emerge is still a matter of debate: some claim that if the magnetic field was absent, extra-energetic particles reacting with the atmosphere might still have encouraged the production of organic molecules.

Plate tectonics is a relatively modern theory that explains Earth's surface composition well. The basic model is that as rocks melt they behave like fluids, and form new mineral compositions. The lighter mineral compositions float to the surface, and stop mixing with the overall system. The creation and movement of these large components of Earth's crust, which make up its surface, may have been needed to produce a stable long-term environment for life. Plate movement is largely driven by radioactive decay in Earth's interior causing heat flow and the convective movement of mantle and crustal rocks. Figure T2.5(a) shows how, as new rocky material is brought to the surface by volcanism and seafloor spreading, older rocks are pulled back into the mantle, where they are melted and recycled. Often, a portion of this material remains buoyant and stays at the surface, forming continental crust and the land on which most of us live.

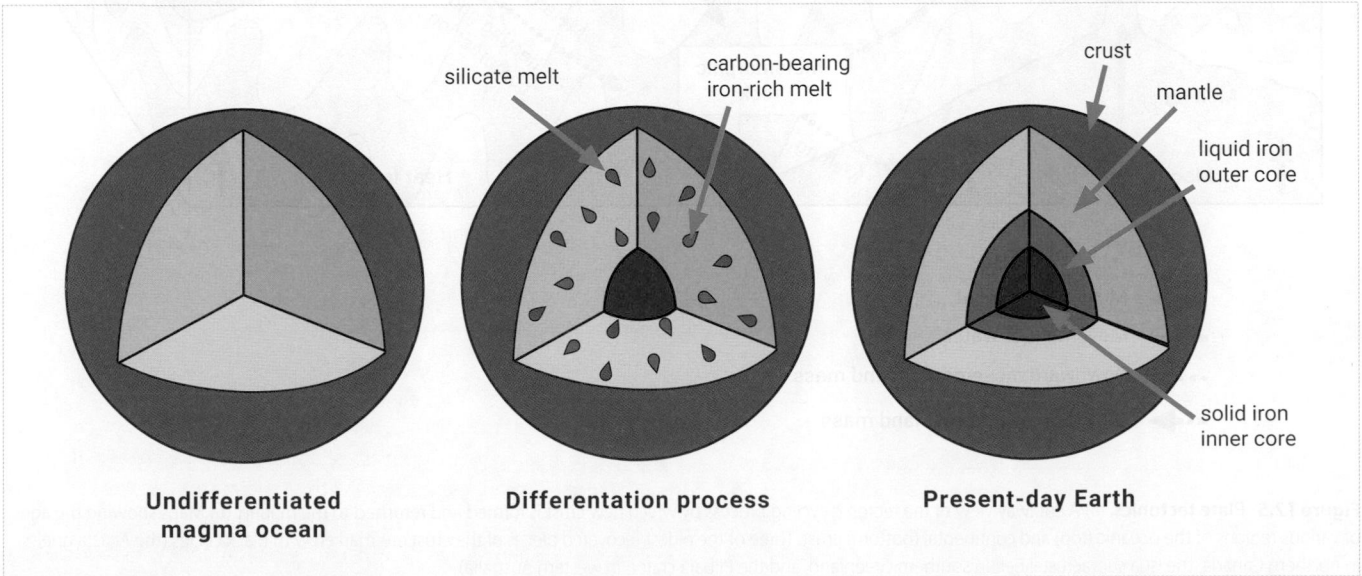

Figure T2.4 The differentiation of Earth. From left to right, the process by which Earth differentiated into its modern layers.
Source: Natalia Solomatova.

It is unclear when this process started and how the volume and area of continental surface has evolved over time, but Earth is the only rocky planet in the Solar System with active and regular tectonic cycling. Plate tectonics may have helped stabilize Earth's surface temperature over geological timescales, but whether it was necessary for life to emerge is still debated.

Presently, the oldest oceanic crust is only about 250 Ma old, with the majority much younger. In contrast, *continental crust* is often much older. In general, the abundance of continental crustal material decreases as it gets older. The oldest crust is found in northern Canada, Greenland, and Western Australia (Figure T2.5b).

Impacts on early Earth

As Earth's surface is so dynamic, little direct evidence remains of its early history, and geochemists must rely on proxies for clues. The moon is smaller and tectonically inactive, so it preserves a record of the early Solar System environment in the form of numerous impact craters. Earth is 81 times bigger than the moon and its gravity would have attracted many more impactors. Collisions would have been intense between 4 and 3.8 Ga and the oceans may have been repeatedly sterilized for long periods of time by large asteroid impacts.

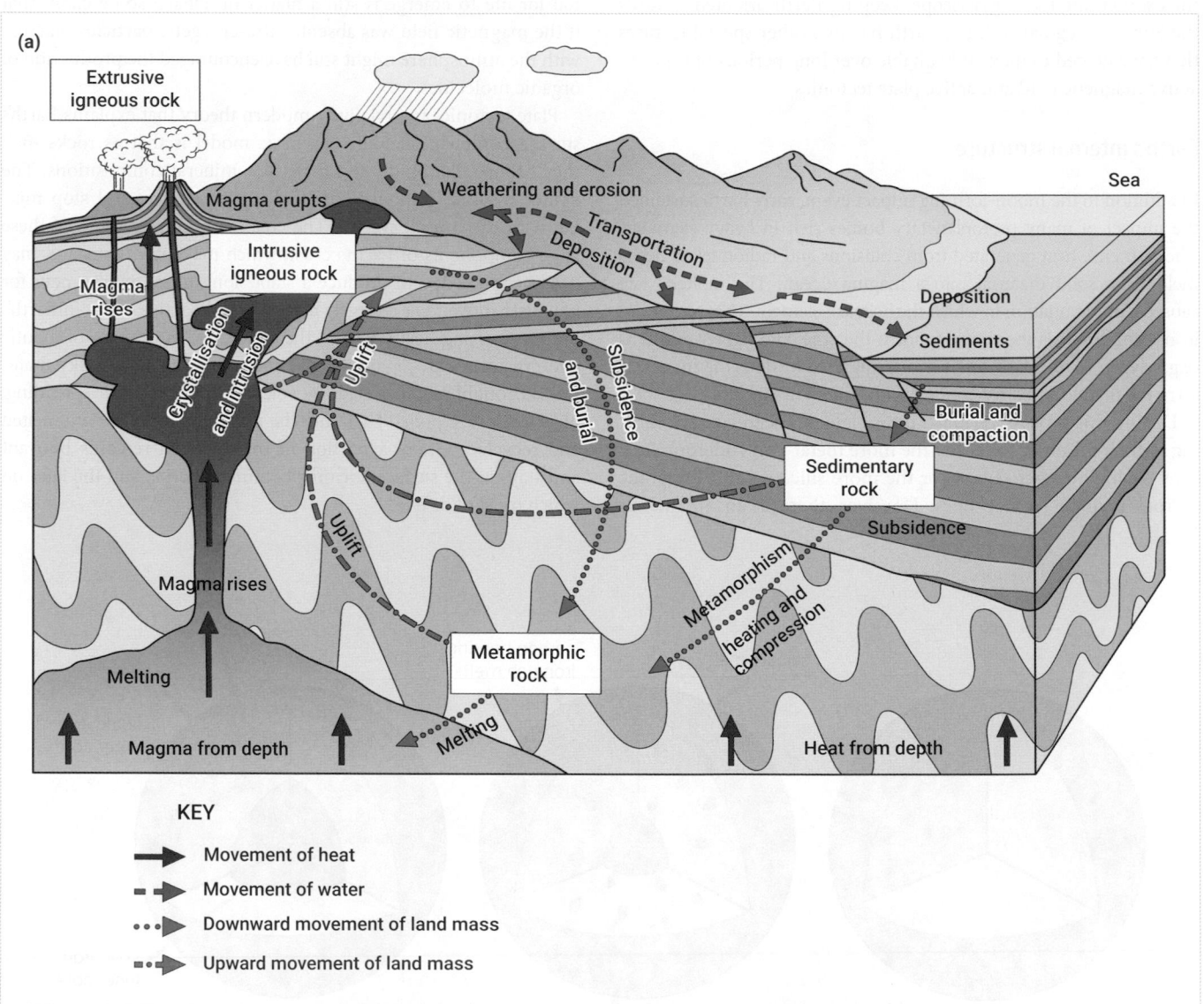

Figure T2.5 Plate tectonics. (a) A cutaway view of the tectonic cycling process by which new crust is formed and returned to the mantle. (b) Maps showing the ages of various regions of the oceanic (top) and continental (bottom) crust. Three of the oldest recovered pieces of the crust are marked with black circles (the Acasta gneiss in northern Canada, the Isua supracrustal belt in southern Greenland, and the Pilbara craton in western Australia).

Sources: (a) 'An introduction to geology', An OpenLearn chunk used/reworked by permission of The Open University copyright © 2021. www.open.edu/openlearn; (b, top) © Mr. Elliot Lim and Mr. Jesse Varner, CIRES & NOAA/NCEI; (b, bottom) Credit: Professor Irina M. Artemieva.

Figure T2.5 (Continued)

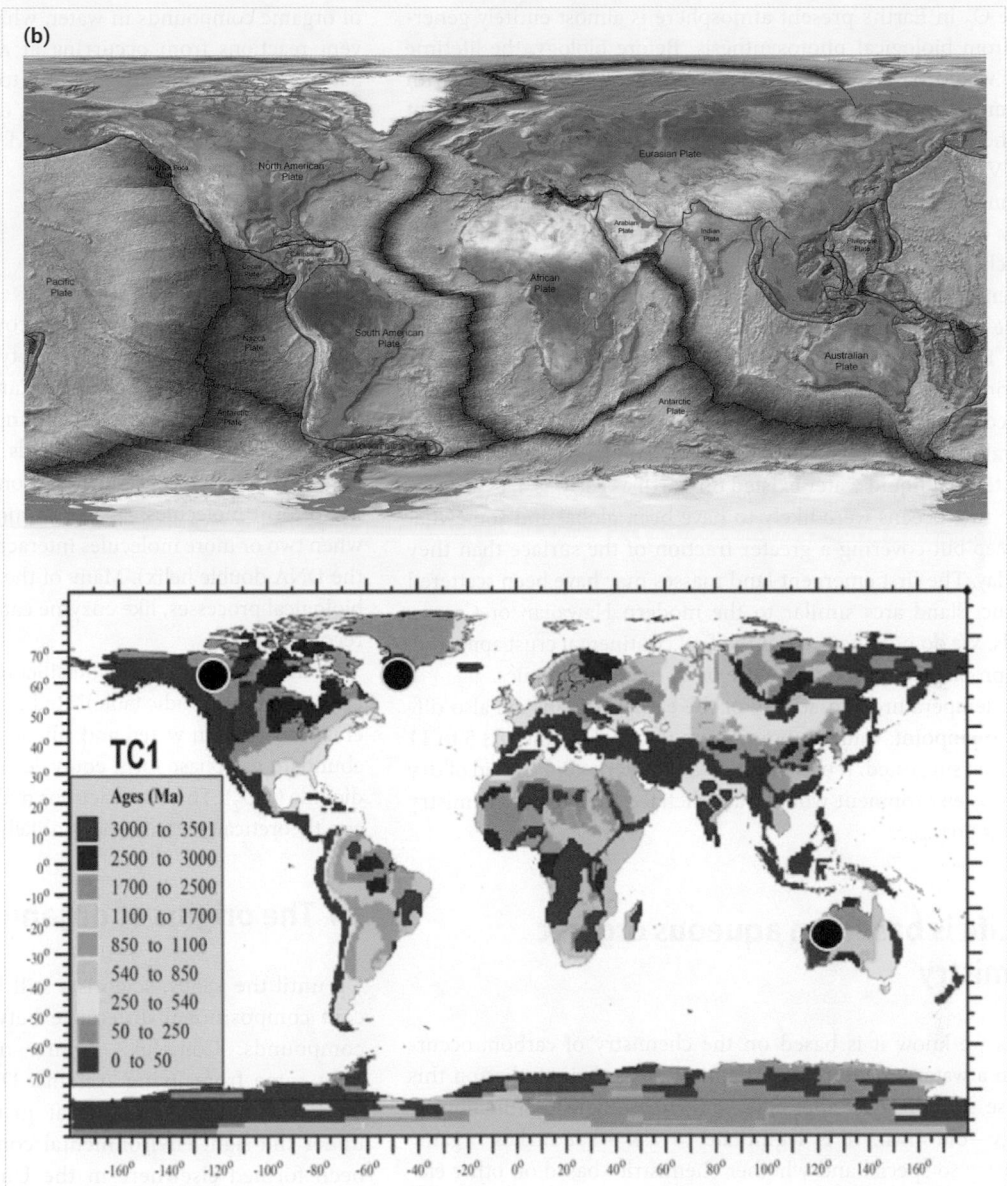

Thus, even if life had arisen repeatedly before 3.8 Ga, only the most recent occurrence could be the parent of the modern biosphere, according to this model. Using the earliest date of 3.8 Ga for the origin of life still leaves a period of about 300 million years for life to emerge. However, there is considerable debate about these arguments.

Early oceans and the atmosphere

If life required an aqueous environment to emerge, and if a large fraction of Earth's early organic molecules were derived from the atmosphere, we need to understand the origin and evolution of Earth's oceans and atmosphere.

In our Solar System, planetary atmospheres vary in their chemical characteristics. Gases like CO_2 and N_2 are oxidized, so atmospheres composed predominantly of these gases are referred to as *oxidizing*. Gases like H_2, NH_3, and CH_4 are reduced and form *reducing* atmospheres. Oxidizing atmospheres include those of Earth, Mars, and Venus, while those of Saturn and Jupiter are highly *reducing*. The redox balance of atmospheres affects the efficiency with which organic compounds can be synthesized by abiotic environmental processes.

Earth's differentiation had a huge effect on its atmosphere. The earliest atmosphere is likely to have formed from gases exiting the early mantle. During differentiation and outgassing, elements such as carbon, nitrogen, and hydrogen rose toward the surface, with their oxidation state determined by the conditions they were exposed to during migration. It is possible that the first atmosphere was reducing (containing abundant CH_4 and CO) when the mantle still contained metallic iron, but it would ultimately have become oxidizing (composed mostly of CO_2 and N_2) as the iron sank to the core.

The O_2 in Earth's present atmosphere is almost entirely generated from biological photosynthesis. Before biology, the lifetime of O_2 would have been extremely short due to its reaction with iron and other elements in the crust. The lack of O_2 also meant that there was little UV-absorbing ozone (O_3), allowing ultraviolet (UV) radiation (which was produced in greater proportions at that time) from the Sun to reach Earth's surface. UV radiation is highly energetic and can break strong chemical bonds and promote chemical reactions.

Scientists are unsure where Earth's water came from, but most believe it was already here by the time of the moon-forming event. Primitive Earth's water probably escaped from the planet's interior as steam, slowly condensing to form surface water as the crust cooled. The presence of zircon silicates (weathering-resistant minerals that are among the oldest known on Earth) on the surface indicates that liquid water existed on Earth as early as 4.4 Ga.

The first oceans were likely to have been global and somewhat less deep but covering a greater fraction of the surface than they do today. The first emergent land masses may have been scattered volcanic island arcs similar to the modern Hawaiian or Canary Islands. We do not know when the first continental crust appeared, but it probably coincided with the onset of plate tectonics.

The temperature and acidity of the earliest oceans are also difficult to pinpoint: values ranging from 0 to 100 °C and pH 5 to 11 have been suggested. If early Earth was a water world devoid of dry land, this environment would have dictated the type of chemistry that was possible.

2.3 Life is based on aqueous organic chemistry

Life as we know it is based on the chemistry of carbon, occurring in a watery medium. Biochemists focus their studies on this because it represents the only known example of functioning biology—our own. It is legitimate to ask what makes organic chemistry so special and whether chemistries based on other elements could also support life.

Why water?

Logic suggests that the complex chemistry necessary for life has to take place in liquid medium. In solids, the diffusion of metabolites occurs too slowly to support significant molecular interactions. Diffusion is much faster in gases, but large molecules are generally non-volatile, making molecular reactions and the creation of stable structures improbable.

Water is among the most abundant liquids in the Universe, remains liquid over a wide temperature and pressure range, and has a high heat capacity. At the atmospheric pressures found on Earth, water has one of the largest liquid stability ranges of any known astrophysically common molecular compound, and can also change into solid (ice) and gas (vapour) phases.

The temperature range over which water remains a liquid is often compatible with the stability and controlled reactivity of organic compounds: high temperatures may cause decomposition of organic compounds in water, while low temperatures may prevent reactions from occurring at all. The lakes of methane on Saturn's moon Titan are too cold to support the sorts of organic reactions we observe in water. So, exploring organic reactivity in water might help us to understand the distribution of life in the Universe.

Why carbon?

Carbon, being able to form four covalent bonds ('tetravalent'), is uniquely able to build up the variety of chemical compounds that allow for diverse biochemistry. Carbon can associate covalently with many other elements (including hydrogen, nitrogen, oxygen, and sulfur) to generate stable bonds, as well as with itself, to give a huge inventory of molecules. Carbon also has the exceptional ability to form molecules capable of molecular recognition, which is when two or more molecules interact to form stable structures (e.g. the DNA double helix). Many of the most important fundamental biological processes, like enzyme catalysis, are based on molecular recognition.

Could life also be based on silicon (which is immediately below carbon in the periodic table)? Reduced silicon compounds are generally unstable in water, and silicon does not have an appreciably abundant gas phase form equivalent to methane (CH_4) or carbon dioxide (CO_2). These restrictions make a metabolism based on silicon theoretically feasible but unlikely.

2.4 The origins of organic molecules

Up until the 1850s, scientists still believed there was a significant compositional difference between organic and inorganic compounds. Complex, organic molecules were thought to only come from living systems. Did they arise spontaneously on Earth from elements that just happened to be available under the right environmental conditions, or could they have been formed elsewhere in the Universe and delivered to the planet?

Delivery from space

It is likely that the arrival of extra-terrestrial material brought carbon-based molecules to early Earth. The comets, meteorites, and interplanetary dust we observe today contain a complex array of organic molecules and the flux of these arriving on Earth's surface was likely higher early in its history.

Comets (Figure T2.6) are mixtures of dust and water ice that formed early in the history of the Solar System. The organic components of several comets have been analysed using spectroscopy. Reactive compounds, including hydrogen cyanide (HCN) and formaldehyde (HCHO), are often abundant. A typical 1-km diameter comet could deliver an amount of HCN comparable to that produced by electric discharges in a reducing atmosphere over the course of an entire year. Experiments also show that the shock pressure and temperature extremes caused by the high velocity entry of extra-terrestrial bodies into Earth's atmosphere

Figure T2.6 Comets. (a) Halley's comet imaged from Earth during its last visit in 1986. (b) Comet 67P/Churyumov–Gerasimenko as imaged from space by the European Space Agency's Rosetta mission. (c) The range of chemical compositions of comets.

Sources: (a) and (b) Halley's Comet: Art Directors & TRIP/Alamy Stock PhotoComet 67P: NASA/ESA/Rosetta/MPS for OSIRIS Team MPS/UPD/LAM/IAA/SSO/INTA/UPM/DASP/IDA; (c) graph © NASA/JPL/SwRI.

can convert their complex organic materials into HCN, among other compounds.

Formation from atmospheric gases

If the initiation of life depended on the *terrestrial* synthesis of organic compounds, we need to understand the source and nature of the molecules that would have been produced. Scientists have explored how the environment, particularly the primordial atmosphere, may have generated them from simple chemical reactions.

HCN is produced more abundantly in relatively reducing (CH_4 or CO, NH_3) than oxidizing (CO_2, N_2) atmospheres, and, in general, it is chemically easier to make organic compounds from reduced gases than from oxidized ones.

Starting in the 1950s, scientists began to test reactions that could produce organic molecules under conditions thought to simulate those present when life emerged. In 1953, an American graduate student named Stanley Miller demonstrated the synthesis of some organic compounds, starting with a reducing gas

mixture comprising CH_4, NH_3, and H_2. (Note that molecular oxygen (O_2), which makes up ~ 20% of Earth's modern atmosphere, was absent.) At the time, CH_4, NH_3, and H_2 were considered to have been components of early Earth's atmosphere. Look at Figure T2.7, which depicts the experiments Miller conducted. Miller passed electrical discharges, mimicking lightning, through this simulated early atmosphere. Water was recirculated as steam and condensed into a flask to simulate rain falling into primitive oceans. Organic molecules would thus 'rain out' and collect in the water flask shown.

After a few days, a complex mixture of organic molecules formed, including several amino acids of biological importance. This experiment is generally considered to be the first successful example of intentional 'prebiotic chemistry' (although Nobel Laureate Melvin Calvin, who coined the term 'chemical evolution', was actively publishing during this time as well). Later studies extended Miller's work and examined how these molecules could self-assemble into more complex compounds relevant to biochemistry.

Figure T2.7 Stanley Miller with his now-famous experimental apparatus.
Source: Stanley Miller Papers, Special Collections & Archives, UC San Diego.

Geochemists now believe, as some did at the time of Miller's experiments, that the early atmosphere is more likely to have been neutral or oxidizing, composed mainly of CO_2 and N_2. Recent experiments also suggest that amino acid yields from neutral atmospheres may be higher than previously thought. Nevertheless, it remains possible that Earth passed through an early stage during which it hosted a reducing atmosphere.

Synthesis in submarine hydrothermal vents

Submarine hydrothermal vents occur at locations where cold sea water comes into contact with hot magma from Earth's interior. They occur at sites called mid-ocean ridges that form the emergent boundaries of tectonic plates in Earth's crust. Geysers and hot springs are comparable structures on land, although the causes of their formation are subtly distinct. The whole ocean circulates through submarine hydrothermal vents approximately every 10^7 years. This circulation process is likely to have begun early in Earth's history when heat flow from the planet's interior was greater than it is now.

The temperatures in hydrothermal vents, such as the one depicted in Figure T2.8a, range from about 4 to 350 °C, with pH values ranging from 0 to 11. Various minerals precipitate as the heated vent water enters the surrounding ocean, leading to the formation of some remarkable rock formations, such as the carbonate chimney shown in Figure T2.8b.

It has been proposed that organic compounds could form in vents as seawater passes through them, and also that conditions would be suitable for such compounds to self-assemble and polymerize into larger molecules. Some vents in active tectonic areas are reducing environments rich in H_2, H_2S, CO, CO_2, and CH_4 and host potentially catalytic minerals.

Unfortunately, it is difficult to test this hypothesis directly by measuring modern vent chemistry because any organic material detected could simply be environmentally processed biological material.

Synthesis on land

Research continues into hydrothermal vent scenarios at the bottom of oceans, but other investigators suggest that the emergence

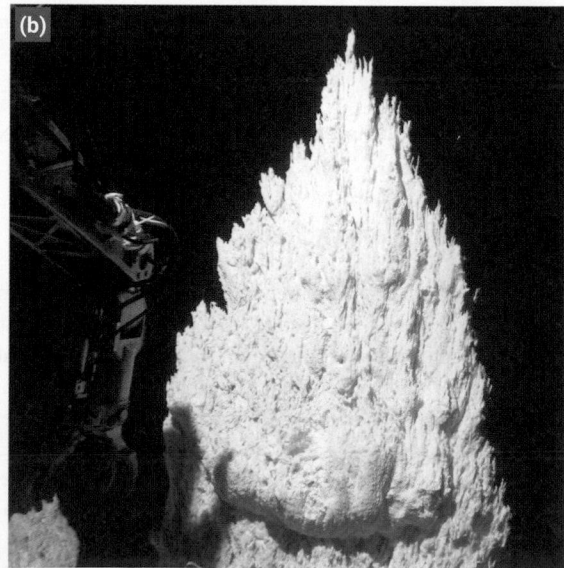

Figure T2.8 Two types of submarine hydrothermal vent systems. (a) Black smokers, which are typically closer to marine spreading centres, and hotter and rich in sulphide (the black discharge is composed of particulate metal sulphides, which form when the hot vent water comes into contact with colder ocean water). (b) A carbonate chimney from a serpentinizing vent system, which are typically somewhat colder and farther from the spreading ridge axes.
Source: 17-MARUM, Center for Marine Environmental Sciences, Universitat Bremen, CC_4.0, 18- Credit NSF, NOAA, University of Washington. Image courtesy National Science Foundation.

of life required much smaller bodies of water, which could only have existed on land. Lakes, rivers, and tidal ponds are more likely to enable the concentrations of organic compounds necessary for the generation of larger molecules and the development of complex biochemical reactions.

Small bodies of water may undergo periodic cycles of wetting and evaporation, which can further drive reactivity and complexification. Polymerization of nucleic acids and peptides, for example, has been demonstrated to occur under such conditions. Small rocky fissures, channels, and beaches might additionally provide mineral surfaces and catalysts conducive to chemical reactions and continuous activation.

2.5 Complex biomolecules

We cannot be sure which compounds were available when life began, and we do not know which were needed for it to be sustained, so we must rely on clues from modern biochemistry.

The absolute minimum required to form a living cell resembling those we see today would appear to be:

- a lipid membrane;
- an information storage mechanism, like RNA and/or DNA;
- encoded catalysts, like peptides; and
- a means of energy storage and transfer.

Modern entities such as viruses and prions possess some of these components, but are notably lacking in their ability to self-generate

them. As such, they may suggest partial steps in the achievement of life-like systems.

Such chemistry might allow the generation of a simple, bounded, self-generating entity or 'protocell', capable of replication. Replication would be necessary to expand the population, and also for continuous change through variation and selection according to the principles of Darwinian evolution. This would be a route to greater complexity and the emergence of the LUCA.

The biochemical components of a protocell would need to be made by simple self-generating processes from materials available in the environment, under environmentally plausible conditions. This may appear to be an unlikely combination of circumstances, but some such materials can be experimentally generated in the laboratory. Their natural chemistry offers some promising clues.

Amino acids

Amino acids are the building blocks of proteins. Several prebiotic processes (such as those modelled in Miller's experiments) can form amino acids, and they are also are found in meteorites and comets. Adolph Strecker, a 19th-century German chemist, discovered that the amino acid alanine could be synthesized by mixing HCN, acetaldehyde (CH_3CHO), and ammonia (NH_3) in water under acidic conditions, as illustrated in Figure T2.9.

Exchanging acetaldehyde for other aldehydes generates other amino acids, while other variations in reaction conditions produce related chemicals called hydroxy acids. These reactions are relatively straightforward and robust, so it seems entirely plausible that protein-building elements were available in Earth's early environments.

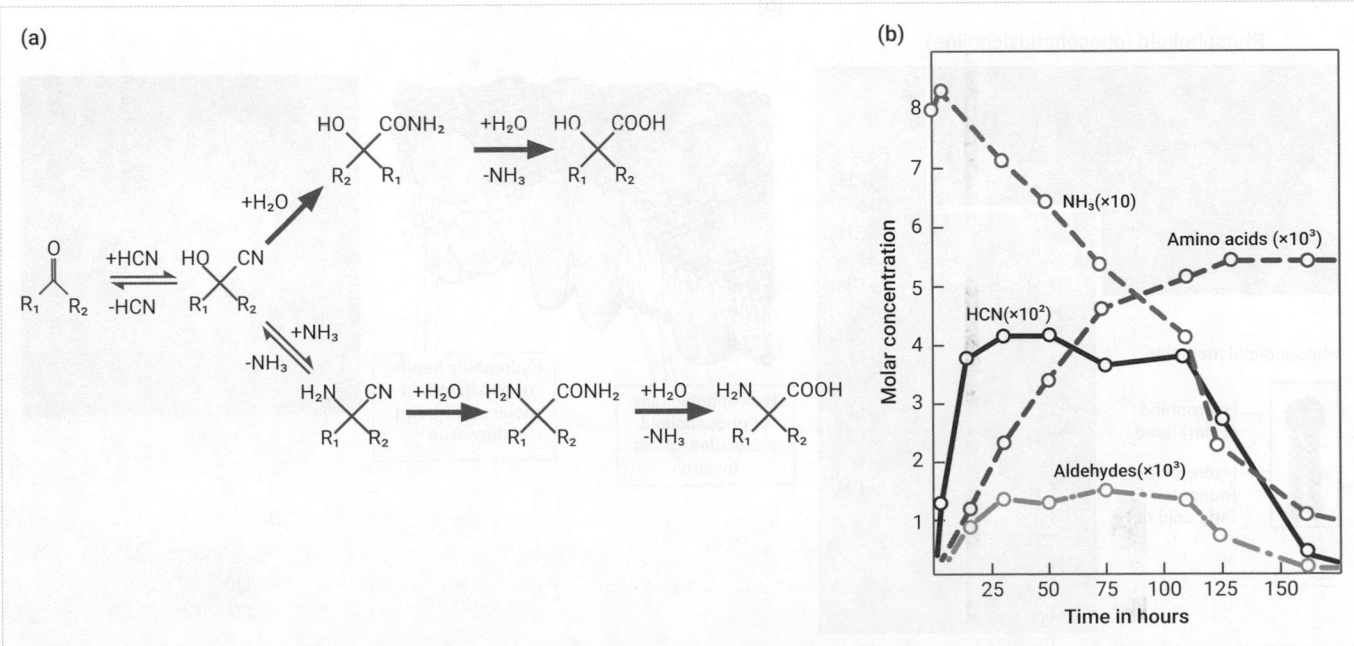

Figure T2.9 The Strecker amino acid synthesis. (a) The chemical mechanism which can result in either the synthesis of hydroxyl acids (top branch) or amino acids (bottom branch). (b) Concentrations of chemicals involved in the Strecker synthesis measured over the course of an electric discharge experiment.

Source: Cleaves, H.J. Prebiotic Chemistry: What We Know, What We Don't. *Evo Edu Outreach* 5, 342–360 (2012). © Springer Nature.

Lipids

Cell membranes are composed of phospholipids, themselves made from long-chain molecules called fatty acids; the general structure of a phospholipid is shown in Figure T2.10a. Fatty acids are amphiphilic (= 'having two loves') because they have a water-soluble part and an oil- or fat-soluble part. Phospholipids in modern cell membranes, such as those depicted in Figure T2.10b, have complex structures, but one can imagine that ancient protocell membranes were formed from simpler amphiphilic building blocks.

Under appropriate conditions, long-chain fatty acids will spontaneously form cell-like structures called vesicles, not unlike those depicted in Figure T2.10c. Despite limited stability, fatty acid vesicles can encapsulate other organic molecules, including RNA. They are also permeable to smaller organic compounds and this may have enabled the uptake of components for complex structures such as transport proteins. Fatty acid vesicles can also maintain proton gradients, a possible basis for energy transfer in early metabolism.

Possible prebiotic sources of fatty acids remain unclear. They can be made in the lab from CO and H_2, but the efficiency of these processes is often low. Small amounts have been discovered in meteorites, but this may be contamination.

Sugars

Sugars (carbohydrates) and their derivatives are crucial to prebiotic chemistry because they (ribose and deoxyribose, respectively) are structural components of RNA and DNA; the structures of RNA and DNA are illustrated in Figure T2.11.

▶ **We learn more about the nucleic acids in Chapter 3.**

Sugars, including ribose, have been detected in meteorites, although their instability over long timescales makes their accumulation in the environment unlikely. They have the empirical formula $(HCHO)_n$ and can be made by linking HCHO molecules under basic conditions (e.g. the formose reaction; Figure T2.12) or by using HCN as a starting material.

Sugar compounds have complicated stereochemistry: they show chirality ('handedness', rather like gloves). The atoms in chiral compounds can be spatially arranged in numerous ways, although biology employs only a few of the possible structures. Biology uses only the right-handed D-sugars (D for *dextro*). Most amino acids are also chiral and biology almost always uses left-handed or *laevo* L-amino acids. The origins of this *homochirality*, the reason why one form predominates, are unknown.

Nucleobases

All contemporary organisms store hereditary information in double-stranded DNA molecules. The process of turning this information into proteins involves several components, including single-stranded messenger RNA (mRNA) molecules, transfer RNA molecules (tRNA), ribosomes, and numerous structural peptides and enzymes. DNA is itself copied using various protein enzymes and small RNA primers.

▶ **We learn more about how the information stored in DNA is used to direct the synthesis of proteins in Chapter 5.**

Nucleobases, along with sugars, are principal components of these nucleic acids and determine the information stored in a genetic code. There are two classes: the purines (adenine and guanine) and the pyrimidines (cytosine, uracil, and thymine).

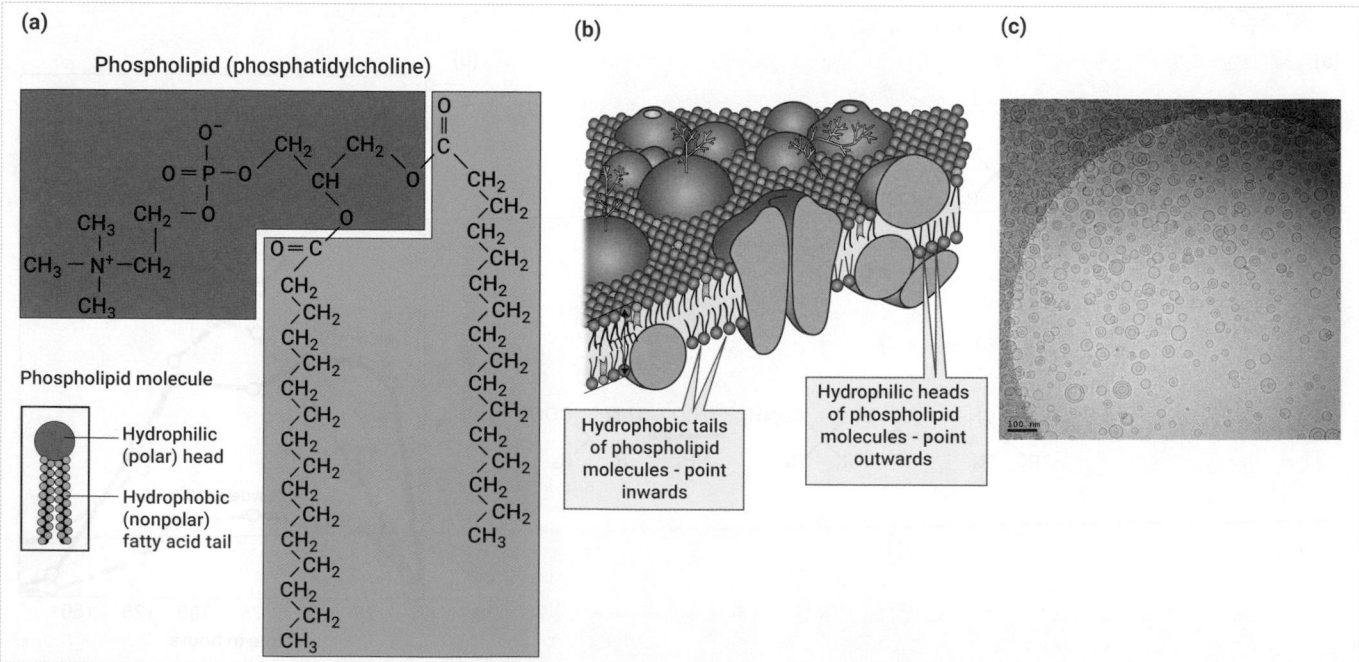

Figure T2.10 (a) The structure of a phospholipid, a modern membrane lipid. (b) Phospholipids aggregate to form a central component of the cell membrane. (c) Cryoelectron micrograph of lipid vesicles formed when phospholipids are agitated.

Sources: (b) Adapted from Figure 2.6D in Bowater, Biochemistry (Oxford Biology Primer). Licensed from PLS Clear; (c) Courtesy of J. Goodwin & H. Khant, NCMI, Baylor College of Medicine, Houston, TX.

Figure T2.11 RNA and DNA. (a) The sugar components of RNA and DNA. The 'B' attached to each sugar group denotes in a general way one of the nucleobases: A, T, G, and C in DNA, and A, U, G, and C in RNA. (b) The structures of polymeric RNA and DNA.

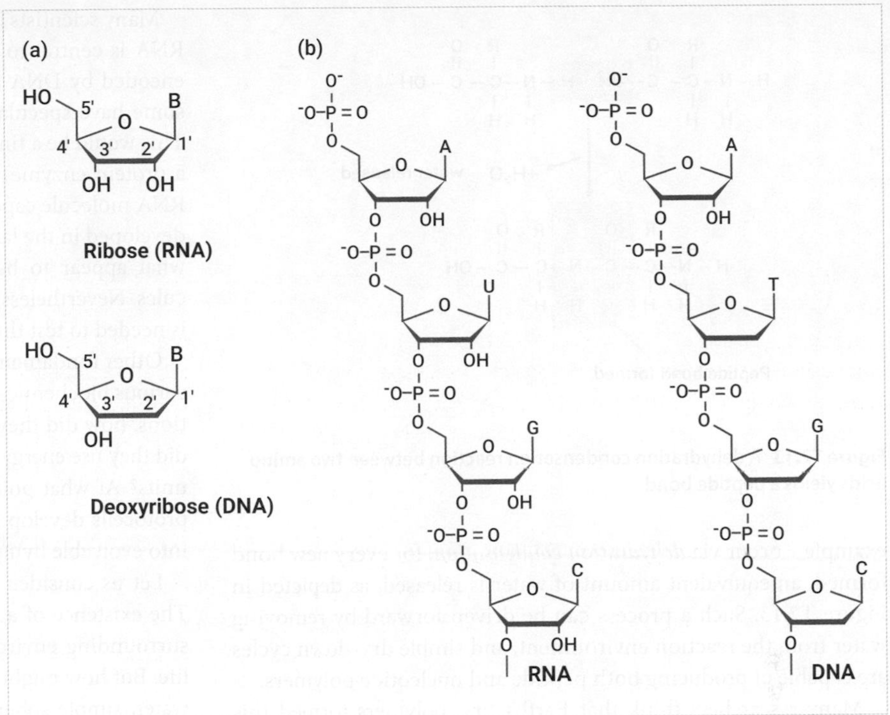

(a)

Ribose (RNA)

Deoxyribose (DNA)

(b)

RNA

DNA

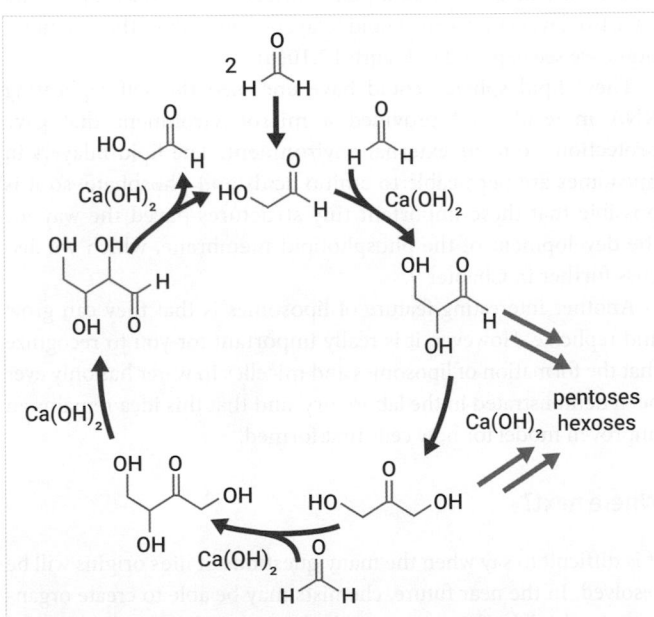

Figure T2.12 The autocatalytic formose reaction. The reaction occurs when formaldehyde (HCHO) is reacted in alkaline water and produces a variety of sugars and related molecules.

Nucleobases have been detected in meteorites, and laboratory experiments have demonstrated potentially prebiotic pathways for their formation. Some researchers suggest that nucleic acids could have arisen from completely different classes of nucleobases. As with most ideas on the origin of life, there is little consensus.

Adenine is perhaps the simplest nucleobase, from a chemical perspective. It is, in terms of its chemical formula, a pentamer of

HCN: $C_5H_5N_5$. Soon after Miller's spark-discharge experiments, adenine was found to form from concentrated aqueous solutions of ammonium hydroxide (NH_4OH) and HCN. Guanine and other purines can also be made by variations of this synthesis. It is intriguing that HCN seems to be such a versatile starting material, allowing for the production of amino acids and sugars, as well as nucleobases, although this simple interpretation does not readily explain the origin of the complex systems which gave rise to biology.

Reactions that produce pyrimidines in abundance have also been discovered, including ones using conditions that simulate drying beaches. For example, cytosine can be made by allowing urea to condense with cyanoacetaldehyde, and uracil can be made directly from cytosine by hydrolysis in water. Simply put, if cytosine can be made, then uracil synthesis is also possible.

Combining nucleobases, ribose (and other types of sugars), and phosphate into nucleotides has proven to be a great challenge, but major research advances in the last decade have shed light on how this problem might be solved. Nevertheless, the complexity of prebiotic nucleotide/nucleoside synthesis means it is likely to remain an active research field for years to come.

Polymerization and synthesis

While scientists can envisage prebiotic routes to the availability of many modern essential biomolecules, it is less easy to imagine how the complex informational process which produce them may have emerged. Scientists assume that the nucleic acid-encoded protein enzymes which underpin modern biochemistry would not have been available initially.

Most biologically important polymerizations—in which amino acids or nucleotides are strung together in linear sequences, for

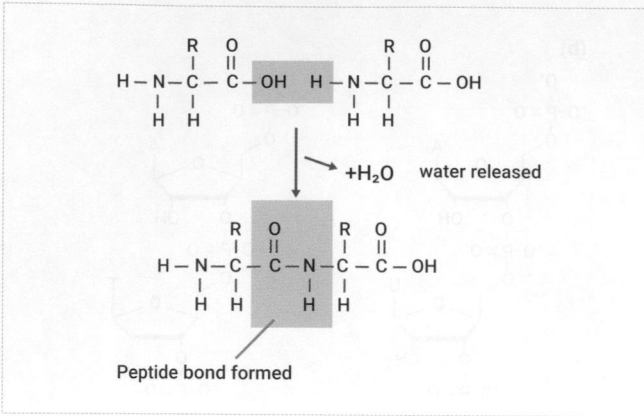

Figure T2.13 A dehydration condensation reaction between two amino acids yields a peptide bond.

example—occur via *dehydration condensation*: for every new bond formed, an equivalent amount of water is released, as depicted in Figure T2.13. Such a process can be driven forward by removing water from the reaction environment, and simple dry-down cycles are capable of producing both peptide and nucleotide polymers.

Many researchers think that Earth's first polymers formed this way, generating an initial mixture of more or less random sequences of polymers. Some of these may have had functional activity, especially when encapsulated in a protocell (vesicle) and subjected to the pressures of selection from which all biological complexity emerges. Lab experiments involving artificial selection support this notion, but success has been limited and there is much that is not understood.

Another question concerns the establishment of the energy-rich molecules on which modern life forms depend. The energy economy of cells, as well as their ability to communicate and react to their environment, involves molecules with high-energy phosphate bonds and it is hard to see how these may have arisen, although several possible theories have been put forward.

2.6 From biomolecules to life forms

In this brief account we have looked at some basic requirements for life and scenarios in which the basic biochemicals of life may have arisen in the absence of pre-existing life. (Topic 3 considers the characteristics of life in more detail.) Now we can ask: Is a combination of the right environmental conditions and necessary chemicals *sufficient* to explain the self-sustaining, continually replicating, evolving phenomenon that we call life?

All forms of life pass on their characteristics from one generation to the next, but this presents a problem. According to what we know, DNA holds the code needed to make functional proteins, but proteins are needed to make DNA. So, even if proteins and nucleotides arose by chance through chemically simple events, which came first?

Many scientists have suggested that RNA can solve this paradox. RNA is central to the expression of genes (enabling information encoded by DNA to give rise to specific protein sequences), and some have speculated that early life emerged in an 'RNA world'. This would be a time when RNA functioned as both a catalyst (like a protein enzyme) and as a self-replicating genetic molecule. An RNA molecule capable of assembling 80% of its sequence has been developed in the laboratory, and other experiments have produced what appear to be self-sustaining, self-propagating RNA molecules. Nevertheless, much more theoretical and experimental work is needed to test this model.

Other fundamental questions about very early life also remain. If various biochemicals emerged under prebiotically plausible conditions, how did they come to be associated with one another? How did they use energy to form large, complex molecules and metabolic units? At what point did a protocell become necessary? How did protocells develop into cells and how did they become integrated into evolvable living systems?

Let us consider one of these questions: the formation of cells. The existence of a cell, the content of which is separate from the surrounding environment, was a critical step in the evolution of life. But how might cells have first formed? When fatty acids are in water, simple spherical structures called **micelles** and **liposomes** form. Micelles are contained by a simple single layer, a monolayer, with a fatty acid core and a polar surface; by contrast, liposomes have two layers that form a lipid bilayer—the basis of the cell membrane we see depicted in Figure T2.10(b).

These lipid spheres could have enclosed the self-replicating RNA molecule and provided a microenvironment that gave protection from the external environment. The lipid bilayers in liposomes are permeable to amino acids and phosphate, so it is possible that these important tiny structures paved the way for the development of the phospholipid membrane, which we discuss further in Chapter 9.

Another interesting feature of liposomes is that they can grow and replicate. However, it is really important for you to recognize that the formation of liposomes and micelles in water has only ever been demonstrated in the laboratory, and that this idea remains an unproven model for how cells first formed.

Where next?

It is difficult to say when the many questions of life's origins will be resolved. In the near future, chemists may be able to create organisms in the lab, *de novo*, entirely from simple materials. Entire synthetic genomes have already been constructed and can be manipulated with great precision. Computer scientists also know how to store and control information in life-like simulations. The creation of artificial life may be a reality in the near future.

A comprehensive understanding of how such systems could have arisen in the environments of primitive Earth still evades us. Despite considerable progress, much work remains to be done and many surprises are likely in store.

FURTHER READING

Here is a range of articles that review in more detail some of the topics we explore in this topic.

Joyce, G. F. & Szostak, J. W. (2018) Protocells and RNA self-replication. *Cold Spring Harb. Perspect. Biol.* **10**: a034801. doi: 10.1101/cshperspect.a034801

Kitadai, N. & Maruyama, S. (2018) Origins of building blocks of life: a review. *Geosci. Front.* **9**: 1117. https://doi.org/10.1016/j.gsf.2017.07.007

Shaw, G. H. (2016) *Earth's Early Atmosphere and Oceans, and The Origin of Life*. Springer International Publishing. doi: 10.1007/978-3-319-21972-1

Zahnle, K., Schaefer, L., & Fegley, B. (2010) Earth's earliest atmospheres. *Cold Spring Harb. Perspect. Biol.* **2**: a004895. doi: 10.1101/cshperspect.a004895

TOPIC 3

LIFE AND ITS EXPLORATION

How Do We Define Life?

Contents

Introduction		30
3.1	Metabolism	31
3.2	Membranes	31
3.3	Defence	32
3.4	Replication and continuity	32
3.5	Variation and selection	32
3.6	Conclusion	33

Introduction

Biological organisms are alive. It is difficult to give a precise, comprehensive, and uncontroversial definition of 'life', even though we usually know what we mean by the word. The distinction between things which are alive and those which are not is usually obvious to us, especially for complex, multicellular plants and animals. The distinction may be less clear when we look at single cells and microbes, and it is especially opaque when we consider entities such as viruses and prions, which have biochemical properties but are not cells.

Our primary wish is to understand life on this planet. Any search for extra-terrestrial life also requires that we have some kind of definition. If we do not, how will we know if and when we have found it? A further problem is that, biochemically speaking, we find it difficult to imagine life that is not built around carbon-containing molecules or that does not depend on a watery environment, for this is the life we know and of which we are a part.

What we can do is list the characteristics that we observe life to have and use this as a guide to what life really is. We can also try to understand the structural units in which life manifests itself.

Lists of such characteristics are easy to find and you might like to begin your study by looking at some of these or creating your own. A problem, however, is the implication that there is a *minimum number of characteristics* that an entity must possess if it is to be regarded as alive. Deeper reflection shows us that life is represented by *a set of five functional qualities*:

1. Metabolism.
2. Membranes.
3. Defence.
4. Replication and continuity.
5. Variation and selection.

The following five sections describe each of these qualities in turn. For clarity, they are approached separately, although in reality they comprise an integrated whole. As we shall see, the units of life that give the biological world its identity exist at many different scales, from organelles and cells, through tissues and organs, to bodies and whole communities.

3.1 Metabolism

Living organisms take up, store, and transmit energy that originates from the Sun. They also process energy-containing molecules so that the energy, generally stored in chemical bonds, can be used in different forms. This overall process is called metabolism and all living organisms carry it out.

To fully appreciate the process of metabolism, we need to draw on some important physical concepts: the law of thermodynamics. The laws of thermodynamics are the most fundamental statements we can make about the Universe, and they are considered to be unbreakable.

They can be expressed in the following statements:

The first law: Energy can be neither created nor destroyed.

The second law: Over time, the entropy of any system increases.

'Entropy' can be understood as a measure of 'disorder'. It describes the extent to which the heat energy of a system is *unavailable* for conversion into mechanical work. The energy that is stored in chemical bonds and mechanical systems is potentially available to do work and change the state of things. The second law is telling us that, over time, the amount of this energy decreases while the free energy of the Universe increases.

These laws, particularly the second, have some important consequences:

1. Energy can only be transferred from entities with high free energy to those with low free energy, not the other way round. (If we find ourselves apparently cooling something—by sweating, say, or with a refrigerator—we must be taking energy from somewhere else.)

2. The transfer of energy from one chemical entity or body to another is never 100% efficient; some free energy is always released.

The laws of thermodynamics are not only useful in defining life, but they will also turn out to be important when we investigate the physiology and biochemistry of living organisms in later modules.

So what has metabolism to do with the laws of thermodynamics? Metabolism can be thought of as temporarily defying the second law of thermodynamics. So long as the organism remains alive, its metabolism causes the interconversion of molecules and facilitates growth. This represents an accumulation of stored energy, decreasing the organism's entropy. Nothing else in the Universe that we know of can achieve this.

(As an aside, physicists will correctly argue that living organisms are not closed systems and therefore that the second law does not strictly apply to them. Living things are, nevertheless, oases of temporarily reduced entropy within the increasing entropy of the expanding Universe.)

From their metabolism, living organisms release free energy in the form of heat (the vibration of molecules). At the same time, they have the remarkable property of continuously regenerating energy-storing materials from food as they live and grow. In other words, while they are alive and active, organisms are in overall net positive energy balance. As soon as they senesce (stop growing or decline in metabolic rate), they undergo a loss of stored energy and increase their entropy. When they die, all remaining energy is released completely to the environment (and often made use of by other organisms). This is why the apparent defiance of the second law is only *temporary*; ultimately, everything tends to a state of disorder.

▶ **We learn more about metabolism in Chapter 7.**

3.2 Membranes

The chemical processes of life take place within enclosed structures. These structures are delimited by sheets of organic material called **membranes**. Membranes may be simple or complex and multi-layered. They define the boundaries of an organism and separate it functionally from its environment.

▶ **We learn more about the composition and function of membranes in Chapters 9, 16, and 28.**

We can envisage each membrane, whether it is the wall of a bacterium, the cellulose cell wall of a plant cell, the phospholipid membrane of an animal cell, the outer wall of a plant stem, or the skin of a complex animal, as delineating and separating off a tiny portion of the matter and energy of the Universe. The membrane contains the organism's shape, form, and structure. It also enables the organism to function biochemically for the duration of its lifespan and to defend itself against invasion by other organisms.

Membranes isolate and restrict the space in which molecular interactions can take place. In an unrestricted, unenclosed space the physical process of **diffusion** will eventually result in an evenly spread, low density, random distribution of molecules, with a corresponding dispersal of free energy and increase in entropy. If the space is enclosed and limited, molecular movements are spatially restricted and local concentrations of energy can be higher.

On the one hand, then, enclosure restricts the freedom with which energy can move, while on the other it concentrates it in local processes. When this happens, large and complex molecules can form and energy can flow in a coordinated and biologically productive manner.

Biologists speak, sometimes loosely, of cells as being the most basic units of life. In **prokaryotic** life forms, a single (but sometimes multi-layered) membrane surrounds the cell, the contents of which are essentially free to move in a non-compartmentalized manner. In **eukaryotic** life forms, besides the external cell membrane, an internal membrane encloses the DNA, forming a nucleus. Other internal membranes delineate other components, creating organelles (mitochondria, endoplasmic reticulum, lysosomes, chloroplasts, etc.).

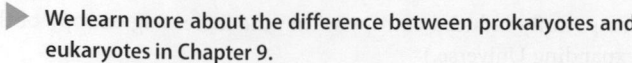

We learn more about the difference between prokaryotes and eukaryotes in Chapter 9.

Membranes enclose biological spaces, but they are far from inert. They enable ionic gradients to be established, control the movement of molecules into and out of the organism or organelle, and determine in several different ways how the organism responds to and interacts with its external environment. This includes responses to chemicals and other external signals, attachment to other cells, anchorage to surfaces, movement, dispersal, and other dynamic processes.

3.3 Defence

Given that life takes place in units separated off from the environment, another quality of living organisms is their ability to defend an enclosed region of the biosphere. A dead animal or plant does not do this. It is quickly invaded by microorganisms and then consumed by opportunistic, scavenging organisms. Its decay releases its component chemicals and stored energy, making them available for other organisms to use.

An alive but inadequately defended organism is similarly susceptible to take-over by other life. This takes the form of predation for food, devastation by disease, or colonization by parasites. At all levels of biology, there is a constant arms race between defence and invasion, between protection and predation. This creates not only an evolutionarily important environmental pressure, but it also produces food chains and the constant recycling of biochemical resources and energy.

The quality of defence can be extended to include homeostasis—the tendency for organisms to counteract the effects of disturbances and maintain the stability of their internal conditions (often called the internal environment or *milieu*, acknowledging the origination of the concept by the 19th-century French physiologist Claude Bernard).

3.4 Replication and continuity

All life has the ability to replicate and is in a state of continual development and growth. These processes have the effect of transmitting life's qualities from one generation to the next. They allow us to conclude that life on Earth arose on a single occasion and has continued expansively ever since. In multicellular organisms, reproduction, growth, and differentiation (the specialization of functions within the organism) all happen at the cellular level, even if the results are most often seen as changes in the whole organism.

Genes are fundamental to all these processes because:

1. they are the units of biological information that are transmitted when an organism reproduces itself; and

2. the information they hold defines an organism's phenotypic potential.

By genotype we mean the complete genetic make-up of an organism. An organism's phenotype is its complete set of physical, biochemical, and behavioural characteristics.

A crucial feature of organisms is that they can replicate their genes with a low error rate. This means that offspring have near-identical genotypes and closely similar phenotypes. All life that we know of uses this process to reproduce, exploiting the particular biochemical properties of DNA, RNA, and protein to do so.

Single-celled and multicellular organisms hold their genetic information in a variety of forms and reproduce themselves using a diverse set of strategies, some of them extremely elaborate. Organisms also vary in the proportions of time and energy they allocate to reproduction and to growth or differentiation. In studying the biological world, we need to look beyond these varied details of process and see that all living things are characterized by their continuity: they transmit through time information about their identity.

We learn more about the replication of biological information in Chapter 3.

3.5 Variation and selection

Living organisms, their structures, and their life processes have evolved from ancestral organisms by a combination of variation and selection. We pointed out that replication has a low error rate, but that rate is not zero. There are also environmental factors which can modify gene structures (cause mutations) or alter their expression as proteins. These effects mean that, in general, offspring are similar but not genetically or phenotypically identical to their parents or siblings.

Selection is the process by which some individuals are able to survive, while others do not. The selection 'pressure' on an organism refers to the total effect of all the forces in the environment which threaten its survival and its ability to reproduce. They include food supply, competition, predation, and disease, as well as human interventions and physical factors. Animals, plants, and microorganisms live or die according to whether their structure and function—their phenotypes—are compatible with the environment they are in. Modern organisms exist precisely and only because their ancestors had phenotypes that allowed them to tolerate their environmental conditions and survive for long enough to reproduce successfully.

It is essential to appreciate that the forces of selection act on the *phenotypes* of individuals. Genes are not under selection pressure directly. They are selected indirectly because they determine phenotype or permit various phenotypes to occur. This is why an understanding of the form and function of organisms and of the ways in which they interact with their environments is of such great importance to biologists.

It is also understood by biologists that phenotypic variations acquired by organisms during non-reproductive periods of life are not transmitted as genetic variations to their offspring: the child of a weightlifter will not inherit the large musculature that its parent developed through training.

The *expression* in offspring of a limited range of genes (called **imprinted genes**) can be altered by parental experiences through a process called **epigenetic modification**, but there is no mechanism for the transmission to the next generation of gene *mutations* resulting from changes in phenotype acquired in adult life.

Variation and selection are the basis of evolution and can explain completely the wondrous diversity of life forms that we find in the biosphere. We do not need to invoke any form of purposeful design or external agency, nor is it necessary to think of evolution as directed towards any form of goal or target of absolute perfection. (Humans may express a concept of 'perfection' in relation to the animals and plants we exploit and control for productive, medical, decorative, or aesthetic purposes. Goal-directed selection, such as that used in breeding pedigree animals or propagating crop plants for food, often results from this and can have a huge influence on species abundance and diversity. This kind of human activity establishes ideals of value and function. It is best seen as an environmental pressure on the selected organisms.)

There is abundant evidence that evolution continues, based solely on the randomness of genetic variations and the opportunism of environmentally driven selection.

▶ **We explore evolutionary principles further in Topic 4, and explore their consequences in Chapter 35.**

3.6 Conclusion

We have identified five qualities of life as we know it:

1. Acquisition, storage, manipulation, and transmission of energy derived from the Sun.

2. Enclosure of these processes within membranous entities.

3. Defence of the enclosed structure, representing a compartmentalization of the biosphere.

4. Replication and transmission of characteristics to subsequent generations.

5. Constant variation, allowing for adaptation and selection.

So far as we know, life is the only thing on our planet that shows all these qualities. We strongly suspect that if life should ever be found on other planets, it will have them too.

Variation and selection are the basis of evolution and have resulted in the wondrous diversity of life on Earth. Its continuity, for at least 3.5 billion years, attests to its success.

Evolutionary Processes

Contents

Introduction   34

4.1   What is evolution?   34

4.2   Seven reasons why the study
       of evolution is important   36

4.3   Conclusion   41

Introduction

Life on Earth as we know it today is the culmination of evolutionary processes that have been occurring across the millennia, since life first emerged. But what are these processes, and why is evolution so important to our understanding—and continued study—of biological science? We explore these questions in this topic, beginning with an important first question: what do we mean by 'evolution'?

4.1 What is evolution?

There are a multitude of definitions of evolution, but perhaps the simplest and most accurate is that proposed by Darwin himself: 'Descent with modification'. This means that evolution is a process whereby life changes over time, and these changes are passed down through the generations (i.e. they are inherited). The theory proposes that all organisms are related in some way, making common ancestry a central property of life on Earth. It also proposes that the immense diversity of life we see today has been produced by the reproduction and diversification of our ancestors. If you think about it, this is an astounding concept: everything that makes you human arose in one of your ancestors, and that ancestor (and all of its descendants) must have survived until reproductive age to pass their features on!

When did evolution begin to take place?

We think that the process of evolution started around 3.5 billion years ago in the oceans, a short while after Earth had formed. As we note in Topic 2, the exact details of the origin of life are still shrouded in uncertainty, but somehow simple organic molecules arose that began to replicate themselves or each other. As time progressed, some of these molecules changed, and those that could copy themselves most effectively spread more quickly than those that could not: life was already evolving.

Over time, more complex molecules evolved that could harvest energy, create new molecules, and speed up chemical reactions. Again these features were inherited, and so passed to successive generations. The complexity of these molecules continued to increase, eventually leading to the formation of simple cells (prokaryotes), and later cells with many compartments (eukaryotes). Multicellular organisms then evolved, with the origin of the animals between 600 and 700 million years ago, and plants at least 470 million years ago. These changes were all caused by the process of evolution.

▶ We learn more about prokaryotes and eukaryotes in Chapter 9.

What causes evolutionary change?

A number of processes cause evolutionary change, including natural selection, mutation (a change in DNA sequence, often caused by copying error), non-random mating, migration, and genetic drift.

▶ We explore these concepts in more detail in Chapter 35.

In brief, mutation is important because it is the ultimate source of the genetic variation that underpins other processes. Genetic drift is a random process that affects a lot of the 'hidden' variation in our genetic sequences and so is of huge interest to geneticists. However, only one process leads to the adaptation of organisms, through which they become better suited to their environment: natural selection. Because of its role in adaptive evolution, natural selection is unquestionably the most important theory in all of biology. It is a process that explains the origin of life, and every critical transition in life over the last 3.5 billion years, from the invasion of the land by plants and animals to the invention of stone tools by the ancestors of humans.

Later in this book, you will become intimately familiar with the process of natural selection. Originally proposed by Charles Darwin and Alfred Russel Wallace in 1858, the theory is elegantly simple: individuals that are successful in the competition to survive and reproduce will tend to pass on to the next generation those 'heritable characteristics' that increase their chances of survival and reproduction. Over time, the 'heritable characteristics' that increase the chances of survival and reproduction will become more common in the population.

▶ We learn more about natural selection in Chapter 34.

Using this theory in its most simple form, it is relatively easy to understand how adaptations like camouflage might evolve to help animals to escape from predators, or how jaws might evolve to allow feeding. However, in the 160 years since natural selection was first proposed, we have discovered a lot more about biological processes, and have been able to refine the theory.

How has our understanding of natural selection been refined over time?

Let's consider some examples of how the theory of natural selection has been refined, while noting the theory has not been fundamentally altered!

1. With our understanding of modern genetics, we can now identify how evolution happens at the molecular level and causes changes in the frequency of alleles of genes in a population over time (evolutionary genetics).

2. We now know that reproduction can sometimes 'trade off' against survival. Sexual selection is a form of natural selection that acts on traits that enhance reproduction, but which may limit survival. For instance, many male birds of paradise have extremely bright and ornate plumage (Figure T4.1a); this not only makes them more vulnerable to predation, but may also act as a signal of genetic quality, or indicate the 'sexiness' of their future offspring to potential mates.

3. We now think that selection does not just happen on 'individuals' in a population. Sometimes, adaptations may arise that allow the genes to be passed to the next generation from outside of the individual that they are in (this is called indirect fitness). This most commonly occurs in situations in which selection acts on traits that improve the overall reproductive success of related members of a group (termed kin selection). This is why, for example, you find that parents will endanger themselves to protect their offspring, and why some ants and termites develop into armoured soldiers (Figure T4.1b) that will sacrifice themselves to defend their queen (who often produces huge numbers of siblings for them).

4. Sometimes, the genes of one organism may be selected through their effects on a different species. For instance, when a fungus, *Ophiocordyceps unilateralis*, infects the brain of the ant species *Camponotus leonardi*, the fungus makes the ant climb up into the canopy, bite onto the vegetation there, and die. The fungus then produces a fruiting body which grows out of the ant's head, before rupturing to release the fungus's spores (Figure T4.1c). The climbing behaviour of the ant ensures that the fungal spores are optimally propagated to be able to infect other ants. The ant behaviour is the result of the genes of the fungus, not the ant. This concept of genes acting outside of their body is known as the 'extended phenotype' and was largely popularized by Richard Dawkins in his 1982 book of the same name.

5. We now think that selection does not necessarily have to act on particular traits, but can act on the ability to create variation in those traits. For instance, we know that organisms can often change their form and behaviour in response to different environmental variables, a phenomenon known as phenotypic plasticity. Selection can act on this ability to change, rather than the specific features themselves. For example, phenotypic plasticity has been seen in tadpoles of the frog species *Rana pirica* where they develop into different forms in the presence or absence of predators (Figure T4.1d).

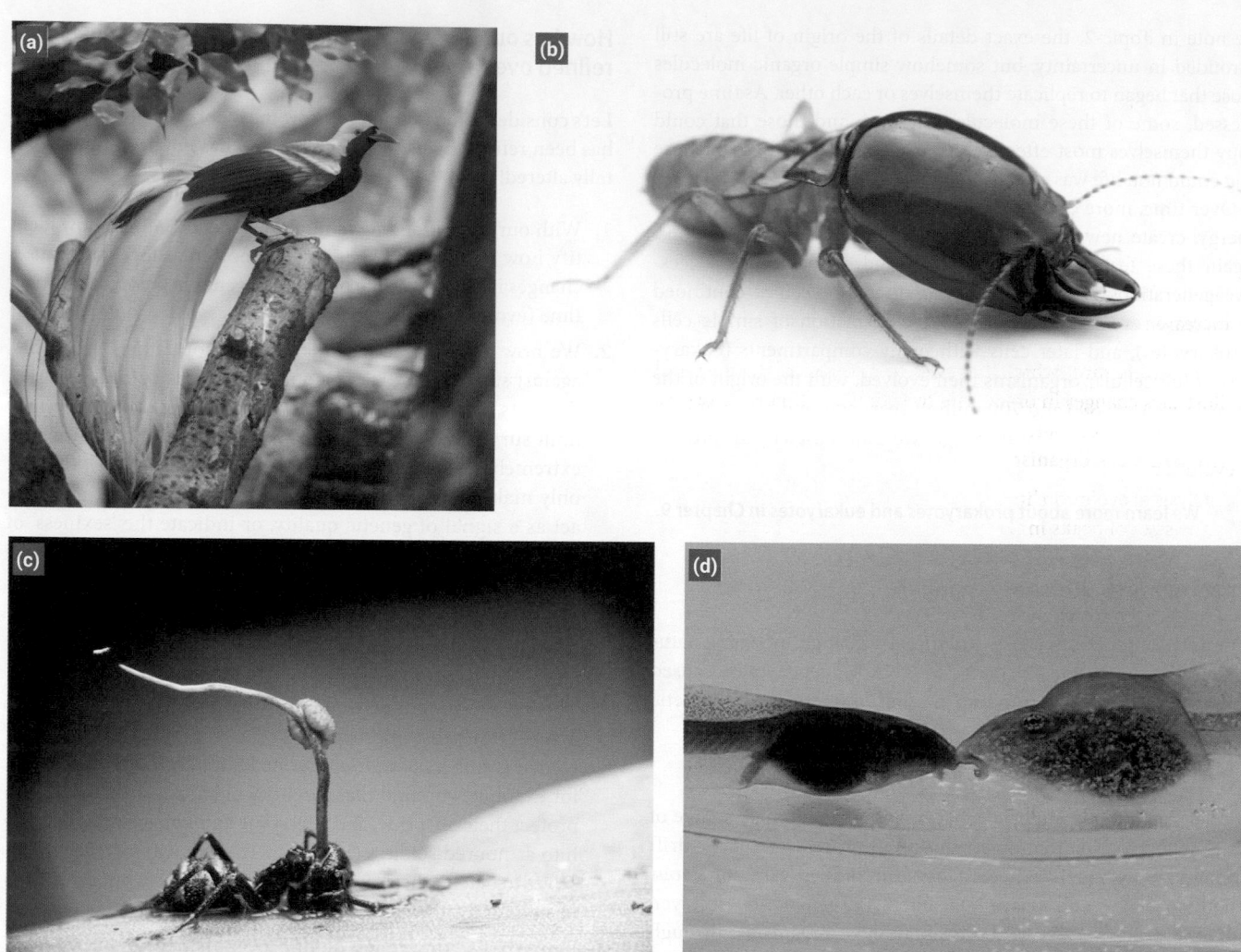

Figure T4.1 Manifestations of our understanding of natural selection. (a) The ornamentation of the greater bird of paradise attracts mates but may decrease survival. (b) Soldier termites develop armour and jaws to defend their queen but will never reproduce. (c) A fungus infects the brain of an ant, changing its behaviour so that fungal spores can be dispersed. (d) The two forms of tadpole in the frog *Rana pirica* are a consequence of phenotypic plasticity.

Sources: (a), (b), (c) Shutterstock; (d) Wikimedia Commons_CC_3.0. Copyright: Tsukasa Mori, Hiroko Kawachi, Chiharu Imai, Manabu Sugiyama, Youichi Kurata, Osamu Kishida, Kinya Nishimura.

4.2 Seven reasons why the study of evolution is important

Although our understanding of the process of evolution has changed slightly in recent years, this does not in any way diminish the central importance of the theory to scientists. For this reason, the rest of this topic focuses on seven key reasons why it is still important to study and understand the process of evolution. Let's now consider each of these reasons in turn.

1. Natural selection is a 'battleground' for scientific thinking

Despite its importance to the vast majority of scientists, the theory of evolution is frequently misunderstood or rejected by large sectors of the world's population. For instance, the Pew Research Center estimates that only around 30% of the adult public in the US accept that unaided evolutionary processes can explain the diversification of all organisms.

There are a number of explanations as to why the theory is disputed by some. First, it is a complex process that requires an integration of knowledge from the life sciences, statistics, and geosciences, and so understanding it does require some scientific literacy. There are also certainly many misconceptions about the process, which are outlined in Topic 1.

Second, there is a growing mistrust of scientific research in general, as seen in attitudes towards the use of genetically modified organisms, on climate change, and on the safety of vaccines. This mistrust has been prompted, in part, by political agendas, and also by difficulties in communicating the findings of advanced scientific research to the general public.

Finally, religious beliefs may influence the acceptance of evolutionary theory; resistance tends to be more visible in societies where the concepts of evolution are at odds with religious ideology (in particular, the speciation, importance, and superiority of humans).

This book will not discuss these issues further, but will instead focus on scientific support for the theory. There is a huge amount of evidence that evolution (by natural selection and other processes) occurs, and the vast majority of scientists around the world accept the theory. Over half a million scientific papers to date have used the word 'evolution' as a keyword, revealing the massive quantity of research that involves the concept. You will find a lot of evidence for evolution later in this book, but sources of evidence include:

1. The fossil record, which provides 'snapshots' of the past and illustrates changes in organisms over time. As time goes on, we are finding more and more transitional fossils (or 'missing links') between organisms.

2. Analysis of evolution 'in action'. Famous studies include changes in the sizes of beaks in finches on the Galapagos islands following a drought, and the evolution of *Escherichia coli* bacteria in the laboratory of Richard Lenski in Michigan State University, but there are many, many more. A fine example is provided by the evolution of stickleback armour, described in Chapter 35.

3. Genetic analysis, which shows us how organisms are related to each other, based upon gradual divergence in their DNA sequences over time.

4. Biogeography, where the global distribution of organisms and the unique features of island species reflect evolutionary and geological changes.

5. The practical uses of evolution in processes such as the development of new drugs, and in conservation programmes.

2. Evolution explains the diversity of life on Earth

Understanding the process of evolution helps us to understand how the diversity of life on Earth arose. The diversity of the life that we know about is astonishing, but most species on Earth remain to be described. Attempts to estimate the number of species have been highly controversial; Camilo Mora and colleagues gave a 'best guess' of around 8.7 million species in 2011, but past estimates have ranged between 3 and 100 million species, excluding viruses and bacteria. However, once microbial biodiversity is taken into account, a recent study estimated that there could be one trillion species, with only one-thousandth of one percent now identified.

New organisms are being discovered all of the time. Two marine phyla have only been discovered in the last 40 years. These are the Loricifera, miniature bullet-shaped organisms (Figure T4.2a), and the Cycliophora, plump symbiotes that attach themselves to the mouths of lobsters and steal their food (Figure T4.2b). Other species have been 'hiding in plain sight': the Soprano pipistrelle (*Pipistrellus pygmaeus*) was only formally separated from the Common pipistrelle (*Pipistrellus pipistrellus*) in 1999 when it was realized that *P. pipistrellus* echolocates at 45 kHz, and *P. pygmaeus* echolocates at 55 kHz.

All of these organisms were produced by the process of evolution, and so knowing how they are related to other animals helps us to understand the processes of evolutionary diversification.

For many years, scientists have been looking for patterns in evolution, and to do this they arranged organisms into hierarchical (or nested) groups, based upon their shared characteristics in a process termed 'classification', 'taxonomy', or 'systematics'. Animals (and other organisms) were assigned to a major group (or phylum) which contains a number of species, all of which share a common

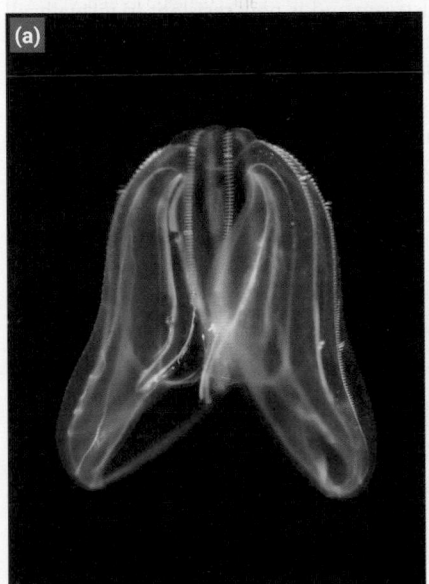
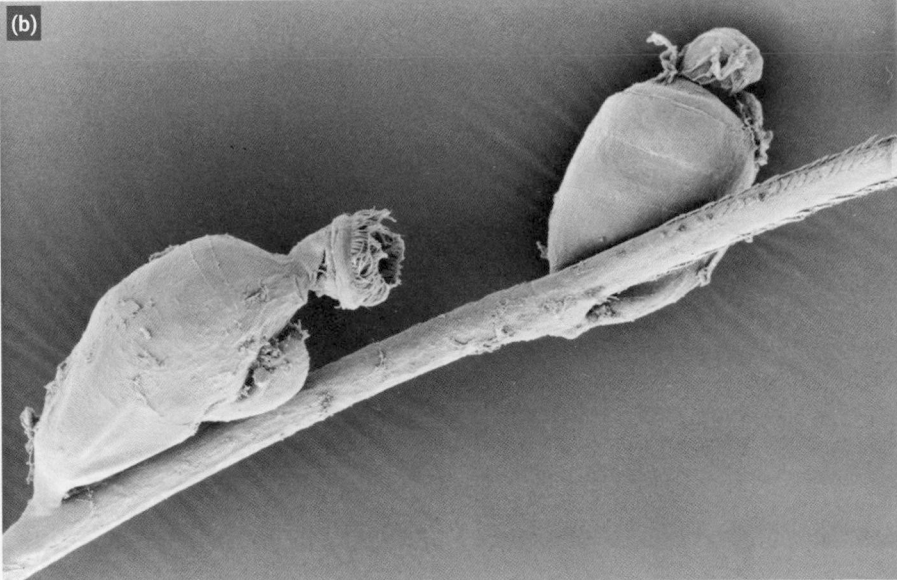

Figure T4.2 Two marine phyla discovered in the past 40 years. (a) A ctenophore, or comb jelly. (b) Cycliophora attached to the mouth of a lobster.

Sources: (a) Shutterstock; (b) electron microscope picture of two pandoras on a lobster whisker. © P. Funch and R. Møbjerg Kristensen.

body plan. Each phylum was subdivided into smaller divisions containing progressively fewer species and defined by an increasing number of important features (or characters).

▶ **We learn more about the classification of organisms in Topic 5.**

This method of classification was useful for quickly learning the characteristics of organisms. However, scientists have also been striving to understand the precise relationships between all organisms on Earth. This is because it is interesting, but more importantly because it allows us to answer interesting questions about living organisms and the evolutionary process itself.

One fascinating example of our changing understanding of animal relationships involves the flatworms (Phylum: Platyhelminthes). Up until around 20 years ago, these were considered to be basal in the evolutionary tree of the animals—that is, they branched from the animal tree before almost every other animal. This was largely because they lack a body cavity (a coelom) and were consequently considered to be relatively simple animals. However, ribosomal gene sequencing then showed us that flatworms are actually part of a higher-level group (called the lophotrochozoans) and lie alongside molluscs and annelids, which do have coeloms. This means that flatworms have almost certainly lost their body cavity over evolutionary time, showing a remarkable trend in evolutionary 'simplification'.

Another example of a changing understanding of relationships comes from the mammals. When the genome of mammals was investigated, we identified a previously unknown grouping of animals, which were later named the Afrotherians. These are hugely diverse in structure, including such animals as elephants, hyraxes, aardvarks, and sea cows (Figure T4.3). The main thing that unites them is that they live in Africa, which indicates that they have diversified separately from the rest of the other mammals. On the anatomical side, we have now managed to identify that their shared characteristics include high vertebral counts, aspects of placental membrane formation, the shape of the ankle bones, and the relatively late eruption of the permanent dentition. However, some members of this group have remarkable similarities with other mammals not in the group, with the golden moles looking very similar to the true moles; they have evolved convergently to look like each other because they live in the same kinds of environment.

These are just a few examples, but they show that knowing the relationships between animals allows us to better understand the evolutionary process. Also, if we can better understand the evolutionary process, we are better able to produce accurate evolutionary trees.

3. Natural selection can kill you!

In the 1920s, Alexander Fleming discovered penicillin, one of the first antibiotics used by doctors. Since then, huge numbers of other chemicals have been discovered that can kill pathogenic microorganisms, and this has led to a huge leap in human life expectancy and the alleviation of much suffering. However, now the evolution of antibiotic resistance in these pathogens means that diseases that were once treatable may now be deadly again; there is a looming antibiotic resistance crisis.

The way in which resistance evolves is now well understood, but it is difficult to prevent. There is often natural variation in resistance: some bacteria can neutralize the chemical drug by making it harmless, some can pump it back outside of the cell, and some can alter their outer structure so that it cannot enter the cell in the first place. After being exposed to antibiotics, sometimes those bacteria with the potential for resistance can survive, and they can then multiply and replace all of the bacteria that were killed off. This is natural selection in action, and we have made the problem worse in recent years by overusing and misusing antibiotics.

In the USA, the Centers for Disease Control and Prevention has estimated that two million people are infected by drug-resistant bacteria each year, and that 23,000 will then die as a result of these infections. A recent report estimates that, in Europe, 33,000 people die each year because of antibiotic resistant infections from eight bacteria. Many of these infections are associated with healthcare environments, with the young and old being particularly susceptible to infection. A particularly serious threat is carbapenem-resistant Enterobacteriaceae (CRE), which allows only a 50% survival rate when it enters the bloodstream and is now resistant to virtually all drugs. Other threats include drug resistance in the causative agent of gonorrhoea, *Neisseria gonorrhoeae*, which can lead to severe complications in reproductive function, and in the 'hospital superbug' MRSA (methicillin-resistant *Staphylococcus aureus*), which is responsible for over 11,000 deaths annually in the USA.

Figure T4.3 A hyrax, a manatee, and an aardvark, all members of the Afrotheria.
Source: Shutterstock.

4. Humans are still evolving

There is a common misconception that humans stopped evolving because of the invention of modern medicine. However, studies comparing the genetics of populations of humans from around the world have found some surprising instances of evolution in our genome, some of it having occurred relatively recently. For instance, a mutation in the *EPAS1* gene has spread rapidly through populations in Tibet over the last 2750 years, making it one of the fastest instances of human evolution ever seen. It is thought that this gene is involved in allowing sufficient oxygenation of body tissues at high altitude without the need for increased numbers of red blood cells.

Extremely fast selection has also been seen in some genes for human diet. For instance, mutations in the gene for the enzyme lactase have spread extremely quickly through European populations, indicating that strong selection has acted on these genes over the last 5000–10,000 years. Lactase is involved in the digestion of milk, and the gene that produces it is normally only expressed (or 'turned on') in infancy. Mutations in the gene for lactase have enabled it to be expressed for longer, allowing humans to digest milk into adulthood; this has been suggested to have evolved alongside the development of dairy farming in Europe.

Many other genes appear to have been under recent and rapid selection. These include genes for detoxification of plant chemicals, immunity, resistance to malaria, hot and cold tolerance, skin pigmentation, language skills, and even tooth enamel. Intriguingly, many of these genes may have been subject to selective pressures that were created by human culture. For example, farming has been implicated as the cause of many blood diseases. This is because the clearing of forests leaves standing pools of water. These pools are ideal breeding grounds for mosquitos, which subsequently spread malarial parasites through human populations. In turn, this leads to the natural selection of different malarial resistance genes (which often have additional effects of causing blood diseases, like sickle cell anaemia).

More and more cases of such evolution are being found, where the selective pressures on humans are now caused by humans themselves. In Chapter 34, you will find a detailed example of how the Baju people of Indonesia have recently evolved a number of specific adaptations linked to diving, which have been caused by their lifestyle.

5. Evolution helps to feed us

An understanding of evolution is also important in agriculture. For thousands of years, humans have been placing artificial selection pressures on plants and animals to help produce food for their populations. However, these pressures have had consequences, and understanding them will be critical for ensuring the future of our species.

We have created numerous crop plants through artificial selection. By artificially selecting those plants that have beneficial characteristics, we have considerably increased the food resources available for our population. For instance, maize (which produces sweetcorn) was produced by artificial selection of its wild grass ancestor, teosinte, in Mexico around 9000 years ago. Teosinte has far

more side branches than maize, and produces fewer female inflorescences (groups of flowers) which develop into ears of corn; this difference is depicted in Figure T4.4. We now know that artificial selection by Mexican farmers simply upregulated the expression of one gene, *TEOSINTE BRANCHED*; this represses side-branches, and allows a bigger investment in growth in the ears of corn.

Artificial selection has also reduced the variation in populations of many of our crop species. Evolutionary theory tells us that populations with low genetic variation are more susceptible to changing environmental conditions than are diverse populations. This was one of the reasons for the Irish potato famine in the 1840s. A specific variety of potato called the 'lumper' had been planted all over Ireland and had been propagated vegetatively, meaning that all of the potatoes were genetically identical to each other. Because of this, a rot caused by *Phytophthora infestans* decimated the potato populations, leading to food shortages that caused the deaths of one in eight Irish people in just three years.

Understanding evolution can help us to mitigate some of these problems. For instance, by understanding the evolutionary history of crops, we can find genes that we can use in genetic engineering. In the 1970s, genes conferring resistance to seven viral diseases that affect domestic corn were found in a previously unknown teosinte species, and subsequently inserted into domestic corn crops through genetic modification. More recently, the genetic engineering of crops has

Figure T4.4 Artificial selection. A comparison of the ears of (a) teosinte and (b) its artificially created descendant, maize.

Sources: Maize: Shutterstock; Teosinte: Matt_Lavin_CC_2.0, Wikimedia Commons.

become more extensively used. This has allowed us to increase crop yields, give resistance to pests and diseases, reduce spoilage, and even improve nutritional value.

Often, genes may be introduced from other species. This is not without controversy, even though there is a scientific consensus that currently available GM food crops pose no greater risk to human health than conventional food. Europe grows relatively few GM crops, but they are far more common in other parts of the world: the US Department of Agriculture estimates that over 90% of the planted areas of soyabeans, corn, and cotton in the USA are genetically modified.

6. Understanding evolution helps in conservation efforts

Biodiversity around the world continues to decline, largely because of changes to the environment caused by humanity. To stop this from happening, we urgently need information not only on the true extent of biodiversity, but also about the potential of species to respond to environmental change. This is why an understanding of evolution can help in conservation.

Variation is critical to the process of evolution, and a loss of variation can cause problems for conservation efforts. A loss of variation can occur when populations are reduced to very low numbers (e.g. through hunting or disease) and can lead to two major issues. The first is called inbreeding depression, and occurs because relatives in small populations tend to mate with each other. This causes alleles that are deleterious and recessive (and so often 'hidden' by the effects of dominant alleles) to come together in offspring, and so be expressed.

▶ We learn more about dominant and recessive alleles in Chapter 4.

One example of inbreeding depression was documented in red deer (*Cervus elaphus*) on the Isle of Rum in Scotland where inbreeding caused a reduction in birth weight and first-year survival. A second issue is that the loss of variation in small populations reduces the ability of populations to cope with environmental change.

Many conservation efforts now consider the importance of genetic variation in their programmes. For instance, the use of wildlife tunnels or bridges in deforested areas can increase community connectivity, increase genetic variation, and improve the long-term persistence of a species. Breeding programmes also seek to maintain the variation of species in captivity. In addition, the identification of particularly varied subpopulations may be important for conservation: in 2014, genetic and morphological evidence for hidden diversity was found in a commercially important sardine that may be critical for their future protection.

Modern conservation initiatives may also take into the account the differences between species when they are deciding how to focus their efforts. One of the best-known examples of this is the EDGE of Existence programme, designed by the Zoological Society of London. The aims of the programme are to identify the most evolutionarily distinct and globally endangered (EDGE) species by identifying how 'isolated' they are on evolutionary trees, and how at risk of extinction they are.

We also now know that species may be lost when they are so similar to others that they can hybridize with them. For instance, the African Rift Valley lakes are well known for the huge diversity of cichlid fish species that are found there. This diversity was partially maintained by sexual selection, where the different forms of fish used visual discrimination to avoid mating with each other. However, recent algal blooms in the lakes have meant that visibility in the lakes has been reduced, and different species of fish have now started to mate with each other, leading to a loss of diversity.

An understanding of evolution is also important when considering how to deal with specific threats to diversity. For example, when trying to assess the potential effects of invasive species, evidence suggests that the best predictor of invasion success is the number and size of introductions. This is because more variation in the invading population allows them to adapt to a new environment.

Evolution is also important in helping organisms to deal with the effects of climate change. When faced with a changing environment, to avoid extinction, organisms must either stay where they are and adapt to the new environment (or show phenotypic plasticity) or move to a more favourable location. In the future, conservation efforts may thus involve moving endangered organisms to different locations, or introducing beneficial variation from other populations that will help them to adapt where they are currently found.

7. Evolutionary theory underpins most of biology (and other subjects)

Finally, it is important to note that evolutionary theory is not only useful in the aforementioned examples, but is actually vital in many biological disciplines. It is often viewed as a recurring theme in biology, which unites the different disciplines. For instance, the theory is vital if we want to understand how gene frequencies change in populations over time (evolutionary genetics). This, in turn, helps us explain differences in the anatomy, physiology, and neurobiology of organisms, as well as the spread of disease and our susceptibility to disease (pathology and immunology).

A particularly exciting (and still relatively new) field of biological research that uses evolutionary theory is called evolutionary developmental biology, or 'evo-devo'. It compares the developmental processes of different organisms to infer how these developmental processes evolved, and how they might evolve in future. Through evo-devo, we have gained remarkable insights regarding evolution. For example, we now know that tiny changes in the expression of genes called *HOX* genes in the leg-bearing segments of crustaceans and insects can lead to large and rapid evolutionary changes, including the growth of extra sets of wings, as depicted in Figure T4.5a.

We also now understand that a lot of the supposedly 'non-coding' DNA in our genome actually contains regulatory elements (or switches) for other genes. Mutations in them cause large changes in development. For example, changes in the regulatory elements of the *PitX1* gene in stickleback fish (but not in the gene itself) can lead to the loss of a set of pelvic fins, as shown in Figure T4.5b.

Figure T4.5 Evolutionary development in action. (a) Mutations in the Ultrabithorax gene (*UBX*) can lead to an extra set of wings in *Drosophila melonogaster*. (b) A mutation in the regulatory element of the *PitX1* gene can lead to a loss of pelvic spines in sticklebacks. *Sources*: (a) David Scharf/Science Photo Library; (b) photo by Mike Shapiro and David Kingsley, Stanford University and Howard Hughes Medical Institute.

Further, we now think that genetic and developmental systems may constrain evolution, limiting the route down which it progresses. This may be part of the reason that we see such striking examples of convergence in many species (where organisms independently evolve similar features).

At the other end of the spectrum of biological disciplines, evolution can be used to explain a lot of animal behaviour. For instance, behavioural ecology seeks to explain animal behaviour by its effects on survival and reproduction. It uses a number of tools including optimality models and game theoretical models. Optimality models consider the benefits and costs of different types of behaviour, attempting to predict the best balance between them. They have been used to explain group sizes, mating strategies, and foraging patterns in many animals. Game theory models populations as a group of competing 'players' in a 'game', each of which has their own strategy in that game. Each strategy will 'win' against some opposing strategies, but may lose against others. This approach has been used to explain the existence of rare cases of altruism towards non-family members (amongst many other things).

4.3 Conclusion

To conclude, evolutionary theory really is critical to understanding much of the biological sciences. This fact is best summed up by the title of a 1973 essay written by Theodosius Dobzhansky:

Nothing in Biology Makes Sense Except in the Light of Evolution.

FURTHER READING

Mora, C., Tittensor, D. P., Adl, S., Simpson, A. G. B., & Worm, B. (2011) How many species are there on Earth and in the ocean? *PLoS Biol.* **9**.
An article that considers the diversity of life on Earth.

Laland, K. N., Odling-Smee, J., & Myles, S. (2010) How culture shaped the human genome: bringing genetics and the human sciences together. *Nat. Rev. Genet.* **11**: 137–48.
An article that considers recent human evolution and how our culture has affected it.

Doebley, J., Stec, A., & Gustus, C. (1995) teosinte branched1 and the origin of maize: evidence for epistasis and the evolution of dominance. *Genetics* **141**: 333–46.
An article that explores the origin of maize.

The Diversity, Organization, and Classification of Life

Contents

Introduction	42
5.1 Taxonomy	43
5.2 What is a species?	44
5.3 Higher taxa: larger groupings, families, and kingdoms	48
5.4 The domains of life: the highest taxa	49
5.5 Systematic interpretation of organismal relationships	52
5.6 Five pitfalls of taxonomic interpretation	57

Introduction

The diversity of the living organisms that surround us is remarkable and endlessly fascinating. Whether we look at plants, animals, bacteria, or any other forms of life, we wonder at its variety, its richness, and its persistence.

It is natural to ask how this diversity came to be, to ponder the possible connections between organisms, and to wonder, perhaps, at the position of humans in the overall scheme of life. The history of biology is a history of attempts to make sense of life's diversity, to rationalize and understand it.

When we think about the world and the objects in it we find it helpful to put things in groups. This does not only apply to living organisms: physicists group atomic particles by energy, charge, and spin; chemists arrange the fundamental elements according to their structure in a table with rows and columns; geologists group rocks and minerals according to their composition and origin; and meteorologists sort clouds into visually and functionally similar types.

Grouping things and identifying them with appropriate labels seems to be a natural human tendency. Besides giving a sense of order to a complex world, it facilitates informed discussion. It is especially useful for biological discourse because it give us an agreed set of names for the organisms we study, based on descriptions of how they look, how they are constructed, and how they work. This allows us to compare findings, to build up a body of information, and to develop a comprehensive view of the biosphere.

The naming and grouping of biological organisms in this way is called **taxonomy** and the resulting collection of names forms a biological **nomenclature**.

5.1 Taxonomy

The science of taxonomy, also called **systematics**, is concerned with establishing the identity of every living organism, naming it and placing it in the most appropriate category for further study. Taxonomy therefore depends on the detailed analysis and precise description of an organism's characteristics.

A **taxon** (plural **taxa**) is any group of organisms which share similar characteristics. Within a taxon, individual types of organisms (usually called species) are identified and named using features that distinguish them from other members of the taxon.

Taxonomy is useful because it means that when we study the group we understand more about its members than we would if we studied them individually. Grouping organisms by the features they have in common implies that they are similar to one another or related in some way. The grouping is informative and we could say that it has heuristic value.

But how should we decide which characteristics to use for grouping? Could it be that some characteristics reflect human concerns (e.g. the usefulness of a plant or the danger posed by a predator) or represent superficial similarities (general shape, say, or colour) while others have a deeper biological significance? Are some groupings more valid than others?

To answer this question, it helps if we recognize two different approaches to taxonomy: **subjective** and **objective grouping**.

Subjective and objective grouping

Taxonomic groupings can be based on whatever feature or features are appropriate to our purpose.

- A nutritionist might separate plants that are edible from those that are not, and then subdivide them according to the energy value of their fruits, seeds, or roots.
- A pharmacologist or herbalist might be more interested in cataloguing those with medicinal properties.
- A marine scientist studying ocean habitats might list all vertebrates that swim but separate those with scales and fins from those with skin and limbs.
- A medical microbiologist studying pathogenic fungi, bacteria, and viruses might find it best to group them according to the types of disease they cause.

Perhaps grouping is most obviously done on the basis of structural or physiological similarities:

- flowering and non-flowering in plants;
- animals which fly and those which crawl;
- insects with one pair of wings or two;
- vertebrates which feed their young with milk;

and there is no reason why other observable features (colour, behaviour, growth rate, chemical composition, habitat, nutritional value . . .) should not be used.

All these examples represent *subjective* grouping because they depend on the interests of the person carrying it out and the particular set of characteristics they have decided to focus on.

Subjective grouping can be useful but it suffers from the disadvantage that one scientist's purpose may be different from another's. It would surely be less confusing and better for scientific progress if there was one system that everyone used. Such a system would be *objective* rather than *subjective* because it would be based on characteristics that are independent of the reason for study.

Objective grouping: Aristotle, Linnaeus, and the origin of names

The first attempt at an objective classification of life forms is usually attributed to the Greek philosopher and scientist Aristotle (384–322 BCE). He distinguished between animals 'with blood' (roughly those we now call vertebrates, possessing back bones) and those 'without blood' (invertebrates). Animals with blood were subdivided into those producing live young (now called mammals) and those laying eggs (fish, birds), while those without blood included the groups we now call insects, crustaceans, and molluscs.

Aristotle's groupings, although based on the structural criteria he saw, were broad and not particularly helpful in identifying individual organisms. He believed there to be a *hierarchy* of functional and structural complexity, extending from plants to humans, and that life forms could be graded according to their degree of ultimate perfection. His scheme was therefore **teleological**: the various forms around us result from nature's purposeful designs towards a final goal of perfection as represented by humankind.

The founder of modern taxonomy was the Swedish physician and botanist Carl Linnaeus (1707–1778). He was a scholar of enormous industry who, with the help of students, correspondents, and travellers, assembled numerous specimens and vast quantities of descriptive information about plants and animals from around the world. He had an orderly approach and an obsessive attention to detail. This allowed him to arrange plants and animals by their physical similarity and to indicate their identities in a system of names.

In his *Systema Naturae* (first edition published 1735, 12th edition 1766) Linnaeus assigned a **binomial** (two-part) name to each organism. This was done using Latin, the language of science and natural philosophy at the time. Each name had a *generic* part and a *specific* part. They were somewhat complex names because he wanted them to provide a unique description of the organism. For example, the common dog-rose was *Rosa sylvestris vulgaris, flore odorata incarnato*, which means 'The common rose of the woods, with a flesh-coloured sweet-scented flower'.

These names soon became unwieldy and were reduced to **trivial names** (*Rosa canina*, which literally means dog-rose) in the manner we are more familiar with today. The usual name for the organism in the speaker's vernacular language (English, Arabic, Chinese . . .) is usually called the **common name**.

Systema Naturae eventually encompassed 7700 plants and 4400 animals and formed the basis for most subsequent attempts to describe the biological world. Each organism was defined by reference to a **type specimen**, and the trivial binomial name carried an abbreviated epithet to indicate the person (the 'authority') who defined the type. It is said that Linneaus' own body is the type specimen for human kind (trivial name: *Homo sapiens* Linn. or *Homo sapiens* L.) because that was the only example he studied in detail. Type specimens (held in museums or other collections) are also called **holotypes**.

Linnaeus' vast collection of specimens, with later additions by colleagues and followers, was assembled in Uppsala after his death before being dispersed in subsequent decades to other centres around the world. Linnaeus also published several treatises on plant structure and identification, and it can be said that his thorough and painstaking work established botany as a science. His method applied equally well to animals and was even adopted by geologists wishing to name and classify rock types. The modern equivalents of Linnaeus' catalogues are the identification keys used by practical biologists, especially those working in the field.

Objective grouping and naming conventions for species

The contemporary nomenclature for individual life forms retains the basic binomial system of species identification established by Linnaeus, including the use of Latin (words with a Greek root are also used). These so-called 'scientific names' are generally written in italics and only the first letter of the generic part is capitalized.

The generic and specific parts of each species name may reflect a functional or morphological feature of the organisms, or they may relate to the name of the discoverer, the place of discovery, or some other unique feature. Some examples are explained in Table T5.1.

Naming organisms in this way is sometimes called **classical** or **numerical taxonomy**. Where there are vast numbers of similar organisms, their characteristics can be scored and analysed by a computer program to assess their degree of similarity. The result is depicted as a dendrogram, which can be helpful in suggesting likely evolutionary relationships. An example based on bacterial strains is given in Experimental Toolkit T5.1.

For familiar genera or where members of the academic community communicate informally, the generic part of the name may be abbreviated. Thus the fruit fly *Drosophila melanogaster* becomes *D. melanogaster*, the nematode roundworm *Caenorhabditis elegans* becomes *C. elegans*, and the bacterium *Escherichia coli* becomes *E. coli*. Sometimes it is handy to refer to familiar organisms by the generic part alone (*Drosophila*, etc.). Where more than one species in a genus is implied, this is indicated by a plural abbreviation, *spp*. For example, *Quercus spp.* signifies the various species of oak tree.

Some taxonomists make use of trinomial names (genus + species + subspecies), for domesticated animals (e.g. *Bos taurus domesticus* for domestic cattle, derived from an ancestral ox called *Bos taurus*) and for subspecies resulting from geographical separation (e.g. *Panthera pardus fusca*, *Panthera pardus melas*, and *Panthera pardus nimr*, three subspecies of leopard found in India, Java, and the Arabian peninsula, respectively). However, this approach is far from consistently applied.

A number of naming conventions are used and these are specified in codes of nomenclature maintained by international organizations. For eukaryotes there are the International Code of Zoological Nomenclature (ICZN) and the International Code of Nomenclature for Algae, Fungi, and Plants (ICN). For prokaryotes there is the International Code of Nomenclature of Bacteria (ICNB) and a code is also maintained by The International Committee on Taxonomy of Viruses (ICTV).

For newly discovered species it is accepted that the discoverer has the right to suggest what the elements of the name should be. Reference lists of species names and descriptions will often include the authority epithet, sometimes with the date of first description, but these details are generally left out in common usage except where it is necessary to avoid confusion.

5.2 What is a species?

The Linnaean method identifies a 'species' on the basis of a detailed analysis of *structure*. This appears to be an objective approach (in the sense explained above), but it is not infallible. Consider, for example, the difference between a German shepherd dog and a Havanese, depicted in Figure T5.1. These are both domestic dogs (*Canis familiaris*), but selective breeding over countless generations has removed any superficial resemblance between them to the extent that an alien observer of Earth's fauna might be surprised to learn that they are the same species. However, the common pipistrelle bat was once believed to be a single species but is now known to comprise at least two. The species are morphologically (structurally) identical and can only be distinguished by the pitch at which they echolocate: the common pipistrelle (*Pipistrellus pipistrellus*) uses 45 kHz, while the soprano pipistrelle (*Pipistrellus pygmaeus*) uses 55 kHz.

Modern taxonomists increasingly rely on information from RNA and DNA sequences, protein structures, and immunological compatibility to identify species. This kind of information is also used to confirm or disprove similarities and differences that were previously based on morphology alone. The information gathered this way is extensive and the quantity of it is increasing rapidly. (We explain how it is obtained in Experimental Toolkit T5.2.)

Such molecular analysis is capable, theoretically, of testing every type description made since Linnaeus began his remarkable work. Where appropriately preserved fossil or other material exists, it is also used to identify extinct life forms and to compare ancient organisms with their contemporary descendants or relatives.

We know, for example, that the genomes of most living humans contain DNA sequences that are similar to those found in well-preserved remains of Neanderthal hominids who became extinct approximately 40,000 years ago. As these hominids were contemporary with the direct ancestors of modern humans, it is concluded that *Homo sapiens* and *Homo neanderthalensis* interbred.

The biological definition and Darwin's legacy

Despite ever-increasing amounts of information and vast molecular datasets, providing an incontrovertible definition of a species can be a problem. For most animals and plants, a good working definition is *an interbreeding group of organisms whose offspring are fertile and closely resemble their parents*. (The words 'breed' and

Table T5.1 Binomial scientific names of some prokaryotic and eukaryotic organisms, and their derivation

Group	Name	Derivation
Prokaryotes	*Escherichia coli*	*Escherichia* = Theodor Escherich who discovered the bacterium in 1888
		coli = bowel or lower gut
	Deinococcus radiodurans	*Deinos* = strange or unusual
		radio = radiation; *durans* = enduring
	Thermus aquaticus	*Thermos* = hot
		aquaticus = found in or by water
	Helicobacter pylori	*Helix* = spiral; *bacter* = rod or staff
		pylorus = gate keeper (lower orifice of stomach)
	Rickettsia prowazekii	*Rickettsia* = named after Howard Taylor Ricketts who first associated this bacterium with typhus
		prowazekii = named after Stanislav von Prowazek, an early investigator of typhus
Eukaryotes: Plants	*Allium cepa* (Onion)	*Allium* = garlic family of plants
		cepa = onion
	Solanum tuberosum (Potato)	*Solanum* = nightshade family of plants
		tuberosum = bulbous or tuberous
	Digitalis purpurea (Foxglove)	*Digitalis* = little glove (from the shape of the flowers)
		purpurea = reddish purple
	Hordeum vulgare (Barley)	*Hordeum* = bristly (the ear has long bristles called awns)
		vulgare = common
	Quercus robur Common oak	*Quercus* = beech and oak family of plants
		robur = hard/strong
Eukaryotes: Animals	*Drosophila melanogaster* (Fruit fly)	*Drosophila* = dew-loving
		melanogaster = dark-bellied
	Hirudo medicinalis (European medicinal leech)	*Hirudo* = adhere (to skin); the genus was named by Linnaeus
		medicinalis = used in medicine
	Haliaeetus albicilla (White-tailed eagle)	*Hali* = sea; *Aetos* = eagle
		albi = white; *cilla* = tail
	Spermophilus tridecemlineatus (Thirteen-lined ground squirrel)	*Spermophilus* = seed-loving
		tridecem = thirteen; *lineatus* = line
	Giraffa camelopardalis rothschildi (One of several sub-species of Northern giraffe, also called the Baringo giraffe or the Ugandan giraffe; see Figure T5.3)	*Giraffe* = giraffe
		camellopardalis = derived from the Greek word for giraffe
		rothschildi = named after British naturalist (Lord) Walter Rothschild (1868–1937)

'variety' are also sometimes used, although with somewhat looser meanings.)

This definition of a species is objective, but it is often challenged. It becomes particularly problematic in the following situations:

- Many plants and some animals reproduce asexually. Individuals do not therefore always interbreed in the sense implied by the definition.

- Many plants, both wild and domestically bred, as well as a few animals, interbreed to form hybrids. The resulting individuals, which may or may not be fertile, may have a mixture of the parental characteristics and are neither one species nor the other. An example is the wild cowslip, *Primula veris*, which hybridizes with other members of the *Primula* genus to produce individuals with multi-headed and multi-coloured flowers, as shown in Figure T5.2.

EXPERIMENTAL TOOLKIT **T5.1**   The classification of bacteria and their relatedness

Strains of bacteria can be grouped according to characteristics of morphology, metabolic processes, and physiology. These traits include cell shape, colour, motility, whether they grow aerobically or anaerobically, cell wall composition, and their ability to metabolize specific chemicals. Each organism can be scored for each trait: 1 for positive and 0 for negative. On a statistical basis, the more characteristics organisms share, the closer their evolutionary relationship is likely to be.

From the scores, a computer algorithm constructs a similarity matrix, as depicted in Figure 1. From this a **dendrogram** can be generated, as shown in Figure 2. A dendrogram shows, by calculation, the most likely pathway for the evolutionary derivation of each strain.

Classical, descriptive taxonomy is still a powerful mechanism for quick identification of bacterial strains. This is carried out in a range of situations

(research laboratories, clinical labs, hospitals, public health labs, field study sites) using API strips, such as those shown in Figure 3.

API strips have small 'wells' into which the unknown bacterial suspension is placed. Each well tests a different biochemical property—for example, whether the bacterium can utilize lactose. A positive result gives a colour change and the results can be represented as a number (0 for no reaction, 1 for a reaction). The resulting binary code can then be matched to an identification database. Different API strips have different combinations of components, to distinguish between specific groups of bacteria.

Although both methods provide identification, comparing morphology to calculated relatedness can be inaccurate as it is not always clear whether a trait is ancient in the ancestry (synapomorphy) or a recent adaptation (apomorphy). The reliability of the calculated relationships is improved with larger quantities of similarity data.

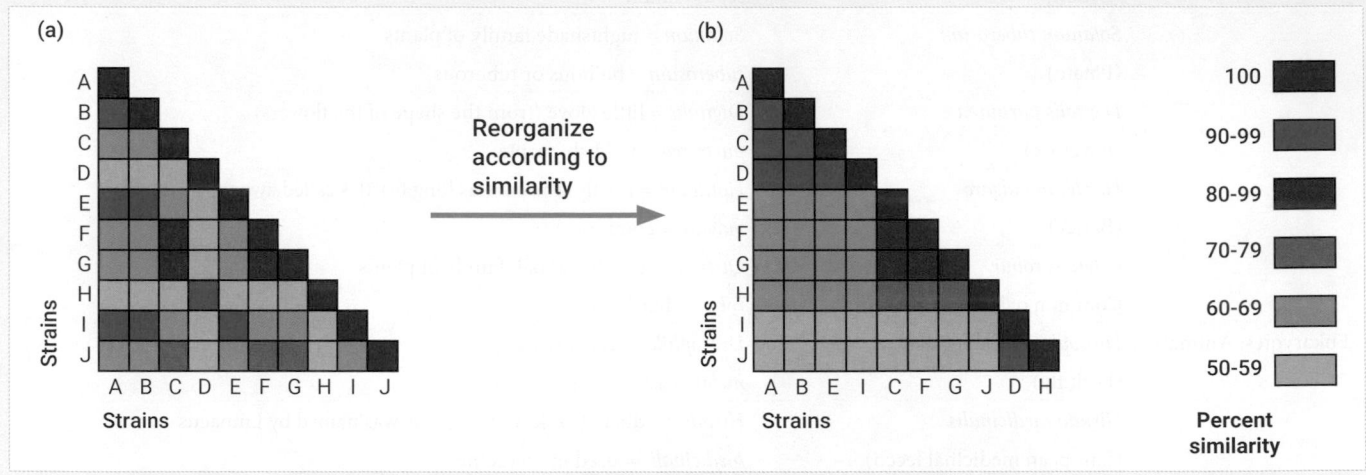

Figure 1 A similarity matrix in which bacterial strains A–J have been compared to estimate their percentage similarity.

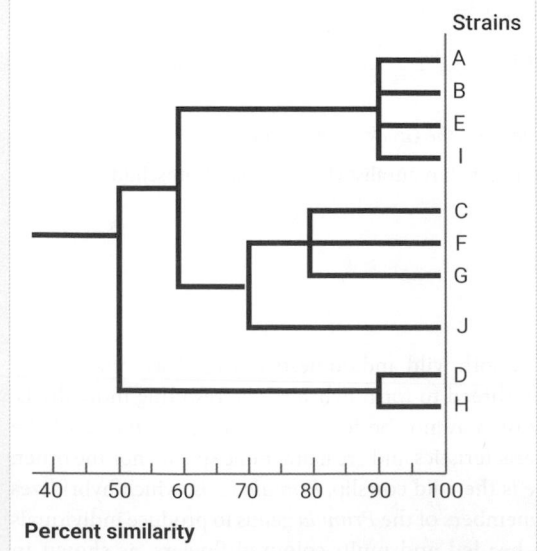

Figure 2 Dendrogram generated from the similarity matrix in Figure 1.

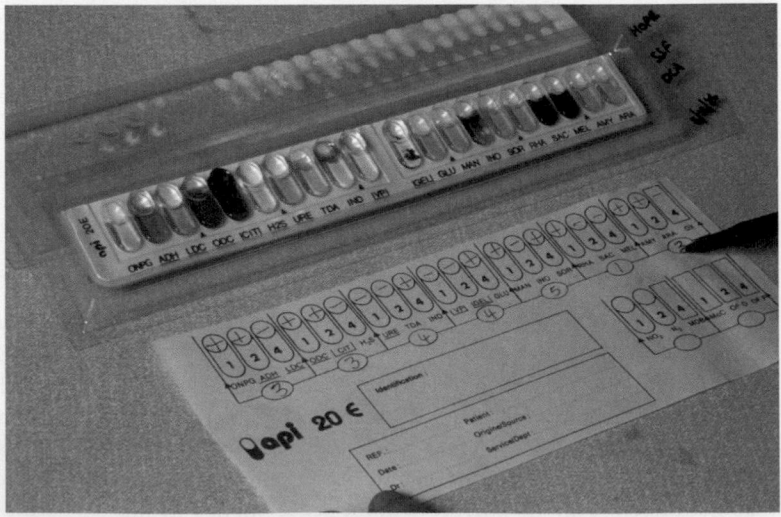

Figure 3 An example of API strips.

Source: Science History Images / Alamy Stock Photo.

Figure T5.1 Contrasting forms in different breeds of a single species. (a) Havanese. (b) German shepherd dog.
Source: (a) Pixabay; (b) Wikimedia Commons/Platyna.

- Two or more populations of individuals may be prevented from breeding by geographical separation. In this process, called **allopatry**, an otherwise homogenous community of organisms becomes divided and the two groups eventually go their separate evolutionary ways under local pressures of climate, predators, resources, or genetic drift. (The opposite process is called **sympatry**: the formation of separate groups in the same location with no apparent physical or other barrier). In allopatric speciation, the point at which the two groups should be considered distinct species is unclear.

Figure T5.2 The formation of hybrids by interbreeding. (a) Cowslip (*Primula veris*) with multi-headed and multi-coloured flowers. (b) *Primula vulgaris* with yellow flowers.
Sources: (a, left) Wikimedia Commons/BerndH; (a, right) Wikimedia Commons/Jasper33; (b) Wikimedia Commons/SvdMolen.

- The domestication and selection of agricultural animals and pets has led to a diversity of breeds that scarcely resemble their ancestors in size, shape, or behaviour. These forms could theoretically breed with their ancestral species or would be capable of breeding with wild individuals but have no opportunity to do so. Should they be identified as different species?

Charles Darwin (1809–1882) considered the effects of hybridization, separation, and selection, both natural and human-driven, in great depth. He realized that, along with the ability of some variants to survive better than others, they contribute significantly to the generation of new species.

Darwin's great insight was to understand that organisms change over time as a result of selection. At the end of *On the Origin of Species* (first published in 1859) he wrote:

> *Thus, from the war of nature, from famine and death, the most exalted object which we are capable of conceiving, namely, the production of the higher animals, directly follows. There is grandeur in this view of life, with its several powers, having been originally breathed into a few forms or into one; and that, whilst this planet has gone cycling on according to the fixed law of gravity, from so simple a beginning endless forms most beautiful and most wonderful have been, and are being, evolved.*

There is no reason to believe that evolution has ceased, and we should really consider the existence of *any* species as a transient state. It is also highly likely that the ability of plant and animal geneticists to make artificial hybrids by molecular manipulation will lead to further debate, especially where fertile and potentially free-living organisms are created.

The phylogenetic definition

An alternative definition of a species is *the smallest group of organisms which share an ancestor and can be distinguished from other groups (by phenotype or genotype)*. This is sometimes called the **phylogenetic definition** for it depends on objective, descriptive criteria related to pedigree (ancestry) rather than on the fortuitous localization of organisms or the breeding opportunities available to them.

By this definition, each identifiably distinct form is taken to have evolved separately and to have its own, unique history. (We shall consider the importance of common ancestry again presently.) The hybridization and group separation problems potentially still remain, however, and it is not always superficially obvious what constitutes a distinct group. The giraffes that range over the savannas of sub-Saharan Africa were for a long time thought to be one species (*Giraffa camelopardalis*) with several subspecies (see Figure T5.3). Modern genetic analysis shows that there are at least four distinct groups that have not interbred for millions of years.

As we can see, the biological and the phylogenetic definitions of species both have limitations. Nevertheless, the word 'species' is meaningful, widely understood, and in very common use. The acceptance of species identities and of precise ways of asserting them remains crucial to studies in all areas of biology.

Figure T5.3 Rothschild's giraffe (*Giraffa camelopardalis rothschildi*), one of several subspecies of giraffe found in sub-Saharan Africa.
Source: Wikimedia Commons/Bernard Dupont.

5.3 Higher taxa: larger groupings, families, and kingdoms

As we have seen, Aristotle attempted to divide life forms into broad divisions and organized them according to their nearness to ultimate perfection. Linnaeus, in contrast, began with individual plants and animals and used his binomial labels to define species (*specific* = species). He refrained from teleological (purposeful or creation-based) interpretations, but he did gather species into larger groupings. He put plants and animals (and minerals) into separate *kingdoms*. Within the kingdoms, he described distinct *families* and *orders* based on broader morphological similarities that could be easily discerned (e.g. the number of stamens and pistils in flowers).

An adaptation of this system is in common use today and, like the system of species names, it follows a set of conventions. The groupings (taxa) above species level are established as *levels of similarity* or *ranks*. As such, the larger groupings—called higher taxa—embrace more species with fewer characteristics in common. The ICN and ICZN specify the higher taxa into which each species falls.

Figure T5.4 shows how this system works in the case of familiar plant and animal examples. The taxa listed represent a generally agreed set of main groupings. Many biologists working with

Figure T5.4 The principal taxonomic divisions of life. Two example organisms: the pea and honey bee.

Sources: Pea: Wikimedia Commons/Rasbak; Honey bee: Wikimedia Commons/Sharp Photography.

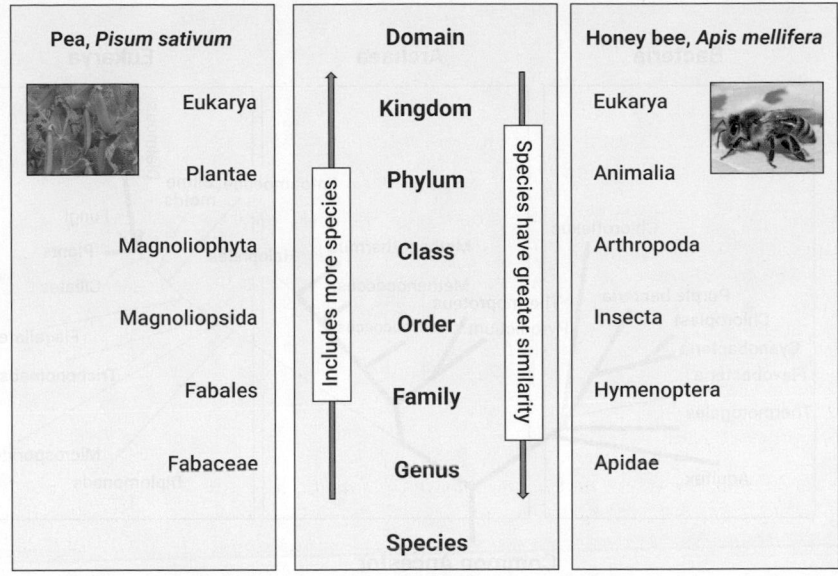

Pea, *Pisum sativum*	Domain	Honey bee, *Apis mellifera*
Eukarya	Kingdom	Eukarya
Plantae	Phylum	Animalia
Magnoliophyta	Class	Arthropoda
Magnoliopsida	Order	Insecta
Fabales	Family	Hymenoptera
Fabaceae	Genus	Apidae
	Species	

(Includes more species → | Species have greater similarity →)

specific types of organism find it convenient to insert additional groupings (e.g. subphylum and subclass) to indicate where other common characteristics are especially important for study.

There are many cases where important group characteristics fail to align conveniently with the overall scheme illustrated in Figure T5.4. Such groupings can be named but are generally specified as 'unranked'. Good examples of this would be the angiosperm (flowering) group of vascular plants and the monocotyledon group of angiosperms: both of these groupings, which are of enormous botanical significance, fall somewhere between the ranks of kingdom and phylum. Similarly, ornithologists commonly refer to birds of prey as *raptors* and to flightless birds without enlarged sternums (emus, ostriches, etc.) as *ratites*. Neither of these groupings, which are informative and widely used, can be positioned conveniently in the list of higher taxa.

What we see from this is that the higher taxa, derived from Linnaeus' system, are groupings of convenience. They result from the human tendency to categorize things and they continue to play an important part in biological discourse. Their meanings can be set out very precisely (in the International Codes or by the specialists who use them), but they are groupings defined by convention: they do not stem from any absolute set of biological characteristics or status, nor do they relate to fixed steps on a developmental ladder.

Furthermore, there is no particular parallelism between taxa in different branches: for example, there is no similarity of meaning between 'phylum' in the plant kingdom and 'phylum' in the animal kingdom.

5.4 The domains of life: the highest taxa

Linnaeus categorized all living organisms into one of two broad kingdoms: Plantae or Animalia. After the discovery of microbial life in the 19th century, Carl von Nägeli proposed (in 1857) that all fungi, bacteria, and protists (other single-celled organisms) were to be placed in the Plantae kingdom. A century or so later, in 1969, Robert H. Whittaker introduced a five-kingdom system consisting of Plantae, Animalia, Fungi, Protista, and Monera (prokaryotes).

In 1977, Carl Woese and George Fox created a *tripartite* phylogenetic tree, and this is the overarching arrangement which is currently in common use. This group's life forms into the three **domains** illustrated in Figure T5.5:

- Bacteria (the majority of prokaryotes);
- Archaea (a group of prokaryotes with distinctive biochemistry);
- Eukarya.

We explain the fundamental differences between prokaryotes and eukaryotes in Chapter 9 but note how plants, animals, and fungi are now all identified as eukaryotes. Figure T5.5 shows that they form a relatively small region of the tree.

The Woese–Fox phylogenetic tree was based on evolutionary relationships rather than on structural or functional resemblances. Points of common ancestry were determined quantitatively by identifying similarities and differences in the genes coding for the small subunits of ribosomal RNA (rRNA; the 16S subunit in prokaryotes and the 18S subunit in eukaryotes) taken from representative organisms. This was the first time that the evolutionary connections between groups of organisms were deduced from an extensive analysis of molecular sequences.

▶ **We learn more about rRNA in Chapter 5.**

The basis for this approach was the realization that the general structure of rRNA is common across all organisms and that small variations in non-essential regions of the molecule (the variable and hypervariable regions) have accrued at a more or less constant rate over evolutionary time. The ribosomes therefore represent a kind of **evolutionary chronometer** or **molecular clock**. Experimental Toolkit T5.2 explains how this type of analysis is carried out.

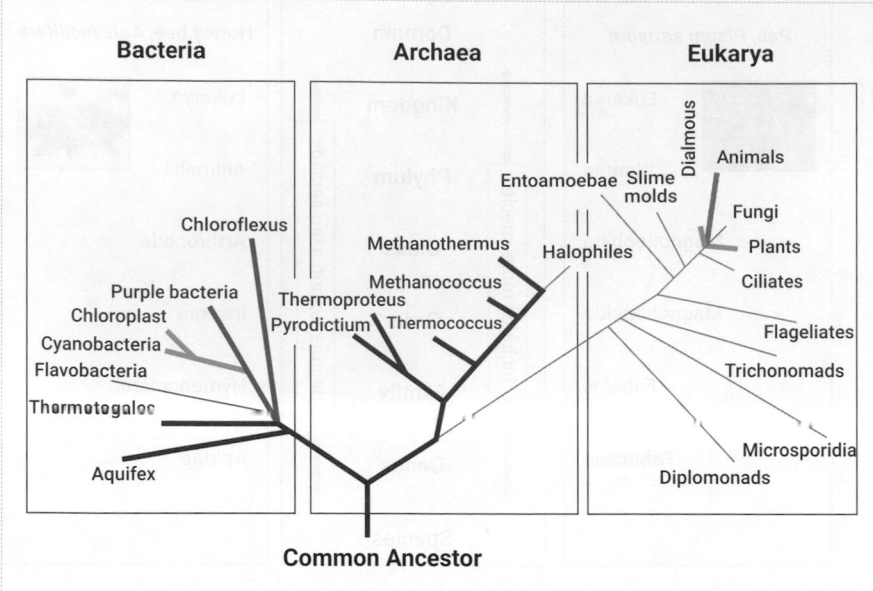

Bacteria   **Archaea**   **Eukarya**

Chloroflexus

Purple bacteria
Chloroplast
Cyanobacteria
Flavobacteria

Thermotogales

Aquifex

Methanothermus
Methanococcus
Thermoproteus
Pyrodictium   Thermococcus

Halophiles

Entoamoebae Slime
molds

Dialmous

Animals
Fungi
Plants
Ciliates
Flageliates
Trichonomads

Microsporidia
Diplomonads

Common Ancestor

Figure T5.5 A phylogenetic tree based on rRNA data. The tree consists of three domains: Bacteria, Archaea, and Eukarya.

Source: Woese, Microbiological Reviews, 1987, 51, 221-271 (p. 231). Licensed from PLS Clear.

EXPERIMENTAL TOOLKIT T5.2   Tracking the molecular evolution of organisms

The underlying structure and function of ribosomal RNA (rRNA) is similar across all organisms. However, small changes (mutations) occur in the sequence of bases (A, C, G, and T) of the genes that code for rRNA at a roughly constant rate over a long period of time. By analysing the number of differences between the rRNA genes of different species, it is possible to calculate the time at which they separated from a common ancestor. As such, the ribosome can be used as an evolutionary chronometer or molecular clock.

The 16S and 18S genes contain both highly conserved regions and non-essential (variable and hypervariable) regions. Mutations in the former will be lethal or cause malfunction and will therefore be rapidly eliminated from the population. Mutations in the latter are likely to be preserved from one generation to the next and subject to further mutations as time goes on. The conserved regions can be used to design PCR primers with which to amplify the gene coding for the rRNA, allowing for gene sequencing and analysis.

Figure 1 illustrates the use of the ribosomal molecular clock. The root sequence represents that of the common ancestor. Over time, mutations in the ancestor sequence occur, which causes a divergence into two groups (2A and 2B). Further accumulations of mutations result in more branching of these groups (3A, 3B, 3C, etc.).

One way to carry out ribosomal DNA analysis is to amplify the gene of interest using the polymerase chain reaction (PCR). This can be done with chromosomal DNA of purified organisms, or it can be done from DNA extracted from different environments. DNA primers are designed to complement small sections of conserved sequence in the 16S or 18S rRNA genes, which are then used in a PCR reaction to amplify a large

quantity of DNA. The DNA yielded can then be purified and used for sequence analysis.

Evolutionary trees are then generated from the sequence data by identifying conserved regions in the genetic code. As not all rRNA genes are the same length, software is used to align sequences so that specific functional regions can be compared according to the maximum likelihood that they are related.

▶ **We learn more about PCR in Chapter 8.**

Molecular classification is not limited to rRNA sequences. Other genes can be used as evolutionary chronometers, but they need to have conserved regions among the group of interest and have homologous functionality. In other words, they need to code for proteins with the same function: comparing a protease enzyme to a lipase enzyme would not work as they are functionally different and would not share conserved regions in their sequences.

Frequently used genes include those for the ATPase and cytochrome oxidases; these genes are found widely found across groups of organisms and were probably crucial in ancestral cells. Comparing the analysed genes to those already listed in databases such as the Ribosomal Database Project (RDP) can provide an insight into what biochemical activity an organism may be capable of, even when this activity cannot be studied directly.

Protein sequences are also used to characterize molecules. Homologous mRNA sequences should create homologous proteins, which will be similar in shape and structure. However, some proteins with similar sequences may incorporate different molecules in their active

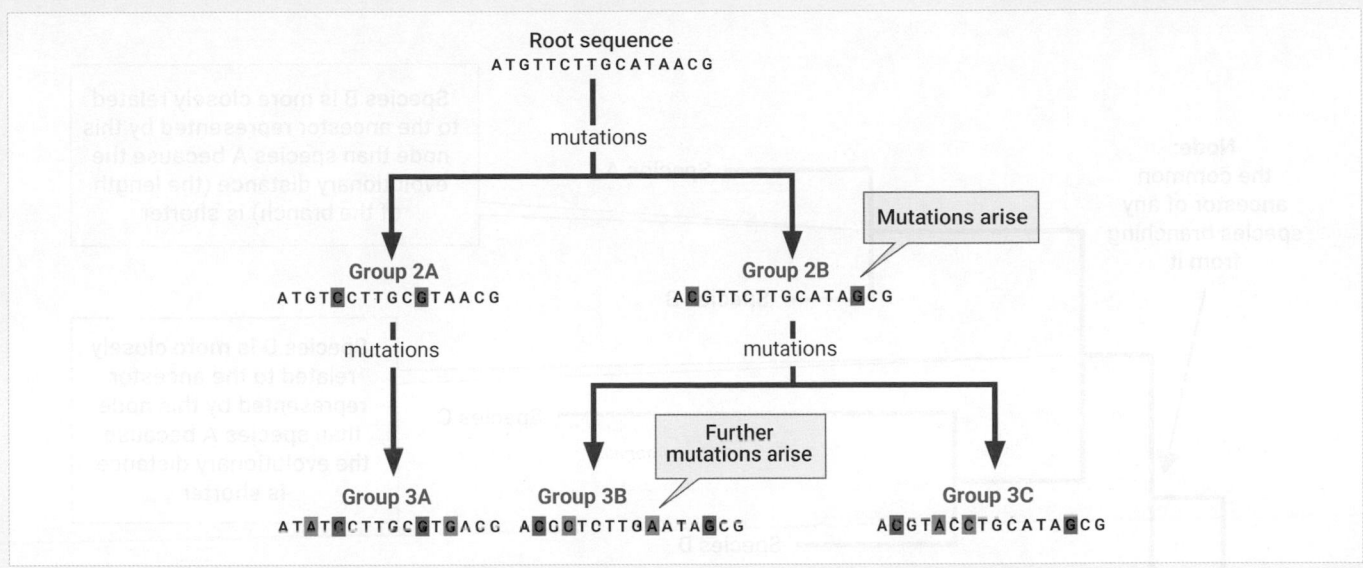

Figure 1 The molecular clock.

sites, so great care must be taken with the analysis and the deductions that follow.

How do we create phylogenetic trees?

A phylogenetic tree, which shows likely evolutionary relationships, is constructed using the aligned sequences by means of a treeing algorithm. This calculates the likely **evolutionary distance** (E_D) between the sequences based on the differences between them.

The treeing program firstly aligns the sequences so that conserved areas can be compared directly, after which the comparative analysis can be performed. An example of such an alignment for two bacterial traits is shown in Figure 2.

The data can be converted into an arrangement of connecting lines called branches, the length of each being proportional to the E_D: a long branch implies a large evolutionary distance, a short branch a small evolutionary distance. The result is a phylogenetic tree or dendrogram that shows the evolutionary distance between the organisms. The higher the homology (sequence similarity), the shorter the evolutionary distance and vice versa. A generalized example of a phylogenetic tree is shown in Figure 3.

Look at this figure, and notice the different branch lengths depicted. Species B has a shorter branch length from its most recent ancestor than species A so is more closely related to it (and therefore shows greater sequence homology with it). Species D has a shorter branch length from the ancestor one step further back on the tree than does species A so, again, is more closely related to that ancestor than is species A. So, we can deduce that it shows greater sequence homology with that ancestor than does species A.

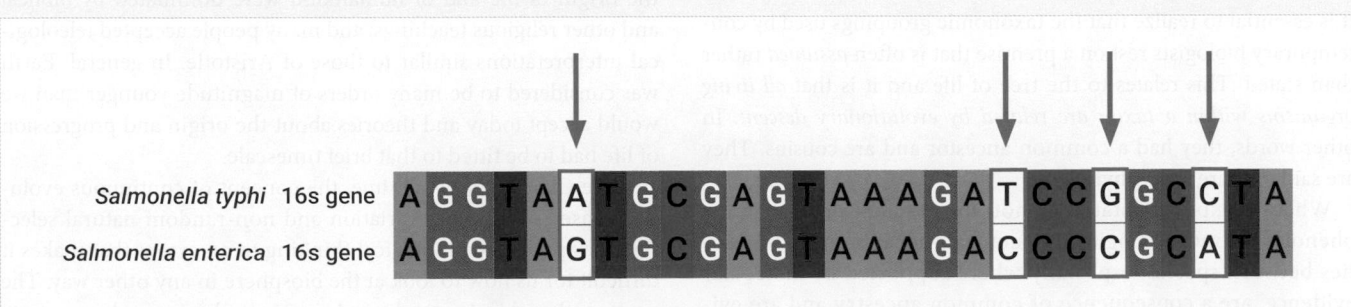

Figure 2 Two sections of the 16S gene sequence from *Salmonella typhi* and *Salmonella enterica* which have been aligned. Highlighted sections show conserved nucleotides.

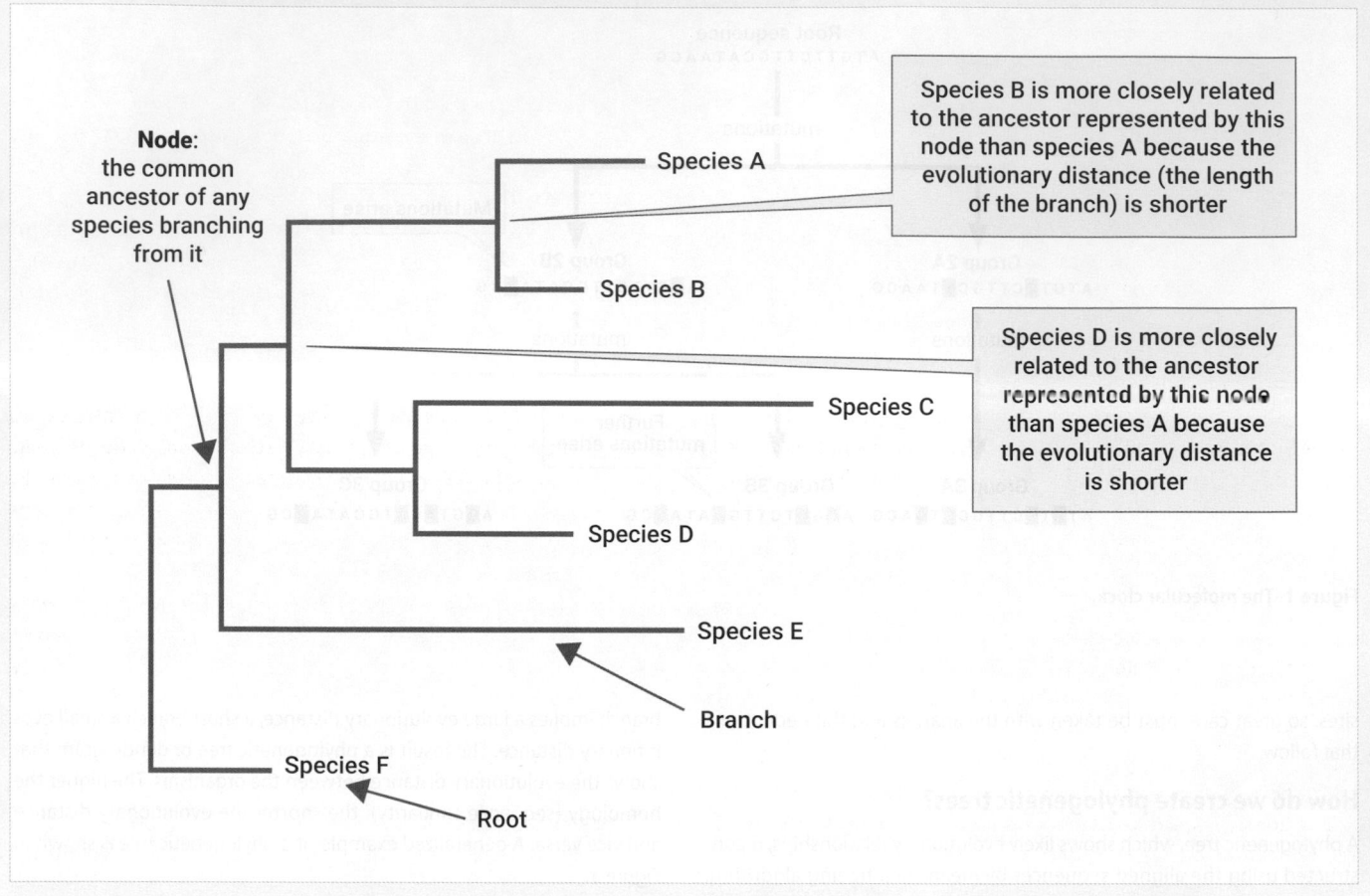

Figure 3 A generalized example of a phylogenetic tree.

Source: © Andy Vierstraete.

5.5 Systematic interpretation of organismal relationships

It is essential to realize that the taxonomic groupings used by contemporary biologists rest on a premise that is often *assumed* rather than stated. This relates to the tree of life and it is that *all living organisms within a taxon are related by evolutionary descent*. In other words, they had a common ancestor and are cousins. They are said to share their **phylogeny**.

Whether explicitly stated or not, the assumption is that any phenotypic (morphological, structural, behavioural) similarities between species, especially where supported by molecular evidence, are a consequence of common ancestry and are evidence of it.

The level of taxonomic grouping shown in Figure T5.4 suggests, in a general way, how far back in evolutionary history one would need to go to find that ancestor: the higher the level, the more ancient the phylogenetic connection.

When Linnaeus laid the foundations of his taxonomy based on the morphology of organisms, he was working at a time when ideas about evolution were essentially vague and fragmented. Views on the origin of life and of humankind were dominated by biblical and other religious teachings, and many people accepted teleological interpretations similar to those of Aristotle. In general, Earth was considered to be many orders of magnitude younger than we would accept today and theories about the origin and progression of life had to be fitted to that brief timescale.

Since Charles Darwin's time, the concept of continuous evolution, based on random variation and non-random natural selection, has permeated biological thinking to an extent that makes it difficult for us now to look at the biosphere in any other way. The geological and palaeontological sciences also now demonstrate the very long timescale over which evolutionary change has taken place.

It is perhaps a tribute to Linnaeus and those who developed his work that his fundamental methods continue to underpin descriptive taxonomy and nomenclature, even if we now add a deeper layer of interpretation of the links between the organisms we see around us.

Evolutionary trees, molecules, and clades

Many biologists find the traditional, Linnaean taxonomic approach restricting and arbitrary, for the reasons discussed above. Even though species can usually be agreed upon, the intermediate and higher taxa may not be as informative or reliable as we would wish. The *evolutionary* history of every species—its pedigree or ancestry—is unique so it would be preferable to use that as the objective basis for organizing our understanding of the biological world.

The ability to analyse connections based on molecular sequences (see Experimental Toolkit T5.2) encourages this evolutionary view and has given rise to complex trees linking all known life forms according to evolutionary principles. A current example is given in Figure T5.6. Diagrams like this, and their fractal equivalents (e.g. OneZoom; http://www.onezoom.org/), present us with a visual sense of how diverse life actually is. They also remind us that the forms we may be most familiar with are far from representative of life as a whole.

The evolutionary relationships between organisms are often described and interpreted using an approach called **cladistics**. The concept of a **clade** was first proposed by a German entomologist, Willi Hennig, in 1950. A clade is defined as '*a branch of the phylogenetic tree (the Tree of Life) which contains an ancestor, all of its descendants, and no other organisms*'.

A clade is a **monophyletic** grouping because all members of the clade have the same evolutionary origin. They are part of the same **lineage** and thus have the same phylogeny, as illustrated in Figure T5.7a. A **synapomorphy** is a phenotypic characteristic shared by all members of a clade.

A clade starts at a common ancestor but can be of any size provided (i) it contains a single common ancestor and all of its descendants; and (ii) that ancestor is not the originator of any organisms outside the clade. If a descendant becomes the ancestor for a branch of the clade, a new clade or subclade can be identified. Clades and subclades can become nested in this way. Besides living descendants, a clade should theoretically include any that may have become extinct, although in practice this is rarely the case.

A grouping that includes members of more than one clade (i.e. one from which an earlier common ancestor is missing) is called **polyphyletic**. One from which some descendants of the common ancestor are missing is called a **paraphyletic** group.

How do we interpret cladograms?

A **cladogram** is a diagrammatic representation of a clade; as such, it is a depiction of the evolutionary history of each organism and the closeness of the links between them. An example is shown in Figure T5.7a. Look at this figure, and note that all the symbols in this figure represent members of a clade whose common ancestor was organism A. Lines indicate the phylogenetic relationships between ancestors and their descendants (as in a family tree).

The rectangular symbols represent species that are currently alive. Species 1 is the only surviving descendant of organism D and may

or may not be taxonomically identical to it. Species 2 and 3 are sister species, as are 4 and 5. Species 3, 4, and 6 are cousins, as are species 1, 2, 4, and 8. The symbols in the green area are monophyletic and form a clade, with common ancestor B. It includes an organism (E) that went extinct without leaving modern descendants.

The symbols in the blue area do not represent a clade because their common ancestor (C) is excluded; the current species in this grouping are thus from different clades and are polyphyletic. The symbols in the orange area do not represent a clade because they exclude a monophyletic group (species 2 and 3 and their ancestor) also descended from the common ancestor; they comprise a paraphyletic group.

The cladogram in Figure T5.7a identifies the positions of individual species and ancestors.

An alternative presentation (Figure T5.7b) concentrates on lineages rather than species. This representation of the Primate clade uses an approach in which new lineages are shown to arise by branching from ancestral ones. By this interpretation, populations are continuous (rather than splitting, as suggested by Figure T5.7a) with new characteristics (**apomorphies**) arising to form distinct subgroups. It is important to understand that no hierarchy is implied by the horizontal arrangement of the current species (e.g. lemurs and lorises are sister groups and their positions could be swapped) and that horizontal distances have no meaning. Ancestry is indicated solely by the vertical position of the branching event (e.g. the lemur/loris group branched away before the haplorrhines appeared).

Many biologists find the form of cladogram shown in Figure T5.7b more acceptable because it suggests that organismal characteristics (apomorphies) are continually changing: it gives a better impression of branching (a subgroup of individuals moving away from the main group and developing their own synapomorphies). It also avoids the implication of sudden evolutionary switches from one ancestral species to another.

All the *living* species in a clade can be described as cousins, except for those which have separated most recently (the closest possible relatives), which are called sister species. (Strangely, they are never called brother species.) Because variation increases through the clade the further away the species get from the common ancestor, the number of synapomorphies *among cousins in subclades* increases. This is also indicated in Figure T5.4.

If that seems counterintuitive, consider a clade comprising the vertebrate animals. All members of this clade, including the original ancestral vertebrate, have a backbone. However, this may be the *only* morphological characteristic they share. For example, while the mammal, amphibian, and bird members of the clade are **tetrapods** (they have four limbs), their common vertebrate ancestor may not have been. All mammals have body hair, feed their young with milk, have red blood cells without nuclei, and are warm-blooded. Amphibians have none of those characteristics. Birds are warm-blooded but have feathers rather than hair, have nucleated red blood cells, and do not feed their young on milk. Thus, the mammal clade, the amphibian clade, and the bird clade are subclades of the vertebrate clade. Each has its own synapomorphies in addition to the vertebrate clade synapomorphy of a backbone.

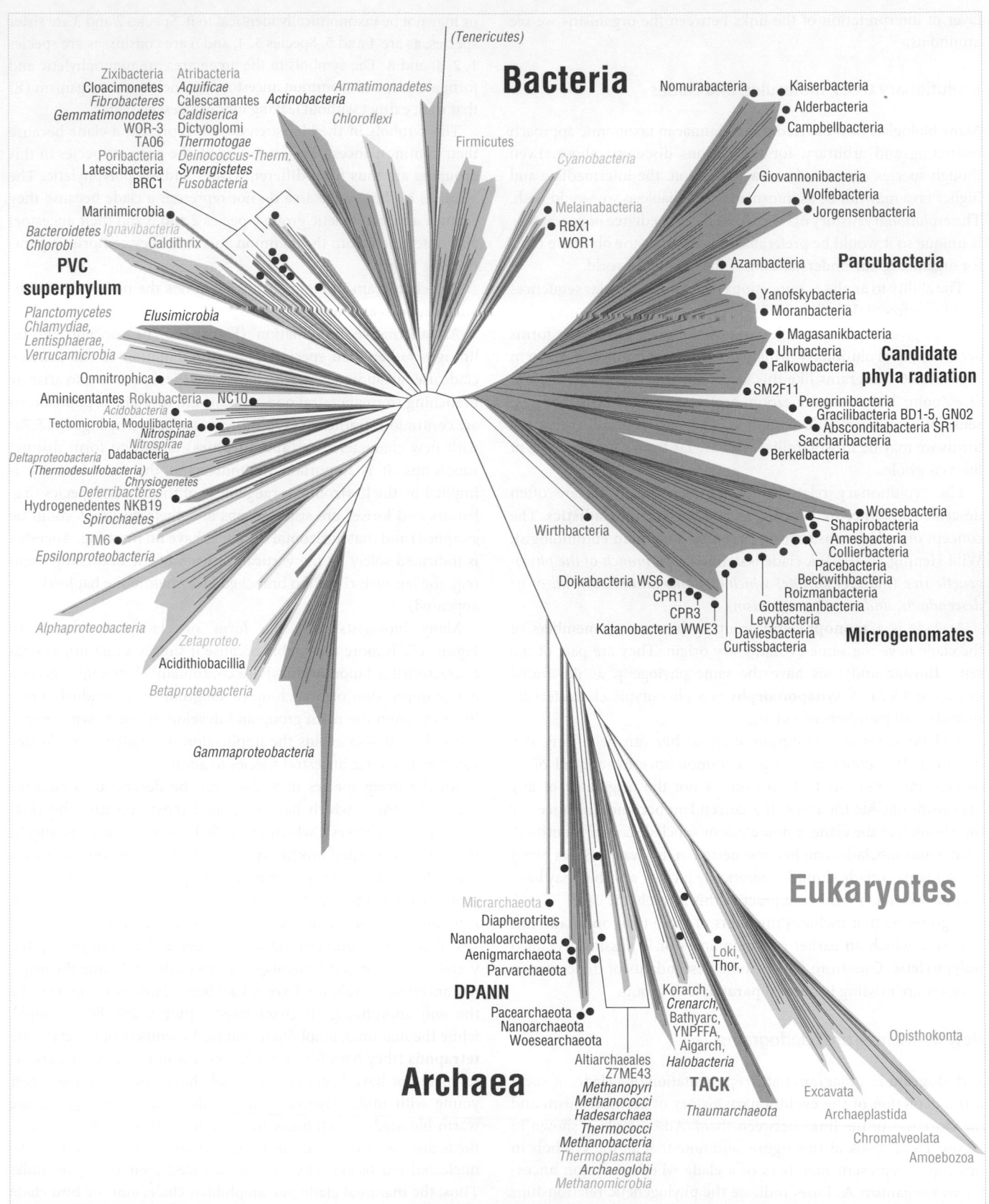

Figure T5.6 Tree of life. A contemporary tree of life, based on molecular data.

Source: Adapted from Hug et al. Nature Microbiology 2016,1:16048.

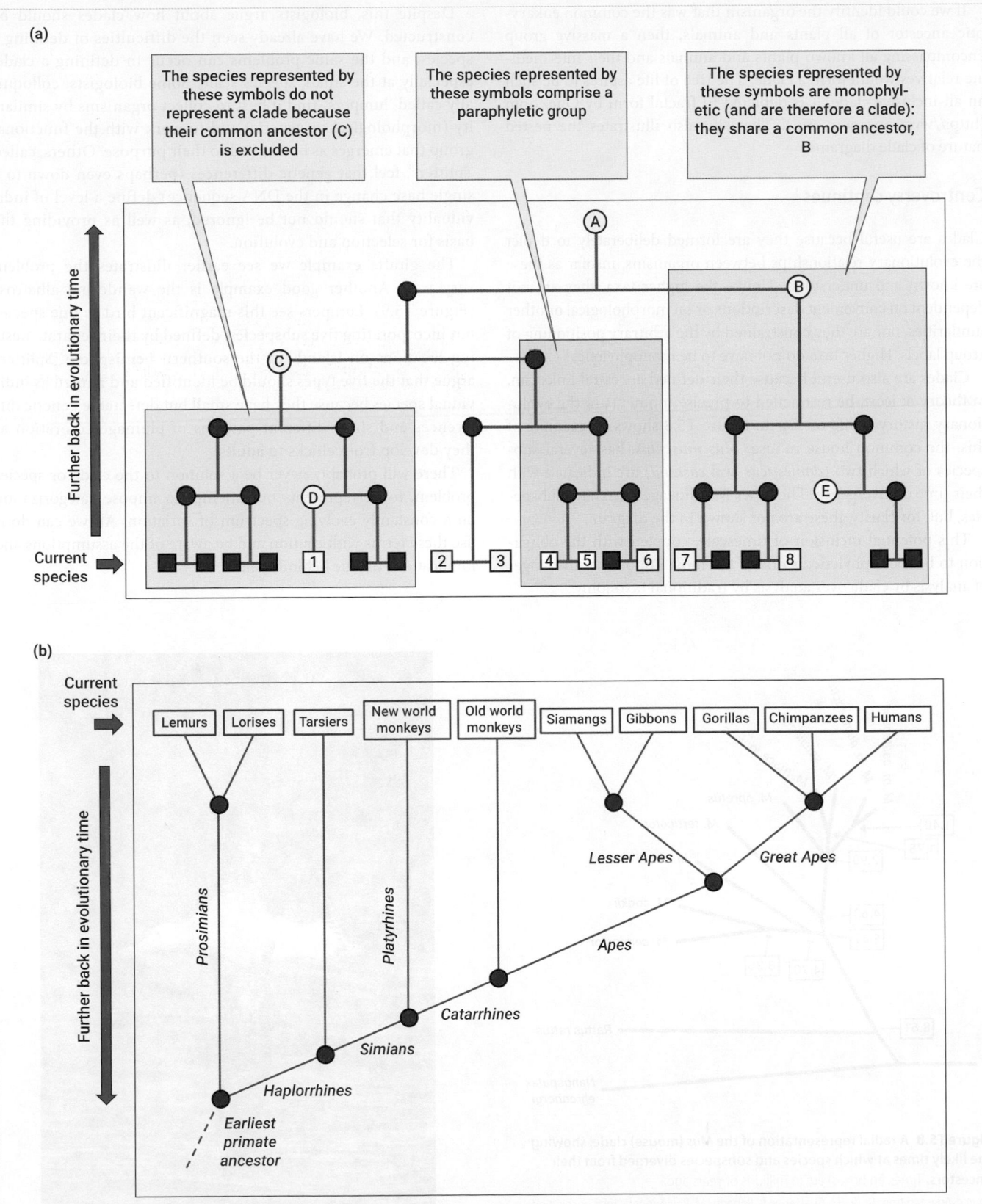

Figure T5.7 Clades and their terminology. (a) Members of a clade whose common ancestor was organism A. (b) An alternative cladogram layout.

If we could identify the organism that was the common eukaryotic ancestor of all plants and animals, then a massive group encompassing all known plants and animals and their intermediate relatives would form a clade. The tree of life aspires to be such an all-inclusive clade. It is depicted in fractal form by OneZoom (http://www.onezoom.org/). That site also illustrates the nested nature of clade diagrams.

Controversy continues

Clades are useful because they are formed deliberately to depict the evolutionary relationships between organisms, insofar as these are known and understood. Unlike the higher taxa, they are not dependent on convenient descriptions or on morphological or other similarities, nor are they constrained by the arbitrary positioning of group labels. Higher taxa do not have to be monophyletic.

Clades are also useful because their defined ancestral links can, in theory at least, be reconciled to precise moments in the evolutionary history of life on Earth. Figure T5.8 shows an example of this: the common house mouse, *Mus musculus*, has several subspecies of which two (*domesticus* and *castanii*) are indicated with their time of divergence. The other *Mus* lineages also have subspecies, but, for clarity, these are not shown in the diagram.

This potential inclusion of timescale, coupled with the obligation to be monophyletic, is one of the most important advantages of analysis by clade over analysis by traditional taxonomy.

Despite this, biologists argue about how clades should be constructed. We have already seen the difficulties of defining a species, and the same problems can occur in defining a clade, especially at the small, species scale. Some biologists, colloquially called 'lumpers', find it best to collect organisms by similarity (morphological or genetic) and to work with the functional group that emerges as best suited to their purpose. Others, called 'splitters', feel that genetic differences (perhaps even down to a single base change in the DNA sequence) define a level of individuality that should not be ignored, as well as providing the basis for selection and evolution.

The giraffe example we see earlier illustrates the problem very well. Another good example is the wandering albatross (Figure T5.9). Lumpers see this magnificent bird as one species but incorporating five subspecies, defined by their separate nesting locations on islands in the southern hemisphere. Splitters argue that the five types should be identified and named as individual species because they have small but detectable genetic differences and show different patterns of plumage coloration as they develop from chicks to adults.

There will probably never be a solution to the clade or species problem, for it represents our attempt to impose categorization on a constantly evolving spectrum of variation. All we can do is use these terms with caution and be aware of the assumptions and implications that lie behind them.

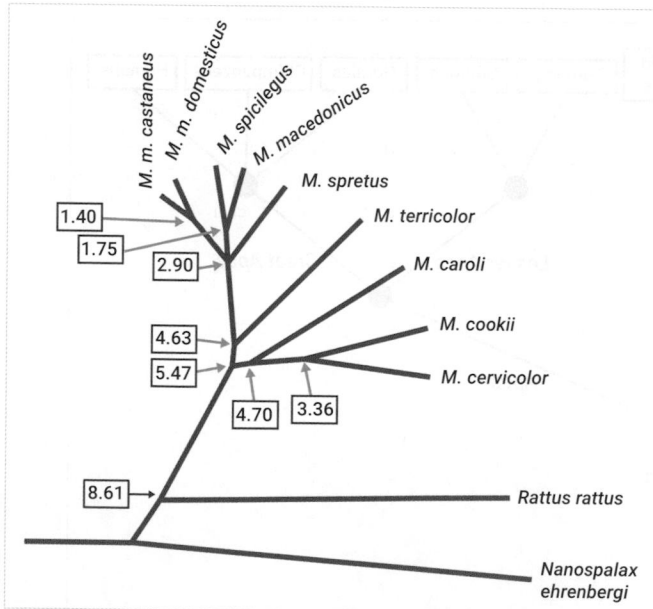

Figure T5.8   A radial representation of the *Mus* (mouse) clade, showing the likely times at which species and subspecies diverged from their ancestors. Times (in boxes) are in millions of years ago.

Source: Adapted from Rudra M, Chatterjee B, Bahadur M. Phylogenetic relationship and time of divergence of Mus terricolor with reference to other Mus species. J Genet. 2016 Jun; 95(2):399–409.

The data were based on genetic analysis of living *Mus* species and two non-*Mus* rodent species representing known ancestral lineages (the common rat, *Rattus rattus,* and the blind mole rat, *Nannospalax ehrenbergi*).

Figure T5.9   The wandering albatross (*Diomedea exulans*).
Source: Wikimedia Commons/JJ Harrison.

5.6 Five pitfalls of taxonomic interpretation

We have discussed some of the assumptions and conventions used in various approaches to biological systematics. We have also indicated that these are not always made explicit when systematic terminology is used. There are several other traps that must be avoided when referring to biological organisms and when interpreting the story of life. Many of these involve the misuse of words. They lead to sloppy and misleading language in colloquial discussion, the popular press, and media reports. We should avoid propagating these mistakes unwittingly.

Trap 1: fixed classification

The existence of reference books, identification keys, international codes, clade diagrams, dendrograms, and taxonomic databases may suggest that organism classifications are fixed and immutable. This is not so. Such resources represent, at any moment, the best interpretation of the evidence currently available. Classifications are provisional: they can and do change as more information becomes available.

Let's consider the two examples depicted in Figure T5.10.

Example 1: Duckweed

Duckweeds (genus *Lemna*; Figure T5.10a) are small aquatic plants consisting of a thallus (a body that is not separated into stem and leaves) of photosynthetic cells from which thin rootlets hang. The plants reproduce mostly by budding, and only produce their tiny flowers extremely rarely. They have no clear morphological similarity to any other flowering plants and were for a long time classified in a family of their own, the 'Lemnaceae'. When the classification was reassessed using molecular methods in 2004 it was found that duckweeds are part of the arum family (Araceae). This is not at all obvious from the plants' outward forms.

Example 2: Cetaceans are artiodactyls

From Linnaeus' time until the start of the 21st century, the 80 or so species of marine mammals, which includes baleen whales (Figure T5.10b), toothed whales, dolphins, and porpoises were grouped into an Order called the Cetacea. DNA analysis, immunological studies, and careful dissection of these animals' hidden hindlimb bones, now shows that their closest relatives are the even-toed ungulates, called the Artiodactyls (cattle, sheep, goats, giraffe, camels, deer, antelope, hippopotamus (Figure T5.10b)). They are thought to be most closely related to hippos, although uncertainty remains. Thus, Cetaceans are no longer considered to be a separate order and are now included in the Artiodactyla. Fortunately, because they represent a monophyletic lineage from a common (artiodactylate) ancestor, it is correct to refer to them as belonging to the cetacean clade.

Trap 2: misleading names

Common names and trivial names (explained above) are usually immutable once they have been agreed and established. The generic part of the trivial name may occasionally change if the members of closely related genera need to be rearranged, regrouped, or reassigned, and specific epithets may be altered if duplications or errors need to be resolved, but otherwise these labels tend to become fixed in the biological nomenclature.

Common names can be very misleading:

- Arums (Figure T5.10a) are often called lilies but are not members of the Liliaceae family;
- Buckwheat (*Fagopyrum esculentum*) is not a wheat or grass but a dicotyledonous plant of the rhubarb family;
- Flying foxes (*Pteropus spp.*) are large, fruit-eating bats, not carnivores;
- Horned toads (*Phrynosoma spp.*) are reptiles (lizards) not amphibians;
- The slow or blind worm (*Anguis fragilis*) is a limbless lizard, not an invertebrate;
- The European water rat (*Arvicola amphibius*) is a vole;
- Cuttlefish (*Sepia spp.*) are cephalopod molluscs, not fish.

Other examples are easy to find. Of course, common names are also language-dependent. Names are a human conceit: they are created by us for our convenience and have no intrinsic validity. It is sensible, therefore, to treat them with caution and avoid being misled.

Trap 3: contemporary ancestors and primitivity

No contemporary organism can be the ancestor of another. Coexisting relatives are either cousin species or sister species, with the degree of separation dependent on the evolutionary position of the common ancestor (Figure T5.7a, b). Thus, chimpanzees, orang-utans, and gorillas, which, like us, are living members of the ape clade, are our cousins and not mysteriously anachronistic remnants of ancestral populations. They have existed for *exactly* the same length of time as we have (since the time of the ancestor we share) and are, like us, the product of continuous evolution.

By the same logic, it is incorrect to describe chimpanzees, orang-utans, gorillas, or any other living species as *primitive*. This descriptor is often carelessly used for organisms that are unusual or that seem superficially to represent ancient forms, as per the examples depicted in Figure T5.11:

- the duck-billed platypus is an egg-laying monotreme mammal which comprises an Order made of a single species, with a few extinct relatives;
- the coelacanths are two rarely observed species of bony fish representing a clade, once thought to be extinct, which diverged very early in the evolution of fishes;
- the tree ferns include several species of non-flowering plants which produce spores on the undersides of fronds and develop layered, trunk-like stems which are actually made of root fibres.

Such organisms may represent either very narrow clades (that is to say, they appear to be the only living descendant, or one of very few descendants, of an ancestor; see species 1, 2, and 3 in Figure T5.7a) or clades which diverged from those of more abundant

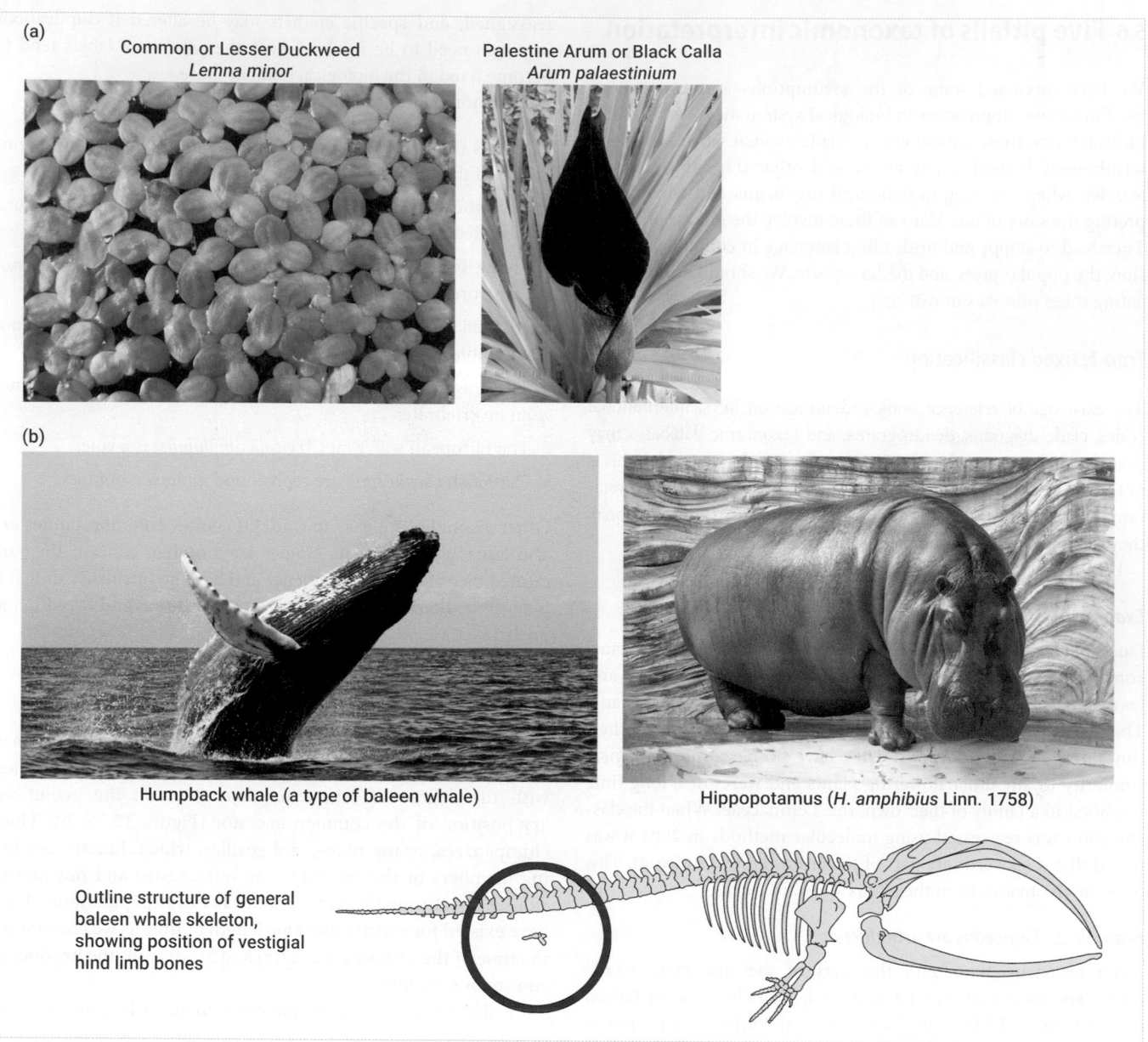

(a)

Common or Lesser Duckweed
Lemna minor

Palestine Arum or Black Calla
Arum palaestinium

(b)

Humpback whale (a type of baleen whale)

Hippopotamus (*H. amphibius* Linn. 1758)

Outline structure of general
baleen whale skeleton,
showing position of vestigial
hind limb bones

Figure T5.10 Examples of updated taxonomy. (a) Duckweed: Lemnaceae or Araceae? (b) Cetaceans are artiodactyls.

Sources: (a) Wikimedia Commons/Barbarossa, Wikimedia Commons/Berichard; (b) Wikimedia Commons/Whit Welles, Wikimedia Commons/Kabacchi, Wikimedia Commons/Andrew Z. Colvin.

species very early in evolutionary time. The unfortunate term 'living fossil' is also used, somehow implying that they remained stubbornly unchanged despite the passage of time, while other organisms have 'progressed'.

A particular phenotypic trait in a living organism may be legitimately described as primitive if it can be shown to have been present in an ancestor as well. It is also true that that rates of genotypic and phenotypic evolution vary between and within clades. Some lineages have evolved more rapidly than others, but that does not make the living species themselves more ancient.

Equally, the absence of sister or close cousin species does not imply a lack of adaptation. On the contrary, such organisms have been adaptively successful, while their extinct evolutionary variants self-evidently were not.

Trap 4: purposeful evolution

As we move from higher taxa to individual species or use a cladogram to follow the phylogenetic trail towards a living organism, we may feel that we are ascending to increased levels of complexity and biological sophistication. This may be so, but we should avoid falling into Aristotle's teleological trap of interpreting it as progress or as development towards ultimate perfection. Evolution is non-directed (unless one invokes religious beliefs) and has no target or endpoint.

Many modern species may indeed be more complex than their ancestors, although this is not necessarily the case. Complexity may result from divergence and specialization based on adaptive variation, but it is not an inevitable progression. The best evidence for this is the *coexistence* of more complex and less complex species:

Figure T5.11 Three species sometimes erroneously described as *primitive*. (a) Duck-billed platypus (*Ornithorhynchus anatinus*). (b) Tree fern (*Dicksonia antarctica*). (c) West Indian Ocean coelocanth (*Latimeria chalumnae*).
Sources: (a) Wikimedia Commons/Stefan Heinrich; (b) Wikimedia Commons/Eliedion; (c) Wikimedia Commons/Citron.

all living organisms are the product of exactly the same length of evolutionary time and they are all, by definition, adapted to their current environment. So there is no sense in which some are more perfectly designed than others.

Trap 5: convergent evolution

Cladograms depict the phylogenetic history of species and identify where lineages have diverged (see Figures T5.7 and T5.8). While the synapomorphies that define a clade are possessed by all its members, it is possible for an apomorphy to exist in more than one subclade but not be present in their common ancestor. In other words, a morphological or any other kind of feature can emerge more than once during the course of evolution. This is known as **homoplasy**, or convergent evolution, and it results in the presence of **homoplastic** or **analogous characters** among distantly related cousins.

Well-known examples are:

- the capacity for muscle-powered flight in insects, birds, and bats;
- the presence of similar enzymatic processes in unrelated organisms;
- multiple cases of C4 photosynthesis in diverse groups of plants;

- similar anticoagulant chemicals in the saliva of unrelated blood-eating animals including vampire bats (mammals), ticks (insects), and leeches (annelid worms);
- the snake-like shape and movements of the limbless lizard called a slow worm;
- carnivory (trapping and consumption of insects and protozoans) in diverse groups of plants.

The trap would be to interpret such apomorphies as evidence of descent from a more recent common ancestor than is, in fact, the case. This danger has been avoided in well-analysed examples like those above, with the presence of homoplastic characters being understood as convergent evolution.

But how can we really tell if similar characters are homoplastic or not? The best evidence usually comes from molecular analysis (detecting **gene mutations** and **duplications** among closely related species and then looking for them among more distantly related members of the clade for which we have independent evidence of distinct lineages) coupled with statistical analysis of the most likely relationships. The evidence can sometimes be incomplete, difficult to decipher, or open to alternative interpretations, and many cases become controversial. Thus, great care needs to be taken to avoid arriving at misleading conclusions.

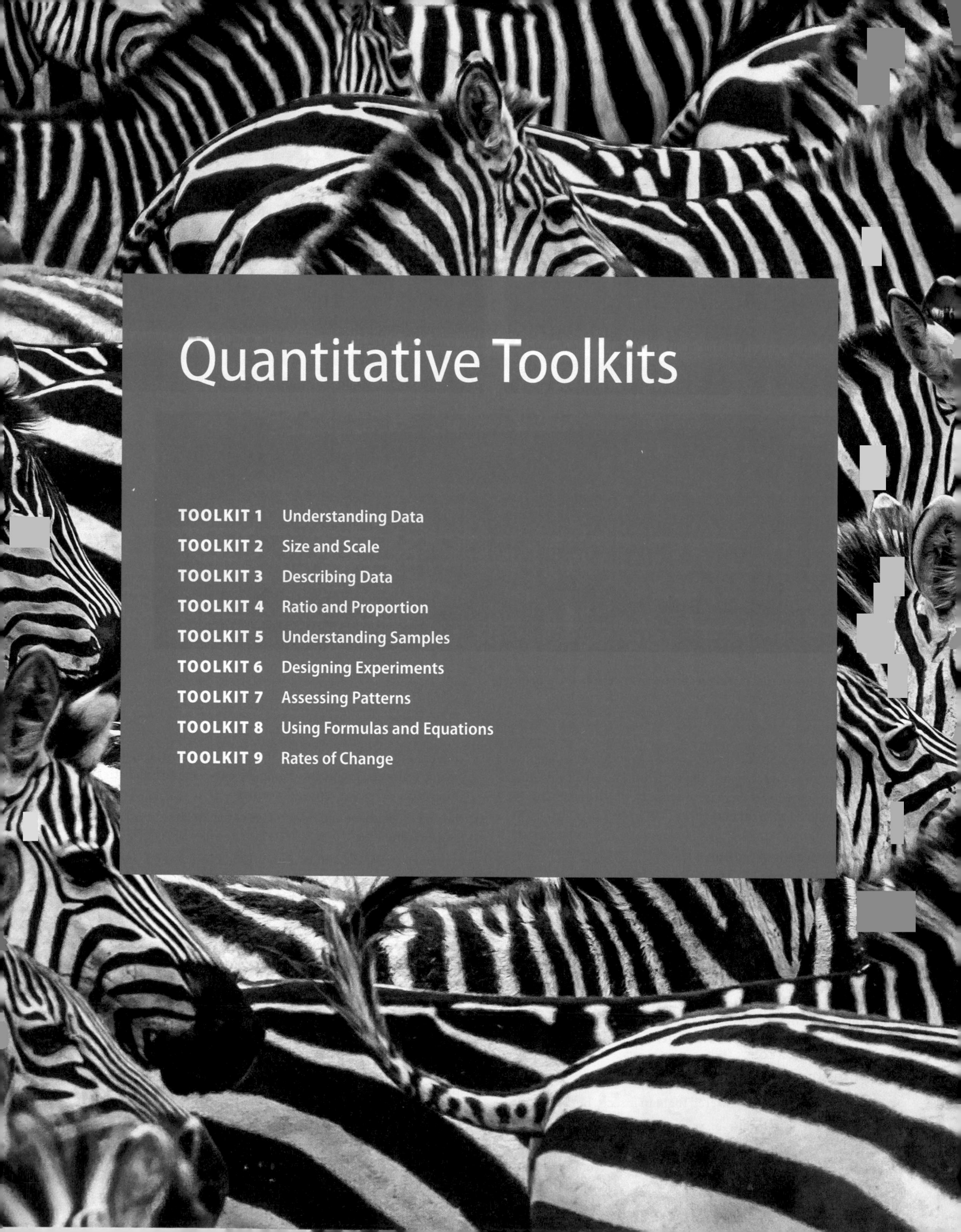

Quantitative Toolkits

TOOLKIT 1   Understanding Data

TOOLKIT 2   Size and Scale

TOOLKIT 3   Describing Data

TOOLKIT 4   Ratio and Proportion

TOOLKIT 5   Understanding Samples

TOOLKIT 6   Designing Experiments

TOOLKIT 7   Assessing Patterns

TOOLKIT 8   Using Formulas and Equations

TOOLKIT 9   Rates of Change

QUANTITATIVE TOOLKITS

Mathematical approaches are revolutionizing biology today, in the way that the microscope did in the past. From high-throughput lab techniques to environmental monitoring, we now generate more data than ever before. The challenge that comes with this is to analyse and make sense of that data, and to communicate its meaning. From climate change to pandemics, data is at the heart of biology: we measure, calculate, estimate, and identify patterns in life from the molecular level right through to populations.

In the Quantitative Toolkits we take the key concepts in mathematics and statistics and apply them in a biological context, focusing on what the maths and statistics allow us to do and to understand.

To describe processes in biology we need to use both very large and very small numbers: in a viral infection an infected person could carry 10^{10} virions, each of which is only 10^{-7} m, or 100 nm, in diameter, and which collectively make up 10 µg in mass. And, of course, these virions and people don't stay still: we can use equations and graphs to describe how a viral infection spreads, calculate an index of biodiversity, or estimate how quickly a substance will diffuse into a cell.

In these Quantitative Toolkits we want you to gain an intuitive sense of the magnitude of numbers written in scientific notation and to be able to look at an equation and picture how the variables relate to each other and imagine how the graph might appear.

The Quantitative Toolkits will also help you to consider how to design an experiment, calculate the concentration of a solution, recognize different types of variables and confounding factors, and think about how to collect samples, whether you are working in the field or in a lab. The use of numbers, units, graphs, and equations in biology involve plenty of conventions; once you know them, you will feel more confident about analysing and reporting your own data.

Once you have collected your data, you will need to make sense of it, starting off by describing it and then going on to look for patterns or changes. Because there is always variation in biology, we need ways of working out just how much variation there is; we then need to determine whether any difference that we see is likely to *really* be a difference or whether it might have happened by chance. Becoming a biological scientist means having the whole quantitative toolbox at your fingertips.

Image: A vivid interpretation of patterns in biology: a zebra herd waiting on the bank of the Mara river, Kenya. *Source:* Manoj Shah/Getty Images.

Understanding Data

LEARNING OBJECTIVES

In this toolkit we will:

- Distinguish different levels of measurement (nominal, ordinal, scale) and between category-like versus continuous data (Section 1.1, Levels of measurement).

- Review the International System of Units (SI units) and other commonly used units in biology (Section 1.1, Units).

- Convert between different prefix–unit combinations for the same units, for example, nm to mm or kg to mg (Section 1.1, Converting between different prefix-unit combinations).

- Recognize different types of variables (independent, dependent, interdependent, and confounding) (Section 1.2, Dependent and independent; Confounding variables).

- Compare when the source of variation in an independent variable is manipulated and when it is natural (Section 1.2, Sources of variation: natural or manipulated).

- Distinguish between different types of study: experimental, quasi-experimental, and observational (Section 1.2, Sources of variation: natural or manipulated).

Contents

1.1 The features of data   64

1.2 Types of variables   66

In biology, as in all sciences, we answer questions by using observations, as we explain in Topic 1. Each record of an observation is a **datum** and together our records constitute **data**. Biological data are variable: observations on the same biological characteristic are almost always different from one another—for example, the length of the necks of giraffes. This variation is fundamental to the process of evolution. A characteristic that varies, like neck length in giraffes, is called a **variable**. Understanding the basic features and types of data and variables will help you plan effective studies (see Quantitative Toolkits 6 and 7), and analyse and present your data appropriately (see Quantitative Toolkits 3 and 7).

1.1 The features of data

Levels of measurement

🔘 Go to the e-book to watch a video that explores levels of measurement.

There are three main levels of measurement: nominal, ordinal, and scale (Figure QT1.1). When you imagine a set of data you probably think of, let us say, a list of numbers representing the level of glucose in blood, the mass of human babies, or the number of eggs in a birds nest. Data like this are **scale level**. These examples illustrate two different ways in which scale data can be recorded: either measured, as in the case of weight, or counted, as in the case of egg numbers. Data that are measured tend to be **continuous**, so, for

example, a baby might weigh 5 kg or 10 kg or anywhere in between. By contrast, data that are counted are **discrete**. For example, you count 1 or 2 eggs but not 1.2 eggs.

However, a datum can just be the name of something, like the name of a base on DNA or the colour of a flower on a plant. Data like this are **nominal level**. You cannot put nominal data into any particular order: they are just categories that can tell whether two observations are the same or different. But you cannot, for example, say if one is heavier, longer, or faster than the other. If you use just two categories the data would also be called **binary**; common examples include infected or not infected, dead or alive, present or absent.

Between nominal and scale levels of measurement is **ordinal level**. Like nominal level, it identifies categories but has the added feature that you can arrange the categories into a logical order that is, you can rank them. The Bristol Stool Scale is a memorable example: medical professionals use the Bristol Stool Scale to rank human faeces from small hard lumps (type 1) to watery with no solid bits (type 7). Other examples include the Beaufort scale for wind speed, the Kendal–Shepherd Ladder of Aggression for measuring dog behaviour, and the Apgar scoring system, which is used to assess the health of newborn babies.

In terms of making appropriate decisions about how to analyse and present data it often boils down to understanding if the data are likely to behave in **categorical** or **continuous** fashion. In general, continuous scale data fall in the continuous part of the divide, while ordinal and nominal fall in the categorical part. However, there is a bit of a grey area around discrete scale data. For example, gulls usually have a clutch size of 2–3 eggs, whereas an emu can lay up to 20 eggs: data on gull-egg counts will behave like categorical data while those from emus will behave like continuous data.

▶ To check you understand levels of measurement, make sure you complete Activity 1 and Activity 2 at the end of this toolkit.

Units

Scale-level data have **units**. For discrete data, these are whatever you are counting: 'eggs' in the last example. For continuous data, examples are seconds, grams, and metres. The **International System of Units (SI Units)**, adopted throughout the world, has seven base units (Table QT1.1). SI units are used to measure things throughout biology but we also use other units (Table QT1.2).

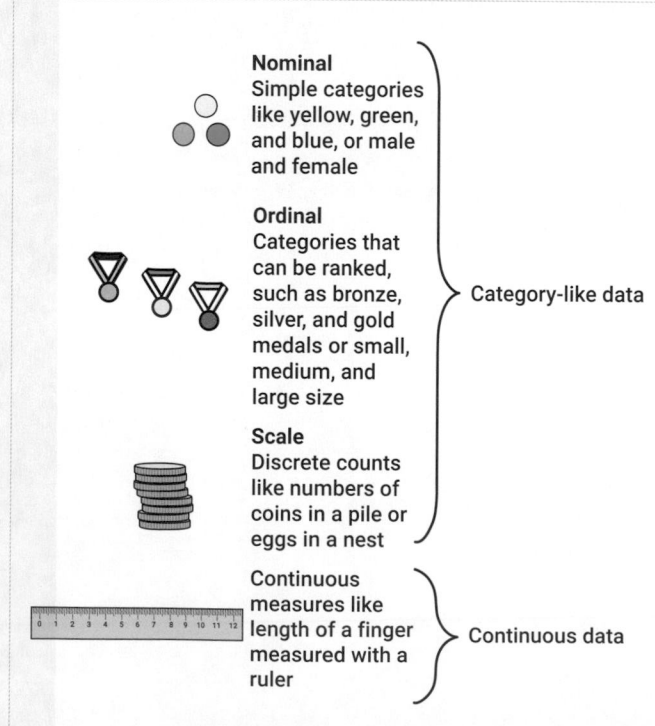

Figure QT 1.1 Levels of measurement and how they relate to recognizing categorical-like and continuous data.

Nominal
Simple categories like yellow, green, and blue, or male and female

Ordinal
Categories that can be ranked, such as bronze, silver, and gold medals or small, medium, and large size

⎱ Category-like data

Scale
Discrete counts like numbers of coins in a pile or eggs in a nest

Continuous measures like length of a finger measured with a ruler ⎱ Continuous data

Table QT1.1 The International System of Units (SI Units)

Unit	Abbreviation	Used to measure
second	s	time
metre	m	length
kilogram	kg	mass
ampere	A	electric current
kelvin	K	thermodynamic temperature
mole	mol	amount of substance
candela	cd	luminous intensity

Table QT1.2 Units commonly used in biology

Unit	Abbreviation	Used to measure	Value in other units
minute	min	time	1 min = 60 s
hour	h		1 h = 60 min = 3600 s
day	d		1 d = 24 h = 86 400 s
hertz	Hz	frequency*	$1 \text{ Hz} = \text{s}^{-1}$ sometimes written as cycles per second
mole	M	concentration	$\text{mol.L}^{-1} = \text{mol.dm}^{-3}$ $= 1000 \text{ mol.m}^{-3}$
katal	kat	catalytic activity	mol.s^{-1}
litre**	L**	volume or capacity	$1 \text{ L} = 1 \text{ dm}^3 = 10^{-3} \text{ m}^3$ note that $1 \text{ mL} = 1 \text{ cm}^3$ and $1 \text{ μL} = 1 \text{ mm}^3$
degree Celsius	°C	temperature	= (temp. in K) – 273.15
becquerel	Bq	disintegrations per second (radioactivity)	s^{-1}
curie	Ci		$1 \text{ Ci} = 3.7 \times 10^{10}$ disintegrations s^{-1}
coulomb	C	electric charge	$A \times s$
newton	N	force, weight	kg.m.s^{-2}
pascal	Pa	pressure	$\text{kg·m}^{-1}\text{·s}^{-2} = \text{N/m}^2$ atm (atmosphere) 1 atm = 101.325 kPa mmHg (millimetres of mercury) 1 mmHg = 0.133 kPa
farad	F	capacitance	$\text{kg}^{-1}\text{·m}^{-2}\text{·s}^4\text{·A}^2 = \text{C/V}$
ohm	Ω	electrical resistance	$\text{kg·m}^2\text{·s}^{-3}\text{·A}^{-2} = \text{V/A}$
siemen	S	electrical conductance	$\text{kg}^{-1}\text{·m}^{-2}\text{·s}^3\text{·A}^2 = \Omega^{-1}$
(not applicable)***		absorbance (at a given wavelength of light)	

* Frequency in this context is a rate of occurrence per unit time and is not to be confused with frequency as used in a statistical context (Quantitative Toolkit 3, Table QT3.7).

** Note that litre can be abbreviated as l or L, but L is often used because of the potential for confusion of l ('ell') and 1 ('one'). Sometimes dm^3 and cm^3 are used instead of L and mL.

*** Absorbance of light in a spectrophotometer is measured by calculating the ratio of the intensity of light through a sample divided by the intensity of the same light source through a reference solution. The ratio is calculated in the spectrophotometer and the output measurement therefore has no units.

There are some **conventions** that will help to avoid confusion. The key ones are:

- Don't make a unit into a plural (i.e. write 5 m not 5 ms to mean 5 metres).
- Don't use p for 'per', e.g. mps (for m/s) or kph (for km/h).
- Leave a space between the numerical value and unit symbol (15 kg but not 15-kg or 15kg).
- g/dm^3 can also be written as g.dm^{-3}.

Converting between different prefix–unit combinations

Go to the e-book to watch a video that explores the conversion of units.

Prefixes, such as kilo and milli, are used in conjunction with units to indicate multiples, or fractions, of these units. For example, a kilogram (kg) indicates a multiple of thousand grams (g) (see Table QT2.1). There are many different approaches to converting between these prefix–unit combinations and most people have their preferred style. The following are two of the most common ways used by biologists.

▶ **To check you understand units, make sure you complete Activity 2 and Activity 3 at the end of this toolkit.**

Remembering (or looking up) the relationships

If you know, or you can look up, the relationship you can simply multiply or divide (Figure QT1.2). For example:

- To convert 57 ng into pg we multiply by 1000. So, 57 ng = 57000 pg.
- To convert 57 ng to μg we divide by 1000. So, 57 ng = 0.057 μg.

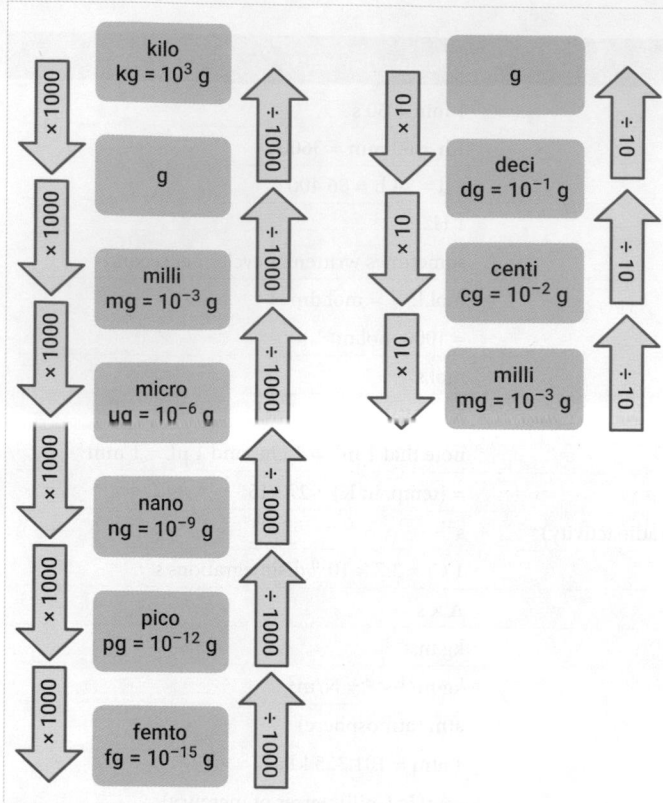

Figure QT1.2 Converting between different prefix-unit combinations using the relationship between them.

Using ratio squares

We explain ratio squares in more detail in Quantitative Toolkit 4. For now, let's go briefly through the process. Write down what you know, i.e. that 1000 pg is 1 ng, then underneath write down what you are trying to find. The pairs of numbers are equivalent ratios (1000 : 1 is the same ratio as 57000 : 57). You can see two examples in Figure QT1.3.

1.2 Types of variables

A characteristic that varies is called a variable. To identify what variables you need to collect data on in any particular study you need to be clear on the purpose of the study. For example, if you were investigating the heart rate of water fleas (*Daphnia* sp.) in relation to ambient temperature, there are two variables of interest: *heart rate* and *temperature*. A **constant**, however, is the same for every observation. In our water flea scenario, the type of animal is a constant (i.e. it is a water flea). In statistics, variables are generally represented by letters from the end of the alphabet (x, y, or z) and constants by letters from the beginning of the alphabet (a, b, or c); see Quantitative Toolkit 8.

▶ **To check you understand types of variables, make sure you complete Activity 4 at the end of this Toolkit.**

Dependent and independent

In studies looking at the patterns between two variables you will often be able to distinguish one variable that is regarded as generating variability in the other one. This is your **independent variable**. The other is your **dependent variable** (Figure QT1.4). Dependent variables are also known as **test**, **data**, or **response variables**, and independent variables are also known by other names, including **predictor**, **grouping**, or **explanatory variables**, or **factors**.

One way of remembering this is to think 'dependent variables depend on independent variables'. In the heart rate in water flea example, *temperature* is the independent variable and *heart rate* the dependent variable as the *heart rate* of water fleas may depend on *temperature*, but the reverse doesn't make sense. Remember, the purpose of your study gives all-important context.

In a study of overnight temperature of babies in babygrows made of different materials, *temperature* would be the dependent variable and *babygrow material* the independent variable.

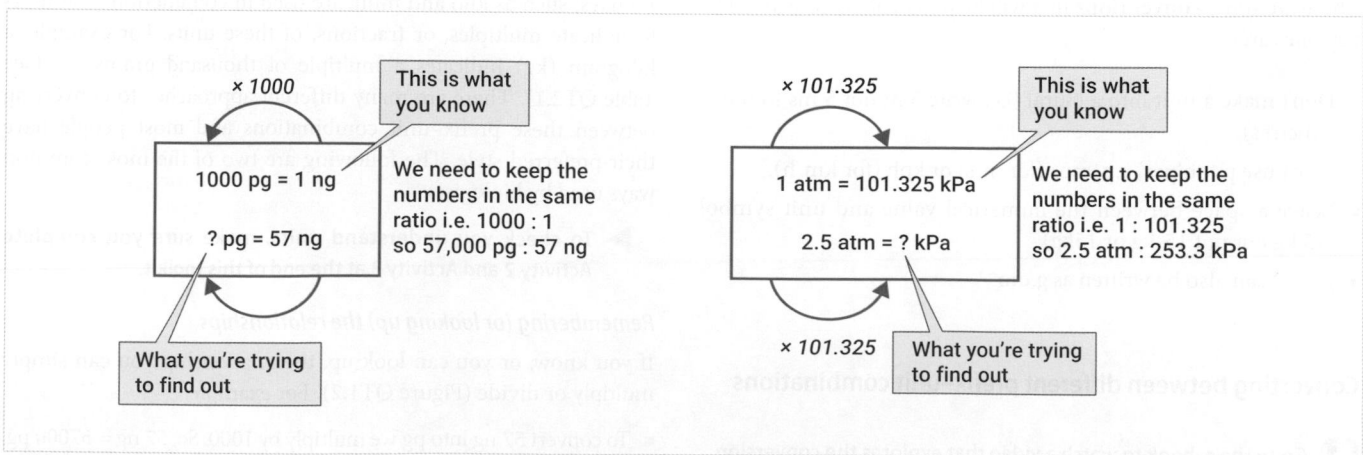

Figure QT 1.3 Converting between different prefix–unit combinations using ratio squares.

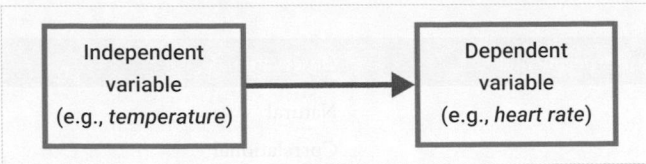

Figure QT 1.4 Types of variable. An independent variable generates variation in the dependent variable (example variable taken from water flea example in text).

Studies can, and often do, involve more than one independent variable and/or more than one dependent variable, as we discuss further in Quantitative Toolkits 6 and 7.

It is not always the case that you will have a dependent–independent variable situation. Studies looking at the patterns between two variables—for example, *arm span* and *height*—may just be looking to see if they are **inter-dependent** (also referred to as **intcr-rclatcd**) or **correlated** (i.e. varying together). The term **orthogonal** is sometimes used in statistics to describe two variables that are not inter-dependent.

More complex study designs with more than two variables can involve variables in dependent–independent and inter-dependent roles (see Quantitative Toolkit 6).

Confounding variables

A **confounding variable**, **confounding factor**, or **third variable** is a variable that is not the focus of your investigation but that can complicate the interpretation of your results. It can do this by causing variability in both the dependent and independent variables. For example, you might be investigating the relationship between *cancer* and *smoking*, thinking that people who smoke are more likely to have cancer. There are a number of potentially confounding variables to consider, including *stress*: perhaps stress causes people to smoke and to get cancer (Figure QT1.5).

Ideally, you would remove the influence of all confounding variables in order to focus on just the variables of interest by controlling them physically (for example, using only *Daphnia* of the same age) or statistically (by collecting data on age and including it in the analyses; see Quantitative Toolkit 7). However, given the complexity of biological systems, you are unlikely to be able to eliminate *all* potentially confounding factors. The important thing is to be

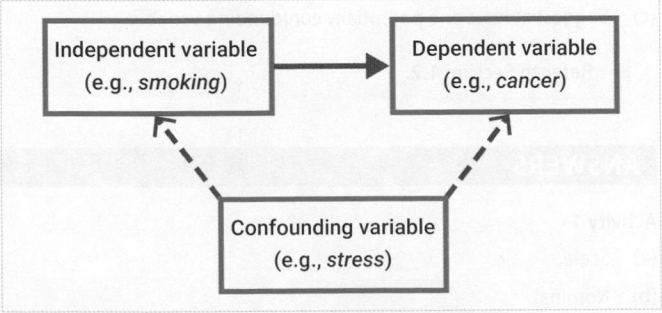

Figure QT1.5 Types of variable. A confounding variable generating variation in both independent and dependent variables complicates interpretation of results (example variable taken from smoking example in text).

aware of potentially confounding variables, to do your best to control them, and to take into account what you have not controlled when interpreting your results.

Sources of variation: natural or manipulated

Returning to our water flea example, in which we look at *heart rate* (dependent variable) in relation to *temperature* (independent variable), you might set up a series of test tubes at different temperatures and measure the heart rates of the water fleas in each one. In doing so, you will have manipulated the variation in the independent variable. Studies where variation in the independent variable is artificially generated like this are often referred to as **experimental**, **manipulated**, **planned**, or **designed studies**.

A handy tip if you are having trouble recognizing which variable is your dependent variable and which is the independent variable is this: if you are manipulating one variable, then this one will be your independent variable.

Sometimes variation cannot, or cannot easily, be artificially generated. In these situations, we must rely on natural variation. But we can manipulate that variation by specifically selecting for it. For example, if you had wanted to look at *heart rate* (dependent variable) in relation to *sex* rather than *temperature*, you might measure the heart rates of 20 male and 20 female water fleas. Studies where variation in the independent variable is manipulated in this way should be referred to as **quasi-experimental studies**, although this is not a commonly used term.

An alternative approach to investigating *heart rate* and *temperature* in water fleas would be to go out into the field and measure the heart rate of water fleas in ponds of different temperatures. In this case, you would be making use of naturally occurring variation in the independent variable. Studies using naturally occurring variation like this are referred to as **natural**, **correlational**, or **observational studies** or **experiments**. The various terms applied to types of study are summarized in Table QT1.3.

Traditionally, the term 'experimental design' has been associated with experimental studies, but the principles of experimental design also apply to quasi-experimental studies and to observational studies.

▶ **For more information on experimental design, see Quantitative Toolkit 6.**

There are often ethical, practical, or financial disadvantages to designs involving manipulated variation and you must also remember to make sure that the variation you generate is typical of biological systems and relevant to your question. However, confounding variables can be more easily controlled in experimental studies.

The source of variation in the independent variable is often associated with the **setting** in which the data are collected. Typically, natural studies are conducted in the **field** and experimental studies are conducted in the **laboratory**. However, this is not fixed: experimental studies, in particular, can be conducted in the field. For example, a study looking at whether fire attracts elephants to woodland could artificially set light to some areas of woodland and artificially exclude fire from other areas. This would be an experimental study in a field setting.

Table QT1.3 Types of study: comparison of three main types

Type of study	Experimental	Quasi-experimental	Observational
Alternative name	Manipulated Planned Designed		Natural Correlational Observational
Source of variation in independent variable	Manipulated	Selected	Natural
Advantages	Physical control of confounding variables easier Cause and effect less ambiguous if designed well	Potential advantages (and disadvantages) of both experimental and observational	Generally fewer ethical, financial, and practical issues Variation definitely within biological appropriate range

ACTIVITIES

Activity 1

For each of the following say whether the data are being measured at nominal, ordinal, or scale level.

(a) A biochemist estimates the concentration of peptides in moles/L.

(b) A public health official records causes of death.

(c) An ecologist counts the number of barnacles in a quadrat.

(d) A behaviourist records the relative social status of adult females in a troop of baboons.

(e) A conservationist records the presence or absence of an endangered frog in fragments of rainforest.

▶ Refer to Section 1.1, Levels of measurement.

Activity 2

For each of the following say whether the data are likely to behave in a continuous or categorical way.

(a) A biochemist estimates the rate of reaction of peptide digestion.

(b) A public health official records cause of death.

(c) An ecologist counts the number of barnacles in a quadrat.

(d) A behaviourist records the relative social status of adult females in a troop of baboons.

(e) A conservationist records the presence or absence of an endangered frog in fragments of rainforest.

▶ Refer to Section 1.1, Levels of measurement.

Activity 3

Convert the following units.

(a) 354 ng to µg

(b) 0.098 nmol to pmol

(c) 43 µL to mL

▶ Refer to Section 1.1, Converting between different prefix–unit combinations.

Activity 4

Multiple sclerosis (MS) is an autoimmune disorder known to cause widespread damage and decreased volume of the central nervous system. However, there has been relatively little study of the effect of MS specifically on the region of the brain called the hippocampus.

Sicotte et al. (2008) thought that it was an area likely to be affected by MS and conducted a study to investigate. They used magnetic resonance imaging to compare hippocampus volume (mm^3) in patients with MS and healthy people. They found that hippocampus size was smaller in patients with MS. (Sicotte, N. L., Kern, K. C., Giesser, B. S., Arshanapalli, A., Schultz, A., Montag, M., & Bookheimer, S. Y. (2008). Regional hippocampal atrophy in multiple sclerosis. Brain **131**: 1134–1141. https://doi.org/10.1093/brain/awn030)

For this study:

(a) Identify the dependent and independent variables and for each indicate:
 – units used, if applicable;
 – level of measurement;
 – whether the data are likely to behave in a continuous or categorical way.

(b) State if the source of variation in the independent variable is natural or manipulated.

(c) Suggest at least one potentially confounding variable.

▶ Refer to Section 1.2.

ANSWERS

Activity 1

(a) Scale.

(b) Nominal.

(c) Scale.

(d) Ordinal.

(e) Nominal.

Activity 2

(a) Continuous.

(b) Categorical.

(c) Probably continuous. It depends on how many—if the maximum is a lot it would be continuous.

(d) Categorical.

(e) Categorical.

Activity 3

(a) 0.354 µg.

(b) 98 pmol.

(c) 0.043 mL.

 Follow the link from the e-book to watch the walk-through video that will help you understand the answer to Activity 3.

Activity 4

(a) Dependent variable = size of hippocampus: mm^3, scale, continuous-like.
Independent variable = health status of patient (measured as healthy v MS) not applicable, nominal, categorical-like.

(b) Natural.

(c) Possible answers include age and gender.

Size and Scale

Contents

2.1   Scientific notation and standard form   71
2.2   Using prefixes   72
2.3   Logarithmic scales   72
2.4   What is a logarithm?   75
2.5   What are natural logarithms?   76

LEARNING OBJECTIVES

In this toolkit we will:

- Convert between numbers in scientific notation, standard form, and normal notation (Section 2.1).

- Compare the magnitude of numbers in scientific notation (Section 2.1).

- Convert numbers from scientific notation to prefixes and vice versa (Section 2.2).

- Become familiar with the terms base, power, index, exponent, and logarithm (Sections 2.1, 2.3).

- Recognize logarithmic scales and distinguish them from linear scales (Section 2.3).

- Define a logarithm and use a calculator to calculate logarithms to the base 10 (Section 2.4).

- Distinguish between logs to the base 10 and natural logarithms (ln), and logs to the base e (Section 2.5).

An *E. coli* bacterium is anywhere between 0.0000007 m and 0.0000014 m long, contains 4,000,000 nucleotide base pairs, and there might be 10,000,000 to 1,000,000,000 bacteria in a millilitre of culture. To make these numbers easier to read, biologists use base numbers with indices (singular = index). The index (also known as 'power' or 'exponent') is written as a superscript to the right of a base.

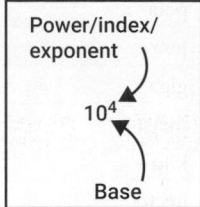

Power/index/exponent

10^4

Base

2.1 Scientific notation and standard form

It is important to gain a sense of the size of numbers when dealing with index notation as biologists rely upon using this as a way of describing and quantifying things that are too small to see. The relationship between numbers in index notation and normal numbers is shown in Figure QT2.1. Going down the table, each number is 10 times smaller than the one above. For numbers greater than 1 the powers are positive and for numbers between 0 and 1 the powers are negative.

'**Standard form**' is when a number between 1 and 9.999 is multiplied by a power to the base 10. For example, 4,000,000 is written in standard form as 4×10^6 and 0.0000014 is written as 1.4×10^{-6}.

'**Scientific notation**' is similar to standard form, but the number does not need to be between 1 and 9.999—for example, 400×10^4 or 0.0014×10^{-3}.

The term **order of magnitude** is used to mean a rough measure of the size of something expressed as a power of 10. Commonly, order of magnitude is used in a relative sense. For example, the diameter of a red blood cell, 10^{-5} m, is an order of magnitude larger than the length of an *E. coli*, 10^{-6} m (as shown in Figure QT2.2). One order of magnitude would be a 10-fold difference, two orders of magnitude would be a 100-fold difference and so on.

Try to get a sense of the size of numbers when they are written in index notation. For example, which is larger, a polio virus at 3×10^{-8} m or a coronavirus SARS-CoV-2 at 1.2×10^{-7} m? Looking at the powers, we are comparing 10^{-8} with 10^{-7}: 10^{-7} is 10 times larger than 10^{-8}. A good way to compare is to convert them to both be in the same power.

$$\div 10 \quad \begin{array}{c} 3 \times 10^{-8} \text{ m} \\ 0.3 \times 10^{-7} \text{ m} \end{array} \quad \times 10$$

Looking at this figure, if the term with the power of 10 is increased 10 times (from 10^{-8} to 10^{-7}) then the multiplier must be decreased by a factor of 10 (from 3 to 0.3) to keep the value of the whole number the same.

Now it is easier to compare the polio virus at 0.3×10^{-7} m diameter with the coronavirus at 1.2×10^{-7} m and it is easier to see that the coronavirus is approximately four times larger in diameter than the polio virus.

Decreasing size					
	÷ 10	1,000,000	10 × 10 × 10 × 10 × 10 × 10	10^6	Approximate radius of the moon in metres
	÷ 10	100,000	10 × 10 × 10 × 10 × 10	10^5	
	÷ 10	10,000	10 × 10 × 10 × 10	10^4	Approximate height of Mount Everest in metres
	÷ 10	1000	10 × 10 × 10	10^3	
	÷ 10	100	10 × 10	10^2	Height of the tallest trees in metres
	÷ 10	10	10	10^1	
	÷ 10	1	1	10^0	Approximate height of a child in metres
	÷ 10	0.1		10^{-1}	Width of the palm of an average adult hand in metres
	÷ 10	0.01		10^{-2}	Approximate length of a grain of rice in metres
	÷ 10	0.001		10^{-3}	Approximate diameter of a grain of salt in metres
	÷ 10	0.0001		10^{-4}	Approximate width of a human hair in metres
	÷ 10	0.000 01		10^{-5}	Approximate diameter of a red blood cell in metres
	÷ 10	0.000 001		10^{-6}	Length of a typical *E. coli* in metres

Figure QT2.1 Ways of writing very large or small numbers.

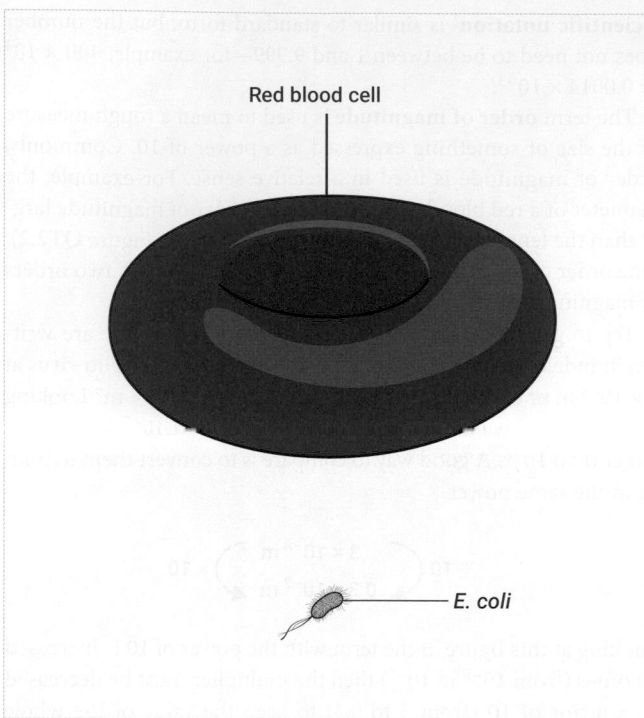

Red blood cell

E. coli

Figure QT2.2 Scale diagram of a red blood cell, 10^{-5} m, alongside an *E. coli*, 10^{-6} m. The blood cell is 10 times longer than the *E. coli*.

Table QT 2.1 Prefixes and their symbols

Power	Prefix	Symbol
10^{24}	yotta	Y
10^{21}	zetta	Z
10^{18}	exa	E
10^{15}	peta	P
10^{12}	tera	T
10^{9}	giga	G
10^{6}	mega	M
10^{3}	kilo	k
10^{2}	hecto	h
10^{1}	deca	da
10^{-1}	deci	d
10^{-2}	centi	c
10^{-3}	milli	m
10^{-6}	micro	μ
10^{-9}	nano	n
10^{-12}	pico	p
10^{-15}	femto	f
10^{-18}	atto	a
10^{-21}	zepto	z
10^{-24}	yocto	y

🌐 Go to the e-book to watch the quantitative skills video that explains how to write large and small numbers in scientific notation, and how to compare their sizes.

▶ Practise using numbers in scientific notation in Activity 1 and Activity 2 at the end of this chapter.

🌐 Go to the e-book to watch the quantitative skills video that will talk you through working with scientific notation and prefixes.

▶ Practise writing numbers in scientific notation with prefixes and get a sense of the relative sizes in Activity 3, Activity 4, and Activity 5 at the end of this chapter.

2.2 Using prefixes

Prefixes such as m (for milli, 10^{-3}), μ (for micro, 10^{-6}), and n (for nano, 10^{-9}) can be used to replace the power of 10. For example, 2×10^{-6} m is 2 μm: the '$\times 10^{-6}$' is replaced by a 'μ'. Table QT2.1 lists the prefixes from 10^{-24} to 10^{24}.

If a number is written using a power that does not correspond to one of the prefixes then it is necessary to convert it so that it *does* match. For example, 2.2×10^{-7} m $= 0.22 \times 10^{-6}$ m $= 0.22$ μm. Alternatively, we could have chosen nm, in which case 2.2×10^{-7} m $= 220 \times 10^{-9}$ m $= 220$ nm.

In the coronavirus and polio virus examples in Section 2.1, we may wish to write both numbers in nanometres (nm):

For polio virus: 3×10^{-8} m $= 30 \times 10^{-9}$ m $= 30$ nm

For coronavirus: 1.2×10^{-7} m $= 120 \times 10^{-9}$ m $= 120$ nm.

2.3 Logarithmic scales

When dealing with measurements that might vary over a very wide range, that is, over many orders of magnitude or powers of 10, it becomes difficult to fit all the measurements onto a linear scale. In these situations, it is more convenient to use a logarithmic scale where the axes go up in multiples of the base rather than by simply adding on. In Figure QT2.4 we can see the same data (taken from Figure QT2.3) plotted on a linear scale (a) where the y-axis tick marks are going up by adding on 1000 each time, and plotted on a logarithmic scale (b) where the number is doubling at each tick mark on the y-axis. It is straightforward to plot graphs on a log or linear scale with a spreadsheet program.

🌐 Go to the e-book to watch the quantitative skills video that will help you master the use of a spreadsheet to draw a graph with linear or logarithmic scales.

CASE STUDY 1 — Using base 2 for describing cell growth

Cells divide into two when they reproduce, so when thinking about cell growth we can use 2 as the base. The powers then reflect the generation number. Each time the population doubles, e.g. from 1 to 2 to 4 to 8, the power increases by 1, that is, 2^0, 2^1, 2^2, 2^3. One way of visualizing this is shown in Figure QT 2.3, which shows how the numbers in index form relate to normal numbers and to the size of the population.

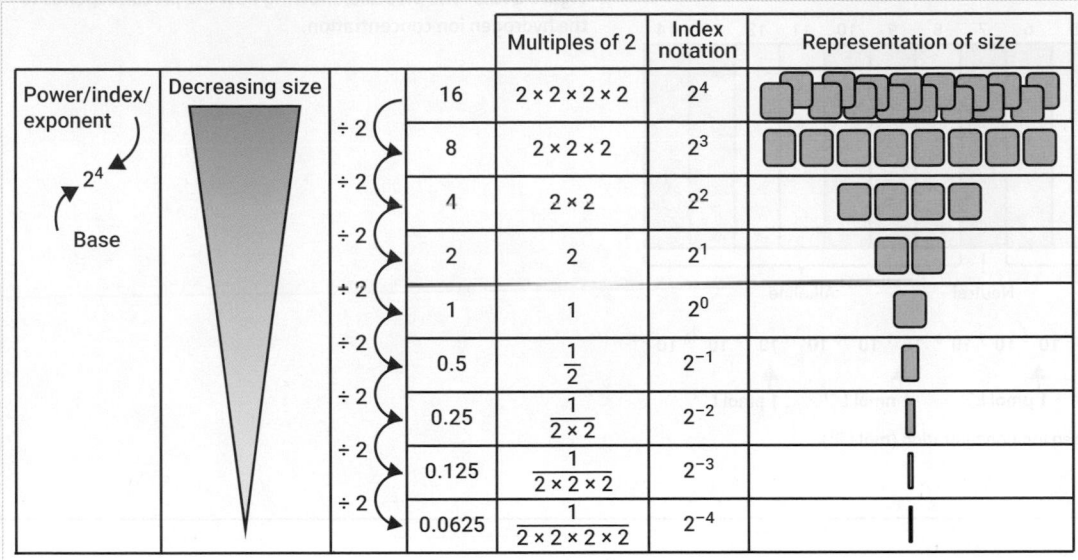

Figure QT 2.3 Visualizing how numbers in index form relate to normal numbers and the size of the population.

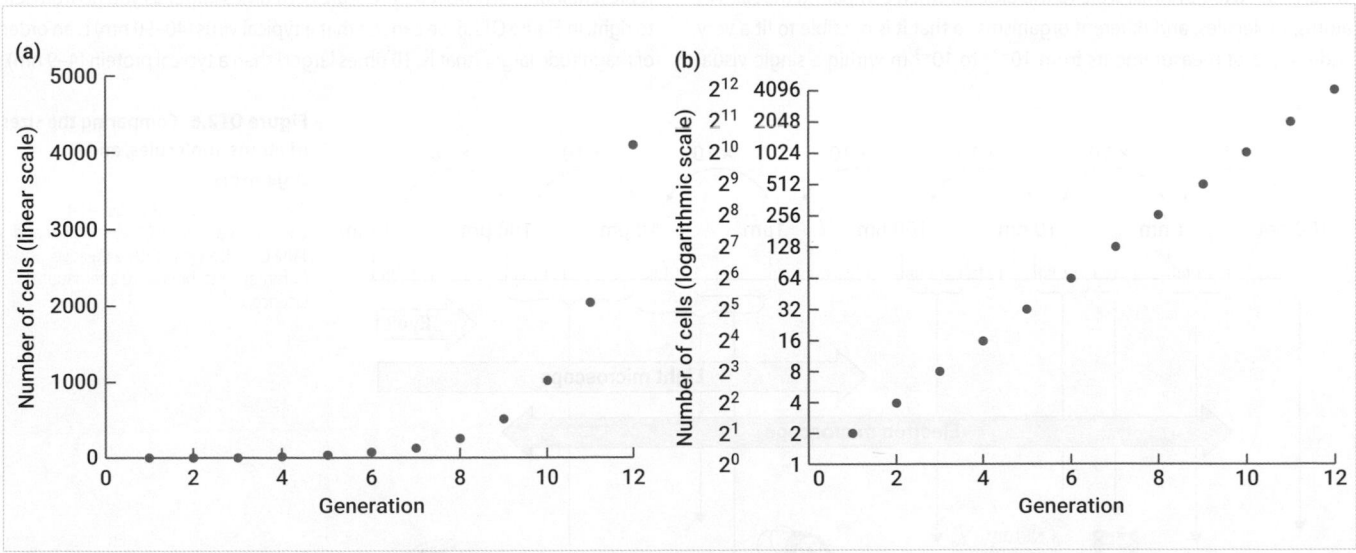

Figure QT2.4 Using linear or logarithmic scales to visualize numbers in index form. Number of cells plotted against generation number. In panel (a) the number of cells is plotted on a linear scale: the number of cells increases by 1000 with each tick mark on the y axis. By contrast, in panel (b) the same data is plotted on a logarithmic scale: in this case the number of cells multiplies by 2 with each tick mark on the y-axis.

CASE STUDY 2 — The pH scale is a logarithmic scale

It is more common to use logarithmic scales based on 10 as in the pH scale, which covers a range of hydrogen ion concentrations as shown in Figure QT2.5. A change in pH from 6 to 7 corresponds to a 10-fold decrease in hydrogen ion concentration from 10^{-6} to 10^{-7} M. Note that as the hydrogen ion concentration goes up, the pH goes down.

Figure QT2.5 The pH scale, showing how the pH corresponds to the hydrogen ion concentration.

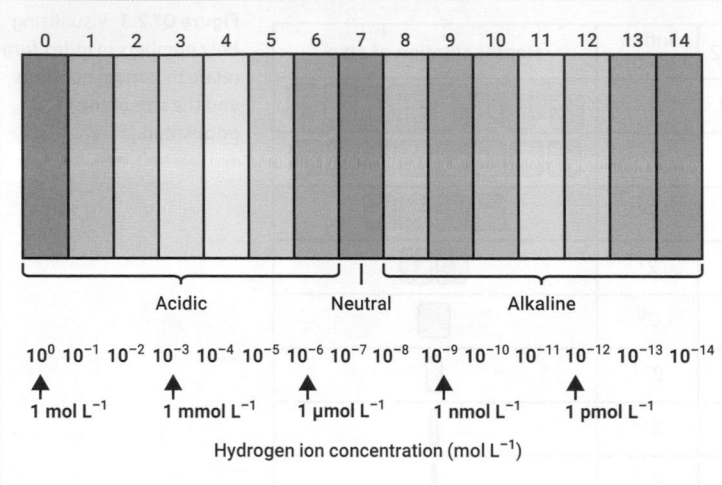

CASE STUDY 3 — A logarithmic scale to compare sizes of atoms, molecules, and organisms

We also use a logarithmic scale to the base 10 when comparing the sizes of atoms, molecules, and different organisms so that it is possible to fit a very wide range of measurements from 10^{-10} to 10^{-3} m within a single visual representation. Each label on the scale is 10 times larger as you go from left to right. In Figure QT2.6 we can see that a typical virus (40–90 nm) is an order of magnitude larger (that is, 10 times larger) than a typical protein (4–9 nm).

Figure QT2.6 Comparing the sizes of atoms, molecules, and organisms.

Source: Concepts of Biology—1st Canadian edition, by Charles Molnar and Jane Gair, licensed under a Creative Commons Attribution 4.0 International Licence.

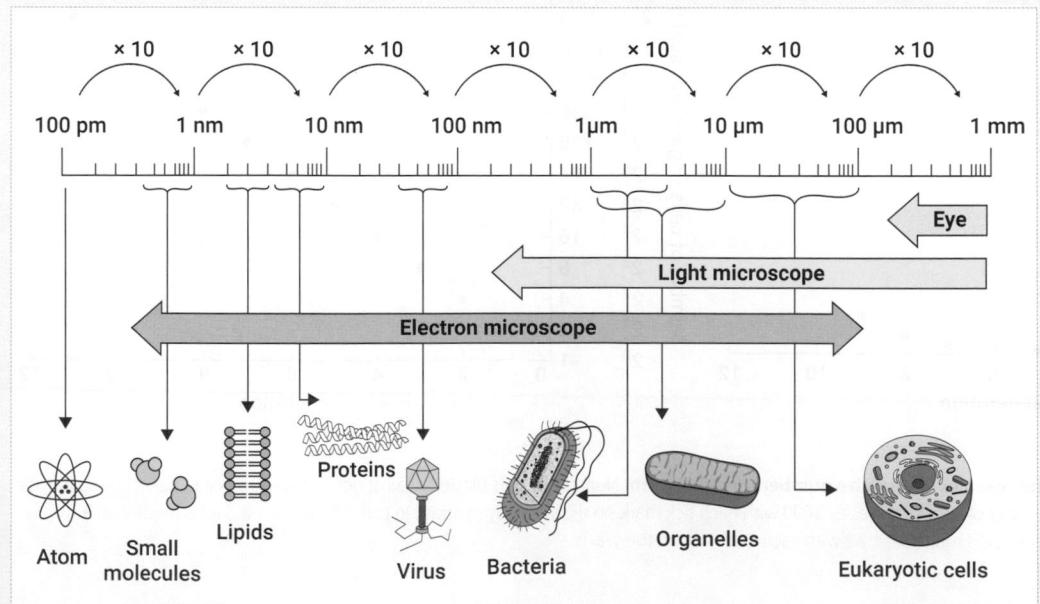

2.4 What is a logarithm?

A logarithm is the power to which the base is raised. For example, the logarithm to the base 10 of 100 is 2 because $100 = 10^2$: the base is 10 and the power is 2 (Figure QT2.7). The log button on a calculator means the logarithm to the base 10.

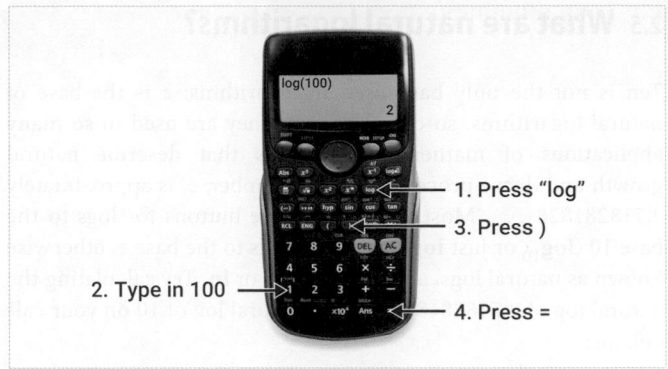

1. Press "log"
3. Press)
2. Type in 100
4. Press =

Figure QT2.7 Using a calculator to evaluate logarithms. Note that other calculators might work differently—check the instruction manual.
Source: Shutterstock.

Try it on your calculator:

EXAMPLE	the log (to the base 10) of 100 is 2	the log (to the base 10) of 1000 is 3	the log (to the base 10) of 1,000,000 is 6
NOTATION $\log_{10}(x) = y$	log(100)　　2	log(1000)　　3	log(1 000 000)　　6
BECAUSE The log is the power to which the base is raised.	$10^2 = 100$	$10^3 = 1000$	$10^6 = 1,000,000$

You can show that 10 to the power of 2 is 100 on your calculator—this is called 'anti-log'

(on most calculators you need to press SHIFT, then log):

10^2　　100	10^3　　1000	10^6　　1 000 000

The log of a number does not have to be a whole number. Try finding the log of 5, 50 and 500 on your calculator.

log(5)　　0.6989700043	log(50)　　1.698970004	log(500)　　2.698970004

This corresponds to: $10^{0.69897} = 5$　　$10^{1.69897} = 50$　　$10^{2.69897} = 500$

If you take the log of a number between 0 and 1, you will get a negative number. To see why this is, think about this example: $0.1 = 10^{-1}$. As the log is the power, then the log of 0.1 is −1. Try a few examples on your calculator.

log(0.1)　　−1	log(0.001)　　−3	log(0.5)　　−0.3010299957

This corresponds to: $10^{-1} = 0.1$　　$10^{-3} = 0.001$　　$10^{-0.301} = 0.5$

Calculating pH from hydrogen ion concentration and vice versa

In the case of the pH scale, it is customary to take the logarithm of the hydrogen ion concentration and multiply it by −1. So if $[H^+] = 10^{-6}$ mol.dm^{-3} then the logarithm is −6 and the pH is 6. If the $[H^+]$ increased to 10^{-5} mol.dm^{-3}, the logarithm would be −5 and so the pH would be 5. A 10-fold increase in the $[H^+]$ corresponds to a one unit decrease in pH.

To calculate the pH from a hydrogen ion concentration of 5×10^{-9} M, type $\boxed{\text{log}}\ \boxed{5}\ \boxed{\times 10^x}\ \boxed{-}\ \boxed{9}\ \boxed{=}$

And the answer given is −8.301 . . . So, the pH is 8.301.

To calculate the hydrogen ion concentration from a pH of 3.4, type $\boxed{\text{SHIFT}}\ \boxed{\text{LOG}}\ \boxed{-}\ \boxed{3}\ \boxed{.}\ \boxed{4}\ \boxed{=}$

And the answer given is 3.981×10^{-4} M.

▶ **Practise calculating pH from hydrogen ion concentrations and vice versa in Activity 6 at the end of this chapter.**

2.5 What are natural logarithms?

Ten is not the only base used in logarithms: e is the base of natural logarithms, so-called because they are used in so many applications of mathematical models that describe natural growth and decay processes. Euler's number, e, is approximately 2.718281828 . . .. Most calculators have buttons for logs to the base 10 ($\log_{10}$ or just **log**), as well as logs to the base e, otherwise known as natural logs, abbreviated $\log_e$ or **ln**. Try calculating the natural log of 2.718281828 and the natural log of 10 on your calculator:

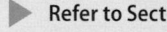

ln(2.718281828)	ln(10)
0.9999999998	2.3025858093

Now try the reverse operations, calculating e to the power of 1 and e to the power of 2.3025858093:

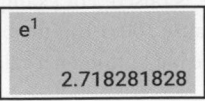

e^1

2.718281828

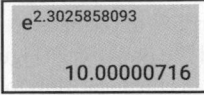

$e^{2.3025858093}$

10.00000716

Natural logarithms and e are found in many biological applications, including bacterial and population growth and changes in substrate concentration during enzyme reactions.

ACTIVITIES

Activity 1

Solutions can be filtered to remove bacteria and a typical pore size of the filter is 2.2×10^{-7} m. Will the following pass through?

▶ **Refer to Section 2.1.**

(a) A bacterium such as *Haemophilus influenzae*, which is 2×10^{-6} m long.

(b) A large virus such as *Megavirus chilensis*, with a diameter of 4×10^{-7} m.

(c) A small virus such as polio, which is 3×10^{-8} m long.

Activity 2

Streptococcus is a spherical bacterium with a diameter of 1.5×10^{-6} m. If *Streptococcus* bacteria were lined up across the head of a pin of diameter 1.5×10^{-3} m, how many would fit?

▶ **Refer to Section 2.1.**

Activity 3

Write the following measurements with an appropriate prefix by filling in the right-hand column. The first one has been done as an example.

▶ **Refer to Section 2.1.**

Diameter of *Streptococcus*	1.5×10^{-6} m	1.5 μm
Diameter of the head of a pin	1.5×10^{-3} m	
Diameter of *Megavirus chilensis*	4×10^{-7} m	
Length of *Haemophilus influenzae*	2×10^{-6} m	
Gap between two neurons in a synapse	3×10^{-8} m	
Concentration of acetylcholine in synaptic vesicles	0.1 M	

Activity 4

Indicate the relative sizes below by filling in the gap between columns with the symbols, $<, >, \approx, =$

30 nm	2 μm
2200 nm	2 μm
5 fs	4 ps
2000 amol	2 fmol
30 nm	3×10^{-8} m
540 μmol	5.5×10^{-7} mol

▶ **Refer to Section 2.2.**

Activity 5

The diagram shows a microscope image of a human hair.

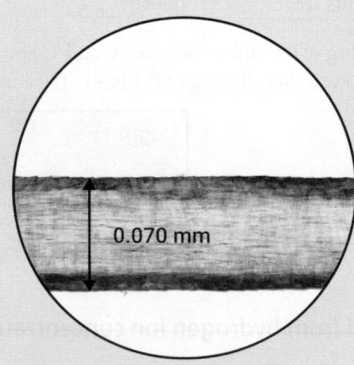

0.070 mm

To get a sense of the relative sizes, make a sketch of this diagram. *To scale*, alongside the hair, sketch *to scale* an outline of a HeLa cell of diameter 50 μm.

Alongside that, sketch *to scale* an outline of a *Staphlococcus* bacterium with diameter 1 μm.

▶ **Refer to Section 2.2.**

Activity 6

Fill out this table to show the relationship between the pH and the hydrogen ion concentration.

pH	[hydrogen ion]/mol.dm⁻³
3	
7.4	
6.8	
	2.3×10^{-12}
	6.31×10^{-4}
	1×10^{-10}

To get the pH when you know the hydrogen ion concentration, on most calculators you press log and then enter the concentration.

To do the reverse process and get the hydrogen ion concentration from the pH, on most calculators you press **SHIFT** then **LOG** then enter a minus sign and the number for the pH.

▶ **Refer to Section 2.3.**

ANSWERS

 Follow the links from the e-book to watch a video walk-through of the answers to these activities.

Activity 1

Haemophilus influenzae, 2×10^{-6} m	2×10^{-6} m = 20×10^{-7} m > 2.2×10^{-7} m ∴ will not pass through
Megavirus chilensis 4×10^{-7} m	4×10^{-7} m > 2.2×10^{-7} m ∴ will not pass through
Polio 3×10^{-8} m	3×10^{-8} m = 0.3×10^{-7} m < 2.2×10^{-7} m ∴ will pass through

Activity 2

$$\frac{1.5 \times 10^{-3}}{1.5 \times 10^{-6}}$$

$$= 1 \times 10^{3}$$

Activity 3

Diameter of *Streptococcus*	1.5×10^{-6} m	1.5 μm
Diameter of the head of a pin	1.5×10^{-3} m	1.5 mm
Diameter of *Megavirus chilensis*	4×10^{-7} m	0.4 μm or 400 nm
Length of *Haemophilus influenzae*	2×10^{-6} m	2 μm
Gap between two neurons in a synapse	3×10^{-8} m	0.03 μm or 30 nm
Concentration of acetylcholine in synaptic vesicles	0.1 mol.dm⁻³	100 mmol.dm⁻³

Activity 4

30 nm = 3×10^{-8} m	<	2 μm = 2×10^{-6} m
2200 nm = 2200×10^{-9} m= 2.2×10^{-6} m	≈	2 μm = 2×10^{-6} m
5 fs = 5×10^{-15} s	<	4 ps = 4×10^{-12} s
2000 amol = 2000×10^{-18} mol = 2×10^{-15} mol	=	2 fmol = 2×10^{-15} mol
30 nm = 3×10^{-8} m	=	3×10^{-8} m
540 μmol = 540×10^{-6} mol = 5.4×10^{-4} mol	>	5.5×10^{-7} mol

Activity 5

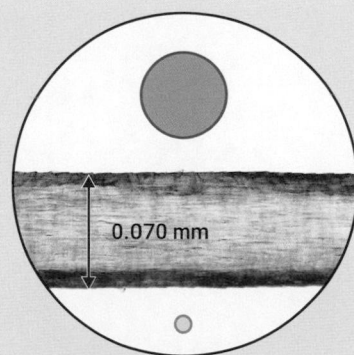

0.070 mm

The hair width is 10 mm on the page.
This corresponds to 0.07 mm, which is 7×10^{-5} m.
A HeLa cell diameter is 50 μm, which is 5×10^{-5} m.

$$10 \text{ mm} \longleftrightarrow 7 \times 10^{-5} \text{ m}$$

$\div 1.24$ ⟶ $\div 1.24$

$$? \text{ mm} \longleftrightarrow 5 \times 10^{-5} \text{ m}$$

10 ÷ 1.4 = 7.1 mm; the HeLa cell is the larger blue circle above the hair.
A 1 μm cell will be 1/50 of the size of the HeLa cell.
7 mm ÷ 50 = 0.14 mm, which is the small green dot below the hair.

Activity 6

pH	[hydrogen ion]/M
3	1×10^{-3}
7.4	4×10^{-8}
6.8	1.58×10^{-7}
11.64	2.3×10^{-12}
3.2	6.31×10^{-4}
10	1×10^{-10}

Describing Data

Contents

3.1	Central tendency	80
3.2	Measures of variability	80
3.3	Frequency distributions	82

LEARNING OBJECTIVES

In this toolkit we will:

- Compare and calculate the mean, median, and mode as measures of central tendency (Section 3.1).

- Compare and calculate the range, interquartile range, variance, and standard deviation as measures of variability (Section 3.2).

- Distinguish between data in raw form and the same data expressed as frequencies (Section 3.3).

- Produce frequency distributions from raw observations and from frequencies for category-like and continuous data (Section 3.3).

- Compare the normal, Poisson, and binomial distributions and where they might be applied (Section 3.3).

Imagine that you have collected data on the length in metres of the necks of eight giraffes and that you know your measurements are continuous scale level (see Quantitative Toolkit 1): what next? If someone asks you to describe your data you could list the numbers one by one—for example, 1.80, 1.86, 1.89, 1.81, 1.77, 1.75, 1.71, 1.68. However, this would be a bit tedious, especially if you had lots of records. It would be better to say how many records you had—in other words, the **sample size (*n*)**—along with a measure of **central tendency** to indicate the length in general, and a **measure of variability**, to indicate if the lengths were all similar or widely different. For our giraffe example you could say the sample size was 8 and the mean was 1.78 m with a standard deviation of 0.07. These are **descriptive statistics**.

▶ The measures of central tendency and variability you should use depend on the features of your data, as we describe in Quantitative Toolkit 1.

An alternative way of communicating the data would be to construct a **frequency distribution** which would show not only the central tendency and variability of the data, but also the 'shape'. This would help you to get to know your data even better and is likely to be useful as you go further with your analyses (see Quantitative Toolkit 7). For our giraffe example you would draw a histogram, but, again, the type of frequency distribution you should use depends on the features of your data (see Quantitative Toolkit 1).

3.1 Central tendency

The three principal ways of measuring central tendency are **mean** (ȳ), **median**, and **mode** (Table QT3.1).

Mean

A mean is calculated by adding up all the values and dividing by the sample size. It works for scale data but is not good for ordinal data and doesn't work at all for nominal data. It's also vulnerable to **outlier** and **extreme values**. Outliers and extremes are variously defined but are basically values that stand noticeably apart from most other observations. They may be a genuine part of the dataset but are often a result of measurement error. In any event, if the purpose of a measure of central tendency is to give you the general flavour of your data then it is not so useful if it changes radically due to one value.

Let us consider a very simple example: 1.4, 1.6, 1.2. You can imagine these numbers representing anything you like—the length of *C. elegans*, for example, or the time dogs spend wagging their

tails each day in hours. The sum of 1.4, 1.6, and 1.2 is 4.2; divide this by 3 and this gives you a mean of 1.4. Change the 1.6 to a 3.4 (an extreme value) and the mean changes from 1.4 to 2.0!

Median

The median is the midpoint of the values—that is, the value with as many values above it as below it.

The median is calculated by putting the numbers in size order, then taking the middle value (half way between the two middle numbers if you have an even number of records).

The median is less sensitive to extremes. In size order 1.4, 1.6, and 1.2 becomes 1.2, 1.4, and 1.6. The median is 1.4: change the 1.6 to 3.4 (an extreme value) and the median is still 1.4.

The median can be used with ordinal, as well as scale, level data. For example, the speed of seven horses measured at ordinal level might look like this: canter, trot, trot, canter, gallop, trot, walk. In rank order this would be walk, trot, trot, trot, canter, canter, gallop. The median would be trot.

Mode

The mode is the most frequently occurring value. It works for scale, ordinal, and nominal data, and is not affected by extreme values. The mode of seven roses coloured yellow, red, yellow, white, red, yellow, and white is yellow—the most frequently occurring value in this data set.

3.2 Measures of variability

Let us go back to the data on the length of the necks of eight giraffe (Figure QT3.1). Imagine that, instead of 1.80, 1.86, 1.89, 1.81, 1.77, 1.75, 1.71, and 1.68 our data set was 1.78, 1.78, 1.78, 1.78, 1.78,

Table QT3.1 Comparison of measures of central tendency

Measure of central tendency	Calculation	Levels of measurement	Sensitivity to extreme values
Mean (ȳ)	Add up the values and divide by the sample size	Scale	Large
Median	Put the data in ascending order and select the middle value	Scale and ordinal	Small
Mode	The most frequently occurring value	Scale, ordinal, and nominal	Small

Figure QT3.1 Giraffe necks of varying length can be summarized by information on sample size (in this illustration, eight), central tendency (for example, mean), and variability (for example, standard deviation).
Source: Shutterstock.

Table QT3.2 Comparison of measures of variability

Measure of variability	Calculation	Levels of measurement	Sensitivity to extreme values
Range	The difference between the minimum and the maximum value	Scale and ordinal	Large
Interquartile range	The difference between the lower and upper quartile	Scale and ordinal	Small
Variance (s^2)	The sum of squares divided by the degrees of freedom (see Table QT 3.3)	Scale	Medium
Standard deviation (s)	The square root of the variance (see Table QT3.3)	Scale	Medium

1.78, 1.78, and 1.78. The sample size would still be 8 and the mean would still be 1.78, but there is clearly something different about this second set of records: they have no variability. Being able to describe the **variability** of your data, as well as its central tendency, is important. Four ways of measuring variability are **range**, **interquartile range**, **variance (s^2)**, and **standard deviation (s)**, as outlined in Table QT3.2.

Range

The range is simply the difference between the **minimum** (smallest) and **maximum** (largest) value. It works for scale level and ordinal data (although you can only state minimum and maximum values for ordinal data; you cannot calculate a difference). The range is no good for nominal data and is very sensitive to extreme values. For example, for 2, 2, 7, 7, 8, 9, 10, 11, 12, 14, 16 the minimum is 2, the maximum is 16, and therefore the range is 2 to 16, or 14. But if the 16 was instead a 22 the range would increase by 6 to 20 (22–2) even though only one value changed.

Interquartile range

The interquartile range is the difference between the middle value of the lower half of the data (known as the **lower quartile**) and the middle value of the upper half of the data (known as the **upper quartile**). For example, for 2, 2, **7**, 7, 8, **9**, 10, 11, **12**, 14, 16, the numbers in bold indicate the lower quartile (7), the median (9), and the upper quartile (12). The interquartile range is 7 to 12, or 5, and would be unchanged if the 16 was a 22. As such, it is less sensitive to extreme values than the range. Again, it works for scale and ordinal data, but the interquartile range is no good for nominal data.

Variance and standard deviation

 Go to the e-book to watch videos that will walk you through the calculation of standard deviation and variance, and introduce you to z-scores and standardizing data.

Unlike for range and interquartile range, the calculation of variance involves all the data points but only works for scale data. Conceptually, you can think of the variance as being a measure of the distance of the data points from the mean. If there is not much variation, and all the values are clustered around the mean, then the variance will be small. It is calculated by dividing the **sum of squares** by the **degrees of freedom**.

To get the sum of squares you find the difference between each value and the mean, square each of these values, then add these up.

(You need to do the squaring: the sum of just the differences would be zero.) Degrees of freedom are related to sample sizes, in this case one less than the sample size (n–1). The formulae to use for these calculations are in Table QT3.3 and a sample calculation is in Table QT3.4.

Table QT3.3 Formulae for calculating variance and standard deviation.
Formulae can be written in different ways. Two different formulae for variance and standard deviation are shown here. You should get the same result with either. Formula A is better for understanding what variance is measuring and formula B is easier to use with larger sample sizes.

	Formula A	Formula B
Variance	$s^2 = \dfrac{\sum(y-\bar{y})^2}{n-1}$	$s^2 = \dfrac{\sum y^2 - \dfrac{(\sum y)^2}{n}}{n-1}$
Standard deviation	$s = \sqrt{\dfrac{\sum(y-\bar{y})^2}{n-1}}$	$s = \sqrt{\dfrac{\sum y^2 - \dfrac{(\sum y)^2}{n}}{n-1}}$

Where n = sample size.

Table QT3.4 Calculating variance and standard deviation using formula A from Table QT3.3

Construct a calculation table with the column headings as shown below:

Column number	1	2	3	4
Column heading	y	$\bar{y}$	$(y-\bar{y})$	$(y-\bar{y})^2$
	3	4	–1	1
	4	4	0	0
	5	4	1	1
Σ				2

Put your data (y) in column 1: in this example y = 3, 4, 5.

- Find the mean of y by dividing the sum of y by the sample size (n): in this example, 12/3 = 4. Put this in column 2 for every value of $\bar{y}$.
- Find the deviation of each value from the mean: see column 3.
- Calculate the squares of the deviations: column 4.
- Add the squares of the deviations together to find the sum of squares: bottom of column 4.
- Divide the sum of squares by the degrees of freedom (sample size minus 1; or n – 1): = 2/2 = 1.
- This is the variance. To find the standard deviation take the square root of this number: = $\sqrt{1}$ = ±1.

3.3 Frequency distributions

It is useful to recognize if the data you are dealing with are already in the form of frequencies or whether they are still in the 'raw observation' form. Units and levels of measurement are most clearly recognizable when data are 'raw' (see Quantitative Toolkit 1). For example, a sample of 13 records of gorilla sex is presented in Table QT3.5 in both its raw form (a) and as frequencies (b), and a sample of 26 records of gorilla sperm length is presented in Table QT 3.6 in both its raw form (a) and as frequencies (b).

Table QT3.5 Frequency distribution table for category-like data. Sex in gorillas: (a) raw observations (male/female); (b) frequency distribution table. This is an example of a frequency distribution table of data measured at nominal level but the approach is similar for ordinal or discrete scale data.

(a)	Male, Male, Female, Male, Female, Female, Female, Male, Male, Male, Male, Female, Female

(b)	Category	Tally	Frequency
	Male	IIIII II	7
	Female	IIIII	6

Table QT3.6 Frequency distribution table for continuous data. Sperm length in gorillas: (a) raw observation (in micrometres, μm); (b) frequency distribution table. This is an example of a frequency distribution table of data measured at continuous scale level.

(a)	60.80, 60.92, 60.51, 60.62, 60.75, 61.19, 61.21, 60.71, 61.25, 60.74, 60.81, 60.83, 60.91, 61.15, 61.22, 60.83, 61.08, 60.93, 61.12, 61.34, 61.01, 61.05, 61.14, 61.44, 61.07, 61.37

(b)	Classes (range of values)	Tally	Frequency
	60.50–60.69	II	2
	60.70–60.89	IIIII II	7
	60.90–61.09	IIIII II	7
	61.10–61.29	IIIII II	7
	61.30–61.49	III	3

In Table QT1.2 Hertz (Hz) are listed as the units used to measure 'frequency'. This is using the word frequency as a physicist would. When we are talking about frequency distributions, we are using it as a statistician would. These two uses of the word frequency are compared in Table QT3.7.

Data in their frequency form indicate how many occurrences there are of each value, or range of values—in other words, its **frequency distribution**. Frequency distributions can be presented as tables (as in Table QT3.5 and Table QT3.6) or as graphs. They not only give us an idea of the central tendency and variability of our data, but also allow us to describe the 'shape' of our data.

To generate a frequency distribution from 'raw observations', we define **classes**, or **class intervals**, and tally up how many records there are in each. Whether we use classes or class intervals depends on whether the raw observation data behave like they are **categorical** or **continuous** (Table QT3.8). For the former (nominal, ordinal variables, and some discrete scale data) we use classes that suggest themselves. For the latter (some discrete and all continuous scale data) we use class intervals that need to be defined. By convention, the bars in a frequency graph for continuous data touch and the graph is known as a **histogram**.

Theoretical frequency distributions

When we generate a frequency distribution from data we have gathered it is empirical—that is, it is based on observation rather than theory. By contrast, a theoretical frequency distribution is one that can be generated by a mathematical equation. It is both remarkable and useful that biological data commonly follow—in other words, can typically be modelled by—theoretical distributions (see Topic 1). It is useful because our ability to assume data follows a distribution opens up a raft of powerful analytical techniques (see Quantitative Toolkit 7). There are three theoretical distributions in particular that are useful to biologists: the normal, Poisson, and binomial distributions (Table QT3.9).

- The normal distribution is typically a good model of continuous scale level data that has been measured.

- The Poisson distribution is typically a good model of discrete scale level data that has been counted.

- The binomial distribution is typically a good model of nominal level data that has been collected using only two categories i.e., binary data.

Table QT3.7 Using the word frequency

	Physics	Statistics
Definition	The number of times an event repeats in a set time	The number of times a value, or value within a specified range, occurs
Units	Hertz (cycles per second)	Not applicable
Examples	The maximum frequency of sound waves that a human can hear is about 20,000 Hertz (Hz)	Mendel found that the frequency of phenotypes in his peas followed a 9:3:3:1 ratio
	If a heart is beating at 60 beats per minute, it has a frequency of 1 Hz	In a sample of the heart rates (in Hz) from seven people of 0.9, **1.2**, **1.4**, **1.4**, 1.6, 1.6, 1.7 the frequency of heart rates between 1 and 1.5 Hz is 3 (values in bold)

Table QT3.8 How to construct a frequency distribution

To generate frequencies starting from the raw observations measured at nominal, ordinal, or discrete scale level with a small range:	To generate frequencies starting from the raw observations measured at continuous scale level, or at a discrete scale level with a large range, is the same as for other types of measurement except, instead using classes suggested by the data, you have to define class intervals:
■ Identify the classes you are going to use (for example, *male* and *female*).	■ Identify the class intervals that you are going to use (e.g., *60.50–60.69, 60.70–60.89*, and so on).
■ Go through your data adding a tally mark to the appropriate class for each observation on your sample.	■ Go through your data adding a tally mark to the appropriate class interval for each observation on your sample.
■ Count the tally marks for each class to get the frequencies.	■ Count the tally marks for each class interval to get the frequencies.
	■ There are no hard-and-fast rules for defining class intervals, but some tips for starting out are:
	● make your classes equal in size;
	● avoid using lots of classes with only a few values in each (no more than 10 categories as a rough guide);
	● avoid using so few categories that you cannot see any detail (no fewer than five categories as a rough guide).

Table QT3.9　Examples of data and the theoretical distribution likely to be the best match

Example variable	Example sample (n=10) raw observations form	Features of data	Example sample (n=10) frequency form (table)	Example sample (n=10) frequency form (graph)	Assumed theoretical distribution
Species of ladybirds	Harlequin, Pine, Harlequin, Cream-spot, Cream-spot, Pine, Harlequin, Harlequin, Pine, 10-Spot	Nominal Category-like	Harlequin = 4 Pine = 3 10-Spot = 1		None
Length of cream-spot ladybirds (mm)	4.1, 4.3, 4.8, 4.5, 4.4, 4.2, 4.9, 4.8, 4.6, 4.4	Scale Measures Continuous	4.00–4.20 = 2 4.21–4.40 = 3 4.41–4.60 = 2 4.80–5.00 = 3		Normal
Number of pine ladybirds	2, 1, 0, 0, 1, 3, 6, 1, 0, 2	Scale Counts Discreet Category-like	0 = 3 1 = 3 2 = 1 3 = 1 4 = 0 5 = 0 6 = 1		Poisson
Occurrence of 22-spot ladybirds	Present, present, absent, absent, present, absent, present, present, absent, present	Nominal Binomial Category-like	Present = 6 Absent = 4		Binomial trial size 1
Number out of six harlequin ladybirds alive, after 3 days, in a Petri dish following exposure to a fungal pathogen	3, 4, 5, 6, 1, 0, 3, 4, 6, 2 (status of the ladybirds in the first Petri dish might be: dead, dead, live, live, dead, live)	Scale Proportion Category-like (Status as dead/live is binary (nominal) binomial)	0 = 1 1 = 1 2 = 1 3 = 2 4 = 2 5 = 1 6 = 2		Binomial trial size >1

ACTIVITIES

Activity 1

The following is a sample of incisor tooth length in humans in millimetres (mm): 18, 19, 21, 21, 13, 12, 16, 19, 21, 20. Are these data in 'raw' or frequency form?

▶ **Refer to Section 3.3.**

Activity 2

Using the data in Activity 1, calculate the following for this sample.

▶ **Refer to Sections 3.1 and 3.2.**

(a) Sample size (n).

(b) Mean.

(c) Median.

(d) Mode.

(e) Range.

(f) Interquartile range.

(g) Variance.

(h) Standard deviation.

Activity 3

If you had constructed a frequency distribution graph of the sample in Activity 1:

(a) Would you draw the columns touching or not?

(b) Would you describe it as a histogram?

▶ **Refer to Section 3.3.**

Activity 4

The Beaufort scale is an ordinal level measuring system, which ranges from 0 (calm) to 12 (hurricane force). The following is a sample of wind speeds measured using this system: 4, 3, 4, 5, 2, 2, 1, 3.

▶ **Refer to Sections 3.1 and 3.2.**

(a) Select the most appropriate measure of central tendency and the most appropriate measure of variability for these data. Justify your choice.

(b) Calculate the measures of central tendency and variability that you selected in (a).

Activity 5

If you had constructed a frequency distribution graph of the sample in Activity 4:

(a) Would you draw the columns touching or not?

(b) Would you describe it as a histogram?

▶ **Refer to Section 3.3.**

Activity 6

For each of the following say which theoretical distribution out of normal, Poisson, and binomial would most likely be usefully modelling the data.

▶ **Refer to Section 3.1.**

(a) The status of 100 cells as dead or alive.

(b) The diameter of 100 cells in micrometres (μm).

(c) The number of cells in each of 100 Petri dishes.

ANSWERS

Activity 1

Raw

Activity 2

(a) 10

(b) 18 (given by 18 + 19 + 21 + 21 + 13 + 12 + 16 + 19 + 21 + 20 = 180 divided by 10)

(c) 19 (the middle of the middle two values, in this case both: 12, 13, 16, 18, **19**, **19**, 20, 21, 21, 21)

(d) 21 (the most commonly occurring value: 12, 13, 16, 18, 19, 19, 20, **21**, **21**, **21**)

(e) 12 to 21 or 9 (the lowest number is 12 and the highest number is 21 and the difference between these is **9**: **12**, 13, 16, 18, 19, 19, 20, 21, 21, **21**)

(f) 16 to 21 or 5 (the middle of the bottom half of the numbers is 16 and the middle of the top half of the numbers is 21).

(g) 10.89 (the sum of squares, 98, divided by the degrees of freedom, 9)

(h) +/−3.30 (the square root of the variance)

🜉 Follow the link from the e-book to watch a walk-through of the answer to this activity.

Activity 3

(a) Yes, touching.

(b) Yes.

Activity 4

(a) Median and interquartile range—should not use mean or variance/ standard deviation with ordinal data (see Table QT3.1 and Table QT3.2 for justification).

(b) Median is 3 and interquartile range is 2 to 4.

🜉 Follow the link from the e-book to watch a walk-through of the answer to this activity.

Activity 5

(a) No, not touching.

(b) No.

🜉 Follow the link from the e-book to watch a walk-through of the answer to this activity.

Activity 6

(a) Binomial.

(b) Normal.

(c) Poisson.

Ratio and Proportion

Contents

4.1 Expressing ratios	86
4.2 Comparing ratios	87
4.3 Relating ratios, fractions, and percentages	89
4.4 Probability	89
4.5 Proportion	90
4.6 Concentrations and dilutions of solutions	90

LEARNING OBJECTIVES

In this toolkit we will:

- Explain the difference between a part-to-part and a part-to-whole ratio (Section 4.1).
- Compare ratios by simplifying to unit ratios (Section 4.2).
- Relate ratios to fractions and percentages (Section 4.3).
- Explain that two quantities which are in direct proportion can be described using ratios (Section 4.5).
- Use the ideas of ratio in calculations of concentrations and dilutions of solutions (Section 4.6).

The ideas of ratio and proportion are used throughout biology and can be expressed using a variety of symbols. The notation of ratios, fractions, and percentages are often used interchangeably when describing these ideas.

4.1 Expressing ratios

A **part-to-part ratio** is the most common way of expressing a ratio and is often used as the default. Take the example of a plant breeding experiment which produces red and white flowers in the ratio of 3 : 1 illustrated in Table QT4.1. This is a part-to-part ratio (sometimes called an internal ratio): for every three red flowers, you'd expect to get one white flower. And you can then scale this up, so for every six red flowers, you'd expect to get two white ones, or for every nine red flowers you'd expect to get three white ones.

A **part-to-whole ratio** (sometimes called an external ratio) is less common. Let's look at this same example. We might say that we get three red flowers out of four. This could be written as a part-to-whole ratio of 3 : 4, although this is uncommon. If spoken aloud, we'd say '3 in 4' or '3 out of 4'.

Part-to-whole ratios are more commonly used when **calculating concentrations** (Table QT4.1). When we talk about a 70 per cent alcohol solution, it means 70 mL ethanol in 100 mL total solution, which actually means 70 mL ethanol plus 30 mL water to make 100 mL in total. You could write it as a 7 : 3 ethanol : water solution, but that would be flying in the face of convention.

▶ Practise using the % concentration notation in Activity 1 at the end of this toolkit.

While a part-to-part ratio is generally the default, it is important to carefully consider the context, especially when working internationally as conventions can vary. Language can be a helpful clue: if the words 'for' or 'to' are used—for example, '2 for 3', or '2 to 3'—then it's likely to be a part-to-part ratio. However, if the words 'in' or 'out of' or 'per' are used—for example, '2 in 3' or '2 out of 3' of '2 per 3'—then it's likely to be a part-to-whole ratio.

🌀 Go to the e-book to watch a video that will help you explore ways of thinking and talking about ratios.

Parts per million (ppm) is a type of part-to-whole ratio: 1 ppm means 1 part of a substance in 1,000,000 parts, while 8.3 ppm is 8.3 parts in 1,000,000 parts. Parts per billion (ppb) is 1 per 1,000,000,000. These are ambiguous units: 'parts' usually means mass in biological examples but there are instances in which it could be used to mean distance or volume. Ppm and ppb are not recommended, but are still sometimes used.

Diagrammatic representations can be very useful tools when thinking about ratios and in interpreting real life problems. In Case studies 1 and 2 we look at using a bar model and a ratio square.

🌀 Go to the e-book to watch a video that explores how the bar model and ratio square methods can be used to support proportional reasoning.

▶ Practise calculating with ratios in Activity 2 at the end of this toolkit.

4.2 Comparing ratios

A 6 : 2 ratio can be simplified to a 3 : 1 ratio: we call these **equivalent ratios**. When using ratios it is common for decimals to be used. For example if we want to simplify the ratio 5 : 4, an equivalent ratio would be 1.25 : 1 (as we would divide both numbers by 4).

In our plant breeding experiment (Table QT4.1) we may have sown lots of seed and discovered that we had 342 red flowers and 98 white flowers. How similar is this to the 3 : 1 ratio? The easiest way to see how similar they are is to simplify the ratio to get a

Table QT4.1 Representations for ratios

Part-to-part (internal)		Part-to-whole (external)	
3 ... 1	3 : 1	3 ... 4	3 : 4 or $\frac{3}{4}$
6 ... 2	6 : 2	6 ... 8	6 : 8 or $\frac{6}{8}$
Language: use 'to' 'for' e.g. 'red *to* white is 3 to 1' or '3 red *for* every 1 white'		**Language:** use 'in' 'out of' e.g. '3 *in* 4 are red' or '3 red *out of* 4 in total'	
Example: Mendelian ratios, nutrient ratios N : P : K = 5 : 7 : 4		**Example:** Diluting solutions, a 1 : 5 dilution means 1 part plus 4 parts to make 5 parts in total	

CASE STUDY 1   Red and grey squirrels

There are two red squirrels for every seven grey squirrels in a forest. If there are 16 red squirrels, how many grey squirrels are there?

A **bar model approach**: the red squirrels are represented by two red squares and the grey squirrels with seven grey squares to show the ratio 2 : 7.

There are 16 red squirrels, so that means each red square represents eight red squirrels. So there must be $7 \times 8 = 56$ grey squirrels.

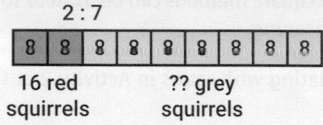

Bar model to represent part-to-part ratios

2 : 7

| 8 | 8 | 8 | 8 | 8 | 8 | 8 | 8 | 8 |

16 red squirrels   ?? grey squirrels

The **ratio square approach**: the aim is to make the scaling up (or down) explicit in the diagram. To get from 2 to 16 for the red squirrels we need to multiply by 8, so the same must be done with the grey squirrels to keep them in the same ratio: $7 \times 8 = 56$ grey squirrels.

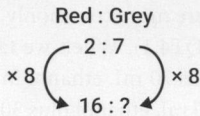

Red : Grey

2 : 7

$\times 8$ $\times 8$

16 : ?

CASE STUDY 2   Phenotype ratios in fruit flies

In a sample of 3000 flies, we might expect the ratios of phenotypes to be 9 : 3 : 3 : 1. The phenotypes being:

- brown body–red eyes (9)
- brown body–brown eyes (3)
- black body–red eyes (3)
- black body–brown eyes (1).

How many flies of each phenotype would we expect to observe?

A bar model approach is shown in Figure QT4.1. The bar illustrates the 9 : 3 : 3 : 1 ratio. If there were these numbers of each, then the total is 16 $(9 + 3 + 3 + 1 = 16)$.

Our actual total is 3000 so we need to scale up from 16 to 3000; in other words, we need to find the number that 16 needs to be multiplied by to get 3000. This is $3000 \div 16 = 187.5$.

Now we can multiply the numbers in the ratio by 187.5 so they all scale up by the same factor. You can check the calculation by ensuring that $1687.5 + 562.5 + 562.5 + 187.5$ add up to 3000.

9		3	3	1	=	16
⊙⊙⊙⊙⊙⊙⊙⊙⊙		○○○	●●●	●		× 187.5
$9 \times 187.5 = 1687.5$		562.5	562.5	187.5		3000

Figure QT4.1 Using the bar model to calculate the ratios of phenotypes of fruit flies. The top row shows the expected ratio of phenotypes 9 : 3 : 3 : 1. The middle row shows the phenotypes, brown body and red eyes, brown body and brown eyes, black body red red eyes, and black body and brown eyes. The bottom row shows the calculation for the expected values in a sample of 3000 flies.

unitary ratio. To do this we divide both numbers in the ratio by the smallest number, in this case 98:

$$322 : 98 = 3.3 : 1$$

Here we can see that it is fairly close to 3 : 1, but we are getting slightly more red flowers for each white flower.

▶ **Practise simplifying ratios in Activity 4 at the end of this toolkit.**

The same principle can be applied to ppm and ppb units. As 1 ppm is equivalent to 1 g in 1,000,000 g we can make the equivalent ratio by dividing both parts by 1000.

1 ppm is the same as . . .	1 g in 1,000,000
Divide both numbers by 1,000 . . .	0.001 g in 1000 g
(this means they are both still in the same ratio)	
This is the same as . . .	1 mg in 1 kg

When talking about a solution it is usual to note that the density of water is 1 kg per litre. In other words, 1 mg in 1 kg of water is equivalent to 1 mg in 1 L of water.

Overall, we have shown that 1 ppm is the same as 1 mg L^{-1}.

We can use the same strategy to show that 1 mg L^{-1} is equivalent to 1 µg mL^{-1}:

1 mg divided by 1000 is 1 µg

1 L divided by 1000 is 1 mL

Because we've divided both elements by 1000, they are both still in the same ratio.

🎥 Go to the e-book to watch a video that will help you convert between ppm and mg/L.

▶ Practise converting from ppm to µg mL^{-1} in Activity 3 at the end of this toolkit

4.3 Relating ratios, fractions, and percentages

A part-to-whole ratio is often expressed as a fraction. The total number of parts (i.e. the whole) is the denominator (the bottom number) and the part that we are interested in is the top number (the numerator). Again, a diagram can be a helpful way to visualize this. Going back to our example of three red flowers for every one white flower, we would say that for every four flowers, three of them are red; writing that as a fraction would be $\frac{3}{4}$.

▶ Practise calculating a fraction from a ratio in Activity 5 at the end of this toolkit.

Language is an important clue to knowing how to represent something mathematically: we would say '3 out of 4' or '3 in every 4' so we would write $\frac{3}{4}$. Fractions are often used when talking about probabilities since we would say the probability of getting one red flower is three out of every four or ¾ or 0.75 or 75 per cent.

A percentage is just a fraction with 100 on the bottom (as the denominator): e.g. 31 per cent is the same as $\frac{31}{100}$, which is the same as 0.31. If we have three red flowers out of every four, then we could use scaling up to say that we would have 75 red flowers out of every 100 (we multiply the top and bottom numbers of the fraction each by 25 to get 75/100):

$$\frac{3\,(\times 25)}{4\,(\times 25)} = \frac{75}{100} = 75\%$$

Once the fraction has been expressed with a denominator of 100, then the percentage is just the top number. In fact the word 'percentage' means 'per hundred'. So we could say we have 75 per cent red flowers. Or we could also say 'we have a 75 per cent chance of getting a red flower'. Or we could also say that we would expect 75 per cent of our flowers to be red.

The word 'per' is often used in the context of part-to-whole ratios. We might say we have 70 mL of methanol *per* 100 mL of our solution, which we could also write as 70 per cent. Or we could talk about a saline solution that is 9 g *per* litre, also written as 9 g/L or 9 g.L^{-1}. Some other examples are:

Energy changes in kJ per mole	kJ/mol	kJ.mol^{-1}
Density is mass per volume	kg/dm^3	kg.dm^{-3}
Speed is distance per time	m/s	m.s^{-1}
Population density is number per unit volume (the number has no units)	/dm^3 or /mL	dm^{-3} or mL^{-1}
Or per unit area	/m^2	m^{-2}

▶ The use of prefixes is covered in Quantitative Toolkit 2 and units are covered in Quantitative Toolkit 1.

4.4 Probability

If you roll a fair/unbiased six-sided die, what is the probability of rolling a 3? The 3 is one side out of six sides so the probability is 1 in 6. Mathematically we would write that as 'the probability of rolling a 3 with a six-sided die is $\frac{1}{6}$'.

To take a more biological example, if 50 per cent of bacteria in a dish are longer than 1 µm, then the probability of randomly picking out a bacterium that is longer than 1 µm is 1 in 2 or 50 per cent or ½—all different ways of describing the same probability (see Table QT4.2). Note the use of the phrase '1 *in* 6' or '1 *out of* 6'—writing that mathematically is $\frac{1}{6}$.

If you were to roll the die 50 times, how many 3s would you expect to get? You would get a 3 one-sixth of the time, so $\frac{1}{6} \times 50 = 8.333$, which is around 8–9 times.

Say you go to a party where there are 250 people and the probability of someone who is infected with a virus being present is 1 in 20 (i.e. $\frac{1}{20}$). How many infected people are likely to be present? It would be one-twentieth of 250, i.e. $\frac{1}{20} \times 250 = 12.5$ people, so 12 or 13 people. Note the use of the word '*of*'—if we say one-twentieth *of* 250 we write $\frac{1}{20} \times 250$.

What would be the probability of getting a 3 on the first roll of the die and a 3 on the second roll? Each event independently has a probability of $\frac{1}{6}$ but the probability of the first being a 3 *and* the second being a 3 is essentially one-sixth *of* one-sixth, i.e. $\frac{1}{6} \times \frac{1}{6} = \frac{1}{36}$.

Table QT4.2 Linking ratios, fractions, percentages, and decimals

Part-to-part ratio	Part-to-whole ratio	Fraction	Decimal	Percentage
3 : 1	3 : 4	$\frac{3}{4} = \frac{75}{100}$	0.75	75%
9 : 1	9 : 10	$\frac{9}{10} = \frac{90}{100}$	0.9	90%
1 : 9	1 : 10	$\frac{1}{10} = \frac{10}{100}$	0.1	10%
1 : 19	1 : 20	$\frac{1}{20} = \frac{5}{100}$	0.05	5%

4.5 **Proportion**

What is the difference between a ratio and a proportion? Proportionality involves making a ratio equal to something—turning it into an equation. If two quantities are in direct proportion, then as one increases the other increases. Mathematically, we would write $y \propto x$ for 'y is proportional to x'.

This kind of relationship can be visualized using a graph (Figure QT4.2). The top black line in the graph shows the y value increasing by 3 for every 1 increase in the x value. This would represent a 3 : 1 ratio. The middle line shows a 2 : 1 ratio and the bottom line shows a 1 : 1 ratio.

There are a number of examples in biology where we might use proportions—for example, in magnification. For a simple 2× magnifying glass, the ratio of the image size to the actual size is 2 so we could write image size : actual size = 2 : 1. Alternatively, we could write it as a fraction:

$$\frac{\text{image size}}{\text{actual size}} = 2$$

Or we could think of the magnification as the 'multiplier' or the 'scale factor' between the actual size and the image size, i.e.:

$$\text{image size} = 2 \times \text{actual size}$$

and we could draw a graph where y is the image size and x is the actual size, which for a 2× magnification would be the middle line in Figure QT4.2.

Interpreting this equation we can say the magnification is the multiplier between the actual size and the image size: Figure QT4.3 shows an example where the actual size is 2 mm; with a magnification of 40× the image size is 80 mm. Or we could equally use the language of ratio: the ratio of the image size to the actual size is 40:80 in this case.

▶ **Practise these ideas of direct proportion in Activity 6 and Activity 7 at the end of this toolkit.**

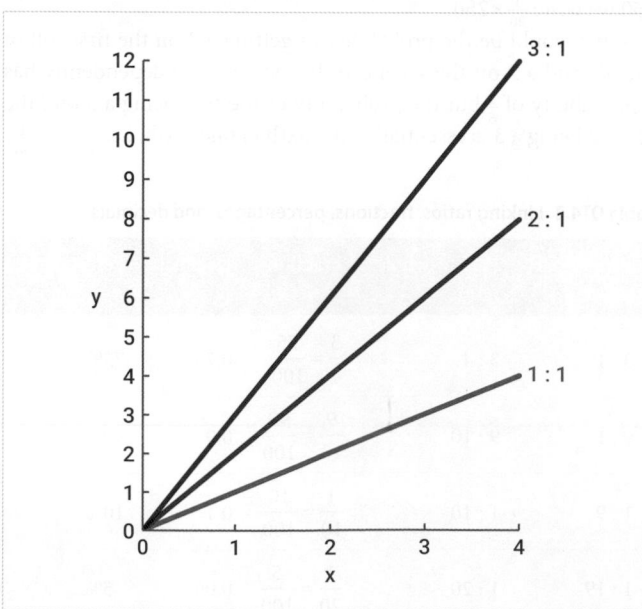

Figure QT 4.2 Directly proportional relationships.

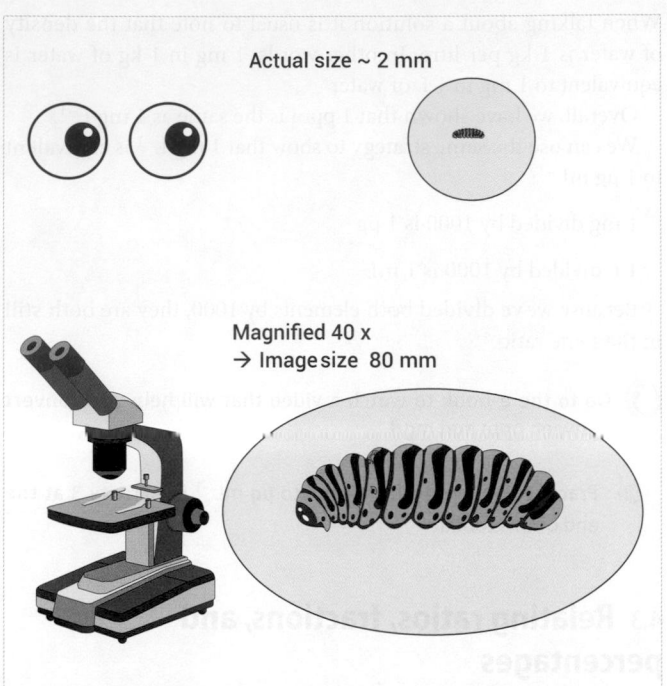

Figure QT4.3 The use of magnification to illustrate the relationship between actual and image size.

🌐 Go to the e-book to watch a video that will help you explore proportion and magnification further.

4.6 **Concentrations and dilutions of solutions**

The definition of the term '**concentration**' in relation to solutions is the mass of substance per volume. Written mathematically, it is:

$$\text{concentration} = \frac{\text{mass}}{\text{volume}}$$

We can therefore think of the concentration as the ratio of the mass to the volume of the solution. This is a part-to-whole ratio and we can use the ideas of ratio to perform calculations. For example, if a solution has a salt concentration of 5 g L^{-1}, what mass would be needed for a solution of 5.3 L? Using the ratio square, we can see that the volume has increased from 1 L to 5.3 L, i.e. multiplied by 5.3. Therefore, we multiply 5 g by 5.3 to get 26.5 g. This shows that 26.5 g in 5.3 L is the same concentration (or same ratio) as 5 g in 1 L.

$$
\begin{array}{c}
\text{g : L} \\
5 : 1 \\
\times 5.3 \overset{\frown}{} \times 5.3 \\
? : 5.3
\end{array}
$$

🌐 Go to the e-book to watch a video that will talk you through concentration calculations.

Sometimes concentrations are expressed in moles per dm^3 (mol.dm^{-3}). To convert between g.dm^{-3} and mol.dm^{-3} it is necessary to convert between g and mol. The relative formula mass of a molecule describes how much mass there is in 1 mole. One mole of a large molecule such as haemoglobin has a mass of 64,000 g, whereas one mole of a small molecule such as water has a mass of 18 g. We can use the ideas of ratios to work out the number of moles in a given mass or the mass in a given number of moles.

Take the example of water: the formula mass of water is 18, which tells us that 1 mole of water has a mass of 18 g. Say we wanted to work out the mass of 5 moles of water; we could use a ratio square. At the top we write down what we know about the ratio of mass to moles for water: 18 g : 1 mol, then underneath we write ? g : 5 mol. We are scaling up the ratio by 5—i.e. multiplying both parts of the ratio by 5. So 18 g per 1 mol is the same as 90 g per 5 mol.

g : mol
18 : 1
× 5   × 5
? : 5

🎧 Go to the e-book to watch a video that will walk you through calculations involving mass and moles.

What if you wanted to know how many moles of water there are in 33.7 g? First, we need to work out what we would multiply 18 by to get 33.7. To do this we do the inverse operation, which is to *divide* 33.7 by 18. This operation gives 1.872. So, if we multiply the left-hand side by 1.872 we get 33.7. We must also multiply the right-hand side by 1.872 and we get 1.872. (Remember: our example is 18 g per 1 mol, so the right-hand side is 1.) Now the bottom ratio is equivalent to the top ratio and we would say '18 to 1 is equivalent to 33.7 to 1.872'.

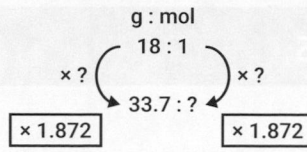

▶ Practise calculating concentrations, amounts, and volumes in Activities 8, 9, and 10 at the end of this toolkit.

Dilutions

If we make a solution 10 times more dilute, that means the concentration is 10 times less. So a 5 g L^{-1} solution would become 0.5 g L^{-1} and we would say the dilution factor is 10. We would call this a '1 in 10 dilution' and it would be written 1 : 10. This is a part-to-whole ratio so it means 1 part of the original solution plus 9 parts of solvent to make 10 parts of the new solution.

🎧 Go to the e-book to watch a video that will walk you through calculating dilutions.

Some examples are shown in Table QT4.3.

Say we need to make 20 mL of this solution: Figure QT4.4 shows how we can use ratios to work out how much of the original solution and how much water would be required.

▶ Practise calculating concentrations, amounts, and volumes in Activities 11, 12, 13, and 14 at the end of this toolkit.

Table QT4.3 Some examples of dilutions

Original solution	Diluted solution	Dilution	Preparation
10 g L^{-1}	5 g L^{-1}	1 : 2	1 part + 1 part → 2 parts
			e.g. 1 mL of original solution + 1 mL water → 2 mL diluted solution
10 g L^{-1}	2.5 g L^{-1}	1 : 4	1 part + 3 part → 4 parts
			e.g. 1 mL of original solution + 3 mL water → 4 mL diluted solution
10 g L^{-1}	2 g L^{-1}	1 : 5	1 part + 4 parts → 5 parts
			e.g. 1 mL of original solution + 4 mL water → 5 mL diluted solution
10 g L^{-1}	1 g L^{-1}	1 : 10	1 part + 9 part → 10 parts
			e.g. 1 mL of original solution + 9 mL water → 10 mL diluted solution
10 g L^{-1}	0.1 g L^{-1}	1 : 100	1 part + 99 part → 100 parts
			e.g. 1 mL of original solution + 99 mL water → 100 mL diluted solution

Concentration = 5 g/L   **Water**   **Concentration = 0.5 g/L**

        1 part of 5 g/L   +   9 parts   10 parts of 0.5 g/L
× 2   × 2   →   × 2
        2 mL   +   18 mL   →   20 mL

Figure QT4.4 Using ideas of ratio to calculate volumes required for a 1 : 10 dilution.

ACTIVITIES

Activity 1

You need to make 500 mL of a 70 per cent ethanol solution. What volumes of ethanol and water are required?

▶ **Refer to Sections 4.1 and 4.2.**

Activity 2

A fertilizer mixture is described as having an N : P : K ratio of 5 : 7 : 4. If the total weight is 500 g, how much phosphorus (P) will it contain?

▶ **Refer to Sections 4.1 and 4.2.**

Activity 3

(a)   Convert 0.48 ppm to mg L^{-1}.

(b)   Convert 0.48 ppm to µg mL^{-1}.

(c)   Convert 3.4 µg mL^{-1} to mg L^{-1}.

▶ **Refer to Sections 4.1 and 4.2.**

Activity 4

In a plant-breeding experiment we might have four phenotypes (A, B, C, and D), which appear in the ratio 106 : 38 : 35 : 12. Simplify this ratio and compare it with the predicted ratio of 9 : 3 : 3 : 1.

▶ **Refer to Sections 4.1 and 4.2.**

Activity 5

If there are two red squirrels for every 15 grey squirrels in a forest, what fraction of squirrels are red?

▶ **Refer to Section 4.3.**

Activity 6

The rate of an enzyme reaction is directly proportional to substrate concentration. If the rate was 30 mol s^{-1} at a concentration of 10 µM, what would the rate be at a concentration of 20 µM?

▶ **Refer to Section 4.5.**

Activity 7

A microscope image shows HeLa cells with a scale bar showing 10 µm. What is the actual approximate length of the cell shown with the ruler?

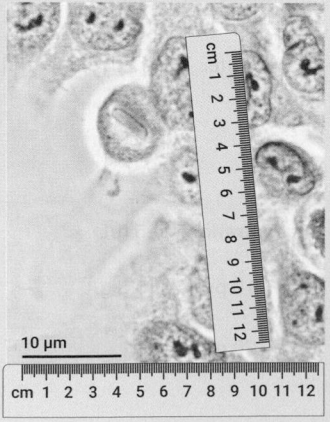

10 µm

Activity 8

You have weighed out 2.5 mg of an enzyme and need to make a 10 mg mL^{-1} solution. How much water do you need to dissolve it in?

▶ **Refer to Section 4.6.**

Activity 9

The concentration of plankton in seawater at a depth of 4 m is roughly 5 mg L^{-1}. What volume of seawater would contain 50 µg?

▶ **Refer to Section 4.6.**

Activity 10

Calculate how much urea is present in the following (in µmol).

▶ **Refer to Section 4.6.**

(a)   10 mL × 0.01 mol.dm^{-3}

(b)   25 mL × 4 mmol.dm^{-3}

(c)   20 µL × 0.5 mol.dm^{-3}

Activity 11

The concentration of a laboratory stock solution of an antibiotic is 25 mg mL^{-1}. How much of the stock solution and how much water would be needed to prepare 100 mL of a working solution with a concentration of 2.5 mg mL^{-1}?

▶ **Refer to Section 4.6.**

Activity 12

ATP is an important molecule required for energy-dependent processes and is often added to *in vitro* experiments for enzymes to function correctly. If you were to add 10 µL of a 0.05 M ATP solution to a final volume of 2 mL, what would be the resulting concentration of ATP? (M is an abbreviation for mol.dm^{-3}.)

▶ **Refer to Section 4.6.**

Activity 13

Describe how you would prepare 1 mL of a 20 nM solution from a 1 mM stock solution.

▶ **Refer to Section 4.6.**

Activity 14

How would you make 10 mL of a 1 µM solution of antimycin from a 1 mM stock solution?

▶ **Refer to Section 4.6.**

Image length from the ruler = 3.4 cm
Scale bar length   = 4.2 cm

▶ **Refer to Sections 4.1 and 4.2.**

ANSWERS

 Go to the e-book to watch a video that will walk you through the answers to these Activities.

Activity 1

70 per cent ethanol is 70 mL ethanol : 30 mL water so 70 mL + 30 mL would give 100 mL total volume. Therefore, we need to scale up by 5 times.

volume (mL) of		
ethanol	water	solution
70	30	100
350	150	500

Activity 2

There are 5 parts of N, 7 parts of P, and 4 parts of K, making 16 parts altogether.
The mixture is 500 g in total so one part would be 500/16 = 31.25 g
There are 7 parts phosphorus so 7×31.25 g = 218.75 g.

5 parts	7 parts	4 parts
500 g		
156.25 g	218.75 g	125 g

Activity 3

From one row to the next, divide each number by 1000.

(a) 0.48 g per 1,000,000 g

   = 0.48 mg per 1000 g
   = 0.48 mg per 1 kg
   = 0.48 mg L^{-1}

(b) 0.48 g per 1,000,000 g

   = 0.48 mg per 1000 g
   = 0.48 µg per 1g
   = 0.48 µg mL^{-1}

(c) 3.4 µg per 1 mL

   = 3.4 mg per 1 L

Activity 4

To simplify to a unit ratio, divide all the numbers by the smallest number, in this case 12.
   106 : 38 : 35 : 12
   = 8.83 : 3.17 : 2.92 : 1 and compare with 9 : 3 : 3 : 1
There are slightly fewer A and C than expected and slightly more B than expected.

Activity 5

There are 17 squirrels altogether so the fraction of red squirrels is $\frac{2}{17}$.

Activity 6

The rate was 30 mol s^{-1} at a concentration of 10 µM. What would the rate be at a concentration of 20 µM?
Clear setting out helps here:
   10 µM gives a rate of 30 mol s^{-1}
   20 µM gives a rate of ? mol s^{-1}
As we have doubled the concentration and the rate is proportional to the concentration, we get double the rate. So the answer is 60 mol s^{-1}.

Activity 7

Image length from the ruler = 3.4 cm
   actual size = x
   Scale bar length = 4.2 cm
   actual length of scale bar = 10 µm (10 µm = 0.01 mm = 0.001 cm)

$$\frac{\text{actual length}}{\text{image length}} = \frac{x}{3.4 \text{ cm}} = \frac{0.001 \text{ cm}}{4.2 \text{ cm}}$$

$$\frac{x}{3.4} = 0.000238$$

$$x = 0.000238 \times 3.4 = 0.00081 \text{ cm} = 0.0081 \text{ mm} = 8.1 \text{ µm}$$

Activity 8

The required volume is 0.25 mL

$$\frac{10 \text{ mg}}{1 \text{ mL}} = \frac{2.5 \text{ mg}}{? \text{ mL}}$$

Activity 9

The required volume is 0.01 L which is 10 mL

$$\frac{5 \text{ mg}}{1 \text{ L}} = \frac{0.05 \text{ mg}}{? \text{ L}}$$

Activity 10

Note that 1 mL = 0.001 dm^3

(a) $$\frac{0.01 \text{ mol}}{1 \text{ dm}^3} = \frac{? \text{ mol}}{0.01 \text{ dm}^3}$$

   The required amount is 0.01 mol $\times$ 0.01 = 10^{-4} mol, which is 100 µmol

(b) $$\frac{4 \text{ mmol}}{1 \text{ dm}^3} = \frac{? \text{ mol}}{0.025 \text{ dm}^3}$$

   The required amount is 4 mmol $\times$ 0.025 = 0.1 mmol = 100 µmol

(c) $$\frac{0.5 \text{ mol}}{1 \text{ dm}^3} = \frac{? \text{ mol}}{20 \times 10^{-6} \text{ dm}^3}$$

   The required amount is $20 \times 10^{-6} \div 2 = 10 \times 10^{-6} = 10$ µmol

Activity 11

The stock solution is 25 mg mL^{-1} and it is being diluted to 2.5 mg mL^{-1}. This is a 1 : 10 dilution.
   For a 1 : 10 dilution you use 1 part of the stock plus 9 parts of water to make 10 parts altogether.
   To get 100 mL altogether we would need 10 mL stock plus 90 mL water.

Activity 12

You are taking 10 µL (0.01 mL) and adding 1.99 mL to make a total volume of 2 mL.

Scale this up: 1 mL + 199 mL gives a total of 200 mL. This is a 1 : 200 dilution.

Therefore, the concentration of ATP will be 200× lower, that is, 0.05 ÷ 200 = 2.5×10^{-4} M.

Activity 13

1 mM : 20 nM

$$= 10^{-3}\ M : 20 \times 10^{-9}\ M$$
$$= 1 : 20 \times 10^{-6}$$
$$= 50,000 : 1$$

volume (mL) of		
1 mM stock	water	20 nM solution
1	49,999	50,000
2×10^{-5}	0.99998	1

Activity 14

1 mM to 1 µM is a 1 : 1000 dilution

Therefore, 0.01 mL stock solution plus 9.99 mL water would give 10 mL of a 1 µM solution.

volume (mL) of		
1 mM stock	water	1 µM solution
1	999	1000
0.01	9.99	10

Understanding Samples

LEARNING OBJECTIVES

In this toolkit we will:

- Distinguish populations and parameters from samples and statistics (Section 5.1).

- Outline the consequences of sampling error (Sections 5.1 and 5.2).

- Define sample size (Section 5.2).

- Describe sample quality in terms of variability and bias (Section 5.3).

- Calculate standard error and 95 per cent confidence intervals of the mean (Section 5.3., Variability, standard error, and confidence intervals).

- Compare related and unrelated samples (Section 5.4).

Contents

5.1	Populations and samples	96
5.2	Sample size	96
5.3	Sample quality	96
5.4	Related and unrelated samples	98

If you wanted to find out how long are the necks of giraffes currently in Serengeti National Park, would you plan to collect data on all the giraffes in the park? Alternatively, if you wanted to know the blood sugar levels of people in Wales, would you take blood samples and measure this for everyone in Wales? The answer, in both cases, is no, you would not. It would not be practical (logistically, ethically, or financially) and, fortunately, it is not necessary: you can get your answer from a subset, or **sample**. Understanding samples, and how you get them through **sampling**, is fundamental to good science.

5.1 Populations and samples

As a biologist the image that comes into your head when you hear the word 'population' is probably a collection of plants or animals of the same species in the same area. However, you need to put this image to one side when you are thinking about data analysis. In this context, the word **population** encapsulates all the possible observations you could make to answer a particular question. For example, if you were interested in the blood sugar level of people in Wales, you would need to measure every person in Wales to get your population. In this case, the population is not the people involved, but their blood sugar levels. If you used *all* the data in your population to calculate a descriptive statistic (a mean, for example; see Quantitative Toolkit 3), it would be an example of what statisticians call a **parameter**.

Your population for any particular analysis is defined by its conceptual framework—that is, the questions you are seeking to answer, and any hypotheses and predictions you develop to answer them (see Topic 1). Populations can be finite or infinite. Compare the questions 'What is the level of blood sugar in people in Wales today?' with 'What is the level of blood sugar of people?'. The former defines a finite population but the latter is infinite because it has no reference to time or space. Generally, a question is more interesting if the population it defines is infinite, but the most important thing is that you are mindful of what your population is when you interpret your results.

While your population is the entire set of observations relevant to your research question, your **sample** is the subset of that population you use to get your answer (Figure QT5.1). If you calculate a descriptive statistic for your sample (for example, a mean), this would be an example of a descriptive **statistic**. In short, a parameter describes a population and a statistic describes a sample.

We do not expect a statistic to be exactly the same as its corresponding parameter, but it can give us a good indication: in more technical parlance we can **infer** the value of a parameter from its corresponding statistic. The difference between a statistic and its corresponding parameter is called **sample error** and it is generated by **sampling error**.

Go to the e-book to watch a video that will help you understand how populations and samples are used to answer questions.

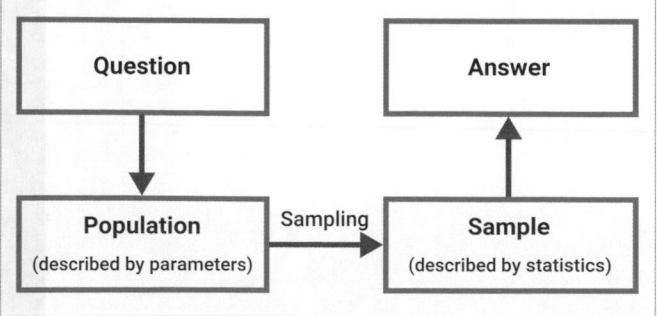

Figure QT 5.1 How populations and samples are used to answer questions.

5.2 Sample size

A fundamental feature of a sample is how big it is—that is, how many data it contains. We have already met the descriptive statistic for this in Quantitative Toolkit 3: it is the **sample size (n)**. Sample size is a simple, but important, bit of information. For example, sample size is a key determinant of statistical power (see Quantitative Toolkit 6 and Quantitative Toolkit 7): the bigger your sample, all other things being equal, the more reliable your analyses will be. If we think about this for a moment, we realize that it is because a larger sample more fully represents its corresponding parameter.

You can see this when looking at frequency distributions of different samples from the same population. Look at Figure QT5.2 and notice how the histograms in each row become more alike as the sample size increases—from 10 to 100 to 1000.

Samples of the same size, from the same population, can appear quite different from each other due to **sampling error**, mentioned in the previous section in relation to the difference between a parameter and a statistic. Bigger samples from the same population tend to look less different from each other, and are also more likely to look like a theoretical distribution (see Quantitative Toolkit 3). Look at Figure QT5.2 and notice how the histograms in the bottom row (n = 1000) look much more like a theoretical distribution than those in the top row.

Just as we do not expect a statistic to be the same as its corresponding parameter (because we are basing it on a subset of the population), we do not expect statistics from different samples of the same population to be identical.

5.3 Sample quality

Variability, standard error, and confidence intervals

Descriptive statistics can be thought of as estimates of their corresponding parameters. Every statistic has an associated **standard error** which we can use to get an idea of how close the statistic is likely to be to its corresponding parameter. You are most likely to have come across the **standard error of the mean**. This type of standard error is so common that if you see standard error written without reference to a particular statistic, then you can assume that this is the type of standard error being referred to.

Standard errors are calculated from the variability of a sample (see Quantitative Toolkit 3): the more variable a sample, the bigger its standard error and the less confidence we can have about how accurately we can infer the value of the parameter from that sample.

The formula for calculating the standard error is:

$$S_{\bar{y}} = \frac{s}{\sqrt{n}}$$

where

- s = standard deviation
- n = sample size.

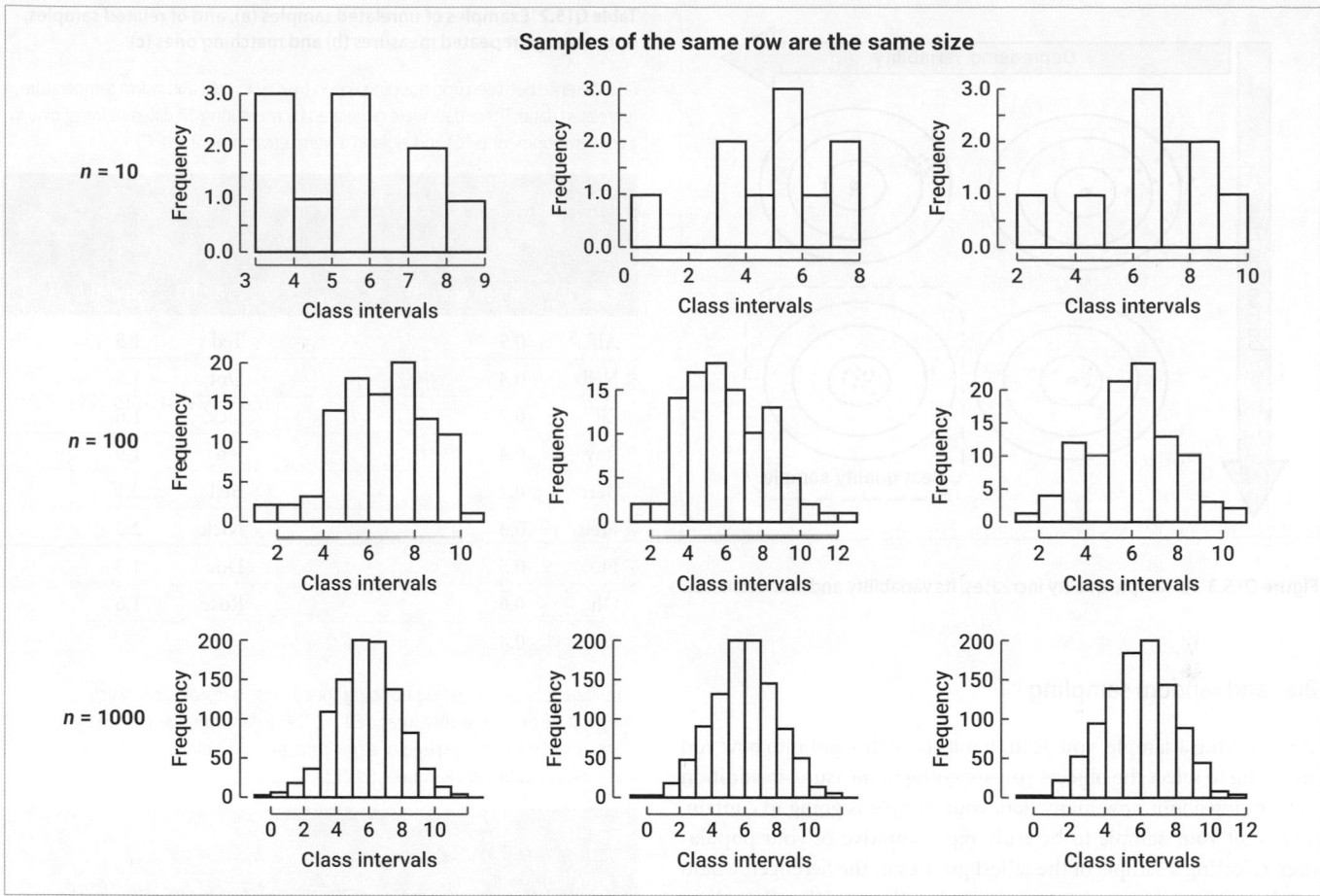

Figure QT5.2 Histograms of nine samples from the same population of scale level continuous data.

🎬 Go to the e-book to watch a video that will help you understand how to calculate sample error and 95 per cent confidence levels.

Standard errors can be used to calculate specific confidence intervals, typically **95 per cent confidence intervals** (Table QT5.1). As for standard errors, confidence intervals can be used to get an idea of how close our statistic is likely to be to our parameter: the bigger the confidence interval, the lower the confidence. A 95 per cent confidence interval is saying that, if we took 100 samples, we would expect 95 of those samples to have confidence intervals that include the population mean. In other words, there is a very good chance that a parameter will fall within its 95 per cent confidence interval.

The confidence interval ranges above the statistic to the **upper confidence limit** and below the statistic to the **lower confidence limit**.

The formulae for calculating the confidence interval and upper and lower confidence limits of the mean are:

- Confidence interval = mean $\pm t S_{\bar{y}}$
- Lower confidence limit = mean $- t S_{\bar{y}}$
- Upper confidence limit = mean $+ t S_{\bar{y}}$

(Degrees of freedom for t = n – 1).

Table QT 5.1 Comparison of error bars for standard deviation, standard error of the mean, and 95% confidence intervals

Standard deviation	Standard error	95 % confidence intervals
Shows the variation within the sample	Shows the variability of the sample in relation to the population	
s	$S_{\bar{y}} = \dfrac{s}{\sqrt{n}}$	= mean $\pm t S_{\bar{y}}$

Although the error bars that you see on graphs in peer-reviewed papers can represent any measure of variability, they are most often standard errors or confidence intervals. Mathematically, all these measures are related (as shown in Table QT5.1) but 95 per cent confidence intervals are most useful for comparing samples and trying to assess patterns in data (see Quantitative Toolkit 7).

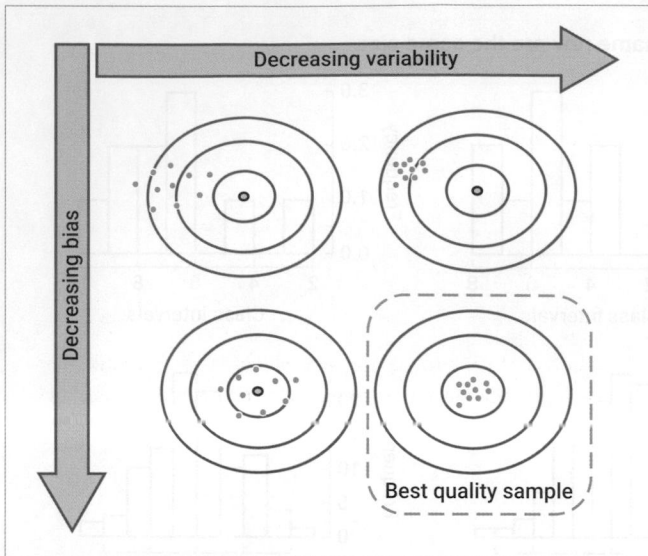

Figure QT5.3 As sample quality increases, its variability and bias decreases.

Bias and random sampling

When taking a sample, you need to put some thought into how you are going to select the objects you are going to measure, in addition to thinking about how many data your sample is going to contain. You want your sample to be truly representative of your population: selecting a sample of the tallest giraffes in the Serengeti would not be a very accurate way of answering the question 'how long are the necks of giraffe in Serengeti National Park?'. Such a sample would be **biased** (Figure QT5.3). The way to avoid bias is to use **random sampling**. This is not as easy as it perhaps sounds, as we discuss further in Quantitative Toolkit 6.

5.4 Related and unrelated samples

If you have more than one sample and are interested in a potential difference between them, it is important to know whether the samples are **related** or **unrelated**.

Samples can be related if they all contain data from the same individuals. This is known as having **repeated measures** (Table QT5.2b). Alternatively, related designs can be based on the similarity of the characteristics of individuals contributing to each sample. This is known as having **matched measures** (Table QT5.2c). When just two samples are involved this is often called **pairing** rather than matching. If the data are not repeated or matched then the samples are unrelated (Table QT5.2a).

Distinguishing related and unrelated samples is just one aspect of understanding the independence of data (see Quantitative Toolkit 6). As long as you are aware of having a related design, and use the appropriate analytical technique, your analyses will remain reliable. For example, paired tests, mixed models, and nested designs can all be used to deal with this kind of lack of independence of data.

Table QT5.2 Examples of unrelated samples (a), and of related samples, both through repeated measures (b) and matching ones (c).

(a) Difference between frog hopping speed (m s^{-1}) in cold and warm temperature (unrelated data). These data were generated by measuring 17 different frogs: nine in a cold environment (5 °C) and eight in a warm environment (20 °C).

Temperature			
Cold		Warm	
Frog ID	Speed (m s^{-1})	Frog ID	Speed (m s^{-1})
Alf	0.9	Ted	1.5
Nell	0.4	Dot	1.5
Lil	0.7	Jack	1.6
Ray	0.4	Jen	1.9
Bert	0.2	Bell	2.1
Fred	0.3	Nick	2.2
Flo	0.5	Doc	1.3
Oli	0.6	Rose	1.6
Sid	0.4		

(b) Differences between frog hopping speed (m s^{-1}) in cold and warm temperature (repeated-measures design). These data were generated by measuring the hopping speed of eight frogs, first in a cold environment (5 °C) and then in a warm environment (20 °C).

Frog ID	Temperature	
	Cold	Warm
Ludwig	0.8	1.9
Blossom	0.6	2.1
John	0.5	1.8
Mona	0.5	1.7
Rollin	0.2	1.6
Ernie	0.0	0.8
Kat	0.7	1.9
Ron	0.9	2.0

(c) Differences between frog hopping speed (m s^{-1}) in cold and warm temperature (matched design). These data were generated from eight pairs of frogs matched for age and sex characteristics. One frog in each pair was in a cold environment (5 °C) and the other in a warm environment (20 °C).

Frog IDs	Temperature	
	Cold	Warm
Kit and Bob	0.8	2.1
Mike and Robbie	0.4	1.8
Louise and Bess	0.9	1.7
Ojo and Laura	0.5	2.2
Will and Bill	0.2	1.6
Harry and Sam	0.1	0.7
Hilary and Jill	0.3	1.5
Louie and Don	0.8	1.3

ACTIVITIES

Activity 1

You are asked to find out the mean age of death of toy poodle dogs in the United Kingdom over a 6-month period. You have access to all the death records for this breed and calculate the mean as 11 years. Your friend is given a similar task but is only provided with a sample of data. They calculate the mean of this sample to be 12 years. What are the general names given to the mean calculated by you and by your friend and what is the name of the difference between the values that you have calculated?

 Refer to Section 5.1.

Activity 2

The values in your friend's sample from Activity 1 were 15, 12, 10, 11, 9, 14, and 13. What is the sample size and standard error of this sample? What else would you need to look up to calculate the 95 per cent confidence intervals?

 Refer to Section 5.3.

Activity 3

The table below presents the values of four samples similar to your friend's in Activities 1 and 2. In this table, which sample is the least biased? Which is the least variable?

 Refer to Section 5.3.

SAMPLE ID	A	B	C	D
Mean	13	10.5	10	12.5
Standard error	0.85	0.80	0.74	0.75
Sample size	7	7	7	7

Activity 4

For each of the following three scenarios, which of the following would appropriately describe the samples produced: related (repeated), related (matched), or unrelated?

 Refer to Section 5.4.

(a) An entomologist working on the transmission of malaria by mosquitoes got 12 volunteers to wash their feet and 12 volunteers not to wash their feet for four days. She then recorded the number of mosquitoes settling on the skin of each volunteer.

(b) An entomologist working on the transmission of malaria by mosquitoes got 12 volunteers to wash their feet for four days and then she recorded the number of mosquitoes settling on the skin of each volunteer. She then asked the same volunteers not to wash their feet for four days and then recorded the number of mosquitoes settling on the skin of each volunteer again.

(c) An entomologist working on the transmission of malaria by mosquitoes got 12 pairs of identical twin volunteers. In each pair, she asked one twin to wash their feet for four days and the other not to wash their feet for four days. Then she recorded the number of mosquitoes settling on the skin of each volunteer.

ANSWERS

Activity 1

Your mean is a parameter as it is generated from all the values in the (statistical) population under study. Your friend's mean is a statistic as it is generated from a sample of values in the population under study. The difference is sample error.

Activity 2

Sample size (n) = 7, Standard error = 0.82. The value of t.

Activity 3

Least biased is B (10.5 is closest to the population mean of 11) and least variable is C (0.74 is the smallest standard error).

Activity 4

(a) Unrelated.

(b) Related repeated.

(c) Related matched.

Designing Experiments

Contents

6.1 The importance of good experimental design	101
6.2 Problems	101
6.3 Solutions	102
6.4 Power analyses	104
6.5 Recognizing lack of independence	104
6.6 Adding variables	105
6.7 Implications for interpretation	105

LEARNING OBJECTIVES

In this toolkit we will:

- Recognize the importance of good experimental design (Section 6.1).
- Summarize the problems caused by variability (noise) and bias (inaccuracy) and distinguish measurement error from biological sources (Section 6.2).
- Identify ways of minimizing noise and inaccuracy (Section 6.3).
- Evaluate the advantages and disadvantages of including more variables (Section 6.4).
- Recommend the use of power analysis to plan for sample sizes for adequate power (Section 6.5).
- Recognize lack of independence in data, including pseudoreplication, and its implications (Section 6.6).
- Highlight four issues to look out for when interpreting your findings: non-representative samples, reverse causation, confounding variables, and placebo effect (Section 6.7).

Various terms are used to describe the things that are measured in a study, be it cells, Petri dishes, leaves, humans, or giraffes. But here we shall call them **subjects**. Experimental design involves deciding what subjects to measure and how to measure them. It also involves deciding when, where, how, and how many of these subjects to sample.

These are really important decisions that need to be made when planning data collection and analyses. Good experimental design helps ensure your work is reliable, valid, and reproducible—three key tenets of good science. No matter how sophisticated the statistical technique or how cool the statistical software, it will not be able to compensate for poor experimental design. Whether it is answering questions about cancer cell growth, wombat physiology, or enzyme activity, if you get the design of your investigation wrong you will waste time and money and not be able to answer your question.

Traditionally, the term 'experimental design' has been associated with studies in which variation in the independent variable is artificially generated (see Quantitative Toolkit 1). Such studies are usually called **experimental studies**. However, the principles also apply to **observational studies** and **quasi-experimental studies** which make use of naturally occurring variation.

While good experimental design is crucial to good science, there is no such thing as a perfect design: you must work ethically, and you will face practical and financial constraints. As long as you are mindful of the limitations of what you have done when interpreting your results, you can learn from the experience, as well as still potentially making a contribution to knowledge. In any event, you can use your understanding of experimental design to critically evaluate the work of others.

6.1 The importance of good experimental design

Investing time in planning your data collection and analyses will help ensure that you:

1. Conduct reliable and valid analyses.
2. Interpret your findings honestly.
3. Work ethically.

Any analyses you use on your data will require it to meet certain standards. If these are not met then, at best, your analyses will be unreliable and, at worst, they will be totally invalid. For example, a t-test, which can be used to assess for statistically significant differences (see Quantitative Toolkit 7), only works if the data are normally distributed. If you conduct a t-test on ordinal data, which cannot be normally distributed, then your results will be completely useless. Considering what analyses you will use on your data at the experimental design stage will help you avoid such disasters.

You are also more likely to be able to use a simple analysis, whose assumptions and limitations you understand, if the data come from a well-planned experiment. Even if your analyses are technically valid, your results will be much more difficult to interpret if you have poor experimental design.

Whether problems stem from analysis, or from interpretation, you can see that poor experimental design can lead to all the resources you put into collecting and analysing the data being wasted. This is unethical in itself, but even more so if this has potential negative impact on animal (including human) welfare or conservation.

6.2 Problems

Two key goals of experimental design are to reduce noise and to improve accuracy—in other words, to improve the quality of your samples as described in Section 5.3 of Quantitative Toolkit 5.

▶ Practise reducing noise and improving accuracy in Activity 1 at the end of this Toolkit.

Variability and noise

The problem with noise is that it makes it harder for you to detect patterns in your data. Imagine if you stood blowing a whistle behind the Royal Philharmonic Orchestra while they played Beethoven's 9th Symphony at the Royal Albert Hall. With all sections of the orchestra playing full blast no one in the audience would be able to hear your whistle. But as each section gradually played more softly, the audience would be increasingly likely to hear your whistle—and if all the musicians stopped playing the audience would hear only you. Using this as an analogy: when designing experiments we are trying to get all sections of the orchestra to be as quiet as possible so that we can hear if anyone is whistling. In other words, reducing noise increases statistical power.

▶ We discuss statistical power in more detail in Quantitative Toolkit 7.

What constitutes noise and what constitutes variation of interest depends on the context. For example, if you want to compare how long it takes salivary amylase to break down starch at body temperature (37 °C) and room temperature (20 °C), then the two variables you are interested in are the time taken for the starch to be broken down and temperature: everything else, such as pH, is a potential source of noise.

Sources of variation can relate to the subject, as well as its environment. The concentration of the solutions in our salivary amylase example could be an issue, but when our subjects are living organisms there is even more potential for those subjects to be a source of noise. For example, if you are studying heart rate in *Daphnia* sp. (water fleas) in relation to the temperature of the water that they are in, the variables you are interested in are heart rate and temperature. Other aspects of the environment, such as the salinity of the water, and characteristics of the individual *Daphnia* sp., such as age and sex, just create noise.

If experiments cannot be carried out at the same time or in the same space, you need to think about the noise implications. Differences in space may be associated with environmental variation—perhaps *Daphnia* kept at one end of the lab are exposed to more light than those at the other end. Differences in time may be associated with subjects getting older or solutions going off.

Noise is not just a consequence of the biological system under investigation. The system of measurement can also generate noise in a range of ways—for example, through shortcomings of the equipment or how it is used, or how observations are recorded.

The difference between the value obtained when a subject is measured and its true value is known as **measurement error**: the smaller the measurement error the greater the **measurement**

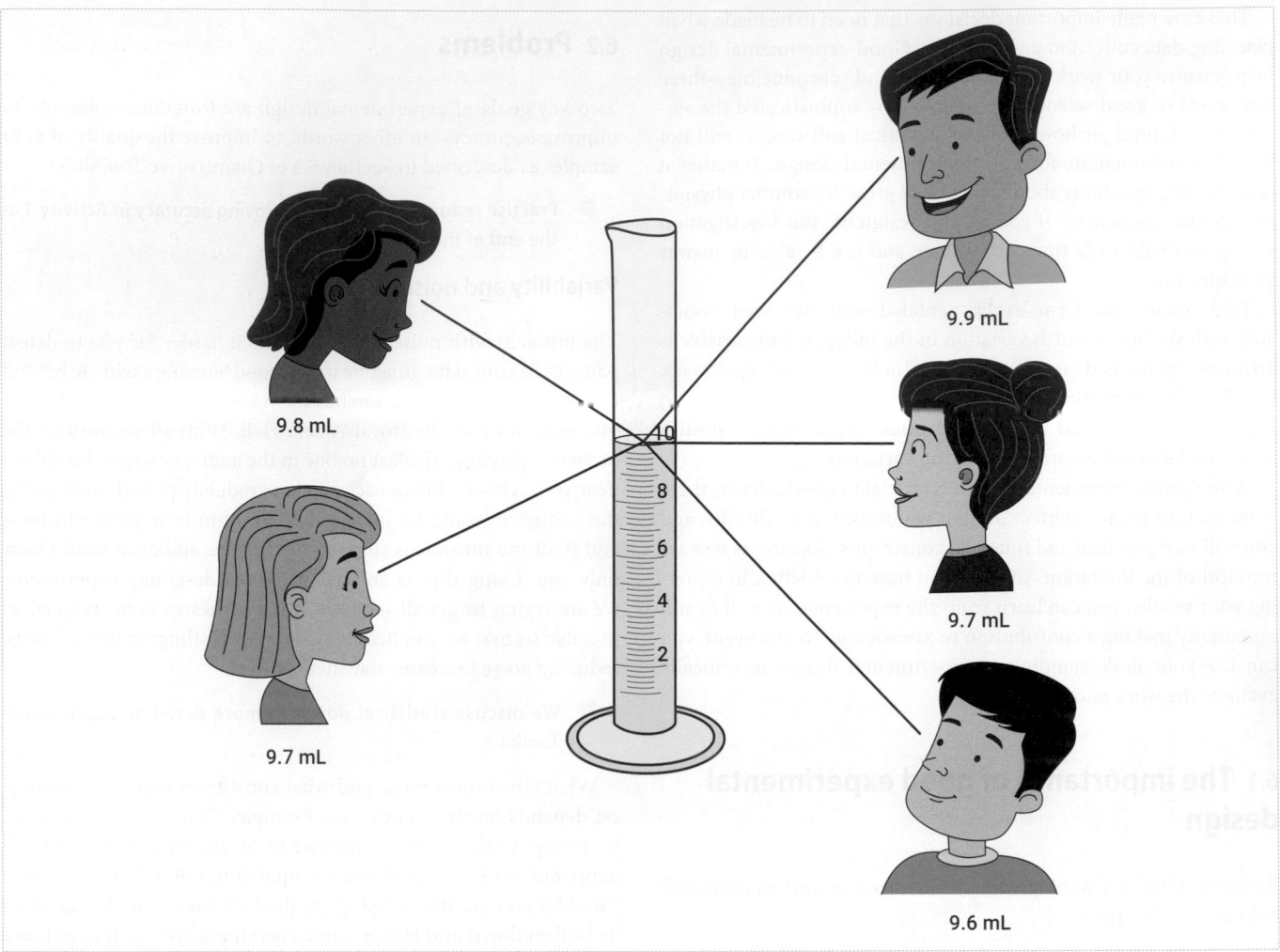

Figure QT6.1 Differences in the way observers use equipment can contribute to measurement error. For example, the angle at which you look at the level of a liquid in a measuring cylinder makes a difference to the measurement you take. This source of noise can be reduced by training observers in standard protocols.

reliability. For example, it is possible to weigh the same animal in the same state 20 times with the same piece of equipment and get different values each time, not because of anything biological but to do with the mechanics of the tool being used. None of these 20 values might be the 'true' weight of the animal. Even something as seemingly straightforward as measuring the amount of liquid in a cylinder can create variation (Figure QT6.1). When the subjects of a study are animals, particularly human animals, the interaction between the experimenter and the subject is another potential source of unwanted variation.

Bias and inaccuracy

The problem with bias is that you are not looking at what you intended to look at. This might be because your sample is not representative of your population (see Quantitative Toolkit 5) or, as we see with noise above, it might be generated by your system of measurement (measurement error). For example, poorly calibrated or otherwise defective equipment can be a source of bias such that the data generated by that equipment will be inaccurate (Figure QT6.2).

6.3 Solutions

Control and randomization

Control and randomization are two key ways of dealing with the problems of noise and bias.

Control

Control can be physical or statistical. Physical control can be achieved by using two approaches, which are often combined. The first is by constraining the variation in the system as much as possible, except for the variables of interest: it gives you what is often called a **controlled experiment**. For example, to look at the heart rate of *Daphnia* in relation to temperature you could set up cold and warm water baths, introduce *Daphnia* and measure their heart rates, while keeping everything else the same as much as possible, including light, time of day, salinity and oxygen content of the water, age of the *Daphnia* sp., and so on.

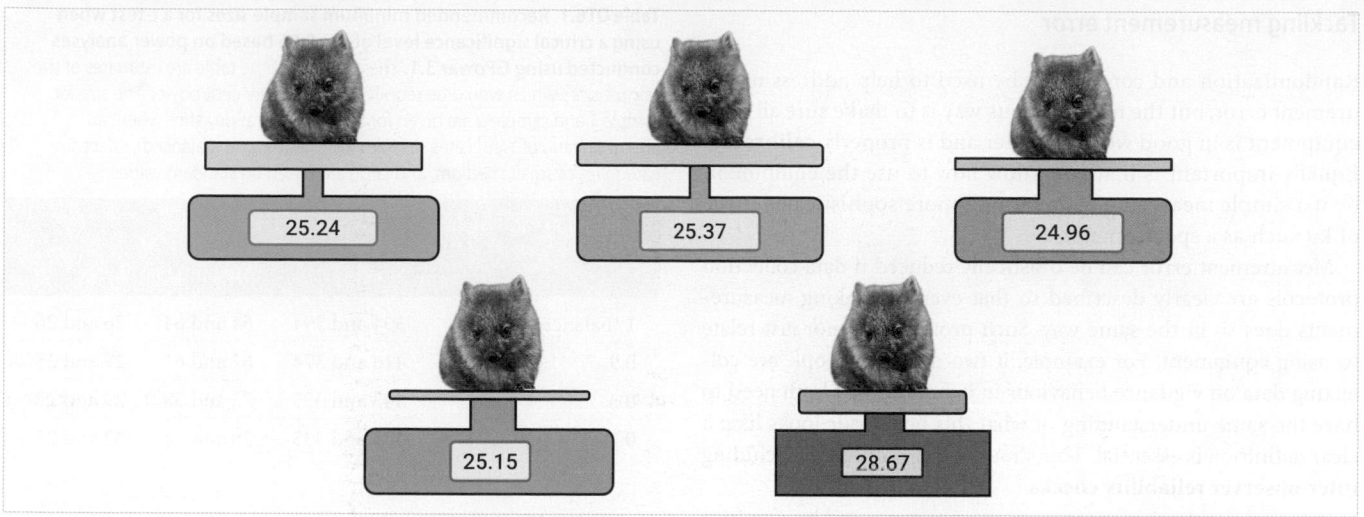

Figure QT6.2 Defects in equipment can contribute to measurement error. For example, if the mass of the same wombat is measured using five different balances you can get five different values. In this illustration, the first four machines have similar values, and the differences are just noise, but the fifth machine is defective and the value is inaccurate. This source of inaccuracy can be reduced by calibrating equipment and zeroing (known as taring) before taking measurements. *Source*: Shutterstock.

The second approach is to use a **control group**. The subjects in a control group are not exposed to the independent variable at all. For example, to look at breast cancer cell growth in relation to oestrogen you could prepare two sets of culture dishes containing skin cells and only add oestrogen to one set. The set without oestrogen added would be your control group. A **placebo** is a type of control group used in medical trials. The placebo is not clinically active at all, perhaps just a tablet made of sugar compared to a tablet containing a new form of pain relief.

Statistical control is achieved by identifying and collecting data on potentially important sources of variation and incorporating these variables into your analysis. For example, with the *Daphnia* heart rate example you could allow light to vary but collect data on it. If you then include light as a variable in your statistical test you can tease out what variation in heart rate you observe is due to temperature and what is due to light. The downside is that you will need more complex statistics, capable of dealing with more than one independent variable, and bigger sample sizes to achieve adequate power (see Section 6.4).

A **within-subject** design, where you measure the same subjects in different conditions, has built into it the control of noise generated by the characteristics of the subjects. For example, if you were to use the same *Daphnia* when measuring heart rate at different temperatures then you minimize the noise generated by the age, sex, etc., of the *Daphnia* in the sample. This contrasts with a **between-subject** design where you measure different subjects in different conditions. A within-subject design might not be feasible or there might be logistical or other difficulties, but it is good to always consider the option (see Section 6.6).

Randomization

If something is chosen, or done, at random it means that it happens by chance alone. In our *Daphnia* heart rate example, by **randomly selecting** each *Daphnia* subject and **randomly allocating**

it to either the cold or the warm temperature, each *Daphnia* has an equal chance of being selected and an equal chance of ending up in either a cold or warm water bath. By using randomization you should avoid, for example, ending up with all the youngest *Daphnia* being allocated to a cold treatment and all the oldest to a warm treatment. Randomization does not get rid of noise but the noise, on average, cancels itself out. Randomization helps ensure that your sample is representative of your population and that the data within the sample are independent.

The example with *Daphnia* that we've just described is an example of a **fully randomized experiment**. You should always carry out a fully randomized experiment if you can. By contrast, quasi-experimental studies do not allow subjects to be allocated to treatments in a fully randomized way. For example, if we wanted to compare heart rate between males and females we could not randomly allocate each *Daphnia* to either male or female. In this case we need to work with what we have. This is OK, but it is good to be aware that you are doing this and consider the consequences when interpreting your results (see Section 6.6).

Randomization can be usefully applied to all aspects of your protocol, some of which might not immediately come to mind. For example, within-subject designs can suffer from **order effects**—that is, the sequence in which subjects receive different treatments can create noise and bias. You can address this issue by randomizing the order in which each subject receives the different treatments.

You might be thinking that designing a randomized experiment is easy; however, humans are notoriously bad at being random in the statistical sense. You can read lots more about this elsewhere, but one very simple example of a way in which you could introduce randomization is as follows. Imagine you had to choose one test tube at random from a row of six. You could randomize this by rolling a dice. If you got the number 3 then you would choose the third one along.

Tackling measurement error

Randomization and control can be used to help address measurement error, but the most obvious way is to make sure all your equipment is in good working order and is properly **calibrated**. Equally important is that you know how to use the equipment, be it a simple measuring cylinder or a more sophisticated piece of kit such as a spectrometer.

Measurement error can be drastically reduced if data collection protocols are clearly described so that everyone taking measurements does so in the same way. Such protocols do not just relate to using equipment. For example, if two different people are collecting data on vigilance behaviour in baboons, they both need to have the same understanding of what this behaviour looks like: a clear definition is essential. This should be monitored by including **inter-observer reliability checks**.

Potential problems of measurement error generated by cognitive biases of an observer can be addressed by making sure the observer knows as little about the subject as possible, known as **blinding**. For example, not knowing which treatment a patient has received will prevent any confirmation bias on the part of the observer (see Topic 1, Table T1.2).

When the subjects of a study are animals, particularly human animals, the interaction between the observer and the subject is another potential source of measurement error. Such **observer effects** can be mitigated for when working with non-human animals by **habituation** and working **double-blind**. Habituation is the process of gradually getting animals used to being observed until they carry out their normal activities as if the observer is an inert part of their environment. (Jane Goodall famously habituated the chimpanzees in Gombe National Park.) In a double-blind study the protocol is designed so that the observer and the subject know as little as possible about each other.

▶ **Apply your learning in Activity 2 at the end of this toolkit.**

6.4 Power analyses

The power of a statistical test is a measure of its ability to detect a pattern in your data if it is there (see Quantitative Toolkit 7). Assuming you have taken steps to minimize the noise in your study, using the procedures we have already discussed, then an increase in your sample size is typically going to be the simplest way for you to increase statistical power. However, returns are diminishing: that is, the more you increase sample size, the less the increase in power you get. Set against this are the practical and ethical costs of increasing your sample size: Is it possible? Is it affordable? Can it be done in the time available? Is it going to raise animal welfare or conservation concerns?

The key question then becomes 'How much is enough?'. The traditional answer is that 80 per cent power is enough. You can use an analytical procedure to estimate what sample sizes would give you this level of power. This procedure is a type of **power analysis**. The minimum information that you will need to conduct this type

Table QT6.1 Recommended minimum sample sizes for a t-test when using a critical significance level of α = 0.05 based on power analyses conducted using GPower 3.1. The numbers in the table are estimates of the sample sizes which would be required to get 80 per cent power. The size for sample 1 and sample 2 are given for different allocation ratios where an allocation ratio of 1 indicates samples are of equal size (balanced). Effect size categories of small, medium, and large are based on standard values.

Allocation ratio (n_2/n_1)	Effect size		
	Small	Medium	Large
1 (balanced)	394 and 394	64 and 64	26 and 26
0.9	416 and 374	67 and 61	27 and 25
0.8	443 and 355	72 and 58	29 and 23
0.7	479 and 335	78 and 54	32 and 22

of power analysis for a **null hypothesis significance test** (NHST) such as a t-test, is:

- the name of the analysis;
- the critical significance level (α) you are going to use;
- the standardized effect size (see Quantitative Toolkit 7).

▶ **We discuss the factors that go into choosing the right NHST, and the basics of how to use them, in Quantitative Toolkit 7.**

If you have a choice between a non-parametric test and its parametric counterpart, then the parametric test will be more powerful, although sometimes only marginally. The critical significance level (α) is whatever you decide, but 0.05 is most common. A standardized effect size incorporates standard deviation and therefore has the variability, or noise, built in. Since the effect size is the thing that you are interested in finding out about you are unlikely to have a value for this at the planning stage unless it comes from previous work. However, there are some standard values you can use for small, medium, and large effect sizes based on your best guess; if you are being cautious, use small. These standard values involve a measure of variability and so have the noise built in. Table QT6.1 shows recommended sample sizes for a t-test.

▶ **Apply your learning in Activity 2 at the end of this toolkit.**

6.5 Recognizing lack of independence

You need to be aware of the independence of data within, and between, samples in order to choose valid analyses (see Quantitative Toolkit 7). For example, the independence of data within samples is an assumption of all NHSTs regardless of whether they are parametric or non-parametric. Data are independent if the selection of one subject you are going to measure does not influence the selection of another, and all subjects remain equally likely to be selected. If your sampling is truly random you will have achieved this independence. Lack of independence of data within samples is often referred to as **pseudoreplication**.

Both space and time can cause issues with independence. If you want to look at the use of shade by elephants and you record that one elephant in a group is under a tree, the elephant next to her might be in the shade just because she is part of her family group. Alternatively, the same elephants might be in the shade an hour later because they have not moved.

In this example, taking one measure per group would be more likely to give you independent measures of shade use by elephants. Alternatively, you could take data from all the elephants in the group and use a measure of central tendency rather than the raw data. In this context this is often called **aggregating** or **condensing** data.

Taking measurements from the same subject, whether it be an elephant or a test tube, is often—but not always—problematic. Independence is relative to what you are investigating. If, for example, you were interested in the reliability of a particular heart pressure gauge, taking multiple readings from the same person could be valid. Taking measurements from the same subject is also okay if it is part of a within-subject design and the data are analysed using the appropriate technique for **related samples** (see Quantitative Toolkit 7).

▶ Apply your learning in Activity 5 at the end of this toolkit.

6.6 Adding variables

Univariate analyses involve one variable. **Bivariate analyses** involve two variables. **Multivariate analyses** involve more than two variables. Our examples so far, such as heart rate in *Daphnia* and temperature, and breast cancer cell growth and oestrogen, have been bivariate.

As already mentioned, you may want to include more than two variables in your analyses in order to statistically control variables that you cannot *physically* control (see Section 6.3). Other potential advantages of including additional variables in your analyses are that you can look at interaction, deal with pseudoreplication (see Section 6.5), and increase the ability of a model to make predictions.

Statistical interaction occurs when two independent variables lack independence from one another in terms of their effect on the dependent variable. For example, you could include both sex and temperature as variables of interest when investigating heart rate in *Daphnia*. An extreme interaction would be that heart rate in males goes up and in females it goes down with increasing temperature. You might need to investigate such interactions in order to be able to interpret your results or because your question is specifically about interaction.

In the context of statistical models like generalized linear models (GZLMs; see Quantitative Toolkit 7), categorical independent variables are known as either **fixed factors** or **random factors**. Fixed factors are generally the key variables in terms of the questions being asked, while random factors are not thought to influence the dependent variable in any systematic way; they just add noise. When you have multiple measurements from the same individual you can address potential issues around pseudoreplication by including subject ID as a random factor. Models that include both random and fixed factors are referred to as **mixed models**.

Another consideration when determining the number of variables to include in your experimental design is that the more variables there are in a model, the better it will be at predicting values of the dependent variable based on novel values of the independent variables.

A disadvantage of a larger number of variables is that the analyses will be more complex and will need to involve more subjects to maintain statistical power. Fully randomized designs are harder to achieve the more variables there are involved. They also require the independent variables to be **fully crossed**; that is, samples are needed for all combinations of conditions. For example, if you are looking at the growth of breast cancer cells in relation to three concentrations of oestrogen and three concentrations of progesterone, then there are nine possible combinations. **Blocking**, **split-plot**, **latin square**, and **nested** designs are examples of alternative designs that maximize randomization within the practical constraints of the study system, but generate data that need more advanced analytical techniques about which you would be wise to consult a statistician for advice at an early stage.

▶ Apply your learning in Activity 4 at the end of this toolkit.

6.7 Implications for interpretation

As we said at the beginning, there is no such thing as a perfect design. The constraints that you will face will vary but, for example, it is easier to use physical control in the laboratory setting than it is in the field setting, and a quasi-experimental or observational study is likely to have more unresolvable design issues than a purely experimental study. What is important is to be aware of the limitations of any particular investigation. Below are some things you need to look out for when interpreting your own findings or assessing the work of others. (Also see Table QT7.2 for additional guidance on interpreting the results of NHSTs.)

▶ Apply your learning in Activity 6 at the end of this toolkit.

Non-representative samples

Random samples are usually representative of the population from which they are drawn, but this is not guaranteed, particularly with small sample sizes (see Quantitative Toolkit 5)—and, of course, full randomization is often not possible.

If your sample is not representative you are not answering the question you thought you were answering. For example, if you are trying to investigate the influence of temperature on the heart rate of *Daphnia* at different temperatures, and all the *Daphnia* in your samples are female, then you are investigating the influence of temperature on the heart rate of female *Daphnia* and you need to have this limitation in mind when interpreting your results.

Cause and effect: reverse causation

As biological scientists we are often aiming to establish cause and effect, but you need to consider the possibility of *reverse* causation, particularly in observational studies. For example, if you find a pattern of higher incidences of smoking in people with cancer it may well be the case that smoking causes cancer. But the reverse is plausible: getting a cancer diagnosis causes people to start smoking.

One way to get a better insight would be to look at *when* people started smoking: before or after diagnosis.

Cause and effect: confounding factors

Interpretation of cause and effect can also be undermined by a **confounding factor**, or **confounding variable** (also known as a **third variable**). For example, what if stress causes cancer and people who are stressed also smoke more? In this scenario stress is a confounding variable (see Figure QT1.5) and a pattern of higher incidence of smoking in people with cancer would be spurious and not causal. If you thought of this at the design stage, you will improve your ability to interpret your findings by collecting data on stress, as well as smoking and cancer, and adding another variable to your analysis (see Section 6.6).

The placebo effect

In clinical trials, subjects in control groups are given a **placebo**, which is just like the treatment under investigation except that it lacks the active ingredient. For example, when the anti-anxiety drug Zoloft was in trials the placebo pills would have contained everything the real pills did except the active ingredient, sertraline.

The use of placebos is good experimental design practice but you need to be aware of the **placebo effect** when interpreting the analyses. The placebo effect is a well-documented phenomenon whereby some subjects receiving the placebo show improvement in their symptoms. In terms of interpreting the results, this means that the treatment is effective but not, or at least not *just*, in the way originally thought.

ACTIVITIES

Activity 1

Identify potential sources of noise and inaccuracy in the scenario below and suggest some design improvement to tackle these.

Question: Does salivary amylase break down starch quicker at body temperature (37 °C) than at room temperature (20 °C)?

Basic design: Set up two test tubes containing starch solution. Put one test tube in a water bath at body temperature (37 °C) and one at room temperature (20 °C). Set up a spotting tile with a drop of iodine in each well to go with each test tube. Use a pipette to take a drop of the solution from each test tube and add it to the first well on each spotting tile and record if starch is present or not (the iodine will turn black if starch is present). Add 1 mL of salivary amylase to each test tube. After 30 seconds use a pipette to take a drop of the solution from each test tube and add it to the next well on the appropriate spotting tile. Continue taking samples every 30 seconds. Record how long it takes for the starch to be absent at each temperature.

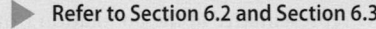 **Refer to Section 6.2 and Section 6.3.**

Activity 2

Why is the planning of sample sizes an important part of experimental design and what analytical tool can be used to do this?

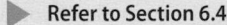 **Refer to Section 6.4.**

Activity 3

What design issue, related to sample size, would have to be considered in a study collecting data from birds in nest boxes to compare fledgling mass in different habitats? Suggest potential solutions.

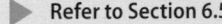 **Refer to Section 6.5.**

Activity 4

Give three reasons you might want to add an independent variable to your design.

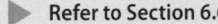 **Refer to Section 6.6.**

Activity 5

An observational study shows that the mass of plastic litter found on beaches with recycling facilities is less than on beaches without such facilities. Do you think recycling bins cause there to be less litter?

 Refer to Section 6.7.

ANSWERS

Activity 1

Potential sources of noise include variation in concentration and volume of starch, amylase, and iodine solutions as well environmental factors such as pH and light. This could be addressed by using standard concentrations and volume for solutions and environment, i.e., physical control. Potential sources of inaccuracy include non-random sampling and poorly used or calibrated equipment. This could be addressed by random sampling and proper maintenance and training in use of equipment as well as setting up multiple tubes at each temperature.

Activity 2

Sample sizes are the way you are most likely to be able to plan for adequate statistical power. Power analysis can be used to estimate adequate sample sizes.

Activity 3

Data points collected from the same nest boxes are not as independent from each other as data points from birds in different next boxes. Potential solutions include using data from just one bird (chosen at random) from each clutch, condensing data from all fledglings for each next box, or including nextbox ID as a random variable in a generalized linear model (GZLM).

Activity 4

To investigate interaction between variables, to statistically control noise, and to deal with pseudoreplication.

Activity 5

You need to be very cautious when interpreting cause and effect. While reverse causation seems unlikely in this example, there are a number of potentially confounding factors, for example, the weather on the day the data was collected or the number of standard bins.

Assessing Patterns

LEARNING OBJECTIVES

In this toolkit we will:

- Identify the main features and basic types of graphs used with biological data (Section 7.1).

- Explain the implication of sampling error for assessing patterns and use 95% confidence intervals to help assess patterns (Section 7.2).

- Explain the need for, and choose between, the basic types of graphs and null hypothesis significance tests (NHSTs) (Section 7.1, Graphs of two variables; Section 7.3, Choosing the right test).

- Use P-values to decide if a pattern is statistically significant or non-significant (Section 7.3, P-values).

- Define Type I and Type II statistical errors and interpret P-values in the context of statistical error and power (Section 7.3, Error and power).

- Describe the use of effect size and distinguish between statistical and biological significance (Section 7.3, Effect sizes).

- Report the outcome of NHSTs (Section 7.3, Reporting).

- Recognize NHSTs in the context of the range of analytical techniques available to biologists (Section 7.4).

Contents

7.1	Using graphs	108
7.2	Implications of sample error	113
7.3	Using null hypothesis significance tests	114
7.4	Other options	117

To answer questions in biology we usually have to assess patterns in data. This is not as straightforward as it might seem. Visually assessing data in **graphs** is a really good starting point, but this can only take us so far because of the complexity of biological systems and the implications of **sampling error**. We need other analytical tools: **null hypothesis significance tests** (NHSTs) have traditionally been used, but an increasing range of other options is available.

7.1 **Using graphs**

We are going to focus here on using graphs to display visually data that you have collected. This contrasts to graphs that are visual displays of mathematical equations, as we see in Quantitative Toolkit 8. However, as we shall see later, these two uses of graph are combined when you add a line to a data graph that represents a model (see Topic 1 and Quantitative Toolkit 8).

Graphing data that you have collected is useful for several reasons. For example, the act of graphing helps you understand and explore your data and allows you to make an initial assessment of potential patterns. Graphs can also help you check if your data meet the assumptions for different analyses and help you communicate your results.

Anatomy of a graph

Graphs are typically called **figures** in books and scientific papers and **charts** in spreadsheet software packages like Excel. The main components of graphs are summarized in Table QT7.1 and Figure QT7.1. Not all these components are found on all graphs: their use depends on the data, context, and, to some extent, preferences of the person creating the graph!

 Go to the e-book to watch a video that will help you understand the 'anatomy' of a graph.

▶ Practise identifying the anatomy of a graph in Activities 1 and 3 at the end of this toolkit.

Graphs of frequency distributions

Frequency distribution graphs are visual depictions of frequency distributions.

▶ We discuss frequency distributions, and the difference between raw observations and frequencies, in Quantitative Toolkit 3.

These graphs involve just one variable and are used to describe a single sample. Frequencies are often converted to percentages when graphs are being used to communicate results: it should be clear from the y-axis label when this has been done. Pie charts can be used as alternatives to frequency distribution graphs, but this only works well for category-like data when there are only a few categories.

Frequency distribution graphs of continuous data have a special name, **histograms**, and their columns should touch. Figure QT7.2a is a histogram of the length of cream-spot ladybird data in Table QT3.9. Frequency distribution graphs of category-like data are not consistently named, but the terms column graph and bar graph are often applied. The columns in such graphs should not touch. Figure QT7.2b is a frequency distribution graph of the ladybird species data in Table QT3.9.

Figure QT7.3 is an example in which a line derived from a mathematical formula has been added to a data graph, in this case a histogram of convict heights: the line is derived from the equation

Table QT7.1 Main components of graphs. Components in bold are marked on Figure QT7.1

Feature	Description
Caption (includes title)	The figure number plus brief title and description of what the graph is about.
Axes	The reference lines against which the data are plotted. Typically, graphs have a horizontal x-axis and a vertical y-axis. Where it applies, an independent variable should be plotted on the x-axis and a dependent variable on the y-axis.
Axis label	A name to indicate what the axis represents, including units in brackets, where appropriate.
Tick marks	Small lines at right angles to an axis representing, or demarcating, a set number of units or category value. On a linear scale they should be equal distances apart, whereas on a logarithmic scale they can present logged values (equidistant) or unlogged values (decreasing distance apart).
Gridlines	Lines that extend out from tick marks and can help with reading values.
Tick label	The value, or category name, of the tick mark.
Data markers	Symbols (crosses, dots, bars) or column/bars representing frequencies, raw data, or descriptive statistics.
Legend	Where different symbols (or colour or shading) are used for different sets of data, this is the key that indicates what the different symbols represent.
Variability markers	Lines or shapes representing variability, which should be used with data markers that represent central tendency. They can extend both above and below the data marker or in one direction only. Lines, called error bars or whiskers, are typically used to show 95% confidence intervals or standard errors around means. Box plots have a rectangle representing interquartile ranges and whiskers representing ranges around medians.
Model line	A line depicting a model of the data. (A model, in this context, is a mathematical statement of how variables are related. For example, the equation $y = mx + c$ is the equation for a straight line which can model variables that have a linear relationship.) Additional lines, or shading, may be used to show error around a model.
Connecting lines	Lines connecting data markers. These are used to help communicate changes from one data point to another and are often useful when the x-axis shows time or distance.

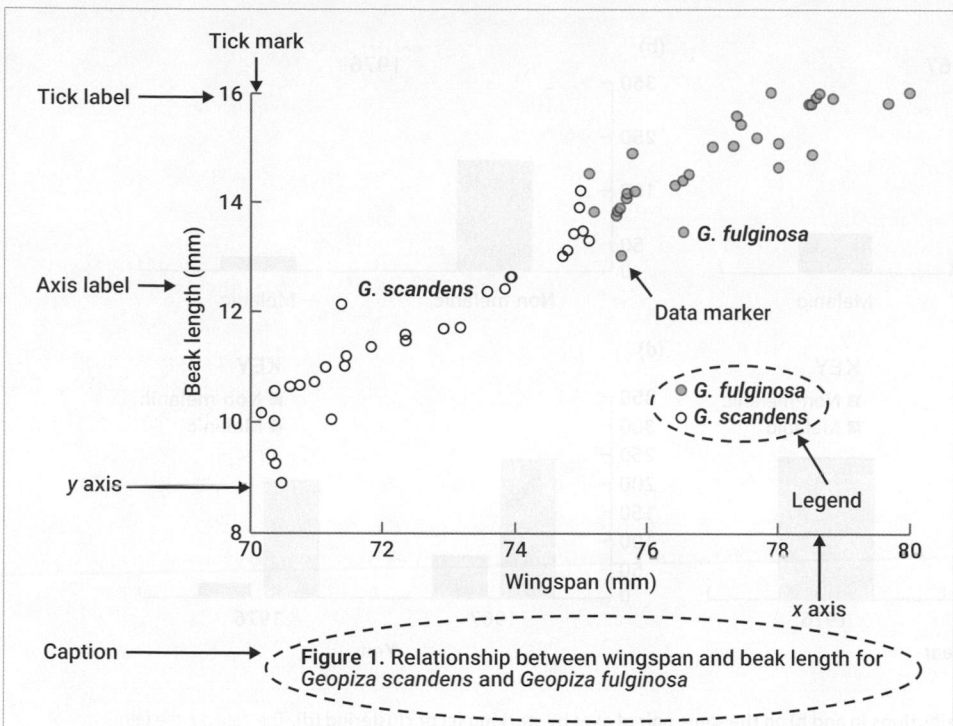

Figure QT7.1 Main components of graphs.

Figure 1. Relationship between wingspan and beak length for *Geopiza scandens* and *Geopiza fulginosa*

for the normal distribution, and you can see it closely fits the data, which would suggest it's a good model.

With appropriate use of colour or shading, multiple frequency distributions can be displayed on the same pair of axes, as illustrated in Figure QT7.4. The peppered moth occurs in melanic (dark) and non-melanic (light) forms known as 'morphs'; Figure QT7.4 displays the frequency of these morphs for two different years (1967 and 1976). In Figure QT7.4a and QT7.4b the frequency distributions for different years are shown on separate axes but are shown on the same axes in (c) and (d) using clustering or stacking: notice the essential use of a legend in (c) and (d). By comparing

these frequency distributions, you can assess if there is an association between the two variables, as we discuss later in this toolkit.

Graphs of two variables

Figure QT7.5 depicts a guide to choosing an appropriate type of graph involving two variables. It is useful to compare frequency distributions when assessing associations between variables

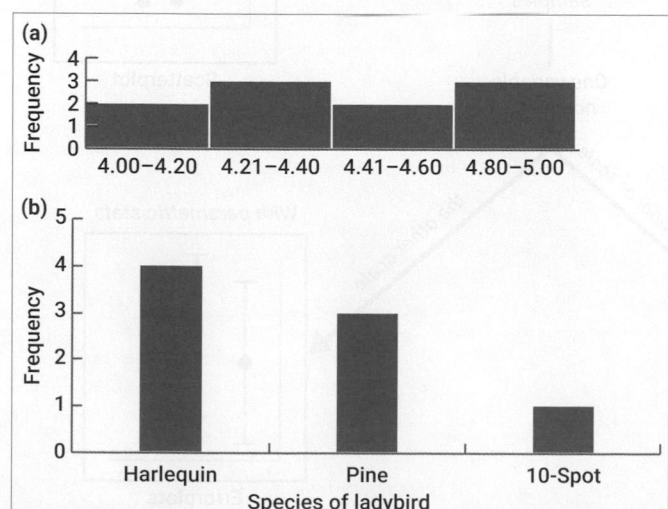

Figure QT7.2 Example graphs of frequency distributions. (a) For continuous scale data, the length of cream-spot ladybirds. (b) For nominal data, the occurrence of ladybirds from different species. Data from Table QT3.9.

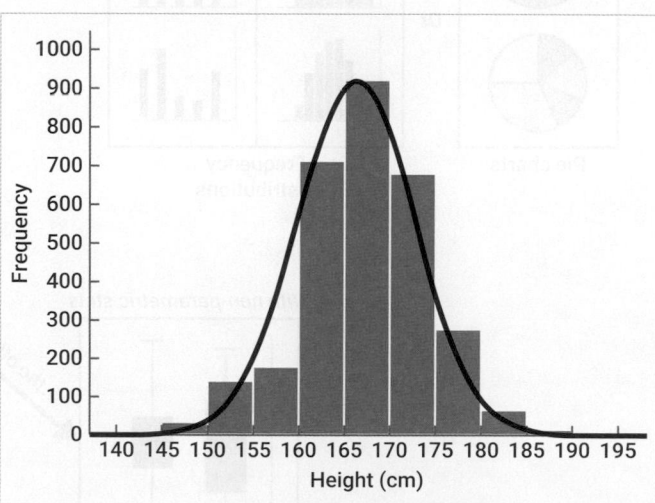

Figure QT7.3 Example of a histogram with a line representing what the predicted values would be if the data followed the normal distribution. The close fit between the data (the columns) and the model (the line) suggests it would be safe to assume these data are normally distributed.

Source of data: MacDonell, W. R. (1902). On criminal anthropometry and the identification of criminals. *Biometrika*, **1**(2), 177–227. doi: 10.2307/2331487. http://www.key2stats.com/data-set/view/361 Key2stats

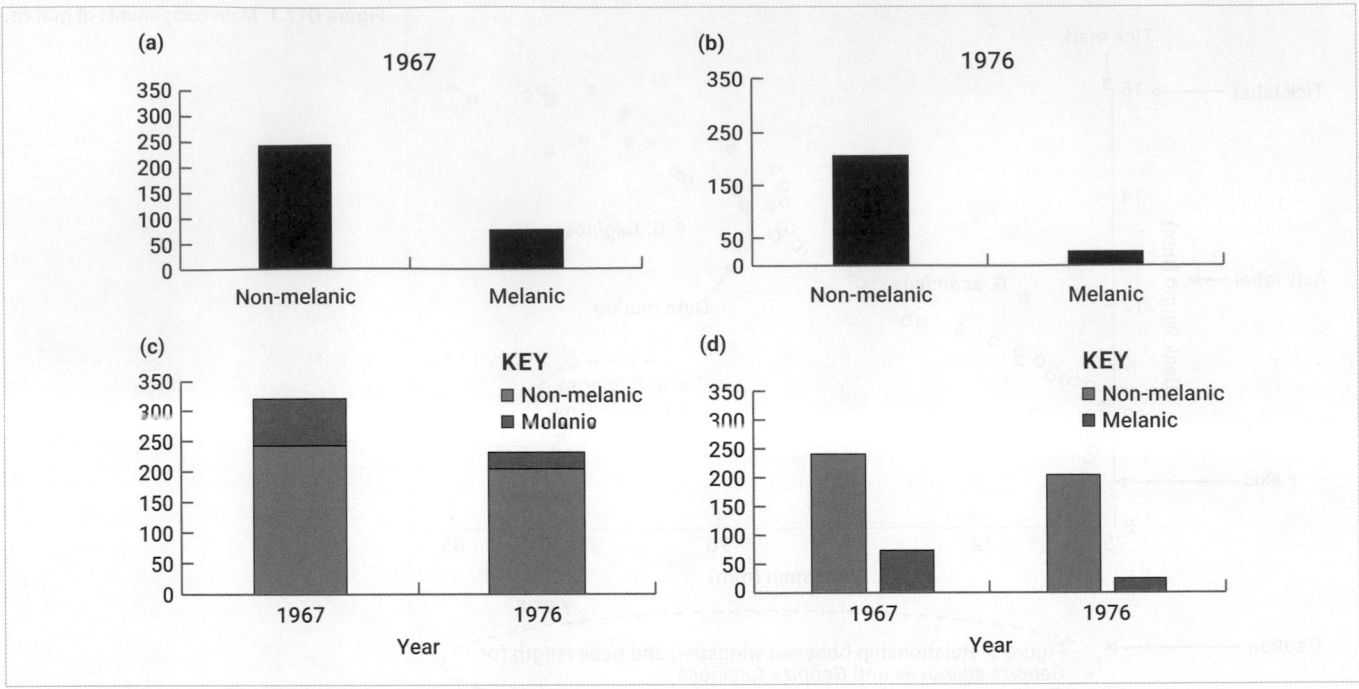

Figure QT7.4 Examples of two frequency distributions (a and b) on the same pair of axes by stacking (c) or clustering (d). The data are the frequency of different moth morphs from two different years.

Source: Brakefield, P. M., and Lees, D. R. (1987) Melanism in *Adalia* ladybirds and declining air pollution in Birmingham. *Heredity*, **59**(2), 273–277.

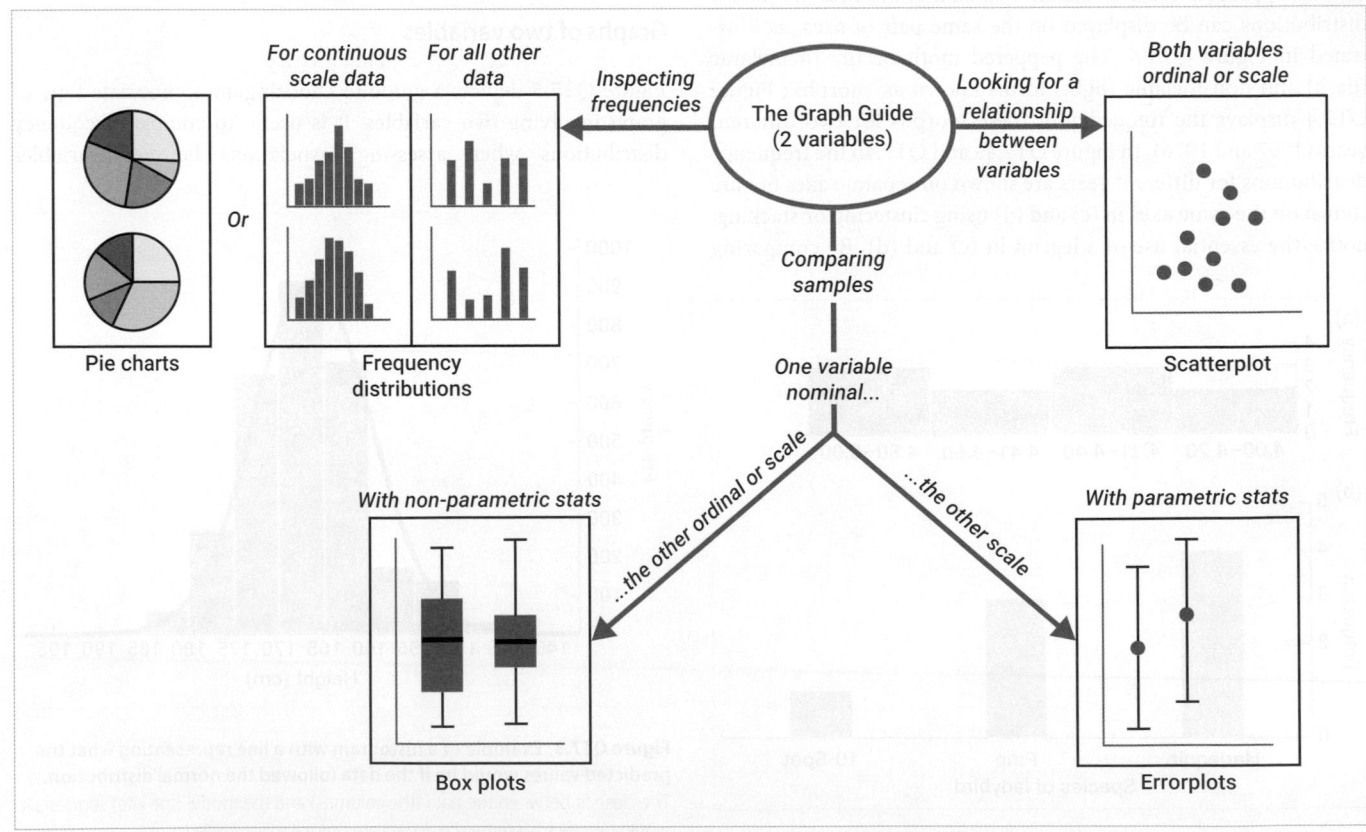

Figure QT7.5 The graph guide. A guide to choosing an appropriate type of graph involving two variables.

(Figure QT7.5, top left). Other types of graph are useful for assessing relationships (Figure QT7.5, top right) and differences (Figure QT7.5, bottom). The most important guiding principle you should follow when choosing a graph type is that it displays your data clearly and fairly.

Although not covered in this toolkit, graphs can potentially portray data on more than two variables, using a second *y*-axis and colour or shading. The primary limitation on the number of variables that you can display is loss of clarity if you try to include too much.

If both the variables are ordinal or scale level and raw observations are plotted, the graph might look like that in Figure QT7.6. Alternatively, the graph may look like that in Figure QT7.7 if measures of central tendency are plotted. This type of graph is typically called a **scatterplot** and helps you assess relationships between variables (see Figure QT7.5, top right). Notice that the data markers in Figure QT7.7 have associated variability markers (Table QT7.1). For example, if the data marker represents means of several data points, rather than a single raw datum, then you should add error bars depicting 95% confidence intervals. Variability markers should be parallel to the axis to which they apply: this is usually the

y-axis, as in Figure QT7.7. It is good practice to include information on the size of the samples used to calculate central tendencies and their associated variabilities.

There are two main types of line that you can add to the graph: connecting lines (Figure QT7.8) and model lines (Figure QT7.9). You would use connecting lines to help communicate changes from one data point to another; this is often useful when *x* is time or distance. The resulting graph is called a **line graph**. You would use a smooth line to depict a model of your data. We have already met the use of a model line on a histogram in Figure QT7.3. Another example that you might come across is a straight line on a scatterplot associated with regression analyses. Model lines on graphs are often referred to as **trend lines**.

> Go to the e-book to watch a video that will help you understand the concept of regression.

If your independent variable is nominal, rather than ordinal or scale, then a graph that looks like Figure QT7.10 (often called an **errorplot**) or Figure QT7.11 (universally called a **box plot**) would likely be a good choice. These types of graph help you assess differences (see Figure QT7.5, bottom).

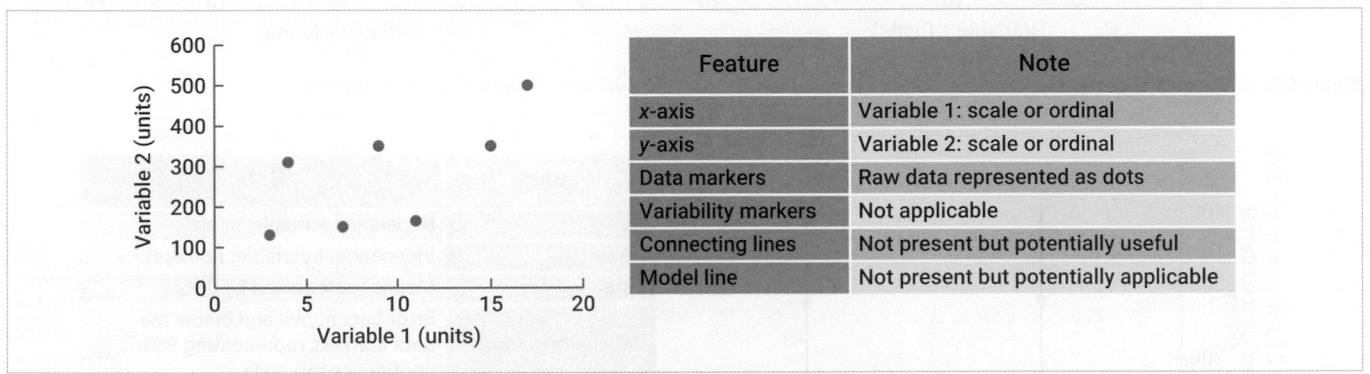

Figure QT7.6 Generalized graph of raw observations for two scale-level variables, often called a scatterplot, with notes on key features highlighted. Variable 1 could be, for example, giraffe neck length (m) and variable 2 giraffe body mass (kg).

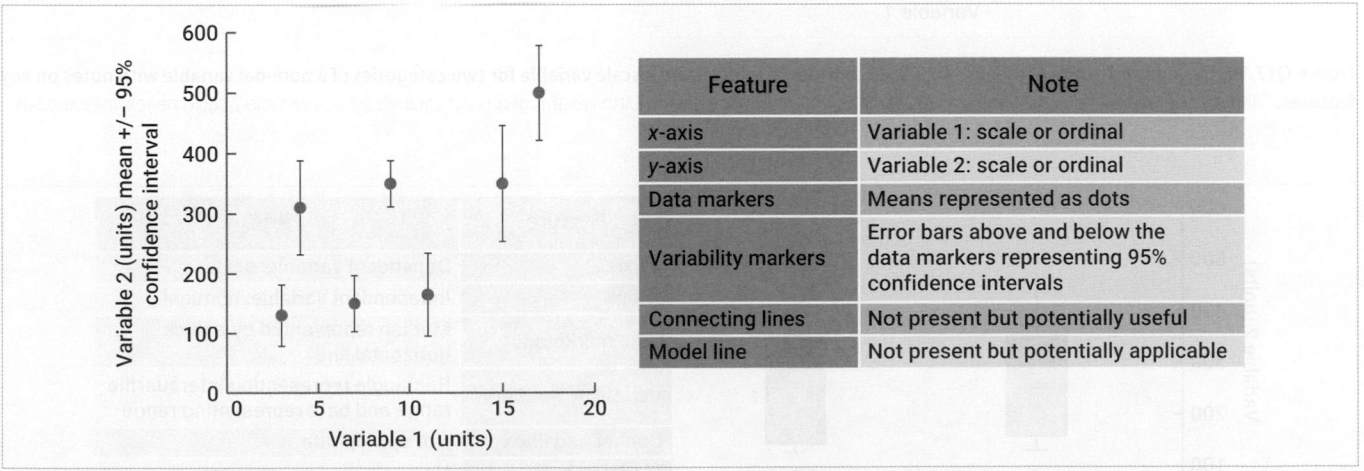

Figure QT7.7 Generalized graph of means with 95% error bars for scale level variables with notes on key features highlighted. It is good practice to also give information, either in the graph or in the caption, on the sample sizes used to calculate each mean and 95% confidence interval. Variable 2 could be heart rate (beats per minute) for water fleas (*Daphnia* sp.) with means of say 10 individuals for each, and variable 1 could be temperature (degrees Celsius).

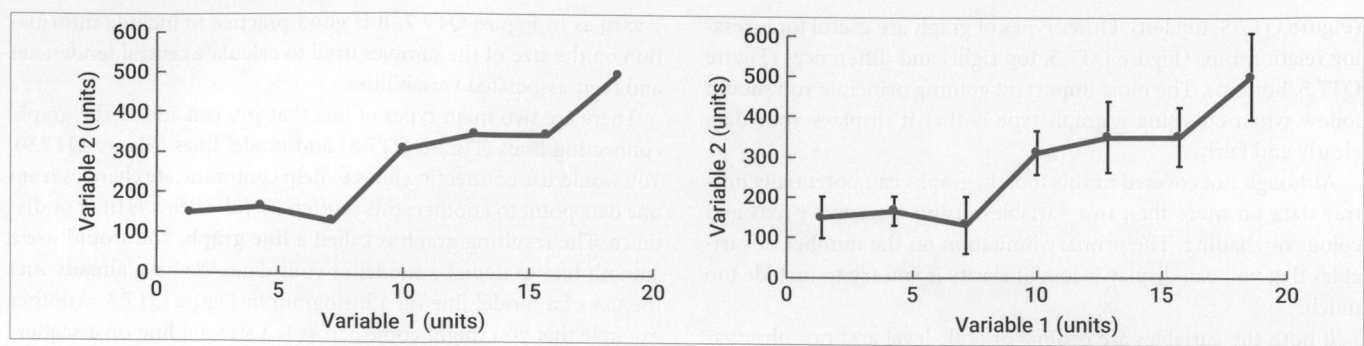

Figure QT7.8 Use of connecting lines. Figures QT7.6 and QT7.7 with connecting lines added; often used when the *x*-axis is time or distance.

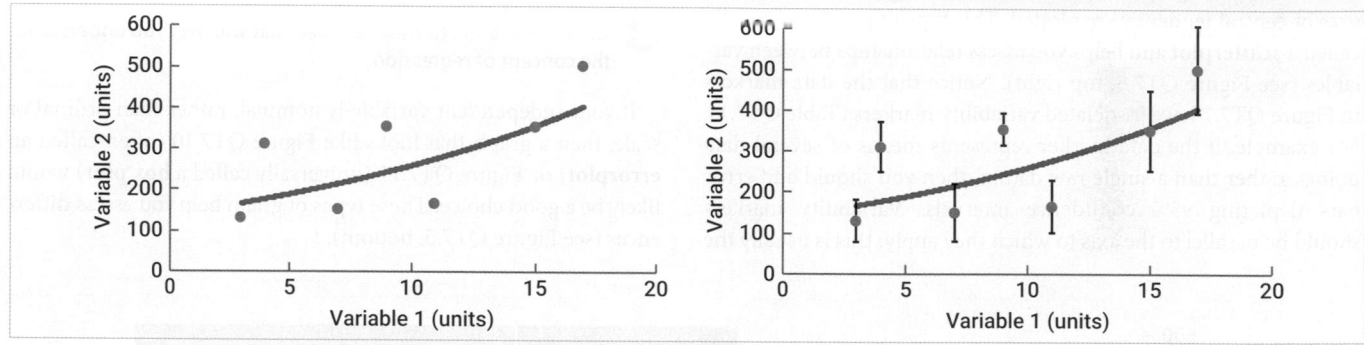

Figure QT7.9 Use of lines to represent models. The data from Figures QT7.6 and QT7.7 with lines representing models of the data.

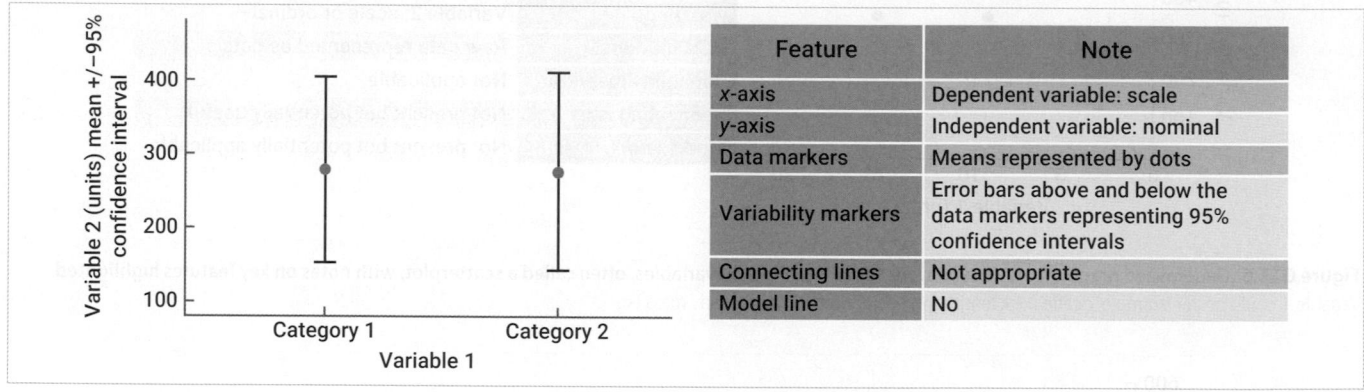

Feature	Note
x-axis	Dependent variable: scale
y-axis	Independent variable: nominal
Data markers	Means represented by dots
Variability markers	Error bars above and below the data markers representing 95% confidence intervals
Connecting lines	Not appropriate
Model line	No

Figure QT7.10 Generalized graph of means (with 95% confidence intervals) of a scale variable for two categories of a nominal variable with notes on key features. This type of graph is often called an errorplot. Variable 1 could be, for example, birth weights of ground squirrels (g) and variable 2 could be sex, measured as female and male.

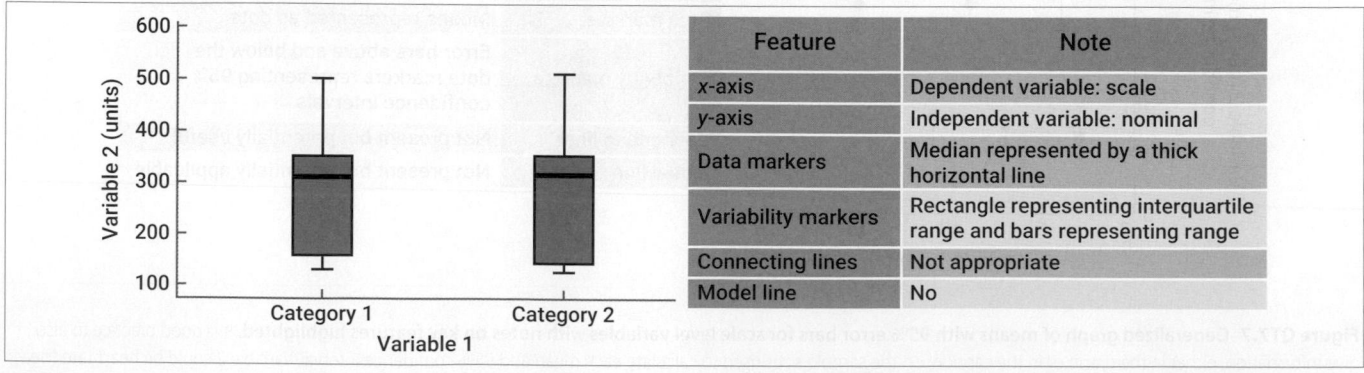

Feature	Note
x-axis	Dependent variable: scale
y-axis	Independent variable: nominal
Data markers	Median represented by a thick horizontal line
Variability markers	Rectangle representing interquartile range and bars representing range
Connecting lines	Not appropriate
Model line	No

Figure QT7.11 Generalized graph of medians (with interquartile ranges and ranges) of an ordinal variable for two categories of a nominal variable with notes on key features. This type of graph is called a box plot. Variable 1 could be, for example, stool consistency (Bristol stool scale) and variable 2 could be time of day, measured as morning and afternoon.

7.2 Implications of sample error

Descriptive statistics (see Quantitative Toolkit 3) and graphs are good ways of exploring data and making an initial assessment of potential patterns, but you need to be cautious. For example, imagine having two samples of blood sugar level with different means, perhaps one from people taking drug A and one from people taking drug B. How can you tell if the two means are different just because of sampling error (see Quantitative Toolkit 5), and in fact come

from the same (statistical) population, or because something else is also going on and they come from different populations? If they came from different populations it would indicate that something of biological interest could be going on with drugs A and B. The same would not be true if the samples came from the same population. Another example is given in Figures QT7.12 and QT7.13.

Measures of variability can be used to help assess if a result is **statistically non-significant** (due to sampling error alone) or **statistically significant** (not just due to sampling error alone). On a graph, the more variability markers overlap, the more likely it is that the samples are from the same population. Confidence intervals (see Quantitative Toolkit 5) are particularly appropriate to use in this respect (Figure QT7.14).

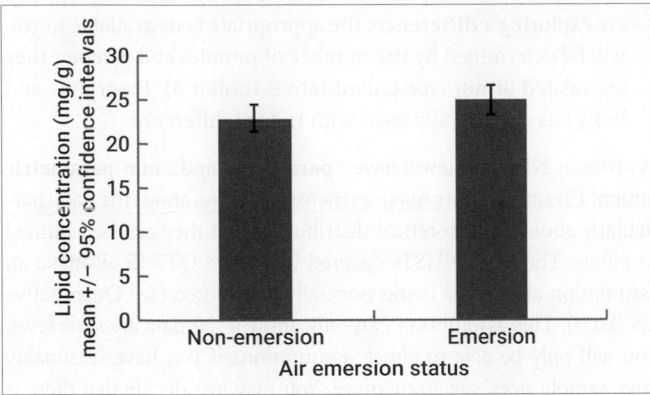

Figure QT7.12 Comparison of lipid levels in the mantles of pimpleback mussels (milligrams of lipid per gram of mussel) kept in water (no air emersion) and emersed in air for 20 minutes. Mean lipid levels are higher in the no-emersion sample, 22.9 mg g^{-1} compared to 25 mg g^{-1}, but is this just sample error?

Source: Greseth, S. L., Cope, W. G., Rada, R. G., Waller, D. L., and Bartsch, M. R. (2003) Biochemical composition of three species of unionid mussels after emersion. *Journal of Molluscan Studies*, **69**(2), 101–106.

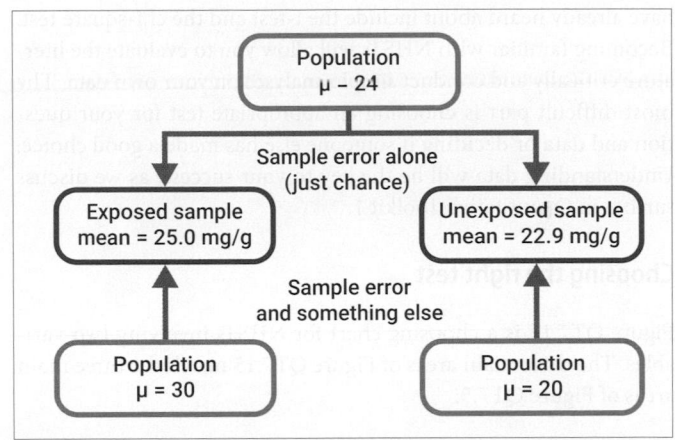

Figure QT7.13 Comparison of example scenarios that could produce the results in Figure QT7.12.

Figure QT7.14 Generalized graphs showing 95% confidence intervals overlapping considerably (a), not overlapping at all (b), and overlapping an intermediate amount (c). The difference in the means is most likely to be sample error only in (a) and least likely in (b).

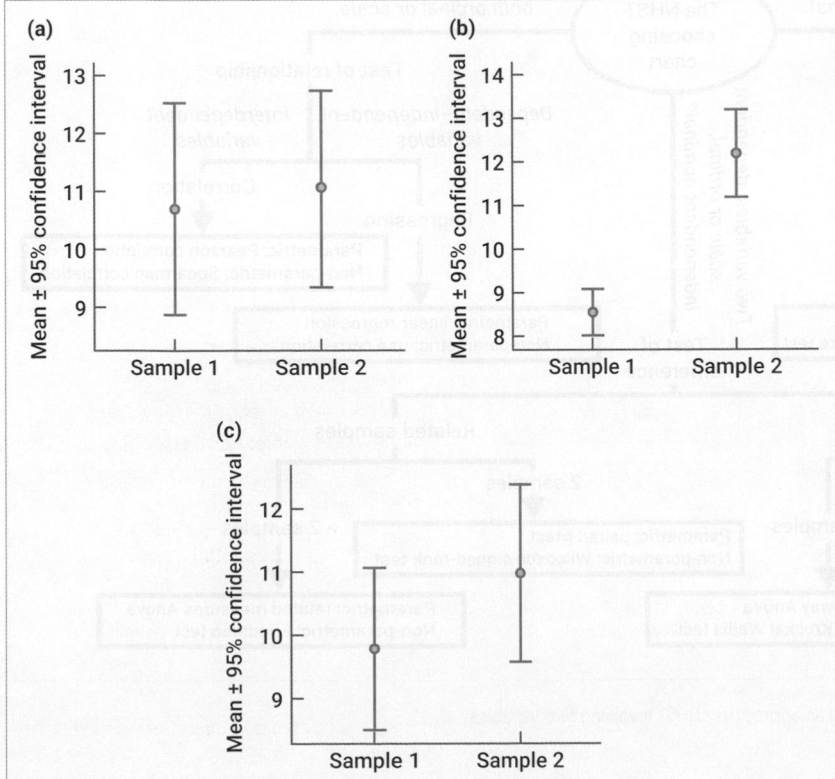

▶ Practise applying the knowledge you gained in this section in Activities 1, 3, 4, and 6 at the end of this toolkit.

7.3 Using null hypothesis significance tests

While inspecting 95% confidence intervals helps you assess whether you have a significant or non-significant result, you will generally also want to use an analysis that leads to more clear-cut decision-making. In other words, you'd want an analysis that, if used by anyone on the same data, using the same criteria, would enable them to come to the same decision about whether any apparent pattern was statistically significant or not as you. Historically, **null hypothesis significance tests** (NHSTs) have been the primary analytical tool used by biologists for this. Examples that you may have already heard about include the t-test and the chi-square test. Becoming familiar with NHSTs will allow you to evaluate the literature critically and conduct simple analyses on your own data. The most difficult part is choosing an appropriate test for your question and data or deciding if someone else has made a good choice. Understanding data will be the key to your success, as we discuss further in Quantitative Toolkit 1.

Choosing the right test

Figure QT7.15 is a choosing chart for NHSTs involving two variables. The three main areas of Figure QT7.15 match the three main areas of Figure QT7.5:

1. Top left: If you have two nominal variables in frequency form (see Quantitative Toolkit 3) then you should look at tests of **frequencies** such as chi-square tests. Frequency distributions, or pie charts, tend to go with tests of frequencies.

2. Top right: If you are interested in a potential **relationship** between two variables that are scale or ordinal then you should look at **correlation** (for inter-dependent variables) or **regression** (when you have a dependent-independent variable situation). Scatterplots are generally used with tests of relationships.

3. Bottom half: If you have a dependent variable that is scale or ordinal and an independent variable that is nominal, and you are exploring a **difference**, the appropriate tests available to you will be determined by the number of samples and whether they are related or not (see Quantitative Toolkit 5). Errorplots and box plots are typically used with tests of difference.

With many NHSTs you will have a **parametric** and a **non-parametric** option. Parametric tests make extra assumptions about the data, particularly about the theoretical distribution that they can be assumed to follow. The basic NHSTs covered by Figure QT7.15 all make an assumption about data being normally distributed (see Quantitative Toolkit 3). This assumption can only apply if the data are scale level. You will only be able to check assumptions if you have reasonably large sample sizes, say 20 or more. You may just decide that there is no reason to think the assumptions are not met and go ahead with a parametric test (Figure QT7.16). The advantage of doing a parametric test, when your data meet the assumptions, is that it has more statistical power, which we will return to later.

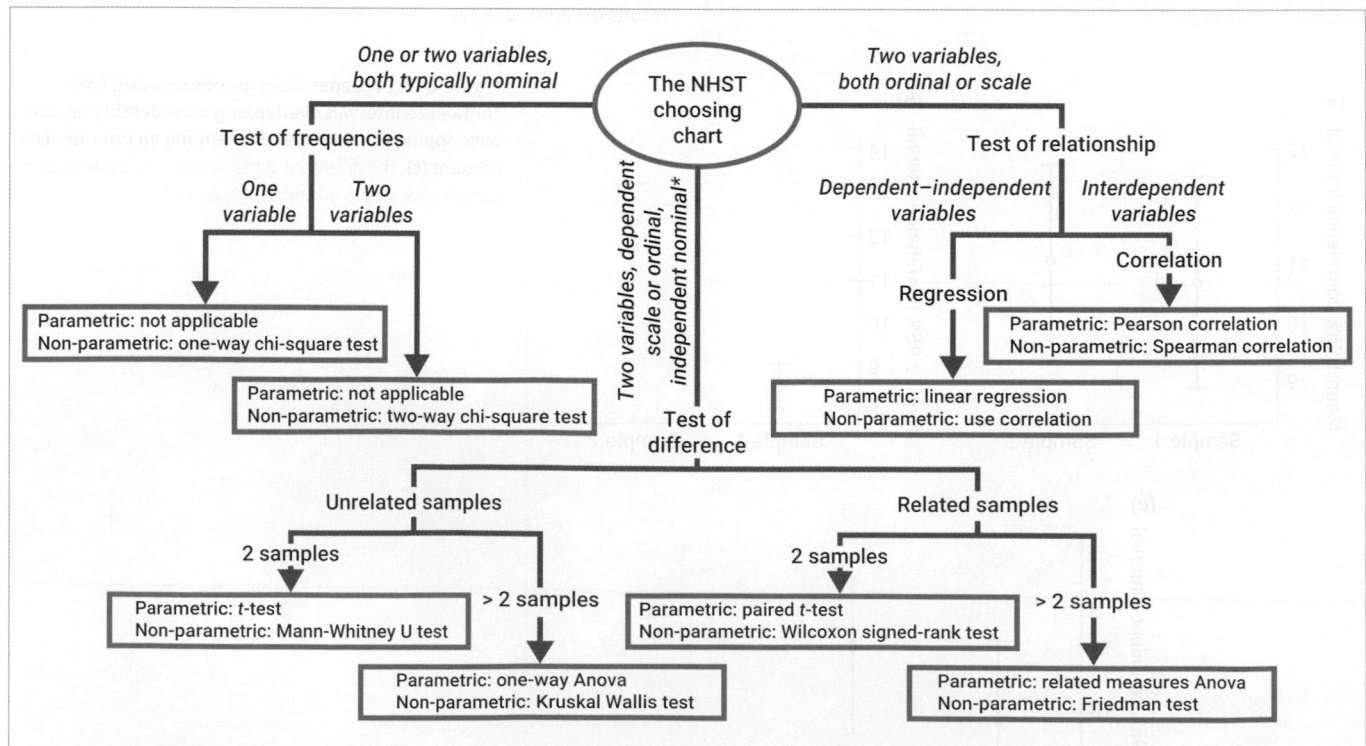

Figure QT7.15 The NHST choosing chart. A guide to choosing an appropriate NHST involving two variables.

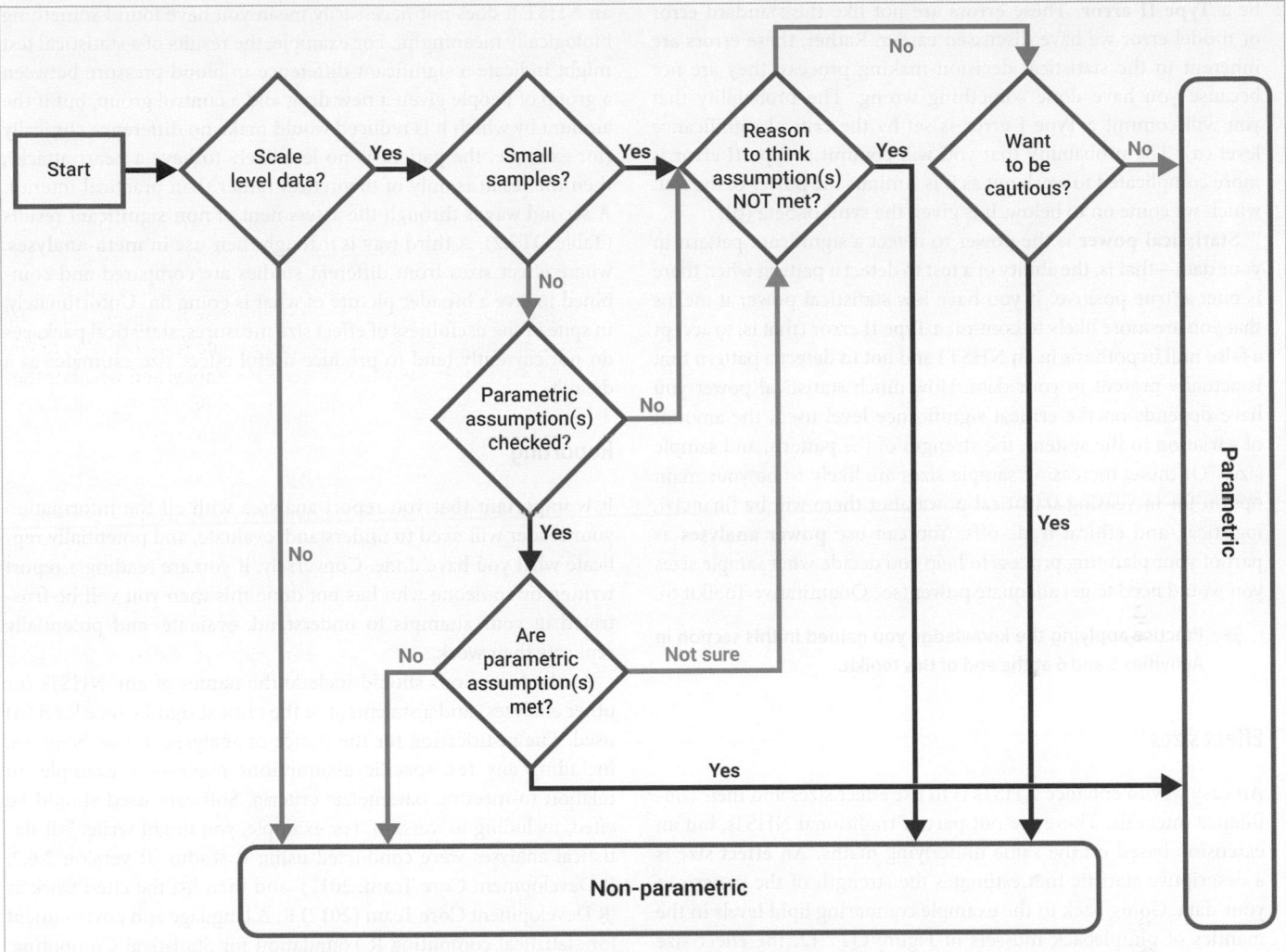

Figure QT7.16 A flow chart to decide between parametric or non-parametric NHST options in Figure QT7.15.

▶ **Practise applying the knowledge you gained in this section in Activities 7 and 8 at the end of this toolkit.**

P-values

These days you will almost always use statistical software to conduct your analyses, of which a freeware package called 'R' has become the most widely used by research biologists. Other well-known examples include SPSS and SAS.

Once you have chosen your test you need to work out how to get the software you are using to conduct the test and to identify the key components of the output. There are lots of sources of assistance available to you for this, including the help that comes with the software, textbooks, and online tutorials. The key components of the output will be the value of the statistic, the sample sizes, degrees of freedom, and the **P-value**. It is this P-value that you use to decide whether a pattern is statistically significant or

non-significant. You compare P to a benchmark probability called a **critical significance level** and given the symbol alpha (α): if P is equal to or less than this, it indicates that you have a statistically significant result. A critical significance level of 0.05 is widely used in biological sciences.

We can write this as:

If P ≤ 0.05 → statistically significant result.

If P > 0.05 → statistically non-significant result.

▶ **Practise applying the knowledge you gained in this section in Activity 5 at the end of this toolkit.**

Error and power

Whenever you conduct a NHST and you decide that you have a statistically significant result this could be a **Type I error** (or **false positive**); if you decide a result is non-significant then this could be a **Type II error**. These errors are not like the standard error or model error we have discussed earlier. Rather, these errors are inherent in the statistical decision-making process: they are not because you have done something wrong. The probability that you will commit a Type I error is set by the critical significance level (α). The probability that you will commit a Type II error is more complicated to work out as it is 1 minus the power of the test, which we come on to below. It is given the symbol beta (β).

Statistical power is the power to detect a significant pattern in your data—that is, the ability of a test to detect a pattern when there is one: a 'true positive'. If you have low statistical power it means that you are more likely to commit a Type II error (that is, to accept a false null hypothesis in an NHST) and not to detect a pattern that is actually present in your data. How much statistical power you have depends on the critical significance level used, the amount of variation in the system, the strength of the pattern, and sample sizes. Of these, increasing sample sizes are likely to be your main option for increasing statistical power, but there will be financial, logistical, and ethical trade-offs. You can use **power analyses** as part of your planning process to help you decide what sample sizes you would need to get adequate power (see Quantitative Toolkit 6).

▶ Practise applying the knowledge you gained in this section in Activities 5 and 6 at the end of this toolkit.

Effect sizes

An easy way to enhance NHSTs is to use effect sizes and their confidence intervals. These are not part of traditional NHSTs, but an extension based on the same underlying maths. An **effect size** is a descriptive statistic that estimates the strength of the pattern in your data. Going back to the example comparing lipid levels in the mantles of pimpleback mussels in Figure QT7.12, the effect size would be the difference between the lipid levels in mussels kept in water (no air emersion) and those emersed in air for 20 minutes. If the two samples came from the same statistical population, and were therefore statistically non-significant, this would be zero. Effect sizes come in unstandardized and standardized versions. The standardized versions are moderated by the sample size and variability to help us compare studies.

Calculated along with their 95% confidence intervals, effect sizes are useful in three main ways. One way is through the assessment of biological significance: If you get a significant result from an NHST it does not necessarily mean you have found something biologically meaningful. For example, the results of a statistical test might indicate a significant difference in blood pressure between a group of people given a new drug and a control group, but if the amount by which it is reduced would make no difference clinically (for example, the patient is no less likely to have a heart attack), then the result is only of theoretical rather than practical interest. A second way is through the assessment of non-significant results (Table QT7.2). A third way is through their use in **meta-analyses**, where effect sizes from different studies are compared and combined to give a broader picture of what is going on. Unfortunately, in spite of the usefulness of effect size measures, statistical packages do not currently tend to produce useful effect size estimates as a default.

Reporting

It is important that you report analyses with all the information your reader will need to understand, evaluate, and potentially replicate what you have done. Conversely, if you are reading a report written by someone who has not done this then you will be frustrated in your attempts to understand, evaluate, and potentially replicate their work.

Methods sections should include the names of any NHSTs (or other analyses) and a statement of the critical significance level (α) used. The justification for the choice of analyses should be given, including any test-specific assumptions made—for example, in relation to meeting parametric criteria. Software used should be cited, including its version. For example, you might write: 'All statistical analyses were conducted using R studio (R version 3.6.2; R Development Core Team, 2017)' and then list the cited work as 'R Development Core Team (2017) R: A language and environment for statistical computing R Foundation for Statistical Computing, Vienna, Austria (2017), RStudio Team, 2016'.

In a Results section, you should cite the following information for each NHST statistical test reported:

1. The name of the test.

2. The value of the test statistic.

3. The degrees of freedom and/or sample size(s). Degrees of freedom are generally best reported as subscribed to the letter or symbol representing the statistic. For non-parametric tests it is good practice to include the sample sizes even if you have

Table QT7.2 Guidance on interpreting the results of NHSTs

Decision	Statistically non-significant	Statistically significant
What might account for this?	The decision is right and the pattern is due to chance only, or it is wrong and a Type II (false negative) error has been committed.	The decision is right and the pattern cannot be accounted for by chance alone, or it is wrong and a Type I (false positive) error has been committed.
Tips for interpretation	Consider your sample sizes and/or the confidence intervals around your effect size. If the sample sizes are lower than recommended by power analyses and/or your effect size confidence intervals are wide you should suggest that further work is needed.	Do not get over-excited by an occasional statistically significant result. If you are using a critical significance level of 0.05 and doing a project which involves 100 tests and you get 5 significant results, that is to be expected.

degrees of freedom, as you cannot fully assess the result without this information.

4. The P-value. Typically, you can report exact P-values if you are using statistical software. Otherwise quote them as being below your critical significance value (e.g. P<0.05). A P-value of 0.000 on a computer printout means the value is less than 0.0005 and therefore rounds up, to three decimal places, to 0.000. You may see non-significant results being reported as NS or P>0.05 but it is better to report exact P-values if you can.

An example of what the statistical citation for a t-test might look like is:

$$\text{t-test}: t_{38} = 2.06, P = 0.046$$

And for a Mann–Whitney U test like his:

$$\text{Mann–Whitney U test}: U = 120.5, n_1 = 20, n_2 = 20, P = 0.032.$$

It is not currently standard to report effect size with your P-value, but you should consider doing so if you can. Effect sizes should be reported with their associated 95% confidence intervals.

▶ **Practise applying the knowledge you gained in this section in Activities 2 and 9 at the end of this toolkit.**

7.4 Other options

Thanks to the dramatic expansion of computing capacity, biological scientists today have a wide range of analytical techniques available to them. For example, **bootstrapping**, **jackknifing**, and the **Monte Carlo method** can free you from the need to make assumptions about the distribution of your data and thus provide an alternative to non-parametric statistics. The **generalized linear model** (GZLM) can be used to undertake analyses equivalent to all the basic parametric NHSTs and much more. You can build and compare different models to help assess what is underlying the variation of interest in your system. **Time series** can be used to look for patterns in data through time—for example, potential seasonal patterns across years. They are often used in association with data which is presented in line graphs. **Multivariate methods**, including principal components analysis (PCA), factor analyses, and cluster analysis, are useful when you have large datasets including more than one dependent variable; they can play an important role in generating research hypotheses and complement NHSTs and modelling techniques. **Bayesian statistics** is a fundamentally different alternative to NHSTs and is being increasingly used by biological scientists. One of the strengths of Bayesian statistics is that it allows you to take into account existing knowledge.

The techniques mentioned above can offer useful alternatives to NHSTs but it is really important to remember that it is all about assessing patterns to explore biological systems and:

- No technique is valid if used inappropriately: you must understand your data.
- All techniques have their limitations: you must interpret your data with these in mind.

ACTIVITIES

Figure QT7.17 is taken from Ripley *et al.* (2021) and presents a study on the effects of temperature on the physiology and behaviour of embryonic catsharks. Catsharks are oviparous (i.e. egg laying) and their embryos develop in pouches (commonly known as 'mermaid purses') in the sea. There are four separate graphs. Each graph presents data on two or three of the following variables:

1. Temperature (15 °C or 20 °C) is the temperature of the water in the tank in which the catshark embryo is developing.

2. Ventilation frequency (Hz) is the number of undulations the embryo makes every second to keep water moving through the pouch.

3. Routine metabolic rate (mg O_2 hour^{-1}) is the amount of oxygen taken up by the embryo measured in milligrams per hour.

4. Freeze response duration (minutes) is how long the embryo stops moving following a mock predator attack.

The research looked at the animal's freeze response following a mock predator attack. The length of time the embryo can remain motionless in this response is limited by the embryo's metabolic rate, which in turn is likely to be limited by oxygen availability. Since oxygen content of the water is related to temperature, with colder water holding more dissolved oxygen than warmer water, you would expect a difference in the freeze response duration of embryonic catsharks kept at different temperatures. In graphs (a), (b), and (c) there are data markers for both the mean values (bars) and individual catshark's values (black dots). The caption mentions a linear model: this means they have used GZLM (see Section 7.4) and P-values can be interpreted as for NHSTs (see Section 7.3). Graphs (a) and (c) include asterisks in association with horizontal square brackets, to indicate statistically significant differences between treatment groups at different critical significance levels.

Study the graphs and then complete the following activities.

1. In graphs (a), (b), and (c) what do the variability markers represent and how do you know? Comment on the use of this measure rather than 95% confidence intervals.

▶ **Refer to Section 7.1, Anatomy of a graph and Section 7.2.**

2. How many catsharks were in the 15 °C sample in (c) and how do you know?

▶ **Refer to Section 7.3, Reporting.**

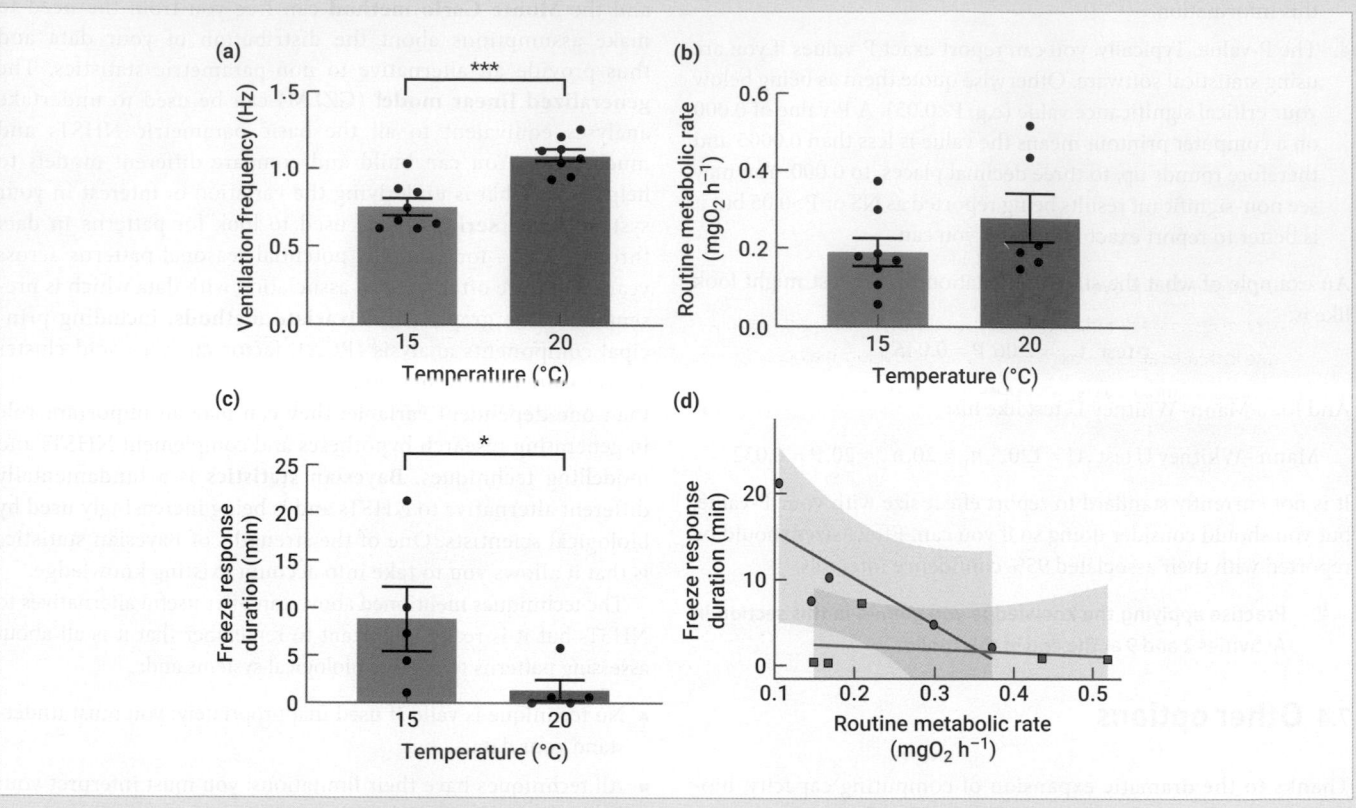

Figure QT7.17 (a) Ventilation frequency, (b) routine metabolic rate, (c) freeze response duration, and (d) the relationship between routine metabolic rate and freeze response duration of *Scyliorhinus canicula* embryos tested at 15 °C (blue) and 20 °C (orange). Symbols denote individuals ± SEM (standard error of mean); (a) n = 8 (15 °C) and 8 (20 °C); (b) n = 8 (15 °C) and 6 (20 °C); (c) n = 5 (15 °C) and 5 (20 °C); (d) n = 5 (15 °C) and 5 (20 °C); the lines plotted were calculated from a linear model (P < 0.05); asterisks denote statistically significant differences between treatment groups (∗P<0.05, ∗∗ P<0.001, ∗∗∗ P<0.0001, Mann–Whitney test).

Source: Ripley, D. M., De Giorgio, S., Gaffney, K., Thomas, L., and Shiels, H. A. (2021) Ocean warming impairs the predator avoidance behaviour of elasmobranch embryos. *Conservation Physiology*, **9**(1).

3. What do the orange and blue lines in (d) represent? And what do the grey areas around these lines represent?

   ▶ **Refer to Section 7.1, Anatomy of a graph.**

4. i. Describe the patterns in means and standard errors in graphs (a), (b), and (c).

   ii. Describe the pattern in the data in graph (d).

   iii. Why do you need to be cautious when interpreting the patterns that you have described in (i) and (ii)?

   ▶ **Refer to Section 7.2.**

5. Referring to the results of the statistical tests, which graphs show patterns that are statistically significant? Justify your answer and say what type of statistical error could have been committed.

   ▶ **Refer to Section 7.3, P-values and Section 7.3, Error and power.**

6. From the results of the Mann–Whitney U test it was concluded that the pattern in (b) was not statistically significant. What then could account for the difference in the means in this graph? What type of statistical error could have been committed?

   ▶ **Refer to Section 7.2 and Section 7.3.**

7. What route would have been taken through Figure QT7.15 to choose to conduct a Mann–Whitney U test in association with Figure QT7.18(c)? Why?

   ▶ **Refer to Section 7.3, Choosing the right test.**

8. The Mann–Whitney U test is a non-parametric test. What route would have been taken through Figure QT7.16 when choosing this test in association with Figure QT7.18(c)?

   ▶ **Refer to Section 7.3, Choosing the right test.**

9. Compared to the list indicating what information should be reported for NHSTs, what information did the authors report for the Mann–Whitney U test in Figure QT7.18(c), and what is missing?

   ▶ **Refer to Section 7.3, Reporting.**

ANSWERS

1. Standard error of the mean (SEM). It tells us in the caption. If 95% confidence intervals had been plotted rather than standard errors, these would have been larger and their overlap (or not) would have been a better indication of whether or not the pattern was statistically significant.

2. Five—it tells us in the caption 'n=5' (Note: lower case 'n' is usually used for sample size rather than upper case 'N').

3. The lines represent models of the data and the greys shaping the model error.

4. i. (a) Mean ventilation frequency is higher at 20 °C than at 15 °C, standard errors do not overlap; (b) mean routine metabolic rate is higher at 20 °C than at 15 °C, standard errors overlap; (c) mean freeze response duration is lower at 20 °C than at 15 °C, standard errors do not overlap.

   ii. Freeze response duration appears to decrease as routine metabolic rate increases at 15 °C, but there is no clear relationship at 20 °C.

   iii. Apparent differences and relationships could be due to sampling error alone.

5. The differences in (a) and (c) are statistically significant based on the information on the Mann–Whitney U test in the caption and the asterisks on the graphs. The relationships in (d) are statistically significant based on the information on the linear model given in the caption. (P-values less than critical significance level of at least 0.05.) When doing an NHST you risk making a Type I error (false positive) every time you conclude a result is significant.

6. Sampling error alone could account for this difference. When doing an NHST you risk making a Type II error (false negative) every time you conclude a result is non-significant.

7. Test of difference in freeze length response of embryos at two different temperatures.

   The dependent variable is freeze response duration (scale level) and the independent variable is temperature (nominal level with two categories: 15 and 20 °C).
   Samples are unrelated because there are different, unmatched, individuals in each sample.
   There are two samples: one of freeze response durations at 15 °C and the other at 20 °C.

8. Scale level? Yes.

   Small samples? Yes.
   Any reason to think parametric assumptions not met? No.
   Want to be cautious? Yes.

9. Reported: name of the test (Mann–Whitney U test), sample sizes (n = 5 (15 °C), n = 5 (20 °C)), P-value (P<0.05).

   Not reported: value of the statistic, degrees of freedom (but not applicable for Mann–Whitney U test).

Using Formulas and Equations

Contents

8.1 Anatomy of an equation   121

8.2 The main types of formulas in biology   122

8.3 Carrying out calculations and keeping track of units   127

8.4 Rearranging formulas   127

LEARNING OBJECTIVES

In this toolkit we will:

- Recognize and use the symbols used in equations, including:
  - $+, -, \times, \div, \Sigma$ act as operators that mean to add ($+$), subtract ($-$), multiply ($\times$), divide ($\div$), or do repeated addition (Σ) (Section 8.1, Symbols for operators).
  - Letters can stand for variables or constants. Constants can be universal or defined for particular experimental conditions (Section 8.1, Letters can stand for variables or constants).
  - Symbols to show relationships, $=, \sim, \propto$ (Section 8.1, Symbols to show relationships).
  - Exponential and logarithm notation (Section 8.1, Exponential and logarithm notation).

- Recognize the main types of equations used in biology, relate the equation to the shape of the graph, and predict how changing the constants affects the graph, including (Section 8.2):
  - direct proportion;
  - linear with positive or negative slope;
  - inverse proportion;
  - rectangular hyperbola; and
  - exponential growth or decay.

- Carry out calculations by substituting numbers for each constant and variable into equations while keeping track of units and prefixes (Section 8.3).

- Be able to rearrange equations to change the subject of the equation (Section 8.4).

When coming across a formula or equation in a text (we are using these terms synonymously), it is important not to jump over it, but rather to think about what it is saying: this means you need to think about what the symbols represent and what it says about the relationship between them. The more experienced you become, the easier it will be for the meaning to jump out from the symbols. At first, though, you will need to make an effort to translate the symbols into meaning. In this Toolkit we will first look at what the symbols mean and what they can represent, and then we will go on to look at some ways to help visualize what an equation is describing.

8.1 Anatomy of an equation

Symbols for operators

Symbols that act as operators include +, −, ×, and ÷, and these tell us to perform a particular action, namely to add, subtract, multiply, or divide, respectively. Letters can be used to stand for numbers that are either constants or variables (more about that later), but if we are using letters then there are some different ways of writing multiplication, division, and repeated addition.

Multiplication: Sometimes a full stop symbol (.) is used instead of a multiplication sign and it is also possible to leave out the multiplication sign.

The formula distance = speed × time

can be written as d = s × t *or* d = st *or* d = s.t

Division: Whereas a word formula for concentration might be written as:

concentration = amount of substance ÷ volume of solution

the same word formula could be written in symbols as a fraction. The line in the fraction means that the top expression is divided by the bottom expression:

$$\text{concentration} = \frac{\text{amount of substance}}{\text{volume of solution}} = \frac{m}{v} = m \div v$$

Summation (repeated addition): The symbol sigma, Σ, represents the sum of terms. If we want to add up the number of individuals of each species in an ecological community, we might call the number of individuals of species 1, x_1 and the number of individuals of species 2, x_2, and so on. x_n is traditionally the last term. x_i stands for any of the terms. The formula below represents the sum of these numbers:

Sigma notation for the sum of numbers:

$$\Sigma x_i = x_1 + x_2 + x_3 + \cdots + x_n$$

This sigma notation is used in Case Study 8.1, the Shannon–Weiner diversity index.

▶ Practise using sigma notation in Activity 1 at the end of this toolkit.

Letters can stand for variables or constants

There are two categories of constant: (1) the universal constants; and (2) constants that are characteristic of a particular molecule. Universal constants always have the values listed in Table QT8.1.

Other constants can vary with different experimental conditions. These are sometimes called coefficients rather than constants. For example, an absorption coefficient, affinity constant, or a rate constant should be cited with reference to the conditions of measurement, usually the temperature (in biology usually either 20 °C or 25 °C or 37 °C) and pressure (usually normal atmospheric pressure, 1 atm, 101.3 kPa).

In an equation there will likely be a variety of letters—some will stand for the constants and others for the variables. For example, as we see in Case Study 8.2, in the Beer–Lambert Law the absorbance, A, of a particular wavelength of light is proportional to the concentration, C, of the substance in solution. We might design an experiment where we prepare a series of test tubes with different concentrations of a coloured solution. Each concentration of the solution will have a different absorbance; these are variables. Here we are choosing to use letters that remind us of the physical quantity that they represent, e.g. A for absorbance and C for concentration.

Symbols to show relationships

The **equals sign**, =, means that one side of an equation is equal to the other side. This is important when we need to rearrange an equation: once we have said that each side of the equation is equal to the other side, if we add a number to one side (for example), we

Table QT8.1 Universal constants used in biology

Name	Symbol	Value	Example of formula
Pi	π	3.14159. . .	The ratio of the circumference (C) to the diameter (D) of a circle $\pi = \dfrac{C}{D}$
Euler's number	e	2.718. . .	e is the base of natural logarithms and is used in exponential growth and decay
Avogadro's number	N_A	6.02×10^{23} mol^{-1}	Converting between mass and amount of substance in moles
Gas constant	R	8.31446 J mol^{-1} K^{-1}	Nernst equation
Boltzmann constant	k_b	1.380649×10^{-23} J K^{-1}	
Faraday constant	F	96 485.33 C mol^{-1}	Nernst equation
Speed of light in a vacuum	c	2.99792458×10^8 m s^{-1}	

also have to add the same number to the other side to keep the two sides equal to each other. This is distinct from the function of the equals sign on a calculator, which is simply a direction to work out the result of a calculation.

▶ We learn more about rearranging equations in Section 8.4.

The **proportional to** symbol, ∝, shows that two variables are related in a particular way. In the Beer–Lambert Law, described in Case Study 8.2, we would say that the absorbance is proportional to the concentration, and would write $A \propto C$, because as the concentration increases so the absorbance increases. In fact, in this instance the relationship is **directly proportional**, which means that when the concentration is zero ($C = 0$), the absorbance is zero ($A = 0$), and the absorbance, A, increases as the concentration, C, increases.

▶ We discuss the concept of proportionality in Quantitative Toolkit 4.

The **approximately equals** sign is also used in biology to describe an estimate. The symbol for this can be one of the following: ≈, ~. For example, the diameter of *Streptococcus* bacteria is ~ 0.8 μm (actually it is usually between 0.6 and 1.0 μm).

Greater than, >, and **less than**, <, are used as descriptors; for example, the diameter of a norovirus is less than 50 nm, so we can write $d < 50$ nm.

Exponential and logarithm notation

Exponential, power, and index notation can appear in a variety of equations. Quantitative Toolkit 2 discusses this notation for numbers so if this is unfamiliar, refer back to this Toolkit. If the base is a number and the exponent is a variable, then it is called an exponential equation

Examples of exponential equations include:

$$y = e^{kx}$$
$$y = 10^{kx}$$
$$y = 2^{kx}$$

includeIf the base is the variable and the exponent is a number, then it is called a power function.

Power functions are usually in the form: $y = Ax^k$, for example, $y = x^{0.75}$

where

- k is a constant
- x is the independent variable
- y is the dependent variable
- e is Euler's number, approximately 2.718.

Exponential terms such as e^{kx} are often used in population growth or when looking at rates of reaction, as we discuss in more detail in Quantitative Toolkit 9.

In exponential growth the dependent variable increases gradually at first and then the curve gets steeper and steeper, as seen in Table QT8.3. An exponential growth equation is one of the form $y = A^{kx}$ where A is a number such as e, 10, or 2 and the exponent or

power is positive. If it is describing exponential decay, the power is negative, i.e. $y = e^{-kx}$ and this is discussed further in Table QT8.4.

▶ **Quantitative Toolkit 2 explains the use of index notation and how to use a calculator to work with the exponential function. A case study showing how exponential notation can describe cell growth can also be found there.**

Multiplying and dividing with expressions in exponential form has important implications for our later sections on keeping track of units and rearranging equations, so let us take a moment to review multiplying and dividing numbers in index form: the left-hand side of Table QT8.2 explores examples of calculations using numbers and the right-hand side generalizes those examples.

▶ **If you would like a recap on logarithms, go to Quantitative Toolkit 2.**

A **logarithm** is the power to which a base number is raised: e.g. for 10^3 the log to the base 10 is 3.

When using the term 'logarithms' we refer to the base of 10. So, if we could write the relationship $y = 10^x$ as $\log(y) = \log(10^x) = x$, we would say 'the log of y equals x'—that is, 'the log of y is the power (x) to which the base (10) is raised'.

Sometimes we use natural logarithms. In this case the base is e and we would write $\ln(x)$. In parallel with the example above, we could write $y = e^x$ as $\ln(y) = \log(e^x) = x$.

We would say 'the natural log of y equals x'—that is, 'the natural log of y is the power (x) to which the base (e) is raised'.

The rules for multiplying and dividing logarithms are summarized in Table QT8.3.

8.2 The main types of formulas in biology

There are a few important types of formulas in biology and Table QT8.4 shows the most common types, relating the formula to the shape of the graph. When looking at an equation in symbols it is important to try to read it—that is, to imagine what the shape of the graph would look like and think about the context. For example, an equation of the form $y = kx$, where k is a positive constant, will form a straight line on a graph because as x increases, y also increases in direct proportion. One way to help visualize this is with the use of a spreadsheet to draw the graphs.

🎥 Go to the e-book to watch a video that demonstrates how to draw graphs from the equation and how to use it to get a feel for how changing the parameters affects the shapes of the graphs.

In addition to being able to relate the equation to the graph, it is also important to be able to see how the equation and graph relate to the biological situation under study. We see an example of this when using a statistical analysis to identify if there is a correlation between two variables; this is explored further in Quantitative Toolkit 7. To illustrate this point, consider Case Study 8.2, the relationship between absorbance, A, and concentration, C.

▶ Practise using these ideas in Activities 5a and 6a at the end of this toolkit.

Table QT8.2 Rules for multiplying and dividing numbers and expressions in exponent form. This table shows how the generalized representations using letters on the right compare with the example calculations shown on the left.

Using numbers. . .		Generalizing with letters instead of numbers. . .
Multiplication: if we multiply two numbers we add the powers		
$8 \times 8 = 64$ could be written	$2^3 \times 2^3 = 2^{3+3} = 2^6 = 64$	$2^a \times 2^b = 2^{a+b}$
$100 \times 10{,}000 = 1{,}000{,}000 =$	$10^2 \times 10^4 = 10^{2+4} = 10^6 = 1{,}000{,}000$	$10^a \times 10^b = 10^{a+b}$

Using numbers. . .		Generalizing with letters instead of numbers. . .
Division: if we divide two numbers we subtract the powers		
$\dfrac{64}{4} = 16$ could be written	$\dfrac{2^6}{2^2} = 2^{6-2} = 2^4 = 16$	$2^a \div 2^b = \dfrac{2^a}{2^b} = 2^{a-b}$
$\dfrac{100{,}000}{1000} = 100$	$\dfrac{10^5}{10^3} = 10^{5-3} = 10^4 = 100$	$10^a \div 10^b = \dfrac{10^a}{10^b} = 10^{a-b}$

Using numbers. . .		Generalizing with letters instead of numbers. . .
If we have the power of a power we multiply the powers		
$(2^3)^4 = 2^3 \times 2^3 \times 2^3 \times 2^3 = 2^{3\times4} = 2^{12}$	Could be written as $8 \times 8 \times 8 \times 8 = 4096$	$\left(2^a\right)^b = 2^{a\times b}$
$(10^2)^3 = 10^{2\times3} = 10^6$	Could be written as $100 \times 100 \times 100 = 1{,}000{,}000$	$\left(10^a\right)^b = 10^{a\times b}$

Using numbers. . .		Generalizing with letters instead of numbers. . .
1 divided by a number in index form		
$\dfrac{1}{100} = 0.01$	Same calculation in index form is $\dfrac{1}{10^2} = 10^{-2}$	$\dfrac{1}{10^a} = 10^{-a}$
$\dfrac{1}{0.01} = 100$	The same calculation in index form is $\dfrac{1}{10^{-2}} = 10^2$	$\dfrac{1}{10^{-a}} = 10^a$

Table QT8.3 Rules for multiplying and dividing numbers and expressions in logarithm form

Using numbers. . .		Generalizing with letters instead of numbers. . .
The log is the power to which the base is raised		
	$\log(1000) = \log\left(10^3\right) = 3$	$\log(10^a) = a$
Multiplication: if we multiply two expressions we add the logs		
$10^{2+4} = 10^2 \times 10^4$	$\log\left(10^2 \times \left(10^4\right)\right) = \log\left(10^2\right) + \log\left(10^4\right) = 6$	$\log(c \times d) = \log(c) + \log(d)$
Division: if we divide two expressions we subtract the logs		
$\dfrac{10^5}{10^3} = 10^{5-3}$	$\log\left(\dfrac{10^5}{10^3}\right) = \log\left(10^5\right) - \log\left(10^3\right) = 2$	$\log\left(\dfrac{d}{c}\right) = \log(d) - \log(c)$
The power of a power becomes repeated addition		
$(10^2)^3 = 10^2 \times 10^2 \times 10^2 = 10^6$	$\log\left(\left(10^2\right)^3\right) = \log\left(10^2 \times 10^2 \times 10^2\right)$ $= \log 10^2 + \log 10^2 + \log 10^2$ $= 3 \times \log\left(10^2\right)$	$\log\left(\left(10^c\right)^d\right) = d \times \log 10^c$

Table QT8.4 Generic forms of equations commonly used in biology

Relationship	Generic form	Description	Example	Typical shape of the graph
y directly proportional to x	$y = kx$	Straight line passing through $(0,0)$	Beer–Lambert Law for absorbance and concentration (Case Study 8.2)	
Straight line	$y = mx + c$	Straight line crossing the y-axis at the point $(c,0)$	Linear regression using statistical analysis	
Straight line, negative slope	$y = -mx + c$	Straight line crossing the y-axis at the point $(c, 0)$ with a negative (downward) slope		
y is inversely proportional to x	$y = \dfrac{k}{x}$	The y value decreases as the x value increases	The relationship between diffusion rate and distance; Fick's Law, see Case Study 8.3	
Rectangular hyperbola	$y = \dfrac{ax}{x + b}$	The y value increases as the x value increases, but then gradually levels off	Michaelis–Menten equation for rate of reaction and substrate concentration, see Case Study 8.4	
Exponential growth	$y = e^{kx}$	For a positive exponent, as the x value increases, the y value also increases, getting steeper and steeper	Bacterial or population growth	

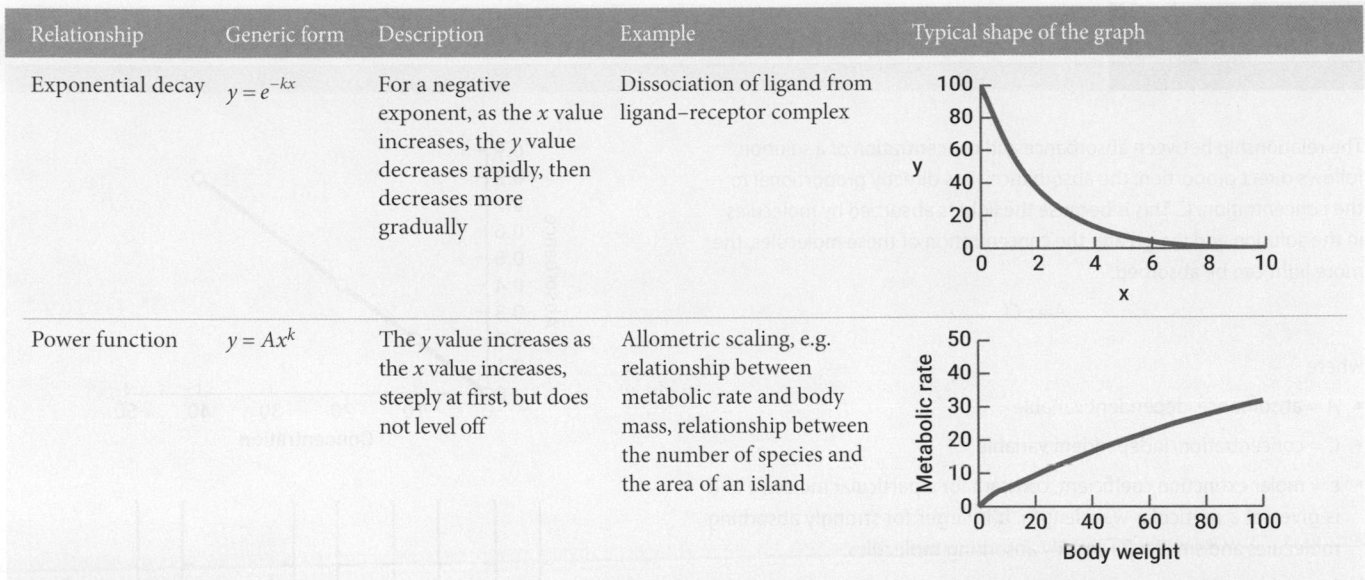

Relationship	Generic form	Description	Example	Typical shape of the graph
Exponential decay	$y = e^{-kx}$	For a negative exponent, as the x value increases, the y value decreases rapidly, then decreases more gradually	Dissociation of ligand from ligand–receptor complex	
Power function	$y = Ax^k$	The y value increases as the x value increases, steeply at first, but does not level off	Allometric scaling, e.g. relationship between metabolic rate and body mass, relationship between the number of species and the area of an island	

CASE STUDY 8.1 Shannon–Weiner diversity index

Some formulas allow you to calculate an index, which is a number used as a descriptor. A good example is the Shannon–Weiner diversity index (H)—the higher the number, the greater the diversity. This index is obtained by first calculating the fractional abundance (p) of each species:

$$\text{fractional abundance of species A} = p = \frac{\text{number of species A}}{\text{number in total population}}$$

The next step is to calculate the natural logarithm of p and multiply it by p.

▶ Refer to Quantitative Toolkit 2 to calculate the natural logarithm of a number.

The reason for doing this is essentially just arbitrary: there is no fundamental rationale for it. It simply gives a number that is usually less than 10 and is greater for more species, each having similar fractional abundance. The index, H, is the sum of $p \times \ln(p)$ for all the species present. And because we tend to prefer a number that is positive, we multiply the answer by -1 to get a positive number.

The sum is indicated by the symbol Σ (capital sigma). The formula for this index is–

Shannon–Weiner diversity index $H = -\Sigma[p_i \times \ln(p_i)]$

If we had, say, three species, we would add up the values for $p \times \ln(p)$ for each of the species]

$$H = -\big[(p \times \ln(p)\,\text{for species 1}) + (p \times \ln(p)\,\text{for species 2})$$
$$+ (p \times \ln(p)\,\text{for species 3})\big]$$

Go to the e-book to watch a video that will help you understand how to use a spreadsheet to do this calculation.

▶ Practise calculating the Shannon–Weiner diversity index in Activity 2 at the end of this toolkit.

Worked example

There are two hypothetical marine lagoons that contain four species of fish: A, B, C, and D. The table shows the data: the number of each species of fish. The column headed p calculates the fractional abundance, i.e. the number of a given species divided by the total number. For example, for species A in lagoon 1 $p = \frac{45}{100} = 0.45$. The next column calculates the natural log of p, $\ln(p)$ and the next column multiplies the value for p by the value for $\ln(p)$. All the values for $p \times \ln(p)$ are added up; for lagoon A the total is -0.893.

H for lagoon 1 = 0.893
H for lagoon 2 = 1.38
Conclusion: lagoon 2 has a greater diversity index.

Species	Lagoon 1				Lagoon 2			
	Number	p	$\ln(p)$	$p \times \ln(p)$	Number	p	$\ln(p)$	$p \times \ln(p)$
A	45	0.45	−0.80	−0.36	25	0.25	−1.39	−0.35
B	50	0.50	−0.69	−0.35	22	0.22	−1.51	−0.33
C	2	0.02	−3.91	−0.078	26	0.26	−1.35	−0.35
D	3	0.03	−3.51	−0.105	27	0.27	−1.31	−0.35
Total	100			−0.893	100			−1.38

CASE STUDY 8.2   Direct proportion: the Beer–Lambert Law

The relationship between absorbance and concentration of a solution follows direct proportion: the absorbance, A, is directly proportional to the concentration, C. This is because the light is absorbed by molecules in the solution and the greater the concentration of these molecules, the more light can be absorbed.

$$A = \varepsilon C l$$

where

- A = absorbance, dependent variable
- C = concentration, independent variable
- ε = molar extinction coefficient, constant for a particular molecule and is given at a particular wavelength. It is larger for strongly absorbing molecules and smaller for weakly absorbing molecules.
- l = pathlength, constant provided the same container is used for all measurements, usually 1 cm.

 Go to the e-book to watch a video that will help you relate an equation to a graph, and the biological context.

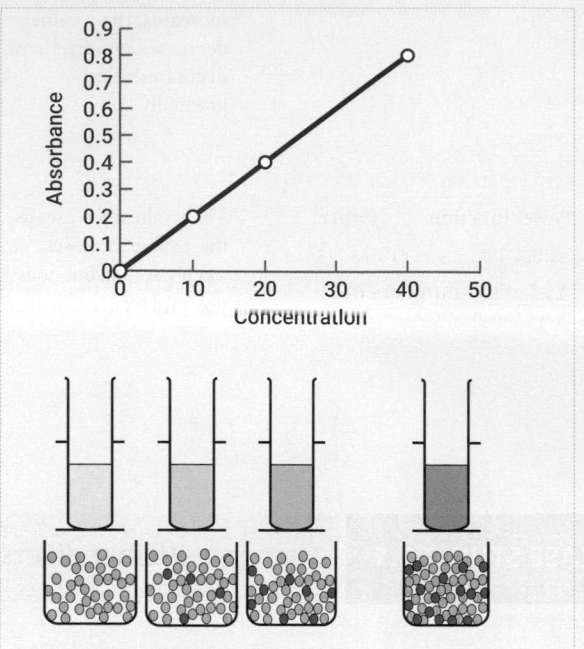

CASE STUDY 8.3   Direct and inverse proportion: Fick's Law for rate of diffusion

 Go to the e-book to watch a video that walks through this case study.

A formula can reflect the relationship between variables derived from experimental observation. For example, as the difference between the concentration (C) of a molecule inside and outside a cell increases, the rate of diffusion, J, increases in direct proportion. We can write the relationship:

The rate of diffusion is proportional to the concentration gradient:

$$J \propto (C_{outside} - C_{inside})$$

where

- J is the rate of diffusion
- $C_{outside}$ is the concentration of molecules or ions outside a membrane
- C_{inside} is the concentration of molecules or ions inside a membrane.

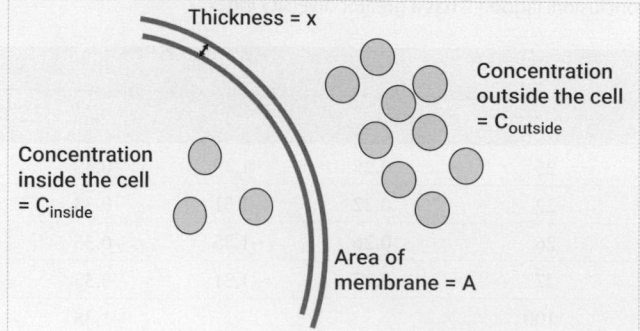

The rate of diffusion is also directly proportional to the area of the membrane, A, which makes sense because the rate of diffusion increases as the area increases.

The rate of diffusion is proportional to the area of the membrane:

$$J \propto A$$

where A is area of the membrane.

The rate of diffusion is **inversely proportional** to the thickness of the membrane, x, which makes sense because the rate of diffusion decreases as the thickness increases.

The rate of diffusion is inversely proportional to the thickness of the membrane: $J \propto \dfrac{1}{x}$

To turn this relationship into a formula we need to introduce the idea of a **constant**. A constant in this case is the constant of proportionality and is different for each type of molecule and for the conditions of the experiment (e.g. temperature and pressure). The constant, D, is called the diffusion coefficient and will have a certain value for a particular molecule and experimental conditions such as temperature and pressure. Fick's Law is written as:

$$\text{Fick's Law } J = \frac{-DA(C_{outside} - C_{inside})}{x}$$

8.3 Carrying out calculations and keeping track of units

Keeping track of the units in a calculation is very important. The units should cancel out appropriately to leave the correct result. For example, when working with the formula for concentration, $C = \frac{m}{v}$, if the mass is given in grams, g, and the volume is given in litres, L, then the resulting concentration will be g L^{-1}.

In a calculation it can be helpful to include the units with the numbers. Let us consider an example: if 0.5 g salt is dissolved in 2 L water, the concentration is $C = \frac{m}{v}$

Assign the numbers to the relevant symbols in the equation, so m = 0.5 g (the mass of salt) and V = 2 L (the volume of water). Then substitute these numbers into the equation and write the units alongside:

$$C = \frac{m}{v} = \frac{0.5\,\text{g}}{2\,\text{L}} = 0.25\,\text{g L}^{-1}$$

Note that we can write either g/L or g L^{-1}.

A more complex example is Fick's Law, as we saw in Case Study 8.3. In a calculation with Fick's Law you might be given typical values, such as the following example of a sucrose solution in dialysis tubing. The thickness of the dialysis tubing is 80 μm; the diffusion coefficient, D, is 520 μm^2 s^{-1}; the area of the membrane is 2500 mm^2 and the difference in concentration is 10 mM.

Fick's Law $\quad J = \dfrac{-DA\left(C_{outside} - C_{inside}\right)}{x}$

The first step is to convert the values so they are in the standard units as given in Table QT8.5.

▶ Refer to Quantitative Toolkit 1 for more information about prefixes.

D, the diffusion coefficient for sucrose is 520 μm^{2}s^{-1}

$= 520 \times (10^{-6}\,\text{m})^2\,\text{s}^{-1}$

$= 5.2 \times 10^{-10}\,\text{m}^2\text{s}^{-1}$

$A = 2500\,\text{mm}^2 = 2500 \times (10^{-3}\,\text{m})^2 = 2.5 \times 10^{-3}\,\text{m}^2$

$C_{outside} - C_{inside} = 10\,\text{mmol L}^{-1} = \dfrac{10 \times 10^{-3}\,\text{mol}}{0.001\,\text{m}^3}$

$= 10\,\text{mol m}^{-3}\,(\text{since } 1\,\text{L} = 0.001\,\text{m}^3)$

Table QT8.5 Symbols and units for Fick's Law

Symbol	Meaning	Units
J	Rate of diffusion in molecules per unit time	mol s^{-1}
D	Diffusion coefficient	m^2 s^{-1}
A	Area of the membrane	m^2
$C_{outside} - C_{inside}$	Concentration gradient across the membrane	mol m^{-3}
x	Thickness of the membrane	

$x = $ thickness of dialysis tubing $= 80$ μm $= 80 \times 10^{-6}$ m $= 8 \times 10^{-5}$ m

Now you have all the numbers with standard units, substitute them into the equation and you will see that the units will simplify.

🔘 Go to the e-book to watch a video that will help you keep track of units in a calculation.

$J = \dfrac{-DA\left(C_{outside} - C_{inside}\right)}{x}$

$= \dfrac{-5.2 \times 10^{-10}\,\text{m}^2\text{s}^{-1} \times 2.5 \times 10^{-3}\,\text{m}^2 \times 10\,\text{mol m}^{-3}}{8 \times 10^{-5}\,\text{m}}$

$= 1.625 \times 10^{-7}\,\text{mol s}^{-1}$

Note that the metres cancel out—looking at the top of the fraction we have m$^2 \times$m$^2 \times$m^{-3} which simplifies to m. If you are not sure about this simplification, refer back to Section 8.1 which shows how to manipulate index notation—the principles that we saw there for numbers and symbols apply equally well to units.

▶ Practise using these ideas in Activities 3, 4, 5, and 6 at the end of this toolkit.

8.4 Rearranging formulas

The equals sign in the formula means that one side of the formula is equal to the other side, so when rearranging a formula it is essential to always do the same thing to both sides. The idea of using the inverse process is very important in working out what to do when rearranging formulas. Inverse processes include:

- addition and subtraction
- multiplication and division
- squaring and taking the square root
- taking the log or raising to a power (called anti-log), as we'll discuss in more detail later in this section.

You can think of an inverse process as one that undoes another, e.g. addition undoes subtraction.

Add and subtract	$3 + 4 = 7$	$7 - 4 = 3$
Multiply and divide	$3 \times 4 = 12$	$12 \div 4 = 3$
Squaring and taking the square root	$3^2 = 9$	$\sqrt{9} = 3$
Raising to a power and taking the log	Using base 10: $10^4 = 10{,}000$	$\log(10000) = 4$
	Using base e: $e^4 = 54.5981\ldots$	$\ln(58.5981\ldots) = 4$

Rearrange a simple equation

 Go to the e-book to watch a video that explains how to rearrange a simple equation using the approach set out here.

Example 1: Concentration (C) is the mass (m) of substance divided by the volume (V) of the solution: $C = m/V$.

We need to rearrange this equation to make m the subject. If we want m to be the subject of the formula, we need to multiply both sides by V. How do we know to do this? It is because m is divided by V in the original formula, so we use the inverse process, i.e. multiplication by V.

Multiply both sides by V:

$$V \times C = \frac{m}{V} \times V$$

V cancels out on the right:

$$V \times C = \frac{M}{\cancel{V}} \times \cancel{V}$$

Resulting in:

$$V \times C = m$$

Or, if you prefer, you can put them the other way round and leave out the multiplication sign:

$$m = VC$$

Example 2: Rearrange $C = \dfrac{m}{V}$ to make V the subject.

Now we need two steps. We start off the same way as in the previous example:

Multiply both sides by V:

$$V \times C = \frac{m}{V} \times V$$

V cancels out on the right:

$$V \times C = m$$

Then divide both sides by C:

$$\frac{m}{C} = \times \frac{V \times C}{C}$$

C cancels out on the right:

$$\frac{m}{C} = V$$

Rearranging the Michaelis–Menten equation

The Michaelis–Menten equation is an important equation in biochemistry as it allows us to compare substrates and enzymes. The Michaelis–Menten equation describes how the rate of an enzyme-catalysed reaction, v, varies with the substrate concentration, $[S]$.

$$v = \frac{V_{\max}[S]}{K_M + [S]}$$

where

- $[S]$ = substrate concentration, the independent variable
- v = initial rate of reaction, the dependent variable
- K_M = the Michaelis–Menten constant
- $V_{\max}$ = maximum rate of reaction.

Rearranging the Michaelis–Menten equation to make $[S]$ the subject is a more complex example than our previous examples and requires more steps, but the principles are the same.

 Go to the e-book to watch a video that works through the example that follows.

Example 3: Rearrange $v = \dfrac{V_{\max}[S]}{K_M + [S]}$ to make $[S]$ the subject.

There are several ways of approaching this and the following is one example.

The first step is to multiply both sides by $(K_M + [S])$ so we do not have a fraction.

First multiply both sides by $(K_M + [S])$:

$$v \times (K_M + [S]) = \frac{V_{\max}[S]}{K_M + [S]} \times (K_M + [S])$$

The $(K_M + [S])$ term cancels out on the right-hand side to give:

$$v(K_M + [S]) = V_{\max}[S]$$

Now expand out the bracket on the left so we have all the terms as separate entitie:

$$vK_M + v[S] = V_{\max}[S]$$

We need to have both terms with $[S]$ on the same side of the equation.

If we subtract $v[S]$ from both sides there will be no term with $[S]$ on the left:

$$vK_M + v[S] - v[S] = V_{\max}[S] - v[S]$$

$$vK_M = V_{\max}[S] - v[S]$$

Now both terms on the right have $[S]$ in them so we can factorize:

$$vK_M = [S](V_{\max} - v)$$

To get $[S]$ on its own we can divide both sides by $(V_{\max} - v)$

$$\frac{vK_M}{(V_{\max} - v)} = \frac{[S](V_{\max} - v)}{(V_{\max} - v)}$$

The terms $(V_{\max} - v)$ on the right-hand side cancel out, so

$$[S] = \frac{vK_M}{(V_{\max} - v)}$$

CASE STUDY 8.4 The Michaelis–Menten equation

This diagram shows a mental model of what is happening inside cells or test tubes when an enzyme is catalysing a reaction to convert substrate to product. Note that while the diagram shows a static image, in reality the enzymes and substrates are constantly moving in a random fashion and occasionally bum into each other.

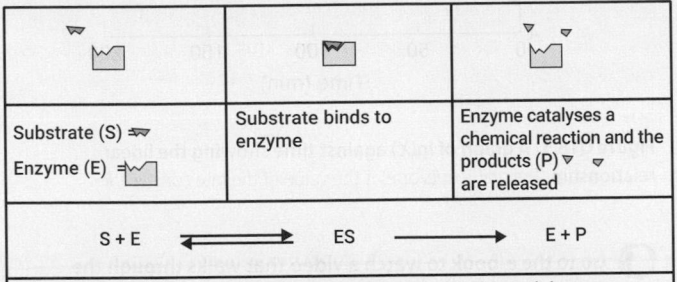

Substrate (S) ☞	Substrate binds to enzyme	Enzyme catalyses a chemical reaction and the products (P) ▽ ▽ are released

$$S + E \rightleftarrows ES \longrightarrow E + P$$

In reality we have many enzymes and substrate molecules and they are constantly moving and bumping into each other in a random manner. As the reaction continues, more product is produced and less substrate remains.

Let us explore the properties of this equation:

$$v = \frac{V_{max}[S]}{K_m + [S]}$$

where

- $[S]$ = substrate concentration
- v = rate of reaction (sometimes called velocity)
- V_{max} = maximum rate (when all enzymes are occupied with substrate)
- K_M = Michaelis–Menten constant.

When $[S]$ is very large compared with K_M, then v approximates V_{max}. Substitute in some numbers to help visualize this. Say K_M is 1 and $[S]$ is 10,000, then

$$v = \frac{V_{max}10,000}{1 + 10,000}$$

$$v = \frac{V_{max}10,000}{10,001}$$

10,000 divided by 10,001 is almost 1 (actually 0.9999) so v approximates to V_{max}

Let us see what happens when $[S]$ and K_M have the same value, i.e. we are saying $[S] = K_M$

$$v = \frac{V_{max}[S]}{[S] + [S]}$$

$[S] + [S]$ is just $2[S]$ so we get:

$$v - \frac{V_{max}[S]}{2[S]}$$

Now let us write in the multiplication signs just to make it a bit clearer:

$$v = \frac{V_{max} \times [S]}{2 \times [S]}$$

Since $[S]$ divided by $[S]$ is just 1, we are left with:

$$v = \frac{1}{2} V_{max}$$

So we can see that when the substrate concentration $[S]$ is equal to the K_M, then the rate of reaction, v, is half the maximum rate of reaction.

Or we can turn it around and say: when $v = \frac{1}{2} V_{max}$ the substrate concentration gives the K_M.

This is useful because we can then compare values for the K_M for different substrates. In some cases, an enzyme may work efficiently with low concentrations of substrate; in other cases, higher substrate concentrations are needed for the enzyme to work efficiently.

 Go to the e-book to watch a video that will help you understand the Michaelis–Menten equation.

Rearranging the exponential growth equation to make time the subject

The exponential growth equation can be used for a number of biological situations such as the growth of bacteria, cells, and whole animals, as well as the number of people infected with a bacteria or virus. In the different biology subdisciplines the symbols used may be different and the words used to describe them may be different, but the underlying mathematics is very similar.

In the following examples we use X for the number of bacteria, cells, animals, or infected organisms, and t stands for time. (We are often looking at the change in these observations over time.) Quantitative Toolkit 9 looks at some of these models in more detail; here we will look at how to manipulate these equations and rearrange them in helpful ways so that we can isolate the time required for a particular amount of growth or the time required for the population to double.

 Go to the e-book to watch a video that walks through the example that follows.

The exponential growth equation

$$X = X_0 e^{kt}$$

where

- X is the number of cells or organisms at any time, t
- X_0 is the number of cells or organisms at the start
- k is the growth rate constant.

Starting with the exponential growth equation, there are a number of ways to proceed. Let us start by rearranging so that only the exponential term is left on one side and all the other terms are on the other side. In this case we need to divide both sides by X_0:

$$X = X_0 e^{kt}$$

$$\frac{X}{X_0} = \frac{X_0}{X_0} e^{kt}$$

X_0 then cancels out on the right:

$$\frac{X}{X_0} = \frac{\cancel{X_0}}{\cancel{X_0}} e^{kt}$$

Now we can take natural logs of both sides:

$$\frac{X}{X_0} = e^{kt}$$

$$\ln\left(\frac{X}{X_0}\right) = kt$$

Then to get t on its own, we divide both sides by k leaving t as the subject of the equation:

$$\frac{1}{k}\ln\left(\frac{X}{X_0}\right) = t$$

Rearranging the exponential growth equation to estimate the rate constant, k

It is possible to show, by taking logs of both sides of this equation, that the log plot against time is a straight line and that the slope of that line gives the value of the rate constant, k.

 Go to the e-book to watch a video that walks through the example that follows.

Starting with the exponential growth equation,

$$X = X_0 e^{kt}$$

Take natural logs of both sides:

$$\ln(X) = \ln\left(X_0 \times e^{kt}\right)$$

Using the first rule from Table QT8.3:

$$\log(a \times b) = \log(a) + \log(b)$$

we write: $\ln(X) = \ln(X_0 + \ln e^{kt})$

Since the log is the power, the term $\ln e^{kt}$ just becomes kt

$$\ln(X) = \ln(X_0) + kt$$

This resembles the equation of a straight line shown in Table QT8.4 and Figure QT8.1

$$y = c + mx$$

where

$$y = \ln(x), c = \ln(X_0), m = \text{slope} = k, x = t$$

Show how the doubling time (t_d) relates to the rate constant

It is also possible to show how the doubling time relates to the rate constant. Using the equation, the doubling time (t_d) occurs when $X = 2X_0$ (i.e. twice the starting concentration).

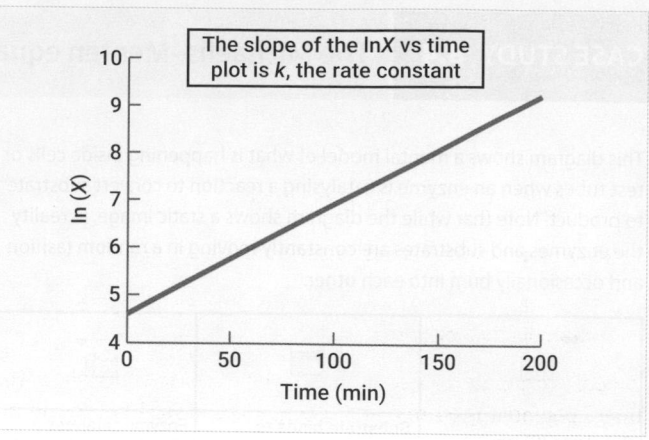

Figure QT8.1 A graph of ln(X) against time showing the linear relationship. The gradient (slope) is the value of the rate constant, k.

 Go to the e-book to watch a video that walks through the example that follows.

Starting with the exponential growth equation:

$$X = X_0 e^{kt}$$

Substitute in $X = 2X_0$, i.e. X is twice its starting value when t is equal to the doubling time i.e. $t = t_d$

$$2X_0 = X_0 e^{kt_d}$$

Now we have X_0 on both sides so we can divide both sides by X_0 to cancel out:

$$\frac{2\cancel{X_0}}{\cancel{X_0}} = \frac{\cancel{X_0} e^{kt_d}}{\cancel{X_0}}$$

$$2 = e^{kt_d}$$

Now it becomes easier to take (natural) logs of both sides:

$$\ln(2) = \ln\left(e^{kt_d}\right)$$

From the first rule in Table QT8.3:

$$\ln\left(e^a\right) = a$$

$$\ln(2) = kt_d$$

Divide both sides by k to leave t_d as the subject of the equation:

$$t_d = \frac{\ln(2)}{k}$$

$$= \frac{0.693}{k} \left(\text{since } \ln(2) = 0.693\right)$$

Rearranging an equation with a log term

As we have seen, the log function is the inverse of an exponential function. Taking logs is the inverse process to antilog. Looking at this with numbers to start with we can see that:

$$\log(1000) = 3$$

Then if we antilog both sides of this equation we get:

$$1000 = 10^3$$

We can use this strategy to rearrange an equation with a log term, for example the Nernst equation, which relates the equilibrium electrical potential of an ion (E_{ion}) to the ratio of concentrations of the ion on the inside and outside of a membrane.

▶ **We learn more about the Nernst equation in Chapter 17.**

🎬 **Go to the e-book to watch a video that walks through the example that follows.**

Take this example for a sodium ion where we know that the membrane potential is 64 mV and $[Na^+]_{out} = 145$ mM but we do not know $[Na^+]_{in}$

$$E_{Na^+} = 2.303 \frac{RT}{zF} \log\left(\frac{[Na^+]_{out}}{[Na^+]_{in}}\right)$$

We can take values for R and F from the table of universal constants (see Table QT8.1).

$R = 8.31447$ J mol^{-1}K^{-1}

$F = 96\ 485.33$ C mol^{-1}

Z = charge on the ion, which is +1 for sodium

T is the temperature in Kelvin; let us take 298 K (which corresponds to 25 °C).

We can substitute this in to the first part of the equation :

$$2.303 \frac{RT}{zF} = 2.303 \times \frac{8.31447\ \text{Jmol}^{-1}\text{K}^{-1} \times 298\ \text{K}}{1 \times 96\ 485.33\ \text{Cmol}^{-1}} = 0.05914\ \text{JC}^{-1}$$

(J C^{-1} means Joules per Coulomb which is the same as Volts (V).)
So, the Nernst equation now becomes:

$$E_{Na^+} = 0.05914 \log\left(\frac{[Na^+]_{out}}{[Na^+]_{in}}\right)$$

We can now substitute in values for $E = +64$ mV ($= 0.064$ V) and $[Na^+]_{out} = 145$ mM

$$0.064 = 0.05914 \log\left(\frac{145}{[Na^+]_{in}}\right)$$

Dividing both sides by 0.05914. . .

$$\frac{0.064}{0.05914} = \frac{0.05914}{0.05914} \log\left(\frac{145}{[Na^+]_{in}}\right)$$

$$1.08217 = \log\left(\frac{145}{[Na^+]_{in}}\right)$$

Now we need to antilog.
Here 1.08217 becomes $10^{1.08217}$

$$\text{And } \log\left(\frac{145}{[Na^+]_{in}}\right) \text{ becomes } \frac{145}{[Na^+]_{in}}$$

$$10^{1.08217} = \frac{145}{[Na^+]_{in}}$$

We ideally want the sodium ion concentration as the subject of the equation so we multiply both sides by $[Na^+]_{in}$

$$[Na^+]_{in} \times 10^{1.08217} = 145$$

$10^{1.08217} = 12.083$ (using a calculator)

$$[Na^+]_{in} \times 12.083 = 145$$

Then divide both sides by 12.083:

$$\frac{[Na^+]_{in} \times 12.083}{12.083} = \frac{145}{12.083}$$

$$[Na^+]_{in} = 12\ \text{mM}$$

So, from knowing that the equilibrium membrane potential was +64 mV and the outside sodium ion concentration was 145 mM, we can work out that the inside sodium ion concentration must be 12 mM.

ACTIVITIES

Activity 1

Five measurements were made of the diameter of coronavirus virions in a sample: 120, 150, 60, 80, 160.
    The formula for the mean, $\bar{x}$, of a set of numbers in $\bar{x} = \dfrac{\sum x_i}{n}$

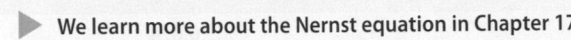
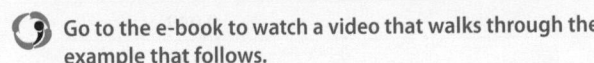

where n is the number in the sample and x_i stands for each of the measurements.

(a) Use the formula to calculate the mean.

The formula for the sample standard deviation is:

$$s = \sqrt{\frac{\sum (x_i - \bar{x})^2}{n-1}}$$

When calculating this by hand, it is helpful to tabulate the data and use columns to work out the intermediate steps.

(b) Fill out the table below to calculate $\sum (x_i - \bar{x})^2$ and then the sample standard deviation.

x_i	$x_i - \bar{x}$	$(x_i - \bar{x})_2$
0.62		
0.67		
0.79		
0.71		
0.62		

▶ **Refer back to Section 8.1, Symbols for operators.**

Activity 2

We have two hypothetical forests which each have three species of bird, species A, B, and C. The populations of these birds are shown in the table.

Species	Forest 1 number of individuals	Forest 2 number of individuals
A	33	3
B	35	5
C	32	92
Total population of birds	100	100

(a)   Looking at the data in the table, which forest do you think has the greater diversity?

(b)   Calculate the Shannon–Weiner diversity index (H) for each forest and compare this with your answer to (a).

▶ **Refer back to Section 8.1, Symbols for operators, and Section 8.1, Exponential and logarithm notation.**

Activity 3

The Beer–Lambert Law is $A = \varepsilon Cd$

A is the absorbance of a particular wavelength of light, no units

ε is the molar extinction coefficient, a constant for a particular molecule, $M^{-1}\ cm^{-1}$

C is the concentration, M

d is the pathlength, cm

For double-stranded DNA, $\varepsilon = 20\ g^{-1}.cm^{-1}.L$. If the absorbance is 0.089 for a 1 cm pathlength, calculate the concentration of DNA in the sample. Ensure that you include the units with your answer.

▶ **Refer back to Section 8.3.**

Activity 4

(a)   Several concentrations of a standard albumin solution were prepared and reacted with the Bradford reagent, which produces a blue-coloured solution that absorbs light of 595 nm wavelength. The absorbance of these solutions was measured with a pathlength of 1 cm and follows the Beer–Lambert Law as described in Activity 3. Plot the data and calculate the extinction coefficient, ε. What are the units for ε? Write the equation that describes this graph.

Albumin standard ($\mu g\ mL^{-1}$)	A_{595}
10	0.024
20	0.049
50	0.125
100	0.246
200	0.509

(b)   A 0.1 mL sample of an unknown protein-containing solution was reacted with Bradford reagent in a similar manner and the absorbance measured was 0.221. Calculate the protein concentration in the unknown sample.

▶ **Refer back to Section 8.3.**

Activity 5

The initial rate of reaction, r, of the hydrolysis of sucrose to form glucose and fructose is proportional to the concentration of sucrose [S]. The constant of proportionality is known as k. The formula is $r = kC$.

At pH 7 the rate constant for this reaction is $2 \times 10^{-11}\ s^{-1}$ at 27 °C and increases to $8.5 \times 10^{-11}\ s^{-1}$ at 37 °C.

(a)   Sketch, on the same axes, graphs to represent how the initial rate of reaction, r, varies with sucrose concentration, [S] at the two different temperatures.

▶ **Refer back to Section 8.2.**

(b)   Calculate the initial rate of reaction at each temperature for a sucrose concentration of 0.15 mol.L^{-1}.

▶ **Refer back to Section 8.3.**

Activity 6

The rate of diffusion, J, of the general anaesthetic propofol in water at 25 °C, 1 atm pressure is related to the area of the membrane, A, the thickness of the membrane, x, and the concentration gradient between the inside and outside of the membrane as niven by Fick's Law.

$$J = \frac{-DA(C_{outside} - C_{inside})}{x}$$

J = rate of diffusion in moles per unit area per unit time, mol.s^{-1}

D = diffusion coefficient, m^2.s^{-1}

A = area of the membrane, m^2

$C_{outside}$ and C_{inside} are the concentrations outside and inside of the membrane in mol.m^{-3}

x = thickness of the membrane, m.

(a)   Draw a sketch to show the shape of a graph which represents the dependence of: (i) J on A; (ii) J on x; (iii) J on $C_{outside} - C_{inside}$

▶ **Refer back to Section 8.2.**

(b) Which of the letters in the formula represent (i) variables and which represent (ii) constants?

▶ Refer back to Section 8.1, Letters can stand for variables or constants.

(c) The diffusion coefficient, D, for propofol in water was given in a scientific paper as 0.02×10^{-6} cm^2.s^{-1}. What is the rate of diffusion of oxygen across a 2 mm^2 section of cell membrane which is 8 nm thick and has a concentration gradient of 0.6 μM?

(**Hint:** It is a good idea to convert all the units into those required by the formula first, before substituting them into the formula.)

▶ Refer back to Section 8.3.

ANSWERS

🎥 Go to the e-book to watch a video that walks through answers to these Activities.

Activity 1

(a) $\sum x_i = 120 + 150 + 60 + 80 + 160 = 570$ nm

$n = 5$ (there are five measurements in the sample

$$\frac{\sum x_i}{n} = \frac{570}{5} = 114 \text{ nm}$$

The mean is 114 nm.

(b)

x_i	$x_i - \overline{x}$	$\left(x_i - \overline{x}\right)^2$
120	6	36
150	36	1296
60	−54	2916
80	−34	1156
160	46	2116

$$\sum \left(x_i - \overline{x}\right)^2 = 36 + 1296 + 2916 + 1156 + 2116 = 7520$$

$$\text{Standard deviation} = s = \sqrt{\frac{\sum \left(x_i - \overline{x}\right)^2}{n-1}} = \sqrt{\frac{7520}{4}} = \sqrt{1800} = 43$$

Activity 2

Forest 1

Species	Number of individuals	p	$\ln p$	$p \times \ln p$
A	33	0.33	−1.1087	−0.3659
B	35	0.35	−1.0498	−0.3674
C	32	0.32	−1.1394	−0.3646

Forest 2

Species	Number of individuals	p	$\ln p$	$p \times \ln p$
A	3	0.03	−3.5066	−0.1052
B	5	0.05	−2.9957	−0.1498
C	92	0.92	−0.0834	−0.0767

Forest 1: $\sum p \times \ln p = -0.3659 - 03674 - 0.3646 = -1.0979$, therefore $H \sim 1.1$.

Forest 2: $\sum p \times \ln p = -0.1052 - 01498 - 0.0767 = -0.3317$, therefore $H \sim 0.33$.

Activity 3

$$A = \varepsilon C d$$

Rearrange to make C the subject:

Divide both sides by εd: $C = \dfrac{A}{\varepsilon d}$

Substitute in the values: $\varepsilon = 20$ g^{-1}.cm^{-1}.L, $A = 0.089$, $d = 1$ cm

$$C = \frac{0.089}{20 \text{ g}^{-1}\text{cm}^{-1}\text{L} \times 1 \text{cm}}$$

$$C = \frac{0.089}{20 \text{ g}^{-1}\text{L}}$$

$$C = 0.00445 \text{ g L}^{-1}$$

Activity 4

Rearreange the equation so that ε is the subject.

$$A = \varepsilon C d$$

$$\frac{A}{C d} = \varepsilon$$

Choosing a point on the line (10, 0.024) and substituting:

$$A = 0.024, C = 10 \text{ μgmL}^{-1}, d = 1 \text{ cm}$$

$$\varepsilon = \frac{0.024}{10 \text{ μgmL}^{-1} \times 1 \text{ cm}} = 0.0024 \text{ μg}^{-1}\text{mL cm}^{-1}$$

If you chose different points you may have calculated 0.0025 μg$^-$ mL cm^{-1}

$A = 0.221$, $\varepsilon = 0.0024$ μg^{-1}mL cm^{-1}, $d = 1$ cm. Substituting in these values:

$$C = \frac{A}{\varepsilon d} = \frac{0.221}{0.0024 \text{ μg}^{-1}\text{mL cm}^{-1} \times 1 \text{ cm}} = 92.08 \text{ μg mL}^{-1}$$

Activity 5

The equation is $r = kC$

(a)

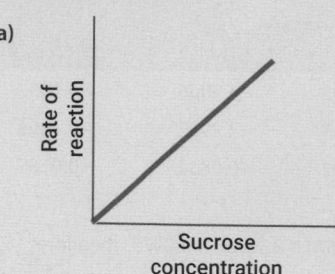

Rate of reaction (vertical axis)

Sucrose concentration (horizontal axis)

(b) When temp = 27 °C, $r = 2 \times 10^{-11}\,s^{-1} \times 0.15\,mol\,L^{-1}$

$$r = 3 \times 10^{-12}\,mol\,L^{-1}\,s^{-1}$$

When temp = 37 °C, $r = 8.5 \times 10^{-11}\,s^{-1} \times 0.15\,mol\,L^{-1}$

$$r = 1.275 \times 10^{-11}\,mol\,L^{-1}\,s^{-1}$$

Activity 6

(a)

   (i)   **(ii)**   **(iii)**

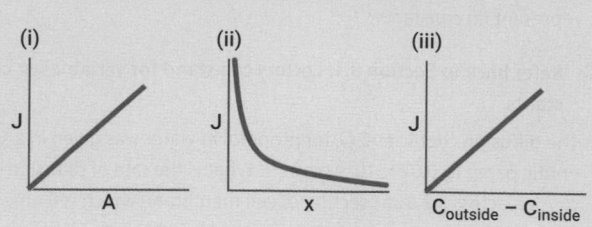

(b) **(i)** Variables are J, A, x, $C_{outside}$, C_{inside}

     (ii) D is a constant

(c) $D = 0.02 \times 10^{-6}\,cm^2\,s^{-1} = 0.02 \times 10^{-6} \times (0.01\,m)^2\,s^{-1} = 2 \times 10^{-12}\,m^2\,s^{-1}$

$A = 2\,mm^2 = 2 \times \left(10^{-3}\right)^2 = 2 \times 10^{-6}\,m^2$

$x = 8\,nm = 8 \times 10^{-9}\,m$

$C_{outside} - C_{inside} = 0.6\,\mu M = 0.6 \times 10^{-6}\,mol\,dm^{-3} = 0.6 \times 10^{-3}\,mol\,m^{-3}$

$$J = \frac{-DA\left(C_{outside} - C_{inside}\right)}{x}$$

$$J = \frac{-2 \times 10^{-12}\,m^2\,s^{-1} \times 2 \times 10^{-6}\,m^2 \times 0.6 \times 10^{-3}\,mol\,m^{-3}}{8 \times 10^{-9}\,m}$$

$$J = 3 \times 10^{-13}\,mol\,s^{-1}$$

Rates of Change

LEARNING OBJECTIVES

In this toolkit we will:

- Interpret graphs to estimate the rate from the slope (gradient) in a variety of biological scenarios (Sections 9.1, 9.2, 9.3, 9.4).

- Calculate average rate and instantaneous rate (Section 9.2).

- Recognize exponential growth and decay on linear and logarithmic scales (Sections 9.3 and 9.4).

- Relate the underlying biology to the multiplicative nature of exponential growth and decay, and gain an intuitive sense of exponential equations (Sections 9.3 and 9.4).

- Know that the ratio of the population size between one generation and the next can be expressed in different ways; examples are the reproduction number R, and the population multiplication rate, λ (Section 9.3).

- Recognize logistic growth and distinguish it from exponential growth (Section 9.5).

Contents

9.1 Rate is the slope (gradient) of a line or curve   **136**

9.2 Quantifying the rate of change   **137**

9.3 Rates involving exponential growth   **139**

9.4 Rates involving exponential decay   **142**

9.5 Logistic growth   **143**

Life never stays still: molecules, subcellular structures, cells, organisms, and populations are constantly changing. Quantifying this change allows biologists to make comparisons and to identify underlying mechanisms. We use equations as mathematical models to describe these changes and make predictions. For example, in the case of bacterial growth we can make predictions about how quickly the bacterial population will increase, and how fast an infection could spread; in the case of the rate of a chemical reaction, we can make predictions about how much of the substrate is metabolized and how quickly this occurs.

There are a number of examples in which the same mathematics underpins observations in many different fields of biology. We aim to help you appreciate these common underlying mathematical structures. The simple models that we introduce here will provide the fundamentals for future exploration of more complex models including the computational approaches that are increasingly being used.

9.1 Rate is the slope (gradient) of a line or curve

The term **rate**, in biology, is generally used to express a ratio of two quantities that usually have different units. For example, heart rate is the number of heart beats per minute; reaction rate is the number of substrate molecules reacted per second. Some of these

rates have special names. For example, speed is the rate of change of distance with time, acceleration is the rate of change of speed with time, and frequency is the numb-r of events per unit time. We often use the term 'per' to relate the two quantities: distance *per* time or molecules *per* time. Occasionally the words 'with' or 'over' can be used (e.g. distance *with* time or distance *over* time). It is common for a rate to be measured against time, and it is also possible to have a rate of change over distance or per cell division or generation, for example.

Let us start by considering a simple example and then see how this can extend to other biological scenarios. Figure QT9.1 shows the distance travelled by a cheetah over time: over the first 5 s it travels 20 m; over the next 5 s it travels 50 m. This information can be displayed graphically as we see in Figure QT9.1a. When graphing distance against time in this way, we see that for the first 5 s the cheetah is travelling more slowly, covering only 20 m, compared with the subsequent 5 s during which it covers 50 m. The **slope** (or **gradient**) shows the rate that the distance is changing with time: if the line is steep, then the distance is changing a

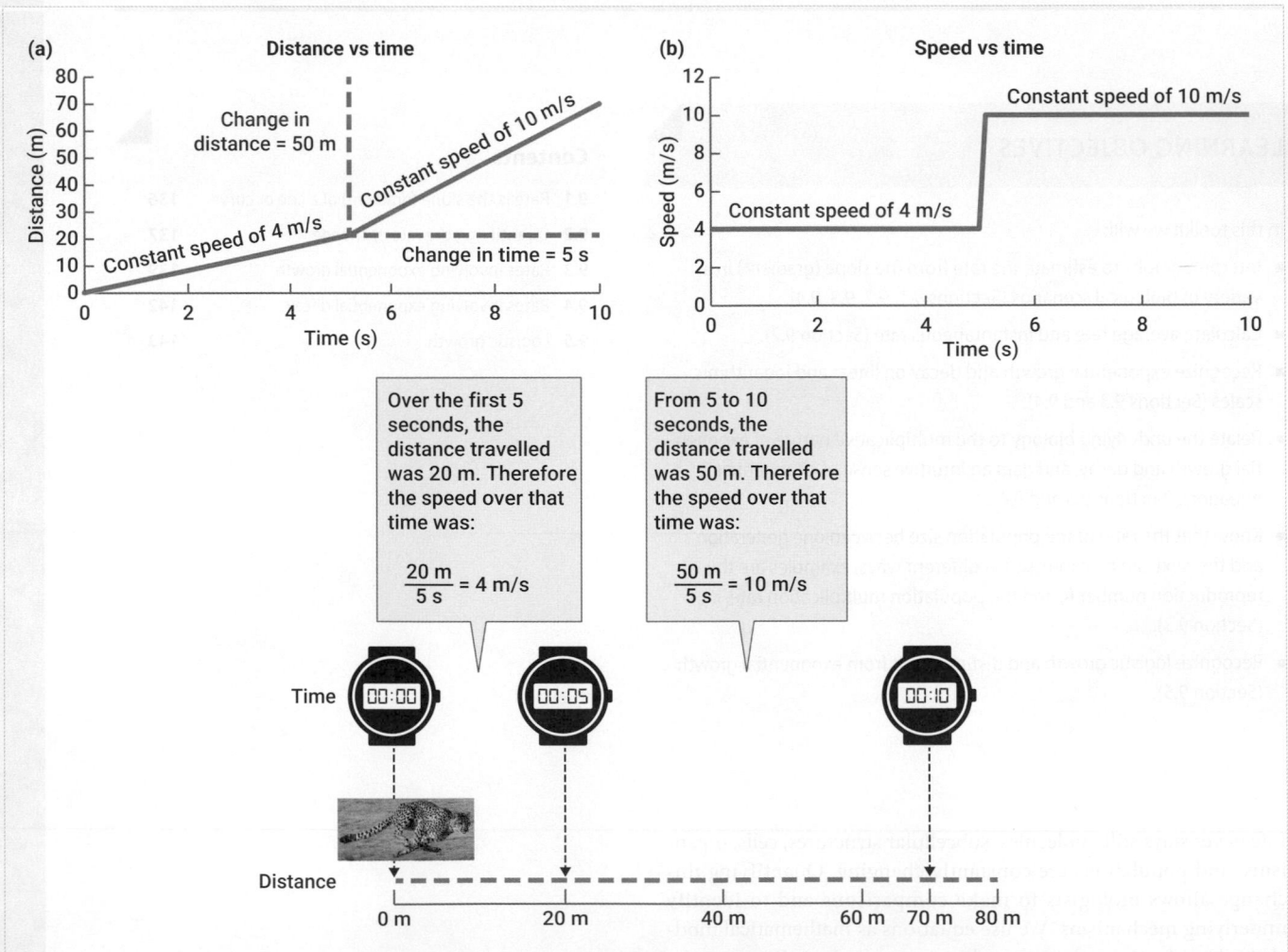

Figure QT9.1 Variable increasing at a constant rate. Diagrammatic representation of a cheetah travelling over 70 m (bottom) and graphical display of the change in distance (a) and speed (b) with time. Speed is rate of change of distance over time; in other words, the slope in (a) is plotted on the *y*-axis in (b).

lot with time. By contrast, if the line is shallow the distance is not changing much with time. The *steepness of the line indicates the rate of change of distance with time*, which is what we call speed. In the first 5 s, the cheetah is travelling more slowly than in the following 5 s and we can see that the line starts off shallower and ends steeper. If the line is flat, there is no change in distance with time (i.e. the object is still and so the speed is zero).

If we plot a graph of *speed* against time we can read off the speed at a particular time just by reading the value from the *y*-axis and we can look at the slope of the line to see how much the speed is *changing* with time (see Figure QT 9.1b). Notice how Figure QT9.1b relates to Figure QT9.1a—the values plotted on the *y*-axis in (b) are the slope of the line at the same time in (a). When the speed is constant, the speed versus time graph is horizontal and the corresponding distance versus time graph has a straight line with positive slope.

It is a little artificial to think that the cheetah will travel at a constant speed for 5 s and then suddenly increase to a higher constant speed. It is more likely that the cheetah will accelerate—that is, its speed will increase gradually with time. Figure QT9.2 shows such a situation in which the steepness of the distance versus time graph (Figure QT9.2a) is gradually increasing, reflecting the increasing speed (Figure QT9.2b). In this case the speed is increasing at a constant rate. In other words, the acceleration is constant (Figure QT9.2c). (The rate of change of speed is the acceleration.)

The slope of the distance versus time graph is the speed and reflects the rate of change of distance with time.

We can extend this way of thinking to a situation in which a quantity is decreasing—for example, a chemical reaction such as a toxin being metabolized in the liver. If the amount of toxin left in the body is decreasing at a constant rate, the concentration versus time relationship will go down at a constant rate. The rate being constant means that the decrease is the same for each time interval—from 0 to 2 min it falls by 1 mg/L (8 down to 7 mg/L) and from 2 to 4 min it also falls by 1 mg/L (from 7 down to 6 mg/L). Since concentration is decreasing at a constant rate, Figure QT9.3a shows a straight line (constant rate) with a negative slope (decreasing concentration). The rate of change of concentration is constant so the graph in Figure QT9.3b is a horizontal straight line.

9.2 Quantifying the rate of change

We have seen that the rate of change is shown by the slope of the line. When the cheetah was travelling with a constant speed, the line on the distance versus time graph was straight and the speed could be calculated by working out the slope (or gradient). This is done by reading off the distance travelled over a certain time, as shown in Figure QT9.1.

A similar approach can be taken with the data for decreasing concentration of toxin shown in Figure QT9.3.

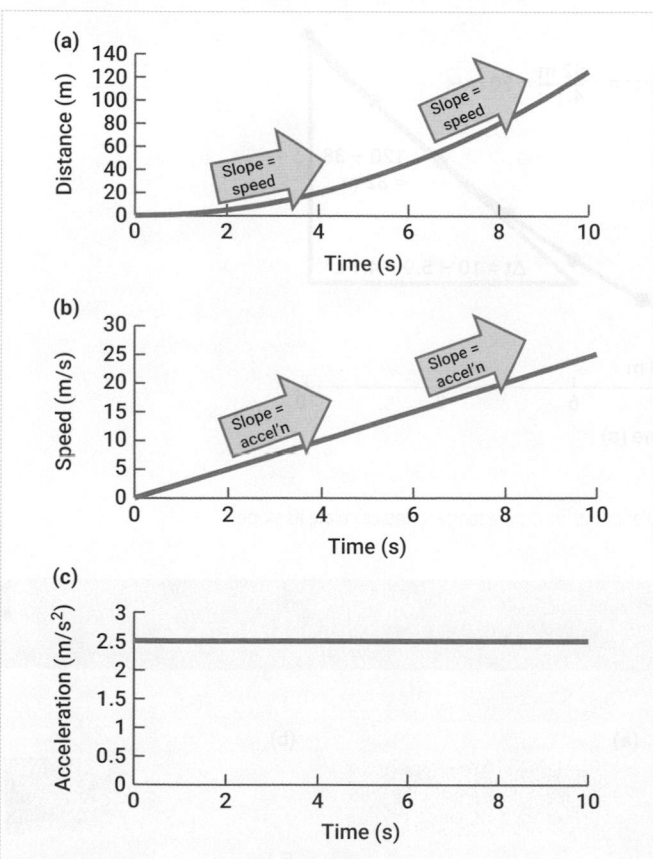

Figure QT9.2 Variable increasing at an increasing rate. Graphical display of the change in distance (a), speed (b), and acceleration (c) with time. Speed is rate of change of distance over time, i.e. the slope in (a) is plotted on the *y*-axis in (b). Acceleration is rate of change of speed over time, i.e. the slope in (b) is plotted on the *y*-axis in (c).

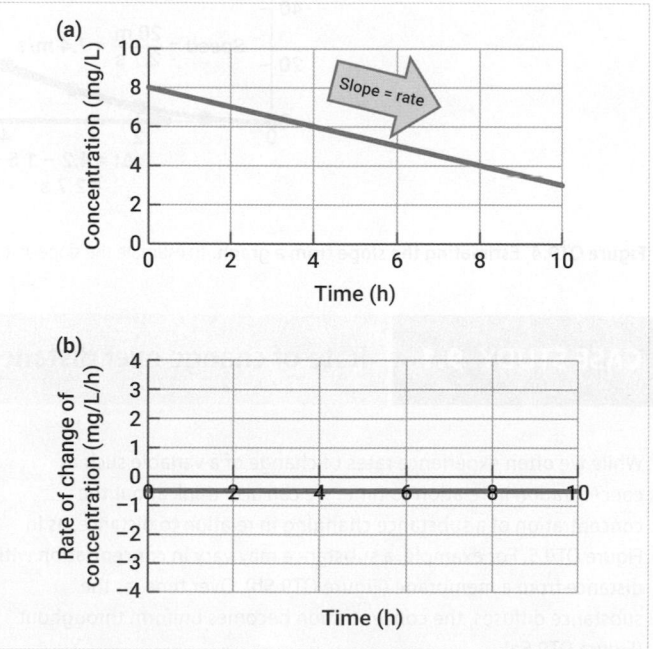

Figure QT9.3 Variable decreasing at a constant rate. Graph of concentration (a) and rate of change of concentration (b) with time. The slope in (a) is plotted on the *y*-axis in (b). The concentration is decreasing at a constant rate.

For example, over a 4-h time period, from 0 to 4 h, the concentration decreased by 2 mg/L so the rate of decrease is:

$$\text{rate of decrease} = \frac{\text{change in concentration}}{\text{difference in time}} = \frac{8-6\,\text{mg/L}}{4\,\text{h}}$$

$$= \frac{2\,\text{mg/L}}{4\,\text{h}} = 0.5\ \text{mgL}^{-1}\text{h}^{-1}$$

We could have chosen any two points on the graph to do this calculation and the resulting slope would be the same. Note that the units have been included in the calculation; refer to Quantitative Toolkit 8 for more information about this.

In the situation in which the cheetah is accelerating—i.e. the speed is increasing, as depicted in Figure QT9.2—if we want to know the speed of the cheetah at 8 s, we could estimate this by drawing a tangent to the curve and then working out the slope of the tangent. A tangent is a straight line that touches the curve at a single point. This is shown in Figure QT9.4: the orange line to the right of the graph shows the tangent to the curve at 8 s, while the blue line to the right of the graph shows the tangent to the curve at 3 s.

Alternatively, we could work out the **average speed** over 10 s by dividing the total distance, 124 m, by the total time, 10 s, to givs

$$\left(\frac{124\,\text{m}}{10\,\text{s}} = \right) 12.4\ \text{m s}^{-1}$$

▶ **Practise applying the concepts introduced in this section by trying Activity 1 at the end of this toolkit.**

It is common to use the symbol Δ (greek capital delta) to indicate the change in a variable. For example, Δx could be used to stand for the change in distance and Δt could be used to stand for the change in time. So the formula for speed would be:

$$\text{speed} = \frac{\Delta x}{\Delta t} = \frac{\text{change in } x}{\text{change in } t}$$

When describing the instantaneous speed (that is, the speed at a given instant), we are thinking about an infinitesimally small change in distance over an infinitesimally small change in time. This is sometimes represented as:

$$\text{speed} = \frac{dx}{dt}$$

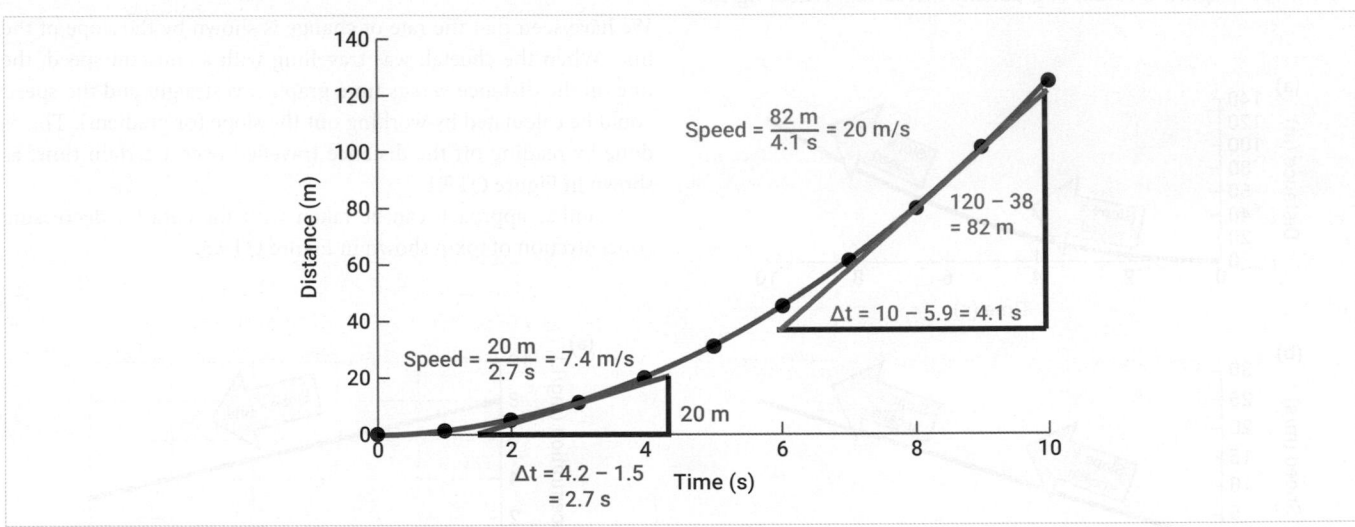

Figure QT9.4 Estimating the slope from a graph. To estimate the slope at a particular point, we draw a tangent and calculate its slope.

CASE STUDY 9.1 | Rate of change over distance

While we often experience rates of change of a variable such as concentration in relation to time, we can also think about the concentration of a substance changing in relation to distance, as in Figure QT9.5. For example, a substance may vary in concentration with distance from a membrane (Figure QT9.5b). Over time, as the substance diffuses, the concentration becomes uniform throughout (Figure QT9.5a).

Figure QT9.5 Rate of change of concentration with distance.

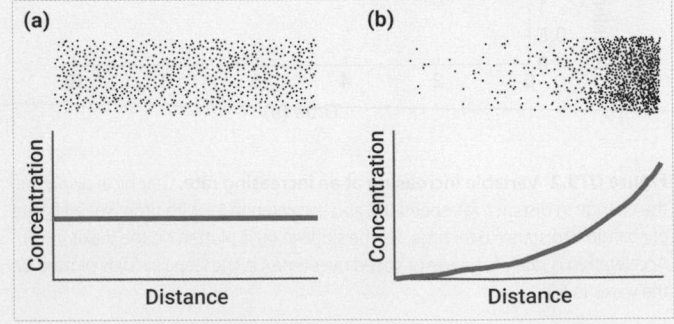

This notation, which you may have seen before, comes from calculus, a mathematical approach to describe the instantaneous rate of change of one variable with another.

9.3 Rates involving exponential growth

Exponential growth can be used to model phenomena such as the replication of DNA, growth of cancer cells, bacteria, a population, or the spread of a disease. These are all situations in which the rate of increase is dependent on what is already there. For example, a larger population has more babies than a smaller one. In these examples the underlying maths is the same, although often the symbols used may be different. The purpose of a quantitative approach is to be able to describe and compare growth rates, and to make predictions. It can also help to identify underlying mechanisms. Let us start by thinking about the simplest example, a bacterium dividing into two. We will then go on to consider more complex scenarios.

If you start with five bacteria and each divides into two, then after one generation there are 10 bacteria (i.e. 5×2). After another generation there are 20 bacteria ($5 \times 2 \times 2$), then 40 ($5 \times 2 \times 2 \times 2$), then 80 ($5 \times 2 \times 2 \times 2 \times 2$), and so on. Table QT9.1 shows the number of bacteria for each generation. Look at this table and notice how the number of bacteria are doubling from one row of the table to the next. The key idea here is that the population size is multiplying with each step, in contrast to linear growth, where the same amount adds on with each step. Exponential growth eventually leads to massive increases over short periods.

Expressing this in symbols, we can write: $N(g) = N_0 \times 2^g$

where

- **g** is the generation number (it is convention to write generation 0 as the starting generation),
- N_0 is the starting number (which is 5 in the example above),
- **N(g)** is the number of bacteria at a particular generation, g (for example N(3) would be the number of bacteria at generation three).

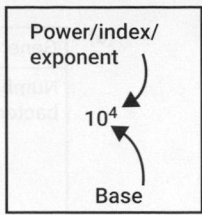

This is where the term 'exponential growth' comes from: the mathematical equations that best model this pattern of rate of change have time, in this case g, as an exponent.

To build up a mental picture of this exponential growth we can display the numbers *diagrammatically*, as shown in Figure QT9.6, and *graphically* using either a linear or logarithmic scale. When plotting these data on a normal linear scale (as in Figure QT9.6b), the numbers rapidly get too large to be able to plot them easily. In this graph, it is very difficult to see the difference between the numbers in the first few generations. To address this issue, we often use a *logarithmic* scale, as in Figure QT9.6c. All this means is that the numbers on the *y*-axis are multiplying by 10 as you go up it rather than them being additive.

An interesting property of exponential growth is that the *rate of increase* (i.e. the growth) also increases exponentially, as seen in Figure QT9.6; Figure QT9.6a shows a diagrammatic representation and Figure QT9.6d shows the rate of change (the growth) increasing exponentially.

Look again at Figure QT9.6. Notice how we can see the exponentially increasing rate when the growth is depicted on a linear scale. When drawn on a logarithmic scale, however, this exponential growth shows a straight line. It is the same data, just displayed on different scales. This straight line on a log plot can be used as a diagnostic for exponential growth. That is, if we have some data and we are not sure if it is following exponential growth, then we plot it on a logarithmic scale: if the data form a straight line, we can infer that the growth was indeed exponential.

▶ **Consolidate the ideas in this section by trying Activities 2, 3, and 4 at the end of this toolkit.**

The time on the horizontal *x*-axis in Figure QT9.6b, QT9.6c, and QT9.6d represents the generation, so a time increase of 1 represents

Table QT9.1 Diagrammatic representation of the calculation of the number of bacteria in each generation

Generation, g	Calculating number of bacteria		Number of bacteria	Change in number of bacteria
0	5	5×2^0	5 (=N_0)	
1	5×2	5×2^1	10	5
2	$5 \times 2 \times 2$	5×2^2	20	10
3	$5 \times 2 \times 2 \times 2$	5×2^3	40	20
4	$5 \times 2 \times 2 \times 2 \times 2$	5×2^4	80	40
5	$5 \times 2 \times 2 \times 2 \times 2 \times 2$	5×2^5	160	80
g		5×2^g	(=N(g))	

This shows how the calculation can be represented using index notation and relates this to a visual representation of the number of bacteria. The final row gives the general relationship.

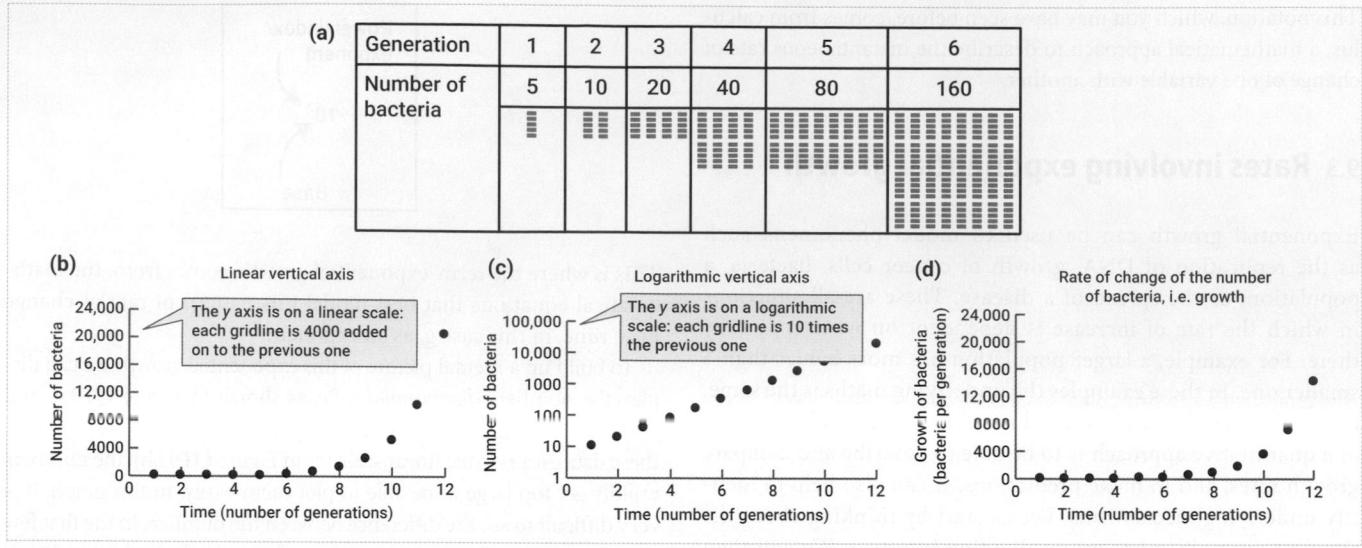

Figure QT9.6 Comparison of linear (b) and logarithmic (c) scales on *y*-axes. The same data (a) for the number of bacteria plotted on both graphs show the increase in cell number at each generation (d).

the time for one generation. We could, however, use minutes, hours, or days—it would just mean scaling the horizontal axis.

In practice, we usually work out the **doubling time** (or **generation time**) from counts of large numbers of bacteria rather than actually timing how long it takes for a single bacterium to divide.

Let us look at an example:

- the initial population, N_0, is 1 million cells
- the population some time later, $N(g)$, is 8 million cells.

There have been three generations, $1 \rightarrow 2 \rightarrow 4 \rightarrow 8$ million cells.

If we know that this happened over a particular length of time, we can calculate the doubling time. So, for example, if we know that this population increase occurred over 5 h, then the doubling time estimate would be:

$$\frac{5\,\text{h}}{3\,\text{generations}} = 1.67\,\text{h per generation}$$

or

$$\frac{300\,\text{min}}{3\,\text{generations}} = 100\,\text{min per generation}$$

 Go to the e-book to watch a video that explores a more complex example where it is not so easy to see how many generations have occurred.

Reproduction number, R

While thinking of growth in terms of generation number and doubling time is useful, it is also possible to consider exponential growth without assuming that the population doubles with each generation. Rather than simply doubling each generation we could say that the population increases R times. This would account for a situation in which, say, two animals mate to produce several offspring, some

of which might die, so that, on average, the second generation has three animals rather than four.

The reproduction number can also be used to represent the spread of infectious diseases, such as COVID-19, where one person might infect several other people. Say, on average, that one person infects six other people: the reproduction number would be 6. When looking at populations we are dealing with averages, so it might be the case that 12 people infect 31 other people. This would be simplified down to one person infecting 2.6 other people ($31 \div 12 = 2.6$) and the reproduction number would be 2.6.

If the population at the start is N_0 and the population size of the first generation is N_1, then:

The population size in generation 1 is

$$N_1 = N_0 \times R$$

The population size in generation 2 is

$$N_1 \times R = N_0 \times R \times R = N_0 \times R^2$$

The population size in generation 3 is

$$N_2 \times R = N_0 \times R \times R \times R = N_0 \times R^3$$

The population size in generation 4 is

$$N_3 \times R = N_0 \times R \times R \times R \times R = N_0 \times R^4$$

In general, the population size in generation g is given by $N_g = N_0 \times R^g$ where

- g is the generation number
- N_g is the population size at a particular generation g
- N_0 is the starting population size.

In infectious disease, R is known as the reproduction number, while in ecology it is called either the 'population multiplication rate' or 'finite rate of increase' and is given the symbol λ.

It is important to remember the underlying maths is the same in all these branches of biology.

When R is greater than 1 the population increases. For example, if R = 1.2, then after 10 generations the population size is just over six times larger than at the start:

$$N_{10} = 1.2^{10} \times N_0 \sim 6.19 \times N_0$$

In contrast, when R is less than 1 the population size decreases. For example, if R = 0.8, then after 10 generations, the population size is approximately 11 per cent of its original size:

$$N_{10} = 0.8^{10} \times N_0 \sim 0.11 \times N_0$$

A useful analogy to this geometric growth rate is compound interest accrued on money in the bank each month or each year.

Continuous growth and e

Another way to describe exponential growth is to use a continuous growth model based on e, Euler's number. You may have seen the equation:

$$N = N_0 e^{\mu t}$$

where

- N = population size (i.e. number of bacteria, cells, or animals)
- N_0 = starting population size
- e = Euler's number ~ 2.71818…
- μ = a constant which is characteristic of the organism under certain conditions
- t = time.

How do we arrive at the exponential growth equation, $N_0 e^{\mu t}$? We can start from the idea that exponential growth occurs when the rate of change of the population size is proportional to the population size. Writing that in symbols:

$$\text{Rate of change of } N = \frac{\Delta N}{\Delta t} \propto N$$

where

- N = population size
- $\dfrac{\Delta N}{\Delta t}$ is the growth
- $\propto$ is the symbol for 'is proportional to'.

Different cells, bacteria, or organisms may double over different times. For example, *E. coli* can double in 20 min, whereas *Vibrio cholerae* (the bacterium responsible for cholera) takes 60 min. This is reflected in the rate constant μ and is characteristic of an organism under certain conditions. The rate of change of N is then:

$$\frac{\Delta N}{\Delta t} = \mu \times N$$

In cell biology μ is called the **specific growth rate** and has units of s⁻¹ or min⁻¹ or h⁻¹ or d⁻¹. If we plot a graph of the rate of change of N, the growth, $\left(\dfrac{\Delta N}{\Delta t}\right)$ on the y-axis and number of bacteria, N, on the x-axis, then we get a straight line with a slope equal to the value of μ.

Essentially what this means is that, when plotting a graph of the number of cells against time, the slope of the line gets steeper as the number of bacteria increases. In Figure QT9.7b we can see that when the population size is about 1000, the slope is shallow (shown by the blue triangle). When the population size is larger, say 12,000, the slope is steeper (shown by the orange triangle). This occurs with bacteria, cells, animals, and infections. The function $e^{\mu t}$ has the unique property that it reflects this characteristic of exponential growth.

Figure QT9.7 shows a comparison between the graph of the exponential growth equation, $N = N_0 e^{\mu t}$ (Figure QT9.7b) and the growth model that we have previously looked at $N = N_0 R^g$ (Figure QT9.7a). They are just two different ways of describing the same underlying biology: one is continuous and the other goes in jumps because you cannot have half a cell or half a bacterium. We often use the continuous model (Figure QT9.7b) when we have very large numbers of cells or bacteria.

A note about symbols: in cell biology we often use the symbol μ, the specific growth rate. The exponential equation is then written:

$$N = N_0 e^{\mu t}$$

In population biology we often use the symbol r, which is called the **intrinsic rate of natural increase** and also has units of s⁻¹, min⁻¹, h⁻¹, or d⁻¹, and the exponential equation is almost the same, except the μ has changed to r:

$$N = N_0 e^{rt}$$

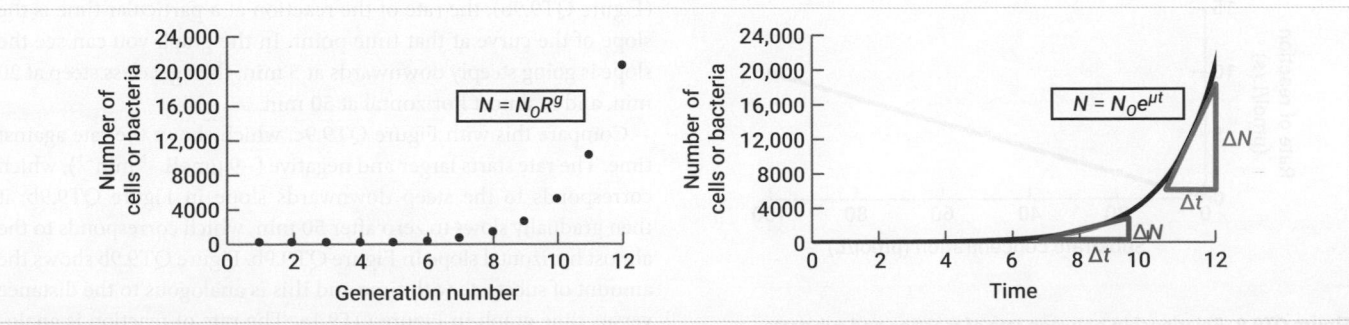

Figure QT9.7 A comparison of discrete and continuous growth models. Both models are shown with a linear scale on the vertical y-axis and the shapes of the graphs are the same, although the equations have a different form.

We have explored several aspects of exponential growth in this toolkit. If you would like to do more in this area, go to the e-book to watch the quantitative skills videos that will help you find out:

- how to estimate the specific growth rate constant, μ, from some data (the same approach is used for estimating r, the intrinsic rate of natural increase);

- how to estimate the doubling time and how the doubling time relates to the specific growth rate μ (or the intrinsic rate of natural increase, r).

9.4 **Rates involving exponential decay**

Now let us look at a scenario in which a variable is *decreasing* exponentially, as in the case of an enzyme-catalysed reaction. Consider a simple example where a substrate is acted upon by an enzyme and is converted to a product. When the substrate molecules are first added to the enzymes there are a lot of substrate molecules so there are many opportunities for collisions to occur and reactions to take place, making the rate of reaction relatively fast. However, as substrate molecules are used up, the concentration of substrate decreases, there are fewer collisions, and the rate of reaction becomes slower.

In this case, we say that the rate of reaction is proportional to the substrate concentration: this is called a first order reaction. The rate of reaction and substrate concentration are related by the rate constant, k, which will have a particular value for that substrate, enzyme, and experimental conditions:

$$\frac{\text{change in substrate concentration}}{\text{time period}}$$

$$= \text{rate of reaction} = k \times \text{substrate concentration}$$

The graph of the initial rate of reaction on the y-axis and substrate concentration on the x-axis will be a straight line, as shown in Figure QT9.8. The slope of the line is k, the rate constant.

Now let us look at how the rate of a chemical reaction can be described graphically and think about how the graph of the amount of substrate versus time compares with the graph

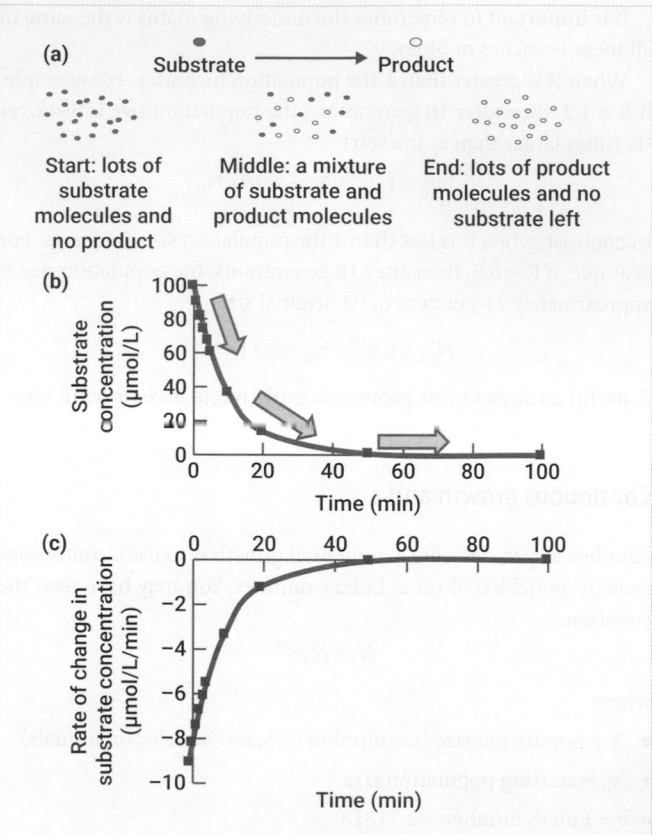

Figure QT9.9 **Substrate concentration and rate of reaction vary with time in a first order reaction.** In this example, the substrate concentration is high at the start. As it is converted to product, however, the substrate concentration decreases (b). The rate of reaction is shown in (c). It starts negative, because the substrate concentration is decreasing, then approaches zero as the rate of reaction slows.

of the rate of reaction versus time. This is an example where the rate is changing.

The rate of a chemical reaction (sometimes called the velocity or speed of reaction) can be quantified by measuring the concentration of substrate or product over time. Figure QT9.9a shows an example of a chemical reaction during which a substrate molecule undergoes an enzyme-catalysed reaction to form a single product. In a plot of the concentration of substrate against time (Figure QT9.9b), the rate of the reaction at a particular time is the slope of the curve at that time point. In the graph you can see the slope is going steeply downwards at 5 min, then gets less steep at 20 min, and is almost horizontal at 50 min.

Compare this with Figure QT9.9c, which shows the rate against time. The rate starts larger and negative ($-9\ \mu\text{molL}^{-1}\ \text{min}^{-1}$), which corresponds to the steep downwards slope in Figure QT9.9b; it then gradually slows to zero after 50 min, which corresponds to the almost horizontal slope in Figure QT9.9b. Figure QT9.9b shows the amount of substrate with time and this is analogous to the distance versus time graph in Figure QT9.1a. The rate of reaction is analogous to the speed, and so Figure QT9.9b is analogous to the speed versus time graph in Figure QT9.1b.

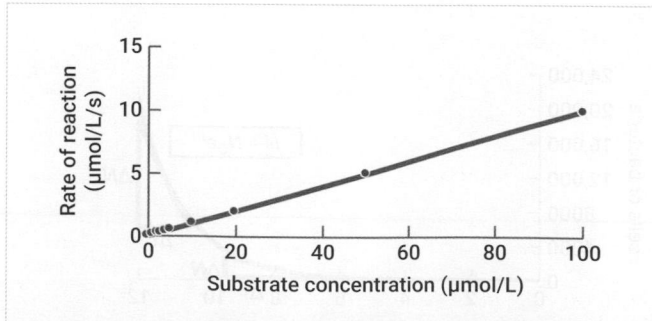

Figure QT9.8 **Relationship between rate of reaction and substrate concentration.** In this example the slope of the line, k, is 0.1 s^{-1}. The rate of reaction is proportional to the substrate concentration.

In Section 9.3 we saw that the exponential growth equation applies whenever we have a situation in which the rate of growth is proportional to the size of the population. Here, we have a situation in which the rate of decay is proportional to the concentration of substrate and so the exponential decay equation applies:

$$C = C_0 e^{-kt}$$

where

- C = concentration at any time
- C_0 = the concentration at the starting time (when $t = 0$)
- k = rate constant for the particular reaction under certain experimental conditions
- t = time (usually in s but sometimes min or h).

The exponential decay equation has a negative exponent, indicating that the concentration is decreasing.

This scenario is relatively straightforward if the reaction proceeds only in the forward direction. However, if the reaction is reversible, the reaction will proceed in the forward direction when only the substrate is present, but it will start to proceed in the backward direction too as soon as product is produced. In this case an **equilibrium** will be reached—the point at which there is no net change in concentration of substrate or product, even though the forward and backward reactions are still occurring. At equilibrium the forward and backward reactions are occurring at the same rate so the amount of product does not change and the overall rate—the net rate—is zero.

Biochemists often measure the **initial rate of reaction** so that they can ignore the problems arising from either running out of substrate or having the product being converted back to substrate. In this case, the rate is calculated from the slope of the line in the first few minutes of the reaction.

9.5 Logistic growth

A logistic growth model is used to describe growth when resources are limiting—that is, where the rate of growth slows as the population gets larger. Looking at the population size in Figure QT9.10a, we can see that when the population size is small, the rate of growth increases. At first it appears exponential, but after about 30 min the rate of growth slows until the population size levels off. (If we compare Figure QT9.10a with QT9.10b, we see how the rate of growth increases to a maximum at about 30 min then falls to zero.) The population size when there is no further growth is called the **carrying capacity**.

If we plot the same data on a logarithmic scale, as in Figure QT9.10c, we can see that the first 20 min or so follows a straight line on the log plot, which indicates exponential growth.

The mathematical equation that models logistic growth is:

$$N(t) = \frac{K}{1 + e^{-r(t-h)}}$$

where

- $N(t)$ is the population size at any time, t
- K = carrying capacity or the equilibrium level, the maximum population size

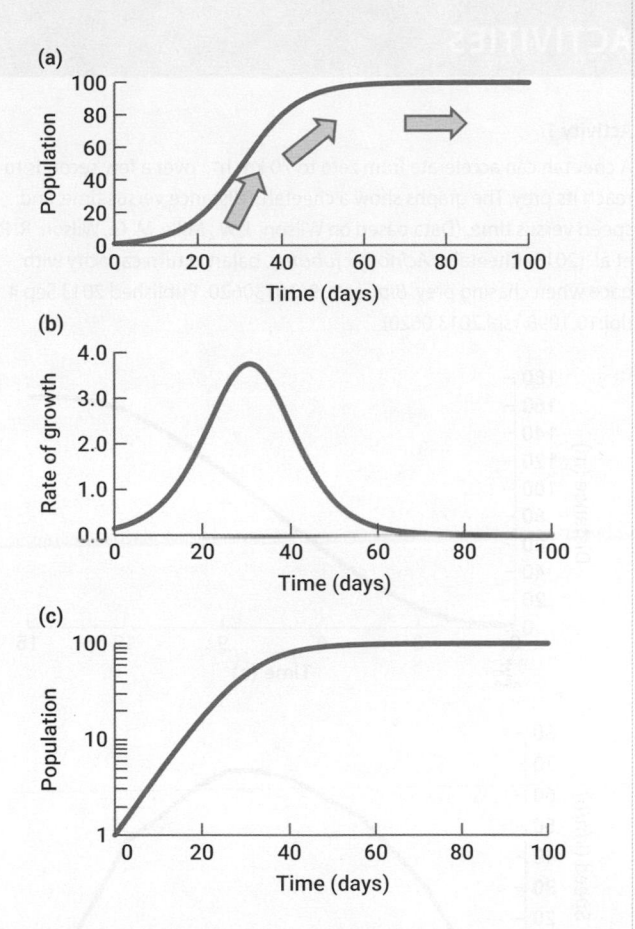

Figure QT9.10 Logistic growth. (a) The population grows rapidly at first, shown by the increasing slope of the line. It is growing fastest (steepest slope) at approximately 30 min, then grows more slowly (less steep) from 30 min until it levels off at just over 60 min. (b) The rate of growth increases to a maximum at about 30 min after which it slows and decreases to zero. (c) The data in panel (a) plotted on a logarithmic scale on the vertical y-axis. In the early phase, from 0 to 30 days, the log plot shows a straight line that indicates the exponential nature of the growth of the population.

- r = rate constant for growth
- t = time
- h = the time when the population size reaches half of the carrying capacity, K.

 To find out more about this topic, go to the e-book to watch a video that explores the logistic growth equation in more detail.

Many other more complex models can be used to describe biological processes—we have only just scratched the surface here to introduce some of the key ideas. It is also common now to use computational models and exploit the potential of computing power to describe more complex systems.

ACTIVITIES

Activity 1

A cheetah can accelerate from zero to 70 km h⁻¹ over a few seconds to reach its prey. The graphs show a cheetah's distance versus time, and speed versus time. (Data based on Wilson, J. W., Mills, M. G., Wilson, R. P., et al. (2013) Cheetahs, *Acinonyx jubatus*, balance turn capacity with pace when chasing prey. *Biol. Lett.* **9**: 20130620. Published 2013 Sep 4. doi:10.1098/rsbl.2013.0620)

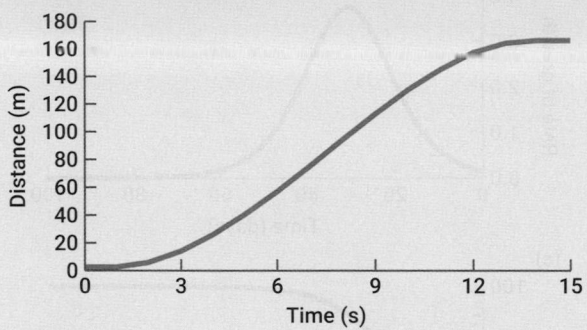

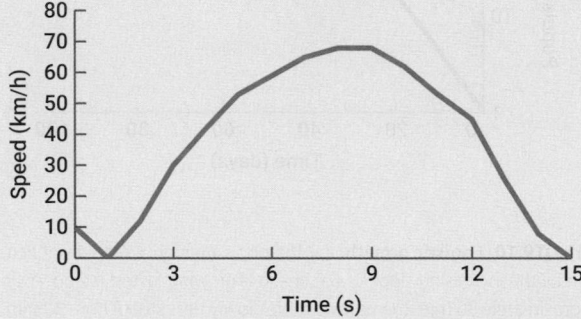

(a) At what time is the cheetah travelling at approximately 68 km h⁻¹?

(b) Use the distance versus time graph to estimate the speed (in m s⁻¹) between 6 and 9 s and compare your answer to the speed (which is given in km h⁻¹) in the speed versus time graph.

(c) At what time is the cheetah accelerating the most?

(d) At what time is the cheetah not accelerating or decelerating?

▶ **Refer to Section 9.1, Section 9.2, and Section 9.3.**

Activity 2

The graph shows data for the population size plotted against time.

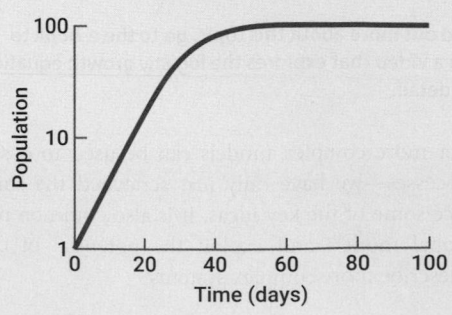

(a) Is the scale on the *y*-axis linear or logarithmic?

(b) Over what timescale is the growth exponential?

(c) Do the data follow a wholly exponential or logistic model?

▶ **Refer to Section 9.4.**

Activity 3

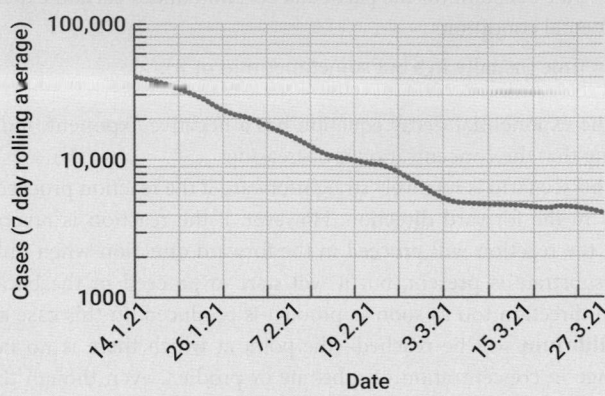

The graph shows the number of COVID-19 cases during the first few months of 2020. (Data from Public Health England (2021) UK Coronavirus Dashboard https://coronavirus.data.gov.uk/details/cases.) During which period(s) is the fall exponential?

▶ **Refer to Section 9.4.**

Activity 4

The following graphs show the rolling 7-day average leading up to the date indicated for the number of new COVID-19 cases and deaths (within 28 days of a positive COVID-19 test) per day in England from September 2020 to March 2021.

(a) What scale is used on the *y*-axis for cases?

(b) How would you describe the change in the number of cases between 12.9.2020 and 10.10.2020?

(c) How would you describe the change in the number of deaths over the period between 2.1.2021 and 13.3.2021?

(d) Calculate an estimate for the percentage drop in deaths per week during the period from 1.2.2021 to 13.3.2021.

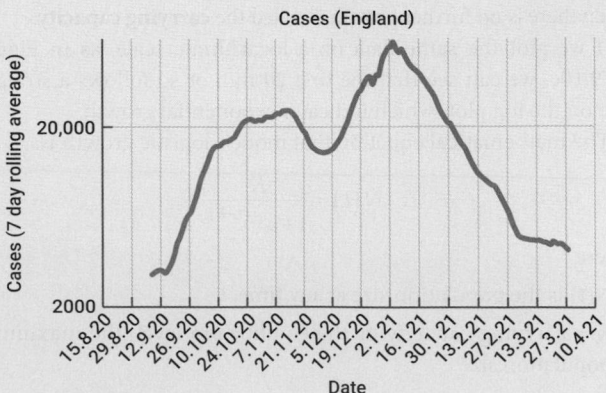

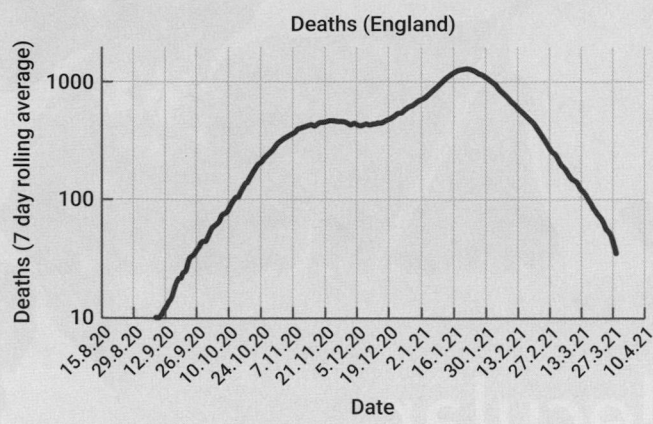

Deaths (England)

(Data from Public Health England (2021) UK Coronavirus Dashboard https://coronavirus.data.gov.uk/details/cases.)

▶ Refer to Section 9.4.

ANSWERS

Activity 1

(a) 8 s.

(b) At 6 s the distance is ~55 m; at 9 s the distance is ~110 m (your values may be slightly different)

$$\text{speed} = \frac{\text{change in distance}}{\text{change in time}} = \frac{110 - 55\,\text{m}}{9 - 6\,\text{s}} = \frac{55\,\text{m}}{3\,\text{s}}$$

We need the answer in km h^{-1} so convert m to km and s to h:
55 m = 0.055 km
3 s = 3 ÷ 60 ÷ 60 = 0.0008333 h

$$\text{speed} = \frac{0.55\,\text{km}}{0.00083333\,\text{h}} = 66\,\text{km h}^{-1}$$

From the speed versus time graph, the speed at 6 s is about 60 km h^{-1} and at 9 s the speed is about 68 km h^{-1}, so the estimated value of 66 km h^{-1} over that range is consistent. Your value for the estimated speed may be slightly different, but it should be within the range of 60–68 km h^{-1}.

(c) The slope of the speed versus time graph indicates the acceleration—it is steepest at about 2 s so the cheetah is accelerating the most at 2 s.

(d) Acceleration is zero when the slope of the speed versus time graph is horizontal (i.e. the speed is not changing) which is at 8–9 s.

Activity 2

(a) Logarithmic.

(b) From 0 to about 25 days (where the data follows a straight line on a logarithmic scale).

(c) From 0 to 25 days it appears exponential, but after 25 days it is clearly being limited and is therefore logistic.

Activity 3

This is plotted on a logarithmic scale for the cases so a straight line shows exponential decay. The period from the start of the graph in early January until approximately February 10 shows exponential decay. Then the last week of February also shows exponential decay. Finally, the part from early March to the end of the data also shows exponential decay.

Activity 4

(a) Logarithmic.

(b) Exponential growth.

(c) Exponential growth from 2.1.21 to 16.1.21 and then exponential decay from 1.2.21 to 20.3.21.

(d) 1.2.21 deaths ~1000
13.3.21 deaths ~100
Decrease = 900
% decrease = 900/1000 = 90%.

MODULE ONE

Life at the Molecular Level

1 Building Blocks: Molecules and Macromolecules

2 Energy: Powering Biochemical Processes

3 Information: Genes and Genomes

4 Mendelian Genetics

5 Reading the Genome: Gene Expression and Protein Synthesis

6 Proteins and Proteomes

7 Metabolism: Energy Capture and Release from Food

8 Molecular Tools and Techniques

MODULE ONE

It is now over 20 years since the draft of the human genome was first published. The intervening years have seen our understanding of life at the molecular level—and our *application* of that understanding—progress hugely. We now understand more fully than ever before how the information in our genomes is used to help organisms grow, survive, and thrive; we understand more about errors in our genome and how they can lead to disease. More than this though, we are beginning to use what we know about these molecular processes to treat and even prevent disease in ways that were previously unimagined. Remarkable experimental approaches, such as gene editing (which we describe in Chapter 8), open up intriguing new possibilities for improving health and wellbeing.

Alongside these strides forward, we are also understanding just how much of life at the molecular level remains constant across the diversity of life. The molecules that maintain life from generation to generation—for example, the nucleic acids that make up the genomes of all living things—have persisted for millennia in the form we see today. It is quite astonishing that such a variety of life continues to be sustained by such conserved molecules and processes.

There is one fundamental prerequisite for life and its continuation: energy. Every living organism must have ready access to a source of energy to survive. For some organisms, including plants and some bacteria, this source of energy is sunlight; however, many other organisms survive by consuming those organisms who have harvested energy from light.

Dependencies such as these give rise to food chains, through which the food of almost any animal can be traced back to plants and sunlight. A food chain consists of producers, consumers, and decomposers. Producers, such as plants, use sunlight and nutrients from the soil to create molecules like starch and sugars. Consumers, such as animals, are unable to harness energy from sunlight in this way and must instead consume producers (or other organisms that themselves consume producers): they obtain energy by metabolizing the starch and sugars the producers contain.

Decomposers feed on dead animals and break down the compounds present in their bodies into simple nutrients that are returned to the soil. From here, these nutrients can be taken up by plants . . . and so the food chain continues.

We begin this module by exploring the concepts from life at the chemical level—the world of atoms and molecules—that underpin the existence of cells, organs, tissues, and organisms. We then go on to explore the concept of energy in Chapter 2 before discovering how organisms liberate energy from food to support their ongoing survival in Chapter 7.

Elsewhere, we see how biological information is stored, retrieved, and used—and how this information is carried from one generation to the next to ensure the continuation of life.

Image: Ultimately, life on Earth is powered by sunlight, harnessed by plants—such as these maple and beech trees. *Source:* Nikada/iStock.

Building Blocks

Molecules and Macromolecules

By the end of this chapter you should be able to:

- Describe atomic structure and covalent bonds in terms of protons, neutrons, and electrons.
- List the major elements that make up biological molecules and explain their bonding properties.
- Compare covalent bonds with different types of non-covalent bonds.
- Explain the nature of biochemical reactions in terms of breaking and making chemical bonds.
- Explain how the chemical and physical properties of water give it a crucial role in life processes.
- Explain the measurement of acidity by the pH scale, distinguish between strong and weak acids, and explain the principle of buffering.
- Discuss the structures and give examples of the functions of the four main classes of biological macromolecules: carbohydrates, lipids, nucleic acids, and proteins.

Chapter contents

Introduction 149

1.1 The elements that make up biological molecules 150

1.2 Formation and properties of biological molecules 151

1.3 Non-covalent bonds allow interactions between molecules 161

1.4 Chemical reactions involve breaking and making bonds 164

1.5 Water plays a key role in supporting life 167

1.6 Acids and bases: molecules that release or bind hydrogen ions 169

1.7 Biological macromolecules: structure and function 172

1.8 Interactions between macromolecules can be specific, flexible, and transient, allowing biological processes to occur 180

Watch the key concepts video in the e-book to prepare yourself for studying this chapter.

Introduction

Despite the complexity of living things and life processes, all are based on chemical compounds and chemical reactions. Thus, to study biology, you will need a basic understanding of some chemical principles, which are covered briefly in this chapter. Chemicals are made up of minute particles called atoms, which are linked together in different combinations to form molecules. Here we explore the structure of atoms, how chemical bonds link them together, and how breaking bonds and making new ones transforms molecules in chemical reactions. We look at the chemical elements that are commonly found in biological molecules and explain some of their important bonding properties. Key properties of molecules, including their size and shape and the reactions they undergo, are determined by the atoms that make them up and the ways these are bonded together into functional groups, which are also explained here. We look the structure and function of some biological molecules, starting with the properties of water, a remarkable small molecule without which life would not be possible. The chapter concludes with a brief review of the main classes of molecules that are found in cells as an introduction to the more detailed coverage of these elsewhere in the book.

1.1 The elements that make up biological molecules

All the substances around us that make up both living and non-living things are comprised of chemical **elements**. An element is a single substance that cannot be broken down by chemical

processes into anything simpler. **Atoms** are minute particles that make up elements, and the characteristics of its atoms determine the identity and properties of each element.

Before exploring the structure of atoms in the next section, we will take a quick look at the periodic table. Chemistry textbooks commonly include this table, which summarizes information about all known elements. The full periodic table is shown in Figure 1.1. Fortunately, there is no need for aspiring biologists to study the entire periodic table, as only a handful of elements (highlighted in Figure 1.1) are commonly found in biological molecules. Nevertheless, before you learn about the nature and behaviour of biological molecules it is essential to understand some basic principles that determine the behaviour of atoms and recall specific information about a few elements. As we discuss in more detail in the next section, atoms can be linked together by chemical bonds. Chemical **compounds** are formed when atoms of more than one element are bonded together.

Atomic structure

The basic structure of an atom consists of a central **nucleus**, containing positive protons and uncharged neutrons, with negatively charged **electrons** around it, as illustrated in Figure 1.2a. The exact size of an atom depends on the element, but atomic radii are of the order of trillionths of a metre (10^{-12} m).

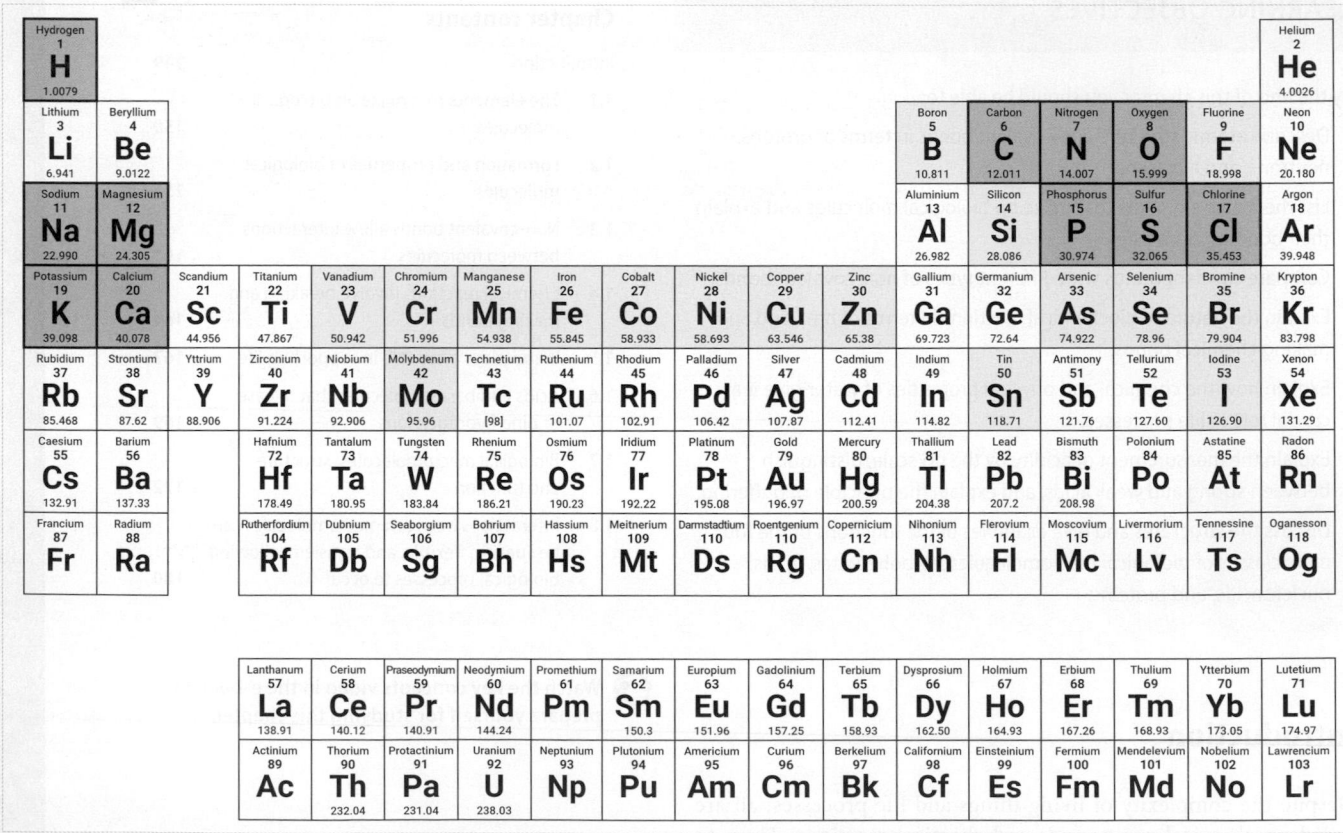

Figure 1.1 Periodic table illustrating the elements essential for life. Vertical columns in the table are known as 'groups' and horizontal rows are 'periods'. The 28 elements essential for animal life are indicated by coloured squares; trace elements are shown in yellow and those present in larger quantities are shown in green. The six most abundant elements in the human body are carbon, hydrogen, oxygen, nitrogen, phosphorus, and calcium, accounting for almost 99 per cent of the mass of an adult human.

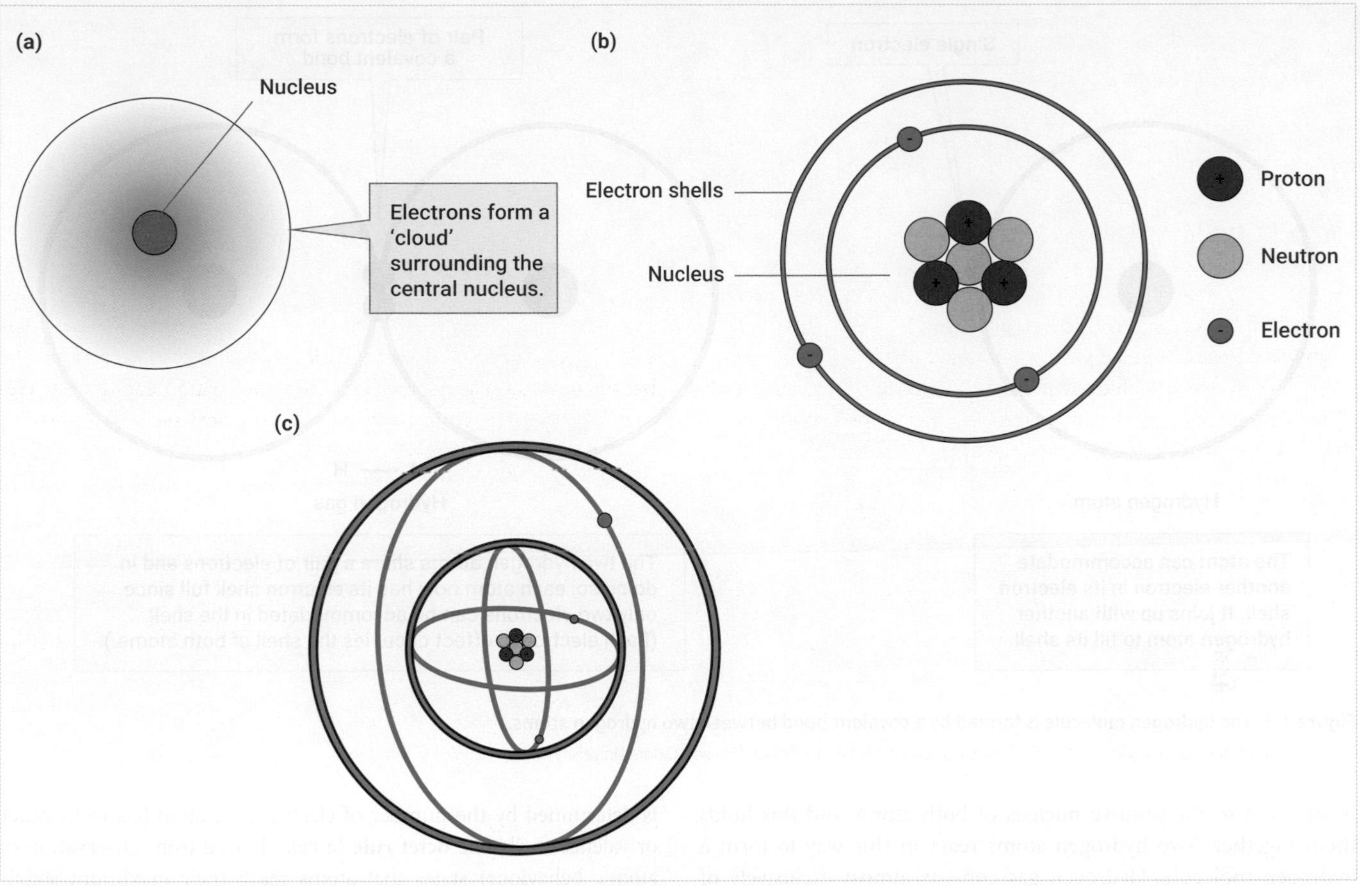

Figure 1.2 Structure of an atom of the element lithium (Li). (a) The nucleus containing protons and neutrons forms a tightly packed structure in the centre, while the electrons form a diffuse cloud around it. (b) The electrons occupy different energy levels or 'shells' around the nucleus. (c) The electron shells can be envisaged as spheres with electrons moving within them.

Source: (a,b) Snape, A. & Papachristodoulou, D. (2018). *Biochemistry and Molecular Biology* (6th ed.). Oxford University Press. (c) Science Photo Library.

The number of protons in the nucleus determines which element the atom belongs to, and is matched by the number of electrons, so that the atom has no overall charge. As described in Section 1.3, atoms can gain or lose electrons, thus becoming negatively or positively charged ions of the same element.

The electrons of an atom occupy different energy levels, or 'shells'. (The concept of energy is explained further in Chapter 2). Although these shells are often depicted as concentric rings, as in Figure 1.2b, the electrons are actually in constant motion within a restricted three-dimensional space around the nucleus, as illustrated in Figure 1.2c.

PAUSE AND THINK

An atom of carbon (chemical symbol C) contains six protons and six neutrons, while an atom of sodium (chemical symbol Na) contains 11 protons and 12 neutrons. How many electrons are there in (a) an atom of carbon (b) an atom of sodium?

Answer:

(a) Six.

(b) 11. The number of electrons is equal to the number of protons.

Check your understanding of the concepts covered in this section by answering the questions in the e-book.

1.2 Formation and properties of biological molecules

Despite containing a relatively small number of different elements, living things contain a huge variety of **molecules**, which are made up of atoms that are bonded together in different numbers and different combinations. Molecules can be formed by two types of chemical bonds that occur between atoms: covalent bonds and ionic bonds. In this section we focus on covalent bonds, while ionic bonds are covered in Section 1.3, as they play a greater part in interactions between biological molecules than in their formation.

Covalent bonds between atoms are formed by electron pair sharing

A covalent bond is formed by two atoms sharing a pair of electrons, a simple example being the formation of a hydrogen molecule from two hydrogen atoms, as shown in Figure 1.3. Each electron

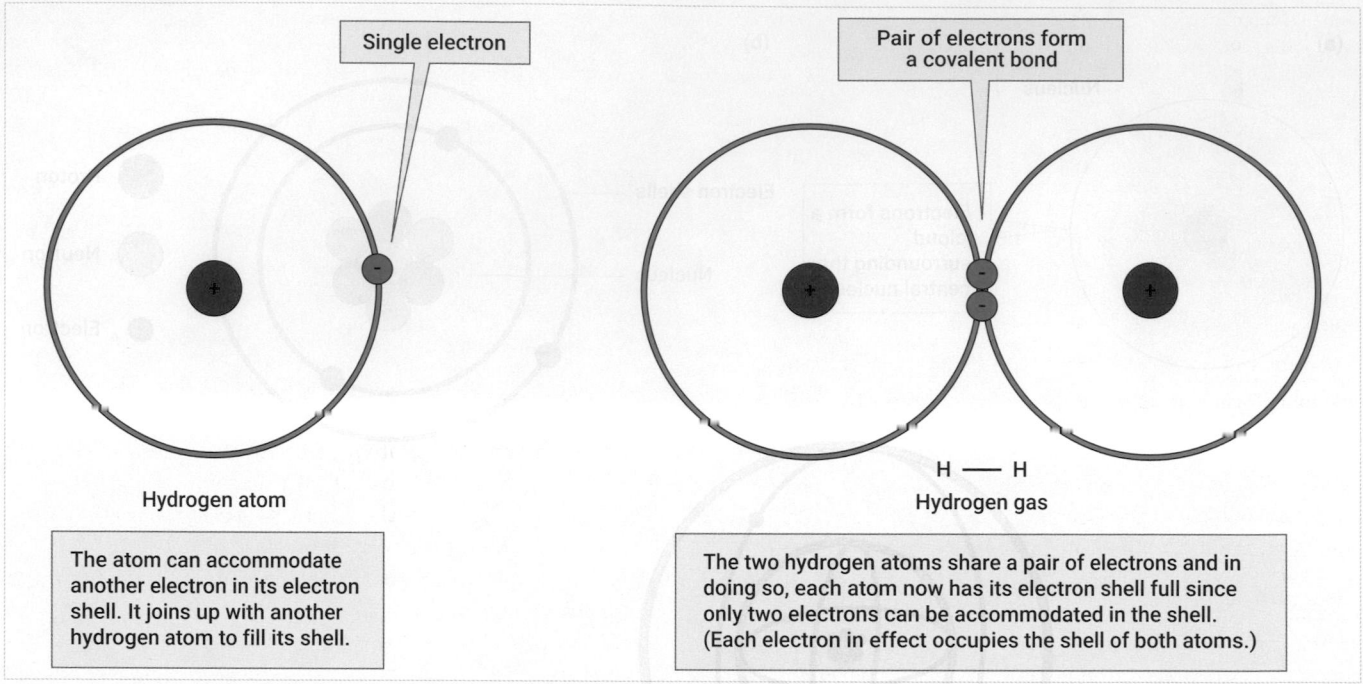

Figure 1.3 The hydrogen molecule is formed by a covalent bond between two hydrogen atoms.

Source: Snape, A. & Papachristodoulou, D. (2018). *Biochemistry and Molecular Biology* (6th ed.). Oxford University Press.

is attracted to the positive nucleus of both atoms and this holds them together. Two hydrogen atoms react in this way to form a hydrogen molecule. Hydrogen gas consists almost exclusively of hydrogen molecules, which is why it is written as H_2; the proportion of hydrogen atoms is negligible. Similarly, oxygen gas (written O_2) consists of oxygen molecules, each comprising two oxygen atoms.

You will note from this that a molecule can consist of two or more atoms of the *same* element bonded together. Many molecules, however, contain atoms of more than one element bonded together, such as carbon dioxide (CO_2) and ammonia (NH_3).

The valence of an atom determines how many covalent bonds it forms

The arrangement of elements in the periodic table places them in groups (vertical columns in the table) that have properties in common. One such property is **valence** or valency, which is the number of covalent bonds an atom of the element can form. Valence is determined by the number of electrons the atom has in its outer or valence shell. The **octet rule** (a rule derived from observation of atomic behaviour) states that atoms reach their maximum stability when the number of electrons in the valence shell is eight. They can reach this number either through sharing electrons with other atoms of the same or different elements (i.e. by covalent bonding), or through gaining or losing electrons to form charged ions as described in Section 1.3. The elements that already have the maximum number of electrons in their valence shell (helium, argon, neon, etc., the so-called 'noble' gases) are extremely non-reactive and rarely bond or ionize. Note that hydrogen, illustrated in Figure 1.3, is somewhat exceptional in that its valence shell holds a maximum of two electrons.

Carbon is a key element in the structure and function of living things. It has four electrons in its valence shell and thus needs to form four covalent bonds to increase the number of electrons to eight and achieve stability. We can succinctly state this fact by saying that carbon 'has a valence of four'. The bonding of a carbon atom is shown in Figure 1.4.

Figure 1.4 Carbon can form four covalent bonds by sharing four electron pairs. (a) Carbon with the four bonds as conventionally drawn for convenience. (b) A more realistic depiction showing the three-dimensional arrangement of the four bonds

Source: Snape, A. & Papachristodoulou, D. (2018). *Biochemistry and Molecular Biology* (6th ed.). Oxford University Press.

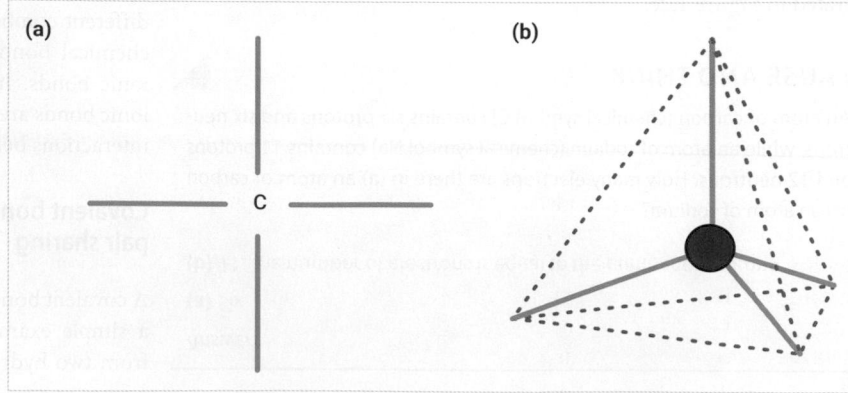

Table 1.1 Electron complements and valences of elements that are commonly found in biological molecules

Element	Electron shell			Number of covalent bonds formed (valence)
	I	II	III	
Hydrogen	1			1
Carbon	2	4		4
Nitrogen	2	5		3
Oxygen	2	6		2
Phosphorus	2	8	5	3 or 5
Sulphur	2	8	6	2, 4, or 6

You will find it useful to memorize the valences of elements that are commonly found in biological molecules (only a very small number of the 118 elements of the periodic table) so that you know how many covalent bonds they need to form to make up biological molecules. The valences of the elements most commonly found in living things—carbon, hydrogen, oxygen, and nitrogen—are shown in Table 1.1.

Sulphur and phosphorus, also shown in Table 1.1, are less common but of crucial importance for life. The valences for these two elements shown in the table need a bit of explanation, as for reasons that will not be explained here sulphur and phosphorus atoms can each achieve a full valence shell in more than one way. Sulphur can achieve a full shell by sharing two, four, or six electrons, while phosphorus can share three or five. In biological molecules sulphur most often has a valence of two, so forms two covalent bonds, and phosphorus has a valence of five and forms five bonds.

Molecules based on carbon are key to life and are termed organic molecules

Carbon atoms can readily form covalent bonds with each other, enabling them to form chains, branched molecules, and even cyclic structures, as shown in Figure 1.5. Carbon can also bond covalently to atoms of other elements, such as hydrogen, oxygen, and nitrogen, which give molecules different chemical properties. The versatility of carbon and its ability to form the framework of molecules that are diverse in size, structure, and chemical function means that it is a key element in living things. Hence, molecules that are built on carbon frameworks are called **organic** molecules. The opposite term, used to describe a molecule that does not contain carbon, is inorganic.

Molecules have characteristic three-dimensional shapes

Chemical structures are often depicted in two dimensions, flat on the page, but it is important to realize that they are actually three-dimensional, with characteristic spatial arrangements of atoms and bonds leading to particular shapes. As you will see later, especially in our discussion of proteins (Chapter 6) the shapes of biological molecules are crucial for determining the interactions that can occur between them. As an example of the reality of molecular shapes, the structure of carbon bonded to four other atoms is shown in Figure 1.4. Figure 1.4a shows the molecule as often depicted for simplicity, while Figure 1.4b shows the true shape of the molecule, a tetrahedron or triangular pyramid.

Differing electronegativities can give bonds polarity

Covalent bonds involve two atoms sharing a pair of electrons, which then occupy the space between the two nuclei, as shown in Figure 1.6. However, atoms of different elements vary in the strength with which they attract electrons, so the electrons in a covalent bond are not necessarily positioned centrally between them. Recall that the electrons are not stationary but move around within a defined space. Figure 1.6a shows that in covalent bonds between oxygen and hydrogen, the space occupied by the shared electron pairs places them closer to oxygen than to hydrogen. This is because oxygen is more **electronegative** than hydrogen, i.e. it is more attractive to electrons.

Figure 1.6b shows the structure of water, and illustrates a further point arising from the differing electronegativities of oxygen and hydrogen. The unequal sharing of electron pairs gives the oxygen atom a slight negative charge, and the two hydrogen atoms a slight positive charge. The covalent bonds formed between atoms of different electronegativities are called **polar** covalent bonds, and as charge is distributed unevenly across the water molecule it is called a polar molecule. The polarity of water and of certain other molecules and chemical groups is of great importance in determining their properties and interactions, as explained later in this chapter.

Table 1.2 shows the electronegativity values of some elements, including those commonly found in biological molecules. The difference in electronegativities of oxygen and hydrogen is clear, and the table also shows that carbon and hydrogen have similar electronegativities, so molecules consisting only of carbons and hydrogens are non-polar.

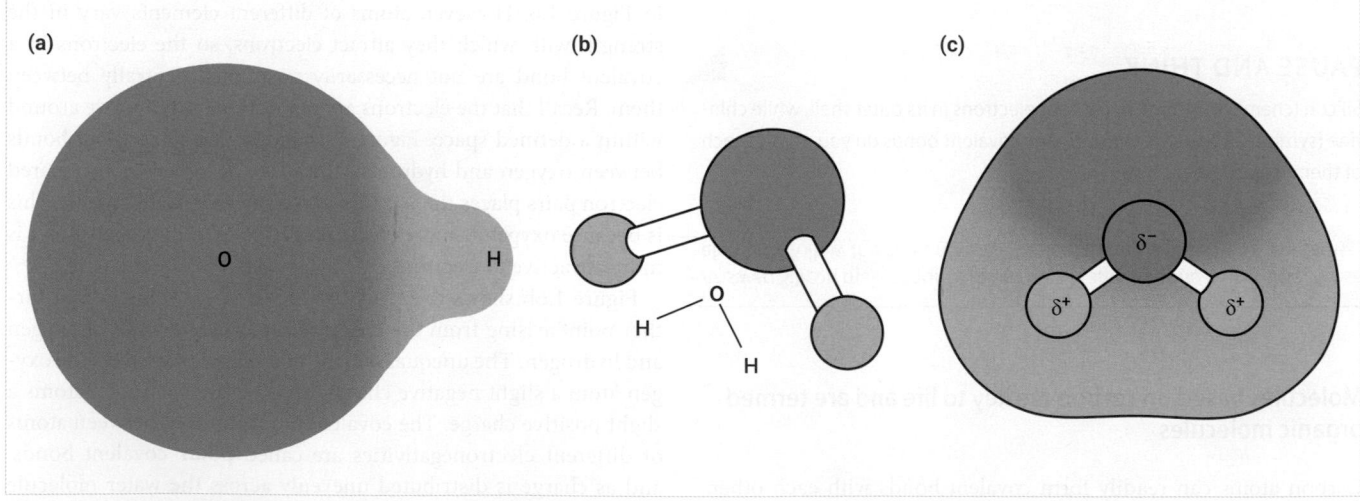

Figure 1.5 Carbon framework. A selection of organic molecules showing how carbon can form the framework for linear, branched, and cyclic structures, by forming covalent bonds between carbon atoms and with atoms of other elements.

Figure 1.6 Covalent bonds in water. (a) The relative distribution of electrons in a covalent bond between oxygen and hydrogen. The greater electronegativity of oxygen compared to hydrogen draws the electrons towards it, forming a polar bond. (b) and (c) The polarity of bonds in the water molecule gives the molecule overall polarity. The symbols δ^- and δ^+ indicate regions of the molecule that are slightly negatively and slightly positively charged.

Source: (b) Crowe, J. & Bradshaw, T. (2021). *Chemistry for the Biosciences: The Essential Concepts* (4th ed.). Oxford University Press.

Table 1.2 Electronegativity values for a set of elements, shown according to their positions in the periodic table

H 2.1							
Li 1.0	Be 1.5		B 2.0	**C** 2.5	**N** 3.0	**O** 3.5	F 4.0
Na 0.9	**Mg** 1.2		Al 1.5	Si 1.8	**P** 2.1	**S** 2.5	**Cl** 3.0
K 0.8	**Ca** 1.0						

Apart from hydrogen, which is a special case, electronegativity increases as we move from left to right across a period (a row of the table) and decreases as we move down a group (a column of the table). Elements that are found frequently in biological molecules are shown in bold.

Source: Adapted from Snape, A. & Papachristodoulou, D. (2018). *Biochemistry and Molecular Biology* (6th ed.). 9780198768111. Oxford University Press.

PAUSE AND THINK

Based on the electronegativity values in Table 1.2, put the following pairs of atoms in order of bond polarization, starting with the most weakly polarized bond and ending with the most strongly polarized bond.

(a) C–H

(b) N–H

(c) O–H

(d) C–O

Answer: Order is: a, b, d, c (increasing difference in electronegativities between the two atoms in the pair).

Single, double, and triple covalent bonds are formed by two atoms sharing one, two, or three pairs of electrons

Figure 1.6 shows covalent bonding in water, where one oxygen atom shares an electron pair with each of two hydrogen atoms. A bond formed by sharing one pair of electrons is called a **single** covalent bond, so we can say that the water molecule contains two single bonds. However, the molecule of oxygen gas (O_2) shown in Figure 1.7a is formed by two oxygen atoms sharing two electron pairs with each other. We can say that the two oxygen atoms are joined by a **double** covalent bond.

Oxygen has a valence of two, so it can form either two single bonds, or one double bond. However, molecules with higher valences, such as carbon with a valence of four, can share up to three electron pairs and hence can form single, double, or triple covalent bonds.

In the carbon-containing molecules ethane, ethene, and ethyne shown in Figure 1.7b, each covalent bond is simply depicted by a line between the two atoms, with each line representing a shared electron pair.

PAUSE AND THINK

1. A double bond can form between which of the following pairs of atoms?

   (a) C, H

   (b) C, O

   (c) O, H

   (d) C, C

2. A triple bond can form between which of the following pairs of atoms?

   (a) C, N

   (b) C, Cl

   (c) C, H

   (d) C, C

Answers:

1. Double bond (b) and (d). Each atom in the pair must have a valence of at least two to form a double bond between them.

2. Triple bond (a) and (d). Each atom in the pair must have a valence of at least three to form a triple bond between them.

Double and triple bonds restrict possible three-dimensional arrangements of atoms within a molecule

Figure 1.8 shows a molecule of 1,2-dichloroethane ($C_2H_4Cl_2$) and one of 1,2-dichloroethene ($C_2H_2Cl_2$). In the names of these molecules, '1,2-' tells us that the chlorines are each bonded to a different carbon, carbon 1 and carbon 2, while 'ethane' and 'ethene'

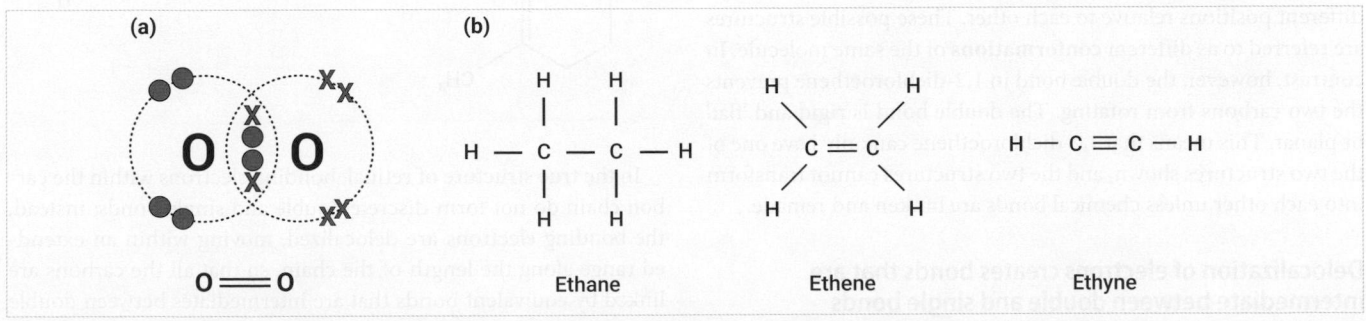

Figure 1.7 Single and multiple covalent bonds. (a) In molecular oxygen, two oxygen atoms share two pairs of electrons, forming a double covalent bond. (b) Carbon can form single (e.g. ethane), double (e.g. ethene), or triple (e.g. ethyne) covalent bonds.

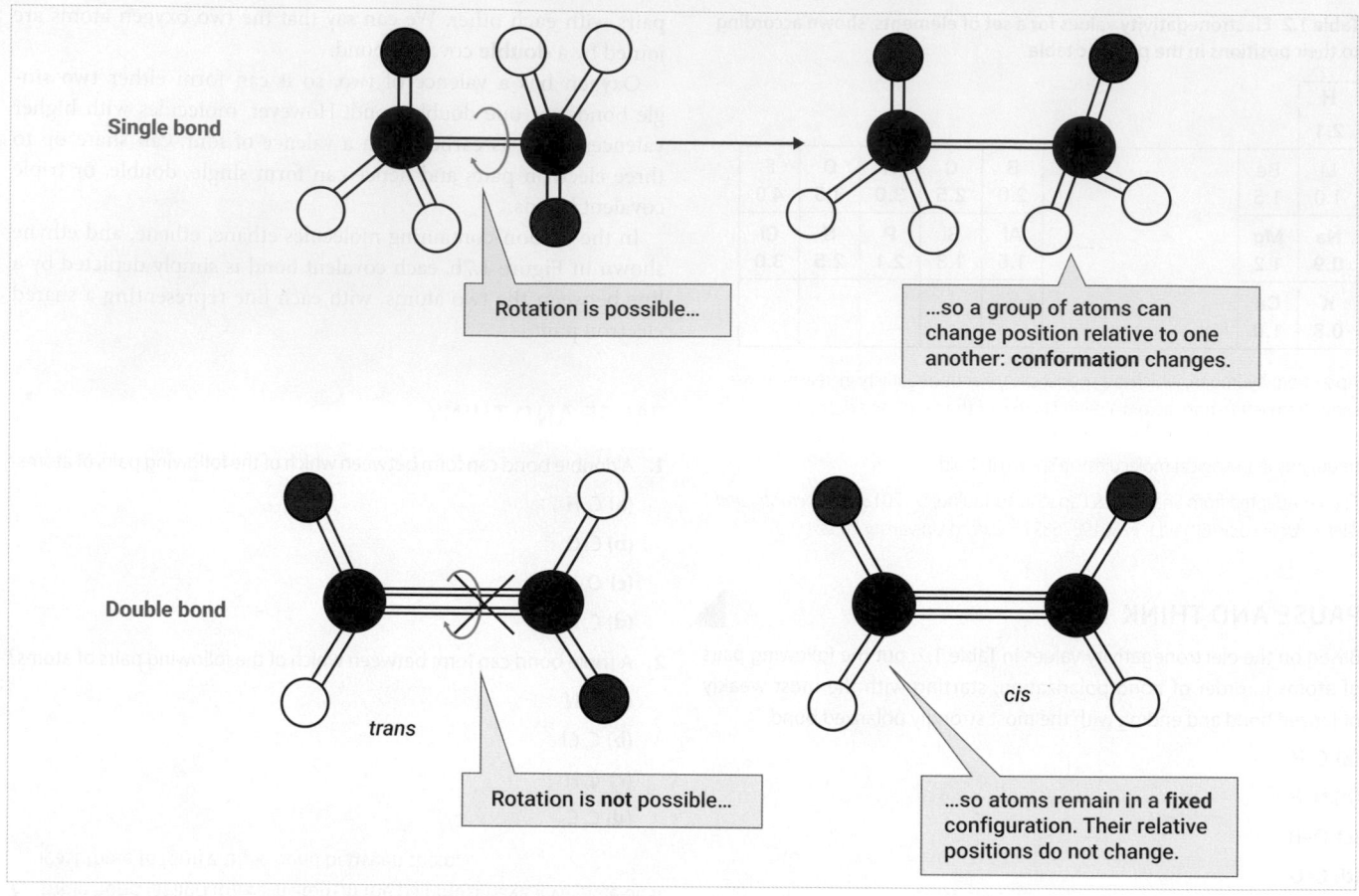

Figure 1.8 Bond rotation about a single bond allows the conformation of atoms joined by the bond to change. This is illustrated by 1,2-dichloroethane, in which the chlorine atoms (red) can change their relative position within the molecule. However, bond rotation about a double bond is not possible. The two 1,2-dichloroethene molecules depicted are different and cannot be transformed into each other unless bonds are broken and reformed. The terms '*trans*' and '*cis*' are explained in the subsection on isomers later in this section.

tell us that one molecule has a single and the other a double bond between the two carbons. We will use these two molecules to illustrate an additional concept that is important for our understanding of the structures and properties of biological molecules, that of **bond rotation**. In the 1,2-dichloroethane molecule, the two carbon atoms can rotate or 'spin' around without disturbing the single bond between them. This means that a single molecule of 1,2-dichloroethane can flexibly adopt different structures in which the hydrogen and chlorine atoms attached to the two carbons are in different positions relative to each other. These possible structures are referred to as different **conformations** of the same molecule. In contrast, however, the double bond in 1,2-dichloroethene prevents the two carbons from rotating. The double bond is rigid and 'flat' or planar. This means that 1,2-dichloroethene can only have one of the two structures shown, and the two structures cannot transform into each other unless chemical bonds are broken and remade.

Delocalization of electrons creates bonds that are intermediate between double and single bonds

Some molecules contain a series of alternating double and single bonds between carbon atoms. This occurs, for example, in the

vitamin A derivative, retinal, which is important in the visual system. The structural formula of retinal is shown here using a 'skeletal structure' convention where carbons bonded to each other are not shown explicitly, but are assumed to be present.

In the true structure of retinal, bonding electrons within the carbon chain do not form discrete double and single bonds; instead, the bonding electrons are delocalized, moving within an extended range along the length of the chain, so that all the carbons are linked by equivalent bonds that are intermediates between double and single bonds. The delocalization of bonding electrons stabilizes the molecule's structure, a phenomenon known as **resonance stabilization**.

Another well-known example of electron delocalization is that of the benzene ring. Benzene is often depicted as if it alternates rapidly between the two structures shown.

However, this is not really the case. An alternative depiction indicates the delocalized electrons by showing a circle within the benzene ring and shows that benzene does not actually contain alternating double and single bonds. All the bonds in the ring are equivalent and intermediate between double and single bonds.

Organic molecules that contain the ring of delocalized electrons found in benzene are termed **aromatic** compounds (as opposed to **aliphatic** compounds, which are organic molecules that do not contain a delocalized electron ring).

A particularly important example of electron delocalization in biological molecules occurs in peptide bonds. These are bonds that link small molecules (amino acids) together to form large protein molecules. Electron delocalization gives peptide bonds a 'partial double bond' character, which prevents rotation around them. As you will see later in this chapter and in Chapter 6, the ability of proteins to adopt specific three-dimensional shapes and to change their shapes by changing conformation is crucial to many aspects of their function. However, the lack of rotation around peptide bonds imposes limitations on protein structures.

Functional groups determine the characteristic reactions of biological molecules

A **functional group** in chemistry is a specific group of atoms and/or bonds. Organic compounds typically consist of a carbon 'skeleton' with one or more functional groups attached. Each functional group can undergo a characteristic set of reactions, for instance a hydroxyl functional group (–OH) can undergo an oxidation reaction to form a carbonyl functional group (–C=O), and a carbonyl functional group in turn be oxidized to form a carboxylic acid group (–COOH). (The nature of oxidation reactions is explained further in Chapter 2). Biological molecules contain characteristic functional groups and it is these that determine important properties of the molecules, such as their solubility in water, the types of bonds and interactions they form, and the reactions they take part in. Table 1.3 shows a summary of the main functional groups found in biological molecules.

PAUSE AND THINK

Identify the three functional groups on this molecule:

Answer: Thiol, amino, carboxylic acid (from top to bottom).

Isomers add to the variety of biological molecules

The identity and properties of a chemical compound depend not only on its composition (i.e. which atoms of which elements it contains), but also on how the atoms are bonded together and how they are arranged in space. For example, the sugars glucose and fructose both have the chemical formula $C_6H_{12}O_6$ but have different structures (as shown in Figure 1.9). **Isomers** is the term used to describe groups of compounds that have the same formula but different arrangements of atoms. We distinguish two main classes of isomer, structural isomers and stereoisomers.

Structural isomers contain the same atoms bonded together in different ways

Glucose and fructose are structural isomers of each other. Some simpler examples, illustrated in Figure 1.9, show that structural isomers can contain the same chemical groups arranged in different ways, or can contain different chemical groups. Where chemical groups differ, for instance between butanal (an aldehyde) and butanone (a ketone) this can have a profound effect on the chemical reactions the molecules can undergo.

Stereoisomers contain the same atoms bonded together in the same ways but arranged differently in space

Figure 1.10 shows two classes of stereoisomer: *cis–trans* isomers and enantiomers or optical isomers. *Cis–trans* isomers arise because of restrictions on the movement of atoms imposed by molecular structure; the example of the double bond was discussed earlier in this section, and Figure 1.10 also shows *cis–trans* isomerism in cyclic structures.

Table 1.3 The main functional groups found in biological molecules

Class of compound	Name of functional group	Structure of functional group	Example
Alkene	Double bond	$>\!\!=\!\!<$	$H_2C\!=\!CH_2$
Alcohol	Hydroxyl	—OH	$H_3C—CH_2—OH$
Ether	Ether linkage	—O—R	
Thiol	Thiol or sulphydryl	—SH	$H_3C—CH_2—SH$
Aldehyde	Carbonyl	$-\overset{\overset{\textstyle O}{\|\|}}{C}-$	$R-\overset{\overset{\textstyle O}{\|\|}}{C}-H$
Ketone	Carbonyl	$-\overset{\overset{\textstyle O}{\|\|}}{C}-$	$R-\overset{\overset{\textstyle O}{\|\|}}{C}-R'$
Carboxylic acid	Carboxyl	$-C\overset{\textstyle O}{\underset{\textstyle OH}{}}$	$H_3C-C\overset{\textstyle O}{\underset{\textstyle OH}{}}$
Ester	Ester linkage	$-C\overset{\textstyle O}{\underset{\textstyle O-R}{}}$	$H_3C-O-\overset{\overset{\textstyle O}{\|\|}}{C}-H$
Amine	Amino	$-NH_2$	$H_3CH_2CH_2C-N\overset{\textstyle H}{\underset{\textstyle CH_2CH_3}{}}$
Amide	Amide	$-C\overset{\textstyle O}{\underset{\textstyle NH_2}{}}$	$H_3C-C\overset{\textstyle O}{\underset{\textstyle NH_2}{}}$
Phosphoric acid ester	Phosphoester	$-\overset{\|}{\underset{\|}{C}}-O--\overset{\overset{\textstyle O}{\|\|}}{\underset{\underset{\textstyle O^-}{\|}}{P}}--O^-$	
Phosphoric acid anhydride	Phosphoanhydride	$-O--\overset{\overset{\textstyle O}{\|\|}}{\underset{\underset{\textstyle O^-}{\|}}{P}}--O--\overset{\overset{\textstyle O}{\|\|}}{\underset{\underset{\textstyle O^-}{\|}}{P}}--O-$	

R denotes a carbon containing group such as an alkyl group. The general formula of an alkyl group is $C_nH_{2n-1}-$: for example, C_2H_5- is an ethyl group. Where a molecule has more than one R group it may have identical or different R groups. R, R', and R'' are used to denote different R groups in a single molecule.

▶ We discuss the importance of *cis* isomerism of fatty acids in cell membrane structure in Chapter 9.

Enantiomers or optical isomers are the second class of stereo-isomer shown in Figure 1.10. This class of isomer occurs where four different atoms or chemical groups are attached to a central atom, forming a **chiral** molecule (one that has no plane of symmetry). As shown, different orders of attachment of the four varying groups around the central, 'chiral' atom lead to there being

two isomers, which are mirror images of each other but cannot be superimposed. In biological molecules the central atom of a chiral centre is usually a carbon atom. There are different conventions for naming enantiomers, but in biology the D- and L- convention is mainly used. Sugars and amino acids, discussed later in this chapter, are example of biological molecules that have enantiomers, with the D-isomers of sugars being most common in living organisms, while the L-isomers of amino acids are the ones found in proteins.

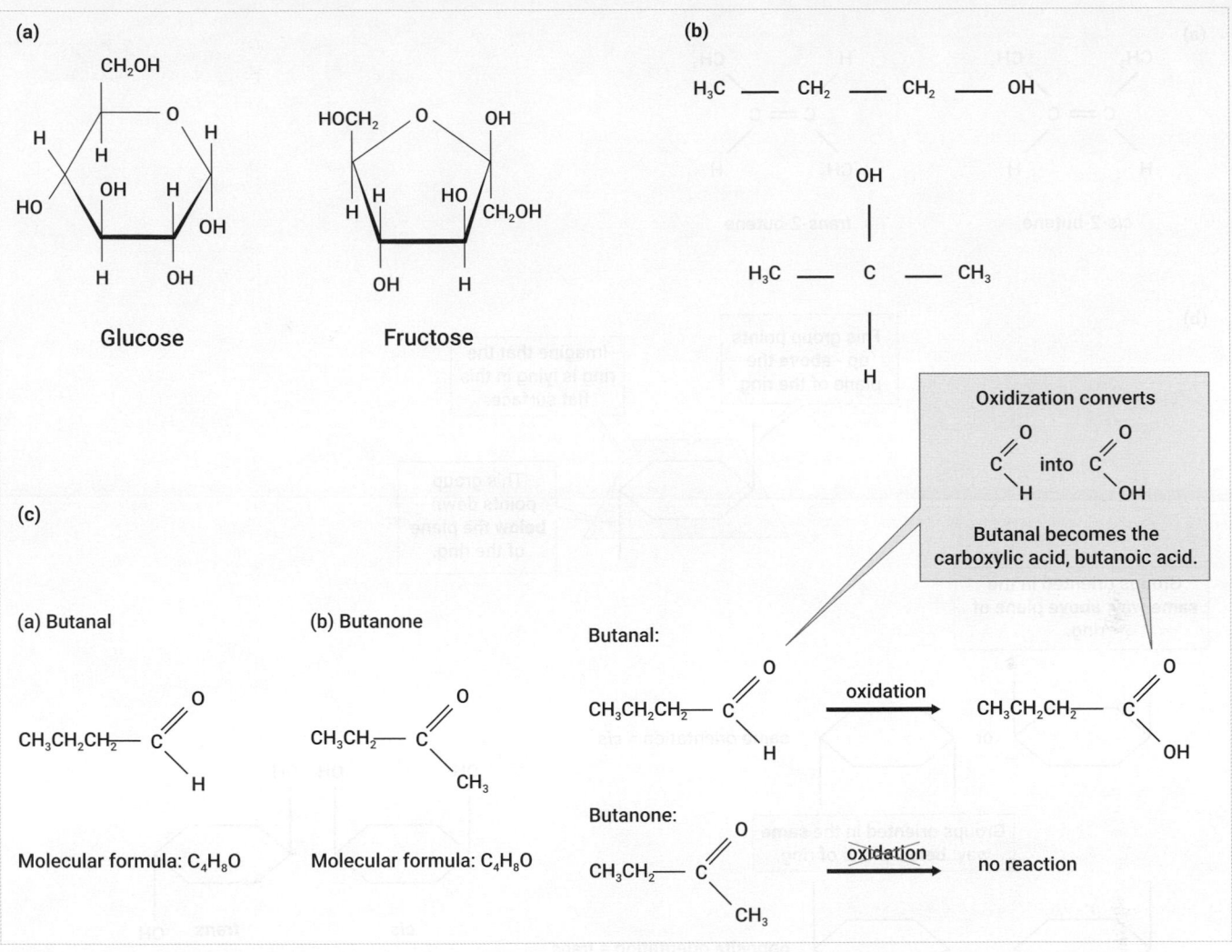

Figure 1.9 Structural isomers. (a) The sugars glucose and fructose are a pair of structural isomers. Both have six carbons, but in glucose five carbons are incorporated into the ring structure, while only four carbons are in the ring of fructose. (b) A simpler example of structural isomers. Both molecules are isomers of propanol. The main structural group, a hydroxyl group (–OH), is the same in both and they belong to the same chemical family, alcohols. (c) An example of structural isomers that share the same chemical formula but belong to different chemical families. Butanal is an aldehyde, while butanone is a ketone. The reactions they undergo will therefore differ.

Source: Crowe, J. & Bradshaw, T. (2021). *Chemistry for the Biosciences: The Essential Concepts* (4th ed.). Oxford University Press.

PAUSE AND THINK

Which of the following compounds can exist as a pair of enantiomers?

Source: Crowe, J. & Bradshaw, T. (2021). *Chemistry for the Biosciences: The Essential Concepts* (4th ed.). Oxford University Press.

Answer: (b) and (d). The other molecules do not have a carbon with four different groups attached.

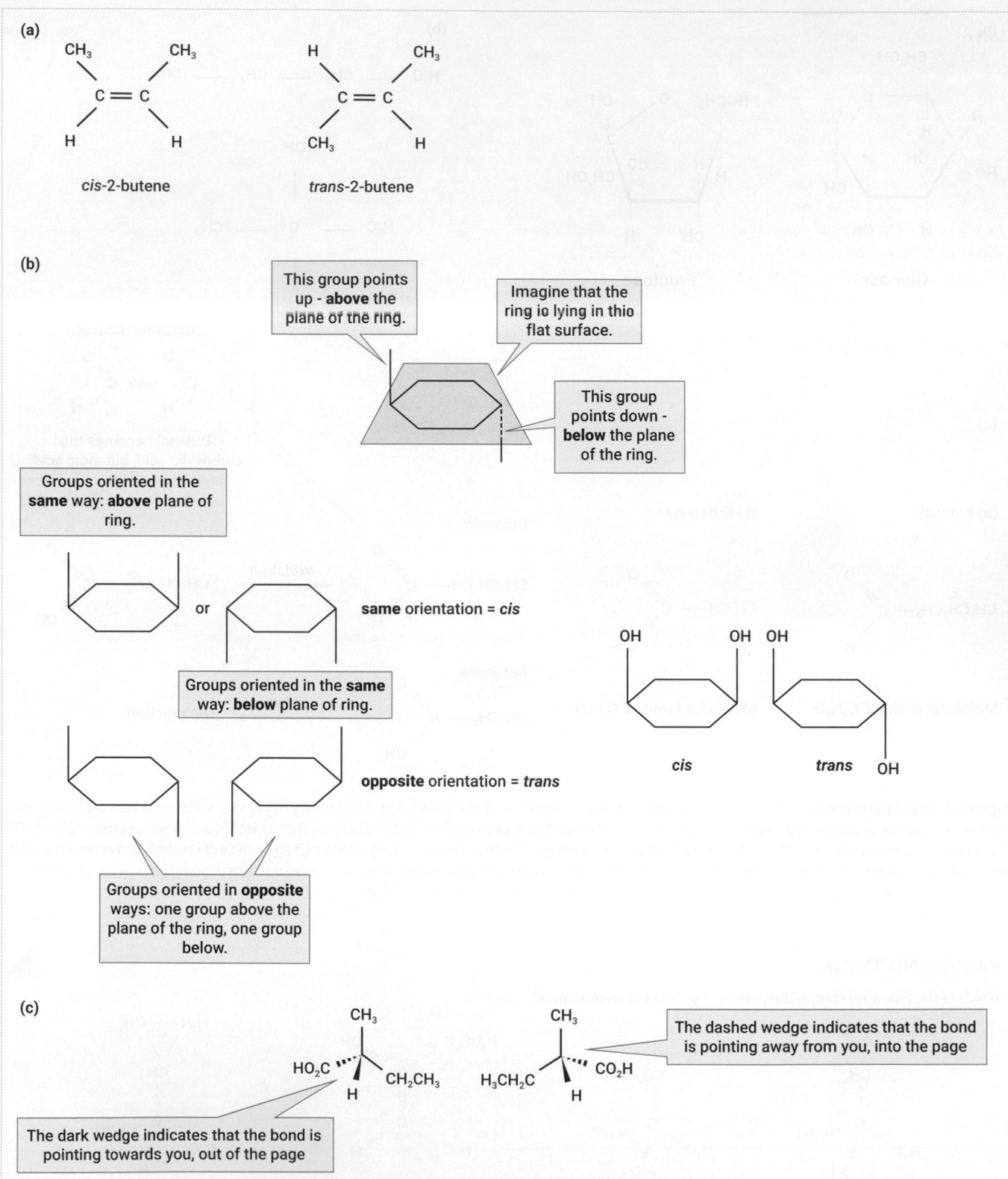

Figure 1.10 Stereoisomers. (a) *Cis–trans* isomerism. In the *cis* isomer the two methyl (–CH₃) groups are on the same side of the double bond, while in the *trans* isomer they are on opposite sides. As there is no rotation about the double bond these are two different molecules. (b) *Cis–trans* isomerism in a cyclic compound. The figure illustrates a convention for drawing cyclic compounds that requires you to envisage the ring as if it were a dinner plate sitting flat on a table, with the lower edge towards you and the upper edge away from you. The functional groups attached to the ring can sit above or below the plane of the ring but cannot interchange. (c) A pair of enantiomers or optical isomers. Both molecules contain the same four functional groups attached to a central carbon, but they are mirror images of each other and not superimposable. The figure illustrates a convention for drawing these molecules to indicate their three-dimensional structure.

Source: Crowe, J. & Bradshaw, T. (2021). *Chemistry for the Biosciences: The Essential Concepts* (4th ed.). Oxford University Press.

Stereochemistry in therapeutic drug development: the tragedy of thalidomide

The activity of many drugs depends on their precise interaction with biological receptor molecules. It is quite common for only one of two enantiomers of a drug to have therapeutic properties, as only one isomer binds the relevant receptor. The problem with the drug thalidomide, which was developed and marketed in the 1950s as a treatment for morning sickness during pregnancy, is that the two enantiomers have different activities, one of which is therapeutic and the other harmful. The drug, the structure of which is shown in Figure 1, contains a chiral carbon and has enantiomers that are named using the terminology used in chemistry, as *R* and *S* isomers.

The drug was withdrawn in 1961, when it was realized that it was causing babies to be born with severe limb malformations and other abnormalities. Over half the affected babies died, and those who lived had significant disabilities. It turns out that the *R* enantiomer of thalidomide acts as a

sedative and is effective against morning sickness, but the *S* form has an entirely different activity, which inhibits the growth of blood vessels in the developing fetus and hence caused birth defects. Even if steps had been taken to separate the enantiomers before giving the drug, this would not have been effective because the two forms can interconvert in the body.

The tragedy of thalidomide prompted many countries to tighten up their drug testing regimes, and the drug remained off the market for many years. However, since the mode of action of the *S* form of thalidomide was elucidated there has been renewed interest in using it as a treatment for cancer. Reducing the blood supply to cancerous cells by blocking their ability to build new blood vessels could be an effective therapy. Obviously, the use of thalidomide and its derivatives against cancer must be strictly controlled.

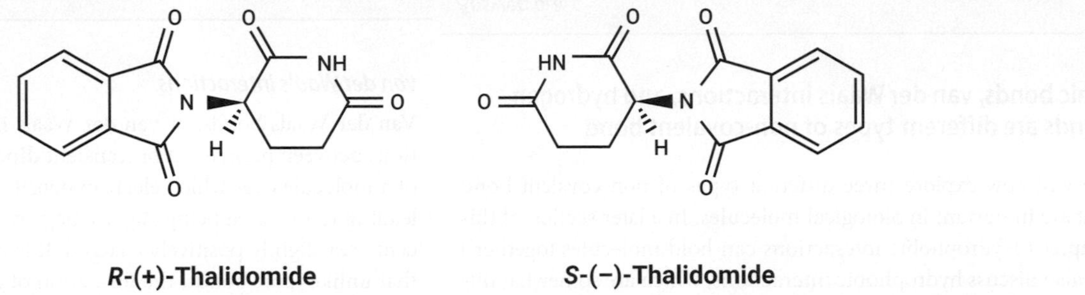

R-(+)-Thalidomide   *S*-(−)-Thalidomide

Figure 1 The two enantiomers of thalidomide.

Despite enantiomers seeming almost identical, the fact that they are not superimposable is important for the function of molecules that need to interact specifically through shape matching, such as drugs and their receptors. Just as you cannot fit your right hand into your left glove, a D-isomer cannot fit into a receptor site that has evolved to fit the L-isomer of the same molecule. The tragic consequences of very different responses to enantiomers of the same drug are explored in Clinical Box 1.1.

 Check your understanding of the concepts covered in this section by answering the questions in the e-book.

1.3 Non-covalent bonds allow interactions between molecules

Molecules consist of atoms joined together by covalent bonds, but other classes of chemical bond exist, enabling molecules to interact with each other without electron-pair sharing. Specific bonding between molecules, which may be transient or long-lasting, is crucial to biological processes, and additionally non-covalent bonding

between different parts of large biological molecules can stabilize their complex three-dimensional structure, which is often crucial for their function. The non-covalent bonds that allow such interactions are weaker than covalent bonds, and before we describe the different types of non-covalent bond, we should briefly explore the concept of bond strength or bond energy. It is important to have a basic knowledge of this and other concepts relating to energy to understand how and why biological reactions occur.

▶ We discuss energy in more detail in Chapter 2.

Bond energies indicate the strength of chemical bonds

When chemical bonds are formed energy is released, and to break them again energy equivalent to that released must be provided. It may help you to understand this concept if you imagine the bond as a cord joining two atoms together, and the energy or work required to break that cord. The stronger the bond, the more energy is needed to break it.

Units of energy are joules (J) or kilojoules (kJ), where 1000 J equals 1 kJ. Bond energies are measured in kJ mol^{-1} where mol refers to a mole (6×10^{23}) bonds of that type. Thus, if we are told that the average bond energy of a covalent bond between one carbon atom and another is 348 kJ mol^{-1}, we know that when 6×10^{23} of these covalent bonds are formed, 348 kJ of energy are released, and that it would require 348 kJ of energy to break the bonds once formed. If we are similarly told that the average bond energy of a covalent bond between a carbon and a hydrogen atom is 412 kJ mol^{-1}, we can tell that carbon–hydrogen bonds are stronger than carbon–carbon bonds, as more energy is released when they are made, and more energy is required to break them.

PAUSE AND THINK

Given the bond energies below, rank the bonds in order from weakest to strongest.

(a) H–H 432 kJ mol^{-1}

(b) H–S 363 kJ mol^{-1}

(c) C=C 602 kJ mol^{-1}

Answer: b, a, c

Ionic bonds, van der Waals interactions, and hydrogen bonds are different types of non-covalent bond

We will now explore three different types of non-covalent bond that are important in biological molecules. In a later section of this chapter ('Hydrophobic interactions can hold molecules together') we also discuss hydrophobic interactions, which are somewhat different in nature from those described here but play an important role in the structures of many molecules found in the cell.

Ionic bonds

Ionic bonds are formed between atoms that have widely different electronegativities (see Table 1.2), to the extent that the more electronegative atom fully gains an electron or electrons, forming a negatively charged anion, while the less electronegative atom loses an electron or electrons and forms a positively charged cation. Generally, an ionic bond forms between atoms whose electronegativities differ by 1.7 or greater. In the example shown in Figure 1.11, a sodium atom donates the single electron in its outer valence shell to a chlorine atom, leaving the sodium atom with one more proton than it has electrons, so it has a single positive charge. The chlorine atom gains one more electron than it has protons, so it becomes a chloride ion, carrying a single negative charge. The oppositely charged ions are bound together by electrostatic forces

of attraction. Although weaker than most covalent bonds, ionic bonds are relatively strong compared to other non-covalent bonds, as shown in Table 1.4.

Ionic bonds are relatively rare in biological molecules, although they play an important role in stabilizing protein structure, where they are often known as 'salt bridges' (see Chapter 6).

PAUSE AND THINK

Refer to Table 1.2 to answer this question.

For which of these pairs of atoms would you predict that they form an ionic bond between them?

(a) C and P

(b) Ca and Cl

(c) O and S

Answer:

b: It is the only pair where the difference in electronegativities is 1.7 or greater.

van der Waals interactions

Van der Waals bonds, or van der Waals interactions, are interactions between permanent or transient **dipoles**. A dipole is a region of a molecule over which electron density is unevenly distributed, leading to one area being slightly negatively charged and an adjacent area slightly positively charged. It is important to understand that unlike ionic bonds, the formation of dipoles does not involve complete loss of an electron by one atom and gain by another, rather it involves the spatial distribution of electrons within a covalent bond, or across a whole or part of a molecule.

Dipoles can be **permanent**, as in the polar water molecule shown in Figure 1.6, or **transient** in non-polar molecules. Hopefully, it is relatively easy to see how polar molecules or polar groups within molecules can bond together through electrostatic attraction. The bonding together of non-polar molecules is harder to envisage. To understand how transient dipoles form in non-polar molecules, it may help to envisage a motorway on which several lanes of vehicles are travelling in the same direction (Figure 1.12). At any one time, the vehicles may be evenly spaced, or a chance occurrence may cause them to bunch together. If you are used to motorway traffic you may well recognize that 'bunching' is a frequent occurrence. Similarly, electrons within a molecule frequently bunch together by chance, transiently creating areas of relative negative charge, and there is a knock-on effect as these negative regions repel electrons

Figure 1.11 Ionic bonds. The formation of the ionic compound sodium chloride, NaCl. Sodium chloride is formed when one electron is transferred from a sodium atom to a chlorine atom to form a sodium and a chloride ion which are held together by the electrostatic attraction between them, an ionic bond.

Source: Crowe, J. & Bradshaw, T. (2021). *Chemistry for the Biosciences: The Essential Concepts* (4th ed.). Oxford University Press.

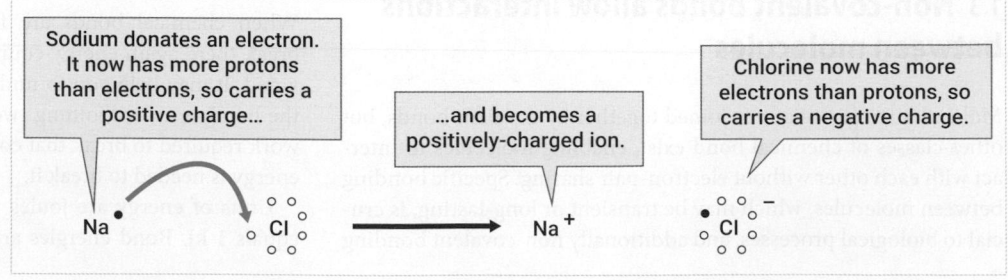

Sodium donates an electron. It now has more protons than electrons, so carries a positive charge…

…and becomes a positively-charged ion.

Chlorine now has more electrons than protons, so carries a negative charge.

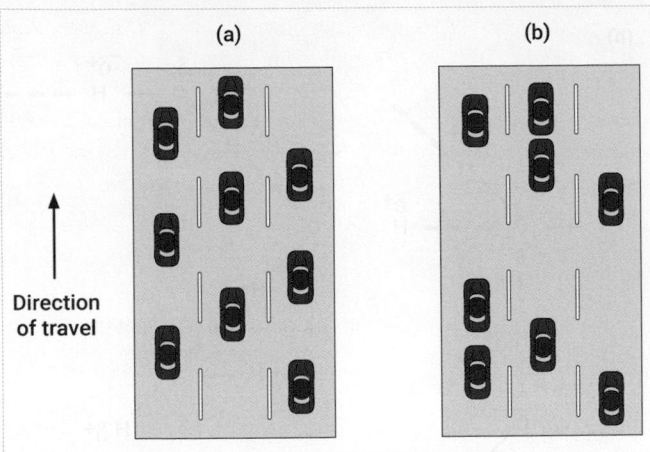

Figure 1.12 Two views of motorway traffic. (a) Traffic in all three lanes is evenly distributed. We rarely see this type of distribution (b) Traffic is unevenly distributed, with vehicles bunched together. We see this type of distribution much more frequently.

Source: Crowe, J. & Bradshaw, T. (2021). *Chemistry for the Biosciences: The Essential Concepts* (4th ed.). Oxford University Press.

Figure 1.13 Transient dipoles. Just as many tiny ropes could pin Gulliver to the ground, multiple weak bonds such as transient dipoles can hold large molecules together.

Source: Alamy.

from adjacent regions, transiently creating areas of relative positive charge called **induced dipoles**.

Both permanent and transient dipoles allow electrostatic attraction and hence bonding to occur between molecules, or between different regions of the same large molecule. As the forces are extremely weak (Table 1.4), molecules must be in close proximity for bonding to occur. It may be hard to envisage how transient dipoles can account for functionally significant associations between large molecules, but the explanation is that the very size of biological molecules allows multiple small interactions to occur between them. You could think of an analogy from Jonathan Swift's book *Gulliver's Travels* where the tiny Lilliputians tie the (to them) gigantic Gulliver to the ground with many small ropes (Figure 1.13). The term van der Waals interactions, rather than van der Waals bonds, is often used as it is descriptive of this type of bonding requiring multiple interactions over a wide area.

Hydrogen bonds

Hydrogen bonds are a specific class of bond formed between permanent dipoles. They are treated separately here both because of their relative strength (Table 1.4) and because of their importance in biological molecules.

Table 1.2 shows the relatively large difference between the electronegativity of hydrogen, and that of oxygen, nitrogen, and fluorine.

In a chemical 'group' (part of a molecule) in which hydrogen covalently bonds to any one of these three atoms, a strongly polar covalent bond is formed, with hydrogen carrying partial positive charge and the other atom (X in Figure 1.14) carrying partial negative charge. As shown in Figure 1.14, this polar covalent bond is *not* the hydrogen bond. The hydrogen bond is formed through the electrostatic interaction between the hydrogen (δ^+H) of one such group and a partially negative atom (δ^-Y) in another group. Hydrogen bonds have variable strength, being strongest when all three nuclei involved are arranged in a straight line, as in the interacting water molecules shown in Figure 1.14.

The hydrogen bond donor atom (X in Figure 1.14) is usually nitrogen, oxygen, or fluorine, and the hydrogen bond acceptor atom can also be nitrogen, oxygen, or fluorine. Fluorine is not found in biological molecules, but hydrogen, oxygen, and nitrogen are common, and groups that can take part in hydrogen bonding, such as the hydroxyl (–OH) group and amino (–NH_2) group, are of great significance in the structure and function of key molecules including proteins and nucleic acids (DNA). The capacity of water to form hydrogen bonds is also important and affects its properties and interaction with other molecules, as discussed later in this chapter.

PAUSE AND THINK

In this chemical structure, identify the atom that is the hydrogen bond donor and the atom that is the hydrogen bond acceptor

Answer: Nitrogen (N) is the donor and oxygen (O) is the acceptor.

Table 1.4 Bond strengths are indicated by bond energy values

	Bond strength (kJ mol^{-1})
van der Waals interactions	0.4–4
Hydrogen bonds	12–30
Ionic bonds	20

The bond energy is the energy needed to break one mole of bonds, or the energy released when one mole of bonds is created.

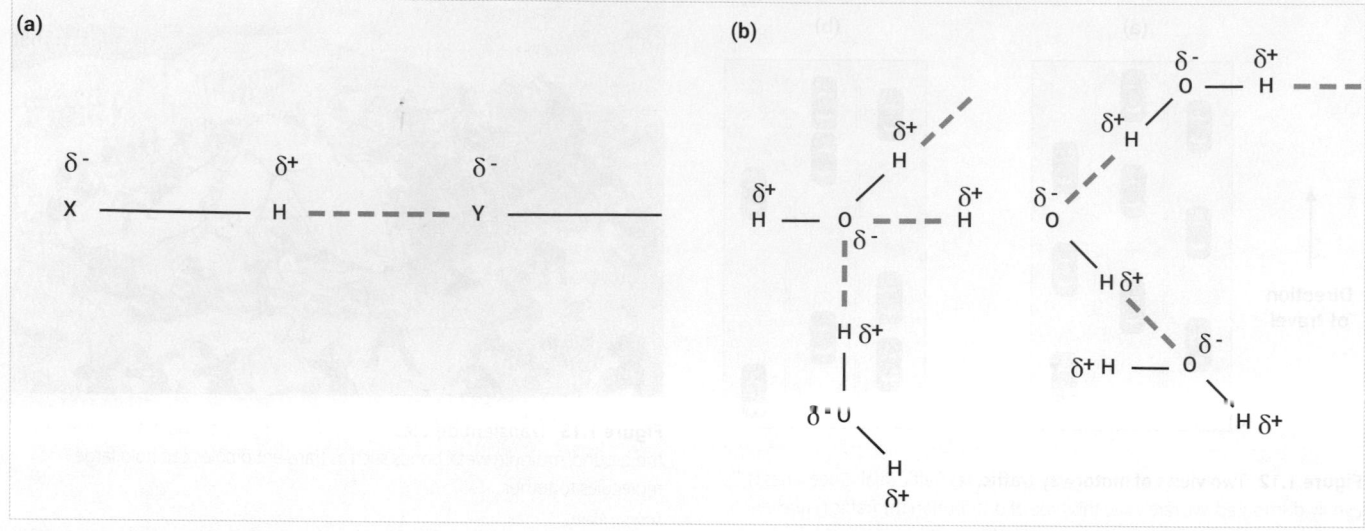

Figure 1.14 Hydrogen bonds. (a) A hydrogen bond (blue dashed line) is formed between a hydrogen atom that is covalently bound to a strongly electronegative atom (X) and therefore has a partial positive charge (δ^+) and an atom (Y) that carries a partial negative charge (δ^-). (b) Hydrogen bonds that form between water molecules are strong because all three atoms involved in each bond (O-H -- O) lie in a straight line.

🌀 **Check your understanding of the concepts covered in this section by answering the questions in the e-book.**

1.4 Chemical reactions involve breaking and making bonds

Chemical reactions occur constantly in all living cells. For example, we describe in Chapter 7 how animals break down large, complex food molecules into simpler ones, which are then used to build new sets of molecules that the organism needs to continue its life processes. Plants use energy from the sun to build large molecules from simpler components. The chemical reaction scheme (or chemical equation), shown below, illustrates the combustion of methane to form carbon dioxide and water. In chemical terminology, methane and oxygen are the **reactants** in this scheme, and carbon dioxide and water are the **products**.

$$CH_4 \quad + \quad 2O_2 \quad \rightarrow \quad CO_2 \quad + \quad 2H_2O$$

methane + oxygen → carbon dioxide + water

This is the reaction that occurs when we burn methane gas as domestic or industrial fuel. It does not take place in living cells, but it provides a simple example to illustrate the point that chemical reactions involve breaking and making chemical bonds. In the reaction, the four covalent bonds linking the hydrogen atoms to carbon are broken and carbon and hydrogen form new bonds with oxygen atoms to make water and carbon dioxide.

Chemical reactions are reversible and proceed to equilibrium

When methane is used as fuel, a spark or flame is needed to start the reaction, but once it gets going all the methane is consumed and converted to carbon dioxide and water. We can say that the

reaction 'goes to completion'. However, most chemical reactions are not so strongly unidirectional. To illustrate this point and explain its significance, we will look at a reaction that does occur in cells, the interconversion of two sugars, glucose 6-phosphate and fructose 6-phosphate.

glucose 6-phosphate $\rightleftharpoons$ fructose 6-phosphate

For now, we will not concern ourselves with the chemical structure of these two molecules. You can read more about them in Chapter 7. What you need to notice is the two half arrows ($\rightleftharpoons$) that link the two sides of the reaction. These arrows tell us that glucose 6-phosphate can be converted into fructose 6-phosphate, and fructose 6-phosphate can also be converted back into glucose 6-phosphate. There is a constant interconversion from reactant to product and from product to reactant.

Figure 1.15 shows what happens if we start with pure glucose 6-phosphate in a container. Some of the glucose 6-phosphate converts to fructose 6-phosphate (the **forward reaction**), but as it builds up, fructose 6-phosphate starts to convert back again (the **reverse reaction**). As the concentration of glucose 6-phosphate decreases, the more slowly the forward reaction proceeds, and conversely the more fructose 6-phosphate builds up, the more quickly the reverse reaction proceeds. When the rate of the forward reaction is equal to that of the reverse reaction, the concentrations of glucose 6-phosphate and fructose 6-phosphate no longer change, and we say the reaction has reached **equilibrium**.

It is important to understand that reactions do not cease when they reach equilibrium. The interconversion of reactants and products continues, but there is no net change in their concentrations. It is also important not to assume that reactions at equilibrium contain a 50:50 mixture reactants and products, as this is rarely the case. In the reaction illustrated in Figure 1.15 the reaction reaches equilibrium when it contains 67% glucose 6-phosphate and 33% fructose 6-phosphate. We can say that the equilibrium of this

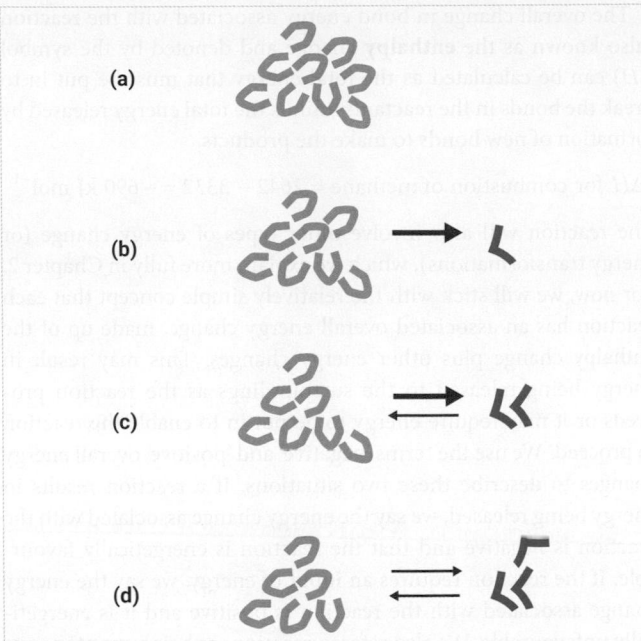

Figure 1.15 Interconversion of glucose 6-phosphate to fructose 6-phosphate. In (a) we have only glucose 6-phosphate, which starts to convert to fructose 6-phosphate as in (b). As fructose 6-phosphate builds up, it begins to convert back to glucose 6-phosphate, as in (c) and (d). The thickness of the arrows denotes the rate of the reaction: the higher the concentration of reactant, the faster the reaction. In (d) the forward and reverse reactions are occurring at equal rates, there is no net change in the concentrations of glucose 6-phosphate and fructose 6-phosphate, and we say the reaction is at equilibrium.
Source: Adapted from Alberts, B. et al. (2019). *Essential Cell Biology*. 4th edn. Garland Science.

reaction favours glucose 6-phosphate, or that the equilibrium of the reaction 'lies to the left'. The position of the equilibrium of a reaction is influenced by the relative energies of the reactants and products, a subject that is discussed further in Chapter 7.

PAUSE AND THINK

The following reaction takes place in the glycolysis pathway:

3-phosphoglycerate ⇌ 2-phosphoglycerate

The equilibrium of this reaction 'lies to the left'. Which compound predominates when the reaction is at equilibrium?

Answer: 3-phosphoglycerate

The direction of a reversible reaction is influenced by the concentrations of reactants and products

The conversion of glucose 6-phosphate to fructose 6-phosphate occurs in cells in the glycolysis pathway, a series of reactions by which glucose is broken down for fuel (see Chapter 7). During glycolysis, a high concentration of glucose 6-phosphate is maintained in the cell as it is formed in the preceding reaction of the

pathway. Conversely, as soon as fructose 6-phosphate is formed it is utilized in the next reaction in the pathway, and this means that its concentration is maintained at a low level, preventing its reconversion to glucose 6-phosphate. The reaction is not at equilibrium and proceeds in one direction.

Thus, if we write the reaction as:

glucose 6-phosphate ⇌ fructose 6-phosphate

We can say that in glycolysis, the reaction goes in the forward direction. To emphasize the point, when considering the glycolysis pathway, we might write the reaction as:

glucose 6-phosphate → fructose 6-phosphate

without the two half arrows.

Although this reaction proceeds in one direction during glycolysis, it can be reversed under different circumstances. For example, in plant cells, the reaction proceeds in the opposite direction during photosynthesis, the series of reactions whereby plants build sugars from carbon dioxide and water. Here, fructose 6-phosphate is the reactant and glucose 6-phosphate the product. Fructose 6-phosphate is rapidly made in the preceding step of photosynthesis and glucose 6-phosphate is rapidly used in the next step of the pathway, causing the reaction to continue to produce more glucose 6-phosphate. So, when considering photosynthesis, we could write our reversible reaction as

fructose 6-phosphate ⇌ glucose 6-phosphate

or as

fructose 6-phosphate → glucose 6-phosphate.

The principle of the reversible reaction proceeding in different directions in different pathways is illustrated in Figure 1.16.

Chemical reactions involve energy changes, which determine whether they will occur in cells

We explore the subject of energy in detail in Chapter 2, but it is useful to consider it briefly at this point, as energy changes play an important role in determining which chemical reactions occur in the cell and the direction of these reactions. Earlier in this chapter we touched on energy changes that occur when chemical bonds are made or broken. We discussed the concept of energy being needed (energy 'put in') to break a chemical bond, and the corollary is that energy is *released* when a bond is made. The sum of all the energy changes associated with bonds being made and bonds being broken contributes to an overall change in energy that occurs when a chemical reaction takes place. We can consider the example of the combustion of methane:

$$CH_4 \quad + \quad 2O_2 \quad \rightarrow \quad CO_2 \quad + \quad 2H_2O$$
methane + oxygen → carbon dioxide + water

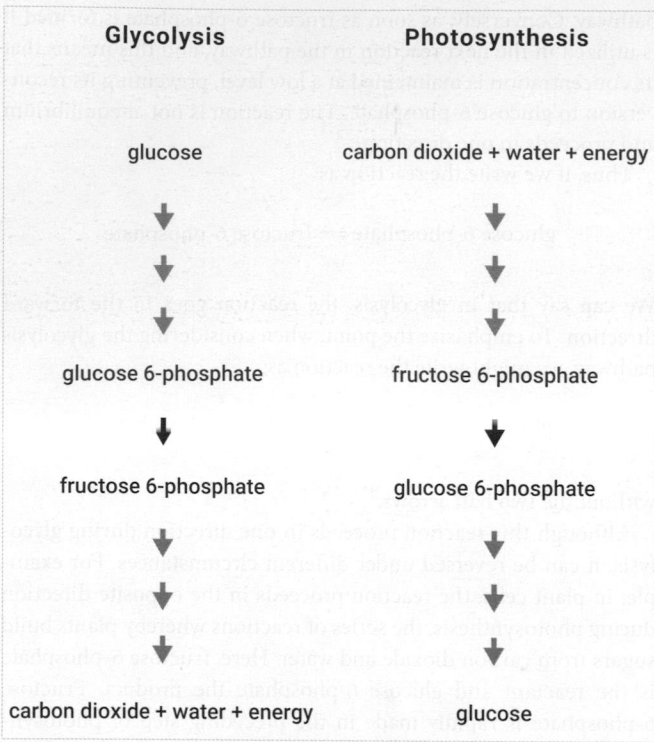

Figure 1.16 The interconversion of glucose 6-phosphate and fructose 6-phosphate proceeds in different directions in two different biochemical pathways. In the glycolysis pathway glucose is used as fuel. It is broken down into smaller molecules, with the eventual production of carbon dioxide and water and release of energy. In photosynthesis, energy harvested from the sun is used to produce glucose from carbon dioxide and water. The reversible reaction that interconverts glucose 6-phosphate and fructose 6-phosphate occurs in both pathways but is driven in opposite directions by the accumulation of different precursor molecules. Note that the pathways are represented here in a highly simplified way, with each blue arrow representing multiple reactions. The pathways are described in detail in Chapter 7.

We can rewrite the reaction to make it easier to keep track of the bonds that are broken and made:

$$H \!-\! \underset{\underset{H}{|}}{\overset{\overset{H}{|}}{C}} \!-\! H + O{=}O + O{=}O \rightarrow O{=}C{=}O + H{-}O{-}H + H{-}O{-}H$$

The balance of energy changes associated with this reaction can be calculated as shown in Table 1.5.

The overall change in bond energy associated with the reaction (also known as the **enthalpy** change and denoted by the symbol ΔH) can be calculated as the total energy that must be put in to break the bonds in the reactants, minus the total energy released by formation of new bonds to make the products.

ΔH for combustion of methane $= 2642 - 3332 = -690$ kJ mol^{-1}

The reaction will also involve other types of energy change (or energy transformations), which we explain more fully in Chapter 2. For now, we will stick with the relatively simple concept that each reaction has an associated overall energy change, made up of the enthalpy change plus other energy changes. This may result in energy being released to the surroundings as the reaction proceeds or it may require energy to be put in to enable the reaction to proceed. We use the terms 'negative' and 'positive' overall energy changes to describe these two situations. If a reaction results in energy being released, we say the energy change associated with the reaction is negative and that the reaction is energetically favourable. If the reaction requires an input of energy, we say the energy change associated with the reaction is positive and it is energetically unfavourable. We also refer to reactions as being **spontaneous** if they release energy and do not require energy input, while reactions that require energy are not spontaneous.

A more technical term used to describe a reaction that releases energy is say that it is **exergonic**, while a reaction that requires an input of energy is **endergonic**. These terms are explained more fully in Chapter 2, but hopefully you can grasp the concept of energetically favourable and unfavourable reactions instinctively, without the need for too much detail at this stage. The reason for introducing the concept here is to explain briefly how energy considerations can determine whether a chemical reaction can occur spontaneously in cells, whether it can proceed in one or in both directions, and how the terms 'reversible' and 'irreversible' reactions are used in this context.

Earlier in this section we emphasized the use of two half arrows to show that the interconversion of glucose 6-phosphate and fructose 6-phosphate can occur in either direction. Technically, all chemical reactions are reversible, with their direction being determined by the concentrations of reactants and products as discussed above, and by other factors such as temperature and pressure. However, energy considerations may mean that one direction is strongly favoured over the other. To illustrate this point, think again about the combustion of methane.

CH$_4$	+	2 O$_2$	$\rightarrow$	CO$_2$	+	2 H$_2$O
methane	+	oxygen	$\rightarrow$	carbon dioxide	+	water

Table 1.5 The balance of energy changes in the reaction

	Reactants		Products	
Molecules	CH$_4$	2 O$_2$	CO$_2$	2 H$_2$O
Bonds	4 × (C–H)	2 × (O = O)	2 × (C = O)	4 × (O – H)
Bond energies	4 × (412 kJ mol^{-1})	2 × (497 kJ mol^{-1})	2 × (740 kJ mol^{-1})	4 × (463 kJ mol^{-1})
	= 1648 kJ mol^{-1}	= 994 kJ mol^{-1}	= 1480 kJ mol^{-1}	= 1852 kJ mol^{-1}
Total energy	1648 + 994 = **2642 kJ mol^{-1}**		1480 + 1852 = **3332 kJ mol^{-1}**	

The reaction results in release of a large amount of energy, as heat. This means that reversing the reaction, to make methane and oxygen from carbon dioxide and water, would require the input of the same large amount of energy. Consequently, the reverse reaction does not occur to any appreciable extent.

Compared to the combustion of methane, the reactions that occur in cells do not result in such large releases of energy. Cells would be consumed by fire and explosions if they did! Consequently, many cellular reactions that involve the release or input of modest amounts of energy can be driven in different directions by changes in the relative concentrations of reactants and products, as in our glucose 6-phosphate $\rightleftharpoons$ fructose 6-phosphate example. (Temperature and pressure changes are generally not a consideration in biological systems.) Some cellular reactions, however, can effectively occur only in one direction, as the input of energy required to drive them in the unfavourable direction is too great to be overcome by achievable changes in the concentrations of reactants and products. Thus, in discussing the reactions that occur in cells, we distinguish between **reversible** and **irreversible** reactions according to energy considerations.

▶ We explain in Chapter 2 how cells can achieve reactions that are energetically unfavourable and hence will not occur spontaneously, by harnessing the energy released by other reactions.

PAUSE AND THINK

The energy changes associated with three biochemical reactions are shown below. Which of these reactions will occur spontaneously in the cell?

(a) sucrose + H_2O $\rightarrow$ glucose + fructose -29 kJ mol^{-1}

(b) glutamate + NH_4^+ $\rightarrow$ glutamine + H_2O $+30$ kJ mol^{-1}

(c) glucose 1-phosphate $\rightarrow$ glucose 6-phosphate -7 kJ mol^{-1}

Answer: (a) and (c) as they have negative energy changes.

Catalysis is required to speed up reactions

While considering that the overall energy change can tell us whether a reaction is likely to occur, it does not tell us anything about the speed of the reaction. The combustion of methane, though leading to a large overall energy release, will not take place at any appreciable rate without an initial energy input, a spark or fire, to get it going. This initial input is termed the **activation energy** of the reaction and can be envisaged as a barrier that needs to be overcome to enable it to occur.

Many biological reactions would not occur fast enough to be useful without the help of a class of molecules called **enzymes**, which act as catalysts. A catalyst is a substance that speeds up a reaction by lowering its activation energy. Enzymes are discussed in more detail in Chapter 6.

 Check your understanding of the concepts covered in this section by answering the questions in the e-book.

1.5 Water plays a key role in supporting life

Around 70% of the weight of a cell is water, and most biochemical reactions take place within this aqueous environment. In this section we will explore the special properties of water, an apparently simple molecule, and explain how these properties have enabled it to play its fundamental role in supporting life.

The polarity of water molecules gives them special properties

As described above, and shown in Figure 1.5, water is a polar molecule. The uneven distribution of charge allows water molecules to interact with each other, via hydrogen bonding, to form a cohesive network (Figure 1.17a). This extensive hydrogen bonding gives water a high boiling point and a high surface tension. The latter enables it to be drawn up from the roots of plants to their foliage, sometimes over great distances. Water also has a high specific heat: it can absorb or release heat through breaking or forming hydrogen bonds without undergoing rapid changes in temperature, and it has a high heat of vaporization, since hydrogen bonds must be broken to allow water molecules to escape the liquid form. As organisms are largely composed of water, it helps to protect them from temperature changes through its high specific heat. Water can also help to cool organisms and their environment by evaporation, due to its high heat of vaporization.

Water is a good solvent

The polarity of water molecules makes it a good solvent. As shown in Figure 1.17(b) charged particles such as sodium and chloride ions can form extensive electrostatic interactions with surrounding water molecules. Molecules that carry positive or negative charge also dissolve readily in water, and so do molecules that are polar, but not charged. As shown in Figure 1.15(c), molecules such as glucose that have polar groups form hydrogen bonds with surrounding water molecules. Even large molecules, such as proteins (discussed in Section 1.7) can dissolve in water through having regions that are charged or polar.

Hydrophilicity and hydrophobicity describe molecules' affinity for water

Molecules that dissolve in water are often termed **hydrophilic**, meaning 'water-loving'. The opposite of hydrophilic is **hydrophobic** ('water-fearing'). Hydrophobic molecules usually contain mainly carbon and hydrogen atoms and they do not have charged or polar groups. They cannot form hydrogen bonds and hence do not dissolve in water. If you have used oils in the kitchen you will know that they do not readily mix with water and tend to form a separate layer. This is due to their chemical structure, which contains a long string of carbons bonded to hydrogen (hydrocarbon chains), as shown in Figure 1.18.

Hydrophobic interactions can hold molecules together

The term hydrophobic forces, or hydrophobic interactions, refer to the forces that impose organization on hydrophobic molecules due to the necessity to minimize their interaction with water. As shown in Figure 1.19, when hydrophobic molecules are placed in water, they prevent some of the water molecules from forming hydrogen bonds. This is an energetically unfavourable situation, which can be minimized if the hydrophobic molecules are held together.

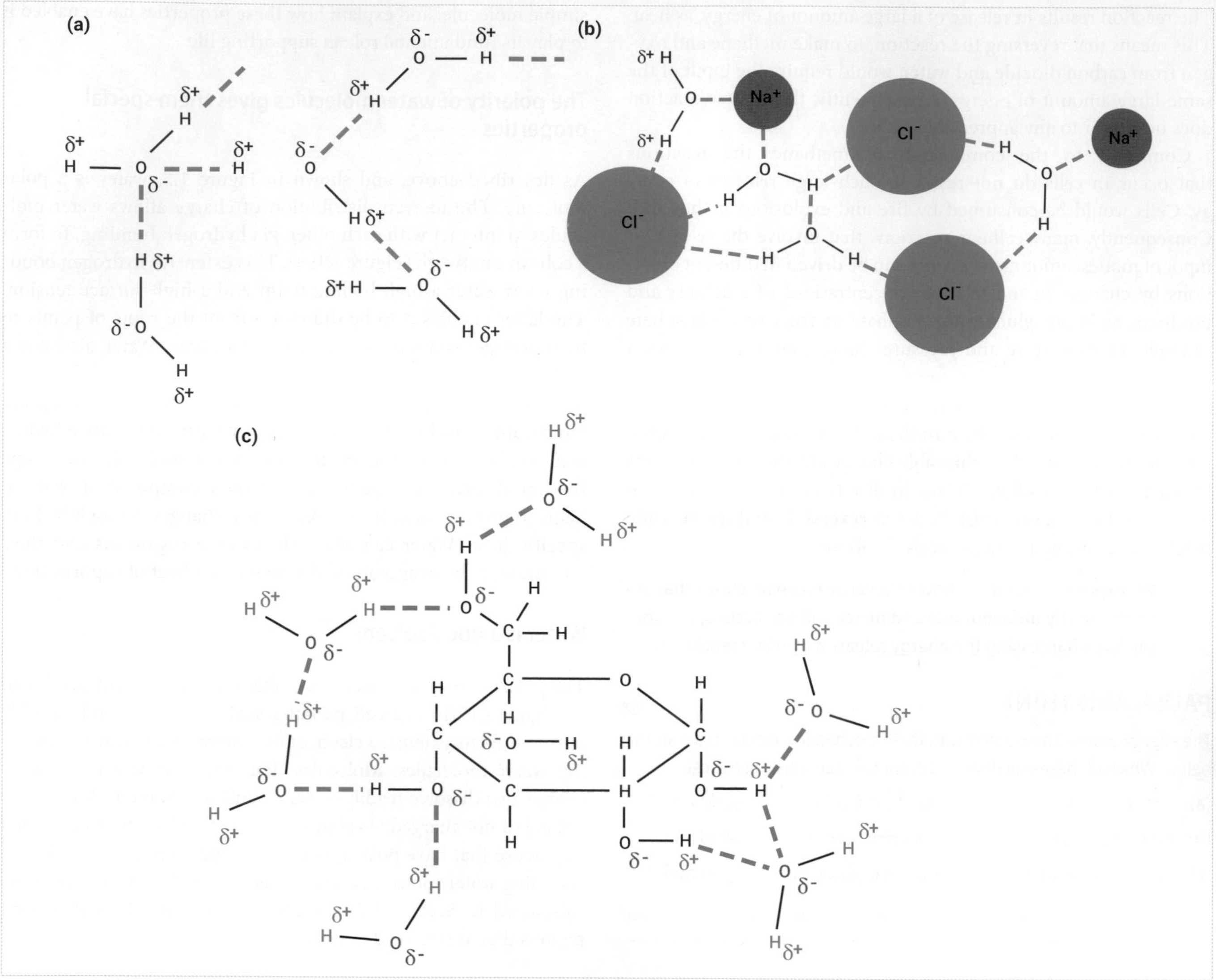

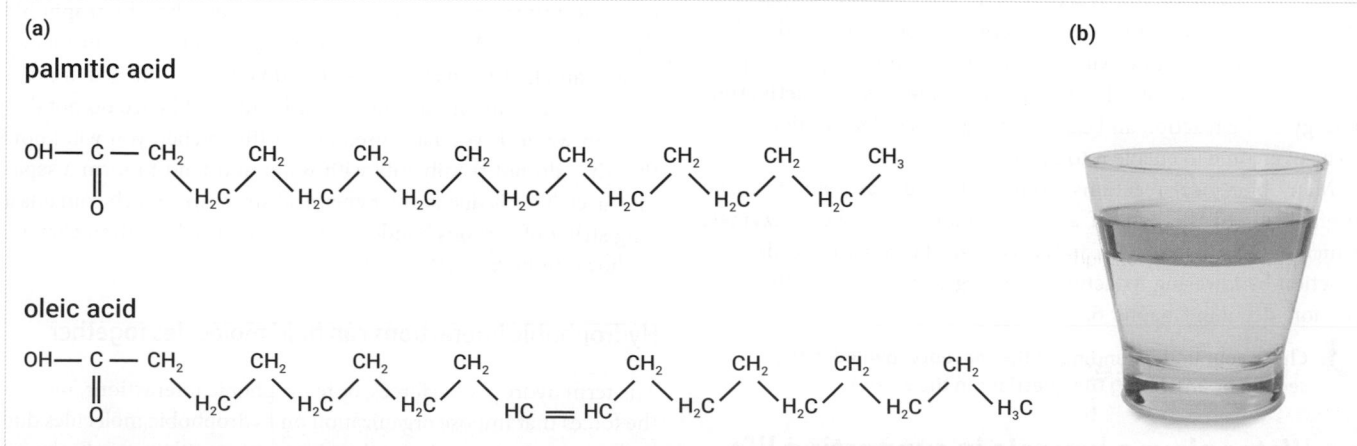

Figure 1.17 Polarity of water molecules. (a) Hydrogen bonding network of water. (b) Solvation of sodium and chloride ions. (c) Glucose is soluble in water as it contains polar groups that can form hydrogen bonds with the water molecules.

(a)

palmitic acid

oleic acid

(b)

Figure 1.18 Structures of fatty acids, the main chemical components of fats and oils. (a) Fatty acids vary in the length of the hydrocarbon chain, and whether they contain carbon–carbon double bonds, as in the examples shown here. In all cases the long hydrocarbon chains make these molecules hydrophobic, so that oils are immiscible with water, as illustrated in (b)

Source: (b) Shutterstock.

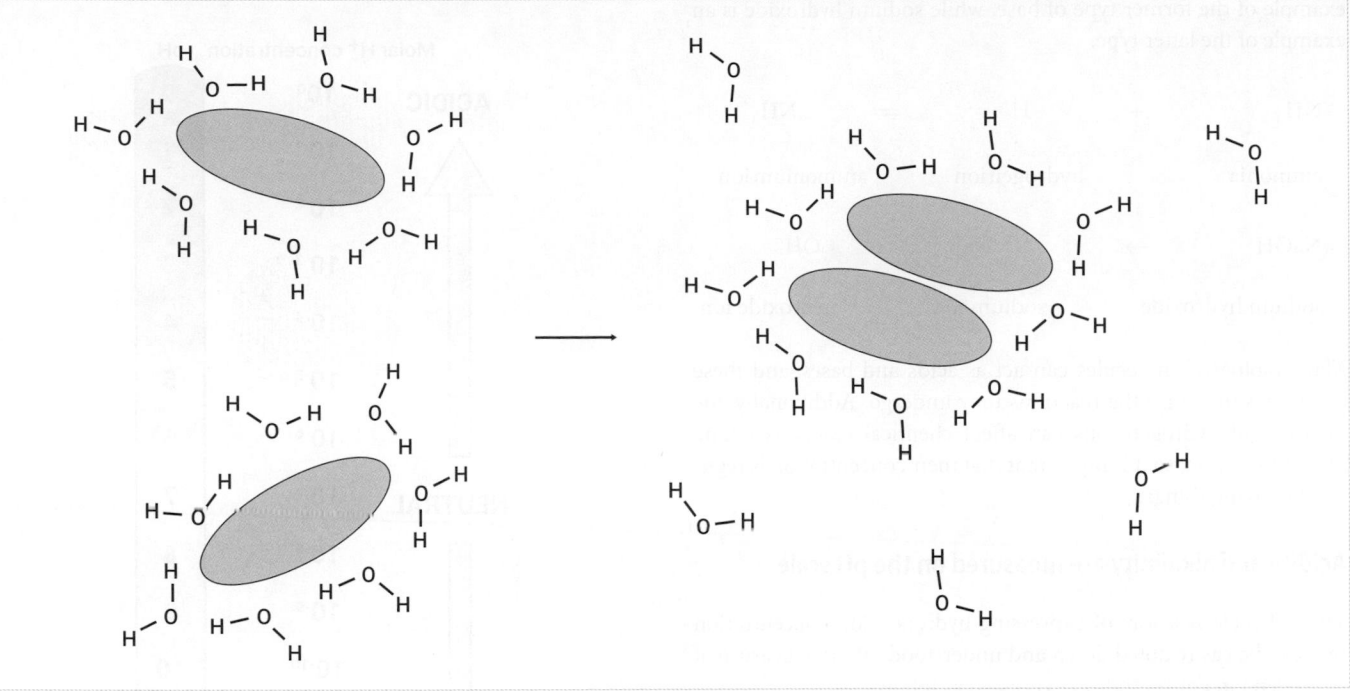

Figure 1.19 Hydrophobic interactions. Water molecules form an ordered structure around hydrophobic molecules (shown in grey), thus disrupting hydrogen bonding between the water molecules. When two hydrophobic molecules aggregate the area exposed to water is reduced. This frees up some of the water molecules that previously formed a 'shell' around them.

Source: A. L. Jonsson, M. A. J. Roberts, J. L. Kiappes, K. A. Scott. Essential chemistry for biochemists. *Essays Biochem* 2017; 61(4): 401–427. doi: https://doi.org/10.1042/EBC20160094.

Hydrophobic interactions are sometimes classed as a type of weak chemical bond. Although no specific links are formed between molecules, these interactions play an important role in the structure of proteins, DNA, and other cellular molecules.

Check your understanding of the concepts covered in this section by answering the questions in the e-book.

PAUSE AND THINK

Which of the molecules below would be hydrophilic, and which hydrophobic?

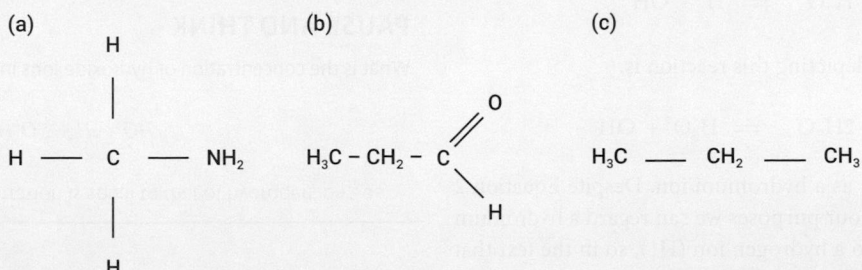

(a)

(b)

(c)

Answer: (a) and (b) Hydrophilic as the NH_2 and C=O groups are electronegative and make the molecules polar. (c) Hydrophobic. This is a non-polar molecule.

1.6 Acids and bases: molecules that release or bind hydrogen ions

An **acid** is a substance that will release hydrogen ions (protons) into solution. A common example is hydrochloric acid (HCl), which breaks down completely to its component hydrogen and chloride ions.

$$HCl \rightarrow H^+ + Cl^-$$

hydrochloric acid   hydrogen ion   chloride ion

Conversely, a **base** is a substance that absorbs hydrogen ions from solution. Bases can either bind hydrogen ions themselves or work more indirectly by releasing hydroxide ions, which can combine with hydrogen ions to form water. As shown below, ammonia is an example of the former type of base, while sodium hydroxide is an example of the latter type.

$$NH_3 \quad + \quad H^+ \quad \rightleftharpoons \quad NH_4^+$$

ammonia   hydrogen ion   ammonium ion

$$NaOH \quad \rightarrow \quad Na^+ \quad + \quad OH^-$$

sodium hydroxide   sodium ion   hydroxide ion

Many biological molecules can act as acids and bases, and these properties influence the reactions they undergo. Additionally, the presence of hydrogen ions can affect chemical reactions taking place in solution, so it is important that their concentration is regulated in living things.

Acidity and alkalinity are measured on the pH scale

The pH scale is a way of expressing hydrogen ion concentration that can be easily noted down and understood. pH is a measure of hydrogen ion concentration expressed as follows:

$$pH = -\log_{10}[H+]$$

The pH scale is shown in Figure 1.20. pH 7 is termed 'neutral' pH, while solutions with pH values below 7 are termed acidic, and solutions with pH values above 7 are termed alkaline. It is important to note that a change of 1 unit on the pH scale represents a 10-fold change in hydrogen ion concentration.

pH 7, or neutral pH, reflects the concentration of hydrogen ions in pure water at 25°C, which is 10^{-7} M. Water dissociates into hydrogen ions and hydroxide ions, as shown in Equation 1 below, only to a very small extent.

Equation 1 $H_2O \quad \rightleftharpoons \quad H^+ + OH^-$

A more accurate way of depicting this reaction is:

Equation 2 $2H_2O \quad \rightleftharpoons \quad H_3O^+ + OH^-$

The H_3O^+ ion is known as a hydronium ion. Despite Equation 2 being more accurate, for our purposes we can regard a hydronium ion as being equivalent to a hydrogen ion (H^+), so in the text that follows we will continue to refer to hydrogen ions in aqueous solution simply as hydrogen ions or protons.

For us, the significance of pH 7 is that the pH within cells, as well as the pH of blood and many other biological fluids, needs to be kept at or near this neutral pH for biological reactions and biological functions to occur. The maintenance of neutral pH is constantly threatened by reactions that produce molecules that can act as acids and bases. Maintenance is achieved by a system known as buffering.

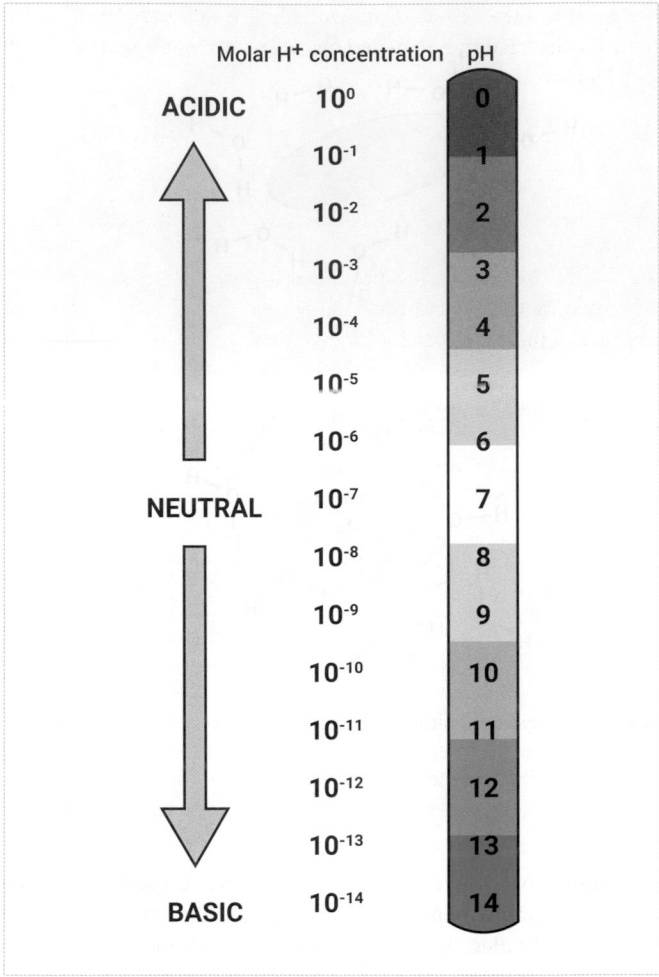

Figure 1.20 The pH scale, showing the relationship between hydrogen ion concentration and pH.

Buffering depends on weak acids and their conjugate bases

To explain how buffering works, we first need to introduce the concept of strong and weak acids. **Strong acids** are those that dissociate completely in solution. Hydrochloric acid, shown at the beginning of Section 1.6, is an example of a strong acid. Many acids, however, dissociate only partially in solution and are termed

weak acids. Ethanoic acid, shown below, is an example of a weak acid. Its dissociation is depicted below as a reversible reaction, with two half arrows.

$$CH_3COOH \quad \rightleftharpoons \quad H^+ \quad + \quad CH_3COO^-$$

ethanoic acid   hydrogen ion   ethanoate ion

As it only partially dissociates, ethanoic acid in solution consists of a mixture of the acid form, shown on the left of the equation above, and the ethanoate ion, shown on the right. As the ethanoate ion can bind a hydrogen ion to reform ethanoic acid, the ethanoate ion can be termed the **conjugate base** of ethanoic acid. Ethanoic acid in solution is therefore a mixture of the acid and its conjugate base, and the position of equilibrium for its dissociation reaction determines the relative proportions of ethanoic acid and ethanoate ions in solution.

For any weak acid, the proportion of acid and its conjugate base can be quantified as its **dissociation constant**, K_a.

$$HA \quad \rightleftharpoons \quad H^+ \quad + \quad A^-$$

acid   conjugate base

$$K_a = \frac{[H^+][A^-]}{[HA]}.$$

Different weak acids each have a characteristic K_a value: the larger the K_a, the further to the right the equilibrium for their dissociation lies, the greater the tendency to dissociate, and the stronger the acid. For ethanoic acid K_a at 25 °C = 1.76×10^{-5}.

However, the K_a value is not much used by biochemists, as there is another way of expressing the strength of an acid by a much more convenient term—the **pK_a** value. The two are related by the equation:

$$pK_a = -\log_{10} K_a$$

Thus, ethanoic acid has a pK_a of 4.76. As pK_a is the negative log of K_a, the stronger the acid the lower its pK_a.

The pH of a solution can be related to the extent of dissociation of a weak acid

If we look again at the scheme for dissociation of ethanoic acid, it tells us that the addition of hydrogen ions to a solution of ethanoic acid would result in more CH_3COOH being formed, as the excess hydrogen ions would react with ethanoate ions. In other words, the reaction would be 'driven' in the direction of the leftwards facing arrow by the addition of hydrogen ions:

$$CH_3COOH \rightleftharpoons H^+ + CH_3COO^-$$

Conversely, the addition of hydroxide ions to the solution would drive the reaction to the right, as the hydroxide ions would react with hydrogen ions to form water and would therefore reduce the concentration of hydrogen ions overall. The further dissociation of ethanoic acid would produce more hydrogen ions to compensate.

The pH (hydrogen ion concentration) of a solution is therefore related to the extent of dissociation of an acid. The relationship can be quantified by the Henderson–Hasselbalch equation:

$$pH = pK_a + \log 10 \left(\left[CH_3COO^- \right] \right) / \left(\left[CH_3COOH \right] \right)$$

As an example, for a solution containing 0.1 M ethanoic acid and 0.1 M sodium ethanoate:

$$pH = 4.76 + \log_{10} \frac{[0.1]}{[0.1]}$$
$$= 4.76 + 0 = 4.76$$

If the mixture contains 0.1 M ethanoic acid and 0.2 M sodium ethanoate:

$$pH = 4.76 + \log_{10} \frac{[0.2]}{[0.1]}$$
$$= 4.76 + \log_{10} 2 = 4.76 + 0.30 = 5.06$$

The Henderson–Hasselbalch equation can be generalized to apply to any weak acid by stating it as:

$$pH = pK_a + \log\ 10 \left(\left[\text{conjugate base} \right] \right) / \left(\left[\text{acid} \right] \right) \quad \text{or as}$$
$$pH = pK_a + \log\ 10 \left(\left[A^- \right] \right) / \left(\left[HA \right] \right)$$

where HA is any acid, and A^- is its conjugate base.

pK_a represents the pH at which a weak acid is half dissociated

It follows from the Henderson–Hasselbalch equation that pK_a is the pH at which a weak acid is half dissociated, i.e. when we have a 50:50 mix of the acid and its conjugate base. This point is illustrated in the example above where the concentrations of ethanoic acid and ethanoate are equal at 0.1 M.

Weak acids buffer at pH values around their pK_a

A **buffer** is a solution that resists changes in pH caused by the addition of acids or bases. A 50:50 mixture of ethanoic acid and ethanoate ions constitutes a buffer solution, because, as described above, the addition of acid is compensated for by the formation of more ethanoic acid, and the addition of base is compensated for by the formation of more ethanoate ions. The overall outcome is that the concentration of hydrogen ions, and hence the pH, remains relatively constant despite the addition of small amounts of acid or base.

There are several examples of weak acid and conjugate base pairs that act as buffers in biological systems. The ammonium ion and ammonia pair, shown below, is one such system and buffers urine.

$$NH_3 \ + \ H^+ \ \rightleftharpoons \ NH_4^+$$

acid   conjugate base

Buffers work best at pH values close to their pK_a, as that is where there is plenty of both the acid and its conjugate base. At low pH (high hydrogen ion concentration), ethanoic acid is almost entirely present as CH_3COOH, so if more hydrogen ions are added there are not enough ethanoate ions to soak them up, and the pH will drop further. At high pH (low hydrogen ion concentration) the solution will contain mainly the ethanoate ion, and there is no 'spare' ethanoic acid to donate additional hydrogen ions on addition of hydroxide ions.

A buffer can be effective within a pH range from one pH point below its pK_a to one pH point above. Maintaining a constant pH within cells and body fluids is vital for the integrity of many biological molecules and for biological reactions to occur. Consequently, several weak acids can act as buffers within living things. In Real World View 1.1 we learn about the devastating consequences for marine animals when the pH of their environment changes. This emphasizes the importance of pH and buffering for maintaining life.

PAUSE AND THINK

Would ethanoic acid make a good buffer within cells? Why or why not?

Answer: No. Ethanoic acid has a pK_a of 4.76, which means it buffers effectively in the pH range 3.76–5.76. Reactions in cells mainly need to take place at or near to pH 7 and ethanoic acid would not be a good buffer to maintain this pH.

 Check your understanding of the concepts covered in this section by answering the questions in the e-book.

1.7 Biological macromolecules: structure and function

While roughly 70% of the weight of a typical bacterial or animal cell is made up of water, its remaining contents are a complex mixture of different ions and chemical compounds. However, despite this complexity, most of the organic molecules in a cell belong to only four main families, with the other components present in only small amounts. The point is illustrated in Table 1.6.

We somewhat arbitrarily divide cellular organic molecules into two categories, small molecules and macromolecules. Small molecules have molecular masses in the range of a few hundred Daltons or less. (A Dalton (Da) is a unit of atomic or molecular mass effectively equal to the mass of a hydrogen atom.) Within cells we find four main classes of macromolecule: carbohydrates, lipids, nucleic acids, and proteins. Although their structure and function will be covered in more detail in later chapters (nucleic

Table 1.6 Approximate chemical composition of a bacterial cell

	Percentage of total cell weight	Number of types of each molecule
Water	70	1
Inorganic ions	1	20
Sugars and precursors	1	250
Amino acids and precursors	0.4	100
Nucleotides and precursors	0.4	100
Fatty acids and precursors	1	50
Other small molecules	0.2	~300
Macromolecules (proteins, nucleic acids, and polysaccharides)	26	~3000

REAL WORLD VIEW 1.1   **Ocean acidification: the other CO_2 problem**

Much attention has been given to the impact of carbon dioxide emissions on Earth's climate, but an additional problem caused by human activity, the effect of CO_2 on the pH of the ocean, is less well known. Before the start of the industrial era the average pH of the ocean at its surface was around 8.2. It is now 8.1. This change may seem trivial, but the fact that it has occurred extremely fast in geographical terms and is likely to continue, and continue at an increasing rate, should be a cause of concern. The delicate balance of life within the ocean could be profoundly upset and mass extinctions would follow.

CO_2 affects the ocean's pH because it dissolves to form carbonic acid, a weak acid that dissociates to form hydrogen and hydrogen carbonate ions.

$$CO_2 + H_2O \ \rightleftharpoons \ H_2CO_3 \ \rightleftharpoons \ H^+ + HCO_3^-$$

Ocean waters are already saturated with carbonate ions (CO_3^{2-}), which are derived from the slow dissolution of rock and sediment. Some, but not all, of the excess hydrogen ions are soaked up by carbonate to produce more hydrogen carbonate, but carbonate cannot be replaced quickly. The net effect is a decrease in pH and in the concentration of carbonate ions, both of which impact on marine life. Acidification affects the respiratory and metabolic functions of animals that have evolved to breathe in water, while reduced carbonate will make it difficult for a wide variety of organisms, including phytoplankton, to build their shells and skeletons. The loss of phytoplankton could have a catastrophic effect on marine life.

acids in Chapter 3, proteins in Chapter 6, and carbohydrates in Chapter 7), it is useful to have an overview here. These macromolecules are formed by linking together subunits consisting of small molecules. When many small molecules that are identical or similar in structure are linked together, the process is called polymerization and the resulting macromolecule can be called a **polymer**. The individual subunits are called **monomers**. Thus, as explained below, carbohydrate macromolecules are polymers of sugars, nucleic acids are polymers of nucleotides, and proteins are polymers of amino acids. The fourth class of macromolecules, lipids, contain small subunits called fatty acids, but these are not joined together in long chains, so lipid molecules are not referred to as polymers.

Carbohydrates

Carbohydrates are polymers of single sugar molecules (monosaccharides) such as glucose. Glucose (shown in Figure 1.21a) has the chemical formula $C_6H_{12}O_6$, and a general formula for many monosaccharides is $C_n(H_2O)_n$, giving rise to the term carbohydrate, from 'carbon' and 'water'.

Sugars are an important source of energy, and polymers of sugars are used as energy stores, but this is not their only function. Glycogen, starch, and cellulose are large polymers known as polysaccharides formed by joining together glucose units in a slightly different manner in each of the three cases. Glycogen and starch function as energy stores in animals and plants, respectively, while cellulose provides structural strength in plants. All three molecules consist only of glucose monosaccharides linked together. More complex polysaccharides exist; for example, chitin, shown in Figure 1.21c, forms the supportive exoskeleton of insects and cell walls in fungi.

PAUSE AND THINK

From looking at Figure 1.21, what can you say is the main structural difference between starch and cellulose and between starch and glycogen?

Answer: Starch consists of glucose monosaccharides all linked together in the same orientation to form a single chain. In cellulose two different orientations of the glucose monosaccharide alternate along the chain. In glycogen the orientation of the glucose monosaccharides is the same as in starch, but the chain is branched.

Lipids

Lipids are biological molecules that are insoluble in water. The two best-known classes of lipids, which have structural features in common, are triglyceride fats or **triacylglycerols**, used in cells for energy storage, and cell membrane lipids. Their small subunits are fatty acids, such as oleic acid and palmitic acid (see Figure 1.22). Fatty acids consist of carbon and hydrogen atoms linked in long hydrocarbon chains, with a carboxylic acid group at one end. The hydrocarbon chains are highly hydrophobic, making lipids insoluble in water.

Unlike the polysaccharides described above, lipid macromolecules do not consist of multiple fatty acid monomers linked to each other. Instead, a glycerol molecule acts as the backbone to which two or three fatty acids are linked, as shown in Figure 1.22. The triacylglycerols, which contain three fatty acids, act as highly concentrated forms of energy storage and are packed into adipose cells as lipid droplets. In membrane **glycerophospholipids**, the third fatty acid is replaced by a polar group, which is linked to glycerol by a phosphate ion. Other membrane lipids have variations of this structure, but crucially for their function, all have a hydrophilic 'head group' linked via a backbone structure to two hydrophobic hydrocarbon tails.

Membrane lipids can be described as **amphipathic** molecules, referring to one end of the molecule (the head group) being hydrophilic, and the other end (the fatty acid hydrocarbon tails) being hydrophobic. When they are placed in water this property enables them to group spontaneously to form small vesicles, each bounded by a lipid bilayer in which the hydrophilic head groups face outward and protect the hydrophobic hydrocarbon 'tails' from the polar water molecules (as shown in Figure 1.23). The lipid bilayer has the same basic structure as the cell membrane, and it is easy to envisage how this capacity of lipids to form vesicles may have aided the evolution of cellular life, by enclosing small pockets of reacting and replicating molecules.

Fatty acids vary in the length of their hydrocarbon chain, and in the number of carbon–carbon double bonds they contain, and these variations affect their specific properties and those of the cell membranes that contain them. Fatty acids containing only single bonds are termed saturated fatty acids (palmitic acid being an example), while those with one double bond are termed monounsaturated (e.g. oleic acid) and those with two or more double bonds are termed polyunsaturated fatty acids.

Before moving on from this section, it is worth noting that the term 'lipid' also covers smaller insoluble biological molecules such as cholesterol (an important component of animal cell membranes) and steroid hormones (as shown in Figure 1.24).

PAUSE AND THINK

At first glance, cholesterol and steroid hormones have very little in common with triacylglycerols. Which shared characteristic leads to all these compounds being classed as lipids?

Answer: All are hydrophobic and hence insoluble in water but soluble in organic solvents.

Nucleic acids

Nucleic acids are the class of molecules that enable cells and hence organisms to replicate. This is because they carry within their chemical structure both the capacity to act as a template for making an exact copy of themselves (i.e. to act as genetic material), and the capacity to act as 'instruction manuals' directing the synthesis of proteins. Proteins enable cells to carry out the multiplicity of functions required for life.

The two major classes of nucleic acid found in cells are **deoxyribonucleic acid** (DNA) and **ribonucleic acid** (RNA).

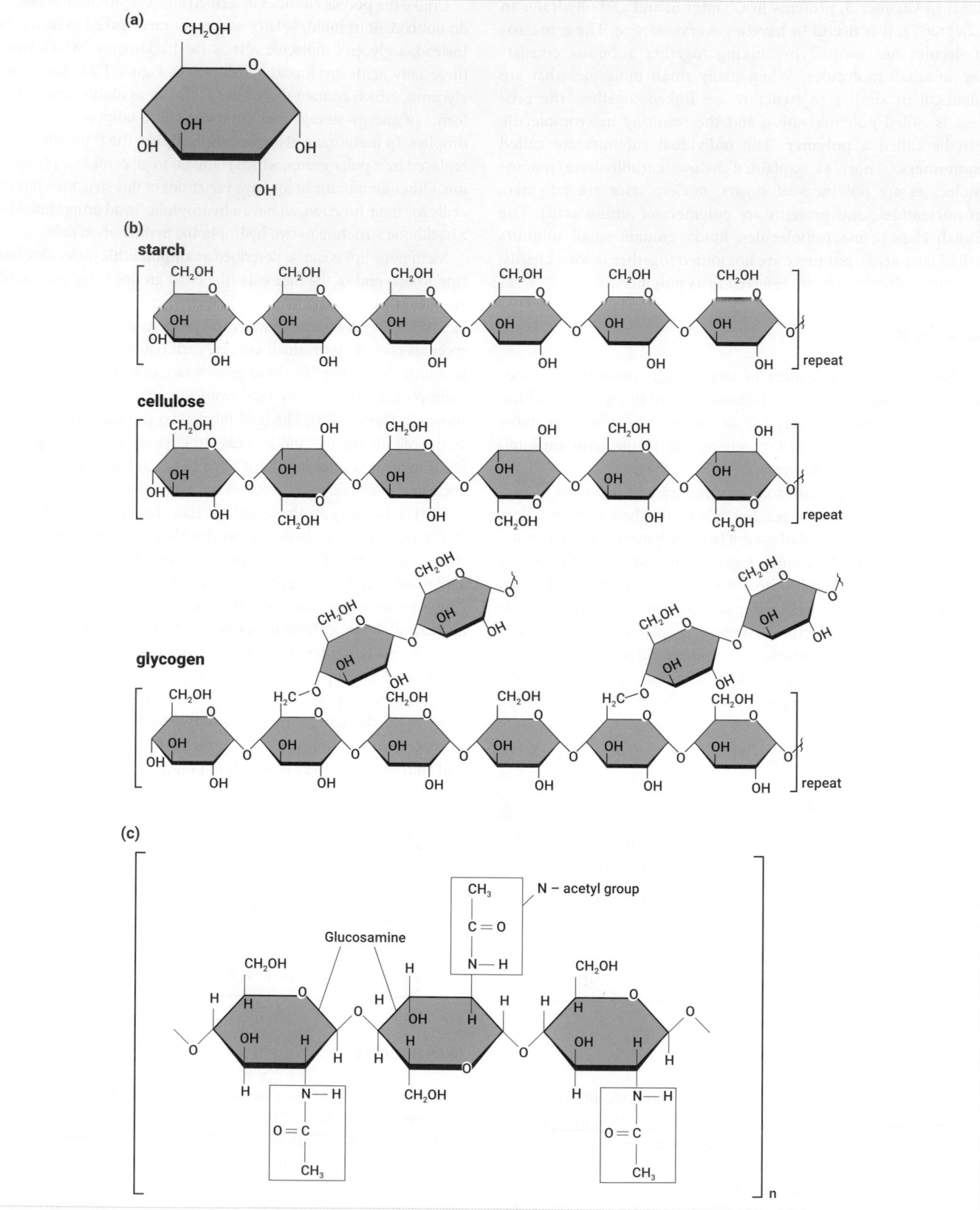

Figure 1.21 Carbohydrate structures. (a) The structure of glucose, a simple sugar. (b) The energy storage molecules glycogen and starch, and the structural plant molecule cellulose, are polymers of glucose, which are linked together in different ways to give polysaccharides with different structures and properties. (c) Chitin, a polymer of a more complex sugar. Chitin is found in the insect exoskeleton and in fungi.

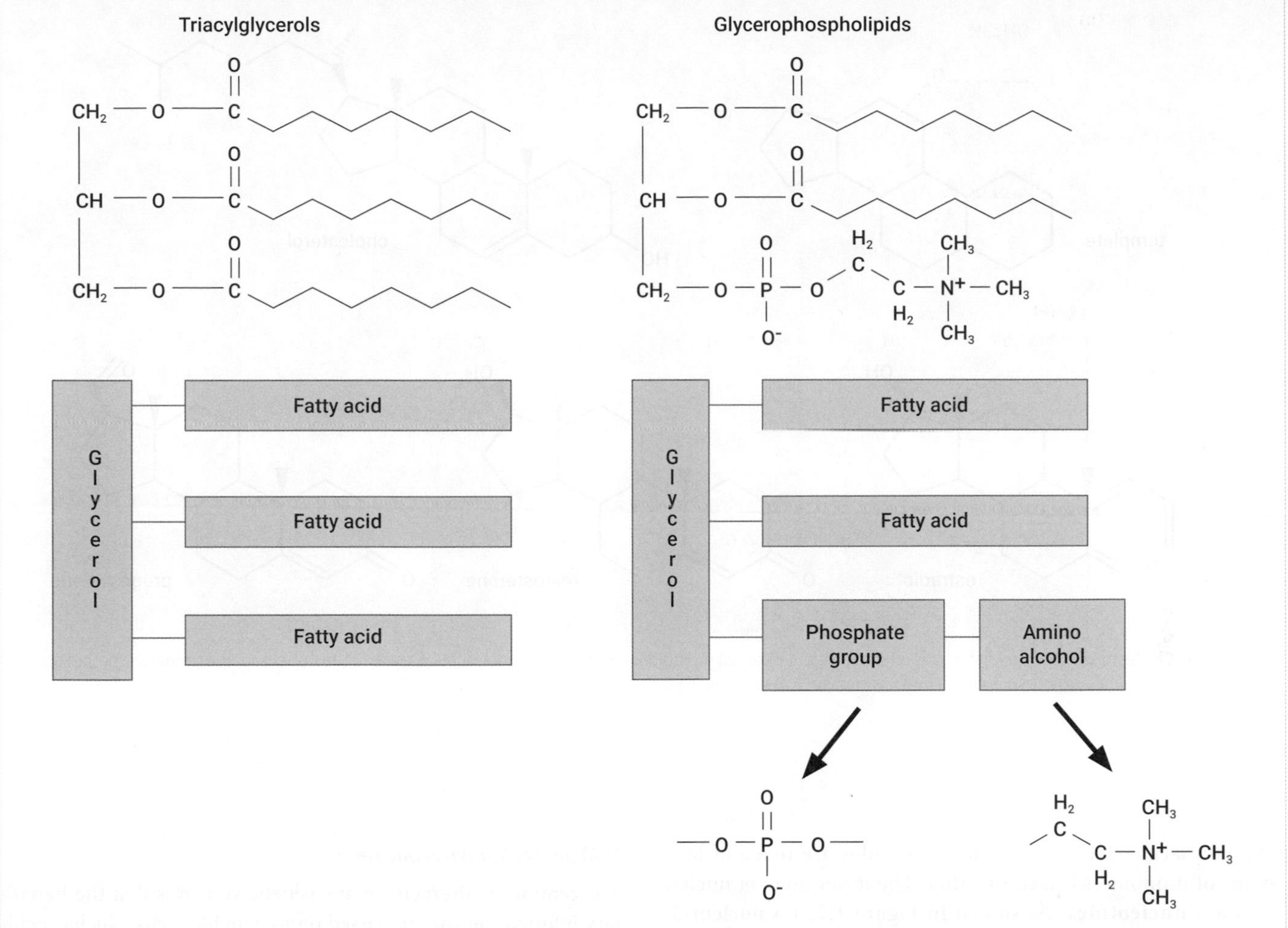

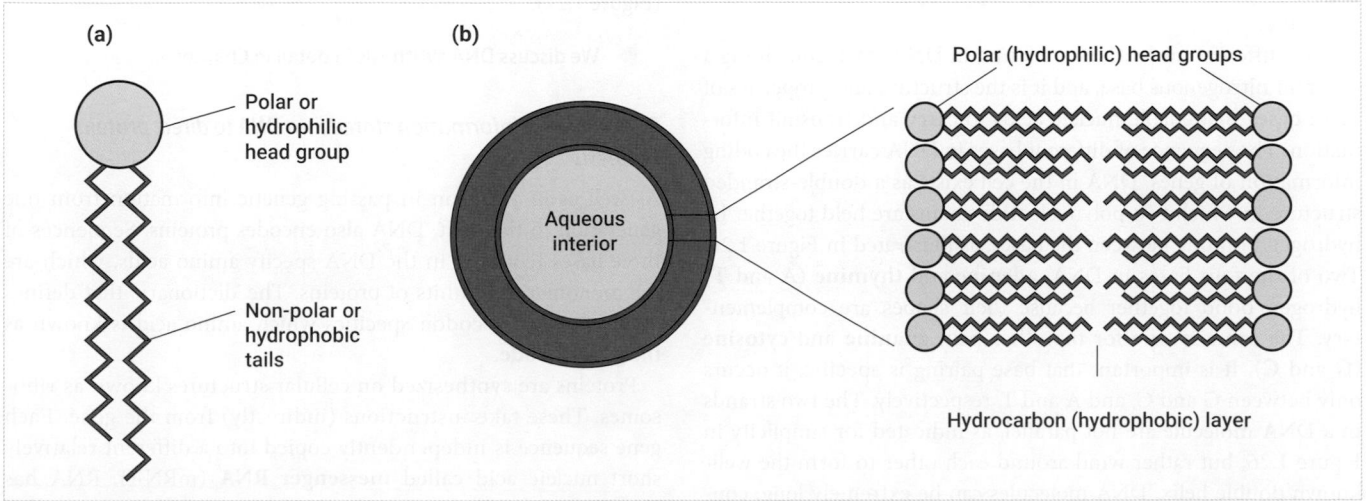

Figure 1.22 A comparison of the general structures of triacylglycerols and glycerophospholipids. The 'amino alcohol' group shown is one example of a range of charged or polar 'head' groups found in membrane lipids.

Source: Crowe, J. & Bradshaw, T. (2021). *Chemistry for the Biosciences: The Essential Concepts* (4th ed.). Oxford University Press.

Figure 1.23 Membrane lipids. These are amphipathic (a) and form vesicles with a bilayer structure in water (b).

Source: Snape, A. & Papachristodoulou, D. (2018). *Biochemistry and Molecular Biology* (6th ed.). Oxford University Press.

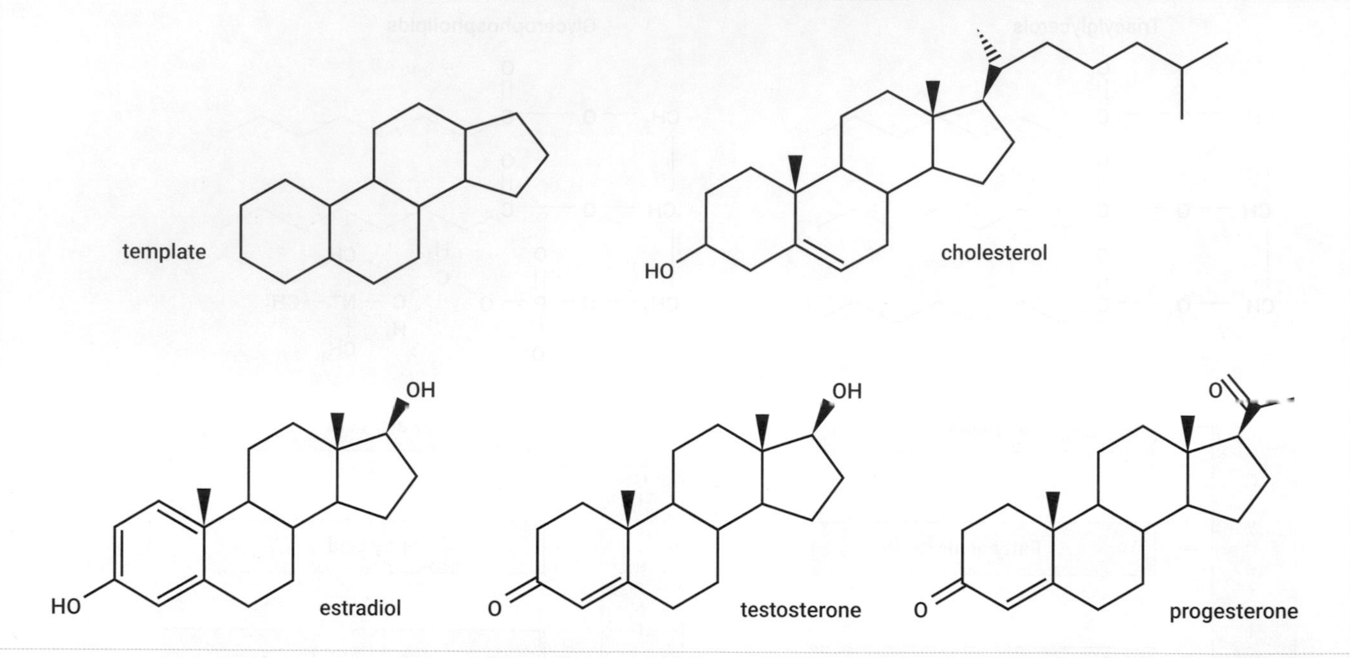

Figure 1.24 Cholesterol. Cholesterol is an important component of cell membranes. Steroid hormones, such as estradiol, testosterone, and progesterone, are lipids with related structures, based on a common template of fused carbon rings.

Like polysaccharides, these macromolecules are made of long chains of monomers linked together. The monomers of nucleic acids are **nucleotides**. As shown in Figure 1.25a a nucleotide has three parts to its structure, a five-carbon sugar (deoxyribose or ribose), a nitrogen-containing ring structure known as a nitrogenous base, and up to three phosphates. Nucleotides can be linked together via phosphate groups, as shown in Figure 1.25b, to create polymers of variable length in which the bases protrude from the sugar–phosphate backbone of the molecule.

Four different nucleotides are found in DNA, each containing a different nitrogenous base, and it is the structure and properties of these bases that enable nucleic acids to carry and transmit information. The sequence of different bases in DNA carries the coding information of genes. DNA in the cell exists as a double-stranded structure in which two polynucleotide chains are held together by hydrogen bonding between the bases, as illustrated in Figure 1.26. Two of the four bases in DNA, **adenine** and **thymine** (**A** and **T**) hydrogen bond together because their shapes are complementary. The same is true for the other pair, **guanine** and **cytosine** (**G** and **C**). It is important that base pairing is specific; it occurs only between G and C, and A and T, respectively. The two strands in a DNA molecule are not parallel, as indicated for simplicity in Figure 1.26, but rather wind around each other to form the well-known double helix. DNA molecules can be extremely long, consisting of millions of base pairs.

▶ **We learn more about the structure of DNA in Chapter 3.**

DNA directs its own replication

The central requirement of any genetic system is that the hereditary information can be passed on to daughter cells. Nucleic acids fulfil this requirement by directing their own replication. The two strands of DNA are separated and each strand acts as a template for the assembly of new partner strands. An A on the template strand matches a T on the new strand, G is matched to a C and vice versa. This results in two new DNA molecules identical to the original (Figure 1.27).

▶ **We discuss DNA synthesis in detail in Chapter 3.**

RNA transmits information stored into DNA to direct protein synthesis

As well as its function in passing genetic information from one generation to the next, DNA also encodes proteins. Sequences of three bases (codons) in the DNA specify amino acids, which are the monomeric subunits of proteins. The 'dictionary' that defines which three-base codon specifies which amino acid is known as the **genetic code**.

Proteins are synthesized on cellular structures known as ribosomes. These take instructions (indirectly) from the gene. Each gene sequence is independently copied into a different relatively short nucleic acid called **messenger RNA** (mRNA). RNA has almost the same structure as a single strand of DNA except that the sugar is ribose rather than deoxyribose, and the base thymine (T) is replaced by uracil (U). The bases in RNA specifically pair

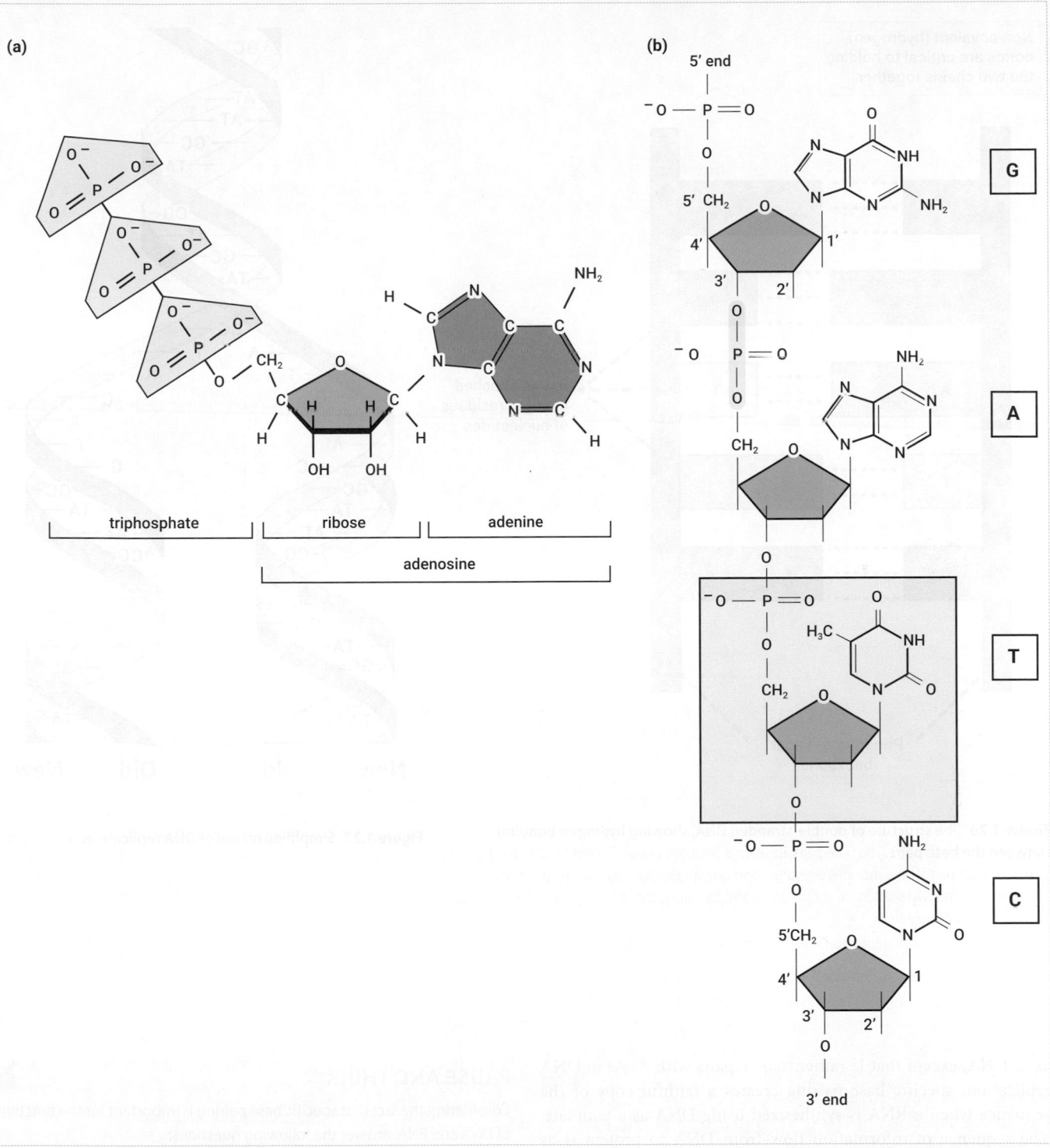

Figure 1.25 Structure of a nucleotide and polynucleotide. (a) This nucleotide example is adenosine triphosphate (ATP). Nucleotides can have up to three linked phosphate ions attached to a ribose or deoxyribose sugar (ribose in ribonucleotides and deoxyribose in deoxyribonucleotides). (b). Structure of a polynucleotide, in this case a short length of DNA. An individual nucleotide is enclosed in a grey box. Links between the nucleotides are via phosphodiester bonds (one is highlighted in yellow). The four nitrogenous bases found in DNA are designated A, C, G, and T, as explained in the text.

Source: Alberts et al. *Essential Cell Biology*, 4th edn. Garland Science.

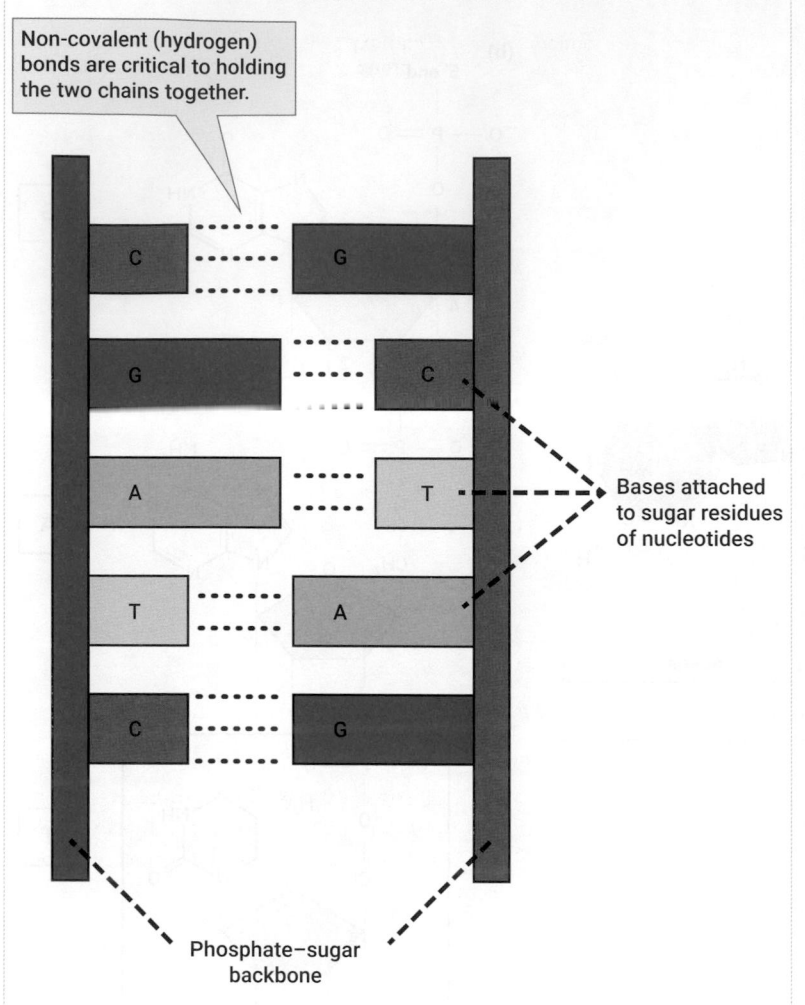

Figure 1.26 The structure of double-stranded DNA, showing hydrogen bonding between the base pairs. The base pairs are always between G and C or between A and T. Note that each base pair always includes one larger and one smaller base so that all base pairs are of the same size. The two strands are shown as parallel for clarity, but in fact they wind around each other to form a double helix.

Source: Snape, A. & Papachristodoulou, D. (2018). *Biochemistry and Molecular Biology* (6th ed.). Oxford University Press.

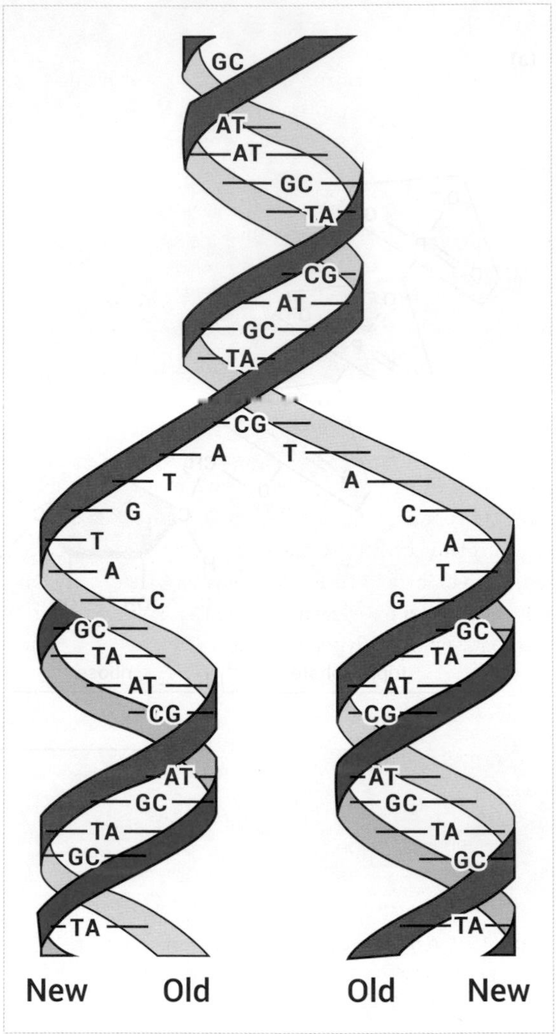

Figure 1.27 Simplified model of DNA replication.

as in DNA, except that U rather than T pairs with A. As in DNA replication, specific base-pairing creates a faithful copy of the sequence when mRNA is synthesized using DNA as a template. The sequence of information flow from DNA to protein is as shown below.

$$\text{DNA of gene 1} \rightarrow \text{mRNA 1} \rightarrow \text{protein 1}$$

$$\text{DNA of gene 2} \rightarrow \text{mRNA 2} \rightarrow \text{protein 2}$$

Details of the process by which mRNA base sequences are interpreted and used to synthesize specific proteins are covered in Chapter 6.

PAUSE AND THINK

Considering the fact that specific base pairing is important for the function of DNA and RNA, answer the following questions:

1. Which type of chemical bond is responsible for base pairing?

2. From looking at Figure 1.26, how many of these bonds form between A and T and between C and G?

3. How many bonds would you predict will form between U and A?

3. As U substitutes for T in RNA it forms two bonds with A.

2. Two bonds between A and T and three bonds between C and G.

1. Hydrogen bonds.

Answer:

Proteins

Proteins are polymeric molecules made up of amino acid monomers. The structure of an amino acid is shown in Figure 1.28. Amino acids contain a central carbon, to which an amino group, a carboxylic acid group, a hydrogen atom, and a side chain or R group are each linked by a covalent bond. Differences in the side chains distinguish amino acids, of which 20 are commonly found in proteins.

As outlined in the section on nucleic acids, the order in which amino acid sequences are linked together to form proteins is directed by the genetic code. Links are formed by peptide bonds between the carboxylic acid group of one amino acid and the amino group of the next, as shown in Figure 1.29. Peptide bonding does not involve the side chains.

The polymer of amino acids is known as a peptide (if short) or a polypeptide (if longer than around 30 amino acids). Polypeptides typically contain 100–1000 amino acids but can be longer or shorter. A protein consists of one or more polypeptide chains, which are folded up to form its three-dimensional structure, as shown in Figure 1.30. In Chapter 6 we learn how the characteristics of the amino acids that make up the polypeptide chain determine the protein folding.

PAUSE AND THINK

With 20 amino acids, how many different five-amino acid peptides can be made?

Answer: 20^5, i.e. 3,200,000 or 3.2×10^6. The number of 100-amino acid polypeptides is 20^{100}—a massive number!

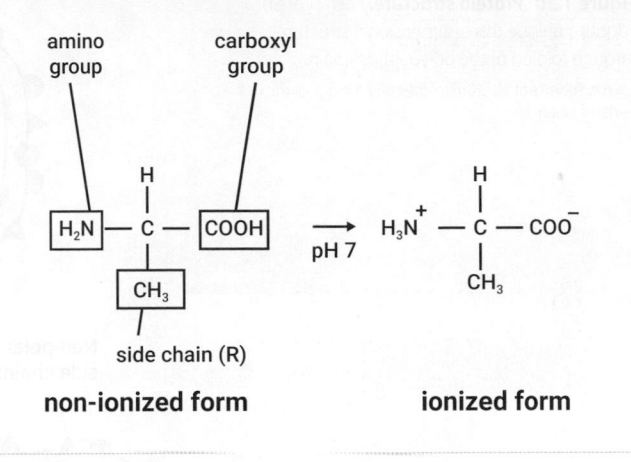

Figure 1.28 Structure of the amino acid, alanine. The central carbon, amino, and carboxyl groups, and hydrogen atom are common to all amino acids. The side chain or R group is variable. At cellular pH the amino group acts as a base and accepts an extra hydrogen ion from water, while the carboxylic acid loses a hydrogen ion, as shown. Thus, the amino acid carries a positive and a negative charge.

Source: Alberts et al. *Essential Cell Biology*, 4th edn. Garland Science.

It is the different properties of its amino acid side chains (charged, polar hydrophilic, or hydrophobic) and the order in which they are linked together that determine the folding, and hence the characteristic structure, of each protein, and it is the enormous variation in protein sizes and shapes that enables them to carry out a

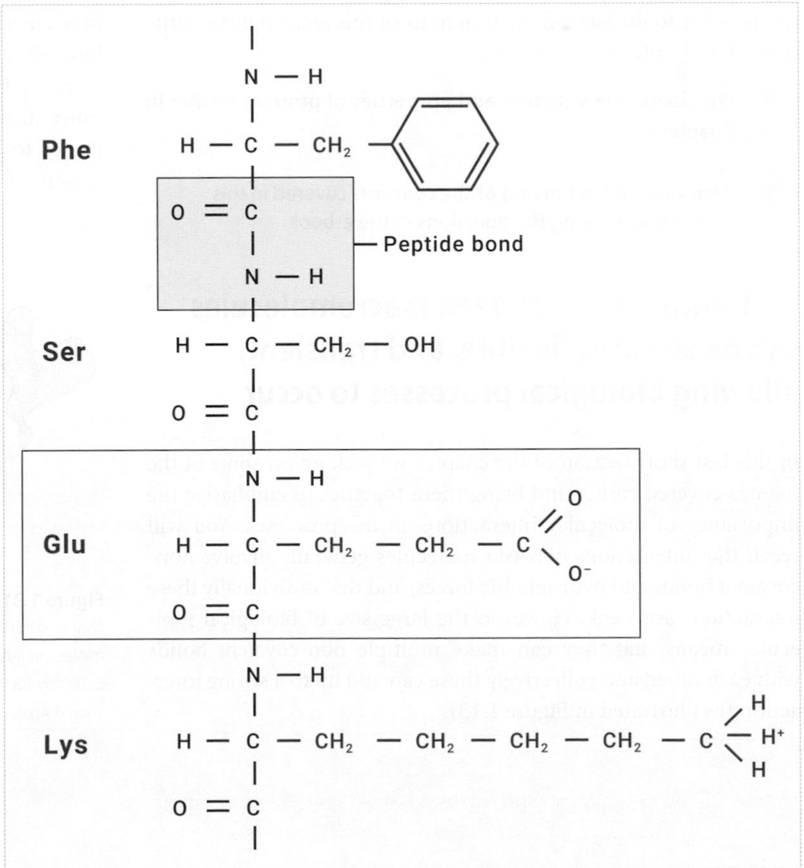

Figure 1.29 Amino acids are linked by peptide bonds (yellow box) to form polypeptides. The side chains of the four amino acids shown, phenylalanine (Phe), serine (Ser), glutamate (Glu), and lysine (Lys), are shown in red.

Source: Alberts et al. *Essential Cell Biology*, 4th edn. Garland Science.

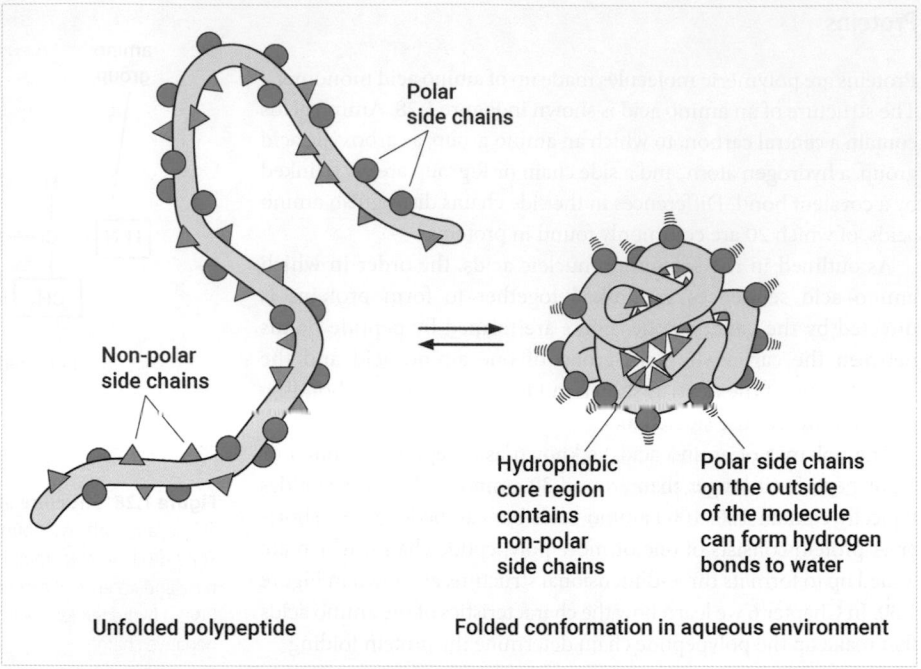

Figure 1.30 Protein structure. Each protein adopts a unique three-dimensional structure through folding of the polypeptide chain.
Source: Alberts et al. (2008). *Molecular Biology of the Cell*. Garland Science.

Polar side chains

Non-polar side chains

Hydrophobic core region contains non-polar side chains

Polar side chains on the outside of the molecule can form hydrogen bonds to water

Unfolded polypeptide

Folded conformation in aqueous environment

multiplicity of roles. Proteins are important structural components of cells, but they also have key functions as enzymes (biological catalysts) that orchestrate complex networks of metabolic reactions, as transport molecules that transmit materials and signals across cell membranes and carry substances around the body, as molecular motors that enable muscles to contract and cells to move, and as protective antibodies that fight infection. In fact, it is not possible to list here all the functions of this amazingly versatile class of molecule.

▶ We discuss the structure and properties of proteins further in Chapter 6.

🕘 Check your understanding of the concepts covered in this section by answering the questions in the e-book.

1.8 Interactions between macromolecules can be specific, flexible, and transient, allowing biological processes to occur

In this last short section of the chapter we pick up on some of the themes covered earlier and bring them together to emphasize the importance of molecular interactions in life processes. You will recall that interactions between molecules generally involve non-covalent bonds and hydrophobic forces, and that individually these interactions are weak. However, the large size of biological molecules means that they can make multiple non-covalent bonds with each other, and collectively these can add up to a strong interaction (as illustrated in Figure 1.13).

Another important point to consider is the requirement for molecules to be in close proximity for non-covalent interactions to occur. If we consider a molecule with a complex shape, as shown in Figure 1.31, it will only be able to come close to another molecule at multiple points, and hence form multiple bonds with that molecule if the second molecule has a similarly complex, but complementary, shape. This requirement for complementarity allows molecular interactions to be highly specific, and this in turn allows biological processes to be tightly regulated. For instance, enzymes work by binding to specific substrate molecules, and antibodies must distinguish exquisitely between 'self' and 'non-self' molecules to carry out their protective role. Because of their great variety of three-dimensional shapes, proteins are the 'masters' of

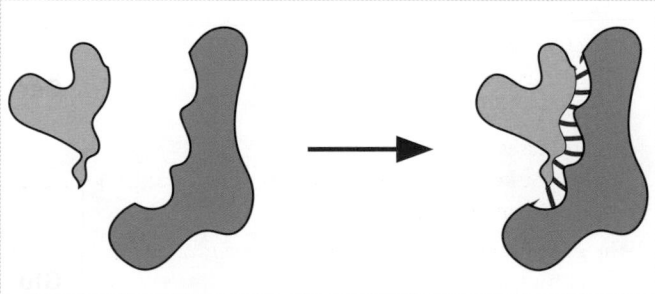

Figure 1.31 Macromolecular interaction. A large molecule with a complex shape, such as a protein, can undergo a very specific interaction with another molecule if it has a complementary shape. This is because non-covalent bonds can only form between regions of the molecules that are very close to each other.
Source: Adapted from Alberts et al. *Essential Cell Biology*, 4th edn. Garland Science.

specific interactions. Many proteins act as receptors by binding to individual smaller molecules such as hormones or other signalling molecules (collectively termed **ligands**), while others combine with multiple other molecules to build up highly complex structures, molecular machines such as the contractile units of muscle or the protein and RNA ribosome, a microscopic factory where new proteins are assembled.

Despite the stability afforded by multiple non-covalent interactions, the fact that individually they can easily be broken and reformed is also important in biological processes. A good example is that of base pairing by hydrogen bonding in DNA and RNA. For DNA replication to occur the two strands of DNA must be split apart by breaking hydrogen bonds, and the bonds are reformed as a new molecule is built on each template strand. Similarly, when RNA is made by copying the DNA sequence, the two strands of DNA are transiently separated over a short stretch, enabling the correct RNA sequence to be assembled by base-pairing with the DNA; the RNA molecule is released and the strands of DNA come together again.

Flexible molecular interactions also take place, and are important, in proteins. As protein folding depends on non-covalent interactions between the side chains of amino acids within the protein sequence, this means that proteins can undergo microscopic changes in their **conformation** (their three-dimensional shape) due to slight changes in these interactions. For instance, such changes take place when muscles contract, when an enzyme binds its substrate, and when a receptor protein binds its ligand. The conformational change of the enzyme enables it to catalyse a reaction, while that of the receptor sets off a chain of responses within the cell. However, these conformational changes must be transient. The muscle must relax again, the substrate is transformed into a product that must be released from the enzyme, and the chain of responses set off by the ligand–receptor interaction must cease when the need for it recedes. The transient and flexible nature of non-covalent interactions allows these dynamic changes to take place.

 Check your understanding of the concepts covered in this section by answering the questions in the e-book.

SUMMARY OF KEY CONCEPTS

- Living things and the world around them are composed of chemicals. Carbon, hydrogen, oxygen, and nitrogen are the most abundant chemical elements found in living things.

- Atoms are the fundamental particles of the chemical elements. An atom has a positively charged nucleus surrounded by negatively charged electrons, with the number of protons determining which element the atom belongs to.

- Atoms can link together via shared electron pairs (covalent bonds) to form molecules.

- Molecules based on carbon skeletons (organic molecules) are of particular importance to life.

- Molecules can link together to form larger structures, through non-covalent bonds: ionic bonds, hydrogen bonds, or van der Waals interactions.

- Life processes depend upon chemical reactions, in which chemical compounds are transformed through breaking and making bonds, with associated energy transformations.

- Water has properties that give it a key role in supporting life. It is a polar molecule that can form hydrogen bonds, and hence dissolve polar, hydrophilic substances, while it repels hydrophobic substances and causes them to cluster together by hydrophobic interactions.

- Acids are substances that dissociate in water to release hydrogen ions, while bases can accept hydrogen ions. Strong acids dissociate fully, while weak acids are only partially dissociated.

- The pH scale is a measure of the hydrogen ion concentration and hence acidity of a solution. Hydrogen ion concentration affects chemical reactions, so maintenance of pH by buffering is important in living systems. Buffering can be achieved by a mixture of a weak acid and its conjugate base at pH values near the pK_a of the acid.

- Cells contain hundreds of different chemical substances but 70% of a cell's content is water, and most of the rest of its constituents belong to four classes of molecule.

- Carbohydrates, lipids, nucleic acids, and proteins are the four main classes of large macromolecules found in cells. Each class is made up of constituent small molecules (sugars, fatty acids, nucleotides, and amino acids respectively), and each has characteristic structural features, properties, and functions.

- Interactions between biological macromolecules are specific, but also flexible and transient—properties that are necessary for the dynamic processes of life.

 Use the flashcards in the e-book to test your recall of key terms introduced in this chapter.

QUESTIONS

Looking for answers? Once you've answered these questions, follow the link in the e-book to the answer guidance and check your work.

Concepts and definitions

1. List the six chemical elements that are most abundant in the human body.

2. Rank these types of chemical bonding in order of strength, from weakest to strongest: (a) ionic bond, (b) hydrogen bond, (c) covalent bond, (d) van der Waals interaction.

3. Name the functional groups in the two molecules below and explain the significance of the circle inside the benzene ring structure in molecule (a).

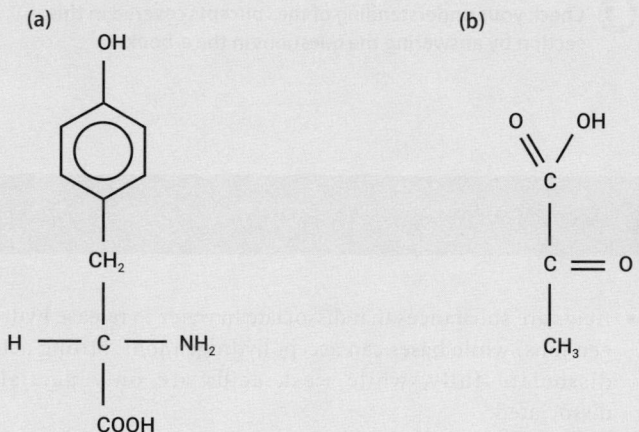

(a)   **(b)**

4. What does the value of its pK_a tell you about the properties of a weak acid?

5. In each case, match the small molecule component with the biological macromolecule:

    Small molecules: fatty acid, monosaccharide, amino acid, nucleotide.
    Macromolecules: ribonucleic acid (RNA), glycogen, membrane phospholipid, protein.

Apply the concepts

6. Explain why carbon plays such a special role in the molecular basis of life on Earth.

7. As well as the concept of a polar bond, certain molecules can be described as polar. Looking at the structures of formaldehyde and carbon dioxide below, explain why formaldehyde is regarded as a polar molecule and carbon dioxide as non-polar.

formaldehyde   carbon dioxide

8. When we consume alcohol it is metabolized by the following reaction:

$$CH_3CH_2OH \; + \; 3O_2 \; \rightarrow \; 2CO_2 \; + 3H_2O$$

ethanol   +   oxygen   →   carbon dioxide + water

Use the bond energies in Table 1.5 to calculate the value of the enthalpy change for the reaction. Is the reaction exothermic or endothermic? The structural formula for ethanol is

9. Which feature of DNA and RNA molecules justify them being classified as acids?

10. For ethanoic acid ($pK_a = 4.75$) show how you would use the Henderson–Hasselbalch equation to calculate the relative amounts of undissociated acid and conjugate base present in a solution at pH values of 4.75, 5.75, 6.75, and 7.75.

Beyond the concepts

11. Scientists researching the chemical basis of life in the early 20th century, prior to the discovery of DNA structure, were reluctant to believe that nucleic acids were the genetic material of cells. Many believed that proteins must be the carriers of genetic information. Can you explain why the scientists held that opinion?

12. It has been argued that hydrogen bonds are particularly important in allowing life to exist and life processes to occur. What features and properties of hydrogen bonds and the molecules in which they are found support this argument?

13. Lipids could be considered the 'odd ones out' among the four major classes of biological macromolecules. Discuss the possible basis for this statement.

FURTHER READING

Jonsson, A. L., Roberts, M. A. J., Kiappes, J. L. & Scott, K. A. (2017) Essential chemistry for biochemists. *Essays Biochem.* **61**: 401–27.
This article provides a brief overview of those areas in chemistry that are most relevant to biochemistry, summarizing basic principles and giving examples on how these principles are applied in biological systems.

Crowe, J. & Bradshaw, T. (2021) *Chemistry for the Biosciences*, 4th edition. Oxford University Press.
A textbook that covers the topics in this chapter and some additional ones in more detail. It is careful to assume very little prior knowledge.

Pace, N. R. (2001) The universal nature of biochemistry. *PNAS* **98**: 805–8.
Explores the reasons why life on Earth is based on carbon-containing molecules.

Ball, P. (2005) Water for life: seeking the solution. *Nature* **436**: 1084–5.
A discussion of water's special role in biology and whether it is right to focus on the presence of water in the search for extra-terrestrial life.

Energy

Powering Biochemical Processes

Chapter contents

Introduction   184

2.1   Living cells transform energy from
      one state to another   185

2.2   Will a reaction happen or not?   186

2.3   Harvesting and releasing free energy
      from food   188

2.4   ATP is the universal energy intermediate
      in all life   189

2.5   Electron transport during metabolism   192

Watch the key concepts video in the e-book to prepare yourself for studying this chapter.

LEARNING OBJECTIVES

By the end of this chapter you should be able to:

- Define the term 'metabolism' and explain the difference between catabolism and anabolism and endergonic and exergonic reactions.

- Explain the terms 'entropy', 'enthalpy', and 'free energy', and the relationship between them, and state the first and second law of thermodynamics.

- Explain the importance of irreversible reactions in a metabolic pathway and give examples.

- Summarize the energy cycle in life and the relationship between light energy and chemical energy in cells.

- Explain the statement 'ATP is a universal energy currency'. Explain what makes ATP a 'high-energy phosphate' compound.

- Outline the principles of food oxidation and harnessing of energy in the form of ATP.

- Explain the importance of redox reactions in metabolism and the importance of coenzymes such as NAD and FAD.

Introduction

Let us start this chapter with a question: Why do you, as a biology student, need to care about energy at all? The simple answer is that living cells resemble chemical factories where thousands of reactions take place. These reactions all involve transformations of matter and energy. We generally believe that we know what energy is, but it is difficult to define. A useful definition is 'the capacity to do work'. An entity with higher energy has more capacity to 'do work' (in essence, to bring about some kind of change) than an entity with lower energy.

In this chapter we will introduce some of the fundamental energy-related concepts that underpin the maintenance of life. We begin by considering how energy is transformed by the biochemical reactions that sustain life.

2.1 Living cells transform energy from one state to another

The reactions that take place in living cells are collectively known as **metabolism**. Some reactions involve the conversion of large and complex molecules into simpler ones and *release* energy in the process. They are known as **catabolic** reactions. Others result in the construction of large and complex molecules from simpler ones such as sugars and nitrogenous compounds. These reactions *consume* energy and are known as **anabolic** reactions. The synthesis of complex molecules from simpler ones involves an increase in the energy content of the cell; an increase in the energy content of a cell requires chemical work to be done.

Most cells get the fuel to power their survival from food molecules; the energy stored in food initially comes from the sun and is converted into chemical energy (in the form of sugars) by photosynthetic plants. For some organisms, such as bacteria living around hydrothermal vents in the ocean, the 'foods' that supply their energy are chemicals such as hydrogen sulphide from Earth's crust.

We learn more about the biochemical reactions through which organisms obtain energy from their food—their metabolism—and the capturing of energy by plants through the process of photosynthesis in Chapter 7. In this chapter, we explore the concept of energy itself—a concept that is fundamental to the maintenance of life.

Energy exists in different forms, of which kinetic, chemical, thermal, and potential are examples.

Kinetic energy is associated with movement, **chemical energy** with molecular composition, **thermal energy** with heat, and **gravitational potential energy** with position and height. A cyclist uses energy derived from food (chemical energy) to move a bicycle (kinetic energy) and in doing so becomes hot (thermal energy). If he cycles to the top of a hill and stops, the kinetic energy is converted into gravitational potential energy and he can use it to go down the hill without pedalling.

Individual living cells also transform energy from one state to another but in a less obvious way. Food molecules are rich in potential chemical energy, while molecules such as water (H_2O) and carbon dioxide (CO_2) have, in this context, none. When glucose is oxidized to carbon dioxide and water during the process of respiration, which we discuss further in Chapter 7, its potential energy is released and is used to drive other reactions in the cell. Plants (and some algae and bacteria), in turn, convert light energy to chemical energy via the reactions of **photosynthesis** (which we also explore in Chapter 7) and utilize this energy to synthesize glucose and oxygen from CO_2 and H_2O.

Free energy

A chemical system such as a living cell contains a huge number of individual molecules, each of which contains a certain amount of energy, dependent on its structure. This energy can be described as the heat content or **enthalpy** of the molecule

When a molecule is converted into a different structure, its energy content (that is, its enthalpy) usually changes. This change in enthalpy is written as ΔH (delta H). The ΔH may be negative (energy is lost from molecules and released during the course of the chemical change) or positive (energy is taken up from the surroundings during the course of the reaction).

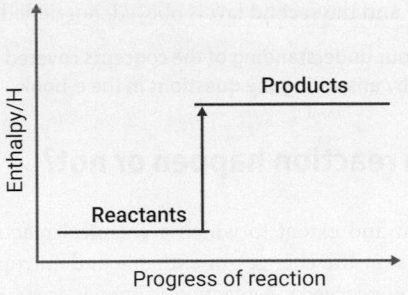

PAUSE AND THINK

Look at this graph. Is ΔH positive or negative? Has energy been taken up from the surroundings or lost to the surroundings?

Answer: ΔH is positive. The product molecules have taken up energy from their surroundings.

The laws governing energy transformations: thermodynamics

The energy changes exhibited by living systems such as cells are governed by certain 'laws'; these laws were determined by the study of energy transformations—a field of enquiry known as 'thermodynamics'.

 You can read more about thermodynamics in Topic 3.

The **first law of thermodynamics** states that energy can be neither created nor destroyed; this means that the total energy content of the universe remains constant. Instead of being created or destroyed, energy is converted from one form to another, as we noted above.

However, not all energy is useful in the sense of being capable of performing work. The heat in a hot car engine is a form of energy, but it cannot be used to propel the car. If you consider the energy released by the oxidation of food, only part of it can be used to do work. The rest of the energy increases the total **entropy** of the universe. Entropy is the degree of randomness or disorder in any system; it is represented by the symbol S. An increase in temperature increases the random motion of molecules and hence increases their entropy. When a molecule breaks down into smaller ones, or anything increases the number of particles in a system, the entropy is also increased. The change in entropy as a process or reaction takes place is written as ΔS (delta S).

The **second law of thermodynamics** specifies that the total entropy of the universe is always increasing—that is, moving towards a state of increasing disorder. Living cells appear to be defying the second law of thermodynamics: they seem to become *more organized*. As they grow and reproduce, they convert small, randomly arranged compounds from the environment, which have high entropy, into the large, highly organized structures from which the cells are constructed, which have lower entropy.

This seeming paradox can be explained by considering where the energy that drives the anabolic processes comes from. Cells oxidize food molecules, releasing energy in the process. This energy release causes a large increase in entropy, as does the breakdown of larger molecules to smaller ones such as CO_2, which escape into the cell's surroundings. This increase in entropy *exceeds* the decrease that occurs during the production of complex and highly organized structures. So, if we consider the cell *plus* its surroundings (known collectively as the 'universe'), we see a net *increase* in total entropy and the second law is obeyed, not defied.

 Check your understanding of the concepts covered in this section by answering the questions in the e-book.

2.2 Will a reaction happen or not?

The direction and extent to which a chemical reaction can proceed depends on the changes in enthalpy and entropy that occur as the reaction proceeds. A reaction is known as spontaneous if it can proceed under a given set of conditions without intervention. Neither enthalpy nor entropy alone can determine whether a reaction can proceed spontaneously in the direction in which it is written, or indeed in the reverse direction. Rather, we need to consider another term known as the '**free energy**' of the reaction, which is denoted by the letter **G**. Free energy is also known as Gibbs' free energy after J. Willard Gibbs who derived the relationship between enthalpy, entropy, and energy. The change in Gibbs' free energy as

a reaction proceeds is related to the change in both enthalpy and entropy as follows:

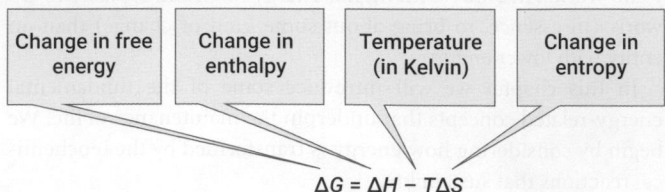

$$\Delta G = \Delta H - T\Delta S$$

The temperature in degrees K is the temperature in degrees Celsius plus 273 (0 K is known as **absolute zero** in terms of temperature. K = °C + 273).

The equation $\Delta G = \Delta H - T\Delta S$ applies to systems in which the temperature and pressure remain constant during a process, which is the case for biological systems.

The **change of free energy** during a reaction is really important. We can use the ΔG of a reaction to predict whether it can take place spontaneously (that is, without the net input of energy) or whether energy must be supplied for the reaction to take place: if the change in Gibbs' free energy is negative, then the reaction is spontaneous.

Figure 2.1 shows the free energy change during the reaction A ⇌ B.

In Figure 2.1a, the product B has a lower energy than the reactant A. ΔG is negative and the reaction can occur spontaneously; it is described as **exergonic**.

In Figure 2.1b, the product B has a higher free energy than the reactant A. ΔG is positive and the reaction will not occur spontaneously. Energy has to be supplied to the system to convert B into A; the reaction is said to be **endergonic**.

The ΔG value of a reaction is not fixed but rather depends on the concentrations of reactants and products (see below). However, the ΔG value of a reaction under specified standard conditions *is* fixed. The definition of 'standard conditions' for biological reactions are those with reactants and products at 1.0 M, 25 °C (298 K), and pH 7.0.

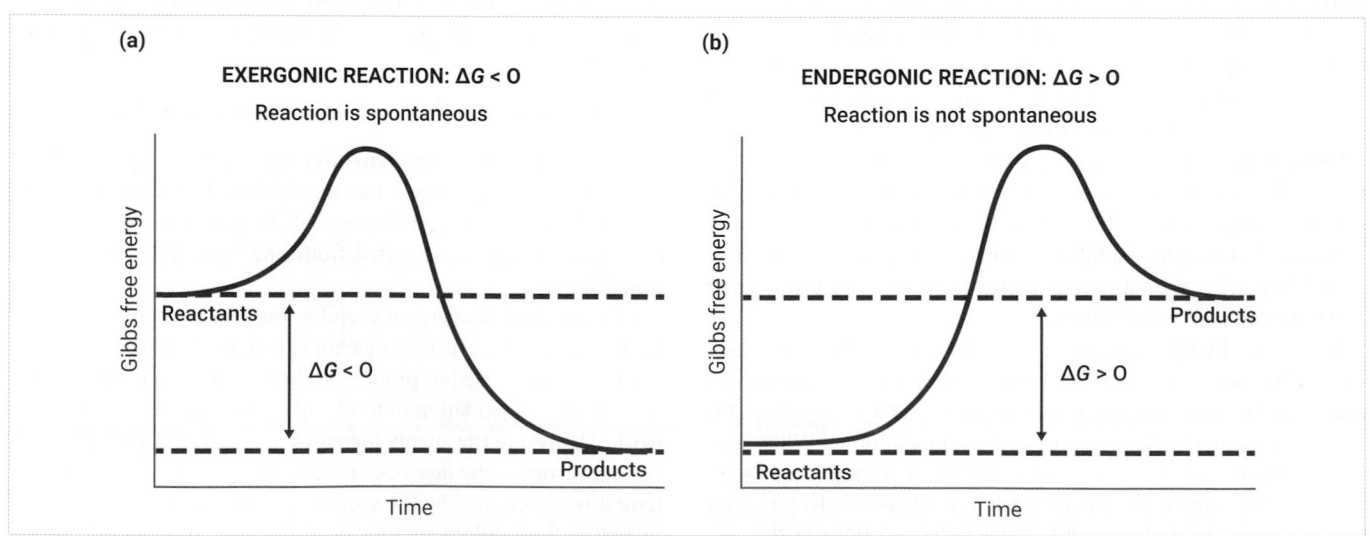

Figure 2.1 Changes in Gibbs' free energy during a reaction. (a) In the first reaction the product has a lower free energy than the reactant so ΔG is negative; the reaction proceeds spontaneously. (b) In the second reaction the product has higher free energy than the reactant so ΔG is positive; the reaction does not proceed spontaneously.

Source: (a,b) OpenStax Biology.

The ΔG value under these conditions is called the **standard free energy change of a reaction**. It is denoted ΔG^0 and its values are a useful guide in understanding metabolic reactions.

Although the concentrations of metabolites in a cell are never 1 M, ΔG^0 values are frequently quoted to explain why certain biological reactions behave as they do. It is a very useful compromise.

PAUSE AND THINK

Do all spontaneous reactions/processes take place at the same speed?

Answer: It depends on the reaction/process. An ice cube will melt faster if placed in hot water than in cold water, but both processes are spontaneous.

Equilibrium

A living cell is at a higher energy level than the random collection of molecules from which it was assembled. As such, it is not in **equilibrium** with its surroundings. Equilibrium is a state of maximum stability where there is no change in energy between the cell and its surroundings. Equilibrium is achieved only when a cell decomposes after its death.

Strictly speaking, *all* chemical reactions are reversible: the reaction A $\rightleftharpoons$ B can also occur in the reverse direction: B $\rightleftharpoons$ A. This might imply that the ΔG value must be negative in both directions, as we said that a reaction cannot occur spontaneously *unless* the ΔG value is negative. The answer to this apparent paradox is that the ΔG of a reaction is not fixed, but varies with the reactant and product concentrations (as mentioned above). In the reaction A $\rightleftharpoons$ B, if A is at a high concentration and B at a low concentration, the ΔG may be negative in the direction A $\rightarrow$ B and, of course, positive from B $\rightarrow$ A. If the concentration of B is high, and the concentration of A is low, however, the ΔG can be negative for the reverse direction, B $\rightarrow$ A.

In general terms, a reaction will proceed to the point at which A and B reach concentrations at which the ΔG is zero in both directions, and so no further net reaction can occur. This is the **chemical equilibrium point**.

There is an important point to note here: systems at equilibrium can do no work, as ΔG is 0 in both directions. However, a system, such as a living cell, *must* do work to stay alive. This tells us that a living cell is not at equilibrium, and only reaches equilibrium when it is dead. The fact that metabolic reactions in a cell are not at equilibrium is therefore a characteristic of life.

If the ΔG for a given reaction is large, for all practical purposes the reaction is irreversible, because the large concentration changes that would be necessary to drive the reaction in the reverse direction are not possible in a cell.

The activation energy

We said above that a negative ΔG is a necessary condition for a reaction to occur spontaneously. But one other consideration is also important: how *fast* is the reaction happening? A reaction might happen spontaneously, but so slowly to be of no practical use to a living cell. Why is this? One reason is because there is an energy barrier to chemical reactions occurring. This energy barrier has a physical form: a transition state through which the substrate of the reaction passes on the way to becoming a product:

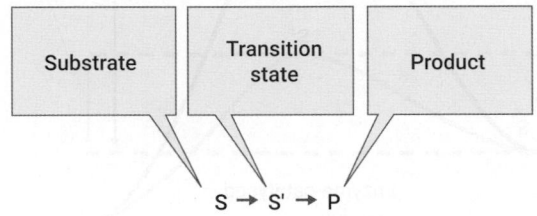

$$S \rightarrow S' \rightarrow P$$

The transition state is at a higher free energy than the substrate, so the free energy change for the conversion of S to S′ is positive. This difference in free energy is called the **activation energy**, and must be overcome before the reaction can proceed. The greater the activation energy, the greater the amount of energy needed to make a reaction happen (and so the slower the reaction will take place without the input of more energy).

But here we face a problem: the energetic cost of overcoming the activation energy for many essential biological reactions is simply too high for life to be possible. Instead, biological systems have developed something of a work-around: they use **enzymes** to lower the activation energy required before a reaction can happen. By lowering the activation energy, enzymes enable reactions to happen more quickly—at the rate required to sustain life. As such, enzymes are biological **catalysts**.

▶ The structure and function of enzymes is described in detail in Chapter 5.

Look at Figure 2.2, which compares the energy profile of an enzyme-catalysed and a non-catalysed reaction. Notice how the activation energy for the non-catalysed reaction is much greater than for the enzyme-catalysed reaction. Also notice that the free energy of the substrate, S, and product, P, remains the same regardless of the activation energy. This leads us to an important point: enzymes do not change the free energy change of a reaction: they merely provide a lower-energy (and therefore faster) path via which the reaction can proceed. They provide the equivalent of striking a match to ignite a mixture of petrol and oxygen, which will not react by themselves as the transition energy is too high.

PAUSE AND THINK

The oxidation of glucose to carbon dioxide and water has a large negative ΔG^0 value but glucose is quite stable in the presence of oxygen. Why is that?

Answer: The ΔG^0 of a reaction determines whether a reaction may proceed but does not determine the rate. The rate is determined by the activation energy of the reaction and the rate at which the transition state is formed.

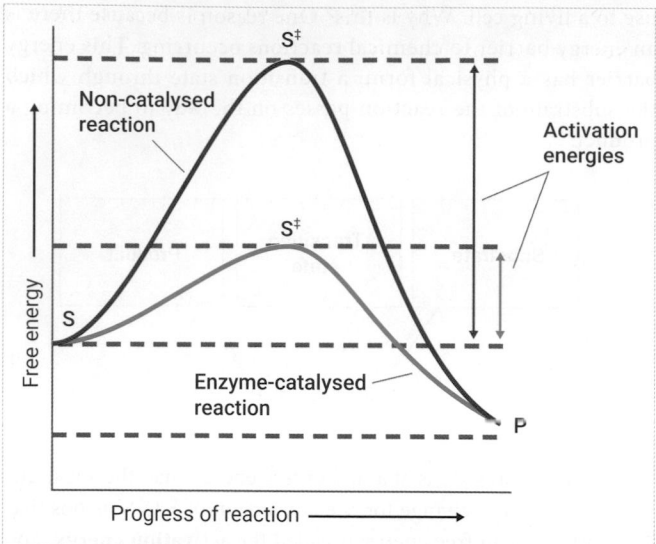

Figure 2.2 Energy profiles of non-catalysed and enzyme-catalysed reactions. S = substrate; S‡ = transition state; P = product.

Source: Snape, A. & Papachristodoulou, D. (2018). *Biochemistry and Molecular Biology* (6th ed.). Oxford University Press.

The importance of reversible and irreversible reactions in metabolism

Chemical reactions in a cell are not single, isolated processes, but are organized into **metabolic pathways**—linked series of reactions occurring within a cell in which the products of the first reaction are the reactants of the next, and so on. For example, about a dozen reactions are involved in the process of converting glycogen (the storage form of carbohydrate in muscle) into lactic acid when a muscle vigorously contracts.

Metabolic pathways are usually considered to flow in one direction. Although all chemical reactions are reversible in theory, conditions in a cell are such that a pathway will flow in one direction at any given time as it would be more favourable thermodynamically for the flux to proceed in one direction only. Many of the individual reactions in a pathway may be freely reversible, but the pathway virtually always contains one or more reactions that are irreversible. Why is this important? If all the reactions in a metabolic pathway were reversible, then the pathway could not proceed in one direction overall. If, for example, glycogen was degraded to lactic acid in muscle, and glycogen was resynthesized at the same time and location, there would be no net change in the amount of glycogen present, and so no yield of energy for muscle contraction. It would resemble an attempt to build the walls of a house with bricks being taken away at the same time as they were being laid: there would be a state of equilibrium during which the wall would fail to increase in height.

From a thermodynamic viewpoint, the only overall requirement is that the free energy change from the first reactant to the last component of the pathway is negative. If this is the case the pathway will proceed to completion, even if the free energy change for some of the individual reactions is positive.

Factors external to the cellular content of metabolites (the intermediates in a metabolic pathway) may determine the direction in

which a pathway will proceed. In the case of glycogen, the release of insulin after a meal will favour glycogen *synthesis*; by contrast, the release of adrenaline into the circulation will favour the *breakdown* of glycogen in muscle, allowing the pathway to provide energy for muscle contraction.

 Check your understanding of the concepts covered in this section by answering the questions in the e-book.

2.3 Harvesting and releasing free energy from food

We will now consider the range of vital biochemical reactions through which energy is made available to the cells of our bodies—reactions that are collectively called metabolic reactions. Without these reactions, and the energy they supply, we would quite simply be unable to exist. But, first, we must introduce two phenomena that are central to biochemical reactions: oxidation and reduction.

Oxidation and reduction reactions

When energy-rich metabolic fuels such as glucose are oxidized, they ultimately produce carbon dioxide and water. But what do we mean by terms such as 'oxidized'? **Oxidation** describes the loss of electrons from one substance to another. The reciprocal acceptance of electrons by a substance from another is known as **reduction.** When a substance is oxidized in a reaction, another substance must be reduced (the two reactions happen in concert, rather than in isolation) so we refer to paired reactions of this kind as '**redox**' reactions.

In the pathway of glucose oxidation, for example, glucose is finally oxidized to produce carbon dioxide and at the same time oxygen is reduced to water:

$$C_6H_{12}O_6 + 6\,O_2 \longrightarrow 6\,CO_2 + 6\,H_2O + energy$$

This process does not happen directly but by a series of coordinated reactions—reactions that we will explore in greater detail in Chapter 7.

Biological oxidations are very rarely additions of molecular oxygen (as the name 'oxidation' might imply). Instead, they are more likely to be transfers of electrons and/or hydrogen ions. For example, oxidation in biological systems often involves enzymatic removal of two hydrogen atoms from a molecule. The following reactions are seen in fatty acid oxidation:

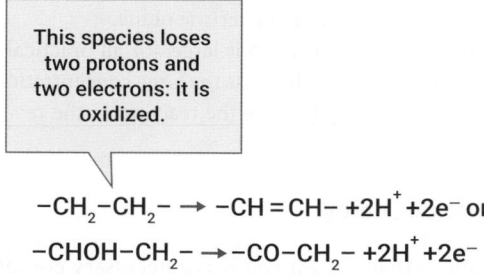

This species loses two protons and two electrons: it is oxidized.

$$-CH_2-CH_2- \rightarrow -CH=CH- +2H^+ +2e^- \text{ or}$$

$$-CHOH-CH_2- \rightarrow -CO-CH_2- +2H^+ +2e^-$$

In such chemical oxidation systems, the electrons must be transferred from the electron donor to an electron acceptor. Depending on the particular electron acceptor, the electron transferred may be accompanied by a proton, in which case a hydrogen atom is transferred, or the proton may be liberated into solution, and only the electron transferred. We learn more about electron acceptors in Section 2.5.

Photosynthesis, catabolism, and anabolism: the energy cycle of life

The conversion of simple precursor molecules to larger cellular molecules (such as DNA and proteins) involves increases in energy, and therefore cannot occur without an input of energy. The required energy comes from the breakdown of food, which also results in the production of carbon dioxide and water.

As already mentioned in this chapter, metabolism as a whole consists of anabolic and catabolic reactions. Anabolic reactions involve positive free energy changes and result in the synthesis of larger molecules from simpler precursors. Catabolic reactions involve negative free energy changes and result in the breakdown of larger molecules to smaller simpler ones. How are these anabolic and catabolic reactions coordinated?

We can summarize the overall situation of the energy cycle in life in Figure 2.3.

Let us start with plants. **Photosynthesis** is the process in plants that converts light energy into chemical energy. Plants convert CO_2 and H_2O (waste products of food breakdown by animals) to food molecules such as glucose or its derivatives, releasing oxygen at the same time. The food molecules produced by plants, such as glucose, are converted into other food molecules such as fats, and all of these foodstuffs are consumed by other organisms such as animals.

▶ **We describe photosynthesis in greater detail in Chapter 7.**

The oxidation of these foodstuffs by animals releases carbon dioxide and water as mentioned above (catabolism); they also release chemical energy, which is captured in the form of ATP (adenosine triphosphate). ATP is then used to drive energetically unfavourable processes in the cell: it is used to convert small, disorganized molecules into large and organized molecules, such as proteins and DNA.

Although the assembly of large cellular structures involves a decrease in entropy (making it energetically unfavourable), the oxidation of food molecules involves a greater increase in entropy (favourable). The entropy change of the total system (cell and surroundings) is positive, and so the second law of thermodynamics is obeyed and energy can both be harvested and released.

The process of food oxidation can be summarized as:

$$\text{Food (organic compounds)} + \text{Oxygen} \rightarrow \text{Carbon dioxide} + \text{Energy (ATP} + \text{heat)}$$

Let us see how ATP is involved in the energy economy of the cell, using Figure 2.4.

Figure 2.4 shows how the oxidation of food molecules is coupled with the synthesis of high-energy phosphoryl groups such as ATP, which are used to provide energy for biological work. We will shortly see what makes ATP such an important and universal energy currency.

The oxidation of energy-supplying food molecules in the cell without it being appropriately coupled to the energy-requiring reactions would simply liberate heat that could not be used to do chemical or other work in the cell. Instead, the free energy change involved in food breakdown must be coupled to the energy-requiring processes. This occurs by converting ADP plus inorganic phosphate to ATP, a high-energy compound that acts as the universal energy intermediate of life. But what do we mean by a high-energy compound? We discuss this next in Section 2.4.

🌀 Check your understanding of the concepts covered in this section by answering the questions in the e-book.

2.4 **ATP is the universal energy intermediate in all life**

ATP is known as a '**high-energy phosphate compound**'. Its chemical structure is shown in Figure 2.5. Notice the three phosphate groups attached to the five-carbon ring.

Figure 2.3 The energy cycle in life.

Source: Snape, A. & Papachristodoulou, D. (2018). *Biochemistry and Molecular Biology* (6th ed.). Oxford University Press.

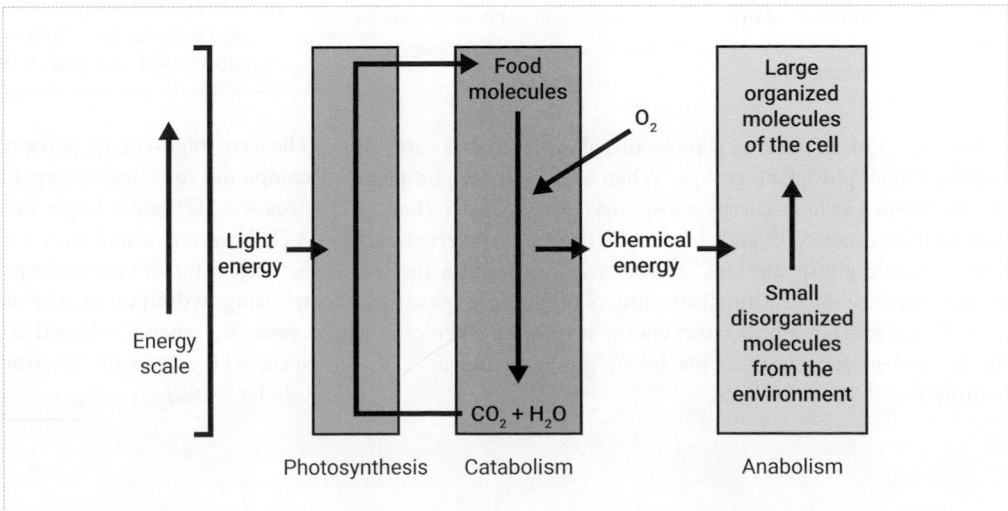

Figure 2.4 The role of ATP in the energy economy of the cell. Note that some types of work involve breakdown of ATP to AMP (adenosine monophosphate) instead of ADP (adenosine diphosphate), but this does not change the concept given here. P_i is inorganic phosphate.

Source: Snape, A. & Papachristodoulou, D. (2018). *Biochemistry and Molecular Biology* (6th ed.). Oxford University Press.

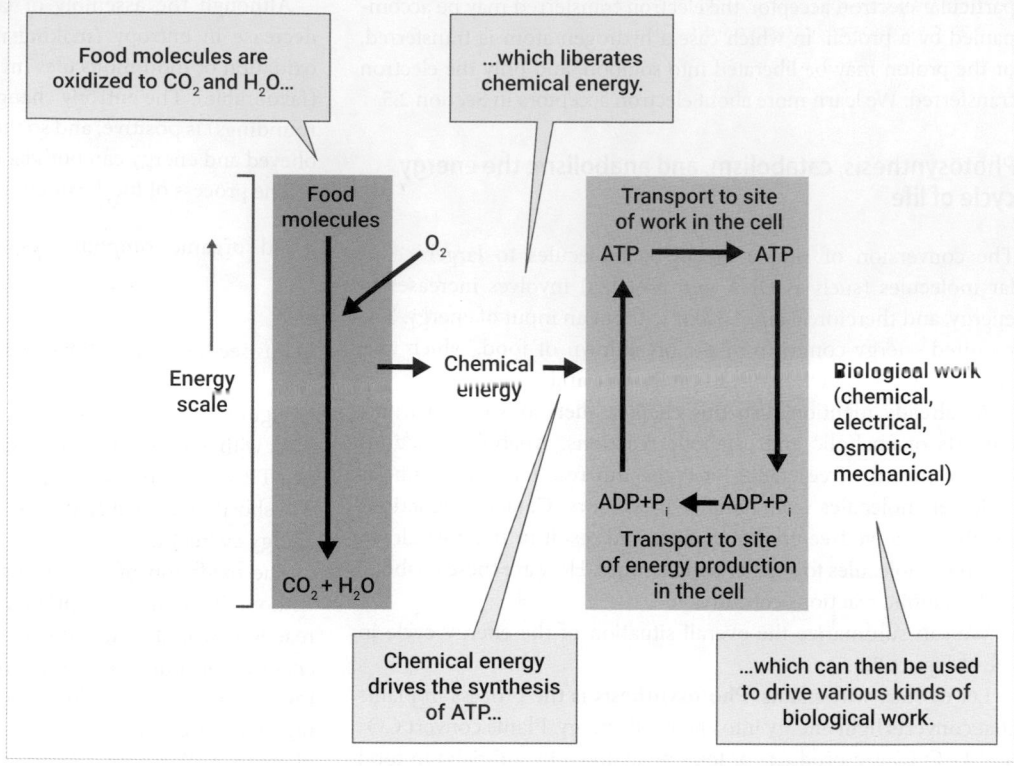

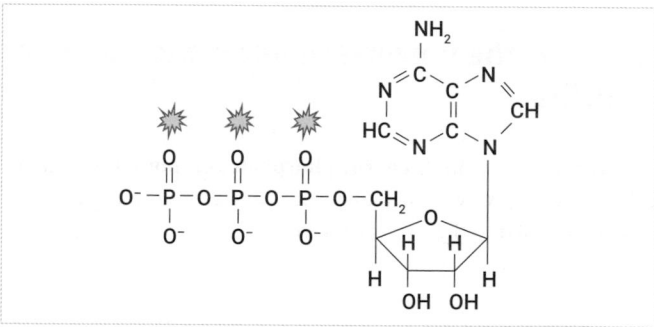

Figure 2.5 The structure of ATP. The stars denote the three phosphate groups.

Source: Snape, A. & Papachristodoulou, D. (2018). *Biochemistry and Molecular Biology* (6th ed.). Oxford University Press.

PAUSE AND THINK

Consider these two statements and notice the difference in the free energy change of the two reactions:

- The ΔG^0 for the hydrolysis of ATP to ADP and P_i is -30.5 kJ mol^{-1}
- The ΔG^0 of hydrolysis of AMP to adenosine and P_i is -14.2 kJ mol^{-1}
  What is the reason for this difference?

Answer: ATP is a high-energy phosphate anhydride, whereas AMP is a low-energy phosphate ester. Release of the phosphate from ATP relieves the strain caused by the proximity of the negatively charged phosphate groups. Hydrolysis of AMP causes little increase in stabilization.

We can think of ATP as a molecule of AMP that is carrying two additional phosphate groups. When each of these phosphate groups (known as high-energy phosphoryl groups) is detached, so that ATP becomes ADP and then AMP, they are converted back into inorganic phosphate ions. This conversion releases the free energy that went into the initial attachment of these groups to form ATP. Figure 2.6 shows how much energy is released when each of the bonds between the three phosphate groups found in ATP is hydrolysed.

The term 'high-energy phosphate' is used when the hydrolysis of a compound to liberate inorganic phosphate (P_i) is associated with negative ΔG^0 values larger than 30 kJ mol^{-1}. Direct hydrolysis of ATP, however, would only release energy as heat, which would be useless for driving endergonic reactions. Instead, the energy-releasing hydrolysis of ATP is coupled to energy-requiring processes: the energy released is harnessed by other processes. As such, ATP acts as the intermediate energy source which powers cellular work.

Figure 2.6 The release of energy when successive phosphate groups are detached from ATP.

Source: Snape, A. & Papachristodoulou, D. (2018). *Biochemistry and Molecular Biology* (6th ed.). Oxford University Press.

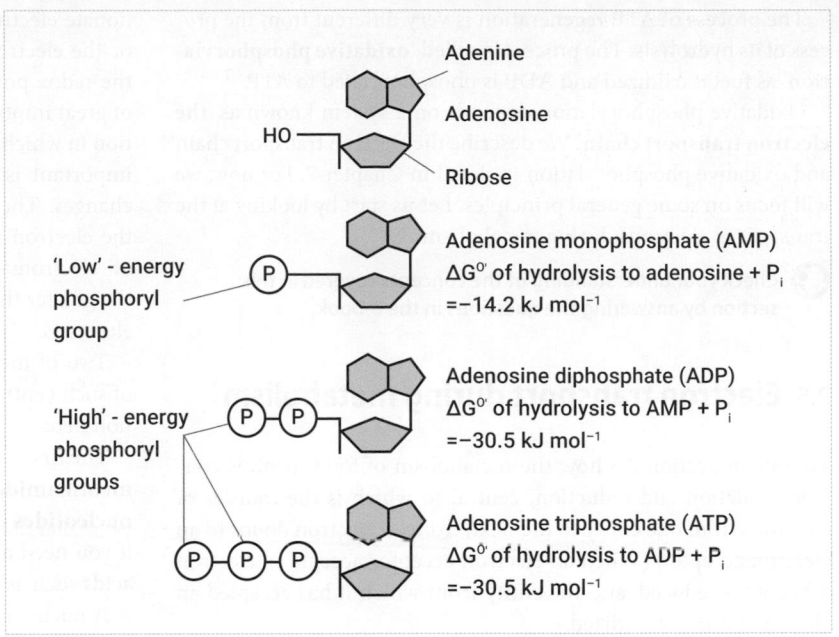

Adenine

Adenosine

Ribose

'Low' - energy phosphoryl group

Adenosine monophosphate (AMP)
$\Delta G^{0'}$ of hydrolysis to adenosine + P_i
=−14.2 kJ mol^{-1}

'High' - energy phosphoryl groups

Adenosine diphosphate (ADP)
$\Delta G^{0'}$ of hydrolysis to AMP + P_i
=−30.5 kJ mol^{-1}

Adenosine triphosphate (ATP)
$\Delta G^{0'}$ of hydrolysis to ADP + P_i
=−30.5 kJ mol^{-1}

Let us now see how ATP is used to generate energy, and how the oxidation of foodstuffs generates ATP.

How does ATP perform work and how is it regenerated?

The cell is able to harness the energy released during ATP hydrolysis to perform chemical, transport, and mechanical work. Let us look at chemical work as an example. The hydrolysis of ATP can be used to drive endergonic reactions, for which the free energy change is positive. As long as the free energy change of the endergonic reaction is *less* than the free energy change of ATP hydrolysis (which is exergonic and negative), the reaction can take place.

Usually, the reaction involves the transfer of a phosphate group from ATP to another molecule: X + ATP ⟶ X–P + ADP.

X–P can then be the reactant for a subsequent reaction in a pathway.

Let us look at the first steps of glucose metabolism in the pathway of glycolysis. Notice how the product of the first reaction becomes the reactant of the second. Such reactions, whereby the product of one becomes the reactant of the second, and so forth, are known as **coupled reactions**.

Reaction 1:

Glucose + ATP ⟶ glucose 6-phosphate + ADP

Reaction 2:

Glucose 6-phosphate ⟶ fructose 6-phosphate

The ΔG^0 for reaction 1 is −17 kJ mol^{-1}. For reaction 2, it is 1.7 kJ mol^{-1}

Therefore, the coupled reaction has a ΔG^0 of −15.3 kJ mol^{-1} and is thermodynamically possible.

Coupled reaction:

Glucose + ATP ⟶ fructose 6-phosphate + ADP

Notice how the free energy change for the coupled reaction is simply the sum of the two individual reactions. In other words, it is additive. The fact that the free energy changes of reactions in a metabolic pathway are additive is very important: it means that pathways can proceed in a particular direction as long as the overall free energy change is favourable.

Note that all the reactions in a cell which involve ATP hydrolysis are enzymatically catalysed. Enzymes which carry a phosphate group from ATP elsewhere are known as **kinases**.

As well as performing chemical work, the breakdown of ATP powers muscle contraction, the generation of electrical signals, and the pumping of ions against concentration gradients, to name but a few examples. The mechanisms are, in principle, the same. Whatever the process, as long as ATP breakdown is coupled to the mechanism, the liberation of free energy that results from the process occurring will drive it forward.

We now come to the question of how ATP is generated or regenerated. Only a small amount of ATP exists in a cell at any one time. Consequently, it needs to be regenerated efficiently. To that end, the ATP 'cycle' indicated in Figure 2.4 takes place very rapidly indeed.

The process of ATP regeneration is very different from the process of its hydrolysis. The process is called 'oxidative phosphorylation' as fuel is oxidized and ADP is phosphorylated to ATP.

Oxidative phosphorylation depends on a system known as 'the electron transport chain'. We describe the electron transport chain and oxidative phosphorylation in detail in Chapter 7. For now, we will focus on some general principles. Let us start by looking at the transport of electrons during metabolism.

 Check your understanding of the concepts covered in this section by answering the questions in the e-book.

2.5 Electron transport during metabolism

We saw in Section 2.3 how the metabolism of food involves coupled oxidation and reduction, central to which is the transfer of electrons from one entity to the next, from an electron donor to an electron acceptor. (When an electron acceptor accepts an electron it becomes reduced, and the entity from which it has accepted an electron becomes oxidized.)

The ultimate electron acceptor in the aerobic cell is oxygen. Oxygen is **electrophilic**, which means it readily attracts or accepts electrons. When it accepts four electrons, it also accepts four protons from the solution and forms water molecules:

$$O_2 + 4e^- + 4H^+ \rightarrow 2H_2O$$

While oxygen may be the ultimate electron acceptor, it is not the *only* electron acceptor in the cell. Other electron acceptors form a relay system, which carries electrons along a 'chain' from metabolites to oxygen. This is how the electron transport chain operates, and it plays a predominant role in ATP generation.

The essential concept, that the transfer of electrons to oxygen is coupled to the generation of ATP, is illustrated in Figure 2.7.

There are a number of electron carriers, which have different affinities for electrons. One with a lesser affinity will tend to donate electrons to one with higher affinity. The electron affinity or the electron donating potential of a redox couple is known as the redox potential value (E_0') and is expressed in volts (V). It is of great importance in biochemistry because it indicates the direction in which electrons will tend to flow between reactants. Equally important is the fact that E_0' values are related to free energy changes. The more negative the redox potential value, the lower the electron affinity, and hence the greater the tendency to pass on electrons—and the greater the tendency to pass on electrons, the greater the reducing potential and the higher the energy of the electrons.

Two of the electron carriers involved in energy production are of such central importance in metabolism that they merit description here.

The first carrier involved in the oxidation of many metabolites is **nicotinamide adenine dinucleotide, NAD⁺**. You have already met **nucleotides** in the form of ATP, ADP, and AMP (look at Figure 2.6 if you need a reminder), and of course in the structure of nucleic acids such as DNA and RNA.

A nucleotide has the general structure:

$$base - sugar - phosphate$$

NAD⁺ is a dinucleotide that comprises two bases: adenine (as in ATP) and nicotinamide. This dinucleotide is formed by linking the two phosphate groups of the two nucleotides (which is unlike the way nucleotides are linked together in nucleic acids). The general structure of NAD⁺ is:

$$Base - sugar - phosphate - phosphate - sugar - base$$

The chemical structure of NAD⁺ is shown in Figure 2.8. Look at this structure, and notice how it follows the base–sugar–phosphate–phosphate–sugar–base structure in terms of the sequence in which its components are joined.

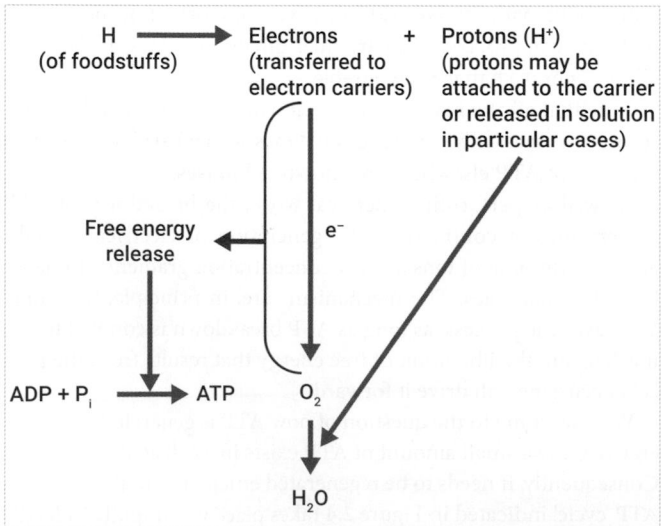

Figure 2.7 The concept of electron transfer to oxygen and coupled ATP generation.

Source: Snape, A. & Papachristodoulou, D. (2018). *Biochemistry and Molecular Biology* (6th ed.). Oxford University Press.

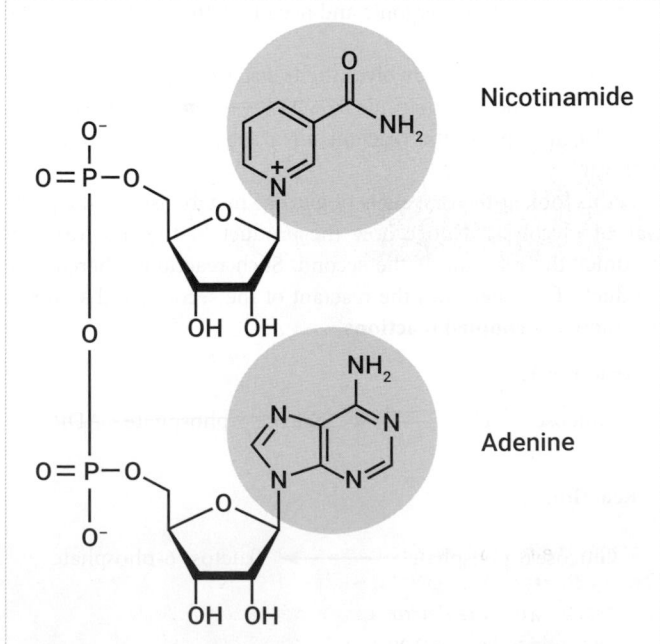

Figure 2.8 The structure of NAD⁺.

NAD$^+$ is a **coenzyme,** a small organic molecule that participates in enzymatic reactions. It differs from an ordinary enzyme substrate in that its reduced form leaves the enzyme and attaches to a second enzyme where it donates its reducing equivalents to a second substrate. (The term 'reducing equivalent' refers to molecules which can transfer the equivalent of one electron in redox reactions. A reducing equivalent can donate a single electron or a hydrogen atom.)

NAD$^+$ acts by being continually reduced and re-oxidized, and in so doing transfers electrons from one molecule to another. Its transfer of electrons is accompanied by the transfer of protons, such that its net effect is to transfer one hydrogen atom for every electron.

NAD+ can be reduced by accepting two electrons from two hydrogens on a metabolite. It is the coenzyme for several **dehydrogenases** which catalyse this type of reaction:

$$AH_2 + NAD^+ \rightarrow A + NADH + H^+$$

The reduced NAD$^+$, in the form of NADH, can then diffuse to a second enzyme and participate in a reaction such as:

$$B + NADH + H^+ \rightarrow BH_2 + NAD^+$$

Notice how, in this instance, NADH is being oxidized to NAD$^+$.

In this way NAD$^+$ acts as the carrier for the transfer of a pair of hydrogen atoms from A to B, even though it carries only one proton:

$$AH_2 + B \rightarrow A + BH_2$$

Two other important hydrogen carriers are **flavin adenine dinucleotid (FAD)** and **flavin mononucleotide (FMN)**. In this case the electrons are transferred as hydrogen atoms. FAD can be reduced to FADH$_2$ by accepting two hydrogen atoms; FMN is reduced in the same way.

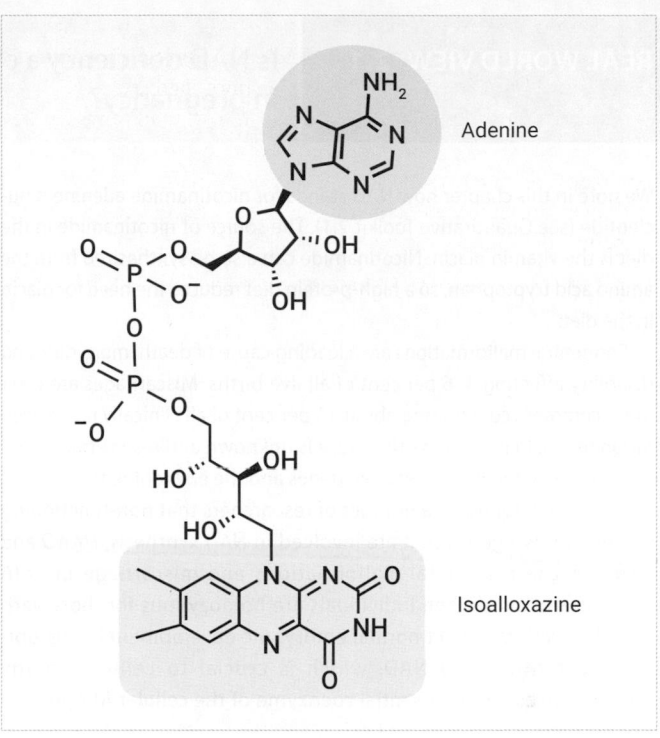

Figure 2.9 The structure of flavin adenine dinucleotide (FAD).

FAD (shown in Figure 2.9) has the structure:

isoalloxazine – ribitol – phosphate – phosphate – ribose – adenine

As shown in Figure 2.10, FMN has the following structure:

isoalloxazine – ribitol – phosphate

When these coenzymes are reduced as a result of fuel oxidation they can donate a pair of electrons to a series of electron carriers

---

QUANTITATIVE TOOLKIT 2.1	Calculating the relationship between the $\Delta G^{0\prime}$ value and the E_0' value in the oxidation of NADH

There is a direct relationship between the $\Delta G^{0\prime}$ value and the E_0' value of a redox reaction. This relationship is quantified by what is called the **Nernst equation**:

$$\Delta G^{0\prime} = -nF\Delta E_0',$$

where

n equals the number of electrons transferred in the reaction, F is the Faraday constant (96.5 kJ V^{-1} mol^{-1}), and E_0' is the difference in redox potential between the electron donor and electron acceptor.

If we consider the oxidation of NADH (see Real World View 2.1):

$$NADH + H^+ + \tfrac{1}{2}O_2 \rightarrow NAD^+ + H_2O$$

The redox potential, $\Delta E_0'$, is +1.136V. Therefore:

$$\Delta G^{0\prime} = -2\left(96.5 \text{ kJ V}^{-1}\text{mol}^{-1}\right)(+1.136\,V) = -219.25 \text{ kJmol}^{-1}.$$

So, a net positive redox potential gives a large and negative free energy change. A spontaneous redox reaction is characterized by a negative value of change in free energy and a positive value of redox potential. Both $\Delta G^{0\prime}$ and $\Delta E_0'$ can be used to predict whether a process is spontaneous.

REAL WORLD VIEW 2.1   **Is NAD deficiency a cause of congenital malformations and miscarriage in pregnancy?**

We note in this chapter how NAD stands for nicotinamide adenine dinucleotide (see Quantitative Toolkit 2.1). The source of nicotinamide in the diet is the vitamin niacin. Nicotinamide can also be synthesized from the amino acid tryptophan, so a high-protein diet reduces the need for niacin in the diet.

Congenital malformations are a leading cause of death, morbidity, and disability affecting 3–6 per cent of all live births. Miscarriages are even more common, constituting about 15 per cent of all clinically recognized pregnancies. In most cases the cause is unknown and results from often complicated interactions between genes and the environment.

It has been found by a number of researchers that non-functioning variants of two genes that are involved in NAD synthesis, *HAAO* and *KYNU*, cause congenital malformations and miscarriage in both humans and mice when individuals are homozygous for those variants. The variants affect normal embryonic development by disrupting the synthesis of NAD, which is crucial to cellular energy metabolism as it is an essential coenzyme of the cellular ATP production system.

Hartmut Cuny and colleagues aimed to find out whether NAD deficiency defects, as seen with homozygous loss-of-function variants in the genes *HAAO* or *KYNU*, can be induced by environmental factors and gene–environment interactions. They investigated whether NAD deficiency-driven embryo loss and congenital malformations occur in wild-type mice or those with heterozygous variants in genes involved in NAD synthesis, if the dietary supply of niacin is deficient.

They found that the malformations and miscarriages were not restricted to the rare homozygous gene variants but could be provoked by reduced dietary intake of niacin in both wild-type mice and those heterozygous for the defective genes. Their study supports the notion that NAD deficiency is clinically relevant, and they are proposing to determine the range of NAD levels during pregnancy in families with a history of miscarriage or congenital malformations and investigate the prevalence of NAD deficiency. Their findings stress the importance of adequate maternal intake of niacin during pregnancy. Niacin is a water-soluble vitamin and as such does not pose problems of toxicity if taken as a supplement.

Read the original work

Cuny, H., Rapadas, M., Gereis, J. *et al.* (2020) NAD deficiency due to environmental factors or gene–environment interactions causes congenital malformations and miscarriage in mice. *Proceedings of the National Academy of Sciences*, 117, 3738–47; DOI: 10.1073/pnas.1916588117.

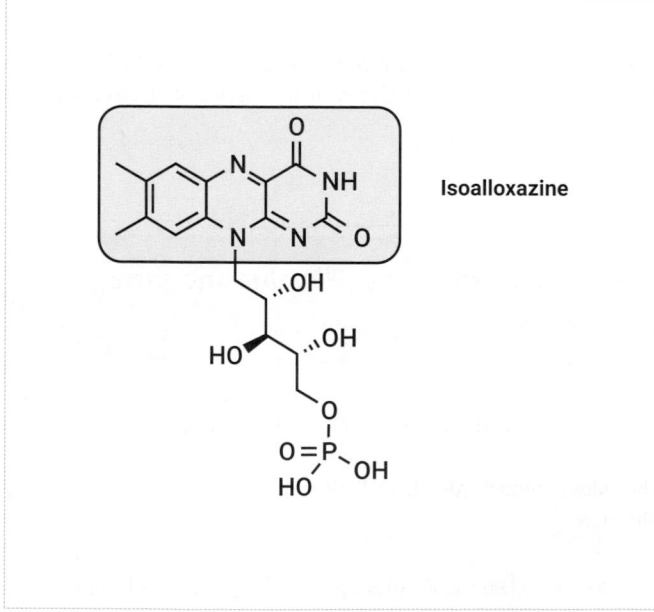

Isoalloxazine

Figure 2.10 The structure of flavin mononucleotide (FMN).

known collectively as the electron transport chain. As electrons are passed from one electron carrier to another, energy can be captured in the form of ATP. This process is known as oxidative phosphorylation, as mentioned earlier; we will discuss it in greater detail in Chapter 7.

PAUSE AND THINK

NAD and FAD accept hydrogen ions when metabolic fuel is oxidized. What would happen if these hydrogen ions were released into the cell or bloodstream instead of being picked up by the coenzymes?

Answer: There would be a number of consequences. The pH of the cell or the blood would drop, causing acidosis, which is pathological. Also, energy would be lost as the electron transport chain, and the process of oxidative phosphorylation that it feeds, does not accept free hydrogen ions but rather reduced coenzymes.

 Check your understanding of the concepts covered in this section by answering the questions in the e-book.

SUMMARY OF KEY CONCEPTS

- ΔG is the free energy change in a reaction. It is an expression of the amount of energy change in a reaction available to perform useful work.

- Free energy is related to enthalpy and entropy by the expression $\Delta G = \Delta H - T\Delta S$ where ΔH is the enthalpy change, T is the absolute temperature and ΔS is the entropy change.

- The ΔG value can be used to determine whether a reaction is likely to be reversible in cells.

- ATP is termed a high-energy phosphate molecule. The release of the two terminal phosphate groups liberates large amounts of free energy, which is used to perform cellular work.

- Enzymes lower the activation energy required before a reaction can happen. In doing so, enzymes enable reactions to happen more quickly: they act as biological catalysts.

- Biological oxidation involves removal of electrons and transfer to another acceptor molecule, which need not be oxygen.

- In biochemical systems, oxidation commonly involves enzymatic removal of two hydrogen atoms from a metabolite molecule.

- A variety of electron/hydrogen carriers transfer electrons to oxygen in the cell.

- The electron affinity of each carrier determines which one is able to donate electrons to another: ones with lesser affinity will tend to donate electrons to ones with higher affinity.

- The electron affinity of carriers is expressed as the redox potential value E'_0 and is expressed in volts.

- The redox potential value associated with a reaction is directly related to the free energy change of that reaction, a relationship represented by the Nernst equation.

- NAD^+ is an important electron carrier: it can accept two electrons and a hydrogen atom to form NADH, and functions as a coenzyme for dehydrogenases.

- FAD is of a similar structure to NAD^+ but the accepting group is the vitamin riboflavin. It is reduced to $FADH_2$.

- FMN is a single-nucleotide form of FAD; it is reduced to $FMNH_2$.

 **Use the flashcards in the e-book to test your recall of key terms introduced in this chapter.**

QUESTIONS

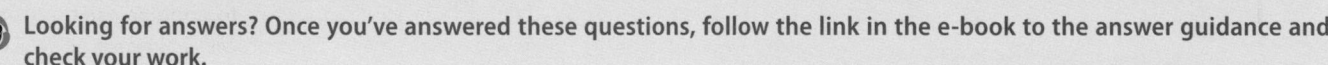 **Looking for answers? Once you've answered these questions, follow the link in the e-book to the answer guidance and check your work.**

Concepts and definitions

1. What is meant by free energy?

2. What is entropy and what is its significance?

3. Which of the following changes in states represent:

    (a) the smallest ΔS ?

    (b) the largest ΔS?

    - Freezing of water to ice.
    - Melting of ice to liquid water.
    - Sublimation of ice to gas.

    (c) How do you know?

4. Which of the following statements correctly describes the process of oxidation?

    (a) The loss (or donation) of electrons to another molecule, atom or ion.

    (b) The acceptance of electrons from another molecule, atom, or ion.

Apply the concepts

5. Does activation energy affect Gibbs free energy in an enzyme-catalysed reaction?

6. The second law of thermodynamics specifies that all processes must increase the total entropy of the universe. Living cells are at a lower entropy than the randomly arranged molecules in their environment. Does that mean that living cells are exempt from the second law? Explain your answer.

7. Calculate the Gibbs free energy change (ΔG) for the following chemical reaction if the reaction occurs at 20 °C, the change in

heat (ΔH) = 19,070 cal, and the change in entropy (ΔS) = 90 cal K^{-1}.

$$ATP \rightarrow ADP + Pi$$

8. Does an increase in reaction temperature make ATP → ADP + P$_i$ more or less likely to occur spontaneously? Explain your answer.

Beyond the concepts

9. In the cell, the ATP synthesizing system converts ADP into ATP, but it does not convert AMP into ATP. How is AMP brought back into the system?

10. The redox couple FAD + 2H$^+$ + 2e$^-$ → FADH$_2$ has an E_0' value of −0.219 V. The redox couple of $\frac{1}{2}O_2 + 2H^+ + 2e^- \rightarrow H_2O$ has a value of -0.816 V.

Calculate the $\Delta G^{0'}$ value for the oxidation of FADH$_2$ by oxygen to water. The Nernst equation is $\Delta G^{0'} = nF\Delta E_0'$ where F = 96.5 kJ V^{-1} mol^{-1}.

FURTHER READING

Otto Meyerhof and the Physiology Institute: the birth of modern Biochemistry.http://www.nobelprize.org/nobelprizes/themes/medicine/states/otto-meyerhof.html
A biography of Otto Meyerhoff on the occasion of receiving the Nobel Prize for Physiology and Medicine in 1922.
Sir Hans Adolf Krebs/German-British biochemist/Brittanica.com:http://www.brittanica.com/biography/Hans-Krebs

A biography of Hans Krebs.
Prebble, J. (2002) Peter Mitchell and the ox phos wars. *Trends Biochem. Sci.* **27**: 209–212.
A paper describing the discovery of the process of oxidative phosphorylation and the arguments around the theory.

Information
Genes and Genomes

Chapter contents

Introduction 197

3.1 The chemical structures of DNA and RNA 198

3.2 Genome organization and packaging 206

3.3 DNA replication 210

LEARNING OBJECTIVES

By the end of this chapter you should be able to:

- Describe the chemical structures of DNA and RNA.

- Compare the sizes and organization of DNA genomes in prokaryotic and eukaryotic cells and outline the organization of DNA in the human genome.

- Describe how eukaryotic genomes are packaged into chromatin.

- Describe the mechanism of DNA replication in prokaryotes.

- Discuss key differences between DNA replication in prokaryotes and eukaryotes.

- Explain the requirement for repair of damaged DNA and outline some repair mechanisms.

 Watch the key concepts video in the e-book to prepare yourself for studying this chapter.

Introduction

The genome is the full genetic material of an organism, and in all cellular organisms the genome is made up of deoxyribonucleic acid (DNA). There can be few more iconic images in the modern world than that of the DNA double helix (Figure 3.1), and in this and the next chapter we will find out how this surprisingly simple molecular structure enables the genome to carry out two vital roles: first, to act as a 'manual' that contains the information necessary for the organism to carry out its cellular functions; and, second, to provide a mechanism that allows organisms to reproduce and pass on their traits to the next generation. The nature of genetic material, the structure and

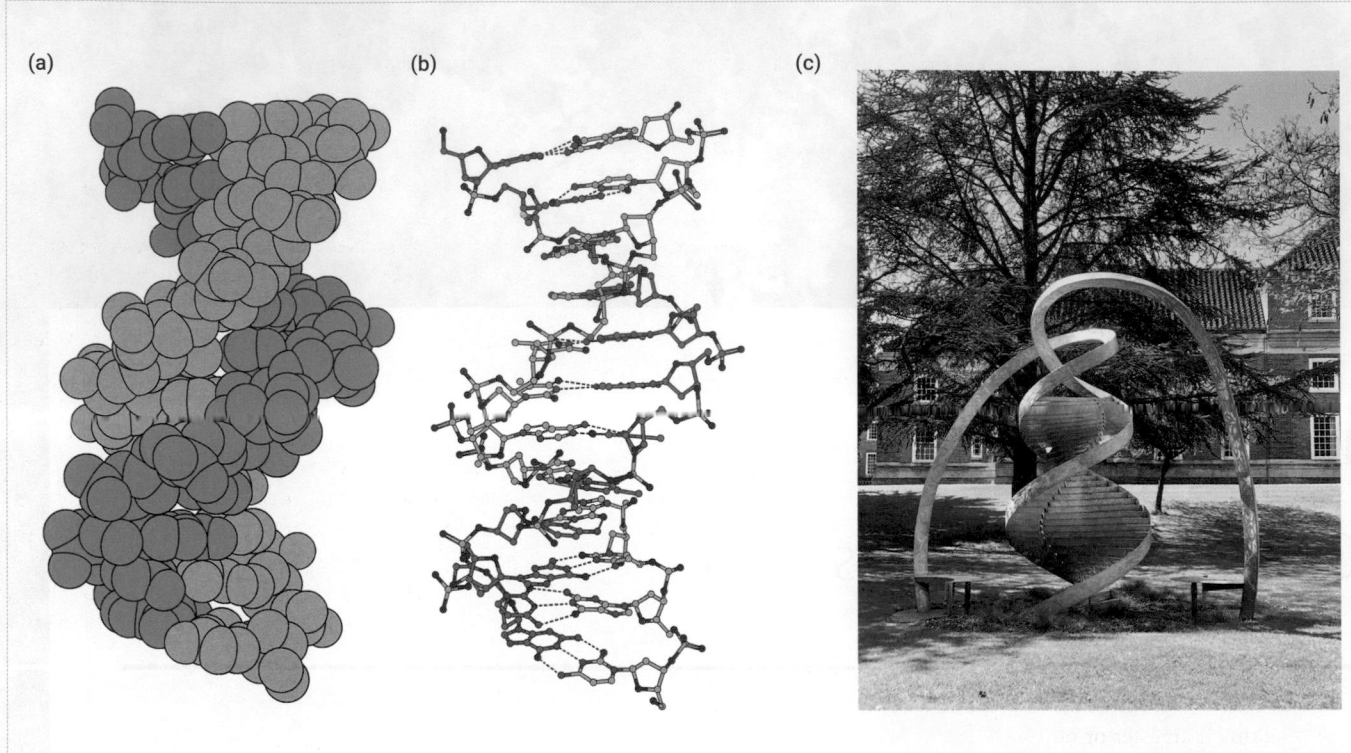

(a)   (b)   (c)

Figure 3.1 The DNA helix. Space-filling (a) and ball-and-stick (b) models of a short section of DNA (Protein Data Bank code 1BNA). The form of DNA shown is the B form, which is most prevalent in cells. B DNA consists of two strands of DNA wrapped around each other to form the famous 'double helix'. The strands are coloured differently here to help you distinguish them. (c) The sculpture 'Mutual Wrap' is situated in the grounds of Clare College, Cambridge. It was commissioned to commemorate the discovery of the double helix structure by James Watson and Francis Crick and is one of a series created by the sculptor Charles Jencks to represent the structure of DNA and show how biological information flows from it.
Source: Shutterstock.

organization of the genome, and its role in reproduction and inheritance are the main subjects of this chapter, while Chapter 5 explains how the information contained in the genome is 'read', directing the synthesis of proteins that carry out cellular functions.

While DNA is the material of genomes, it is thought that at the origin of life genetic information was carried in the closely related molecule ribonucleic acid (RNA). Despite being superseded by the DNA genome as cellular life evolved, RNA retains great importance as an intermediate in genome-directed protein synthesis. For this reason, in this chapter we describe the basic molecular structure of RNA alongside that of DNA.

Protein synthesis is a two-stage process. The first stage is **transcription**, in which the information held in the DNA is copied to make a corresponding messenger RNA molecule, and the second stage is **translation**, in which the information held in the messenger RNA is used to link amino acids together in a specific order to make a protein. Transcription and translation are described in detail in Chapter 5, but you will find it useful to be aware of these two stages of protein synthesis while studying the material in this chapter.

It is also worth noting that certain classes of virus have an RNA genome, including human immunodeficiency virus (HIV) and

coronaviruses, one of which, SARS-CoV-2, is the cause of the Covid-19 pandemic. Viruses are not considered to be cells as they cannot replicate independently and rely on some functions of their hosts to do so, so they do not violate the rule that all cells have a DNA genome.

3.1 The chemical structures of DNA and RNA

DNA and RNA are termed nucleic acids because DNA was originally isolated from cell nuclei. Their chemical compositions are surprisingly simple, to the extent that for decades after the discovery of nuclear DNA in 1869 scientists had difficulty believing that it could constitute the genetic material, instead focusing their attention on proteins. Acceptance of the DNA genome came from experiments such as those by Avery, MacLeod, and McCarty, and by Hershey and Chase in the 1940s and 1950s. We explore the work of Avery and his colleagues in Scientific Process 3.1.

Nucleotides are the building blocks of DNA and RNA

Nucleic acids are polymers of **nucleotides**, which each consist of a pentose sugar, a nitrogenous base, and one or more phosphates, as shown in Figure 3.2. A nucleoside is a sugar plus base, while the addition of phosphate makes it a nucleotide.

Figure 3.3 shows the structures of the pentose sugars found in DNA and RNA in more detail. The carbons are numbered 1′, 2′, 3′, etc. using the ′ or 'prime' symbol to distinguish them from carbons in the nitrogenous bases, which are numbered simply 1, 2, 3, etc. The pentose sugar is 2′-deoxyribose in deoxyribonucleotides or ribose in ribonucleotides. The difference is that 2′-deoxyribose (often simply termed deoxyribose) lacks a hydroxyl group on the 2′ carbon.

We can envisage each of the bonds that link the phosphate and the base to the pentose sugar as being formed through the loss of a molecule of water, as shown in Figure 3.4. The actual biochemical reactions by which nucleotides are synthesized in the cell are a lot more complex than this, but the simplification is helpful to understand the structure of the nucleotide and the nomenclatures used for the bonds. Phosphates are salts of phosphoric acid. Therefore, the bond that links the phosphate to the sugar in nucleic acids forms between an acid (phosphoric acid) and an alcohol (the hydroxyl group attached to the 5′ carbon of the pentose ring), making it an ester or phosphoester bond (Figure 3.4). Note that the bonds between linked phosphates are termed phosphoanhydride bonds (Figure 3.2). The bond between the sugar and the base is a bond between a carbohydrate (the sugar) and another molecule, making it a glycosidic bond.

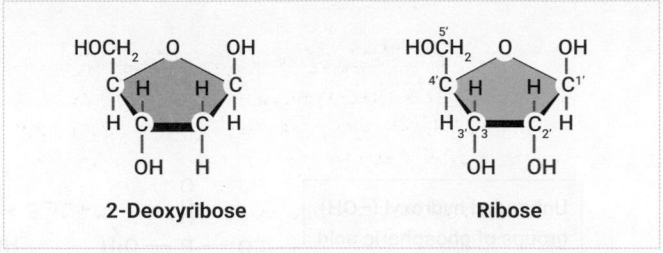

Figure 3.3 The structures of 2′-deoxyribose and ribose, the pentose sugars found in DNA and RNA, respectively. The carbons are numbered 1′, 2′, 3′, etc. Note that the structures are identical except that 2′-deoxyribose (often simply termed deoxyribose) lacks a hydroxyl group on the 2′ carbon.

As also shown in Figure 3.4, at cellular pH the phosphates are negatively charged through the loss of protons from their unbonded hydroxyl groups. This ability to donate protons explains the acidic nature of DNA and RNA.

Nitrogenous bases are ringed structures, which are classified as **purines** if based on a double-ringed structure, and **pyrimidines** if they are single-ringed. Each nucleotide of DNA contains one of four bases, as shown in Figure 3.5. RNA also contains three of these four bases, but it contains uracil as a fourth base instead of thymine. Additional unusual and modified bases with specific functions do occur in cellular DNA and RNA, but adenine (A), guanine (G), cytosine (C), thymine (T), and uracil (U) are the fundamental five.

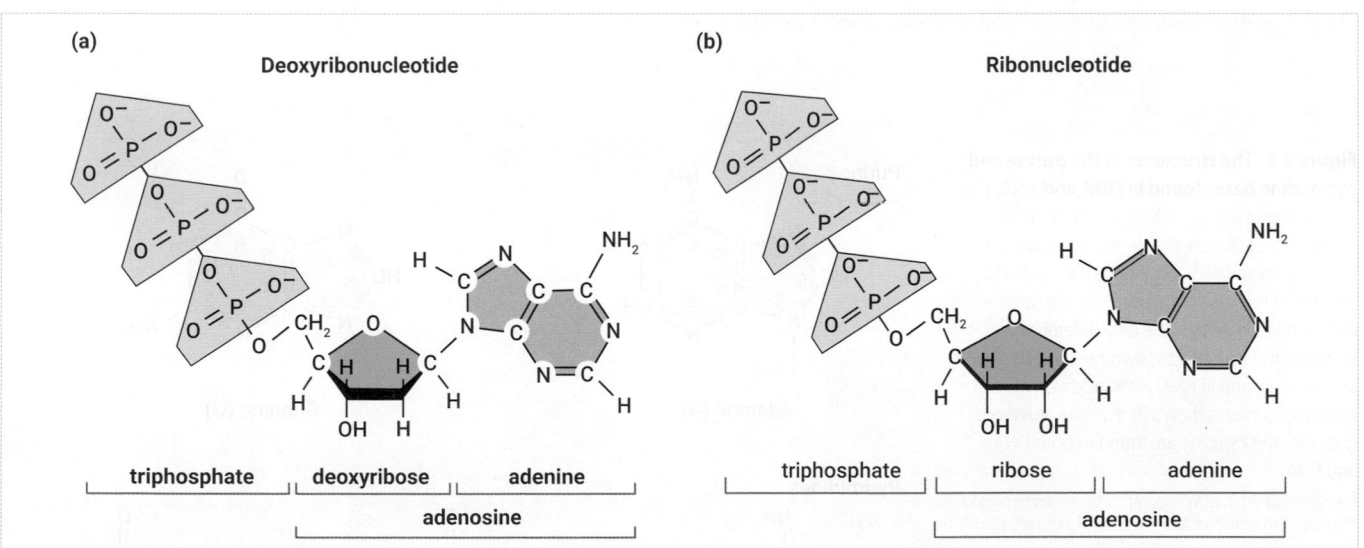

Figure 3.2 The structural formula of a nucleotide. (a) A deoxyribonucleotide (component of DNA). (b) A ribonucleotide (component of RNA). Each nucleotide contains a pentose (five carbon) sugar (blue), a nitrogenous base (green), and one to three phosphates (yellow). In these examples the base is adenine and there are three phosphates, making these nucleotides deoxyadenosine triphosphate (dATP) and adenosine triphosphate (ATP) respectively. The difference between the deoxyribonucleotide and the ribonucleotide is in the structure of the sugar, highlighted in red. Further details of the structures and nomenclature of nucleotides are explained in the text.

Source: Alberts et al. *Essential Cell Biology*, Edn. 4. Garland Science.

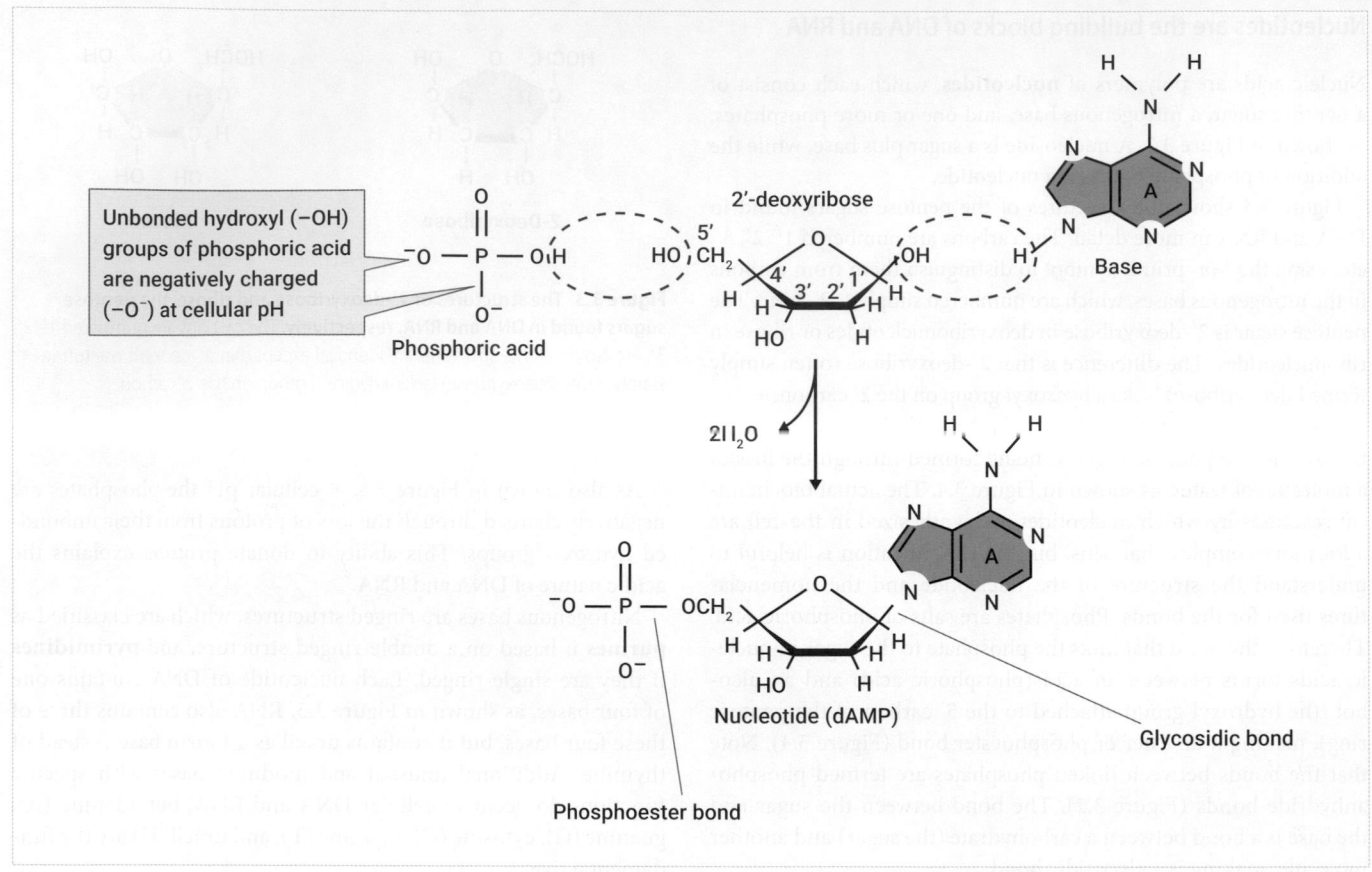

Figure 3.4 Bonding of the phosphate and base to 2'-deoxyribose to form a nucleotide, deoxyadenosine monophosphate (dAMP). Note that while this figure gives useful information on nucleotide structure, simple condensation reactions with loss of water are not used to synthesize nucleotides in the cell. The figure also shows that nucleotides are negatively charged through loss of protons from the phosphates.

Source: adapted from Watson et al. *Molecular Biology of the Gene*, 7th edn. Pearson. 2014.

Figure 3.5 The structures of the purine and pyrimidine bases found in DNA and RNA. The numbering of atoms making up the rings is shown. For each base the dotted line indicates the position at which formation of a glycosidic bond links the base to carbon 1' of ribose or deoxyribose. Note that the only difference between the structure of thymine (found in DNA) and uracil (found in RNA) is the addition of a methyl group to carbon 5 in thymine. Adenine, guanine, and cytosine are found in both DNA and RNA.

Source: adapted from Watson et al. *Molecular Biology of the Gene*, 7th edn. Pearson. 2014.

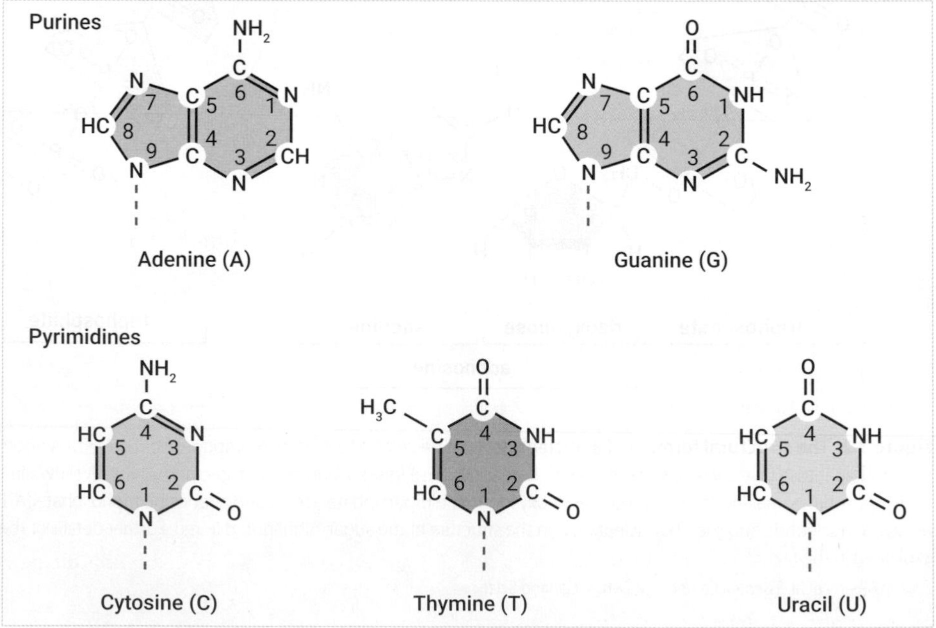

Table 3.1 Nomenclature of the major bases and nucleosides found in DNA and RNA

Base	Deoxyribonucleoside	Ribonucleoside
Adenine	Deoxyadenosine	Adenosine
Guanine	Deoxyguanosine	Guanosine
Cytosine	Deoxycytidine	Cytidine
Thymine	Deoxythymidine *or* Thymidine	–
Uracil	–	Uridine

The nomenclature of nucleosides is derived from the names of the bases as shown in Table 3.1. Deoxythymidine is often simplified to thymidine, since the ribonucleotide equivalent is rarely found in cellular nucleic acids.

Nucleotides are in turn named from the nucleoside, for example adenosine monophosphate (AMP), adenosine diphosphate (ADP), and deoxyadenosine triphosphate (dATP). The names are usually abbreviated, but it is worth learning the full versions from the table.

PAUSE AND THINK

Give the full names of the nucleotides that have the abbreviations dCTP, GDP, and TMP.

Answer: Deoxycytidine triphosphate, guanosine diphosphate, and thymidine monophosphate.

DNA and RNA are linear polynucleotides in which an ester bond links the phosphate group of one nucleotide to the 3′ hydroxyl group of the next. Thus, the 5′ carbon of one sugar is linked to the 3′ carbon of the next by a phosphate and two ester bonds, a structure termed a **phosphodiester bond** or phosphodiester linkage. Figure 3.6 shows a short section of DNA consisting of four nucleotides in which the phosphodiester linkage is labelled. It also illustrates the polarity of polynucleotides. At one end of the molecule there is a 5′ phosphate that is not linked to another sugar, while at the other end there is a free 3′ hydroxyl group. The two ends are thus termed the **5′ and 3′ ends**, respectively.

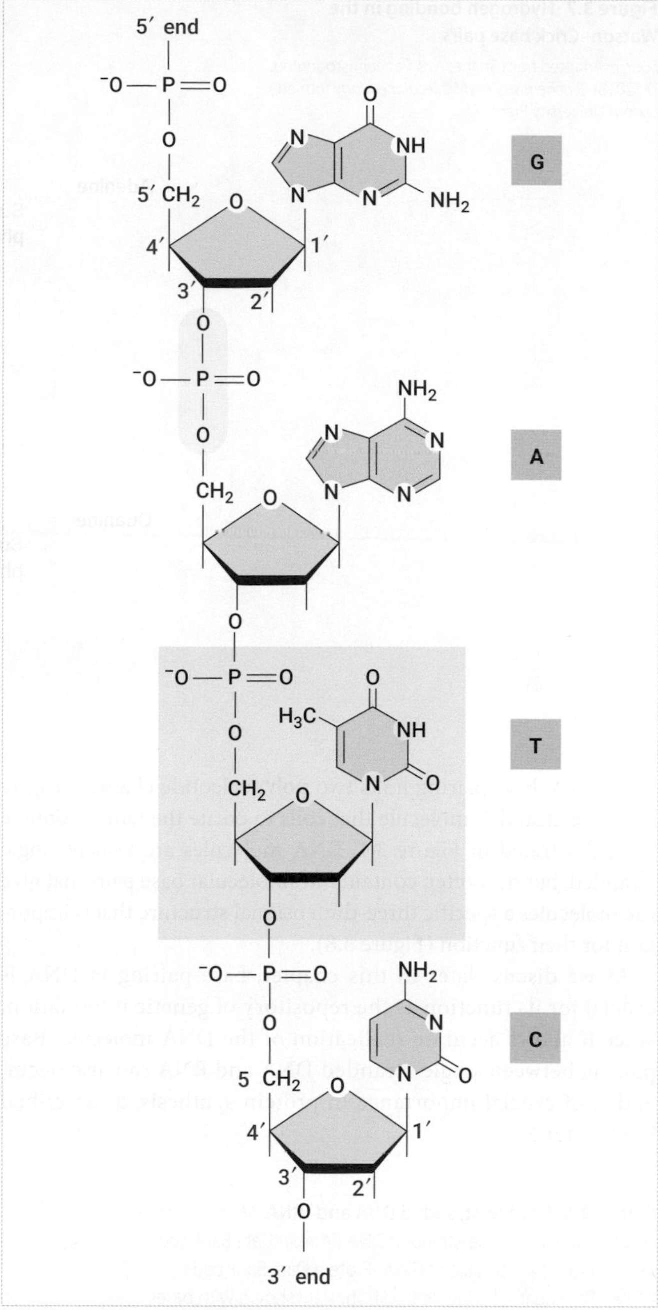

Base pairing is crucial for the function of DNA and RNA

Cellular DNA and RNA molecules vary greatly in length but can contain thousands or even, in the case of genomic DNA, millions of nucleotides. It is the order of the bases in DNA and RNA molecules that constitutes their information content. The feature of base structure that enables them to carry the information for coding proteins and have the capacity to direct their own replication is their capacity for **complementary base pairing**. Erwin Chargaff, working in the 1940s, established that the four nucleotides are not necessarily present in DNA in equal amounts, but in any DNA sample the ratio of purines to pyrimidines is approximately 1:1. Moreover, he found that the number of adenine molecules is equal to the number of thymine molecules and the number of guanine

Figure 3.6 A short length of a single strand of DNA, showing the 5′ and 3′ ends. An individual nucleotide, thymine (T), is in enclosed in a box and one phosphodiester bond is highlighted.

Source: Alberts et al. *Essential Cell Biology*, Edn 4. Garland Science, 2014.

molecules is equal to the number of cytosine molecules. Chargaff's 'rules' provided an important clue to the scientists who determined the three-dimensional structure of DNA. The ratios observed by Chargaff are explained by the observation that pairs of bases have the capacity to hydrogen bond with each other, forming specific purine–pyrimidine pairs: A with T (or with U in RNA) and G with C (see Figure 3.7). These hydrogen-bonded structures are known as Watson–Crick base pairs.

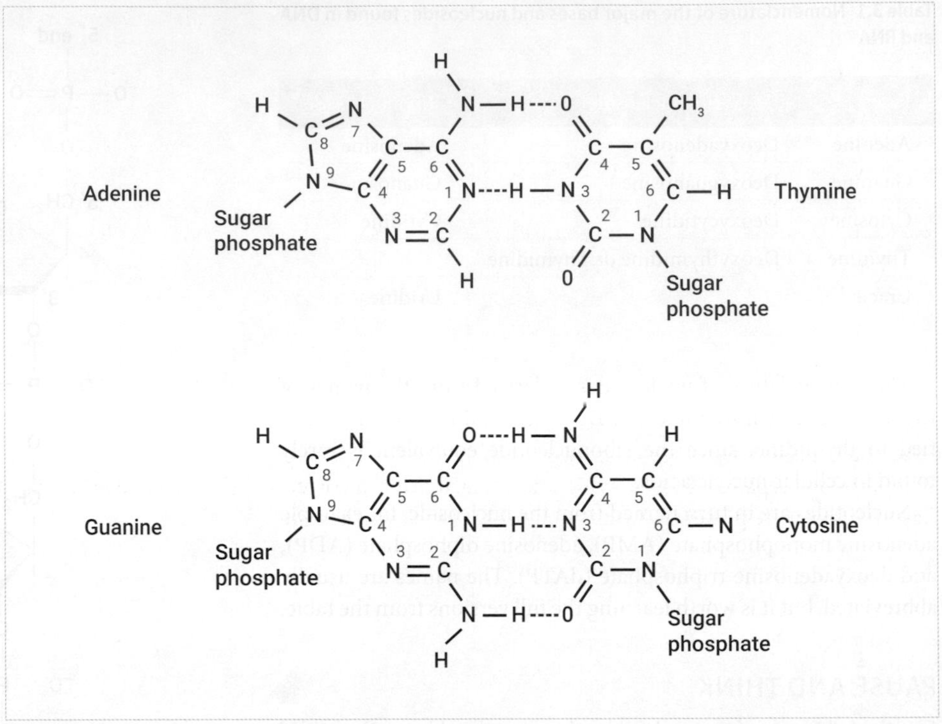

Figure 3.7 Hydrogen bonding in the Watson–Crick base pairs

Source: Adapted from Snape, A. & Papachristodoulou, D. (2018). *Biochemistry and Molecular Biology* (6th ed.). Oxford University Press.

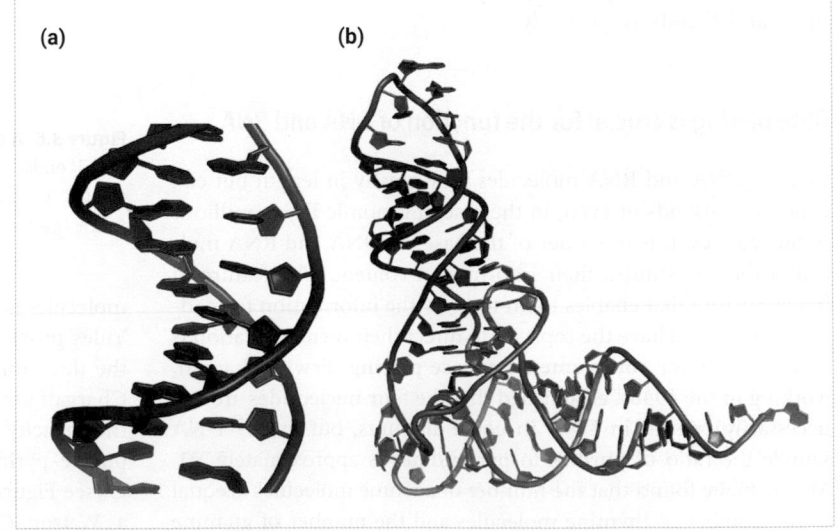

In DNA, base pairing links two polynucleotide chains, creating a double-stranded molecule that coils to create the famous double helix illustrated in Figure 3.1. RNA molecules are usually single stranded, but they often contain intramolecular base pairs that give the molecules a specific three-dimensional structure that is important for their function (Figure 3.8).

As we discuss later in this chapter, base pairing in DNA is crucial for its function as the repository of genetic information, since it allows accurate replication of the DNA molecule. Base pairing between single-stranded DNA and RNA can also occur, and is of crucial importance in protein synthesis, as described in Chapter 5.

PAUSE AND THINK

The percentage of adenine bases in the human genome is approximately 30. What are the approximate percentages of each of the other three bases?

Answer: Thymine 30%, guanine 20%, and cytosine 20%.

Structural differences between DNA and RNA reflect their functions

Before moving on to discuss the three-dimensional structure of DNA, it is worth noting two further points of interest. First, DNA

Figure 3.8 Double-stranded DNA and tRNA. Models of (a) a short section of double-stranded DNA (Protein Data Bank code 1BNA) and (b) a molecule of tRNA (Protein Data Bank code 4TNA). The sugar–phosphate backbones are shown with bases protruding from them. (a) The two strands of the DNA double helix, here coloured grey and pink, are linked together by base pairing. (b) The tRNA molecule consists of a single strand, which folds to create a complex three-dimensional shape stabilized by intramolecular base pairing. This shape includes regions of double helix, including that shown here in blue. We discuss the function of tRNA in Chapter 5.

is chemically a much more stable molecule than RNA. This is because the 2′ hydroxyl group on the ribose sugar of RNA is capable of interacting with the adjacent phosphorus attached to the 3′ carbon, and in doing so it breaks the link between the phosphate group and the 5′ carbon of the next sugar. This is probably the reason why through evolution DNA has replaced RNA as the material of the genome, while, as explained in Chapter 5, RNA retains a role as a more transient 'messenger' molecule in protein synthesis.

Second, the presence of thymine rather than uracil in DNA is explained by the need to repair damaged DNA to maintain the integrity of the genome as discussed later in this chapter. A careful look at the base structures in Figure 3.6 will show you that loss of the amino group from cytosine would convert it into uracil. Spontaneous deamination of cytosine occurs in cellular DNA, and this would cause a mutation (a change in the genome sequence) if it were not repaired by removing and replacing the incorrect base. If uracil were present naturally in DNA the repair mechanism

would be unable to distinguish between normal uracil bases and those that arise through cytosine deamination. The occurrence of thymine in DNA allows recognition and replacement of uracil but not thymine, so that repair can take place without accidental alteration of the normal base sequence.

Before moving on to discuss the structure of DNA in more detail, in Scientific Process 3.1 we explore the crucial experiments that provided evidence of DNA's role as the genetic material of the cell.

SCIENTIFIC PROCESS 3.1   Evidence that DNA is the genetic material

Research question

In the early part of the 20th century scientists debated the nature of genetic material. Many argued that nucleic acids, with their relatively simple chemical composition, could not be the repository of all the information required to direct the development and functions of a complex organism. Instead, they proposed that protein was the substance of genes. A number of experiments, however, led to the conclusion that the genome is indeed composed of DNA. Important among them was the work of Oswald Avery and co-workers, published in 1944.

Prediction

Avery and his team built upon the work of Frederick Griffiths, who had shown that a virulent strain of the bacterium *Streptococcus pneumoniae*, termed the S strain due to its smooth outer capsule, could 'transform' bacteria of a non-virulent R or rough strain so that they gained its virulent characteristics. Mice injected with either live R strain bacteria, or heat-killed bacteria from the S strain, did not develop disease. However, co-injection of the live R strain and heat-killed S strain caused the mice to die. The experiments designed by Avery and his co-workers predicted that they could isolate the 'transforming substance' from bacteria and analyse its chemical composition. They also decided to treat the purified substance with enzymes. If it was DNA, then an enzyme that could break down DNA should destroy its transforming activity.

Methods

Rather than injecting bacteria into mice, the team used an assay that allowed them to detect whether transformation of the R strain occurred by observing the bacteria in culture (Figure 1). They experimented with a variety of methods for purifying crude extracts of heat-killed S strain bacteria

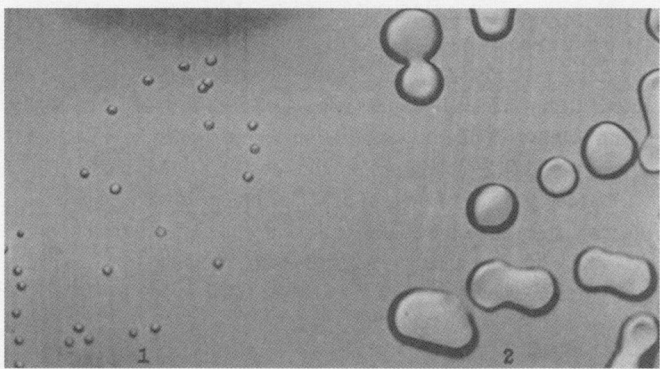

Figure 1 *Streptococcus pneumoniae* **experiment.** Colonies of R strain *S. pneumoniae* grown on a culture plate are shown on the left, while colonies of S strain bacteria that have arisen from transformation of the R strain are shown on the right. Magnification 3.5×

Source: Avery, O. T. et al. (1944) Studies on the chemical nature of the substance inducing transformation of pneumococcal types. *J. Exp. Med.*, 79, 137–157.

and established a protocol that gave them a relatively pure extract that was still able to transform the R strain when mixed with it. They subjected the purified material to chemical analysis and they also treated it with a panel of enzymes. At that time, they were unable to buy pure DNase enzyme 'off the shelf' so they used extracts of enzymes prepared from a number of different animal tissues. They tested the ability of these extracts to break down DNA by using them on pure DNA samples from other sources.

Results

Table 1 shows the results of chemical analysis of four preparations of the transforming substance, and compares them to the known composition of the sodium salt of DNA. The similarity in chemical composition to DNA

Table 1 Elementary chemical analysis of purified preparations of the transforming substance

Preparation No.	Carbon (%)	Hydrogen (%)	Nitrogen (%)	Phosphorus (%)	N/P ratio
37	34.27	3.89	14.21	8.57	1.66
38B	–	–	15.93	9.09	1.75
42	35.50	3.76	15.36	9.04	1.69
44	–	–	13.40	8.45	1.58
Theory for sodium desoxyribonucleate	34.20	3.21	15.32	9.05	1.69

Table 2 The inactivation of transforming principle by crude enzyme preparations

Crude enzyme preparation	Phosphatase	Tributyrinesterase	Depolymerase for desoxyribo-nucleate	Inactivation of transforming principle
Dog intestinal mucosa	+	+	+	+
Rabbit bone phosphatase	+	+	–	–
Swine kidney phosphatase	+	–	–	–
Pneumococcus autolysates	–	+	+	+
Normal dog and rabbit serum	+	+	+	+

provided evidence that at least the majority of the substance in the purified extract was DNA. This was consistent with the results of tests of its physical characteristics, which showed that it was a large molecule and absorbed ultraviolet light of the same wavelength as does DNA.

The results in Table 2 show that enzyme extracts are only able to destroy the activity of the transforming substance if they possess DNase activity (described in the table as 'depolymerase for desoxyribonucleate'). The researchers also showed that treatment with protease and ribonuclease enzymes did not destroy the activity of the substance.

Conclusion

Avery and his colleagues point out in the discussion section of their paper that DNA, a nucleic acid, has produced a change in *S. pneumoniae* bacterial cells that enables them to produce an entirely different substance, the smooth bacterial capsule. They also note that production of the capsule

must involve the transformed calls carrying out a series of enzyme-catalysed reactions. At that stage the structure of DNA was unknown, and the researchers had no idea of the mechanism by which DNA could enable this to occur. We now understand how DNA encodes proteins, as described in Chapter 5.

The researchers also noted that the transformation, once it had occurred, was 'stable': that is, the newly acquired characteristics were passed on as the bacteria divided. They went so far as to state that 'The inducing substance has been likened to a gene, and the capsular antigen which is induced in response to it has been regarded as a gene product'. However, it would take several more years and further research before it was fully established and widely accepted that DNA is the genetic material.

Read the original work

Avery, O. T. *et al.* (1944) Studies on the chemical nature of the substance inducing transformation of pneumococcal types. *J. Exp. Med.* 79: 137–157.

The DNA double helix

The story of the race to discover the structure of DNA has been told many times. Key evidence came from X-ray diffraction images produced by Rosalind Franklin at King's College London, including the famous 'Photo 51' (as shown in Figure 3.9a). The cross shape in the image is characteristic of a helical structure. In 1953, James Watson, who was working on the structure of DNA at the University of Cambridge, was shown the unpublished X-ray photograph. Watson and his collaborator Francis Crick were quickly able to create a model of the DNA molecule that incorporated previously established data and was compatible with the X-ray image. Their structure for DNA was published in April 1953 and in the decades since it has proved substantially correct.

The DNA double helix is shown in Figure 3.9b and 3.9c. Its key features are:

1. It consists of two polynucleotide chains or strands, linked together by hydrogen bonds between the bases (complementary base pairing).

2. Adenine is linked to thymine by two hydrogen bonds, while guanine is linked to cytosine by three hydrogen bonds.

3. The hydrophilic sugar–phosphate backbones of the two strands are on the outside of the molecule, in contact with its aqueous environment.

4. The two strands are antiparallel: that is, they are arranged in different directions with respect to the 5′ to 3′ polarity of the sugar–phosphate backbone.

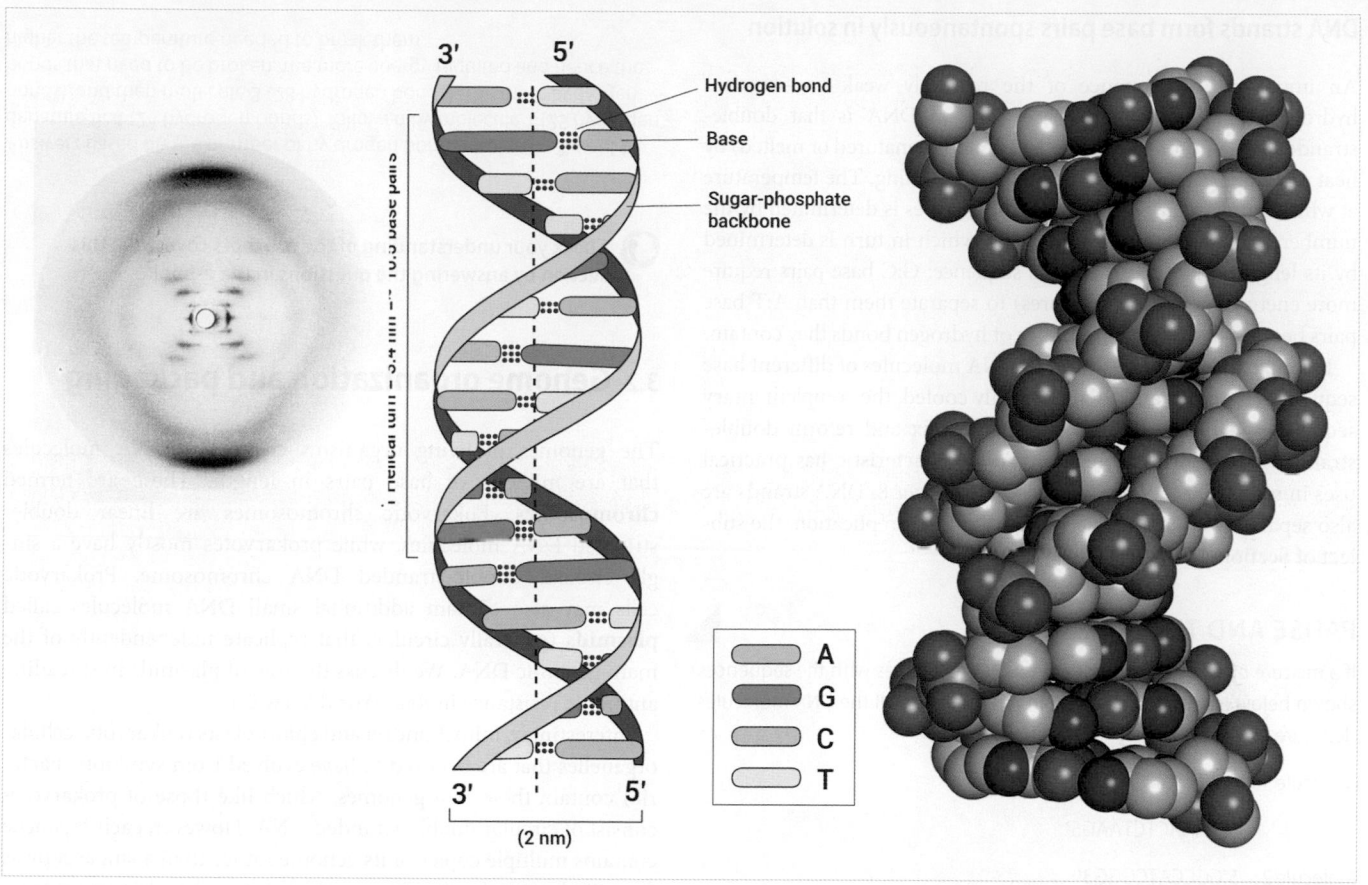

Figure 3.9 The double helix structure of DNA. The form of DNA shown is the B form, which is most prevalent in cells. (a) 'Photo 51', an X-ray diffraction image taken by Rosalind Franklin and Ray Gosling at King's College London in May 1952, provided confirmation of the helical structure of DNA. The photograph (together with their own deductions) enabled James Watson and Francis Crick of the University of Cambridge to build the first correct model of the DNA molecule. A schematic structure based on the Watson–Crick model is shown in (b). The space-filling model in (c) illustrates the two grooves of different sizes in which the edges of the base pairs are exposed to the exterior of the molecule. Atom colours are: C, grey; N, blue; O, red; P, orange.

Source: (a) King's College London Archives/Science Photo Library. (b) Adapted from Watson et al. (2014) *Molecular Biology of the Gene*, 7th edn. Pearson. (c) Shutterstock.

5. Complementary base pairing, with purine–pyrimidine pairs, means that the two strands are evenly spaced throughout the length of molecule.

6. The base pairs are tucked away inside the molecule, so that their hydrophobic faces are not exposed to the aqueous environment. However, the edges of the base pairs are accessible in the major and minor grooves, as shown in Figure 3.9b. We explain the significance of this for gene regulation in Chapter 5.

7. The base pairs are arranged like the rungs of a ladder, but each successive base pair is slightly rotated compared to the one below it, giving the helical structure.

8. The helices are right-handed. Imagine walking up a strand as if it were a spiral staircase—you will continually turn clockwise. Alternatively, imagine a right-handed person driving in a standard screw—their wrist rotates in the same direction as the helix.

9. The dimensions of the DNA double helix vary slightly depending on its base sequence and whether it is in its crystallized form or in solution. Approximate dimensions are:

- 2 nm diameter
- 10–10.5 base pairs per 360° turn
- A 'pitch' (vertical length of one 360° turn) of 3.4 nm and hence a distance from one base pair to the next of approximately 0.34 nm.

PAUSE AND THINK

Nucleic acid sequences are, by convention, written in the 5' to 3' direction from left to right.

(a) What would be the complementary sequence to a single strand of DNA that has the sequence 5' GATCGGA 3'?

(b) What would be the sequence of an RNA molecule that is complementary to the DNA sequence 5' GATCGGA 3'?

(b) 5' UCCGAUC 3'

(a) 5' TCCGATC 3'

Answer:

DNA strands form base pairs spontaneously in solution

An important consequence of the relatively weak non-covalent hydrogen bonding between base pairs in DNA is that double-stranded DNA in solution can be separated (denatured or melted) by heat and then renatured (or annealed) on cooling. The temperature at which a particular DNA molecule denatures is determined by the number of hydrogen bondsmit forms, which in turn is determined by its length in base pairs and its sequence: G:C base pairs require more energy (higher temperatures) to separate them than A:T base pairs because of the higher number of hydrogen bonds they contain.

If a mixture of double-stranded DNA molecules of different base sequences is heated and then gradually cooled, the complementary sequences will 'find' each other in the mix and reform double-stranded base-paired molecules. This characteristic has practical uses in molecular biology as we see in Chapter 8. DNA strands are also separated by enzyme action during DNA replication, the subject of Section 3.2.

PAUSE AND THINK

If a mixture of short double-stranded DNA molecules with the sequences shown below is gradually heated, in which order will the DNA molecules denature, and why?

Molecule 1:   5'ATTTGAGATTC3'

3'TAAACTCTAAG5'

Molecule 2:   5'GGCCATCGGG3'

3'CCGGTAGCCC5'

Molecule 3:   5'TCGGTTTAGA3'

3'AGCCAAATCT5'

Answer: Based on the number of hydrogen bonds, molecule 3 would denature first (24 hydrogen bonds), followed by molecule 1 (25 hydrogen bonds), and then molecule 3 (28 hydrogen bonds). The more hydrogen bonds that need to be broken, the more energy required and hence the higher the temperature needed to break them.

 Check your understanding of the concepts covered in this section by answering the questions in the e-book.

3.2 Genome organization and packaging

The genomes of living organisms consist of DNA molecules that are millions of base pairs in length. These are termed **chromosomes**. Eukaryotic chromosomes are linear double-stranded DNA molecules, while prokaryotes mostly have a single circular double-stranded DNA chromosome. Prokaryotic cells may also contain additional small DNA molecules called **plasmids** (generally circular) that replicate independently of the main genomic DNA. We discuss the role of plasmids in spreading antibiotic resistance in Real World View 3.1.

Interestingly, mitochondria and chloroplasts (eukaryotic cellular organelles that are believed to have evolved from symbiotic bacteria) contain their own genomes, which like those of prokaryotes consist of circular double-stranded DNA. However, each organelle contains multiple copies of its genome rather than a single copy as found in bacterial cells. The genomes of mitochondria and chloroplasts encode some of the specialized proteins they require to function, while other proteins they need are encoded by the nuclear genome.

The organization of the genome in prokaryotic and eukaryotic cells is shown schematically in Figure 3.10.

REAL WORLD VIEW 3.1   **Plasmids and the spread of antibiotic resistance**

Antibiotics have saved countless lives since penicillin was first used to treat bacterial infections in the 1940s. Yet in the early 21st century we are in danger of a returning to a world where deadly diseases such as tuberculosis and bacterial sepsis cannot be cured. The reason is the spread of antibiotic resistance. Antibiotics work by inhibiting processes in bacteria that are essential for their life or growth, such as making cell walls or synthesizing proteins. Bacteria develop resistance through gaining new or altered genes, which may chemically inactivate the antibiotics or pump them out of the bacterial cells. Frequently, these resistance genes are carried on plasmids. Plasmids help to spread antibiotic resistance very rapidly, because copies of their DNA can be made and transferred from one bacterium to others in its vicinity—a process known as 'horizontal transfer' that occurs alongside 'vertical transfer' from one generation to the next.

Widespread overuse of antibiotics has fuelled the spread of resistance at an alarming rate. Doctors are often under pressure to prescribe them, even for viral illnesses against which they are ineffective. They are also used widely in agriculture where they are fed routinely to livestock animals to promote growth and are sprayed on to crops. As the antibiotics kill bacteria, including those that are non-pathogenic, bacterial strains that are resistant grow rapidly with reduced competition and can pass on their resistance genes.

Very few new antibiotics have been developed since the 1980s. The race is now on to discover new drug targets in bacteria and agents that may work against them. There is also a great deal that we can and should be doing to change our behaviour to stop the further spread of resistance to existing antibiotics.

You can find out more on this topic from the website of *Antibiotic Action* (http://antibiotic-action.com/), a global initiative established by the British Society for Antimicrobial Chemotherapy in 2011.

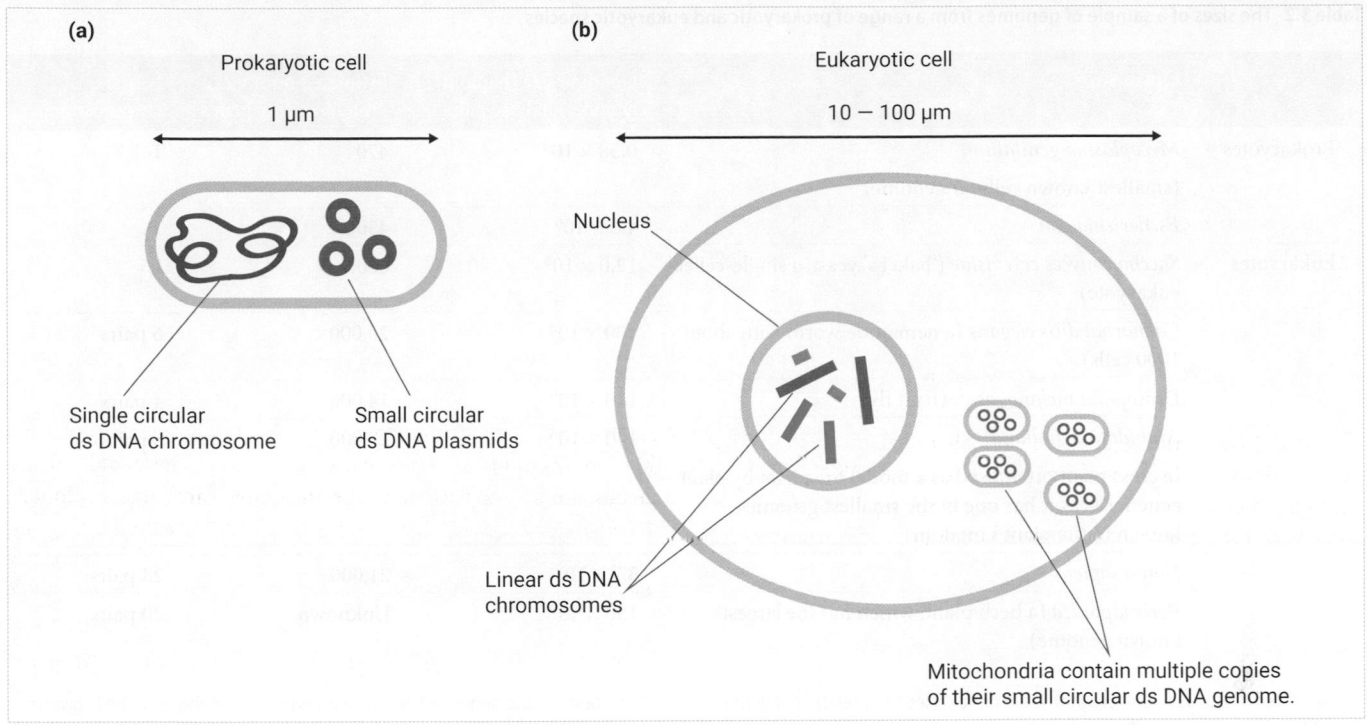

Figure 3.10 Organization of the genome in prokaryotic and eukaryotic cells. All cellular genomes consist of double-stranded (ds) DNA. (a) Prokaryotic cells each contain a single circular genome, which is highly folded and compacted to fit it in the small space. They may also contain small double-stranded circular plasmids that can replicate independently. (b) The genome of eukaryotic cells is contained in the nucleus and is organized as linear double-stranded chromosomes, the number and sizes of which vary between species. Eukaryotic cells contain homologous pairs of chromosomes, denoted here by colour coding. We explain the significance of homologous pairs later in Chapter 4. The eukaryotic cell contains mitochondria, which have multiple copies of their own double-stranded circular genome. In plant cells, chloroplasts also have their own genome.

Genome sizes vary between species

The size of an organism's genome (and hence the number and length of its chromosomes) generally, although not invariably, reflects the complexity of the organism. Genome sizes of a range of organisms are shown in Table 3.2.

Table 3.2 also indicates the number of protein-coding **genes** found in the genomes of these organisms. Like genome size, the number of genes an organism contains in its genome generally increases with complexity. However, the genomes of many eukaryotes (including humans) contain much more DNA than can be accounted for by their genes alone. The nature of these **intergenic sequences** in the human genome is discussed later in this section, but before this we consider two additional points regarding the genes themselves, firstly the question of what actually constitutes a gene, and secondly, the existence of 'split genes' in eukaryotes.

What is a gene?

It is surprisingly hard to define exactly what we mean by a 'gene'. For many purposes it is sufficient to think of the DNA sequence that encodes a particular protein. However, some genes encode RNA sequences that are not translated into proteins, for example tRNAs and rRNAs (discussed further in Chapter 5). Additionally, DNA sequences that flank coding sequences but are not themselves transcribed into RNA are involved in regulating gene

function and may therefore be considered an integral part of the gene. The regulatory sequences of genes are also considered further in Chapter 5.

Eukaryotes contain split genes consisting of introns and exons

In prokaryotes the protein-coding region of a gene is continuous, but in eukaryotes the coding region is typically interrupted by segments of DNA that do not code for amino acids. These 'intervening' sequences are termed **introns,** while the coding or 'expressed' sequences are termed **exons.** In the human genome exons total only about 1.6 per cent of the DNA, while the transcribed sequences of genes, including introns, make up around 25 per cent of the genome. When split genes are transcribed into RNA the sections of the primary transcripts that correspond to the introns are removed and the remaining sequences are joined together to make a continuous messenger RNA, a process known as **splicing.**

▶ We discuss splicing in more detail in Chapter 5.

The human genome contains transposable elements and other repetitive DNA sequences

Figure 3.11 illustrates the composition of the human genome and shows that about half of our DNA is made up of repetitive sequences. A surprisingly large proportion of these derive from sequences

Table 3.2 The sizes of a sample of genomes from a range of prokaryotic and eukaryotic species

	Organism	Genome size (haploid* genome) in base pairs	Estimated number of protein-coding genes	Number of chromosomes
Prokaryotes	*Mycoplasma genitalium* (smallest known cellular genome)	0.58×10^6	470	1
	Escherichia coli	4.6×10^6	4300	1
Eukaryotes	*Saccharomyces cerevisiae* ('baker's' yeast, a single-celled eukaryote)	12.0×10^6	6600	16
	Caenorhabditis elegans (a nematode worm with about 1000 cells)	100×10^6	20,000	6 pairs
	Drosophila melanogaster (fruit fly)	140×10^6	14,000	4 pairs
	Arabidopsis thaliana (a cress plant often used as a model organism by plant geneticists as it has one of the smallest genomes known in the plant kingdom)	140×10^6	27,000	5 pairs
	Homo sapiens	3.2×10^9	21,000	23 pairs
	Paris japonica (a herb plant, which has the largest known genome)	150×10^9	Unknown	20 pairs

* The cells of most eukaryotic organisms are diploid (see Chapter 4). That is, they contain two copies of the genome. The size of a single copy of the genome is given in the table.

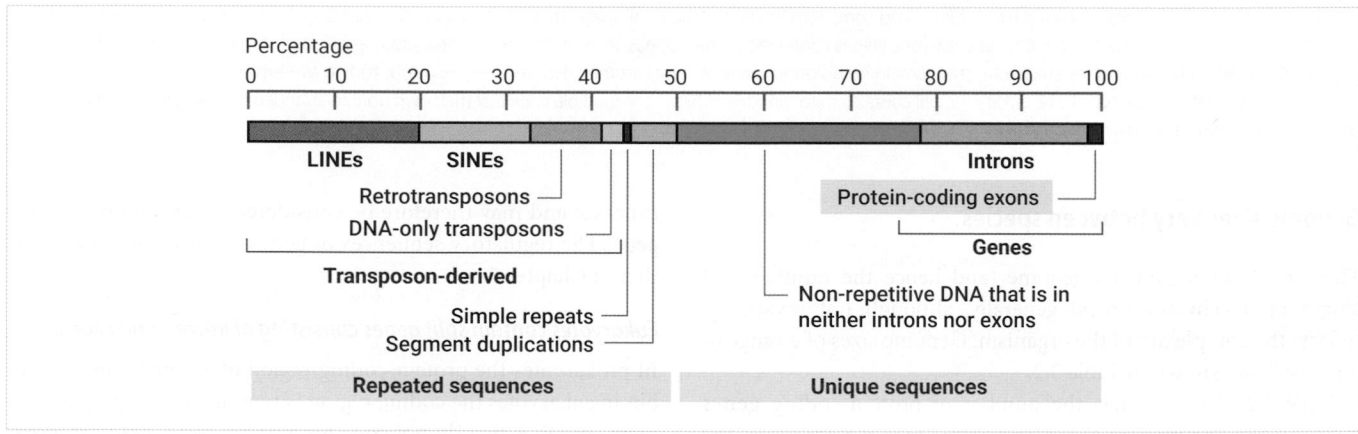

Figure 3.11 Composition of the human genome. Half of the genome is made up of repetitive sequences, many of which derive from transposable elements or transposons. The main classes of transposons (LINEs, SINEs, other retrotransposons, and DNA-only transposons) are described in the text. Non-transposon repeats include simple repeats, also described in the text, and segment duplications, which are large blocks of sequence (1000–200,000 base pairs) that are found at two or more places in the genome. The non-repetitive unique sequences that are not introns or exons include genes that encode functional RNAs (such as tRNA and rRNA) and regulatory sequences, as well as sequences with no known function.

Source: Matylla-Kulinska, K., Tafer, H., Weiss, A. and Schroeder, R. (2014) Functional repeat-derived RNAs often originate from retrotransposon-propagated ncRNAs. *WIREs RNA*, 5: 591–600. https://doi.org/10.1002/wrna.1243

that can (or could in the past) move from one part of the chromosome to another. These transposable elements, or **transposons**, popularly known as 'jumping genes', are classified according to the mechanism by which they move around. **DNA transposons** are those in which the sequence is cut out of the chromosome and inserted in a new position. This type of transposon is also found in bacterial genomes. **Retrotransposons** are found only in eukaryotes. They are not cut out of the DNA, but instead replicate by being transcribed into RNA copies, which are then copied back into DNA by a retrotransposon encoded enzyme, **reverse transcriptase**. The DNA copies insert themselves into new positions in the genome.

Some retrotransposons within the human genome have sequence similarities to **retroviruses**, but retrotransposons can only reproduce their DNA and move around within the genome of the cell, they cannot become fully infectious viral particles.

Reproduction and movement of transposable elements have played a major role in the evolution of the human genome, but most of the transposons in our present-day genome have undergone mutations that mean they are no longer mobile. The small proportion that can still move around contribute to human genome variation and occasionally cause disease by 'jumping' into the middle of protein-coding genes or into gene regulatory sequences.

Repetitive DNA in our genome can be classified according to whether it is derived from transposable elements. Two major types of retrotransposon-derived repeats are **LINEs (long interspersed elements)** and **SINEs (short interspersed elements)**. LINEs are a few thousand base pairs long and together make up around 21 per cent of the genome, while SINEs are 100–500 base pairs long and make up 13 per cent of the genome. The main class of SINEs is called the *Alu* element, of which there are more than one million in the genome. LINEs and SINEs are found as single or clustered sequences in many genomic locations, and comparison of their locations in different mammalian species suggests that they have proliferated and moved around relatively recently in our evolutionary history. However, the vast majority are no longer mobile.

The other main category of repetitive DNAs, which are not transposon-derived, are termed **simple sequence repeats**. These make up at least three per cent of the genome and consist of short nucleotide units, 14 base pairs or less in length, arranged in head to tail 'tandem' arrays, for example TTCCA/TTCCA/TTCCA repeated dozens or thousands of times. Such DNA sequences are found particularly around the **centromeres** (the region of the chromosome that attaches to the spindle at cell division) and **telomeres** (ends of the chromosomes), but also at other places in the genome. The smallest repeat unit found is the dinucleotide repeat, for example CACACACA. These **microsatellite** sequences are found in multiple positions (**loci**) on the chromosomes and are **polymorphic** in that the number of repeats at any particular locus varies between individuals. This gives them some important practical applications, for example in forensic science where they are used in DNA profiling, as described in Real World View 3.2.

REAL WORLD VIEW 3.2 DNA profiling

Polymorphic microsatellite loci provide a method that can be used for forensic testing of DNA, perhaps at a crime scene or in paternity testing. The principle is illustrated in Figure 1, where details are shown only of 'Locus 1', a region of the genome where there is a microsatellite consisting of the sequence GATA repeated a variable number of times. Copies of the region are made using the polymerase chain reaction (PCR) which gives

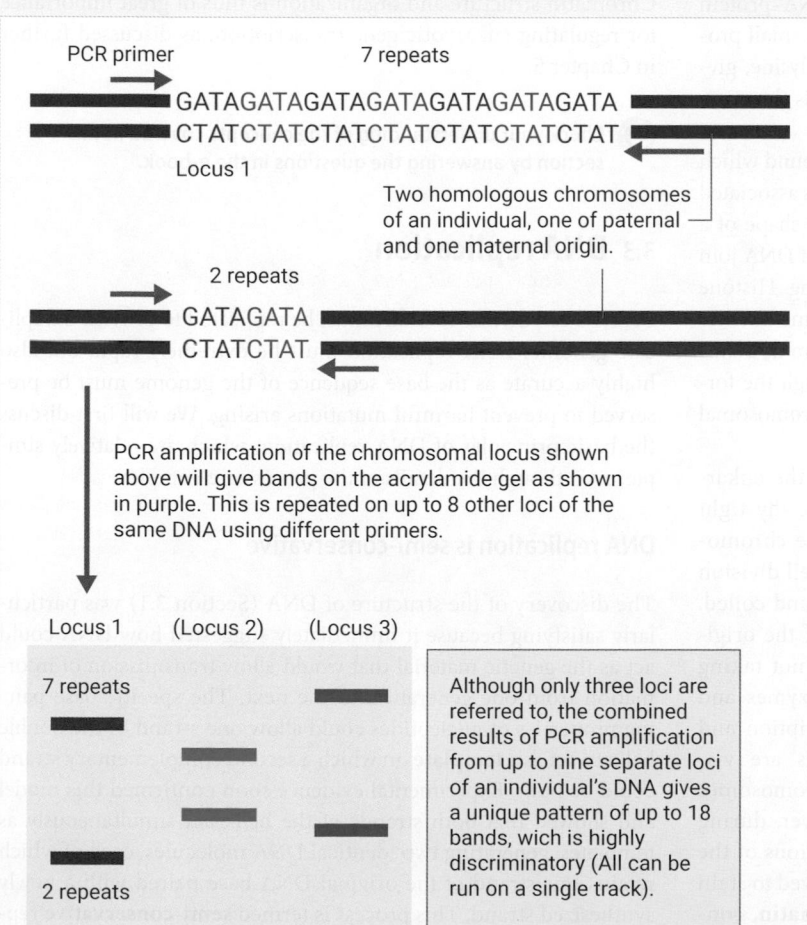

Figure 1 Diagram illustrating the principle of DNA fingerprinting in forensic science. A microsatellite locus on homologous chromosomes of an individual is shown. PCR products give two bands, with the lower repeat number giving the smaller product that runs faster than the larger one. This process is repeated on several more selected loci on the DNA of the same individual so that altogether a pattern of bands will result that is unique to the individual. The process is repeated on the forensic sample of DNA to see if the patterns match. (Note that loci 2 and 3 are not illustrated but the bands that might be produced from them are shown.)

Source: Snape, A. & Papachristodoulou, D. (2018). *Biochemistry and Molecular Biology* (6th ed.). Oxford University Press.

different-length products depending on the number of repeats present. (PCR is covered in more detail in Chapter 8.)

The individual whose DNA is being analysed in Figure 1 has inherited a chromosome containing seven GATA repeats at Locus 1 from one of their parents, and a chromosome containing two repeats at Locus 1 from the other parent. As the locus is highly polymorphic, a different person could have chromosomes containing 2, 3, 4, 5, 6, or 7 GATA repeats at that point. In reality, at least 10 different loci at different positions around the genome are analysed in any one test, though results from only three loci are shown here. If multiple highly variable loci are analysed, the odds that another individual will have the same number of repeats at all loci as the individual whose DNA is being tested, become vanishingly small. If the DNA has been collected from a crime scene it can be matched with DNA from a suspect, and so long as the profiling has been carried out under strictly controlled conditions it gives evidence of identity that is acceptable in court.

The DNA genome is packaged with proteins

One of the most intriguing problems in biology is that of packaging the long DNA genome into the microscopic cell. For example, the human cell nucleus contains about 2 m of chromosomal DNA, and this is packed into a sphere about 10 μm in diameter. To put this in perspective, if the nucleus were the size of a tennis ball, the genome would be the length of 500 tennis courts! In both prokaryotic and eukaryotic cells, DNA is bound by positively charged proteins that counteract the negatively charged phosphate groups of the DNA, allowing it to be compacted into a small space.

DNA in eukaryotic cells exists as **chromatin**—a DNA–protein complex. The main proteins of chromatin are **histones**, small proteins that are rich in the amino acids arginine and/or lysine, giving them positive charges. The structure of chromatin is shown in Figure 3.12.

Eight histone proteins form an '**octamer**' complex around which are wrapped 146 base pairs of DNA. The octamer and its associated DNA form a unit called a **nucleosome**, which has the shape of a disc. Short linker sequences of about 30–40 base pairs of DNA join successive nucleosomes, arranged like beads on a string. Histone H1 binds the DNA as it enters and leaves the nucleosome, further condensing the nucleosomes together into a fibre 30 nm in diameter. Further packaging of chromatin takes place through the formation of long loops that are attached to a central chromosomal non-histone protein scaffolding.

The tightness of DNA packaging changes during the eukaryotic cell cycle. As shown in Figure 3.13, it is especially tight during cell division (**mitosis** and **meiosis**) when the chromosomes must be separated without tangling. During cell division the looped structure of chromatin is further folded and coiled, achieving a 10,000-fold compaction of the length of the original DNA. During **interphase**, when cell division is not taking place, chromatin is less tightly packed to allow enzymes and regulatory molecules to access the DNA for transcription and DNA replication. Therefore, mitotic chromosomes are visible using the light microscope, while individual chromosomes are not distinguishable in interphase nuclei. However, during interphase, chromatin packing varies in different regions of the genome, with less compact chromatin, which is observed to stain lightly in microscopy studies and is termed **euchromatin,** contrasting with tightly compacted chromatin identifiable as dark staining regions known as **heterochromatin**. Coding regions of the genome, which are potentially transcriptionally active, are generally less compacted, while repetitive DNA sequences in regions of the chromosomes that have no coding function, such as the centromeres and telomeres, remain compacted.

 We explore the eukaryotic cell cycle in more detail in Chapter 10.

The exact degree of packing of transcriptionally active chromatin is highly dynamic, perhaps alternating between 10 nm and 30 nm fibre structures with short sequences of the DNA becoming transiently free of nucleosomes to allow access by other factors. Chromatin structure and organization is thus of great importance for regulating eukaryotic gene transcription, as discussed further in Chapter 5.

Check your understanding of the concepts covered in this section by answering the questions in the e-book.

3.3 DNA replication

Before a cell divides its DNA must be replicated to provide a duplicate genome. DNA replication must be extremely rapid but also highly accurate as the base sequence of the genome must be preserved to prevent harmful mutations arising. We will first discuss the basic principles of DNA replication, which are relatively simple, even though the details of the process are complex.

DNA replication is semi-conservative

The discovery of the structure of DNA (Section 3.1) was particularly satisfying because it immediately suggested how DNA could act as the genetic material that would allow transmission of information from one generation to the next. The specific base pairing properties of nucleotides could allow one strand of the double helix to act as a **template** on which a second complementary strand could be built. Experimental evidence soon confirmed this model and showed that both strands of the helix act simultaneously as templates, generating two identical DNA molecules, each of which retains one strand of the original DNA base paired with a newly synthesized strand. This process is termed **semi-conservative** replication and is illustrated in Figure 3.14.

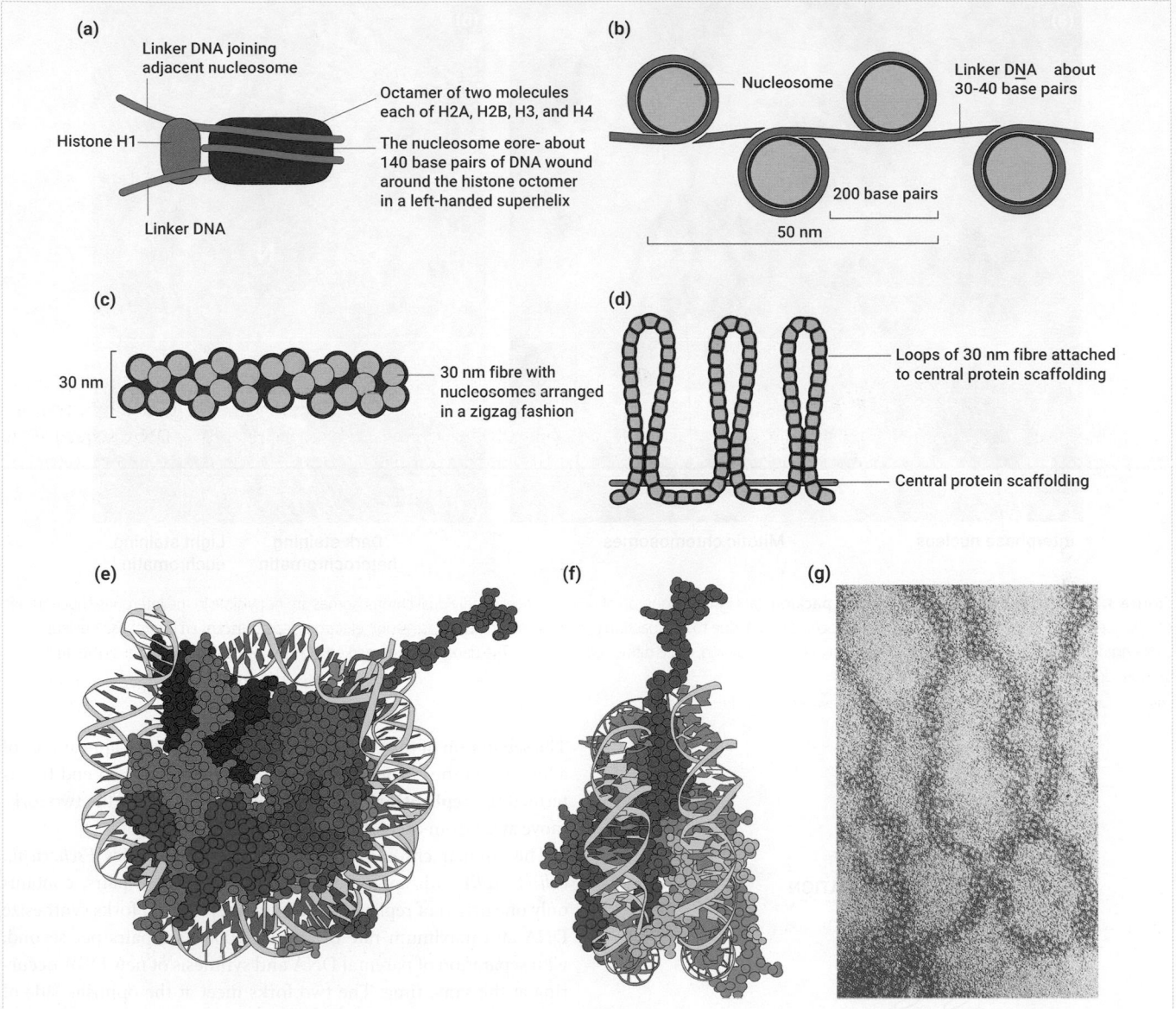

Figure 3.12 Chromatin packing in eukaryotes. (a) Diagram of a nucleosome. (b) Beads on a string form, the 10 nm fibre. (c) A 30 nm fibre of chromatin. (d) Loops of the 30 nm fibre are attached to a central protein scaffold in a 360° array. Further looping and coiling then takes place to compact the chromatin into a chromosome. (e) and (f) Two views of the structure of the nucleosome core particle showing the DNA (ribbon structure) wrapped around the octamer histone core (space-filling model). Each histone protein is shown in a different colour, while DNA is grey. The histone proteins are tightly packed into the core of the octamer, as shown, but each histone also has a flexible 'tail' section that protrudes from the nucleosome. Only the tail of one of the histone H3 molecules is shown here. The flexible histone tails are involved in regulating transcription of genes, as discussed in Chapter 5. (g) Electron micrograph of a 30 nm fibre of chromatin.

Source: (a–d) Snape, A. & Papachristodoulou, D. (2018). *Biochemistry and Molecular Biology* (6th ed.). Oxford University Press. (g) Lewin B. (1994). Genes V. Oxford University Press, Oxford. Photograph was provided by Prof. B. Hamkalo.

Replication is initiated at specific chromosome sequences

For replication to occur the two strands of the DNA double helix must be separated. This does not occur simultaneously across the entire chromosome. Instead, the strands are initially separated at short sequences called **replication origins**, which are typically rich in A:T base pairs. Specific proteins aid in the process, which requires energy from ATP hydrolysis. Enzymes that unwind the DNA double helix are called **helicases**.

PAUSE AND THINK

How does the presence of multiple A:T base pairs at replication origins aid the process of initiation?

Answer: As A is linked to T by two hydrogen bonds rather than three, having A:T rather than G:C base pairs means it is easier to separate the two strands of DNA, as required for replication.

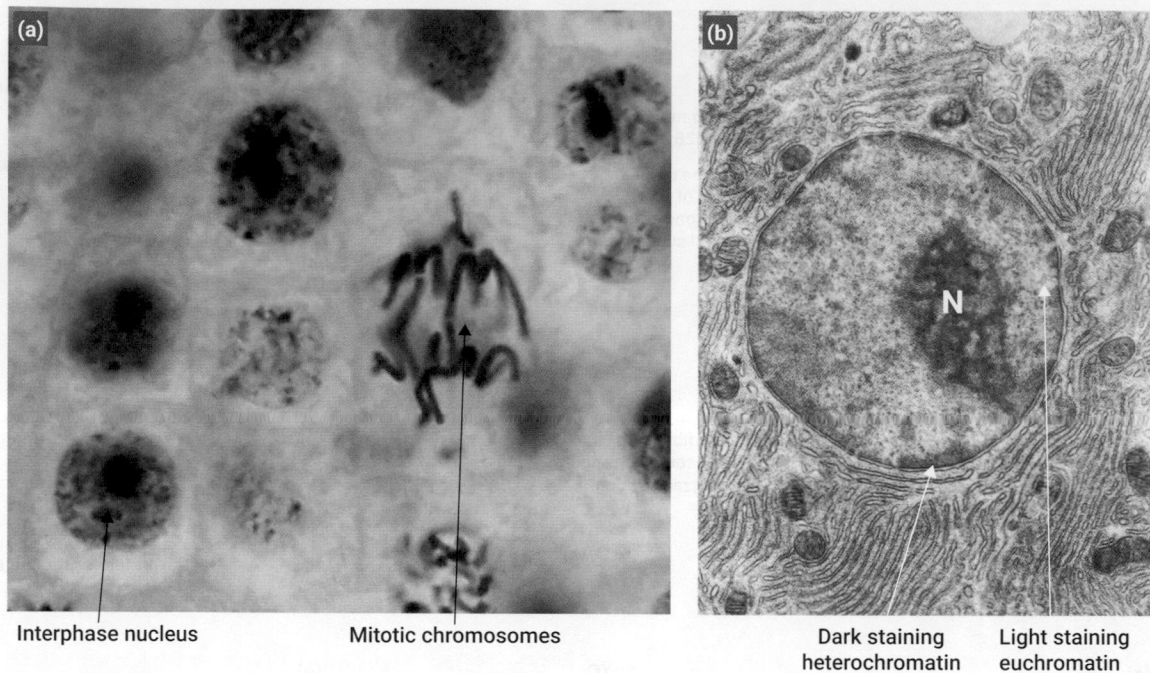

Interphase nucleus   Mitotic chromosomes   Dark staining   Light staining
                                                          heterochromatin   euchromatin

Figure 3.13 Different degrees of chromatin packing. (a) Light micrograph of onion root cells. Individual chromosomes are not visible in the interphase nucleus but can be seen in the cell that is undergoing mitotic cell division, due to compaction of the chromatin. (b) Transmission electron micrograph of an interphase nucleus, showing dark-staining compact heterochromatin and light-staining, more diffuse euchromatin. The dark structure marked 'N' is the nucleolus. (We discuss the nucleolus in more detail in Chapter 9.)

Source: (a) agefotostock/Alamy Stock Photo. (b) Don Fawcett/Science Photo Library.

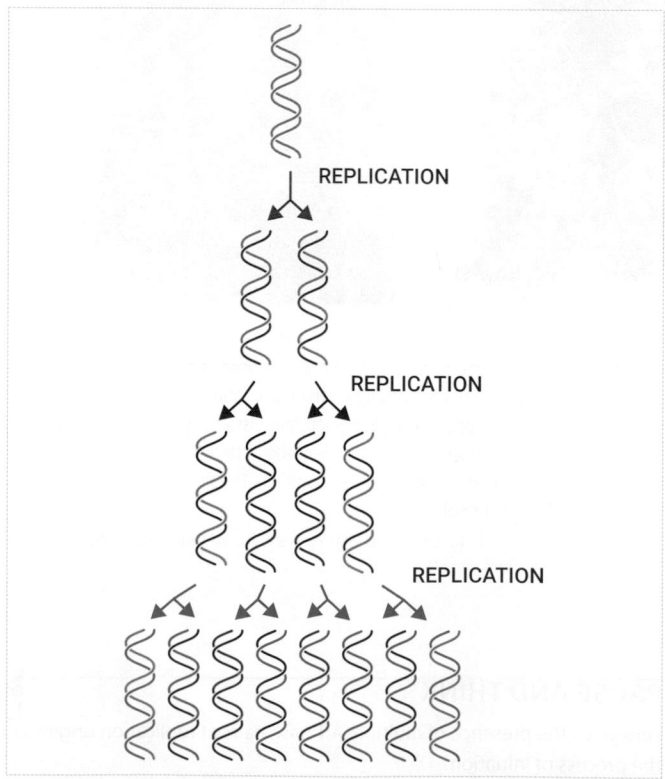

REPLICATION

REPLICATION

REPLICATION

Figure 3.14 DNA replication is semi-conservative. Each of the two strands of the double helix acts as a template on which a new strand is synthesized, so the 'daughter' DNA molecules consist of one strand conserved from the previous generation and one newly made strand.

Source: Cooper. *The Cell*, 8th edn. Oxford University Press/Sinauer Associates.

The separation of the two strands of DNA causes the appearance of a 'bubble' in the chromosome, with a structure at each end that is termed the **replication fork**. As replication proceeds the two forks move away from each other as shown in Figure 3.15.

The circular chromosome of the bacterial species *Escherichia coli* (*E. coli*) with approximately 4.6 million base pairs, contains only one origin of replication. The two replication forks synthesize DNA at a maximum rate of around 1000 base pairs per second, with separation of parental DNA and synthesis of new DNA occurring at the same time. The two forks meet at the opposite side of the circle (Figure 3.15a).

Eukaryotic chromosomes are linear and longer than the *E. coli* chromosome. In eukaryotes the rate of DNA synthesis is slower than in prokaryotes, perhaps because of their more complex chromatin structure. Rates vary but are estimated to be 4–40 base pairs per second. Typically, the replication of an animal cell genome takes around 8 hours, and this could not be achieved if each chromosome had only a single origin of replication. Therefore, multiple origins are spaced at intervals along the chromosome (Figure 3.15b).

PAUSE AND THINK

Human chromosome 1 consists of around 249 million base pairs. Assuming a replication rate of 40 base pairs per second, how many hours would it take to completely replicate the chromosome if there were only one replication origin placed centrally in the chromosome?

Answer: 1729.16 hours.

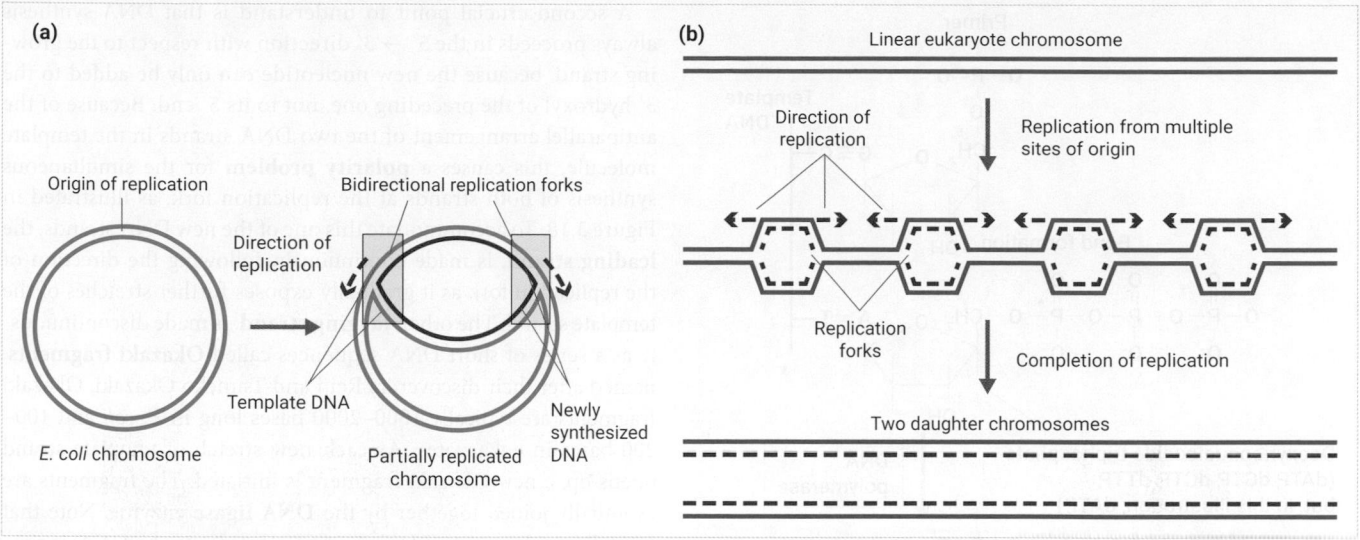

Figure 3.15 DNA replication. (a) Bidirectional replication of the *E. coli* chromosome. Template strands are blue; newly synthesized strands are red. (b) Multiple bidirectional replication forks in a eukaryotic chromosome.

Source: Snape, A. & Papachristodoulou, D. (2018). *Biochemistry and Molecular Biology* (6th ed.). Oxford University Press.

Topoisomerase enzymes relieve supercoiling ahead of the replication fork

DNA strand separation by helicase presents problems due to the introduction of additional **supercoiling** into the DNA. You can illustrate this problem for yourself using twisted cord or rope as shown in Figure 3.16. The cord is held or clamped at one end and this prevents it from rotating and relaxing as the strands are separated.

DNA in the cell is similarly not free to relax ahead of the replication fork. In *E. coli* the closed-circle chromosome effectively 'clamps' the DNA, while in eukaryotes the length of the DNA and

its attachment to protein structures restrict it from rotating. The resulting supercoils will oppose the separating process, bringing replication to a halt. To prevent this, enzymes called **topoisomerases** act on the DNA. They transiently break the sugar–phosphate backbone of either one or both strands and allow the supercoils to relax before rejoining the strands with no loss of or change to the base pair sequence.

Replication machinery of prokaryotic cells

The basic process of DNA replication is similar in prokaryotes and eukaryotes, but there are some differences, for example in the enzymes involved. We will look in some detail at prokaryotic replication using the well-studied bacterial species *Escherichia coli* (*E. coli*) as an example and will review important differences between prokaryotes and eukaryotes later in this section (see Comparison of DNA replication and repair in prokaryotes and eukaryotes).

The enzymes that catalyse replication are called **DNA polymerases**; they polymerize nucleotides into DNA, using deoxynucleoside triphosphates (**dNTPs**) as substrates. The reaction is illustrated in Figure 3.17.

The incoming deoxynucleoside triphosphate (specifically dATP in Figure 3.17) base pairs with the template strand through hydrogen bonding, which is spontaneous and energetically favourable. The enzyme-catalysed reaction is the formation of a new phosphodiester bond between the 3' hydroxyl of the preceding nucleotide in the new DNA chain, and the α-phosphate of the new dNTP. The energy for this reaction is provided by release of inorganic pyrophosphate (PP_i) from the incoming dNTP and its subsequent hydrolysis into two phosphate ions.

There are three major DNA polymerases in *E. coli*, **Pol I, II, and III**—named in order of their discovery. DNA synthesis at the replication fork is mainly catalysed by **Pol III**, but **Pol I** plays an

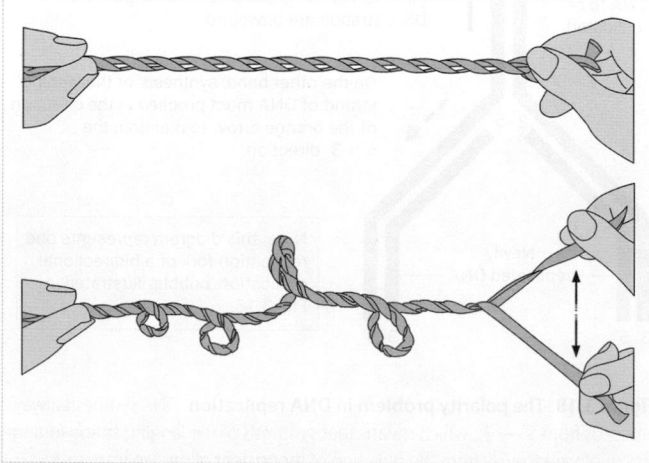

Figure 3.16 Supercoiling of DNA. Supercoiling ahead of the replication fork is analogous to the coiling and twisting of a cord as you try to separate the strands. Topoisomerase enzymes cut and rejoin the DNA, allowing the supercoils to relax so that strand separation and replication can continue.

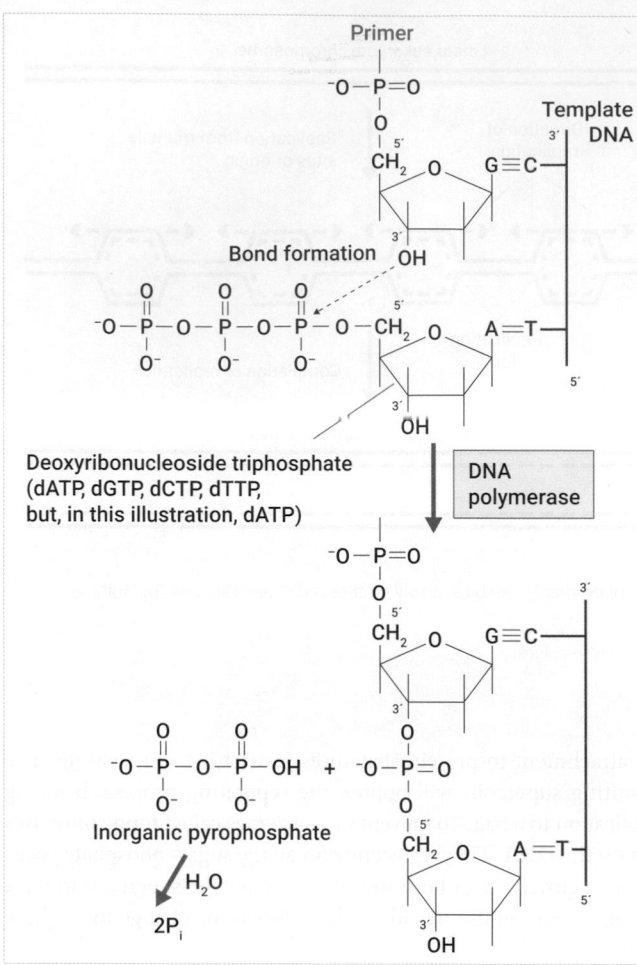

Figure 3.17 The elongation reaction catalysed by DNA polymerase. The diagram shows the addition of an adenine deoxynucleotide from dATP to the 3′ end of the primer DNA strand. The base selected for addition is determined by the base on the template strand, which would be thymine (T) in this example. Note that the synthesis is in the 5′→3′ direction; the chain is being lengthened in the 5′→3′ direction. The dotted line with arrow shows the attack of the 3′-OH on the α-phosphate.

Source: Snape, A. & Papachristodoulou, D. (2018). *Biochemistry and Molecular Biology* (6th ed.). Oxford University Press.

essential role in DNA replication (described below), as well as in repair. Less is known of Pol II, but it is believed to be associated with certain types of DNA repair.

An important point to appreciate is that DNA polymerases cannot initiate a new DNA chain. They can only elongate (add to) a pre-existing polynucleotide strand called a **primer**. In contrast, new RNA molecules can be initiated 'from scratch'. Therefore, the process of new DNA synthesis actually starts with the synthesis of a short RNA sequence by the enzyme **primase**. The RNA primer is synthesized by copying the template DNA, and the primase reaction is like that catalysed by DNA polymerase, except that the substrates are nucleoside triphosphates (ATP, CTP, GTP, and UTP) instead of dNTPs. Once an RNA primer of 10–20 nucleotides has been made, DNA polymerase takes over and extends the chain. The primers are later removed and replaced by DNA, as described later in this section.

A second crucial point to understand is that DNA synthesis always proceeds in the 5′ → 3′ direction with respect to the growing strand, because the new nucleotide can only be added to the 3′ hydroxyl of the preceding one, not to its 5′ end. Because of the antiparallel arrangement of the two DNA strands in the template molecule, this causes a **polarity problem** for the simultaneous synthesis of both strands at the replication fork, as illustrated in Figure 3.18. To accommodate this one of the new DNA strands, the **leading strand**, is made continuously, following the direction of the replication fork as it gradually exposes further stretches of the template strand. The other, **lagging strand**, is made discontinuously as a series of short DNA sequences called **Okazaki fragments**, named after their discoverers Reiji and Tsuneko Okazaki. Okazaki fragments are typically 1000–2000 bases long in *E. coli* and 100–200 bases in eukaryotes. As each new stretch of template strand opens up, a new Okazaki fragment is initiated. The fragments are eventually joined together by the **DNA ligase** enzyme. Note that DNA ligase does not add any new nucleotides to the DNA; it simply catalyses formation of a phosphodiester bond between adjacent Okazaki fragments.

The synthesis of leading and lagging strands is illustrated in Figure 3.19.

A more detailed view of the replication fork including the enzymes and proteins involved is shown in Figure 3.20. Among the key enzymes shown are helicase to unwind the double helix, attached to which is primase, which synthesizes a single RNA primer for the leading strand and a series of RNA primers at intervals on the lagging strand. SSB, which has a high affinity for single-stranded DNA, but with no base sequence specificity, binds to the separated DNA strands and prevents them from reannealing prematurely.

As indicated, in *E. coli* there are two connected molecules of Pol III at the replication fork, one synthesizing the leading strand

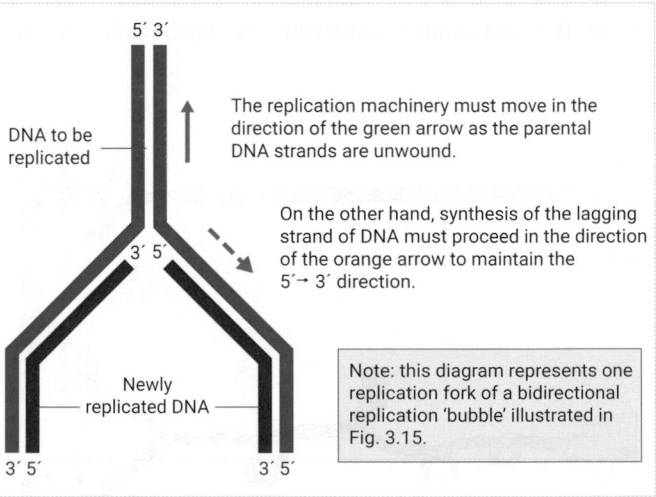

Figure 3.18 The polarity problem in DNA replication. DNA synthesis always proceeds from 5′ → 3′, which means that synthesis of the lagging strand (orange arrow) proceeds away from the direction of movement of the replication machinery (green arrow).

Source: Adapted from Snape, A. & Papachristodoulou, D. (2018). *Biochemistry and Molecular Biology* (6th ed.). Oxford University Press.

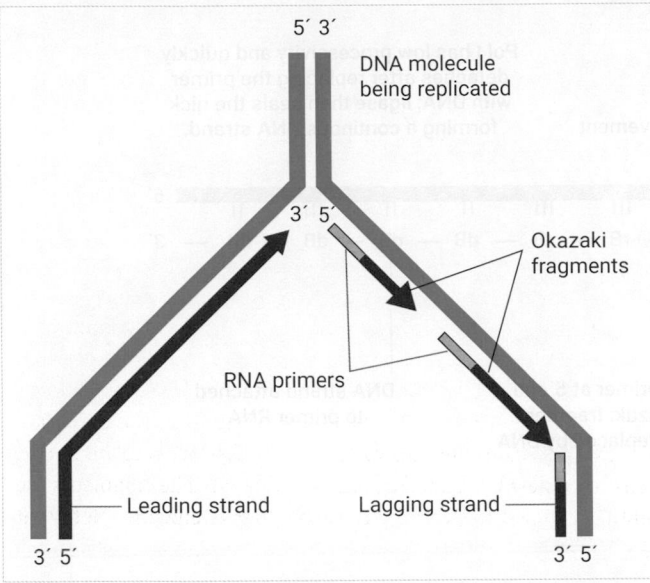

Figure 3.19 Diagram of a replication fork. The leading strand is synthesized continuously, while the lagging strand is synthesized as a series of short (Okazaki) fragments.

Source: Snape, A. & Papachristodoulou, D. (2018). *Biochemistry and Molecular Biology* (6th ed.). Oxford University Press.

towards the replication fork, and the other synthesizing the lagging strand away from the replication fork, but both moving in the same overall direction. To allow this the lagging template strand is looped, so that for a short distance it is oriented with the same polarity as the leading-strand template. The replication machinery can therefore proceed in the direction of the fork and synthesize both new strands. The **sliding clamp** protein complex shown in

Figure 3.20 stops Pol III dissociating prematurely, especially when it has temporarily to detach from the template for the lagging strand, when synthesis of a new Okazaki fragment begins.

We turn now to the question of how the Okazaki fragments are linked up into continuous DNA. In *E. coli*, when Pol III reaches the RNA primer of the preceding Okazaki fragment, it disengages from the DNA, leaving a **nick** (a break in the sugar–phosphate backbone of one strand of double-stranded polynucleotide) at the DNA/RNA junction. This is where Pol I comes in. Pol I is an astonishing enzyme with three separate catalytic activities on the same molecule: DNA polymerase, **5′→3′ exonuclease**, and **3′→5′ exonuclease**. Exonuclease activity refers to removal of nucleotides by hydrolysis from the end of a polynucleotide chain ('exo' as the nucleotides are 'nibbled' away from the end). The action of Pol I in processing Okazaki fragments is illustrated in Figure 3.21.

In summary:

- Pol I attaches to the nicks between successive Okazaki fragments and uses its DNA polymerase activity to add nucleotides to the 3′-OH of the preceding fragment, moving in the 5′ → 3′ direction.

- Since, as Pol I moves, it encounters the RNA primer of the next Okazaki fragment, the nucleotides of this are hydrolysed off by its 5′→3′ exonuclease activity and replaced by the new DNA sequence it is synthesizing.

- Pol I (unlike Pol III) has *low processivity*—it does not hold on to the DNA template strand firmly and detaches relatively soon after the RNA has been replaced. It does not have the sliding clamp to hold it on to the DNA.

- When Pol I detaches a nick is left in the chain, which is healed by DNA ligase. DNA ligase catalyses formation of a phosphodiester bond between the 3′-OH of one DNA fragment and the 5′-phosphate of the next.

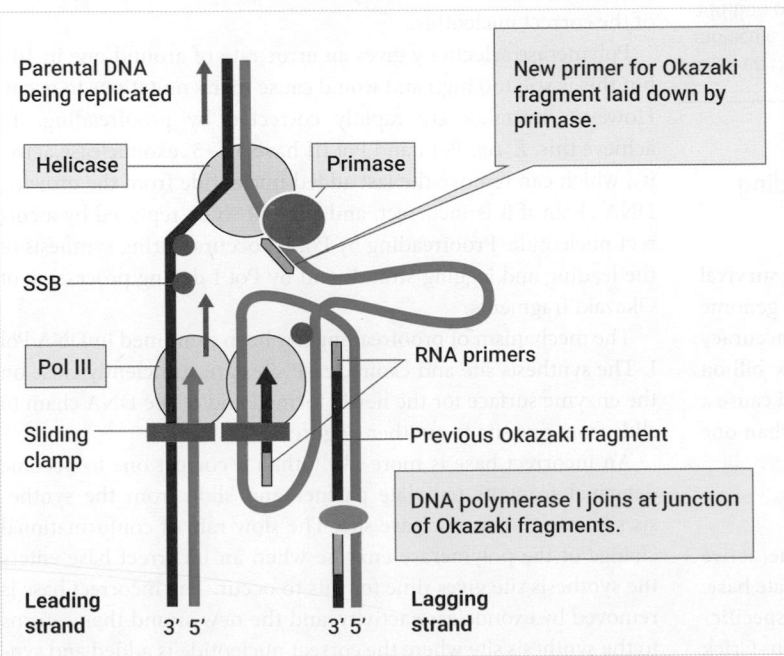

Figure 3.20 Enzymes and proteins at the replication fork of *E. coli*. Functions of the enzymes are described in the text. **SSB** (single-strand binding protein) stops the template strands from reannealing. The loop in the lagging strand solves the problem of how DNA polymerase III can continue to move forward (upwards on the page) but synthesizes DNA in the required 5′→3′ direction on both strands.

Source: Snape, A. & Papachristodoulou, D. (2018). *Biochemistry and Molecular Biology* (6th ed.). Oxford University Press.

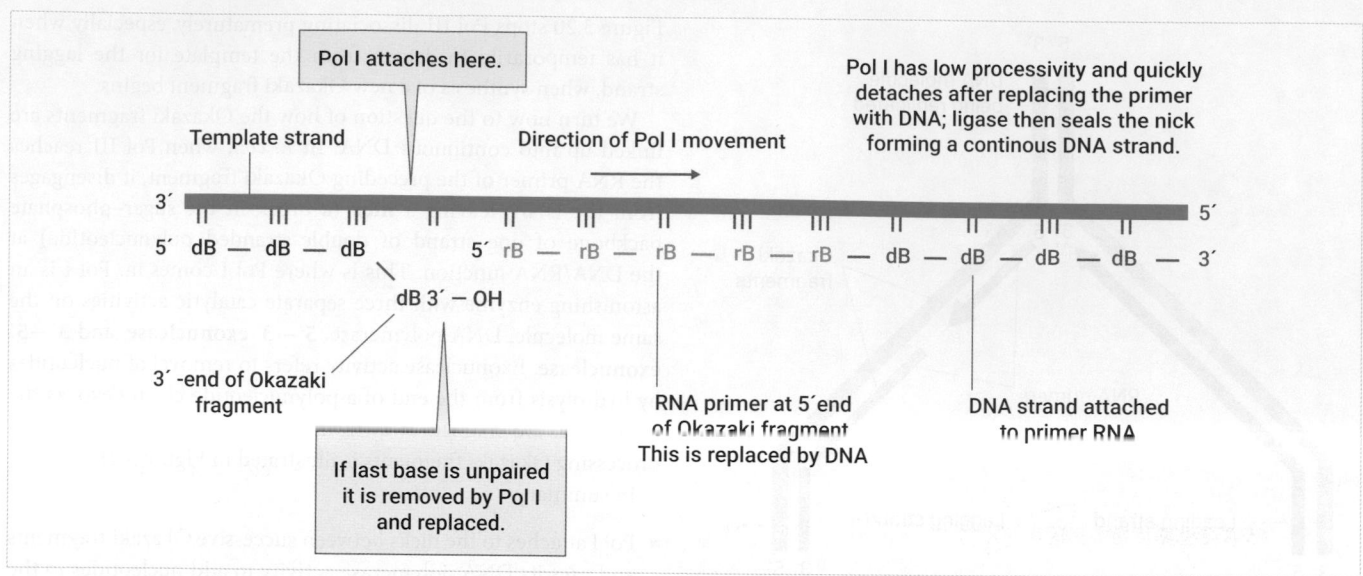

Figure 3.21 **DNA polymerase I (Pol I) actions in processing Okazaki fragments.** As well as replacing RNA primers with DNA, Pol I can remove an unpaired (incorrect) base as part of the proofreading process that increases the accuracy of DNA replication. dB, deoxynucleotide; rB, ribonucleotide.
Source: Snape, A. & Papachristodoulou, D. (2018). *Biochemistry and Molecular Biology* (6th ed.). Oxford University Press.

As well as 5′→3′ exonuclease activity, which allows it to remove RNA primers ahead of new DNA synthesis, Pol I has 3′→5′ exonuclease activity, which is involved in **proofreading**, described in the next subsection.

PAUSE AND THINK

DNA polymerase and DNA ligase both catalyse formation of a phosphodiester bond in the DNA backbone, but what is the key difference between the two classes of enzyme?

Answer: DNA polymerase adds a new nucleotide to the growing DNA molecule. DNA ligase links two existing DNA molecules together without addition of a new nucleotide.

Accuracy of DNA synthesis is improved by proofreading and mismatch repair

Accurate replication of its DNA sequence is critical for the survival of any organism. Bacteria such as *E. coli* replicate their genome extremely rapidly, within 40 minutes, but must achieve accuracy at the same time. A human cell must replicate over six billion base pairs for each cell division and a single mistake could cause a genetic disease. The error rate of DNA replication is less than one in a billion. How is this achieved?

Proofreading corrects the new strand as it is made

When a deoxynucleotide triphosphate (dNTP) enters the active site of a DNA polymerase, it must base pair with the template base. Despite the emphasis we have previously placed on the specificity of base pairing, hydrogen bonding between non-Watson–Crick base pairs can sometimes occur. Unusual 'wobble' base pairing is,

in fact, useful in protein synthesis (as discussed in Chapter 5), but it must be avoided in DNA replication. For the most part, this is achieved by DNA polymerase. Watson–Crick base pairs (A:T and G:C) are almost identical to each other in their size and shape, but 'illegitimate' base pairs have a different geometry and therefore do not fit well into the polymerase active site. Once a correct dNTP has entered the active site and base pairs with the template, the polymerase very quickly closes around it and places it in the appropriate position to make a phosphodiester bond. This rapid conformational change does not occur if an incorrect dNTP enters the active site. Thus, DNA polymerase is very selective for addition of the correct nucleotide.

Polymerase selectivity gives an error rate of around one in 10^5, but this is still too high and would cause many mutations to occur. However, mistakes are rapidly corrected by proofreading. To achieve this, *E. coli* Pol I and Pol III have 3′→5′ exonuclease activity, which can remove the last added nucleotide from the growing DNA chain if it is incorrect, and allow it to be replaced by a correct nucleotide. Proofreading by Pol III occurs during synthesis of the leading and lagging strands and by Pol I during processing of Okazaki fragments.

The mechanism of proofreading has been examined in DNA Pol I. The synthesis site and exonuclease sites are sufficiently close on the enzyme surface for the newly formed end of the DNA chain to slide from one site to another (Figure 3.22).

An incorrect base is more likely than a correct one to become detached from its template partner and slide from the synthesis site into the exonuclease site. The slow rate of conformational change of the polymerase enzyme when an incorrect base enters the synthesis site gives time for this to occur. The incorrect base is removed by exonuclease activity, and the new strand then returns to the synthesis site where the correct nucleotide is added and synthesis resumes.

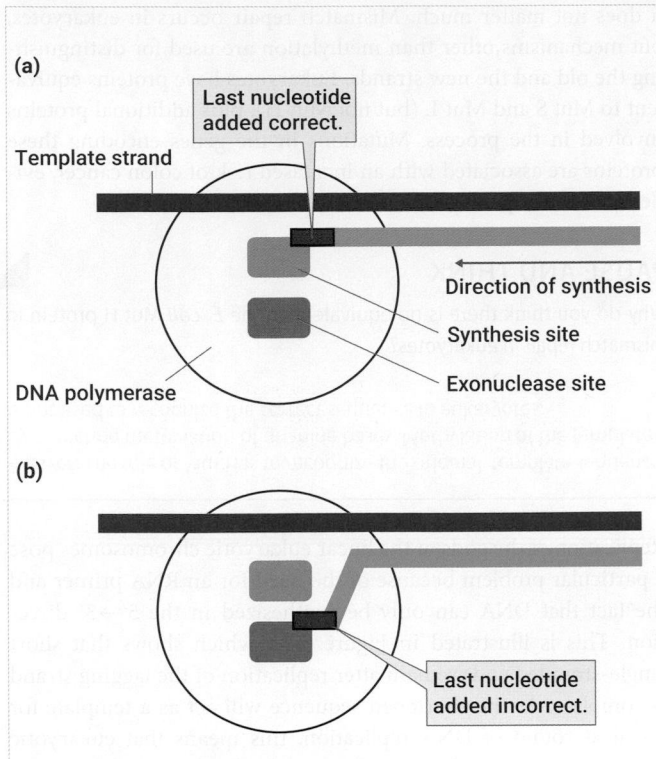

Figure 3.22 Simplified diagram of proofreading by DNA Pol I. (a) The situation if the last addition was correct. (b) The situation if the last addition was not correct. The incorrect nucleotide in (b) is more likely to detach from the template strand and slide to the exonuclease site, where it is removed. The newly synthesized chain will then slide back to the synthesis site, where the correct nucleotide is added.

Source: Snape, A. & Papachristodoulou, D. (2018). *Biochemistry and Molecular Biology* (6th ed.). Oxford University Press.

Methyl-directed mismatch repair corrects errors after replication

Despite proofreading, the number of mismatches that slip through into the newly synthesized DNA would still give an unacceptable rate of mutation. The cell can therefore replace incorrect bases even after the newly synthesized DNA has been released from the polymerase. In *E. coli* this is achieved by **methyl-directed mismatch repair**, which is illustrated in Figure 3.23.

Figure 3.23 illustrates what happens when a mismatch of adenine (A) in the newly synthesized strand with guanine (G) in the template strand escapes proofreading correction, causing a distortion in the DNA double helix. The repair system can recognize the distortion, but it must also discriminate so that it removes the A base from the new strand and replaces it, rather than replacing the correct G base in the template, as that would perpetuate the mutation.

How is this strand discrimination made? It depends on DNA methylation. In *E. coli* some adenine bases in the DNA sequence are methylated by a specific enzyme. This does not affect DNA structure and function. After DNA replication there is a short period when the template strand is methylated but the newly synthesized strand has not yet been methylated. This allows the repair system to identify the correct, template sequence. The repair system involves proteins designated Mut S, Mut L, and Mut H, which recognize the distortion in the double helix and remove a stretch of

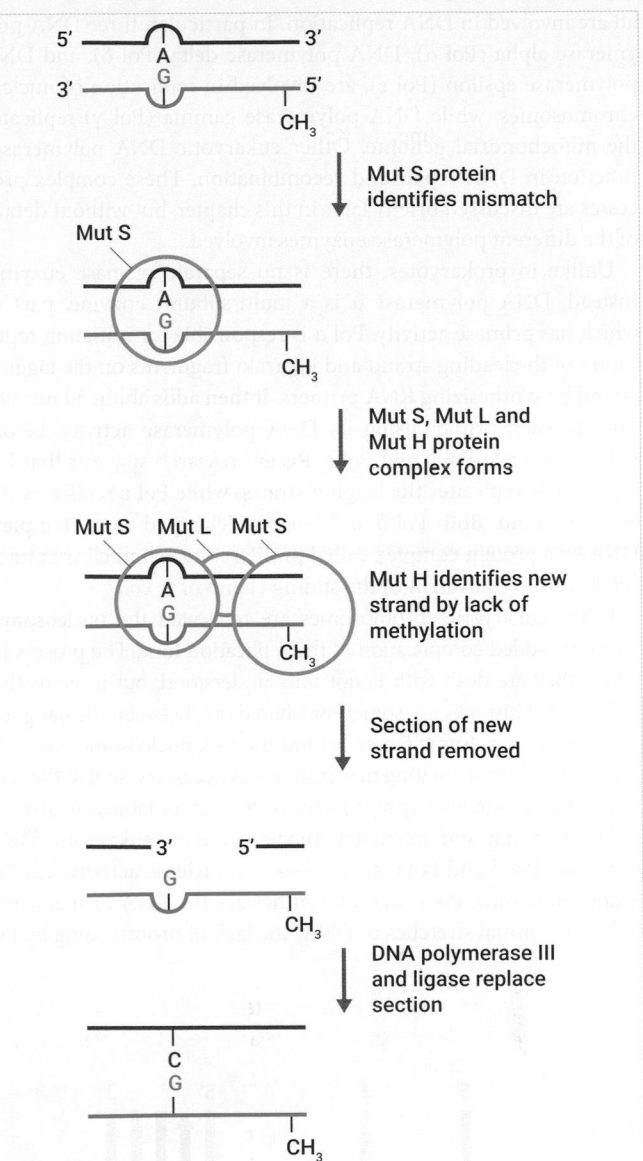

Figure 3.23 Mismatch repair in *E. coli*. The repair system includes three proteins, Mut S, Mut L, and Mut H, which recognize the distortion in the double-stranded DNA, identify the new (incorrect) strand because it does not contain a methylated base nearby, and drive removal and replacement of the incorrect sequence.

Source: Snape, A. & Papachristodoulou, D. (2018). *Biochemistry and Molecular Biology* (6th ed.). Oxford University Press.

newly replicated DNA containing the mismatch. DNA polymerase III synthesizes a replacement, thereby correcting the error. DNA ligase completes the repair by sealing the nick. This system increases the fidelity of replication so that there is a final error rate of less than or equal to one in 10^{10}.

Comparison of DNA replication in prokaryotes and eukaryotes

The general principles of DNA replication in *E. coli* also apply to eukaryotes, but there are differences in detail. Many eukaryotic DNA polymerases have been identified (15 in humans), but not all are involved in DNA replication. In particular, three DNA polymerase alpha (Pol α), DNA polymerase delta (Pol δ), and DNA polymerase epsilon (Pol ε), are involved in replication of nuclear chromosomes, while DNA polymerase gamma (Pol γ) replicates the mitochondrial genome. Other eukaryotic DNA polymerases function in DNA repair and recombination. These complex processes are discussed briefly later in this chapter, but without details of the different polymerase enzymes involved.

Unlike in prokaryotes, there is no separate primase enzyme. Instead, DNA polymerase α is a multi-subunit enzyme, part of which has primase activity. Pol α is responsible for initiating replication of the leading strand and Okazaki fragments on the lagging strand by synthesizing RNA primers. It then adds about 30 nucleotides to each primer using its DNA polymerase activity, before handing over to Pol δ and Pol ε. Recent research suggests that Pol δ primarily replicates the lagging strand, while Pol ε replicates the leading strand. Both Pol δ and Pol ε are clamped to the template DNA by a protein complex called proliferating cell nuclear antigen (PCNA), the equivalent of the sliding clamp of *E. coli.*

When eukaryotic chromosomes are replicated the nucleosomes create an added complication at the replication fork. The process by which they are dealt with is not fully understood, but it seems that the parental histones are somehow 'shared out' between the daughter DNA molecules. Immediately behind the fork nucleosomes are fully reassembled, incorporating new histones as necessary, so that the replicated DNA immediately regains the normal chromatin structure.

Proofreading and mismatch repair occur in eukaryotic DNA synthesis. Pol δ and Pol ε have 3'→5' exonuclease activity, but Pol α does not. Since Pol α mainly synthesizes the RNA primers, and only short initial stretches of DNA, the lack of proofreading by Pol α does not matter much. Mismatch repair occurs in eukaryotes, but mechanisms other than methylation are used for distinguishing the old and the new strands. Eukaryotes have proteins equivalent to Mut S and Mut L (but not Mut H), plus additional proteins involved in the process. Mutations in the genes encoding these proteins are associated with an increased risk of colon cancer, evidence of the importance of mismatch repair in humans.

Replication of the ends of the linear eukaryotic chromosomes pose a particular problem because of the need for an RNA primer and the fact that DNA can only be synthesized in the 5'→3' direction. This is illustrated in Figure 3.24, which shows that short single-stranded ends remain after replication of the lagging strand is complete. As the shortened sequence will act as a template for the next round of DNA replication, this means that eukaryotic chromosomes would become shorter at each successive round.

The problem is solved by the existence of a specialized enzyme, telomerase, which can add multiple short repeated sequences to the ends of linear chromosomes, creating the structures known as telomeres. Telomerase is especially active in actively dividing cells such as embryonic cells, and then becomes less active as the

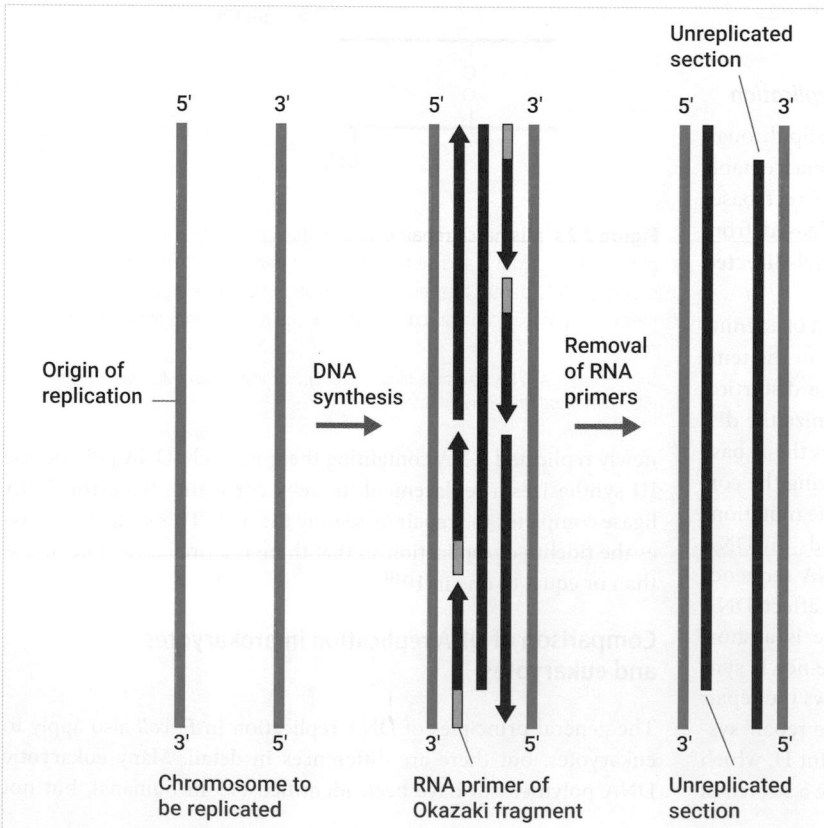

Figure 3.24 The 'end replication' problem of linear chromosomes. Removal of the RNA primer leaves a short section (around 100 base pairs) unreplicated at the 5' end of each new strand. At the next round of replication, the new, shorter strands will be replicated and through the same problem the chromosome will shorten again. (Note that for this illustration a very short chromosome is shown. In reality, a single chromosome would have multiple origins of replication, but the ends of the chromosome would still be truncated as shown here).

Source: Snape, A. & Papachristodoulou, D. (2018). *Biochemistry and Molecular Biology* (6th ed.). Oxford University Press.

organism ages. This means that telomeres shorten gradually during successive cell divisions, but important sequences at the ends of chromosomes are protected. It is suggested that shortening of our telomeres may be connected to the ageing process.

DNA suffers frequent damage, but can be repaired

The mechanisms described above ensure that DNA is replicated accurately. However, chemical changes that damage DNA occur at a rate that would result in each cell accumulating many mutations per day if there were not constant repair. Some types of damage occur spontaneously, while other types are caused or accelerated by external factors, such as chemicals and radiation, or internal factors, such as reactive oxygen species (also called free radicals) generated during cellular respiration. Some examples of DNA damage are illustrated in Figure 3.25.

Such damage can interfere with transcription (copying the DNA sequence to RNA for protein synthesis) and with DNA replication.

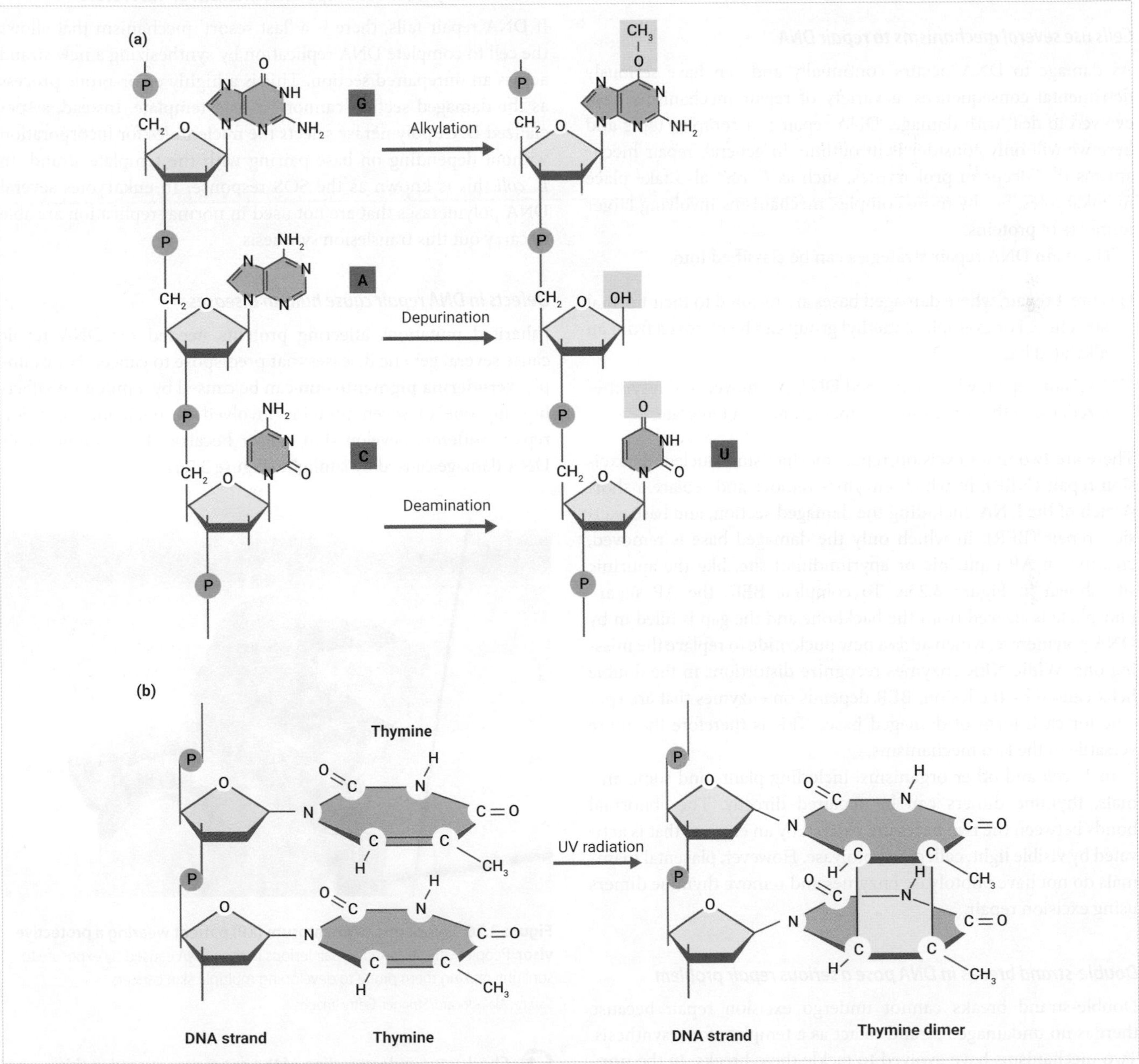

Figure 3.25 Examples of DNA damage. (a) A sequence of undamaged DNA is shown on the left. On the right, each base has undergone a different form of damage. In reality, adjacent bases are unlikely to undergo different types of damage at the same time: this is shown for illustration. **Alkylation** is the addition of a methyl or ethyl group to the base by a chemical agent. Here, guanine (G) is methylated (yellow highlight). The resulting O6-methylguanine mispairs with thymine during DNA replication, causing a mutation. **Depurination**, through cleavage of the glycosidic bond that links adenine or guanine to the deoxyribose sugar, and **deamination** of cytosine, adenine, or guanine are common types of spontaneous DNA damage. Deamination of cytosine creates uracil (U), as shown. (b) Exposure to ultraviolet (UV) light can cause fused structures of adjacent pyrimidine bases such as the thymine dimer shown.

Source: (a) Adapted from Cooper G. M. (2000) *The Cell: A Molecular Approach*, 2nd edn. Sinauer Associates, Sunderland (MA); and *Essential Cell Biology*, 4th edn. (b) Alberts et al. *Essential Cell Biology*, 5th edn. W.W. Norton and company, 2019.

During replication damaged bases may 'mispair', introducing mutations. For example, deamination of cytosine forms uracil, which pairs with adenine at the next round of replication, causing conversion of a C:G base pair to A:T. Thymine dimers, on the other hand, completely block replication as the DNA polymerase cannot get past them.

Some DNA-damaging agents attack the sugar–phosphate backbone of the DNA rather than the bases. For instance, X-rays can cause double-stranded breaks in the DNA molecule, a particularly dangerous form of damage that can be lethal to the cell.

Cells use several mechanisms to repair DNA

As damage to DNA occurs continually and can have seriously detrimental consequences, a variety of repair mechanisms have evolved to deal with damage. DNA repair is a complex topic and here we will only consider it in outline. In general, repair mechanisms that occur in prokaryotes, such as *E. coli,* also take place in eukaryotes, but by more complex mechanisms involving larger numbers of proteins.

The main DNA repair strategies can be classified into:

(1) Direct repair, where damaged bases are restored to their normal structure. For example, a methyl group can be removed from an alkylated base.

(2) Excision repair, where damaged DNA is removed and resynthesized, using the remaining normal strand as a template.

There are two main excision repair mechanisms: nucleotide excision repair (NER), in which enzymes remove and replace a short stretch of the DNA, including the damaged section, and base excision repair (BER), in which only the damaged base is removed, creating an AP (apurinic or apyrimidinic) site, like the apurinic site shown in Figure 3.25a. To complete BER, the AP sugar–phosphate is cleaved from the backbone and the gap is filled in by DNA polymerase, which adds a new nucleotide to replace the missing one. While NER enzymes recognize distortions in the double helix caused by the lesion, BER depends on enzymes that are specific for each form of damaged base. NER is therefore the more versatile of the two mechanisms.

In *E. coli* and other organisms, including plants and some animals, thymine dimers can be repaired directly. The abnormal bonds between the two bases are cleaved by an enzyme that is activated by visible light, called a photolyase. However, placental mammals do not have photolyase enzymes and remove thymine dimers using excision repair.

Double-strand breaks in DNA pose a serious repair problem

Double-strand breaks cannot undergo excision repair because there is no undamaged strand to act as a template for resynthesis. Two mechanisms have evolved to tackle these breaks. In the **non-homologous end-joining** mechanism, the two cut ends are simply

joined by ligation. This can change the DNA sequence if nucleotides have been trimmed off the cut ends before they rejoin. The other mechanism, repair by **homologous recombination**, is more accurate. It uses a second copy of the double-stranded sequence, if one is available (for example, if it has been synthesized by DNA replication), to direct the repair. The details of the end-joining and recombination mechanisms are complex and are not described here.

Translesion synthesis is a repair mechanism of 'last resort'

If DNA repair fails, there is a 'last resort' mechanism that allows the cell to complete DNA replication by synthesizing a new strand across an unrepaired section. This is a highly error-prone process as the damaged section cannot act as a template. Instead, a specialized DNA polymerase selects the nucleotides for incorporation without depending on base pairing with the template strand. In *E. coli* this is known as the SOS response. In eukaryotes several DNA polymerases that are not used in normal replication are able to carry out this translesion synthesis.

Defects in DNA repair cause human diseases

Inherited mutations affecting proteins needed for DNA repair cause several genetic diseases that predispose to cancer. For example, **xeroderma pigmentosum** can be caused by a mutation affecting any one of seven proteins involved in nucleotide excision repair. Sufferers develop skin cancer because they cannot repair DNA damage caused by sunlight (Figure 3.26).

Figure 3.26 Xeroderma pigmentosum (XP) patient wearing a protective visor. People with XP cannot repair lesions in their DNA caused by exposure to sunlight, making them prone to developing multiple skin cancers.
Source: Alain Jocard/Stringer, Getty Images.

 Check your understanding of the concepts covered in this section by answering the questions in the e-book.

SUMMARY OF KEY CONCEPTS

- All cellular organisms have a genome that is made up of DNA.

- DNA and RNA are polymers of nucleotides (deoxyribose or ribose sugar, base and phosphate) that are linked to each other by phosphodiester bonds.

- The four bases in DNA are adenine, cytosine, guanine, and thymine (A, C, G, and T), while in RNA uracil (U) replaces thymine.

- DNA forms a double helix of two antiparallel strands, in which the sugar phosphate backbone gives 5′ to 3′ polarity, and complementary bases are paired by hydrogen bonds (A:T and G:C).

- Prokaryotes have a single circular double stranded DNA chromosome, while eukaryotic chromosomes are long linear double-stranded DNA molecules, which are packaged with histones and other proteins to form compact chromatin.

- Eukaryotic genes are split into non-coding introns and coding exons. Exons make up only around 1.6 per cent of the human genome sequence.

- Around 25 per cent of the human genome sequence consists of introns and exons—perhaps 21,000 protein-coding genes. The rest consists of a variety of non-coding sequences, some regulatory, much of viral origin, and much of it repetitive and of unknown function.

- DNA replication is semi-conservative: each strand acts as a template for polymerization of complementary nucleotides (catalysed by DNA polymerase) so that two identical molecules are made simultaneously at each replication fork.

- The main polymerase enzymes in prokaryotic DNA replication are DNA Pol III and DNA Pol I. DNA Pol I plays a specific role in replacing RNA primers, which are necessary to start off the new strands.

- New DNA strands are synthesized in the 5′ to 3′ direction, so at each replication fork the leading strand is made continuously and the lagging strand is made discontinuously as Okazaki fragments, which are then linked together by DNA ligase.

- DNA replication is extremely accurate, and fidelity is improved by proofreading. Incorrectly incorporated nucleotides may also be removed by mismatch repair, to achieve a final error rate of less than or equal to one in 10^{10}.

- DNA replication, proofreading, and mismatch repair processes are essentially similar in prokaryotes and eukaryotes, with differences in the detail.

- DNA that is damaged can be repaired by direct repair and excision repair mechanisms. Double-stranded breaks in DNA can also be repaired but the repairs are error prone. Defects in DNA repair can lead to cancer.

 Use the flashcards in the e-book to test your recall of key terms introduced in this chapter.

QUESTIONS

 Looking for answers? Once you've answered these questions, follow the link in the e-book to the answer guidance and check your work.

Concepts and definitions

1. Name the bases found in DNA and RNA that are purines, and those that are pyrimidines.

2. The sequence of one strand of a DNA molecule is:

    5′ – ATTGATCCGTGA – 3′

    What would be the sequence of the complementary strand in the double helix, and how many full turns of the double helix would this sequence make?

3. Approximately how many nucleosomes would be formed by a DNA molecule that is 2000 base pairs in length? Assume the linker sequence between nucleosomes is 35 base pairs.

4. Which enzyme is responsible for each of the following roles in prokaryotic DNA replication?

    (a) Unwinding the two strands of the DNA double helix.

    (b) Synthesis of the RNA primer.

    (c) Synthesis of the leading strand.

(d) Synthesis of the lagging strand.

(e) Replacing RNA primers in Okazaki fragments.

(f) Sealing 'nicks' in the sugar phosphate backbone to link Okazaki fragments.

5. What is the accuracy of DNA replication with and without proof-reading and mismatch repair?

Apply the concepts

6. Compare and contrast the structures of DNA and RNA.

7. James Watson and Francis Crick's 1953 publication *A Structure for Deoxyribose Nucleic Acid* states: 'It has not escaped our notice that the specific pairing we have postulated immediately suggests a possible copying mechanism for the genetic material.' Explain what this statement refers to.

8. Explain why looping of the lagging strand template is needed for DNA replication.

9. While DNA Pol I and III in prokaryotes can both catalyse the synthesis of DNA, DNA Pol I also has both $5' \rightarrow 3'$ exonuclease, and $3' \rightarrow 5'$ exonuclease activity, whereas DNA Pol III has only $3' \rightarrow 5'$ exonuclease activity. Explain how the different exonuclease activities reflect the functions of the two enzymes.

10. What are the main classes of repetitive DNA sequences found in the human genome? Why are some of these sequences referred to as 'fossils'?

Beyond the concepts

11. Figure Q3.11 illustrates the edges of C:G and A:T base pairs when viewed from the major and the minor grooves of a DNA molecule.

DNA-binding proteins involved in gene transcription can bind to specific DNA sequences by recognizing these characteristic chemical groupings of different base pairs.

For each base pair, identify chemical groups that proteins can recognize in the major and minor groove.

You should be able to identify at least one example of each of the following features:

(a) Hydrogen bond acceptor.

(b) Hydrogen bond donor.

(c) Non-polar hydrogen atom that cannot participate in hydrogen bonds.

(d) Methyl group.

What impact could the degree of chromatin compaction (euchromatin versus heterochromatin) have on accessibility of these groups to proteins?

12. Meselson and Stahl carried out the following experiment that demonstrated that DNA replication is a semiconservative process:

The DNA of *E. coli* was 'labelled' by growing the bacteria in a medium in which the nitrogen source was ^{15}N so that both DNA strands were 'heavy'. The bacteria were then transferred to a medium containing ^{14}N so that all subsequent DNA chains would be 'light'. After three generations (i.e. three rounds of replication) the DNA was analysed by density gradient centrifugation, which separates the DNA molecules according to their weight.

Which of the following density gradient patterns shown in Figure Q3.12 would be observed? Explain your answer.

More information on Meselson and Stahl's experiment can be found in Further reading.

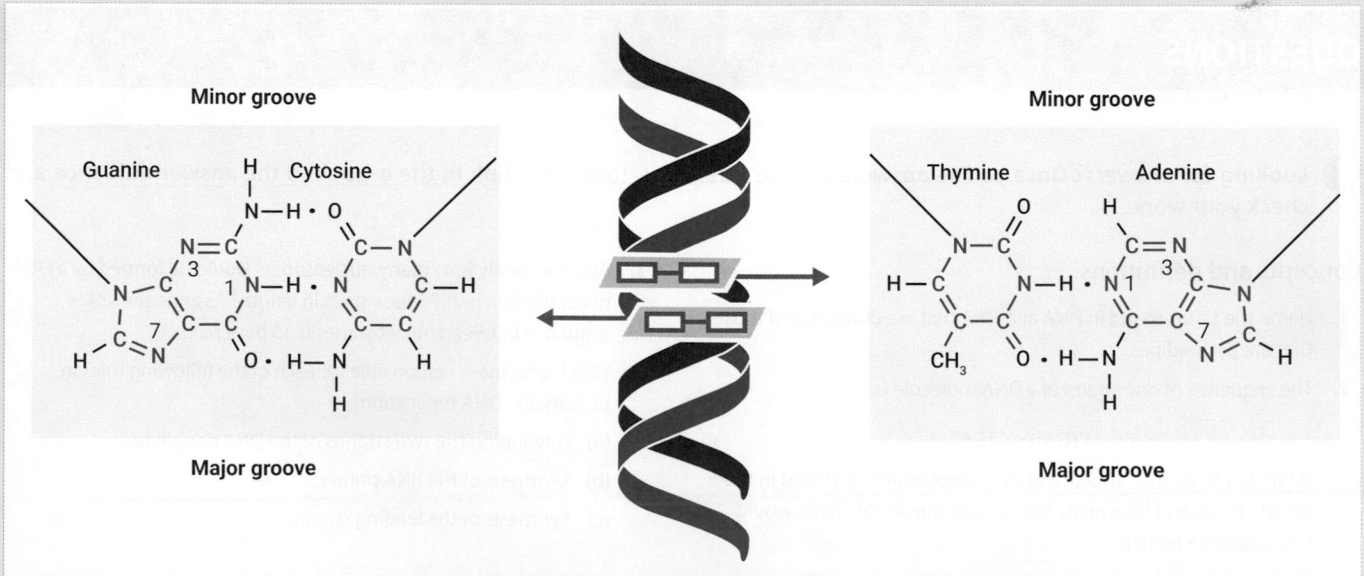

Figure Q3.11 Characteristic chemical groups are seen when DNA base pairs are viewed from either the major or the minor groove. For example, the minor groove of the G:C base pair contains N-3 of guanine, O-2 of cytosine and the amino group ($-NH_2$) attached to C-2 of guanine.

Source: D. Papachristodoulou, A. Snape, W. H. Elliott and D. C. Elliott. *Biochemistry and Molecular Biology*.

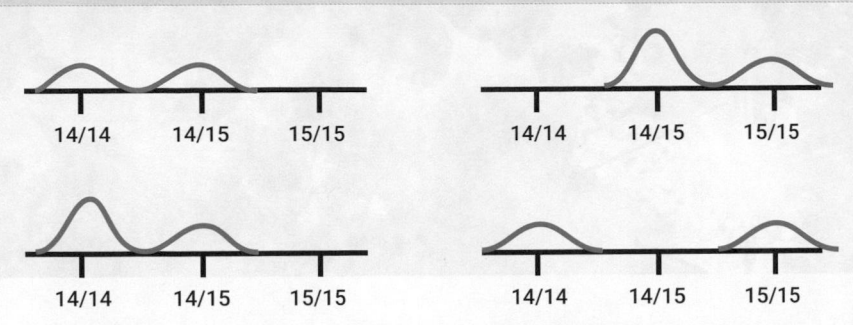

Figure Q3.12 Possible density gradient patterns of DNA after bacteria are *transferred* from ¹⁵N- to ¹⁴N-containing medium and allowed to undergo three subsequent rounds of replication. Distance along the horizontal axis reflects the weight of the DNA, with the lightest DNA on the left and the heaviest on the right. The area under the curve is a measure of the amount of double-stranded DNA with a particular composition of ¹⁵N- and ¹⁴N-labelled chains.

13. The enzymes topoisomerase and telomerase have both been proposed as targets for anti-cancer drugs. What is the role of each of these enzymes in DNA replication, and why might inhibiting their activity be therapeutic strategies to treat cancer?

14. It is often suggested that the size of the genome reflects the complexity of the organism. Discuss this with reference to the data in Table 3.2 that gives the genome sizes and gene numbers of humans and the nematode worm, *Caenorhabditis elegans*.

15. The genetic disease xeroderma pigmentosum (XP) is caused by defects in proteins required for NER. People with XP suffer from skin cancer caused by exposure to sunlight. Outline the chain of events within skin cells that leads to cancer in these individuals.

FURTHER READING

Chambers, D. A., Reid, K. B. M., & Cohen, R. L. (1994). DNA: the double helix and the biomedical revolution at 40 years. *FASEB J.*, **8**: 1219–26.
Reviews a meeting to mark the 40th anniversary of the double helix. Despite being somewhat out of date it provides a good overview of work that followed on from the publication of the structure of DNA.

Hanawalt, P. C. (2004) Density matters: the semiconservative replication of DNA. *Proc Natl Acad Sci USA* **101**: 17889–94.
A perspective piece on the elegant original experiments that demonstrated that DNA replication is semiconservative, and some of the follow up.

Lander, E. S. (2011) Initial impact of the sequencing of the human genome. *Nature* **470**: 187–97.
An exploration of the impact of the human genome sequence in the decade after its publication, and the road ahead in fulfilling the promise of genomics for medicine.

Fuss, J. O. & Cooper, P. K. (2006) DNA repair: dynamic defenders against cancer and aging. *PLoS Biol.* **4**: e203. https://doi.org/10.1371/journal.pbio.0040203
An accessible summary of DNA repair mechanisms focusing on humans and the consequences of defects in repair systems.

Levy, S.B. (1998) The challenge of antibiotic resistance. *Sci. Am.* **278**: 46–53. doi: 10.1038/scientificamerican0398-46.
While this is an older article, it clearly sets out the challenges posed by antibiotic resistance in accessible language.

Mendelian Genetics

Chapter contents

Introduction   224

4.1   Chromosomes and genes are carried
      in pairs   225

4.2   Mendel's experiments established the
      laws of inheritance   227

4.3   The behaviour of chromosomes as cells
      divide can explain the basis of inheritance   231

4.4   Extensions and refinements of
      Mendel's laws   239

4.5   Genetic disease   241

Watch the key concepts video in the e-book to
prepare yourself for studying this chapter.

LEARNING OBJECTIVES

By the end of this chapter you should be able to:

- Explain the organization of genes and chromosomes in sexually reproducing eukaryotes.

- Explain the chromosomal basis of mammalian sex determination.

- Describe how the experiments of Gregor Mendel led to our modern
  understanding of the 'laws' of inheritance.

- Define the terms genotype, phenotype, allele, homozygote, and
  heterozygote.

- Discuss some extensions and refinements of Mendel's laws.

- Relate our understanding of inheritance to the behaviour of chromosomes at meiosis.

- Explain the inheritance of human autosomal dominant, autosomal
  recessive, and X-linked recessive disorders using simple pedigree
  diagrams.

Introduction

For thousands of years people have recognized that inherited characteristics (also called traits) are passed on from parents to their
children, and farmers have known how to increase desirable traits
in plants and animals through selective breeding. Despite this
ancient knowledge, scientists struggled to understand the basis of
heredity. Many scientists, including Charles Darwin, believed in

a 'blending' mechanism in which the traits of both parents were mixed in the child; for example, according to this hypothesis, the offspring of a red and a white-flowered plant might have pink flowers. However, the blending hypothesis could not explain common observations, such as the disappearance of a trait in one generation of a family, followed by its reappearance in subsequent generations. Through his experiments on garden peas, Gregor Mendel (1822–1884), a relatively obscure scientist working in his monastery garden, established laws of inheritance that apply to all sexually reproducing plants and animals, including humans.

In this chapter we explain how genetically determined traits are transmitted in sexually reproducing eukaryotes from one generation to the next. This field of study is known as **Mendelian genetics**, after Gregor Mendel (see Figure 4.1). It may also be called **classical genetics** or **transmission genetics**. Although Mendel's work preceded our understanding of DNA, genes, and chromosomes, you will see that his results can now be interpreted in the light of our current knowledge, and indeed be explained by that knowledge.

Before exploring Mendel's experiments in detail, we will briefly review the organization of genes and chromosomes in sexually reproducing eukaryotes. Although Mendel's work pre-dates their discovery, later in this chapter we will interpret his results by considering the behaviour of genes and chromosomes during cell division; it is useful to have some background information at this stage.

4.1 Chromosomes and genes are carried in pairs

As described in Chapter 3, eukaryotic chromosomes are double-stranded linear DNA molecules. Each species has a characteristic complement of chromosomes in its cells, and the cells are **diploid** in most stages in the life cycle of sexually reproducing eukaryotes. This means that they have two sets of chromosomes, one derived

from each parent. Humans, for example, have 46 chromosomes consisting of 22 **homologous pairs** known as **autosomes** and two **sex chromosomes**, X and Y in the male and X and X in the female (see Figure 4.2). The Y chromosome is notable for carrying very few genes, although in Scientific Process 4.1 we look at an experimental study that showed that a single gene on the Y chromosome determines male development.

▶ We discuss the function of this *SRY* gene further in Chapter 33.

The diploid chromosome number is often designated '2n', so in humans we can express the diploid chromosome number by writing '2n = 46'. Gametes are **haploid**: they contain only one copy of each chromosome pair. We can express the haploid chromosome number for humans by writing 'n = 23'.

During interphase (the phase of the cell cycle in which cells are not dividing) the DNA of the chromosomes is packaged with proteins into chromatin, but in a relatively diffuse manner, which does not allow us to distinguish individual chromosomes. During cell division (mitosis and meiosis, described further in Section 4.3) the chromosomes become compacted. Mitotic chromosomes can be stained and then visualized using a light microscope, as shown in Figure 4.2. The full set of chromosomes found in an individual is referred to as its **karyotype**.

Different variants of the same gene are called alleles

The two members of a homologous chromosome pair each have the same genes in the same order, so eukaryotes have two copies of each autosomal gene. The base sequences of the two homologous chromosomes are therefore almost the same, but the two copies of an individual gene may differ slightly in base sequence. Different variants of the same gene are known as **alleles**, and the inheritance of different alleles contributes to genetic variation. For example, a gene on human chromosome 4 encodes a protein, glycophorin A, that spans the membrane of red blood cells (a blood group antigen).

Figure 4.1 Heredity. Although farmers have used selective breeding for thousands of years, to increase desirable traits in plants and animals, the basis of heredity was not understood until the 19th century. (a) A Peruvian farmer selling different varieties of potatoes that were produced by selective breeding. (b) Gregor Mendel, the 'founding father' of modern genetics. (c) *Pisum sativum*, the garden pea used by Mendel in his experiments.

Source: (a) Shutterstock. (b) Bateson, William, Public domain, via Wikimedia Commons. (c) Kilom691, Public domain, via Wikimedia Commons.

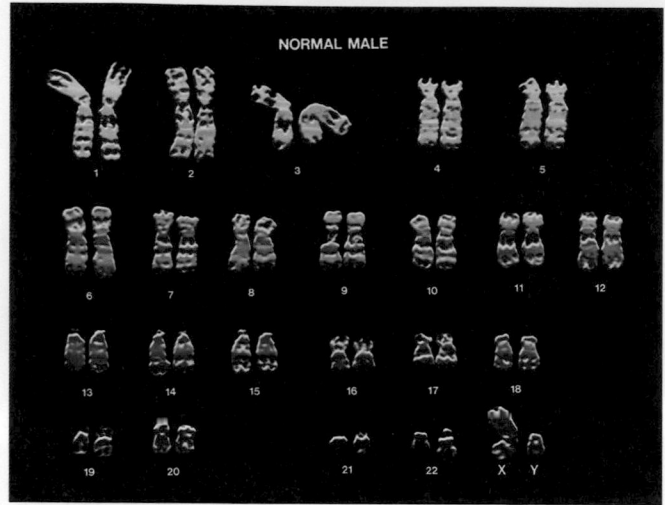

Figure 4.2 A karyogram (a photograph illustrating the karyotype) showing human chromosomes. The chromosomes of a mitotic cell have been stained using Giemsa dye to produce a 'G-banded' preparation. The chromosomes are numbered in order of size, with chromosome 1 being the largest. Chromosome 22 was thought to be the smallest human chromosome, but modern methods have shown that chromosome 21 is actually slightly smaller. Chromosomes 1–22 are autosomes. X and Y are sex chromosomes. *Source*: Biophoto Associates/Science Photo Library.

Two alleles of the gene, the M and N alleles, encode proteins that differ at just two out of 131 amino acids, leading to individuals having different MN blood groups depending on which alleles they inherit.

PAUSE AND THINK

Genes encoded by the X chromosome are called X-linked genes. How many copies of each X-linked gene are found in females and in males?

Answer: Females have two copies of each X-linked gene. Males have only one.

Later in this chapter we explore how alleles are distributed and passed on from one generation to the next. Before we look at the role of cell division in this process, in Section 4.2 we will explore the basic 'laws' of inheritance, discovered by Gregor Mendel in the 19th century.

Gene dosage is important: extra or missing chromosomes cause abnormalities

As we will see in Section 4.3, meiotic cell division ensures that gametes (eggs and sperm) each receive just one member of each autosomal homologous chromosome pair and one sex chromosome.

| SCIENTIFIC PROCESS 4.1 | Randy the XX male mouse |

Research question

From the 1950s it was known that a gene, or genes, on the Y chromosome in mammals initiates the development of the testes, which in turn secrete all the hormones needed for male development. Researchers wished to know which Y chromosome gene(s) were the determinants of sex. Koopman and colleagues tested the hypothesis that the *Sry* gene, and not another gene called *Zfy*, encodes the 'testis determining factor'.

Materials and methods

A fragment of DNA containing the *Sry* gene was injected into fertilized mouse eggs, which were cultured until they became two-cell embryos. These were implanted into host mother mice. Some of the transgenic embryos were analysed after 14 days of development, when the phenotypic sex and chromosome makeup of the embryos can be determined. Others were allowed to develop to term before being examined.

Results

Out of 741 embryos examined after 14 days development, two were found to be phenotypically male and had two X chromosomes. These

embryos were shown to contain the injected *Sry* gene but not *Zfy*. Out of 93 experimental animals that were born after being allowed to develop to term, five were found to be transgenic (i.e. the injected *Sry* gene had integrated into their own genome). While some of these happened to be male mice who also had a whole Y chromosome, one mouse (m33.13) had *two* X chromosomes but appeared to be fully male. This mouse (nicknamed Randy and shown in Figure 1) showed complete male development and male copulatory behaviour when paired with female mice. Randy was sterile because the possession of two X chromosomes interferes with sperm development, but in all other respects he was a normal male mouse.

Conclusion

This study demonstrated that a single gene on the Y chromosome, *Sry*, is sufficient to determine male development in mammals. Around the same time, very rare cases were found of human males who had two X chromosomes but whose genome also contained a short sequence of DNA that included the *SRY* gene (the human equivalent of *Sry*). These findings provided additional evidence that the mechanism of sex determination is the same in mice and humans.

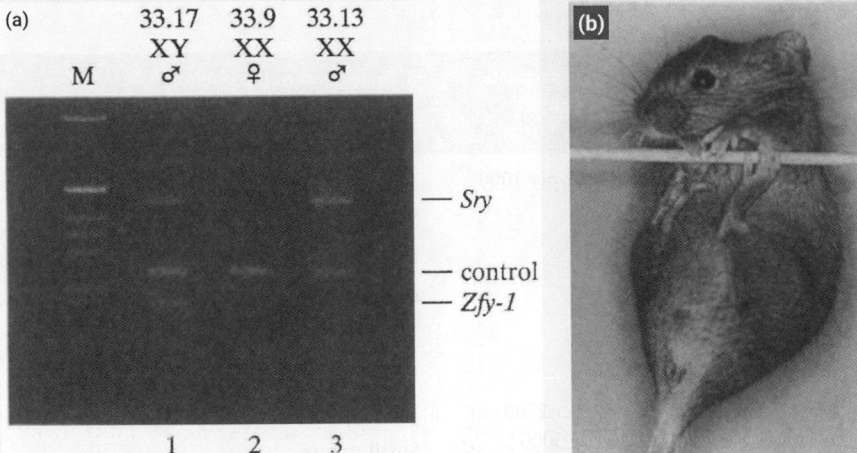

Figure 1 (a) Polymerase chain reaction (PCR) analysis of DNA of three mice from the same litter. All three mice are positive for 'control DNA', showing that the PCR reaction has worked. Mouse 33.17 has two Y chromosome genes, *Sry* and *Zfy*, as expected since it has an X and a Y chromosome. Mouse 33.9 has two X chromosomes and, as expected, has neither the *Sry* nor the *Zfy* gene. Mouse 33.13 ('Randy') has two X chromosomes but is transgenic for the *Sry* gene. He does not have the *Zfy* gene. (b) Mouse 33.13 ('Randy') has the fully male external phenotype, conferred by the presence of the *Sry* transgene.

Source: (a) Koopman P, Gubbay J, Vivian N, Goodfellow, P, Lovell-Badge R. Male development of chromosomally female mice transgenic for Sry. *Nature* 351, 117–121 (1991). (b) Koopman P., Sinclair A., Lovell-Badge R. Of sex and determination: marking 25 years of Randy, the sex-reversed mouse. *Development*. 2016 May 15;143(10):1633-7. doi: 10.1242/dev.137372. PMID: 27190031.

Further reading

Koopman, P., Gubbay, J., Vivian, N., Goodfellow, P., & Lovell-Badge, R. (1991) Male development of chromosomally female mice transgenic for Sry. *Nature* **351**: 117–121.

Koopman, P., Sinclair, A., & Lovell-Badge, R. (2016) Of sex and determination: marking 25 years of Randy, the sex-reversed mouse. *Development* **14**: 1633–7. doi: 10.1242/dev.137372. PMID: 27190031

Occasionally, however, errors give rise to gametes with incorrect numbers of chromosomes. If these are fertilized the resulting embryos are usually so abnormal that they are spontaneously aborted, causing a miscarriage, but a few of these **aneuploidies** (numerical chromosomal abnormalities) are less severe in their impact and the fetus may survive to term.

In humans, several medical syndromes are characterized by an individual having an extra chromosome (a trisomy) or a missing chromosome (a monosomy); these syndromes are summarized in Table 4.1. As you can see, the phenotype of each syndrome is complex. This is because each chromosome carries many genes and having the correct number of copies of a gene (what we call the **gene dosage**) affects how much of the gene product is made. Making too much or too little of multiple gene products affects many aspects of our development.

 Check your understanding of the concepts covered in this section by answering the questions in the e-book.

4.2 Mendel's experiments established the laws of inheritance

Gregor Mendel studied seven traits of the garden pea, *Pisum sativum*, as illustrated in Figure 4.3. For each of these, he could easily distinguish two forms—for instance purple or white flowers, yellow or green peas. In modern terminology, we would call these different phenotypes. A **phenotype** is a measurable characteristic of an organism arising from its genetic makeup, its **genotype**. (The formal definition of a phenotype states that it depends on both the organism's genes and its environment, and we will explore the effect of environment further in Section 4.5. For the purpose of understanding Mendel's experiments, however, we only need to consider the genotypes of his peas.)

> **PAUSE AND THINK**
>
> Having an extra or missing autosome is usually lethal. Only three autosomal trisomies support development of a fetus to term, and of these only Down syndrome (trisomy 21) is compatible with survival to adulthood. Looking at Figure 4.2, can you suggest why trisomy 21 causes less severe medical problems than other autosomal trisomies?
>
> *Answer:*
> Chromosome 21 is the smallest human chromosome. Therefore, it carries fewer genes than other autosomes, and the levels of fewer gene products are disrupted than in other autosomal trisomies.

Mendel was a careful scientist. For each trait he studied, he used pea plants that were true-breeding. That is, when he self-fertilized a purple-flowered plant or fertilized it with pollen from another purple-flowered plant he obtained only purple-flowered offspring, plants with round peas produced offspring only with round peas, those with wrinkled peas produced offspring with only wrinkled peas (see Figure 4.4), and so on.

Having established this, Mendel went on to carry out **monohybrid cross** experiments, in which he crossed plants that

Table 4.1 Human syndromes caused by numerical chromosomal abnormalities. Note that while each syndrome has a characteristic set of signs and symptoms, the incidence and severity of many of these varies between individuals.

Syndrome	Trisomy/ monosomy	Karyotype	Approximate incidence (live births)	Main features of the phenotype
Klinefelter syndrome	Trisomy (sex chromosome)	47, XXY	1–2 per 1000	Male development
				Small testes
				Infertility
				Gynaecomastia (development of breast tissue)
				Tall stature
				Learning difficulties
Turner syndrome	Monosomy (sex chromosome)	45, X	1 per 2500–1 per 3000	Female development
				Small stature
				Ovaries degenerate before birth
				Girls do not undergo puberty and are mainly infertile
				Some individuals have skin flaps on neck
				Swelling of hands and feet
				Kidney or heart defects
				Most have normal intelligence
Down syndrome	Trisomy (autosome, chromosome 21)	47, XY, +21 or 47, XX, +21	1 per 1000	Typical facial characteristics with upward slant to eyes
				Small stature
				Poor muscle tone
				Intellectual disability
				Congenital heart defects
				High incidence of leukaemia and early-onset Alzheimer disease
Edwards syndrome	Trisomy (autosome, chromosome 18)	47, XY, +18 or 47, XX, +18	1 per 5000	Abnormalities affecting many organs
				Congenital heart defects
				Small head
				Clenched fists
				Severe intellectual disability
				Most babies die before their first birthday
Patau syndrome	Trisomy (autosome, chromosome 13)	47, XY, +13 or 47, XX, +13	1 per 16,000	Abnormalities affecting many organs
				Congenital heart defects
				Severe intellectual disability
				Micropthalmia (abnormally small eyes)
				Deafness
				5–10% of children live past their first birthday

differed in just one of the seven characteristics. He may have been surprised to see that when he crossed purple-flowered plants with white-flowered plants, all the offspring had purple flowers, and when he crossed plants with yellow peas with plants with green peas, all the offspring had green peas. These results continued for all seven traits: in each cross, the offspring in this **F₁ generation** consistently displayed the phenotype of just *one* of the parent plants, the one shown on the top row in Figure 4.3. We call this phenotype **dominant**, while the phenotype that is not seen in the F₁ generation is **recessive**.

The results of one of these monohybrid cross experiments are illustrated in Figure 4.4. It shows the outcome in the F₁ generation, where crossing plants with round peas with plants with wrinkled peas gave plants only with round peas. It also shows what happened when Mendel self-fertilized plants from the F₁ generation. In the F₂ generation, there was a mixture of plants with round peas and plants with wrinkled peas. Mendel counted the F₂ offspring and found that 75 per cent of the plants had round peas, while 25 per cent had wrinkled peas, giving a 3:1 ratio of plants with the dominant, round pea phenotype to plants with the recessive, wrinkled pea phenotype.

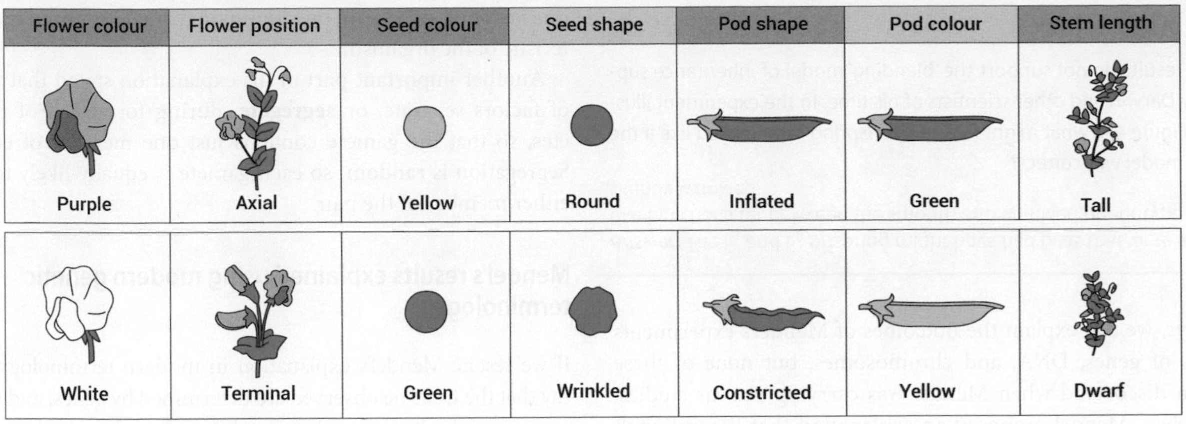

Flower colour	Flower position	Seed colour	Seed shape	Pod shape	Pod colour	Stem length
Purple	Axial	Yellow	Round	Inflated	Green	Tall
White	Terminal	Green	Wrinkled	Constricted	Yellow	Dwarf

Figure 4.3 The seven traits of garden peas studied by Mendel. For each of these, there are two easily distinguished phenotypes. For each trait, the dominant phenotype is the one shown on the top row.

Source: Meneely, P. M. et al. (2017). *Genetics* (1st edn). Oxford University Press.

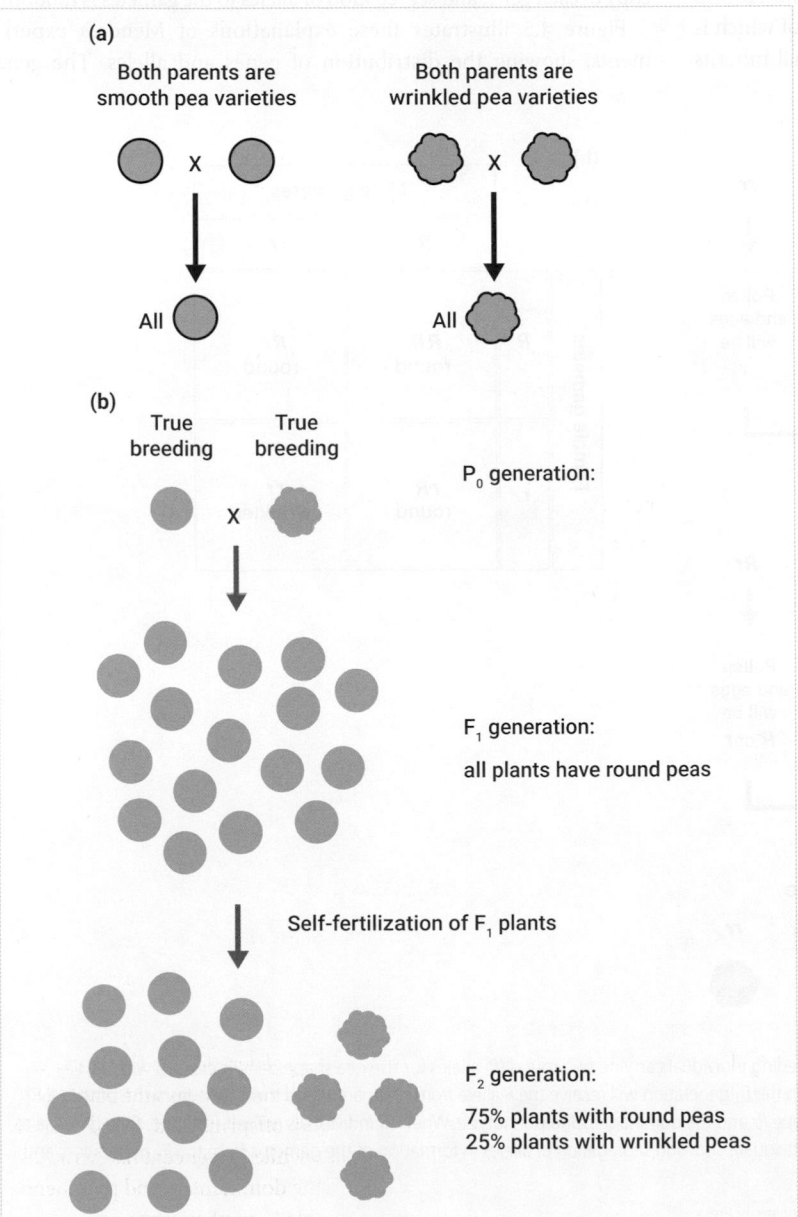

(a)

Both parents are
smooth pea varieties

Both parents are
wrinkled pea varieties

All

All

(b)

True
breeding

True
breeding

P_0 generation:

F_1 generation:

all plants have round peas

Self-fertilization of F_1 plants

F_2 generation:

75% plants with round peas
25% plants with wrinkled peas

Figure 4.4 Mono-hybrid cross experiment in peas. (a) In a true-breeding variety, all the offspring of self-fertilization are identical to each other and to the parent. (b) Results of a monohybrid cross between true-breeding round and wrinkled pea plants. In the F_1 generation, all plants have round peas. Self-fertilization of plants of the F_1 generation gives a mixture of the two phenotypes in the F_2 generation: a 3:1 ratio of plants with round peas to plants with wrinkled peas.

Source: Adapted from Meneely, P. M. et al. (2017). *Genetics* (1st edn). Oxford University Press.

PAUSE AND THINK

Mendel's results do not support the 'blending' model of inheritance supported by Darwin and other scientists of his time. In the experiment illustrated in Figure 4.4, what might F₁ and F₂ offspring have looked like if the blending model was correct?

Answer: The F₁ and F₂ offspring might have had peas that were intermediate in character between the smooth and wrinkled phenotype—just slightly wrinkled.

Nowadays, we can explain the outcomes of Mendel's experiments in terms of genes, DNA, and chromosomes, but none of these had been discovered when Mendel was carrying out his studies. Nevertheless, Mendel proposed an explanation that fits well with our current knowledge. He suggested that the traits he studied are controlled by inherited 'factors' and that the organism inherits a pair of factors that determine each trait, with one of the pair inherited from the male parent and one from the female parent. He also suggested that each factor has two different 'forms', one of which is dominant and one recessive, so that if a single individual inherits one factor of each form, the dominant form determines the characteristic of the organism.

Another important part of his explanation stated that the pairs of factors separate, or **segregate**, during formation of the gametes, so that the gamete contains just one member of each pair. Segregation is random, so each gamete is equally likely to receive either member of the pair.

Mendel's results explained using modern genetic terminology

If we restate Mendel's explanation in modern terminology, we can say that the traits he observed are determined by genes, and that each organism inherits two copies of each gene, one from each parent. The genes he studied each have two alleles, one of which is dominant and one recessive. If an individual inherits one copy of each allele, the dominant allele determines its phenotype. Gametes contain only one copy of each gene, and segregation of alleles to the gametes is random.

Figure 4.5 illustrates these explanations of Mendel's experiments, showing the distribution of genes and alleles. The gene

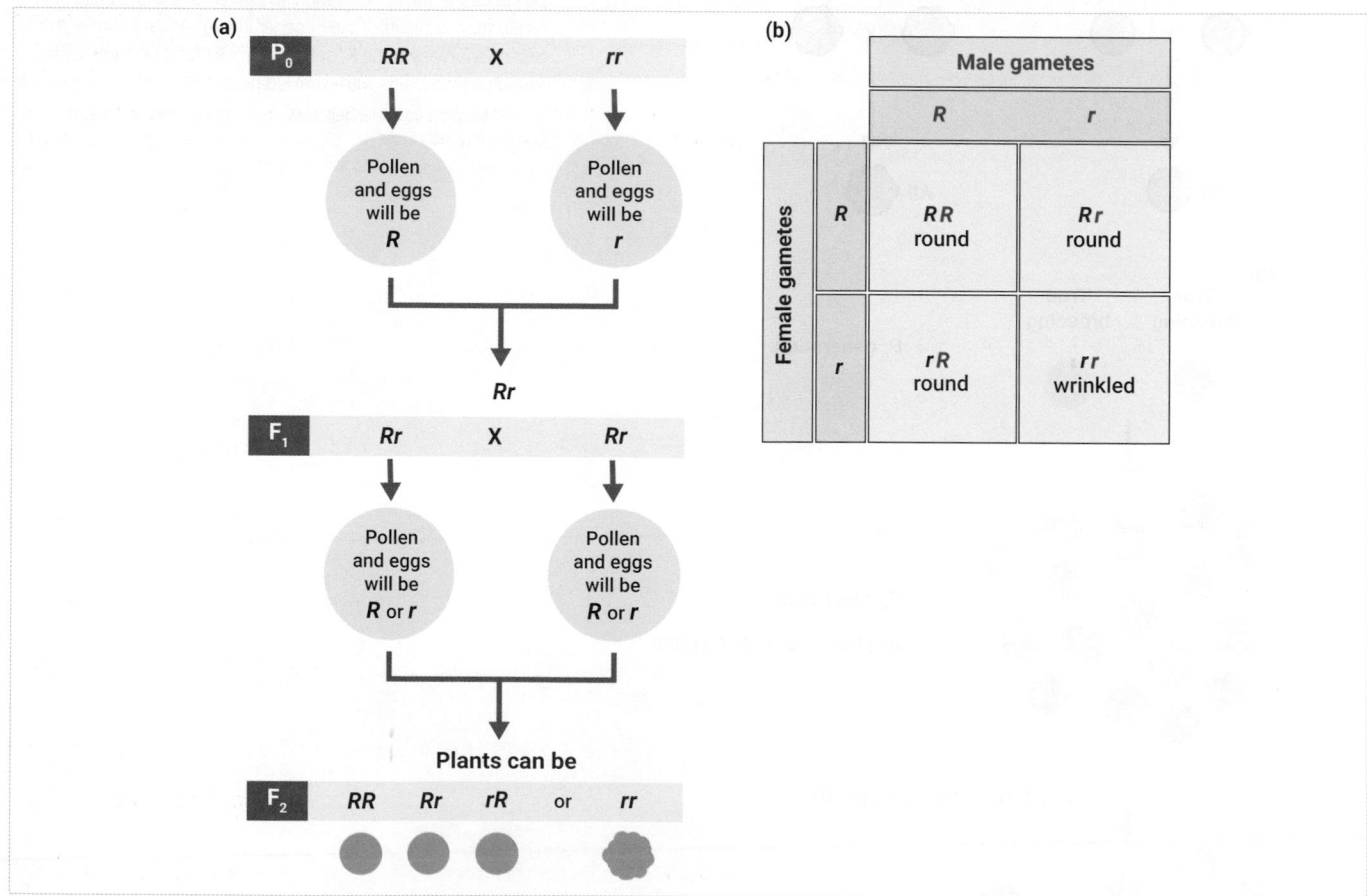

Figure 4.5 Distribution of genes and alleles. (a) When a true-breeding individual carrying two round (*R*) alleles for the pea shape gene is crossed with a true-breeding individual carrying two wrinkled (*r*) alleles, all the offspring in the F₁ generation will receive the *R* allele from one parent and the *r* allele from the other. All F₁ individuals are heterozygous for the pea shape gene, have the genotype *Rr* and the dominant round phenotype. When F₁ individuals are self-fertilized, they give rise to the F₂ generation with genotypes and phenotypes shown. (b) Punnett square showing segregation of alleles in formation of the gametes from the F₁ generation, and how they combine at fertilization.

Source: Adapted from Meneely, P. M. et al. (2017). *Genetics* (1st edn). Oxford University Press.

that determines pea shape is denoted by the letter R or r, with (by convention) capital R denoting the dominant (round) allele and lower-case r denoting the recessive (wrinkled) allele. Individuals who carry two identical alleles (RR or rr) are described as being **homozygotes** or as being **homozygous** for the pea shape gene, while those who carry two different alleles (Rr) are **heterozygotes** or are **heterozygous** for the gene.

We can see in Figure 4.5a that there are three different genotypes in the F_2 generation (RR, Rr, and rr) but because R is dominant to r, these give rise to only two phenotypes (round and wrinkled). Figure 4.5b shows a Punnett square, a convenient way of setting out the outcome of the F_2 self-fertilization. The Punnett square shows the gametes produced by each parent. One parent's set of gametes is shown across the top of the square, while the other parent's set of gametes is shown down the side. The completed square shows the genotypes of all possible progeny produced by combining one gamete from each parent. In Figure 4.5b note that completion of the square predicts the 3.1 ratio of plants with round peas to plants with wrinkled peas.

Mendel extended his experiments to show that genes for different traits segregate independently

Having looked separately at seven different traits, Mendel continued his experiments by crossing pea plants that differed in two or more traits. One such experiment is illustrated in Figure 4.5. The gene for pea colour has two alleles, Y (dominant, yellow) and y (recessive, green). When Mendel crossed true-breeding plants with yellow, round peas with true-breeding plants with green, wrinkled peas, all the F_1 offspring were yellow and round.

Figure 4.6 shows how the parental plants were homozygous for both genes (as they are true breeding). By contrast, the F_1 plants were *heterozygous* for both genes. When Mendel self-fertilized plants of the F_1 generation, four different phenotypes appeared in the F_2 offspring, in the ratios shown (9:3:3:1). The Punnett square shows how this result comes about: the different alleles of the two genes segregate independently to the gametes, so that four classes of gamete occur with equal probability.

The conclusion Mendel derived from experiments with two or more genes is often called the 'Law of Independent Assortment'. This law states that the inheritance of one gene does not affect the inheritance of another. This holds true for all the seven genes he studied, and for many other genes. However, as we will see later, it is not *universally* true: genes that are close together on the same chromosome can be inherited together.

In the next section, we consider how our knowledge of chromosomes, and particularly of their behaviour during cell division, helps us to interpret Mendel's experiments and to understand and extend his laws of inheritance.

 Check your understanding of the concepts covered in this section by answering the questions in the e-book.

4.3 The behaviour of chromosomes as cells divide can explain the basis of inheritance

DNA replication occurs as a prelude to cell division. It is worth considering how chromosomes behave when cells divide, as it helps us to understand how genetically determined traits are passed from generation to generation. Cell division is covered in greater detail in Chapter 10. Here we focus on its role in inheritance, but we will start with an outline of the process.

There are two types of cell division in sexually reproducing eukaryotes: **mitosis**, which occurs in the majority of cells (the somatic cells); and **meiosis**, which is the specialized cell division that occurs during formation of the gametes. While meiosis is of particular importance in understanding inheritance we will describe both mitosis and meiosis so that we can compare the two processes.

Cell division by mitosis produces diploid 'daughter' cells that are genetically identical to each other and to their 'parent' cell

In all cells except the germ line cells that give rise to gametes, eukaryotic cell division involves mitosis, the stages of which are illustrated in Figure 4.7.

▶ **Mitosis is explained in detail in Chapter 10.**

As the cell enters the first stage of mitosis (**prophase**), duplicated chromosomes become condensed. At this stage each duplicated chromosome consists of two 'sister' **chromatids**, as depicted in Figure 4.8.

The chromatids are identical double-stranded DNA molecules produced by semi-conservative DNA replication (discussed in detail in Chapter 3): in each chromatid one of the strands is newly synthesized, and the other is the parental strand. At this stage the pair of chromatids is held together by specialized proteins.

The cell now undergoes a major structural change: during **prometaphase** the nuclear membrane disintegrates, allowing the **mitotic spindle** to access the chromosome. The spindle is an arrangement of microtubules (see Figure 6.19), originating at the **centrosomes**, a complex of proteins located at each end or **pole** of the cell. At **metaphase** (Figure 4.7, step 3) the chromosomes are arranged at the **equator** (that is, the centre) of the cell, with each chromatid attached to spindle fibres via the **centromere** of each chromatid. In **anaphase** (Figure 4.7, step 4) the chromatids separate. The separated chromatids, now called **daughter chromosomes**, move towards the poles and the poles move further apart (Figure 4.7, step 5), all the movement being brought about by the spindle fibres. This process is referred to as **segregation** of the daughter chromosomes.

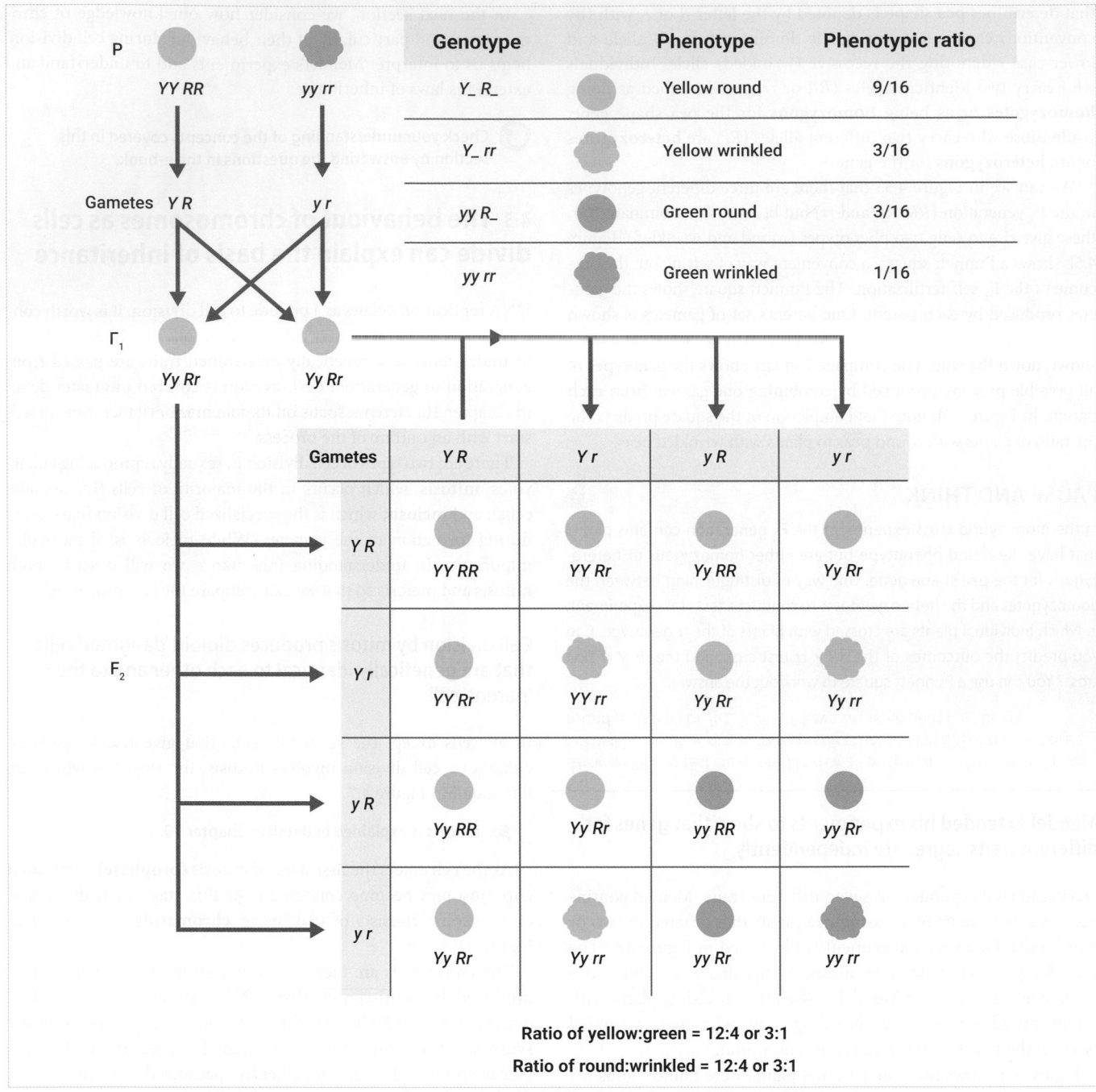

Figure 4.6 The outcome of Mendel's experiments involving two genes shown using a Punnett square. Gametes from a cross of heterozygotes (*YyRr × YyRr*) are combined in a Punnett square to predict the resulting offspring. The phenotypic ratio of the offspring is 9:3:3:1. Considering each gene alone shows the expected 3:1 ratio of dominant to recessive phenotype.

Source: Meneely, P. M. et al. (2017). *Genetics* (1st edn). Oxford University Press.

Segregation is completed at **telophase** when the chromosomes reach the poles. The mitotic apparatus is disassembled and nuclear membranes reform. **Cytokinesis** or cytosolic division is now completed, with each daughter cell receiving one full set of chromosomes. The chromosomes decondense into the interphase state.

Cell division by meiosis produces haploid gametes containing a single copy of each gene

Germ line cells divide to produce gametes (eggs and sperm in mammals). The crucial genetic difference between gametes and the other cells in the body (somatic cells) is that gametes are

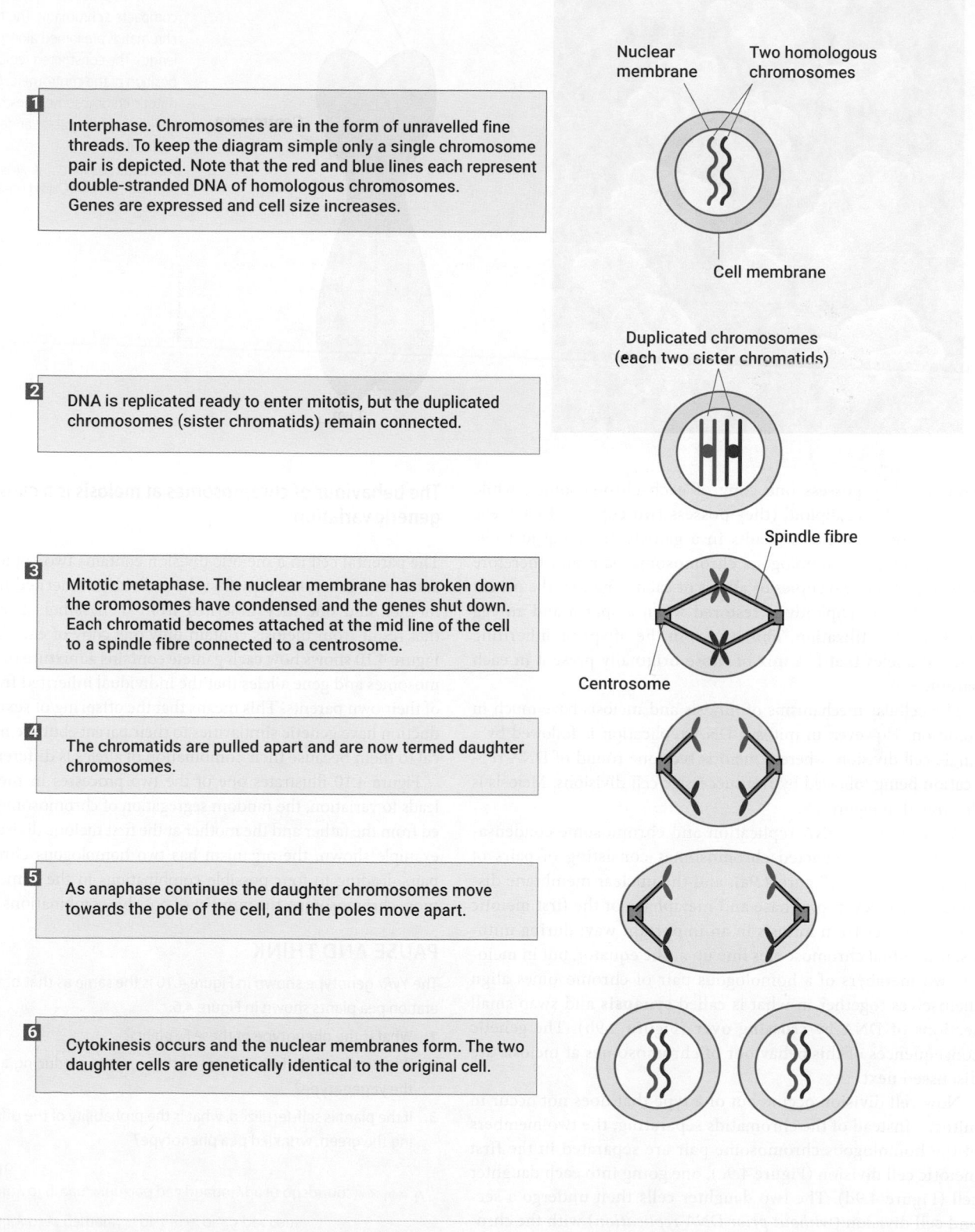

1 Interphase. Chromosomes are in the form of unravelled fine threads. To keep the diagram simple only a single chromosome pair is depicted. Note that the red and blue lines each represent double-stranded DNA of homologous chromosomes. Genes are expressed and cell size increases.

2 DNA is replicated ready to enter mitotis, but the duplicated chromosomes (sister chromatids) remain connected.

3 Mitotic metaphase. The nuclear membrane has broken down the cromosomes have condensed and the genes shut down. Each chromatid becomes attached at the mid line of the cell to a spindle fibre connected to a centrosome.

4 The chromatids are pulled apart and are now termed daughter chromosomes.

5 As anaphase continues the daughter chromosomes move towards the pole of the cell, and the poles move apart.

6 Cytokinesis occurs and the nuclear membranes form. The two daughter cells are genetically identical to the original cell.

Figure 4.7 A simplified diagram of mitosis. Only one pair of homologous chromosomes is shown. As the focus is to explain the behaviour of chromosomes, not all intermediate stages are shown.

Source: Adapted from Snape, A. & Papachristodoulou, D. (2018). *Biochemistry and Molecular Biology* (6th edn). Oxford University Press.

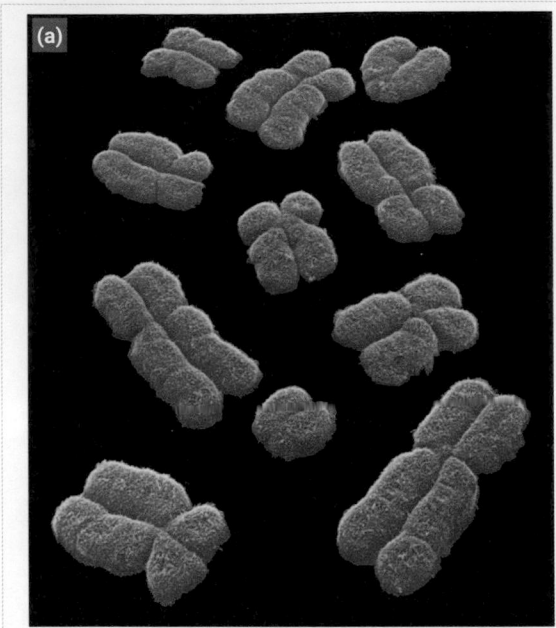

(a)

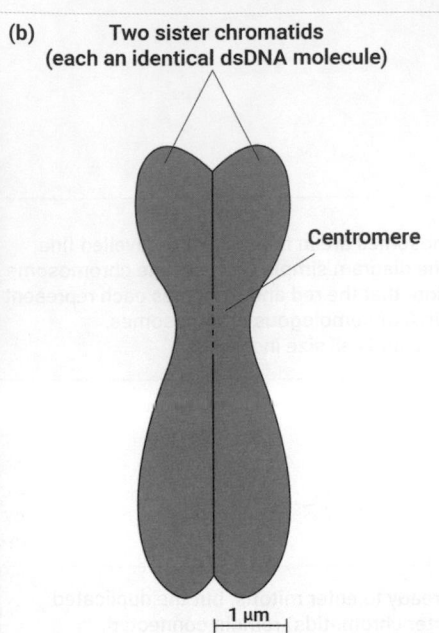

(b) **Two sister chromatids**
(each an identical dsDNA molecule)

Centromere

1 µm

Figure 4.8 Chromatids. (a) A scanning electron micrograph of mitotic chromosomes, showing tightly compacted chromatin. The two chromatids are joined along their length. The constricted region is at the position of the centromere. (b) A mitotic chromosome. The chromatids are identical double-stranded DNA molecules.

Source: Cooper. *The Cell, A Molecular Approach*, 8th edn. Oxford University Press.

haploid (they possess one copy of each chromosome), while somatic cells are diploid (they possess two copies). To achieve this difference, meiosis results in a gamete receiving just *one* member of each homologous chromosome pair, and therefore only *one* of the two copies or alleles of each gene that the parent cell contained. Diploidy is restored when a sperm and an egg cell fuse at fertilization. This results in the offspring inheriting a set of alleles that is a mix of those originally present in each parent.

The cellular mechanisms of mitosis and meiosis have much in common. However, in mitosis, DNA replication is followed by a single cell division, whereas meiosis sees one round of DNA replication being followed by *two* successive cell divisions. Meiosis is illustrated in Figure 4.9.

As in mitosis, DNA replication and chromosome condensation produce compacted chromosomes consisting of pairs of sister chromatids (Figure 4.9a), and the nuclear membrane disappears. However, prophase and metaphase of the first meiotic division differ from mitosis in an important way: during mitosis, individual chromosomes line up at the equator, but in meiosis two members of a homologous pair of chromosomes align themselves together in what is called **synapsis** and swap small sections of DNA by **crossing over** (Figure 4.9b). The genetic consequences of this behaviour of chromosomes at meiosis are discussed next.

Now cell division occurs but of a type that does not occur in mitosis. Instead of the chromatids separating, the two members of the homologous chromosome pair are separated in the first meiotic cell division (Figure 4.9c), one going into each daughter cell (Figure 4.9d). The two daughter cells then undergo a second cell division (*without prior DNA replication*) with the chromatids now separating (Figure 4.9e) as in mitosis. This gives four haploid cells (gametes) from the original diploid parent cell (Figure 4.9f).

The behaviour of chromosomes at meiosis is a cause of genetic variation

The parental cell in a meiotic division contains two copies of each autosomal gene, one copy that the individual inherited from their mother, and one they inherited from their father. The gametes that result from meiosis contain only one copy of each gene, and Figure 4.10 shows how each gamete contains a mixture of the chromosomes and gene alleles that the individual inherited from either of their own parents. This means that the offspring of sexual reproduction have genetic similarities to their parents but are not identical to them because their combination of alleles is different.

Figure 4.10 illustrates one of the two processes in meiosis that leads to variation, the random segregation of chromosomes inherited from the father and the mother at the first meiotic division. In the example shown, the organism has two homologous chromosome pairs, leading to four possible combinations in the gametes. With more chromosomes, the number of possible combinations is greater.

PAUSE AND THINK

The *YyRr* genotype shown in Figure 4.10 is the same as that of the F$_1$ generation pea plants shown in Figure 4.6.

1. What is the phenotype of these F$_1$ plants?
2. What is the probability of one of these plants producing a gamete of the *yr* genotype?
3. If the plant is self-fertilized, what is the probability of the offspring having the green, wrinkled pea phenotype?

Answer:

1. F$_1$ phenotype: yellow, round peas.
2. Probability of yr gamete: ¼ or 1 in 4, or 25 per cent.
3. Probability of green, wrinkled pea phenotype in offspring: ¼ × ¼ = ¹/₁₆ or 1 in 16.

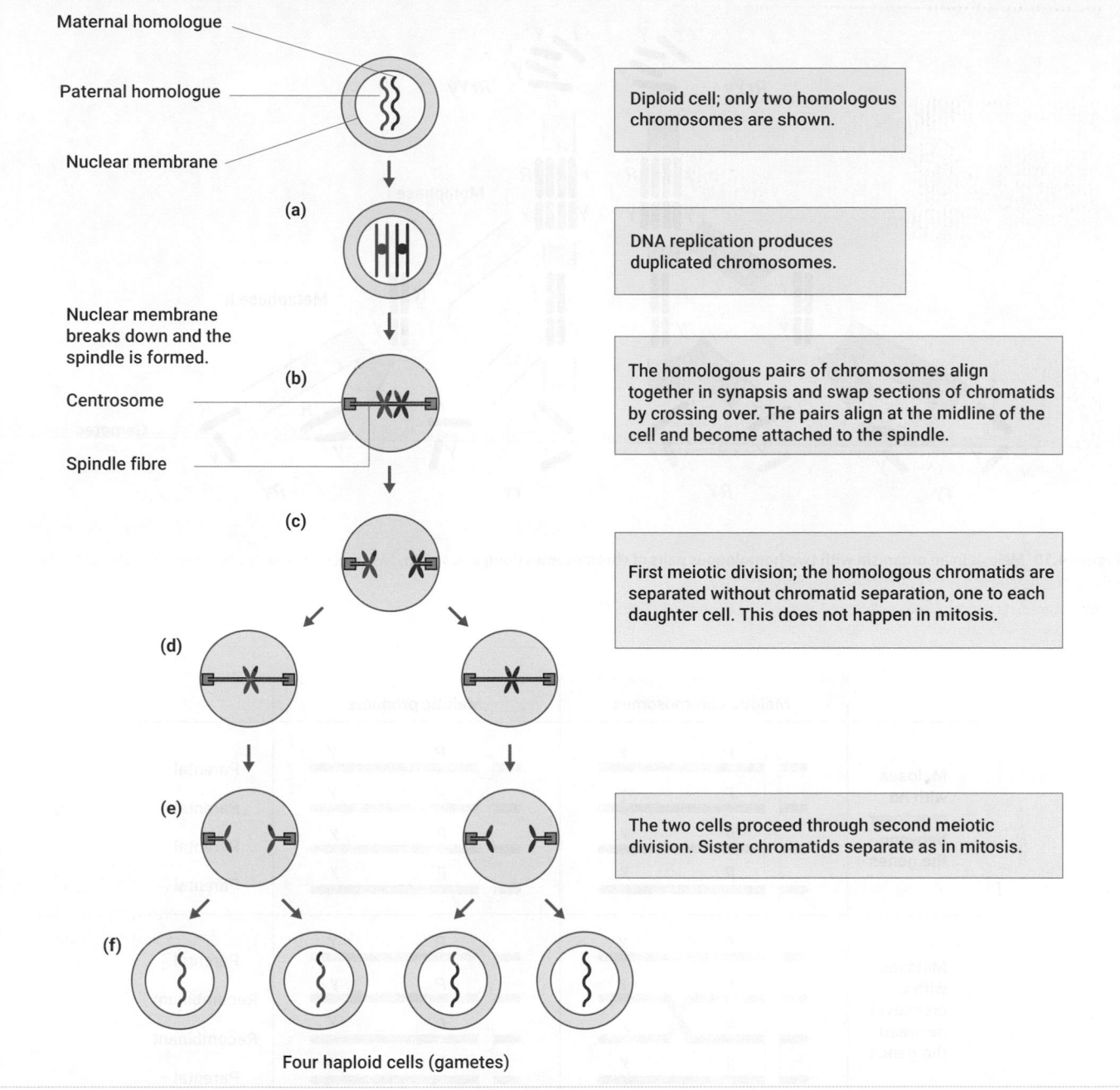

Figure 4.9 A simplified diagram of meiosis. As in mitosis, chromosome replication occurs prior to cell division. Unlike in mitosis, however, meiosis sees two cell divisions following a single round of replication. As a result, the daughter cells (the gametes) each receive only one member of each homologous chromosome pair. In the figure only a single pair of homologous chromosomes is shown for simplicity. A single crossing over event is shown. Crossing over and its genetic consequences are described in the text.

Source: Snape, A. & Papachristodoulou, D. (2018). *Biochemistry and Molecular Biology* (6th edn). Oxford University Press.

The second meiotic process that yields yet more variation is crossing over, a process that leads to genetic **recombination** between genes that are **linked** on the same chromosome. Figure 4.11 shows how breakage and rejoining of sections of chromatids during prophase of meiosis I lead to the production of **recombinant chromosomes**, which have a set of alleles that matches that of neither parent.

Linkage and crossing over distort Mendelian ratios

To understand how linkage and crossing over affect the outcomes of breeding experiments such as those carried out by Gregor Mendel, we will consider inheritance of two pea genes that are both on the same chromosome. These are the genes illustrated in Figure 4.10. We encountered the gene labelled *Y* or *y* earlier: it is the gene that

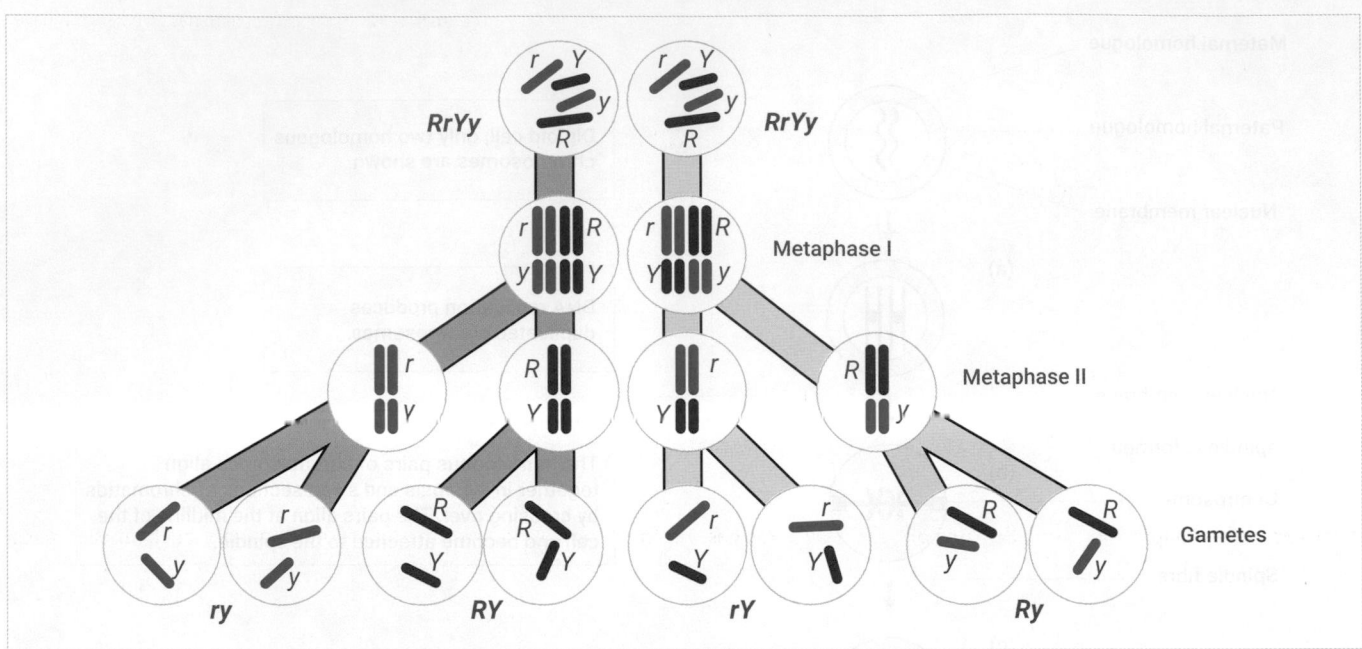

Figure 4.10 Meiosis in an organism with two homologous pairs of chromosomes (long and short). Two genes are shown, one on each chromosome. The organism is heterozygous for both genes: its genotype is *YyRr*.

Source: Wikimedia Commons. OpenStax Biology, Copyright Rice University, All Rights Reserved.

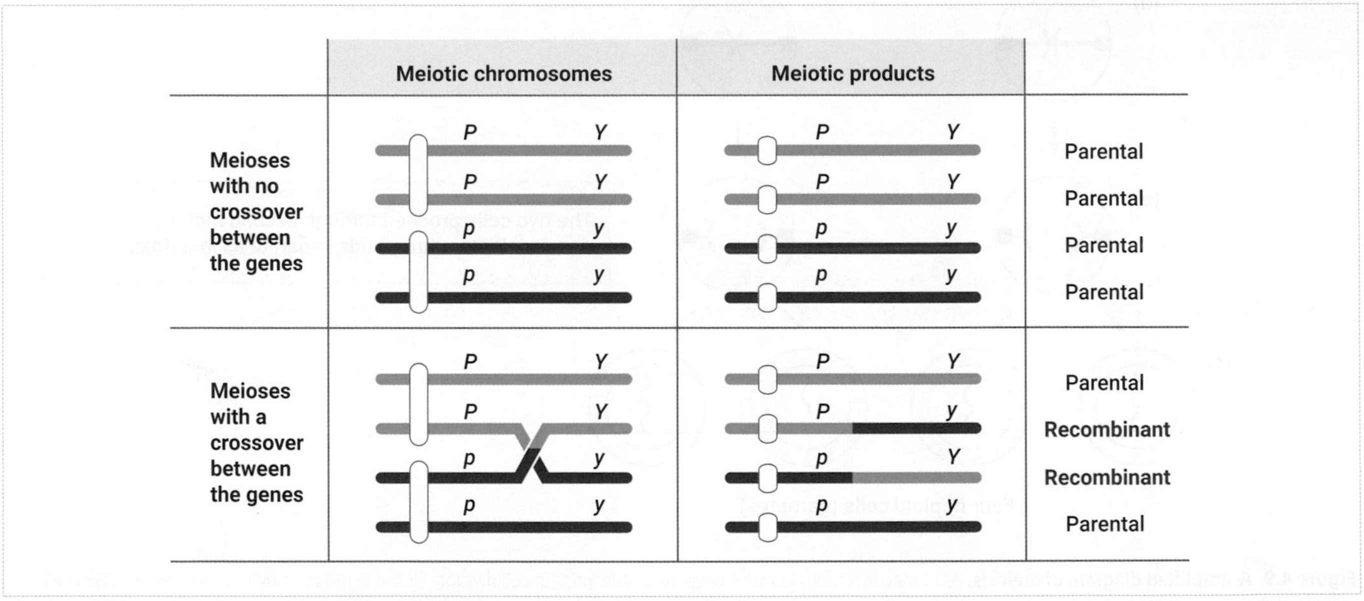

	Meiotic chromosomes		Meiotic products		
Meioses with no crossover between the genes	*P*	*Y*	*P*	*Y*	Parental
	P	*Y*	*P*	*Y*	Parental
	p	*y*	*p*	*y*	Parental
	p	*y*	*p*	*y*	Parental
Meioses with a crossover between the genes	*P*	*Y*	*P*	*Y*	Parental
	P	*Y*	*P*	*y*	Recombinant
	p	*y*	*p*	*Y*	Recombinant
	p	*y*	*p*	*y*	Parental

Figure 4.11 Outcome of meiosis when two genes are linked on the same chromosome. The parent organism has the genotype *PpYy*, with the *P* and *Y* alleles on one member of the homologous chromosome pair and the *p* and *y* alleles on the other. When no crossing over occurs, gametes will have the genotype *PY* or *py*. Crossing over occurs when one chromatid from each member of the homologous pair breaks and rejoins with a chromatid of the other member, as shown. Gametes produced by a meiotic division where crossing over has taken place between the two genes may have the parental genotypes *PY* and *py* if they receive a chromosome that was not involved in the crossover, or the recombinant genotypes *Py* and *pY* if they receive a chromosome that was involved. As explained in the next section, the probability of a crossover taking place between two genes and therefore the frequency of recombination depends on how far apart they are on the chromosome.

Source: Adapted from Griffiths A. J. F., Miller J. H., Suzuki D. T., et al. *An Introduction to Genetic Analysis*. 7th edn. New York: W.H. Freeman; 2000.

determines whether the plants produce peas that are yellow or green in Mendel's experiments. The *P* gene was not one of those that Mendel investigated. It determines whether the peas have purple or green pods (Figure 4.12a). The *P* allele, which gives purple pods, is dominant over the *p* allele, which gives green pods.

Let us consider what would happen in a cross between a plant that is heterozygous for both genes and one that is homozygous for the recessive alleles of both genes. We can use Punnett squares to compare what we expect to happen if the cross involves two genes that are on *different* chromosomes (the *Y* and *R* genes we looked

(a)   **(b)**

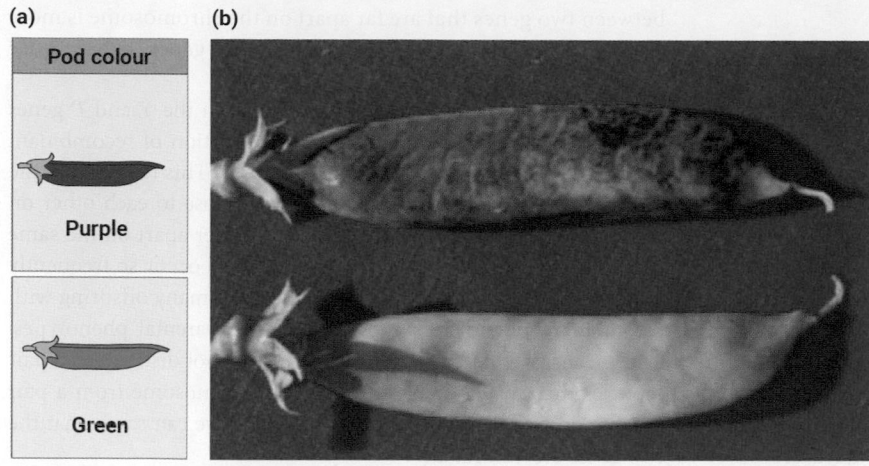

Figure 4.12 The P gene determines whether peas have purple or green pods, with the *P* allele (purple) being dominant over *p* (green).

Source: (a) Adapted from Meneely et al. *Genetics; Genes, Genomes, and Evolution* (2017). Oxford University Press.

at earlier in this chapter) with one that involves the *Y* and *P* genes, which are on the *same* chromosome.

In Figure 4.13a we see that, when the two genes under consideration are unlinked on two different chromosomes, the heterozygous plant produces four classes of gamete, each with equal probability. By contrast, the homozygous plant can only produce one class of gamete. The pea colour and pea shape genes on different chromosomes follow Mendel's law of independent assortment, so offspring of this cross will have four different phenotypes: Yellow, round; Yellow, wrinkled; Green, round; Green, wrinkled. The predicted ratio of phenotypes is 1:1:1:1, so we will see roughly equal numbers of offspring of each phenotype. Look at Figure 4.13a and also note that two of the phenotypes (Yellow, round and Green, wrinkled) are those of the parent plants, while the other two (Yellow, wrinkled and Green, round) are not.

Figure 4.13b shows an equivalent cross where the two genes are linked on the same chromosome. Note that this prevents the two genes segregating independently. We are considering the situation illustrated in Figure 4.11, where one copy of the chromosome in the heterozygous parent carries the dominant allele of each gene, and the other copy carries the recessive allele of each gene.

If there is no crossing over between the two genes, the heterozygous parent plant can only produce two classes of gamete, each with equal probability. The offspring will have two different phenotypes: Yellow, purple and Green, green, which match the phenotypes of the parents. The predicted ratio of phenotypes is 1:1, so we would expect to see roughly equal numbers of offspring of each of the two phenotypes.

Table 4.2 shows the results of a cross in which 100 offspring were counted. In each case, we can compare the predicted number of offspring of each class with the number seen. As you can see, there is a striking mismatch between our prediction and what is actually observed, and this can be explained by crossing over.

Offspring with yellow peas and green pods and those with green peas and purple pods are the results of crossovers between the two genes that took place during meiosis. We can describe these offspring as **recombinants** or say that they have recombinant phenotypes as opposed to parental phenotypes.

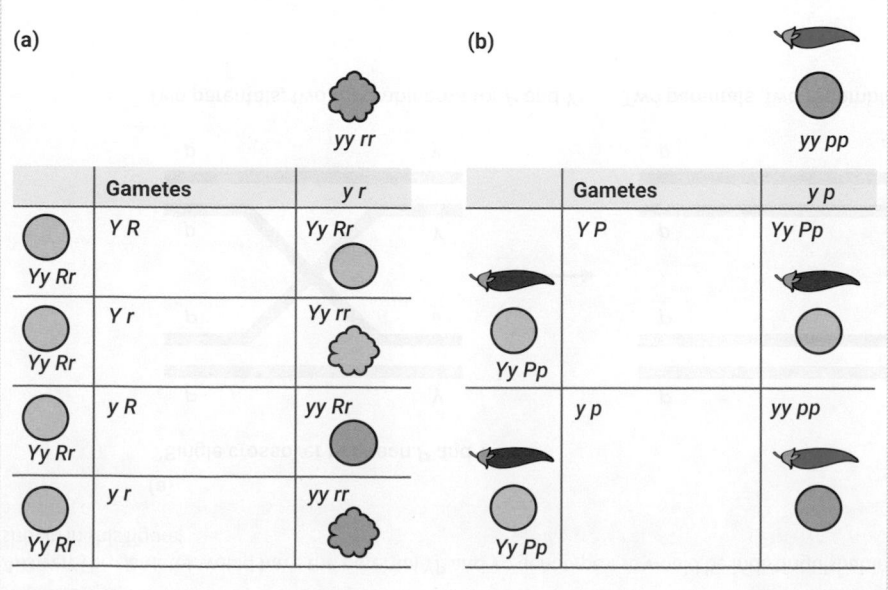

Figure 4.13 Punnett square predictions of the outcome of two crosses. In each case a parent plant that is heterozygous for each of two genes is crossed with a parent that is homozygous for the recessive alleles of the two genes. (a) The two genes are unlinked on different chromosomes. (b) The two genes are linked on the same chromosome. In the heterozygous parent the dominant alleles of both genes are on one chromosome, and the recessive alleles are on the other. Note that here we are assuming there is no crossing over between the two linked genes.

Table 4.2 Results of *YyPp* × *yypp* cross

Offspring phenotype	Offspring genotype	Predicted number	Observed number
Yellow, purple	*YyPp*	50	31
Yellow, green	*Yypp*	0	19
Green, purple	*yyPp*	0	17
Green, green	*yypp*	50	33

The probability of crossing over occurring at any position along the length of a chromosome is influenced to some extent by factors such as the base pair sequence of the DNA, but it is largely determined by chance. This means that a crossover between two genes that are far apart on the chromosome is more likely to occur than a crossover between two genes that are close together.

Table 4.2 shows that crossing over between the *Y* and *P* genes is a frequent event, as there is a high proportion of recombinant offspring (19 + 17 out of 100, or 36 per cent). This tells us that the *Y* and *P* genes are linked but are not very close to each other on the chromosome. If two genes are even further apart on the same chromosome crossing over between them will occur so frequently that in experiments of this type we will see as many offspring with recombinant phenotypes as offspring with parental phenotypes. This means that experiments of this type cannot distinguish a pair of genes that are far apart on the same chromosome from a pair that are on two different chromosomes. All we can say is that the two genes are 'unlinked'.

PAUSE AND THINK

During a meiotic division we generally see more than one crossover between each homologous pair of chromosomes. In the cross illustrated in Figure 4.13b, what gametes would be produced by a meiotic division in which two crossovers take place between the *Y* and *P* genes?

Answer: The gametes would have the parental YP and yp genotypes, so would be indistinguishable from those produced if no crossover takes place, as shown in this figure:

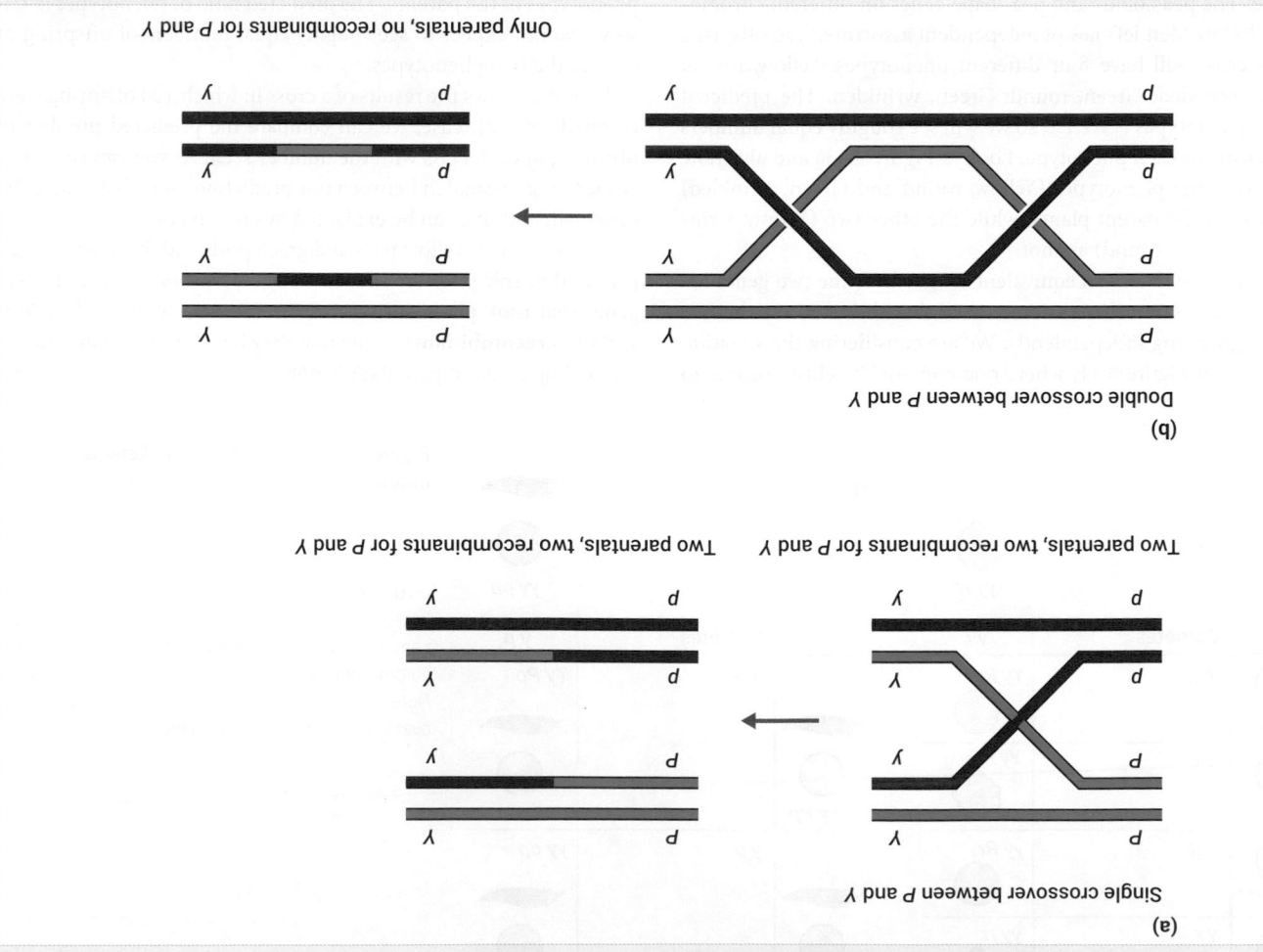

(a) The single crossover between *P* and *Y* produces two recombinant and two parental chromosomes. (b) The double crossover between *P* and *Y* restores the original relationship of the *P* and *Y* alleles. This would not be detected as a crossover in an experiment investigating the frequency of recombination between *P* and *Y*.

Recombination frequencies have been used to construct genetic maps

Before the availability of large-scale DNA sequencing and computer-based analysis, researchers were able to investigate the locations of genes on chromosomes by using experimental data such as those shown in Table 4.2. The recombination frequency between two genes gives an approximate measure of the distance between them, which is expressed in 'map units' or sometimes in centiMorgans (cM), a unit named after the distinguished geneticist Thomas Hunt Morgan. The data shown in Table 4.2 give a recombination frequency of 36 per cent for the *P* and *Y* genes, which places them 36 map units or 36 cM apart. An example of this type of genetic map (often called a linkage map) can be found in the 'Further reading' article by Susan Offner about Mendel's peas.

Genetic maps generated using recombination frequencies have some limitations. As we discussed earlier in this chapter, crossing over between two genes that are far apart on the same chromosome is inevitable. This produces a recombination frequency of 50 per cent, which cannot be used to infer distance. Even with genes that are closer together, double crossovers can cause complications in measuring recombination frequencies. Also, map unit distances do not correlate perfectly with physical distances measured in base pairs, because the recombination rate varies somewhat in different regions of the genome.

Despite these limitations, genetic mapping has provided a great deal of useful information on the organization of genes and genomes. Mapping was originally limited to experimental organisms such as peas and fruit flies, which can be used for breeding experiments in large numbers; such organisms also have many genes with multiple alleles that produced easily recognizable phenotypes. Later, mapping was extended using genetic 'markers' such as small sequence variations that could be recognized by molecular methods, enabling the production in the 1990s of a genetic map of the human genome using data from a panel of large families. (A genetic map of just one chromosome is shown in Figure 4.14.) The genetic map of the human genome was used by researchers working on the Human Genome Project; it helped them to place sections of DNA sequence data in the correct order to give the overall sequence. The human genome sequence was published in 2001.

 Check your understanding of the concepts covered in this section by answering the questions in the e-book.

4.4 Extensions and refinements of Mendel's laws

Each of the seven traits that Mendel chose to study showed two distinct and easily distinguishable phenotypes. He found that each trait was determined by a single gene with one dominant and one recessive allele. This simplicity helped Mendel to determine the basic laws of inheritance. However, many inherited traits exhibit greater complexity. For example, we may find multiple alleles of the same gene, or that more than one gene determines a single characteristic. In this section we briefly consider some examples of genes

that illustrate the need for extensions and refinements of Mendel's laws to fully explain their inheritance.

Genes may have several alleles

Some genes have multiple alleles, and in some cases there is no clear dominance of one allele over another. In the example of human MN blood group alleles, described earlier, individuals who inherit one copy of each allele make both the M and the N form of the glycophorin A protein and hence their phenotype (red blood cells expressing both the M and N antigen) equally reflects both inherited alleles. The alleles are said to be **co-dominant**. In the ABO blood group system, there are three alleles of a single gene encoding an enzyme that synthesizes molecules (different from glycophorin A) that are also found on the surface of red blood cells. The alleles of this gene are designated A, B, and O, so individuals can have the following genotypes: AA, AO, BB, BO, AB, or OO. A and B are co-dominant, while O is recessive to both A and B. This means there are four possible ABO blood group phenotypes: blood group A (AA and AO genotypes), blood group B (BB and BO genotypes), blood group AB (AB genotype), and blood group O (OO genotype).

Genetic heterogeneity and pleiotropy illustrate the complexity of phenotypes

We frequently come across situations in which mutations in several different genes can cause the same phenotype. Hereditary deafness is a good example. If you think about the process of hearing, which depends on both the complex structures of the ear and the proper functioning of the nervous system, it should be no surprise to find out that hereditary deafness can be caused by mutations in many different genes. This **genetic heterogeneity** can cause complications in genetic analysis.

Hereditary deafness also provides a good example of **pleiotropy**, a situation in which a single mutation affects multiple phenotypes. Many genes and mutations are known that cause non-syndromic deafness, where no effects are observed other than loss of hearing. But there are also multiple examples of syndromic deafness—for example, Alport syndrome, in which a single mutation causes hearing loss, eye abnormalities, and kidney disease; and Usher syndrome, in which there is loss of both vision and hearing. You can find out more about the genetics of hearing loss in Clinical Box 4.1.

Phenotypes can be influenced by the environment

Earlier, we defined a phenotype as a measurable characteristic of an organism arising from its genetic makeup. Many of the phenotypes we have discussed can be observed directly (flower colour, pea shape) while others (blood groups) require biochemical assays to detect them. Strictly speaking, phenotypes are determined by a combination of genetic and environmental factors, and in some cases the contribution of the environment can complicate genetic analysis. Environmental factors that can affect a phenotype include diet and temperature. To take one example, the *agouti* gene affects coat colour in several mammalian species, including mice, dogs, and horses. Several mutations of the *agouti* gene are known

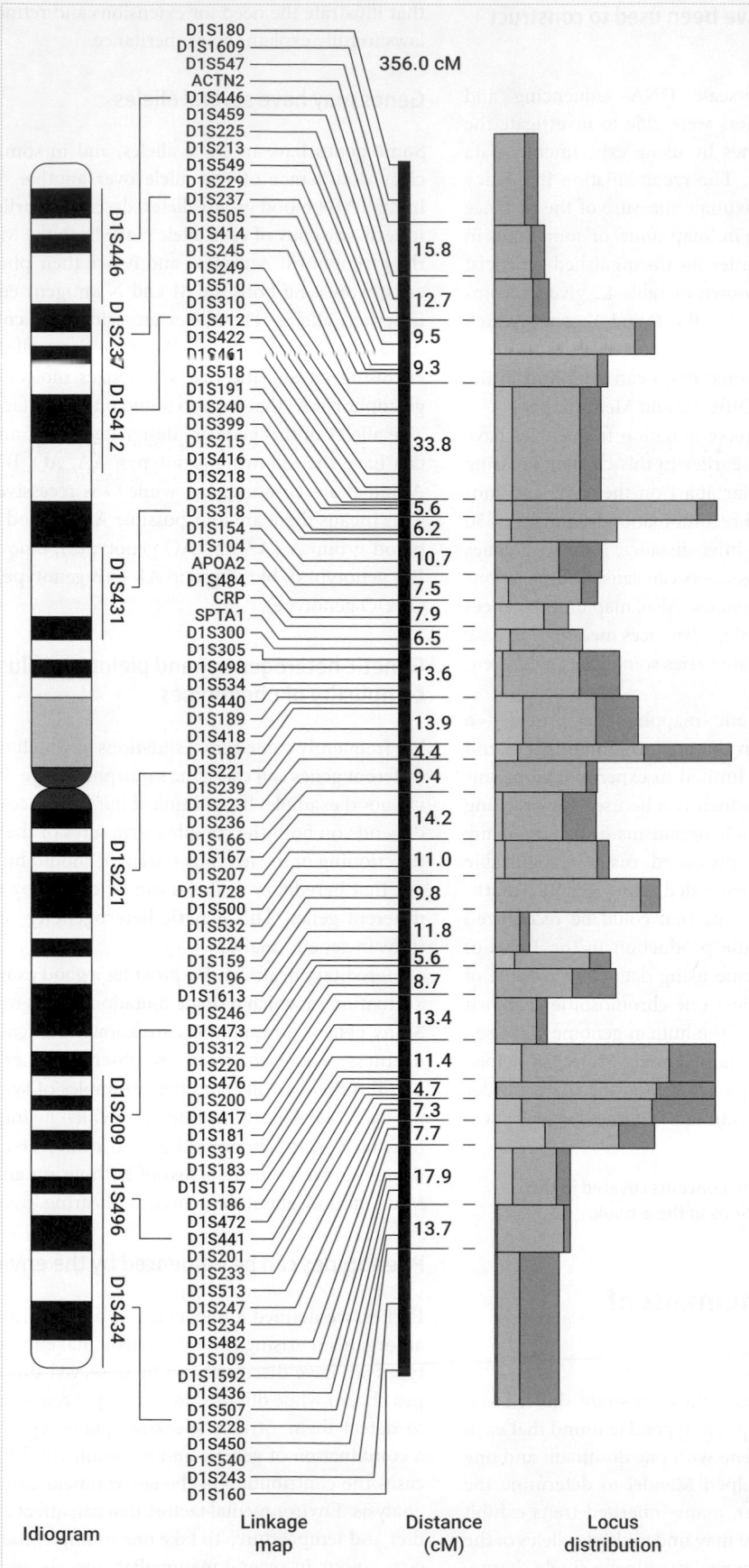

Figure 4.14 Linkage map of human chromosome 1, correlated with the chromosome banding pattern. Most of the map shows the positions of markers based on sequence variation, while genes giving a known phenotype are shown in green. When this map was produced only a small proportion of the markers could be placed on the banding pattern diagram (ideogram).

Source: B. R. Jasney et al. (1994) The Genome Maps 1994. *Science, 265,* 2055–2070. https://science.sciencemag.org/content/265/5181/2055/tab-pdf.

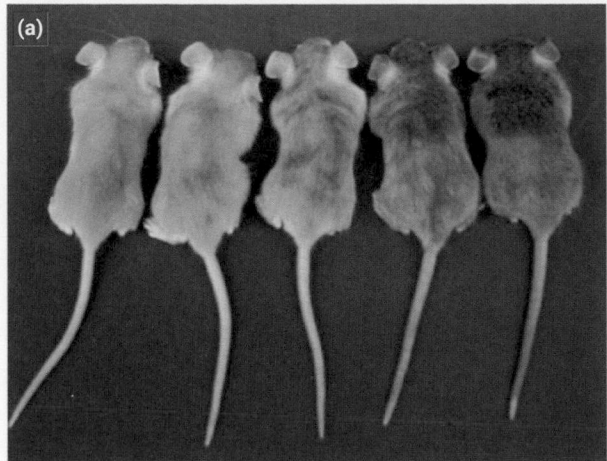

Figure 4.15 Environmental effects on phenotype. (a) The five mice shown all have the same agouti genotype. However, their coat colours—the phenotype they display—differ due to different levels of expression of the agouti A^{vy} allele, which is influenced by their mothers' diet. (b) The Himalayan rabbit breed carries a mutation in a gene affecting coat colour, which leads to temperature sensitive pigmentation. Dark pigment is made in cooler tissues such as the ears and tail, but the mutation makes the required enzyme inactive at warmer temperatures in more central body parts.

Source: (a) Dolinoy, D. C., Huang, D., & Jirtle, R. L. (2007) Maternal nutrient supplementation counteracts bisphenol A-induced DNA hypomethylation in early development. *Proc Natl Acad Sci USA*, 104(32), 13056–61. https://doi.org/10.1073/pnas.0703739104. Copyright (2007) National Academy of Sciences, USA. (b) Grant Heilman Photography/Alamy Stock Photo.

in mice, including one that gives a range of coat colours from yellow to brown, as depicted in Figure 4.15a. It has been shown that differences in the diet of mother mice during gestation affect gene expression of the mutant allele in their offspring, with the highest levels of expression giving the fully yellow coat colour.

Another intriguing example is that of Siamese cats and Himalayan rabbits, which carry a mutation in a gene that determines their coat colour. The mutated gene encodes an enzyme, tyrosinase, which is required for pigment production. The mutation makes the enzyme temperature sensitive, so it is mostly inactive at the animal's body temperature. At the slightly cooler extremities the enzyme functions, producing darker fur (see Figure 4.15b). Himalayan rabbits raised at low temperatures have enlarged areas of dark fur and may even become fully pigmented over their whole body.

Many heritable traits show polygenic or multifactorial inheritance

While the traits we have looked at so far are each determined by a single gene, many heritable traits are affected by multiple genes and some are also influenced by environmental factors. Human height is a good example. We observe that humans cannot simply be divided into two classes: 'tall' or 'short'. Nevertheless, we have evidence that height is at least partially inherited, since tall parents tend to have tall children, and we see trends in height according to people's ethnicity or geographic origin. Height variation is continuous, as shown in Figure 4.16, and this is typical of traits that are **polygenic** (influenced by many genes). Analysis of a large amount of data on height and genetic variation suggests that several hundred genes are involved.

▶ If you need help with reading these graphs, read through Quantitative Toolkit 6 and in particular the section on theoretical frequency distributions.

Although height is mainly genetically determined, we know that it is also influenced by environmental factors, such as nutrition of the mother during pregnancy and of the individual during childhood and adolescence. As shown in Figure 4.16b, average human heights increased during the 20th century due to better nutrition and a reduction in disease. A trait such as height that is determined by many genes and the environment is termed a **multifactorial** trait.

 Check your understanding of the concepts covered in this section by answering the questions in the e-book.

4.5 Genetic disease

One impetus for studying human inheritance is the incidence of diseases that clearly run in families. Many of these, such as heart disease and diabetes, are multifactorial: the probability of any individual developing the disease depends on a combination of several genes and their environment. However, our focus here will be on disorders in which a mutation in a single gene causes the disease. Studying the patterns of inheritance of single-gene disorders can help to unravel their causes and enable patients and their families to receive appropriate advice about their medical care and reproductive choices. Modern techniques have allowed us to identify the genes associated with almost all known single-gene disorders, leading to further understanding of the diseases, more rapid and accurate diagnosis, and sometimes the development of new therapies.

Single-gene disorders are classified by their mode of inheritance

Single-gene disorders are classified according to whether the gene is carried on an autosome or a sex chromosome, and whether inheritance of the disease is dominant or recessive. It is useful to remind yourself at this point that alleles of genes are described as dominant or recessive based on how they affect the phenotype. Genes on autosomal chromosomes and on the X chromosome in

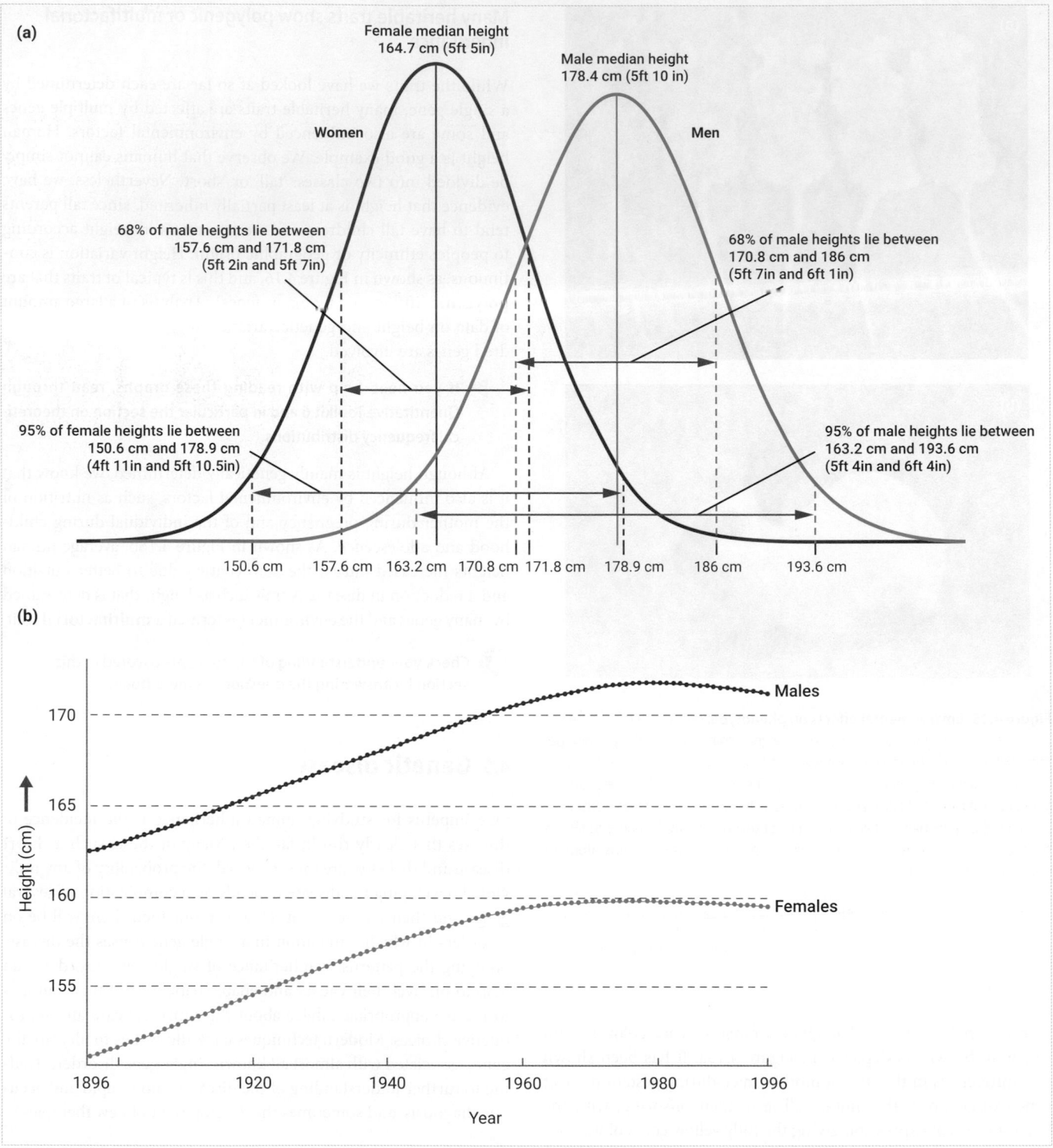

Figure 4.16 Height distributions. (a) The distribution of adult heights for men and women based on large cohort studies across 20 countries in North America, Europe, East Asia, and Australia. Shown is the sample-weighted distribution across all cohorts born between 1980 and 1994 (so reaching the age of 18 between 2008 and 2012). As height is influenced by many genes, it is a continuous phenotype, showing that both males and females show a normal distribution. (b) Average adult height by year of birth, world, 1896–1996.

females are inherited in pairs, one from each parent. A dominant allele manifests itself in the phenotype whether it is inherited on both members of a homologous chromosome pair or only on one member. A recessive allele, however, is masked by a dominant one. This means that a recessive allele only manifests itself in the phenotype if it is inherited from both parents, or if the individual inherits only one copy of the gene.

PAUSE AND THINK

Under what circumstances would someone normally inherit only one copy of a gene?

Answer: Males inherit only one copy of genes on the X chromosome (X-linked genes) as they have only one X chromosome. The Y chromosome carries very few genes.

We will now look at some examples of single-gene disorders that illustrate the different modes of inheritance.

Autosomal recessive inheritance

Oculocutaneous albinism type 1 (OCA1) is an example of a condition that is inherited in autosomal recessive fashion. The gene involved is on chromosome 11 and encodes an enzyme, tyrosinase, that is required to produce the pigment melanin. Individuals who have two mutated copies of the gene (i.e. are homozygous for the recessive alleles) lack pigment in their skin, hair, and eyes (as depicted in Figure 4.17). By contrast, heterozygous individuals who have one normal and one mutated copy of the gene make enough of the necessary enzyme to have normal pigmentation. They are unaffected by the condition but have the potential to pass it on to their children, so are known as heterozygous **carriers**.

The inheritance pattern of an autosomal recessive condition such as OCA1 is shown in Figure 4.18.

Autosomal dominant inheritance

Another inherited condition caused by a mutation in a single autosomal gene is Huntington disease (HD), a degenerative condition that affects movement, emotion, and cognition. Unlike OCA1, the inheritance of HD is dominant. The gene involved is on chromosome 4 and encodes a protein called huntingtin. The disease-causing mutation creates a version of the huntingtin protein that includes additional amino acids. The normal function of huntingtin is not fully understood but it is important in the nerve cells in the brain. It appears that the mutant protein acquires some abnormal functions that cause the disease to manifest even when the individual has only inherited a single mutated allele and their other, recessive, allele produces normal protein.

The inheritance pattern of an autosomal dominant condition such as HD is shown in Figure 4.19.

Figure 4.17 Albinism. Kelly Gallagher is an alpine ski racer from Northern Ireland with oculocutaneous albinism, which causes her to be visually impaired. Skiing with Charlotte Evans as a guide she won a gold medal at the Winter Paralympic Games in Sochi in 2014. Kelly is also an advocate for the rights of people with albinism worldwide.
Source: Ian Walton/Staff, Getty Images.

PAUSE AND THINK

For some autosomal dominant conditions, it is relatively common to find a single affected child in a family where neither of the parents has the condition. Can you explain how this might occur?

Answer: This can occur where a new mutation occurs during formation of the egg or sperm cell that gave rise to the affected child.

Sex-linked inheritance

As there are so few genes on the Y chromosome we can generally assume that a disorder referred to as 'sex-linked' is associated with a gene on the X chromosome. Therefore, the terms 'sex-linked' and 'X-linked' inheritance are often used interchangeably. X-linkage gives rise to unusual inheritance patterns because males only have one copy of the gene. We will only consider an example of X-linked recessive inheritance because X-linked dominant conditions are rare.

Figure 4.18 Autosomal recessive inheritance. Here, both parents are unaffected by the condition but are heterozygous carriers. Their children will be affected if they inherit the mutated, recessive allele from both parents. By probability, ¼ (25%) of the couple's children are likely to be affected, while ½ (50%) of their children are likely to be heterozygous carriers, and ¼ (25%) are likely to inherit two normal copies of the gene and be neither affected nor carriers. You should note that these numbers are based on probabilities. The actual proportions of affected and unaffected children observed in a single family may well differ from these predictions, particularly when families are small.

Source: Courtesy of the US National Library of Medicine.

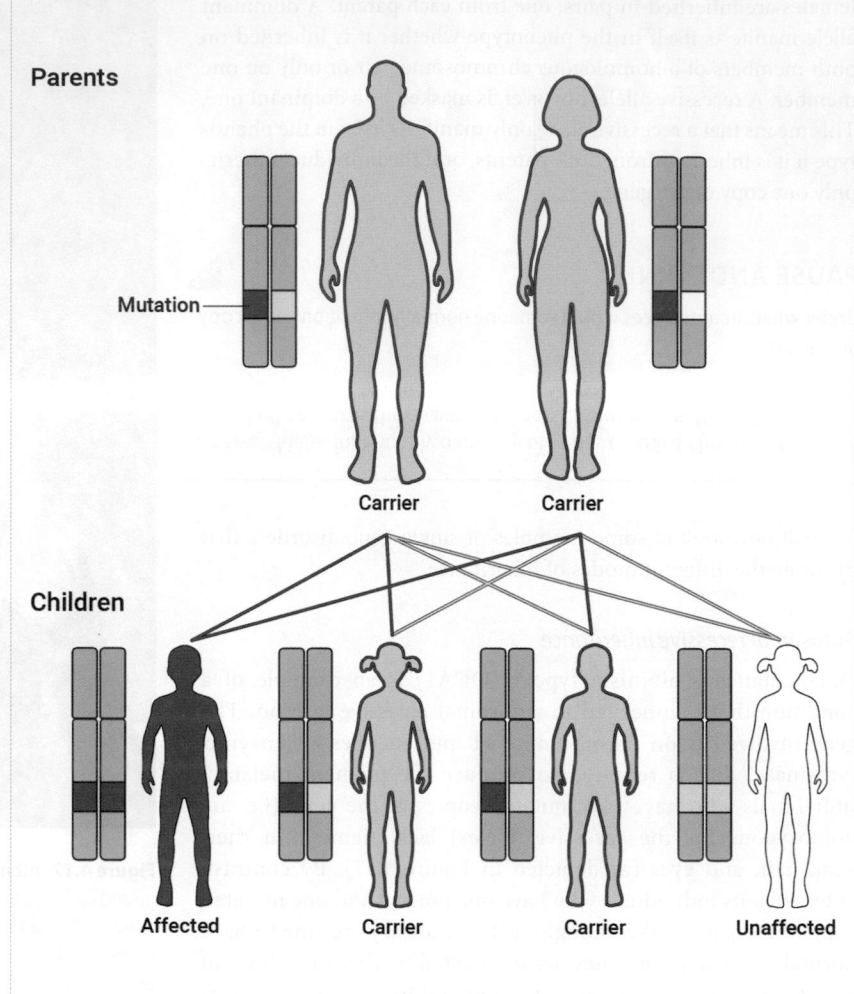

Haemophilia A is one of the best-known X-linked disorders. The causative gene encodes a protein called factor VIII, which is required for blood clotting. People with haemophilia suffer from prolonged bleeding following injury or surgery, and they may have complications arising from internal bleeding. The condition can be treated by regular injections of the missing clotting factor. Haemophilia A is a recessive disorder: heterozygous 'carrier' females typically have reduced levels of factor VIII and their blood may be slower to clot than normal, but usually there is no clinical impact.

Figure 4.20 shows inheritance patterns for X-linked recessive disorders. A key point is that an affected father cannot pass the condition to his son, as males must inherit the Y chromosome from their fathers. All daughters of an affected father will inherit a mutated copy of the X chromosome from their father and will therefore be heterozygous carriers.

X-linked recessive disorders are observed more frequently in males than in females. This is because men who inherit a single X chromosome carrying the mutation will have the disorder, but females will only be affected if they inherit a mutated X chromosome from their father and one from their mother. The chance of a female inheriting an X chromosome carrying the mutation

from both parents depends on how common the mutated gene is in the population. For instance, one estimate for a relatively common X-linked recessive disorder, deuteranopia (also known as red–green colour blindness), is that around one in 12 males and one in 200 females in populations of northern European origin have the condition. Some other X-linked disorders are rare and are only very rarely seen in females. This is especially the case for severe diseases for which affected males are unlikely to have children.

Pedigrees are used to study human genetics

When studying human genetics we obviously cannot carry out large breeding experiments of the type used by Mendel. However, we often construct family trees or pedigrees to show the inheritance of a disease or trait within a family. Analysing these pedigrees can give useful information about the likely mode of inheritance of a condition, and genetic counsellors can use their knowledge to give advice to individuals and families, for instance on the probability of a disease being passed on to their children. Pedigree analysis can also be combined with modern molecular genetic

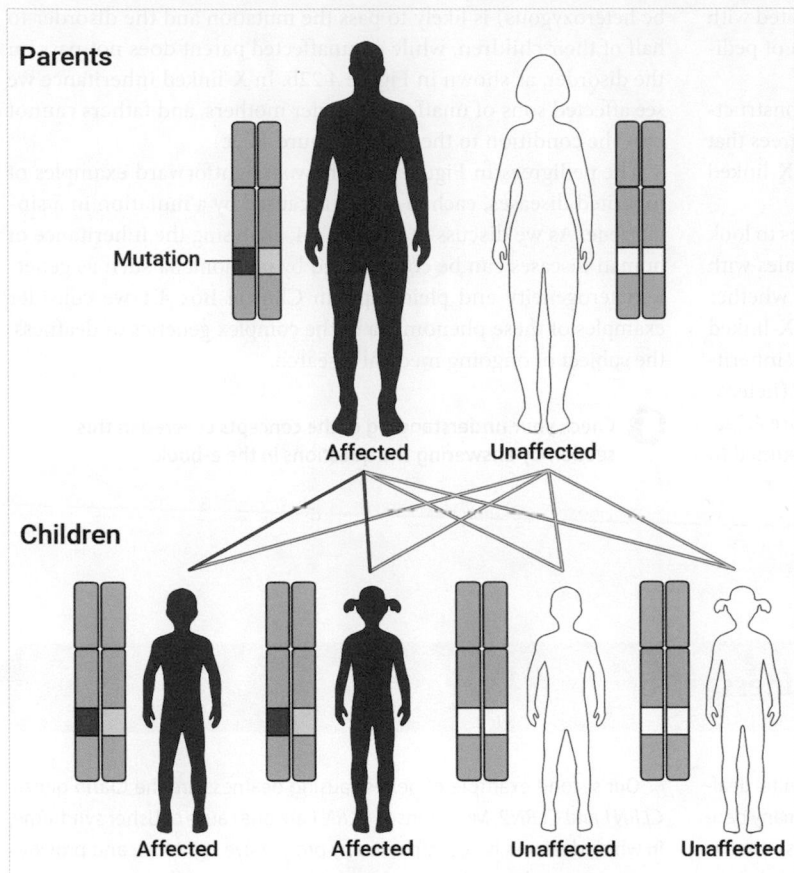

Parents

Mutation

Affected   Unaffected

Children

Affected   Affected   Unaffected   Unaffected

Figure 4.19 Autosomal dominant inheritance. Here the father has one mutated copy of the gene. As the mutated allele is dominant, he is affected by the condition. The mother has two normal copies of the gene and is unaffected. By probability, ½ (50%) of this couple's children are likely to inherit the mutated allele from their father and be affected. Note that in autosomal dominant inheritance we do not see unaffected heterozygous carriers, as a single mutated allele causes the disease. As previously, you should note that the predicted numbers are based on probabilities. The actual proportions of affected and unaffected children observed in a single family may well differ from these predictions, particularly when families are small.

Source: Courtesy of the US National Library of Medicine.

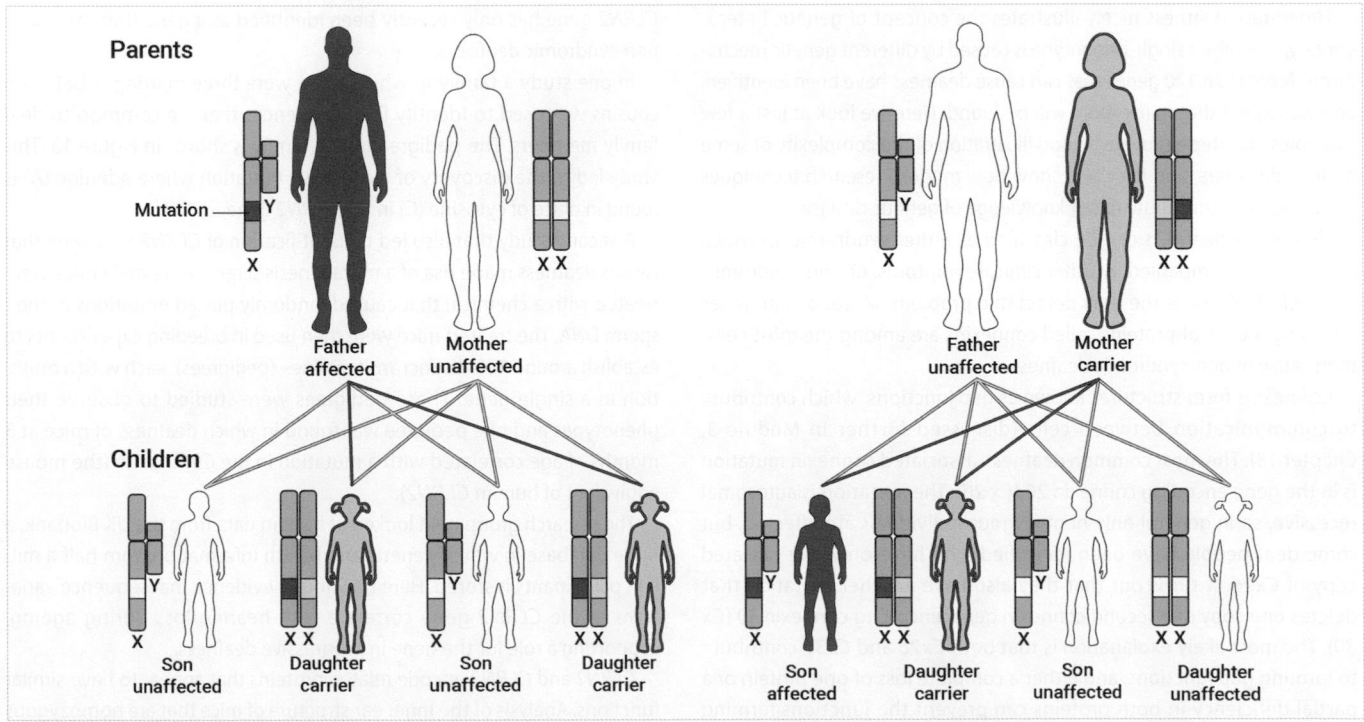

Figure 4.20 X-linked recessive inheritance. Two possible scenarios are shown. Where the father is affected and the mother is unaffected, all sons will be unaffected but all daughters will be heterozygous carriers of the condition. Where the father is unaffected and the mother is a heterozygous carrier, by probability ½ (50%) of the couple's sons are likely to inherit the mutated allele from their mother and be affected, and ½ (50%) of the couple's daughters are likely to be heterozygous carriers.
Source: Courtesy of the US National Library of Medicine.

techniques to identify genes and analyse mutations associated with genetic disease. We explore an example of this application of pedigree analysis in Clinical Box 4.1.

Figure 4.21 shows the most common symbols used in constructing human pedigrees, while Figure 4.22 shows three pedigrees that illustrate autosomal recessive, autosomal dominant, and X-linked recessive inheritance.

When analysing human pedigrees, there are two features to look out for: whether the disorder affects both males and females with similar frequencies, indicating autosomal inheritance; or whether the disorder is seen only or mainly in males, indicating X-linked inheritance. In pedigrees showing autosomal recessive inheritance, it is common to see two unaffected individuals (hetero zygous carriers) having affected children, as shown in Figure 4.22a. In autosomal dominant inheritance an affected parent (assumed to

be heterozygous) is likely to pass the mutation and the disorder to half of their children, while an unaffected parent does not pass on the disorder, as shown in Figure 4.22b. In X-linked inheritance we see affected sons of unaffected carrier mothers, and fathers cannot pass the condition to their sons (Figure 4.22c).

The pedigrees in Figure 4.22 show straightforward examples of inherited diseases, each of which is caused by a mutation in a single gene. As we discuss in Section 4.4, analysing the inheritance of human diseases can be complicated by phenomena such as genetic heterogeneity and pleiotropy. In Clinical Box 4.1 we consider examples of these phenomena in the complex genetics of deafness, the subject of ongoing medical research.

 Check your understanding of the concepts covered in this section by answering the questions in the e-book.

CLINICAL BOX 4.1   The genetics of deafness

It has been estimated that around 50 per cent of cases of congenital deafness (deafness that is present at birth) are genetic. In addition, many people progressively lose their hearing as they age, a process that is influenced by their genes. The genetics of deafness is therefore an important research topic.

Hereditary deafness nicely illustrates the concept of genetic heterogeneity, whereby a single phenotype is caused by different genetic mechanisms. More than 120 genes that can cause deafness have been identified, and we expect that many more will be found. Here we look at just a few examples, but these provide a good illustration of the complexity of some genetic disorders, and they also show how modern research techniques can powerfully contribute to our knowledge of genetic disease.

Hereditary deafness can be classified as either syndromic, in which deafness is accompanied by other clinical symptoms, or non-syndromic, in which deafness is the only detectable problem. Mutations in genes encoding a class of proteins called connexins are among the most common cause of non-syndromic deafness.

Connexins form structures known as gap junctions, which contribute to communication between cells (discussed further in Module 3, Chapter 18). The most common deafness-associated connexin mutation is in the gene encoding connexin 26 (Cx 26). The mutation is autosomal recessive, so in general only homozygous individuals are affected, but some deaf people have been identified who have only one mutated copy of Cx26. It turns out that they also have another mutation that deletes one copy of a second connexin gene, encoding connexin 30 (Cx 30). The most likely explanation is that both Cx26 and Cx30 contribute to forming gap junctions, and either a complete loss of one protein or a partial deficiency in both proteins can prevent the junctions forming correctly.

Our second example of genes causing deafness are the Clarin genes, CLRN1 and CLRN2. Mutations in CLRN1 are one cause of Usher syndrome, in which deafness is accompanied by progressive sight loss, and provides a good example of pleiotropy (a single gene that affects two or more phenotypes, in this case hearing and sight). The related but distinct CLRN2 gene has only recently been identified as a gene that can cause non-syndromic deafness.

In one study a family in which there were three marriages between cousins was used to identify DNA sequences that are common to deaf family members. The pedigree of this family is shown in Figure 1a. This study led to the discovery of a causative mutation where adenine (A) is found in place of cytosine (C) in the CLRN2 gene.

A second study that also led to identification of CLRN2 as a gene that causes deafness made use of a mutagenesis screen. Here male mice were treated with a chemical that caused randomly placed mutations in their sperm DNA. The treated mice were then used in breeding experiments to establish around 100 distinct 'mutant lines' (pedigrees), each with a mutation in a single gene. These pedigrees were studied to observe their phenotypes and one pedigree was found in which deafness of mice at 3 months of age correlated with a mutation in the Clrn2 gene (the mouse equivalent of human CLRN2).

The research group next looked at human data from the UK Biobank, a huge database in which genetic and health information from half a million participants is stored. Here they found evidence that sequence variations in the CLRN2 gene correlate with hearing loss during ageing, supporting a role for the gene in progressive deafness.

CLRN1 and CLRN2 encode related proteins that appear to have similar functions. Analysis of the inner ear structure of mice that are homozygous for a mutation in Clrn2 showed gradual loss of stereocilia from the hair

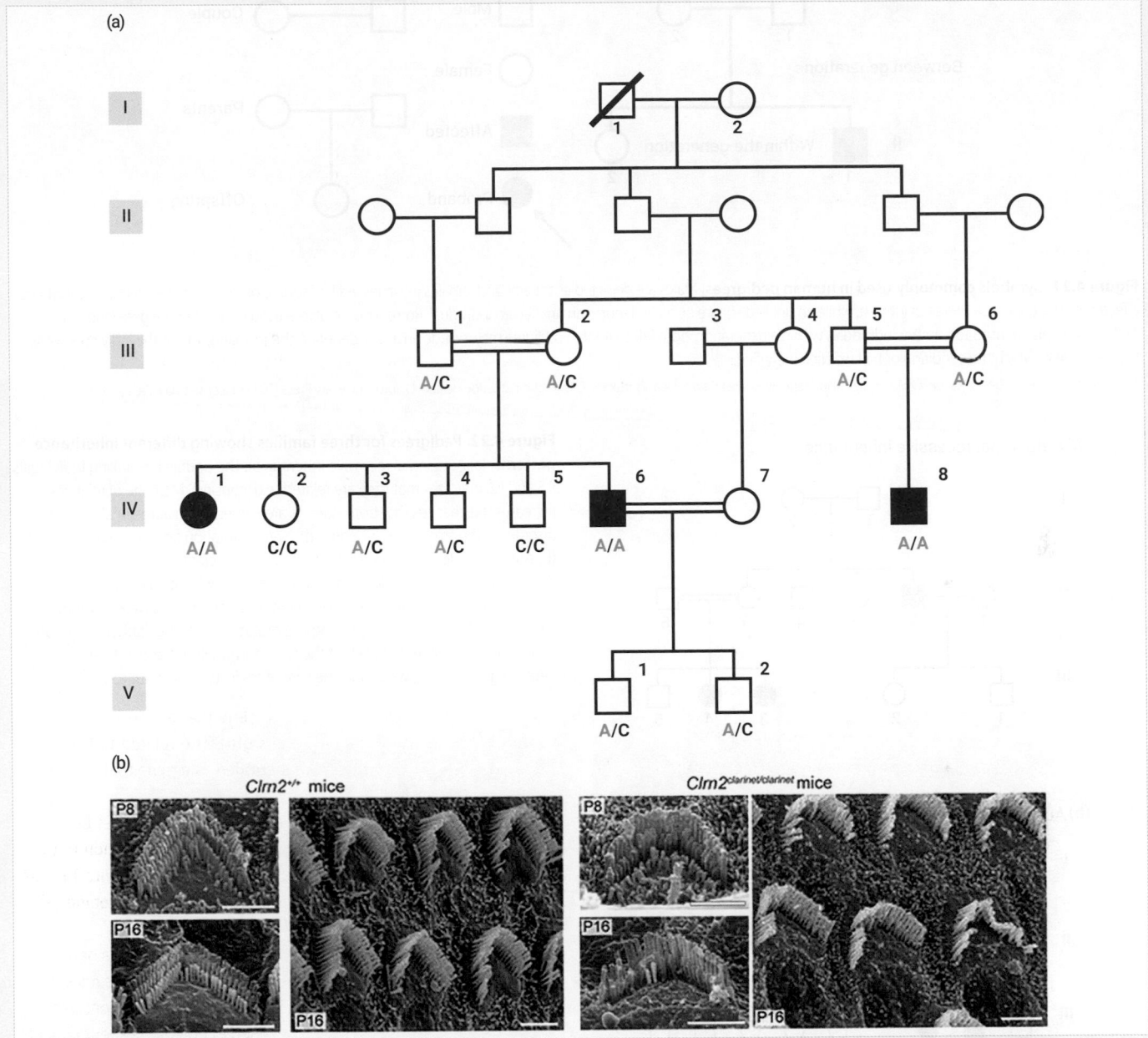

Figure 1 Autosomal recessively inherited deafness. (a) Pedigree of a family showing autosomal recessively inherited deafness. Individual I.1 is deceased. Double horizontal lines indicate marriage between cousins. 'A' or 'C' denotes the base found at a particular position within the *Clarin2* gene. 'A' is a mutation that causes deafness. Consanguineous marriages (between cousins) in the pedigree have caused all individuals with the mutation to inherit it from a single common ancestor. (b) Stereocilia on hair cells of the mouse inner ear, viewed with an electron microscope. P8: 8 days after birth, P16: 16 days after birth. On the left, Clrn2⁺/⁺ denotes cells from mice with no *Clarin-2* mutation. On the right, Clrn2^clarinet/clarinet denotes cells from mice that are homozygous for the *Clarin-2* mutation. Mice that are homozygous for the *Clarin-2* mutation have fewer stereocilia, with increased loss of stereocilia at P16. Scale bars: 2 μm.

Source: (a) Vona, B., Mazaheri, N., Lin, S. J. et al. A biallelic variant in CLRN2 causes non-syndromic hearing loss in humans. Hum Genet 140, 915–931 (2021). (b) Dunbar, L. A. et al. Clarin-2 is essential for hearing by maintaining stereocilia integrity and function. EMBO Mol Med 11(9): e10288 (2019). © 2019 the authors. Published under the terms of the CC BY 4.0 licence.

cells (Figure 1b), suggesting that the protein has a role in maintaining these structures. Further research is needed to fully understand the functions of *CLRN1* and *CLRN2* and to explain how *CLRN1* mutations affect vision as well as hearing.

Further reading

Nance, W. E. (2003) The genetics of deafness. *Ment. Retard. Dev. Disabil. Res. Rev.* **9**: 109–119.

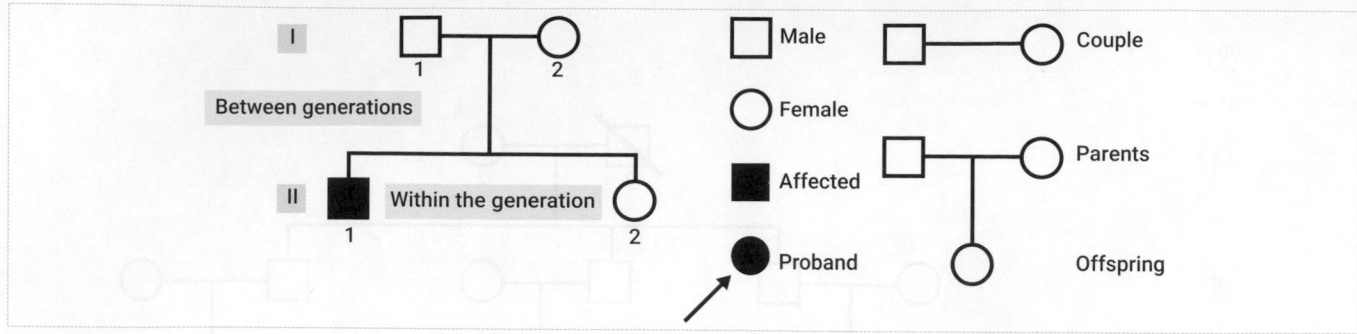

Figure 4.21 Symbols commonly used in human pedigrees. Males are depicted as squares and females as circles. A filled square or circle denotes an individual who is affected by the condition or trait of interest, while an unfilled square or circle denotes an unaffected individual. Roman numerals are used to number the generations, and Arabic numerals are used to number individuals within a generation, from left to right. A pedigree may include an arrow denoting the 'proband', who is the individual within the pedigree in which the condition or trait was first identified.

Source: Genetics: Genes, Genomes and Evolution by Philip Meneely, Rachel Dawes Hoang, Iruka N. Okeke, and Katherine Heston. Oxford University Press (2017). Licensed via PLSclear.

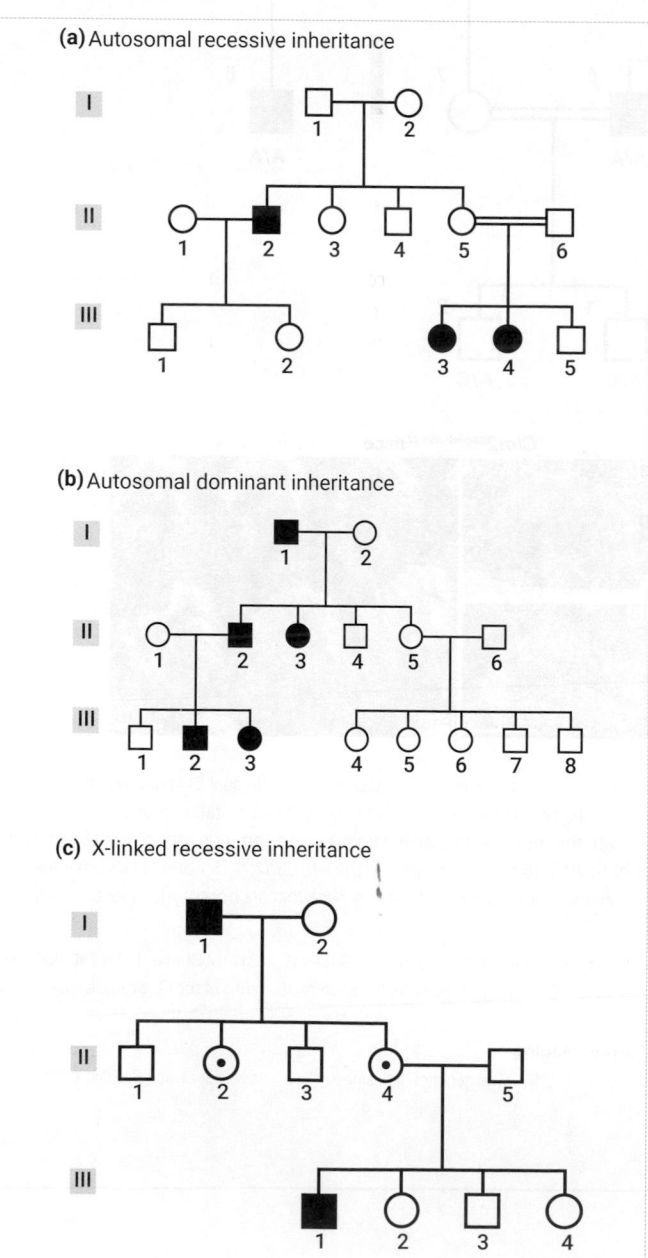

Figure 4.22 Pedigrees for three families showing different inheritance patterns. (a) Autosomal recessive inheritance. The double line linking individuals II.5 and II.6 indicates that they are related (e.g. cousins). Marriage of relatives increases the chances that both parents are carriers of an autosomal recessive disorder as they may have inherited the same mutation from a common ancestor. (b) Autosomal dominant inheritance. (c) Three generations of a family illustrating X-linked inheritance. Individuals II.2 and II.4 are shown with a dot in the middle of the symbol, indicating that they are 'obligate carriers'; that is, we know they have inherited an X chromosome carrying the mutation from their father. Although there is a 50% chance that each of the two individuals III.2 and III.4 are heterozygous carriers, we cannot be sure of their status without further testing.

PAUSE AND THINK

In the autosomal recessive pedigree shown in Figure 4.22a, calculate the probability that individual II.3 is a heterozygous carrier of the condition. You can assume the condition has an early onset (i.e. we know that II.3 will not develop it later in life).

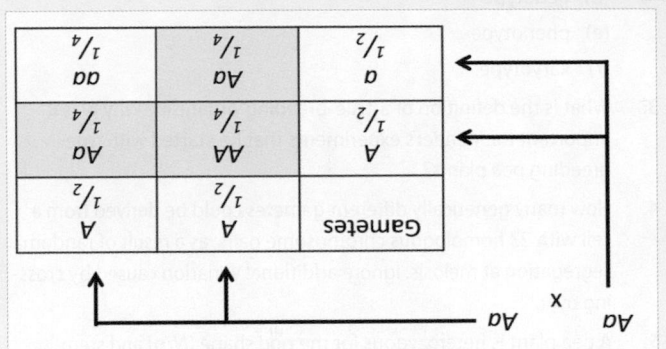

Answer: $2/3$ or 0.67%. Using a Punnett square, we can depict the first two generations of this pedigree as:

The overall probability of a child of two heterozygous carrier parents being a carrier is $1/4 + 1/4$ (= $1/2$), but in this case we know that individual II.3 is not homozygous recessive. Her possible genotypes are therefore confined to the shaded cells in the Punnett square, giving a 2 out of 3 chance that she is heterozygous.

SUMMARY OF KEY CONCEPTS

- Eukaryotic cells contain homologous pairs of chromosomes. In humans there are 22 pairs of autosomes and two sex chromosomes (XX in females and XY in males).

- Cells therefore carry two copies of each autosomal gene, which may be present as different alleles.

- Alleles of a gene may be described as dominant or recessive, depending on how they are expressed in the phenotype (measurable characteristics) of the organism.

- Gregor Mendel's experiments established the basic laws of inheritance, which state that genes are inherited in pairs, one copy from each parent, and that genes for different traits segregate independently, provided they are not linked on the same chromosome.

- Meiosis is a special form of cell division that occurs during formation of gametes, in which haploid daughter cells each receive only one member of each homologous chromosome pair.

- Random segregation of homologous chromosomes at meiosis and recombination between linked genes by crossing over increase genetic diversity.

- Many characteristics are not simply determined by a single gene with one dominant and one recessive allele. Genes with multiple alleles, co-dominance, genetic heterogeneity, and pleiotropy are just some examples that give more complex inheritance patterns.

- Phenotypes can be affected by both genes and the environment. Characteristics such as human height that show continuous variation are polygenic and multifactorial (affected by many genes and a combination of genes and the environment).

- Human genetic diseases that are determined by a single gene show characteristic patterns of inheritance, which may be autosomal dominant, autosomal recessive, or X-linked recessive. These can be studied using pedigrees.

Use the flashcards in the e-book to test your recall of key terms introduced in this chapter.

QUESTIONS

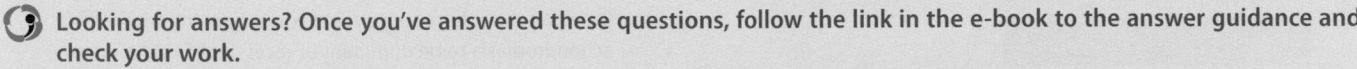

Looking for answers? Once you've answered these questions, follow the link in the e-book to the answer guidance and check your work.

Concepts and definitions

1. A female fruit fly of the species *Drosophila melanogaster* has a diploid chromosome number of eight, including two X chromosomes. How many pairs of autosomes does this species have?

2. Give definitions of the following terms used in genetics:
   (a) allele
   (b) homozygote
   (c) heterozygote

(d) genotype

(e) phenotype

(f) karyotype.

3. What is the definition of a 'true-breeding' organism? Why was it important for Mendel's experiments that he started with true-breeding pea plants?

4. How many genetically different gametes could be derived from a cell with 23 homologous chromosome pairs, as a result of random segregation at meiosis. Ignore additional variation caused by crossing over.

5. A pea plant is heterozygous for the pod shape (N/n) and stem length (l/t) genes, which are linked on the same chromosome. One chromosome carries the N and the T alleles of the genes, while its homologue carries the n and the t alleles. What allele combinations can be found in the gametes of this plant? Which are parental and which are recombinant allele combinations and what determines the frequency with which recombinant gametes arise?

Apply the concepts

6. How many copies of an X-linked gene would you find in the following human cells from a normal female?

(a) A somatic cell.

(b) A germ cell immediately prior to the first meiotic division.

(c) A cell that has completed the first meiotic division.

(d) A mature gamete.

7. In the fruit fly, *Drosophila melanogaster*, you are studying a body colour gene that has two alleles, grey and ebony, and a wing length gene that has two alleles, long and vestigial. You carry out a cross between true-breeding flies with the following phenotypes.

grey, vestigial X ebony, long

All the F_1 offspring have the following phenotype: grey, long

(a) For each of the two genes, which allele is dominant?

(b) Using suitable symbols and a Punnett square, work out the genotypes and phenotypes of the F_2 generation from this cross. Assume the two genes are not linked.

8. A cross was carried out between a pea plant that is heterozygous for two linked, genes, D and H, and a plant that is homozygous recessive for both these genes. The number of offspring of different genotypes are shown below. Use these data to calculate the distance between the D and H genes.

Cross: Dd Hh × dd hh
Offspring

Genotype	Number
Dd Hh	94
dd Hh	4
Dd hh	6
dd hh	96

9. Coat colour in shorthorn cattle is determined by a gene which has two alleles, red and white. When a true-breeding red female is crossed with a true-breeding white male, all the F_1 offspring are an intermediate colour, roan. Close inspection of roan cattle shows that the intermediate coat colour is caused by them having a mixture of red hairs and white hairs.

(a) What type of inheritance is shown here?

(b) What would be the coat colour phenotypes and genotypes of the F_2 cattle, in what proportions?

10. David and Sangita marry and have two sons followed by three daughters. Their second son and their second and third daughters have a condition that causes joint pain in their hips and knees. Neither David nor Sangita has this condition.

You suspect a single-gene disorder. Construct a pedigree for this family and explain what type of inheritance you suspect. Give your reasons.

Beyond the concepts

11. As well as a dihybrid cross, Mendel carried out a trihybrid cross involving three unlinked genes. The genotype of the F_1 offspring was *AaBbCc*. What are the possible genotypes of the gametes the F_1 offspring could produce? How many different phenotypes would be seen in the F_2 offspring?

12. In the worm species, *C. elegans*, crossing over at meiosis is highly regulated so that one (and only one) crossover occurs for each homologous chromosome pair. How does this differ from the situation in other organisms such as garden peas? How does the inheritance of two traits that are determined by genes that are at each end of the longest chromosome differ between *C. elegans* and the garden pea?

13. Mendel investigated flower colour in garden pea plants, looking at a single gene in which the dominant allele gives purple flowers and the recessive allele gives white flowers. Later, geneticists investigated flower colour genetics in other plants, with different results. In the four o'clock plant (*Mirabilis jalapa*) crossing true-breeding white-flowered plants with true-breeding pink-flowered plants gives pale pink flowers in the F_1 generation. The F_2 generation gives 25% pink, 50% pale pink, and 25% white-flowered plants. Suggest an explanation for these results in terms of genes and alleles, and in terms of the biochemical reactions leading to pigment formation.

14. Achondroplasia is a form of short-limbed dwarfism. It is a single gene disorder caused by a mutation in the gene that encodes a receptor protein. During normal development the receptor is activated when the long bones of the limbs have grown, and it prevents them from growing further. Mutations that cause achondroplasia make the receptor protein active at all times. Based on this information would you expect inheritance of achondroplasia to be dominant or recessive? Explain your answer.

15. A deaf man who is homozygous for a recessively inherited deafness gene marries a deaf woman who is also homozygous for a recessively inherited deafness gene. The couple have six children, all of whom can hear. Can you explain this family pedigree?

FURTHER READING

Offner, S. (2011) Mendel's peas and the nature of the gene: genes code for proteins and proteins determine phenotype. *Am. Biol. Teach.* **73**: 382–7.

An interesting update on Mendel's experiments, explaining how some of the genes he studied lead to the phenotypes observed.

Gleason, K. (2017) The linear arrangement of six sex-linked factors in *Drosophila*, as shown by their mode of association (1913), by Alfred Henry Sturtevant. *Embryo Project Encyclopedia.* ISSN: 1940-5030. Available at: https://embryo.asu.edu/pages/linear-arrangement-six-sex-linked-factors-drosophila-shown-their-mode-association-1913-alfred

An account of a classic study in which Alfred Sturtevant showed that the arrangement of genes on a chromosome could be established from recombination frequencies. This article also highlights some modern applications of gene mapping.

Jackson, M., Marks, L., May, G. H. W., & Wilson, J. B. (2018) The genetic basis of disease. *Essays Biochem.* **62**: 643–723.

A wide-ranging review, aimed at undergraduate students, that explores and takes further many of the topics covered in this chapter, as well as covering some more advanced topics. A useful resource.

Yamamoto, F., Clausen, H., White, T., Marken, J., & Hakomori S. (1990) Molecular genetic basis of the histo-blood group ABO system. *Nature* **345**: 229–33.

Explains the molecular basis for the three alleles, two co-dominant and one recessive.

McEvoy, B.P. & Visscher, P.M. (2009) Genetics of human height. *Econ. Hum. Biol.* **7**: 294–306.

An accessible explanation of multifactorial inheritance and the methods used to identify the many genes involved in determining human height.

Steel, K. (2011) Mouse genetics for studying mechanisms of deafness and more: an interview with Karen Steel. *Dis. Models Mechan.* **4**: 716–18.

An interview with Karen Steel explaining her motivation for studying the mechanisms of deafness and introducing a project that aims to mutate every mouse gene and study the resulting phenotypes.

Reading the Genome

Gene Expression and Protein Synthesis

Chapter contents

Introduction	252
5.1 Gene expression: overview of the process	253
5.2 The genetic code	254
5.3 RNA synthesis: transcription	255
5.4 Protein synthesis: translation	264
5.5 Control of gene expression	271

Watch the key concepts video in the e-book to prepare yourself for studying this chapter.

LEARNING OBJECTIVES

By the end of this chapter you should be able to:

- Describe in outline the two main phases of gene expression and protein synthesis: transcription and translation.
- Describe the structures of three main classes of RNA (mRNA, tRNA, and rRNA) and their roles in protein synthesis.
- Explain how the genetic code specifies the amino acid sequences of proteins.
- Describe the enzymes and processes involved in transcription, post-transcriptional modification of RNA, and translation. Compare transcription and translation in eukaryotes and prokaryotes.
- Discuss the regulation of transcription by DNA binding proteins in prokaryotes and eukaryotes.
- Outline how chromatin conformation, cell signalling, and DNA methylation contribute to gene regulation in eukaryotic cells.
- Discuss some mechanisms of post-transcriptional gene regulation.

Introduction

In Chapter 3 we described the DNA genome as the 'instruction manual' that enables an organism to carry out its cellular function. In this chapter we explain how cells utilize the information stored in their genome. The overall process is often referred to as gene expression. More specifically, in the first part of this chapter we describe the two main phases of gene expression: transcription, which leads to the synthesis of RNA molecules; and translation, where the information encoded in the RNA is used to synthesize proteins. Proteins are the ultimate products encoded by most genes, but, as explained in this chapter, some genes encode RNA molecules that are not subsequently translated into proteins, but themselves have specific cellular functions.

It is important that cells regulate which genes they express and how much of each gene product they make. This regulation enables them to respond appropriately to changes in their environment. Additionally, in multicellular organisms where cells have specialized functions, different cell types express different genes. For example, muscle cells must synthesize actin and myosin required for muscle contraction, while developing red blood cells need large amounts of globin proteins to make haemoglobin. In the second part of the chapter we explore how this differential gene expression is achieved through regulatory mechanisms.

5.1 Gene expression: overview of the process

Before looking in detail at the molecules, reactions, and enzymes involved in each step of gene expression it is worth gaining an overview of the process. As mentioned earlier, there are two main phases: transcription and translation. Transcription involves copying the nucleotide sequence of DNA to make an RNA molecule, often called a 'transcript'. In the case of protein-coding genes, the RNA molecule acts as an intermediate or 'messenger' that carries the information required to link amino acids together in the appropriate sequence to make a protein. Translation is the process whereby the nucleotide sequence of the messenger RNA (mRNA) is used to direct protein synthesis.

The order in which transcription and translation occur in protein synthesis is illustrated in Figure 5.1. If you have difficulty recalling which term refers to which process, it may help to remember that to *transcribe* a section of text is to copy it, while to *translate* is to convert the text into a different language. As DNA and RNA are both polymers of nucleotides and are closely related in structure, mRNA can be regarded as a *transcribed* copy of DNA in which the information content is preserved in the sequence of bases. Proteins, however, are polymers of amino acids, a completely different type of molecule, so utilizing the nucleotide sequence of RNA to make a protein requires *translation* of the information from the 'language' of nucleic acids to the 'language' of proteins.

Francis Crick, one of the discoverers of DNA structure, introduced the term 'the central dogma' to describe the principle that genetic information flows only in the direction shown in Figure 5.1, from DNA to RNA to protein. A dogma is a principle that is held to be universally true. Nowadays the universality of the central dogma requires some qualification, as we know that retroviruses, a class of virus that includes the human immunodeficiency virus (HIV), can copy their RNA genomes to make DNA in a process known as reverse transcription. This qualification of Crick's central dogma is of interest because of the medical importance of retroviruses and because reverse transcription has proved a useful tool for genetic engineering.

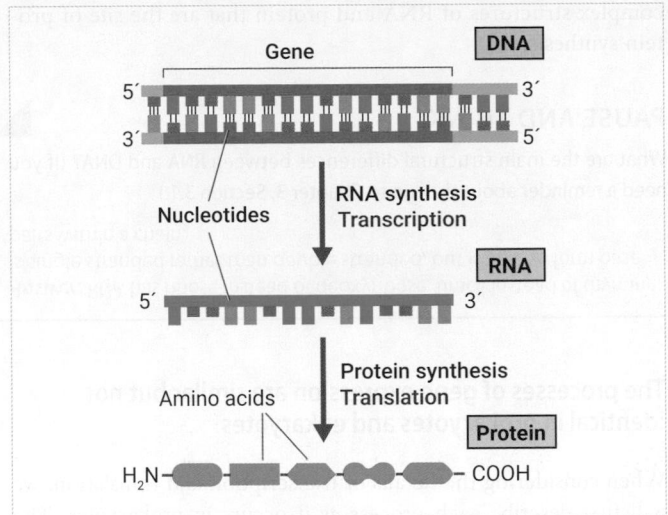

Figure 5.1 The order of events in gene expression. Genetic information stored as DNA is first transcribed into an intermediate 'messenger' RNA molecule containing a copy of its nucleotide sequence. The nucleotide sequence of the RNA is the translated into the amino acid sequence of a protein.

▶ We discuss genetic engineering in more detail in Chapter 8.

Gene expression involves three major classes of RNA

Most genes encode proteins, which are synthesized via expression of an intermediate RNA molecule called a messenger RNA (mRNA). However, some genes encode other types of RNA that are not destined to be translated into protein. As shown in Table 5.1, cellular RNA molecules can be classified according to their different functions. mRNA is one of the three major classes of RNA that pla;tkey roles in gene expression. The othe two are transfer RNA (tRNA) and ribosomal RNA (rRNA). These will be covered in more detail when translation is discussed later in the chapter, but for now it is enough to know that tRNAs transport amino acids and place them in the correct order as directed by the mRNA, while rRNAs are the main components of ribosomes,

Table 5.1 Classes of RNA that are found in cells; their characteristics and functions are described in more detail later in the chapter

Class of RNA	Function
messenger RNA (mRNA)	Codes for proteins
ribosomal RNA (rRNA)	Contributes to the structure and function of the ribosome: a molecular 'factory' for protein synthesis.
transfer RNA (tRNA)	Delivers amino acids for protein synthesis
Other non-coding RNAs	Multiple functions, including mRNA splicing and gene regulation

complex structures of RNA and protein that are the site of protein synthesis.

PAUSE AND THINK

What are the main structural differences between RNA and DNA? (If you need a reminder about this, go to Chapter 3, Section 3.1.)

Answer: RNA has ribose instead of deoxyribose, uracil instead of thymine, is single stranded rather than double stranded, but is able to form base pairs within a chain.

The processes of gene expression are similar but not identical in prokaryotes and eukaryotes

When considering the details of transcription and translation, we will first describe each process as it occurs in prokaryotes. The fundamentals of each process are the same in eukaryotes but there are also some differences that we will describe. Before going into detail, it is useful to consider how the structures of prokaryotic and eukaryotic cells affect the main sequence of events involved in gene expression. In prokaryotic cells, transcription and translation both take place in the cytosol. This means that as soon as a molecule of prokaryotic mRNA is made, tRNA and ribosomes assemble on it, and translation begins. In eukaryotes, transcription occurs in the

nucleus, and RNA molecules are then exported to the cytosol. As explained later, eukaryotic mRNA undergoes several modifications in the nucleus before it is exported and translated.

 Check your understanding of the concepts covered in this section by answering the questions in the e-book.

5.2 The genetic code

As soon as the structure of DNA was known, scientists were eager to work out how a code consisting of four bases could be used to encode the 20 different amino acids that make up proteins. Clearly, there could not be a one to one correspondence with one base encoding one amino acid. If two bases were read together this would give 16 (4×4) possible combinations—still not enough—but reading the base sequence in groups of three would give 64 ($4 \times 4 \times 4$) combinations, more than enough for 20 amino acids. This system of reading three bases at a time is the one that is used. Each group of three bases is called a **codon**, and the identification of which codons correspond to which amino acids constitutes the **genetic code**. The code is virtually universal in that all organisms use it, although minor variations are found in mitochondria and some protozoans.

The assignment of mRNA codons to amino acids, the genetic code, is shown in Table 5.2. Three codons have been reserved as

Table 5.2 The genetic code

5′ base	Middle base				3′ base
	U	C	A	G	
U	UUU Phe	UCU Ser	UAU Tyr	UGU Cys	U
	UUC Phe	UCC Ser	UAC Tyr	UGC Cys	C
	UUA Leu	UCA Ser	UAA Stop*	UGA Stop*	A
	UUG Leu	UCG Ser	UAG Stop*	UGG Trp	G
C	CUU Leu	CCU Pro	CAU His	CGU Arg	U
	CUC Leu	CCC Pro	CAC His	CGC Arg	C
	CUA Leu	CCA Pro	CAA Gln	CGA Arg	A
	CUG Leu	CCG Pro	CAG Gln	CGG Arg	G
A	AUU Ile	ACU Thr	AAU Asn	AGU Ser	U
	AUC Ile	ACC Thr	AAC Asn	AGC Ser	C
	AUA Ile	ACA Thr	AAA Lys	AGA Arg	A
	AUG Met†	ACG Thr	AAG Lys	AGG Arg	G
G	GUU Val	GCU Ala	GAU Asp	GGU Gly	U
	GUC Val	GCC Ala	GAC Asp	GGC Gly	C
	GUA Val	GCA Ala	GAA Glu	GGA Gly	A
	GUG Val	GCG Ala	GAG Glu	GGG Gly	G

Codons are given here as RNA sequences. The codon sequences in the coding strand of DNA are the same, except that T replaces U. Each amino acid is designated by a three-letter code, which is a shortened form of its name. The 20 amino acids are: alanine (Ala), arginine (Arg), asparagine (Asn), aspartic acid (Asp), cysteine (Cys), glutamic acid (Glu), glutamine (Gln), glycine (Gly), histidine (His), isoleucine (Ile), leucine (Leu), lysine (Lys), methionine (Met), phenylalanine (Phe), proline (Pro), serine (Ser), threonine (Thr), tryptophan (Trp), tyrosine (Tyr), and valine (Val).

* Stop codons have no amino acids assigned to them.

† The AUG codon is the initiation codon, as well as that for other methionine residues.

'stop' signals that indicate to the protein-synthesizing machinery that the protein is complete. These three codons (UAA, UAG,vand UGA) have no amino acids assigned to them. Of the remaining 61 codons, all code for amino acids, which means that most of the amino acids are encoded by more than one codon, giving what is known as a **degenerate code**.

PAUSE AND THINK

Which amino acid is encoded by each of the following codons?

(a) ACU in RNA

(b) GUG in RNA

(c) CTT in DNA

Answer:

(a) Thr (threonine)

(b) Val (valine)

(c) Leu (leucine)

 Check your understanding of the concepts covered in this section by answering the questions in the e-book.

5.3 RNA synthesis: transcription

In this section we look in more detail at the first stage of gene expression, transcription or RNA synthesis. In Section 5.1 we mention three main types of RNA that are found in cells: mRNA, tRNA, and rRNA. Whereas mRNA is an intermediate that then undergoes translation to make the final product of protein-coding genes, tRNA and rRNA are the final products of the genes that encode them. All types of RNA are synthesized by transcription, but we will concentrate on the details of mRNA synthesis. In later sections of this chapter we find out how the three main classes of RNA are involved in translation.

RNA polymerase catalyses transcription and copies a section of DNA

The building-block reactants for RNA synthesis are the ribonucleotides ATP, CTP, GTP, and UTP. In *E. coli*, all RNA is synthesized from these components by a single enzyme called **RNA polymerase**. In eukaryotes three RNA polymerases synthesize different types of RNA, discussed in more detail later in this section. Eukaryotic mRNA is synthesized by RNA polymerase II.

The process of transcription is illustrated in Figure 5.2. It first requires that the DNA strands are separated to provide a single-stranded template for directing the sequence of nucleotides to be assembled into RNA. The two strands are transitorily separated over a short sequence at the site of RNA synthesis, and then come together again after the polymerase has passed. In effect, a separation 'bubble' moves along the DNA. The basic process is much the same as DNA synthesis in that the base of the incoming ribonucleotide is complementary to the base on the DNA template (Figure 5.2) but, unlike DNA synthesis, only one strand of the DNA is copied and only one RNA strand is formed. The strand of DNA that is copied

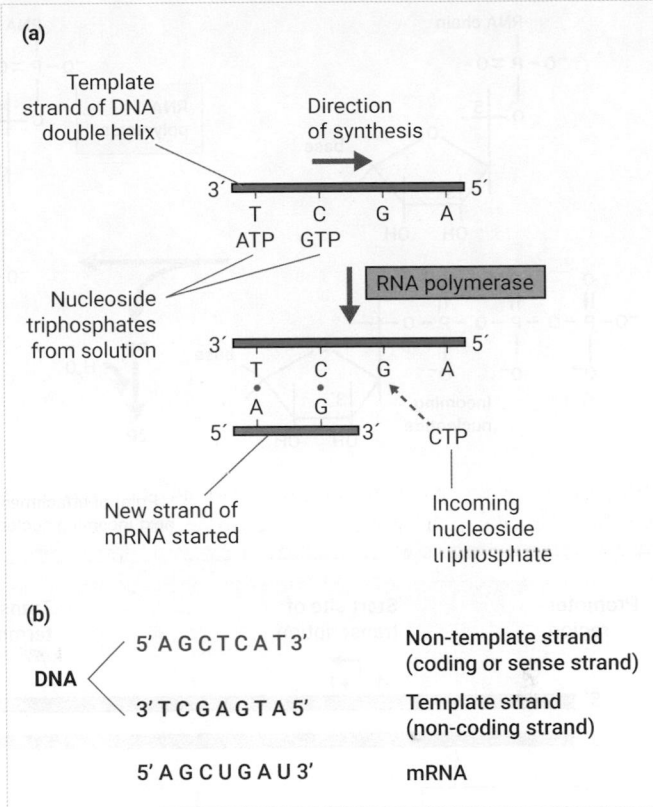

Figure 5.2 Transcription. (a) Copying mRNA from a DNA template strand. Note that the separation of the two DNA strands is transitory. (b) Relationship of transcribed mRNA to template and non-template strands of DNA.

Source: Snape, A. & Papachristodoulou, D. (2018). *Biochemistry and Molecular Biology* (6th edn). Oxford University Press.

is called the **template strand**, or sometimes the **non-coding** strand. The DNA strand that is not copied has the same base sequence as the newly made RNA (except that the DNA has T where the RNA has U), so it is called the **coding strand** of the DNA, or sometimes the **sense** strand. The two strands of DNA and their relationship to the mRNA sequence are illustrated in Figure 5.2.

PAUSE AND THINK

(a) What is the RNA sequence that would be made by transcribing a section of DNA that has the following sequence on the coding strand: 5′ ATTCGC 3′?

(b) What RNA sequence would be made by transcribing DNA that has the following sequence on the non-coding (template) strand: 5′ GTCCAT 3′?

Answer:

(a) 5′ AUUCGC 3′.

(b) 5′ AUGGAC 3′.

The RNA polymerase works its way along the template, joining together the nucleotides in the correct order as determined by the DNA template. Like DNA synthesis, *RNA synthesis is always in the 5′ → 3′ direction*. That is, new nucleotides are added to the 3′-OH

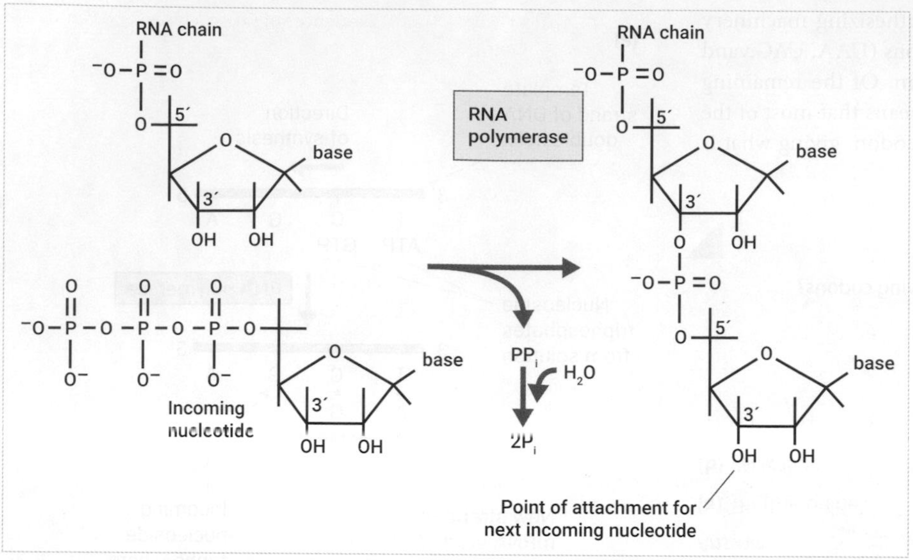

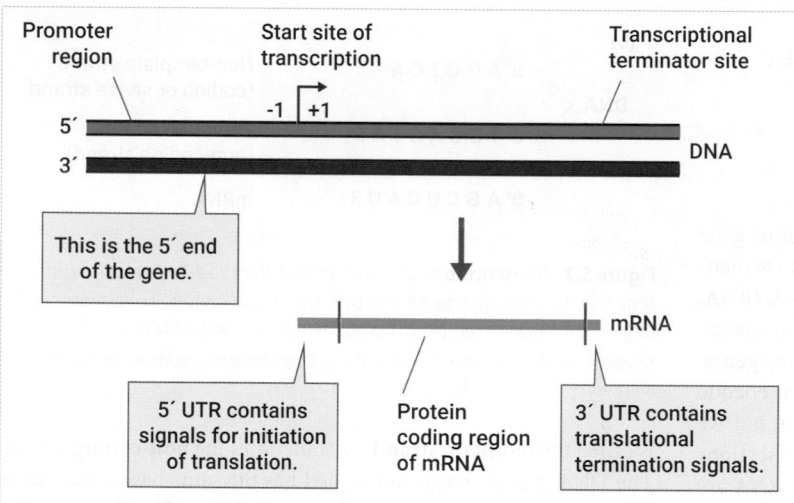

Figure 5.3 The reaction catalysed by RNA polymerase.

Source: Snape, A. & Papachristodoulou, D. (2018). *Biochemistry and Molecular Biology* (6th edn). Oxford University Press.

Figure 5.4 Geography of a prokaryotic gene and its mRNA. The '5' end' of a gene refers to the coding or sense strand of the DNA. Note that the figure is not to scale as the transcribed sequence of a gene is usually much longer than the promoter. The 5' and 3' UTRs are untranslated regions of the mRNA.

Source: Snape, A. & Papachristodoulou, D. (2018). *Biochemistry and Molecular Biology* (6th edn). Oxford University Press.

and so the chain elongates in the 5'→3' direction. The chemical reaction catalysed by the polymerase (Figure 5.3) is very much like that of DNA synthesis in that it involves the attachment of the α-phosphoryl group (the first one attached to the ribose) of the nucleoside triphosphates to the 3'-OH of the preceding nucleotide, splitting off inorganic pyrophosphate (PP_i). Splitting off PP_i and its subsequent hydrolysis to two P_i molecules provides the energy to drive the synthesis reaction, again as in DNA synthesis.

An important point of difference, however, is that RNA polymerase *can initiate new chains*—it does not need a primer; it can synthesize the entire mRNA molecule from the four nucleoside triphosphates, provided a DNA template is there. This is quite different from DNA polymerase, which can only elongate existing chains.

mRNA molecules are short-lived copies of DNA

DNA is immortal in cellular terms, but mRNA is short-lived, with a typical half-life of 20 minutes to several hours in mammals and about two minutes in bacteria. Thus, when a protein is needed in the cell a continuous stream of mRNA molecules must be produced

to maintain its synthesis. This might seem wasteful, but it gives the important benefit of permitting control of the expression of individual genes. Once mRNA synthesis ceases and the mRNA already made breaks down, synthesis of that protein stops.

DNA sequences adjacent to the transcribed sections play important roles in transcription

DNA adjacent to the transcribed sequence of a gene, while not itself copied into RNA, plays an important role in positioning RNA polymerase at the start site for transcription and in ensuring that transcription ends at the correct point. Figure 5.4 illustrates a typical prokaryotic gene, including these adjacent sequences. It is important to understand the terminology that is used to describe the location of these in relation to the transcribed sequences. In Figure 5.4, both strands of DNA are shown, but when we talk about the '5' and 3' ends' of a gene we do so with reference to the coding strand, which has the same polarity as the RNA transcript.

The 5' **promoter** region is essential for positioning RNA polymerase at the correct start site. At the opposite (3') end is a terminator region necessary for termination of transcription. The first

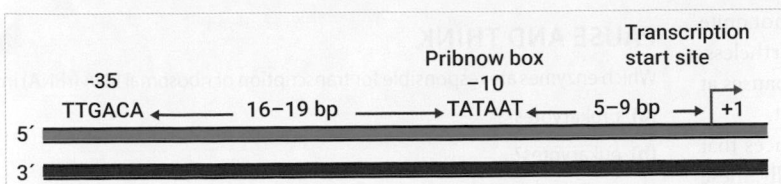

Figure 5.5 Typical sequences of *Escherichia coli* promoter elements.
Source: Snape, A. & Papachristodoulou, D. (2018). *Biochemistry and Molecular Biology* (6th edn). Oxford University Press.

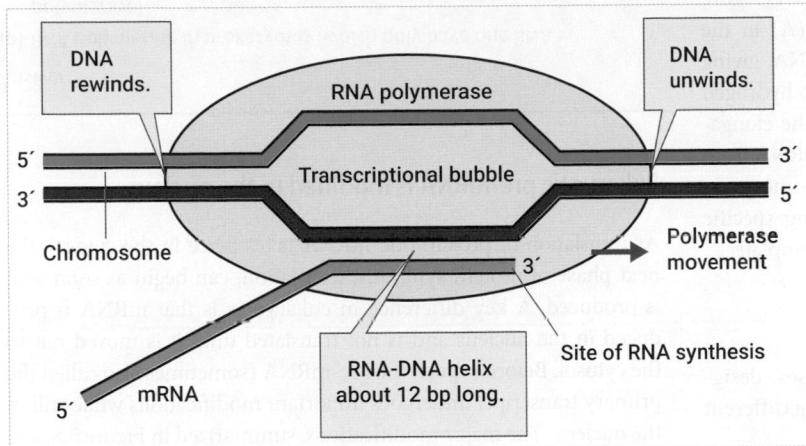

Figure 5.6 DNA transcription by *E. coli* RNA polymerase.
The polymerase transiently unwinds about 17 base pairs of DNA, forming a transcription 'bubble' that progresses along the DNA. The newly formed RNA forms an RNA–DNA double helix about 12 base pairs in length.

Source: Snape, A. & Papachristodoulou, D. (2018). *Biochemistry and Molecular Biology* (6th edn). Oxford University Press.

nucleotide that appears in the RNA sequence is given the number +1 and the nucleotide 5′ to this −1 and so on. The start site is illustrated by an arrow (→) that indicates the direction of transcription. By analogy to the flow of a river, DNA sequences 5′ to this are referred to as 'upstream' and 3′ to this as 'downstream'.

A final point to note from Figure 5.4 is that an mRNA molecule has sections at each end that are not translated into protein. The 5′ untranslated region (UTR) contains sequences necessary for initiation of translation and the 3′ UTR signals for its termination.

While the basic chemistry of RNA synthesis is the same in prokaryotes and eukaryotes, there are differences in the details of the transcription process, with that in prokaryotes being simpler. Consequently, we will first explore transcription in prokaryotes (specifically *E. coli*) and then in eukaryotic cells.

There are three phases of transcription in *E. coli*: initiation, elongation, and termination

Initiation takes place when RNA polymerase locks on to the gene at the promoter. The polymerase binds specifically to short DNA sequences that are similar in many promoters. These are often called 'boxes'. In a typical *E. coli* promoter, there are two boxes: the Pribnow box (named after its discoverer, David Pribnow), which is centred at nucleotide −10, and a second box centred at −35. Typical sequences of the boxes are shown in Figure 5.5. The sequence of both strands of the DNA is shown to emphasize the fact that the polymerase and other proteins required for transcription recognize and bind to the double-stranded DNA. They do so because each base pair presents characteristic chemical groupings in the major and minor grooves of the DNA so the sequence of base pairs creates a specific binding site (you can read more about this in Chapter 3). This ability of enzymes and other proteins to recognize and bind to specific DNA sequences is also important when considering regulation of gene expression (Section 5.5).

Correct initiation of transcription is obviously important. Synthesis of an mRNA needs to commence at the correct nucleotide on the template and on the correct strand. The −35 and Pribnow boxes are the signals for positioning the RNA polymerase. RNA polymerase of *E. coli* is a large complex of several protein subunits. The 'core' enzyme containing five subunits has an affinity for DNA, but it cannot recognize the correct initiation site unless it is joined by an additional subunit called the sigma (σ) subunit. With this attached, the polymerase binds to the −35 and Pribnow boxes giving it the correct starting point and correct orientation, and initiation of transcription can start.

The polymerase can now separate the DNA strands as it moves, thus making the template strand bases available for pairing with incoming bases of NTPs. The enzyme synthesizes the first few phosphodiester bonds from nucleoside triphosphates and initiation is thus achieved. At this point sigma factor protein detaches (to be used again) and the polymerase, now released, moves down the gene, synthesizing mRNA. The polymerase moves at the rate of 50–100 nucleotides per second (compare with DNA polymerase at up to 1000 nucleotides per second) and unwinds the DNA ahead. The DNA rewinds behind it forming a temporary unwound 'bubble', which passes along the gene with the polymerase, as depicted in Figure 5.6.

PAUSE AND THINK

Why is the orientation of RNA polymerase on the DNA important, and how is it achieved?

Answer: RNA polymerase binds to double-stranded DNA, but once bound it must move in the correct direction and use the correct strand of DNA as a template. The enzyme is oriented by its sigma subunit, which binds to the −35 and Pribnow boxes.

An incorrect base in an mRNA molecule could lead to synthesis of an abnormal protein but as mRNA molecules are short-lived, and many copies are transcribed from a single gene, fidelity is not quite so crucial in transcription as it is in DNA replication. Nevertheless, RNA polymerase does carry out some proofreading, as it pauses at and can remove the latest nucleotide added if it is incorrect.

At the ends of prokaryotic genes are termination sequences that cause the mRNA to detach from the DNA template. Typically, these encode a sequence of guanines (G) and cytosines (C) that base pair with each other to form a 'stem-loop' structure in the mRNA (Figure 5.7). The stem-loop is followed by string of adenines (A) in the DNA template, which are copied as uracil (U) in the mRNA, giving weak bonding of the RNA to DNA (as there are only two hydrogen bonds between A and U). This weak bonding disrupts the elongation process and facilitates complete detachment of the mRNA from the DNA template, hence terminating transcription. In some genes alternative mechanisms of termination operate, involving specific proteins that actively unwind the mRNA from the DNA template.

Eukaryotes have three RNA polymerase enzymes

In eukaryotes there are three different RNA polymerases, designated I, II, and III, which are responsible for transcribing different classes of gene.

- RNA polymerase I (Pol I) transcribes a large ribosomal RNA (rRNA) molecule that is processed by splitting into shorter sections to make most types of rRNA.
- RNA polymerase II (Pol II) transcribes mRNA. It also transcribes some types of small RNAs that do not code for proteins but have various functions including RNA splicing (see later in this chapter).
- RNA polymerase III (Pol III) transcribes a variety of small RNAs including tRNAs and 5S rRNA.

The basic enzymatic reaction by which RNA is synthesized by all three RNA polymerase enzymes is the same as in prokaryotes (Figure 5.3). In the following sections we focus on transcription of mRNA by RNA polymerase II.

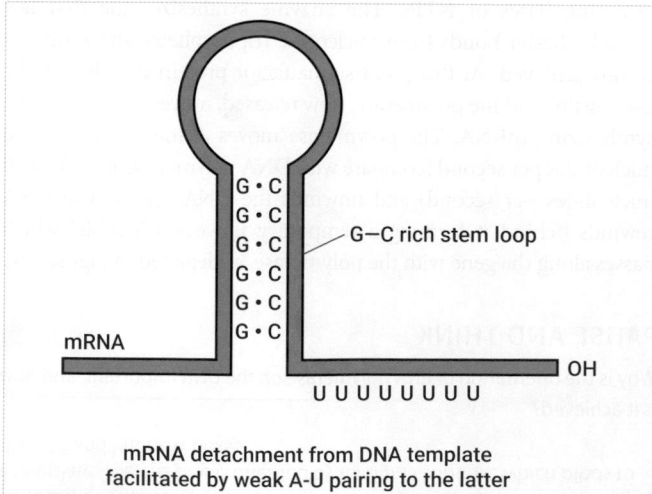

mRNA detachment from DNA template facilitated by weak A-U pairing to the latter

Figure 5.7 The stem-loop structure. A stem-loop structure in an RNA transcript, followed by a string of uracils, facilitates complete detachment of the mRNA from the DNA template and terminates transcription.

Source: Papachristodoulou, D., Snape, A., Elliott, W. H., Elliott, D. C. *Biochemistry and Molecular Biology*.

Eukaryotic pre-mRNA is modified in the nucleus

As translation of prokaryotic mRNA takes place in the cytosol, the next phase of protein synthesis, translation, can begin as soon as it is produced. A key difference in eukaryotes is that mRNA is produced in the nucleus and is not translated until it is moved out to the cytosol. Before export, the pre-mRNA (sometimes also called the primary transcript) undergoes important modifications while still in the nucleus. The major modifications, summarized in Figure 5.8, are:

- addition of a modified 'cap' nucleotide to the 5′ end;
- splicing to remove introns; and
- addition of a poly(A) tail to the 3′ end.

Each of these modification steps is described in more detail later in this chapter.

Transcription initiation in eukaryotes requires general transcription factors

Figure 5.9 shows the typical promoter of a 'type II' gene, i.e. one that is transcribed by RNA polymerase II. It contains DNA sequences that are required to initiate transcription. The **basal promoter** contains the **initiator (Inr)**, a short pyrimidine-rich sequence at the start site, and the **TATA box**, centred usually around 25 base pairs upstream of the start site, with a typical sequence TATAAAA. Many, but not all, type II genes have the TATA box. Its role is to position the RNA polymerase correctly at the promoter. In genes that do not have a TATA box its role is taken by other sequences.

The **upstream control elements** are found at variable positions within about 50 to 200 base pairs upstream of the start site. Those most commonly found are the CAAT (pronounced CAT) box with a typical sequence GGCCAATCT, and the GC box (GGGCGG). Eukaryotic genes usually have at least one of these common control elements. In addition to the CAAT and GC boxes there may be any number of additional control elements. Their function is discussed in Section 5.5 on control of gene expression.

Unlike prokaryotic RNA polymerase, which is positioned by its sigma subunit, eukaryotic RNA polymerases require additional protein factors called **general transcription factors** to position them on the promoter. General transcription factors also help to unwind the DNA to allow transcription to begin. The factors associated with RNA polymerase II are designated TFIIA, TFIIB, etc. The exact functions of all the TFII proteins and the sequence of events when they assemble with the RNA polymerase on the DNA is not known

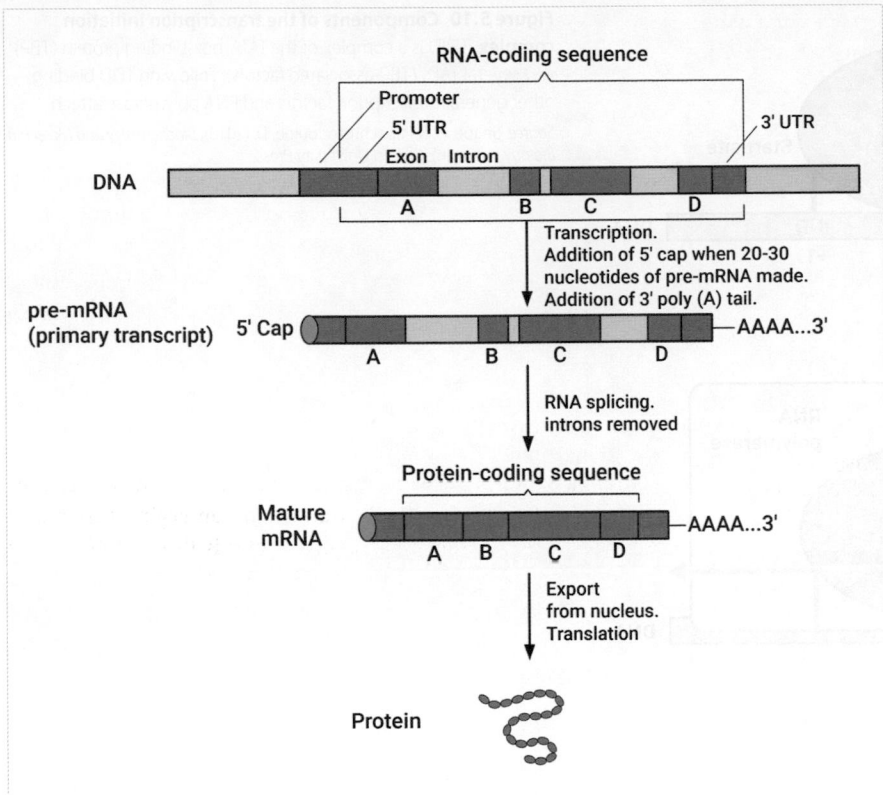

Figure 5.8 Post-transcriptional modifications of eukaryotic mRNA. Addition of the 5′ cap and splicing to remove introns take place during transcription. Once these processes are complete and the poly(A) tail is added, the mature mRNA is exported from the nucleus for translation. Note that different genes contain different numbers of introns and these are of variable length: in reality, intron sequences are often much longer than exons.

Source: A Primer of Human Genetics by Gibson, OUP.

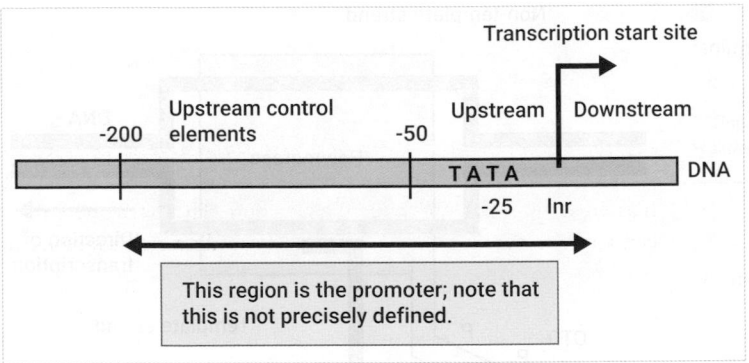

Figure 5.9 Eukaryotic type II gene control elements. Examples of upstream common control elements are the CAAT box and the GC box. Inr, initiator (a pyrimidine-rich stretch on one of the strands).

Source: Snape, A. & Papachristodoulou, D. (2018). *Biochemistry and Molecular Biology* (6th edn). Oxford University Press.

for certain: the order of assembly may vary. However, a large complex called **TFIID** is a key component in committing a gene to transcription. The heart of the TFIID complex is the **TATA box-binding protein (TBP)**, which attaches TFIID to the TATA box (Figure 5.10). TBP binding causes localized bending and distortion of the DNA, which opens it up for interactions with other transcription factors and RNA polymerase. The position of the TATA box relative to the start site enables TFIID to position RNA polymerase II at the start site and ensures that it is pointing in the right direction.

Elongation of the transcript requires RNA polymerase II modification

After the initiation complex is formed, RNA polymerase II must travel along the DNA template. For this to occur, modification

of the enzyme is needed, specifically phosphorylation of amino acids in part of its structure called the carboxy terminal domain (CTD). The CTD structure contains multiple repeats of a seven-amino acid sequence and is special to eukaryotic RNA polymerase II as there is no equivalent in prokaryotic RNA polymerase. One of the general transcription factors in the initiation complex, TFIIH, phosphorylates several serine amino acids in the CTD. This enables the polymerase to leave the initiation complex and bind a new set of factors associated with elongation and processing of the RNA transcript.

As RNA polymerase II moves away from the promoter, many of the initiation factors are removed, but TFIID remains bound to the TATA box, thus facilitating further rounds of initiation. TFIID will leave the promoter only after the need for transcription is over.

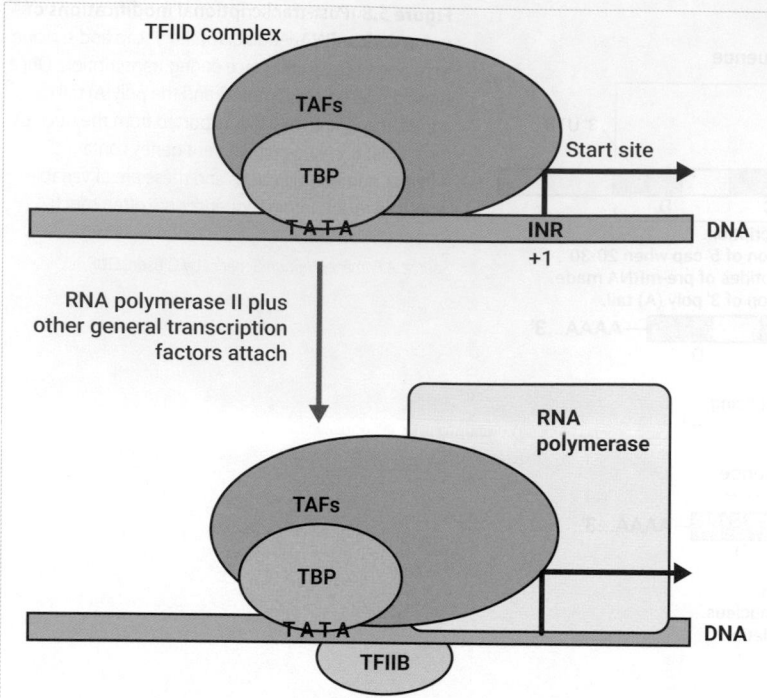

Figure 5.10 Components of the transcription initiation complex. TFIID is a complex of the TATA-box-binding protein (TBP) and several TAFs (TBP-associated factors). Following TFID binding, other general transcription factors and RNA polymerase attach.

Source: Snape, A. & Papachristodoulou, D. (2018). *Biochemistry and Molecular Biology* (6th edn). Oxford University Press.

PAUSE AND THINK

Which chemical group is added to amino acids in the carboxy terminal domain of RNA polymerase II to enable elongation to begin?

Answer: Phosphate groups are added to serine amino acids by an enzyme activity of TFIIH.

Figure 5.11 illustrates the elongation phase of transcription by RNA polymerase II.

Capping and splicing occur simultaneously with transcription

The RNA of the eukaryotic gene primary transcript immediately undergoes a modification at its 5′ end, called capping. As shown in Figure 5.12, the 5′ end of the primary RNA transcript has the triphosphate group from the ribonucleoside triphosphate that began the chain. The terminal phosphate of this group is removed and a GMP residue is added from GTP, forming an unusual 5′–5′ triphosphate linkage. The G is then methylated in the N-7 position, and often there is additional methylation of ribose sugars as shown in Figure 5.12. The cap stabilizes the mRNA by protecting its 5′ end from breakdown by nuclease enzymes, and it is involved in initiation of translation as described in Section 5.5.

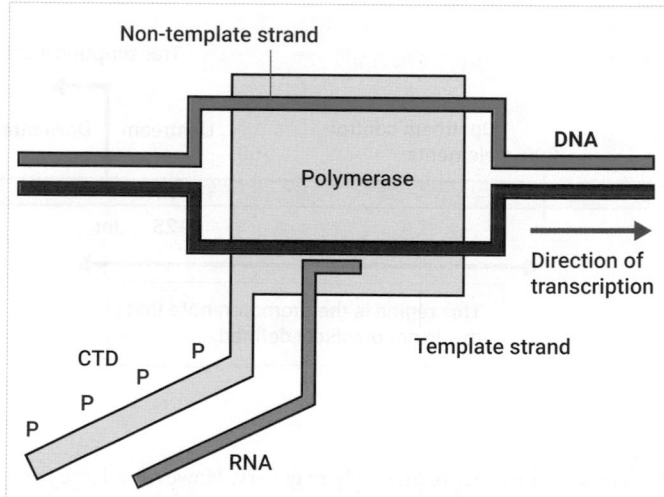

Figure 5.11 RNA polymerase II transcribing a gene. This diagram is designed to represent the enzyme unwinding the DNA as it progresses; the template strand is in a groove of the enzyme containing the catalytic site while the other strand takes a separate path. The two reassociate behind the enzyme. The RNA forms an eight-base pair hybrid with the template strand but is then diverted to a separate exit from the enzyme in the direction of the phosphorylated carboxy terminal domain (CTD). The CTD has enzymes attached for capping and splicing the RNA transcript as it is synthesized, as well as packing the mRNA for transport into cytosol.

Source: Snape, A. & Papachristodoulou, D. (2018). *Biochemistry and Molecular Biology* (6th edn). Oxford University Press.

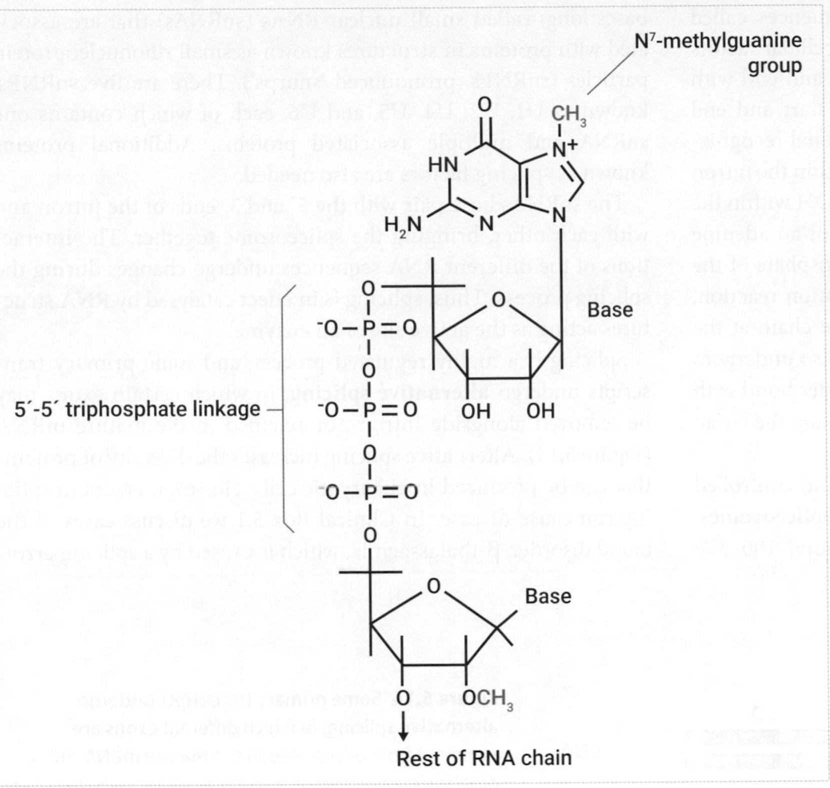

Figure 5.12 Structure of the 5′ cap in eukaryotic mRNA.
The terminal nucleoside triphosphate of the primary RNA transcript is converted to a diphosphate followed by a reaction with GTP in which pyrophosphate is eliminated. This is followed by methylation of the added guanosine and of the ribose of the first nucleotide in the primary transcript, as shown. In some cases, a third methyl group may be added to the 2′-OH of the next nucleotide in the primary transcript.

Source: Snape, A. & Papachristodoulou, D. (2018). *Biochemistry and Molecular Biology* (6th edn). Oxford University Press.

Figure 5.13 A summary of the mechanism of RNA splicing.

Source: Snape, A. & Papachristodoulou, D. (2018). *Biochemistry and Molecular Biology* (6th edn). Oxford University Press.

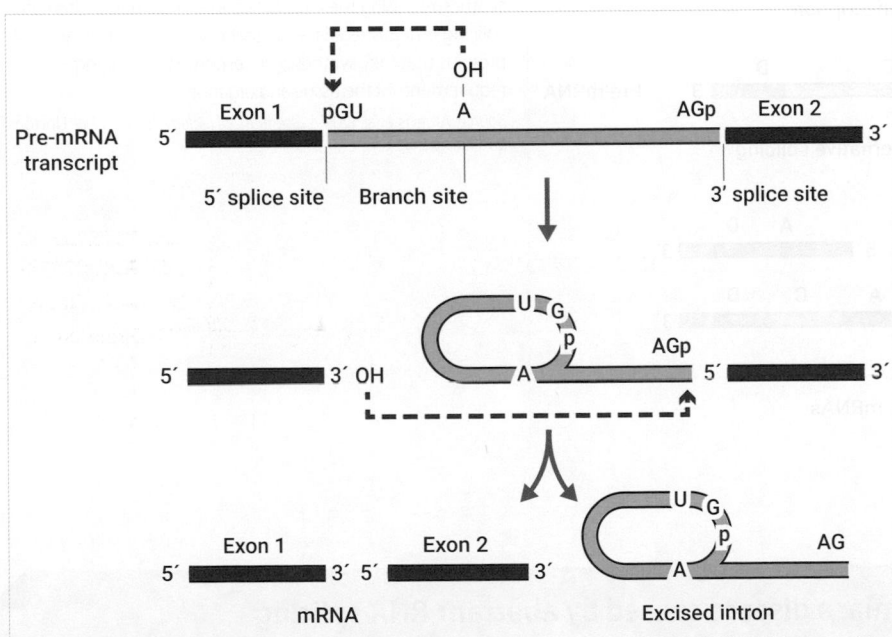

Splicing is the removal of introns and joining of exons

Introns are the 'intervening sequences' within eukaryotic genes that do not code for polypeptide chains. They are removed from the primary transcript, as transcription proceeds, by the process of RNA splicing. Genes vary in the number of introns they contain, which can be anything from one to 500 or more, and the lengths of the introns also varies, with anything between 50 and 20,000 base pairs being common. Exons also vary in size but usually are smaller than introns, around 150 base pairs on average.

A summary of the splicing process is shown in Figure 5.13. The key reaction is the transesterification reaction, in which a phosphodiester bond is transferred from a hydroxyl (–OH) group on one nucleotide to a hydroxyl group on a different nucleotide within the same RNA chain.

The exon–intron junctions are 'marked' by sequences called the 5′ and 3′ splice sites that are recognized by the cellular factors involved in the process. All introns begin with GU and end with AG, although the full lengths of the recognizable start and end sequences are longer than just two bases. An additional recognizable site is a short sequence of around seven bases within the intron and close to its 5′ end, called the branch site. The A–OH within the branch site in Figure 5.13 denotes the 2′-hydroxyl of an adenine nucleotide. The 2′-hydroxyl group links to the 5′ phosphate of the G nucleotide at the 3′ splice site by a transesterification reaction, forming a **lariat** (looped) structure. This breaks the chain at the end of exon 1, thus producing a free 3′-OH, which also undergoes a transesterification reaction, forming a phosphodiester bond with the 5′ end of exon 2, joining the two exons and releasing the lariat, which is subsequently broken down.

The splicing reaction in the nucleus is catalysed and controlled by very complex protein–RNA structures called **spliceosomes**. These are complexes of small RNA molecules, around 100–300 bases long, called small nuclear RNAs (snRNAs) that are associated with proteins in structures known as small ribonucleoprotein particles (snRNPs, pronounced 'snurps'). There are five snRNPs, known as U1, U2, U4, U5, and U6, each of which contains one snRNA and multiple associated proteins. Additional proteins known as splicing factors are also needed.

The snRNAs base pair with the 5′ and 3′ ends of the intron and with each other, bringing the spliceosome together. The interactions of the different RNA sequences undergo changes during the splicing process. Thus, splicing is in effect catalysed by RNA structures acting as the active site of an enzyme.

Splicing is a highly regulated process, and some primary transcripts undergo **alternative splicing**, in which certain exons may be removed alongside introns, or retained in the mature mRNA (Figure 5.14). Alternative splicing increases the diversity of proteins that can be produced in eukaryotic cells. However, errors in splicing can cause disease. In Clinical Box 5.1 we discuss cases of the blood disorder, β-thalassaemia, which is caused by a splicing error.

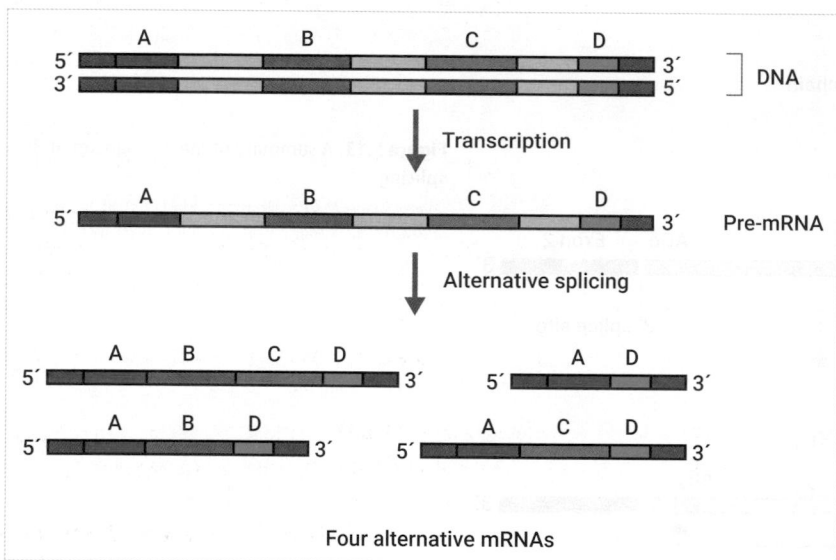

Figure 5.14 Some primary transcripts undergo alternative splicing, in which different exons are removed from or retained in the mature mRNA. This process increases the number of different proteins that can be synthesized using the information in a single gene. Alternative splicing is highly regulated so that cells can alter the range of proteins that they synthesize in response to changing requirements of the tissue and organism.

Source: Alberts et al. (2019) *Essential Cell Biology*, Edn 5. W.W. Norton and company.

CLINICAL BOX 5.1   β-Thalassaemia: a disease caused by aberrant RNA splicing

The thalassaemias are a family of genetic diseases caused by mutations that affect haemoglobin production. The haemoglobin protein consists of α and β subunits, which are encoded by separate genes, and the disease is known as α- or β-thalassaemia depending on which gene, and therefore which protein subunit, is affected. The main health problems associated with thalassaemia are anaemia caused by reduced or no production of haemoglobin, and overload of iron in the body caused by the regular blood transfusions that are given as treatment. More than 200 different mutations in the β-globin gene can cause β-thalassaemia, and some of the most common mutations affect splicing of the primary transcript. Figure 1 illustrates one of these mutations, which produces an extra splice site in an intron. This results in retention of part of the intron sequence in the spliced mRNA as shown in Figure 1, preventing full translation of the β-globin protein.

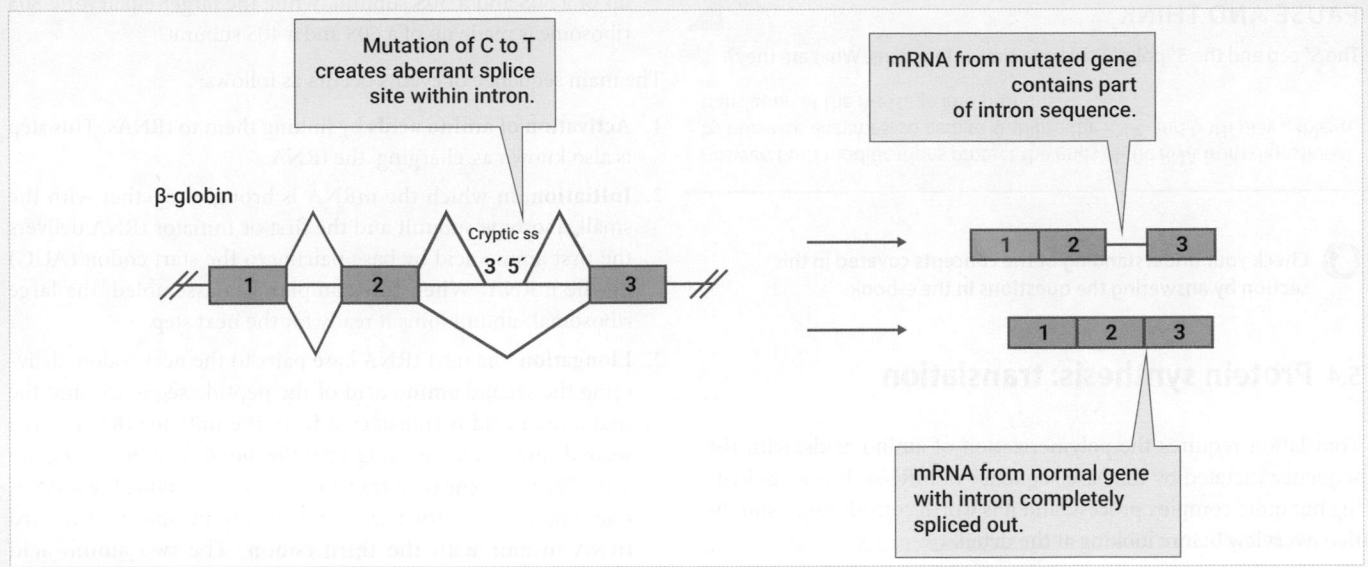

Figure 1 A mutation causes aberrant splicing of the β-globin gene. The mutation of a C to a T nucleotide within intron 2 creates a 5' splice site within the intron. When the mutation is present the splicing machinery also uses a 'cryptic' 3' splice site within the intron that is not normally used. The result is that part of the intron is retained in the mature mRNA. An individual who has two copies of the mutated gene will not make β-globin protein and will suffer from β-thalassaemia. It may be possible to develop therapies that restore normal splicing in people carrying this mutation.

Source: Adapted from Genetic therapies for RNA mis-splicing diseases. Suzan M. Hammond Matthew J. A. Wood.

Eukaryotic mRNAs end with a poly(A) tail

Most eukaryotic mRNAs end in a string of up to 250 adenine residues known as a 3' poly(A) tail. The poly(A) tail is not directly encoded by the gene, but its position is directed by a polyadenylation signal (AAUAAA) that is encoded by the gene and transcribed in the primary transcript. The polymerase transcribes a short distance beyond the polyadenylation signal and then terminates transcription.

Although the signal and process for the actual termination is less well understood than in prokaryotes, the process of polyadenylation has been studied extensively and is shown in Figure 5.15.

The RNA is cleaved near the polyadenylation signal by a specific endonuclease (an enzyme that cuts within a polynucleotide sequence). Then another enzyme, poly(A) polymerase, which is not dependent on a DNA template, uses ATP as the source of adenine nucleotides to form a poly(A) tail. The poly(A) tail increases the stability of the mRNA and increases the efficiency of translation.

Polyadenylation of the primary transcript can occur after the transcript has been capped and spliced, or it can take place before splicing is complete. After the post-transcriptional modifications are complete, mature, processed mRNAs are exported from the nucleus through the nuclear pores.

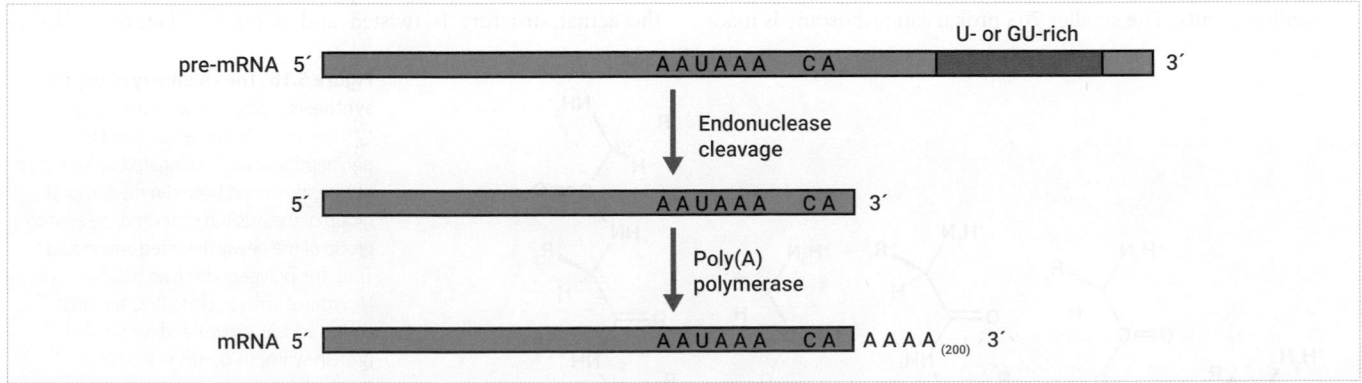

Figure 5.15 Polyadenylation involves the binding of protein factors to the AAUAAA signal upstream of the poly(A) site and to a G/U-rich sequence downstream of it. The two factors interact to form an active endonuclease complex, which cleaves the transcript in between them at the polyadenylation site (CA). The free 3' end of the upstream region is polyadenylated by the enzyme poly(A) polymerase, while the downstream sequence that has been cleaved away is degraded.

Source: Adapted from Fig. 6.9 in Latchman, D. (2015) *Gene Control* 2nd Edn, Garland.

PAUSE AND THINK

The 5′ cap and the 3′ poly(A) tail have similar functions. What are they?

Answer: Both modifications protect the ends of the RNA from digestion by nuclease enzymes, so help to stabilize the RNA, and both play a role in translation of the message into proteins.

Check your understanding of the concepts covered in this section by answering the questions in the e-book.

5.4 Protein synthesis: translation

Translation requires the polymerization of amino acids, with the sequence dictated by the base sequence of mRNA. It is a fascinating but quite complex process, and it is worth considering a simplified overview before looking at the detail.

A simplified overview of translation

The three main players in translation are listed below.

1. **mRNA** contains the sequence of bases that are read in groups of three (codons) to give the amino acid sequence of the protein.
2. **tRNAs** are small 'adaptor' RNA molecules that each deliver a specific amino acid to the mRNA. Each tRNA contains a three-base anticodon sequence that pairs with the corresponding codon on the mRNA.
3. **Ribosomes** are complex subcellular particles consisting of rRNA and proteins (see Figure 5.19). They move along the mRNA providing sites at which the mRNA and the tRNAs (carrying amino acids) interact. The ribosome also catalyses formation of a peptide bond between the incoming amino acid and the one next to it in the growing peptide sequence. Prokaryotic and eukaryotic ribosomes have different sizes, which are designated by their sedimentation coefficient (a measure of their behaviour when subject to centrifugation). The units of ribosome size are S or Svedberg units. The smaller 70S prokaryotic ribosome is made up of a 50S and a 30S subunit, while the larger eukaryotic 80S ribosome is made up of a 60S and a 40S subunit.

The main sequence of events occurs as follows:

1. **Activation of amino acids** by linking them to tRNAs. This step is also known as 'charging' the tRNA.
2. **Initiation**, in which the mRNA is brought together with the small ribosome subunit and the first or initiator tRNA delivers the first amino acid by base pairing to the start codon (AUG) on the mRNA. When this complex has assembled, the large ribosomal subunit joins it ready for the next step.
3. **Elongation**: the next tRNA base pairs to the next codon, delivering the second amino acid of the peptide sequence, and the first amino acid is transferred from the initiator tRNA to the second amino acid, forming a peptide bond, as shown in Figure 5.16. The ribosome then translocates (moves) along the mRNA, releasing the initiator tRNA and freeing up space for a third tRNA to pair with the third codon. The two amino acid sequence is transferred to the third amino acid on its tRNA, and the ribosome translocates again. This elongation process keeps repeating until the ribosome reaches a stop codon.
4. **Termination** occurs when the ribosome reaches a stop codon. Instead of a tRNA, a protein release factor binds the stop codon sequence, and catalyses release of the completed protein sequence from the last tRNA. The rest of the translation complex disassembles, and its components can be reused in another round of translation.

We will now look at the process in more detail, describing translation in prokaryotes. We will then highlight some key points about eukaryotic translation where it differs from the process in prokaryotes.

Transfer RNAs act as adaptor molecules through codon : anticodon base pairing

Transfer RNAs (tRNAs) are small RNA molecules, less than 100 nucleotides in length. When depicted diagrammatically they have a cloverleaf structure (Figure 5.17a), although folding means that the actual structure is twisted and narrower (Figure 5.17b), to

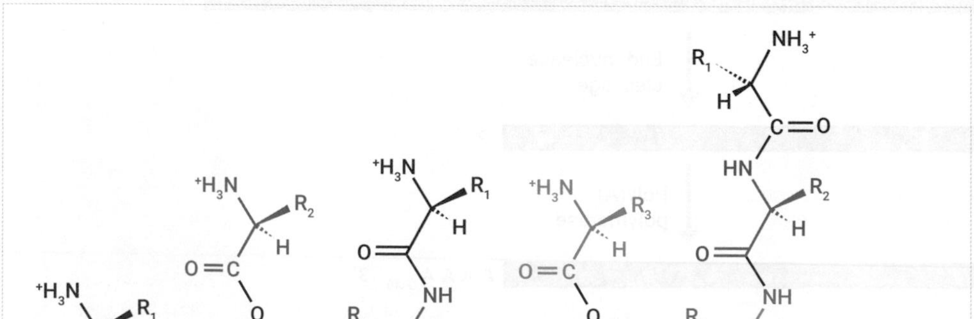

Figure 5.16 The chemistry of peptide synthesis. Successive amino acids are delivered by tRNA molecules, and the polypeptide chain is elongated by formation of a peptide bond between the carboxyl group of the existing chain and the amino group of the newly delivered amino acid. Thus, the polypeptide chain grows from its N-terminal amino acid to its C-terminal amino acid, as shown, and we say that protein synthesis occurs in the N to C terminal direction.

Source: Berg et al. *Biochemistry*, Edn 5. Macmillan.

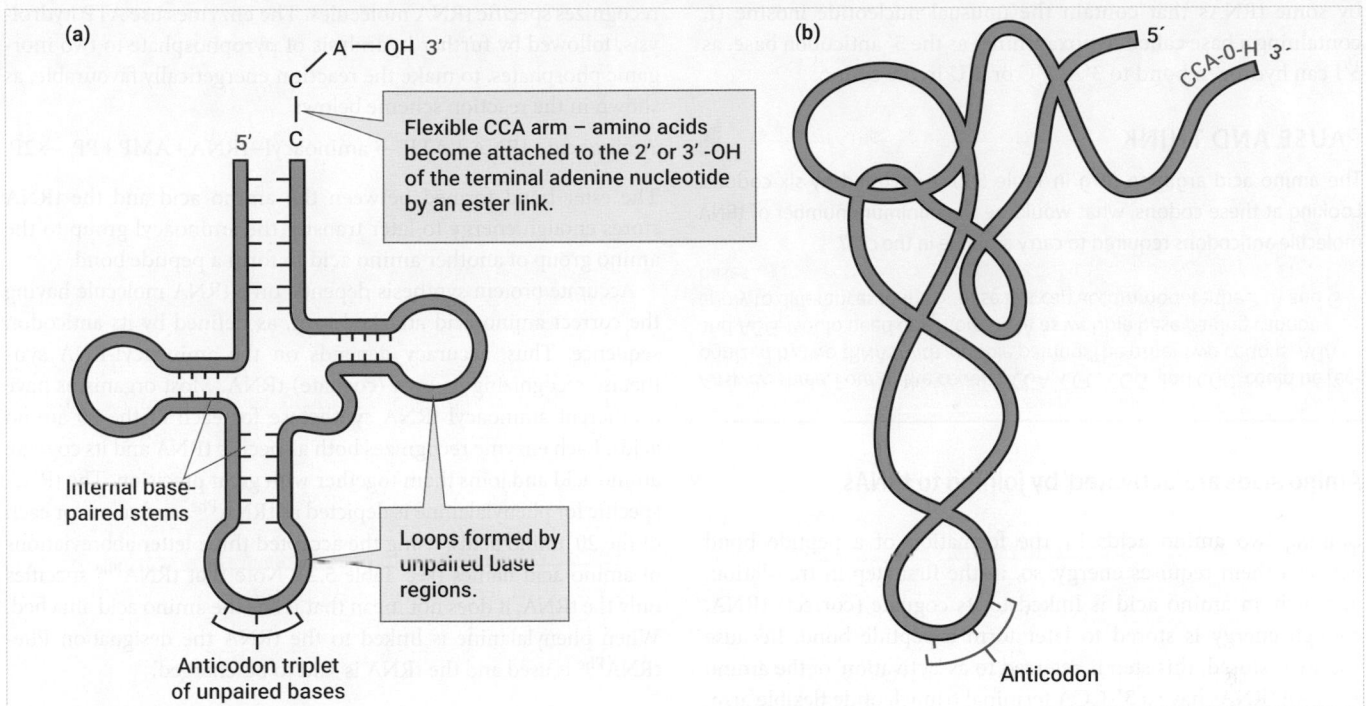

Figure 5.17 Transfer RNA. (a) The cloverleaf structure of tRNA. (b) The folded structure is a more accurate depiction of the three-dimensional shape of the tRNA molecule. *Source*: (a) Snape, A. & Papachristodoulou, D. (2018). *Biochemistry and Molecular Biology* (6th edn). Oxford University Press.

allow them to pack close to each other during translation. Internal base pairing forms the stem-loops, with one loop containing the three bases that form the anticodon, a triplet of bases complementary to a codon. The anticodon is located at a hairpin bend of the tRNA so that the three bases are unpaired and available for hydrogen bonding. Thus, if a codon on the mRNA is 5′UUC3′ (coding for phenylalanine), the anticodon corresponding to this on a tRNA molecule will be 5′GAA3′.

Wobble base pairing reduces the number of tRNA molecules required to recognize 61 different codons

Since the genetic code has 61 codons that each represent an amino acid, it might be expected that cells contain 61 different tRNA molecules, each with its own anticodon complementary to one codon and each accepting the one amino acid represented by its codon. In fact, there are fewer than 61 tRNA species—the exact number varies in different organisms. There must be at least one tRNA for each of the 20 amino acids, but some tRNA molecules can recognize more than one codon differing from each other in the 3′ base. For example, in the genetic code there are two codons that encode phenylalanine, 5′UUC3′ and 5′UUU3′, but yeast cells have only one corresponding tRNA, with the anticodon sequence 5′GAA3′. The single anticodon can recognize and base pair to either codon through an unusual type of base pairing called **wobble pairing**.

The wobble mechanism allows the 5′G of the anticodon to pair with either 3′C or 3′U in the codon as shown in Figure 5.18. In other tRNAs, a 5′U in an anticodon will pair with either 3′A or 3′G in the codon. Importantly, this departure from Watson–Crick base pairing is only possible at the 3′ base of the codon, where a little flexibility can be accommodated. Additional flexibility is provided

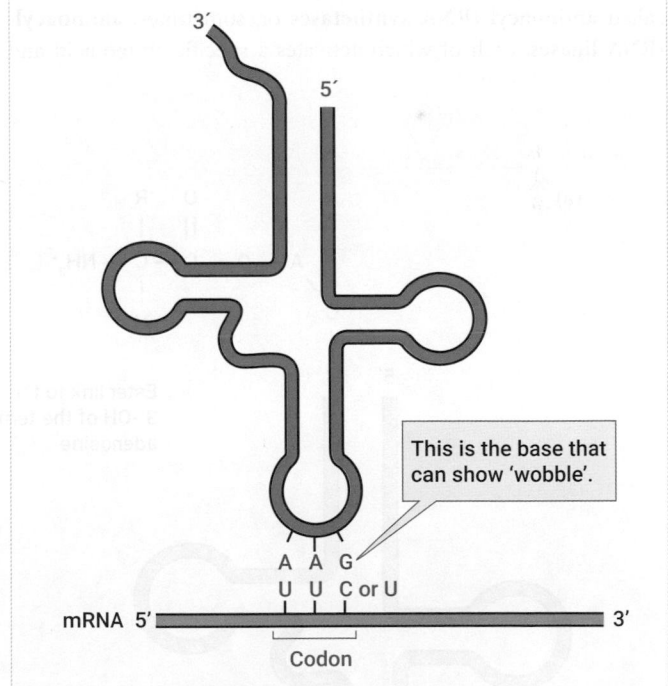

Figure 5.18 Base pairing of an anticodon of a tRNA molecule to an mRNA codon. Note that in this figure the tRNA is drawn so that it is 'flipped' compared to the usual convention of depicting RNA molecules with the 5′ end to the left and the 3′ end to the right (see Figure 5.17) to show the antiparallel base pairing of the anticodon with the codon. In the example shown, the anticodon sequence 5′GAA3′ in the tRNA can base pair with codons 5′UUC3′ and 5′UUU3′, both of which code for phenylalanine.

Source: Snape, A. & Papachristodoulou, D. (2018). *Biochemistry and Molecular Biology* (6th edn). Oxford University Press.

by some tRNAs that contain the unusual nucleotide inosine (I, containing a base called hypoxanthine) as the 5′ anticodon base, as 5′I can hydrogen bond to 3′A, 3′C or 3′U in the codon.

PAUSE AND THINK

The amino acid arginine (Arg in Table 5.2) is encoded by six codons. Looking at these codons, what would be the minimum number of tRNA molecule anticodons required to carry arginine in the cell?

Answer: Three. Four of the codons, CGA, CGC, CGG, and CGU, could be recognized by two tRNAs with wobble pairing. The other two codons, AGA and AGG, would need one more tRNA as wobble base pairing cannot apply to differences in the 5′ base but can accommodate the 3′ A and G bases.

Amino acids are 'activated' by joining to tRNAs

Linking two amino acids by the formation of a peptide bond between them requires energy, so, in the first step in translation, in which an amino acid is linked to its cognate (correct) tRNA, enough energy is stored to later form a peptide bond. Because energy is stored, this step is referred to as 'activation' of the amino acid. All tRNAs have a 3′-CCA terminal trinucleotide flexible arm, to which the amino acid is attached by an ester bond between a hydroxyl group on the ribose at the 3′ terminus of the tRNA and the carboxyl group of the amino acid as shown in Figure 5.19.

Ester bond formation is catalysed by a family of enzymes called **aminoacyl-tRNA synthetases** or, sometimes, **aminoacyl-tRNA ligases**, each of which activates a specific amino acid and

recognizes specific tRNA molecules. The enzymes use ATP hydrolysis, followed by further hydrolysis of pyrophosphate to two inorganic phosphates, to make the reaction energetically favourable, as shown in the reaction scheme below:

$$\text{Amino acid} + \text{tRNA} + \text{ATP} \rightarrow \text{aminoacyl} - \text{tRNA} + \text{AMP} + PP_i \rightarrow 2P_i$$

The ester bond formed between the amino acid and the tRNA stores enough energy to later transfer the aminoacyl group to the amino group of another amino acid to form a peptide bond.

Accurate protein synthesis depends on a tRNA molecule having the correct amino acid attached to it, as defined by its anticodon sequence. Thus, accuracy depends on the aminoacyl-RNA synthetase recognizing its own (cognate) tRNA. Most organisms have a different aminoacyl-tRNA synthetase for each of the 20 amino acids. Each enzyme recognizes both a specific tRNA and its cognate amino acid and joins them together with great precision. The tRNA specific for phenylalanine is depicted as tRNA^{Phe}, and so on for each of the 20 amino acids, using the accepted three letter abbreviations of amino acid names (see Table 5.2). Note that tRNA^{Phe} specifies only the tRNA; it does not mean that it has the amino acid attached. When phenylalanine is linked to the tRNA the designation Phe-tRNA^{Phe} is used and the tRNA is said to be 'charged'.

Ribosomes provide the structures on which translation takes place

Ribosomes derive their name from the content of RNA, which accounts for about 60 per cent of the dry weight. As illustrated in Figure 5.20, a ribosome consists of two subunits; in prokaryotes

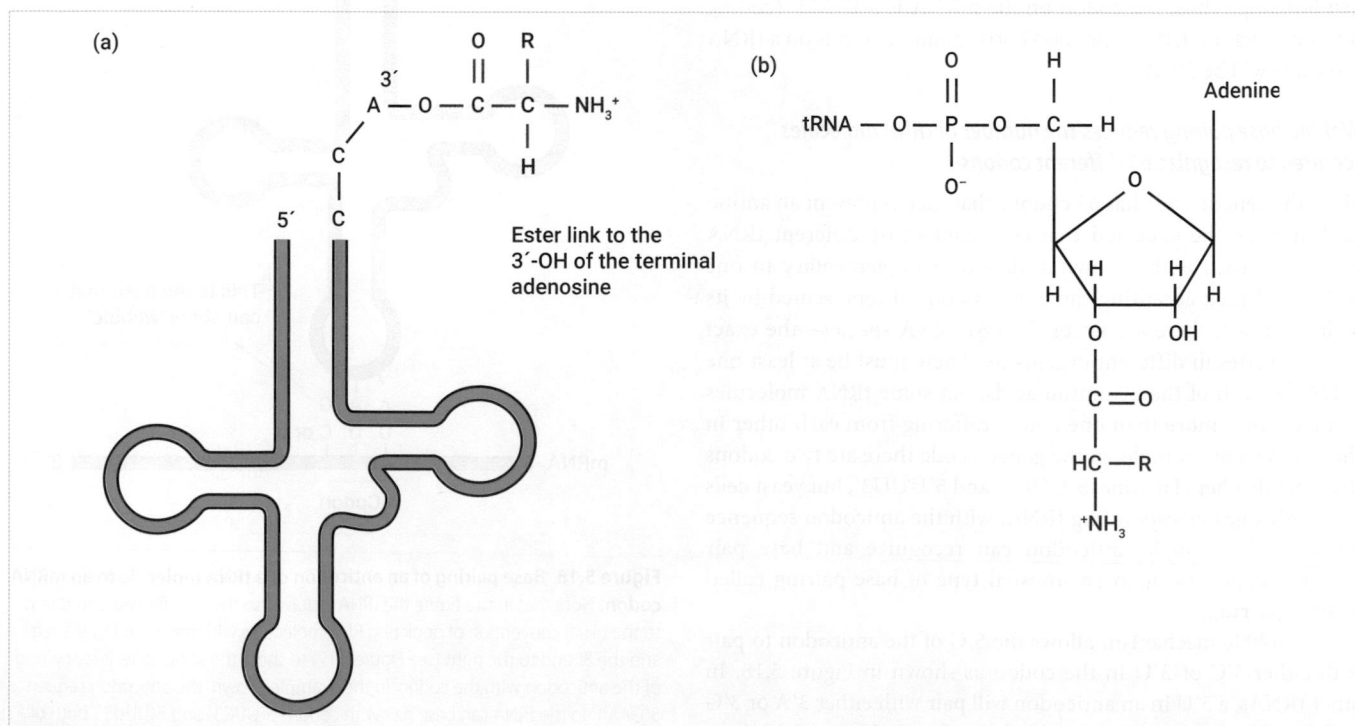

Figure 5.19 (a) Diagram of tRNA molecule showing the CCA base sequence at the 3′ end, where the amino acid is attached by an ester link. (b) Structure of the terminal nucleotide with the attached amino acid. In some cases, the ester link is on the 3′-OH of the ribose.

Source: (a) Snape, A. & Papachristodoulou, D. (2018). *Biochemistry and Molecular Biology* (6th edn). Oxford University Press.

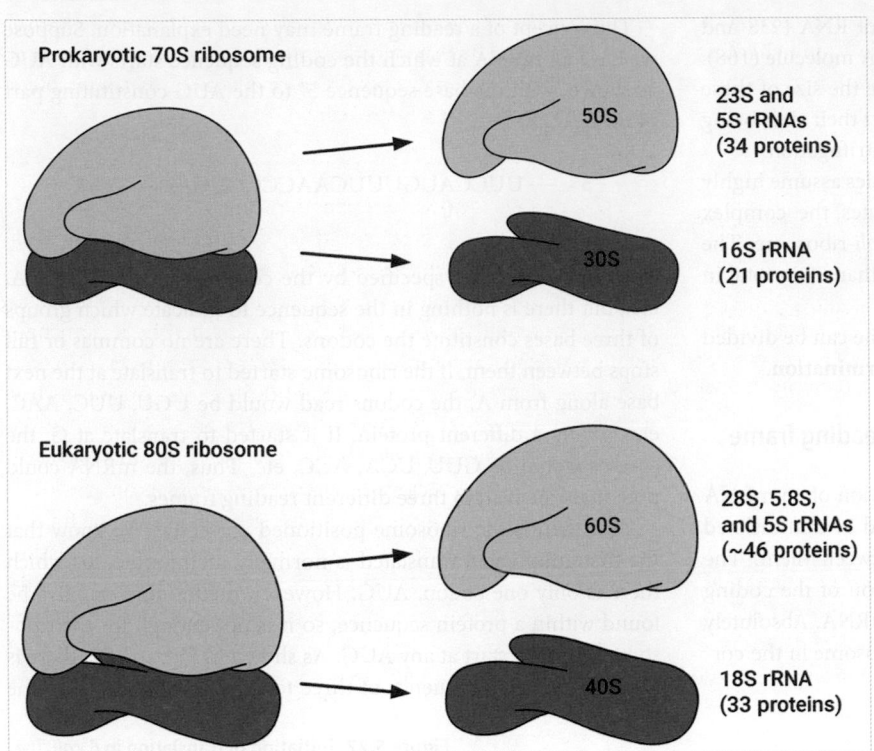

Prokaryotic 70S ribosome

50S — 23S and 5S rRNAs (34 proteins)

30S — 16S rRNA (21 proteins)

Eukaryotic 80S ribosome

60S — 28S, 5.8S, and 5S rRNAs (~46 proteins)

40S — 18S rRNA (33 proteins)

Figure 5.20 Overview of the structure of prokaryotic and eukaryotic ribosomes. The two subunits contain different sets of rRNA molecules and proteins. All ribosomes contain the two major rRNA molecules (23S and 16S in prokaryotes, 28S and 18S in eukaryotes) and one or two smaller RNAs, as shown. Ribosomal proteins are numbered and designated 'L' or 'S' depending on whether they are found in the large or small subunit.

Source: Lodish et al. *Molecular Cell Biology*. Edition 4. Macmillan.

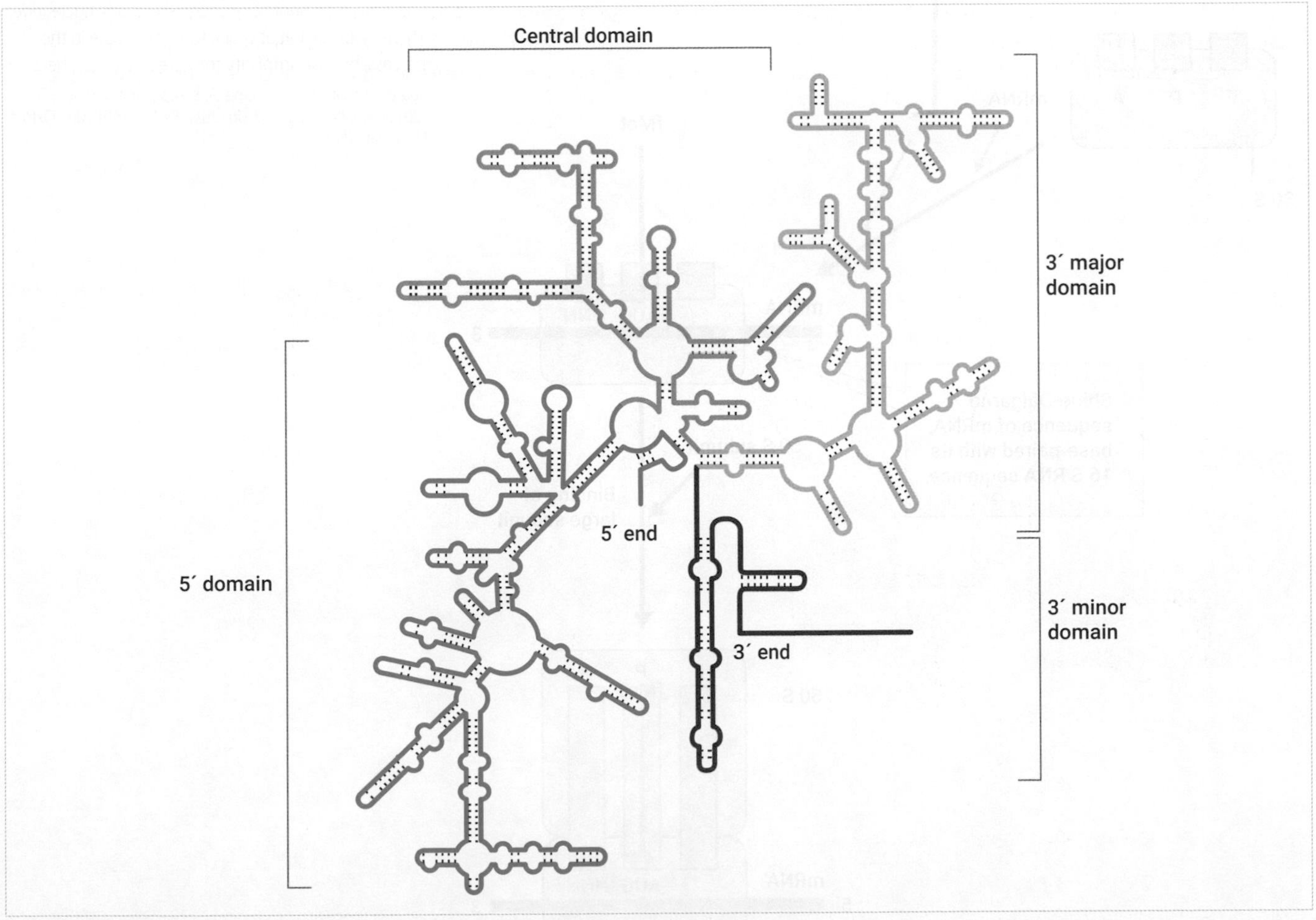

Central domain

3′ major domain

5′ domain

5′ end

3′ end

3′ minor domain

Figure 5.21 16S rRNA. The many regions of internal base pairing are typical of ribosomal RNAs and give the molecules complex three-dimensional structures.

Source: Fig 9.5 from Lewin, B. (1994) *Genes V*. Oxford University Press, Snape, A. & Papachristodoulou, D. (2018). *Biochemistry and Molecular Biology* (6th edn). Oxford University Press.

the larger 50S subunit contains two molecules of RNA (23S and 5S) and the smaller 30S subunit contains one RNA molecule (16S). The S or Svedberg units that are used to describe the size of these RNA molecules and the ribosomal subunits reflect their sizes being determined by measuring their behaviour on centrifugation.

Because of internal base pairing, rRNA molecules assume highly folded compact structures. Figure 5.21 illustrates the complex folded structure of an RNA molecule of an *E. coli* ribosome. The rRNAs are associated with many proteins (more than 50 in total in prokaryotes), forming solid particles.

Synthesis of a protein molecule on the ribosome can be divided into three phases—**initiation**, **elongation**, and **termination**.

Initiation puts the ribosome in the correct reading frame

It is important that a ribosome begins its translation of an mRNA at exactly the correct point. mRNAs have 5′ and 3′ untranslated regions (UTRs) and the coding region lies between them. The ribosome therefore must recognize the *first* codon of the coding sequence, which is not at the beginning of the mRNA. Absolutely precise initiation is essential, for this puts the ribosome in the correct **reading frame**.

The concept of a reading frame may need explanation. Suppose we have an mRNA at which the coding sequence starts with AUG as shown, with the base sequence 5′ to the AUG constituting part of the 5′ UTR:

$$5'---UUCCAUGUUUCAACCCCUG------3'$$
$$\Uparrow$$

The amino acids are specified by the codons AUG, UUU, CAA, etc., but there is nothing in the sequence to indicate which groups of three bases constitute the codons. There are no commas or full stops between them. If the ribosome started to translate at the next base along from A, the codons read would be UGU, UUC, AAC, etc., giving a different protein. If it started to translate at G, the codons would be GUU, UCA, ACC, etc. Thus, the mRNA could potentially be read in three different reading frames.

How then is the ribosome positioned correctly? We know that the first amino acid translated is normally methionine, for which there is only one codon, AUG. However, methionine can also be found within a protein sequence, so it is not enough for the ribosome simply to start at any AUG. As shown in Figure 5.22, there is also a purine-rich sequence of three to eight nucleotides 5′ to the

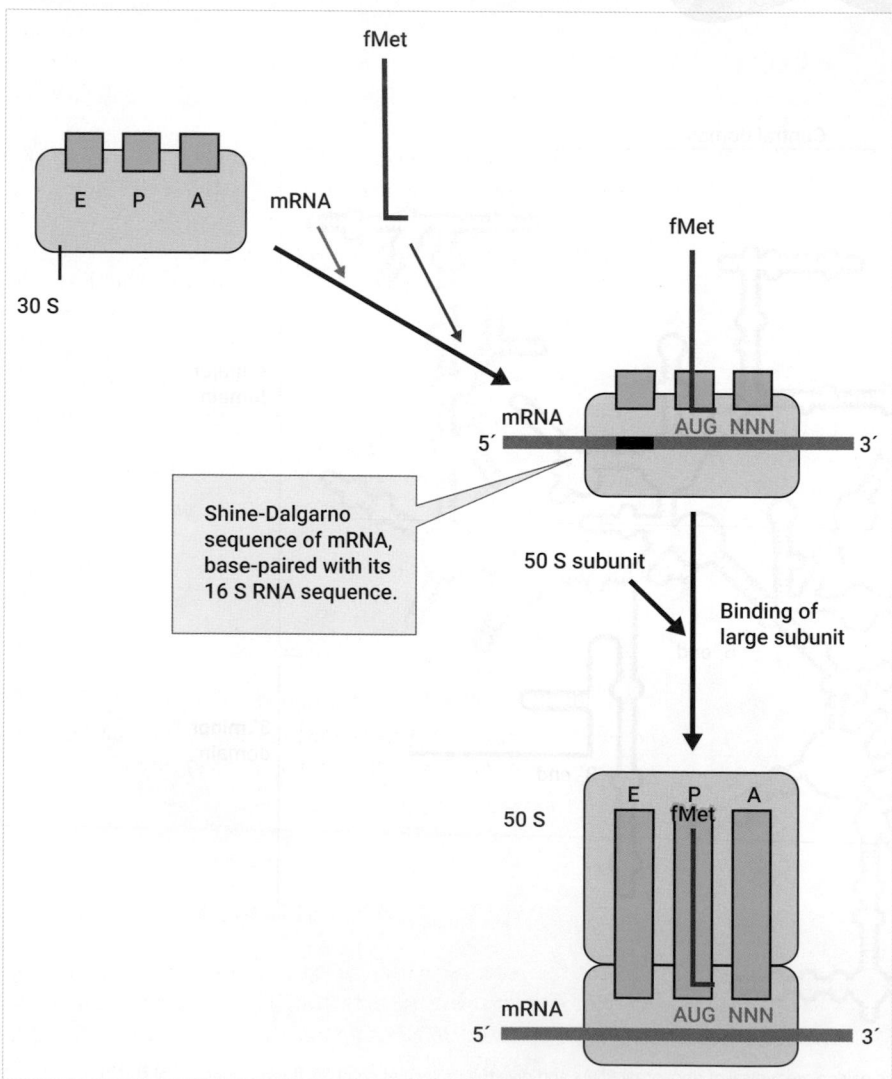

Figure 5.22 Initiation of translation in *E. coli*. The initiating transfer RNA, tRNA^fMet, is represented by the purple line, the anticodon being the horizontal short line. NNN represents any codon (N for any nucleotide). Three cytosolic initiation factors participate in the process, but for simplicity they are not shown here.

Source: Adapted from Snape, A. & Papachristodoulou, D. (2018). *Biochemistry and Molecular Biology* (6th edn). Oxford University Press.

start site on the mRNA known as the **Shine–Dalgarno sequence**, which base pairs to a section of the 16S rRNA of the small 30S ribosomal subunit. Thus, the recognition signal for initiation is effectively longer than just an AUG as it consists of the AUG with the Shine–Dalgarno sequence upstream of it.

A ribosome has three sites on it, each of which can accommodate a tRNA molecule. These are the A, P, and E sites (for **acceptor** or **aminoacyl**, **peptidyl**, and **exit** sites), the roles of which will be explained further in the subsection on elongation. At initiation, the mRNA is positioned such that the first and second codons on the mRNA are aligned with the P and A sites, respectively (Figure 5.22).

The first amino acid in prokaryotic translation is a modified methionine

Once the AUG is in the correct position in the P site, the initiating tRNA is placed so that its anticodon can base pair with the AUG initiation codon on the mRNA. There are two different types of tRNAs specific for methionine, both with the same anticodon, but one is used exclusively for initiation and the other exclusively for inserting methionine internally into the growing polypeptide during the elongation process. The methionyl-tRNA that is specific for initiation is the only aminoacyl-tRNA that can directly enter the P site of the ribosome; all others must enter the A site first.

In prokaryotes there is another difference between the two methionyl-tRNAs. The initiator methionine is modified by addition of a formyl group as shown in Figure 5.23. Thus, prokaryotic proteins are synthesized with *N*-formylmethionine (fMet) as the first amino acid residue, although the formyl group, and frequently the methionine also, are removed before completion of the synthesis.

The initiating tRNA can be called tRNA^{fMet} (f for formyl), and the charged version fMet-tRNA^{fMet}, while the tRNA for methionine involved in elongation is called tRNA^{Met}.

Once the charged initiator tRNA is positioned in the P site of the 30S subunit, the large 50S joins the complex. We now have a complete 70S ribosome positioned on the mRNA with the fMet-tRNA^{fMet} in the P site with its anticodon base paired with the initiating AUG codon. The A site is vacant, awaiting delivery of the second amino acid on its tRNA. Initiation is complete.

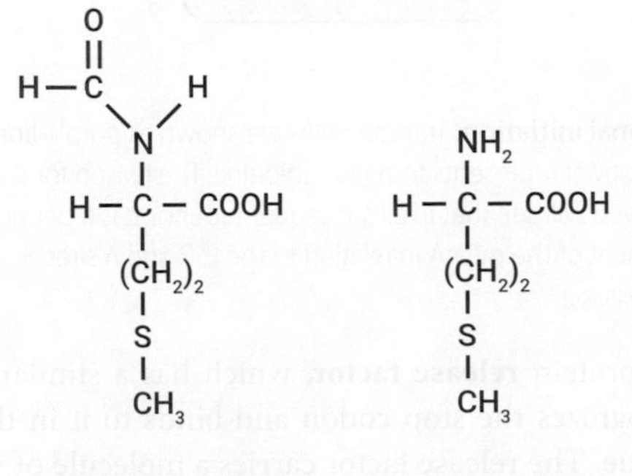

Figure 5.23 Comparison of the structures of *N*-formylmethionine and methionine. The formyl group is shown in red. '*N*' refers to the formyl group being bonded to the nitrogen of the amino group. As *N*-formylmethionine is the first, *N*-terminal, amino acid of the polypeptide, its modification has no impact on peptide bond formation.

PAUSE AND THINK

List the components required for initiation of translation in *E. coli*.

Answer: mRNA, charged fMet-tRNA^{fMet}, initiation factors, 30S and then 50S ribosomal subunit.

Once initiation is achieved, elongation is the next step

We have already given the chemistry of elongation (Figure 5.16). We suggest that you follow the steps in Figure 5.24 as you read the next part. Starting with the initiation complex (state a), we have an fMet-tRNA^{fMet} in the P site, and the A site is vacant. It might be worth re-emphasizing that *only* in the initiation process does the P site accept a tRNA charged with an amino acid—in this case *N*-formylmethionine; *all* subsequent aminoacyl-tRNAs enter the A site. Delivery of the second aminoacyl-tRNA to the A site with its anticodon positioned at the mRNA codon (Figure 5.24, state b) is therefore the next step.

The aminoacyl groups on the two tRNA molecules on the P and A sites are in the vicinity of the catalytic site of **peptidyl transferase**, which, after a pause for proofreading, transfers the fMet group from the tRNA in the P site to the free amino group of the incoming aminoacyl-tRNA in the A site, producing a dipeptide attached to the second tRNA (state d). Despite its name, peptidyl transferase is not a protein enzyme. It is an RNA molecule with catalytic properties, part of the 23S RNA of the large ribosomal subunit; it is a **ribozyme**.

The aminoacyl group on the tRNA in the P site is transferred to the free amino group of the aminoacyl-tRNA in the A site. As the synthesis of the polypeptide proceeds, the group attached to the tRNA in the P site is the partially completed polypeptide chain, which is *transferred* to the incoming amino acid on the tRNA in the A site. This is why the process is called the **peptidyl transferase reaction**.

Adjacent to the peptidyl transferase in the ribosome there is the opening to a tunnel through the large ribosomal subunit. As the polypeptide is synthesized it is fed into this tunnel to emerge from the subunit, amino terminal end first.

The next step is translocation: movement of the ribosome relative to the mRNA. This precise movement shifts the now discharged initiator tRNA into the E site, from which it exits the ribosome, while the growing peptide chain occupies the P site, and the A site aligns to the next codon (state e), ready to receive the third aminoacyl-tRNA and repeat the process.

PAUSE AND THINK

In order to synthesize a polypeptide 100 amino acids in length, how many times would the ribosome need to translocate along the mRNA.

Answer: Ninety-nine times. The first two aminoacyl-tRNAs are accommodated in the P and A sites without translocation needing to take place.

Initiation and elongation factors keep translation on track

Besides the proteins that are integral to the ribosome, cytosolic proteins known as **initiation factors** (IFs) and **elongation factors** (EFs) play key roles in ensuring that the stages of translation proceed correctly. We will not discuss their roles in detail, but some

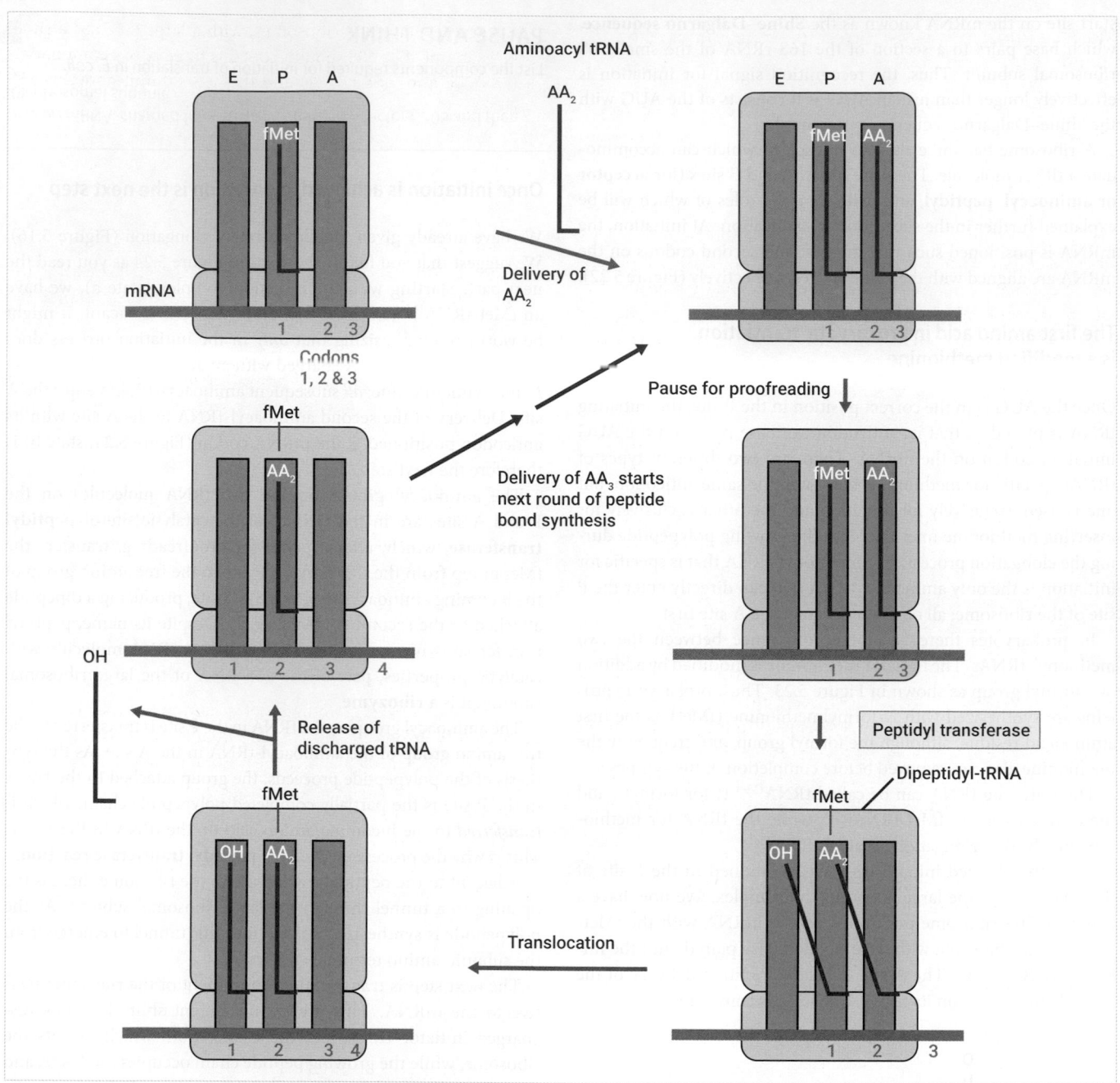

Figure 5.24 The elongation process in protein synthesis in prokaryotes following translational initiation. Transfer RNAs are shown as purple lines, the anticodon being represented by the short horizontal section; AA$_2$ and AA$_3$ represent amino acids and fMet represents formylmethionine. The reason for the naming of the enzyme peptidyl transferase may not be evident, but if you consider the next round of synthesis you will see that in all subsequent rounds it is a peptide that is transferred to the incoming aminoacyl-tRNA. E, exit site; P, peptidyl site; A, acceptor site. Note movement of the mRNA in relation to the E, P, and A sites.

Source: Snape, A. & Papachristodoulou, D. (2018). *Biochemistry and Molecular Biology* (6th edn). Oxford University Press.

are involved in delivering aminoacyl-tRNAs to the ribosome, while others bind the ribosome, enable proofreading, and physically prevent the next stage of translation from starting until the previous one is complete.

Termination takes place at a stop codon

At the end of each mRNA coding section there is a stop codon (Table 5.2). These codons do not have corresponding tRNAs.

Instead, a protein **release factor,** which has a similar shape to a tRNA, recognizes the stop codon and binds to it in the A site of the ribosome. The release factor carries a molecule of water to the ribosome that hydrolyses the ester bond between the now completed polypeptide chain and the final tRNA. This releases the polypeptide from the ribosome.

After this the release factor dissociates, the remaining ribosome complex is disassembled into the two subunits, and the remaining bound tRNAs and mRNA are released.

Multiple ribosomes translate a single mRNA molecule, speeding up translation

It takes approximately 20 s for a ribosome to synthesize a polypeptide of average length (about 400 amino acids) in the bacterium *E. coli*. However, protein synthesis can proceed much faster than this because multiple ribosomes can move along the mRNA at any one time. As soon as a ribosome has got under way and has moved along about 30 codons, another initiation can occur. The ribosomes follow one another down the mRNA, each independently synthesizing a protein molecule. This greatly increases the rate of protein synthesis. The term **polysome** (a shortened version of polyribosome) refers to the complex of multiple ribosomes associated with an mRNA molecule in the process of translation. As prokaryotic mRNA does not undergo post-transcriptional modification and does not have to be exported from a nucleus, translation can begin on an mRNA molecule even before its synthesis is complete, speeding up protein synthesis still further.

Prokaryotic mRNAs are typically polycistronic, encoding more than one protein

An important point that has not yet been explained is that prokaryotic mRNA molecules typically encode more than one protein. They are therefore described as **polycistronic**. On the bacterial chromosome, genes encoding proteins that all take part in a single cellular process are typically adjacent to each other, forming a functional unit called an **operon**. The advantage of operons in coordinating gene regulation will be explained in Section 5.5. For now, it is enough to appreciate that all the genes in an operon can be transcribed together, forming a single mRNA with successive start codons, coding sequences, and stop codons arrayed along its length (Figure 5.25). This means that ribosomes must also initiate translation at multiple points along the mRNA, directed by multiple Shine–Dalgarno sequences.

Translation in eukaryotes and prokaryotes differs mainly at the initiation stage

Like transcription, the basic process of translation is the same in eukaryotes and prokaryotes, but there are some differences:

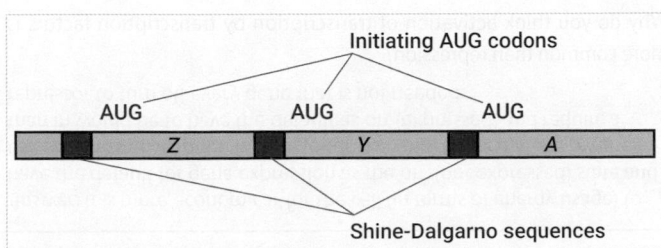

Figure 5.25 The structure of a polycistronic prokaryote mRNA. In this example, the three genes, Z, Y, and A encode three enzymes that function in the metabolism of the sugar, lactose. Hence, when the bacterium needs to use lactose as an energy source, the enzymes required are all made in a coordinated process, from a single mRNA. This system, the *lac* operon, is discussed later in this chapter.
Source: Snape, A. & Papachristodoulou, D. (2018). *Biochemistry and Molecular Biology* (6th edn). Oxford University Press.

1. Eukaryotes have 80S ribosomes, with a large 60S and a small 40S subunit. The 60S subunit contains three rRNA components (28S, 5S, and 5.6S rRNA) and the 40S subunit contains 18S rRNA.

2. Eukaryotes have more and different initiation factors and elongation factors, although several of them have equivalent functions to those in prokaryotes.

   Initiation of translation is similar in eukaryotes in that it requires the initiating methionyl-tRNA to bind into the P site of the small ribosomal subunit that positions itself correctly at the initiating AUG. Following the formation of this complex the large subunit joins and elongation proceeds.

3. A difference in initiation in eukaryotes compared to prokaryotes is that the initiating amino acid is methionine, not *N*-formylmethionine, although it still has a special initiator tRNA, designated tRNAi^{Met}, rather than the regular tRNA^{Met}, to bind to.

4. Another difference is that there is no equivalent to the Shine–Dalgarno sequence to direct the ribosome to the first AUG. Instead, the small (40S) ribosomal subunit, already complexed with Met-tRNAi^{Met} in the P site, recognizes and binds with a complex of proteins near the cap structure at the 5′ end of the mRNA to form a 'pre-initiation complex'. This complex scans the message by moving along it until it finds the correct AUG (usually the first one encountered).

This process of 5′ cap binding and scanning for the first AUG explains why eukaryotic mRNAs are usually monocistronic, only encoding a single protein. Once the pre-initiation complex reaches the first AUG the large 60S ribosomal subunit joins to the small subunit forming the 80S complex.

Elongation and termination of translation in eukaryotes are essentially the same as in prokaryotes. The elongation and release factors are somewhat different but carry out the same functions.

PAUSE AND THINK

Explain why the use of the 5′ cap to direct ribosome binding means that eukaryotic mRNAs are monocistronic and not polycistronic.

Answer: The ribosome attaches at the 5′ cap and moves along the mRNA towards the 3′ end until it finds the first AUG, where translation begins. That means there is no opportunity for additional AUG codons in the mRNA to be used as start codons as the first protein being translated would get in the way of ribosomes scanning for more downstream start codons.

 Check your understanding of the concepts covered in this section by answering the questions in the e-book.

5.5 Control of gene expression

Gene control is necessary for several reasons. Firstly, it enables cells to conserve energy and cellular resources that are used up by protein synthesis. For example, it would not be favourable for a bacterium to constantly synthesize all the proteins encoded by

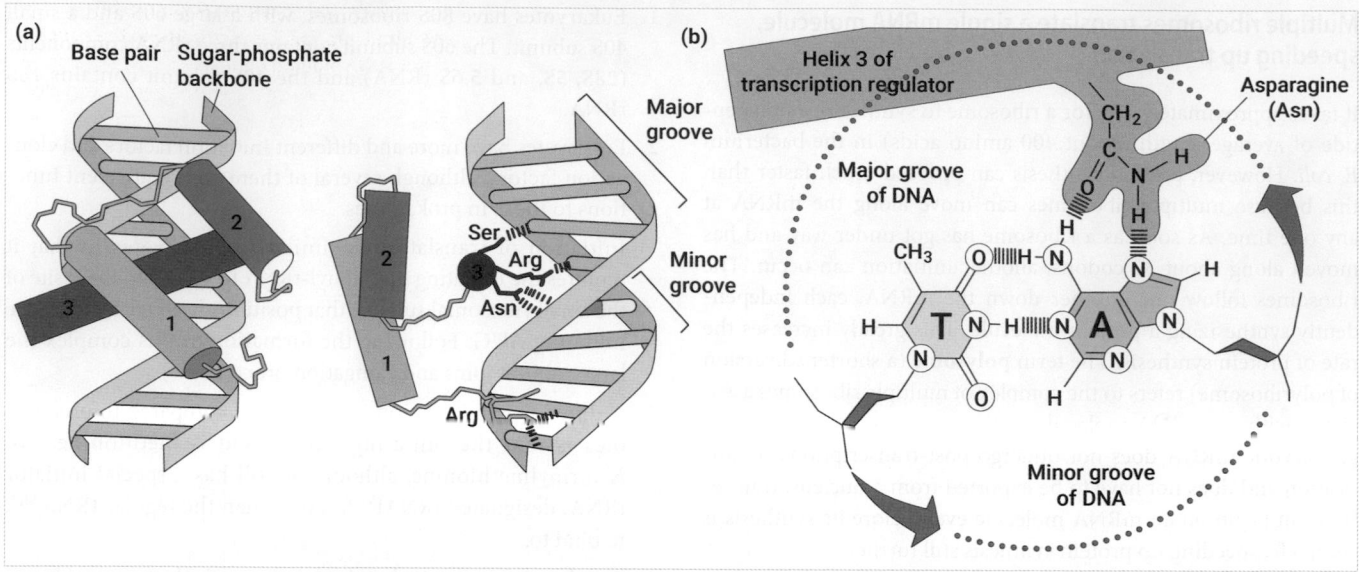

Figure 5.26 A transcription factor interacts with the major groove of the DNA double helix. (a) The transcription factor contains a characteristic arrangement of three α helices, called the homeodomain, that is found in many eukaryotic DNA-binding proteins. (b) Amino acid residues in the third helix (shown here end on) make contact with the edges of the base pairs, without disrupting hydrogen bonding between the bases. (c) An example of an amino acid–base pair interaction: the side chain of asparagine forms hydrogen bonds with an adenine base in an A:T base pair. Typically, 10–20 such interactions provide sequence-specific DNA binding by the transcription factor.

Source: Alberts et al. (2019) *Essential Cell Biology*, Edn 5. W.W. Norton and company.

its genome. It can conserve its energy and enhance its chances of survival by synthesizing only the enzymes needed to metabolize nutrients available in its environment and by changing its pattern of gene expression in responses to changes in the environment. Eukaryotic cells must also respond to environmental changes, such as hormonal signals or the availability of particular metabolites. A second driver for gene regulation is that multicellular eukaryotes consist of multiple cell types. This cell specialization also requires controlled gene expression so that the appropriate proteins are made in each cell type—globin in developing red blood cells, and contractile actin and myosin in muscle cells, for example.

As protein synthesis takes place in multiple stages there are many levels at which it is regulated: mRNA transcription, processing, and nuclear export of mRNA (in eukaryotes) and translation. However, most energy saving is achieved by regulation early in the process, so the bulk of regulation takes place at the level of transcription, and mostly at the initiation step. It is this transcriptional regulation that will be our main focus, with a brief exploration of other mechanisms at the end of the chapter.

DNA binding proteins act as transcription factors that regulate RNA synthesis

Gene regulation at the level of transcription is carried out by proteins that can recognize and bind to specific DNA sequences. These proteins are known as **transcription factors**, and selective DNA binding is crucial to their function. They bind to double-stranded DNA in which the bases are already paired, so must recognize the edges of the bases that are exposed in the grooves of the double helix (you can read more about this in Chapter 3). Depending on the base pair, these edges present different chemical groups for

non-covalent interaction with amino acid side chains, particularly in the major groove (Figure 5.26). The DNA sites that the transcription factors recognize are usually about 10–20 base pairs in length.

Examples of DNA binding domains in transcription factors are shown in Figure 5.27.

Transcription factors generally have a domain structure, with one domain of the protein that binds the DNA, as shown in Figure 5.27, and another that exerts its effect on transcription. Transcription factors often activate transcription, but in some cases their effect is repressive. We will explore some of the mechanisms by which they regulate transcription in the following sections.

The *E. coli lac* operon as an example of gene control in prokaryotes

The *lac* operon is a cluster of three genes that encode enzymes required for the cellular import and metabolism of the disaccharide, lactose. Regulation of the *lac* operon was the first example of

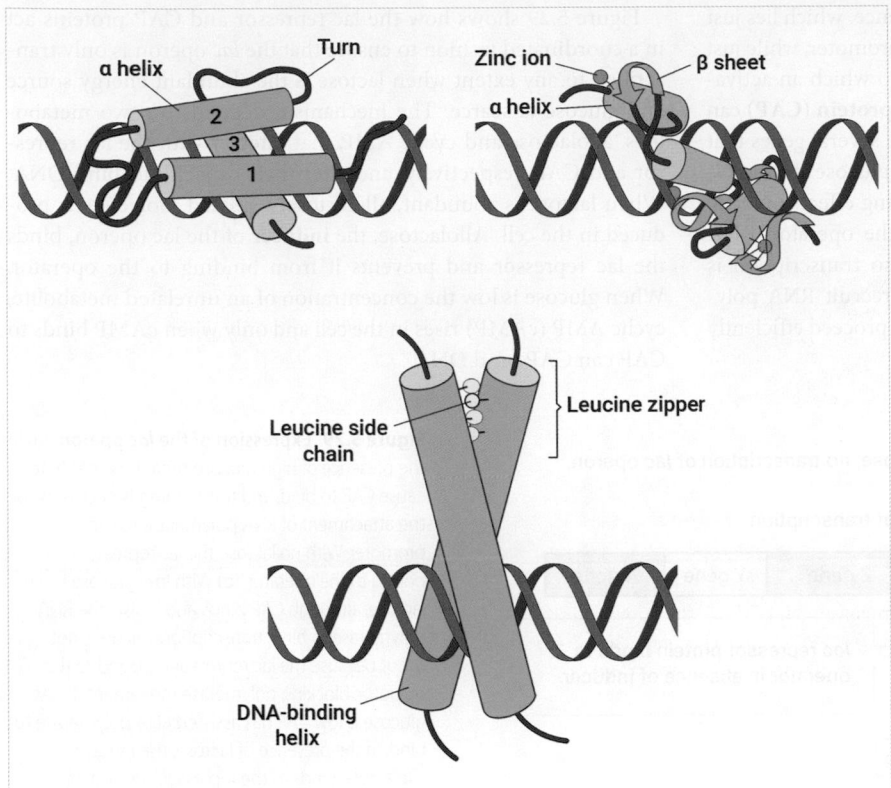

Figure 5.27 Transcription factors can be organized into 'families' according to the structure of their DNA binding domain. Examples from some of the main families are shown. The 'zinc finger proteins' contain zinc ions (green spheres) that coordinate the structure of their DNA binding domains: finger-like elements that reach around the DNA molecule. The 'helix-turn-helix' domain is found in many other transcription factors, including forming part of the homeodomain shown in Figure 5.26.

Source: From McKee & McKee, *Biochemistry*, 7e. Oxford University Press.

gene control to be elucidated, and it neatly illustrates the principle of DNA-binding proteins that can activate or repress transcription and respond to environmental signals. Regulation of the *lac* operon reflects *E. coli*'s preference for glucose as its energy and carbon source. Hence, its glucose-metabolizing enzymes are always made and the genes encoding them are constantly ('constitutively') transcribed. In contrast, the enzymes required to utilize lactose are inducible; that is, they are synthesized only when glucose, the preferred source, is scarce and lactose is abundant. Regulation is at the level of transcription initiation, and the arrangement of the three genes as a single operon that is transcribed into a polycistronic mRNA encoding all three enzymes allows coordinated regulation of their expression.

Figure 5.28 shows the organization of the *lac* operon. The key enzyme needed to digest lactose is β-galactosidase, while lactose permease is needed to transport lactose into the cell, and there is a third enzyme, a transacetylase, whose role is less well understood. These three enzymes are encoded by the *lacZ*, *lacY*, and *lacA* genes, respectively. The promoter, where RNA polymerase binds to begin transcribing the operon, is flanked by two regulatory DNA sequences. Each of these is recognized and bound by a different regulatory protein, one of which represses, and the other activates, transcription. The **lac repressor** protein is encoded by the *lacI* gene, which is adjacent to the operon but has a separate promoter and is constitutively expressed regardless of the presence or absence of lactose.

Figure 5.28 Diagram of the *lac* operon. Note that the *lacI* gene is an independent gene that codes for the lac repressor protein.

Source: Snape, A. & Papachristodoulou, D. (2018). *Biochemistry and Molecular Biology* (6th edn). Oxford University Press.

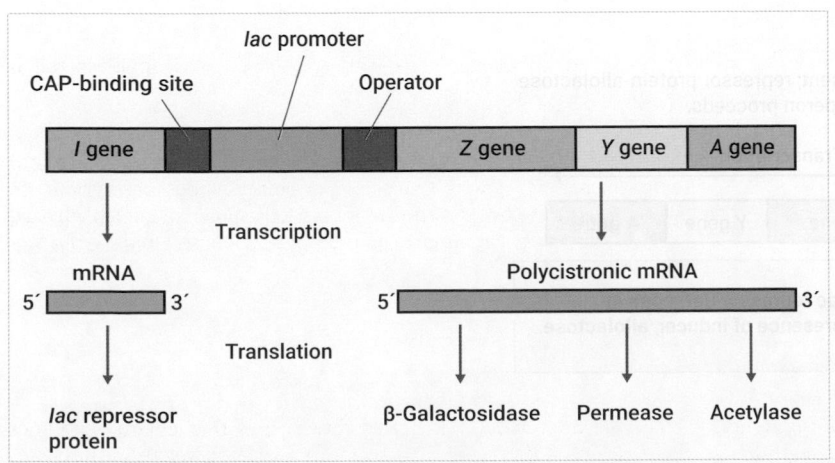

The lac repressor binds to the **operator** sequence, which lies just downstream of (and partially overlapping) the promoter, while just upstream of the promoter there is a sequence to which an activator protein called the **catabolite gene activator protein (CAP)** can bind. CAP is given its name because it activates several genes that enable *E. coli* to metabolize sugars other than glucose. In the *lac* operon, the lac repressor and CAP have opposing effects on gene expression. When the repressor is bound to the operator, RNA polymerase cannot proceed along the DNA, so transcription is blocked. CAP binding, however, is needed to recruit RNA polymerase to the promoter, so transcription cannot proceed efficiently without it.

Figure 5.29 shows how the lac repressor and CAP proteins act in a coordinated fashion to ensure that the *lac* operon is only transcribed to any extent when lactose is the abundant energy source and glucose is scarce. The mechanism depends on two metabolites, allolactose and cyclic AMP, that interact with the lac repressor and CAP, respectively, and alter their capacity to bind DNA. When lactose is abundant, allolactose, a related molecule, is produced in the cell. Allolactose, the **inducer** of the lac operon, binds the lac repressor and prevents it from binding to the operator. When glucose is low the concentration of an unrelated metabolite, cyclic AMP (cAMP) rises in the cell and only when cAMP binds to CAP can CAP bind DNA.

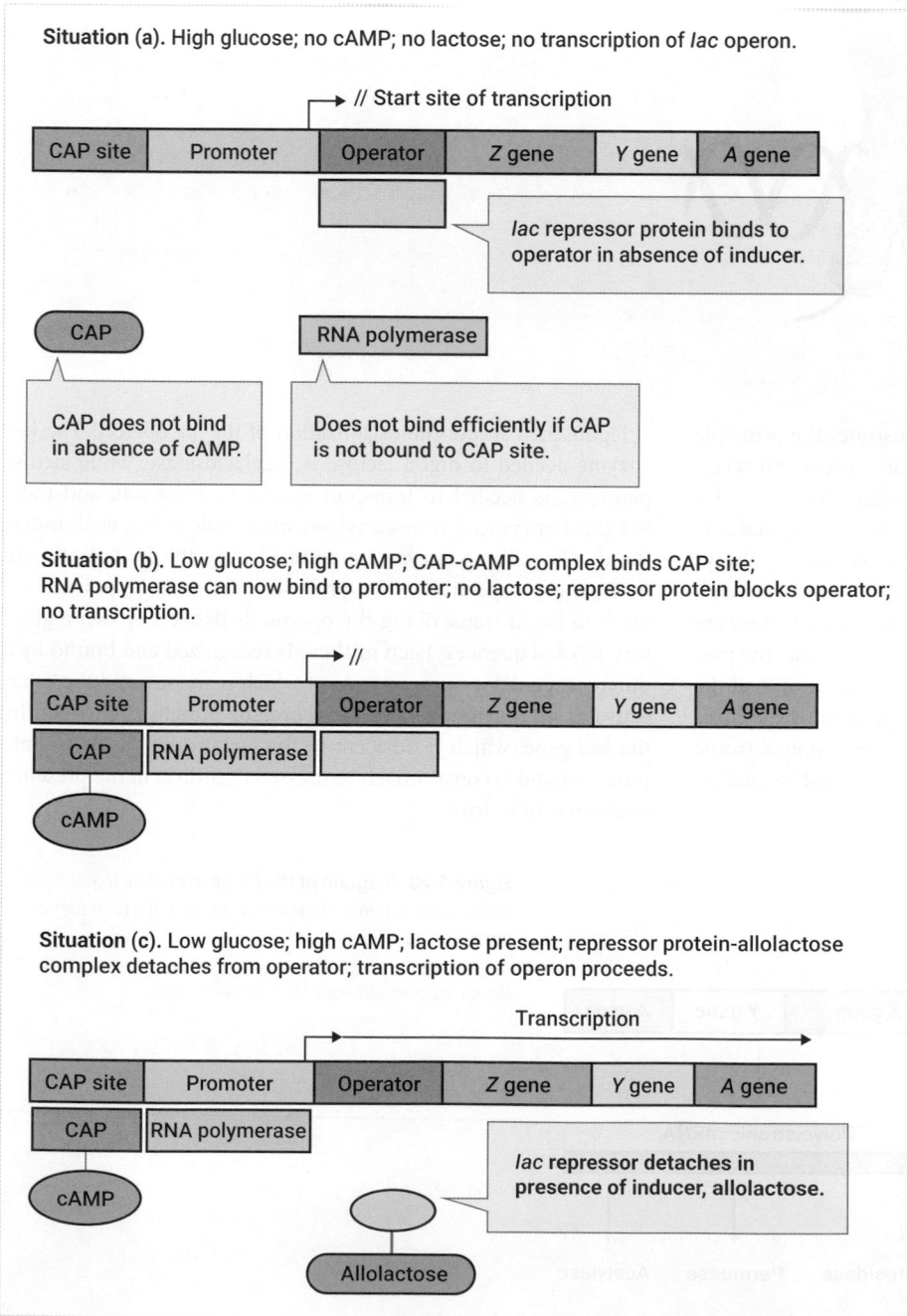

Situation (a). High glucose; no cAMP; no lactose; no transcription of *lac* operon.

// Start site of transcription

| CAP site | Promoter | Operator | Z gene | Y gene | A gene |

lac repressor protein binds to operator in absence of inducer.

CAP

RNA polymerase

CAP does not bind in absence of cAMP.

Does not bind efficiently if CAP is not bound to CAP site.

Situation (b). Low glucose; high cAMP; CAP-cAMP complex binds CAP site; RNA polymerase can now bind to promoter; no lactose; repressor protein blocks operator; no transcription.

//

| CAP site | Promoter | Operator | Z gene | Y gene | A gene |

| CAP | RNA polymerase |

cAMP

Situation (c). Low glucose; high cAMP; lactose present; repressor protein-allolactose complex detaches from operator; transcription of operon proceeds.

Transcription

| CAP site | Promoter | Operator | Z gene | Y gene | A gene |

| CAP | RNA polymerase |

cAMP

lac repressor detaches in presence of inducer, allolactose.

Allolactose

Figure 5.29 Expression of the *lac* operon. (a) In the presence of high glucose there is no cAMP to cause CAP to bind, and this binding is necessary for the attachment of RNA polymerase to the promoter. With no lactose the lac repressor is bound to the operator. (b) With low glucose but no lactose, although CAP binds and assists the RNA polymerase to bind, transcription still does not occur because the lac repressor is bound to the operator, blocking polymerase movement. (c) As glucose is low CAP has assisted RNA polymerase to bind. In the presence of lactose, the inducer allolactose binds to the repressor, causing its release from the operator and transcription can proceed.

Source: Snape, A. & Papachristodoulou, D. (2018). *Biochemistry and Molecular Biology* (6th edn). Oxford University Press.

The *lac* operon was the first understood example of prokaryotic gene control, but similar mechanisms involving DNA binding repressor and activator proteins were later found to apply to operons associated with other metabolic pathways. Moreover, similar mechanisms operate in transcriptional regulation in eukaryotes, as discussed below.

PAUSE AND THINK

What would be the consequence, in terms of *lac* operon gene expression, of a mutation in the *lacI* gene that results in no functional lac repressor protein being made.

Answer: The lac operon genes would be expressed at all times (constitutively) regardless of the presence or absence of lactose in the growth medium.

Transcriptional regulation in eukaryotes involves promoter and enhancer DNA sequences

Regulation of transcription in eukaryotes involves several classes of DNA element that are illustrated in Figure 5.30 and described in more detail below.

We have already introduced eukaryotic gene promoters in Section 5.3. As a reminder, the region around the start site of transcription contains sequences that bind general transcription factors such as the TFIID complex, which are required to position RNA polymerase II correctly. These sequences make up the **basal** or **core promoter**; the minimal sequence of DNA that can correctly initiate transcription. However, with only the core promoter sequences,

transcription is slow and inefficient. Additional **upstream control elements** such as the CAAT and GC boxes (see Figure 5.9) are also present. These and other common control elements are present in the promoters of many genes, in variable numbers and at various positions. They are recognized and bound by transcription factors that are common to many cell types, and generally increase the rate of transcription.

In addition, there may be control elements that mediate many aspects of gene regulation, for example cell-type specific and hormonal control of gene expression. To give a single example, the promoters of genes encoding globin proteins have multiple regulatory elements containing the base sequence GATA (Figure 5.31). A specific transcription factor found only in developing red blood cells binds the GATA sequence and is required for activation of the globin genes. This factor is known as GATA-1, named for the DNA sequence it binds. Other cell types such as muscle do not contain GATA-1, and therefore globin genes are not expressed there.

The eukaryotic gene promoter is usually considered to be in the region of 200 base pairs upstream (5′) of the start site of transcription. However, fully regulated gene expression often involves much more distant sequences, known as **enhancers**, which can greatly increase the expression of the gene. Like promoters they contain clusters of short control sequences, many of which bind the same transcription factors that bind upstream control elements in promoters. The remarkable thing is that a given enhancer may be thousands of base pairs distant from the gene that is affected and may be upstream or downstream of the gene. For example, enhancers both upstream and downstream of the globin genes contain GATA-1 binding sites, as does the promoter, and all these control sequences work together to give a very high level of globin gene expression in developing red blood cells.

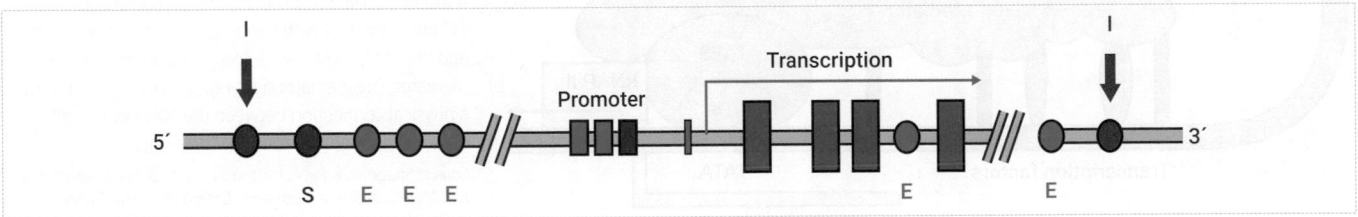

Figure 5.30 DNA elements involved in eukaryotic gene control. Transcription is indicated by the green arrow. The promoter contains both positive (green) and negative (red) regulatory elements. Enhancer elements (E) are typically thousands of base pairs away from the transcription start site and may be upstream (5′) or downstream (3′) of, or even within an intron of, the transcribed sequence. Distant regulatory elements that repress transcription are termed silencers (S). Insulators (I) are sequences that demarcate the regulatory unit and prevent the regulatory sequences within it from influencing adjacent genes.

Source: Snape, A. & Papachristodoulou, D. (2018). *Biochemistry and Molecular Biology* (6th edn). Oxford University Press.

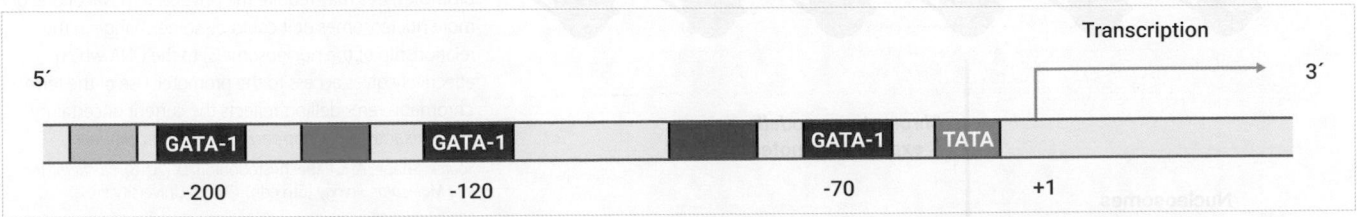

Figure 5.31 The promoter of the human β-globin gene contains binding sites for the red blood cell-specific transcription factor, GATA-1. The position of the TATA box is shown in green, while other coloured boxes represent binding sites for additional transcription factors that increase the rate of transcription.

Source: Adapted from B. Emerson. *Gene Expression: General and Cell-Type Specific* [M. Karin, ed.], pp. 116–161. Boston: Birkhauser, 1993.

Some distant regulatory sequences do not activate transcription, but instead repress it in circumstances where the gene product is not required. It is obviously not appropriate to call these sequences enhancers, so an alternative term, **silencers**, is used.

Eukaryotic transcription factors work via interactions with the transcription apparatus and with chromatin

Eukaryotic transcription factors recognize and bind promoter and enhancer sequences via their DNA-binding domains. As mentioned earlier, they also contain activation (or sometimes repression) domains that exert their effects on transcription once they are bound to their target gene. There are two main mechanisms by which these domains influence transcription, firstly by interaction with the transcription initiation complex (RNA polymerase and its associated general transcription factors) as shown in Figure 5.32, and secondly by chromatin modification.

Transcriptional activators can increase the efficiency with which RNA polymerase and its associated factors bind the promoter, or the efficiency with which they initiate transcription. When bound to more distant sequences in enhancers, they achieve this by looping of the DNA between the promoter and its enhancer so that the two can be brought into proximity with one another, as shown in Figure 5.32.

Chromatin modification is the other major mechanism by which transcription factors influence gene expression. As discussed in Chapter 3, eukaryotic genes are complexed with histones and other chromatin proteins. It is not difficult to envisage that these reduce the accessibility of DNA to RNA polymerase and its associated factors and, therefore, the 'default' state of chromatin is a 'shut-down' condition—the genes are inactive. Transcriptional activation in eukaryotes involves 'opening up' or unblocking the promoters. It requires modification of the nucleosome structure, a process known as **chromatin remodelling** (Figure 5.33). It is not known whether a nucleosome physically leaves the DNA or just changes its attachment to permit the transcription complex to assemble on the promoter. It is, however, known that nucleosome remodelling is carried out by protein factors and is an energy-requiring ATP-dependent process.

Another form of chromatin modification is covalent modification via enzymes that add chemical groups such as acetyl (ethanoyl), methyl, or phosphate groups to amino acid residues in the histone proteins. Histones contain many lysines and arginines, as these amino acids have basic, positively charged side chains that interact with the negative sugar–phosphate backbone of DNA. Addition of extra groups alters these interactions, making the DNA more (or less) accessible for transcription. Histone acetylation by **histone acetyl transferase** enzymes (HATs) negates the positive charge on lysines and arginines, so it is generally associated with

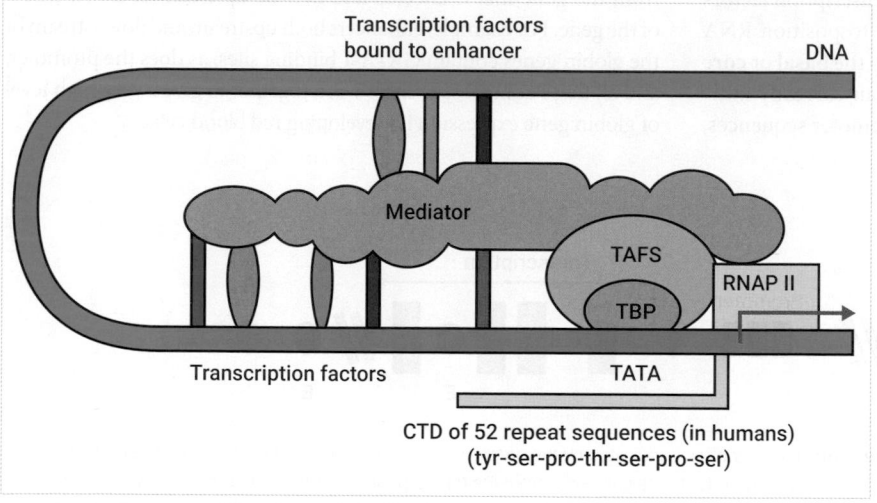

Figure 5.32 Illustration of the transcription initiation complex. The sizes, positions, and interactions of the various components are conjectural. They represent the binding of the general transcription factors (TBP and associated TAFS of the TFIID complex) to the TATA box, which positions the RNA polymerase II (RNAPII) at the correct site. The upstream sequence-specific transcription factors bound to the promoter and enhancer interact with the general transcription factors and RNA polymerase via a complex of proteins called the mediator. The mediator does not bind to DNA but forms a physical connection between the polymerase and other components of the initiation complex.

Source: Snape, A. & Papachristodoulou, D. (2018). *Biochemistry and Molecular Biology* (6th edn). Oxford University Press.

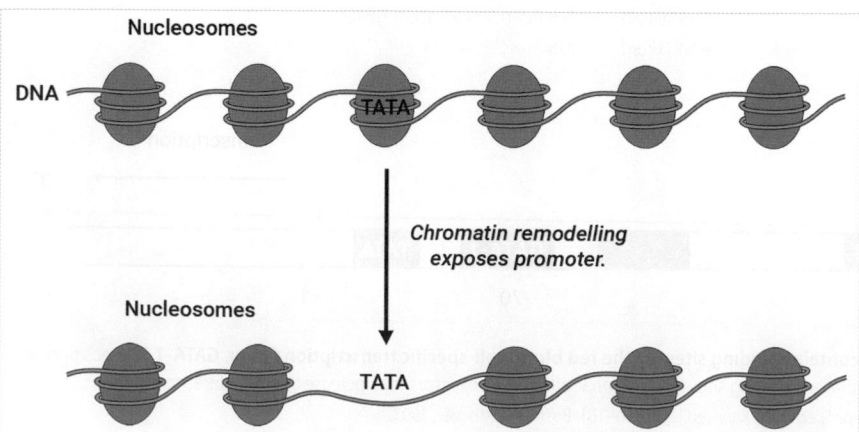

Figure 5.33 Chromatin remodelling. The principle is that the promoter in chromatin is blocked by nucleosomes. Gene activation requires exposure of the promoter; this may require the physical removal of one or more nucleosomes or it could be some change in the relationship of the nucleosome(s) to the DNA which effectively gives access to the promoter. Use of the term 'chromatin remodelling' reflects the current uncertainty about exactly what happens at the molecular level.

Source: Snape, A. & Papachristodoulou, D. (2018). *Biochemistry and Molecular Biology* (6th edn). Oxford University Press.

'loosely wound' chromatin and therefore with transcriptionally active genes. Removal of acetyl groups by **histone deacetylase** enzymes (HDACs) has the opposite effect, shutting down transcription. Several transcription factors work by recruiting these and other histone-modifying enzymes to the chromatin at the region where they bind to the DNA.

We have so far described eukaryotic transcription factors mainly as activators of transcription. However, eukaryotic gene transcription is typically regulated by multiple signals to a given promoter, and these signals may be either negative or positive. The combination of multiple signals gives finely balanced regulation that can respond to complex situations. Negative control is achieved by transcription factors acting as repressors. Repression by transcription factors can occur through a wide variety of mechanisms, some essentially equivalent to those operating in activation, that is by interaction of a repression domain with the initiation complex, or by repressive chromatin modification such as histone deacetylation. However, refinement of eukaryotic gene control is often achieved by transcriptional activators and repressors competing for the same or overlapping binding sites on DNA, and in this case the repressor may act simply by blocking access of an activator to the promoter. Whether the gene is activated or repressed will therefore depend on the relative levels or activities of the repressor and the activator in the cell (Figure 5.34).

Gene control is required for cell differentiation and regulating cell function

Gene control in eukaryotes is based on the same principles as in prokaryotes, but it is more complex, partly because of packaging of eukaryotic genomes into chromatin, as discussed above. A second complication is the large number of different cell types in multicellular organisms. An *E. coli* cell has about 4000 genes, each of which may need to be transcribed at some point. For the most part what is required is an 'on-off' switch for regulated genes, such as those of the *lac* operon, or, for constitutive genes, a permanent 'on' condition. A human cell has about 21,000 protein-coding genes and in humans there are many different types of cell, such as liver, muscle, brain, epithelial, blood, and bone cells, all of which arise through differentiation of embryonic cells that originate from a single fertilized egg. Many proteins are common to all cell types, for example the enzymes required for glycolysis, and these are encoded by what are known as housekeeping genes. However, each type of cell also has its own cohort of proteins needed for specific cell functions, such as globin required in red blood cells, for which cell-type specific gene regulation is required.

Even when differentiation into cell types has been achieved, another layer of gene control is required. The activities of each cell must be such that they correspond to the needs of the whole organism. An obvious example is that cell division should not proceed independently (as happens in cancer): the cell must receive one or more signals from other cells before it proceeds to division. Additionally, the rate of synthesis of individual proteins varies over time according to the organism's needs. For example, after a meal, the level of enzymes required for breakdown or storage of foodstuffs increases through hormonal activation of specific genes. The activities of many cells are controlled by a whole battery of hormones, **growth factors**, and **cytokines** which regulate appropriate genes. A given gene in a cell may be simultaneously regulated by a multiplicity of signals from hormones or other factors.

A consequence of this complexity is that there can be many short regulatory DNA sequences associated with each eukaryotic gene. A plethora of different transcription factors in eukaryotic cells interact with these sequences. It has been estimated that more than 5 per cent of human genes code for transcription factors. Cell-type specific gene expression often depends on the presence of

Figure 5.34 Some eukaryotic transcriptional repressors work by blocking access of activator proteins to DNA. Gene expression is regulated by modulating the relative levels of activator and repressor proteins in the cell.

Source: Lodish et al. *Molecular Cell Biology*, Edition 4, Figure 10-62. Available on NCBI bookshelf: https://www.ncbi.nlm.nih.gov/books/NBK21677/#A2649.

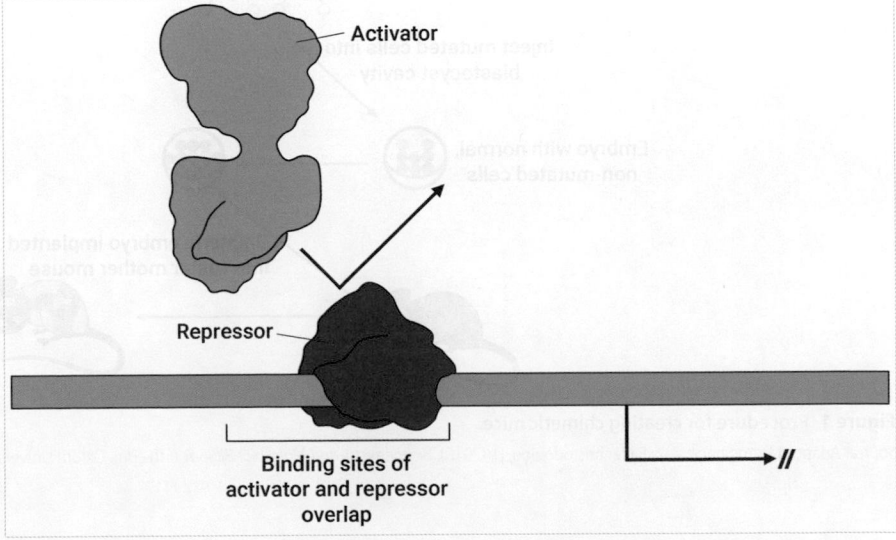

particular transcription factors in the cells; for instance, the GATA-1 factor is only present in developing red blood cells, as described in Scientific Process 5.1. Some transcription factors are activated only when the cell receives the appropriate signal such as from a hormone. This system gives some of the very great flexibility needed for gene control in eukaryotes.

SCIENTIFIC PROCESS 5.1 — Transcription factor GATA-1 is required for red blood cell development

Research question

Based on previous evidence that showed that the transcription factor GATA-1 binds to regulatory sequences of globin genes and other genes that are specifically expressed in erythroid cells (developing red blood cells), Pevny and colleagues tested the hypothesis that GATA-1 is required for red blood cell differentiation.

Materials and methods

The study made use of mouse embryonic stem cells (ESCs), which are cells derived from embryos at a very early stage in their development. These cells are pluripotent; that is they are not yet specialized and should be able to develop into various cell types. In one set of ESCs, DNA manipulation was used to introduce a targeted mutation into the gene that encodes GATA-1. As the GATA1 gene is on the X chromosome and the ESCs used in this study were derived from male embryos, it was only necessary to mutate a single copy of the gene to produce ESCs that could not make any GATA-1 protein. These GATA-1-deficient ESCs were then used to create chimeric embryos (Figure 1) containing a mixture of cells with a normal functioning GATA1 gene, and those in which GATA1 was mutated. The chimeric embryos were transferred into foster mother mice and allowed to develop.

Results

Tissues from fetal and neonatal mice derived from chimeric embryos were analysed to test whether they contained cells derived from both the normal and GATA-1-deficient embryonic cells. The derivation of cells in each tissue could be distinguished because they contained slightly different forms of a protein, GPI, that are separated from each other by electrophoresis (a method for separating proteins according to their charge). A representative set of results from one of nine mice tested is shown in Figure 2a and demonstrates that each of the tissues tested contains cells derived from both normal and GATA-1-deficient embryonic cells, apart from blood, which has contributions only from normal embryonic cells. More detailed analysis, shown in Figure 2b, showed that GATA-1-deficient ESCs can make a small contribution to blood, but only to white blood cells and not to red blood cells. The researchers noted that many of the chimeric mice were not healthy, and analysis of the liver of one of the fetal

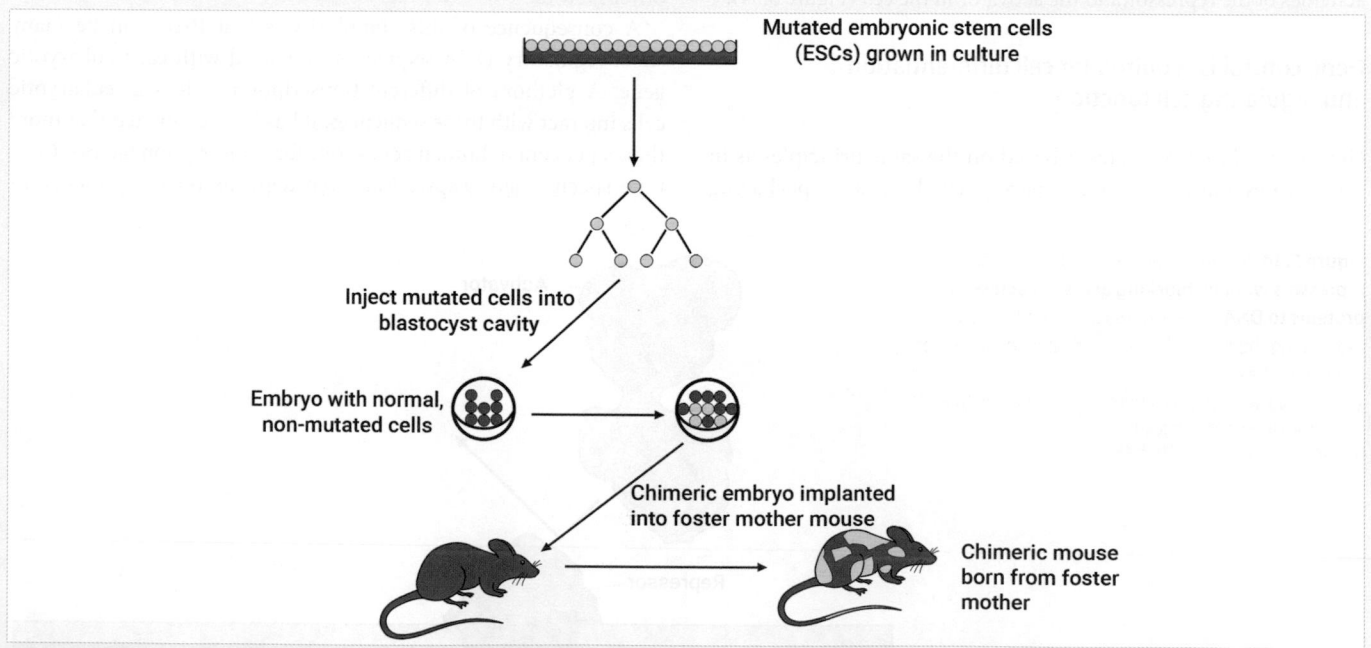

Figure 1 Procedure for creating chimeric mice.

Source: Adapted from Snape, A. & Papachristodoulou, D. (2018). *Biochemistry and Molecular Biology* (6th edn). Oxford University Press.

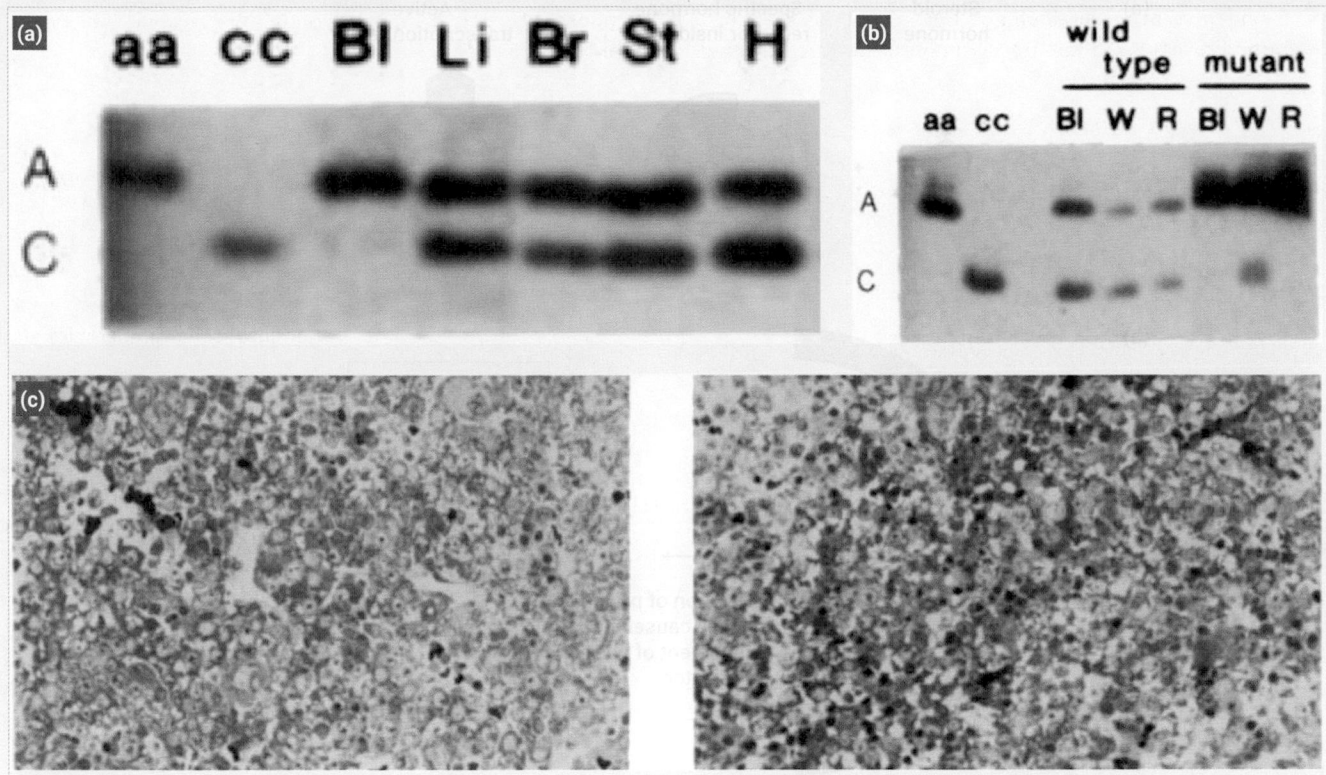

Figure 2 Analysis of tissues and blood from a chimeric mouse. (a) Analysis of tissues from a chimeric mouse. The figure shows electrophoretic analysis of the GPI protein. GPI$_A$ comes from normal cells, and GPI$_C$ comes from cells derived from GATA-1-deficient ESCs. Lanes marked aa and cc are samples of blood from control mice containing only GPI$_A$ (aa) or GPI$_C$ (cc) cells. Tissues tested are: Bl (blood), Li (liver), Br (brain), St (stomach), H (heart). (b) Analysis of blood from a chimeric mouse. aa and cc are controls as in (a). 'Wild type' is blood from a control mouse that has only normal non-GATA-1-deficient cells, some of which make GPI$_A$ and some GPI$_C$. This shows that both GPI$_A$ and GPI$_C$ cells can normally make white and red blood cells. 'Mutant' is from a chimeric mouse that has both normal and GATA-1-deficient cells. GPI$_A$ comes from normal cells, and GPI$_C$ comes from cells derived from GATA-1-deficient ESCs. 'Bl' is whole blood, 'W' is white blood cells, 'R' is red blood cells. (c) Samples of liver from a fetal chimeric and a control fetus. The liver is the site of red blood cell development at the fetal stage. The sample on the right is from the control, normal fetus, and contains many erythroid precursor cells, recognizable by their darkly staining nuclei. The sample on the left is from the chimeric fetus, and erythroid precursor cells are relatively scarce.

Source: Reprinted by permission of Springer Nature. Pevny, L., Simon, M. C., Robertson, E., Klein, W. H., Tsai, S. F., D'Agati, V., Orkin, S. H., and Costantini, F. Erythroid differentiation in chimaeric mice blocked by a targeted mutation in the gene for transcription factor GATA-1. *Nature* 349, 257–60 (1991).

chimeric mice showed that it was deficient in erythroid cells (Figure 2c), i.e. the normal embryonic cells had not completely compensated for failed development of GATA-1-deficient red blood cells.

Conclusion

This study demonstrated the importance of a transcription factor, GATA-1, in the differentiation of red blood cells. Similar studies have also shown

the key role of transcriptional regulation and specific transcription factors in the development of specialized cell types, such as nerve and muscle, in multicellular eukaryotic organisms.

Read the original paper

Pevny, L., Simon, M. C., Robertson, E., Klein, W. H., Tsai, S. F., D'Agati, V., Orkin, S. H., & Costantini, F. (1991) Erythroid differentiation in chimaeric mice blocked by a targeted mutation in the gene for transcription factor GATA-1. *Nature* **349**: 257–60.

Cell signalling contributes to gene control

Transcription factors that are involved in complex gene control often exist in an inactive form in the cell and cannot stimulate transcription until they are activated. Activation may be by phosphorylation or another modification that causes a conformational change in the protein, allowing it to bind its target DNA sequence. Activation may also be associated with the movement of the transcription factor from the cytosol to the nucleus where it can then bind to DNA.

Activation is usually the result of signals arriving at the cell from other cells. Figure 5.35 gives two examples, in outline, of the activation mechanisms involved: in Figure 5.35a a steroid hormone is shown to enter the cell directly (due to its lipid solubility) and on binding to a soluble receptor protein causes a conformational change in the latter so that it is now an active transcription factor. Figure 5.35b shows the general concept of the way in which many hormones bind to membrane receptors and induce a signal cascade inside the cell. This results in activation of specific transcription factors, often by their phosphorylation.

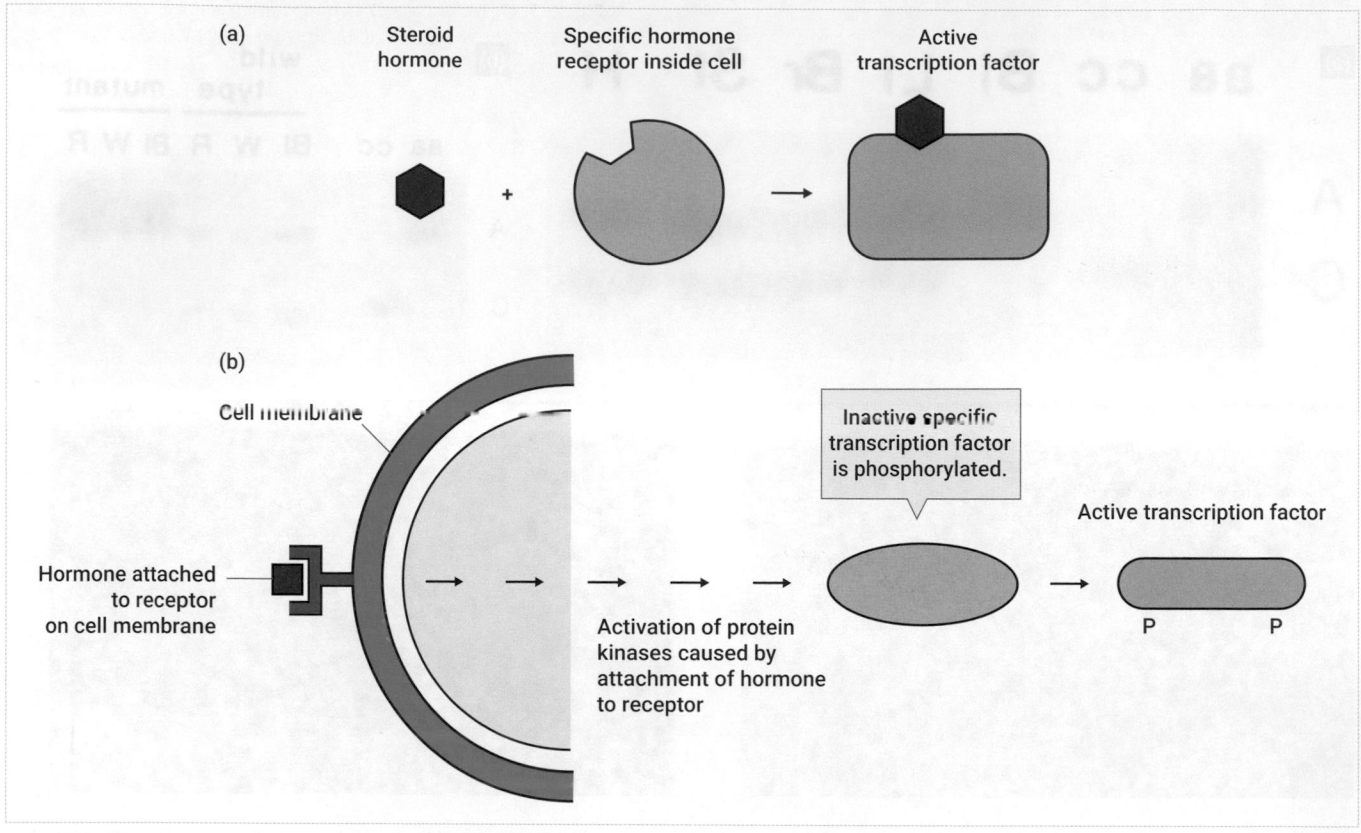

Figure 5.35 Examples of transcription factor activation. (a) A steroid hormone enters the cell; it attaches to a receptor specific for that hormone and causes a conformational change in the receptor protein, which is now an active transcription factor. This activates the gene(s) that are controlled by the hormone. (b) Protein hormones such as insulin do not enter the cell but bind to receptors on the cell surface. This results in a sequence of events that ends in the phosphorylation of the appropriate inactive transcription factors and activates them. Note that there are many different hormones that bind to specific receptors and activate different transcription factors by varying mechanisms. Often activation causes transport of the transcription factors from the cytosol into the nucleus. The transcription factors bind to specific response elements of different genes. Thus, each hormone can exert control over appropriate genes.
Source: Snape, A. & Papachristodoulou, D. (2018). *Biochemistry and Molecular Biology* (6th edn). Oxford University Press.

DNA methylation maintains gene control through cell division, giving cells 'memory' of their status

Methylation of bases of DNA occurs in both bacteria and eukaryotes. In bacteria, as described in Chapter 3, it distinguishes the 'old' from the 'new' DNA strand immediately following DNA replication. In mammals, methylation of cytosine in so-called 'CpG' sequences has a different role, transcriptional regulation. The presence of 5-methylcytosine, shown in Figure 5.36, is associated with transcriptionally inactive genes.

A key feature of mammalian cytosine methylation is that the methylation pattern of a DNA sequence can be preserved through DNA replication and cell division. To understand this, we must consider the responsible enzymes, DNA methyltransferases. There are two classes of methyltransferases: *de novo* **methyltransferases** methylate previously unmethylated DNA, while **maintenance methyltransferases** recognize methylated cytosine in CpG on the template DNA strand and add a methyl group to the cytosine paired with the guanine of the CpG on the newly replicated strand (Figure 5.37). The importance of this is that CpG methylations associated with particular patterns of gene expression, and hence with particular differentiated cell types, are preserved when the cell divides.

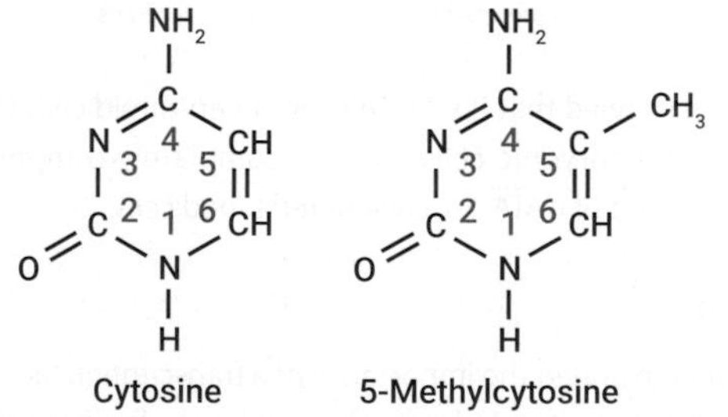

Figure 5.36 Cytosine and 5-methylcytosine.
Source: *Biochemistry and Molecular Biology*, 6th Edn. Despo Papachristodoulou, Alison Snape, William H. Elliott, and Daphne C. Elliott.

This kind of change in DNA, which is heritable in the sense of being preserved through cell division but does not involve a mutation or change of base sequence, is termed an **epigenetic** modification.

Post-transcriptional gene regulation allows fine tuning of gene expression

Although gene control mainly occurs at the first stage of gene expression, there are many examples of regulation at later stages.

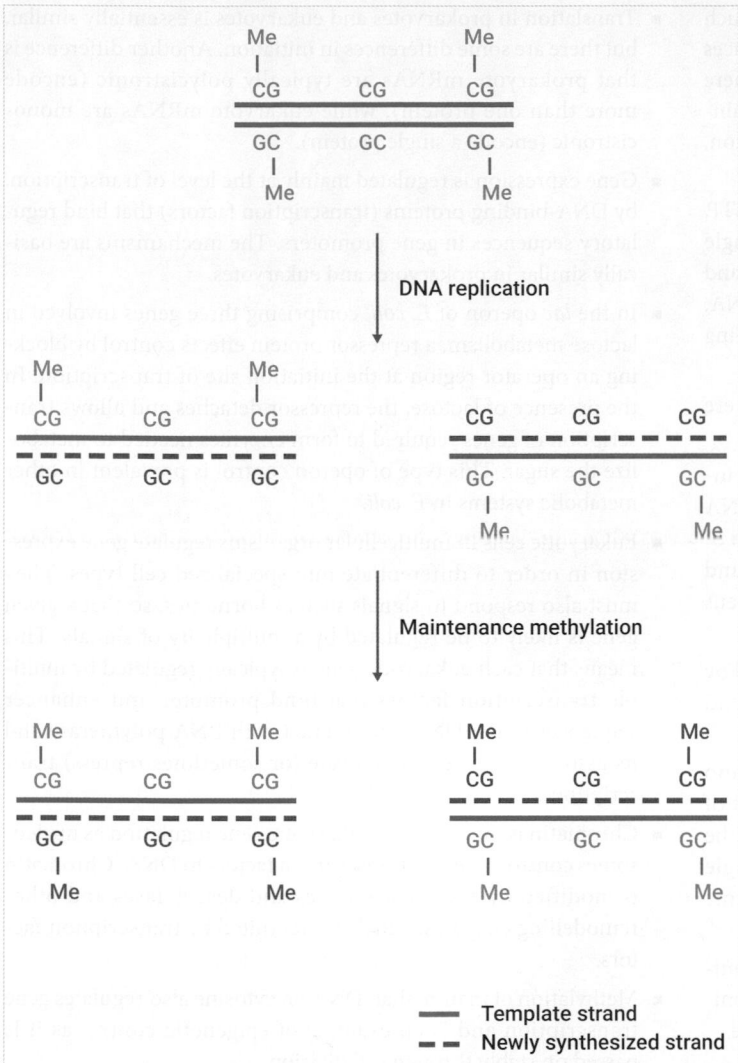

Figure 5.37 Differential DNA methylation is preserved after DNA replication by the action of maintenance methyltransferases. The methyltransferase enzyme recognizes 5-methylcytosine in CpG islands on the template DNA strands, and methylates corresponding cytosines on the newly synthesized strands.

Source: Snape, A. & Papachristodoulou, D. (2018). *Biochemistry and Molecular Biology* (6th edn). Oxford University Press.

Once transcription is initiated it is sometimes terminated before synthesis of the RNA is complete. This mechanism, known as **attenuation**, is used in prokaryotes to regulate transcription rapidly in response to fluctuating levels of key metabolites in the cell. In eukaryotes, alternative splicing of mRNA is also a potential control point: the relative quantities of differently spliced mRNAs (and thus different proteins) produced from a single gene may vary in different cell types or in response to changed conditions. The stability of mRNAs is also regulated; the longer the lifetime of a message the more rounds of translation it can undergo, and hence the more protein is made. Translation may also be regulated directly, and the breakdown of translated proteins is another regulated process.

A relatively recent discovery has been that of widespread eukaryotic gene regulation by several classes of small RNA. Small regulatory RNAs exert their effect through base pairing with mRNAs, either causing them to be broken down or interfering with their translation. This phenomenon of gene silencing by **RNA interference (RNAi)** has been the focus of recent attention, not least because it offers a powerful experimental and potentially therapeutic tool for controlling gene expression.

 Check your understanding of the concepts covered in this section by answering the questions in the e-book.

SUMMARY OF KEY CONCEPTS

- Gene expression involves two processes: transcription (synthesis of RNA using a DNA template) and translation (synthesis of protein by linking amino acids in a sequence determined by the base sequence of mRNA).

- Three main types of RNA are synthesized by transcription: mRNA, rRNA, and tRNA. Each has a distinct function in protein synthesis.

- The genetic code defines how the base sequence of DNA, which is copied into mRNA, directs the synthesis of proteins. Sequences of three bases (codons) each encode a single amino acid. There are 64 codons, of which 61 encode amino acids, with the remaining three being stop codons. AUG is the initiator or start codon, encoding methionine.

- Transcription is catalysed by RNA polymerase, using ATP, CTP, GTP, and UTP as building blocks. Prokaryotes have a single RNA polymerase, while eukaryotes have three (Pol I, Pol II, and Pol III). Pol II transcribes protein-coding genes to make mRNA, while Pol I and Pol III transcribe other types of RNA, including rRNAs and tRNAs.

- Promoter and terminator sequences in the DNA direct where RNA polymerase should begin and end transcription.

- In prokaryotes mRNA can be translated as soon as it is transcribed in the cytosol, while in eukaryotes the pre-mRNA undergoes post-transcriptional modifications: addition of a 5′ cap and a 3′ poly(A) tail and splicing to remove introns and retain exons. The mature mRNA is exported from the nucleus for translation in the cytosol.

- The sequence of codons in an mRNA molecule is translated by cytosolic ribosomes; large complexes containing rRNA and many proteins.

- Transfer RNAs (tRNAs) act as adaptors between the amino acids and the mRNA. Each tRNA has an unpaired triplet of bases known as the anticodon, which hydrogen bonds to the codon on the mRNA. Wobble base pairing allows a single tRNA to bind to more than one codon that encodes the same amino acid.

- Translation has three phases: initiation, elongation, and termination. Initiation involves assembly of the ribosome in a complex with mRNA and the initiator tRNA carrying methionine.

- Elongation involves delivery of aminoacyl-tRNAs one at a time to the ribosome, which moves along the mRNA in the 5′ to 3′ direction. Each incoming amino acid is added to the C-terminus of the growing peptide chain by formation of a peptide bond. When the ribosome reaches a stop codon the protein is released.

- Translation in prokaryotes and eukaryotes is essentially similar, but there are some differences in initiation. Another difference is that prokaryote mRNAs are typically polycistronic (encode more than one protein), while eukaryote mRNAs are mono-cistronic (encode a single protein).

- Gene expression is regulated mainly at the level of transcription, by DNA-binding proteins (transcription factors) that bind regulatory sequences in gene promoters. The mechanisms are basically similar in prokaryotes and eukaryotes.

- In the *lac* operon of *E. coli*, comprising three genes involved in lactose metabolism, a repressor protein effects control by blocking an operator region at the initiation site of transcription. In the presence of lactose, the repressor detaches and allows transcription of genes required to form enzymes needed to metabolize the sugar. This type of operon control is prevalent in other metabolic systems in *E. coli*.

- Eukaryotic cells in multicellular organisms regulate gene expression in order to differentiate into specialized cell types. They must also respond to signals such as hormones, so that a given gene is likely to be regulated by a multiplicity of signals. This means that each eukaryotic gene is typically regulated by multiple transcription factors that bind promoter and enhancer sequences in the DNA and interact with RNA polymerase and its associated protein to activate (or sometimes repress) transcription.

- Chromatin is important in eukaryotic gene regulation as nucleosomes control access of transcription factors to DNA. Chromatin is modified by histone acetylases and deacetylases and other remodelling enzymes, which are recruited by transcription factors.

- Methylation of mammalian DNA on cytosine also regulates gene transcription and is an example of epigenetic control as it is passed on stably through cell division.

- Gene expression can be fine-tuned by post-transcriptional regulation. A recent discovery is RNA interference (RNAi): a natural mechanism of silencing protein-coding genes in eukaryotes, which has potential as an experimental and therapeutic tool.

 Use the flashcards in the e-book to test your recall of key terms introduced in this chapter.

QUESTIONS

 **Looking for answers?** Once you've answered these questions, follow the link in the e-book to the answer guidance and check your work.

Concepts and definitions

1. Explain the difference between transcription and translation.

2. Give two ways in which RNA synthesis differs from DNA synthesis.

3. What is the role of the sigma (σ) subunit in initiation of prokaryote transcription?

4. In the genetic code 61 codons are used to specify the 20 amino acids, but there are fewer than 61 tRNA molecules. Explain how this is so.

5. Describe how the *E. coli lac* operon is controlled in response to the presence or absence of lactose in the growth medium.

Apply the concepts

6. In what ways does eukaryotic transcription differ from that in prokaryotes?

7. If you are given the base sequence of the coding region of an mRNA, can you deduce the amino acid sequence of the protein it codes for? If you are given the amino acid sequence of a protein, can you deduce the sequence of the coding region of the mRNA that directed its synthesis? Explain your answers.

8. Studies have indicated that in *E. coli* tRNA molecules (with their aminoacyl or peptidyl attachments) straddle A, P, and E sites on the ribosome. Explain why this is advantageous.

9. The mechanism of initiation of translation in eukaryotes is not compatible with polycistronic mRNA. Explain why.

10. List two mechanisms by which eukaryotic transcription factors can repress rather than activate gene expression.

Beyond the concepts

11. Proofreading of RNA synthesis does not occur to the same extent as proofreading of DNA synthesis. Why do you think this is the case?

12. What possible biological significance does the existence of introns have?

13. Consider an mRNA that codes for a protein 200 amino acid residues in length. What would the resultant polypeptide be from translation of this messenger if codon number 100 was mutated so that its first base was deleted or if the first and second bases were deleted? What if all three bases were deleted? Explain your conclusions.

14. Mitochondria contain their own DNA genome, which encodes some of the proteins required for their structure and function (others are encoded in the nucleus). In order to express these proteins, mitochondria utilize their own ribosomes, which are 70S rather than 80S ribosomes. The initiator amino acid in mitochondria is *N*-formylmethionine rather than unmodified methionine. Suggest an explanation for these observations.

15. What is one medical potential of RNAi?

FURTHER READING

Woychik, N. A. & Reinberg, D. (2001) RNA Polymerases: Subunits and Functional Domains. eLS http://www.els.net DOI: 10.1038/npg.els.0003302
A review of the structure, function and evolution of both prokaryotic and eukaryotic RNA polymerases in an online resource: eLS (Encyclopedia of Life Science).

Ast, G. (2005) The alternative genome. *Sci. Am.* **292**: 58–65.
A readable discussion of alternative splicing.

Nirenberg, M. (2004) Historical review: deciphering the genetic code—a personal account. *Trends Biochem. Sci.* **29**: 46–54.

A personal history describing the experiments that enabled the genetic code to be worked out, by one of the 1968 Nobel prize winners.

Yonath, A. (2005) Ribosomal crystallography: peptide bond formation, chaperone assistance and antibiotics activity. *Mol. Cells* **20**: 1–16.
A 'mini-review' by a Nobel prizewinner, relating findings on the structure of ribosomes to various aspects of protein synthesis and folding, and explaining how antibiotics can interfere with ribosome function in a selective fashion.

Struhl, K. (1999) Fundamentally different logic of gene regulation in eukaryotes and prokaryotes. *Cell* **98**: 1–4.
Mini-review giving a good overview of the principles of gene regulation, with additional information on chromatin remodelling in eukaryotes.

Proteins and Proteomes

Chapter contents

Introduction   284

6.1   Amino acids are the building blocks
of proteins   285

6.2   There are four levels of protein structure   291

6.3   Protein structures reflect their function   296

6.4   Enzymes: proteins as biological catalysts   306

6.5   The proteome is the full set of proteins
expressed in a cell, tissue, or organism   315

Watch the key concepts video in the e-book to
prepare yourself for studying this chapter.

LEARNING OBJECTIVES

By the end of this chapter you should be able to:

- Classify the 20 amino acids that are commonly found in proteins according to their structures and properties.

- Define the four levels of protein structure.

- Explain that the amino acid sequence of a protein determines its three-dimensional structure and outline how proteins fold in the cell.

- Give examples of different types of proteins and explain how their structures reflect their function.

- Explain how proteins function as enzymes (biological catalysts) and describe key properties of enzymes.

- Compare enzymes that display Michaelis–Menten kinetics with allosteric enzymes and outline different ways in which enzymes are regulated.

- Define the terms proteome, genome, proteomics, genomics, and bioinformatics, and describe sample applications of genomics and proteomics in science and medicine.

Introduction

In Chapter 5 we described how the sequence of nucleotides in the DNA genome directs the synthesis of proteins. While DNA and RNA can be regarded as molecules that store and transmit information to support life, proteins carry out an astonishingly diverse range of functions. Proteins can be hormones that transmit signals from one cell to another or they can transport other molecules such as oxygen around the body by carrying them in the blood. They can transport molecules in and out of cells, or to different regions within cells. The three-dimensional structure of proteins allows them to interact with each other with exquisite precision, enabling them to act as receptors on cell surfaces, and to build stable complexes that provide structure and strength to cells, tissues, and organs.

Yet proteins are also flexible molecules that can change their arrangements ('conformations') in three-dimensional space, enabling them to contribute to dynamic processes such as muscle contraction and cell movement. It is remarkable that the microscopic movements of protein molecules can be multiplied and organized to allow our muscles to contract with enormous force. Large assemblies of proteins such as those found in ribosomes or in skeletal muscle, which come together to carry out complex but precise processes, can be even be regarded as 'molecular machines'.

The dynamic nature of proteins also allows them to play a key role in the biochemical reactions of cellular metabolism through their function as enzymes (biological catalysts), a specialized function that we consider in Section 6.4.

This incredible diversity of protein function is reflected in their huge range of shapes and sizes, some of which are illustrated in Figure 6.1. The building blocks of proteins are amino acids, and just 20 of these can be linked together in different numbers and different sequences to make the thousands of proteins that give such variety to life processes.

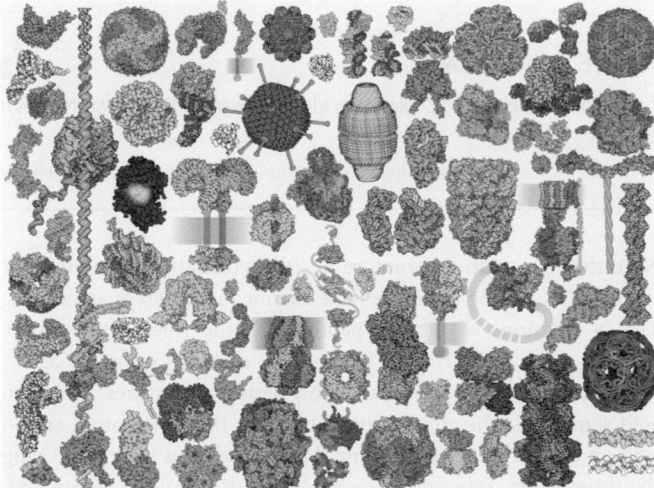

Figure 6.1 The diversity of protein structures. An illustration created by molecular biologist and artist David Goodsell for the Research Collaboratory for Structural Bioinformatics Protein Data Bank (RCSB PDB), showing the diversity and beauty of protein structures. The Protein Data Bank is a repository for the three-dimensional structural data of large biological molecules and is available worldwide as a free resource for research and education. Goodsell's illustrations appear regularly in the popular 'Molecule of the Month' feature in the PDB.
Source: Copyright David S. Goodsell and the RCSB PDB. Licensed via CC-BY-4.0 licence.

Much of the work of protein scientists is devoted to solving the structures of proteins and trying to understand the 'rules' that enable cells to fold a linear string of amino acids into exactly the right shape. Solving protein structures can help investigators to understand exactly how proteins work, for instance by clarifying how protein antibodies react with antigens and thereby trigger an immune response to infection. In turn this understanding can be of medical significance; for example, it enables researchers to design drugs that block the antibody–antigen interaction when it takes place inappropriately, as in allergies. Understanding protein structure can lead researchers to devise their own 'engineered' proteins that are of medical or industrial significance, such as gene therapy proteins that work within cells to repair faulty DNA, or enzymes that break down plastic bottles to reduce pollution.

In this chapter we first explain the basics of protein structure, including the properties of the 20 amino acids that form the building blocks of proteins. We then look in detail at some examples that illustrate key points about the structure and function of proteins.

6.1 Amino acids are the building blocks of proteins

Let us start by discussing some terminology, to make sure you understand how the terms **peptide**, **polypeptide**, and **protein** are used when describing protein structure, and explain how we describe the sizes of protein molecules.

- A peptide is a molecule that consists of amino acids linked together by peptide bonds. Strictly speaking, the term could be used for a molecule containing any number of amino acids, but in practice it tends to be used for a sequence of 30 or fewer amino acids, while the term 'polypeptide' is used for longer chains.

- A protein is a functional molecule made up of one or more polypeptide chains. The largest known protein is titin, an abundant protein in muscle, which has between 27,000 and 33,000 amino acids in a single polypeptide chain. The enormous (for a protein) size of titin reflects the requirement for it to link different parts of the contractile apparatus of muscle cells together and to stretch and contract, like a molecular spring, as the muscle contracts and relaxes. Most proteins do not have to be anything like this size in order to carry out their function: sizes of a hundred to a thousand amino acids are more common. You can read more about titin and its role in muscle contraction in Chapter 20.

When we define the size of a polypeptide or protein, we often do so by stating how many amino acids it contains. However, another convention is to quote its molecular mass in daltons (Da) or kilodaltons (kDa), where a dalton is approximately equal to the mass of a hydrogen atom. For example, adult human haemoglobin contains 574 amino acids and has a molecular mass of approximately 64,000 Da or 64 kDa.

So, what does the structure of a peptide or polypeptide look like?

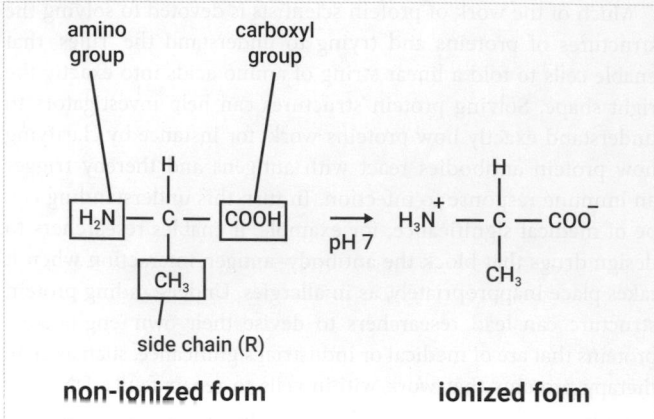

Figure 6.2 Structure of the amino acid, alanine. The central carbon, amino, and carboxyl groups, and the hydrogen atom are common to all amino acids. The side chain or R group is variable. At pH 7 (the pH of the intracellular environment) the amino group acts as a base and accepts an extra hydrogen ion from water, while the carboxylic acid loses a hydrogen ion, as shown. Thus, the amino acid carries a positive and a negative charge.
Source: Alberts et al. *Essential Cell Biology*, 4th edn. Figure 2.22 A.

Amino acid structure and peptide bonding

Firstly, let us look at the basic structure of an amino acid, which you can see in Figure 6.2.

The central carbon of the amino acid is termed the α-carbon; every amino acid has the same three groups bonded to the α-carbon:

- an amino group;
- a carboxyl group; and
- a hydrogen.

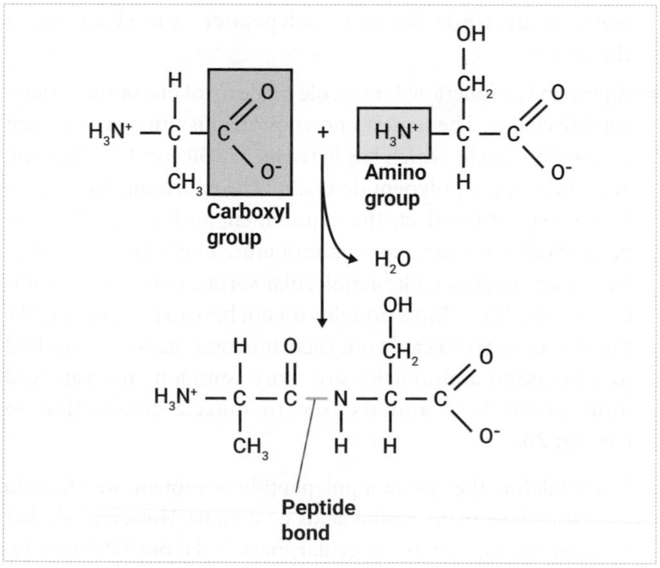

Figure 6.3 Formation of a peptide bond between two amino acids, alanine on the left and serine on the right. The amino acids differ from each other only in the structure of their side chains, which are not involved in peptide bond formation. The peptide bond is formed by a condensation reaction between the carboxyl group of one amino acid and the amino group of another, as shown.

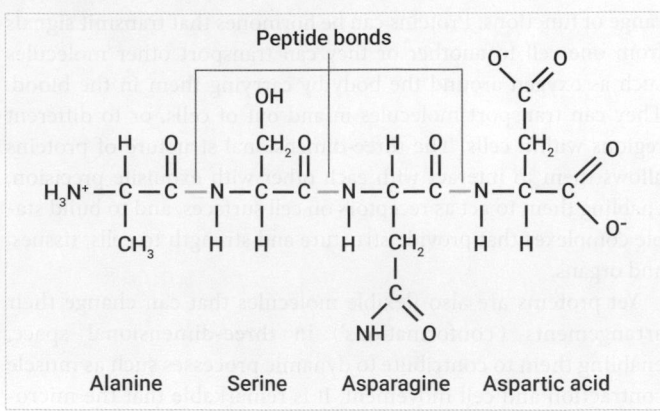

Figure 6.4 A tetrapeptide consisting of four amino acids linked by three peptide bonds. The names of the amino acids are shown below the structure.

The fourth bond of the α-carbon attaches the R group or side chain, which is the part of the structure that varies between amino acids.

Amino acids can be linked together by a peptide bond. A peptide bond is formed by a condensation reaction that occurs between the free amino group ($-NH_3^+$) of one amino acid and the free carboxyl group ($-COO^-$) of another, with the removal of H_2O. The result is **a dipeptide**, as shown in Figure 6.3.

The dipeptide is the simplest form of a 'peptide unit'. As multiple identical peptide bonds are formed, a peptide structure emerges as shown in Figure 6.4.

It is worth a brief reminder here that, as described in Chapter 5, cellular protein synthesis is not simply a series of condensation reactions, as each amino acid is first bonded to a tRNA and then 'handed over' to the growing peptide chain. However, the resulting peptide structure is the same as that shown in Figure 6.4.

PAUSE AND THINK

If a polypeptide contains 80 amino acids, how many peptide bonds does it contain?

Answer: 79 (one less than the number of amino acids).

Note that the peptide structure has **directionality**, meaning that the two ends of the molecule have chemically different structures. The N terminus of the polypeptide has the free amino group, $-NH_3^+$, and the C terminus has the free carboxyl group, $-COO^-$. When writing out peptide and polypeptide structures as amino acid sequences, the convention is to write the N-terminal amino acid on the left and the C-terminal amino acid on the right. Using the three-letter code for amino acids (given in full in Figure 6.6) the sequence of the peptide shown in Figure 6.4 could be written as Ala-Ser-Asn-Asp.

It is useful to explain another piece of terminology, the term '**residues**', before we turn our attention to the properties of the individual amino acids. Amino acids that are linked together by peptide bonds are often termed residues. We may talk about 'a 20-residue peptide', a '400-residue protein', or 'the N-terminal and C-terminal residues' rather than 'the N-terminal and C-terminal amino acids'. The term is used as it reflects the loss of some atoms of each amino acid (to make water) when the peptide bond is formed. What is left of each amino acid in the peptide is termed a residue.

PAUSE AND THINK

Name the N- and C-terminal amino acids for the following peptides, which are written using the one letter code shown in Figure 6.6.

(a) ANTGH

(b) SPCR

Answer:

(a) N terminal: alanine; C-terminal: histidine.

(b) N terminal: serine; C terminal: arginine.

Structure and properties of the 20 amino acids commonly found in proteins

A functional protein is formed when a polypeptide 'folds', as shown in Figure 6.5. The stable three-dimensional shape of the folded protein is crucial for its function and depends on specific interactions (mainly non-covalent bonds) between the amino acids, and between the amino acids and molecules in the protein's environment—for example, water molecules in the cytosol of a cell. The structure and properties of the individual amino acids and the order in which they are linked together in the chain determine which interactions will occur and therefore how the protein folds. This means it is important to have some knowledge of the properties of the amino acids that make up proteins.

While more than 500 different amino acids have been found in nature, only 20, shown in Figure 6.6, are used in all organisms for protein synthesis by translation. We will focus on these 20, exploring their structures and properties and their role in supporting life.

Look at Figure 6.6 and notice the two types of abbreviation that are routinely used to identify each amino acid: the three-letter and single-letter codes. While the three-letter versions are easier to remember, if you are going to study proteins in any depth, it is worth learning the single-letter abbreviations as they are more often used in computer programs that are used to store and analyse protein sequences and structures.

Notice from Figure 6.6 how these 20 amino acids have the same underlying structure shown in Figure 6.2. How, then, do we classify them into groups? Amino acids can be classified according to the structures and properties of their side chains. In Figure 6.6 the classifications are 'polar', 'non-polar', and 'charged'.

▶ If you are not sure about the meanings of these terms you should refer to Chapter 1, where they are explained.

Charged side chains may be basic (proton acceptors and hence positively charged at pH 7) or acidic (proton donors and hence negatively charged at pH 7). The reason for using this classification system is that a key determinant of how the amino acid will interact with other chemical groups is the hydrophobic or hydrophilic ('water hating' or 'water loving') nature of their side chain. This characteristic strongly influences how a polypeptide chain will fold in three dimensions. When a protein that is found in the cytosol of a cell folds, hydrophilic side chains will be on the outside of the structure, in contact with the aqueous environment, while hydrophobic side chains will be tucked away inside the structure to 'hide' them from the water molecules.

Non-polar side chains are hydrophobic, while polar and charged side chains are hydrophilic. However, it is important to appreciate that some amino acids are not easily defined as having hydrophobic or hydrophilic side chains. Three amino acids fall into this category: tyrosine, cysteine, and glycine.

Tyrosine (Tyr or Y) not only has a polar hydrophilic hydroxyl group, but also a large non-polar, hydrophobic benzene ring. So, there is no right or wrong answer to the question of whether tyrosine has a hydrophilic or hydrophobic side chain, and its position in a folded protein depends on factors such as the amino acid residues in the protein that interact with it.

Cysteine (Cys or C) is also neither strongly hydrophilic nor hydrophobic, but it has a unique role in protein structure because

Figure 6.5 Proteins fold into three-dimensional shapes that are determined by the identity and order of the amino acids that form the polypeptide chain. Interactions between the amino acids, and between the amino acids and their environment (usually the aqueous environment of the cell's interior) determine how the protein folds and stabilize its folded structure. In the figure, the 'main chain' of the polypeptide consisting of central carbons linked by peptide bonds (see Figure 6.4) is coloured grey and the side chains of the amino acids are in different colours.
Source: Cooper, *The Cell*, 8e. Oxford University Press.

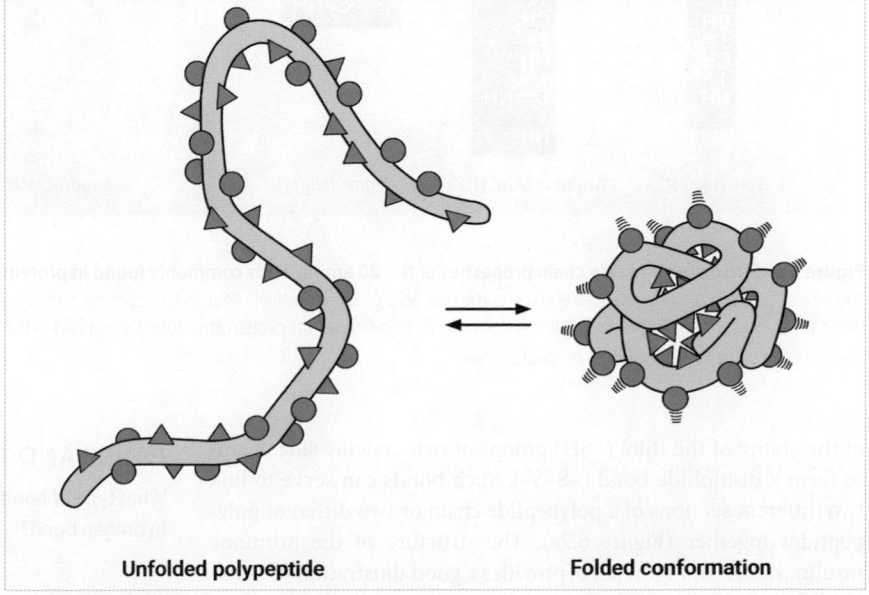

Unfolded polypeptide   **Folded conformation**

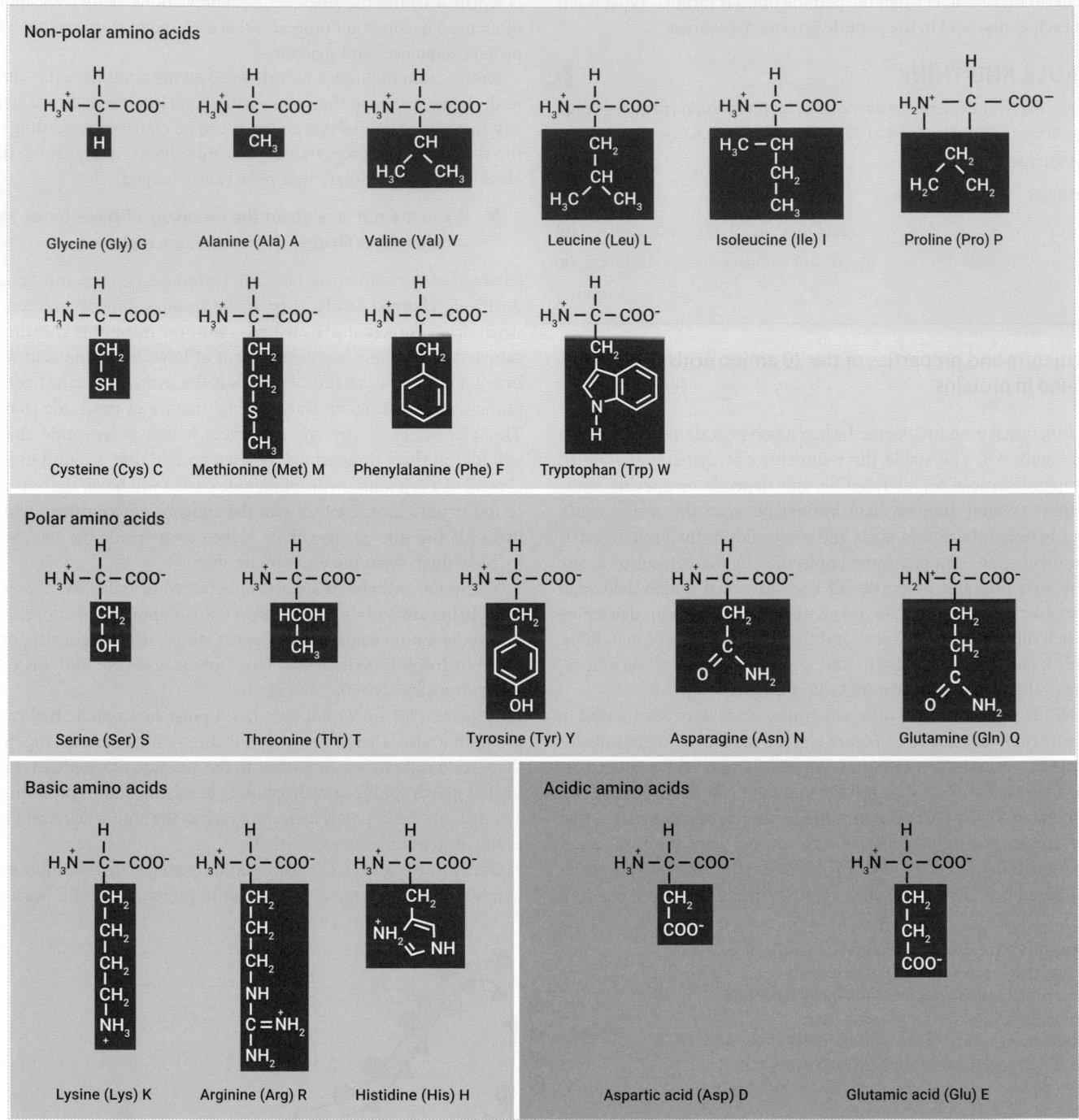

Non-polar amino acids

Glycine (Gly) G   Alanine (Ala) A   Valine (Val) V   Leucine (Leu) L   Isoleucine (Ile) I   Proline (Pro) P

Cysteine (Cys) C   Methionine (Met) M   Phenylalanine (Phe) F   Tryptophan (Trp) W

Polar amino acids

Serine (Ser) S   Threonine (Thr) T   Tyrosine (Tyr) Y   Asparagine (Asn) N   Glutamine (Gln) Q

Basic amino acids

Lysine (Lys) K   Arginine (Arg) R   Histidine (His) H

Acidic amino acids

Aspartic acid (Asp) D   Glutamic acid (Glu) E

Figure 6.6 Structures and side chain properties of the 20 amino acids commonly found in proteins. The structures are depicted as they would be at pH 7, with the amino group positively charged and the carboxylic acid group negatively charged. The amino acids are classified according to whether their side chains are non-polar, polar, or charged. This classification system and its significance for protein structure is explained in the text.
Source: From Cooper, *The Cell*, 8e. Oxford University Press.

of the ability of the thiol (–SH) groups of two cysteine side chains to form a disulphide bond (–S–S–). Such bonds can serve to link two different sections of a polypeptide chain or two different polypeptides together (Figure 6.7a). The structure of the hormone insulin, shown in Figure 6.7b, provides a good illustration of disulphide bonding.

PAUSE AND THINK

What type of bond is a disulphide bond? Is it a (a) covalent, (b) ionic, or (c) hydrogen bond?

Answer: It is (a) a covalent bond.

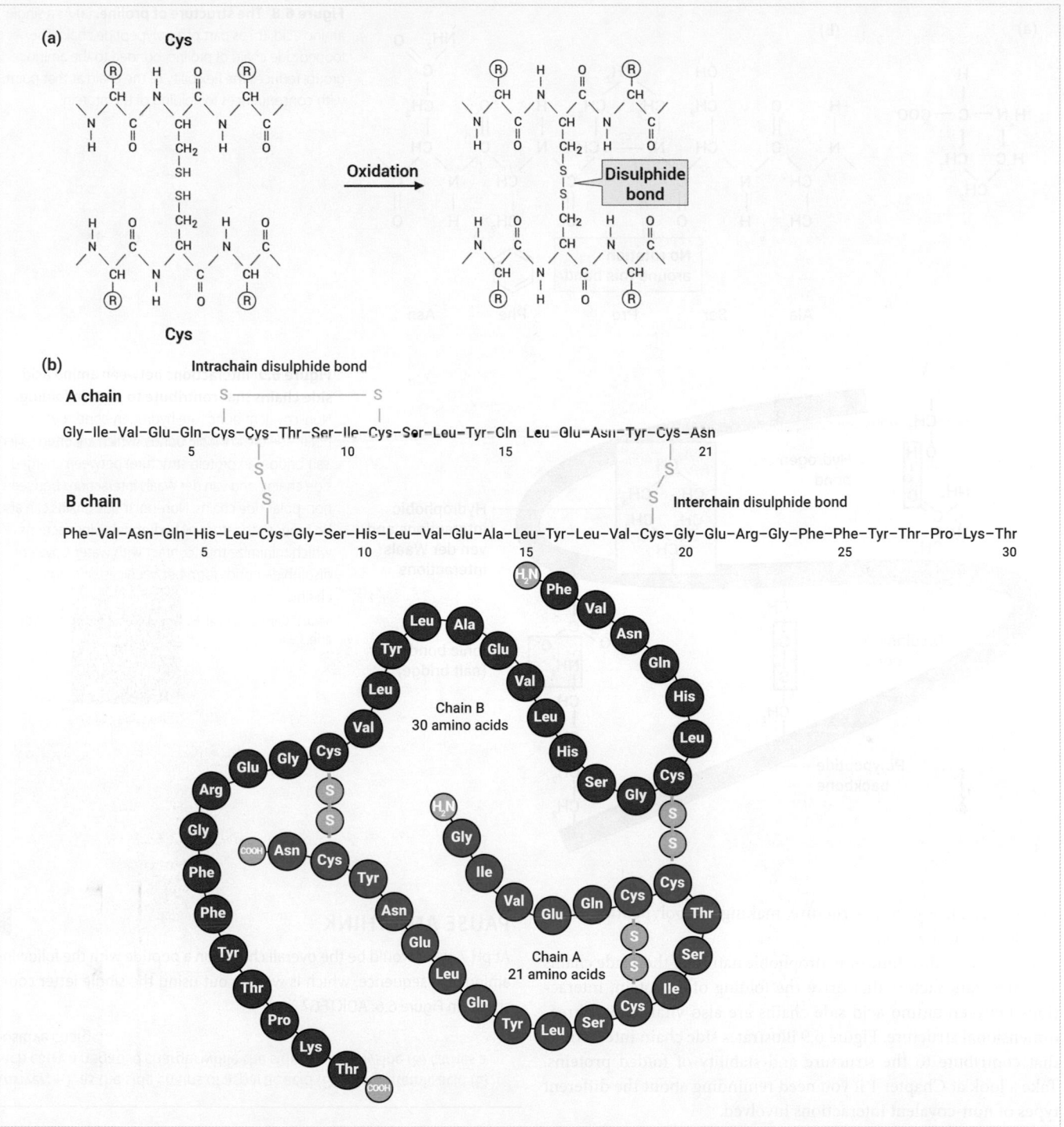

Figure 6.7 Disulphide bonds. (a) Formation of a disulphide bond between two cysteine residues in a polypeptide chain. (b) Disulphide bonds stabilize the structure of the hormone, insulin. Insulin is synthesized initially as a single polypeptide chain: the amino acid sequence that links the A and B chains is cleaved off within the cell, before the hormone is secreted into the bloodstream. Disulphide bonds are formed before cleavage takes place, so that the two chains remain together.

Glycine (Gly or G) is the third amino acid that is not easily designated hydrophilic or hydrophobic, as its R group is simply a second hydrogen. It is therefore small, and neither polar, charged, nor strongly hydrophobic. In the folded protein, glycine can tuck flexibly into small spaces.

Proline (Pro or P) is another amino acid with distinctive properties. Its side chain can be unambiguously classified as hydrophobic, but it has an unusual structure that constrains the flexibility of polypeptide chains. The structure of proline in Figure 6.8 may look odd, but it is an ordinary amino acid except that the side chain forms a loop by bonding at one end to the central carbon and at the other end to the nitrogen, giving it an $-NH_2^+$ rather than an $-NH_3^+$ group. This looped side chain of proline prevents rotation about the bond between the nitrogen and

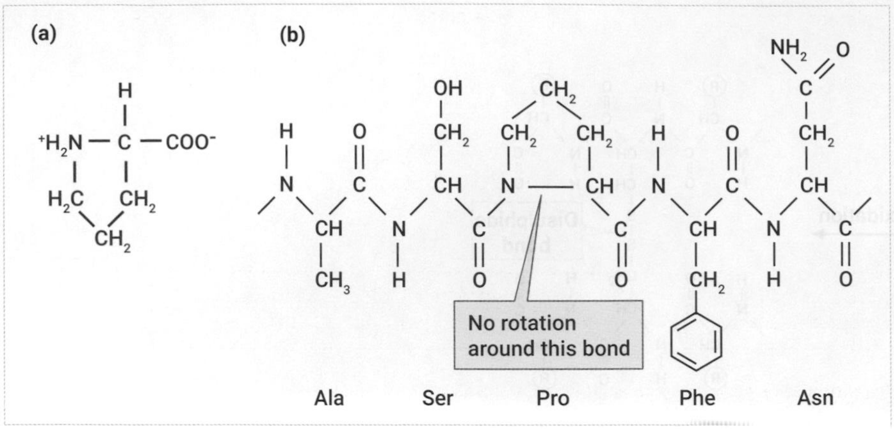

(a)   **(b)**

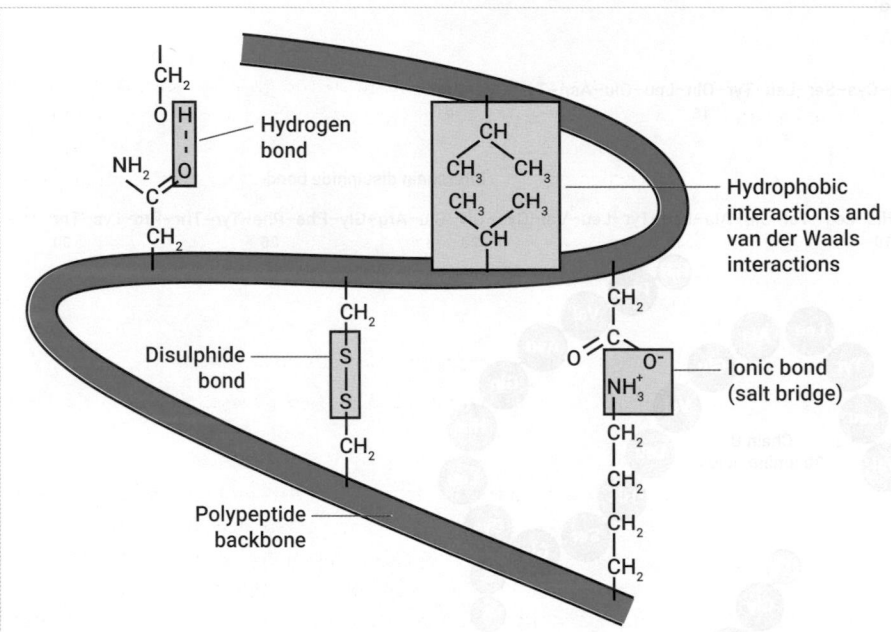

Figure 6.8 The structure of proline. (a) As a single amino acid; (b) as part of a polypeptide chain. The looped side chain of proline, bonded to the amino group, reduces the flexibility of the chain at that point, with consequences for folding of the protein.

Figure 6.9 Interactions between amino acid side chains that contribute to protein folding. Non-covalent bonds are hydrogen bonds between polar side chains, ionic bonds (which are often called 'salt bridges' in protein structure) between charged side chains, and van der Waals interactions between non-polar side chains. Non-polar side chains can also be brought together by hydrophobic interactions which minimize their contact with water. Covalent disulphide bonds form between cysteine side chains.

Source: Campbell et al. *Biology, A Global Approach*. 10th edn. Pearson.

the central carbon in its structure, making the polypeptide chain less flexible at that point.

While the hydrophilic or hydrophobic nature of their side chains are important factors that drive the folding of a protein, interactions between amino acid side chains are also vital for its three-dimensional structure. Figure 6.9 illustrates side chain interactions that contribute to the structure and stability of folded proteins. Take a look at Chapter 1 if you need reminding about the different types of non-covalent interactions involved.

When looking at the depictions of single amino acids in Figure 6.6 you will notice that the amino groups and carboxyl groups are charged, as indeed they would be at pH 7. However, whether a polypeptide is charged or uncharged overall is determined solely by the side chains of its component amino acids. Why is this? Recall from Figures 6.2 and 6.3 that when amino acids polymerize, most of the amino and carboxyl groups are involved in uncharged peptide bonds, leaving only a single free amino group at one end, (positively charged), and a single free carboxyl group at the other, (negatively charged). Thus, the overall charge of a peptide or protein is determined by the side chains of any charged amino acids in its sequence, *not* by the amino and carboxyl groups.

PAUSE AND THINK

At pH 7, what would be the overall charge on a peptide with the following amino acid sequence, which is written out using the single letter codes shown in Figure 6.6: ADKTEG?

Answer: −1, as the side chains of aspartic acid (D) and glutamic acid (E) each carry a negative charge, while the side chain of lysine (K) carries a positive charge.

Figure 6.10 illustrates another important feature of amino acid structure: the existence of **optical isomers** (enantiomers). The amino acid shown is alanine, in which the presence of four different groups attached to the central carbon make it an 'asymmetric' carbon atom. Two different forms of alanine are therefore possible, D- and L-alanine, in which the groups attached to the central carbon are differently configured, making the two structures mirror images of each other. All amino acids in proteins, except for glycine, have an asymmetric central carbon and are of the L-configuration. D-amino acids are only rarely encountered in nature and they are not

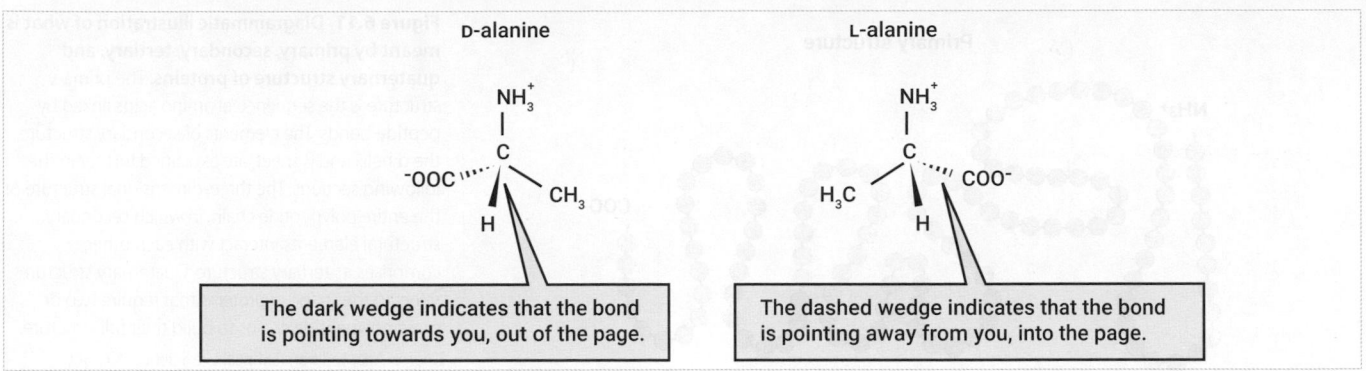

Figure 6.10 **D- and L-alanine are optical isomers (or 'enantiomers').** All amino acids apart from glycine have an asymmetric central carbon and hence can exist as D- and L-isomers. L-isomers are the ones found in proteins.

Source: Crowe, J., & Bradshaw, T. (2021) *Chemistry for the Biosciences: The Essential Concepts* (4th edn). Oxford University Press.

found in proteins. We do not know why life evolved in this way, but it has consequences for the structures that proteins can adopt.

PAUSE AND THINK

Explain why glycine has no L- and D-isomers.

Answer: Glycine has a second hydrogen attached to the central carbon instead of a side chain. Therefore, it does not have an asymmetric carbon with four different groups attached.

 Check your understanding of the concepts covered in this section by answering the questions in the e-book.

6.2 There are four levels of protein structure

Now that we understand the chemical nature of amino acid residues, we can examine how their interactions influence the folding of the linear polypeptide chain into a three-dimensional protein.

The basic concept of protein folding seems quite simple: the polypeptide chain is flexible and by twisting and bending it allows different parts of the chain to interact with each other to achieve a three-dimensional shape, as illustrated in Figure 6.5. In reality, this is a complex process that is still not fully understood, not least because each polypeptide has a multiplicity of ways it could fold, only one of which gives the correct structure for the protein to carry out its function.

Proteins can be thought of as having four levels of structure, which are called the primary, secondary, tertiary, and quaternary structures. The sequence of amino acids that are linked together covalently in a polypeptide chain is the **primary** structure (Figure 6.11). When we considered the joining together of amino acids to form a polypeptide in the previous section, we were describing the primary structure of a polypeptide. However, the primary structure says nothing about how that polypeptide is arranged in a three-dimensional space, just the order of the amino acids. **Secondary** structure refers to the conformation of the polypeptide 'backbone', which includes sections of regular repeating structures known as α helices (helices being the plural of helix) and β sheets (Figure 6.11). The secondary structure then folds on itself to give the **tertiary** structure (Figure 6.11).

The folded molecule formed by the primary, secondary, and tertiary structures may be the final functional protein or it may be a **monomer** or **subunit**, which associates with other monomers (which may be the same or different) to form a functional protein. The interaction of two or more separate polypeptide chains forms a protein that has **quaternary** structure (Figure 6.11).

PAUSE AND THINK

Why is it that not all proteins have quaternary structure?

Answer: Many proteins consist of only one polypeptide chain.

We described the primary structure of proteins earlier, in the section 'Amino acid structure and peptide bonding'. In the following sections we discuss secondary, tertiary, and quaternary protein structure, but first we will look at an important feature of the peptide bond, its 'partial double-bond' character, which has an impact on the three-dimensional structure of proteins.

The peptide bond constrains protein folding

If you look again at the polypeptide structure in Figure 6.4 you will see that the structure has a 'backbone' consisting of alternating central α-carbons and peptide bonds. The folding of the chain into a three-dimensional conformation requires rotation or 'twisting' around bonds in this polypeptide backbone. However, you will recall from Chapter 1 that rotation can occur around single covalent bonds, but not around double bonds. Although the peptide bond is written as an ordinary single bond (about which rotation might be expected), delocalization of electrons (also described in Chapter 1) means that the peptide bond behaves as a hybrid between the two structures shown in Figure 6.12: (1) in which the bond between the carbon and nitrogen atom is a single bond; and (2) in which it is a double bond.

The situation is often described as the C–N bond having 'partial double-bond' character. This character is sufficient to prevent rotation about it, making the polypeptide chain more rigid. Folding the polypeptide involves rotation only around the other bonds in the polypeptide chain (those shown in blue in Figure 6.12).

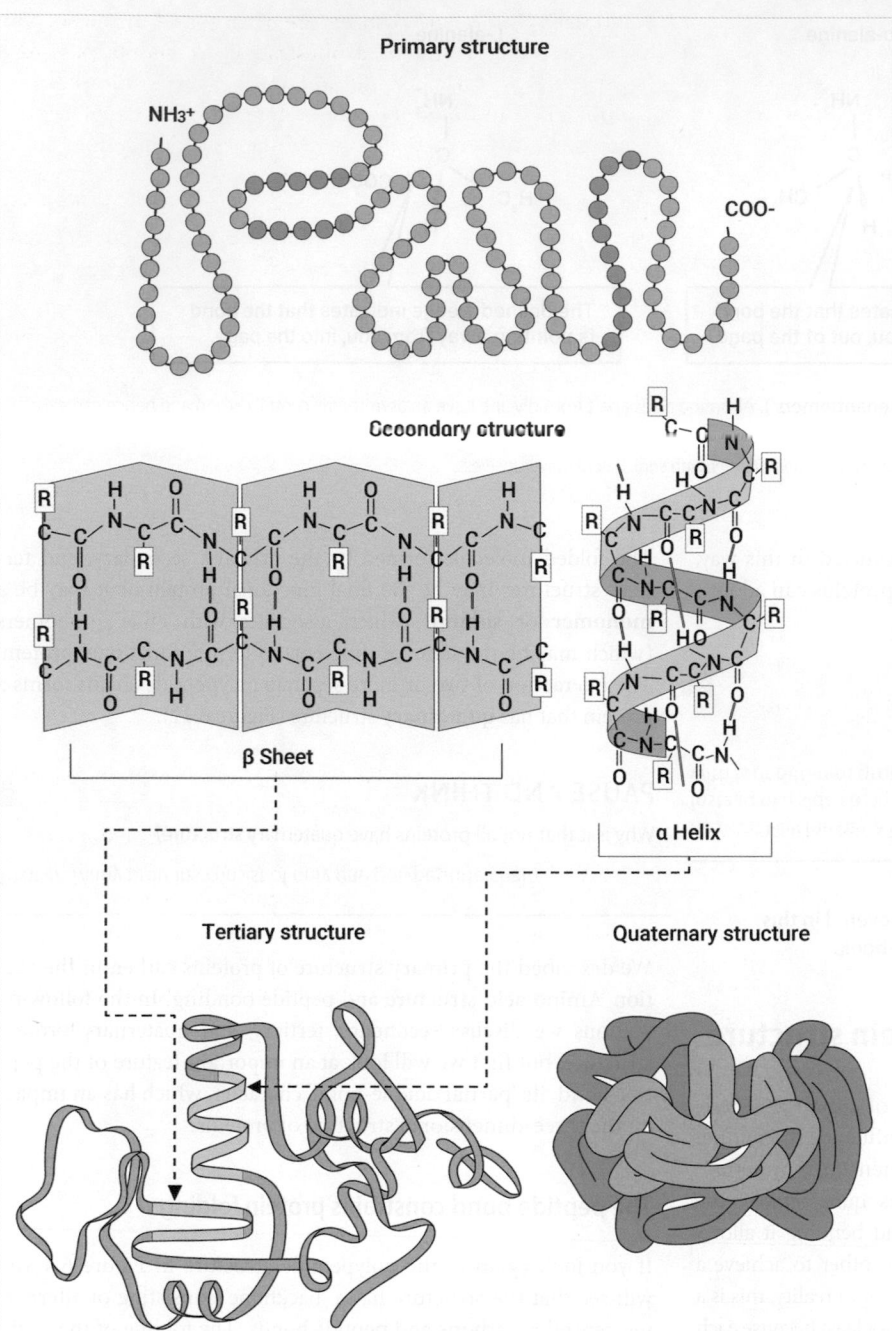

Primary structure

NH3+

COO-

Secondary structure

β Sheet

α Helix

Tertiary structure

Quaternary structure

Figure 6.11 Diagrammatic illustration of what is meant by primary, secondary, tertiary, and quaternary structure of proteins. The primary structure is the sequence of amino acids linked by peptide bonds. The elements of secondary structure, the α helix and β sheet, are explained further in the following sections. The three-dimensional structure of the entire polypeptide chain, in which secondary structural elements interact with each other, comprises its tertiary structure. Quaternary structure refers to the shape of proteins that require two or more polypeptide chains to build their full structure.

Source: Adapted from Lubrizol Life Science. All rights reserved.

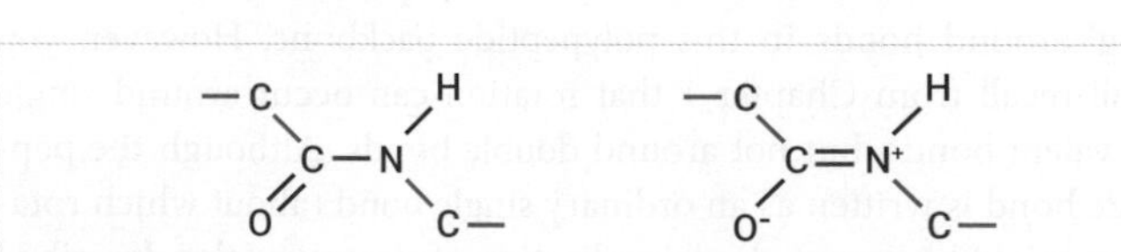

Figure 6.12 The peptide bond is hybrid between the two structures shown. When a protein folds, rotation is not possible around the peptide bond (shown in red) but rotation can occur around the other bonds in the main peptide backbone (shown in blue).

Source: Snape, A. & Papachristodoulou, D. (2018). *Biochemistry and Molecular Biology* (6th edn). Oxford University Press.

In theory, there are two possible configurations in which the peptide bond could form, *cis* and *trans* as shown in Figure 6.13. In practice, peptide bonds in proteins are almost always in the *trans* configuration as the side chains prevent *cis* peptide bond formation by getting in each other's way. (This phenomenon is called steric hindrance.)

Secondary structure depends on hydrogen bonding between polar groups in different peptide bonds

We will now discuss protein secondary structure in more detail. It comes about because the backbone of a polypeptide contains multiple polar groups—the C–O and N–H of the peptide bond—which are capable of hydrogen bonding with each other. Two main classes of secondary structures are formed by hydrogen bonding between peptide bonds: the **α helix** and the **β sheet**. These structures are shown in Figure 6.14. The α helix arises when the backbone is

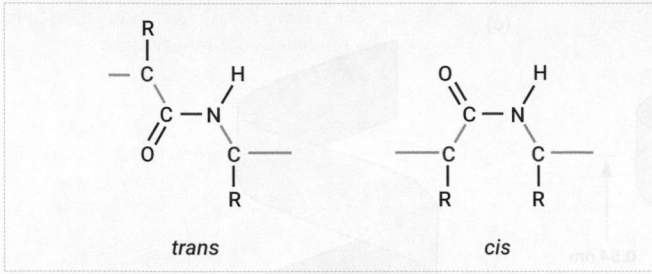

Figure 6.13 The peptide bond (red) almost always forms in the *trans* configuration, as shown on the left. Formation of the *cis* conformation, shown on the right, is rare because in this configuration the R groups, being on the same side of the peptide bond, get in each other's way, a phenomenon known as steric hindrance.

Source: Snape, A. & Papachristodoulou, D. (2018). *Biochemistry and Molecular Biology* (6th edn). Oxford University Press.

arranged in a spiral; the β sheet occurs when extended polypeptide backbones are side by side. The regular pattern of hydrogen bonding between elements of the backbone means that both these structures are stable. In both cases the side chains extend outwards from the structure made by the peptide backbone, allowing them to interact with other amino acid side chains as the protein folds further.

In the α helix, the polypeptide backbone is twisted into a right-handed helix which, for L-amino acids, is more stable than a left-handed one. You can visualize the direction of twist of the right-handed helix; if you look down the axis either way the helix turns clockwise. You can also imagine tightening a conventional screw with your right hand to give you the direction. The α helix structure has 3.6 amino acid units per turn, which results in the C=O of each peptide bond being aligned to form a hydrogen bond with the peptide bond N–H of the fourth distant amino acid residue. All the C=O and N–H groups of the polypeptide backbone are hydrogen bonded in pairs, as shown in Figure 6.14a, forming a cylindrical, rod-like structure.

The β sheet also forms a stable structure in which the polar groups of the polypeptide backbone are hydrogen bonded to one another. As shown in Figure 6.14, the polypeptide chain is extended, with the C=O and N–H groups hydrogen bonded to those of a neighbouring section of the chain. It is 'pleated' because successive α-carbon atoms of the amino acid residues lie alternately slightly above and below the plane of the β sheet. The side chains also alternate on either side of the plane of the sheet. The adjacent polypeptide chains bonded together can run in the same direction (parallel) or opposite directions (antiparallel). In the latter case, the polypeptide makes tight 'β turns' or hairpin loops to fold the chain back on itself. Proline residues are often found in β turns because of their ability to introduce a kink in the polypeptide chain.

PAUSE AND THINK

What would be the approximate length of an α helix consisting of 18 amino acids?

Answer: 2.7 nm, since 18 amino acids would give five turns of the helix, and the 'pitch' (vertical length of the helix) is 0.54 nm as shown in Figure 6.14.

Multiple α helices and β sheets are often found in the secondary structure of a protein, as illustrated in Figure 6.15. This is because they do not depend on bonding of the side chains of amino acids, and therefore can be formed by different peptide sequences. However, the nature of the side chains does influence the likelihood of a sequence forming a particular secondary structure, with proline being known as a 'helix breaker' because its kinked side chain prevents rotation about the bond between the nitrogen and the α-carbon in its structure (as we discuss in Section 6.1).

Sections of the polypeptide chain that do not form α helices and β sheets adopt less regular secondary structures, termed random coils or loops, that link the α helices and β sheets together (Figure 6.15).

PAUSE AND THINK

Are the β sheet structures in the protein shown in Figure 6.15 parallel or antiparallel?

Answer: They are antiparallel, as shown by the arrow heads.

Tertiary structures of proteins are stabilized mainly by non-covalent bonds

The various secondary structures of a polypeptide chain undergo further folding and packing together to form its tertiary structure. As shown in Figure 6.9, the bonds involved are predominantly non-covalent—hydrogen bonds between polar amino acid side chains or between polar side chains and peptide bonds, van der Waals interactions between hydrophobic side chains, and some salt bridges between charged side chains. In globular proteins that are found in the cytosol, hydrophobic interactions that bury non-polar and non-charged side chains in the interior away from the aqueous environment are also important.

PAUSE AND THINK

Which type of covalent bond contributes to protein tertiary structure?

Answer: The disulphide bond.

Most folded proteins are only marginally stable

The folding of a protein is a chemical reaction, so it has an associated free energy change, ΔG, which must have a negative value if folding is energetically favourable and therefore spontaneous. As discussed in Chapter 2, the value of ΔG depends on the values of ΔH (enthalpy or bond energy changes) and ΔS (entropy changes) that occur as the protein folds. Although a folded protein is evidently more ordered than an unfolded polypeptide, which can have multiple conformations, the entropy change of protein folding is somewhat favourable. This is because an unfolded polypeptide exposes a large surface area, around which water molecules create an ordered shell. The protein is more compact once folded, disrupting the ordered shell of water, and therefore allowing the entropy of the water to increase.

It can be calculated that the overall outcome of combined enthalpy and entropy changes when a protein folds is that the overall free

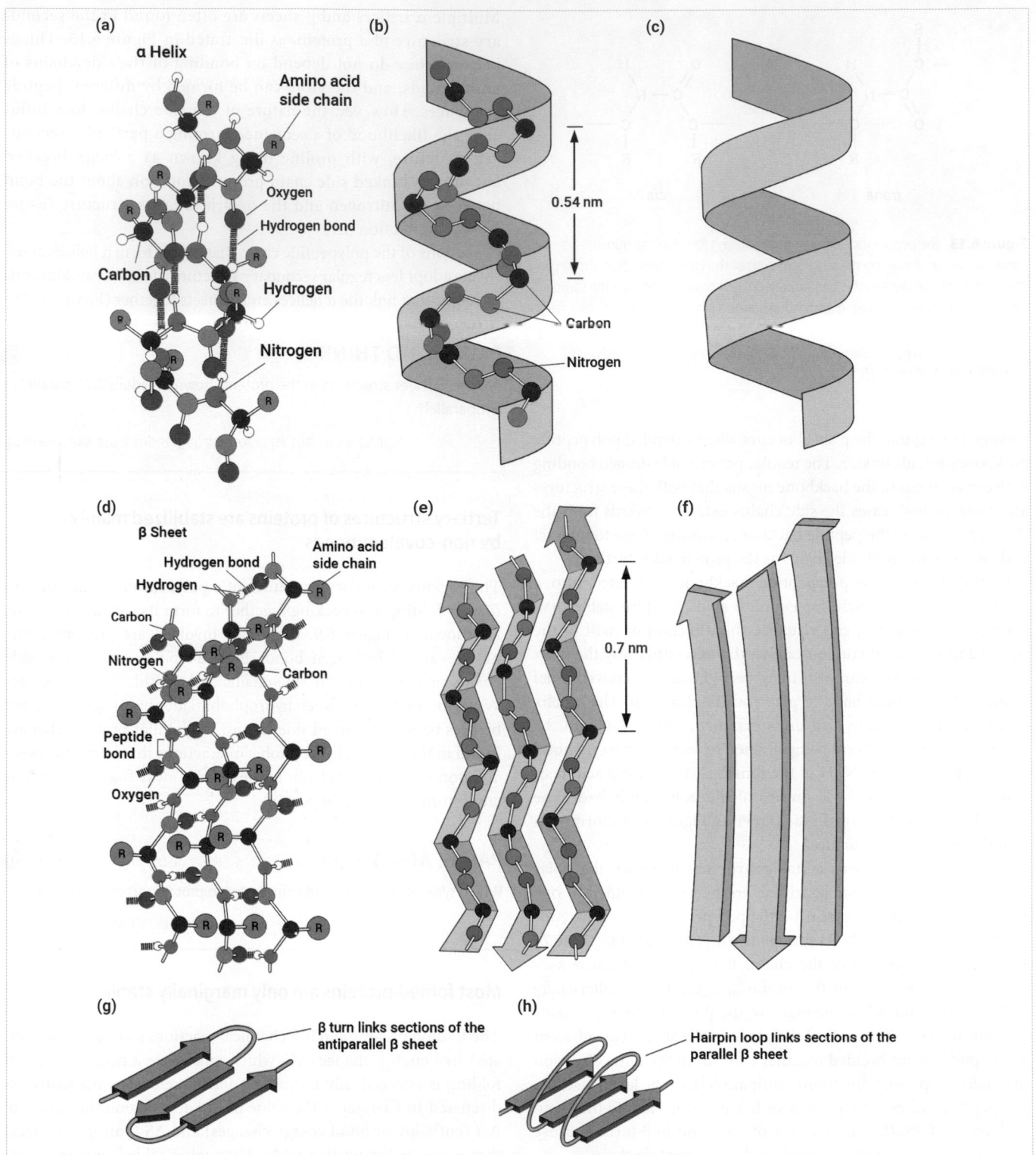

Figure 6.14 Protein secondary structures. (a), (b), and (c) show visualizations of the α helix, while (d), (e), and (f) show the β sheet. (a) and (d) show the full molecular structure of the polypeptide backbone, with the hydrogen bonds between the C=O and N–H groups of the polypeptide bond in red, and with side chains denoted by 'R'. (b) and (e) focus on the main atoms that form the polypeptide backbone, with the α-carbons shown in black, peptide bond carbons in grey, and peptide bond nitrogens in blue. Superimposed on the atomic structures in (b) and (e) are 'ribbon diagrams' that illustrate a convention used to depict the α helix and β sheet in graphic representations of protein structure, while (c) and (f) show the ribbon diagrams alone. In the β sheet the arrow denotes the direction of the polypeptide chain from N to C terminal; (e) also shows why the β sheet is also called the 'β-pleated sheet', because successive α-carbon atoms of the amino acid residues lie alternately slightly above and below the plane of the sheet. (g) and (h) emphasize that β sheets are formed by interactions of components of the same polypeptide chain, with loops linking the sections that are hydrogen bonded to each other to form the sheet. (g) is an antiparallel β sheet in which alternate hydrogen-bonded sections are oriented in opposite directions with respect to the N and C terminal of the polypeptide chain, while (h) is a parallel β sheet in which the hydrogen-bonded sections all run in the same direction. (d)–(f) show details of an antiparallel β sheet.

Source: Alberts et al. *Essential Cell Biology*, Edn 4. Garland Science, 2014.

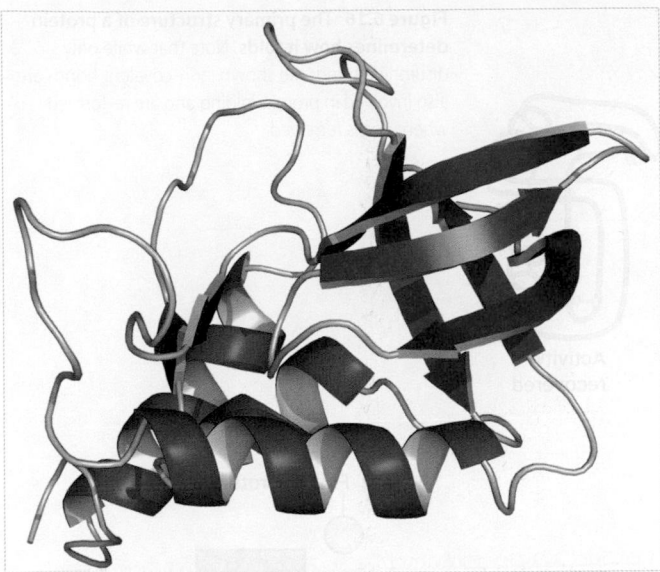

Figure 6.15 Structure of staphylococcal nuclease. The figure shows how multiple elements of secondary structure, α helices, and β sheets, are linked together by less structured sections of the polypeptide chain.

Source: Snape, A. & Papachristodoulou, D. (2018). *Biochemistry and Molecular Biology* (6th edn). Oxford University Press.

energy change associated with protein folding is small and negative (i.e. most folded proteins are only marginally stable). You can show this for yourself by frying an egg: the heat is enough to denature the egg-white albumin protein (i.e. break the non-covalent bonds), creating a tangled insoluble mass as opposed to the clear, ordered and soluble protein that is seen prior to heating.

Quaternary structure brings different polypeptide chains together

Many proteins are fully functional once they are folded into their tertiary structure, but others require two or more subunits, each of which consists of a separately synthesized polypeptide chain or monomer. Proteins consisting of two, three, and four subunits are called dimers, trimers, and tetramers respectively. Multimeric or multisubunit proteins are said to have quaternary structure.

In quaternary structure, protein subunits are linked together by non-covalent bonds of the same types that stabilize tertiary structures. Non-covalent bonds form between complementary regions of the subunits, which must be closely aligned by their three-dimensional shapes: for example, a hydrophobic bulge on the surface of one subunit will interact with a hydrophobic groove on the surface of another, while in an adjacent region polar amino acid side chains may allow hydrogen bonding between the two subunits. The combination of different interactions allows the subunits to bind together in a very specific way.

Polypeptides linked together to form proteins with two or more subunits may be the same as or different from each other. Haemoglobin (see Figure 6.21) is an example of a protein with quaternary structure. It consists of two identical 'α' subunits and two identical 'β' subunits (not to be confused with α helices

and β sheets) with the order of events during assembly being that an α and a β subunit first form a dimer, and then two dimers come together to form the fully functional protein.

The primary structure of a protein determines how it folds

A famous study carried out in the laboratory of Christian Anfinsen established the principle that the primary amino acid sequence of a protein is sufficient to direct its folding into a functional three-dimensional shape. He 'denatured' (unfolded) an enzyme, ribonuclease, by exposing it to high concentrations of urea, which broke down the non-covalent bonds involved in its tertiary structure. He also exposed it to a reducing agent, which broke the disulphide bonds (by reducing –S–S– to –SH HS–). The denatured protein showed no enzyme activity, but when the urea was removed by dialysis and it was reoxidized its activity was restored. Anfinsen's explanation was that the polypeptide chain of the enzyme had spontaneously refolded itself into the correct conformation. The experiment is illustrated in Figure 6.16.

Protein folding in the cell requires chaperones

The outcome of Anfinsen's famous experiment, published in 1961, may seem obvious to us now. Yet, despite its apparent simplicity, we still do not understand exactly how proteins are correctly folded in the cell. One problem is that a peptide chain may start to fold while it is being synthesized despite being incomplete, and another is that many of the myriad incorrect conformations a protein can adopt are only slightly less stable than the correct one.

A solution to these problems has evolved in the shape of **molecular chaperones**, which are themselves a class of proteins. This term stems from a tradition of having someone, known as a 'chaperone', to accompany and look after young people while they socialize. Chaperone proteins are so named because they prevent premature interactions and promote 'proper' ones. The function of one such chaperone, a protein complex termed the GroEL/GroES system, is illustrated in Figure 6.17. The system is sometimes more catchily called an Anfinsen cage, because the newly made protein is sequestered within it, away from the crowded cellular environment, until it is properly folded.

Despite the existence of molecular chaperones it is quite common for proteins not to fold properly or to partially unfold during their lifetime. Misfolded proteins can be refolded or removed by various cellular quality control mechanisms, but if these fail, accumulation of misfolded proteins can cause disease (Clinical Box 6.1).

 Check your understanding of the concepts covered in this section by answering the questions in the e-book.

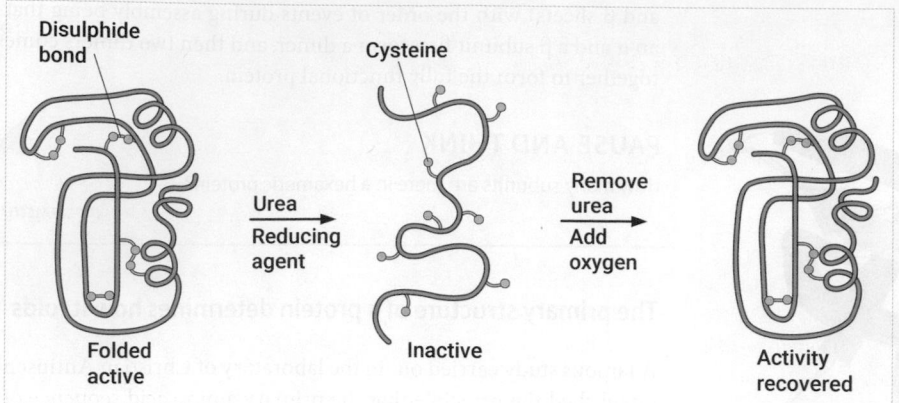

Figure 6.16 The primary structure of a protein determines how it folds. Note that while only disulphide bonds are shown, non-covalent bonds are also involved in protein folding and are re-formed when urea is removed.

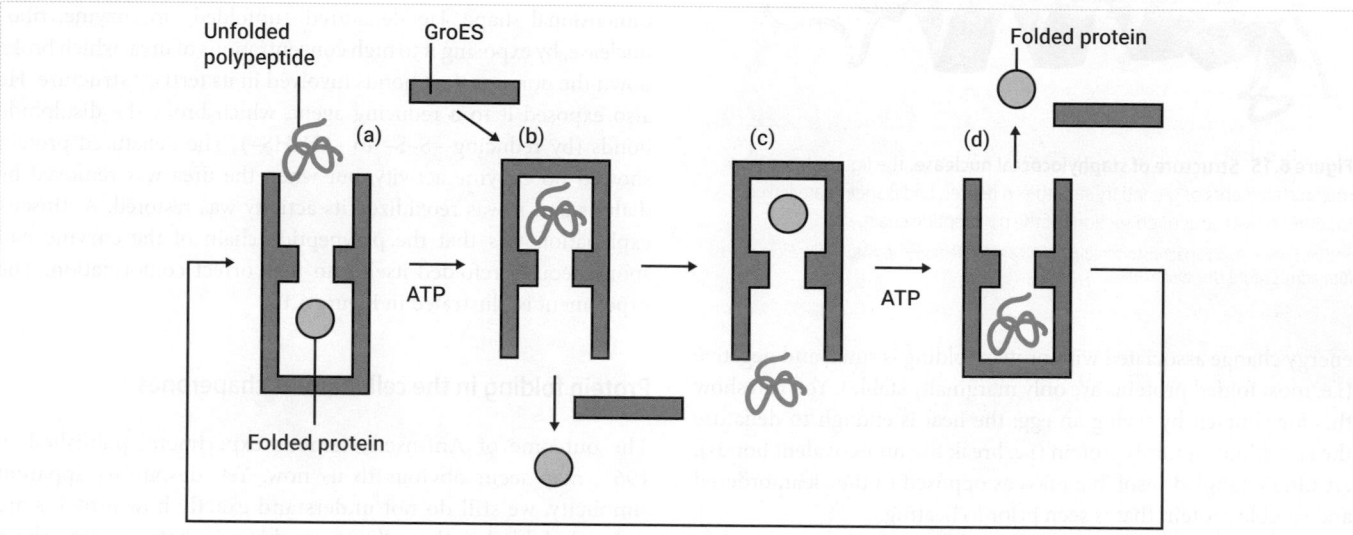

Figure 6.17 Simplified diagram to illustrate the principle of GroEL action in *E. coli*. The 'lid' structure is known as GroES. There are two folding chambers, which in this model are postulated to work alternately. In (a) an unfolded polypeptide with hydrophobic groups exposed has attached to GroEL by hydrophobic interactions. The lower cavity has a folded polypeptide represented as a solid circle waiting for release. In (b) the unfolded polypeptide has entered the cavity, the 'lid' seals it in, and the cavity undergoes a conformation change that creates a hydrophilic environment for it. Meanwhile, the lower cavity has opened, undergoing a conformation change that switches the environment from hydrophilic to hydrophobic, and hence it expels the folded protein which now has hydrophilic amino acid residues on the outside. If the protein had not properly folded and still had hydrophobic residues exposed it would 'stick' in the cavity which would close again and give it another go. In (c) and (d), the situation is just the same as (a) and (b) but upside down. The mechanism requires energy from the hydrolysis of ATP at the steps indicated.

Source: Snape, A. & Papachristodoulou, D. (2018). *Biochemistry and Molecular Biology* (6th edn). Oxford University Press.

6.3 **Protein structures reflect their function**

Figure 6.1 illustrates the great diversity of protein sizes and shapes that reflect their range of functions. To remind you of just a few of the roles they carry out, proteins can be:

- **Structural proteins**, providing mechanical support and strength both inside and outside the cell. Examples are collagen and elastin, both found in extracellular matrix, tendons, and ligaments; and keratin, the protein that forms hair, nails, claws, and horns. Within the cell, proteins of the cytoskeleton such as actin and tubulin create a supporting network (see Figure 6.19), but also provide 'tramlines' along which substances can be transported around the cell.

- **Enzymes**, catalysts that enable biological reactions to take place in the cellular environment.

- **Transport proteins**: these can be proteins that carry molecules around the body in the circulation system (for example, haemoglobin, discussed in detail later in this section), or proteins involved in movement of small molecules and ions in and out of the cell. The latter are membrane-embedded proteins, which form a structurally distinct class of proteins as discussed later in this chapter.

- **Signalling proteins**, including some regulatory hormones that travel in the blood from one part of the body to another. For instance, insulin, produced in the pancreas when blood glucose is high after a meal, travels to other tissues such as liver, muscle, and adipose tissue, where it regulates their metabolism appropriately. Intracellular signalling proteins, such as the 'G proteins' described in Section 17.7, transduce signals arriving from outside a cell and regulate reactions within it.

- **Receptors**, which bind with exquisite sensitivity and specificity to signalling molecules, which may be other proteins or small

CLINICAL BOX 6.1   Misfolded proteins can cause disease

Prion diseases are an unusual group of fatal neurological degenerative diseases that affect humans and animals. In sheep, the disease is known as 'scrapie' because the animals scrape off their wool by rubbing against fence posts; cattle are affected by bovine spongiform encephalopathy (BSE), commonly called 'mad cow disease'. Human prion diseases include Creutzfeldt–Jakob disease (CJD) and kuru. The diseases can be transmitted when infected tissue is consumed or, rarely, can be an inherited trait. The only known case of a prion disease being transmitted from animals to humans is BSE, which when transmitted to humans is known as new-variant CJD (nvCJD). An outbreak of nvCJD in the 1990s, mainly in the United Kingdom, was caused by the consumption of meat products from animals that had been fed infected material. The spread of disease has since been controlled by stricter legislation.

The cause of prion diseases was very difficult to track down and was initially controversial. It is now known that the diseases are caused by an abnormal form of a normal protein found in the brain. The function of this normal protein is unknown. The term prion derives from 'proteinaceous infectious particle' and the protein itself is called PrP^c (for the normal prion protein) or PrP^{Sc} (for the abnormal version). Surprisingly, the two proteins have identical amino acid sequences and are coded for by the same gene, but their folded conformations are different, as shown in Figure 1. PrP^c is soluble and protease sensitive, while PrP^{Sc} is less soluble, tends to aggregate into long fibrils, and is protease resistant.

The controversy about prion diseases centred on the question of how an improperly folded protein can be infectious. No mechanism is known by which a protein molecule can direct its own replication. Instead, PrP^{Sc} somehow causes PrP^c to convert to the abnormal form. This has been demonstrated *in vitro* by incubating the two together. It is believed that it occurs by a 'seeding' mechanism where PrP^{Sc} proteins form aggregates to which PrP^c proteins attach and refold, giving more PrP^{Sc}. The conversion *in vivo* of PrP^c to PrP^{Sc} is a rare event in the absence of infection by PrP^{Sc}, so the spontaneous occurrence of prion disease is rare. It is believed that mutations in the gene for the normal PrP^c may increase the probability of incorrect folding, which might explain the hereditary origin of some cases of the disease. Once some PrP^{Sc} is formed, it would then trigger the formation of more.

It is not known with certainty how prions cause disease. It is thought to be associated with accumulation in the brain of the long aggregate fibrils which are known as amyloids. It has been found that other proteins can form similar β strand-rich aggregates, also often called amyloids, which are associated with diseases such as Alzheimer disease (Figure 2). In the case of Huntington disease, protein aggregates are formed by an abnormal protein encoded by an inherited mutated gene.

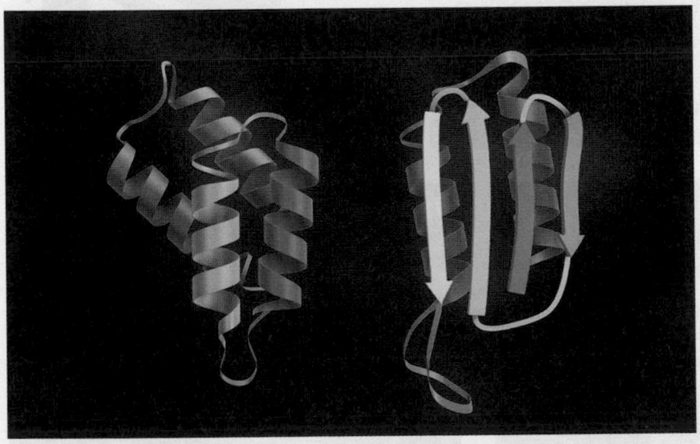

Figure 1 Normally folded prion protein PrP^c on the left, with the abnormally folded PrP^{Sc} on the right.
Source: Shutterstock.

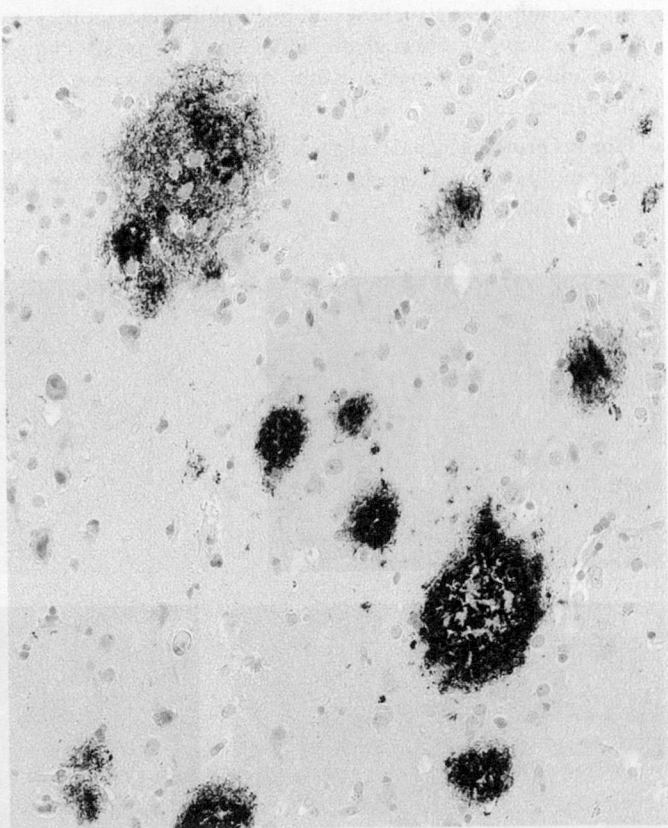

Figure 2 Brain tissue from an Alzheimer patient stained with an antibody that reacts with amyloid protein aggregates (dark brown).
Source: Nephron, CC BY-SA 3.0 <https://creativecommons.org/licenses/by-sa/3.0>, via Wikimedia Commons.

molecules. Examples are the insulin receptor and the acetylcholine receptor. A generic term for the molecule that binds a receptor is a ligand. Thus, insulin is the ligand for the insulin receptor (described in Section 19.3) and acetylcholine is the ligand for the acetylcholine receptor (described in Chapter 17).

- **Motor proteins**: these remarkable proteins, which include myosin, dynein, and kinesin, transform energy derived from the hydrolysis of ATP into kinetic energy, thus enabling cells to move, muscles to contract, and substances to be transported within cells.

Figure 6.18 illustrates some interesting examples of proteins carrying out different functions.

Proteins mainly fall into three structural classes

If we look at the list of proteins above, we can consider that they mainly fall into three structural classes: globular, fibrous, and membrane-embedded proteins.

- **Globular proteins** are found in aqueous environments, such as the cytosol or blood. Their sizes vary, but they mainly form compact shapes with the hydrophobic amino acid residues buried within the structure, and hydrophilic ones on the outside. Examples are myoglobin and haemoglobin (see Figures 6.20 and 6.21), and most enzymes, such as the nuclease shown in Figure 6.15.
- **Fibrous proteins** include collagen, found in skin, tendons, cartilage, and bone, and keratin, the main component of hair and

nails. They have elongated shapes and polymerize to give them strength and enable them to span long distances. Their structures involve helices coiled around each other. While keratin contains coiled α helices, the helical structure of collagen, shown in Figure 6.25, is not an α helix.

The cytoskeletal proteins actin and tubulin form extended structures in a slightly different way, by polymerization of globular proteins (G actin, α and β tubulin) to form microfilaments (also called F actin) and microtubules (as you can see in Figure 6.19).

- **Membrane-embedded proteins** (or simply membrane proteins) are structurally distinct because they function in a hydrophobic environment, the lipid bilayer. The arrangement is therefore opposite to that of globular proteins, with hydrophobic residues facing outward. Membrane transport proteins such as the bacterial porin illustrated in Figure 6.27 have a central channel lined with hydrophilic residues, to allow the movement of polar molecules and ions in and out of the cell.

PAUSE AND THINK

The amino acid lysine is generally found on the outside of folded globular proteins, but on the inside of folded membrane proteins. Why is this?

Answer: Lysine has a positively charged, hydrophilic side chain, so is on the outside of globular proteins where it interacts with their aqueous environment but is hidden away from the hydrophobic lipid environment of the membrane proteins.

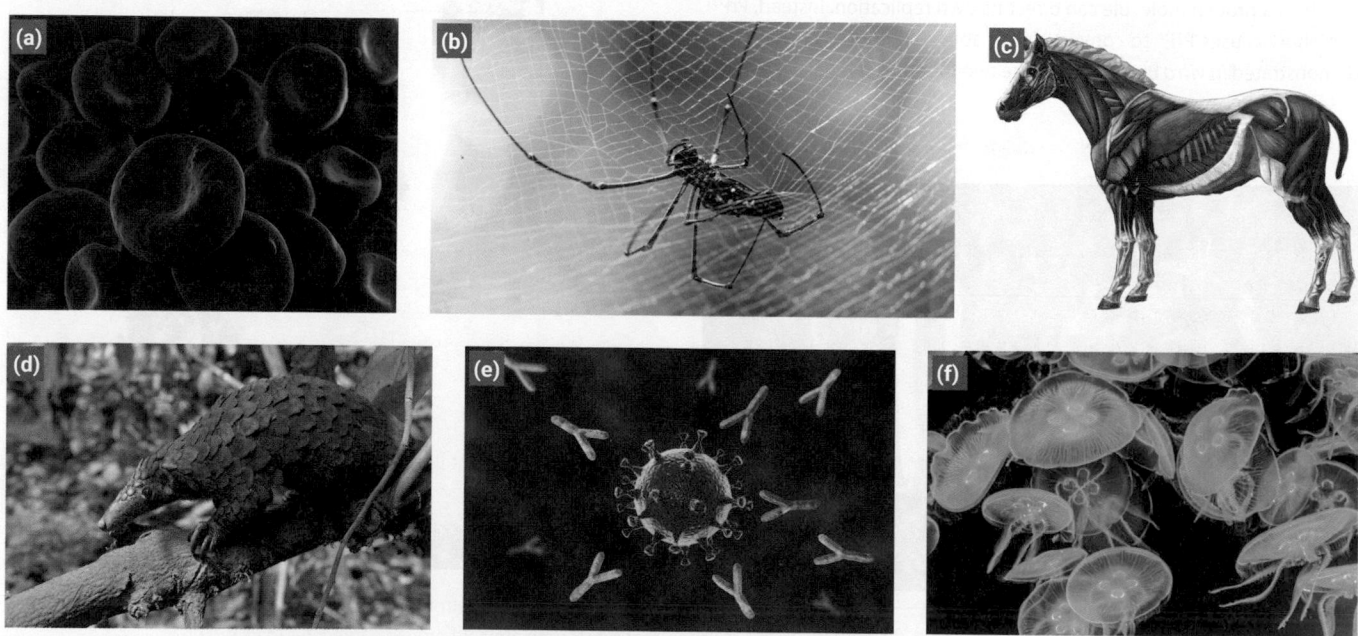

Figure 6.18 Proteins have diverse functions. (a) Red blood cells contain the protein haemoglobin, which enables them to transport oxygen around the body. (b) Spider silk is a strong protein that traps prey in the spider's web. (c) Muscle proteins provide strength and allow movement. (d) Pangolin scales are made of keratin, the same protein that makes up hair and nails. Pangolins are the only scaly mammals. (e) Protein spikes on the coronavirus allow it to bind to receptors and invade cells, unless the Y-shaped antibody proteins produced by our immune system attack and prevent it. (f) Jellyfish produce a green fluorescent protein that has many applications in biological research.
Source: (a, b, d–f) Alamy. (c) Shutterstock.

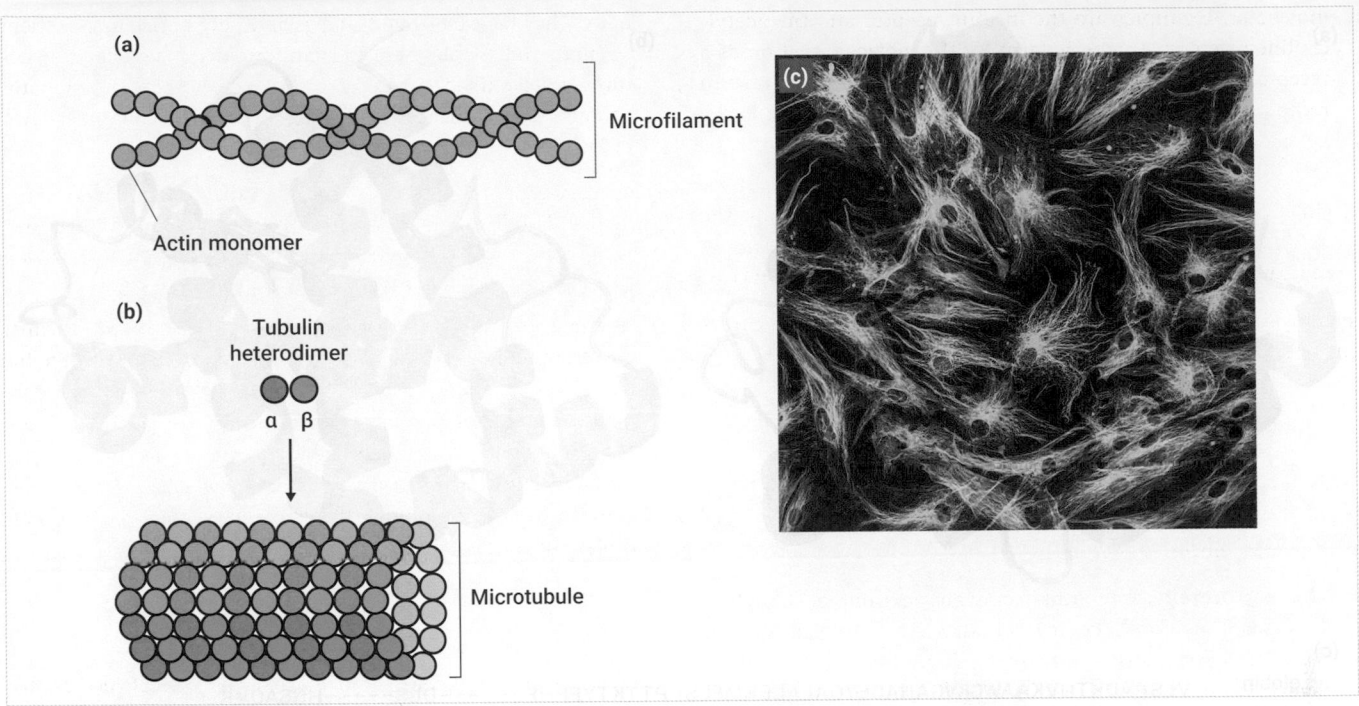

Figure 6.19 Actin and tubulin are examples of fibrous proteins formed by polymerization of globular monomers. (a) Actin microfilaments are polymers of single actin monomers that form two strands, which wind around each other. The monomers are all identical despite being shown here in different colours to differentiate the two strands. The strands are tightly apposed forming a rod-like structure but are shown in a more open form for clarity. (b) α and β tubulin form heterodimers (dimers in which the two protein monomers are different), which self-assemble into microtubules. (c) Human skin cells in culture viewed using fluorescence microscopy. Nuclei are labelled blue, actin microfilaments are pink, and tubulin microtubules are green.

Source: (a, b) Adapted from Snape, A. & Papachristodoulou, D. (2018). *Biochemistry and Molecular Biology* (6th edn). Oxford University Press. (c) Shutterstock.

We will now discuss examples of proteins in each of these structural classes in more detail.

The oxygen-binding proteins, haemoglobin and myoglobin, are examples of globular proteins

We are considering haemoglobin and myoglobin as examples of globular proteins that illustrate the relationship of protein structure and protein function. Haemoglobin is the oxygen carrier in blood, while myoglobin acts as an oxygen store in striated muscle, a tissue that requires additional oxygen during periods of intense activity. The two proteins are closely related, both in structure and evolutionary ancestry. However, myoglobin is a relatively simple molecule, while haemoglobin has evolved into a sophisticated molecular machine superbly adapted for its functions.

The myoglobin protein is monomeric, consisting of a single polypeptide chain (as illustrated in Figure 6.20a), whereas haemoglobin is a tetramer, consisting of two α and two β polypeptide chains (as illustrated in Figure 6.21). The two α subunits of haemoglobin are identical, as are the two β chains. Each of the subunits of haemoglobin has close similarity to myoglobin, both in its amino acid sequence (Figure 6.20c) and in its folded structure (Figure 6.20b).

In both myoglobin and haemoglobin, a molecule of haem is stably bound into a pocket in the folded polypeptide chain. Haem (Figure 6.22a) is an example of a **prosthetic group**, a non-peptide structure that is required for a protein to fully function. Haem has

an intense red colour that is responsible for the colour of red blood cells. The ferrous iron in the centre of haem can form six bonds; four bonds hold it in place by binding to the nitrogen atoms of the haem group, the fifth attaches to a histidine residue of the protein, and the sixth is available for the reversible attachment of oxygen (Figure 6.22b). As each of the four subunits of haemoglobin contains a haem group, a single molecule of haemoglobin can bind up to four molecules of oxygen, while a molecule of myoglobin can only accommodate one. Oxygen can be termed the **ligand** of both haemoglobin and myoglobin. In this context a ligand is a substance that binds to a specific site on a protein.

Figure 6.23 shows that myoglobin and haemoglobin have different abilities to bind oxygen at varying oxygen concentrations. Myoglobin is almost fully saturated, even at the low oxygen concentration of 20 mmHg present in exercising muscle. This indicates that myoglobin has a high affinity for oxygen, enabling it to extract oxygen from the blood and hold it in the muscle tissue. Myoglobin only gives up oxygen to muscle when intense activity greatly lowers the oxygen concentration in the tissue.

In contrast to the relatively simple function of myoglobin, haemoglobin needs to function in a more complex way as is needs to pick up as much oxygen as it can in the lungs but then deliver it to other tissues. The oxygen saturation curve of haemoglobin is sigmoid (S-shaped), in contrast to the hyperbolic curve that rises smoothly and then levels off seen with myoglobin (see Figure 6.23). Haemoglobin must pick up as much oxygen as it can in the

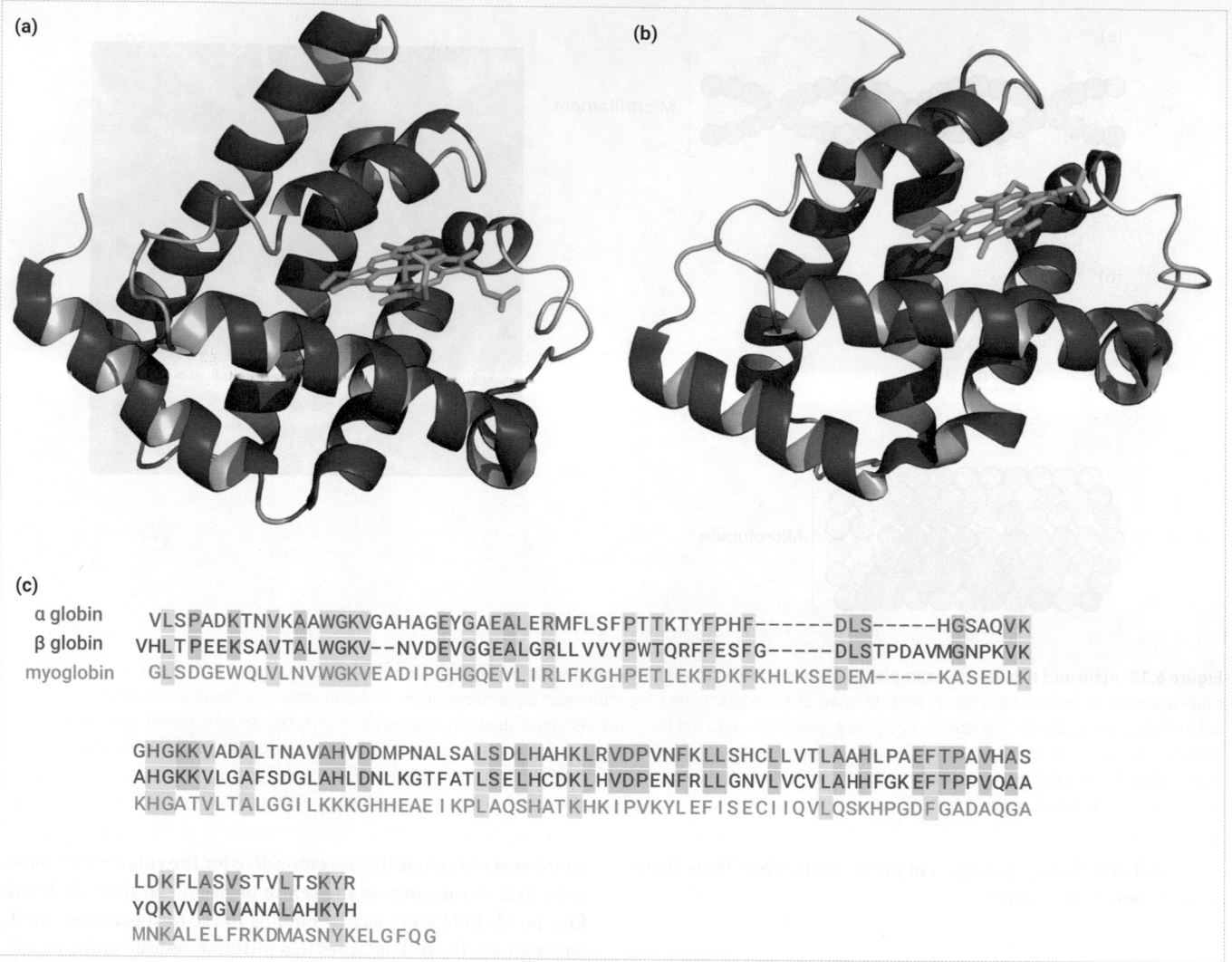

Figure 6.20 Myoglobin and haemoglobin. Models of (a) myoglobin (PDB: 1MBO) and (b) a single chain (β chain) of haemoglobin (PDB: 1HHO). Computer-generated diagrams showing the folding of the polypeptide in myoglobin and haemoglobin, and the positioning of haem (green) in the molecule. (c) Amino acid sequence alignment of human α and β globin and myoglobin. Some 'gaps' (dashes) have been introduced to maintain the alignment. Amino acids that are identical at the equivalent position in all three proteins are highlighted in yellow, while those that are identical in α and β globin but different in myoglobin are highlighted in blue. The evolution of these three proteins is discussed in Section 6.5.

Source: Snape, A. & Papachristodoulou, D. (2018). *Biochemistry and Molecular Biology* (6th edn). Oxford University Press.

lungs but then surrender it in the tissue capillaries. Its sigmoid oxygenation curve indicates that haemoglobin releases some oxygen to tissues even when they are at rest, with oxygen concentrations around 40 mmHg, because it has a lower oxygen affinity than does myoglobin. Nevertheless, it does retain most of its bound oxygen in resting tissue and releases it very efficiently in exercising tissue where the oxygen concentration is lower, around 20 mmHg.

The differences in oxygen binding properties of myoglobin and haemoglobin that are required for their distinct functions are achieved by their protein structures exhibiting key differences, as explained in the next section.

How are the differences in oxygen saturation curves achieved?

The simple hyperbolic oxygen saturation curve of myoglobin reflects the fact that it is single subunit protein. Oxygen binding is

a chemical reaction in which each molecule of myoglobin is either bound or not bound by oxygen, and as myoglobin has a high affinity for oxygen, we can say that the equilibrium of the binding reaction shown below 'lies to the right'. (You should consult Chapter 1 if you need reminding about equilibria).

$$Mb + O_2 \rightleftharpoons MbO_2$$

The sigmoid curve for oxygen binding by haemoglobin comes about because of cooperative oxygen binding. This cooperative binding is made possible by the protein having quaternary structure with four subunits. Each of haemoglobin's four haem groups, one per subunit, provides a binding site for an oxygen molecule; as each site is occupied, it becomes easier for another oxygen molecule to bind. Thus, we see a progressive increase in affinity of haemoglobin for oxygen as more sites become occupied. This

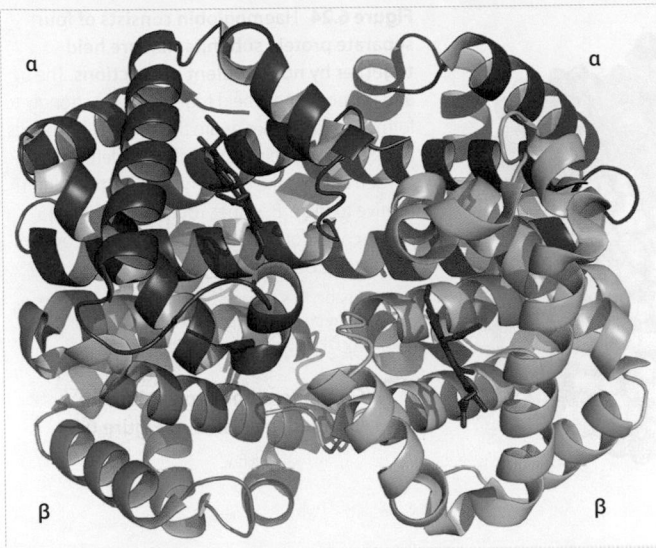

Figure 6.21 Quaternary structure of haemoglobin (PDB: 1A3N). The two α and two β polypeptide chains are labelled. Each chain contains a haem prosthetic group (shown here in pink). Oxygen binds iron in the haem group, as described in the text. The figure shows haemoglobin with no oxygen bound.

Source: Snape, A. & Papachristodoulou, D. (2018). *Biochemistry and Molecular Biology* (6th edn). Oxford University Press.

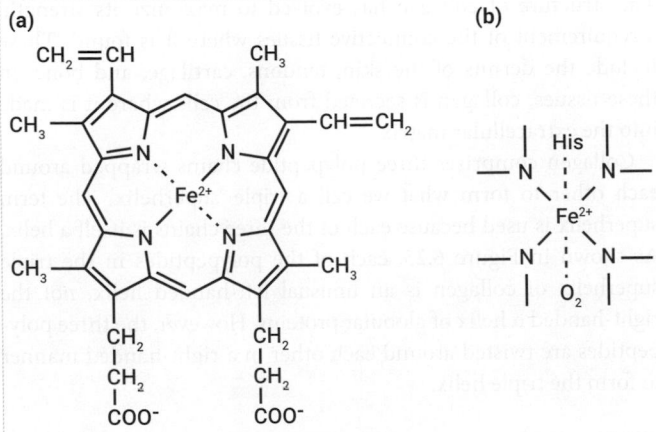

Figure 6.22 The structure and binding ability of haem. (a) Haem structure. Note that, at one side of the molecule, there are two hydrophilic propionate groups ($-CH_2CH_2COO^-$) while the remaining side chains are all hydrophobic. In myoglobin, haem sits in a cleft of the molecule with the hydrophilic side pointing out towards the aqueous environment and the hydrophobic groups buried into the non-polar interior of the protein. (b) Binding ability of Fe^{2+} in haem. The iron atom in haem form six bonds in total: four bonds to the nitrogen atoms in the flat ring structure of haem, as shown, and the other two above and below the plane of the page (denoted here by the dotted pink lines). One of the perpendicular bonds is bound to the nitrogen atom of a histidine residue, the other is the binding site for an oxygen molecule.

Source: Snape, A. & Papachristodoulou, D. (2018). *Biochemistry and Molecular Biology* (6th edn). Oxford University Press.

explains why, as shown in Figure 6.23, the oxygen saturation of haemoglobin initially increases gradually as oxygen concentrations increase above zero, but then the curve gets steeper (oxygen binds more readily as more sites become occupied) before levelling out as full saturation is approached.

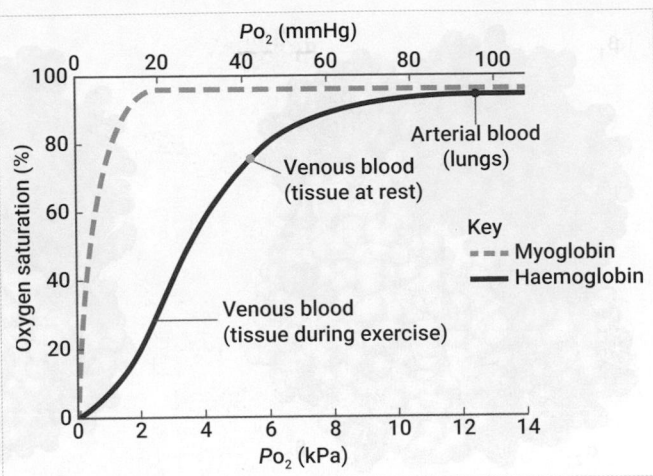

Figure 6.23 Oxygen saturation curve for haemoglobin and myoglobin. Haemoglobin in red blood cells is fully saturated with oxygen at the concentration of oxygen (c. 100 mmHg) in the lungs, but gradually releases it at oxygen concentrations found in both resting and exercising tissue (40 mmHg and 20 mmHg). The higher affinity for oxygen of myoglobin compared with haemoglobin means that myoglobin in the muscles readily accepts oxygen delivered to it by haemoglobin in the blood, and only gives up oxygen for use in muscle tissue respiration at the very low oxygen concentrations reached in muscle during intense exercise.

Source: Berg, Tymocczko and Stryer. *Biochemistry*, 7th edn. (2012) W. H. Freeman and Company.

Haemoglobin is an allosteric protein

The cooperative binding of oxygen by haemoglobin is an example of a phenomenon exhibited by many proteins, called **allostery**. Haemoglobin can be termed an allosteric protein. Allosteric proteins are proteins that have more than one binding site for ligands and in which binding of a ligand (oxygen in the case of haemoglobin) at one site influences the way in which the protein interacts with a further ligand (an additional oxygen molecule in haemoglobin) at another site. Allostery has been extensively studied because it allows fine-tuning and regulation of protein function.

So how does the binding of oxygen to one subunit on a haemoglobin molecule make it easier for another oxygen molecule to bind to another subunit? The answer is in a structural change that occurs when oxygen binds. To understand the structural change it is helpful to view haemoglobin as a molecule made up of two protein dimers, one formed from the $α_1$ and $β_1$ subunits, which are firmly bound together, and the other similarly formed from the $α_2$ and $β_2$ subunits. The two dimers associate to form the complete quaternary structure. Oxygen binding causes a relative rotation between the two dimers as shown in Figure 6.24.

In Figure 6.24a haemoglobin is shown in the so-called 'Tense' or 'T' state, in which it has a low affinity for oxygen, while in 6.24b it is in the 'Relaxed' or 'R' state, in which it has high oxygen affinity. This structural change to the haemoglobin molecule is caused by binding of oxygen to haem, which makes the iron atom in the haem group move slightly. The protein rearranges itself because of the movement of the iron atom, causing a relative movement of one αβ dimer with respect to the other. This movement results in the T→R change.

We will see a further example of allostery in Section 6.4, where we discuss its importance in regulating enzyme activity.

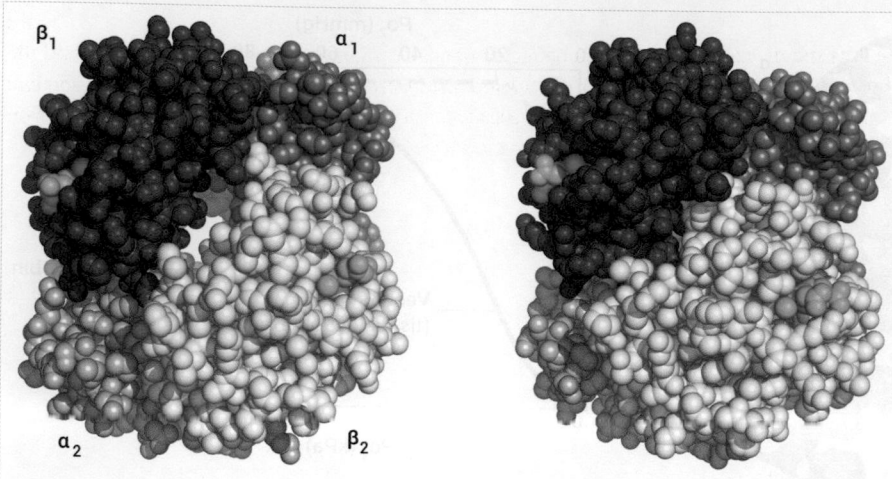

Figure 6.24 Haemoglobin consists of four separate protein subunits that are held together by non-covalent interactions. The α_1 and β_1 subunits are held together quite strongly to form a protein dimer, as are the α_2 and β_2 subunits. Deoxyhaemoglobin is shown on the left. On oxygen binding the $\alpha_2\beta_2$ dimer shifts its position relative to the $\alpha_1\beta_1$ dimer, rotating by about 15° with a corresponding change in the non-covalent interactions between the two dimers.

Source: Snape, A. & Papachristodoulou, D. (2018). *Biochemistry and Molecular Biology* (6th edn). Oxford University Press.

 Go to the e-book to explore an interactive version of Figure 6.24.

PAUSE AND THINK

Haemoglobin in the fetus has a different structure from adult haemoglobin; the two β globin subunits are replaced by related but slightly different γ (gamma) subunits. Why is this difference necessary?

Answer: Fetal haemoglobin needs to have a higher affinity for oxygen than adult haemoglobin, so that oxygen is transferred from the mother to the fetus. This higher affinity is provided by the slight structural difference provided by the γ subunits.

Mutations affecting the haemoglobin protein can have clinical consequences

The protein structure of haemoglobin has evolved so that it is superbly adapted to deliver oxygen to our tissues. The importance of this is brought home by the observation that mutations that change its amino acid sequence, and consequently change the three-dimensional structure of haemoglobin, can have devastating consequences. More than 1000 naturally occurring human haemoglobin variants are known that differ from the normal haemoglobin molecule in a single amino acid. Many of these variants have little impact, but others cause diseases of varying severity. In Clinical Box 6.2 we discuss sickle cell disease, which is caused by a change of just one amino acid in the sequence of β globin.

Collagen is a fibrous protein built for strength

We are looking at collagen as an example of a fibrous protein because it is the most plentiful protein in the mammalian body. The structure of collagen has evolved to maximize its strength, a requirement of the connective tissues where it is found. These include the dermis of the skin, tendons, cartilage, and bone. In these tissues, collagen is secreted from the cells where it is made into the extracellular matrix.

Collagen comprises three polypeptide chains wrapped around each other to form what we call a triple 'superhelix'. The term superhelix is used because each of the three chains is itself a helix. As shown in Figure 6.25, each of the polypeptides in the triple superhelix of collagen is an unusual left-handed helix, not the right-handed α helix of globular proteins. However, the three polypeptides are twisted around each other in a right-handed manner to form the triple helix.

CLINICAL BOX 6.2   Sickle cell disease

Sickle cell disease was one of the first genetic diseases to be understood at the molecular level. It is caused by a mutation, a change in the DNA sequence encoding the β globin chain, and illustrates how as little as just one amino acid change in a protein can have a profound effect.

In the β chains of the normal human haemoglobin tetramer (haemoglobin A), amino acid number 6 is glutamic acid, which has a side chain that is negatively charged and highly hydrophilic. In the haemoglobin of people with sickle cell disease (haemoglobin S), this glutamic acid is replaced by a hydrophobic valine residue. The change requires only a single base mutation: from A to T in the DNA sequence (Figure 1).

The hydrophobic valine on the haemoglobin S binds to a hydrophobic pocket on another haemoglobin tetramer and so on, resulting in the formation of long rigid rods. In oxygenated haemoglobin, because of its different conformation, the hydrophobic pocket is not exposed, and the haemoglobin tetramers do not bind to each other. In deoxygenated blood of people with sickle cell disease long deoxyhaemoglobin rods build up and distort the normal biconcave red blood cells into sickle shapes that tend to block capillaries, causing tissue damage (Figure 1). The abnormal red cells also break up, causing anaemia.

Despite being fatal if untreated, sickle cell disease is prevalent in geographical areas where malaria is, or was, common and in the descendants of people who migrated from those areas, such as African Americans. This high incidence can be explained by the mutated gene undergoing positive selection. The haemoglobin abnormality inhibits the development of the malarial parasite, even in individuals who carry one normal and one mutated copy of the gene and therefore do not suffer from sickling of their red blood cells. This so-called heterozygote advantage protects unaffected genetic carriers against death from malaria and the mutation is therefore preserved in the population by natural selection: carriers are more likely to survive long enough to reproduce (and pass on the mutated gene to their offspring).

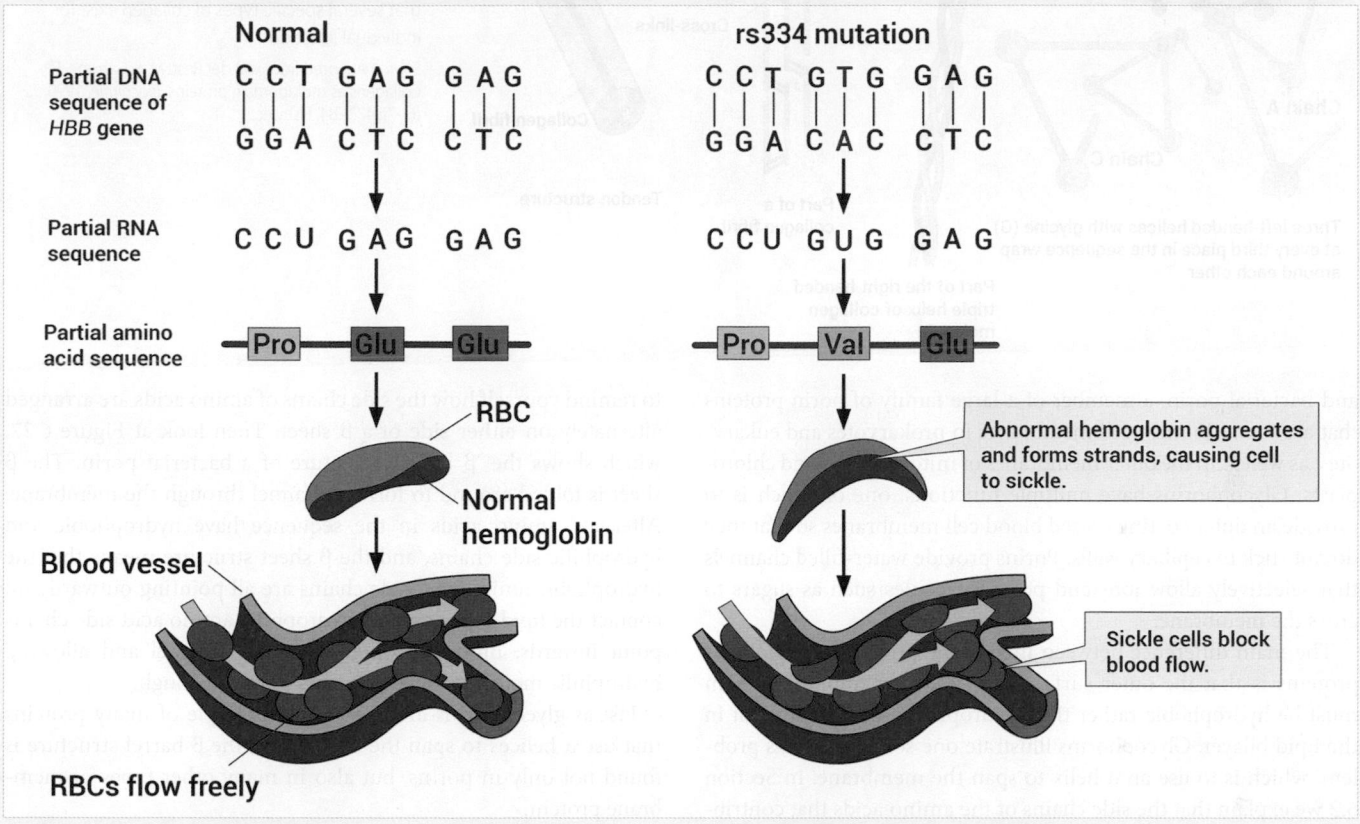

Figure 1 Sickle cell disease. A single base mutation from A to T in the DNA sequence encoding β globin causes the replacement of glutamic acid by valine in the protein sequence. The abnormal haemoglobins that result (HbS) stick together in low oxygen conditions, causing sickling of the red blood cells (RBCs) and blockage of capillaries. *Source*: Adapted from van der Rest, M. & Garrone, R. Collagens as multidomain proteins. *Biochimie*, 1990; 72 (6–7): 473–84. Elsevier.

The three strands of the triple superhelix are in close association, forming a very strong structure. The side chains of the amino acids in polypeptide chains would normally prevent such close association, but the structure of collagen allows this. The left-handed helix of collagen has a distinctive structure, with three amino acid residues in each turn of the helix, rather than 3.6 as in the α helix. Every third amino acid is glycine, and the small structure of glycine, in which the 'side chain' is just a hydrogen atom, allows the close packing of the three chains.

Another distinctive feature of collagen is that approximately one in every three amino acids in its sequence is proline, which you may recall is not found in an α helix (see Section 6.2). Proline contributes to the strength of collagen in two ways. First, the occurrence of proline causes the collagen chain to wind up into the left-handed helix with three amino acids per turn. Second, the side chains of proline in collagen are modified by the addition of a hydroxyl group; this hydroxyl group is added by an enzyme-catalysed reaction in the endoplasmic reticulum after synthesis of the protein. The hydroxyl group is polar and allows the formation of hydrogen bonds that hold the three helices together in the superhelix, contributing to the strength of the structure. Interestingly, our requirement for ascorbic acid (vitamin C) in our diet arises because it is needed as a cofactor of the enzyme that hydroxylates proline in collagen.

PAUSE AND THINK

Scurvy, caused by lack of vitamin C in the diet, is characterized by bleeding gums and failure of wound healing. How do these symptoms arise from vitamin C deficiency?

Answer: In scurvy, connective tissue is weakened because collagen cannot form properly.

Glycophorins and porins are membrane proteins and have distinctive structures due to their hydrophobic environment

We are now going to look at two examples that illustrate typical structures of membrane-embedded proteins: glycophorins, which are a major component of erythrocyte (red blood cell) membranes,

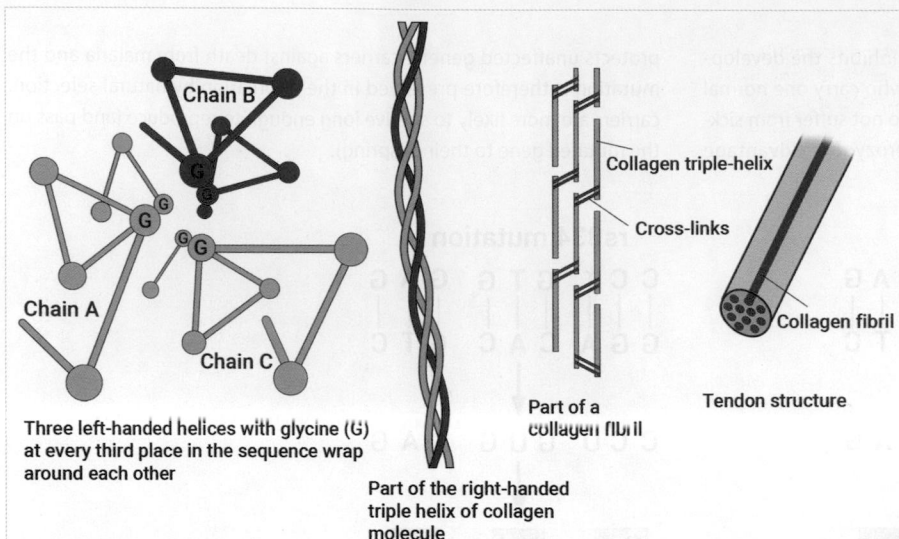

Figure 6.25 The structure of collagen. Three left-handed helices are wound around each other to form a right-handed triple helix. The 'ropes' of collagen so formed are lined up in parallel as shown and cross-linked to each other to form collagen fibrils in tendons, cartilage, and bone, or they form a network in the dermis of the skin. Note that several specific types of collagen exist for individual functions.

Source: Adapted from van der Rest, M. & Garrone, R. Collagens as multidomain proteins. *Biochimie*, 1990; 72 (6–7): 473–84. Elsevier.

and bacterial porin, a member of a large family of porin proteins that are found in most cell membranes in prokaryotes and eukaryotes, as well as in the outer membranes of mitochondria and chloroplasts. Glycophorins have multiple functions, one of which is to provide an outer coating on red blood cell membranes so that they do not stick to capillary walls. Porins provide water-filled channels that selectively allow ions and polar molecules such as sugars to cross the membrane.

The main difference between membrane proteins and globular proteins is that the outer surface of a folded membrane protein must be hydrophobic rather than hydrophilic, to allow it to sit in the lipid bilayer. Glycophorins illustrate one solution to this problem, which is to use an α helix to span the membrane. In Section 6.2 we explain that the side chains of the amino acids that contribute to an α helix point outwards, away from the axis of the helix. (See Figure 6.14a if you need reminding of this.) An α helix can therefore be used to cross the membrane, if the amino acids that make up the helix have hydrophobic side chains.

Figure 6.26 illustrates the structure of glycophorins, in which a single hydrophobic α helix crosses the membrane. The protein also has hydrophilic extracellular and intracellular domains. As shown in Figure 6.26a, the extracellular domain of glycophorins is modified by glycosylation (the addition of sugars that are covalently linked to certain amino acids), which makes it even more strongly hydrophilic.

Many other membrane proteins besides glycophorins use α helices with hydrophobic side chains to cross the membrane. In some proteins the polypeptide chain crosses the membrane several times, as shown in Figure 6.26b, with hydrophilic extra- and intracellular loops linking the α helices.

PAUSE AND THINK

Which structural features of the sugar molecules that are added to glycosylated proteins make them hydrophilic?

Answer: Sugars such as glucose and galactose contain multiple polar hydroxyl groups, which are hydrophilic.

Rather than using an α helix to cross the membrane, porins use the other type of secondary structure, the β sheet. Look at Figure 6.14d

to remind yourself how the side chains of amino acids are arranged alternately on either side of a β sheet. Then look at Figure 6.27, which shows the 'β barrel' structure of a bacterial porin. The β sheet is folded around to form a channel through the membrane. Alternate amino acids in the sequence have hydrophobic and hydrophilic side chains, and the β sheet structure means that the hydrophobic amino acid side chains are all pointing outward and contact the lipid bilayer, while hydrophilic amino acid side chains point inwards, lining the interior of the channel and allowing hydrophilic molecules such as sugars to pass through.

Just as glycophorins are simply one example of many proteins that use α helices to span the membrane, the β barrel structure is found not only in porins, but also in many other types of membrane protein.

Many proteins have two or more structural domains

If we consider a protein molecule that comprises a single polypeptide chain, we often see two or more regions that form compact domains of folded structure linked together by a less structured section of the chain. Often each structural domain makes a distinct contribution to the overall function of the protein. An example is seen in the enzyme pyruvate kinase, which has three such domains, as illustrated in Figure 6.28.

▶ **Enzymes are introduced briefly in Chapter 2, and we explore their structure and function as biological catalysts in detail in Section 6.4.**

The central domain of the pyruvate kinase binds its substrate and has catalytic properties, while the other two domains have separate functions. The 'cap' domain is attached to the catalytic domain by a flexible loop, which allows it to shut down over the catalytic domain once the substrate is bound. The third domain is the regulatory domain, which enables the enzyme to be active only when required. Proteins that are made up of two or more distinct domains are termed modular proteins.

▶ **We explore the evolutionary significance of modular proteins in Section 6.5.**

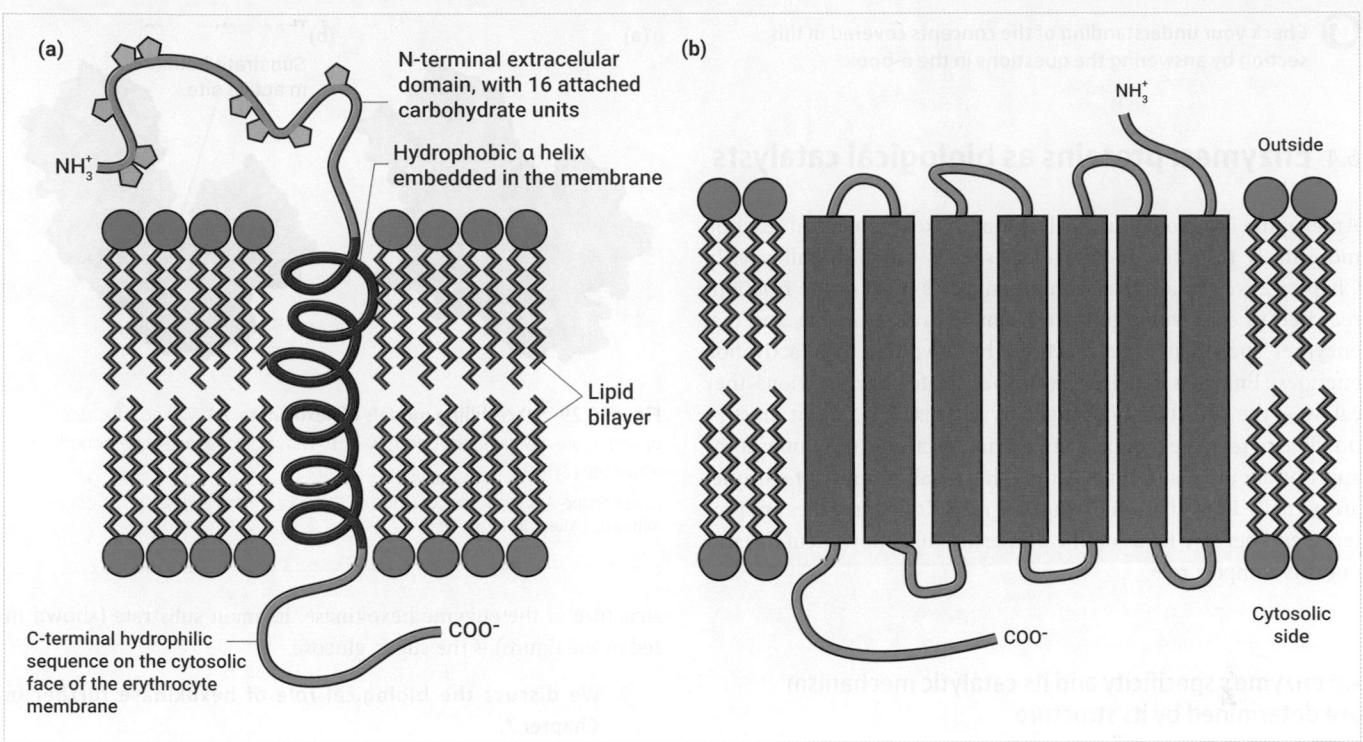

Figure 6.26 Glycophorin and transmembrane proteins. (a) The structure of glycophorin, a protein of the erythrocyte membrane. The protein consists of a single polypeptide chain. The α helical portion of the protein (shown in red) consists of a chain of 19 hydrophobic amino acids that form a helix approximately 3 nm in length, sufficient to span the non-polar interior of the lipid bilayer. The extracellular and intracellular domains (shown in green) contain hydrophilic amino acids. The extracellular domain is modified by the addition of sugars, which make it even more hydrophilic. (b) Representation of a transmembrane protein with seven α helices spanning the lipid bilayer. There are many examples of proteins with this 'seven transmembrane' structure. The α helices are not actually arranged in a row, as shown here for convenience, but are clustered compactly together.

Source: Snape, A. & Papachristodoulou, D. (2018). *Biochemistry and Molecular Biology* (6th edn). Oxford University Press.

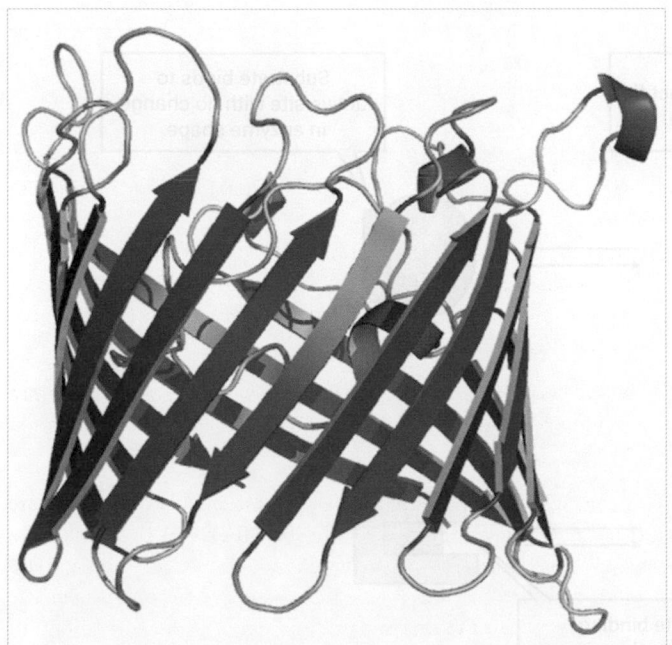

Figure 6.27 The β barrel structure of a bacterial porin (Protein Data Bank code 1BH3). β sheet strands are purple. Short α helical sections are pink, and other parts of the structure are grey.

Source: Snape, A. & Papachristodoulou, D. (2018). *Biochemistry and Molecular Biology* (6th edn). Oxford University Press.

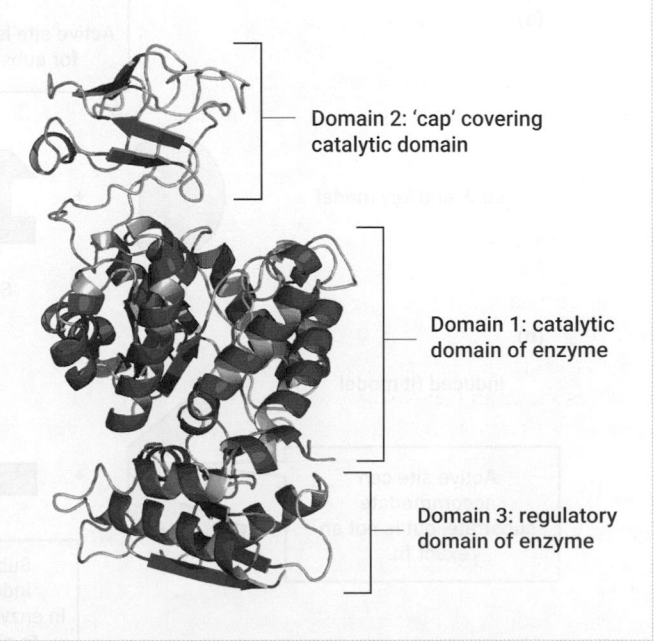

Domain 2: 'cap' covering catalytic domain

Domain 1: catalytic domain of enzyme

Domain 3: regulatory domain of enzyme

Figure 6.28 Structure of the enzyme pyruvate kinase. The single polypeptide chain folds into three linked structural domains, which each makes a distinct contribution to the enzyme's overall function.

Check your understanding of the concepts covered in this section by answering the questions in the e-book.

6.4 Enzymes: proteins as biological catalysts

An enzyme is a biological catalyst that makes a chemical reaction more likely to occur, and therefore speeds up its overall rate. In Chapter 2 we explain that without enzymes most of the reactions required in cells would occur too slowly to support life, and that enzymes speed up these reactions by lowering their activation energies. Enzymes must be very specific for the reactions they catalyse, they must be unchanged by the reaction, and it must be possible to regulate their activity so that reactions occur only at the appropriate place and time. Although a small number of enzymes are actually RNA molecules, the vast majority of enzymes are proteins, and here we focus on the properties that allow them to carry out this complex role.

An enzyme's specificity and its catalytic mechanism are determined by its structure

Enzyme specificity can be achieved because proteins can adopt particular three-dimensional shapes. A crucial feature of enzyme structure is that the polypeptide chains are folded in such a way that an 'active site' is formed on the surface of the enzyme. The active site is a three-dimensional pocket or cleft into which the substrates of the enzyme (the molecules that take part in the reaction) fit very precisely. This is illustrated in Figure 6.29, which shows the

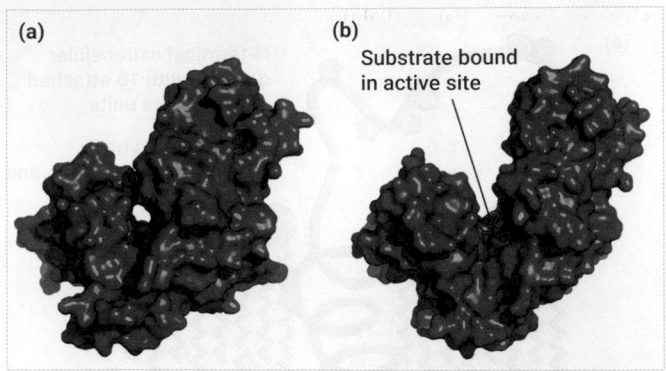

Figure 6.29 Space-filling models of hexokinase. (a) With no substrate bound (Protein Data Bank Code 1HKG). (b) With a glucose analogue bound in the active site (2YHX).

Source: Snape, A. & Papachristodoulou, D. (2018). *Biochemistry and Molecular Biology* (6th edn). Oxford University Press.

structure of the enzyme hexokinase. Its main substrate (shown in red in the figure) is the sugar, glucose.

▶ **We discuss the biological role of hexokinase further in Chapter 7.**

Early explanations of enzyme interactions with their substrates suggested a 'lock-and-key' model in which the active site was envisaged to be a rigid structure (the lock) with the substrate (the key) fitting into it (as illustrated in Figure 6.30a). A more up-to-date concept, however, is the 'induced-fit' mechanism, which is based on the view that the enzyme is not a rigid structure analogous to a lock, but rather is a flexible structure capable of changing

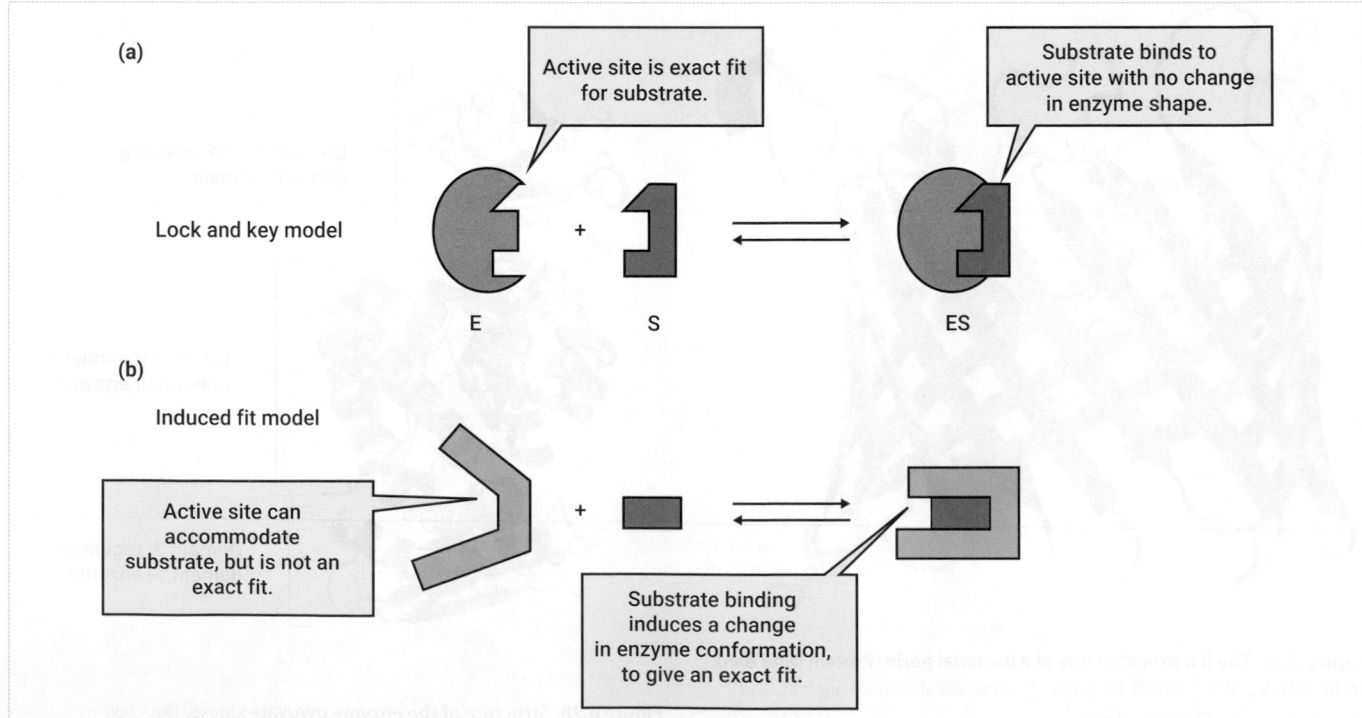

Figure 6.30 Models of enzyme action. (a) Lock-and-key model of enzyme mechanism. E, enzyme; S, substrate. (b) Induced-fit model of the hexokinase mechanism.

Source: Adapted from Snape, A. & Papachristodoulou, D. (2018). *Biochemistry and Molecular Biology* (6th edn). Oxford University Press.

its conformation (shape) slightly in an interactive way when its substrate binds. This change has the important effect of altering the spatial arrangements of groups on the molecule. Proteins can change their conformation, as already described for haemoglobin, through changes in the interactions between amino acid side chains, and such changes may be triggered by substrate binding.

The induced-fit mechanism was first established for hexokinase. As shown in the space-filling model in Figure 6.29, and more schematically in Figure 6.30b, the enzyme has two 'wings' to its structure. In the absence of glucose, these have an 'open' conformation, but on binding of glucose, the wings close in a jaw-like movement, which results in the creation of the catalytic site.

As with all specific protein–ligand binding, the substrate attaches to the active site of an enzyme reversibly by weak non-covalent bonds, but specificity is achieved because *several* weak bonds are formed—literally a case of strength in numbers. Unless there is a precise fit between the interacting groups on the substrate and enzyme, the attachment will not occur.

So how does the attachment of substrates to the active site of an enzyme lead to catalysis? In Chapter 2, we explain that enzymes lower the activation energy of the reaction they catalyse, but how does this occur? The answer varies, but some common mechanisms are illustrated in Figure 6.31. In 6.31a, the reaction involves two substrate molecules, which are brought into close alignment

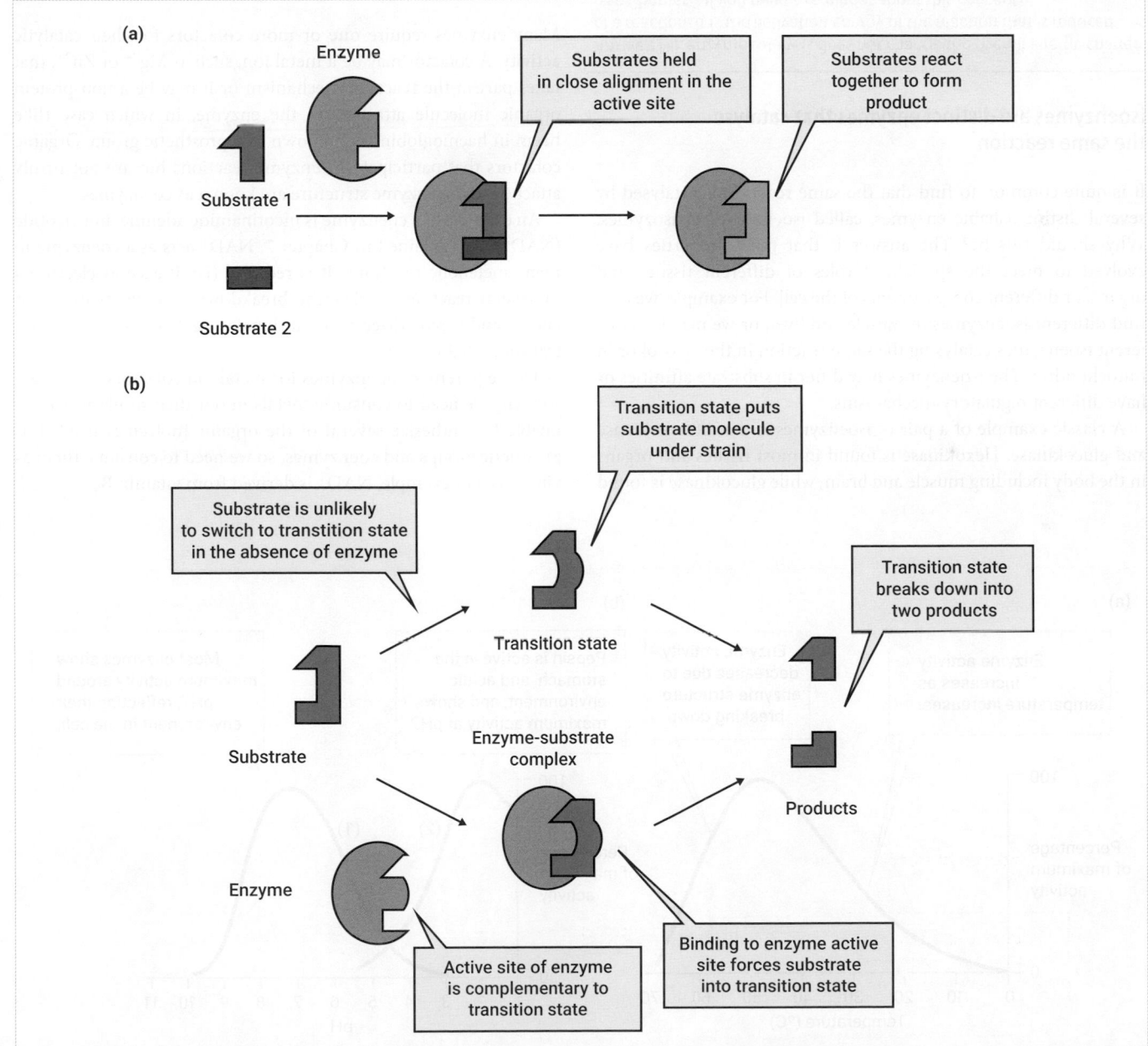

Figure 6.31 Mechanisms of enzyme action. In (a) two substrate molecules are brought into close alignment at the active site, allowing them to react together and form a single product. In (b) the enzyme active site is complementary to the substrate in its transition state. In the transition state the substrate molecule is under strain and is likely to break down to two product molecules. The transition state can form spontaneously in the absence of enzyme, but this happens only rarely so the reaction proceeds very slowly unless enzyme is present.

Source: Adapted from Snape, A. & Papachristodoulou, D. (2018). *Biochemistry and Molecular Biology* (6th edn). Oxford University Press.

through binding to the active site, facilitating their interaction. In Figure 6.31b, binding to the active site puts strain on a substrate molecule, forcing it into an intermediate state called a transition state, which then leads to the substrate splitting into two products. Because the active site is complementary in shape to the transition state, it is easier for the substrate to achieve this state when the enzyme is present than when it is not, so the frequency of the reaction is increased.

PAUSE AND THINK

Would the mechanisms illustrated in Figure 6.31 alter the overall energy change associated with the reactions catalysed?

Answer: No, enzyme catalysis does not change the overall energy change of a reaction. It is the activation energy of the reaction that is reduced (see Chapter 2 if you need a reminder about this concept).

Isoenzymes are distinct enzymes that catalyse the same reaction

It is quite common to find that the same reaction is catalysed by several distinguishable enzymes, called isoenzymes or isozymes. Why should this be? The answer is that their properties have evolved to meet the specialized roles of different tissues and organs, or different compartments of the cell. For example, we may find different isoenzymes in muscle and liver, or we may find different isoenzymes catalysing the same reaction in the cytosol or in mitochondria. The isoenzymes may differ in substrate affinities or have different regulatory mechanisms.

A classic example of a pair of isoenzymes is that of hexokinase and glucokinase. Hexokinase is found in most tissues and organs in the body including muscle and brain, while glucokinase is found specifically in the liver. Both isoenzymes catalyse the addition of a phosphate group to glucose. However, the two isoenzymes have different affinities for glucose, reflecting the different ways that phosphorylated glucose is used in the different organs. Hexokinase has a high affinity for glucose and is found in organs that constantly need to metabolize glucose for energy, with phosphorylation of glucose being the first step in the glycolysis pathway (as we explore further in Chapter 7). In contrast, glucokinase has a relatively low affinity for glucose, as the liver only carries out this reaction when glucose levels are high as a precursor to converting glucose to glycogen, an energy storage molecule.

Many enzymes require non-protein cofactors

Many enzymes require one or more cofactors for their catalytic activity. A cofactor may be a metal ion, such as Mg^{2+} or Zn^{2+}, that takes part in the reaction mechanism or it may be a non-protein organic molecule attached to the enzyme, in which case (like haem in haemoglobin) it is known as a prosthetic group. Organic cofactors that participate in enzyme reactions but are not firmly attached to the enzyme structure are known as coenzymes.

An example of a coenzyme is nicotinamide adenine dinucleotide (NAD^+). As explained in Chapter 7, NAD^+ acts as a coenzyme in many metabolic reactions. It is reduced (i.e. it accepts electrons) in several reactions during the breakdown of carbohydrates for energy and is reoxidized when it donates electrons to the electron transport chain.

The requirement of enzymes for metal ion cofactors is one reason why we need to consume metals in our diet; similarly, we are unable to synthesize several of the organic molecules needed as prosthetic groups and coenzymes, so we need to consume them as vitamins. For example, NAD^+ is derived from vitamin B_3.

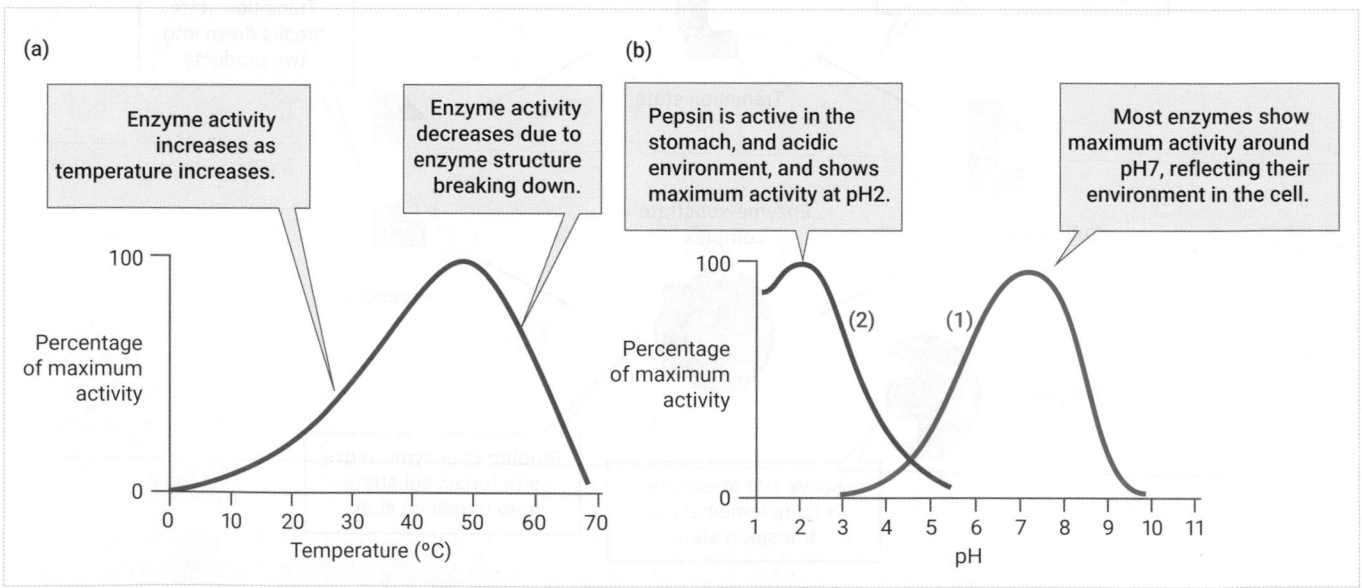

Figure 6.32 Effect of pH and temperature on enzyme activity. (a) Effect of temperature on a typical enzyme. The precipitous drop in activity at high temperatures is due to breakdown of the enzyme structure. (b) Effect of pH on enzyme activity. Curve 1 is typical of most enzymes with maximal activity near physiological pH. Curve 2 represents pepsin, an exceptional case, since this digestive enzyme functions in the acidic stomach contents.

Source: (a) Adapted from Snape, A. & Papachristodoulou, D. (2018). *Biochemistry and Molecular Biology* (6th edn). Oxford University Press.

PAUSE AND THINK

Enzymes are not altered by the reactions they catalyse. What is distinctive about the role of coenzymes such as NAD+ in enzyme reactions?

Answer: Molecules such as NAD+ are altered by the enzyme-catalysed reactions they take part in. An NAD+ molecule acts as an electron acceptor in one reaction, then carries the electrons to a different enzyme for a reaction in which it acts as an electron donor.

pH and temperature affect enzyme activity

As enzymes are proteins and must maintain a specific three-dimensional structure to operate as they should, their activity is affected by temperature and pH, as shown in Figure 6.32.

Increasing temperature generally increases the rate of a chemical reaction, and this is observed for enzyme-catalysed reactions up to a point. However, at higher temperatures the non-covalent bonds that maintain the enzyme structure break down, leading to reduced activity and ultimately complete inactivation. A typical enzyme optimum temperature plot would appear as shown in Figure 6.32a.

The activity of an enzyme is influenced by pH in several ways. As explained in Section 6.1, ionic bonds between positively and negatively charged amino acid side chains contribute to the structure of the enzyme. The charge on the side chains, and hence the ionic bonds, can be disrupted by pH changes, which are changes in the hydrogen ion concentration of the surrounding solution.

▶ We discuss pH further in Chapter 1.

pH changes can also affect the interaction of the substrate with the enzyme and the catalytic mechanism of the enzyme. Enzyme pH activity profiles vary, but the optimum is often around pH 7, a typical pH for the interior of a cell. Exceptions include the digestive enzyme, pepsin, which functions in the acidic stomach contents, and which has a pH optimum near 2.0.

PAUSE AND THINK

By what mechanism does increasing temperature initially increase the rate of an enzyme-catalysed reaction?

Answer: Increased temperature increases the movement of molecules so that more collisions occur between them, including between enzymes and substrate molecules.

In Real World View 6.1, we look at an example of 'enzyme engineering' where the structure of an enzyme is altered to stabilize it and enable it to maintain activity outside its normal temperature and pH environment.

Enzyme kinetics: investigating the rates of enzyme-catalysed reactions

As illustrated in Real World View 6.1, we can gain useful information about enzymes by investigating their protein structure. A complementary approach to investigating enzymes is to measure the rate of an enzyme-catalysed reaction under experimentally controlled conditions. These studies of **enzyme kinetics** provide data that enable us to compare different enzymes and gain insight into their activities.

Figure 6.33a illustrates the results of a typical experiment in which the rate of an enzyme-catalysed reaction is measured in a series of reaction tubes. Substrate concentration is increased in successive tubes, while enzyme concentration remains the same. Initially the rate of reaction (V_0) increases almost linearly as substrate concentration increases, but at high substrate concentrations the reaction rate levels off and approaches a maximum rate (V_{max}), which is reached when all enzyme active sites are bound by substrate. At V_{max} the reaction continues, but new substrate can only enter the active site of an enzyme when it releases the product, so further increases in substrate concentration cannot increase the reaction rate any further. We say that the reaction is at V_{max} when the enzyme is **saturated** with substrate.

The hyperbolic shape of the curve shown in Figure 6.33a is typical of many enzymes. It can be described by an algebraic equation named after researchers Leonor Michaelis and Maud Menten. The equation, shown here, is known as the Michaelis–Menten equation, and enzymes that give curves of this type are said to display **Michaelis–Menten kinetics**.

$$V_0 = \frac{V_{max}[S]}{K_M + [S]}$$

V_0 is the initial rate or initial velocity of the enzyme reaction and is determined experimentally. It is important that it is the *initial* rate of reaction at each substrate concentration, measured immediately after mixing enzyme and substrate, because the substrate concentration changes over the course of the reaction as it is converted to product. Measuring the initial rate avoids this complication. The value of [S] for each measurement is set by the experimenter. The values of V_{max} and K_M can be estimated from the graph and are useful for comparing the catalytic efficiencies of different enzymes.

As we explained earlier, V_{max} is the maximum rate of reaction and is reached when the enzyme is saturated with substrate. It gives a measure of how fast the enzyme converts substrate to product when it is working at full capacity. The value of V_{max} is specific to a particular enzyme and substrate, and it is dependent on enzyme concentration: increasing enzyme concentration increases V_{max}.

K_M is a constant known as the Michaelis constant and is defined as the concentration of substrate that gives half the maximum rate of reaction (0.5 V_{max}). The value of K_M is also specific to a particular enzyme and substrate, but unlike V_{max} it is unaffected by changes in enzyme concentration. For many enzymes the value of K_M indicates how tightly the enzyme binds its substrate. An enzyme with a low K_M value binds its substrate efficiently and needs relatively little substrate to reach half its maximum rate, compared to an enzyme with a high K_M value, which binds its substrate only weakly.

PAUSE AND THINK

The rate of an enzyme-catalysed reaction can often be measured by tracking the appearance of a reaction product. In an experiment where [S] is measured in mmoles per litre (mM) and V_0 is measured in moles of product produced per minute (moles min^{-1}), what would be the units of K_M and V_{max}?

Answer: K_M units are units of substrate concentration, mM. V_{max} units are units of V_0 (moles min^{-1}).

REAL WORLD VIEW 6.1   Engineering enzymes for industrial applications

In Chapter 1 we explained that triglyceride fats consist of three fatty acid molecules, each linked to a glycerol backbone (see Figure 1.22). Lipases are enzymes that release fatty acids from glycerol, allowing the fatty acids to be used as energy sources, as described in Chapter 7. Lipases have many industrial applications, perhaps the most familiar being their inclusion in detergents to break down fat and oil stains on our clothes. They can also be used in the synthesis of biodiesel, the production of drugs and agrochemicals, and in food processing and the production of flavour compounds, to name just few additional examples.

For many of these applications, it is advantageous to make the enzymes more thermostable (i.e. to stop them breaking down at the higher temperatures typically used in industrial processes). An example of how this can be done using knowledge of protein structure and protein engineering is seen in the work of Han and colleagues.

They based their work on structural studies that showed that many lipases have a 'lid' that covers the active site, as shown in Figure 1a. The lid remains closed when the enzyme is in the aqueous environment but when it is placed in an organic solvent, as would happen when it encounters a potential substrate, the lid opens to give the hydrophobic substrate access to the active site.

Han and colleagues aimed to stabilize the lid of a lipase produced by the fungus *Rhizomucor miehei* by changing two amino acids in the enzyme from proline and lysine to cysteines, which formed a disulphide bond between them. They showed that the engineered enzyme retained its full catalytic activity significantly longer than the original non-engineered version when heated to 60° (Figure 1b).

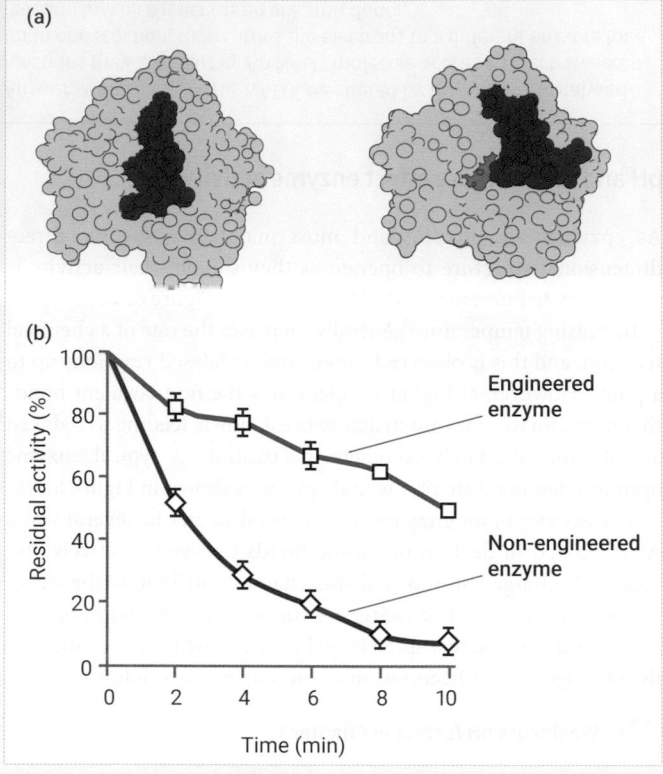

Figure 1 (a) Surface view of the structure of lipase from the fungus *Rhizomucor miehei*, on the left in an aqueous solution, and on the right in an organic solvent. The 'lid' structure, shown in red, is closed in aqueous solution and open in the organic solvent, exposing the active site (blue) to the lipid substrate. (b) Lipase engineered to make the lid more stable retains its catalytic activity at 60° for longer than the non-engineered enzyme.

Source: (a) Adapted from Rehm et al. (2010) *Protein Science*, 19: 2122–2130. doi:10.1002/pro.493, Figure 5. (b) Adapted from Han et al. *Appl Microbiol Biotechnol* (2009) 85:117–126 DOI 10.1007/s00253-009-2067-8, Figure 5.

While we can obtain estimated values of V_{max} and K_M from a graph such as the one shown in Figure 6.33a (a 'Michaelis–Menten plot'), there is an issue that reduces the accuracy of these estimates. We are looking for the point at which the curve levels off to get a reading of V_{max}, and we use that to read off the value of [S] that gives $0.5\,V_{max}$. However, a careful look at the graph will show you that in the experiment, the value of V_0 never quite reaches V_{max}, so the value of V_{max} has to be estimated by drawing a horizontal line above the top of the curve. A simple 'fix' often used to give more accurate estimates is to use a double reciprocal plot, as shown in Figure 6.33b. Here $\frac{1}{V_0}$ is plotted against $\frac{1}{[S]}$, which gives a straight line that can be extrapolated to give values for $\frac{1}{V_{max}}$ at the intercept with the vertical axis, and $-\frac{1}{K_M}$ at the intercept with the horizontal

axis. The double reciprocal plot, often called the 'Lineweaver–Burk' plot is also used to give a quick visual analysis of the effects of enzyme inhibitors, the subject of our next section.

Toxins and drugs can inhibit enzymes

Many chemicals can specifically inhibit the activity of a particular enzyme. Since blocking a reaction is likely to be harmful to the organism, some naturally occurring and man-made enzyme inhibitors are classed as toxins. However, some of these chemicals can be repurposed, or even designed, to act as therapeutic drugs. An example is the class of drugs known as ACE inhibitors, which act on angiotensin-converting enzyme (ACE), an enzyme involved in regulating blood pressure. A natural ACE inhibitor identified in the venom of the Brazilian pit viper, *Bothrops jararaca* (Figure 6.34)

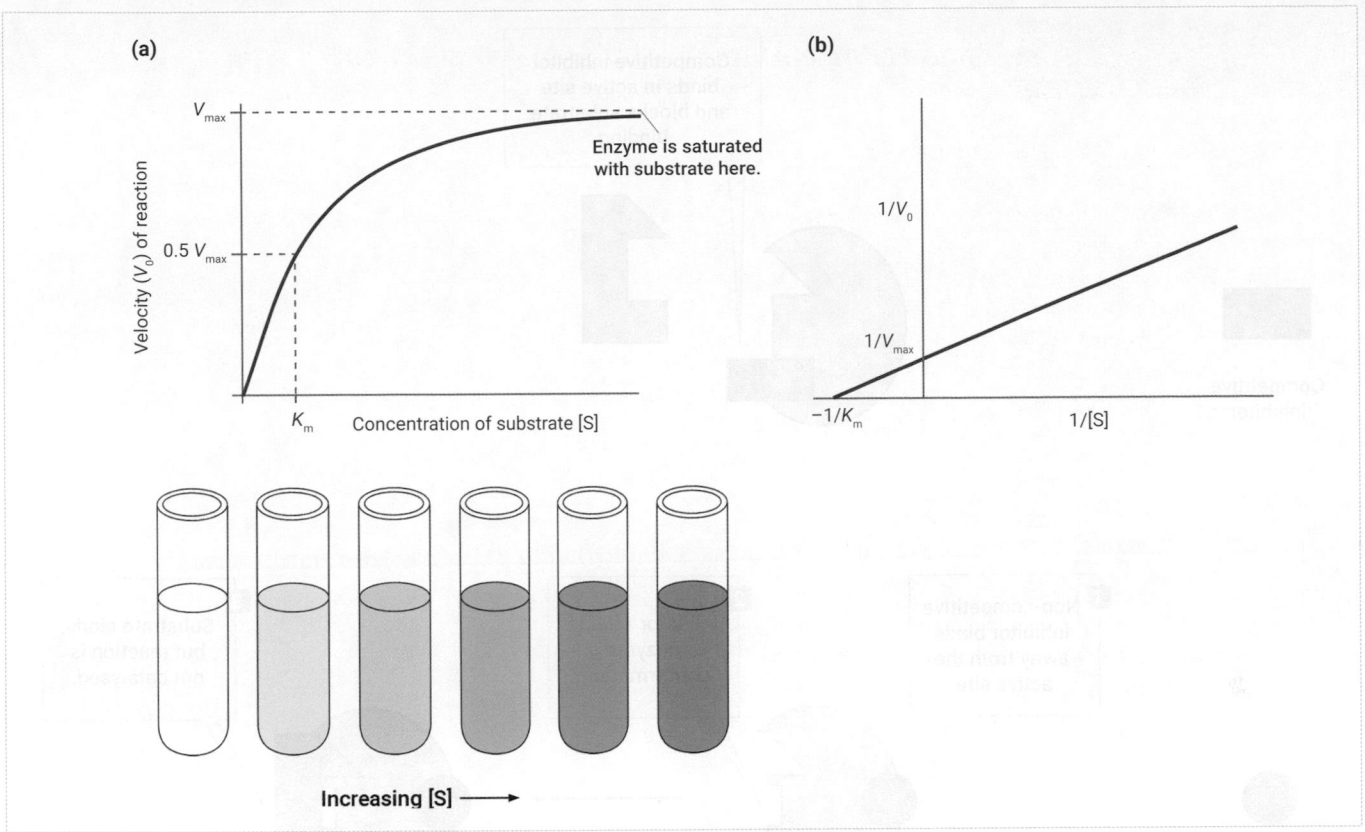

Figure 6.33 Enzyme kinetics. (a) Experimental measurement of the effect of substrate concentration on the rate of an enzyme-catalysed reaction. Initial rate of reaction (V_0) is measured in a series of reaction tubes (shown below the graph). Enzyme concentration is the same in each tube, while the substrate concentration [S] is increased. V_{max} is the maximum rate of reaction. K_M is the substrate concentration at which the initial rate of reaction is half V_{max} and is explained further in the text. (b) Data from the experiment can be plotted as a double reciprocal plot ($\frac{1}{V_0}$ vs $\frac{1}{[S]}$), so that values for K_M and V_{max} can be obtained from the intercepts on the x- and y-axes.

Source: (a, top) Snape, A. & Papachristodoulou, D. (2018). *Biochemistry and Molecular Biology* (6th edn). Oxford University Press; (a, bottom) Figure 4–36a in Alberts et al. *Essential Cell Biology* (5th edn), W. W. Norton and Co. (b) Snape, A. & Papachristodoulou, D. (2018). *Biochemistry and Molecular Biology* (6th edn). Oxford University Press.

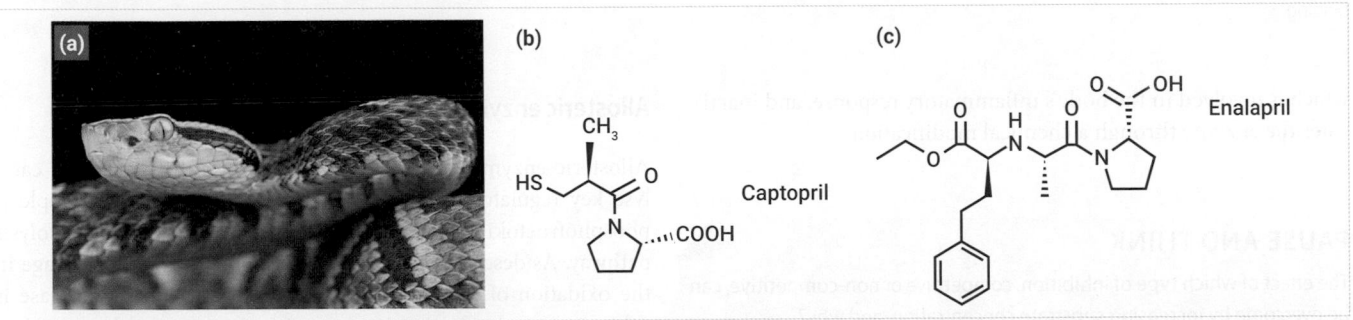

Figure 6.34 Toxins can inhibit enzymes. The venom of (a) the Brazilian pit viper, *Bothrops jararaca*, contains a molecule that can kill its victim by inhibiting angiotensin-converting enzyme (ACE), causing a fatal drop in blood pressure. The molecule has been adapted to make therapeutic ACE inhibitors such as (b) captopril and (c) enalapril, used for the treatment of hypertension (high blood pressure) and heart failure.

Source: (a) reptiles4all/Shutterstock.

was adapted to make the first therapeutic ACE inhibitor, captopril, which was designed to treat hypertension (high blood pressure) and heart failure.

These ACE inhibitors are **competitive inhibitors** of enzyme activity. As illustrated in Figure 6.35, a competitive inhibitor is similar enough in shape to the substrate of an enzyme to bind its active site, preventing the actual substrate from binding. In contrast, **non-competitive inhibitors** do not interfere with substrate binding, but interfere with the enzyme's activity in some other way, so that although the substrate binds there is no catalysis and the substrate is not converted to product. The drug aspirin is an example of a non-competitive inhibitor. It binds to the enzyme cyclooxygenase,

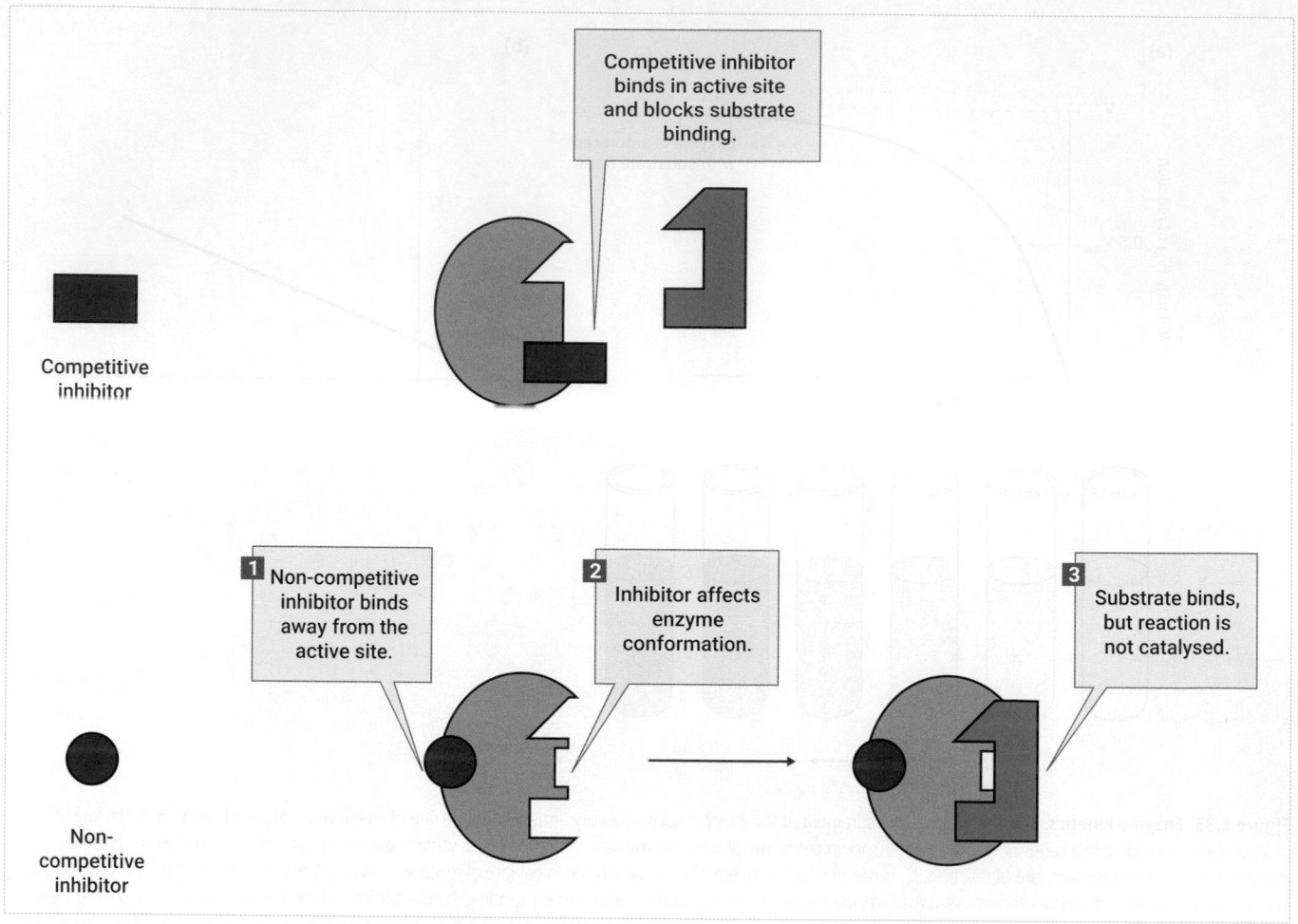

Figure 6.35 Competitive and non-competitive enzyme inhibitors have different modes of action. While competitive inhibitors mimic the active site and prevent substrate binding, non-competitive inhibitors inhibit catalysis through other mechanisms. In the example shown, the non-competitive inhibitor binds at a position distant from the active site, but in doing so it induces a change to the enzyme's overall structure that affects its catalytic mechanism despite allowing substrate binding.

which is involved in the body's inflammatory response, and inactivates the enzyme through a chemical modification.

PAUSE AND THINK

The effect of which type of inhibition, competitive or non-competitive, can be overcome by increasing substrate concentration, and why?

Answer: Competitive inhibition can be overcome by increasing substrate concentration. This is because when the substrate concentration is high enough it will compete' the inhibitor and all enzyme active sites will be bound by substrate. Increasing substrate concentration will not overcome non-competitive inhibition because the inhibitor does not bind to the active site.

The effects of inhibitors can be investigated using enzyme kinetic studies of the type described in the previous section. Figure 6.36 compares the effects of a competitive and a non-competitive inhibitor on the values of K_M and V_{max}, using a double reciprocal plot.

Allosteric enzymes regulate metabolic pathways

Allosteric enzymes are an important class of enzymes that catalyse key regulatory steps in metabolic pathways. An example is phosphofructokinase, which catalyses a reaction in the glycolysis pathway. As described in Chapter 7, glycolysis is the first stage in the oxidation of glucose for energy, and phosphofructokinase is subject to regulation that increases its activity—and hence drives the pathway forward—when energy levels in the cell are low. Conversely, phosphofructokinase is inhibited—and hence glycolysis is slowed down—when cellular energy levels are high. We will come back to the specific regulators of phosphofructokinase at the end of the section, but first we need to understand more about the nature and properties of allosteric enzymes.

At this point it may be helpful to remind yourself of the definition of an allosteric protein, discussed earlier in the context of haemoglobin (see Section 6.3). As stated there, allosteric proteins are proteins that have more than one binding site for ligands, whereby the binding of a ligand at one site influences the way in which the protein interacts with a further ligand at another site.

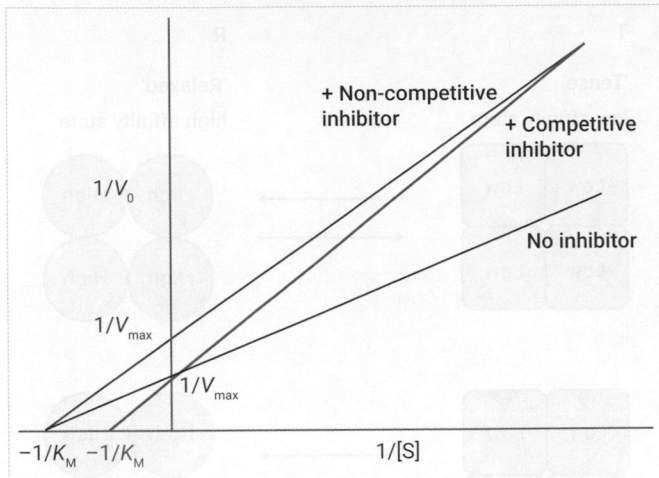

Figure 6.36 Effects of a competitive and a non-competitive inhibitor on enzyme kinetics, illustrated using a double reciprocal plot. The competitive inhibitor interferes with substrate binding, and this causes an apparent decrease in the strength of substrate binding and an increase in the K_M value. However, V_{max} is not changed, as increasing the substrate concentration will eventually saturate the enzyme with substrate and 'out compete' the competitor. The non-competitive inhibitor does not affect substrate binding, so K_M is unchanged. V_{max} is reduced since adding more substrate cannot overcome the effect of the non-competitive inhibitor.
Source: Snape, A. & Papachristodoulou, D. (2018). *Biochemistry and Molecular Biology* (6th edn). Oxford University Press.

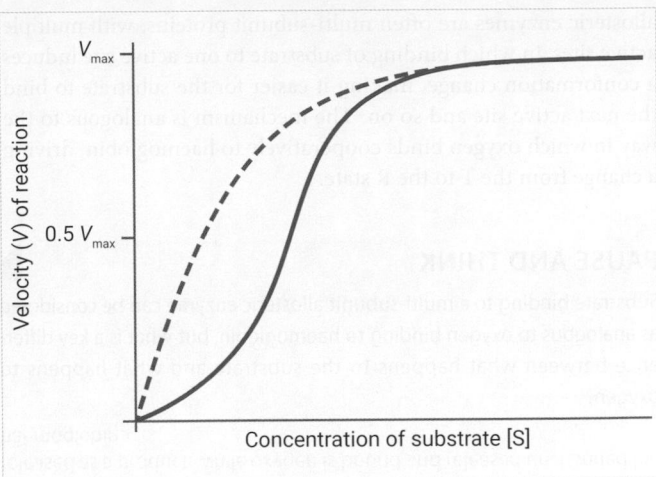

Figure 6.37 Effect of substrate concentration on the rate of an enzyme-catalysed reaction. The dashed line shows a reaction catalysed by a non-allosteric enzyme, while the solid line shows a reaction catalysed by an allosteric enzyme.
Source: Adapted from Snape, A. & Papachristodoulou, D. (2018). *Biochemistry and Molecular Biology* (6th edn). Oxford University Press.

Before we explore how this definition applies to allosteric enzymes, we will look at a graph showing how the rate of an enzyme-catalysed reaction changes as we increase substrate concentration, comparing a non-allosteric and an allosteric enzyme (Figure 6.37). Understanding this graph helps us to understand what distinguishes allosteric enzymes from other enzymes and makes them suitable for their regulatory role.

You will notice that the shape of the curve differs depending on whether the reaction measured is catalysed by a non-allosteric enzyme (showing Michaelis–Menten kinetics) or an allosteric enzyme. If you compare Figure 6.37 with Figure 6.23, which shows the oxygen saturation curves for myoglobin and haemoglobin, you will see a similarity. The non-allosteric enzyme acts like myoglobin: the rate of reaction initially rises steadily as substrate concentration

is increased and then levels off, giving a hyperbolic curve that is similar to the oxygen saturation curves of myoglobin. However, both the allosteric enzyme and haemoglobin give a sigmoid curve, in which there is an initial gradual increase in reaction rate or oxygen binding, followed by an abrupt change to a more rapid increase before the curve levels off.

Figure 6.37 illustrates one reason why allosteric enzymes are good at regulating metabolic reactions: they are sensitive to small changes in substrate concentration, switching from a very slow to a rapid rate of catalysis. It is like having an on/off switch for the reaction.

To understand how this is achieved, we can take the analogy between the allosteric enzyme and the allosteric oxygen-binding protein haemoglobin further. Like haemoglobin, allosteric enzymes have two different protein conformations, designated the 'Tense' or 'T' state and the 'Relaxed' or 'R' state. While the T and R states of haemoglobin have low and high affinities for their ligand, oxygen, the T and R states of allosteric enzymes have low and high affinities for their substrate. As illustrated in Figure 6.38,

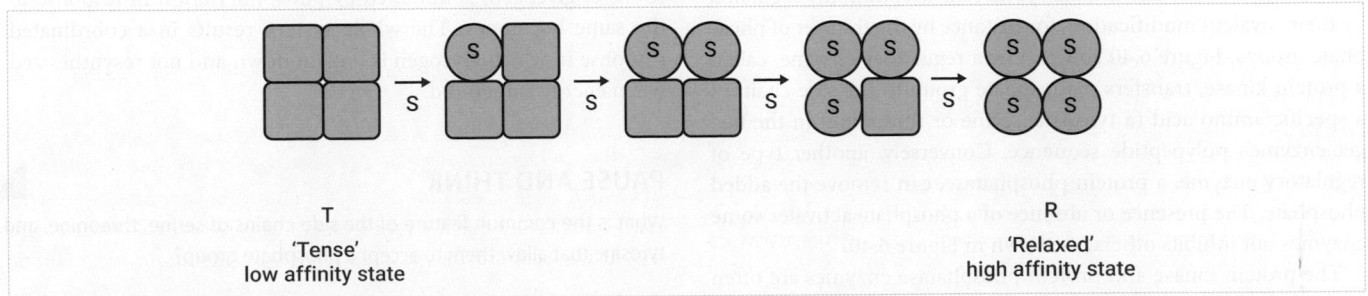

Figure 6.38 A model illustrating how cooperative binding of its substrate to a multi-subunit allosteric enzyme drives the change from the enzyme's 'tense' low affinity state to its 'relaxed' high affinity state. The binding of a single substrate molecule to one active site causes a conformational change in the subunit. This facilitates the conformational change of the second subunit when its active site is bound by a substrate molecule, and so on for the next subunit. The net effect is to make it 'easier' for successive subunits to undergo the conformational change, which is seen as increasing affinity for the substrate by successive subunits as the whole enzyme transitions from the T to the R state.
Source: Adapted from Snape, A. & Papachristodoulou, D. (2018). *Biochemistry and Molecular Biology* (6th edn). Oxford University Press.

allosteric enzymes are often multi-subunit proteins, with multiple active sites, in which binding of substrate to one active site induces a conformation change, making it easier for the substrate to bind the next active site and so on. The mechanism is analogous to the way in which oxygen binds cooperatively to haemoglobin, driving a change from the T to the R state.

PAUSE AND THINK

Substrate binding to a multi-subunit allosteric enzyme can be considered as analogous to oxygen binding to haemoglobin, but what is a key difference between what happens to the substrate and what happens to oxygen?

Answer: The substrate is altered by the enzyme-catalysed reaction and is released as a product, while oxygen is bound and released unchanged by haemoglobin.

An additional important feature of allosteric enzymes is that, besides the active site, they contain specific binding sites for other molecules, termed allosteric regulators. Allosteric regulators act by stabilizing the enzyme in either its T or its R state. Figure 6.39 shows how regulators that stabilize the T state decrease the enzyme's affinity for its substrate and therefore inhibit the enzyme, while those that stabilize the R state increase the enzyme's affinity for its substrate and activate it.

Phosphofructokinase, our example of an allosteric enzyme, is activated by AMP (adenosine monophosphate) and inhibited by ATP (adenosine triphosphate). Research has shown that when ATP binds to phosphofructokinase at its allosteric regulatory site the enzyme undergoes a change in conformation and has decreased affinity for its substrate. As explained in Chapter 2, ATP acts as an energy storage molecule in cells, so it makes sense that when ATP is abundant it acts as a feedback inhibitor to slow the glycolysis pathway and stop unnecessary breakdown of glucose. Conversely, AMP is a metabolite of ATP which acts as a signal that the cell is short of ATP and glycolysis is needed to make more.

Covalent modification regulates enzymes in response to the needs of the organism

The activity of enzymes in eukaryotic cells is commonly regulated by their covalent modification, for instance by the transfer of phosphate groups. Figure 6.40 shows how a regulatory enzyme, called a protein kinase, transfers a phosphate group to the side chain of a specific amino acid (a tyrosine, serine or threonine) in the target enzyme's polypeptide sequence. Conversely, another type of regulatory enzyme, a protein phosphatase, can remove the added phosphate. The presence or absence of a phosphate activates some enzymes but inhibits others, as shown in Figure 6.40.

The protein kinase and protein phosphatase enzymes are often under hormonal control, allowing the enzymes they target to be regulated according to the needs of the organism. For example, key enzymes involved in the metabolism of glycogen (the storage form of glucose, which we discuss in Chapter 2) respond to the hormones glucagon and adrenaline which are released when energy

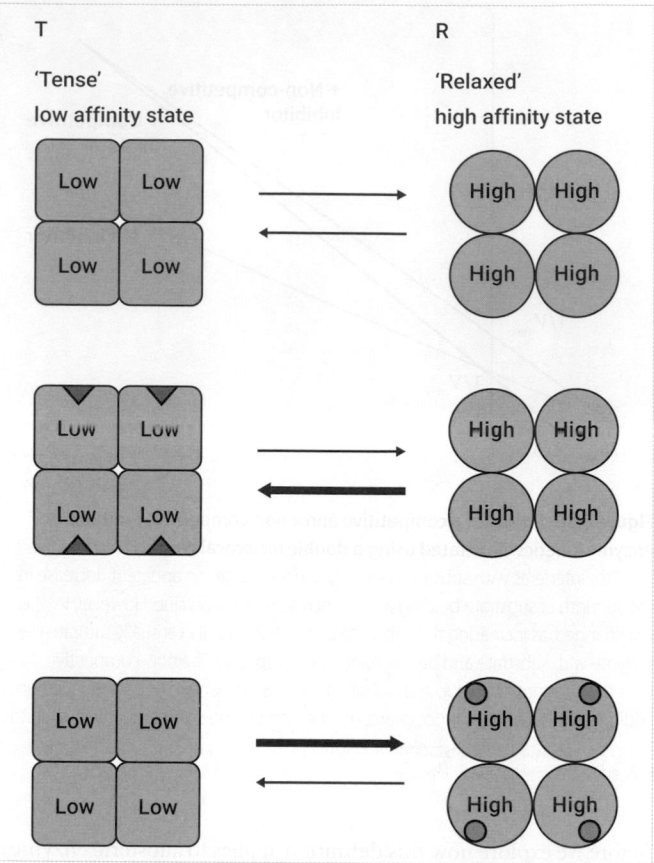

Figure 6.39 A model illustrating the effect of allosteric regulators on enzyme activity. In the absence of a regulator the enzyme exists as a mixture of the T and the R forms, which can interconvert through a change in the enzyme's conformation. An allosteric inhibitor binding to the enzyme (triangles) stabilizes it in the T form, thus reducing its affinity for the substrate, while an allosteric activator binding to the enzyme (circles) stabilizes it in the R form, thus increasing its affinity for the substrate.

Source: Adapted from Snape, A. & Papachristodoulou, D. (2018). *Biochemistry and Molecular Biology* (6th edn). Oxford University Press.

is required. As illustrated in Figure 6.41, glycogen is broken down by an enzyme that is activated by phosphorylation in response to these hormones. However, a second enzyme, required for synthesis of glycogen, is inhibited by phosphorylation in response to the same hormones. The whole system results in a coordinated response in which glycogen is broken down and not resynthesized when energy is required.

PAUSE AND THINK

What is the common feature of the side chains of serine, threonine, and tyrosine that allow them to accept a phosphate group?

Answer: They all have a hydroxyl group (–OH) in the side chain.

 Check your understanding of the concepts covered in this section by answering the questions in the e-book.

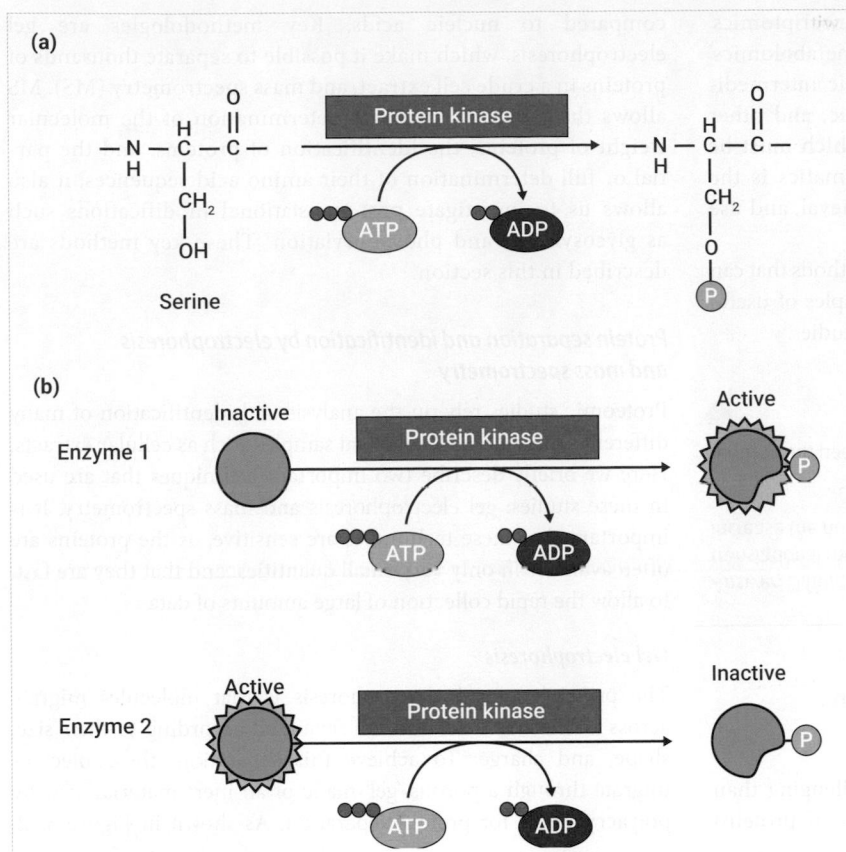

Figure 6.40 Covalent modification of enzymes can regulate their activity. (a) An enzyme has a phosphate group transferred from ATP to a specific serine side chain by a regulatory protein kinase. The phosphate can subsequently be removed by a regulatory protein phosphatase enzyme. (b) Some target enzymes are activated by phosphorylation, while others are inactivated.

Source: From Cooper, *The Cell*, 8e, Oxford University Press.

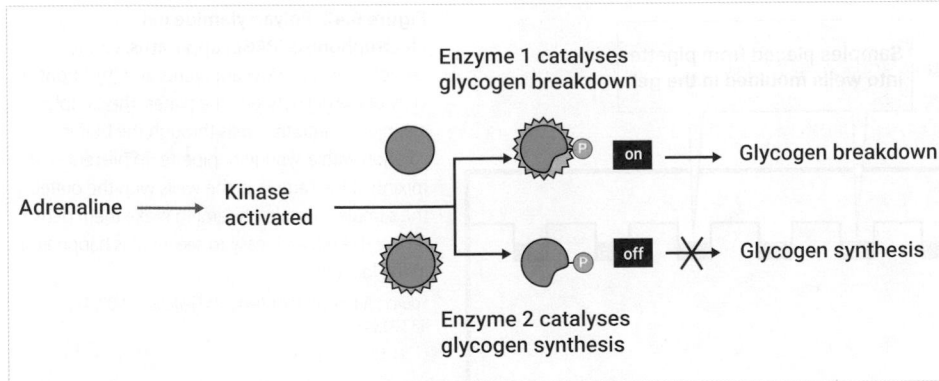

Figure 6.41 In mammals the hormone adrenaline is secreted when energy is needed, for example in a 'fight or flight' situation. Adrenaline has multiple effects, including activation of a protein kinase that stimulates the breakdown of glycogen in muscle cells. The protein kinase phosphorylates two target enzymes, one of which is activated and catalyses glycogen breakdown in response to the need for energy. The other enzyme, which would catalyse glycogen synthesis, is inhibited by phosphorylation.

Source: Based on elements of Figure 4-46 in Alberts et al. *Essential Cell Biology* (5th edn), W. W. Norton and Co.

6.5 The proteome is the full set of proteins expressed in a cell, tissue, or organism

A living cell or organism contains a large number of different proteins, the entirety of which is referred to as the 'proteome'. This is analogous to the term 'genome', which refers to the entire DNA sequence of an organism. However, with a tiny number of exceptions, the genome of all cells in an organism is the same, whereas the proteome varies from cell type to cell type. For example, the proteomes of liver, brain, and muscle cells overlap, as all will contain essential structural proteins and metabolic enzymes, but they also differ to a considerable extent as each cell type makes proteins specific to its specialized function.

The types of proteins found in a cell may change over time, for instance as specialized cell types differentiate during embryonic development. They may also change in response to physiological needs; for example, in a liver cell some enzymes are needed specifically after eating to facilitate the breakdown of food molecules. A comparison of the proteomes of normal and diseased cells may also reveal differences that can be attributed to the disease state. These examples illustrate why it may be useful to study the proteome as it varies in development, in response to physiological needs, and in diseases such as cancer.

The study of large collections of proteins is known as 'proteomics'. Proteomics differs from conventional 'protein chemistry' because it involves the investigation of multiple proteins in a single experiment. By analogy, the large-scale study of genes is called 'genomics'. Analogous terms such as 'transcriptomics' (studying the full set of RNA transcripts) and 'metabolomics' (studying the full set of small molecule metabolic intermediates) have also been coined. Genomic, proteomic, and other '-omic' studies generate vast amounts of data, which must be stored and made available for analysis. Bioinformatics is the branch of science that deals with the storage, retrieval, and use of biological data.

In the sections that follow we will outline some methods that can be used to investigate the proteome and give examples of useful information that can be gained through proteomic studies.

PAUSE AND THINK

The number of proteins an organism can produce may exceed the number of its protein-coding genes. How is this possible?

Answer: Differential splicing of mRNAs (see Chapter 5) and post-translational modifications such as removal of part of the protein can increase the number of different proteins made in a cell.

Modern technology allows the rapid acquisition and analysis of proteomic data

Proteomic studies are technically much more challenging than genomic studies due to the chemical complexity of proteins compared to nucleic acids. Key methodologies are gel electrophoresis, which make it possible to separate thousands of proteins in a crude cell extract, and mass spectrometry (MS). MS allows the rapid and accurate determination of the molecular weight of proteins, the identification of proteins, and the partial or full determination of their amino acid sequences; it also allows us to investigate post-translational modifications such as glycosylation and phosphorylation. These key methods are described in this section.

Protein separation and identification by electrophoresis and mass spectrometry

Proteomic studies rely on the analysis and identification of many different proteins from biological samples such as cellular extracts. Here we briefly describe two important techniques that are used in these studies: gel electrophoresis and mass spectrometry. It is important that these techniques are sensitive, as the proteins are often available in only very small quantities, and that they are fast, to allow the rapid collection of large amounts of data.

Gel electrophoresis

The principle of gel electrophoresis is that molecules migrate across an electric field, and are separated according to their size, shape, and charge. To achieve this separation, the molecules migrate through a porous 'gel' made of an inert material, usually polyacrylamide for protein separation. As shown in Figure 6.42,

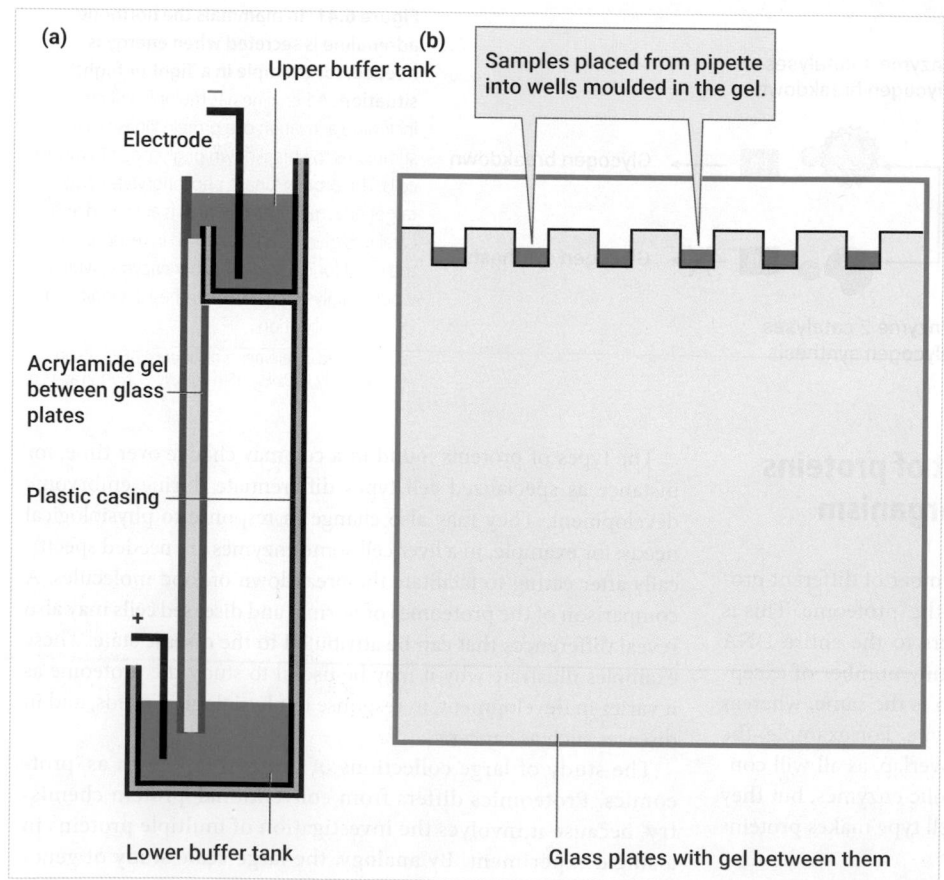

Figure 6.42 Polyacrylamide gel electrophoresis (PAGE) apparatus. (a) Cross-sectional view of PAGE apparatus and (b) a front view of the gel between the plates. The samples are injected into the wells through the buffer solution with a syringe or pipette. To prevent mixing of the sample in the wells with the buffer, the samples contain glycerol to make them dense. A blue dye makes it easy to see what is happening in the loading.

Source: Adapted from *Trends in Genetics*, 2002; 18, 433–434.

the gel is held in place between vertical plates, with an aqueous buffer solution in the reservoir tanks at the top and bottom. The buffer also fills the pores in the gel, so that an electric field is established across it when a voltage is applied via electrodes attached to the top and bottom tanks.

It is generally most useful to separate proteins according to their size while eliminating the effects of their different charges and three-dimensional shapes. In order to do this the proteins are denatured (i.e. their secondary, tertiary, and quaternary structures are broken down) by treating them with a reducing agent that breaks disulphide bonds, and by heating them in a buffer that contains a strong detergent called sodium dodecyl sulphate (SDS). SDS has a hydrophobic tail and a negatively charged sulphate group. It inserts its hydrophobic tail into the proteins, breaking down their structures, and coats them with negative charge. Large amounts of the SDS attach, roughly one molecule per two amino acid residues, swamping whatever charge the native protein had, so that all proteins have a strong negative charge proportionate to their size.

This separation technique is known as SDS polyacrylamide gel electrophoresis (SDS-PAGE). The method is illustrated in Figure 6.42.

The separated proteins can be visualized as 'bands' in the gel by staining, for example with a dye called Coomassie blue. An example is shown in Figure 6.43, which illustrates the use of SDS-PAGE to check stages in the purification of a protein.

Isoelectric focusing

A useful variant on electrophoresis is isoelectric focusing. The isoelectric point (pI) of a molecule with different ionizing groups ($-COOH$, $-NH_2$, and certain amino acid side chains in the case of proteins) is the pH at which the positive and negative charges exactly balance so the net charge on the molecule is zero. The principle is illustrated in Figure 6.44a. The gel first has a stable pH gradient established across it and the native non-denatured proteins migrate across it towards the point where the pH matches the pI. The proteins then remain at that point.

Isoelectric focusing is combined with SDS-PAGE in two-dimensional (2-D) gel electrophoresis, as illustrated in Figure 6.44b. A

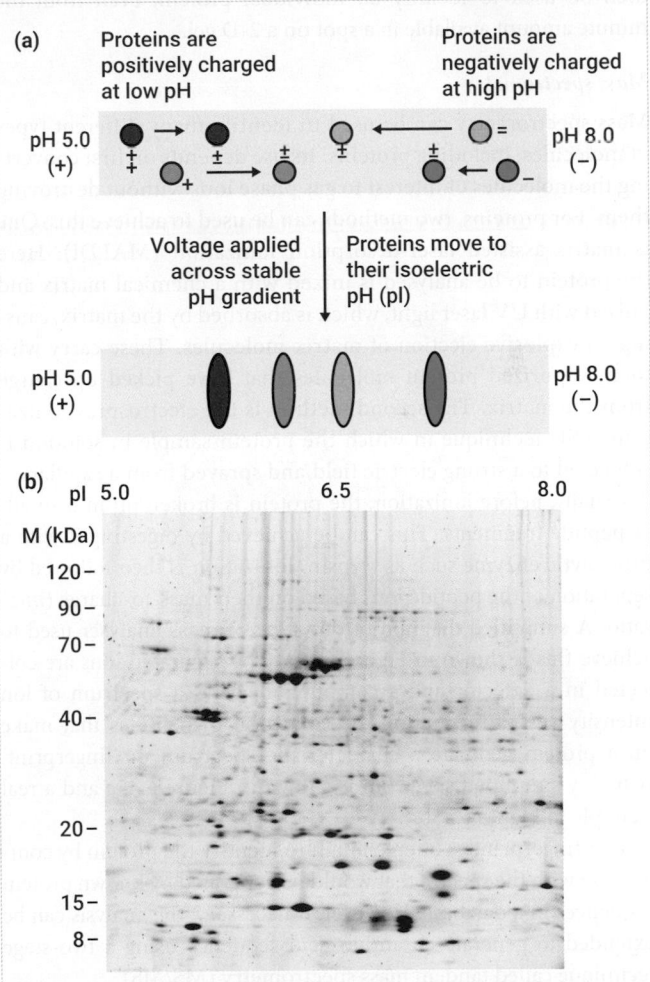

Figure 6.44 Isoelectric focusing. (a) Principle of isoelectric focusing. A stable pH gradient is established in a narrow tube of polyacrylamide gel. Proteins subjected to electrophoresis in the gel move according to their charge. The charge on amino acid side chains alters with varying pH and when each protein reaches the pH at which its overall charge is zero (its pI) it stops moving. The figure shows four proteins, but often a complex mix is analysed of which several proteins will have the same pI. For 2-D electrophoresis the narrow tube of gel is then subjected to SDS-PAGE with the electric field applied at right angles to that used for the first separation, giving further separation by size. (b) Representative 2-D electrophoresis gel of whole-cell proteins from the gut pathogen *Helicobacter pylori*. Proteins were separated using immobilized pH gradient isoelectric focusing in the first dimension and a second-dimension slab gel containing SDS buffer. Gels were stained with fluorescent Sypro Ruby. pI is the isoelectric point (see text). M is the molecular weight.

Source: Adapted from Koehorst, J. J., Saccenti, E., Schaap, P. J. et al. *F1000Research*, 2017, 5:1987 (doi: 10.12688/f1000research.9416.3).

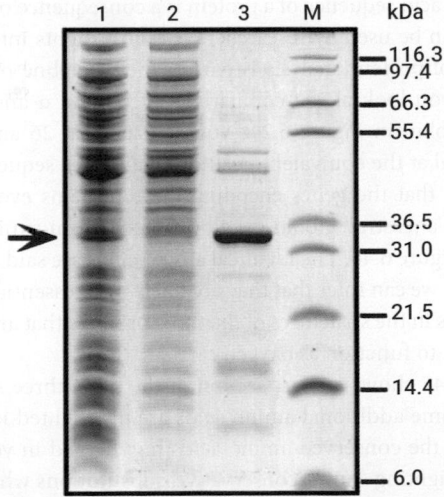

Figure 6.43 SDS-PAGE used to check the purification of a protein from *Escherichia coli*. Lane 1 shows all the proteins extracted from the *E. coli* bacteria. The arrow indicates the protein of interest. The purification was carried out using a chromatographic technique. Lane 3 shows the protein of interest, now almost completely separated from other proteins in the sample, while lane 2 shows the unwanted proteins that were not captured by the purification. Lane M (for markers) shows proteins of known molecular weights. By comparison with the position of these on the gel, we can see that the protein of interest has a molecular weight of 35.3 kDa. (kDa is the molecular weight unit, kiloDaltons.)

Photograph courtesy Dr Anne Chapman-Smith, Department of Molecular Biosciences, University of Adelaide, Australia.

sample is first separated by isoelectric focusing on a gel strip with an established pH gradient and the 'strip' is transferred to an SDS gel and electrophoresed at right angles to the first direction. The first separation is by pI and the second by size. A crude cell extract analysed by 2-D electrophoresis can give rise to many hundreds of separate protein spots as illustrated in Figure 6.44b. The result looks complex, but the method is powerful because it shows up differences in the whole collection of proteins found in a cell or tissue type, for instance by comparing the patterns produced by a normal versus a cancer cell population. Mass spectrometry can then be used to identify an individual protein, even from the minute amount available in a spot on a 2-D gel.

Mass spectrometry

Mass spectrometry can be used to identify many different types of molecules, including proteins. Its use depends on first converting the molecules of interest to gas phase ions without destroying them. For proteins, two methods can be used to achieve this. One is matrix-assisted laser-desorption ionization (MALDI). Here the protein to be analysed is mixed with a chemical matrix and pulsed with UV laser light, which is absorbed by the matrix, causing an explosive ejection of matrix molecules. These carry with them vaporized protein molecules that have picked up charge from the matrix. The second method is the electrospray ionization (ESI) technique in which the protein sample in solution is subjected to a strong electric field and sprayed from a capillary.

Usually, before ionization, the protein is broken up into smaller peptide fragments. This can be achieved by digesting it with a proteolytic enzyme such as trypsin. Ionization is then followed by separation of the peptide ions based on their mass-to-charge (m/z) ratio. A simplified diagram of one type of mass analyser used to achieve this is shown in Figure 6.45a. The separated ions are collected in a detector and a computer displays a spectrum of ion intensity versus m/z ratio. The collection of peptides that make up a protein generate a characteristic spectrum or 'fingerprint'. A highly simplified spectrum is shown in Figure 6.45a and a real example in Figure 6.45b.

The 'fingerprint' is often enough to identify the protein by comparison with the spectra that would be generated by known protein sequences in a database. However, if necessary, the analysis can be extended to generate an amino acid sequence using a two-stage technique called tandem mass spectrometry (MS/MS).

Proteomic and genomic studies can shed light on evolution

The acquisition of data from millions of proteins has shown that many of them can be assigned to 'families' based on similarities in their amino acid sequences. In Section 6.4 we explore how a protein kinase can regulate enzyme activity by transferring a phosphate group from ATP to the target enzyme. We discuss a single example there, but in fact protein kinases comprise a very large protein family. The human genome encodes more than 400 protein kinases, which can be recognized by their characteristic amino acid sequences and hence by similarities in their protein structure. All protein kinases contain a recognizable catalytic domain, while variation in other parts of their structure allows them to transfer

phosphates to different target proteins and to take part in a wide range of cellular processes.

Besides the protein kinase family, the human genome contains many other recognizable families of proteins that are related by structure and function. Examples include DNA-binding proteins that regulate gene expression, proteins forming channels to transport different ions across membranes, and proteins involved in cell signalling pathways.

Proteins that share similar amino acid sequences have often evolved from a common ancestral protein. Such proteins are said to share sequence **homology**, or to be **homologous** to each other. How does the evolution of homologous proteins occur? It is believed that a common mechanism, illustrated in Figure 6.46, involves the duplication of gene sequences, followed by mutation. Gene duplication should not be confused with DNA replication, although it can result from errors in the replication process. It refers to the addition of an extra copy of a DNA sequence into the genome of an organism. Often the duplicate sequence is arranged 'in tandem' with the original sequence on the chromosome, as shown in Figure 6.46.

PAUSE AND THINK

What is the origin of the phrase 'in tandem' used to refer to the relative positions of Genes A1 and A2 in Figure 6.41?

Answer: It has the same origin as the term used for a tandem bicycle, where two riders sit one behind the other. The two genes are next to each other on the same chromosome.

The amino acid sequence of a protein is a consequence of its evolution and can be used by researchers to gain insights into the past. This type of protein analysis is part of the discipline of bioinformatics. If you look at the comparison of human α and β globin and myoglobin in Figure 6.20c you can see that 26 amino acids are identical at the equivalent position in all three sequences. This is evidence that the genes encoding these proteins evolved from a common 'ancestral' globin gene by a mechanism similar to that shown in Figure 6.46. The identical amino acids are said to be 'conserved', and we can infer that they are likely to be essential to maintain features in the structures of the three proteins that are required for them all to function as oxygen carriers.

Figure 6.47 shows a short section of the same three sequences, in which some additional amino acids are highlighted in green in addition to the conserved amino acids highlighted in yellow. The green highlighting shows 'conservative' substitutions where amino acids have been replaced with others in which the side chains have similar physical and chemical properties. For example, leucine (L), isoleucine (I), and valine (V) all have hydrophobic side chains of similar size, while aspartic acid (D) and glutamic acid (E) both have negatively charged side chains. These conservative substitutions are less likely to alter the structure and function of the protein than are 'non-conservative' or 'radical' substitutions such as the replacement of an uncharged, hydrophobic leucine with a charged, hydrophilic lysine.

Systems of comparison have been devised, which can be used to evaluate the likelihood that aligned protein sequences are true

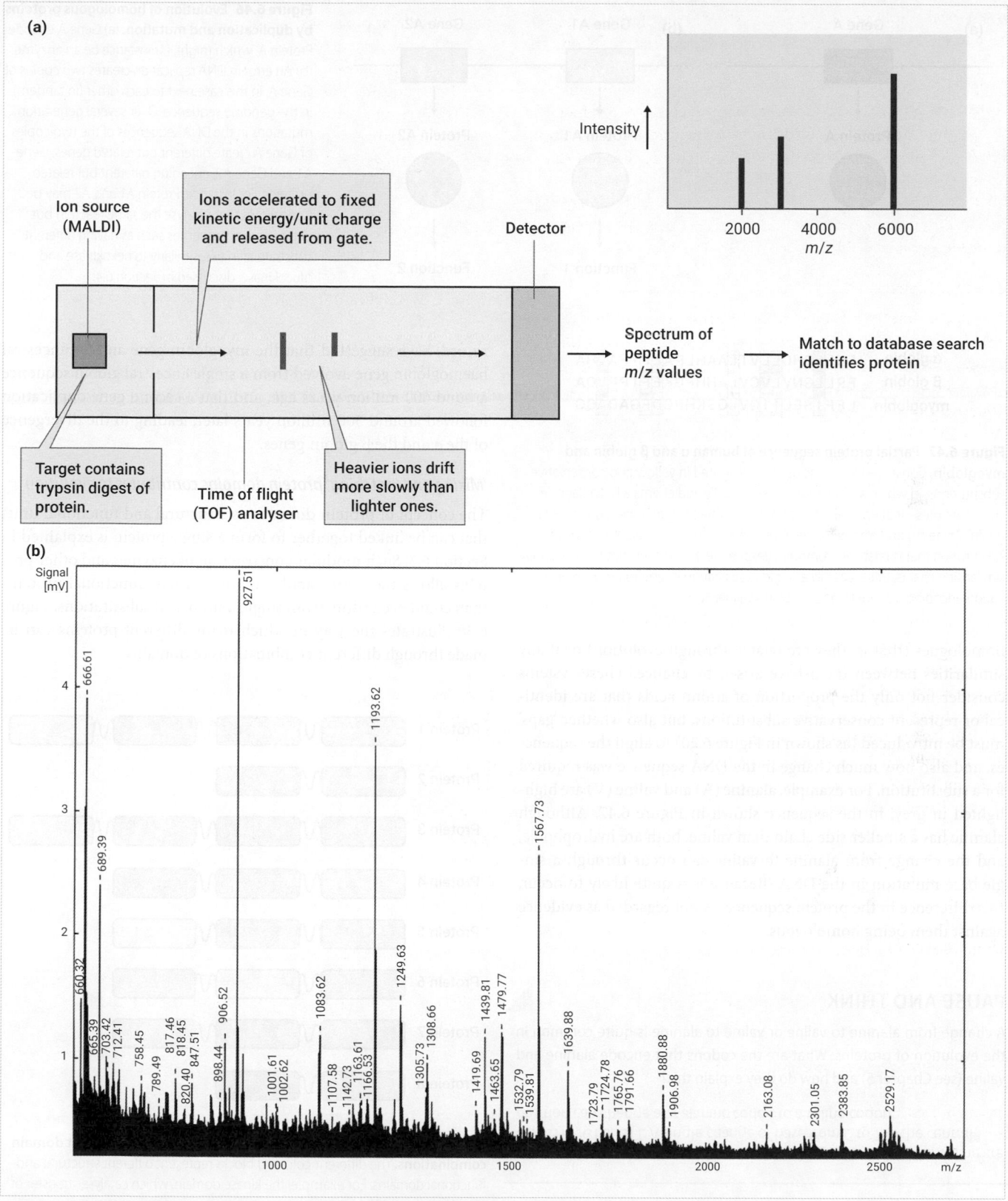

Figure 6.45 (a) Simplified diagram of peptide analysis by matrix-assisted laser-desorption ionization time of flight (MALDI-TOF) 'fingerprinting'. The protein is digested with trypsin, and the resulting peptides are ionized. A high voltage is then used to propel them through a grid into a flight tube. The flight tube has no electrical or magnetic field: the peptides simply drift passively through the tube, with the heavier ones taking longer to reach the detector. Matching the mass analysis spectrum to databases can identify a protein. (b) Mass spectrometric analysis of the protein bovine serum albumin (BSA). The protein was digested with trypsin and a MALDI spectrum generated. The mass charge (m/z) ratios are indicated on the x-axis and the relative intensity is plotted on the y-axis.

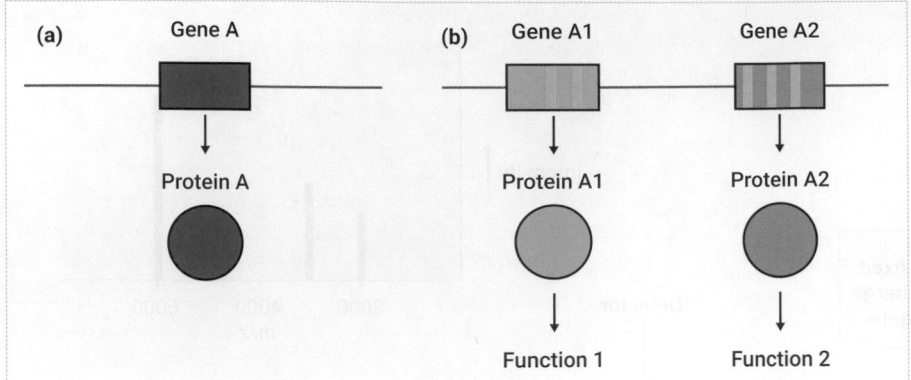

Figure 6.46 Evolution of homologous proteins by duplication and mutation. (a) Gene A encodes Protein A, which might, for instance be an enzyme. (b) An error in DNA replication creates two copies of Gene A, in this case next to each other (in tandem) in the genome sequence. Over several generations, mutations in the DNA sequences of the two copies of Gene A create different but related genes, Gene A1 and Gene A2, encoding different but related proteins. For instance, Protein A1 and A2 may be isoenzymes that catalyse the same reaction but differ in other properties such as having different substrate affinities, similarly to hexokinase and glucokinase, discussed in Section 6.4.

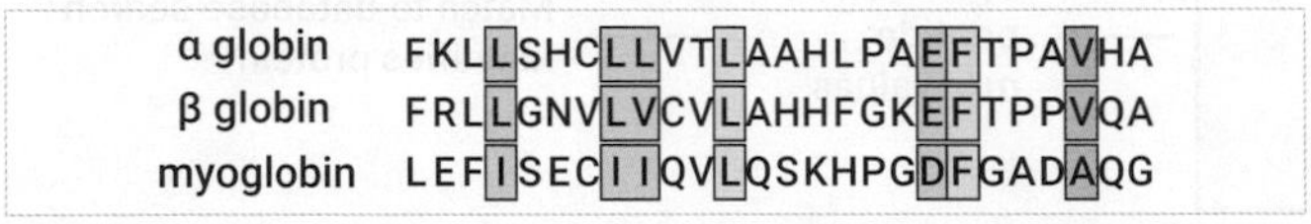

α globin	FKLLSHCLLVTLAAHLPAEFTPAVHA
β globin	FRLLGNVLVCVLAHHFGKEFTPPVQA
myoglobin	LEFISECIIQVLQSKHPGDFGADAQG

Figure 6.47 Partial protein sequence of human α and β globin and myoglobin. Conserved amino acids are highlighted in yellow and conservative substitutions, in which amino acids with physically and chemically similar side chains are present at equivalent positions in all three proteins are highlighted in green. Valine (V) and alanine (A) are highlighted in grey to illustrate a type of substitution that is relatively common, despite the amino acids not having closely similar side chains, because only a single base change is required to change a codon encoding alanine to one encoding valine.

homologues (that is, they are related through evolution) or if any similarities between them have arisen by chance. These systems consider not only the proportion of amino acids that are identical or represent conservative substitutions, but also whether 'gaps' must be introduced (as shown in Figure 6.20) to align the sequences, and also how much change in the DNA sequence was required for a substitution. For example, alanine (A) and valine (V) are highlighted in grey, in the sequence shown in Figure 6.47. Although alanine has a smaller side chain than valine, both are hydrophobic, and the change from alanine to valine can occur through a single base mutation in the DNA. Because it is quite likely to occur, this difference in the protein sequences is not regarded as evidence against them being homologous.

PAUSE AND THINK

A change from alanine to valine or valine to alanine is quite common in the evolution of proteins. What are the codons that encode alanine and valine (see Chapter 5) and how do they explain this?

Answer: Codons for alanine are GCA, GCC, GCG, GCT. Codons for valine are GTA, GTC, GTG, GTT. A single change of base from C to T in the central position can change any alanine codon to a valine codon.

Sequence comparison of the three members of the globin family illustrated in Figures 6.20c and 6.42, coupled with comparisons with related proteins from other organisms and the fossil

record, have suggested that the myoglobin gene and an ancestral haemoglobin gene evolved from a single ancestral globin sequence around 600 million years ago, and that a second gene duplication followed around 300 million years later, leading to the divergence of the α and the β globin genes.

'Mixing and matching' protein domains contributes to evolution

The concept of protein domains as structural and functional units that can be linked together to form a single protein is explained in Section 6.3. Such modular construction of enzymes and other proteins allows the more rapid evolution of new functional proteins than could occur only from single amino acid substitutions. Figure 6.48 illustrates the way in which many different proteins can be made through different combinations of domains.

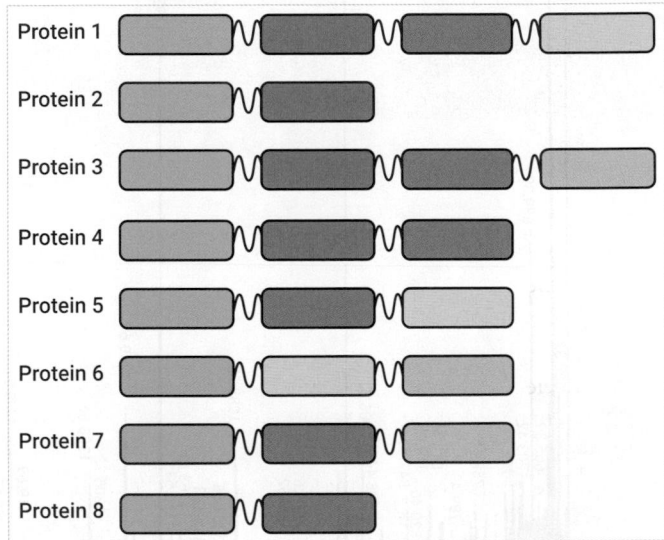

Figure 6.48 Modular construction of proteins through different domain combinations. The different coloured blocks represent different structural and functional domains. For example, the kinase domain, which catalyses transfer of phosphate from ATP to a substrate molecule, is one of the most common domains in the human genome. If all eight proteins here are enzymes that share a (blue) kinase domain, they could have different substrates or different regulation, depending on the other domains that make up the complete enzyme.

Domain shuffling (or domain swapping) is the name given to the evolutionary process in which new genes are assembled from sections of DNA that code for pre-existing protein domains. In eukaryotes a protein domain is often (although not always) encoded by a specific exon.

▶ **Exons are described in Chapter 5.**

During evolution, exons can be duplicated and recombined in different combinations and orders in different parts of the genome, thus building new genes through a 'mix and match' process.

PAUSE AND THINK

DNA sequence analysis suggests that a common occurrence in the evolution of new proteins in eukaryotes is an existing exon being copied and inserted in a new position in the genome. What aspects of the intron/exon structure of eukaryotic genes favour this as a mode of protein evolution?

Answer: A single exon frequently encodes a protein domain, which can be inserted as a new functional unit in an existing protein structure. Additionally, a new exon can be inserted within an intron without disrupting existing protein sequence, and provided the new exon carries the sequences required for splicing, it will be incorporated into the mRNA when the gene is transcribed.

Proteomics and bioinformatics have multiple applications in science and medicine

It is impossible here to give more than a flavour of the increasing contribution being made to science and medicine by proteomic and other 'omics' studies, coupled with the use of bioinformatics to analyse the vast amount of data they generate. Below we give just a few examples.

Besides information on evolution, the comparison of novel protein sequences and structures with those already in the databases can suggest that they belong to particular classes, such as membrane proteins or kinase enzymes, and thus can help to attribute functions to these new discoveries. Further knowledge of the relation of protein structures to their functions can be used to engineer proteins for specific medical and industrial applications.

'Omic' data can allow detailed molecular characterization of disease states, which can be used to identify both 'biomarkers' of disease for diagnostic purposes and potential targets for drug treatment. Knowledge of the proteins involved in disease processes can be used for the rational design of drugs that will interfere with their function.

Genomic data can also be used to compare individuals who have different observed susceptibilities to common diseases, allowing genetic risk factors to be identified. These can then be used to tailor medical advice and treatment to the individual, giving the possibility of personalized medicine.

Much of the data that are generated and analysed in proteomic and other 'omic' studies are stored on publicly accessible databases. In the 'Further reading' for this chapter we introduce some of these and give information on training and educational resources they provide that are accessible for undergraduate students.

 Check your understanding of the concepts covered in this section by answering the questions in the e-book.

SUMMARY OF KEY CONCEPTS

- Amino acids are the building blocks of proteins.
- Proteins are made up of one or more polypeptide chains (amino acids linked by peptide bonds) constructed from 20 species of amino acids.
- The 20 amino acids differ in their side chains or R groups, which have different sizes and degrees of hydrophobicity, hydrophilicity, and electrical charge. The interactions between side chains are important determinants of proteins' three-dimensional structure.
- There are four levels of protein structure. The amino acid sequence of its polypeptide chain determines the folded structure of a protein.
- The primary structure is the linear amino acid sequence.
- Secondary structure involves folding of the polypeptide backbone. The main secondary structure motifs are the α helix and the β pleated sheet. Proteins are built up of various combinations of these structures linked by connecting loops.
- Tertiary structure involves the further folding of the secondary structure motifs into the three-dimensional form of the protein,

which is stabilized by non-covalent interactions between amino acid side chains, and in some cases by covalent disulphide bonds between cysteines.
- The association of protein molecules by non-covalent interactions to form multi-subunit proteins produces the quaternary structure.
- Protein structures reflect their function.
- Proteins mainly fall into three structural classes: globular proteins that are found in aqueous environments, fibrous proteins such as actin, tubulin, and collagen, and membrane proteins.
- In globular proteins, hydrophobic residues are mainly inside the molecule and hydrophilic ones outside in contact with water. Membrane proteins have hydrophobic amino acids on the outside in contact with the hydrophobic membrane interior.
- Myoglobin and haemoglobin are related globular proteins that bind oxygen in muscle and red blood cells respectively. Haemoglobin is an allosteric protein, in which binding of oxygen to one subunit alters the protein structure and facilitates additional oxygen binding.

- Collagen has a strong triple superhelix structure. The individual helices have three amino acids per turn (tighter than an α helix) and are left-handed but form a right-handed superhelix.

- Glycophorins and porins represent two common types of membrane protein. Glycophorins use an α helix with hydrophobic amino acid side chains pointing outwards to cross the cell membrane, while porins use a β sheet folded around to form a channel, with hydrophobic side chains pointing outwards into the membrane, and hydrophilic ones pointing inwards.

- Larger proteins contain domains, which are sections of the polypeptide chain folded into compact structures and linked together by less structured sections. Each domain has a distinct role in the overall function of the protein.

- An enzyme is a protein that lowers the activation energy of the reaction it catalyses, making the reaction more likely to occur, and therefore speeding it up.

- On binding of substrates to enzymes the protein often changes shape, a process known as induced fit.

- Catalysis occurs at the active site of the enzyme, which is often complementary in shape to the transition state of the substrate.

- Different enzymes catalysing the same reaction are known as isoenzymes or isozymes. They have characteristics tailored to their particular roles usually in different tissues.

- Many enzymes require non-protein cofactors such as metal ions, tightly attached organic prosthetic groups, or organic coenzymes such as NAD^+ that take part in the reaction.

- Enzymes are affected by temperature, pH, and inhibitors. Inhibitors may compete with the substrate for the active site or be non-competitive and inhibit by binding elsewhere.

- Many enzymes show Michaelis–Menten kinetics. V_{max} and K_M are useful values derived from kinetic studies of these enzymes.

- Allosteric enzymes do not show Michaelis–Menten kinetics. They are particularly sensitive to changes in substrate concentration and have binding sites for regulatory molecules. They often regulate metabolic pathways.

- The proteome is the full set of proteins that are expressed in a cell, tissue, or organism.

- Modern protein technology methods such as electrophoresis and mass spectroscopy have given us the ability to rapidly identify minute amounts of proteins. They have led to the study of proteomics in which large numbers of proteins are studied at once.

- Protein and DNA databases, which record vast amounts of information on all aspects of proteins, now complement these technologies.

- Proteomic and genomic studies can shed light on evolution. Sequence comparisons discover homologous proteins that share similar amino acid sequences and have evolved from a common ancestral protein.

- Proteins have evolved via mutations and by domain shuffling, in which the coding regions of genes have been reassembled to code for new combinations of domains to produce new proteins.

- Bioinformatics, which involves the computer-assisted use and analysis of database information, is complementary to proteomics and genomics and has multiple applications in science and medicine.

 Use the flashcards in the e-book to test your recall of key terms introduced in this chapter.

QUESTIONS

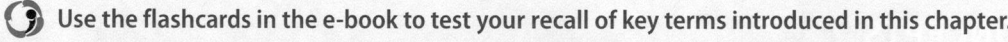 Looking for answers? Once you've answered these questions, follow the link in the e-book to the answer guidance and check your work.

Concepts and definitions

1. Write down the structure and name of an amino acid with each of the following side chains:

   (a) H

   (b) Aliphatic hydrophobic

   (c) Aromatic hydrophobic

   (d) Acidic (lose a proton at pH 7)

   (e) Basic (gain a proton at pH 7)

2. What is meant by denaturation of a protein?

3. Compare the α helix and the collagen triple helix.

4. Explain how an enzyme catalyses a reaction.

5. What is meant by the term proteome?

Apply the concepts

6. The peptide bond is said to be planar. Explain briefly what is meant by this, give the structural basis, and state the consequences of the peptide bond being planar.

7. The α helix and β sheet structures are prevalent in proteins. What is the common feature that makes them suitable for this role?

8. In a globular cytosolic protein would you statistically expect to find most of the residues of each of these amino acids on the inside or

the outside of the folded protein? Where would you expect to find them in a membrane-embedded protein? (a) phenylalanine, (b) aspartic acid, (c) arginine, (d) isoleucine.

9. Using the Michaelis–Menten equation, calculate the values of V_0 (as a fraction of V_{max}):

   (a) when $[S] = 2 \times K_M$

   (b) when $[S] = 5 \times K_M$

   (c) when $[S] = 10 \times K_M$

10. If a typical allosterically controlled enzyme is exposed to saturating levels of substrate, what would be the effects of allosteric activators on the reaction velocity?

11. Suppose you have a minute amount of an unidentified protein as a spot- on a gel. How could it be sufficiently characterized rapidly to identify a corresponding protein in a protein database?

Beyond the concepts

12. In the context of protein tertiary structures it has been commented that when nature is onto a good thing it sticks with it. Discuss this briefly.

13. Anfinsen's experiment might suggest that protein folding is a simple process, yet we still do not fully understand how it takes place in the cell. Discuss this statement.

14. Hexokinase is found in most tissues in the body, including skeletal muscle. It catalyses the conversion of glucose to glucose 6-phosphate, a first step in glucose metabolism. The K_M of hexokinase with respect to glucose is 0.15 mM. Glucokinase is a liver-specific isoenzyme of hexokinase that catalyses the same reaction, but the K_M of glucokinase with respect to glucose is 20 mM.

    (a) Which enzyme, hexokinase or glucokinase, binds glucose more strongly?

    (b) Blood glucose concentration is around 5 mM but rises to 10 mM or higher after a carbohydrate-rich meal. What are the implications for metabolism of glucose by skeletal muscle and the liver?

15. A competitive inhibitor for a specific enzyme works by combining with its active site and blocking access to it by its substrate. Transition state analogues have been found to be very effective in some cases. Why would you expect such a molecule to be more effective than a competitive analogue of the substrate of the same enzyme?

16. What is meant by the term proteomics? It has come into research prominence only relatively recently. What has been a major factor in this?

17. Protein databases have assumed great importance. Briefly explain their relevance and use.

FURTHER READING

Pace, C. N., Grimsley, G. R., & Scholtz, J. M. (2009) Protein stability. In: eLS. John Wiley & Sons Ltd, Chichester. http://www.els.net
A discussion of the thermodynamics of protein folding and stability in an on line resource: eLS (Encyclopedia of Life Science).

Minor, D. L. Jr (2007) The neurobiologist's guide to structural biology: a primer on why macromolecular structure matters and how to evaluate structural data. *Neuron* **54**: 511–33.
A review of the methodology of protein structure determination, and the value of structural knowledge, for a scientifically literate but non-specialist audience.

Robinson, P. K. (2015) Enzymes: principles and biotechnological applications. *Essays Biochem* **59**: 1–41. doi: https://doi.org/10.1042/bse0590001

Cravatt, B. F., Simon, G. M., & Yates, J. R. III (2007) The biological impact of mass-spectrometry-based proteomics. *Nature* **450**: 991–1000.
Gives an overview of mass spectrometry proteomic techniques, and then gives examples of how they have been used to increase understanding of biological pathways and processes, often with the potential for medical applications

You can also explore publicly available websites and databases that support proteomic and bioinformatic research in the e-book.

CHAPTER 7

MODULE ONE

Metabolism

Energy Capture and Release from Food

LEARNING OBJECTIVES

By the end of this chapter, you should be able to:

- State the three stages of complete oxidation of glucose and outline the events in each one, and outline the pathway of glycolysis in terms of ATP production and reduced coenzymes.

- Describe the conversion of pyruvate to acetyl-CoA and explain the importance of this reaction.

- Outline the tricarboxylic acid (TCA) cycle, concentrating on yield of ATP and reduced coenzymes.

- Describe the function of the electron transport chain and how the electron transport chain results in a proton gradient across the mitochondrial or cell membrane.

- Explain the concept of chemiosmosis and how the proton gradient results in synthesis of ATP by ATP synthase.

- Describe briefly the principles and processes of anaerobic metabolism and fermentation.

- Outline the 'light' and 'dark' reactions in photosynthesis and explain the importance of photosynthesis in sustaining life on Earth.

Chapter contents

Introduction   324

7.1   The main stages of glucose oxidation   325

7.2   Energy release from oxidation of fats and amino acids   328

7.3   Glucose oxidation: a deeper look   329

7.4   Anaerobic respiration and fermentation: production of ATP without the use of oxygen   337

7.5   Photosynthesis   338

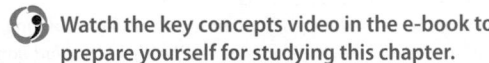 Watch the key concepts video in the e-book to prepare yourself for studying this chapter.

Introduction

The release of energy from glucose, fats, and amino acids involves long and somewhat complicated metabolic pathways. In Chapter 2 we considered the central role of energy in biology: how energy is transformed, and why that energy transformation is so important to the biochemical reactions that sustain life. In this chapter we will look in more detail at the pathways that release energy from metabolic fuels in the form of ATP. We will also consider the synthesis of metabolic fuels by plants—that is, the capture of energy by the process of photosynthesis, which we mentioned in Chapter 2.

7.1 The main stages of glucose oxidation

We begin our exploration of the release of energy from foods by looking at the oxidation of glucose.

Respiration, the overall process by which aerobic organisms derive energy from foods, amounts to the oxidation of glucose. We can represent this overall process as follows:

$$C_6H_{12}O_6 + 6O_2 \rightarrow 6CO_2 + 6H_2O$$

The $\Delta G^{0'}$ for this reaction is 2820 kJ mol^{-1}. it liberates large amounts of energy.

In the cell, this oxidation process is accompanied by the synthesis of more than 30 molecules of ATP from ADP and P_i.

The entire process of glucose oxidation to CO_2 and H_2O can be divided into three stages:

- **Stage 1: glycolysis:** Glycolysis results in the splitting ('lysis') of glucose into two 3-carbon fragments (C_3) ultimately yielding **pyruvate,** accompanied by reduction of nicotinamide adenine dinucleotide (NAD$^+$). This occurs in the cytosol of cells. No oxygen is involved.

- **Stage 2: the tricarboxylic acid (TCA) cycle,** also known as the **Krebs cycle** (after its discovery by Hans Krebs, in the 1930s). For consistency, we will refer to it as the TCA cycle throughout the chapter. In mitochondria, the carbon atoms of pyruvate are converted into an acetyl group and CO_2; in the cycle, electrons from the acetyl groups are transferred to electron carriers such as NAD$^+$ and FAD. No molecular oxygen is involved at this stage either. Carbon atoms are released as CO_2, the oxygen derived largely from water. The cycle is located inside mitochondria in eukaryotes and in the cytosol of prokaryotes.

- **Stage 3: the electron transport system (or chain):** Electrons are transported from the electron carriers to oxygen, where, with protons from the solution, water is formed. It is in stage 3 that most of the ATP is generated. This occurs in the **inner mitochondrial membrane** in eukaryotes and in the cell membrane in prokaryotes.

What follows is a summary of the events in the three stages and then a more detailed account of the processes. It is important to get the overall picture and not to lose perspective by diving into the molecular detail straight away.

Stage 1 in the release of energy from glucose: glycolysis

Glycolysis does not involve oxygen, and only two ATP molecules per molecule of glucose lysed are produced from ADP. The end products are pyruvate and NADH, as shown in Figure 7.1.

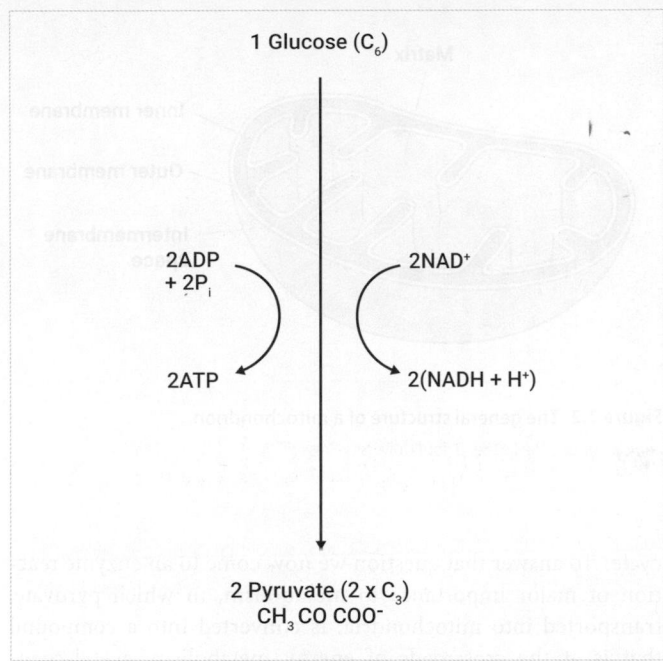

Figure 7.1 The net result of glucose oxidation. NAD$^+$ is reduced in the process of glycolysis to yield NADH. NADH is subsequently reoxidized to NAD$^+$ by mitochondria in eukaryotes and on the cell membrane in prokaryotes.

Source: Snape, A. & Papachristodoulou, D. (2018). *Biochemistry and Molecular Biology* (6th edn). Oxford University Press.

Stage 2 in the release of energy from glucose: the TCA cycle

The **mitochondria** are small organelles located in the cytosol of the cell in eukaryotes; look at Figure 7.2 for a general depiction of their structure. The inner membrane of a mitochondrion is the site of ATP generation. Its area is increased by being invaginated into compartments known as **cristae**. The interior of the mitochondrion is called the **matrix** and is filled with a concentrated solution of enzymes. It is here that stage 2 of glucose metabolism—the TCA cycle—mainly occurs, only one reaction being located in the inner mitochondrial membrane.

▶ We learn more about mitochondria in Chapter 9.

What links glycolysis to the TCA cycle? The oxidative decarboxylation of pyruvate

We saw above how glycolysis results in the oxidation of glucose to yield pyruvate. But how does this pyruvate enter the TCA

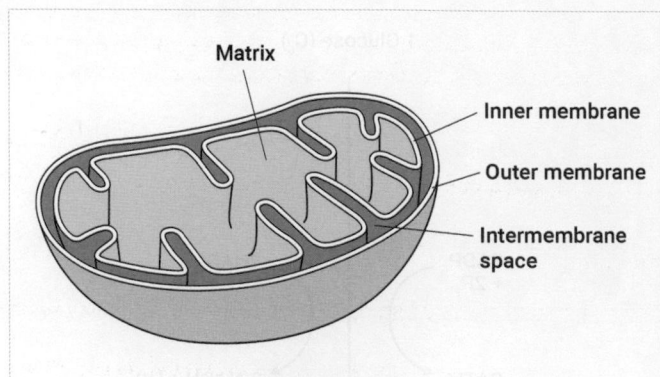

Figure 7.2 The general structure of a mitochondrion.
Source: Cooper, *The Cell* 8e, Oxford University Press.

cycle? To answer that question we now come to an enzyme reaction of major importance to metabolism, in which pyruvate, transported into mitochondria, is converted into a compound that is at the crossroads of energy metabolism: acetyl-coenzyme A (acetyl-CoA). This reaction is called the **oxidative decarboxylation** of pyruvate.

As the name implies, the substantive part of acetyl-CoA is the molecule **coenzyme A**. Coenzyme A is usually referred to as CoA for short, but is written in equations as CoA–SH because its thiol group is the reactive part of the molecule. Its structure is shown in Figure 7.3.

Unlike NAD^+ and FAD, CoA is not an electron carrier, but an acyl group carrier (A for acyl). Like NAD^+ and FAD, it is a dinucleotide that includes the vitamin **pantothenic acid** in its structure.

Acetyl-CoA is formed from CoA when an acetyl group is attached to the rest of the CoA molecule via a sulphur atom to form what is called a thiol ester. As such, acetyl-CoA can be written as CH_3CO–S–CoA. The thiol ester is a high-energy compound.

So, what happens when pyruvate is subjected to oxidative decarboxylation? We see CO_2 released (this is the decarboxylation), a pair of electrons transferred to NAD^+ (oxidation), and an acetyl group transferred to CoA. The reaction, catalysed by the enzyme **pyruvate dehydrogenase**, is as follows:

$$Pyruvate + CoA-SH + NAD^+ \rightarrow Acetyl-S-CoA + NADH + H^+ + CO_2$$
$$\Delta G^{0'} = -33.5 \text{kJmol}^{-1}$$

The large negative free-energy change means that the oxidative decarboxylation reaction is irreversible.

Acetyl-CoA enters the TCA cycle

The acetyl group of the acetyl-CoA produced from pyruvate is now fed into the TCA cycle. The essential point of the cycle is that the carbon atoms of the acetyl group are converted into CO_2 while NAD^+ is reduced to NADH. In addition, a molecule of FAD is reduced to $FADH_2$, the electrons coming indirectly, in part, from water. The cycle also generates one 'high-energy' phosphoryl group from P_i (the energetic equivalent of ATP) for each acetyl group fed in—so, therefore, *two* for each glucose molecule being oxidized. (Remember: one molecule of glucose generates two molecules of acetyl Co-A.) Stage 2 of glucose oxidation is illustrated in Figure 7.4.

In summary, a molecule of pyruvate from the cytosol is converted in the mitochondria by pyruvate dehydrogenase and the TCA cycle into three molecules of CO_2; in the process, three molecules of NAD^+ and one molecule of FAD are reduced. We have hardly started to make ATP: only two molecules in glycolysis and two in the TCA cycle (for each glucose molecule oxidized). Almost 30 are still to be made!

So far, glucose oxidation has mainly involved the preparation of fuel. The big return in the form of ATP generation comes in stage 3.

Stage 3 in the release of energy from glucose: electron transport to oxygen

The oxidation of the NADH and $FADH_2$ generated from the TCA cycle takes place in the inner mitochondrial membrane in eukaryotes and in the cell membrane in prokaryotes. It involves electrons being transferred from NADH and $FADH_2$ to oxygen with the formation of water:

$$NADH + H^+ + \frac{1}{2}O_2 \rightarrow NAD^+ + H_2O$$

This transfer does not happen in one step. Rather, the electrons pass along a chain of electron carriers, which collectively form the

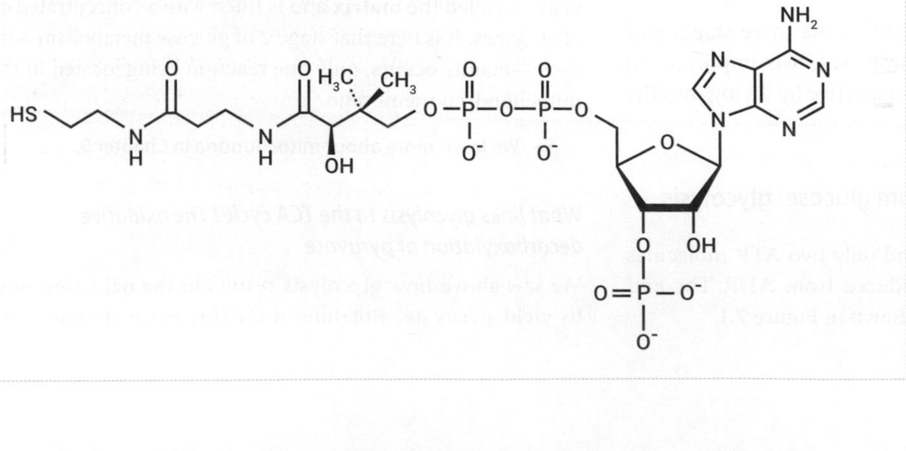

Figure 7.3 The structure of coenzyme A.

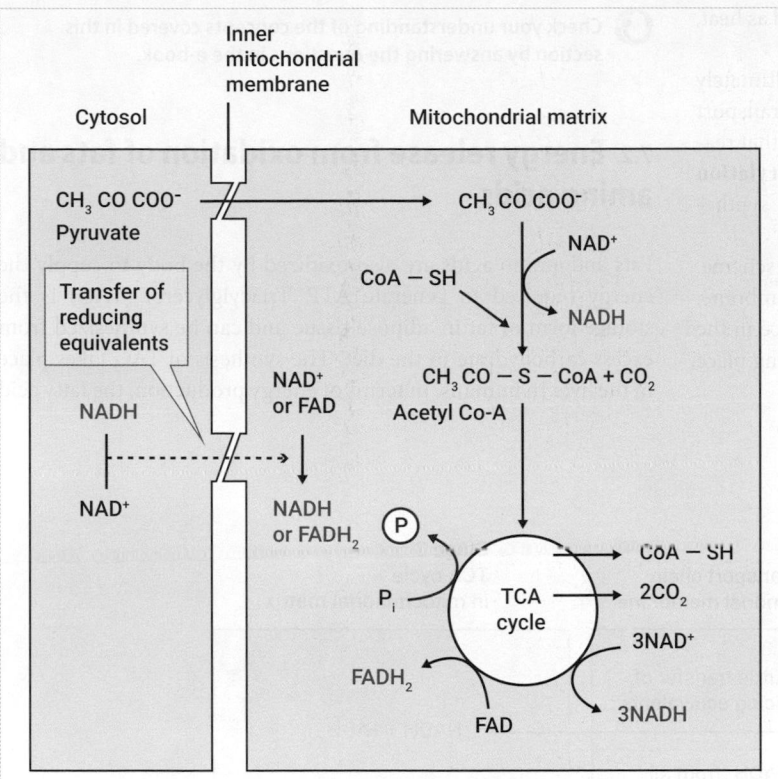

Figure 7.4 Stage 2 of glucose oxidation. Pyruvate, from glycolysis, is converted into acetyl-CoA, which enters the TCA cycle. Notice how NADH and FADH$_2$ are generated by the TCA cycle, when NAD$^+$ and FAD are reduced. Notice also the generation of a high-energy phosphate group, equivalent energetically to ATP.

Source: Snape, A. & Papachristodoulou, D. (2018). *Biochemistry and Molecular Biology* (6th edn). Oxford University Press.

electron transport chain. The electron carriers alternate between the reduced and oxidized state as they accept and donate electrons. Each component of the chain accepts electrons from another component, which has a lower affinity for electrons (that is, it is less electronegative). It then donates the electrons to the next component, which is *more* electronegative. Overall, then, electrons from NADH and FADH$_2$ descend an electron 'staircase' as illustrated in Figure 7.5. Look at this figure and notice how, as electrons descend the staircase, they are handed from electron carriers with low electron affinity to carriers with relatively higher electron affinity—as indicated by the increase in redox potential as we proceed down the staircase.

In this way the free energy is liberated in manageable parcels, in the sense that it can be harnessed by mechanisms that result in

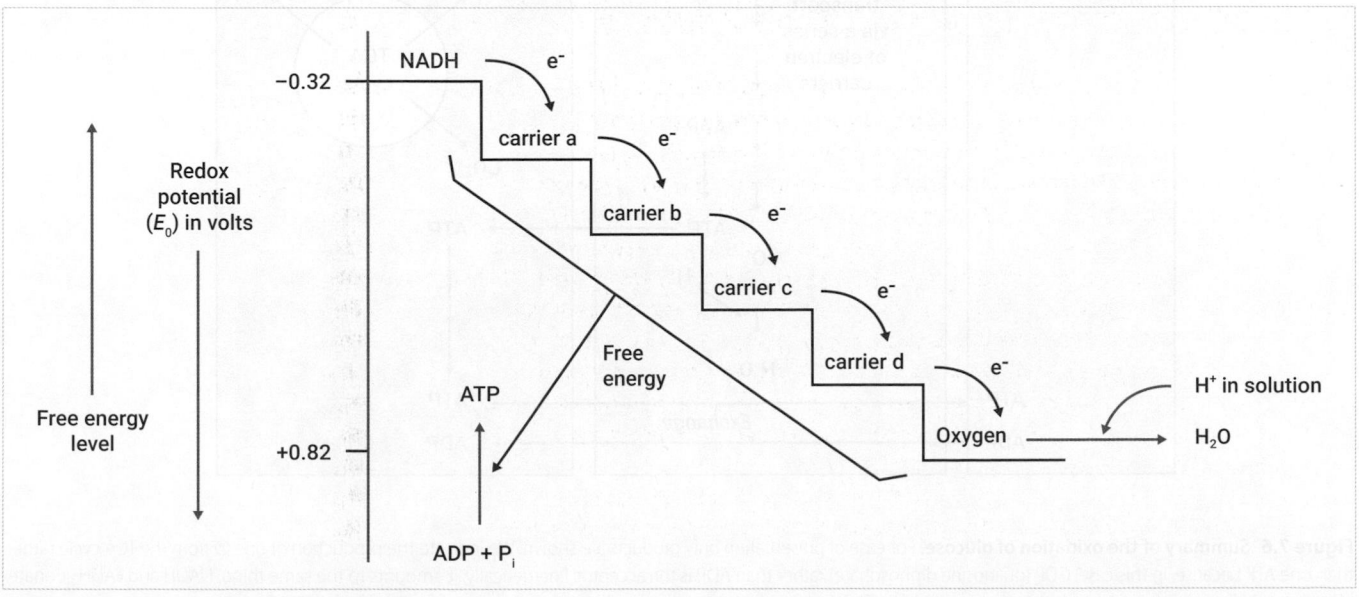

Figure 7.5 Principle of the electron transport chain. The number of carriers shown is arbitrary; each is a different carrier. Some carriers only accept electrons, with protons being liberated into solution; other carriers accept protons along with the electrons. The final reaction with oxygen involves protons from solution.

Source: Snape, A. & Papachristodoulou, D. (2018). *Biochemistry and Molecular Biology* (6th edn). Oxford University Press.

ATP generation from ADP and P_i rather than being wasted as heat, as would occur in the simple burning of glucose.

But to what use is the oxidation of NADH and $FADH_2$ ultimately put? Their oxidation during the course of the electron transport chain drives the conversion of ADP and P_i to ATP. It is for that reason that the complete process is called **oxidative phosphorylation** as mentioned before. Thirty or more ATP molecules are synthesized from the oxidation of one molecule of glucose.

Figure 7.6 shows all three stages put together in one scheme. Notice how these three stages span the mitochondrial membrane, with stage 1 taking place in the cytosol, stage 2 taking place in the mitochondrial matrix, and stage 3 straddling the two, taking place across the mitochondrial membrane.

 Check your understanding of the concepts covered in this section by answering the questions in the e-book.

7.2 Energy release from oxidation of fats and amino acids

Fats and amino acids are also oxidized by the body to supply the energy required to generate ATP. Triacylglycerol (TAG) is the storage form of fat in adipose tissue and can be synthesized from excess carbohydrate in the diet. The synthesis of TAG takes place in the liver in humans. In terms of energy production, the fatty acid

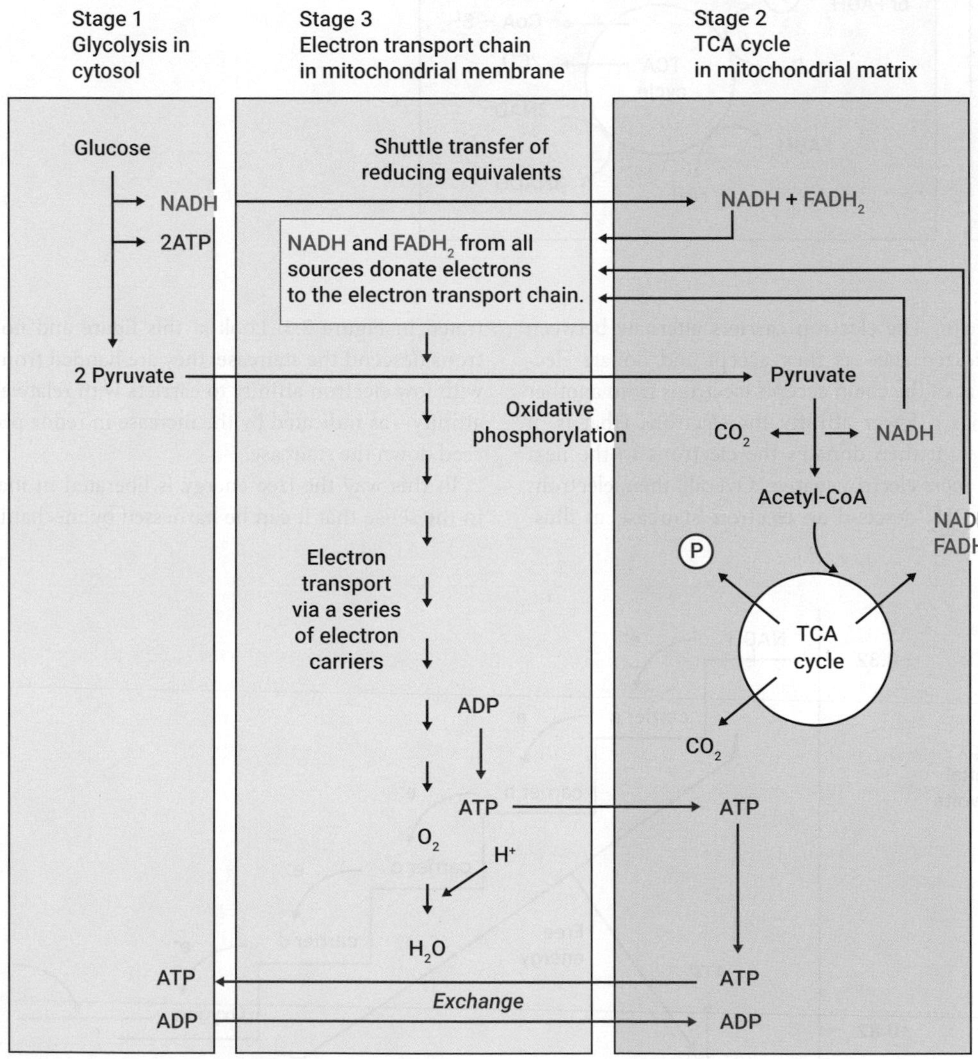

Figure 7.6 Summary of the oxidation of glucose. For ease of presentation only products are shown. We indicate the production of one ℗ from the TCA cycle rather than one ATP because, in this case, GDP (guanosine diphosphate) rather than ADP is the acceptor. Energetically, it amounts to the same thing. NADH and $FADH_2$ donate electrons to different points in the electron transport chain. $FADH_2$ donates electrons to the chain at a point lower down than does NADH.

Source: Snape, A. & Papachristodoulou, D. (2018). *Biochemistry and Molecular Biology* (6th edn). Oxford University Press.

components of TAG are important, the glycerol portion being less significant:

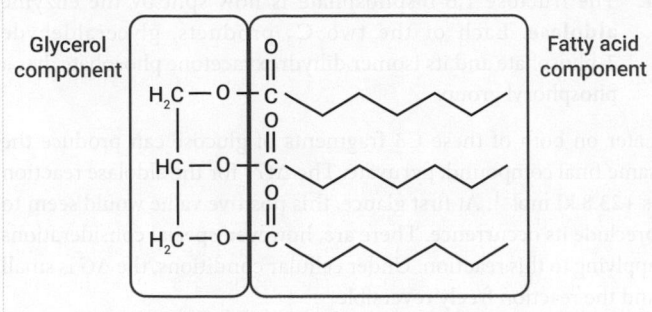

Glucose, as described, is oxidized and the acetyl-CoA so formed is fed into the TCA cycle. The storage form of glucose is glycogen, which is formed in all tissues but mainly in the liver and skeletal muscle. Glycogen is formed when glucose is plentiful (the 'fed state'). Liver glycogen is then degraded to free glucose when glucose is in short supply (the 'fasting state'). This glucose enters the blood stream and supplies the brain and erythrocytes. Muscle glycogen is not broken down to free glucose but to glucose 6-phosphate, which enters the glycolytic pathway and provides energy for muscle contraction.

Glycogen synthesis and degradation are under hormonal control: insulin stimulates glycogen synthesis, catalysed by the enzyme glycogen synthase, while adrenaline and glucagon stimulate degradation through the action of kinase and phosphatase enzymes. Glycogen synthase is activated by dephosphorylation and glycogen phosphorylase is activated by phosphorylation, as we discuss further in Section 7.3. It is important to distinguish between a phosphatase and a phosphorylase. A phosphatase is an enzyme that removes a phosphate group from a given molecule. A phosphorylase is a hydrolytic enzyme that adds a phosphate group using inorganic phosphate.

Fatty acids are also metabolized by detaching two carbon atoms at a time as acetyl-CoA, which is also fed into the TCA cycle. In the preliminary metabolic steps, NAD$^+$ and FAD are reduced in both systems, and the electrons they carry are fed into the electron transport chain, as we see for glucose oxidation.

Amino acids, which are only stored in the body in the form of muscle protein, can be metabolized to pyruvate or to acetyl-CoA and join stage 2, the TCA cycle.

Figure 7.7 illustrates how glucose, fat, and amino acid oxidation relate to one another. Look at this figure and notice how fats (in the form of fatty acids) and amino acids feed into the same metabolic pathways as those used during glucose oxidation, but at later stages in the overall process.

PAUSE AND THINK

Fatty acids and glucose have very different structures but have a common pathway of complete oxidation. What is the first common metabolite in their oxidation pathways?

Answer: Acetyl-CoA. Glucose is oxidized by glycolysis to pyruvate which is converted into acetyl-CoA. Fatty acids are oxidized to acetyl-CoA. The acetyl-CoA enters the TCA cycle regardless of whether it has been produced by glucose or by fatty acid oxidation.

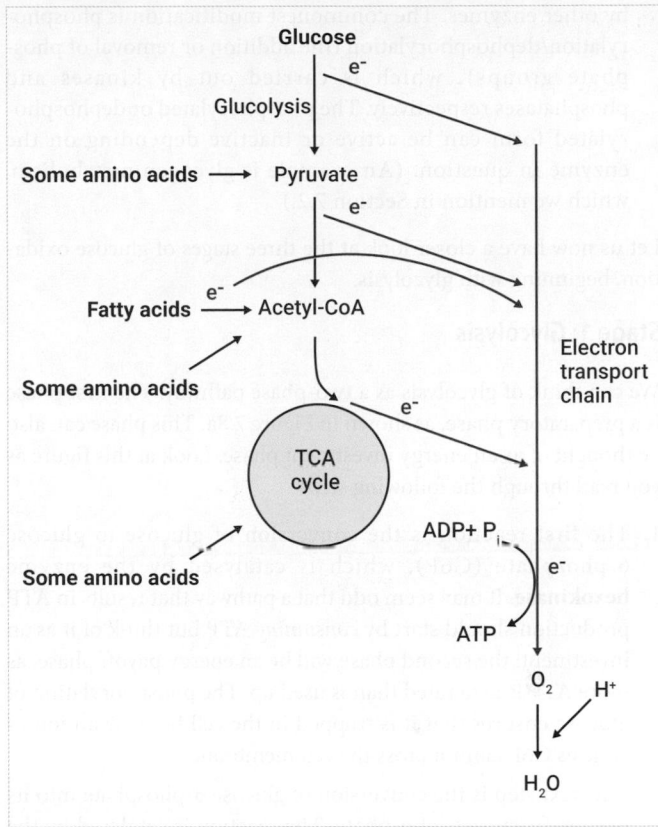

Figure 7.7 The relationship between glucose, fat, and amino acid oxidation for energy generation.

Source: Snape, A. & Papachristodoulou, D. (2018). *Biochemistry and Molecular Biology* (6th edn). Oxford University Press.

 Check your understanding of the concepts covered in this section by answering the questions in the e-book.

7.3 Glucose oxidation: a deeper look

Metabolic pathways are regulated by activating and/or inhibiting key enzymes. So, before we look at the individual reactions seen during the course of glucose oxidation, let us remind ourselves of the different strategies of enzyme regulation, which these reactions employ.

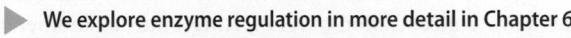 We explore enzyme regulation in more detail in Chapter 6.

1. **Allosteric interactions:** Allosteric enzymes have two binding sites: one for the substrate and one for an effector, which can be an activator or an inhibitor. The binding of the effector is non-covalent and reversible (e.g. phosphofructokinase, which we mention shortly)

2. **Feedback inhibition:** In this type of regulation, the enzyme is inhibited by the product of the reaction when the concentration is higher than the demands of the cell (e.g. hexokinase, which we also mention shortly)

3. **Covalent modification:** In this type of regulation the enzymes are activated or inhibited by covalent modification carried out

by other enzymes. The commonest modification is phosphorylation/dephosphorylation (the addition or removal of phosphate groups), which is carried out by kinases and phosphatases respectively. The phosphorylated or dephosphorylated form can be active or inactive depending on the enzyme in question. (An example is glycogen metabolism, which we mention in Section 7.2.)

Let us now have a closer look at the three stages of glucose oxidation, beginning with glycolysis.

Stage 1: Glycolysis

We can think of glycolysis as a two-phase pathway. The first phase is a preparatory phase, as shown in Figure 7.8a. This phase can also be thought of as an energy investment phase. Look at this figure as you read through the following steps

1. The first reaction is the conversion of glucose to glucose 6-phosphate (G6P), which is catalysed by the enzyme **hexokinase**. It may seem odd that a pathway that results in ATP production should start by *consuming* ATP but think of it as an investment: the second phase will be an energy payoff phase, as more ATP is generated than is used up. The phosphorylation of glucose ensures that it is trapped in the cell because an ion as large as G6P cannot cross the cell membrane

2. The next step is the conversion of glucose 6-phosphate into its isomer, fructose 6-phosphate. The reaction is catalysed by the enzyme **phosphoglucose isomerase**

3. The next step is another energy investment reaction catalysed by **phosphofructokinase**. Another molecule of ATP is used to phosphorylate fructose 6-phosphate, yielding fructose 1, 6-bisphosphate. This reaction is a key step in glycolysis as phosphofructokinase is allosterically controlled by ATP and its metabolite AMP.

   ▶ **We discuss phosphofructokinase and its allosteric control in Chapter 6.**

ATP is a substrate for the reaction, but it can also bind to a site other than the active site of the enzyme (when ATP is present in high concentrations). In this way it inhibits the enzyme and therefore inhibits glycolysis. As a result, when the energy level of the cell is high, glycolysis will be inhibited as there is no need for more ATP to be generated. Conversely, if levels of AMP are high (which indicates a low concentration of ATP), AMP binds to another site on the enzyme and activates it, stimulating glycolysis (and therefore energy production in the form of ATP).

It might seem paradoxical that ATP is both the substrate of the reaction and its inhibitor, but this paradox is resolved when we consider the relative affinities of the enzyme active site and the allosteric sites for ATP. ATP binds the active site (the substrate site) at low concentrations, but when the concentrations rise—indicating that the energy level of the cell is high—it binds the allosteric site and inhibits the enzyme. The control is even finer because the inhibition of phosphofructokinase means that fructose 6-phosphate concentrations increase. As fructose 6-phosphate is in equilibrium with glucose 6-phosphate, glucose 6-phosphate levels increase and inhibit hexokinase—an example

of product inhibition. In this way, glycolysis is inhibited by the lower activity of both enzymes.

4. The fructose 1,6-bisphosphate is now split by the enzyme **aldolase**. Each of the two C_3 products, glyceraldehyde 3-phosphate and its isomer dihydroxyacetone phosphate, has a phosphoryl group.

Later on both of these C3 fragments of glucose can produce the same final compound, pyruvate. The $\Delta G^{0'}$ for the aldolase reaction is +23.8 kJ mol^{-1}. At first glance, this positive value would seem to preclude its occurrence. There are, however, special considerations applying to this reaction. Under cellular conditions, the ΔG is small and the reaction freely reversible.

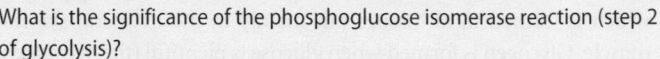

PAUSE AND THINK

What is the significance of the phosphoglucose isomerase reaction (step 2 of glycolysis)?

Answer: For the 6-carbon sugar to be split it needs to be in the aldose form (i.e. the fructose isomer) so that the enzyme aldolase can act on it.

5. An enzyme, triose phosphate isomerase, catalyses the interconversion of the isomers glyceraldehyde 3-phosphate and dihydroxyacetone phosphate. The two compounds are in equilibrium. However, glyceraldehyde 3-phosphate is continually removed by the next step in glycolysis. As a result, all of the dihydroxyacetone phosphate is progressively converted into glyceraldehyde 3-phosphate.

We now enter the energy payoff phase. Look now at Figure 7.8b as we walk through the remaining steps.

6. The next reaction is the conversion of glyceraldehyde 3-phosphate into 1,3 bisphosphoglycerate. The aldehyde group of glyceraldehyde 3-phosphate is oxidized by glyceraldehyde 3-phosphate dehydrogenase, using NAD$^+$ as an electron acceptor and inorganic phosphate is added, forming 1,3 bisphosphoglycerate

7. The $RCO-O-PO_3^{2-}$ group on 1,3 bisphosphoglycerate is a high-energy compound and so its phosphoryl group can be transferred to ADP, forming ATP and leaving us with 3-phosphoglycerate (3-PGA). The enzyme responsible is phosphoglycerate kinase. It is so named because, in the reverse direction, it transfers a phosphoryl group from ATP *to* 3-PGA. (By convention, kinases are always given their name in reference to the side of the reaction that involves ATP, whether the reaction can proceed in that direction or not.)

The phosphoryl group generated in this process is attached to the actual substrate (1,3-bisphosphoglycerate). For this reason, the reaction is described as substrate-level phosphorylation, as opposed to oxidative phosphorylation, which we will describe in more detail shortly.

8. The phosphoryl group remaining on 3-phosphoglycerate is transferred from the 3- to the 2- position as shown in Figure 7.8b, catalysed by the enzyme **phosphoglycerate mutase**.

9. The next step in glycolysis is the removal of a molecule of water from 2-phosphoglycerate to form phosphoenolpyruvate.

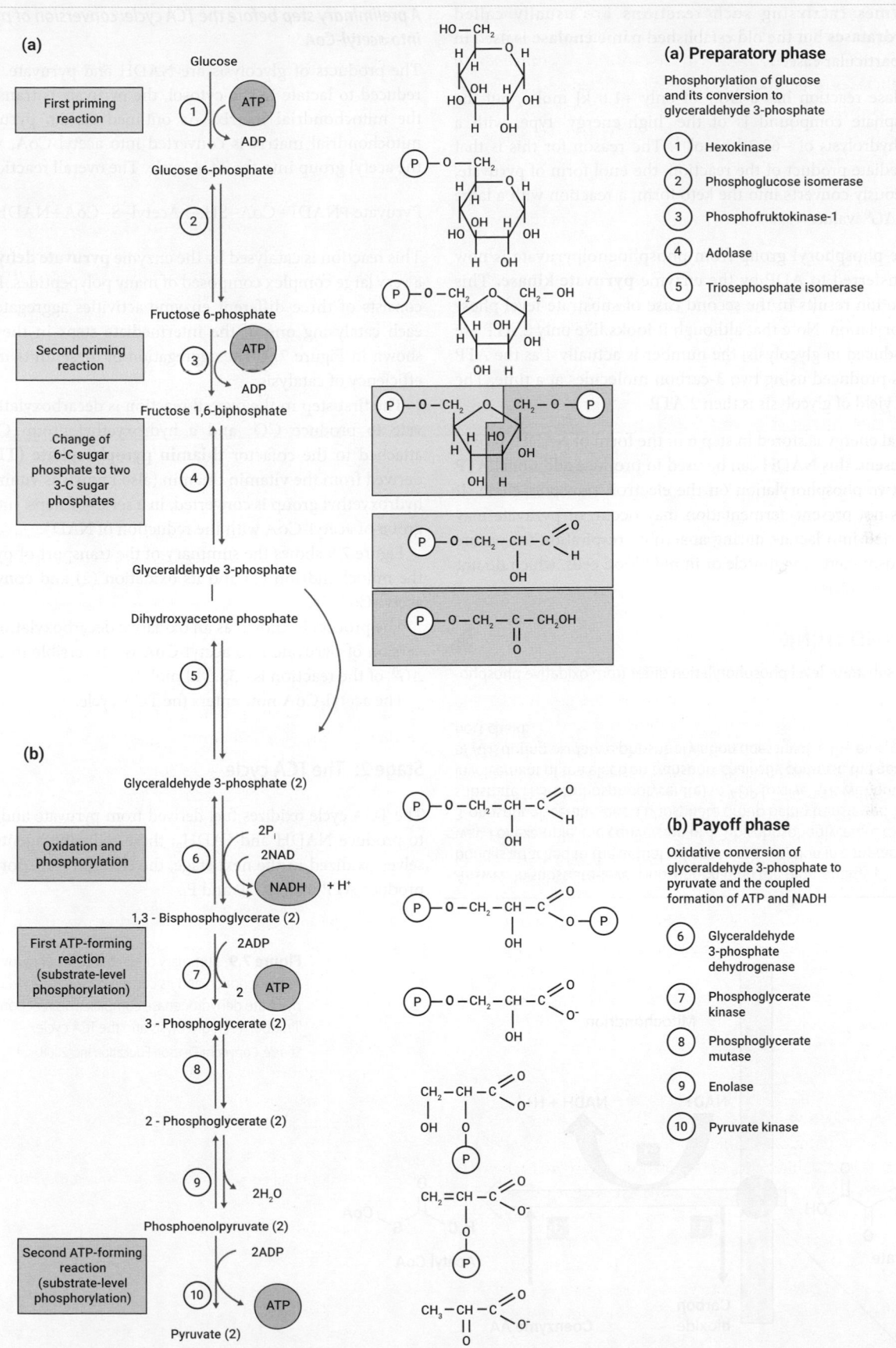

Figure 7.8 The two stages of the glycolytic pathway. (a) The preparatory phase; (b) the payoff phase.

Source: Nelson, D. L. and Cox , M. M. from Lehninger, *Principles of Biochemistry*, 4th Edition. W. H. Freeman. Figure 2.

Enzymes catalysing such reactions are usually called **dehydratases** but the old established name **enolase** is used in this particular case.

The enolase reaction has a $\Delta G^{\circ\prime}$ of only +1.8 kJ mol^{-1}, but the enolphosphate compound is of the 'high-energy' type, with a $\Delta G^{\circ\prime}$ of hydrolysis of −62.2 kJ mol^{-1}. The reason for this is that the immediate product of the reaction, the enol form of pyruvate, spontaneously converts into the keto form, a reaction with a large negative $\Delta G^{0\prime}$ value.

10. The phosphoryl group from phosphoenolpyruvate is now transferred to ADP by the enzyme **pyruvate kinase.** This reaction results in the second case of substrate-level phosphorylation. Note that although it looks like only 2 ATP are produced in glycolysis, the number is actually 4 as the ATP was produced using two 3-carbon molecules at a time. The net yield of glycolysis is then 2 ATP.

Additional energy is stored in step 6 in the form of NADH. If oxygen is present, this NADH can be used to produce additional ATP by oxidative phosphorylation on the electron transport chain. If oxygen is not present, fermentation may occur, or pyruvate may be converted into lactate during anaerobic respiration (as occurs in vigorously exercising muscle or in red blood cells, which do not have mitochondria).

PAUSE AND THINK

How does substrate-level phosphorylation differ from oxidative phosphorylation?

Answer: In substrate-level phosphorylation the high-energy phosphoryl bond is attached to the actual substrate of a reaction in a metabolic pathway. For example, the conversion of 1,3-bisphosphoglycerate to 3-phosphoglycerate sees a phosphate group being transferred from the substrate (1,3-bisphosphoglycerate) to ADP to give ATP, without the involvement of the electron transport chain. By contrast, the generation of ATP during oxidative phosphorylation does involve the electron transport chain.

A preliminary step before the TCA cycle: conversion of pyruvate into acetyl-CoA

The products of glycolysis are NADH and pyruvate. Unless it is reduced to lactate in the cytosol, the pyruvate is transported into the mitochondrial matrix. As outlined earlier, pyruvate in the mitochondrial matrix is converted into acetyl-CoA, which feeds the acetyl group into the TCA cycle. The overall reaction is:

$$Pyruvate + NAD^+ + CoA-SH \rightarrow Acetyl-S-CoA + NADH + H^+ + CO_2$$

This reaction is catalysed by the enzyme **pyruvate dehydrogenase**, a very large complex composed of many polypeptides. It essentially consists of three different enzyme activities aggregated together, each catalysing one of the intermediate steps in the process, as shown in Figure 7.9. The aggregation of these units increases the efficiency of catalysis.

The first step in the overall reaction is decarboxylation of pyruvate to produce CO_2 and a hydroxyethyl group CH_3CHOH– attached to the cofactor **thiamin pyrophosphate (TPP)**. TPP is derived from the vitamin thiamin (also known as vitamin B_1). The hydroxyethyl group is converted, in a series of steps, into the acetyl group of acetyl-CoA with the reduction of NAD^+.

Figure 7.9 shows the summary of the transport of pyruvate into the mitochondrion (1) and its oxidation (2) and conversion into acetyl-CoA(3).

The process is known as an oxidative decarboxylation. The conversion of pyruvate into acetyl-CoA is irreversible in animals; the $\Delta G^{0\prime}$ of the reaction is −33.5 kJ mol^{-1}.

The acetyl-CoA now enters the TCA cycle.

Stage 2: The TCA cycle

The TCA cycle oxidizes fuel derived from pyruvate and acetyl-CoA to produce NADH and $FADH_2$; these reducing agents are themselves oxidized in the next stage, the electron transport system, to produce ATP from ADP and P_i.

Figure 7.9 Summary of the transport of pyruvate into the mitochondrion and its conversion into acetyl-CoA by the pyruvate dehydrogenase complex. This reaction forms the 'link' between glycolysis and the TCA cycle.
Source: Copyright Pearson Education Inc, 2008.

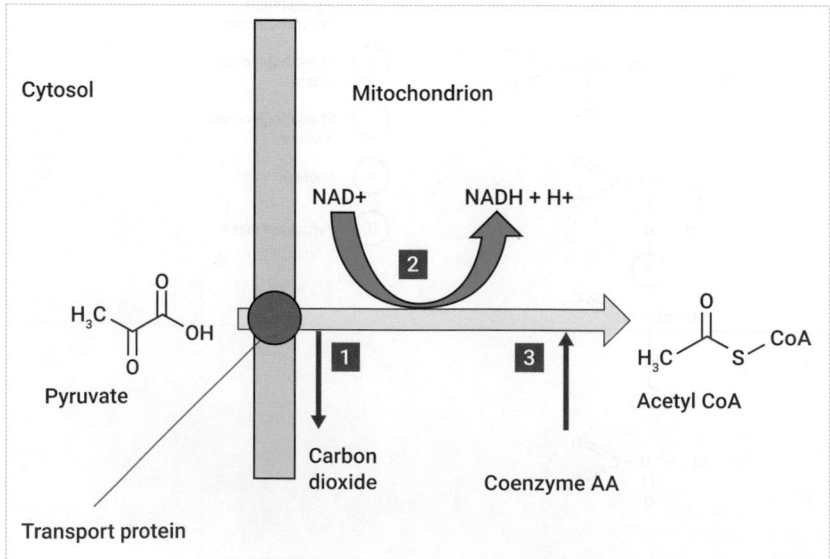

A summary of the TCA cycle is shown in Figure 7.10. The TCA cycle is thermodynamically feasible because it is driven by the free energy made available from the elimination of the acetyl group of acetyl-CoA. In addition to the production of NADH and $FADH_2$, the cycle produces one molecule of GTP (guanosine triphosphate), which is equivalent to one ATP, for every turn of the cycle, by substrate-level phosphorylation.

Acetyl-CoA enters the cycle by reacting with oxaloacetate to produce citrate. This reaction involves the input of a molecule of water. With one complete 'turn' of the cycle, oxaloacetate is reformed, and the acetyl group of acetyl-CoA is consumed. Look again at Figure 7.10 and notice how the overall output from one complete cycle, with one acetyl group as the input, is as follows:

- two molecules of CO_2
- the reduction of three molecules of NAD^+ to three molecules of NADH
- the reduction of one molecule of FAD to $FADH_2$
- the production of one molecule of GTP, which is equivalent to one molecule of ATP.

Now look at Figure 7.11, which shows the complete TCA cycle. The complete cycle consists of nine enzyme-catalysed reactions; the enzymes themselves are named above the reaction arrow for each step. In eukaryotic cells, the TCA cycle enzymes are

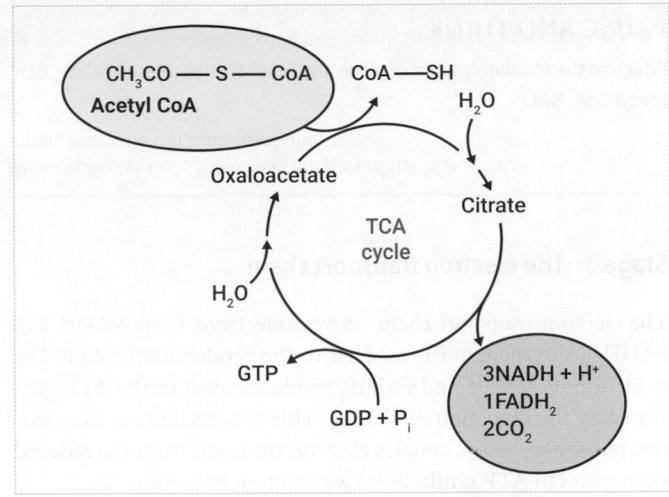

Figure 7.10 The inputs and outputs of the TCA cycle. Individual cycle reactions are not shown.

Source: Snape, A. & Papachristodoulou, D. (2018). *Biochemistry and Molecular Biology* (6th edn). Oxford University Press.

located in the mitochondrial matrix; the exception is succinate dehydrogenase, which is located in the inner mitochondrial membrane.

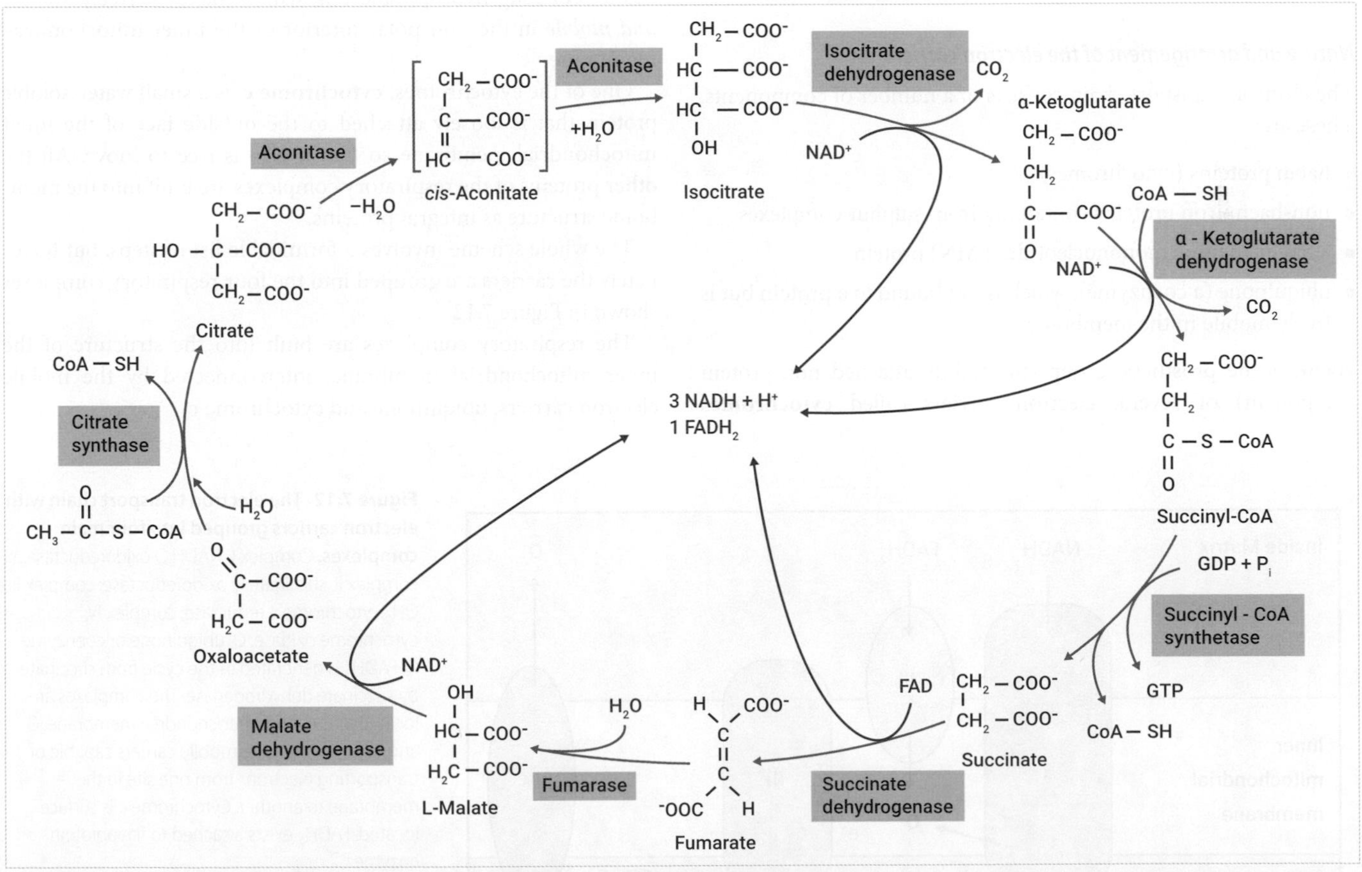

Figure 7.11 The complete TCA cycle. Red highlights the production of reducing agents from the cycle. Purple highlights supply of the elements of H_2O to the cycle. (The conversion of citrate to isocitrate involves removal and addition of H_2O but there is no net gain.)

Source: Snape, A. & Papachristodoulou, D. (2018). *Biochemistry and Molecular Biology* (6th edn). Oxford University Press.

Stage 3: The electron transport chain

The electron transport chain conveys electrons from NADH and FADH₂ to oxygen and is coupled to the production of ATP. The oxidation of NADH and FADH₂ produces most of the ATP generated by the oxidation of glucose. This process, known as oxidative phosphorylation, couples electron transport from the reduced coenzymes to ATP synthesis, as we mention in Section 7.1.

We explored the principle of electron transport earlier in the chapter. To remind you, the electron transport chain comprises electron carriers, located in the inner mitochondrial membrane in eukaryotes and in the cell membrane in prokaryotes, which alternate between the reduced and oxidized state. Each component of the chain accepts electrons from a carrier that has a lower affinity for electrons than itself; in so doing it becomes reduced. It then donates the electrons to another carrier with *greater* affinity for electrons (and becomes oxidized in the process). Let us now consider the components and operation of the electron transport chain in a little more detail.

Nature and arrangement of the electron carriers

The electron transport chain consists of a number of components. These are:

- haem proteins (cytochromes)
- non-haem iron proteins containing iron–sulphur complexes
- a flavin adenine mononucleotide (FMN) protein
- ubiquinone (a coenzyme), which is not bound to a protein but is freely mobile in the membrane.

Haem is the prosthetic group (the tightly attached non-protein component) of several electron carriers called **cytochromes** (so called because of their red colour, which is caused by the iron they contain). The different cytochromes are called c_1, c, a, and a_3 (in order of their participation in the chain). In cytochromes, the Fe atom oscillates between the Fe^{2+} and Fe^{3+} states as it accepts an electron from the preceding carrier and donates it to the next carrier in the chain. So, different cytochromes can have different electron affinities even though they all have haem as their prosthetic group.

Non-haem iron proteins are also electron carriers. Here the iron is bound to the thiol side group of the amino acid cysteine of the protein; it is also bound to inorganic sulphide ions to form **iron–sulphur complexes**. As with the cytochromes, the iron atom in these complexes can accept and donate electrons in a cyclical fashion, oscillating between the ferrous and ferric state. These iron–sulphur centres are associated with flavin enzymes—FAD-containing enzymes such as succinate dehydrogenase—in which they can accept electrons from FAD.

FMN (flavin adenine mononucleotide) consists of the flavin half of FAD. We have seen its structure in Figure 2.10. It carries electrons from NADH to an iron–sulphur centre. All the iron–sulphur centres transfer electrons to ubiquinone.

Ubiquinone is the electron carrier that is not bound to a protein. Its name implies that it can exist as a quinone and is found ubiquitously. It is often referred to as coenzyme Q (CoQ), UQ, or Q. It is an electron carrier as it can accept protons. The molecule has a very long hydrophobic tail, which makes it freely soluble *and mobile* in the non-polar interior of the inner mitochondrial membrane.

One of the cytochromes, **cytochrome c**, is a small water-soluble protein that is loosely attached to the outside face of the inner mitochondrial membrane so that it also is free to move. All the other proteins of the respiratory complexes are built into the membrane structure as integral proteins.

The whole scheme involves a formidable list of steps, but fortunately the carriers are grouped into the four respiratory complexes shown in Figure 7.12.

The respiratory complexes are built into the structure of the inner mitochondrial membrane, interconnected by the mobile electron carriers, ubiquinone and cytochrome c.

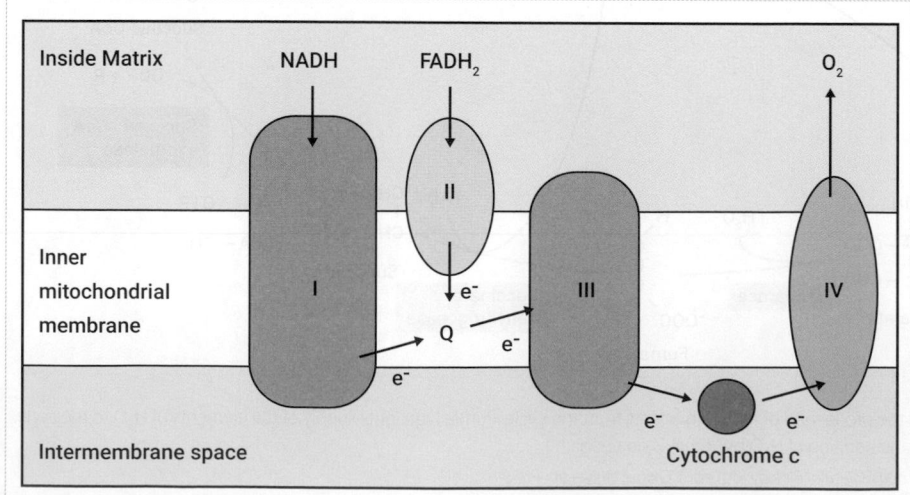

Figure 7.12 The electron transport chain with electron carriers grouped into four main complexes. Complex I, NADH:Q oxidoreductase; complex II, succinate:Q oxidoreductase; complex III, QH₂:cytochrome c reductase; complex IV, cytochrome oxidase. Q, ubiquinone or coenzyme Q. FADH₂ is generated in the cycle from succinate by succinate dehydrogenase. The complexes are located in the inner mitochondrial membrane. Q and cytochrome c are mobile carriers capable of transporting electrons from one site in the membrane to another. Cytochrome c is surface-located. FADH₂ exists attached to flavoprotein enzymes.

Source: Snape, A. & Papachristodoulou, D. (2018). *Biochemistry and Molecular Biology* (6th edn). Oxford University Press.

- Q takes electrons from complexes I and II and delivers them to complex III. Cytochrome c is the intermediary between complexes III and IV.

- Complex I carries electrons from NADH to Q.

- Complex II carries electrons from succinate and other substrates (e.g. fatty acids) via $FADH_2$ to Q.

- Complex III uses QH_2 to reduce cytochrome c.

- Complex IV transfers electrons from cytochrome c to oxygen.

Complexes I, III, and IV are, for convenience, referred to as NADH:Q oxidoreductase, QH_2:cytochrome c oxidoreductase, and cytochrome oxidase, respectively. Complex IV, cytochrome oxidase, is also a multi-subunit structure; electrons are donated to it by cytochrome c on the outer face of the inner mitochondrial membrane. Cytochrome oxidase contains the haem proteins cytochromes a and a_3, and copper centres, which participate in the final transfer of electrons to oxygen.

The final reduction catalysed by cytochrome oxidase is:

$$O_2 + 4e^- + 4H^+ \rightarrow 2H_2O.$$

But *how* does the transfer of electrons from NADH and $FADH_2$ to oxygen during the course of the electron transport chain produce ATP? ATP isn't produced directly. Instead, the mitochondrion couples the process of electron transport and energy release to the synthesis of ATP by a mechanism known as **chemiosmosis**.

PAUSE AND THINK

What is the immediate role of electron transfer in the respiratory chain?

Answer: To pump protons from the mitochondrial matrix to the outside of the inner mitochondrial membrane, thereby generating a proton gradient that can be used to drive ATP synthesis.

Oxidative phosphorylation: the generation of ATP coupled to electron transport

In 1961 Peter Mitchell produced a theory of how electron transport causes ATP synthesis that was so novel that it was at first hardly taken seriously by most scientists and was strongly opposed by many. His proposed mechanism was finally accepted and he received the Nobel Prize for it in 1978. (We learn more about Mitchell's work in Scientific Process 7.1.)

The concept, known as chemiosmosis, is based on the simple notion that chemical gradients have the ability to do work. Energy stored in the form of a hydrogen concentration gradient is used to drive the synthesis of ATP. 'Osmos' is the ancient Greek word meaning 'push' or 'thrust'. Molecules or ions will migrate from a high concentration to a low concentration and, if a suitable energy-harnessing device spans this concentration gradient, useful work can be done as the molecules or ions migrate.

What is needed for ATP generation to be coupled to electron transport? The electron transport must create a gradient, and the gradient should be allowed to flow back through a device which uses the energy of the gradient to synthesize ATP from ADP and P_i. This requires the existence of an intact membrane; without it, a gradient cannot be formed. It explains the observation that a damaged mitochondrial membrane can allow electron transfer but no harnessing of the energy in the form of ATP.

The inner mitochondrial membrane is virtually impermeable to protons, but special proton-conducting channels are inserted into it. Protons flow from the outside through these channels back into the mitochondrial matrix and the energy of this flow is harnessed to the formation of ATP from ADP and P_i.

The proton-conducting channels are **ATP synthase** complexes: knob-like structures which completely cover the inner surface of the cristae. ATP synthase complexes convert ADP and P_i into ATP, the process being energetically driven by the proton flow. Figure 7.13 shows a simplified version of the generation of ATP in mitochondria by the chemiosmotic mechanism.

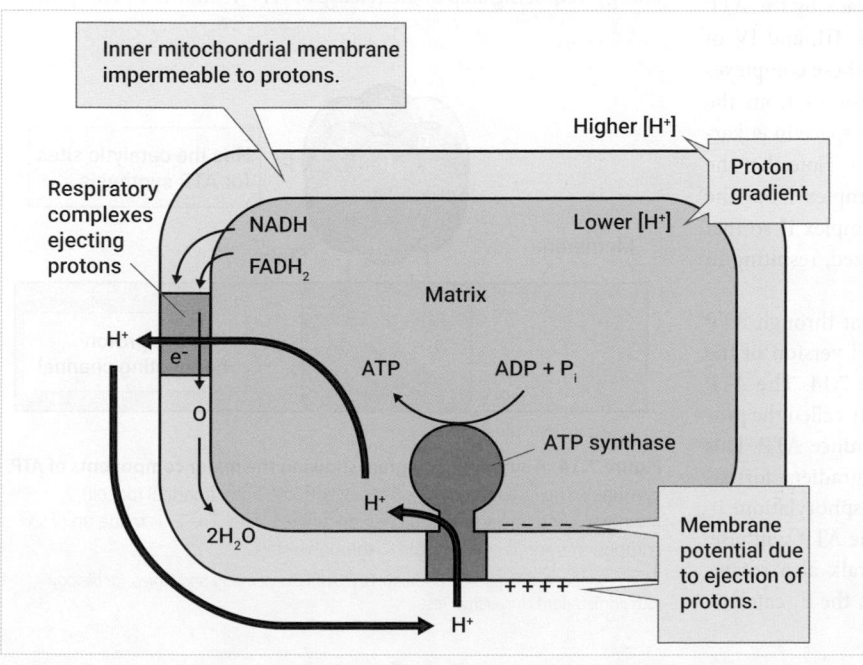

Figure 7.13 Generation of ATP in mitochondria by the chemiosmotic mechanism. Protons are ejected by the electron transport chain into the intermembrane space and establish a proton gradient, which is used to synthesize ATP. This ATP synthesis is catalysed by ATP synthase in the mitochondrial matrix.

Source: Snape, A. & Papachristodoulou, D. (2018). *Biochemistry and Molecular Biology* (6th edn). Oxford University Press.

SCIENTIFIC PROCESS **7.1**   **Mitchell's chemiosmotic hypothesis**

The hypothesis

The chemiosmotic hypothesis was proposed by Peter Mitchell in 1961. He based the concept on the simple notion that gradients have the ability to do work. The hypothesis states that the synthesis of ATP is driven by a proton motive force. As electrons go through the electron transport chain, protons would be pumped across the inner mitochondrial membrane. This movement would result in a proton concentration gradient across the inner mitochondrial membrane with a lower pH (high proton concentration) in the intermembrane space and a higher pH (lower proton concentration) in the mitochondrial matrix. The proton gradient and the membrane potential would be the proton motive force that would drive the synthesis of ATP.

The evidence

The hypothesis was originally rejected by many scientists. It was, however, accepted in the face of accumulating evidence that supported the hypothesis. Here is some of the evidence:

(a) Electron transport does generate a proton gradient. The pH measured on the outside is lower than that inside the mitochondria.

(b) Damaged mitochondria in which a proton gradient cannot be established do not synthesize ATP.

(c) The enzyme ATP synthase was discovered on the inner mitochondrial membrane.

(d) ATP can be synthesized by generating a proton gradient by means other than electron transport. Racker and Stoeckenius showed that

generation of a proton gradient in an artificial lipid membrane using bacteriorhodopsin, a light-driven bacterial proton pump, resulted in the setting up of a proton gradient that enabled a mitochondrial ATPase from beef heart to synthesize ATP (*J. Biol. Chem.* 1974 **25**: 662–3).

(e) Substances such as dinitrophenol (1), which destroy the proton gradient by shuttling protons back across the membrane, result in the functioning of the electron transport chain without generation of ATP.

2,4-Dinitrophenol (1)
DNP

Conclusion

It is now firmly established that the chemiosmotic theory is the explanation for most of the ATP production in mitochondria. The importance of the membrane potential as part of the proton motive force was challenged in an article published by the Royal Society in 2019, but the fact remains that a proton gradient across the mitochondrial inner membrane (and in plants and prokaryotes) is responsible for the synthesis of ATP.

Further reading

For more on the challenge to the importance of the membrane potential, see:

Morelli, A. M., Ravera, S., Calzia, D., and Panfoli, I. (2019) An update of the chemiosmotic theory as suggested by possible proton currents inside the coupling membrane. *Open Biol.* **9**: 180221. http://doi.org/10.1098/rsob.180221.

How are the protons ejected?

How are the protons ejected into the cytosolic side of the inner mitochondrial membrane before being pumped back by the ATP synthase complex? The answer is by complexes I, III, and IV of the electron transport chain (see Figure 7.12). As these complexes accept and donate electrons, they also pump protons from the mitochondrial membrane into the intermembrane space in eukaryotes or outside the cell membrane in prokaryotes. Note that the oxidation of NADH pumps protons through complex I but the oxidation of $FADH_2$ pumps electrons through complex II so that fewer protons are pumped when $FADH_2$ is oxidized, resulting in fewer molecules of ATP being produced.

The protons then flow back down their gradient through ATP synthase, which is in the membrane. A simplified version of the components of ATP synthase is shown in Figure 7.14. The ATP synthase uses the flowing of protons (more formally called the proton motive force) to phosphorylate ADP and produce ATP. This combined process of electron transport, proton gradient formation, and ATP synthesis is known as oxidative phosphorylation.

Protons are taken up by the F_0 component of the ATP synthase, which actually rotates in the membrane. The stalk also rotates and in doing so causes conformational changes in the F_1 catalytic

subunits α and β. These conformational changes catalyse the formation of ATP from ADP and inorganic phosphate. The actual energy-requiring step is the release of ATP from the F_1 component;

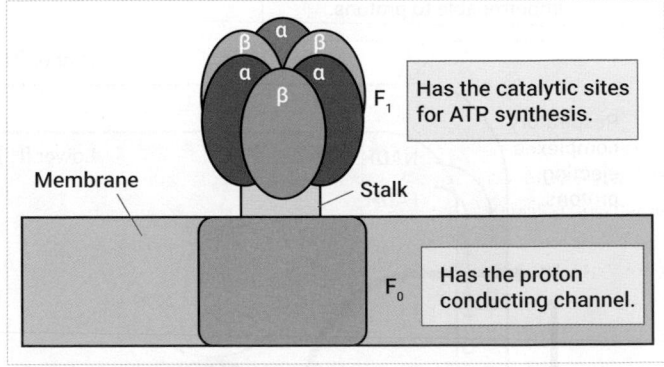

Figure 7.14 A simplified diagram showing the major components of ATP synthase. The F_0 component is integral with the mitochondrial (or cell) membrane and forms the proton conducting channel. The F_1 is made up of six components and is linked to the F_0 through a stalk.

Source: Snape, A. & Papachristodoulou, D. (2018). *Biochemistry and Molecular Biology* (6th edn). Oxford University Press.

the conformational changes that occur when the protons are travelling through the F_0 and the stalk provide this energy.

Knowing this makes it clear why ATP cannot be formed if the membrane is damaged: a proton gradient would be impossible to set up, there would be nothing to drive rotation of the ATP synthase, and so nothing to catalyse the formation of ATP. A number of compounds, known as uncouplers of oxidative phosphorylation, have the same effect: they inhibit the coupling between the electron transport and phosphorylation reactions, in so doing inhibiting the synthesis of ATP without affecting the respiratory chain. Examples include dinitrophenol (mentioned in Scientific Process 7.1) and the potent antibiotic valinomycin. Healthy mitochondria are essential for survival and correct function of the body. Defects in mitochondrial function have detrimental effects on cell function (Real World View 7.1).

The yield of ATP from the complete oxidation of glucose is approximately 30 ATP. Two come directly from glycolysis by substrate-level phosphorylation, two from the TCA cycle via GTP, and the rest from oxidation of cofactors NADH and $FADH_2$. These values are estimates as there is no exact whole number relationship between electron transport, proton ejection, and ATP generation.

PAUSE AND THINK

What causes the F_0 part of the ATP synthase to rotate?

Answer: When negatively charged residues on the F_0 are protonated they seek a hydrophobic environment instead of the hydrophilic environment in which they are stable in the unprotonated state. This means that they move towards the interior of the synthase and away from the outside. This causes a rotation of the F_0 component.

Check your understanding of the concepts covered in this section by answering the questions in the e-book.

7.4 Anaerobic respiration and fermentation: production of ATP without the use of oxygen

Oxygen is not always plentiful in animals as the delivery of oxygen by the blood to tissues may not be adequate. This is especially so in vigorously exercising muscle. So how does glucose oxidation, which normally terminates with the reduction of oxygen, proceed under these circumstances? The answer is that NADH must be reoxidized to NAD^+. This is achieved by the conversion of pyruvate to lactate and the concomitant oxidation of NADH to NAD^+, a process catalysed by the enzyme lactate dehydrogenase, as shown in Figure 7.15. This process is termed anaerobic respiration—that is, the oxidation of glucose in the *absence* of oxygen.

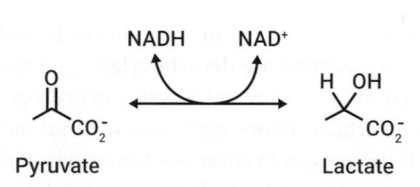

Figure 7.15 The conversion of pyruvate to lactate.

Mitochondria are responsible for generating 90 per cent of the energy needed to sustain life and support organ function. Defects in mitochondrial function can cause cell injury and cell death—with those parts of the body that require the greatest amount of energy being most affected. They are the heart, brain, skeletal muscles, and lungs.

It is often difficult to diagnose mitochondrial disease as signs and symptoms differ from individual to individual. Symptoms can include seizures, strokes, inability to walk, talk, and see, and other developmental delays. Mitochondrial disease primarily affects children but appearance in adulthood is also possible.

Mitochondrial diseases arise from errors in mitochondrial DNA—DNA that is specific to the mitochondria, and distinct from the DNA stored in a cell's nucleus (what we call nuclear DNA). The *inheritance* of mitochondrial DNA is also different from that of nuclear genes. It is described as 'maternal inheritance' because mitochondria (and their DNA) are passed on to offspring from the egg and not the sperm.

While most mitochondrial function is controlled by nuclear DNA (most of the 1500 or so mitochondrial proteins are encoded in nuclear DNA), mitochondrial DNA encodes 13 proteins of the respiratory chain. Mutations in these genes have serious consequences for ATP production.

Treatment options are currently limited. Vitamins and pyruvate have been tried as therapeutic options with little evidence of effectiveness. However, help is now available in the form of mitochondrial replacement therapy. Researchers at Newcastle University, led by Douglas Turnbull, transplanted mitochondrial DNA from healthy female donors into the eggs of women with mitochondrial disease. This procedure led to ethical questions about biological motherhood as the child receives genes from two females. However, in 2013 the UK government agreed to develop legislation that could legalize this so-called 'three-person IVF', and in 2015 the UK became the only country to give regulatory approval for germline modification, in so doing making an exception to its wider prohibition on human germline modification. The issue remains controversial in most other countries.

To find out more about the research into mitochondrial replacement therapy pioneered by Douglas Turnbull and his colleagues, see Craven, L., Tuppen, H., Greggains, G. *et al.* (2010) Pronuclear transfer in human embryos to prevent transmission of mitochondrial DNA disease. *Nature* **465**: 82–85. https://doi.org/10.1038/nature08958.

The production of lactate during anaerobic respiration ensures that there is an adequate supply of NAD⁺ for glycolysis to occur even when the supply of oxygen is inadequate. We experience the side effects of the production of lactate during vigorous exercise. During such exercise our muscles are forced to operate anaerobically (because their need for oxygen outstrips its supply); as a result, they start to produce lactate. It is this lactate that causes the burning sensation in our muscles that is a common (unpleasant!) feature of strenuous exercise.

Erythrocytes absolutely depend on anaerobic respiration because they lack mitochondria. As a result, the TCA cycle and oxidative phosphorylation do not occur. Instead, erythrocytes also convert pyruvate into lactate and in this way continue the production of ATP via glycolysis.

Figure 7.16 illustrates the net result of the anaerobic glycolysis of glucose. Cancerous cells metabolize glucose anaerobically at high rates. Detection and possible inhibition of this process are important in the diagnosis and treatment of malignancies (Clinical Box 7.1).

Fermentation is another example of glycolysis proceeding in the absence of oxygen. Yeast lives entirely on anaerobic glycolysis and uses a mechanism analogous to that seen in erythrocytes and anaerobically respiring muscle cells to reoxidize NADH, as shown in Figure 7.17.

In this case, the pyruvate is converted into acetaldehyde and carbon dioxide by the enzyme **pyruvate decarboxylase**, and the acetaldehyde is converted into ethanol by alcohol dehydrogenase, regenerating NAD⁺, which allows glycolysis to continue. Pyruvate decarboxylase is not present in animals—fortunately perhaps, as vigorous activity would result in alcohol intoxication!

 Check your understanding of the concepts covered in this section by answering the questions in the e-book.

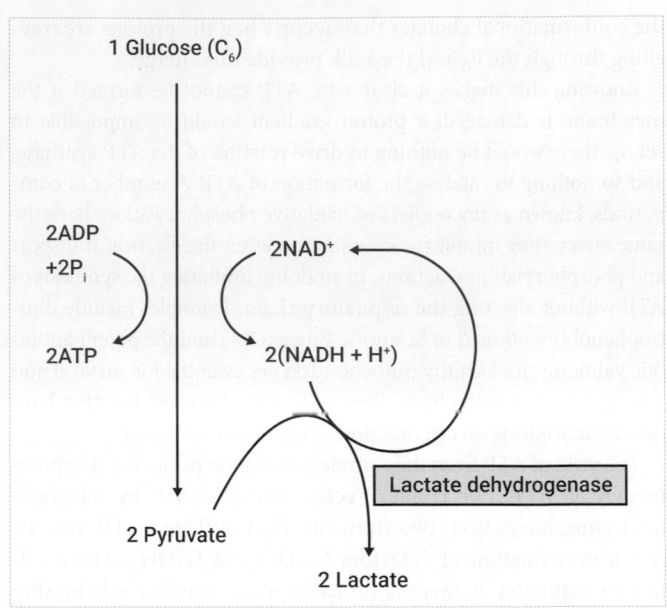

Figure 7.16 The net result of anaerobic metabolism (glycolysis) of glucose.

Source: Snape, A. & Papachristodoulou, D. (2018). *Biochemistry and Molecular Biology* (6th edn). Oxford University Press.

7.5 Photosynthesis

We have seen in previous sections how ATP generation in aerobic cells depends on the electrons present in food, with high energy potential, being transported down the energy scale to end up as the electrons present in the hydrogen atoms in water. If life on Earth

---

CLINICAL BOX 7.1   The Warburg effect: metabolism in cancerous cells

The Warburg effect is a form of modified metabolism in cancer cells, which favours an anaerobic form of metabolism over the normal aerobic metabolism seen in most other cells in the body.

Otto Warburg was awarded the Nobel Prize in Physiology in 1931 for his 'discovery and nature and mode of action of the respiratory enzyme'. He observed that tumour cells mainly generated energy by the conversion of glucose into pyruvate and subsequently into lactate; this was in contrast to 'healthy' cells, which would produce energy by the conversion of glucose into pyruvate and its subsequent aerobic oxidation. This observation is referred to as the 'Warburg effect'.

Warburg's hypothesis—that cancer is caused by non-aerobic metabolism of glucose by tumour cells—was postulated in 1924 and articulated later in a paper entitled 'The prime cause and prevention of cancer', which he presented at a meeting of Nobel laureates in 1966. He proposed that cancer was a result of mitochondrial dysfunction which would not allow the tumour cells to function normally.

While this view has changed in the intervening years, the Warburg effect has re-attracted attention as an approach to cancer detection

and monitoring of treatment, particularly in solid tumours. We now know that the difference in metabolism in tumour cells—anaerobic rather than aerobic—is not the cause of cancer but rather reflects the effects on metabolism of mutations causing cancer. Malignant cells can carry out glycolysis at rates up to 200 times those of normal cells. If glycolysis can be inhibited in these cells, then cancer growth may be halted.

Scientists have been investigating the therapeutic value of Warburg's observations since 2013. Studies using glycolytic inhibitors, such as dichloroacetic acid (DCA) and 2-deoxy-D-glucose (2DG), have shown promising results, destroying cancer cells *in vitro* and also in some animal studies. Alpha-cyano-4-hydroxycinnamic acid has been found in pre-clinical trials to inhibit the activity of proteins called monocarboxylate transporters (MCT), which prevent lactic acid accumulation in tumours. Allowing lactic acid to accumulate in these cells inhibits glycolysis and halts cancer cell growth. Some higher-affinity MCT inhibitors are currently being tested in clinical trials.

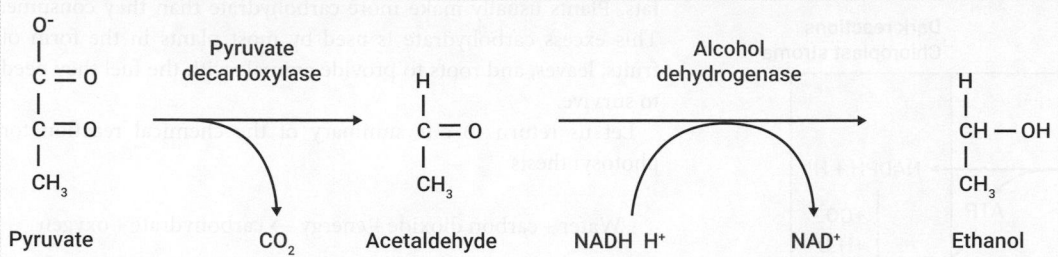

Figure 7.17 The reoxidation of NADH by yeast, through which ethanol is produced from pyruvate.

is to continue, a way must exist to raise these electrons back up the energy scale; if this was a one-way process—an inexorable slide towards low-energy electrons—the continuation of life would be impossible. Apart from life forms found in deep oceans, which use hydrogen sulphide to release energy, **photosynthesis** occurring in plants is the biological process by which electrons from water are 'recycled', producing oxygen and carbohydrate.

Figure 7.18 illustrates the relationship between oxidative phosphorylation and photosynthesis.

We can represent the process of photosynthesis by the following reaction scheme:

$$6CO_2 + 6H_2O \rightarrow C_6H_{12}O_6 + 6O_2$$

Notice how this is the reverse of respiration (summarized in Section 7.1): carbon dioxide and water are consumed to generate glucose and oxygen.

There are two essential requirements for the synthesis of glucose from carbon dioxide and water: a reducing agent of sufficiently high energy, and ATP. The source of reducing power in photosynthesis is NADPH.

Photosynthesis occurs in the **chloroplasts** of the cells of green plants. (In fact, it is the pigment chlorophyll in chloroplasts that give green plants their colour.) Light energy is directly involved

only in transferring electrons from H_2O to $NADP^+$ and in the generation of the proton gradient which drives ATP production.

Chloroplasts resemble mitochondria in that they are membrane-bounded organelles found in the cytosol of cells. Unlike mitochondria, however, they contain another type of membrane-bounded structure: **thylakoids**. They, in turn, contain the light-harvesting pigment **chlorophyll**. Outside the thylakoids is the chloroplast stroma as shown in Figure 7.19.

What follows is an overview of the process of photosynthesis. It is, in fact, not one process but two, each with a number of steps. The two processes are called the '**light reactions**' and the '**dark reactions**', as summarized in Figure 7.20. Look at the figure and notice how the two reactions occur in distinct locations within the chloroplast: 'light reactions' occur in the thylakoid membrane, while 'dark reactions' take place in the chloroplast stroma.

In the light reactions, H_2O is split by the light absorbed by chlorophyll, producing electrons and protons (i.e. hydrogen ions and molecular oxygen). The hydrogen ions are transferred to $NADP^+$, which is closely related to our familiar NAD^+. It is in fact NAD^+ with an additional phosphate group. The light reactions use solar energy—that is, energy from sunlight—to form NADPH. They also result in the generation of ATP by chemiosmosis. This time, the proton gradient is formed across the thylakoid membrane in the process of proton transfer from water to $NADP^+$. This proton gradient results in the production of ATP.

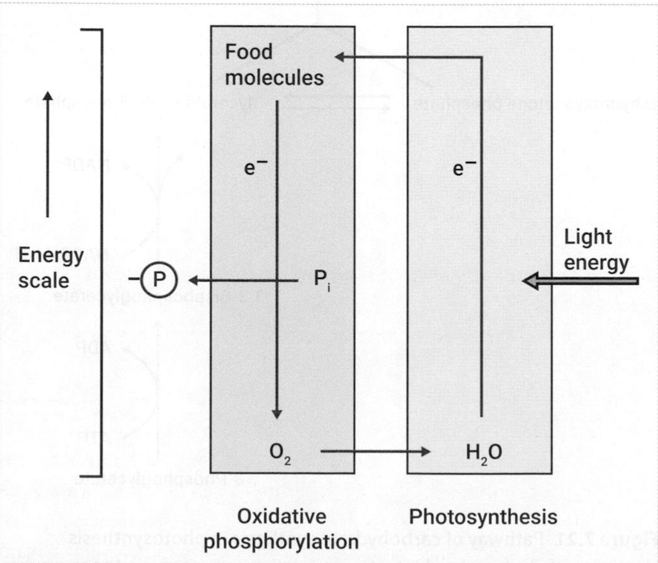

Figure 7.18 Electron cycling between high- and low-energy states by oxidative phosphorylation and photosynthesis.

Source: Snape, A. & Papachristodoulou, D. (2018). *Biochemistry and Molecular Biology* (6th edn). Oxford University Press.

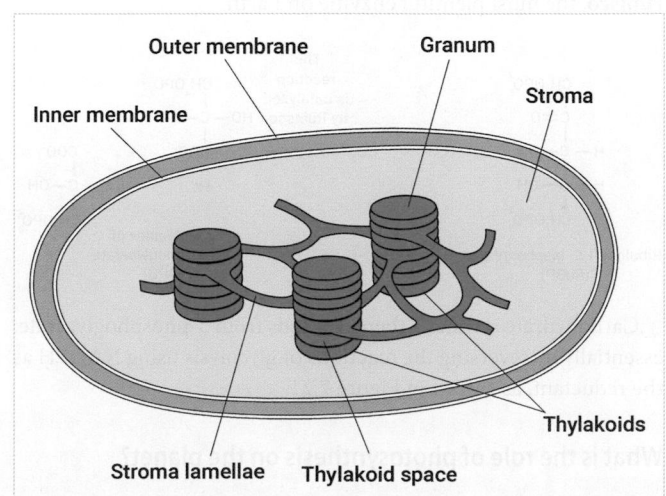

Figure 7.19 A schematic representation of a chloroplast. Grana are stacks of thylakoids.

Source: Snape, A. & Papachristodoulou, D. (2018). *Biochemistry and Molecular Biology* (6th edn). Oxford University Press.

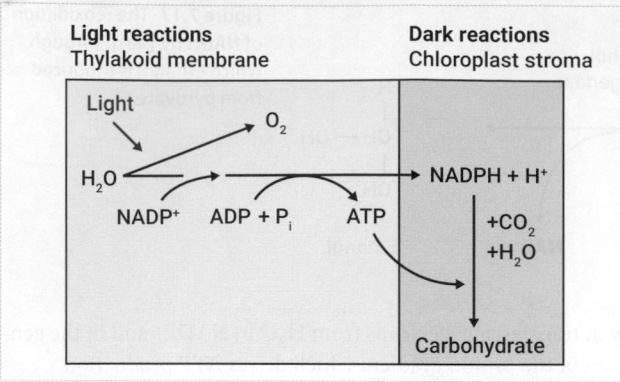

Figure 7.20 The processes of photosynthesis. A summary of the light and dark reactions, and their locations within the chloroplast.

Source: Snape, A. & Papachristodoulou, D. (2018). *Biochemistry and Molecular Biology* (6th edn). Oxford University Press.

So far, no carbohydrate has been produced, but this is driven by the series of reactions known as the 'dark reactions' or the **Calvin cycle**. 'Dark' refers to the fact that light is not needed for these reactions, which result in the conversion of carbon dioxide and water into glucose. This series of reactions takes place in the chloroplast stroma.

PAUSE AND THINK

Do the dark reactions only take place in the absence of light?

Answer: No, dark reactions do take place in bright sunlight. However, the term 'dark' denotes the fact that light is not essential for the reactions to take place, in contrast with the light reactions, which cannot proceed in the absence of sunlight.

The key reaction in the synthesis of carbohydrate from carbon dioxide and water is the production of 3-phosphoglycerate from the sugar ribulose 1,5-bisphosphate by the action of the enzyme **ribulose 1,5-bisphosphate carboxylase/oxygenase**, known as **rubisco**, the most plentiful enzyme on Earth:

CH$_2$OPO$_3^{2-}$
|
C=O
| This reaction is catalyzed by Rubisco.
H—C—OH   +   H$_2$O   +   CO$_2$ ⟶
|
H—C—OH
|
CH$_2$OPO$_3^{2-}$

Ribulose 1,5–bisphosphate
(RuBP)

CH$_2$OPO$_3^{2-}$
|
HO—C—H   H$^+$
|
COO$^-$ +   COO$^-$
|
H$^+$ H—C—OH
|
CH$_2$OPO$_3^{2-}$

2 molecules of
3-phosphoglycerate
(3PG)

Carbohydrate synthesis then proceeds from 3-phosphoglycerate, essentially by reversing the reactions of glycolysis using NADPH as the reductant, as shown in Figure 7.21.

What is the role of photosynthesis on the planet?

The carbohydrate that is synthesized by photosynthesis can supply the plant with chemical energy and the carbon 'skeletons' it needs to produce other materials; these include cellulose, proteins, and

fats. Plants usually make more carbohydrate than they consume. This excess carbohydrate is used by most plants in the form of fruits, leaves, and roots to provide animals with the fuel they need to survive.

Let us return to our summary of the chemical reaction for photosynthesis:

Water + carbon dioxide + energy → carbohydrate + oxygen

It shows that, apart from harnessing the power of the sun to provide animals with food (in the form of carbohydrate), plants are responsible for providing oxygen in the atmosphere, without which life as we know it would not be possible.

⚙ **Check your understanding of the concepts covered in this section by answering the questions in the e-book.**

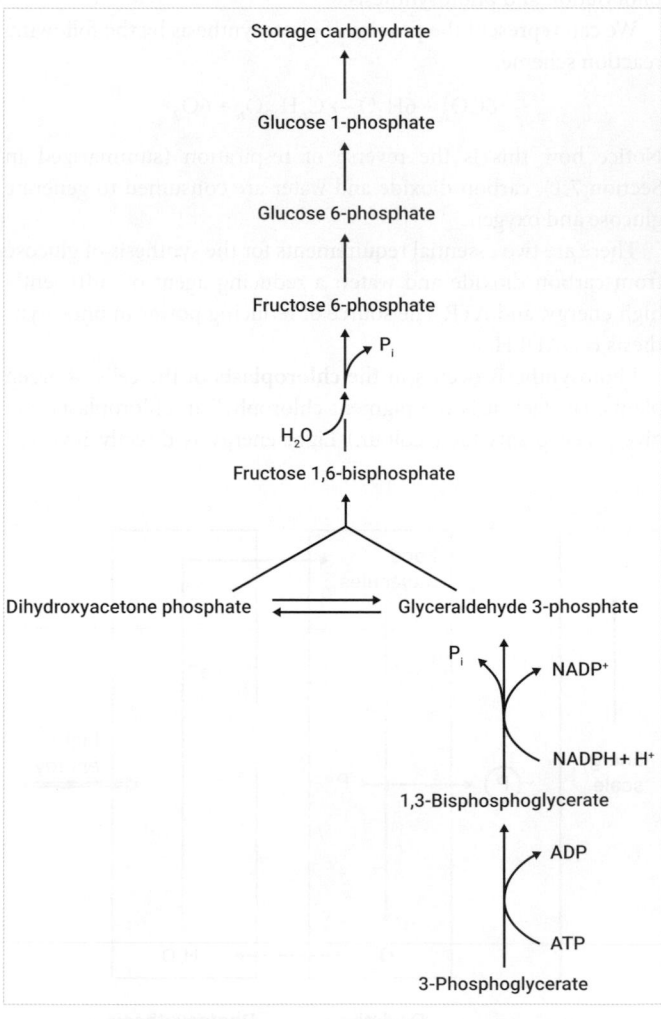

Figure 7.21 Pathway of carbohydrate synthesis in photosynthesis starting with 3-phosphoglycerate. The reactions are essentially the reversal of glycolysis, but using NADPH rather than NADH.

Source: Snape, A. & Papachristodoulou, D. (2018). *Biochemistry and Molecular Biology* (6th edn). Oxford University Press.

SUMMARY OF KEY CONCEPTS

- Biological oxidation involves the removal of electrons and their transfer to another acceptor molecule, which need not be oxygen.

- Biological oxidation commonly involves the enzymatic removal of two hydrogen atoms from a metabolite molecule.

- A variety of electron/hydrogen carriers participate in the transfer of electrons to oxygen.

- NAD^+ is an important electron carrier. It can accept two electrons and a hydrogen atom to form NADH.

- FAD is of a similar structure to NAD^+, but the accepting group is the vitamin riboflavin. It is reduced to $FADH_2$.

- FMN is a single-nucleotide form, which is reduced to $FMNH_2$.

- Glucose oxidation is the complete oxidation of glucose or glucosyl units of glycogen. It occurs in three stages:
  - stage 1, glycolysis, occurs in the cytosol;
  - stage 2, the TCA cycle, occurs in the mitochondrial matrix;
  - stage 3, the electron transport system, is in the inner mitochondrial membrane, the main site of ATP generation.

- Glycolysis causes the lysis of the C_6 glucose molecule into the two C_3 molecules of pyruvate. It occurs in the cytosol.

- Glycolysis produces a net gain of only two ATP molecules but prepares the glucose for the next stage, the TCA cycle. There is also one oxidation step, which reduces NAD^+ to $NADH + H^+$.

- NAD^+ must be reoxidized via the mitochondrion or glycolysis would halt.

- As a pre-step to the TCA cycle, pyruvate is transported into the mitochondrial matrix where it is converted by pyruvate dehydrogenase into acetyl-CoA, which then enters the TCA cycle.

- In a single turn of the TCA cycle, electrons from the acetyl group (plus extra ones originating from water) are transferred to NAD^+ and FAD, and the carbon atoms are removed as CO_2.

- During the cycle only two ATP molecules are produced per molecule of glucose (in the equivalent form of GTP), but fuel is generated in the form of NADH and $FADH_2$, which are used in stage 3.

- The electron transport system is stage 3 of the oxidation of glucose, and generates most of the ATP.

- During stage 3, electrons move along the hierarchy of electron carriers from NADH and $FADH_2$, releasing free energy, which is used to generate a proton gradient across the inner mitochondrial membrane by pumping protons out of the mitochondria.

- The electron carriers are grouped into four complexes. Proton pumping occurs in complexes I, III, and IV but not II.

- The proton gradient is used to drive ATP synthesis by the molecular machine known as ATP synthase in the inner mitochondrial membrane.

- ATP synthase is a minute rotating motor driven by proton flow. The rotation causes conformational changes in its subunits, the energy of which drives the condensation of $ADP + P_i$ to ATP.

- The yield of ATP per molecule of glucose cannot be calculated with absolute precision but approximately 30 molecules are produced from ADP + phosphate.

- There are no mitochondria in prokaryotes, but the cell membrane operates in an equivalent way to the inner mitochondrial membrane for the purposes of ATP synthesis.

- Photosynthesis occurs in plant cell chloroplasts. It comprises light-dependent ('light') reactions and light-independent ('dark') reactions.

- The part dependent on light is the splitting of water to generate NADPH.

- NADPH is used for the reductive synthesis of carbohydrate from CO_2 and water.

- Chlorophyll is a green pigment, which receives light energy. It is present in the membrane of organelles called thylakoids.

- When activated by photons, chlorophyll molecules donate electrons to chains of electron carriers arranged in two photosystems (PSI and PSII). The electrons are finally used to reduce $NADP^+$.

- During passage of electrons from one photosystem to the other, ATP is generated by the chemiosmotic mechanism.

- Carbohydrate is synthesized using NADPH and ATP in the Calvin cycle.

- The key reaction in this synthesis is catalysed by ribulose 1,5-bisphosphate carboxylase/oxygenase (rubisco).

- Rubisco generates 3-phosphoglycerate from which carbohydrate synthesis proceeds by reversal of glycolytic reactions, but using NADPH as reductant.

Use the flashcards in the e-book to test your recall of key terms introduced in this chapter.

QUESTIONS

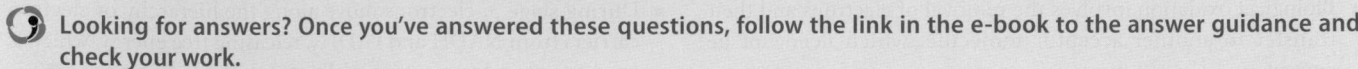

Looking for answers? Once you've answered these questions, follow the link in the e-book to the answer guidance and check your work.

Concepts and definitions

1. What are the three major phases involved in the oxidation of glucose and where do they occur?

2. What is FAD and what is its role?

3. Glycolysis and the TCA cycle produce NADH and FADH₂. What happens to them?

4. What normally happens to acetyl-CoA generated by the pyruvate dehydrogenase reaction?

5. What is the main source of acetyl-CoA apart from the pyruvate dehydrogenase reaction?

Apply the concepts

6. What is achieved by anaerobic glycolysis in muscle? Under what circumstances does it occur?

7. Give a brief account of the complexes that constitute the electron transport system of the inner mitochondrial membrane with particular reference to the creation of the proton gradient across the membrane.

8. The reaction for which the enzyme pyruvate kinase is named never occurs in the cell, i.e. it does not use ATP to phosphorylate pyruvate. Discuss this.

9. The pyruvate dehydrogenase reaction is of great importance in metabolism. Explain why.

10. Can fatty acid be converted into glucose in animals?

Beyond the concepts

11. In calculating how many molecules of ATP are produced as a result of the oxidation of cytosolic NADH we cannot be sure of the exact answer in eukaryotes. Why is this?

12. Proton pumping due to electron transport in photosystem II causes movement of protons from the outside of thylakoids to the inside. In mitochondria, protons are pumped from the inside to the outside. Comment on this.

13. The active sites in the F₁ subunits of ATP synthase are said to be cooperatively interdependent. What does this mean?

14. What is the difference between an inhibitor of the electron transport chain and an uncoupler of oxidative phosphorylation? Give examples.

15. Brown adipose tissue has a naturally occurring uncoupler. Name it and explain its function. And what could be the use of an artificially made uncoupler?

FURTHER READING

Mitchell, H. (1998) The mechanism of proton pumping by cytochrome c oxidase. *Proc. Natl. Acad. Sci. USA* **95**: 12819–24.

The is article outlines a detailed model of the coupling of individual electron transfer steps to proton pumping. It analyses the steps involved in the reduction of oxygen to water that is accompanied by pumping of four protons across the mitochondrial or bacterial membrane.

Fillingame, R. H. (1999) Molecular rotary motors. *Science* **286**: 1687–8. DOI: 10.1126/science.286.5445.1687.

This article discusses an elegant structural model of how ATP synthase works: it behaves like a tiny molecular rotary motor, coupling the mechanical force of an electrochemical proton gradient to the formation of the chemical bond between ADP and P.

Itoh, H., Takahashi, A., Adachi, K., et al. (2004) Mechanically driven ATP synthesis by F₁-ATPase. *Nature* **427**: 465–8.

This article describes an elegant experiment showing how the rotation of the 'stalk' of the F₁ component using a magnetic bead could lead to ATP synthesis without the need for a proton gradient.

Molecular Tools and Techniques

LEARNING OBJECTIVES

By the end of this chapter, you should be able to:

- Describe the use of restriction enzymes to cut DNA at defined sequences.
- Explain how electrophoresis is used to analyse DNA.
- Describe how the polymerase chain reaction (PCR) is used to amplify selected DNA sequences.
- Describe how DNA is ligated into vectors and how DNA libraries are produced.
- Explain how DNA is sequenced using the chain-termination method and next-generation sequencing methodology.
- Describe the use of cDNA and RT-PCR to analyse transcripts.
- Describe how cloned cDNA is used for large-scale protein production in bacterial systems.
- Describe how gene targeting is used to produce knockout mice.
- Describe how genes are edited using the CRISPR-Cas9 system.
- Outline how genomics and bioinformatics approaches using DNA sequence data are applied.

Chapter contents

Introduction — 343

8.1 Basic methodologies using DNA — 344

8.2 Applications of recombinant DNA technology — 353

8.3 Genomics and bioinformatics complement recombinant DNA technology — 359

Watch the key concepts video in the e-book to prepare yourself for studying this chapter.

Introduction

From the early days of the Covid-19 pandemic in the spring of 2020 we became used to daily updates on the number of individuals testing positive for the Covid-19 virus. Graphs of these positive cases quickly became a powerful visual indicator of the progress of the pandemic. Before the introduction of lateral flow tests in the UK in late 2020, PCR tests were the mainstay of the government's testing regime. The 'PCR' in PCR tests is a reference to the polymerase chain reaction—a method introduced during the 1980s that allows the rapid *in vitro* production of billions of copies of a specific-DNA sequence.

The PCR tests that became so commonplace during the Covid-19 pandemic are just the latest in a series of tools that have been developed to harness our understanding of molecular biology. Since the 1970s, important advances in many areas of biological and medical science have sprung from scientists' ability to manipulate DNA. Following the publication of DNA structure in 1953, researchers made rapid progress in discovering and purifying naturally occurring enzymes such as DNA polymerase and DNA ligase that can be co-opted for use in the laboratory. We learn more about DNA polymerase and ligase in Chapter 3. Significant papers published during the 1970s demonstrated the capacity of viruses and plasmids to incorporate DNA from a range of sources, replicate it within bacteria, and even produce functional proteins from 'foreign' DNA.

After a short pause to establish the safety of this recombinant DNA technology, the discipline of gene cloning took off. **Recombinant DNA** refers to DNA that is made by covalently linking DNA sequences from different sources, while **gene cloning** or molecular cloning refers to the ability to make multiple copies of defined DNA sequences.

Other advances that contributed to the rapid expansion and increasing power of DNA technology and genetic engineering include the use of restriction enzymes to cut DNA at defined sequences, and the ability to acquire and analyse large quantities of DNA and RNA sequence data rapidly and with a high degree of accuracy.

You will find references to the use of DNA technology and DNA sequence data throughout this book and increasingly in your everyday life. In this chapter we provide an overview of the techniques that are widely used to analyse and manipulate DNA. We help you to understand how some of the key methodologies work and how they can be used experimentally in the biotechnology industry and in medical science.

8.1 Basic methodologies using DNA

Cellular DNA molecules are extremely large. The *E. coli* chromosome contains 4.6 million base pairs, while the smallest human chromosome is more than 10 times larger: it contains 48 million base pairs. These sizes are too large for intact chromosomes to be handled easily in the laboratory. Instead, a key requirement for working with DNA is to obtain fragments that are small enough to handle—generally a few thousand base pairs. A second requirement is to replicate these DNA fragments to obtain multiple copies for analysis and manipulation. In this section we explore some basic techniques that enable scientists to achieve these first steps.

Restriction enzymes cut DNA at specific sequences

Restriction enzymes, also known as restriction endonucleases, are bacterially produced enzymes that cut double-stranded DNA at specific sequences by cleaving the phosphodiester bond between nucleotides. These enzymes form part of the bacterium's defence system against 'foreign' DNA, such as viral genomes, but we can also use them in the laboratory to cut cellular DNA into pieces of a manageable size, a process often called restriction digestion.

We discuss the structure of nucleic acids, including the location of phosphodiester bonds, in Section 1.7.

Restriction enzymes generally recognize sequences four to eight base pairs in length, which are typically palindromic (i.e. they read the same 'forwards' on one strand as 'backwards' on the complementary strand). Figure 8.1 shows the DNA recognition sequences of some restriction enzymes. Each enzyme is derived from a distinct bacterial species, and this is reflected in the naming convention for the enzymes. For example, *Eco*R1 (pronounced 'Ee-co-are-one') is derived from *Escherichia coli*.

A feature of many restriction enzymes is the production of a 'staggered' cut in the DNA, leaving short overhanging single-stranded ends. Of the examples shown in Figure 8.1b three enzymes produce overhanging ends and one, *Hpa*I, produces 'blunt' ends. As we will see later, the overhanging ends give a useful degree of control when ligating (joining) DNA from different sources.

PAUSE AND THINK

A scientist wishes to cut a bacterial genome into fragments for further study. Which restriction enzyme do you predict would give longer fragments, *Alu*I, which recognizes and cuts a four base pair sequence, or *Not*I, which recognizes and cuts an eight base pair sequence?

Answer: Not I. A specific four-base sequence will be present by chance more frequently in a DNA molecule than a specific eight base sequence, so Not I will have fewer target sites in the bacterial genome and will therefore cut it into longer fragments.

DNA fragments can be separated by electrophoresis

Once molecules of DNA of a suitable size have been produced, it is often useful to separate them according to size, either as part of further analysis or to obtain a pure sample of a particular fragment for further manipulation. Electrophoresis separates molecules by exploiting the fact that their rate of movement across an electric field depends on their size and charge. In the case of DNA, each phosphate group carries a negative charge and the regular repeat of the phosphate groups means that all DNA molecules have the same charge to mass ratio. Therefore, separation by size is achieved by the movement of the molecules through what we call a 'gel'— an inert porous material. Small molecules can migrate quickly through the pores in the gel, while larger ones are impeded and move more slowly. This phenomenon is often described as 'molecular sieving'.

For larger DNA molecules the gel is made from agarose, a polysaccharide that is purified from seaweed. For molecules of less than

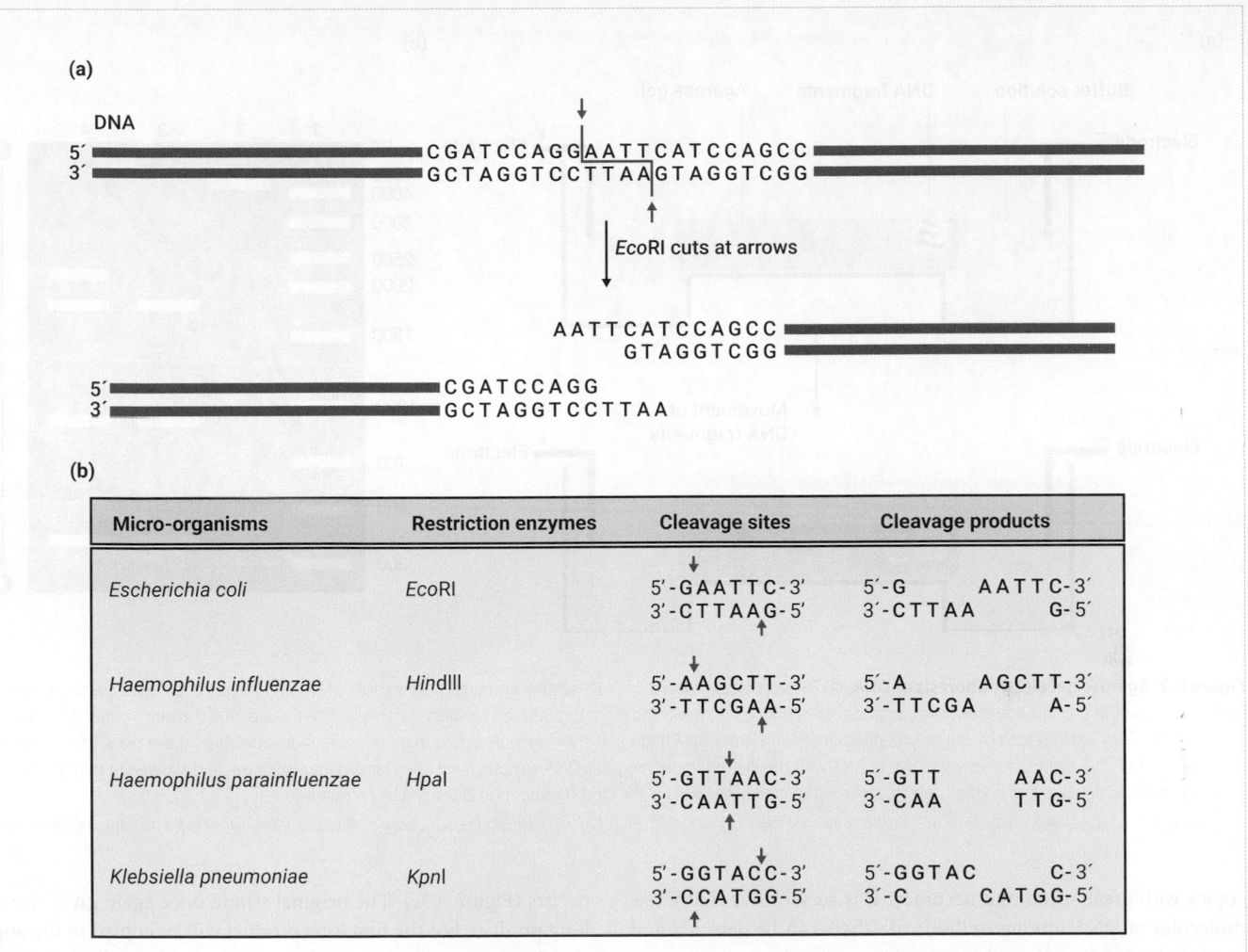

Figure 8.1 DNA recognition sequences. (a) Restriction enzymes recognize specific DNA sequences and cut them at defined points. In the example shown, *Eco*R1 recognizes the palindromic sequence 5'GAATTC3' and cuts immediately after the G on both strands of the DNA, leaving 5' overhanging ends on both strands. (b) Examples of restriction enzymes produced by different bacteria. The enzymes recognize different palindromic sequences and leave different types of overhang. *Eco*R1 and *Hind*III both leave 5' overhangs, *Kpn*I leaves 3' overhangs, and *Hpa*I leaves blunt ends.

Source: Craig et al. (2021). *Molecular Biology* (3rd edn). Oxford University Press.

1000 base pairs, polyacrylamide is used. Polyacrylamide forms smaller pores than agarose and can achieve separation of small fragments that differ in length by just one base pair.

The principle of separation by gel electrophoresis is illustrated in Figure 8.2a, while Figure 8.2b shows an image of an agarose gel after electrophoresis. Here the fragments have been visualized using a dye, ethidium bromide, which makes the DNA fluoresce under ultraviolet (UV) light. As well as obtaining an estimate of the size of the DNA fragments in Figure 8.2b, it is possible to obtain a pure sample of a particular fragment for further use by cutting out of the gel the section (the 'band') containing the fragment of interest and purifying the DNA away from the agarose.

PAUSE AND THINK

Estimate the sizes (to within 100 base pairs) of the DNA fragments in lane 3 of the gel shown in Figure 8.2b.

Answer: 1800 bp, 1100 bp, 900 bp, and 400 bp.

DNA sequences can be amplified using PCR

The PCR technique was developed during the 1980s. It has assumed enormous importance because it allows a sequence of DNA to be selected and replicated *in vitro*, making millions of

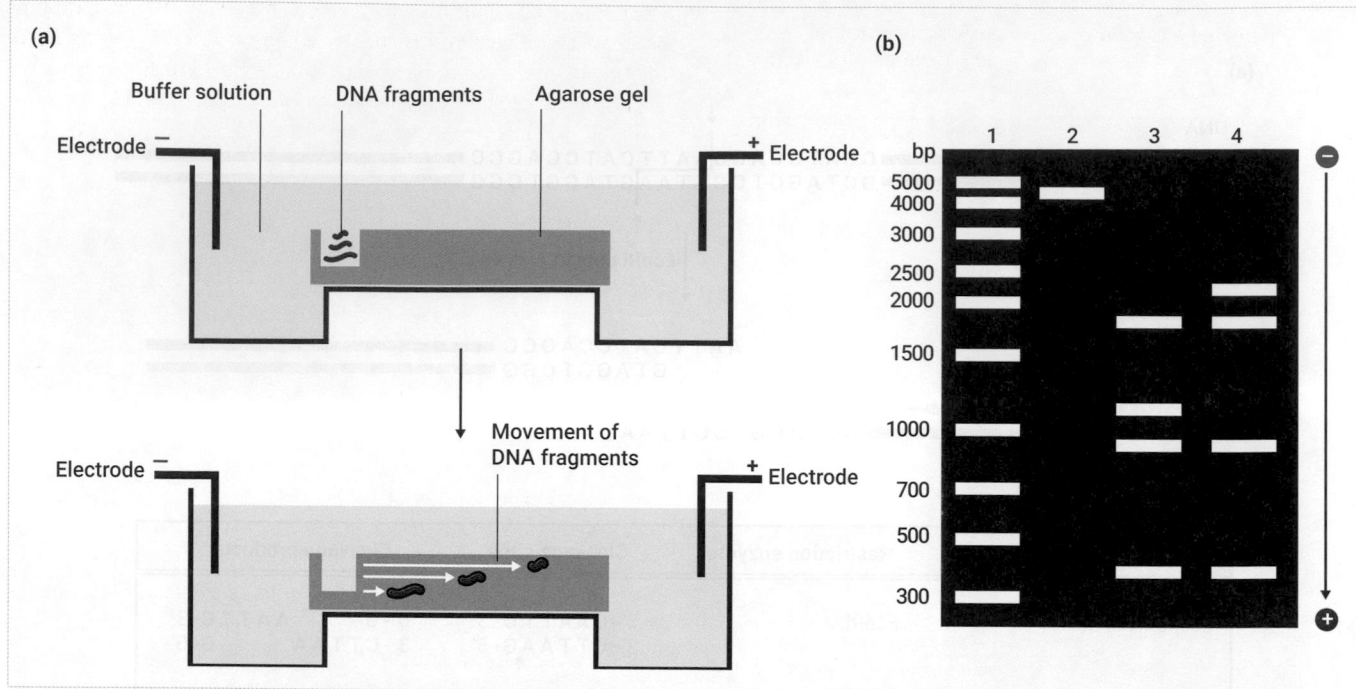

Figure 8.2 Agarose gel electrophoresis of DNA. (a) The apparatus is shown in cross-section and side on. A mixture of DNA fragments of different sizes has been loaded into a well at the cathode (negative electrode) end and the fragments move through the gel towards the anode (positive electrode) when a voltage is applied across the gel. They are impeded by the polymeric agarose, with larger DNA fragments moving more slowly than the more manoeuvrable smaller ones. (b) Illustration of a typical result of DNA electrophoresis, visualizing DNA with the dye, ethidium bromide. DNA samples have been loaded into lanes 1–4 and allowed to migrate. Lane 1 contains DNA fragments of known sizes, which allows estimation of the sizes of the DNA fragments in lanes 2–4 by comparison.

Source: Snape, A. & Papachristodoulou, D. (2018). *Biochemistry and Molecular Biology* (6th edn). Oxford University Press. (b) Craig et al. (2021). *Molecular Biology* (3rd edn). Oxford University Press.

copies with great speed and accuracy. It is so sensitive that a few molecules of DNA among millions of others can be detected and amplified exponentially within a few hours.

The basic PCR process is illustrated in Figure 8.3. An essential requirement of this process is that sequences flanking each end of the required section are known, so that complementary single-stranded DNA **primers** can be made. Primers are synthesized chemically and are typically around 20 nucleotides in length, sufficiently long to enable them to base pair (anneal) via hydrogen bonding with great specificity to the complementary sequence in the target DNA.

To describe the method, we will consider a stretch of DNA that we want to amplify. This is shown in yellow in Figure 8.3. First, the parent strands are separated by heating (Figure 8.3b). For simplicity, in Figure 8.3b onwards we only show what happens for *one* of the two parent strands since the same process occurs for both.

After strand separation, the two chemically synthesized DNA primers are added in large excess compared with the number of template molecules, because each round incorporates the primers into a newly made strand. The primers anneal to the complementary sequence on the template strand when the mixture is cooled (Figure 8.3c), and then replication to the end of the template occurs, producing a 'long' product (Figure 8.3d). That is the end of the first cycle.

The next cycle is started by heating to separate the strands again. After cooling, primers attach to both strands and replication occurs (Figure 8.3e). The original strand once again gives rise to a long product, but the first long product will be copied in the opposite direction to produce a 'short product' (Figure 8.3f) which is the sequence we wanted to amplify. At the end of the second cycle we therefore have the original strand, two long products, and one short product.

The third cycle is started by heating, and primers again attach to the separated strand on cooling (Figure 8.3g). Replication now produces a double-stranded version of the short product (Figure 8.3h). From now on the number of short products increases exponentially with each cycle but the long products increase only arithmetically. The long products therefore become a negligible proportion of the total and do not interfere with any further use of the amplified section of DNA.

The entire process is automated with the heating and cooling performed for multiple cycles (typically 20–30), each cycle taking only a few minutes. Amplification by millions-fold is easily achieved as 2^n copies are made, where n = number of cycles.

As well as the template DNA and primers, the incubation mixture contains the deoxynucleoside triphosphates (dATP, dCTP, dGTP, and dTTP) that are required to synthesize the new DNA molecules, and DNA polymerase, the enzyme that catalyses DNA replication. The DNA polymerase must be heat stable so that it can remain active throughout multiple cycles of heating and cooling and does not need replenishing. Heat-stable DNA polymerases are derived from bacterial species that survive in high

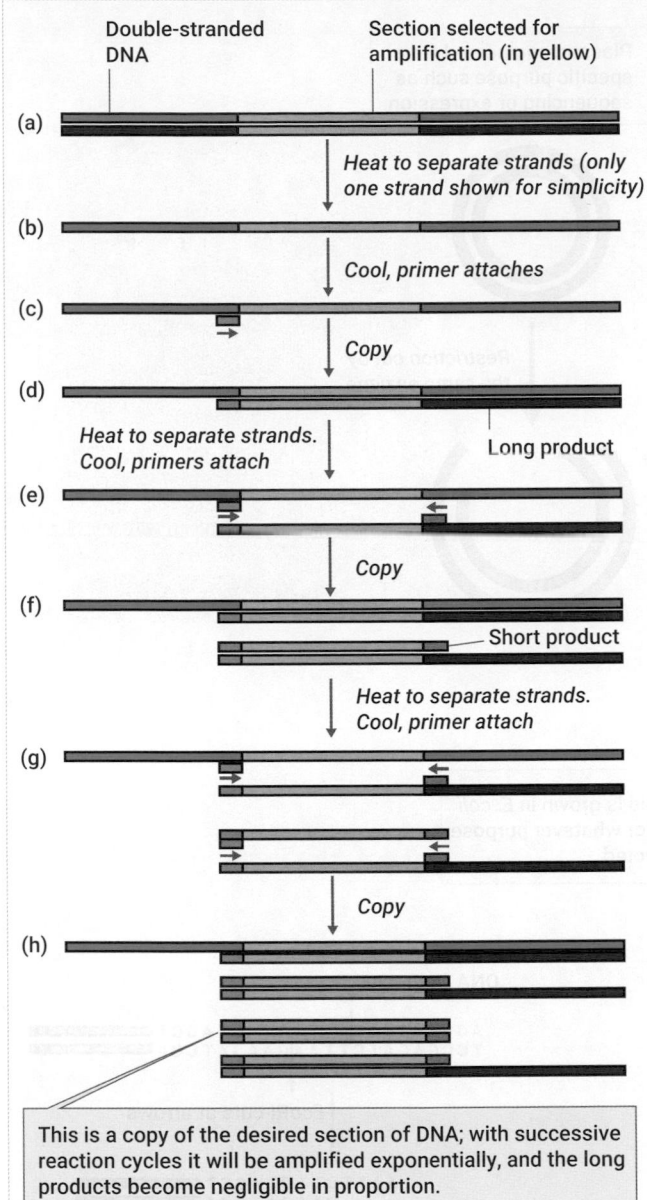

Figure 8.3 Amplification of a DNA section by PCR. Incubations contain four dNTPs, a heat-stable DNA polymerase, and primers. (a) The yellow bars represent the section chosen for amplification by using appropriate primers. (b) Heating separates the strands. From this point, to keep the diagram to a manageable size, amplification of only one strand is shown, although the process amplifies both strands. (c) Cooling allows the primer appropriate for this strand to anneal. (d) DNA is replicated (in this case beyond the end of the desired piece). (e) Heating separates the strands and after cooling new primers anneal. (f) With this round of replication, the desired short product is made. (g) and (h) With further rounds of heating, cooling, and replication, many double-stranded copies of the desired short product are made.

Source: Snape, A. & Papachristodoulou, D. (2018). *Biochemistry and Molecular Biology* (6th edn). Oxford University Press.

temperature environments, such as volcanic hot springs. For example, Taq polymerase from *Thermus aquaticus* is often used. Primers and dNTPs are also provided in excess so that they do not need adding during the reaction cycles.

PCR has multiple uses in molecular biology. We will explore some of them later in this chapter when we discuss cloning DNA and next-generation sequencing, while in Real World View 3.2 we describe the PCR-based amplification of polymorphic genome sequences in DNA profiling.

PAUSE AND THINK

How many copies of a DNA sequence will be made after 25 PCR cycles?

Answer: $2^{25} = 3.3 \times 10^7$ *or 33 million.*

Creating recombinant DNA molecules: joining DNA by ligation

Creating DNA molecules of a manageable size through restriction digestion or PCR is often the first step in creating recombinant DNA by joining together DNA from different sources. For example, Figure 8.4 shows how a DNA sequence may be inserted into a plasmid vector, enabling it to be replicated within bacterial host cells. The circular plasmid has been linearized by cutting with a restriction enzyme, and the DNA being added (the 'insert') has overhanging ends that are complementary to those created in the linearized plasmid. The overhanging ends base pair (anneal) with each other via hydrogen bonding, while the enzyme **DNA ligase** must be added to the reaction to catalyse the formation of phosphodiester bonds to covalently link the sugar–phosphate backbones. The reaction that joins the DNA molecules is called a ligation, after the ligase enzyme.

Once a section of DNA has been ligated into a plasmid vector the plasmid can be inserted into an *E. coli* host cell, through a process called transformation. The plasmid replicates within the host cell and the bacterial host cell also replicates rapidly by cell division, greatly increasing the number of copies of the DNA sequence. We refer to the production of multiple copies of a DNA sequence as 'cloning', and the replicated DNA sequence of interest as a 'clone'.

Cloned DNA is stored in libraries

Plasmid vectors containing inserts provide a convenient way to store cloned DNA, as they can be frozen for long-term storage or propagated within bacterial host cells. A collection of DNA clones is referred to as a **library**. Figure 8.5 outlines the steps involved in constructing a genomic library: a collection of clones that include all the sequences in an organism's genome.

PAUSE AND THINK

If the average size of a cloned insert is 4000 base pairs, how many clones would a genomic library need to contain to hold the full genome of the yeast species, *Saccharomyces cerevisiae* (genome size 12×10^6 base pairs)? How many clones would be needed to hold the full human genome (3.2×10^9 base pairs)?

Answer: The *Saccharomyces cerevisiae* genomic library would require 3000 clones. The human genomic library would require 800,000 clones—too many to be practical.

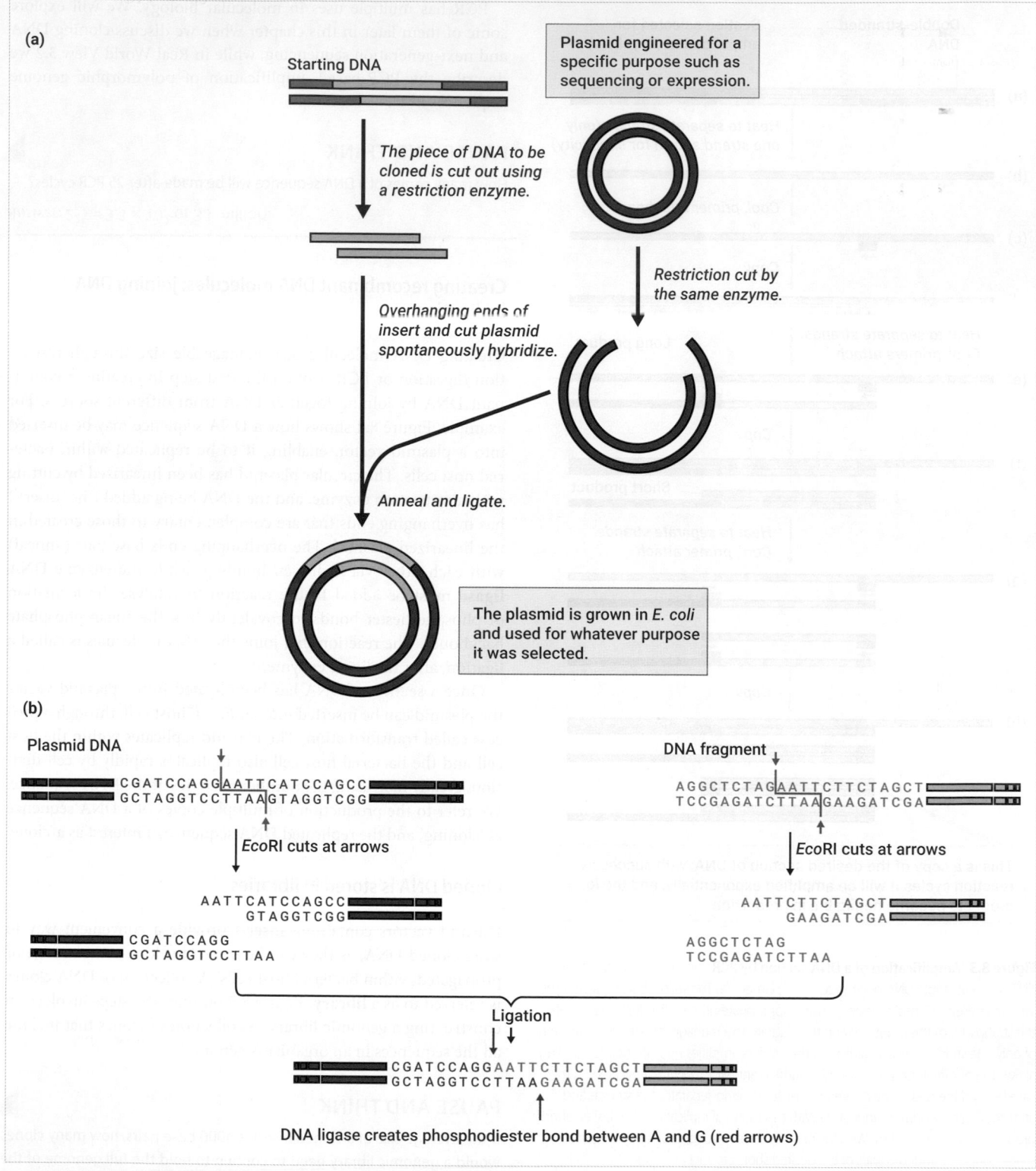

(a)

Starting DNA

Plasmid engineered for a specific purpose such as sequencing or expression.

The piece of DNA to be cloned is cut out using a restriction enzyme.

Restriction cut by the same enzyme.

Overhanging ends of insert and cut plasmid spontaneously hybridize.

Anneal and ligate.

The plasmid is grown in *E. coli* and used for whatever purpose it was selected.

(b)

Plasmid DNA

    CGATCCAGGAATTCATCCAGCC
    GCTAGGTCCTTAAGTAGGTCGG

*Eco*RI cuts at arrows

    AATTCATCCAGCC
        GTAGGTCGG

    CGATCCAGG
    GCTAGGTCCTTAA

DNA fragment

    AGGCTCTAGAATTCTTCTAGCT
    TCCGAGATCTTAAGAAGATCGA

*Eco*RI cuts at arrows

    AATTCTTCTAGCT
        GAAGATCGA

    AGGCTCTAG
    TCCGAGATCTTAA

Ligation

    CGATCCAGGAATTCTTCTAGCT
    GCTAGGTCCTTAAGAAGATCGA

DNA ligase creates phosphodiester bond between A and G (red arrows)

Figure 8.4 Insertion of DNA into a plasmid vector. (a) Method of constructing a recombinant plasmid for cloning a DNA insert. (b) Detail of annealing and ligation of the plasmid and insert DNA. In this example both the plasmid and the insert have been cut with the restriction enzyme *Eco*R1, creating complementary overhanging ends that anneal by base pairing. DNA ligase creates the phosphodiester bonds that link the sugar–phosphate backbones.

Source: (a) Snape, A. & Papachristodoulou, D. (2018). *Biochemistry and Molecular Biology* (6th edn). Oxford University Press. (b) Craig et al. (2021). *Molecular Biology* (3rd edn). Oxford University Press.

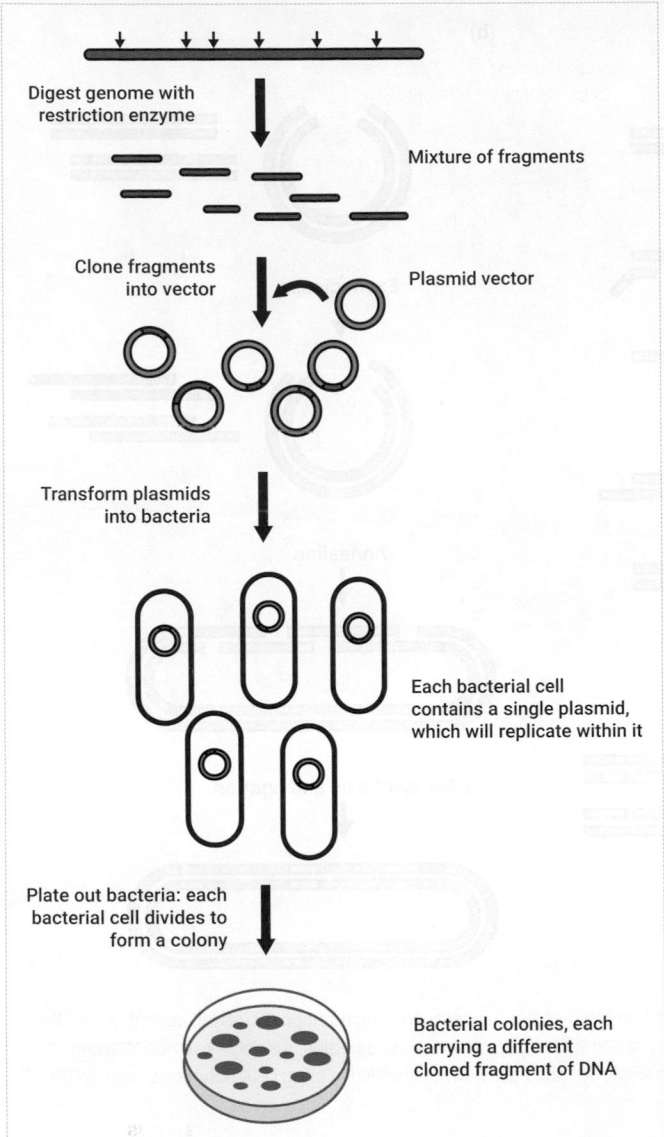

Figure 8.5 Steps in constructing a genomic library. The genomic DNA is cut into fragments of a few thousand base pairs by using a restriction enzyme. The fragments are then ligated into a plasmid vector, and the plasmids are transformed into *E. coli* bacterial host cells, with each cell picking up a single plasmid. The transformed *E. coli* can be propagated on nutrient plates as shown. Each bacterial colony on the plate results from division of a single transformed *E. coli* cell, so represents a single clone. The aim is for every sequence in the genome to be represented in the total library of clones.

Source: Cooper, *The Cell*, 8th edn. Oxford University Press.

Plasmid vectors can accommodate inserts up to a maximum size of approximately 10,000 base pairs. This limit on insert size means that libraries with an unmanageable number of clones would be needed to accommodate large genomes such as the human genome. Fortunately, alternative vectors that allow larger insert sizes can be used. For example, viral vectors based on bacteriophage lambda can hold up to 20,000 base pairs, while for even larger inserts of 500,000 base pairs or longer, yeast artificial chromosomes (YACs) and bacterial artificial chromosomes

(BACs) can be used. Libraries produced in BACs were used to clone the human genome in preparation for its sequencing.

PCR can be used to adapt DNA for cloning

In recent years, PCR-based techniques have added flexibility and efficiency to methods for cloning DNA. An example is illustrated in Figure 8.6. In Figure 8.6a additional short 'adapter' sequences are added to the ends of two DNA molecules using PCR primers. In Figure 8.6b a technique called 'Gibson assembly' (after one of the inventors) is used to simultaneously insert both DNA molecules in a defined order into a plasmid vector. The exonuclease step in Gibson assembly uses an enzyme that digests a few nucleotides DNA from the 3′ ends of the DNA inserts and vector to create complementary overhanging ends.

DNA sequencing

The ability to 'read' the base sequence of DNA is crucial to our growing understanding of the structure and function of genes and genomes. Sequence data are used in many fields, from medical science to biotechnology and evolutionary biology. The full genome sequences of humans, and an increasing number of other species, have been determined and are freely available to researchers. Following the publication of the 'reference' human genome sequence in 2003, we have been able to use genome sequence information and the comparison of genomes from many individuals to understand more about our human ancestry and the impact of genetic variation on many aspects of our health.

For many years, the standard method for determining DNA sequence was based on the method developed by Frederick Sanger and colleagues in the 1970s. This is the **chain-termination method**, often referred to as **Sanger sequencing**. The Human Genome Project, which led to the publication of the first full draft of the human genome in 2001, was a major impetus for increasing the speed and decreasing the cost of sequencing DNA. Many of these 'next-generation sequencing' technologies were based on the same principles as Sanger sequencing, but increasingly novel methods have been developed. We cannot cover all these here but will explain the basis of sequencing using the chain-termination method and give a short account of some of the new techniques. We will also consider strategies for compiling sequence data to give full genome sequences.

The principles of sequencing by the chain-termination method

As illustrated in Figure 8.7, the chain termination method depends on the DNA sequence of interest being copied by DNA polymerase to synthesize a new DNA chain. The reaction contains a mixture of the four deoxynucleotides that are normally required for DNA synthesis (dATP, dGTP, dCTP, and dTTP) plus a small proportion of dideoxynucleotides (ddATP, ddGTP, ddCTP, and ddTTP). The structure of a dideoxynucleotide (ddNTP) differs from a deoxynucleotide (dNTP) in that it has a hydrogen, rather than a hydroxyl group, linked to the 3′ carbon of the sugar ring, as illustrated in Figure 8.7a. This means that no further elongation is possible once DNA polymerase has incorporated a ddNTP into a growing DNA

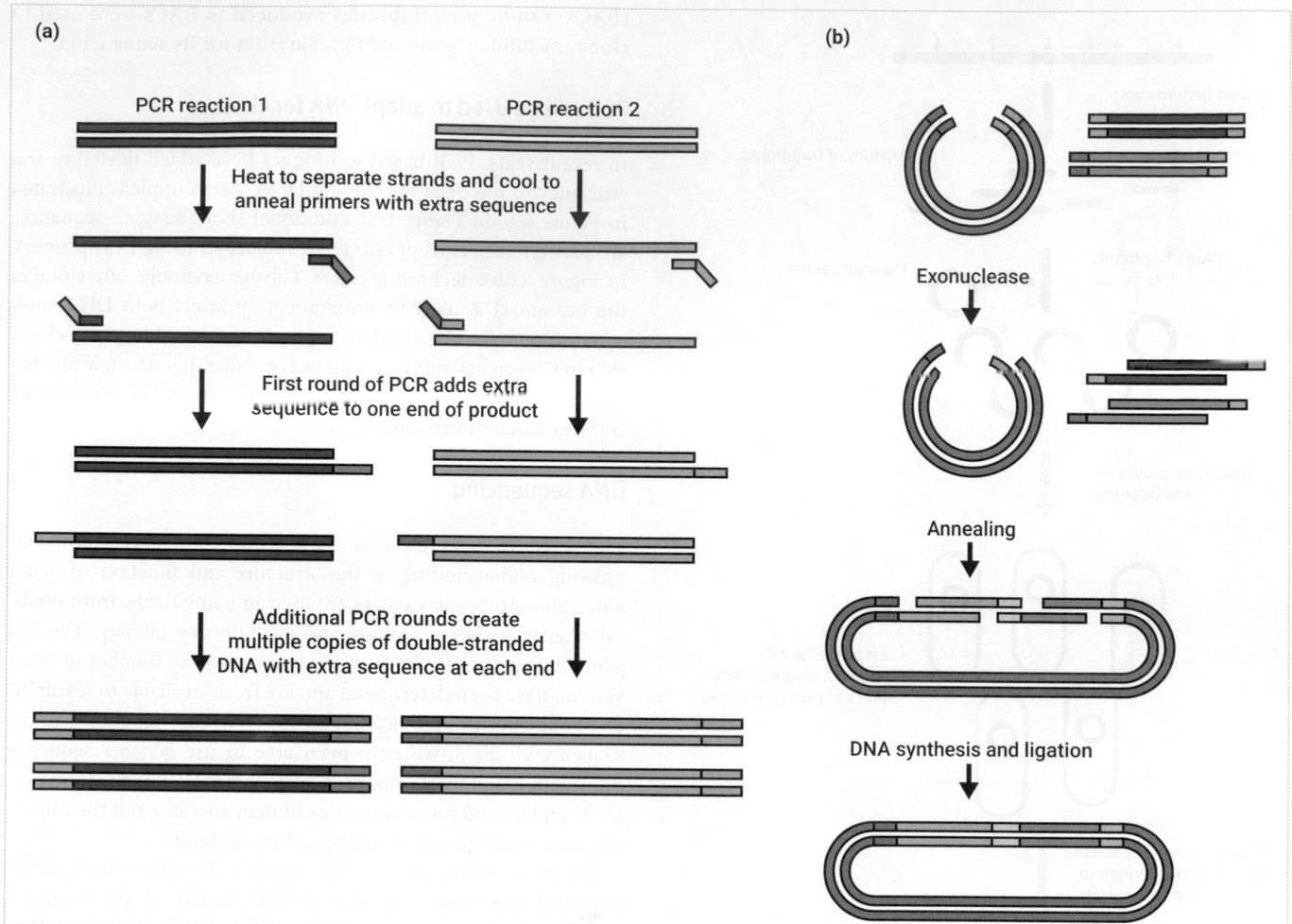

Figure 8.6 Cloning DNA. (a) PCR can be used to add defined 'adapter' sequences to DNA molecules to aid subsequent cloning steps. (b) Gibson assembly: the DNA molecules created in (a) are inserted into a plasmid vector. All the DNA molecules are mixed and subjected to exonuclease digestion in a single reaction, creating overhanging ends. The DNA fragments are then allowed to anneal to each other, leaving single-stranded gaps, which are filled in using DNA polymerase and dNTPs. DNA ligase then creates phosphodiester bonds.

Source: (b) Craig et al. (2021). *Molecular Biology* (3rd edn). Oxford University Press.

chain because the next incoming nucleotide cannot bond to the hydrogen on the 3′ carbon.

The products of a sequencing reaction are shown in Figure 8.7c. Multiple copies of the template strand are made by the DNA polymerase. If we consider the addition of the first nucleotide after the primer sequence (a G added opposite a C in the template strand), ddGTP will be added in a proportion of the reactions, and chain elongation will proceed no further. In other reactions, however, dGTP will be added, allowing elongation to proceed with addition of the next nucleotide, which again may be a dNTP, allowing further elongation or a ddNTP leading to termination, and so on. The result is a set of replicated molecules with lengths progressively growing in a step of one nucleotide.

Figure 8.7d shows the results of a manual sequencing experiment analysed using an electrophoresis gel. While manual sequencing is rarely carried out these days, being able to interpret the gel helps to understand the methodology as a whole. To produce these results, four separate sequencing reactions were carried out using the same template sequence that is shown in Figure 8.7c, although the

primer sequence is not represented in Figure 8.7d. Each sequencing reaction contains all four dNTPs but contains just one of the four ddNTPs. The sequencing reaction containing ddATP produces DNA fragments of three different sizes, which are seen in the left-hand lane on the electrophoresis gel. In the next lane we see the products of the reaction containing ddGTP, four fragments, and so on for the reactions containing ddCTP and ddTTP.

In manual sequencing, the DNA products can be visualized in the electrophoresis gel by incorporating radioactive versions of the nucleotides. Advances that made more rapid, automatic sequencing possible are illustrated in Figure 8.7b, e, and f. Here the ddNTPs are each 'tagged' with a different coloured fluorescent dye, so that all four ddNTPs can be added to a single sequencing reaction. The products are analysed by gel electrophoresis as before, but this time the gel is held in a narrow capillary tube and each product is detected using a laser beam, which gives a distinguishable signal for each fluorescent molecule that passes through it. The base sequence is displayed as a series of coloured peaks with the written sequence above, as shown in Figure 8.7f.

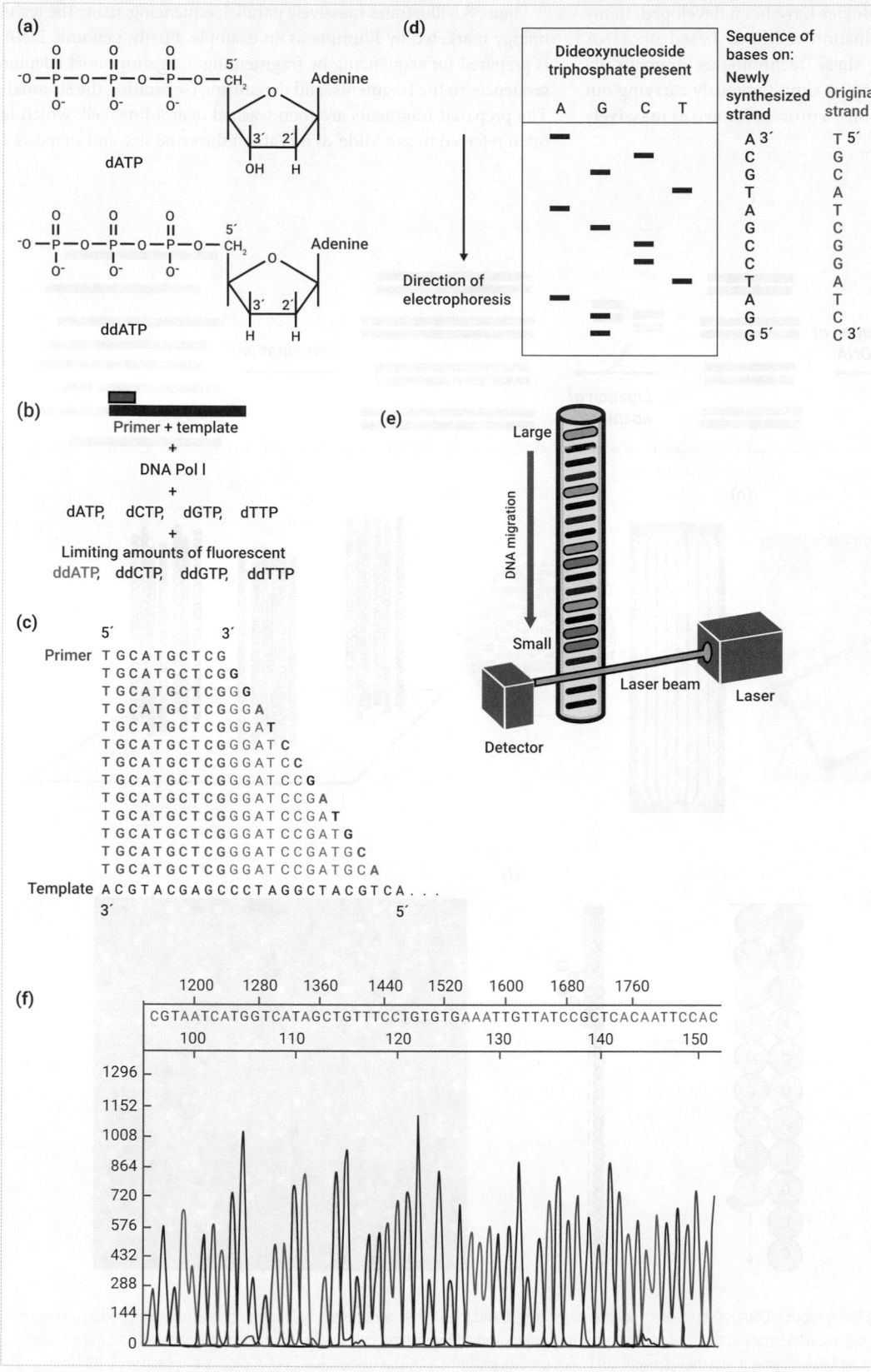

Figure 8.7 Chain-termination DNA sequencing using ddNTPs. (a) Comparison of the structures of dATP and ddATP. ddATP has a hydrogen atom rather than a hydroxyl group attached to the 3′ carbon. (b) A sequencing reaction. A primer is annealed to the DNA molecule whose sequence is being determined, and DNA synthesis is carried out by the DNA polymerase enzyme. dNTPs are provided for new DNA synthesis, together with a small proportion of ddNTPs. (c) Products of DNA synthesis, each terminated by one of the ddNTPs. (d) Products of a manual sequencing experiment. Four separate synthesis reactions are carried out, with a different ddNTP in each reaction. The reactions also contain all four dNTPs, one or more of which is radioactive to enable detection of the products. The products are separated by size using gel electrophoresis, allowing the sequence to be 'read' from the gel. (e) Products of automated sequencing are analysed using capillary electrophoresis. A single sequencing reaction contains all four ddNTPs, distinguishable through 'tagging' with fluorescent dyes, which are detected as they pass a laser beam.
(f) An example of the output of an automated sequencing reaction using fluorescent ddNTPs. Note that this is unrelated to the example sequence shown in (b)–(e).

Source: (a, d, f) Snape, A. & Papachristodoulou, D. (2018). *Biochemistry and Molecular Biology* (6th edn). Oxford University Press. (b, c, e) Craig et al. *Molecular Biology* (2nd edn). Oxford University Press.

Next-generation sequencing technologies make sequence acquisition faster and cheaper

The development of automated sequencing made it possible for the international Human Genome Project consortium to complete the sequence of the entire 3.2×10^9 bases of human DNA in 2003. Since then, the focus of human genomic studies has switched to sequencing individual genomes. For this to be a practical proposition faster and cheaper methods were needed.

Several next-generation technologies have been developed, many of them (like the chain-termination method) based on DNA synthesis reactions. However, these technologies dramatically increase the speed of sequencing by simultaneously carrying out and analysing millions of reactions, a process known as **massively parallel sequencing**.

Figure 8.8 illustrates massively parallel sequencing, using the technology marketed by Illumina as an example. Firstly, genomic DNA is prepared for sequencing by fragmenting it, ligating short adapter sequences to the fragments, and denaturing (separating the strands). The prepared fragments are then washed over a flow cell, which is often referred to as a 'slide' as it is about the same size and shape as a

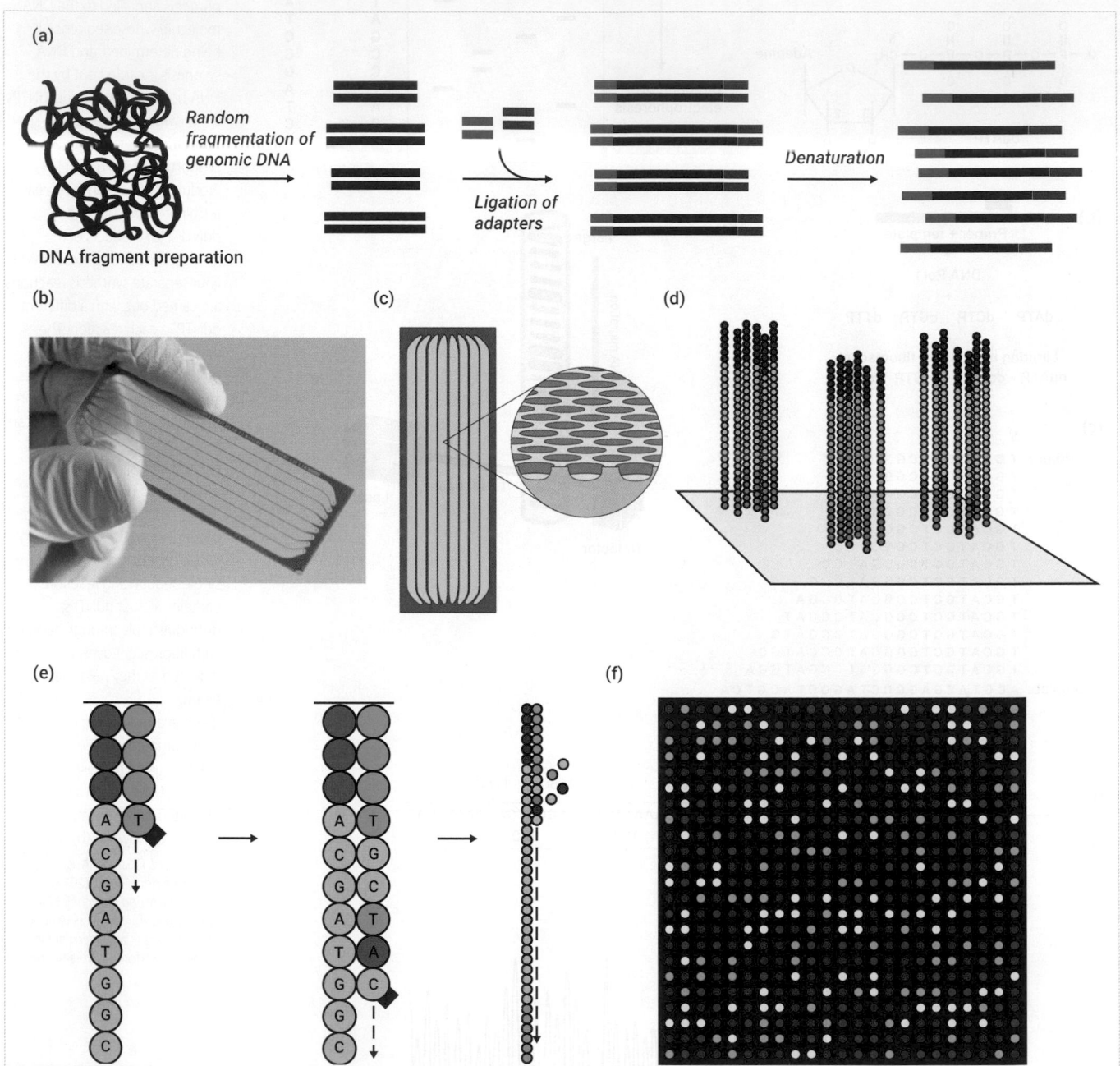

Figure 8.8 Illumina sequencing. (a) The genomic DNA is fragmented, adapters are ligated, and the DNA is denatured. (b) Fragments are then attached to a flow cell via short primer DNA sequences that are complementary to the ligated adapters. The primers and adapters base pair. (c) The flow cell contains billions of microscopic nanowells. The system is set up so that one DNA fragment attaches per well. (d) PCR is used to produce a cluster of identical DNA molecules in each well. (e) The molecules in each cluster are then sequenced simultaneously. A specific primer and four 3′ end blocked nucleotides, each with a different fluorescent label, are used for DNA synthesis. After each addition, the added nucleotide is determined by its fluorescence (f). The chemical block to 3′ addition is then removed, and another round of synthesis takes place.

Source: (a, d, e) Snape, A. & Papachristodoulou, D. (2018). *Biochemistry and Molecular Biology* (6th edn). Oxford University Press. (b) Bainscou, CC BY 3.0 <https://creativecommons.org/licenses/by/3.0>, via Wikimedia Commons. (f) Thomas Shafee, CC BY 4.0 <https://creativecommons.org/licenses/by/4.0>, via Wikimedia Commons.

standard glass microscope slide (see Figure 8.8b and c). The flow cell contains billions of microscopic nanowells and the DNA molecules are captured by short DNA sequences that have been fixed to the nanowell surface and are complementary to the adapters. The aim is for one DNA fragment to be tethered within each well.

Following the capture stage, multiple rounds of PCR produce a cluster of molecules with the same sequence within each nanowell (see Figure 8.8d) and the sequencing reactions are started. All four nucleotides are provided, but each nucleotide not only has a fluorescent 'tag' attached that allows it to be detected, but also prevents elongation. This means that after the first synthesis reaction each of the DNA molecules is elongated by the addition of just one nucleotide. A 'snapshot' image of the flow cell is taken (see Figure 8.8f) in which each nanowell is detectable as a coloured dot that shows which nucleotide was added to the DNA sequence it contains.

The fluorescent tags are then removed using a chemical reaction so that a second synthesis reaction can take place, adding one more nucleotide to each sequence, and a second snapshot is taken. This process is repeated many times very rapidly using fully automated methods.

Sequencing technologies are developing at a very rapid pace. Many, like the Illumina method, rely on PCR to generate multiple copies of the DNA, and DNA synthesis in millions of separate reactions. They use a variety of methods to detect the addition of nucleotides. For instance, some rely on the release of a pyrophosphate ion during the synthesis reaction, while others measure changes in pH caused by the release of hydrogen ions. Even newer methods dispense with the need for PCR amplification and DNA synthesis, instead reading the sequence of a single DNA molecule by forcing it, nucleotide by nucleotide, through a microscopic pore (a nanopore) in a membrane and measuring electrical differences across the membrane, which change as the DNA emerges. Nanopore methods allow long stretches of DNA to be read in a single reaction, using sequencing machines the size of a USB flash drive.

Assembling whole-genome sequences

Traditional sequencing methods produce sequencing 'reads' of a few hundred base pairs, while next-generation sequencing methods typically produce shorter reads of only tens of base pairs. Assembling these short sequence reads in the correct order to produce a full genome sequence is a challenge. Figure 8.9 compares the strategies used to generate the first 'reference' human genome sequence in 2003 with those now available to sequence the genomes of individuals.

To start the work of sequencing the reference human genome, genomic DNA was first broken down into large fragments 100,000–200,000 base pairs in size. These fragments were then cloned into bacterial artificial chromosome vectors (BACs). Preparatory work had created a 'map' of the genome, in which the positions of certain landmark sequences were established, and the correct order of BACs was worked out by comparing them with the map. Each BAC clone was then broken down further into smaller fragments, and the full sequences of these smaller fragments were determined and ordered against the BAC clone before the overlapping BAC clone sequences were finally put together.

Sections of highly repetitive DNA sequences provided a challenge as it is hard to read them accurately, and it is also hard to

order them accurately if they occur at several locations in the genome. Some repeat sequences in the human genome were published for the first time as late as 2021. However, the reference sequence was deemed to be 'complete' in 2003, after 13 years of work when 99 per cent of the sequence was ready for publication.

To generate an individual's genome sequence using modern methods it is no longer necessary to go through the stage of generating clones. The genomic DNA is broken down into fragments that are sequenced directly using one of the next-generation technologies. Assembling multiple short sequence reads without any prior ordering of clones is known as 'shotgun' sequencing. Doing this on the massive scale required to generate a large genome sequence has only become possible recently with the assistance of powerful computers. The availability of a reference sequence for alignment is also helpful, as illustrated in Figure 8.9. A major aim of sequencing individual genomes is to investigate human variation, and comparison with the reference sequence allows us to rapidly identify individual differences. An individual's genome sequence can now be 'read' within a day, although the analysis may take longer.

 Check your understanding of the concepts covered in this section by answering the questions in the e-book.

8.2 Applications of recombinant DNA technology

Recombinant DNA technology has multiple applications in many fields, including agriculture, archaeology, biological and medical research, forensic science, and the pharmaceutical industry. Here we explore just a few key examples and you will find others mentioned in various places throughout the book.

Working with RNA and cDNA

So far, we have focused on genomic DNA as the source material for cloning, creating libraries, and generating DNA sequence data. However, it is often desirable to work with messenger RNA (mRNA) instead. mRNA is the first product made during the process of gene expression and it represents sections of the genome that encode proteins. Analysing mRNA can give information about which genes are expressed in different cell types, at different stages of development, and under different conditions. For example, looking for differences between the genes expressed by a normal cell and a cancer cell derived from the same tissue can give useful information on how the cancer developed and on possible targets for treatment.

cDNA is a copy of mRNA

Like DNA, mRNA can be purified from cells, but RNA is much less stable than DNA and cannot be ligated into cloning vectors and propagated in host cells. mRNA molecules can instead be copied in the laboratory to make complementary DNA (**cDNA**). mRNA is copied using an enzyme called **reverse transcriptase**, which is derived from retroviruses. The process is summarized in Figure 8.10.

cDNA molecules can be ligated into plasmid vectors to create cDNA libraries. The inserts found in a cDNA library are copies of the mRNA sequences found in the source tissue.

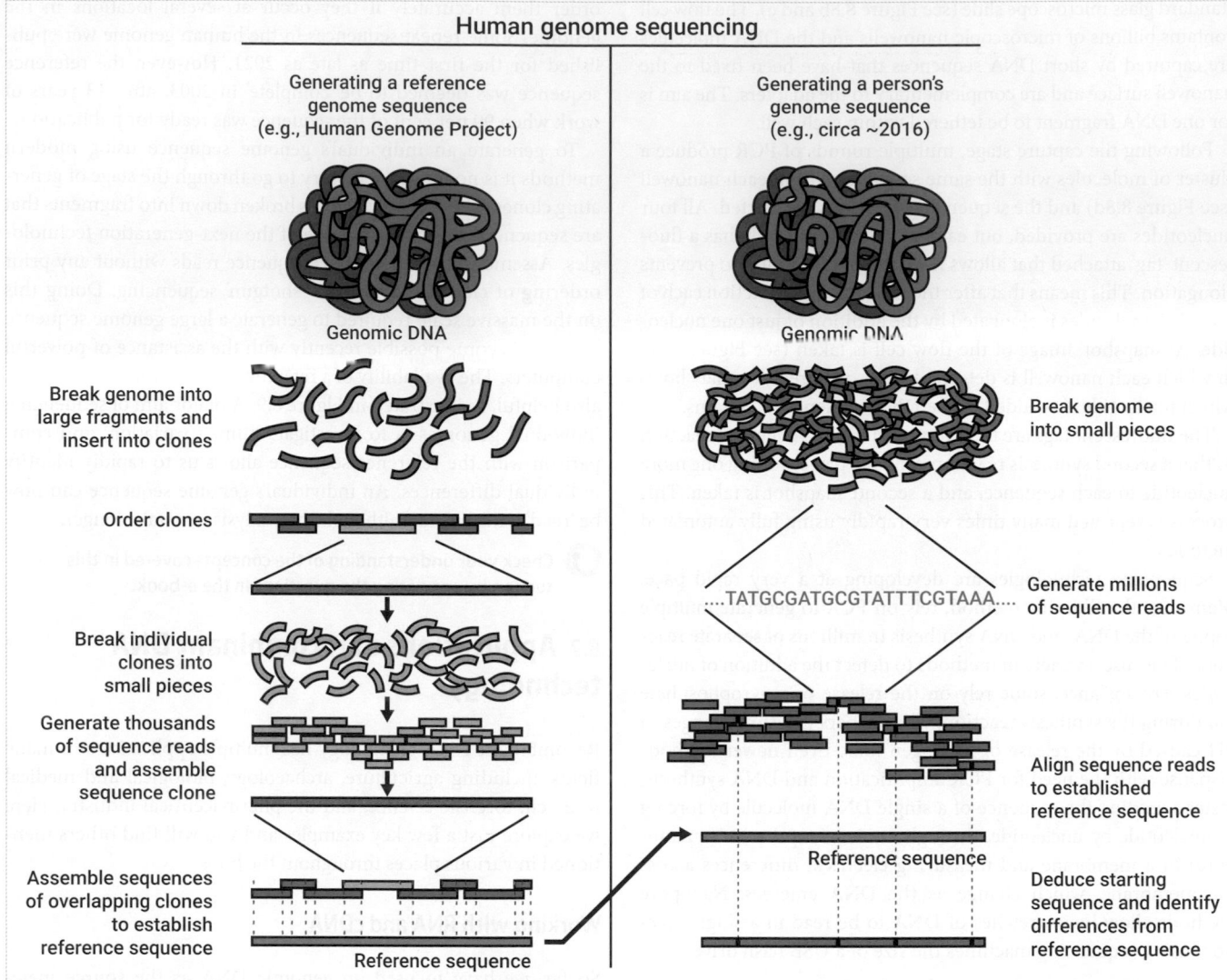

Figure 8.9 Strategies for assembling whole-genome sequences, illustrated by the human genome. Assembling the 'reference' human genome sequence published in 2003 involved multiple cloning, sequencing, and assembly steps. By 2016, modern sequencing methods allowed an individual's genome sequence to be established rapidly and cheaply, bypassing the cloning steps. The individual genome is broken into short fragments which are sequenced many times over, and the overlapping sequence reads are then aligned with the reference sequence.

Source: National Human Genome Research Institute: genome.gov

PAUSE AND THINK

A genomic library is prepared using genomic DNA from human liver cells, and a cDNA library is prepared using mRNA from human liver cells. Will the two libraries contain the same DNA sequences?

Answer: All the DNA sequences found in the cDNA library should also be found in the genomic library, since cDNAs are copies of mRNA molecules that are made using protein-coding genes of the genome as templates. However, the genomic library will contain many additional sequences, since a large proportion of the genome does not encode proteins, and not all the protein-coding genes are expressed in liver cells.

cDNA can be used to quantify gene expression

mRNA expression can be quantified directly using a variety of techniques. However, these are not described here because they have largely been replaced by methods where cDNA is used as a more robust surrogate for mRNA. RT-PCR is a commonly used technique that combines reverse transcription (RT) with the polymerase chain reaction (PCR) to detect a particular mRNA transcript present in a cell or tissue type.

During the process of RT-PCR, the mRNA content of a cell is first copied using reverse transcriptase to make single-stranded cDNA, which is then subjected to PCR using primers specific for the transcript of interest. If the transcript was present in the cell an amplified double-stranded DNA PCR product will be made; if it was absent, no PCR product will be made.

RT-PCR methodology has been successfully adapted to make it quantitative (qPCR or real-time PCR) so that the level of gene expression, not just the presence or absence of a transcript, is measured. The results from a typical qPCR experiment are shown in Figure 8.11. Here four sets of primers were used to detect four different mRNA transcripts.

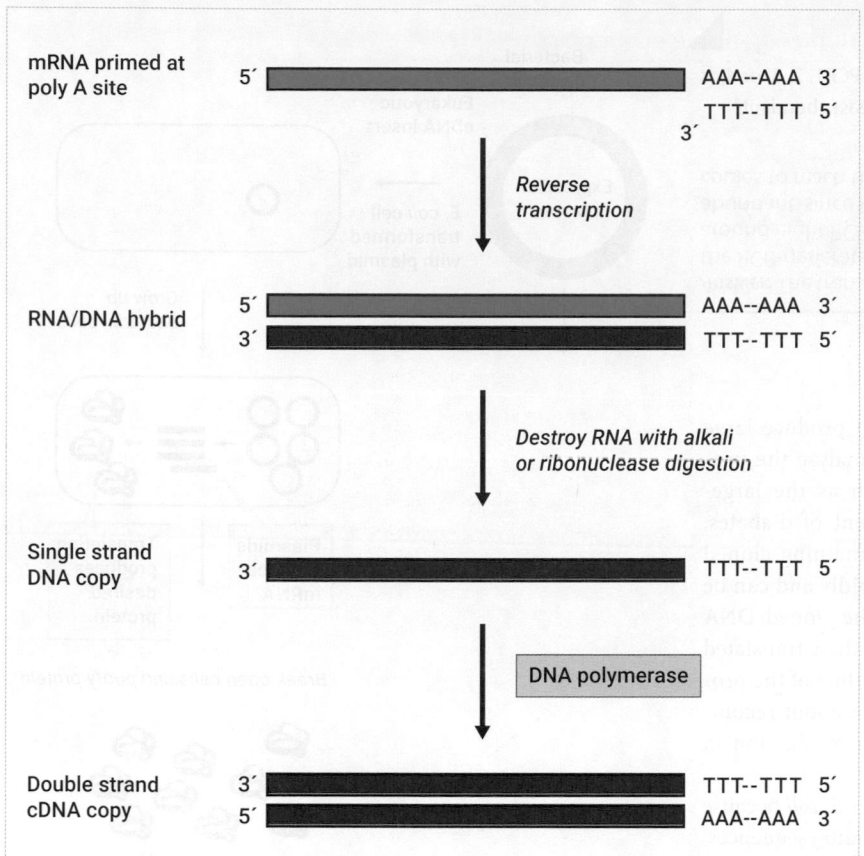

mRNA primed at poly A site

5′ [   ] AAA--AAA 3′

TTT--TTT 5′

3′

Reverse transcription

RNA/DNA hybrid

5′ [   ] AAA--AAA 3′

3′ [   ] TTT--TTT 5′

Destroy RNA with alkali or ribonuclease digestion

Single strand DNA copy

3′ [   ] TTT--TTT 5′

DNA polymerase

Double strand cDNA copy

3′ [   ] TTT--TTT 5′

5′ [   ] AAA--AAA 3′

Figure 8.10 Preparation of complementary DNA (cDNA) from eukaryotic mRNA. This method makes use of the poly A tail, a common feature of most eukaryotic mRNAs. A poly T primer anneals to the poly A tail and the reverse transcriptase enzyme can synthesize the cDNA by adding dNTPs to the primer, using the mRNA as a template. The mRNA is then digested away using alkali or a ribonuclease enzyme, and DNA polymerase is used to make the second DNA strand.

Source: Snape, A. & Papachristodoulou, D. (2018). *Biochemistry and Molecular Biology* (6th edn). Oxford University Press.

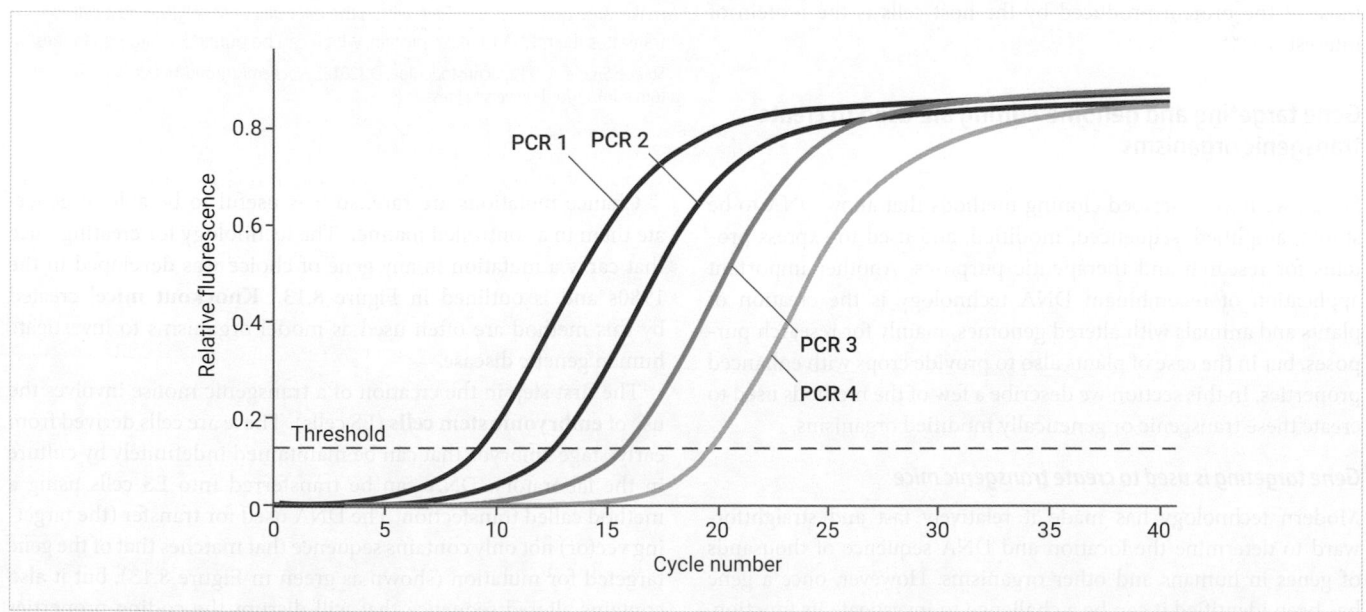

Figure 8.11 Amplification plot from qPCR reactions. Each curve shows the progress of a PCR reaction measured by incorporation of a fluorescent molecule into the DNA products. Each of the four PCR reactions uses a different pair of primers to detect four different mRNA transcripts. During the first few PCR cycles the amount of DNA doubles with each cycle but is below the level of detection. The number of cycles at which the amount of product reaches a detectable threshold level is used for quantification, as it is proportional to the amount of template present at the start of the reaction. The reactions plateau when one more of the components becomes limiting.

Source: Snape, A. & Papachristodoulou, D. (2018). *Biochemistry and Molecular Biology* (6th edn). Oxford University Press.

Producing proteins from cDNA clones

Recombinant DNA technology is often used to produce large amounts of a specific protein, for example to analyse the protein structure or for industrial purposes, such as the large-scale production of human insulin for treatment of diabetes. *Escherichia coli* bacteria or other host cells containing cloned DNA encoding the protein of interest divide rapidly and can be grown in large quantities in liquid cultures. The cloned DNA sequence is transcribed into mRNA, which is then translated into protein within the host cell. A simplified outline of the process is shown in Figure 8.12. You can read more about recombinant DNA technology and recombinant drug production in microbes in Chapter 15.

cDNA is used to produce eukaryotic proteins in *E. coli* because genomic DNA contains additional non-protein-coding sequences, called introns, that bacteria cannot deal with during gene expression. The cDNA is cloned into a plasmid, called an expression plasmid or expression vector, which contains a DNA sequence (the promoter) that is required to direct transcription of the cDNA insert within the bacterial cell. The system can be designed so that most of the protein produced by the host cells is the protein of interest.

Gene targeting and genome editing are used to create transgenic organisms

So far, we have described cloning methods that allow DNA to be stored, amplified, sequenced, modified, and used to express proteins for research and therapeutic purposes. Another important application of recombinant DNA technology is the creation of plants and animals with altered genomes, mainly for research purposes, but in the case of plants also to provide crops with enhanced properties. In this section we describe a few of the methods used to create these transgenic or genetically modified organisms.

Gene targeting is used to create transgenic mice

Modern technology has made it relatively fast and straightforward to determine the location and DNA sequence of thousands of genes in humans and other organisms. However, once a gene has been identified it can be a challenge to investigate its function. Organisms carrying a mutation that inactivates the gene are often useful in this respect. For instance, excessive eating in a mutant mouse suggests that the mutated gene has a role in appetite control.

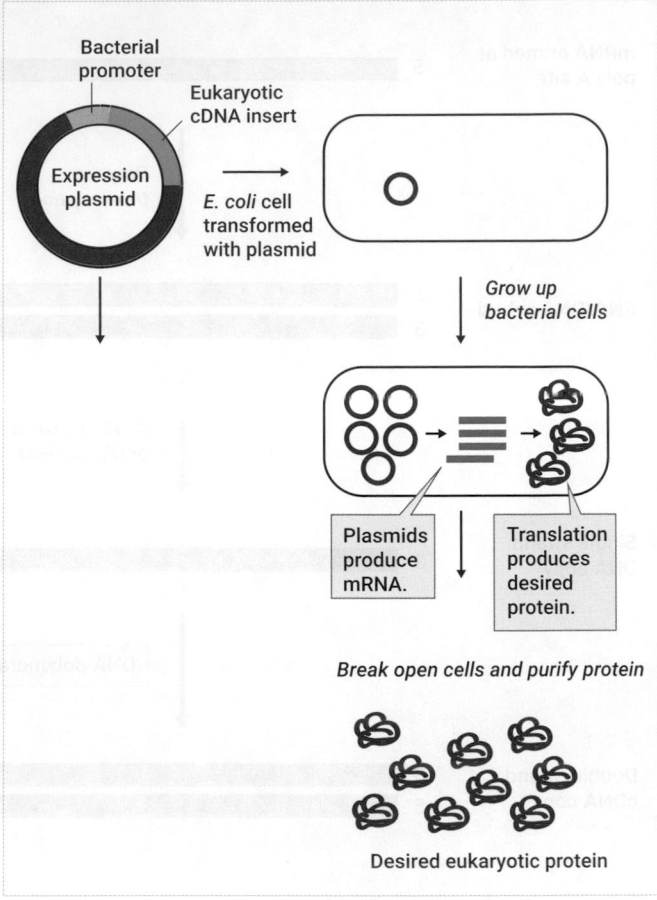

Figure 8.12 Expression of a eukaryotic protein in *E. coli*. The plasmid vector is of a particular type called an 'expression plasmid', which contains a DNA sequence (the promoter) that drives transcription of the cloned DNA insert to make large quantities of mRNA within the bacterial host cell. The host cell translates the mRNA to make protein, which can be purified in large quantities.
Source: Snape, A. & Papachristodoulou, D. (2018). *Biochemistry and Molecular Biology* (6th edn). Oxford University Press.

Chance mutations are rare, so it is useful to be able to generate them in a controlled manner. The technology for creating mice that carry a mutation in any gene of choice was developed in the 1980s and is outlined in Figure 8.13. '**Knockout mice**' created by this method are often used as model organisms to investigate human genetic disease.

The first step in the creation of a transgenic mouse involves the use of **embryonic stem cells** (ES cells). These are cells derived from early-stage embryos that can be maintained indefinitely by culture in the laboratory. DNA can be transferred into ES cells using a method called transfection. The DNA used for transfer (the targeting vector) not only contains sequence that matches that of the gene targeted for mutation (shown as green in Figure 8.13), but it also contains altered sequence that will disrupt the coding properties of the gene (red in Figure 8.13). In a small number of transfected ES cells, the altered sequence inserts into the targeted gene by a process called homologous recombination. Cells in which this has

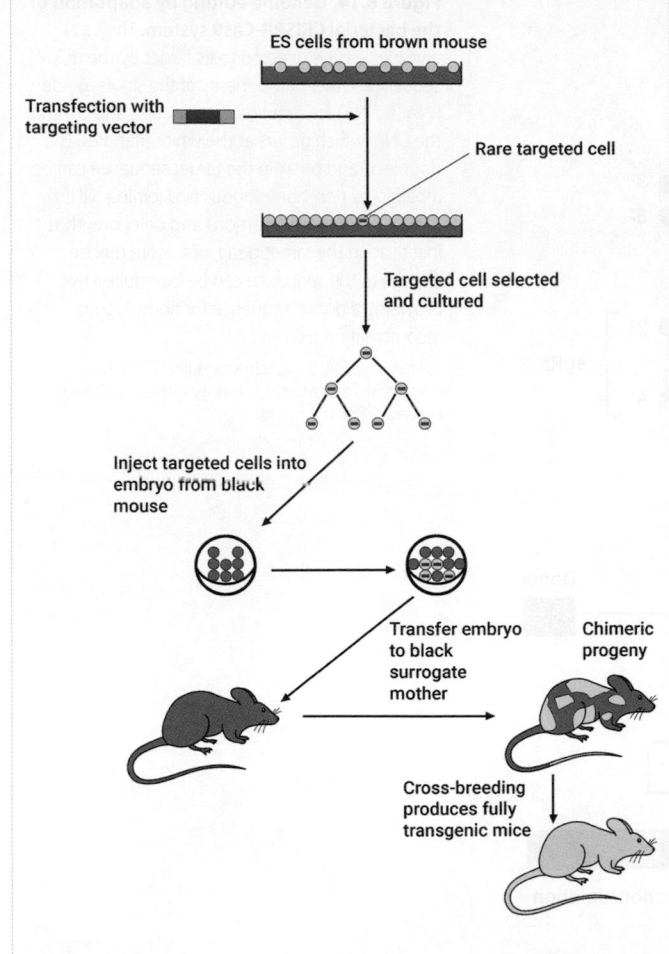

Figure 8.13 An outline of the procedure for generating knockout mice by gene-targeted mutation. A targeting vector is used to disrupt the sequence of a specific gene in mouse embryonic (ES) cells. ES cells containing the mutated gene are injected into host embryos, which are then transferred to a foster mother and allowed to develop to term. Offspring derived from the injected embryos are chimeras, containing cells from the black host embryo and the brown ES cells. The different coat colours provide a useful method of recognizing chimeric offspring, which are then used for crossbreeding to produce fully transgenic mice.

Source: Snape, A. & Papachristodoulou, D. (2018). *Biochemistry and Molecular Biology* (6th edn). Oxford University Press.

happened can be selected, grown in culture, and then injected into host embryos.

The injected embryos are transferred into a 'foster mother' mouse, who gives birth to chimeric offspring (i.e. offspring containing two genetically distinct cell types, some derived from the host embryo and some derived from the injected, genetically altered ES cells). Breeding from the chimeric offspring is used to establish a line of mice in which all cells have the gene of interest 'knocked out' by incorporation of the altered sequence into both copies of the gene.

The methodology used to create knockout mice has been adapted and extended, for example to create 'conditional' knockouts in

which the gene of interest is mutated only in a specific tissue, organ, or stage of development. However, there are a number of reasons why this system is not easily adaptable to other organisms: the dependence on ES cells, which are not available for many species; the need for an *in vitro* selection system to select rare targeted cells; and the need for extensive crossbreeding to create fully transgenic animals. Consequently, much effort has gone into developing new technologies that allow the genome of an organism to be altered precisely at a specific location. These technologies are referred to collectively as gene editing or genome editing.

While several methods are available, one in particular has gained prominence because of its flexibility, adaptability to many species, and ease of use. The CRISPR-Cas9 system, which we describe next, led to the award of the 2020 Nobel Prize in Chemistry to Emmanuelle Charpentier and Jennifer Doudna.

Genome editing using CRISPR-Cas9

The CRISPR-Cas9 system is based on a defence system that bacteria use against viruses. CRISPR stands for clustered regularly interspersed short palindromic repeats, referring to short sequences of viral DNA that are incorporated into the bacterial genome. These sequences are transcribed into RNA transcripts called 'guide sequences', which form a complex with a CRISPR-associated (Cas) protein. The function of a Cas protein varies depending on the bacterial species. The system used for genome editing uses Cas9, an endonuclease enzyme that cuts the sugar–phosphate backbone of DNA within the sequence. When used by bacteria against viruses, the guide RNA pairs with the invading viral DNA, which is cleaved and destroyed by the Cas9 endonuclease.

The power of this system when used for genome editing, as illustrated in Figure 8.14, is that the Cas9 nuclease can be directed against any genome sequence by designing a guide RNA that is complementary to the target. Figure 8.14 shows how Cas9 introduces a double-stranded break in the genome. This allows the genome sequence to be altered by two possible routes. One uses a natural DNA repair system called non-homologous end joining (NHEJ), in which the broken strands of DNA are reattached. NHEJ is an imprecise process in which random sequences are added to the end of the DNA before they are rejoined. The extra sequences disrupt the coding sequence of the target gene, creating a 'knockout'. The other route sees more precise changes being introduced by providing DNA that is partly a match for the target gene (homologous DNA) and also containing the desired sequence change. Homologous recombination takes place during the repair, introducing the desired change to alter the coding sequence of the gene.

The CRISPR-Cas9 method could potentially be used to correct mutations in the human genome, offering the possibility of its use in gene therapy. The use of such a process on human embryos is of course controversial. In 2016 the UK Human Fertilisation and Embryology Authority (HFEA) granted permission for the use of editing on normal human embryos, for research purposes only. The embryos are to be destroyed after seven days. In Clinical Box 8.1 we discuss the progress and promises of gene therapy since it was first trialled in 1990.

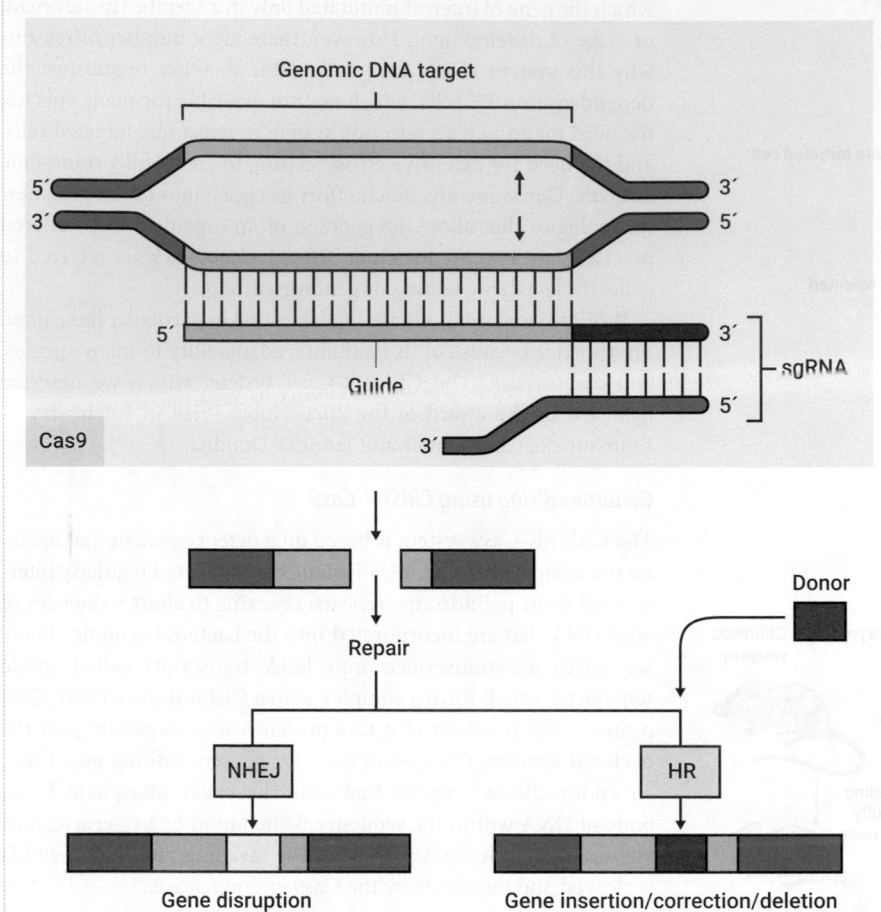

Figure 8.14 Genome editing by adaptation of the bacterial CRISPR-Cas9 system. The Cas9 endonuclease is directed to its target by the guide sequence. Other components of the single-guide RNA (sgRNA) are required for targeting and cutting the DNA, which occurs at the white triangles. The double-strand break in the target sequence can be repaired by non-homologous end joining (NHEJ), producing random insertions and deletions, thus inactivating the target gene, or a more precise change to the sequence can be introduced by providing a 'donor' sequence for homologous recombination (HR).

Source: Snape, A. & Papachristodoulou, D. (2018). *Biochemistry and Molecular Biology* (6th edn). Oxford University Press.

CLINICAL BOX 8.1   ## Gene therapy: pioneers, pauses, and progress

In the late 20th century the development of sophisticated genetic engineering techniques raised hopes that gene therapy for human diseases would soon be on the horizon. The aim was to treat diseases caused by mutation in a single gene, either by adding a correct copy of the gene sequence to the recipient's cells, or in a more sophisticated but more challenging scenario, by correcting the error in the recipient's own gene sequence.

Ethical considerations generally rule out changing the genetic makeup of all an individual's cells, as this would alter their germ line and would lead to genetic changes being passed to the next generation. Instead the cells and tissues most affected by the disease are targeted. For example, trials of gene therapy for cystic fibrosis made use of a modified virus that would target cells in the respiratory system and carry a corrected copy of the *CFTR* gene to the patient's lungs.

Disorders of the haematopoietic (blood cell) system were common early targets for gene therapy, as it was established that haematopoietic

stem cells can be isolated from the bone marrow of an individual, before being cultured, treated, and used to replace their bone marrow, which has meanwhile been destroyed by irradiation. The reimplanted stem cells can reconstitute all blood cell types.

The world's first gene therapy clinical trial began in 1990, when 4-year-old Ashanti de Silva was treated for a form of severe combined immunodeficiency disease (SCID), in which a mutation in a gene called adenosine deaminase (*ADA*) caused toxicity to her immune cells. Her cultured haematopoietic stem cells were treated with a functional *ADA* gene carried in a vector based on a retrovirus (a type of virus that inserts its DNA into the host cell genome, which is discussed in Chapter 13). This gave encouraging results with Ashanti and several other children.

Gene therapy suffered a setback with the death of 18-year-old Jesse Gelsinger in 1999. Jesse suffered a severe inflammatory response to the viral gene therapy vector that was used to treat him for a genetic liver disorder. As Jesse had suffered from a relatively mild and already treatable form of the disorder questions were asked about the ethics of the trial, and the case led to many trials being paused or halted.

In 2001 a French group used gene therapy to treat children with a very severe and fatal immune defect called X-linked, severe combined immunodeficiency (SCID-X1), which was then treatable only by isolation in a 'bubble' to prevent infection. This disease, found almost exclusively in boys, is different from that caused by *ADA* deficiency. The researchers used a retroviral vector to transfer a corrected version of the causative *IL2RG* gene into the patient's cultured haematopoietic stem cells. Ten out of 11 patients benefited significantly from the therapy but, unfortunately, within three years of the treatment, two of the patients had developed leukaemia. This was caused by the retroviral vector being inserted into the patient's genome in a position that activated a cancer-causing gene. After this setback, work focused on modifying the retroviral vector to avoid it causing leukaemia and now many children are successfully treated for SCID-X1 using gene therapy.

Despite several successes, gene therapy has proved more difficult than was anticipated. Many gene therapy trials have been disappointing not because of harmful side effects but through lack of efficacy. Nevertheless, hundreds of clinical trials are in progress, many of them aimed at treating cancer, where the unusual genetic makeup of tumour cells provides hope that they can be targeted without harming the patient. Gene editing methods such the CRISPR-Cas9 system raise the possibility of precise targeting and correction of faulty gene sequences.

Another key development that may enhance the success of gene therapy is the development of induced pluripotent stem cells (iPSCs). These are cells that can be isolated from individuals and induced to form many different cell types. Such cells could undergo genetic correction outside the body and then be replaced in the patient.

You can read more about the impact of Jesse Gelsinger's death and recent advances in gene therapy in this online article: Rinde, M (2019) *The death of Jesse Gelsinger, 20 years later*, https://www.sciencehistory.org/distillations/the-death-of-jesse-gelsinger-20-years-later

---

PAUSE AND THINK

Figure 8.14 shows two possible ways in which the CRISPR-Cas9 system can alter the genome sequence. Which of these could potentially be used to treat sickle cell disease, where there is a change in the sequence of the β globin gene resulting in one amino acid being changed in the protein?

Answer: It would be necessary to use homologous recombination, not non-homologous end joining, as the requirement is to introduce a precise change to correct the sequence of the gene, not to produce a random change that would inactivate it.

Transgenic plants can be produced using a plasmid vector or a 'gene gun'

The potential value of genetic modification is not just limited to therapeutic applications in humans. Genetic modification has also been widely used in agricultural plants. Foreign genes can be inserted successfully into plant chromosomes by using, as a cloning vector, the Ti (or tumour-inducing) plasmid, which is normally found in the pathogenic soil bacterium *Agrobacterium tumefaciens*. The plasmid is altered to make it harmless to the plants and to provide a way for plant cells that have incorporated the new gene to be selected in culture. Entire new plants can be generated from these modified cells. Alternatively, DNA molecules may be literally shot into plant cells, with a gun-type instrument that fires a cloud of fine shot loaded with DNA.

Crop plants are being engineered with specific characteristics, such as resistance to herbicides. The purpose is to control weeds by blanket spraying with the herbicide so that only the resistant crop plant survives. Other plants have been genetically modified to alter their nutritional properties. For example, 'golden rice' is

designed to produce β-carotene to combat vitamin A deficiency. The introduction of genetically modified (GM) crops has met with some opposition because of environmental concerns and questions about their ownership by for-profit companies, but they are gradually gaining acceptance.

 Check your understanding of the concepts covered in this section by answering the questions in the e-book.

8.3 Genomics and bioinformatics complement recombinant DNA technology

In Chapter 6 we introduced the disciplines of genomics, proteomics, and bioinformatics and described some of their many applications in the study of proteins and in medical science. The use of publicly accessible databases such as GenBank (http://www.ncbi.nlm.nih.gov/genbank/), containing large amounts of sequence data, is also vital for researchers studying DNA.

One of the major challenges arising from the Human Genome Project was the need to identify protein-coding genes, given that they make up only a small proportion of the genome. Identification can be automated, using computer-based methods to search for:

- open reading frames (sequences that code extended amino acid sequences that are not interrupted by stop codons);
- a 'signature' sequence such as a TATA box, indicating the existence of a gene promoter;
- splice donor and acceptor sites that identify introns.

The ENCODE project (https://www.encodeproject.org/) is a publicly funded project with the goal of building a list of functional elements in the human genome, including protein-coding genes and regulatory sequences.

Once protein-coding genes are identified, further studies and experiments are needed to establish their biological functions. Research in this area is often termed **functional genomics**. As described in Chapter 6 with respect to amino acid sequences, automated searches for DNA sequences that show homology with other known genes can help to shed light on gene function. Homology may be found with known genes in the same species or with equivalent or similar genes in other species. Homology in parts of their sequence can indicate that genes belong to a

'family' that share some aspects of their function, such as DNA-binding proteins that act as transcription factors (discussed in Chapter 5).

In Chapter 4 we explore examples of naturally occurring mutations in plants, animals, and humans that may enable scientists to identify the normal function of the affected genes. Functional genomics often involves the use of model organisms, in which a gene sequence can be deliberately altered by gene editing and the effects observed. Such experiments cannot be carried out in humans, but Figure 8.15 shows an example where combining homology searching with the use of a mouse model

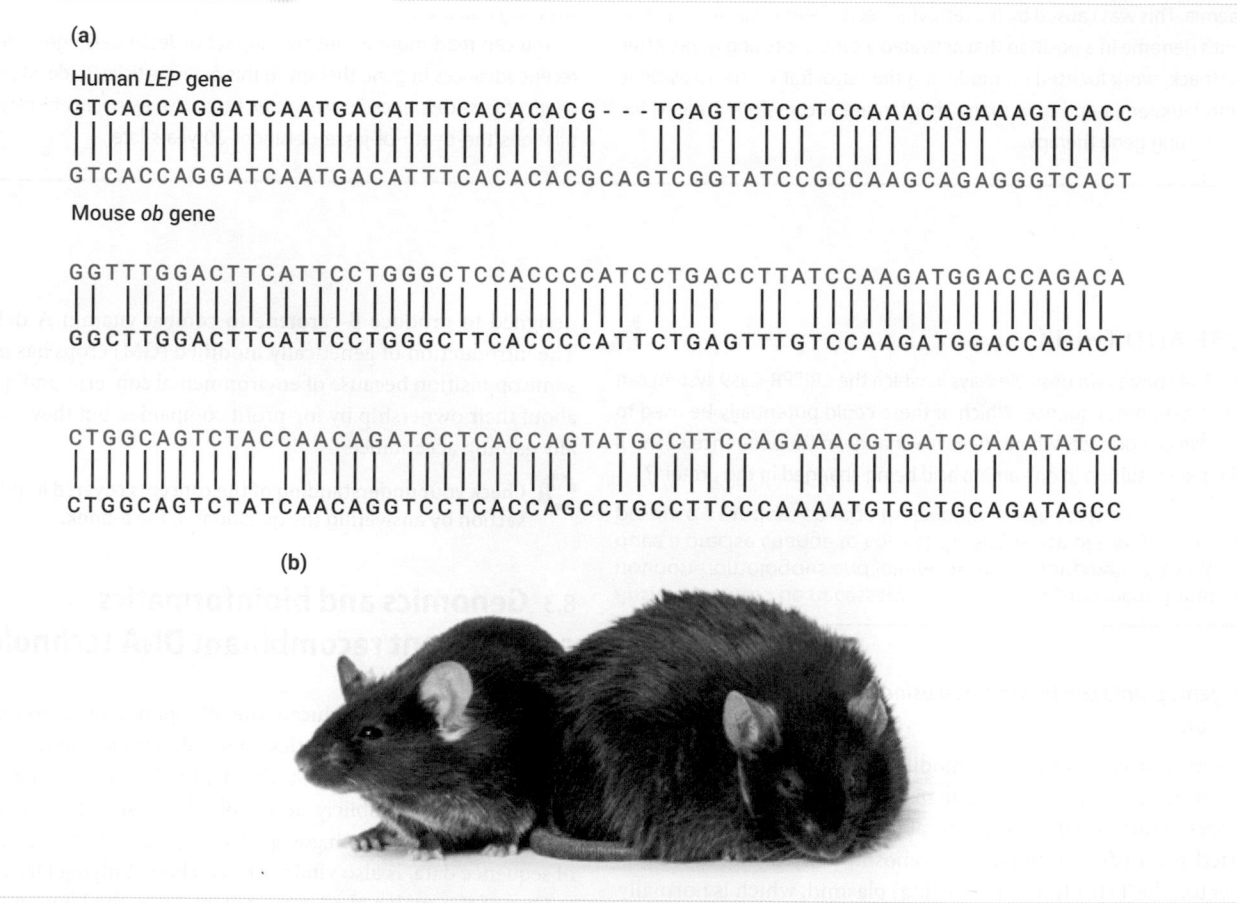

Figure 8.15 Discovery of leptin. (a) Comparison of partial DNA sequences of the human *LEP* gene and the mouse *ob* or *lep* gene, both encoding the hormone leptin. The gene was first discovered in mice, and the homologous human gene was identified through sequence similarity. Vertical lines link nucleotides that are identical in the mouse and human genes. The three dots in the human gene sequence indicate a 'gap' where three nucleotides found in the mouse gene are missing from the human sequence. This difference does not affect the gene function. (b) The mouse on the right carries a mutation in both copies of its *ob* gene and is obviously obese in comparison to the normal mouse on the left. Knowledge of this 'obese' mouse mutant strain assisted researchers to identify the *ob* gene and assign a function to the previously unknown leptin hormone, in both mice and humans.

Source: (a) Klug et al. *Essentials of Genetics*, Edition 9. Pearson. (b) © 2021 The Jackson Laboratory.

led to the discovery of the human hormone, leptin, which regulates hunger and fat storage. Mutations that make the leptin gene dysfunctional cause extreme obesity but are very rare in humans. More commonly, obese individuals develop leptin resistance, which stops them feeling full and contributes to them continuing to overeat despite high leptin levels. Treatments for obesity are proposed that would reduce leptin levels and restore leptin sensitivity.

The leptin story is only one example of the many uses of DNA sequence analysis and bioinformatics. A few other examples that illustrate the broad reach of this field of study include: from medicine, genetic profiling of cancer cells to identify suitable treatments; from conservation biology, studying genetic variation to avoid inbreeding in programmes to save endangered species; from agriculture, clarifying relationships among varieties of fruits and vegetables; and from ancestry research, the growing trend for individuals to trace their genetic origins and identify their close and distant relatives through DNA comparison. An important field of study is that of the microbiome, the full set of genetic material from micro-organisms in a specific niche or habitat, such as the human gut or a soil sample from a particular area. You can read more about genomic studies of the microbiome in Chapter 15.

 Check your understanding of the concepts covered in this section by answering the questions in the e-book.

SUMMARY OF KEY CONCEPTS

- Genetic engineering, encompassing recombinant DNA technology and gene cloning, has become a powerful technology that is widely used in many fields.

- Recombinant DNA refers to DNA that is made by covalently linking DNA sequences from different sources, while gene cloning or molecular cloning refers to the ability to make multiple copies of defined DNA sequences.

- Traditional cloning methods involve cutting DNA at known sequences using restriction enzymes. The DNA fragments produced are analysed using electrophoresis.

- PCR can amplify a selected stretch of DNA millions of fold in a few hours. The sequence to be amplified is delimited by the PCR primers.

- The ends of DNA molecules or fragments are covalently joined in precise ways using DNA ligase. Sections of 'foreign' DNA ligated into plasmid vectors are amplified within bacterial host cells, producing cloned DNA, which is stored in 'libraries'. A genomic library contains all the sequences in an organism's genome.

- DNA sequencing for many years used the chain-termination method, using dideoxynucleotides (ddNTP). Next-generation sequencing technologies, often PCR-based, have greatly increased the speed and reduced the cost of generating massive amounts of sequence data.

- The original determination of the human genome sequence took 13 years, but modern, fully automated sequencing and associated powerful computer analysis can now sequence an individual's genome within a day.

- DNA technology can be used to analyse RNA sequences if they are first copied into cDNA. RT-PCR combines reverse transcription, used to make cDNA, with PCR to quantify RNA transcripts. cDNA can be cloned into plasmid vectors and used to produce proteins for medical or industrial purposes.

- Gene targeting combined with stem cell technology has been harnessed to generate genetic knockout mice, which can provide animal models of human disease.

- Genome editing using the CRISPR-Cas9 system is a powerful new technology that allows precise changes to be introduced into genome sequences. It has many applications, including the possibility of gene therapy.

- Transgenic crop plants (GM crops) have been produced that have enhanced properties such as increased nutrient value or herbicide resistance.

- Genomics and bioinformatics complement recombinant DNA technology, using publicly available databases that store sequence data.

 Use the flashcards in the e-book to test your recall of key terms introduced in this chapter.

QUESTIONS

 Looking for answers? Once you've answered these questions, follow the link in the e-book to the answer guidance and check your work.

Concepts and definitions

1. The recognition sequences of five restriction enzymes are shown in Figure Q8.1. Which type of 'end' (5′ overhang, 3′ overhang, or blunt end) does each of the enzymes produce?

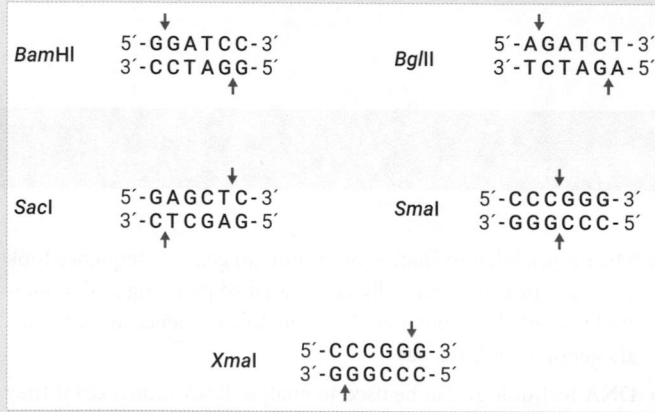

Figure Q8.1 Recognition sequences of five restriction enzymes. Red arrows show where the enzymes cut DNA.

2. Which enzyme is used in the polymerase chain reaction (PCR)? What distinctive feature of the enzyme makes it suitable for use in PCR?

3. What is the natural function of the DNA ligase enzyme, and how is it used in recombinant DNA technology?

4. What is the difference between a dideoxynucleotide triphosphate (ddNTP) and a deoxynucleotide triphosphate (dNTP) and how does it allow the ddNTP to play a key role in chain-termination sequencing?

5. In the context of CRISPR-Cas9 genome editing, what is the function of the guide RNA?

Apply the concepts

6. Two of the five restriction enzymes shown in Figure Q8.1 produce the same overhanging ends, despite having different recognition sequences. This allows DNA fragments produced by cutting with either enzyme to be ligated together. Which two enzymes are these?

7. The restriction enzymes *Sma*I and *Xma*I are isoschizomers (i.e. different enzymes that cut the same recognition sequence). Why might you choose to use either *Sma*I or *Xma*I in a DNA cloning project?

8. A PCR reaction is carried out using repeated cycles of heating and cooling. In step 1 of each cycle the temperature is 95 °C, in step 2 the temperature is 55 °C, and in step 3 the temperature is 72 °C. Explain what happens in each of the three steps.

9. Figure Q8.9 shows the sequences of forward and reverse PCR primers designed to amplify a section of the 100 base pair DNA molecule shown (note that only one strand of the target DNA is shown, but you can assume the existence of the complementary strand). Locate the sequences the primers will anneal to and give the size (in base pairs) of the PCR product.

Hint: When looking at the PCR primers, remember to consider 5′ and 3′ and whether they are a match for the DNA sequence shown or for its complement.

Forward primer	5′	CGGTAATCCGACGAGTCAGT 3′
Reverse primer	5′	ACCGCTAAGTAACAGAGGGC 3′

```
5′  1  atccctgggg cggtaatccg acgagtcagt aataggattc ctgcactccg  50

   51  ggctagttcg gaacctgagc gccctctgtt acttagcggt cctagtgaca 100  3′
```

Figure Q8.9 Forward and reverse PCR primers and the 100 base pair DNA molecule containing the target sequence for PCR amplification. Note that only one strand of the target DNA molecule is shown, but you can assume the existence of the complementary strand. Additionally, by convention, the DNA sequence is divided into stretches of 10 base pairs and split across more than one line to make it easier to read, whereas in reality it is continuous.

10. Many of the methods described in this chapter make use of annealing or hybridization (sequence-specific base pairing) by nucleic acids. What examples can you find?

Beyond the concepts

11. In 1990 the United States Congress committed $3 billion dollars to fund the Human Genome Project. Many argued that this large sum of money would be better spent on other projects, but Congress considered that the potential technological and medical benefits of the project made the expenditure worthwhile. How many benefits and applications of the project can you list?

12. As shown in Figure 8.12, human proteins can be made in large quantities in *E. coli* bacteria for therapeutic purposes. Why is a cDNA clone encoding the desired protein used in preference to a genomic clone?

13. The CRISPR-Cas9 system could potentially be used therapeutically to correct errors in human genome sequences. However, 'off-target' effects, where the guide RNA binds to sequences other than the intended target, are a potential barrier to its use in gene therapy. Briefly outline how CRISPR-Cas9 could be used for gene therapy. What could be the problems caused by these 'off-target' effects?

FURTHER READING

Roberts, M. A. J. (2019) Recombinant DNA technology and DNA sequencing. *Essays Biochem.* **63**: 457–68.
A concise and up-to-date review, aimed at undergraduate students, which covers many of the techniques discussed in this chapter, plus some extras.

Hood, L. & Rowen, L. (2013) The Human Genome Project: big science transforms biology and medicine. *Genome Med.* **5**: 79. https://doi.org/10.1186/gm483
A useful account of the origins, realization, and impact of the Human Genome Project.

Berlec, A. & Strukelj, B. (2013) Current state and recent advances in biopharmaceutical production in *Escherichia coli*, yeasts and mammalian cells. *J. Ind. Microbiol. Biotechnol.* **40**: 257–74.
A general update on the expression of recombinant proteins.

Terns, R. M. & Terns, M. P. (2014) CRISPR-based technologies: prokaryotic defense weapons repurposed. *Trends Genet.* **30**: 111–18.
A fairly concise review that covers both the biological function of CRISPR systems and their adaptation for experimental and therapeutic purposes.

Life at the Cellular Level

9   Characteristics of Prokaryotic and Eukaryotic Cells

10   Cell Division in Prokaryotes and Eukaryotes

11   Microbial Diversity

12   The Growth, Measurement, and Visualization of Cells

13   Microbes in Life: Harnessing Their Power

14   Microbes as Agents of Infectious Disease

15   Viruses

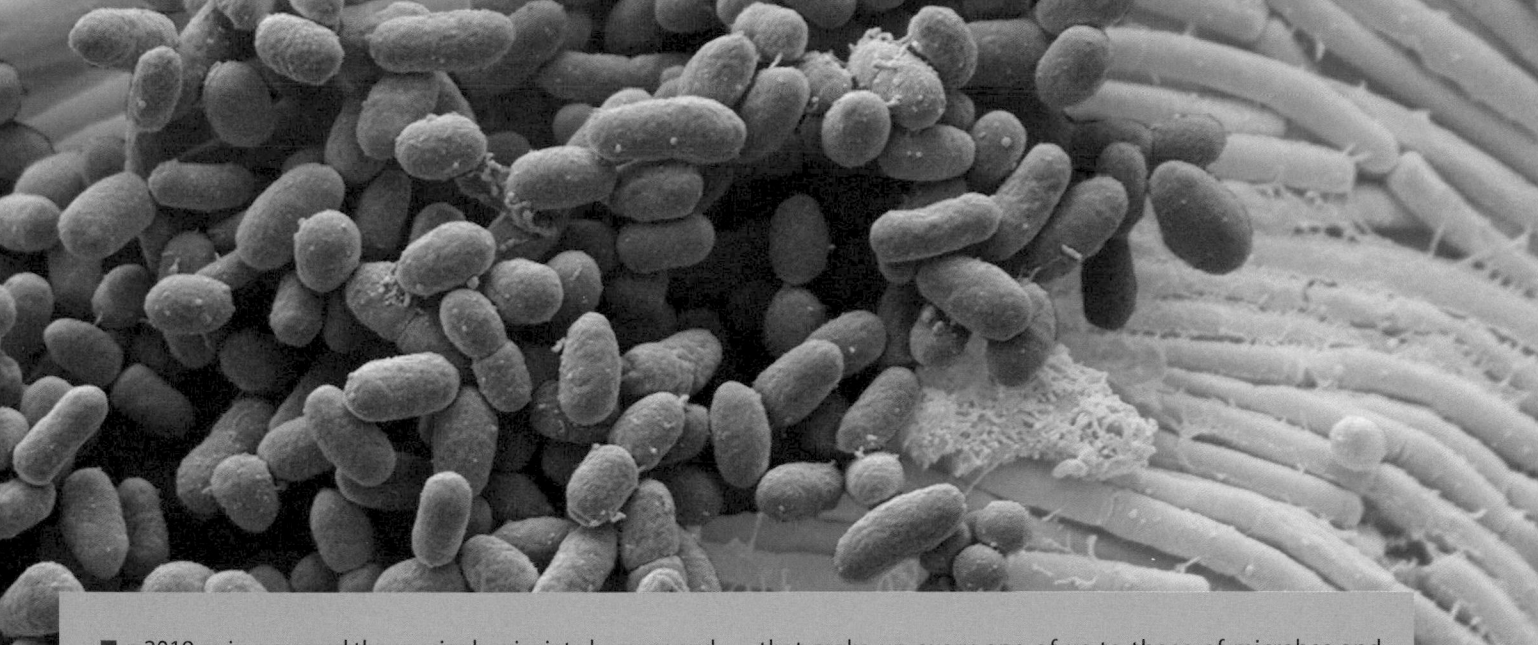

MODULE TWO

In 2019 a virus crossed the species barrier into humans and caused a global pandemic that, at the time of writing, the world is still attempting to control. But global pandemics are not a new phenomenon: the flu pandemic of 1918 bears witness to the fact that they have happened before, and they will happen again.

SARS-CoV-2, the virus that causes Covid-19, has led to tens of millions of infections and millions of deaths worldwide. This disease will almost certainly never be eradicated by vaccination in the way that smallpox disease was; the only certainty seems to be that we will be living with it for many years to come.

As the coronavirus pandemic has made so clear, our continued health and the health of the planet depends on our understanding of life at the cellular level, which includes the world of microbes, and our understanding of viruses, where the cellular and acellular collide.

In this module, we take you on a fascinating journey into the fundamental unit of life: the cell. We look at the characteristics of individual cells—be they cells that come together to form multicellular organisms or those single cells that exist as organisms in their own right—and we look at the features that differentiate them, from the cells that make up every one of us to those of microbes and through to the cells of plants.

Our planet is dominated by microbes: they are on us and in us and they occupy most environmental niches on Earth. They predated humans as occupants of our planet and they will remain long after our species has gone. During our exploration of microbial life, we will uncover a fascinating web of interdependence between microbes and multicellular organisms—for example, the beneficial relationships between certain species of fungi and terrestrial plants. We also discover how microbes can pose a major threat to our wellbeing—as agents of disease—but equally, how microbial activity can be harnessed for our benefit—to generate medical products such as antibiotics, in the production of food and drink, and even to protect our environment by helping to tackle pollution.

Our journey then ends where this opening began: an exploration of viruses—entities that have evolved to exploit life at the cellular level to their advantage, with consequences for us all.

Image: Microbes are everywhere: a coloured scanning electron micrograph (SEM) of bacteria cultured from a smartphone. *Source:* Steve Gschmeissner/Science Photo Library.

MODULE TWO

Characteristics of Prokaryotic and Eukaryotic Cells

LEARNING OBJECTIVES

By the end of this chapter, you should be able to:

- Describe the key characteristics shared by all cells, and the ways in which cells can vary.

- Describe the numerous components and structures found as part of different cells types and their roles.

- Recall the similarities and differences between the cells of the Bacteria, the Archaea, and the Eukarya

- Explain that many organisms exist as single cells, whereas others are complex combinations of a large numbers of cells and cell types

- Recall that the extracellular matrix is as important as the resident cells for the structure and function of tissues and organs.

Chapter contents

Introduction   367
9.1 The diversity of cells   368
9.2 The characteristics of the Bacteria   373
9.3 The characteristics of the Archaea   383
9.4 The characteristics of eukaryotic cells   384
9.5 The key eukaryotic cellular structures   386
9.6 From cells to tissues   401

🕐 Watch the key concepts video in the e-book to prepare yourself for studying this chapter.

Introduction

Cells are the fundamental units of life. Some living organisms comprise just a single cell, while the human body is thought to contain around 30 trillion, working in exquisite harmony to keep us alive. Cells also display remarkable diversity—a diversity that makes possible the range of living organisms that populate our planet. At a fundamental level, however, all cells fall into one of two types: prokaryotic or eukaryotic.

In this chapter we consider both **prokaryotic** and **eukaryotic** cells, and the similarities and differences between them. We take you on a journey around the interior and exterior components of these cells, exploring both the structure of these cell components as well as their functions. We will then explore the diversity in both structure and function of different groups of cells, and consider how both prokaryotic and eukaryotic cells duplicate (or replicate).

We begin by considering how cells were first identified, and the nature of a structural feature that is common to all life: the phospholipid membrane.

9.1 The diversity of cells

Cells have existed for millennia, but the word 'cell' is a relatively recent invention. Robert Hooke (1635–1703) first coined the term 'cell' in 1665. Hooke, the Curator of Experiments for the Royal Society of London, viewed a thin slice of a piece of cork with a microscope. He thought the regular appearance of the shapes he observed—depicted in Figure 9.1—looked like monks' cells, which led to the term he used to describe these structures: from the Latin *cella* meaning 'small room'.

However, what he viewed was not actually cells as we know now, but the empty cell walls of dead plant tissue. At a similar time, Dutch textile merchant, Anton van Leeuwenhoek (1632–1723)

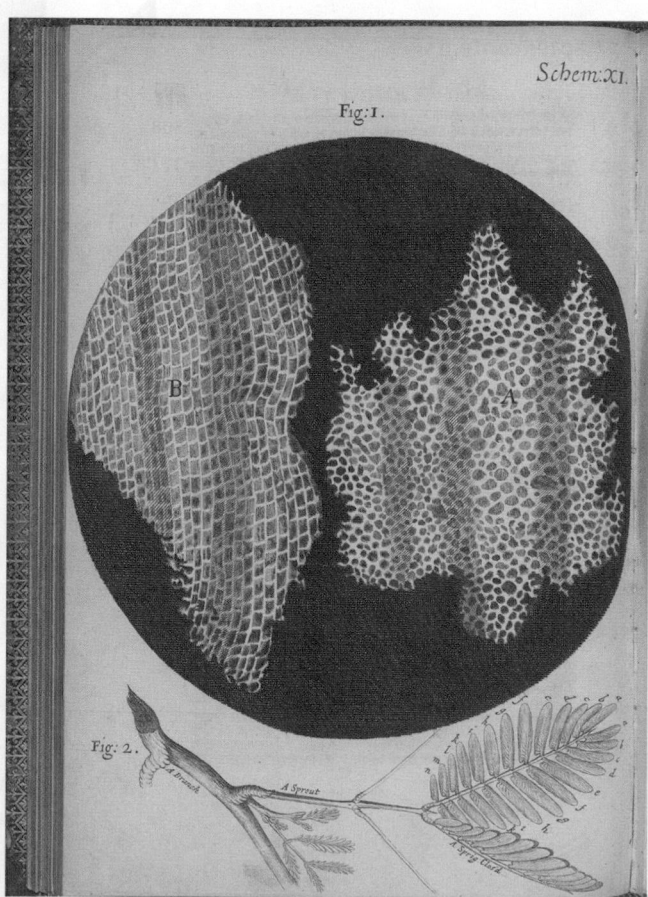

Figure 9.1 Robert Hooke's drawings of the cellular structure of cork and a sprig of sensitive plant.

Source: 4.0 International (CC BY 4.0), engraving from *Micrographia*, 1665, by Robert Hooke. Wellcome Collection.

was also experimenting with microscopes. His microscopes were capable of a higher magnification than Hooke's, and Leeuwenhoek went on to discover and visualize a variety of different cells, including those of some micro-organisms, red blood cells, and spermatozoa.

Nearly two centuries later, Matthias Schleiden (1838) and Theodore Schwann (1839), in looking at plant and animal tissues respectively, developed the first two tenets of 'cell theory':

1. All living organisms are composed of one or more cells.
2. The cell is the (smallest) basic or fundamental unit of structure and organization in organisms.

A few years later, in 1855, Rudolf Virchow added the third tenet, recognizing that cells do not just spontaneously generate:

3. Cells arise from pre-existing cells ('*omnis cellula ex cellula*').

A cell can be described as a 'sack' or 'ball' that is mostly composed of water (water being approximately 70% of the total cell mass).

As we learn in Chapter 3, there are two main types of cell: those that are prokaryotic (the archaea and bacteria) and those that are eukaryotic. Eukaryotes comprise organisms such as protists, fungi, plants, and animals.

While there are many similarities between these two cell types there are also a number of differences, which are summarized in Table 9.1.

The single cells of unicellular prokaryotes and eukaryotes have a variety of different dimensions and shapes while being responsible for all functions necessary for those cells to remain viable and to reproduce. In multicellular eukaryotes, the various cells are again of a variety of different dimensions and shapes, and carry out the necessary processes required for them to be viable, and also carry out a number of specialized functions. The relative sizes of different cells are depicted in Figure 9.2.

The cells of unicellular organisms are responsible for *all* functions. By contrast, being part of a multicellular organism allows for cell differentiation whereby different cells have specialized functions. As a result, cells displaying the same function can build tissues, and tissues can build organs, all of which serves to increase the complexity of the organism.

Why are cells limited to the sizes illustrated in Figure 9.2? There are a number of reasons. Prokaryotes are limited by how efficiently they can carry out the metabolism sufficient to maintain viability. Their surface area to volume ratio is crucial. The ratio is the surface area divided by the volume and is important because it represents the area of the cells that is in contact with the environment over which they can acquire nutrients necessary for growth and metabolism. The larger the cell gets, the smaller the ratio, and the fewer opportunities there are to gain nutrition. This is even more vital as they have no **endocytosis** or **phagocytosis** to acquire nutrients. (Endocytosis and phagocytosis are methods by which many eukaryotic cells can take up nutrients.)

Many eukaryotic cells are also limited in size by their surface area to volume ratio, the nucleo-cytoplasmic ratio (that is, how big the nucleus is in relation to the cell), and the fragility of their membrane (since they have no cell wall for protection). Plant cells can be larger in size due to a large central vacuole, which is responsible for growth, and the support provided by the cell wall.

Table 9.1 Comparison of prokaryotic and eukaryotic cells

Similarities	Possess a lipid membrane (semi-permeable barrier, selective transport)
	Contain cytosol (with amino acids and proteins, fats, carbohydrates, and other molecules)
	Have DNA and RNA
	Undertake metabolic (chemical) reactions
	Undergo growth
	Can reproduce (but note: prokaryotes—binary fission; eukaryotes—mitosis and meiosis)
	Experience ageing and death
	Perform mechanical activities (movement of molecules, structures or the whole cell)
	Respond to stimuli
	Capable of self-regulation
	Capable of cellular evolution
Differences	Prokaryotic cells have no nucleus or membrane-bound organelles (prokaryote means 'before nut or kernel')
	Eukaryotic cells ('true nut or kernel') have a nucleus plus organelles such as endoplasmic reticulum, Golgi, lysosome, vacuole (plant cells only), for example
	Prokaryotes have a cell wall (Gram-positive or Gram-negative)
	Eukaryotes are mixed: animal cells do not have a cell wall, but plant (cellulose) and fungi (chitin) cells do
	Prokaryotes and eukaryotes occupy different ecological niches (particularly the Archaea)
	Prokaryotes and some eukaryotes exist as single-celled organisms
	Other eukaryotes exist as multicellular organisms
	Prokaryotes show biochemical diversity (organotrophic, phototrophic, lithotrophic)
	Prokaryotes have circular DNA
	Eukaryotes have linear DNA ('small' vs 'large' genomes)
	Prokaryotic organisms (and thus cells) can form colonies
	Eukaryotic organisms can be unicellular, colony forming, or cells can exist as part of multicellular organisms with tissues and organs

Figure 9.2 The relative sizes of cells and their components.

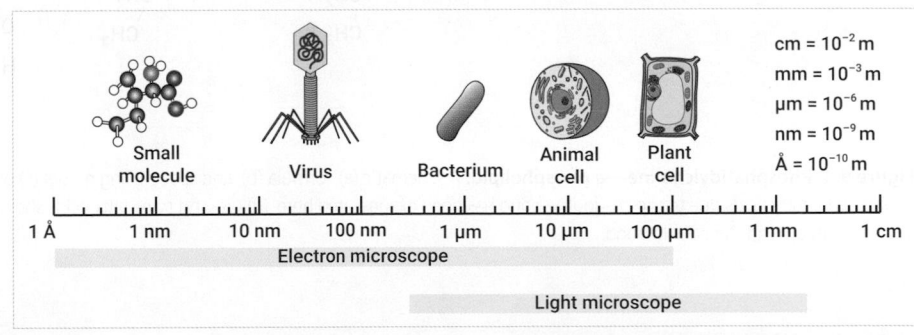

The phospholipid membrane: a unifying feature

The structure and function of any cell is critically dependent upon the plasma membrane that defines the cell's outer boundary and separates the contents of the cell from the external environment. In addition, for eukaryotic cells, the intracellular membranes also define and contribute to the structure of various organelles such as the nucleus and mitochondria. A membrane's functions depend upon not only the phospholipid bilayer but also the various proteins and other molecules found within, and associated with, the bilayer.

Let us build up a picture of the phospholipid membrane, starting with the 'building blocks' from which it is formed. The three main classes of membrane lipid that make up the bilayer are the phospholipids, glycolipids, and sterols.

The structure of a phospholipid is illustrated in Figure 9.3. Phospholipids are the major class of membrane lipids and are derived from either glycerol (the phosphoglycerides or glycerophospholipids) or sphingosine (the sphingolipids). Figure 9.3a shows how glycerol acts as a scaffold for a phosphorylated head group (predominantly ethanolamine, choline, or serine) and two fatty acid chains bound to a glycerol molecule with ester linkages (C–O–C=O). Sphingosine is a more complex structure than glycerol, having one hydrocarbon chain itself. However, it acts in a similar fashion to glycerol: as a scaffold for the addition of a fatty acid chain, as well as a head group such as ethanolamine, serine, or choline.

The glycolipids are sugar-containing lipids, which are found in the outer leaflet of the plasma membrane. The generalized structure

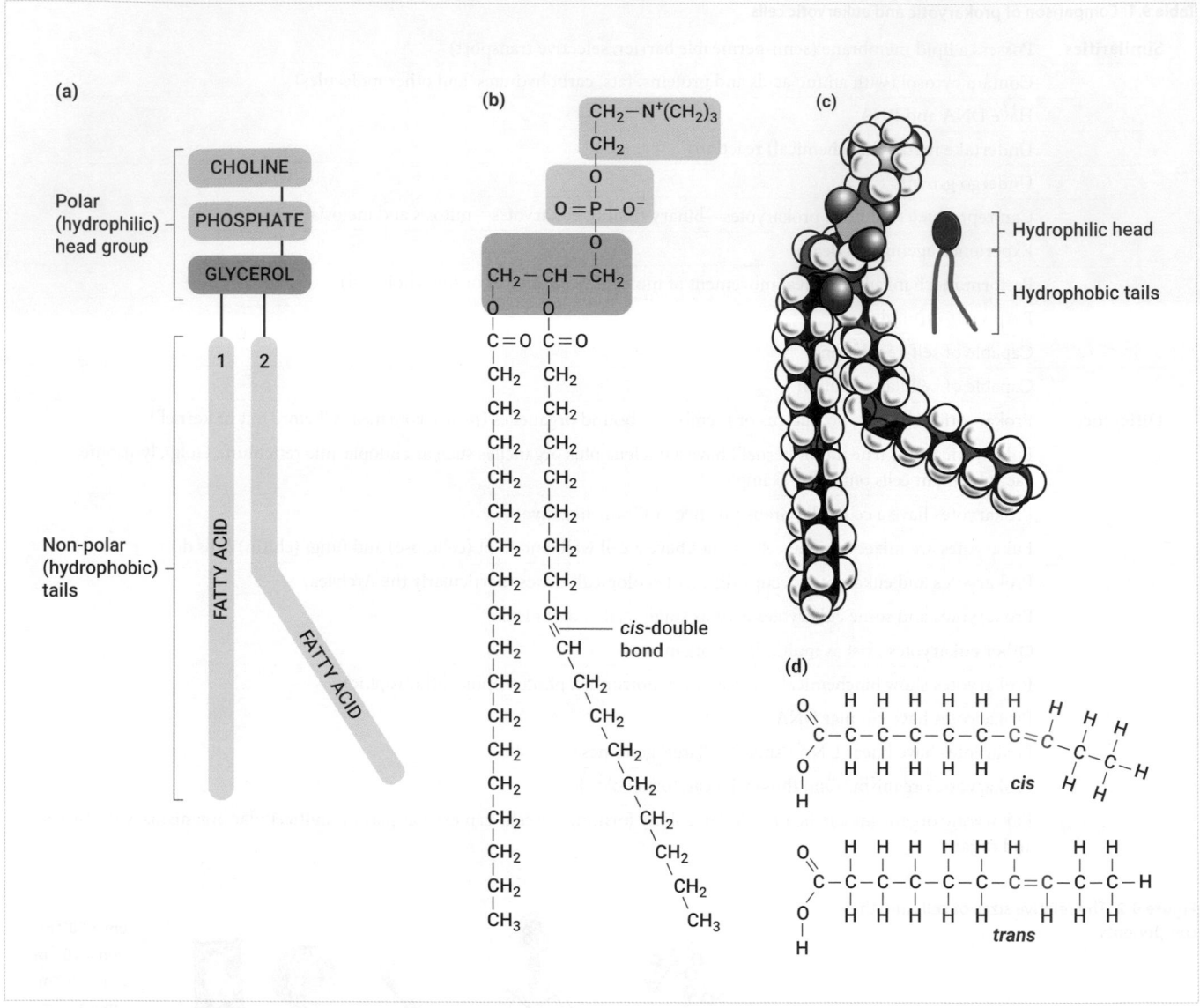

Figure 9.3 Phosphatidylcholine—a phospholipid. A schematic (a), formula (b), and space-filling model (c) of the phospholipid. The 'kink' in the hydrocarbon chain of one of the fatty acids due to the *cis*-double bond has been exaggerated here. (d) *Cis*- and *trans*-fatty acids showing the consequence of the hydrogen atoms' arrangement around the double bond.

of a glycolipid is shown in Figure 9.4. Notice how the structure of a glycolipid is broadly similar to that of a phospholipid, with the exception that a glycolipid has a sugar group attached as a head group, in contrast with the phosphorylated group present in phospholipids.

Sterols are present in eukaryotic plasma membranes but not in most prokaryotes. The plasma membrane of animal cells is rich in cholesterol, whose structure is shown in Figure 9.5. Typically, however, the membranes of organelles have smaller amounts than cell

membranes. In plant and fungal plasma membranes, the sterol is sitosterol and ergosterol, respectively.

The lipid bilayer spontaneously aligns in an aqueous environment to have a **hydrophobic** area on the inside and two **hydrophilic** surfaces exposed to the environment; this structure is depicted in Figure 9.6. The formation of this structure is a consequence of the **amphipathic** nature of the lipid components and hydrophobic interactions. 'Amphipathic' refers to the dual chemical nature of the membrane lipids: the fatty acid hydrocarbon chains

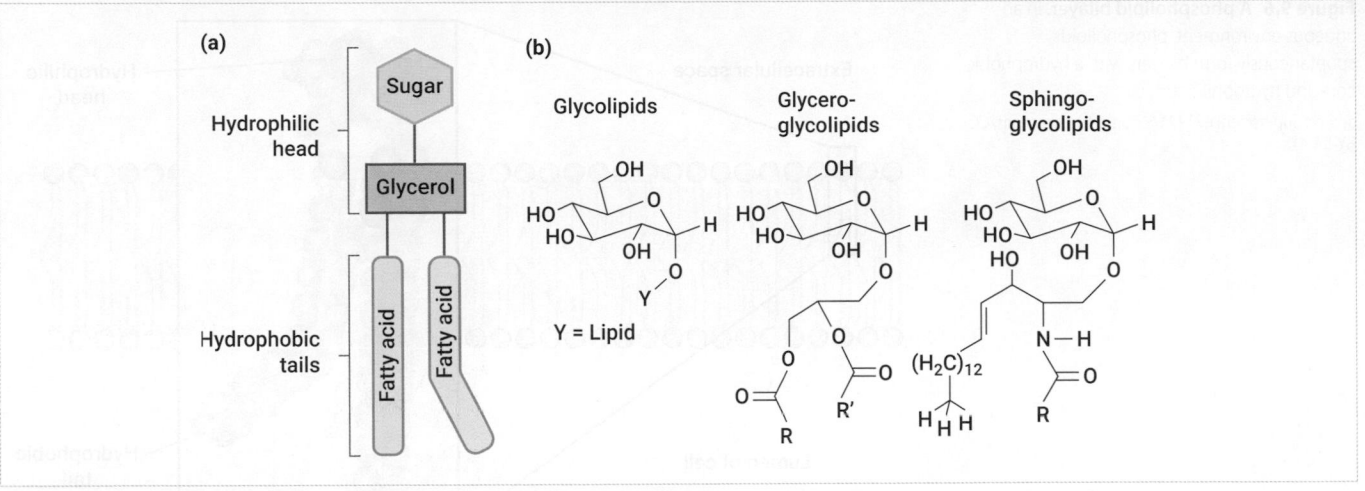

Figure 9.4 Glycolipids. (a) A schematic of a glyceroglycolipid and (b) the molecular structures of glycolipids, glyceroglycolipids, and sphingoglycolipids.

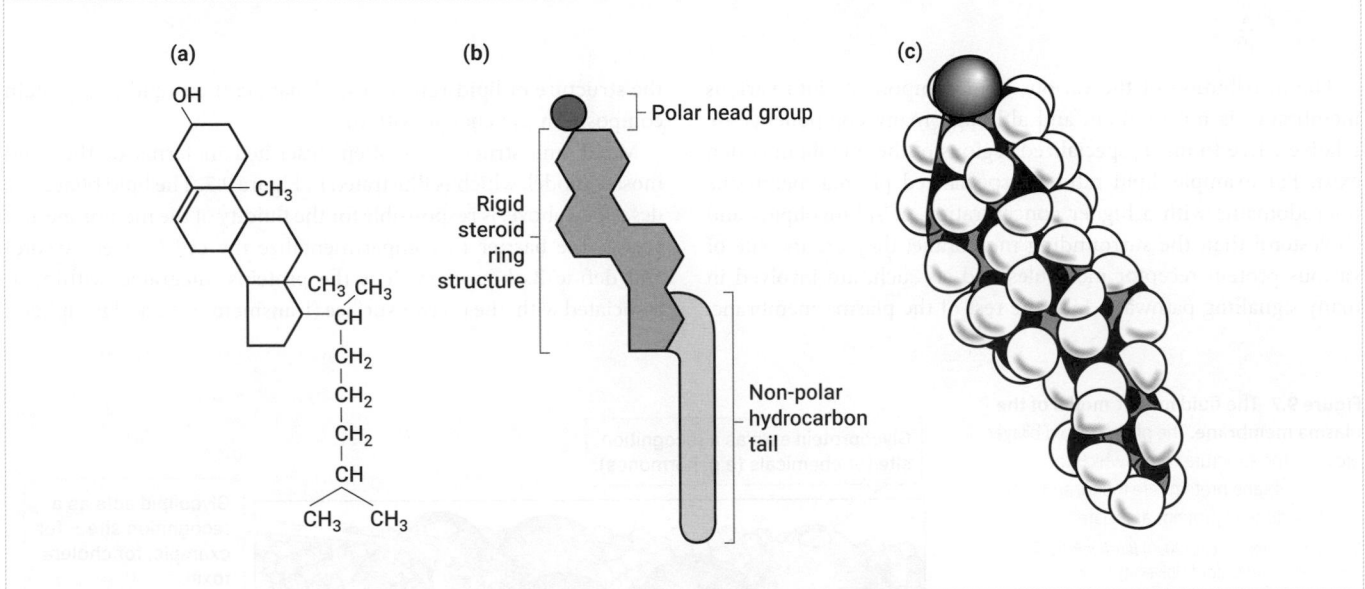

Figure 9.5 Cholesterol. (a,b) Schematic/formula and (c) space-filling models of cholesterol.

of phosphoglycerides are hydrophobic ('water-fearing'), whereas the phosphorylated head groups are hydrophilic ('water-loving').

Membranes, or lipid bilayers, rely on being fluid in nature to function as they should. Many of their components, such as the phospholipids, are able to move relative to one another across the surface of the bilayer. In particular, the fatty acids of the phospholipids are vital to maintaining the fluidity of the bilayer. Fatty acids, which are linear chains of carbon atoms, can differ in length—from 14 to 24 carbon atoms—and in the degree of saturation of the carbon–carbon bonds. A saturated fatty acid contains only single carbon–carbon bonds. However, most fatty acids are unsaturated: one or more of the carbon–carbon bonds along the fatty acid

chain are *cis*-double bonds (with only one hydrogen joined to each carbon). This conformation is shown in Figure 9.3d.

Why is the double bond important? Each double bond introduces a kink, or bend, into the chain, which affects the ability of the phospholipids to pack together, thereby influencing the fluidity of the bilayer.

Cholesterol also plays an important role in regulating the fluidity of the membrane. The cholesterol molecules orient themselves within the membrane, predominantly between the fatty acid chains, which they interact with and partially immobilize (thus reducing fluidity). At high concentrations, cholesterol can prevent the fatty acid chains packing together and therefore increase membrane fluidity.

Figure 9.6 A phospholipid bilayer. In an aqueous environment, phospholipids spontaneously form bilayers with a hydrophobic core and hydrophilic surfaces.

Source: Superscience71421/Wikimedia Commons/CC BY-SA 4.0.

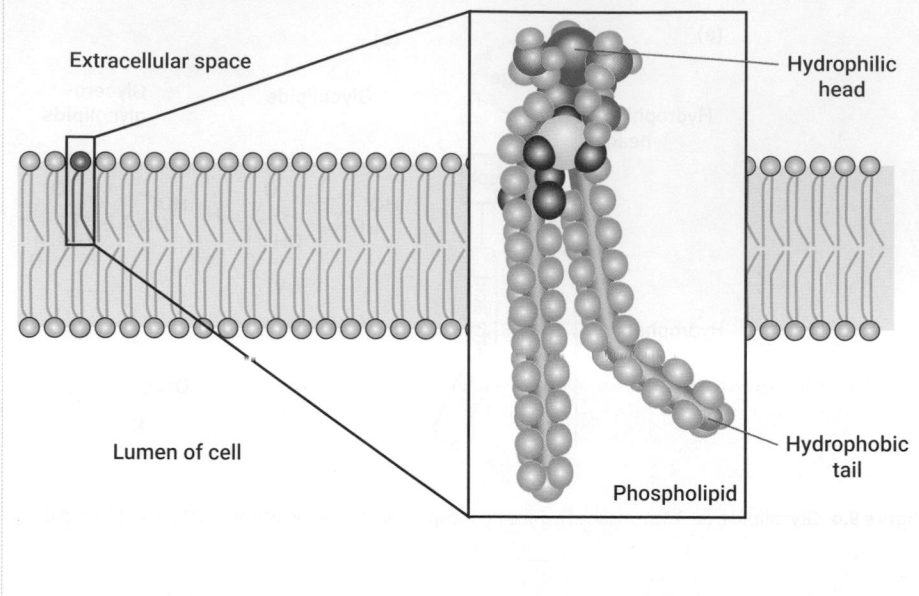

Extracellular space

Hydrophilic head

Lumen of cell

Hydrophobic tail

Phospholipid

The distribution of the various lipid components into various membranes is not random, and although many components are relatively free to move, specialized regions of the membrane often exist. For example, lipid rafts are specialized plasma membrane microdomains with a higher concentration of sphingolipids and cholesterol than the surrounding membrane; they are the site of various protein receptor molecules and, as such, are involved in many signalling pathways. Like the rest of the plasma membrane,

the structure of lipid rafts is also dynamic: their lipid and protein composition can change with time.

Membrane structure is often described in terms of the fluid mosaic model, which is illustrated in Figure 9.7. The lipid bilayer, as described above, is responsible for the fluidity of the membrane and acts as the barrier to compartmentalize the cell (if a eukaryote) and define its boundary. It is the proteins integrated within, or associated with, the bilayer surface (transmembrane and peripheral

Figure 9.7 The fluid mosaic model of the plasma membrane. The phospholipid bilayer provides the structure within which the transmembrane proteins are found, and with which peripheral proteins associate.

Source: Bowater R. et al. (2020) *Biochemistry: The Molecules of Life.* Oxford University Press.

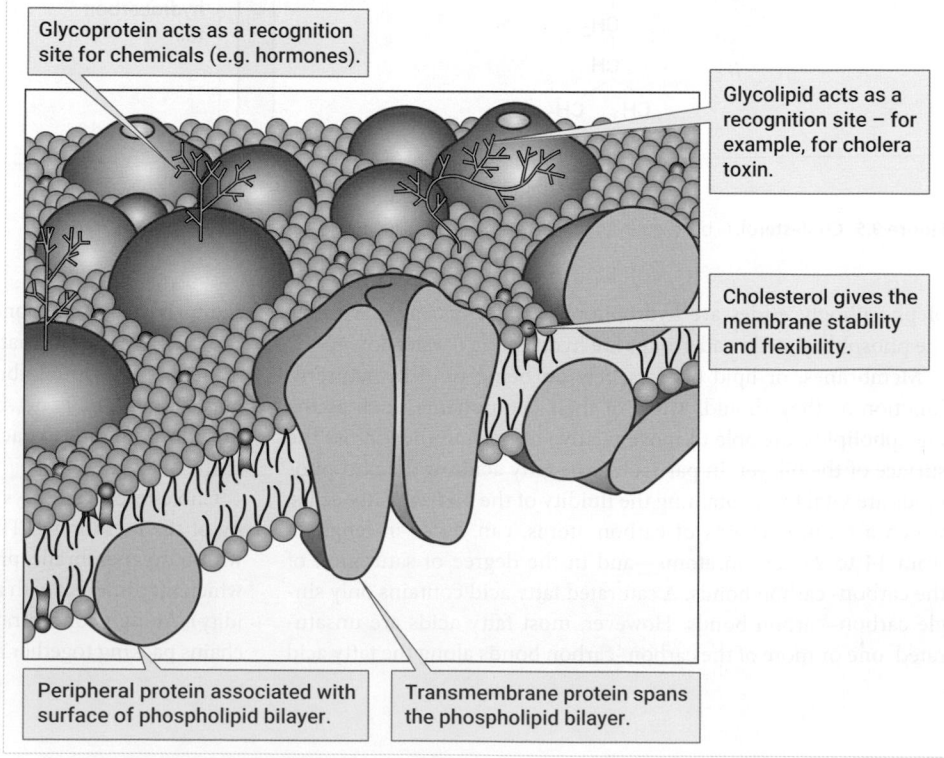

Glycoprotein acts as a recognition site for chemicals (e.g. hormones).

Glycolipid acts as a recognition site – for example, for cholera toxin.

Cholesterol gives the membrane stability and flexibility.

Peripheral protein associated with surface of phospholipid bilayer.

Transmembrane protein spans the phospholipid bilayer.

proteins, respectively) that are responsible for the mosaic pattern of membranes. These integral and membrane-associated proteins are critical for many roles of the membrane—roles that differ depending upon which membrane is being considered—and for the successful functioning of the cell itself.

Integral proteins can serve as:

- channels that allow selected ions to pass from one side of the membrane to another (by diffusion along a concentration gradient);
- transporters or pumps of ions or molecules (often by active transport against a concentration gradient);
- receptors that bind to specific molecules (ligands) and send a signal across the membrane (often from outside to inside the cell allowing the cell to respond to information in its environment); and
- enzymes that catalyse reactions at the membrane surfaces.

Some are also important for cell–cell recognition, and for attaching a cell to another cell or to the surrounding extracellular matrix, through which cells can be built into structures or tissues. Integral membrane proteins are frequently the targets for many therapeutic drugs.

Peripheral proteins, often linked to the integral proteins, play important functions in the passage of signals within and between cells, for example, or function as enzymes. Specific examples of various membrane proteins, both integral and membrane-associated, are discussed at various points throughout this book.

Archaeal phospholipids are structurally slightly different to the eukaryotic and prokaryotic phospholipids. The hydrophobic tails consist of isoprene units rather than fatty acids; these are bound to the glycerol with **ether linkages** (C–O–C), as illustrated in Figure 9.8. Side branches off the isoprene can be joined together to form carbon rings, which help to increase the stability of the membrane.

Despite these differences, archaea still exhibit hydrophilic surfaces and a hydrophobic internal environment in what is known as a lipid monolayer. The monolayer membrane acts similarly to the phospholipid layer and carries out the same functions; the main

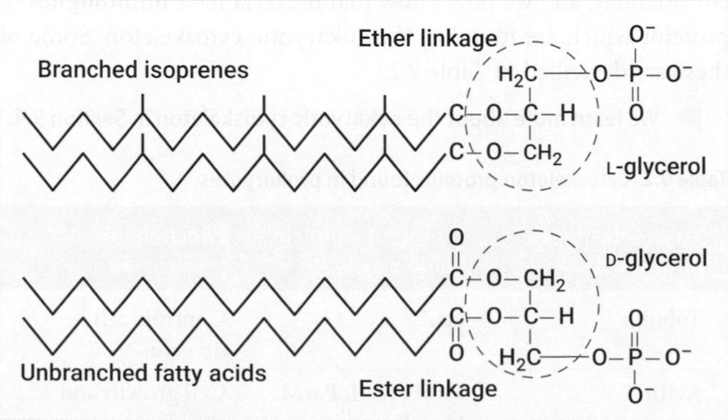

Figure 9.8 Archaeal phospholipids. The phospholipids in archaea (top) differ from phospholipids from other prokaryotes and from eukaryotes (bottom) in the structure of the glycerol molecule, the type of bond used to join the hydrocarbon chains to the glycerol, and the nature of the hydrocarbon chain.

difference is that the monolayer is structurally more stable, and therefore can be subjected to higher temperatures, vital for the survival of thermophiles, which live in very hot environments.

PAUSE AND THINK

Why are archaeal phospholipids different from eukaryotic and prokaryotic phospholipids?

Answer: Archaeal phospholipids have similar functions and properties to those of prokaryotes and eukaryotes, but the monolayer is structurally stable at extremes of temperature.

Go to the e-book to complete an activity to reinforce the relative sizes of cells and their components.

Check your understanding of the concepts covered in this section by answering the questions in the e-book.

9.2 The characteristics of the Bacteria

Bacteria are prokaryotic cells. They are unicellular organisms and belong to their own domain. As you can see in Figure 9.9, bacteria have a variety of cell morphologies, the main shapes being coccus (round), bacillus (rods), filament, and spirochete (spiral). They can be between 0.5 and 5 μm in diameter.

We learn more about the remarkable diversity of the bacteria in Chapter 11.

Despite their diminutive size, bacteria have a huge impact on our ecosystem and how we live. For example, photosynthetic bacteria produce the majority of oxygen we breathe, and then respire. In this section, we will examine a range of structures and processes, some of which are common to all bacteria, and some specialized structures that are not found in all bacteria. A simple diagram of a bacterial cell is shown in Figure 9.10.

During the remainder of this section, we will consider each of the following structures:

- nucleoid;
- plasmids;
- cytoskeleton;
- ribosomes;
- reproduction;
- cell wall;
- specialized structures.

The nucleoid

Bacteria are prokaryotes and their genetic material—their DNA—is free floating in the cytoplasm rather than enclosed in a nucleus, like that found in eukaryotic cells. Despite being free floating, bacterial DNA is not distributed evenly throughout the cell interior. Rather, it is concentrated in a region called the **nucleoid**. The DNA forms a single, unbroken ring called a chromosome, which has an origin of replication.

We learn more about origins of replication in Section 3.3.

Name	Shape	Example
Coccus		*Paracoccus denitrificans, Micrococcus luteus*
Diplococci		*Streptococcus pneumoniae, Neisseria meningitidis*
Streptococci		*Streptococcus thermophillus, Streptococcus pyogenes*
Bacillus		*Pseudomonas aeruginosa, Escherichia coli*
Vibrio		*Vibrio cholerae*
Spirochete		*Treponema pallidum, Borellia burgdorferi*
Helical		*Helicobacter pylori*
Filamentous		*Penicillium roqueforti*

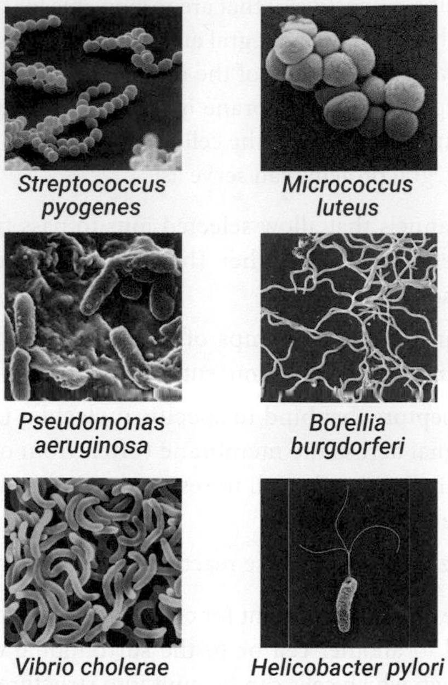

Streptococcus pyogenes

Micrococcus luteus

Pseudomonas aeruginosa

Borellia burgdorferi

Vibrio cholerae

Helicobacter pylori

Figure 9.9 Bacterial cell morphologies.

Sources: Eye of Science/Science Photo Library; AMI Images/Science Photo Library; CDC/Science Photo Library; Science Source/Science Photo Library; Dennis Kunkel Microscopy/Science Photo Library.

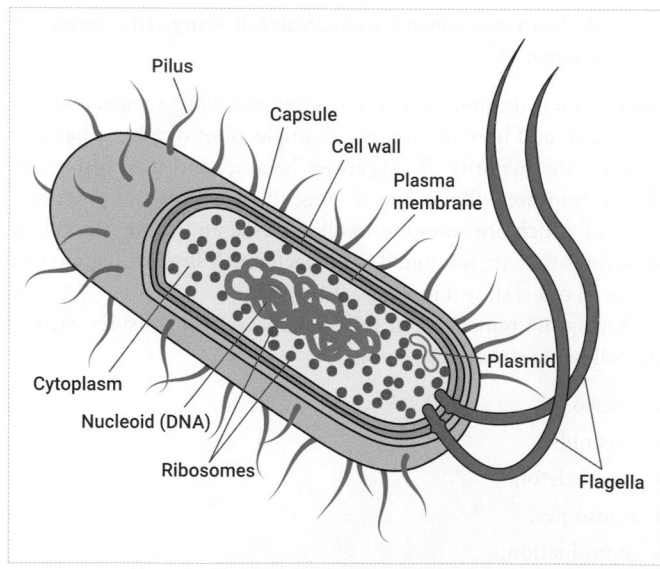

Figure 9.10 Schematic diagram of a bacterial cell.

Source: https://www.yourgenome.org/sites/default/files/downloads/activities/genomics-and-microbes-pack/yourgenome-genomicsandmicrobespack.pdf yourgenome, Genome Research Limited CC BY 4.0.

Plasmids

In addition to the nucleoid, smaller rings of DNA can be found within bacterial cells; these smaller rings, called **plasmids**, are capable of self-replication. Plasmids often contain genes that are not essential for life but can confer an advantage—for example, resistance to antibiotics or **virulence factors** for pathogenicity. If plasmids are removed from the bacterium, the cell can still replicate.

PAUSE AND THINK

Why do some bacteria have plasmids while others do not?

Answer: Many plasmids encode proteins that are important for a bacterial cell when they encounter particular situations. For example, plasmids can encode heavy metal resistance genes, useful if a bacterium is in a contaminated environment. Bacterial cells will pass plasmids between them using a process called conjugation. Plasmids place an extra metabolic burden on the cell, and therefore plasmids can easily be lost during reproduction if a selection pressure to keep them is not maintained.

The cytoskeleton

For many years it was thought that bacteria did not have a cytoskeleton. However, our understanding around this has changed considerably, and we now know that bacteria have **homologues** of proteins which are found in the eukaryotic cytoskeleton. Some of these are described in Table 9.2.

▶ We learn more about the eukaryotic cytoskeleton in Section 9.4.

Table 9.2 Cytoskeleton proteins found in prokaryotes

Eukaryotic protein or filament	Prokaryotic protein homologue	Function in prokaryotes
Tubulin	FtsZ	Controls cell division
Actin	MreB, MbI, ParM, MamK	Cell growth and shape
Intermediate filaments	CreS	Controls a curved shaped
Molecular motors	Not found yet	

PAUSE AND THINK

Why was it that for many years scientists thought that bacteria did not have a cytoskeleton?

Answer: For many years it was thought that only eukaryotic cells had a cytoskeleton, and that it evolved after the first eukaryotic cell. Scientists were not really looking for these structures in bacteria as bacterial cells were so much smaller, and their cell division process less complex. Bacteria also have a tough cell wall, so it was thought a cytoskeleton was unnecessary. The development of better imaging technology allowed scientists to get a much closer look inside the tiny cells. Sequencing technology has also revealed prokaryotic homologues of eukaryotic cytoskeleton proteins.

Ribosomes

The cytoplasm of a bacterium is far from empty; it is packed with ribosomes for protein synthesis, together with various metabolic enzymes. Figure 9.11 shows how prokaryotic ribosomes consist of two subunits: these are named the 30S and 50S subunits. The S refers to their sedimentation value in Svedberg units, which is a measure of how fast the molecule settles during centrifugation, the means through which the two subunits were first isolated.

The smaller 30S subunit contains 21 ribosomal proteins and also a 16S rRNA molecule. The gene for this rRNA is often sequenced and used in the identification and classification of prokaryotes. The larger 50S ribosomal subunit contains 34 proteins and two rRNA molecules, the 5S and the 23S. Like those of the eukaryotes, ribosomes are essential in protein biosynthesis, as we discuss further in Chapter 5, Section 5.4.

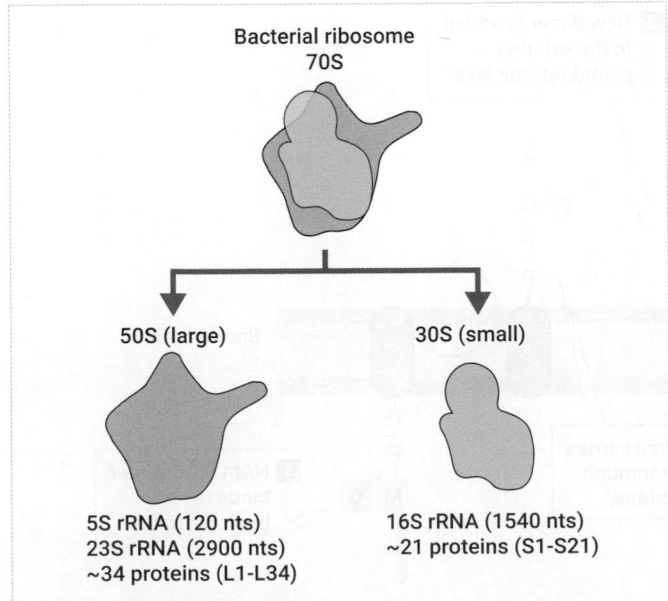

Figure 9.11 Structure of the bacterial ribosome.

Source: Nancy Craig et al., *Molecular Biology, Principles of Genome Function*, 3rd edn, 2021, Oxford University Press.

The bacterial cell wall

Bacteria can be classified into two large groups: Gram-positive and Gram-negative. This classification is based on their reaction to a procedure called the Gram stain, which distinguishes bacteria based on the structure and composition of their cell walls. Let us now explore the Gram stain and the bacterial cell wall in more detail, and investigate the characteristics of Gram-positive and Gram-negative bacteria.

Bacteria have a rigid cell wall which helps determine their shape. It also acts as a protective barrier by preventing water from entering the cell via osmosis. The bacterial cell wall can be broken down by the enzyme lysozyme. If this occurs, then the influx of water into the cell that results will cause it to expand and burst, a process called **lysis**.

The molecule responsible for the rigidity of the cell wall is the molecule **peptidoglycan**, which consists of a polysaccharide component and a peptide component. The layer of peptidoglycan in the cell wall is made from chains of the polysaccharide that lie adjacent to each other, which in turn form sheets that are able to stack up onto each other.

The components and overall structure of peptidoglycan chains are illustrated in Figure 9.12. Look at this figure and notice how the peptidoglycan chains have a polysaccharide backbone, consisting of alternating *N*-acetylglucosamine (NAG) and *N*-acetylmuramic

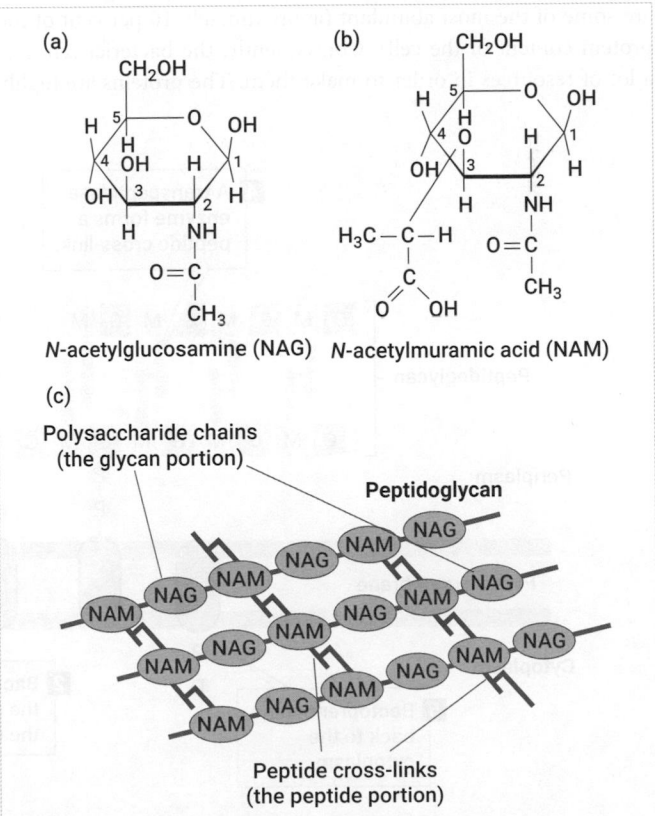

Figure 9.12 Peptidoglycan chains. Structure of (a) NAG and (b) NAM residues that make up the polysaccharide component of peptidoglycan, and (c) how the monomers are joined together with short peptide chains.

acid (NAM) monosaccharides connected with glyosidic bonds. The polysaccharide chains are joined together by short polypeptides that are always attached to the NAM monosaccharide. These peptide chains consist of the amino acids L-alanine (L-Ala), D-glutamic acid (D-Glu), meso-diaminopimelic acid (m-DAP), and D-alanine (D-Ala). This cross-linking gives the peptidoglycan layer its strength.

Growth of the cell wall

The NAG and NAM residues are made in the cytoplasm of the cell. The pentapeptide is joined to NAM. They then become attached to a membrane carrier molecule called **bactoprenol** before being transported across the cell membrane (see Figure 9.13). In order to insert new NAG and NAM residues, both glycosidic and peptide links have to be broken, and this is achieved with enzymes called autolysins. The NAM–NAG dimer is inserted into the peptidoglycan layer using a transglycosylase enzyme, and the final step is the re-sealing of the peptide cross-links using transpeptidase enzymes. Where new cell wall is inserted will determine the cell shape.

S-layers

Many bacteria also have an S-layer (surface layer)—a two-dimensional crystalline layer of protein or glycoprotein that surrounds the entire bacterial cell. The proteins comprising the S-layer are some of the most abundant (approximately 10 per cent of the protein content of the cell). Consequently, the bacterial cell uses a lot of resources in order to make them. The proteins are highly ordered across the surface and form a lattice of hexagonal, oblique, or square units. As you can see in Figure 9.14 they can look a bit like Roman mosaic floor tiles! This layer is thought to add extra protection against osmotic stress or predation.

The S-layer is not a universal trait across *all* bacteria. For example, both the model organisms *Escherichia coli* and *Bacillus subtilis* lack this layer. Indeed, it is interesting to note that research into the S-layer was probably hindered, as these model organisms lacked them.

The Gram stain

Now that we have considered the components of the cell wall, we turn our attention towards the iconic Gram stain, which is used to differentiate bacteria based on the structure of their cell wall.

Gram staining was developed in 1884 by Danish scientist Hans Christian Gram; the procedure is discussed in more detail in Experimental Toolkit 9.1. Over 130 years after its development the Gram stain is still used frequently by microbiologists as it is cheap, quick, easy to perform, and provides useful information to aid in identification.

The procedure has not changed much from the original, but now uses two stains, crystal violet and the counter-stain safranin, as shown in Figure 9.15. Initially, bacterial cells are heat-fixed onto a glass microscope slide. The cells are then flooded with crystal violet, which enters the cell wall. Iodine, which acts as the **mordant**, is then added; this attaches the crystal violet to the peptidoglycan. Ethanol is used to decolourize the cells, and then a counter stain of safranin is used.

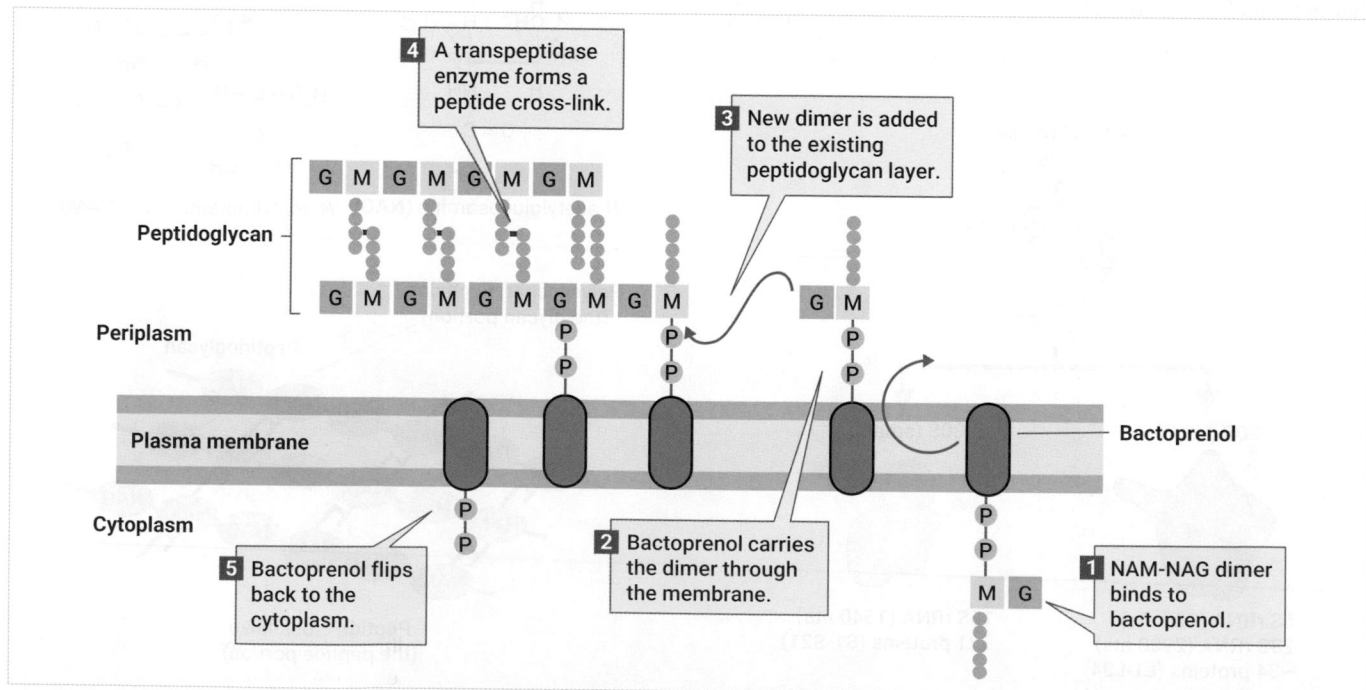

Figure 9.13 Growth of new bacterial cell wall. The NAG residues are denoted as (G) and the NAM residues as (M). NAM–NAG dimer is bound to bactoprenol (shown in blue) and carried through the plasma membrane. NAM–NAG residues in the peptidoglycan layer are broken using autolysins and the new NAM–NAG dimer inserted using transglycosylase.

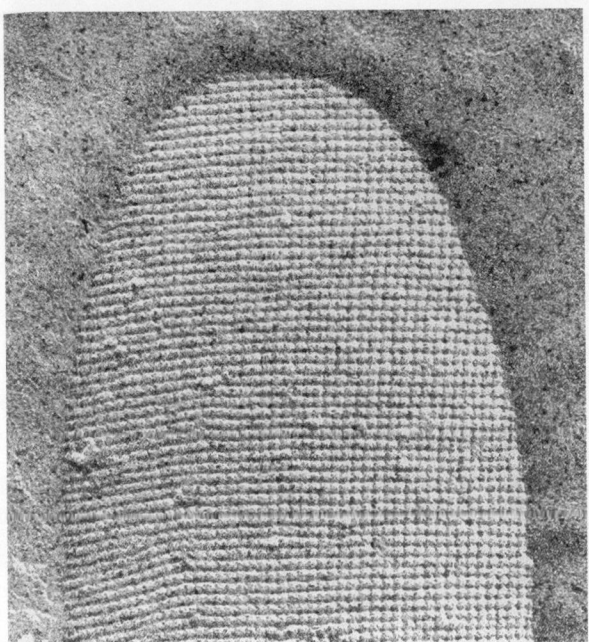

Figure 9.14 The S-layer seen in many bacteria.

Source: Reprinted by permission from Springer. Uwe B. Sleytr et al. Structural and chemical characterization of S-layers of selected strains of *Bacillus stearothermophilus* and *Desulfotomaculum nigrificans*. *Archives of Microbiology*, 146, 19–24. © 1986.

If the bacterial cell wall contains a lot of peptidoglycan, the crystal violet dye is retained within the cell wall and the cells are observed as purple under a light microscope: they are considered Gram-positive. If there is little peptidoglycan in the cell wall, then the crystal violet is washed away with the ethanol. When the cells are then stained with safranin they appear to be red under a light

microscope and are considered Gram-negative (also referred to as **diderms**).

PAUSE AND THINK

How has the Gram stain been important for clinical microbiology?

Answer: The Gram stain is used in clinical microbiology laboratories not only to distinguish between Gram-positive and Gram-negative bacteria, but also to examine cell morphology. It is a stain that is cheap and easy to perform without the use of expensive equipment.

Gram-positive bacteria

Gram-positive bacteria possess a rigid peptidoglycan layer within the cell wall. This polysaccharide layer increases the strength of the cell wall and helps to protect the cell. In Gram-positive bacteria, up to 90 per cent of the cell wall consists of peptidoglycan and the layer is 20–80 nm thick. The peptidoglycan is reinforced with teichoic acid—chains of glycerol or ribotol linked with phosphodiester bonds. These teichoic acid molecules can extend beyond the cell wall (see Figure 9.16), and could act as an attachment for infection by **bacteriophages**, viruses that attack bacterial cells.

Gram-positive bacteria are capable of secreting enzymes, molecules, and proteins, including **exotoxins**, which are vital for protection and survival; they are excreted by **exocytosis** into their immediate environment. We will learn more about exotoxins in Chapter 14.

Gram-negative bacteria

Gram-negative bacteria have a double plasma membrane separated by a periplasmic space (or **periplasm**). In contrast to Gram-positive cell walls, Gram-negative cell walls consist of a thin layer

---

EXPERIMENTAL TOOLKIT 9.1   **The 'perfect' imperfect stain and exceptions to the rule**

Staining was a really important technique for early microbiologists, and many different stains were used. One problem when dealing with pathological specimens was how to differentially stain the bacteria, and not the surrounding tissue. Mordants were one way this could be achieved: mordants fix a stain within the cell, and other chemicals can then be used to decolourize.

Hans Christian Gram wanted to stain lung tissue from patients who had died from pneumonia. He used crystal violet and then fixed this stain with the mordant iodine, but he found that this then precipitated the crystal violet. In an attempt to clean up the slide, he used ethanol to decolourize, and when he examined the slides under the microscope he found that the lung tissue had been decolourized due to the alcohol, but that the iodine had fixed the crystal violet in the bacteria.

Gram published his observation, stating '*I have therefore published the method, although I am aware that as yet it is very defective and*

imperfect; but it is hoped that also in the hands of other investigators it will turn out to be useful'. Gram was not aware that his stain would form the basis of distinguishing between the cell walls of Gram-positive and Gram-negative bacteria, and that his stain would still be in use over 130 years later!

Later methods used safranin to counterstain Gram-negative bacteria, so that they would appear red.

As enduring as this stain has become, there are complexities to Gram status. Some bacteria, such as the *Mycoplasma*, lack a cell wall and so cannot be stained at all. Other bacteria, such as *Deinococcus*, which can survive millions of rads of radiation, have an incredibly thick layer of peptidoglycan, and also possess two membranes, more associated with Gram-negative bacteria.

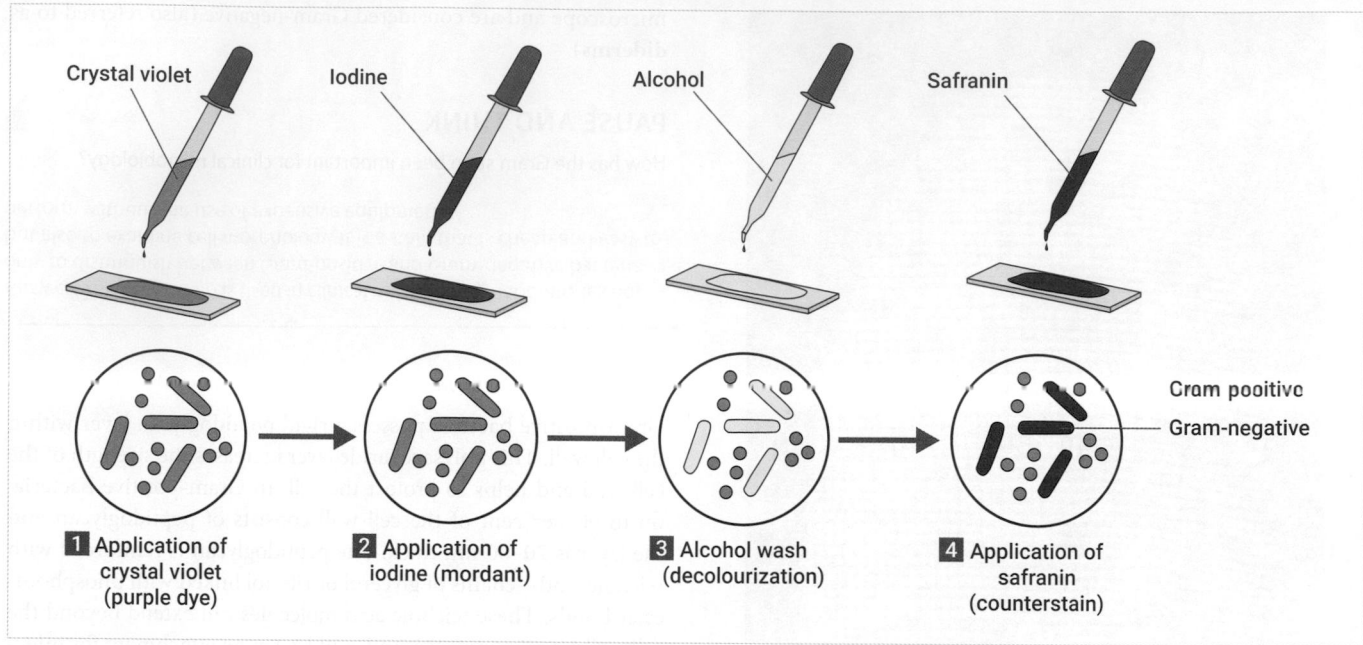

Figure 9.15 Steps in the procedure of the Gram stain.

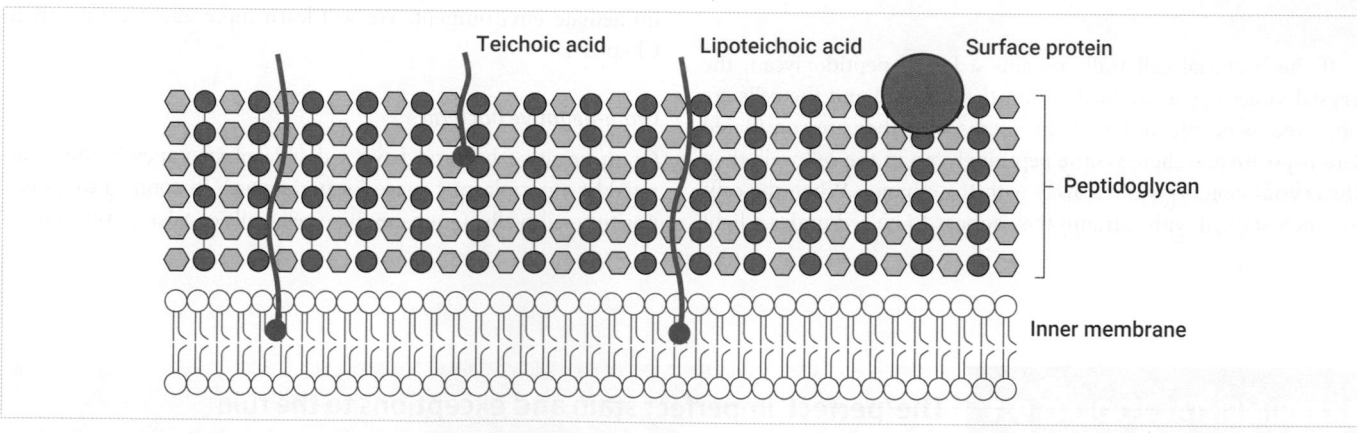

Figure 9.16 Gram-positive cell wall showing a phospholipid membrane layer (blue) with a thick peptidoglycan layer. Hexagons represent the polysaccharide backbone of the peptidoglycan layer and the orange circles represent the amino acids cross-linking. Surface proteins are integrated into the peptidoglycan layer, as well as teichoic and lipoteichoic acids.

of peptidoglycan of only one or two sheets, constituting less than 10 per cent of the cell wall overall, and being only 2–10 nm thick.

In contrast to the Gram-positive cell wall, the peptidoglycan layer of Gram-negative bacteria is covered by the outer membrane, as illustrated in Figure 9.17. The outer membrane contains lipo-polysaccharide (LPS) and porins. Porins are a class of proteins that form channels in this outer membrane; these channels allow the movement of small molecules, such as ions, in and out of the peri-plasmic space via passive diffusion. The periplasm is a very active area for metabolism: it is the location for many enzymes and trans-port proteins.

Integrated in the outer membrane are complex lipopoly-saccharide (LPS) molecules. LPS, also known as **endotoxins**, play a vital role in toxicity as these molecules are capable of inducing an immune response in higher organisms. We will learn more about the role of endotoxins in bacterial infection in Chapter 14. LPS are also involved with various cell interactions, surface adhesion, bacteriophage sensitivity, and they also increase the stability of the outer cell membrane.

The differences between the cell walls, illustrated in Figure 9.18, result in bacterial cells with quite different properties. These differ-ences are summarized in Table 9.3.

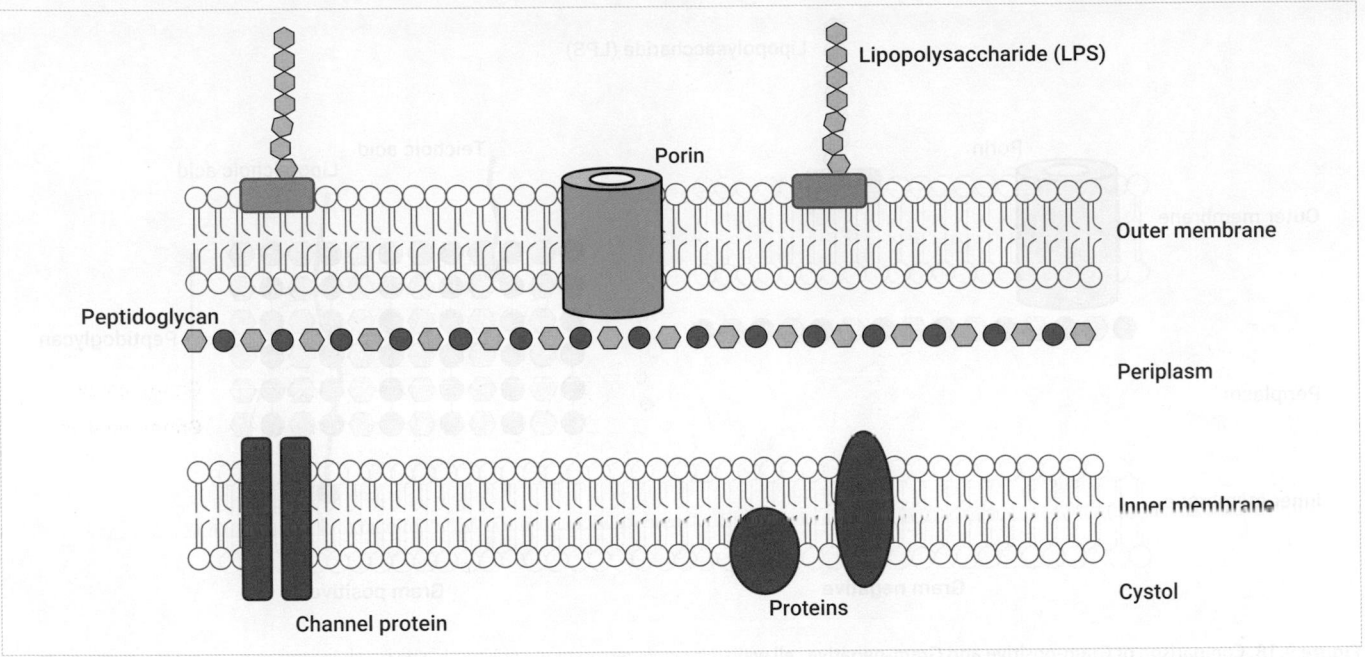

Figure 9.17 Gram-negative cell wall showing the two phospholipid membrane layers (blue) with a thin peptidoglycan layer. Hexagons represent the polysaccharide backbone of the peptidoglycan layer and the orange circles represent the amino acids cross-linking. The membrane includes proteins integrated into both the inner and outer membrane, and the LPS present in the outer membrane.

Table 9.3 Summary of key differences between Gram-positive and Gram-negative bacteria

Gram-negative	Gram-positive
Gram stain—observed as red due to the counterstain safranin	Gram stain—observed as purple due to retention of the crystal violet stain
<10% peptidoglycan	Thick peptidoglycan layer (90%), multiple layers
Periplasm	No periplasm
Flagella—4 rings	Flagella—2 rings
Primarily endotoxins (LPS)	Primarily exotoxins
Lower resistance to drying	Higher resistance to drying
Increased susceptibility to streptomycin, chloramphenicol, and tetracycline	Increased susceptibility to penicillin
Endospores	

Bacteria with no cell wall

The *Mycoplasma* and *Phytoplasma* are genera of bacteria that lack a cell wall. Instead, the plasma membrane is stabilized by the presence of sterols. Despite the lack of a cell wall they are very closely related to bacteria that *do* have a cell wall. This suggests they lost the cell wall during their evolution. Bacteria that belong to the genus *Phytoplasma* are obligate plant parasites that are limited to the phloem of the plant and are transmitted by insects.

Mycoplasmas are all pathogens of plants, animals, and humans; a good example is *Mycoplasma pneumonia*, which causes a respiratory tract infection.

L-forms are bacteria that can lose their cell wall. L-forms have the intriguing capacity, having lost their cell walls, to abandon fission as a means of reproduction. Instead, they use membrane blebbing, tubulation, or vesiculation, which does not require cytoskeletal proteins. These strategies are summarized in Figure 9.19.

Bacteria without cell walls can also be made artificially by digesting the peptidoglycan. Gram-positive bacteria form **protoplasts** and Gram-negative produce **spheroplasts** when the cell wall is removed in this manner. These are different to L-forms as they cannot replicate.

> **PAUSE AND THINK**
>
> Why have some bacteria evolved to lose their cell wall?
>
> *Answer:* Cell walls were not needed for bacteria such as *Mycoplasma* because they live within other cells where the osmotic environment is controlled.

Other specialized structures

We learned earlier in this section about the features shared by the majority of bacteria—their nucleoid, cell wall, S-layers, and mode of reproduction. In this section we discover that bacteria also have a number of other structures, both internal and external, which

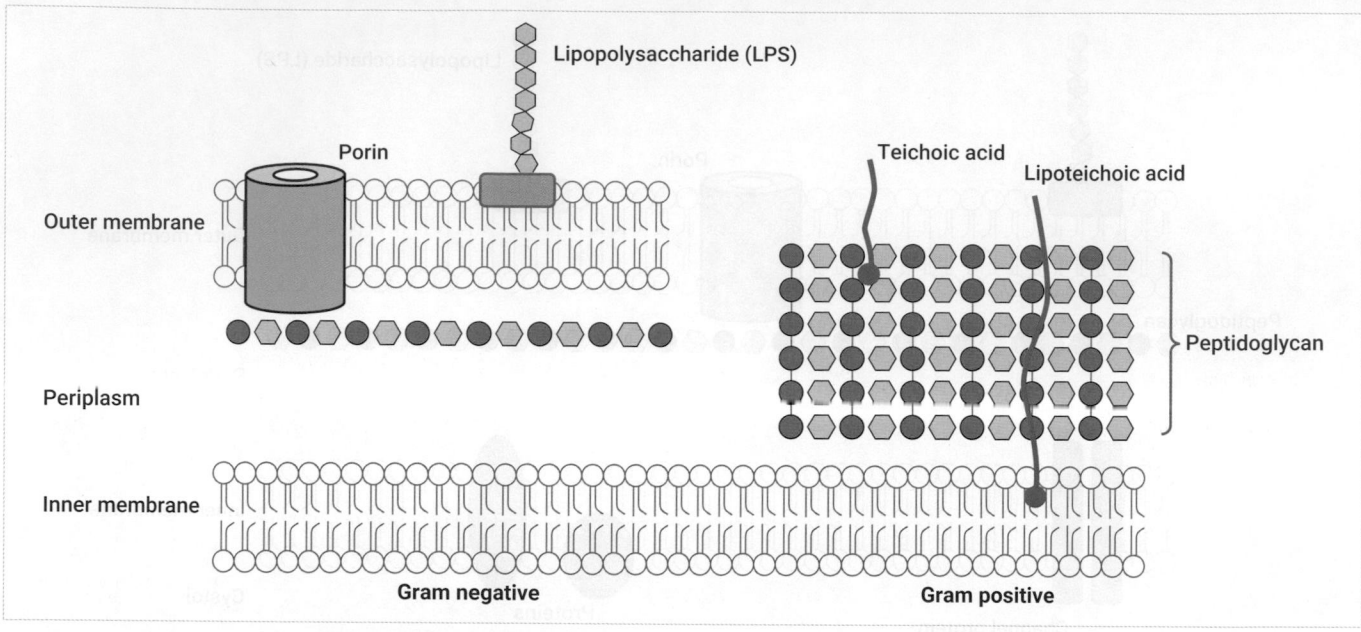

Figure 9.18 Comparison of Gram-positive and Gram-negative cell walls. Hexagons represent the polysaccharide backbone, circles represent the amino acids cross-linking. Note the absence of the periplasm in Gram-positive cell walls.

Figure 9.19 L-form reproduction.

Source: Errington, Jeff. L-form bacteria, cell walls and the origins of life. *Open Biology* vol. 3,1 120143. 8 Jan. 2013, doi:10.1098/rsob.120143 CC BY 3.0 http://creativecommons.org/licenses/by/3.0/.

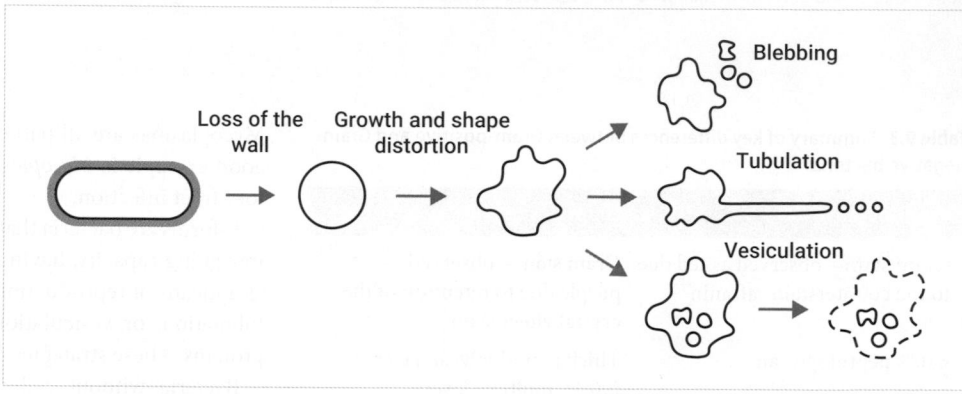

enable their survival in different environments. However, these structures are not shared by all bacteria.

The glycocalyx

The **glycocalyx** consists of polysaccharides and/or polypeptides that form a viscous and sticky layer surrounding the external cell wall of the bacterium. When the structure is organized and attached to the cell wall, the glycocalyx layer is described as a **capsule**; if it is loosely attached and disorganized it is referred to as a **slime layer**.

The glycocalyx layer provides protection from desiccation, and also has a vital role in protecting pathogenic bacteria from being phagocytosed. An example of this is *Streptococcus pneumoniae*, a Gram-positive bacterium, which can infect the upper respiratory tract and cause pneumonia. It has a polysaccharide glycocalyx capsule that prevents it being destroyed by the host's immune system. We will learn more about this virulence factor in Chapter 14.

Flagella

Many, but not all, bacteria are capable of moving between different environments. Bacterial movement is referred to as **taxis**. There are many reasons why bacteria may need to move; for example, anaerobic bacteria can move towards an environment with lower oxygen present. The movement towards or away from a chemical is called **chemotaxis** and the movement towards or away from light is **phototaxis**.

Flagella (singular flagellum; also known as axil filaments) play a vital role in movement and virulence in bacteria. These hair-like structures made from the protein flagellin are anchored in the inner membrane using the basal body, which is composed of rings. The hook is located at the base of the filament and is wider than the filament itself; this attaches the filament to the motor protein at the base, as illustrated in Figure 9.20. The rotation of the motor protein drives the rotation of the filament, which propels the bacteria.

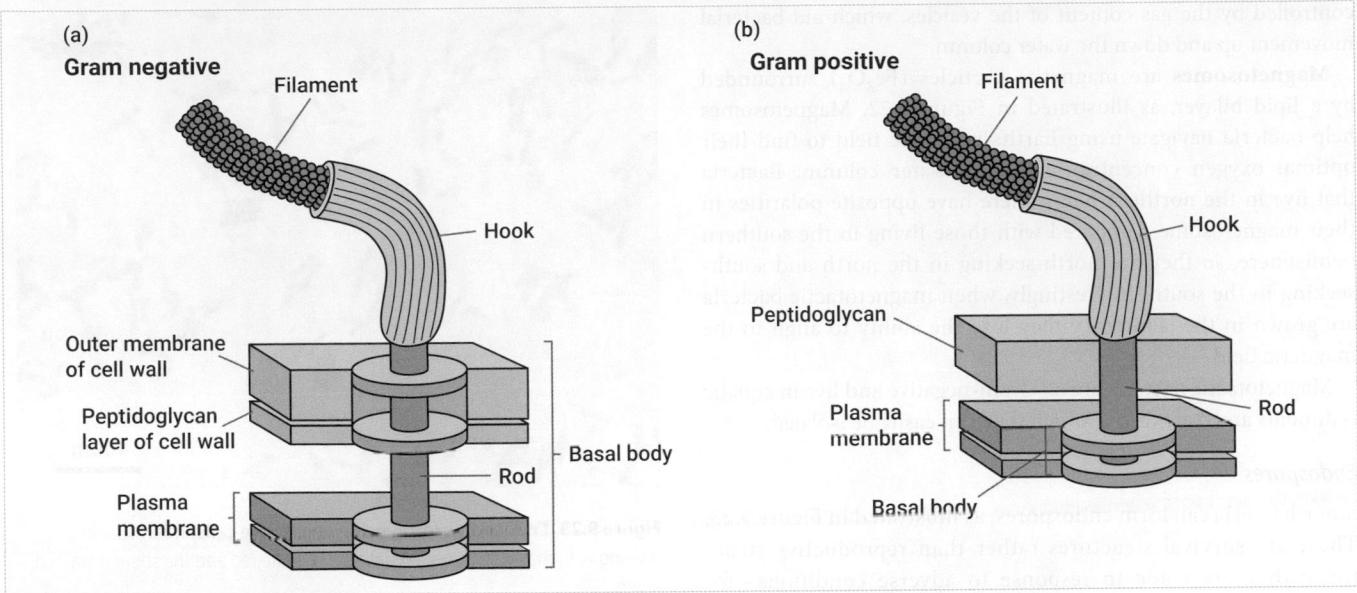

Figure 9.20 Parts and attachment of flagellum. (a) The flagellum in a Gram-negative bacterium. (b) The flagellum in a Gram-positive bacterium.
Source: Clinicalgate.com.

The structure of the flagellum differs depending upon the type of bacteria: Gram-negative bacteria have four rings in the basal body, whereas Gram-positive bacteria have only two.

Flagella can be arranged differently over the surface of bacteria, and these arrangements can be seen in Figure 9.21. The arrangement of flagella all over the surface of a bacterium is known as **peritrichous** arrangement; this arrangement is common in *Salmonella* and *Escherichia* species. A **lophotrichous** arrangement sees flagella bundled at one end of the cell. An **amphitrichous** arrangement occurs when one flagellum, or a bundle of flagella, is found at either pole of a bacterium, common in Spirochaetes. Finally, a **monotrichous** arrangement describes an arrangement whereby a singular flagellum is located at one end of the cell.

PAUSE AND THINK

Why don't all bacteria have flagella?

Answer: Flagella are important for movement, to find new sources of nutrition, and to adhere to other cells. So, not to possess them might seem to be a disadvantage. But some bacteria, such as *Myxococcus xanthus*, have the ability to move without flagella and glide over surfaces. The production and then use of flagella is expensive in terms of energy. Interestingly, it has recently been shown that some non-motile species can cheekily hitchhike a ride on other motile species.

Pili

Pili (singular pilus) are found on the surface of bacteria. They are long, thin, filamentous structures, made from a protein called pillin, that extend from the surface of bacteria. (Look back at Figure 9.10 to see pili represented in the context of other bacterial cell features.)

There are two types of pili. One type is involved in **conjugation**, whereby genetic material (for example, plasmids) can be exchanged between bacterial cells. This is sometimes termed as a 'sex pilus'. The other type of pili are called **fimbriae**. These pili help bacteria to adhere to surfaces and other cells and therefore also play an important role in forming **biofilms**. The ability to attach to host cells and form biofilms means the bacteria are capable of colonizing an environment. Biofilms can be useful in bioremediation. However, in a clinical setting they can cause many issues, such as leading to infections of prosthetic implants.

▶ We learn more about biofilms when we consider quorum sensing later in this section.

Gas vesicles and magnetosomes

Gas vesicles are hollow structures made of protein that are permeable to gases but not to water. They control the floating properties of bacteria that live in water; they are buoyancy regulators that are

Figure 9.21 Different arrangements of flagella across the surface of the bacterial cell.
Source: Openstax CC BY 4.0 Access for free at https://openstax.org/books/microbiology/pages/1-introduction.

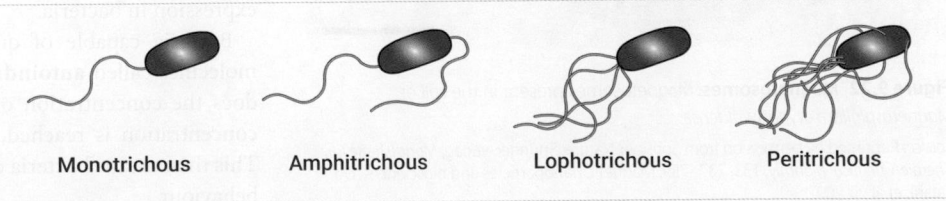

Monotrichous   Amphitrichous   Lophotrichous   Peritrichous

controlled by the gas content of the vesicles, which aid bacterial movement up and down the water column.

Magnetosomes are magnetite particles (Fe_3O_4) surrounded by a lipid bilayer, as illustrated in Figure 9.22. Magnetosomes help bacteria navigate using Earth's magnetic field to find their optimal oxygen concentration in the water column. Bacteria that live in the northern hemisphere have opposite polarities in their magnetosome compared with those living in the southern hemisphere, so they are north-seeking in the north and south-seeking in the south. Interestingly, when magnetotactic bacteria are grown in the laboratory they lose the ability to align to the magnetic field.

Magnetotactic bacteria are all Gram-negative and live in aquatic sediments and water, from which they can easily be isolated.

Endospores

Some bacteria can form endospores, as illustrated in Figure 9.23. These are survival structures rather than reproductive structures: they are made in response to adverse conditions—for example, a lack of nutrients. One bacterium will produce a single endospore within its cell. Endospores do not metabolize, contain very little water, and are extremely resistant to extreme environments, such as heat, desiccation, radiation, acids, and chemical disinfectants.

Endospores have a unique structure, shown in Figure 9.24. They have an outer protein-based coat, beneath which is a thick layer of peptidoglycan called the cortex. Under the cortex is a germ cell wall that will produce the new wall of the bacterium when the endospore germinates. The core contains the DNA, a few ribosomes, and dipicolinic acid, which stabilizes and protects the spore DNA.

Endospore formation is a feature of soil-dwelling organisms such as *Bacillus* spp. Endospores can lay dormant for thousands of years before germinating in the presence of water and nutrients.

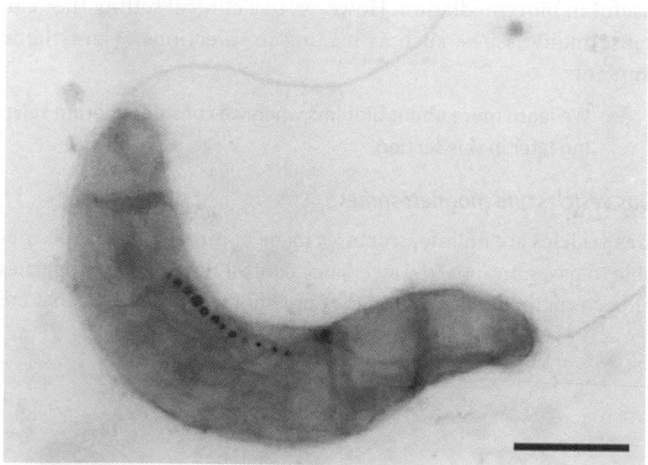

Figure 9.22 Magnetosomes. Magnetosomes present in the cell of *Magnetospirillum gryphiswaldense*.

Source: Reprinted by permission from Springer Nature: Springer-Verlag, *Monatshefte für Chemie/Chemical Monthly*, 133, 737–759. Magnetic nanoparticles and biosciences. Ivo Šafařík et al., © 2002.

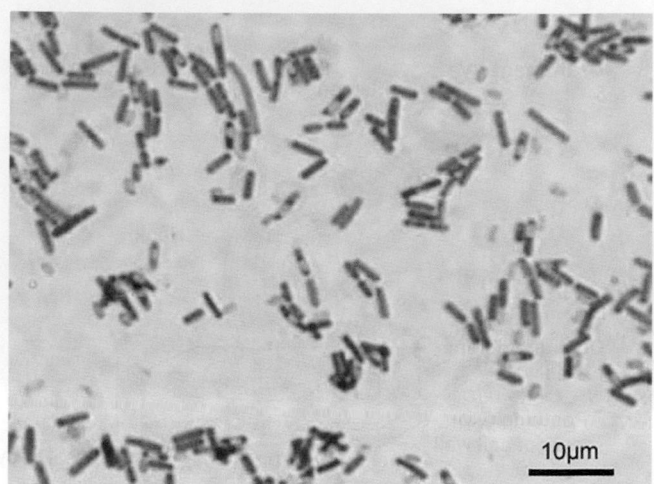

Figure 9.23 Endospores form inside vegetative cells. This is shown by staining with malachite green. Vegetative cells are red and the spore is stained green.

Source: Wikimedia Commons/Y tambe/CC BY-SA 3.0.

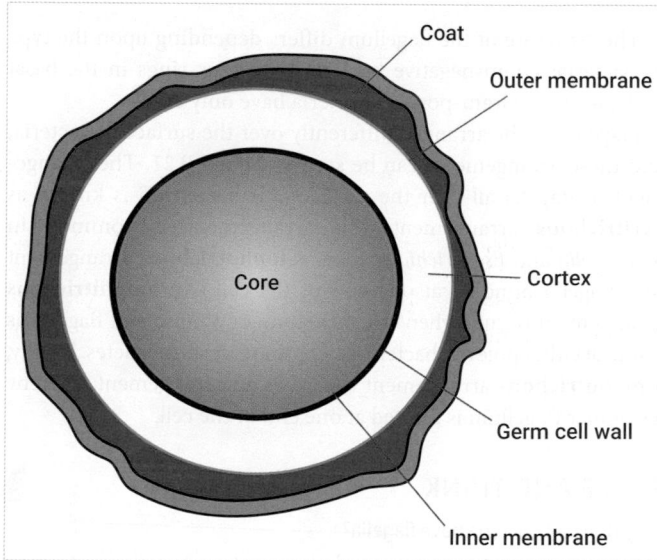

Figure 9.24 The structure of a bacterial endospore.

Biofilms

Prokaryotes have evolved, over billions of years, a method of communication called **quorum sensing**. Although it is not as advanced as the communication exhibited by multicellular organisms, quorum sensing is a vital process and has a huge impact on gene expression in bacteria.

Bacteria capable of quorum sensing release chemical signal molecules called **autoinducers**. As a cell population increases, so does the concentration of these autoinducers. Once a threshold concentration is reached, gene expression in bacteria is altered. This then means bacteria can coordinate gene expression and their behaviour.

The evolution of quorum sensing is thought to be one of the early developments in the formation of multicellular organisms, as different bacteria can co-inhabit a single environment, coordinate, and communicate with each other. Gram-positive and Gram-negative bacteria are both capable of quorum sensing, but they use different mechanisms and molecules.

Quorum sensing is involved in a whole range of prokaryotic (and archaeal) activities, such as the formation of symbiotic relationships, motility, sporulation, conjugation, production of antibiotics, and virulence. It is also central to the formation of biofilms. A biofilm is a population of bacteria that, having originally been free-floating (planktonic), subsequently stick together or aggregate on a surface to form a film or mat. The ability to form biofilms has long been a feature of prokaryotes. Stromatolites dating back 3.7 billion years were formed from layers of biofilms that trapped sediments; some stromatolites are shown in Figure 9.25.

Biofilm-forming bacteria produce an extracellular protective layer called **extracellular polymeric substance** (EPS), which consists of polysaccharides and proteins. This helps protect the community from external influences such as desiccation, antibiotics, and host immune systems. A biofilm can consist of many species of bacteria. However, they share nutrients and communicate using quorum sensing to coordinate gene expression, especially the genes involved in producing EPS.

Biofilms are found in a wide range of environments, such as in wastewater treatment plants and on root nodules, as well as in hydrothermal vents and hot springs. Biofilm formation can have devastating effects when they develop in a clinical setting, however; for example, biofilms on catheters can lead to urinary tract infections, while prosthetic joints and heart valves can provide surfaces for biofilm formation, along with plaque forming on teeth. All biofilms formed in a clinical setting help increase the bacteria's virulence, as their protective layer shields them from antibiotics and from being attacked by the immune system, making treatment of infections significantly more complicated.

 Check your understanding of the concepts covered in this section by answering the questions in the e-book.

9.3 The characteristics of the Archaea

Having considered the features of bacterial cells, let us now focus on the characteristics of the Archaea.

Archaea (from the Greek for 'ancient things') are sometimes confused for bacteria: they are also prokaryotic but are structurally different to bacteria and often inhabit harsh or extreme environments. Archaea belong to their own domain, which is extremely diverse and includes many species that are capable of tolerating a wide range of niches, from high saline lakes to hydrothermal vents. They are also capable of utilizing various energy sources, such as metal ions and organic compounds; some can even use hydrogen gas. We will learn more about the diversity of the Archaea in Chapter 11; in this section, however, we will explore the structure of the cells of the Archaea, and how their characteristics differ from those of the Bacteria and the Eukarya.

Cell morphology and structure

The Archaea have some traits that are similar to bacteria in that they have no organelles and their genetic material is 'free-floating' in the cytoplasm. They range in size from 0.1 to 15 µm in length and reproduce by binary fission. Their cell morphology is also similar to bacteria: they can exist as rods, cocci, and spirals. Some archaea can have an unusual morphology such as the halophile *Haloquadratum walsbyi*; Figure 9.26 shows how it has a flat, square shape.

▶ We learn more about binary fission in Chapter 10.

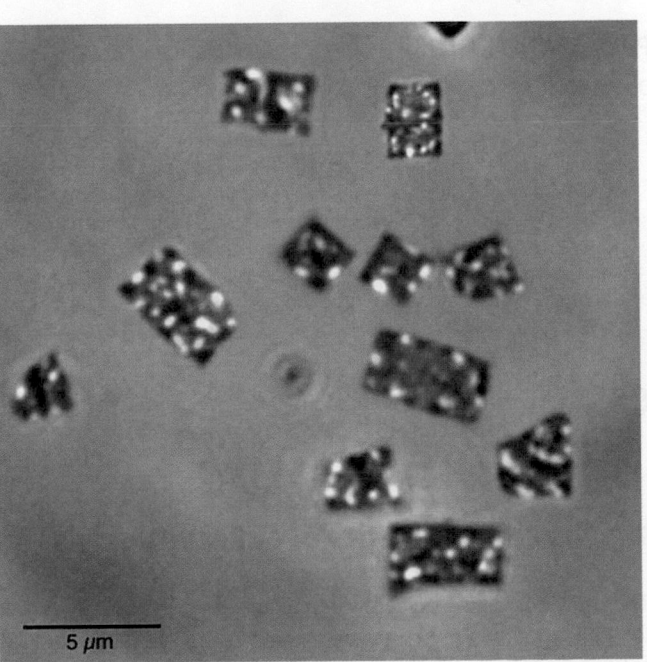

5 µm

Figure 9.26 Cells of the halophile *Haloquadratum walsbyi*.
Source: charvannoort/Shutterstock.com.

Figure 9.25 Stromatolites in Shark Bay, Australia.

Haloquadratum walsbyi contains gas-filled vesicles to increase its buoyancy, which is thought to help increase the amount of light captured by the cells. This archaeon has been found in hyper-saline environments where sea water evaporates and where ultraviolet (UV) light is also very high.

The cell membrane of archaea is different to that of the bacterial membrane discussed in Section 9.1. Their membrane consists of phospholipids containing glycerol-ether linkages (see Figure 9.8), which increase their stability. In many archaea the fatty acids fuse to form one molecule with two polar (hydrophilic) phosphate heads; this increases the stability of the membrane even further and helps them to tolerate extreme environments.

The cell wall of archaea is also different to that of bacteria, and contains pseudopeptidoglycan. We already know from Section 9.2 that bacterial peptidoglycan contains alternating residues of *N*-acetylglucosamine (NAG) and *N*-acetylmuramic acid (NAM) residues joined together with β 1–3 linkages. However, the Archaea feature pseudopeptidoglycan that has *N*-acetyltalosaminuronic acid instead of NAM, joined with β 1–4 linkages. The cell wall of archaea is resistant to digestion by lysozyme, and it is not susceptible to the antibiotic penicillin.

Although there is still debate, it is thought that the Eukarya branched from the Archaea early in evolution. Indeed, the Archaea share a number of similarities with the Eukarya, which might point to a common ancestor:

- they have histone-like proteins associated with their DNA, as do eukaryotes;

- their DNA is also organized into a chromatin-like structure;

- their DNA and RNA polymerases are very similar to those in eukaryotes.

PAUSE AND THINK

Do you think the Archaea are suitably named?

Answer: Archaea were first discovered from extreme environments, which many scientists thought were similar to those on ancient Earth where life first evolved. Thus, they considered them to be ancient ancestors, and named them Archaea, which comes from the Greek and means 'ancient things'. Modern phylogenetic analysis, however, points to evidence that the Eukarya branched early from the Archaea, and that there is likely to be an even more ancient ancestor.

 Check your understanding of the concepts covered in this section by answering the questions in the e-book.

9.4 The characteristics of eukaryotic cells

The activities taking place inside eukaryotic cells are responsible for driving almost every process going on in the body. Indeed, eukaryotic cells and the multitude of processes taking place within them drive much of life on Earth. By the end of this section we will have helped you to understand how.

Eukaryotic cell structure and function

All organisms within the domain Eukarya consist of the same basic type of cell: the eukaryotic cell. This means that, at the cellular level, all known animals, plants, and fungi are really very similar. For example, various metabolic and other processes (such as glycolysis and transcription) will be the same whatever organism a particular eukaryotic cell is found in. However, the precise genes being expressed—and therefore the functions of different eukaryotic cells—can vary considerably within cells of an organism and between cells from different organisms. For example, more than 200 different cell types have been found within the human body, each with its own shape, size, and function.

Eukaryotic cells differ vastly in their structure from that of prokaryotic cells but, as we have learnt, all cells—whether eukaryotic or prokaryotic—share some basic common features. To remind you, these shared features are:

- being enclosed by a selectively permeable barrier known as a plasma membrane;

- having internal cellular components suspended within a fluid-like substance known as the cytosol;

- the presence of protein-building components known as ribosomes; and

- the presence of genetic material in the form of DNA, coiled up to form chromosomes.

This final similarity is also linked to one of the major differences between eukaryotic and prokaryotic cells: the location of those chromosomes within a cell. A eukaryotic cell has its chromosomes located within a membrane-bound structure known as a nucleus. (Indeed, the name 'eukaryotic' comes from the Latin meaning 'true (*eu*) nucleus (*karyon*)'.) Prokaryotic (meaning 'before (*pro*) nucleus') cells have their chromosome(s) located in an area known as the **nucleoid**—a region that is not enclosed by a membrane and is therefore not physically separated from the rest of the cell's internal environment. So, the presence of internal membrane-bound organelles within eukaryotic cells (and their absence from prokaryotic cells) forms the fundamental difference between these two types of cell.

Membranes found within eukaryotic cells divide them into different compartments. These membrane-bound compartments, known as organelles, are suspended within the cytosol. The cytosol plus the organelles (with the exception of the nucleus) constitutes the cytoplasm, the region enclosed within the plasma membrane of a eukaryotic cell. Multiple types of organelle are found within each eukaryotic cell, some numbering just one per cell and others numbering in the hundreds, or even thousands. Each organelle has a specific function; the concerted processes carried out by all these organelles constitutes a cell's metabolism.

 We learn more about a cell's metabolism in Chapter 7.

Endosymbiosis: the evolution of organelles

Eukaryotic organelles are believed to have originated from ancestral prokaryotic organisms as a result of **endosymbiosis**.

Endosymbiosis is seen when two species live together, one inside the other.

The idea of endosymbiosis being linked to evolution was first proposed by Lynn Margulis. She was interested in living organisms that exhibit endosymbiotic relationships. Such relationships are common in nature; for example, the bacterium *Rhizobium* can form bacteroids inside the roots of leguminous plants, where it fixes atmospheric nitrogen. Another example is the algae *Chlorella*, which lives inside the ciliate *Paramecium bursaria*. *Chlorella* fixes carbon through photosynthesis, providing the *P. bursaria* with sugar in the process. In return, the *P. bursaria* provides a sheltered environment for the *Chlorella*.

In order for endosymbiosis to occur, the early ancestral eukaryote first had to lose its cell wall. This change enabled the processes of **exocytosis** and **endocytosis**—the movement of substances such as macromolecules out of and into the cell—to develop. This cell, with its ability to move large molecules across its membrane, started to act as a predator towards other single-celled prokaryotic organisms.

Approximately 2 billion years ago, a small prokaryotic cell capable of producing energy through **oxidative phosphorylation** was engulfed by a larger prokaryotic cell—the early ancestral eukaryote. Instead of being digested, the smaller cell was retained intact inside the larger predatory cell. The relationship was mutually beneficial: the larger cell, thought to be related to *Lokiarchaeota* (from the Archaea domain), provided nutrients and shelter, while the smaller, engulfed cell was able to produce energy that the larger cell was able to utilize, giving it a selective advantage. Over time, the smaller engulfed cell lost the genes needed for its independent survival through **reductive evolution** and evolved into the organelle that we recognize today as the mitochondrion.

▶   We learn more about oxidative phosphorylation in Chapter 7.

There are several lines of biological evidence for this theory:

- mitochondria retain some of their own genome, which on analysis appears to be closely related to free-living aerobic bacteria such as *Paracoccus* species;
- mitochondria are able to replicate themselves via binary fission;
- mitochondria possess a double membrane; and
- remnants of transcription and translation apparatus exist within mitochondria.

Plant cells evolved from a similar endosymbiotic event that happened a lot later, about 1.2 billion years ago. It is proposed that early cyanobacteria were engulfed by an amoeboid or a flagellated cell to give rise to a **plastid**. Primary plastids are located in the cytosol of their hosts and have two membranes and a peptidoglycan layer, all of which indicate that they originated from cyanobacteria. A second endosymbiotic event then saw an algal cell containing plastids being engulfed by a non-photosynthetic eukaryote. Evidence for this secondary endosymbiosis takes the form of plastid-derived genetic information, located within a small vestigial nucleus. It was not until 1980 that 16S rRNA analysis from plastids confirmed that cyanobacteria are indeed ancestors of the chloroplast.

Two main theories seek to explain how the nucleus of eukaryotic cells evolved. One is a symbiotic theory: that an archaea cell engulfed a proteobacterium forming the nucleus. The other theory is that the plasma membrane invaginated and formed the endoplasmic reticulum and nucleus. As we see elsewhere in this chapter, many organelles retain a small chromosome; it is thought that the remaining genome from the engulfed prokaryotes was involved in the transfer of genes to the nucleus of the host cell.

What are the benefits of having organelles?

The compartmentalization of different areas of the cell into organelles enables eukaryotic cells to be larger and more complex than prokaryotic cells. A typical eukaryotic cell might be 10–100 μm in diameter, whereas a prokaryotic cell might be 100 times smaller than that. However, the organisms that contain these cells can vary hugely in size: one eukaryote may be thousands of times larger than another. The cells of a blue whale are really very similar in size to those of a shrimp; the key difference is the sheer number of cells that come together to form each animal.

At a fundamental level, the size of a cell is dictated by a combination of its function and shape. The more metabolic processes taking place inside a cell, the more organelles (such as mitochondria) it will likely contain, and the larger it will be. To ensure the cell can conduct all these activities efficiently, it needs to be able to exchange substances with the extracellular world effectively: materials such as nutrients and waste products must pass into and out of the cell in sufficient quantity sufficiently quickly. So what limits how big a cell can be?

To understand why there *are* limits on cell size, we need to appreciate the relationship between surface area (i.e. the amount of plasma membrane available) and volume (i.e. how big it is inside the cell)—a relationship that also explains why larger organisms are multicellular. As cells increase in size, both their internal volume and the metabolic processes occurring within them will also increase. As a result, the cell will consume nutrients and produce waste substances at a faster rate. The surface area will also increase but at a different rate to that of the volume: if the radius of a cell increases, it will result in a *cubed* (x^3) increase in cell volume but only a *squared* (x^2) increase in surface area.

So how does this relationship affect the ratio of cell surface area to volume? Let us consider the answer, assuming a perfectly spherical cell (which is rarely true in nature!).

The relationship between surface area and volume: the surface area to volume ratio

Table 9.4 shows that as cells get larger the ratio of surface area to volume decreases. That is, as the volume of a cell gets incrementally larger, its surface area increases at a slower rate. A very small cell has a large surface area relative to its volume; a very large cell has a small surface area relative to its volume. Look at the fourth column of Table 9.4 to verify this for yourself.

This decrease in the ratio is significant when considering an increase in overall organism size, the total number of cells within

Table 9.4 The relationship between cell size and surface area to volume ratio assuming a spherical shape

Cell radius (arbitrary units)	Cell surface area ($4\pi r^2$) (units^2)	Cell volume ($4/3\pi r^3$) (units^3)	Ratio of surface area : volume of the cell
1	12.57	4.19	3
2	25.13	33.51	0.75
3	37.70	113.10	0.33
4	50.27	268.10	0.19
10	125.66	4188.79	0.03

the organism, and the proper functioning of the cells. Remember that the amount of surface area available for the exchange of materials with the extracellular environment is vital for efficient cell function: if there is not enough surface area, exchange cannot happen efficiently enough to support life.

It is this relationship between cell surface area and cell volume that is believed to be the main reason why cells are small: cells are small so that they can efficiently exchange nutrients and waste materials between the intra- and extracellular environments. This also explains why larger organisms consist of an increasing number of cells, as illustrated in Figure 9.27. Look at this figure and notice how an increase in the number of cells as an organism gets larger enables it to retain a steady surface area to volume ratio (meaning that it can continue to exchange nutrients and waste in an efficient way).

PAUSE AND THINK

How might the differing rates of change between volume and surface area affect the efficiency—the acquisition of required nutrients and the removal of waste products—of a cell?

Answer: Assuming a cell is a sphere, then

volume $= 4/3\pi r^3$ and SA $= 4\pi r^2$

so SA/V ratio $= 3/r$

As a cell increases in size, its volume and surface area both increase. However, likely do not do so at the same rate because the volume increases at a faster rate, and so the surface area to volume ratio decreases. In very broad terms, the larger a cell is, the more nutrients it requires and the more waste products it needs to remove; the rates of these processes will be affected by the surface area. If a cell becomes too large it will not have a large enough surface area to support the required rate of diffusion of nutrients and waste products. A cell can increase its surface area without changing its volume by changing its shape from a sphere to one where the membrane has lots of folds to create structures such as microvilli, for example.

🌀 **Check your understanding of the concepts covered in this section by answering the questions in the e-book.**

9.5 The key eukaryotic cellular structures

Having reviewed the general characteristics of eukaryotic cells, let us now examine in more detail the features they share, namely the nucleus, ribosomes, endomembrane system, Golgi apparatus,

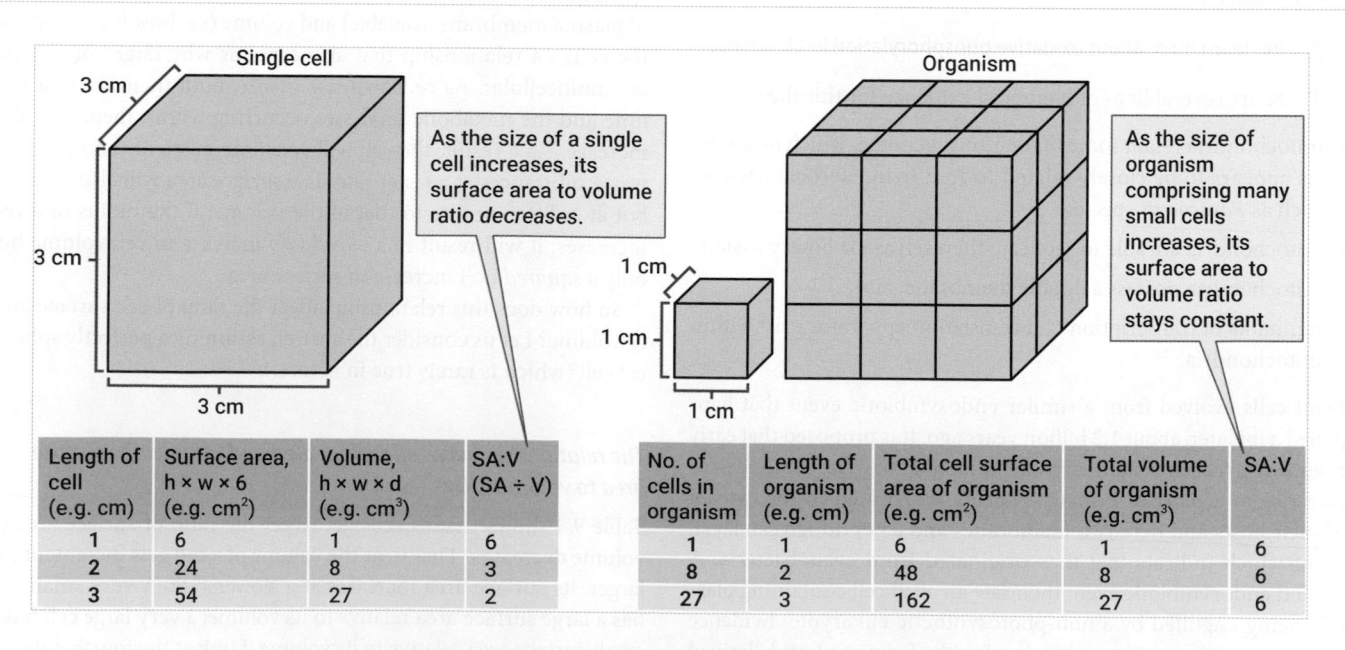

Figure 9.27 Surface area to volume ratio. A schematic indicating the difference in surface area (SA) to volume (V) ratio between single cells and multicellular organisms.
Source: © Dr Mike Dyall-Smith.

lysosomes, cytoskeleton, and mitochondria. We will then examine those structures that are not universal: the cell wall, chloroplasts, and peroxisomes.

The nucleus

A eukaryotic cell contains a single nucleus and it is here that the majority of genetic material within the cell can be found. Additional genes are also found in two different organelles: mitochondria and chloroplasts. The nucleus is a relatively large organelle, about 5 μm in diameter, making it one of the most noticeable organelles within the cell, and one which is easily identifiable with the use of a light microscope, as demonstrated in Figure 9.28a.

As shown in Figures 9.28b and 9.28c the structure of the nucleus is one of layers, starting with the outermost layer comprising two membranes, which form the nuclear envelope. The two membranes lie parallel to each other around 25–40 nm apart; the space in between is called the **perinuclear** space and is continuous with the lumen of the rough endoplasmic reticulum. However, there are multiple points throughout the membrane where these two membranes are joined together by pore complexes, known as **nuclear pores**. These pores create channels that span both the membranes and enable the controlled movement of substances into and out of the nucleus. Nuclear pores are crucial for effective cell function as they regulate the movement of substances, such as proteins, into the nucleus as well as the movement out of substances produced and/or assembled inside the nucleus, such as RNA and ribosomes.

Within the nuclear envelope lies the genetic material of the cell in a form known as **chromatin**. Chromatin is made up of a cell's genetic material (that is, deoxyribonucleic acid (DNA)), wound around a series of histone-packaging proteins. It is the job of some of these packaging proteins to encourage the DNA to coil up into a more compact shape that eventually forms the structure known as a chromosome: one long DNA molecule wound around a series of packaging proteins. Unwound, a single molecule of human DNA from a single cell could be 2–3 m long. When we consider that each typical human cell contains 46 chromosomes, it is easy to see why the role of the packaging proteins is so important: the DNA coiled

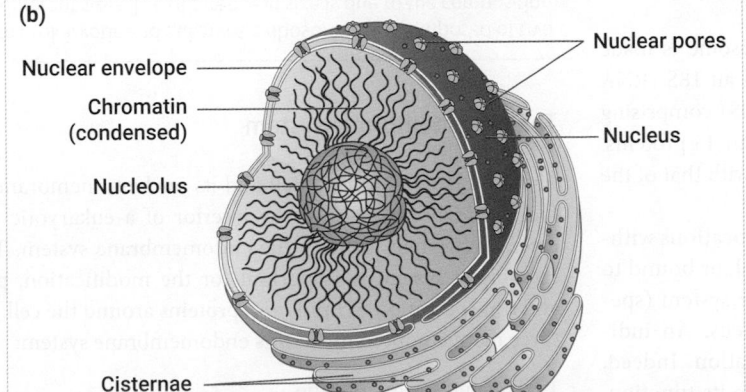

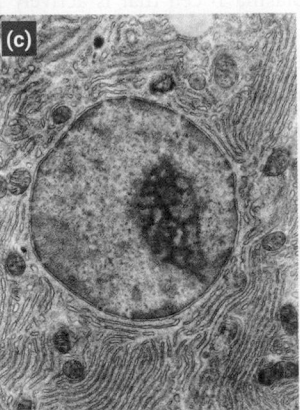

Figure 9.28 The eukaryotic cell nucleus. (a) Light microscope image of cheek cells that have been stained with methylene blue. The nucleus is clearly visible as a darker mass within the cell. (b) Diagram of the nucleus of a eukaryotic cell showing its key components. (c) Transmission electron micrograph of the nucleus.

Source: (a) Joseph Elsbernd/Flickr CC BY-SA 2.0 https://creativecommons.org/licenses/by-sa/2.0/; (b) Openstax CC BY 4.0 Download for free at http://cnx.org/contents/14fb4ad7-39a1-4eee-ab6e-3ef2482e3e22@22.42; (c) Don Fawcett/Science Photo Library.

around the packaging proteins means that a huge amount of genetic material can be packaged inside the nucleus of each individual cell.

▶ We learn more about the cell's genetic material in Chapter 3.

Nestled in the centre of all the genetic material is a structure known as the **nucleolus**. Only visible in cells which are not undergoing division, the nucleolus is a dense region of mainly proteins whose primary function is to produce ribosomal ribonucleic acid (rRNA), which is then assembled with ribosomal proteins into the mature 40S and 60S ribosome subunits. These are then able to leave the nucleolar region (and the nucleus itself) through the pores in the nuclear envelope.

PAUSE AND THINK

What are the advantages of having a membrane-bound nucleus?

Answer: A membrane-bound nucleus helps with the packaging of DNA into one location; it provides extra protection for the DNA from the many cellular reactions occurring within the cytoplasm or other parts of the cell. The compartmentalization allows the specialism of the nucleus, increasing the efficiency of related processes (such as DNA replication and production of RNA); it also allows a level of regulation of gene expression by being able to control the rate of RNA leaving the nucleus via the nuclear pores.

Eukaryotic ribosomes

Ribosomes are small structures made up of a combination of proteins and rRNA. Their primary function, like those ribosomes in prokaryotic cells, is the synthesis of proteins. Unlike the nucleus, ribosomes are not enclosed by a membrane and therefore they are not considered to be organelles. However, ribosomes and other large supramolecular complexes (e.g. the spliceosome and proteosome) are sometimes referred to as non-membrane-bound organelles. Also, while there is only one nucleus located inside each cell, a eukaryotic cell that is producing large quantities of proteins—for example, a plasma B cell that is actively producing antibodies—may contain many millions of individual ribosomes.

As you can see from Figure 9.29, each 80S ribosome is made up two subunits: a small subunit (40S) comprising an 18S rRNA molecule and 30 proteins; and a large subunit (60S) comprising three rRNA molecules, 28S, 5.8S, and 5S, along with 45 proteins. Look back at Figure 9.11 to compare this structure with that of the prokaryotic ribosome.

Ribosomes are able to function at two different locations within a cell: either freely suspended within the cytosol, or bound to the outer membrane of either the endomembrane system (specifically the endoplasmic reticulum) or the nucleus. An individual ribosome is able to function at either location. Indeed, it can alternate between locations and still perform its function. Ribosomes that are bound to the endomembrane system are well placed to produce proteins that are destined to be secreted by the cell: they are operating directly next to the specific organelle, the Golgi body, that will then package and transport those proteins out of the cell.

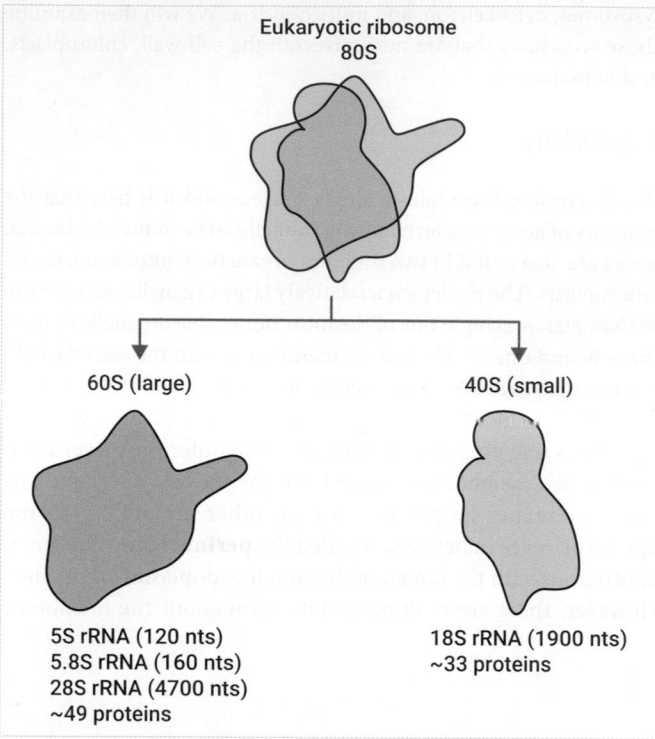

Figure 9.29 The structure of the eukaryotic ribosome.

Source: Nancy Craig et al., *Molecular Biology, Principles of Genome Function*, 3rd edn, 2021, Oxford University Press.

PAUSE AND THINK

Why are ribosomes different sizes in prokaryotic and eukaryotic cells?

Answer: Both prokaryotic and eukaryotic ribosomes are composed of two subunits of different sizes. The difference in size is due to the composition of the subunits. In particular, there are more proteins in both the small and large subunits of the eukaryotic ribosome compared with the subunits of the prokaryotic ribosome. Interestingly, eukaryotic mitochondrial and chloroplast (plants) ribosomes are more like prokaryote ribosomes, being smaller in size. This is one of the pieces of evidence that supports the endosymbiotic theory, which we discuss further in Section 9.4.

The endomembrane system

As well as surrounding a cell and its nucleus, membranes are also found widely throughout the interior of a eukaryotic cell, most extensively when forming the endomembrane system. The endomembrane system is important for the modification, packaging, and transportation of lipids and proteins around the cell. There are three main components to this endomembrane system:

1. The nuclear membrane.
2. The rough and the smooth endoplasmic reticulum (ER).
3. The Golgi apparatus.

In addition, other membrane-bound structures, such as lysosomes, vesicles, endosomes, and the cell membrane itself, which are involved in the trafficking and exchanging of material throughout and into or out of the cell, are also part of the endomembrane system.

The ER is an extensive network of membranes that extends from the nuclear membrane to form tubular compartments throughout the cell. The smooth ER lacks ribosomes along its membranes, whereas the rough ER is studded with ribosomes on its outer surface membrane, giving it a 'rough' appearance when viewed through a microscope. Look at Figure 9.30 to see how the rough and smooth ER differ in appearance.

The interior of the compartments formed by the membrane of the ER is known as the lumen. This is the site of many important metabolic processes. The lumen of the smooth ER is a hive of metabolic activity, dominated by the production and modification of lipids, such as the steroids and phospholipids used in the building of membranes, as well as the breakdown of carbohydrates to release glucose, and the storage of calcium ions for use during processes such as muscle contraction. In some cells, particularly those in the liver, the lumen of the smooth ER is also home to many important enzymes, including those needed to detoxify harmful chemicals such as ethanol and barbiturate drugs.

Activity of the rough ER is focused mainly around proteins. As noted above, the outer membrane of the rough ER is studded with ribosomes, which are the sites of protein synthesis. The proteins produced by the ribosomes associated with the rough ER may be destined for use within the cell or transported beyond the cell membrane. Either way, the proteins produced by these ribosomes enter the lumen of the rough ER and are modified, re-folded, and packaged prior to the ER releasing them for transportation to other parts of the endomembrane system for use. During this packaging, the rough ER wraps the proteins in a membrane-bound transport vesicle. While some of these vesicles transport their contents along components of the cytoskeleton to secrete them across the plasma membrane and out of the cell, many vesicles are actually destined for another component of the endomembrane system, the Golgi apparatus.

Golgi apparatus

The Golgi apparatus is also referred to as the Golgi complex or the Golgi body. It is a series of many (sometimes numbering into the hundreds) individual, flattened, membrane-enclosed discs, known as **cisternae**, that are stacked adjacent to each other, as illustrated in Figure 9.31. The arrangement of the discs within each stack results in a receiving side of the stack and a dispatching side of the stack.

The receiving side of the stack is known as the *cis* side and is oriented towards the ER so that it can receive the substances produced by the ER. The dispatching side of the stack is known as the *trans* side and is oriented towards the plasma membrane of the cell. Associated with these cisternae are many hollow vesicles, which

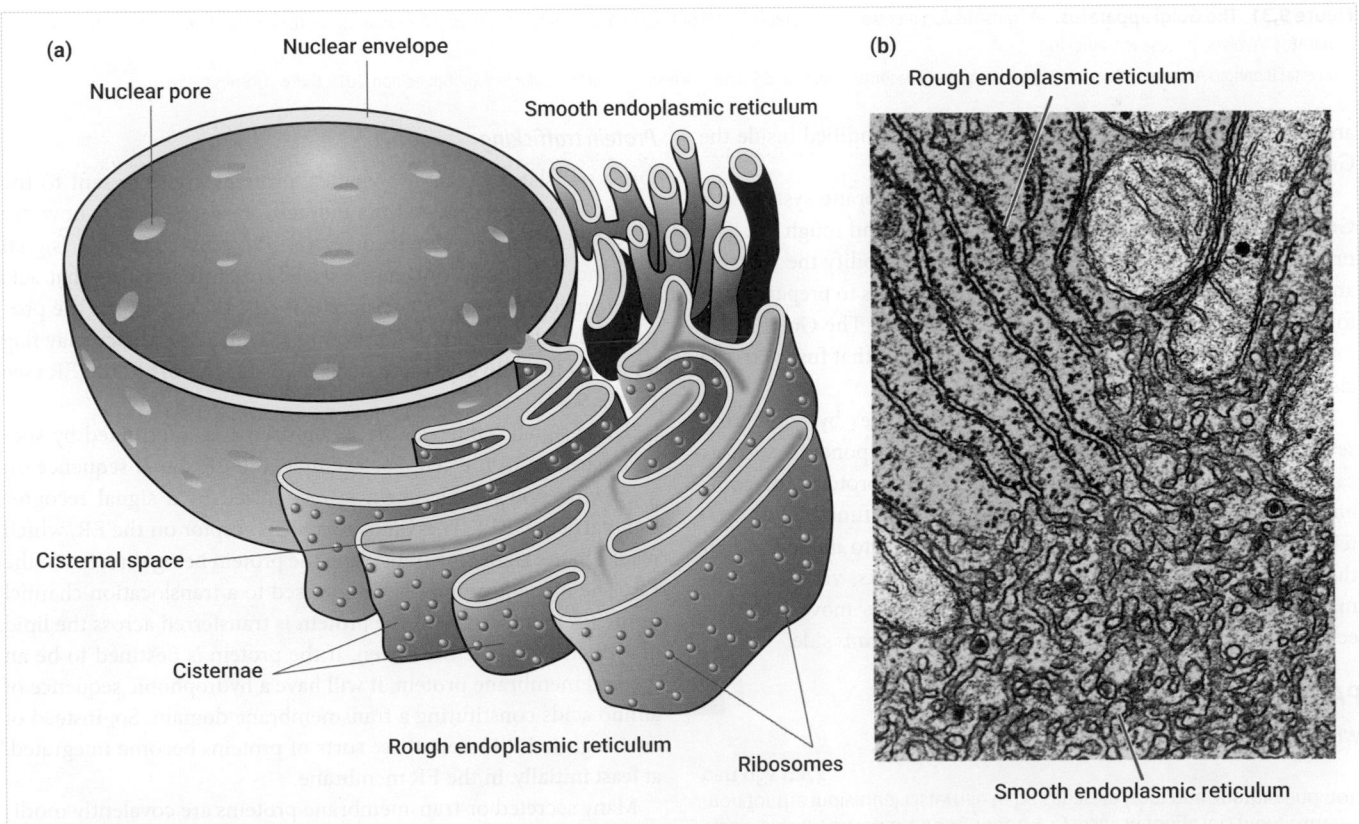

Figure 9.30 The endoplasmic reticulum. (a) A schematic representation of the endoplasmic reticulum and (b) a transmission electron micrograph image showing the contrasting appearance of the smooth and rough ER.

Source: (b) Don Fawcett/Science Photo Library.

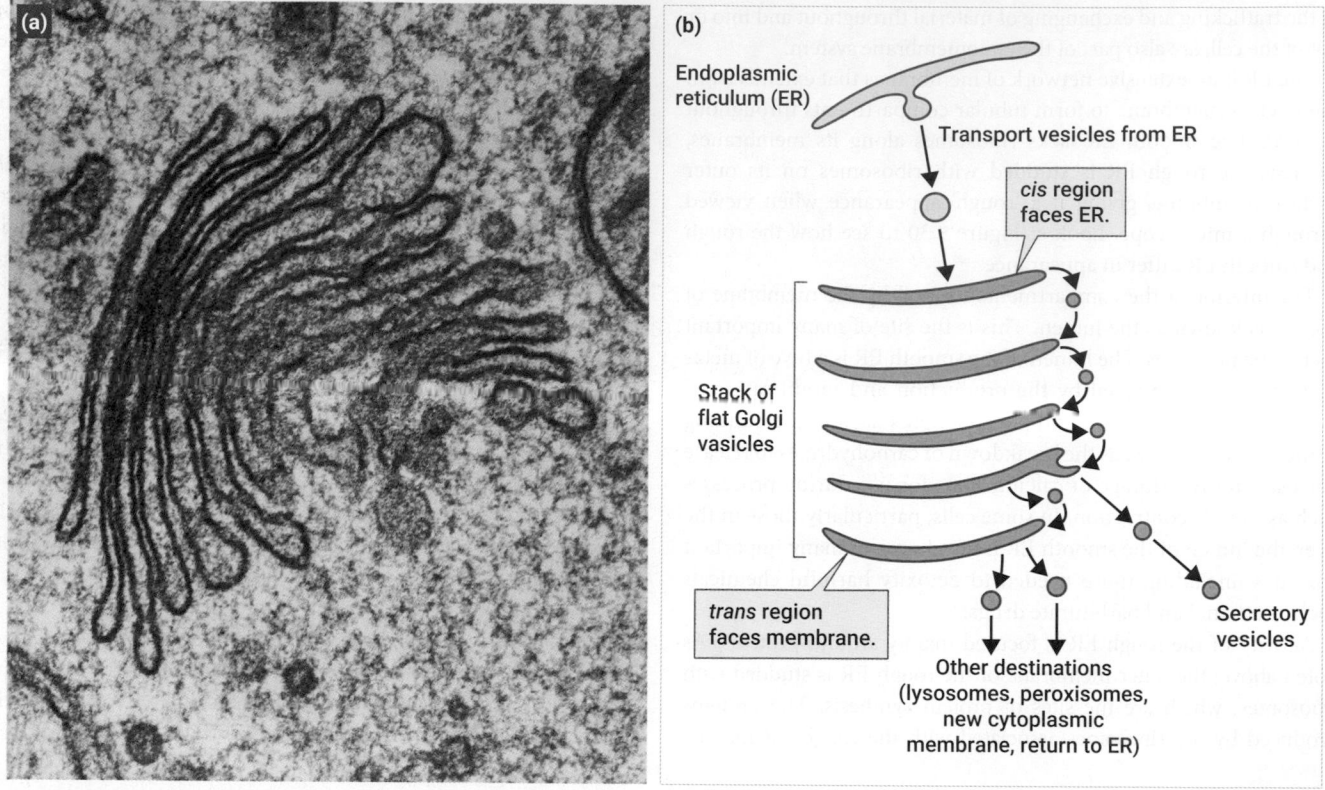

Figure 9.31 The Golgi apparatus. (a) Transmission electron micrograph (TEM) of the Golgi apparatus of *Chlamydomonas* algae. (b) *Cis* and *trans* faces of the Golgi apparatus involved in vesicle trafficking.
Source: (a) Biophoto Associates/Science Photo Library; (b) Papachristodoulou and Snape, *Biochemistry and Molecular Biology*, 6th edition, 2018, Oxford University Press.

are ready to transport the substances that are modified inside the Golgi apparatus.

As one of the components of the endomembrane system, the Golgi apparatus works closely with the smooth and rough ER. The primary function of the Golgi apparatus is to modify the proteins and lipids produced by the rough and smooth ERs to prepare them for transportation within, or outside of, the cell. The Golgi apparatus receives the products of the ER via vesicles that fuse with the cisternae located on the *cis* side of the stack.

The Golgi apparatus modifies these molecules by combining several together or by attaching additional components, such as carbohydrate chains to proteins to form glycoproteins, or combining polypeptide chains together to build a functioning protein. These molecules and components are able to move through the Golgi apparatus, between the different stacks, via a series of membrane-bound vesicles. These vesicles always move the molecules from the *cis* side of the stack towards the *trans* side.

PAUSE AND THINK

Where is the Golgi apparatus found within cells?

Answer: In the case of mammalian cells, it is generally accepted that the Golgi is located near the nucleus of the cell and close to the centrosome. In muscle and in most invertebrates and plants, the Golgi exists as individual stacks scattered throughout the cytosol. In fungi, it is more common for the individual cisternae to be separate from one another and not part of a stack.

Protein trafficking

Once translated by the ribosomes, proteins are then sent to the correct compartment within the cell or secreted. But how are these proteins processed correctly? Proteins contain a 'signal sequence'—a short continuous stretch of amino acids—that acts in an analogous way to a postcode: it tells the cell where the protein should be delivered. For example, a signal sequence may flag a protein for delivery to the nucleus, or for import to the ER (see Figure 9.32).

These signal sequences are recognized and interpreted by specific particles in the cell. For example, the ER signal sequence on a growing polypeptide chain is recognized by a signal recognition particle (SRP). This binds to a SRP receptor on the ER, which results in the ribosome translating the protein being directed to the ER. The nascent protein is then passed to a translocation channel in the ER membrane and the protein is transferred across the lipid bilayer and into the ER lumen. If the protein is destined to be an integral membrane protein, it will have a hydrophobic sequence of amino acids constituting a transmembrane domain. So, instead of entering the ER lumen, these sorts of proteins become integrated, at least initially, in the ER membrane.

Many secreted or transmembrane proteins are covalently modified (such as by *N*-linked glycosylation, the attachment of sugars via a particular covalent bond) once in the ER, or later in the Golgi, before they are trafficked to their ultimate destination via membrane-bound vesicles. These vesicles, carrying their cargo of

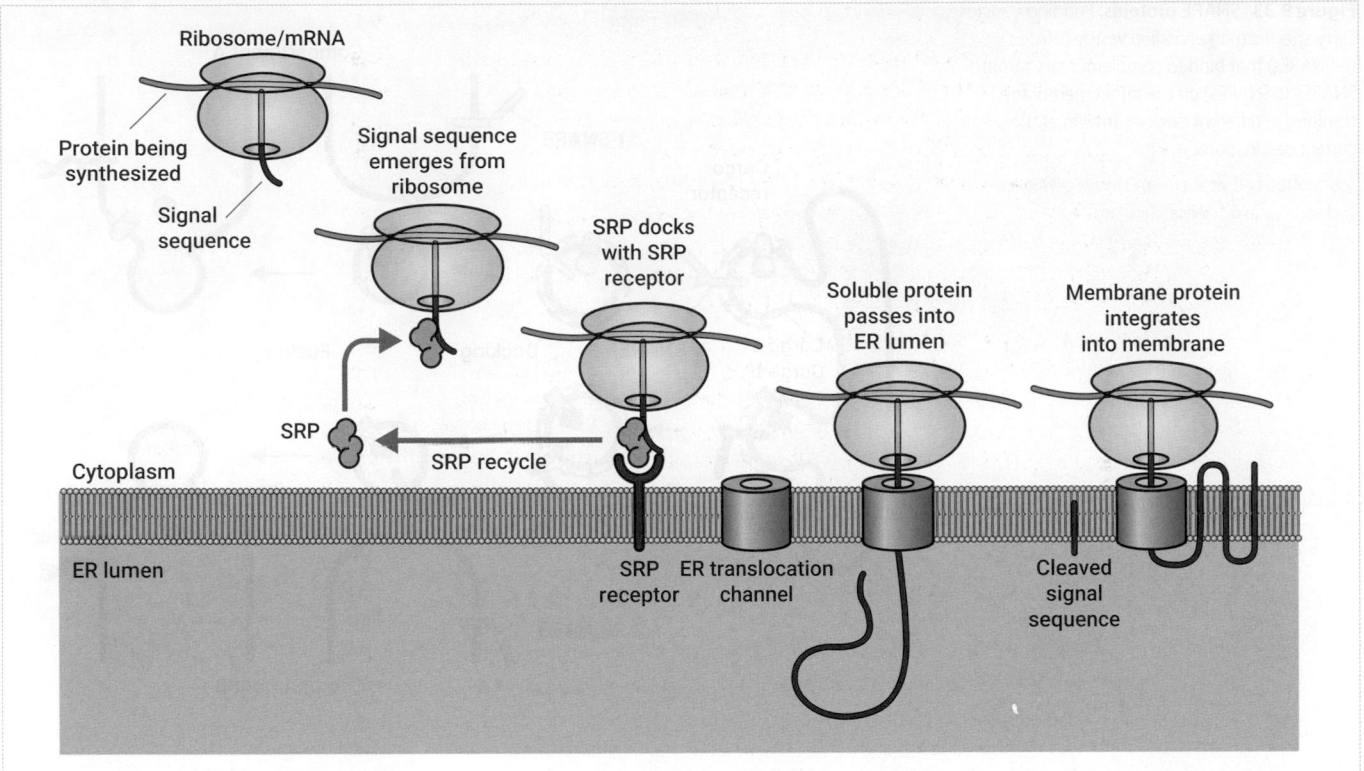

Figure 9.32 Protein targeting to the endoplasmic reticulum. As the new protein is made by the ribosome, the signal sequence is recognized by the SRP, which in turn binds to its receptor on the ER. Soluble proteins pass through the ER translocation channel (or transport complex) into the lumen of the ER, whereas proteins destined to be membrane-bound are incorporated initially into the ER membrane. The signal sequence is then removed from all soluble proteins and most integral membrane proteins.

Source: Adapted from Papachristodoulou and Snape, *Biochemistry and Molecular Biology*, 6th edition, 2018, Oxford University Press.

proteins or other material, bud from one membrane such as the ER, move through the cytosol, and fuse with another membrane (such as the *cis* side, or receiving surface, of the Golgi). Different proteins (called SNARE proteins) on both the vesicle (v-SNAREs) and target membranes (t-SNARES) help the vesicles fuse with the correct destination, as shown in Figure 9.33. Vesicles can then bud from the *trans*, or dispatching side, of the Golgi as part of the exocytosis pathway of secretion. Cells also have similar processes for the process of endocytosis (the specialized process of taking material into the cell).

In addition to these important roles in modifying, secreting, and transporting substances made inside the cell, the Golgi apparatus also has a very important role in producing specific vesicles that contain important enzymes used in the breakdown of damaged cell components: these are the lysosomes.

Lysosomes

Lysosomes are essentially membrane-enclosed fluid pockets containing a broad range of digestive enzymes; they can be considered the cell's own digestive or removal/recycling system. The lysosomes are produced by the Golgi apparatus while the digestive enzymes contained within them are produced by the rough ER. Each lysosome may contain dozens of different enzymes, which include glycosidases (break bonds between sugar groups), proteases (break down proteins), and lipases (break down fats). These enzymes are able to speed up the rate at which macromolecules are broken down within cells. As a result, one of the roles of the lysosome is to protect cells from the digestive enzymes they contain! If these enzymes were to escape from the lysosome, they would be able very quickly to digest and destroy many important cells and tissues.

All enzymes have conditions under which they work best. For lysosomal enzymes this is typically towards the more acidic end of the pH spectrum: the pH of the interior of a lysosome is around 5. We have previously mentioned that the contents of a lysosome are potentially dangerous to the cell, but the cytosol of most cells is of a more neutral pH, one that is beyond the optimal range of many lysosomal enzymes. As such, many of the enzymes contained within lysosomes are activated when in the acidic conditions inside a lysosome, but not at the more neutral pH of the cytosol. To maintain their slightly acidic internal environment, the membrane of lysosomes contains proton pumps that actively pump H^+ ions (protons) from the cytosol into the lysosome.

Figure 9.33 SNARE proteins. Budding vesicles carry specific markers called vesicle SNAREs (v-SNAREs) that bind to complementary target SNAREs (t-SNAREs) on the target membrane, resulting in different cargoes arriving at the correct destinations.

Source: Alberts, B. et al. (2004) *Essential Cell Biology*, 2nd edn. Garland Science.

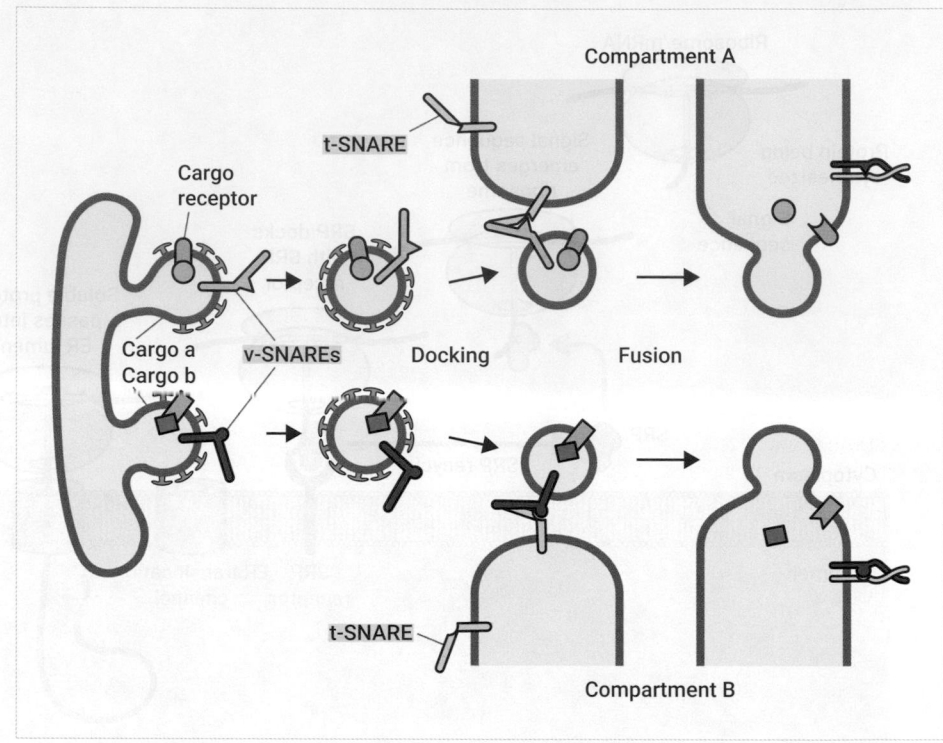

Figure 9.34 The role of lysosomes in endocytosis, phagocytosis, and autophagy. Primary lysosomes fuse with membrane-bound vesicles derived from endocytosis, phagocytosis, or autophagy. The digestive enzymes found within the lysosomes are then able to break down the contents of the endosome, phagosome, or autophagosome.

Source: Lodish, H. et al. (2008) *Molecular Cell Biology*, 6th edn. WH Freeman and Company.

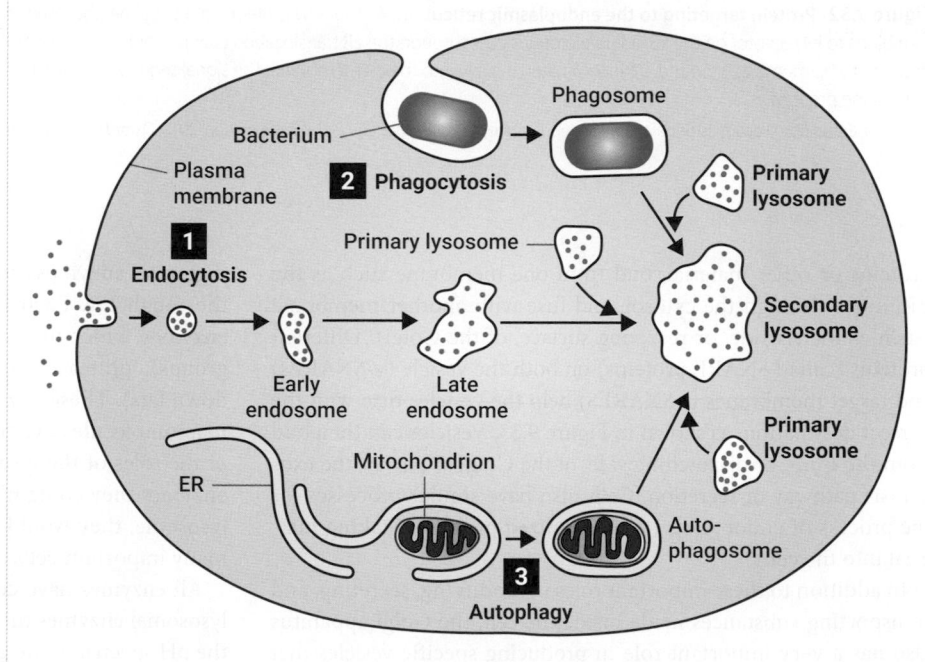

▶ **We learn more about the activity of enzymes in Chapter 6.**

Lysosomes degrade and digest materials present in the cell as a result of three processes: endocytosis, phagocytosis, and autophagy, as illustrated in Figure 9.34:

■ Endocytosis is the process by which a cell brings molecules into the cell from the external environment.

■ Phagocytosis involves large cell fragments or particles being engulfed and taken into a cell—for example, particles of pathogens which have been broken down after invading the body.

■ Autophagy is the process by which the internal components of a cell are broken down with the help of the lysosomes and replaced as they cease to function effectively or pose a threat to the overall functioning of the cell.

PAUSE AND THINK

What might be the consequences if lysosomes were not functional in some manner?

Answer: Think about the functions of the lysosome and what would therefore be lost if it could not work properly, or at all. Molecules brought into the cell by endocytosis would not be properly broken down, which may result in a shortage of precursors for various metabolic processes. Perhaps more importantly, large cell fragments or pathogenic micro-organisms, as well as damaged or unnecessary internal cellular compo-nents, would not be dismantled and the components removed or recycled by phagocytic and autophagic processes. A subsequent build-up of large molecules and damaged cellular components could eventually result in the death of the cell. A rare group of typically inherited diseases, called lysosomal storage diseases, are caused by lysosomal dysfunction, usually as a consequence of the malfunction of a single enzyme critical for the breakdown of the macromolecules.

Cytoskeleton

The cytoskeleton is a complex network of protein filaments that extend throughout the cytoplasm of all cells to form a protein fibre-based scaffold. This scaffold provides both support for the cell and its contents as well as a form of transport network: the internal components of a cell can move along these specific protein fibres.

All cells, whether prokaryotic or eukaryotic, contain some form of cytoskeleton (as we note in Table 9.1). Before the application of electron microscopy there were sizeable gaps in our understanding of the structure and function of the cytoskeleton as well as its significance for overall cell function. We now know that the cytoskeleton is a dynamic, constantly changing network of multiple different protein fibres, all working together to give a cell its shape and to facilitate various cell activities, such as movement and division.

▶ We discuss the mechanical properties of the cytoskeleton further in Chapter 28.

In this section we will focus on the eukaryotic cytoskeleton, which is composed of three main protein fibres: microtubules, actin or microfilaments, and intermediate filaments. These three fibres are illustrated in Figure 9.35.

Microtubules

Microtubules are the thickest of the three fibre components and are made up of α and β tubulin proteins that coil around to form a rigid, hollow tube (see Figure 9.35a and notice the light and dark green strands). These fibres are found in all known eukaryotic cells and can perform a range of functions, from giving a cell its shape or providing a transport pathway within a cell, to being involved in the separation of chromosomes during cell division.

Microtubules are dynamic structures: they are continually being assembled and reassembled in response to changing conditions within the cell, with a fast-growing plus end and a slow-growing minus end, as illustrated in Figure 9.36. The microtubule assembly centre within animal cells is called the centrosome (or **microtubule organizing centre**, MTOC), an organelle typically formed by two centrioles situated perpendicular (at a 90° angle) to each other, located in the cytoplasm close to the nucleus in non-dividing cells; the centrosome duplicates when the cell enters active cell division. During cell division, each centrosome moves towards

Figure 9.35 The components of the cytoskeleton. (a) schematic of microtubules, intermediate filaments, and microfilaments (left to right); (b) cells fluorescently stained to show the microtubules (green), intermediate filaments (yellow), and microfilaments (red).

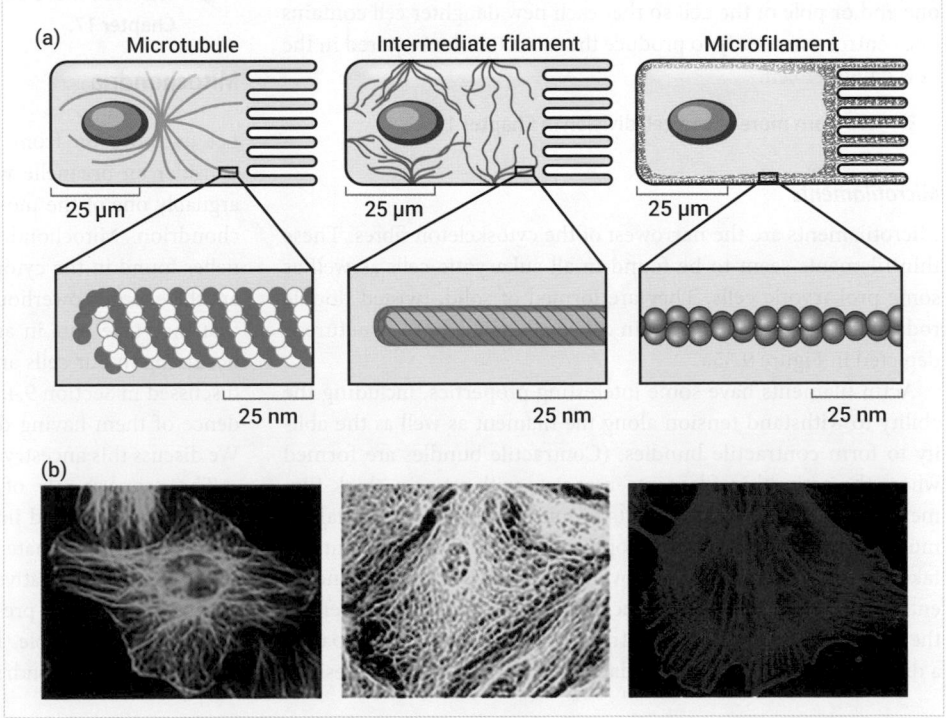

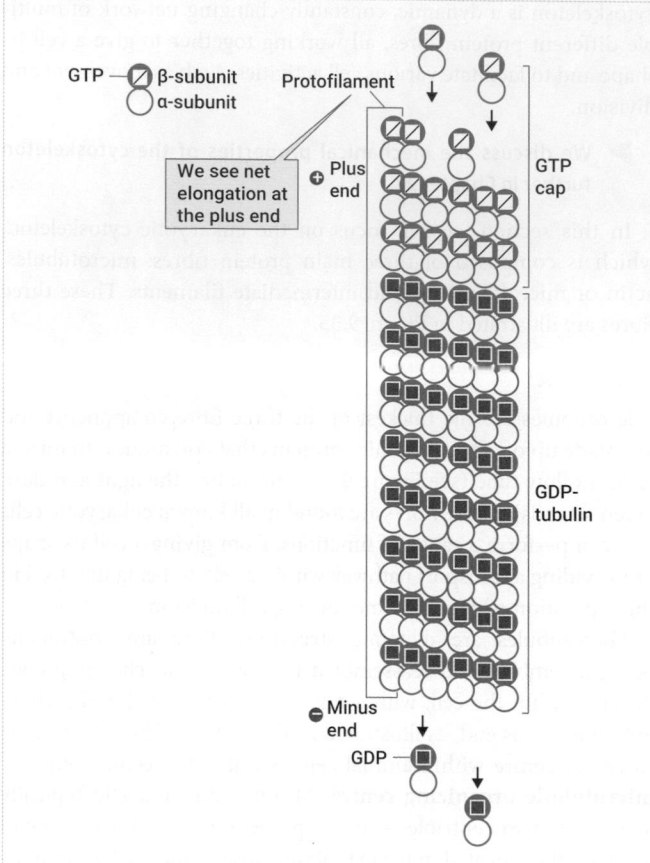

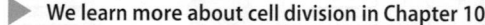

Figure 9.36 Dynamic nature of microtubules. Tubulin dimers can be added or removed from the existing microtubule.

Source: Craig et al., *Molecular Biology* (3rd edn, 2021), Oxford University Press.

one end or pole of the cell so that each new daughter cell contains one centrosome, ready to produce the microtubules required in the new daughter cell.

▶ **We learn more about cell division in Chapter 10.**

Microfilaments

Microfilaments are the narrowest of the cytoskeleton fibres. These thin filaments seem to be found in all eukaryotic cells as well as some prokaryotic cells. They are formed of solid, twisted double rods made of a globular protein called **g-actin**, whose structure is depicted in Figure 9.35a.

Actin filaments have some interesting properties, including the ability to withstand tension along the filament as well as the ability to form contractile bundles. (Contractile bundles are formed where the actin 'thin filaments' associate with myosin 'thick filaments'; they are key components of muscle cells and essential for muscle contraction.) These properties enable microfilaments to take on important roles in movement, both within cells and of entire cells, as well as on the much larger scale of the movement of the skeleton when muscles contract. Microfilaments also separate a dividing animal cell into two daughter cells during cytokinesis.

▶ **We learn more about cytokinesis in Chapter 10.**

Microfilaments share with microtubules the characteristic of having polarity: they have plus and minus ends, and are assembled and disassembled to meet the needs of the cell. This behaviour enables cells to move and respond to changes in their environment. One such example is when cells of the immune system need to move to the site of an infection to fight the invading pathogen. After the cells have arrived at the site of infection they leave the circulatory system's capillaries and migrate through the tissues to deal with the pathogen.

Intermediate filaments

Intermediate filaments derive their name from their diameter, which is smaller than that of the microtubules but larger than that of the microfilaments. Contrary to the previous two filaments, intermediate filaments are not found within all eukaryotic cells, but seem only to be present in some animal cells, including our own and those of some invertebrates.

The main role of the intermediate filaments is structural: these diverse filaments provide integral mechanical strength to a range of cells and tissues, particularly tissues such as skin. Unlike the microtubules (which are made of the protein tubulin) and the microfilaments (which are made of the protein actin), intermediate filaments are made of many different proteins: more than 50 have been identified to date. Those you might have heard of include keratin (the main protein found in hair and nails and in animal horns) and neurofilament proteins (which make up a key component of the axons in nerve cells). Other intermediate filaments also perform roles in the nuclear envelope, within our white blood cells, and in forming epithelial tissues, which line all the surfaces of the body, including skin.

▶ **We learn more about the structure of the nervous system in Chapter 17.**

Mitochondria

Let us move on from the cytoskeleton and its components to consider an organelle whose role in energy generation makes it arguably one of the most important cellular organelles: the mitochondrion. Mitochondria are double-membrane enclosed organelles found in the cytoplasm of almost all eukaryotic cells. They are the energy powerhouse of the cell and the metabolic processes that they take part in are essential to many of the key processes occurring in our cells and the cells of our eukaryotic relatives. As discussed in Section 9.4, they possess many traits that provide evidence of them having descended from ancient prokaryotic cells. We discuss this ancestry further in Real World View 9.1.

The primary role of mitochondria is to convert the chemical energy from food into a form that cells can utilize, known as adenosine triphosphate (ATP). This complex and efficient process, known as oxidative phosphorylation (aerobic cellular respiration), requires the presence of oxygen. Cells with a high energy demand—for example, muscle cells and sperm cells—have high numbers of mitochondria, often in excess of several thousand per

REAL WORLD VIEW 9.1 | Mitochondrial Eve

Mitochondria have retained many characteristics from the ancestral cells from which they evolved by virtue of possessing their own genes, which are stored in strands known as mitochondrial DNA (mtDNA). Mitochondrial DNA contains 37 genes, which are only ever passed on along the maternal line. This means that every person on Earth (aside from those individuals who have been conceived through mitochondrial replacement therapy) has inherited their mtDNA from their mother; there is no recombining of a mother's and father's mtDNA to introduce variation between generations.

The mutation rates of mtDNA are around 10 times higher than those seen for nuclear DNA. This is likely due to damage to mtDNA being repaired less efficiently, the presence of oxygen radicals in mitochondria causing a more mutagenic environment, and that mitochondria have an increased number of replications each time cells divide. This high mutation rate, along with the maternal inheritance, presents evidence to help us use mtDNA genes to answer questions about our evolution. Analysis has found that human mtDNA has descended from a common ancestor known as mitochondrial Eve, who originated in Africa. This analysis provides a strong body of evidence to support the 'Out of Africa' hypothesis, which proposes that the main group of ancestors for modern-day humans came from Africa approximately 200,000 years ago. This emergence of humans from Africa, as deduced from mtDNA evidence, is illustrated in Figure 1.

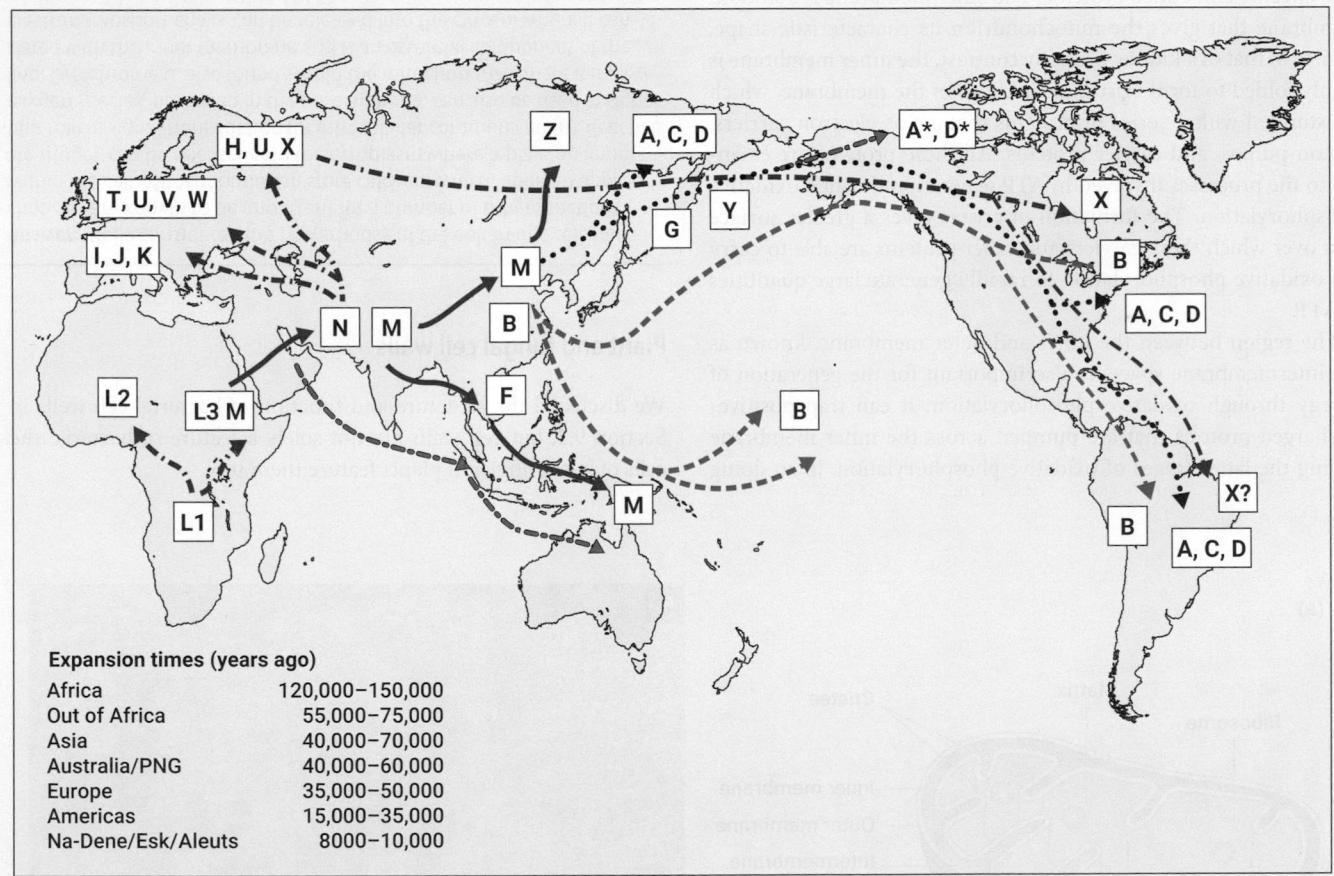

Expansion times (years ago)

Africa	120,000–150,000
Out of Africa	55,000–75,000
Asia	40,000–70,000
Australia/PNG	40,000–60,000
Europe	35,000–50,000
Americas	15,000–35,000
Na-Dene/Esk/Aleuts	8000–10,000

Figure 1 A map of the world showing the migration of human population groups out of Africa, as deduced from female mtDNA. Each population group (called a haplogroup) is denoted by a letter, and subgroups are denoted by numbers. Migration starts from the population group denoted L1.
Source: FamilyTreeDNA.

cell, so that they have available the useable energy (in the form of ATP) they need to survive and thrive.

▶ **We discuss oxidative phosphorylation in more detail in Chapter 7.**

Owing to their ancestral prokaryotic origins, mitochondria retain some of their own genetic material. As such, they are able to control their own division by binary fission to produce the huge numbers of mitochondria needed in the most active eukaryotic cells.

In addition to acting as energy converters, mitochondria also help to regulate one of the processes of programmed cell death, known as apoptosis, and in the production of cellular components such as heme, found in the blood gas transport protein haemoglobin, and in producing the plasma membrane component and hormone precursor, cholesterol.

Figure 9.37 shows how the structure of a mitochondrion is well adapted to support its function. The two membranes that surround the mitochondrion are both composed of a phospholipid bilayer with specific embedded proteins. The outer membrane is a smooth membrane that gives the mitochondrion its characteristic shape, similar to that of a kidney bean. By contrast, the inner membrane is highly folded to form cristae, or foldings of the membrane, which are studded with a series of proteins that act as electron carriers, proton pumps, and carrier proteins. All these proteins are essential to the processes involved in ATP generation through oxidative phosphorylation. The formation of cristae gives a greater surface area over which these carriers and other proteins are able to carry out oxidative phosphorylation to rapidly generate large quantities of ATP.

The region between the inner and outer membrane, known as the intermembrane space, is also important for the generation of energy through oxidative phosphorylation: it can trap positively charged protons that are pumped across the inner membrane during the latter stages of oxidative phosphorylation, in so doing

building up a powerful source of potential energy. The final stage of oxidative phosphorylation is the use of this potential energy to drive the generation of ATP by the enzyme ATP synthase.

▶ **We learn more about the operation of ATP synthase in Chapter 7.**

The inner membrane surrounds the central region of the mitochondrion, known as the mitochondrial matrix, where mtDNA is located. It is also home to many of the ribosomes and enzymes that are necessary for the mitochondrion to perform its many different functions.

PAUSE AND THINK

Is energy production the only function of mitochondria?

Answer: While considered the powerhouse of the eukaryotic cell, mitochondria are known to be important for a number of other functions within the cell. Mitochondria can store calcium, which, upon an appropriate trigger, can be released into the cytoplasm where it plays an important role in signal transduction (or intracellular communication). Reactive oxygen species, produced in the mitochondria, can also be used in signalling. Cytochrome C, also found within the mitochondria, where it is associated with the inner membrane and is an essential component of the electron transport chain, can be released into the cytosol where it can combine with a protein called APAF-1 to trigger apoptosis (a type of cell death). There is also some evidence to show that mitochondria are also involved in the regulation of cellular metabolism and the synthesis of various macromolecules (such as heme and steroid hormones).

Plant and fungal cell walls

We discussed the structure and function of bacterial cell wells in Section 9.2, but cell walls are not solely a feature of bacteria: the cells of both fungi and plants feature them too.

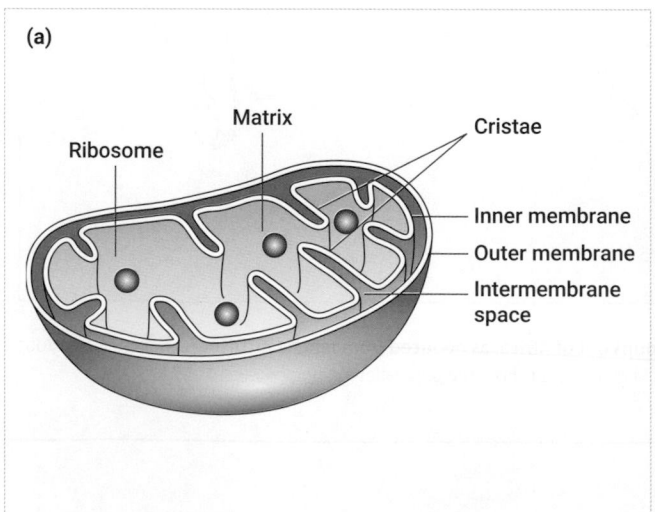

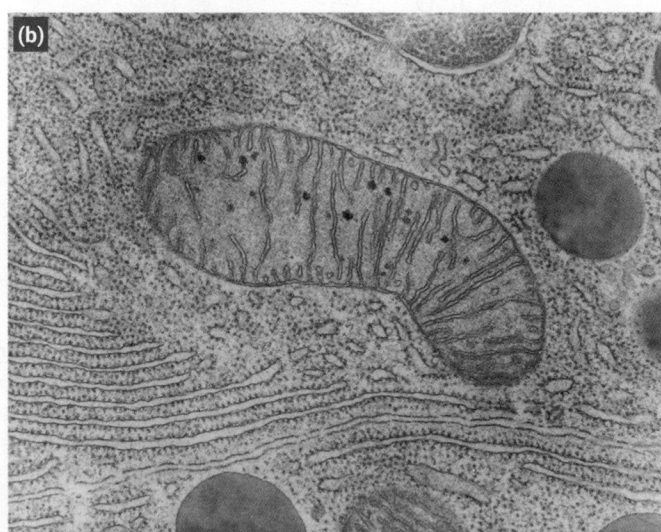

Figure 9.37 The mitochondrion. (a) A schematic illustration of the key components and structural features of a mitochondrion. (b) A transmission electron micrograph image of a mitochondrion that clearly shows the double membrane that surrounds this organelle.

Source: (a) Geoffrey Cooper, *The Cell, A Molecular Approach*, 8th edition, 2018, Oxford University Press; (b) Science History Images/Alamy Stock Photo.

Plant cell walls

The cell wall of a plant cell is its outermost layer. Primarily made of cellulose, the most abundant macromolecule on Earth, it offers protection from some forms of damage (including mechanical damage), helps to maintain cell shape, and prevents a cell from taking in too much water or drying out due to water loss. While plant cell walls are primarily composed of cellulose, a polysaccharide, there are many other molecules involved. These include proteins such as cell wall proteins (CWPs), the complex organic polymer lignin, and the polysaccharide pectin, all of which combine to form a cell wall matrix. CWPs play vital roles in the organization and rearrangement of a plant cell wall while polysaccharides typically perform a more structural function.

As you can see in Figure 9.38, a plant cell wall is arranged in layers on the extracellular side of the plasma membrane; this is the primary cell wall. Some plant cells also have a secondary cell wall. Neighbouring plant cells are connected to each other by a middle lamella, which attaches the cell walls to each other.

Growing plant cells are surrounded by a primary cell wall that is rich in polysaccharides. This layer dynamically responds to the changing needs of the organism: as the cell grows and divides, this

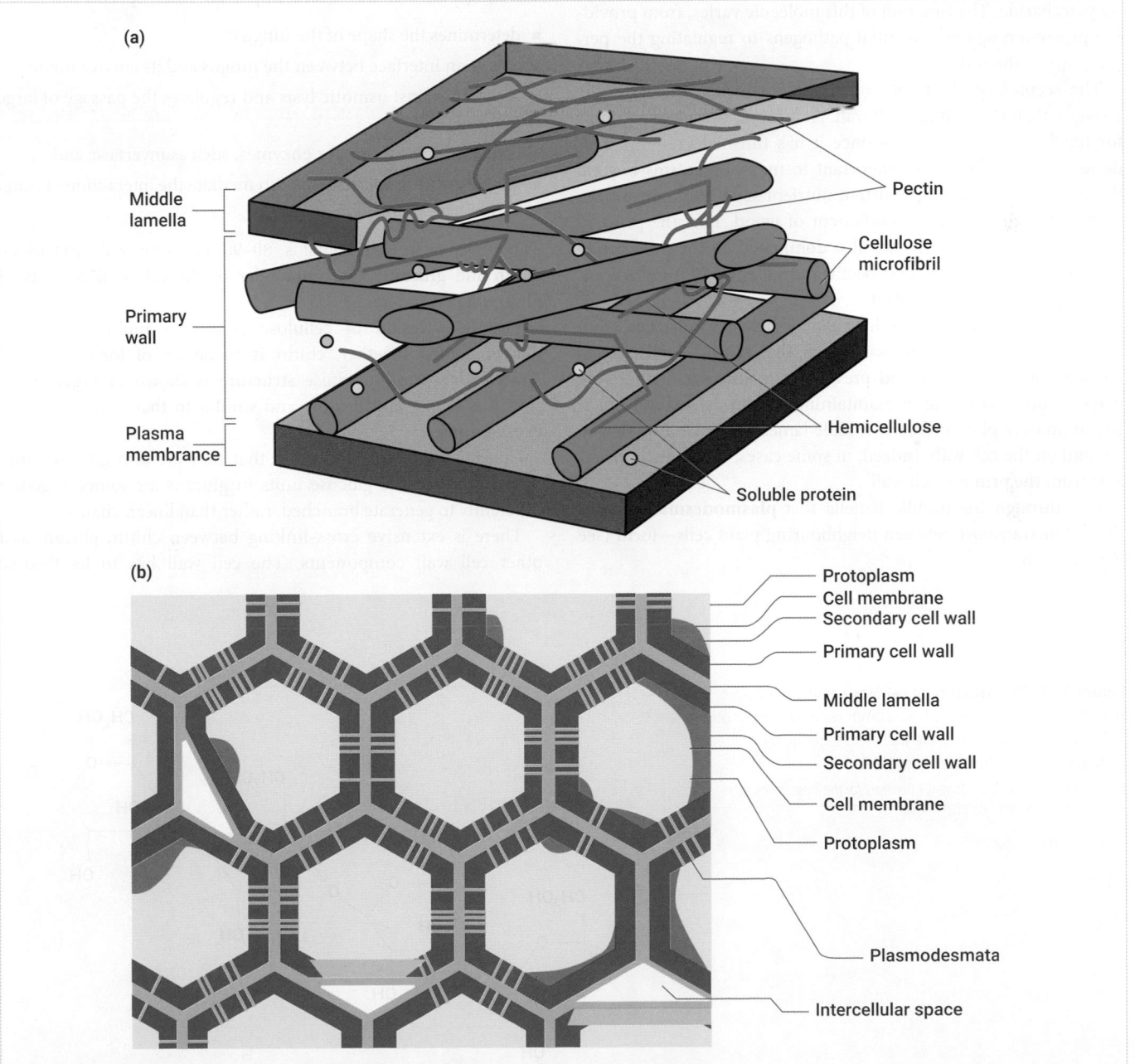

Figure 9.38 A plant cell wall is arranged in layers on the extracellular side of the plasma membrane. (a) Here you can see the plasma membrane and the primary cell wall with the middle lamella (the secondary cell wall is not pictured). (b) A stylized cartoon showing the arrangement of cell wall and middle lamellae between cells.

Source: (a) Reprinted by permission from Springer Nature, Nature Publishing Group. Sticklen, M. Plant genetic engineering for biofuel production: towards affordable cellulosic ethanol. *Nature Reviews Genetics*, 9, 433–443; © 2008; (b) Rit Rajarshi/Wikimedia Commons/CC BY-SA 4.0 https://creativecommons.org/licenses/by-sa/4.0/deed.en.

layer changes accordingly. Understanding the precise organization of the molecules within this layer is complex and not known with certainty. What *is* known, however, is that this layer of the cell wall is typically thin, flexible, and well hydrated, and its primary constituent molecule is cellulose, a strong polysaccharide made from β-glucose monomers. Its structure is shown in Figure 9.39. Hydrogen bonds link neighbouring glucose chains to form cellulose fibres. Working with, and often binding to, the cellulose fibres are xyloglucan molecules, one of the hemicellulose molecules. The network formed by cellulose and xyloglucan bears the majority of the load distributed throughout the primary cell wall.

In addition, many primary cell walls also contain pectin, another polysaccharide. The function of this molecule varies, from providing protection against potential pathogens to regulating the permeability of the wall.

The secondary plant cell wall is much thicker and generally stronger than the primary cell wall. As a result, this layer accounts for much of a cell's biomass once it has finished growing. This dense layer of biomass is important to many organisms beyond the plant as it forms an important nutrient source for consumers, as well as being a major component of wood. Like the primary layer, secondary cell walls are also dominated by the presence of cellulose, hemicellulose, and pectin but in differing proportions, depending on the type of plant.

The middle lamella acts like a glue between adjacent plant cells. It is a pectin-rich polysaccharide that can help cells stay in contact with each other and prevent them moving. It therefore plays an important role in maintaining a plant's structural integrity. In mature plant cells, the middle lamella can also function to strengthen the cell wall. Indeed, in some cases, it is indistinguishable from the primary cell wall.

It is through the middle lamella that **plasmodesmata**—pores that allow transport between neighbouring plant cells—form (see Figure 1.39b).

PAUSE AND THINK

Why do plants need a strong cell wall?

Answer: The plant cell wall is made up of cellulose, a structural carbohydrate that has great tensile strength. There is no internal skeletal structure inside plants to hold them up, so the cell wall provides the strength that enables some plants, such as giant redwood trees, to reach heights of over 90 m.

The fungal cell wall

The fungal cell wall has several important roles. Specifically, it:

- determines the shape of the fungus;
- acts as an interface between the fungus and its environment;
- protects against osmotic lysis and regulates the passage of large molecules;
- contains binding sites for enzymes, such as invertase; and
- has antigenic properties that can mediate the interaction of fungi with other organisms.

The fungal cell wall contains 80–90 per cent polysaccharides: **chitin** and **glucans**. The remainder of the cell wall consists of glycoproteins and lipids.

Just as a strand of cellulose comprises multiple units of glucose joined together, chitin is made up of long chains of *N*-acetylglucosamine, whose structure is shown in Figure 9.40. Chitin is strong and flexible and similar to that found in insect exoskeletons.

Glucans resemble cellulose in that they are also glucose polymers. However, the glucose units in glucans are joined together differently to generate branched, rather than linear, chains.

There is extensive cross-linking between chitin, glucan, and other cell wall components. The cell wall has to be flexible

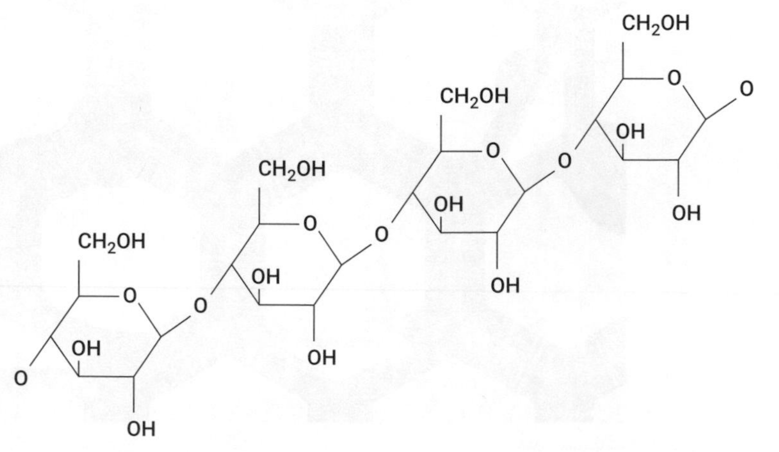

Figure 9.39 The structure of cellulose. A single cellulose strand, such as that depicted here, comprises multiple glucose units. Neighbouring strands are then held together by hydrogen bonds to form cellulose fibres.

Source: J. Crowe and T. Bradshaw, *Chemistry for the Biosciences*, 4th edition, 2021, Oxford University Press.

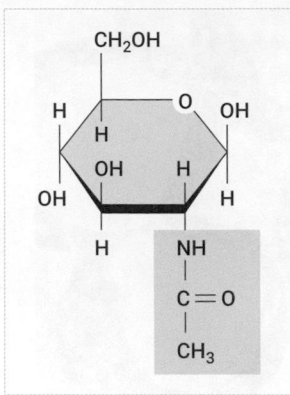

Figure 9.40 The structure of the chitin monomer, *N*-acetylglucosamine.

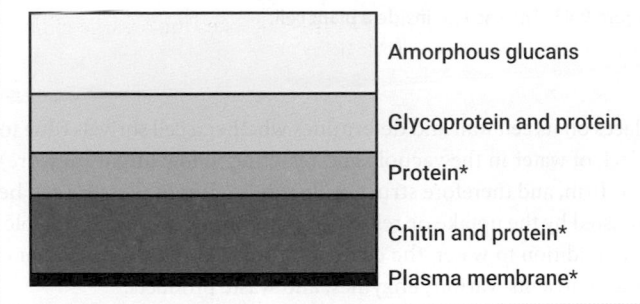

Amorphous glucans

Glycoprotein and protein

Protein*

Chitin and protein*

Plasma membrane*

Figure 9.41 Composition of the fungal cell wall. Those layers marked with * are the only layers present at the apex (apical tip) of the hyphae.

enough to cope with changes in cell morphology, such as hyphal branching and the formation of reproductive spores. Enzymes such as hydrolases have chitinase and glucanase activity, and this contributes to the breaking and re-forming of bonds to keep the cell wall flexible.

Fungal cell walls have four zones, which blend into one another; these zones are depicted in Figure 9.41. The outermost zone consists of amorphous glucans. Beneath this layer lies a network of glyco-protein within a protein matrix. There follows a layer of protein, with an innermost layer of chitin microfibrils embedded in protein. At the apex (or tip) of the hyphae the cell wall is thinner, and just has an outer layer of protein and an inner layer of chitin and protein.

Chloroplasts

Chloroplasts are one member of a group of specialized plant organelles known as plastids. Like mitochondria, chloroplasts are also double-membrane-enclosed organelles, but they are found exclusively in cells of green plants and algae and not in other eukaryotic cells. Chloroplasts convert light energy to chemical energy, in the form of sugars, through a series of reactions known as photosynthesis. These sugars can then either be used by the plant or stored within its cells. This has the consequence that plants form a food and energy source for organisms further along a food chain.

▶ We learn more about the biochemical basis of photosynthesis in Chapter 7.

Chloroplasts contain pigments that are able to absorb light at particular wavelengths. The most prevalent of these pigments is usually a group known as the chlorophylls. When present, chlorophylls result in the leaves, and many other parts of the plant, being green. Look at Figure 9.42a, which depicts the internal structure of these organelles. Notice how both the membranes forming the double membrane are smooth. The flattened discs that you can see in Figure 9.42b are a membranous network known as the **thylakoids**; when stacked on top of each other these form the **grana** (singular granum). Many of the reactions

(a)

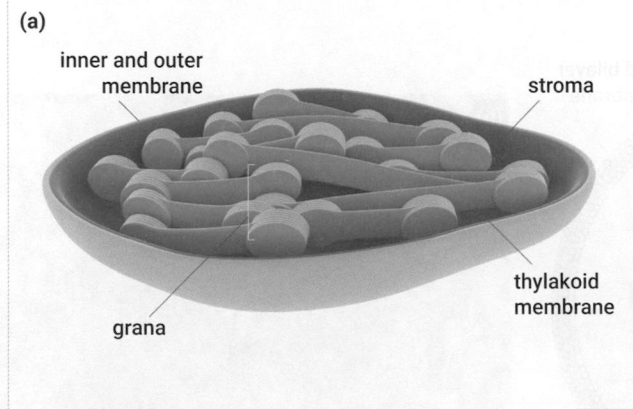

inner and outer membrane

stroma

grana

thylakoid membrane

(b)

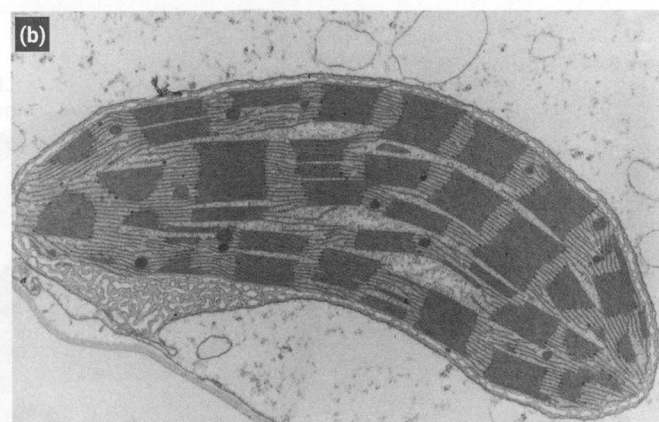

Figure 9.42 The chloroplast. (a) A schematic diagram of the structural components of a chloroplast. (b) A transmission electron micrograph image of a chloroplast, which clearly shows the double membrane that surrounds the organelle along with the flattened thylakoid membranes where many of the reactions of photosynthesis take place.

Source: (a) Science Photo Library; (b) Science History Images/Alamy Stock Photo.

of photosynthesis take place on these thylakoid membranes. Other reactions also take place in the intermembrane space, the region between the inner and outer membrane, and also in the stroma, the fluid region of the chloroplast that surrounds the thylakoids.

Other plastids include the amyloplast, which gains its name from the sugar amylose (a form of starch) that it stores, typically in the underground storage tissues of plants such as roots and tubers; and the gerontoplast, a modified chloroplast that disassembles components of other plastids as plant cells age.

PAUSE AND THINK

What were the early origins of the chloroplast?

Answer: Scientists think that the chloroplast had its ancient origins in an endosymbiotic event where an ancient ancestor of the eukaryotic cell engulfed a photosynthetic cyanobacterium, but instead of ingesting it, retained it within its cell.

Central vacuole

Vacuoles are found inside many cells, some prokaryotic and many eukaryotic. The name comes from the Latin word *vacuus* meaning 'a cell space empty of cytoplasm'. Typically small and numbering into the tens, or even 100s, inside some cells, these membrane-bound storage compartments serve many purposes depending upon the cell type. However, the role of vacuoles is particularly significant for plant cells.

Plant cells contain one super-sized vacuole known as the central vacuole, as depicted in Figure 9.43. This is a unique feature of plant cells and can occupy as much as 95 per cent of a plant cell's volume. The main role of the central vacuole is to maintain turgor pressure within the cell. Turgor pressure is the pressure that a cell

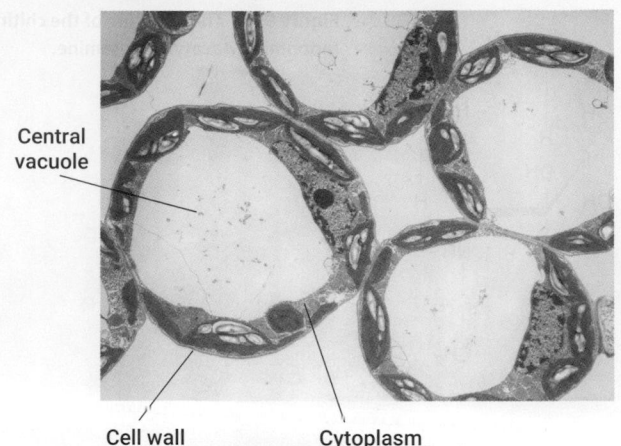

Figure 9.43 The vacuole inside a plant cell.

Source: Dr Jeremy Burgess/Science Photo Library.

places on its cell wall and determines whether a cell shrivels (due to a lack of water in the vacuole and resulting in low turgor pressure) or is firm, and therefore structurally robust. Turgor pressure can be adjusted by the uptake or release of water from the central vacuole.

In addition to water, the central vacuole can also store food and other nutrients, various enzymes, and waste products.

Peroxisomes

Peroxisomes are small organelles found in nearly all eukaryotic cells, but they vary in number and size depending upon the needs of the cell. The structure of a peroxisome is depicted in Figure 9.44. The primary function of these small organelles is to oxidize various biomolecules, including fatty acids, and to synthesize a type of membrane lipid called plasmalogens, in which one of the hydrocarbon chains is joined to the glycerol by an ether bond instead of the usual ester bond.

Figure 9.44 Peroxisome. Diagram (a) and transmission electron micrograph (b) of peroxisomes.

Source: (a) Qef/Wikimedia Commons/Public Domain; (b) Figure 282 from *The Cell* (2nd Edition) by Don W. Fawcett Md, 1981, Elsevier B.V.

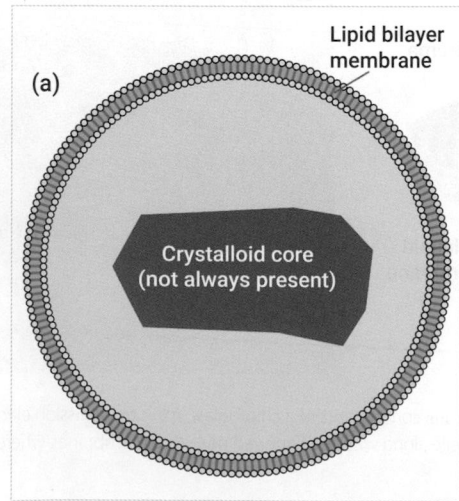

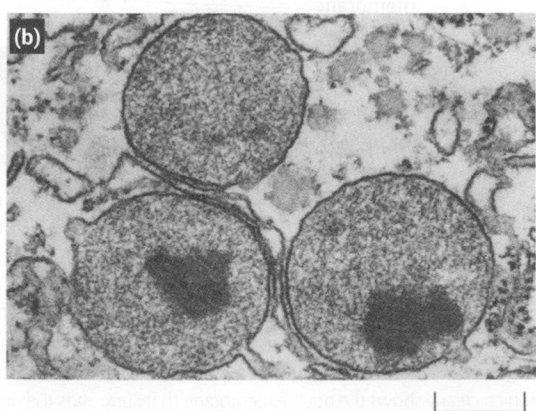

Peroxisomes have additional functions in plants: they are involved in the conversion of fatty acids to carbohydrates via the glyoxylate cycle in seeds, and are involved in photorespiration (in the recycling of carbon from phosphoglycolate) in leaves.

 Check your understanding of the concepts covered in this section by answering the questions in the e-book.

9.6 From cells to tissues

All the cells in any one organism share the same genetic material in their nuclei, but they are not identical. Specialized cells undertake specific functions due to cell-specific gene expression and consequent protein production. In multicellular organisms the majority of specialist cell types are components of various tissues and organs. The overall arrangement of these structures relies not only on the specialist cells themselves, but also on a variety of cell–cell and cell–extracellular matrix (ECM) interactions and connections.

The ECM (such as the basal lamina shown in Figure 9.45) is a complex three-dimensional structure comprising various proteins and long linear polysaccharides, which are made of repeating two-sugar units called glycosaminoglycans.

In higher eukaryotes (for example, mammals) there are four major tissue types: epithelial, connective, nervous, and muscle.

Variations in tissues are due to:

- the different cell types and their functions, shapes, and spatial arrangements;

- how the cells interact with, and react to, their neighbouring cells and extracellular surroundings (which also vary in composition);

- how cells respond to signalling molecules such as hormones.

Cell–cell and cell–matrix interactions involve molecular structures such as tight, gap, and adherens junctions, desmosomes, hemidesmosomes, and focal adhesions involving integrins. Each of these structures is illustrated in Figure 9.45, and all are explored in more detail in Chapter 28.

Taken together, these cell–cell and cell–matrix structures can help form barriers and define tissue compartments, add mechanical strength to a collection of cells in a tissue, or—in the case of gap junctions—provide a channel for the selective transit of molecules between cells.

The extracellular matrix

The ECM, consisting of water, polysaccharides, and fibrous proteins, contributes to the structure and stability of tissues. Its components are produced by the numerous resident cells, especially fibroblasts, although other cells have this function in various tissues.

The fibrous proteins (collagen, elastin, fibronectin, and laminin) are embedded in a hydrated 'gel', creating an extensive meshwork as illustrated in Figure 9.46. The gel is formed from proteoglycans, carbohydrate polymers (glycosaminoglycans, GAGs) attached to core ECM proteins, which have a net negative charge. This negative

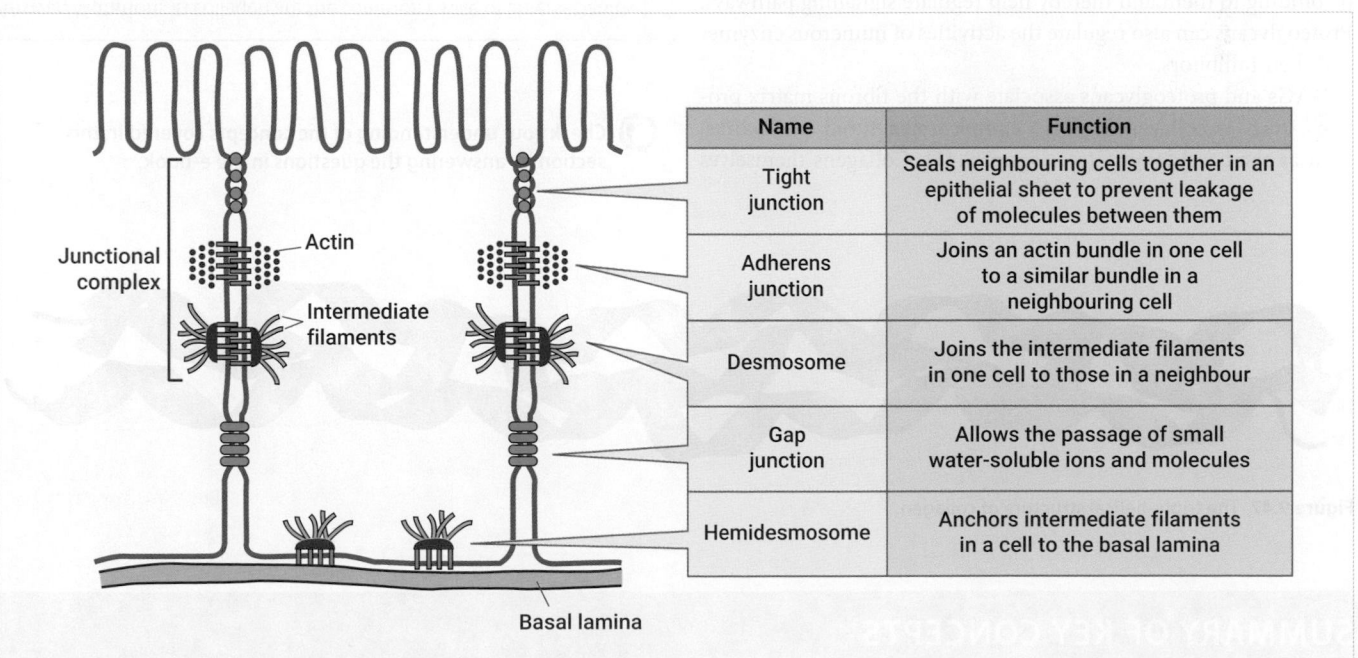

Figure 9.45 Types of cell–cell and cell–extracellular matrix contacts.
Source: Alberts, B. et al. (2004) *Essential Cell Biology*. Garland Science.

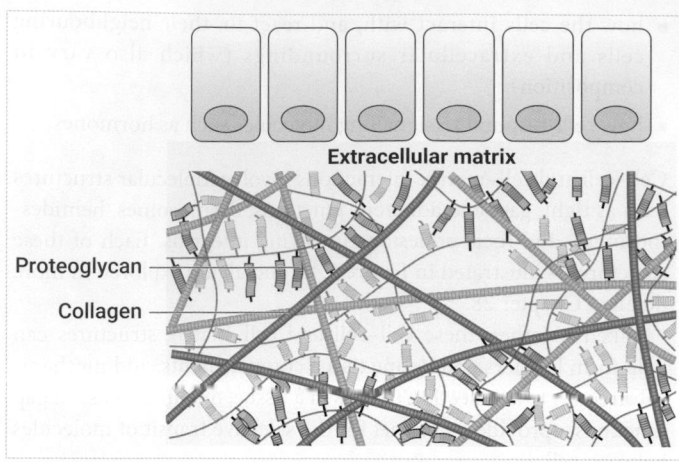

Figure 9.46 A representation of some of the key components of the extracellular matrix.

Source: (top) Peter Takizawa; (bottom) from T. D. Pollard and W. C. Earnshaw, *Cell Biology* 2nd Edition, 2008, Elsevier B.V.

charge attracts positively charged sodium ions, and subsequently water, to create the gel.

A huge variety of proteoglycans exists, with different ones being found in various ECMs. For example, aggrecan (a major component of cartilage) has over 100 GAG chains added.

As well as providing structural support, proteoglycans can regulate a variety of other activities. Depending on the composition of the gel, they can serve as sieves to control the movement of molecules within the tissue or structure. They can store growth factors by binding to them and thereby help regulate signalling pathways. Proteoglycans can also regulate the activities of numerous enzymes and their inhibitors.

GAGs and proteoglycans associate with the fibrous matrix proteins, such as collagen, to form complex structured meshworks, such as the basal lamina (see Figure 9.45). Collagens themselves

make up ~25 per cent of the total protein mass in mammals. Collagen is composed of a triple helical structure, as depicted in Figure 9.47, and is found in skin and bone. The collagens assemble into fibrils, which in turn often aggregate into fibres several micrometres in diameter.

Different collagens have different functions, one being to anchor the basal lamina of epithelial layers to the underlying connective tissue. They also help the tissue resist tensile forces. Elastin, another fibrous matrix protein, has a different function: it gives tissues such as skin and lungs the ability to stretch and return to their original shape.

Collectively, the particular cellular and molecular composition of tissues, and how all the components interact and respond to each other, result in the unique and vital functions of the various tissues and organs. These then contribute to the overall functioning and well-being of the organism itself.

PAUSE AND THINK

Individuals with Ehlers–Danlos syndrome have 'stretchy' or hyperextensible skin. With your knowledge of the extracellular matrix and its components, what could be the underlying reason for these clinical symptoms?

Answer: Mutations in collagen are the common cause of Ehlers–Danlos syndrome. One of collagen's functions is to help a tissue resist tensile forces as collagen is normally a tough, fibrous protein. The problems seen in patients with this syndrome can either be due to the resulting poor strength of collagen or due to the absence of sufficient amounts of structurally normal collagen within the tissue. While many types of Ehlers–Danlos syndrome commonly result from mutations in collagen, mutations in other genes related to collagen processing or proteins that interact with collagen can also cause the syndrome.

Check your understanding of the concepts covered in this section by answering the questions in the e-book.

Figure 9.47 The triple-helical structure of collagen.

SUMMARY OF KEY CONCEPTS

- All life is made up from cells, which share a number of key characteristics (such as being enclosed by a lipid membrane) but can vary extensively in shape, size, complexity, and function.

- Many organisms exist as single cells, whereas others are complex combinations of a large number of cells and cell types.

- Prokaryotic cells lack a nucleus, while eukaryotic cells possess one.

- There are two types of cell wall found in bacteria, Gram-positive and Gram-negative. Gram-positive bacteria possess a rigid peptidoglycan layer within their cell wall, while Gram-negative bacteria have much less peptidoglycan and have a periplasm separating their inner and outer membranes.

- Bacteria do not all share the same internal and external cell structures. Those exhibited by only some bacteria include flagella, gas vacuoles, and magnetosomes.

- Like bacteria, archaea are also prokaryotic but exhibit various structural differences, which enable them to inhabit often extreme environments.

- Eukaryotic cells have double-membraned organelles that enable different processes and cellular functions to take place in different cellular regions.

- The cytoskeleton is a complex network of protein filaments that extend throughout the cytoplasm of all cells to form a scaffold. This scaffold provides both support for the cell and its contents, as well as a form of transport network.

- Unlike other eukaryotes, the cells of plants and fungi have cell walls. A key component of plant cell walls is cellulose, while the key components of fungal cell walls are chitin and glucans.

- The chloroplast is an organelle unique to plant cells and is the site of photosynthesis.

- Specialist cells come together to form organs and tissues, mediated by interactions between different cells and between cells and the extracellular matrix.

 Use the flashcards in the e-book to test your recall of key terms introduced in this chapter.

QUESTIONS

Looking for answers? Once you've answered these questions, follow the link in the e-book to the answer guidance and check your work.

Concepts and Definitions

1. Who coined the term 'cell', and what inspired the choice of name?

2. What are the three tenets of cell theory?

3. What is the fluid mosaic model?

4. What characterizes the Archaea?

5. Describe the structure of the eukaryotic cell nucleus.

6. What is the function of the Golgi apparatus?

7. What is the cytoskeleton?

8. What are the four main tissue types in higher eukaryotes?

Apply the concepts

9. Why are cells limited in size?

10. What are the key differences between prokaryotic and eukaryotic cells?

11. Name and describe three features of bacterial cells that are *not* shared by all bacteria.

12. How are Gram-positive bacteria different from Gram-negative bacteria?

13. Why do bacteria use quorum sensing?

14. What are the key differences between plant and fungal cell walls?

15. What role do mitochondria play in the cell?

16. What are the functions of ribosomes? Where are ribosomes located and how do they differ in eukaryotes compared with prokaryotes?

Beyond the concepts

17. Do all bacteria have a cell wall?

18. What are the key characteristics of the phospholipid membrane components that are critical for membrane functions?

19. What is peptidoglycan and how is it assembled?

20. How does a newly synthesized protein end up being secreted?

21. What are the similarities and differences between mitochondria and chloroplasts?

22. Why is the extracellular matrix important for the structure and function of tissues?

FURTHER READING

https://theconversation.com/first-animal-cells-could-have-been-created-by-viruses-71202

Read this to find out more about how viruses can hijack bacterial cells and reprogramme them to behave more similarly to some eukaryotic cells.

https://theconversation.com/strange-microorganism-under-the-sea-may-be-missing-link-in-evolution-41445

Read this to find out more about the evolution of early life and the links between prokaryotic and eukaryotic life.

https://theconversation.com/weve-been-wrong-about-the-origins-of-life-for-90-years-63744

Read this to find out more about an alternative hypothesis for the origin of life on Earth; that focused around deep-sea hydrothermal vents.

Koonin, E. V. (2010) The two empires and three domains of life in the post-genomic age. *Nat. Educ.* **3**: 27.

Simpson, A. G. B. & Roger, A. J. (2004) The real 'kingdoms' of eukaryotes. *Curr. Biol.* **14**: R693–6.

Walsh, D. A. & Doolittle, W. F. (2005) The real 'domains' of life. *Curr. Biol.* **15**: R237–40.

van der Gulik, P. T. S., Hoff, W. D., & Speijer, D. (2017) In defence of the three-domains of life paradigm. *BMC Evol. Biol.* **17**: 218.

Cavicchioli, R. (2011) Archaea—timeline of the third domain. *Nat. Rev. Microbiol.*, **9**: 51–61.

A few examples of articles discussing the classification of life into domains and kingdoms, and how the classification of life is not a simple process, particularly with advances in genomic analysis.

Scheffers, D.-J. & Pinho, M. G. (2005) Bacterial cell wall synthesis: new insights from localization studies. *Microbiol. Mol. Biol. Rev.* **69**: 585–607.

A detailed overview of the components of bacterial cell walls and how they are made, with an emphasis on how various techniques such as fluorescence microscopy have helped our understanding of the processes.

Murat, D., Byrne, M., & Komeili, A. (2010) Cell biology of prokaryotic organelles. *Cold Spring Harb. Perspect. Biol.* **2**: a000422.

A review looking at lipid- and protein-bounded organelles in prokaryotes; prokaryotes are not as simple as once thought!

Sallman-Almen, M., et al. (2009) Mapping the human membrane proteome: a majority of the human membrane proteins can be classified according to function and evolutionary origin. *BMC Biol.* **7**: 50.

A very comprehensive article where the authors describe their bioinformatics study of membrane proteins.

Mohandas, N. & Gallagher, P. G. (2008) Red cell membrane: past present, and future. *Blood* **112**: 3939–48.

A review looking specifically at the red blood cell membrane, and how alterations in various components can result in a number of diseases.

Histology Guide. Virtual histology laboratory. Available at http://histologyguide.org/index.html

An online resource of a collection of images showing the microanatomy of human cells, tissues, and organs.

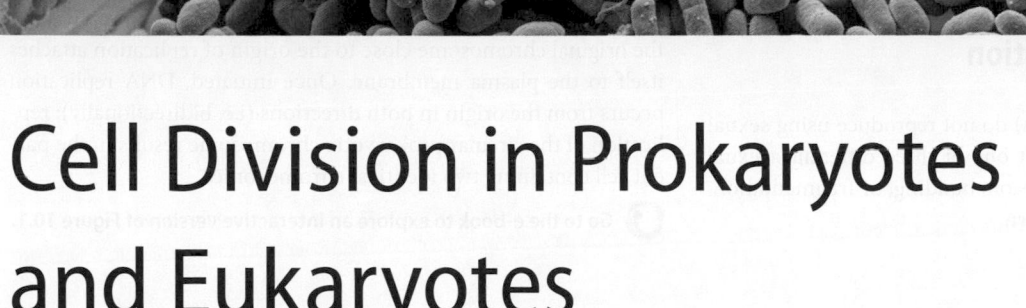

Cell Division in Prokaryotes and Eukaryotes

LEARNING OBJECTIVES

By the end of this chapter, you should be able to:

- Describe how prokaryotes reproduce primarily by binary fission, and recall that budding and fragmentation are also possible.
- Explain how eukaryotes produce two identical daughter cells by the process of mitosis.
- Recall that meiosis of eukaryotic cells results in four non-identical daughter cells with half the DNA complement of the original parental cell.

Chapter contents

Introduction		405
10.1	Bacterial reproduction	406
10.2	Eukaryotic cell division	412
10.3	Mitosis: a deeper look	415
10.4	Meiosis	418

 Watch the key concepts video in the e-book to prepare yourself for studying this chapter.

Introduction

We saw in Chapter 1 how all living organisms are composed of one or more cells, which represent the fundamental unit of structure. Collectively, these two concepts capture the first two tenets of 'cell theory'. All cells arise from the division of existing cells, with every living cell thought to have descended from an ancestral cell 3–4 billion years ago (see Topic 2). Life on Earth relies on cell division and the transfer (or transmission) of genetic information from parental cell to daughter cell. These processes drive the propagation of individuals and the evolution of species.

It is remarkable how the fusion of an egg and sperm cell can give rise to a multicellular individual; that bacterial cells can rapidly grow and double their population every 20 minutes is equally impressive. This chapter will look in detail at the third tenet of cell theory—that cells arise from pre-existing cells—by considering various types of cellular reproduction. In single-celled organisms cell reproduction (or division by binary fission, for example) results in new individual organisms. In multicellular organisms cell reproduction (such as mitosis) allows collections of cells to form tissues and organs in the processes of homeostasis, growth, and repair. A specialized type of cell division, meiosis, allows the production of gametes for sexual reproduction.

10.1 **Bacterial reproduction**

Prokaryotes (archaea and bacteria) do not reproduce using sexual reproduction. Instead, they adopt one of three different asexual reproduction strategies: binary fission, budding, or fragmentation. Let us consider each of these in turn.

Binary fission

Binary fission is the most common form of asexual reproduction exhibited by prokaryotes and produces two genetically identical daughter cells. This is **vertical gene transfer** whereby DNA is inherited by one generation from the previous generation.

In order for cells to reproduce both the DNA (chromosome) and cellular content has to be duplicated. Look at Figure 10.1 and notice how the circular chromosome has a region called the origin of replication. Prior to the prokaryote (and its genetic material) replicating, the original chromosome close to the origin of replication attaches itself to the plasma membrane. Once initiated, DNA replication occurs from the origin in both directions (i.e. bidirectionally); replication of the circular prokaryotic chromosome results in the parent cell containing two identical chromosomes.

🜊 Go to the e-book to explore an interactive version of Figure 10.1.

Figure 10.1 The process of reproduction as exhibited by prokaryotic cells. This process is known as binary fission.
Source: © OpenStax. Textbook content produced by OpenStax is licensed under a Creative Commons Attribution License 3.0 licence.

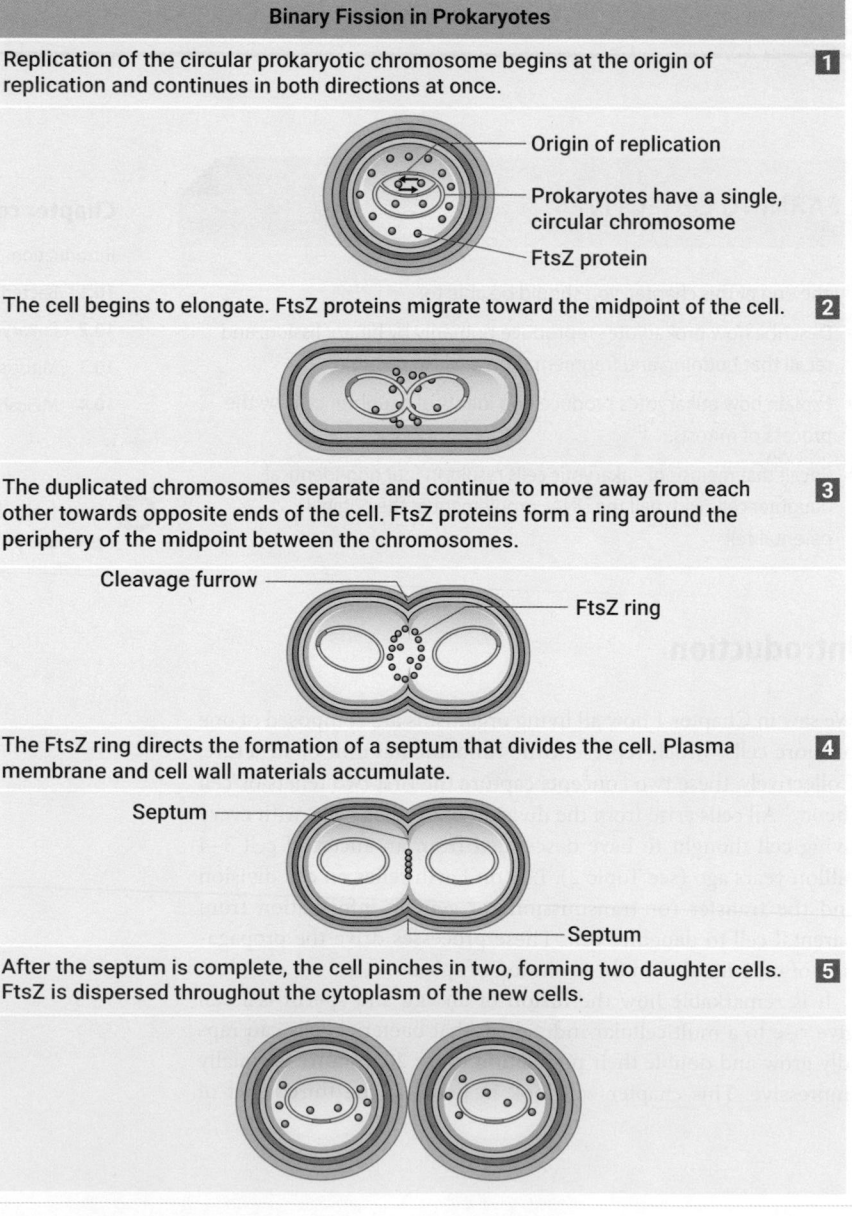

Binary Fission in Prokaryotes

Replication of the circular prokaryotic chromosome begins at the origin of replication and continues in both directions at once. **1**

Origin of replication

Prokaryotes have a single, circular chromosome

FtsZ protein

The cell begins to elongate. FtsZ proteins migrate toward the midpoint of the cell. **2**

The duplicated chromosomes separate and continue to move away from each other towards opposite ends of the cell. FtsZ proteins form a ring around the periphery of the midpoint between the chromosomes. **3**

Cleavage furrow

FtsZ ring

The FtsZ ring directs the formation of a septum that divides the cell. Plasma membrane and cell wall materials accumulate. **4**

Septum

Septum

After the septum is complete, the cell pinches in two, forming two daughter cells. **5**
FtsZ is dispersed throughout the cytoplasm of the new cells.

Once replication of the chromosome is initiated several other processes within the cell are also triggered, including the activation of cytoskeletal FtsZ proteins, as well as elongation and an increase in volume of the prokaryotic cell (see steps 2 and 3 in Figure 10.1). Fts proteins are found in all prokaryotes and the Archaea, and are similar to the tubulin proteins of eukaryotes. 'Fts' is an abbreviation for 'filamentous temperature sensitive', so named because *fts* genes were discovered when temperature-sensitive mutants were being studied. This is explored further in Scientific Process 10.1. The Fts proteins interact to form the **divisome**, the apparatus that orchestrates not only division of the cell but also synthesis of the new cytoplasmic membrane and cell wall.

During elongation (step 2 in Figure 10.1) the bacterial cell needs to increase metabolism in order to make more cell wall and peptidoglycan. Elongation of the cell enables the two identical chromosomes to separate and move away from each other towards opposing ends of the cell.

Once the two chromosomes have moved away from the midpoint of the cell, the FtsZ proteins can begin to form a ring of new plasma membrane and cell wall, called the Z ring, around the midpoint of the cell. This process also involves the protein FtsI, which is also called the penicillin-binding protein because its activity is blocked by the antibiotic penicillin.

The Z ring causes constriction at the midpoint of the cell through which a septum forms (Figure 10.1, step 4). The formation of the septum ultimately results in the cytoplasm of the parent cell dividing to form two separate, identical daughter cells (Figure 10.1, step 5).

To ensure that the Z ring only forms in the midpoint of the cell, another group of proteins called Min proteins depolymerize FtsZ at the poles of the cell. This activity stops the Z ring from forming in these areas of the cell. Min proteins also ensure the equal division of the two daughter cells. We discuss the discovery of the Min proteins in Scientific Process 10.2.

SCIENTIFIC PROCESS 10.1   The use of conditional lethal mutants to examine the process of binary fission

Research question

How do you study a fundamental concept such as cell division, when mutations in the genes that code for vital proteins necessary for the process are lethal?

Materials and methods

In order to study binary fission, scientists isolated conditional lethal mutations—mutations that are lethal under one condition (the restrictive condition) but not under another condition (the permissive condition).

Cell division was examined in the model bacterium *Escherichia coli*. The *E. coli* mutants with impaired cell division were made by treating the bacterial cells with a mutagen. Scientists then selected mutants that could divide normally at 30 °C (the permissive condition) but had impaired cell division at 42 °C (the restrictive condition).

Results

Scientists found that rather than dividing normally at 42 °C, these mutants formed long filaments that contained many nuclei, as illustrated in Figure 1. These long filaments formed because the cells were unable to form a septum and therefore failed to divide.

Conclusion

This behaviour was found to be caused by mutations in the *fts* genes, which were temperature sensitive: abnormal behaviour was only seen when the temperature exceeded the permissive temperature.

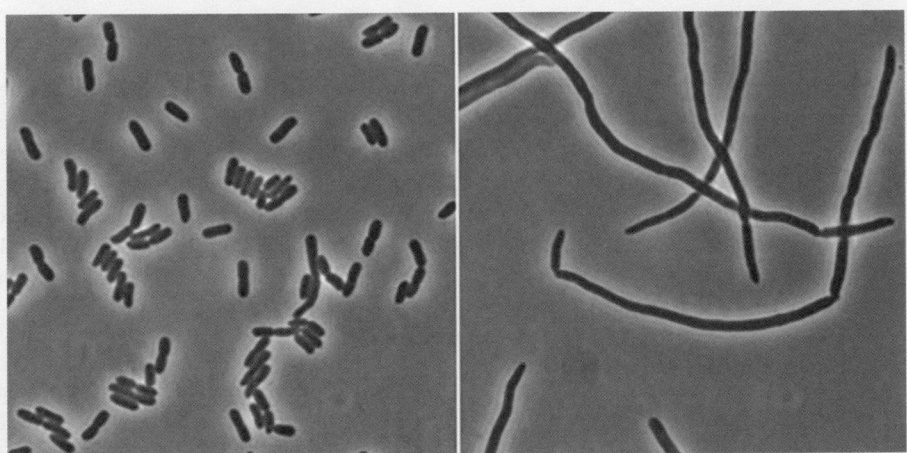

Figure 1 *Escherichia coli* **cell division.** Left: normally dividing *E. coli* cells. Right: a mutation in the *FtsB* gene causes the cells to divide improperly under a restrictive condition.
Source: Alessandro Senes, Senes Lab, Dept. of Biochemistry, University of Wisconsin—Madison.

SCIENTIFIC PROCESS 10.2   Where's my middle? The accidental discovery of mini-cells

Research question

This research is a good example of where a discovery can lead to the answering of a question that was never actually posed! In this case the question would have been: why does the Z ring form only in the middle of the bacterial cell?

Materials and methods

Over 50 years ago, in 1967, Adler and his colleagues published a paper in *The Proceedings of the National Academy of Sciences* that described their discovery of an *E. coli* mutant that could make mini-cells that lacked DNA. They found the mutant as part of a screening programme to find *E. coli* mutants that were resistant to ionizing, but not ultraviolet, radiation. Interestingly, these mini-cells remained metabolically active, although they could not divide because they lacked the bacterial chromosome.

Results

The researchers found that the mutant bacteria divided near the pole of the cell, effectively pinching off a piece of the pole, which became the mini-cell. The formation of a mini-cell is shown in Figure 1.

It was an intriguing discovery: while the mutant did not have the properties being looked for in the original screening, the mini-cells allowed scientists to discover the Min proteins that are required if a cell is to successfully undergo binary fission to yield two daughter cells, and which effectively allow the bacterium to know where the middle of the cell is.

Conclusion

Adler reported in his paper that mini-cells may have a useful function in understanding biology. Indeed, they are used today to examine cell surface structures and are investigated as nanoparticles for the purposes of drug delivery.

Read the original work

Adler, H., Fisher, W. D., Cohen, A. and Hardigree, A. A. (1967) Miniature *Escherichia coli* cells deficient in DNA. *Proc. Natl. Acad. Sci. USA* 57(2): 321–326.

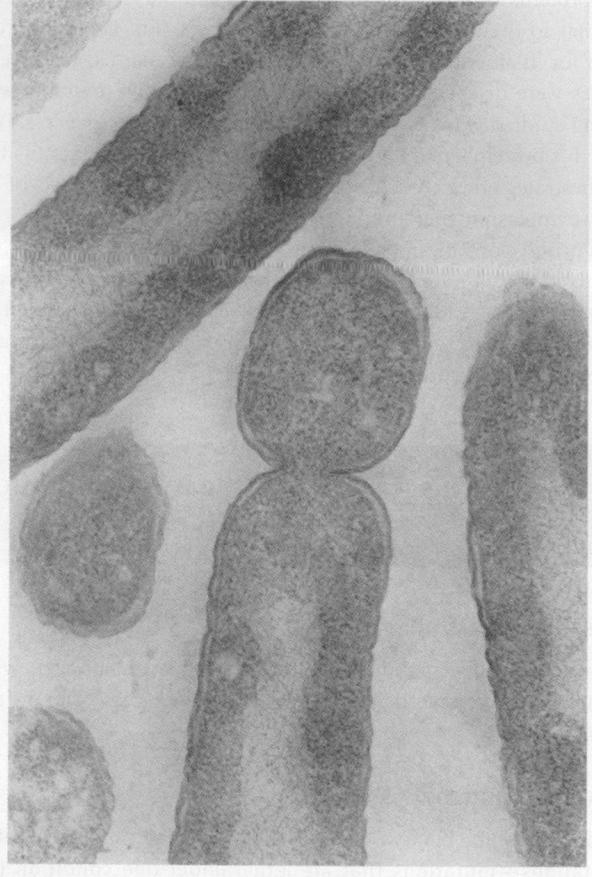

Figure 1   Thin section electron micrograph of a cell producing a mini-cell.

Source: Reproduced from H. I. Adler, W. D. Fisher, A. Cohen, and Alice A. Hardigree (1967) Miniature *Escherichia coli* cells deficient in DNA. *Proc. Natl Acad. Sci. USA* 57(2): 321–326.

It is also interesting to note that the genes for Fts and Min proteins are found in plant genomes, as they have a role in chloroplast division. This is really strong evidence for the prokaryotic origin of the chloroplast through the endosymbiosis of cyanobacteria as discussed in Chapter 9.

The formation of two cells from one is referred to as a generation. So, the rate at which bacteria duplicate is referred to as their doubling time or generation time. This rate is highly variable and depends upon a number of nutritional and genetic factors. We discuss the growth of cell populations in more detail in Chapter 12.

Budding

Some bacteria are known to reproduce by a process called budding, where a daughter cell is produced by the mother cell at a particular site. The bacterial nucleoid divides and one moves to the young bud. Then, when the bud is mature enough, it is released from the mother cell as a daughter cell (see Figure 10.2). Budding is also commonly seen in eukaryotic yeast cells.

Figure 10.2 shows how budding sees the formation of a bud, with an outward growth at the surface of the cell. The bud grows until it is a similar size to the parent cell, at which point it breaks away.

Fragmentation

A third type of reproduction, called fragmentation, sees fragments of filamentous cells break off. These fragments then grow into full-sized cells. This process is fairly common in filamentous cyanobacteria. By contrast, unicellular cyanobacteria divide by binary fission.

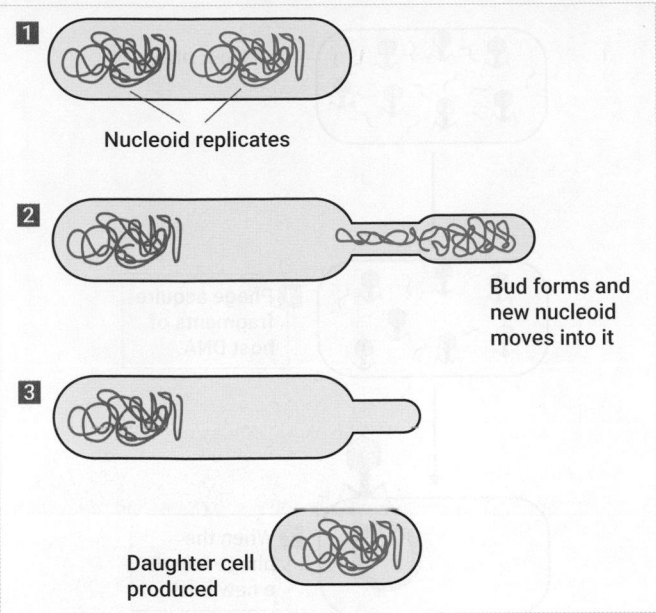

1 Nucleoid replicates

2 Bud forms and new nucleoid moves into it

3 Daughter cell produced

Figure 10.2 Bacterial reproduction via budding.

Horizontal gene transfer

Cell division results in vertical gene transfer, whereby the genome of a parent cell is replicated and passed in its entirety to its daughter cell. However, the genetic diversity of prokaryotes is also increased through **horizontal gene transfer**. During this process, chromosomes and segments of DNA can be replicated and exchanged between different bacterial cells as a result of transposition, transformation, conjugation, and transduction. Let us briefly consider each mechanism in turn.

Transposition

Figure 10.3 shows how transposition is mediated by **transposons**, which are mobile genetic elements that can hop from one location on the chromosome to another. During this movement, transposons can hop into areas of the bacterial chromosome and disrupt the function of genes found in that area, leading to a mutation.

When a transposon hops elsewhere in the chromosome it can sometimes take with it a small piece of the bacterial DNA from the previous location; this DNA can be left behind when the transposon moves on.

Transposons can also hop into plasmids and these can then be passed to another bacterial cell through conjugation.

Transformation

Transformation describes the uptake of exogenous DNA from the environment, as illustrated in Figure 10.4. This DNA often comes from dead cells that are located close to living cells.

Once DNA has been taken up by the cell it can be either broken down or integrated into the bacterial chromosome. In instances where DNA has been taken up as a plasmid, the plasmid can exist

Figure 10.3 An illustration of transposition. Transposition can follow one of two mechanisms. A cut-and-paste mechanism sees the transposon excised from one part of the chromosome and reintegrated at a different location. A copy-and-paste mechanism sees the transposon duplicated in a second location, while remaining at its initial location.

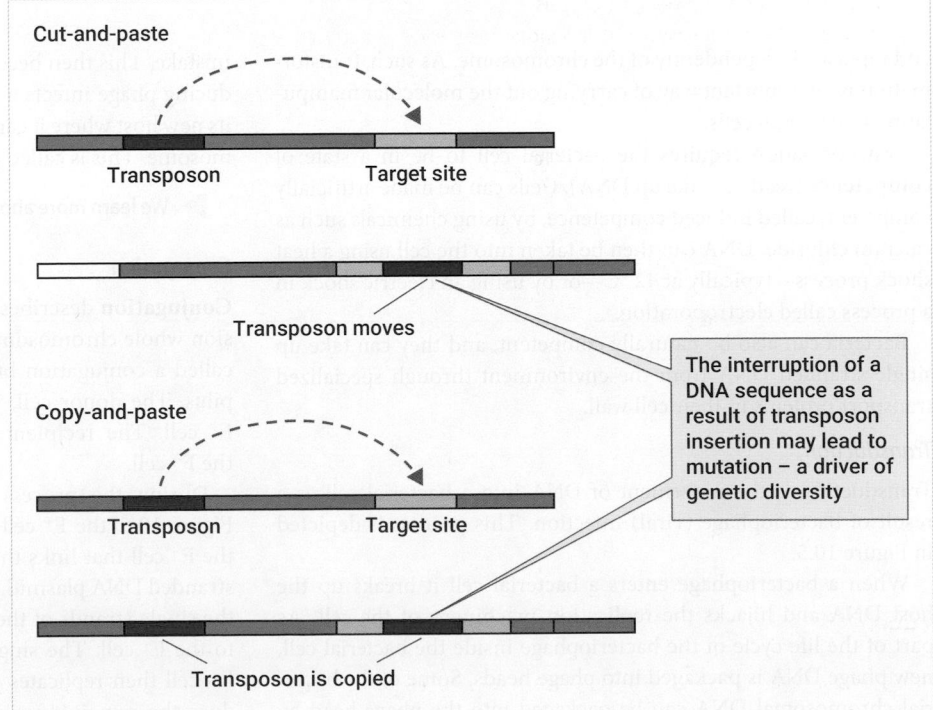

Cut-and-paste

Transposon   Target site

Transposon moves

Copy-and-paste

Transposon   Target site

Transposon is copied

The interruption of a DNA sequence as a result of transposon insertion may lead to mutation – a driver of genetic diversity

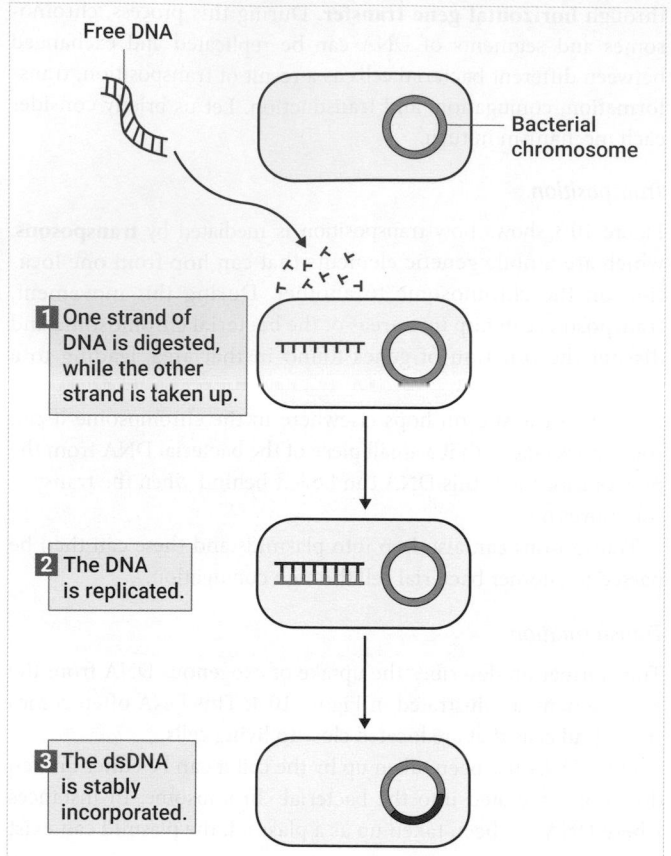

Figure 10.4 The process of transformation. During transformation, genetic material from the external environment is taken up by a bacterial cell and may become integrated into its genome.

Source: Philip Meneely et al. (2017) *Genetics*. Oxford University Press.

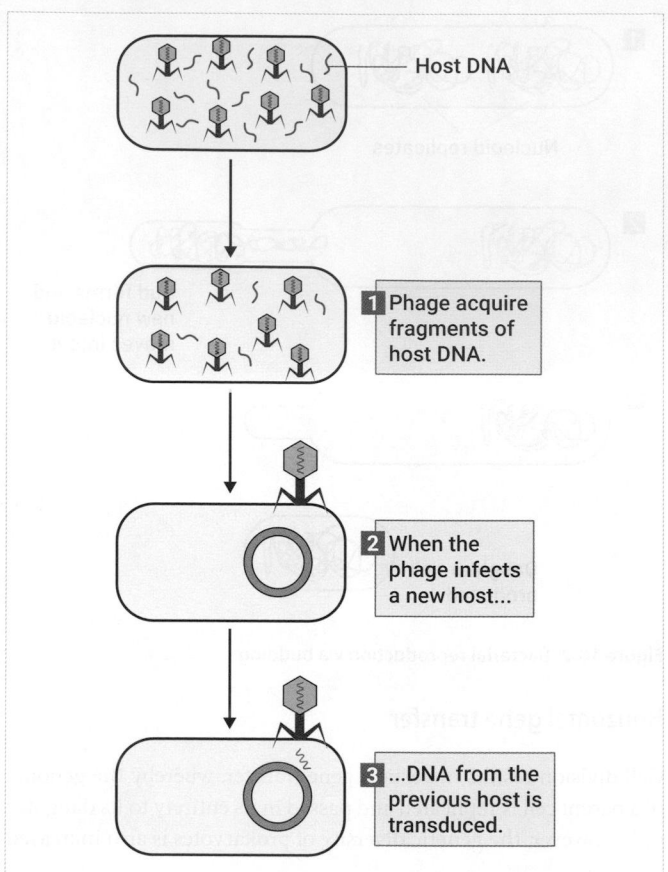

Figure 10.5 A schematic illustration of generalized transduction. Notice how transduction sees a bacteriophage introducing exogenous DNA into a host cell.

Source: Philip Meneely et al. (2017) *Genetics*. Oxford University Press.

and replicate independently of the chromosome. As such, transformation is an important way of carrying out the molecular manipulation of bacterial cells.

Transformation requires the bacterial cell to be in a state of **competence** (ready to take up DNA). Cells can be made artificially competent, called induced competence, by using chemicals such as calcium chloride. DNA can then be taken into the cell using a heat shock process—typically at 42 °C—or by using an electric shock in a process called electroporation.

Bacteria can also be naturally competent, and they can take up single-stranded DNA from the environment through specialized transport proteins in their cell wall.

Transduction

Transduction is the movement of DNA into a bacterial cell as a result of bacteriophage (viral) infection. This process is depicted in Figure 10.5.

When a bacteriophage enters a bacterial cell it breaks up the host DNA and hijacks the replication machinery of the cell. As part of the life cycle of the bacteriophage inside the bacterial cell, new phage DNA is packaged into phage heads. Some of the bacterial chromosomal DNA can be packaged into the phage head by

mistake. This then becomes a transducing phage. When the transducing phage infects a new bacterial cell, it injects this DNA into its new host where it can become integrated into the bacterial chromosome. This is called generalized transduction.

 We learn more about bacteriophages in Chapter 15.

Conjugation

Conjugation describes the movement of plasmids (and on occasion whole chromosomes) between two cells through a structure called a conjugation bridge, which is formed by a bacterial sex pilus. The donor cell, which contains the plasmid, is called the F+ cell. The recipient cell, which lacks the plasmid, is called the F− cell.

During the process of conjugation, which is illustrated in Figure 10.6, the F+ cell produces the sex pilus, which attaches to the F− cell that links the two bacterial cells together. The double-stranded DNA plasmid is nicked (cut) at a specific point and one of the single strands of the plasmid DNA crosses through the bridge to the F− cell. The single-stranded DNA plasmid inside the new F+ cell then replicates to produce a double-stranded plasmid, as does the remaining single-stranded DNA plasmid in the donor

Figure 10.6 An overview of conjugation. Notice how both cells involved in the conjugation become donor (F⁺) cells by the end of the process.

Source: Philip Meneely et al. (2017) *Genetics*. Oxford University Press.

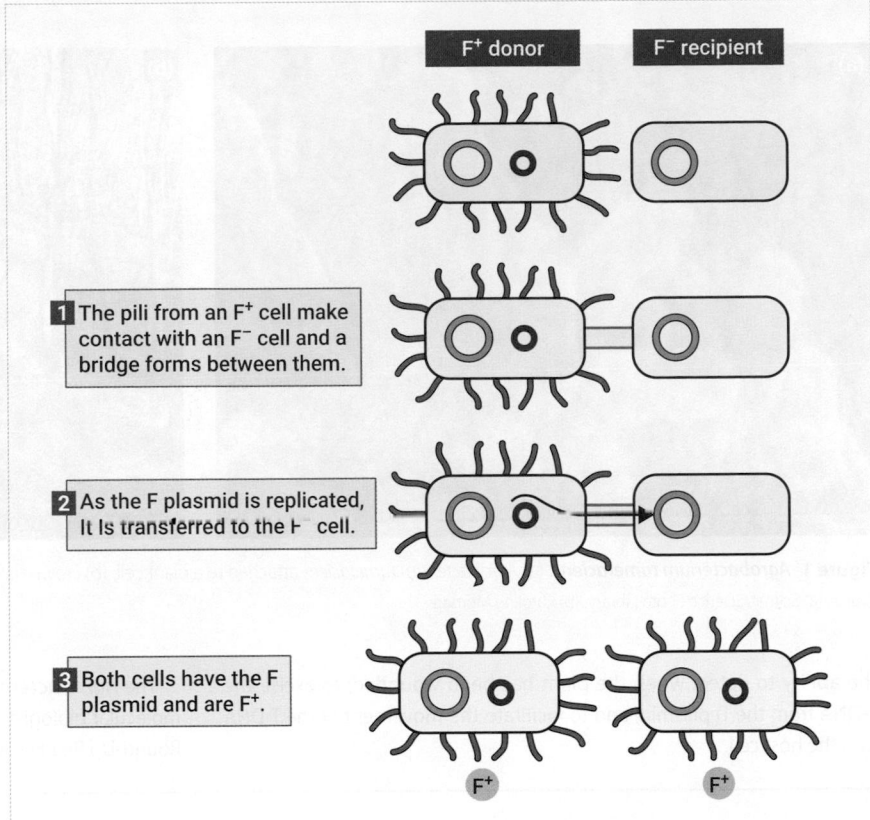

F⁺ donor   F⁻ recipient

1 The pili from an F⁺ cell make contact with an F⁻ cell and a bridge forms between them.

2 As the F plasmid is replicated, it is transferred to the F⁻ cell.

3 Both cells have the F plasmid and are F⁺.

F⁺   F⁺

cell. After the plasmid has been taken up (once conjugation is complete), the F⁻ cell becomes an F⁺ cell: it is now a donor cell too.

Look back at Figures 10.3 to 10.6 and notice how these four mechanisms of horizontal gene transfer increase genetic diversity: the movement of DNA and its incorporation into the host cell change the host cell's genetic make-up. This is an important driver of biodiversity: natural selection acts on the resulting organisms, and new species may emerge as they are selected for over time.

We see an excellent example of naturally occurring horizontal gene transfer within the bacterium *Agrobacterium tumefaciens*, which causes crown gall disease (the growth of a type of tumour) in plants. This is described in more detail in Real World View 10.1.

REAL WORLD VIEW 10.1   Horizontal gene transfer in nature and its use for genetic engineering

Agrobacterium tumefaciens is a rod-shaped Gram-negative bacterium that belongs to the alpha proteobacterium. Its morphology is shown in Figure 1a. *Agrobacterium tumefaciens* is the causative agent of crown gall disease, the consequences of which are shown vividly in Figure 1b.

Agrobacterium tumefaciens can be found in the area of soil, called the rhizosphere, directly adjacent to the plant root, and it is chemically attracted towards substances exuded by plant roots. Pathogenic strains of *A. tumefaciens* contain a large plasmid (approximately 200 kb in size), called the Ti (tumour-inducing) plasmid. Remarkably, this bacterium can transfer a segment of DNA from the Ti plasmid, called T-DNA, into the nucleus of the infected plant cell, where it is incorporated into the host genome. The subsequent expression of the T-DNA by the infected host ultimately leads to the onset of crown gall disease. It is fascinating not

only because the bacterium transfers its DNA to a plant, which is a different domain of life, but also because it can be harnessed as a tool for genetic engineering.

The Ti plasmid encodes the genes needed for its transfer into the plant cell to occur. Specifically, it contains:

1. Oncogenic genes that encode enzymes involved in the synthesis of auxins and cytokines, which drive tumour formation.

2. Opine genes that encode proteins for the synthesis of opines. The opines are produced inside the tumour cells and act as both a carbon and nitrogen source to fuel the growth of *A. tumefaciens*.

Transfer of the T-DNA into the plant host cell is mediated by the *vir* region of the Ti plasmid, which contains a number of different genes that encode

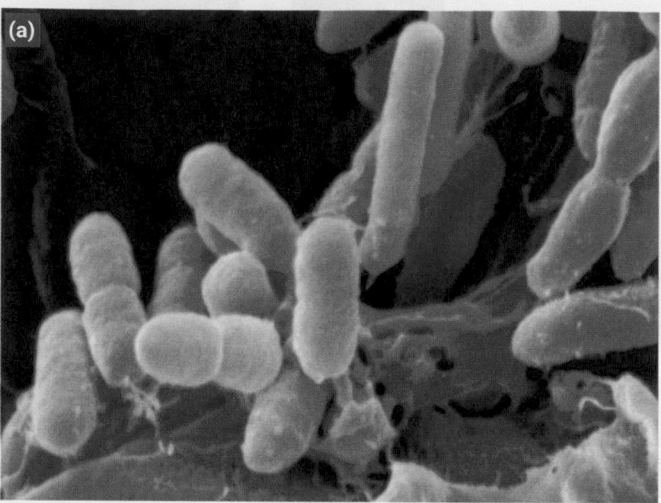

Figure 1 *Agrobacterium tumefaciens.* (a) *Agrobacterium tumefaciens* attached to a plant cell; (b) crown gall disease.
Source: (a) Scimat/Science Photo Library; (b) Caroline Morgan.

the ability to detect when the plant has been wounded; to excise the T-DNA from the Ti plasmid; and to facilitate the movement of the T-DNA into the host cell.

The *Agrobacterium* system of T-DNA transfer has been harnessed by molecular biologists to produce genetically modified plant crops, such as Round-Up Ready soya engineered by the company Monsanto.

PAUSE AND THINK

Do you think that bacterial cells are at an evolutionary disadvantage as they can only reproduce asexually?

Answer: Bacteria are asexual, which means their progeny are genetically identical to the parent cell. As a result, genetic diversity through reproduction is limited. However, bacteria are very good at taking up DNA from their environment and through other processes, such as transformation and conjugation, which is genetic exchange between cells. This means that they can react to changes in their environment, which to a great extent replaces the need for sexual reproduction. Bacteria are highly successful organisms that occupy a wide variety of environmental niches.

 Check your understanding of the concepts covered in this section by answering the questions in the e-book.

10.2 Eukaryotic cell division

All cells have a **cell cycle**, which represents the life of a cell from the moment it is created from a parent cell until it divides to produce daughter cells (or dies). Look at Figure 10.7 and notice how the cell cycle consists of two consecutive, continuous phases: the **mitotic (M) phase** and **interphase**. Also notice how interphase involves three different stages: G_1, S, and G_2.

Interphase

When a cell is in the mitotic phase it is undergoing cell division. By contrast, interphase is not considered a part of cell division. Instead, interphase sees the cell engaged in its normal metabolic activities while also preparing for its subsequent division by activities such as DNA replication and organelle duplication. Approximately 90 per cent of a cell's cycle is spent in interphase; for some adult mammalian cells this phase lasts for about 20 hours. It is not usually possible to see the individual chromosomes within the nucleus during interphase; this is only possible in the early mitotic stages once the genetic material has condensed.

PAUSE AND THINK

We have just noted that interphase in some adult mammalian cells lasts for about 20 hours. But how long is the whole of the cell cycle, and do all eukaryotic cells undergo division in the same way?

Answer: The length of the cell cycle is actually quite variable in different kinds of cell. In the example above, the full cell cycle would likely take 21–24 hours. Other types of cell, such as budding yeasts, have a cell cycle as short as 1.5 hours. Some adult cell types (such as some liver cells) divide only very occasionally in response to injury, whereas others (termed post-mitotic) stop dividing altogether.

There are three stages to interphase; G_1, S, and G_2. The activities that take place during each of these stages are outlined in Table 10.1.

During the S phase of interphase DNA molecules replicate. As they are synthesized, the duplicated DNA molecules are linked together with protein complexes called cohesins. Once a DNA

Figure 10.7 The cell cycle. (a) The mitotic (M) phase of the cell cycle and the three separate stages of interphase; G₁, S, and G₂; and (b) the number of chromosomes and chromatids at various stages of the cell cycle.

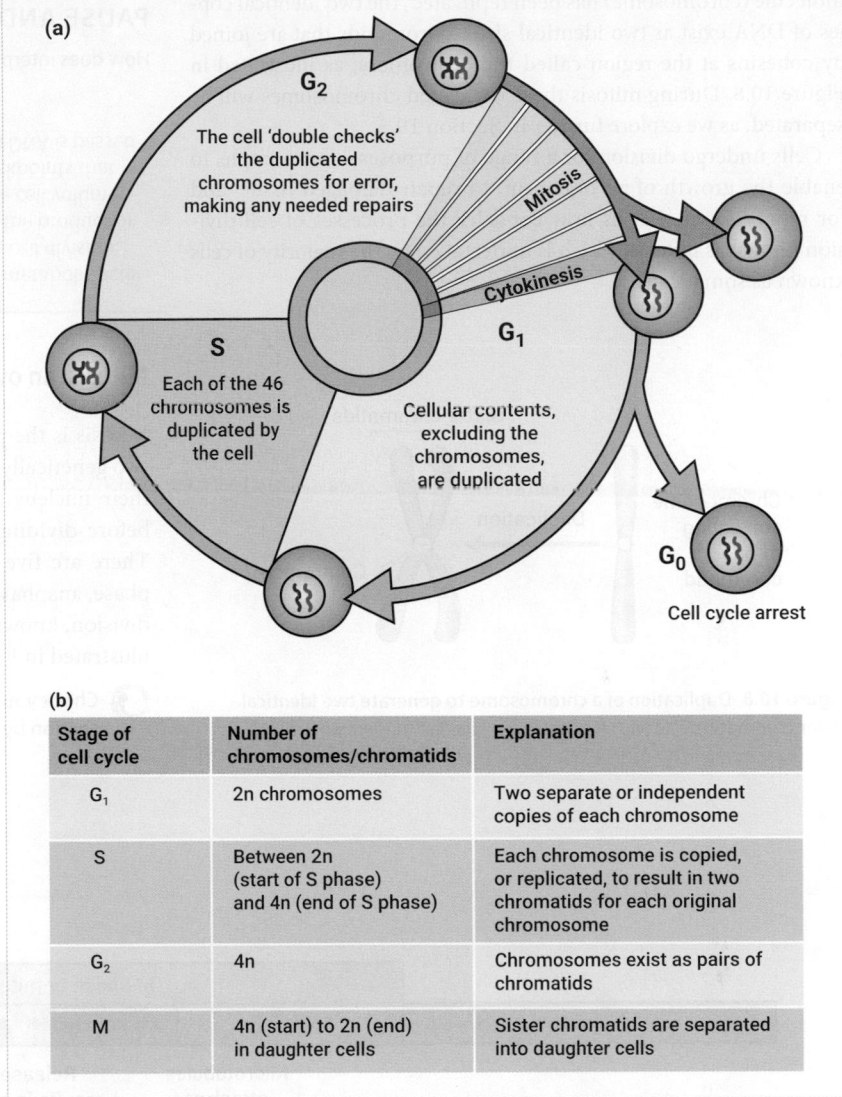

(a)

G₂

The cell 'double checks' the duplicated chromosomes for error, making any needed repairs

Mitosis

Cytokinesis

G₁

S

Each of the 46 chromosomes is duplicated by the cell

Cellular contents, excluding the chromosomes, are duplicated

G₀

Cell cycle arrest

(b)

Stage of cell cycle	Number of chromosomes/chromatids	Explanation
G₁	2n chromosomes	Two separate or independent copies of each chromosome
S	Between 2n (start of S phase) and 4n (end of S phase)	Each chromosome is copied, or replicated, to result in two chromatids for each original chromosome
G₂	4n	Chromosomes exist as pairs of chromatids
M	4n (start) to 2n (end) in daughter cells	Sister chromatids are separated into daughter cells

Table 10.1 The different activities that take place during the three different stages of interphase of the cell cycle; G₁, S, and G₂

G₁ (1st gap)	S (synthesis)	G₂ (2nd gap)
Cell recovers from previous division	DNA is replicated, resulting in two identical copies (or sister chromatids) of each chromosome joined at the centromere. Another name for replicated is synthesized, hence the name of this stage	Cell continues to grow and increase in volume; proteins required for chromosome manipulation are produced
Cell grows and increases in volume; high amount of protein synthesis	Cell continues to grow and replicate organelles; the centrosome (in animal cells) is also duplicated	Cytoskeletal filaments change; these will later aid in the movement of chromosomes during the mitotic phase and cell shape changes
Organelles, such as the mitochondria and ribosomes, are duplicated	Specialized chromatin structures (such as sister-chromatid cohesions made from cohesin protein complexes) are constructed to prepare chromosomes for separation in M phase	
Materials needed for DNA replication are accumulated	S checkpoint to ensure DNA replication is complete	G₂ checkpoint takes place to check that DNA has replicated correctly
G₁ checkpoint takes place to check that DNA is suitable for replication		

molecule (chromosome) has been replicated, the two identical copies of DNA exist as two identical sister chromatids that are joined by cohesins at the region called the centromere, as illustrated in Figure 10.8. During mitosis these replicated chromosomes will be separated, as we explore further in Section 10.3.

Cells undergo division for a range of purposes—for example, to enable the growth of an organism, to repair damaged tissue, and for reproduction. Let us now consider the processes of cell division known as mitosis, which is undertaken by the majority of cells known as somatic cells.

Mitosis: an overview

Mitosis is the process by which one single cell divides to produce two genetically identical daughter cells. During mitosis, cells divide their nucleus and replicated chromosomes and their organelles before dividing their cytoplasm to form the two daughter cells. There are five stages to mitosis: prophase, prometaphase, metaphase, anaphase, and telophase. An additional stage of cytoplasmic division, known as **cytokinesis**, also occurs. The overall process is illustrated in Figure 10.9.

Check your understanding of the concepts covered in this section by answering the questions in the e-book.

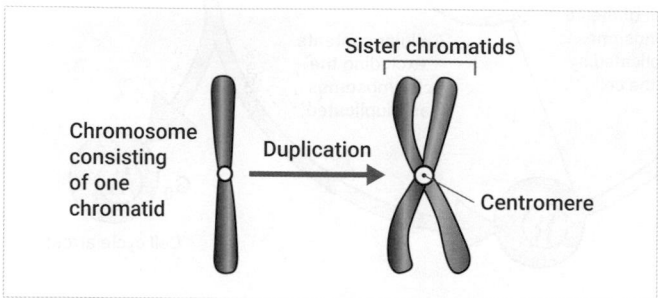

Figure 10.8 Duplication of a chromosome to generate two identical sister chromatids. The two chromatids are joined at the centromere.

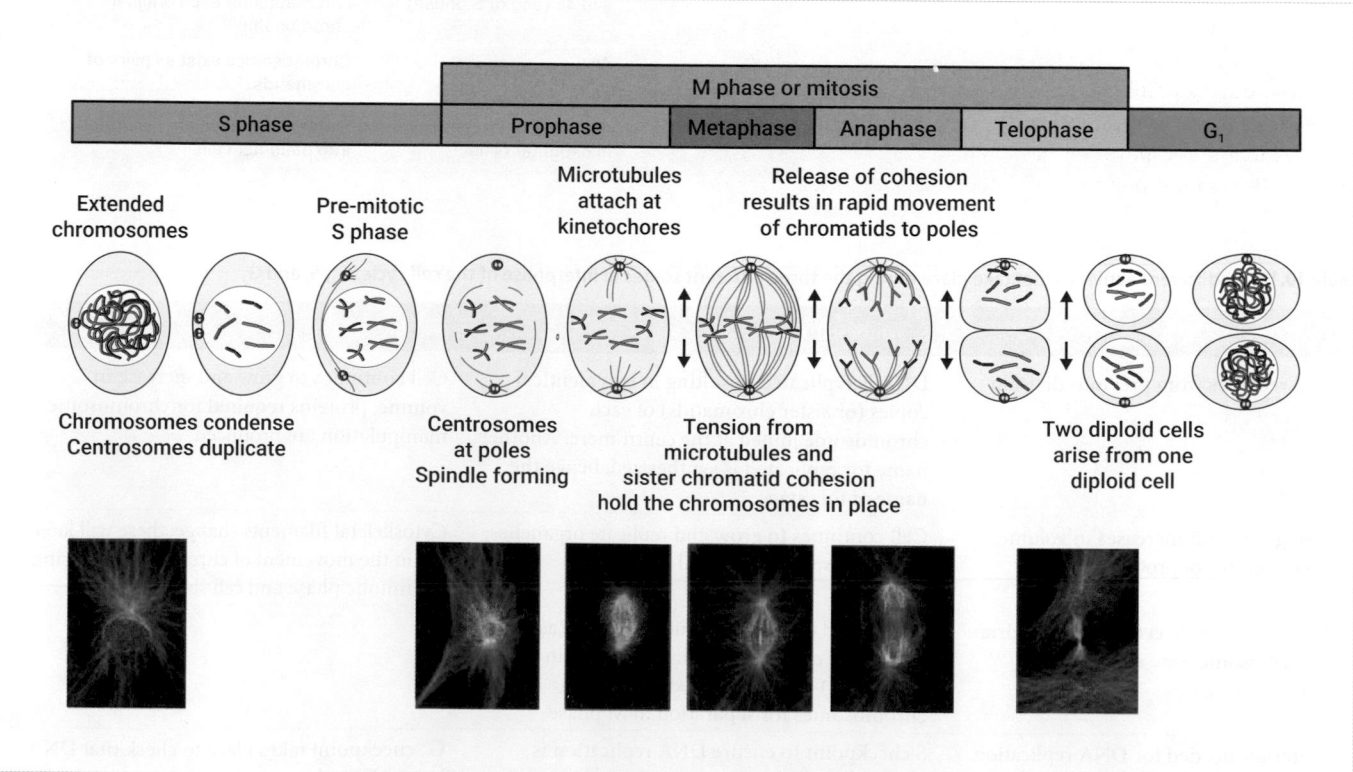

Figure 10.9 An overview of mitosis. For simplicity, this overview demarks four of its five stages. In practice, metaphase is preceded by prometaphase, as explained in the text.

Source: Philip Meneely et al. (2017) *Genetics*. Oxford University Press; and Nancy Craig et al. (2021) *Molecular Biology, Principles of Genome Function*, Third Edition. Oxford University Press.

10.3 **Mitosis: a deeper look**

Having looked at mitosis in overview, let us now consider each stage in turn.

Prophase

Prophase is the first stage of mitosis. During prophase, the nucleolus disappears, the nuclear envelope starts to fragment, and the DNA and associated proteins found within a cell, known as chromatin, start to coil up and condense. The process of chromatin coiling up makes the DNA molecules more compact until they have condensed to a point at which they become visible through a light microscope.

During prophase, organelles called **centrosomes** (in animal cells) move to the two opposite poles of the cell and make a series of long tubulin filaments, a form of microtubule. These microtubules are polar, with the minus ends embedded in the spindle pole in the centrosome, and the plus ends pointing outward. Collectively, these filaments form the **mitotic spindle**, the structure that will eventually be responsible for separating and moving the chromosomes into the new daughter cells. Microtubules are dynamic polymers that constantly grow and shrink: this lengthening and shortening of the mitotic spindle is an essential element of cell division, as we will discover shortly.

Prometaphase

Prometaphase follows prophase. During prometaphase, in cells other than fungal cells, the nuclear membrane breaks down fully, releasing the DNA into the cytoplasm of the cell. The microtubules produced during prophase now extend and attach to the chromosomal centromeres at regions called the **kinetochores** in a process called search and capture; this attachment is illustrated in Figure 10.10. Although not yet fully understood, the kinetochores feature molecular motor proteins that, along with changes to the microtubules themselves, provide the pulling force to separate and move the sister chromatids.

The size and complexity of kinetochores varies among different species; in higher eukaryotic cells they are multi-layered protein complexes that form at the centromere and can each bind 20–40 microtubules. Once these microtubules have attached to the kinetochores, the chromosomes can then begin the movement needed for the subsequent stages of mitosis.

During both prometaphase and metaphase (which follows) the chromosomes undergo dramatic structural changes as they condense, which help with their later separation. We noted earlier how sister chromatids are held together by a type of protein called cohesin. During these stages, however, most cohesin complexes are lost from the arms, but they remain at the centromeres. As a result, the centromeres become the only locations at which the chromatids are joined, as illustrated in Figure 10.10.

Metaphase

Metaphase is the middle stage of mitosis, during which the replicated chromosomes line up along the equator of the dividing cell,

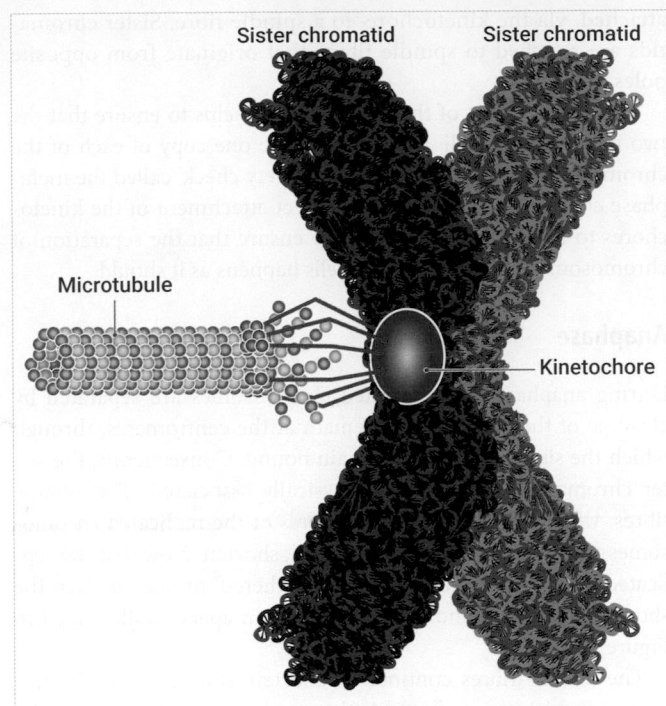

Figure 10.10 The attachment of microtubules at the kinetochore. The kinetochore is found at the centromere.
Source: Philip Meneely et al. (2017) *Genetics*. Oxford University Press.

as depicted in Figure 10.11. The spindle fibres, which attached to the kinetochores of the replicated chromosomes during prophase, contract and relax; in doing so, they draw the replicated chromosomes to the centre of the cell so that they line up along a region known as the metaphase plate. Each sister chromatid is

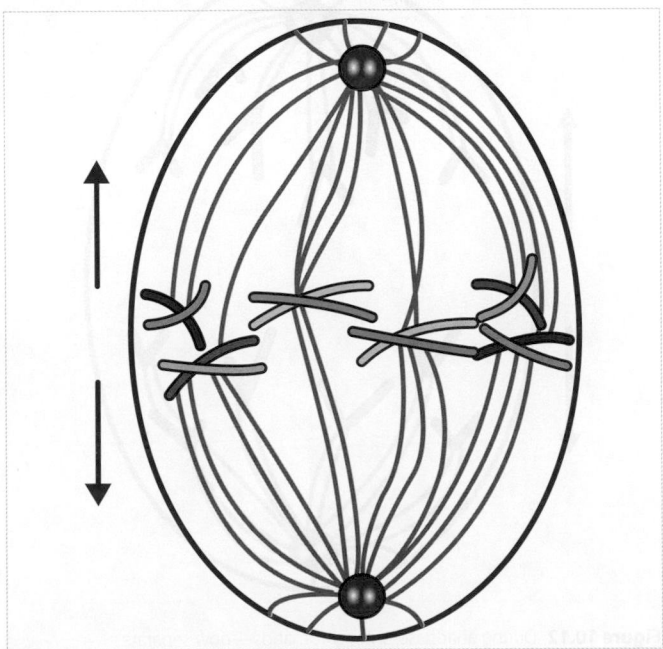

Figure 10.11 Metaphase. Chromosomes line up at the centre of the cell along a region called the metaphase plate.
Source: Adapted from Philip Meneely et al. (2017) *Genetics*. Oxford University Press.

attached, via the kinetochore, to a spindle fibre. Sister chromatids are attached to spindle fibres that originate from opposite poles of the cells.

This arrangement of the chromosomes helps to ensure that the two new daughter cells will each receive one copy of each of the chromosomes. A form of molecular 'safety check' called the metaphase checkpoint also confirms correct attachment of the kinetochores to the spindle, again to help ensure that the separation of chromosomes into the daughter cells happens as it should.

Anaphase

During anaphase, the replicated chromosomes are separated by cleavage of the cohesins that remain at the centromeres, through which the sister chromatids remain bound. Consequently, the sister chromatids are no longer physically associated. The spindle fibres, which connect the kinetochores of the replicated chromosomes with one of the poles of the cell, shorten. Now that the replicated chromosomes are no longer 'tethered' to one another, the shortening of the spindle fibres pulls them apart, as illustrated in Figure 10.12.

The spindle fibres continue to shorten, drawing the chromosome to which they are attached towards one of the poles of the cell. This ensures that each pole receives the same kind and number of chromosomes as the original parental cell.

Telophase

By the time telophase is reached the two complete sets of genetic information are situated at opposite poles of the dividing cell.

During telophase these individual, separated chromosomes begin to disperse and decondense, reforming the relaxed chromatin; at this point they are no longer visible under a light microscope. A new nuclear membrane forms around each set of chromosomes and the spindle fibres disappear as they are dismantled. It may also be possible to see the cytoplasm of the cell starting to prepare for division during this stage.

Cytokinesis

Cytokinesis is the final stage of cell division, during which the cytoplasm is divided to produce the daughter cells. Cytokinesis proceeds differently in plant cells compared with animal cells. In animal cells, a contractile ring of actin and myosin filaments forms around the centre of the cell. Figure 10.13a shows how this gradually contracts to form a **cleavage furrow**, which becomes smaller in diameter until the cytoplasm of the cell is eventually pinched into two, forming the two daughter cells. Each daughter cell contains a full complement of organelles and one nucleus with a complete set of chromosomes.

Plant cells need to separate and rebuild a section of their cell wall, in addition to the cell membrane, to create daughter cells. This rebuilding happens when modified vesicles from the Golgi apparatus form across the previous cell plate until they eventually fuse together and with the plasma membrane. A new cell wall forms between these new membranes and division of the parent cell occurs, as illustrated in Figure 10.13b.

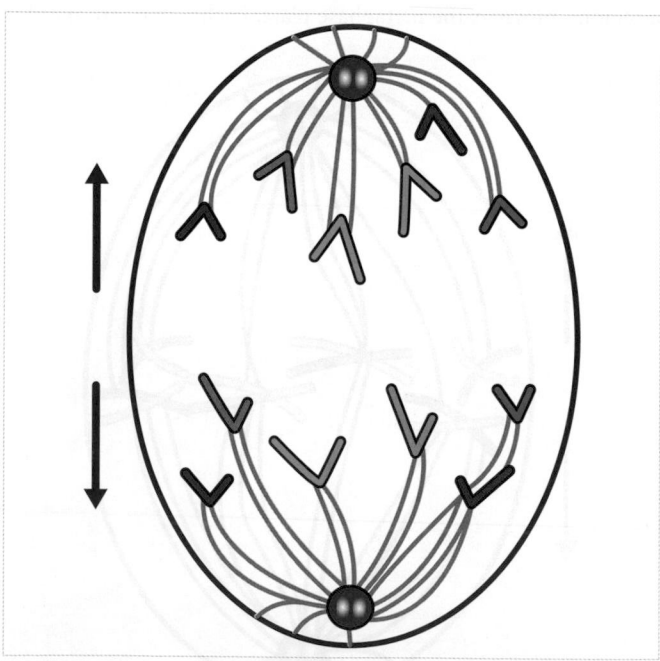

Figure 10.12 During anaphase, sister chromatids—now separate chromosomes—are pulled to the poles of the cell.
Source: Adapted from Philip Meneely et al. (2017) *Genetics*. Oxford University Press.

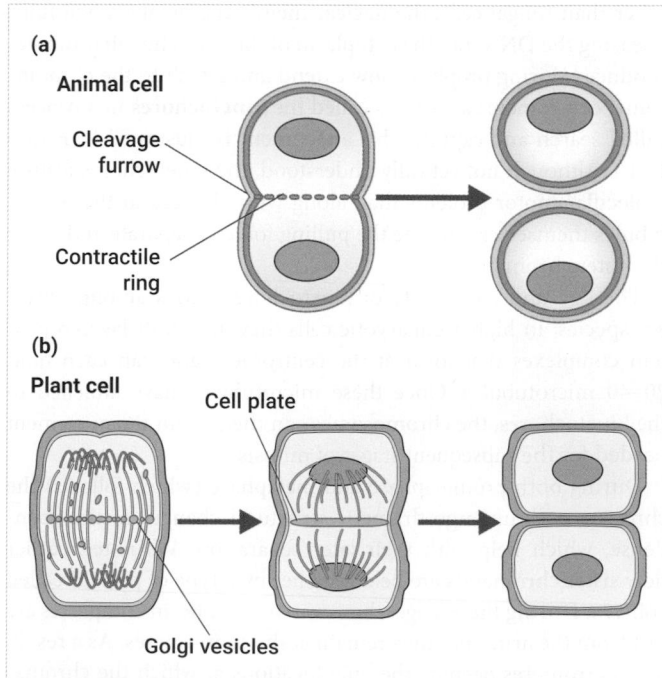

Figure 10.13 Cytokinesis. The process of cytokinesis within (a) animal and (b) plant cells.
Source: Openstax CC BY 4.0 C Rye et al. *Biology*, 2016. Section URL: https://openstax.org/books/biology/pages/10-2-the-cell-cycle.

Figure 10.14 Key cell cycle checkpoints and the cyclin–CDK complexes that regulate them.

Source: Ee Phie Tan, Francesca E. Duncan, Chad Slawson. The sweet side of the cell cycle. *Biochem. Soc. Trans.* 2017; 45 (2): 313–322. doi: https://doi.org/10.1042/BST20160145 Biochemical Society.

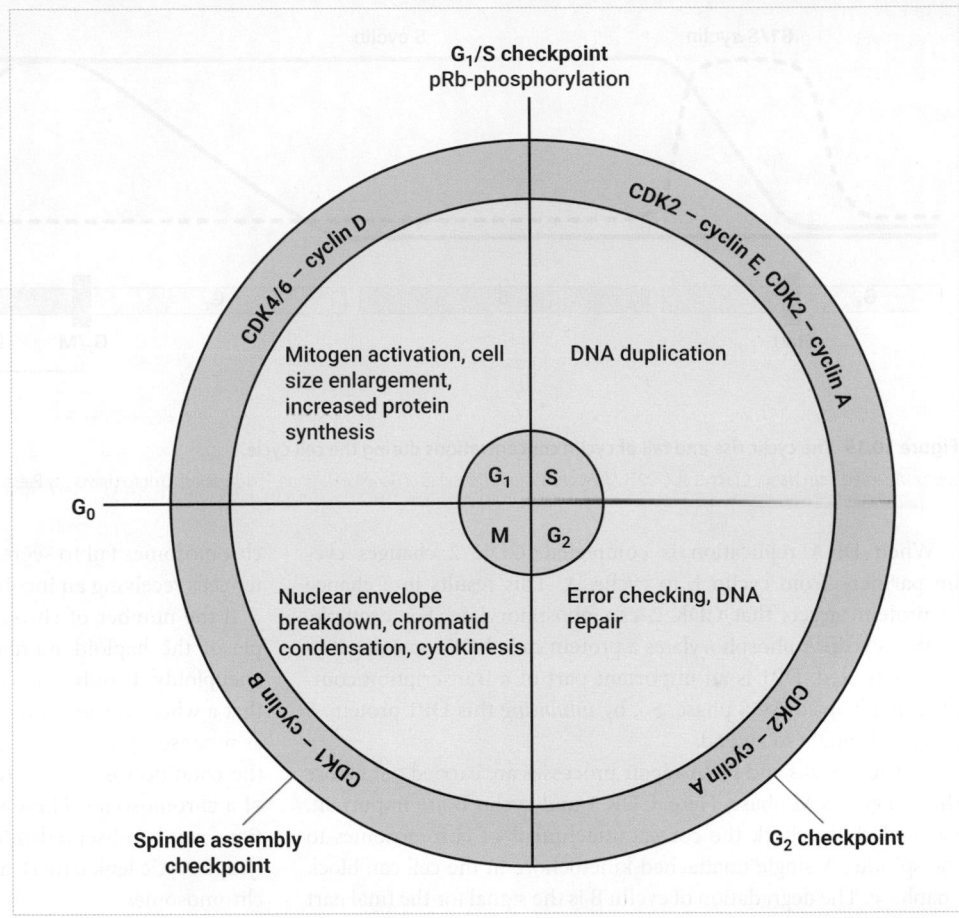

▶ We learn more about the Golgi apparatus in Chapter 9.

How is mitosis controlled?

It is important that a cell undergoes mitosis only when new daughter cells are required. In other words, mitosis should occur at the right time and in the right place. Normal, healthy eukaryotic cells will not undergo mitosis unless they receive signals, such as growth factors, telling them to grow and divide. When they are not undergoing mitosis, many cells exist in a non-dividing, but viable, state called G_0. Mitosis can then be initiated through the binding of growth factors to cell-surface receptors, which span the plasma membrane. This binding triggers the activation of cell signalling pathways, which in turn trigger an increase in the amount of cyclin D proteins present.

The D-type cyclins are members of one of two key groups of proteins that control the various stages of the cell cycle, and which also ensure that the stages occur in the correct order. The key regulatory proteins are the cyclin dependent kinases (CDKs or CDCs) and the cyclins.

Control is exerted at certain 'checkpoints' within the cell cycle. These checkpoints are illustrated in Figure 10.14. The purpose of each checkpoint—and the regulatory proteins involved in imposing control—is described in Table 10.2. Look at this table and notice that each checkpoint involves a combination of cyclins and CDKs, although the identity of the proteins is specific to each checkpoint.

Table 10.2 Main cell cycle checkpoints and the CDK/cyclin heterodimers involved

Checkpoint	Purpose of checkpoint	CDK	Cyclin
Restriction (R) or G_1 to S	Checks for: cell size, availability of nutrients, signals (positive growth cues), DNA damage	4 and 6 2	D E
G_2	Checks for: DNA damage, DNA replication completeness	2	A
M	Checks for chromosome attachment to spindle at metaphase plate	1	B

The cyclin-dependent kinases are inactive without a cyclin binding partner. So, when growth factor-induced expression of the D cyclins raises their concentration, they can then bind with either CDK 4 or 6. CDK 4 and 6 are then able to phosphorylate their target proteins, ultimately leading to the transcription and expression of proteins required for DNA replication during S phase. Cyclin E is also expressed at this point; together with its partner CDK 2 it can help the cell to progress into S phase.

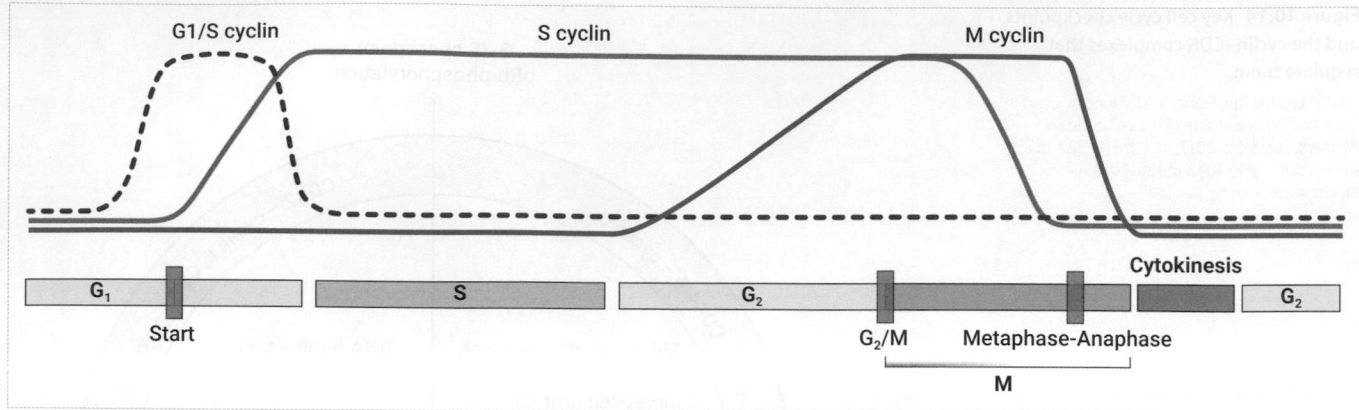

Figure 10.15 The cyclic rise and fall of cyclin concentrations during the cell cycle.
Source: Adapted from Nancy Craig et al. (2021) *Molecular Biology, Principles of Genome Function*, Third Edition. Oxford University Press.

When DNA replication is completed, CDK 2 changes cyclin partner—from cyclin E to cyclin A. This results in a change of protein targets that CDK 2 can phosphorylate. For example, CDK 2/cyclin A phosphorylates a protein called DP1, causing it to be deactivated. DP1 is an important part of a transcription complex that helps drive S phase. So, by *inhibiting* this DP1 protein, S phase is brought to an end.

Further checks and DNA repair processes are carried out before the cell enters M phase. Here, CDK 1 and cyclin B are important, particularly to check the correct attachment of chromosomes to the spindle. A single unattached kinetochore in the cell can block anaphase. The degradation of cyclin B is the signal for the final part of mitosis to occur.

The various CDKs are present throughout the cell cycle, whereas the cyclins change in concentration, as their name implies. It is the sequential and cyclic increase and decrease in different cyclin proteins, and their binding to the CDKs, that ensures the various stages of the cell cycle occur in the correct order—for example, that M phase or cell division only occurs after DNA replication in S phase, and G2, has taken place. The cyclic rise and fall of cyclin concentration during the course of mitosis is illustrated in Figure 10.15.

As we have seen, the cyclins control the activity of their CDK binding partners. However, the progression and control of the cell cycle is not as simple as this. The CDKs need activating by being phosphorylated, and for any inhibitory phosphates to be removed, before they can act on their targets. They are also subject to inhibition by families of proteins collectively known as the cyclin-dependent kinase inhibitors (CDKIs, also sometimes known as CKIs). Mutations in any of these important families of proteins (such as amplification or too much cyclin or CDK, or loss of one of the CDKIs) could result in mitosis occurring when it should not. This could lead to too many cells being produced, one key characteristic of a group of diseases known as cancers.

Errors in mitosis

Despite there being a number of checks to ensure that cell division occurs correctly, the process of mitosis can still go wrong. One problem that can occur is called non-disjunction, whereby the chromosomes fail to segregate properly. This results in the daughter cells receiving an incorrect number of chromosomes.

If the number of chromosomes in a cell is not an exact multiple of the haploid number, the cell exhibits a condition called aneuploidy. If only one copy of a chromosome is present—such that a whole chromosome is missing—the condition is referred to as monosomy. By contrast, if an additional chromosome is present the condition is termed trisomy, whereby a cell has three copies of a chromosome. These sorts of numerical chromosome aberrations can be observed in cancer cells (e.g. trisomy 12 in chronic lymphocytic leukaemia), alongside other structural changes to the chromosomes.

▶ **We learn more about haploid numbers and the condition of aneuploidy in Section 10.4.**

PAUSE AND THINK

What could be the consequences of cells undergoing mitosis at inappropriate times and places?

Answer: Inappropriate or uncontrolled mitosis can result in an increase in cell numbers, termed a neoplasm. This unwanted increase may cause problems for the tissue within which the cells are found. For example, the increased cell mass could exert pressure on a nerve or blood vessel, blocking or impairing its function. Alternatively, the mass could disrupt the normal functioning of the tissue. This sort of neoplasm—in which the effects are mostly localized—can be referred to as a benign neoplasm or tumour. However, mutations in a variety of genes can cause some neoplasms to be more dangerous: the cells may be able to invade local and distant tissues. These neoplasms are then termed malignant and more commonly known as cancers.

 Check your understanding of the concepts covered in this section by answering the questions in the e-book.

10.4 Meiosis

The majority of animal and plant cells contain two compete sets of chromosomes within their nucleus, known as the diploid number of chromosomes. In humans this diploid number is 46 chromosomes, organized as 23 chromosome pairs. One of each of these chromosome pairs is inherited from the mother and one is inherited from the father. This means that for each gene that a diploid cell contains there are two copies of that gene.

All organisms also contain some cells that possess only *one* copy of each chromosome, known as the haploid number of chromosomes. In animal cells, the only haploid cells are the gametes: the sperm and egg cells. When a sperm cell fuses with an egg cell during fertilization, the two haploid gametes each pass on their single set of chromosomes so that the fertilized egg contains two full sets of chromosomes: it is diploid.

Meiosis is a special form of cell division that occurs during the formation of gametes; the overall process is illustrated in Figure 10.16. Gametes are sex cells (e.g. sperm, eggs, and pollen).

The cells destined to become gametes (germline stem cells) undergo a different form of cell division from somatic cells (which only undergo mitosis) as they need to contain *half* the amount of genetic information that somatic cells contain: they need to be haploid rather than diploid. Why is this difference in chromosome number important? As we note above, when fertilization occurs during sexual reproduction, the two sets of haploid chromosomes—one set from the sperm, and one from the egg (in the case of non-plant reproduction)—come together so that the fertilized egg has a full quota of chromosomes: it is diploid.

So, how does the process of meiosis bring about a reduction in chromosome number, from diploid to haploid? The reduction happens because meiosis comprises *two* nuclear divisions (meiosis I and meiosis II) to produce the haploid daughter cells. This is in contrast to the one division seen in mitosis.

Look at Figure 10.17, which compares the processes of mitosis and meiosis. Mitosis sees one diploid parent cell undergoing one nucleic division to produce two genetically identical daughter cells. This process of cell division is usually undertaken by somatic cells for growth of the organism, repair of damaged tissues, or for asexual reproduction.

> Go to the e-book to explore an interactive version of Figure 10.17.

By contrast, the process of meiosis involves one diploid parent cell undergoing two nuclear divisions, the first being a reduction

division, the final objective being to produce four genetically different haploid daughter cells. This process is undertaken by eukaryotic organisms during the production of gametes.

Meiosis I is a reduction division: it separates the homologous pairs of chromosomes and, in doing so, halves the number of chromosomes within the resulting daughter cells. It is also during this division that recombination (crossing over) occurs, leading to an increase in genetic variation.

By contrast, meiosis II occurs in a similar manner to mitosis: replicated chromosomes (sister chromatids) separate and move into the final daughter cells (in this case, the gametes).

The process of meiosis also differs from mitosis because it allows alleles to be swapped between similar chromosomes during a process known as **crossing over**. This introduces genetic variation during the process of gamete production and is central to the differences exhibited by offspring compared with their parents. We learn more about crossing over in the section that follows.

The stages of meiosis

Before meiosis can begin the meiotic S phase takes place, during which the DNA is replicated and the chromosomes are loaded with a meiosis-specific cohesin complex, which keeps the sister chromatids together until anaphase of meiosis II. Let us now consider the stages of meiosis that follow.

PAUSE AND THINK

Why does a cell need to complete the S phase before it can enter meiosis?

Answer: Meiosis results in a diploid cell undergoing two nuclear divisions to produce four haploid daughter cells. DNA replication occurs during the S phase and is essential to ensure the parent cell contains sufficient DNA to be able to provide copies for the four resultant daughter cells.

Prophase I

As with the prophase seen during mitosis, prophase I of meiosis sees the nucleolus disappear, microtubules that will form the spindle begin to form, and there is coiling up and condensing

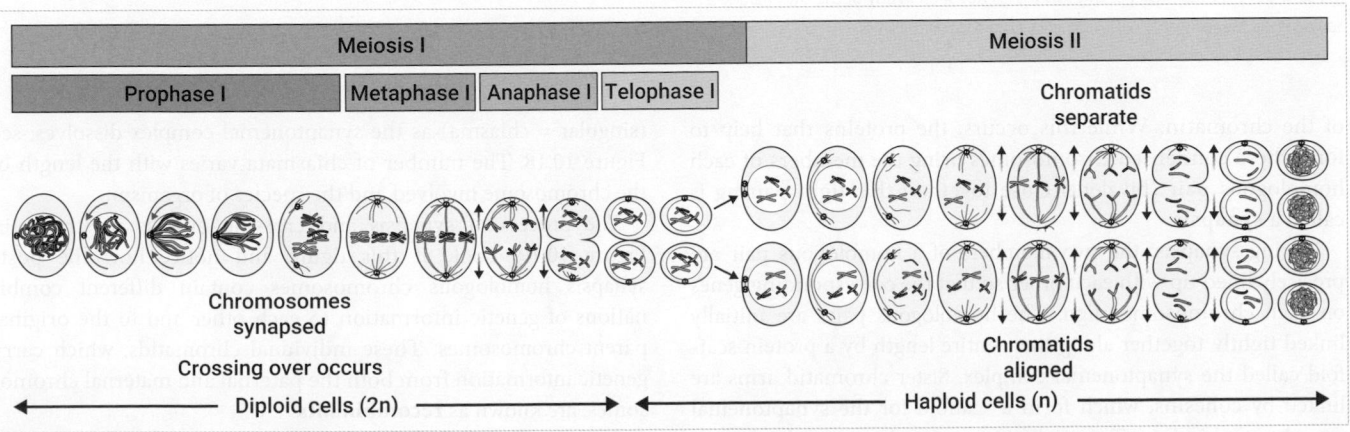

Figure 10.16 The stages of meiosis.

Source: Philip Meneely et al. (2017) *Genetics*. Oxford University Press.

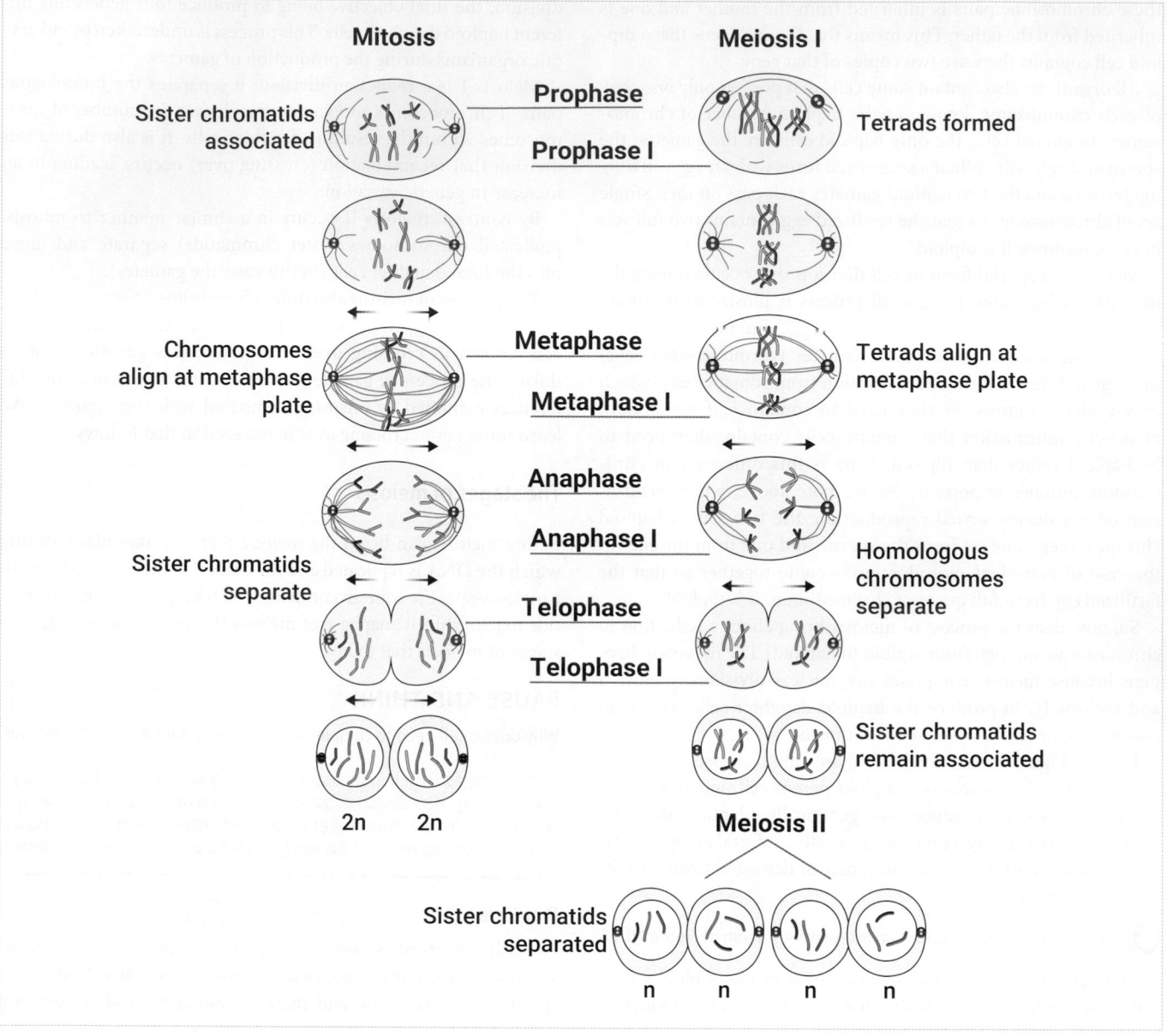

Figure 10.17 An overview of the movement of chromosomes during the processes of mitosis and meiosis. Notice how there is an exchange of genetic material during prophase I of meiosis that is not seen in any of the other stages.

Source: Philip Meneely et al. (2017) *Genetics*. Oxford University Press.

of the chromatin. While this occurs, the proteins that help to form these condensed chromosomes bring the members of each homologous pair (bivalent) close together; this tight pairing is called a **synapsis**.

During synapsis, the two members of a homologous pair are precisely lined up with each other so that the corresponding genes on each chromosome align. The homologous pairs are initially linked tightly together along their entire length by a protein scaffold called the synaptonemal complex. Sister chromatid arms are linked by cohesins, which form a scaffold for the synaptonemal complex to bind to. This tight interaction between the homologous pairs means that sections of non-sister chromatids may exchange with each other through a process known as crossing over. These sites of crossing over are seen visually as **chiasmata**

(singular = chiasma) as the synaptonemal complex dissolves; see Figure 10.18. The number of chiasmata varies with the length of the chromosome involved and the species of organism.

The process of synapsis and crossing over is illustrated in Figure 10.19. Look at this figure and notice how the post-synapsis homologous chromosomes contain different combinations of genetic information to each other and to the original parent chromosomes. These individual chromatids, which carry genetic information from both the paternal and maternal chromosomes, are known as **recombinants**.

Crossing over recombines the genetic information found in the homologous pairs of chromosomes in many different ways. As such, it is a fundamental way of introducing diversity to any population of sexually reproducing organisms.

Figure 10.18 Chiasmata showing crossing over.

Source: Philip Meneely et al. (2017) *Genetics*. Oxford University Press.

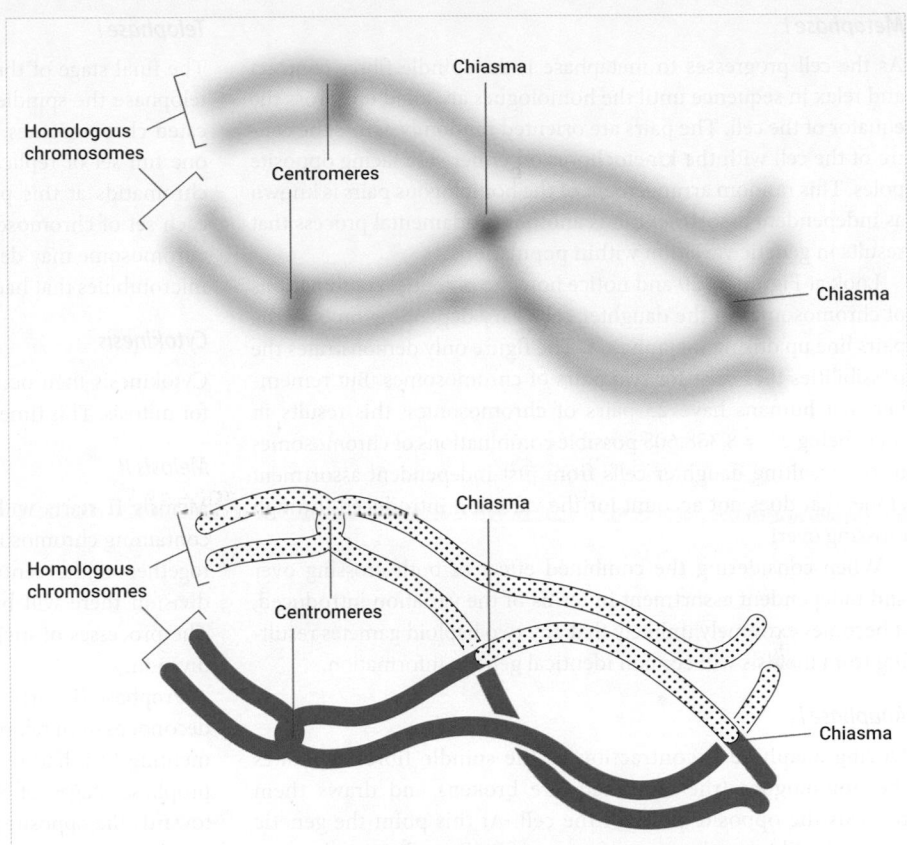

Figure 10.19 The process of crossing over between homologous pairs of chromosomes during prophase I of meiosis. The four daughter chromosomes that result are known as a tetrad.

Source: Philip Meneely et al. (2017) *Genetics*. Oxford University Press.

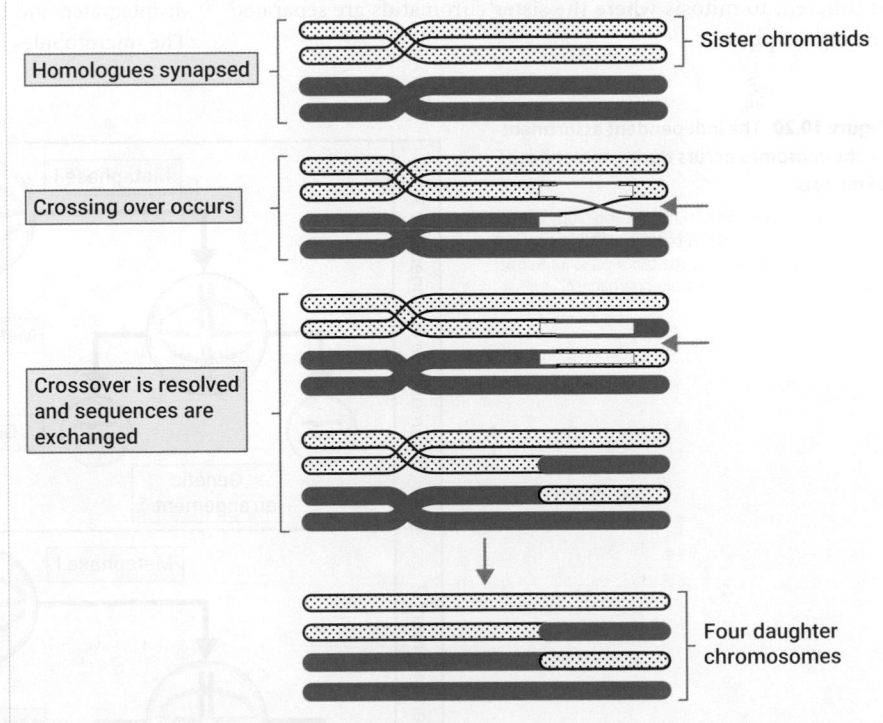

Prometaphase I

As the stage of prophase I continues into prometaphase I, the homologous pairs of chromosomes are no longer as closely associated to each other: they are held together only by the chiasmata.

The microtubules that form the spindle fibres connect to the kinetochores of the homologous chromosomes, as well as to the poles of the cell. By this stage, the nuclear membrane of the cell has fully broken down so that the spindle fibres are able to move the chromosomes within the cell.

Metaphase I

As the cell progresses to metaphase I, the spindle fibres contract and relax in sequence until the homologues are lined up across the equator of the cell. The pairs are oriented randomly across the centre of the cell with the kinetochores on either side facing opposite poles. This random arrangement of the homologous pairs is known as independent assortment; it is another fundamental process that results in genetic variation within populations.

Look at Figure 10.20 and notice how the potential combinations of chromosomes in the daughter cells vary depending on how the pairs line up during metaphase I. The figure only demonstrates the possibilities that exist for two pairs of chromosomes. But remember that humans have 23 pairs of chromosomes: this results in there being $2^{23} = 8,388,608$ possible combinations of chromosomes in the resulting daughter cells from just independent assortment alone; that does not account for the variation introduced through crossing over!

When considering the combined effect of both crossing over and independent assortment in terms of the variation introduced, it becomes extremely unlikely that any two haploid gametes resulting from meiosis will contain identical genetic information.

Anaphase I

During anaphase I, contraction of the spindle fibres separates the homologues (the chiasmata are broken) and draws them towards the opposite poles of the cell. At this point the genetic material still takes the form of a pair of replicated sister chromatids joined together at the centromere by cohesin proteins. This is different to mitosis where the sister chromatids are separated during anaphase.

Telophase I

The final stage of the first meiotic division is telophase I. During telophase the spindle fibres complete the movement of the replicated chromosomes so that each pole of the parent cell contains one full set of replicated chromosomes; although there are sister chromatids at this point, it is considered a haploid set. Around each set of chromosomes a nuclear envelope may reform and the chromosome may decondense (depending upon the species). The microtubules that had formed the spindle disappear.

Cytokinesis

Cytokinesis then occurs in the same manner as described earlier for mitosis. This time, two non-identical daughter cells are formed.

Meiosis II

Meiosis II starts with each of the daughter cells from meiosis I containing chromosomes made up of two sister chromatids joined together at the centromere. By the end of this second meiotic division there will be four non-identical haploid daughter cells. The processes of meiosis II are, overall, very similar to a mitotic division.

Prophase II starts with chromosomes condensing again (if they decondensed in telophase I) and with the nuclear membrane fragmenting (if it had reformed during telophase I). As seen with the prophase stages of other cell divisions, the centrosomes move towards the opposite poles of the cell and the spindle fibre microtubules reform.

By prometaphase II, the nuclear envelopes have completely disintegrated and the meiotic spindle in each cell is fully formed. The microtubules of the spindle attach to the kinetochores of each

Figure 10.20 The independent assortment of chromosomes occurs during metaphase I of meiosis.

Source: Openstax CC BY 4.0 C Rye et al. Biology, 2016. Section URL: https://openstax.org/books/biology/pages/11-1-the-process-of-meiosis. Access for free at https://openstax.org/books/biology/pages/1-introduction.

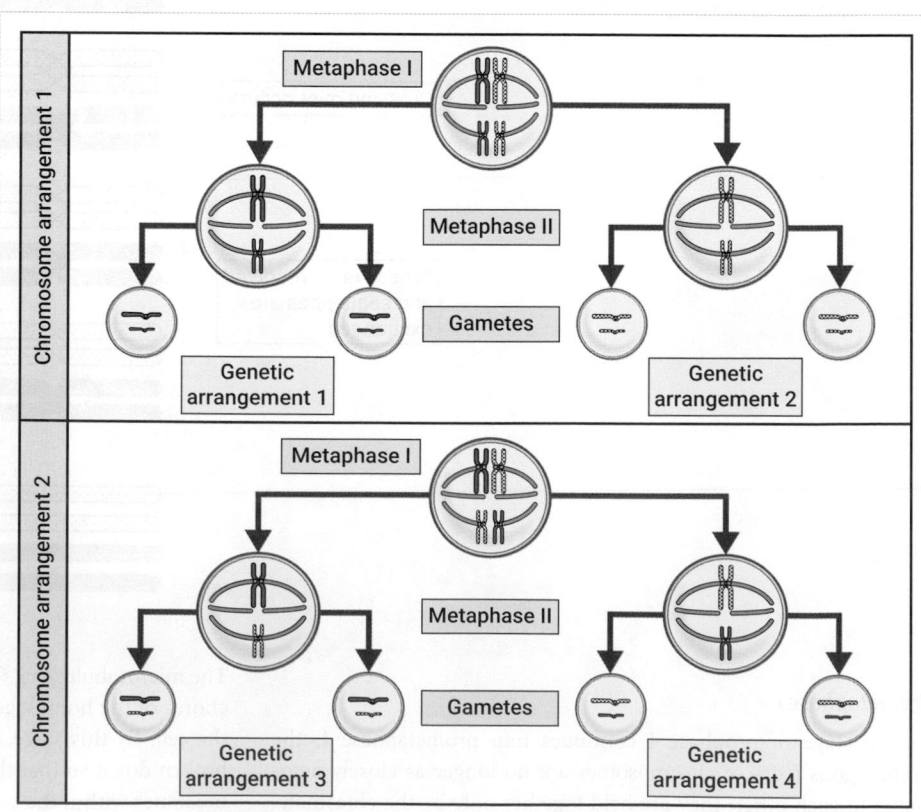

of the sister chromatids to link them to the centrosomes at each pole of the cell. Once again, the spindle fibres contract and relax to manoeuvre the sister chromatids to the equator of the cell, where they line up prior to being pulled apart by the action of the spindle fibres in anaphase II. The sister chromatids, which are now separate, migrate towards opposite poles of the cell such that, by telophase II, each pole of the cell contains a single haploid set of chromosomes.

Once again, a nuclear envelope forms around the haploid set of chromosomes and cytokinesis takes place. Each of the four daughter cells produced as a result of the two divisions of meiosis is genetically different to the others and to the original parent cell.

How is meiosis controlled?

As we discussed for mitosis, a number of regulatory control points also operate during the production of the haploid gametes to both initiate meiosis and to regulate the various steps through the two cell divisions. The germline stem cell first has to leave the mitotic cell cycle and enter the meiotic cell cycle (a step called the mitosis/meiosis decision). Interestingly, in the female of mammalian species (including humans), meiosis starts just a few months after conception in the fetal ovary. Meiosis is then halted at metaphase II until the individual reaches puberty, when one (or a few) of the oocytes are released from the ovary at a time. If the oocyte then meets a sperm, meiosis is completed and fertilization occurs. In males, in contrast, meiosis is all postnatal.

The basic cell cycle machinery responsible for the control of the mitotic cell cycle (the CDKs and cyclins) is also in control of the main steps of the meiotic cell cycle. However, the lack of an S phase between meiosis I and II, and the recombination events, require specific regulation. The overall frequency of recombination,

as well as crossing over at particular regions of the chromosomes, is regulated.

Checkpoints are also as crucial to meiosis as they are to mitosis, and again similar ones are in control. Additional checkpoints, such as the meiotic recombination checkpoint, are present to ensure recombination is complete. If not, the meiotic division is arrested or delayed, thus preventing the formation of defective gametes. Other steps of meiosis, such as chromosome separation, are also regulated.

PAUSE AND THINK

How does the reduction division of meiosis help to produce haploid, rather than diploid, daughter cells?

Answer: Meiosis involves two nuclear divisions. The first of these is the reduction division, which sees one diploid parent cell dividing to form two haploid daughter cells. During the first nuclear division of meiosis, the homologous pairs of chromosomes are separated and haploid daughter cells produced, which then proceed to undergo the second meiotic nuclear division. Without this initial reduction division, haploid cells would not form during meiosis.

Errors in meiosis

Errors that arise during meiosis can result in inherited diseases. These can be a result of numerical aberrations (too many or too few chromosomes—that is, aneuploidy) or structural changes to the chromosomes (including partial duplications, deletions, inversions, and translocations).

Aneuploidy is often the result of a failure of the chromosomes to segregate properly during meiosis I or II, a phenomenon known as non-disjunction. Non-disjunction is illustrated in Figure 10.21. Aneuploidy is thought to be quite common, with about 5 per cent of

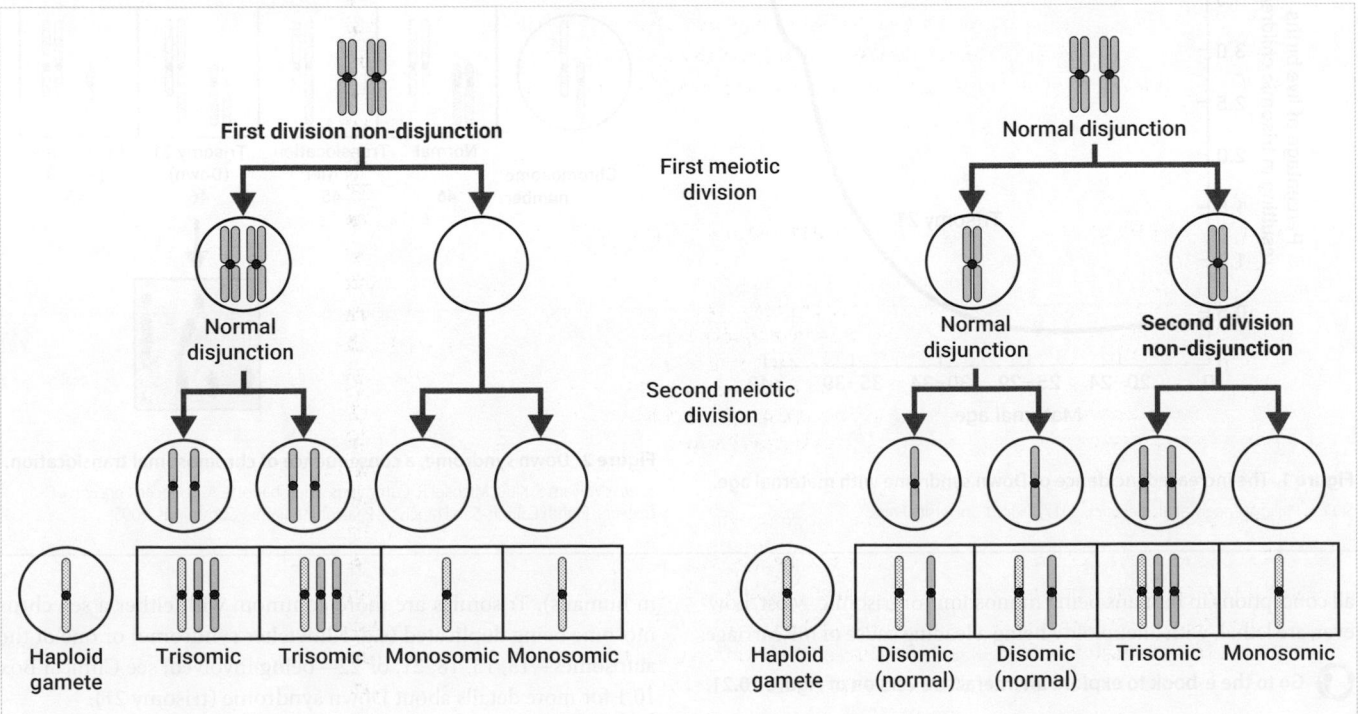

Figure 10.21 Non-disjunction. Non-disjunction during either meiosis I or II can result in monosomy or trisomy.

| CLINICAL BOX 10.1 | Down syndrome: the most common human trisomy |

Down syndrome is most commonly caused by an extra copy of chromosome 21 (termed trisomy 21). It is a condition associated with characteristic facial appearance, intellectual disability, and weak muscle, although there are a number of other phenotypic features that an individual can display. It is thought that the extra copies of the genes found on chromosome 21, and the resulting proteins expressed from the genes, disrupt normal fetal development and result in the characteristic features that are seen.

Individuals with Down syndrome have a higher risk of early death than the general population, most commonly due to heart problems or infections. However, life expectancy for Down syndrome individuals is now typically 50–60 years in developed countries owing to improved medical care.

In most cases the parents of individuals with Down syndrome are genetically normal. However, the occurrence of non-disjunction during meiosis during the development of the ovum is thought to cause up to 95 per cent of cases. The incidence is, on average, about 1 in 1000 births, but the age of the mother can greatly affect the chance of conceiving an individual with Down syndrome, as depicted in Figure 1.

Meiosis in a female actually starts when she herself is a fetus: cells destined to become eggs start the process of meiosis but become arrested in meiosis I during synapsis of the homologues. Meiosis restarts when the female reaches puberty but is again arrested in meiosis II and not fully completed until fertilization takes place. Consequently, the ovum released each month has been arrested for one month longer than the ovum

released the previous month. As a result, ova released when the woman is 35–45 years are significantly older than those produced 20 years earlier. As the woman ages, the meiotic machinery also ages and is not as efficient or effective; mistakes such as non-disjunction of chromosome 21 are more likely.

Far less commonly, Down syndrome can occur when part of chromosome 21 (the Down syndrome critical region (DSCR)) becomes translocated to another chromosome (often 14) during the formation of either the egg or sperm. After fertilization, affected individuals have two normal copies of chromosome 21 plus the extra material from chromosome 21 attached to the other chromosome, the consequence being three copies of genetic material from chromosome 21, as illustrated in Figure 2.

It is also possible that Down syndrome results from an extra copy of chromosome 21 in only some of the body's cells. In affected individuals the condition is called mosaic Down syndrome.

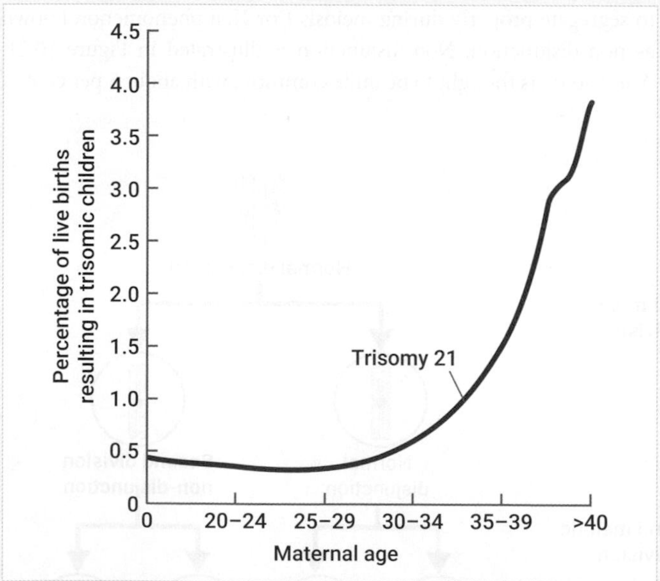

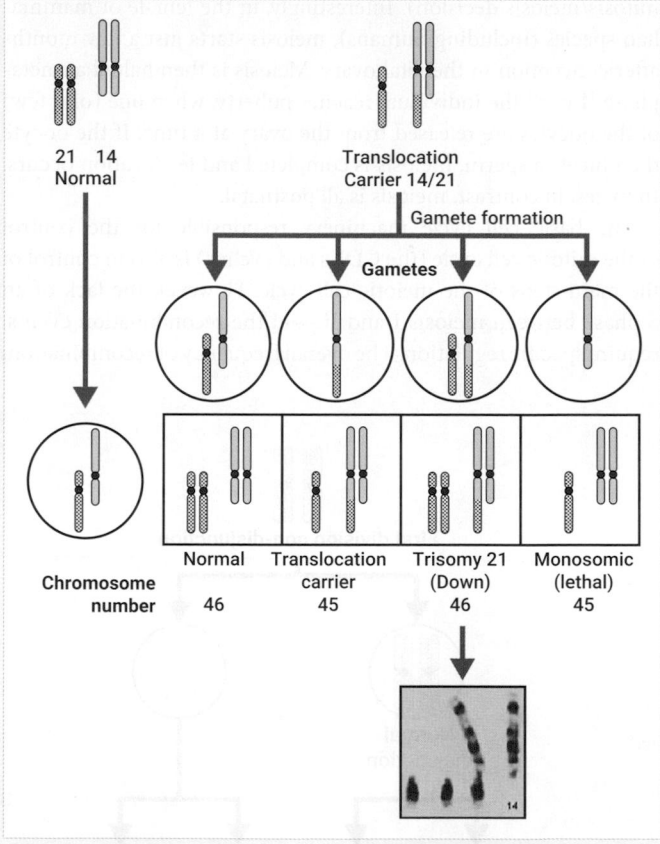

Figure 2 Down syndrome, a consequence of chromosomal translocation.
Source: William S. Klug, Michael R. Cummings, and Charlotte A. Spencer. *Concepts of Genetics*, Eighth Edition. San Francisco: Pearson Benjamin Cummings, 2005.

Figure 1 The increased incidence of Down syndrome with maternal age.
Source: Philip Meneely et al., *Genetics*, 2017, Oxford University Press.

---

all conceptions in humans being monosomic or trisomic. Most, however, are lethal, with aneuploidy being a leading cause of miscarriage.

 Go to the e-book to explore an interactive version of Figure 10.21.

In animals, monosomies only result in live births when one of the sex chromosomes is lost (an example being Turner syndrome

in humans). Trisomies are more common, with either a sex chromosome being duplicated (e.g. Klinefelter syndrome) or one of the autosomes—13, 15, 18, 21, or 22—being involved; see Clinical Box 10.1 for more details about Down syndrome (trisomy 21).

 Check your understanding of the concepts covered in this section by answering the questions in the e-book.

SUMMARY OF KEY CONCEPTS

- Cellular reproduction occurs primarily via binary fission for prokaryotes, although budding or fragmentation are also seen as alternative strategies.

- Genetic variation in prokaryotes is enhanced by transposons, transformation, transduction, and conjugation.

- Cellular reproduction occurs via mitosis for eukaryotes, and yields diploid daughter cells.

- Meiosis is the process of producing haploid gametes for sexual reproduction of eukaryotic organisms.

- Crossing over and independent assortment during meiosis increase genetic variation in eukaryotes.

Use the flashcards in the e-book to test your recall of key terms introduced in this chapter.

QUESTIONS

Looking for answers? Once you've answered these questions, follow the link in the e-book to the answer guidance and check your work.

Concepts and definitions

1. What is the process for cell division in bacteria?

2. Apart from binary fission, name two other ways in which bacteria can reproduce?

3. What are FtsZ proteins, and during cell division what do they form?

4. Name four ways through which horizontal gene transfer could occur.

5. Name the phases of the eukaryotic cell cycle.

6. What is the result of mitotic cell division?

7. Name the stages of mitosis in order.

8. What is non-disjunction?

9. Name two trisomy syndromes.

Apply the concepts

10. What role does the sex pilus play in conjugation?

11. What role do spindle fibres play in eukaryotic cell division?

12. What type of errors can be seen in both meiosis and mitosis?

13. In what way does mitosis differ from meiosis?

Beyond the concepts

14. What role does vertical gene transfer play in bacterial genetic diversity?

15. Why is it important to have control processes that regulate the cell cycle in multicellular organisms?

16. What are the purposes of G1 and G2 phases of the cell cycle?

FURTHER READING

https://theconversation.com/mitotic-spindles-could-help-develop-better-chemo-drugs-16919
An interesting article on how mitotic spindles could help to develop better chemotherapy drugs.

Divan, A. & Royds, J. (2020) *Cancer Biology and Treatment.* Oxford University Press: Oxford.
A more in-depth explanation of how errors in cell division can lead to cancer.

Microbial Diversity

Chapter contents

Introduction	426
11.1 Classification and taxonomy	427
11.2 Microbial habitats	429
11.3 Bacterial diversity	432
11.4 Archaeal diversity	439
11.5 Fungal biodiversity and reproduction	445
11.6 Protist diversity	458

🌀 Watch the key concepts video in the e-book to prepare yourself for studying this chapter.

LEARNING OBJECTIVES

By the end of this chapter, you should be able to:

- Explain why a common classification system is needed.
- Recall the differences between classical and molecular taxonomy.
- Explain how sequencing and metagenomics has opened up our understanding of unculturable micro-organisms.
- Describe the diverse environments microbes can occupy.
- Describe the vast diversity of microbes across all three domains of life.

Introduction

The diversity of microbes is staggering. They can occupy virtually every conceivable environmental niche that Earth has to offer. If you consider the three domains of life, two of them, the Bacteria and the Archaea, are entirely composed of microbes, and the Eukarya also have microbial Kingdoms such as the Fungi and Protista. So, it is no exaggeration to say that we truly live in and on a microbial world; without them, we would cease to exist.

This chapter will take you on a journey through this microbial world, examining the diversity of species across the three domains of life. We will also explore how technologies such as sequencing and metagenomics have opened our eyes to the hidden world of unculturable microbes. First though, we need to consider how organisms are classified and the current methods used for this purpose.

11.1 Classification and taxonomy

As we see in Topic 5, scientists need a system of classification in order to make sense of the huge number of living organisms that populate our planet. Biological classification is called taxonomy, and it involves the grouping together of organisms that share properties. The way in which we categorize organisms has changed over time, not least because we now know far more about how organisms evolved from each other.

In 1977 Carl Woese and George Fox created the tripartite **phylogenetic tree** which remains in use today. In this tree, life is grouped into three domains—Bacteria, Archaea, and Eukarya—based on the small subunits of rRNA. These three domains are illustrated in Figure 11.1.

We also see in Topic 5 that the names we use for different species must be the same if we are to share knowledge. Agreed naming conventions are captured by what we call biological nomenclature. Biological nomenclature—a species name, essentially—consists of the genus followed by the specific epithet (species). The names of micro-organisms can seem very arbitrary, but they can tell us something about the organism's history and perhaps where it can be found.

Table 11.1 shows some examples of named bacteria. Look at this table and notice how the names have two components: the first word tells us the genus; the second word identifies the specific species.

Figure 11.1 Phylogenetic tree based on rRNA data.
The tree consists of three domains: Bacteria, Archaea, and Eukarya.

Source: Woese, Microbiological Reviews, 1987, 51, 221–271 (p. 231). Licensed from PLS Clear.

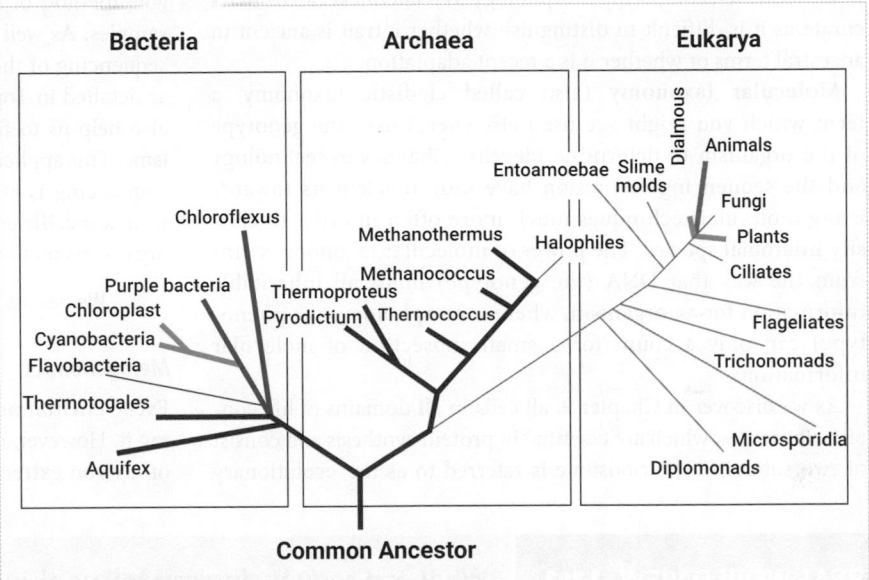

Table 11.1 Names of prokaryotic microbes and their derivation

Name	Derivation
Escherichia coli	Named after Theodor Escherich who found the bacterium in 1888
Deinococcus radiodurans	*deinos*: strange or unusual
	radio: radiation
	durans: enduring
Thermus aquaticus	*thermos*: hot
	aquaticus: found in or by water
Helicobacter pylori	*helix*: -spiral
	bacter: rod or staff
	pylorus: gate keeper (lower orifice of stomach)
Rickettsia prowazekii	*Rickettsia*: named after Howard Taylor Ricketts who first associated this bacterium with typhus
	prowazekii: named after Stanislav von Prowazek, an early investigator of typhus

Microbial taxonomy

As we discovered in Topic 5 there are two main ways in which we can gather information to determine the classification of an organism: (i) classical and (ii) molecular (cladistic) taxonomy.

Classical taxonomy (also called numerical taxonomy) is based on phenotypic characteristics such as shape, metabolic activity, pathogenicity, and nutritional requirements. Scored characteristics can be built up into a dendogram.

 We learn more about dendrograms in Topic 5, Experimental Toolkit T5.1. Look at that panel for an example.

In microbiology, classical taxonomy is still a powerful mechanism for quick identification, and this is illustrated by the use of API strips, which we discuss in Experimental Toolkit T5.1. API strips are used for bacterial identification in a wide range of situations. However, comparing morphology to relatedness can be inaccurate as it is difficult to distinguish whether a trait is ancient in ancestral terms or whether it is a recent adaptation.

Molecular taxonomy (also called cladistic taxonomy, a term which you might see used elsewhere) uses the genotype of the organism to determine identity. Changes in technology and the sequencing revolution have moved scientists towards using molecular techniques much more often in order to classify microbial species. The power of molecular taxonomy stems from the way that DNA (the genotype) holds all inheritable information for an organism, whereas morphology (the phenotype) can only account for a small subsection of molecular information.

As we discover in Chapter 9, all cells in all domains of life contain ribosomes, which are essential in protein synthesis and consist of two subunits. The ribosome is referred to as the 'evolutionary chronometer' or molecular clock (as we explore further in Topic 5). But why is this?

As we discuss in Chapter 9, ribosomal subunits contain characteristic, conserved rRNA molecules. The rRNA sequence is therefore very useful as its structure and function is similar across all organisms: any differences in sequence between species can be instructive from the point of view of identification and classification. A common method of identifying bacterial species involves isolating and sequencing the gene for the 16S ribosomal subunit; for eukaryotic organisms the gene for 18S rRNA is used. We discuss this sequencing further in Topic 5.

Sequencing and its impact

The development of sequencing has led to many great scientific discoveries. Sequencing is now a very available and affordable method of finding new microbial species in environmental samples. As well as helping us to classify micro-organisms by the sequencing of the genes that encode for small ribosomal submits, as detailed in Topic 5, Experimental Toolkit T5.2, sequencing can also help us to find out more about the biology of micro-organisms. This application is explored further in Real World View 11.1. Sequencing is also used in **metagenomics**, where DNA is taken from a specific environment and sequenced to discover the micro-organisms that live there.

 We discuss sequencing in more detail in Chapter 8.

Metagenomics

Every environment on Earth has some type of microbe inhabiting it. However, many of these habitats are either difficult to get to or are too extreme for humans and higher organisms to tolerate.

REAL WORLD VIEW 11.1   **What can sequencing can tell us about the biology of micro-organisms?**

Thermotoga maritima is a thermophilic prokaryote that was isolated from geothermic soil, whose temperature ranges from 60 to 90 °C. As such, *T. maritima* is capable of growing over a large range of temperatures. The reason for sequencing the genome of this organism is that it may hold vital information about evolution, as *T. maritima* is thought to harbour traits that would have existed in early micro-organisms.

Temperature can have a huge effect on proteins and DNA and so it is interesting to find out how enzymes are able to work at high temperatures without denaturing. On sequencing *T. maritima* it was found that they have very strong promoter regions and ribosome binding sites on their DNA, which are involved in transcription and translation respectively. These properties help to reduce the chances of errors occurring during protein synthesis, even at high temperatures. Sequencing this bacterium has given insights into how genes are expressed at high temperatures, which helps to picture what life would have been like during the beginnings of early life, and suggests the mechanisms employed by archaea and prokaryotes for their survival.

Recent years have seen a rise in multi-drug resistance (MDR) in bacteria. One major pathogen that is becoming increasingly problematic in the UK is the multi-drug resistant *Mycobacterium tuberculosis* (MDR-TB). In 2014, MDR-TB accounted for 1.4 per cent of all UK cases of tuberculosis. The sequencing of this pathogen has provided vital information about the drugs MDR-TB is resistant to.

In order to identify antibiotics that will kill the infection in a clinical setting, the culture has to be grown and then tested against a range of antibiotics. Results for this can take up to 2–3 weeks. By contrast, using a molecular test to determine the resistance genes present in the genome of MDR-TB (for example, by carrying out a polymerase chain reaction (PCR) on sputum collected from an individual carrying an infection to look for target resistance genes) could save time and money by helping to predict antibiotic resistance much more quickly.

▶ **We learn more about PCR in Chapter 8.**

Nonetheless, many microbes make these inhospitable areas their home. When we want to study such organisms, a few hurdles might hinder us; for example, high temperatures and pressures might be needed for cultivation.

It is standard procedure to grow microbes by emulating their natural environment. This might include growing them in oxygen-poor environments, with different carbon sources, or with other microbes. Some microbes grow slowly and need specialized equipment, such as bioreactors, which can keep cultures going for long periods of time. For example, a group of bacteria that can perform anammox (anaerobic ammonium oxidation), which is part of the nitrogen cycle, grow really slowly and the population only doubles in size once every 2–3 weeks. As a result, it can take many months to get enough cells to study these organisms.

▶ **We learn more about the nitrogen cycle in Chapter 13.**

No matter how specific your media is, however, the vast majority of microbes cannot be grown in the lab: they are unculturable. This can be for a whole host of reasons: exposure to oxygen, lack of essential minerals or vitamins, media composition, and so on. (After all, it is hard to recreate hydrothermal vents in a laboratory setting!)

These organisms do not go unstudied just because we are unable to grow them, however. One thing that all these microbes have in common is that they contain DNA. Therefore, if we isolate the DNA from the environment and use PCR to amplify the rRNA genes, we can then sequence them, and compare those sequences to genome databases. By doing so, we can obtain a lot of information about these unculturable organisms. This approach is called **metagenomics**. Samples taken from sites ranging from hydrothermal vents to ancient glacial layers have yielded DNA and have given us a picture of what organisms exist in these extreme environments. The total genetic material from a given microbial population is termed the **microbiome**.

PAUSE AND THINK

What is the main difference between classical taxonomy and molecular taxonomy?

Answer: Classical taxonomy uses the characteristics of what an organism looks like (i.e. its phenotype) to group organisms together. Molecular taxonomy uses the genotype (i.e. its DNA and RNA sequences) to group organisms together in a phylogenetic tree.

 Check your understanding of the concepts covered in this section by answering the questions in the e-book.

11.2 Microbial habitats

The incredible diversity of microbial life reflects the seemingly limitless places that microbes can call home. Microbes are everywhere: they live on us, in us, and all around us. They can live in the most

inhospitable of environments, including deep in the ocean where light never reaches and pressures are immense, as well as in the frozen conditions of the Arctic and on the inside walls of nuclear reactors. The study of microbes that live in different habitats and the interactions they have with each other is called microbial ecology.

Microbes can occupy different environmental niches within habitats and each type of microbe can have more than one niche. A niche might feature different microenvironments. As the name suggests, a microenvironment can occupy a very small area: there can be many thousands of different microenvironments within just a few millimetres because conditions such as nutrient levels, moisture, pH, and oxygen levels can change within the space of a few micrometres (μm).

Microbes often form biofilms within their environment. Biofilms are essentially microbial communities surrounded by a slime layer of polysaccharide. Often the communities are mixed: they may comprise different bacterial species as well as eukaryotic microbes such as fungi and protists. Biofilms can be complex structures containing clumps of bacteria, and different organisms can live in different parts of the biofilm.

Before exploring microbial diversity in more detail we will now look at three habitats that support a vast majority of this diversity: soil, freshwater, and seawater habitats.

Soil environment

All soils might seem to you to be the same, but they actually represent highly complex and diverse environments. Soil composition is very much dependent on the geology of the land underneath the soil, and the climate conditions. Microbial populations in different soils will change depending on organic content, and water content. Table 11.2 provides further information on soil types and their characteristics.

Various properties of soil, including the availability of water and nutrients, are determined by the nature of its aggregation—that is, the way in which the main soil particles are held together with organic matter and other particles. The microbial inhabitants also have a significant effect on the physical, chemical, and biological characteristics of the soil. Clay particles have the highest surface area for interactions with water molecules; they can also adsorb organic material and provide a surface for microbes to colonize.

Humic material, such as that shown in Figure 11.2, is made up of undefined polymers and simple organic compounds. It is formed from the partial decomposition of plants, animals, and microbes. Humic material can bind small organic molecules, and this makes them unavailable for use by microbes. It can also cause soil minerals to aggregate into particles, termed microaggregates, which are <50 μm in size and contribute to the soil structure. Macroaggregates are larger than >50 μm and are stabilized by polysaccharides and micro-organisms.

What characterizes the soil microbial population?

Microbes such as bacteria, fungi, protists, and archaea constitute only a fraction of the soil mass (0.5%), but they have a major impact on the properties of the soil. The microbial biomass comprises a mix of growing cells and resting cells, though any growth of

Table 11.2 Types of soil and their characteristics

Soil type	Water retention	Nutrient levels
Sandy	Fine grain, retain very little water, large air spaces, can be acidic	Poor level of nutrients
Clay	Little air space, poor drainage, retains water	High level of nutrients, trapped onto the clay particles
Peaty	High water content but is acidic	Rich in organic matter and high level of nutrients
Chalky	Alkaline soil, retains little water. Contains calcium carbonate or lime	High level of nutrients, but these might not always be available
Silty	Intermediate particle size. Can retain water and drain poorly	Medium level of nutrients
Loam	A perfect soil mix of 40% sand, 40% silt, 20% clay, fertile and well draining	Good nutrient levels

Figure 11.2 Humic material.

Source: Onur Ersin/Shutterstock.com.

microbial cells in the soil is quite slow. Within different soil types, such as those described in Table 11.2, the population of microbes will vary in both their number and their diversity. Typically, soil contains 10^8 to 10^{10} cells per gram, and the number of species can range from just a few up to 10,000 species per gram. This may still be an underrepresentation as we begin to know more about the viable but unculturable microbes within these environments.

Many microbes in the soil environment occupy the area of soil adjacent to the plant roots. This area is called the **rhizosphere**, and is rich in plant root exudates, such as amino acids, sugars, and vitamins. We encounter several examples of microbes that interact with the roots of plants later in this chapter.

Freshwater environments

There are two main freshwater environments; these are characterized by lakes, and by rivers and streams.

Lakes

The microbial content of fresh water in a lake is influenced by available light and oxygen. A freshwater lake environment comprises different zones, as shown in Figure 11.3. These zones have different levels of light and nutrients, and different microbes live in these niches.

The **photic zone** exists between the water surface and the region where the light can penetrate the furthest. This zone also possesses the greatest amount of oxygen, due to the presence of photosynthetic microbes.

The **littoral zone** is the part of the photic zone closest to land; here, sunlight can penetrate to the bottom of the zone.

The **limnetic zone** exists at the same depth, with the same level of light penetration, but it is open water away from the shore. In these zones you will find primary producers such as cyanobacteria and microalgae, and bacteria such as *Pseudomonas* spp., *Cytophaga* spp., and *Caulobacter* spp.

The **profundal zone** lies in the deepest water and contains little oxygen. It is home to purple and green sulphur bacteria—anaerobic photosynthetic organisms that use wavelengths of light for photosynthesis that are different to those used by the cyanobacteria and microalgae in the photic zone. They can also metabolize H_2S to sulphur and sulphate.

 We learn more about the photosynthetic bacteria in Section 11.3.

The **benthic zone** is the sediment at the bottom; it is rich in nutrients, but anoxic so here you will find microbes that are anaerobic and lithotrophic—for example, *Clostridium* spp.

Rivers and streams

Rivers and streams often feature running water, which promotes greater oxygenation of the water than occurs in lakes. Rivers and streams can also be shows, meaning that light can penetrate to the bottom. Micro-organisms will be found growing in biofilms which are attached to rocks and other hard surfaces.

Marine environment

Seawater has a 3.5 per cent weight by volume (w/v) concentration of salt, and light can only penetrate to around 100 m. Marine systems have less variation in pH and temperature compared to freshwater habitats, but pressure increases as the depth increases. The nutrient levels of open marine water (known as the pelagic zone) can be low compared to freshwater habitats. Despite these characteristics, a wide diversity of microbes live in marine environments; these include bacteria, archaea, protists, and unicellular fungi. The cells are small, however, due to the low nutrient levels.

Primary production in the open ocean is conducted by prokaryotic phototrophs, the predominant organisms being

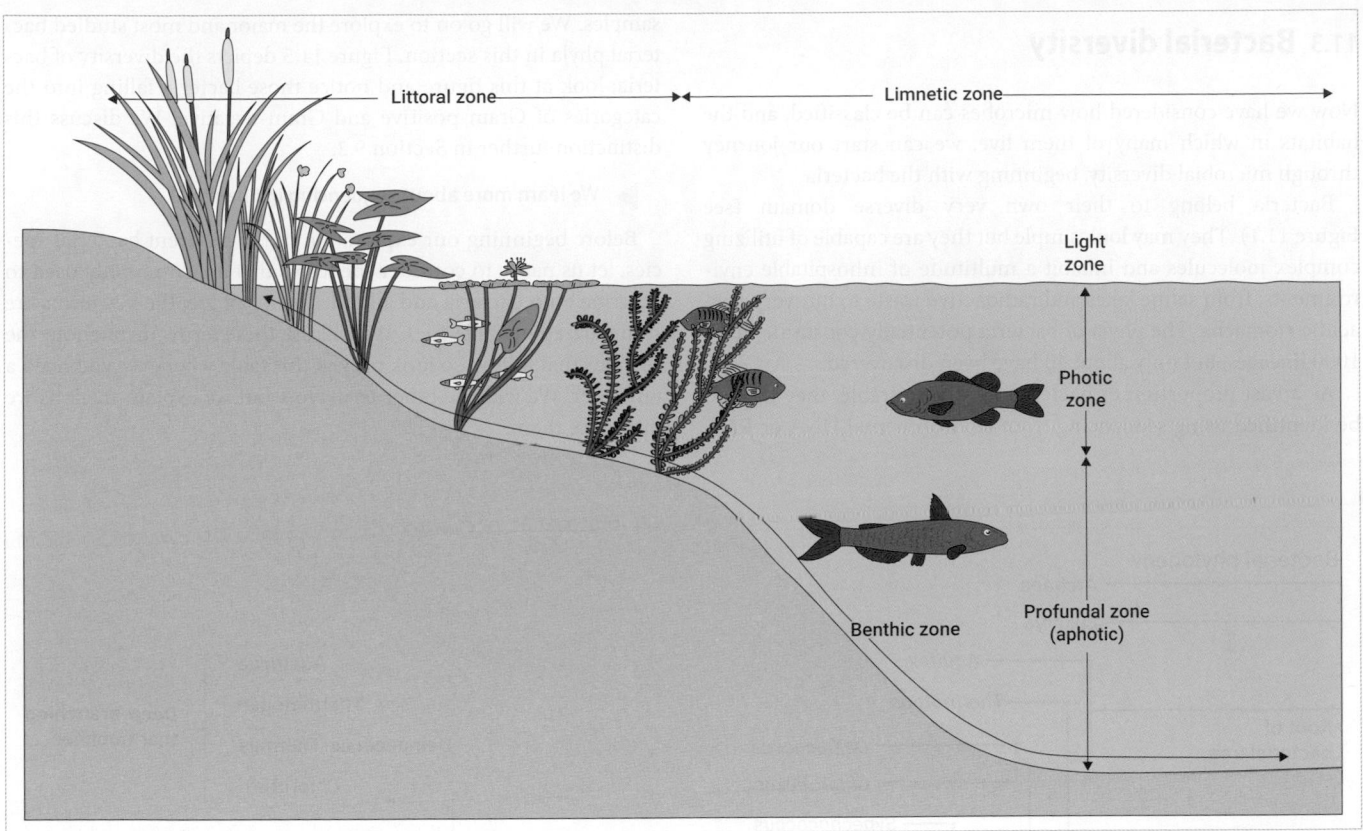

Figure 11.3 **The zones associated with a freshwater lake.**

Source: Adapted from *Texas Aquatic Science Textbook*, Rosen 2014, courtesy of the Texas Aquatic Science partners: Texas Parks and Wildlife, The Meadows Center for Water and the Environment, and the Harte Research Institute for Gulf of Mexico Studies.

Prochlorococcus and *Synechococcus*, both of which are members of the Prochlorophytes (shown in Figures 11.4a and 11.4b, respectively). These two species are responsible for 25 per cent of photosynthesis globally. Despite being very similar organisms, they occupy different environment niches: *Prochlorococcus* is found in low-nutrient, warm conditions, whereas *Synechococcus* is found in nutrient-rich and coastal environments.

🌀 Check your understanding of the concepts covered in this section by answering the questions in the e-book.

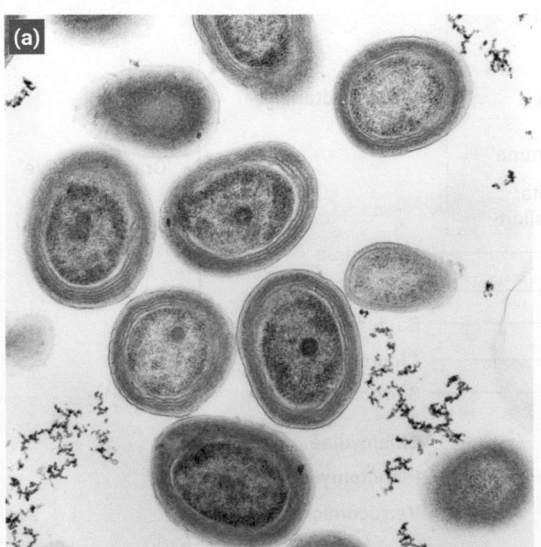
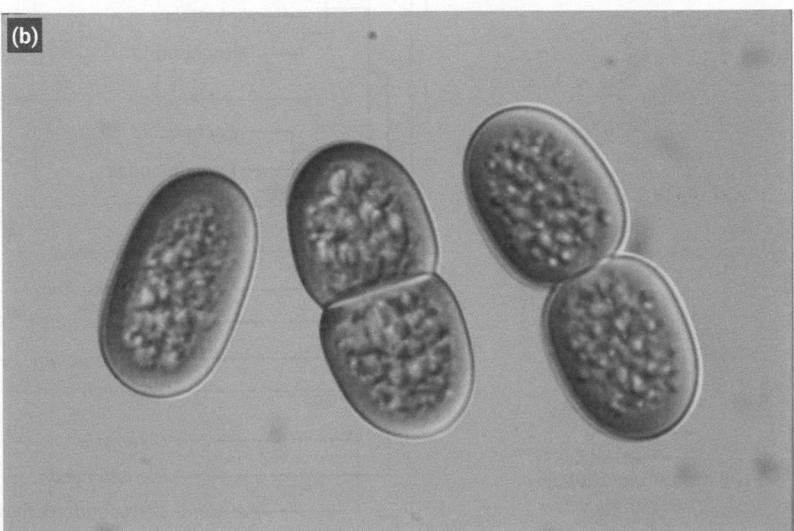

Figure 11.4 **Prokaryotic phototrophs.** (a) *Prochlorococcus* and (b) *Synechococcus*.

Source: (a) Chisholm Lab/Flickr/Public Domain; (b) Photo by Y. Tsukii http://protist.i.hosei.ac.jp/.

11.3 Bacterial diversity

Now we have considered how microbes can be classified, and the habitats in which many of them live, we can start our journey through microbial diversity, beginning with the bacteria.

Bacteria belong to their own very diverse domain (see Figure 11.1). They may look simple but they are capable of utilizing complex molecules and inhabit a multitude of inhospitable environments, from saline lakes and radioactive waste to our very own acidic stomachs. The phyla of bacteria potentially constitute up to 1000 lineages, but only about 40 have been discovered.

As a vast proportion of bacteria are unculturable, they have to be identified using sequencing from environmental DNA or RNA

samples. We will go on to explore the major and most studied bacterial phyla in this section. Figure 11.5 depicts the diversity of bacteria; look at this figure and notice those bacteria falling into the categories of Gram-positive and Gram-negative. We discuss this distinction further in Section 9.2.

▶ **We learn more about sequencing in Chapter 8.**

Before beginning our exploration of the different bacterial species, let us pause to consider the terms that are commonly used to describe both bacteria and all other forms of life; the key terms are summarized in Table 11.3. We will use these terms throughout the sections that follow, so look back at this table whenever you need a reminder. We will use other terms too, but we explain them as we introduce them.

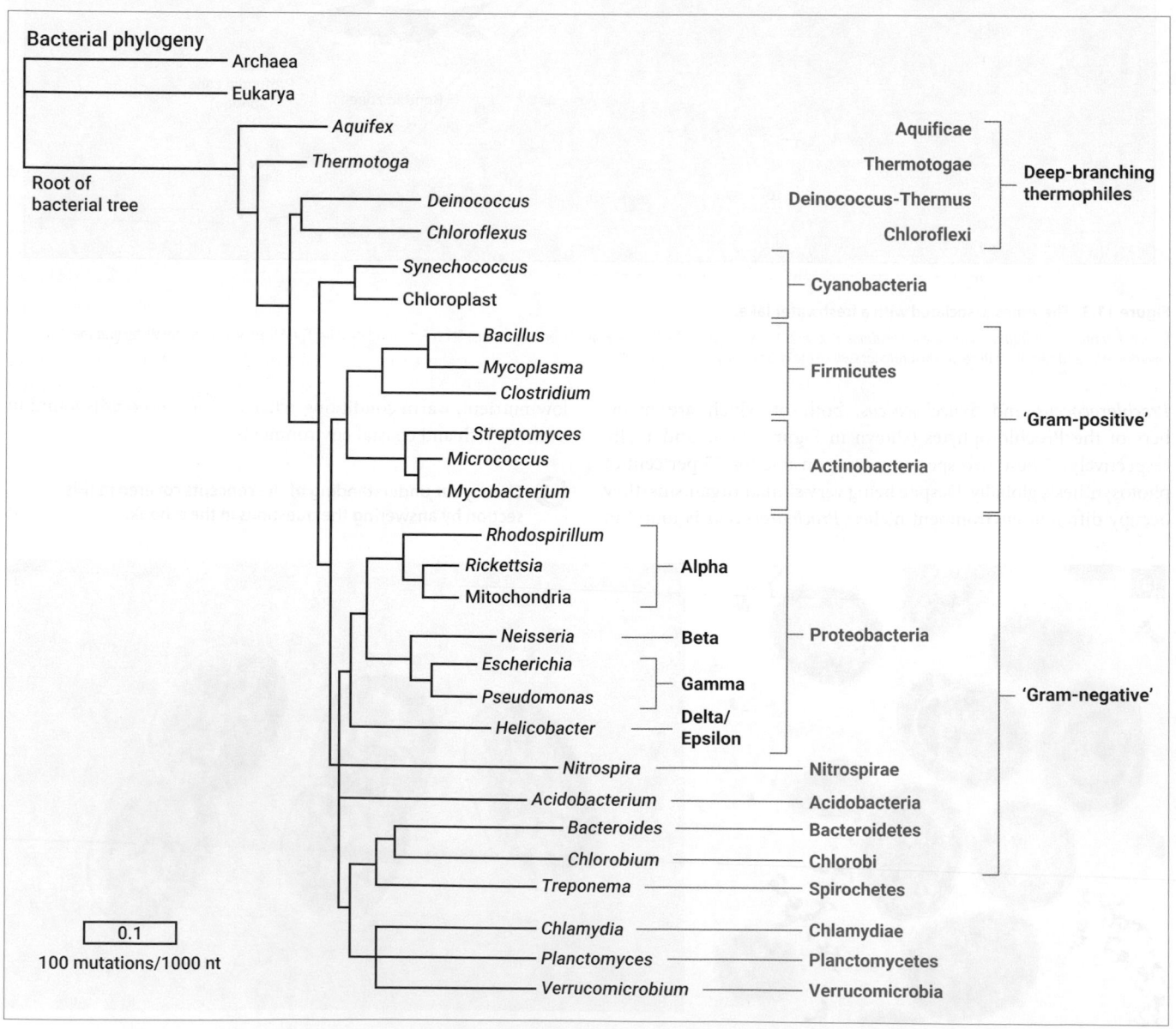

Figure 11.5 A phylogenetic tree depicting the diversity of bacteria.

Table 11.3 Terms commonly used to describe the contrasting characteristics of microbes

Aerobic	Undertakes metabolic processes that require oxygen
Anaerobic	Undertakes metabolic processes that do not require oxygen
Commensal	Inhabits a host, from which it derives food, but does so without harming the host
Parasite	Inhabits or otherwise interacts with a host in a way that is detrimental to the wellbeing of that host
Facultative	May exhibit a behaviour under certain conditions, but such behaviour is not obligatory for its survival
Obligate	Must exhibit a certain behaviour: it is obligatory for the survival of that organism
Autotroph	An organism that can produce its own food
Photoautotroph	An organism that produces its own food using light as an energy source
Heterotroph	An organism that obtains energy by consuming other plants or animals

Aquificales and Thermotogales

The Aquificales and Thermotogales are termed 'deep-branching', as scientists consider them to have evolved first from the last universal common ancestor (LUCA) (see Topic 2). Look at Figure 11.5, and notice how they are close to the root of the tree. They are all **hyperthermophiles**, which means that they grow at extremely high temperatures, up to 95 °C. They are also both hydrogentrophs (they oxidize hydrogen gas with molecular oxygen to make water), and obligate **autotrophs** (they fix O_2 into biomass using the reverse TCA (tricarboxylic acid) cycle).

Thermotogales grow at 50–80 °C. An example is *Thermotoga maritima*, which has been isolated from a geothermal vent. Cells have a loosely bound sheath or 'toga' which can be seen in Figure 11.6.

Another interesting example is *T. subterranea*, which is found in underground oil reservoirs. They use anaerobic respiration to survive, reducing elemental sulphur, thiosulphate, and sulphite to hydrogen sulphide.

Thermus and *Deinococcus*

Thermus and *Deinococcus* are also deep-branching bacteria, which share a unique structural trait: instead of having diaminopimelic acid in the peptidoglycan cross bridge of their cell wall, they have L-ornithine.

▶ We learn more about the structure of the bacterial cell wall in Chapter 9.

Thermus spp. grow at 70–75 °C. They are heterotrophs and can be isolated from hot springs such as those at Yellowstone National Park (Figure 11.7). *Thermus aquaticus* has contributed significantly to the development of techniques in molecular biology: its DNA polymerase (Taq polymerase) is a vital component in PCR (see Chapter 8).

Deinococcus spp. are not thermophiles, but they can withstand extremely high doses of radiation and can resist desiccation. They have a thick cell wall which stains Gram-positive. *Deinococcus radiodurans*—affectionately known as 'Conan the Bacterium' (Figure 11.8)—is a very interesting organism to study: being able to survive high doses of radiation means that it must have incredible DNA repair processes. (Radiation causes DNA damage, which—if extensive and left unchecked—would be lethal to an organism.) Indeed, scientists have found that it can reassemble a functioning genome from hundreds of DNA fragments.

Chloroflexi: thermophilic green photoheterotrophs

The Chloroflexi, which can be known as the 'green non-sulphur' bacteria, are heat-loving (thermophilic) green bacteria and are also considered to be deep-branching. They exist as long filaments, as illustrated in Figure 11.9, and are photoheterotrophs: they absorb light and use that energy to split organic molecules. They

Figure 11.6 Ultrathin section of *Thermotoga maritima*.
Source: Stothard, P., et al. (2005) BacMap: an interactive picture atlas of annotated bacterial genomes. *Nucleic Acids Res* 33:D317–D320. Image Source: K. O. Stetter.

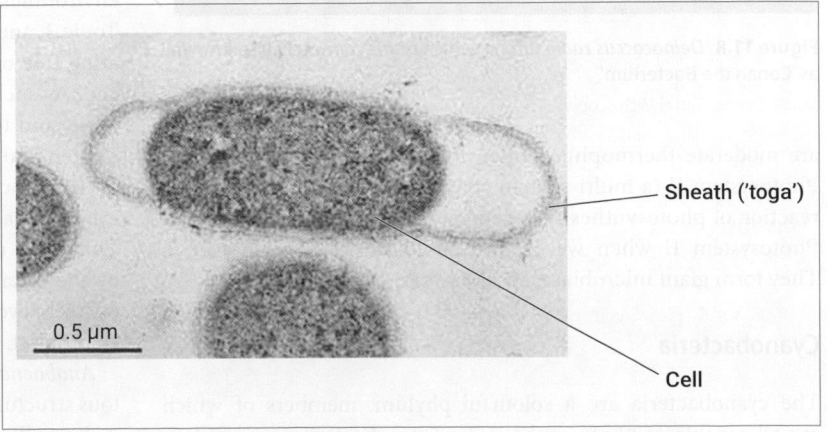

0.5 μm

Sheath ('toga')

Cell

Figure 11.7 A hot spring at Yellowstone National Park, from which *Thermus* spp. have been isolated.

Source: Lane V. Erickson/Shutterstock.com.

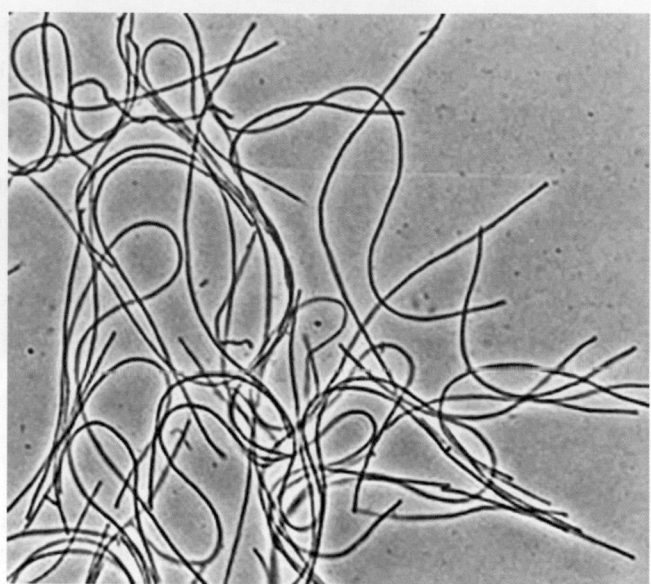

Figure 11.9 Long filaments of *Chloroflexus aurantiacus*.

Source: The genome portal of the Department of Energy Joint Genome Institute: 2014 updates. Nordberg, H., Cantor, M., Dusheyko, S., et al. *Nucleic Acids Res.* 2014, 42(1): D 26–31. Photo: Sylvia Herter.

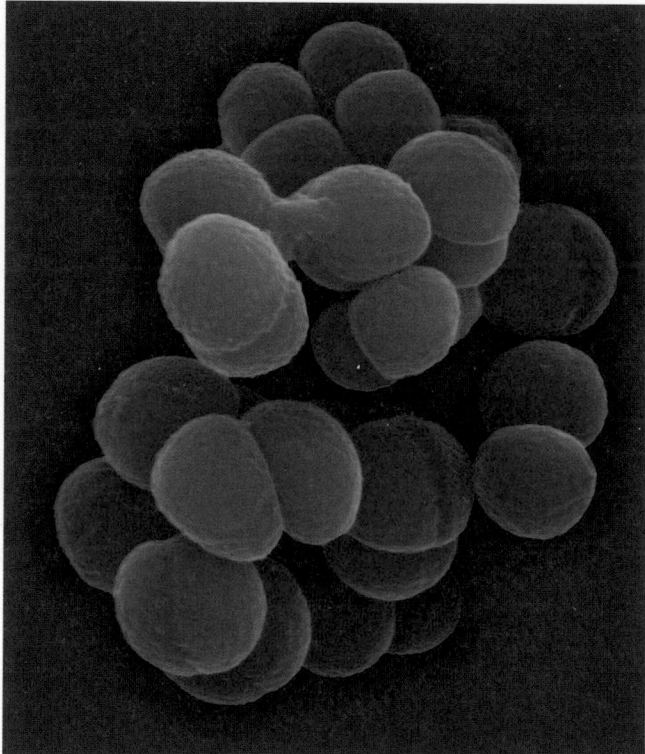

Figure 11.8 *Deinococcus radiodurans*, which is affectionately also known as 'Conan the Bacterium'.

Source: Dennis Kunkel Microscopy/Science Photo Library.

are moderate thermophiles, often found in hot springs, and use Photosystem II (a multi-protein complex that performs the initial reaction of photosynthesis) to generate ATP. We learn more about Photosystem II when we explore photosynthesis in Chapter 7. They form giant microbial mats along with cyanobacteria.

Cyanobacteria

The cyanobacteria are a colourful phylum, members of which are all capable of photosynthesizing. Cyanobacteria can be blue-green or green-brown in colour, the result of blue and red light

being absorbed by chlorophylls *a* and *b*. Some can appear red due to the presence of phycoerythrin. Chlorophylls and phycoerythrin are pigments, which is why the cyanobacteria are so colourful. Photosynthesis in plants uses both Photosystem I and Photosystem II; the cyanobacteria are the only prokaryotes also to use both of these protein complexes for their photosynthesis.

Cyanobacteria have **carboxysomes** inside their cells, which are the site of CO_2 fixation. Carboxysomes contain enzymes and are thought to concentrate CO_2 in order to overcome the inefficiency of the Rubisco enzyme (which is involved in the first major step in photosynthesis, as we see in Chapter 7). They store energy-rich compounds in their lipid bodies. They also have **thylakoids**, similar to those found in the chloroplasts of plant cells, which contain the photosynthetic apparatus.

Cyanobacteria are found in terrestrial and aquatic environments and often get a bad press for being involved in algal blooms, which can cause a huge amount of damage to aquatic environments as they can produce toxins. As we discovered in Topic 2, ancient cyanobacteria were the cells responsible for creating free oxygen in the atmosphere about 3.5–2.5 billion years ago. As such, they have had a significant impact on the oxygen cycle, and today are responsible for a significant amount of the oxygen in our atmosphere.

Ancient cyanobacteria were also involved in an endosymbiotic event with an ancient eukaryote, which gave rise to the chloroplast. This event essentially involved the cyanobacteria being engulfed by the eukaryote, and from then on existing within the eukaryotic cell, evolving to become a distinct organelle in the form of a chloroplast.

Anabaena is a genus of cyanobacteria that form large filamentous structures, which are clearly seen in Figure 11.10.

As well as being capable of photosynthesis, *Anabaena* spp. can also fix nitrogen (along with many other cyanobacteria) in

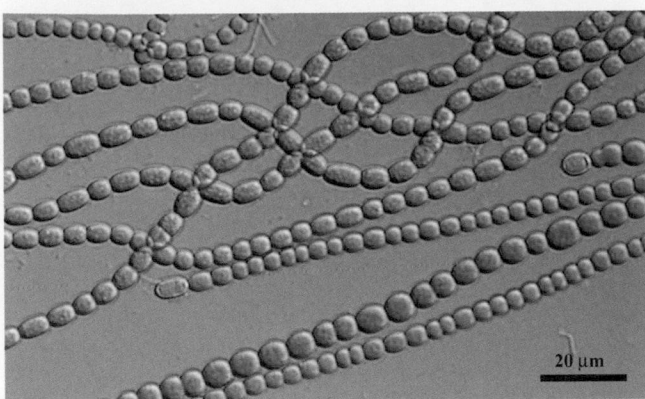

Figure 11.10 Filaments of *Anabaena*.

Source: Image © CCALA Culture Collection of Autotrophic Organisms.

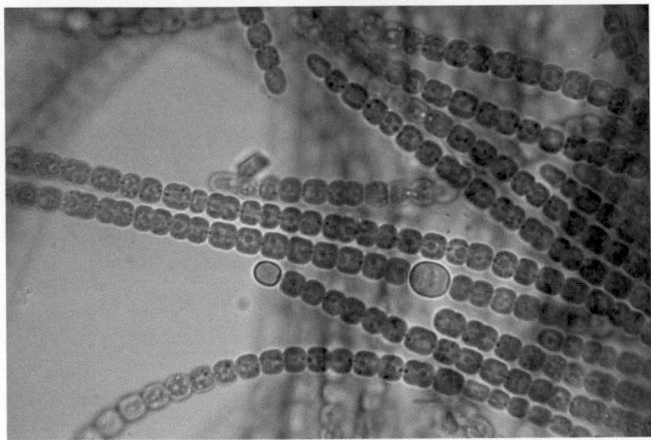

Figure 11.11 An *Anabaena* filament showing a heterocyst, the large, round cell in the middle right of the image.

Source: John Walsh/Science Photo Library.

specialized cells called **heterocysts**, which occur every 10–15 cells along a filament. A heterocyst can be seen in Figure 11.11 as a larger, round cell. Heterocysts are thick-walled cells, which are not capable of photosynthesis, but do have a membrane system.

PAUSE AND THINK

Why was the evolution of eukaryotes dependent on cyanobacteria?

Answer: Cyanobacteria are capable of photosynthesis, and the ancient cyanobacteria were responsible for increasing the levels of oxygen in Earth's atmosphere. This increase in oxygen was required for aerobic metabolism, which led to the evolution of more complex life. Aerobic respiration produces more ATP than anaerobic respiration, which means that more metabolic reactions can take place within cells. In addition, it is thought that the chloroplast, the organelle in plants responsible for photosynthesis, evolved from an ancient endosymbiotic event.

The *Oscillatroia* genus are filamentous blue-green cyanobacteria that reproduce by fragmentation. They get their name for their oscillating movement, which moves the filaments towards light sources.

Another common cyanobacterium can be found in health stores, marketed as a super food under the name Spirulina. Spirulina is made from *Arthrospira platensis* and *A. maxima*, which are both filamentous and blue-green in colour.

Proteobacteria

Proteobacteria is the largest phylum of the prokaryotes. They are all Gram-negative, but have a range of different cell morphologies. As the phylum is so large, it is split into five subgroups depending on their phylogenetic relationships as deduced from analysis of their rRNA:

- α (alpha) proteobacteria;
- β (beta) proteobacteria;
- γ (gamma) proteobacteria;
- δ (delta) proteobacteria;
- ε (epsilon) proteobacteria.

Let us now consider each of these five groups in turn.

α (Alpha) proteobacteria

The α proteobacteria contain the genera *Rickettsia*, an **intracellular** pathogen that causes typhus. This group also contains environmentally important bacteria such as *Rhizobium* spp., which live in symbiosis with legume-producing plants (plants that produce pods, such as peas, beans, chickpeas, and lentils): they associate with root nodules where they fix nitrogen to nitrate. An example of such a root nodule is shown in Figure 11.12.

Another important genera is *Agrobacterium*. A soil-dwelling bacterium, *A. tumefaciens* causes crown gall disease: its DNA genome includes a tumour-inducing plasmid (Ti plasmid). A small part of this plasmid enters the plant, and when expressed by the plant cell genetic machinery it disrupts the regulation of cell division, causing tumours to form.

Agrobacterium tumefaciens is not completely bad news for plants, however: it is used as a genetic engineering tool for plants. Specifically, genes can be inserted into the Ti plasmid, which are then integrated into the plant genome. We learn more about *A. tumefaciens* and its role in genetic engineering in Real World View 10.1.

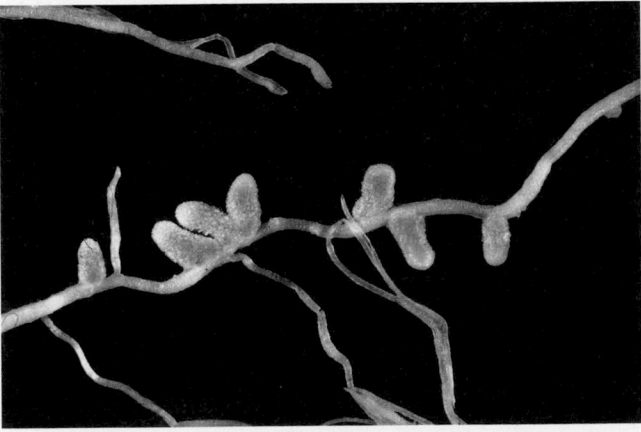

Figure 11.12 Legume-producing plants. Pink-tinged root nodules on leguminous plants inside of which live the *Rhizobia* spp.

Source: Wikimedia Commons/Ninjatacoshell/CC BY-SA 3.0.

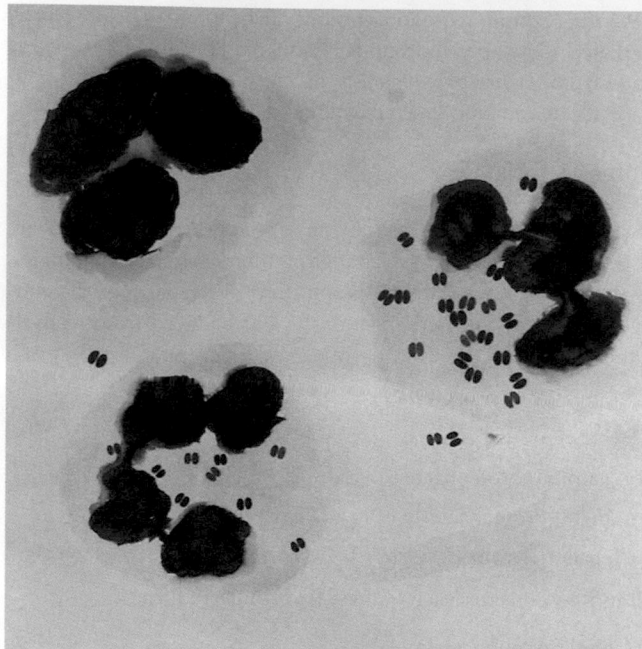

Figure 11.13 *Neisseria gonorrhoeae.* The image shows neutrophils with ingested bacteria.
Source: Mediscan/Alamy Stock Photo.

β (Beta) proteobacteria

The β proteobacteria are a diverse group and comprises everything from human pathogens to environmentally important nitrifying bacteria. The human pathogens include *Neisseria meningitidis*, which can cause sepsis, and *N. gonorrhoeae*, which causes gonorrhoea (see Figure 11.13). The bacteria involved in nitrification, such as *Nitrosomonas*, are responsible for some of the most important environmental impacts of this group: they are able to oxidize ammonium to nitrite, and therefore play a huge role in wastewater treatment.

γ (Gamma) proteobacteria

The γ proteobacteria as a group is home to many common opportunistic pathogens that can grow aerobically and anaerobically. One of the most famous examples is *Escherichia coli*, shown in Figure 11.14, which is a gut microbe. *Escherchia coli* has been used as a **model organism** to study bacteria for many years, and has been described as a 'work horse' in molecular biology as it is often used as a host organism for genetic transformation. (You will almost certainly use *E. coli* in the laboratory as part of your undergraduate practicals.)

Some *Escherichia* are notoriously pathogenic: the *E. coli* strain O157:H7, found in bovine products, causes foodborne illness and enterohemorrhagic diarrhoea. In 2011 the strain of *E. coli* denoted O104:HB4 caused an outbreak in Germany which originated from either contaminated cucumbers from Spain, or sprouts from Italy. However, not all *E. coli* strains are so dangerous: many are commensals—that is, they use their host as a source of food, but without causing overt harm to that host. Other pathogens in this group are *Salmonella* strains, which are often implicated in food poisoning from undercooked poultry, and *Yersinia pestis*, the causative agent of the Black Death.

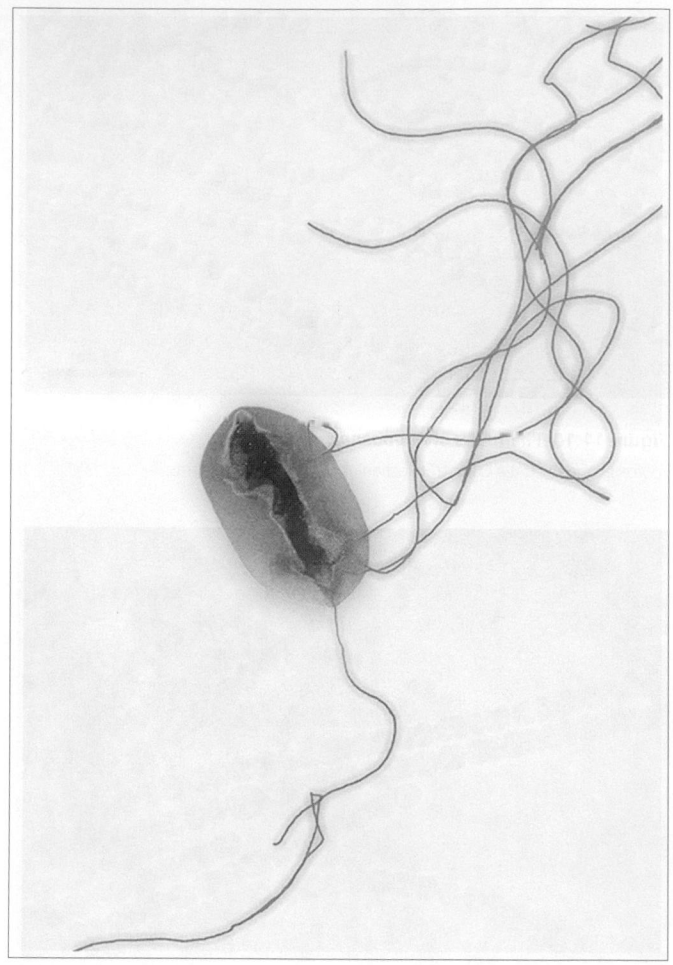

Figure 11.14 A colour-enhanced transmission electron micrograph (TEM) of *Escherichia coli* 0157:H7 strain. (Magnification 6836×)
Source: Phanie/Alamy Stock Photo.

δ (Delta) proteobacteria

This group tends to consist of aerobic bacteria. (That is, they carry out aerobic respiration, as we describe in Chapter 7.) The most well-known bacteria in this group are in the *Desulfovibrio* genus. These organisms are sulphate-reducing bacteria and typically reside in aquatic environments and in the intestines of animals. They are becoming more widely studied due to their bioremediation (decontamination) properties, especially the strains capable of reducing toxic metals such as uranium.

ε (Epsilon) proteobacteria

The ε proteobacteria have very few known genera, and are often found to colonize the gut. The majority of those that have been identified are capable of nitrate reduction, a pathway of the nitrogen cycle. The ε proteobacteria that have been isolated from hydrothermal vents have genes that are highly conserved with other members of the group such as *Wolinella succinogenes* and the pathogens *Helicobacter hepaticus* and *Campylobacter jejuni*. Many species of this group, like those from the *Sulfurimonas* genus, are also capable of metabolizing sulphur and hydrogen (as well as nitrate).

▶ We learn more about the nitrogen cycle in Chapter 13.

Spirochaetes

Spirochaetes are all Gram-negative bacteria, which exhibit a unique coiled cell morphology that gives them their name, as illustrated in Figure 11.15. Another unique feature of this phylum is the presence of axial filaments at either pole of the cell. These filaments run lengthwise in the periplasmic space of the cell wall, and can rotate and move the cell through liquids in a corkscrew-like manner. Look at Figure 11.15, and note the axial filament running along the centre of the spiral structure. Spirochetes can range in size from 3 to 250 μm, *Spirochaeta plicatilis* being one of the longest.

The genus *Spirochaeta* are often found in aquatic environments; they are anaerobic and facultative anaerobes and able to ferment carbohydrates.

Many species in the Spirochaetes phyla are pathogenic. The genera *Treponema*, *Borrelia*, and *Leptospira* all contain human or animal pathogens. *Treponema pallidum* is an anaerobic parasite of humans that causes the sexually transmitted infection (STI) syphilis. Another human pathogen is *Borrelia burgdorferi*, which is the causative agent of Lyme disease, a disease transmitted by infected ticks.

Chlamydiae

Chlamydiae is a diverse and interesting phylum: they are intracellular, aerobic pathogens that stain Gram-negative. The cell wall contains lipopolysaccharide and many species lack a peptidoglycan layer.

▶ We learn more about the structure of the bacterial cell wall in Chapter 9.

Organisms of the *Chlamydia* and *Chlamydophila* genera are obligate intracellular parasitic bacteria and cause disease in humans and animals. *Chlamydia* have a very small genome and are deficient in many metabolic activities. As a result, they need to parasitize resources from the host cells, and are only able to replicate within them.

The life cycle of *Chlamydia* is unique to the genus. An **elementary body** is a small, dense, non-growing cell that is fairly resistant to desiccation (that is, it can be dried out and still survive). This is the infective form of the cell that can exist outside of the host. When the elementary body is engulfed by a host cell it is converted to a **reticulate body**—a larger, non-infectious cell that replicates by binary fission in the host cell to form new elementary bodies. These elementary bodies are then released when the host cell dies, and go on to infect other cells. And so the cycle continues.

▶ We learn about binary fission in Section 10.1.

The most commonly known species of this phylum is *Chlamydia trachomatis*, which causes the STI chlamydia and the eye disease trachoma in humans. A cell infected with *Chlamydia trachomatis* is shown in Figure 11.16.

Another pathogen is *Chlamydophila psittaci*, which causes the disease psittacosis, also known as parrot fever. This zoonotic disease is transmitted from infected birds (from Psittacidae) and causes a disease with similar symptoms to pneumonia.

Firmicutes and actinobacteria

All Gram-positive bacteria were once classified as Firmicutes. Today, though, we have Firmicutes, which have a low 'GC' content, and Actinobacteria, which have a high 'GC' content. (The GC content refers to the relative number of guanine and cytosine residues in their chromosomal DNA.)

▶ We learn more about the composition of DNA in Section 1.7.

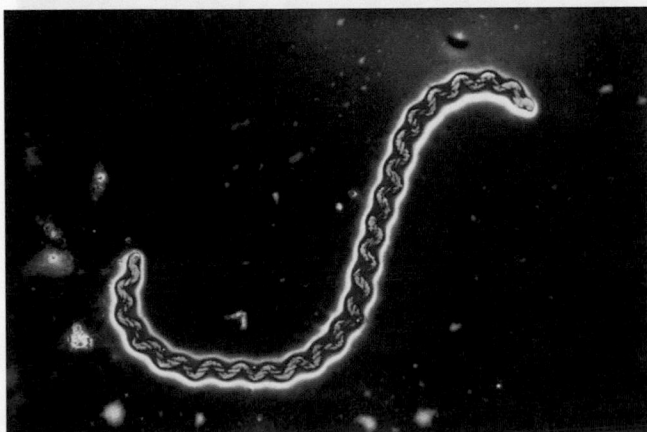

Figure 11.15 Spirochaetes. A colour-enhanced scanning electron microscope image of the coiled cell morphology in the spirochaete *Leptospira interrogans*.
Source: CNRI/Science Photo Library.

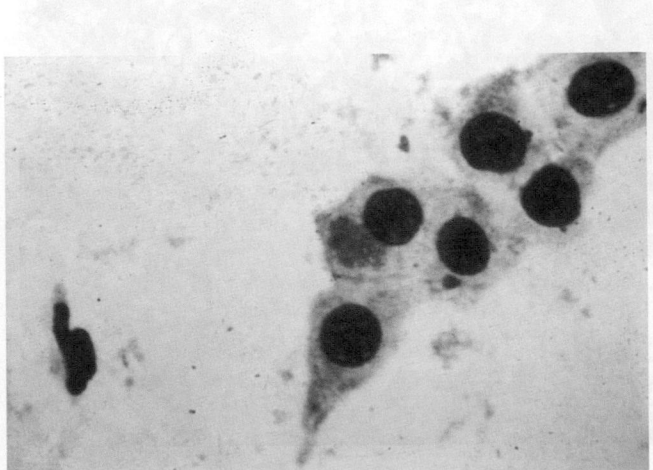

Figure 11.16 A cell infected with *Chlamydia trachomatis*.
Source: Getty Images.

Firmicutes

Bacilli is a class of bacteria that include the genera *Bacillus*, *Lactobacillus*, and *Staphylococcus*. These are all Gram-positive bacteria with low GC, so they belong to the Firmicutes. *Lactobacillus* and *Staphylococcus* are both unable to form spores, unlike *Bacillus*, which we discuss later in this section.

Lactobacillus are rod-shaped bacteria, as illustrated in Figure 11.17. They inhabit low pH environments (some as low as pH 4). They are facultative anaerobes that are also capable of fermenting glucose. Their ability to ferment glucose to lactate has been used by humans to help preserve food and they are vital in the production of cheese, yoghurt, and pickled vegetables. They are also cultured in sourdough 'starter cultures'.

However, *Lactobacillus* are not only involved in food preservation, but are also found within the human digestive, urinary, and genital systems. *Lactobacillus* spp. have a sizeable presence in our gut microbiota and are thought to play vital roles in how humans digest food. They are even included in probiotic drinks and supplements.

Lactobacillus are very rarely pathogenic. By contrast, *Listeria* spp., which are closely related to *Lactobacillus*, can cause major foodborne illnesses. For example, *L. monocytogenes* is often found in unpasteurized dairy products, soil, and sewage. It can cause the infection listeriosis in those who have ingested food contaminated with it, causing serious illness in immunocompromised people, such as pregnant women, newborns, and the elderly. The ability of *L. monocytogenes* to grow at low temperatures (4 °C) is one reason that pregnant women are told to avoid unpasteurized cheese, unless the cheese is heated thoroughly to kill the pathogen.

Staphylococcus are non-motile, coccus-shaped cells that are commonly found among human and animal microflora. Many species of *Staphylococcus* are opportunistic pathogens and can therefore not only cause infections and disease in immunosuppressed people, or in patients after operations, but can also be found in more minor infections such as skin lesions and boils. *Staphylococcus aureus* has received a lot of attention in recent years as many isolates of the species have been found to be resistant to methicillin (a β-lactam antibiotic), which is usually used to treat *S. aureus* infections. These isolates are commonly referred to as MRSA: methicillin-resistant *Staphylococcus aureus*.

The *Streptococcus* genus consists of cocci bacteria that often form chains. They are facultative anaerobes that are able to ferment sugars into lactic acid. This characteristic has proven useful in the production of fermented products such as buttermilk. Many species are commensals of human and animal microflora, and have been found on skin and in intestines. However, some species that are pathogenic can cause a condition that we often refer to as a 'strep throat' or streptococcal pharyngitis, as well as meningitis, bacterial pneumonia, and pink eye.

The genus *Streptococcus* is subdivided into two groups depending on their ability to lyse red blood cells, a process called **β-haemolysis**. Strains that *are* capable of β-haemolysis, including *S. pyogenes*, produce the enzyme streptolysin. To test for the presence of this enzyme, cultures are grown on blood agar. A zone of clearing appears around the colony, as depicted in Figure 11.18.

Strains that are not capable of β-haemolysis do not produce haemolysins and so do not cause red blood cell lysis: they carry out α-haemolysis instead. Rather than colonies appearing with a clear zone around them, the colour around the colony is green/brown—a result of the haemoglobin inside the red blood cells being oxidized to methemoglobin (which is a green colour), by the action of hydrogen peroxide, which is produced by the bacteria.

The Firmicutes also include the endospore-forming, low GC bacteria, which includes the *Bacillus* and *Clostridium* genera.

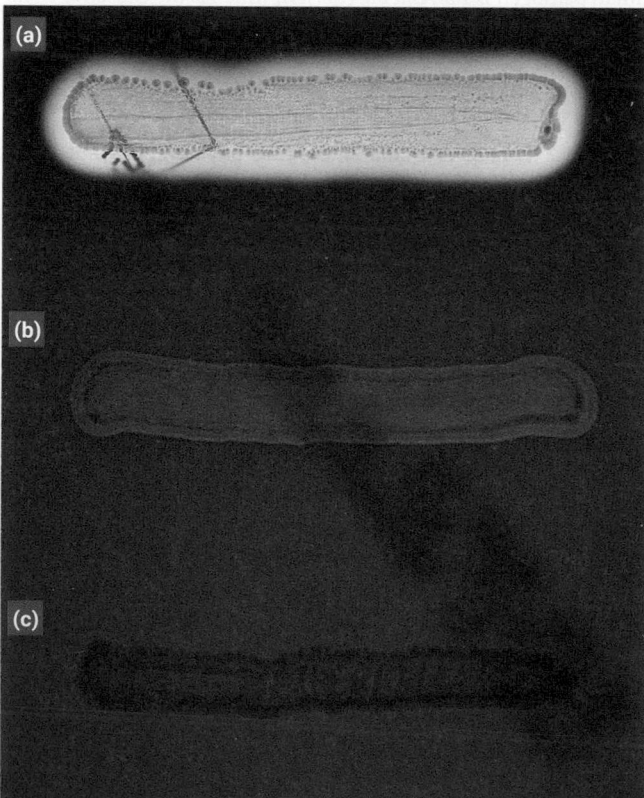

Figure 11.18 Colonies of *Streptococcus* spp. on blood agar showing β-haemolysis zone of clearing. (a) This zone appears yellow against the red of the agar, (b) α-haemolysis, which appears green on the media, and (c) no haemolysis. *Source*: © Tasha Sturm.

Figure 11.17 Scanning electron micrograph of *Lactobacillus casei*. *Source*: Power and Syred/Science Photo Library.

CLINICAL BOX 11.1   Bacterial toxins

Typically, bacterial toxins have a detrimental effect on humans. *Clostridium difficile* produces toxins that increase its virulence and cause changes in human intestinal epithelial cells. Another species, *C. botulinum*, produces a very potent neurotoxin called botulinum, which causes muscle weakness and paralysis in humans and animals, if ingested. This can ultimately be fatal if respiratory muscles are affected.

However, the fact that it causes muscle paralysis is a sought-after attribute in a therapeutic context despite the fact that botulinum can be fatal: the toxin can be injected in low concentrations into overactive muscles, such as spasming muscles, to reduce their activity; it can also be injected into hyperactive nerves, such as those related to excessive sweating.

The botulinum toxin is also used cosmetically, and commonly goes under the name Botox® (Figure 1). In this process the toxin is injected into facial muscles, which subsequently relax, reducing the appearance of wrinkles.

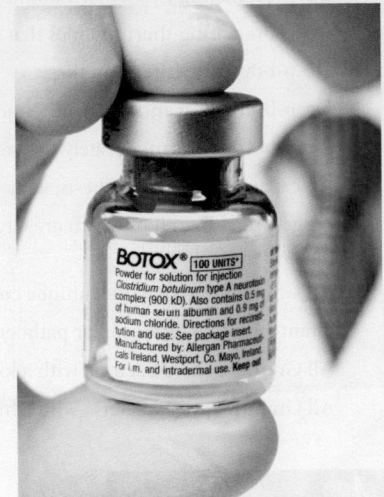

Figure 1 A cosmetic use of the botulinum toxin.

A Botox® product.

Source: Saturn Stills/Science Photo Library.

Bacillus are facultative aerobic, rod-shaped bacteria capable of forming spores. They are found in a multitude of environments: they are commensals on our skin and are very common in soils. They are capable of forming biofilms in soil and have been found to associate with plant root systems. Many *Bacillus* species produce antibiotics—for example, polymixin and bacitracin—which helps to protect them in mixed microbial communities.

 We learn more about biofilms in Section 9.2.

Clostridium are strictly anaerobic bacteria, which are capable of fermenting sugars into butyric acid or buterol. It is these compounds that give cultures of *Clostridium* their characteristic foul smell. Typically, *Clostridium* are found in environmental soil samples, but some are pathogenic: *C. perfringens*, which can be found in soils and on organic decaying material, is responsible for causing gas gangrene in humans, along with bacteraemia and food poisoning. This is due to the production of toxins, which increases the virulence of the organism. We learn more about bacterial toxins in Clinical Box 11.1.

Another pathogen is *C. tetani*, which is responsible for causing tetanus. This environmental species can infect humans from animal bites and wounds. The bacteria divide and produce a toxin that causes severe muscle spasms. There are very few cases of tetanus in the UK today due to the vaccination programme, which protects people from contracting the fatal disease.

The formation of endospores is highly advantageous to *Bacillus* and *Clostridium* species. The spore-like structures are formed within the cell and contain the cell's chromosomes. These dormant forms of the bacterium are resistant to heat, desiccation, and UV light, and are capable of surviving for hundreds of years.

 We learn more about endospores in Section 9.2.

Bacillus anthracis is well known for forming endospores: it causes the infection anthrax, which is a common problem for livestock. Humans can also contract the infection from working in high-risk environments, such as working with livestock, dead animals, and animal products such as wool. The resistance of the spores to desiccation means they have also been used in biological warfare.

Actinobacteria

The Actinobacteria include the *Streptomyces*, which are filamentous spore-forming bacteria. They play a major role in the soil ecosystem, as well as giving us the majority of our clinically useful antibiotics, as we discuss further in Chapter 13.

Summary of the characteristics of the main bacterial phyla

The characteristics of the different bacterial phyla are complex, but are summarized in Table 11.4.

Check your understanding of the concepts covered in this section by answering the questions in the e-book.

11.4 Archaeal diversity

The Archaea are extremely diverse. However, their ability to live in extreme environments makes many of them unculturable. As you can see from Figure 11.19, the environments in which these organisms can be found appear to be very alien—from hydrothermal vents (**thermophiles**), and highly acidic environments such as the gut or sulphuric pools (**acidophiles**) to beneath the Arctic ice and subglacial lakes (**psychrophiles**). The Archaea can be free-living or found in association with other organisms.

Table 11.4 Summary of the characteristics of the main bacterial phyla

Phylum	Brief description
Aquificales	Deep-branching hyperthermophiles that are autotrophic
Thermotogales	Deep-branching thermophiles that have a sheath surrounding their cells
Thermus	Deep-branching thermophiles
Deinococcus	Deep-branching and resistant to radiation
Chloroflexi	Deep-branching, moderately thermophilic, and photosynthetic
Cyanobacteria	All capable of photosynthesis; some species are capable of nitrogen fixation
Proteobacteria	Largest phylum, all of which are Gram-negative bacteria; there are five sub-groups: alpha, beta, gamma, delta, and epsilon
Spirochaetes	All Gram-negative with a unique coiled cell morphology
Chlamydiae	Gram-negative intracellular pathogens
Firmicutes	All Gram-positive bacteria with a low GC content in their DNA
Actinobacteria	All Gram-positive bacteria with a high GC content in their DNA

Figure 11.19 **Some of the extreme environments that the Archaea call home.** (a) Hydrothermal vents; (b) sulphuric ponds; (c) subglacial lakes.

Source: (a) Sully Vent in the Main Endeavour Vent Field, NE Pacific, Courtsey of NOAA PMEL EOI Program https://www.pmel.noaa.gov/eoi/gallery/smoker-images.html; (b) SFM Titti Soldati/Alamy Stock Photo; (c) Johann Helgason/Shutterstock.com.

As it is so hard to culture these extremophiles in the laboratory, a lot of work on the Archaea has been carried out using metagenomics, a technique that enables scientists to identify genes and proteins that are involved with metabolism and other processes such as reproduction. Their genomes are of great interest to many as they often provide us with information about how life evolved.

Figure 11.20 shows how the Archaea domain consists of four phyla. The first to branch off near the root are the Nanoarchaeota, closely followed by the Korarchaeota. The third group are called the Crenarchaeota and the largest phyla are the Euryarchaeota. We will now consider each of these in turn.

Nanoarchaeota

Nanoarchaeota is the newest addition to the Archaean phylogenetic tree. This phylum was added in 2002 after the discovery in Iceland of *Nanoarchaeum equitans* in submarine hot rocks. It was found to have a symbiotic relationship with another archaean, *Ignicoccus hospitalis* (from the Crenarchaetoa phylum), as illustrated in Figure 11.21.

Nanoarchaeum equitans is a hyperthermophile: it can withstand temperatures between 75 and 95 °C and is strictly anaerobic. The existence of only one known species means that little is known about the phylum. However, metagenomics has been used to reveal that genes from this organism have been found in high-temperature environments across the world.

Korarchaeota

Korarchaeota are deeply rooted in the archaeal phylogenetic tree. One of the most well-known species to be identified was *Korarchaeum cryptofilum*, which was discovered in Yellowstone National Park in 2008. Since its discovery its genome has been sequenced and it has shed light on many ancestral genes, found to be conserved in Crenarchaeota and Euyarchaeota. It has a small

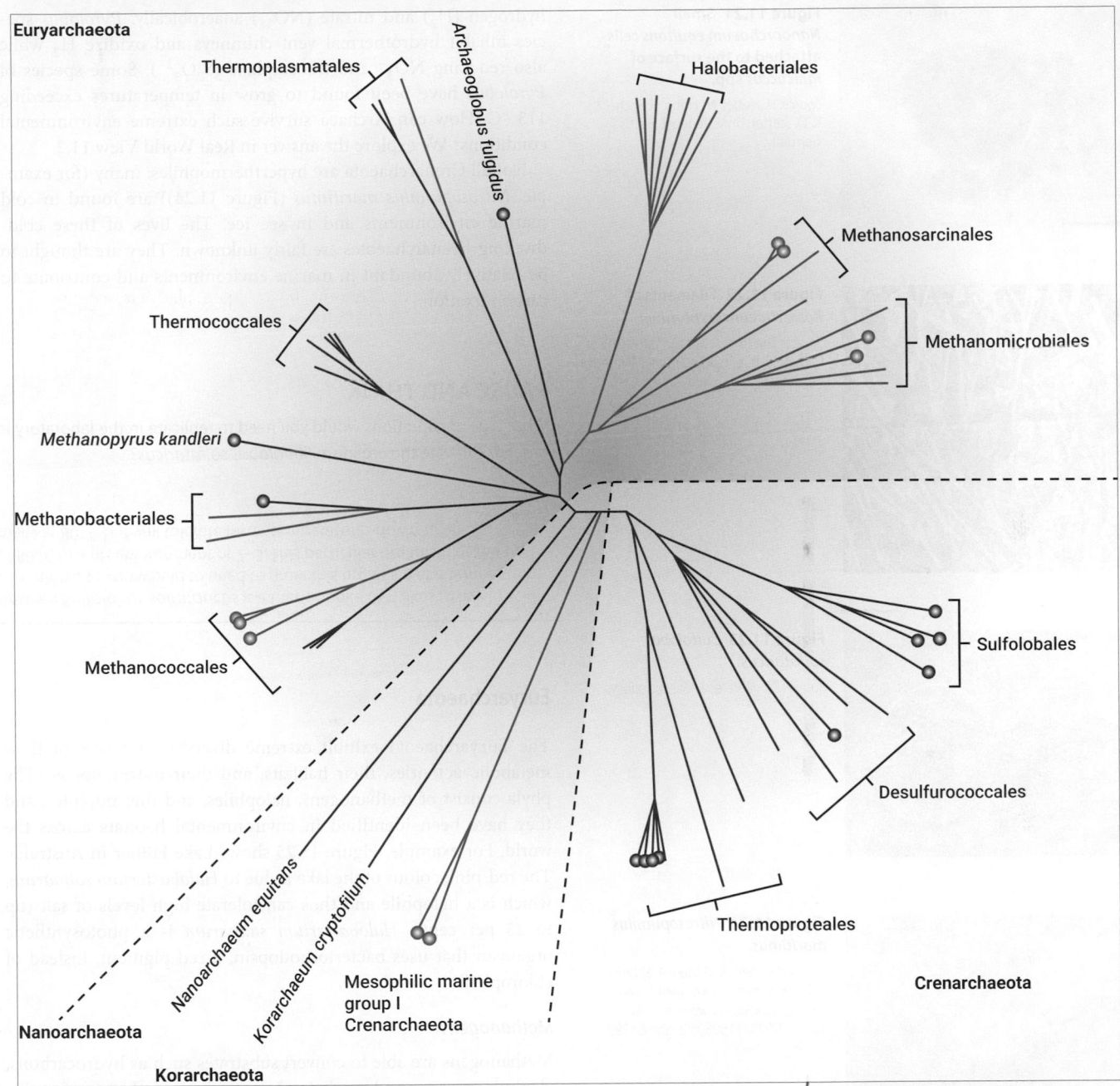

Figure 11.20 A phylogenetic tree of the Archaea domain.

Source: Reprinted by permission from Springer Nature; Figure 1 from Berg, I., Kockelkorn, D., Ramos-Vera, W. et al. Autotrophic carbon fixation in archaea. *Nature Review Microbiol* 8, 447–460 ©2010. https://doi.org/10.1038/nrmicro2365.

genome and the cells have a filamentous morphology, as illustrated in Figure 11.22.

Crenarchaeota

Crenarchaeota inhabit a variety of hot and cold habitats in terrestrial and marine environments. The hyperthermophillic Crenarchaeota have been found in hot springs and hydrothermal vents, which can also be rich in sulphur, as well as artificial habitats such as outflows of geothermal power plants.

Sulfolobus solfataricus, which is capable of metabolizing sulphur, is one of the original Crenarchaeota to have been discovered; it is depicted in Figure 11.23. It is classed as a thermoacidophile as not only can it survive high temperatures, but it is also capable of surviving in acidic environments: its optimum pH is 2–3.

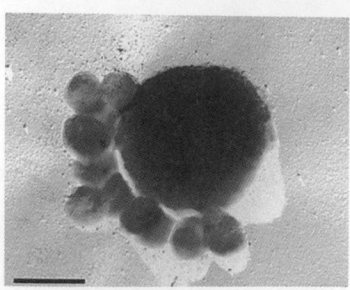

Figure 11.21 Small *Nanoarchaeum equitans* cells attached to the surface of *Ignicoccus* spp.

Source: H. Huber, M. Hohn, R. Rachel & K. O. Stetter, Univ. Regensburg, Germany.

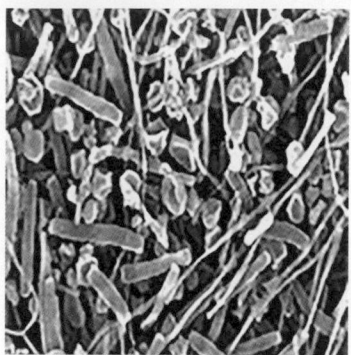

Figure 11.22 Filaments of *Korarchaeum cryptofilum*.

Source: The Regents of the University of California, Public domain, via Wikimedia Commons.

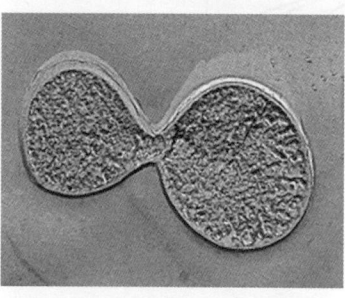

Figure 11.23 *Sulfolobus solfataricus*.

Source: Sonja-Verena Albers; Sulfosys.com.

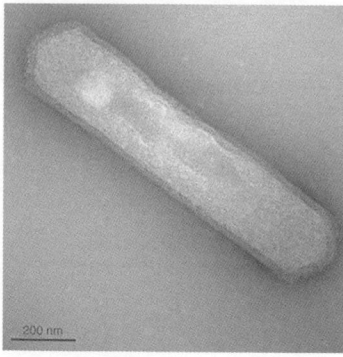

Figure 11.24 *Nitrosopumilus maritimus*.

Source: Qin, Wei & Martens-Habbena, Willm & Kobelt, Julia & Stahl, David. (2016). *Candidatus nitrosopumilus*. 10.1002/9781118960608.gbm01290.

Since its discovery, *S. solfataricus* has been found in the hot springs of Yellowstone National Park and hot mud pools in volcanic craters near Naples.

It was originally thought that Crenarchaeota all metabolized sulphur. However, many genera have been found to metabolize

hydrogen (H_2) and nitrate (NO_3^-) anaerobically. *Pyrolobus* species inhabit hydrothermal vent chimneys and oxidize H_2 while also reducing NO_3^- and thiosulphate ($S_2O_3^{2-}$). Some species of *Pyrolobus* have been found to grow in temperatures exceeding 113 °C. How can Archaea survive such extreme environmental conditions? We explore the answer in Real World View 11.2.

Not all Crenarchaeota are hyperthermophiles: many (for example, *Nitrosopumilus maritimus* (Figure 11.24)) are found in cold marine environments and in sea ice. The lives of these cold-dwelling Crenarchaeotes are fairly unknown. They are thought to be relatively abundant in marine environments and contribute to carbon fixation.

PAUSE AND THINK

What type of conditions would you need to replicate in the laboratory in order to cultivate the organism *Sulfolobus solfataricus*?

Answer: Sulfolobus solfataricus is a thermophile that likes to live in acidic conditions, so you would need to grow this organism at a temperature of 80 °C in a pH environment of 2–3. This particular organism grows optimally at 80 °C but will tolerate lower temperatures down to 60 °C.

Euryarchaeota

The Euryarchaeota exhibit extreme diversity in terms of their metabolic activities, their habitats, and their morphologies. The phyla consist of methanogens, halophiles, and thermophiles, and they have been identified in environmental habitats across the world. For example, Figure 11.25 shows Lake Hillier in Australia. The red/pink colour of the lake is due to *Halobacterium salinarum*, which is a halophile and thus can tolerate high levels of salt (up to 25 per cent). *Halobacterium salinarum* is a photosynthetic organism that uses bacteriorhodopsin, a red pigment, instead of chlorophyll.

Methanogens

Methanogens are able to convert substrates such as hydrocarbons, alcohols, acetate, and methyl substrates into methane, anaerobically. They are obligate anaerobes and are found in habitats such as in the guts of ruminants and termites, wetlands (paddy fields), and even in deserts.

Methanogens have become of more interest recently due to methane being a greenhouse gas. Archaea helped to create the atmosphere we have today. However, there are gases that are damaging to our atmosphere: greenhouse gases such as methane, nitrous oxide, and carbon dioxide all contribute to climate change. As such, the methane-producing methanogens have become the focus of more interest in recent years in the context of controlling climate change.

REAL WORLD VIEW 11.2   How do the Archaea survive extreme environments?

Archaea are very well adapted to extreme environments and many—such as the halophiles (which live in saline environments) discussed earlier—have evolved mechanisms to survive, such as using ion pumps to reduce the osmotic potential across the membrane. These pumps work to ensure the halophiles have an intercellular environment that is the same as their external environment: hypersaline. Halophiles also possess unique proteins that retain their shape despite the salinity, ultimately enabling the organism to survive in ion-rich saline conditions that would result in the desiccation of many other organisms.

Another adaptation exhibited by thermophiles is that the lipids forming the outer membrane are far more heat stable than those found in other prokaryotes. Unlike bacteria, whose phospholipids comprise fatty acid groups, Figure 1 shows how the phospholipids of archaea contain isoprenoid groups. These groups can form cross-links and carbon rings, which make the phospholipid monolayer in archaea more heat stable.

It is well known that high temperatures denature proteins and melt DNA. So, how do archaea solve this problem?

DNA denaturation typically occurs at about 94–95 °C (the temperature commonly used in PCR). That these thermophiles can survive above these temperatures demonstrates how their DNA has evolved to be more heat stable than that found in most other organisms.

DNA binding proteins in the Euryarchaetoa, which we explore in the next subsection, have some homology to histones in eukaryotes and are involved in the compaction of double-stranded DNA into nucleosome-like structures; they also help to regulate gene expression. These archaeal histones are not only thought to help increase the stability of DNA in high temperatures, but they are also found in (saline-tolerant) halophiles too.

▶ We explore the role of histones in the packaging—and therefore stability—of DNA in Chapter 3.

In hyperthermophiles, a reverse DNA gyrase enzyme creates a particular type of twisting in the DNA called positive supercoiling, which is thought to further increase its stability. (Think about how a rope comprises multiple strands twisted together: this gives the rope strength and

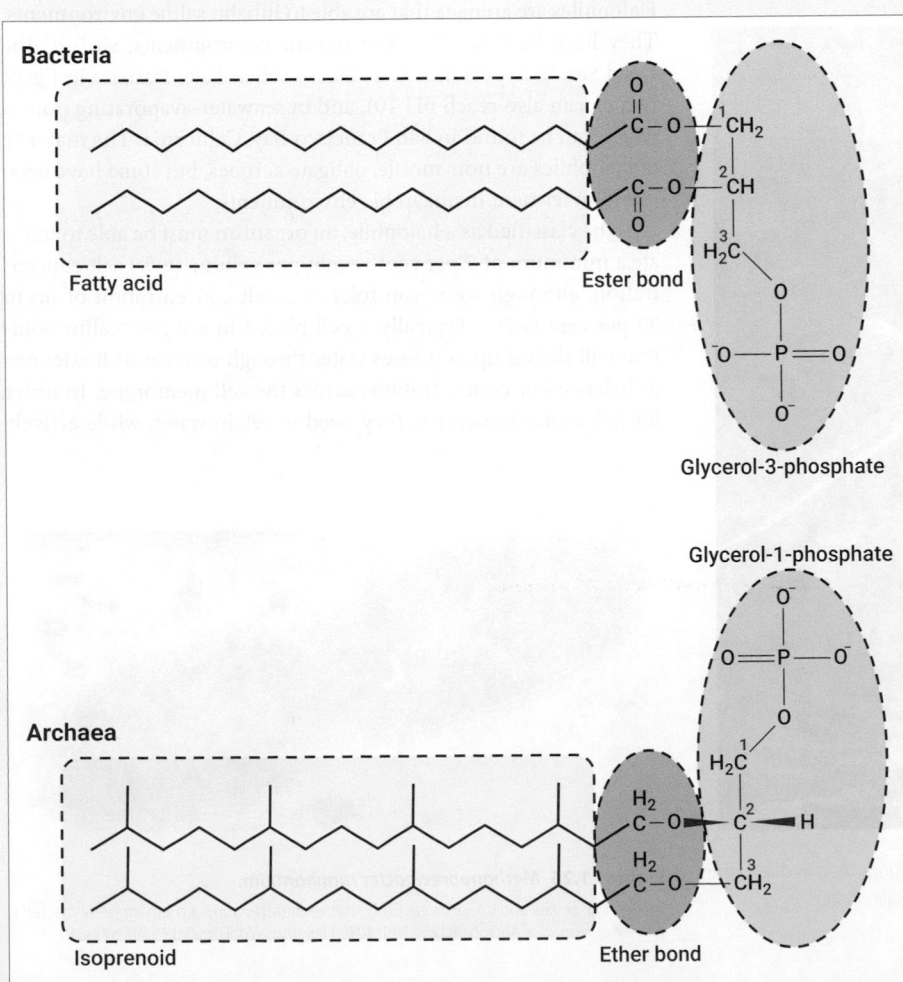

Figure 1 Isoprenoid groups. While bacterial phospholipids contain fatty acid groups, the phospholipids of archaea contain isoprenoid groups, which can cross-link and form carbon rings to make the structures they are part of more heat stable.

Source: Reprinted from Antonella Caforio, Arnold J. M. Driessen, Archaeal phospholipids: Structural properties and biosynthesis. *Biochimica et Biophysica Acta* (BBA), Molecular and Cell Biology of Lipids, 2017, 1862(11) 1325–1339. © 2017 with permission from Elsevier.

stability.) This enzyme has only been found in archaea that can grow above 80 °C.

A range of forces stabilize the folding of proteins. These forces, such as hydrogen bonds, cysteine bridges, and so on, all contribute to a protein's stability. It should therefore be no surprise that proteins that have been found to function in high temperatures have particularly strong bonds involved in their folding. A large number of ionic bonds operating across the surface of a protein and the presence of a highly hydrophobic core all help to increase the stability of proteins at high temperatures, and therefore make them less likely to unfold/denature.

The presence of heat shock proteins is also useful: they help to fold proteins that have been partially denatured.

Why have methanogens come to have such an impact? The increase in human agricultural activity has seen an increase in the farming of ruminants—for example, cattle, sheep, deer, and buffalo. Emissions from these ruminants are a significant source of methane. For example, the methane-producing *Methanobrevibacter ruminantium* (Figure 11.26) is a strictly anaerobic euryarchaeota and accounts for nearly 62 per cent of all ruminant archaea.

Research into the metabolic activities of *Methanobrevibacter* spp. can help scientists understand and alter the amount of methane emitted into the atmosphere by this organism: if they can identify conserved genes and proteins that are known to be important in methane production from ruminants it could be possible to alter their expression, limiting methane synthesis in the process. Further, by understanding the metabolic processes involved—the chemical compounds used by the organism to synthesize methane—it is possible to alter the diet of ruminants such that archaea receive compounds that reduce the amount of methane produced.

Halophiles

Halophiles are archaea that are able to inhabit saline environments. They have been identified in marine environments, such as the Dead Sea, in hypersaline soda lakes such as Lake Hamara in Egypt (which can also reach pH 10), and in seawater-evaporating ponds, which can be found in San Francisco Bay, California. The majority of halophiles are non-motile, obligate aerobes, but some have been found to ferment in anaerobic environments.

To be classified as a halophile, an organism must be able to tolerate a minimum of 9 per cent weight per volume (w/v) salt concentration, although some can tolerate a salt concentration of up to 30 per cent (w/v). Typically, a cell placed in a highly saline solution will shrivel up as it loses water through osmosis as it attempts to balance salt concentrations across the cell membrane. In order for halophiles to survive, they need to retain water, while actively

Figure 11.25 The red Lake Hillier in Australia. Its striking colour is due to the presence of the photosynthetic halophile *Halobacterium salinarum*.
Source: EyeEm/Alamy Stock Photo.

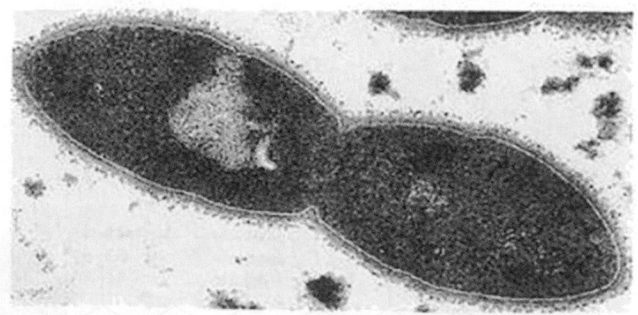

Figure 11.26 *Methanobrevibacter ruminantium*.
Source: J. G. Zeikus and V. G. Bowen. Comparative ultrastructure of methanogenic bacteria. *Canadian Journal of Microbiology*. 21(2): 121–129. https://doi.org/10.1139/m75-019.

maintaining a salt concentration imbalance across the membrane. (We say that they need to maintain their osmotic potential.) *Halobacterium* are extreme halophiles, and they maintain a salt concentration imbalance by pumping potassium (K^+) into their cytoplasm from the environment. This ensures that the concentration of K^+ inside the cell is greater than sodium (Na^+) outside the cell, with the result that the cell maintains a positive osmotic potential.

Halophiles are found in a number of supermarket products: they are used to make salted and fermented food products such as soy sauce and salted fish.

Some species of halophile have also been found to carry out light-driven synthesis of ATP. Despite being light-driven it is not photosynthesis as it does not involve chlorophyll. Rather, *Halobacterium salinarum* (Figure 11.27a) and a few other species are able to synthesize a purple-coloured protein called bacteriorhodopsin, which is incorporated into the membrane. Bacteriorhodopsin absorbs light energy, which then catalyses the activity of a proton pump. This mechanism is employed under oxygen-limiting situations.

▶ **We learn more about photosynthesis and ATP synthesis in Chapter 7.**

Another halophilic archaea capable of utilizing the sun's energy is *Haloquadratum walsbyi*, which has an unusual flat, square morphology as depicted in Figure 11.27b. It contains gas-filled vacuoles that are thought to increase its buoyancy, and along with the flat surface are thought to increase the area for light harvesting.

Having explored the range of species encapsulated by the bacteria and archaea, let us continue on our journey of microbial diversity by considering the fungi.

 Check your understanding of the concepts covered in this section by answering the questions in the e-book.

11.5 Fungal biodiversity and reproduction

The study of fungi is termed **mycology**, a term that comes from the Greek words *mykes* ('cap') and *logos* ('discourse'). Look at Figure 11.28. This depicts the kind of image you will probably have in your mind when fungi are spoken about (i.e. a large fleshy fungus with a cap and a stipe (stem)). However, this image is misleading: most fungi actually appear as moulds, as in Figure 11.29, where the fungus *Penicillium camemberti* is growing on agar.

Fungi are fascinating organisms: as well as having intriguing means of growth and reproduction, they are steeped in folklore and feature in many fairy stories.

As a reminder, fungi are eukaryotic: they have a nuclear membrane and complex cells. Fungi are one of the three major groups of eukaryotic organisms, along with plants and animals, and have significant environmental and economic impact, as we discuss further in Chapter 13.

▶ **We learn more about the characteristic features of eukaryotes in Chapter 9.**

The key structural features of fungi

If you were to examine the *P. camemberti* shown in Figure 11.29 under a microscope you would see the 'body' of the fungus, which consists of long thin filaments called **hyphae**. Hyphae are effectively the cells of a fungus. You can see hyphae in Figure 11.30a. Look at this figure 11.30a, and notice that the hyphae can branch to form a network of filaments. When hyphae grow and branch they can form a visible **mycelium** which you can see on the surface of the plate in Figure 11.29, or sometimes on your bread if you've left it out too long!

Figure 11.27 Electron scanning micrographs of halophilic bacteria. (a) *Halobacterium salinarum* and (b) *Haloquadratum walsbyi*.

Source: (a) Figure 2 from Stan-Lotter, H., Fendrihan, S. Halophilic Archaea: life with desiccation, radiation and oligotrophy over geological times. *Life* 2015, 5, 1487–1496. https://doi.org/10.3390/life5031487 CC BY 4.0; (b) Figure 1 from Sublimi Saponetti, M., Bobba, F., Salerno, G., et al. (2011) Morphological and structural aspects of the extremely halophilic Archaeon *Haloquadratum walsbyi*. *PLoS ONE* 6(4): e18653. https://doi.org/10.1371/journal.pone.0018653.

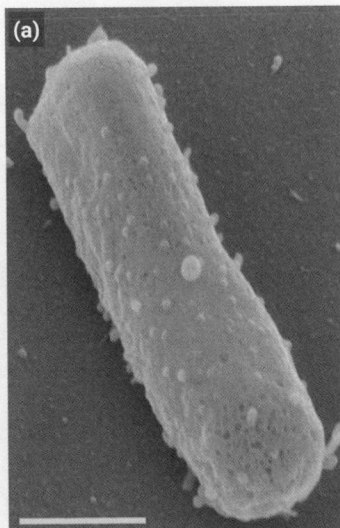

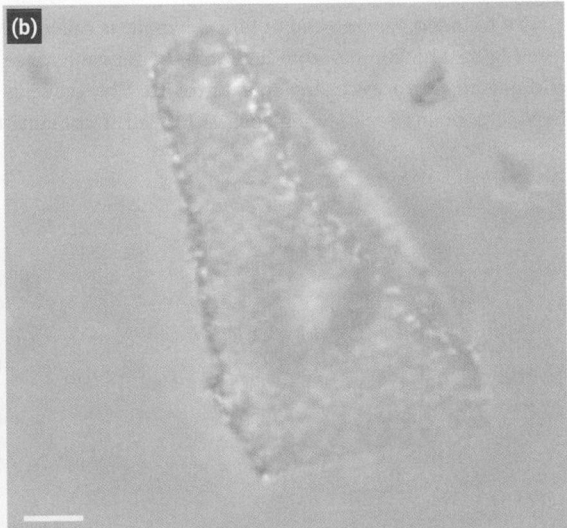

Figure 11.28 A typical image used when fungi are discussed. The 'death cap' *Amanita phalloides*.

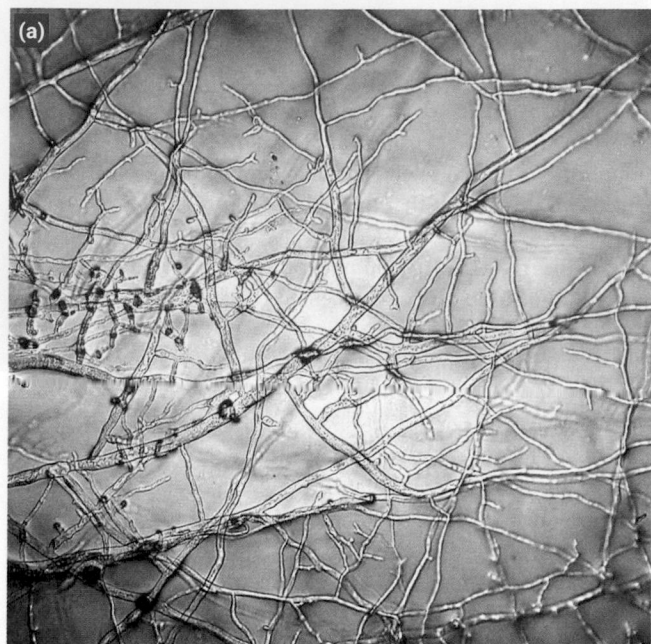

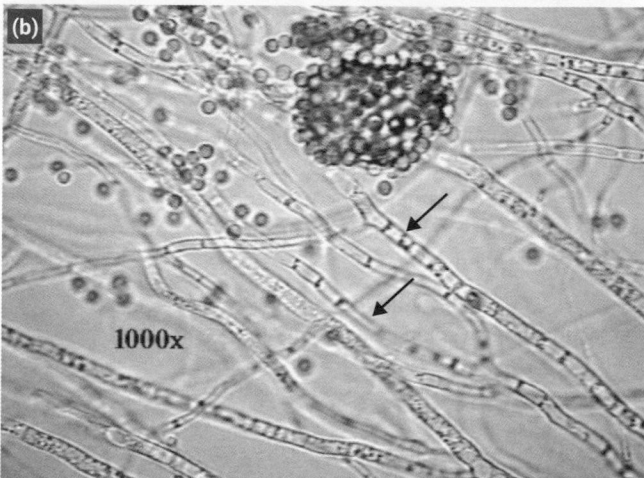

Hyphae are filled with protoplasm, which contains the organelles that are familiar to use from other eukaryotes, including the nucleus, mitochondria, and Golgi body. This overall structure is illustrated in Figure 11.31. The majority of fungal species have a structure in their hyphae called a septum (plural septa), depicted in Figure 11.30b. The septum is a wall that divides the fungal hyphae into compartments. The septa usually have holes in them, which allow the cytoplasm and sometimes organelles to move between different parts of the hyphae (Figure 11.31). Behind each septum are protein lattices called **Woronin bodies**. These can plug the holes in the septa if the hyphae become damaged and when they get older.

The rounded, tapered end of fungal hyphae is called the apical tip (see Figure 11.30c). It is from here that the fungal hyphae can grow and extend, a process called **apical growth**. The growing tip of the hyphae contains very few organelles; instead it contains the apical

Figure 11.29 Most fungi actually appear as moulds. *Penicillium camemberti* growing on potato dextrose agar.

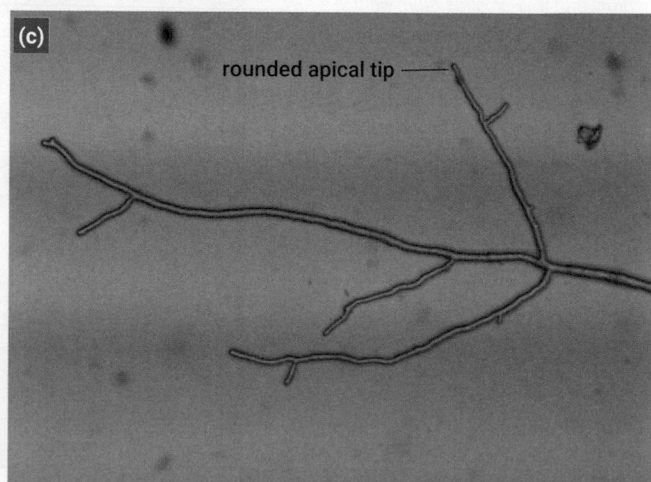

rounded apical tip

Figure 11.30 Hyphae. (a) Branching fungal hyphae; (b) septa in hyphae indicated by arrows; (c) rounded apical tip.

Source: (a) Wikimedia Commons/Bob Blaylock/CC BY-SA 3.0; (b) Roberts Lab, School of Aquatic and Fishery Sciences, University of Washington.

Figure 11.31 A diagrammatic representation of the fungal hyphae with organelles.

Source: Adapted from Playfair & Bancroft, *Infection & Immunity*, 4th edition, 2013, Oxford University Press.

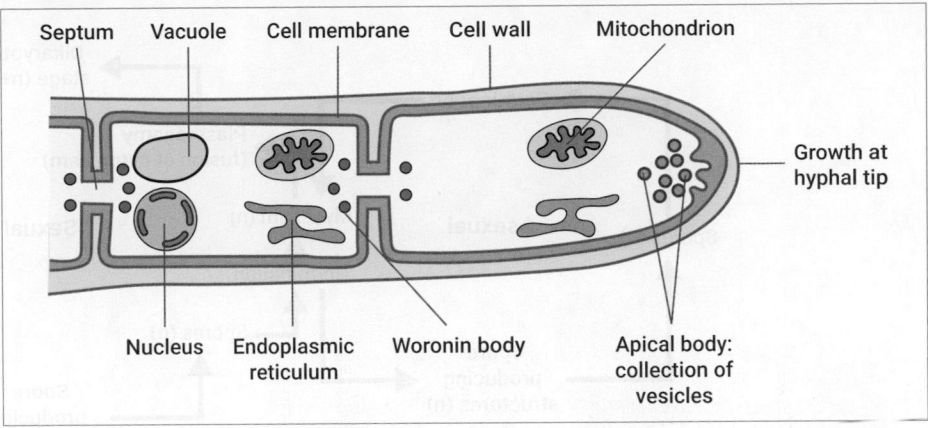

body, which is a collection of membrane-bound vesicles. These vesicles arise from the Golgi body and contain components that are needed to both break down and rebuild the cell wall at the growing tip. The vesicles fuse with the plasma membrane at the apical tip and form new cell wall material, resulting in apical growth. Apical growth is a distinguishing feature of fungi, as it enables fungal species to very effectively push their way through soil particles. It also contributes to their success as plant pathogens by enabling them to penetrate plant surfaces.

Not all fungi exist as hyphae. Instead, some exist as single cells called **yeasts**, as depicted in Figure 11.32. Two very well-known yeasts you will have come across are *Saccharomyces cerevisiae*, better known as baker's yeast, and *Candida albicans*, the causative agent of an infection called thrush. We learn more about yeasts in Sections 13.4 and 14.4.

Fungal habitats

Fungi can gain their nutrition in a variety of ways—ways that often determine the habitat in which they can be found. In order to obtain their nutrition, fungi can be:

- **saprophytes**: they obtain their nutrition from dead organic material;
- **biotrophs**: they obtain their nutrition from living host tissue (examples include lichens, mycorrhizal fungi, and endophytes);
- **necrotrophs**: they kill the host cell then feed on the dead tissue (examples include some plant pathogens).

About 90,000 species of fungi have been described, but as many as 1.5 million may exist and new species are being continually discovered. This makes mycology a fascinating topic.

Mycorrhizal relationships

Mycorrhizae are mutualistic symbiotic associations that form between the roots of most species of plants and fungi. The relationship develops and is maintained within the soil environment. A bidirectional movement of nutrients occurs, whereby fixed carbon from photosynthesis by the plant flows to the fungus, and inorganic nutrients from the soil move to the plant. Mycorrhizal fungi form a crucial link between the plant root and the soil.

As much as 20 per cent of the total carbon fixed by the plant is transferred to the fungus and the plant increases photosynthetic

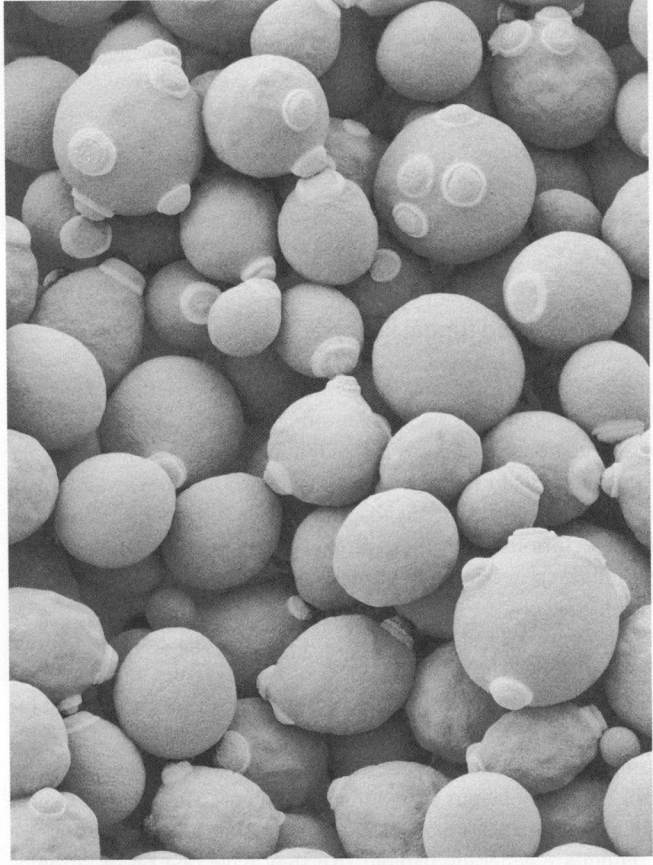

Figure 11.32 Yeast cells of *Candida albicans*.

Source: Dennis Kunkel Microscopy/Science Photo Library.

activity to compensate for the loss. The flow of carbon to the soil mediated by mycorrhizal fungi serves several important functions:

1. Fungal hyphae produce hydrolytic enzymes, which increase nutrient availability by breaking down complex organic compounds.

2. Fungal hyphae bind soil particles together, thereby improving soil aggregation and characteristics such as drainage properties.

3. The development of a unique rhizosphere microbial community called the **mycorrhizosphere**, which is the area of soil adjacent to the fungal hyphae.

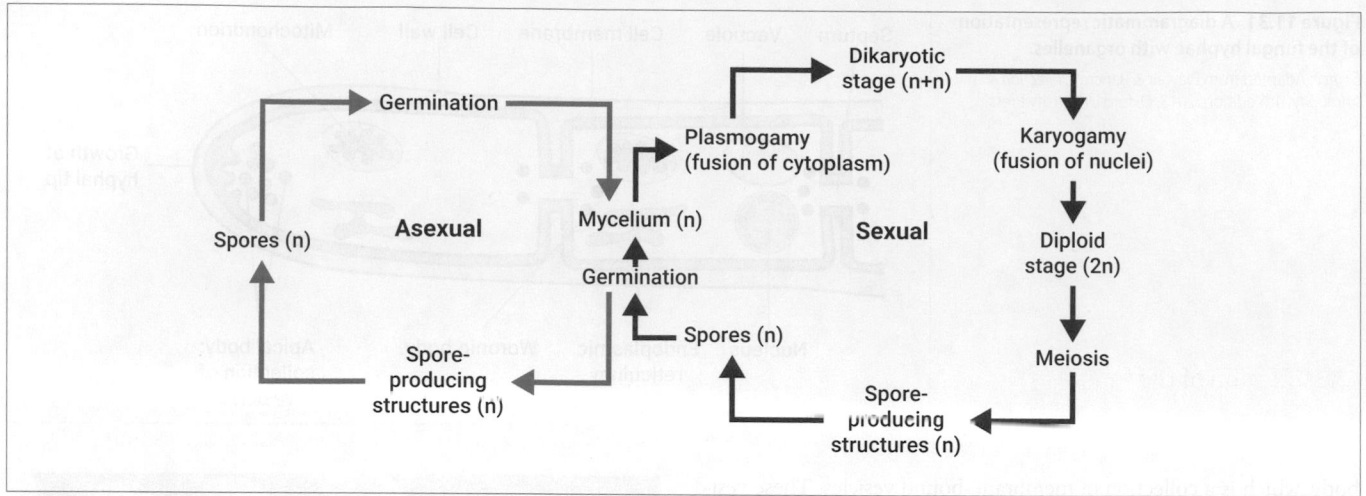

Figure 11.33 Fungal life cycle. A generalized fungal life cycle showing both asexual and sexual cycles.

We will shortly consider fungal biodiversity, and how fungi are classified in phyla according to their method of sexual reproduction. First, though, we need to consider a generalized fungal life cycle.

The generalized fungal life cycle

Individual fungi are able to reproduce using both asexual and sexual reproduction. Asexual reproduction allows for the dispersal of a large number of genetically identical progeny, whereas sexual reproduction allows for genetic diversity in the progeny and therefore the potential ability to adapt to new environments. The essential structure for both these processes is called the **fungal spore**. All fungal spores are microscopic, but they vary in size, structure, shape, and colour.

A generalized fungal life cycle, showing both asexual and sexual cycles, is shown in Figure 11.33. Look at this figure and notice the generation of spores as part of the both the sexual and asexual life cycles. It is important to note that fungal reproduction is a highly complex topic, and there are many variations. For example, not all fungi have an obvious sexual stage as part of their life cycle, which can make classification challenging.

Fungi are generally haploid (n): they have one set of chromosomes. During their life cycle, however, many fungi can move from haploid to dikaryotic (n+n) to diploid (2n). The dikaryotic condition describes the situation in which we see two nuclei existing in the same portion of the fungal hyphae prior to nuclear fusion (karyogamy). A dikaryotic condition is maintained in part by the septa in the fungal hyphae.

The asexual cycle is shown as the left-hand cycle in Figure 11.33. During this cycle, the fungal hyphae can differentiate to produce specialized reproductive structures upon which spores—that are genetically identical to the parent—are produced. These spores are haploid and can be dispersed by a number of different mechanisms, including wind, insects, and animals. The spores germinate to produce more fungal hyphae, which are also haploid.

With sexual reproduction, shown in the right-hand cycle in Figure 11.33, many (but not all) fungi are required to have different mating types. In humans, the mating types are male and female, and we see obvious differences in the secondary sexual characteristics between the two. By contrast, the mating type in fungi is not so obvious as hyphae look very much the same.

If hyphae of compatible mating types meet within an environment such as the soil then fusion of the cytoplasm (a process called **plasmogamy**) can occur. This gives rise to a dikaryotic stage (n+n). Fusion of the nuclei then occurs, which is called **karyogamy**. This gives rise to the diploid stage (2n).

The diploid stage can then undergo meiosis, and the fungi will produce the specialized reproductive structures upon which spores—which are genetically *different* from the parent—are produced. Just as with the asexual cycle, these spores are haploid, and can be dispersed by a variety of mechanisms. Spores germinate to produce new hyphae that are also haploid. And so the cycle continues.

We will provide more specific information about fungal reproduction when we examine each of the major fungal phyla. Refer back to this section to see how the specific strategies relate to the general cycles described here.

Dual nomenclature

As we have just seen, the classification of fungi depends on their method of sexual reproduction. However, there are some fungi for which no sexual stage has been found, and therefore classification of them can be hard. For classification purposes, these fungi are dumped temporarily into the artificial group, the deuteromycetes, which are also known as the fungi imperfecti.

However, it has been accepted more recently that these fungi belong to the Ascomycota, and the sexual stages have simply not yet been discovered. As soon as the sexual stage of a deuteromycete has been found it is reclassified, which means quite a bit of moving around in terms of fungal classification. In addition, fungi were often given two names: one name for the asexual stage (the anamorph) and the other for the sexual stage (the telomorph). The giving of two names is called **dual nomenclature**.

A good example is the causative agent of the ash dieback disease. The asexual stage is called *Chalara fraxinea*, and the sexual stage is called *Hymenoscyphus pseudoalbidus*. As you can imagine, this made things very confusing and does not take into account how our understanding has evolved in the light of molecular phylogeny. In 2011 changes were made to the International Code of Nomenclature for Algae, Fungi, and Plants which meant that from 2013 a fungus could only have one name.

Classification of the fungi

It is currently accepted that fungi can be grouped into five major phyla according to their method of sexual reproduction:

- Chytridiomycota;
- Zygomycota;
- Glomeromycota;
- Ascomycota;
- Basidiomycota.

A phylogenetic tree showing these phyla is illustrated in Figure 11.34.

Chytridiomycota

The Chytridiomycota (chytrids) was the first fungal phylum to diverge from a common ancestor: it appeared in the late Cambrian, more than 500 million years ago. The chytrids can be found in moist soil and freshwater habitats, where they decompose cellulose. The majority are filamentous and grow as hyphae, but they do not contain any septa (cross walls). Some are unicellular.

The Chytridiomycota are the only fungal phylum to produce motile spores called **zoospores** as a means of both asexual and sexual reproduction. These zoospores have a single whiplash flagellum to power movement, as depicted in Figure 11.35.

Chytrids live as saprophytes, but they are also facultative parasites. A specialized group of chytrids grow in the rumen of herbivorous animals where they degrade cellulose. An example of these rumen fungi is *Neocallimastix* spp. This fungus is unusual as it is an obligate anaerobe: it can only survive in the absence of oxygen. Its mitochondria have evolved into organelles called hydrogenosomes, which ferment carbohydrate, generating hydrogen gas in the process.

Other Chytridiomycota species can be plant pathogens. These include *Synchytrium endobioticum*, which causes black wart of potato, as depicted in Figure 11.36. Chytrids can also be pathogens of amphibians, as we explore further in Real World View 11.3.

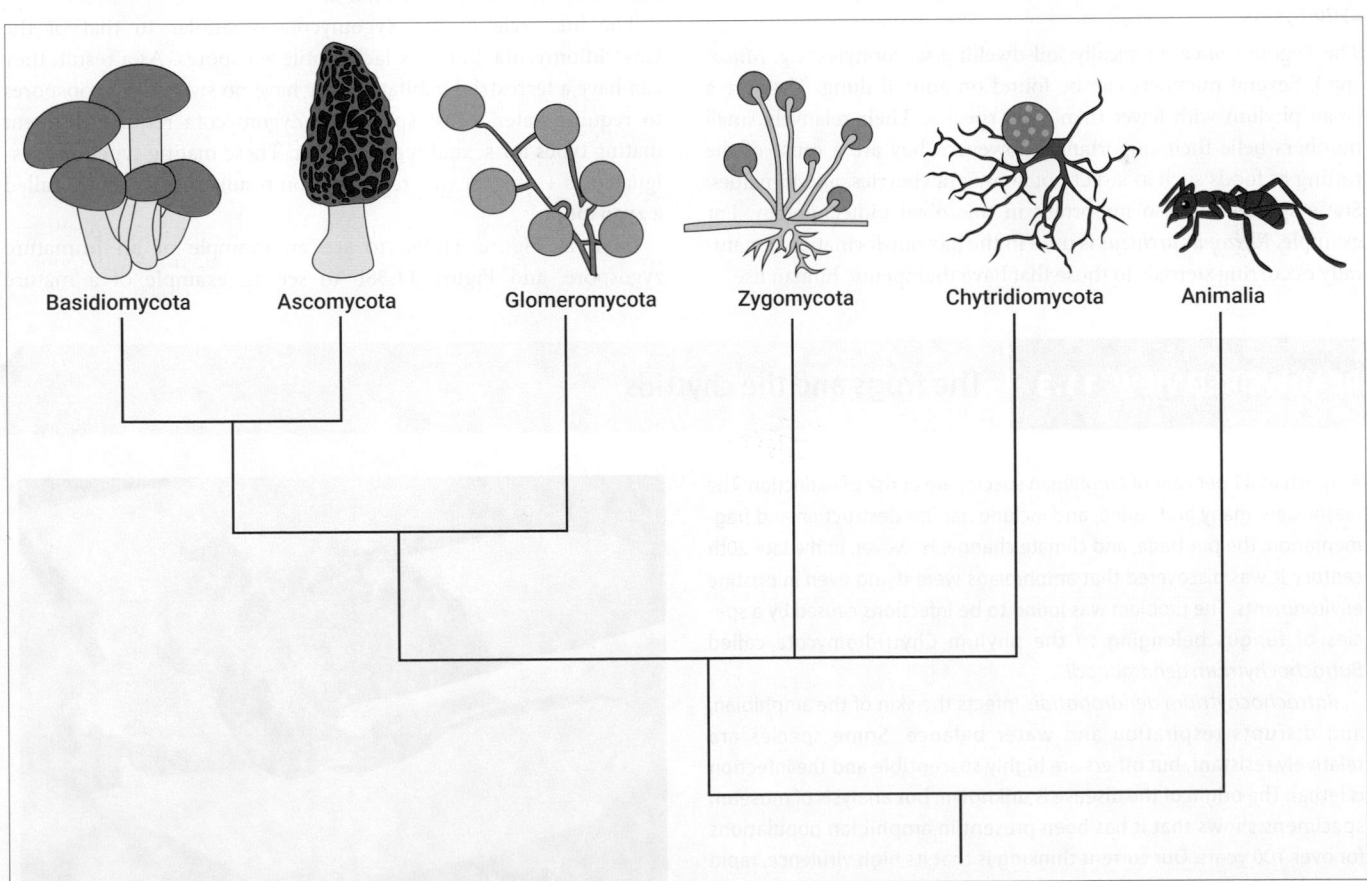

Figure 11.34 The five major phyla of fungi. Fungal phylogenetic tree showing the five main phyla: Chytridiomycota, Zygomycota, Glomeromycota, Ascomycota, and Basidiomycota.

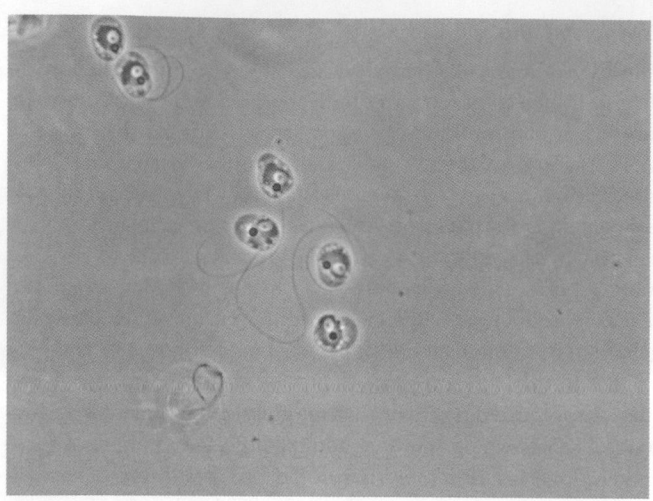

Figure 11.35 Zoospore from *Batrachochytrium dendrobatidis*. Notice the single flagellum extending from the cells in the middle of the image.
Source: Joyce E. Longcore, School of Biology and Ecology, University of Maine.

Figure 11.36 Black wart of potato caused by *Synchytrium endobioticum*.
Source: UK Crown Copyright, courtesy of Fera.

During sexual reproduction, a structure called a zoosporangium is produced, within which the motile zoospores are formed.

Zygomycota

The Zygomycota are typically soil-dwelling saprophytes (e.g. *Mucor* spp.). Several members can be found on animal dung. They are a small phylum with fewer than 1000 species. Their relatively small numbers belie their importance, however: they are a cause of the rotting of foods such as sweet potatoes, strawberries, and tomatoes. Some species are also important in microbial biotechnology. For example, *Rhizopus arrhizus* is used in the biotransformation of naturally occurring steroids to those that have therapeutic human use.

The Zygomycota have a crucial environmental role because of their involvement in the decomposition of plant material. They are common as bread moulds (e.g. *Rhizopus stolonifer*) and are often termed 'pin-head' fungi, due to their long, thin hyphae, which end in the asexual reproductive structures, the **sporangia**, which contain the spores (see Figure 11.37).

The life cycle of the Zygomycota is similar to that of the Chytridiomycota, but they lack motile zoospores. As a result, they can have a terrestrial habitat as they have no swimming zoospores to require water. Most species of Zygomycota require different mating types for sexual reproduction. These mating types are designated as + or −. Sexual reproduction results in a structure called a **zygospore**.

Look at Figure 11.38a to see an example of an immature zygospore, and Figure 11.38b to see an example of a mature

---

REAL WORLD VIEW 11.3 | **The frogs and the chytrids**

As much as 41 per cent of amphibian species are at risk of extinction. The reasons are many and varied, and include habitat destruction and fragmentation, the pet trade, and climate change. However, in the late 20th century it was discovered that amphibians were dying even in pristine environments. The problem was found to be infections caused by a species of fungus belonging to the phylum Chytridiomycota called *Batrachochytrium dendrobatidis*.

Batrachochytrium dendrobatidis infects the skin of the amphibian, and disrupts respiration and water balance. Some species are relatively resistant, but others are highly susceptible and the infection is lethal. The origin of the disease is unknown, but analysis of museum specimens shows that it has been present in amphibian populations for over 100 years. Our current thinking is that its high virulence, rapid spread, and broad host range will make this disease a major cause of extinction in amphibians. More research is needed to understand the spread and the effects of the disease so that we can better plan to mitigate for it.

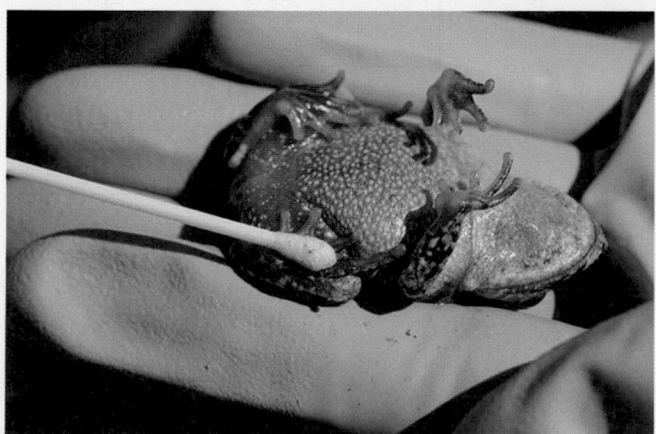

Figure 1 Toad being tested for the presence of *Batrachochytrium dendrobatidis*.
Source: Nature Picture Library/Alamy Stock Photo.

Figure 11.37 Sporangia. Bread mould (*Mucor* spp.) under a microscope.
Source: Getty Images.

zygospore. Notice how the immature zygospore has a thin wall, while the mature zygospore has a much thicker wall, as well as cell wall protrusions, giving it a hairy appearance.

The zygospore opens and a hypha is produced. The hypha forms a sporangium at the end, as illustrated in Figure 11.39, which can then release spores. Spores released from the sporangium germinate into new hyphae, which can also reproduce asexually by the production of sporangia.

Spore release from the Zygomycota is one of the fastest active expulsions in nature. For example, the name of *Pilobolus*, which can be found on horse manure, means 'hat thrower'. *Pilobolus* has evolved an immensely powerful mechanism of active spore liberation to enable its spores to travel over the horse dung, known as the 'zone of repugnance', and to reach fresh grass beyond.

Glomeromycota

The Glomeromycota are the newest phylum: they used to be part of the Zygomycota. The fungi in this phylum are fascinating. They associate with the root cortical cells of woody plants, as illustrated

in Figure 11.40; in so doing, the fungal hyphae massively extend the root network of the plant, so that more water and essential minerals for growth can be transported. In return, the fungus receives sugar from the plant, which increases its photosynthetic capacity to compensate.

▶ We learn more about associations between fungi and plants in Chapter 14.

Ascomycota

The Ascomycota is the largest fungal phylum with over 15,000 species. Many of our industrially important fungi are in this phylum; these include *Saccharomyces cerevisiae* (baker's and brewer's yeast) and *Penicillium chrysogenum*, which plays an important role in the production of the antibiotic penicillin. We also rely heavily on the Ascomycota for food production and the development of flavour in such products as the cheeses stilton and camembert.

The Ascomycota have numerous environmental habitats—terrestrial, aquatic, and marine—where they live as saprophytes, biotrophs, necrotrophs, and parasites. Most Ascomycota are either filamentous moulds, or single-cell yeasts, but some species can form larger fruiting bodies that can be seen with the naked eye. Examples of fungi that produce these larger fruiting bodies include the highly prized culinary truffles, which you can see in Figure 11.41a.

Truffles are fascinating as they can produce chemicals called pheromones, which enable pigs and dogs to sniff them out in woodland. You can also find hard fruiting structures such as *Daldinia concentrica*, which are also known as cramp balls or King Alfred's cakes (Figure 11.41b). More information about this fungus can be found in Real World View 11.4.

Some species of Ascomycota are pathogens of animals and plants. This includes the most common blood-borne fungal pathogen in humans, the yeast *Candida albicans*, which causes candidiasis, an infection of warm mucosal surfaces, such as the mouth and vaginal region. We explore fungal toxins and fungal pathogens of animals and humans further in Chapter 14.

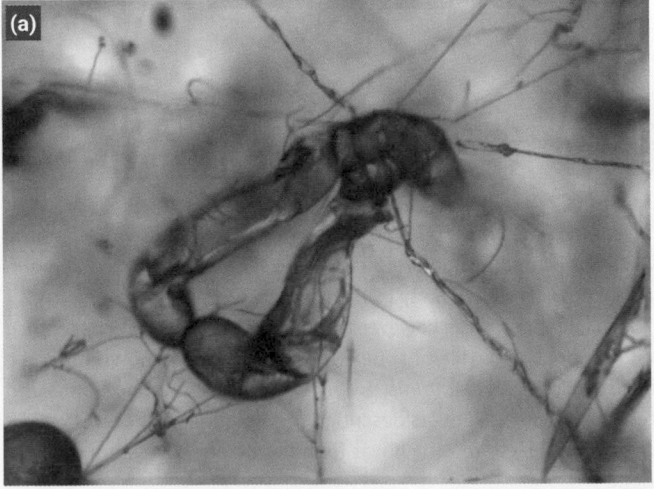

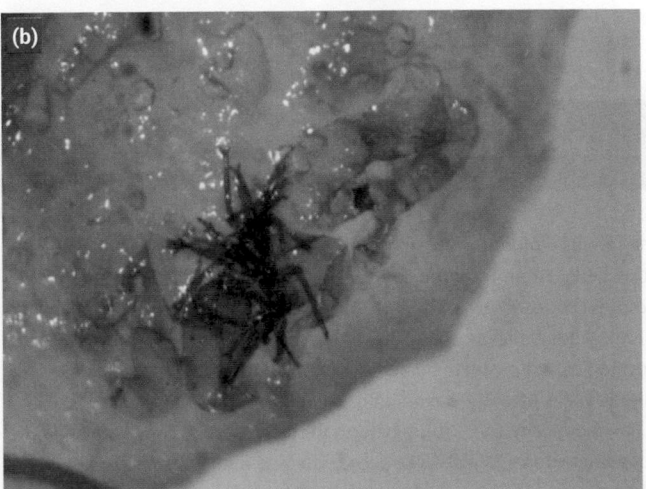

Figure 11.38 Zygospores of *Phycomyces blakesleeanus*. (a) Immature zygospore; (b) mature zygospore.

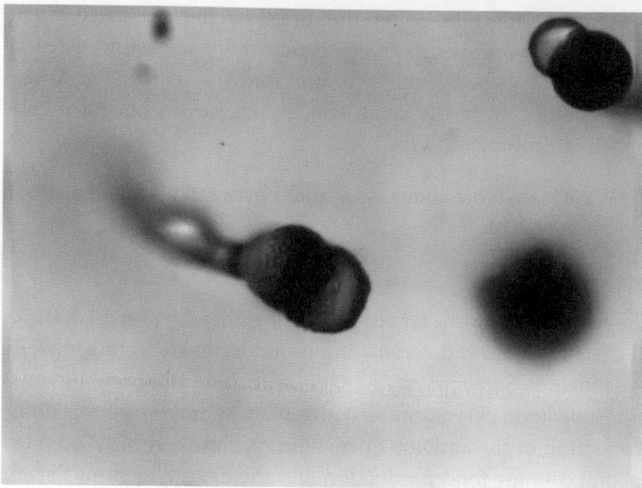

Figure 11.39 Sporangium of *Phycomyces blakesleeanus*.

Figure 11.41 Truffles. (a) Black truffle and (b) *Daldinia concentrica*.
Source: (a) Image by WikiImages from Pixabay.

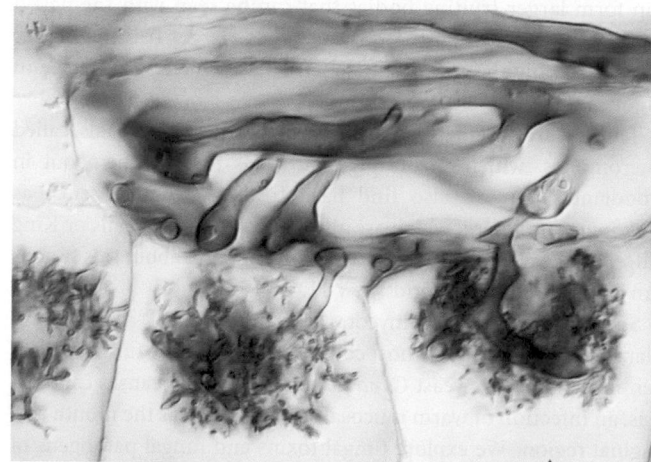

Figure 11.40 Fungi associating with the root cortical cells of Canada wild ginger (*Asarum canadense*).

Source: Image © Mark Brundrett.

Fungal plant pathogens include the agent of Dutch elm disease, *Ophiostoma novo-ulmi*, which is spread to these trees by wood-boring insects. The fungus interferes with the movement of water in the elm which leads to wilt, as shown in Figure 11.42.

Asexual reproduction in the Ascomycota involves the production of specialized hyphae called **conidiophores**, upon which the

---

REAL WORLD VIEW 11.4   Cramp balls and cakes: *Daldinia concentrica*

The fungus *Daldinia concentrica* is an interesting example of a superstitious belief in an intervention that actually has no therapeutic value. *Daldinia concentrica* is also known as 'cramp balls' because people in the Middle Ages believed that putting these hard-fruiting bodies into your clothing would protect you against cramp. They are also known as 'King Alfred's cakes' after the Saxon king.

Taxonomically, these fungi belong to the phylum Ascomycota and produce their spores in a sac-like ascus. The asci and spores are quite large, which makes them easy to see under a light microscope.

Daldinia concentrica has an intriguing spore dispersal mechanism, which is inhibited by light. As a result, the spores are released at night, and are possibly carried away by night insects such as moths.

In an interesting twist, however, the therapeutic value of *D. concentrica* is not entirely a matter of superstition: it has been known for some time that species of *Daldinia* produce antimicrobial compounds and steroids. In 2006, scientists described the isolation of concentricolide, a benzofuran lactone, which was found to be effective as an anti-HIV agent.

Figure 11.42 Fungal plant pathogens cause disease. Dutch elm disease showing distinct wilting.

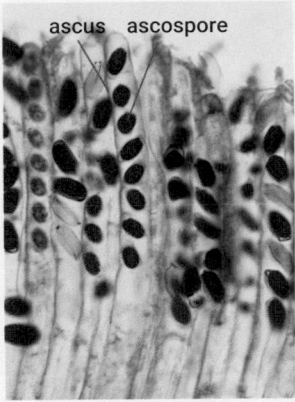

Figure 11.44 Ascus. Eight ascospores inside the sac-like ascus.

haploid spores, called **conidia**, are formed. The conidiophores can be clearly seen on *Penicillium camemberti* in Figure 11.43.

Sexual reproduction in many of the species of Ascomycota requires different mating types, known as + and − strains. As a result of sexual reproduction an **ascus** is formed. This structure looks like a sac, and often the Ascomycota are referred to as the sac-fungi (see Figure 11.44). Eight haploid ascospores are produced inside the ascus; look at Figure 11.44 and notice how obvious these eight structures are.

The spores are shot out of the ascus under pressure. This is called the water cannon mechanism. But how does this mechanism operate? A precisely timed increase in molecules called osmolytes occurs in the ascus, which draws in water by osmosis to increase turgor pressure. The turgor pressure provides the force for stretching the ascus and ultimately for firing out the spores, which can achieve massive velocities. An example is shown in Figure 11.45.

Yeasts

Yeast are not a taxonomic group but are, in fact, a growth form comprising a single cell (i.e. they are unicellular). The majority of fungi that form yeast cells are found in the Ascomycota, but there are some examples in the Basidiomycota and Zygomycota.

Some fungi can exist in two forms: they can differentiate from a hyphae form to a yeast form, or vice versa. This behaviour is called **dimorphism**. Dimorphism is known to be common in many human fungal pathogens; we will return to this concept in Chapter 14.

Unicellular yeasts reproduce asexually using either **fission** or **budding**. In fission, two daughter cells are produced from a mother cell, with new cell wall material being laid down between the two cells (see Figure 11.46a). In budding, a new daughter cell arises as a protrusion from the mother cell. The nucleus of the mother cell is duplicated and the duplicate elongates into the daughter cell. When the daughter cell separates away from the mother cell, it

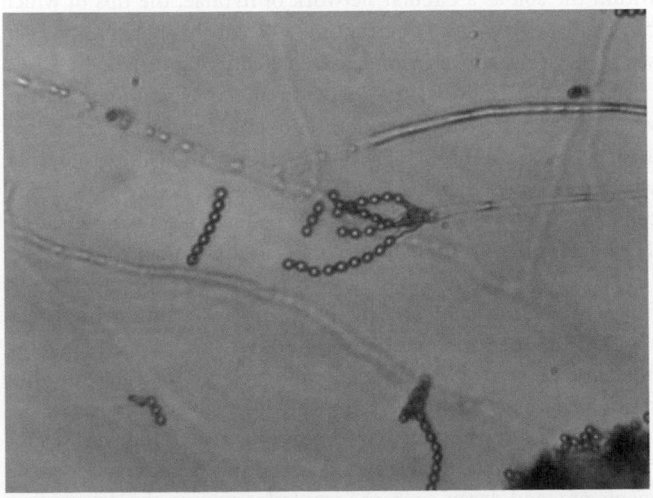

Figure 11.43 Conidiophores. Conidiophores of *Penicillium camemberti*.

Source: William Jacobi, Colorado State University, Bugwood.org.

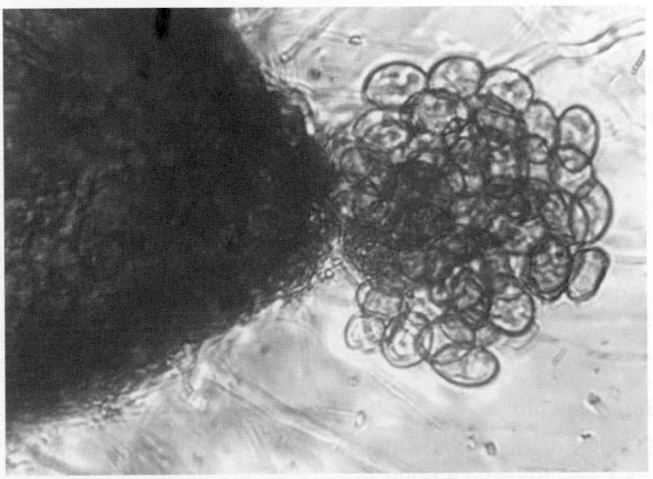

Figure 11.45 The ascus of *Sordaria fimicola* showing ascospores being forcibly discharged via water pressure from within the fruiting body.

Source: © Visuals Unlimited/naturepl.com.

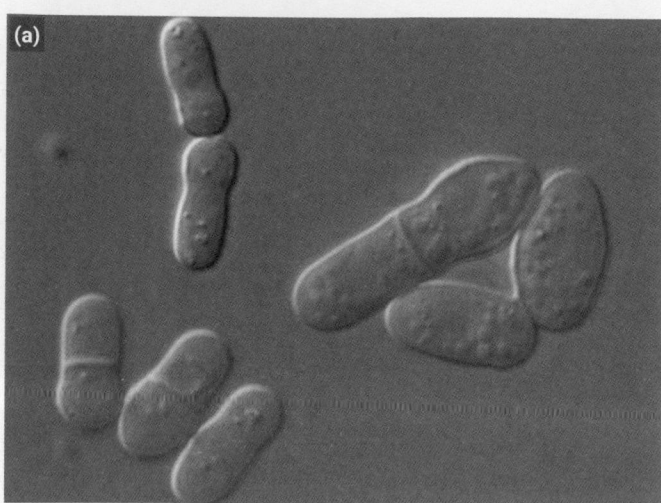

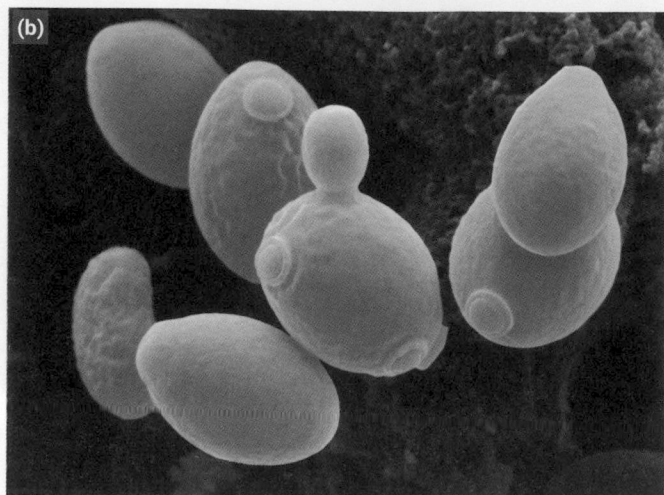

Figure 11.46 Yeast reproduction. (a) Fission sees two daughter cells being produced from a parent cell. (b) Budding sees a daughter cell emerge from an initial protrusion. The bud leaves a scar on the mother cell, seen here as the prominent circles on the cell surface.
Source: (a) Image from Wiley Online Library, Yeast 31: 83–90 (2014), Niki, H., '*Schizosaccharomyces japonicus*: the fission yeast is a fusion of yeast and hyphae' https://doi.org/10.1002/yea.2996. (b) Mediscan/Alamy Stock Photo.

leaves behind a bud scar, which can be easily seen in Figure 11.46b. Daughter cells can only bud from unscarred regions of the mother cell, so when these areas are no longer available, the mother cell stops reproduction and dies.

> ▶ We learn more about fusion and budding when we explore cell division in Chapter 10.

Yeasts can also exhibit sexual reproduction, which relies on there being two different mating types. In *Saccharomyces cerevisiae*, for example, these two cell types are called 'a' and 'α'. Plasmogamy is controlled by chemical pheromones: the 'a' cell produces 'a' factor, and the 'α' cell, 'α' factor. These pheromones are detected by the opposite mating type, and act as signals towards which the cells grow, until contact is achieved. The cells then undergo cell fusion-plasmogamy to form an a/α diploid, and then nuclear fusion occurs to form the zygote.

PAUSE AND THINK

How might fungal dimorphism benefit the fungus?

Answer: The term fungal dimorphism describes how some fungi can exist in two growth forms, either as hyphae (long filaments) or as a yeast (single cell). Hyphal forms of fungi are very good at growing through soil: their mechanism of apical growth means they can push their way through soil particles. Yeasts are very good at dispersing in fluids. Many pathogenic fungi are dimorphic and we will learn more about them in Chapter 14.

Basidiomycota

The Basidiomycota are the second largest phylum, comprising approximately 13,000 species, including the mushrooms, bracket fungi, fairy clubs, puffballs, stinkhorns, birds nest fungi, and jelly fungi. The reproductive structures of the familiar larger, fleshy fungi of the Basidiomycota have an amazing diversity of colours and shapes, a few of which you can see in Figure 11.47. While the reproductive structures are very visible, the majority of the fungus body (thallus) is in the soil as microscopic hyphae.

The fruiting body contains hyphae that can fuse to form **basidia**, which are diploid cells (2n). As you can see in Figure 11.48, the basidia are club shaped and they can undergo meiosis to produce haploid **basidiospores**.

Look closely at Figure 11.48 and notice how each spore develops on a small stalk called a **sterigma**. The basidiospore develops from the sterigma asymmetrically; at maturity, a drop of fluid, called the Buller's drop, forms on the apiculus and fluid also builds up on the adjacent spore surface, as shown in Figure 11.49. When the Buller's drop touches the spore surface fluid the two coalesce and the momentum discharges the spore from the sterigma.

After the basidiospores are released they can germinate in the soil to produce a mycelium. The mycelium can grow outwards within the soil as a circular network of hyphae, the tips of which can produce new fruiting structures. When this happens, we can see a 'fairy ring' above ground, as depicted in Figure 11.50.

Many of the fleshy Basidiomycota fungi that we commonly see in woodlands are the fruiting structures of mycorrhizal fungi (i.e. they associate with plant roots). These differ from the Glomeromycota, which we discussed earlier, because they surround the root cortical cell in a sheath called the **Hartig net** instead of growing inside it. The benefits of this mutualistic symbiosis are the same.

A good example of a Basidiomycota is one of the most deadly fungi, *Amanita phalloides*. This fungus forms a mycorrhizal relationship with oak and beech trees. Another species of Amanita, *A. muscaria*, known for its classic fairytale look of a red cap with white spots, forms a mycorrhizal relationship with silver birch trees.

Most of the fungi in the Basidiomycota are saprophytes, which cause the decay of leaf litter, wood, and dung. Some can be serious agents of wood decay, such as *Serpula lacrymans*, the dry rot fungus. Others are severe parasites, such as *Armillariella mellea*, the honey agaric, which destroys a wide range of woody plants. Two other important groups of plant pathogens, the rusts and smuts, are also classified within this phylum.

Figure 11.47 Examples of the Basidiomycota. (a) *Amanita muscaria* (fly agaric); (b) *Geastrum triplex* (common earth-star); (c) *Phallus impudicus* (common stinkhorn); (d) *Crucibulum laeve* (common birds-nest fungus).

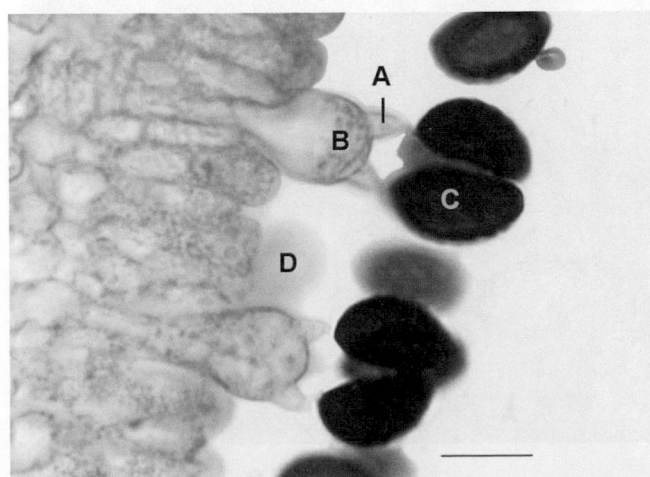

Figure 11.48 Basidiospores developing on a basidium. A, Sterigma; B, basidium; C, basidiospore; D, immature basidium.

Source: Jon Houseman and Matthew Ford/Wikimedia Commons/CC BY-SA 3.0.

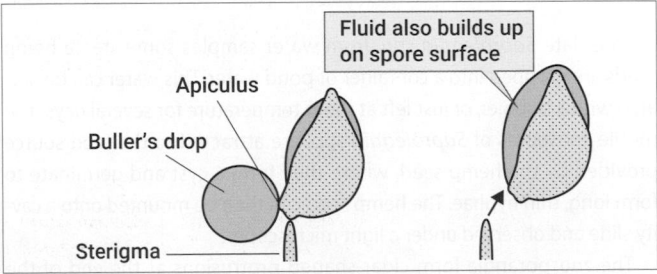

Figure 11.49 The formation of the Buller's drop on a basidiospore.

Source: Reprinted from J. Webster, M. C. F. Proctor, R. A. Davey, G. A. Duller. (1988) Measurement of the electrical charge on some basidiospores and an assessment of two possible mechanisms of ballistospore propulsion. *Transactions of the British Mycological Society*, 91(2), 193–203, with permission from Elsevier.

PAUSE AND THINK

Which fungal phylum has the largest number of macroscopic fungi?

Answer: While the Ascomycota is the largest fungal phylum, these fungi are mostly microscopic. It is the Basidiomycota that has the largest number of macroscopic fungi; these are the larger fleshy fungi you can see in the environment.

Figure 11.50 Fairy ring. A fairy ring of *Clitocybe nebularis* (the clouded funnel mushroom).

Source: Josimda/Wikimedia Commons/CC BY-SA 3.0.

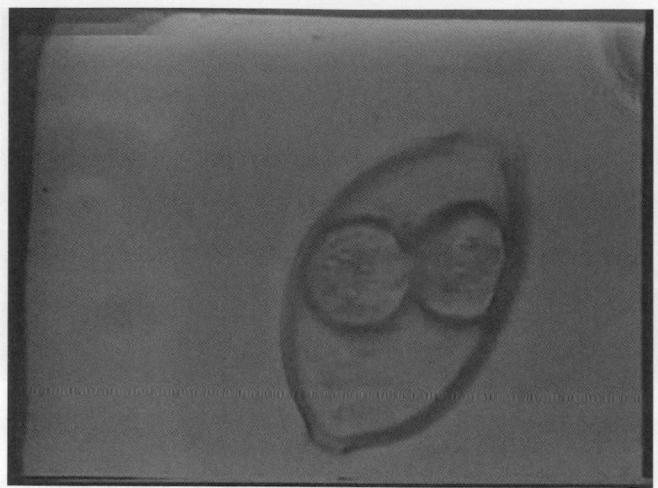

Figure 11.51 Zoosporangium. Lemon-shaped zoosporangium of *Phytophthora infestans*.

Oomycetes

Oomycetes are curious organisms: they are not true fungi, as they have cellulose in their cell walls rather than chitin, and neither do they synthesize ergosterol. However, they behave in a similar way to fungi and for a long time were classified with them. Recently, more modern taxonomy has shown that these organisms do not belong in the kingdom Fungi, leading to them being moved to the kingdom Protoctista.

The oomycetes produce motile zoospores as a means of asexual reproduction in a similar way to the Chytridiomycota and therefore require watery environments or moist soils for these swimming cells to move within. The motile zoospores are formed inside a lemon-shaped zoosporangium, the shape of which can easily be seen in Figure 11.51.

Oomycetes are able to infect a number of different hosts, including algae, plants, protists, fungi, and arthropods. One example of an oomycete that has had devastating social consequences is the causative agent of potato blight, *Phytophthora infestans*; this pathogen was responsible for the Irish potato famine of 1845–1849.

Another oomycete of interest is *Saprolegnia* spp. If you have ever kept an aquarium, you will be familiar with this organism as it is a common fish pathogen. William Arderon, a Norfolk naturalist, first observed it as a causative agent of vertebrate disease in 1748. A really nice experiment can be done to bait *Saprolegnia* spp. from water samples; this experiment is described in in Experimental Toolkit 11.1.

EXPERIMENTAL TOOLKIT 11.1   Baiting for *Saprolegnia* spp.

To isolate *Saprolegnia* spp. from water samples some sterile hemp seeds are dropped into a container of pond water. This water can be aerated with a bubbler, or just left at room temperature for several days. The motile zoospores of *Saprolegnia* spp. are attracted to the food source provided by the hemp seed, where they form a cyst and germinate to form long, thin hyphae. The hemp seed can then be mounted onto a cavity slide and observed under a light microscope.

The zoosporangia form cigar-shaped protrusions at the end of the hyphae. If you are lucky you will be able to see the zoospores hatching from the zoosporangium as in Figure 1.

The hyphae of *Saprolegnia* provide an ideal habitat for other single-celled protozoa, such as *Vorticella* spp., which can anchor its stalk to the *Saprolegnia* hyphae, as shown in Figure 2a. The *Paramecium* spp. in Figure 2b simply cannot resist the potential food source surrounding the hyphae!

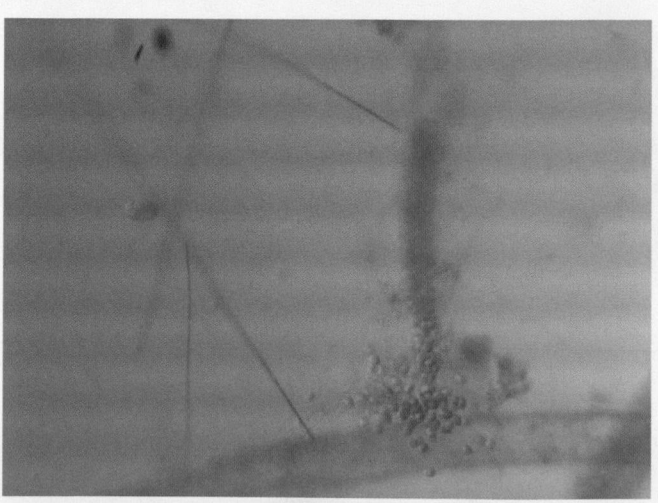

Figure 1 Zoospores of *Saprolegnia* spp. hatching from a zoosporangium.

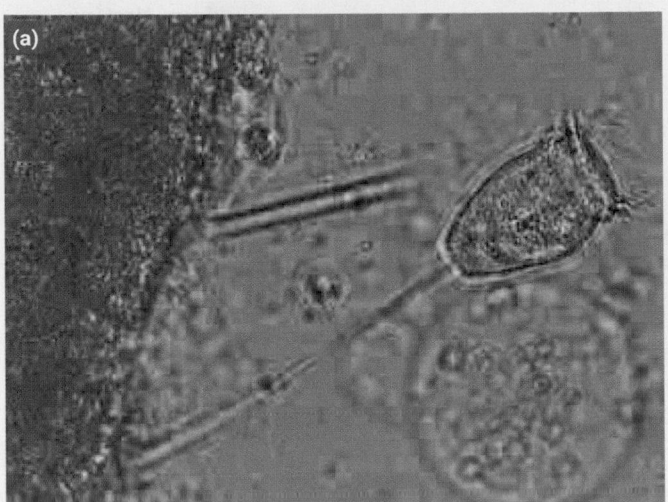

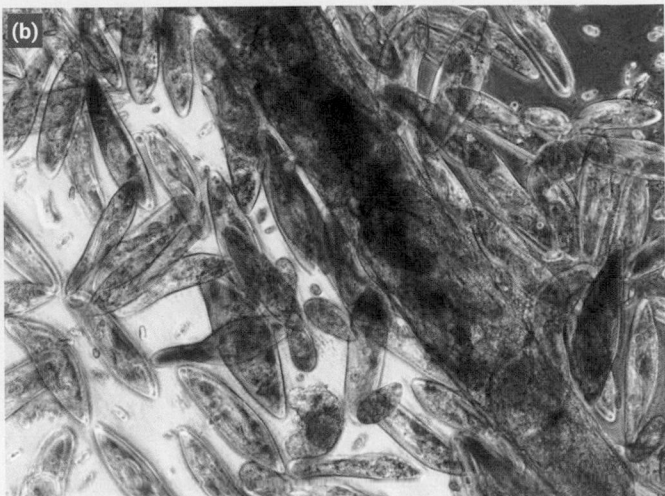

Figure 2 (a) *Vorticella* spp. attached to the *Saprolegnia* hyphae. (b) *Paramecium* spp. attracted to the fungal hyphae as a food source.

PAUSE AND THINK

Why are the Oomycetes not considered to be true fungi?

Answer: For many years mycologists considered the Oomycota to be fungi, and studied them extensively. However, the Oomycota have cellulose in their cell walls, as opposed to chitin. Their cell membranes also lack ergosterol, which is the main sterol in fungal plasma membranes. More recent molecular taxonomy has also placed them much closer to plants than true fungi.

Lichens

Lichens are an example of a mutualistic symbiotic relationship between a fungus and a simple photosynthetic autotroph, which is most commonly an alga. The two organisms are so intertwined that they produce a single **thallus** (body).

For many years, lichens were considered to be a separate taxonomic group, until work, perhaps most notably done by the author Beatrix Potter, showed them to be two dissimilar organisms living together. The fungal partner is termed the **mycobiont**, and the algal partner is the **phycobiont**.

More than 15,000 lichens are known, and they are classified according to the fungal partner. Lichens are able to survive extreme conditions, and are found in virtually every environment. The largest number of lichenized fungi occur in the Ascomycota (98 per cent), although some Basidiomycota are also lichenized. The alga symbiont can be either:

- prokaryotic: 8 per cent of cases are cyanobacteria (blue green algae);
- eukaryotic: 92 per cent of cases are Chlorophyta (green algae) or Xanthophyta (yellow green algae).

Fungi and algae associations are not species specific: one fungus can form lichens with many different algae.

Lichens can be divided into three distinct groups according to their thallus morphology, as illustrated in Figure 11.52:

1. **Crustose**, a flat and crusty thallus which is firmly fixed to a surface.
2. **Foliose**, a leaf-like thallus which is loosely attached to the surface.

Figure 11.52 Lichen thallus morphology. (a) Crustose; (b) foliose; (c) fruticose.

Source: (a) Longklong/Shutterstock.com; (b) Przemyslaw Muszynski/Shutterstock.com; (c) Phagalley/Shutterstock.com.

Figure 11.53 Lichen thallus. Lichen section stained with methylene blue, showing the algal layer, medulla layer, and cortex.
Source: iStock.

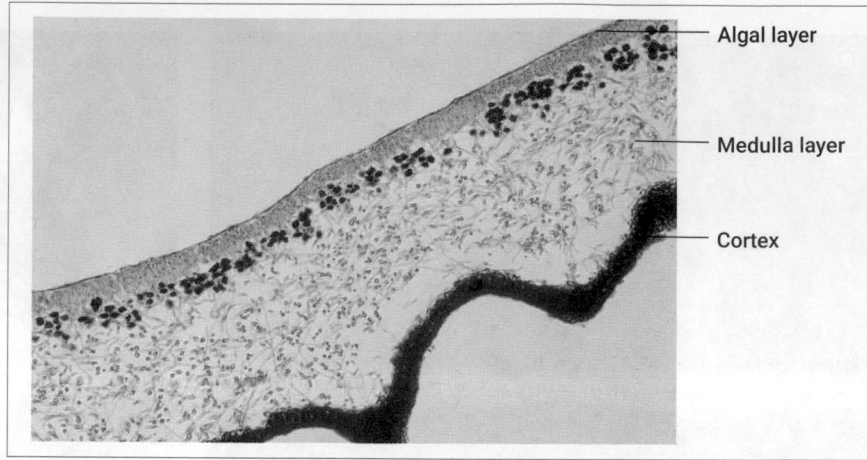

Algal layer

Medulla layer

Cortex

3. **Fruticose**, a bushy or shrubby thallus which is highly branched and attached to the surface with hyphae which are called rhizoids.

The thallus, of the lichen can be divided into tissue layers, as illustrated in Figure 11.53. The most distinct is the **algal layer**, where algae cells lie among loosely intertwined hyphae. The **medulla layer** consists of intertwined hyphae, and no algae cells. This layer lies next to, and supports, the algal layer. The cortex is a surface layer that becomes an epidermis-like tissue.

The symbiosis between the mycobiont and the phycobiont is **mutualistic**, so both partners benefit. The fungus delivers water and minerals to the alga, and protects it from adverse environmental conditions. It also provides the structural support for the algal cells. In return, the algae photosynthesize and supply sugar to the fungus. The fungus removes carbon from the algal cells in the form of sugars by means of **haustoria** (suction pipes) which penetrate the algal cell.

▶ **We encounter haustoria again in relation to fungal plant pathogens in Chapter 14**

Asexual reproduction results in the production of **soredia** and **isidia**. These structures are dispersed by splashing water, wind, insects, or animals. Asexual reproduction can also occur when the lichen is decayed or damaged. Small fragments are readily dispersed, while large fragments can germinate *in situ*. The majority of lichenized fungi belong to the phylum Ascomycota, and thus sexual reproduction would follow the mechanisms as previously described with the formation of a sac-like ascus.

Lichens have a range of uses in our society, as we explore in Real World View 11.5.

PAUSE AND THINK

Lichens are an example of what type of symbiosis and what benefits does this bring?

Answer: Lichens are an example of mutualistic symbiosis—that is, both partners in the relationship benefit from being together. The fungal partner benefits from the photosynthetic algae, as fixed carbon from photosynthesis in the form of sugar is passed to them for growth. The algae benefit as they obtain a sheltered environment, as well as water and minerals from the fungus.

 Check your understanding of the concepts covered in this section by answering the questions in the e-book.

11.6 Protist diversity

It is currently accepted that the current number of protists that have been described only reflects a small proportion of the total number of protists that exist in nature. This is largely due to the very diverse range of habitats that protists can occupy, from soils through to aquatic ecosystems and even inside our own gut.

Scientists disagree on the best way to classify the protists, but we have opted to follow the system that considers the phyla of the Euglenozoa, the Alveolates, the Stramenopiles, the Cercozoans, the Amoebozoa, and, lastly, the group covered under the term of the Algae. Figure 11.54 shows the phylogenetic relationship between these phyla. Look at this figure, and notice how the protists are not all on one branch within the Eukarya.

In the remainder of this section we discuss the characteristics and features that some of these phyla share, as well as those that are unique to each phylum.

Phylum Euglenozoa

The Euglenozoa are single-celled flagellate organisms. They possess one or two flagella, which enable these protists to move. As a result, the Euglenozoa exhibit a range of feeding modes, which can include being predatory, parasitic, or **photoautotrophs** (which carry out photosynthesis).

There are two main subphyla for the Euglenozoa: the kinetoplastids and the euglenids.

Kinetoplastids

The kinetoplastids are a group of biologically important obligate parasites: they are not free-living. This group includes some pathogenic species: the leishmaniasis-causing *Leishmania* species, and

The many uses of lichen

Lichens are used by humans in diverse ways. In the Arctic region, lichens are harvested as fodder for reindeer (Figure 1a). They are also used to flavour food in India and Nepal. Traditionally, lichens were used to prepare dyes (blue, red, brown, yellow) for cloth (Figure 1b). However, the major (and perhaps unexpected) use of lichens today is to help prolong the aromas in perfumes, cosmetics, and lotions.

Lichens are also exquisitely sensitive to sulphur dioxide levels in the atmosphere. This trait means that they are useful as biological indicators of atmospheric pollution. The more polluted an environment, the less diverse the lichen population becomes. Fruticose lichens, in particular, disappear in polluted areas.

Figure 1 (a) Lichen as a foodstuff: a reindeer eats lichen on the Arctic tundra. (b) Lichen as a dye: Harris Tweed® cloth, handwoven in the Outer Hebrides, Scotland, used to be dyed using lichen.
Source: (a) Arterra Picture Library/Alamy Stock Photo; (b) Toby Adamson/Alamy Stock Photo.

the protists responsible for causing African sleeping sickness and Chagas' diseases, the trypanosomes.

Kinetoplastid cells contain a full complement of the organelles typically associated with the eukaryotes but have one unique feature: a kinetoplast. The kinetoplast is an extensive network of mitochondrial DNA. An example is shown in Figure 11.55.

Euglenids

Euglenids possess chloroplasts, which means they are mostly autotrophic: they can produce their own sugars by photosynthesizing. (We discuss photosynthesis further in Chapter 7.) Linked to this, they also contain an eyespot, which functions as a light shield and a light detector that enables them to respond to light by using their flagellum to move towards it. Unusually, though, some euglenids are able to switch between being autotrophs and being heterotrophs: when light is available they use it to produce their own organic molecules but they consume nutrients from their environment when light is limiting.

Phylum Alveolates

The alveolates are a group of protists that possess small membrane-enclosed sacs known as alveoli under their plasma membrane. The alveolates are a common and abundant group of organisms that exhibit a diverse range of nutrition strategies. They include many photosynthetic species, as well as pathogenic species such as the malaria-causing protist, *Plasmodium*. The alveolates comprise three groups: the dinoflagellates, the apicomplexans, and the ciliates.

Dinoflagellates

The dinoflagellates are a group of organisms commonly found in aquatic ecosystems; an example is shown in Figure 11.56. Each of these species possesses two flagella, which enable dinoflagellates to move through their aquatic environment, typically in a whirling or spinning motion.

Approximately half of the almost 5000 known species of dinoflagellates are photosynthetic, making dinoflagellates important

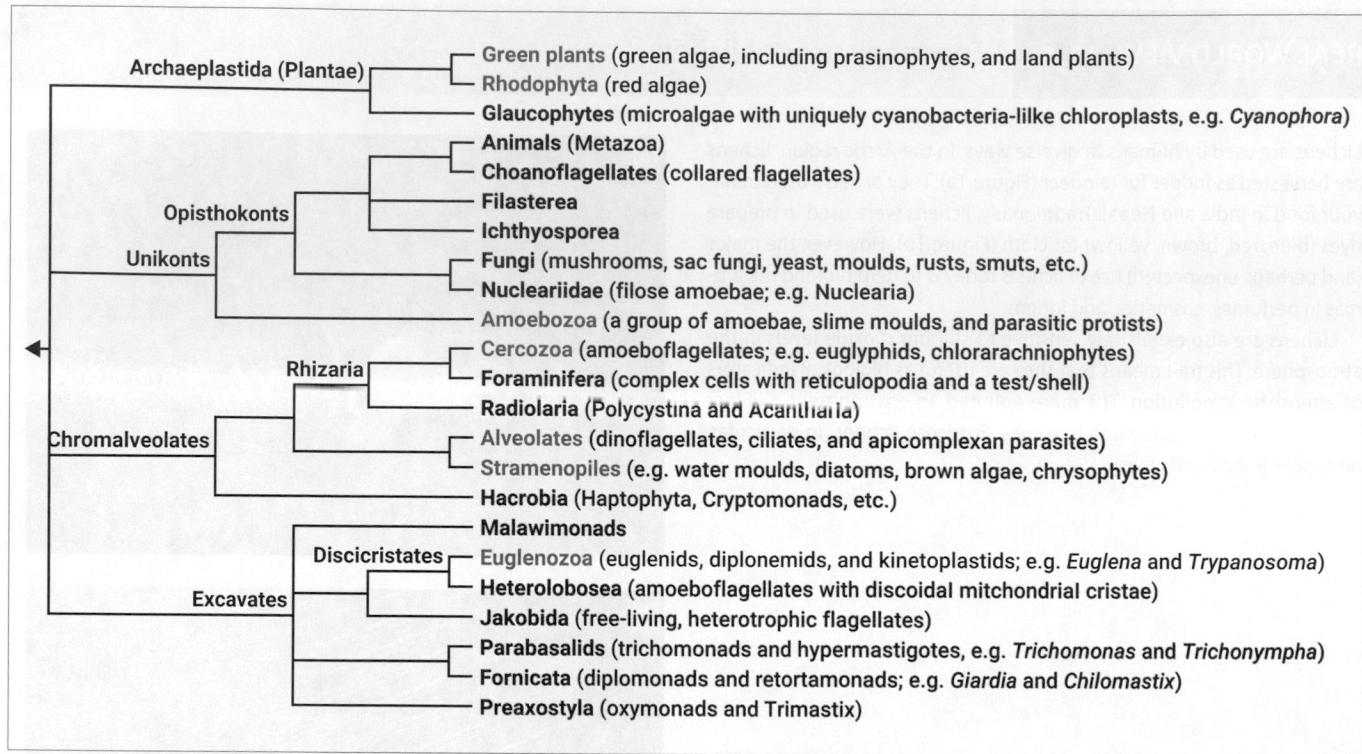

Figure 11.54 Eukarya. A phylogenetic tree for some of the major groups of known eukaryotic organisms from the kingdom Eukarya.

Source: From Keeling, P., Leander B. S., and Simpson, A. 2009. Eukaryotes, Eukaryota, Organisms with nucleated cells. Version 28 October 2009. http://tolweb.org/Eukaryotes/3/2009.10.28 in The Tree of Life Web Project, http://tolweb.org/.

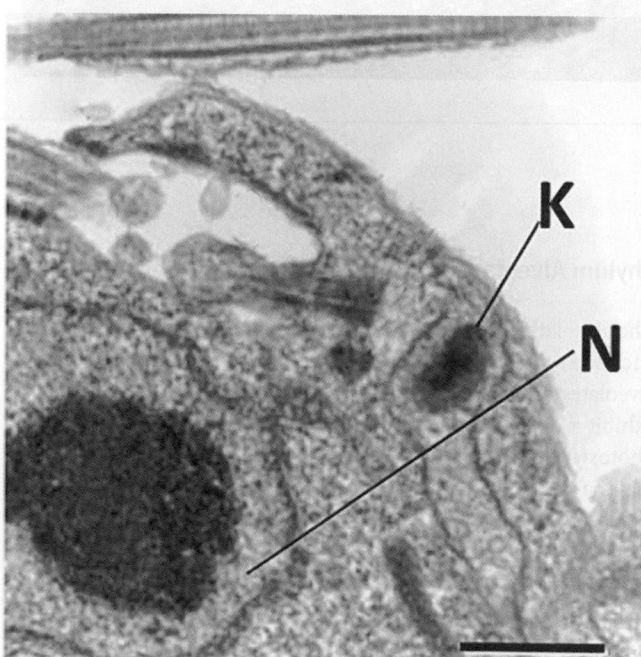

Figure 11.55 Kinetoplasts. A kinetoplast (K) in *Trypanosoma brucei*. N denotes the nucleus.

Source: Selvapandiyan, A., Kumar, P., Salisbury, J. L., Wang, C. C., and Nakhasi, H. L. (2012) Role of centrins 2 and 3 in organelle segregation and cytokinesis in *Trypanosoma brucei*. *PLoS ONE* 7(9): e45288. https://doi.org/10.1371/journal.pone.0045288.

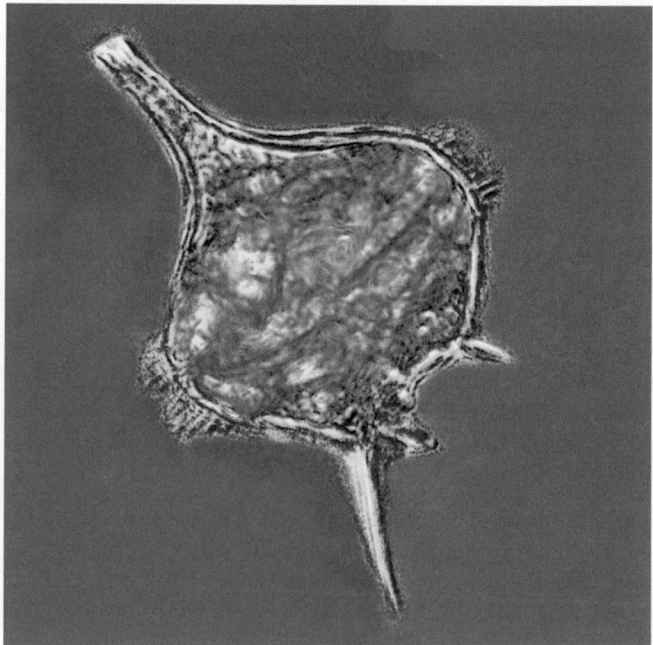

Figure 11.56 Dinoflagellates. A dinoflagellate viewed under the light microscope.

Source: Tintinnidguy/Wikimedia Commons/CC BY-SA 4.0.

producers in aquatic food chains, with most of the rest of this group being strict heterotrophs.

Populations of some species of dinoflagellate can undergo periods of rapid population growth, which produces algal blooms sometimes also known as red tides. These red tides can cause the death of large numbers of aquatic organisms as some dinoflagellate species produce toxins that are lethal to invertebrates and fish and, in some cases, to humans.

Some dinoflagellates are capable of producing bioluminescence, which can result in spectacular displays during a period of algal bloom. Figure 11.57 shows a red tide algal bloom by day and another by night where the bioluminescence of the dinoflagellates is very clear.

Apicomplexans

Estimates suggest that there could be as many as 10 million apicomplexan species but little more than 10,000 have been described. Of those known species, almost all are parasites of animals and the vast majority of these have a negative impact on their host. The most widely known apicomplexans are those that cause the disease malaria. These are a group of species within the *Plasmodium* genus, which collectively result in more than 1 million human deaths each year.

Other apicomplexans that have a significant impact on humans are *Toxoplasma gondii*, which is transmitted from cat faeces and can causes brain cysts in a human host, and *Eimeria* spp., which cause coccidiosis disease in a range of livestock, costing the global agricultural industry in excess of $2 billion each year.

Apicomplexans typically have complex life cycles, which may involve a series of intermediate hosts as well as those within which the protist reproduces. Figure 11.58 shows some of the characteristic features of these organisms seen at two life stages, including the unique apical complex, a feature that is used to classify organisms into this Apicomplexa group.

Ciliates

The ciliates are named for their possession of thousands of **cilia**: small hair-like projections, which cover part or all the surface of the organism. Some ciliates are large enough to be seen with the naked eye. For example, *Stentor* spp. can reach 2 mm in length. Movement of these cilia enables the ciliate to move and, in some cases, to feed. Many cilia are heterotrophs: they feed on other smaller protists or on bacteria.

Figure 11.59 shows some of the characteristic features of the ciliates, including the presence of two types of nuclei: tiny micronuclei and large macronuclei. Each ciliate will have one or more nuclei of each of the two types.

Phylum Stramenopiles

The Stramenopiles is a large group of organisms, mostly algae, characterized by the presence of a 'hairy' flagellum. In many species this hairy flagellum is also accompanied by a smooth, hairless flagellum (as shown in Figure 11.60).

The Stramenopiles include some of the most important photosynthetic organisms on Earth, particularly within aquatic systems where Stramenopiles can range from microscopic unicellular diatoms through to the huge macroscopic multicellular brown algae. Let us now continue our exploration of this group of organisms by focusing on the diatoms, brown algae, and golden algae.

Diatoms

Diatoms are photosynthetic algae. As such, they are responsible for carrying out one of the fundamental processes that enable the survival of life on Earth: carbon fixation through photosynthesis. They are an abundant group of organisms, present in almost every environment that is both wet and receives sunlight. If you were to take a sample of just 20 litres of surface seawater, it could contain millions of these microscopic

Figure 11.57 Algal blooms. (a) A red tide seen off the coast of San Diego, USA, during the day; (b) a separate bloom, also off San Diego, taken at night in May 2018 shows the bioluminescence of the dinoflagellates. This red tide is the result of an explosion in the population numbers of dinoflagellates, including the species *Lingulodinium polyedra*.

Source: (a) James R. D. Scott/Getty Images.

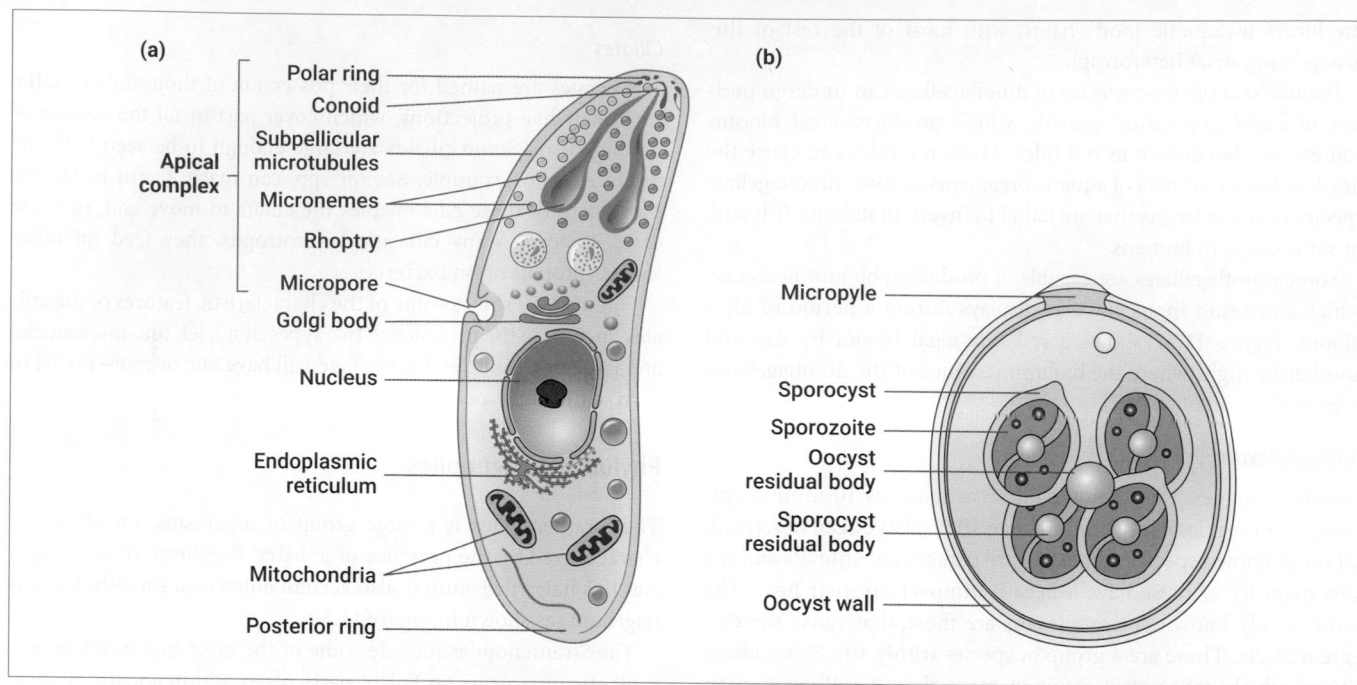

Figure 11.58 Two of the life stages of a typical apicompolexan cell. (a) The merozoite (sexual form) stage and (b) the oocyst.

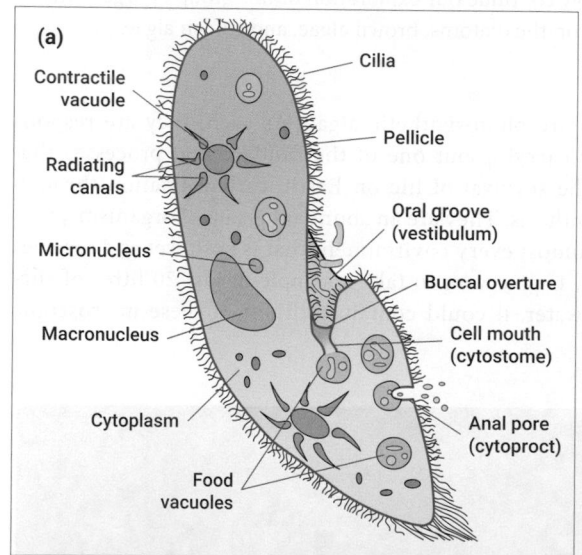

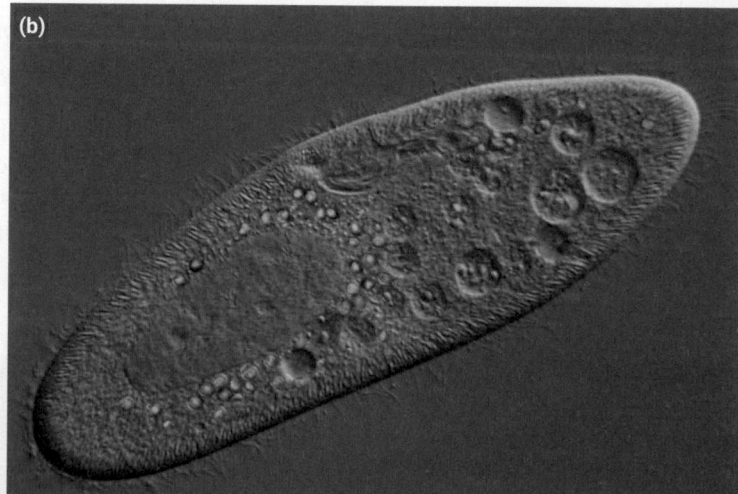

Figure 11.59 Characteristic features of ciliates. (a) The key internal features of a ciliate organism known as a *Paramecium*; and (b) a *Paramecium* as seen under a light microscope.

Source: (a) Dorling Kindersley Ltd/Alamy Stock Photo; (b) Deuterostome/Wikimedia Commons/CC BY-SA 4.0.

aquatic organisms. Diatoms depend on light to drive photosynthesis and they are vital to life on Earth in this regard: they fix more atmospheric carbon than all of the world's tropical forests combined!

Look at Figure 11.61 and notice how diatoms are diverse in structure. An unusual feature that all diatoms have in common, however, is that they possess a cell wall that is rich in silica, known as a frustule. Silica is the main component of glass and so diatoms are sometimes referred to as 'algae living in a glass house'. The frustule is composed of two halves that fuse together. Each half contains many perforations that enable the exchange of materials across this otherwise impermeable layer of protection.

Brown algae

Also known as Phaenophytes, brown algae are photosynthetic Stramenopiles that range in size from the microscopic through to the giant bladder kelp *Macrocystis pyrifera* that can be found across the eastern Pacific and southern oceans, as depicted in Figure 11.62.

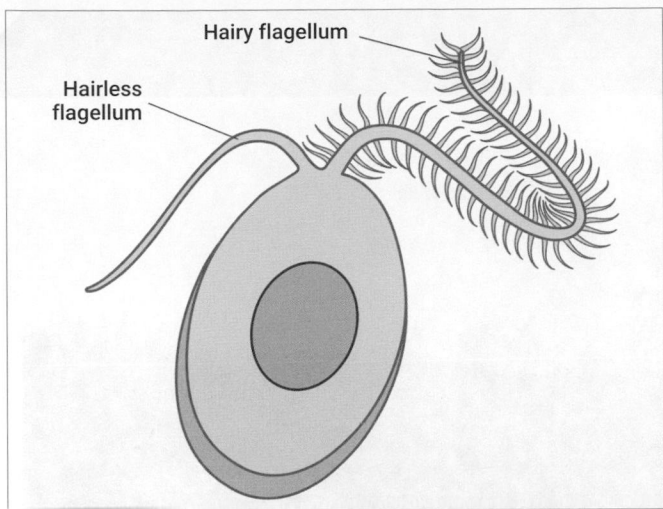

Figure 11.60 The typical arrangement of flagella within Stramenopiles.

Source: C. Rye, R. Wise, V. Jurukovski, et al. OpenStax, 'Biology' 2016. © Sep 15, 2020 OpenStax. Textbook content produced by OpenStax is licensed under a Creative Commons Attribution License 4.0 licence. Access for free at https://openstax.org/books/biology/pages/1-introduction.

Macrocystis pyrifera is one of the fastest-growing and largest organisms on Earth: it can grow up to 60 cm in a single day to reach an overall length of 45 m or more.

All brown algae species are multicellular, although many species have unicellular spores in their lifecycle. The plastids found in brown algae species originated from a secondary endosymbiosis event that involved an ancestral red algal cell. (We learn more about red algae later in this section.) These plastids, descended from red algae, possess carotenoids as their photosynthesizing pigments. These pigments reflect orange and red light strongly, which explains why these photosynthesizing organisms are brown rather than the green we might usually associate with this process.

Golden algae

The approximately 1200 known species of golden algae can be found across both marine and freshwater habitats. The majority of these species are unicellular organisms that possess two flagella

Figure 11.62 The giant bladder kelp, *Macrocystis pyrifera*.

Source: Velvetfish/istockphoto.com.

for movement. However, some species may have just the one flagellum, live as colonies, or be filamentous, while some are even encased in a silica cyst. These characteristics can be used to distinguish between different species.

Golden algae have two *unique* characteristics: the possession of a yellow pigment known as fucoxanthin, which gives many of these organisms their golden hue, and the use of oil droplets as a store of energy.

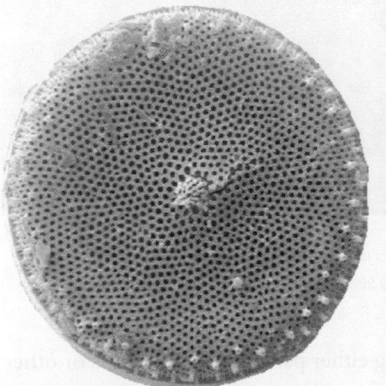 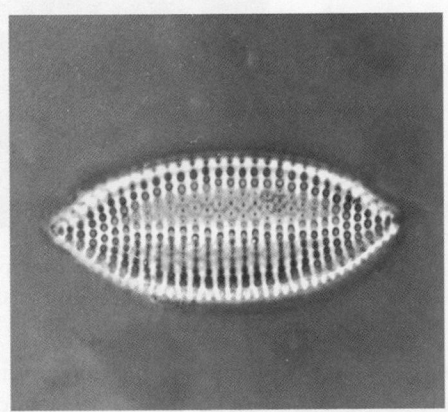 

Figure 11.61 Diatoms. Examples of the structural diversity of diatoms with the hard, porous frustule visible in all images. In the right-most image, both halves of the frustule can be clearly seen.

Source: University College London 2002. All rights reserved.

Tackling toxic algae

Scientists from the University of East Anglia and the John Innes Centre in Norwich, UK, have found a simple solution to tackle the problem of algal blooms caused by a species of golden algae, *Prymnesium parvum*. This particular species of algae does not produce a bloom that is obviously visible. However, over a time period of a few days or less, they can produce sufficient toxins to kill large numbers of the fish who happen to occupy the same waterways as the algae—in this case, brackish habitats such as that found within the Norfolk Broads National Park (see Figure 1)

After a bloom of *P. parvum* in 2015, nearly 750,000 fish were rescued from one region of this National Park and released into an area that was not experiencing a bloom. Following this bloom and subsequent rescue of fish, scientists started to conduct experiments to see if hydrogen peroxide could be used in sufficiently low quantities to be able to prevent the algal blooms but without impacting on the fish or macro-invertebrate populations. So far the results have been promising, with the hydrogen peroxide proving to be a labour-efficient breakthrough and local anglers reporting that fish populations are also recovering.

Figure 1 Toxic algae. The devastating effect on wildlife of a *Prymnesium parvum* bloom. Upper Thurne, Norfolk, 2015.
Source: Photo taken by M. Rejzek.

Golden algae can use their flagellum/flagella to move within their aqueous habitat to stay within the light-penetrating zone, enabling them to maximize the opportunity for photosynthesis.

As we have seen with the dinoflagellates, some species of golden alga can reproduce rapidly to form an algal bloom under the right environmental conditions. We discuss recent research around these golden algae blooms in Real World View 11.6.

What phylum of protists includes a group of organisms that possess a frustule? What is the frustule made of?

Answer: Frustules are primarily made of silica and are seen in the diatoms, which are organisms within the Stramenopiles phylum.

Phylum Cercozoa

The cercozoa are a group of protists with a diverse range of forms, with some having developed an outer shell made of silicone-containing scales (see Figure 11.63). All of this group appear to have descended from a predatory amoeboflagellate, and have evolved into species consisting primarily of amoeboflagellates and

Figure 11.63 Cercozoans. A species from the Imbricatea group on the surface of which the silica scales are clearly seen.
Source: NEON/Wikimedia Commons/CC BY-SA 2.5.

flagellates, which move using either pseudopodia, flagella, or other forms of podia such as reticulopodia.

Some cercozoans underwent an endosymbiotic event whereby they engulfed a photosynthesizing algae. Subsequently, they retained the photosynthetic abilities of their intracellular companion.

Phylum Amoebozoa

There are close to 2500 known species of amoebozoa. These organisms share common features that include the **endoplasm** and **ectoplasm**. The endoplasm is the fluid, granular interior of the organism—effectively the densest region of a cytoplasm. The ectoplasm is the more watery region of the cytoplasm and is found adjacent to the plasma membrane. These two regions of the cytoplasm work together to facilitate locomotion: the endoplasm flows forwards and the ectoplasm flows backwards to generate a form of intercellular propulsion.

Amoebozoans also have tubular or lobed pseudopodia, which further facilitate movement and also feeding. For most amoebozoa this feeding involves the pseudopodia engulfing food particles, which the organism then takes in through a series of vacuoles in a process known as phagocytosis.

There are three main groups of amoebozoans: the slime moulds, tubulinids, and entamoebas. Let us now consider each of these in turn.

Slime moulds

The slime moulds are a small group or organisms, comprising approximately 700 species. (They are not related to fungi despite their name.) These organisms have no commercial significance, but scientists study them as they undergo fascinating morphological changes from single cell to multicellular forms. They are saprophytes and live on rotting plant material, such as leaves and decaying wood.

Slime moulds fall into two groups: the acellular slime moulds and the cellular slime moulds. Acellular slime moulds, such as *Physarum*, exist as a mass of protoplasm called a **plasmodium**, which in some ways resembles an amoeba. The plasmodium flows over surfaces (e.g. woody branches), as can be seen in Figure 11.64, and engulfs food such as bacteria and fungal spores. They can be very brightly coloured and spread over large areas. The amoeboid motion comes from streaming of the cytoplasm, which is facilitated by the protein actin.

The plasmodium mass is diploid and there are no separate cells. Under adverse environmental conditions, such as drying or nutrient starvation, a structure called a sporangium can be produced, inside of which haploid spores are made via meiotic division. These spores can then be released and germinate to form swarm cells. When swarm cells fuse, the diploid plasmodium re-forms.

The cellular slime moulds include *Dictyostelium discoideum*, which is a popular model organism as it has genuine multicellularity as part of its life cycle and could shed light on the evolution of multicellular organisms. Cellular slime moulds exist as single cell amoeba-like organisms. However, they can aggregate and then move as a cell mass, which is called the **pseudoplasmodium**. This is different to the plasmodium of acellular slime moulds as the individual cells within the pseudoplasmodium have cell membranes.

Starved cells produce signal molecules that attract other cells, which then aggregate to form a large slug-like mass. The slug forms a fruiting body called a **sorocarp**, as depicted in Figure 11.65. The sorocarp releases spores that then germinate into amoeboid-like cells.

Figure 11.64 Slime mould. The acellular slime mould *Physarum* spreading over dead wood.
Source: Nature Picture Library/Alamy Stock Photo.

Tubulinids

Tubulinids are a broad, diverse group of amoebozoans that are most commonly present in soil, but they are also found in aquatic environments, both freshwater and marine. Tubulinids have either lobe- or tube-shaped pseudopodia, which they tend to use to help them locate, move towards, and engulf their prey. Most tubulinids, such as the tubulinid amoeba, shown in Figure 11.66, are heterotrophs, and eat other protists or bacteria.

 Go to the e-book to watch a video that illustrates amoeba movement.

Archamoebae

The Archamoebae is a group of protists that are often internal parasites or are commensal within other, larger-bodied, animals. The most widely known archamoebae are probably the *Entamoeba* spp., shown in Figure 11.67. The *Entamoeba* are parasites of the human gut and can cause the disease amoebic dysentery.

Non-parasitic, non-commensal archamoebae typically possess a single nucleus and flagellum (unlike their parasitic cousins). However,

Figure 11.65 Sorocarp. Sorocarps being produced from the slug mass of *Dictyostelium discoideum*, a cellular slime mould.

Source: Usman Bashir/Wikimedia Commons/CC BY-SA 4.0.

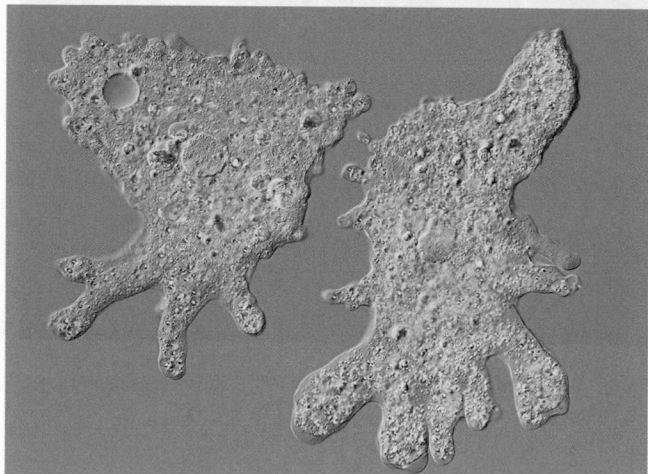

Figure 11.66 Tubulinids. Tubulinid amoebas.

Source: Lebendkulturen.de/Shutterstock.com.

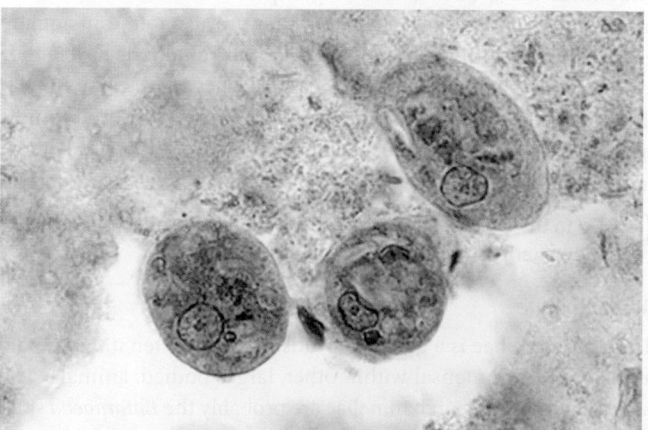

Figure 11.67 Archamoebae. *Entamoeba coli*, a common non-pathogenic protist found within the human intestinal tract.

Source: Image from www.atlas-protozoa.com, courtesy Stefano Laganà © medical-itc.it.

the very large amoeba *Pelomyxa* goes against this: each of these unicellular organisms possesses many nuclei and multiple flagella.

Algae

The term 'algae' is not a true phylogenetic group. Organisms referred to as algae come from multiple lineages so they do not share a common ancestor. However, they do all share the presence of plastids which enable these organisms to photosynthesize. In this regard, some species of algae are quite closely related to land plants. Others, such as the brown algae we have explored already, are genetically very different from land plants, and experienced a secondary endosymbiosis event in their evolutionary past that gave rise to the presence of plastids within their cells.

We will conclude this section by considering the closest algal relatives of land plants, the species of red and green algae.

Red algae

Red algae are a group of around 7000 known species of multicellular algae that solely exist in aquatic environments. The vast majority (over 6500 species) live in the marine environment and the remaining 5 per cent live in freshwater habitats. Red algae depend on their aqueous surroundings to transport their immobile sperm to the female organs to facilitate their reproduction.

Red algae are also known as the rhodophytes (from the Greek *rhodos*, meaning red): they derive their name from phycoerythrin, one of the photosynthetic pigments that they contain, which strongly reflects red light. In addition to phycoerythrin, red algae possess other photosynthetic pigments such as phycocyanin. Together, the pigments enable these organisms to absorb light across the blue and green portions of the visible spectrum, a region that is able to penetrate deeply into water.

The quantities of phycoerythrin present vary with habitat: species that occupy the deeper aquatic environments possess greater quantities of this red pigment and thus appear a much darker shade of red. In fact, some species appear almost black.

Figure 11.68 Nori. *Porphyra* spp.
Source: Nature Picture Library/Alamy Stock Photo.

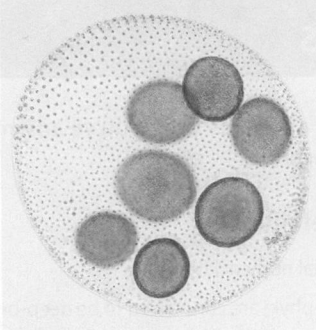

Figure 11.69 A colonial algae. *Volvox globator*, a colony of alga cells, each of which possess two flagella
Source: Nnehring/istockphoto.com.

Some species of red algae have a long tradition of being consumed by human societies across much of the world. For example, a group of red algae known as nori (*Porphyra* spp.), depicted in Figure 11.68, has been cultivated in Japan for many centuries and is one of the main 'seaweeds' used in wrapping sushi rolls, as well as being a traditional snack food.

Green algae—the chlorophytes and charophytes

There are two main groups of green algae: the chlorophytes and the charophytes. In common with the land plants, charophytes and chlorophytes both possess the photosynthetic pigment chlorophyll plus other accessory pigments such as carotenoids. They also produce and store sugars as starch granules, and possess cell walls comprising primarily cellulose fibres.

The charophytes are complex algae, which are now known to be the ancestors of land plants, with some species being more closely related to some of the Plantae species than they are to other charophytes!

Charophytes often inhabit the fringe regions of aquatic environments, such as along rocky shores or around the edges of freshwater bodies such as lakes, all of which experience dry periods. As a result, individuals that can tolerate these dry spells will survive better than those who cannot. Over time, this can lead to organisms being able to survive for longer periods in drier conditions and eventually can lead to life as a land plant.

Charophytes are able to tolerate these dry periods as a result of the presence of **sporopollenin**, a tough polymer found in the walls of plant spores and pollen grains, which can protect charophyte gametes from drying out.

The chlorophytes, on the other hand, can be unicellular or multicellular and, in the case of *Volvox* (Figure 11.69), can even be a colonial algae. Found only in aquatic environments and occasionally in damp environments such as the soil, these organisms may live independently or symbiotically within other organisms. Like all of the groups of algae, they are important producers within their environments. Unlike many other algae species, however, some chlorophytes have formed close symbiotic relationships with other organisms such as hydra and fungi-forming lichens.

 Check your understanding of the concepts covered in this section by answering the questions in the e-book.

SUMMARY OF KEY CONCEPTS

- Microbes exhibit vast diversity across all three domains of life.
- Bacteria belong to their own diverse domain, and are capable of utilizing complex molecules.
- Bacteria can inhabit a wide variety of different habitats, including seemingly inhospitable environments.
- Sequencing and metagenomics has opened up our understanding of unculturable micro-organisms.

- The Archaea are extremely diverse, but their ability to live in extreme environments makes many of them unculturable.
- Fungi are eukaryotic organisms. They can grow in a filamentous hyphal form or they can exist as single cells (yeasts).
- Fungi can undergo both asexual and sexual reproduction and can form a range of symbiotic relationships with other organisms.
- Protists are eukaryotic organisms, which occupy a diverse range of habitats.

 Use the flashcards in the e-book to test your recall of key terms introduced in this chapter.

QUESTIONS

Looking for answers? Once you've answered these questions, follow the link in the e-book to the answer guidance and check your work.

Concepts and definitions

1. What is binomial nomenclature?

2. What bacterial phyla are considered to be deep-branching?

3. Bacteria of what phylum are capable of photosynthesis?

4. What is the largest bacterial phylum?

5. What is the difference between the Firmicutes and the Actinobacteria in terms of their GC content?

6. Name the five fungal phyla. Which phylum contains the largest number of species, and which is the newest phylum?

7. Describe some of the habitats where protists are known to live.

Apply the concepts

8. Explain how Woese and Fox created the tripartite phylogenetic tree and why it is still used today.

9. Why are many Archaea unculturable?

10. Why do fungi have both asexual and sexual reproduction?

Beyond the concepts

11. Do you agree that we live on a planet that is dominated by microbes?

12. Do you think there should be an effort to identify all microbial species?

13. Why do you think fungi feature so heavily in folklore?

14. How are slime moulds able to contribute to our understanding of complex multicellular life?

FURTHER READING

https://theconversation.com/the-cities-of-the-future-could-be-built-by-microbes-63545
An interesting read on how the cities of the future could be built by microbes.

https://theconversation.com/strange-microorganism-under-the-sea-may-be-missing-link-in-evolution-41445
An exploration of how the microbes being discovered can shed light on evolution

Locey, K. J. & Lennon, J. T. (2016) Scaling laws predict global microbial diversity. *PNAS* **113**: 5970–5.
A paper that looks at scaling laws which can help to predict global microbial diversity. Their modelling suggests that Earth is home to more than 1 trillion microbial species.

Lopez, P., Halary, S., & Baptiste, E. (2015) Highly divergent ancient gene families in metagenomic samples are compatible with additional divisions of life. *Biol. Direct* **10**.
This paper looks at how current techniques of using phylogenetic trees might be missing novel lineages of organisms in the 'microbial dark universe'.

The Growth, Measurement, and Visualization of Cells

LEARNING OBJECTIVES

By the end of this chapter, you should be able to:

- Explain why sterility is important to ensure that your culture only contains the cells under investigation, and why a septic technique is an important part of maintaining sterility.

- Explain that many experiments require us to culture cells in the laboratory in order to understand them.

- Recall that cells can be grown in either complex or defined media, which at a minimum provide the essential nutrients needed for growth and reproduction.

- Describe the different phases of growth that cells undergo when grown in batch culture.

- Recall that a cell population can be measured using a variety of techniques, such as the use of optical density, dry weight, or the use of a haemocytometer.

- Explain that cells are generally so small that they can only be visualized by the use of microscopes.

Chapter contents

Introduction	469
12.1 Maintaining a sterile environment	470
12.2 Obtaining and using microbial culture collections	473
12.3 Culturing cells from multicellular organisms	474
12.4 Measuring the growth of microbial cells	479
12.5 Visualizing cells	488

Watch the key concepts video in the e-book to prepare yourself for studying this chapter.

Introduction

As we have seen in previous chapters in this module, diverse cell types exist across all the domains of life. They vary in size, shape, complexity, composition, and function. But how do we know about cells and what they are made of? While cells can be studied *in vivo* (within a whole living (multicellular) organism), much of what we have learnt has come from observations and experiments on cells carried out *in vitro* (separate from the body or in laboratory conditions). This chapter explores these laboratory techniques and equipment, and how we use them to study cells and cellular processes.

12.1 Maintaining a sterile environment

Before cells of any type can be cultured in a laboratory, it is important to understand why sterility is important and how to maintain sterility through **aseptic technique**. A sterile environment is one that contains no living organisms; sterilization is the process by which all living organisms are either killed or removed in order to render an environment sterile.

Good aseptic technique is crucial to master, as it enables you to work safely and protect yourself and others, and to ensure that you do not contaminate your cultures with unwanted microbes.

At a minimum, aseptic technique requires you to work around a Bunsen burner flame. The flame heats the air above and creates an updraft which microbes cannot enter. This enables the working area to remain sterile. Bottlenecks of culture flasks are flamed prior to and post inoculation as shown in Figure 12.1a. Sterile loops are heated until red-hot before inoculating (Figure 12.1b), or glass spreaders are initially dipped in 70% (v/v) ethanol before passing through a flame to sterilize. Some processes require working in a biological safety cabinet.

Before you inoculate growth media with a bacterium (or other cell types), you need to be sure that your culture flask, the media you are using, and any additives to your media are sterile. Sterilization can be achieved in a number of ways:

1. Using heat (as we see in the examples above).
2. Filter sterilization of heat labile liquids.
3. Irradiation.
4. Detergents.

Let us now consider each of these in turn.

Sterilization with heat

The technique of pasteurization uses heat to kill unwanted organisms, but it is not sterilization, as pasteurization only reaches a temperature of 72 °C for 15 s. Pasteurization is performed on milk as the temperature is enough to kill organisms such as *Brucella abortus* and *Mycobacterium tuberculosis* but not high enough to alter the taste of the milk.

Most bacterial cells can be killed at temperatures of around 70 °C, but some bacteria can produce endospores which can germinate to form new bacterial cells, and these are really tolerant to heat. So how are these endospores destroyed? To do so we need to heat under pressure. In the laboratory this is done using an autoclave, which is like a big pressure cooker; you can see an autoclave in Figure 12.2.

▶ **We learn more about endospores in Section 9.2.**

Typically, culture vessels and media are heated in an autoclave to 121 °C at a pressure of 15 pounds per square inch (PSI) for 20 min. Before equipment and media are sterilized, the air is removed from the autoclave, so that the atmosphere is just composed of steam. A wet, steamy environment is important as water penetrates the 'contaminating' cells.

To ensure the correct temperature and pressure has been reached, culture vessels can be marked with autoclave tape, which will change colour if the correct conditions have been applied. Some materials that you want to use for growth, such as soil, might need several rounds of sterilization, with resting in between; this process allows endospores to germinate into vegetative cells, which can then be killed.

Sterilization with irradiation

Ionizing and non-ionizing radiation can both be used for sterilization. The most common non-ionizing radiation used is ultraviolet (UV) light. Absorption of the energy present in UV light by the cells causes chemical bonds to rupture so critical cellular functions cannot take place. UV lamps can be found in food

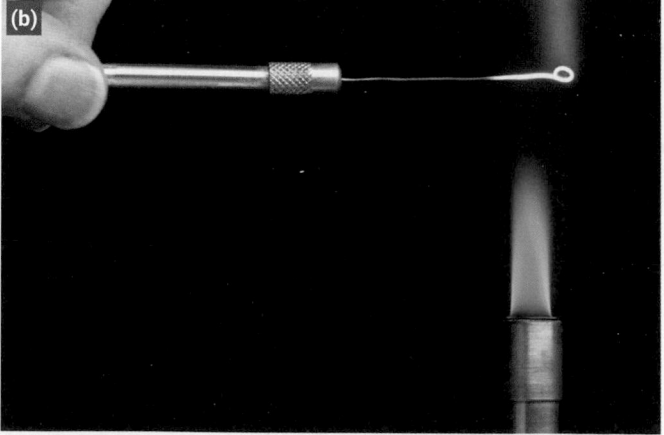

Figure 12.1 Aseptic technique. (a) The sterilization of a culture flask neck using a Bunsen burner flame. (b) A sterile loop is heated until red-hot before use.
Source: (a) Trevor Clifford Photography/Science Photo Library; (b) Martyn F. Chillmaid/Science Photo Library.

Figure 12.2 A typical autoclave used to sterilize solutions and media.

Source: Timof/Shutterstock.com.

preparation areas, operating theatres, and tissue culture facilities. Ionizing radiation has a shorter wavelength and, therefore, more energy, which results in greater penetrating power. For example, gamma radiation is used to sterilize surgical equipment, syringes, and catheters.

Sterilization by filtration

Filtration can sterilize liquids or gases. Filtration is often carried out when the liquid or gas has heat-sensitive components, which would be destroyed if subjected to the conditions in an autoclave. A good example would be an antibiotic solution which needs to be added to a microbial growth medium.

Filters of different pore sizes can be used and are usually made from nitrocellulose or polycarbonate. A pore size of 0.22 μm is commonly used as this will remove bacteria and also larger cells, such as yeasts. Filters with smaller pore sizes than this are needed to remove mycobacteria and viruses. Sterilized filter units can be purchased quite cheaply and used with sterile syringes. You can see a typical filtration unit in Figure 12.3.

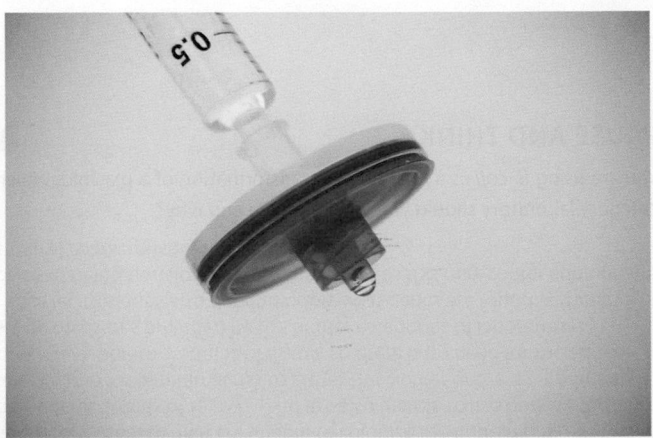

Figure 12.3 Filtration unit and syringe.

Source: DarwelShots/Shutterstock.com.

Other filters such as high efficiency particulate air (HEPA) filters are used in biological safety cabinets and laminar flow hoods to sterilize air coming into and out of the cabinets.

Sterilization with chemicals

The gas ethylene oxide can be used to sterilize large items of medical equipment (also dressings and mattresses). It is also used in the food industry as an antifungal fumigant for items such as dried fruits, nuts, and spices. The ethylene oxide is used in a specialized chamber filled with the gas in a humid atmosphere at 40–50 °C.

Disinfection

While sterilization kills all living organisms, some organisms are able to survive treatment with chemical disinfectants. Disinfectants, which are used to treat work surfaces and floors, can contain a variety of chemicals such as alcohols, halogens, phenolics, and surfactants. Common disinfectants that you might find in the laboratory as an undergraduate include the phenolic-based product Hibiscrub.

Microbial safety cabinets

The UK Health and Safety Executive define a micro-organism as '*a microbiological entity, cellular or non-cellular, which is capable of replication or of transferring genetic material*'. It is important to recognize that this definition includes viruses, which are non-cellular and can transfer genetic material.

Micro-organisms are grouped according to their pathogenicity and risk of infection to humans. These groups, together with representative examples, are described in more detail in Table 12.1. Working with Group 2 organisms and above requires the use of a microbial safety cabinet, as shown in Figure 12.4a. Different types of cabinet are used according to the organism group being handled. Microbial safety cabinets and laminar flow cabinets are equipped with high-efficiency particulate-absorbing (HEPA) filters. In laminar flow cabinets (Figure 12.4b), the air is drawn through the filter in the back, enabling sterile air to flow through to the front, as depicted in Figure 12.4c. They are useful for pouring petri dish plates in sterile conditions, but must not be used when handling microbes because microbial aerosols could be pushed towards the user. In the microbial safety cabinet, air is taken out of the cabinet,

Table 12.1 Biological hazard as defined by the UK Health and Safety Executive

Hazard level	Description	Example
Group 1	Unlikely to cause human disease	*Lactobacillus casei* (used to ferment foods)
Group 2	Can cause human disease and may be hazardous to employees; it is unlikely to spread to the community and there is usually effective prophylaxis or treatment available	*Bordetella pertussis* (causes whooping cough)
Group 3	Can cause severe human disease and may be serious hazard to employees; it may spread to the community, but there is usually effective prophylaxis or treatment available	*Bacillus anthracis* (causes anthrax)
Group 4	Causes severe human disease and is a serious hazard to employees; it is likely to spread to the community and there is usually no effective prophylaxis or treatment available	Zaire ebolavirus (causes ebola)

Source: Health and Safety Executive.

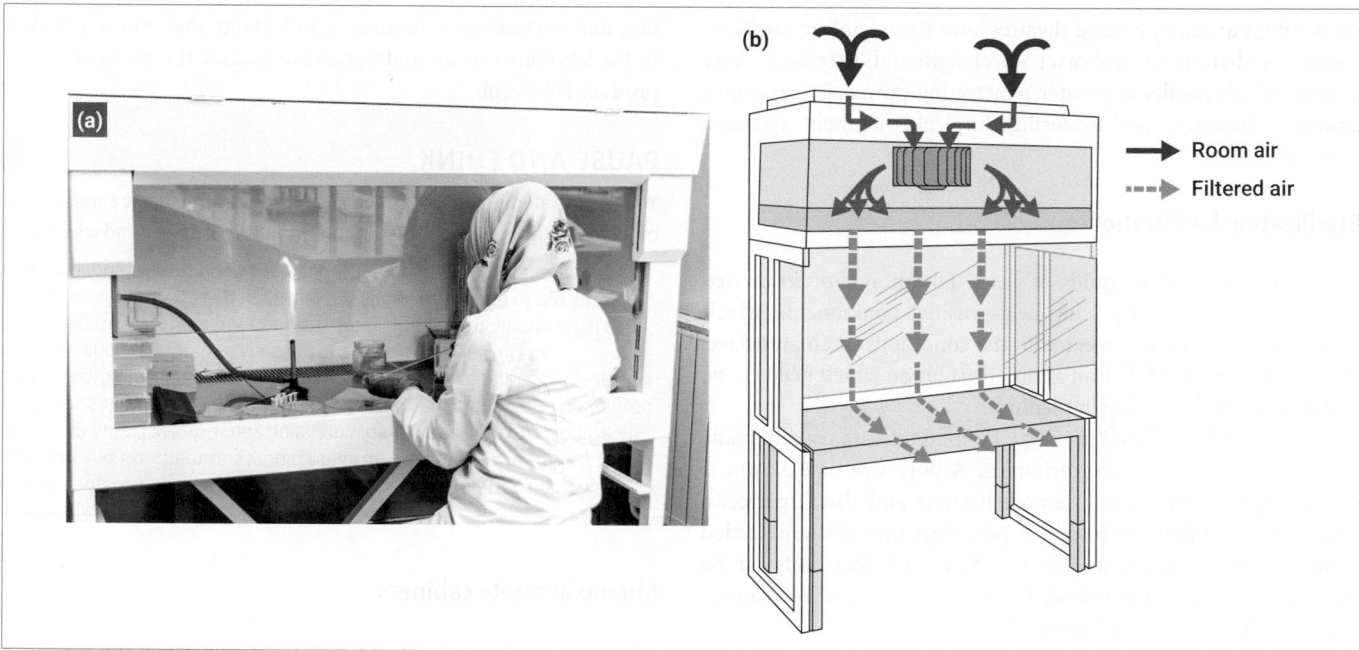

Figure 12.4 Laminar flow cabinet. (a) Laminar flow cabinet and (b) the direction of air flow in a laminar flow cabinet. Notice how filtered air is pulled towards the user.
Source: (a) Muamer Osmanbegovic/Alamy Stock Photo.

which is then passed through the HEPA filter, trapping microbial and viral particles.

Laboratory categories

Professional microbiologists and scientists working with genetically modified organisms work in different levels of containment depending on what it is they are working with: the higher the level, the greater the risk to health posed by the organisms being handled. These are known as Laboratory categories 1–4 and are described in Table 12.2. It is important to know what category level you are working in; as an undergraduate student you will be typically be working in either category 1 or 2 laboratories.

PAUSE AND THINK

You are using *E. coli* as a host for the transformation of a plasmid. Which category laboratory should you be working in and why?

Answer: Biosafety level 1, providing the strain of *E. coli* being used is a laboratory strain (which are unable to colonize the human gut—unlike pathogenic strains of *E. coli*—and are also unable to live outside the laboratory). There is a minimal risk to either the worker or the environment posed by laboratory strains of *E. coli*, so there is no need for the added safety measures provided by the higher categories of laboratories. However, this does also depend upon what genes are found within the plasmid being transformed into the *E. coli*, as this may confer additional survival characteristics to the bacteria.

Table 12.2 Different laboratory category levels

Laboratory category (risk group)	Biosafety level	Laboratory type	Laboratory practices	Safety equipment
1	Basic Biosafety Level 1	Basic teaching, research	Good Microbial Technique (GMT)	None, open bench work
2	Basic Biosafety Level 2	Primary health services; diagnostic services, research	GMT plus protective clothing and biohazard sign	Open bench plus biological safety cabinet (BSC) for potential aerosols
3	Basic Biosafety Level 3	Special diagnostic services, research	As level 2 plus special clothing, controlled access, directional airflow	BSC and/or other devices for all activities
4	Maximum containment	Dangerous pathogen units	As level 3 plus airlock entry, shower exit, special waste disposal	Class 111 BSC or positive pressure suits in conjunction with Class 11 BSC, double-ended autoclave (through the wall), filtered air

Source: Reprinted from *Laboratory Safety Manual*, Third Edition, 1. General Principles, Table 2. World Health Organization, 2004.

 Check your understanding of the concepts covered in this section by answering the questions in the e-book.

12.2 Obtaining and using microbial culture collections

Having now considered good laboratory technique and other important factors that relate to handling biological materials in the lab, let us now explore one context in which such techniques and factors are important: when wishing to grow (or 'culture') micro-organisms for further study: the production of microbial culture collections.

Microbial culture collections represent a community of microbes that have been isolated from a particular environmental niche, and subsequently grown (that is, cultured) in the lab to allow the individual species present to be identified, and be subsequently studied in more detail (or used for different purposes). As such, microbial culture collections are not just static archives that exist purely for curiosity: they are an incredibly important resource for furthering research and understanding.

A number of microbial culture collections are housed around the world. These include the National Collection of Type Culture (NCTC) in London and the National Collection of Industrial and Marine Bacteria in Aberdeen. Scientists internationally can deposit samples into these collections, and withdraw samples from them for use in their own research.

How are microbial culture collections produced?

In addition to using national culture collections, scientists may also want to isolate their own culture collections from specific environmental niches. A good example is the collection of *Streptomyces* spp. bacteria from the leaf cutter ants (see Clinical Box 13.1). But how would you go about doing this?

Firstly, you need to have an environment from which you want to isolate the microbes, and you need to decide whether you want to attempt to gather a representation of the overall **culturable** microbial population. (It is important to remember at this point that the majority of microbes are '**unculturable**', as we note in Chapter 11, so you are not going to be able to collect every single type of microbe from your environment.)

As an example, let us imagine that you want to obtain a culture collection of fungi that can trap nematode worms. Nematode-trapping fungi are found in many different environments, but a woodland soil would be a good starting point. You need to decide on the medium for the cultivation: different media will select for different organisms. In this case, a good starting medium would be potato dextrose agar (PDA) as this supports the growth of a wide range of fungal species.

A range of woodland soil samples of known locations would be collected and weighed. A 1 g sample of soil would be put into a known volume of sterile distilled water, or perhaps a buffer solution such as Ringers. A dilution series would be prepared and then a known volume of selected dilutions spread across the surface of the PDA plates. (We discuss how to perform a dilution series in Section 12.4.)

You would incubate your fungal isolation plates at 25–30 °C. Colonies of fungi would start to appear after approximately 48 h. When the fungal colonies have been established, they need to be purified using a method called hyphal tipping, a technique that involves isolating the tips of growing hyphae and culturing them on further PDA plates.

You might need to do several cycles of hyphal tipping until you are confident that you have obtained a pure isolate. Your purified isolates are now your own culture collection, much like the one shown in Figure 12.5.

You would then put your isolates through a screen to see if any of them were able to trap nematode worms. Those that could would then be examined in more detail—for example, to characterize the type of trap used, and environmental conditions that induce trap formation.

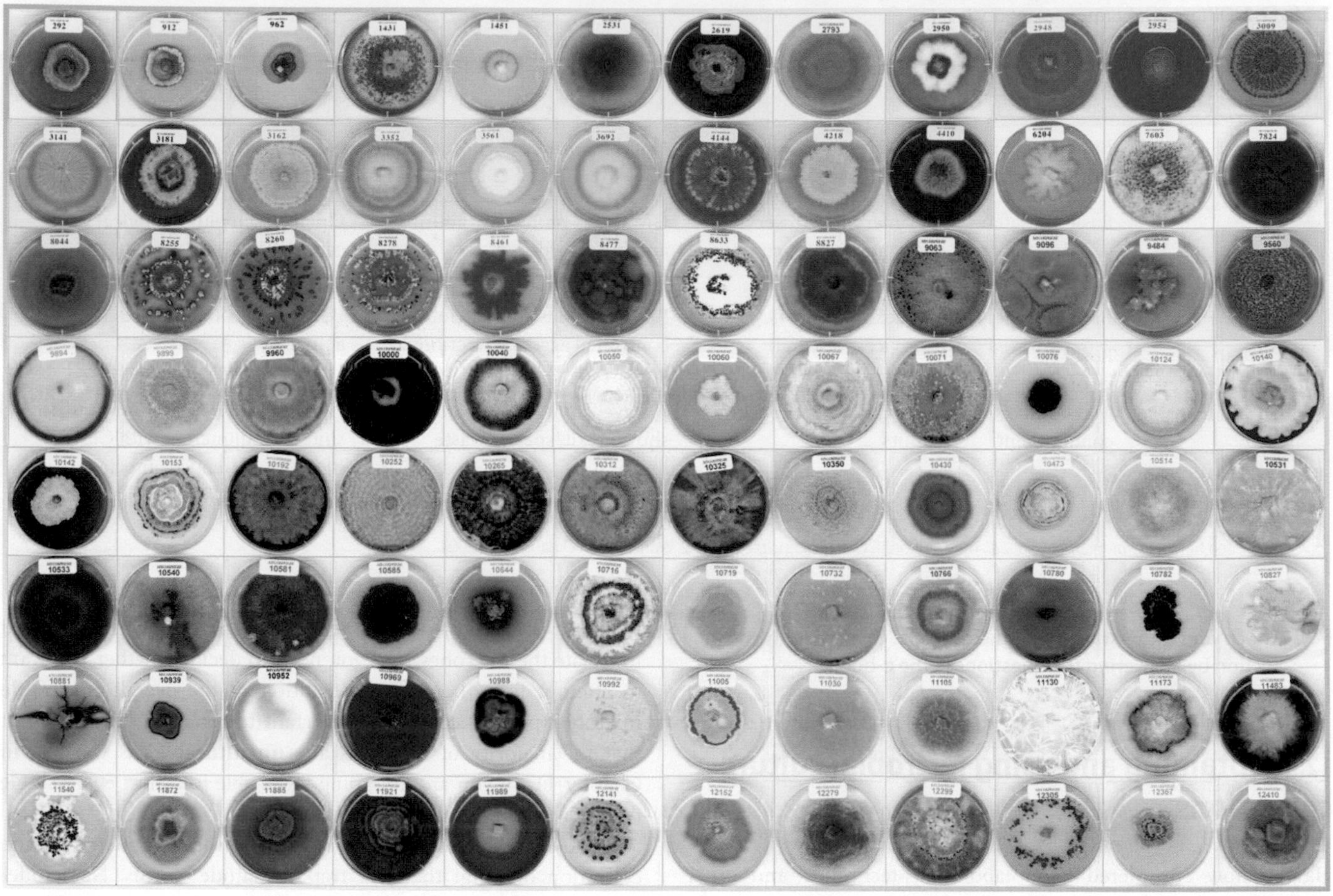

Figure 12.5 A fungal culture collection.
Source: Copyright © 2001–2018 Mycosphere. All rights reserved.

You would want to store your culture collection. For fungi, slopes of media can be made and inoculated with each isolate. They can then be covered with mineral oil and kept at 4 °C for long-term storage. Fungi can also be freeze-dried for storage, which brings with it the advantage that a large number of samples can be stored in a small space. You would number each isolate and keep a computer record of the isolate, where it was from, and the date it was isolated, together with the date that it was stored and the storage location.

If any of your isolates proved to be very interesting, you would seek to identify them. This is best done through ribosomal analysis, as we discuss in Section 11.1.

Check your understanding of the concepts covered in this section by answering the questions in the e-book.

12.3 Culturing cells from multicellular organisms

Many of the concepts and techniques we explored in Section 12.1 apply to the growth of cells from multicellular organisms.

Eukaryote cell culture

In order to study cells from a multicellular organism, it is important to be able to culture them *in vitro*—that is, to be able to grow them

Figure 12.6 A cell culture flask.
Source: Jens Goepfert/Shutterstock.com.

outside the body in some sort of petri dish or flask (as depicted in Figure 12.6).

A wide variety of animal and plant cells are routinely grown and manipulated in culture. These include:

- two-dimensional monocultures—cultures of a single cell type;
- three-dimensional monocultures in various matrices (a matrix is usually a gel-like substance containing one or more proteins that

mimic, to a greater or lesser extent, the extracellular matrix or surroundings that cells are situated in within tissues—that is, the cells' environmental 'niche');

- organoids: miniaturized and often simplified organs formed *in vitro* in three dimensions, from one or a few cells from a tissue, or from embryonic or pluripotent stem cells;

- co-cultures—cultures containing more than one cell type;

- tissue **explants**: small pieces of tissue that can be been transferred from the original host organism to a new place or into culture media.

Both organ and cell culture can be referred to as **tissue culture**. Indeed, these terms are often used interchangeably. Primary cells are derived directly from tissue explants by using enzymes to digest the extracellular matrix to free the cells. These cells usually have finite lifespans *in vitro*. By contrast, cell lines are usually able to grow indefinitely (or continuously) and are abnormal in some manner. For example, they are usually originally derived from tumours, or they may be transformed and immortalized deliberately.

Why are cells grown in culture?

Cells are grown in culture for a number of reasons. In the case of cancer cells, they are grown to try to understand how they are different from their normal counterparts, and how and why they behave in the ways they do. To achieve this, genomic DNA and RNA populations within the cells can be isolated for various genetic analyses. Proteins can be studied *in situ* by using various staining techniques, or they can be extracted and analysed by western blotting, for example. Cell behaviour, such as proliferation, differentiation, migration, invasion, or tubule and organoid formation can all be investigated using cell cultures.

How do culture cells grow?

Many mammalian cells adhere to the surface of the vessel as they grow. These are called adherent or anchorage-dependent cell lines, as illustrated in Figure 12.7. By contrast, blood cells (cells that we refer to as being haematopoietic) usually grow in suspension; we say these are anchorage-independent because they do not have to be attached to a surface to be able to grow: they can float freely in suspension. A suspension cell ine is depicted in Figure 12.8.

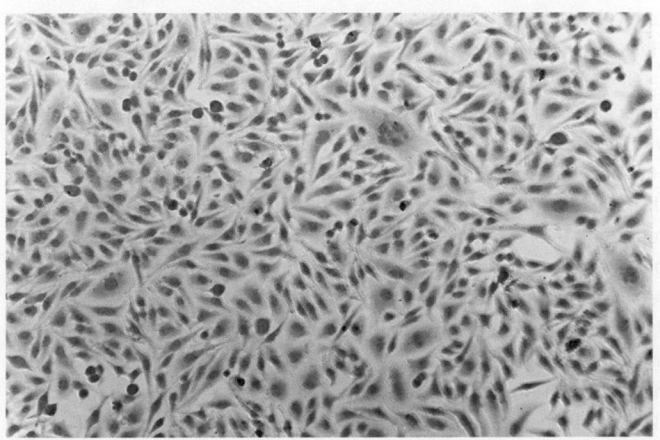

Figure 12.7 HeLa cells—an adherent cell line.

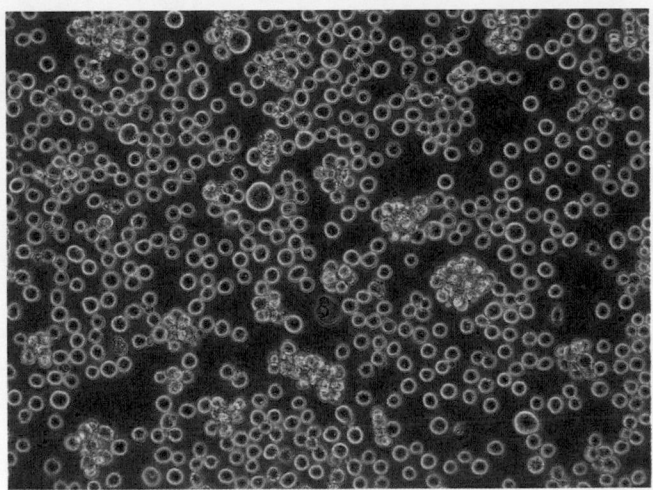

Figure 12.8 K562 cells—a suspension cell line.
Source: AddexBio.

The first cell line to be successfully cultured in the early 1950s was the HeLa cell line by cell biologist George Otto Gey (see Figure 12.7). The cell line originated from cervical carcinoma cancer cells obtained via an explant taken from a patient called Henrietta Lacks. As was the custom of the time, she was not asked for permission to harvest the cells. Unfortunately, Henrietta died only a few months after the surgery, but her cells live on to this day. They have played key roles in the development of vaccines, the testing of drugs, and research into cancer to name but a few. To learn more about the ethics of cell lines read Real World View 12.1.

The ability to grow cells *in vitro* like this was greatly aided by the development of media that had been tailored for the cells' requirements. Harry Eagle carried out a systematic analysis of the nutrients required to support the growth of cells in culture. His basal medium led the way for the development of other media, which have subsequently allowed the growth, and therefore study, of a wide variety of cell types. Examples of media developed for culturing mammalian cells are given in Table 12.3.

What does successful cell culture require?

A number of requirements must be met in order to successfully culture mammalian cells *in vitro*. These requirements vary according to what is being cultured but include sterile conditions, a source of nutrients, and the correct temperature and pH.

As we note in Section 12.1, it is important to understand aseptic technique (to maintain sterility) before attempting to culture cells as it is important to culture only the cells required. The various media that eukaryotic cells are grown in are very rich sources of nutrients. The minimum composition of a culture medium is listed in Table 12.4. It is also important to recognize that these same nutrients also make ideal nutrient sources for bacterial or fungal contaminations, making an aseptic environment even more important.

Look at Table 12.4 and notice that the culture medium contains animal products, which can raise ethical concerns. As such, scientists are working to develop non-animal derived alternatives.

REAL WORLD VIEW 12.1   Ethics of cell lines

Henrietta Lacks (Figure 1), born Loretta Pleasant on 1 August 1920, went to the Johns Hopkins Hospital in Baltimore on 29 January 1951 with abdominal pain and abnormal bleeding. It was there that Dr Howard Jones biopsied and subsequently diagnosed a cervical cancer (later identified as an adenocarcinoma). Henrietta received treatment in the way of radium tube inserts but died from the disease on 4 October 1951 at the age of 31.

Unbeknown to her or her family, tissue and cells from the biopsies were retained and passed to physician and cancer researcher George Otto Gey. This was common practice at the time: Gey had been receiving cells from cervical cancer biopsies for several years in the hope of keeping the cells alive and growing them in the laboratory in order to study them. However, he could only keep such cells alive for a few days at most. That was until he attempted to grow cells from Henrietta's cancer in the laboratory.

For some reason, her cells, which he called HeLa (using the first two letters from her two names), were different from all the previous samples he had collected: they kept dividing and were essentially immortal. This was the breakthrough he, and others, had hoped for.

The immortality of these cells has allowed numerous studies to be undertaken on them in the intervening years in order to try to understand the cancer. They have also been used in a variety of other medical research projects. For example, Jonas Salk used HeLa cells in the development of the polio vaccine, and HeLa cells have had a pivotal role in the investiga-

tion of the effects of radiation and chemotherapy, for undertaking gene mapping, and for developing *in vitro* fertilization, to name a few important advances.

It has been estimated that over 20 tonnes of HeLa cells have been grown since 1951, and there are ~11,000 patents and ~80,000 scientific publications using these cells. HeLa cells have even been sent into space to study the effects of microgravity.

All of this was achieved without the original consent of Henrietta or her family. As was the custom of the time, consent to keep and study the cells in research was not required nor sought by her doctors. Despite the fact that her cells have been used commercially, neither she nor her family received any form of compensation. Indeed, it was not until the 1970s that her family found out, by chance, that her cells had been kept, and used extensively. Later, in the 1980s, family medical records were published, also without consent, and then in 2013 the HeLa genome was published without permission.

Henrietta Lacks' case, and the subsequent use of her cells and genetic information, raises a number of questions about how an individual's samples should be used for research and other purposes. In February 2010, Johns Hopkins released the following statement concerning the cervical samples that were taken from Henrietta:

Johns Hopkins Medicine sincerely acknowledges the contribution to advances in biomedical research made possible by Henrietta Lacks and HeLa cells. It's important to note that at the time the cells were taken from Mrs. Lacks' tissue, the practice of obtaining informed consent from cell or tissue donors was essentially unknown among academic medical centers. Sixty years ago, there was no established practice of seeking permission to take tissue for scientific research purposes. The laboratory that received Mrs. Lacks's cells had arranged many years earlier to obtain such cells from any patient diagnosed with cervical cancer as a way to learn more about a serious disease that took the lives of so many. Johns Hopkins never patented HeLa cells, nor did it sell them commercially or benefit in a direct financial way. Today, Johns Hopkins and other research-based medical centers consistently obtain consent from those asked to donate tissue or cells for scientific research.

In 2013, Johns Hopkins and the National Institutes of Health (NIH) worked with the Lacks family to formulate an agreement that requires scientists to receive permission to use Henrietta Lacks' genetic blueprint, or to use HeLa cells in NIH-funded research.

Of course, HeLa cells are now not the only cell lines used in cancer and other areas of research: many other cell lines originally derived from individuals are available for use in research. Now, legislation such as the Human Tissues Act 2004 (UK) regulates the 'removal, storage, use, and disposal of human bodies, organs, and tissue'. As such, a patient (or their representative) at many hospitals worldwide (including in the UK) must give informed consent for any samples from biopsies or blood tests to be retained for research or other purposes.

Henrietta Lacks' legacy is enormous: not only were her cells instrumental in numerous biomedical studies and important findings, but her case also highlighted the importance of informed consent and the need for doctors to discuss with their patients the purposes for which their tissue will be used.

Figure 1 Henrietta Lacks.
Source: Science History Images/Alamy Stock Photo.

Table 12.3 Example media used for mammalian cell cultures

Media type	Examples	Uses
Balanced salt solutions	PBS, Hanks' BSS, Earle's salts DPBS HBSS EBSS	Form the basis of many complex media
Basal media	MEM	Primary and diploid culture
	DMEM	Modification of MEM containing increased level of amino acids and vitamins. Supports a wide range of cell types, including hybridomas
	GMEM	Glasgow's modified MEM was defined for BHK-21 cells
Complex media	RPMI 1640	Originally derived for human leukaemia cells. It supports a wide range of mammalian cells including hybridomas
	Iscoves DMEM	Further enriched modification of DMEM which supports high density growth
	Leibovitz L-15	Designed for CO_2-free environments
	TC 100	Designed for culturing insect cells
	Grace's insect medium	
	Schneider's insect medium	
Serum-free media	CHO	For use in serum-free applications
	HEK293	
	Ham F10 and derivatives Ham F12 DMEM/F12	Note: these media must be supplemented with other factors such as insulin, transferrin, and epidermal growth factor. These media are usually HEPES buffered
Insect cells	Serum-free insect medium 1 (Cat no. 53777)	Specifically designed for use with Sf9 insect cells

Source: Public Health England.

Table 12.4 Minimum medium requirements for eukaryotic cell culture

Basic constituent	Examples	Why added?
Bulk ions	Na^+, K^+, Ca^{2+}, Mg^{2+}, Cl^-	Osmotic balance To regulate membrane potential To act as enzyme cofactors
Carbohydrates	Glucose	Source of energy
Amino acids	Essential	Building blocks of proteins
Vitamins	B, etc.	Precursors for numerous co-factors
Fatty acids and lipids	Cholesterol	Membrane function
Proteins and peptides	Albumin, transferrin, hormones, and growth factors	Promote cell growth
Serum	Fetal calf (or bovine) serum	Contains growth factors to support growth Binds toxins Neutralizes trypsin and other proteases (which can damage cell membrane proteins)
Trace elements	Iron, zinc, and selenium	Selenium is a detoxifier
Buffering system	CO_3/HCO_3 or HEPES	Maintain pH
Antibiotics	Penicillin	Not required, but often added to control the growth of bacterial contaminants

Figure 12.9 Tissue culture hood. This is a laminar flow type cabinet where the air is drawn through the filter in the back enabling sterile air to flow through to the front.

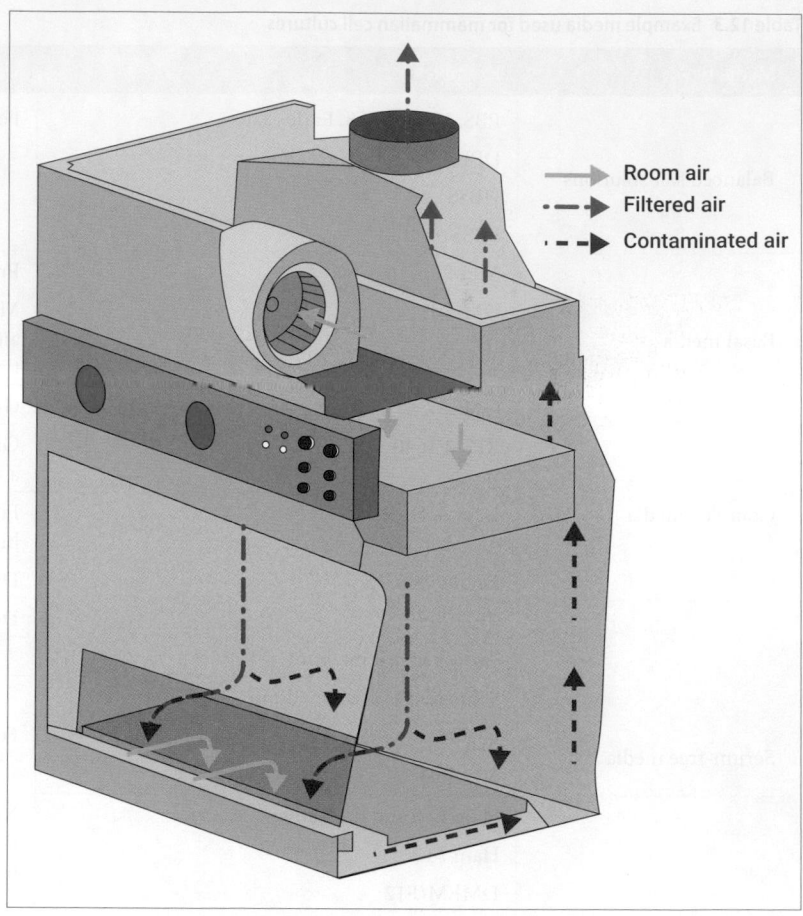

Room air
Filtered air
Contaminated air

The handling of mammalian cell cultures is routinely undertaken in specialized tissue culture hoods of the appropriate biohazard category. Most cell lines require laboratories to be designated as Biohazard Category 2 (Advisory Committee on Dangerous Pathogens, 1995)). A schematic diagram of a tissue culture hood is shown in Figure 12.9.

Any solutions or reagents added to the cell cultures should also be sterilized. In fact, many such solutions are often pre-bought as sterile or are sterilized via autoclaving or by filter sterilization (the latter for heat- and/or pressure-sensitive compounds and solutions). Antibiotics are also added to prevent contaminating bacterial growth (a penicillin/streptomycin mix is common).

How are cells grown?

The cells can be grown in a variety of flasks, plates, or dishes, with the cells grown directly on to the plastic or with a coating (such as gelatin or extracellular matrix protein) applied. Culture flasks usually have a filter in the lid to allow gases to pass in and out.

The cells are grown in an incubator with a humidified atmosphere to prevent evaporation of the medium, and with 5–10%

CO_2; this is part of the buffering system of the media (comprising sodium bicarbonate/carbonic acid) to maintain the correct pH. Human cells are usually grown at 37 °C, but this may be altered if, for example, they have been modified with a temperature-sensitive mutation.

Mammalian and other cells can be stored in liquid nitrogen for preservation, either in the liquid phase (−196 °C) or in the vapour phase (−156 °C). A cryoprotective agent (such as glycerol or DMSO) is added and the cells are slowly cooled to prevent potentially damaging ice crystals from forming.

How can cells in culture be viewed and measured?

Cells in culture can be viewed using a range of different microscopes, a phase contrast microscope being common. Haemocytometers (or counting chambers) and automatic cell counters, such as a Coulter counter, can be used to determine the concentration of cells within a culture. The inclusion of a dye such as trypan blue, which live cells exclude, can be added if a measure of the number of live cells (cell viability) is required. Fluorescent-activated cell sorter (FACS) machines, such as the one shown in Figure 12.10, can not only be

Figure 12.10 A close-up of the laser in a fluorescence-activated cell sorting (FACS) machine. FACS provides a way of sorting a mixture of cells into two or more containers, one cell at a time, based upon the specific light scattering and fluorescent characteristics of each cell.

used for counting, but they can also be used for identifying and sorting cells into different types by using fluorescently tagged antibodies specific for cellular proteins.

▶ We learn more about light microscopy, including phase contrast microscopy, in Section 12.5.

Plant cell culture

The same principles apply to plant cell cultures as they do to mammalian cell cultures. Cells from the appropriate part of the plant need to be obtained and grown in sterile conditions with appropriate medium and supplements. Plant cells can be grown as an embryo, organ, callus (a mass of undifferentiated cells), and in cell culture, as depicted in Figure 12.11a and b.

The culturing of undifferentiated plant cells requires the inclusion of a careful balance of plant hormones such as **auxins** and **cytokinins**. Some plant cells have the potential to regenerate a whole plant; by altering the growth conditions plant cell cultures can be manipulated to grow into any of the cell types that make up a plant. The ability to introduce genetic modifications introduces a range of options for generating genetically manipulated crops.

PAUSE AND THINK

Why was the development of basal media important in cell culture?

Answer: The development of basal media allowed cells to be grown *in vitro* as they were provided with all the nutrients to support their survival and growth. This then led to the development of more complex media specifically tailored to support the growth of a wide variety of different cells, which allowed more areas of research to be possible.

 Check your understanding of the concepts covered in this section by answering the questions in the e-book.

12.4 Measuring the growth of microbial cells

Microbial cells can be grown in liquid and solid media. There are two types of media: defined and complex. A chemically defined medium is where all the components are known and defined: they

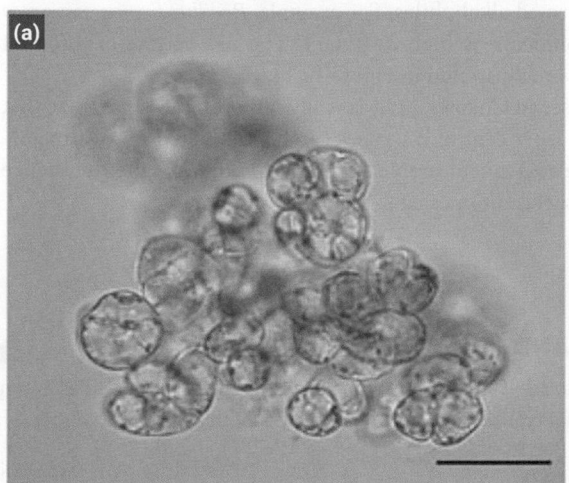

Figure 12.11 Plant cell culture. (a) Cluster of single plant cells in a suspension culture (400× magnification). (b) Callus grown from *Nicotiana tabacum*.

Source: (a) Reprinted by permission from Springer Nature. Figure 1 h from: Moscatiello, R., Baldan, B., and Navazio, L. (2013) Plant cell suspension cultures. In: Maathuis, F. (ed.) *Plant Mineral Nutrients. Methods in Molecular Biology (Methods and Protocols)*, vol 953. Humana Press, Totowa, NJ; (b) Igge/Wikimedia Commons/CC BY-SA 3.0.

are in a pure form and exact amounts of chemicals are added to give known concentrations. Chemically defined media perform in a consistent way. This type of medium is therefore important when microbial cells are used to produce therapeutic substances, such as human factor 8, which is used in the treatment of haemophilia. It is essential that such therapeutic substances can be produced in a reliable, controlled way to ensure they behave as expected when administered in the body.

An example of a chemically defined medium is given in Table 12.5. In contrast to defined media, a complex medium has components within it that have not been totally characterized, and the concentration of components such as sugars and amino acids are not accurately known. A good example of a complex medium is Luria-Bertani (LB) broth, which is used for the routine cultivation of *Escherichia coli*; its components are shown in Table 12.6.

Solid media is made from liquid media by the addition of a setting agent, most commonly agar. Agar, which comes from seaweed, is a complex polysaccharide. Its value as a setting agent stems from the way that it does not melt until the temperature reaches near boiling point. This means that microbes can be cultured at higher temperatures than 50 °C. After autoclaving but prior to setting, the media can be poured into containers such as petri dishes, which can come in a variety of different shape and sizes as depicted in Figure 12.12. Such containers are often used for spread-plating and obtaining pure cultures. Other containers include plastic or glass tubes for making slopes, which are used for longer-term storage.

Figure 12.12 A Petri dish used for cultivation.
Source: Hilal Korkut/Shutterstock.com.

Cultivation requires a basic understanding of the physiology of the organism that you want to culture. Figure 12.13 shows how different microbes have different temperature optima:

- psychrophiles can grow between −5 and 20 °C;
- mesophiles grow between 15 and 45 °C;
- thermophiles grow between 45 and 80 °C;
- hyperthermophiles grow between 65 and 105 °C.

Microbes also grow optimally at different pHs. Those that grow at less than pH 5.5 are called acidophiles. For example, the bacterium *Sulfolobus solfataricus* that grows in Yellowstone National Park is an extreme acidophile. Most microbes are neutrophiles and grow between pH 5.5 and 8.5. Those capable of growing above pH 7.5 are termed alkaliphiles. For example, *Bacillus firmus* can grow in environments as high as pH 11. The pH optima of different microbes are illustrated in Figure 12.14.

As we see in Chapter 11, it is really important to emphasize that less than 1 per cent of microbes can be cultured. The vast majority are termed 'unculturable' and we only know of their existence through metagenomics analysis of environments.

Table 12.5 A chemically defined medium for the growth of *Bacillus megaterium*

Component	Amount (g L^{-1})	Function
Sucrose	10	Carbon and energy source
K_2HPO_4	2.5	pH buffer; P and K sources
KH_2PO_4	2.5	pH buffer; P and K sources
$(NH_4)2HPO_4$	1.0	pH buffer; N and P sources
$MgSO_4.7H_2O$	0.2	S and Mg^{2+} sources
$FeSO_4.7H_2O$	0.01	Fe^{2+} source
$MnSO_4.7H_2O$	0.007	Mn^{2+} source
pH 7.0		

Table 12.6 The components of LB broth or agar

Component	Amount (g L^{-1})
Tryptone	10
Yeast extract	5
NaCl	0.5
For solid media add agar	15

PAUSE AND THINK

You want to see if a bacterium can grow using sucrose as the sole carbon source. What type of medium would you use for this study and why?

Answer: You would use a chemically defined medium, including sucrose as the only possible source of carbon the bacterium could use. If another carbon source, such as glucose, or a complex component such as yeast extract was included in the medium, and you found your bacterium was able to grow, you would not be able to determine if it was capable of using just sucrose.

Figure 12.13 Temperature optima of different micro-organisms.

Source: N. Parker, M. Schneegurt, A.-H. Thi Tu, et al. OpenStax, 'Microbiology' 2016. © May 3, 2021 OpenStax. Textbook content produced by OpenStax is licensed under a Creative Commons Attribution License 4.0 licence. Access for free at https://openstax.org/books/microbiology/pages/1-introduction.

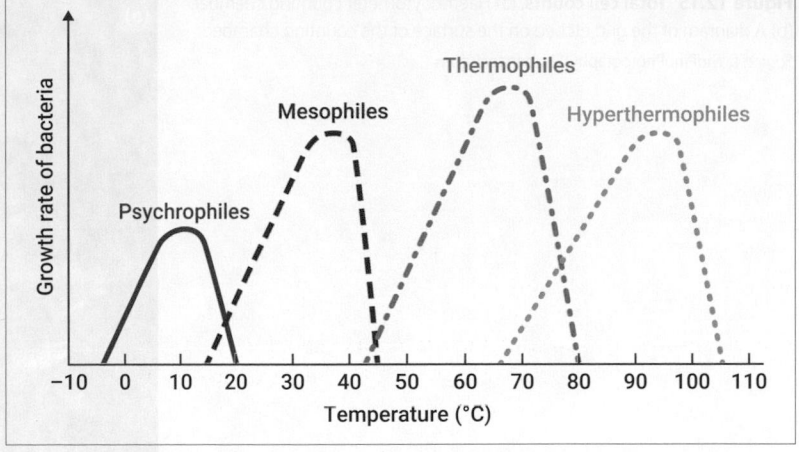

Figure 12.14 pH optima for different micro-organisms.

Source: N. Parker, M. Schneegurt, A.-H. Thi Tu, et al. OpenStax, 'Microbiology' 2016. © May 3, 2021 OpenStax. Textbook content produced by OpenStax is licensed under a Creative Commons Attribution License 4.0 licence. Access for free at https://openstax.org/books/microbiology/pages/1-introduction.

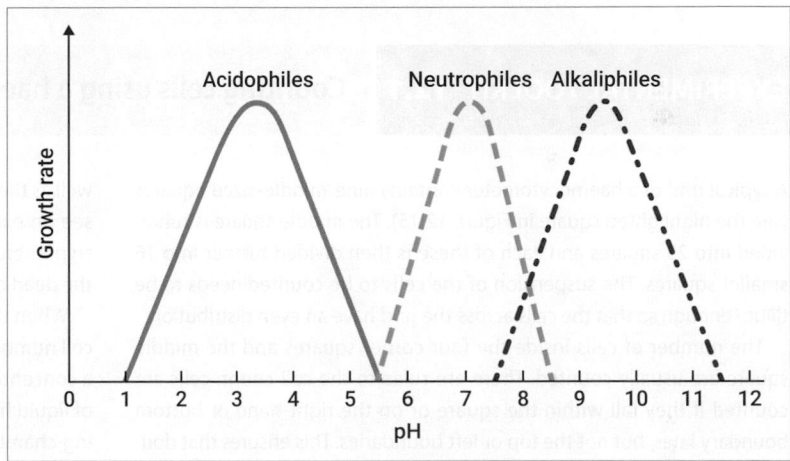

How do we count the number of cells in a population?

The number of cells in a population can be measured in a variety of ways, with each method having distinct advantages and disadvantages. Some methods are more suitable for some organisms than others. Let us now examine these methods in more detail.

Total cell counts

Total cell counts use microscopic examination with a specialized slide or counting chamber—for example, a Neubauer chamber, one type of haemocytometer counting chamber, which has a grid etched on the surface. The slide is made of glass and the etched grid can only be seen clearly under the microscope. Look at Figure 12.15 and notice how the etched grid is not clearly visible when the slide is viewed with the naked eye.

A known volume of liquid can be trapped over the grid enabling cells to be counted per unit volume. Yeasts cells, algal cells, and other eukaryotic cells such as mammalian cells can be counted in this way, but bacterial cells are generally too small. Counting is quick, and this method can distinguish between dead and live cells if a stain such as trypan blue is used. Read Experimental

Toolkit 12.1 to get an understanding of how a haemocytometer is used.

Viable counts

Viable counts allow us to determine the number of living cells in a solid such as soil, or a liquid such as sea water. They can also be used to monitor the growth of cells in broth culture. They rely on the use of a serial dilution series and plate culture.

A serial dilution describes the stepwise dilution of a sample over a number of steps; the dilution factor at each step is the same. For example, in a 10-fold dilution series, such as that shown in Figure 12.16, the sample is diluted by a factor of 10 at each step such that a large dilution of the original sample can be achieved within a few steps. Samples (of a known volume) of each dilution can then be plated onto growth media. Each individual microbial cell when on the plate will divide to form a visible colony which can be counted, as shown in Figure 12.16. The cell number is then expressed as a colony forming unit (cfu).

However, there are problems with this method: cells can clump together; the method is quite time consuming; and it also requires an understanding of the microbe being counted, as a suitable

Figure 12.15 Total cell counts. (a) Haemocytometer counting chamber. (b) A diagram of the grid etched on the surface of the counting chamber. *Source:* DavidPinoPhotography/Shutterstock.com.

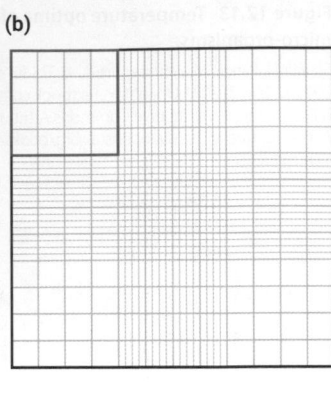

EXPERIMENTAL TOOLKIT 12.1   Counting cells using a haemocytometer chamber

A typical grid of a haemocytometer contains nine 'middle-sized' squares (see the highlighted square in Figure 12.15). The middle square is subdivided into 25 squares and each of these is then divided further into 16 smaller squares. The suspension of the cells to be counted needs to be dilute enough so that the cells across the grid have an even distribution.

The number of cells inside the four corner squares and the middle square are usually counted. There are rules to the cell count: cells are counted if they fall within the square or on the right-hand or bottom boundary lines, but not the top or left boundaries. This ensures that double counting does not occur.

Cells can be stained with trypan blue; as only dead cells take up this stain, you could determine the total number of cells (blue and clear), as well as the number of dead (blue) or viable (clear) cells. Look at Figure 1 to see an example of a haemocytometer square to which a sample of a trypan blue-stained yeast cell suspension has been added. Notice how the dead cells are a prominent dark blue colour.

When the cells have been counted, it is possible to calculate the mean cell number per one of the 'middle-sized' squares. To turn this count into a concentration (a cell count per mL) we need to know the total volume of liquid held within the chamber we are working with. (Different counting chambers have different volumes.) Let us say that the volume over our haemocytometer is 100 nL (or 0.1 μL or 0.0001 mL). To obtain a count per mL, we multiply the mean by 10^4. (0.0001 mL $\times 10^4 = 1$.) If the cells have been diluted with a stain prior to counting, then this dilution factor also has to be taken into account when calculating the original cell count per mL.

Example

Cells in a suspension were diluted 1:5 with trypan blue. (So, if 1 mL of cell suspension was used, it would have been added to 4 mL of trypan blue to achieve a final volume of 5 mL: a five-fold dilution.) A sample was placed in a haemocytometer and the four outside squares and the middle square were counted. The counts were: 50, 48, 54, 45, and 51.

1. Calculate the mean number of cells across the five counts:

$$(50+48+54+45+51) \div 5 = 49.6$$

2. Multiply by 10^4 to determine the number of cells per mL of trypan blue solution:

$$49.6 \times 10\,000 = 4.96 \times 10^5 \text{ cells } per\ mL$$

3. Multiply by the dilution factor to determine the number of cells per mL in the original suspension. The cells were diluted 1:5 with trypan blue, so multiply by 5:

$$4.96 \times 10^5 \times 5 = 2.48 \times 10^6 \text{ cells } per\ mL \text{ in the original suspension.}$$

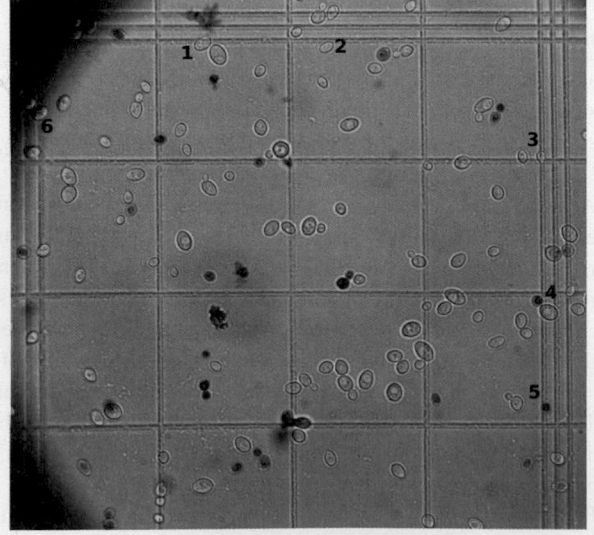

Figure 1 A haemocytometer square showing a trypan blue stained yeast cell suspension.
Source: Copyright by Eureka Brewing Blog.

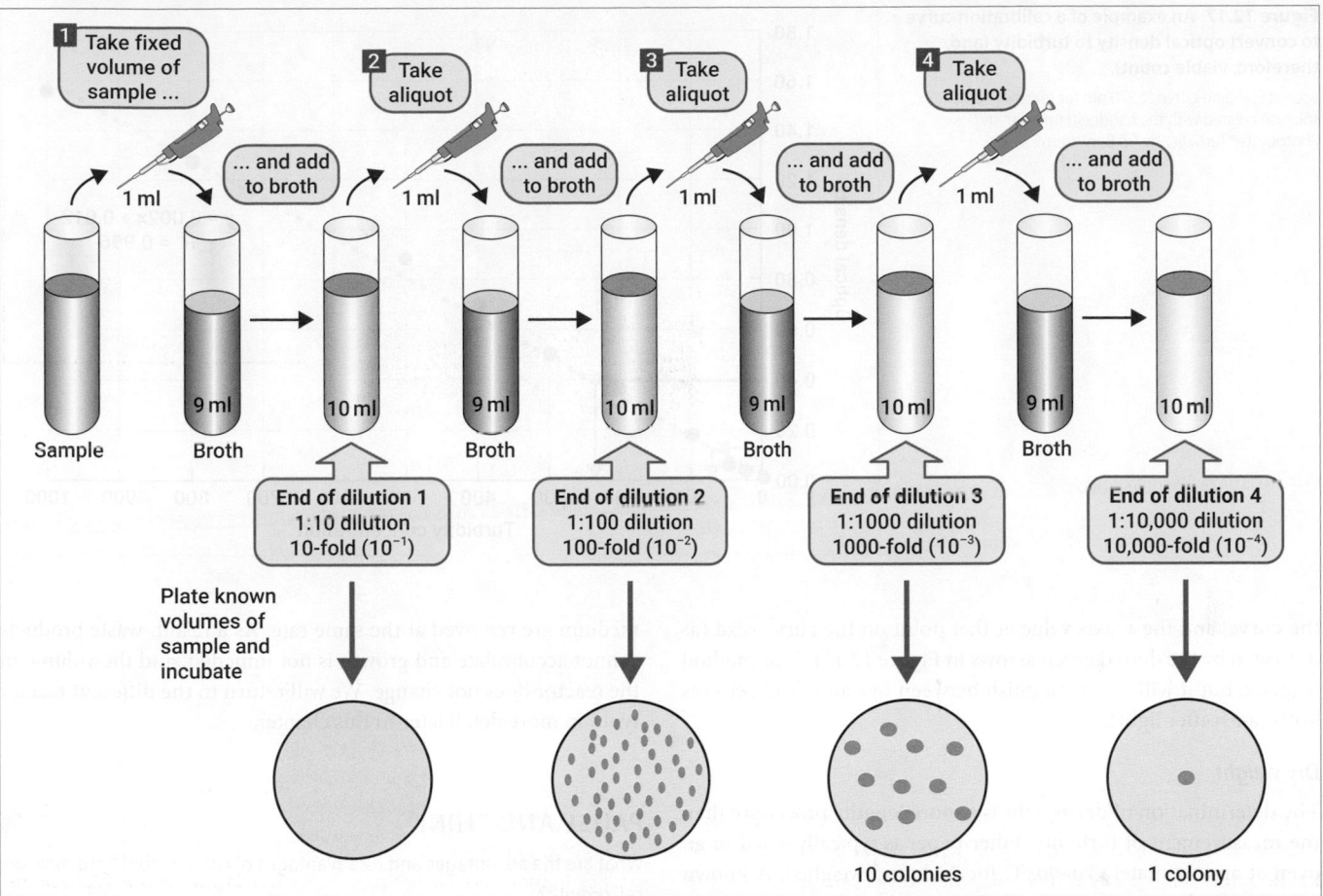

Figure 12.16 Viable counts. The use of a serial dilution to obtain a viable count.

growth medium needs to be used. Another limitation is that it will only determine the number of culturable cells present in the sample (and, as we know, only a small fraction of the microbial community can be cultured).

Example: Calculating a viable count

Let us now work through an example of a viable count calculation, drawing on the approach we have just outlined.

Imagine that you are carrying out an experiment to measure the number of *Streptomyces* bacteria in soil. You add 1 g of soil to 10 mL of sterile distilled water and mix. You then perform a 1,000,000-fold serial dilution (diluting from 10^{-1} to 10^{-6}) and plate 100 µL of each dilution onto selective medium for the growth of *Streptomyces* spp. bacteria. You then incubate at 30 °C and record the number of colonies on each plate. You find there are 60 colonies on the 10^{-3} dilution.

1. What is the colony forming unit (cfu) per mL of the *Streptomyces* bacteria in the sample?

2. How many *Streptomyces* are there in 1 g of the soil sample?

The 10^{-3} dilution is a 1000-fold dilution of the original sample. Therefore, we multiply by 1000 to determine the number of colonies in the original sample:

$$60 \times 1000 = 6.0 \times 10^4 \text{ colonies (cfu)}$$

We know that 100 µL was spread onto the plate, so to get the number of bacteria in 1 mL (1000 µL) we multiply by 10:

$$6.0 \times 10^4 \times 10 = 6.0 \times 10^5 \text{ cfu/mL}$$

The second question asks how many *Streptomyces* bacteria there were in the 1 g of soil. We know that 1 g of soil was placed into 10 mL of water, and we know that there are 6.0×10^5 cfu per mL in that original sample.

So, the total number of cfu in the original 10 mL of water = $\left(6.0 \times 10^5 \text{ cfu/mL}\right) \times 10 \text{ mL} = 6.0 \times 10^6 \text{ cfu}$.

Turbidity

The change in optical density of a clear liquid medium can also be used to measure growth. Optical density can be measured using a spectrophotometer, usually at 600 nm. A beam of light is shone through the liquid and the amount of light that travels through the liquid is recorded. This amount of light depends on how much the light has been scattered by cells suspended in the liquid: the higher the number of cells present, the greater the scattering (the higher its optical density), and the lower the amount of light that passes through to be detected.

Optical density readings can be linked to viable counts using a calibration curve such as that shown in Figure 12.17. The optical density reading is found on the y-axis; that y-axis value is found on

Figure 12.17 An example of a calibration curve to convert optical density to turbidity (and, therefore, viable count).

Source: Standard curve at 570 nm for turbidity standard solution created with the handheld turbidimeter Photopette® Turbidity by Tip Biosystems.

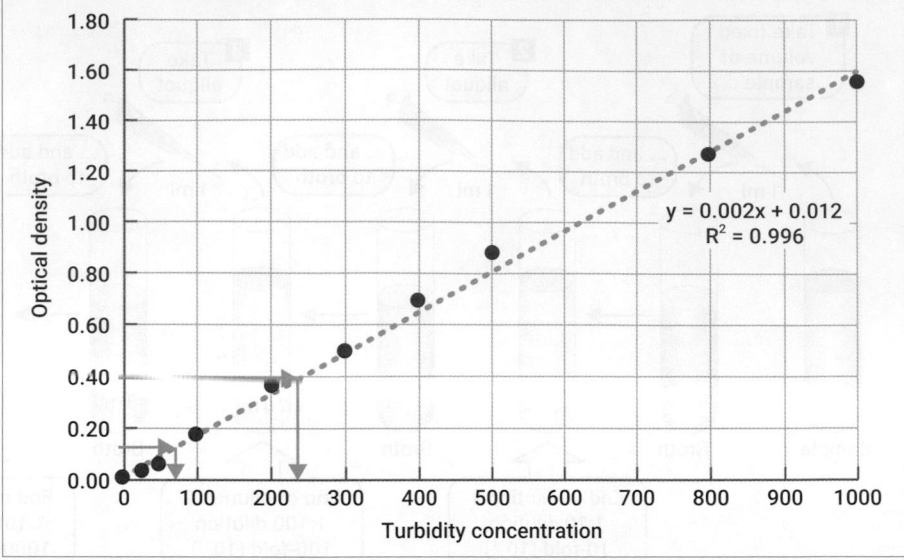

the curve, and the *x*-axis value at that point on the curve read (as indicated by the dotted green arrows in Figure 12.17). This method is quick, but it will not distinguish between live and dead cells (as both can scatter light).

Dry weight

The determination of dry weight is a more lengthy procedure than the measurement of turbidity. Filter paper is typically dried in an oven at approximately 70–90 °C for 24 h and weighed. A known volume of culture is then filtered through the paper and the paper then re-dried at 70 °C and weighed. The difference in weights then provides a dry weight measurement of the microbial cells captured on the filter paper. This method will not distinguish between live and dead cells. Neither is it useful if the media used for the cultivation has suspended particulate matter: this matter will also be captured on the filter paper, and will artificially inflate the dry weight obtained.

Growth of microbial cells in a bioreactor

Microbial cells can be grown in both the laboratory and on an industrial scale in three main ways:

1. Batch culture.
2. Fed-batch culture.
3. Continuous culture.

There are distinct differences between these approaches, as illustrated by the very simplified reactor diagrams in Figure 12.18.

Batch culture is a closed system. Medium is not added or removed and so the volume inside the reactor stays the same. Nutrients will be used up as growth occurs and waste products will accumulate.

In fed-batch culture new nutrients in the form of fresh medium can be added at specific time intervals, but waste products will still accumulate as culture and spent medium is not removed. This is still a closed system, but the volume in the reactor does increase.

Continuous culture is an open system. New nutrients are added in the form of fresh medium at a specific rate, and culture and spent

medium are removed at the same rate. As a result, waste products cannot accumulate and growth is not impeded, and the volume in the reactor does not change. We will return to the different reactor types in more detail later in this chapter.

PAUSE AND THINK

What are the advantages and disadvantages of using turbidity to measure cell growth?

Answer: The main advantage of using turbidity to measure cell growth is that this method is quick. A small sample of the liquid culture is taken and the turbidity can be measured within a minute or two. This process can be repeated at various times as the liquid culture is incubated to determine the rate of growth. Undertaking a colony count, or measuring the dry weight, to determine cell number (and repeating this over time to determine cell growth) can be time consuming, as these techniques usually take an overnight or 24-h step. Using a calibration curve for the turbidity measurements allows a quick determination of the approximate cell number. The disadvantage of this method, however, is that the turbidity measurement does not tell you whether the cells are alive or dead. As turbidity measurements are usually undertaken at a wavelength of 600 nm, if there is something else in the culture that absorbs light at 600 nm, the resulting turbidity reading could be inaccurate.

Growth phases in batch culture

When microbes grow in batch culture they pass through a number of distinct growth phases. For example, if you were to add 1 mL of a pure culture of *Escherichia coli* that had been grown in a larger volume of nutrient-rich medium into a conical flask containing the same medium (a batch culture closed system), the *E. coli* would undergo a series of four growth phases:

1. Lag phase.
2. Exponential (log) phase.
3. Stationary phase.
4. Death phase.

Figure 12.18 Different types of bioreactor systems. (a) Batch. (b) Fed-batch. (c) Continuous.

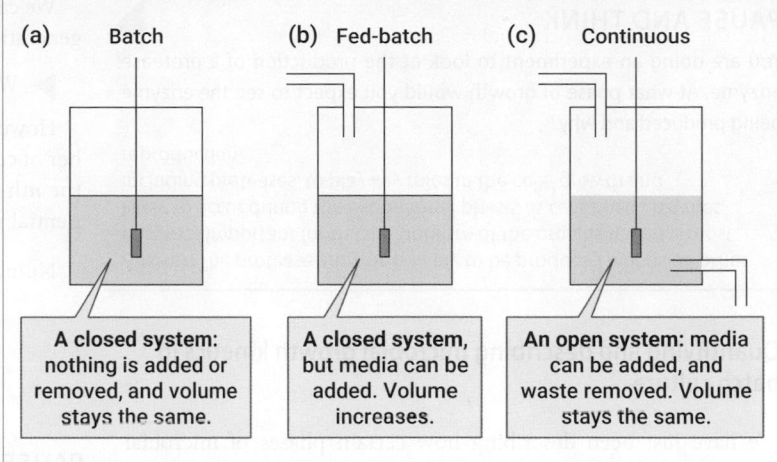

(a) Batch	(b) Fed-batch	(c) Continuous
A closed system: nothing is added or removed, and volume stays the same.	A closed system, but media can be added. Volume increases.	An open system: media can be added, and waste removed. Volume stays the same.

Look at Figure 12.19, which shows these different phases, with time on the *x*-axis and the number of cells as a log value on the *y*-axis. Notice how there are two phases during which the number of cells does not change (as indicated by the horizontal portions of the curve): the lag and stationary phases. By contrast, there are two phases when the number of cells does change (as indicated by the curve having either a positive (upward-sloping) or negative (downward-sloping) gradient): the exponential and death phases.

 Go to the e-book to watch a video that shows the nature of microbial growth over a period of four hours.

Let us now consider each of these phases in turn.

Lag phase

When cells are placed into fresh medium (a process called inoculation), they need time to adjust to the new conditions. Cells might be old or damaged, or lack enough ATP or ribosomes in order to start growing straight away, for example. Such components must be made before the cells can start reproducing. Cells will start detecting their new environment, prompting the expression of specific genes—for example, genes encoding for enzymes needed for growth—which will enable the cells to move into the next phase of growth.

The length of this lag phase is very much dependent on the type of change to which the cells have been subjected: they may have been placed into different media, or the temperature and pH might be different. For example, cells which had been growing in a complex, nutrient-rich medium, but which had then been placed into a minimal medium, would have a longer lag phase. Alternatively, cells which are fresh and inoculated into the same medium may have a very short lag phase.

It is important to note that while the number of cells does not increase in lag phase (as indicated in Figure 12.19), the cells remain metabolically active.

Exponential (log) phase

During exponential phase, the cells grow and divide at their maximum rate (called μ_{max}). The cell population doubles in a specific length of time, which is dependent on the microbe itself, on the medium, and also on the growth conditions provided. For example, a population of *E. coli* grown with shaking at 37 °C in nutrient-rich medium will double in size every 20 min.

Cells are also typically at their largest during early exponential growth. Primary metabolites, such as enzymes, are being produced, which aid in the growth and reproduction of the organism. In late exponential phase, however, the doubling rate slows and new genes needed for survival will be expressed.

Stationary phase

The increase in cell number stops during the stationary phase, either because of a lack of a key nutrient or because of the build-up of waste products. The physiology of the cells changes as well: they are now not as able to adapt to new conditions. Some bacteria will make endospores, others reduce their cell size, and enzymes that confer a resistance to stress are produced. Stationary cells are typically more tolerant to changes in environmental conditions (e.g. heat and osmotic pressure). Secondary metabolites such as antibiotics are produced during this phase.

Death phase

As no new nutrients are being provided (remember: batch culture is a closed system), the cells will die as waste products build up. The rate at which the cells die increases exponentially with time: death is slow to start with, and then increases. This may seem like a minor detail but death rates are really important as they help to determine the effectiveness of both food preservation and antibiotics.

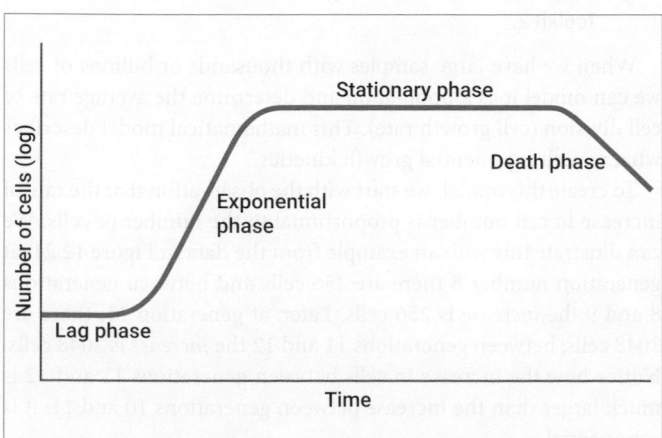

Figure 12.19 The microbial growth phases seen in batch culture.

PAUSE AND THINK

You are doing an experiment to look at the production of a protease enzyme. At what phase of growth would you expect to see the enzyme being produced and why?

Answer: The protease enzyme is likely to be produced alongside other enzymes important for the metabolism of the organism. This is most likely to occur during the exponential phase, as cells need enzymes, including proteases, to play key roles in the cells' growth and reproduction.

Quantifying and describing microbial growth kinetics in batch culture

We have just been describing how certain phases of microbial growth are exponential in nature. But how do we quantify this kind of growth?

Single-celled organisms such as bacteria reproduce by binary fission: each cell replicates its DNA, then a septum forms around the middle of the cell, which separates it into two daughter cells. By contrast, eukaryotic single cells reproduce by mitosis, which involves the segregation of chromosome pairs. Despite these differences, we can use the same mathematical tools to describe the growth of both prokaryotic and eukaryotic single cells in batch culture.

▶ We learn more about binary fission in Section 10.1 and explore mitosis in more detail in Section 10.3.

Look at Figure 12.20. Imagine you have one cell in a volume of liquid. This cell reproduces, generating an identical daughter cell. So, there are now two cells in the liquid. Both cells now reproduce and each produces another daughter cell: the population is now four cells. The population then continues to double with each generation: 2, 4, 8, 16, 32, 64, 128 . . . .

Alternatively, we can express the number of cells using what we call index notation: 2^1, 2^2, 2^3, 2^4, 2^5, 2^6, 2^7. Here we can see that the first generation has 2^1 cells ($2^1 = 2$), the second generation has 2^2 cells ($2^2 = 2 \times 2 = 4$), the third generation has 2^3 cells ($2^3 = 2 \times 2 \times 2 = 8$) and so on . . . . (Notice how 2^1, 2^2, 2^3 etc. are acting as mathematical 'shorthand' for our original sequence of numbers: 2, 4, 8 etc.)

Generation	Number of cells	Number of cells expressed in index notation	Diagrammatic representation
0	1	2^0	
1	2	2^1	
2	4	2^2	
3	8	2^3	
4	16	2^4	
n		2^n	

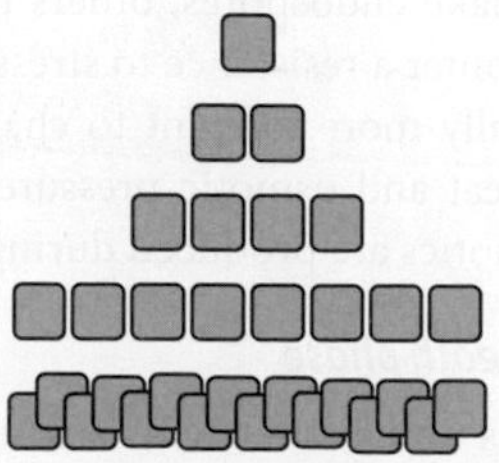

Figure 12.20 Representing exponential growth using mathematical notation.

We can say that, in general terms, the number of cells in the *n*th generation is 2^n cells.

▶ We learn more about powers in Quantitative Toolkit 2.

However, we rarely start with just one cell. If we call the number of cells at the start X_0, then we can find the number of cells in the *n*th generation by multiplying X_0 by 2^n. This is called exponential growth.

Number of cells in the nth generation = starting number x 2^n

$$X_n = X_0 2^n$$

> **Equation 1**
> X_n = number of cells in the n
> X_n = start number of cells

PAUSE AND THINK

Complete this table to show the number of cells present after 10 generations given the starting number of cells shown:

Generation number *n*	Starting number of cells X_0	Number of cells $X_n = X_0 \times 2n$
3	1	$1 \times 2^3 = 8$
10	1	
3	150	$150 \times 2^3 = 1200$
10	150	

Answer: Generation number 3: 1024, Generation number 10: 153,600.

Graphical representations of these growth kinetics are shown in Figure 12.21. Figure 12.21a shows the number of cells with a standard linear scale; this scale dramatically shows the exponential nature of the growth: notice the sharp rise in the plot towards the right of the graph. In contrast, the *y*-axis scale in Figure 12.21b is logarithmic (base 2)—that is, the tick marks increase in multiples of 2: note that when the data are plotted in this way the line is straight. This is an important diagnostic indicator of exponential growth: if data are plotted on a logarithmic scale and the line is straight, that indicates the underlying mechanism is one where growth is exponential, that is, the number of cells will double in a certain time.

▶ We learn more about logarithmic scales in Quantitative Toolkit 2.

When we have large samples with thousands or billions of cells we can model it as a population and determine the average rate of cell division (cell growth rate). This mathematical model describes what we call exponential growth kinetics.

To create this model, we start with the observation that the rate of increase in cell number is proportional to the number of cells. We can illustrate this with an example from the data in Figure 12.21: at generation number 8 there are 256 cells and between generations 8 and 9 the *increase* is 256 cells. Later, at generation 11, there are 2048 cells; between generations 11 and 12 the *increase* is 2048 cells. Notice how the increase in cells between generations 11 and 12 is much larger than the increase between generations 10 and 11: it is exponential.

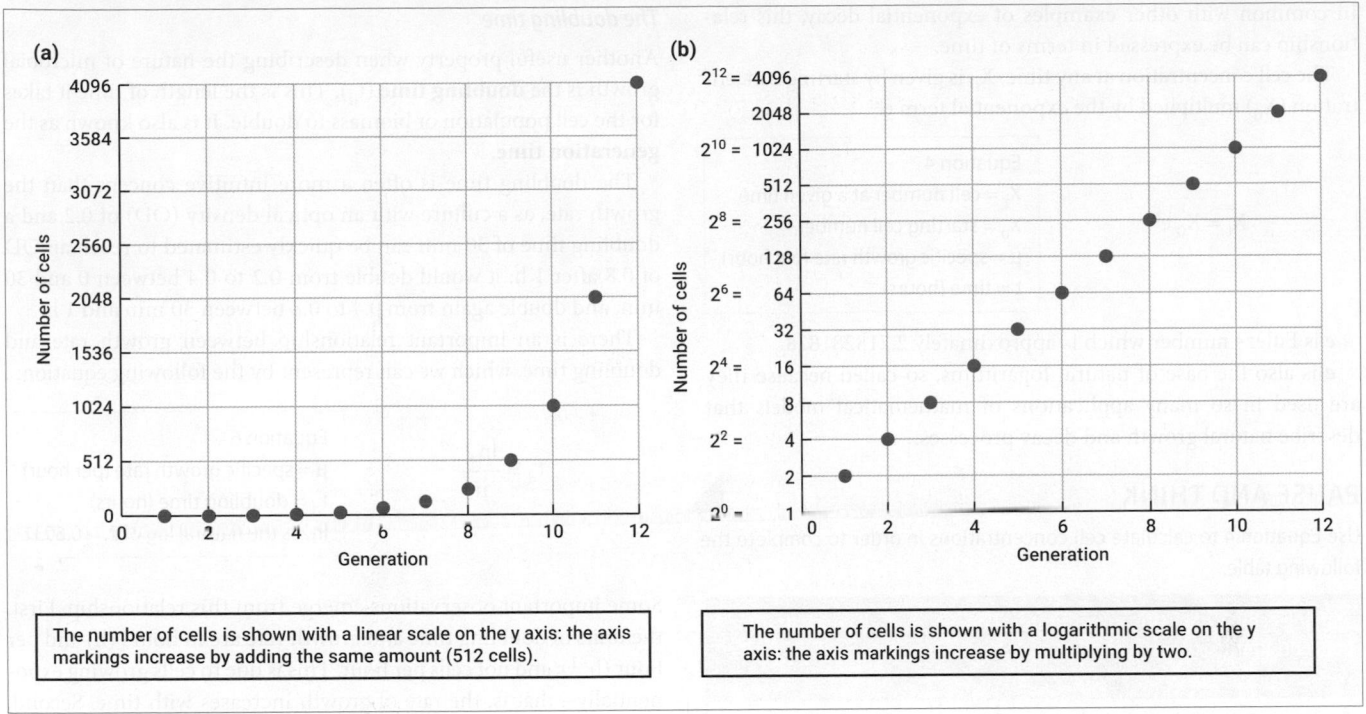

(a)

The number of cells is shown with a linear scale on the y axis: the axis markings increase by adding the same amount (512 cells).

(b)

The number of cells is shown with a logarithmic scale on the y axis: the axis markings increase by multiplying by two.

Figure 12.21 Graphical representations of growth kinetics. (a) The number of cells is shown with a standard linear scale. Notice the rapid increase in the number of cells in later generations. We call this type of increase exponential. (b) The y-axis scale is logarithmic (base 2). Although we are plotting the same data as in (a), the logarithmic scale means that the plot gives a straight line.

We construct our model as follows:

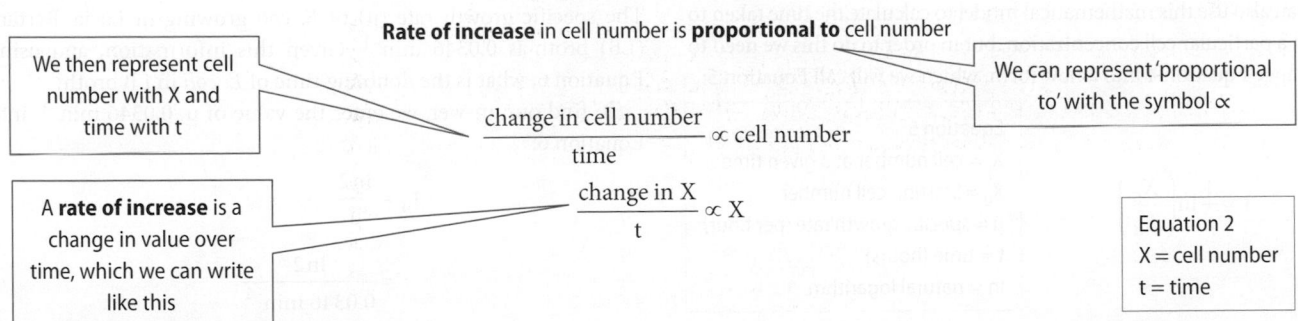

We then represent cell number with X and time with t

Rate of increase in cell number is **proportional to** cell number

We can represent 'proportional to' with the symbol ∝

$$\frac{\text{change in cell number}}{\text{time}} \propto \text{cell number}$$

A **rate of increase** is a change in value over time, which we can write like this

$$\frac{\text{change in X}}{t} \propto X$$

Equation 2
X = cell number
t = time

This relationship is found in many aspects of science, physics, chemistry, and biology, when we describe continuous growth or decay. One example is radioactive decay where the rate of radioactive decay is proportional to the number of radioactive isotopes present. Another example is a chemical reaction during which the rate of reaction depends upon the concentration of one of the reactants.

Each type of microbe has a characteristic specific growth rate under particular conditions. This specific growth rate is usually given the symbol μ and is constant for a given microbe under given conditions.

Rate of increase in cell number = specific growth rate × cell number

$$\frac{\text{change in cell number}}{\text{time}} = \text{specific growth rate} \times \text{cell number}$$

$$\frac{\text{change in X}}{t} = \mu X$$

Equation 3
X = cell number
t = time interval (h)
μ = specific growth rate (per hour)

In common with other examples of exponential decay, this relationship can be expressed in terms of time.

The cell concentration at any time, X_t, is given by starting concentration (X_0) multiplied by the exponential term $e^{\mu t}$.

$$X_t = X_0 \, e^{\mu t}$$

> **Equation 4**
> X_t = cell number at a given time
> X_0 = starting cell number
> μ = specific growth rate (per hour)
> t = time (hours)

e is Euler's number which is approximately 2.718281828.

e is also the base of natural logarithms, so-called because they are used in so many applications of mathematical models that describe natural growth and decay processes.

PAUSE AND THINK

Use Equation 4 to calculate cell concentrations in order to complete the following table.

Starting cell concentration X_0	Specific growth rate (per hour) μ	Time (hours) t	Cell concentration at time, t X_t
1	0.3	10	20
100	0.3	10	2009
100	0.3	100	
100	0.9	10	
100	0.9	100	

Answer: 1.07×10^{15}; 810,308; 1.22×10^{41}

We can also use this mathematical model to calculate the time taken to reach a particular cell concentration, but in order to do this we need to rearrange Equation 4 into a new form, which we will call Equation 5:

$$t = \frac{1}{\mu} \ln\left(\frac{X_t}{X_0}\right)$$

> **Equation 5**
> X_t = cell number at a given time
> X_0 = starting cell number
> μ = specific growth rate (per hour)
> t = time (hours)
> $\ln$ = natural logarithm

PAUSE AND THINK

Use Equation 5 to calculate the time taken to reach a particular cell concentration so that you can complete empty cells in the following table.

Starting cell concentration X_0	Specific growth rate (per hour) μ	Time (hour) t	Cell concentration at time, t X_t
1	0.3	9.99	20
100	0.3		2009
100	0.3		1,000,000
100	0.9		1,000,000
100	0.9		2009

Answer: 10; 30.7; 10.2; 3.3

The doubling time

Another useful property when describing the nature of microbial growth is the **doubling time** (t_d). This is the length of time it takes for the cell population or biomass to double. It is also known as the **generation time**.

The doubling time is often a more intuitive concept than the growth rate, as a culture with an optical density (OD) of 0.2 and a doubling time of 30 min can be quickly estimated to reach an OD of 0.8 after 1 h: it would double from 0.2 to 0.4 between 0 and 30 min, and double again from 0.4 to 0.8 between 30 min and 1 h.

There is an important relationship between growth rate and doubling time, which we can represent by the following equation:

$$t_d = \frac{\ln 2}{\mu}$$

> **Equation 6**
> μ = specific growth rate (per hour)
> t_d = doubling time (hours)
> $\ln 2$ is the natural log of 2: ~ 0.6931

Some important observations emerge from this relationship. First, the units of doubling time and growth rate are in hours (h) and per hour (h^{-1}), and not cells per hour. This is due to cells growing exponentially—that is, the rate of growth increases with time. Second, the overall time for the biomass to double (t_d) is ~ 0.69 times the reciprocal of the average growth rate ($1/\mu$). What does this tell us? If cells divided uniformly as shown in Figure 12.21, then the growth rate and doubling time would be equal. But because cells behave differently and divide at different times, the doubling time for each cell is *smaller* than the overall rate of growth.

Example: using the specific growth rate to determine doubling time

The specific growth rate (μ) of *E. coli* growing in Luria–Bertani (LB) broth is 0.0346 min^{-1}. Given this information, and using Equation 6, what is the doubling time of *E. coli* in LB broth?

To find our answer, we enter the value of μ, 0.0346 min^{-1}, into Equation 6:

$$t_d = \frac{\ln 2}{\mu}$$

$$= \frac{\ln 2}{0.0346 \text{ min}^{-1}}$$

$$= 20 \text{ min}$$

 Check your understanding of the concepts covered in this section by answering the questions in the e-book.

12.5 Visualizing cells

The majority of cells, whether animal, plant, or microbial, are far too small to be seen with the naked eye. In general, bacteria are in the order of 2–5 microns in size and eukaryotic cells perhaps ten times larger, at around 20–50 μm across. By contrast, a fine human hair (which a good human eye can just about see) is around 0.04 mm (or 400 μm) wide.

As we learnt in Chapter 9, it was not until the mid-17th century that it began to be possible to visualize cells using microscopy.

Ever since Robert Hooke published his *Micrographia* in 1665, and his Dutch contemporary Antonie van Leeuwenhoek observed the structure of spermatozoa, protozoa, and bacteria using his custom-made single-lens instruments, microscopy has provided biologists with a window on cell structure and function. Indeed, microscopes of one sort or another have played a central role in biological science ever since.

How do microscopes work?

The central components of any optical microscope are lenses, which are usually made of glass. Lenses have been used in eye-glasses since the 13th century, and it is likely that this use led to the development of single-lens magnifying glasses to view small objects. In fact, it was Leeuwenhoek's use of magnifying glasses in his draper's shop to examine the quality of thread that encouraged him to make a better device. This device—his microscope—is shown in Figure 12.22.

Using his improved single-lens instruments, van Leeuwenhoek was able to obtain far better resolution than anyone had achieved previously. No one knows how he created his lenses—he was notoriously secretive. But it is believed that he fused together small threads of glass to form extremely small 'balls' of glass—the smallest balls having the largest magnification. His lenses enabled him to see protozoa in water samples, which he called 'animalcules' (Figure 12.23).

By contrast, Hooke obtained better magnification by placing a row of ground-glass lenses in a long tube—an instrument that we would recognize today as a microscope. Hooke's microscope is shown in Figure 12.24. These early microscopes worked by shining light either through or onto the specimen, and relied upon the diffraction of light by the sample to obtain a useable image.

A modern light microscope such as those you will come across in your undergraduate practicals is shown in Figure 12.25.

We learn more about how to use the light microscope in Experimental Toolkit 12.2.

This type of microscopy is referred to as **brightfield** because the sample generally appears dark against a bright background.

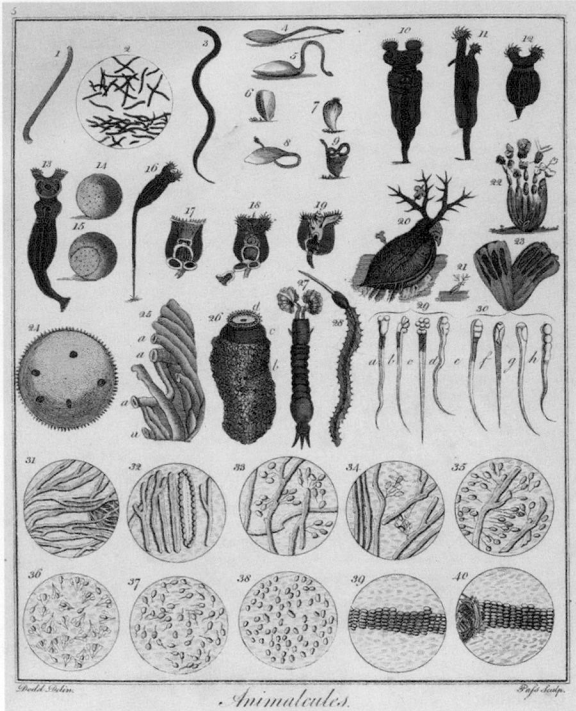

Animalcules.

Figure 12.23 Animalcules observed through a microscope by Antonie van Leeuwenhoek in *c*.1795. These include human spermatozoa from a warm cadaver (31–40).
Source: Ann Ronan Picture Library/Heritage Images/Science Photo Library.

Figure 12.24 Hooke's compound microscope. Hooke obtained better magnification by placing a row of ground-glass lenses in a long tube, in an instrument recognizable today as a microscope.
Source: © Science Museum Group.

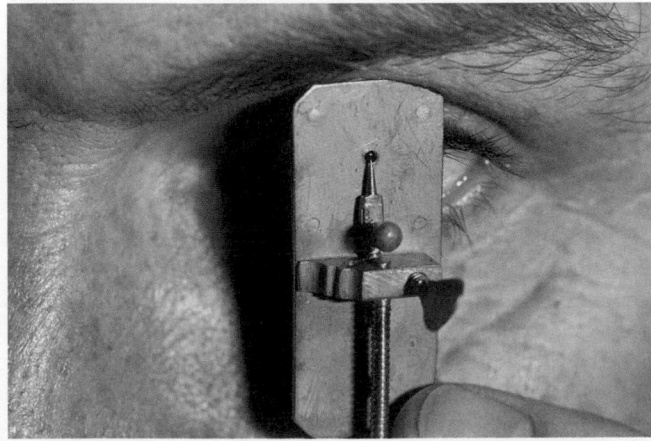

Figure 12.22 An early light microscope. Antonie van Leeuwenhoek's microscope.
Source: Biophoto Associates/Science Photo Library.

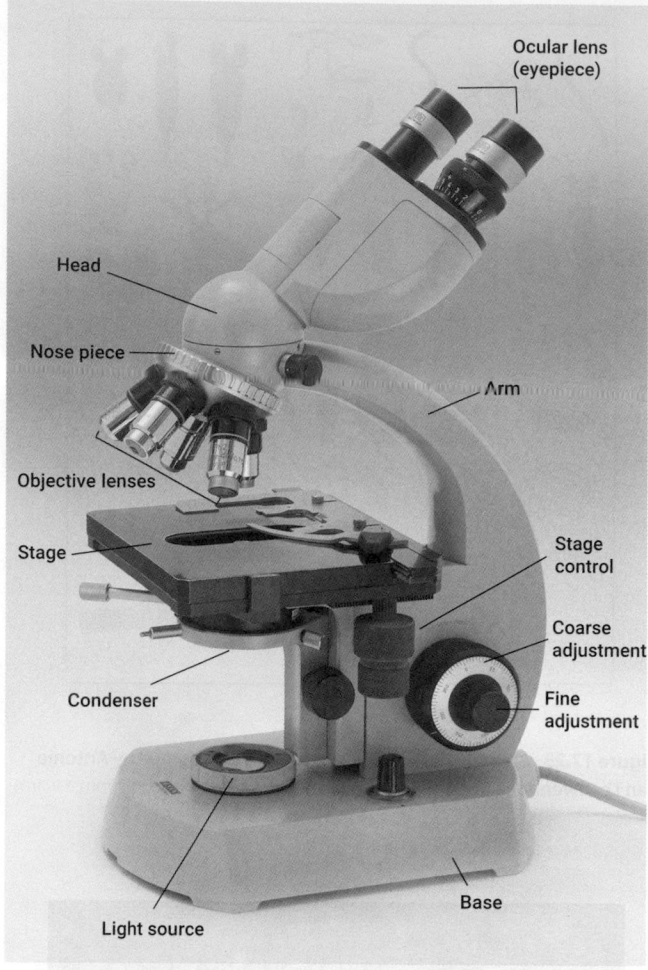

Figure 12.25 A standard compound light microscope. There are key components to the light microscope, but the important elements that determine the overall magnification of a specimen is the magnification of the eye piece and the magnification of the objective lens.

Source: Scenics & Science/Alamy Stock Photo.

Resolution: the limitation of using visible light

Any visualization technique that makes use of visible light is limited by the properties of that light. Of particular concern is the resolution that can be achieved. The **resolution** describes the shortest distance between two very close objects that still allows us to discern those objects as being separate, rather than being joined together—a concept illustrated in Figure 12.26. The resolution is dependent on the wavelength of light and the numerical aperture (NA) of the objective lens. The best resolution is approximately equal to half the wavelength of the illuminating light, or 200–350 nm in the visible range.

Let us pause to think a little more about how the resolution is influenced by the wavelength of light and the numerical aperture of the lens being used.

Look at Figure 12.27. 'A' is the angular aperture of the lens and θ is half of the angular aperture (also known as the acceptance angle). The numerical aperture (NA) of a lens is a measure of its light gathering capacity, and is given by the relationship:

$$NA = n(\sin \theta)$$

where n is the refractive index of the medium through which the sample is viewed. The larger the NA, the better the resolving power.

The resolution (D) of the lens is given by what is called the **Rayleigh criterion**:

$$D = 0.61 \times \lambda / NA$$

where λ is the wavelength of the light used for illumination.

Suppose we want to know the resolution of a light microscope lens that has a numerical aperture of 0.45 given that the wavelength of white light is approximately 0.52 μm. We would enter the wavelength and the value of the NA into the above expression:

$$D = 0.61 \times \frac{0.52 \text{ μm}}{0.45}$$

$$= 0.7 \text{ μm}$$

So, our lens has a resolution of 0.7 μm.

Staining cells to improve visibility

Most biological samples do not diffract light very well and so the resulting images have very poor contrast—that is, there is relatively little difference in how light or dark the sample is relative to the background. Various techniques have been developed to improve contrast, however. The simplest of these methods involves the use of coloured dyes to stain biological samples. For bacteria, the most well-known staining procedure is Gram's method, which we first learnt about in Chapter 9. To remind you, this technique relies upon the formation of a purple precipitate of iodine and crystal violet within the bacterial cells. Crystal violet is a basic molecule that is taken up by microbes and is thought to bind to acidic components of the cell (e.g. nucleic acids). The crystal violet is then precipitated inside the cells when iodine (the mordant) is added.

All bacteria have peptidoglycan cell walls, but of varying thickness. In those species with thin walls (Gram-negative), the purple precipitate is washed out of the cells during the decolorization step of the procedure. By contrast, it is trapped within those species with very thick walls (Gram-positive). A red counterstain (safranin) is then applied; like crystal violet, this is a basic molecule that binds to acidic components. Because the safranin is added *after* the decolorization step, it stains *all* the bacteria. As a result, the Gram-negative bacteria appear reddish-pink and the Gram-positive bacteria dark blue/purple. Look at Figure 12.28a.

Perhaps the most frequently used staining technique for animal cells is the haematoxylin–eosin method (H&E). In this procedure, a complex of oxidized haematoxylin (haematein) and aluminium is used to stain cell nuclei blue. A basic counterstain (eosin) is then applied which stains acidic structures—in particular, the cytoplasm—pink, as illustrated in Figure 12.28b. This method can be applied to virtually any animal tissue.

By contrast, many staining protocols have been developed to highlight particular structures or cellular components—for example, Golgi's silver stain for neurones, Van Gieson's stain for

EXPERIMENTAL TOOLKIT 12.2 — How do you use a light microscope?

The microscope shown in Figure 12.25 may not be exactly the same as the ones you will use in the laboratory, but they have the same set of parts!

The specimen is placed onto a **stage**. A high-intensity bulb generates light, which is focused onto the specimen with a **condenser lens**. Light passing through the specimen is collected by an **objective lens**, and magnified and focused to form an image (within the microscope). An **ocular (eyepiece) lens** carries out further magnification and forms an image that you position your eye to see.

Your microscope will likely have three magnifications for the objective lens: scanning (4×), low (10×), and high (40×). It may also have an oil immersion lens (100×) to view stained bacterial cells. As well as this, the eyepiece (ocular lens) will also magnify the sample, usually 10×. So, the total magnification of the specimen is given by the **ocular magnification** × the **objective magnification**. Table 1 shows how much larger the samples will appear under the magnifications used.

Tips for focusing on specimens

1. **Always start with the lowest magnification power** (scanning objective lens). You may need to move the specimen around to get it into the field of view, but you should be able to see something on this setting. Use the *coarse knob* to focus on the specimen. Don't worry if the image appears to be tiny: you start with the lowest power to enable you to locate the specimen, and can move onto higher powers once the specimen is in the field of view.

2. **If you can see your image under the scanning objective, switch to the low-power lens**. Do not move the slide at this stage or you will likely need to return to the first step again. If you cannot see the specimen clearly, use the *coarse knob* to refocus. Again, if you have not focused on the image at this level, you will find it very difficult to move to the next level.

3. **If you can see your image under the low power, switch to the high-power lens, if required**. (If you have a thick slide, or a slide without a cover, do not use the high power objective lens: it is longer than the others and you may damage the slide or microscope as you try to turn it into place.) At this point, only use the fine adjustment knob to focus the image of the specimen.

4. As you increase the magnification, you may find that the specimen appears to be darker or lighter. If this happens, try adjusting the diaphragm to change the amount of light reaching the specimen.

5. If you see a line in your viewing field, try twisting the eyepiece. The line should move. It might be that your microscope has a pointer (to point at parts of your specimen) or an eyepiece graticule (used for measuring specimens).

Table 1 Using the magnification of the microscope to obtain the size of a specimen

Original size of specimen	Objective lens magnification	Ocular lens magnification	Size of specimen through microscope
0.2 mm	10× (low power)	10×	20 mm
2 mm	4× (scanning)	10×	80 mm
0.05 mm	40× (high power)	10×	20 mm
2 µm	100× (oil immersion)	10×	2000 µm (2 mm)

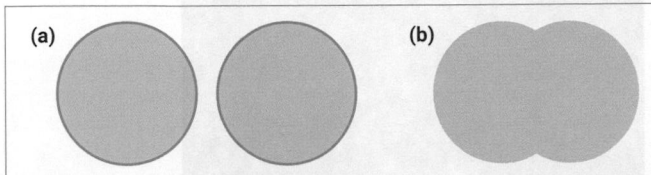

Figure 12.26 The concept of resolution. (a) These two objects are resolved. (b) These objects are not resolved and appear to be joined together.

connective tissue, and periodic acid–Schiff's stain (PAS) for sugar-containing components such as mucins (see Figure 12.28c). Although very powerful, these coloured staining techniques have now largely been superseded by the use of fluorescent probes.

Staining of cells and tissues with fluorescent molecules provides much better contrast and specificity, as can be seen in Figure 12.29.

Fluorescent chemicals absorb short-wavelength (high-energy) photons and emit longer-wavelength (lower-energy) photons. This property means that, in the visible part of the electromagnetic

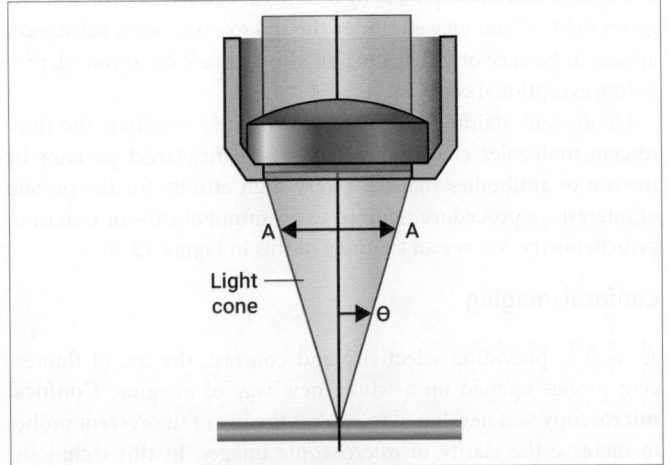

Figure 12.27 The light cone produced by a lens.

Source: Olympus Corporation, 'Angular Aperture', via https://www.olympus-lifescience.com/ja/microscope-resource/primer/anatomy/numaperture/.

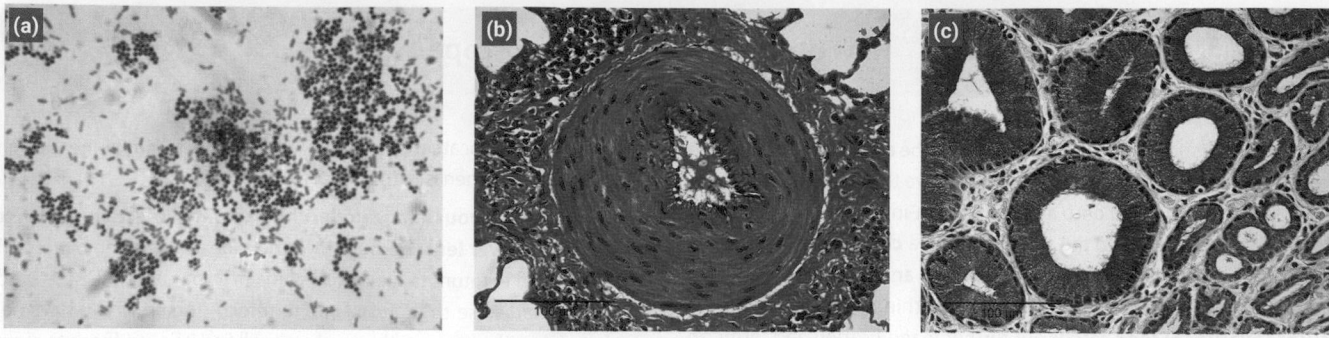

Figure 12.28 **The addition of stain to specimens can help us to visualize them** (a) Gram's stain showing a Gram-positive bacterium, *Staphylococcus aureus* (dark blue/purple) and a Gram-negative bacterium, *Escherichia coli* (reddish-pink). (b) Haematoxylin–eosin (H&E) stain applied to a sample of rat lung artery. (c) Periodic acid–Schiff's stain (PAS) applied to a sample of human stomach.

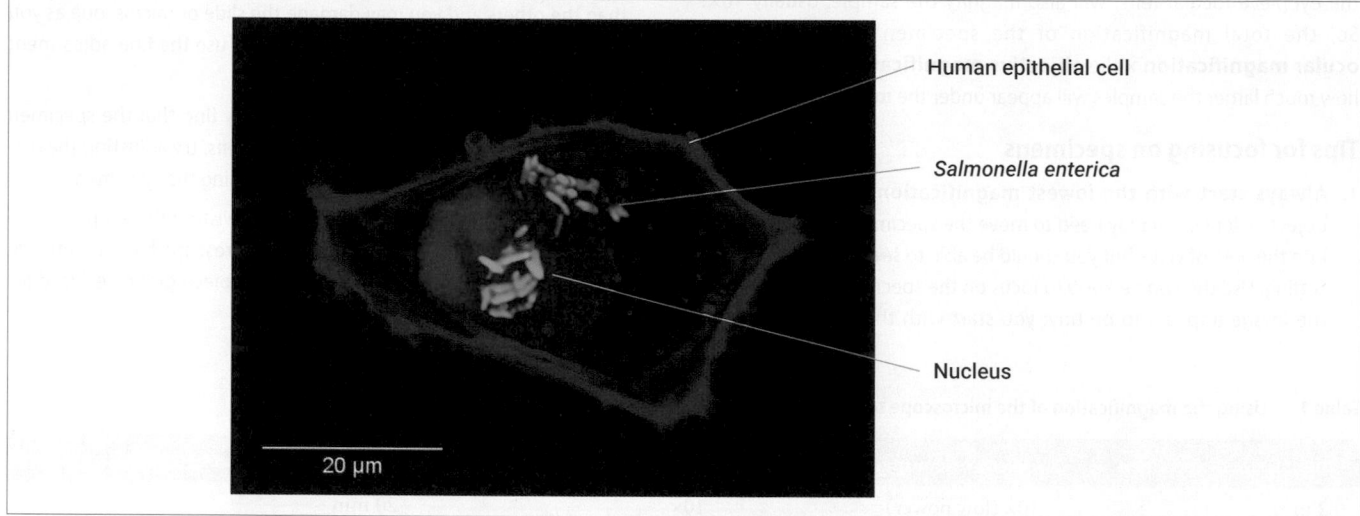

Figure 12.29 **Fluorescent staining.** A human epithelial cell (red) infected with *Salmonella enterica* expressing green fluorescent protein (GFP, green). Cell nucleus stained blue.

spectrum, the emitted light has a different colour than the light which is absorbed. Typically, blue fluorescent probes are excited by UV light, green fluorophores by blue light, and red fluorophores by green light. When viewed under the microscope, such substances appear to be very bright against an almost black background, providing exceptional contrast.

Fluorescent staining can also be extremely selective: the fluorescent molecules can be directed to specific, target proteins by the use of antibodies that have very high affinity for the protein of interest—a procedure referred to as immunohisto- or immunocytochemistry. We see an example of this in Figure 12.30.

Confocal imaging

As well as providing selectivity and contrast, the use of fluorescent probes opened up a whole new way of imaging. **Confocal microscopy** was developed to exploit the use of fluorescent probes to increase the clarity of microscopic images. In this technique, emitted light reaching the detector is restricted by a small aperture or pinhole. By strategically placing the aperture in a particular

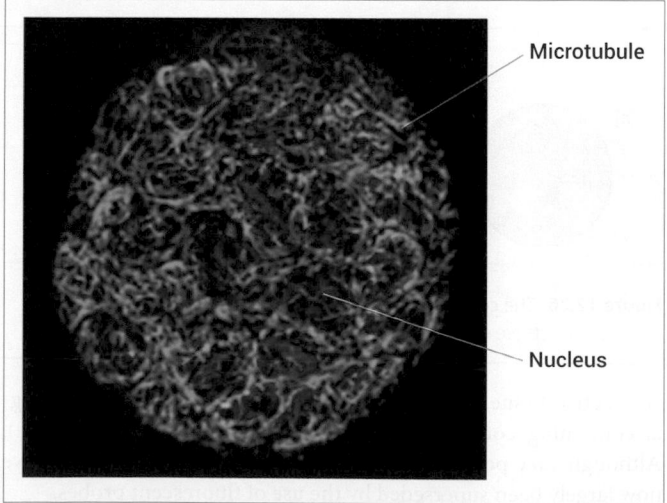

Figure 12.30 **Immunohistochemistry.** Fluorescent image of a cluster of kidney cells stained with an antibody directed against microtubules (green), with nuclei stained blue.

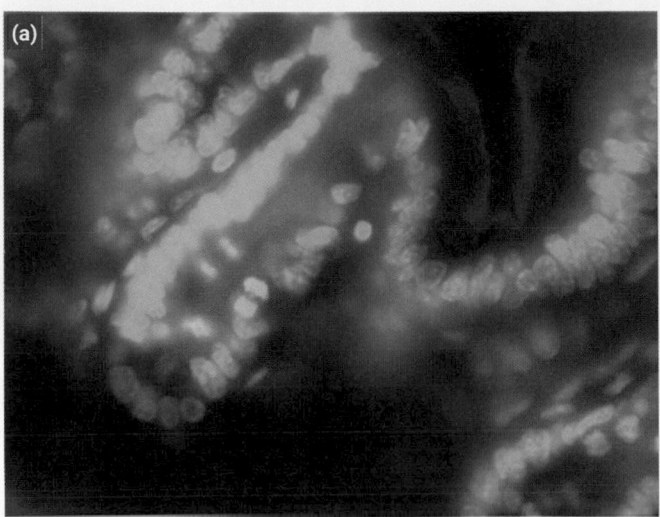
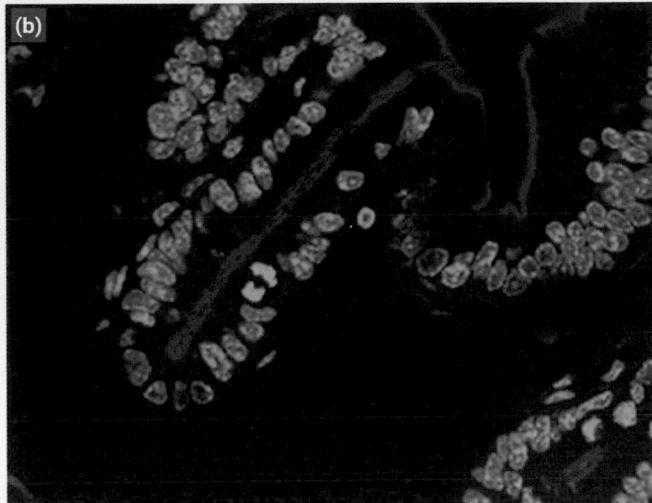

Figure 12.31 Confocal imaging. Standard fluorescence image (a) and confocal image (b).

location within the light path of the microscope, only light from the point at which the objective lens is focused (its **focal point**) reaches the detector, and any out-of-focus light is prevented from contributing to the image. As a result, a confocal image is actually an optically generated 'slice' or section through the specimen at the level of the focal plane. Look at Figure 12.31 to see the improvement in image quality achieved with a confocal image.

As well as achieving wonderful optical clarity, the confocal microscope also creates optical sections, and by moving the focus of the objective through the sample, a series of these sections can be captured. These slices can then be combined digitally to create a three-dimensional model of the sample.

Viewing live samples

The staining procedures described earlier in this section are only useable on 'fixed' (that is, dead) samples. In order to obtain high-contrast images of live specimens biologists need to use different approaches.

Phase contrast and differential interference contrast microscopy

The first attempts to improve contrast in the microscopy of live specimens involved the use of specialized optics in the microscope itself. When light passes through a specimen in brightfield microscopy, the amplitude (or brightness) of the light is changed by the sample due to the way it diffracts the light. This diffraction causes the specimen to appear dark against a bright background. As well as the amplitude of the light, what is known as the phase of the light is also changed. However, this phase change is invisible to the eye.

In the 1930s, Frits Zernike, a Dutch physicist, developed a way to make these phase changes visible by modifying the microscope to create destructive interference between the deviated and the undeviated light passing through the sample. This interference has the effect of making the specimen appear much darker than

in brightfield. Because of its use of phase changes, this method is termed **phase-contrast microscopy**.

A comparison of the kind of image captured by phase-contrast and brightfield microscopy is shown in Figure 12.32.

Differential interference contrast (DIC) microscopy uses interference with polarized light to give samples an almost three-dimensional appearance (see Figure 12.32c). Nevertheless, the contrast achieved with these methods remains inferior to that which can be attained with fluorescence microscopy.

Fluorescence microscopy

In the cold waters of the Pacific Ocean off the coast of North America lives a jellyfish by the name of *Aequorea victoria*, an example of which is shown in Figure 12.33.

This hydromedusa produces a protein, called **green fluorescent protein** (GFP), which is naturally fluorescent. This protein has been isolated and cloned, allowing the development of powerful molecular biological methods that make it possible to express GFP in virtually any living organism. Furthermore, biologists have managed to create chimeric genes that attach gene sequences that code for proteins of interest to the gene for GFP. As a result, when a cell transcribes such chimeric genes, it produces the protein of interest along with its own fluorescent tag—a tag that can be tracked wherever the protein goes.

Other fluorescent proteins have since been discovered and, through mutation, GFP itself has been modified so that we see different colours when it fluoresces—colours that fall in different parts of the electromagnetic spectrum, from blue to far-red (see Figure 12.32d). These genetic tools have allowed us to investigate intracellular protein dynamics in unprecedented ways.

However, proteins are not the only cellular components of interest. Indeed, one of the simplest cellular components—the calcium ion (Ca^{2+})—was the first to be investigated by fluorescent methods in live cells.

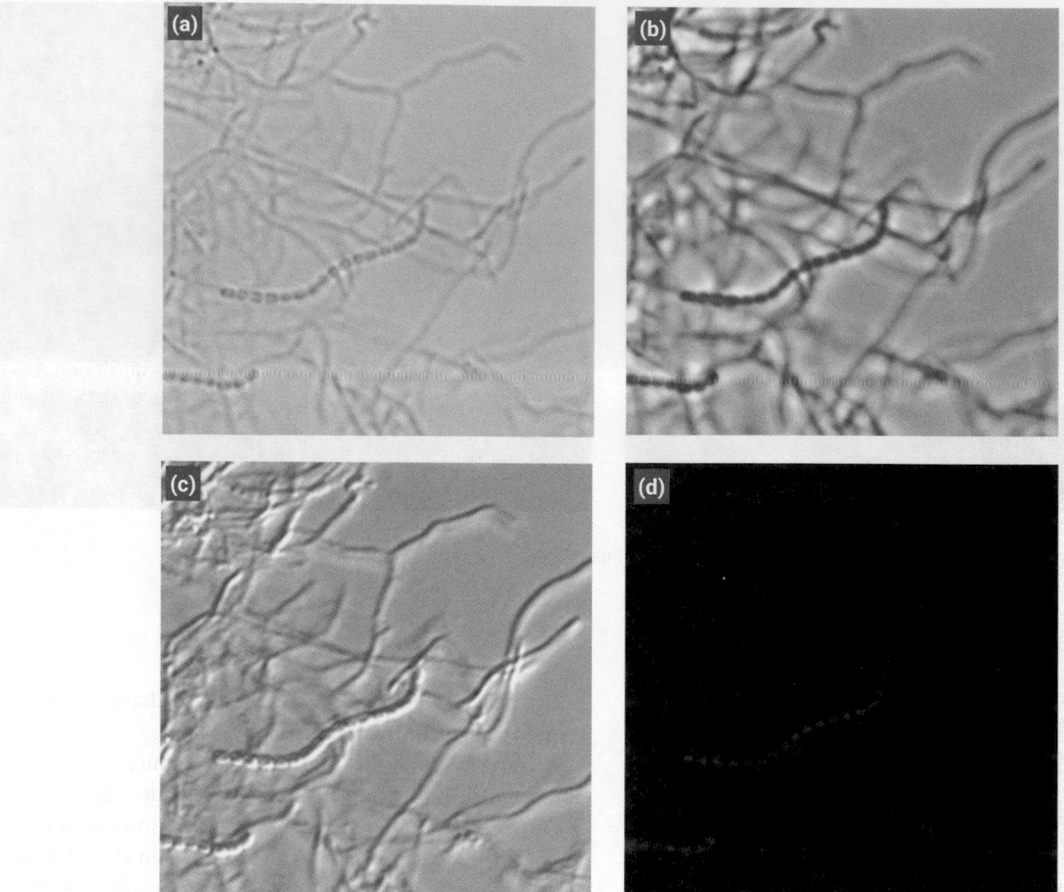

Figure 12.32 Different types of microscopy. Examples of (a) brightfield, (b) phase-contrast, (c) DIC, and (d) fluorescence microscopy.

Research carried out in the late 1960s and early 1970s first made it clear that Ca^{2+} is fundamental to the regulation of numerous biological processes, from muscle contraction to synaptic transmission. However, monitoring the dynamics of this small ion inside living cells was extremely difficult. Then, in around 1980, a PhD student at the University of Cambridge synthesized the first fluorescent probe that could be used to measure intracellular Ca^{2+} concentration: quin2. The student's name was Roger Tsien and he would later be awarded the Nobel prize for his work on GFP.

Tsien's invention of quin2 and its derivatives fura-2 and indo-1 revolutionized the study of Ca^{2+} by enabling levels of Ca^{2+} in cells to be monitored, as illustrated in Figure 12.34. Since then, numerous probes for other ions (e.g., H^+, Na^+, Cl^-, Mg^{2+}) have been developed by Tsien and other workers.

Fluorescence imaging techniques have also been extended to much larger structures than small ions, namely cellular organelles. Probes to explore particular characteristics of these organelles have been designed. For example, active mitochondria have an interior that is highly oxidative, so probes that become negatively charged in an oxidizing environment have been developed. Because charged molecules cannot pass through membranes, the negatively charged probes become trapped inside the mitochondria where they can be visualized (Figure 12.35).

Similarly, weak bases that acquire a proton in low pH environments become passively trapped in acidic organelles such as lysosomes. Consequently, fluorescent weak bases have been developed to specifically label lysosomes.

Smaller still: how electrons help us to see

We saw earlier in this section how visualization techniques that rely on visible light are limited by the resolution that can be achieved. In order to obtain better resolution, biologists must make use of something that has a much shorter wavelength than the photons that make up visible light—namely the electron.

The effective wavelength of an electron is in the order of 100,000× shorter than the photons comprising visible light, and this pushes the resolution of the electron microscope (EM) down to around 0.2 nm—that is 1000× better than the optical microscope. The major disadvantage of the EM is that biological samples are invisible to electrons. The only way to make them visible is to coat them (or 'infiltrate' them) with heavy metals, which absorb and reflect electrons. However, doing so means fixing the sample, and so limits EM to dead specimens.

In optical microscopy, brightfield imaging is referred to as 'transmitted-light' imaging because the light is transmitted through the sample. By contrast, fluorescence microscopy is referred to as 'reflected-light' microscopy because the light is projected onto the sample via the objective lens. Similarly, electron microscopy also has two modes of image acquisition: transmission electron

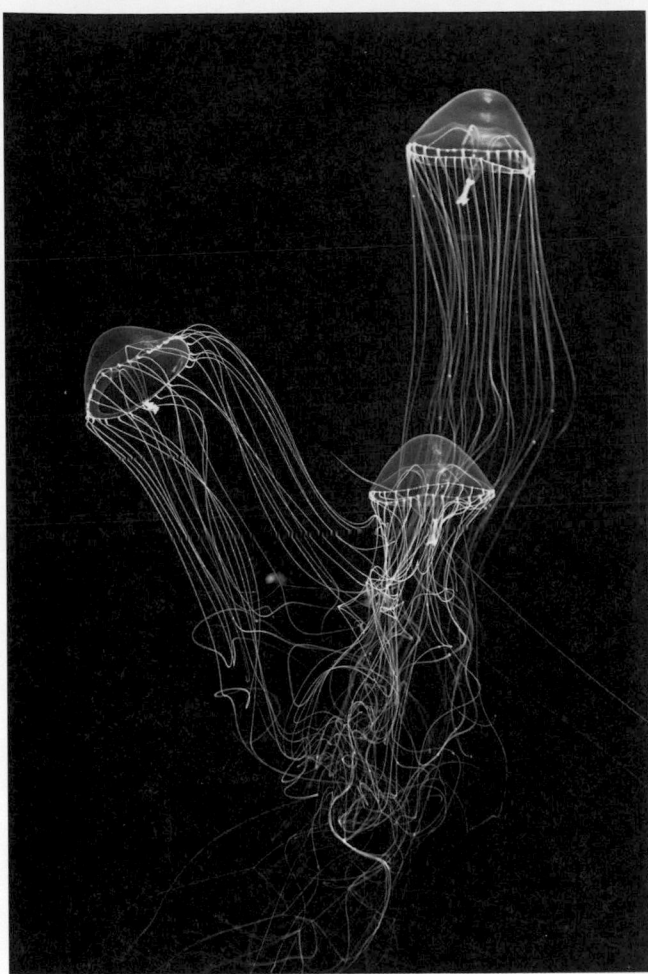

Figure 12.33 The jellyfish *Aequorea victoria*. This jellyfish produces green fluorescent protein, which is naturally fluorescent.

Source: Alex Archontakis/Alamy Stock Photo.

microscopy (TEM) and scanning electron microscopy (SEM), which rely on transmitted and reflected electrons, respectively. Both use an electron 'gun' to generate electrons that are focused onto the specimen using electromagnetic, as opposed to glass, lenses. Figure 12.36 shows images taken using TEM and SEM.

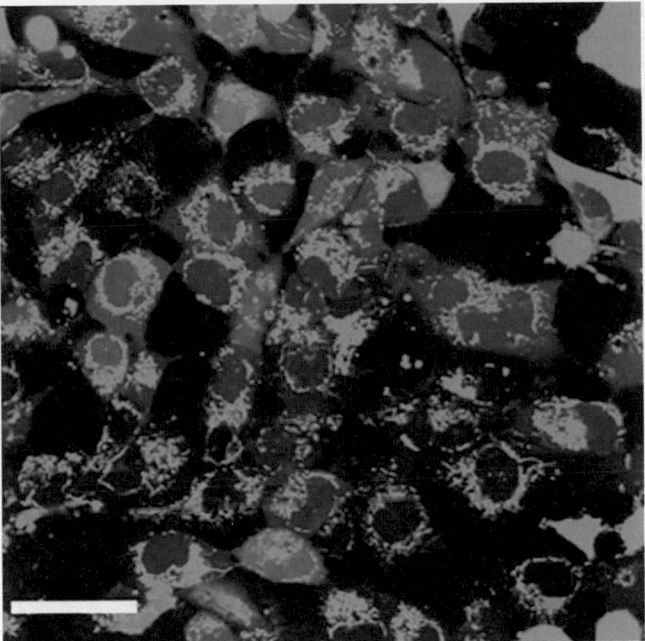

Figure 12.35 The use of a fluorescent probe to visualize mitochondria.

TEM relies upon the scattering of electrons as they pass through the sample. If the electrons encounter a continuous film of metal they will be reflected or 'back-scattered'. So the specimen must be very thin if electrons are to pass through it. Sections of fixed material, stained with heavy metals such as osmium or uranium, are cut to about 50 nm thickness and are then placed in the TEM. An image is formed from the electrons transmitted by the specimen by their conversion to photons when they strike a phosphorescent screen; the light emitted is captured by a camera. The combination of the thinness of the section, heavy-metal staining, and exquisite resolution of the TEM allows the investigation of the internal structures of even the smallest microbes.

In SEM, after fixation and dehydration, the sample is coated with a thin layer of gold or platinum, and a beam of electrons is then scanned across the specimen. The electrons reflected from the surface of the sample are captured by a detector within the microscope. This detector contains a scintillant that converts the electrons into photons. These photons are then counted by a

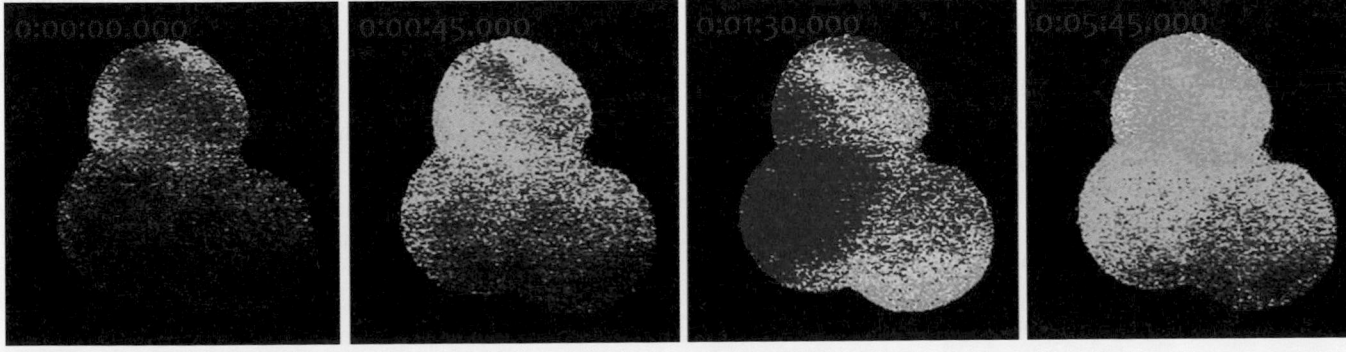

Figure 12.34 Calcium dynamics in three pancreatic acinar cells stimulated with acetylcholine revealed using fura-2. Cool colours: low Ca^{2+} concentration; warm colours: high Ca^{2+} concentration.

Figure 12.36 Transmission (TEM) and scanning (SEM) electron microscopy. Sporulating *Streptomyces coelicolor* viewed with TEM (left), which allows the visualization of internal structures, or SEM (right), which allows for the visualization of the surface of the cell or spore.

Source: Cottet-Rousselle, C., Ronot, X., Leverve, X., and Mayol, J.-F. (2011) Cytometric assessment of mitochondria using fluorescent probes. *Cytometry*, 79A: 405–425. https://doi.org/10.1002/cyto.a.21061.

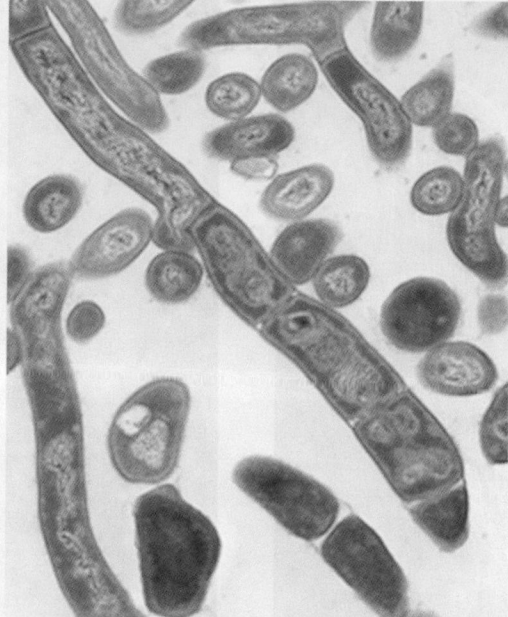
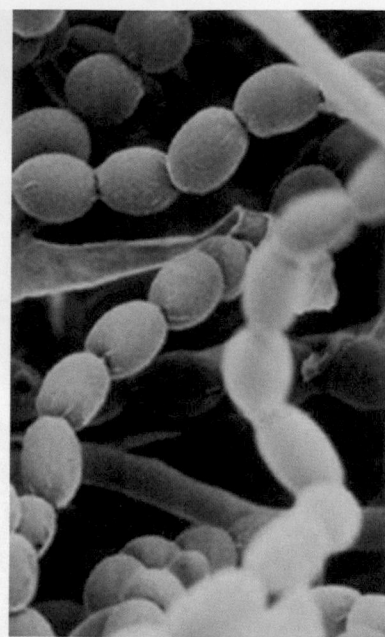

photomultiplier tube, as in the scanning confocal, and an image showing the fine details of the surface of the sample is computer-generated pixel by pixel.

Ultimately, biologists would like to image live cells with the resolution of the EM, and techniques have recently been developed to break the 'diffraction barrier' in optical microscopes, pushing the resolution to below 50 nm. However, these 'super-resolution' methods are still fairly limited with regard to the types of samples they can be used on, the numbers of fluorescent colours that can be imaged in the same sample, and the time required to acquire the images.

 Check your understanding of the concepts covered in this section by answering the questions in the e-book.

SUMMARY OF KEY CONCEPTS

- Sterilization is the removal of all contaminating micro-organisms. It can be achieved using heat, filtration, irradiation, and detergents.

- The media used for the growth of cells is generally sterilized using an autoclave. Heat-sensitive chemical solutions are sterilized using filtration.

- Microbes are grouped according to their pathogenicity and risk of infection to humans. Microbes in high-risk categories have to be handled in category 4 laboratories, which have high levels of containment.

- The ability to culture eukaryotic cells *in vitro* helps us develop our understanding of cell biology.

- Organ and cell culture is referred to as tissue culture.

- The development of media was important for the culture of both eukaryotic and prokaryotic cells. Cells can be grown in both minimal and complex media.

- Micro-organisms have different temperature and pH optima; these optima are important to recognize if microbes are to be grown successfully in the laboratory.

- The growth of cells can be measured in a number of ways including using a counting chamber, viable count, turbidity, and dry weight.

- Cells grown in batch culture go through four distinct growth phases: lag, log, stationary, and death.

- Cell growth can be described mathematically using growth kinetics.

- Cells can be visualized using microscopes, the most common being a light microscope.

- Cells can be stained to improve their visibility and allow us to see different cell structures.

- Confocal imaging increases the clarity of microscopic images.

- Phase-contrast microscopy is used to view live specimens.

- Electron microscopy allows for better resolution than techniques that use visible light as the electron has a much shorter wavelength than the photon.

Use the flashcards in the e-book to test your recall of key terms introduced in this chapter.

QUESTIONS

Looking for answers? Once you've answered these questions, follow the link in the e-book to the answer guidance and check your work.

Concepts and definitions

1. Why is aseptic technique important in maintaining sterility?
2. What are the four main ways in which sterilization can happen?
3. Draw a diagram of the growth phases seen in cell batch culture, and label each phase.
4. Who invented the first microscope, and what did he see?
5. What lenses can be typically found on a light microscope?
6. What staining method is most often used for animal cells?
7. What method would you use to view live samples?

Apply the concepts

8. Explain why the Zaire ebolavirus is classed as hazard level Group 4.
9. What are the key differences between a complex medium and a defined medium?
10. Describe the advantages and disadvantages of using dry weight to determine cell growth.
11. The doubling time of *Bacillus subtilis* is 120 minutes. Starting with one cell, how many cells would be present after 24 hours of growth?
12. Why was confocal microscopy developed?

Beyond the concepts

13. How do electrons help us with resolution?
14. The data in Table 1 are from a microbial growth curve experiment using batch culture where growth was measured as dry weight (in g). Using graph paper, plot the data and fully annotate the graph.

Table 1 Microbial growth data

Time (min)	Dry weight (g)
0	0.1
20	0.1
40	0.15
60	0.2
80	0.4
100	0.8
120	1.4
140	1.6
160	1.6
180	1.61
200	1.3
220	1
240	0.7

15. Using your graph from question 14 calculate the specific growth rate of the microbe with the unit h^{-1}. The specific growth rate equation is:

$$\mu = \frac{\ln 2}{t_d}$$

16. On your graph from question 14 label each of the growth phases 1 through to 4 and describe what is happening to the microbial growth at each phase.

FURTHER READING

Yeoman, K., Fahnert, B., Clarke, T., & Lea-Smith, D. (2020) *Microbial Biotechnology* (Oxford University Press).
  To learn more about microbial growth kinetics.
Skloot, R. (2010) *The Immortal Life of Henrietta Lacks* (Picador).
  To learn more about Henrietta Lacks.

'Bendy laser beams can examine human tissue like never before'.
  https://theconversation.com/bendy-laser-beams-can-examine-human-tissue-like-never-before-95488
  To find out more about how we are advancing microscope imaging read this article from The Conversation.

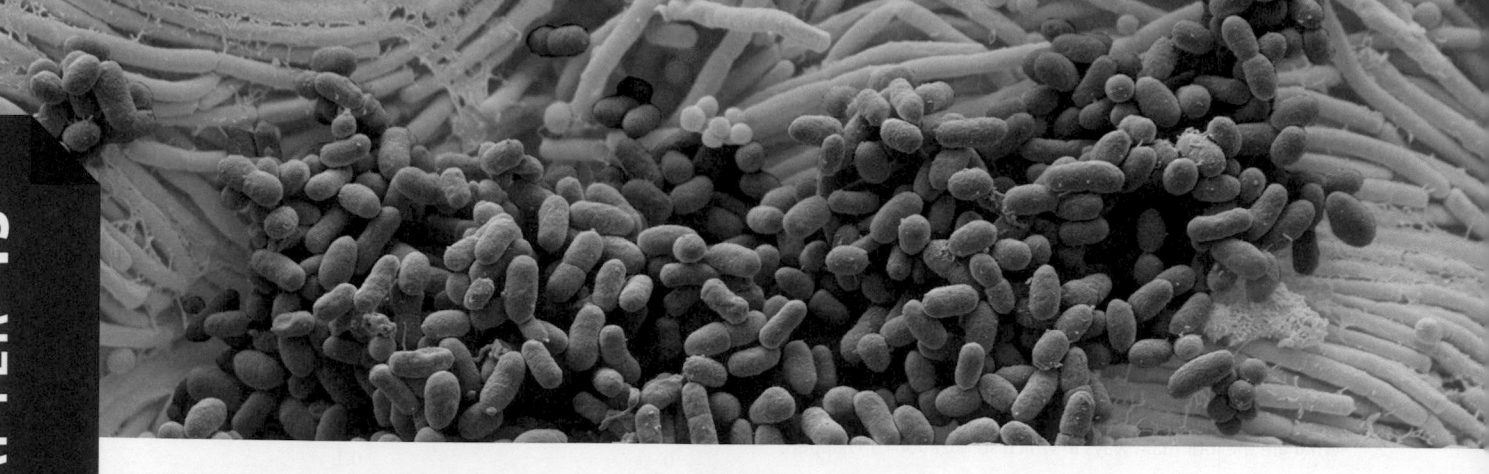

Microbes in Life

Harnessing Their Power

Chapter contents

Introduction   498

13.1   Microbial biotechnology   499

13.2   The pharmaceutical industry   499

13.3   The production of chemicals and fuels   504

13.4   Biotechnology and food and drink production   507

13.5   Microbial polysaccharides   513

13.6   Biological control agents and bioremediation   514

13.7   Biogeochemical cycles   517

🌀 Watch the key concepts video in the e-book to prepare yourself for studying this chapter.

LEARNING OBJECTIVES

By the end of this chapter, you should be able to:

- Explain that biotechnology is not a new concept, and that humankind has been using ancient biotechnology processes for several thousand years.

- Evaluate microbes for their usefulness within modern, novel biotechnology processes.

- Describe the modern molecular biology techniques that can manipulate the microbial host to produce new products for industry.

- Name pharmaceutical products and explain the roles microbes play in their production.

- Explain that microbes can also be used to make fine chemicals such as enzymes, which are then used in other processes.

- Recall that microbes have been used for the preservation and production of food for thousands of years.

- Discuss that microbes can be found in virtually every environment, and often have complex mutualistic symbiosis with other organisms.

- Evaluate the roles that microbes play in geochemical cycling.

Introduction

Despite being the smallest forms of life, microbes have a huge global impact. They existed on Earth millions of years before more complex multicellular eukaryotes. Indeed, without these early prokaryotic microbes pumping oxygen into the atmosphere, complex eukaryotic life would never have evolved. Remarkably, the fossil remains of these early microbes can still be found in the form of stromatolites—the fossils of photosynthetic cyanobacteria, which we see in Figure 9.25.

We are still heavily dependent on microbes for many processes that are central to life. For example, they carry out the chemical transformations crucial to the geochemical cycling of elements such as nitrogen and carbon, as we discover in Section 13.7. More intimately, we carry with us a vast number of diverse microbes on our skin and in our bodies, which contribute significantly to our health and well-being.

This chapter is very wide-ranging and will explore how we have put microbes to use as chemical factories to produce pharmaceutical drugs, and chemicals for industry, and how they are involved in food and drink production. We will also delve into different microbial habitats and the role the microbes play in the environment. While very few micro-organisms are agents of disease in plants, animals, and humans, their impact in this area cannot be underestimated, as we will explore in Chapter 14.

13.1 Microbial biotechnology

Humankind has a long history of using microbes to its advantage, albeit for the most part unintentionally. **Fermentation** sees microbes being used to convert sugars into alcohol or organic acids, and we will see examples of its use later in the chapter. For now, we note that its use as a means of preserving food, and as a means to produce foods and beverages, has been well documented. The manufacture of beer is thought to stretch back to 7000 BCE and the Babylonians; wine was produced by Assyrians in 3500 BCE and records show that the ancient Egyptians made bread. These records can be seen in depictions of ancient Egyptians making bread and filling bread moulds as shown in Figure 13.1.

The deliberate use of microbes by humans began with Louis Pasteur, who proved in 1851 that alcoholic fermentation was the result of microbial activity. The use of a single microbe,

Figure 13.1 Depiction of bread-making in ancient Egypt.

Source: © DEA/G Dagli Orti/De Agostini Editore/agefotostock.

Saccharomyces carlsbergensis, for brewing—a process called pure strain brewing—was carried out for the first time in 1883.

The use of any living organism in an industrial process is called **biotechnology**. The large-scale industrial use of microbes began in the 1900s with the production of acetone and butanol by fermentation, a process that uses the bacterium *Clostridium acetobutylicum* to produce acetone followed by citric acid production by the filamentous fungus *Aspergillus niger*. Since then the biotechnology industry has exploded, yielding a vast array of products that impact on every aspect of our lives.

Micro-organisms are extremely useful in biotechnology for a number of reasons:

1. Microbes are relatively easy to cultivate and can grow rapidly when the right conditions are provided.

2. Many microbes can grow on cheap waste substrates, e.g. bagasse, which is a waste product from the sugar production industry.

3. A hugely diverse range of potential microbial products are of use to humankind.

4. Many microbes can readily undergo genetic manipulation to modify and improve strains, or to produce genetically modified organisms.

> **PAUSE AND THINK**
>
> Do you consider processes such as the making of cheese and beer to be examples of ancient biotechnology?
>
> *Answer:* The definition of biotechnology is the use of any organism in an industrial process. Hundreds of years ago both beer and cheese production were often local processes, and the use of microbes in the production of these products was not through a deliberate understanding of microbiology. However, we know that fermented products were traded, and papers referencing 'ancient biotechnology', which describes the production of beer and cheese.

In later sections, we will look at how microbes are used in a variety of different biotechnology processes, from the production of pharmaceuticals, to organic acids and alcohols for the chemical industry, to enzymes and fine chemicals. Microbes can be used as food, but also to *make* food. Microbes also have a role in the treatment of our waste material, as biological control agents, and in bioremediation.

 Check your understanding of the concepts covered in this section by answering the questions in the e-book.

13.2 The pharmaceutical industry

The production of pharmaceuticals is a big business, worth many billions of pounds each year across the world. Microbes are hugely important in the pharmaceutical industry: not only do they produce an amazing array of diverse chemicals with therapeutic benefits, but they can also be used to transform existing chemical compounds into therapeutically useful ones, and to act as mini factories for drug production.

Figure 13.2   Alexander Fleming in his laboratory at St Mary's Hospital.
Source: World History Archive/Alamy Stock Photo.

Antibiotics

Look at Figure 13.2. You might recognize this scientist as Alexander Fleming. It is well known that Fleming, working at St Mary's Hospital in 1928, first noted the effect that the filamentous fungus, *Penicillium notatum*, had on certain bacteria: the penicillin produced by this fungus inhibited bacterial growth. What is not quite so well-known is what happened next. Penicillin might have remained a scientific curiosity had it not been for two other scientists, Howard Florey and Ernst Chain (Figure 13.3), who pioneered work to produce pure penicillin on a large scale. Howard Florey was the head of a laboratory at the University of Oxford and Ernst Chain was an immigrant, who had escaped to England from Nazi Germany.

Florey employed Chain in his laboratory and together they began to investigate penicillin in more detail, both men being motivated by casualties from the Second World War and the deaths seen among those casualties from infection. Chain came up with

Figure 13.3   Nobel Prize laureates of 1945. From left: Artturi Virtanen, Chemistry; Alexander Fleming, Medicine; Ernst. B Chain, Medicine; Gabriela Mistral, Literature; and Howard Florey, Medicine.
Source: TT News Agency/Alamy Stock Photo.

a method of purifying and concentrating penicillin, enough for it to be tested on mice, and eventually humans. However, sufficient quantities were unable to be made in the UK because manufacturing plants were being taken over by the war effort. Florey used his extensive contacts in the USA to get large quantities made in that country and the first doses of penicillin were taken over on the D-day landings.

All three men—Fleming, Florey, and Chain—won the Nobel Prize for Physiology or Medicine in 1945. Penicillin started the so-called 'golden age' of antibiotics and has contributed to increased life expectancy.

Despite penicillin coming from a filamentous fungus, most clinically useful antibiotics do not actually come from fungi, but from bacteria, especially species belonging to the genus *Streptomyces*. *Streptomyces* are filamentous bacteria that primarily live in soil. The first of these antibiotics to be discovered was streptomycin, from the bacterium *Streptomyces griseus*, in the early 1940s. There are several different groups of antibiotics, which have different modes of action. These are shown in Table 13.1.

New groups of antibiotics are still being discovered; take a look at Clinical Box 13.1 to find out more.

Antibiotic resistance

One of the biggest challenges humans now face is antibiotic resistance. You are likely to have heard of the hospital-acquired infection methicillin-resistant *Staphylococcus aureus* (MRSA), which can cause serious infection. The only antibiotic that can treat this infection now is vancomycin, although it is worrying that some MRSA strains are developing resistance to this, too.

Antibiotic resistance is not a new phenomenon. Indeed, it was not long after penicillin started to be used that the first cases of resistance were reported. New antibiotics are vital if we are to overcome the threat from antibiotic resistance and to prevent deaths from infection, but it might surprise you to know that the last new group of antibiotics to reach the market was the lipopeptides, back in 2003. The problem is that it takes a long time for a new antibiotic to be discovered or synthesized, and subsequently approved for use in humans. However, it takes comparatively little time for bacteria to become resistant.

The process of drug discovery is also extremely expensive, and pharmaceutical companies are unwilling to take the financial risk involved in developing new antibiotics which are then only used by patients for a short length of time, or are only used as a last resort. For more information on the discovery of new antibiotics read Clinical Box 13.1.

Steroid production

Steroids are used to treat a variety of illnesses, such as rheumatoid arthritis, asthma, and eczema. They are also used as oral contraceptives. The production of steroids used to be a costly business, as the hormones had to be extracted from the adrenal glands of animals. However, less cost-intensive methods have since been developed.

For example, the mould *Rhizopus arrhizus* (shown in Figure 13.4) is able to convert a steroid called diosgenin, which is found in Mexican yams, into an intermediate compound that can then be

Table 13.1 Antibiotic groups, modes of action and the microbes producing them

Antibiotic group	Examples	Producing organism	Mode of action
β-lactams	Penicillin G	*Penicillium chrysogenum*	Inhibit cell wall synthesis
	Methicillin		
	Oxacillin		
	Ampicillin	Semi-synthetic penicillin	
Cephalosporins	Cephalothin	*Cephalosporium acremonium*	Inhibit cell wall synthesis
	Cefotetan		
	Cefepime		
Aminoglycosides	Gentamicin	*Micromonospora purpurea*	Inhibit protein synthesis
	Kanamycin	*Streptomyces kanamyceticus*	
	Neomycin	*Streptomyces fradiae*	
	Streptomycin	*Streptomyces griseus*	
Fluoroquinolones	Ciprofloxacin	Synthetic antibiotic	DNA synthesis inhibitors
	Moxifloxacin		
Monobactams	Aztreonam	*Chromobacterium violaceum*	Inhibit cell wall synthesis
Carbapenems	Ertapenem	*Streptomyces* spp.	Cell wall synthesis
	Meropenem		
Polypeptide antibiotics	Bacitracin	*Bacillus subtilis*	Inhibit cell wall synthesis
Macrolides	Erythromycin	*Saccharopolyspora erythraea*	Inhibit protein synthesis
	Clarithromycin	Semi-synthetic	
	Vancomycin	*Streptomyces orientalis*	Inhibit cell wall synthesis
Other	Tetracycline	*Streptomyces* spp.	Inhibit protein synthesis
	Trimethoprim	Synthesized from gallic acid	Folic acid synthesis inhibitor

---

CLINICAL BOX 13.1   The war against antimicrobial resistance: the race to find new antibiotics

Since the discovery and purification of penicillin we have been in a 'golden age' of antibiotics where deaths from infections caused by minor wounds are a feature of our past. Yet many scientists feel that we are now at the end of this golden period, and that the phenomenon of antimicrobial resistance is making us vulnerable to infectious disease once again. We have had no new antibiotics since the lipopeptides in 2003, and infections from so called 'super-bugs' such as MRSA are becoming harder to treat.

We can combat this lack of new antibiotics to some extent by ensuring hospitals are clean, by washing our hands, and by being responsible in our prescribing and use of antibiotics. However, we urgently need to discover and develop new antibiotics. So how can this be done?

One avenue of discovery is to investigate novel ecological niches that may be home to microbes that produce previously undiscovered antibiotics. Two research groups working in collaboration, one at the University of East Anglia, led by Professor Matt Hutchings, and the other at the John Innes Centre, led by Professor Barrie Wilkinson, have done exactly that.

Leaf cutter ants (Figure 1) belong to the tribe Attini and have a mutualistic relationship with fungi: the ants collect plant material, which they chew up and then feed to a symbiotic fungus called *Leucoagaricus*

Figure 1   The leaf cutter ant *Acromyrmex* sp.
Source: Matt Hutchings/Norwich Research Park Image Library/CC BY 4.0.

gongylophorus. The fungus is then used as the sole source of food for the growing ant larvae. This is an ancient relationship, having evolved between 50 and 60 million years ago.

The environment created by the ants to grow their fungal food is both warm and moist. These are ideal growing conditions not only for the fungus itself, but also for other microbes, including other fungi, which could potentially infect and damage their fungal gardens. With this in mind, the ants constantly care for their fungal gardens, removing any unwanted interlopers.

In addition to this careful weeding, the leaf cutter ants have evolved another remarkable symbiotic relationship to combat these pathogens: they carry Actinobacteria on their bodies, which include bacteria from the genus *Streptomyces*. These bacteria produce antibiotics, which the ants use to suppress incoming pathogens that could otherwise infect their fungal gardens.

Scientists are very interested in these *Streptomyces* bacteria, as currently 70 per cent of our clinically useful antibiotics come from this genus.

Professors Hutchings and Wilkinson have been successful in isolating these *Streptomyces* and other Actinobacteria from leaf cutter ants. Indeed, they recently isolated an entirely new species of *Streptomyces*, called *S. formicae*, from the African ant *Tetraponera penzigi*, which harvests the acacia plant. They have found that *S. formicae* produces a new antibiotic that no-one has seen before. The antibiotic—a type of chemical called a pentacyclic polyketide—belongs to a new group of antibiotics called the formicamycins. Preliminary results have shown that it is active against both MRSA and vancomycin-resistant enterococci.

Figure 13.4 *Rhizopus arrhizus*, used for the biotransformation of the steroid diosgenin.

Source: Science History Images/Alamy Stock Photo.

chemically transformed into cortisone for human use. The Upjohn chemical company commercialized this process, after which the cost of cortisone production reduced dramatically.

Steroid production is an example of **biotransformation**: the steroid is not produced directly by the microbe, but instead the microbe is used to modify a naturally occurring steroid.

Hypocholesterolaemic drugs

Cholesterol is the main sterol found in the plasma membranes of animals, where it plays an important role in stabilizing the membrane. Around 30 per cent of the cholesterol in your body comes from your diet; the rest is synthesized by your liver. High levels of cholesterol in the blood can cause atherosclerosis, whereby a build-up of plaque in the arteries can lead to coronary heart disease. Statins are drugs that reduce cholesterol levels by reducing the synthesis of cholesterol by the liver. They do this by inhibiting the activity of an enzyme called 3-hydroxy-3-methylglutaryl-coenzyme A reductase.

Naturally occurring statins can be obtained from different genera and species of filamentous fungi. For example, lovastatin is mainly produced by *Aspergillus terreus* strains, and mevastatin by *Penicillium citrinum*. One type of statin, atorvastatin, is the most commercially successful drug in history, with revenue from sales exceeding $120 billion worldwide.

Immunosuppressants

Immunosuppressants, which suppress the activity of the immune system, have been highly successful in preventing the rejection of transplanted organs. Ciclosporin A was first discovered as an anti-fungal agent in the filamentous fungus *Tolypocladium inflatum*. Two other products, rapamycin and tacrolimus, are also on the market; both come from actinomycete bacteria.

All these drugs work by binding to an immunophilin protein, a type of intracellular protein that is important in the immune system. The complex produced when the drug binds to the immunophilin protein interferes with activation of lymphocytes, a type of blood cell. Interestingly, tacrolimus has also proven successful as a treatment for atopic dermatitis, a common skin disease.

▶ **We learn more about how the immune system works in Chapter 27.**

Recombinant DNA technology and drug production in microbes

The development and application of recombinant DNA technology, which we explore further in Chapter 8, has enabled the production of a wide range of pharmaceutical products. Human insulin was the first product to be made in this way and to be approved for use in humans. The insulin protein consists of two polypeptide chains: the A chain is 21 amino acids in length and the B chain is 30 amino acids in length; its structure is shown in Figure 13.5. Recombinant DNA technology sees the genes coding for the A and B chains being cloned into separate plasmids, as shown in Figure 13.6. The plasmids are transformed into the host bacterium *Escherichia coli*. The two chains are produced by the *E.coli* and the A and B chains are then mixed to produce functional insulin.

▶ **We learn more about transformation in Section 10.1.**

Other medicinal products produced by recombinant DNA technology include human growth hormone and Factor VIII for the treatment of haemophilia. Table 13.2 shows some further examples.

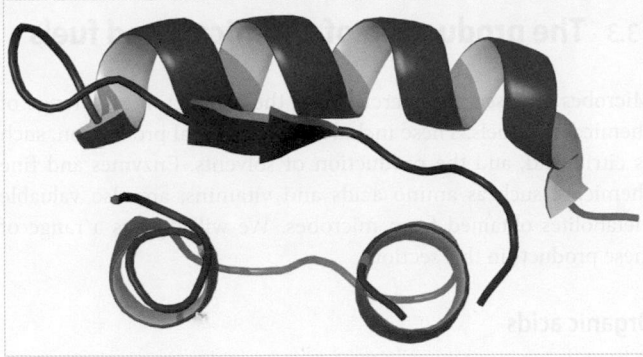

Figure 13.5 The structure of insulin. This side-on view shows how insulin comprises two separate polypeptide chains.

Source: Crowe & Bradshaw, *Chemistry for the Biosciences* 4th edn, 2021, Oxford University Press.

Table 13.2 Examples of recombinant proteins used in treatments

Disease	Example protein
Cancer treatment	Interferons
	Interleukins
Cardiovascular disease	Erythropoietin
	Urokinase
Endocrine (hormonal) disorders	Human growth hormone
	Insulin
Neurological disorders	Endorphins
	Neuropeptides
Vaccines	Hepatitis B
Blood clotting	Factor VIII

Figure 13.6 Recombinant insulin. Steps in the production of recombinant insulin.

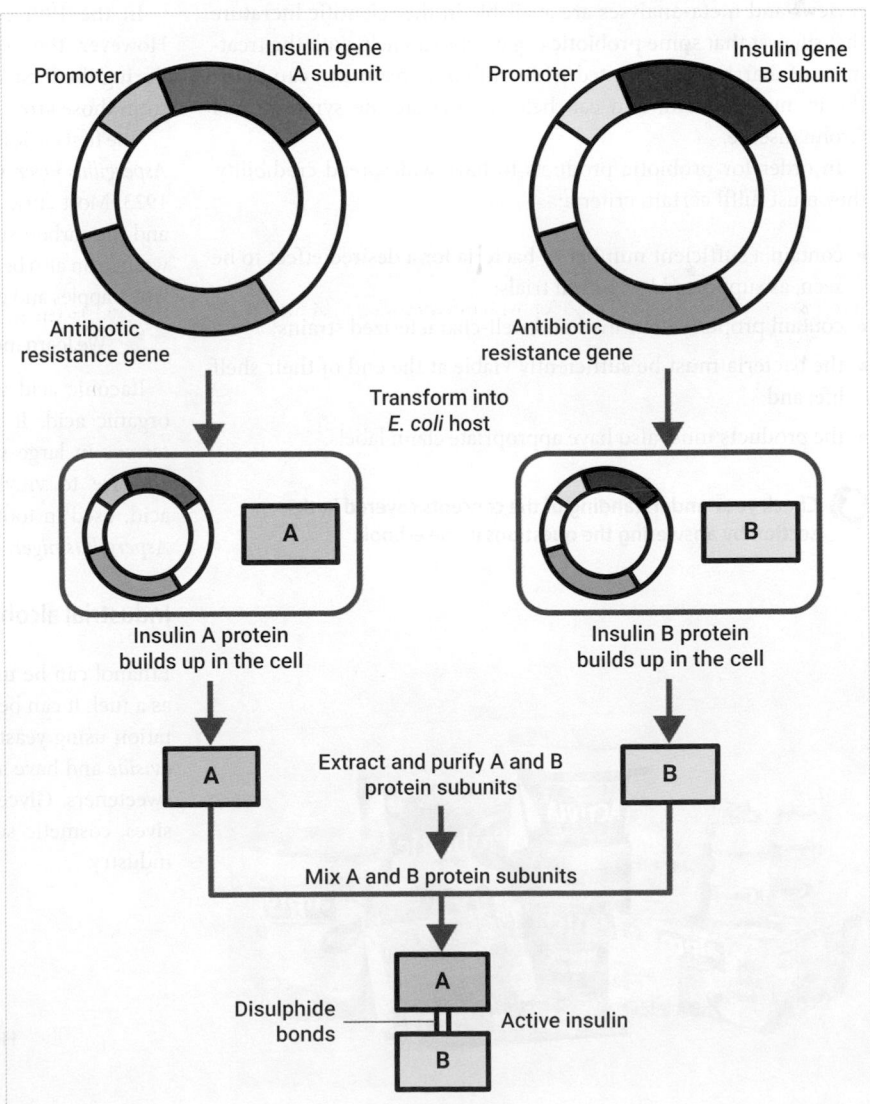

Probiotics

Probiotics are often called 'friendly' bacteria. The popularity of probiotic products has increased dramatically in recent years, with a whole range of products now available, from drinks and chocolate to tablets. Some of these products are shown in Figure 13.7.

Probiotics are often portrayed in advertisements as 'helping to speed digestive transit' and we spend millions on them in the hope that they will improve our health. The worldwide market for probiotics is thought to be in excess of $44 billion. Probiotics are defined by the World Health Organization as 'living micro-organisms that, when administered in adequate amounts, confer a health benefit on the host'. Probiotics must be able to tolerate and survive stomach gastric acid in order to colonize the small and large intestines.

Probiotics are not genera or species; a particular type of bacterium does not become a probiotic until it is shown by evidence to confer a specific health benefit. Typical organisms associated with probiotic products include *Lactobacillus* spp. and *Bifidobacterium* spp. There have been many trials conducted, and numerous reviews and meta-analyses are available in the scientific literature that suggest that some probiotic organisms can help with the treatment of diarrhoea associated with antibiotic treatment, can boost the immune system, and can help to alleviate the symptoms of Crohn disease.

In order for probiotic products to have widespread credibility, they must fulfil certain criteria:

- contain a sufficient number of bacteria for a desired effect to be seen, as supported by clinical trials;
- contain properly identified and well-characterized strains;
- the bacteria must be sufficiently viable at the end of their shelf life; and
- the products must also have appropriate claim labels.

 Check your understanding of the concepts covered in this section by answering the questions in the e-book.

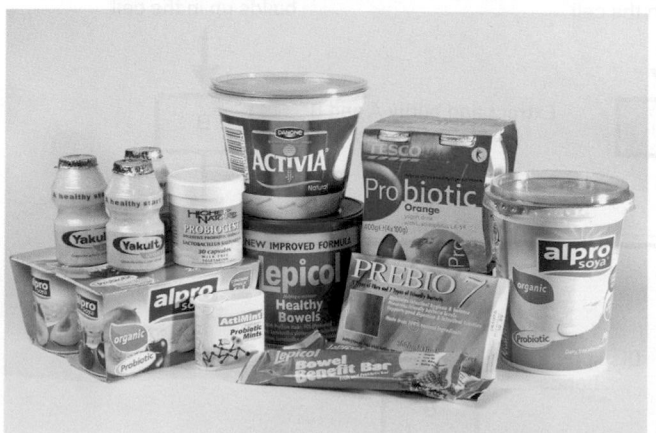

Figure 13.7 Probiotics. A range of commercially available probiotic products.
Source: Cordelia Molloy/Science Photo Library.

13.3 The production of chemicals and fuels

Microbes are used commercially for the production of a range of chemical and fuels. These include bulk chemical production, such as citric acid, and the production of solvents. Enzymes and fine chemicals, such as amino acids and vitamins, are also valuable metabolites obtained from microbes. We will discuss a range of these products in this section.

Organic acids

Citric acid is a metabolite of both plants and animals; its structure is shown in Figure 13.8. It is a remarkable chemical. it is safe, as well as biodegradable, and has a whole range of different uses. For example, citric acid is used extensively in the food industry where its addition to canned products prevents the loss of vitamin C. It can also be found in soft drinks, jams, and confectionery. Non-food uses include cleaning metals, maintaining metals in solution for electroplating, and calico printing.

In the 19th century, most citric acid was produced in Italy. However, the poor maintenance of the lemon and lime groves during the First World War resulted in prices increasing as output from those groves fell.

The first citric acid production plant to use the filamentous fungus, *Aspergillus niger*, was established by Pfizer in Brooklyn, New York, in 1923. Most citric acid is made using a **batch fermentation process**, and the carbon substrate generally used is molasses, although other wastes can also be used, such as apple and grape pomace (the pulp left when apples and grapes are pressed), and carrot waste.

▶ **We learn more about batch fermentation in Section 12.4.**

Itaconic acid is another example of a commercially produced organic acid. It is made by the filamentous fungus *Aspergillus terreus* in large stainless steel tanks. Itaconic acid is used as an additive to vinyl polymers, such as emulsion paint. Gluconic acid, used in toothpaste manufacture, is produced by the fungus *Aspergillus niger*.

Industrial alcohols

Ethanol can be used as a solvent in other industrial processes, or as a fuel. It can be made chemically from petroleum or by fermentation using yeasts. Glycerols are produced by *Saccharomyces cerevisiae* and have a wide range of uses, as solvents, plasticizers, and sweeteners. Glycerols are also used in the manufacture of explosives, cosmetic soaps, and anti-freeze, as well as in the printing industry.

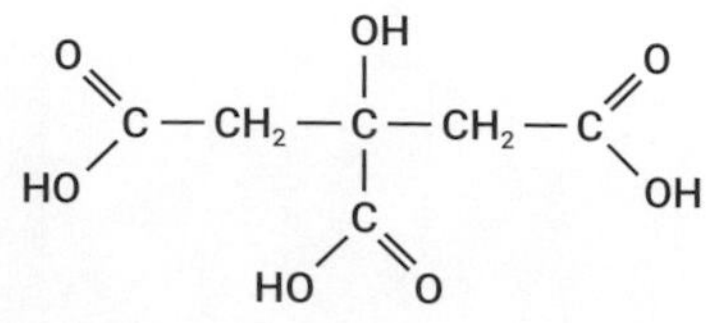

Figure 13.8 The structure of citric acid.

Enzymes and fine chemicals

Of the enzymes produced on an industrial scale, over half of them come from fungi, and one-third from bacteria. Enzymes are important for many different purposes. For example, proteases produced by *Bacillus* spp. (e.g. *Bacillus licheniformis*) are used in washing powders. Enzymes are also used in the food industry. Table 13.3 shows some more examples of enzymes, the organisms that produce them, and their uses. You can see from this table that the uses are very wide-ranging; some may even surprise you!

PAUSE AND THINK

Why are fungi so good at producing commercially useful enzymes?

Answer: Obviously, fungi do not produce enzymes just for our convenience. Many fungi live saprophytic lives in the soil where they come across complex organic molecules, which they can use for growth. As fungi do not have a gut, they produce a wide range of extracellular enzymes that can break down complex organic compounds into smaller molecules which can be taken into the fungal hyphae by passive diffusion.

Amino acids are produced commercially for use in a wide range of products. For example, aspartic acid is used to manufacture aspartame, a low-calorie sweetener. However, L-glutamic acid (GA) production represents the largest industry, with over 200,000 tonnes produced annually. The main use of L-glutamic acid is as a seasoning, better known as monosodium glutamate (MSG). MSG used to be produced by the hydrolysis of wheat gluten or soybean protein. However, it is now produced by fermentation.

The organisms used in fermentation to produce GA are Gram-positive, non-spore forming, and non-motile bacteria. Examples include *Corynebacterium glutamicum* and *Brevibacterium flavum*. The carbon source for the fermentation is molasses or sugar cane juice; more complex organic wastes tend not to be used as it is hard to purify the glutamic acid post-fermentation.

As populations become more health conscious, the market for vitamin supplements has grown dramatically. In the UK we spent nearly £500 million on vitamins in 2020,[1] and it is estimated that 50 per cent of us take them on a daily basis. Vitamins used to be made from both plant and animal biomass, but microbes are now used as the major source. For example, strains of the single-celled fungus *Saccharomyces cerevisiae* are used to make a range of B vitamins. Table 13.4 shows some more examples.

Microbial fuels

Biofuels are defined as gas, liquid, and solid fuels generated from renewable materials that come from plants and animals, more generally called biomass. Examples of these fuels include

Table 13.3 Microbial production of enzymes and their uses

Enzyme	Producing organism(s)	Use
Amylase	*Bacillus licheniformis*	Hydrolysis of starch in the bread and beer industry
	Aspergillus oryzae	Textile paper and detergent industries
Catalase	*Aspergillus* spp.	Generation of oxygen from peroxide to convert latex to rubber
Cellulase	*Trichoderma reesei*	Production of fruit drinks
	Bacillus spp.	Used in the textile industry to finish denim fabric
	Clostridium spp.	
Invertase	Yeasts, *Aspergillus* spp.	Hydrolysis of sucrose
		Used to make soft centres in chocolates
Lactase	*Aspergillus niger* and *Aspergillus oryzae*	Hydrolysis of lactose in milk products
Lipases	*Bacillus* spp.	Detergents, textile, and cosmetic industries
	Candida rugosa	Biodegradation
Pectic enzymes	*A. niger, P. notatum*	Pre-treatment of fresh juices
Proteases	*Aspergillus* spp.	Tenderize meat
	Bacillus spp.	Rennin replacement
		Washing powder enzymes
Phytase		Producing animal feed stocks
Rennet	*Cryphonectria parasitica*	Milk coagulation in cheese manufacture
Restriction enzymes and ligase enzymes	Various bacterial species	Use in recombinant DNA technology
		Restriction enzymes used to cut DNA at specific recognition sequences
		Ligase enzymes join cut DNA ends back together

---

[1] Source: https://store.mintel.com/uk-vitamins-and-supplements-market-report.

Table 13.4 Microbial production of vitamins and pigments and their uses

Product	Producing organism	Use
Vitamin C	Biotransformation process	Nutritional supplement
B vitamins	Yeasts: *Saccharomyces cerevisiae*	Nutritional supplement
B$_{12}$	*Pseudomonas denitrificans* and *Propionibacterium shermanii*	Nutritional supplement
β-carotene	*Phycomyces blakeleeanus*	Precursor of vitamin A; colouring agent for margarine and baked goods
Ergosterol	*Saccharomyces cerevisiae*	Source of vitamin D
Riboflavin	*Ashbya gossypii*	Nutritional supplement

Table 13.5 Oil content of micro-organisms

Micro-organism	Species	Oil content (% dry weight)
Microalgae	*Botryococcus braunii*	25–75
	Schizochytrium spp.	50–77
Yeasts	*Lipomyces starkeyi*	64
	Rhodotorula glutinis	72
Bacteria	*Arthrobacter* spp.	>40
	Acinetobacter calcoaceticus	27–38
	Umbelopsis isabellina	86
Filamentous fungi	*Thermomyces lanuginosa*	75

ethanol, methanol, and biodiesel. Microbial biofuels are seen as an alternative to fossil fuels. They have advantages in that they do not contribute to greenhouse gases, and they do not compete for agricultural land. Despite these benefits, they have not yet reached their potential as viable alternatives to petrol or diesel as the technology to produce them is still being developed.

Microbes such as fungi (both yeasts and filamentous fungi) and microalgae could be used as biomass for the production of biodiesel as they have a rich oil content. For example, microalgae can have an oil content of greater than 80 per cent by weight of dry biomass. Table 13.5 shows more microbes which have the potential to be used for biodiesel production.

The production of bioalcohols uses fermentation. The substrate for fermentation is usually corn starch or cane sugar, but other plant materials can be used, providing they have been pre-treated to release utilizable sugars. Organisms such as the fungus *Saccharomyces cerevisiae* or the bacterium *Zymomonas mobilis* are then grown in these substrates, producing ethanol as they grow. Methanol can be produced by methanotrophic bacteria such as *Methylosinus trichosporium* using methane as a carbon source.

Biohydrogen is another example of a potential biofuel as it produces large amounts of energy when combusted and can be converted to electricity by fuel cells. These fuel cells can have a number of applications, including the powering of vehicles (Figure 13.9).

Figure 13.9 Biofuels. The refuelling of a car with a hydrogen fuel cell.
Source: Sunpix Environment/Alamy Stock Photo.

Biohydrogen is a form of clean energy as the product of combustion is water with no associated gas emissions. The industrial development of biohydrogen is currently focusing on photosynthetic bacteria such as *Cyanobacteria* spp. and also green algae.

 Check your understanding of the concepts covered in this section by answering the questions in the e-book.

13.4 Biotechnology and food and drink production

Microbes can be used within the manufacturing process to make different foods and drinks, but they can also be used directly as food in the form of **single-cell protein**. Let us now explore how microbes are used within the food and drink industry in both these ways.

Microbes as food: single-cell protein

Single-cell protein (SCP) is the outcome of the production of biomass from microbes, including single-celled and filamentous fungi, bacteria, and algae. SCP is a bit of a misleading term, as it is not a pure protein, but is in fact the whole microbial biomass, which also contains carbohydrate, lipid, nucleic acids, minerals, and vitamins.

The production of SCP was prompted by the two world wars, during which there was a shortage of meat and fish protein. This shortage stimulated interest in using microbial biomass as a source protein for both human and animal consumption to try to boost protein supplies. For example, strains of *Saccharomyces cerevisiae* and *Candida utilis*, which was grown to produce Torula yeast, were used to supplement other sources of protein. The population expansion after the Second World War made protein production an even more urgent issue, leading to a number of different microbial protein production ventures, especially in the 1960s and 1970s.

There are distinct advantages to using microbial protein over plant and animal protein. These include:

- microbes grow rapidly and are highly productive;
- microbial cells have a high protein content (30–80 per cent on a dry weight basis);
- microbes can use a wide range of substrates for growth;
- microbe strains can be selected and further developed;
- microbial biomass production occupies little land area when compared to plant crops and animal production;
- microbial biomass production can occur all year round as it is independent of season;
- product quality is consistent.

Given all these advantages, why don't we all eat microbial protein? Well, there are some issues of safety and societal acceptability.

One major issue with SCP is the elevated nucleic acid levels they exhibit, especially RNA. These high levels of RNA are due to high growth rates and high protein content. Unfortunately, the digestion of RNA leads to the generation of purine compounds, which lead to an excess of uric acid build up. This excess uric acid can crystallize in the joints to give gout-like symptoms or kidney stones. RNA can be reduced to acceptable levels by using heat, although this adds to the cost of SCP production.

So what of societal acceptability? While most people are happy to eat mushrooms, the thought of eating bacteria and mould is not palatable to some. The manufacturers of Quorn™ have overcome this to a large extent by promoting its health benefits, and the growing market in probiotic products has made the idea of consuming bacteria as part of the diet more acceptable.

Costs of SCP production: the carbon substrate

The overall cost of SCP depends on many variables, but the major cost of any SCP process is the carbon substrate, which can represent up to 50–60% of the total cost. Renewable sources are desirable, including in particular agricultural, dairy, and wood-processing wastes. However, some substrates require more pre-treatment than others.

The more chemically reduced the substrate (e.g. oils and hydrocarbons), the better the biomass yield of the microbe. Reduced substrates are more expensive, but the advantage is that they do not need any pre-treatment. By contrast, oxidized substrates such as cellulosic wastes are cheaper, but they have to be pre-treated to release more simple carbohydrate molecules such as glucose. The cost of the pre-treatment adds to the cost of the SCP production, and these oxidized substrates are not converted into microbial biomass as efficiently.

Table 13.6 shows some of the carbon substrates that are used to produce microbial biomass.

How is SCP produced?

Several SCP ventures failed mainly due to the cost of the substrate used for microbial growth. In 1971 British Petroleum joined with the Italian company ANIC to manufacture Toprina, a product made by growing yeast on the residues of crude oil. Its market was as an alternative to soybean cake in the animal feed industry. It was also proposed as a supplement to the diet of the developing world. But Toprina was a spectacular failure: it suffered heavy financial

Table 13.6 Carbon substrates used to produce microbial biomass

Carbon substrate	Micro-organism
Carbon dioxide	*Spirulina*
	Chlorella
Liquid hydrocarbons	*Saccharomyces lipolytica*
	Candida tropicalis
Methane	*Methylomonas methanica*
	Methylococcus capsulatus
Methanol	*Candida boidinii*
Ethanol	*Candida utilis*
Glucose	*Fusarium venenatum*
Inulin	*Candida* spp.
Molasses	*Candida utilis*
	Saccharomyces cerevisiae
Spent sulphite waste	*Paecilomyces variotii*
Whey	*Kluveromyces marxianus*
	Penicillium cyclopium
Lignocellulose waste	*Agaricus bisporus*
	Cellulomonas spp.

Source: Waites, M. J. et al. (2001) *Industrial Microbiology: An Introduction*, Table 14.3, p. 221. Wiley-Blackwell.

losses, mainly due to the oil crisis in the 1970s. Safety also became an issue when it was shown that Toprina was high in RNA, which can cause gout, as discussed at the start of this section.

Pruteen was an animal feed for chickens, pigs, and veal calves; it was made by ICI and used the bacterium *Methylophilus methylotrophus*, which was grown on methanol. Methanol mixes completely with water, and as it is available in pure form there was no need to further purify the protein. While Pruteen production was once the largest continuous aerobic process, manufacture stopped due to the rising cost of methanol.

An SCP production success story: Quorn™

Despite the failure of some SCP production, one product has been a massive success. In 1991, Marlow Foods (a subsidiary of ICI) isolated a species of *Fusarium graminearum*, a filamentous fungus. We know this fungus as QUORN™, which contains 12 per cent protein and has no cholesterol or animal fat. These qualities mean that it is marketed as a health food product.

Quorn™ generated sales of nearly £230 million in 2019, with 1000 tonnes being produced each year in the 40 m³ airlift fermenter in the northeast of England. The fermenter is operated continuously at 30 °C and pH 6.0. The fungus, now called *Fusarium venenatum*, is grown on food-grade glucose syrup as the carbon source, and ammonia is used both as the nitrogen source and to control the pH. Growth of the fungus is supplemented with biotin and mineral salts.

The fungal mycelium is harvested by a process called **vacuum filtration**—a technique for separating liquids and solids. The 'filter cake' produced is a mat of intertwined fungal hyphae, which has a meat-like texture; this feature is used to make products that are familiar to us such as burgers, chicken pieces, sausages, and mince. A range of these products is shown in Figure 13.10.

The final biomass contains 10 per cent RNA, which is too high for humans to eat (as discussed at the start of this section), so RNA levels have to be reduced by exposing the biomass to a thermal shock of 64 °C

for 30 min. This kills the fungus and activates RNases, enzymes that break down the RNA into nucleotides. These nucleotides then diffuse out of the cells and this method reduces the RNA levels to 2 per cent.

The company that now owns the Quorn™ brand, Philippines-based Monde-Nissin, is set to expand production with a recently announced £150 million investment, and sales of Quorn™ are projected to steadily increase.

Dihé

The production of SCP does not have to be highly technical and nor does it need to use specialized equipment, as clearly depicted in Figure 13.11. The cyanobacterium *Spirulina platensis* is used to make a food called Dihé in the African regions of Kanem and Lake Chad. *Spirulina platensis* is taken from Lake Chad and poured into a depression created in the ground. When dried, the *S. platensis* is cut into chunks and sold. *Spirulina platensis* is also sold as a health food supplement.

Mushroom production

Mushroom production is a substantial industry. The major cultivated mushroom (37 per cent) is *Agaricus bisporus* and is worth approximately £1 billion per annum. Other commercial species include the oyster mushroom, *Pleurotus ostreatus*, and the shiitake mushroom, *Lentinula edodes*, which is grown on logs.

Mushroom production makes very efficient use of plant waste material: the product is directly edible and harvesting is easy. Protein conversion efficiency per unit of land and per unit of time is good compared with animal protein production.

Figure 13.10   Some products made using the mycoprotein Quorn™.

Source: SarahJaneJ/Shutterstock.com.

Figure 13.11   The production of Dihé in Africa.

Source: © FAO/Marzio Marzot.

Figure 13.12 Mushroom production in trays using solid state fermentation.
Source: Dexter McMillan.

Mushroom growth uses a **solid-state fermentation** process, a process that does not involve water or any other liquid. It can be carried out in small spaces, as shown in Figure 13.12. The solid substrate used in the fermentation can be made from a variety of different materials depending on the country of production; typically this is

composting straw, manure, and fertilizer. For example, in Europe the mix is usually wheat straw (40–50%), horse manure or stable bedding (20–25%), poultry manure (10–15%), and gypsum (5–10%).

The production of the compost has four phases. In phase I, the components are mixed together and the mesophilic micro-organisms—those that grow between 20 °C and 45 °C—develop. The carbohydrates and the proteins within the mix are converted to heat and ammonia. As the temperature rises the thermophilic micro-organisms develop; they feed on the straw, which softens it. In phase II the compost is conditioned for 8 h at 56 °C and then kept at 45 °C. The micro-organisms used at this stage are actino-mycete bacteria and the thermophilic fungus *Humicola insolens*.

The preparation of the *A. bisporus* inoculum starts with the growth of the spawn on sterilized cereal grains such as millet or rye. The spawn is added to the compost at the end of phase II. The mycelium grows through the compost for 12–16 days. The compost then enters phase III, whereby temperatures are reduced to 20–25 °C. The fruiting bodies of *A. bisporus* then start to appear in phase IV and can be harvested after 18–21 days. The fruiting bodies are picked, but they continue to appear in cycles of 7–8 days (a phenomenon known as flushing). The yield of mushrooms and its biological efficiency is explored further in Real World View 13.1.

REAL WORLD VIEW 13.1   Mushroom yields and biological efficiency

The commercially cultivated oyster mushroom, *Pleurotus ostreatus*, is considered a restaurant delicacy and can command a higher retail price than the normal button mushroom, *Agaricus bisporus*. It has historically been cultivated mainly on sawdust, but with this substrate becoming more uncommon, other agricultural waste products are now being investigated to see if they can provide good yields of the mushroom.

Table 1 shows data from a study conducted in 2003 by Obodai *et al.* This team used a known quantity of different waste products, sterilized

them, and then inoculated them with spores of the oyster mushroom. The mushrooms went through three flushes of growth. The total fresh weight was then calculated together with the biological efficiency of conversion of the substrate into mushroom biomass.

Look at this table and notice how the substrate giving the *best* biological efficiency is composted sawdust; by contrast, fresh sawdust gives the worst biological efficiency (other than elephant grass, which gives no fungal growth at all). Also notice how we see a reduction in fresh weight of mushrooms across the three flushes.

Table 1   Cumulative mushroom yields and biological efficiency (BE) on different substrates

Substrate (1 kg dry weight)	Fresh weight of mushroom by flushes (g)				BE (%)
	First	Second	Third	Total fresh weight (g)	
Fresh sawdust	13.0	n.f.*	n.f.	13.0	1.3
Composted sawdust	83.6	79.2	20.4	183.2	18.32
Rice husk	23.3	n.f.	n.f.	23.3	2.33
Corn husk	25.2	14.2	10.1	49.5	4.95
Banana leaves	58.2	43.1	10.2	111.5	11.15
Maize stover	50.7	37.1	n.f.	87.8	8.78
Rice straw	50.2	49.2	12.2	111.6	11.16
Elephant grass	n.f.	n.f.	n.f.		

* n.f. no flush, i.e. no fungal growth was observed.

Use of microbes to produce food and beverages

Beyond being foodstuffs in their own right—be that as animal protein substitutes or mushrooms—microbes are widely used in the production of other foodstuffs. Let us now consider some of the most prominent examples.

Bread

Bread is a staple part of the diet of most cultures. The origins of bread may reach back as far as 30,000 years. Archaeologists working on the shore of Lake Zurich have found bread-like objects in late Neolithic layers, as depicted in Figure 13.13. When analysed, these bread-like objects were found to contain barley and wheat remains.

Bread-making was refined by the Ancient Greeks and Romans, and became a normal part of household activity. Unleavened bread is made from flour with no raising agent, but the bread we are most familiar with is leavened: the single-celled fungus *Saccharomyces cerevisiae*, as also known as baker's yeast, is added to form dough, and causes that dough to rise. (Early bread makers would have relied on naturally occurring environmental yeasts to get the dough to rise.)

The rising of dough during the proving and baking stages is the result of the release of carbon dioxide from fermentation: the *S. cerevisiae* ferments the sugars present in the dough, releasing carbon dioxide as it does so. The carbon dioxide gives the bread a fluffy texture, but the yeast also imparts other properties to the dough (including flavour) as a result of the release of metabolites such as organic acids, glycerol, and aroma compounds.

Bread can be made from different types of flour, including wheat, rye, and corn. Different flours give rise to different types of dough because they contain different sugars, which the yeast can ferment. These doughs are then used to produce distinctive breads, which can be standout features of food cultures. These include naan, an oven-baked flat bread from India, and the pretzel from Germany.

Bacteria can also be used in the making of dough. For example, *Lactobacillus sanfrancisco* is used in the making of sourdough; its production of lactic acid gives this dough its characteristic flavour.

Vinegar (acetic acid)

Vinegar is made when aerobic acetic acid bacteria convert ethanol to acetic (ethanoic) acid. These bacteria belong to the genera *Acetobacter* and *Gluconobacter*.

Acetic acid fermentation proceeds according to the following reaction:

$$C_2H_5OH + O_2 \rightarrow CH_3COOH + H_2O$$

Ethanol   Acetic (ethanoic) acid

Vinegar can be made from any drink containing alcohol, but it is usually produced from wine, beer, or cider. Food-grade vinegar is most often used as a food preservative and as a condiment. Artisan vinegars are becoming more sought after and can command higher retail prices.

Large-scale commercial vinegar production is done by **semi-continuous submerged fermentation**. This is essentially a repeated fed batch culture, which we explore in Chapter 12. It is quite difficult to grow aerobic acetic acid bacteria in the laboratory, so the submerged fermentations are inoculated with unselected cultures from previous fermentations.

Figure 13.13 Bread-like objects found in Neolithic layers at Lake Zurich.

Source: Öai-Öaw, Vias/A. G. Heiss. Figure 4 from Heiss, A. G., Antolín, F., Bleicher, N., et al. (2017) State of the (t)art. Analytical approaches in the investigation of components and production traits of archaeological bread-like objects, applied to two finds from the Neolithic lakeshore settlement Parkhaus Opéra (Zürich, Switzerland). *PLoS ONE* 12(8): e0182401. https://doi.org/10.1371/journal.pone.0182401.

Cheese production

Cheese production is another very old process, reaching back as far as 400 BCE. Indeed, Homer's *Iliad* mentions the use of a goat kid's stomach to make cheese. Cheese is made from milk using an enzyme called rennin (also called chymosin). The commercial form of rennin is rennet, a protease enzyme that is used to coagulate the milk into curds and whey.

While traditionally sourced from calf stomachs, rennin can now be obtained by other means. For example, the plant pathogen *Endothia parasitica* (previously *Cryphonectria parasitica*) produces a rennin with very similar characteristics to calf rennin. Fungal rennins have become increasingly important in cheese manufacture as they can be produced more cheaply. Rennin can also be created using genetically modified *E. coli*.

Microbes are also used in other ways—for example, to add flavour to cheese. These flavours come from the ripening process and are due to the production of chemical compounds such as acids, esters, and diacetyl and sulphur-containing compounds. For example, the white mould *Penicillium camemberti* (shown in Figure 13.14) is used to flavour camembert cheese.

In the production of the blue cheese roquefort, curds produced from sheep's milk are inoculated with spores of *Penicillium roqueforti* and are loosely packed together. The spores give the 'bubbles' of mould throughout the cheese, as depicted in Figure 13.15.

Bacteria such as *Propionibacteria freudenreichii* are important in the flavour of continental cheeses such as Edam and Gouda; the bacteria also produce the holes that are characteristic of these types of cheese.

Asian foods

Species of the filamentous fungi *Rhizopus* are used to make tempeh from soybeans. Tempeh, shown in Figure 13.16, is the major source of protein for Indonesians.

Shoyu (or soy sauce) is made in East and Southeast Asia. It has a long history and has been made in China for 2000 years. Soy sauce is produced when the filamentous fungus *Aspergillus oryzae* ferments a mixture of soybeans, wheat kernels, and raw or roasted wheat flour.

Figure 13.15 Roquefort cheese. The mould that makes this cheese a blue cheese is caused by spores of *Penicillium roqueforti*.
Source: Picture Partners/Shutterstock.com.

Figure 13.16 The filamentous fungi *Rhizopus* are used to make tempeh from soybeans.
Source: Ika Hilal/Shutterstock.com.

Figure 13.14 *Penicillium camemberti* growing on an agar plate.

Japan leads the world in soy sauce production: its 27,000 manufacturers make their soy sauce from equal amounts of soybeans and wheat. In contrast, Chinese soy sauce uses soybeans with little or no wheat.

The first step in production is to make the koji—the product of the solid state fermentation of soybeans and wheat kernels with *Aspergillus oryzae*. The next step is brine fermentation (called moromi in Japan), which uses high concentrations of salt. The fermentation is carried out by halophilic (salt-loving) bacteria and osmophilic yeasts (yeast that tolerate high osmotic pressure).

Beverages

Alcoholic beverages also illustrate how different cultures have developed different products, depending on the availability of the necessary starting materials for production. Table 13.7 provides some examples; in the following sections we consider beer and wine production in more detail.

Table 13.7 Types of alcoholic beverages and their country of origin

Substrate	Non-distilled beverage	Product of alcoholic fermentation distilled to form	Location
Apples	Cider	Cider brandy	Northern Europe and North America
Cacti/succulents	Pulque	Tequila	Mexico, Central America
Grapes	Wine	Cognac	Southern Europe, North and South America, Australia, New Zealand
Pears	Perry	Pear brandy	Northern Europe
Honey	Mead	–	UK
Sugar cane	–	Rum	West Indies
Barley and other cereals	Beer	Whisky	UK mostly
Rice	Sake	Shochu	Japan

Source: Waites, M. J. et al. (2001) *Industrial Microbiology: An Introduction*, Table 12.1, p. 180. Wiley-Blackwell.

Brewing beer

Brewing is one of the oldest biotechnology processes. The first records of brewing date to 6000 BCE, in Egypt. Beer was considered a 'safe' drink when water supplies were contaminated with animal and human sewage.

The first micro-organism to be grown in pure culture (a culture that only contains the organism required) was *Saccharomyces carlsbergensis* (Carlsberg Yeast number 1) and pure strain brewing was carried out for the first time in 1883 at the Carlsberg brewery in Denmark.

There are four main types of beer, as shown in Figure 13.17: ales, porters, lagers, and stouts. Lager accounts for 90 per cent of the beer produced by the brewing industry. The type of beer depends on the yeast used during the fermentation process.

Beer production involves four main stages.

1. **Malting:** during this stage, the barley grains are partially germinated in a controlled way. Brewer's yeast cannot produce the enzyme amylase (which breaks down starch), so the starch within the grain cannot be used until the seed has germinated (at which point hydrolytic enzymes such as amylases and proteases are released). After germination, the grains are dried in a kiln at 50–60 °C and then cured at 80–110 °C. This produces the malt.

2. **Wort preparation:** the wort is made from the malt, although other ingredients can also be added; these include unmalted barley and other cereals. The malt and the additives are milled and mixed with hot water. The liquid is separated from the solids to form the wort, which is then stabilized by boiling. Hops can be added at this point, which gives the beer its bitter flavour.

3. **Yeast fermentation:** the yeasts ferment the wort to produce the beer. The yeasts must be able to use the wort sugars for growth and be able to tolerate high levels of alcohol.

Two types of yeast are used to produce different beers: top-fermenting yeasts are used to make ales and stouts, while bottom-fermenting yeasts are used to produce lager.

Figure 13.17 The four types of beer. From left to right: ale, porter, lager, and stout.

Source (left to right): stevepb/Pixabay; stevepb/Pixabay; ariprodz/Pixabay; Ernest_Roy/Pixabay.

'Top-' or 'bottom-fermenting' describes how a yeast behaves after it has formed flocs. Large enough to be visible, flocs comprise thousands of individual yeast cells and form after the wort sugars have been used. Top-fermenting yeasts float when they form flocs, while bottom-fermenting yeasts sink: they form a sediment.

Beer flavours are developed partly as a result of aromatic alcohols, which are made from the wort amino acids. Esters are also made; these include ethyl acetate, which has a solvent aroma, and phenyl ethyl acetate, which smells of roses and honey.

4. **Post-fermentation:** during post-fermentation, the majority of yeast is removed, but the 'green' beer has to be matured. The nature of this maturation differs according to the product being made.

For cask and bottled ales, priming sugar is added and there is a secondary fermentation which carbonates the beer to produce the fizz. For lager, the beer is held at 8 °C for several weeks and no sugar is added, but the remaining yeast will continue to ferment the wort sugar. This also carbonates the lager.

Large-scale beer production relies on storage ageing, where no further sugar is added and the beer is stored at 1–4 °C for 7–10 days. A chill haze forms, which is then removed by filtration.

The majority of beers are then pasteurized or filtered and packaged into kegs, bottles, cans, or tanks.

The use of fungi in wine production

Wine making has existed in the Mediterranean region for over 7000 years. The Roman philosopher Pliny the Elder, who lived between 23 and 79 CE, referred in his book *Naturalis Historia* to '*in vino veritas*': 'the truth is in the wine'. Ancient Greeks even had a god of wine, Dionysus.

Wines are non-distilled alcoholic drinks made from a variety of fruits, including grapes, peaches, plums, apricots, and bananas. Many wines are made using yeasts that are present naturally on the surface of the fruit.

When wine is produced from grapes, the grapes are initially crushed to produce a liquid called the 'must'. Many genera and species of yeast are found in the grape must. These include both *Saccharomyces* and non-*Saccharomyces* species.

There are three types of non-*Saccharomyces* groups:

1. Yeasts that are largely aerobic.
2. Yeasts with low fermentative activity.
3. Yeasts with fermentative metabolism.

Non-*Saccharomyces* groups have been considered problematic due to their volatile acidity, sensitivity to SO_2, poor fermentation ability, and intolerance to ethanol. However, it is now thought that these yeasts could impart regional flavour characteristics.

Noble rot is caused by the filamentous fungus *Botrytis cinerea* (Figure 13.18); it infects the grape, causing it to dehydrate. However, this process is not as detrimental to the wine-making process as it may first seem: it concentrates the sugars, giving sweeter flavours such as honey and ginger. The fungus also produces a chemical called phenylacetaldeyde, which gives a fruity aroma to the wine at low levels.

Figure 13.18   Noble rot caused by the filamentous fungus *Botrytis cinerea*.
Source: blickwinkel/Alamy Stock Photo.

PAUSE AND THINK

What are the key differences between beer and wine production?

Answer: The substrate for beer production is mainly malted barley; for wine, it is grapes (although other fruits can be used). Whilst *Saccharomyces cerevisiae* is important for both processes, the addition of brewer's yeast to the fermentation of the wort is a more controlled process. Different types of yeast are used—e.g. top- and bottom-fermenting yeasts—to produce either ales or lagers, respectively. In wine production, the use of yeasts for the fermentation is a less well-controlled process. Many different species of yeasts are present in the grape 'must', and contribute to the flavour of the wine. Other fungi such as *Botrytis cinerea* also have a role to play.

 Check your understanding of the concepts covered in this section by answering the questions in the e-book.

13.5 Microbial polysaccharides

Polysaccharides—carbohydrates comprising many small sugar units—can be produced by a range of different microbes, including bacteria, yeasts, moulds, and algae. For example, Figure 13.19 shows the yellow shiny colonies of *Xanthomonas campestris* that produce xanthan gum.

Many (but not all) microbial polysaccharides are hydrophilic: they are water loving and disperse easily in water. They can be made up multiple units of one monosaccharide, in which case they are called homopolysaccharides. Alternatively, they can comprise multiple units of several different monosaccharides; these are called heteropolysaccharides.

Microbial polysaccharides are classified into three classes depending on where they are located in the cell:

1. Intracellular: located on part of the cytoplasmic membrane.
2. Cell wall polysaccharides: forming a structural part of the cell wall itself.
3. Extracellular: loosely attached to the cell surface in the form of slime, or covalently bound to the cell surface, forming a capsule.

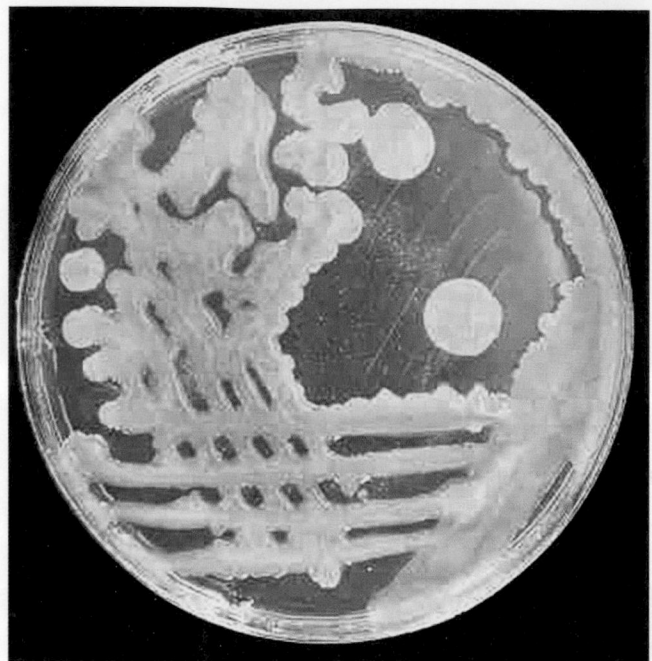

Figure 13.19 Microbial polysaccharides. *Xanthomonas campestris* growing on an agar plate.

Source: Jim Deacon, Institute of Cell and Molecular Biology, University of Edinburgh.

Microbial polysaccharides have an important role in the development of biofilms, where they help the cells to adhere to surfaces and enhance resistance to environmental conditions. They are also used in a number of different industries, including food production, pharmaceuticals, and engineering. (Most microbial polysaccharides have to be slightly modified, either chemically or physically, before they can be used, however.) Table 13.8 shows the

bacteria that produce commercial polysaccharides, their classification, and how they are used.

▶ We learn more about biofilms in Section 9.2.

🌀 Check your understanding of the concepts covered in this section by answering the questions in the e-book.

13.6 Biological control agents and bioremediation

Microbes can be deliberately used in a variety of ways environmentally, to either control pests or clean up contaminated land through a process of bioremediation. Specific microbes are also used as soil inoculants to promote nitrogen fixation or phosphate mobilization. Microbes have also been used very effectively for wastewater treatment and the production of compost. Let us now consider some of these uses.

Biological control

Biological control can be defined as the use of one organism to control another. It is an attractive alternative to chemical pesticides as it is considered to be safer. Microbes, mainly bacteria and fungi, have been used as biological control agents for nearly 40 years. However, their use has failed to keep pace with new chemical pesticides: their success has been limited by their ability to control a narrow range of pests, by their slow action, and their short field life. *Bacillus thuringiensis* (Bt) is the most successful bioinsecticide, but it is not very persistent in the environment, and the Bt toxin is sensitive to solar UV.

Mycoparasites are fungi that parasitize other fungi: they obtain all their nutrition from their living host—a characteristic that has been used for biological control. Let us consider a couple of examples:

Table 13.8 Commercial microbial polysaccharides, their classification, and how they are used

Bacteria	Polysaccharide	Class	Group	Use
Agrobacterium spp.	Curdlan	Exopolysaccharide	Homopolysaccharide	Food industry, pharmaceutical industry, heavy metal removal, concrete additive
Halomonas eurihalina	Levan	Exopolysaccharide	Homopolysaccharide	Prebiotic, pharmaceutical, and cosmetic industries
Leuconostoc dextranicum	Glucan	Cell wall	Homopolysaccharide	Pharmaceutical and food industries
Leuconostoc mesenteroides	Dextran	Exopolysaccharide	Homopolysaccharide	Pharmaceutical industry and as a thickener in food
Gluconacetobacter xylinum	Cellulose	Cell wall	Homopolysaccharide	Pharmaceutical in wound healing, tissue engineering, and audio speaker diaphragms
Pseudomonas putida	Alginate	Cell wall	Heteropolysaccharide	Food industry, textiles, and paper making
Sphingomonas paucimobilis	Gellan	Exopolysaccharide	Heteropolysaccharide	Confectionery and jams
Xanthomonas spp.	Xanthan gum	Exopolysaccharide	Heteropolysaccharide	Food industry as a stabilizer and thickener. Ceramic, cosmetic, and textile industry. Also used as a lubricant for oil drills

Figure 13.20 Mycopesticide products from *Metarhizium anisopliae*.
Source: © Agriplex India.

- *Paecilomyces lilacinus* is registered in the Philippines as Biocon®. It is a soil-borne fungus that parasitizes the eggs and larvae of some nematodes that cause significant damage to plant roots.

- A mycopesticide called Green muscle® is based on the spores of the insect pathogen fungus *Metarhizium anisopliae*. Green muscle® is available as a spore powder or an oil-miscible concentrate, and is sprayed using normal equipment. Examples of pesticides yielded from *Metarhizium anisopliae* are depicted in Figure 13.20.

The fungus *Heterobasidion annosum* is the causative agent of butt rot, which damages timber at the base of the tree trunk, as shown in Figure 13.21. It is a serious pathogen of conifers in Great Britain, continental Europe, and North America. It spreads slowly by mycelial growth along the roots of diseased trees, and infects healthy trees by root-to-root contact. This disease can be controlled by the fungus *Peniophora gigantea*. Spores are sold as suspensions in sachets, with the inclusion of a dye so that the forester can ensure the stump is completely covered.

PAUSE AND THINK

Why are there not more commercially available biological control agents?

Answer: It takes many years to develop biological control agents, and many of those that show promise in the laboratory and in small field trials fail when used on a larger scale. Biological control agents are also limited by their ability to control a narrow range of pests, by their slow action, and their short field life.

Bioremediation

Bioremediation is the use of living organisms or their products to metabolize or detoxify environmental pollutants. The use of biological processes is not a novel concept: wastewater treatment and

Figure 13.21 Butt rot. Damage at the base of a tree trunk caused by *Heterobasidion annosum*.
Source: Strobilomyces/Wikimedia Commons/CC BY-SA 3.0.

composting have long been used as biological approaches to handling human and other organic wastes. For example, bioremediation was used to treat the Exxon Valdez oil spill in Alaska in 1989, which polluted 800 km of intertidal shoreline. An oleophilic (oil-tolerant) fertilizer was added to the shoreline to act as stimulant to enhance the growth of hydrocarbon-degrading bacteria.

Most of the research on fungal bioremediation has centred around the lignin-degrading enzyme system of white rot fungi. White rot fungi degrades the lignin in wood, causing it to decay.

Mycofiltration is the impregnation of fungal spores and hyphae into fabric landscaping cloth. These fabrics are overlaid onto contaminated ground, and the fungal mycelium acts as a filter, trapping and degrading the contaminants. These technologies are new, and are being used on a small scale. Widespread adoption depends on the acceptance of the technology by the public and by regulators.

Wastewater treatment

Humans have always needed to manage water. The Babylonians developed storm water drainage systems and larger homes had cesspools below ground. The technology for dealing with wastewater was developed further by the Romans, whose latrines were essential parts of buildings even in remote settlements. More recently, London's sewerage system was designed as a result of the 'Big Stink' of 1858 and persistent outbreaks of cholera. Joseph Bazalgette, chief engineer to the Metropolitan Board of Works, conceived the underground infrastructure; at the time, it was the biggest civil engineering project in the world.

Wastewater can come from many different places, domestic and industrial, and some water can be highly contaminated. The term sewerage is used for material which contains human or animal faecal waste. The aim of wastewater treatment is to reduce organic and inorganic materials to a level that no longer supports microbial growth, and to eliminate other potentially toxic materials and microbial pathogens. Bacteria are the most numerous pathogens in wastewater; most common are *Salmonella* spp., *Vibrio cholerae*, and *Shigella* spp., although it is worth bearing in mind that high numbers of these organisms are generally required to cause illness.

The most frequent waterborne illnesses are caused by enteric viruses, which are shed in high numbers in faecal material. These virus particles, such as norovirus and rotavirus, are not removed by filtration. However, non-viral pathogens are also an issue for water quality: eukaryotic microbes such as protozoa produce cysts and oocysts that can withstand harsh environments and are chlorine resistant. Examples include *Giardia* and *Cryptosporidium*, which is the most common waterborne infection in UK.

▶ **We learn more about viruses in Chapter 15.**

If the treatment of the wastewater has been successful, the **biological oxygen demand** (BOD) of the water will have been reduced. BOD is a measure of the amount of oxygen needed by micro-organisms to oxidize the organic matter in the water and is an indicator of pollution. A high BOD implies that there is a lot of organic matter (i.e. pollution) to be oxidized; it translates into a high level of oxygen removal from the water, which can cause the death of freshwater animals.

The process of wastewater treatment is divided into four main parts, as shown in Figure 13.22. Let us consider each of these in turn.

Primary treatment and primary sedimentation

Primary treatment involves the filtering out of solids, debris, and coarse material using screens. The wastewater then moves to sedimentation tanks for primary sedimentation, during which fine particles sediment out to form sludge. In this phase the BOD is reduced by 30–40%. The sludge from primary treatment can be taken to landfill, incinerated, or further treated.

Secondary treatment

During secondary treatment, the wastewater is treated using both biological and physical processes. The object is to speed up biological degradation using microbes, reducing the BOD by 80–90%.

Different types of secondary treatment are available. Trickling filters use beds of stones or moulded plastic which the water trickles over. Microbial biofilms develop on the surfaces of the stones, creating complex communities of bacteria, fungi, and protozoa.

Another secondary process is activated sludge treatment. This process is aerobic and needs continual agitation to maintain a ready supply of oxygen. The wastewater is brought into contact with a mixed microbial population in the form of flocs. These flocs are formed when bacteria such as *Zoogloea ramigera* and species of *Acinetobacter*, *Flavobacterium*, and *Pseudomonas* produce microbial polysaccharides. The bacteria are prey for eukaryotic microbes such as ciliates and amoebas.

▶ **We learn more about floc production in Section 13.4.**

Secondary treatment also produces sludge which can be further treated using anaerobic digestion. Anaerobic digestion produces methane and carbon dioxide. The methane can be used to power the sewerage plant and the remaining organic material can be used as fertilizer for crops.

Tertiary treatment

During tertiary treatment the secondary effluent is further treated using chemicals such as chlorine, or with UV light. The water is then released into the freshwater system.

Composting

We considered some aspects of composting when we looked at the production of mushroom compost in Section 13.4. Many thousands of households across the UK do their own composting with kitchen and garden waste. There are also municipal composting sites, which use aerated piles, tunnel systems, and rotating drums.

All composting relies on the action of a diverse array of microbes, and different populations will develop at different stages of the composting process. **Mesophiles**—organisms that grow at moderate temperatures—will inhabit the compost at the start of the process. As heat is generated through microbial degradation of the biomass, the thermophilic microbial population will take over. These microbes thrive at elevated temperature (>45 °C), although

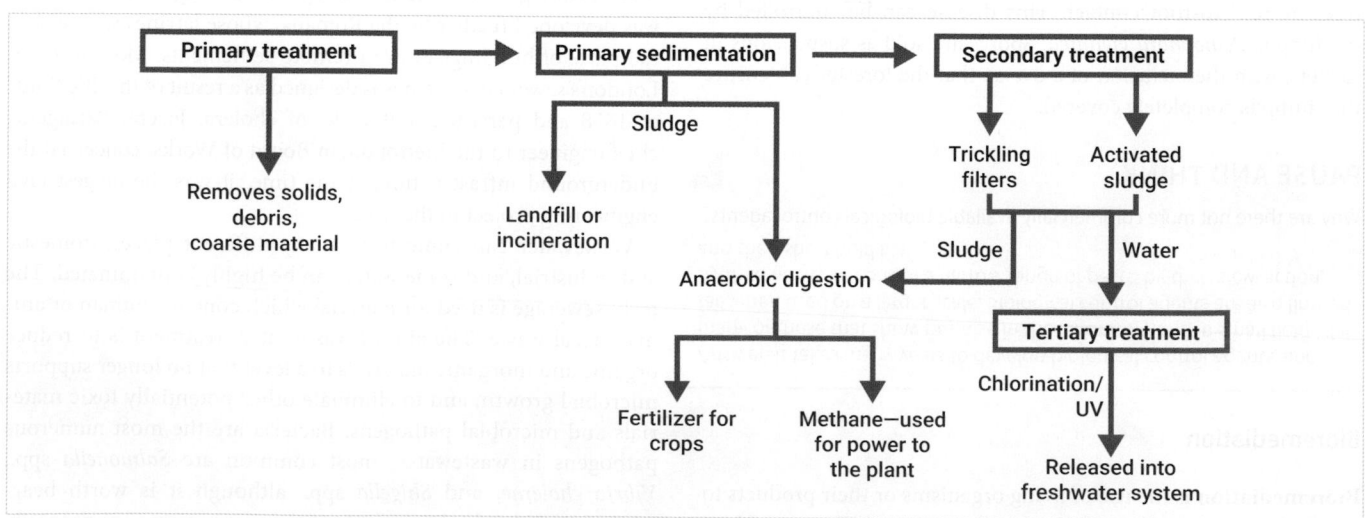

Figure 13.22 The four stages of wastewater treatment.

garden compost heaps often do not reach the really high temperatures of municipal compost, which can reach 60–70 °C.

The compost should contain a mix of green and brown material, green being discarded vegetables, flowers, stems, and cut grass, and brown being leaves and woody material such as branches. Green material is high in nitrogen, while brown material is rich in carbon but low in nitrogen. Ideally, there should be a carbon to nitrogen ratio of 30:1. Good aeration and moisture are both needed for effective composting to take place.

 Check your understanding of the concepts covered in this section by answering the questions in the e-book.

13.7 Biogeochemical cycles

All living organisms need the chemical elements carbon, hydrogen, oxygen, nitrogen, phosphorus, and sulphur for survival as they are essential for building cell biomass. All the elements from which organisms are formed are cycled between organic or inorganic, water soluble or insoluble, gaseous, liquid, or solid forms. Such cycling is called a **biogeochemical cycle**.

A biogeochemical cycle encapsulates the processes involved in cycling an element through various biological, chemical, and geological forms in air (biosphere), water (hydrosphere), and soil (lithosphere) systems. Microbes play an essential role in these chemical transformations in a variety of different habitats. Though there are four key cycles—the nitrogen cycle, carbon cycle, phosphorus cycle, and the sulphur cycle—we will focus on the first two of these here.

The nitrogen cycle

Nitrogen gas (N_2) is the major component of the atmosphere (78%). The nitrogen cycle has major economic importance and is essential for living organisms. Nitrogen is a component of a variety of macromolecules, including amino acids, proteins, and nucleic acids. It has also been suggested that the oxidation of ammonic nitrogen (NH_4^+) in Earth's mantle was a source of liquid water for early Earth. Microbial transformations play a major role in the nitrogen cycle and they involve important interactions between microbes and plants. The nitrogen cycle is shown in Figure 13.23 and we will be referring back to this figure over the next few sections.

Biological nitrogen fixation

We will begin looking at the cycle at the point at which atmospheric nitrogen (N_2) is converted to ammonium compounds. This process is called biological nitrogen fixation (BNF) and is highlighted in green in Figure 13.23. Nitrogen fixation can occur during a lightning strike, which produces nitrogen oxides (NO_x), but this fixes nitrogen in much smaller amounts than biological nitrogen fixation at only five teragrams of nitrogen per year (that is, 5 Tg N yr^{-1}, where 1 Tg = 1,000,000,000,000 g). (By contrast, biological nitrogen fixation fixes around 260 Tg N yr^{-1}.)

Ammonia can be man-made through the Haber–Bosch process. Ammonia produced in this way is used in the production of fertilizers and also explosives. The amounts of nitrogen produced (Tg N yr^{-1}) naturally and by man-made processes are summarized in Figure 13.24.

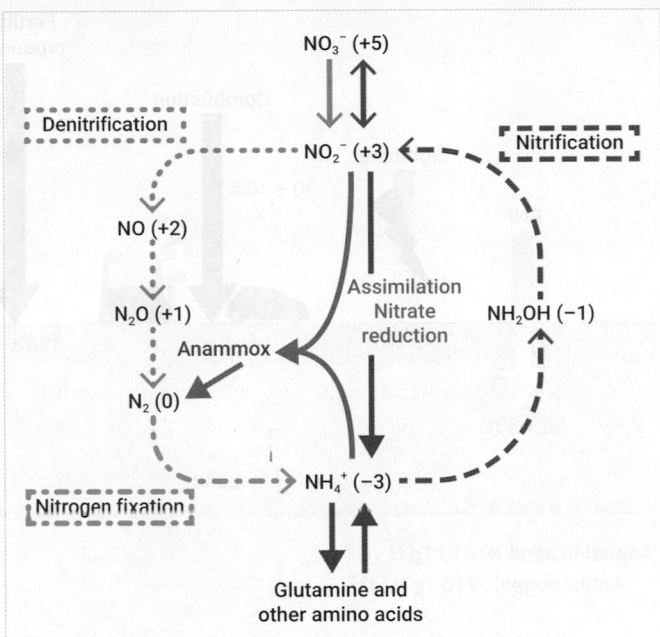

Figure 13.23 The nitrogen cycle. The nitrogen cycle showing nitrogen fixation in blue, nitrification in purple, and denitrification in green. The anammox reaction is shown in orange and assimilatory nitrate reduction in red.

Biological nitrogen fixation is only performed by prokaryotes; such prokaryotes are called diazotrophs. There are many different types of diazotroph, but all belong in one of two groups. Symbiotic diazotrophs form close symbiotic associations with leguminous plants. Free-living diazotrophs are not associated with plants, but rather live in the bulk soil; they can fix nitrogen in either aerobic or anaerobic conditions. The different types of diazotroph are shown in Table 13.9.

All diazotrophs use the same enzyme to fix nitrogen; this enzyme is called nitrogenase. It requires a lot of energy to drive the reaction as the triple bond in N_2 makes it very stable: 16 ATP molecules are required to break the triple bond, as represented by the following equation:

$$N_2 + 8H^+ + 8e^- + 16ATP \rightarrow 2NH_3 + H_2 + 16ADP + 16P_i$$

The nitrogenase enzyme is inactivated by the presence of oxygen, which means that the reaction is anaerobic. This is clearly a problem for aerobic organisms that require oxygen. As a consequence, they have evolved mechanisms for protecting their nitrogenase enzyme from oxidative damage.

Free-living diazotrophs

Free-living diazotrophs have evolved the following mechanisms to protect their nitrogenase enzyme from oxidative damage:

1. Production of an extracellular polysaccharide slime layer which limits O_2 diffusion.

2. High rates of respiration and excessive oxidization to use up oxygen inside the cell.

Cyanobacteria are phototrophs, meaning that they can photosynthesize. Unicellular cyanobacteria, such as *Gloeotrichia* spp., have a membrane system within their cells to protect the nitrogenase

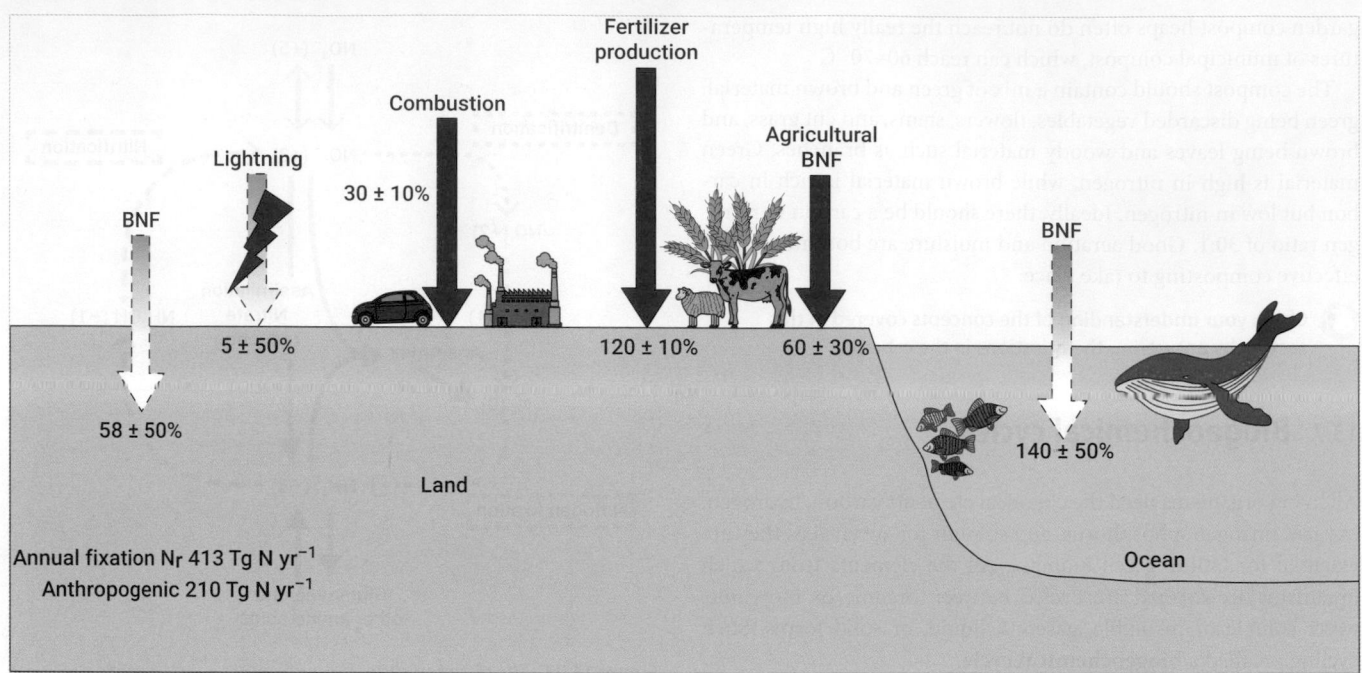

Figure 13.24 Global nitrogen fixation. Global nitrogen fixation from natural, biological nitrogen fixation (BNF) and anthropogenic sources. The unit Tg is a teragram (1 Tg is 10^9 kg).

Source: Fowler, D., Steadman, C., Stevenson, D. et al. (2015) Effects of global change during the 21st century on the nitrogen cycle. *Atmospheric Chemistry and Physics Discussions*. 15, 1747–1868. 10.5194/acpd-15-1747-2015. CC BY 3.0.

Table 13.9 Examples of diazotrophs

Free-living	Aerobes	Anaerobes
Chemo-organotrophs	*Azotobacter* spp.	*Clostridium* spp.
Phototrophs	Cyanobacteria	
Chemolithotrophs	*Thiobacillus* spp.	

Symbiotic	Bacteria	Plant
Chemo-organotrophs	*Rhizobium* spp.	Peas, beans, clover
	Bradyrhizobium spp.	Soya
	Frankia spp. (actinomycete)	Alder
	Sinorhizobium spp.	Alfalfa

enzyme from oxygen. They also carry out nitrogen fixation at night when they are not producing O_2 through photosynthesis.

Multicellular cyanobacteria—for example, *Nostoc* spp. in grasslands and *Anabaena* spp. in rice paddies—are filamentous and have specialized cells for N_2 fixation called heterocysts. These can be seen as light brown structures in Figure 13.25. Heterocysts are thick-walled cells, which occur after every 10–15 cells. These heterocysts do not contain photosynthetic systems and have specialized membranes with low O_2 levels and high N_2 diffusion.

Species such as *Anabaena* are essential for the long-term stability of rice paddies. Because they produce NH_3, additional nitrogen in the form of fertilizer is not required.

Anaerobes, such as the bacteria *Klebsiella* spp. and *Clostridium* spp., do not require mechanisms to protect their nitrogenase from

oxygen, but have less energy available to drive nitrogen fixation. This is because anaerobic respiration does not produce as much ATP as aerobic respiration.

Symbiotic diazotrophs

Symbiotic diazotrophs form a mutualistic symbiotic relationship with plants—a relationship from which both partners benefit. They often produce root nodules as part of the symbiosis. The bacteria get carbon from the plant through photosynthesis and the plant gets ammonium compounds from the bacteria through nitrogen fixation.

Frankia are actinomycete bacteria. They are filamentous and form root nodules in a symbiotic relationship with woody plants, called actinorhizal plants, such as alder trees.

Rhizobium–*legume symbiosis*

One of the best-studied symbiotic relationships occurs between *Rhizobium* spp. and leguminous plants. The relationship can be quite specific, so *Rhizobium* spp. bacteria will form nodules on the roots of peas, beans, vetch, and clover, while *Bradyrhizobium* spp. will form nodules on the roots of soya plants, and *Sinorhizobium* on alfalfa. Again, these are mutualistic relationships: the plants provide carbon to the bacteria in the nodules, and in return the bacteria provide the plants with ammonium.

Steps in the development of the symbiosis

Symbiotic nitrogen-fixing bacteria such as *Rhizobium* spp. are often present in the rhizosphere, the thin layer of soil directly adjacent to the plant roots. The steps in the symbiosis are shown in Figure 13.26.

Figure 13.25 **Heterocysts.** *Nostoc* spp. cyanobacterium with heterocysts.

Source: Dr Robert Calentine/Visuals Unlimited, Inc.

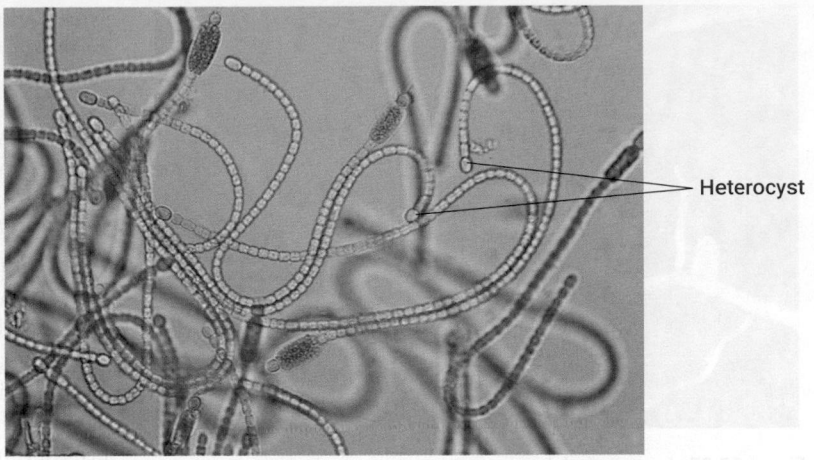

Heterocyst

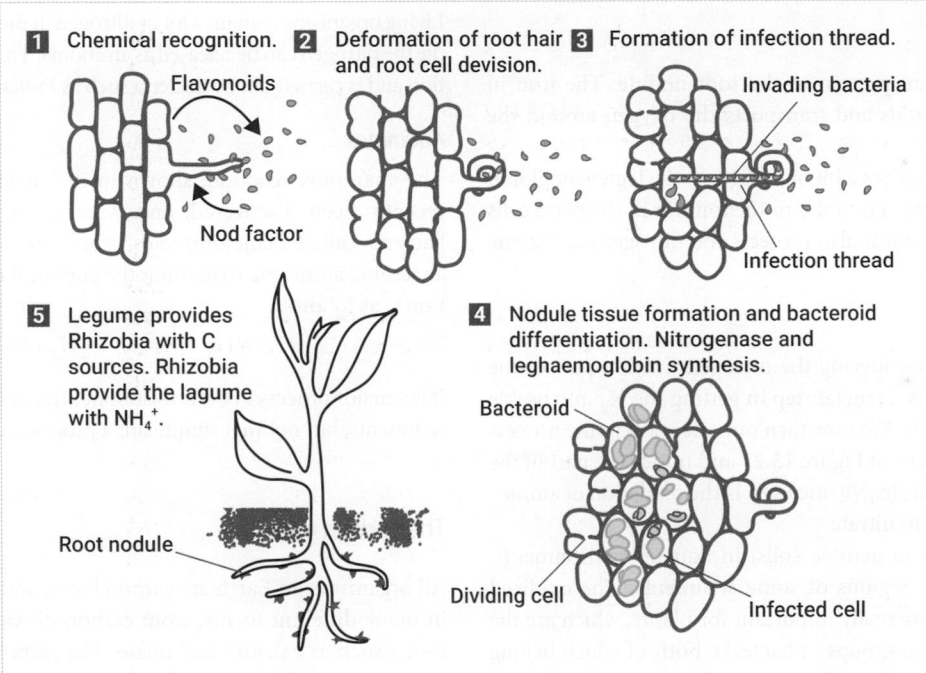

1 Chemical recognition.

Flavonoids

Nod factor

2 Deformation of root hair and root cell devision.

3 Formation of infection thread.

Invading bacteria

Infection thread

5 Legume provides Rhizobia with C sources. Rhizobia provide the lagume with NH₄⁺.

Root nodule

4 Nodule tissue formation and bacteroid differentiation. Nitrogenase and leghaemoglobin synthesis.

Bacteroid

Dividing cell

Infected cell

Figure 13.26 **Symbiosis.** The development of root nodules on a leguminous plant.

The development of the symbiotic relationship starts when the legume plant produces chemical signals called flavonoids in low nitrogen conditions (Figure 13.26, step 1). These flavonoids induce the transcription of a series of genes in the rhizobia bacteria called *nod* genes. Flavonoids can be very specific and only induce *nod* genes in compatible rhizobia. For example, *Rhizobium leguminosarum* will only induce its *nod* genes in the presence of specific flavonoids produced by plants such as peas, clover, and vetch. The *nod* genes can be found clustered together on a giant plasmid, called the Sym plasmid, inside the bacterial cell. Many of the *nod* genes code for enzymes that synthesize nod factors.

Nod factors are signalling molecules that cause a root hair to deform and start to curl (Figure 13.26, step 2). This action traps the rhizobia, which can then form an infection thread that runs through the root cortical cells (Figure 13.26, step 3). The nod

signalling factors also cause the root cortical cells to swell and divide, and this produces the root nodule itself.

When inside the cytoplasm of the root cortical cells, the rhizobia differentiate into structures called bacteroids (Figure 13.26, step 4). This is a permanent differentiation process: the bacteroids can never return to being free-living bacteria. These bacteroids then carry out nitrogen fixation through the induction of another cluster of genes called *nif* genes, which code for the nitrogenase enzyme, and the *fix* genes, which are needed for the nitrogen fixation process.

A compound called leghaemoglobin, which is produced by the plant, protects the nitrogenase enzyme, which is susceptible to oxidative damage within the root nodule. Leghaemoglobin is almost identical to our own haemoglobin molecule and is responsible for the pinky tinge of the root nodules (shown in Figure 13.27).

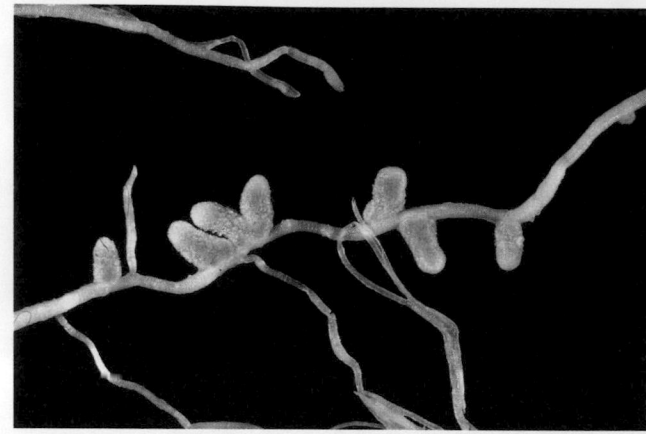

Figure 13.27 Leghaemoglobin. Root nodules showing the pink tinge of leghaemoglobin.
Source: Ninjatacoshell/Wikimedia Commons/CC BY-SA 3.0.

It is the most abundant protein in the root nodule. The iron in the leghaemoglobin binds and transports the oxygen around the plant.

Nitrogen fixation requires a lot of energy, so the leghaemoglobin not only delivers the oxygen to the mitochondria in the plant cells and to the bacteroids, but it also protects the nitrogenase enzyme from oxidative damage.

Nitrification

We have spent time considering the nitrogen fixation part of the nitrogen cycle, as this is a crucial step in getting the N_2 into usable ammonium compounds. We now turn our attention to the process of nitrification: look back at Figure 13.23 and notice the part of the cycle highlighted in purple. Nitrification is the oxidation of ammonia to nitrite and then to nitrate.

Nitrification occurs in aerobic soils, in aquatic environments, and also in the upper regions of some sediments. The oxidized nitrogen compounds are really important for plants, which are the primary producers. Two groups of bacteria, both of which belong to the family Nitrobacteraceae, carry out the process in concert. The ammonium oxidizing bacteria (AOB), such as *Nitrosomonas* spp., turn ammonium into nitrite before nitrite oxidizing bacteria (NOB), such as *Nitrobacter* spp., turn nitrite into nitrate.

The process occurs in three separate reactions.

Firstly, the enzyme ammonia monooxygenase from the AOB catalyses the synthesis of hydroxylamine (NH_2OH) from ammonia:

$$NH_3 + 2H^+ + O_2 + 2e^- \rightarrow NH_2OH + H_2O$$

Secondly, hydroxylamine is then oxidized to nitrite (NO_2^-):

$$NH_2OH + H_2O \rightarrow NO_2^-$$

Finally, the nitrite is oxidized further to nitrate by the NOB:

$$NO_2^- + H_2O \rightarrow NO_3^- + 2H^+ + 2e^-$$

Nitrification can be a problem in agricultural soils, as nitrate is easily leached (washed away), which means that the primary producers such as plants then cannot use it. If the nitrate goes into freshwater systems it can also cause **eutrophication**, which is where nutrient build-up in water causes excessive algal growth (called algal blooms), which is discussed further in Chapter 11.

Denitrification

Let us return to Figure 13.23 and look at the process of denitrification, shown in dark blue. Denitrification turns NO_2^- and NO_3^- anions back to N_2. This can be an anaerobic process, when it takes place in water-logged soils, or it can be aerobic. Bacteria capable of performing this reaction exist in soil, marine, and freshwater environments. Examples include *Bacillus licheniformis*, *Paracoccus denitrificans*, and *Pseudomonas stutzeri*.

Ammonification

Living organisms contain a lot of nitrogen in their biomass. When they die the nitrogen can be released as ammonia. This is called ammonification and is carried out by bacteria such as *Proteus* and *Clostridium* spp.

Anammox

The anammox reaction, shown in red in Figure 13.23, has only recently been discovered, and is carried out by Gram-negative bacteria called Planctomycetes. The term 'anammox' stands for anaerobic ammonia oxidation; the chemical equation for the reaction is as follows:

$$NH_4^+ + NO_2^- \rightarrow N_2 + H_2O$$

It is a major process in both wastewater treatment and anoxic marine sediments, but not that significant a process in aerobic soils.

The carbon cycle

All organisms on Earth are carbon based and carbon can be found in many different forms, from carbon dioxide, as gas, to mineral forms such as calcium carbonate. The carbon cycle is depicted in Figure 13.28.

Most of the carbon on Earth is found in carbonate rocks such as limestone, but it is the carbon in the atmosphere that is cycled most rapidly, with the atmosphere acting as both source and sink of carbon. Carbon dioxide from the atmosphere enters the system through photosynthesis, a process carried out not only by plants, but also by cyanobacteria and other photosynthetic prokaryotes.

Other organisms (consumers) then gain their carbon by eating the biomass of plants and/or cyanobacteria. The carbon consumed is used for the production of biological macromolecules such as DNA, proteins, carbohydrates, and lipids. Some carbon is lost back to the atmosphere through respiration.

Primary consumers (the organisms that eat plants and/or cyanobacteria) can be eaten by secondary consumers (the organisms that eat the primary consumers), so carbon can pass through the food

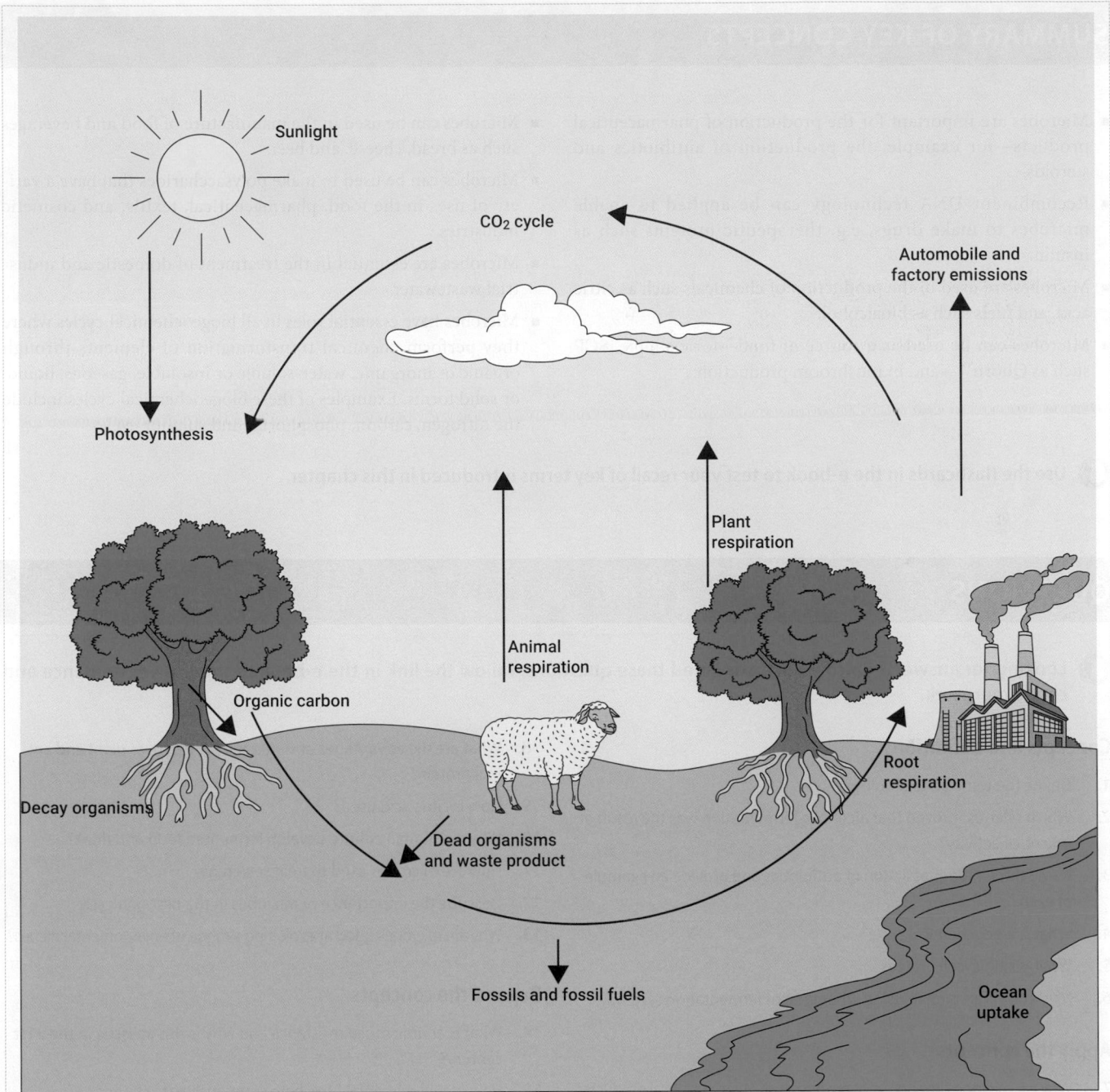

Figure 13.28 The carbon cycle.
Source: © UCAR.

chain. Those plants, animals, and microbes that are not eaten will eventually die and their biomass is decomposed by bacteria and fungi, enabling the carbon to be released into the habitat.

Fungi are especially good at decomposing: they possess a wide variety of enzymes, such as cellulases, xylanases, and lignases, which can break down complex organic molecules into simpler molecules.

The burning of fossil fuels—be that in power stations or in the combustion engines of cars, aeroplanes, and other vehicles—has added considerably to the carbon dioxide in the atmosphere, and this is contributing towards the current climate emergency we discussed in Topic 1 at the start of this book.

⊙ **Check your understanding of the concepts covered in this section by answering the questions in the e-book.**

SUMMARY OF KEY CONCEPTS

- Microbes are important for the production of pharmaceutical products—for example, the production of antibiotics and steroids.

- Recombinant DNA technology can be applied to enable microbes to make drugs, e.g. therapeutic proteins such as insulin.

- Microbes are used in the production of chemicals such as citric acid, and fuels such as bioalcohol.

- Microbes can be used as a source of food—for example, SCP such as Quorn™—and in mushroom production.

- Microbes can be used in the manufacture of food and beverages such as bread, cheese, and beer.

- Microbes can be used to make polysaccharides that have a variety of uses in the food, pharmaceutical, textile, and cosmetic industries.

- Microbes are essential in the treatment of domestic and industrial wastewater.

- Microbes have essential roles in all biogeochemical cycles where they perform chemical transformation of elements through organic or inorganic, water-soluble or insoluble, gaseous, liquid, or solid forms. Examples of these biogeochemical cycles include the nitrogen, carbon, phosphorus, and sulphur cycles.

Use the flashcards in the e-book to test your recall of key terms introduced in this chapter.

QUESTIONS

Looking for answers? Once you've answered these questions, follow the link in the e-book to the answer guidance and check your work.

Concepts and definitions

1. Define the term 'biotechnology'.

2. Which scientist proved that alcoholic fermentation was the result of microbial activity?

3. Name three modes of action of antibiotics and provide an example of each.

4. What is a geochemical cycle?

5. What is single-cell protein?

6. Mushrooms are grown using what type of fermentation system?

Apply the concepts

7. Why are microbes important as producers of pharmaceutical products?

8. What are the advantages of single cell protein over plant and animal protein?

9. How is citric acid used?

10. Why did human culture develop fermented food and drink?

11. How are microbes used to treat sewerage?

12. Describe the importance of microbes in the nitrogen cycle.

13. Why are microbes good at exploiting a range of environmental niches?

Beyond the concepts

14. What is antimicrobial resistance and why is this an issue in the 21st century?

15. How would microbial fuels help combat climate change?

FURTHER READING

Yeoman, K., Lea-Smith, D., Fahnert, B., & Clarke, T. (2020) *Microbial Biotechnology*. Oxford: Oxford University Press.
   For an introduction to how microbes are used in biotechnology.
Money, N. P. (2015) *Microbiology: A Very Short Introduction*. Oxford: Oxford University Press.
   For broader information on microbes in general, as well as a look at environmental microbiology.

https://theconversation.com/why-bacteria-could-be-the-answer-to-a-future-without-oil-35443
   To find out more about how bacteria could be the answer to a future without oil, this article from The Conversation is an interesting read.

Microbes as Agents of Infectious Disease

LEARNING OBJECTIVES

By the end of this chapter, you should be able to:

- Recall that a microbial organism must express or exhibit molecules or characteristics that are called virulence factors in order to cause a disease.

- Recall Koch's postulates and describe how they can be used to identify which microbial pathogen is the causative agent of a particular disease.

- Recall that microbial pathogens can be bacterial, fungal, or protozoan.

- Explain that infectious microbial diseases can affect many millions of people, often from the poorest communities.

- Explain that the problem of drug resistance is becoming a major problem despite many therapeutic approaches to treat microbial infections having been developed.

- Recall that plants can also be subject to infectious disease and describe some key ways in which they can be affected.

Chapter contents

	Introduction	523
14.1	Koch's postulates: how can we tell if something is a true pathogen?	524
14.2	Bacterial pathogens	526
14.3	Antibacterial drugs	531
14.4	Fungal pathogens	534
14.5	Protozoan diseases	539
14.6	Plant pathology	547

Watch the key concepts video in the e-book to prepare yourself for studying this chapter.

Introduction

In Chapter 11 we discovered that microbes can live in many different environments, from hot springs to salty lakes. Human, animal, and plant bodies (called hosts) are no exception, and all provide suitable living places for microbes. Many microbes live on and in animals, plants, and us quite harmlessly. But sometimes microbes can gain entry to a host and cause that host harm. These microbes are then deemed to be pathogens.

In this chapter we discuss a range of microbial pathogens from the bacteria, fungi, and protozoa, how they invade organisms, and how they produce toxins. We discuss another group of important pathogens, the viruses, in Chapter 15.

14.1 Koch's postulates: how can we tell if something is a true pathogen?

A **pathogen** is defined as any organism or molecule that is capable of causing a disease. Pathogens range from prokaryotes such as *Salmonella typhimurium* (which causes food poisoning), eukaryotes like *Trypanosoma brucei* (sleeping sickness), viruses such as human immunodeficiency virus (acquired immunodeficiency syndrome), and even small proteins called prions that can cause diseases such as bovine spongiform encephalopathy (BSE, which is also sometimes known as 'mad cow disease'). As shown in Figure 14.1, infectious diseases are a significant cause of death worldwide, particularly in low-income countries where living conditions—including poor sanitation and poor water quality—put the population at particular risk.

In order to cause an infection, micro-organisms must establish themselves either by attaching to or penetrating one of the surfaces that line the body, as illustrated in Figure 14.2. In order to achieve this, microbial pathogens exhibit or produce a variety of **virulence factors** that help them acquire nutrition and avoid the host's immune system, for example.

We have not always known the pathogens or causative agents that cause certain diseases. In the 19th century, Frenchman Louis Pasteur (1822–1895), German Robert Koch (1843–1910), and other microbiologists first started to make a connection

between pathogens and the diseases they cause when they proposed the **germ theory of disease**, also sometimes called the pathogenic theory of medicine. (The term 'germ' in this context

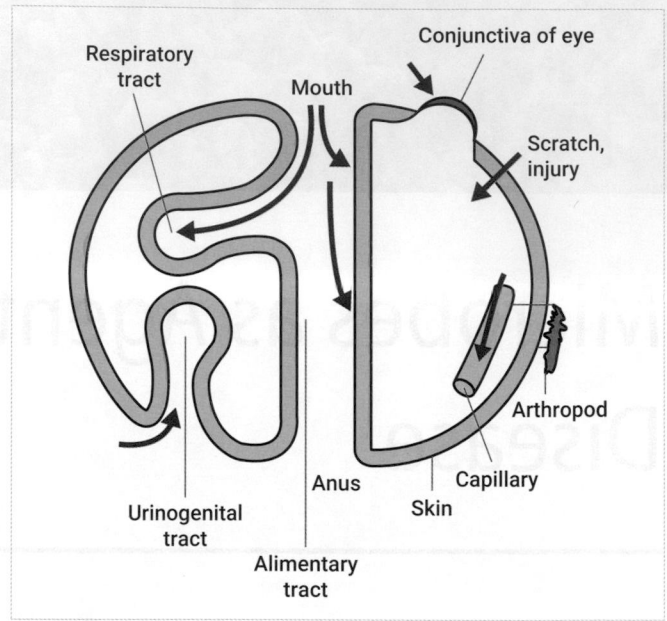

Figure 14.2 Body surfaces and systems that are sites of microbial infection.
Source: basicmedicalkey.com.

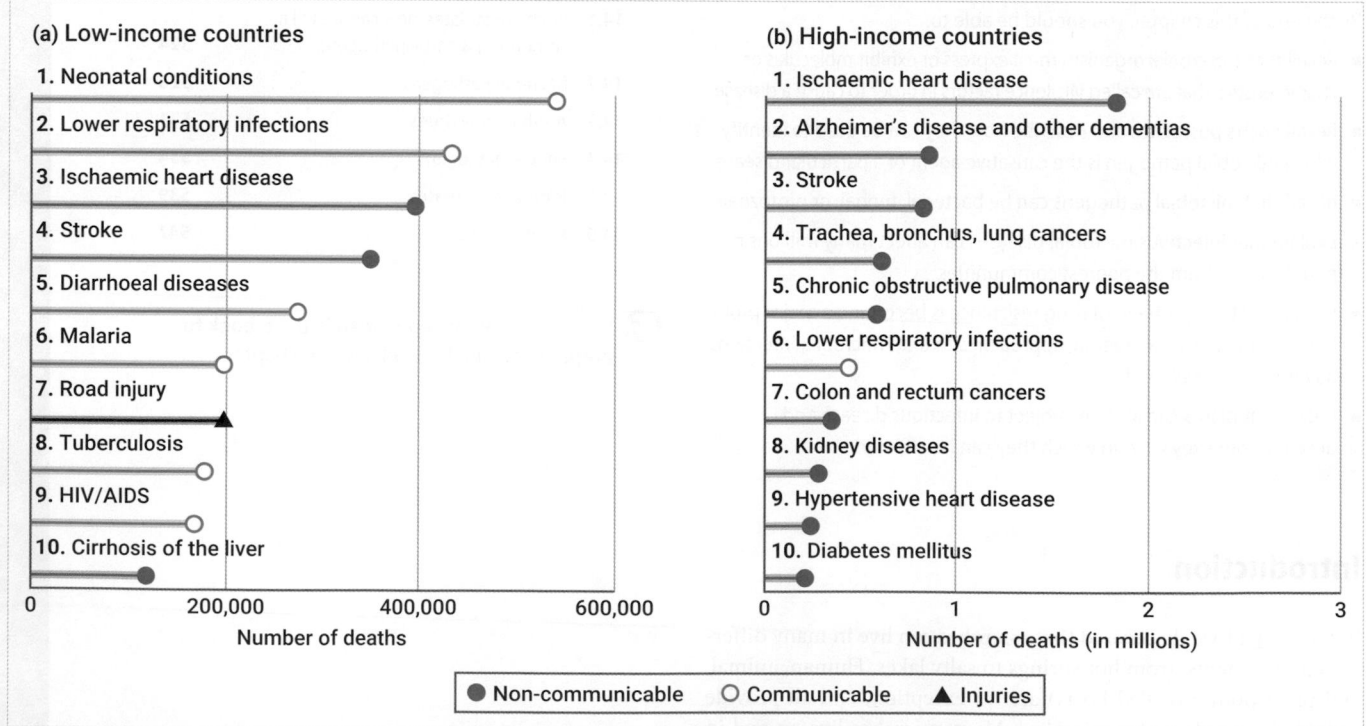

Figure 14.1 Infectious diseases are a significant cause of death. Top 10 causes of death in (a) low-income and (b) high-income countries in 2016. Communicable diseases—those diseases that occur due to transmission of an infectious agent from an infected animal or reservoir to a susceptible animal, either directly or indirectly through a vector or the environment—are responsible for a significant proportion of deaths, particularly in low-income countries.

Source: Reproduced from WHO, The top 10 causes of death, published 9th December 2020; https://www.who.int/news-room/fact-sheets/detail/the-top-10-causes-of-death, accessed 27/12/2021.

applies not just to various bacteria, but also to viruses, fungi, protozoa, and prions.)

But how do we know which pathogen causes a particular disease? Koch developed a series of essential conditions, or postulates, which must be satisfied to prove a particular microbe causes a particular disease. These postulates, which are summarized in Figure 14.3, are:

1. The suspected pathogen must be present in abundance in every case of the disease (but not found in healthy individuals).

2. That pathogen must be isolated and grown in pure culture.

3. The cultured pathogen must cause the disease when it is inoculated into a healthy, susceptible experimental host.

4. The same pathogen must be re-isolated from the diseased experimental host.

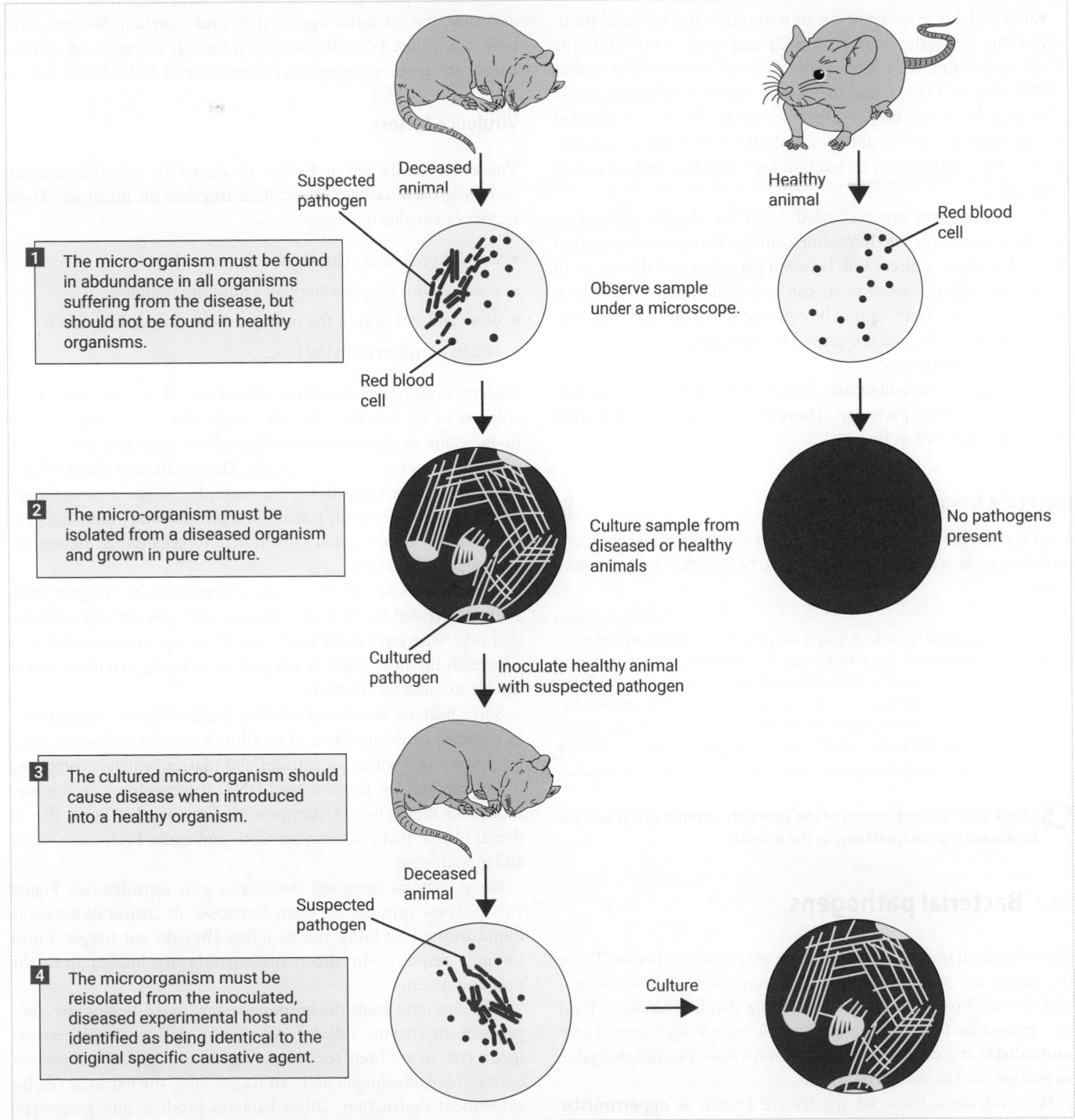

Figure 14.3 A visual representation of Koch's postulates.

Source: [mike jones]/Wikimedia Commons/CC BY-SA 3.0.

Koch used these rules to demonstrate in 1876 that the bacterium *Bacillus anthracis* was the cause of anthrax. He originally isolated the bacterium from farm animals that had died of anthrax. After growing *B. anthracis* in the laboratory, he introduced it into a mouse that subsequently developed the disease. Koch reisolated the bacterium from the diseased mouse and repeated the infection–isolation cycle many times, in so doing demonstrating that the bacterium caused the same disease on each occasion.

Koch and his team went on to determine the bacterial pathogens that cause tuberculosis (1882) and cholera (1883). Using Koch's postulates, the bacteria that cause diseases such as typhus (1880), tetanus (1884), and the plague (1894) were subsequently identified by a number of other scientists. Koch was awarded the Nobel Prize in Physiology or Medicine in 1905 in recognition of his contribution to bacteriology and the understanding of disease.

These postulates are still used today to identify pathogenic agents, although they do have shortcomings that mean they cannot be used to prove a direct link between pathogen and disease in all cases. For example, some hosts can be infected but are asymptomatic; some pathogens cannot be cultured under laboratory conditions; and some diseases are caused by a combination of pathogen, host, and the environment.

It is unethical to deliberately infect a healthy human host with a suspected disease pathogen. However, we discuss a self-testing experiment in Clinical Box 14.1.

PAUSE AND THINK

A micro-organism has been isolated from a skin infection. Apply Koch's postulates to show how the organism would be identified as the causative agent.

Answer: Having isolated the micro-organism from the infected skin, it should then be grown in a pure culture in the laboratory. A healthy, susceptible host organism should then be infected with the micro-organism using this laboratory culture. If the pathogen causes a similar skin infection to that which was observed initially, and the same micro-organism can again be isolated from the new infection, then it is likely to be the causative agent.

🌀 **Check your understanding** of the concepts covered in this section by answering the questions in the e-book.

14.2 **Bacterial pathogens**

Many bacterial species live in harmony on and in our bodies. These organisms are part of our normal microbiota. The human gut and skin are home to millions of bacteria that help us digest food and protect us from various pathogenic micro-organisms. These **mutualistic** and **commensal** relationships have evolved alongside us and are vital to our everyday life.

However, some bacterial species are known as **opportunistic pathogens**. Typically, these micro-organisms do not harm their host but can invade the organism's system and cause a disease given the correct environment and stresses. This is a huge problem in particular for people with suppressed immune systems, such as those going through chemotherapy, as an otherwise harmless bacterium can cause an infection in such individuals. Other bacteria are parasites: their interaction with the host benefits them while the host is harmed.

In this section we explore some of the characteristics of bacterial pathogens that enable them to cause infection, and how the bacterial agents that cause certain diseases have been identified. We will also take a look at examples of specific pathogens such as *Staphylococcus aureus* and *Helicobacter pylori*.

Virulence factors

Virulence factors are molecules produced by micro-organisms, including bacteria, that allow them to cause an infection. These factors can enable the pathogen to:

- attach to cells and colonize a niche in the host;
- evade and/or suppress the host's immune response;
- enter and exit cells (if the pathogen is an intracellular one);
- obtain nutrition from the host.

Bacteria must be able to adhere to host cells if they are to establish colonies or an infection. In order to do this they use specialized lipoproteins or glycoproteins called adhesion factors (also called adhesins) or structures such as pili. These adhesion factors determine the host cell specificity. For example, *Neisseria gonorrhoeae*, which causes the sexually transmitted genitourinary infection gonorrhoea, has adhesins that only adhere to cells lining the urethra and vagina of humans.

A pathogen's specificity is also determined by receptor molecules expressed by the host cell; these are typically glycoproteins that otherwise have other functions. If the bacterium is unable to adhere to the host cell it is referred to as being **avirulent** and is unable to cause an infection.

Some bacteria attach and colonize the host by way of a biofilm, as depicted in Figure 14.4a. A biofilm is a sticky web of bacterial cells embedded within an extracellular matrix, which is composed of polysaccharides, proteins, and DNA; it is produced by the bacterial cells themselves. A common example of a biofilm is that of dental plaque that can form on teeth and cause both tooth decay and gum disease.

Some bacteria surround their cells with capsules (see Figure 14.4b). These capsules are often composed of chemicals normally found in the host body and therefore they do not trigger a host immune response. In effect, the bacteria are hidden from the immune system.

Some bacteria evade the immune system in a different way: they produce an enzyme called IgA protease. IgA protease destroys IgA, a type of antibody (or 'immunoglobulin', Ig) that can prevent bacterial cell attachment and can flag or label the bacterial cell for subsequent destruction. Other bacteria produce anti-phagocytic chemicals that prevent lysosomes, which contain digestive enzymes, from fusing with those phagocytic vesicles that contain the bacterial cell. Free of the risk of digestion, the bacterium can then escape from the vesicle and hide from the immune response within the cell that phagocytosed it.

▶ We learn more about phagocytosis in Section 27.2.

Bacterial cells often require a source of iron, but it can be in short supply within the host environment because it is insoluble. To address this issue, some bacteria produce various iron-binding proteins, called siderophores, which scavenge and bind ferric iron (Fe^{3+}) and make the mineral available to the bacterial cell.

Many bacteria are able to produce toxins, an activity that contributes in part to the ability of the bacteria to cause disease. There are two broad types of toxin: exotoxins and endotoxins. Exotoxins are secreted molecules, and endotoxins are components of

---

CLINICAL BOX 14.1   Determining the cause of disease: Koch's postulates and *Helicobacter pylori*

Prior to 1982 it was thought that spicy food, stress, and lifestyle were among the main factors to cause the development of peptic ulcers and gastritis (inflammation of the stomach). However, pivotal research by Barry Marshall and Robin Warren demonstrated that *Helicobacter pylori* (Figure 1) was responsible instead. This discovery resulted in these two scientists being awarded the Nobel Prize for Physiology or Medicine in 2005.

It was generally considered that no microbial species would be able to live within the acidic conditions found within the stomach (at least not for very long). However, building on a few, mostly forgotten, reports of spiral-shaped bacteria being found in stomach linings or stomach washings over the previous 100 years, the pathologist Robin Warren observed small curved or spiral bacterial cells in stomach biopsies taken from ~50 per cent of cases. He also found that inflammation was always seen in the tissue at the site of the infection.

Barry Marshall was able to culture these then-unidentified bacteria from some of the biopsies. Together, Warren and Marshall were then able to demonstrate that the bacteria were present in almost all patients with gastritis or some sort of gastric ulcer and inflammation.

Their theory that the bacteria caused these diseases was not well received by the scientific and medical community, however. So, in an experiment to demonstrate 'cause and effect', Marshall deliberately drank a live culture of *H. pylori*. It was shown that he subsequently developed signs of gastritis, which were then cured with a course of antibiotics. (This experiment obviously went against health and safety practices but followed one of Koch's postulates.)

Much more ethically designed experiments have since shown that *H. pylori* causes >90 per cent of duodenal ulcers and up to 80 per cent of gastric ulcers, and that the antibiotic clarithromycin, used as part of a triple therapy (i.e. used alongside two other medications), is an effective treatment for many gastritis cases.

Despite the success of the antibiotic therapy—and in line with many other bacterial pathogens—it is possible for *H. pylori* to develop resistance to clarithromycin. This is the main reason for treatment failure.

The identification of *H. pylori* as a causative agent has revolutionized the treatment of many thousands of individuals for whom gastritis and ulcers were a long-term chronic illness. The identification of the role of *H. pylori* in inflammatory conditions of the stomach has also changed our view of the causes of some cancers too. The widespread inflammation seen in some individuals with *H. pylori* infection predisposes them not

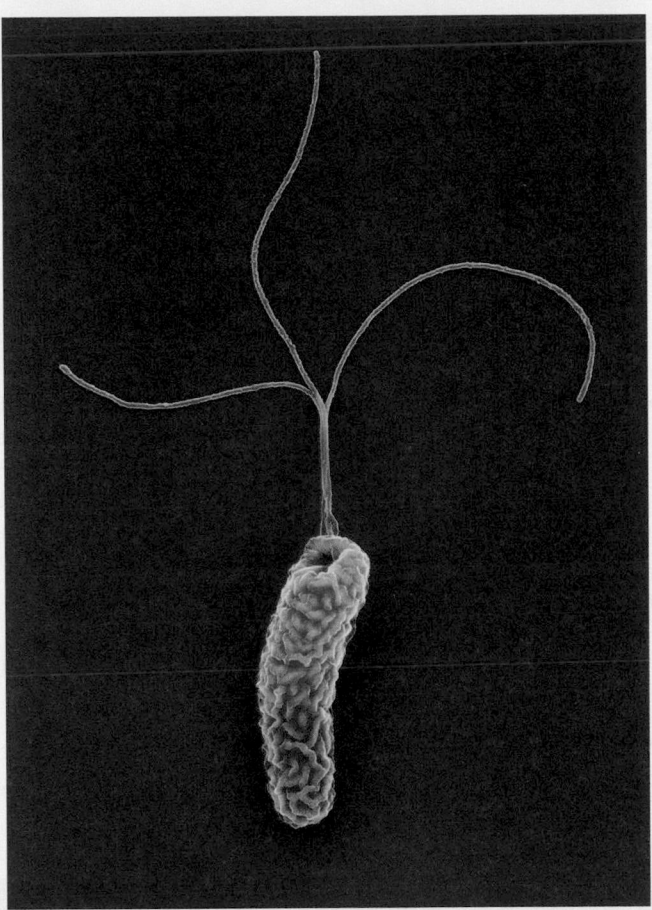

Figure 1 *Helicobacter pylori.*
Source: Science Photo Library.

only to ulcers, but also to stomach cancers, and is a risk factor for mucosa-associated lymphoid tissue (MALT) lymphoma.

A couple of caveats should be noted, however. Many people with *H. pylori* infections are asymptomatic and do not develop gastritis, ulcers, or more serious conditions such as cancers. A small proportion of gastritis and ulcer patients also show no signs of infection with *H. pylori*: other factors can and do contribute to the development of gastric illnesses.

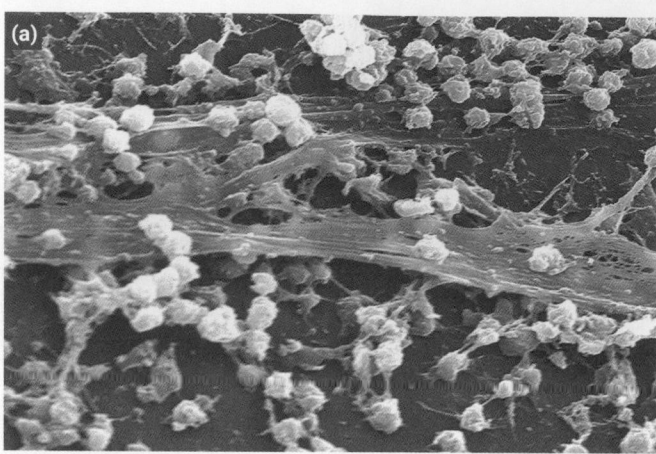
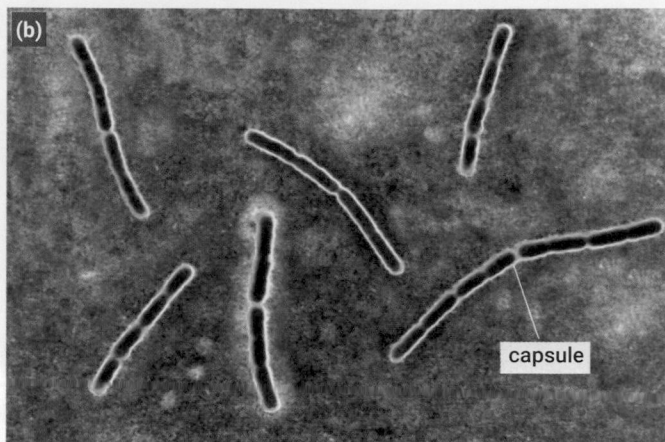

Figure 14.4 Biofilm. (a) Biofilm formed by *Staphylococcus aureus*. (b) Capsule surrounding *Bacillus anthracis* cells.
Source: (a) Science History Images/Alamy Stock Photo; (b) BSIP SA/Alamy Stock Photo.

Gram-negative bacterial cell walls. Look at Table 14.1 for a further comparison of the two.

Look at Table 14.1 and notice how exotoxins fall into three categories: neurotoxins, cytotoxins, and enterotoxins.

- Neurotoxins affect the nervous system. For example, *Clostridium botulinum* releases the botulinum toxin that specifically attacks motor neurones, preventing muscle excitation and therefore causing paralysis.

- Cytotoxins, the largest group of toxins, have a huge range of direct effects on specific cells in an organism: they can inhibit protein synthesis, destroy cellular membranes, and can cause destruction of cells. The diphtheria toxin is an example. Produced by *Corynebacterium diphtheriae*, diphtheria toxin blocks protein synthesis by inhibiting the elongation of polypeptides.

- Enterotoxins are protein exotoxins released by bacteria that target the intestines. Staphylococcal enterotoxin B (SEB) is produced by *Staphylococcus aureus* and is a common cause of food poisoning.

Many virulence factors that enable bacteria to colonize and/or harm the body are the products of quorum sensing (QS) genes.

Quorum sensing is a cell-to-cell communication process which involves the production, release, and community-wide sensing of molecules called autoinducers. Autoinducers modulate gene expression in response to the density and species diversity of a bacterial population. In *S. aureus*, QS controls a variety of virulence factors such as adhesin molecules, which enable the pathogen to adhere to cell surfaces, and haemolysins—molecules that break open red blood cells.

PAUSE AND THINK

Why are virulence factors important for pathogens?

Answer: Virulence factors are molecules produced by micro-organisms, including bacteria, that allow them to cause an infection—for example, adhesion molecules such as specialized lipoproteins or glycolipids; factors that help the pathogen evade the host immune response (e.g. capsules and IgA protease); and factors that help the pathogen obtain nutrients (such as siderophores). Without appropriate virulence factors, the micro-organisms are either unable to cause infection (and are called avirulent) or have reduced effectiveness at causing infection and its subsequent symptoms and/or the immune system is able to eliminate the pathogen.

Table 14.1 A comparison of exotoxins and endotoxins

	Exotoxins	Endotoxins
Source	Gram-positive and Gram-negative bacteria	Gram-negative bacteria
Relation to bacteria	Secreted metabolic product	Portion of outer cell wall released upon cell death
Chemical nature	Protein or short peptide	Lipid (A) portion of lipopolysaccharide
Toxicity	High	Low, may be fatal in high doses
Heat stability	Unstable >60 °C	Stable for up to 1 h at autoclave temperatures (121 °C)
Effect on host	Variable: acts as a cytotoxin, neurotoxin, enterotoxin	Fever, lethargy, malaise, shock, blood coagulation
Fever producing	No	Yes
Antigenicity	Strong: stimulates antibody production	Weak
Representative diseases	Botulism, tetanus, gas gangrene, diphtheria, cholera, plague, staphylococcal food poisoning	Typhoid fever, tularaemia, endotoxic shock, urinary tract infections, meningococcal meningitis

What are the sources of infectious diseases and how are they transmitted?

Many bacterial (and other) pathogens cannot exist for long outside of the host organism. A range of 'reservoirs' can harbour the pathogen, with or without causing disease; the pathogen can be transmitted from these reservoirs to a susceptible host.

Many human diseases involve the pathogen being transmitted from an animal host; these are referred to as zoonotic diseases. Humans can also act as carriers: they do not exhibit symptoms (they are 'asymptomatic') but can infect others, sometimes for years. For example, 'Typhoid Mary' (Mary Mallon) is considered to be the first person identified in the USA as being an asymptomatic carrier of *Salmonella typhi* and was responsible for infecting over 50 individuals, who subsequently developed typhoid fever. Much more recently, asymptomatic carriers have also been a concern during the Covid-19 pandemic, when it was found that both children and adults alike could carry the virus without themselves exhibiting symptoms. Individuals with tuberculosis and cholera may also be asymptomatic.

Various pathogens are also found in non-living reservoirs such as soil, water, and food.

Depending upon the source and nature of the pathogen, the mode of transmission can vary. A bacterial pathogen can be transmitted directly, via activities such as kissing, hand shaking, and sex. Indirect routes include via toothbrushes, toys, and crockery. These indirect routes involve a **fomite**: a non-living object that may be contaminated with an infectious pathogen and be involved in its transmission.

Fomites are associated particularly with infections acquired in hospitals, also known as **nosocomial** infections. These objects enable pathogens to be passed between doctors, visitors, and patients, and between patients themselves. Fomites include objects such as clothing, as well as hospital equipment—for example, stethoscopes, catheters, and life support equipment. Fomites are one reason why the NHS dress code states that doctors should wear short-sleeved shirts: they reduce the risk of long sleeves coming into contact with patients, potentially spreading infection in the process.

Careful sterilization can prevent cross-infection, but this can be hindered by the generation on the objects of biofilms by bacterial pathogens, which can provide some protection against sterilization.

A number of pathogens are transmitted with the help of some form of vehicle. These include dust particles for airborne transmission, and water (such as streams or swimming pools). Pathogens can also be transmitted by some form of vector, including the surface of insect bodies or via insect bites.

Bacterial pathogens: some examples

Many different types of bacterial pathogen exist. To provide an exhaustive list would take a whole book. Let us now focus on some specific examples that infect the skin, lungs, digestive system, and the genitourinary tracts. The pathogens involved—*Staphylococcus aureus*, *Streptococcus pneumoniae*, *Helicobacter pylori*, and *Neisseria gonorrhoeae*—are illustrated in Figure 14.5.

Staphylococcus aureus is a normal member of our microbiota but can become an opportunistic pathogen and there are now strains that have resistance to various antibiotics. *Streptococcus pneumoniae* is a bacterium that can cause infection of the lungs with potentially fatal consequences. *Helicobacter pylori* was found to cause many gastric ulcers; this realization revolutionized the treatment of thousands of people. *Neisseria gonorrhoeae* colonizes mucosal surfaces of the urethra, uterine cervix, anal canal, throat, and conjunctiva. It is often transmitted through sex and causes the sexually transmitted disease gonorrhoea.

Let us now consider each of these examples in more detail.

Staphylococcus aureus

Staphylococcus aureus (*S. aureus*) is a Gram-positive, round-shaped (coccal) bacterium that is a member of the normal microbiota of the body. It is frequently found in the nose, respiratory tract, and on the skin. However, it is an opportunistic pathogen and infections can occur when the skin or mucosal barrier is breached. In addition to causing minor skin infections (impetigo and folliculitis), *S. aureus* infection can result in more serious, sometimes life-threatening, conditions, including pneumonia, toxic shock syndrome, and sepsis. It can also cause food poisoning via ingestion of enterotoxin-contaminated food.

Staphylococcus aureus exhibits a number of virulence factors that help it cause disease. It can evade the host immune system process of phagocytosis by expressing a protein called protein A. This bacterial protein binds to a host antibody called IgG, which then cannot trigger the destruction of the bacterium in the way it normally would.

▶ We learn more about types of antibody in Chapter 27.

Staphylococcus aureus also produces a number of enzymes that help it spread during infection. Two examples are staphylokinase and hyaluronidase.

Staphylokinase catalyses the conversion of plasminogen to plasmin within the host. Plasminogen and plasmin form part of the body's mechanism for inducing and controlling the clotting of blood; when produced from plasminogen, plasmin digests the long filaments of fibrin that help to give blood clots their structure and stability. By inhibiting the clotting of blood in this way, staphylokinase increases the spread of *Staphylococcus* and therefore promotes infection.

Hyaluronidase breaks down hyaluronic acid, which is found in connective, epithelial, and neural tissues. It has been suggested that the breakdown of hyaluronic acid by hyaluronidase leads to a degradation of connective tissue, again promoting the spread of *Staphylococcus* within the degraded tissue, and thereby promoting infection (although there is little evidence to support this role). The breakdown of the hyaluronic acid by hyaluronidase has been shown to provide a source of carbon for the pathogen.

Staphylococcus aureus also produces toxins: superantigens that can trigger toxic shock syndrome; exfoliative toxins, which cause peeling of the skin; and other toxins that act on cell membranes.

Some *Staphylococcus* express β-lactamase, an enzyme that can break down penicillin, resulting in resistance to that antibiotic.

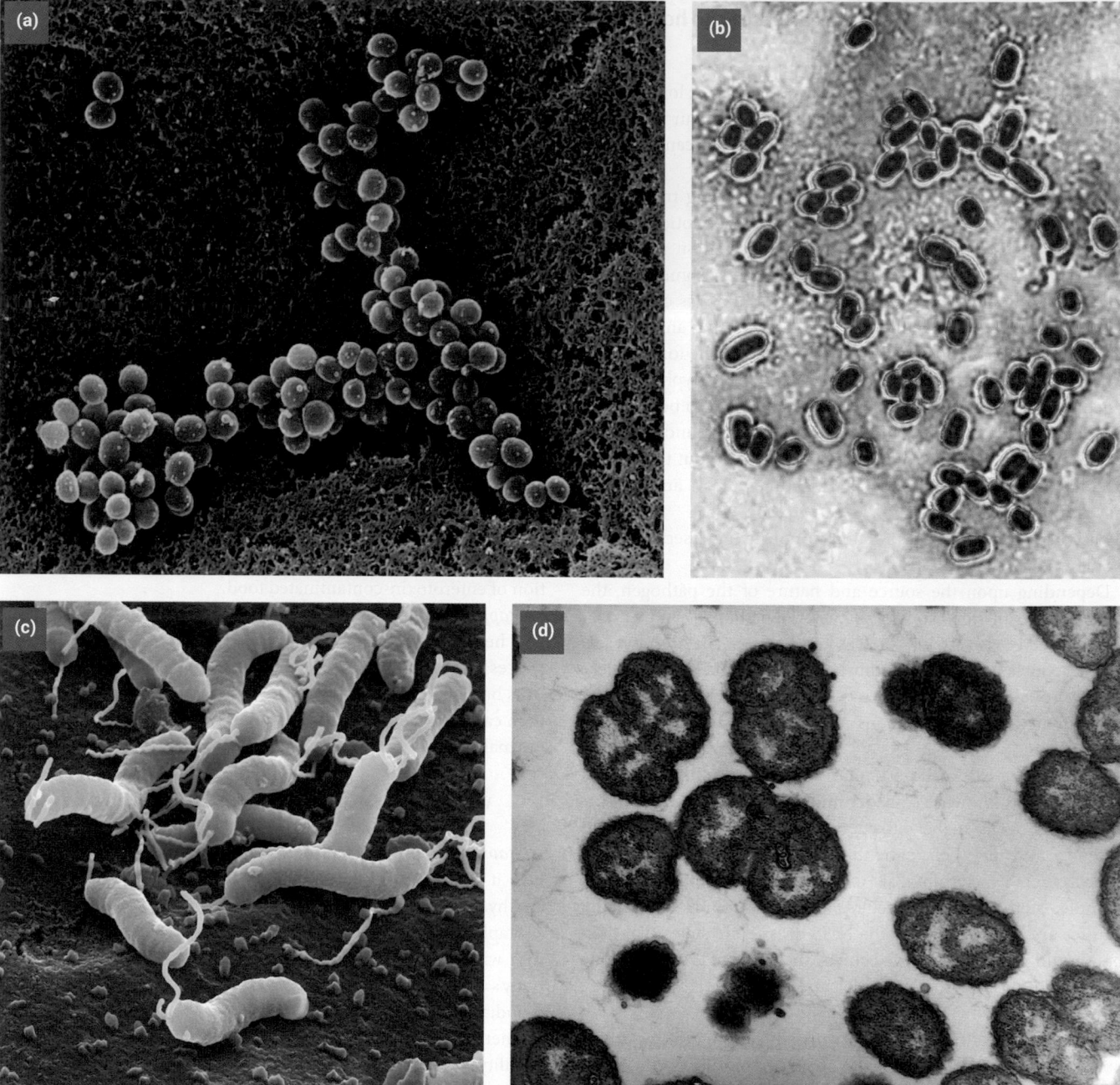

Figure 14.5 Examples of bacterial pathogens. (a) *Staphylococcus aureus*; (b) *Streptococcus pneumoniae*; (c) *Helicobacter pylori*; and (d) *Neisseria gonorrhoeae*.
Source: (a, b) Steve Lowry/Science Photo Library; (c) Juergen Berger/Science Photo Library; (d) Phanie/Alamy Stock Photo.

Penicillin is the antibiotic of choice for treating *S. aureus* infections. However, it is of no use if the bacteria express the β-lactamase enzyme. In these situations, other β-lactam-based antibiotics (which have the same mode of action as penicillin but are β-lactamase resistant) are prescribed instead.

Resistance to penicillin was relatively rare when the antibiotic was first introduced, but many strains of *S. aureus* are now resistant to this and other antibiotics. For example, we now have methicillin-resistant *Staphylococcus aureus* (MRSA), as we discuss further in Chapter 13.

Streptococcus pneumoniae

Streptococcus pneumoniae (*S. pneumoniae*) is a Gram-positive cocci that was a major cause of pneumonia in the late 19th century. Since that time, several vaccines have been developed and the World Health Organization (WHO) recommends routine childhood pneumococcal vaccination. Despite this progress, pneumococcal disease remains relatively common in developing countries where vaccinations and treatments are not always readily available, and where a higher prevalence of human immunodeficiency virus

(HIV) infection makes individuals more susceptible. It is also more common during winter and early spring.

Healthy carriers can harbour *S. pneumoniae* in their respiratory tracts as part of their normal microbiota without displaying any symptoms. However, the bacteria can become pathogenic in susceptible individuals, such as those with a weakened immune system. When the bacterium colonizes the alveoli of the lungs the host responds with inflammation, causing the alveoli to fill with plasma, blood, and leukocytes, which contribute to the symptoms of pneumonia.

Streptococcus pneumoniae can also cause a number of other diseases such as sinusitis (infection of the sinus), otitis media (ear infection), and pneumococcal meningitis (infection of the meninges, the membranous lining of the brain).

Streptococcus pneumoniae uses a number of virulence factors to contribute to infection and disease symptoms. The pore-forming toxin pneumolysin is released by the bacterium and aids colonization by helping it to adhere to the host cell. Other virulence factors include the enzyme IgA protease, which degrades the host immunoglobulin IgA; and a polysaccharide capsule, which provides protection once the bacterium is phagocytosed by the host. Both therefore contribute to evasion by the bacteria of the host immune response.

Streptococcus pneumoniae infections can be treated by a number of antibiotics, including penicillin. However, the 1990s saw an increase in the levels of multidrug-resistant *Streptococcus pneumoniae* (MDRSP), which remains a concern to this day.

Helicobacter pylori

Helicobacter pylori (*H. pylori*) is a highly motile Gram-negative bacterium that is usually found in the stomach. It was originally identified in the early 1980s, as discussed further in Clinical Box 14.1 The discovery of the role the bacterium plays in causing gastric ulcers has revolutionized the treatment of such ulcers. The majority of infected individuals are asymptomatic, however, and the bacterium may be an important part of our normal microbiota.

Symptoms of *H. pylori* infection can include gastritis with abdominal pain and nausea. Infection leads to an increased risk of developing ulcers and even gastric cancers.

Helicobacter pylori has four to six flagella, which are responsible not only for movement, but also for the burrowing of the bacterium into the mucosal lining of the stomach. (See Clinical Box 14.1, Figure 1, and notice the flagella that are clearly visible.) This burrowing helps the bacterium avoid the acidic conditions found in the stomach.

The bacterium attaches to the epithelial cells using adhesin molecules, and the bacterium is occasionally found inside these cells. *Helicobacter pylori* produces large amounts of the enzyme urease, which helps to neutralize the acid in the surrounding environment. Urease breaks down urea found in the stomach into carbon dioxide and ammonia; these react with and neutralize the acid. Unfortunately, the ammonia that is produced is toxic to the epithelial cells and causes their apoptotic death. Infection with *H. pylori* also triggers chronic inflammation. This, alongside the damage to the epithelial cells, ultimately results in the development of ulcers.

Antibiotics, such as clarithromycin and amoxicillin, can be used to treat *H. pylori* infection. However, as with other bacterial pathogens, antibiotic resistance is increasingly a problem. Prior to the discovery of *H. pylori*'s role in gastric ulcers, the only treatment option to relieve the symptoms was to reduce the acidity of the stomach by using therapeutics such as antacid tablets.

Neisseria gonorrhoeae

Neisseria gonorrhoeae (*N. gonorrhoeae*) is a Gram-negative diplococcus that colonizes mucosal surfaces of the urethra, uterine cervix, anal canal, throat, and conjunctiva. It is often transmitted through sex and causes the sexually transmitted disease gonorrhoea. If left untreated it can result in pelvic inflammatory disease.

Neisseria gonorrhoeae has a typical Gram-negative bacterium cell wall and membrane but with a unique lipopolysaccharide called lipooligosaccharide (LOS). The bacterium has fimbriae, which play a major role in its adherence to the host's non-ciliated epithelial cells.

▶ We learn more about the structure of the Gram-negative bacterial cell wall in Section 9.2.

Neisseria gonorrhoeae is exclusively a human pathogen. This specificity may be because the bacteria obtain the iron they require only by binding to the human iron-chelating molecules transferrin and lactoferrin.

During infection, bacterial LOS and peptidoglycan are released when the cells are lysed. The LOS stimulates the production of a molecule called tumour necrosis factor, which can damage the cells. The damage attracts cells of the immune system, including neutrophils, which can then engulf the bacteria. Many of the bacteria can then survive intracellularly. The bacterial LOS, and resultant inflammation, is thought to be responsible for the symptoms of infection.

As with other bacterial pathogens, infection with *N. gonorrhoeae* can be treated with antibiotics. However, antibiotic resistance is also common, particularly to penicillins. This has led to the emergence of the term 'super gonorrhoea'.

PAUSE AND THINK

How can the spread of bacterial pathogens between individual members of a human population be reduced?

Answer: As bacteria can be transmitted through direct person-to-person contact, via the air, water, or food, or by living vectors, the ways to reduce the transmission varies. Such measures include water treatment, immunization of animals and humans, good personal and food hygiene measures, and safer sex practices.

 Check your understanding of the concepts covered in this section by answering the questions in the e-book.

14.3 Antibacterial drugs

After a bacterial pathogen has infected a host organism a disease arises only if the pathogen multiplies sufficiently to adversely affect the host body. However, as we note in Section 14.1, not all infections have noticeable symptoms: some hosts may be infected but without symptoms. These infections are referred to as asymptomatic or subclinical. However, when an infection is recognized (whether symptomatic or not) a number of drugs can be used to treat the pathogenic bacterial agent; some of these drugs and their targets are summarized in Figure 14.6.

As the name implies, antibacterial drugs are used to combat infections caused by bacteria. These drugs fall into two types: bactericidal and bacteriostatic. Bactericidal drugs kill the bacteria, while bacteriostatic drugs prevent them from multiplying. Once rendered unable to replicate, the bacteria become a ready target for the host's immune system, which can then destroy them.

The largest group of antimicrobial drugs are called **antibiotics** and come in two forms: broad spectrum and narrow spectrum. Broad-spectrum antibiotics affect a large number of both Gram-positive and Gram-negative bacteria; by contrast, narrow-spectrum antibiotics affect a more specific range of bacteria.

The relatively non-specific action of broad-spectrum antibiotics can cause problems: they may also destroy the host's harmless symbiotic bacteria (e.g. those found in the gut), which can have a negative effect on the host.

How do antibacterial drugs operate?

Let us now consider the main ways in which antibacterial drugs operate.

Cell wall inhibition

We saw in Section 9.2 how the cell walls of both Gram-positive and Gram-negative bacteria include peptidoglycan, although it is present in much lower amounts in Gram-negative bacteria. Therefore, antibiotics that disrupt the synthesis of peptidoglycan are broad spectrum: they can affect a wide range of bacteria.

Peptidoglycan plays a huge role in the rigidity of the cell wall: if it is destroyed, the cell loses its structural integrity. The class of antibiotics that affect cell wall synthesis is called β-lactams; they include penicillin and its derivatives (e.g. methicillin and cefoxitin). As β-lactams inhibit the synthesis of new peptidoglycan, only *growing* cells are affected: their cell walls become weak and fragile and the cells eventually lyse.

Protein synthesis inhibition

Protein synthesis is a complicated, multi-step process, as we discover in Section 5.4. Protein synthesis inhibitors can have effects on many of the enzymes and other proteins involved in the process, resulting in its disruption. The main complex involved in protein synthesis is the ribosome; the inhibition of its activity can have a huge impact.

We saw in Chapter 9 how bacteria are prokaryotes and humans are eukaryotes. The ribosomes of prokaryotes and eukaryotes comprise different components. Consequently, it is possible to specifically inhibit the activity of a pathogen's ribosomes without impeding the activity of the host's ribosomes. Examples of antibiotics that target bacterial ribosomes are the tetracyclines, aminoglycosides, and others such as erythromycin and chloramphenicol.

Figure 14.6 Antibiotics. Types of antibiotics and their bacterial targets. *Source*: oxfordmedicine.com.

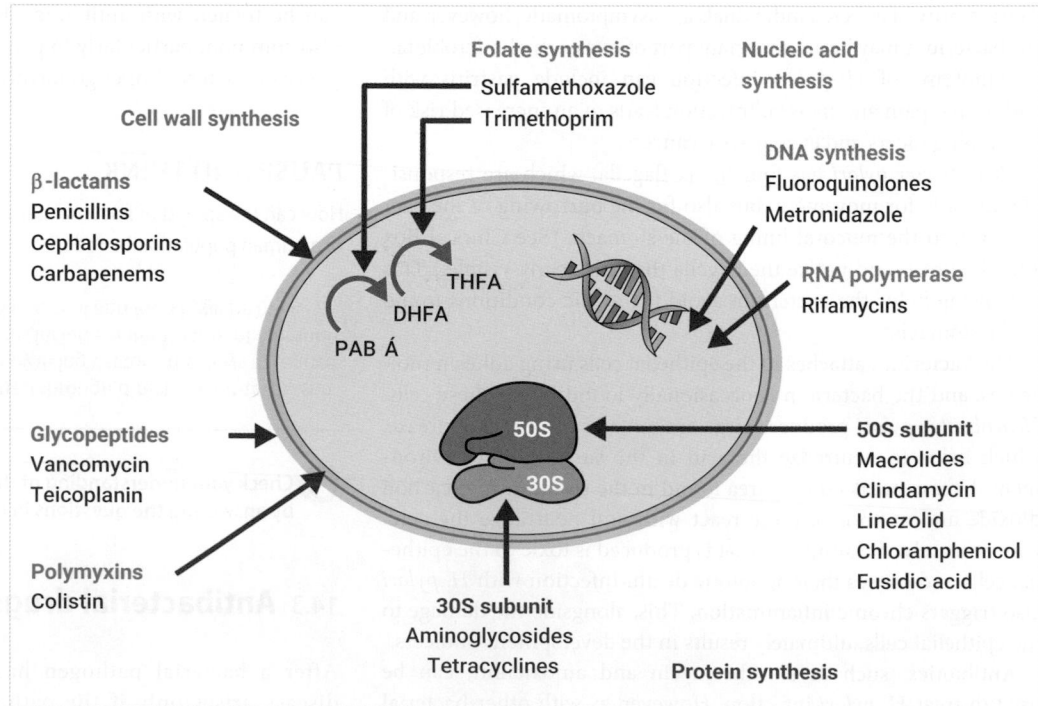

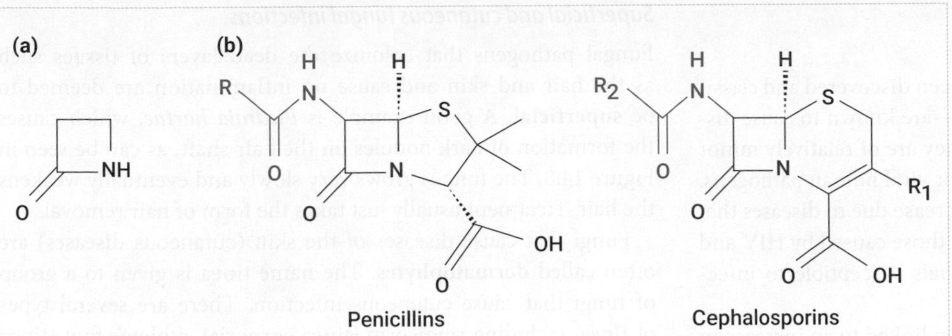

(a)   **(b)**

Penicillins

Cephalosporins

Figure 14.7 β-lactam antibiotics. (a) The structure of the β-lactam ring. (b) Examples of classes of antibiotic that include the β-lactam ring structure.

Nucleic acid inhibition

Antibacterial drugs that interfere with the DNA and/or RNA of a bacterial pathogen can have a devastating effect on them. However, these drugs can also interfere with the host's cells if they are not specific enough.

DNA and RNA synthesis can be disrupted at many stages. The antibiotics metronidazole and nitrofurantoin are commonly used to treat pelvic inflammatory disease and bladder infections, respectively. They act by generating metabolites that damage DNA strands as they are being synthesized. These drugs are selectively toxic to anaerobic organisms but *can* affect human cells.

Disruption to the plasma membrane

Changes in the permeability of the plasma membrane can also have devastating effects on the bacterial cell, which can lose vital metabolites, proteins, and water as a result. Polymyxin B attaches to the lipopolysaccharide (LPS) of the bacterial cell and interrupts the inner and outer membrane in Gram-negative bacteria. Daptomycin targets membrane function and peptidoglycan synthesis and is often used in soft tissue staphylococcal infections.

Disruption to metabolite synthesis

Bacteria are typically unable to utilize free folic acid and therefore have to synthesize their own. Sulphonamides and trimethoprim are broad-spectrum antibiotics that block specific steps in bacterial folic acid metabolism. These bacteriostatic drugs stop the growth of the bacteria, enabling the host to destroy the infection; they are commonly used to treat urinary tract infections.

Bacterial antibiotic resistance

Since the discovery of penicillin in 1928, antibiotics have been a vital weapon in the fight against bacterial infections. However, the resistance of bacteria to penicillin-based, and many other, antibiotics is now becoming a huge issue: it threatens the ability of healthcare providers to tackle bacterial pathogens in an effective way. So how do bacteria manage to resist such antibiotics?

Many bacteria have evolved enzymes called β-lactamases, or penicillinase. These enzymes are able to hydrolyse the β-lactam ring that is part of the chemical structure of many antibiotics (as illustrated in Figure 14.7). Once the β-lactam ring has been hydrolysed, the antibiotic becomes deactivated.

Bacteria have also developed resistance by evolving to produce altered versions of proteins to which the β-lactams would normally bind. With the β-lactams unable to bind, the antibiotic is no longer effective. This mechanism of resistance is exhibited by MRSA.

It is fairly common for a given bacterium to be resistant to one antibiotic. In fact, some species of bacteria have always contained genes that confer resistance in their genomes. For example, *Paracoccus* is resistant to rifampicin. However, it only really becomes an issue when bacteria cause an infection in higher organisms and when communities or organisms gain resistance genes.

Resistance genes are often found on plasmids and therefore are capable of being transferred to other bacteria through horizontal gene transfer. Many types of MRSA, for example, are becoming resistant to other antibiotics in addition to methicillin, resulting in multidrug resistant (MDR) bacteria.

▶ **We learn more about horizontal gene transfer in Section 10.1.**

So, how can such resistance be overcome? One approach is to give more than one antibiotic at once. Independent resistance to sulphonamide antibiotics and trimethoprim is becoming more common. Therefore, two drugs are often used together to disrupt two different steps in the metabolism of folic acid in a bid to overcome an infection (and avoid resistance).

PAUSE AND THINK

Why does antibiotic resistance matter?

Answer: Antibiotic-resistant infections are considered to be one of the leading threats to human health. Without effective antibiotics that kill the pathogenic bacteria many routine treatments—such as basic operations or setting broken bones—will become increasingly dangerous. When bacterial infections do not respond to frontline antibiotics, then more expensive medicines may have to be used. It may even be the case that no alternative medicines can be used. This could then result in a longer duration of illness and treatment, often increasing healthcare costs, as well as the economic burden on families and societies. Unfortunately, the increase in antibiotic-resistant infections will result in a rise in mortality, too.

Ⓖ **Check your understanding of the concepts covered in this section by answering the questions in the e-book.**

14.4 **Fungal pathogens**

Of the 70,000 fungi that have currently been discovered and classified, only a very few—some 400 species—are known to cause disease in animals and humans. As such, they are of relatively minor importance when compared to bacteria or viral human pathogens. However, fungal infections are on the increase due to diseases that can weaken the immune system, such as those caused by HIV and cancer. (This weakening leaves individuals susceptible to infections in general, including fungal ones.)

A rise in fungal infections has also been linked to an increase in the number of organ transplants being performed. Again, this is a consequence of individuals being immunocompromised: to avoid transplanted organs being rejected, those receiving the transplants are given immunosuppressive drugs. While this might reduce the likelihood of the immune system attacking the transplanted organ, it has the unavoidable consequence of leaving the individual more prone to infection.

The natural environment for disease-causing fungi is the soil, where they live as saprophytes: they live on and break down dead organic matter. However, they can become parasitic when the opportunity arises. As such, they are **facultative** parasites. Very few fungi are obligate parasites.

It is interesting to note that fungi, being larger than bacteria, were recognized first as being agents of disease. In 1841 David Gruby isolated a fungus from a ringworm infection, and experimentally showed that this fungus could cause the disease on healthy skin. This procedural method followed Koch's postulates, which we learnt about in Section 14.1.

Fungal infections

In this section we will consider the different levels of fungal infection: superficial, cutaneous, subcutaneous, and systemic. We will look at the ways in which fungal infections can be identified and treated and will also examine the role played by fungi as allergens in the environment.

Superficial and cutaneous fungal infections

Fungal pathogens that colonize the dead layers of tissues such as the hair and skin and cause no inflammation are deemed to be **superficial**. A good example is *Piedraia hortae*, which causes the formation of dark nodules on the hair shaft, as can be seen in Figure 14.8. The fungus grows very slowly and eventually weakens the hair. Treatment usually just takes the form of hair removal.

Fungi that cause diseases of the skin (cutaneous diseases) are often called **dermatophytes**. The name tinea is given to a group of fungi that cause cutaneous infection. There are several types of tinea, including ringworm (tinea corporis), athlete's foot (tinea pedis), and jock itch (tinea cruris). These fungal infections lead to obvious damage to tissues such as hair, nails, and skin, but this usually involves the dead layers only.

The inflammatory response to cutaneous disease is intense but is localized at the site of infection. In tinea corporis (ringworm), lesions are red, raised, and itchy; they also tend to be circular, as you can see in Figure 14.9. It is this characteristic shape that gives ringworm its name.

Athlete's foot is the most common dermatophyte. Most cases of athlete's foot are caused by three fungi: *Trichophyton rubrum*, *T. mentagrophytes*, and *Epidermophyton floccosum*. The infection is found between the toes (in the toe webbing), where the epidermis becomes red, cracked, weeping, and itchy, as depicted (somewhat unpleasantly) in Figure 14.10. Athlete's foot is often caused by wearing socks and shoes that can build up moisture and heat. The infection can be easily picked up in gyms, swimming pools, and dormitories. Athlete's foot is quite hard to treat, as all the fungi involved in the infection must be removed for treatment to be successful.

Subcutaneous infection

In subcutaneous fungal infections the fungi gain entry to the subcutaneous tissue—the tissue found under the skin—via wounds that penetrate the skin and into the subcutaneous tissue beneath. These wound typically take the form of scratches caused by thorny plants. These diseases therefore tend to be more prevalent in rural and tropical regions.

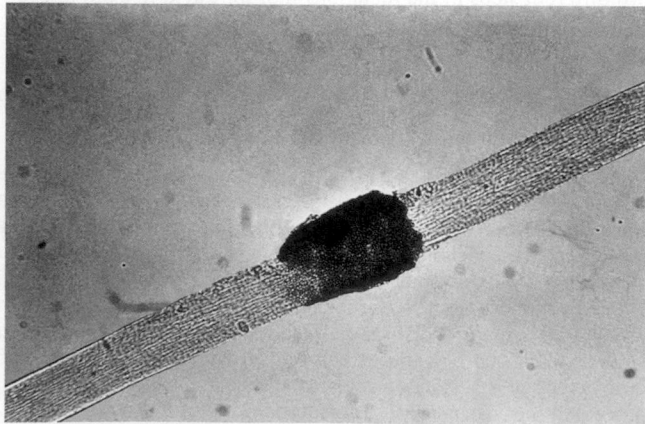

Figure 14.8 Fungal colonization of the hair. *Piedraia hortae* growing on a hair shaft.
Source: Courtesy Public Health Image Library.

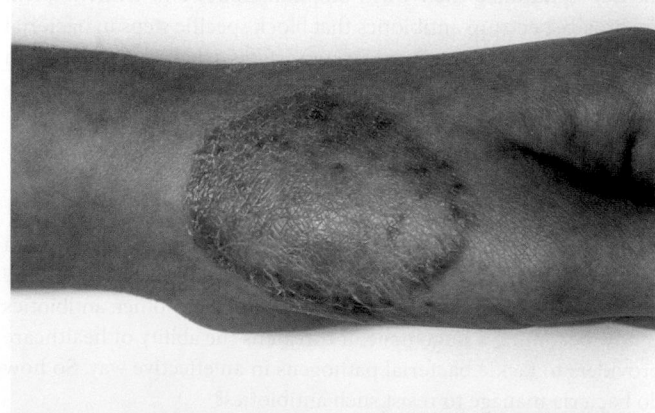

Figure 14.9 Ringworm. The red, circular lesion of tinea corporis.
Source: Mediscan/Alamy Stock Photo.

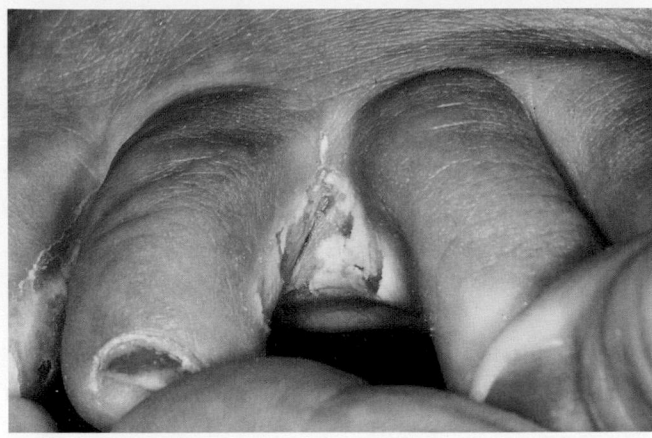

Figure 14.10 Athlete's foot. Athlete's foot infecting the webbing between the toes.

Source: Dr P. Marazzi/Science Photo Library.

Infections are localized, but they can form abscesses called **mycetomas**. Tissue damage is variable with these abscesses but can be very disfiguring.

One fungus that can cause serious deformities is *Madurella mycetomatis*, which is most common in tropical regions and can be picked up by walking barefoot on contaminated soil. Another example is the fungus *Sporothrix schenckii*, which enters the body through damage caused by rose and barbary thorns. Infection with *Sporothrix schenckii* has also occurred though interaction with plant debris such as baled hay and pine-bark mulch. There have also been outbreaks among building workers exposed to heavily infected timbers.

Systemic infection

The most serious level of fungal infection is seen when it becomes systemic—that is, when the fungus spreads from one internal organ to another. In many (but not all) cases, systemic fungal infections originate in the lungs after the inhalation of fungal spores. (As we see in Section 11.5, spores are used by fungi as a means of dispersal after both asexual and sexual reproduction.) If a vital organ such as the brain becomes a target of systemic infection, it frequently leads to death.

Before we examine examples of systemic fungal infection in more detail, we need to reintroduce the concept of **dimorphism**, which we first encountered in Section 11.5. Dimorphism describes the way in which fungi can exist in tissues as either single cells (yeast) or a mycelial (multicellular) form. Dimorphic fungi, which are pathogenic in animals, are more likely to exist outside the host in a mycelial form. Once fungal pathogens have gained entry to the host organism, however, conditions promote the conversion of the mycelium form to the yeast form. These conditions include elevated temperature, high carbon dioxide concentrations, low oxygen concentrations, and the presence of different sugars, such as glucose.

The formation of yeast cells helps the fungus to spread throughout the host body via the blood and lymphatic systems. The infection then becomes systemic (body wide). The hyphal form, due to the mechanism of apical growth (see Section 11.5), can quickly invade solid tissue.

Dimorphism is an important virulence factor in many systemic fungal infections. However, not all fungi that cause systemic infections are dimorphic, and many fungi that display dimorphism are not pathogenic.

Let us now turn our attention to four examples of systemic fungal infection: aspergillosis, candidiasis, histoplasmosis, and coccidioidomycosis.

Aspergillosis

Aspergillosis is caused by the filamentous fungus *Aspergillus fumigatus*, which is shown in Figure 14.11. *Aspergillus fumigatus* belongs to the Ascomycota.

This disease most often occurs in farmers and people who have been handling decaying organic matter. Individuals suffering from leukaemia and immunosuppressive disorders are also at risk.

Aspergillus fumigatus lives as a saprophyte in soil and decaying vegetation. Unusually for a fungus, it is very tolerant of elevated temperatures and can live at temperatures up to 65 °C. This means that it can live in, and produce a lot of spores in, environments such as composting plant material. *Aspergillus fumigatus* reproduces asexually to produce spores called **conidia**. These conidia have a diameter of 2–3 μm. This small size means that they can be dispersed easily in the air. Every day we breathe in several hundred conidia: there are approximately 10–400 colony forming units (cfu) in every cubic metre of air.

Typically, infection occurs when conidia are inhaled into the lungs: they can bypass the defences of the nasal and bronchial cavities and reach the lung alveoli. The conidia then germinate to form hyphae, which can penetrate the tissue and invade blood vessels, where thrombosis (blood clotting) can occur.

One form of aspergillosis is often referred to as 'fungus ball' or aspergilloma. In this disease, *A. fumigatus* occupies previously existing cavities in the lung, particularly old tuberculosis lesions, where it forms a compact ball of mycelia surrounded by a dense fibrous wall. The fungus balls are solitary and vary in size but are generally less than 8 cm in diameter. Perhaps surprisingly, patients with aspergilloma often display no symptoms or have only a moderate cough. However, haemorrhage may occur if the infection

Figure 14.11 Stained asexual sporing structures of *Aspergillus fumigatus*.

Source: William.W.Mangin-1 (talk | contribs), Microbe Wiki.

reaches a blood vessel. Treatment takes the form of surgical removal of the fungus ball.

Candidiasis

Candida albicans is a yeast that belong to the Ascomycota. *Candida albicans* is an opportunistic pathogen, which lives as a **commensal** organism in a healthy host on moist mucosal surfaces such as the mouth and vagina. However, under certain conditions it can cause cutaneous infections (candidiasis) in the same locations. These appear as soft grey/white lesions of fungal mycelium over a red mucosa. Other areas of infection include the digestive tract, bones, and the brain.

It is estimated that 70 per cent of the female population will at some point have vaginal candidiasis. Infection often occurs in pregnancy due to an increase in glycogen in the vaginal mucous membranes which promotes the growth of *C. albicans*. Oral thrush is also common in newborn babies as a result of infection during a vaginal birth, as can be seen in Figure 14.12. Other contributing issues to infection are other diseases present in the host, malnutrition, and the overuse of antibiotics.

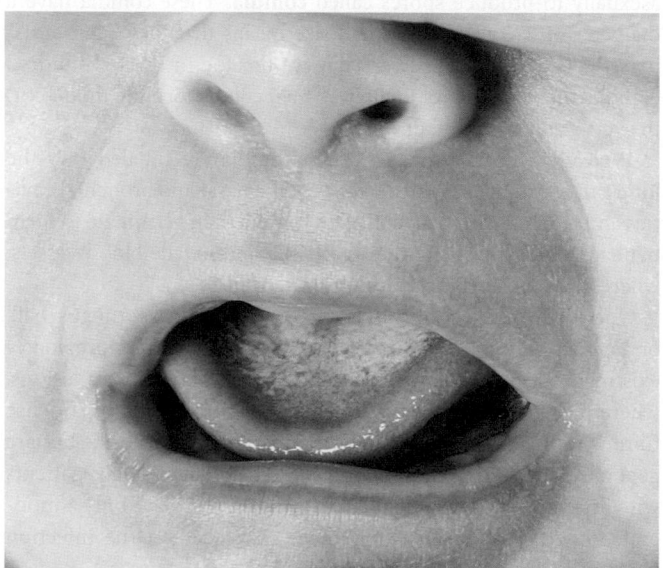

Figure 14.12 Candidiasis. Newborn suffering from oral candidiasis
Source: RioPatuca/Alamy Stock Photo.

Candidiasis can also occur in other animals such as reptiles and birds. The severity of the candidiasis is often related to other underlying health conditions. Systemic infection is rare, but the transition from harmless commensal to unrelenting pathogen is a fine line and depends on a myriad of virulence factors. The virulence factors that are important in *C. albicans* infections are detailed in Table 14.2.

▶ **We discuss virulence factors in association with bacterial disease in Section 14.2.**

Histoplasmosis

Histoplasmosis is also called Darling's disease. It occurs throughout the world but is endemic in some areas such as Ohio and the lower Mississippi river valley of the USA. Symptoms can vary, although can include fever, headache, muscle ache, and dry cough. The causative agent is the ascomycete fungus *Histoplasma capsulatum* (shown in Figure 14.13), which can commonly be found in the soil and also in bat and bird faeces.

Histoplasma capsulatum is a thermal dimorph—that is, it changes form as the temperature changes. At room temperature it has a mycelial form, but at body temperature (37 °C) it takes on a yeast form. Strains of *H. capsulatum* that are unable to carry out this dimorphic switch do not cause infection: we say they are avirulent.

The yeast form replicates inside macrophages, which can be carried from the lungs to sites all over the body. (Macrophages are part of our immune system and are involved in the phagocytosis and destruction of pathogenic bacteria, as we discover in Chapter 27.) Most people who contract histoplasmosis will recover without any treatment, but for some patients who have compromised immune systems the fungus can become systemic and cause severe illness.

Coccidioidomycosis

Coccidioidomycosis is also known as San Joaquin Valley fever and is caused by *Coccidioides immitis*. This fungal pathogen will cause disease in a normal healthy individual. It occurs mainly in the South Western part of the USA, where the dusty, dry conditions enable the spores to easily disperse. Skin tests have shown that almost everyone who lives in this part of the world contracts the disease but its symptoms may be so mild as to go unnoticed. Where symptoms *are* noticed, they are similar to those of tuberculosis or pneumonia.

Table 14.2 Virulence factors of *C. albicans*

Virulence factor	Description
Dimorphism	The change from a hyphal (mycelial) to a yeast form triggered by environmental conditions
Adhesion factors	Adherence to cell surfaces is achieved by phospholipases and proteases which break down the host cell membranes. There are also non-specific mechanisms—including electrostatic charge and van der Waals forces—which allow the yeast to attach to a wide range of tissue types and inanimate surfaces, e.g. plastics
Phenotypic switching	The switch from white cells to opaque cells. White cells are the classic yeast shape, but opaque yeast cells are elongate and they have a pocked surface. The switch from white to opaque cells is needed for sexual reproduction
Protease enzymes	There are 10 different secreted aspartyl proteases, which have a variety of functions. They are involved in adhesion, dimorphism, phenotypic switching, escaping host immune cells, and nutrient acquisition
Lipase enzymes	There are 10 different lipase enzymes involved in adhesion and dimorphism

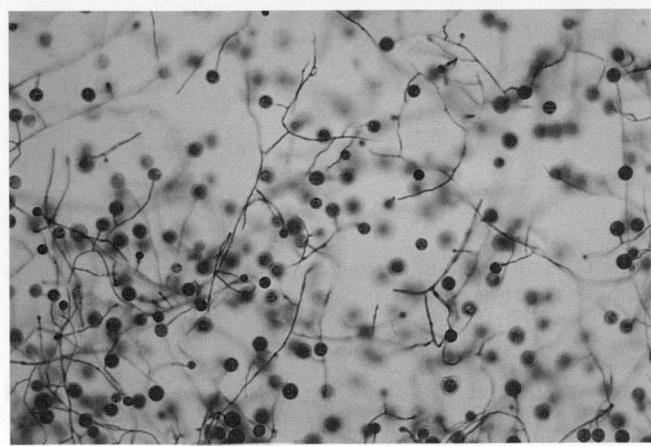

Figure 14.13 Fungal hyphae and sporing structures of *Histoplasma capsulatum*.
Source: Gado Images/Alamy Stock Photo.

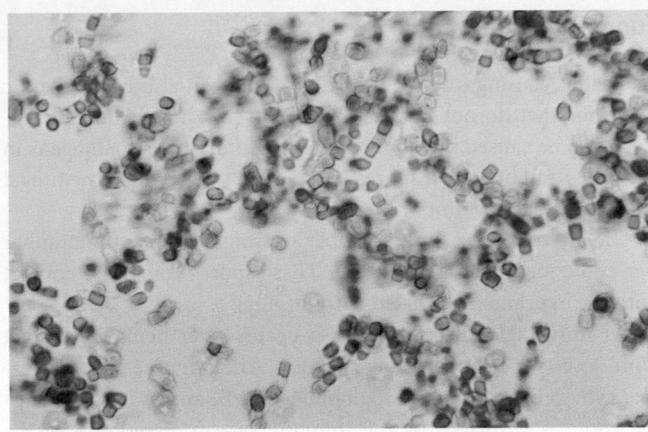

Figure 14.14 Stained *Coccidioides immitis* arthrospores.
Source: Courtesy Public Health Image Library

Once an individual has contracted and recovered from the disease they usually experience lifelong immunity.

Coccidioidomycosis has an interesting demographic: it occurs most frequently among dark-skinned males aged between 30 and 50 years, while light-skinned females have the greatest immunity to the disease.

Coccidioides immitis produces **arthrospores** (illustrated in Figure 14.14) through asexual reproduction. After being inhaled into the lungs the arthrospores become round and split to form endospores, which then spread throughout the body.

The disease may become systemic in immunocompromised hosts, leading to the formation of lesions of the skin and subcutaneous tissue, bones, joints, and brain. When the infection becomes systemic, death usually occurs.

Treatment of fungal infections

The treatment of fungal infections is often difficult: as both the host and the parasite are eukaryotic, it is hard to attack the fungus without also damaging the host cells. One of the most effective ways to treat a fungal infection is to administer an antifungal antibiotic (also known as an antimycotic). However, very few antimycotic compounds can be used to treat humans. Of those that are available, the polyene macrolide antibiotics are used most frequently.

Polyene macrolide antibiotics are produced by bacteria of the *Streptomyces* genus and include amphotericin B, nystatin, and pimaricin. They all have a large ring structure, as depicted in Figure 14.15, and bind specifically to ergosterol, which is the sterol in the fungal membrane. The antibiotics work by increasing the permeability of the plasma membrane, which allows the cellular contents to leak out. Lysis and death of the cell follows.

▶ We learn more about the structure of the fungal membrane in Chapter 9.

Amphotericin B is administered intravenously and has a broad spectrum of activity against most fungi. It is used to treat systemic infections and severe fungal infections of the skin, hair, and nails. At low concentrations this antibiotic can also boost the immune system. Unfortunately, it is toxic to the kidneys in higher concentrations.

Nystatin is too toxic to be administered intravenously and cannot be absorbed by the digestive tract. It is therefore applied to the skin in what are called **topical** preparations along with pimericin to treat superficial infections such as thrush (*Candida albicans*).

PAUSE AND THINK

Why are fungal infections of humans hard to treat?

Answer: Fungi are eukaryotic and so are humans. This means that many of our cellular processes are very similar. Drugs that target metabolic pathways that are the same in both humans and fungi will not only damage the fungus, but damage us as well. With this in mind, drugs that are good at treating fungal infections target the differences between humans and fungi. One of those differences is the sterol in the plasma membrane. We have cholesterol, but fungi have ergosterol. So, drugs that specifically target the biosynthesis of ergosterol will only harm fungal cells.

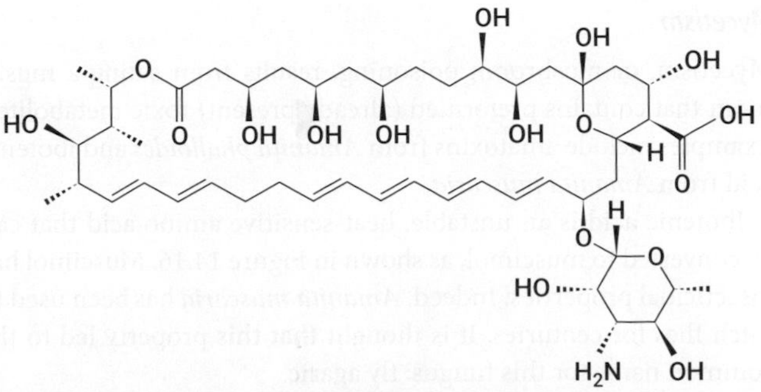

Figure 14.15 The chemical structure of the polyene macrolide antibiotic nystatin. Notice the large ring structure, which is characteristic of this group of antibiotics.

Fungal allergens

As well as causing infectious disease, fungi can also cause allergic reactions that do not involve invasion of the host. Allergic reactions can be caused by fungi ingested in food, fungal pathogens in the body, and by fungal spores in the air. Of these, airborne fungal spores are the most important.

The air contains a huge number of fungal spores: the number per cubic metre can be $1–2\times10^4$. These spores come into contact with the eyes and are inhaled with the air.

An immediate allergic response occurs after being in contact with dust which contains fungal spores. The immune system over-reacts by producing immunoglobulin E (IgE) antibodies. The fungal spore is destroyed by the immune system. Symptoms of the allergy include itchy eyes, running nose, and sore throat, as well as asthma. Exposure to fungal spores over a long period of time can cause allergic alveolitis, the symptoms of which include chills, fever, and a dry cough.

▶ We learn more about the different types of antibody in Chapter 27.

Encounters with massive spore numbers are often associated with specific occupations. Agriculture is one such occupation, and, in particular, the processes of harvesting and making hay. Individuals working in feed mills or food-processing plants can also be exposed to a large number of spores from *Aspergillus*, *Penicillium*, and *Mucor*.

The processing of municipal waste as compost or sludge can encourage the growth of *Aspergillus fumigatus*, which can cause an allergic reaction or infective mycosis.

Fungal poisoning

Fungi do not only harm humans by triggering allergic reactions or by causing infection, as we have seen in the previous two sections, but they can also be poisonous and inflict harm in this way.

There are two ways in which you can fall victim to fungal poisoning. These are:

1. Mycetism.

2. Mycotoxicosis.

Let us now explore what each of these terms mean.

Mycetism

Mycetism, or mushroom poisoning, results from eating a mushroom that contains preformed (already present) toxic metabolites. Examples include amatoxins from *Amanita phalloides* and ibotenic acid from *Amanita muscaria*.

Ibotenic acid is an unstable, heat-sensitive amino acid that can be converted to muscimol, as shown in Figure 14.16. Muscimol has insecticidal properties. Indeed, *Amanita muscaria* has been used to catch flies for centuries. It is thought that this property led to the common name for this fungus: fly agaric.

In humans who have ingested *A. muscaria*, muscimol alters neuronal activity, resulting in the symptoms of *A. muscaria* poisoning, which include euphoria, hallucinations, nausea, and depression. Muscimol has this effect by activating the receptors for the brain's

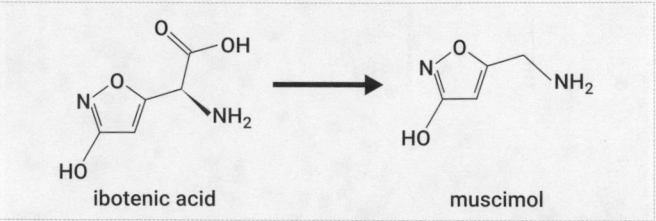

Figure 14.16 The amino acid ibotenic acid can be converted to muscimol, which can affect neural activity in humans.

principal inhibitory neurotransmitter, gamma aminobutyric acid (GABA), found in numerous parts of the brain.

Amatoxins and phallotoxins are some of nature's most deadly poisons. They occur in a number of fungal genera, but species in the genus *Amanita* are the most poisonous. The mortality rate after eating these mushrooms can be as high as 90 per cent.

The amatoxins are cyclopeptides that comprise eight amino acids. They cause cell destruction in the liver, kidney, and gastrointestinal tract. Phallotoxins are cyclopeptides comprising seven amino acids; they act quickly, but their toxicity is limited to liver cells. The phallotoxin phalloidin has an interesting use in microscopy, which is explored further in Real World View 14.1.

Given the serious damage that mushroom poisons can potentially cause, treatment for mushroom poisoning needs to be administered quickly, even before the mushroom has been identified. The gastrointestinal tract is emptied by inducing vomiting, pumping the stomach, or performing an enema. The patient's kidneys and liver are then monitored.

Mycotoxicosis

The term mycotoxicosis describes the intoxication experienced after eating fungi that have produced toxins while growing on foodstuffs. Mycotoxicosis is associated with a specific food that shows signs of fungal growth and is seasonal, non-transmissible, and unresponsive to drug and antibiotic treatment.

Over 100 species of filamentous fungi can produce mycotoxins, including species of the genus *Claviceps*. *Claviceps purpurea* is a fungus that infects crops such as wheat, barley, and rye, causing a disease called ergot. Kernels of wheat infested with the fungus develop into light-brown curved pegs called sclerotia, which can be seen in Figure 14.17. The sclerotia contain a mix of alkaloids that can be harmful if consumed.

In the Middle Ages, before people knew about fungal toxins, these sclerotia were harvested with the wheat grains and turned into bread. When contaminated bread was consumed, a condition called ergotism (also known as St Anthony's Fire) would develop. Ergotism has two forms: one causes gangrene (the death of body tissue as a result of blood supply loss) and the other causes convulsions. Ergotism used to be common in human populations but has now been virtually eradicated due to better grain harvesting and storage, both of which limit the risk of the fungal infection giving rise to ergot.

A more recent example of mycotoxicosis is the discovery of aflatoxins in the 1960s. Aflatoxin is produced by a range of filamentous

Using phalloidin as a stain for F-actin

The phallotoxin phalloidin has an interesting use in microscopy. Phalloidin binds to the cytoskeletal protein F-actin inside both living and fixed cells and prevents its depolymerization. Tagging phalloidin with a fluorescent marker means that F actin can be visualized under a microscope, as shown in Figure 1. The fluorescent marker used is eosin.

Not only can F-actin be seen using this technique, but the level of fluorescence can be used to determine the *amount* of F-actin present in the cell. The ability to tag F-actin in this way has helped us to study the cell processes that involve it, including the modification of cellular shape, cell migration, and cell division.

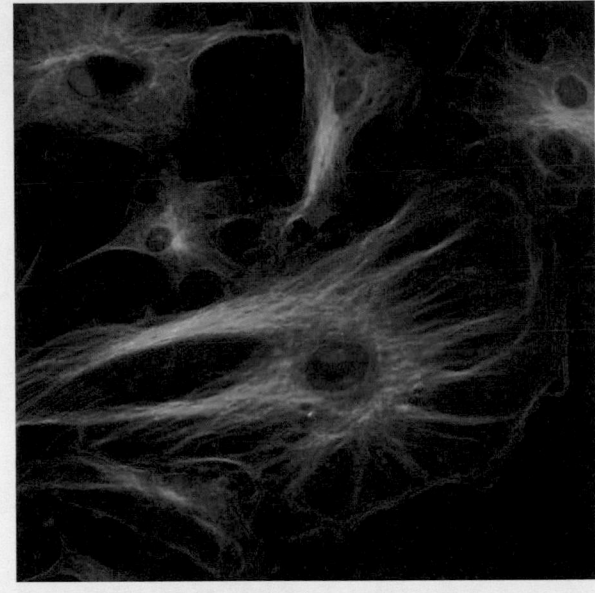

Figure 1 Cells stained with phalloidin that has been tagged with a fluorescent marker. F-actin is shown in red.
Source: Wikimedia Commons.

fungi, including *Aspergillus parasiticus*, *Aspergillus ostianus*, and some species of *Penicillium*. Aflatoxins cause liver damage, particularly in livestock. Livestock are exposed to aflatoxins when they consume contaminated feed such as mouldy peanut meal and corn.

Public health officials are concerned about being able to detect aflatoxins in foods, given that the lethal dose for adults is only 10–20 mg. It is virtually impossible to eradicate aflatoxins completely, so the challenge is to ensure levels are acceptably low. Once detected, batches of food with high levels of aflatoxin are discarded.

Check your understanding of the concepts covered in this section by answering the questions in the e-book.

14.5 Protozoan diseases

When considering microbes and their impacts on human health, our journey is not complete until we have spent time contemplating protozoan diseases. This group of diseases is caused by a diverse group of organisms known as protists.

The protozoa are an almost entirely unicellular group of eukaryotes. Like other eukaryotes, protozoa have a nucleus that contains multiple chromosomes, organelles that perform specific functions, and a cytoskeleton that includes microtubules. The protozoa reproduce either asexually by mitosis, sexually by meiosis, or by both modes of reproduction: a characteristic that is less common for most eukaryotes, although is similar to fungi.

▶ We learn more about the characteristics of eukaryotic cells in Chapter 9 and learn about different modes of reproduction in Chapter 10.

This fascinating group of organisms comprises some of the most important organisms on Earth in terms of their impact on human health, food production, and ecosystem functioning. In this section we explore ways in which protozoa are impacting human health, first by examining their capacity to cause disease and then by focusing on some examples of infectious diseases with protozoa as their causative agent, namely leishmaniasis and giardiasis.

The capacity of protozoans to cause disease

There are over 30,000 protozoan species, including those examples depicted in Figure 14.18. Not all are parasites, however. Protozoa are small (typically 10–100 µm long) and are visible under the microscope. They possess a range of virulence factors that include **adhesins** (which provide the ability to adhere to a host cell or surface), the ability to produce/secrete toxins, and an ability to survive inside the **phagocytic** vesicles of a host organism. Typically, protozoa have extracellular features such as cilia or flagella that can enable them to actively move.

The taxonomy of the protozoa is complex. Rapid advances in molecular taxonomy in recent years (see Topic 5) make it a fluid

Figure 14.17 Sclerotia. Black curve-shaped pegs of sclerotia of *Claviceps purpurea* on the ears of wheat.

Source: Stefan Dinse/istockphoto.com.

and changing area of biology. For the purposes of this section we will consider the protozoa that cause disease in humans to be organized into four different categories based on how they move. You should be aware that there are other approaches to protozoan taxonomy involving molecular techniques and you may well encounter a different approach in other literature sources.

The four groups that we will consider are:

1. The Sarcodina, which include the amoebae. The amoebae are represented by organisms such as *Entamoeba*, which use extensions from their cytoplasm, known as **pseudopodia**, to move.

2. The Mastigophora, which include the flagellates. Examples of the flagellates include *Trypanosoma* spp., which cause African sleeping sickness and Chagas disease. Trypanosomes move by beating or rotating their flagella. We discuss two examples of protozoa from this group in more detail below: *Leishmania* sp. and *Giardia lamblia*.

3. The ciliates, which include ciliates such as *Balantidium* (causes dysentery). These organisms move by beating the tiny cilia found on their outer surface in a coordinated way.

4. The Sporozoa, which comprise organisms such as *Plasmodium* that are non-motile in their adult form.

Some protozoan parasites have different forms during their life cycle. Cysts, for example, are multi-layered shells that can endure hostile conditions for long periods of time. They are passive and cannot move or eat. When cysts are returned to their feeding environment they turn back into the active stage called the **trophozoite**. This process is known as **excystation** and is illustrated in Figure 14.19. Conversely, the transformation from the active trophozoite into a cyst is called **encystation**.

The diseases that result from infection with protozoans cause a substantial burden to healthcare systems across many parts of the

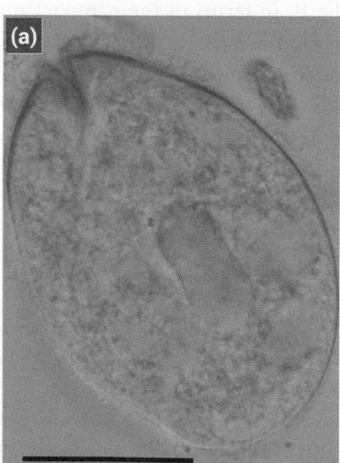
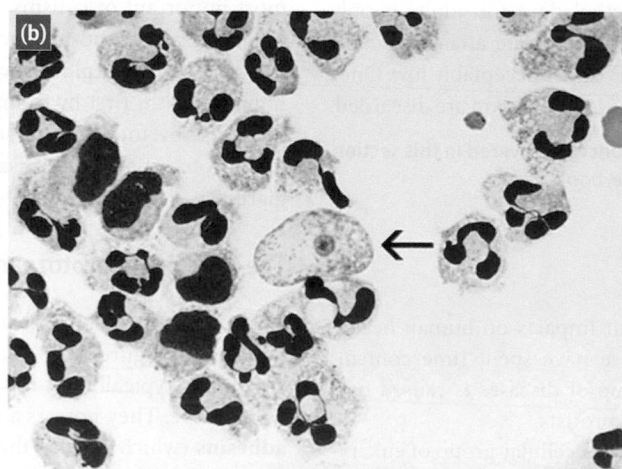
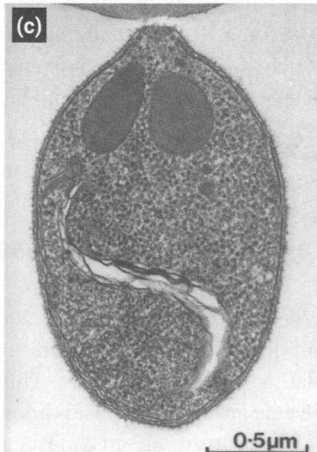

Figure 14.18 Protozoa. (a) The only ciliate known to be pathogenic to humans, *Balantidium coli*. Image is of a trophozoite, the active stage of the protozoan life cycle, as explained further in the text. Scale bar: 50 μm. (b) The arrow indicates a trophozoite of the amoeba *Naegleria fowleri* at ×600 magnification. (c) *Plasmodium knowlesi*, a species of apicomplexans. This image shows a merozoite, a motile stage of this parasite's life cycle.

Source: (a) Ponce-Gordo, F., Jirků-Pomajbíková, K. (2017). *Balantidium coli*. In: J. B. Rose and B. Jiménez-Cisneros, (eds) Water and sanitation for the 21st century: Health and microbiological aspects of excreta and wastewater management. Michigan State University, UNESCO. https://doi.org/10.14321/waterpathogens.30; (b) L. G. Capewell, A. M. Harris, J. S. Yoder, et al. Diagnosis, clinical course, and treatment of primary amoebic meningoencephalitis in the United States, 1937–2013. *Journal of the Pediatric Infectious Diseases Society*, 4 (4), 2015, e68–e75, https://doi.org/10.1093/jpids/piu103; (c) E. Moles, J. J. Valle-Delgado, P. Urbán, et al. Possible roles of amyloids in malaria pathophysiology. *Future Science* OA 2015 1:2/CC BY 4.0.

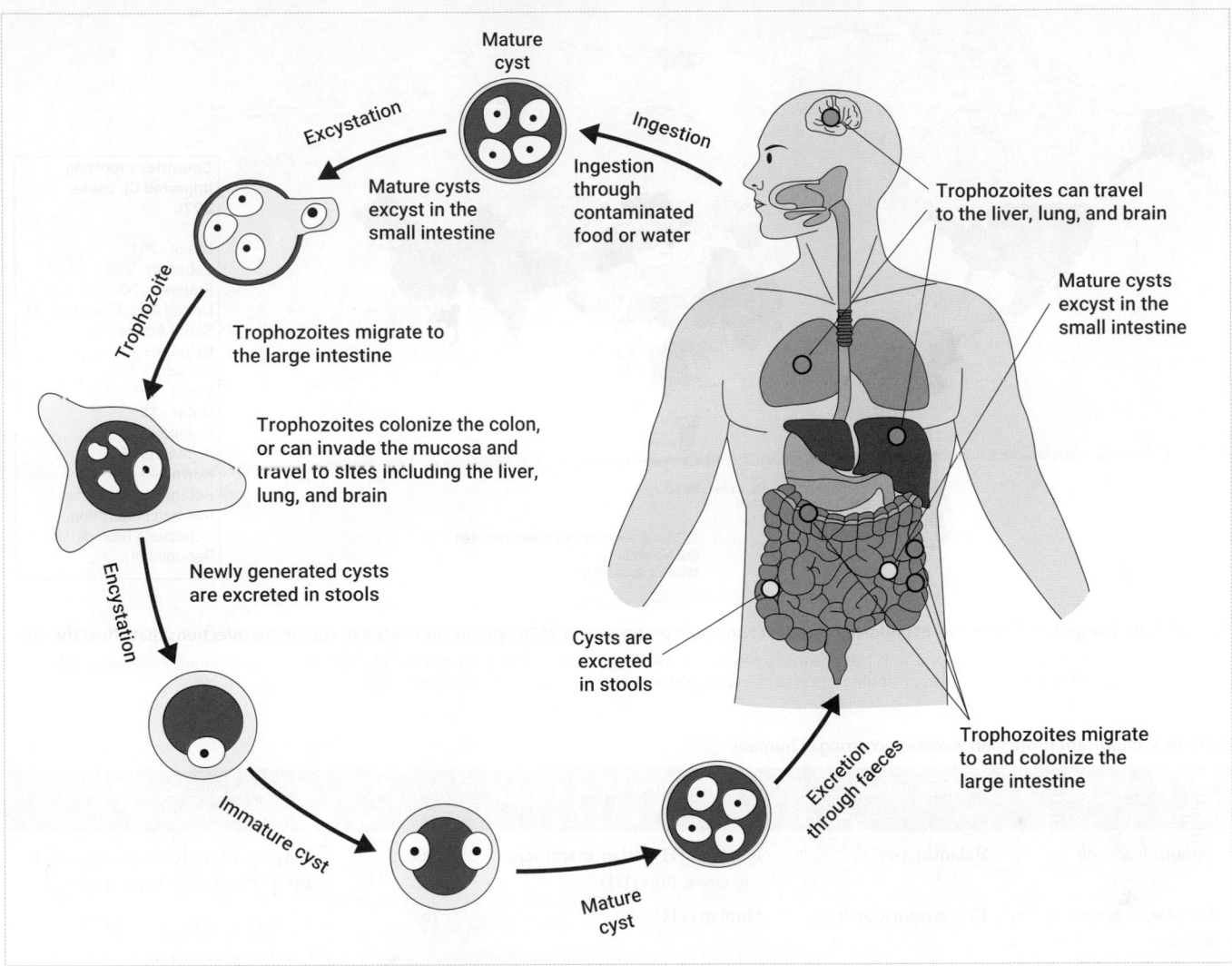

Figure 14.19 Protozoan life cycle. The different forms seen during the life cycle of *Entamoeba histolytica*.

Source: Reprinted from S. Begum, H. Gorman, A. Chadha, K. Chadee. Entamoeba histolytica. *Trends in Parasitology*, 2021; 37(7), 676–677. With permission from Elsevier.

world. While many protozoan diseases are associated with poverty and with developing countries, they also cause problems for developed countries as travellers returning from overseas bring protozoans with them (usually unintentionally!). These protozoans may then go on to cause additional infections in the person's home country.

An example of this can been seen in Figure 14.20, which shows the countries reporting imported cases of *Leishmania* protozoans during 2020, according to a 2017 report by the World Health Organization (WHO). These protozoans can go on to cause the disease leishmaniasis.

In general, protozoan diseases have, until recently, received little research and public health attention, resulting in some of these diseases being included under the umbrella term of 'neglected tropical diseases' (NTDs). Table 14.3 illustrates some of the important protozoan diseases found in humans. We discuss NTDs more fully in Real World View 14.2.

Of the protozoan diseases listed in Table 14.3, malaria causes the most deaths annually: according to the WHO's World Malaria Report 2020 there were approximately 219 million cases in 2019, of which approximately 409,000 were fatal.

Many protozoans that go on to cause disease in humans have life cycles involving multiple organisms and multiple life-cycle stages. The two protozoan diseases leishmaniasis and giardiasis are examples of how these microscopic organisms can have a substantial negative impact on their human host and wider human populations in different ways. Let us now consider each of these in turn.

Leishmaniasis

There are more than 20 species of protozoan within the *Leishmania* genus. A person who is infected with these parasites and is exhibiting symptoms of infection is said to be suffering from **leishmaniasis**, although the majority of people infected with these protozoans do not actually develop symptoms.

Leishmania parasites are transmitted to humans by means of a vector, a small blood-consuming female sandfly. More than 90 species of sandfly, an example of which is shown in Figure 14.21, are

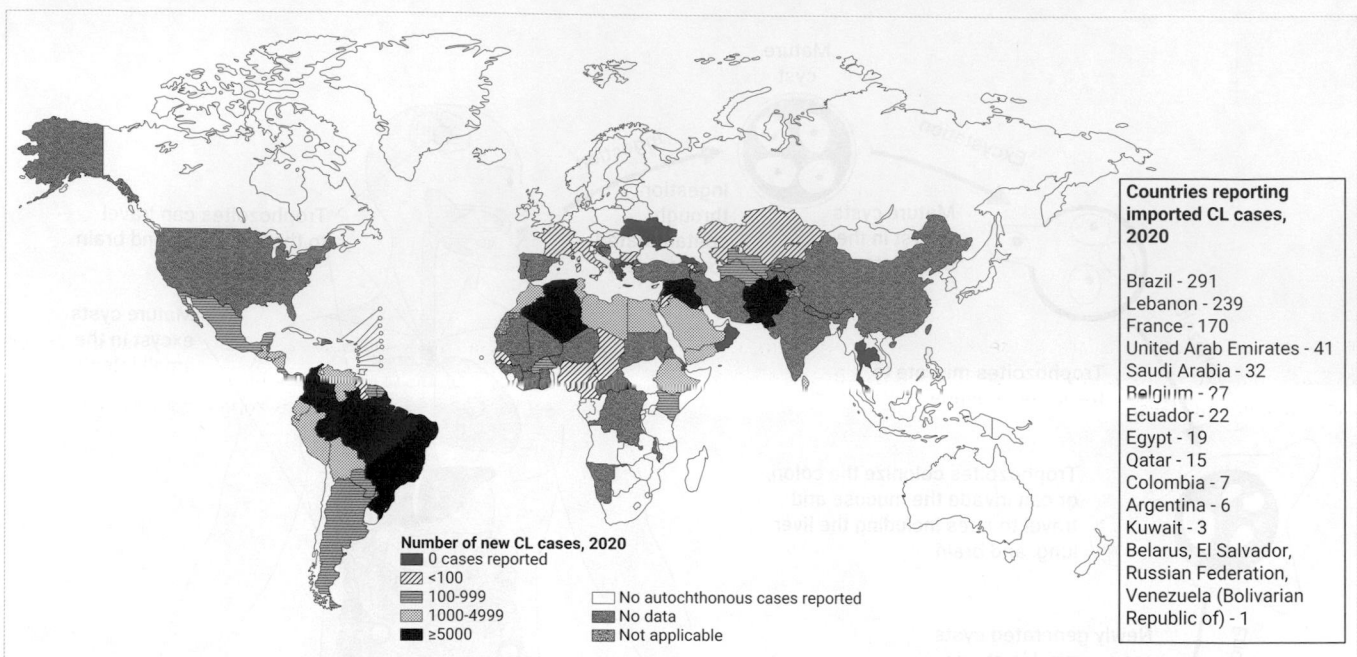

Countries reporting imported CL cases, 2020

Brazil - 291
Lebanon - 239
France - 170
United Arab Emirates - 41
Saudi Arabia - 32
Belgium - 27
Ecuador - 22
Egypt - 19
Qatar - 15
Colombia - 7
Argentina - 6
Kuwait - 3
Belarus, El Salvador, Russian Federation, Venezuela (Bolivarian Republic of) - 1

Number of new CL cases, 2020
- 0 cases reported
- <100
- 100-999
- 1000-4999
- ≥5000
- No autochthonous cases reported
- No data
- Not applicable

Figure 14.20 The global distribution of imported cases of *Leishmania* protozoans in 2020, specifically related to cutaneous infections that affect the skin.

Source: Reproduced from: J. A. Ruiz-Postigo et al. Global leishmaniasis surveillance: 2019–2020, a baseline for the 2030 roadmap. Weekly Epidemiological Record, 3 September 2021 url: https://www.who.int/images/default-source/maps/leishmaniasis_cl_2020.png?sfvrsn=81df4387_5 accessed 14 December 2021.

Table 14.3 Important protozoan diseases occurring in humans

Protozoan pathogen	Disease caused	Organisms involved in life cycle	Multiple life cycle stages	Estimated number of affected people worldwide
Balantidium coli	Balantidiasis	Humans (H), other mammals, e.g. cows, pigs (IH)	Yes	Largely asymptomatic, estimated up to 1% of population infected
Cryptosporidium parvum	Cryptosporidiosis	Humans (H)	Yes	
Cyclospora cayetanensis	Cyclosporidiosis	Humans (H)	Yes	
Entamoeba histolytica	Amoebic dysentery (amoebiasis)	Humans (H)	Yes	~500 million infections annually, approx. 10% are symptomatic
Giardia lamblia	Giardiasis	Humans (H)	Yes	200 million symptomatic cases in low- and middle-income countries alone
*Leishmania spp.	Leishmaniasis	Mammals (H), female sandfly (V)	Yes	12 million people infected and 200,000 new cases each year
Naegleria fowleri	Amoebic meningoencephalitis	Humans (H)	No	Sources agree cases are rare
Plasmodium spp.	Malaria	Mammals (H), female *Anopheles* mosquitoes (V)	Yes	445,000 deaths in 2016
Toxoplasma gondii	Toxoplasmosis	Felids (H), humans (H), birds (IH), rodents and other small mammals (IH)	Yes	Over 6 billion people infected globally, majority asymptomatic
Trichomonas vaginalis	Vaginal trichomoniasis	Humans (H)	Yes	More than 160 million annually
*Trypanosoma brucei gambiense	African sleeping sickness	Mammals (H), tsetse fly (*Glossina* spp.) (V)	Yes	70,000–300,000 across both forms of African sleeping sickness

* Denotes a disease considered a neglected tropical disease (NTD).

H, host; IH, intermediate host; V, vector.

REAL WORLD VIEW 14.2   Neglected tropical diseases

Neglected tropical diseases (NTDs) are a group of diseases that cause a significant disease burden for more than 1.5 billion people living across some of the world's most deprived communities. NTDs include diseases caused by parasitic worms, such as lymphatic filariasis and onchocerciasis (also known as river blindness); diseases caused by protozoan parasites, such as trypanosomiasis (African sleeping sickness) and leishmaniasis: diseases caused by bacterial pathogens, such as trachoma and leprosy; and diseases caused by viruses, such as rabies and dengue fever.

These diseases cause a range of symptoms that can lead to social exclusion and isolation and can prevent sufferers from being able to walk or earn a living. For 170,000 people each year these diseases result in death. NTDs trap some of the poorest people in an unending cycle of disease and poverty.

NTDs impair physical and cognitive development, contribute to mother and child illness and death, make it difficult to farm or earn a living, and limit productivity in the workplace. However, many of these diseases are easily treated. For just 50 US cents (around 40p), a rapid-impact treatment can deliver medication that treats one person for the five most common NTDs (trachoma, intestinal worms, lymphatic filariasis, onchocerciasis, and schistosomiasis, shown in Figure 1), making the treatment of these conditions a very cost-effective public health solution. For more information on these diseases and the work to treat them look at the work of the END Fund (https://end.org/).

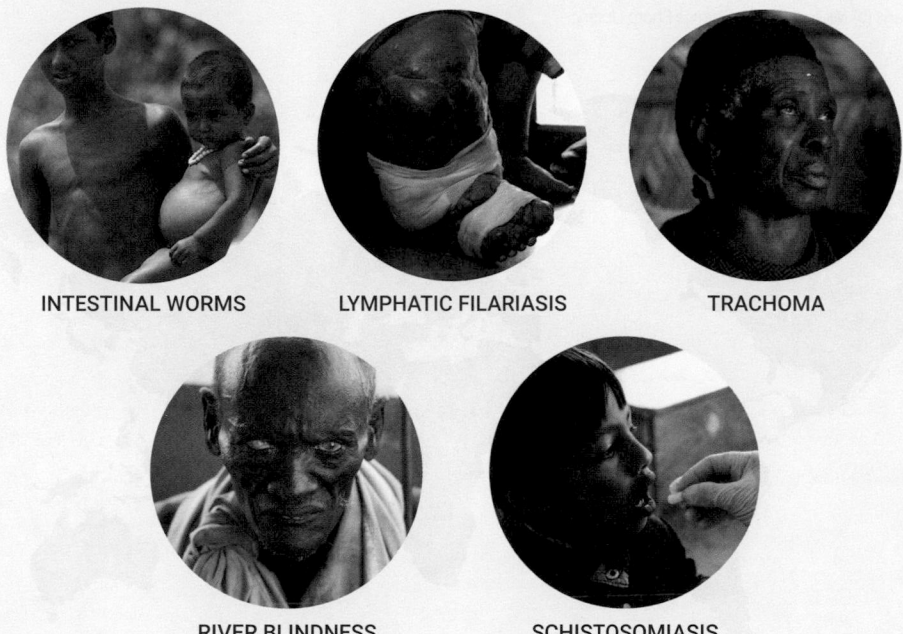

INTESTINAL WORMS   LYMPHATIC FILARIASIS   TRACHOMA

RIVER BLINDNESS   SCHISTOSOMIASIS

Figure 1 The five most common neglected tropical diseases. Together these five diseases affect more than 1.5 billion people across the world.

Source (clockwise from top left): Shafiqul Islam Kajol/Majority World/age fotostock; MintZaa/Shutterstock.com; DMA/Alamy Stock Photo; Mike Goldwater/Alamy Stock Photo; Xinhua/Alamy Stock Photo.

---

known to be able to transmit the *Leishmania* parasites. This transmission has resulted in leishmaniasis being endemic in close to 100 countries worldwide as shown in Figure 14.22.

Leishmaniasis can occur in three main forms:

- cutaneous—the most common form, involving infections of the skin;
- visceral—involving infections of internal organs;
- mucocutaneous—involving infections of the mucosal lining of the nose and throat.

All these forms can cause life-altering consequences for a sufferer, ranging from the destruction of the mucous membranes of the nasal passage (mucocutaneous), debilitating long-term skin lesions and ulcers (cutaneous), and even death (visceral).

The symptoms associated with each form of the disease are due to the parasites' relationship with the human host cells they infect. *Leishmania* are trypanosomal obligate intracellular parasites. This means that, for at least part of their lifecycle, they must live inside the cell of another organism. These particular parasites require a vertebrate host, such as a human, and an invertebrate vector, a sandfly.

Look at Figure 14.23, which shows the infection lifecycle of *Leishmania* parasites within the human and sandfly vector. When inside the sandfly, the *Leishmania* parasites live outside of the

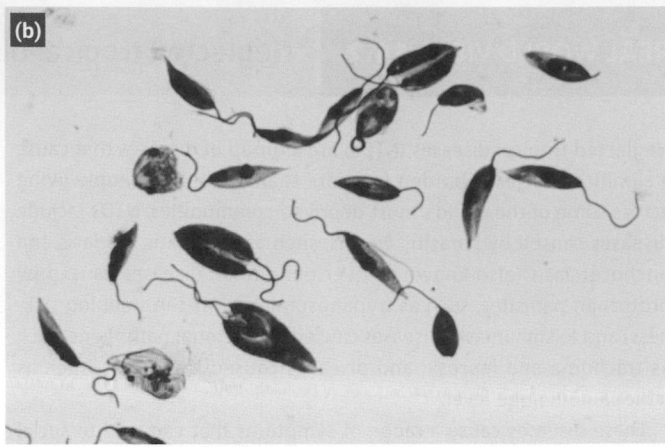

Figure 14.21 *Leishmania* **parasites are transmitted by means of a small fly**. (a) An example of one of the sandfly (*Phlebotomus papatasi*) vectors of *Leishmania*. (b) A section from a culture of *Leishmania* promastigotes, a stage of development for some protozoa (including *Leishmania*).

Source: (a) CDC/Science Photo Library; (b) Michael Abbey/Science Photo Library.

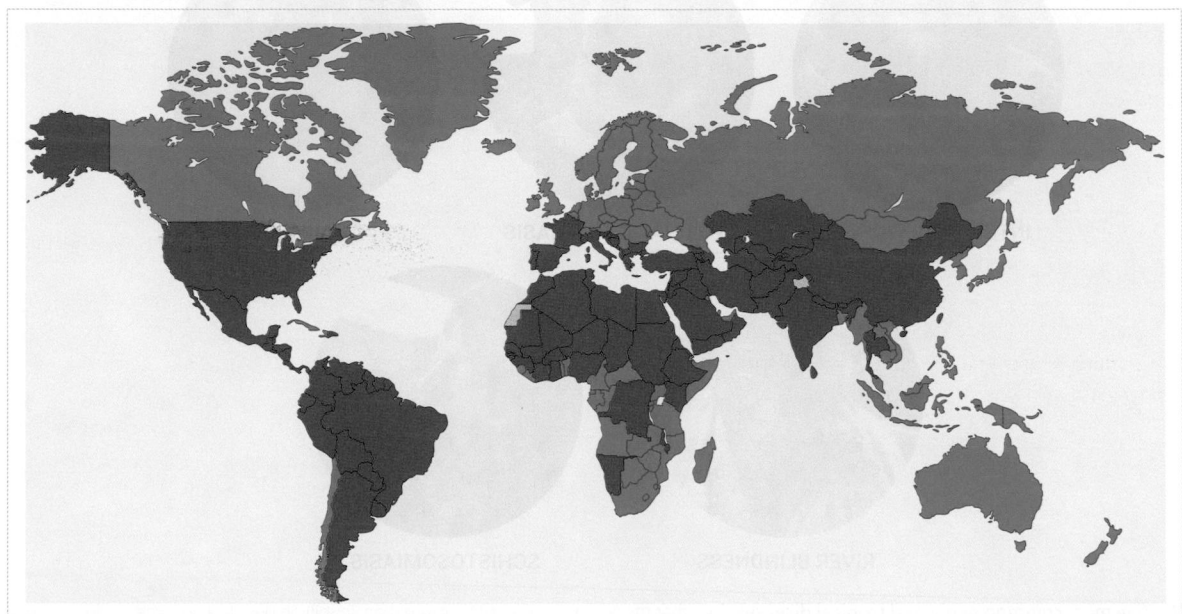

Figure 14.22 **Leishmaniasis is endemic in almost 100 countries.** A map showing those countries around the world reporting endemic cutaneous leishmaniasis infections in 2020.

Source: Reproduced from Leishmaniasis, Status of endemicity of cutaneuos leishmaniasis, map at the following URL: https://apps.who.int/neglected_diseases/ntddata/leishmaniasis/leishmaniasis.html, accessed 14 December 2021.

cells of the vector within the alimentary canal (step 8 in Figure 14.23). While in the alimentary canal they go through a series of developmental changes, waiting for the opportunity to leave and enter the skin of their vertebrate host; this can happen when the sandfly feeds on a suitable host mammal (step 1 in Figure 14.23).

Once within the cells of the human host, the parasites go through further developmental changes, including rapid multiplication and subsequent infection of some of the host's white blood cells, specifically macrophages (steps 2–4 in Figure 14.23). When a sandfly then feeds on the infected host, they also ingest macrophages that contain the parasites as part of their blood meal. Consequently, the infection and re-infection cycle continues (steps 5–8 in Figure 14.23).

What are the impacts of leishmaniasis?

Leishmaniasis is a disease that affects the poorest of the poor and is linked to socio-economic factors such as illiteracy, poor housing, social inequality, and malnutrition, as well as environmental factors such as deforestation and urbanization. The WHO estimates that more than one billion people are at risk of contracting leishmaniasis, with 12 million people currently infected and more than 200,000 new cases reported each year.

The range of symptoms that can be caused by the different forms of the disease means that individuals who have suffered disfigurement as a result of mucocutaneous lesions can experience social exclusion and isolation, while those with multiple cutaneous

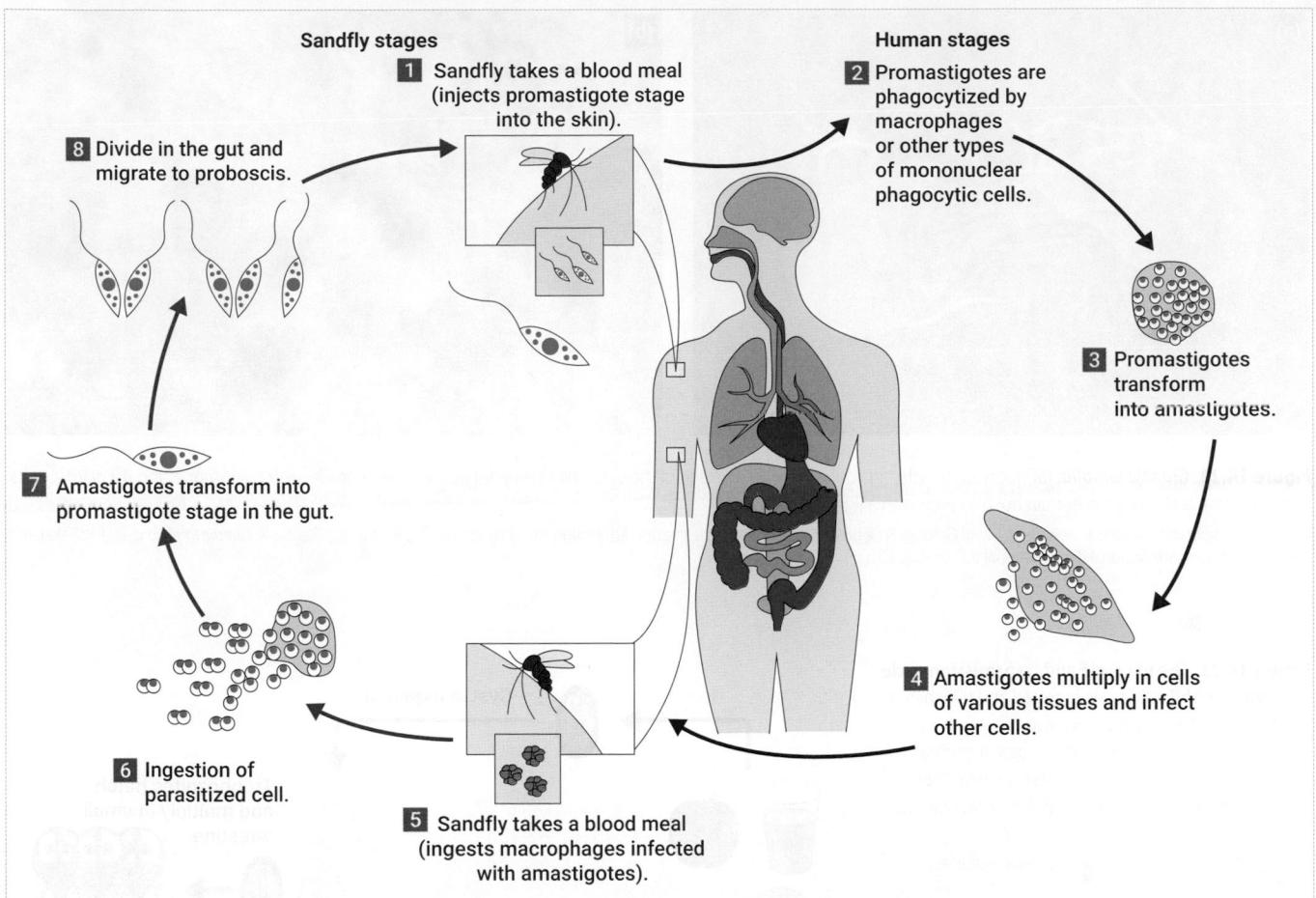

Sandfly stages

1 Sandfly takes a blood meal (injects promastigote stage into the skin).

8 Divide in the gut and migrate to proboscis.

Human stages

2 Promastigotes are phagocytized by macrophages or other types of mononuclear phagocytic cells.

3 Promastigotes transform into amastigotes.

7 Amastigotes transform into promastigote stage in the gut.

4 Amastigotes multiply in cells of various tissues and infect other cells.

6 Ingestion of parasitized cell.

5 Sandfly takes a blood meal (ingests macrophages infected with amastigotes).

Figure 14.23 *Leishmania* **life cycle.** The life cycle of *Leishmania* parasites within the human–sandfly vector infection cycle.
Source: CDC. Life cycle image and information courtesy of DPDx.

lesions may be unable to walk or care for themselves. (Such cutaneous lesions can also cause disfigurement and consequent social exclusion: one person may have more than 100 lesions across exposed areas of skin on their body.) More generally, those suffering from this disease almost always suffer reduced productivity, loss of earnings, and social stigmatization, with those most affected coming from some of the most deprived communities.

The treatment and prevention of leishmaniasis can also be costly. In Nepal, a study found that the average cost of treatment for visceral leishmaniasis (taking into account loss of earnings, as well as treatment itself) was more expensive than the annual household income per capita, while in Bangladesh the total median expenditure on treatment and lost income was 1.2 times the annual income per capita.

Giardiasis

Giardiasis is one of the leading causes of diarrhoeal disease in humans and other mammalian vertebrates. The disease results from infection with a eukaryotic flagellated protozoan known as *Giardia lamblia* (also known as *Giardia intestinalis*), which you can see in Figure 14.24.

In contrast to the *Leishmania* parasites, *G. lamblia* is not an intracellular parasite but instead lives in the intestinal tract of the host. The intestines of a mammal are a hostile environment for many organisms, but *G. lamblia* is well adapted for survival here by being able to enclose itself in a cyst, which protects it from harsh environments and enables it to survive for many months. While in the intestinal tract of the host, these parasites undergo important developmental changes before continuing their life cycle.

Look at Figure 14.25, which shows the life cycle and transmission cycle of *G. lamblia*. The life cycle of *G. lamblia* begins when a mammalian host comes into contact with a contaminated object (called a fomite) or ingests food or drink that is contaminated with faeces containing dormant *G. lamblia* cysts. When these cysts reach the intestines of their new host they hatch and the **trophozoite** life stage emerges.

The trophozoites divide rapidly and embed themselves in the epithelial lining of the duodenum, which triggers the host's immune system to initiate an inflammatory response. This inflammatory response brings about the majority of symptoms exhibited by infected hosts; these include discomfort, nausea, vomiting, diarrhoea, and the characteristic 'eggy burps', to name just a few.

After multiplying, the trophozoites are able to form cysts. They are then eliminated from their host in this form in the host faeces. These eliminated cysts then persist in the environment until a suitable new host comes into contact with them, at which point their

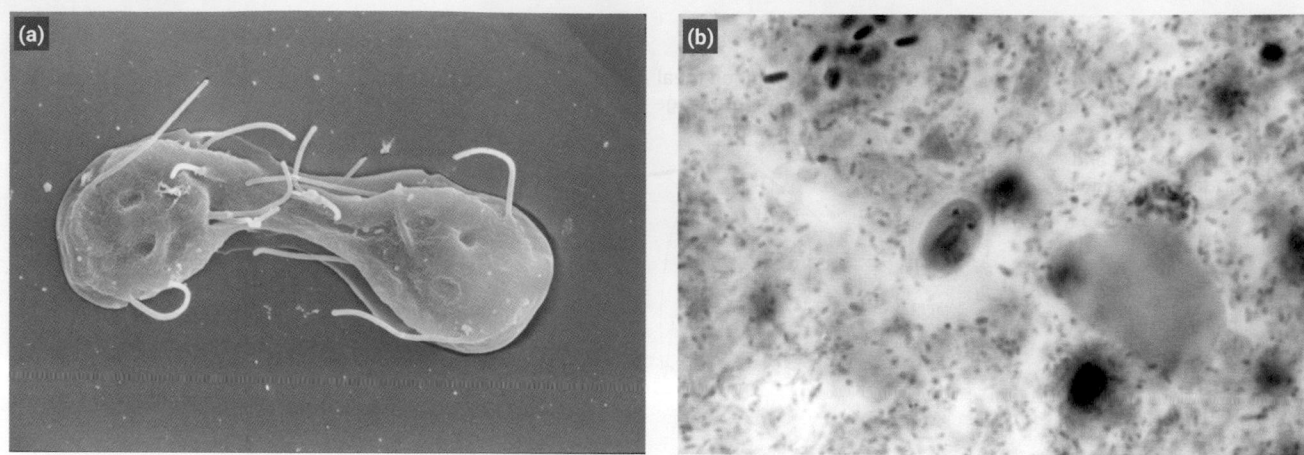

Figure 14.24 _Giardia lamblia._ (a) Trophozoites (the reproductively active stage of the _G. lamblia_ life cycle) as seen under an electron microscope, and (b) giardia cysts (the stage that is shed by the human host) as seen with a light microscope.

Source: (a) _Giardia lamblia_, dorsal view, SEM. David Gregory & Debbie Marshall. Wellcome Images. Attribution 4.0 International (CC BY 4.0); (b) Giardiasis: _Giardia lamblia_ cyst. J. R. Baker. Wellcome Images. Attribution 4.0 International (CC BY 4.0).

Figure 14.25 The life cycle and transmission cycle of _Giardia lamblia._ Cysts are ingested and they hatch to form trophozoites, which divide in the small intestine, but go on to encyst in the large intestine. Cysts are expelled in the faecal material where they contaminate the environment and also other objects (fomites).

Source: CDC. Life cycle image and information courtesy of DPDx.

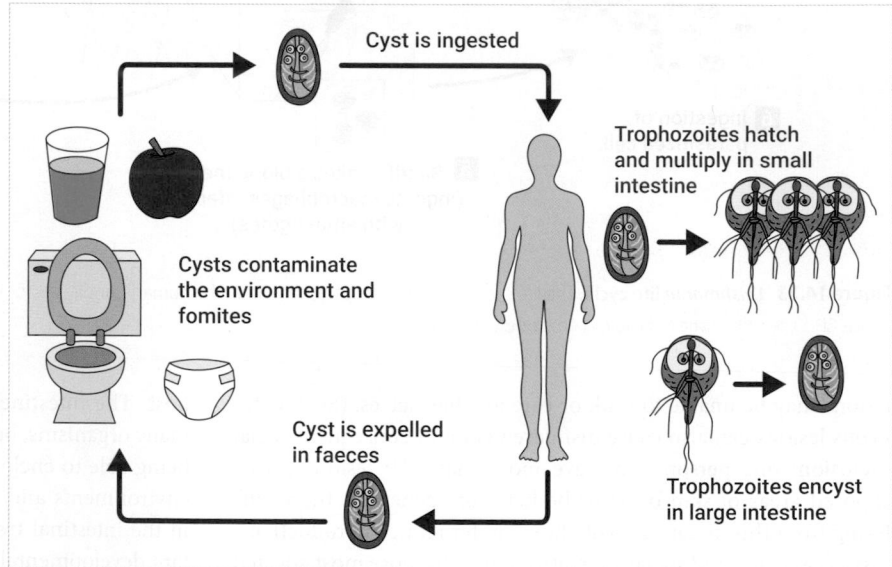

Cyst is ingested

Trophozoites hatch and multiply in small intestine

Trophozoites encyst in large intestine

Cyst is expelled in faeces

Cysts contaminate the environment and fomites

life cycle continues. As few as 10 cysts can cause infection in a new human host.

What are the impacts of Giardia?

Despite being one of the main causes of gastroenteritis worldwide, with an estimated 200 million infections occurring across Africa, Asia, and Latin America each year, _Giardia_ tends to have short-term impacts on those suffering from this disease, without there being lasting consequences. (This is quite different from leishmaniasis, which we have just discussed.) Indeed, many people infected with _Giardia_ protozoans are asymptomatic and can continue to go about their daily activities as usual.

Because its mode of transmission involves contact with faecally contaminated food or objects known as fomites (as illustrated in

Figure 14.25), outbreaks of _Giardia_ are particularly common in areas with poor sanitation. Children are often affected and suffer from persistent diarrhoea, which can then lead to other conditions such as dehydration, malnutrition, and weight loss.

Treatment of an individual exhibiting symptoms is usually straightforward: a single dose of the drug tinidazole is often enough to cure the patient. However, more widespread epidemics can be much harder to bring under control due to the ability of the parasites to survive for long periods outside of their host.

How are protozoan diseases treated?

As we discussed earlier in this section, many protozoans that cause disease in humans have complex life cycles. Treatment for protozoan diseases usually involves killing the protozoan or altering the environment of the human host so that the protozoan is not able to reproduce within the host, therefore limiting its spread.

For some protozoan pathogens—for example, the malaria-causing *Plasmodium*—the vectors that carry the pathogens (in the case of *Plasmodium*, female *Anopheles* mosquitoes) are controlled to limit pathogen transmission and therefore the spread of disease. Vector control methods that limit transmission of biting insects have the additional benefit that they can also help to control the spread of other diseases transmitted by biting insects, such as chikungunya and dengue fever.

There are challenges surrounding the treatment of protozoan infections, however. These include making the treatment accessible and available to those affected, given that some of the most affected communities are in remote areas. Another challenge surrounds the fact that existing treatments are often most effective when administered during the early stages of infection, but—for some diseases, at least—few patients are diagnosed until later stages of the disease.

One example of a disease for which this is a problem is Chagas disease, caused by *Trypanosoma cruzi* and affecting approximately 8 million people across the world, primarily in Latin America. The current treatment focuses on the use of two drugs, benznidazole and nifurtimox, both of which are very effective when administered during the early stages of infection. However, the effectiveness of these drugs reduces with time: the longer a person is infected, the less effective these drugs are in treating the *T. cruzi* infection and the more toxic the treatment becomes to the patient. These drugs are also not recommended for use in pregnant women. Consequently, no effective treatment for pregnant women infected with *T. cruzi* currently exists.

 Check your understanding of the concepts covered in this section by answering the questions in the e-book.

14.6 Plant pathology

Throughout this chapter we have focused on microbes as agents of disease in humans. However, microbial pathogens can also have a devastating effect on the health and survival of plants. In this closing section of the chapter, we consider the role of microbes in plant disease.

Plant diseases can be caused by a range of different organisms: fungi, bacteria, algae, nematodes, viruses, and insects. However,

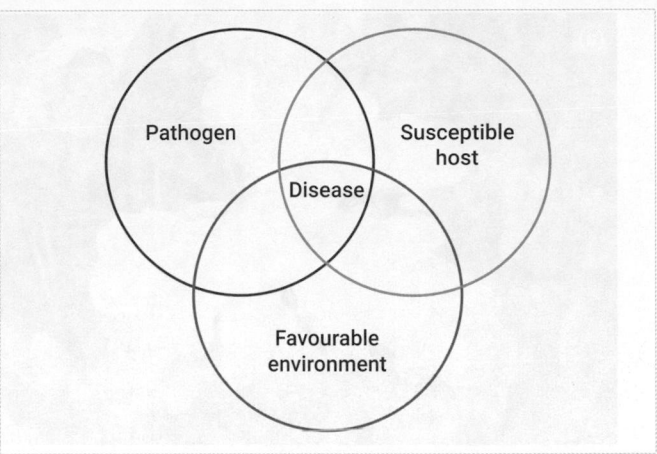

Figure 14.26 The disease triangle. This depicts how disease will only develop when there is a pathogen, a susceptible host plant, and favourable environmental conditions.

the number of plant parasitic fungi is greater than all the others put together. Indeed, fungi cause 70 per cent of all known plant diseases. As you can see from Figure 14.26, plant diseases develop when three conditions overlap: when there is a susceptible host, a pathogen, and favourable conditions of humidity, moisture, etc. This situation is called the disease triangle.

All plants are susceptible to disease, and diseases can wipe out entire crops if conditions are favourable. In the developed world, plant diseases do not now lead to starvation. But the sudden loss of an entire crop in other parts of the world can still be catastrophic. Most crop losses tend to be due to small, regular outbreaks of disease, which reduce the yield of the crop and require expensive treatments. However, some plant diseases have a much greater impact. For example, wheat rust can reduce yield by up to 80 per cent and threaten worldwide production.

Plant disease can spread through a number of different mechanisms:

1. Spores from fungi and bacterial cells on plant surfaces being dispersed by the wind and rain, and by insect activity.
2. Human activity.
3. Poor disposal of diseased plant material.
4. Using contaminated water for irrigation.

As with human and animal pathogens, plant pathogens have a range of virulence factors that aid in disease progression in the host.

In the following sections we will explore examples of plant diseases caused by bacteria, fungi, and viruses. But first, let us consider the types of parasitism displayed by different pathogens, and how those pathogens gain entry to the host plant.

Types of plant parasitism and lifestyles

Organisms that cause plant disease vary in the type of parasitism displayed. The **obligate parasites** are the most specialized and live their entire life cycle on the host. Obligate parasites—including those classed as obligate intracellular parasites—do not often kill the host plant. Why is this? Fungal and bacterial obligate

Figure 14.27 Fungal obligate parasites. (a) Powdery mildew seen as a white growth across the leaf surface. (b) The yellowing of leaves due to phytoplasma infection.

Source: (a) PaulMaguire/istockphoto.com; (b) rdonar/Shutterstock.com.

parasites are biotrophs: they gain their nutrition from living plant tissue. As such, they depend on the living plant for their ongoing survival.

Fungal obligate parasites include the powdery mildew fungi (Figure 14.27a) and the rust fungi. An example of a bacterial obligate parasite is phytoplasma, which causes jujube witches' broom disease.

Obligate pathogens tend to show very narrow host specificity. The species of powdery mildew fungi, for example, are only capable of infecting a particular species of host plant.

Facultative parasites infect the plant tissue when the opportunity becomes available. They grow on dead organic matter in soil or plant debris (i.e. they are saprophytic). Some of the facultative parasites are biotrophs, but others are necrotrophs: they kill the host plant and then obtain their nutrition from dead plant tissue. Others still fall between these two extremes: they are known as hemibiotrophs and feed first on live tissue and then later on dead tissue as the disease progresses.

Facultative parasites tend to infect a wide range of hosts. For example, *Verticillium albo-atrum* has been isolated from more than 70 host genera. The bacterial plant pathogen *Ralstonia solanacearum* can infect a range of plants, including potatoes, tobacco, and bananas.

Some common plant pathogens and their associated lifestyles are summarized in Table 14.4.

The plant host–pathogen relationship

The infection of a plant by a pathogen involves four stages:

1. Penetration of the host.
2. Establishment of the infection within the host plant.
3. Pathological effects on the host plant.
4. Host resistance.

Let us consider each of these in turn.

Penetration of the host

Pathogenic organisms can enter the host plant in different ways, but there are three main entry methods: wounds, natural openings, and direct penetration.

1. *Wounds:* plant wounds can be caused by insects or larger browsing animals. They can also be caused by cultivation methods, especially the use of farm implements. Wounds can also arise as result of natural growth as seen with leaf scars, for example. The damaged cells within a wound site provide a source of nutrients that can be used by the parasite before it then invades adjacent cells. All types of plant pathogen can enter plants via wounds.

2. *Natural openings:* Of the openings that occur naturally on plants, the stomata provide the most important mode of entry for pathogens. Stomata are found on the underside of the leaf and are

Table 14.4 Common plant pathogens and their life style

Pathogen	Disease	Pathogen type	Lifestyle
Sclerotinia sclerotiorum	Soft rot	Fungus	Necrotroph
Ophiostoma novo-ulmi	Dutch elm disease (wilt)	Fungus	Necrotroph
Blumeria graminis	Powdery mildew	Fungus	Biotroph
Cladosporium fulvum	Tomato leaf mould	Fungus	Biotroph
Phytoplasma spp	Jujube witches' broom disease	Bacterium	Biotroph
Agrobacterium tumefaciens	Crown gall disease	Bacterium	Biotroph
Phytophthora infestans	Potato blight	Oomycete	Hemibiotroph

essential for gas and water exchange. They are closed at night (when the plant is not photosynthesizing) but open during the day (when photosynthesis is occurring and the plant needs to efficiently exchange carbon dioxide, oxygen, and water with its surroundings).

All types of plant pathogen can enter plants via natural openings. For example, when the spores of pathogenic fungi germinate, the germ tubes of the fungi are chemically attracted to the stomatal opening and grow through them. Look at Real World View 14.3 to find out more about the relationship between stomata and bacterial plant pathogens.

3. *Direct penetration:* this method of entry is used only by fungal pathogens and is the most common route of entry for these organisms. Fungi that penetrate the epidermis frequently form

REAL WORLD VIEW 14.3 Closing the gap

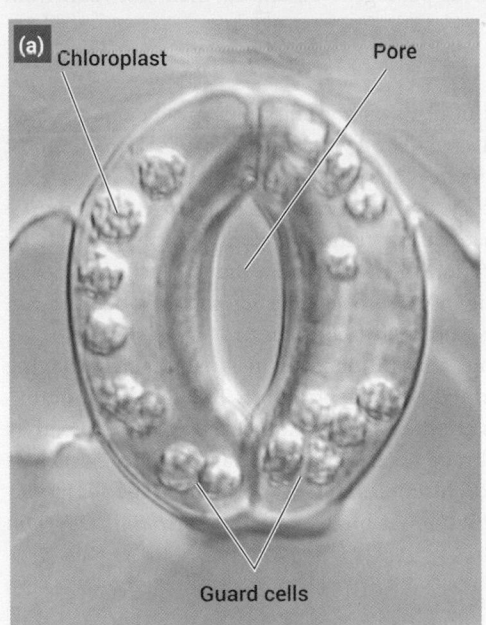

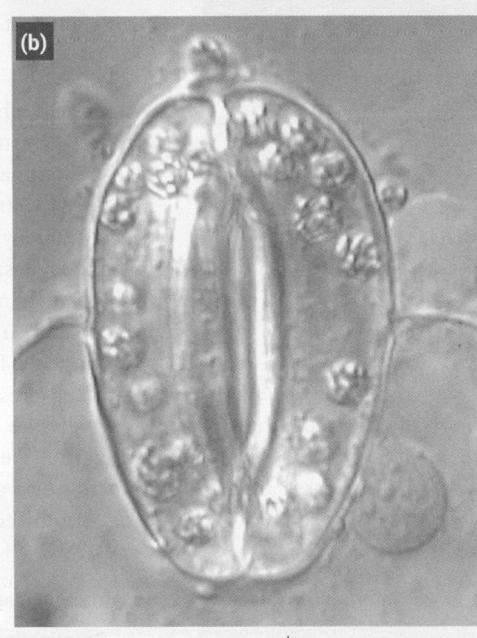

(a) Chloroplast Pore Guard cells

(b)

20 μm

Figure 1 Stomata. The opening (a) and closing (b) of stoma, mediated by swelling and shrinking of guard cells.

Source: Taiz et al. *Plant Physiology & Development*, 6th edn. Sinauer, Oxford University Press.

The stomata are important entry points of pathogens into the host plant. Figure 1 shows how a pair of guard cells surrounds each stomatal opening; their role is to open and close the stomata. This opening and closing is mediated by osmosis: in the light, the guard cells take up water and become turgid, sealing the gap between them, while in the dark, they lose water and become flaccid, and an opening forms between them.

Studies using *Arabidopsis* as a model plant have found that guard cells have evolved to prevent entry of pathogens into the plant: stomata close when plants are under attack by microbial pathogens.

But how do the plants detect that they are under such attack? Many bacterial plant pathogens have flagella, which enable their movement through water. The guard cells can sense the presence of flagellin protein, a structural component of flagella. Such detection induces the closure of the stomata to prevent further entry of the pathogen.

However, studies of the plant pathogen *Pseudomonas syringae*, which causes a disease of leaves (see Figure 2), has shown that bacteria can fight back: *P. syringae* produces a toxic compound called coronatine, which triggers the re-opening of the stomata within 3 hours of their closure.

Bacteria are not the only organisms to do this: fungi can also force stomata to stay open by releasing compounds such as fusicoccin and oxalic acid.

Figure 2 Necrotic regions in a leaf caused by *Pseudomonas syringae*.

Source: Photo courtesy of G. J. Holmes, Cal. Poly. State University, San Luis Obispo.

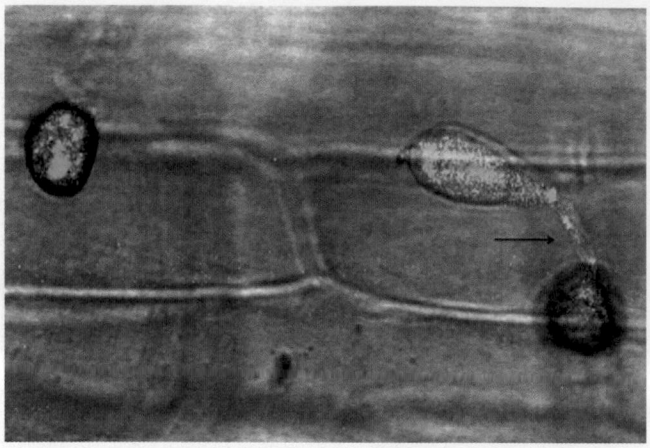

Figure 14.28 The rice blast fungus *Magnaporthe oryzae*. This fungus infects rice plants through a structure called an appressorium. It exerts pressure on the plant leaf until it punctures it.

Source: Research image courtesy of Bais Laboratory/University of Delaware.

an enlarged, flattened 'appressorium' that adheres firmly to the cell wall, preventing the entry of other pathogenic organisms. Penetration of the host plant occurs via a slender hyphal outgrowth called the infection peg, which forces its way into the cell, as depicted in Figure 14.28.

Fungi also produce enzymes that partially degrade the cell wall, aiding in direct penetration. The most important of these enzymes is cutinase, although pectinases and cellulases are also produced.

How is infection within the host established?

After penetration, the parasite requires nutrients to grow and reproduce. Successful pathogens obtain all their nutrition from their plant host.

Fungi

When the hyphae of plant pathogenic fungi invade and grow, an infection is established. The hyphae can grow *between* the host plant cells (in an intercellular manner), or they grow *through* the plant cells (intracellular). In addition, some fungi that produce intercellular hyphae absorb nutrients from living plant cells by forming a haustorium, which sits within the host cell. This structure is shown in Figure 14.29. We come across this structure again in Chapter 37 in association with lichens.

Pathogenic fungi also produce toxins that can harm living cells when present in low concentrations. The structures of toxins are diverse, as are their metabolic effects. For example, they can disrupt membrane function and damage organelles such as mitochondria and chloroplasts. One example is the toxin victorin, produced by the oat blight pathogen *Helminthosporium*

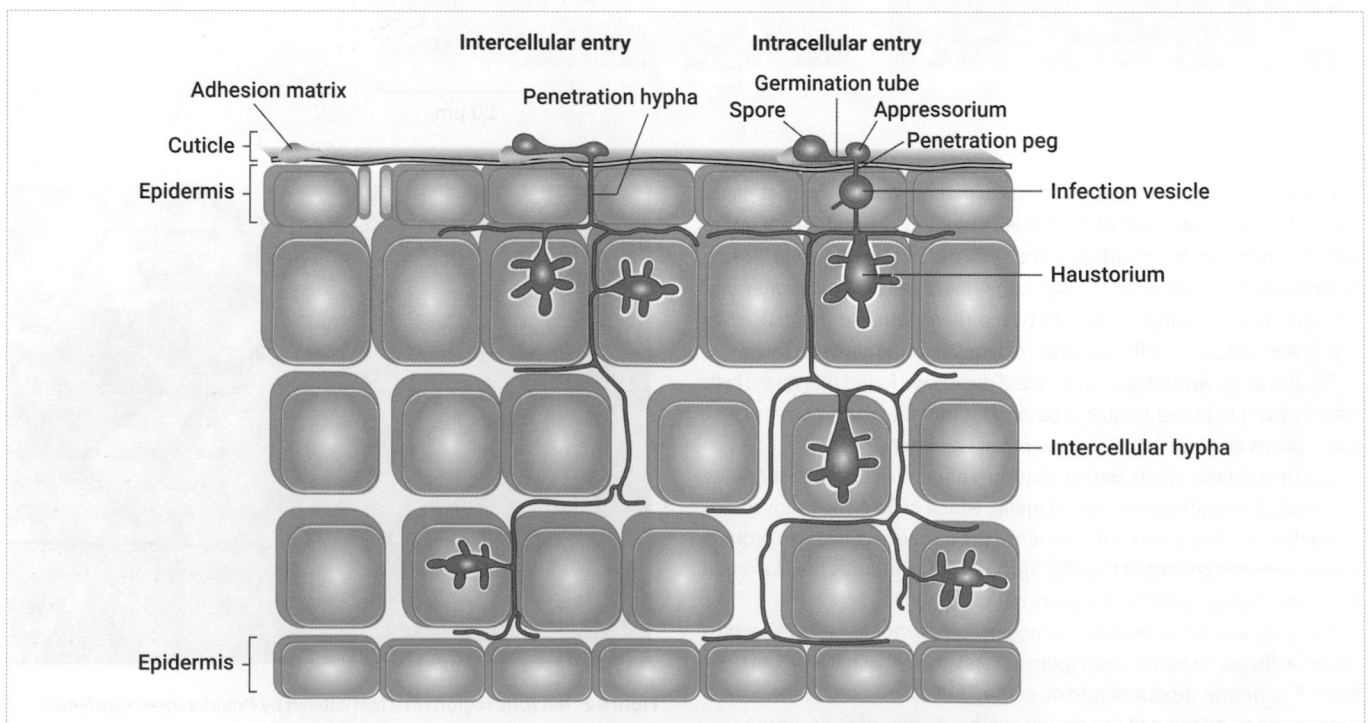

Figure 14.29 Haustorium. The production of haustoria by fungal pathogens. The haustorium sits inside the plant cell and has finger-like projections.

Source: Adapted from T. Torto-Alalibo, S. Meng, R. A. Dean. Infection strategies of filamentous microbes described with the Gene Ontology. *Trends in Microbiology*, 2009; 17(7), 320–327. With permission from Elsevier.

Table 14.5 Bacterial secretion systems and their role in plant pathogenicity

Secretion system	Example organism	Effect
Type I (T1SS)	*Pectobacterium carotovorum*	Binding of pathogen to plant cells
Type II (T2SS)	*Dickeya dadantii*	Production of enzymes (e.g. pectate lyase) to break down cell wall, which leads to soft rot diseases
Type II (T2SS)	*Xanthomonas fastidiosa*	Blockage of xylem
Type III (T3SS)	*Erwinia* spp.	Supress host immunity and increase nutrient acquisition
Type IV (T4SS)	*Agrobacterium tumefaciens*	DNA translocation
Type V (T5SS)	*Dickeya dadantii*	Adhesion to the surface of leaves

victoriae. Victorin has a variety of effects, including the inhibition of protein synthesis and the destruction of the plasmalemma (the plasma membrane that lines the inside of the plant cell wall).

Bacteria

Some bacteria can move between plant cells using their flagella, multiplying rapidly in different areas of the plant they reach. Bacteria also use secretion systems to help establish infections in animals and plants; examples of their effects can be seen in Table 14.5. These systems deliver proteins into the plant cells that enable the bacteria to produce toxins, enzymes, and specific structures which help with adhesion and mobility within the plant host.

Viruses

Plant virions spread to uninfected cells through the plasmodesmata, the membrane channels that connect cells together. This movement requires the expression of proteins that are controlled by the viral genome. Long distance movement within the plant occurs through the phloem.

▶ **We learn more about plant cell structure in Chapter 28.**

How do diseases affect the plant host?

The effects of disease can range from reduction in growth to death of the entire plant. These effects can be seen in different ways, and can be caused by fungi, bacteria, or viruses. Let us now consider some examples.

Rot and soft rot diseases

The production and activity of cell wall-degrading enzymes can result in the destruction of host tissue. Such destruction is a common feature of soft rot diseases; these are caused by pathogens that include the fungus *Sclerotinia fructigena* (soft rot of apples), *Rhizopus* (soft rot of sweet potatoes), and *Pectobacterium carotovorum* (bacterial soft rot of potato).

Soft rot diseases are characterized by extensive watery disintegration of the tissue as cell death and extensive tissue damage ensues, as depicted in Figure 14.30. Many diseases seen in stored foods are of this soft rot type and are recognizable from the necrotic symptoms they produce in leaf stems and roots.

A rot involving seeds or seedling is known as damping off, while rots of roots and stems are known as root rots or stem rots, respectively. Plants exhibiting root rots generally have retarded growth and their leaves may become yellow in dry

Figure 14.30 Soft rot diseases. (a) The soft rot of apples caused by *Sclerotinia fructigena*. (b) The soft rot of potato by *Pectobacterium carotovorum*.
Source: (a, b) Nigel Cattlin/Alamy Stock Photo.

Figure 14.31 Apple scab caused by *Venturia inaequalis*.
Source: Nigel Cattlin/Alamy Stock Photo.

Figure 14.32 Dutch elm disease caused by *Ophiostoma novo-ulmi*. Disease results in wilting of the tree.
Source: Minnesota Department of Natural Resources - FIA, Bugwood.org CC BY 3.0 US.

weather, and subsequently fall off. Examples of fungi causing rots can be found in various species of *Fusarium*: *F. solani* and *F. oxysporum* affect a great variety of vegetable, flower, and field crops. *Fusarium* root rots are common in beans, peanuts, and asparagus.

Spots and blights

Leaves and fruits can also exhibit symptoms of dead or dying (necrotic) tissue. A localized area of such tissue, which appears brown in colour, is called a spot. An example of a spot-causing pathogen is *Venturia inaequalis*, which causes apple scab, one of the most important diseases of apples (Figure 14.31). Another example is *Xanthomonas perforans*, which causes spots on field tomatoes.

Necrotic areas can be more extensive than those occurring in spots. One such symptom is known as blight, an example of which is the raspberry cane blight caused by *Leptosphaeria coniothyrium*.

Wilting

Many plant diseases also disrupt water movement. These are called wilt diseases (because of the visible wilting caused by a lack of water). Examples include *Fusarium* wilt of tomatoes and Dutch elm disease caused by *Ophiostoma novo-ulmi* (Figure 14.32).

Blast diseases

Blast diseases cause the sudden death of young seedlings and buds. For example, rice blast caused by the fungus *Magnaporthe oryzae* results in 10–30 per cent of global rice crop loss.

Dieback

Dieback is a symptom of disease often seen in woody plants in which twigs, branches, shoots, or roots die from the tips downwards. The example with which you are perhaps most familiar is ash dieback caused by the fungus *Hymenoscyphus fraxineus*. Figure 14.33 shows the dark lesions that appear on the bark of the ash tree following infection.

Figure 14.33 Ash dieback. The dark lesions of the ash dieback disease are caused by *Hymenoscyphus fraxineus*.
Source: PJ Photography/Shutterstock.com.

Hypoplastic effects

Hypoplastic effects describe the failure of plants or organs to develop properly. Examples include dwarfing (a failure to grow to

Figure 14.34 Tobacco mosaic virus causing areas of chlorosis on the leaf.
Source: Nigel Cattlin/Alamy Stock Photo.

normal size) and chlorosis (a failure to develop normal green coloration). An example of a plant disease that causes chlorosis is the tobacco mosaic virus; look at Figure 14.34 and notice the areas of yellowing caused by viral infection.

Hyperplastic effects

While hypoplastic effects describe the *under*-development of plants or organs, hyperplastic effects describe the *over*-development of plant parts. These effects include leaf curls, scabs, and tumours (e.g. crown galls). Crown gall is caused by *Agrobacterium tumefaciens*; we learn more about it in Chapter 10.

How can the host fight back?

Most plants have structural barriers that help prevent infection by plant pathogens. These include thick leaf cuticles and corky layers of stems. Some plants can even produce pre-formed fungitoxic compounds to combat fungal infections. These toxins are called phytoanticipins. Examples include the saponins of oats and tomatoes.

Infection can also induce the production of fungitoxic compounds called phytoalexins. One example of a phytoalexin is pisatin from pea plants.

The most common type of structural response to infection, however, is the development of papilla, a localized thickening of the host cell wall where the hypha penetrates the cell. We learn more about the diverse defence mechanisms adopted by different organisms, including plants, in Chapter 32.

 Check your understanding of the concepts covered in this section by answering the questions in the e-book.

SUMMARY OF KEY CONCEPTS

- Microbial pathogens express a variety of virulence factors that enable them to infect a host and cause disease.

- Robert Koch developed a series of tests (or 'postulates') that can be used to determine whether or not a particular microbial species causes a disease.

- While many bacterial species live in harmony on and in our bodies, and form a natural part of our normal microbiota, some can become opportunistic pathogens: they invade the organism and cause disease.

- As bacterial pathogens are prokaryotes (and the infected human hosts are eukaryotic) it has been possible to develop a number of antibiotic drugs that target and inhibit key molecules and processes unique to the pathogen.

- Antibiotic resistance, whereby bacterial pathogens are no longer inhibited or killed by the application of the drugs, has become a major problem in dealing with bacterial infectious diseases.

- Very few fungi cause disease in humans and animals but fungal infections can be common.

- There are four levels of fungal infection: superficial, cutaneous, subcutaneous, and systemic. Systemic fungal diseases are the most dangerous.

- Treatment of fungal infections can be difficult because fungal eukaryotic cells are similar to animal cells: it is difficult to make therapeutic strategies specific to fungi without also causing harm to animal cells.

- Some disease-causing fungi are capable of a differentiation process called dimorphism.

- Fungal spores can cause allergic reactions.

- Fungal poisoning can occur through the ingestion of pre-formed fungal toxins (mycetism) or the ingestion of foodstuffs that have been contaminated with fungi (mycotoxicosis).

- Protozoans—a diverse group of more than 30,000 mostly unicellular eukaryotic organisms—may be parasites, which depend on a host for their survival, and may have different forms during their life cycle.

- Protozoan diseases impact billions of people globally and are typically associated with deprived areas and poor sanitation. As such, these diseases often affect communities already enduring hardship.

- The impacts of protozoan diseases can range from minor illness through to life-altering disabilities and even death.

- All plants are susceptible to disease, which can be caused by a range of different organisms, but fungi cause the most damage.

- To cause an infection a pathogen must enter and then establish itself in the plant.

- Plants can protect themselves against disease by having tough waxy cuticles, corky stems, and some can also produce toxins.

- Plant diseases can be caused by a range of organisms, but the most common organisms to be pathogenic are fungi.

- Plant pathogens have a range of virulence factors that aid in disease progression within the host.

- Some plant pathogens have a narrow host specificity, while others can parasitize a wide range of host plant species.

- Plant pathogens enter the host through natural openings, natural wounds, and wounds inflicted by farm machinery or predation.

- Plant pathogens can cause a range of symptoms such as rots and wilts.

- Plants can prevent infection through structural barriers, such as bark, and also the production of toxins.

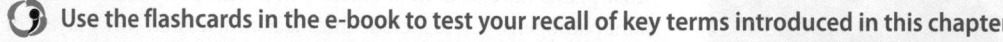 Use the flashcards in the e-book to test your recall of key terms introduced in this chapter.

QUESTIONS

 Looking for answers? Once you've answered these questions, follow the link in the e-book to the answer guidance and check your work.

Concepts and definitions

1. Define a pathogen.
2. State Koch's postulates.
3. What are zoonotic diseases?
4. What is a fomite and what environment are they particularly associated with?
5. What are protozoan cysts?

Apply the concepts

6. Which aspects of infection do virulence factors help with?
7. Describe one mechanism of antibiotic resistance.

8. Describe the differences between endotoxins and exotoxins.
9. Why are fungal infections of humans hard to treat?
10. What is a fungal thermal dimorph and name one disease-causing fungus that is a thermal dimorph?
11. What is the difference between mycetism and mycotoxicosis?
12. How are Protozoa that cause infections in humans categorized?

Beyond the concepts

13. Why are there not more drugs which can treat fungal infections?
14. Why are protozoan diseases associated with developing countries?

FURTHER READING

Waskito, L. A. & Yamaoka, Y. (2019) The story of *Helicobacter pylori*: depicting human migrations from the phylogeography. *Adv. Exp. Med. Biol.* **1149**: 1–16.
An interesting article on how Helicobacter pylori *could be used to examine ancient human migrations.*
Kim, J. (2016) Human fungal pathogens: Why should we learn? *J. Microbiol.* **54**: 145–148.
An editorial piece on the importance of more research into human fungal pathogens.

https://theconversation.com/how-we-can-use-light-to-fight-bacteria-73036
To find out more about how light can be used to fight bacteria, read this article from The Conversation.

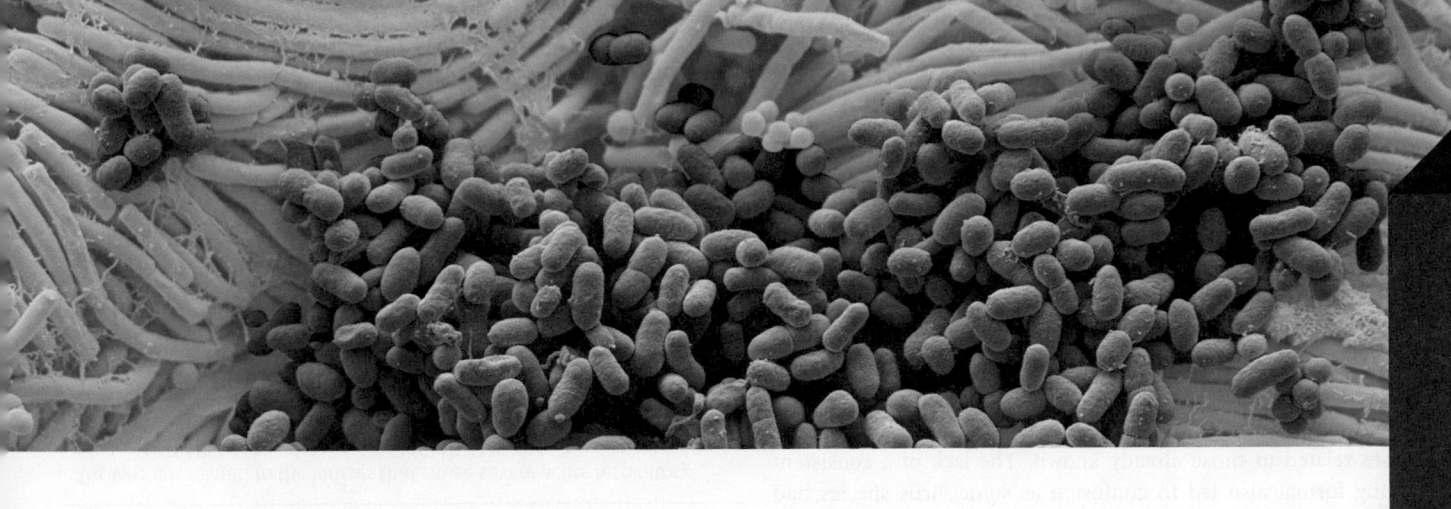

Viruses

Chapter contents

Introduction   555
15.1   How do we name and classify viruses?   556
15.2   The structure of viruses   556
15.3   How do viruses replicate?   559
15.4   The diversity of viral replication strategies   564
15.5   Bacteriophages   567
15.6   Antiviral therapy   569
15.7   Viruses beyond disease   572

LEARNING OBJECTIVES

By the end of this chapter, you should be able to:

- Describe the basic structure of viruses.
- Recall the rules regarding the taxonomy of viruses.
- List the various strategies for transcription among different classes of virus.
- Explain the steps involved in virus replication.
- Describe how antiviral therapies work.
- Appraise the various uses of viruses.

🔊 Watch the key concepts video in the e-book to prepare yourself for studying this chapter.

Introduction

At just a few tens of nanometres in diameter, viruses are one of the smallest agents of disease. However, what they lack in size they make up for in number and impact: it has been estimated that the world's oceans alone contain around 10^{30} individual virus particles, or **virions**. If these virions were laid end to end, they would reach to a distance of 10 million light years away.

Viruses do not just cause disease in humans. All forms of life—animals, plants, fungi, protozoa, bacteria, and archaea—are known to be susceptible to viral infection. As you might expect, viruses have a significant effect on many ecosystems, influencing nutrient availability, species diversity, and even climate change via the release of carbon from dead hosts. From the perspective of human disease—and as illustrated so plainly and unrelentingly throughout the Covid-19 pandemic—viruses can not only cause suffering and death on a global scale, but can also be the cause of considerable economic losses.

15.1 How do we name and classify viruses?

Until 1966, the naming of viruses was in chaos. There was no universal taxonomic scheme, and people who discovered viruses named them however they liked, without considering how those viruses related to those already known. The lack of a consistent naming format also led to confusion as some virus species had multiple names, having been 'discovered' more than once by different researchers. The International Union of Microbiological Societies agreed that something must be done and so convened the International Committee on Taxonomy of Viruses (ICTV) to tackle the problem. The ICTV put together a scheme that placed viruses into groups based on their genetic relatedness, and provided a set of rules for the naming of new viruses.

Under this system, the highest taxon in virus taxonomy is the realm; this taxon is given a single word name that ends in the suffix –viria (e.g. *Riboviria*). Table 15.1 shows the rules for naming of all taxa, with examples.

Look at Table 15.1 and notice how the names of the taxa are always written in italics, and the first word begins with a capital letter. This is always the case when discussing virus taxonomy. When discussing viruses themselves, however, the species name is usually not used: a familiar name is used instead, and this is not written in italics. Capitals are only used when the name contains a proper noun or begins a sentence. For example, *Zaire ebolavirus* is more commonly known by its familiar name Ebola virus (named after the Ebola river in the Democratic Republic of Congo). Virus names may also be abbreviated, in this case to EBOV.

Virus taxonomy is a frequently changing field, with yearly updates that often bring extensive changes. It is likely that there are many more changes to come in the near future, as at the time of writing 46 virus families have still not been assigned to higher taxa. However, the rules of virus taxonomy are constant, and so it

is arguably more important to remember these rules than it is to commit the classification of individual viruses to memory.

 Check your understanding of the concepts covered in this section by answering the questions in the e-book.

15.2 The structure of viruses

As we noted in the Introduction, 'virion' is the name given to the complete, infectious virus particle. Before we look at the structure of viruses, it is important to recognize that virions are not cells—they have no organelles, no membranes or cell wall, and at their simplest they can be composed of just nucleic acids and a protein coat, known as a **capsid**. Figure 15.1 shows a generic virion structure.

Virions typically range in size from 17 nm for circoviruses, to 300 nm for poxviruses. (By contrast—and to give a sense of scale— a typical eukaryotic cell nucleus has a diameter of 6 μm—that is, 6000 nm.) However, recently discovered giant viruses such as pandoravirus and pithovirus can be up to 1.5 μm (i.e. 1500 nm) in diameter, which is larger than the smallest bacterium.

The small size of most viruses precludes the use of light microscopy to view them, as the shortest wavelength of visible light is around 400 nm; as a result, all images of viruses are either computer generated or electron micrographs.

Table 15.1 Naming of virus taxa

Taxon	Suffix	Example
Realm	-viria	*Riboviria*
Phylum	-viricota	*Negarnaviricota*
Subphylum*	-viricotina	*Haploviricotina*
Class	-viricetes	*Monjiviricetes*
Order	-virales	*Mononegavirales*
Suborder*	-virineae	
Family	-viridae	*Filoviridae*
Subfamily*	-virinae	
Genus	-virus	*Ebolavirus*
Species	-virus	*Zaire ebolavirus*

* Not all subtaxa are used for every virus species.

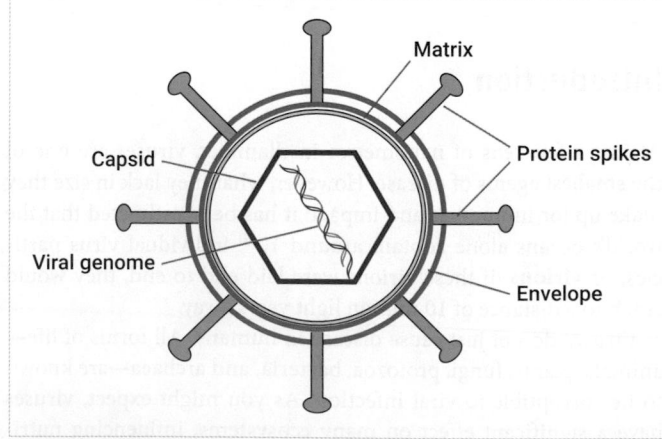

Figure 15.1 Structure of a virion. This is a generic illustration of a virion. Specific structural features vary between virus species and may even be absent in some cases.

We learn more about the use of microscopy in Chapter 12.

Let us now consider the main components of a virion in more detail.

Capsid

The capsid is a protein shell made up of repeating units known as **capsomeres**. One or more different types of protein may make up the capsid, as shown in Figure 15.2. Most virus capsids have either an icosahedral (Figure 15.3a) or a helical structure (Figure 15.3b), or both in the case of some **bacteriophages** (viruses which infect bacteria, Figure 15.3c). The capsid may also be directly associated with the nucleic acids, in which case the two may be referred to together as the **nucleocapsid**. The helical structure shown in Figure 15.3b denotes a nucleocapsid. Look at this figure, and notice how the protein subunits associate with the nucleic acid, in this case RNA.

Icosahedral capsids are given a triangulation number, or T number, based on the size and complexity of the capsid: the higher the T number, the larger and more complex the capsid, and the greater the variety of capsomeres. Look at Figure 15.4, which depicts capsids with increasing T numbers. All three capsids comprise 60 repeating units, but as the complexity of the units increases, so the complexity of the capsid (and the T number) increases too. The icosahedral form is extremely common among animal viruses. Herpesviruses are an example.

Helical capsids are generally rod-shaped or filamentous. The latter vary in terms of size and shape and may also be branched. Most plant viruses are rod-shaped; these include the tobacco mosaic virus (TMV; see Figure 15.3).

The capsid may also contain viral enzymes that are needed to replicate the virus in the host cell if the structure of the viral genome prohibits transcription using host enzymes (see Sections 15.6 and 15.7).

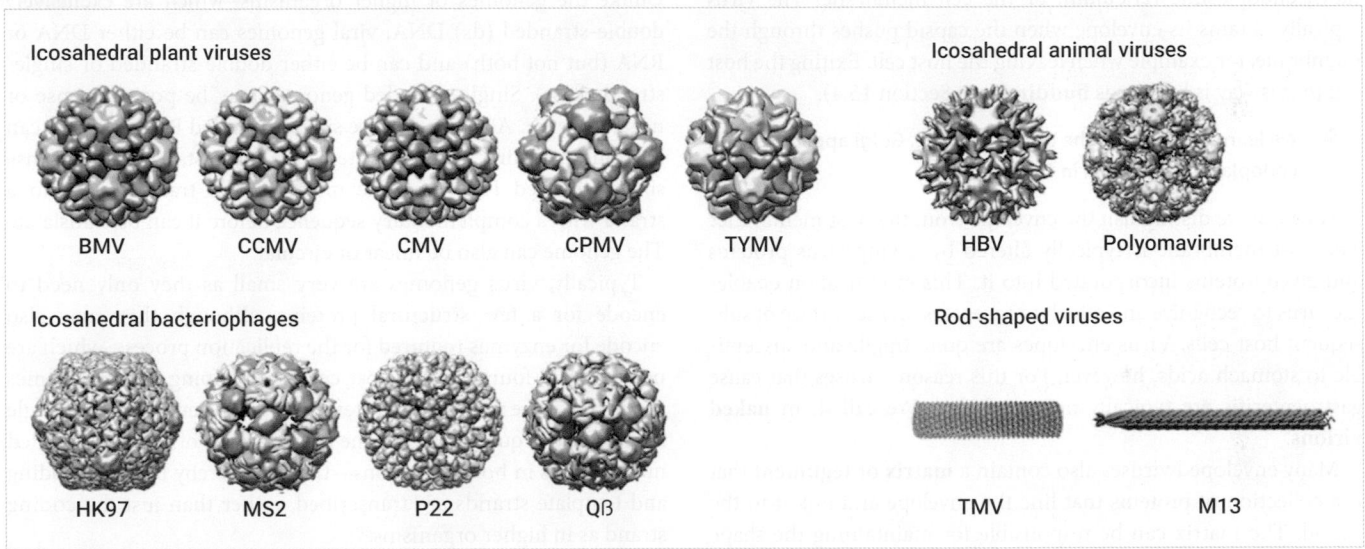

Figure 15.2 Virus capsids. Capsids vary widely in size, shape, and symmetry. The colours used in these images are to distinguish between the different types of capsomere and are not representative of any capsid colour.

Source: Lise Schoonen and Jan C. M. van Hest (2014) Functionalization of protein-based nanocages for drug delivery applications. *Nanoscale* 6: 7124–7141. 09 May 2014 DOI: 10.1039/C4NR00915K http://pubs.rsc.org/en/content/articlehtml/2014/nr/c4nr00915k.

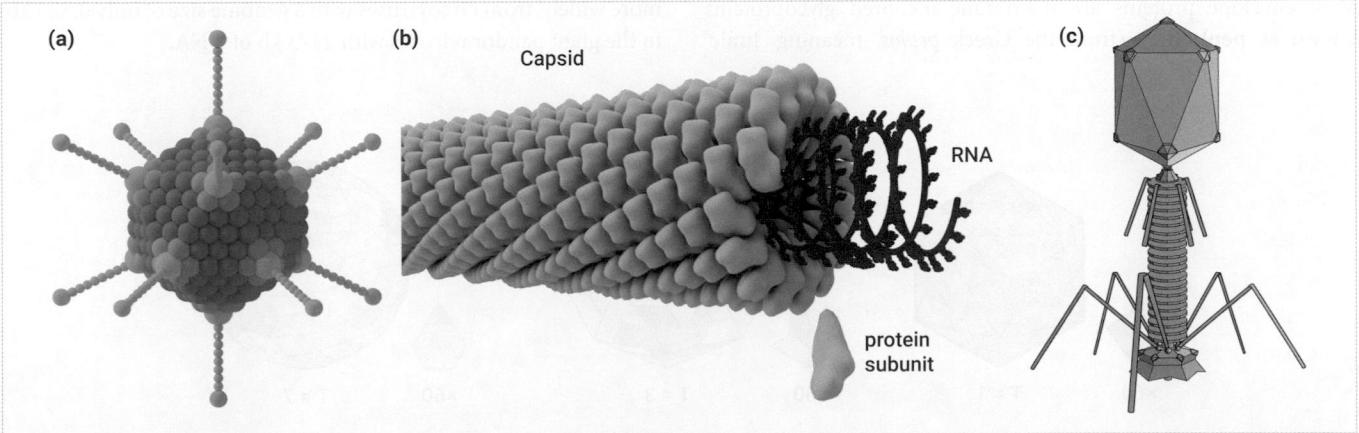

Figure 15.3 Computer-generated images of different capsid symmetries. (a) Icosahedral (adenovirus); (b) helical (tobacco mosaic virus); (c) a typical head-tail bacteriophage.

Source: (a, b) Thomas Splettstoesser (www.scistyle.com)/Wikimedia Commons/CC BY-SA 4.0; (c) Adenosine/Wikimedia Commons/CC BY-SA 3.0.

Envelope

Enveloped viruses are surrounded by a lipid bilayer, usually derived from one of the membranes of its host cell—that is, the Golgi apparatus, endoplasmic reticulum, or the cell membrane. The virus typically obtains its envelope when the capsid pushes through the membrane, for example when leaving the host cell. Exiting the host cell in this way is known as **budding** (see Section 15.4).

> ▶ We learn more about the cell membrane, Golgi apparatus, and endoplasmic reticulum in Chapter 9.

How can we distinguish the envelope from the host membrane? The host membrane is typically altered by having virus proteins and glycoproteins incorporated into it. This modification enables the virus to recognize and attach to receptors on the surface of subsequent host cells. Virus envelopes are quite fragile and susceptible to stomach acids, however. For this reason, viruses that cause gastroenteritis are typically not enveloped. We call them **naked virions**.

Many enveloped viruses also contain a **matrix** or **tegument** that is a collection of proteins that line the envelope and link it to the capsid. The matrix can be responsible for maintaining the shape of the virion, entry into the host cell, gene expression, or immune evasion.

Protein spikes

Most envelope proteins are membrane-anchored glycoproteins known as **peplomers** (from the Greek *peplos*, meaning 'tunic' or loose garment). They resemble spikes, which is what they are more commonly known as. For example, the trimeric spikes on the coronavirus envelope are so large and numerous they are easily visible under an electron microscope, where they resemble a crown or solar corona (from which the virus gets its name). Figure 15.3a shows the long glycoprotein spikes of adenovirus, which project directly from the surface of the naked capsid.

The function of spikes is typically to interact with host membrane receptors, usually for attachment and entry into the cell or in some cases to enable exit of progeny viruses from the cell. These interactions are very specific and determine the preference of a virus for a particular host cell, known as host **tropism**. We learn more about spikes, and the important part they can play in vaccine development, in Clinical Box 15.1.

The virus genome

Unlike the genomes of higher organisms, which are exclusively double-stranded (ds) DNA, viral genomes can be either DNA or RNA (but not both) and can be either double-stranded or single-stranded (ss). Single-stranded genomes may be positive sense or negative sense. A positive-sense single-stranded RNA genome can be translated directly into protein. By contrast, a negative-sense single-stranded RNA genome must first be transcribed into a strand with a complementary sequence before it can be translated. The genome can also be linear or circular.

Typically, virus genomes are very small as they only need to encode for a few structural proteins, although they may also encode for enzymes required for the replication process, which are not typically found in the host cell. Overlapping reading frames can increase the number of proteins that can be encoded by a single nucleic acid sequence, as can the transcription of double-stranded nucleic acids in both directions—that is, whereby both the coding and template strands are transcribed, rather than just the coding strand as in higher organisms.

> ▶ We learn more about the process of transcription in Chapter 5.

RNA viruses typically have smaller genomes than DNA viruses. For example, the largest RNA genome is that of coronaviruses (ssRNA viruses) at up to 32 kb. The size of DNA genomes varies more widely, from circoviruses with a genome size of only 0.825 kb, to the giant pandoraviruses with 2473 kb of DNA.

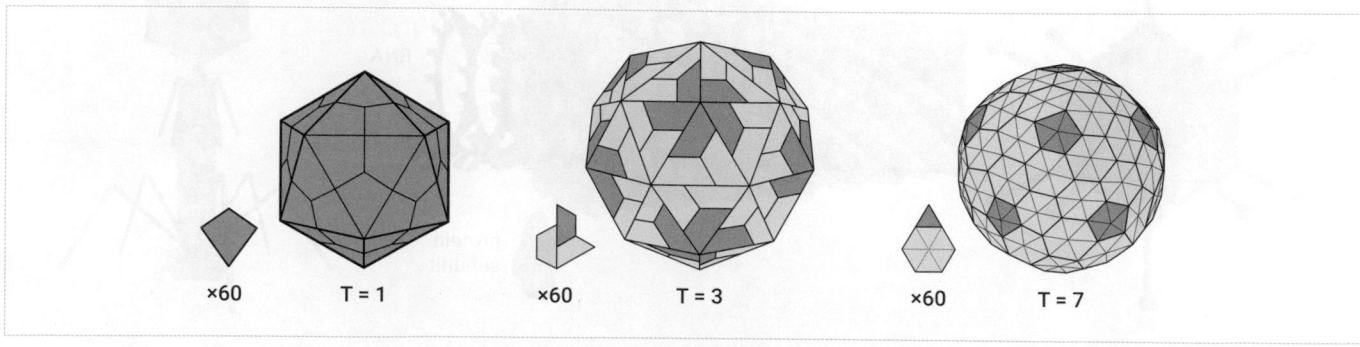

Figure 15.4 Capsids with increasing *T* numbers. The complexity of icosahedral capsids increases as the *T* number increases, in this case from *T* = 1 to *T* = 7.
Source: ViralZone, SIB Swiss Institute of Bioinformatics/CC BY 4.0.

CLINICAL BOX 15.1   The Covid-19 spike protein: the target of vaccine development

The spike protein of the SARS-CoV-2 virus (better known to many of us as Covid-19) was the focal point of research and development activities that sought to develop vaccines against this virus.

The spike protein of viruses such as SARS-CoV-2 acts as the point of initial contact and binding between virus and target cell before infection takes place. If the spike protein is somehow blocked to prevent this initial binding, infection is no longer possible.

The blocking of spike proteins to prevent infection is precisely the role served by antibodies produced by our immune systems: specific antibodies bind selectively to the spike protein, and in so doing inhibit its interaction with host cells, preventing the virus of which it is part from mounting an infection.

Covid vaccines stimulate our bodies to produce antibodies against the spike protein; they do this by mimicking the spike protein itself. When our immune systems subsequently encounter the actual virus, they have already been primed (by the action of the vaccine) to mount a rapid secondary response: to quickly release a wave of virus-specific antibodies, which inhibit infection.

The spike protein of SARS-CoV-2—more properly called the S-glycoprotein—has a trimeric structure: it comprises three subunits, as illustrated in Figure 1. The sequence of this protein was published on 10 January 2020; phase 1 clinical trials—a significant step in vaccine research and development—commenced just six weeks later.

With the sequence of the spike protein in hand, a host of different academic and industrial vaccine groups then sought to develop different Covid vaccines that emulated the structure of that protein. Although they used different molecular approaches, they were united behind a common strategy: to develop a vaccine that stimulates the immune system to behave as if exposed to the SARS-CoV-2 spike protein. The administration of Covid vaccines has rapidly become commonplace in many parts of the world, and the speed with which those vaccines were developed remains quite remarkable: a true testament

Figure 1 SARS-CoV-2 spike protein. A side-on view of the three-dimensional molecular structure of the SARS-CoV-2 S glycoprotein ('S-protein').

Source: Martínez-Flores, D., Zepeda-Cervantes, J. et al. (2021) SARS-CoV-2 vaccines based on the spike glycoprotein and implications of new viral variants. *Frontiers in Immunology*, **12**, 2774.

to the power of our current understanding and application of molecular biology.

PAUSE AND THINK

Distinguish between a positive- and negative-strand RNA virus. Why are these terms not applicable for DNA viruses?

Answer: For a positive (+) strand RNA virus, the RNA is the same as an mRNA and can be translated directly. For a negative (−) strand RNA virus, the RNA is the complement of the mRNA so the other RNA strand has to be synthesized before it is translated. As DNA is not translated, these terms are not applicable even with single-stranded DNA viruses.

 Check your understanding of the concepts covered in this section by answering the questions in the e-book.

15.3 How do viruses replicate?

Viruses do not reproduce in the same way that higher organisms do, neither do they divide like bacteria. Instead, the infecting virus is dismantled and its genome is used to instruct the production of progeny viruses.

The virus replication cycle can be divided into seven main stages:

1. Attachment and fusion.
2. Entry and uncoating.
3. Transcription.
4. Translation.
5. Assembly.
6. Formation of virus envelope.
7. Release.

We will explore each stage in turn later in this section.

The whole of the virus replication cycle takes place inside a host cell and typically requires the normal host cell functions to be suppressed. Much of our knowledge of virus replication in animal viruses is a result of studying animals under experimental conditions and, more recently, comes from the study of cell cultures consisting of monolayers of immortalized or semi-continuous cells. Our understanding of the replication of viruses that do not grow in such conditions is limited.

> ▶ **We learn more about cell culture in Chapter 12.**

An early culture method was to inoculate the chorioallantoic membrane of embryonated eggs. This membrane functions in a similar way to the human placenta in that it supplies the developing embryo with oxygen. Indeed, the inoculation of eggs is still used today as a way of culturing influenza viruses as a part of vaccine development.

Plant viruses and bacteriophages are more easily studied because there are no ethical barriers to cultivating and infecting the host organism; for human virus research, we must be satisfied with whatever information we can derive from cultured cells.

Let us now consider each stage of the virus replication cycle, starting with the point at which the virus first attaches to the host cell it is about to infect.

Attachment and fusion

Viruses are non-motile and encounter host cells completely by chance. The virion must therefore become securely attached to the cell surface to avoid missing the opportunity for infection, to trigger the process of fusion of membranes (in the case of enveloped viruses), and ultimately to gain entry into the cell.

The process of attachment is complex, and varies greatly even among viruses in the same genus. In general terms, however, viruses attach to host cells by using host cell surface molecules as receptors for viral glycoproteins.

One well-studied example of attachment is exhibited by human immunodeficiency virus (HIV), which attaches to the CD4 receptor on macrophages and helper T lymphocytes via a protein named gp120—a glycoprotein that is found on the surface of the virus and is a subunit of the Env protein. This process is illustrated in Figure 15.5.

The initial binding triggers a conformational change in the gp120 protein, drawing the virion closer to the host cell membrane and allowing the gp120 to bind with a second receptor (called a co-receptor), such as CCR5 or CXCR4. Another part of the viral Env protein called gp41 is then exposed. This part of the protein is called the fusion peptide. It inserts into the host cell membrane and folds to create a six-helix bundle that disrupts the host cell membrane and prompts it to fuse with the viral envelope. Once a fusion pore is created, the virus capsid can be delivered into the host cell cytoplasm.

Entry and uncoating

Following attachment, viruses enter the host cell via one of two processes—membrane fusion or receptor-mediated endocytosis; these processes are illustrated in overview in Figure 15.6. However, most mammalian viruses enter the host cell via receptor-mediated endocytosis (RME), a more detailed view of which is given in Figure 15.7a. RME is normally used by the cell to take in nutrients,

metabolites, and hormones, and requires a protein called clathrin. So how does RME proceed?

Clathrin moves to the plasma membrane when a receptor on the membrane surface is activated. This activation initiates the formation of a clathrin-coated pit, which forms the basis of a vesicle. The virion is then drawn into the vesicle and becomes internalized. This process is illustrated in Figure 15.7b.

Influenza viruses enter cells via this method and escape the vesicle once it has become acidic due to the influx of hydrogen ions. The low pH triggers a conformational change in the virus glycoprotein haemagglutinin, which causes the virus envelope to fuse with the endosomal membrane. This fusion discharges the nucleocapsid into the host cell cytoplasm.

For ssRNA viruses, the capsid only needs to be partially removed before transcription of the viral genome can begin. For most viruses, however, such uncoating must be complete before transcription can start.

The strategies employed for uncoating are very diverse. Adenovirus, for example, uncoats with the assistance of viral proteases, activated by the reducing conditions of the host cytoplasm. Polyomaviruses first enter the host endoplasmic reticulum, whereupon the viral capsid destabilizes due to the conditions found there.

By contrast, viruses that enter the host cell by endocytosis must employ different strategies. For example, influenza virus uncoats in the endosome, when fusion with host lysosomes causes the pH to fall (see Figure 15.7a).

Uncoating may occur in the cytoplasm, or in the nucleus; the location depends on the site of virus replication, which is itself dictated by the composition of the virus genome. RNA virus replication is usually in the cytoplasm, and DNA viruses replicate in the nucleus. (One notable exception is the dsDNA poxviruses, which replicate in the cytoplasm.)

Transcription

We see in Chapter 5 how the process of the one-way flow of genetic information is known as 'the central dogma' of molecular biology, which states that in all living things DNA is transcribed to RNA, which is translated to protein.

The processes by which viruses approach the transcription of messenger RNA (mRNA) from their nucleic acids varies depending on the composition of the virus genome. For example, as all living things have dsDNA genomes, they do not synthesize enzymes capable of catalysing the transcription of mRNA from ssRNA. It follows therefore that some classes of virus must be capable of expressing their own polymerase enzymes, which are packaged into the capsid during assembly of the progeny virions, ready to initiate transcription during the next replication cycle.

In 1971 David Baltimore sought to simplify our understanding of viral transcription by describing the five different means by which the transcription of mRNA can be accomplished. These became known as Baltimore classes, and a further two classes were added later with the discovery of **retroviruses** and **pararetroviruses**. The classes and their approach to transcription are given in Table 15.2.

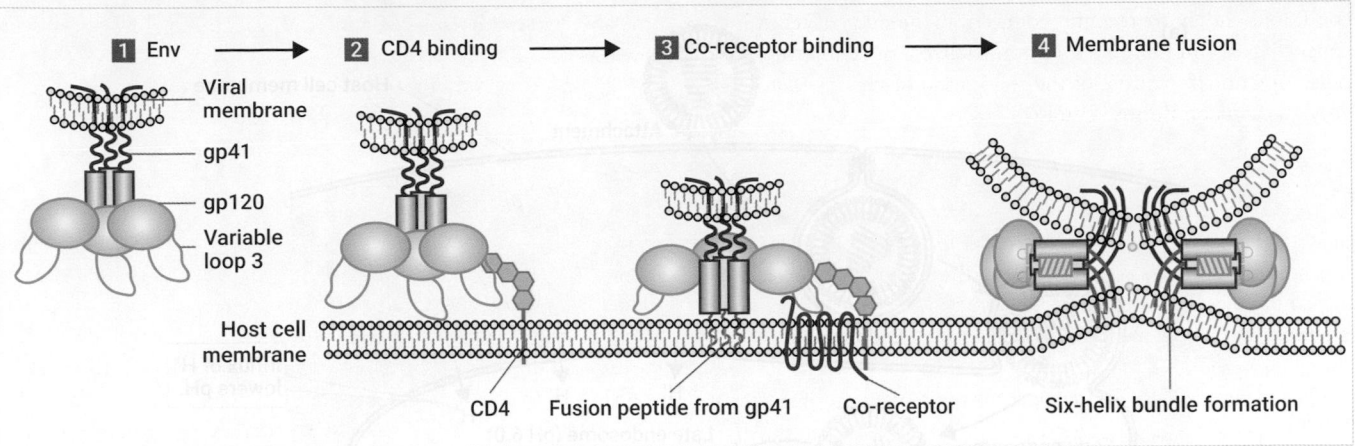

Figure 15.5 Attachment and entry of HIV to a host cell. The gp120 subunit of HIV Env (1) first attaches to the CD4 surface molecule of the host cell (2). Subsequent conformational changes in Env then allow co-receptor binding (3). Membrane fusion is initiated and the fusion peptide of gp41 inserts into the target membrane. The formation of a six-helix bundle facilitates complete membrane fusion (4).

Source: Wilen, C. B. et al. (2012) HIV: cell binding and entry. *Cold Spring Harb. Perspect. Med.* 2012;2:a006866.

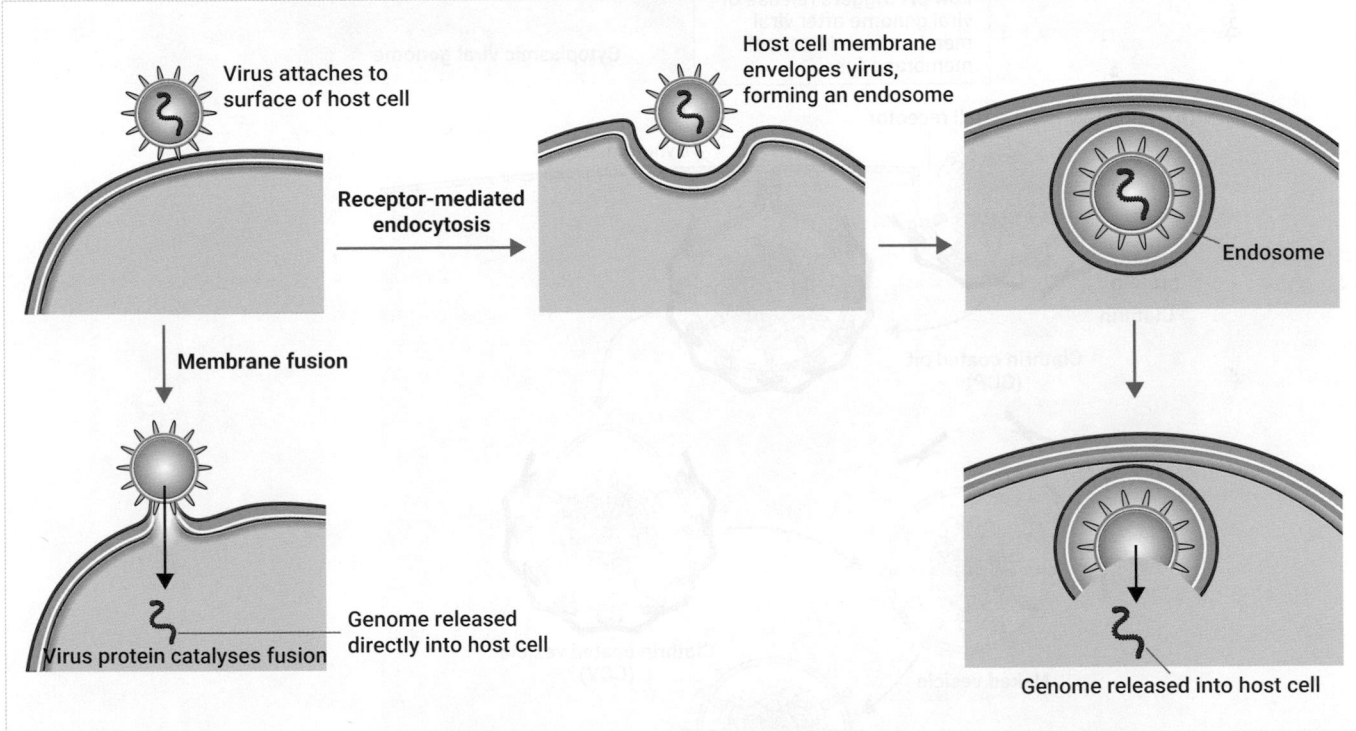

Figure 15.6 Virus entry. A virus can enter a host cell via one of two pathways: membrane fusion or receptor-mediated endocytosis. Both pathways culminate in the release of the viral genome into the host cell.

Source: John Oxford; Paul Kellam; Leslie Collier, *Human Virology*, 2016, Oxford University Press.

Viruses may also undergo two distinct cycles of transcription, known as early and late. Early proteins are generally functional—that is, they are involved in the inhibition of the host cell transcription, prevention of apoptosis, immune evasion, and the assembly of virus particles. Late proteins are more usually structural; they include capsid and matrix components, as well as any enzymes that require packaging into the capsid.

PAUSE AND THINK

Which category of viruses is the most likely to encode an RNA-dependent RNA polymerase in its genome?

Answer: A minus or negative (−) strand RNA virus as it has to synthesize the other RNA strand and uses the − strand for its template.

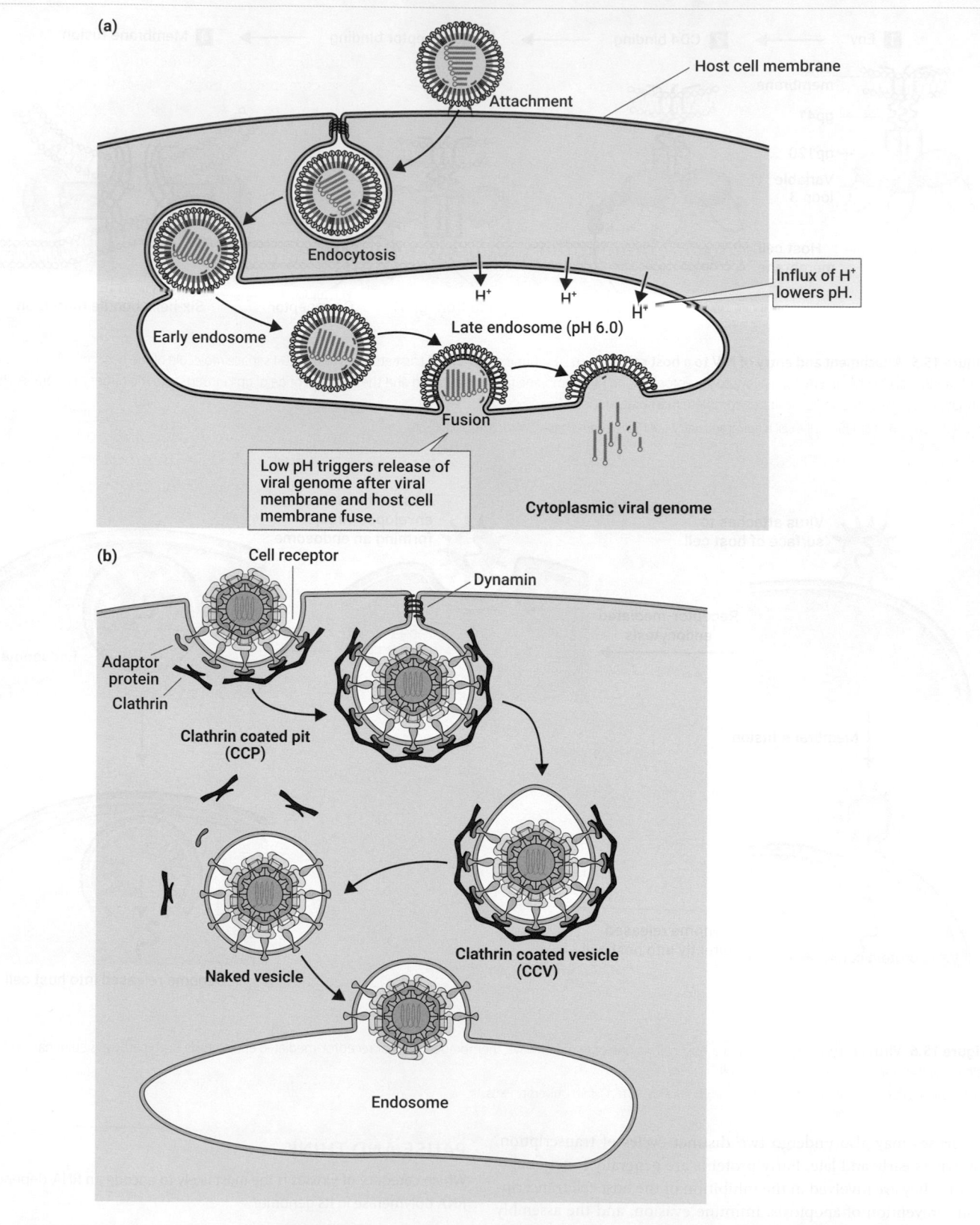

Figure 15.7 The entry of a virus into a host cell via receptor-mediated endocytosis. (a) Influenza viruses enter host cells via endocytosis and escape the endosome so formed when the viral and cellular membranes fuse. This process is triggered by the lowering pH of the endosome. The viral nucleocapsid, which in influenza A virus is in eight distinct sections, is discharged into the host cytoplasm. (b) RME involves the formation of a clathrin-coated pit, which mediates the encircling of the virus with the host cell membrane to form the endosome.

Source: (a, b) ViralZone, SIB Swiss Institute of Bioinformatics.

Translation

The process of translating mRNA into protein utilizes the host cell ribosomes, transfer RNA, and amino acids. Most viral proteins undergo some form of post-translational modification such as phosphorylation (in the case of nucleocapsid proteins) or glycosylation (membrane glycoproteins), or post-translational cleavage. Some viral mRNAs encode more than one polypeptide (they are called **polycistronic**) and possibly the entire genome is transcribed into one mRNA molecule, as in flaviviruses. The subsequent polypeptide must therefore be cleaved into individual proteins, usually by virus-encoded proteases. In some cases, the polycistronic mRNA is cleaved into **monocistronic** mRNA prior to translation.

Genome replication

Replication of the virus genome has two outcomes: first, the number of nucleic acid templates available for transcription is increased; second, there are genomes available for the progeny virions.

Genomes can be replicated by using the genome of the infecting virus directly, or less commonly via an RNA or DNA intermediate. In general, double-stranded genomes are copied directly, and single-stranded genomes are copied via a complementary intermediate. The exceptions here are viruses in classes VI and VII, the retroviruses and pararetroviruses. We discuss the replication strategies of the different classes further in Section 15.4.

Assembly

Once sufficient numbers of viral genomes and proteins have been synthesized, these components must be assembled into progeny virions. As with other stages of replication, this process can occur in the nucleus or the cytoplasm, depending on whether the virus has a DNA or RNA genome, respectively. Simple icosahedral capsids may self-assemble spontaneously, due to the electrostatic and hydrophobic interactions operating between them. Such a process is energetically favourable. By contrast, more complex capsids may require scaffolding proteins—proteins that serve only to assist in virus assembly, before being removed from the finished capsid. Scaffolding proteins may be inside or outside the assembling capsid (procapsid); the scaffolding proteins of herpesviruses are internal, for example. The assembly of a herpesvirus capsid is illustrated in Figure 15.8.

Herpesvirus capsids are composed of four structural proteins and the matrix component may comprise upwards of 23 different proteins. Scaffolding proteins are directly attached to capsomeres on the inside of the procapsid. Once the procapsid is assembled, the scaffolding proteins are cleaved by a protease called VP24, which is itself a component of the capsid. This cleavage allows the procapsid to take on its mature icosahedral formation, as shown in the last step of Figure 15.8.

Once the capsid has assembled, the viral DNA is then packaged into it via a channel at one of the vertices of the icosahedron, a process which alone requires seven different packaging proteins to

Table 15.2 The seven Baltimore classes of viral transcription

Class	Genome	Transcription strategy	Example
I	dsDNA	These viruses use the host cellular machinery for transcription of the (−) strand by DNA-dependent RNA polymerase, in the same way as the host	Herpesviruses
II	ssDNA	The genome may be either the (+) or (−) strand and is first copied to dsDNA by the host DNA-dependent DNA polymerase. The (−) strand is then transcribed	Parvoviruses
III	dsRNA	The (−) strand is transcribed using a viral RNA-dependent RNA polymerase. This enzyme must be packaged into the capsid in order for transcription to be possible in subsequent hosts	Rotavirus
IV	(+)ssRNA	The genome of these viruses is identical to mRNA, and thus can be used directly in translation	Hepatitis C virus
V	(−)ssRNA	The (−) strand is transcribed using the viral enzyme RNA-dependent RNA polymerase, as for dsRNA viruses.	Measles virus
VI	(+)ssRNA	Retroviruses; the ssRNA is first *reverse* transcribed into dsDNA by the viral enzyme RNA-dependent DNA polymerase, or **reverse transcriptase**. The dsDNA is then integrated into the host genome, from where it can be transcribed using the host cellular machinery	Human immunodeficiency virus
VII	dsDNA	Pararetroviruses; transcription is the same as for Class I viruses and also the host cell; the reverse transcription step is involved in replication of the genome (see 'Replication in retroviruses and pararetroviruses' in Section 15.4)	Hepatitis B virus

achieve. The successful packaging of the DNA leads to expulsion of the cleaved scaffolding proteins from the capsid and the assembly is then complete.

The process of helical nucleocapsid assembly is usually different from that for icosahedral capsids. Instead of capsid assembly being followed by genome packaging, the capsomeres assemble *around* the nucleic acids, directly associating with them. One of the best understood helical capsids is that of TMV, in which each capsomere is associated with three nucleotides (see Figure 15.3b).

Formation of the virus envelope

Virus envelopes are usually derived from the host cell membrane, although they can be newly synthesized specifically for the virus. However, such *de novo* synthesis of viral envelopes is a much less common strategy.

Membrane-derived envelopes are acquired when the virion extrudes through a cellular membrane, either an internal membrane (associated with a cellular organelle) or during exit from the cell by budding. Herpesviruses obtain their envelope from the nuclear membrane, for example, and the envelope of hepatitis B virus comes from the endoplasmic reticulum. Influenza virus and HIV both obtain their envelope from the plasma membrane during egress.

Release

Viruses may leave the host cell by essentially destroying it (cell lysis) or by budding from the cell surface. During budding, glycoproteins first modify the host membrane, displacing host membrane proteins and providing a site to which the capsid can bind. This binding of the capsid is often done via M protein, which attaches to both the capsid and the cytoplasmic portions of the glycoproteins that extend through the membrane. Binding of the capsid to M protein triggers the extrusion of the capsid from the host cell. Budding as an exit strategy leaves the plasma membrane intact and an infected cell can produce virions continuously for hours or days.

The surface of the host cell from which the virus buds is crucial as it influences both onwards transmission and the ability of a virus to cause systemic (body-wide) infection. Respiratory and gastro-intestinal viruses, for example, bud from the surfaces of endothelial cells that are exposed to the external environment (apical surfaces), whereas HIV buds from the membrane adjacent to other tissues (the basolateral membrane), from where it can access the interior of the host.

Most non-enveloped viruses are released from the host by cell lysis. In contrast to budding viruses, which can be produced continuously, the production of naked viruses must be coordinated so that all virions mature simultaneously or can be held within the host cell until released. Hundreds or thousands of virions may be released from a single infected cell. Once the host cell has lysed, no more infectious virions will be produced.

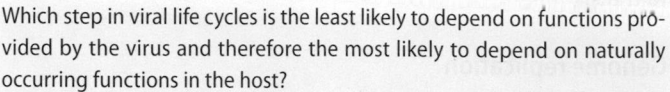

PAUSE AND THINK

Which step in viral life cycles is the least likely to depend on functions provided by the virus and therefore the most likely to depend on naturally occurring functions in the host?

Answer: Entry and uncoating is almost entirely dependent on naturally occurring processes in the host.

 Check your understanding of the concepts covered in this section by answering the questions in the e-book.

15.4 The diversity of viral replication strategies

Having considered the general stages in the replication of a virus, let us now focus on the ways in which replication unfolds in DNA and RNA viruses specifically, beginning with DNA viruses.

Replication in DNA viruses

Before the replication of a DNA genome can begin, the DNA needs a primer. A primer takes the form of a short strand of RNA with a sequence that is complementary to that of the site of the origin of replication, and is the point at which the first nucleotide can attach to the template strand. For dsDNA viruses, including herpesviruses and papillomaviruses, the primer is synthesized by a primase enzyme.

Figure 15.8 The assembly of a herpesvirus capsid.

Source: Reprinted from *Current Opinion in Virology*, Volume 1, Issue 2, Jay C. Brown, William W. Newcomb, Herpesvirus capsid assembly: insights from structural analysis, Pages 142–149, Copyright 2011, with permission from Elsevier.

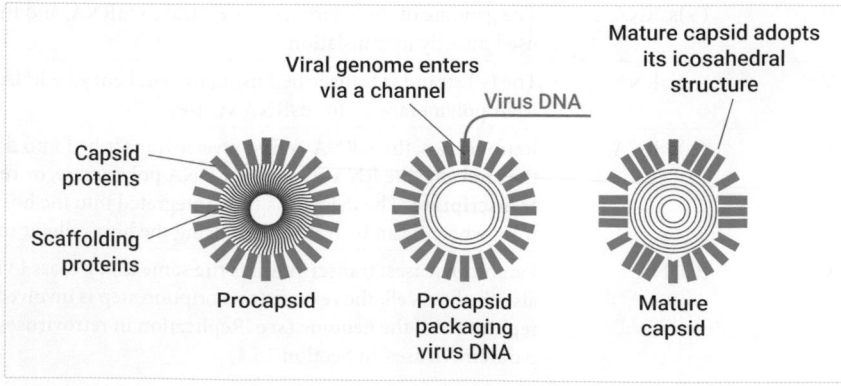

▶ **We learn more about DNA replication in Chapter 3.**

By contrast, the genome of some ssDNA viruses such as parvoviruses can self-prime at the 3′ end by folding over into a hairpin structure, provided the nucleotides are complementary. A complementary strand is then synthesized to convert the single-stranded genome into a double-stranded one before genome replication can begin in earnest.

A third means by which replication can be initiated is seen in adenoviruses, whereby a protein primer is used.

A number of viral enzymes are required to initiate replication: a helicase, to unwind the dsDNA helix; a protein which keeps the two unwound DNA strands separate; an RNase for degrading the RNA primers; a ligase for joining **Okazaki fragments** (see 'Replication fork' section); and a DNA polymerase, which generates the daughter strands. Usually, the DNA polymerase required for the replication of dsDNA is provided by the host cell, although in some cases it is viral in origin. For example, poxviruses encode all their own enzymes for DNA replication.

Figure 15.9 shows how there are two mechanisms for double-stranded DNA replication: via formation of a **replication fork**, as occurs in eukaryotes, and via **strand displacement**. The latter is possible in linear genomes as the parent strands can be copied from both ends simultaneously.

PAUSE AND THINK

DNA replication always requires a primer.

(a) How is DNA replication primed in dsDNA viruses?

(b) How is DNA replication primed in some ss DNA viruses?

a hairpin structure, which can serve as a primer.
(b) In some ssDNA viruses, the genome can base-pair internally and make
serve as a primer, similar to what occurs in most other organisms.
(a) An enzyme known as primase synthesizes the short RNA strand to
Answer:

Replication fork

Virus families with circular dsDNA genomes synthesize daughter strands via a replication fork. Viruses adopting this strategy include polyomaviruses and papillomaviruses. Some viruses (herpesviruses, for example) may circularize their linear genomes prior to replicating them in this manner.

Replication begins with the unwinding of the DNA by the helicase enzyme. Once unwinding has happened, DNA synthesis proceeds in both directions simultaneously. However, as both daughter strands can only be synthesized in the 5′→3′ direction, the synthesis of a single, continuous daughter strand can only occur on the leading template strand. (This is called continuous DNA synthesis.) By contrast, synthesis on the lagging strand must be discontinuous: repeated RNA primers are synthesized and DNA synthesis from these primers gives rise to short strands of newly synthesized DNA known as Okazaki fragments. The RNA primers are degraded by RNase enzymes and the Okazaki fragments are joined by ligases to form a continuous strand that is complementary to that synthesized from the leading strand.

▶ **We learn more about Okazaki fragments when we consider DNA replication in Chapter 3.**

Circular dsDNA genomes can adopt one of two different modes of replication. These are named theta and sigma replication due to the shapes made by the genomes as they are replicated: they resemble the Greek symbols θ and σ as illustrated in Figure 15.10.

During theta replication (Figure 15.10a) two replication forks move away from each other, as the entire genome is duplicated to give two nascent genomes: each one contains one parent and one daughter strand. By contrast, in sigma (or 'rolling circle') replication, one DNA strand is nicked and displaced and becomes the lagging strand, whereas the leading strand remains whole and may become the template for a very long linear strand of DNA known as a **concatemer**. The concatemer is then cut into individual genomes.

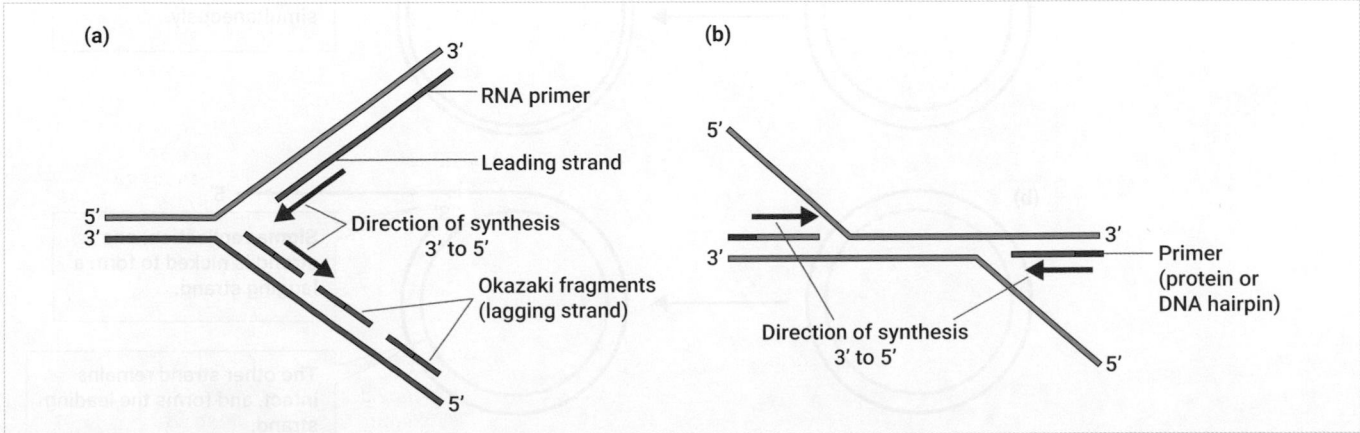

Figure 15.9 Replication of dsDNA virus genomes. (a) Via a replication fork; (b) via strand displacement.

Strand displacement

In some viruses with linear dsDNA genomes (e.g. adenoviruses and poxviruses), replication occurs via strand displacement, as illustrated in Figure 15.9b. The 5′ ends of adenovirus genomes have identical but inverted sequences and are linked to proteins that act as the primers for replication. Replication therefore proceeds from both ends of the genome, via continuous synthesis only, with no need for a replication fork or the synthesis of Okazaki fragments.

Replication in RNA viruses

Viral RNA genomes may be double-stranded, as in the reoviruses; single-stranded positive-sense, as in coronaviruses and picorna-viruses (a family that includes poliovirus); or single-stranded negative-sense, as in filoviruses (a family that includes Ebola virus). As we note earlier, positive-sense RNA genomes can be translated directly, in the same manner as mRNA. By contrast, negative-sense RNA genomes need to be copied to give a complementary strand that can then be translated. Not all RNA viruses require a primer to initiate replication and some use a protein primer. Orthomyxoviruses use oligonucleotide caps taken from host mRNAs to serve as primers.

The RNA transcribed from dsRNA genomes can be either translated into protein by the host ribosomes or used as a template for genome replication. The synthesis of RNA daughter strands from an RNA template only ever occurs in nature as a feature of viral transcription and genome replication, however. Therefore, the RNA polymerase enzyme required for this process must be encoded by the viral genome: it will not be synthesized by the host cell because the host cell simply does not need it.

▶ **We learn more about the process of translation in Chapter 5.**

The replication of ss(+)RNA viruses occurs via a negative-sense strand, transcribed from the positive-sense genome and then used as a template for the synthesis of further genomes. While many of the copies of the ss(+)RNA genomes that result from replication are used to construct progeny viruses, some may be used to scale-up protein synthesis—a phenomenon aided by the fact these genomes can be translated directly, as we note above. Some daughter strands are also used as further templates for synthesis of the ss(−)RNA intermediate.

Replication of ss(−)RNA genomes occurs via a ss(+)RNA intermediate, and therefore may be performed by the same viral RNA polymerase as is used for transcription. In some viruses, however, the processes of transcription and replication are distinct and are catalysed by different polymerases.

Replication in retroviruses and pararetroviruses

The most well-known of the retroviruses are undoubtedly the immunodeficiency viruses, in particular HIV. These viruses have two identical copies of a ss(+)RNA genome, but mRNA is transcribed via a dsDNA intermediate.

Replication is preceded by a reverse transcription step whereby retroviral ss(+)RNA is converted into dsDNA. The newly reverse-transcribed genome then migrates to the nucleus where it is inserted into the host genome via the viral enzyme integrase. At this point the retrovirus is known as a **provirus**. The provirus is then transcribed into mRNA and new viral genomes using the same molecular mechanisms as the host cell uses when transcribing its own genes. This overall process is outlined in Figure 15.11.

The only two pararetrovirus families are *Hepadnaviridae* (which includes hepatitis B virus) and *Caulimoviridae*. Hepadnaviruses have partially double-stranded DNA genomes, the positive-sense strand being only about 50–85 per cent complete. Figure 15.12 illustrates what happens when these viruses replicate; look at this figure as we go on to explore the key steps now.

Figure 15.10 Theta and sigma modes of replication. (a) Theta replication; (b) sigma or rolling circle replication.

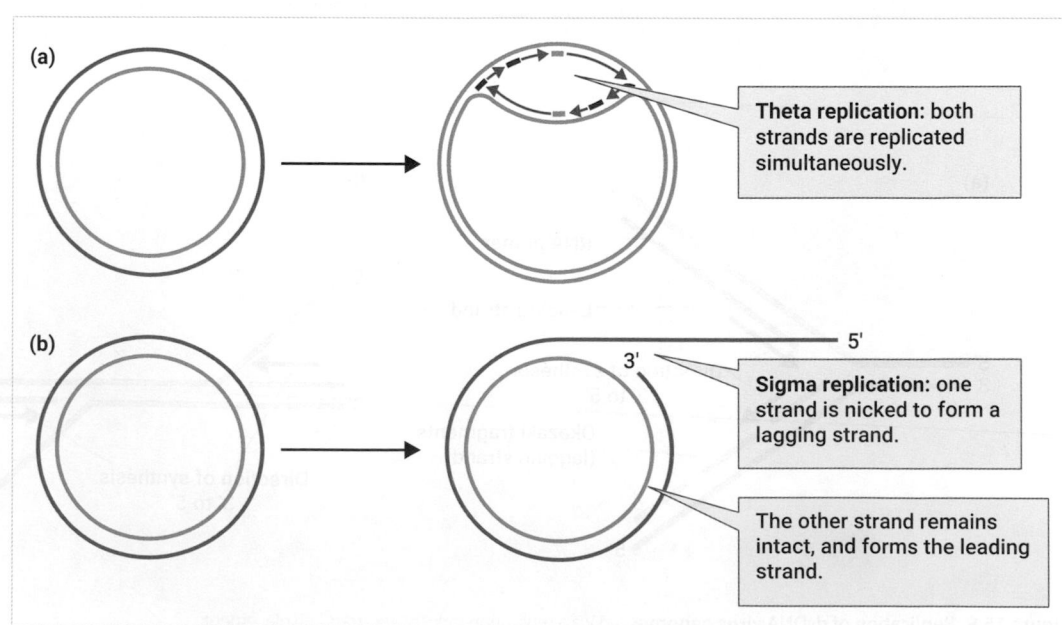

(a)

Theta replication: both strands are replicated simultaneously.

(b)

5′
3′

Sigma replication: one strand is nicked to form a lagging strand.

The other strand remains intact, and forms the leading strand.

Shortly after entering the host cell, the gaps in the positive strand are filled in, creating dsDNA. A protein primer from the negative-sense strand and an oligonucleotide primer from the positive-sense strand are then removed, and the ends are joined to give a covalently closed circular DNA structure (cccDNA).

mRNA transcripts and an mRNA 'pregenome' are produced from the cccDNA. The transcripts are translated into protein and the pregenome serves as an intermediate for genome replication. Viral-encoded reverse transcriptase synthesizes a complementary ssDNA strand, whereupon the ssRNA template is degraded. The nascent DNA strand then serves as the template for a complementary positive-sense DNA strand, producing a new dsDNA genome. This genome may be used as a template for further mRNA transcription, or may be packaged into progeny virions for release.

 Check your understanding of the concepts covered in this section by answering the questions in the e-book.

15.5 Bacteriophages

Bacteriophages (or phages for short) are viruses that exclusively infect bacteria; in fact, their name literally means 'bacteria eaters'. Phages most commonly exhibit what we call a **prolate** structure: an elongated icosahedron and a helical tail (see Figure 15.3c), although simple icosahedral and filamentous forms also exist. Enveloped phages are rare, as the rigid cell wall of host cells means that virions are released by lysis rather than budding.

Upon entering the host cell, a phage can enter one of two cycles, depending on whether it is a virulent or temperate phage. These two cycles are the lytic and lysogenic cycles, shown in Figure 15.13.

The **lysogenic** cycle is typically followed if environmental conditions do not favour release of the phage (e.g. if there is poor nutrient availability and hence few potential hosts). To enter this cycle, the phage may insert its DNA into the host cell genome via the action of viral integrase enzymes, whereupon it becomes known as

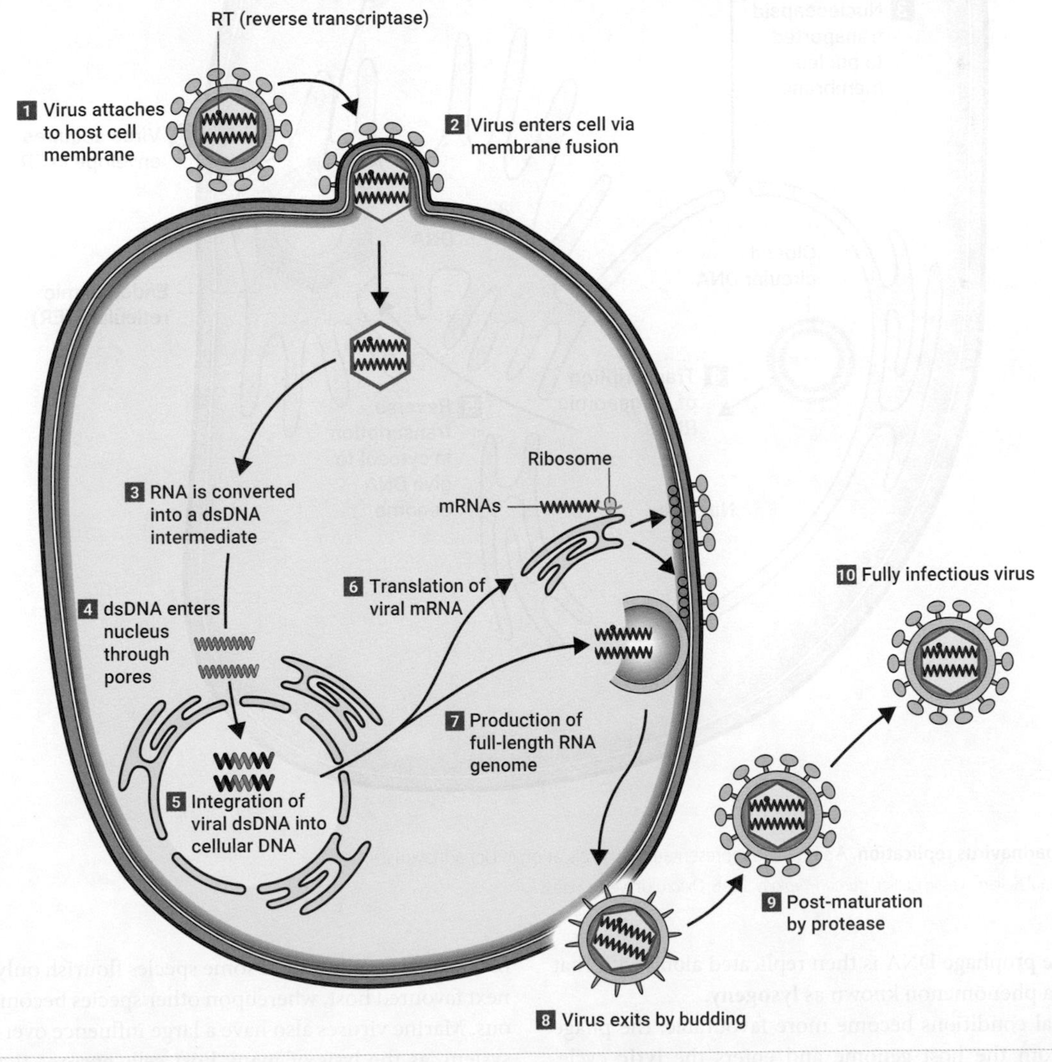

Figure 15.11 Replication in retroviruses. Notice how the viral genome is converted into a DNA intermediate before being integrated into the host cell genome.

Source: John Oxford, Paul Kellam, Leslie Collier, *Human Virology*, 2016, Oxford University Press.

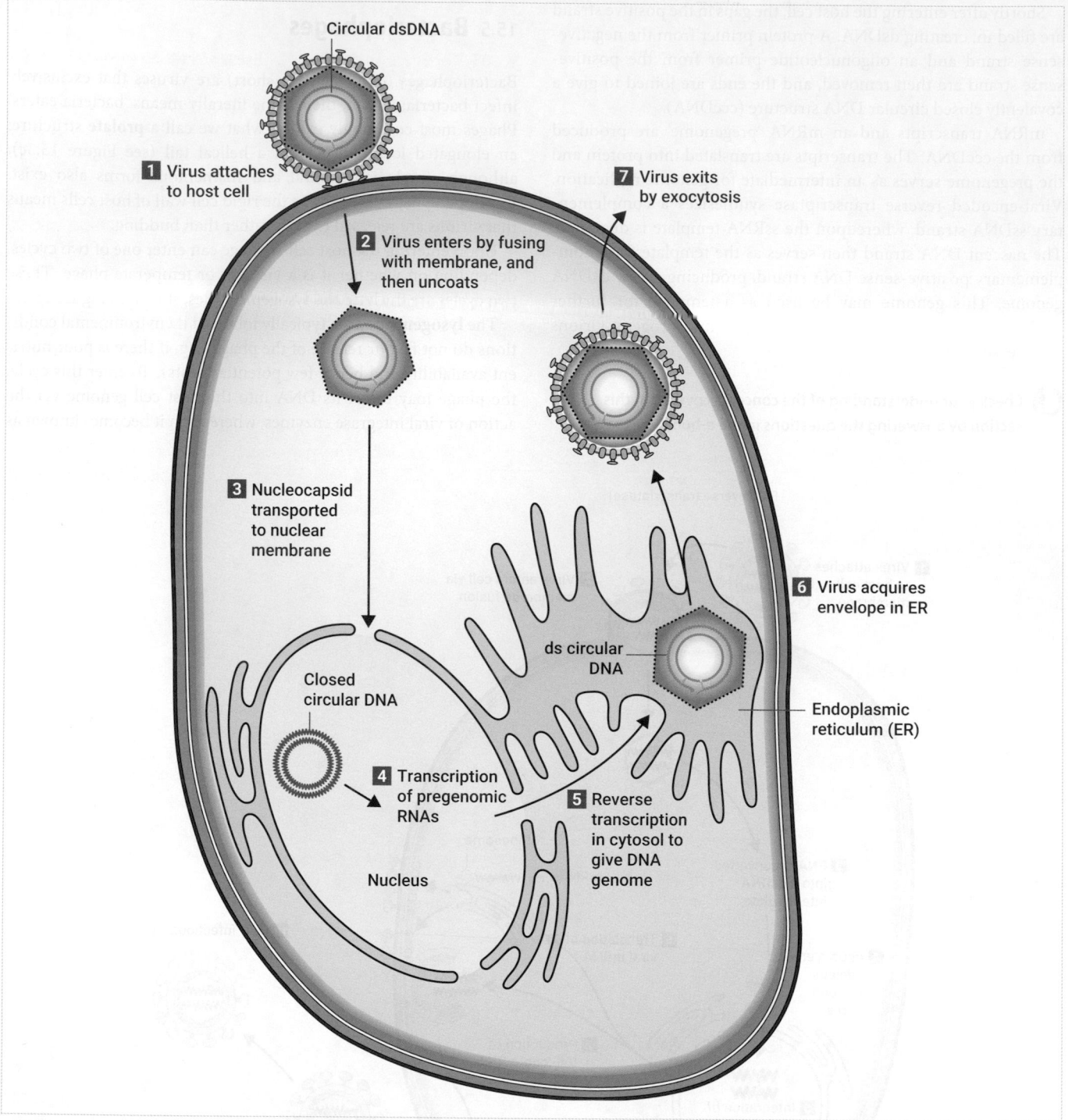

Figure 15.12 Hepadnavirus replication. A simplified representation of replication in Hepadnaviruses.
Source: John Oxford, Paul Kellam, Leslie Collier, *Human Virology*, 2016, Oxford University Press.

a **prophage**. The prophage DNA is then replicated along with that of the host cell, a phenomenon known as **lysogeny**.

When external conditions become more favourable, the phage DNA excises from the host genome and enters the **lytic** cycle, whereby phage genes are expressed and new virions are synthesized, culminating in the death of the host cell and the release of phage progeny. In this way, phages have control over the diversity

of bacterial communities: some species flourish only to become the next favoured host, whereupon other species become more numerous. Marine viruses also have a large influence over the whole ecosystem, as the lysis of many host cells ensures that nutrients are available for other organisms and that the carbon, nitrogen, and phosphorus cycles are maintained for higher life forms in what is essentially a nutrient-limited environment.

Host specificity varies greatly between phage species and some promiscuous phages can even make the jump to different biomes.

The effects of lysis can be observed on an agar plate culture of bacteria, in the form of **plaques**, as illustrated in Figure 15.14. These are zones of clearing in the lawn of bacterial growth where the agar below can be seen, representing the presence of virulent bacteriophages.

Lysogeny is a form of gene transduction. As such, it can have advantageous effects on the host bacterium. For example, phages are known to carry virulence factors between bacterial hosts; *Vibrio cholerae* is only capable of causing disease in humans after it has been infected with a phage carrying the cholera toxin gene.

PAUSE AND THINK

Some phages have a temperate phase in their life cycle, but all phages have a lytic phase in the life cycle. What are these phases?

Answer: Temperate phages can insert their DNA genomes into the genomes of bacteria as a quiescent resident of the genome. Nearly all phage functions are shut off during the temperate phase. When a temperate phage is induced, it enters the lytic phase. During the lytic phase, the phage destroys the host cell and infects nearby host cells. For lytic phage, they lyse the host cell upon infection and cannot insert into the host genome.

🕮 **Check your understanding of the concepts covered in this section by answering the questions in the e-book.**

15.6 **Antiviral therapy**

We have now considered how viruses infect their hosts and subsequently replicate themselves. But how can the impact of such infections on the host organism be minimized? We see in Chapter 14 how antibiotics are used to tackle bacterial infections. Can equivalent therapeutics be used to limit the effects of viral infection? To answer that question, we will now go on to explore the topic of antiviral therapy.

Most viral infections are mild and self-limiting and so do not require treatment but some infections are severe or cause debilitating illness. As such, recent years have seen an increase in research into novel therapeutic agents in an attempt to combat or control these diseases. Viral infections typically responsive to treatment include those caused by herpesviruses, hepatitis B virus, hepatitis C virus, influenza virus, and HIV.

As viruses are essentially inert outside the host cell, the only suitable targets for therapy are actively replicating viruses. The main replicative targets are shown in Figure 15.15. A key priority in the development of any therapeutic agent is to specifically target the harmful agent (in this case the virus), not the host. With this in mind, the more a viral process differs from that of the host, the more specifically a drug can be targeted to that process and the less likely the drug is to be toxic to the patient.

Let us now consider each of these potential targets in more detail.

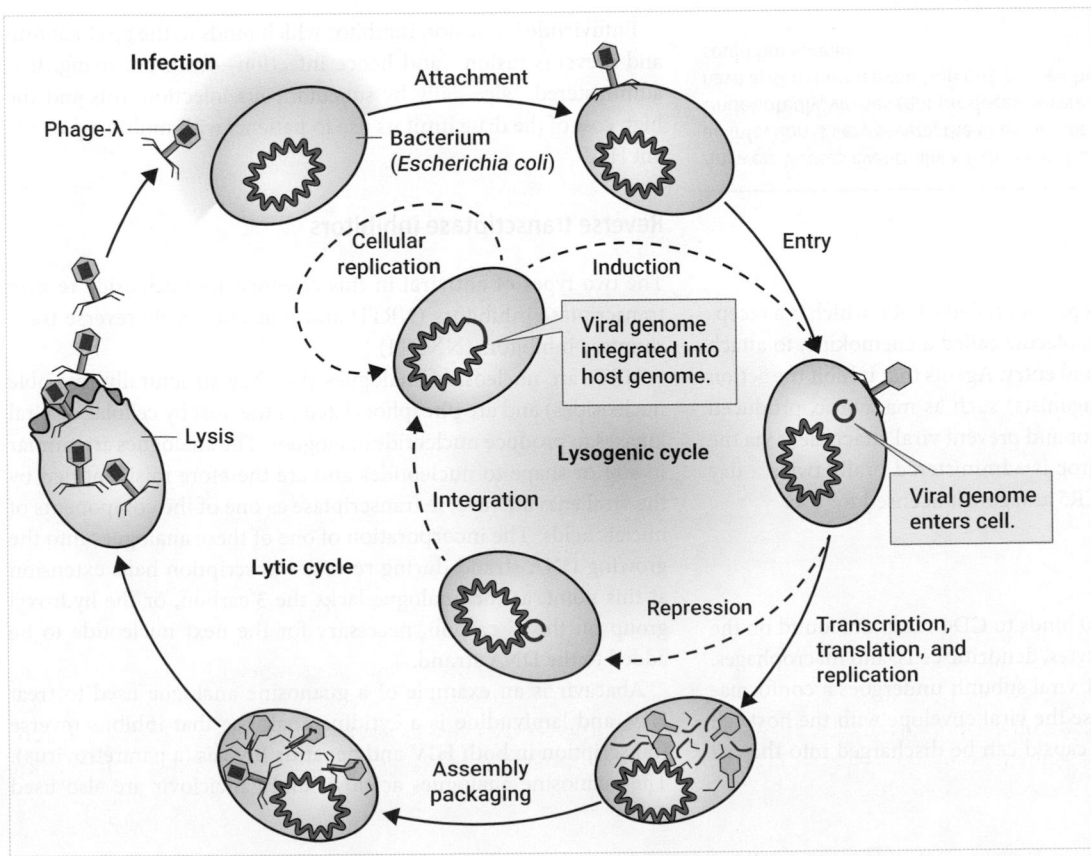

Figure 15.13 The lytic and lysogenic cycles exhibited by bacteriophages.

Source: Reprinted by permission from Springer, *Nat. Rev. Genet.* 4, 471–477, Campbell, A. The future of bacteriophage biology. Copyright 2003.

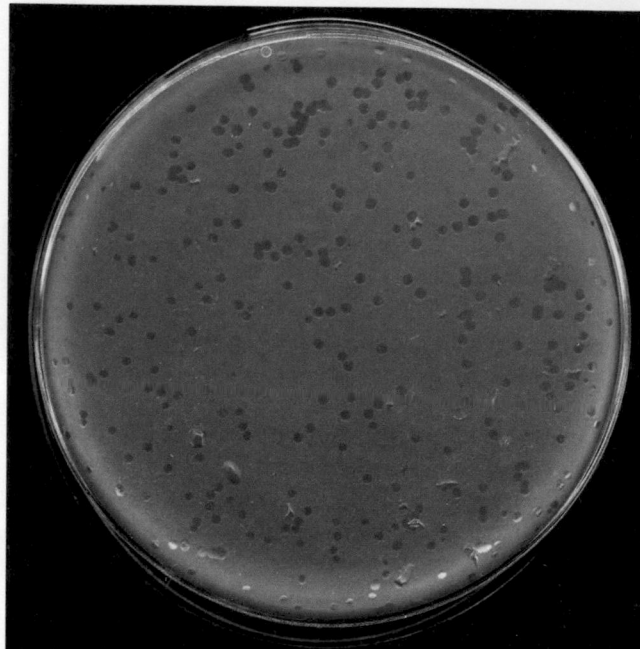

Figure 15.14 Bacteriophage plaques on a lawn of bacterial growth. The growth of the bacterial cells is inhibited where there are bacteriophages present; these manifest as 'holes' in the lawn of growth where the agar below can be seen.
Source: Madboy/Wikimedia Commons/CC BY-SA 3.0.

PAUSE AND THINK

Why are most viral infections 'mild and self-limiting'?

Answer: Viruses eventually kill the cells that they infect. If the outcome of an infection is very severe, the virus cannot spread and infect a new host. Undoubtedly, viruses that produced severe and immediate cell death have arisen during evolution but, as their hosts quickly died, the viruses could not spread.

Attachment inhibitors

In humans, HIV virus uses a protein called CCR5, which is a receptor for a type of signalling molecule called a chemokine, to attach to host cells prior to fusion and entry. Agents that inhibit the action of CCR5 (called CCR5 antagonists) such as maraviroc, produced by Pfizer, bind to this receptor and prevent viral attachment via the glycoprotein gp120. Maraviroc is administered orally twice a day and is currently the only CCR5 antagonist licensed for use.

Fusion inhibitors

The HIV glycoprotein gp120 binds to CD4 receptors found on the surface of helper-T lymphocytes, dendritic cells, and macrophages. Following binding, the gp41 viral subunit undergoes a conformational change in order to fuse the viral envelope with the host cell membrane so that the viral capsid can be discharged into the cell interior.

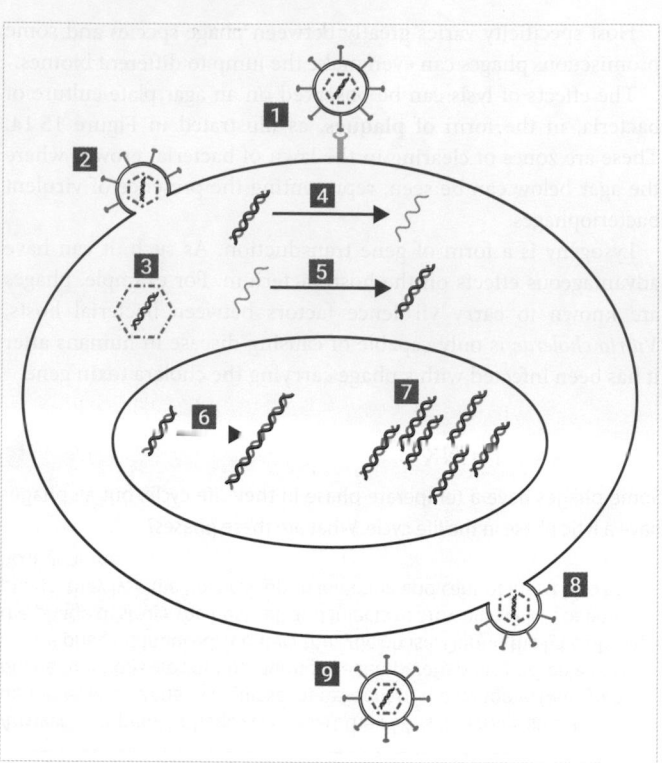

Figure 15.15 Potential targets for antiviral drugs. (1,8) Attachment; (2,9) fusion; (3) uncoating; (4) transcription; (5) reverse transcription; (6) integration; (7) genome replication.
Source: From Carter & Saunders (2013) Virology: *Principles and Applications*. John Wiley & Sons.

Enfuvirtide is a fusion inhibitor which binds to the gp41 subunit and prevents fusion—and hence infection—from occurring. It is administered twice daily by subcutaneous injection; this and the high cost of the drug limit its use to patients with multidrug resistant HIV.

Reverse transcriptase inhibitors

The two types of antiviral in this category are nucleoside reverse transcriptase inhibitors (NRTI) and non-nucleoside reverse transcriptase inhibitors (NNRTI).

NRTI are nucleoside analogues (i.e. they structurally resemble nucleosides) and are phosphorylated in the host by cellular or viral kinases to produce nucleotide analogues. The analogues are similar in size or shape to nucleotides and are therefore misidentified by the viral enzyme reverse transcriptase as one of the components of nucleic acids. The incorporation of one of these analogues into the growing DNA strand during reverse transcription halts extension at this point, as the analogue lacks the 3′carbon, or the hydroxyl group on the 3′ carbon, necessary for the next nucleotide to be added to the DNA strand.

Abacavir is an example of a guanosine analogue used to treat HIV, and lamivudine is a cytidine analogue that inhibits reverse transcription in both HIV and hepatitis B virus (a pararetrovirus). The guanosine analogues aciclovir and ganciclovir are also used to inhibit genome replication in herpesviruses, and ribavirin can be used to treat hepatitis C virus by inhibition of RNA replication.

NNRTI may work in synergy with NRTI to prevent reverse transcription as they work at a site on the enzyme that is distant from the active site. Nevirapine is an NNRTI used in antiretroviral therapy but side-effects are common and include skin rashes, nausea, and hepatotoxicity.

Integrase inhibitors

Viral integrases are enzymes that catalyse the insertion of viral dsDNA into the host genome, thereby creating a provirus. The final step in the insertion sequence is strand transfer, whereby the viral DNA 3′ ends are covalently linked to the host DNA strands. Dolutegravir is an HIV integrase strand transfer inhibitor: it blocks the enzyme active site and prevents this final step from taking place. This drug is used for first-line treatment of HIV in the UK,

as recommended by the World Health Organization, and is also available as a single-pill combination drug along with abacavir and lamivudine.

Protease inhibitors

Following capsid assembly, HIV virions undergo a maturation step that takes place outside the host cell following budding from the surface. Two polyproteins, Gag and Pol, are cleaved via proteases to produce active forms of the proteins; Gag comprises the four structural capsid proteins and Pol comprises the three enzymes reverse transcriptase, integrase, and protease. Inhibition of the maturation step by protease inhibitors such as ritonavir gives rise to non-infectious virions.

We learn more about the treatment of HIV in Real World View 15.1.

 Check your understanding of the concepts covered in this section by answering the questions in the e-book.

REAL WORLD VIEW 15.1 | Treating HIV

HIV is primarily a sexually transmitted virus that is found in a variety of bodily fluids, such as semen, vaginal secretions, blood, and breast milk. It can also be transmitted via contaminated needles, for example in people who inject drugs, and from mother to child during childbirth or breast-feeding. The virus infects cells with a cluster of differentiation 4 (CD4) receptor on their surface, most of which are cells of the immune system such as helper T lymphocytes.

After initial infection, infected cells are destroyed and replaced without there being any outward symptoms. This is described as a period of clinical latency and is a period that may last for up to 10 years. Eventually, however, the CD4 count of infected individuals begins to fall, with a synchronous rise in viral load—the amount of virus present in the body.

Damage to the immune system as a result of the decline in the CD4 cell population leaves the individual open to opportunistic infections and certain cancers, at which point they are said to have developed acquired immune deficiency syndrome (AIDS). Without appropriate therapy the weakening of the immune system has a devastating effect and AIDS is inevitably fatal.

The weakening of the immune system caused by HIV infection makes individuals particularly vulnerable to certain diseases, such as *Pneumocystis jirovecii* pneumonia, tuberculosis, Kaposi's sarcoma, and oesophageal candidiasis. These are known as AIDS-defining illnesses and treatment of these is separate from the treatment of the underlying viral infection.

Antiretroviral therapy (ART) is the therapeutic regimen used for the treatment of HIV. Highly active antiretroviral therapy (HAART), also often known as combination antiretroviral therapy (cART), sees the administration of three or more antiretroviral drugs to prevent the onset of resistance. This resistance is a consequence of the fact that HIV replicates very quickly: it produces billions of new viruses every day in each infected individual and has a high mutation rate. In fact, the mutation rate is so high that most viral progeny are non-infectious, but among the many viable offspring there is a high chance of drug resistance arising.

In the early days of ART the use of one drug quickly led to resistance, and so now three drugs are used in combination. The chance of a mutation occurring that provides resistance to all three drugs simultaneously is extremely small. However, the recent introduction of two-drug combination regimens and single-pill dual therapy drugs means that the terms HAART and cART are being used less frequently.

The timing of ART is down to local policy. For example, treatment may be made available to individuals when their CD4 count drops to below 350 cells/mm³, or when they develop an AIDS-defining illness. Serodiscordant couples—those in which one partner has the virus (seropositive) and the other does not (seronegative)—may both be offered ART to prevent onward transmission to the seronegative partner.

Some exciting research over the past decade, known as the PARTNER studies, has shown that the risk of a seropositive individual who has been virally suppressed through taking ART transmitting the virus to their partner is effectively zero. In the seronegative partner, preventative drug therapy is known as pre-exposure prophylaxis (PrEP).

NHS England now recommends that all seropositive individuals begin ART immediately following diagnosis so that viral replication is suppressed as early in infection as possible to avoid damage to the immune system. Such early treatment prevents the onset of AIDS and, increasingly, people with HIV can expect to have a normal life expectancy. Unfortunately, ART must be lifelong as the patient cannot be cured of HIV: the viral genome integrates into the human host cell's genome as a provirus that cannot be expunged.

Despite this, there have been two reports of individuals with HIV being virally suppressed *without* ART, following bone marrow transplants; these exciting findings provide hope that ongoing research may give rise to a cure. Until such a time, combating the disease depends on education and ART to prevent transmission; it is hoped that one day a vaccine will be available but this goal has thus far proven elusive due in part to the mutable nature of the virus.

15.7 Viruses beyond disease

As viruses are by their very nature exclusively pathogenic it is difficult to think of them in a positive light. However, humankind has exploited viruses for centuries in the form of vaccines, using both killed and attenuated viruses to elicit an immune response that can then provide lifelong immunity to infection. In this section we consider examples of how viruses have been used other than as agents of disease.

Phage typing

Phage typing is the testing of bacterial strains against different bacteriophages to identify which specific hosts the phages can infect. In this way, bacterial strains can be given a 'phage type' based on their susceptibility to different phages and relatedness between strains can be determined. This is important in investigating outbreaks of bacterial infections, such as food poisoning.

To determine the phage type, a suspension of the bacterium in question is spread over the surface of an agar plate, but a series of phage suspensions are dropped onto the plate in a predetermined order before incubation (see Figure 15.16). After the bacteria have been allowed to grow overnight, plaques in the culture identify which phages have successfully infected the bacterial cells and the pattern of infection gives the phage type.

Phage therapy

Antimicrobial resistance is on the rise and the search for alternative therapies is a current hot topic in biomedical research. The bactericidal activity of virulent phages has the potential to be harnessed as a valid treatment option, as the host specificity of the phages means that there is no risk of disease in the patient, and the disruption of a person's normal microbiota (e.g. their gut microbiome) as a result of treatment is likely to be minimal. Popular targets include gut microbes, antibiotic-resistant species, and lung infections in individuals with cystic fibrosis.

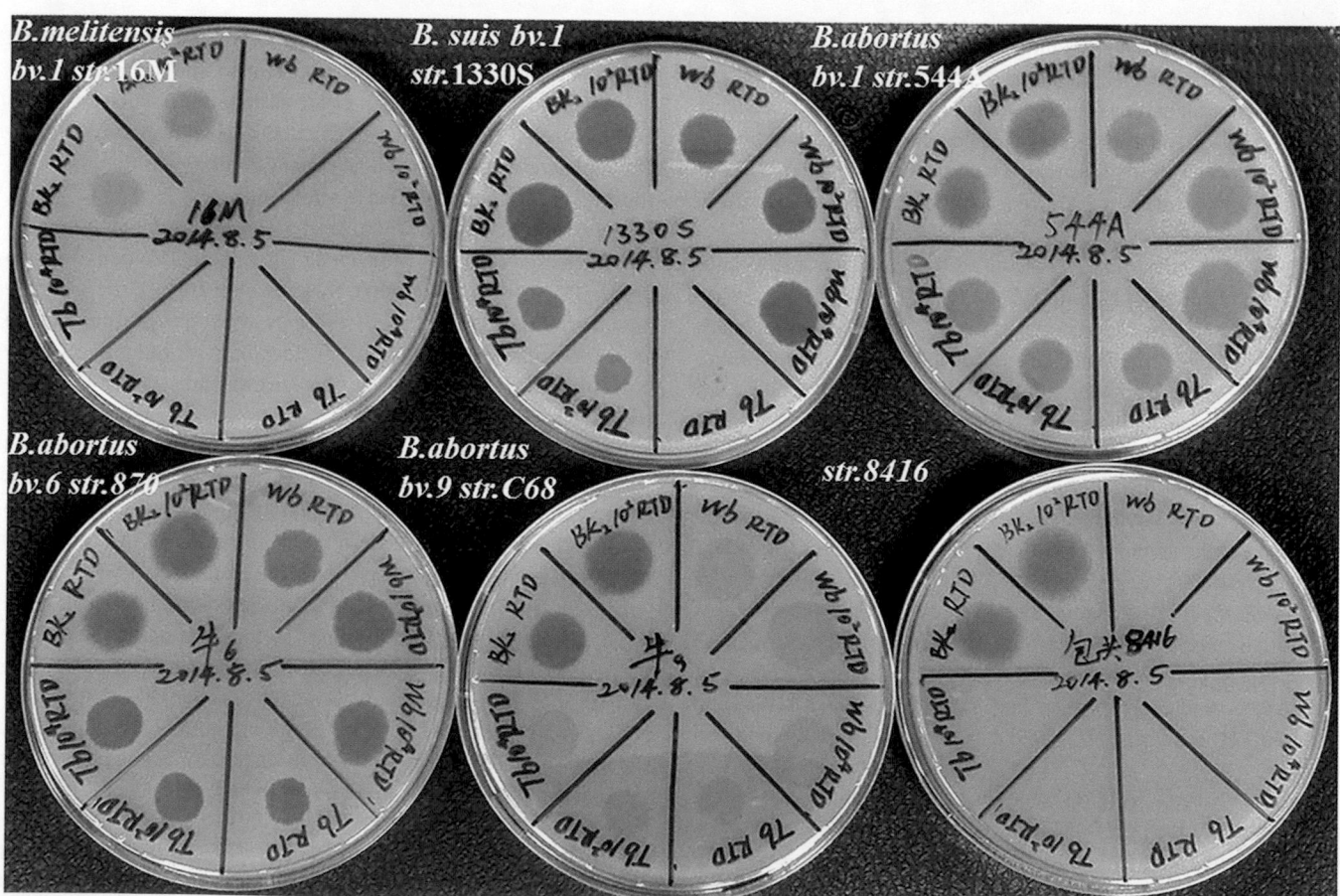

Figure 15.16 Phage typing. A series of different phage suspensions have been dropped into a bacterial culture before incubation. The pattern of plaque formation gives the phage type for this particular bacterial strain.

Source: Frontiers in Microbiology, volume 6 2015, p. 1375. Kang Yao-Xia et al. Typing Discrepancy Between Phenotypic and Molecular Characterization Revealing an Emerging Biovar 9 Variant of Smooth Phage-Resistant *B. abortus* Strain 8416 in China, © 2015 Kang, Li, Piao, Tian, Jiang, Jia, Lin, Cui, Chang, Guo and Zhu, CC BY 4.0.

Therapeutic doses can be low, as the phage will multiply in accordance with the number of bacterial cells present and there is low toxicity to the patient as the phage is composed of biological macromolecules.

Recent research has shown that the lysis of bacteria can further stimulate the immune system and enhance clearance of the pathogen. However, the phage itself will also elicit an immune response, which could prevent future treatment from being successful if used systemically.

Pest control

Viruses can be used for biological pest control (e.g. as viral insecticides) and there are a number of species in the family *Baculoviridae* that in some countries are already being used for this purpose. Again, the host specificity of viruses prevents the loss of insect species that we do not want to target, and there is minimal impact on the environment (unlike with the use of chemical pesticides). Around one-fifth of the world's crop losses every year is due to insects, making this research another currently important and expanding area.

The use of viruses for biological control raises certain ethical issues, however—both for the culling of pests and also for the deliberate release of a (possibly genetically modified) virus into the environment. Real World View 15.2 discusses this issue further.

PAUSE AND THINK

What is the common and shared property of viruses that has made them potentially useful for phage typing, phage therapy, and pest control?

Answer: All three of these applications rely on the specificity between a virus and its host.

Molecular biology

A virus that integrates its nucleic acids into the host's genome could in theory be used as a vehicle to deliver novel genes into that host. This type of genetic engineering is known as **gene therapy**, and could be used to replace faulty genes such as *CFTR* in patients with cystic fibrosis.

As with phage therapy, a way must be found to minimize the immune response to the virus or the therapy will quickly become ineffective. In addition, the placement of the gene within the host genome may be problematic: it must be inserted into an area which is easily accessible to the transcription machinery of the cell, and not in a location where it will do harm—for example, in the middle of another gene.

Many assays used in DNA research and diagnostics require enzymes that are not found in eukaryotic and prokaryotic cells. For example, anyone wishing to synthesize ds DNA from ss RNA *in vitro* will need an RNA-dependent DNA polymerase to do so. This enzyme is rare in nature but exists in retroviruses in the form of reverse transcriptase. Experimental Toolkit 15.1 discusses this further.

Other virally derived enzymes that are commonly used in molecular biology research include restriction enzymes, which cut nucleotide sequences, and ligases, which repair cuts and join sequences together. One of the best characterized RNA ligases is the T4-bacteriophage RNA ligase, although other thermostable phage ligases are also popular for experiments which require high temperature conditions.

 Check your understanding of the concepts covered in this section by answering the questions in the e-book.

REAL WORLD VIEW 15.2 — Myxomatosis

European rabbits were first introduced to Australia in the 18th century, primarily for food. However, they were later released into the wild for sport. Rabbits breed prolifically and their numbers were soon out of control, causing huge crop losses and devastation to the local ecology due to overgrazing. Numerous measures were employed in an attempt to control the population size, including building a 2000-mile-long rabbit-proof fence across Western Australia in 1901. Unfortunately the fence barely slowed the advancing rabbit population, and towards the end of the late 19th century a biological means of control was sought.

In 1950 the myxoma virus was introduced into Australia; this virus was first identified in Uruguay and causes myxomatosis in rabbits. The symptoms of this disease are primarily tumours and blindness, followed by death in around 4–14 days. Initially, the rabbit population was devastated by the disease, which has a case fatality rate of up to 90 per cent. However, the remaining rabbits had some partial immunity due to genetic resistance, and these rabbits now make up most of the (almost completely recovered) population.

The virus was also introduced to Europe, reaching the UK in 1953. Despite an initial reduction in rabbit numbers of around 99 per cent, the population has now reached its previous level again.

Reverse transcription polymerase chain reaction

The polymerase chain reaction (PCR) is a technique that was developed in 1983 by Kary Mullis for the amplification of DNA. The technique, which we discuss in Chapter 8 and which is illustrated in Figure 1, involves denaturation of the dsDNA at around 94–98 °C, followed by an annealing step at 50–65 °C during which single-stranded primers are added to the 3′ ends of the target sequence, and finally an elongation step at 72 °C whereby thermostable polymerases synthesize a DNA strand complementary to the template strand. Each step takes only a few seconds and leads to an exponential increase in the amount of target DNA sequence present.

Before this technique was developed, large amounts of starting material were needed to obtain large amounts of DNA for study—something that is not always possible with non-culturable viruses. So revolutionary

was this technique that it took only 10 years for Mullis' contribution to be recognized with a Nobel Prize in Chemistry; today, PCR is used in research and diagnostic laboratories the world over.

The polymerase used in the elongation step, however, is DNA-dependent, meaning that PCR cannot be used for the compilation of gene expression libraries from mRNA or the amplification of viral RNA genomes. This limitation led to the development of reverse transcription PCR (RT-PCR). This technique features an initial reverse transcription step whereby ss RNA is copied into complementary DNA (cDNA) before entering the thermal cycling of PCR proper. In this way, we see virally derived enzymes being used in the research and diagnosis of viral disease.

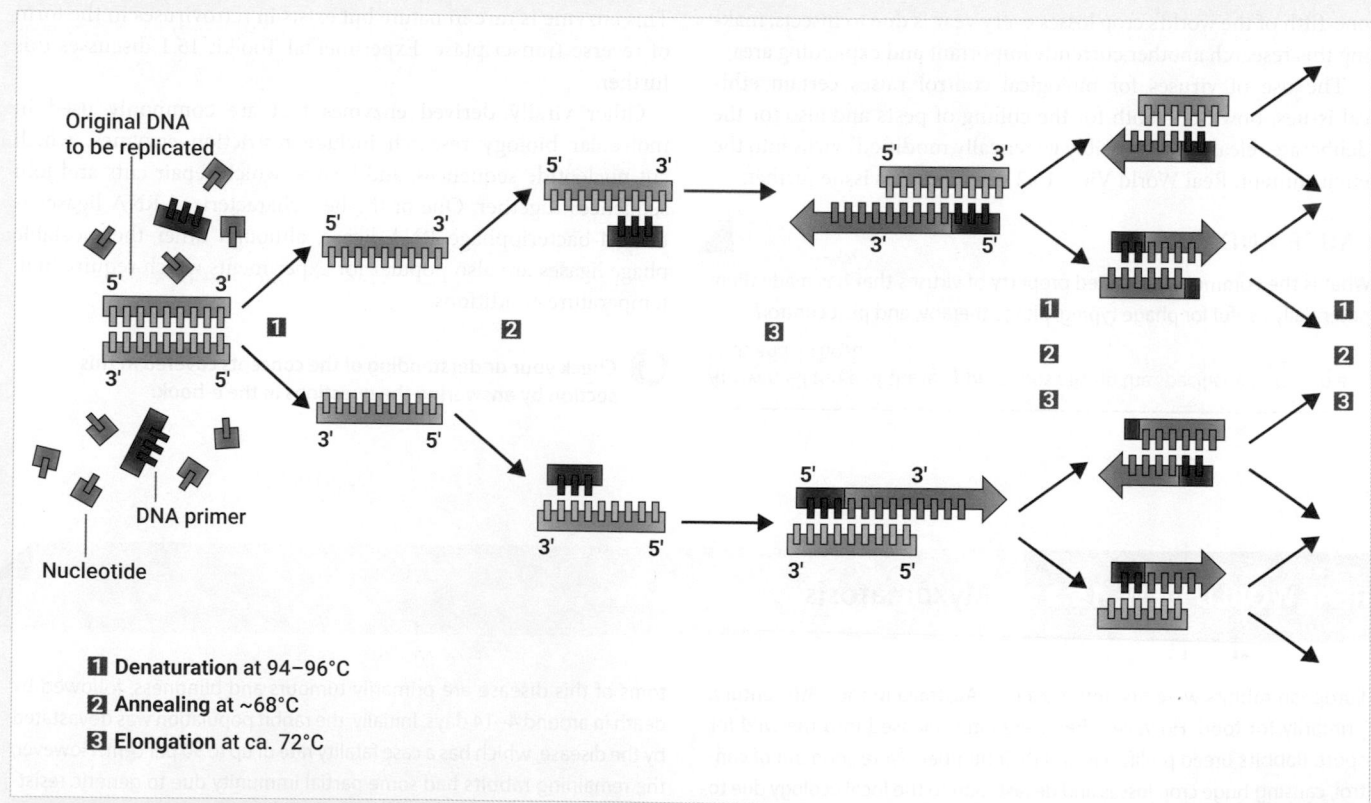

1 Denaturation at 94–96°C

2 Annealing at ~68°C

3 Elongation at ca. 72°C

Figure 1 The polymerase chain reaction. Cycling between the denaturation, annealing, and elongation steps—all at different temperatures—leads to the synthesis of millions of copies of the target DNA sequence.

Source: Enzoklop/Wikimedia Commons/CC BY-SA 3.0.

SUMMARY OF KEY CONCEPTS

- Viruses cannot be said to be living things, as they do not comply with any of the definitions of life. Nevertheless, viruses are many times more abundant than any living organism in the world.

- Viruses are entirely parasitic and are dependent on a host cell for replication. They have no metabolism and are completely inert outside of the host cell.

- Viruses contribute to a considerable disease burden, as well as economic losses.

- In its simplest form, a virion may consist solely of a protein capsid and nucleic acids. However, many viruses have an envelope derived from host cell membranes or internal enzymes to assist with establishing infection in a new host.

- Capsid symmetry is generally either icosahedral or helical. Helical capsids can be rod-shaped or filamentous.

- Viruses commonly have membrane- or capsid-associated glycoprotein spikes, which facilitate attachment and entry into host cells. These glycoproteins also determine directionality by attaching only to certain cell surface receptors.

- Viral genomes can be either DNA or RNA (but not both) and can be either double-stranded or single-stranded. Single-stranded genomes may be positive sense or negative sense. The genome can also be linear or circular.

- Entry into the host cell is usually by receptor-mediated endocytosis in mammalian viruses, but can also be by membrane fusion.

- Transcription of viral nucleic acids defies the 'central dogma' of molecular biology and the process varies according to the genome composition.

- Viruses may leave the host cell via lysis, which kills the host cell, or budding from the cell surface, which does not kill the host cell.

- DNA genomes may be replicated via formation of a replication fork, as occurs in eukaryotes, and strand displacement. RNA viruses replicate via a DNA intermediate and retroviruses replicate via an integrated provirus.

- Bacteriophages are viruses that infect bacteria. They can undergo a lytic or lysogenic cycle depending on environmental conditions.

- Antiviral therapy targets steps involved in the replication of viruses, although it is not commonly used.

- Despite the pathogenic nature of viruses, we have many uses for them—for example, in phage typing, phage therapy, pest control, gene therapy, molecular research, and diagnostic assays.

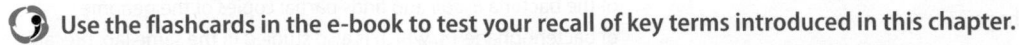 Use the flashcards in the e-book to test your recall of key terms introduced in this chapter.

QUESTIONS

Concepts and definitions

1. Define capsomere.

2. What is reverse transcriptase? Which categories of viruses are the most likely to encode a reverse transcriptase and which are the least likely?

3. Define tropism.

4. What is a provirus?

5. What is a bacteriophage and is it different from a virus?

6. Define what is meant by a nucleoside analogue and give an example.

Apply the concepts

7. Chemical analysis of the base composition of the genomes of the isolates of unknown viruses yielded the following results. Based on these base compositions, what type of virus is each one likely to be?

Isolate	A (%)	C (%)	G (%)	T (%)	U (%)	Type of virus
1	29	23	27	21	0	
2	24	28	26	0	22	
3	28	22	22	28		
4	21	29	29	0	21	

8. What is the function of the spike protein?

9. What is the difference between an RNA-dependent RNA polymerase and a reverse transcriptase? What types of viruses produce each type of enzyme?

10. Although viruses have small genomes, they often make more polypeptides than expected. What are some evolutionary strategies seen in viruses such as HIV to encode numerous different polypeptides from the same genome, that is, to increase the coding capacity of the viral genome?

11. Multiple choice: Figure 15.14 shows a lawn of bacteria with plaques or clear areas in it. If one wanted to isolate copies of the phage in order to perform electron microscopy, which is the best location from which to draw a sample?

    (a) The intact lawn.

    (b) From the centre of an area with large plaques.

    (c) Either the lawn or the plaque.

    (d) From the edge of an area with a few widely separated small plaques.

12. The bacteriophage lambda is a temperate phage. What does this imply about the lambda life cycle?

13. Many enzymes widely used in experimental molecular biology laboratories are derived from viruses. What is the function of these enzymes in the virus?

    (a) DNA ligase

    (b) Integrase

    (c) RNA polymerase from the phage T7

Beyond the concepts

14. Molecular changes occur constantly in viruses, as in other organisms, and can affect any part of the virion. For SARS-CoV-2, the virus responsible for Covid-19, most of the public attention has been focused on changes that occur in the spike protein. As described in Clinical Box 15.1, vaccines have been directed specifically against parts of the spike protein.

    (a) What are the likely outcomes of changes in a spike protein?

    (b) Why is the spike protein such a good target for vaccine development?

    (c) Do you think that additional changes in the spike protein may result in different or 'booster' vaccines that perhaps recognize somewhat different targets within the spike protein?

15. Multiple choice: You may wish to review Chapter 9. Which of these macromolecules is the most likely to be found in a viral envelope? Be sure to explain your reasoning.

    (a) RNA polymerase

    (b) Sphingolipid

    (c) Protein kinase

    (d) Glucose 6-phosphate dehydrogenase.

16. Bacteriophage T4 is a lytic virus that infects the bacteria *E. coli* and has been extensively studied. One of the major structural proteins in the capsid or head of the virus T4 is called gp23. Is gp23 expected to be transcribed and translated as an early protein or a late protein in the T4 life cycle and why?

17. Many viruses depend on the presence of host cell proteins on the cell surface for attachment. As viruses kill the infected cell, why are these receptor proteins present on the host cell?

18. Sequence analysis of the genomes of nearly all multicellular organisms have found many copies of what appear to be viral genomes, which may or may not be active viruses. What might be an example of a coding region that serves as a signature of a viral genome?

19. DNA replication of circular genomes in dsDNA viruses can occur by two mechanisms, known as theta and sigma for the shape of the replicating molecule. Which of these is more similar to what occurs in an organism with a linear genome?

20. Monique is a student who is isolating DNA from a culture of the bacteria *E. coli* and finds partial copies of the genome of bacteriophage T4, which is also studied in the same lab. (Recall that T4 is a lytic phage that infects *E. coli*.) Her lab mates offer different hypotheses about the unexpected presence of the viral genome in her bacterial DNA preparation. Anna wonders if a piece of lab glassware or a plastic tip was contaminated and Monique might have been careless. Brian thinks that T4 might have inserted into the *E. coli* genome. Charlene wonders if some T4 infected the growing *E. coli* culture that Monique used. Which of these suggestions is the least likely and why?

21. Phage therapy is considered a less expensive alternative to antibiotics as a means to combat bacterial infections. What is the primary limitation in the widespread use of phage therapy?

FURTHER READING

Carter, J. & Saunders, V. (2013) *Virology: Principles and Applications*. 2nd edition. John Wiley & Sons.
A good general introduction to undergraduate virology.

Flint, S. J., Racaniello, V. R., Rall, G. F., & Skalka, A. M. (2015) *Principles of Virology*. 4th edition. ASM Press.
An extensive two-volume set, covering both molecular biology (volume 1) and pathogenesis and control (volume 2).

Knipe, D. M. & Howley, P. M. (2013) *Fields Virology*. 6th edition. Lippincott Williams and Wilkins.
Probably the most comprehensive virology reference book, suitable for both undergraduate and postgraduate students.

Burrell, C. J., Howard, C. R., & Murphy, F. A. (2017) Chapter 38: Prions. In *Fenner and White's Medical Virology*. 5th edition. Academic Press.
At the start of this chapter we said viruses were 'one of the smallest agents of disease'. Infectious proteins, known as prions, are actually smaller than viruses and have been linked to diseases such as bovine spongiform encephalopathy (BSE) and Creutzfeldt-Jakob disease (CJD). You can find out more about them in this book.

The Human Organism
Tissues, Organs, and Systems

16   An Overview of Physiology

17   Communication and Control 1: The Nervous and Endocrine Systems

18   Communication and Control 2: Sensory Systems

19   Communication and Control 3: Controlling Organ Systems

20   Muscle and Movement

21   Cardiovascular System

22   Respiratory System

23   Exercise Physiology

24   Renal System

25   Digestive System

26   Reproductive System

27   Immune System

MODULE THREE

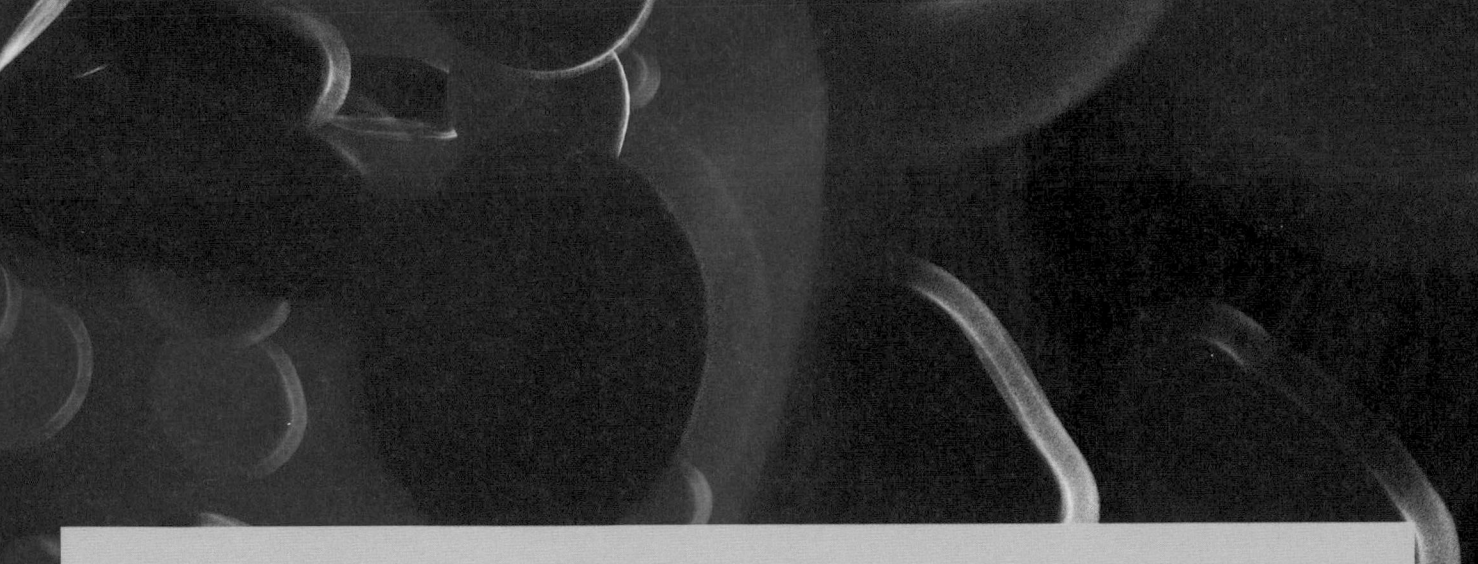

Blood pressure is of major clinical importance: the British Heart Foundation reported in 2019 that an estimated 14.5 million adults aged 16 or older—that is, about 26% of the UK's adult population—had high blood pressure (known more formally as hypertension) and with it an increased risk of potentially fatal conditions such as heart disease and stroke.

The regulation of blood pressure is a classic example of the integration of a whole range of organ systems. Your heart rate and the force of each heartbeat have a major influence upon blood pressure and these are tightly regulated in the short term via monitoring sense-receptors located in key arteries such as the aorta and the carotid. These receptors feed into the autonomic nervous system.

But blood pressure is also affected by how easily or otherwise blood is able to flow through the blood vessels— something that can be controlled by the contraction or relaxation of the smooth muscle walls of the blood vessels. This state of contraction or relaxation is under both local and body-wide control. However, this flow can be compromised by the narrowing of the blood vessels as a result of a disease called atherosclerosis, which in turn can add to the loading on the heart.

Other factors are at play too. For example, our kidneys play an important role in regulating fluid volumes which, in turn, affect blood pressure.

The control of blood pressure throughout the body and the local regulation of blood flow to different body regions involves the continual integration of a complex variety of body systems, regulated by both the autonomic nervous system and the endocrine system. (We only have to imagine the potential changes in demand placed on the body during exercise to appreciate the complexity of this regulation.)

The understanding of how this body system—and all the other body systems— operate together to maintain the processes essential for life is what physiology is all about, and is what we will be exploring in this module.

Physiology is fascinating because it takes us all the way from molecular biology to the whole animal—for example, how the properties of proteins are important in determining the processes of signalling between cells, or why genetics is significant when it comes to explaining why some people might be at higher risk of developing hypertension than others. In this module we will focus on the physiology of mammals, predominantly humans, but the topics interleave closely with the other modules of this book.

Image: Blood cells are pumped round our bodies under carefully regulated pressure. *Source:* Sebastian Kaulitzki/Science Photo Library.

An Overview of Physiology

LEARNING OBJECTIVES

By the end of this chapter you should be able to:

- Explain the range of physiology as an area of study.
- Discuss the concept and significance of homeostasis.
- Describe the structure of the cell membrane.
- Understand the different routes by which substances can be transported across the cell membrane.
- Explain the principles of diffusion and factors affecting the rate of diffusion.
- Describe the mechanisms of active transport across the cell membrane.
- Understand the role of transport mechanisms in cystic fibrosis.

Chapter contents

Introduction   581
16.1   Homeostasis   582
16.2   The cell membrane   584
16.3   Conclusion   592

Introduction

Physiology is a broad subject area that, as with much of biology, does not have clearly defined boundaries. The *Oxford Dictionary of English* (3rd edition) defines physiology as:

> The branch of biology that deals with the normal functions of living organisms and their parts.

and

> The way in which a living organism or bodily part functions.

Physiology therefore incorporates the study of all living organisms, but this module will specifically focus on human physiology. There will, though, be clear links with Module 4 on organismal biology as we will be exploring common themes, but Module 4 will take a much broader perspective.

In this module we will build on the material covered in Module 1 on molecular science and Module 2 on cellular science to take us from the subcellular through to organ systems and set the scene for integration in the whole organism. We focus especially on communication and control as the building blocks for the systems of the body. This allows us to look at how the different systems and mechanisms in the body interact and integrate to maintain a state of dynamic equilibrium. Maintenance of this dynamic equilibrium engages all the body systems enabling key variables such as temperature, pH, intracellular (inside the cell) and extracellular (outside the cell) concentrations of ions, water balance, and so on, all to be maintained within tight limits. There will also be links back to the molecular science and cellular science modules through the discussion of the role of the cell membrane as both the gatekeeper and substrate of communication mechanisms between cells.

The understanding of human physiology is vital to medical science, for example our appreciation of how blood pressure is regulated and how the body responds to the different demands of activity and exercise. Physiology reflects much of our conscious perception of the internal and external world: we are often aware of changes taking place within our bodies, for example the increases in heart rate and in breathing when we exercise, or of the pounding of our heart when we are scared. We are also very aware of our external environment: changes in temperature that make us feel hot or cold, for example, or our special senses that allow us to see, smell, and taste the food we eat, or to hear each other speak.

16.1 Homeostasis

One of the fundamental concepts that underpin our exploration of physiology is that of **homeostasis**: the maintenance of the internal environment within very tight boundaries.

Walter Cannon worked as a military physician in World War I. Based on his experiences he coined the phrase 'fight or flight' to describe the responses an animal makes to life-threatening challenges (see Section 19.1). He also built on the work of the 19th-century French physiologist, Claude Bernard. Bernard stated that 'a free and independent existence is possible only because of the stability of the internal milieu' (see Further reading, Modell, H., et al. (2015)). Cannon developed this view into the concept of homeostasis.

In physiological terms, homeostasis operates at a range of levels: the internal environment of a cell, of an organ system, and of the body as a whole. As you may know from molecular biology, many processes are very sensitive to changes in the local environment. For example, many enzymes will cease to function effectively if the pH or temperature change by more than a small margin (see Section 6.4). This regulation of the internal environment of the body is achieved by the operation of a remarkable variety of systems and processes that continually monitor the state of the body. If that state changes, the systems act to reverse that change in the internal environment, restoring the original condition.

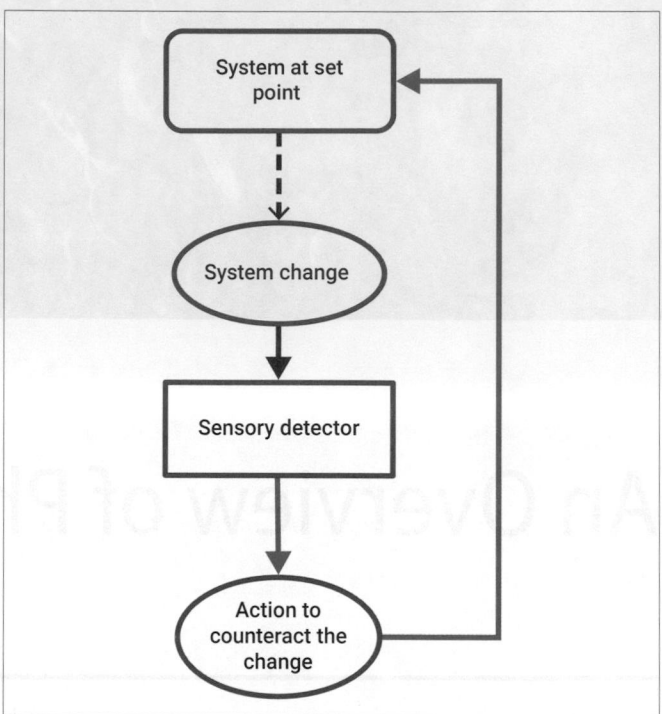

Figure 16.1 Flow diagram illustrating negative feedback. The system is monitored by a sensory detector. When there is a disturbance such that the system moves away from the set point, that change is detected by the sensory detector and an action is triggered to counteract the change and restore the set point.

The core process underpinning homeostasis is a **negative feedback** mechanism (Figure 16.1). For each set variable that is controlled there will be a **set point**, which will be the target for that variable. For example, the core temperature of our bodies is normally maintained within tight boundaries, close to a set point of 37 °C (Figure 16.2), although this set point does vary over any given 24-hour period, showing a **circadian rhythm**. For example, the core body temperature does drop slightly, by about 0.5–1 °C during sleep. For adults, the normal range of resting body temperatures, as measured in the mouth, is taken as 36.4–37.6 °C. The NHS defines a fever as a resting temperature of >38 °C.

🌀 **Go to the e-book to explore an interactive version of Figure 16.2.**

If our core temperature rises, then the body responds by triggering a range of processes to cool the body down: some are physiological, such as increasing sweating; others are behavioural, such as removing clothing. These and related responses are all examples of negative feedback processes, whereby the body's systems act to oppose the rise in temperature by increasing cooling.

Why is it so important to maintain core temperature within such tight boundaries? Many biological processes are affected by temperature, such as the rate of chemical reactions. For example, temperatures above 41 °C can lead to changes in the tertiary structure of proteins ('denaturation') causing enzymes to stop working.

What other body parameters are tightly regulated by feedback mechanisms? There are many at all organizational levels! Examples include blood pH; blood glucose levels; the distribution of ions inside and outside cells; metabolic rate; cardiac output and

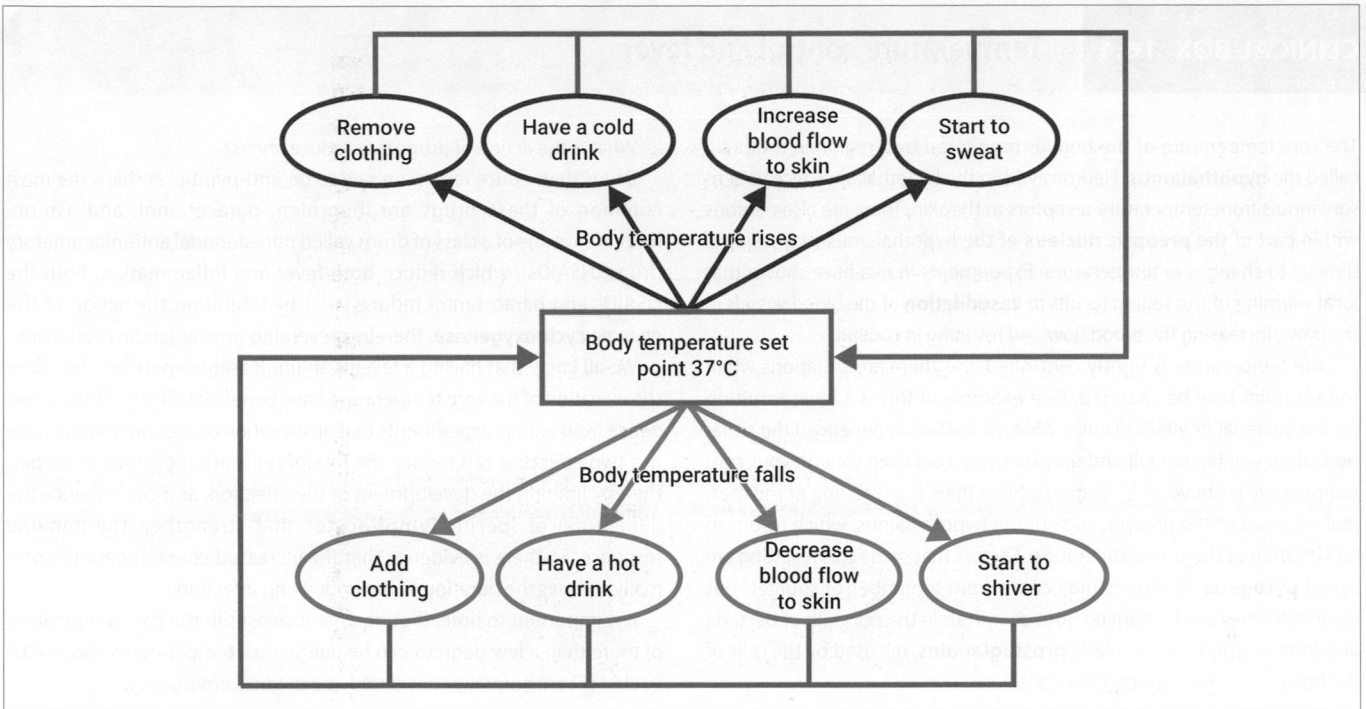

Figure 16.2 Diagram of a control system regulating temperature. Note that there can be more than one type of action in response to the change in the set point of the temperature. These actions can include, for example, physiological responses such as changing the flow of blood to the skin, sweating/shivering (on the right of the diagram), and behavioural responses such as putting on/removing clothing or having a hot/cold drink (on the left).

blood pressure; and control of body posture. We will come across all these, and more, in later sections of this book. Indeed, we can observe negative feedback in operation in almost every aspect of physiology.

It is important to note that negative feedback is not absolutely exact: the aspect being controlled will still vary a little, but normally this is within strictly defined limits. In part, this is due to the inertia in the system: when you start to get warm and your body temperature rises, there is a threshold level of change before that change is detected and the response is triggered. It will then take time for the response to have an effect, so the body temperature will continue to rise for a while before it starts to fall back down.

There are situations where the set point may be varied: for example, the core temperature of the body fluctuates on a daily cycle, falling from a daytime norm of 37 °C to around 36 °C during some phases of sleep. Another example of a change in the temperature set point is the response to infection: the generation of a fever, which we explore further in Clinical Box 16.1.

Thinking about homeostatic control for a complex system such as the body is made more complex because there are two types of variable that we need to consider:

- **Regulated variables**: these are variables that are directly controlled or regulated and for which there are specific sensors that monitor any changes. An example of this, as we have already seen, is body temperature, which is specifically monitored by peripheral and central temperature receptors.

- **Unregulated variables**: these are variables that contribute to the control process but do not have specific sensors. An example of this would be the rate of breathing when we exercise: breathing rate is increased in response to metabolic demands such as an increase in the production of carbon dioxide by the tissues (see Chapter 22). The carbon dioxide levels are specifically monitored, so this is a regulated variable. The rate of breathing is changed in response to this, but the rate of breathing itself is not directly monitored. So, rate of breathing is an unregulated variable, but one that plays an important part in the control of the regulated variable (carbon dioxide levels).

The opposite of negative feedback is **positive feedback**. In positive feedback systems a change in one parameter gives rise to further change such that the amount of change progressively increases. In physiology, such systems are rare because they bring about the opposite of homeostasis. In effect these could cause a 'run-away' phenomenon: taking our consideration of temperature as an example, if an increase in temperature resulted in a response that further increased the temperature, the core temperature would very rapidly rise to lethal levels.

Positive feedback systems do occur in very specific situations, for example in the generation of the nervous impulse, which we will discuss in Chapter 17, and the maintenance of uterine contractions during childbirth (Chapter 26). A very important feature of all positive feedback systems is that there has to be an end point at which the system is switched off and homeostasis is restored.

CLINICAL BOX 16.1   Temperature control and fever

The core temperature of the body is monitored by a region in the brain called the **hypothalamus**. Neurons within the hypothalamus receive sensory inputs from temperature receptors in the skin. There are also neurons within part of the **preoptic nucleus** of the hypothalamus that respond directly to changes in temperature. Experiments in rats have shown that local warming of this region results in **vasodilation** of the blood vessels in the paws, increasing the blood flow and resulting in cooling.

Core temperature is tightly controlled, but there are situations when the set point may be changed. One example of this is a fever resulting from a bacterial or viral infection. Most of us have experienced the situation when you become ill and may feel very cold even though your core temperature is above 37 °C. During a fever there is a resetting of the thermal set point of the preoptic area of the hypothalamus, which results in an elevation of the core temperature. Factors that cause this resetting are called **pyrogens**. Pyrogens may come from a number of sources, but commonly they can be derived from chemicals in the cell walls of bacteria and from chemicals, particularly **prostaglandins**, released by the cells of the body's immune system (Chapter 27).

What is the action of drugs that reduce fevers?

Drugs that reduce fevers are said to be 'anti-pyretic'. Perhaps the most common of these drugs are ibuprofen, paracetamol, and aspirin. Ibuprofen is one of a class of drugs called non-steroidal anti-inflammatory drugs (NSAIDs), which reduce both fever and inflammation. Both the NSAIDs and paracetamol reduce fever by inhibiting the action of the enzyme **cyclooxygenase**, thereby preventing prostaglandin production.

We all know that having a fever is an unpleasant experience, but does the elevation of the core temperature have beneficial effects? There is evidence from animal experiments that an elevation of the core temperature has two effects: it can reduce the rate of replication of some microbes, thereby limiting the development of the infection, and can enhance the generation of specific **lymphocytes** that strengthen the immune response. So there is evidence that the increased core temperature normally reduces the duration of the underlying infection.

It is important to note, though, that increases in the core temperature of more than a few degrees can be dangerous: temperatures above 40.0 °C (104 °F) are normally considered as a medical emergency.

PAUSE AND THINK

Thinking about homeostasis, try drawing a diagram like that in Figure 16.2 to illustrate some examples of negative feedback control of water balance in the body.

Answer: In your diagram you could include behavioural responses such as drinking more and physiological responses such as reductions in urine production (you can refer to Chapter 24 for some more ideas).

 Check your understanding of the concepts covered in this section by answering the questions in the e-book.

16.2 The cell membrane

The common structure that is found in all the systems we will be investigating in this review of physiology is the cell. The cells found in different organs and tissues vary, both in structure and function, depending on the function of the organ system itself: for example, a nerve cell is different in many ways to a cell in the wall of the stomach. Nonetheless, many aspects—such as the presence of a nucleus and organelles such as mitochondria—are common features. These features have been considered in detail in Chapter 9, and so will not be reviewed here. However, we will consider some of the specific features of the plasma membrane because its properties underpin many processes in physiology.

The **plasma membrane** (Figure 16.3) encloses the cell and plays a vital role in the regulation of cell function by: (a) controlling what substances can enter or leave the cell; and (b) controlling the communication between the cell and the rest of the body, for example through the interaction of hormones or nervous impulses.

The plasma membrane allows each cell to maintain a very different composition of its internal cytoplasm compared with the extracellular fluid. This is because the membrane is readily permeable to only a few substances and is either impermeable or shows regulated permeability to other substances.

The main constituents of the plasma membrane are phospholipids and proteins. The phospholipid molecules comprise a polar head that is **hydrophilic** (literally 'hydro'—water, 'philic'—loving) and two non-polar tails that are **hydrophobic** (water 'hating') (Figure 16.3). In an aqueous solution the phospholipids naturally come together as a bilayer with the hydrophilic heads on the outer surface and the hydrophobic tails in the middle. The hydrophobic core of the lipid bilayer makes it impermeable to most water-soluble, polar molecules and to charged ions, such as sodium ions. The bilayer of the membrane is also impermeable to large molecules such as amino acids and proteins. The bilayer is, however, permeable to very small polar modules such as O_2, CO_2, and water, and also to fat-soluble substances such as steroid hormones.

The proteins within the membrane can be divided into two main groups depending on their location within the membrane:

1. Extrinsic proteins are linked to the surface of the membrane. These can be facing inwards—for example, the proteins that make up the cytoskeleton, some membrane-bound enzymes, and G-proteins (Section 17.7). These extrinsic proteins interact either with other proteins of the cytoskeleton or with the polar heads of the lipids. Other extrinsic proteins are outward facing, attached to the external surface of the membrane.

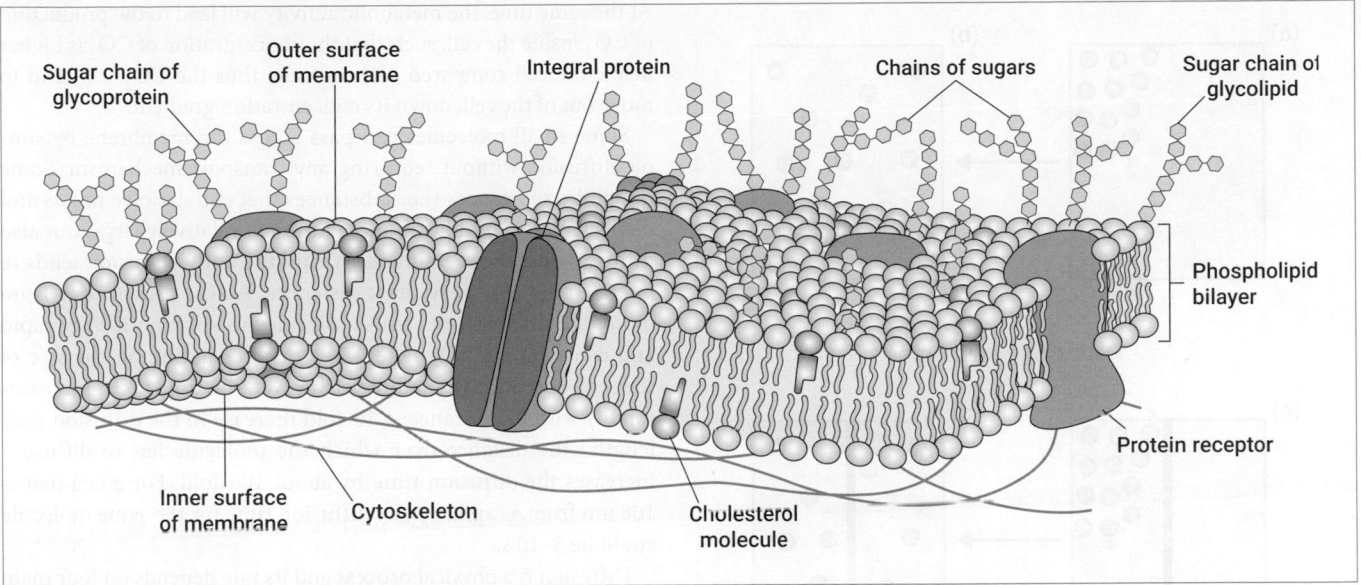

Figure 16.3 The plasma membrane. The core structure is the phospholipid bilayer in which are inserted proteins, many of which play a key role in the communication systems and in regulating the transport of substances across the membrane.

Source: Open University and Krane, C. M. and Kishore, B. K. (2003). Aquaporins: the membrane water channels of the biological world. *Biologist*, 50(2).

2. Integral (intrinsic) proteins are embedded in the bilayer and may span the width of the membrane so they are exposed on both the intra- and extracellular surfaces. These different proteins perform a number of roles. For example, some of the integral, membrane-spanning proteins provide means of transport across the membrane, for example through membrane pores. These enable transport of a variety of substances, most of which cannot cross the membrane through the lipid bilayer because they are too large or are polar molecules.

Transport across the plasma membrane

A whole range of substances are transported across the cell membrane all the time. These range from ions, such as Na^+, K^+, and Cl^-, through very small molecules such as oxygen and water, to the very large macromolecules such as proteins. The effective regulation of the movement of substances into and out of the cell is vital to physiological function and we will be discussing the different mechanisms of transport of different molecules in almost every chapter of this module!

The transport of substances across the plasma membrane can be by either **passive** or **active transport** mechanisms.

- Active transport requires the expenditure of energy to move substances across the membrane, whereas passive transport, as its name suggests, does not require energy to enable movement of substances.

- Passive transport can only occur where there are differences in the concentration or electrical gradients across the cell membrane and it results in a reduction in those gradients, for example by leading to an equalization of the concentration of the substance either side of the membrane.

There are four main ways by which substances can pass across the plasma membrane:

1. Diffusion.

2. Passive protein-mediated transport.

3. Active protein-mediated transport.

4. Endo/exocytosis.

Diffusion

Diffusion is a passive process by which molecules move from a region of high concentration to a region of lower concentration, referred to as moving down a **concentration gradient**. The end-point of diffusion is the even distribution of the molecules (Figure 16.4). For example, as illustrated in Figure 16.4, when a number of molecules of a substance are dissolved in a beaker of water they will initially be concentrated together where they were put in the beaker (Figure 16.4a). Due to **Brownian motion**, the molecules will move randomly within the water such that they end up being distributed evenly throughout the beaker of water (Figure 16.4b).

Brownian motion is described as the random movement of particles suspended in a liquid or gas. The particles are constantly being collided with by other molecules that are continually moving. The energy of the movement increases with increasing temperature.

 Go to the e-book to watch a video that illustrates the process of diffusion.

If a membrane that is permeable to the molecules is placed across the beaker, the process of diffusion will result in the movement of the molecules to and fro across the membrane with the result that there will be an even distribution of the molecules either side of the membrane (Figure 16.4c and d). In other words, in Figure 16.4c

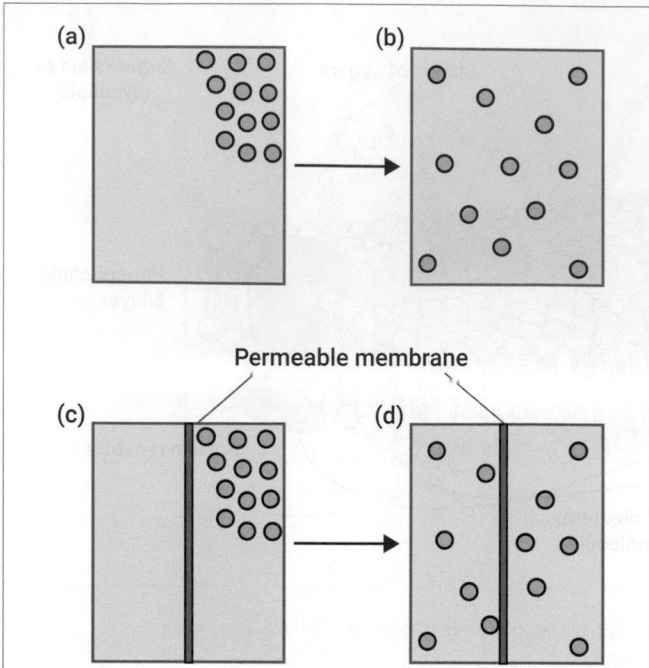

Figure 16.4 The process of diffusion. (a, b) Molecules in solution will move randomly by Brownian motion until they are evenly distributed within the solution. At this stage the individual molecules continue to move, but there is no net change in their distribution. (c, d) When the solution is divided by a membrane that is permeable to the molecules, the molecules will pass to and fro across the membrane, establishing an equal concentration on either side.

there is initially a concentration gradient between the right- and left-hand sides of the beaker. The molecules will tend to move from the area of high concentration down the concentration gradient to the area of low concentration until there is an even distribution (Figure 16.4d).

Note that, once there is an even distribution, the individual molecules do not stop moving, but their random motion means that, as a whole, they will remain evenly distributed. There will therefore be an equilibrium, but it is a **dynamic equilibrium** in that each individual molecule is still free to move across the membrane.

What happens if the membrane is impermeable to the molecules? If the membrane is impermeable, then the molecules will diffuse so that they are evenly distributed within the one-half of the beaker, but they will not cross the membrane. As a result, there will be a concentration gradient across the membrane, with a high concentration of the molecule on one side and none on the other.

Diffusion across the plasma membrane of the cell is therefore a passive process driven by the relative concentration gradient of the substance across the plasma membrane.

Where the membrane is permeable to a molecule, the net movement of that molecul across the membrane will depend on the relative concentrations of the molecule either side of the membrane. Thus, if a cell is metabolizing and using O_2 then the O_2 concentrations within the cell will be lower than those in the adjacent capillaries, so O_2 will tend to diffuse into the cell, down its concentration gradient, thereby supplying the metabolic needs of the cell.

At the same time, the metabolic activity will lead to the production of CO_2 inside the cell, such that the concentration of CO_2 is higher inside the cell compared with outside, thus the CO_2 will tend to move out of the cell, down its concentration gradient.

Some small molecules can pass across the membrane by simple diffusion without requiring any transport mechanism. Some lipophilic substances (i.e. substances that can dissolve in fats and oils) such as steroid hormones, although relatively large, can also diffuse across the membrane. Within the body, diffusion tends to operate only over very short distances. Across distances of 1 μm or so, the diffusion of small molecules in solution is very rapid and can typically only take 0.5–1 ms, depending on the size of the molecule and the medium. However, the time taken increases rapidly with the distance: a 10-fold increase in the diffusion path length—the distance over which the molecule has to diffuse—increases the diffusion time by about 100-fold. For a cell that is 100 μm from a capillary, the diffusion time for the same molecule could be 5–10 s.

Diffusion is a physical process and its rate depends on four main factors:

1. **Temperature**: this affects the kinetic energy of the molecules, so the higher the temperature, the faster the movement of the molecules. Within the body, the temperature is well-regulated and therefore changes in kinetic energy are unlikely to be significant.

2. **Density**: the higher the density, or the higher the viscosity of the medium through which the diffusion is taking place, the slower will be the rate of diffusion because the greater will be the frequency of collisions between the diffusing particles and the medium within which they are diffusing, thereby leading to a slower rate of diffusion.

3. **Molecular/particle size**: the larger the molecule or particle that is diffusing, the greater will be the frequency of collisions between the diffusing particles and the medium within which they are diffusing, thereby leading to a slower rate of diffusion.

4. **Concentration gradient**: the greater the difference in concentration between the two regions, the more rapid will be the rate of diffusion from the higher to the lower concentration region. As the system moves towards equilibrium, so the rate of diffusion will slow down.

Points 2 and 3 above can be combined in a value termed the **diffusion coefficient** (D). The diffusion coefficient depends on the size of the diffusing molecule and the density and viscosity of the medium through which it is passing: the larger the molecule and the greater the viscosity, the smaller will be the value of the diffusion coefficient, and the lower will be the rate of diffusion for a given temperature and concentration gradient. This is defined in Graham's law of diffusion, which states that the rate of diffusion is inversely proportional to the square root of the molecular masses of the two substances.

To calculate the rate of diffusion, we can use the work of mathematics student-turned-doctor Adolf Fick. Fick's first law of diffusion relates to the movement of particles, the **diffusion flux**, such that the net movement of particles is proportional to the steepness of the concentration gradient. Fick's second law incorporates the distance over which the diffusion is taking place in that the time taken for diffusion from one region to another to occur increases in proportion to the square of the distance between the two regions.

We can combine Fick's laws to allow the calculation of the rate of movement of a substance across a cell membrane. At a given temperature, the rate of diffusion (J) across a membrane is *proportional* to the diffusion coefficient (D), the area of the membrane (A), and the concentration gradient (Δc), and the rate is inversely proportional to the thickness of the membrane (x):

$$J = -DA\frac{\Delta c}{x}$$

where

- J = the rate of diffusion in molecules per unit time crossing the membrane, mol m^{-2} s^{-1}
- D = the diffusion coefficient, m^2 s^{-1}
- A = the area of the membrane, m^2
- Δc = the concentration gradient across the membrane, the difference in the numbers of particles, m^{-4}
- x = the thickness of the membrane, m.

Why is there a minus sign in the equation? J is a positive value because there is a net movement of molecules across the membrane, but the concentration gradient reflects the difference between the starting point (initially high concentration) and the end point (initially low concentration) and so has a negative value. The inclusion of the minus sign therefore gives a positive overall value.

Osmosis is a special form of diffusion involving the movement of water. Water will diffuse from an environment where the solute is relatively dilute to that where it is more concentrated. For example, if there are two solutions of different concentrations of sodium chloride either side of the plasma membrane, the sodium and chloride ions would not be able to pass through the membrane (because the membrane is impermeable to ions) but the water will flow from the dilute (**hypotonic**) solution to the more concentrated (**hypertonic**). This is still the same as diffusion because water is moving from its region of high concentration to the region where it is at a lower concentration. The flow of water across the cell membrane plays an important part in the regulation of cell volume (see Chapter 9).

Go to the e-book to watch a video that illustrates the process of osmosis.

If the core of the plasma membrane is made up of the *hydrophobic* tails of the phospholipid molecules, how can water pass through? Water is a very small molecule; the hydrophobic tails of the phospholipid molecules are spaced apart and are often kinked, creating larger gaps, so the water can pass through the spaces between the molecules. There are also integral protein channels, called **aquaporins**, that allow water to pass through (see 'Passive protein-mediated transport', later in this section).

Remember that, as a passive process, diffusion can only lead to an equalization of concentrations either side of the membrane—it cannot generate increases in the concentration difference.

PAUSE AND THINK

For a given molecule such as CO_2, what will be the main variable affecting its diffusion from a metabolizing cell to the blood stream?

Answer: The diffusion coefficient and diffusion path length will all remain constant. Local tissue temperature may vary slightly depending on the level of activity, but the main variable will be the concentration gradient of CO_2 between the cell and the capillary. As the cell's metabolic activity increases, so its production of CO_2 will increase.

Passive protein-mediated transport

Passive protein-mediated transport involves the integral proteins. The proteins involved can take the form of **channels** and **carrier proteins**.

Channels are protein structures that create a pore that spans the membrane, so the substance being transported is separated from the phospholipid bilayer. In their simplest form, these channel proteins form a simple tube allowing the substance to pass through. The channels will be selective for the substances they allow through based on the size of the molecule and the charge associated with it.

One of the most widespread of these channels is the potassium (K^+) channel (Figure 16.5), which plays a key role in the maintenance of the resting potential of neurons (Section 17.3). The K^+ channel comprises four subunits (shown as different colours in Figure 16.5) that create the pore structure. The width of the pore and the location of the four K^+ binding sites (S1–S4) create a selectivity filter that restricts the passage of other ions.

There are also channels specific for water, called aquaporins, which are widely distributed and are involved in the regulation of water movement across the cell membranes. They also have specific roles in regulating water balance in the body as a whole via key structures such as the collecting ducts of the kidney (Chapter 24).

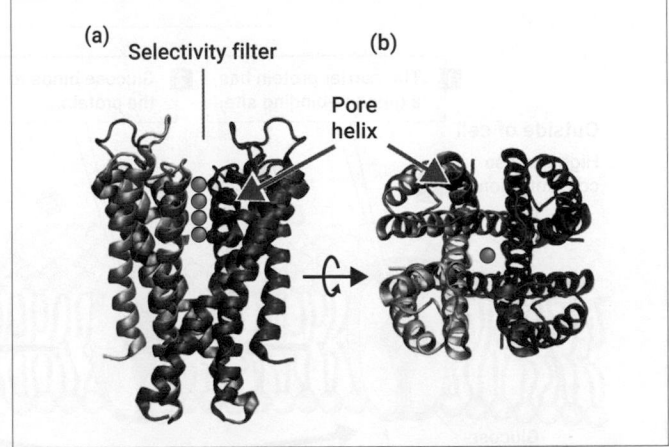

Figure 16.5 Structure of a potassium leakage channel shown in (a) lateral view and (b) surface view. The channel comprises four subunits (shown in different colours) that form the pore. The selectivity filter is on the extracellular surface of the channel and filters ions based on charge and size, such that only K^+ ions can pass through (shown as the green spheres).

Source: Figure 1 from D. Naranjo, H. Moldenhauer, M. Pincuntureo, I. Díaz-Franulic (2016). Pore size matters for potassium channel conductance. *J Gen Physiol*, 148(4), 277. Permission conveyed through Copyright Clearance Center, Inc.

Other types of channels, called **gated channels**, can be open or closed. These will typically be closed at rest but can be opened, for example by a change in the electrical gradient across the membrane which causes a small change in the tertiary structure of the protein allowing the channel to be opened. These **voltage-gated channels**, such as the voltage-gated sodium (Na⁺) channel, are integral to the generation of the action potential (Section 17.4), the process whereby electrical signals are passed along neurons. **Ligand-gated channels** also show a change in conformation to allow passage of substances such as ions. In this case, the channel protein has a specific binding site to which a transmitter molecule can attach, causing the change in conformation, thereby opening the channel. Examples of such channels are commonly found on the post-synaptic membrane of neural **synapses**, which we will return to in Section 17.6.

These channels can therefore allow the passage of substances through the membrane that cannot normally pass through the bilayer or, in the case of the aquaporins and water, allow passage at a much higher rate than by simple diffusion. Although these channels facilitate the diffusion process, enabling faster transport and the transport of substances that could not normally pass through the lipid bilayer, the mechanism is still passive and so does not require energy and can only function in the presence of a concentration or electrical gradient for the substance being transported.

> Go to the e-book to watch a video that illustrates membrane transport in more detail.

Carrier proteins (Figure 16.6a) can also facilitate the passive transport of substances across the membrane. Similar to channel proteins, the substance will pass through the membrane with the protein forming the channel. However, unlike channel proteins, the substance *binds* to a site on the surface of the protein in the same way as a substrate will bind to an enzyme. The binding

process leads to a change in the shape of the protein that results in the transport of the substance across the membrane. As with the channel proteins, carrier proteins are often highly specific for the substance being transported.

Because the transport depends on binding, the process shows the same kinetics as for enzyme-mediated reactions (Chapter 6) and the rate of transport can be limited by the number of carriers: if the concentration of the substance being transported rises above a certain level the carriers can become saturated and the rate of transport will not increase irrespective of further increases in the concentration of the substance (see Figure 16.6b). An example we will explore in Chapter 24 is that of the re-absorption of glucose in the tubules of the kidney. If the blood glucose levels rise too high, as can occur in type 1 diabetes, then the glucose carrier proteins will be saturated and glucose will appear in the urine. Indeed, this phenomenon is the origin of the name for type 1 diabetes, 'diabetes mellitus'. 'Diabetes' means pass through (urinate), and mellitus means 'sweet', as the urine of the patient tastes sweet because of the sugar content.

Active protein-mediated transport

Active protein-mediated transport requires energy to transport the substance and it can transport the substance against its concentration gradient and lead to a steepening of the concentration gradient. The most widespread active transport mechanism is the Na⁺ K⁺ ATPase pump (Figure 16.7). This pump directly uses energy from ATP (adenosine triphosphate—Chapter 2) to pump the ions and so is referred to as a primary active transport mechanism.

> Go to the e-book to watch an animation that depicts the functioning of the Na⁺ K⁺ ATPase pump.

The Na⁺ K⁺ ATPase pump has its evolutionary origins in the maintenance of osmoregulation but in the mammalian cell it has become the driver for a wide range of processes, including the generation of

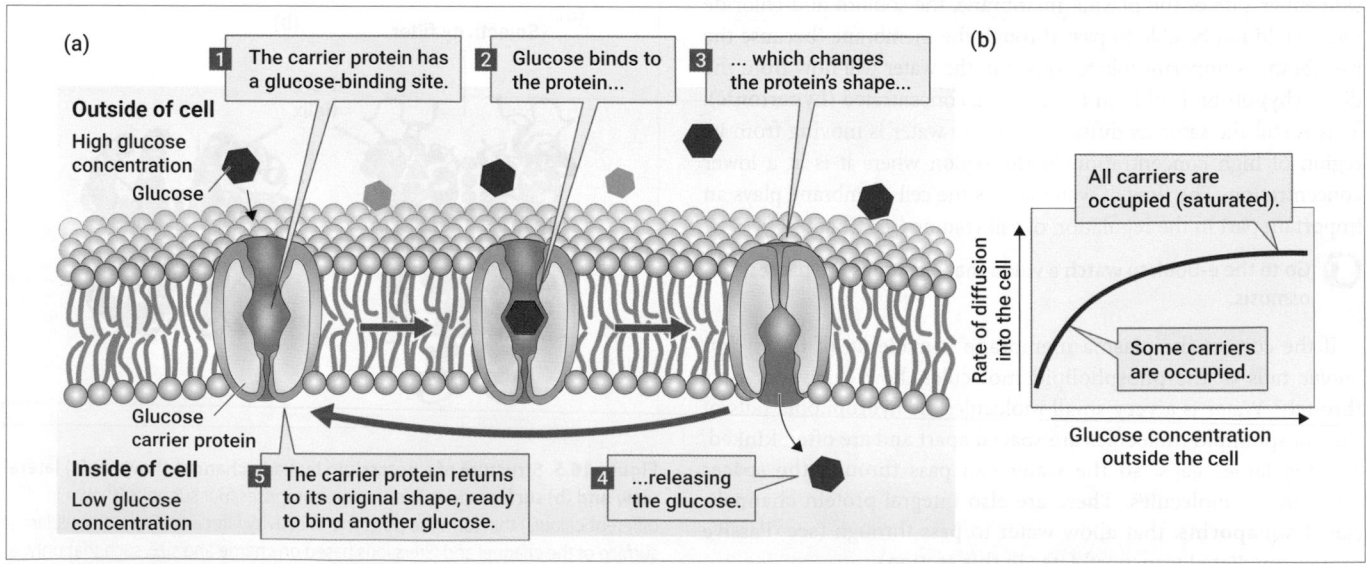

Figure 16.6 Facilitated diffusion of glucose via an integral carrier protein. (a) The steps of facilitated diffusion. (b) The rate of diffusion into the cell is limited by the number of carrier proteins in the membrane: once the glucose concentration exceeds the number of carriers available, the rate of diffusion will reach a maximum, irrespective of further increases in glucose concentration.

Source: Hillis, Sadava, Hill, Price, 2014. *Principles of Life*, 2nd edn. Oxford Publishing Limited.

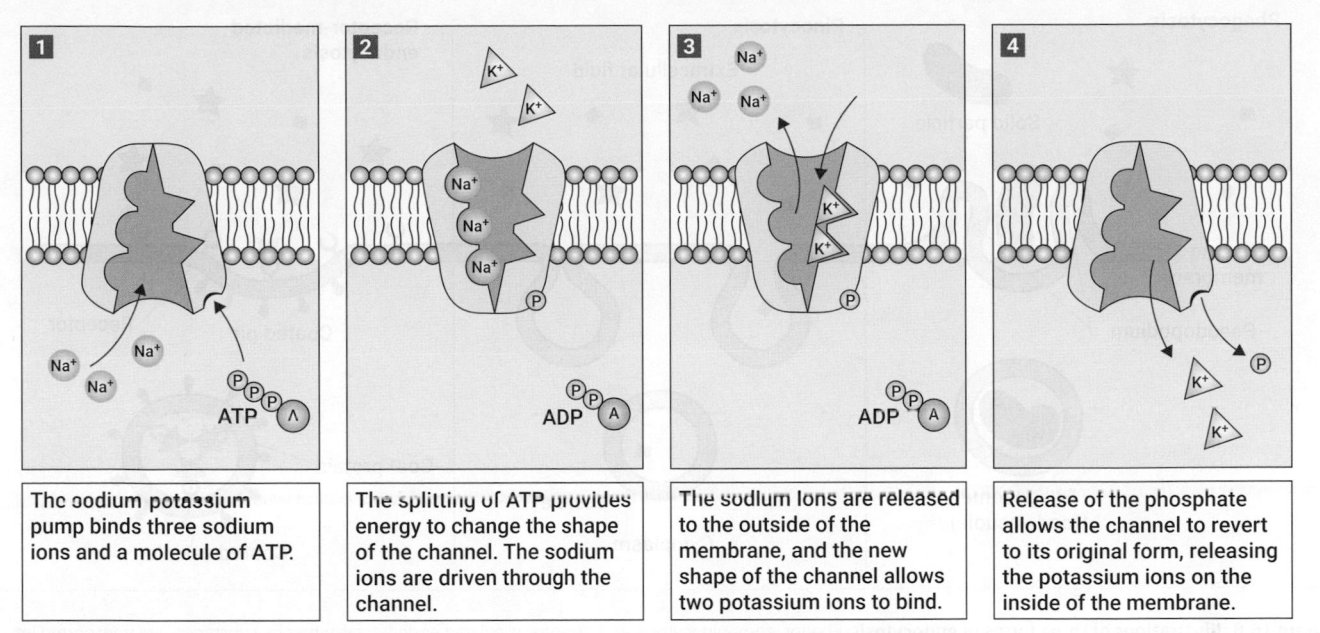

Figure 16.7 The Na$^+$ K$^+$ ATPase pump. The Na$^+$ K$^+$ ATPase uses ATP to pump three sodium ions from the cytoplasm across the membrane to the extracellular fluid against the sodium concentration gradient. In doing so, the ATP is converted into ADP and phosphate and two potassium ions enter the cell. **Stage 1:** the three Na$^+$ ions and one ATP molecule bind to the membrane-spanning protein. **Stage 2:** ATP phosphorylates (adds a phosphate group to) the pump protein, releasing ADP. **Stage 3:** phosphorylation of the pump protein leads to a conformational change releasing the 3 Na$^+$ ions into the extracellular space and allowing 2 K$^+$ ions to bind. **Stage 4:** the pump protein dephosphorylates, reverting to its original shape and releasing the 2 K$^+$ ions into the intracellular space.

Source: Crepalde, M.A., Faria-Campos, A.C. & Campos, S.V. (2011). Modeling and analysis of cell membrane systems with probabilistic model checking. *BMC Genomics*, 12, S14. https://doi.org/10.1186/1471-2164-12-S4-S14/CC BY 2.0.

the nervous impulse (Section 17.4), the absorption of substances by the gut (Chapter 25), and the re-absorption and excretion of substances by the kidney (Chapter 24). These processes are enabled because the Na$^+$ K$^+$ ATPase pump uses energy to create concentration gradients across the membrane: the energy from the splitting of ATP is used to pump sodium ions out of the cell in exchange for potassium ions (Figure 16.7), so that the inside of the cell is maintained with a low concentration of sodium and a relatively high concentration of potassium ions compared with the extracellular fluid. In fact, the pump is asymmetric in that in each cycle it pumps three sodium ions out in exchange for two potassium ions.

The plasma cell membrane typically has within it potassium channels that are ungated and therefore allow the free passage of potassium ions in either direction (a type of passive protein-mediated transport). As a consequence, the potassium ions can flow by diffusion, whereas the sodium gradients are maintained because the membrane is impermeable to sodium ions. These sodium gradients therefore represent a source of potential energy that can be used to drive other processes including the transport of substances that cannot pass through the membrane; this is termed secondary active transport.

There are two types of secondary active transport: **cotransport** (or symport) and **exchange** (or antiport). In the case of cotransport, the substance being transported moves in the same direction as the sodium ion: an example of secondary active cotransport is the absorption of glucose and some amino acids by the small intestine (Chapter 25). In the case of exchange, the substance moves

in the opposite direction: an example of exchange transport is the movement of calcium out of the cell in exchange for sodium, as occurs in cardiac muscle cells.

The epithelial cells lining the small intestine have Na$^+$ K$^+$ ATPase pumps on the basolateral membranes (the membranes which are located on the sides and base of the cell, away from the lumen of the intestine) which pump sodium ions out of the cells into the extracellular fluid. This pumping maintains a low concentration of sodium within the cells. On the apical membrane of the cells are carrier proteins that cotransport sodium and glucose, driven by the movement of sodium ions down their concentration gradient into the cell.

What will happen to glucose absorption if metabolic activity in the cells is stopped? Although the carrier proteins themselves do not make direct use of metabolic energy to transport the substances into the cells, they can only function in the presence of a sodium gradient which is maintained by the operation of the Na$^+$ K$^+$ ATPase pump on the basolateral membrane, hence the term secondary active transport. In the absence of metabolic activity, the sodium gradient would decline and so the intestine would no longer be able to absorb glucose.

There are other examples of primary active transport mechanisms, for instance the Ca^{2+} ATPase pump that plays an important role in the regulation of calcium concentrations within the sarcoplasmic reticulum of muscle. We will return to the Ca^{2+} ATPase pump in Section 20.1.

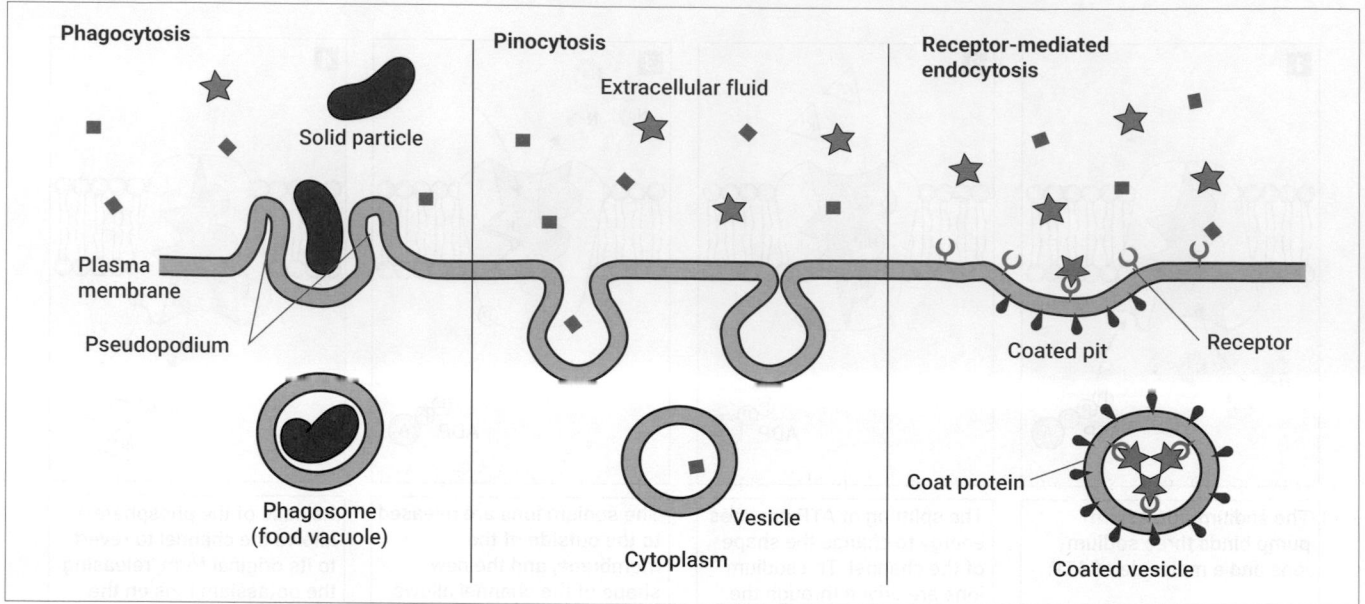

Figure 16.8 Illustrations of three forms of endocytosis. Phagocytosis, pinocytosis, and receptor-mediated endocytosis, whereby substances are transported into the cell without crossing the cell membrane.

Source: Wikimedia Commons/Mariana Ruiz Villarreal LadyofHats/Public Domain.

PAUSE AND THINK

What is the main operation of the Na⁺ K⁺ ATPase pump?

Answer: The pump uses energy to transport Na⁺ ions out of the cell in exchange for K⁺ ions. This maintains a concentration gradient for Na⁺ across the cell membrane of all cells. This concentration gradient can be used to facilitate the transport of other substances across the cell membrane or to enable the electrical activity of nerve cells.

Endocytosis and exocytosis

Endocytosis is the transport into the cell of substances enclosed in membrane-bound vesicles. Endocytosis is the process whereby substances which cannot normally pass through the plasma membrane can be transported into the cell. The substance to be transported is engulfed by an invagination of the membrane to create a vesicle that is then internalized within the cell, as illustrated in Figure 16.8. This process also requires energy.

Endocytosis occurs in three main ways:

- **Phagocytosis** (Figure 16.8a) is a very important element of the immune response, the process whereby cells such as macrophages engulf and subsequently break down pathogens (Chapter 27).

- **Pinocytosis** (Figure 16.8b) is a process whereby fluids containing solutes may be taken up into cells. This process occurs, for example, in the small intestine to absorb fat droplets (Chapter 25).

- **Receptor-mediated endocytosis** (Figure 16.8c), as its name suggests, involves the binding of the substance to be transported to receptors on the surface of the cell. The receptors are located in pits in the surface of the membrane that are coated with a protein complex called **clathrin**, so-called **coated pits** (Figure 16.9).

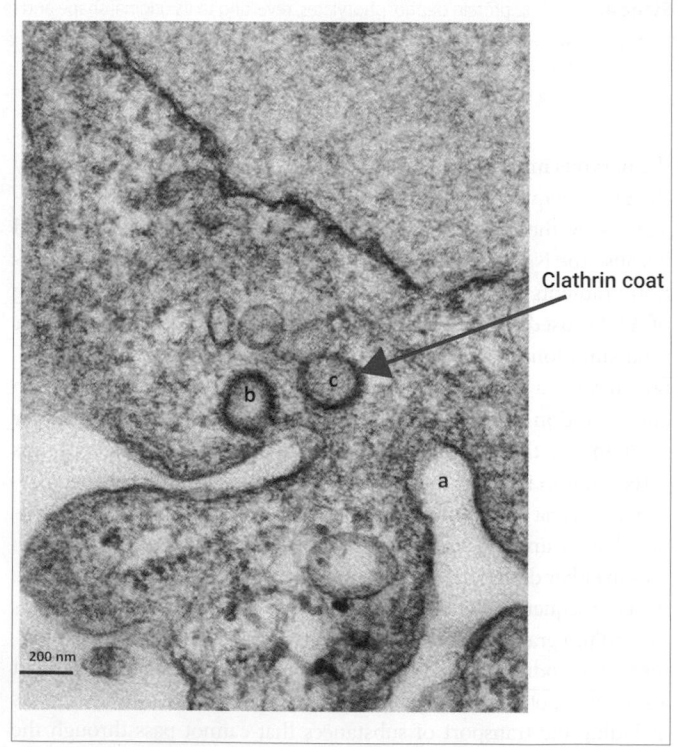

Figure 16.9 Microscope image showing stages of formation of clathrin-coated vesicles. (a) Initial invagination; (b) an almost complete vesicle with the clathrin coat in evidence; (c) a fully formed, clathrin-coated vesicle.

Source: Figure 1 from Johnson, F.B., Dudleenamjil, E., 2012. Clathrin-associated endocytosis as a route of entry into cells for parvoviruses. In: B. Ceresa (ed.) *Molecular Regulation of Endocytosis*. IntechOpen, London. 10.5772/48510.

When the substance is bound to the receptor, this triggers the formation of a clathrin-coated vesicle of membrane that is then internalized within the cell. An example of this process is the recycling of membrane-bound receptors: the membrane protein is first tagged with a protein called **ubiquitin** followed by formation of the clathrin-coated vesicle allowing the receptor protein to be internalized into the cytoplasm of the cell. This is then transported to the lysosomes for recycling.

Substances can also be transported out of the cell by the reverse process, **exocytosis**. In this case, vesicles are assembled within the cell containing the substance to be transported. The vesicles then assimilate into the cell membrane, releasing the substance into the extracellular space. Exocytosis is the mechanism whereby neurotransmitters are released from nerve terminals (Section 17.1) and secretory proteins such as enzymes can be released, for example in the intestine.

CLINICAL BOX 16.2   Cystic fibrosis

Given the importance of membrane transport mechanisms in the regulation of cell function, it is not surprising that there are numerous clinical disorders associated with malfunctioning transport systems. A well-known condition that results from a genetic mutation affecting channel proteins is cystic fibrosis. Cystic fibrosis is characterized by the production of a very thick, viscous mucus that blocks the narrower airways in the lungs and also affects the pancreas, and causes blockage of the intestines through the production of thickened faeces (Figure 1a). Patients particularly show significant symptoms of respiratory difficulty due to airway

blockage and also repeated bacterial respiratory tract infections (Figure 1b and c). The impact of this mucus production significantly affects patients' quality of life, and the median age of death is 31 years (UK Cystic Fibrosis Registry Annual Data Report 2017 (2018)), although this is progressively increasing.

Cystic fibrosis is an inherited disorder, caused by a mutation in the gene cystic fibrosis transmembrane conductance regulator (*CFTR*). The condition is an autosomal-recessive disease, so both parents have to be carriers of the defective gene for the offspring to be homozygous and

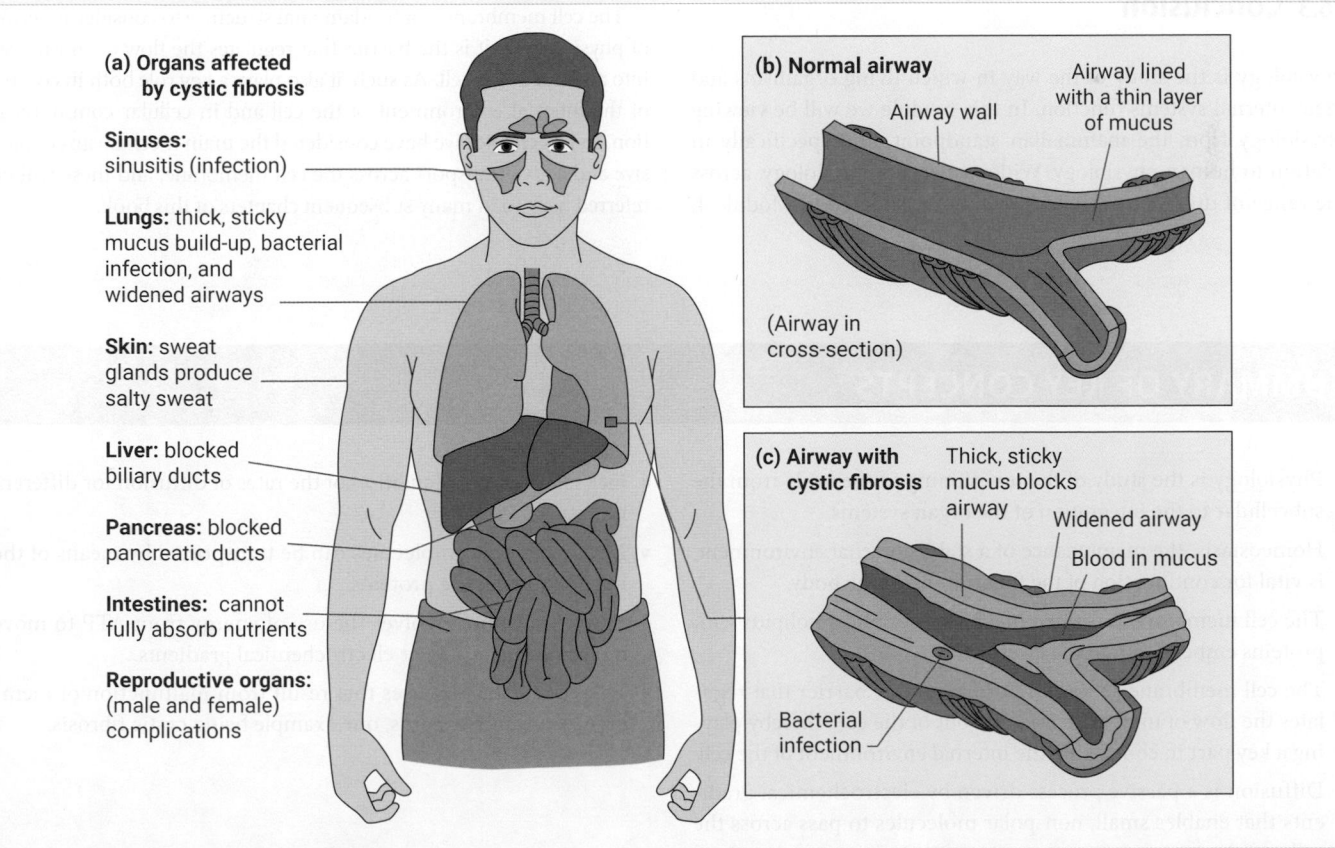

Figure 1 (a) Main organs affected by cystic fibrosis; (b) and (c) effect of cystic fibrosis on the airways.

Source: Wikimedia Commons/National Heart Lung and Blood Institute (NIH)/Public Domain.

so be affected by the disease. All babies can be tested a few days after birth via the heel blood spot test. The heterozygote frequency is about 1 in 20 in the UK, giving an incidence of about 1 in every 2500 births and so there are about 10,500 people living in the UK with cystic fibrosis.

In this reflection on cystic fibrosis we will focus on the impacts on lung function.

In healthy lungs, the airways are lined with fluid—the airway surface liquid (ASL)—on top of which is a layer of mucus that traps bacteria and particles that enter the respiratory tree. Cilia that line the airways continually sweep the mucus and trapped bacteria up away from the lungs.

The mutation in *CFTR* results in changes in the channel protein of the epithelial cells of the airways which affects the movement of chloride ions out of the cell into the ASL. The malfunction of *CFTR* also leads to failure of regulation of the epithelial sodium channel (ENaC). The net effect is an increase in sodium resorption from the airways, which in turn causes water to be drawn from the mucus layer into the cell, following the sodium. The result is a much thicker, viscous mucus layer. This leads to narrowing of the airways and makes it more difficult for cilia to sweep effectively. This means the airways are not cleared, allowing a build-up of bacteria and other particles, leading to repeated infections and local inflammation.

Currently, there is no cure for cystic fibrosis, with most approaches being directed towards limiting the effects of the disease. In relation to lung function, these include, in particular, physiotherapy to help move the mucus and clear the airways. Drug treatments include drugs that reduce the viscosity of the mucus, thereby making it easier to clear the airways. Recent developments include drugs that enhance the opening of the chloride channels, which can be beneficial in patients who have a mutation that affects the rate of production of the channel proteins, although this only represents a small proportion of patients. Patients are likely to require frequent treatment for bacterial infections with antibiotics. Some patients are treated with lung transplants when the lungs become too damaged by repeated infections.

 Check your understanding of the concepts covered in this section by answering the questions in the e-book.

16.3 Conclusion

Physiology is the study of the way in which living organisms and their internal systems function. In this module we will be viewing physiology from the mammalian standpoint, and specifically in relation to human physiology. Wider aspects of physiology, across the range of different organisms, will be considered in Module 4.

The core principles we are considering relate to homeostasis, the maintenance of a near-constant internal environment. This relies on a variety of sensory systems that detect changes and initiate responses to oppose those changes—negative feedback.

The cell membrane is a fundamental structure to consider in terms of physiology as it is the barrier that regulates the flow of substances into and out of the cell. As such, it also plays a key role both in control of the internal environment of the cell and in cellular communication. In this chapter we have considered the main mechanisms of passive and active transport across the cell membrane, and these will be referred back to in many subsequent chapters of this book.

SUMMARY OF KEY CONCEPTS

- Physiology is the study of the functioning of the body from the subcellular to the integration of the organ systems.

- Homeostasis, the maintenance of a stable internal environment, is vital for continuation of the functioning of the body.

- The cell membrane is formed of a bilayer of phospholipids with proteins embedded into the layers.

- The cell membrane acts as a semi-permeable barrier that regulates the flow of molecules into and out of the cell, thereby playing a key part in controlling the internal environment of the cell.

- Diffusion is a passive process driven by electrochemical gradients that enables small, non-polar molecules to pass across the cell membrane.

- Fick's laws allow calculation of the rates of diffusion for different molecules.

- Larger and polar molecules can be transported by means of the integral membrane proteins.

- Active transport involves the use of energy from ATP to move molecules against their electrochemical gradients.

- There are many diseases that result from malfunction of membrane transport systems, one example being cystic fibrosis.

 Use the flashcards in the e-book to test your recall of key terms introduced in this chapter.

QUESTIONS

 Looking for answers? Once you've answered these questions, follow the link in the e-book to the answer guidance and check your work.

Concepts and definitions

1. Why is it important that core body temperature is maintained within tight boundaries?

2. What are the two main components of the plasma membrane?

3. Explain the term 'semi-permeable membrane' in relation to the cell membrane.

4. How does water pass through the cell membrane?

5. How is the intracellular concentration of sodium actively regulated?

Apply the concepts

6. Draw a flow diagram showing the general format of a negative feedback loop.

7. Compare the operation of negative and positive feedback loop pathways.

8. Describe the processes involved in generating a fever and explain why this may be beneficial.

9. Compare the location and roles of integral (intrinsic) and extrinsic proteins.

10. Explain the role of the Na^+K^+ ATPase pump in active transport mechanisms.

Beyond the concepts

11. Discuss the role of the cell membrane in regulating the movement of substances in and out of the cell.

12. Explain the term homeostasis. Identify a physiological parameter that is regulated homeostatically and discuss the mechanisms involved in that regulation.

13. Research a disease that results from cell membrane dysfunction and explain how the pathology of the disease results from that dysfunction.

14. Explain the process of diffusion and discuss the factors that affect the rates of diffusion.

15. Compare and contrast the mechanisms of active transport across cell membranes.

FURTHER READING

Bernardino de la Serna, J., Schütz, G. J., Eggeling, C., & Cebecauer, M. (2016). There is no simple model of the plasma membrane organization. *Front. Cell Develop. Biol.* **4**: 106.
Provides a good overview of the different models of membrane structure and function.

Lombard, J. (2014). Once upon a time the cell membranes: 175 years of cell boundary research. *Biol. Direct* **9**: 32.
A good overview of the development of ideas regarding the plasma membrane.

Modell, H., Cliff, W., Michael, J., McFarland, J., Wenderoth, M. P., & Wright, A. (2015). A physiologist's view of homeostasis. *Adv. Physiol. Educ.* **39**: 259–66.
As its title suggests this provides a perspective on homeostasis that is relevant to physiologists and is easy to read.

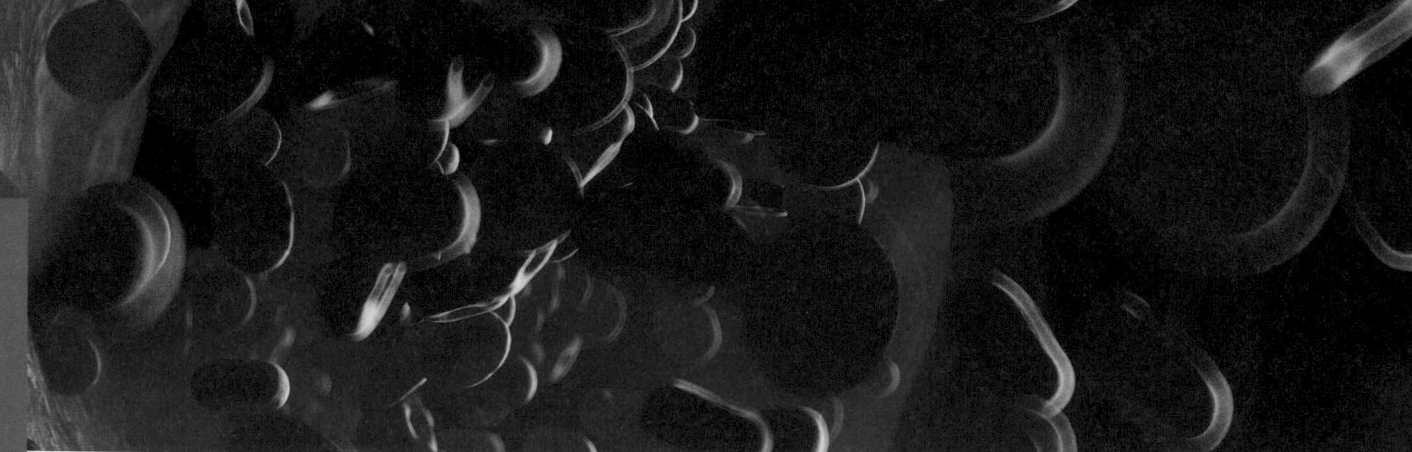

Communication and Control 1

The Nervous and Endocrine Systems

Chapter contents

Introduction		594
17.1	The nervous system	595
17.2	Signal conduction within the nervous system	595
17.3	Resting potential	596
17.4	Action potentials	600
17.5	Action potential conduction	604
17.6	Transmission between neurons	605
17.7	Cellular signalling	614

Watch the key concepts video in the e-book to prepare yourself for studying this chapter.

LEARNING OBJECTIVES

By the end of this chapter you should be able to:

- Describe the general structure of a neuron.
- Explain the processes underlying the generation of the resting membrane potential.
- Explain the processes underlying the generation of the action potential.
- Compare the different types of synapse and compare transmission between neurons.
- Compare the different mechanisms of cell signalling.

Introduction

You will be aware that multicellular organisms are highly complex. The bodies of mammals are made up of many billions of cells that are specialized for different functions and which are often co-located with similar cells in the form of organ systems. These organ systems communicate with each other and with the central control systems to maintain the overall functioning of the body. The flow of information takes the form of continuous feedback from sense organs within the body that signals how that system is operating, input from receptors that monitor the external environment, and activating signals that change the activity of the different body systems.

In this chapter we will mainly focus on the processes of signalling in the nervous system but will also consider some of the other forms of signalling, for example, via hormones, which we will discuss in more detail in later chapters.

17.1 The nervous system

In humans, we can think of the nervous system as providing our conscious and unconscious sensations from internal and external environments, as being the source of our consciousness and generation of our behaviours. It is also the controller of specific actions, such as voluntary movement, or the regulation of core functions, such as heart rate or breathing rate.

The nervous system is a complex organization of specialized cells that can be broadly grouped into having three functions:

- sensory feedback;
- integration;
- command output.

The mammalian nervous system is divided into two main sections:

- **central nervous system** (CNS)—comprising the brain and spinal cord;
- **peripheral nervous system** (PNS)—comprising the sensory and motor neurons that connect the CNS to all the different regions of the body.

The specialized cells that convey information within the CNS and PNS are the nerve cells or **neurons**. There are billions of neurons in the mammalian nervous system; it is estimated that there are 10^{12} neurons in the human brain alone. The main supporting cells are called **glial cells**, and in the brain there are even more of these than there are neurons. As we will see in Section 17.5, glial cells have different roles and play a key role in the functioning of the nervous system.

The PNS can be divided into the **somatic** and the **autonomic** nervous systems. The somatic component of the PNS comprises sensory or **afferent neurons** that carry information towards the CNS, and motor or **efferent neurons** that carry signals from the CNS to the skeletal muscles controlling movement.

The autonomic nervous system is further divided into **sympathetic**, **parasympathetic**, and **enteric** neurons (Section 19.1). The *sympathetic* nervous system prepares the body for action, for example, by increasing cardiac output and the rate and depth of breathing, and by diverting blood flow to the muscles. The *parasympathetic* nervous system maintains core physiological functions such as digestion and does have some opposite actions to the sympathetic nervous system—for example, by slowing the heart rate. The *enteric* nervous system comprises sensory and motor pathways and is specifically associated with control of the digestive system (Section 19.1 and Chapter 25).

The neuron

Neurons are cells that are specialized for signal transmission and, although they take up a variety of different shapes, they normally comprise a common set of structures, as you can see in Figure 17.1:

- **Soma**—the cell body, containing the nucleus and other organelles.
- **Axon**—a single process, extending from the soma, carrying the signals to other cells. This may vary in different parts of the body from being fractions of a millimetre to over a metre in length.
- **Dendrites**—a set of branching processes arising from the soma that receive input connections from other neurons.
- **Terminals**—the end(s) of the axon connecting with other cells to transmit the signal.

How does the nervous system know what is being signalled, for example the sense of touch on the thumb as opposed to the little finger? Or the sense of temperature as opposed to touch? During the development of the nervous system, specific neurons connect with specific regions of the brain creating a so-called **labelled line**. This means that the neurons conveying the sense of touch from the thumb terminate in one part of the sensory cortex of the brain, whereas those from the little finger terminate in a different, although close-by, region. Likewise, the senses to touch and temperature are routed to different areas of the sensory cortex.

▶ We learn more about the sensory systems in Chapter 18.

Signal transmission through the nervous system involves both electrical and chemical signals. The axon carries electrical signals from one cell to another in the form of nerve impulses or **action potentials**. Where the axon contacts the dendrites of another neuron, transmission from one cell to the other occurs at a specialized junction called a **synapse**. In most synapses the signal is transmitted via a chemical messenger called a **neurotransmitter**. There are also some **electrical synapses**, where the two neurons come into direct contact and the electrical signal can be passed directly from one neuron to another without the mediation of a neurotransmitter. For most of this chapter we will focus on chemical synapses, as these represent the most common form of transmission.

 Check your understanding of the concepts covered in this section by answering the questions in the e-book.

17.2 Signal conduction within the nervous system

Now that we have been introduced to the basic structure of the neuron, we can focus on how neurons communicate. Information is conducted along neuronal axons by means of action potentials. These electrical impulses are all-or-nothing events; this means that the electrical impulse is not graded in amplitude and remains the same amplitude along the length of the axon.

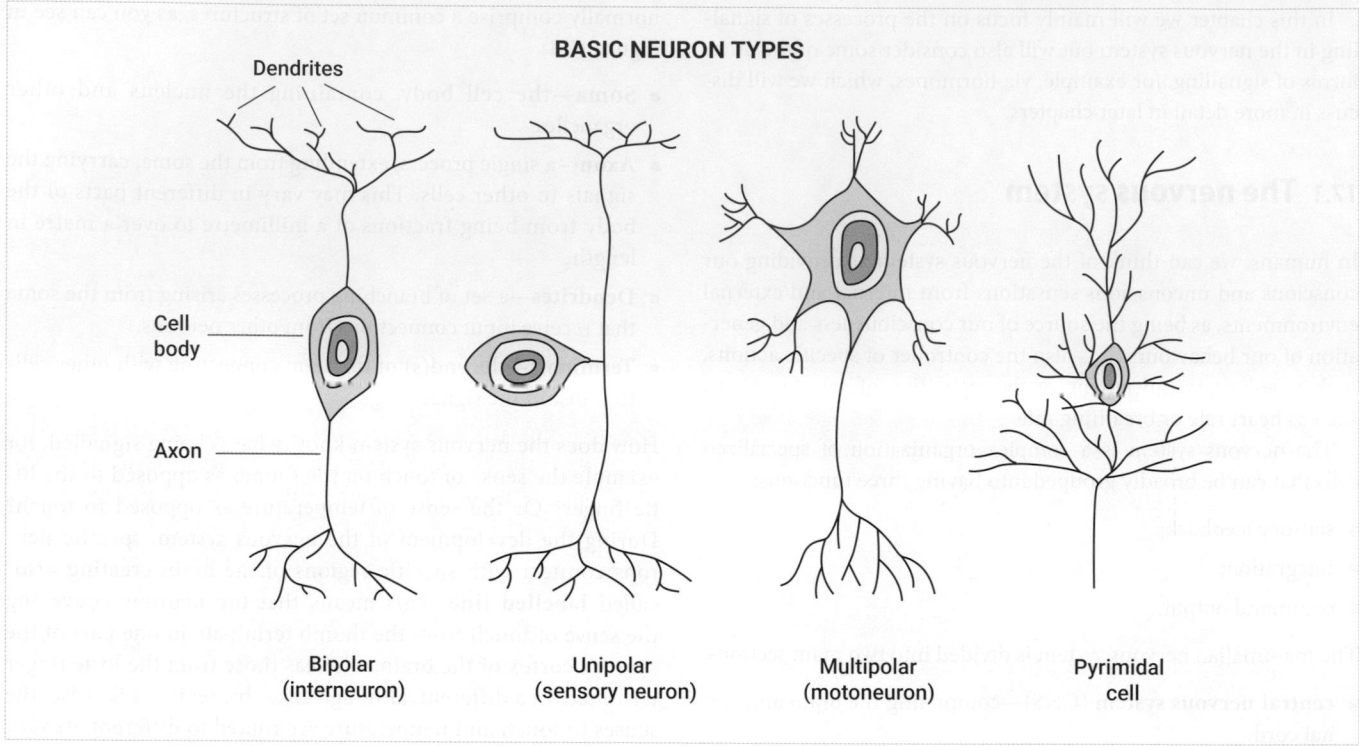

BASIC NEURON TYPES

Dendrites

Cell body

Axon

Bipolar (interneuron)

Unipolar (sensory neuron)

Multipolar (motoneuron)

Pyrimidal cell

Figure 17.1 Different types of neurons. Diagrams of the main types of neurons within the CNS and PNS. Note that, although the appearances differ markedly, the common structures of the cell body (soma), dendrites (input connections), and axon (output process) are present in the different neurons, although a small subset of neurons, such as the amacrine cells of the retina, do not have an axon.

Source: Blamb/Shutterstock.com.

How is information encoded? Take a moment to think about it. Information takes the form of a **frequency code**. This refers to the number of action potentials that the neuron generates per second, often referred to as the action potential frequency or rate of firing. The stronger the stimulus, the higher is the rate of action potential firing.

Action potentials are therefore electrical signals that are generated in neurons and conducted along the axon. Depending on the neuron, the action potential may travel only a millimetre before reaching a synapse, or it may travel well over a metre without being degraded: security of information flow is clearly an important feature of signal transmission.

The key questions are how is the action potential generated and how is it maintained along the length of the axon? These can be answered by considering the specific properties of the cell membrane of the neuron.

PAUSE AND THINK

What is the form of the signal within a neuron and how is the strength of the stimulus encoded?

Answer: The signal within each neuron is conducted in the form of an all-or-nothing action potential. Changes in stimulus strength are encoded not in terms of action potential size, but in the frequency of action potential discharge: the frequency code.

 Check your understanding of the concepts covered in this section by answering the questions in the e-book.

17.3 Resting potential

Resting potential overview

The easiest way to think about generating action potentials is to use the metaphor of a battery. If we take a standard 1.5 V battery and connect a voltmeter to the positive and negative terminals we should record a voltage, or potential difference, of +1.5 V, as you can see in Figure 17.2a.

We then connect our voltmeter across a nerve axon, with one electrode inserted into the **axoplasm** (axoplasm is the term often used to describe the intracellular contents (cytoplasm) found in neuronal axons) and the other in the extracellular fluid, demonstrated in Figure 17.2b. We will record a voltage that will be typically about −70 mV, with the inside of the axon being at a negative potential compared to the extracellular fluid. This voltage is referred to as the **membrane potential**. All cells in the body have a membrane potential across the cell membrane, although this varies in amplitude depending on the cell type. In the case of neurons, this membrane potential is referred to as the **resting potential**, which distinguishes it from the action

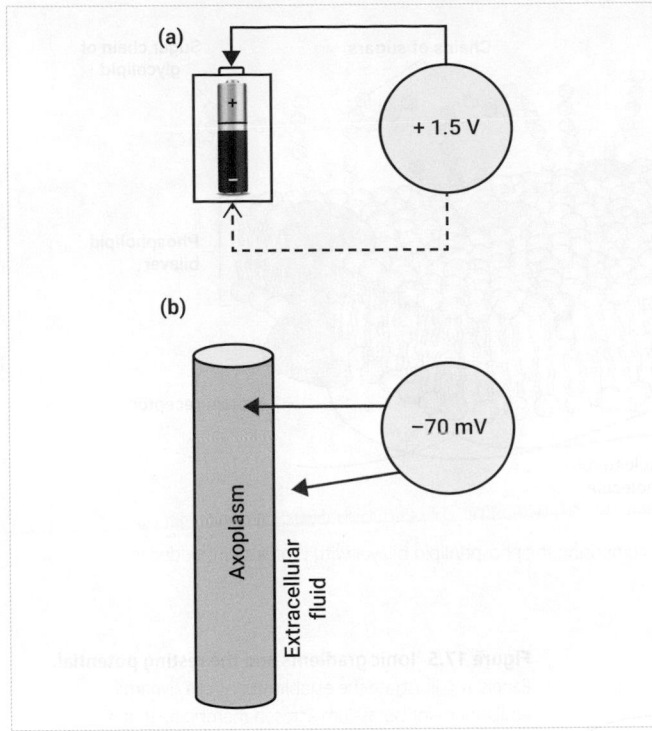

Figure 17.2 Measuring voltage. (a) This shows the terminals of a standard 1.5 V battery connected to a voltmeter, recording a voltage of +1.5 V. (b) This shows an axon connected to a voltmeter, where the 'terminals' are the axoplasm and the extracellular fluid. In this case, the recorded voltage is −70 mV (0.07 V) with the axoplasm at a negative potential with respect to the extracellular fluid.

Source: MarySan/Shutterstock.

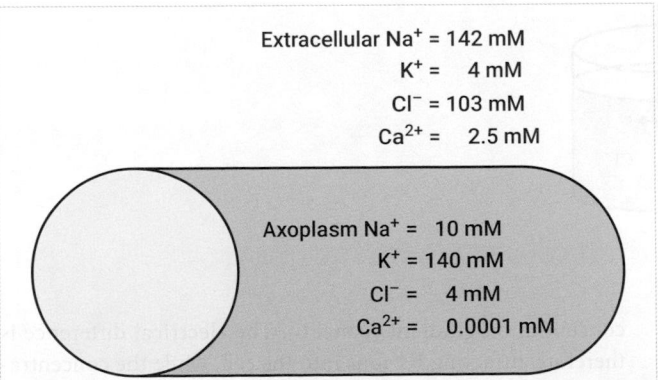

Figure 17.3 The relative concentrations of some key ions in the axoplasm and extracellular fluids.

potential that represents the signal transmission. The existence of the resting potential is fundamental to the physiology of signal transmission within the neurons.

The resting potential arises as a result of the specific properties of the axonal membrane itself and a difference in the distribution of ions either side of the membrane, as shown in Figure 17.3. The ionic composition of the axoplasm is characterized by relatively high concentrations of potassium ions and low concentrations of

sodium and chloride ions. The opposite is true of the extracellular fluid, where there is a low concentration of K^+ ions and relatively high concentrations of Na^+ and Cl^-.

Cell membrane

The properties of the cell membrane are discussed in detail in Module 2 and in Section 16.2 so will only be summarized in this chapter in relation to neuronal signalling. Figure 17.4 shows a diagram of the cell membrane.

The basic structure of the cell membrane comprises a bilayer of phospholipid molecules with associated proteins, some of which are integral within the bilayer, spanning between the intra- and extracellular surfaces. There are other molecules associated with the membrane structure, including glycolipids and glycoproteins, which are lipids and proteins that have sugars attached to them. Some of these play an important role as receptor molecules in neuronal and hormonal signalling (Section 17.7 and Section 19.3).

The phospholipid molecules have two distinct regions, as you can see in Figure 17.4, and as illustrated in more detail in Figure 1.23a: the head, which is polar, and the tail, composed of fatty acids, which is non-polar. The polar head is hydrophilic, whereas the non-polar tail is hydrophobic. When the bilayer is formed, the hydrophilic regions are arranged on the external surfaces, where they come into contact with the aqueous intra- or extracellular fluids. The hydrophobic tails of the two layers come together in the centre of the membrane sheet where they are not exposed to water.

What is the significance of the phospholipid bilayer for the movement of substances across the membrane? The lipid bilayer limits the type of substance that can cross the membrane by diffusion: such substances have to be fat soluble, of small molecular size, and not charged. For example, small molecules such as oxygen and carbon dioxide can diffuse through the membrane bilayer, but the charged ions such as potassium, sodium, calcium, or chloride, cannot (Chapter 16).

The lipid bilayer therefore prevents the movement of ions across the cell membrane. This movement can only occur via the proteins that are inserted into the bilayer. These proteins can enable passive movement of ions, for example through **ion channels**, or active transport of ions, for example by the Na^+K^+ ATPase pump (see Chapter 16 and later in this section).

Equilibrium potential

As we noted, the resting potential across the neuronal membrane is dependent on the distribution of the ions either side of the membrane and the permeability of the membrane to those ions. How these relate together is discussed in the model below.

1. If there is an aqueous solution of potassium chloride (KCl) in a beaker, then the positively charged ions (**cations**), K^+, and the negatively charged ions (**anions**), Cl^-, will be distributed evenly throughout the solution (Figure 17.5, panel 1).

2. Placing a membrane across the beaker, thereby dividing the solution in half, will create two equal solutions (panel 2).

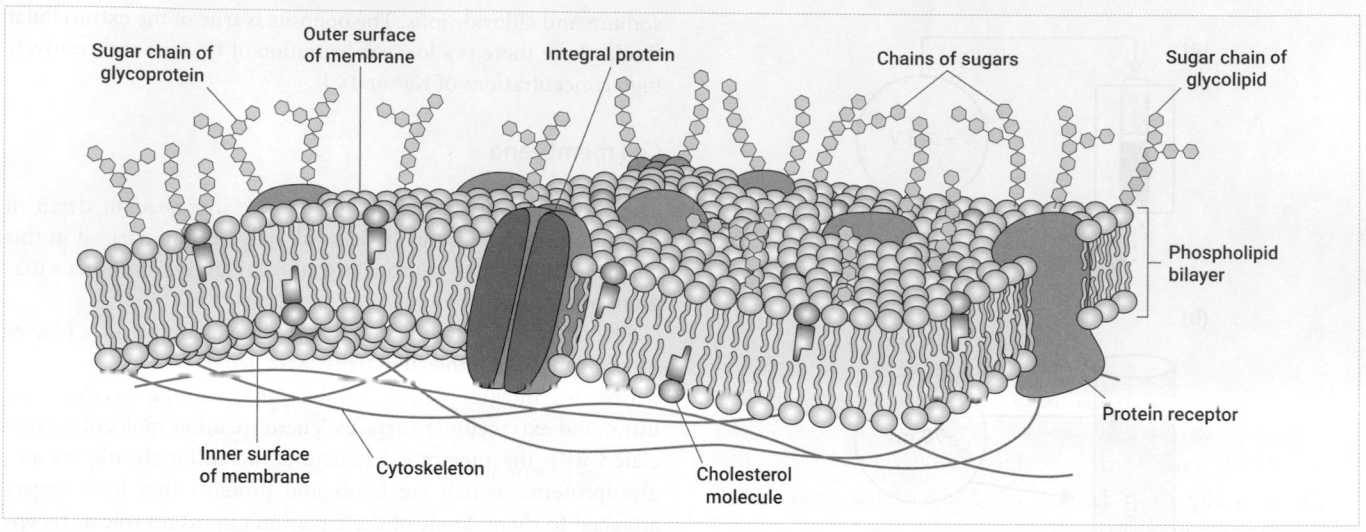

Figure 17.4 The cell membrane. This shows the basic structure of the cell membrane, comprising the phospholipid bilayer with proteins embedded in it. *Source*: © The Open University.

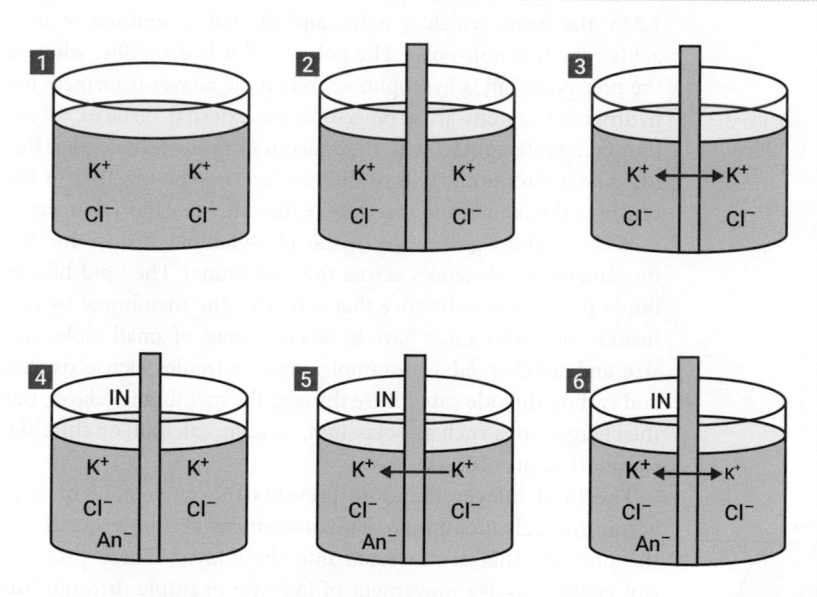

Figure 17.5 Ionic gradients and the resting potential.
Panels 1–6 illustrate the establishment of a dynamic equilibrium for potassium across a membrane that is permeable only to K+ ions. See text for more detail.

3. If the membrane is only permeable to K+ ions, then those ions are free to move in either direction, but the net effect will remain as an equal distribution of the ions (panel 3).

4. If we think about the inside and the outside of a cell, then the cytoplasm of cells contains proteins and amino acids, which are usually negatively charged (shown as An− in panel 4). These molecules are too large to move across the membrane and so there is an unequal distribution of charges with the inside of the cell having more negative charges than the outside.

5. As a consequence, K+ ions will move into the cell to rebalance the difference in charge (panel 5).

6. As this happens, the concentration of K+ ions inside the cell becomes higher than that outside the cell, creating a concentration gradient (panel 6). The electrical difference is therefore attracting K+ ions into the cell, while the concentration gradient is tending to move K+ ions out of the cell. The two directions of movement therefore oppose each other: as more K+ ions move into the cell, so the electrical gradient becomes smaller, while the concentration gradient becomes larger until a point is reached when they are equal and opposite in force. At this point, an equilibrium is reached when there will be no more net movement of K+ ions in either direction, but the inside of the cell will still be at an electrical potential that is negative with respect to the outside, the **equilibrium potential**. This is also referred to as a **dynamic equilibrium**, because there is ionic movement in both directions across the membrane, but no net change.

Ion channels

In the neuron, the passage of charged ions across the membrane is made possible by the presence of passive protein-mediated pathways in the form of ion channels. These ion channels are formed of integral proteins that span the membrane, creating a pore through which ions can pass. There are two main types of channel:

- Leakage or passive channels that are always open.
- Gated channels, which are normally closed but can be opened in response to specific stimuli:
  - Voltage-gated channels open or close in response to changes in the membrane potential.
  - Ligand-gated channels open or close in response to binding of specific chemicals such as neurotransmitters.
  - Mechanically gated channels respond to mechanical stimuli such as stretching.

In the neuronal membrane at rest, the gated channels are closed, so any ion movements are through the leakage channels. These are predominantly permeable to potassium ions and relatively impermeable to other ions such as sodium, chloride, or calcium.

Why are channels permeable to different ions? The ion channels are proteins made up of several subunits that come together to form a pore or tube through the membrane. The leakage channels that allow potassium to travel through are mainly **tetrameric** (comprised of four subunits) proteins where the four subunits come together to create the pore. The channel that is created contains a **selectivity filter** that limits ion passage based on the charge associated with the ion and its diameter. Therefore, these leakage channels only allow K^+ ions through, because the pore of the channel contains charged segments that repel negatively charged ions (e.g. Cl^-) and which create a pore diameter that does not allow larger ions to pass through (Figure 17.6).

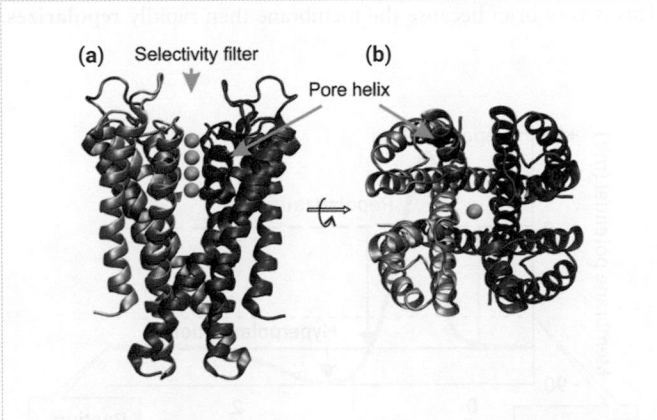

(a) Selectivity filter **(b)** Pore helix

Figure 17.6 Structure of a potassium leakage channel shown in (a) lateral view and (b) surface view. The channel comprises four subunits that form the pore. The selectivity filter is on the extracellular surface of the channel and filters ions based on charge and size, such that only K^+ ions can pass through.

Source: Figure 1 from D. Naranjo, H. Moldenhauer, M. Pincuntureo, I. Díaz-Franulic (2016). Pore size matters for potassium channel conductance. *J Gen Physiol*, 148(4), 277. DOI: 10.1085/jgp.201611625.

Calculating the equilibrium potential

We have seen that there is an unequal distribution of ions on either side of the cell membrane and, as a result, there are both concentration and electrical gradients acting on the ions (Figures 17.3 and 17.5). There is therefore a relationship between the relative concentrations of the ions and their equilibrium potentials. This relationship is expressed in the form of the **Nernst equation**:

$$E_{ion} = 2.303 \frac{RT}{zF} \log \frac{[ion]_{out}}{[ion]_{in}}$$

where

- E_{ion} = the equilibrium potential for a given ion
- R = gas constant
- T = absolute temperature in K (kelvin)
- z = charge on the ion
- F = Faraday's constant
- log = logarithm to base 10.

For a given ion, the charge (z) is constant, and R and F are constants; therefore, we can summarize the relationship in words as:

The equilibrium potential for a given ion is proportional to the logarithm of the ratios of the concentrations of that ion either side of the membrane and to the temperature.

At a body temperature of 37 °C, which is an absolute temperature of 310.15 K, the equation for potassium simplifies to:

$$E_K = 61.54 \log \frac{[K^+]_{out}}{[K^+]_{in}} \text{ mV}$$

▶ **You can find a detailed guide to the Nernst equation in Quantitative Toolkit 8.**

We saw in Figure 17.3 that the relative concentrations of potassium were 4 mM and 140 mM outside and inside the neuron, respectively. If we insert these values into the Nernst equation, this gives us an equilibrium potential for potassium of:

$$E_K = -95 \text{ mV}$$

What does an equilibrium potential of −95 mV mean? The equilibrium potential is the electrical potential across the membrane that would exactly balance the concentration gradient. So, with the inside of the neuron at −95 mV compared to the extracellular fluid, the electrical attraction of the positive charge on the potassium ions (which draws the ions into the neuron) exactly balances the concentration gradient (which moves the potassium ions out of the neuron).

Figure 17.2 showed that the membrane potential for a neuron at rest is approximately −70 mV. If the membrane potential is dependent on K^+, why is the resting potential not the same as the equilibrium potential for potassium? There is a difference between the membrane potential that is measured and the equilibrium potential for K^+ because the membrane is not perfectly impermeable to other ions that make up the extracellular fluid, and because the membrane potential is also partially dependent on the actions of the Na^+K^+ ATPase pump.

The Nernst equation can be used to calculate the equilibrium potentials for each of the ions present in the intra- and extracellular fluids, the most important ions being potassium, chloride, sodium, and calcium.

$$E_{Cl} = -61.54 \log \frac{\left[Cl^- \right]_{out}}{\left[Cl^- \right]_{in}} mV$$

Note that chloride is negatively charged, so z = −1.

$$E_{Cl} = -86.8\,mV$$

$$E_{Na} = 61.54 \log \frac{\left[Na^+ \right]_{out}}{\left[Na^+ \right]_{in}} mV$$

$$E_{Na} = 70.9\,mV$$

$$E_{Ca} = 30.77 \log \frac{\left[Ca^{2+} \right]_{out}}{\left[Ca^{2+} \right]_{in}} mV$$

Note that calcium has two positive charges, so z = 2.

$$E_{Ca} = 135.3\,mV$$

So, the equilibrium potentials for sodium and calcium are both positive compared with the negative potentials for potassium and chloride. In other words, if the membrane was permeable to sodium alone, the membrane potential would move to approximately 71 mV. If there is slight leakage of sodium into the cell, then the resting membrane potential will still be dominated by potassium but will be slightly less negative than the potassium equilibrium potential because of the small leakage of positive sodium ions into the neuron.

The Na^+K^+ ATPase pump

The sodium–potassium ATPase pump is the key to the operation of many cellular transport systems within the body, as we shall see when we explore the operation of systems such as the kidney and the gastrointestinal tract. The pump also has a key role in the maintenance of the resting membrane potential.

The pump actively pumps Na^+ ions out of the cell against their electrical and concentration gradients, and pumps K^+ ions into the cell. The actual process involves the **hydrolysis** of ATP:

$$ATP \rightarrow ADP + Pi$$

For each molecule of ATP that is hydrolysed, three Na^+ ions are pumped out in exchange for every two K^+ ions pumped in. Therefore, there is a net loss of positive ions from the inside of the cell and so the pump contributes to the generation of the negative resting potential. This action of the Na^+K^+ ATPase pump is termed **electrogenic**, because it generates electrical potential difference. As well as maintaining the resting potential during inactivity, the

pumping action is also important in pumping out Na^+ ions following the generation of action potentials.

 Check your understanding of the concepts covered in this section by answering the questions in the e-book.

17.4 Action potentials

The action potential is the signalling event: the transmission of a brief electrical pulse along the length of an axon. This is defined as an **all-or-nothing** event: this means that if the stimulus exceeds the threshold level an action potential is generated and the amplitude of the action potential will always be the same, irrespective of the strength of the stimulus above the threshold level. As described in Section 17.2, the strength of the stimulus is signalled by the frequency of the action potentials (frequency code), not by their amplitude.

If we record the electrical activity in an axon by inserting a microelectrode into the axoplasm at a specific point and measuring the membrane potential relative to the extracellular fluid (Figure 17.2), we will record the resting membrane potential of −70 mV. If we carry on recording while an action potential passes along the axon, the membrane potential will, for a very brief period of time, change to about +30 mV before reverting to the resting potential, as shown in Figure 17.7.

At rest, the axonal membrane is said to be **polarized** because, as with the terminals of a battery, there is an electrical potential existing across the membrane. At the start of the action potential (Figure 17.7), there is a gradual change in the membrane potential towards zero. Such a change is referred to as **depolarization** because the membrane is becoming less polarized—i.e. it is moving towards being electrically neutral.

When the extent of depolarization exceeds a certain threshold level, the rate of depolarization suddenly increases and the membrane potential changes to the point where the inside of the axon actually becomes positive with respect to the extracellular fluid. This is very brief because the membrane then rapidly **repolarizes**,

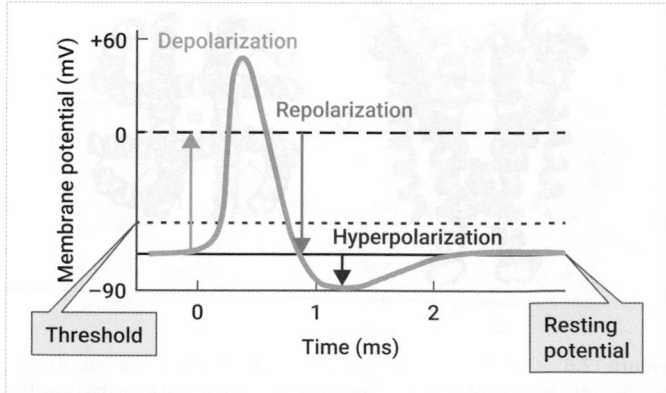

Figure 17.7 The action potential. The graph shows the change in membrane potential at a specific point on the axon as the action potential passes along it. Brief opening of the Na^+ channels causes the potential to be positive before reverting to the negative resting potential.

restoring the negative membrane potential. Indeed, the membrane potential will actually go more negative than the resting potential: it will **hyperpolarize**, approaching −90 mV before recovering to the normal resting membrane potential. As shown in Figure 17.7, all of this is very rapid, taking place in just a few milliseconds.

What is the significance of the threshold level? If there is a very weak stimulus, such that the membrane depolarization does not exceed the threshold level, then the action potential will not be triggered and the membrane potential will return to the resting state. Such subthreshold stimuli may occur, for example, at synapses or in sensory systems when the sensory stimulus is large enough to be detected by the sense organ but not large enough to generate a signal that is transmitted to the brain.

The threshold depolarization for generating an action potential is therefore a critical point in the signalling process. At rest, the membrane is permeable to K⁺ ions through the potassium channels that are always open, but it is effectively impermeable to other ions such as Na⁺, Cl⁻, and Ca²⁺. The resting membrane therefore has a high **conductance** to K⁺ but almost zero conductance to the other ions. Conductance (measured in Siemens) is a measure of the ease with which a structure conducts electricity and is the opposite of resistance, so a substance that has high conductance has low resistance.

During the action potential, a group of voltage-gated Na⁺ channels open: as their name suggests, these channels open in response to changes in the membrane potential and are specific for sodium ions. In this case, depolarization of the membrane causes a change in the folding of the protein pores, causing them to open. The opening of the voltage-gated Na⁺ channels greatly increases the sodium conductance as shown in Figure 17.8. As we saw from the Nernst equation calculations, the equilibrium potential for Na⁺ is about 70 mV because there is a high concentration of Na⁺ ions outside the neuron compared with inside, and the inside of the axon, at rest, is at a negative potential, so both the concentration and the electrical gradients will act to draw Na⁺ ions into the axon. Therefore, opening of the voltage-gated Na⁺ channels will allow Na⁺ ions to enter the axon and the influx of the positively charged ions will make the inside of the cell positive with respect to the outside.

Why does the membrane potential only change to about 30 mV rather than reaching the equilibrium potential for Na⁺ of 70 mV? During the action potential, the K⁺ leakage channels remain open. As the membrane depolarizes and moves further away from the K⁺ equilibrium potential, so potassium ions will tend to move out of the axon, therefore the final level reflects the balance of the inward movement of Na⁺ and the outward movement of K⁺ ions.

The open Na⁺ channels are unstable and remain open for only a very brief period, about 1 ms, before they inactivate, shutting off the inward flow of sodium ions. In their inactive state, the Na⁺ channels cannot be re-opened until the membrane potential returns to the resting state.

If the inward flow of Na⁺ ions is stopped, then the membrane will start to repolarize as the K⁺ ions will continue to flow outwards because the concentration and electrical gradients drive K⁺ out of the cell. This repolarization would be a relatively slow process, but it is accelerated by the opening of another group of channels,

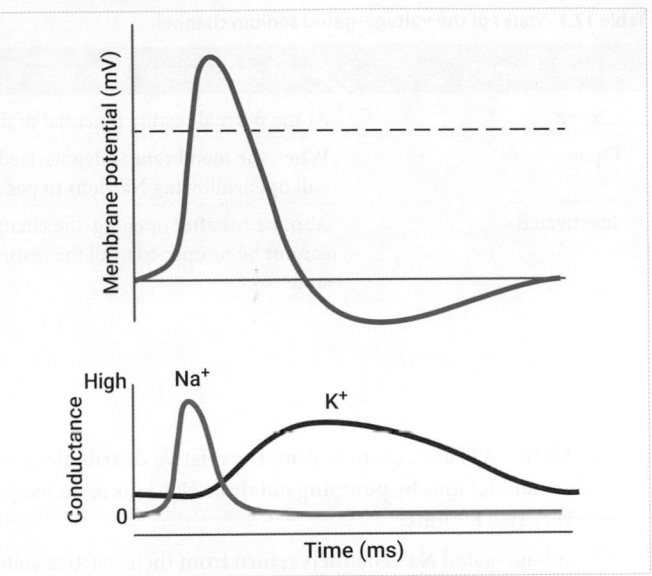

Figure 17.8 Changes in ionic conductance leading to the action potential. Opening of the voltage-gated sodium channels increases the conductance to sodium ions (lower panel, orange line) allowing them to move in down their concentration and electrical gradients. This movement gives rise to the positive upswing in the membrane potential (upper panel). The sodium channels only open briefly and then close, terminating the action potential and repolarization is aided by the outflow of K⁺ ions through the passive K⁺ channels and opening of voltage-gated K⁺ channels (lower panel, red line).

the voltage-gated K⁺ channels, which increase the K⁺ conductance (Figure 17.8) and the rate of loss of K⁺ ions. As positive ions flow out of the axon, the membrane repolarizes and, because the conductance for K⁺ is higher than at rest, the membrane potential actually moves closer to the equilibrium potential for K⁺, so at this stage the membrane is hyperpolarized. As the voltage-gated K⁺ channels close in turn, the membrane potential gradually recovers to the normal resting level.

As the resting state of the membrane is restored, the Na⁺K⁺ ATPase pumps the Na⁺ ions out of the axon and K⁺ ions back in, restoring the respective concentration gradients and contributing to the restoration of the resting membrane potential.

So, to summarize, the stages of the generation of the action potential are:

1. Initial depolarization of the membrane exceeds the threshold level, opening the voltage-gated Na⁺ channels.

2. Na⁺ enters the neuron, moving the membrane potential to a positive level, approaching the equilibrium potential for Na⁺.

3. Voltage-gated Na⁺ channels inactivate, shutting off the influx of Na⁺ ions.

4. Voltage-gated K⁺ channels open, increasing potassium conductance and facilitating the repolarization.

5. The increased K⁺ conductance leads to an initial hyperpolarization as the membrane potential actually becomes more negative than at rest.

Table 17.1 States of the voltage-gated sodium channel

State of channel	Electrical field across the membrane
Closed	At the normal resting potential of the membrane
Open	When the membrane is depolarized beyond the activation threshold, which is about -40 mV, the channels will open, allowing Na^+ ions to pass through
Inactivated	About 1 ms after opening, the channels spontaneously shut and enter an inactive state from which they cannot be re-opened until the resting membrane potential is restored and the channels revert to their closed state

6. The Na^+K^+ ATPase acts to restore the relative distributions of the Na^+ and K^+ ions by pumping out three Na^+ ions in exchange for every two K^+ ions.

7. The voltage-gated Na^+ channels return from their inactive state to their resting state, ready to open again for the next action potential, and the voltage-gated K^+ channels close.

Voltage-gated sodium channel

The voltage-gated sodium channel, like the K^+ leakage channel, forms a pore in the membrane that allows a specific ion to pass through: in this case Na^+ ions. However, unlike the leakage channel, the voltage-gated sodium channel exists in three states, depending on the electrical field across the membrane, as shown in Table 17.1.

The channel is formed from a single polypeptide chain, the α-subunit, and one or two associated β-subunits. The α-subunit, which forms the structure of the pore, is divided into four groups, or domains, each of which in turn comprises six transmembrane alpha-helices that extend the full width of the membrane (Figure 17.9).

Depolarization of the membrane leads to a **conformational** change in the channel protein due to movement of the S6 segments (Figure 17.9) and which results in channel opening.

There are numerous different types of voltage-gated sodium channels, which differ in terms of sections of the α-subunit polypeptide sequences. These channels are found in different locations in the CNS, PNS, and muscles. These differences offer the potential for targeted therapies, for example, the Na_v 1.8 and 1.9 channels are involved in the conduction of pain signals, and therefore offer potential targets for pain relief.

PAUSE AND THINK

When the neuron is at rest, why does sodium not flow into the axon given its concentration and electrical gradients?

Answer: At rest the axonal membrane is impermeable to sodium. The sodium channels are closed. When the neuron is depolarized, the change in electrical field alters the shape of the sodium channel proteins, opening the voltage gate and allowing sodium ions to enter the neuron.

Refractory periods

As mentioned before, the frequency or rate of action potential discharge increases with the increasing strength of the stimulus. However, it will reach a point where the axon cannot respond any faster. The maximum frequency is about 1000 impulses s^{-1}, or one impulse every ms, although most neurons do not discharge at such high rates.

The rate of firing of action potentials is limited by the **refractory period** of the axon. This can be divided into two periods, one following the other:

- **absolute refractory period**: the period following one action potential during which the axon cannot fire another action potential regardless of how strong the stimulus;
- **relative refractory period**: the period following the absolute refractory period during which the axon can fire another action potential but requires a stronger than normal stimulus to do so.

After the opening of the voltage-gated Na^+ channels, the channels spontaneously close and become inactive. This is the absolute refractory period, when the channels cannot be re-opened and therefore the axon cannot fire another action potential.

As the membrane repolarizes, the voltage-gated Na^+ channels change from the inactive state to their resting state and so they return to a state from which they can be made to open again. During this relative refractory period some of the voltage-gated K^+ channels are still open and the membrane is hyperpolarized, therefore it requires a stronger than normal stimulus to bring the membrane potential to the threshold and to re-open the voltage-gated Na^+ channels.

How many ions actually move during the action potential?

The action potential occurs as a result of the opening of voltage-gated Na^+ channels, which allow Na^+ ions to enter the axon down their concentration and electrical gradients, and which causes the membrane potential to change from -70 mV to about $+30$ mV. In many descriptions of the action potential you will read that this change in potential is brought about by 'the sodium ions rushing in…', the implication being that that there is a large movement of Na^+ ions. In reality, the number of ions that move is very small by comparison with the total number of ions in the intra- and extracellular fluids. In other words, a single action potential results in a negligible change in the concentrations of the ions.

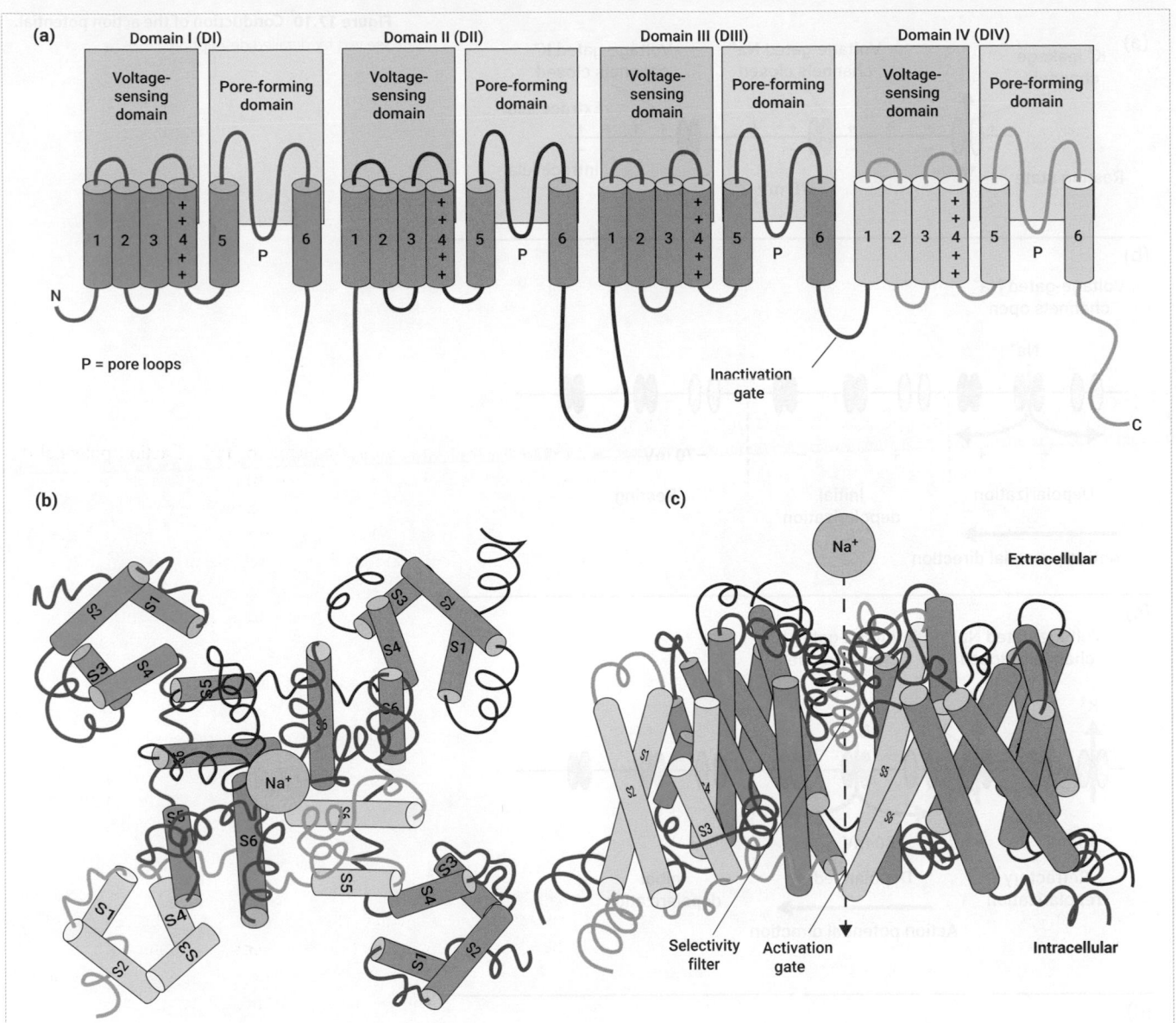

Figure 17.9 Structure of the voltage-gated sodium channel. (a) The four domains of the α-subunit (D I – IV) that are linked by intracellular chains. The domains are shown spread out linearly. Each domain comprises six transmembrane helices: S1–4 of each domain form the voltage sensor, while S5 and S6 come together, along with the pore loops, to form the channel pore. Movement of the S6 segment, the activation gate, causes the channel to open. The link between domains III and IV forms the inactivation gate that closes the pore following channel opening. (b) Configuration of the channel from the extracellular surface with a Na+ ion. (c) Side view of the channel with the selectivity filter, which is the narrowest part of the pore, that acts to prevent ions other than sodium from passing through, and the activation gate, formed of the segments (S6) that move during depolarization to open the channel.

Source: de Lera Ruiz, M., Kraus, R.L. Voltage-gated sodium channels: structure, function, pharmacology, and clinical indications. *J Med Chem* 2015;24;58(18):7093–118. https://pubs.acs.org/doi/10.1021/jm501981g. Further permission related to this material excerpted should be directed to the ACS.

This might at first seem difficult to understand: in order to do so, we need to go into a little more depth regarding the electrical properties of the plasma membrane.

The axoplasm and the extracellular fluids are both aqueous solutions containing charged ions that can move freely in those solutions. The solutions therefore act as good electrical conductors that are separated by the cell membrane that acts as an insulator. In such a system, the membrane is separating the two solutions that have different amounts of electrical charges, so it is acting as a **capacitor**. Positive and negative charges attract each other, so when the membrane is at rest positive charges accumulate on the external surface of the membrane, while negative charges accumulate on the axoplasmic surface. The more effective the capacitor is, the greater the number of electrical charges that can be stored either side of the membrane. It is this separation of charge that is fundamental to the resting membrane potential.

During the action potential, the influx of Na+ ions results in dissipation of the charges accumulated across the membrane, thereby depolarizing the axon. This means that the membrane potential changes without there being a significant change in the overall ion

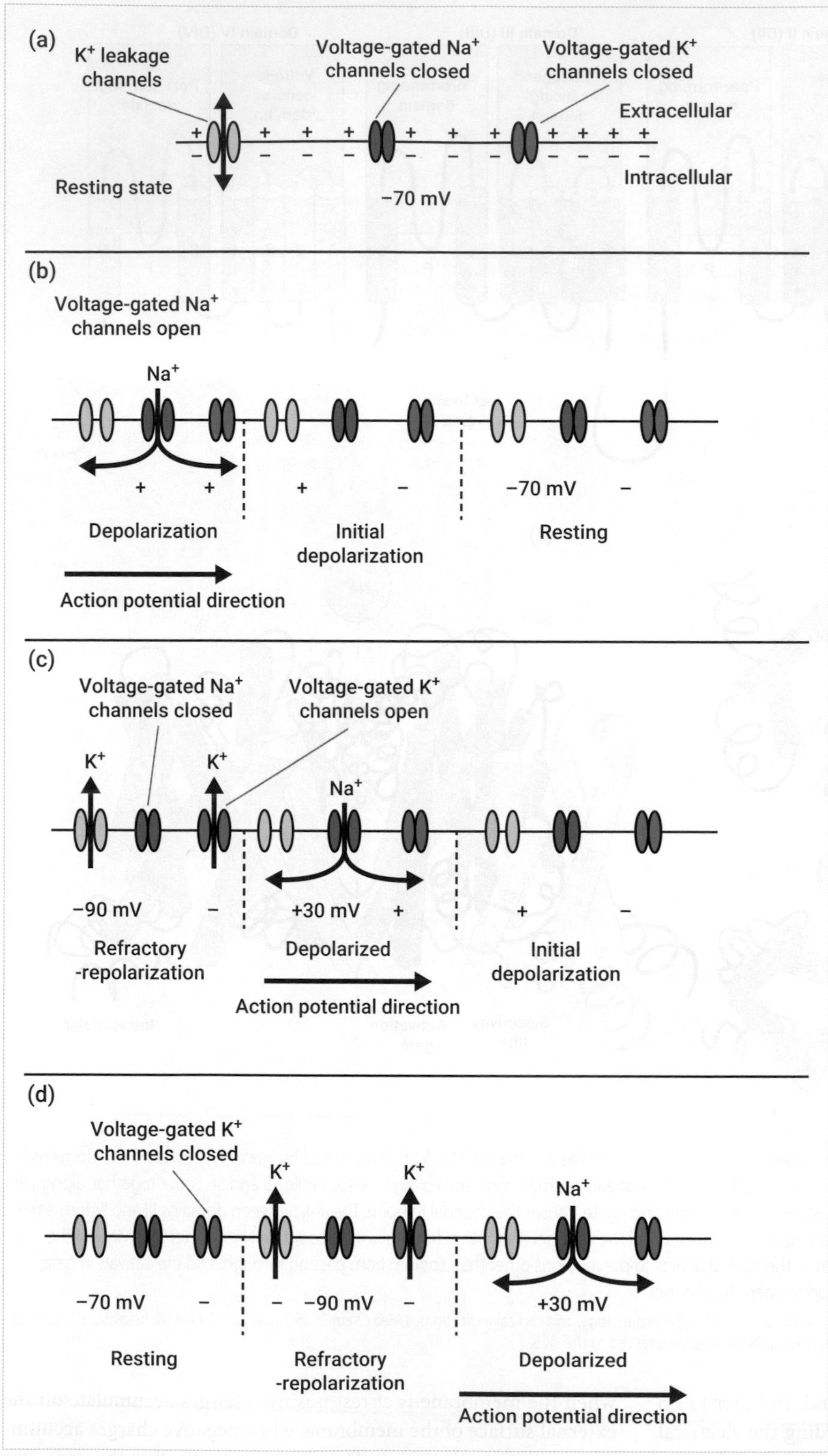

(a)

K+ leakage channels

Voltage-gated Na+ channels closed

Voltage-gated K+ channels closed

Extracellular

Intracellular

Resting state

−70 mV

(b)

Voltage-gated Na+ channels open

Na+

Depolarization

Initial depolarization

Resting

−70 mV

Action potential direction

(c)

Voltage-gated Na+ channels closed

Voltage-gated K+ channels open

K+

K+

Na+

−90 mV

Refractory -repolarization

+30 mV

Depolarized

Initial depolarization

Action potential direction

(d)

Voltage-gated K+ channels closed

K+

K+

Na+

−70 mV

Resting

−90 mV

Refractory -repolarization

+30 mV

Depolarized

Action potential direction

Figure 17.10 Conduction of the action potential.
See text for detailed description.

concentrations. A very rough calculation indicates that the total number of Na+ ions in an axon increases by about 0.05% as a result of the Na+ influx during an action potential.

 Check your understanding of the concepts covered in this section by answering the questions in the e-book.

17.5 Action potential conduction

So far, we have seen how an action potential is generated at a specific location on the axon. We next need to consider how the action potential moves along the axon and how fast it can be conducted.

1. Panel (a) of Figure 17.10 shows the resting state with the voltage-gated Na⁺ and K⁺ channels closed and the only ion flow being the bidirectional flow of K⁺ ions through the leakage channel.

2. Panel (b) shows an action potential at the left-hand end of the axon, with the voltage-gated Na⁺ channels open and the inward current carried by sodium ions reversing the membrane potential, such that the inside of the axon is positive with respect to the outside. The right-hand end of the axon is still at its resting potential. The influx of positive ions through the voltage-gated Na⁺ channels creates an imbalance within the axon with the left-hand section positive but the rest of the axon negative. This sets up a local current, which initiates depolarization of the section of axon next to the part that is carrying the action potential.

3. The local depolarization brings the voltage-gated Na⁺ channels up to threshold, so that they then open, as shown in the middle section of panel (c), allowing sodium ion influx. This, in turn, establishes the local currents that lead to depolarization of the next section of axon, at the right-hand end of the panel. Meanwhile, the left-hand end of the panel shows that the voltage-gated Na⁺ channels have closed and the voltage-gated K⁺ channels have opened, so this section of the axon is in its refractory state and is repolarizing.

4. In panel (d), the action potential has moved further along the axon, so the right-hand end of the section is now at a positive potential, the left-hand end is back in its resting state, and the central section is refractory.

 Go to the e-book to watch an animation that illustrates the movement of an action potential along an axon.

PAUSE AND THINK

Why does the action potential travel in only one direction?

Answer: The action potential cannot travel backwards because the section of axon immediately behind the action potential is refractory and so cannot fire another action potential until it has recovered. If, however, we carried out an experiment of electrically stimulating an axon mid-way along its length, the action potential could travel in both directions because there would initially be no refractory regions.

So, the process of opening and closing the voltage-gated Na⁺ and K⁺ channels along with the local current flow in the axon leads to the action potential moving along the axon, and it remains the same amplitude because it is continually being regenerated by the ion fluxes.

Conduction velocity

Mammalian axons conduct action potentials at velocities that range from 0.5 to 120 ms⁻¹. The two main factors that affect conduction velocity are the diameter of the axon and the presence of a **myelin sheath**: axons may therefore be classified as myelinated and unmyelinated.

Diameter of the axon

As we have seen, there are two routes for current flow during an action potential: across the membrane through the ion channels, and via local currents through the axoplasm. The thicker the axon, the lower the axoplasmic resistance relative to the resistance of the membrane, so the local currents spread further in advance of the action potential,

depolarizing the membrane, before decaying due to loss through leakage through the membrane. As a consequence, in thicker axons the conduction velocity is greater than in thin axons.

Myelin sheath

Special types of glial cells, the support cells that are present in huge numbers in the nervous system, are associated with the production of myelin. In the PNS these are the **Schwann cells**, while in the CNS they are the **oligodendrocytes**. These cells wrap their plasma membranes around the axons (Figure 17.11), with the phospholipid membranes acting as an electrical insulator around the axon, greatly increasing the relative electrical resistance compared with that of the axoplasm.

As seen in Figure 17.11, each segment of myelin is separated by a gap, the **node of Ranvier**, where the axonal membrane is exposed to the extracellular fluid. In an unmyelinated axon ion channels are distributed all along the length, whereas in a myelinated axon the voltage-gated Na⁺ channels are mainly confined to the nodes of Ranvier. As a consequence, the ion movements associated with the action potential occur only at the nodes with local currents flowing between the nodes, so the action potential effectively 'jumps' from node to node, a process called **saltatory conduction**. Saltatory conduction therefore enables much higher conduction velocities than occur in unmyelinated axons. Furthermore, the gaps between nodes are further apart in thicker axons, where the axoplasmic resistance is relatively lower, contributing to the higher velocities seen with the increase in axonal diameter.

The slowest conducting nerve fibres are therefore the unmyelinated axons, those that lack a myelin sheath. These C fibres also typically have the smallest diameters and conduct at velocities below 2 ms⁻¹. In C fibres, the voltage-gated channels are spread along the entire length of the axon and the axons also have small diameters.

 Check your understanding of the concepts covered in this section by answering the questions in the e-book.

17.6 Transmission between neurons

So far, we have considered the conduction of an action potential along the length of the axon of a neuron. Signalling within the body also involves communication between neurons, often on a very large scale: for example, a neuron in the cerebral cortex of your brain might receive inputs from several million other neurons. These interconnections between neurons are called synapses.

There are two main types of synapse, depending on the mechanism of transmission involved:

- Electrical synapses, which are relatively rare, involve the direct electrical transmission of the action potential from one neuron to the next.

- Chemical synapses involve the transmission of the signal between the two neurons by means of diffusion of a chemical, the neurotransmitter.

Electrical synapses

Because there is direct electrical connection between the two neurons, transmission across an electrical synapse is very rapid. The two

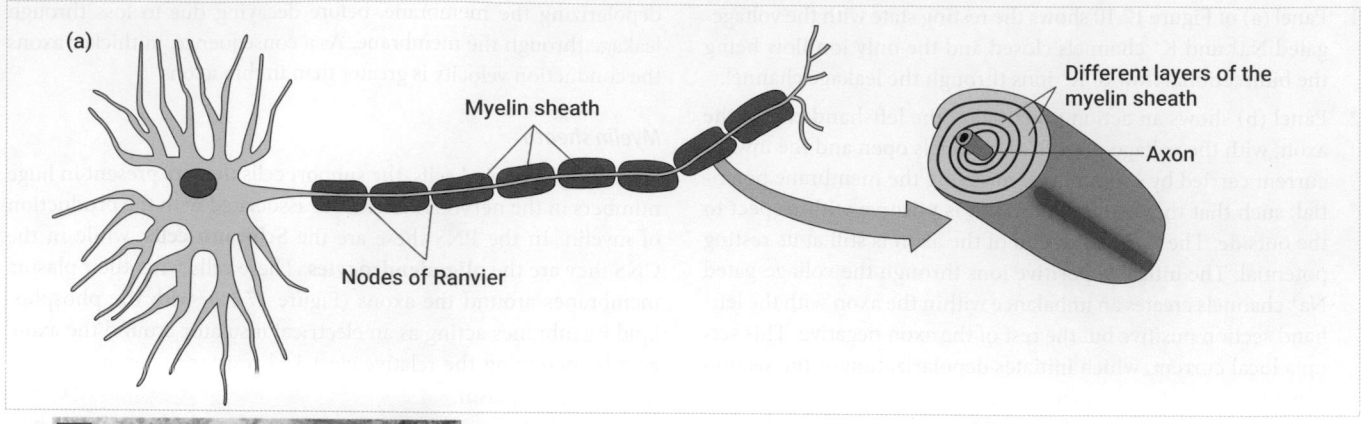

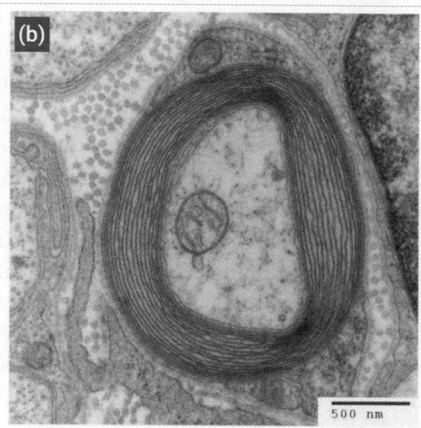

Figure 17.11 Myelin sheath. (a) The arrangement of the myelin sheath around the axon of a peripheral neuron. Each section of myelin is separated by a node of Ranvier. (b) A cross-section of a myelinated axon; the individual layers of plasma membrane making up the myelin sheath are clearly visible.
Source: (a) R. Joseph, 2011. (b) Wikimedia Commons/Mariana Ruiz Villarreal LadyofHats/Public Domain.

neurons may be only 3 nm or less apart, with transmission occurring across so-called **gap junctions** (Figure 17.12a and b). Where the gap junction is formed, there are matching channels, called **connexons**, in each membrane, creating a channel between the two. Each connexon is formed of six protein subunits, called **connexins**, that join to form the pore of the channel such that ions can pass through the channel, enabling transmission between the two neurons. Although there is direct connection, the strength of the connection may be such that the electrical potential generated in the second neuron (the post-synaptic potential) may not be large enough to generate an action potential in the second cell on its own, and may require many of these 'subthreshold' potentials added together to generate an action potential (see 'Synaptic integration' later in this section).

 Go to the e-book to watch an animation that illustrates synaptic transmission

Chemical synapses

By far the most common mechanism of transmission of signals between neurons and between neurons and other target structures, such as muscle fibres, is by chemical transmission. In this system, the arrival of the action potential at the synaptic terminal of the axon causes the release of a neurotransmitter that diffuses across the **synaptic cleft**, which is a gap of 20–50 nm between the two cells (Figure 17.12c and d). On reaching the other side, the neurotransmitter will bind with receptors, specific for that

neurotransmitter, leading to a response in the post-synaptic cell. At rest, the molecules of neurotransmitter are stored in synaptic vesicles in the pre-synaptic terminals of the axon.

Let us look in a little more detail at the process of a neurotransmitter being released. Depolarization of the pre-synaptic terminal causes opening of voltage-gated calcium channels. These are similar to the voltage-gated Na^+ channels but are specific for calcium ions. As with Na^+, Ca^{2+} is a positively charged ion with a very low intracellular concentration and relatively high extracellular concentration. As a consequence, opening of Ca^2-specific channels leads to an influx of calcium ions. The influx of calcium triggers the fusion of the membrane of the synaptic vesicle with that of the pre-synaptic terminal so that the neurotransmitter is released by exocytosis (Section 16.2).

The release of the neurotransmitter is a rapid process. One reason for this is that at the active site of the pre-synaptic terminal the vesicles are already attached to the terminal membrane, a process referred to as **docking** (Figure 17.13). Docking occurs due to the interaction of two groups of so-called **SNARE** proteins, vesicle-SNARE (v-SNARE) and target-SNARE (t-SNARE). These bind the vesicle to the active site of the pre-synaptic membrane. When Ca^{2+} enters the terminal, it interacts with a second vesicular protein called **synaptotagmin**. This interaction triggers fusion of the vesicular membrane with that of the pre-synaptic terminal, releasing the neurotransmitter. The vesicular membrane is therefore incorporated into the terminal membrane, and subsequently recycled by endocytosis.

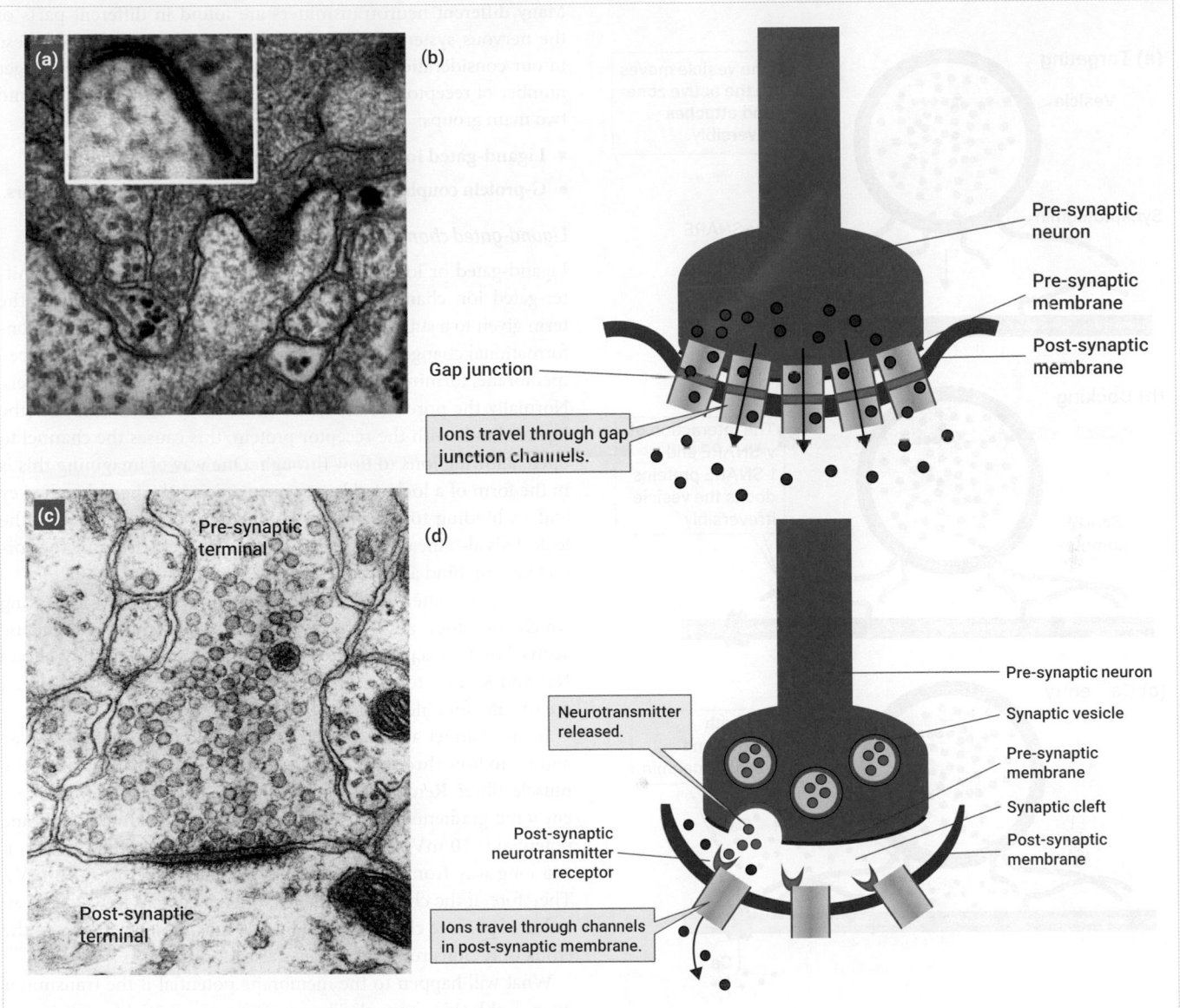

Figure 17.12 Comparison of electrical (a and b) and chemical (c and d) synapses. (a) and (c) show micrographs of the synapses, while (b) and (d) are diagrammatic representations of the synapses. Note the gap junction of the electrical synapse, which brings the processes of the two neurons into very close proximity (about 3 nm apart). The gap in the chemical synapse is wider and the pre-synaptic terminal is characterized by the presence of large numbers of synaptic vesicles.

Source: (a) Rogachevsky et al. Optical nano-control. FENS Forum 2010. http://www.synapsis.ru/publications/posters/poster2010fens.jpg; (b,d) Purves et al., *Neuroscience*, 6th edn, 2017. Oxford University Press; (c) Figure 21.4 from Darnell, J.E., Lodish, H., Berk, A., Zipursky, L. *Molecular Cell Biology*, 4th Edn. W.H.Freeman & Co Ltd (1999).

Research over many years has shown that there are many different types of neurotransmitter. While some have an excitatory action on the post-synaptic cell, others can be inhibitory. Furthermore, some neurotransmitters can have an excitatory action in one part of the nervous system and an inhibitory action elsewhere: an example of this is acetylcholine, which acts as an excitatory transmitter at the neuromuscular junction but is inhibitory when released from the parasympathetic neurons at the pacemaker site of the heart. This is because the action of neurotransmitters is dependent on the properties of the post-synaptic receptor to which they bind.

Neurotransmitters can be divided into three main groups, based on their structure:

- Amines—e.g. acetylcholine (ACh), dopamine (DA), noradrenalin, serotonin (5-HT).
- Amino acids—e.g. gamma-aminobutyric acid (GABA), glutamate (Glu), glycine (Gly).
- Peptides—e.g. encephalins, substance P, vasoactive intestinal peptide (VIP), neuropeptide Y.

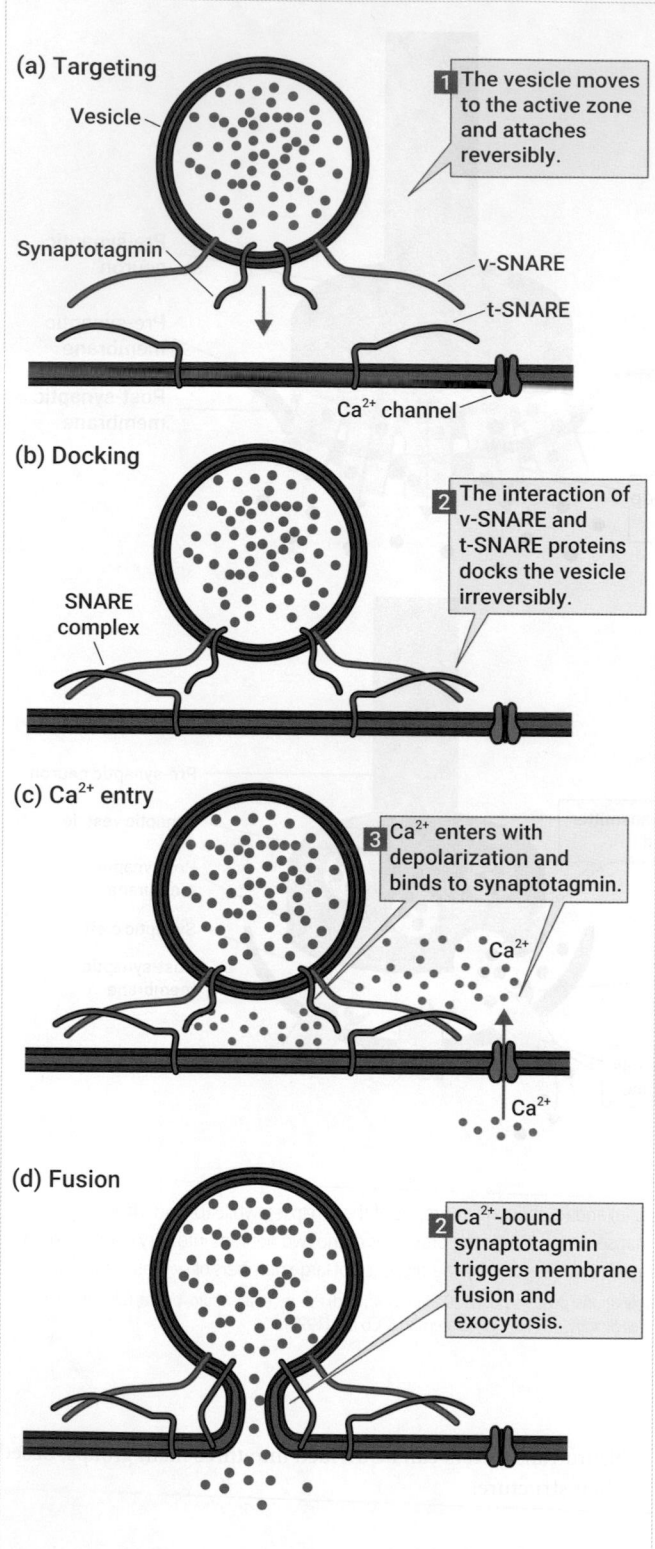

(a) Targeting

Vesicle

Synaptotagmin

v-SNARE

t-SNARE

Ca²⁺ channel

1 The vesicle moves to the active zone and attaches reversibly.

(b) Docking

SNARE complex

2 The interaction of v-SNARE and t-SNARE proteins docks the vesicle irreversibly.

(c) Ca²⁺ entry

3 Ca²⁺ enters with depolarization and binds to synaptotagmin.

Ca²⁺

Ca²⁺

(d) Fusion

2 Ca²⁺-bound synaptotagmin triggers membrane fusion and exocytosis.

Figure 17.13 SNARE proteins and neurotransmitter release. (a, b) The v-SNARE (vesicle-SNARE) proteins interact with the t-SNARE (target-SNARE) proteins 'docking' the vesicle in place on the post-synaptic membrane. (c) When Ca²⁺ enters the terminal, it binds to the vesicle protein synaptotagmin, triggering fusion of the vesicular and membrane proteins and releasing the neurotransmitter (d).

Source: Purves et al., *Neuroscience*, 6th edn, 2017. Oxford University Press.

Many different neurotransmitters are found in different parts of the nervous system and we will come across a number of these in our consideration of different systems. There is an even larger number of receptors, well over 100, but they can be classified into two main groups:

- **Ligand-gated ion channels—ionotropic receptors**.
- **G-protein coupled receptors** (GPCRs)—**metabotropic receptors**.

Ligand-gated channels

Ligand-gated or ionotropic channels are also known as transmitter-gated ion channels or ligand-gated receptors ('ligand' is the term given to a substance that binds to another, resulting in a conformational change). These are integral proteins that span the cell membrane, forming a pore in the same way as other ion channels. Normally the pore is closed, but when the neurotransmitter (the ligand) binds with the receptor protein, this causes the channel to open, allowing ions to flow through. One way of imagining this is in the form of a lock and key to open a door: the ligand is the key and its binding to the protein is the key's action in opening the lock. This also means that there is a strong specificity: only the correct key can bind and lead to channel opening.

The ligand-gated ion channels also show specificity regarding which ions they allow through: for example, the post-synaptic acetylcholine receptor at the neuromuscular junction only allows Na⁺ and K⁺ ions to pass through, whereas other types of channels are specific for chloride ions.

If the channel at the neuromuscular junction allows both Na⁺ and K⁺ to flow through, how does it result in depolarization of the muscle fibre? Remember back to our calculations of the different ionic gradients at rest (Section 17.4): whereas the membrane potential (−70 mV) is close to the equilibrium potential for K⁺, it is a long way from the equilibrium potential for Na⁺ (+70 mV). Therefore, if the channel is opened, the net current flow will be an influx of positive charge carried by Na⁺ ions, which will exceed the efflux of K⁺ ions resulting in depolarization.

What will happen to the membrane potential if the transmitter opens a chloride channel? The equilibrium potential for Cl⁻ ions is more negative than the resting membrane potential, therefore Cl⁻ ions will flow into the axon carrying negative charges, making the inside of the cell more negative—hyperpolarization.

Changes in the membrane potential of the post-synaptic cell that lead to depolarization are termed **excitatory post-synaptic potentials (EPSPs)**.

Changes in the membrane potential of the post-synaptic cell that lead to hyperpolarization are termed **inhibitory post-synaptic potentials (IPSPs)**.

The neurotransmitters bind reversibly to the ionotropic receptors and so, when they detach, the channel closes. If the transmitter is not removed from the synaptic cleft, it could continue to bind and be released, thereby maintaining the cycle of channel opening and closing. Normally the transmitter is removed from the synaptic cleft, which can happen in three ways:

1. Diffusion via the extracellular fluid.

2. Reuptake into the pre-synaptic terminal via specific transporter proteins, which is the process for most amino acid and amine neurotransmitters.

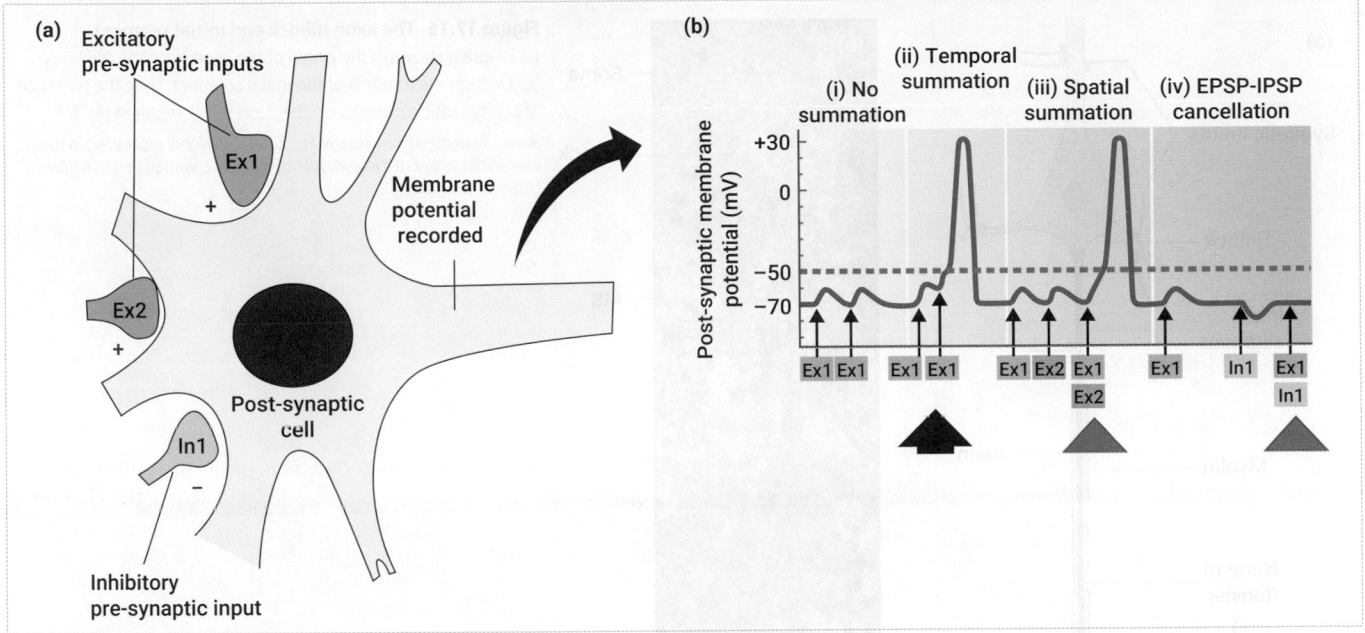

Figure 17.14 Synaptic integration. In (a) Ex1 and Ex2 are two excitatory synapses acting on the post-synaptic neuron; In1 is an inhibitory synapse. In (b), the left-most pair of EPSPs (i), both generated by the same synapse, are too far apart in time to summate, as the first one has decayed away before the second one is initiated. The second pair (ii—black arrow) are closer together and there is summation, leading to an action potential. In the case of the third set (iii), the first pair of EPSPs, generated by Ex1 and Ex2, are again too far apart in time, but the second pair are very close together and generate an action potential (orange arrow). When In1is also active, it leads to hyperpolarization and so can prevent the generation of an action potential (iv, right-hand record).

Source: Purves et al., *Neuroscience*, 6th edn, 2017. Oxford University Press.

3. Enzymatic breakdown in the synaptic cleft. This is the process for removal, for example, of acetylcholine in the neuromuscular junction, where the transmitter is broken down by the enzyme acetylcholinesterase.

What would happen if the transmitter was not removed? The transmitter would continue to be active. An example of this is the highly dangerous nerve gas sarin, which inhibits acetylcholinesterase. This causes death very rapidly, usually due to asphyxia but accompanied by autonomic symptoms such as drooling and loss of control of the sphincter muscles.

Synaptic integration

Although the neuromuscular junction is one of the best-studied synapses (mainly due to its accessibility to experimentation), it is also unusual in that the action potential in the pre-synaptic neuron and associated release of the neurotransmitter acetylcholine result in a sufficiently large EPSP to trigger an action potential in the muscle cell, i.e. there is a 1:1 relationship. At most synapses, the EPSP is relatively small, typically only a few millivolts, and so is well below the threshold for triggering an action potential.

Why do EPSPs vary in size? One reason is that the number of vesicles released varies at different synapses: each vesicle contains several thousand molecules of neurotransmitter, all of which are released together when the vesicle fuses with the pre-synaptic membrane. The actual number varies significantly: at the neuromuscular junction this is typically several hundred, whereas in some CNS synapses it may be as few as only one or two. Therefore, one factor affecting the strength of the synaptic connection is the number of vesicles of neurotransmitter that are released.

If each EPSP is relatively small, then generating an action potential in the post-synaptic cell requires summation of a number of EPSPs to bring the cell membrane to threshold. For example, at synapses in the CNS the EPSPs may only be 1–4 mV in size, therefore it may require the summation of 10 or 20 EPSPs to bring the post-synaptic neuron to threshold to generate an action potential.

Unfortunately, the picture is more complicated than adding together all the depolarizations. There are four factors to consider:

1. The timing of the EPSPs—**temporal summation**.

2. The location of other synapses on the dendritic tree of the post-synaptic cell—**spatial summation**.

3. The size and effect of the EPSPs generated by each synapse—synaptic strength.

4. Whether there are also IPSPs that will reduce the depolarizing current.

Whereas each action potential lasts only about 1 ms, the post-synaptic potentials are longer lasting, reaching a peak and then decaying over a period of about 5 ms; therefore, temporal summation can occur when the pre-synaptic neuron is discharging at a high rate and the EPSPs generated at the post-synaptic site can add together (Figure 17.14bii): this summation may be sufficient to bring the post-synaptic cell to threshold through the repeated excitation occurring at a single synapse.

If we return to a consideration of the structure of the neuron (Figure 17.1), we can see that the cell body often has arising from it numerous processes called dendrites. The synaptic connections from inputting neurons are located at different sites, from the cell body itself to the tip of the dendrites. If a number of these synapses

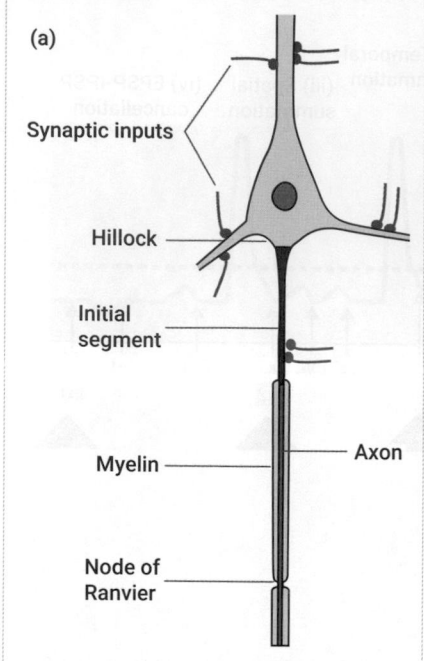

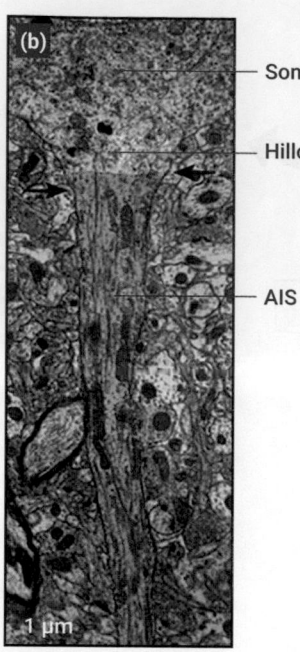

Figure 17.15 The axon hillock and initial segment.
(a) Diagram showing the origin of the axon from the cell body.
(b) Electron micrograph of the initial segment. Note the presence of pre-synaptic terminals on the axon initial segment (AIS).

Source: Reprinted from Kole, M.H., Stuart, G.J. Signal processing in the axon initial segment. *Neuron*, 26;73(2). © 2012, with permission from Elsevier.

are activated at the same time, the EPSPs generated at the different locations on the dendrites can add together to bring the post-synaptic neuron to threshold for generating an action potential (Figure 17.14biii). This is referred to as spatial summation.

We have already noted that the strength of individual synapses may vary as a consequence of the amount of neurotransmitter that is released. The location of the synapse on the dendritic tree is also important. If each post-synaptic potential is below threshold for generating an action potential, then as the depolarization spreads along the dendrite it decays away. The farther away a synapse is located from the cell body, then the weaker will be the effect of the EPSP because it will have reduced in amplitude by the time it reaches the point of integration. So, synapses that are found towards the ends of dendrites will have less effect than those that are nearer.

So far, we have only considered EPSPs, but IPSPs also contribute to synaptic integration: the effect of the IPSP is to reduce the likelihood of the post-synaptic neuron reaching threshold (Figure 17.14b(iv)) as it acts to hyperpolarize the membrane, moving the potential further away from the threshold for generating an action potential.

So synaptic integration is a complex process: whether the post-synaptic neuron reaches threshold for generating an action potential depends on the rate of firing of the inputting neurons, the number of neurons firing, the strength and location of their synapses, and the balance between excitatory and inhibitory inputs.

Where does integration take place to generate an action potential? At the point where the axon originates from the cell body, the cell body funnels down into the relatively thin, longitudinal axon. This funnel region is termed the **axon hillock** (Figure 17.15a). From there arises an unmyelinated length of axon called the **initial segment**. This initial segment is characterized by a high density of voltage-gated sodium channels and it is here that all the EPSPs

and IPSPs are integrated; if there is a depolarization that exceeds threshold an action potential will be generated that is then conducted along the length of the axon.

Note that in the electron micrograph in Figure 17.15 there are synaptic terminals on the initial segment itself. These then are likely to have a significant influence in determining whether an action potential is generated.

Metabotropic receptors

The second main class of post-synaptic receptors are the metabotropic receptors, also known as **G-protein coupled receptors (GPCRs)**. The binding of the transmitter to the GPCR does not result in direct action on an ion channel, as is the case for the ionotropic receptors. Instead, they act by an indirect mechanism through an intermediate transducing molecule called a GTP-binding protein (or **G-protein** for short). The binding of the transmitter molecule to the GPCR causes it to undergo a conformational change that allows it to interact with a G-protein in the plasma membrane.

The best way to consider G-proteins is as a sort of 'go between' molecule linking the signal of ligand binding to a receptor to the activation of the downstream signalling pathways that mediate the actual cellular response to the signal.

G-proteins are referred to as **heterotrimeric proteins** because they contain three different subunits ('hetero' roughly translates to 'difference', while 'trimeric' refers to 'three units'): alpha, beta, and gamma (α, β, and γ) (Figure 17.16). In the inactive state, the alpha subunit of the G-protein is bound to guanosine diphosphate (GDP). When the ligand molecule binds to the receptor, the GPCR activates the G-protein, causing the Gα subunit to exchange GDP for GTP (guanosine triphosphate).

The Gα subunit dissociates from the Gβγ dimer (two units) and these are free to diffuse away from the receptor complex and

Figure 17.16 G-protein coupled receptor activation. Binding of the transmitter activates the G-protein (middle panel), which exchanges GTP for GDP. This leads to dissociation of the G-protein into its Gα and Gβγ subunits.

Source: Pocock et al. *Human Physiology*, 5th edn, 2017. Oxford University Press.

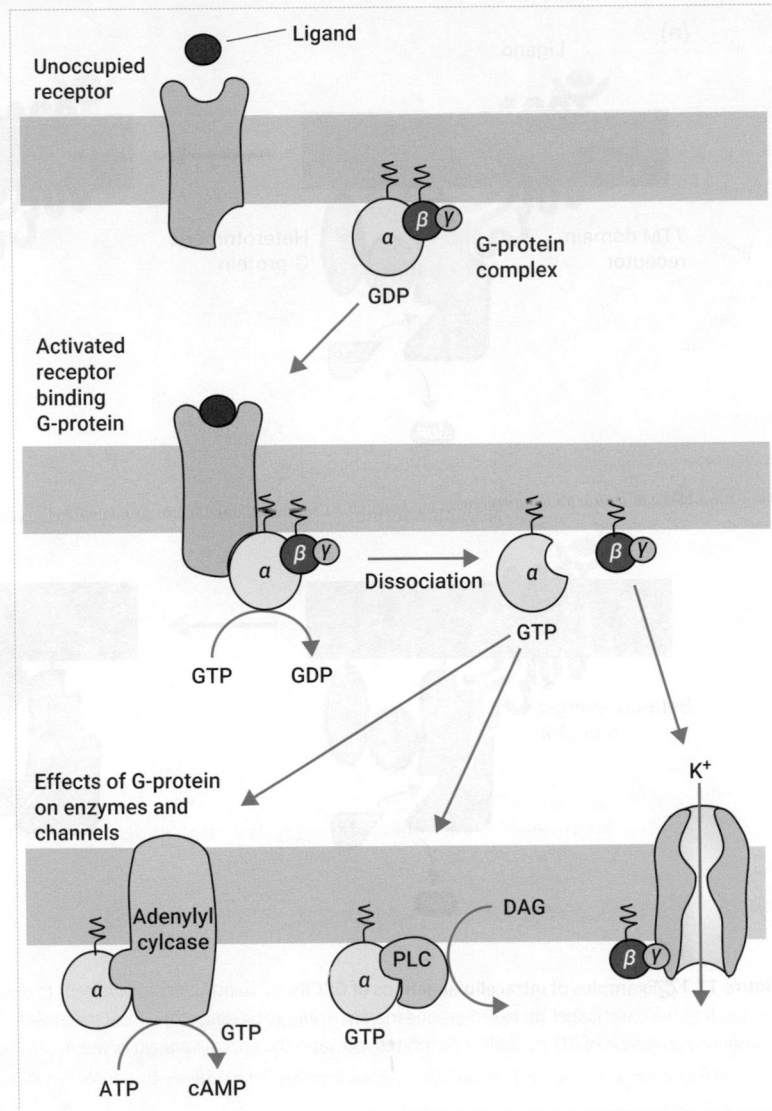

activate cellular mechanisms. The Gα subunit has enzymatic properties that can activate second messenger systems and also cause the breakdown of GTP to GDP + P$_i$, so limiting the duration of activity by rendering the subunit inactive. The inactivated Gα subunit also then rebinds to the Gβγ dimer, reforming the inactive G-protein.

The activation of the GPCRs can be effective in the cell through two main mechanisms: G-protein-gated ion channels and G-protein-activated enzymes.

The action on the G-protein-gated ion channels is mediated through the Gβγ dimer, which can move in the membrane and bind to a nearby ion channel (Figure 17.17a). An example of action through a G-protein-gated ion channel is the effect of acetylcholine, released from the vagus nerve of the parasympathetic nervous system, on heart rate (Chapter 21). In this case, the released Gβγ dimer binds to a nearby K$^+$ channel in the membrane of the sino-atrial node of the heart. Opening the K$^+$ channel allows an increase in outward flow of K$^+$ ions as the pacemaker spontaneously depolarizes. The outflow of K$^+$ will therefore reduce the rate of rise of the pacemaker depolarization, leading to slowing of the heart rate. Whereas the action

of the G-protein-gated ion channel is localized within the membrane, a special feature of the G-protein-activated enzyme systems is that, although they are slower acting than the ionotropic receptors or the G-protein-gated ion channels, they can also give rise to an enzyme cascade that amplifies the response and acts more widely within the cell, often including the nucleus. These systems involve the activation of **second messenger systems** (the original transmitter that bound to the GPCR is the first messenger) (Figure 17.17b). There is a number of different second messenger systems, one of the most common being the cAMP-protein kinase cascade.

When the ligand molecule binds to the receptor the GPCR activates the G-protein, causing the Gα subunit to exchange GDP for GTP (guanosine triphosphate). The Gα subunit then dissociates from the Gβγ dimer and binds to the enzyme **adenylyl cyclase** (Figure 17.16b). The activated adenylyl cyclase catalyses the conversion of ATP to cyclic adenosine monophosphate (cAMP). cAMP is then free to bind to, and so activate, the enzyme phospho-kinase A (PKA). PKA then acts within the cell by phosphorylating target proteins, for example, ion channel proteins. Each molecule of adenylyl cyclase can produce many molecules of cAMP and

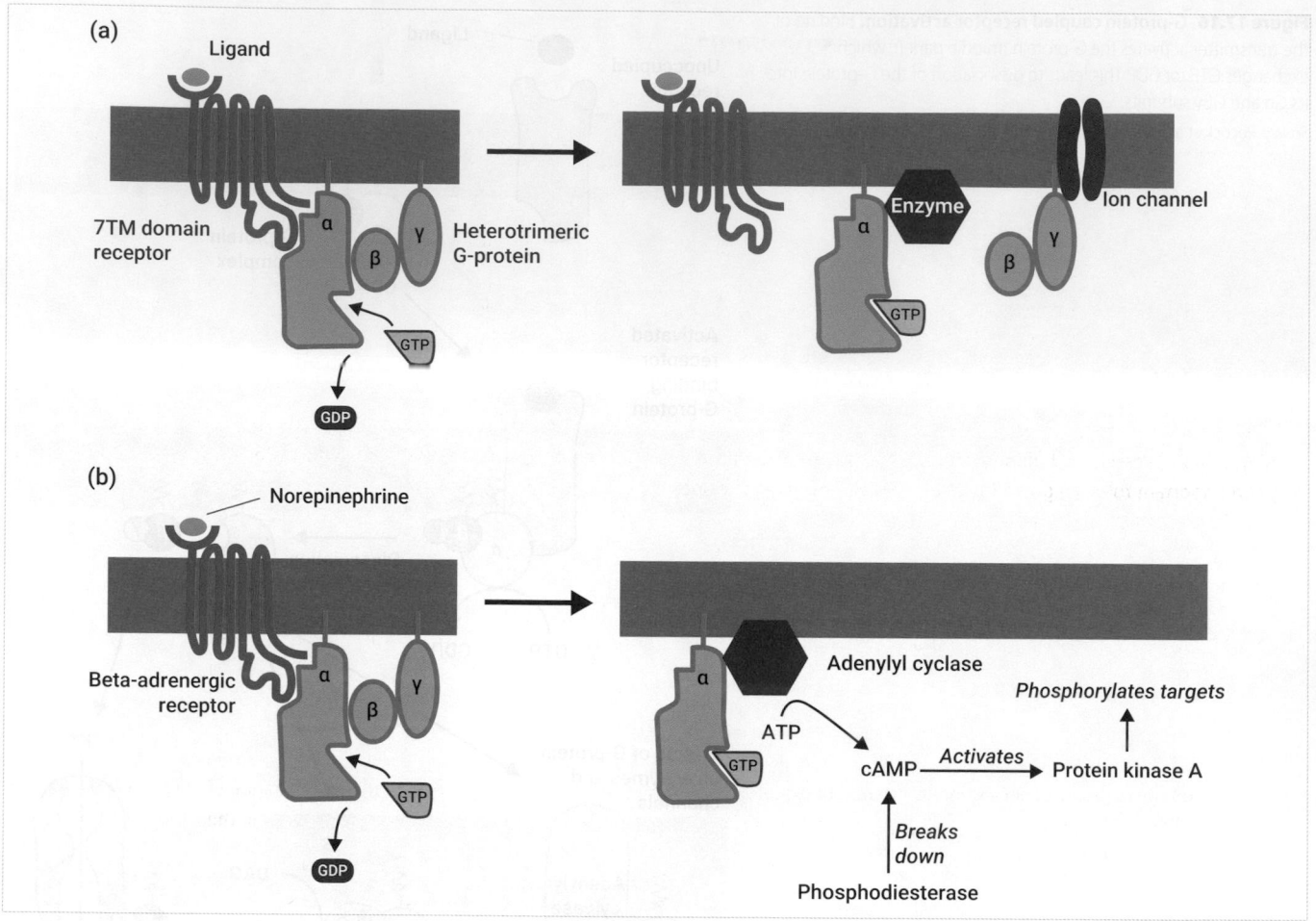

Figure 17.17 Examples of intracellular actions of GPCRs. (a) Transmitter binding leads to dissociation of the G-protein. The Gβγ dimer migrates along the membrane and binds to the ion channel. (b) Noradrenaline (norepinephrine) binding to the receptor releases the activated Gα subunit that binds to the enzyme adenylyl cyclase, promoting conversion of ATP to cAMP, which in turn activates the enzyme protein kinase A. cAMP is rapidly broken down by the enzyme phosphodiesterase.

Source: M.F. Bear, B.W. Connors, M.A. Paradiso (2015). *Neuroscience: Exploring the Brain*, 4th edn. Wolters Kluwer.

each PKA can phosphorylate a number of molecules of target protein, therefore the second messenger system can lead to significant amplification of the response compared with the single channel actions seen for the ligand-gated ion channels and the G-protein-gated ion channels.

There are many hundreds of GPCRs, all of which utilize G-proteins to convey their signals, so you might be wondering how specificity is accomplished in such a system. For example, pacemaker cells in the heart contain G-protein coupled receptors for both adrenaline and acetylcholine. Both these receptors convey the signal of ligand binding by activating a G-protein, yet the effects on heart rate are in opposite directions with adrenaline increasing heart rate and acetylcholine decreasing it. How can such opposing signals be conveyed by the same signalling mechanism? The answer lies in the fact that not all G-proteins are the same. In humans there are 21 different Gα subunits, six different Gβ subunits, and 12 Gγ subunits. We don't yet fully understand the full significance of all the different possible combinations by which this multitude of subunits can come together, but the vast majority of GPCRs act through three main classes of G-protein signalling pathway based on sequence similarities within the Gα subunit: these are the Gαs, Gαi, and Gαq-linked pathways.

Gαs and Gαi have opposing effects at the same effector protein: adenylyl cyclase. The effects of these two G-proteins on adenylyl cyclase activity is easily remembered as 's' stands for stimulatory and 'i' stands for inhibitory. The example shown in Figure 17.17b is of a Gαs pathway, which activates adenylyl cyclase, whereas the action of Gαi would suppress the enzyme.

Rather than cAMP, the second messengers downstream of Gαq-linked receptor activation are inositol 1,4,5-trisphosphate (IP_3) and diacylglycerol (DAG). IP_3 is a water-soluble molecule and diffuses through the cytoplasm to bind to a specific receptor (the IP_3 receptor) located on the endoplasmic reticulum or, in the case of muscle, the sarcoplasmic reticulum. The IP_3 receptor is a ligand-gated ion channel and once activated allows Ca^{2+} to be released from intracellular stores. The subsequent rise in cytoplasmic Ca^{2+} concentration affects a range of cellular processes, for example, in smooth muscle cells a rise in intracellular Ca^{2+} would lead to smooth muscle contraction.

DAG is highly lipophilic and remains within the plasma membrane. The main role of DAG is to activate *protein kinase C* (PKC).

SCIENTIFIC PROCESS 17.1   Expression cloning of a GPCR

Introduction

Platelets are small fragments of cytoplasm derived from cells in the bone marrow and which circulate in the bloodstream where they function to stop bleeding by clumping together, facilitating clot formation at blood vessel injuries. Inappropriate activation of platelets, however, can lead to pathological thrombus (clot) formation, blocking small blood vessels and potentially resulting in stroke and heart attack. There is therefore much interest in understanding the receptors and signalling pathways controlling the normal activation of platelets.

It had been known for many years that extracellular ADP released from damaged blood vessels, red blood cells, and the platelets themselves plays an important role in regulating platelet activity by activating specific GPCRs present in the platelet plasma membrane. One of these receptors, the P2Y1 receptor that acts to mobilize intracellular Ca^{2+}, was already known, but a second ADP receptor coupled to the inhibition of adenylyl cyclase remained elusive.

This unidentified second ADP receptor is of particular importance as it is the target of the widely used antithrombotic drug, clopidogrel. Determining the molecular identity of this second ADP receptor was therefore an important goal as this would allow the development of more specific antiplatelet drugs to treat cardiovascular disease. In their study, Hollopeter and colleagues used an ingenious method of expression cloning to identify this second ADP receptor, which they named P2Y12.

Methods

In order to identify the P2Y12 receptor, Hollopeter and co-workers used the technique of expression cloning from a platelet cDNA library. cDNA libraries consist of hundreds of thousands of individual DNA clones and are made by reverse transcribing the mRNA isolated from cells into complementary DNA, followed by cloning of the cDNAs into an expression vector.

As the library used in this study had been made from mRNA isolated from platelets, a small number of clones amongst the hundreds of thousands present in the library must correspond to the P2Y12 receptor. The trick is finding one of these P2Y12 clones and this is a little like searching for the proverbial needle in a haystack! Pools of cDNA clones from the library (around 7000 clones per pool) were expressed in *Xenopus* oocytes (cells from the ovaries of the clawed frog, Figure 1). As *Xenopus* oocytes do not express ADP receptors, any responses to ADP observed must have come from expression of one of the cDNA clones present in the pool expressed in the oocyte rather than from one of the oocyte's own genes.

The problem Hollopeter and co-workers faced is that activation of the P2Y12 receptor does not produce a signal that can be readily measured in a *Xenopus* oocyte expression system.

Why would it be difficult to record a response from the P2Y12 receptor? This receptor is linked to inhibition of the enzyme adenylyl cyclase and the response is a decrease in the concentration of the intracellular second messenger, cAMP. It is difficult to determine the concentration of cyclic

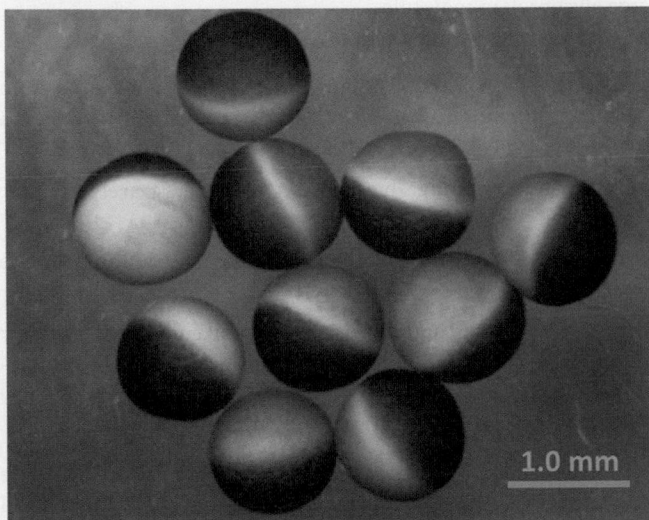

Figure 1 *Xenopus* oocytes. Oocytes isolated from *Xenopus* (commonly known as the clawed frog) are a useful experimental model for ion channel and receptor research. The cells are large so can be easily injected with RNA or DNA, which they readily express into protein.

Source: Cristofori-Armstrong, B., Soh, M., Talwar, S. et al. *Xenopus borealis* as an alternative source of oocytes for biophysical and pharmacological studies of neuronal ion channels. *Sci Rep*, 5, 14763 (2015). https://doi.org/10.1038/srep14763.

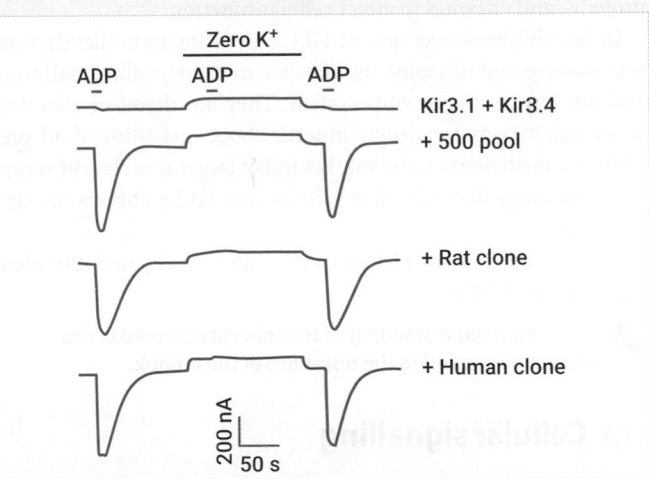

Figure 2 P2Y12 is a G-protein coupled receptor that responds to ADP. Activation of Kir3.1 and 3.4 potassium currents in response to ADP (10 mM) (short bars) in the presence or absence (long bar) of extracellular potassium (70 mM). Oocytes injected with Kir3.1 and 3.4 RNA alone (top trace) do not exhibit a significant current in response to ADP. Oocytes injected with Kir3.1 and 3.4 RNA and a positive cDNA pool (+500 pool) do, however, respond to ADP, showing that one of the 500 clones injected was P2Y12. ADP induced currents from isolated rat and human P2Y12 clones co-injected into oocytes with Kir3.1 and 3.4 RNA are shown as controls (lower traces).

Source: Hollopeter, G., Jantzen, H.M., Vincent, D. et al. Identification of the platelet ADP receptor targeted by antithrombotic drugs. *Nature*, 409, 202–207. Reprinted by permission from Springer Nature: Springer, © 2001.

nucleotides and detecting a tiny decrease in cAMP, resulting from expression of a single cDNA clone amongst a pool of ~7000 cDNA clones, would have been impossible.

This problem was overcome by also injecting the oocytes with mRNA encoding two specific forms of potassium channel, called Kir 3.1 and Kir 3.4. When activated, these ion channels produce a readily measurable potassium current through the plasma membrane (Figure 2). Thus, if a P2Y12 clone was present in one of the pools of clones expressed in an oocyte, applying ADP to the oocyte would now result in a membrane current from the Kir 3.1 and 3.4 ion channels, and this current could be readily detected by electrophysiological techniques.

Results and discussion

After screening 48 pools of ~7000 clones each, one pool was identified that gave a signal to ADP. This positive pool was further subdivided into a number of new pools and the process of injecting oocytes and testing for responses to ADP repeated. Over several rounds of this process, with the number of clones decreasing in each round (a process called sib selection), eventually the size of the pool giving a response to ADP was reduced to 96 clones. DNA sequencing of all 96 of these clones showed that one of the clones corresponded to a novel GPCR, which was then selected for further analysis.

Hollopeter and colleagues showed that this novel GPCR is present on platelets and that its pharmacological and signalling profile matched that of the previously unidentified platelet ADP receptor, in that it was activated by extracellular ADP, leading to $G_{\alpha i}$ activation and the inhibition of cAMP formation. In order to strengthen their findings, Hollopeter and co-workers also investigated a previously described patient with a bleeding disorder characterized by platelets that fail to respond to ADP. Sure enough, this patient was shown to lack a normal form of the gene encoding the P2Y12 receptor, confirming without doubt that P2Y12 does indeed correspond to the previously unidentified platelet adenylyl cyclase-inhibiting ADP receptor. This discovery represented a milestone as it opened the door for the subsequent development of more specific antithrombotic drugs, with fewer side effects compared to their predecessors, by defining the molecular target for drug design.

Read the original work

Hollopeter, G., et al. (2001). Identification of the platelet ADP receptor targeted by antithrombotic drugs. *Nature* **409** (6817): 202–7. doi:10.1038/35051599

Similar to protein kinase A, which is activated by the second messenger cAMP, PKC phosphorylates a range of specific target proteins such as transcription factors, ion channels, cytoskeletal proteins, and enzymes to affect cellular function.

In just this brief overview of GPCRs it is apparent that they are very diverse and of major significance in cell signalling pathways and not just in the nervous system. They are therefore also very important targets for drugs; indeed, about one-third of all prescription medicines on the market today target this class of receptor, from drugs that treat heart disease and schizophrenia through to viagra.

In Scientific Process 17.1 we explore an investigation into identification of a GPCR.

 Check your understanding of the concepts covered in this section by answering the questions in the e-book.

17.7 Cellular signalling

So far in this chapter we have focused on the nervous system, but the processes of cell signalling are much more diverse. Cells respond to changes in their environment, as well as communicating with each other to coordinate processes such as growth, differentiation, metabolism, and even cell death. To achieve this, a cell must integrate a multitude of signals received from neighbouring cells, blood-borne hormones, and neurotransmitters released from nerve terminals, as well as physical cues such as stretching of the plasma membrane or changes in temperature. The response that the cell makes depends on the nature of the signal and typically involves modulation of the activity of existing enzymes and/or changes in the repertoire of genes expressed by the cell.

Chemical messengers

Most of the signals a cell receives are in the form of chemical messengers termed **signalling molecules**. Many different types of signalling molecule are used to transmit information between cells and these range in complexity and size from a single proton to large glycoprotein molecules.

There are several different forms of cell signalling:

- **Paracrine signalling**: the signalling molecule is released from one cell in response to a particular stimulus and travels to a nearby target cell by simple diffusion through extracellular fluid.
- **Endocrine signalling**: the signalling molecules travel to remote target cells via the bloodstream.
- **Autocrine signalling**: the signalling molecules act on the same cell that produced them.
- **Juxtacrine signalling**: the signalling molecules never actually leave the cell of origin but are instead expressed on the cell surface, where they come into contact with touching target cells by direct cell-to-cell communication.

Despite this range of different transport mechanisms, all signalling molecules, or ligands, act at target cells by essentially the same overall mechanism as we have seen with the neurotransmitter: they bind to and activate a specific receptor protein expressed either on the surface or inside the target cell. The binding of the ligand to its corresponding receptor causes a conformational change in receptor structure, which triggers a series of molecular events that elicit the cellular response. An important point to appreciate here is that the complement of receptors expressed differs between different cell types. Thus, cells do not respond to every signalling molecule they encounter; they can only respond to a signalling molecule if the specific receptor for that signalling molecule has been expressed by the cell.

Types of receptor

There are literally hundreds of different signalling molecules utilized for cellular communication and an even greater number of corresponding receptors, as many signalling molecules have more than one type of receptor. Thankfully, based on their molecular structure and the downstream signalling pathways that are initiated by receptor activation, the vast majority of this seemingly impenetrable array of different signalling molecule–receptor combinations can be simplified into just four 'super families' of receptors that share similar characteristics:

- Ligand-gated (ionotropic) ion channels
- G-protein coupled receptors (GPCRs)
- Kinase-linked receptors
- Steroid receptors.

The first three of these receptor super families are activated by signalling molecules that are unable to pass through the lipid bilayer of the plasma membrane (i.e. hydrophilic signalling molecules). The receptors that make up these three families therefore typically span the plasma membrane by one or more transmembrane domains so that their ligand-binding site is exposed to the extracellular environment.

Steroid receptors, on the other hand, are activated by lipid-soluble (hydrophobic) signalling molecules that readily pass through the lipid bilayer of the plasma membrane into the cytosol. The steroid receptors are therefore intracellular proteins that reside in the cytosol or nucleus of the cell.

We have already discussed the ionotropic receptors and the GPCRs in Section 17.6. In this section we will briefly consider the kinase-linked and the steroid receptors.

Kinase-linked receptors

Kinase-linked receptors are integral membrane proteins. The common theme and key mechanism with this class of receptor is that the catalytic activity of the kinase is regulated by the binding of an extracellular ligand. The kinase phosphorylates specific amino acid residues either on itself, by a process known as autophosphorylation, or on intracellular target proteins to create docking sites for intracellular signalling proteins. These newly formed docking sites allow a signalling complex to be assembled at the activated receptor and the signal of extracellular ligand binding to the receptor is then typically transmitted to the nucleus to modulate gene expression. The main families of kinase-linked receptor are:

- Receptor tyrosine kinases (RTKs)
- Receptor serine–threonine kinases
- JAK-binding receptors.

Examples of the messengers for kinase-linked receptors are the polypeptide growth factors, such as fibroblast growth factor and platelet-derived growth factor, as well as the hormones insulin and erythropoietin, which stimulate erythrocyte production.

Steroid receptors

So far, we have looked at cell surface receptors for water-soluble chemical messengers that are unable to freely cross the lipid barrier of the plasma membrane. However, some messengers, such as the steroid hormones, thyroid hormone, retinoic acid, and vitamin D, are lipophilic and can therefore readily cross the plasma membrane to enter the cell (Figure 17.18). As these messengers are insoluble in water they must be transported in blood bound to carrier proteins. While there are some emerging cases of cell surface receptors for lipophilic hormones, for example the G-protein coupled receptor GPR30 is activated by oestrogen, the vast majority of the

Figure 17.18 The pathways of action of the steroid hormones. Type I: the receptor is located in the cytosol where it binds the hormone. The receptor–hormone complex then enters the nucleus where it binds to DNA and alters gene transcription. Type II: the receptor is bound to the DNA and when the hormone binds as well this results in changes in gene transcription.

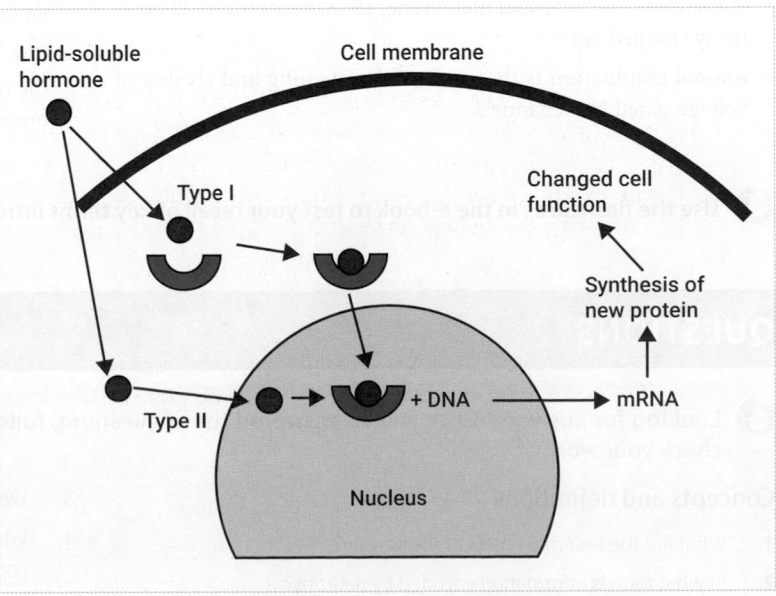

receptors for lipophilic messengers reside inside the cell and are referred to as the steroid hormone super family of receptors.

The steroid hormone receptor super family of receptors are transcription factors that directly regulate the expression of target genes. While thyroid hormone, retinoic acid, and vitamin D are structurally distinct from the steroid hormones, their receptors share a common mechanism of action with steroid hormone receptors and are therefore part of the same super family. All the steroid hormone receptors share three related domains, a hydrophobic hormone-binding region, a DNA-binding region rich in cysteine and basic amino acids, and a variable region involved in transcriptional activation.

Type I steroid hormone receptors, such as the glucocorticoid receptor, reside within the cytosol of the cell (Figure 17.18). Binding of hormone induces a conformational change in receptor structure that reveals a nuclear localization signal. The receptor–hormone complex then translocates to the nucleus where it binds specific regions of DNA and alters gene transcription. However,

most steroid hormone receptors, permanently reside in the cell nucleus and these are referred to as type II receptors. For example, the thyroid hormone receptor is constitutively bound as a dimer to the DNA regulatory region of target genes, regardless of whether or not hormone is present. In the absence of hormone this interaction usually represses gene expression. When thyroid hormone enters the nucleus and binds to the receptor a conformational change occurs in the receptor that switches it from a transcriptional suppressor to a transcriptional activator. With both type I and type II steroid hormone receptors it is the expression of new genes leading to new protein production that facilitates the cellular response to the hormonal chemical signal. As this takes time (minutes to hours), the response to steroid hormones is usually much slower than those mediated through cell surface receptors and ion channels.

 Check your understanding of the concepts covered in this section by answering the questions in the e-book.

SUMMARY OF KEY CONCEPTS

- Almost all neurons comprise a common set of structures: soma—the cell body; axon—that conducts the action potentials; and dendrites—that act as input terminals.

- Signal transmission takes the form of a frequency code, where the rate of action potential discharge relates to the strength of the input stimulus. The maximum frequency of action potential discharge is determined by the refractory period of the neuron.

- All neurons have a resting membrane potential that is determined by the relative distributions of the ions in the intra- and extracellular fluids and the relative permeability of the neuronal membrane to those ions.

- The action potential is the result of a change in the ionic conductance of the neuronal membrane, allowing an influx of positively charged ions.

- Axonal conduction is the result of the opening and closing of voltage-gated Na$^+$ channels.

- Axonal conduction velocity depends on the axonal diameter and the presence of a myelin sheath.

- Signal transmission between neurons is most commonly by means of the release of a chemical neurotransmitter at the synaptic terminal.

- Neurotransmitters may have an excitatory or inhibitory action depending on the properties of the postsynaptic receptor.

- There are four main types of receptors: ionotropic and metabotropic receptors, kinase-linked receptors, and steroid receptors.

- Ionotropic receptors are transmitter-gated ion channels, whereas metabotropic receptors involve indirect action mediated by a G-protein. These can give rise to response amplification through an enzyme cascade.

- The majority of steroid receptors are located within the cell as the transmitter is lipophilic and can pass through the cell membrane.

 Use the flashcards in the e-book to test your recall of key terms introduced in this chapter.

QUESTIONS

 Looking for answers? Once you've answered these questions, follow the link in the e-book to the answer guidance and check your work.

Concepts and definitions

1. What are the four main parts of the neuron?

2. In what form is information encoded by neurons?

3. Which ion is the main determinant of the resting potential?

4. What are the two main structural features that determine the conduction velocity along an axon?

5. What are the four main factors that affect post-synaptic summation?

6. What are the two main classes of post-synaptic receptors?

7. What are the four main forms of chemical signalling?

Apply the concepts

8. Explain the concept of dynamic equilibrium.

9. How does the structure of ion channels affect their permeability to different ions?

10. Describe the process of transmission at chemical synapses.

11. What determines the maximum rate of action potential discharge?

12. How are metabotropic receptors linked to intracellular mechanisms?

Beyond the concepts

13. Compare and contrast metabotropic and ionotropic mechanisms of cell signalling.

14. Describe the properties of axons that determine the conduction velocity of the action potential.

15. Describe the structure of the voltage-gated sodium channel and explain its role in the conduction of the nerve impulse.

16. What is the resting membrane potential? Explain the neuronal properties that underpin the establishment and maintenance of the resting potential and the role of the $Na^+K^+ATPase$ pump.

17. Compare and contrast the different mechanisms of cellular signalling.

FURTHER READING

Bear, M. F., Connors, B., & Paradiso, M. (2015). *Neuroscience: Exploring the Brain*. 4th Edition. Wolters Kluwer.
This is a detailed text covering all the nervous system and can take you at least up to second year and into third year levels. The content of this chapter is addressed in more detail in Chapters 2–6.

De Lera Ruiz, M. & Kraus, R. L. (2015). Voltage-gated sodium channels: structure, function, pharmacology and clinical indications. *J. Med. Chem.* **58**: 7093–118.
This is a very good overview of the sodium channel and its function.

Kress, G. J. & Mennerick, S. (2009). Action potential initiation and propagation; upstream influences on neurotransmission. *Neuroscience* **158**: 211–22.
This article provides additional detail on the mechanisms of generation of the action potential and of conduction.

Pocock, G., Richards, C. D., & Richards, D. A. (2017). *Human Physiology*, 5th edition. Oxford University Press.
Another very readable account of neuronal signalling but in slightly less advanced detail than is given in the two neuroscience texts.

Purves, D., et al (2018). *Neuroscience*, 6th edition. Oxford University Press.
Chapters 2–7 provide a more detailed account of the processes of neuronal signalling covered in this chapter.

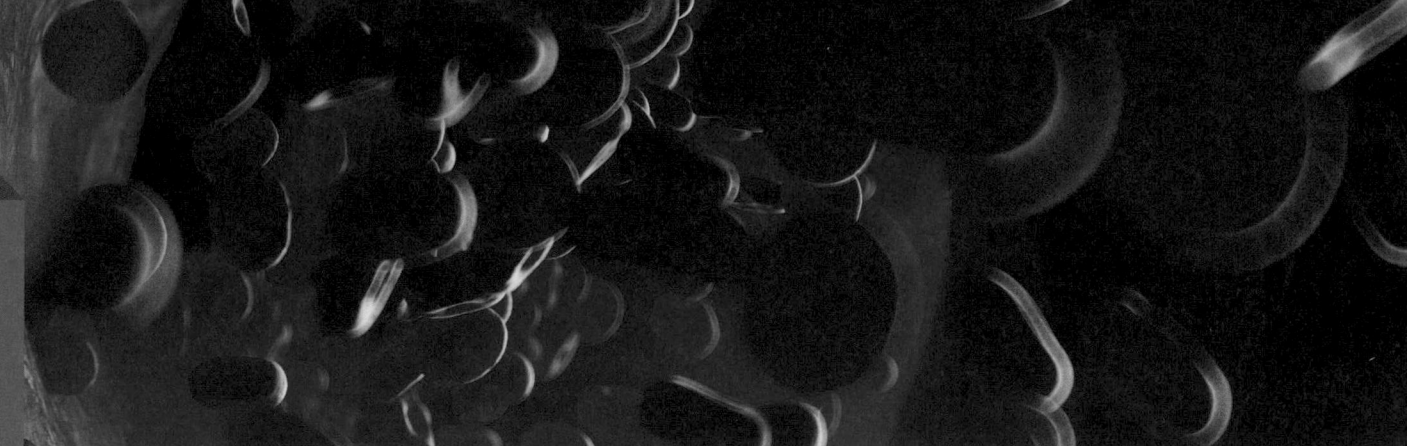

Communication and Control 2

Sensory Systems

Chapter contents

Introduction   618

18.1   General sensation   619

18.2   Touch   620

18.3   Pain and temperature   626

18.4   Proprioception   629

18.5   Special senses   632

 Watch the key concepts video in the e-book to prepare yourself for studying this chapter.

LEARNING OBJECTIVES

By the end of this chapter you should be able to:

- Identify the roles of the different systems that contribute to sensation and their organization.
- Explain the nature of different stimuli and the associated processes of transduction from stimulus to neural response.
- Describe the different types of sensory systems contributing to touch and pain and their neural pathways.
- Compare the physical and neural mechanisms of stimulus processing in the special senses.

Introduction

Our sensory system continually provides us with a remarkably rich supply of information about our internal and external environments. These sensations tell us about the world around us: for example, what is immediate to us—what we are touching or smelling, or what the temperature is, or what is further away—to a view of what is in the far distance. They also provide information about what is going on within our bodies, including when things go wrong, as indicated by pain.

It is important to remember that our conscious awareness of sensation is only part of the story: some elements of the sensory system contribute to both conscious and unconscious systems, such as the information from muscles and joints that informs the control of movement. Other elements of the sensory system never reach our consciousness but play an important role in regulating physiological systems, for example those that monitor blood pressure and therefore are important in regulating cardiac output and peripheral resistance (Section 21.2).

As we discuss the operation of the sensory system, it is very important to remember that our conscious perception of the world is constructed by our brains from the huge variety of sensations that are being detected at any one time. This perception depends very much on the neural processes by which sensory stimuli are converted into patterns of action potentials and transmitted to the brain. However, our perception of the sensory world also depends on a variety of other factors, which may change from minute to minute, for example if we are focusing very hard on solving a complicated problem we may be oblivious of other sensations, such as the noise in the room around us, even though our ears are still detecting it.

A further complicating factor is the way the conscious brain processes the sensory information to create the perception we have. Our assumption may be that these perceptions are an accurate and constant reflection of the physical world: we often may, mistakenly, think of the eye as being like a camera, but the games we can play with visual illusions quickly show us that this is not the case (Figure 18.1).

One further aspect of our perception of the sensory world that we should remember is that our sensory systems are particularly responsive to change or contrast. This can be evidenced in two ways:

- First, our sensory systems adapt to maintained stimuli. When we go into a room where there is a particular smell, we are very aware of it at first but, over time, our sense of smell adapts and we are no longer aware of the sensation.

- Second, when there is a gradual change in the environment, we may not be aware of that change for some time. For example, if we sit in a room where the temperature is very gradually decreasing, we won't become aware of the cooling down until the temperature had changed by a significant amount. Conversely, if we went from one room to another with the same difference in temperatures, we would be aware of the difference straight away.

18.1 General sensation

The sensory system can be categorized in a number of ways, one way is to separate it into the general sensory system and the **special senses** (Figure 18.2).

The general sensory system refers to the sensations that derive from the body, both external (e.g. touch, skin pain, and temperature sense) and internal (e.g. pain from muscles or internal organs and the sensory systems that monitor blood pressure or the pH of the blood). The general sensory system may be further divided into **somatic** and **visceral** sensations.

Somatic sensation includes:

- external (skin): e.g. touch, pressure, temperature, pain;
- internal (muscles and joints): proprioception (knowing where your body is in relation to itself), pain.

Visceral sensation:

- internal organs: pain, pressure.

The neurons that carry the sensory information from the receptors to the central nervous system (CNS) are referred to as **afferent neurons**. There are two main groups of afferent neurons in the peripheral nervous system (PNS):

- Dorsal root ganglion neurons carry information from the receptor to the spinal cord and have their cell bodies in the dorsal root ganglia of the spinal segments (Figure 18.3).

- Cranial nerve afferents carry information from the head and face to the cranial nerve nuclei in the brainstem.

The special senses, as indicated in Figure 18.2, are those specifically associated with the head and special organs of sensation, such as the eyes for vision and the ears for hearing. These will be discussed in Section 18.5.

Sensory stimulus

The sensory system responds to four main types of stimulus:

- mechanical
- chemical
- light
- thermal.

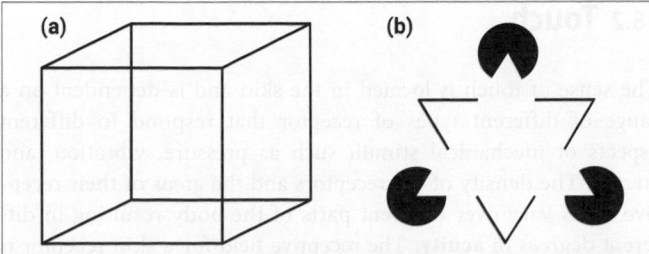

Figure 18.1 Visual illusions illustrating changes in our perceptions of objects. (a) The 'Necker cube'. Look at the cube and determine its orientation: you may perceive the front panel to be to the upper right or the lower left and it can switch between the two. (b) The 'Kanizsa triangle': our brains create a very strong perception of a triangle even though we 'know' there isn't really one there.

Source: (a) BenFrantzDale/ Wikimedia Commons/ CC BY-SA 3.0; (b) Kanizsa, G. (1955). Margini quasi-percettivi in campi con stimolazione omogenea. *Rivista di Psicologia*, 49(1): 7–30.

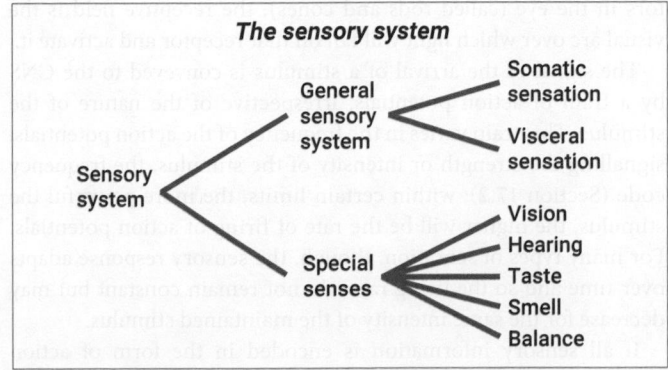

Figure 18.2 Organization of the sensory system.

Some of these, such as light, only contribute to a single perception, which is vision. Others, such as mechanical stimuli, give rise to a variety of sensations. You may be surprised to see that pain is not listed as a stimulus: we will consider pain specifically in Section 18.3.

What examples of mechanically based sensory systems can you think of? Take a moment before reading on. Touch, pressure, vibration, stretch, joint movement, and hearing are all examples of systems in which a mechanical stimulus is converted into a sensory perception.

Each of these sensory processes is underpinned by specialized sensory receptors which respond to these specific stimuli. Each receptor, therefore, has a specific stimulus to which it is particularly sensitive, which is termed the **adequate stimulus**. Irrespective of the different nature of these stimuli, all of them interact with the specific sensory receptor cells resulting in a change in action potential discharge in the sensory pathways that carry information towards the brain. The process of converting the stimulus into action potentials is termed **transduction**.

Transduction involves the interaction between the stimulus and the sensory receptor which results in a change in the membrane potential of the receptor, referred to as the **receptor potential**. In most cases this receptor potential will take the form of depolarization (one key exception to this is vision, see Section 18.5). Where the sensory receptor is itself a neuron, the receptor potential may directly give rise to a change in action potential discharge. In other systems, the sensory receptor may connect with a separate sensory neuron which generates the action potentials. The action potentials are then conveyed to the CNS via the afferent (sensory) pathways. Those signals that reach our conscious attention then give rise to a sensory percept, which is our conscious interpretation of the signal.

Not all sensory stimuli that give rise to a receptor potential will actually result in a change in action potential discharge. In the same way that at synapses some excitatory post-synaptic potentials (EPSPs) may cause the post-synaptic neuron to depolarize but not sufficiently to generate an action potential (see Figure 17.14), so some stimuli may be too small to exceed the threshold for generating action potentials (Figure 18.3). So, although the receptor itself may respond to the stimulus we will not be aware of it.

The area over which arrival of a stimulus activates a specific receptor defines the **receptive field** for that receptor. For a touch receptor in the skin, for example, this is the area of the skin which, if touched, will activate that specific receptor; for the vision receptors in the eye (called rods and cones), the receptive field is the visual arc over which light will fall on that receptor and activate it.

The signal of the arrival of a stimulus is conveyed to the CNS by a train of action potentials, irrespective of the nature of the stimulus. This train varies in the frequency of the action potentials, signalling the strength or intensity of the stimulus, the frequency code (Section 17.2): within certain limits, the more powerful the stimulus, the higher will be the rate of firing of action potentials. For many types of sensation, though, the sensory response adapts over time and so the firing rates do not remain constant but may decrease for the same intensity of the maintained stimulus.

If all sensory information is encoded in the form of action potentials, how do we know, for example, which sense is which? Or

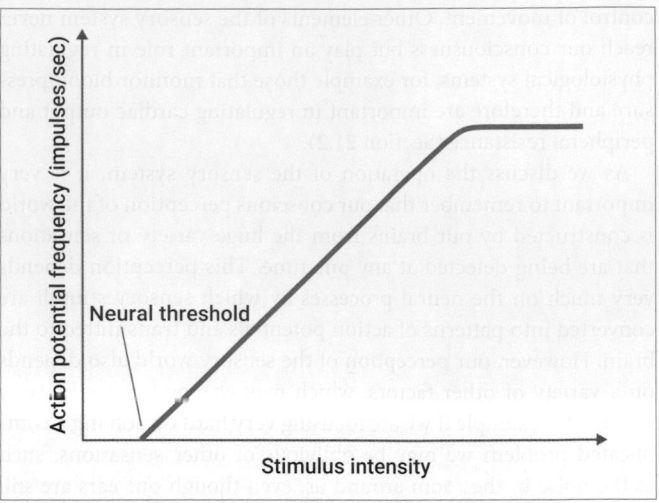

Figure 18.3 The relationship between action potential frequency and stimulus intensity.

which part of the body has been touched? The sensory pathways all involve a **labelled line** system: what this means is that specific sensory information is relayed to specific areas of the brain. In the case of senses such as touch, those areas are laid out as maps of the body (Figure 18.7), so the sensation of being touched on our index finger is relayed to a specific region of the brain concerned with touch and which is close to, but different to, that which detects being touched on the thumb.

PAUSE AND THINK

When we enter a building, we may be very aware of a specific smell. Why do we no longer notice that smell after a while?

Answer: The sensory response adapts over time so that unchanging stimuli no longer trigger a strong response and we become less aware of them. A feature of the sensory system, therefore, is that we are most sensitive to changes in our environment.

 Check your understanding of the concepts covered in this section by answering the questions in the e-book.

18.2 Touch

The sense of touch is located in the skin and is dependent on a range of different types of receptor that respond to different aspects of mechanical stimuli such as pressure, vibration, and stretch. The density of the receptors and the areas of their receptive fields vary over different parts of the body resulting in different degrees of **acuity**. The receptive field for a skin receptor is the area of skin which, when stimulated, will trigger a response in that receptor. The regions of the body where the receptors are most densely distributed and the receptive fields are smallest are, therefore, the areas where the acuity is greatest, regions such as the fingertips and the lips (Figure 18.4). The measurement of this acuity is referred to as **two-point discrimination**, which is the

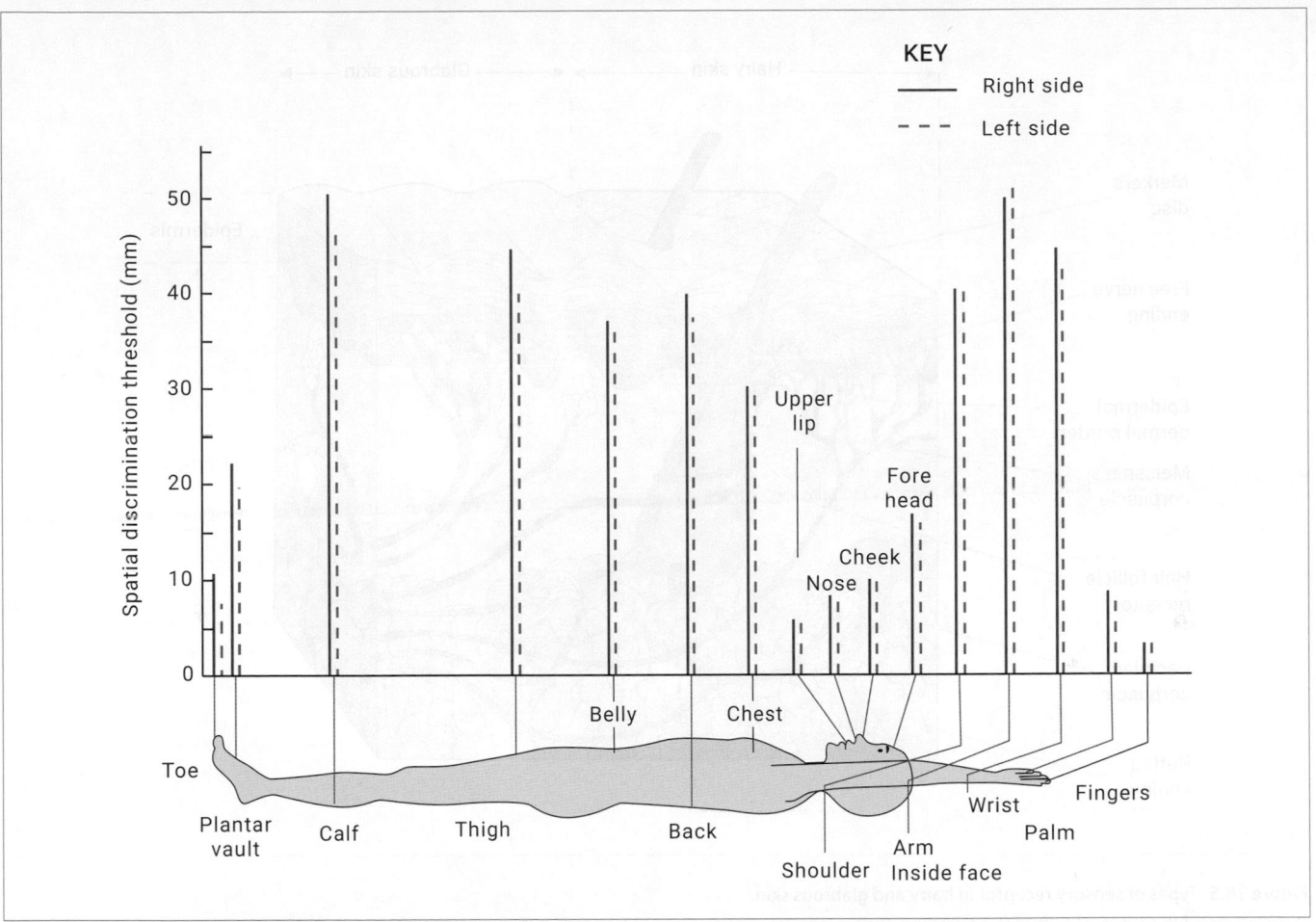

Figure 18.4 Variation in acuity of touch across different areas of the skin. Variation is shown by two-point discrimination: the minimum distance two stimuli have to be apart to be distinguishable as two separate stimuli.

Source: © 2011 L. Schacher, S. Bensaid, S. El-Ghezal Jeguirim, D. Adolphe. Adapted from: Sensory and physiological issues, advances in modern woven fabrics technology, Savvas Vassiliadis; originally published under CC BY 3.0 license. Available from: DOI: 10.5772/25244.

minimum distance two stimuli have to be apart to be distinguishable as separate stimuli rather than merging into one. As can be seen in Figure 18.4, those distances vary very significantly, from about 2 mm on the fingertips and mouth area to more than 50 mm on areas such as the back.

There are two types of skin surface: hairy skin, which covers most of the body, and glabrous (hairless) skin, which covers regions such as the palms of the hands. Within the layers of dermis, underlying the epidermis, are located a variety of different receptor types, which respond to the different types of mechanical stimulus (Figure 18.5, Table 18.1).

In the previous section we noted that the response of sensory receptors to unchanging stimuli adapts over time. This sensory adaptation can be due both to neural mechanisms and also due to the properties of the receptor. In the case of each of the receptors in Table 18.1, the afferent nerve ending terminates in a specialized structure which acts as a mechanical filter that confers the response properties of the ending. For example, the neurons that end in the Meissner's corpuscle and the Ruffini ending are similar but show different response properties because of the mechanical filtering

afforded by the accessory structure. So, the Meissner's corpuscle only responds briefly to the onset and offset of the stimulus (i.e. it adapts very quickly), whereas the Ruffini ending adapts slowly and will tend to fire action potentials throughout the contact with the stimulus.

Sensory pathways

Figure 18.6 illustrates the sensory pathway for touch, which is also called the **dorsal column-medial lemniscal pathway**. The pathway to the somatosensory cortex comprises three different classes of afferent neurons.

First-order afferents—the peripheral sensory neurons that conduct the action potentials from the receptors up to the region of the brainstem called the medulla oblongata. These afferent neurons are classified as Aβ neurons that have relatively large-diameter myelinated axons (6–12 µm) with conduction velocities of 35–75 m s^{-1}. Their cell bodies are located in the dorsal root ganglion of the spinal cord and their axons ascend in the white matter dorsal columns of the **ipsilateral** (same side) cord. The neurons are arranged in

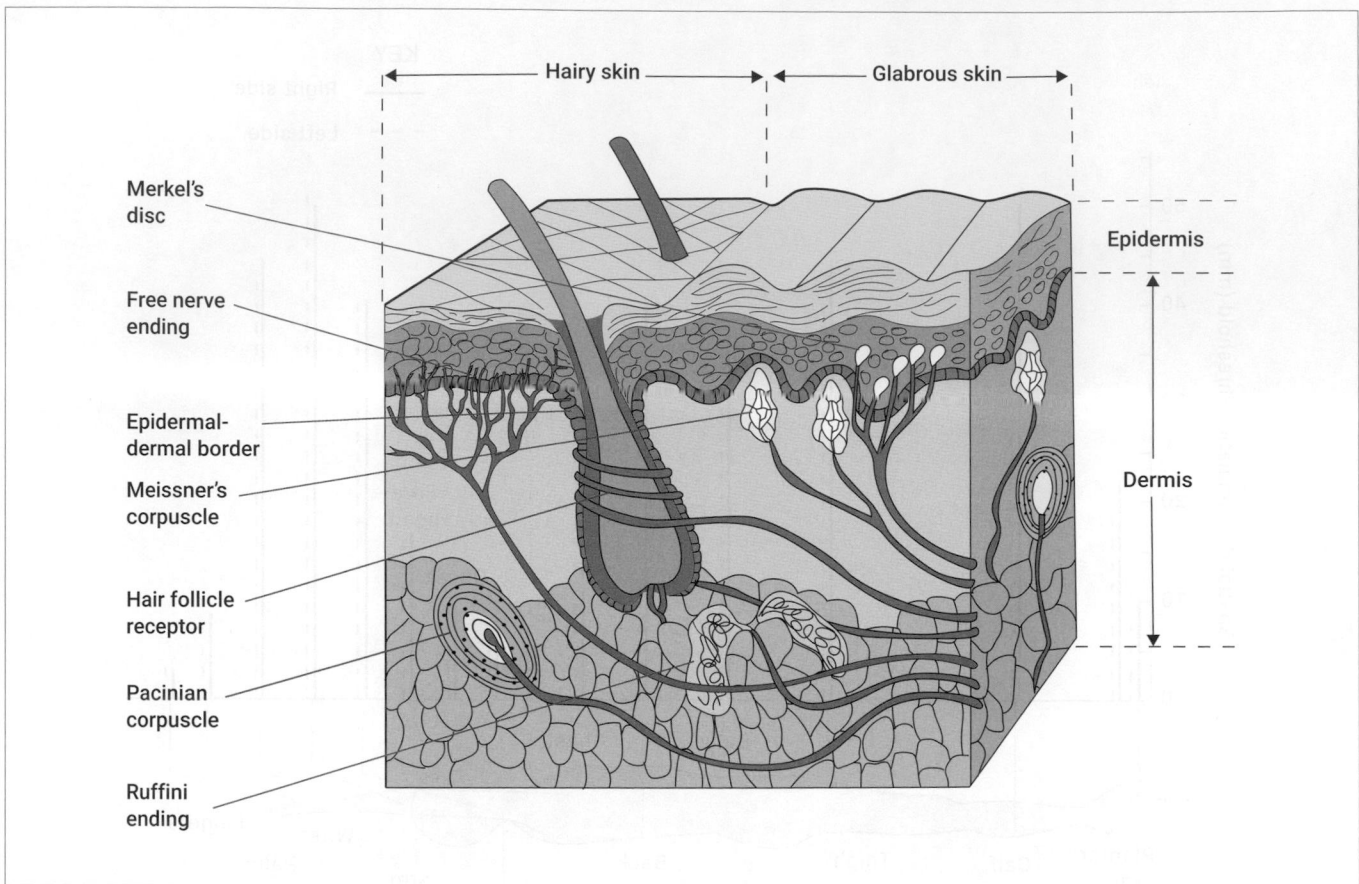

Figure 18.5 Types of sensory receptor in hairy and glabrous skin.

Source: M.F. Bear, B.W. Connors, M.A. Paradiso. *Neuroscience: Exploring the Brain*, 3rd edn, Figure 12.1. Lippincott Williams & Wilkins, Wolters Kluwer, 2006.

Table 18.1 Cutaneous receptor types and their response properties

Receptor	Stimulus	Rate of adaptation	Receptive field size
Hair follicle receptors	Stroking	Fast	Small
Meissner's corpuscle	Stroking	Fast	Small
Pacinian corpuscle	Vibration	Fast	Large
Merkel's disc	Pressure	Slow	Small
Ruffini ending	Stretch	Slow	Large

order, so that the afferents from the lower parts of the body (i.e. the feet and legs) lie closest to the midline of the cord and those from the upper parts of the body are located more laterally.

How long is a neuron? Think about the different parts of your body: neurons vary greatly in length, some in the brain have axons that may be only 1 mm in length. The first-order afferents of the dorsal column–medial lemniscal pathway are among the longest in the body: they can stretch from the big toe to the brainstem, so they can be more than 1.5 m in length in adults!

Second-order afferents—the cell bodies of the second-order afferents are located in the nuclei in the lowest part of the brainstem, the medulla oblongata. Here, there are synaptic connections between the first- and second-order afferents. The second-order afferents cross over (**decussate**) to the other side—contralateral—of the spinal cord and ascend in a white matter tract called the medial lemniscus.

Third-order afferents—the cell bodies of the third-order afferents are located in the thalamus: this is a very complex group of nuclei that process and route the different forms of conscious sensation to the different areas of the cerebral cortex. The third-order afferents of the touch pathway ascend to the **primary somatosensory cortex**, or S1.

The dorsal column–medial lemniscal pathway carries information from the body up to the somatosensory cortex. Sensory information from the face and head is mainly carried via the **trigeminal pathways** which are also routed through the thalamus to S1.

The S1 is located on the postcentral gyrus (Figure 18.7a). Note that because all the sensory fibres decussate (cross over) the representations of the body are also crossed, so the left side of the brain

Figure 18.6 The dorsal column–medial lemniscal pathway mediating touch and conscious proprioception. The name of the pathway reflects the two main white matter segments where the afferent axons are located.

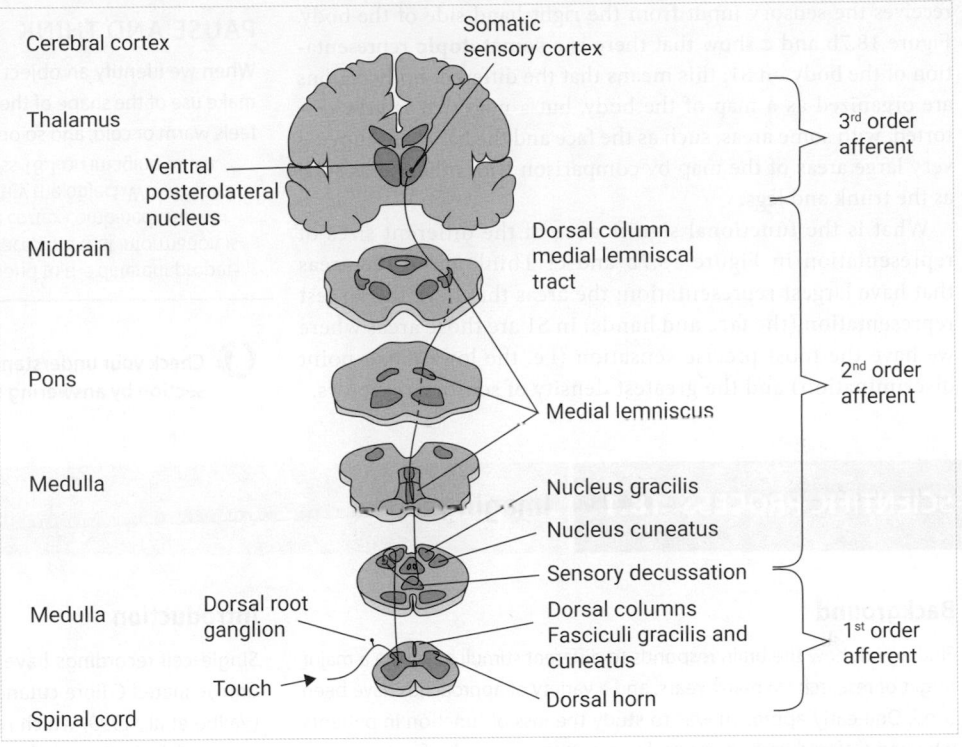

Figure 18.7 Somatosensory (S1) cortex.
(a) Location of S1 on the postcentral gyrus of the parietal lobe. (b, c) Illustrations to show how various body parts would look if their size was proportionate to: (b) the amount of tissue they take up in the somatosensory cortex and (c) the acuity of sensation.

Source: (a, b) Jarvis, M., Okami, P. (2020) *Principles of Psychology*. Oxford University Press; (c) Figure 9.11 from Purves et al., *Neuroscience*, 6th edn, 2017. Oxford University Press.

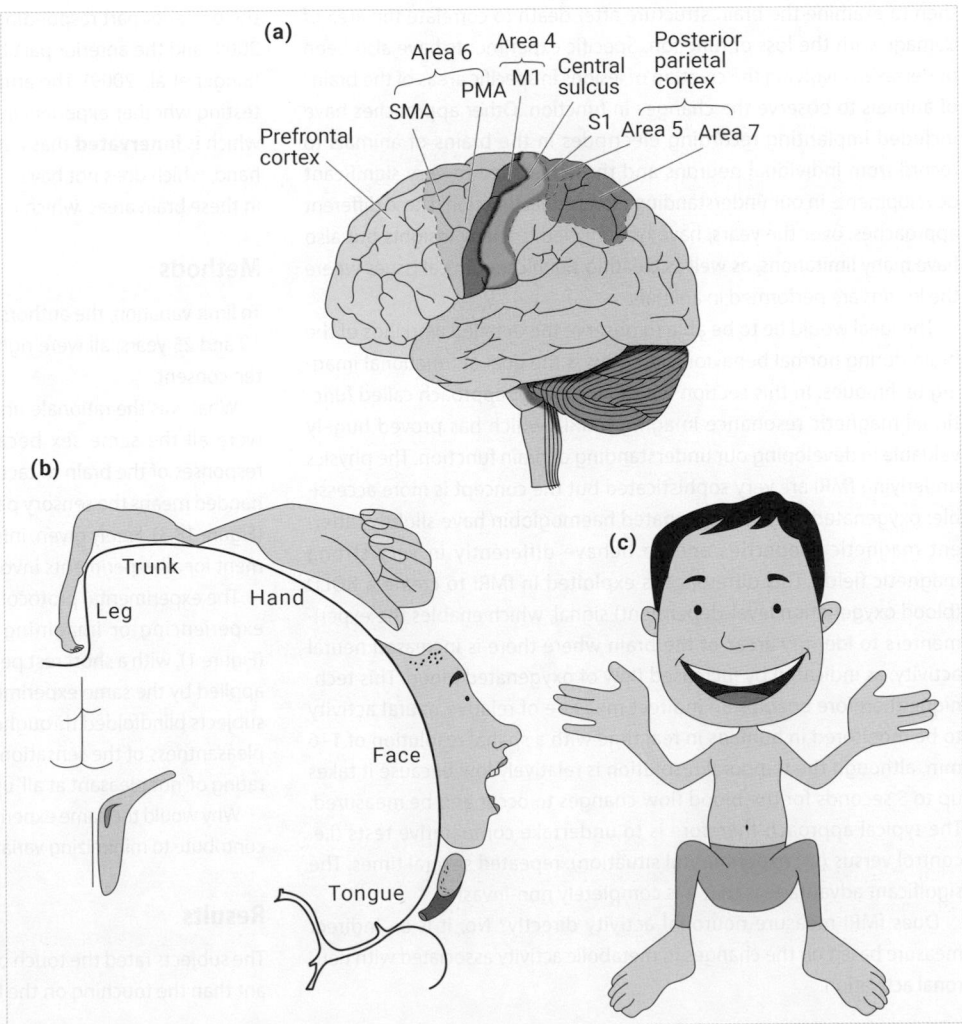

receives the sensory input from the right-hand side of the body. Figure 18.7b and c show that there is a **somatotopic** representation of the body on S1: this means that the different body regions are organized as a map of the body, but a map that is very distorted, with some areas, such as the face and the hands, occupying very large areas of the map by comparison with other areas such as the trunk and legs.

What is the functional significance of the different sizes of representation in Figure 18.7b and c? Think about the areas that have largest representation: the areas that have the largest representation (the face and hands) in S1 are those areas where we have the most precise sensation (i.e. the lowest two-point discrimination) and the greatest density of sensory receptors.

 Check your understanding of the concepts covered in this section by answering the questions in the e-book.

SCIENTIFIC PROCESS 18.1 | Imaging touch

Background

Finding out how the brain responds to different stimuli has been a major target of research for many years, and a variety of approaches have been used. One early approach was to study the loss of function in patients who had suffered brain damage, for example as a result of a stroke, and then to examine the brain structure after death to correlate the area of damage with the loss of function. Specific experiments have also been undertaken involving the creation of lesions in specific areas of the brains of animals to observe the changes in function. Other approaches have included implanting recording electrodes in the brains of animals to record from individual neurons and these have led to very significant developments in our understanding of neuronal function. These different approaches, over the years, have not only led to many insights but also have many limitations, as well as creating significant ethical issues where the lesions are performed in animals.

The ideal would be to be able to observe the detailed workings of the brain during normal behaviours and this is the goal of functional imaging techniques. In this section we will look at an approach called functional magnetic resonance imaging (fMRI), which has proved hugely valuable in developing our understanding of brain function. The physics underlying fMRI are very sophisticated but the concept is more accessible: oxygenated and de-oxygenated haemoglobin have slightly different magnetic properties and so behave differently in very strong magnetic fields. This difference is exploited in fMRI to create a BOLD (blood oxygenation level-dependent) signal, which enables the experimenters to identify areas of the brain where there is increased neural activity, as indicated by increased flow of oxygenated blood. This technique therefore enables an indirect measure of relative neural activity to be monitored in humans in real time with a spatial resolution of 1–6 mm, although the temporal resolution is relatively low because it takes up to 5 seconds for the blood flow changes to occur and be measured. The typical approach therefore is to undertake comparative tests (i.e. control versus (vs) experimental situation), repeated several times. The significant advantage is that it is completely non-invasive.

Does fMRI measure neuronal activity directly? No, it is an indirect measure based on the changes in metabolic activity associated with neuronal activation.

Introduction

Single-cell recordings have indicated that there is a specific group of unmyelinated C fibre cutaneous neurons called C-tactile (CT) afferents (Vallbo et al., 1999) which respond to gentle stroking. It appears that a region of the brain, called the insula, is divided into function regions with the posterior part responding to C-tactile stimulation (Björnsdotter et al., 2009) and the anterior part being responsive to subjective feeling states (Singer et al., 2009). The authors set out to explore these differences by testing whether experiencing or imagining gentle touching of the arm, which is **innervated** (has a nerve supply) by CT afferents, or palm of the hand, which does not have CT afferents, would evoke different responses in these brain areas, which could be recorded using fMRI.

Methods

To limit variation, the authors selected a group of 17 males aged between 19 and 25 years; all were right-handed. All subjects gave informed, written consent.

What was the rationale underlying the subject selection? The subjects were all the same sex because there may be sex differences in the responses of the brain to tactile stimulation. Having them all being right-handed means the sensory pathways project to the same side of the brain (Figure 18.3). Freely given, informed, written consent is an ethical requirement for all experiments involving human subjects.

The experimental protocol involved four procedures, with the subjects experiencing or imagining gentle touch to the right arm or palm (Figure 1), with a short rest period in between each trial. The touching was applied by the same experimenter in all trials, using a soft brush with the subjects blindfolded throughout. The subjects were also asked to rate the pleasantness of the sensation using a 5-point 'Likert' scale, with 1 being a rating of 'not pleasant at all' up to 5, which was 'extremely pleasant'.

Why would the same experimenter carry out all the touching? This would contribute to minimizing variation within the experimental protocol.

Results

The subjects rated the touch on the arm as being significantly more pleasant than the touching on the hand (paired-sample *t*-test, $t=2.75$, $P=0.01$).

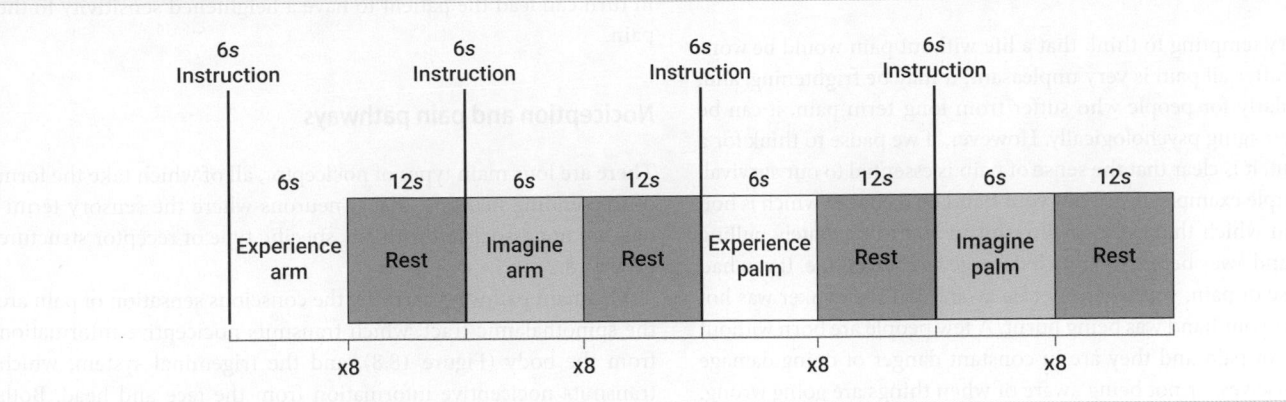

Figure 1 Experimental protocol for fMRI trial.

Source: Lucas, M.V., Anderson, L.C., Bolling, D.Z., Pelphrey, K.A. & Kaiser, M.D. (2015) Dissociating the neural correlates of experiencing and imagining affective touch. *Cerebral Cortex*, 25, 2623–2630. doi:10.1093/cercor/bhu061 Reproduced by permission of Oxford University Press.

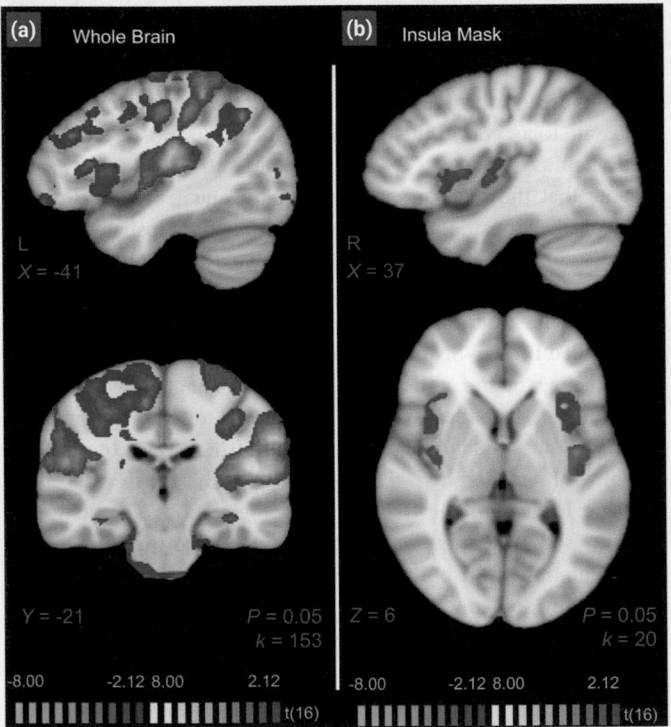

Figure 2 Coloured fMRI displays showing the brain regions where the BOLD signal showed greater activation for experienced vs imagined. Red shows the most increase in activation, green would show a decrease in activation. The left-hand panels (a) show sagittal (upper) and coronal (lower) sections of the brain. The right-hand panels (b) show sagittal (upper) and horizontal (lower) sections with specific activation of the two regions of the insula where the response was greater for experienced vs imagined conditions.

Source: Lucas, M.V., Anderson, L.C., Bolling, D.Z., Pelphrey, K.A. & Kaiser, M.D. (2015) Dissociating the neural correlates of experiencing and imagining affective touch. *Cerebral Cortex*, 25, 2623–2630. doi:10.1093/cercor/bhu061 Reproduced by permission of Oxford University Press.

Analysis of the fMRI showed that the left postcentral gyrus and the posterior insula on both sides showed positive responses to the experienced condition but not to the imagined condition (Figure 2).

Why would the left postcentral gyrus be active during actual touching but not during imagined touching? The postcentral gyrus is the location of the somatosensory cortex (Figure 18.7), S1, which is the first cortical region to receive projections from the ascending sensory pathways.

The anterior insula on both sides responded to both conditions but showed greater activation during the imagined vs the experienced conditions. Furthermore, within each of these anterior and posterior regions of the insula were more discrete regions that showed greater responses to arm (CT) vs palm (non-CT) for both experienced and imagined conditions.

Conclusions

In these experiments, the authors used functional brain imaging techniques to identify specific regions of the brain that respond to different patterns of cutaneous stimulation. They were able to show that the posterior insula only responded to the physical experience of touch, whereas the anterior insula showed a response to both the physical and imagined touch conditions. Furthermore, there were specific subregions that responded more to the CT than non-CT inputs for both imagined and experienced conditions. These findings made an important contribution to the understanding of the processing of different forms of touch and its social/emotional significance within the brain.

Read the original work

Björnsdotter, M., Löken, L., Olausson, H., Vallbo, Å., & Wessberg, J. (2009) Somatotopic organization of gentle touch processing in the posterior insular cortex. *J. Neurosci.* **29**: 9314–20.

Lucas, M. V., et al. (2015) Dissociating the neural correlates of experiencing and imagining affective touch. *Cereb. Cortex* **25**: 2623–30. doi:10.1093/cercor/bhu061

Singer, T., Critchley, H. D., & Preuschoff, K. (2009) A common role of insula in feelings, empathy and uncertainty. *Trends Cog. Sci.* **13**: 334–40.

Vallbo, Å. B., Olausson, H., & Wessberg, J. (1999) Unmyelinated afferents constitute a second system coding tactile stimuli of the human hairy skin. *J. Neurophysiol.* **81**: 2752–63.

18.3 Pain and temperature

It is very tempting to think that a life without pain would be wonderful: after all pain is very unpleasant, it may be frightening, and, particularly for people who suffer from long-term pain, it can be very damaging psychologically. However, if we pause to think for a moment, it is clear that the sense of pain is essential to our survival: as a simple example, if you put your hand on a cooker which is hot, the pain which that causes will result in you immediately pulling your hand away before too much damage has been done. If you had no sense of pain, you would not be aware that the cooker was hot and that your hand was being burnt. A few people are born without a sense of pain and they are in constant danger of doing damage to themselves, or not being aware of when things are going wrong. Such people often die quite young.

Pain, therefore, has a very important protective aspect, warning of damage and evoking responses to remove the body from harm's way.

What is the initial response you make if you accidentally put your hand on something very hot or sharp (apart from yelling!)? Think about what happened last time you experienced this. What was your response? There is a reflex response—the **withdrawal reflex**. This is a spinal reflex mediated by the pain fibre afferents activating the motor neurons of the flexor muscles to pull your arm away from the source of the damage. As a spinal reflex it is very rapid, thereby minimizing the extent of damage.

We also learn very quickly from the results of painful experience what things to avoid in future, and indeed may well become fearful of the cause of the pain: people who have been stung by wasps as children may develop a genuine fear of them that persists throughout their lives.

Before considering the processes involved, it is important to distinguish between two different aspects: pain and **nociception**. Nociception is the physiological process whereby signals of damage to tissues are generated and conducted to the CNS, whereas pain is the actual perception of the associated discomfort. The same nociceptive stimulus may give rise to very different sensations of pain depending on the context; for example, the soldier fighting for life in battle may be unaware of quite significant injuries, whereas relatively minor injuries suffered during normal life may cause the same person to feel a lot of pain. In some cases, pain may be completely separated from nociception, where the pain is generated with the brain itself. Pain may also have significant impact on the state of mind of the sufferer; for example, pain may give rise to anxiety or fear, especially if the cause is unknown or

is associated with specific conditions, such as heart disease. This in turn can lead the patient to have a heightened sensitivity to the pain.

Nociception and pain pathways

There are four main types of nociceptor, all of which take the form of free-ending neurons, that is neurons where the sensory terminals are not associated with any specific type of receptor structure (Table 18.2).

The main pathways carrying the conscious sensation of pain are the spinothalamic tract, which transmits nociceptive information from the body (Figure 18.8), and the trigeminal system, which transmits nociceptive information from the face and head. Both of these pathways have relays in the thalamus and project to the somatosensory cortex of the brain.

First-order afferents—these afferent neurons carry the nociceptive information from the periphery to the spinal cord where they terminate in the dorsal horn, in a region called the substantia gelatinosa.

Second-order afferents—the cell bodies of the second-order afferents are located in the spinal cord and their axons immediately decussate to ascend in the contralateral cord (opposite side) up to the thalamus.

Note that the two sets of sensory pathways travel in different parts of the cord: the pathway for touch remains ipsilateral in the dorsal columns (Figure 18.6), whereas the pathway for pain crosses over and ascends in the ventro-lateral part of the contralateral cord.

If a patient suffers an injury that cuts one side of the spinal cord (a so-called Brown–Séquard lesion), what will happen to their sensation below the site of the injury? If the right-hand side of the cord has been cut, the patient will lose the sense of touch from the right-hand side, but the sense of pain from the left-hand side of the body.

Third-order afferents—the cell bodies of the third-order afferents are located in the thalamus and their axons ascend to the primary somatosensory cortex (S1) to map onto the somatotopic representation of the body (Figure 18.7).

Why can someone who has had a limb amputated still feel pain from that limb? Before moving on, think about the neural pathways. Phantom limb pain is a well-recognized phenomenon whereby patients may feel pain that seems to originate from a limb that has been lost. Although the limb and its nerve endings have been removed, the limb is still represented in the central pathways and on the somatosensory cortex, therefore spontaneous discharge from these neurons can elicit different sensations, including pain.

Table 18.2 Types of nociceptor

Nociceptor	Activating stimulus	Afferent fibre
Mechanical	Intense mechanical stimuli, e.g. pinching, cutting	Aδ: small diameter, myelinated afferent neurons
Thermal	Extreme cold or heat (typically < 5 °C or > 45 °C)	Aδ: small diameter, myelinated afferent neurons
Chemical	Chemicals released as a result of tissue damage, e.g. histamine, bradykinin, potassium	C: small diameter, slow conducting (< 1 ms^{-1}) unmyelinated afferent neurons
Polymodal	A range of stimuli including chemical, thermal, or mechanical	C: small diameter, unmyelinated afferent neurons

Figure 18.8 The spinothalamic tract.

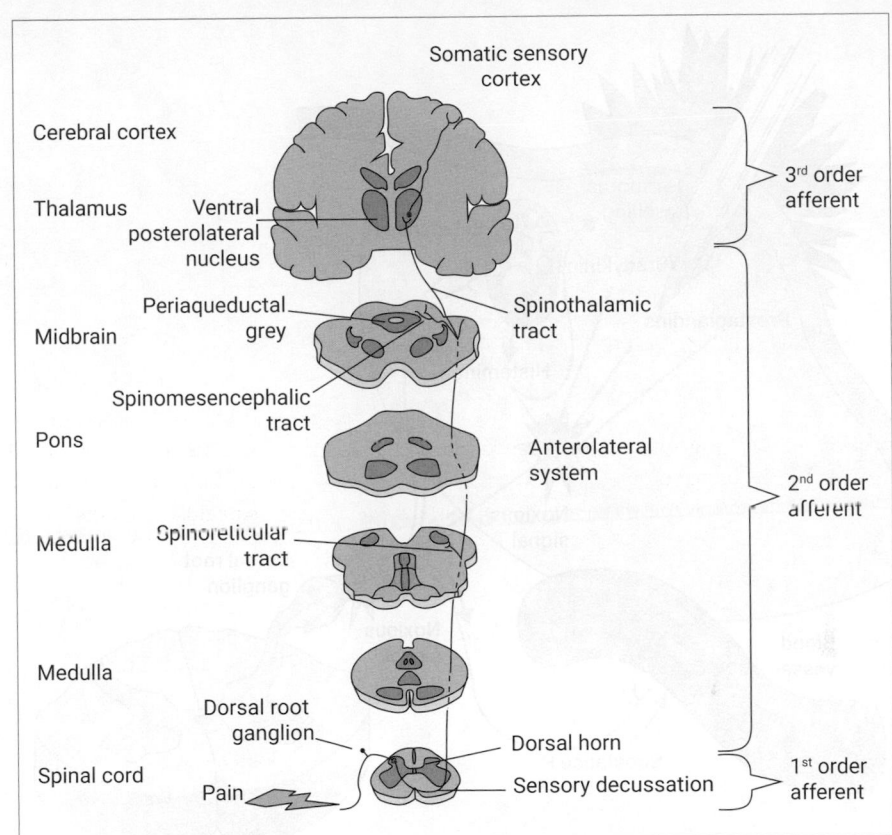

Pain sensation and tissue damage

The sensation of pain is triggered by noxious (harmful) stimuli that cause, or have the potential to cause, tissue damage. Associated with this is the release of a number of chemical mediators that can activate the free nerve endings that signal pain (Figure 18.9).

The local mediators include bradykinins, prostaglandins, Substance P, and histamine, and elicit a range of responses, including the well-recognized **primary hyperalgesia**, characterized by increased pain sensitivity and inflammation (associated tissue swelling and redness). These responses both help to promote healing and also elicit a change in behaviour: the release of the mediators and associated hyperalgesia means that the pain may be ongoing after the removal of the stimulus itself (think of how a burn can be painful for hours or days after the removal of the heat) and this also changes behaviour because we protect the region from further damage. Where the tissue damage is more significant, there may be **secondary hyperalgesia**, which is also mediated in part by CNS mechanisms and results in a more widespread increase in pain sensitivity whereby non-noxious stimuli such as light touch can also evoke pain.

As shown in Figure 18.8 the ascending pain pathways also terminate in various regions of the brainstem. Some of the connections from these pathways impact on the arousal systems of the brain, which focus attention on specific stimuli and pain is often one such stimulus, particularly when the cause of the pain is unknown. These mechanisms are also associated with affective responses, such as anxiety and depression which are commonly associated with pain particularly where it is of long duration.

Controlling pain

The market for pharmaceutical agents controlling pain runs into billions every year. Many of these agents can be purchased over the counter and you may well have taken some in the last few weeks. Many common **analgesics** (pain killers) act to block the production of the local mediators, for example the non-steroidal anti-inflammatory drug (NSAID) analgesics, such as salicylic acid (aspirin) and ibuprofen, act to block the cyclooxygenase (COX) enzymes that are involved in the production of prostaglandins. These drugs therefore have a local action, as do the local **anaesthetics** (agents which block all sensation), which are also commonly used to control pain: anyone who has had dental treatment will be greatly appreciative of the efficacy of local anaesthetics which block sodium channels and therefore prevent neural transmission.

More potent analgesics such as the opioids act on central synaptic mechanisms in the pain pathways in the spinal cord and the brain. These drugs, such as morphine, are also commonly associated with generating feelings of euphoria and their use has to be tightly controlled because they are often highly addictive.

The body, though, also has its own pain regulation systems: as we noted earlier, soldiers in battle may suffer horrific injuries and

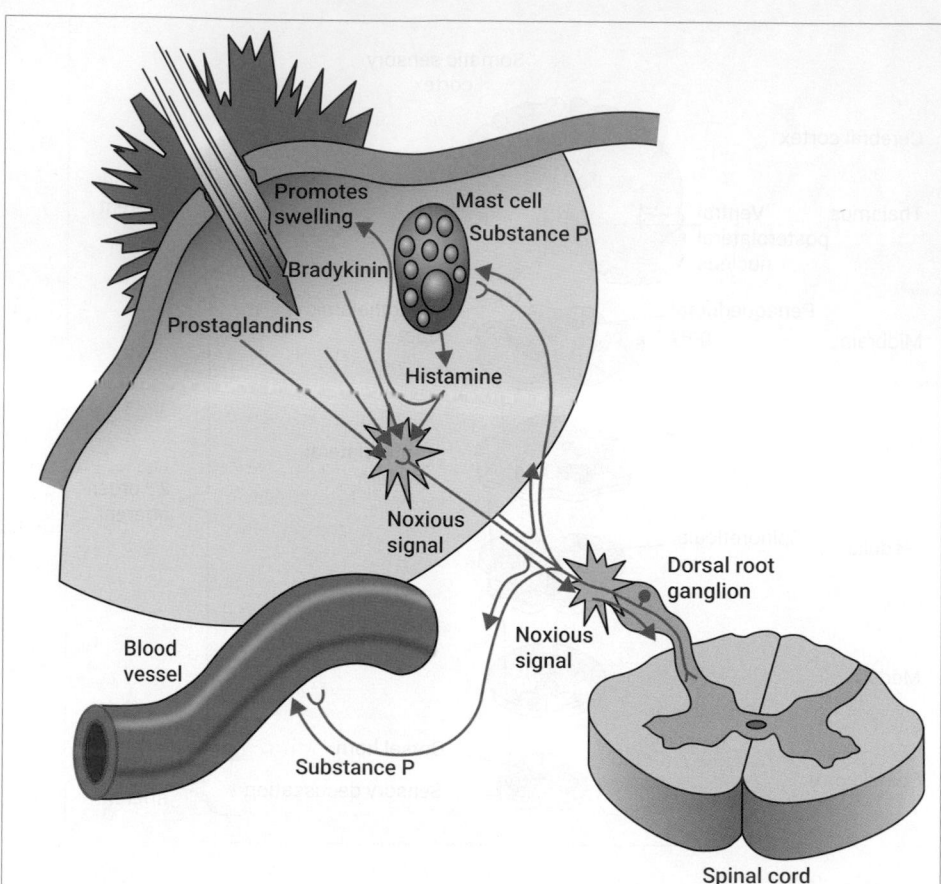

Figure 18.9 Release of local mediators as a result of tissue damage.

Source: Figure 10.7 from Purves et al. *Neuroscience*, 6th edn, 2017. Oxford University Press.

be unaware of them while they are fighting for their lives. Pain regulation can occur at various levels within the nervous system. At the level of the spinal cord, there is evidence of the operation of a '**pain gate**'. This hypothesis was originally developed to explain the phenomenon that 'rubbing makes it better': for example, when you have bruised yourself, an automatic reaction is to rub the area of skin around the site of the injury and it often makes it feel better. The concept of the pain gate is that the activity in large diameter sensory afferents from the skin (e.g. those mediating touch) can block the transmission of the pain signal by the inhibitory action of interneurons in the spinal cord.

A second, powerful site of inhibition of pain is a region of the brainstem called the periaqueductal grey (PAG). Experiments have shown that stimulation of this site activates descending neurons that block the transmission of pain at the level of the spinal cord. The body also produces its own opioids, called **endorphins**. These endorphins have a variety of actions, including inhibiting the second-order pain fibres. Injections of endorphins into the PAG have also been shown to be very effective in producing analgesia by activating the descending pain control system.

What is the placebo effect? The placebo effect occurs when a patient believes that a treatment will be effective even if that 'treatment' actually contains no active ingredients. In the case of pain, for example, giving a patient a pill and telling them it will reduce the pain may be effective, even though the pill contains no active ingredients. The belief may be sufficiently strong to activate the release of endorphins: some experiments have shown that giving patients drugs that block the action of endorphins prevent the placebo effect.

Temperature

The body has temperature receptors in the skin and in central structures such as the hypothalamus and the spinal cord. The central receptors play an important role in the regulation of the core temperature of the body, which is normally maintained within very tight bounds. There are separate receptors in the skin for detecting cold and hot and both types of receptor show adaptation, so if you are sitting in a room that is gradually cooling down, you may well not be aware of the temperature change until it has reached several degrees centigrade, whereas you would detect a much small change in temperature if it took place very quickly.

An interesting example of the adaptive process is shown by Weber's three bowl experiment: three bowls are filled with water, one with cold water, one at room temperature and one with moderately warm water. The subject holds one hand in the cold bowl and one in the warm one for 2–3 min. Then places both hands in the room-temperature bowl. At this moment, the two hands will feel as though they are at different temperatures: for the hand that was previously in the cold bowl, the new one will feel very warm but cold for the hand that was previously in the warm bowl.

The pathways for temperature are almost exactly the same as those for pain (Figure 18.8). Cold receptors are associated with Aδ and C fibres, whereas non-painful hot receptors are linked to C fibres only.

 Check your understanding of the concepts covered in this section by answering the questions in the e-book.

18.4 Proprioception

Proprioception is the sense of the position of the body and the relative position of the body parts. For example, the sense of proprioception enables us to close our eyes and touch our nose without seeing our hands.

This proprioceptive information is provided by several different types of sensory receptor, including:

- Feedback from muscle—the muscle spindle and tendon organ.
- Feedback from skin—the stretch receptors in skin, especially over joints.
- Feedback from joint receptors associated with the capsules of synovial joints.

We also receive important information from the visual system by looking at where our limbs are, but we are perfectly capable of performing very precise movements without watching what our limbs are doing: for example a touch typist or a skilful pianist can perform rapid, precision movements without watching what their hands are doing. However, that ability is lost if the sensory feedback from the skin and muscles is lost. Patients who are affected by a neurological disorder termed large-fibre sensory neuropathy lose the function of the large-diameter sensory neurons from the affected part of the body, which will include loss of proprioception.

These patients are typically able to perform accurate movements while they are watching the affected limb, but in the absence of vision there is postural drift and the movements, especially small movements, are grossly inaccurate.

Proprioceptive information is both conscious and unconscious. The unconscious pathways carry the information to the spinal reflex pathways and to structures such as the cerebellum (Section 20.4) that have important roles in controlling movements. The conscious pathways, as with other conscious sensations, relay the information to the cerebral cortex where it is integrated to generate our awareness of how our body is moving.

The processes involved in proprioception are considered in greater detail in Chapter 20.

How do we perceive the proprioceptive information? Think for a moment before moving on. Although some of the feedback from muscles is based on whether the muscle is shortening or lengthening, we don't think about movement in that way. For example, if the elbow is flexed, the sense organs from the muscle feedback information that the biceps brachii is shortening while the triceps is lengthening, but the actual conscious sensation we have is not that the muscles are changing length but that the elbow is flexing (Scientific Process 18.2).

Muscle spindle—responding to muscle length

Muscle spindles are located in all of the skeletal muscles, the muscles that are attached to bones and generate movement of the body, and respond to stretch, thereby signalling movement. The muscle spindle is one of the most complex of the sense organs in the body. The relative density of spindles in each muscle is related to the precision of control that we have over that muscle, for example the small muscles in the hand can have up to 90 muscle spindles per gram of muscle tissue compared with only 5 spindles per gram in

---

SCIENTIFIC PROCESS 18.2	Proprioception

Introduction

Proprioception is provided by a number of different receptor types. For a number of years there was a debate as to whether or not muscle spindles contribute to the conscious sensation of movement.

Methods

The study was undertaken on four patients who were undergoing surgery at the wrist under local anaesthesia. A second stage involved exposing and cutting the tendon of one of the muscles that extends the big toe in the foot of one of the paper's authors (DIM).

In all the subjects, tendons were pulled upon to stretch the muscles and simulate movement although the joints themselves were prevented from moving.

What was the significance of using a local anaesthetic to block sensation? The local anaesthetic was required for the operating procedure

to prevent pain, but it also meant that there was no associated sensation from the skin of the subjects that could signal movement was taken place.

Results and discussion

All subjects were able to detect the stretching of the muscles which they reported as rotation of the relevant joint to which the muscle was attached. Movements of less than 1 mm could be detected. Therefore, the researchers were able to conclude that the muscle spindles do indeed make a major contribution to the conscious sensation of joint movement.

Read the original work

McCloskey, D. I., Cross, J., Honner, R., & Potter, E. K. (1983) Sensory effects of pulling or vibrating exposed tendons in man. *Brain* **106**: 21–37, https://doi.org/10.1093/brain/106.1.21

muscles such as the calf muscle, gastrocnemius, which is used for walking and running.

Figure 18.10 shows a schematic diagram of the muscle spindle and also a micrograph of a spindle from a human hand. The spindle is made up of highly specialized muscle fibres, which are much smaller than the typical skeletal muscle fibres and are called **intrafusal muscle fibres**. In their central regions the intrafusal muscle fibres are innervated by large diameter sensory (afferent) nerve axons, called Ia axons, which wrap around the individual muscle fibres to form the primary ending.

Alongside the primary ending there are usually one or more secondary endings formed by small-diameter group II afferent axons. The terminals of the Ia and II afferent axons spiral around the intrafusal fibres.

When the muscle is lengthened, the intrafusal fibres are stretched, which also stretches out the spirals of the sensory endings. This mechanical stretching opens Na$^+$ channels that are mechanically sensitive, leading to depolarization of the axons and the generation of action potentials which are conducted up to the spinal cord via the Ia and II afferents.

The two types of ending convey slightly different types of information:

- the Ia afferents, forming the primary endings, are sensitive to both the length and the dynamic changes in length of the muscle, that is the velocity of the movement
- the II afferents, forming the secondary endings, mainly respond to the actual length changes.

Therefore, the CNS receives information about both the change in length of the muscle and the rate at which that is happening. As we discussed above, in Scientific Process 18.2, the conscious perception we have is of the movement of the joint, not of the changes in muscle length themselves.

The muscle spindle also has another special feature: just like skeletal muscle fibres, the intrafusal muscle fibres are innervated by motor neurons. In this case, these are smaller-diameter neurons compared with the α-motor neurons (Section 20.2) and are called γ (gamma)-motor neurons (Figure 18.1). These γ-motor neurons cause the intrafusal fibres to contract and so they can change the

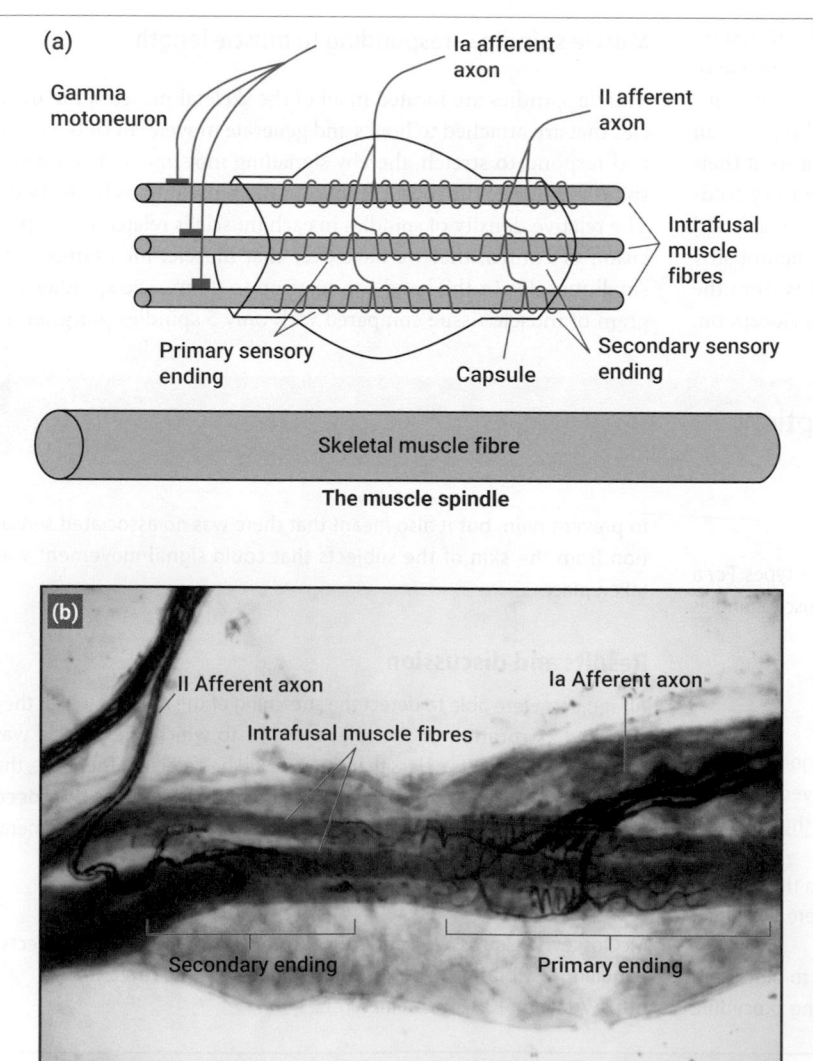

Figure 18.10 The muscle spindle. (a) Diagram illustrating the composition of the muscle spindle. In (b), the micrograph shows a muscle spindle from a human hand muscle, which has been stained with silver nitrate to show the axons. Image length ~ 1 mm.

sensitivity of the muscle spindle to the changes in muscle length. In particular, they can bias the nature of the response, to make the muscle spindle more sensitive either to the actual change in length, or to the velocity of the stretch.

How does this compare with other sensory systems? For some of the other complex sensory systems the overall sensitivity can be modified, for example for both vision and hearing (Section 18.5); however, it is not possible to control the nature of the response. Thus, with hearing the overall sensitivity can be reduced by action of the muscles in the middle ear (Section 18.5), but we cannot make our ears specifically more sensitive to higher or lower frequencies. This ability of the CNS to control the nature of the response from the muscle spindle is very special and is very important in the control of movement.

The Golgi tendon organ—responding to muscle contraction

The tendon organ, or to give it its full title, the Golgi tendon organ, was first described by the famous scientist Camillo Golgi (who also gave his name to a number of other structures, including the Golgi apparatus in the cell (Section 10.4)). Tendon organs are present in all our skeletal muscles and, like the muscle spindle, their

density is a reflection of the level of precision control we have over the muscle.

As its name suggests, the tendon organ is located at the junction between the muscle fibres and the collagen fibres of the tendon (Figure 18.11). The sensory innervation is provided by a large diameter, afferent axon, called the Ib axon. The terminals of the afferent are wrapped around the strands of collagen that make up the body of the organ.

When the muscle fibres contract the tendon is stretched. As a consequence, the collagen strands forming the body of the tendon organ are stretched, leading to stretching of the spirals of the sensory ending. As with the muscle spindle, this stretching opens the Na^+ stretch-activated channels, leading to depolarization and the generation of action potentials. The frequency of action potential discharge reflects the tension being generated within the muscle.

The sensory feedback from the muscle spindle and the tendon organ therefore provides the motor control system with the information needed to enable planning and monitoring of ongoing movements: position, change in position, and the active force being generated.

 Check your understanding of the concepts covered in this section by answering the questions in the e-book.

Figure 18.11 The Golgi tendon organ.
(a) Diagram of the structure of the tendon organ.
(b) Histological preparation of a tendon organ from a human hand muscle. The organ shown is about 0.5 mm in length.

Source: (b) Figure 7.1 from Scott, J.J.A. (2005). The golgi tendon organ. In: Dyck, P.J. & Thomas, P.K. (eds) *Peripheral Neuropathy*, 4th edn. Elsevier.

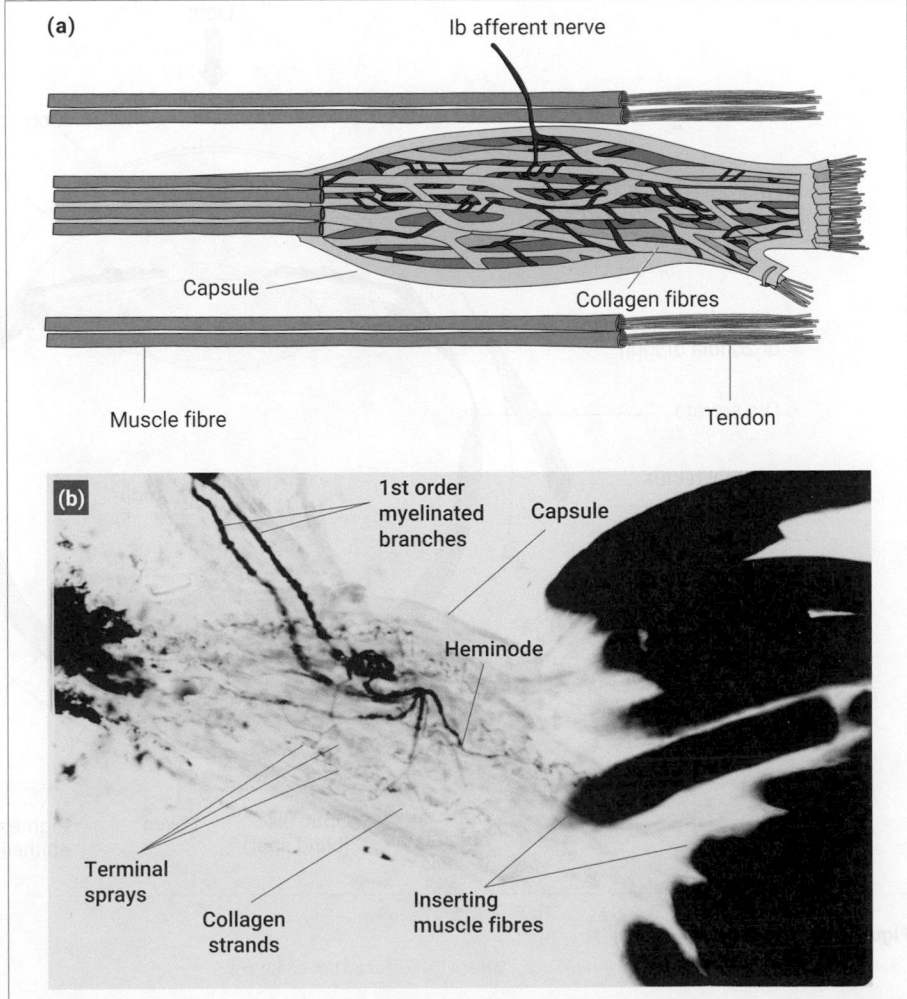

18.5 Special senses

The special senses comprise the group of senses that are specifically associated with the head and comprise vision, hearing, balance, taste, and smell. Vision and hearing have particular roles in that they enable us to explore the external world at a distance from us and to communicate over large distances.

Vision

The visual system comprises the eye and extensive regions of the brain that are involved in processing the wealth of information. The visual system is often compared to a camera, but it does not create a faithful image of the visual world because the neural processing accentuates features such as contrast in the visual field (Figure 18.12).

Figure 18.12 Visual contrast. The two inner squares are actually the same shade of grey but appear to be different because of the relative contrast against the darker or lighter outer squares.

Source: Purves et al. *Neuroscience*, 6th edn, 2017. Oxford University Press.

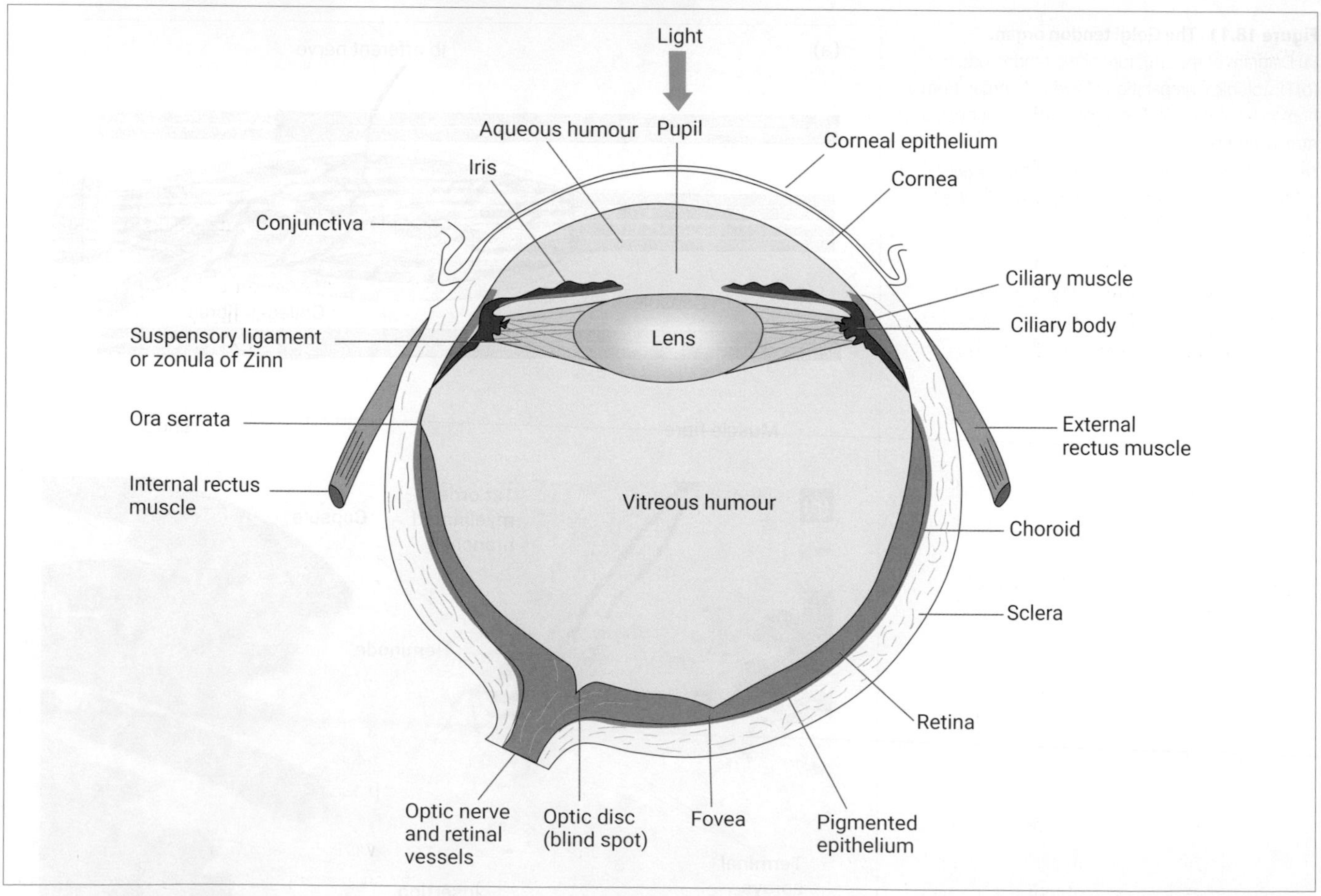

Figure 18.13 The eye.

Source: Pocock, Richards & Richards. *Human Physiology*, 5th edn (2017). Oxford University Press.

Figure 18.14 Focusing the image. Focusing is a two-stage process: the initial stage is refraction by the cornea, which acts as a fixed refractive device—the greater the curvature of the cornea, the shorter is the focal distance. The second stage is the lens, which can be adjusted in its power (accommodation): when the image is distant (a), the lens is relatively flat, with lower refractive power; when the image is close to (b), the lens is much more convex and so can focus the nearby image on the retina. The near point represents the closest point an image can be to the eye and still be focused onto the retina. With age, the near point gradually moves further from the eye as the capacity to adjust the lens decreases, a condition called presbyopia.

Source: M.F. Bear, B.W. Connors, M.A. Paradiso. *Neuroscience: Exploring the Brain*, 4th edn. Wolters Kluwer, 2015.

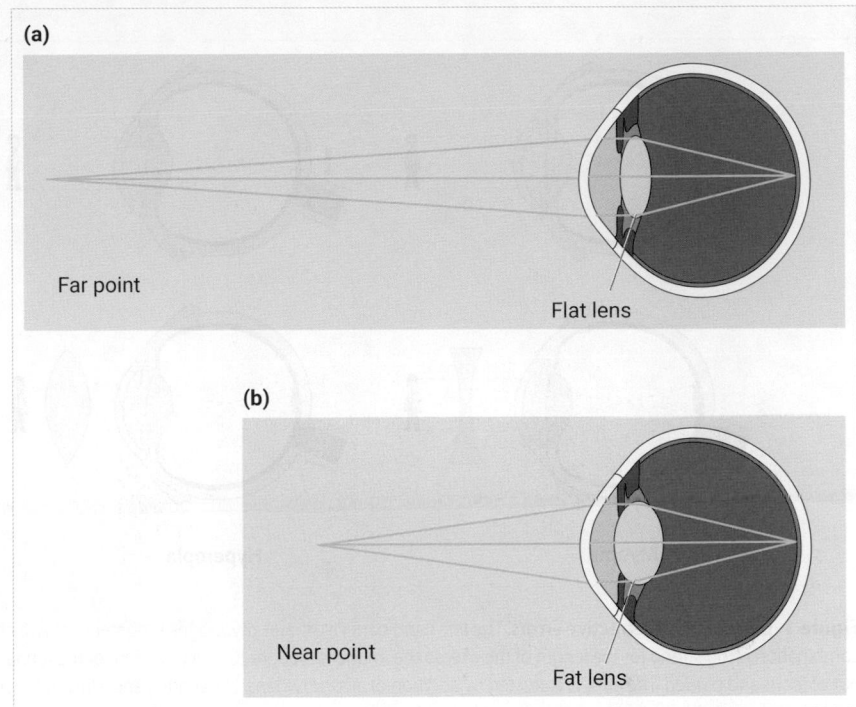

(a)

Far point

Flat lens

(b)

Near point

Fat lens

The cornea and lens

Vision begins with the eye (Figure 18.13), with the first stages being the focusing of the image on the retina by the cornea, which is fixed, and the lens, which can be adjusted depending on how near or far away the image is (Figure 18.14).

Most of the refraction is caused by the cornea, while the lens can be adjusted in convexity to allow the focusing of images that are at different distances from the eye, a process termed **accommodation**. The lens is controlled by the ciliary muscles: for images that are far away, the ciliary muscles are relaxed and the lens is flat. For images that are close to, the ciliary muscles contract, increasing the convexity of the lens so that the light rays can be refracted further to bring them to focus on the retina.

 Go to the e-book to watch a video that explains how the eye focuses an image.

Many people are affected by optical disorders, whereby the image is not focused correctly on the retina, so that vision is blurred. In the vast majority of these cases, the vision can be corrected by the addition of lenses in front of the eye to increase or decrease the amount by which the light is refracted (see Clinical Box 18.1).

CLINICAL BOX 18.1 | Lenses and correcting refractive errors

The most common visual disorders are those affecting the focusing of the visual image on the retina, so-called refractive errors. These take the form of the image:

a) being brought to a focus in front of the retina—near sightedness or myopia;

b) being brought to a focus behind the retina—far sightedness or hyperopia;

c) being focused unevenly across the retina—astigmatism;

d) not being focused when relatively close to the eye due to stiffening of the lens, particularly affecting activities such as reading as people age—presbyopia.

All these refractive errors can be rectified by the use of lenses to bring the image to a focus correctly onto the retina (Figure 1).

In the case of presbyopia, which is a common feature of ageing, the near-point moves further away from the eye. As people develop presbyopia, they start to realize they are having to hold books further away from their eyes in order to be able to read. They will also have increasing difficulty with small print. This can again be corrected by addition of convex lenses, as in the case of myopia in Figure 1, that increase the refractive power of the eye.

Corrective lenses are often worn in the form of spectacles ('glasses') or contact lenses. Many people now also opt for laser surgery where the surface of the cornea can be slightly reshaped to alter the refractive index.

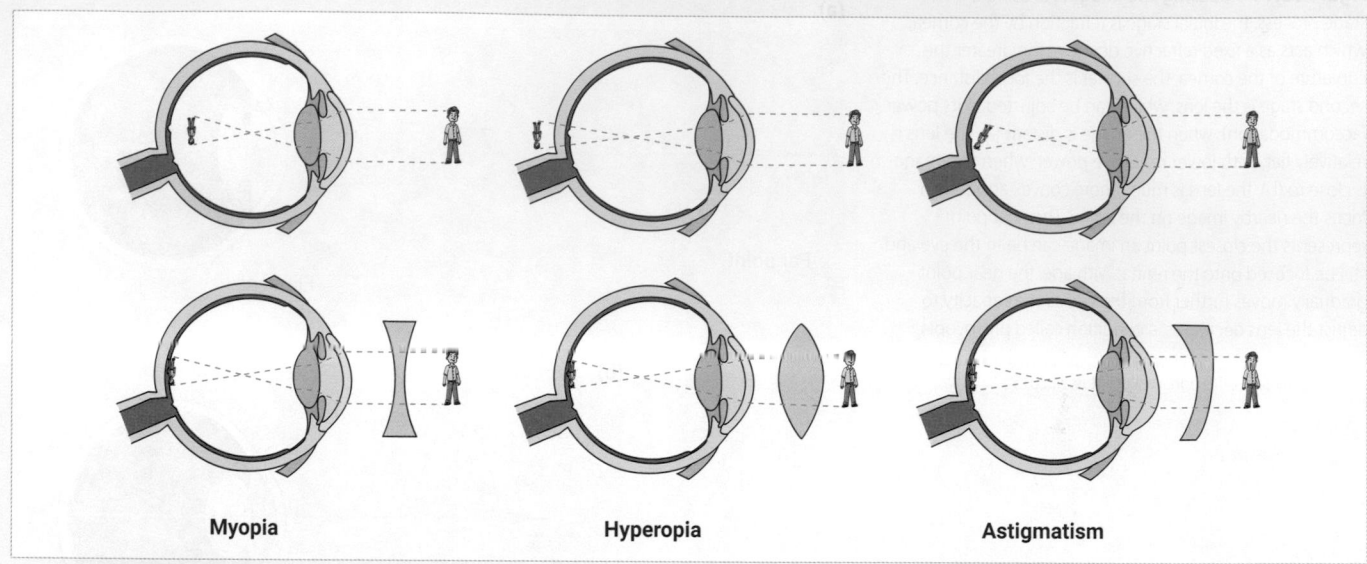

Myopia   **Hyperopia**   **Astigmatism**

Figure 1 Correction of refractive errors. The left-hand panel illustrates myopia (near sightedness) which occurs when the refractive index of the cornea and lens combination is too strong for the length of the eye, so the image is brought to a focus in front of the retina. As a result, the image that actually appears on the retina is out of focus and blurred. The error is corrected by addition of a concave lens. The centre panel illustrates hyperopia (far sightedness). In this case the refractive index is too low and so the image is brought to correct focus by addition of a convex lens. In astigmatism the refractive power of the cornea is uneven so the image is blurred and can be corrected by using a non-spherical lens, the refractive power of which varies across the surface to compensate for the variation across the cornea.
Source: Peter Hermes Furian / Shutterstock.com.

Between the cornea and the lens lies a ring of muscle, the iris, which is pigmented and gives us our characteristic eye colour (Figure 18.13). The iris controls the amount of light passing through the pupil, which appears as the black circle in the centre of the eye. The **pupillary reflex** is the constriction or dilation of the pupil by the iris in response to changes in light level, so that in dim light the pupil is dilated to allow more light to enter. However, this also reduces the depth of field of the image. The pupillary reflex is a bilateral reflex, so that the two pupils are always the same diameter. For example, if you shine a torch into one eye, both pupils will constrict, even though the other eye may be in dim light.

The retina

The image we are looking at is focused on the retina, which is a thin sheet of cells that perform the initial neural processing of the visual stimulus (Figure 18.15).

At the very back of the retina is a layer of photoreceptors comprising the **rods** and the **cones**. The cones are responsible for colour vision when light levels are relatively high, whereas the rods only function in very low light levels and give greyscale vision. Although we are typically much more aware of colour vision, there are about 90 million rods in the retina compared with 5 million cones. The cones are very densely located in a specific region of the retina called the **fovea** (Figure 18.13) which lies at the centre of a pigmented region called the **macula** (Figure 18.16). The fovea is only about 1.5 mm across and is the region of our visual field which we associate with the centre of our vision and where we have the greatest visual acuity. As can be seen from Figure 18.15, with the photoreceptors at the back of the retina, the light actually has to pass through the inner layers of the retina before reaching the

rods and cones, thereby reducing acuity because of some scattering of the light.

How is acuity preserved in the fovea? At the fovea, the inner cell layers are actually displaced to the sides so that light in this region reaches the photoreceptors directly rather than passing through the other cell layers

The outer segments of the rods and cones (Figure 18.17) comprise numerous membranous discs that contain photopigments. In the rods, the photopigment is **rhodopsin**, which is made up of opsin and retinal, which is derived from vitamin A.

When light is absorbed by the rhodopsin it sets in motion the transduction of the light stimulus into an electrical signal (Figure 18.18). The activation of opsin sets in train an enzyme cascade that leads to very large numbers of cyclic guanosine monophosphate molecules (cGMP) being broken down for each molecule of rhodopsin that is activated. In the dark, the cGMP molecules are bound to the Na^+ channels keeping them open. When the cGMP is broken down to GMP by the enzyme phosphodiesterase, the GMP detaches from the Na^+ channels and so they close, leading to hyperpolarization.

Is there anything unusual in the response of the photoreceptors to light compared with other sensory systems? When the photoreceptors are stimulated by light, this results in the closure of Na^+ channels so the cell hyperpolarizes and reduces the release of neurotransmitter, whereas in most sensory systems, the arrival of the stimulus causes depolarization and increased release of neurotransmitter. In the dark, there is a continuous influx of Na^+ into the photoreceptors, the so-called **dark current**.

The process is similar in the cones, except they have three different types of opsins, which absorb light of red, green, or blue

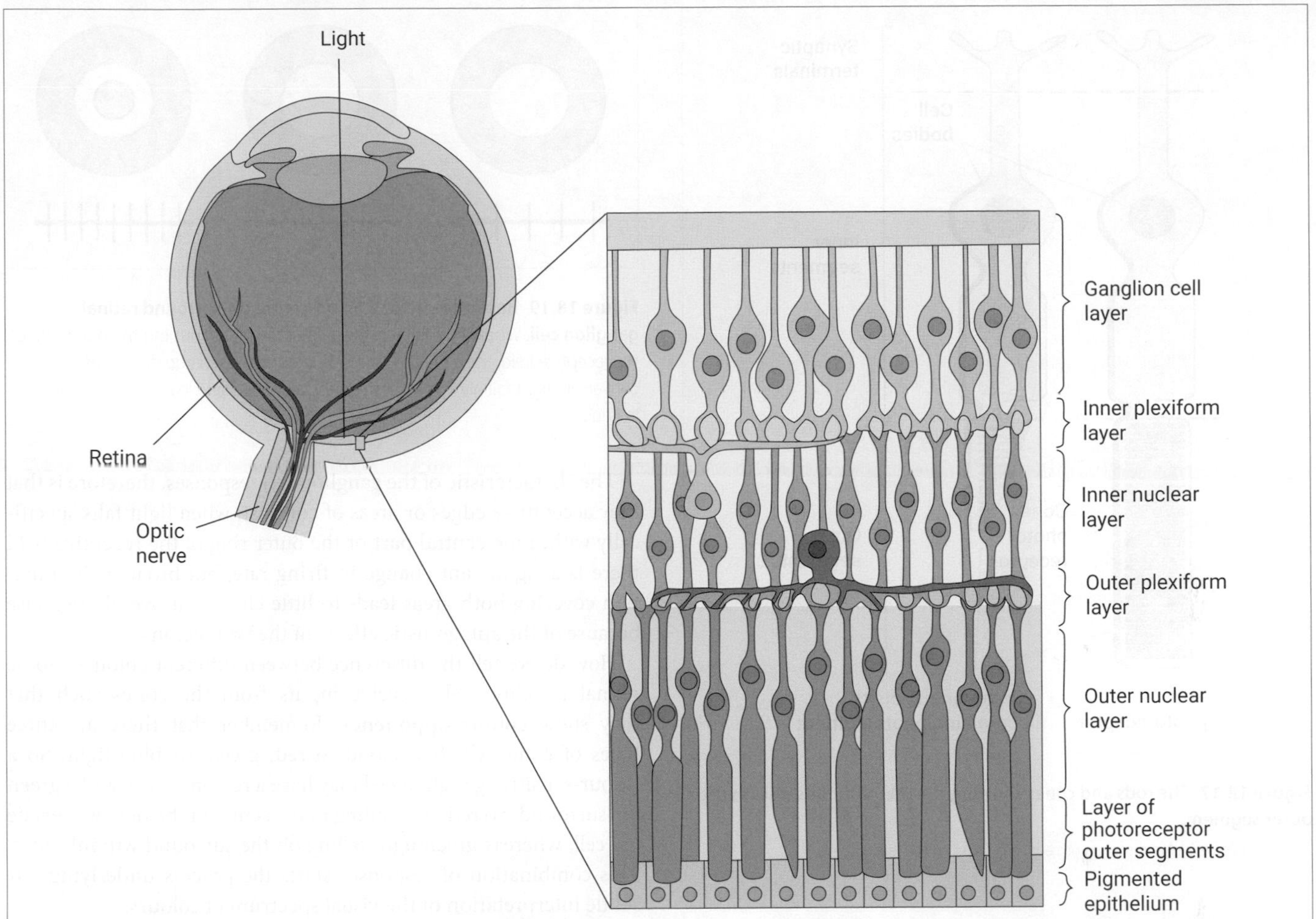

Figure 18.15 The cellular structure of the retina.

Source: Purves et al. *Neuroscience*, 6th edn, 2017. Oxford University Press.

[Figure 18.15 labels: Light; Retina; Optic nerve; Ganglion cell layer; Inner plexiform layer; Inner nuclear layer; Outer plexiform layer; Outer nuclear layer; Layer of photoreceptor outer segments; Pigmented epithelium]

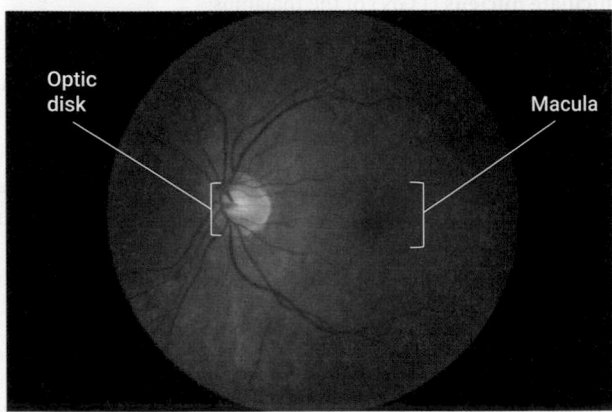

Figure 18.16 The author's retina showing the retinal blood vessels and the macula.

[Figure 18.16 labels: Optic disk; Macula]

colours (this is the case in humans—different species of mammals have different types of opsins). The combination of these three cone systems enables us to see different colours.

Why does it take a long time for our vision to adjust when going into a dark place? Full dark adaptation takes about 20 min. In normal daylight, all the rhodopsin in the rods is activated (the pigment is said to be 'bleached'). In the dark the rhodopsin is reconstituted into its inactivated form by activity in the pigmented epithelium (Figure 18.15), which takes time. As the pigment is restored, so you gradually become aware that your vision is increasing: during this time the sensitivity to light increases by about a million-fold!

The neural response to light stimuli is processed by the layers of cells within the layers of the retina (Figure 18.15). The photoreceptors connect with the bipolar and the horizontal cells and there is processing by these and by the **amacrine** and **ganglion** cells. The axons of the ganglion cells form the optic nerve, which leaves the retina at the optic disc (Figure 18.16).

The retinal ganglion cells that generate the neural output from the retina have very characteristic **centre-surround receptive fields**. These receptive fields comprise an inner circle and an outer ring (Figure 18.19) which show opposing responses. Thus, a dot of light falling on the area of the retina that maps onto the centre of the receptive field of a specific ganglion cell may increase the firing rate of that cell, whereas if the light falls on the outer ring of the receptive field, it would cause the cell to decrease its firing rate. This is therefore termed an on-centre off-surround cell. There are also off-centre on-surround cells.

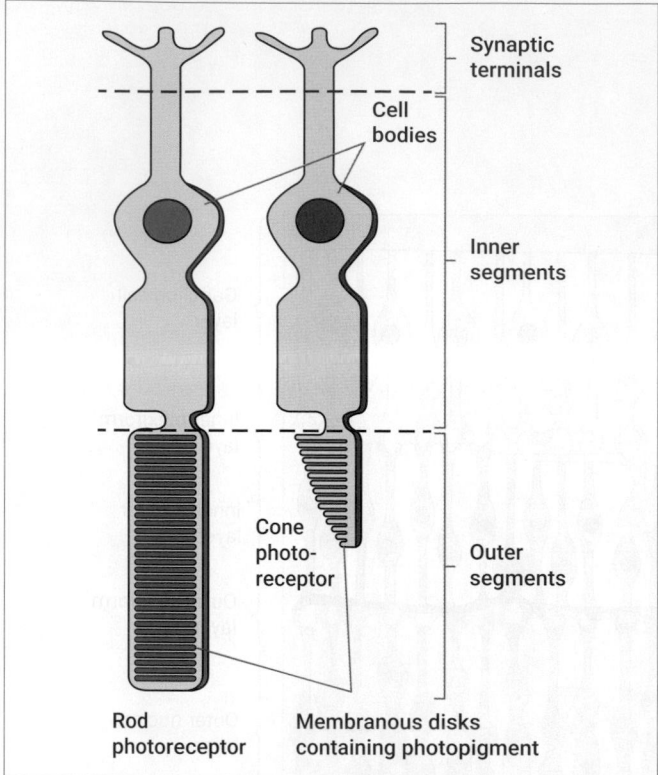

Figure 18.17 The rods and cones showing the membranous discs in the outer segment.

Source: Purves et al. *Neuroscience*, 6th edn, 2017. Oxford University Press.

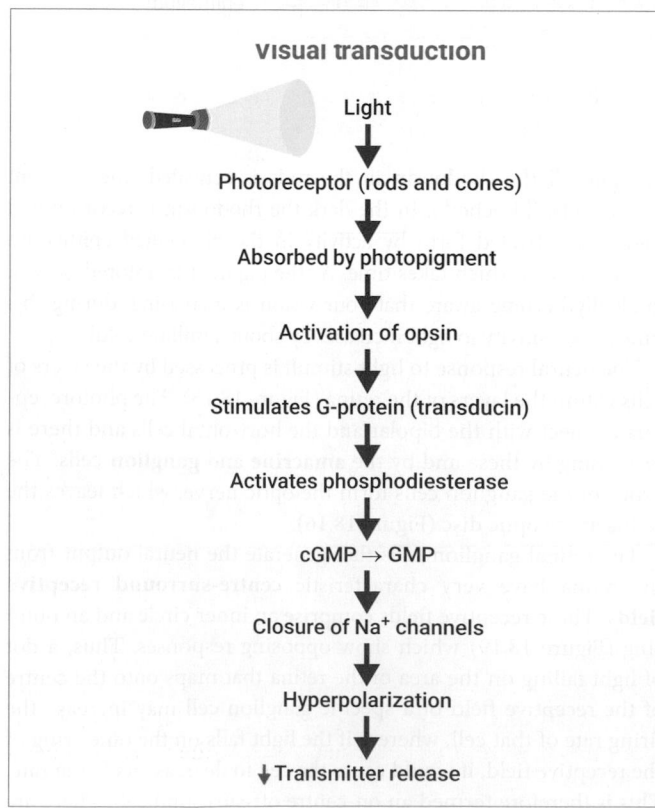

Figure 18.18 Flow diagram illustrating the transduction process.

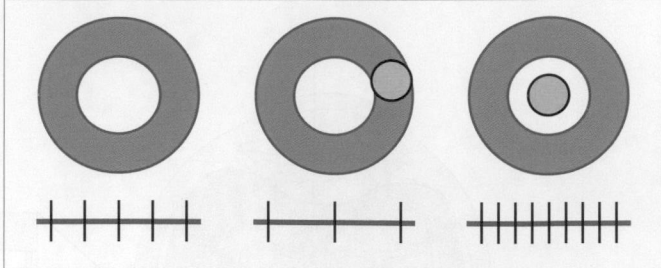

Figure 18.19 Receptive field of an on-centre off-surround retinal ganglion cell. When light falls on the area of the retina forming the outer ring of the receptive field, the cell's rate of action potential discharge decreases; conversely, light falling in the centre of the receptive field leads to an increase in firing.

The characteristic of the ganglion cell responses, therefore is that they accentuate edges or areas of contrast: when light falls specifically within the central part or the outer ring of the receptive field there is a significant change in firing rate, but broader illumination covering both areas leads to little change in overall response because of the antagonistic effects of the two regions.

How do we tell the difference between different colours? Some retinal ganglion cells receive inputs from the cones such that they show **colour opponency**. Remember that there are three types of cone, which respond to red, green, or blue light. So a colour-sensitive ganglion cell may have a red on-centre and a green off-surround. So red light falling in the centre of the field will excite the cell, whereas green light falling on the surround will inhibit it. This combination of responses starts the process underlying our subtle interpretation of the visual spectrum of colours.

It is evident that there are many different types of ganglion cell, based on their response properties and the connectivity within the retina and beyond. One way in which they may be categorized is based on their connections beyond the retina, to the **lateral geniculate nucleus** (LGN) of the thalamus:

- P-type (*parvo*, meaning small) cells make up about 80 per cent of the ganglion cell population. These cells have relatively small receptive fields and project to the parvocellular layers of the LGN.

- M-type (*magno*, meaning large) cells comprise about 10 per cent of the population. These cells have larger receptive fields and project to the magnocellular layers of the LGN.

- Non-M–non-P-type cells make up the remainder of the population.

The visual pathway

Each retina receives light from the visual scene and the axons of the retinal ganglion cells form the two optic nerves that conduct information towards the brain. The visual fields of each eye are divided into two regions: nasal (on the side of the nose) and temporal (on the side of the temple) (Figure 18.20).

The optic nerves from the two eyes come together at the **optic chiasm**. Here the axons from the cells of the nasal fields cross over.

What is the effect of the crossing-over of some of the axons at this point? Before moving on, think about how the visual images may be brought together for binocular vision. This means that the

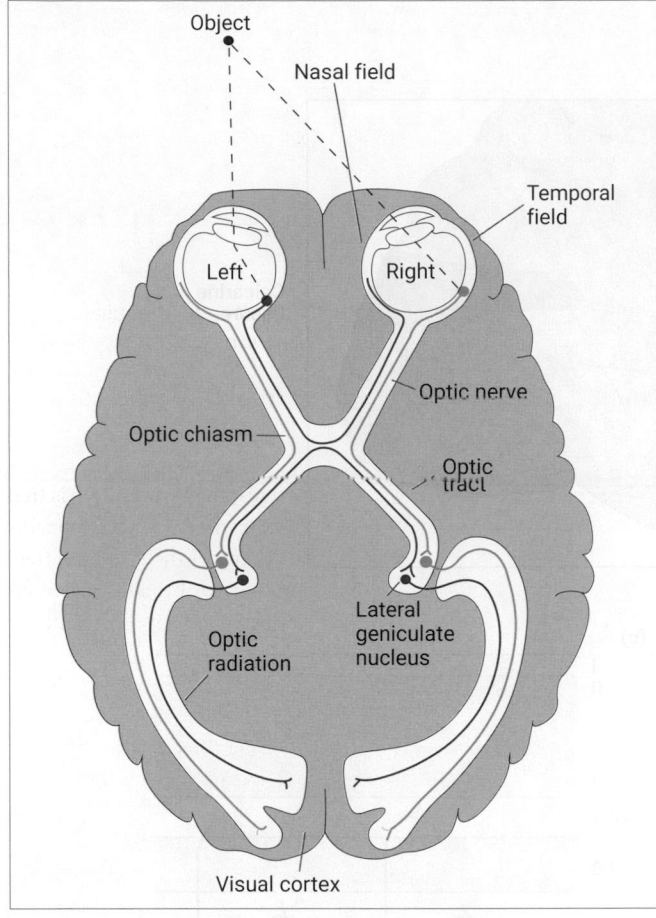

Figure 18.20 The visual pathway from the retina to the visual cortex.

Source: Purves et al. *Neuroscience*, 6th edn, 2017. Oxford University Press.

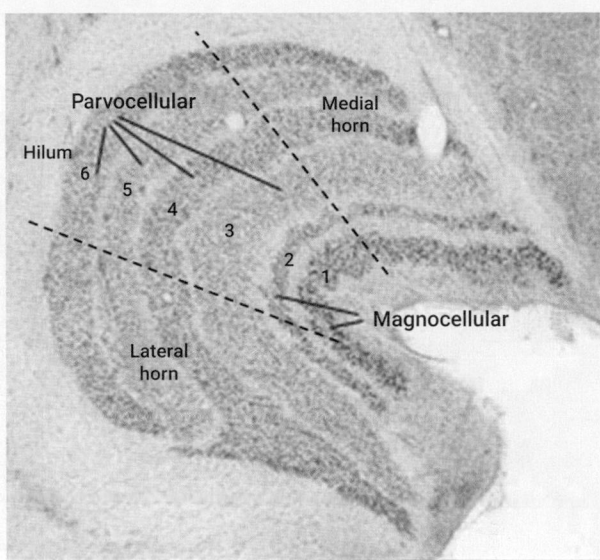

Figure 18.21 Organization of the lateral geniculate nucleus.

Source: Reprinted from C. Kennard, R. John Leigh (2011). *Handbook of Clinical Neurology*, Volume 102, 3–19, with permission from Elsevier.

corresponding images are now brought together: the object that is seen to the left in Figure 18.20 projects onto the nasal field of the left eye and the temporal field of the right eye. At the optic chiasm, the axons from the nasal field of the left eye cross over to join the axons from the temporal field of the right eye, so the two projections of the left visual field are brought together in the right optic tract that enters the right LGN; the two projections of the right visual field go to the left LGN.

The lateral geniculate nucleus

The LGN is one of the nuclei of the thalamus in which sensory information is processed before being passed to the cerebral cortex. The LGN on each side is divided into six layers of cells (Figure 18.21), with layers 1–4 being the parvocellular pathway and 5–6 the magnocellular pathway. The axons from each eye remain separate at this stage, so for the left thalamus, layers 2, 3, and 5 receive inputs from the temporal retina of the left eye and layers 1, 4, and 6 receive inputs from the nasal retina of the right eye (Figure 18.21). For the right thalamus, therefore, layers 2, 3, and 5 receive inputs from the temporal retina of the right eye and layers 1, 4, and 6 receive inputs from the nasal retina of the left eye.

In between each strip of parvocellular or magnocellular layers, is a thin koniocellular strip which receives inputs from the non-M–non-P cells.

The receptive fields of the neurons of the LGN are similar to those of the retinal ganglion cells and the cells retain their relative location, such that each layer of the LGN is laid out as a retinotopic map. But the LGN also receives inputs from other areas of the thalamus and a very significant input from the visual cortex itself. Therefore, the LGN plays a role in processing the visual signal, rather than being just a relay station as would first appear.

Visual cortex

The primary visual cortex (V1) is located along the calcarine fissure of the occipital lobe of the brain, as you can see in Figure 18.22a. There are six main layers to the visual cortex, with the inputs from the LGN entering into layer IV (Figure 18.22c). At this stage the inputs from the M- and P-pathways project onto separate levels of layer IV. The inputs representing the receptive fields of the two eyes project onto the same layers of layer IV, but they connect with different cells, which make up the **ocular dominance columns**. If the columns are seen as being oriented vertically to the surface of the cortex, cells in one column will be driven preferentially by one eye, whereas the cells in the adjacent column will be driven preferentially by the input from the other eye. From layer IV, there are projections to the other layers of the visual cortex, where the neurons receive input from both eyes, thereby enabling binocular processing of the neural input.

As with the other sensory areas of the cerebral cortex, the visual cortex retains a topographic organization, that is, it is laid out as a map of the retina, although the map is distorted in the same way as the somatotopic map of S1 is distorted (Figure 18.7) with different regions having greater representation. In the case of the **retinotopic** map, the representation of the fovea occupies a much larger area of the map than do the more peripheral regions of the retina.

As we saw, the receptive fields of the retinal ganglion cells and the cells of the LGN are circular, with a centre-surround opponent

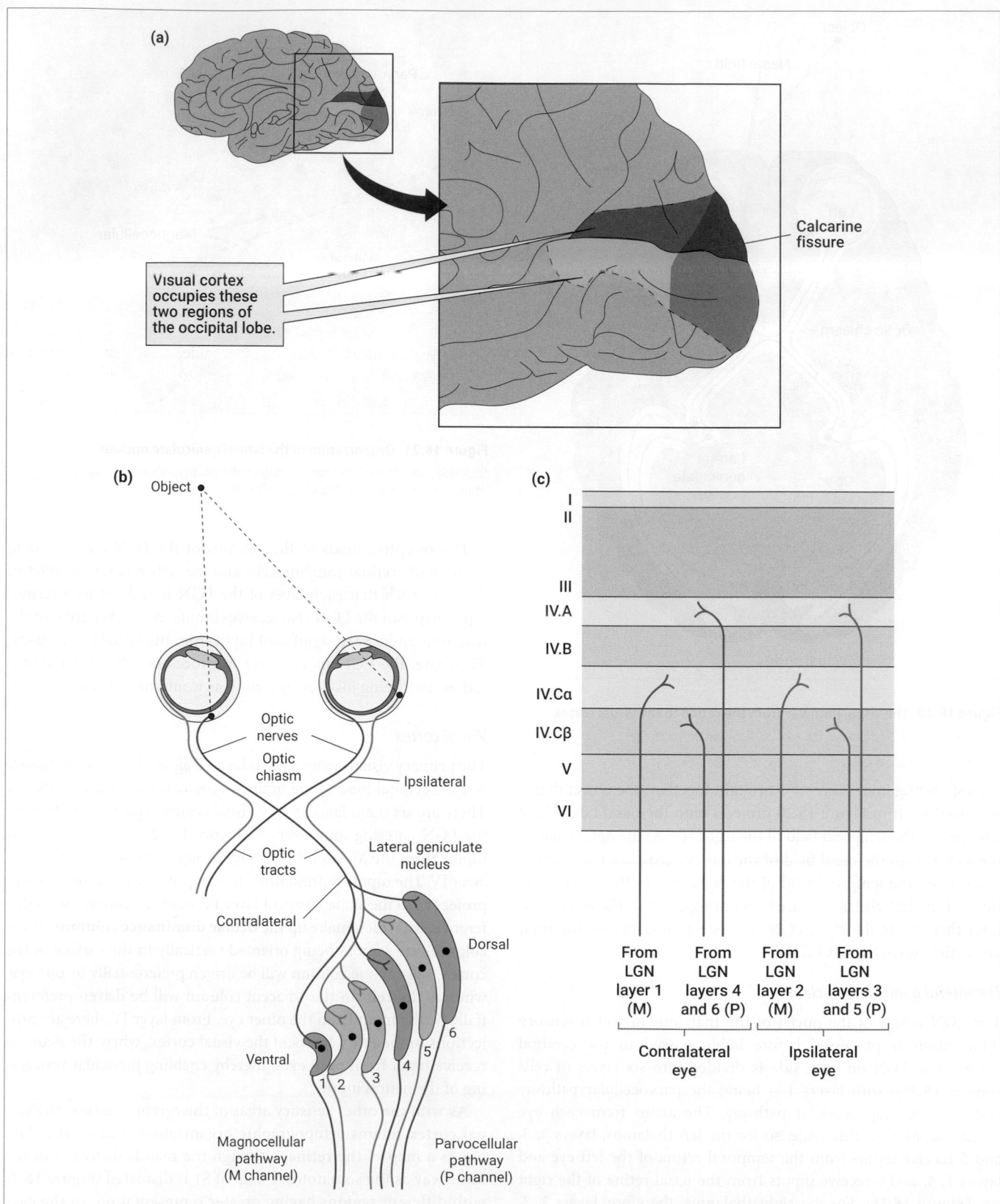

Figure 18.22 The visual cortex. (a) Location of the visual cortex in the occipital lobe. (b) The LGN and its projections to the primary visual cortex. (c) The layers of the primary visual cortex showing the inputs from the LGN. The ipsilateral eye relates to the eye for which the neural projection remains on the same side (see the pathway in (b)), whereas the contralateral eye is the one from which the fibres have crossed over.

Source: (a) Purves et al. *Neuroscience*, 6th edn, 2017. Oxford University Press; (b, c) © The Open University.

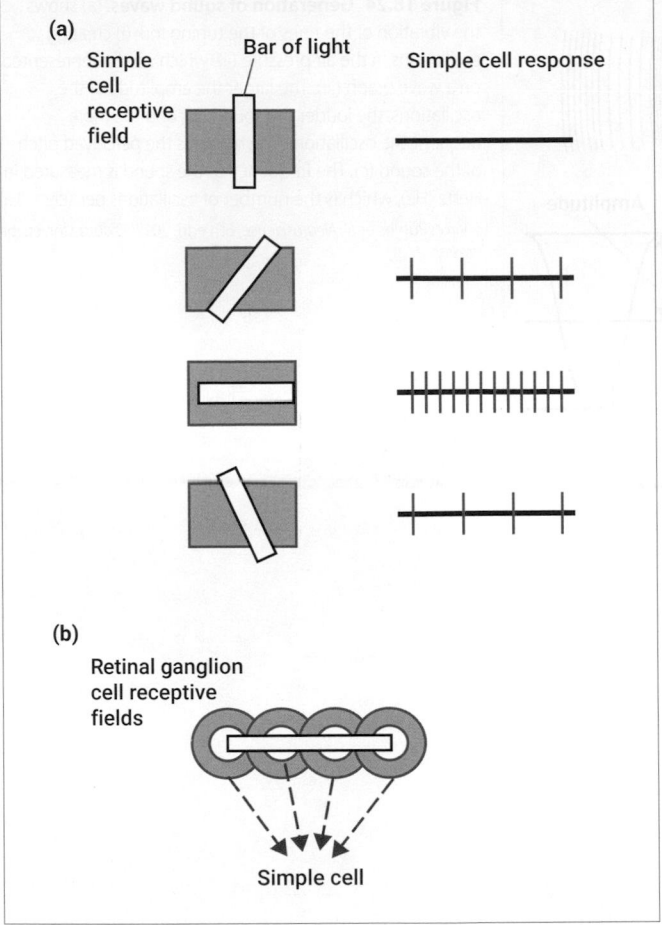

(a)

Simple cell receptive field   Bar of light   Simple cell response

(b)

Retinal ganglion cell receptive fields

Simple cell

Figure 18.23 (a) Receptive field of a visual cortical simple cell responding to a bar of light shone across its receptive field at different orientations; (b) building up the simple cell response from convergence of inputs from a row of retinal ganglion cells.

Hearing

The sense of hearing, like vision, enables us to have a very rich appreciation of our environment and it also underpins our major means of social communication, speech, whereby we interpret different patterns of sound waves as having specific meanings.

Sound is a mechanical stimulus created by rapid changes in air pressure. If you strike a tuning fork, or use a tuning app on your phone, you can create a pure sound wave. In each case, this is created by oscillations in the air pressure: as the prongs of the tuning fork vibrate, or the cone of the speaker in your phone vibrates they cause the air molecules to move to and fro, creating regions of higher and lower pressure, called compressions and rarefactions, respectively (Figure 18.24). The greater the amplitude of movement of the tines of the tuning fork, the greater will be the difference between the peaks and troughs in the air pressure (the amplitude of the wave), and so the louder the sound will be perceived to be.

Why are tuning forks of different lengths? Using tuning forks of different lengths will change the pitch of the sound produced because the tines of the tuning fork will vibrate at different frequencies: short tuning forks will vibrate more rapidly (higher frequency) than those with long tines. The higher the frequency of the vibration, the higher is the perceived pitch of the sound.

The sound waves are detected by the ear (Figure 18.25). The pinna acts to direct the sound wave towards the auditory canal where it strikes the **tympanum** (ear drum), which forms the boundary between the outer and the middle ear. The tympanum is a tightly stretched membrane: when the sound wave hits the tympanum it is made to vibrate at the same frequency as the oscillations of the sound wave. The movements of the tympanum, in turn, cause the three **ossicles** ('little bones', the malleus, incus, and stapes (hammer, anvil, and stirrup)) of the middle ear to move to and fro, again at the same frequency (Figure 18.26), and to transmit the mechanical stimulus to the inner ear, where we find the **organ of Corti** (Figure 18.27), which is the sense organ itself, inside the cochlea.

The outer and middle parts of the ear are both air filled, but the inner ear is fluid filled. The fluid has much greater inertia than does air: if the sound waves hit the surface of the fluid directly, almost all the energy of the wave would be reflected back from the surface of the fluid and very little would be transmitted into the inner ear itself (if you have ever put your head under the surface of the water in a noisy swimming pool, you will know how little of the sound is still audible because most of the sound is reflected back from the surface of the water).

arrangement (Figure 18.19). In the visual cortex, the receptive fields of the neurons become more complicated, responding preferentially to specific shapes. The so-called simple cells respond preferentially to a bar of light of specific orientation; Figure 18.23 shows how such a response could be built up from the convergence of inputs from a group of retinal ganglion cells. As with the ocular dominance columns, so the simple cells are organized in columns such that as one tracks a line tangential to the surface of the visual cortex, so the preferred orientations change in sequence.

Other visual cortical neurons have more complex receptive fields that may be generated by the convergence of inputs from different simple cells. Further complex processing occurs within the other areas of the visual cortex: neurons from the primary visual cortex (V1) project to further visual cortical regions (V2–5). In these different regions, there is evidence of parallel processing of different features of the image, for example motion in the visual field is processed through one route, whereas visual perception and object recognition are processed by different cortical regions, further down the line; neurons in regions of the temporal lobe have been shown to be specifically responsive to image recognition, such as recognizing someone's face.

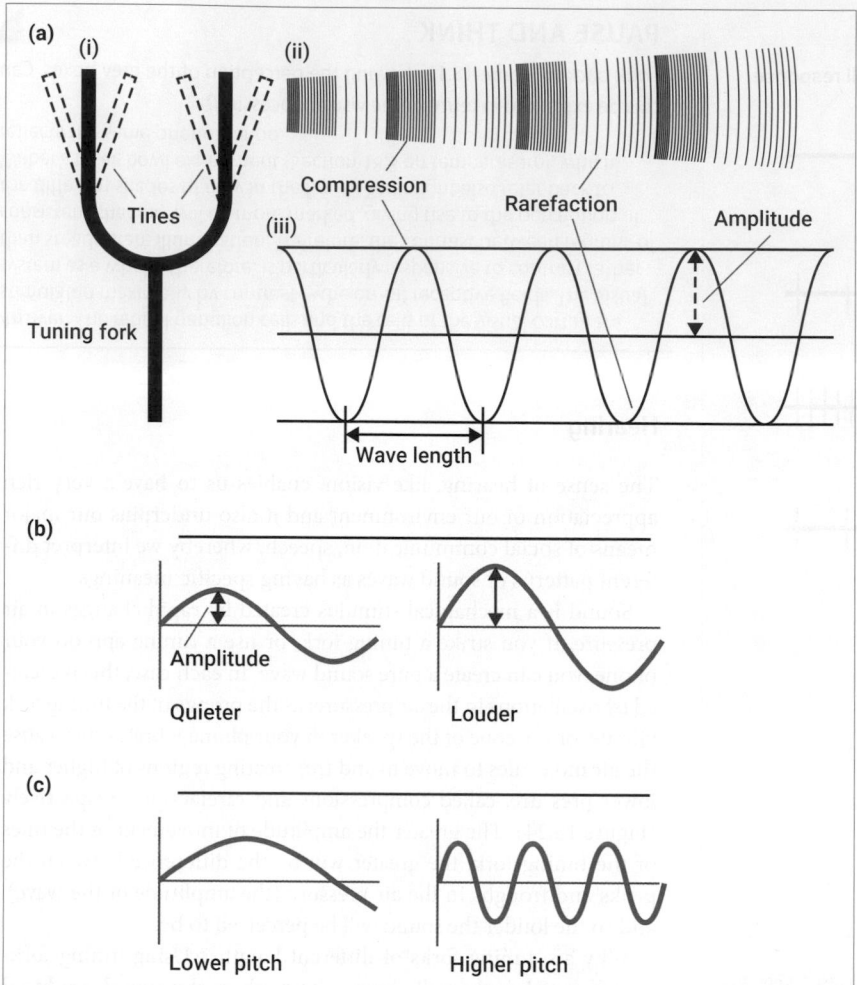

Figure 18.24 Generation of sound waves. (a) shows the vibration of the tines of the tuning fork (i) creating oscillations in the air pressure (ii), which can be represented on a wave graph (iii). The larger the amplitude of the oscillations, the louder the sound (b); and the more frequent the oscillations, the higher is the perceived pitch of the sound (c). The frequency of the sound is measured in Hertz (Hz), which is the number of oscillations per second.

Source: Purves et al. *Neuroscience*, 6th edn, 2017. Oxford University Press.

How do the sound waves entering the inner ear overcome this greater inertia? This is where the middle ear plays an important role in two ways: by area ratio and by lever action (for information about levers see Figure 20.18).

1. The membrane covering the opening to the inner ear, the **oval window** (Figure 18.26), has an area about 17 times smaller than that of the tympanum. The pressure exerted on a surface is the force per unit area; therefore, if all the force that was exerted on the tympanum by the sound waves was transmitted to the oval window, the pressure would be 17 times as much (someone accidentally standing on your foot while wearing stiletto-heeled shoes will exert much more pressure, and therefore cause you much more pain, than the same person would if they were wearing flat-heeled shoes!).

2. The ossicles (Figure 18.26) act as a lever system to magnify the force transmission by about 1.2-fold.

Taking these two systems together results in the pressure applied to the oval window being increased by about 20 times, by comparison with that applied at the tympanum. As a result, the sound waves can be transmitted effectively from the air to the fluid of the inner ear.

The middle ear is air-filled and has a connection to the nasal passages, and therefore the outside air, via the **Eustachian tube** (Figures 18.25 and 18.26), which is normally closed by a valve. When you experience changes in air pressure, for example when taking off in a plane, or driving up a steep hillside, the air pressure difference between the outside air and the middle ear can press on the eardrum and be quite painful. Normally, this can be easily rectified by swallowing or yawning, both of which open the valve and allow the pressures to equalize, reducing the discomfort: you will often also be aware of a clicking in the ear as this happens.

Also in the middle ear are two tiny muscles: the tensor tympani and the stapedius muscle (Figure 18.26). As their name suggests, these muscles attach to the ear drum and the stapedius bone, respectively. When you are subjected to very loud noises, these muscles contract and reduce the amplitude of movement in the middle ear and so help protect the delicate inner ear from the damaging effects of very loud sounds.

Inner ear

The inner ear is where the sound energy transmitted from the middle ear is transduced into neural signals. The middle ear converts the soundwaves into the movements of the ossicles which are then transmitted to the cochlea of the inner ear via the oscillations of

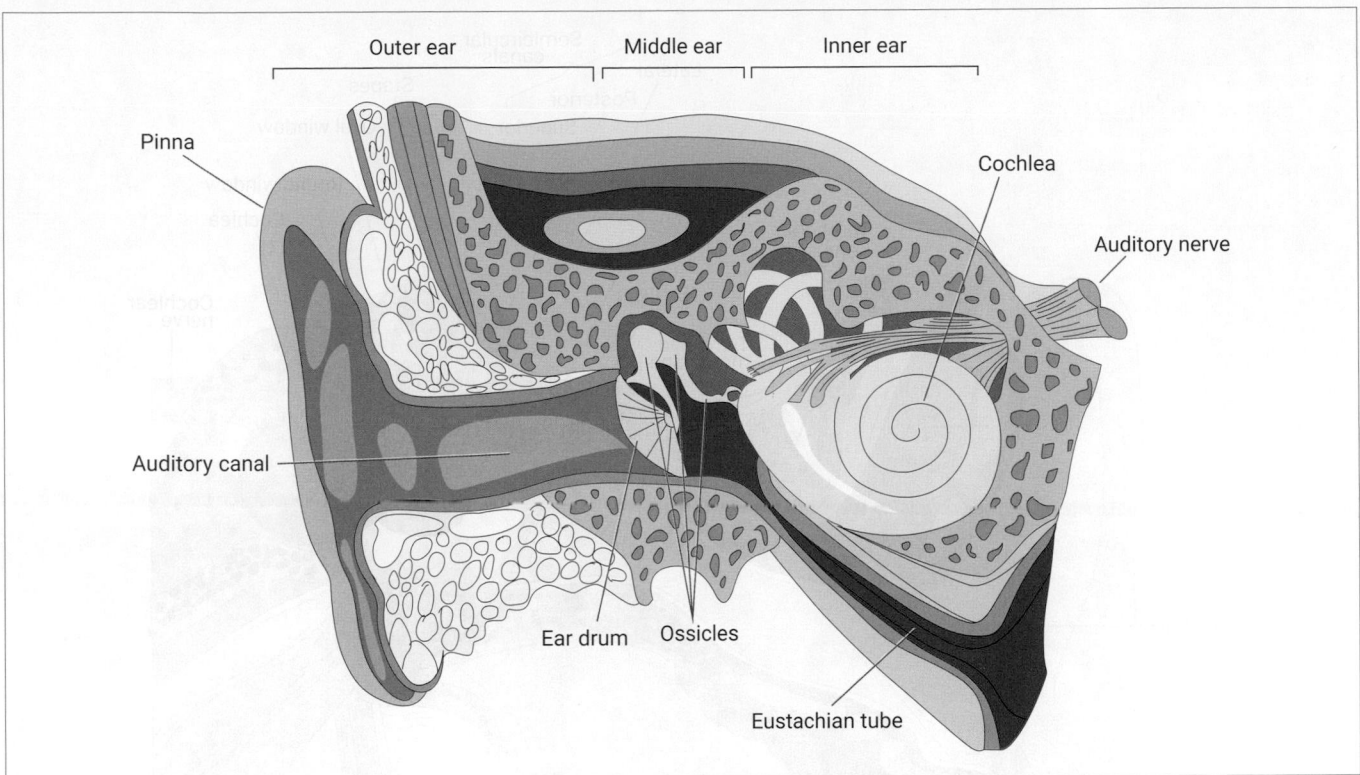

Figure 18.25 Anatomy of the human ear.

Source: Purves et al. *Neuroscience*, 6th edn, 2017. Oxford University Press.

Figure 18.26 The middle ear.

Source: Purves et al. *Neuroscience*, 6th edn, 2017. Oxford University Press.

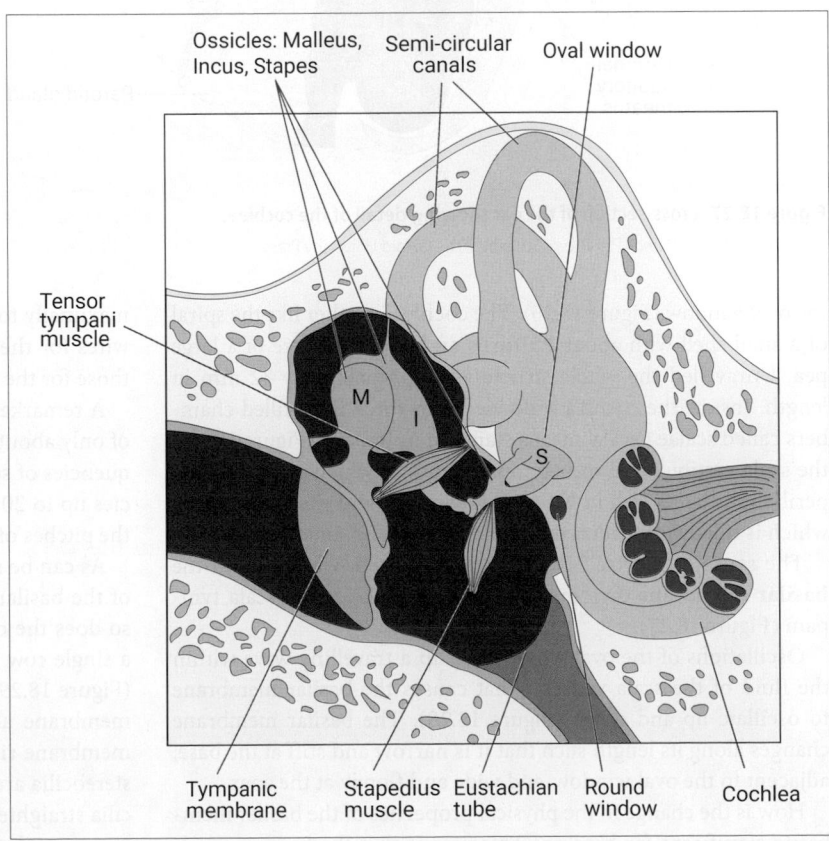

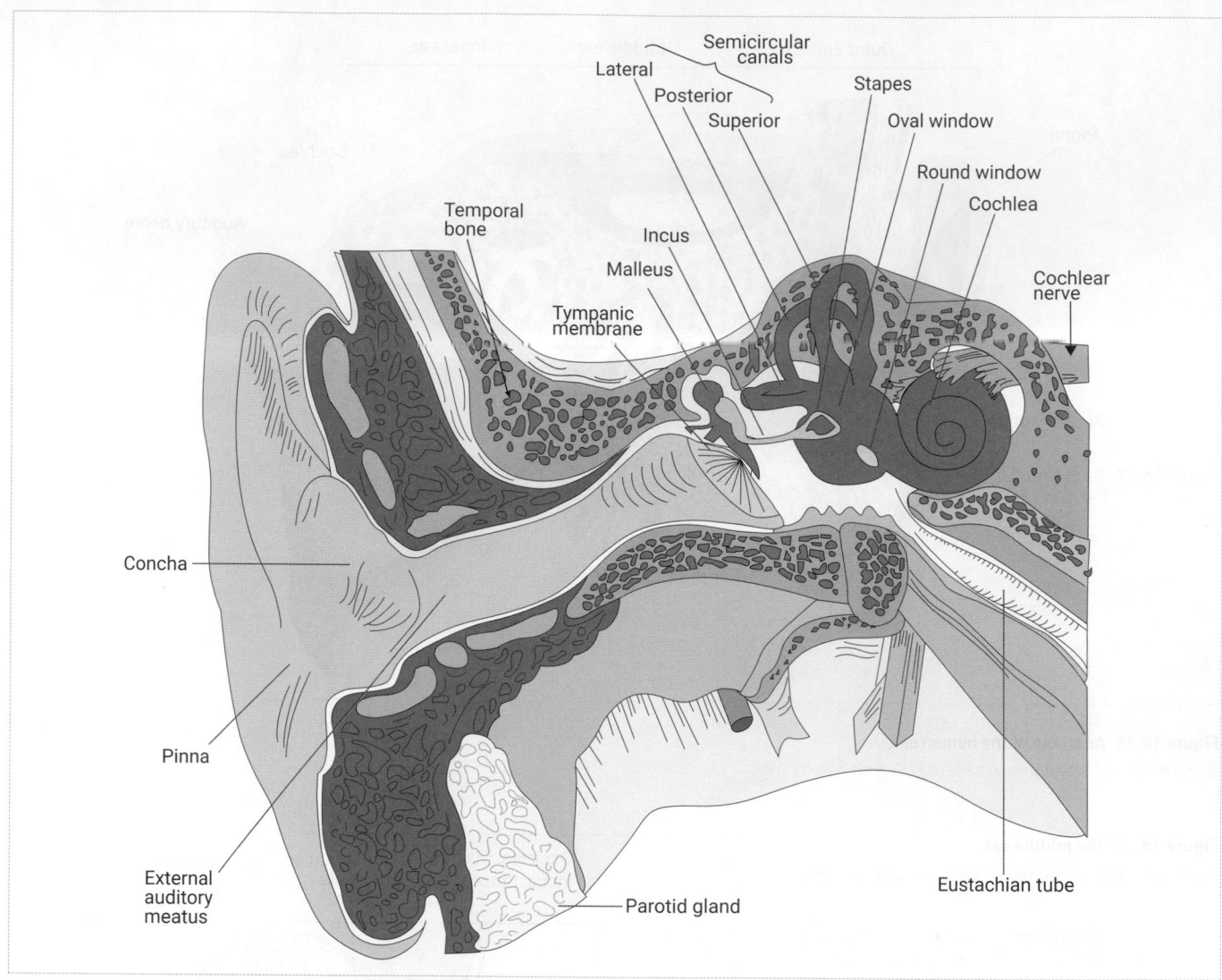

Figure 18.27 Cross-section of the ear showing detail of the cochlea.

Source: Pocock et al. *Human Physiology*, 5th edn, 2017. Oxford University Press.

the oval window (Figure 18.26). The cochlea is rather like the spiral of a snail shell with about 2.5 turns and about the size of a large pea. Unravelled the whole structure is approximately 32 mm in length. Inside, the cochlea is divided into three fluid-filled chambers called scalae ('scala' means staircase in Italian) (Figure 18.27): the scala vestibuli and scala tympani, both of which are filled with perilymph that is rich in Na^+ and low in K^+, and the scala media, which is filled with endolymph that is rich in K^+ and low in Na^+.

The sense organ itself is the **organ of Corti**, which sits in the **basilar membrane** dividing the scala media from the scala tympani (Figure 18.27).

Oscillations of the oval window set up a travelling wave within the fluid of the scala vestibuli that causes the basilar membrane to oscillate up and down (Figure 18.28). The basilar membrane changes along its length such that it is narrow and stiff at the base, adjacent to the oval window, and wide and floppy at the apex.

How is the change in the physical properties of the basilar membrane significant for hearing? This means that the base responds maximally to high frequency sounds, whereas the apex responds

maximally to low frequency sounds: if you look inside a piano, the wires for the very highest notes are short and very taut, whereas those for the lowest notes are long and much thicker.

A remarkable feature of the basilar membrane is that a structure of only about 30 mm in length can encompass all the different frequencies of sounds we can hear, from 20 Hz at the lowest frequencies up to 20 kHz at the top end, and we can distinguish between the pitches of sounds that are only a few Hertz apart.

As can be seen from Figure 18.27, the organ of Corti sits on top of the basilar membrane: as the membrane moves up and down, so does the organ. The organ comprises two groups of hair cells: a single row of inner hair cells and three rows of outer hair cells (Figure 18.29). The stereocilia ('hairs') of the hair cells contact the membrane above them, the tectorial membrane. As the basilar membrane rises, the tectorial membrane moves outwards and the stereocilia are bent. As the basilar membrane descends, the stereocilia straighten up. Therefore, the hair cells will be bending in synchrony with the frequency of the sound waves: up to 20,000 times a second.

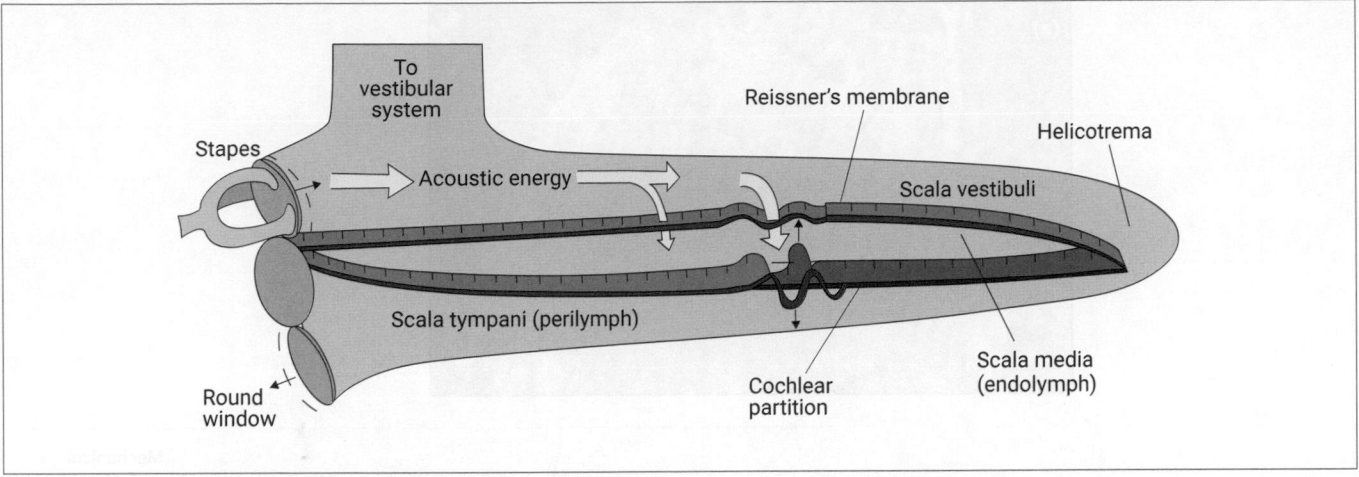

Figure 18.28 Travelling waves within the cochlea driven by oscillations of the oval window. A travelling wave, as its name suggests, is one that moves along, a good example being the wave on the seashore (as opposed to a stationary wave, which does not move, an example being the wave seen on a violin string).

Source: Purves et al. *Neuroscience*, 6th edn, 2017. Oxford University Press.

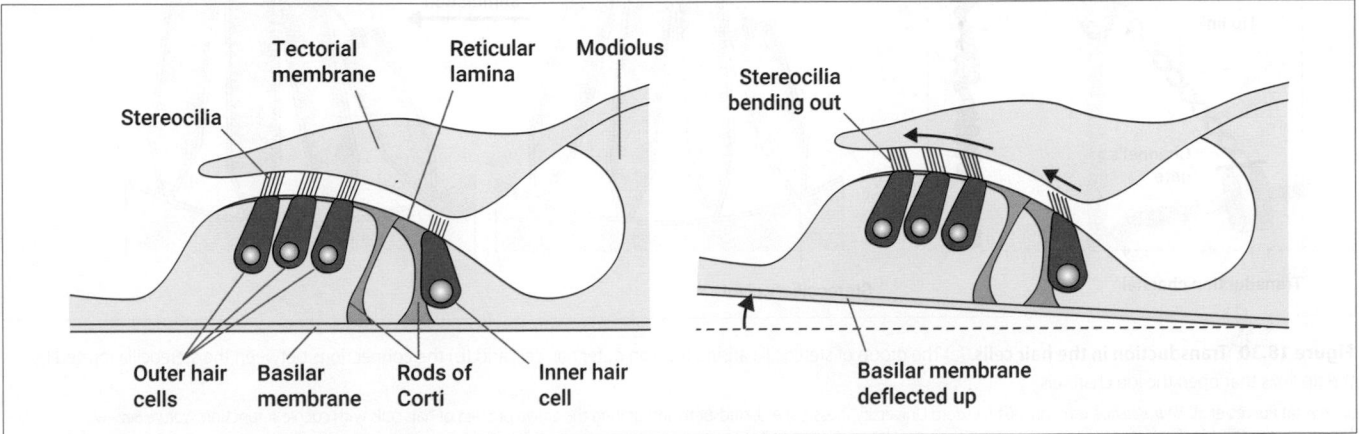

Figure 18.29 Movement of the hair cells in response to the flexion of the basilar membrane.

Source: Purves et al. *Neuroscience*, 6th edn, 2017. Oxford University Press.

There is a puzzle, though: how can we hear quiet sounds that might only move the ear drum by 10 pm (10^{-12} m) and discriminate between frequencies that are only 0.2 per cent apart? Observations of the movement of the basilar membrane *in vitro* showed that the spread of movement was much too broad and the amplitude too small to enable us to hear the range and variety of frequencies we can hear. The key to the story was the discovery of two force-generating systems in the outer hair cells: a very special 'motor' protein called **prestin** and an active process triggered by Ca^{2+} influx. These two systems together act to focus and amplify the movements of the basilar membrane by about 50 times.

The outer hair cells act to amplify the movements of the basilar membrane, while the inner hair cells provide the main neural transduction in hearing.

When the basilar membrane rises, the stereocilia are bent in the direction of the longest stereocilia, which stretches the tip links that connect the stereocilia (Figure 18.30). As the tip links are stretched so they pull on the ion channels in the tip of the adjacent

stereocilia causing them to open. The endolymph that surrounds the stereocilia is very rich in K^+ to the extent that there is a gradient for K^+ to enter the hair cell and cause it to depolarize. The actual movements involved are tiny: the maximal movement of the tips is about 20 nm (1 nm = 10^{-9} m).

How does this compare with the process of depolarization of typical neurons? Before moving on, think about the process of depolarization in a nerve axon. This is a different ionic mechanism to the depolarization of neurons, where the inward current is normally carried by Na^+.

So, as the stereocilia are bent in one direction the hair cells depolarize; as they return in the reverse direction the channels close again and the flow of K^+ is halted. This means that the receptor potential in the hair cells depolarizes and repolarizes at the same frequency as the incoming sound wave.

Hearing loss is a very common disorder that often results from a decreased efficiency of sound transmission or long-term damage to the delicate structures of the inner ear (see Clinical Box 18.2).

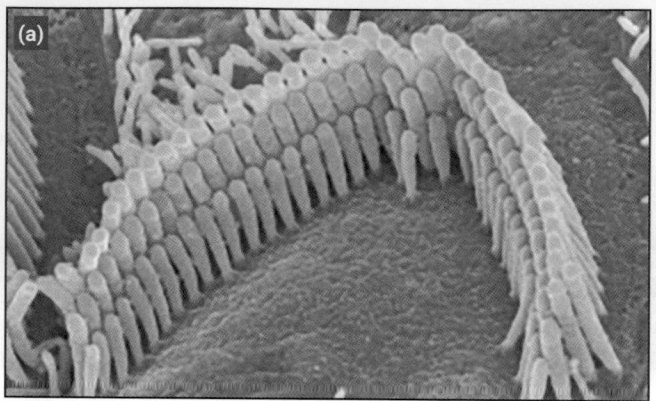

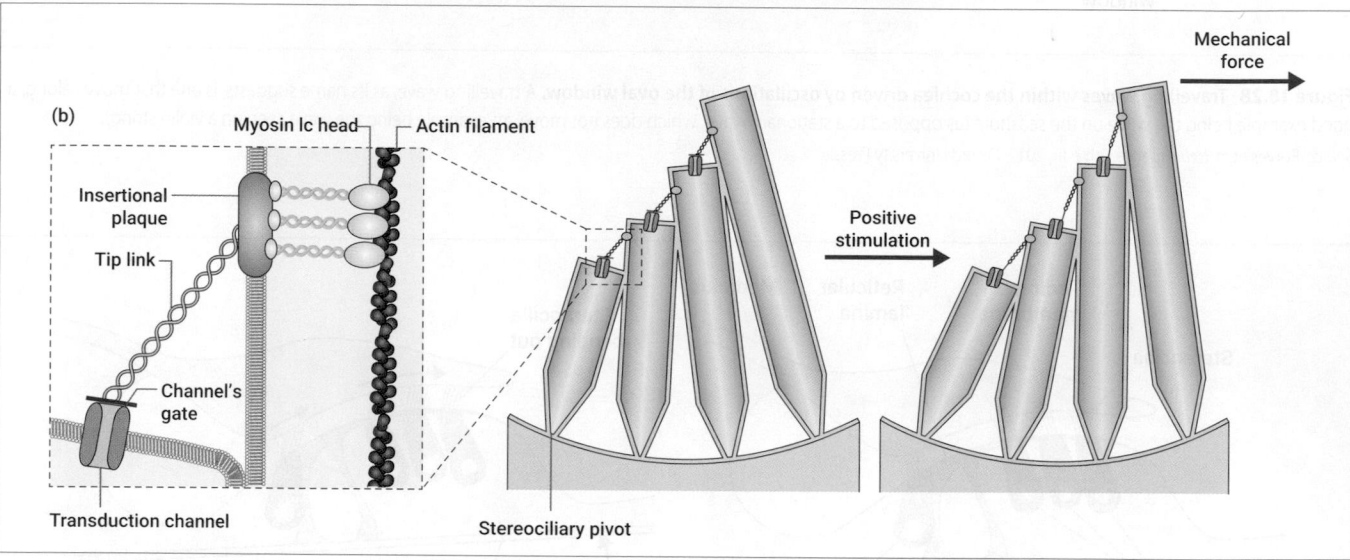

Figure 18.30 Transduction in the hair cells. (a) The group of stereocilia arising from an outer hair cell; and (b) the connections between the stereocilia formed by the tip links that open the ion channels.

Source: (a) Purves et al. *Neuroscience*, 6th edn, 2017. Oxford University Press; (b) A. J. Hudspeth. Integrating the active process of hair cells with cochlear function. *Nature Reviews Neuroscience*, 600–614. Reprinted by permission from Springer Nature: Nature © 2014.

Central pathways

The hair cells synapse with neurons of the auditory nerve (the vestibulo-cochlear nerve), which convey the auditory signals to the brainstem (Figure 18.31). The first set of synapses in the brainstem, in the cochlear nuclei, are with neurons that project to nuclei called the **superior olivary nuclei**. At this stage the signals from the two ears come together and are transmitted to the **medial geniculate nucleus** of the thalamus (recall that the visual pathway is routed to the lateral geniculate nucleus) and then to the auditory cortex of the temporal lobe (Figure 18.31).

A key question is how are the different frequencies of sound encoded in the neural pathway?

There are two main ways in which this can happen:

Tonotopic representation: we have already seen how the S1 cortex is laid out as a map of the body such we know which part of our body is being touched because that specific body region feeds into a specific part of that map (Figure 18.7); and in the same way, we know where objects are in our visual field because of the location the ganglion cells project to in the LGN and consequently the visual cortex. In the case of hearing, the basilar membrane is represented on the auditory cortex, so the pitch of a sound is represented in terms of the location of the projection on the tonotopic map.

Phase locking: the receptor potentials of the hair cells respond in phase with the frequency of the sound wave. Therefore, at low frequencies, up to about 200 Hz, the neurons in the auditory pathway are able to fire action potentials at the same frequency, in phase with the sound. At higher frequencies, the refractory periods (Section 17.4) of the neurons prevent the neurons from following the cycles of the sound wave on a 1:1 basis but they can remain in phase by responding, for example, to every fourth cycle of the sound wave. Given that there will be more than one auditory neuron responding to each frequency, the group of neurons will represent the frequency as a whole, a process referred to as **volley coding**. For frequencies above 5 kHz, the tonotopic representation is the only system for encoding the pitch of the sound.

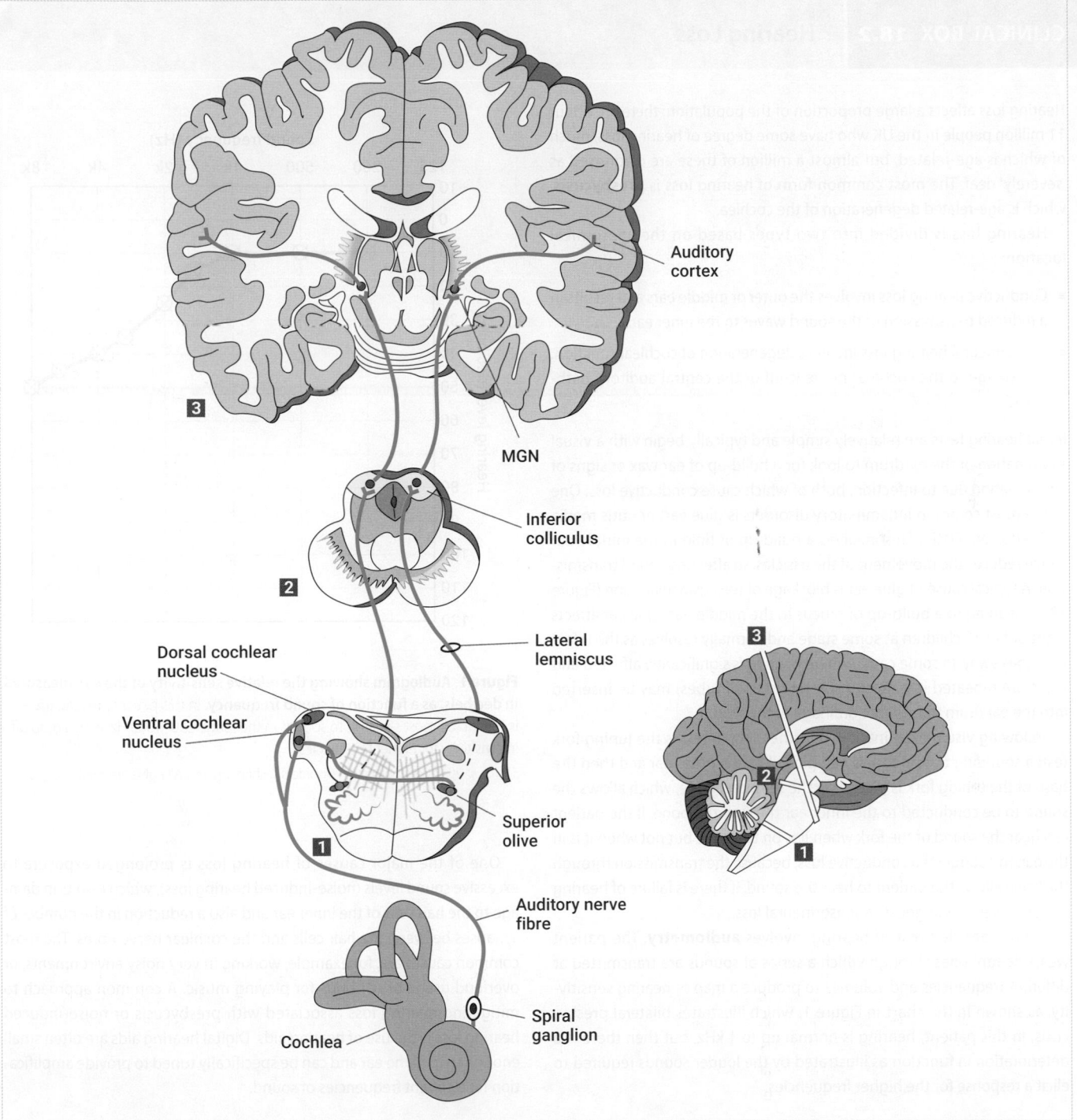

Figure 18.31 The auditory pathway.

Source: Figure 13.13 from Purves et al. *Neuroscience*, 6th edn, 2017. Oxford University Press.

PAUSE AND THINK

How does the mechanism of depolarization of the hair cells compare with that seen in other mechanoreceptors such as touch receptors?

Answer: The extracellular fluid around the stereocilia, the endolymph, is very rich in K⁺ to the extent that there is a gradient for K⁺ to enter the hair cell and cause it to depolarize. In comparison, other mechanoreceptors depolarize when the channels open and allow Na⁺ ions to enter.

Balance

The sense of balance, rather like proprioception (Section 18.4), is one of the senses that we are not normally aware of unless it is perturbed. However, like proprioception it is vital for our ability to control movement and, if it is disturbed or is in conflict with our visual system, it can lead to very unpleasant sensations of nausea such as motion sickness.

CLINICAL BOX 18.2   Hearing Loss

Hearing loss affects a large proportion of the population: there are about 11 million people in the UK who have some degree of hearing loss, much of which is age-related, but almost a million of these are diagnosed as 'severely' deaf. The most common form of hearing loss is **presbycusis**, which is age-related degeneration of the cochlea.

Hearing loss is divided into two types based on the anatomical location:

- Conductive hearing loss involves the outer or middle ears and results in a reduced transmission of the sound waves to the inner ear.
- Sensorineural hearing loss involves degeneration of cochlear function, or damage to the cochlear nerve itself or the central auditory pathways.

Initial hearing tests are relatively simple and typically begin with a visual examination of the eardrum to look for a build-up of ear wax or signs of inflammation due to infection, both of which cause conductive loss. One of the most common inflammatory disorders is 'glue ear', or otitis media with effusion (OME). This involves a build-up of fluid in the middle ear which reduces the movement of the ossicles, so affecting sound transmission. A typical cause of glue ear is blockage of the Eustachian tube (Figure 18.26) leading to a build-up of mucus in the middle ear. Glue ear affects eight out of 10 children at some stage and normally resolves as the infection goes away. In some cases, where hearing is significantly affected and there are repeated instances, grommets (small tubes) may be inserted into the ear drum to facilitate drainage of the fluid.

Following visual examination, a further simple test is the tuning fork test: a sounding tuning fork is first placed by the outer ear and then the base of the tuning fork is placed on the temporal bone, which allows the sound to be conducted to the inner ear through the bone. If the patient can hear the sound of the fork when it is on the bone but not when it is in the ear, this suggests a conductive loss, because the transmission through the bone allows the patient to hear the sound. If there is failure of hearing in both cases, this suggests a sensorineural loss.

A more detailed test of hearing involves **audiometry**. The patient wears headphones through which a series of sounds are transmitted at different frequencies and volumes to produce a map of hearing sensitivity, as shown in the chart in Figure 1, which illustrates bilateral presbycusis. In this patient, hearing is normal up to 1 kHz, but then there is a deterioration in function as illustrated by the louder sounds required to elicit a response for the higher frequencies.

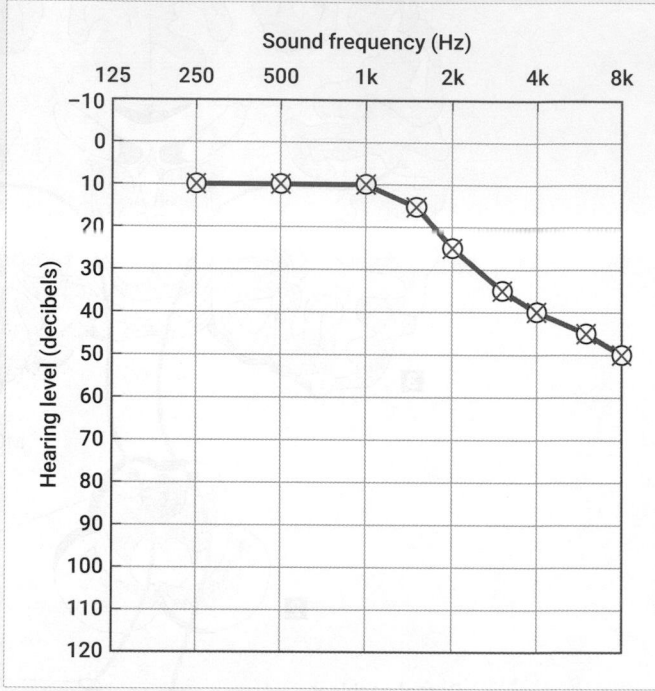

Figure 1 Audiogram showing the relative sensitivity of the ear, measured in decibels, as a function of sound frequency. In this patient's record, the ear is seen to be most sensitive to sounds in the range 250–1000 Hz, with a drop off in sensitivity for higher frequencies.

Source: © 2017 Communication Disorders Technology, Inc. All rights reserved.

One of the major causes of hearing loss is prolonged exposure to excessive sound levels (noise-induced hearing loss), which results in damage to the hair cells of the inner ear and also a reduction in the number of synapses between the hair cells and the cochlear nerve fibres. The most common causes are, for example, working in very noisy environments, or overloud usage of ear buds for playing music. A common approach to mitigating hearing loss associated with presbycusis or noise-induced hearing loss is the use of hearing aids. Digital hearing aids are often small enough to fit in the ear and can be specifically tuned to provide amplification for different frequencies of sound.

The sense of balance is mediated by the **vestibular system** (Figure 18.32), which is linked to the inner ear. There are two main sets of sensory receptors:

- **Semicircular canals**—three on each side of the head.
- **Otolith organs**—**utricle** and **saccule**.

The sensory mechanisms are very similar to those in the cochlea, in both the semicircular canals and the otolith organs are hair cells

with stereocilia that are linked to gated ion channels. The three semicircular canals on each side of the head are oriented at right angles to each other, so they monitor movement in the three planes of movement of the head. At the base of each canal is a swelling, called the **ampulla** (Figure 18.33) containing hair cells, the stereocilia of which project into a gelatinous structure, the cupula. The canals are filled with endolymph. When you turn your head, the bony structure of the head moves, but the endolymph has

Figure 18.32 Vestibular system showing the semicircular canals and otolith organs.

Source: Purves et al. *Neuroscience*, 6th edn, 2017. Oxford University Press.

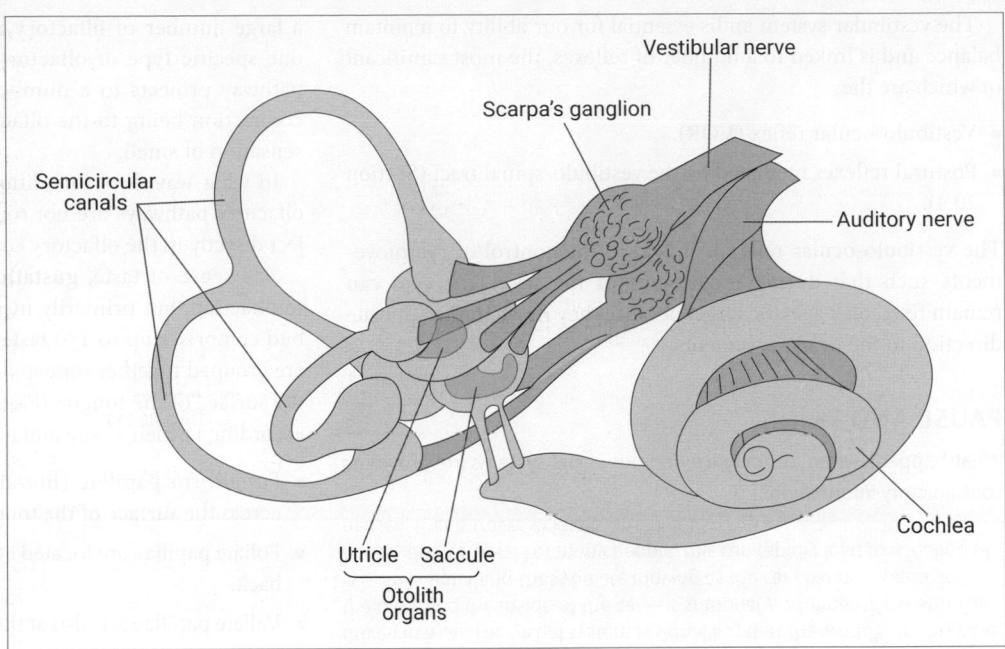

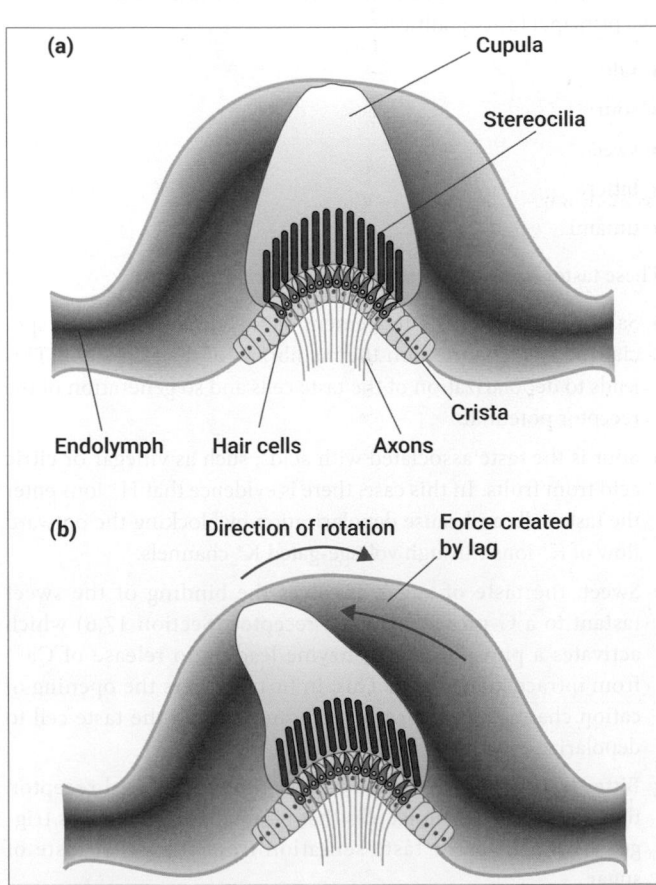

Figure 18.33 Ampulla of a semicircular canal. (a) At rest; and (b) the onset of rotation of the head.

Source: Purves et al. *Neuroscience*, 6th edn, 2017. Oxford University Press.

inertia and so its movement lags behind that of the head as a whole. As a result, the endolymph presses against the cupula, bending the stereocilia, which results in an increase in firing from the sensory neurons. Meanwhile, the neurons in the ampulla of the semicircular canal that mirrors this one on the opposite side of the head will reduce their firing.

What happens if we keep turning at the same speed? The semicircular canals actually respond to rotational acceleration of the head, not movement itself. When you stir a drink in a cup, initially the spoon is rotating but the fluid takes time to catch up due to its inertia. Once you have been stirring for a short time, the fluid is turning at the same speed as the spoon. In the semicircular canals, if you keep rotating the head at the same speed (e.g. if you stand on a playground roundabout), the endolymph 'catches up' with the angular rotation and so no longer exerts significant pressure on the cupula and the afferent firing rate drops back to its resting level. If you suddenly stop the roundabout, the vestibular system will respond as if the head was being accelerated in the opposite direction: if you have ever tried jumping off a spinning roundabout, you will know it is very hard to keep your balance!

The otolith organs respond to linear acceleration and, in particular, are responsible for our sense of whether our heads are up or down in relation to gravity. In this case, the body of the organ comprises a flat sheet of hair cells with their stereocilia projecting into the gelatinous otolithic membrane. On top of the membrane are crystals of calcium carbonate. The saccule is predominantly oriented horizontally, while the utricle is mainly vertical. When you tilt your head in a given direction, the otolith membrane moves slightly, bending the hairs and so stimulating the otolith organs.

The vestibular system andis essential for our ability to maintain balance and is linked to a number of reflexes, the most significant of which are the:

- Vestibulo-ocular reflex (VOR).
- Postural reflexes mediated by the vestibulo-spinal tract (Section 20.4).

The vestibulo-ocular reflex is linked to the control of eye movements such that during movements of the head, the eyes can remain fixed on a specific target because they move in the opposite direction to the head movements.

PAUSE AND THINK

What happens when our balance and our visual systems are providing contradictory information?

Answer: A good example of this is when you try to read while being driven in a car. The visual system is signalling that the world—in this case the book and the inside of the car—is stationary, whereas the vestibular system is signalling that you are moving as the car changes speed and goes round corners. For many people, this can rapidly lead to a feeling of nausea, that is travel sickness.

Taste and smell

Taste and smell are chemical senses that provide us with important information about the outside world and, in particular, about our food. The two senses both involve the detection of chemicals and often operate in conjunction: for example, if you have a 'blocked' nose, so you cannot smell, your appreciation of the taste of food is also significantly diminished.

The sense of smell is based on the detection of a range of odorants, which are typically small molecules that are gaseous at room temperature. Some smells—often the less pleasant ones (!)—can be very potent and we can detect tiny amounts of them in the air, such as ammonia (NH_3), hydrogen sulphide (H_2S)—the smell of rotten eggs—and acetic acid (CH_3COOH)—vinegar.

As we breathe through the nose, air is drawn up into the nasal cavity and across the surface of the **olfactory** epithelium, which contains the olfactory neurons (Figure 18.34). The surface of the olfactory epithelium is covered with a layer of mucus. Cilia project from the olfactory neurons into the mucous layer. The odorants in the air dissolve in the mucous layer and then bind to the olfactory receptor molecules in the cell membranes of the cilia.

The olfactory receptors are linked to G-proteins. Binding of an odorant activates the enzyme adenylyl cyclase, which creates cAMP from ATP. The cAMP binds to ligand-gated channels causing them to open to allow an influx of Na^+ and Ca^{2+} leading to depolarization and action potential generation. There is more information on G-proteins and signalling in Section 17.6.

The axons of the olfactory neurons are routed through small holes in the skull at the top of the nasal cavity, called the **cribriform plate**, and enter the olfactory bulb where they synapse with the dendrites of the second-order axons, the mitral cells (Figure 18.34). Each mitral cell receives converging inputs from a large number of olfactory neurons, all of which only express one specific type of olfactory receptor molecule. The olfactory pathway projects to a number of areas of the brain, one direct connection being to the olfactory cortex, creating the conscious sensation of smell.

In what way does this differ from other sensory systems? The olfactory pathways are not routed through the thalamus but project directly to the olfactory cortex.

The sense of taste, **gustation**, is also a chemical sense linked to olfaction, but primarily mediated by the taste buds. Each taste bud comprises up to 150 taste cells and the taste buds themselves are grouped together on papillae, which are projections mainly on the surface of the tongue (Figure 18.35). The papillae are grouped according to their shape and location:

- Fungiform papillae (literally 'mushroom-like') are scattered across the surface of the tongue.
- Foliate papillae are located at the sides of the tongue, towards the back.
- Vallate papillae are also at the back in the V-formation across the midline.

The taste cells have small microvilli that project into the opening of the taste bud to the surface of the mouth, through which the chemicals in the mouth can interact with the taste cells. There are five principal taste qualities:

- salt;
- sour;
- sweet;
- bitter;
- umami.

These tastes have different transduction mechanisms:

- Salt taste involves the movement of Na^+ directly through specialized Na^+ channels on the membrane of the taste cells. This leads to depolarization of the taste cells and so generation of the receptor potential.
- Sour is the taste associated with acids, such as vinegar or citric acid from fruits. In this case, there is evidence that H^+ ions enter the taste cells and cause depolarization by blocking the outward flow of K^+ ions through voltage-gated K^+ channels.
- Sweet, the taste of sugar, involves the binding of the sweet tastant to a G-protein coupled receptor (Section 17.6) which activates a phospholipase enzyme leading to release of Ca^{2+} from intracellular stores. This, in turn, triggers the opening of cation channels in the cell membrane causing the taste cell to depolarize.
- Bitter taste is also mediated by a G-protein coupled receptor, though clearly the molecules that bind to these receptors trigger a very different taste sensation from the sweet taste of sugar.
- Umami may be a taste name that you have not come across before but it is certainly one you will have tasted as it is the characteristic taste of 'savoury' foods. The most potent tastant is

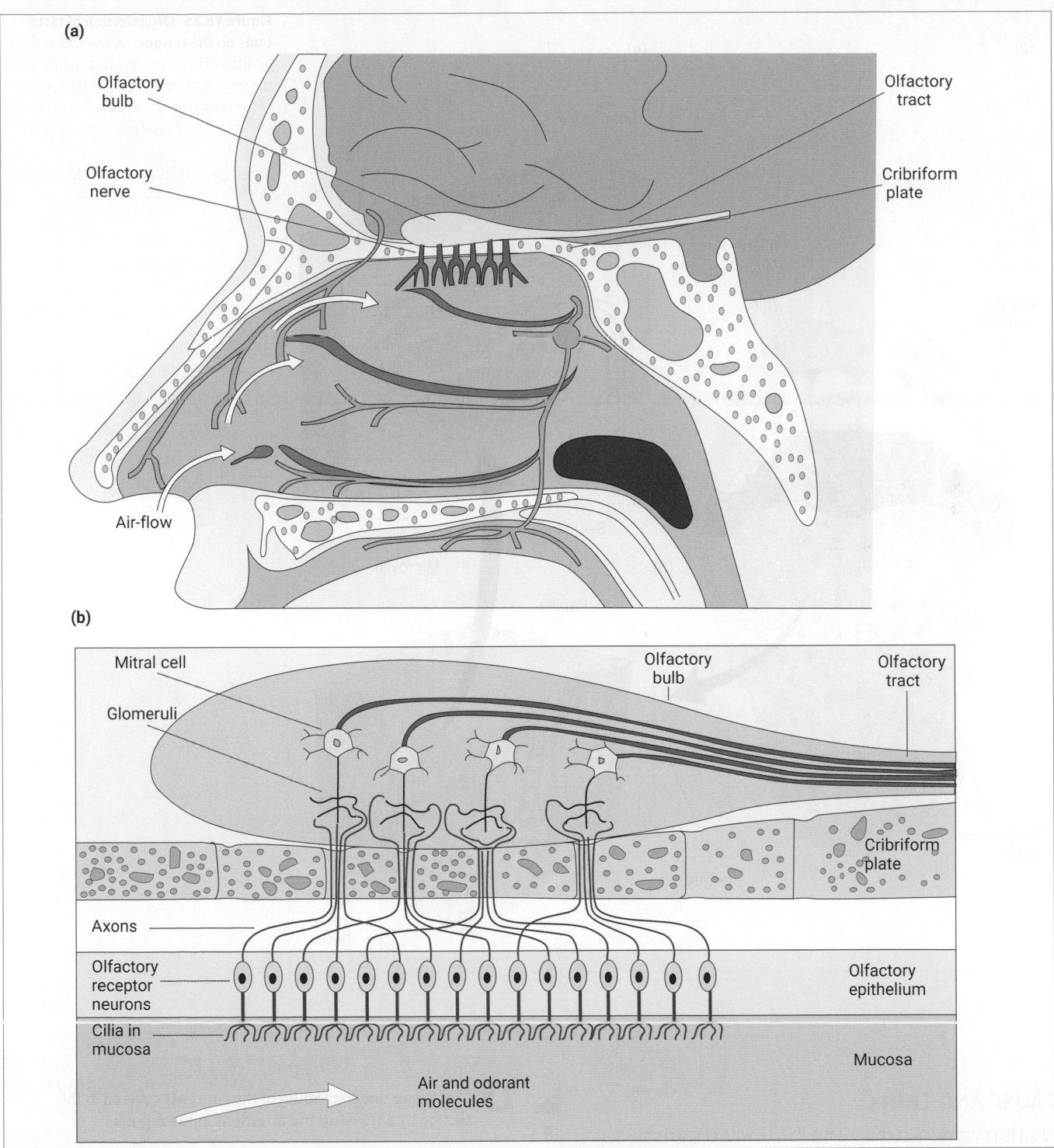

Figure 18.34 The nose and olfaction. (a) The passage of air through the nasal cavity and the location of the organ of smell; (b) the location of the olfactory neurons and the olfactory bulb.

Source: Purves et al. *Neuroscience*, 6th edn, 2017. Oxford University Press.

glutamate, the excitatory neurotransmitter, and many processed foods have their flavours enhanced by the addition of MSG—monosodium glutamate. Again, the transduction mechanism involves a G-protein coupled receptor, activation of which leads to depolarization of the taste cell.

The axons from the gustatory afferents transmit the information to the gustatory nucleus in the medulla of the brainstem. From here the information is relayed to the thalamus and then on to the primary gustatory cortex: note that taste, unlike olfaction, follows the common route of being processed through the thalamus before reaching the cortex.

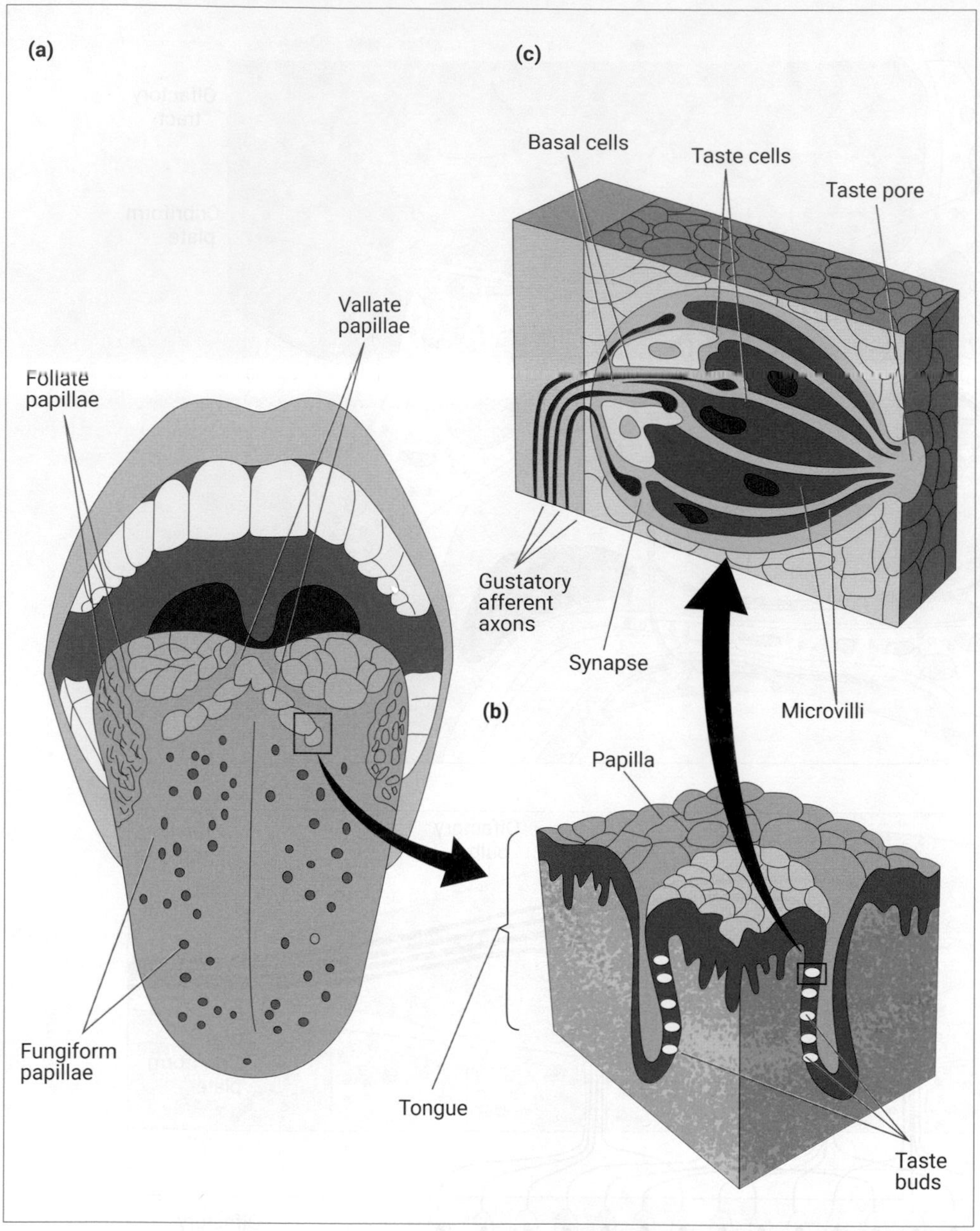

(a)

(c)

Basal cells

Taste cells

Taste pore

Vallate papillae

Foliate papillae

Gustatory afferent axons

Synapse

Microvilli

(b)

Papilla

Fungiform papillae

Tongue

Taste buds

Figure 18.35 Organization of taste buds on the tongue. (a) This shows the location of the three groups of papillae on the surface of the tongue; (b) the location of the taste buds in the pits of a papilla; and (c) the group of taste cells making up a taste bud.

Source: Purves et al. *Neuroscience*, 6th edn, 2017. Oxford University Press.

PAUSE AND THINK

What happens to your sense of taste if you have a blocked nose from a cold?

Answer: Without the olfactory input, the sense of taste is reduced, indicating that the two systems are closely linked in terms of generating the full sense of taste.

Check your understanding of the concepts covered in this section by answering the questions in the e-book.

SUMMARY OF KEY CONCEPTS

- Each sensory modality responds to a specific type of stimulus that generates a neural response.
- Sensory systems respond strongly to change or contrast rather than to steady state conditions.
- During maintained stimulation there is adaptation of the response.
- The transduction processes (with the exception of vision) lead to depolarization of the receptors, generating a receptor potential which, if large enough, will in turn lead to action potential transmission.
- Topographic representation in different forms is a consistent feature: cutaneous sensation retains its mapping through the ascending pathways and up to the S1 cortex; likewise, the visual cortex is laid out as a map of the retina (retinotopic) and the auditory cortex is a map of the cochlea (tonotopic).
- Acuity of each sense is inversely related to the size of the receptive fields of the receptors and directly related to the size of the map in the specific sensory cortex.
- The sensory systems (with the exception of olfaction) are all routed through nuclei of the thalamus before being distributed to the different primary cortical regions.

 Use the flashcards in the e-book to test your recall of key terms introduced in this chapter.

QUESTIONS

 Looking for answers? Once you've answered these questions, follow the link in the e-book to the answer guidance and check your work.

Concepts and definitions

1. Define the term 'receptive field'.
2. What are the four main types of stimulus? Give an example of each one.
3. Why does the left sensory cortex detect sensation in the right side of the body?
4. Compare the responses of the muscle spindle and the tendon organ to ongoing movements.
5. What are the stages of focusing an image on the retina?
6. Name the three ossicles in the middle ear. What is the role in hearing?
7. What are the five qualities of taste detected by the tongue?

Apply the concepts

8. What is the significance of the different sizes of representation of the body and the somatosensory cortex?
9. Describe the process of transduction of the light stimulus within a photoreceptor. What is unusual about the receptor potential?

10. Draw a diagram to illustrate the receptive field of an on-centre, off-surround retinal ganglion cell and show the responses to a dot of light shone on the centre and on the surround.
11. Explain why you are likely to lose your balance if you step off a rapidly rotating playground roundabout.
12. What are the main forms of hearing loss? Give an example of each. How can the form of hearing loss be diagnosed?

Beyond the concepts

13. Define the term 'receptive field' and discuss how the topographic representation of receptive fields is maintained throughout the neural system.
14. Compare and contrast how sound waves of very low and very high frequencies are transduced into neural signals.
15. Describe the process of contrast detection in the visual system.
16. Discuss the significance of adaptation within sensory systems.
17. Discuss the mechanisms involved in controlling pain. In your answer consider both the body's own pain regulation systems and the action of pharmaceutical interventions.

FURTHER READING

Bear, M. F., Connors, B., & Paradiso, M. (2015). *Neuroscience: Exploring the Brain*. 4th edition Wolters Kluwer.
This text gives an advanced, detailed view of the sensory systems and associated mechanisms in Chapters 8 to 12. This is useful for more detailed additional information.

Heinbockel, T. (ed) (2018). Sensory Nervous System. *IntechOpen*. doi: 10.5772/68128.
This is a very readable, online text of sensory systems.

Pocock, G., Richards, C. D., & Richards, D. A. (2017). *Human Physiology*. 5th edition. Oxford University Press.
This is a more general physiology text that provides more detail on sensory systems than the current chapter but is not at an advanced level.

Purves, D., et al (2018). *Neuroscience*. 6th edition. Oxford University Press.
This is another advanced text giving a detailed view of the sensory systems through Chapters 9 to 15.

Communication and Control 3

Controlling Organ Systems

By the end of this chapter you should be able to:

- Describe, in general terms, the organization of the autonomic nervous system and the roles of the sympathetic and parasympathetic branches in the control of organ systems.

- Describe the roles of acetylcholine and noradrenaline as neurotransmitters in the autonomic nervous system and the receptors and ion channels they activate.

- Define the term 'hormone', list the classes of chemical substances that can act as hormones, and describe the features of communication processes involving hormones.

- Understand the relationship between the hypothalamus and the pituitary in the regulation of the endocrine system, and describe the negative feedback pathways involved in their control.

- Discuss the endocrine control of growth, the thyroid, adrenal glands, pancreas, and calcium homeostasis.

Chapter contents

Introduction	653
19.1 The autonomic nervous system	654
19.2 Examples of autonomic nervous system function	662
19.3 The endocrine system	665
19.4 Conclusion	683

 Watch the key concepts video in the e-book to prepare yourself for studying this chapter.

Introduction

In addition to performing their own unique specific functions, organs and tissues also work together as groups to facilitate the various physiological processes necessary for life. These groups of organs and tissues acting together in concert are referred to as **organ systems**. The circulatory system, for example, includes the heart, blood, and blood vessels, which together supply tissues with oxygenated blood and nutrients, and remove the waste products of metabolism. The urinary system, composed of the kidneys, ureters, bladder, and urethra, filters waste products from blood and excretes them from the body. While there is some overlap between the components of different organ systems (for example, the heart is part of the circulatory system as a pump, the muscular system as a type of muscle, and the endocrine system by virtue of its release of hormones such as atrial natriuretic peptide), the organs and tissues of the body are typically classified into 10 different systems, as shown in Table 19.1.

The organ systems themselves also work together in concert to achieve specific goals and maintain **homeostasis** (for more detail, see Chapter 16). Let us, for example, consider the body's response to a perceived harmful event or threat to survival; the so-called 'fight or flight' response. During this response:

- The activity of our cardiovascular and respiratory systems rapidly increases, supplying our skeletal muscles with the additional oxygen required to support an impending bout of physical activity.
- Our digestive system is temporarily put on hold because reacting to the immediate threat is more important than digesting food.
- Our energy stores of glycogen and fat start to be mobilized in anticipation of an increased demand from contracting skeletal muscle.

- The hairs on our skin stand up to conserve heat.
- Our peripheral vision is decreased to allow us to focus on the immediate threat.

For these changes, and indeed the changes associated with virtually every other physiological process in our body to occur, communication between the different organ systems is required. Each organ system receives instructions that allow it to respond to the requirements of the other systems at any moment in time by adapting activity accordingly. This control of activity of our organ systems is what underlies homeostasis and is facilitated by two different modes of communication: the **autonomic nervous system** and the **endocrine system**. In this chapter we will consider the components and functions of these two communication systems and how they control the activity of the different organ systems.

19.1 The autonomic nervous system

We are often completely unaware of the activities of the autonomic nervous system (ANS) as it works autonomously in the background without our conscious effort. We do not, for example, have to consciously 'think' to make our heart beat faster when we start to run or make our pupils change diameter to adapt to varying levels of light. These and many other physiological processes just happen 'automatically' by the actions of the ANS.

Table 19.1 Major organ systems of the body

System	Organs/tissues	Major function(s)
Circulatory system (Chapter 21)	Heart, blood, blood vessels, lymph, lymphatic vessels	Delivery of oxygen and nutrients to tissues Removal of waste products of metabolism from tissues
Digestive system (Chapter 25)	Mouth, stomach, intestines, anus, and associated organs such as liver and pancreas	Processing of food to release of nutrients
Endocrine system (Chapter 17)	Endocrine glands, some tissues such as adipose, and some organs such as heart, liver, stomach, and kidney	Synthesis and release of hormones for communication between systems
Integumentary system	Skin, hair, nails	Barrier between internal and external environments, retention of body fluids, temperature regulation
Muscular system (Chapter 20)	Skeletal, smooth, and cardiac muscle	Generating movement, maintaining posture, heat generation, gut movements, blood flow
Nervous system (Chapter 17)	Brian, spinal cord, nerves, and sensory organs	Collection, processing, and dissemination of information from internal and external environments
Reproductive system (Chapter 26)	Ovaries, fallopian tubes, uterus, cervix, and vagina in females Testes, epididymis, vas deferens, urethra, and penis in males	Gamete production and delivery Development of embryos in females
Respiratory system (Chapter 22)	Nose, larynx, trachea, bronchi, lungs, diaphragm	Gaseous exchange between blood and air
Skeletal system	Bones, cartilage, ligaments	Structural support and protection Production of red and white blood cells
Urinary system (Chapter 24)	Kidneys, ureters, bladder, urethra	Filtering of blood to remove excess fluid and waste products

The ANS is a division of the peripheral nervous system and consists of efferent motor neurons that control the activities of smooth muscle, cardiac muscle, and various glands. Sensory (afferent) neurons also arise from the target organs, tissues, and glands of the ANS to provide the central nervous system (CNS) with information concerning the state of organs; these afferent pathways are also considered by some to be part of the ANS. However, autonomic pathways are modulated by a diverse range of other sensory inputs such as those from the visual, auditory, olfactory, and vestibular systems so it makes sense to consider the ANS as just the efferent motor part of the system.

Classically, the ANS consists of two distinct branches: the **sympathetic branch** and the **parasympathetic branch**. The nerves of the sympathetic branch leave the spinal cord in the thoracic region (chest), whereas the parasympathetic nerves leave the cord either side ('para') of the sympathetic branch at the cranial and sacral regions. The sympathetic and parasympathetic divisions of the ANS are often presented as binary opposing systems with the sympathetic division facilitating '**fight or flight**' and the parasympathetic division facilitating '**rest and digest**'-type responses. While this concept provides a useful generalization to illustrate the nature of the actions of the two divisions, the idea that one division is activated in its entirety in preference to the other is misleading as each division consists of discrete pathways that can be activated in isolation or in patterns depending on the specific response required. The sympathetic division is not just involved in emergency 'flight or fight' responses. Our organ systems are regulated by a tonic input from both sympathetic and parasympathetic divisions for normal daily functions. It is therefore the *balance* between sympathetic and parasympathetic inputs to effector organ systems that is crucial for maintaining homeostasis.

Organization of the autonomic nervous system

The ANS receives inputs from areas in the brain such as the amygdala, hippocampus, hypothalamus, and olfactory cortex. These brain areas integrate information received from both the internal and external environments and use the ANS to relay information regarding the appropriate responses to be made to organ systems. The pathways that make up the ANS are composed of two distinct sets of neurons: preganglionic fibres and postganglionic fibres (Figure 19.1).

Preganglionic fibres are small-diameter myelinated neurons that originate in the CNS and go on to make connections (synapses) with nerve cell bodies located in clusters called **ganglia** (singular ganglion) that reside outside the CNS. The role of the preganglionic fibres, therefore, is to relay action potentials from the CNS to the ganglia. In the sympathetic division these preganglionic neurons are relatively short as the ganglia they synapse with are located adjacent to the spine. In the parasympathetic division, however, the preganglionic neurons are much longer as the ganglia they connect with tend to be located close to the target organ.

The efferent neurons that arise from ganglia are called postganglionic fibres and, not surprisingly, tend to be long in the sympathetic division, as they travel from near the vertebral column to the target organ, and short in the parasympathetic division as the

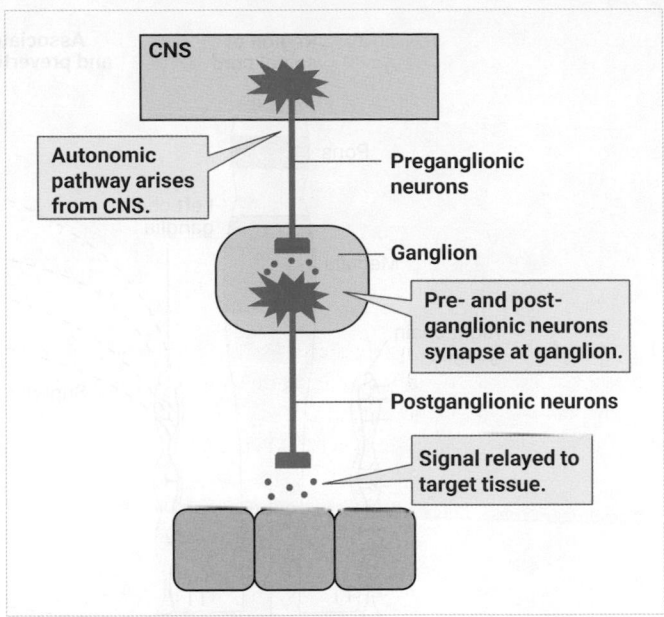

Figure 19.1 Autonomic pathways arising from the CNS consist of pre- and postganglionic fibres that synapse in a ganglion.

parasympathetic ganglia are close to their target organ. These postganglionic fibres are small-diameter unmyelinated neurons that relay signals in the form of action potentials from the ganglia to effector organs. The neuronal connections between these efferent postganglionic neurons and their effector organ consist of multiple swellings called **varicosities**. These varicosities facilitate the release of neurotransmitters over a large surface area of the effector organ enabling many cells to be stimulated simultaneously.

Organization of the sympathetic division

The preganglionic neurons of the sympathetic system originate within the lateral horn of the spinal cord between the second thoracic vertebra (T_2) and lumbar vertebrae L_2–L_3. The majority of sympathetic ganglia that these preganglionic neurons connect with are located in the sympathetic trunks, which consist of two chains of 22 interconnected ganglia that lie close to and either side of the vertebral column, as shown in Figure 19.2. However, sympathetic prevertebral ganglia such as the celiac, superior mesenteric, and inferior mesenteric also lie outside the sympathetic trunks.

Each preganglionic sympathetic neuron may form synapses with multiple postganglionic neurons located in several ganglia and the average ratio between pre- and postganglionic neurons is around 1:20. This allows a coordinated sympathetic stimulation of several target organs at once. Long postganglionic neurons emerge from the sympathetic trunk and prevertebral ganglia to directly innervate effector organs such as the heart, lungs, glands, bladder, stomach, intestines, and radial muscles in the iris of the eye. One exception, however, is in the adrenal gland where the preganglionic sympathetic neurons from the splanchnic nerve innervate modified neuroendocrine cells in the adrenal medulla called **chromaffin cells**. These chromaffin cells have an analogous function to postganglionic sympathetic neurons, but instead of releasing neurotransmitter locally into a synapse with the effector organ,

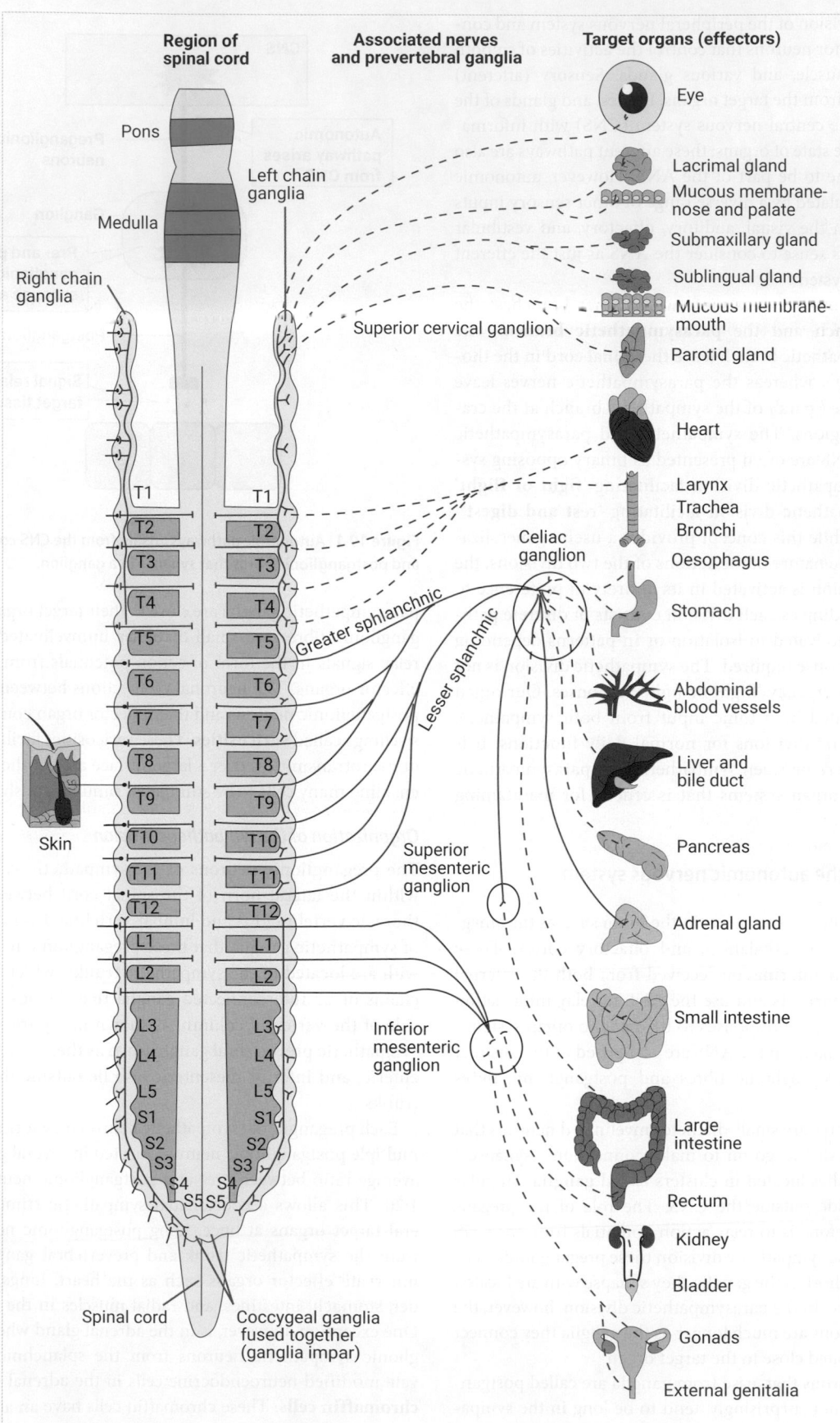

Figure 19.2 Connections in the sympathetic nervous system.

Source: Pocock et al. *Human Physiology*, 5th edn, 2017. Oxford University Press.

the 'transmitter' is the hormone **adrenaline**, which is released into the bloodstream and transported to remote sites around the body.

Organization of the parasympathetic division

The preganglionic neurons of the parasympathetic system originate from within the brainstem and sacral region of the spinal cord either side of the sympathetic origins. The fibres arising from the brainstem exit from the 3rd, 7th, 9th, and 10th cranial nerves. The 10th cranial nerve, the **vagus nerve**, is of particular importance as it contains around three-quarters of all preganglionic parasympathetic neurons, as shown in Figure 19.3. The preganglionic parasympathetic neurons arising from the sacral region of the spinal cord exit via spinal nerves S2, S3, and S4, commonly referred to as the **pelvic splanchnic nerves**. The preganglionic neurons of the parasympathetic division are typically much longer than those in the sympathetic division because the parasympathetic terminal ganglia are located close to or sometimes even within (termed **intramural ganglia**) the effector organ.

The postganglionic neurons of the parasympathetic division therefore only project a short distance, typically 1 or 2 mm, to connect to effector organs which include lacrimal (tear) and salivary glands, the pancreas, heart, and sphincter muscles in the iris of the eye. The ratio of pre- to postganglionic neurons in the parasympathetic system is much smaller (~1:1) than in the sympathetic system. This means the effects of the parasympathetic system tend to be more discrete and localized compared to the sympathetic system, where several effector organ systems can be stimulated at once.

The enteric nervous system

The enteric nervous system (ENS) consists of a highly organized system of neurons located in the walls of the gastrointestinal tract and controls motility of the smooth muscle and secretion (see Section 25.1). The ENS can function relatively independently from both the CNS and the ANS through local neuronal circuits. However, the ENS also connects with ganglia of the ANS and with the CNS itself, and receives motor inputs from parasympathetic neurons in the vagus nerve (also termed the 10th cranial nerve) and sympathetic neurons from prevertebral ganglia. For this reason the ENS is considered by some to be a third division of the ANS.

PAUSE AND THINK

Compare the specificities of the sympathetic and the parasympathetic nervous systems.

Answer: The parasympathetic nervous system has short postganglionic neurons that tend to have very localized actions. The preganglionic neurons of the sympathetic nervous system may synapse with a number of much longer postganglionic neurons, enabling a coordinated response of several target organs or tissues at once.

Neurotransmission in the autonomic nervous system

The two primary neurotransmitters utilized in the ANS are **acetylcholine** (ACh) and **noradrenaline**. Several neuropeptides such as vasoactive intestinal peptide (VIP) and neuropeptide Y (NPY) can also act as co-transmitters, as well as the gaseous substance nitric oxide (NO) and the purine adenosine triphosphate (ATP).

- Neurons that utilize acetylcholine as their primary transmitter are referred to as **cholinergic**.

- Neurons that utilize noradrenaline or adrenaline are referred to as **adrenergic**.

- Autonomic neurons that utilize neurotransmitters other than acetylcholine, adrenaline, or noradrenaline are often referred to as non-adrenergic, non-cholinergic (NANC) neurons.

The actions of transmitters and transmitter analogues are described through a number of terms that have specific meanings as set out below.

- **Agonist**—a chemical that binds to a receptor and generates a response; for example, this may be a neurotransmitter or a hormone.

- **Antagonist**—a chemical that binds to a receptor and blocks the action of the agonist; they may therefore also be called blockers.

- **Competitive antagonist**—a chemical that binds to the same site as the agonist and blocks its action.

If an agonist and a competitive antagonist are both present, what will determine the size of the response generated by the receptor? The agonist and antagonist compete for binding at the same site, so the size of the response generated will depend on (i) the relative affinities of the agonist and the antagonist; and (ii) their relative concentrations.

- **Non-competitive antagonist**—a chemical that binds to a receptor and blocks the action of the agonist. This may be by binding at the same site as the agonist in such a way that the agonist cannot bind irrespective of the relative concentrations, or by binding at a different site on the receptor, which then prevents the receptor being activated even if the agonist binds to it.

- **Inverse agonist**—a chemical that binds to a receptor and generates a response that is opposite to that of the agonist.

- **Ligand**—a chemical that binds to a biological molecule. In neurotransmission, the transmitter is the ligand that binds to the receptor. A ligand may be an agonist, antagonist, or inverse agonist.

Cholinergic transmission

All preganglionic neurons of both the sympathetic and parasympathetic systems are cholinergic as they employ acetylcholine as the neurotransmitter at their synapse with the ganglionic junction (Figure 19.4). The post-synaptic receptor activated by the acetylcholine released from preganglionic neurons is a ligand-gated ion channel called nAChR.

Neurotransmission in preganglionic cholinergic synapses

ACh is synthesized from choline and acetyl coenzyme A (acetyl-CoA) by the enzyme choline acetyltransferase (CAT), which is

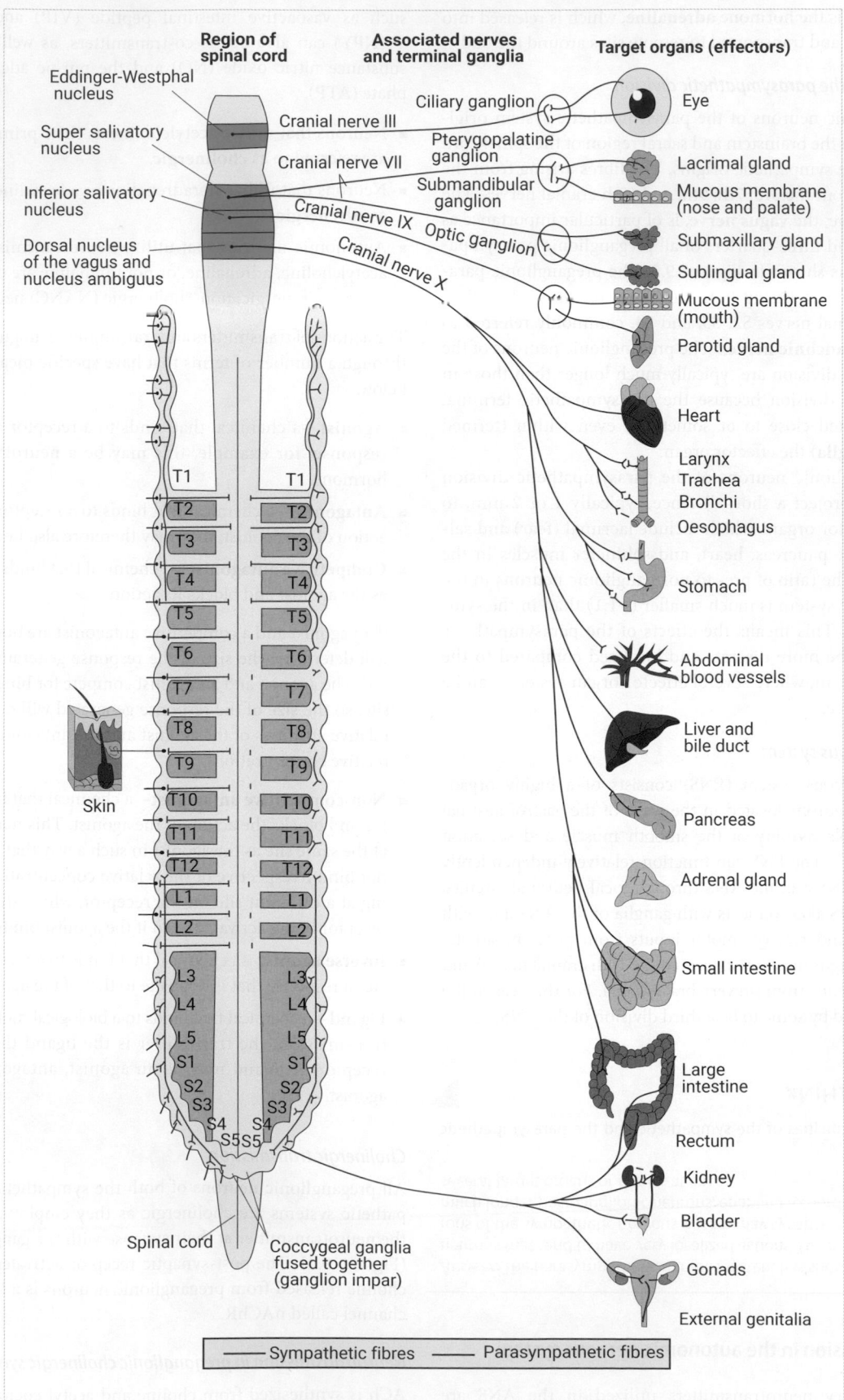

Figure 19.3 Connections in the parasympathetic nervous system.

Source: Pocock et al. *Human Physiology*, 5th edn, 2017. Oxford University Press.

Figure 19.4 Neurotransmission in preganglionic cholinergic synapses.

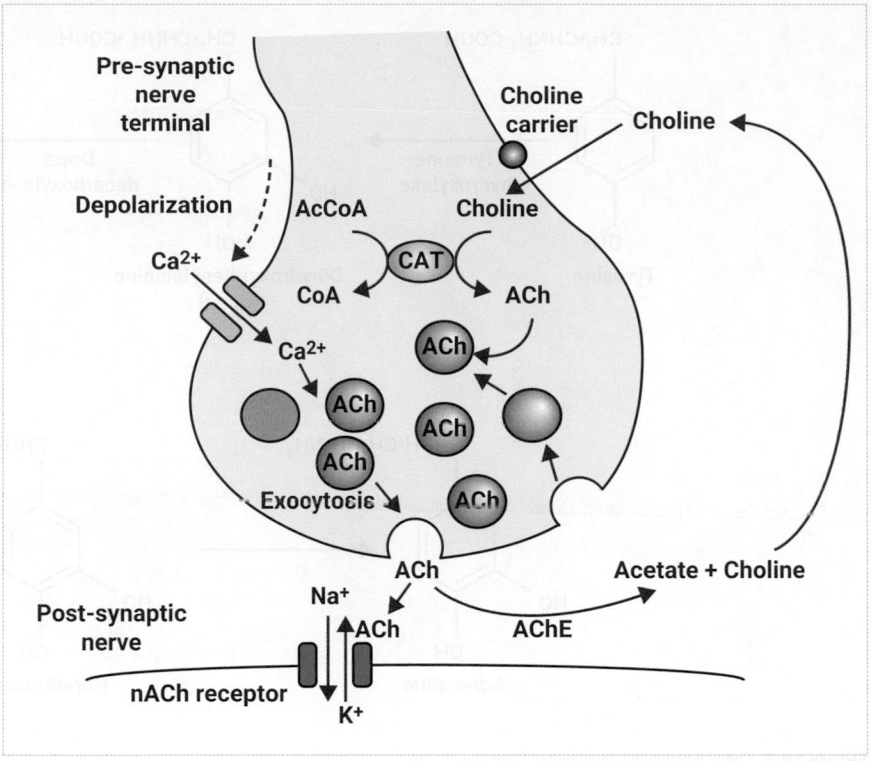

present at high concentration in the cytoplasm of cholinergic nerve terminals. This enzyme is first synthesized in the cell body of the neuron and then transported to the nerve terminal by a process known as anterograde axonal transport. ACh in the nerve terminal is packaged at high concentration into synaptic vesicles by an active transport system. The preganglionic cell body, which lies within the CNS, receives inputs that, if of sufficient strength, cause the generation of an action potential that travels down the axon to the preganglionic nerve terminal. Membrane depolarization caused by arrival of an action potential at the nerve terminal results in the opening of voltage-operated Ca²⁺ channels (VOCCs). This calcium entry causes the synaptic vesicles to fuse with the neuronal cell membrane at active zones within the synapse and release ACh into the synaptic cleft where it can activate nAChRs on the post-synaptic membrane and propagation of an action potential to the postganglionic neuron. The enzyme acetylcholinesterase (AChE) bound to the basement membrane of the nerve terminal facilitates the very rapid (within milliseconds) removal of ACh by hydrolysis to acetate and choline. Acetate is metabolized following uptake into either the pre- or post-synaptic neuron, whereas choline is recycled by transporter proteins so it can be reused for ACh synthesis.

The postganglionic fibres of the parasympathetic system are also cholinergic; however, the receptors activated at the effector side of the postganglionic neuron–effector synapse are G-protein coupled receptors rather than nicotinic ligand-gated ion channels. While nearly all postganglionic neurons in the sympathetic system use noradrenaline as a neurotransmitter, there are some notable exceptions that, like the parasympathetic postganglionic fibres, use acetylcholine. These include sympathetic postganglionic cholinergic neurons that innervate the sweat glands, the piloerector

muscles that make our body hair stand on end resulting in 'goose bumps', and some blood vessels.

Adrenergic transmission

Adrenergic transmission in the ANS is restricted to the sympathetic division with no neurons in the parasympathetic system using noradrenaline or adrenaline as a transmitter. Noradrenaline and adrenaline (called norepinephrine and epinephrine, respectively, in the USA) are referred to as **catecholamines** and, together with the neurotransmitter dopamine, are synthesized from the amino acid tyrosine by a series of enzymatic reactions. You can see the stages of the reaction from tyrosine to adrenaline in Figure 19.5.

The key rate-limiting enzyme in this series of reactions is tyrosine hydroxylase, which is present in high concentrations in sympathetic nerve terminals. The enzymes DOPA decarboxylase and dopamine β-hydroxylase are less specific in their substrate specificity and are not restricted to catecholamine-synthesizing cells. Similar to the choline acetyltransferase enzyme utilized in cholinergic transmission, these enzymes are synthesized in the cell body and transported to the nerve terminal by anterograde transport. The first two enzymes in the pathway, tyrosine hydroxylase and DOPA decarboxylase, are found only in the cytoplasm, whereas dopamine β-hydroxylase, which converts dopamine to noradrenaline, is present only within the synaptic vesicles. Thus, it is the presence of the enzyme dopamine β-hydroxylase within the vesicles of sympathetic neurons that ensures that noradrenaline rather than dopamine is the neurotransmitter that accumulates. Similarly, chromaffin cells of the adrenal medulla contain phenylethanolamine N-methyltransferase. This enzyme methylates noradrenaline to

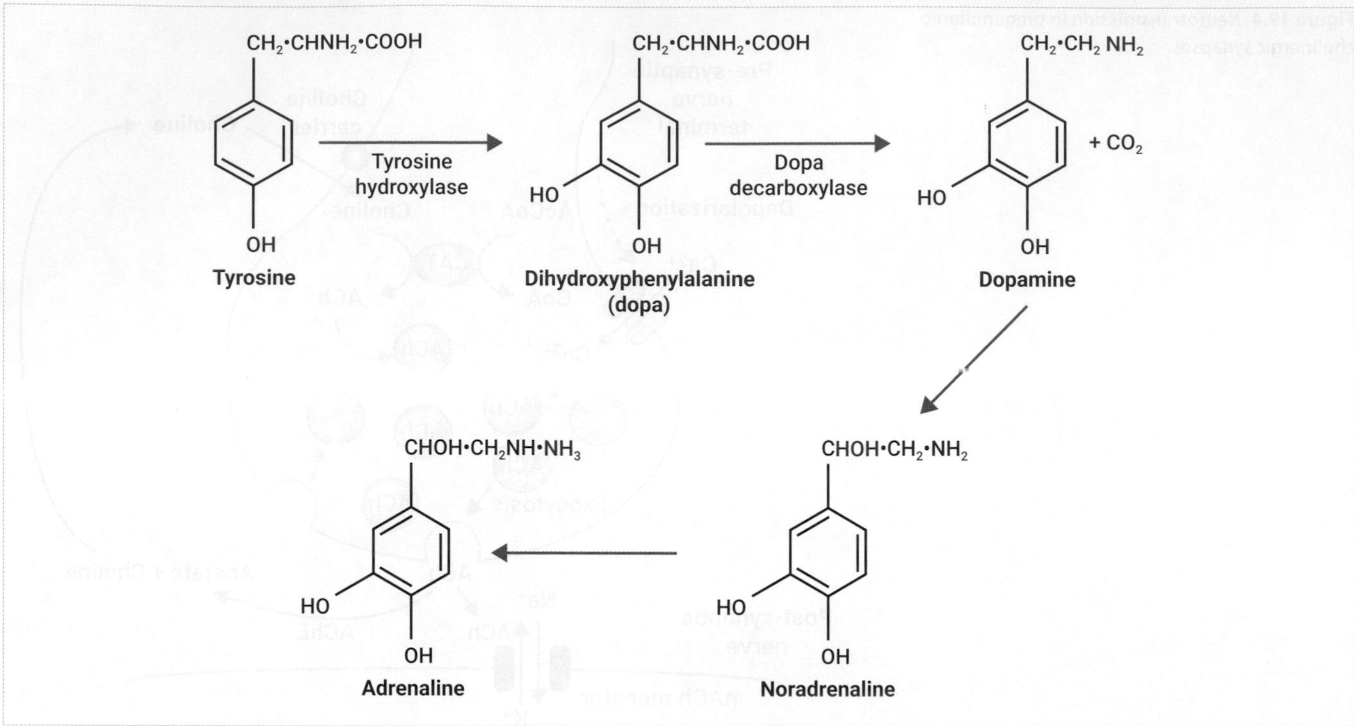

Figure 19.5 Catecholamine synthesis from tyrosine.

Source: D. Papachristodoulou, A. Snape, W.H. Elliott, & D.C. Elliott. *Biochemistry and Molecular Biology*, 6th edn, 2018. Oxford University Press.

form adrenaline, and the hormones released into the bloodstream after sympathetic stimulation consist of ~20 per cent noradrenaline and ~80 per cent adrenaline (as shown in Figure 19.5). These circulating catecholamines are eventually inactivated by the enzyme catechol-*O*-methyltransferase (COMT) in the liver.

Neurotransmission at adrenergic synapses

Arrival of an action potential at the nerve terminal leads to depolarization of the membrane potential and opening of VOCCs, as shown in Figure 19.6. Calcium entry causes the fusion of synaptic vesicles with the neuronal cell membrane and release of noradrenaline into the synaptic cleft, where it activates G-protein coupled adrenergic receptors on the post-synaptic membrane to initiate an intracellular signalling cascade. The signal is terminated by removal of noradrenaline from the synaptic cleft by two distinct transport mechanisms, termed uptake 1 and uptake 2. Once back inside the neuron, noradrenaline is metabolized by the enzyme monoamine oxidase (MAO).

PAUSE AND THINK

What are the main neurotransmitters found in the sympathetic and parasympathetic nervous systems?

Answer: In both the sympathetic and parasympathetic nervous systems the preganglionic neurons are cholinergic, releasing ACh at the synapses. The postganglionic neurons of the parasympathetic nervous system are also cholinergic, but the postganglionic neurons of the sympathetic nervous system are adrenergic, releasing the catecholamines adrenaline or noradrenaline.

Receptors in the autonomic nervous system

It is remarkable that just three primary neurotransmitters—acetylcholine, noradrenaline, and adrenaline—initiate the wide range of effector organ responses facilitated by the ANS. Furthermore, the same neurotransmitter may stimulate activity in some tissues but inhibit activity in others. For example, acetylcholine released from parasympathetic vagal neurons innervating the sino-atrial node in the heart decreases heart rate by suppressing the pacemaker current, whereas acetylcholine released from parasympathetic neurons innervating the lungs causes contraction of airways smooth muscle and bronchoconstriction. How can such a wide range of actions be facilitated by such a limited number of neurotransmitters? The answer lies in the diversity of neurotransmitter receptor types that are expressed by the different organ and tissue types and the range of subsequent downstream signalling pathways that are activated by these receptors.

Acetylcholine receptors

Acetylcholine receptors are split into two classes based on their ability to respond to the drugs **nicotine**, a plant alkaloid, and **muscarine**, a compound isolated from some types of mushroom. The modes of action and molecular structures of these two classes of acetylcholine receptor are fundamentally different. The receptors that respond to the drug nicotine, as well as acetylcholine, are ligand-gated ion channels (see Section 17.6) and are referred to as **nicotinic receptors,** while the receptors that respond to muscarine, as well as acetylcholine, are a type of G-protein coupled receptor called **muscarinic receptors**.

Figure 19.6 Neurotransmission at adrenergic synapses.

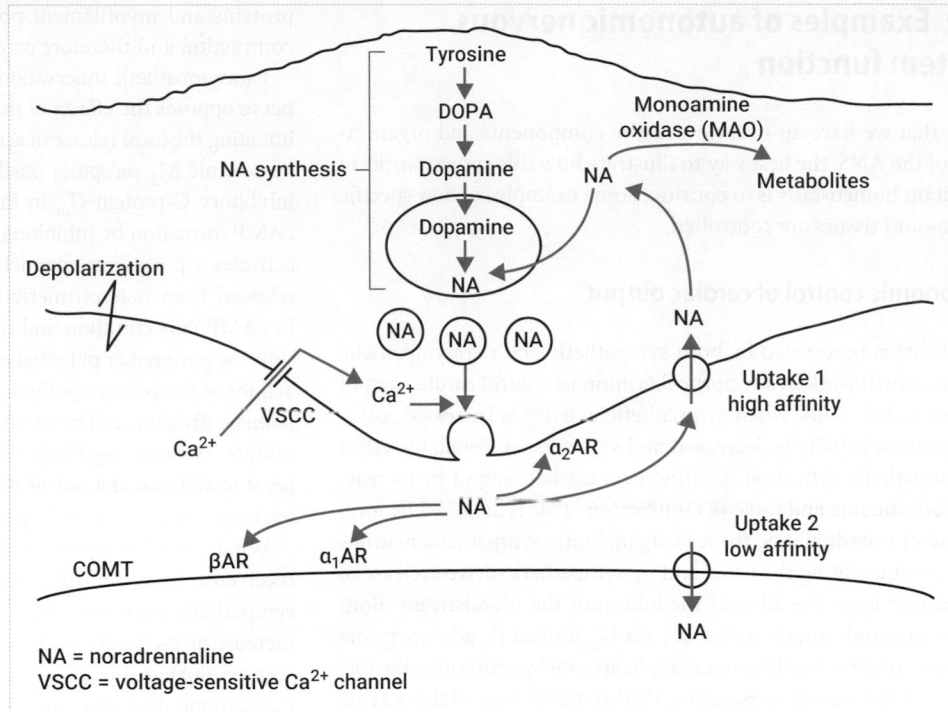

NA = noradrennaline
VSCC = voltage-sensitive Ca²⁺ channel

Nicotinic cholinergic receptors

Nicotinic acetylcholine receptors (nAChRs) are present in all post-ganglionic neurons in both the sympathetic and parasympathetic divisions. Similar to the nicotinic acetylcholine receptors found at the neuromuscular junction (Chapter 20), the ganglionic nAChRs are ligand-gated ion channels permeable to the cations Na⁺ and K⁺. However, the subunit composition of the ganglion type nAChR differs from that of the skeletal muscle type. Binding of acetylcholine (the 'ligand') results in opening of the channel pore and flow of Na⁺ and K⁺ ions down their electrochemical gradients. Na⁺ ions enter the cell and K⁺ ions leave through the open pore, but because more Na⁺ ions enter than K⁺ ions leave, there is a net inward flow of positive charge into the cell. This influx of positive charge depolarizes the membrane potential (i.e. the membrane potential becomes less negative), resulting in an excitatory response which, if of sufficient magnitude, will propagate an action potential in the postganglionic neuron leading to the effector organ.

Muscarinic cholinergic receptors

Muscarinic acetylcholine receptors are members of the transmembrane family of receptors also known as G-protein coupled receptors and are present in the plasma membrane of all effector organ cells innervated by parasympathetic postganglionic neurons. As mentioned previously, sympathetic postganglionic neurons that innervate sweat glands, piloerector muscles, and some blood vessels are cholinergic rather than adrenergic, and these tissues therefore also express muscarinic acetylcholine receptors. Whether the response is excitatory or inhibitory depends on the type of muscarinic receptor expressed by the target tissue. There are five subtypes of muscarinic receptor termed M_1–M_5. The odd numbered receptors (M_1, M_3, and M_5) couple to the $G_{\alpha q}$ G-protein and therefore cause an increase in intracellular Ca^{2+} via inositol 1,4,5-trisphosphate (IP_3) receptor-mediated release of Ca^{2+} from intracellular stores (see Section 17.6). The even-numbered subtypes (M_2 and M_4) couple to $G_{\alpha i}$ causing an inhibition of the production of the second messenger cAMP (again, see Section 17.6 for a description of second messenger pathways).

Adrenergic receptors

All adrenergic receptors are G-protein coupled and respond to the agonists noradrenaline and adrenaline. Most postganglionic neurons of the sympathetic division release noradrenaline as a neurotransmitter, which activates adrenergic receptors in post-synaptic effector organ cells. In addition to this neuronal stimulation of adrenergic receptors, adrenaline released from sympathetic stimulation of the chromaffin cells in the adrenal medulla also travels through the bloodstream to stimulate adrenergic receptors in target tissues. Similarly to the muscarinic acetylcholine receptors, the type of response produced by noradrenaline or adrenaline binding depends on the type of adrenergic receptor present in the plasma membrane of the effector organ cell. There are two main types of adrenoceptor, termed α and β. The α type has two subtypes: α_1 receptors facilitate an increase in intracellular Ca^{2+} via coupling to $G_{\alpha q}$; and α_2 receptors facilitate a decrease in the intracellular second messenger cAMP via coupling to the inhibitory G protein $G_{\alpha i}$. There are three subtypes of β adrenoceptor, termed β_1, β_2, and β_3, and all of them promote an increase in cAMP by coupling to the stimulatory G protein $G_{\alpha s}$.

Check your understanding of the concepts covered in this section by answering the questions in the e-book.

19.2 Examples of autonomic nervous system function

Now that we have an overview of the components and organization of the ANS, the best way to illustrate how this system works to maintain homeostasis is to consider some examples of how specific organs and tissues are controlled.

Autonomic control of cardiac output

The heart is innervated by both sympathetic and parasympathetic inputs, which work in a reciprocal fashion to control cardiac output (Chapter 21). Thus, when sympathetic activity is increased, parasympathetic activity is decreased and vice versa. Overall, the effect of sympathetic activation is to increase cardiac output by increasing both the rate and force of contraction. This is achieved by local release of noradrenaline from postganglionic sympathetic neurons directly innervating the heart and by sympathetic-driven release of adrenaline from the adrenal medulla into the bloodstream. Both adrenaline and noradrenaline act via G_{as}-linked β_1 adrenoceptors on sino-atrial node cells to increase heart rate by promoting the formation of the second messenger cAMP and activation of the enzyme protein kinase A (PKA) (Figure 19.7). cAMP has direct stimulatory effects on the cyclic nucleotide-gated ion channel responsible for initiating the pacemaker current and PKA phosphorylates various target proteins, further enhancing the current. These changes decrease the time taken for the pacemaker potential to reach threshold for action potential generation and therefore increase heart rate.

In addition to the pacemaker cells, the cardiac muscle cells themselves also express β_2 adrenoceptors, and sympathetic-driven PKA activation here results in phosphorylation of various regulatory proteins and myofilament proteins, leading to increased force of contraction and therefore cardiac output.

Parasympathetic innervation of the sino-atrial node from the vagus nerve opposes the effects of the sympathetic system on heart rate by initiating the local release of acetylcholine, which acts on G_{ai}-coupled muscarinic M_2 receptors on the pacemaker cells. Activation of the inhibitory G-protein G_{ai} in the pacemaker cells not only decreases cAMP formation by inhibiting the enzyme adenylyl cyclase, but also activates a potassium channel via the $G_{\beta\gamma}$ subunits, which are also released from heterotrimeric G-protein dissociation. The decrease in cAMP concentration and the increased potassium current act to slow the pacemaker potential and therefore heart rate. At rest it is the actions of the parasympathetic division that dominate over the sympathetic division, and heart rate is maintained at around 70 beats per minute. Without vagal input from the parasympathetic division the heart would beat at a rate of 100–110 beats per minute, driven by the intrinsic automaticity of the pacemaker cells in the sino-atrial node.

What would happen to the heart rate in a patient who has received a heart transplant? Think about the action of the parasympathetic nervous system on the heart. The heart rate would increase as the vagal nerve connection to the heart will have been severed during transplantation. With the loss of this parasympathetic input, the heart rate would default to the intrinsic automaticity of the pacemaker cells in the sino-atrial node.

Regulation of the bladder

While there is some voluntary control over micturition (urination) (see Section 24.8), the sympathetic and parasympathetic divisions of the ANS also play essential roles at an involuntary level.

During the urinary storage phase, noradrenaline released from sympathetic neurons acts at β_3 adrenoceptors in the detrusor

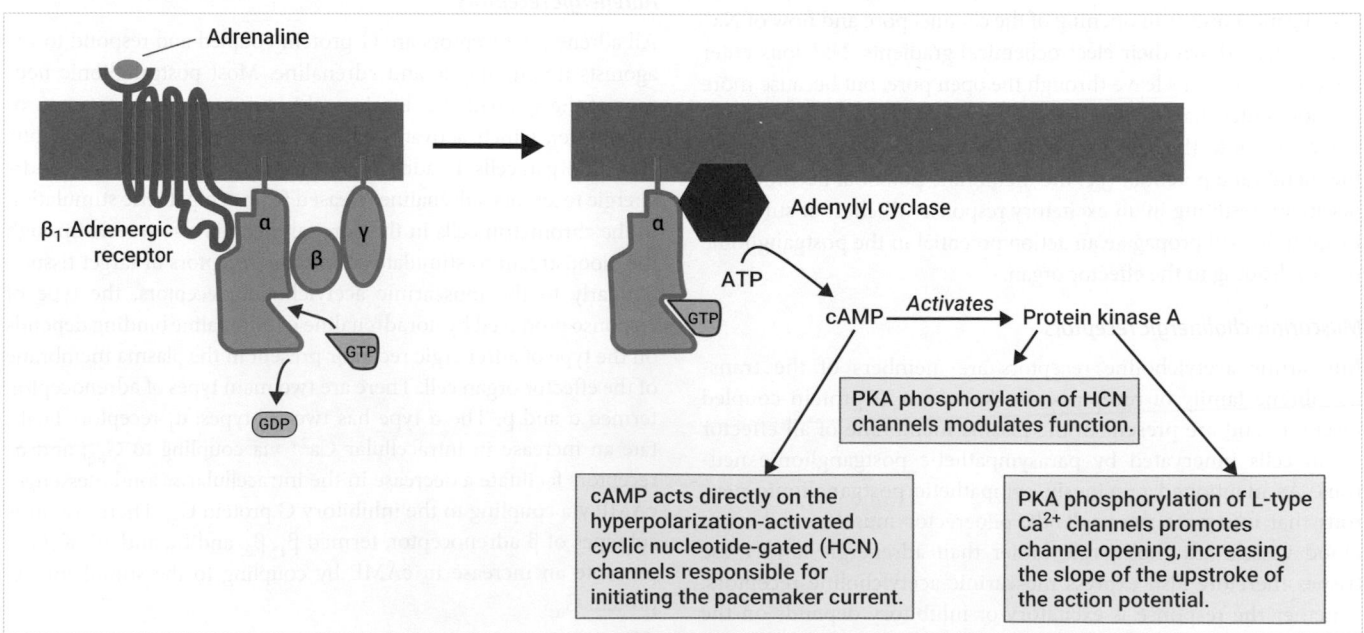

Figure 19.7 Actions of adrenaline on the pacemaker current in the heart.

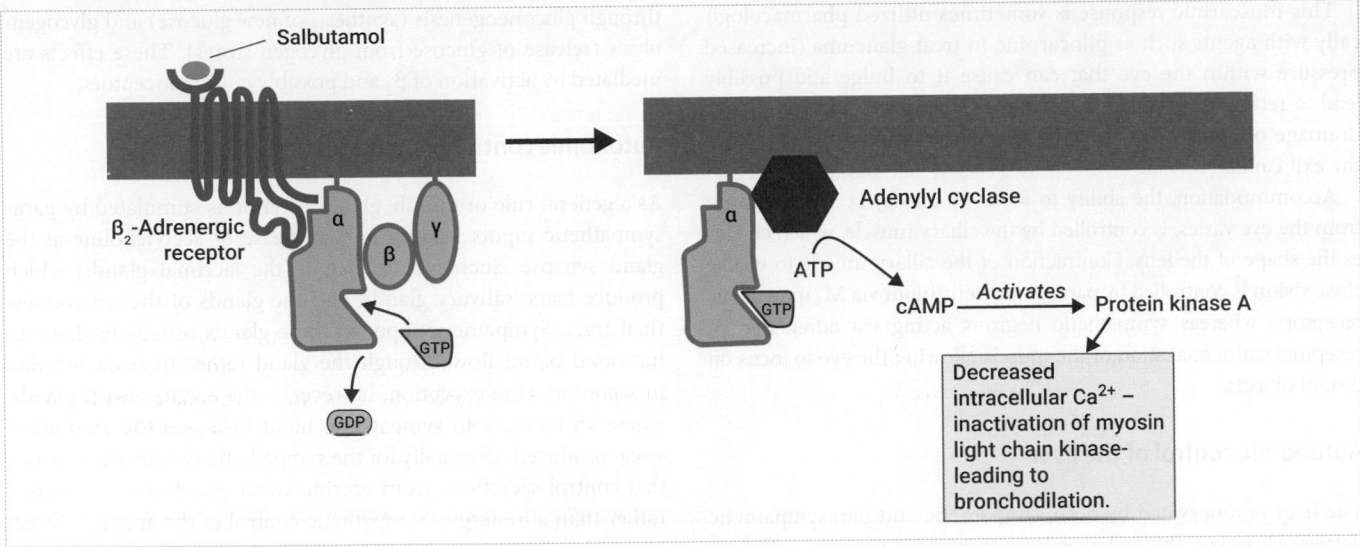

Figure 19.8 Actions of salbutamol at the β_2 adrenergic receptor in airway smooth muscle.

muscle of the bladder wall promoting relaxation and on α_1 adrenoceptors in the outflow tract to increase outlet resistance, allowing the bladder to fill.

Bladder emptying is facilitated by smooth muscle contraction initiated by parasympathetic release of acetylcholine acting mainly at muscarinic M_3 receptors in the detrusor muscle. Parasympathetic postganglionic neurons innervating the bladder also release ATP, which acts in concert with acetylcholine to contract the detrusor muscle, and nitric oxide, which relaxes urethral smooth muscle. The epithelial cells lining the bladder are also a potential source of acetylcholine, the release of which increases on stretching of the bladder during filling. This stretch-induced release of acetylcholine increases with age, which possibly explains why over-active bladder, a condition where a person regularly gets a sudden and compelling need to pass urine, is more common in older people.

Autonomic control of the airways

The tone, and therefore calibre, of the airways is regulated by smooth muscle in the trachea and bronchial tree up to the level of terminal bronchioles (Chapter 22). Airway smooth muscle is tonically activated by parasympathetic neurons from the vagus nerve, which project to ganglia associated with the larger airways. Postganglionic cholinergic neurons from these ganglia innervate the airways smooth muscle and produce a baseline tone by activating muscarinic G_{aq}-linked M_3 receptors on the airways smooth muscle cells. While there is sparse-to-no direct sympathetic innervation of airways smooth muscle, sympathetic neurons do innervate the airways vasculature, submucosal glands, and parasympathetic ganglia to modulate output. Despite a lack of direct sympathetic innervation, β_2 adrenergic receptors are present on the airways smooth muscle cells and activation of these G_{as}-linked receptors by circulating adrenaline and noradrenaline facilitates a decrease in intracellular Ca^{2+} leading to relaxation of smooth muscle tone and bronchodilation. This bronchodilatory response

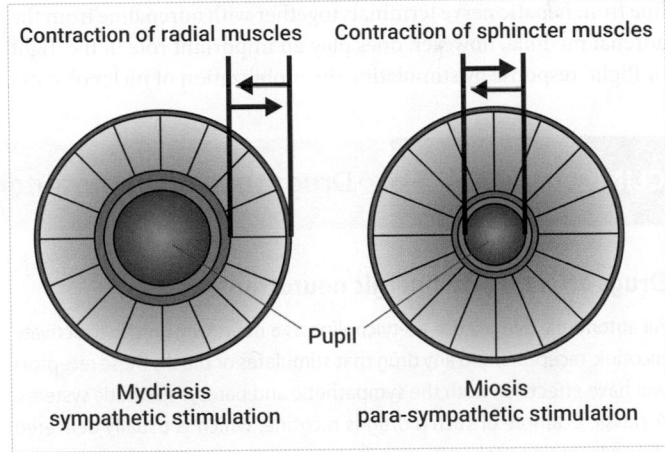

Figure 19.9 Radial and sphincter muscles of the iris.

Source: www.basicmedicalkey.com.

is utilized by β_2 agonist drugs, such as salbutamol, to treat the condition of asthma (see Clinical Box 22.1), a mechanism you can see outlined in Figure 19.8.

Autonomic control of the eye

The iris of the eye has radial and sphincter smooth muscles that regulate pupil diameter and therefore the amount of light reaching the retina (Figure 19.9). Muscle contraction in both types of muscle is initiated by an increase in intracellular Ca^{2+}, resulting from G_{aq}-linked receptor activation, but the type of receptor involved differs. In low light, sympathetic stimulation of the radial muscles activates α_1 adrenoceptors causing the muscle to contract and the pupil to dilate. Conversely, in bright light conditions parasympathetic stimulation of the sphincter muscles causes contraction and pupil constriction by activation of muscarinic M_3 receptors.

This muscarinic response is sometimes utilized pharmacologically with agents such as pilocarpine to treat glaucoma (increased pressure within the eye that can cause it to bulge and possibly lead to retinal damage), as sphincter muscle contraction improves drainage of aqueous humour by relieving folding of the iris over the exit canal.

Accommodation, the ability to focus on an object as its distance from the eye varies, is controlled by the ciliary muscle, which changes the shape of the lens. Contraction of the ciliary muscle to enable close vision is controlled by parasympathetic input via M_3 muscarinic receptors, whereas sympathetic neurons acting via adrenergic β_2 receptors cause relaxation of the muscle allowing the eye to focus on distant objects.

Autonomic control of the liver

The liver is innervated by both sympathetic and parasympathetic nerves. However, the fact that liver transplant patients suffer no life-threatening metabolic disturbances calls into question the importance of autonomic control in regulating whole-body metabolism. Sympathetic stimulation of the liver by release of noradrenaline from hepatic nerve terminals together with adrenaline from the adrenal medulla, however, does play an important role in the 'fight or flight' response by stimulating the mobilization of fuel molecules through gluconeogenesis (synthesis of new glucose) and glycogenolysis (release of glucose from glycogen stores). These effects are mediated by activation of β_2 and possibly α_1 adrenoceptors.

Autonomic control of gland secretion

As a general rule of thumb, gland secretion is stimulated by parasympathetic inputs leading to the release of acetylcholine at the gland synapse. Such glands include the lacrimal glands, which produce tears, salivary glands, and the glands of the gastrointestinal tract. Sympathetic input to these glands usually leads to an increased blood flow through the gland rather than an increase in secretion. One exception, however, is the eccrine sweat glands, where an increase in sympathetic input increases the amount of sweat produced. Unusually for the sympathetic system, the neurons that control secretions from eccrine sweat glands are cholinergic rather than adrenergic. Sympathetic control of the apocrine sweat glands found in the armpit, perineum, ear, and eyelids, however, is mediated by adrenergic postganglionic neurons.

In addition to the endogenous neurotransmitters and enzymes we have discussed in the text, the activity of the ANS can be modulated by a range of exogenous chemicals. These may be of natural origin, such as plant extracts, or man-made in the form of pharmaceutical agents (see Clinical Box 19.1)

CLINICAL BOX 19.1   ## Drugs that affect the autonomic system

Drugs affecting cholinergic neurotransmission

All autonomic ganglia use acetylcholine as a neurotransmitter to activate nicotinic receptors, and any drug that stimulates or blocks these receptors will have effects on both the sympathetic and parasympathetic systems. A classic example of such a drug is nicotine, which is usually delivered through smoking tobacco or, in more recent times, vaping e-cigarettes. Some drugs, however, preferentially affect particular autonomic pathways depending on which receptors or enzyme is or are affected by the drug. *Pilocarpine*, for example, is a non-selective agonist at all muscarinic acetylcholine receptors and is often used to treat disorders of the eye such as glaucoma (increased intraocular pressure) through contraction of the ciliary muscles, allowing drainage of aqueous humour. *Atropine* is a non-selective muscarinic antagonist derived from a number of plants of the nightshade family, such as deadly nightshade and mandrake. The berries of these plants are highly toxic if ingested due to their high atropine content, but atropine in smaller doses can have a number of medical uses related to the inhibition of parasympathetic pathways. These include temporary paralysis of the accommodation reflex and pupil dilation in the eye, the treatment of bradycardia (heart rate <60 beats per minute), and in combating the effects of organophosphate poisoning. *Pirenzepine* is a muscarinic antagonist with selectivity for M1 receptors and can be used to inhibit gastric secretions to treat gastric ulcers. *Darifenacin* is a selective antagonist at the M3 receptor and is used to treat urinary incontinence. *Neostigmine* also affects cholinergic transmission, but this drug works by inhibiting the enzyme acetylcholinesterase rather than affecting receptor function and enhances cholinergic transmission by prolonging the life of acetylcholine in the synapse.

Drugs affecting adrenergic neurotransmission

Adrenergic neurotransmission predominates in the sympathetic division of the ANS and many drugs have been developed that either target adrenoceptors, block the reuptake of noradrenaline from the synapse, or increase the production of noradrenaline. Many of these drugs are commonly used to modulate cardiac output and blood pressure. The so-called **sympathomimetic** drugs mimic the actions of noradrenaline at sympathetic synapses and some are specific for a particular α or β adrenoceptor subtype. *Phenylephrine*, for example, is a selective α1 adrenergic receptor agonist commonly used as a decongestant, and can also be used to increase blood pressure. By activating α_1 adrenoceptors in bronchiole smooth muscle, phenylephrine causes the airways to dilate, allowing mucus to be cleared. The effect of phenylephrine on the α_1 adrenoceptors present in blood vessels, however, is to cause vasoconstriction, thereby raising blood pressure, which can be useful in cases of septic shock.

Clonidine is often used in anaesthesia as a sedative and can also be used to treat high blood pressure. It is classed as a centrally acting α_2 adrenoceptor agonist due to its actions in the CNS; however, it also has actions in the periphery by reducing noradrenaline release through activation of presynaptic α_2 adrenoceptors. Drugs that block rather than activate adrenoceptor activity include *prazosin*, a selective α_1 adrenoceptor antagonist; *atenolol*, a selective β_1 adrenoceptor antagonist; and *propranolol*, a non-selective β-adrenoceptor antagonist. The β-adrenoceptor antagonists, such as propranolol and atenolol, are commonly called '**beta-blockers**' and are primarily used to manage cardiac arrhythmias and to protect the heart from a second myocardial infarction after a heart attack has occurred. They can also be used to treat hypertension, although they are no longer the first choice for most patients.

Check your understanding of the concepts covered in this section by answering the questions in the e-book.

19.3 The endocrine system

The second communication system used by the body to control organ systems and maintain homeostasis is the endocrine system. Similar to the neuron-to-neuron and neuron-to-effector cell connections that we have seen in the ANS, the endocrine system also uses chemical agents to facilitate communication between organ systems. However, rather than local release of neurotransmitter into a synapse, the chemical agents (hormones) in the endocrine system are released into the bloodstream to exert effects on distant organs. Any given hormone released into the bloodstream may act on the cells of several organ systems. As such, a hormone may not have a specific 'target' organ or tissue: it will have its action on each organ system, the cells of which express the receptors for the hormone (i.e. the responsive cells).

Classically, the field of endocrinology concerned the secretion and effects of hormones released from the endocrine glands (Figure 19.10). However, hormones are now known to be released into the bloodstream by a range of other organs and tissues in addition to the endocrine glands. For example, the heart releases natriuretic peptides, the liver insulin-like growth factor, the stomach gastrin and ghrelin, fat cells release leptin, and the kidney renin and calcitriol.

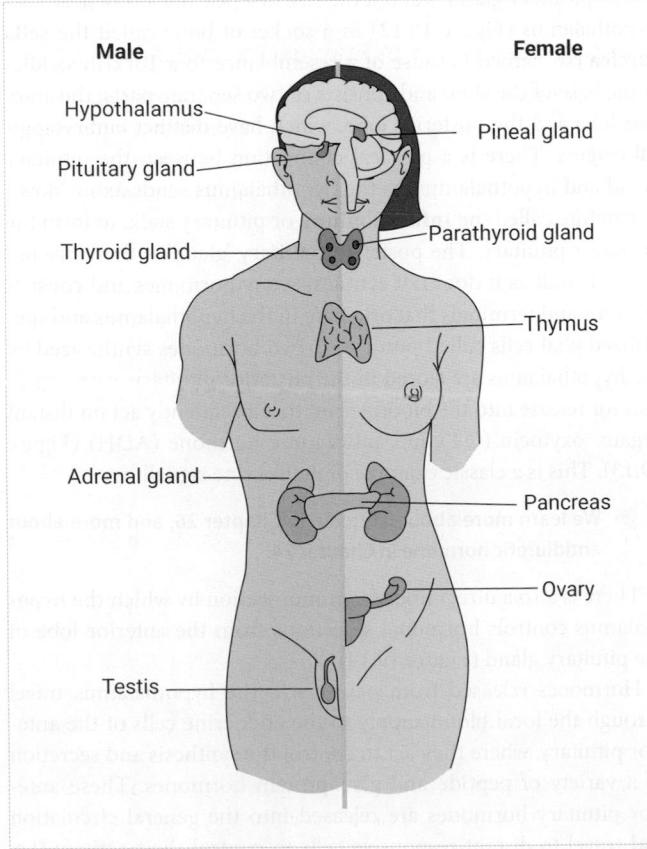

Male   **Female**

Hypothalamus
Pineal gland
Pituitary gland
Parathyroid gland
Thyroid gland
Thymus
Adrenal gland
Pancreas
Ovary
Testis

Figure 19.10 The major endocrine glands.

Source: Jarvis, M. & Okami, P. *Principles of Psychology*, 2020. Oxford University Press.

Endocrine communication via the bloodstream is similar to synaptic transmission in the nervous system in that responses are determined by the type of receptors expressed by target cells. The difference between a neurotransmitter and a hormone therefore lies in the *mode of transport* rather than mode of action at the effector cell.

The term '**endocrine communication**' is used where an endocrine gland or tissue secretes a hormone into the bloodstream to affect distant cells bearing receptors that bind that hormone. However, the endocrine system also includes situations where hormones are synthesized by the cell bodies of specialized neurons and, after transport down the axon of the neuron, are released into the bloodstream to affect distant responsive cells (Figure 19.11). This type of communication is called **neurocrine signalling** and is particularly relevant to the hypothalamic–pituitary axis, the master control centre in the brain for much of the endocrine system. There are also situations where secreted hormones travel only a short distance through interstitial fluid rather than blood and affect neighbouring cells. This type of intra-organ or intra-tissue communication where the hormone does not enter the bloodstream is called **paracrine signalling**. Furthermore, some cells release a hormone which acts back at receptors on the surface of the same cell that produced the hormone. This type of ultra-local communication is called **autocrine signalling**; one example of this is the release of the cytokine interleukin-1 from monocytes to regulate immune responses.

Types of hormone

Hormones can be classified on the basis of either their chemical nature or their solubility in water. There are four different chemical classes of hormones:

- *Peptide/polypeptide hormones*—short or long chains of amino acids (e.g. insulin, glucagon, and growth hormone).
- *Glycoprotein hormones*—large protein molecules, often made up of different subunits, with carbohydrate side chains (e.g. luteinizing hormone (LH), follicle stimulating hormone (FSH) (Chapter 26), and thyroid stimulating hormone (TSH), which are secreted by the anterior pituitary gland).
- *Amino acid derivatives (amines)*—small molecules synthesized from the amino acids tyrosine (e.g. adrenaline and the thyroid hormones), or from tryptophan (e.g. melatonin).
- *Steroid hormones*—these are all derived from cholesterol (e.g. cortisol, aldosterone, testosterone, and oestrogen).

It is also sometimes useful to classify hormones according to their solubility as this gives an indication of their mode of action.

- *Water-soluble hormones*—all the peptide and glycoprotein hormones as well as the catecholamines (adrenaline, noradrenaline,

Figure 19.11 Types of hormonal communication.

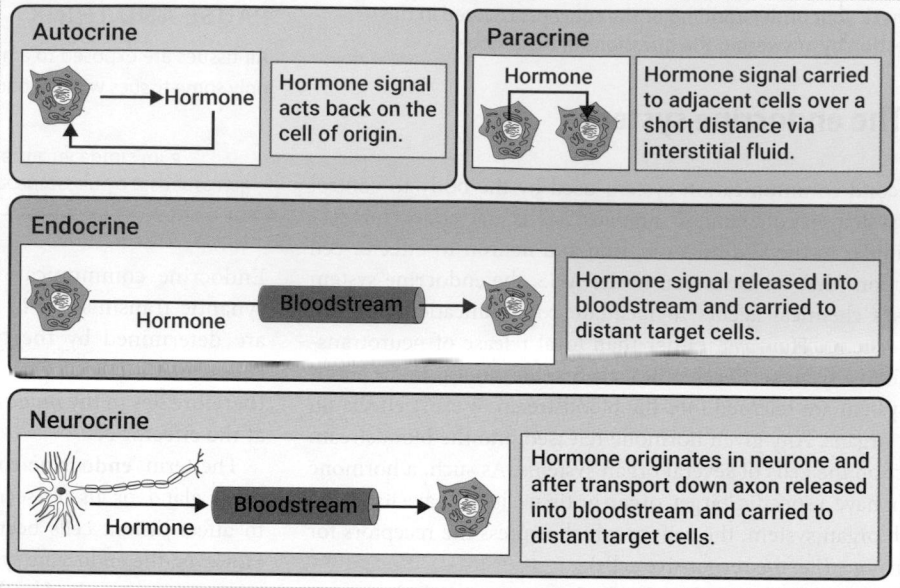

and dopamine) are water-soluble. As these hormones are hydrophilic they are unable to pass across the lipid bilayer of cell membranes. Receptors for this class of hormone must therefore be present at the cell surface and are typically G-protein coupled receptors or tyrosine kinase receptors.

- *Lipid-soluble hormones*—all steroid hormones and the thyroid hormones are lipid-soluble and can therefore pass across the lipid bilayer of cell membranes. Receptors for this class of hormone reside inside the cell in the cytoplasm or nucleus rather than on the cell surface. Binding of hormone to receptor typically allows the hormone–receptor complex to interact with DNA to modify the expression of specific genes that contain hormone response elements in their regulatory region.

Prostaglandins

It is also important to recognize that other signalling agents are released by tissues. An example of such agents is the **prostaglandins**, which is a group of lipid mediators called eicosanoids; these have hormone-like actions and can act via autocrine and paracrine routes. Prostaglandins are synthesized from arachidonic acid via the cyclooxygenase (COX-1 and COX-2) enzymes.

There are different forms of prostaglandins and a variety of receptor types, so the prostaglandins have a wide range of actions that include stimulating both smooth muscle contraction and relaxation, depending on the receptor sub-type. Thus, prostaglandins can significantly affect blood flow and airway flow. They are also involved in mediating allergic responses and pain mechanisms (Section 18.3). In the case of pain, inhibition of prostaglandin synthesis by the COX enzymes is the target of the pain-relieving analgesic non-steroidal anti-inflammatory drugs (NSAIDs).

The hypothalamic–pituitary axis

It has long been known that the pituitary gland secretes several hormones that control the function of other endocrine glands and

is therefore sometimes referred to as the 'master' endocrine gland. The activity of the pituitary gland is regulated by the hypothalamus and, together, the hypothalamus and pituitary gland constitute the major link between the nervous and endocrine systems. Situated, as its name suggests (*hypo* = below), beneath the thalamus in the brain the hypothalamus monitors many aspects of the state of the organ systems by integrating input from a wide range of sensory pathways.

The pituitary gland is about the size of a pea and sits beneath the hypothalamus (Figure 19.12) in a socket of bone called the **sella turcica** (so named because of a resemblance to a Turkish saddle) at the base of the skull and consists of two separate parts; the anterior lobe and the posterior lobe, which have distinct embryological origins. There is a physical connection between the pituitary gland and hypothalamus, as the hypothalamus sends axons down a structure called the **infundibulum,** or pituitary stalk, to form the posterior pituitary. The posterior pituitary 'gland' is therefore not a gland at all as it does not synthesize any hormones and consists of axons and terminals that originate in the hypothalamus and specialized glial cells called pituicytes. Two hormones synthesized by the hypothalamus are stored in the posterior pituitary axon terminals for release into the bloodstream to subsequently act on distant organs: oxytocin (OT) and antidiuretic hormone (ADH) (Figure 19.13). This is a classic example of neurocrine signalling.

▶ We learn more about oxytocin in Chapter 26, and more about antidiuretic hormone in Chapter 24.

There is also a direct line of communication by which the hypothalamus controls hormonal secretions from the anterior lobe of the pituitary gland (Figure 19.14).

Hormones released from neurons in the hypothalamus travel through the local blood supply to the endocrine cells of the anterior pituitary, where they act to control the synthesis and secretion of a variety of peptide and glycoprotein hormones. These anterior pituitary hormones are released into the general circulation and travel to distant responsive cells to control their activity (i.e. classic endocrine function). The anterior pituitary hormones may

Figure 19.12 Location of the hypothalamus and pituitary gland.

Source: Openstax CC BY 4.0 C J. Gordon Betts et al. *Anatomy and Physiology*, 2013. Access for free at: https://openstax.org/books/anatomy-and-physiology/pages/1-introduction Section URL: https://openstax.org/books/anatomy-and-physiology/pages/17-3-the-pituitary-gland-and-hypothalamus.

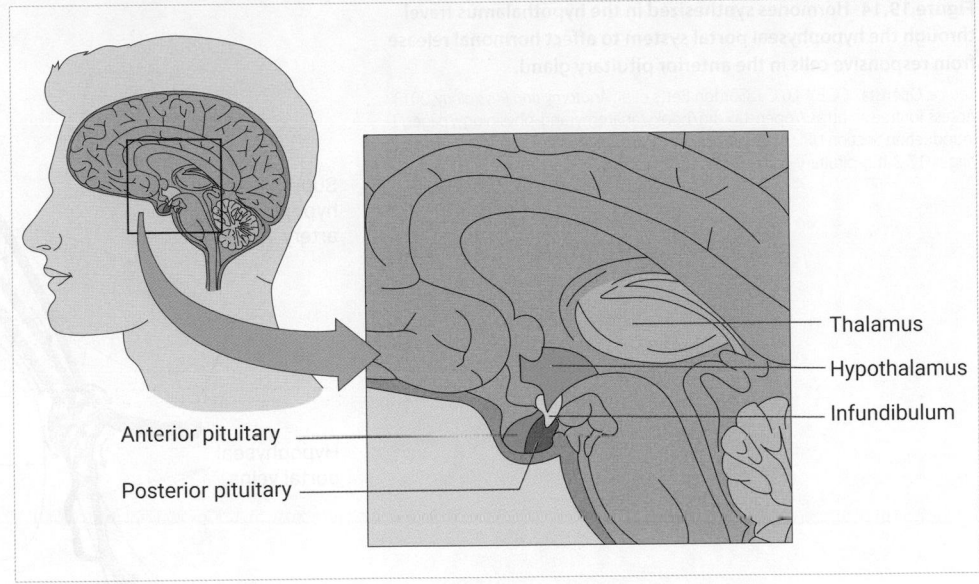

Thalamus
Hypothalamus
Infundibulum

Anterior pituitary

Posterior pituitary

Figure 19.13 Oxytocin (OT) and antidiuretic hormone (ADH) are synthesized by neurons in the hypothalamus and transported down axons before release from the posterior pituitary.

Source: Openstax CC BY 4.0 C J. Gordon Betts et al. *Anatomy and Physiology*, 2013. Access for free at: https://openstax.org/books/anatomy-and-physiology/pages/1-introduction Section URL: https://openstax.org/books/anatomy-and-physiology/pages/17-3-the-pituitary-gland-and-hypothalamus

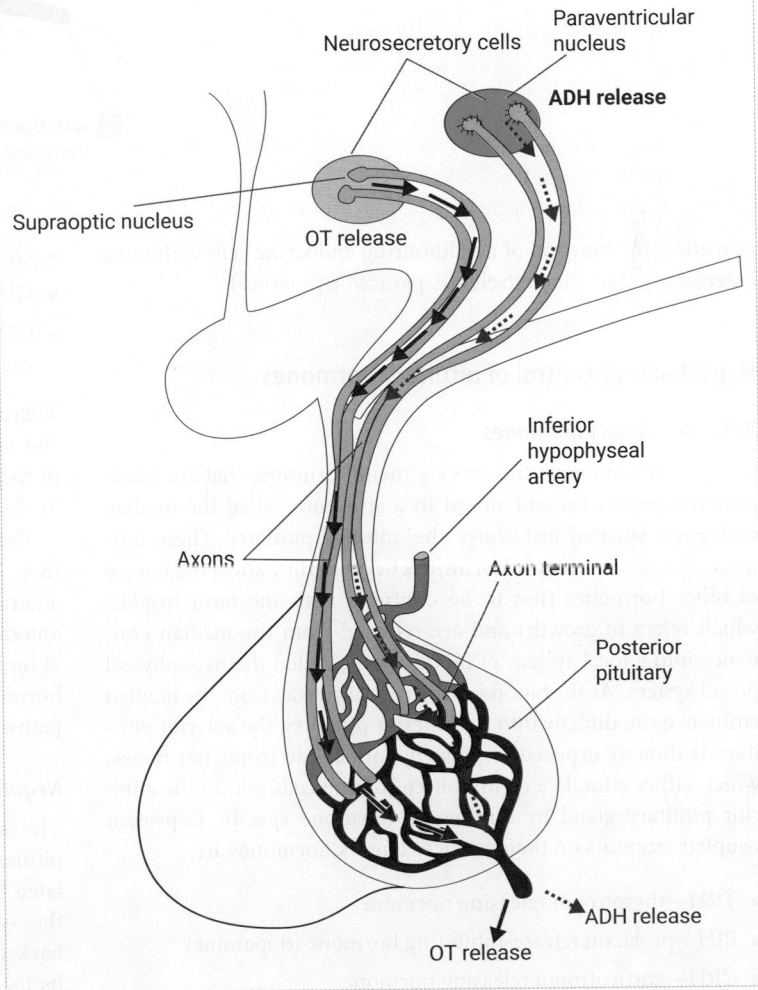

Neurosecretory cells

Paraventricular nucleus

ADH release

Supraoptic nucleus

OT release

Inferior hypophyseal artery

Axons

Axon terminal

Posterior pituitary

ADH release

OT release

Figure 19.14 Hormones synthesized in the hypothalamus travel through the hypophyseal portal system to affect hormonal release from responsive cells in the anterior pituitary gland.

Source: Openstax CC BY 4.0 C J. Gordon Betts et al. *Anatomy and Physiology*, 2013. Access for free at: https://openstax.org/books/anatomy-and-physiology/pages/1-introduction Section URL: https://openstax.org/books/anatomy-and-physiology/pages/17-3-the-pituitary-gland-and-hypothalamus

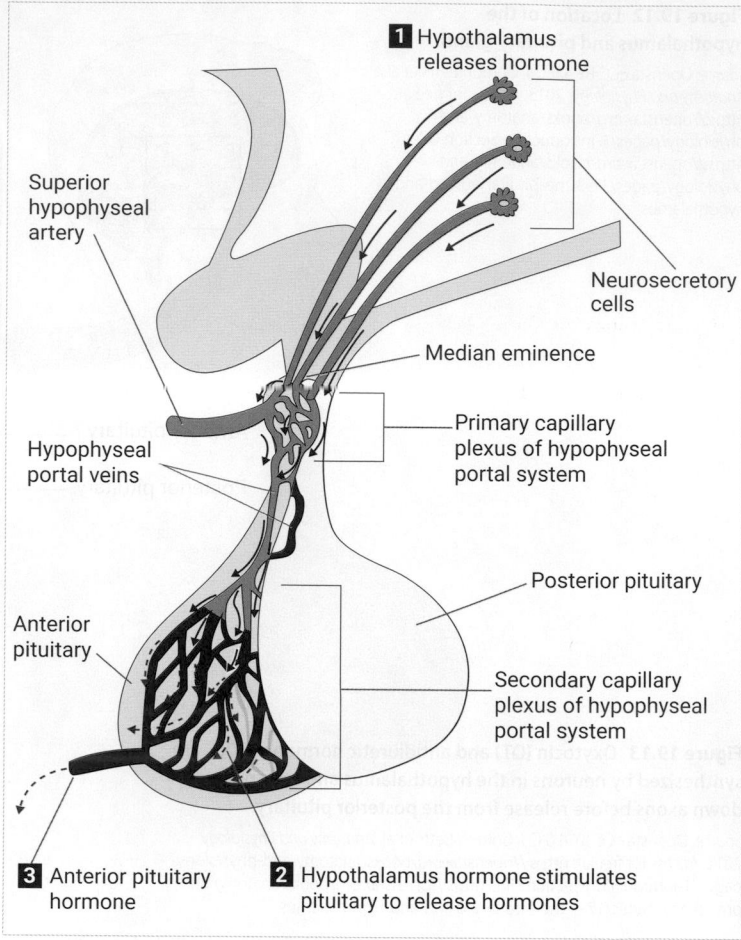

also affect the function of neighbouring endocrine cells within the anterior pituitary gland itself (i.e. paracrine function).

Hypothalamic control of pituitary hormones

Anterior pituitary hormones

The hypothalamus synthesizes six main hormones that are transported down axons and stored in a structure called the median eminence, situated just above the anterior pituitary. These hormones are termed **tropic hormones** because they affect the release of other hormones (not to be confused with the term trop*hic*, which refers to growth) and are released from the median eminence into a local system of blood vessels called the hypophyseal portal system. As the blood vessels running away from the median eminence run directly into the anterior pituitary, the anterior pituitary is directly exposed to these hypothalamic tropic hormones, which either stimulate or inhibit endocrine cells within the anterior pituitary gland by binding to hormone specific G-protein coupled receptors on their surface. The six hormones are:

- TRH—thyrotropin releasing hormone
- PIH—prolactin release-inhibiting hormone (dopamine)
- CRH—corticotropin releasing hormone

- GnRH—gonadotropin releasing hormone
- GHRH—growth hormone releasing hormone
- GHIH—growth hormone inhibitory hormone (also called somatostatin).

There are five types of responsive endocrine cells within the anterior pituitary that produce six major hormones under the control of tropic hormones released from the hypothalamus, as shown in Table 19.2.

The endocrine-responsive cells in the anterior pituitary release their hormonal products into the capillary bed surrounding the anterior pituitary that drains into the systemic circulation, and the anterior pituitary hormone is then transported to distant peripheral responsive tissues to stimulate the production of a final effector hormone, which ultimately facilitates the physiological effect of the pathway (Figure 19.15).

Negative feedback

The secretion of hypothalamic releasing hormones, anterior pituitary hormones, and peripheral effector hormones are regulated by negative feedback loops, which act at different levels in the system (Figure 19.16). In **ultra-short-loop** negative feedback the hypothalamic releasing factor itself (hormone 1) limits its own production in an autocrine/paracrine fashion within

Table 19.2 Endocrine cell types in the anterior pituitary gland

| Cell type | Hypothalamus | | Anterior pituitary | |
	Stimulating hormone	Inhibitory hormone	Hormonal product(s)	Main actions
Thyrotropes	TRH	GHIH	Thyroid stimulating hormone	Stimulates secretion of thyroid hormones from thyroid gland
Lactotropes	TRH (minor +ve control on prolactin)	PIH and GHIH	Prolactin	Stimulates milk production in mammary glands
Corticotropes	CRH	–	Adrenocorticotropic hormone (ACTH)	Stimulates glucocorticoid (mainly cortisol) secretion from adrenal cortex
Somatotropes	GHRH	GHIH	Growth hormone	Stimulates insulin-like growth factor (IGF) production by liver
				Direct growth effects on responsive cells
Gonadotropes	GnRH	–	Luteinizing hormone	Stimulates progesterone and oestrogen production in females and testosterone in males
				Initiates ovulation
	GnRH	–	Follicle stimulating hormone	Stimulates gamete production (eggs in females, sperm in males)

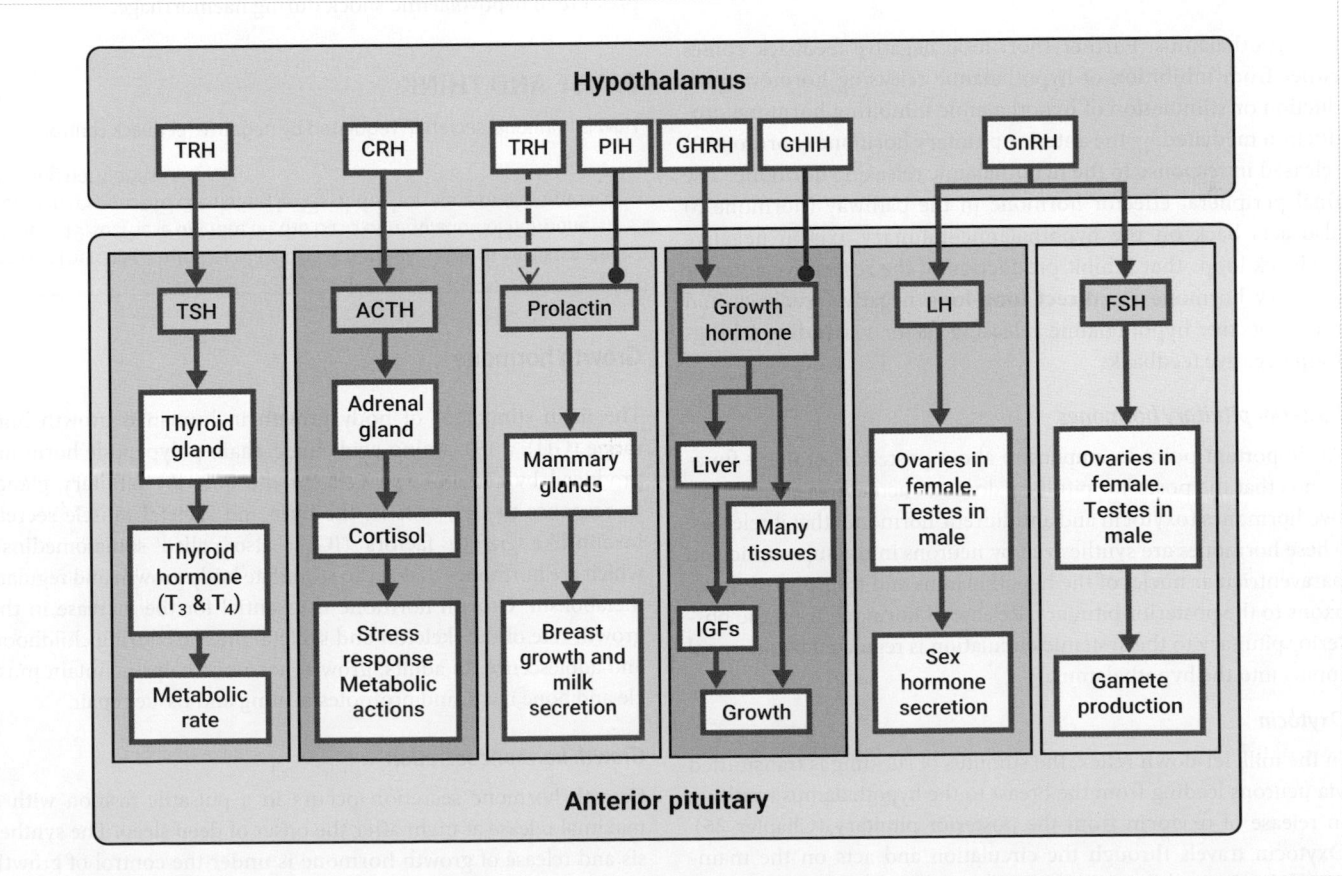

Figure 19.15 Summary of the relationships between the hormonal products of the hypothalamus and the anterior pituitary gland.

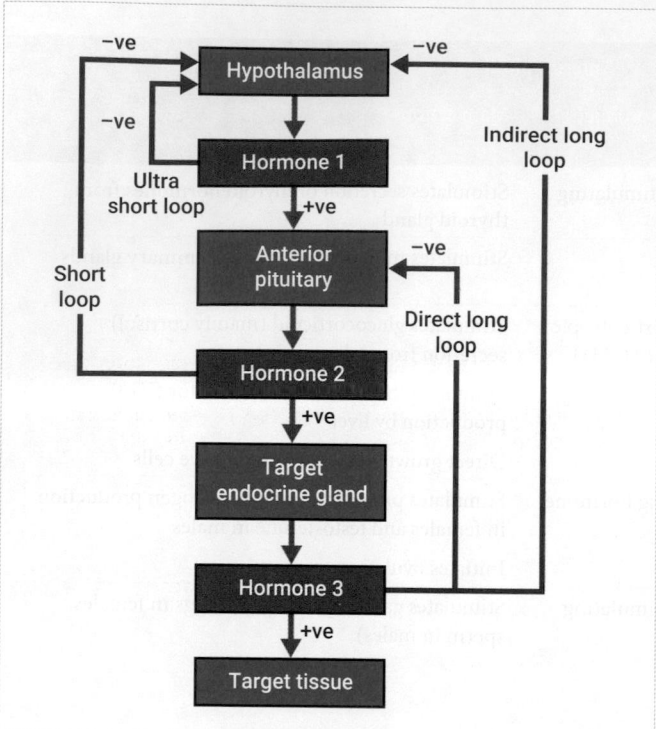

Figure 19.16 Negative feedback pathways in the hypothalamic–pituitary axis.

the hypothalamus. Further short-loop negative feedback comes either from inhibition of hypothalamic releasing hormone production or stimulation of hypothalamic inhibiting hormone production mediated by the anterior pituitary hormone (hormone 2) released in response to the hypothalamic releasing hormone. The final peripheral effector hormone in the pathway (hormone 3) also acts back on the hypothalamic–pituitary axis in negative feedback loops that inhibit production of the respective anterior pituitary hormone via **direct long-loop** negative feedback and the respective hypothalamic releasing factor via **indirect long-loop** negative feedback.

Posterior pituitary hormones

An important point to emphasize about posterior pituitary function is that the posterior pituitary does not actually synthesize the two hormones (oxytocin and antidiuretic hormone) that it releases. These hormones are synthesized by neurons in the supraoptic and paraventricular nuclei of the hypothalamus and transported down axons to the posterior pituitary. Release of hormone from the posterior pituitary to the systemic circulation is regulated by neuronal inputs into the hypothalamus.

Oxytocin

In the milk let-down reflex, the stimulus of suckling is transmitted via neurons leading from the breast to the hypothalamus resulting in release of oxytocin from the posterior pituitary (Chapter 26). Oxytocin travels through the circulation and acts on the mammary glands to cause milk release by contracting the myoepithelial cells surrounding alveoli, thereby squeezing milk into the duct

system. During childbirth the stimulus of pressure on the cervix and uterine wall is again transmitted to the hypothalamus via neuronal input and the release of oxytocin from the posterior pituitary into the general circulation initiates powerful uterine contractions by activation of oxytocin receptors on the uterine smooth muscle cells. Synthetic oxytocin (pitocin) is often administered to increase uterine tone and control bleeding just after birth.

Antidiuretic hormone

ADH, as its name suggests, causes a reduction in urine production (Chapter 24). Receptors for ADH are present on the distal tubular epithelium of the collecting ducts in the kidneys and when activated by ADH facilitate an increase in permeability by inducing translocation of **aquaporin** water channels in the plasma membrane of collecting duct cells, allowing more reabsorption of water and electrolytes back into the blood. Drinking alcohol inhibits ADH release from the posterior pituitary, explaining the increased urination and ultimately dehydration often experienced with drinking alcohol to excess. Osmoreceptors in the hypothalamus detect changes in plasma osmolality and control the amount of ADH released and also the feeling of thirst (Figure 19.17). An alternative name for ADH is vasopressin and this reflects the ability of ADH to also increase peripheral vascular resistance by activating ADH receptors on the smooth muscle cells of blood vessels, causing vasoconstriction and an increase in arterial blood pressure. Vasoconstriction mediated by ADH is particularly important for restoring blood pressure in hypovolaemic shock during haemorrhage.

PAUSE AND THINK

How is hormonal secretion regulated by negative feedback control?

Answer: There are a number of feedback pathways, which are illustrated in Figure 19.16. These operate on the basis of regulation of hormone release in response to circulating levels of the hormone or of hormone releasing hormones.

Growth hormone

The main stimulator of body growth in human is growth hormone (GH), a 191-amino acid single-chain polypeptide hormone produced by somatotrope cells in the anterior pituitary gland. In response to GH, cells in the liver and skeletal muscle secrete insulin-like growth factors (IGFs; also called somatomedins), which are hormones that act to stimulate body growth and regulate metabolism. Growth hormone is essential for the increase in the growth rate of the skeleton and skeletal muscles during childhood and adolescence. In adults, growth hormone helps maintain muscle and bone mass, and promotes healing and tissue repair.

Growth hormone secretion

Growth hormone secretion occurs in a pulsatile fashion with a maximal release at night after the onset of deep sleep. The synthesis and release of growth hormone is under the control of growth hormone releasing hormone (GHRH) and somatostatin released from the hypothalamus. GHRH stimulates the production and

Figure 19.17 Control of plasma osmolality by antidiuretic hormone.

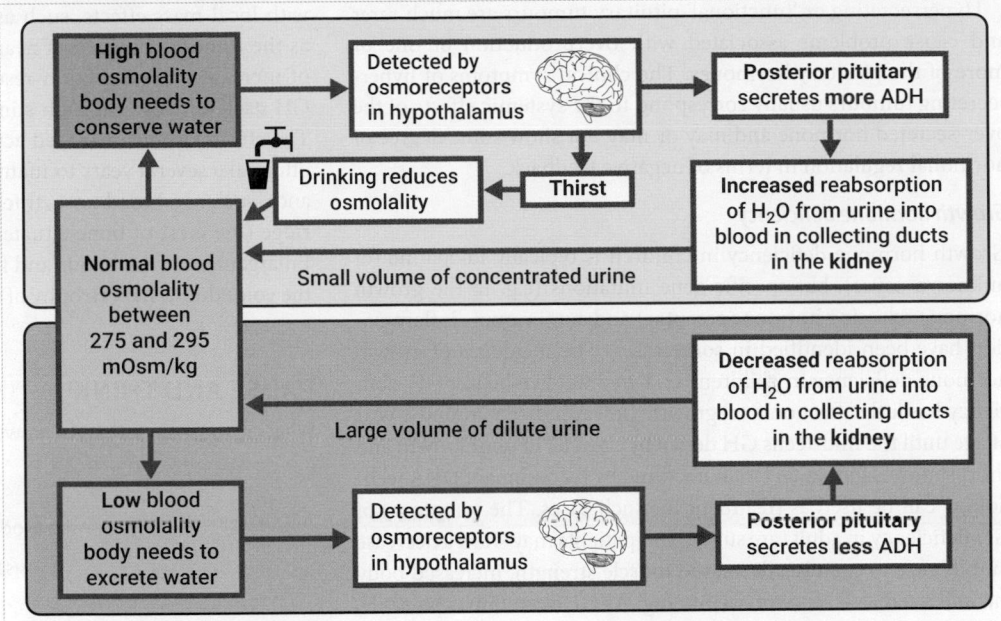

release of GH, while somatostatin inhibits production and release. The secretion of GH is also regulated by plasma glucose and free fatty acid concentrations:

- A decrease in glucose or free fatty acid leads to an increase in GH secretion.
- An increase in glucose or free fatty acid leads to a decrease in GH secretion.
- Fasting increases GH secretion whereas obesity leads to a reduction in GH secretion.

The CNS also regulates GH secretion via inputs into the hypothalamus, effecting GHRH and somatostatin levels:

- There is a surge in GH secretion after the onset of deep sleep.
- Light sleep (rapid eye movement (REM) sleep) inhibits GH secretion.
- Stress (e.g. trauma, surgery, and fever) increases GH secretion.
- Exercise increases GH secretion.

Regulation of GH secretion occurs via long- and short-loop negative feedback. Long-loop negative feedback is mediated by IGFs that inhibit the release of GHRH from the hypothalamus, stimulate somatostatin release from the hypothalamus, and inhibit the action of GHRH in the anterior pituitary. Short-loop negative feedback is mediated by GH itself via the stimulation of somatostatin release from the hypothalamus.

How does GH exert its effects on cells?

GH acts both directly through its own receptor and indirectly by stimulating the production of insulin-like growth factor 1 (IGF-1). The GH receptor is a member of the cytokine receptor superfamily and is coupled to an intracellular enzyme called Janus kinase (JAK). One of the downstream effects of JAK, subsequent to GH binding to its receptor, is activation of a transcription factor that

turns on production of IGFs. There are two forms of IGFs in mammals (IGF-1 and IGF-2) and these are mainly produced in the liver (~75%) and skeletal muscle in response to GH, although many other tissues, such as bone, kidney, and the CNS, also respond to GH by producing IGFs. The IGFs circulate in the blood bound to specific binding proteins that modulate their availability. The IGF receptors present on responsive cells are members of the tyrosine kinase family of receptors and show some similarities to the insulin receptor (also a tyrosine kinase). IGF-1 mediates the majority of the effects of GH in adults, including:

- an increase in cell size (hypertrophy);
- an increase in cell number (hyperplasia);
- an increase in the rate of protein synthesis;
- an increase in the rate of lipolysis in adipose tissue (fat);
- a decrease in glucose uptake.

IGF-2 appears to be more important during growth and development before birth. The actions of both IGFs can be paracrine and autocrine, as well as endocrine. In some tissues IGF-1 inhibits apoptosis (programmed cell death), and some types of tumour express abundant IGF-1 receptors, which inhibit apoptosis.

Pituitary disorders

The most common cause of pituitary malfunction is a benign tumour (adenoma). Such adenomas affect ~70,000 patients in the UK and manifest in either over- or under-secretion of pituitary hormones. Most pituitary tumours are non-functional in that the tumour cells themselves do not produce any hormone. Such non-functioning pituitary tumours can result in inadequate production of one or more of the pituitary hormones due to physical pressure from the growing tumour on glandular tissue. Pressure on surrounding structures in the vicinity of the tumour can also result in headaches, visual problems (by compression of the nearby optic nerve), vomiting, and nausea.

Hypersecreting or 'functional' pituitary tumours are much rarer and cause problems associated with overproduction of one or more of the pituitary hormones. The clinical symptoms of hypersecreting tumours usually correspond to the systemic effects of the over-secreted hormone and may or may not show some degree of hormonal regulation in terms of negative feedback.

Growth hormone deficiency

Growth hormone deficiency in children is typically idiopathic (of unknown cause), but specific gene mutations (e.g. in the growth hormone releasing hormone receptor) and autoimmune inflammation have been identified in some cases. The incidence of growth hormone deficiency in children is ~1 in 3800 live births. GH deficiency has little effect on fetal growth; however, from around 1 year of age until the mid-teens GH deficiency results in poor growth and short stature. Human GH manufactured by recombinant DNA technology can be used as treatment for such cases. The symptoms of GH deficiency in adults are subtle, and patients may show a decrease in tolerance to exercise, decreased muscle strength, increased body fat, and a reduced sense of 'well-being'. As GH secretion is pulsatile, deficiency is often difficult to diagnose and a combination of direct and indirect measurements is required. GH deficiency in adults is usually due to the 'mass effects' of a pituitary adenoma.

Growth hormone excess

Excessive growth hormone secretion in childhood, before fusion of the epiphyseal plates in the long bones, results in the condition of gigantism. Indeed, the world's tallest man ever recorded (Robert Wadlow 1918–1940) suffered from a GH secreting pituitary adenoma. In more recent times, however, gigantism rarely develops due to early diagnosis and treatment.

Excessive growth hormone secretion in adulthood results in the condition of acromegaly. Growth hormone secreting pituitary adenomas are typically large and therefore are also associated with local mass effects, such as headache and visual field defects, as the tumour compresses nearby structures. The systemic effects of increased GH secretion result from both the direct actions of GH itself and also through stimulation of local IGF-1 production. The effects of this increased activation of GH and IGF-1 receptors often take several years to manifest as a change in physical appearance, such as a broad nose, thick lips, and a prominent supraorbital ridge (the crest of bone situated on the frontal bone of the skull). Enlargement of the hands and feet also occur, as does deepening of the voice due to hypertrophy of the soft tissues of the upper airways.

PAUSE AND THINK

What are the main actions of growth hormone in stimulating growth?

Answer:
- Increase in cell size (hypertrophy).
- Increase in cell number (hyperplasia).
- Increase in the rate of protein synthesis.
- Increase in the rate of lipolysis in adipose tissue (fat).
- Decrease in glucose uptake.

The thyroid gland

The thyroid hormones **triiodothyronine** (T3) and **thyroxine** (T4) exert effects on virtually every cell in the body by regulating the transcription of specific genes. Unlike the pituitary and hypothalamic hormones we have looked at previously, the thyroid hormones are lipid-soluble and their receptors reside inside the cell functioning as hormone-regulated transcription factors, as you can see in Figure 19.18 (see also Scientific Process 19.1). Thyroid

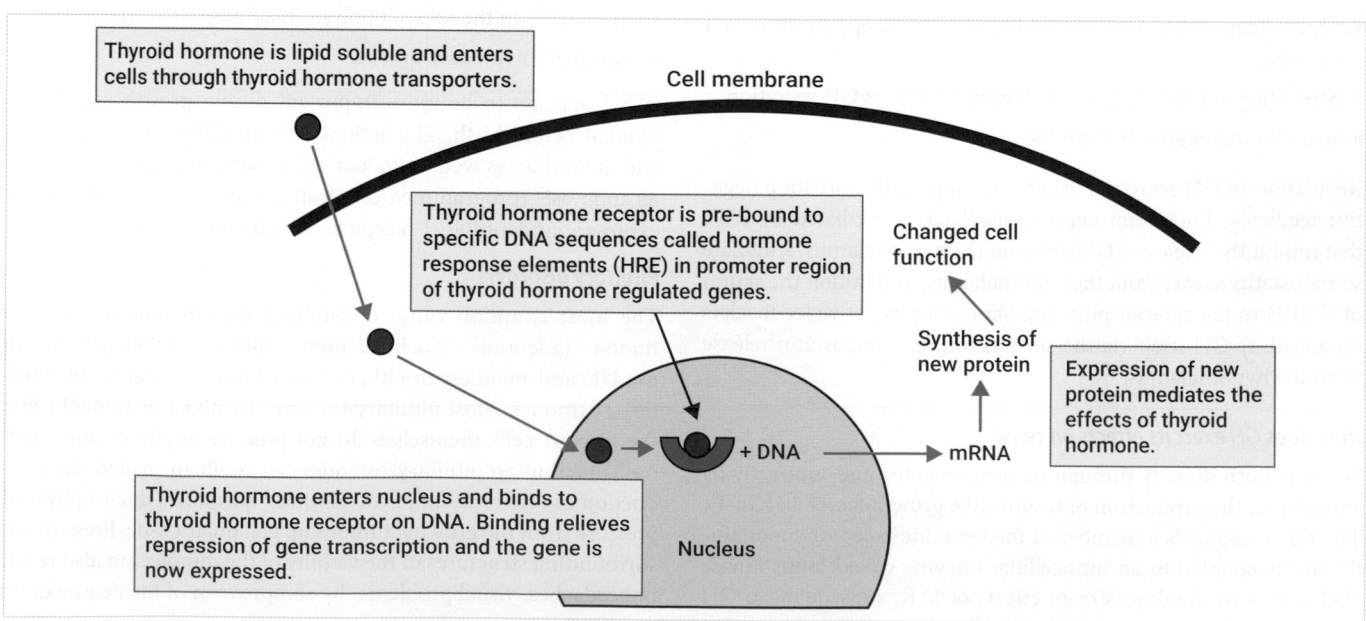

Thyroid hormone is lipid soluble and enters cells through thyroid hormone transporters.

Cell membrane

Thyroid hormone receptor is pre-bound to specific DNA sequences called hormone response elements (HRE) in promoter region of thyroid hormone regulated genes.

Changed cell function

Synthesis of new protein

Expression of new protein mediates the effects of thyroid hormone.

+ DNA → mRNA

Thyroid hormone enters nucleus and binds to thyroid hormone receptor on DNA. Binding relieves repression of gene transcription and the gene is now expressed.

Nucleus

Figure 19.18 Thyroid hormone receptors regulate gene transcription.

SCIENTIFIC PROCESS 19.1 — Thyroid hormone receptors

Research question

By the 1970s, around 80 years after the British doctor George Murray first showed that hypothyroid patients could be treated by injecting sheep thyroid gland extract, it was known that thyroid hormone induces an increase in mRNA synthesis and that high-affinity specific binding sites for T3 existed within the nucleus of some cell types. The research question of how thyroid hormone exerts its effects on cells, however, had still to be answered.

Hypothesis

Given that high-affinity specific binding sites for T3 had previously been detected within the nucleus of some cell types that are known to respond to thyroid hormone such as heart, liver, and kidney, It was hypothesized that these specific binding sites mediate the actions of T3.

Prediction

The prediction from the above hypothesis was that high-affinity specific T3 binding sites would only be found in cell types that respond to thyroid hormone and would be absent in tissues, such as brain, that did not respond to thyroid hormone in terms of an increased oxygen consumption and mitochondrial function.

Methods

To test their hypothesis, Oppenheimer and colleagues looked for high-affinity nuclear T3 binding sites in a range of rat tissues, some of which were believed to be T3 responsive (liver, heart, and kidney) and some of which were not (brain, testes, and spleen). Radioactive T3 (^{125}I-T3) was first injected into rats in order to determine the equilibrium time point for each tissue tested. This point is when the specific activity of nuclear ^{125}I-T3 can be assumed to be the same as that in plasma as the rate of entrance into the nucleus is the same as the rate of exit from the nucleus. The equilibrium time point is measured as the peak time point on a nuclear accumulation curve. Once the equilibrium time point for each tissue was known, a different set of experiments co-injected unlabelled T3 with a fixed amount of ^{125}I-T3 and the animals were sacrificed at the equilibrium time point for each particular tissue of interest. By progressively increasing the concentration of unlabelled T3 in each experiment it was then possible to calculate the nuclear binding capacity of each particular type of tissue.

Results

The surprising result from these experiments was that all tissues were shown to contain at least some nuclear T3 binding sites, even those tissues that were previously thought to be unable to respond to thyroid hormone, such as the brain, testis, and spleen. Furthermore, the apparent number of binding sites varied widely from one tissue type to another (Table 1).

The highest T3 binding capacity of 0.8 ng mg^{-1} DNA was found in the pituitary. Heart tissue had about half this number of binding sites (0.4 ng mg^{-1} DNA), with liver and kidney tissues somewhere in between (0.61 and 0.53 ng mg^{-1} DNA, respectively). A very low binding capacity was observed in testis and spleen tissue, which was consistent with the lack of response of these tissues to T3 by the conventional criteria of enhanced oxygen consumption. However, brain, a tissue that also does not respond to T3 with a measureable increase in oxygen consumption, showed a binding capacity of 0.27 ng mg^{-1} DNA.

Table 1 Nuclear T3 binding characteristics of rat tissue

Tissue	DNA recovery (%)	Total DNA/g tissue (mg)	Nuclear T3* (% total tissue)	Binding capacity ng/mg DNA	Binding capacity Normalized to liver (=1)	Binding capacity ng/g tissue	Binding capacity Normalized to liver (=1)	k_i/k_h**
Liver	59	2.90	12.9	0.61	1.0	1.77	1.00	1.0
Brain	31	1.55	13.5	0.27	0.44	0.42	0.24	0.7
Heart	26	2.01	15.4	0.40	0.65	0.80	0.45	1.0
Spleen	56	17.27	13.0	0.018	0.03	0.31	0.18	1.20
Testis	23	9.56	3.0	0.0023	0.004	0.022	0.01	—
Kidney	42	4.93	9.0	0.53	0.87	2.61	1.47	0.6
Anterior pituitary	84	8.33	52.6	0.79	1.30	6.58	3.72	1.0

* Corrected for DNA losses.

** Ratio of nuclear association constant of given tissue (i) to association constant of liver (h).

Data for testis are not sufficiently precise to allow calculations.

Source: J. H. Oppenheimer, H. L. Schwartz, M. I. Surks. (1974) Tissue differences in the concentration of triiodothyronine nuclear binding sites in the rat: liver, kidney, pituitary, heart, brain, spleen and testis. *Endocrinology*, 95, 897–903. By permission of Oxford University Press.

Conclusion

This study was the first to quantify thyroid hormone binding sites in different tissue types and the finding that T3 binding site number differed between tissue types and did not correspond to the ability of thyroid hormone to increase oxygen demand in that particular tissue laid the grounds for several future studies that established the fundamental mechanism of thyroid hormone receptor action. Firstly, the fact that brain had a relatively high number of binding sites called into question the previous assumption that increased oxygen consumption was an obligate effect of thyroid hormone; we now know that thyroid hormone can affect various cellular pathways independent of oxygen consumption. These studies also demonstrated for the first time that the response to thyroid hormone may differ between different tissues at both the biochemical level, in terms of the amount of thyroid hormone able to bind, and also at the physiological level, in terms of the ultimate response, and this set the stage for the subsequent cloning of the actual thyroid hormone receptors

in the 1980s. Finally, the various tissues investigated all showed that approximately 50 per cent of the binding sites present are occupied at endogenous levels of thyroid hormone and this ultimately led to the idea and discovery that thyroid hormone receptors also have biological activity in the thyroid hormone unbound state.

Read the original works

Tata, J. R. (1963) Inhibition of the biological action of thyroid hormones by actinomycin D and puromycin. *Nature* **197**: 1167–8.

Oppenheimer, J. H., Koerner, D., Schwartz, H. L., & Surks, M. I. (1972) Specific nuclear triiodothyronine binding sites in rat liver and kidney. *J. Clin. Endocrinol. Metab.* **35**: 330–3.

Oppenheimer, J. H., Schwartz, H. L., & Surks, M. I. (1974) Tissue differences in the concentration of triiodothyronine nuclear binding sites in the rat: liver, kidney, pituitary, heart, brain, spleen and testis. *Endocrinology* **9**: 897–903. doi: 10.1210/endo-95-3-897

hormone receptors can bind to DNA in the absence of hormone, typically leading to transcriptional repression, and binding of thyroid hormone to the receptor causes a conformational change in the receptor that changes its function from a transcriptional repressor to a transcriptional activator.

The thyroid hormones are released into the general circulation from the thyroid gland, which is situated in the neck in front of the lower larynx and upper trachea. The gland has a bowtie shape with two lateral lobes joined by a central isthmus and is one of the larger endocrine glands in the body at around 25–30 g in mass. There are two major cell types making up the gland: follicular and parafollicular cells. Follicular cells are arranged in numerous microscopic functional units called follicles, which are spherical sacs lined with follicular cells surrounding a central lumen. The lumen acts as a store for the substance colloid, which is rich in thyroglobulin, the protein scaffold on which the thyroid hormones are made. Parafollicular cells lie within the connective tissue surrounding the follicles and are much fewer in number than the follicular cells. These parafollicular cells (sometimes called C-cells) produce the hormone calcitonin involved in the regulation of calcium homeostasis.

Major physiological actions of thyroid hormones

Thyroid hormones have general effects on the metabolic activity of most tissues and, in general, the response occurs slowly over a period of days to weeks. In most tissues (exceptions include the brain), thyroid hormones increase metabolic rate by increasing the number and size of mitochondria and by activating catabolic pathways for fat, carbohydrate, and protein metabolism. Thyroid hormones are therefore important for normal growth and development due to their general effects on metabolism, but they also have specific effects on certain tissues such as directly affecting bone mineralization and the synthesis of cardiac muscle proteins. The CNS is particularly sensitive to T3 and T4 during development and the absence of thyroid hormones from birth to puberty results in a condition known as congenital hypothyroidism, characterized by mental, as well as physical, retardation. The thyroid hormones also have sympathomimetic effects in that they increase the response of

responsive cells to catecholamines, such as adrenaline, by increasing the number of receptors on the responsive cell.

Thyroid hormone synthesis

The thyroid hormones are small lipid-soluble molecules derived from the amino acid tyrosine with the addition of three atoms of iodine in the case of triiodothyronine (T3) and four atoms of iodine in the case of thyroxine (T4). Most (~90 per cent) of the thyroid hormone secreted by the thyroid gland is in the form of T4, which is subsequently converted to T3 in tissues by removal of the 5′-iodide. This represents an important regulatory mechanism to control the amount of biologically active hormone in cells as T3 has ~10 times the activity of T4. Removal of the 3′-iodide from T4 produces an inactive form of T3 called reverse T3 (rT3), which can block the effect of T3 (Figure 19.19).

The basic steps of thyroid hormone synthesis in the thyroid follicles are as follows (Figure 19.20):

1. Active transport of iodide ions from blood into the follicular cells against a concentration gradient.

2. Synthesis of the glycoprotein thyroglobulin by follicular cells. Thyroglobulin is rich in tyrosine residues, but only a handful of these are used for thyroid hormone synthesis.

3. Secretion (exocytosis) of thyroglobulin into the lumen of the follicle.

4. Oxidation of iodide to iodine. Iodide must first be oxidized to iodine before it can iodinate tyrosine.

5. Iodination of the side chains of tyrosine residues in thyroglobulin to form MIT (mono-iodotyrosine) and DIT (di-iodotyrosine).

6. Coupling of DIT with MIT or DIT with DIT to form T3 and T4, respectively, within the thyroglobulin protein scaffold.

7. Droplets of colloid are 'pinched' into the follicular cell by pinocytosis.

8. Colloid droplets fuse with lysosomes.

9. Lysosomes break down the thyroglobulin by proteolytic cleavage, releasing T3 and T4.

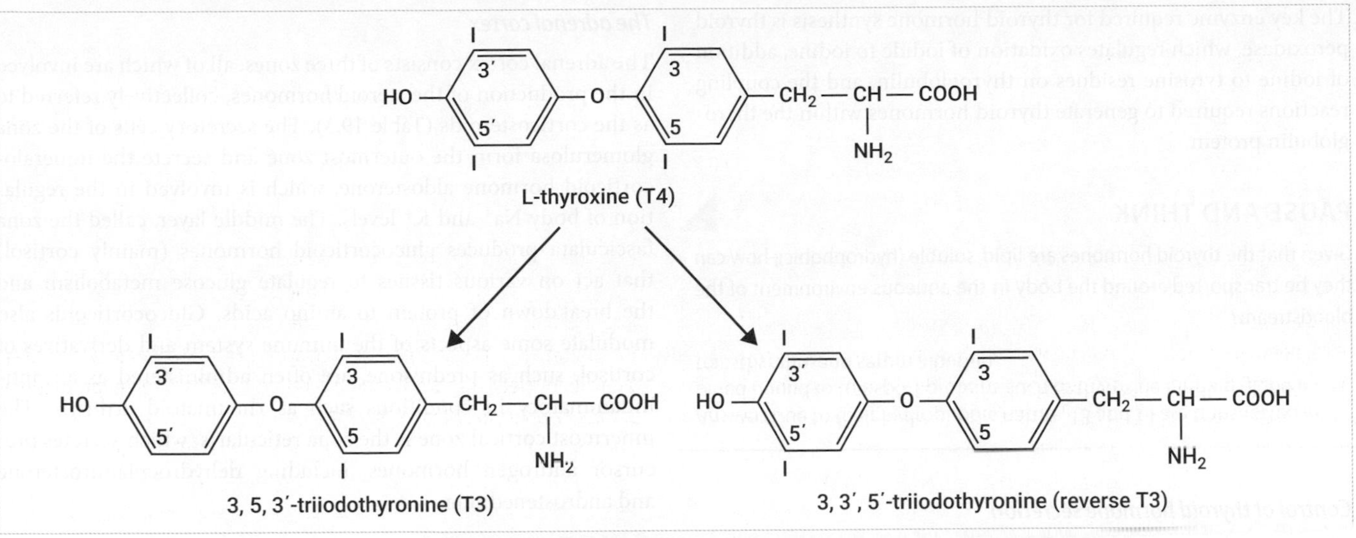

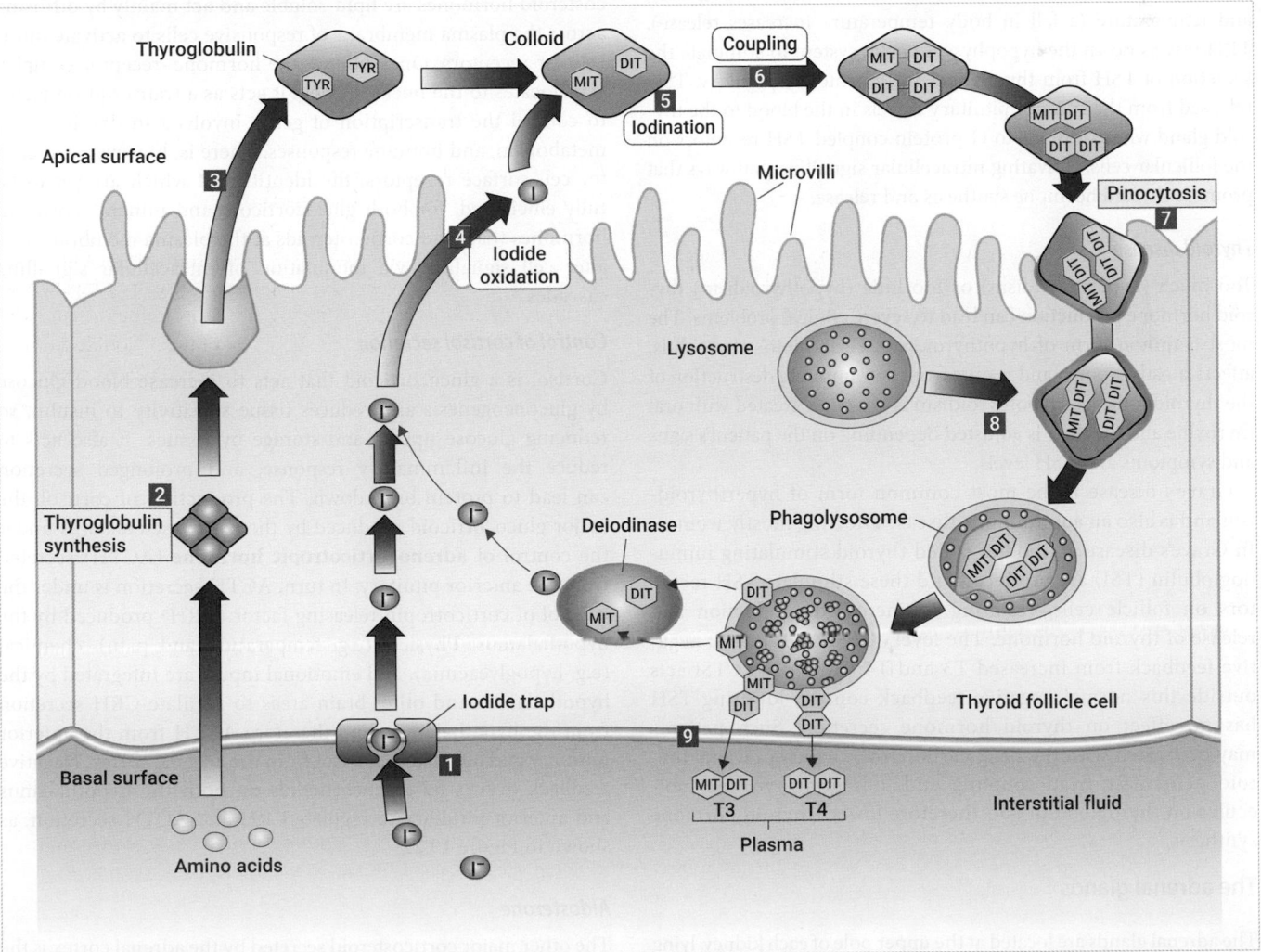

Figure 19.19 Relationship between thyroxine (T4), 3,5,3´-triiodothyronine (T3) and 3,3´,5´-triiodothyronine (reverse T3).

Source: The Global Library of Women's Medicine, Welfare of Women, Global Health Programme.

Figure 19.20 Thyroid hormone synthesis.

Source: R. Arrangoiz et al. Comprehensive review of thyroid embryology, anatomy, histology, and physiology for surgeons. Published by *International Journal of Otolaryngology and Head & Neck Surgery*, 7(4) 2018. CC BY 4.0.

The key enzyme required for thyroid hormone synthesis is thyroid peroxidase, which regulates oxidation of iodide to iodine, addition of iodine to tyrosine residues on thyroglobulin, and the coupling reactions required to generate thyroid hormones within the thyroglobulin protein.

PAUSE AND THINK

Given that the thyroid hormones are lipid-soluble (hydrophobic), how can they be transported around the body in the aqueous environment of the bloodstream?

Answer: Due to their hydrophobic nature, T3 and T4 are transported in blood bound to transport proteins such as thyroxine binding globulin, transthyretin, and serum albumin.

Control of thyroid hormone secretion

The synthesis and secretion of thyroid hormone is under the control of the hypothalamic–pituitary axis. TRH is synthesized and released by the hypothalamus under the negative feedback influence of the circulating levels of T3 and T4, stress (increases release), and temperature (a fall in body temperature increases release). TRH travels down the hypophyseal portal system to stimulate the secretion of TSH from thyrotropes in the anterior pituitary. TSH released from the anterior pituitary travels in the blood to the thyroid gland where it binds to G-protein coupled TSH receptors on the follicular cells, activating intracellular signalling pathways that promote thyroid hormone synthesis and release.

Thyroid disease

Too much (hyperthyroidism) or too little (hypothyroidism) thyroid hormone production can lead to severe clinical problems. The most common form of hypothyroidism, Hashimoto's thyroiditis, affects mostly women and results from autoimmune destruction of the thyroid follicles. Hypothyroidism is generally treated with oral thyroxine and the dose is adjusted depending on the patient's signs and symptoms and TSH levels.

Grave's disease is the most common form of hyperthyroidism and is also an autoimmune disease affecting mostly women. In Grave's disease antibodies called thyroid stimulating immunoglobulin (TSI) are produced and these stimulate TSH receptors on follicle cells, resulting in increased production and release of thyroid hormone. The level of TSH falls due to negative feedback from increased T3 and T4. However, as TSI acts outside this normal negative feedback control, lowering TSH has no effect on thyroid hormone secretion. Such patients may be treated with the drug carbimazole, which prevents thyroid peroxidase from coupling and iodinating tyrosine molecules on thyroglobulin and therefore lowers thyroid hormone synthesis.

The adrenal glands

The adrenal glands are located at the upper pole of each kidney, lying against the diaphragm. The two glands have a combined weight of ~10 g and each consists of two functional regions, an outer cortex and an inner medulla, shown in more detail in Figure 19.21.

The adrenal cortex

The adrenal cortex consists of three zones, all of which are involved in the production of the steroid hormones, collectively referred to as the corticosteroids (Table 19.3). The secretory cells of the zona glomerulosa form the outermost zone and secrete the mineralocorticoid hormone aldosterone, which is involved in the regulation of body Na^+ and K^+ levels. The middle layer, called the zona fasciculata produces glucocorticoid hormones (mainly cortisol) that act on various tissues to regulate glucose metabolism and the breakdown of protein to amino acids. Glucocorticoids also modulate some aspects of the immune system and derivatives of cortisol, such as prednisone, are often administered as an anti-inflammatory in conditions such as rheumatoid arthritis. The innermost cortical zone is the zona reticularis, which secretes precursor androgen hormones, including dehydroepiandrosterone and androstenedione.

Corticosteroid synthesis and mode of action

All the steroid hormones are synthesized from cholesterol via progesterone in a series of enzyme-catalysed reactions. The corticosteroid hormones are lipid-soluble and act mainly by diffusing across the plasma membrane of responsive cells to activate intracellular receptors. Once bound, the hormone–receptor complex translocates to the nucleus where it acts as a transcription factor to control the transcription of genes involved in development, metabolism, and immune responses. There is, however, evidence for cell surface receptors, the identities of which are yet to be fully elucidated, for both glucocorticoid and mineralocorticoid hormones that bind corticosteroids at the plasma membrane and alter cell signalling via modulation of intracellular signalling cascades.

Control of cortisol secretion

Cortisol is a glucocorticoid that acts to increase blood glucose by gluconeogenesis and reduces tissue sensitivity to insulin, so reducing glucose uptake and storage by tissues. It also acts to reduce the inflammatory response, and prolonged secretion can lead to protein breakdown. The production of cortisol, the major glucocorticoid produced by the zona fasciculata, is under the control of **adrenocorticotropic hormone** (ACTH) secreted from the anterior pituitary. In turn, ACTH secretion is under the control of corticotropin releasing factor (CRH) produced by the hypothalamus. Physical (e.g. temperature and pain), chemical (e.g. hypoglycaemia), and emotional inputs are integrated by the hypothalamus and other brain areas to regulate CRH secretion from the hypothalamus, and therefore ACTH from the anterior pituitary and ultimately cortisol from the adrenal cortex. Negative feedback occurs by glucocorticoids on both the hypothalamus and anterior pituitary to regulate CRH and ACTH secretion, as shown in Figure 19.22.

Aldosterone

The other major corticosteroid secreted by the adrenal cortex is the mineralocorticoid aldosterone, which is involved in the regulation of mineral balance. Aldosterone stimulates Na^+ reabsorption from urine in the kidney in exchange for K^+ or H^+ ions. Over-secretion

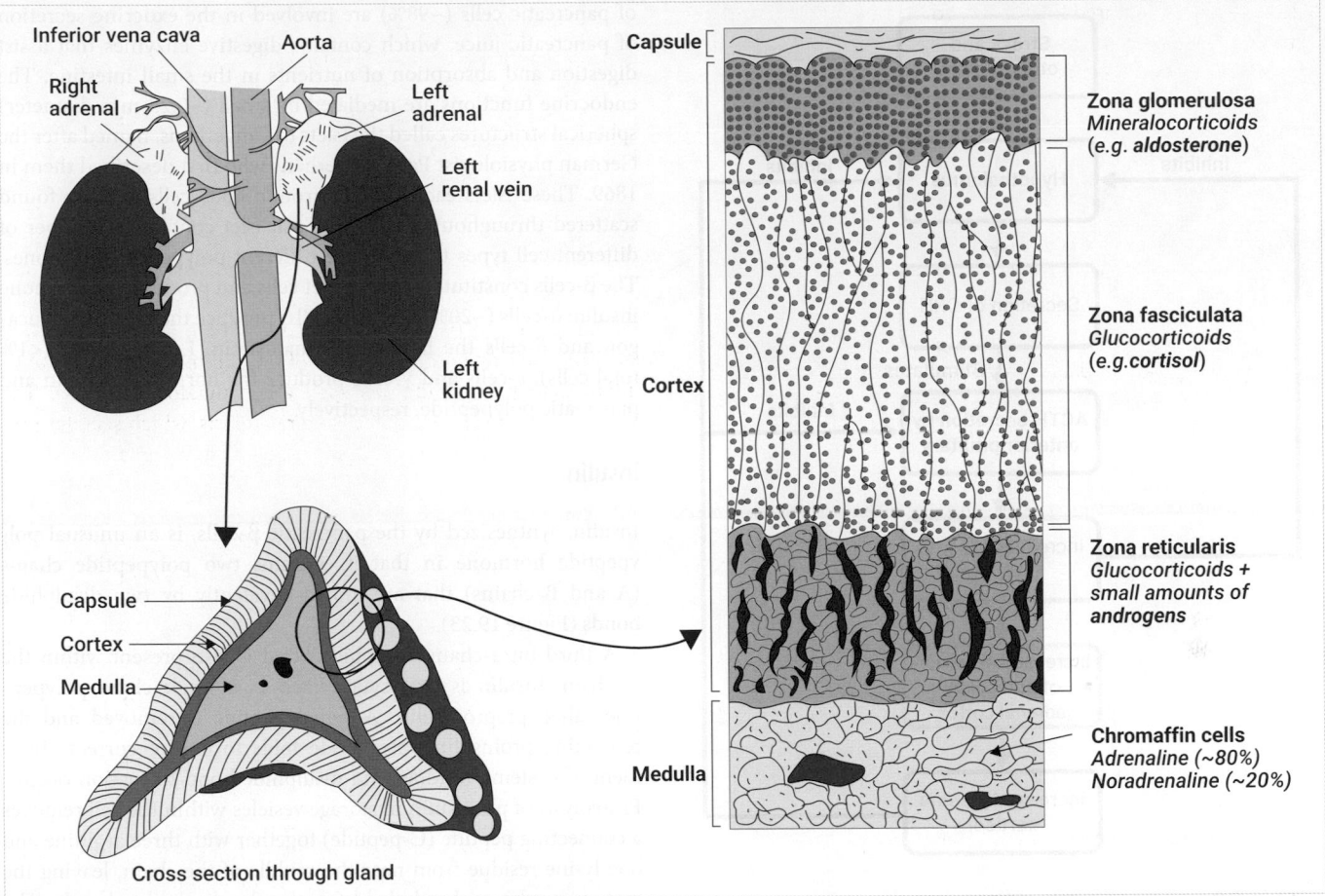

Figure 19.21 Anatomical location and layers of the adrenal glands.

Source: Pocock et al. *Human Physiology*, 5th edn, 2017. Oxford University Press.

Table 19.3 Comparison of the corticosteroid and catecholamine hormones of the adrenal gland

	Corticosteroids	Catecholamines
Synthesized in	Cortex	Medulla
Hormones	Cortisol, aldosterone, androgens	Adrenaline, noradrenaline
Derived from	Cholesterol	Tyrosine
Mode	Endocrine	Neurocrine
Storage	Synthesized and released	Stored in vesicles before release
Solubility	Lipid-soluble	Water-soluble
Receptors	Nuclear receptors	GPCRs (α and β adrenergic receptors)
Typical effect on enzymes	Regulates amount by gene expression	Regulates activity of existing enzymes
Speed of response	Slow (several minutes to hours)	Fast (seconds)

of aldosterone therefore increases Na⁺ and water retention in blood and a loss of K⁺ ions resulting in hypertension and muscle weakness. Too little aldosterone secretion results in the opposite, causing hypotension. The production and secretion of aldosterone from the adrenal cortex is primarily regulated by the hormone angiotensin II as part of the renin–angiotensin aldosterone system (Chapter 24).

The adrenal medulla

In functional terms, the adrenal medulla is essentially a modified ganglion of the ANS and receives neuronal input from sympathetic preganglionic neurons. The adrenal medulla synthesizes the catecholamine hormones adrenaline and noradrenaline and releases these into the circulation under the control of the ANS. Adrenaline is released in response to stressful situations and acts to increase

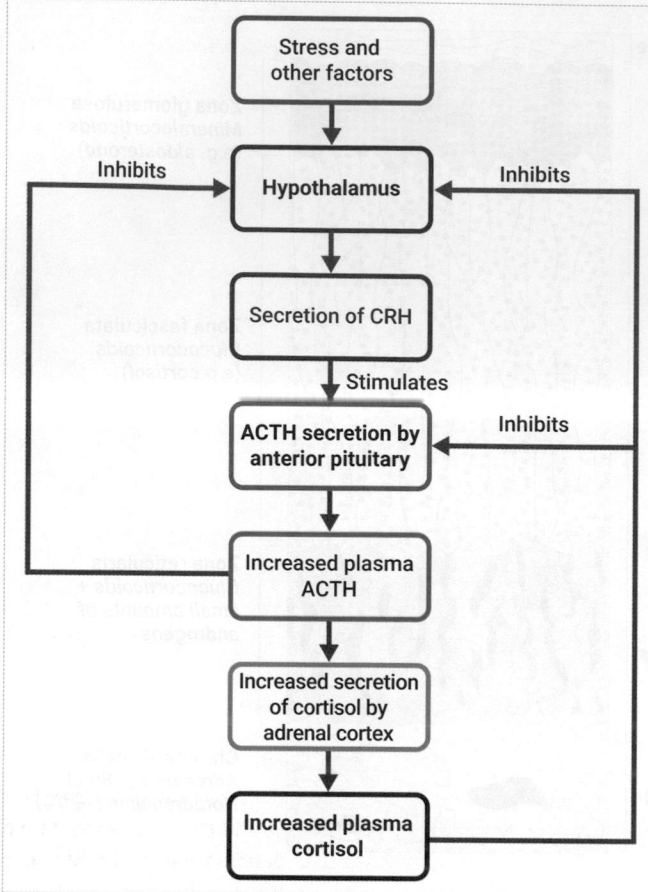

Figure 19.22 Negative feedback pathways in the hypothalamic–pituitary–adrenal axis.

heart rate, increase blood pressure, mobilize glycogen stores from liver and muscle to increase plasma glucose, reduce blood flow to the skin and digestive tract, and to dilate bronchioles in the lung. Overproduction of adrenaline by the adrenal medulla, usually due to a type of tumour called phaeochromocytoma, may be associated with symptoms similar to a 'panic-attack', such as hypertension, anxiety, palpitations, pallor, and sweating.

PAUSE AND THINK

What is the relationship between the adrenal medulla and sympathetic nervous system?

Answer: The adrenal medulla receives neural input from the sympathetic nervous system and secretes adrenaline and noradrenaline as hormones. These then act in the same way as the sympathetic nervous system, for example increasing cardiac output and mobilizing glucose stores while also reducing blood flow to the skin and digestive system.

The endocrine pancreas

The pancreas, located in the abdominal cavity adjacent to the stomach, has both endocrine and exocrine functions. The vast majority

of pancreatic cells (~98%) are involved in the exocrine secretion of pancreatic juice, which contains digestive enzymes that assist digestion and absorption of nutrients in the small intestine. The endocrine functions are mediated by small (~0.25 mm diameter) spherical structures called the Islets of Langerhans, named after the German physiologist Paul Langerhans who first described them in 1869. These islets each contain around 6000 cells and are found scattered throughout the gland. Each islet contains a number of different cell types that produce different polypeptide hormones. The β-cells constitute ~75% of islet cells and produce the hormone insulin: α-cells (~20% of all islet cells) produce the hormone glucagon and δ-cells the hormone somatostatin. Less frequent (<1% total cells), ε-cells and γ-cells produce the hormones ghrelin and pancreatic polypeptide, respectively.

Insulin

Insulin, synthesized by the pancreatic β-cells, is an unusual polypeptide hormone in that it contains two polypeptide chains (A and B chains) that are linked covalently by two disulphide bonds (Figure 19.23).

A third intra-chain disulphide bond is also present within the A chain. Insulin is first synthesized as a single-chain polypeptide called preproinsulin. A signal peptide is removed and the remaining proinsulin polypeptide folds to ensure correct alignment of cysteine residues and disulphide bond formation occurs. Proteolysis of proinsulin in storage vesicles within the cell removes a connecting peptide (C-peptide) together with three arginine and one lysine residue from near the middle of the chain, leaving the mature insulin molecule held together by disulphide bonds. The storage vesicles contain both insulin and C-peptide in equimolar amounts and C-peptide is therefore released into the bloodstream together with insulin. As C-peptide has a longer half-life than insulin, plasma C-peptide concentration is a useful clinical marker of endogenous insulin release. However, recent evidence also suggests that rather than being an inert by-product, C-peptide itself may also act as a hormone with actions in addition to those of insulin.

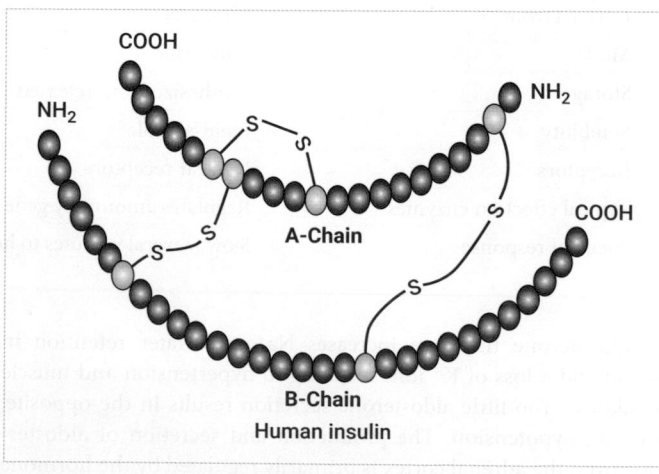

Figure 19.23 Human insulin.

Insulin-responsive tissues and action

The major responsive tissues for insulin are the liver, skeletal muscle, and adipose tissue. The insulin receptor is a member of the tyrosine kinase family of receptors, although it is slightly unusual in that ligand binding does not induce receptor dimerization as the receptor already pre-exists in a dimerized form prior to ligand binding. Binding of insulin to the receptor activates intracellular signalling pathways that regulate carbohydrate, amino acid, and lipid metabolism. A major function of insulin is to clear absorbed glucose from blood following a meal. This effect occurs rapidly as it involves changes in the activities of pre-existing functional proteins, such as enzymes and transport molecules in responsive tissues. In addition, insulin also has long-term effects on cell growth and division by stimulating the synthesis of new protein and DNA replication.

The major metabolic actions of insulin are to (Figure 19.24):

- increase the transport of blood glucose into adipose and skeletal muscle cells;
- increase the synthesis of glycogen form glucose in skeletal muscle and liver;
- decrease the breakdown of glycogen to glucose in skeletal muscle and liver;

- decrease the formation of new glucose by gluconeogenesis in the liver;
- increase the utilization of glucose as an energy source by glycolysis in liver and adipose tissue;
- decrease the mobilization of fat stores (lipolysis) in adipose tissue;
- increase the formation of fat (lipogenesis) in liver and adipose tissue;
- increase amino acid uptake and protein synthesis in liver, muscle, and adipose tissue.

Control of insulin secretion

As insulin is very efficient in lowering blood glucose concentration its secretion must be tightly controlled to ensure that the plasma glucose concentration stays within the normal physiological range (see Clinical Box 19.2). Maintaining plasma glucose concentration at around 5 mM is particularly important for tissues that use glucose as an obligate energy source, such as red blood cells and the brain. Insulin secretion is therefore controlled by a number of factors. An increase in the plasma concentration of metabolites, such as glucose, amino acids, and fatty acids, all act to increase insulin

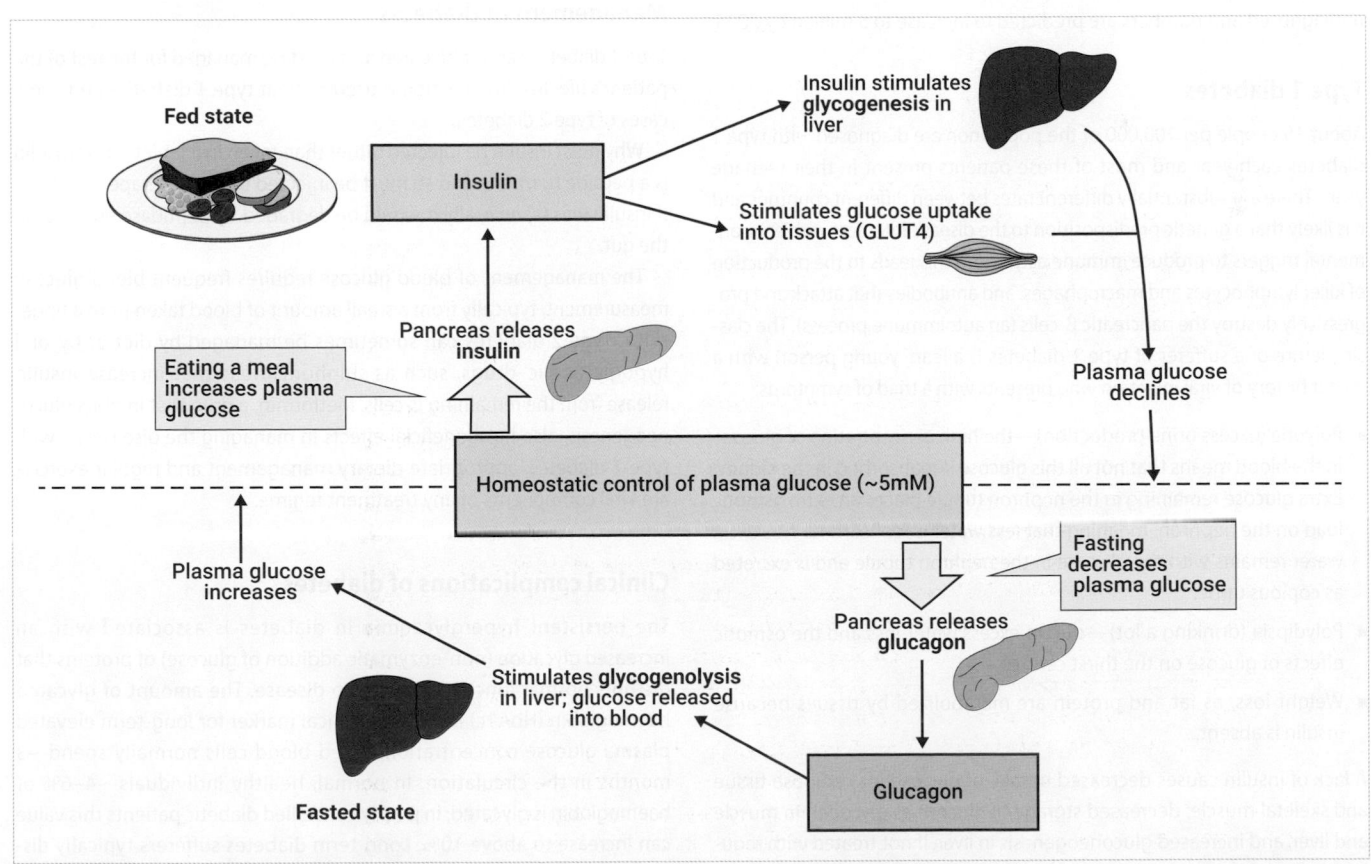

Figure 19.24 Plasma glucose homeostasis.

Source: Ryzhkov Photography/ Shutterstock.com.

CLINICAL BOX 19.2   Diabetes mellitus

Diabetes mellitus is a metabolic disease characterized by chronic hyper-glycaemia (elevated blood glucose concentration). The disease can be due to deficiency in the hormone insulin, resistance to the actions of insulin, or both. There are two major types of the disease.

Type 1 diabetes

- Most common in the young.
- Characterized by the progressive loss of all or most of the pancreatic β-cells.
- Rapidly fatal if not treated.
- Must be treated with insulin.

Type 2 diabetes

- Affects a large number of usually older individuals.
- Characterized by the slow progressive loss of β-cells along with disorders of insulin secretion and tissue resistance to insulin.
- May be present for a long time before diagnosis.
- May not initially need treatment with insulin, but sufferers usually progress to a state where they eventually do.

Approximately 3.9 million people in the UK have diabetes, the majority (~90%) with type 2 disease. A further 0.59 million are believed to be undiagnosed and numbers are predicted to increase to 5 million by 2025.

Type 1 diabetes

About 15 people per 100,000 of the population are diagnosed with type 1 diabetes each year and most of these patients present in their teenage years. There are substantially different rates between different countries and it is likely that a genetic predisposition to the disease interacts with environmental triggers to produce immune activation. This leads to the production of killer lymphocytes and macrophages, and antibodies that attack and progressively destroy the pancreatic β-cells (an autoimmune process). The classic picture of a sufferer of type 1 diabetes is a lean, young person with a recent history of viral infection who presents with a triad of symptoms:

- Polyuria (excess urine production)—the high concentration of glucose in the blood means that not all this glucose is reabsorbed in the kidney. Extra glucose remaining in the nephron tubule places an extra osmotic load on the nephron, meaning that less water is reabsorbed. This extra water remains with the glucose in the nephron tubule and is excreted as copious urine.
- Polydipsia (drinking a lot)—due to excess water loss and the osmotic effects of glucose on the thirst centre.
- Weight loss, as fat and protein are metabolized by tissues because insulin is absent.

A lack of insulin causes decreased uptake of glucose into adipose tissue and skeletal muscle, decreased storage of glucose as glycogen in muscle and liver, and increased gluconeogenesis in liver. If not treated with regular insulin injection, these individuals will progress to a life-threatening crisis called diabetic ketoacidosis. The high rate of fat oxidation in the liver

coupled with a low insulin/glucagon ratio leads to the production of large amounts of ketone bodies producing a metabolic acidosis—ketoacidosis. The features of ketoacidosis are prostration, hyperventilation, nausea, vomiting, dehydration, and abdominal pain.

Type 2 diabetes

Type 2 diabetes is relatively common in populations with an affluent lifestyle. Typically, patients are older and often overweight, and the disease has often been present for some time before diagnosis. While there is good evidence for a genetic predisposition to type 2 diabetes, there is also recent evolving evidence of the involvement of the immune system. At diagnosis, patients typically retain ~50% of their β-cells and their plasma glucose is raised due to an insensitivity of cells to the actions of insulin. However, as the number of β-cells falls, as the disease progresses, a lack of insulin in addition to insulin resistance also contributes to the raised blood glucose. Patients with type 2 disease may also present with the classic triad of symptoms but are more likely to show a variety of problems, such as lack of energy, persistent infections, slow healing minor skin damage, and visual problems.

Management of diabetes

Type 1 diabetes cannot be cured and must be managed for the rest of the patient's life. Insulin injection is used to treat type 1 diabetics and some cases of type 2 diabetes.

Why must insulin be injected rather than taken in a tablet form? Insulin is a peptide hormone and so must be injected to have therapeutic effect. If insulin was taken orally it would be degraded by peptidase enzymes in the gut.

The management of blood glucose requires frequent blood glucose measurement, typically from a small amount of blood taken from a finger prick. Type 2 diabetes can sometimes be managed by diet or by 'oral hypoglycaemic' drugs, such as sulphonylureas that increase insulin release from the remaining β-cells. Metformin, a drug that inhibits gluconeogenesis, also has beneficial effects in managing the disease. As with type 1 diabetes, appropriate dietary management and regular exercise are vital components of any treatment regime.

Clinical complications of diabetes

The persistent hyperglycaemia in diabetes is associated with an increased glycation (non-enzymatic addition of glucose) of proteins that disrupts normal function leading to disease. The amount of glycated haemoglobin (HbA1c) is a useful clinical marker for long-term elevated plasma glucose concentration as red blood cells normally spend ~3 months in the circulation. In normal, healthy individuals ~4–6% of haemoglobin is glycated; in poorly controlled diabetic patients this value can increase to above 10%. Long-term diabetes sufferers typically display a number of vascular complications linked to the non-specific glycation of proteins, including increased risk of stroke, myocardial infarction, and poor circulation to the periphery (particularly the feet). Specific problems include:

- Diabetic eye disease: visual problems arise from changes in the lens due to the osmotic effects of glucose, but the most important problem is diabetic retinopathy—damage to blood vessels in the retina, which can lead to blindness.

- Diabetic kidney disease (nephropathy): the kidney is affected by damage to the glomeruli, poor blood supply because of changes in kidney

blood vessels, or damage from infections of the urinary tract, which is more common in people with diabetes.

- Diabetic neuropathy: diabetes damages peripheral nerves in a number of ways, producing a variety of effects that include changes to, or a loss of, sensation, and changes due to alteration in the function of the ANS.

- Diabetic feet: poor blood supply, damage to nerves, and increased risk of infection all conspire to make the feet of diabetic people particularly vulnerable and loss of the feet through gangrene is not uncommon.

secretion as do the hormones glucagon, gastrin, secretin, and glucose-dependent insulinotropic peptide (GIP). Conversely, a decrease in glucose, amino acids, or free fatty acids will also decrease insulin secretion. Acetylcholine released from parasympathetic nerve terminals in the pancreas stimulates insulin production, whereas adrenaline and noradrenaline from sympathetic innervation and circulatory adrenaline decrease insulin secretion.

Glucagon

Glucagon is a single-chain polypeptide hormone lacking any disulphide bonds and is synthesized by the pancreatic α-cells. Glucagon also has a larger precursor molecule (preproglucagon) that undergoes post-translational processing to produce the biologically active hormone.

The major actions of glucagon are to:

- increase the breakdown of glycogen to glucose (glycogenolysis) in liver;

- decrease the synthesis of glycogen (glycogenesis) in liver;

- increase the synthesis of glucose (gluconeogenesis) in liver;

- increase the breakdown of fat (lipolysis) in adipose tissue.

As listed above, the major responsive tissues for glucagon are the liver and adipose tissue. Skeletal muscle cells lack glucagon receptors as the glycogen stores located here are a local reserve of energy for muscle contraction rather than a source of glucose for other tissues. The glucagon receptor is a $G_{\alpha s}$ coupled G-protein coupled receptor and therefore activates the enzyme adenylyl cyclase to increase the second messenger cAMP, thereby activating protein kinase A (PKA). Phosphorylation of responsive enzymes by PKA alters their activity to mediate the effects of glucagon on glucose metabolism. A decrease in the blood glucose concentration is the major stimulus for the pancreas to release glucagon, whereas secretion is inhibited by insulin and an increase in blood glucose concentration (Figure 19.24).

Endocrine control of calcium and phosphate homeostasis

Calcium ions (Ca^{2+}) play a central role in many cellular processes, including neuromuscular excitability, synaptic transmission, intracellular signalling, bone formation, and the activation

and inactivation of many enzymes. It is not surprising therefore that plasma Ca^{2+} concentration is one of the most tightly controlled variables in the body, with levels kept within a narrow range between 1.0 and 1.3 mM. The consequences of plasma calcium levels straying outside this range are significant. Too little calcium (hypocalcaemia) results in hyper-excitability in the nervous system, leading ultimately to paralysis and even convulsions, whereas chronic hypercalcaemia (too much calcium) may result in kidney stones, constipation, dehydration, tiredness, and depression. The major store of calcium in the body exists as hydroxyapatite crystals ($Ca_{10}(PO_4)_6(OH)_2$) in bone. Phosphate (HPO_4^{2-}) homeostasis is therefore intimately linked with calcium homeostasis: first, because the mobilization of calcium from bone also involves the release of phosphate, and, second, because calcium and phosphate levels are regulated by the same hormones. Unlike calcium, however, the plasma phosphate concentration is not tightly regulated and its level fluctuates throughout the day, particularly after meals.

Parathyroid hormone

The parathyroid glands are located in the neck behind the thyroid grand. Humans usually have four parathyroid glands, each about size of a pea. Although they are in close physical association and both part of the endocrine system, the thyroid and parathyroid glands are otherwise unrelated. Chief cells (sometimes called principal cells) within the parathyroid glands produce the peptide hormone parathyroid hormone (PTH), which is the major acute regulator of calcium and phosphate ions in the blood. The overall effect of PTH is to increase plasma calcium level and decrease plasma phosphate by acting on three major organs (Figure 19.25).

- Bones: PTH increases the number and activity of osteoclasts cells whose role is to break down bone matrix and release Ca^{2+} and phosphate into the blood.

- Kidneys: PTH slows the rate at which Ca^{2+} is lost from blood into urine and increases the loss of phosphate from blood in urine. As more phosphate is lost from the urine than is gained from the bones, there is a net loss of phosphate. Also, PTH promotes the formation of the hormone calcitriol (the active form of vitamin D) in the kidney.

- Gastrointestinal tract: The effects of PTH on the gastrointestinal tract are indirect by promoting the formation of calcitriol by the kidneys. Calcitriol increases the rate of Ca^{2+} and phosphate absorption from food in the gastrointestinal tract.

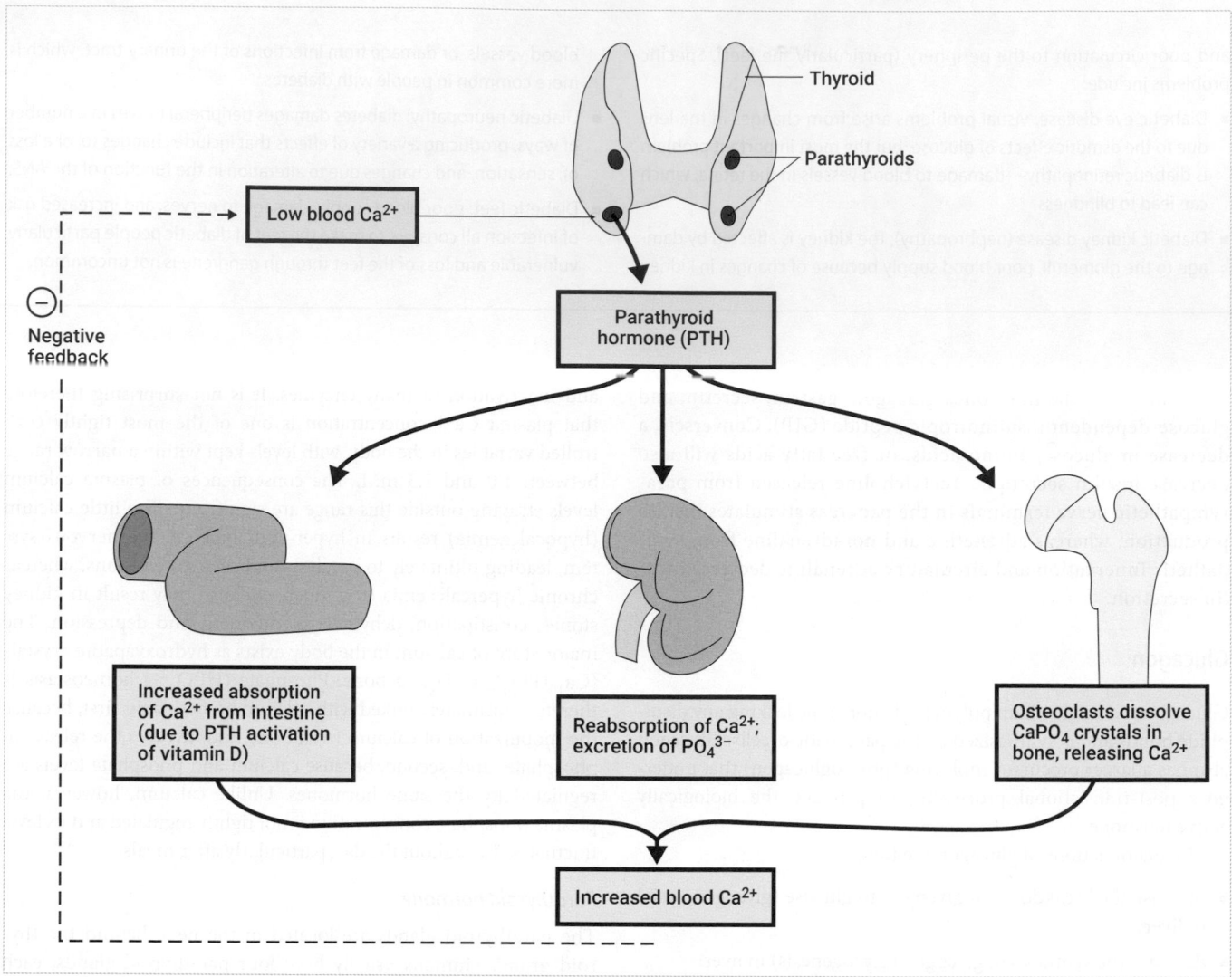

Figure 19.25 Control of plasma calcium homeostasis.

Source: Dave Klemm.

The parathyroid gland expresses a G-protein coupled receptor called the calcium-sensing receptor which, when activated by an increased plasma calcium concentration, results in negative feedback to reduce PTH secretion.

Calcitriol

Vitamin D is the collective term for a group of prohormones, the two major forms of which are vitamin D_2 (ergocalciferol) and vitamin D_3 (cholecalciferol). Vitamin D is absorbed from the diet and also produced from cholesterol in the skin in response to sunlight. Vitamin D is biologically inactive and must undergo two hydroxylation reactions to be converted into its active form calcitriol (1,25-dihydroxycholecalciferol). The first of these hydroxylation reactions occurs in the liver, while the second occurs in the kidney. Parathyroid hormone activates the kidney enzyme responsible for hydroxylation and therefore regulates the amount of calcitriol produced.

Calcitriol acts to increase plasma calcium levels by promoting absorption of dietary calcium from the gastrointestinal tract and also by reducing the loss of calcium from urine by increasing renal reabsorption of calcium. Calcitriol also stimulates release of calcium from bone by promoting the actions of parathyroid hormone in activating osteoclasts.

The effects of calcitriol are mediated by the vitamin D receptor which functions as a hormone-activated transcription factor to regulate gene expression. Calcitriol is therefore more involved in the long-term regulation of calcium rather than the short-term effects mediated by parathyroid hormone.

Calcitonin

The third hormone involved in calcium homeostasis is the peptide hormone calcitonin, which is produced by the parafollicular cells of the thyroid gland. In animal models calcitonin lowers serum calcium levels. However, lack of calcitonin in humans from removal or destruction of the thyroid gland has no apparent effects on calcium homeostasis, suggesting it has little function in humans. There is some suggestion, however, that during pregnancy calcitonin may serve to preserve the maternal skeleton.

Measurement of hormones by immunoassay

Immunoassay is a widely used analytical technique that provides a sensitive and powerful method to quantify substances such as peptides, proteins, organic molecules, pollutants, and drugs in biological samples. Immunoassays are widely used in the clinical setting to determine the plasma level of a particular hormone, such as cortisol, IGF-1, growth hormone, TSH, or thyroid hormone in patients. An immunoassay, as the name suggests, is based on the ability of an antibody to recognize and selectively bind to an antigen. In this scenario the antigen is the hormone of interest that the investigator wishes to quantify. The earliest use of immunoassays to quantify hormones involved labelling the antigen or antibody in the assay with a radioisotope, a technique known as radioimmunoassay (RIA). The potential harmful effects of handling radioisotopes on a daily basis in a routine assay laboratory, however, has meant that RIAs have largely been replaced with the use of fluorescent, rather than radioactive, tags. The principles of the assay, however, remain the same—just a different detection technique is required to detect the label. There are two main types of immunoassay employed: competitive and non-competitive.

Competitive assays

In a competitive immunoassay, samples containing the hormone of interest are mixed with a purified antibody specific for that particular hormone together with a low fixed concentration of labelled hormone. To construct a calibration curve for each set of measurements, known amounts of the unlabelled hormone are added to aliquots of the antibody and labelled hormone mixture. The unlabelled antigen molecules in both test and standard curve samples will compete with the labelled antigen for binding sites of the antibody molecules. Therefore, increasing the amount of

unlabelled hormone added to the standard curve samples will decrease the amount of labelled hormone bound to antibody as the labelled and unlabelled hormone molecules are competing for the same fixed number of binding sites available from the antibody in the sample. Antibody-bound hormone is then separated from free antigen and the amount of fluorescence in each sample is measured. A common way to achieve this is to use an assay format where the hormone-specific antibody is bound to the inner walls of a microtitre plate. Free unbound antigen can therefore be conveniently washed leaving the bound hormone stuck to the walls of the assay plate. The standard curve allows the concentration of antigen in the unknown test samples to be calculated from the amount of fluorescence they contain.

Non-competitive assays

Non-competitive assays can be used where the hormone of interest is large enough to have two non-overlapping epitopes for two different antibodies. One of these antibodies is bound to a physical support, such as the walls of a microtitre plate, and the sample containing the hormone of interest added and allowed to equilibrate. A second labelled antibody is then added to the sample and a 'sandwich' complex between the first immobile antibody, the hormone, and the second labelled antibody is formed. Again, a standard calibration would be produced with known quantities of hormone, allowing the amount of hormone in the test samples to be determined. The advantage of this non-competitive format is that it produces a straight-line calibration curve over several orders of magnitude and has increased sensitivity, since every molecule of hormone in the sample has the ability to produce a signal once bound.

PAUSE AND THINK

Why is it important that plasma calcium concentrations are tightly regulated within the body?

Answer: Calcium ions (Ca^{2+}) play a central role in many cellular processes, including neuromuscular excitability (Chapter 20), synaptic transmission (Section 17.6), intracellular signalling, and, in the longer term, bone formation. For example, lowered plasma concentration of calcium can result in neuronal hyper-excitability and paralysis.

As we have mentioned, disorders of hormone secretion can have serious clinical implications, so it is important to be able to measure the circulating levels of different hormones within the bloodstream (see Experimental Toolkit 19.1).

 Check your understanding of the concepts covered in this section by answering the questions in the e-book.

19.4 Conclusion

From this overview of the ANS and the endocrine system it should now be obvious that the actions of these two systems are fundamental in maintaining homeostasis, as it is these two systems that coordinate the functions of the other organ systems in the body. Homeostasis occurs in many forms, such as the regulation of body temperature, blood pressure, blood volume, acid–base balance, and glucose concentration, and it is the communication provided by the ANS and endocrine systems that allows for these parameters to be kept within their set limits. The hypothalamus is the major control and integration centre for both systems and provides a functional link between the two. A range of sensory inputs from the internal and external environments are received by the hypothalamus, which then coordinates the hormonal and nervous responses to these stimuli through the endocrine and ANSs.

SUMMARY OF KEY CONCEPTS

- Communication between the CNS and the organ systems is essential for homeostasis. Two systems facilitate this communication: the ANS and the endocrine system.

- The ANS is a division of the peripheral nervous system that works autonomously without conscious effort and consists of sympathetic and parasympathetic branches.

- In general, the sympathetic branch facilitates responses required for 'fight or flight', whereas the parasympathetic branch facilitates 'rest and digest' type responses. Tonic input from both branches, however, is required for normal function of most organ systems and it is the balance between these inputs that is key.

- Pathways that make up the ANS are composed of preganglionic and postganglionic fibres.

- All preganglionic fibres use acetylcholine as a neurotransmitter to activate post-synaptic nicotinic ligand-gated ion channels. Postganglionic fibres of the parasympathetic branch also use acetylcholine as a neurotransmitter, but the post-synaptic receptors in this case are muscarinic G-protein coupled receptors.

- Postganglionic fibres in the sympathetic branch use noradrenaline and adrenaline as neurotransmitters to activate adrenergic G-protein coupled receptors.

- Chromaffin cells in the adrenal medulla are modified neuroendocrine cells with analogous function to postganglionic sympathetic neurons. Instead of releasing neurotransmitter locally into a synapse, the chromaffin cells release adrenaline into the bloodstream.

- Hormones are chemical messengers that facilitate intercellular communication through paracrine, autocrine, neurocrine, or endocrine pathways. Classically, hormones were considered as the chemical messengers produced by the endocrine glands and transported in blood. We now know that hormones can be also be produced by a range of cell and tissue types.

- The hypothalamic–pituitary axis constitutes the major link between the nervous and endocrine systems. Six hormones made by the anterior pituitary gland under the control of local hypothalamic hormones are released into the general circulation to regulate the activity of the other endocrine glands. The two hormones released from the posterior pituitary are made in the hypothalamus and stored in the posterior pituitary before release.

- Hormone secretions are regulated by negative feedback loops, which act at different levels in the system.

- Growth hormone produced by the anterior pituitary is the main stimulator of body growth. It has direct effects in responsive cells through the growth hormone receptor and indirect effects by stimulating the production of insulin-like growth factor 1 (IGF-1).

- Thyroid hormone receptors are hormone activated transcription factors. Thyroid hormones (T3 and T4) are synthesized from the tyrosine residues present in the protein thyroglobulin located in the colloid of thyroid follicles.

- The adrenal cortex consists of the zona glomerulosa, zona fasciculata, and zona reticularis, which synthesize mineralocorticoid, glucocorticoid, and androgen hormones, respectively.

- The β-cells in the islets of Langerhans within the pancreas produce the hormone insulin, whereas the α-cells produce the hormone glucagon. Insulin promotes the transport and utilization of blood glucose into adipose and skeletal muscle cells. Glucagon opposes these effects.

- Plasma Ca^{2+} concentration is one of the most tightly controlled variables in the body and is primarily controlled by the actions of parathyroid hormone.

- Parathyroid hormone promotes the release of calcium from bone, slows the rate of calcium loss in urine, and, indirectly through promoting the production of calcitriol, increases the rate of Ca^{2+} absorption from the gastrointestinal tract.

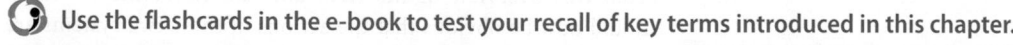 Use the flashcards in the e-book to test your recall of key terms introduced in this chapter.

QUESTIONS

 Looking for answers? Once you've answered these questions, follow the link in the e-book to the answer guidance and check your work.

Concepts and definitions

1. Outline the neuronal pathways that make up the ANS.

2. What are the neurotransmitters of the sympathetic and parasympathetic nervous systems?

3. What is the role of antidiuretic hormone?

4. What are the main effects of thyroid hormone on body tissues?

5. Where is aldosterone secreted from and what is its main action?

Apply the concepts

6. Compare the anatomical organization of the sympathetic and parasympathetic nervous systems.

7. What are the two main types of cholinergic receptors and how do they differ?

8. Compare the different pathways involved in the feedback control of hormone secretion and give an example of hormonal regulation via these pathways.

9. What are the main actions of insulin and how is its secretion controlled?

10. Compare and contrast type 1 and type 2 diabetes.

Beyond the concepts

11. Give an overview of the organization and functions of the sympathetic and parasympathetic branches of the nervous system and explain how the balance between the outputs of these two systems regulates cardiac output.

12. What are the anatomical, neurocrine, and endocrine relationships between the hypothalamus and the anterior/posterior pituitary glands?

13. How is the synthesis and release of thyroid hormone regulated? Describe two disease states that result from malfunction of these processes.

14. What factors control the secretion of human growth hormone and how does growth hormone exert its effects on cells?

15. How are plasma calcium and phosphate levels regulated by the endocrine system?

FURTHER READING

Oakes, P. C., Fisahn, C., Iwanaga, J., DiLorenzo, D., Oskouian, R. J., & Tubbs, R. S. (2016) A history of the autonomic nervous system: part I: from Galen to Bichat. *Childs Nerv. Syst.* **32**: 2303–2308.

Oakes, P. C., Fisahn, C., Iwanaga, J., DiLorenzo, D., Oskouian, R. J., & Tubbs, R. S. (2016) A history of the autonomic nervous system: part II: from Reil to the modern era. *Childs Nerv. Syst.* **32**: 2309–2315.

The two papers listed above give a very useful narrative of our understanding of the ANS. In particular, they provide a readable account of the underlying research and development of current ideas.

Endotext (www.endotext.org/)

This is an online resource focused on clinical endocrinology and which also gives details of the underlying physiological mechanisms.

Bear, M. F., Connors, B., & Paradiso, M. (2015) *Neuroscience: Exploring the Brain.* 4th edition. Wolters Kluwer.

This textbook provides a more advanced account from a neural perspective; in particular, Chapters 15 and 16 consider the hypothalamic–pituitary axis.

Ritter, J., Flower, R., Henderson, G., Loke, Y. K., MacEwan, D., & Rang, H. P. (2019) *Rang & Dale's Pharmacology.* 9th edition. Elsevier.

This is one of the core textbooks on pharmacology and is very useful for providing more detail regarding the basic principles of the ANS and endocrine functions and the ways in which drugs interact with them.

Hinson, J. & Raven, P. (2019) *Hormones.* Oxford Biology Primers, Oxford University Press.

This text provides an excellent and very readable overview of hormones, their regulation, and modes of action.

Muscle and Movement

Chapter contents

Introduction 686

20.1 Skeletal muscle fibres and contraction 687

20.2 Organization of skeletal muscle 694

20.3 Reflexes and posture 701

20.4 Overview of the motor system 705

20.5 Conclusion 715

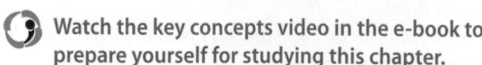 Watch the key concepts video in the e-book to prepare yourself for studying this chapter.

LEARNING OBJECTIVES

By the end of this chapter, you should be able to:

- Describe the structure and organization of skeletal muscle.
- Explain the sliding-filament theory of muscle contraction.
- Describe the organization of motor units.
- Explain the mechanisms for controlling the force of muscle contraction.
- Explain the role of spinal reflexes in control of posture.
- Describe the organization of the descending motor pathways.
- Explain the role of the basal ganglia and cerebellum in controlling movement.
- Describe the organization and role of the motor cortex in the control of voluntary movement.

Introduction

In 1924 the British neurologist Sir Charles Sherrington stated: 'To move things is all that mankind can do... for such the sole executant is muscle whether in whispering a syllable or felling a forest'. At the current time, despite all the technological advances that facilitate our interaction with the world, this statement still holds true: every interaction we have with the external world is mediated through muscle contraction. Controlling movement is also one of the most complex functions we perform: we have over 600 skeletal muscles that are adapted and organized in different ways for the various functions they perform. Muscles range from the very small, high-precision muscles that control the movements of the eyeball to the very powerful muscles of the thighs that power us in walking and running.

When we make a movement, for example to pick something up, we have to identify the location of the target and coordinate a sequence of muscle contractions to reach out to the object, grip it, and lift it. All of these processes require precision control of the different muscles involved based on sensory information, previous learning, and detailed planning. At the core of this process is the skeletal muscle, which generates the required contractile forces under high-level control from the motor areas of the brain.

The specialized contractile cells that make muscle are called **muscle fibres** and their purpose is to transform the chemical energy stored in ATP into mechanical energy in order to generate force, perform work, and produce movement. There are three distinct types of muscle:

- skeletal—muscles under voluntary control generating body movements;

- cardiac—the muscle of the heart;

- smooth—the muscle lining blood vessels and the gastrointestinal system—these are under involuntary control by the autonomic system.

These muscle types are distinguishable on the basis of their appearance when viewed under a microscope (**striated** or **non-striated**), the ability to contract a muscle at will (**voluntary** or **involuntary**), and location and function within the body. Skeletal and cardiac muscle are classed as striated because of the ordered repeating arrangement of the contractile units within the cells called **sarcomeres**. When viewed under the microscope the arrangement of sarcomeres produces a characteristic banding pattern called striations (see Figure 20.2). Smooth muscle cells, however, are so called because they lack an ordered repeating arrangement in their contractile units and therefore appear non-striated or smooth when viewed under the microscope. Skeletal muscle is considered voluntary because we can make it contract or relax under conscious control. Cardiac and smooth muscles, however, are not under conscious control and so are classed as involuntary muscles. Skeletal muscle is also to some extent controlled on an involuntary basis; for example, we do not have to consciously think to contract and relax (i) our diaphragm when breathing or (ii) muscles to maintain posture when sitting or standing.

20.1 Skeletal muscle fibres and contraction

Skeletal muscle, as its name suggests, is associated with the bones of the skeleton where it attaches by bundles of collagen fibres called **tendons**. Skeletal muscle generates movement of the body. However, even when there is no obvious movement, contraction of skeletal muscle also helps to maintain body posture by stabilizing joints.

A layer of dense, fibrous connective tissue called **deep fascia** divides groups of muscles with similar functions into **fascial compartments** (Figure 20.1). Three further layers of connective tissue surround, protect, and support the fibres. The outermost layer is the **epimysium,** which encloses the entire surface of each individual muscle. Continuous with this epimysium is a layer

Figure 20.1 Skeletal muscle structure.

Source: Butler et al. *Animal Physiology*, 2021. Oxford University Press.

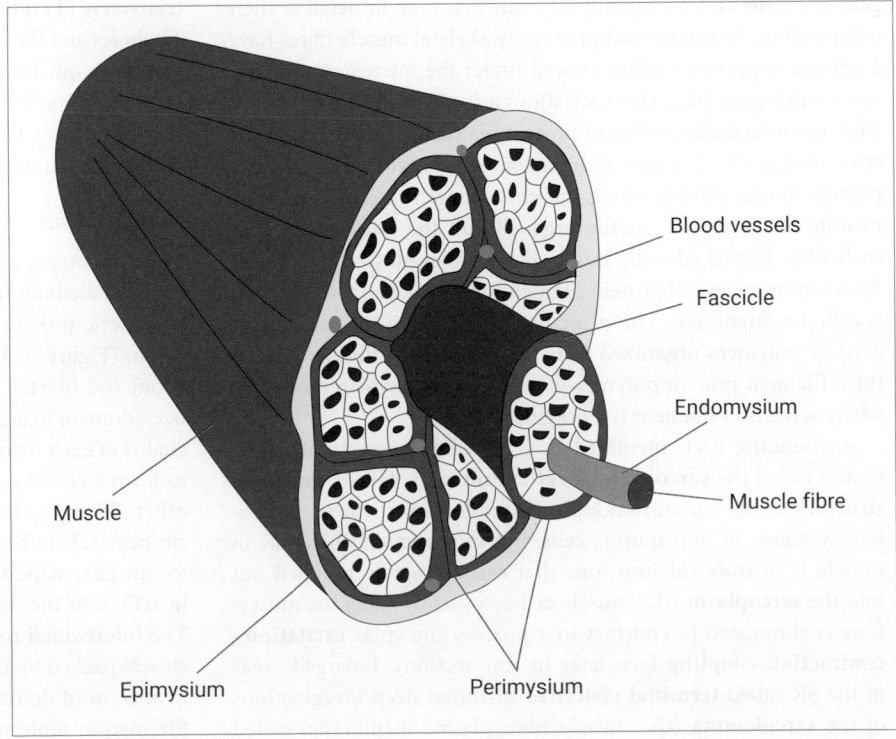

called the **perimysium,** which encloses groups of 10 to 100 or more individual muscle fibres into bundles called **fascicles**. Fascicles are large enough to be seen by the naked eye and if you have ever looked closely at a slice of roast beef, you may have noticed the fascicles as the small cell-like structures covering the surface of the slice. The final layer is the **endomysium** that surrounds each individual muscle fibre.

Skeletal muscle fibres are unusual cells in that each contains up to several thousand peripherally located nuclei. This is because skeletal muscle fibres originate from the fusion of thousands of individual myoblast cells during embryonic development to form a syncytium of cells with a shared cytoplasm, nuclei, and the many mitochondria required to supply the ATP required for contraction. The number of muscle fibres present in each muscle is therefore fixed shortly after birth as mitosis is no longer possible in a syncytium containing thousands of nuclei.

When we strengthen our muscles by exercising and they become bigger, how does this happen? Take a moment to reflect on what you already know about this. Skeletal muscles can only grow larger in size by an increase in the diameter of the existing fibres (**hypertrophy**) rather than an increase in the number of fibres (**hyperplasia**).

A small finite population of quiescent **myosatellite cells**, however, do remain associated with skeletal muscle fibres after embryonic development and these cells can be reactivated upon muscle damage to re-enter the cell cycle and fuse with one another or with existing fibres to provide additional nuclei to facilitate regeneration and repair.

Structure of the skeletal muscle fibre

In order to understand how skeletal muscle fibres contract to generate force and movement we must first look in detail at their composition. As mentioned previously, skeletal muscle fibres have a striated appearance when viewed under the microscope, as you can see in Figure 20.2. These striations arise from the highly organized repeating arrangement of protein filaments within thread like structures, each ~2–3 μm in diameter, called **myofibrils**, that run parallel to one another along the entire length of each fibre. The myofibrils are the contractile organelles of the muscle fibre and each fibre is packed with hundreds to thousands of myofibrils. Each repeating unit of protein filaments that make up the myofibril is called a sarcomere. The proteins that make up the sarcomeres exist as polymers organized into **thick** and **thin filaments**. The thick filament protein polymer is composed of **myosin** molecules whereas the thin filament is a polymer of **actin**.

Surrounding each myofibril is a fluid-filled system of membranes called the **sarcoplasmic reticulum** (SR). This is similar in structure to the smooth endoplasmic reticulum involved in protein synthesis in non-muscle cells, but the function of the SR in muscle is to store calcium ions that can be rapidly pumped out into the **sarcoplasm** (the muscle cell cytoplasm) when the muscle fibre is stimulated to contract in a process known as **excitation–contraction coupling** (see later in this section). Enlarged areas of the SR called **terminal cisternae** surround deep invaginations of the **sarcolemma** (the muscle fibre plasma membrane) called

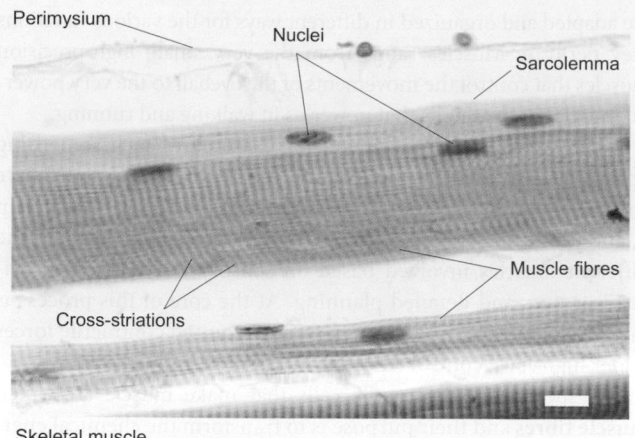

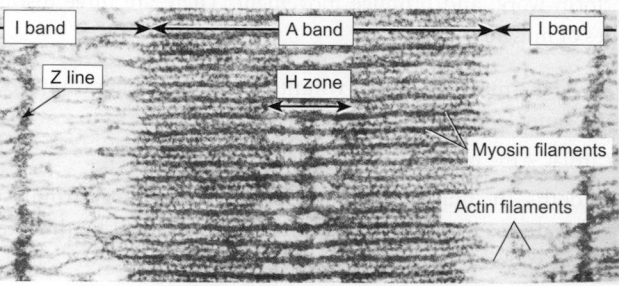

Figure 20.2 Striated muscle. The upper panel shows a light microscope image of myofibrils, with the characteristic striated appearance of skeletal muscle. The lower panel shows a high-magnification view of a sarcomere, as viewed under the electron microscope.

Source: Pocock et al. *Human Physiology*, 5th edn, 2017. Oxford University Press.

transverse (T) tubules (see Figure 20.6). Each fibre has thousands of T tubules and their function is to allow depolarization of the sarcolemma to quickly penetrate the interior of the cell. The structure formed by one T tubule and the terminal cisterna on either side of it is known as a **triad**, which forms the anatomical basis of excitation–contraction coupling.

Thick filaments

Thick filaments are placed centrally within the sarcomere structure and are built from myosin. In its native state, myosin is a large hexameric protein consisting of two heavy chains and four light chains (Figure 20.3). Each myosin heavy chain molecule consists of a long rod-like tail region at one end linked by a flexible hinge-like neck domain to a globular head region at the other. The two heavy chains of each myosin molecule intertwine along their rod regions to form a coiled coil rather like two snakes wrapped around each other. Emerging from the end of this coiled coil are the two myosin heavy chain heads. It is these head regions that are responsible for the enzymatic activity that converts the chemical energy stored in ATP into the mechanical work that drives muscle contraction. The intertwined rod regions of around 300 myosin molecules are closely packed to form the backbone of the thick filament with the myosin head domains projecting from its surface. The packing of the myosin molecules in the thick filaments is such that they are

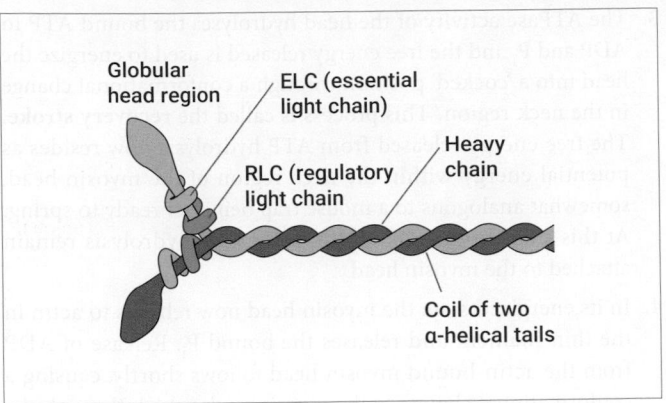

Figure 20.3 The myosin thick chain.

Source: Cooper et al. *The Cell: A Molecular Approach*, 3rd Edn, 2003. Oxford University Press.

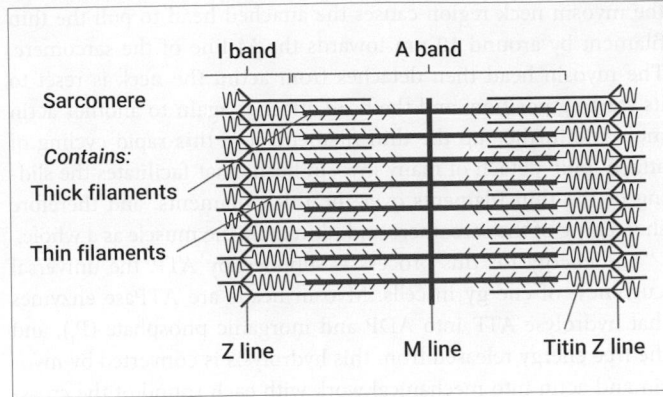

Figure 20.4 Structure of the sarcomere.

Source: England, J., Loughna, S. (2013) Heavy and light roles: myosin in the morphogenesis of the heart. *Cell Mol Life Sci*, 70, 1221–1239. https://doi.org/10.1007/s00018-012-1131-1.

in a tail-to-tail orientation in the centre extending outward from a central structure called the **M-line** (Figure 20.4). When viewed under the microscope the entire length of the thick filaments extending in each direction from the M-line appears as a dark region termed the **A-band**. The central region of the thick filament is devoid of myosin heads creating a bare zone (the **H-zone**) which appears as a paler region within the A-band. Each myosin heavy chain molecule has two much smaller myosin light chain proteins attached around the neck domain lying between the globular head and rod domain. These myosin light chains are composed of one essential light chain, required for structural stability of the neck region, and a regulatory light chain, which is phosphorylated to modulate activity of the myosin head.

Thin filaments

Overlapping each end of the thick filaments within the sarcomere are thin filaments (Figure 20.4). These originate from before the overlap region from a dense structure called the **Z-disc**. The Z-discs are seen as dark lines when striated muscle is viewed under the electron microscope (see Figure 20.2) and define the boundaries of the sarcomere. The region of thin filament between the Z-disc and the overlapping region appears lighter when viewed under the electron microscope and is called the **I-band**. The backbones of the thin filaments are built from a helical assembly of the globular protein actin. Actin monomers join in a head-to-tail fashion to form a helix which has a groove along its length where long thin strands of the rod-shaped protein **tropomyosin** sit associated with a complex of proteins called the **troponin complex**. As we shall see later, both tropomyosin and troponin are involved in regulating the interaction of the myosin heavy chain heads with the actin filaments, the key step in the initiation of muscle contraction.

Structural filaments

Besides thick and thin filaments, sarcomeres also contain structural filaments composed of the protein **titin**. Titin, as its name suggests (titan = gigantic), is the largest known protein at around 30,000 amino acids long and is the third most abundant protein in skeletal muscle after actin and myosin. Titin forms a continuous link between the M-line and Z-disc and is thought to function as a

sort of molecular spring which contributes to the passive elasticity of muscle and limits the range of sarcomere motion under tension. Other structural proteins within the sarcomere include **myomesin**, which forms the M-line, **nebulin**, which wraps around the thin filaments, and **α-actinin**, which helps anchor actin to the Z-discs.

 Go to the e-book to complete an activity related to Figure 20.4.

The sliding filament mechanism of striated muscle contraction

Now we have seen the components that make up a sarcomere, we can move on to understand how this structure relates to the molecular basis of muscle contraction. Muscle contraction is brought about as a result of the sliding past each other of the thick and thin filaments within the sarcomere, producing a greater overlap of the filaments without a change in the length of the filaments themselves. During this sliding the thick filaments remain stationary and the thin filaments slide inwards over the thick filaments towards the M-band of the sarcomere. As the thin filaments slide they pull on the Z-discs they are attached to, causing the sarcomere to shorten in length. As each muscle fibre contains many sarcomeres in series, all of which are shortening simultaneously, the whole muscle contracts as a result of this sliding motion.

The crossbridge cycle

How does sliding of the thick and thin filaments over each other occur? The answer lies in the cyclic interactions made between the heads of the myosin molecules in the thick filaments and actin molecules within the thin filaments.

Essentially, movement in the neck region between the myosin head and body acts as a sort of ratchet mechanism which, through the interaction between the myosin head and actin, pulls on the thin filaments causing them to slide over the thick filaments. The two myosin heads of each myosin molecule act independently from one another with only one head interacting with actin at any given time. Each head attaches to an actin molecule in an adjacent thin filament forming a **crossbridge**. A conformational change in the myosin neck region causes the attached head to pull the thin filament by around 10 nm towards the M-line of the sarcomere. The myosin head then detaches from actin, the neck is reset to its starting position, and the head attaches again to another actin molecule further up the thin filament. It is this rapid cycling of attach–pull–detach of many myosin heads that facilitates the sliding of the thin filaments over the thick filaments, and therefore shortening of the sarcomere and ultimately the muscle as a whole.

The energy for this process is supplied by ATP, the universal 'currency' of energy in cells. Myosin heads are ATPase enzymes that hydrolyse ATP into ADP and inorganic phosphate (P_i), and the free energy released from this hydrolysis is converted by myosin and actin into mechanical work with each round of the crossbridge cycle. To more fully understand the sequence of molecular events that facilitate this conversion of energy, let us now follow what happens to a single myosin head during each crossbridge cycle (Figure 20.5):

1. At the end of the previous crossbridge cycle the myosin head is in a nucleotide-free state (i.e. there is no ATP, ADP, or P_i bound) and is tightly attached to an actin molecule in an adjacent thin filament. This state of myosin head attachment to actin is called the **rigor state** and the myosin head can only detach if it binds ATP.

2. ATP binds to the myosin head causing it to detach from the actin filament.

3. The ATPase activity of the head hydrolyses the bound ATP to ADP and P_i and the free energy released is used to energize the head into a 'cocked' position through a conformational change in the neck region. This process is called the **recovery stroke**. The free energy released from ATP hydrolysis now resides as potential energy within the neck region of the myosin head, somewhat analogous to a mouse trap being set ready to spring. At this stage both ADP and P_i from ATP hydrolysis remain attached to the myosin head.

4. In its energized state the myosin head now rebinds to actin in the thin filament and releases the bound P_i. Release of ADP from the actin-bound myosin head follows shortly, causing a conformational change in the myosin neck, which through the attached head pulls on the actin thin filament, causing it to slide over the thick filament. This process is called the **power stroke**. Release of ADP from the myosin head can therefore be thought of as the trigger that sets off the mouse trap, converting the potential energy stored from the previous recovery stroke into mechanical work. The myosin head is now again in the rigor state of actin attachment and the cycle can recommence.

Go to the e-book to explore an interactive version of Figure 20.5.

Pause here to think about what would happen to the crossbridge cycle if ATP was unavailable. The myosin heads would be unable to detach from the actin filaments and would be locked in the rigor

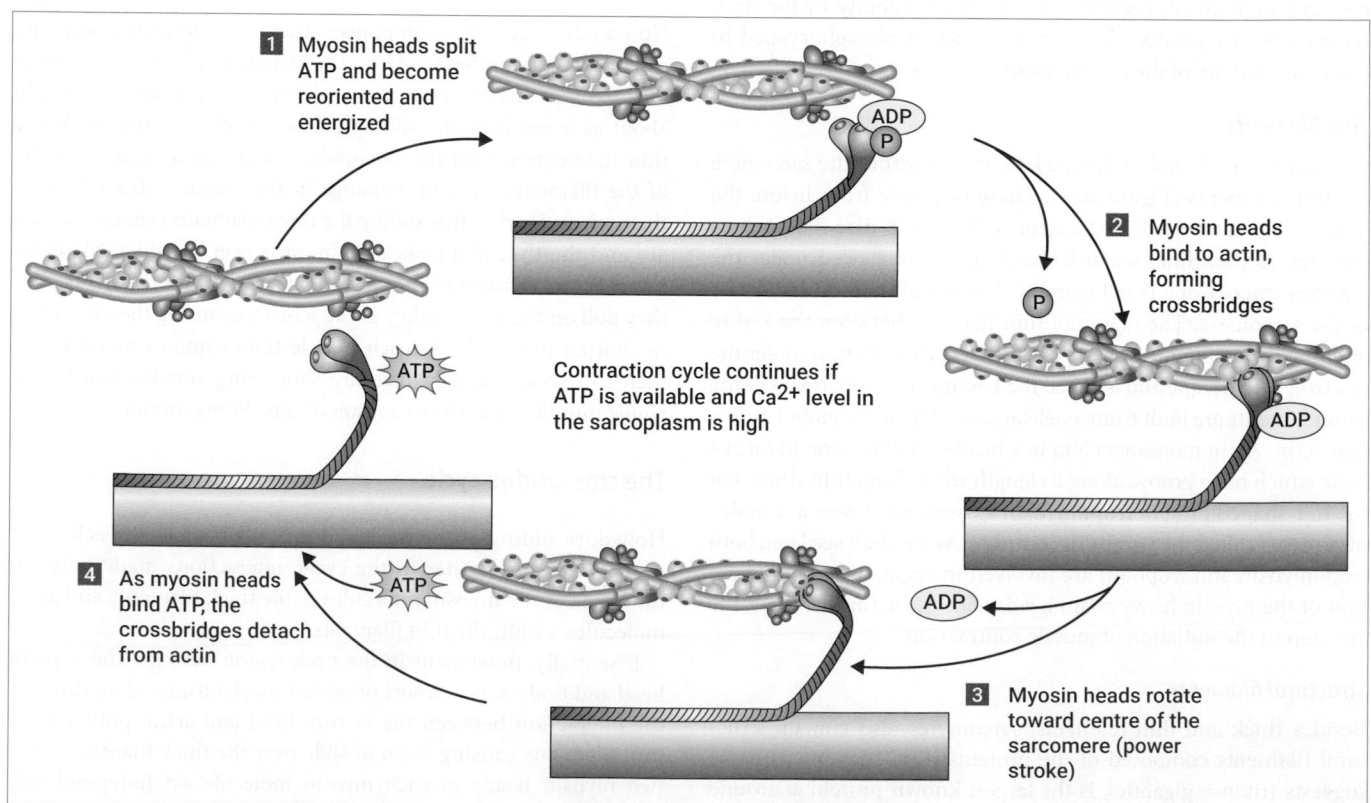

1 Myosin heads split ATP and become reoriented and energized

ADP
P

2 Myosin heads bind to actin, forming crossbridges

P

ATP

Contraction cycle continues if ATP is available and Ca^{2+} level in the sarcoplasm is high

ADP

4 As myosin heads bind ATP, the crossbridges detach from actin

ATP

ADP

3 Myosin heads rotate toward centre of the sarcomere (power stroke)

Figure 20.5 Crossbridge cycle of muscle contraction.

Source: Pocock et al. *Human Physiology*, 5th edn, 2017. Oxford University Press.

state of attachment. This is essentially what happens in *rigor mortis* after death. Around 2–6 hours after death the muscles of the corpse become very stiff as ATP is no longer being produced from metabolism. With no ATP the thick and thin filaments become rigidly linked as ATP is required to release the myosin heads from their rigor-attached state.

A key feature of the crossbridge cycle is that the formation of the numerous myosin to actin attachments within each sarcomere is asynchronous; if all the myosin heads attached and detached at the same time then there would be nothing to prevent the thin filaments from sliding back to their starting position once the myosin heads detach. Instead, at any one time during a muscle contraction there will be some heads attached and undergoing a power stroke while others will be detached and undergoing their recovery stroke.

PAUSE AND THINK

At what stage in the crossbridge cycle is the ATP hydrolysed to ADP and P_i?

Answer: ATP binds to the myosin head, causing it to detach from the actin. The ATPase activity of the myosin head then hydrolyses the bound ATP to ADP and P_i, and the free energy released is used to energize the head into a 'cocked' position through a conformational change in the neck region (the recovery stroke) prior to rebinding to the actin.

Regulation of sarcomere shortening

We now have a basic understanding of the mechanism by which sarcomeres shorten, but how is this mechanism controlled? There must be a way of switching crossbridge cycling on and off on demand; otherwise, we would not be able to control our skeletal muscles. The 'on–off' switch for crossbridge cycling lies at the level of the thin filament proteins troponin and tropomyosin.

When skeletal muscle is not contracting the myosin binding sites on the actin thin filaments are blocked by troponin and tropomyosin. This means that the myosin heavy chain heads in the thick filaments are unable to bind to actin because tropomyosin and troponin are blocking the binding site. Therefore, crossbridge cycling cannot occur and the muscle is in a relaxed state. Associated with each tropomyosin molecule is the troponin complex, which is made up of three different types of troponin:

- troponin-T is responsible for the interaction with tropomyosin;
- troponin-I can inhibit the myosin interaction on its own;
- troponin-C is capable of reversibly binding calcium.

For the myosin heads to be able to bind actin in the thin filaments, tropomyosin must first shift its position slightly on the thin filament in order to uncover the myosin binding sites. The stimulus for this is Ca^{2+} binding to troponin-C, which causes the rest of the troponin complex to induce a rotatory shift in the position of tropomyosin in the actin helix, thereby exposing the myosin binding sites. With the myosin binding sites now revealed, providing ATP is present, the myosin heads can now bind actin and crossbridge cycling begins. Ca^{2+} is therefore the key initiator of muscle contraction and changes in the concentration of Ca^{2+} within the sarcoplasm determine the amount of crossbridge cycling and therefore ultimately the rate and strength of muscle contraction.

Skeletal muscle excitation–contraction coupling

Given that the onset and extent of crossbridge cycling are determined by Ca^{2+}, the cytoplasmic Ca^{2+} concentration must be tightly regulated to control the rate and force of contraction. Ca^{2+} is stored at high concentration in a network of interconnected compartments called the sarcoplasmic reticulum surrounding each myofibril. In order to regulate the concentration of cytoplasmic Ca^{2+}, and therefore the amount of crossbridge cycling, the muscle fibre can release Ca^{2+} from this store into the cytoplasm to initiate sarcomere contraction and also remove Ca^{2+} from the cytoplasm by pumping it back into the store to initiate relaxation.

The signal for a muscle to contract comes from the arrival of an action potential causing depolarization of the sarcolemma membrane potential. What is needed is a way of linking this signal of membrane depolarization to the release of Ca^{2+} from the sarcoplasmic reticulum. This link is known as the process of excitation–contraction coupling.

In addition to the sarcoplasmic reticulum, another membranous system within the muscle fibre, the T-tubules are also key to facilitating excitation–contraction coupling. These tube-shaped perpendicular invaginations of the sarcolemma are present at each junction of the A-band and I-band in the sarcomere and penetrate deep into the muscle fibre allowing action potentials to be conducted into the interior of the cell (Figure 20.6). Surrounding each T-tubule are swellings of the sarcoplasmic reticulum called terminal cisternae. One T-tubule and the two terminal cisternae on each side form a structure known as a triad, which is the site at which excitation–contraction coupling occurs. The very close proximity of the T-tubule and sarcoplasmic reticulum membranes in the triad facilitates the formation of supramolecular complexes consisting of **dihydropyridine receptors** (DHPRs), present on the T-tubule membrane side of the triad, and **ryanodine receptors**, present on the terminal cisternae side (Figure 20.7). Both these proteins are types of calcium channel. Ryanodine receptors are so called because they can be locked into an open position by the plant chemical ryanodine.

The DHPRs on the T-tubule membrane side of the triad are voltage-gated Ca^{2+} channels that sense the membrane depolarization caused by the arrival of an action potential and change their shape as a result. The conformational change in protein structure of the DHPR in response to membrane depolarization is transmitted to the ryanodine receptors present in the terminal cisternae membrane side of the triad through their close physical interaction. This causes the ryanodine receptors to open, allowing the release of Ca^{2+} from the intracellular store into the sarcoplasm, thereby triggering contraction of the muscle fibre.

Relaxation

We have seen how Ca^{2+} release from the sarcoplasmic reticulum initiates skeletal muscle contraction by the process of excitation–contraction coupling. It should come as no surprise, therefore,

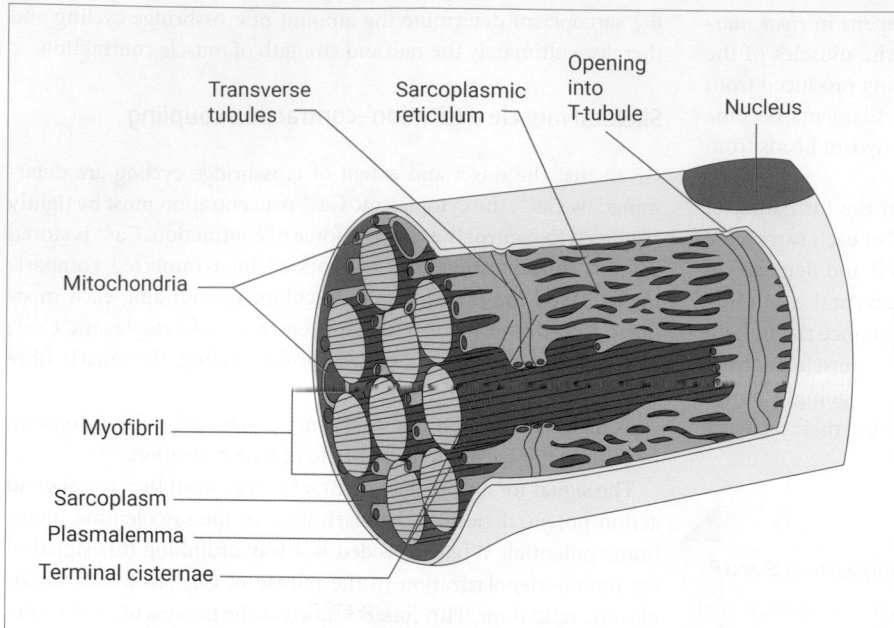

Figure 20.6 Organization of the T-tubule system around the myofibrils.

Source: Frontera, W.R., Ochala, J. Skeletal muscle: a brief review of structure. *Calcified Tissue International and Musculoskeletal Research*. Reprinted by permission from Springer: Springer. © 2015.

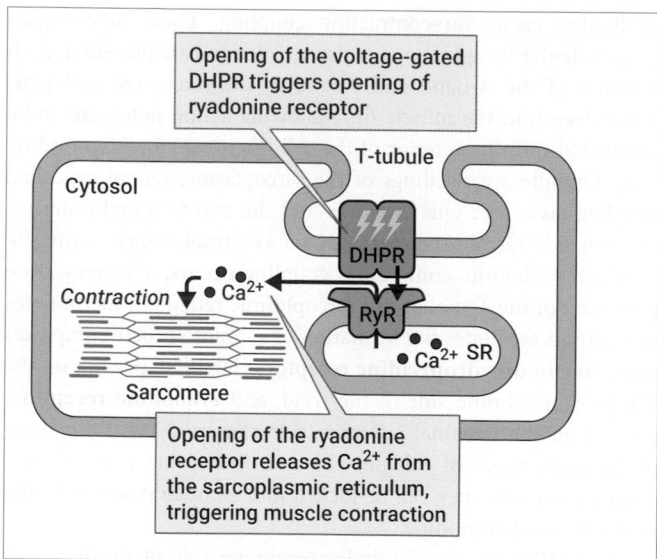

Figure 20.7 Diagram of a T-tubule showing a dihydropyridine receptor (DHPR) and associated ryanodine receptor (RyR). Opening of the voltage-gated DHPR leads to opening of the RyR and release of Ca^{2+} from the sarcoplasmic reticulum to trigger muscle contraction.

Source: Jurkat-Rott K., Lehmann-Horn F. (2005) Muscle channelopathies and critical points in functional and genetic studies. *J Clin Invest*, 115(8), 2000–9. Permission conveyed through Copyright Clearance Center, Inc.

that for muscle relaxation to occur the cytoplasmic Ca^{2+} concentration needs to be reduced. This is achieved by a Ca^{2+} pump located in the membrane of the sarcoplasmic reticulum. This pump consumes energy in the form of ATP as it actively transports Ca^{2+} into the sarcoplasmic reticulum from the cytoplasm and is called the **sarcoplasmic reticulum Ca^{2+} ATPase** (or SERCA for short). Once the stimulus for muscle contraction is removed, Ca^{2+} is no longer released and the ongoing activity of SERCA lowers

cytoplasmic calcium by pumping it back into the sarcoplasmic reticulum store. Once the concentration has fallen sufficiently, Ca^{2+} is released from troponin-C and tropomyosin shifts back to its 'off' position in the groove of the actin thin filament helix. This blocks the myosin binding sites on the thin filament and prevents crossbridge cycling from occurring. The thin filaments, freed from their attachment to the thick filaments, can now passively return to their resting position and the muscle is in a relaxed state.

The neuromuscular junction

The neurons that stimulate skeletal muscle to contract are called **somatic motor neurons** and originate in the central nervous system. At the site of its junction with a muscle, the axon of a motor neuron forms a connection with a muscle fibre called the **neuromuscular junction** (Figure 20.8), which is a specialized type of synapse (Chapter 2). The muscle cell side of the synapse is called the **motor end plate** and contains numerous invaginations of the sarcolemma, which serve to increase the surface area of membrane exposed to the synaptic cleft. The pre-synaptic axon of the motor neuron terminates at a bulge called the **terminal bouton**, which projects into the folds of the muscle cell sarcolemma. Like all synapses, there is no actual physical contact between the neuron and the target cell (muscle fibre). This means that action potentials travelling down the motor neuron cannot simply pass straight from the nerve axon to the muscle fibre as the electrical signal is unable to 'jump' across the gap (**synaptic cleft**) between the nerve terminal and the muscle fibre. Instead, the neurotransmitter acetylcholine, which is stored in vesicles in the pre-synaptic nerve terminal, is released from the vesicles into the synapse and used to pass the signal across the synaptic cleft to the muscle fibre. If sufficient neurotransmitter is released, a new action potential is generated in the muscle fibre itself and this initiates the process of excitation–contraction coupling.

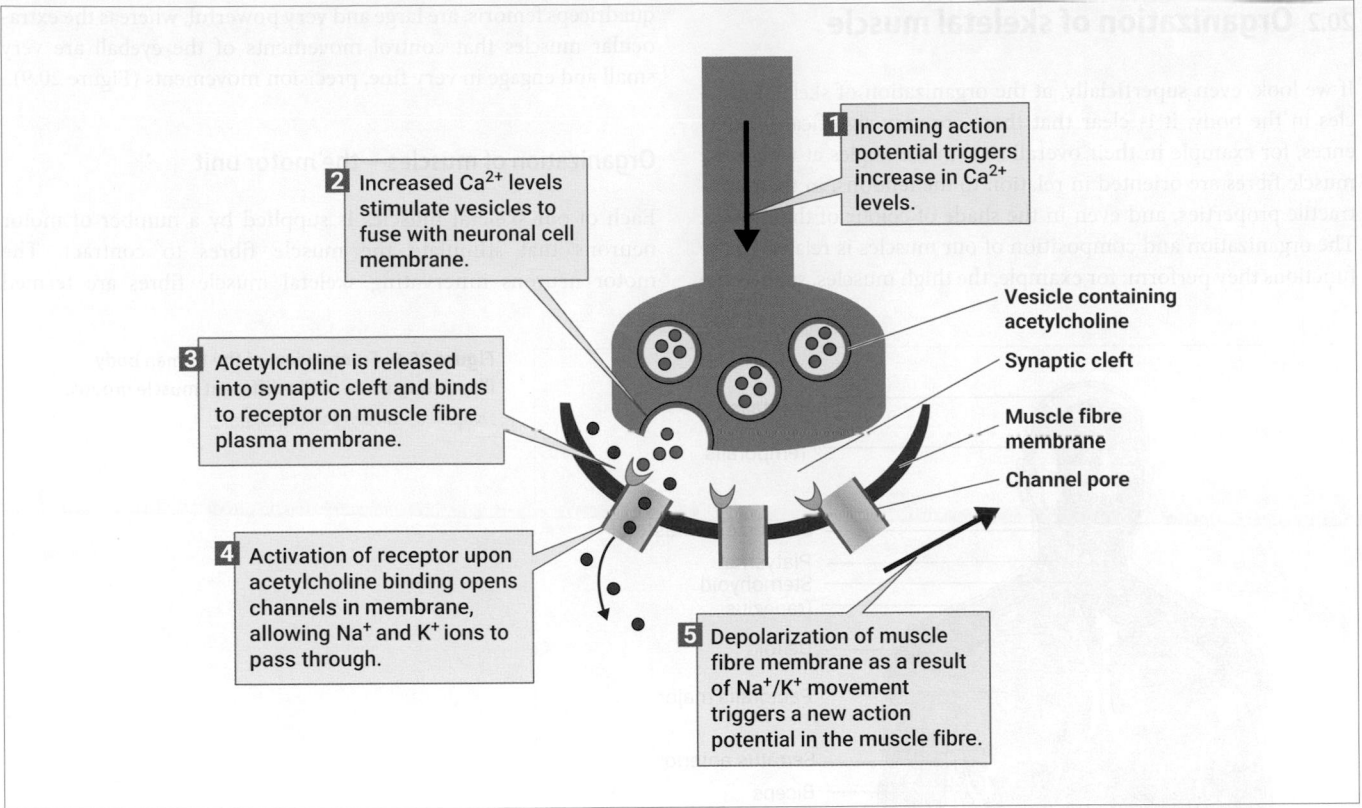

1 Incoming action potential triggers increase in Ca²⁺ levels.

2 Increased Ca²⁺ levels stimulate vesicles to fuse with neuronal cell membrane.

3 Acetylcholine is released into synaptic cleft and binds to receptor on muscle fibre plasma membrane.

4 Activation of receptor upon acetylcholine binding opens channels in membrane, allowing Na⁺ and K⁺ ions to pass through.

5 Depolarization of muscle fibre membrane as a result of Na⁺/K⁺ movement triggers a new action potential in the muscle fibre.

Vesicle containing acetylcholine

Synaptic cleft

Muscle fibre membrane

Channel pore

Figure 20.8 The neuromuscular junction.

Let us now take a closer look at the sequence of events by which an action potential arriving at a motor neuron terminal elicits an action potential in the muscle fibre.

1. Membrane depolarization caused by a neuronal action potential arriving at the pre-synaptic terminal bouton of the neuromuscular junction triggers the opening of **voltage-gated calcium ion channels** in the nerve terminal membrane and an influx of extracellular Ca²⁺ through the open channels.

2. The resulting increase in intracellular Ca²⁺ causes vesicles containing the neurotransmitter acetylcholine to fuse with the neuronal cell membrane and release their contents into the synaptic cleft.

3. Acetylcholine diffuses across the synaptic cleft and binds to **nicotinic acetylcholine receptors** (a type of ligand-gated ion channel) present in the muscle fibre motor end plate plasma membrane.

4. Activation of the motor end plate nicotinic acetylcholine receptors opens a channel pore in the receptor allowing sodium and potassium ions to flow down their electrochemical gradients. As more Na⁺ ions enter the cell through the open nicotinic acetylcholine receptors than K⁺ ions leave, the net result is a depolarization of the muscle fibre membrane potential (i.e. it becomes less negative).

5. The local depolarization of the muscle fibre membrane potential activates voltage-sensitive sodium channels in the muscle membrane, resulting in the generation of a new action potential

in the muscle fibre. This action potential passes along the sarcolemma and T-tubule system initiating the process of excitation–contraction coupling.

6. Acetylcholine in the synaptic cleft is rapidly broken down by the enzyme acetylcholinesterase.

Several toxins can selectively block events at the neuromuscular junction and therefore result in muscle paralysis. Botulinum toxin is an extremely potent neurotoxic protein produced by the bacterium *Clostridium botulinum* and also produced commercially for medical and cosmetic use under the trade name Botox. This toxin inhibits the release of acetylcholine at the neuromuscular junction by interfering with vesicle release. A lack of acetylcholine in the synaptic cleft results in a transient paralysis of the affected muscle.

 Check your understanding of the concepts covered in this section by answering the questions in the e-book.

PAUSE AND THINK

Explain the role of calcium ions in excitation–contraction coupling.

Answer: Ca²⁺ binds to troponin-C, which causes the rest of the troponin complex to induce a rotatory shift in the position of tropomyosin in the actin helix, thereby exposing the myosin binding sites. The myosin heads can now bind actin and crossbridge cycling begins. Ca²⁺ is therefore the key initiator of muscle contraction and changes in the concentration of Ca²⁺ within the sarcoplasm determine the amount of crossbridge cycling and therefore ultimately the rate and strength of muscle contraction.

20.2 Organization of skeletal muscle

If we look, even superficially, at the organization of skeletal muscles in the body, it is clear that there are very significant differences, for example in their overall size, in the angles at which the muscle fibres are oriented in relation to the tendons, in their contractile properties, and even in the shade of colour of the muscle. The organization and composition of our muscles is related to the functions they perform; for example, the thigh muscles, such as the quadriceps femoris, are large and very powerful, whereas the extraocular muscles that control movements of the eyeball are very small and engage in very fine, precision movements (Figure 20.9).

Organization of muscles—the motor unit

Each of our skeletal muscles is supplied by a number of motor neurons that stimulate the muscle fibres to contract. The motor neurons innervating skeletal muscle fibres are termed

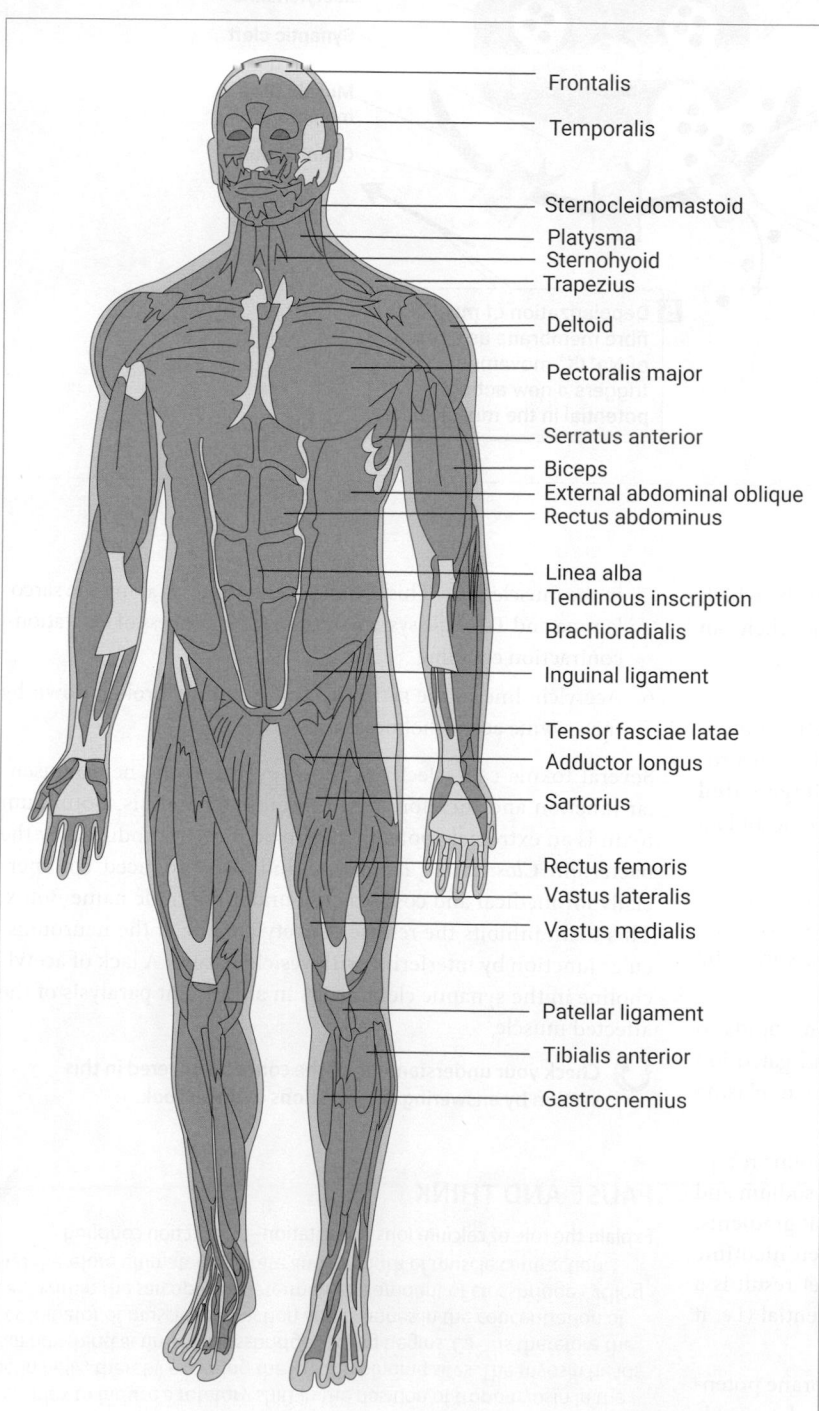

Figure 20.9 Frontal view of the human body illustrating some of the different muscle groups.

Source: Hank Grebe/istockphoto.com.

Frontalis
Temporalis
Sternocleidomastoid
Platysma
Sternohyoid
Trapezius
Deltoid
Pectoralis major
Serratus anterior
Biceps
External abdominal oblique
Rectus abdominus
Linea alba
Tendinous inscription
Brachioradialis
Inguinal ligament
Tensor fasciae latae
Adductor longus
Sartorius
Rectus femoris
Vastus lateralis
Vastus medialis
Patellar ligament
Tibialis anterior
Gastrocnemius

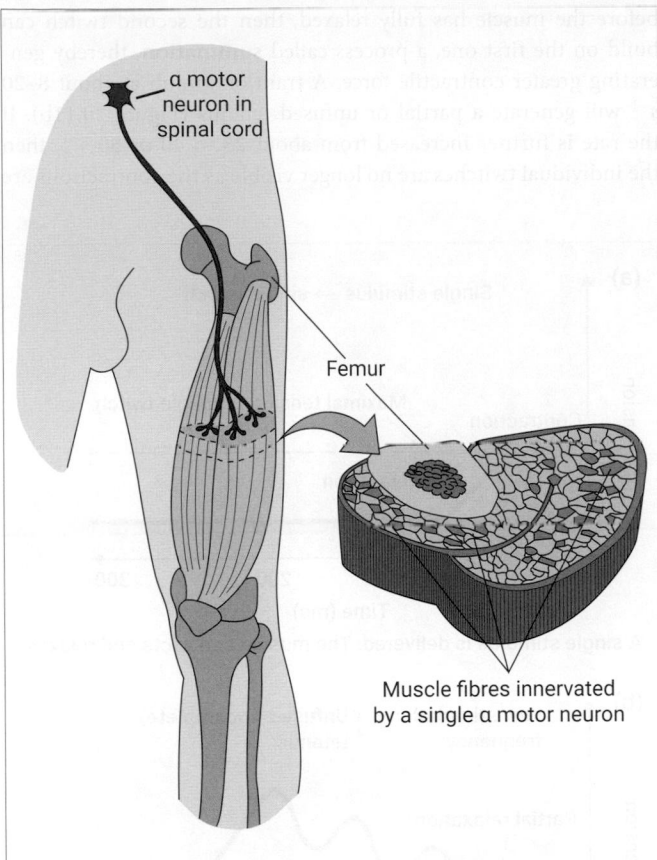

Figure 20.10 Cross-section of the spinal cord, illustrating two α-motor neurons. The cell bodies of the neurons are in the ventral horn of the spinal cord and their axons exit via the ventral root, forming part of a peripheral nerve. Within the muscle the axons of the motor neurons terminate in the neuromuscular junction.

Source: Purves et al. *Neuroscience*, 6th edn, 2017. Oxford University Press.

alpha-motor neurons (α-motor neurons). These α-motor neurons have their cell bodies located in the ventral horn of the spinal cord (Figure 20.10), or in the motor nuclei of the brainstem. The motor neurons from the spinal cord control the muscles of the body (the limbs and trunk), while those from the brainstem control the muscles of the head and face. The axons of these neurons are myelinated (see Section 17.5) and are relatively large in diameter, and therefore have fast conduction velocities of between 50 and 70 m s^{-1}.

The axons of the α-motor neurons leave the central nervous system, forming part of a peripheral nerve, to supply the muscle fibres of the skeletal muscle. The connection with the individual muscle fibres is through the neuromuscular junction, where acetylcholine is released as the transmitter.

The functional unit of each muscle is termed the **motor unit**, which is defined as an α-motor neuron and the group of muscle fibres that it innervates.

Each skeletal muscle fibre is innervated by only one α-motor neuron, but each α-motor neuron innervates a number of muscle fibres. The muscle fibres of each motor unit all function together, so when there is an action potential in the α-motor neuron, all the muscle fibres that it innervates will contract at the same time.

The number of muscle fibres in a motor unit, the **innervation ratio**, varies between muscles, depending on the function of the muscle (Table 20.1).

How might the size of a motor unit affect the precision control of a muscle? Take a moment to think about this question. Typically, those muscles that perform precise movements, such as those in the hand or controlling eye movements, may only have a few tens of muscle fibres in each unit. Conversely, large and powerful muscles, such as the thigh or calf muscles (Table 20.1), may have several thousand muscle fibres in each unit. In the case of the precision muscles, adding the contractions generated by individual motor units together will lead to small gradations in force by comparison with the more powerful muscles that control much less precise movements.

Muscle fibres differ in the speed at which they can contract and in the amount of force they produce:

- slow muscle fibres (Type I) contract relatively slowly and produce low amounts of force;
- fast muscle fibres (Type II) contract relatively quickly and can produce large amounts of force (Table 20.2).

In human muscles the Type II, fast muscle fibres can be further subdivided into Type IIA and Type IIX: Type IIX muscle fibres are classified as fast glycolytic, while IIA are fast oxidative/glycolytic. Type IIA fibres therefore represent an intermediate form of muscle fibre as they can metabolize oxidatively, but, when under exercise stress, with insufficient oxygen supply, they can metabolize glycolytically. The skeletal muscles of smaller mammals such as rats and cats also possess a third, faster contracting type of muscle fibre called IIB.

The group of muscle fibres that makes up each motor unit are all of the same contractile type, so each motor unit is fast or slow

Table 20.1 Comparison of the numbers of muscle fibres per motor unit for different muscles

Muscle	Average number of muscle fibres per motor unit
Inferior rectus (moves the eyeball)	9
First interosseus (hand)	108
Tibialis anterior (front of lower leg)	562
Gastrocnemius (calf muscle)	1934

Table 20.2 Comparison of the properties of the two main types of muscle fibre

Muscle fibre properties	Type I	Type II
Contractile speed	Slow	Fast
Force produced	Low	High
Fibre diameter	Small	Large
Metabolism	Oxidative	Glycolytic
Myosin ATPase rate	Slow	Fast
Resistance to fatigue	High	Low
Myoglobin levels	High	Low
Colour	Red	White
Density of capillaries	High	Low

contracting. The slow motor units are very well-vascularized with a high **myoglobin** content; therefore, they have a rich supply of O_2. They also show an oxidative metabolism, so, although they contract relatively slowly and produce low amounts of force, they can maintain contractions for a long period of time. The fast muscle fibres have a glycolytic metabolism and can not only produce ATP quickly to enable a fast, powerful contraction, but they also fatigue quickly (Table 20.2).

Do you think the proportion of fast and slow units vary between muscles? It does, and this also reflects the function of the muscle, so muscles that have a mainly postural role will have a much higher proportion of Type I muscle compared with those that are involved in fast or powerful movements.

Take a moment to think about examples of these differences. If you have ever looked at the muscles of a chicken, you may see that the leg muscles are much darker in colour than the breast muscles and the fibres are thinner: chickens use their legs for running around but only flap their wings vigorously in very short bursts. The leg muscles have a very high proportion of Type I and IIA muscle fibres, whereas the breast muscles are mainly Type IIB.

Controlling muscle force

Each muscle is made up of a large number of muscle fibres that are grouped together as motor units. The number of muscle fibres per unit and the proportion of the different types of unit vary between muscles depending on their function. The contractile force produced by a muscle therefore depends on:

- the number of muscle fibres in the motor units;
- the type of motor units in the muscle;
- the number of motor units being stimulated;
- the rate of stimulation.

The force generated by each motor unit can be controlled by varying the frequency of stimulation. If the motor neuron fires a single action potential, this will evoke a single twitch contraction of the muscle fibres (Figure 20.11a); often, this will be insufficient to generate any significant movement. If a second stimulus is given

before the muscle has fully relaxed, then the second twitch can build on the first one, a process called **summation**, thereby generating greater contractile force. A train of stimuli at about 8–20 s^{-1} will generate a partial or unfused tetanus (Figure 20.11b). If the rate is further increased from about 25 to 40 or 50 s^{-1}, then the individual twitches are no longer visible as the contractions are

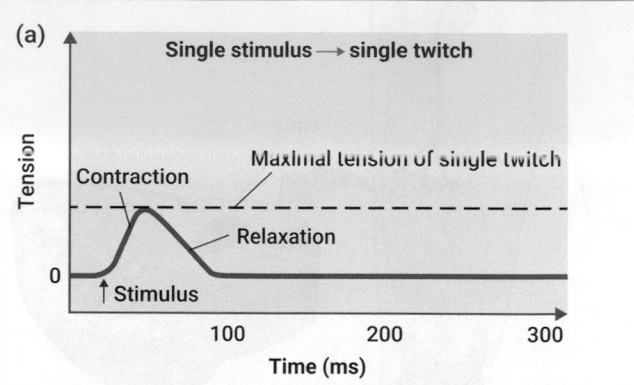

A single stimulus is delivered. The muscle contracts and relaxes.

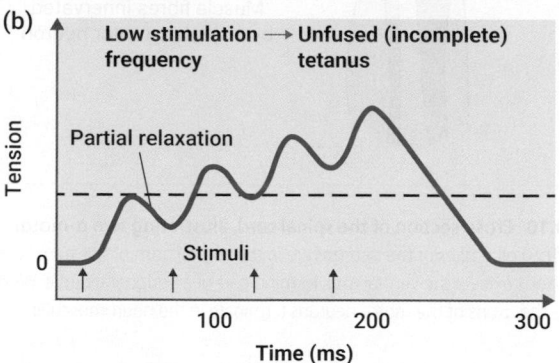

If another stimulus is applied before the muscle relaxes completely, then more tension results. This is wave (or temporal) summation and results in unfused (or incomplete) tetanus.

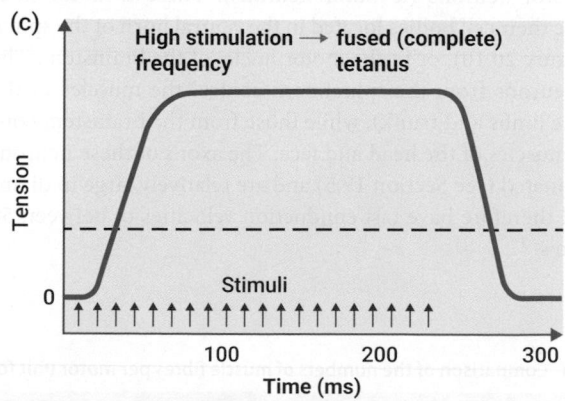

At higher stimulus frequencies, there is no relaxation at all between stimuli. This is fused (complete) tetanus.

Figure 20.11 Contractile tension generated by a single motor unit in response to stimulation at different frequencies.

Source: Pocock et al. *Human Physiology*, 5th edn, 2017. Oxford University Press.

completely fused (Figure 20.11c). This is referred to as a **tetanic contraction**, or tetany, which produces much greater forces than those of the single twitch.

During a single twitch the amount of force and movement developed is small because:

1. Much of the force of the twitch is expended in stretching the elastic tissue of the muscle and so does not produce actual movement.

2. During the contraction, the maximum force is not attained before the Ca^{2+} is pumped back into the SR and so the muscle fibres relax.

During stimulation at higher frequencies the elastic structures (including the tendon) are stretched and so the muscle as a whole shortens and generates movement. However, there is also an increase in the concentrations of Ca^{2+} in the sarcoplasm because it is not all being pumped back as fast as it is being released, enabling more crossbridge interactions and therefore greater force production.

Go to the e-book to watch a video that depicts how contractile tension is generated in response to stimulation.

If you make a muscle contraction of steadily increasing force there is a typical pattern of activation:

- The first motor units to be activated are the small Type I, slow motor units, being stimulated at relatively low action potential rates (8–10 impulses s^{-1}).

- As more force is required, the rate of stimulation increases and more motor units are activated. As each new unit starts to contract, it will begin at 8–10 impulses s^{-1} and gradually increase up to 40–50 impulses s^{-1}, going from an incomplete tetanus (Figure 20.11) to a complete tetanus when the muscle fibres are maximally contracted.

- With more force being required, the larger Type II motor units will start to be activated, again beginning at low rates and gradually increasing until all the motor units in the muscle are contracting.

During typical activities the different motor units are being activated at different rates to suit the type of movement. During standing or walking slowly, the leg muscles will typically be showing contractions of the slow (Type I) units. If the speed of walking increases then there is gradual increase in firing rate and also recruitment of the smaller Type II motor units. When running fast, all the Type II motor units will be involved in powering the movements.

If you try to maintain powerful contractions for a long time, then the muscle will start to fatigue quite quickly: when mitochondrial respiration becomes insufficient to supply the ATP required to maintain the levels of muscle contraction, then ATP regeneration occurs via glycolysis with associated proton production. As a consequence, there is a fall in pH and the amount of force the muscle can generate will decrease. Lactate production within the muscle fibres increases under these conditions and this slows the overall development of acidosis, by supplying the NAD^+ that

Figure 20.12 Different forms of training: (a) distance running for endurance and (b) weight lifting for strength.
Source: (a) Maxisport/Shutterstock.com; (b) Rawpixel.com/Shutterstock.com.

is important in glycolysis and therefore the regeneration of ATP. Lactate accumulation does not therefore appear to be the cause of the acidosis occurring during fatigue, although its presence is an indicator of the development of fatigue and there is some evidence that the levels of lactate in the blood may contribute to the mental feelings of fatigue during intensive exercise. Exercise is covered in more detail in Chapter 23.

Although they are markedly more resistant to fatigue because of their greater vascularization and slower, oxidative metabolism, the Type I muscle fibres will also start to fatigue as the contraction progresses.

What is the effect of training on muscles? Take a moment to think about this question. The effect of training depends to some extent on the type of training (Figure 20.12)! There are two main aspects that can vary: the density of the blood capillaries and the diameter of the muscle fibres. Endurance training will result in muscles that have a richer blood supply, larger mitochondria, and are more resistant to fatigue. Strength training will lead to an increase in fibre diameter (hypertrophy) and so increase the amount of force that can be produced. Note that there is not an increase in the number of muscle fibres (hyperplasia). However, if you sit still day after day, the muscle fibres will gradually get thinner (**atrophy**) and weaker.

PAUSE AND THINK

What factors control the force being generated during an ongoing muscle contraction?

■ The rate of stimulation of each of the units.
■ The number of motor units being stimulated.
■ The type of motor units in the muscle.
■ The number of muscle fibres in the motor units.

Answer:

Muscle architecture

The fibres within muscles are arranged in different ways, again reflecting the function of the muscle. There are two main patterns of arrangement:

■ **longitudinal** architecture;
■ **pennate** ('feather') architecture.

In a longitudinal muscle, the muscle fibres lie in parallel with the direction of movement of the muscle, whereas in a pennate muscle the fibres are orientated at an angle to the direction of movement (Figures 20.13 and 20.14).

In a longitudinal muscle, each muscle fibre may extend for most, if not all, of the length of the muscle, so the fibres tend to be relatively long with a large number of sarcomeres in series. By comparison, in a pennate muscle the muscle fibres tend to be relatively short but there are more fibres for the same volume of muscle (Table 20.3). As a result, longitudinal muscles tend to have a large range of movement and can shorten rapidly but generate lower amounts of force compared with pennate muscles, which have more muscle fibres.

Pennate muscles can vary in the range of angles at which the muscle fibres are oriented and can also vary in the complexity of the architecture; for example, some muscles are bipennate or multipennate (e.g. the deltoid muscle of the shoulder) with several different orientations of the fibres within the body of the muscle.

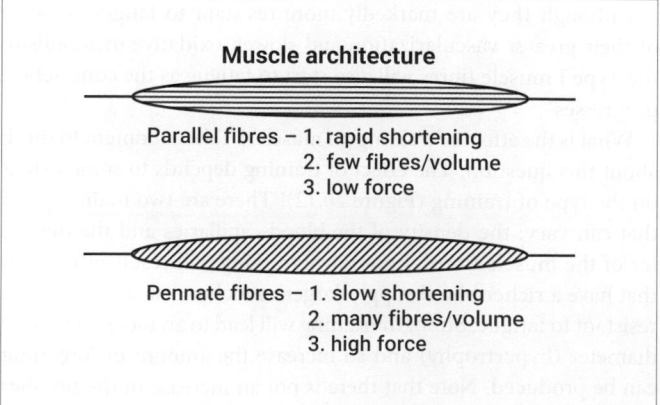

Muscle architecture

Parallel fibres – 1. rapid shortening
2. few fibres/volume
3. low force

Pennate fibres – 1. slow shortening
2. many fibres/volume
3. high force

Figure 20.13 The two basic types of muscle architecture.

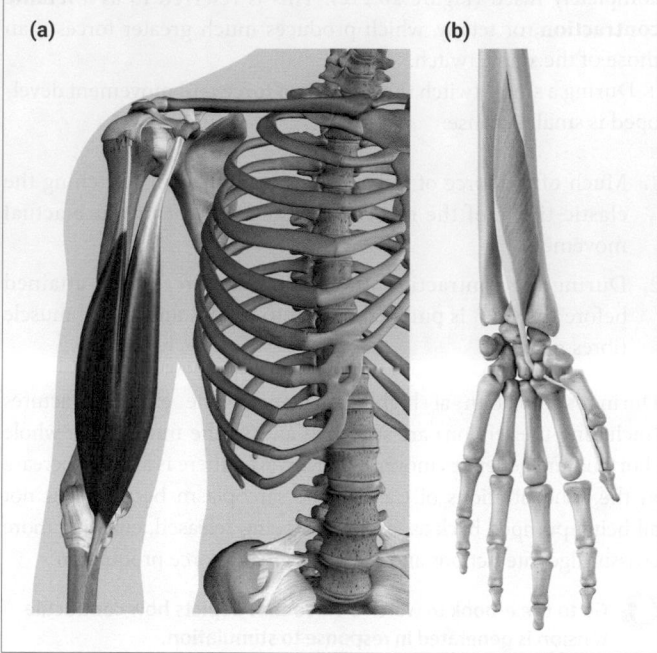

Figure 20.14 Examples of (a) longitudinal (biceps brachii) and (b) pennate (flexor pollicis longus) muscles. Note that in the longitudinal muscle the muscle fibres attach to the tendon at each end, whereas in the pennate muscle the distal tendon runs almost the full length of the muscle, providing attachment for the muscle fibres.

Source: (a) Hank Grebe/istockphoto.com; (b) SciePro/istockphoto.com.

Table 20.3 Comparison of the properties of muscles with different arrangements of muscle fibres

Muscle properties	Longitudinal	Pennate
Fibre length	Long	Short
Number of sarcomeres in series	High	Low
Range of movement	Large	Small
Velocity of shortening	High	Low
Number of fibres in cross-section	Small	Large
Contractile force	Low	High

Tendons and muscle contraction

Tendons form the mechanical link between the muscle fibres and the bones of the skeleton, and so they transmit the movement force of muscle shortening to move the bones of the skeleton. As such, they could be viewed simply as ropes that pull on the bones when the muscles contract, but they also play an important role in transmitting the force of movement from bones and storing the energy as a result of stretching.

Tendons are mainly made up of tightly packed parallel fibres of the connective tissue protein, type 1 **collagen**, along with a small amount of elastin. Where the muscle fibres join the tendon, there is a gradual merging between muscle tissue and collagen. At the other end, the tendon extends into the **periosteum**, the connective tissue membrane that covers the bone. The length and degree of elasticity of the tendon varies significantly, with some tendons being short and stiff, whereas others, such as the Achilles tendon of the heel (Figure 20.15), are long and relatively elastic.

Stretching long tendons stores potential energy within the collagen fibres that can be released to aid subsequent movements. For example, during walking, when the foot hits the ground, the calf muscles are contracted and therefore only lengthen (eccentrically)

a small amount, so the weight of the body on the foot as it flattens onto the ground leads to stretching of the Achilles tendon (Figure 20.16). When the foot then pushes against the ground to take the next step (just before the start of the swing phase), some of the potential strain energy stored in the stretched Achilles tendon is returned as the tendon elastically shortens. This storage and release of elastic energy thereby increases the efficiency of movement; it also means the tendon can act like a shock absorber. During running the amount of stretching increases with the increase in the force being applied to the tendon.

Can you think of an animal that has a very long Achilles tendon that is stretched during locomotion? Take a moment to think before reading on. The kangaroo stores significant amounts of potential strain energy as it stretches its Achilles tendons during hopping and the elastic recoil helps power the next bounce.

The muscle–tendon unit acts rather like a weight on a spring: if the weight is pulled down and released it will bounce up and down on the spring at a specific frequency, the **resonant frequency**, which is determined by the mass of the weight and the stiffness of the spring. The frequency increases with increasing stiffness of the spring and decreases with increasing mass.

How does this relate to locomotion? Again, take a moment to think before reading on. If we assume that the body and the Achilles tendon act as a mass–spring system, then there will be an optimal frequency of running at which the elastic properties of the tendon store and return the highest amounts of energy, and therefore locomotion is most efficient (at least in terms of this system). Experiments of people hopping on a treadmill show that over a range of treadmill speeds, the subject maintains the same frequency of hopping but varies the hop distance. This frequency coincides with the optimum recovery of potential energy from the Achilles tendon. When the subject is forced to hop faster or slower, then the energy recovery is less.

When a muscle is contracting it is not necessarily shortening. That might seem strange at first but the term 'contraction' means the active cycling of the crossbridges between the actin and myosin

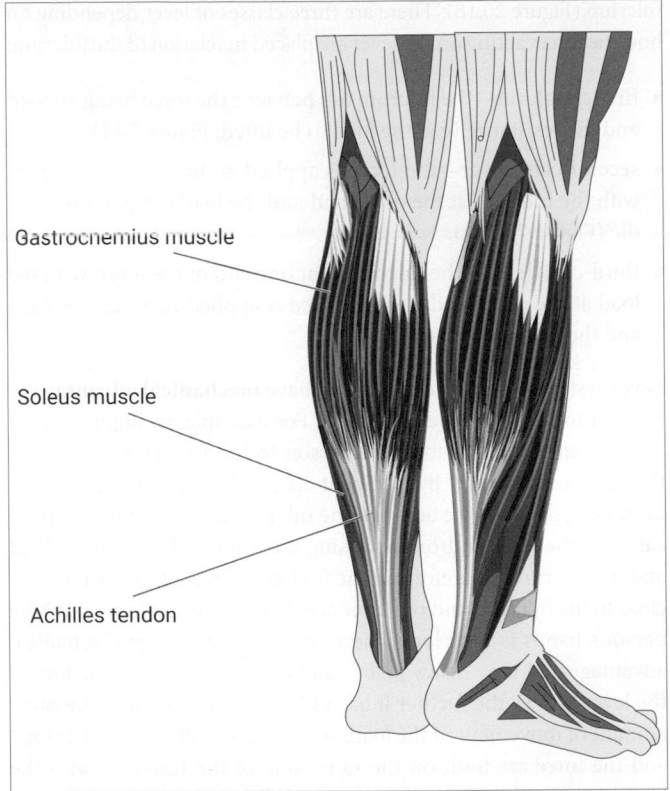

Figure 20.15 The Achilles tendon.

Figure 20.16 Changes in length of the human Achilles tendon during (a) walking and (b) running.

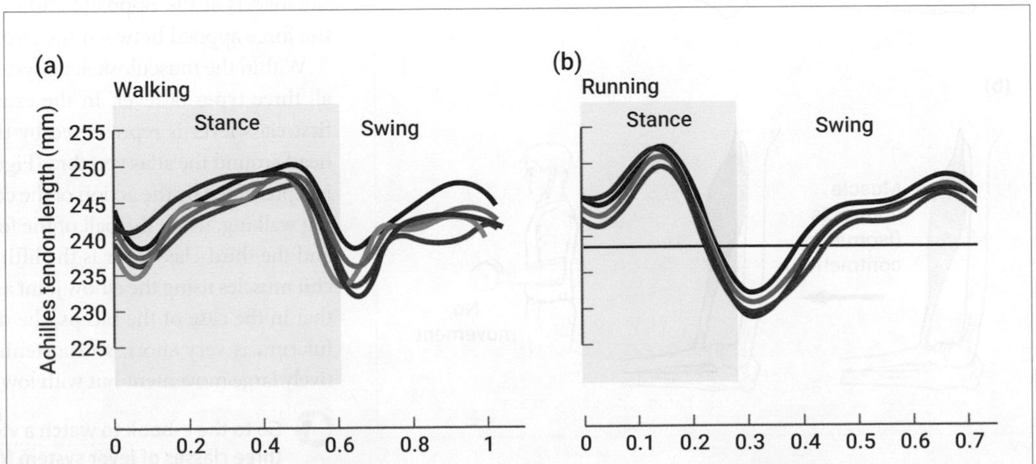

filaments (see Figure 20.5). Muscle contractions can be defined mechanically in several ways:

- **Isotonic contractions**: 'equal tension'—this is a contraction where the tension within the muscle remains the same during the shortening process, for example lifting a fixed weight at a constant velocity.

- **Isometric contractions**: 'equal length'—this is a contraction where the load against the muscle equals or exceeds the contractile force being generated, for example holding a weight in a fixed position, or trying to lift a weight off the ground that is too heavy and so does not move; in both cases, the muscle is generating tension but is not shortening (Figure 20.17b).

- **Concentric contractions**: contractions during which the muscle is shortening; the tension being developed may change or not during the contraction. For example when raising a weight or when standing up from a squatting position (Figure 20.17a).

- **Eccentric contractions**: contractions during which the muscle is actually lengthening, for example when lowering a weight to the ground or when lowering oneself from standing to squatting. When you run downhill, the quadriceps muscles of the thighs are typically undergoing eccentric contractions because you are using them to slow your descent.

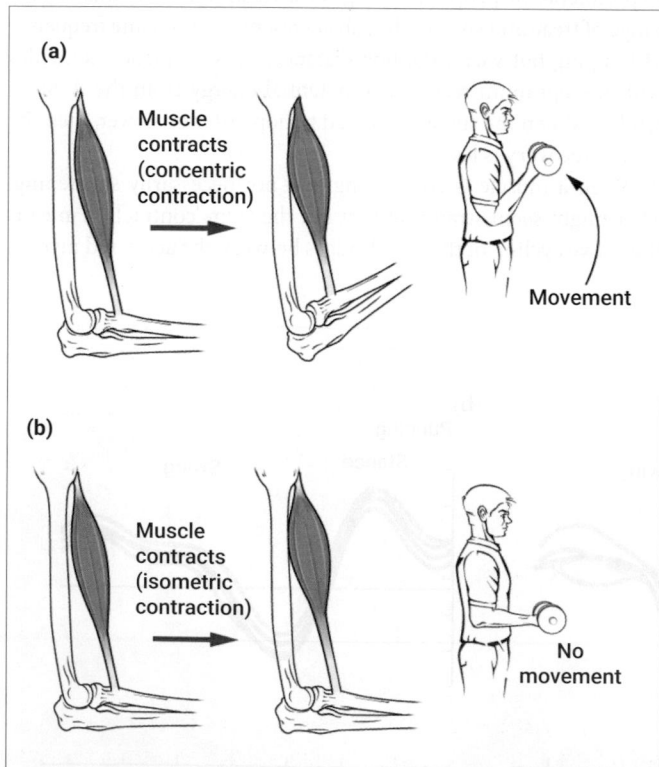

Figure 20.17 (a) Concentric and **(b)** isometric contractions.

The limb as a lever

In order for us to make a movement, the muscles and tendons must act on the bones of the skeleton. The bones are covered by the periosteum, which is a thin layer of very tough connective tissue. The collagen of the tendons and of the ligaments and joint capsules merges with the periosteum creating a continuous set of linkages.

The bones of the skeleton act as levers: the structure of the bone acting as the lever arm and the joint between the bones acting as the fulcrum (Figure 20.18). There are three classes of lever depending on how the forces acting on the lever are placed in relation to the fulcrum:

- first-class lever—the fulcrum lies between the force being applied and the resistance (e.g. the load to be lifted; Figure 20.18a);

- second-class lever—the force is applied at one end of the lever, with the fulcrum at the other end and the load lying in the middle (Figure 20.18b);

- third-class lever—the fulcrum is at one end of the lever, with the load at the other end and the force is applied between the load and the fulcrum (Figure 20.18c).

Lever systems enable the muscles to have **mechanical advantage** in relation to the object being moved. For example, in Figure 20.18d the first-class lever enables the person to lift a much heavier load than they could simply by picking it up. The leverage is improved by increasing the distance between the fulcrum and the point of application of the force and/or decreasing the distance between the load and the fulcrum. Therefore, in the first-class lever, if the load is very close to the fulcrum and the lever arm between the fulcrum and the person's hands is very long, there will be a very large mechanical advantage and very heavy loads can be lifted. However, the longer the lever arm is, the further it has to be moved to achieve the same amount of movement of the load. In the second-class lever, the load and the force are both on the same side of the fulcrum, with the load in the middle, as in the example of the wheelbarrow (Figure 20.18e). Again, the lever enables heavy loads to be moved: the closer the load is to the fulcrum, the heavier the load that can be lifted but the height of the lift is reduced. For the third-class lever, the load is at the opposite end of the lever from the fulcrum with the force applied between the two (Figure 20.18f).

Within the musculoskeletal system of the human body we can find all three types of lever. In the examples shown in Figure 20.18, the first-class lever is represented by the up or down movements of the head around the atlas vertebra (Figure 20.18g). The second-class lever is represented by the action of the calf muscles in raising the body during walking, using the ball of the foot as the fulcrum (Figure 20.18h), and the third-class lever is the lifting of the weight by the biceps brachii muscles using the elbow joint as the fulcrum (Figure 20.18i). Note that in the case of the biceps, the distance between the force and the fulcrum is very short, so shortening of the muscle produces a relatively large movement but with low mechanical advantage.

Go to the e-book to watch a video depicting examples of the three classes of lever system found within the body.

Check your understanding of the concepts covered in this section by answering the questions in the e-book.

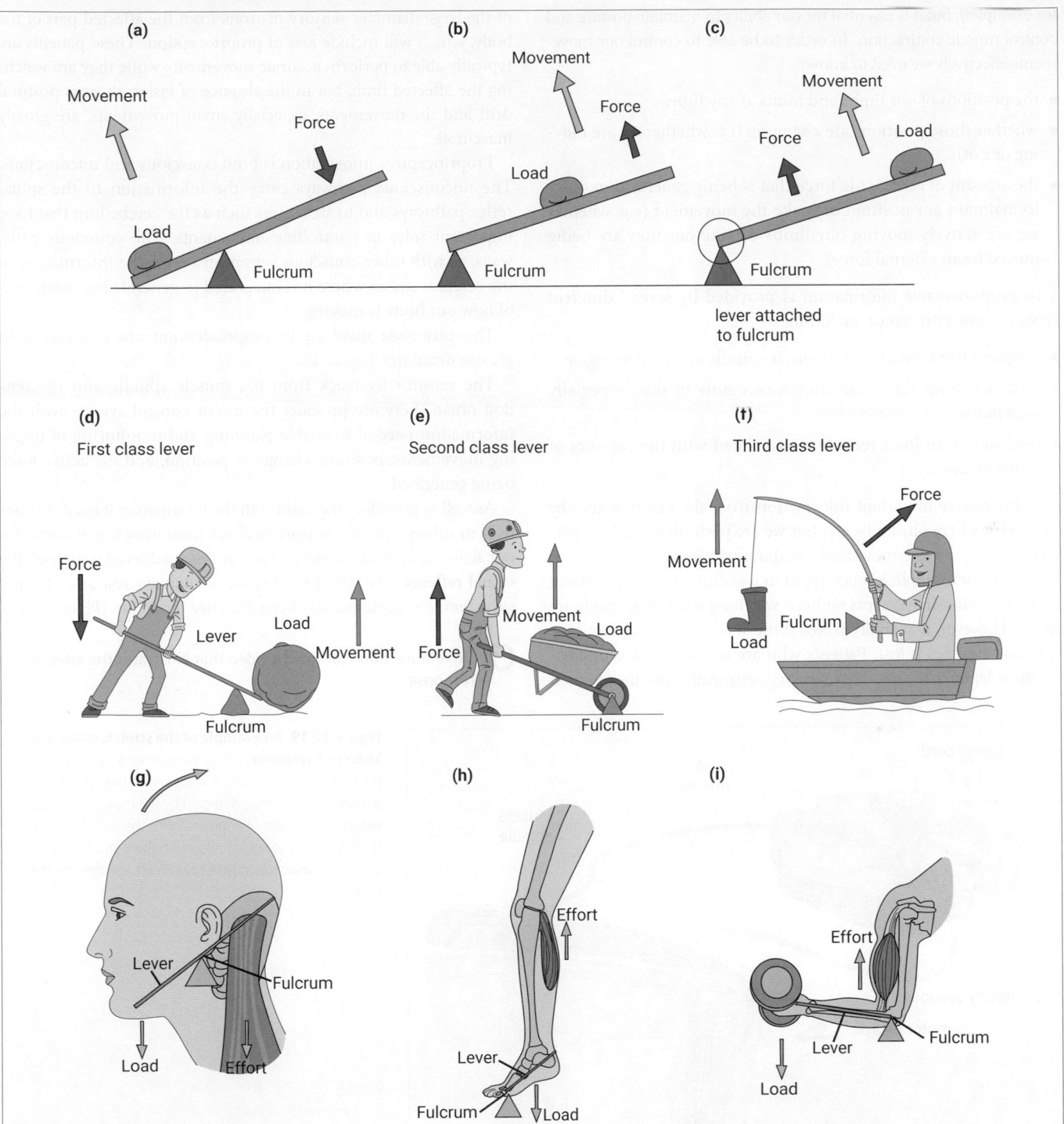

Figure 20.18 Examples of the three classes of lever system found within the body. (a, b, c) The mechanical arrangements; (d, e, f) examples of how we might use these systems externally; (g, h, i) examples of where they are found in the body.
Source: udaix/Shutterstock.com.

20.3 Reflexes and posture

If we look at the composition of a peripheral nerve supplying a muscle, about two-thirds of the nerve fibres are sensory neurons carrying information from the muscle back to the central nervous system. As well as the pain fibres, there are also large numbers of sensory fibres conveying information about the activity of the muscle: whether it is lengthening or shortening, and the contractile tension being generated. This feedback contributes to the sense of **proprioception** (Section 18.4): we don't normally think consciously about proprioception (unlike pain, for example!), but it is essential for our ability to maintain posture and control muscle contraction. In order to be able to control our movements effectively we need to know:

■ the positions of our limbs and joints at any time;

■ whether those positions are changing (i.e. whether we are moving or not);

■ the amount of contractile force that is being generated in order to maintain our position or make the movement (e.g. whether we are actively moving our limbs or whether they are being moved by an external force).

This proprioceptive information is provided by several different types of sensory receptor, including:

■ feedback from muscle—the muscle spindle and tendon organ;

■ feedback from skin—the stretch receptors in skin, especially over joints;

■ feedback from joint receptors associated with the capsules of synovial joints.

We also receive important information from the visual system by looking at where our limbs are, but we are perfectly capable of performing very precise movements without watching what our limbs are doing; for example, a touch typist or a skilful pianist can perform rapid, precision movements without watching what their hands are doing. However, that ability is lost if the sensory feedback from the skin and muscles is lost. Patients who are affected by a neurological disorder termed large-fibre sensory neuropathy lose the function of the large diameter sensory neurons from the affected part of the body, which will include loss of proprioception. These patients are typically able to perform accurate movements while they are watching the affected limb, but in the absence of vision there is postural drift and the movements, especially small movements, are grossly inaccurate.

Proprioceptive information is both conscious and unconscious. The unconscious pathways carry the information to the spinal reflex pathways and to structures such as the cerebellum that have important roles in controlling movements. The conscious pathways, as with other conscious sensations, relay the information to the cerebral cortex where it is integrated to generate our awareness of how our body is moving.

The processes involved in proprioception are considered in greater detail in Chapter 18.

The sensory feedback from the muscle spindle and the tendon organ therefore provides the motor control system with the information needed to enable planning and monitoring of ongoing movements: position, change in position, and the active force being generated.

As well as providing the brain with the information it needs to control our movements, the sensory feedback from muscle is also vital for our ability to maintain posture. One way this is achieved is through the **spinal reflexes**. The spinal reflexes are involuntary responses to specific stimuli, a classic example being the knee-jerk reflex (Figure 20.19).

Go to the e-book to watch a video that illustrates the knee-jerk response.

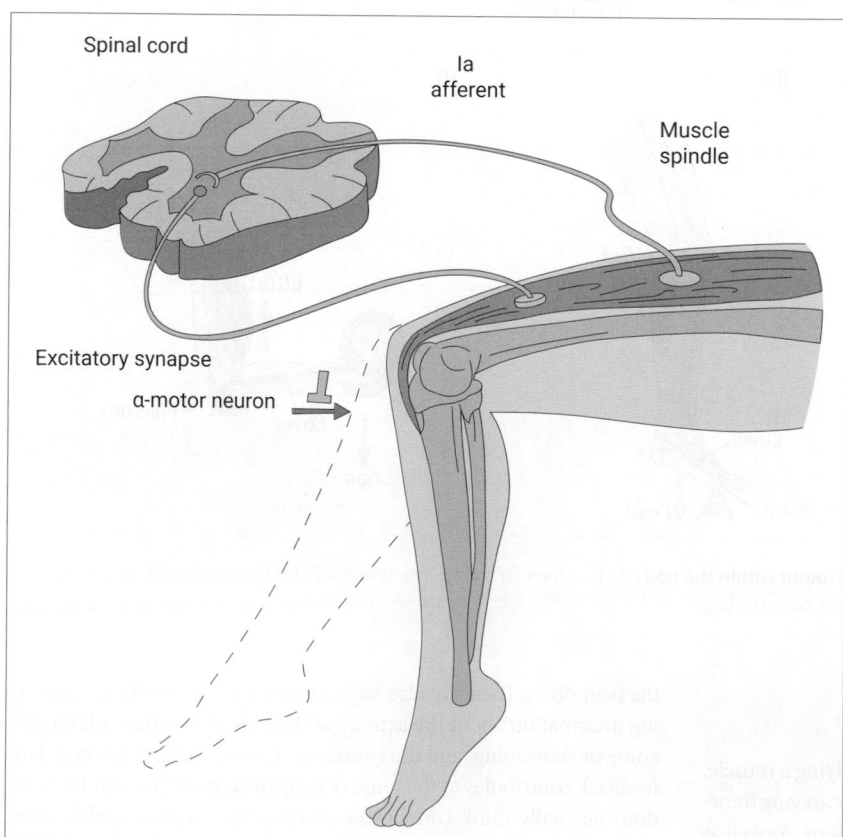

Figure 20.19 An example of the stretch reflex: the knee-jerk response. When the patellar tendon is struck by the hammer, a brief stretch is applied to the quadriceps femoris muscle. This is detected by the muscle spindles, which send a burst of impulses up to the spinal cord. The Ia (and II) afferent axons synapse with the α-motor neurons, which stimulate the muscle to contract and oppose the stretch.

Spinal cord

Ia
afferent

Muscle
spindle

Excitatory synapse

α-motor neuron

The knee-jerk reflex is a specific example of the monosynaptic **stretch reflex**. In terms of their neural circuitry, the stretch reflexes are the simplest of all the reflex arcs and can be generated by almost all of the skeletal muscles. As shown in Figure 20.19, the stretch reflex is mediated by the muscle spindle: when the muscle is stretched (in this case by the tap from the tendon hammer), the muscle spindle signals the stretch with a burst of action potentials in the Ia (and II) afferent. This is relayed to the spinal cord where the Ia afferent branches extensively; one set of synaptic connections is made with the α-motor neurons that innervate the same muscle, thereby stimulating the muscle to contract.

The stretch reflex is therefore an example of a **negative feedback loop**: stretch of the muscle gives rise to a contraction that counteracts the stretch, restoring the muscle to its original length. The stretch reflex operates in combination with a second reflex: **reciprocal inhibition** (Figure 20.20).

In the case of the knee-jerk, the muscle spindle afferents from the quadriceps femoris muscle excite the α-motor neurons to that muscle, causing it to contract. At the same time, they also synapse with inhibitory **interneurons** that inhibit the α-motor neurons of the antagonist muscle, biceps femoris, causing it to relax. The reciprocal inhibition reflex thereby facilitates the action of the stretch reflex.

What is the significance of these reflexes to our control of posture? Take a moment to think about this. The stretch reflex and associated reciprocal inhibition are fundamental to our ability to maintain posture because they enable very rapid responses to postural disturbances.

As an extreme example, think about a ballet dancer standing on the tips of their toes, also called 'en pointe' (shown in Figure 20.21).

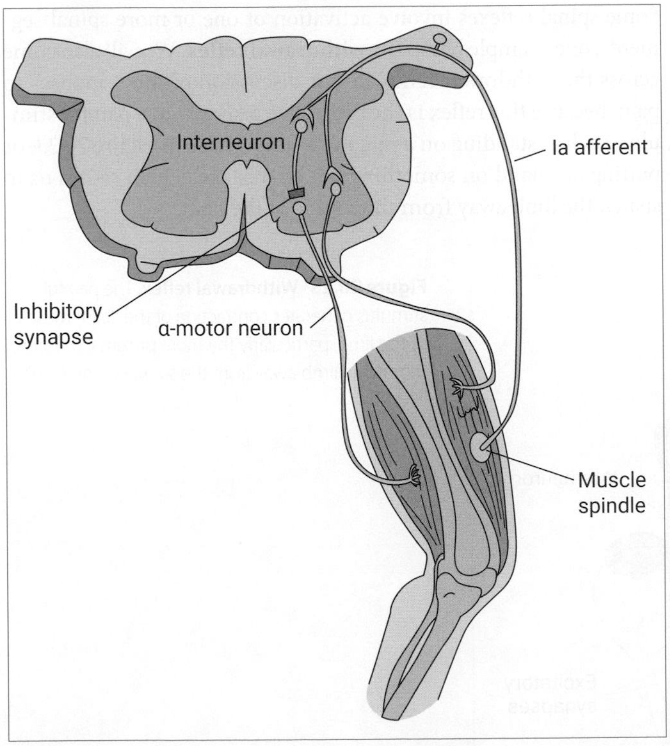

When we are standing, we are unstable because our centre of gravity (which is approximately around the navel) is a long way above the points of articulation such as the ankle. Stability around the ankle joint is maintained by the contraction of the muscles acting on the joint. As the body sways, muscles around the joint will be stretched, or will shorten accordingly. If one of the muscles is stretched, due to sway, the stretch reflex will be activated, leading to increased contraction of that muscle and relaxation of the antagonist, until the correct posture is restored. The correction is very rapid because the spinal reflex pathway only extends as far as the specific segment of the spinal cord; in the case of the ankle, the time for detecting the stretch and stimulating the muscle to increase contraction is about 30 ms.

You can feel your stretch reflexes in operation if you stand on one leg and feel how the muscles around the ankle are spontaneously contracting and relaxing to maintain your balance. This is even more pronounced if you try closing your eyes—but make sure you have someone to catch you if you fall!

The strength of the reflexes can be controlled depending on the conditions; for example, during walking, the size of the stretch reflexes varies depending on the phase of the movement. During the swing phase (i.e. when the foot is clear of the ground and swinging forward), the reflexes are relatively weak, but at the start of the stance phase, when the foot makes contact with the ground and the leg is taking the body weight, the strength of the reflex is greatly increased.

How might the CNS control the strength of the reflexes? Experimental evidence shows that this is a complex process involving

Figure 20.20 Reciprocal inhibition acts to complement the stretch reflex. The muscle spindle response to stretch causes the agonist muscle to contract and also, via the inhibitory interneuron, causes the antagonist muscle to relax.

Figure 20.21 The stretch reflex in action. A ballet dancer standing en pointe is inherently unstable and relies on stretch reflexes to help maintain balance.
Source: Ievgen Repiashenko/Shutterstock.com.

a number of systems, but one way is by modifying the sensitivity of the muscle spindle by stimulating the γ-motor neurons (see Figure 18.10).

Why might a clinician test the stretch reflexes, such as the knee-jerk? Take a moment to think about your answer to this question. It is because the amplitude of the stretch reflexes is under CNS control, so hypo- or hyperactive active reflexes can be indicative of neurological damage.

The monosynaptic, spinal stretch reflex provides a very rapid response to disturbances to posture or ongoing muscle contractions, but this is supplemented by the 'long-latency' stretch reflex, which is also partly initiated by activation of the muscle spindles. The long-latency stretch reflex, as its name suggests, takes longer to have effect than the spinal stretch reflex, with onset times of about 50 ms for the arms and 100 ms for the lower legs. There is evidence that these reflexes may be routed via the motor cortical areas of the brain and that there are very different patterns of modulation of the reflex depending on the nature of the movement or postural task being performed.

So, given what you have just read, what is wrong with the common term 'knee-jerk reaction'? Pause while you consider this question. When people use the term knee-jerk reaction, they take it to mean an unthinking, stereotyped response to a stimulus. Actually, as we have seen, the knee-jerk, like the other stretch reflexes, provides a very rapid involuntary response, but one that can be finely modulated to suit the specific situation.

The muscle spindle is not the only sensor to be involved in eliciting reflexes. The Ib afferent of the tendon organ (Section 18.4) also contributes to a spinal reflex, **autogenic inhibition**. Autogenic inhibition, as illustrated in Figure 20.22, involves the inhibition of ongoing muscle contraction as a result of the feedback from the tendon organ Ib sensory afferent neuron.

The Ib afferents from the tendon organ branch on entering the spinal cord and one set of synaptic connections is with inhibitory interneurons that synapse with the α-motor neurons innervating the same muscle. As a result, the feedback from the tendon organ results in inhibition of the ongoing contraction. At first it might seem strange for the reflex to inhibit the muscle contraction, but

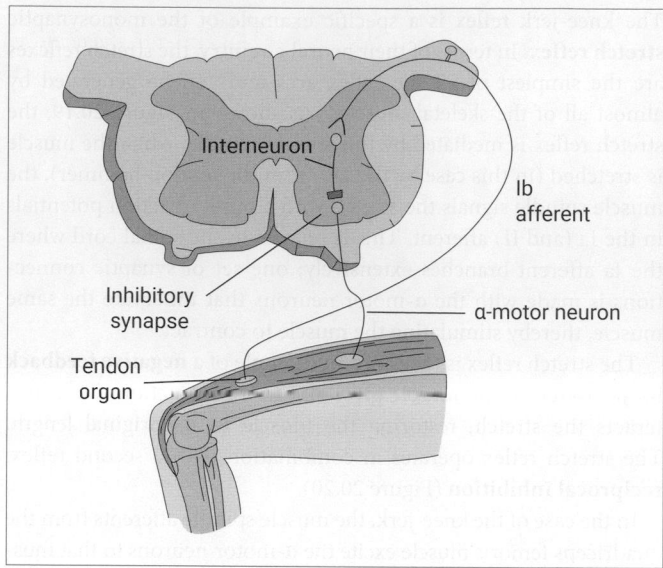

Figure 20.22 Autogenic inhibition. This reflex is mediated by sensory feedback from the Ib afferent of the tendon organ, resulting in inhibition of the α-motor neurons innervating the same muscle.

there is evidence that the action of the feedback is to balance out the level of activity across the muscle, thereby assisting with the generation of an even contraction.

The spinal reflexes we have considered so far are termed **segmental reflexes**, because the interconnections involved do not ascend or descend in the spinal cord, but remain at the same level. Some spinal reflexes involve activation of one or more spinal segments, one example being the **withdrawal reflex**. We will also come across the withdrawal reflex in our discussion of the responses to pain, because this reflex is the protective response to a painful stimulus, such as standing on a pin, the example used in Figure 20.23, or putting our hand on something hot by mistake, which causes us to snatch the limb away from the source of the pain.

Figure 20.23 Withdrawal reflex. The painful stimulus generates contraction of the flexor muscle of the limb, particularly the more proximal muscles, to pull the limb away from the source of the pain.

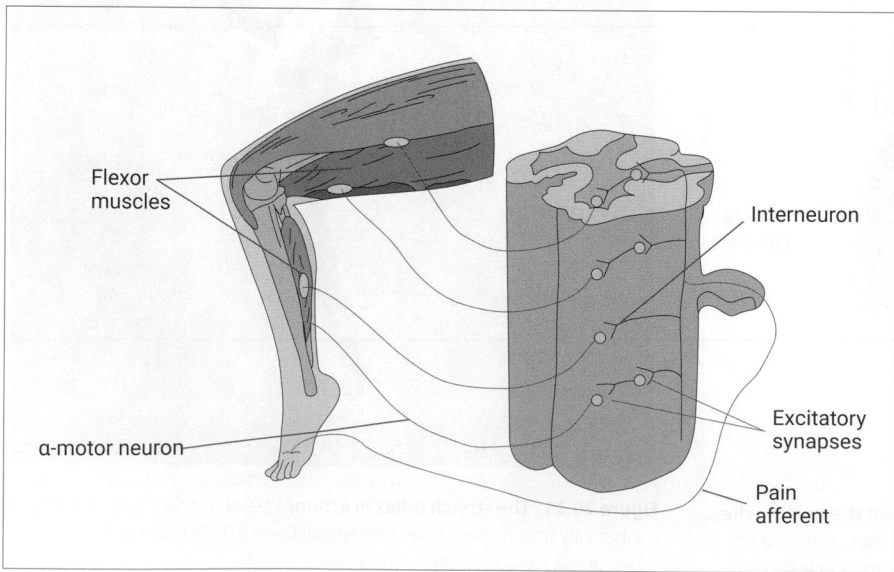

The pain afferents from the skin ascend several segments within the spinal cord and make excitatory connections with the α-motor neurons that innervate the flexor muscles of the limb, particularly those of the proximal muscles (i.e. those closer to the trunk of the body). When there is a painful stimulus, which therefore is likely to cause damage to the body, the withdrawal reflex causes the flexor muscle to contract and so rapidly pull the limb away from the source of the pain.

 Check your understanding of the concepts covered in this section by answering the questions in the e-book.

20.4 Overview of the motor system

The control of movement is one of the most complex functions that is coordinated by the CNS and involves a number of areas of the brain, in particular the motor areas of the cerebral cortex: primary motor cortex, pre-motor cortex, and supplementary motor area; and the sub-cortical structures: the cerebellum, the basal ganglia, and the brainstem (all marked on Figure 20.24). Learning how to control movement is a lengthy process: the baby batting at the mobile in the cot is learning how to control his or her arm and that learning process continues well into adolescence as we develop expertise at the most complex precision movements. This learning process requires continued practice: at its highest level, think of the tennis star who can precisely control the speed, direction, and spin on the serve, or the concert pianist who can play the most demanding concert repertoire.

Controlling each movement depends on the development and execution of a coordinated movement plan that is built up from a wide range of sensory information derived from external sensory information, such as vision and touch, and internal information such as proprioception, the feedback from the muscles and joints. This sensory information is used in conjunction with motor memory, the learnt motor patterns that have been built up through practice of different types of movement (Figure 20.25).

During the course of execution of the movement the sensory feedback is continuously being updated. If the movement is relatively slow, this feedback can be used to correct for errors during the execution of the movement and also used to improve the plan for next time as we learn from mistakes.

PAUSE AND THINK

Why can't we correct fast, so-called 'ballistic' movements during their execution?

Answer: Ballistic movements, such as throwing a ball, are executed as a single performance because they are too fast for us to enable the use of feedback.

Descending motor pathways

The motor pathways that convey the motor commands from the brain to the α-motor neuron populations of the spinal cord and brainstem originate from two main areas: the motor areas of the cerebral cortex and the brainstem (see Figure 20.24). The neurons that make up these pathways are referred to as **upper motor neurons**, distinguishing them from the **lower motor neurons** (e.g. the α- and γ-motor neurons) that innervate the skeletal muscles (Section 20.2) (see also Clincal Box 20.1). There are two major sets of pathways descending from the motor cortical areas:

- corticospinal tract;
- corticobulbar tract.

The names of the descending pathways identify their routes, so the corticospinal pathway descends from the cerebral cortex to the spinal cord and the corticobulbar pathway descends from the cerebral cortex to the 'bulb', which is the old name for the brainstem.

There are several motor pathways that originate in the brainstem and descend to the spinal cord, the major ones being:

- vestibulospinal tract;
- tectospinal tract;
- reticulospinal tract;
- rubrospinal tract.

We will come back to these brainstem pathways shortly, but for now let us focus on examining the corticospinal and corticobulbar tracts in more detail.

CLINICAL BOX 20.1   Upper and lower motor neuron lesions

In clinical terms, the lower motor neurons are those neurons that originate within the spinal cord and brainstem and directly innervate the skeletal muscles. The upper motor neurons are those that form part of the descending motor system and are restricted to the CNS.

Lesions of the lower motor neurons will therefore leave the respective skeletal muscle(s) without a nerve supply. As a result, voluntary control of the muscle will be lost, stretch reflexes will be absent and, over time, the muscle will atrophy as a result of disuse.

Lesions of the upper motor neurons, for example as a result of a spinal injury or a stroke, tend to be widespread and may result in paralysis of significant regions of the body, such as hemiplegia, which is paralysis of one side of the body. Note that because the motor pathways cross over, strokes affecting regions such as the motor cortex will result in deficits in the opposite side of the body. Unlike lower motor neuron lesions, upper motor neuron lesions will result in exaggerated stretch reflexes.

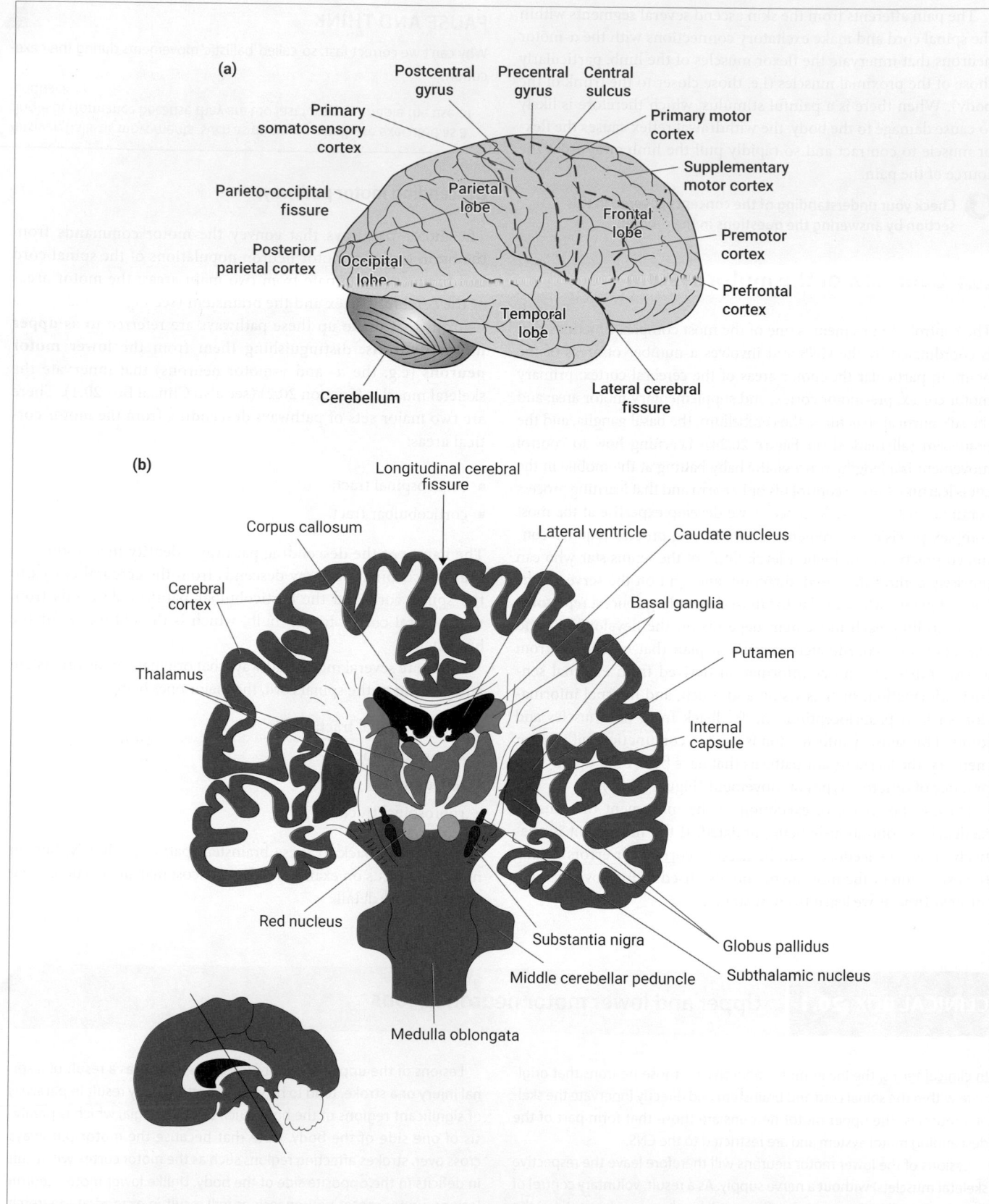

Figure 20.24 Motor areas of the brain. (a) Lateral view of the brain; (b) a coronal section to illustrate the location of the nuclei making up the basal ganglia.

Source: Pocock et al. *Human Physiology*, 5th edn, 2017. Oxford University Press.

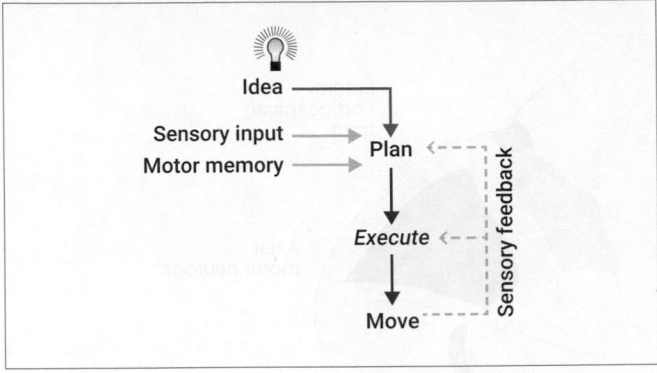

Figure 20.25 Schematic showing the development of a conscious movement.

Corticospinal and corticobulbar tracts

The corticospinal tract originates mainly from the motor cortical areas, with 30 per cent of the neurons originating in the primary motor cortex, 30 per cent from the pre-motor cortex and supplementary motor area, and the remaining 40 per cent from the somatosensory cortex and parietal lobe (Figure 20.26).

The neurons of the corticospinal tract descend from the motor cortical areas through the **internal capsule** and then down through the **brainstem**, where they lie on the anterior surface of the midbrain, pons, and medulla oblongata (shown in Figure 20.26). In the medulla oblongata the neurons are located in swellings on the anterior surface of the medulla, which are referred to as the **medullary pyramids**. As a consequence, these tracts are also sometimes termed the **pyramidal tracts**. At the lowest level of the medulla, about 85 per cent of the fibres cross over to the other side of the brainstem, a process referred to as **decussation**. The fibres that have crossed over comprise the **lateral corticospinal tract** and travel in the lateral part of the contralateral (opposite side) spinal cord (Figure 20.27).

The neurons terminate in the ventral horn of the spinal cord, in the mseeotor neuron pools of the lower motor neurons that innervate the muscles of the limbs, particularly the distal muscles: a much greater proportion of the corticospinal tract neurons supply the more distal muscles of the limbs, especially those of the hand, compared with the more proximal muscles, which corresponds with the relative areas of the motor cortex given over to controlling these different regions (Figure 20.28). As a consequence of the fibres crossing over in the medulla, the left side of the brain controls the right side of the body and vice versa.

PAUSE AND THINK

What is the consequence of the different sizes of representation of the body areas?

Answer: The areas that have a larger representation, in particular the hands and the face, are those parts of the body over which we have the most precise control. Also, think back to the equivalent sensory representations (see Figure 18.7).

Figure 20.26 Pathways of the corticospinal tract.

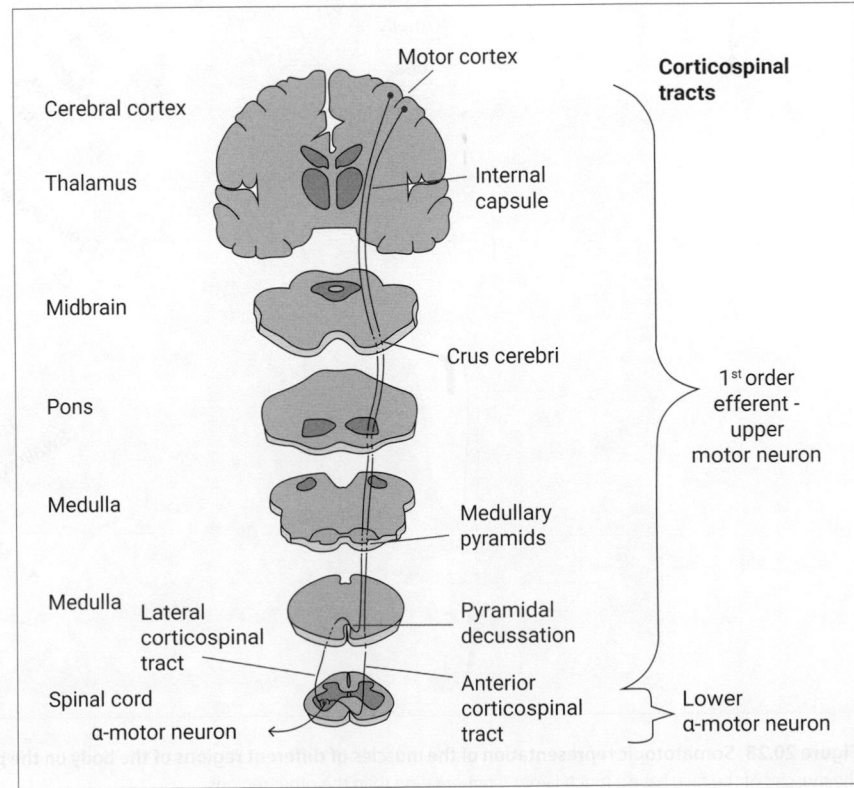

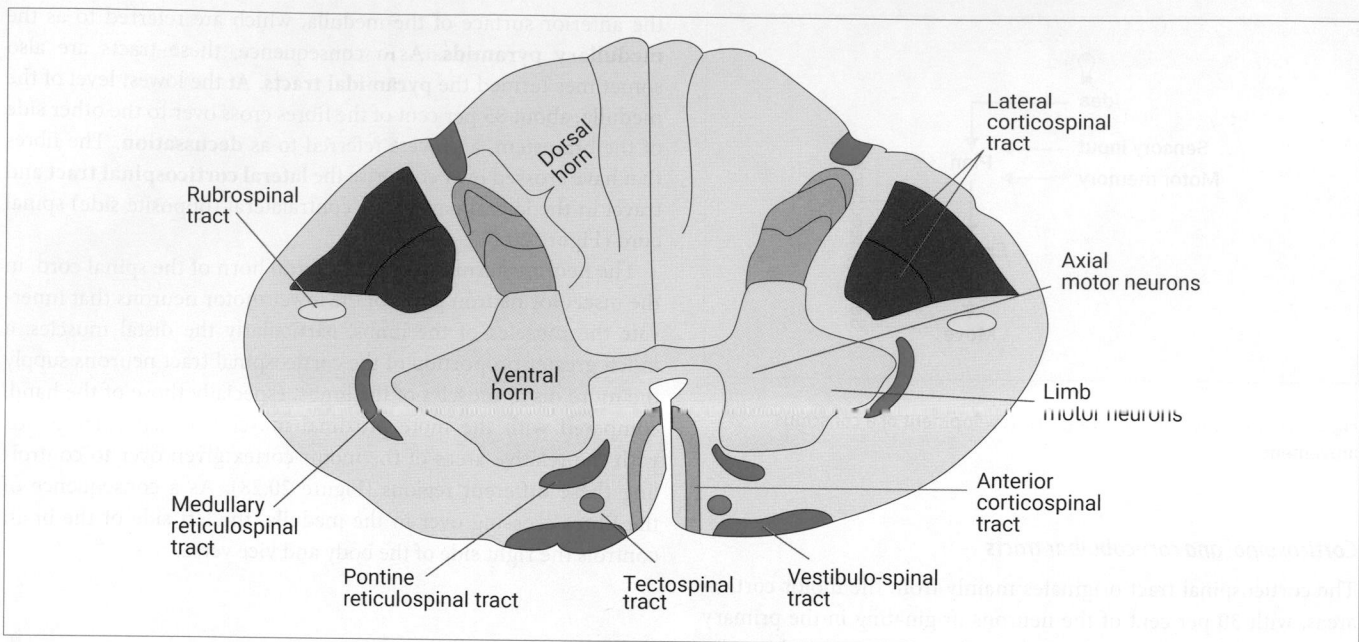

Figure 20.27 Location of the descending motor pathways in a cross-section of the spinal cord.

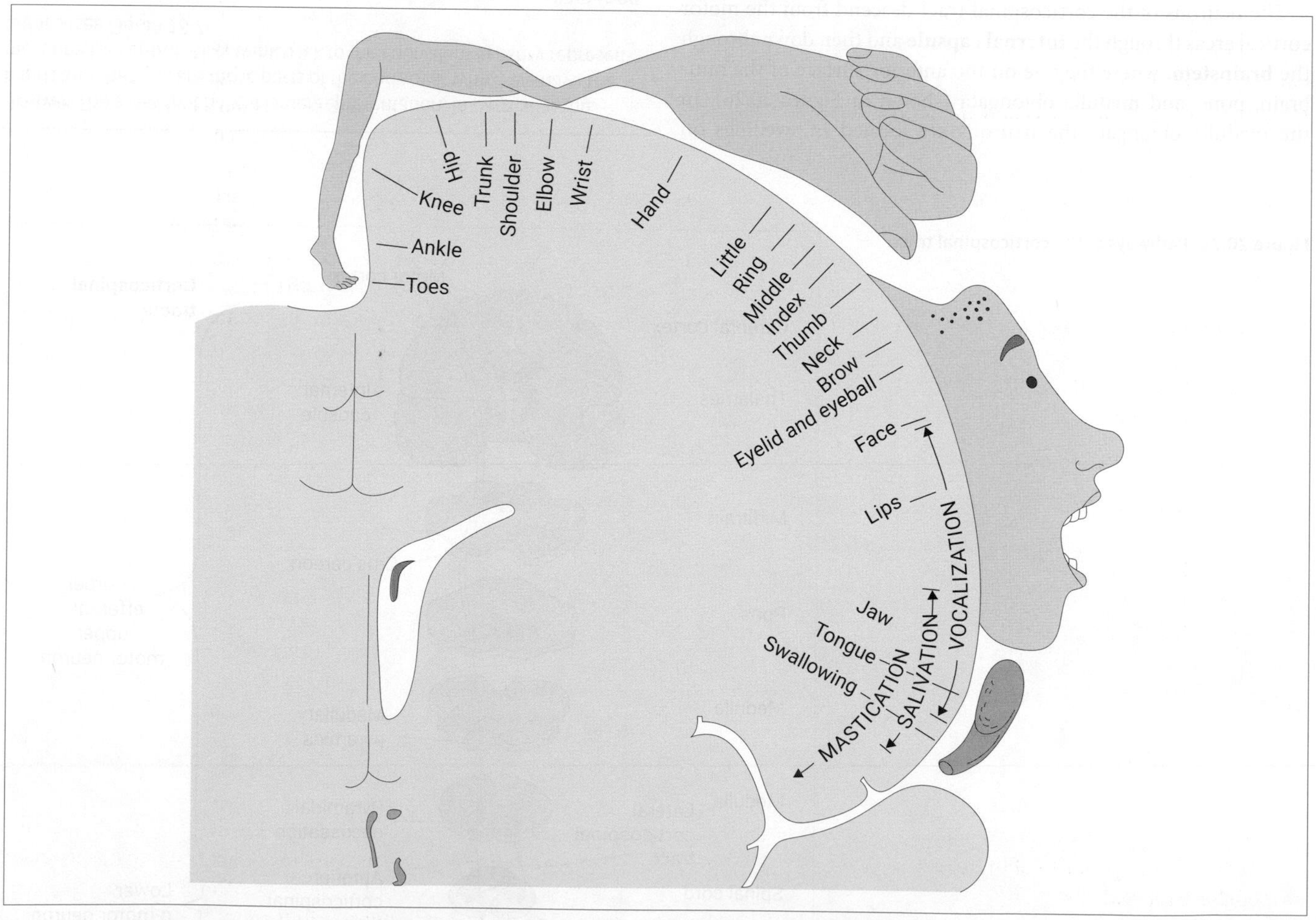

Figure 20.28 Somatotopic representation of the muscles of different regions of the body on the primary motor cortex. The distal muscles of the arm and the muscles of the face have a much larger representation than the other regions.

Source: Pocock et al. *Human Physiology*, 5th edn, 2017. Oxford University Press.

Figure 20.29 Approximate number of neurons in the corticospinal tracts of the rat, monkey, and human with the relative indexes of dexterity.

Source: Figure 1 from Courtine et al. (2007) Can experiments in nonhuman primates expedite the translation of treatments for spinal cord injury in humans? *Nature Medicine*, 2007, 13(5), 561–6. Permission conveyed through Copyright Clearance Center, Inc.

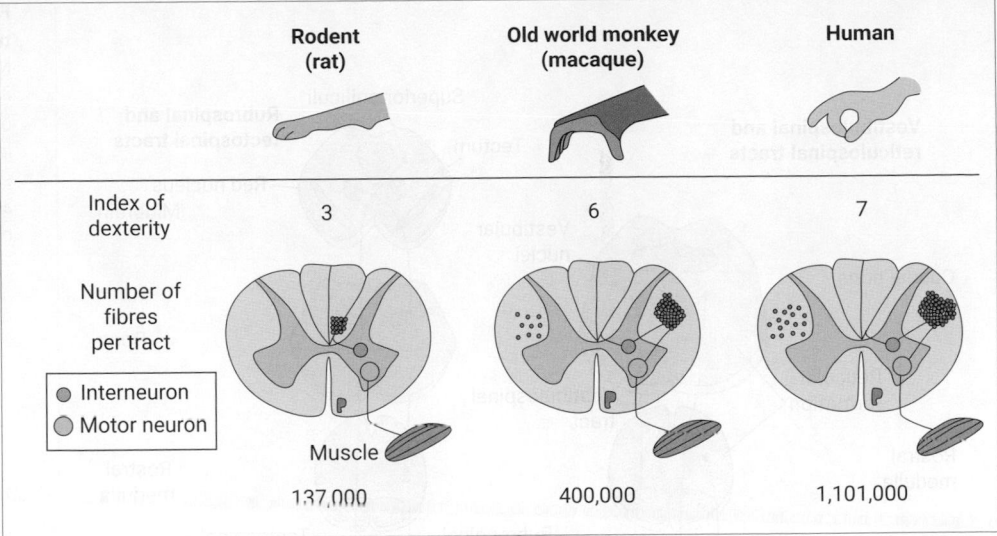

While the majority of the corticospinal neurons synapse with spinal interneurons, about one-third synapse directly with the lower motor neurons. The lateral corticospinal tract therefore provides a route whereby the motor cortex can directly control the movements of the limbs, especially the precision movements of the hands.

The corticospinal tract shows significant differences in scale across the different species of mammals and this is directly related to the degree of dexterity shown by those species, in particular the ability to oppose the thumb and other digits as humans can (Figure 20.29). Thus, humans have more than double the number of corticospinal neurons as macaque monkeys and almost 10 times as many as the rat.

The 15 per cent of fibres that do not decussate in the medulla remain on the same side (ipsilateral) and form the **anterior corticospinal tract**. The anterior corticospinal tract lies in the anterior part of the spinal cord, close to the midline (see Figure 20.26) and some of its axons project bilaterally, to control the lower motor neurons that innervate the muscles of the body axis and limb girdles.

What is the result of having bilateral projections? Take a moment to think before reading on. The motor neurons branch and innervate the same muscle group on each side of the body; therefore, there is synchronous activation of the corresponding muscles.

The neurons of the corticobulbar pathway also originate in the motor regions of the cerebral cortex, with a very large representation of the musculature of the face in the primary motor cortex (see Figure 20.28). This pathway therefore has a very important role in the voluntary control of the muscles of speech, facial expression, and eye movements.

Brainstem pathways

The motor pathways that originate in the nuclei of the brainstem, also referred to as the **extrapyramidal tracts**, synapse with interneurons in the lower motor neuron pools; they do not make any direct connections with the lower motor neurons.

The **vestibulospinal tract** starts in the vestibular nuclei, which receive sensory input from the organs of balance in the vestibular system (see Section 18.5, Balance). One part of the tract projects to both sides of the spinal cord in the neck and is associated with the maintenance of head stability through the control of the neck muscles. The other part of the tract descends on each side of the spinal cord (Figure 20.30), but the fibres remain ipsilateral: that is, they do not cross over (Figure 20.30). These neurons project mainly to the motor neuron pools of the anti-gravity muscles and play an important part in the postural reflexes that maintain balance in response to movements of the head.

The **reticulospinal tract** has its origin in the reticulospinal nuclei of the pons and medulla and it too remains ipsilateral (Figure 20.30). This tract has a role in regulating the strength of the spinal reflexes involved in postural regulation.

The **tectospinal tract** originates in the tectum ('roof') of the midbrain in association with the superior colliculi (Figure 20.30). The superior colliculi receive inputs from branches of the optic tract and the tectospinal tract has a role in controlling head movements, particularly in association with visual stimuli. This is important in enabling us to coordinate movements of the head and eyes to fix our gaze on specific visual targets. The fibres of the tectospinal tract decussate at the level of the midbrain and terminate mainly in the contralateral motor neuron pools of the muscles controlling movements of the neck.

The **rubrospinal tract** also originates in the mid-brain, in a small area called the red nucleus. This tract also crosses over and descends in the spinal cord close to the lateral corticospinal tract (Figure 20.30) and it terminates in the motor neuron pools of the distal muscles. This tract has a limited role in humans compared with many non-primate mammalian species, having been superseded by the corticospinal tract.

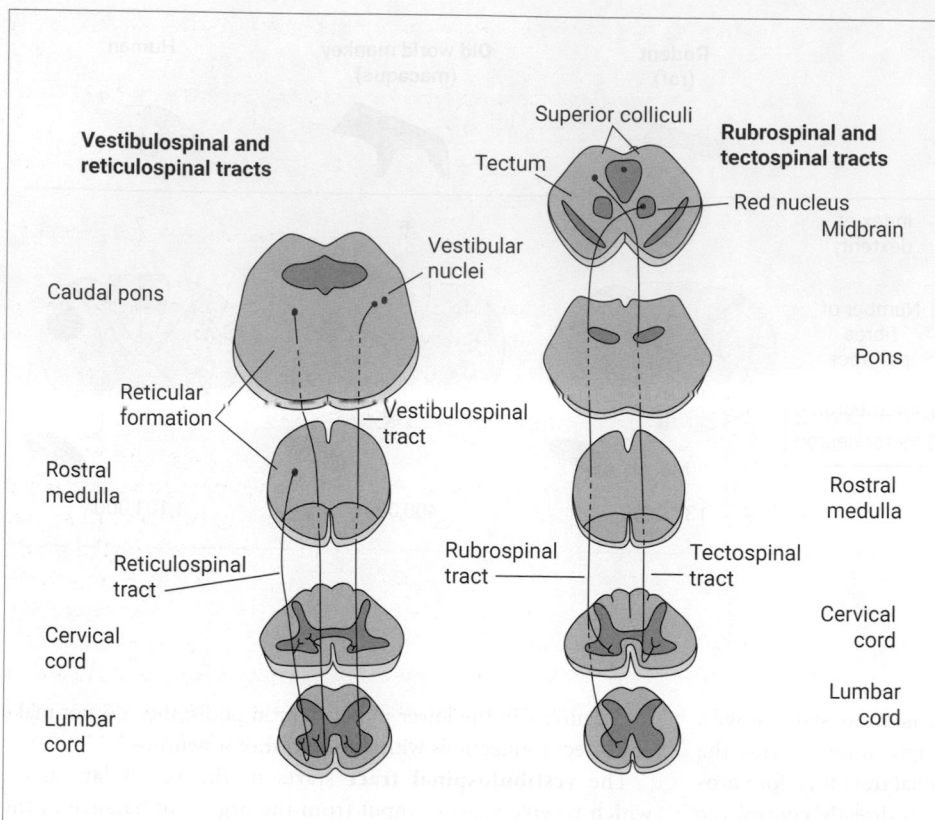

Figure 20.30 The pathways of the major tracts that originate in the brainstem. Note that the vestibulospinal and reticulospinal tracts do not cross over to the other side of the cord (i.e. they remain ipsilateral and so act on muscles on the same side of the body), whereas the tectospinal and rubrospinal tracts cross over and so act on contralateral muscles.

Motor cortex

The term motor cortex refers to three main areas of the frontal lobe (see Figure 20.24):

- primary motor, cortex (M1);
- premotor area (PMA);
- supplementary motor areas (SMA).

The **primary motor cortex** is located on the **pre-central gyrus** and is laid out as a somatotopic map of the body (see Figure 20.28), which is very similar to the sensory map seen on the somatosensory cortex (S1) on the post-central gyrus (Chapter 18). The main descending output pathways from the primary motor cortex are the corticospinal and corticobulbar tracts (see Figure 20.26). Stimulation of specific regions of the primary motor cortex results in contraction of specific muscles and, as we have seen above, the regions of the body over which we can exert the finest control also have the largest representation in the primary motor co,rtex.

For a long time it was thought that this map was totally specific, in other words each output neuron in M1 only connected with the lower motor neurons of a single muscle. However, more detailed examination has shown that each of the M1 neurons actually interact with the motor neuron pools of several muscles and, indeed, that the actions can be excitatory or inhibitory (Figure 20.31).

This evidence suggests that activation of each cortico-motor neuron generates a specific vector of movement, which represents the sum of the contractile activity of the group of muscles which it affects. The overall direction of the voluntary movement of a body region therefore represents the sum of the different vectors of all the neurons that have been selected and the level of activation of each of those neurons.

Figure 20.32 shows the relative levels of activity within the primary motor cortex during performance of a simple motor task, a-repetitive finger-to-thumb opposition movement, compared with imagining performing the task and a control visualization task that was not related to movement. Although actual performance of the task leads to a significant increase in activity in M1, mental rehearsal of the task also leads to increased activity, even though, there is no resultant output in terms of movement gen.ration, which indicates a role for the primary motor cortex in movement planning, particularly in relation to simple movements.

The planning of complex movements is particularly associated with the other two motor cortical regions, the premotor area and the supplementary motor area, and recordings of the activity within these regions shows that they begin to be active up to a second before the movement actually begins. One way of exploring the roles of different areas of the brain is to stimulate the neurons: this can be done using a technique called transcranial magnetic stimulation (TMS). Using this technique it is possible to stimulate areas of the brain without causing significant discomfort to the subject (clearly an important criterion!). Using TMS it is possible to explore the effects of applying stimulation during the performance of movements of varying complexity (see Scientific Process 20.1). If a subject is asked to perform a complex sequence of finger movements, applying a burst of repetitive stimulation to the primary motor cortex interrupts the

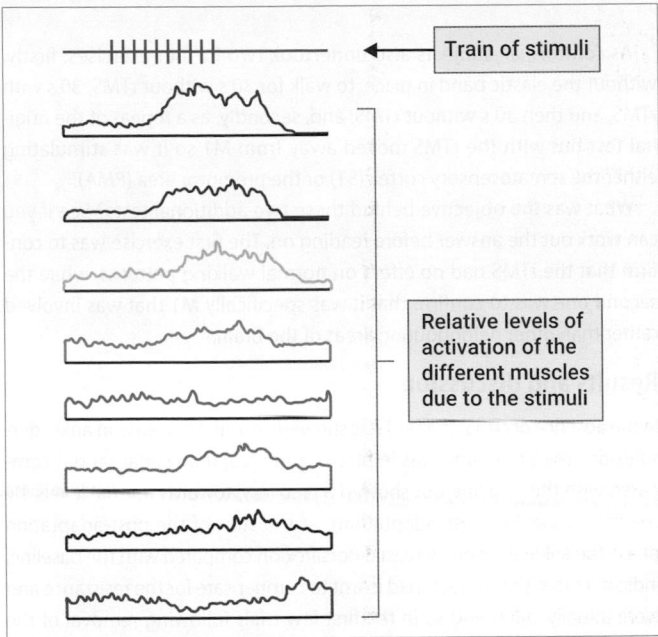

Figure 20.31 Averaged post-synaptic potentials recorded in the motor neuron pools of muscles acting around the thumb in response to a train of stimuli to a single cortico-motor neuron. The top trace shows the train of stimuli, the lower traces show the relative levels of activation of the different muscles: note that, as well as generating different levels of excitation, the motor neurons may also give rise to inhibitory activity, as shown by the depression seen in the bottom trace.

Source: Reprinted from: Lemon R. The output map of the primate motor cortex. *Trends in Neurosciences*, 11, 501–506. © 1988, with permission from Elsevier.

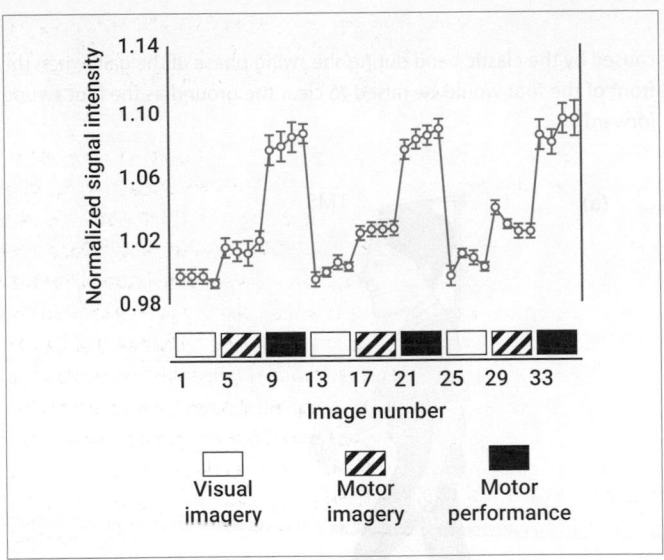

Figure 20.32 Relative levels of activity in the primary motor cortex, measured using functional magnetic resonance imaging (fMRI) during three tasks: performing a motor task using the hand (motor performance), imagining performing the task (motor imagery), and, as a control, visualizing a natural scene (visual imagery).

Source: Figure 2 from C.A. Porro et al. (1996) Primary motor and sensory cortex activation during motor performance and motor imagery: a functional magnetic resonance imaging study. *J Neurosci*, 16, 7688–98. Permission conveyed through Copyright Clearance Center, Inc.

performance of the task but does not cause it to break down. By contrast, if the same stimulation is applied to the supplementary motor area (SMA), this leads to a breakdown in the performance of the task. This suggests that the SMA has a particular role in the planning of complex movements. This idea is further supported by the evidence that patients with lesions affecting the SMA have great difficulty in performing more complex motor tasks, especially those requiring bimanual coordination, such as doing up buttons.

The premotor area appears to have an important role in movement timing and also in the directional planning of movements.

SCIENTIFIC PROCESS 20.1 — Research into the control of walking by the motor cortex

Introduction

Although walking is a highly automated activity, gait also has to adapt to the prevailing conditions, for example when going from a smooth surface to a rough or slippery one (think of how you change gait when walking on an icy surface). These adaptations not only occur rapidly, but also have to be de-adapted on return to a normal surface. One hypothesis is that these adaptations involve corticospinal mechanisms. In these experiments this hypothesis was tested using repetitive transcranial magnetic stimulation (rTMS) to disrupt the cortical processing.

TMS is a non-invasive stimulation technique that allows focal stimulation of regions of the brain by means of a brief, high-intensity magnetic pulse, which stimulates the neurons. Use of repetitive stimulation can be used to test the effect of disrupting cortical processing (Chen et al., 1997) or motor learning (Lundbye-Jensen et al., 2011). The experiments described in this paper involved testing the role of the primary motor cortex (M1) in the storage of walking adaptation in response to changes in walking conditions.

Methods

The authors selected 38 volunteers: 21 females and 17 males, aged 24 ± 4 years, with no known neurological disorder. As these were experiments on humans the study had to be approved by the local ethics committee and conform to the principles of the Declaration of Helsinki.

The subjects were tested by walking on a treadmill, the gait disturbance being provided by an elastic band attached to the ankle to resist dorsiflexion of the ankle (i.e. raising of the foot). The subjects walked normally for 30 s, then for 30 s with the band in place, followed by 120 s post-adaptation (Figure 1). Each subject completed two trials, one with rTMS and one without, the order of testing being randomized across the subjects.

Why were the trials randomized? This was to make sure that there was no sequence effect. For example if the subjects had all received the rTMS on the first trial this might have led to a possible longer-term effect on the second trial.

The rTMS was applied in bursts synchronized to the walking cycle so that it coincided with the disturbance to the normal ankle movement caused by the elastic band during the swing phase of the gait, when the front of the foot would be raised to clear the ground as the foot swung forward.

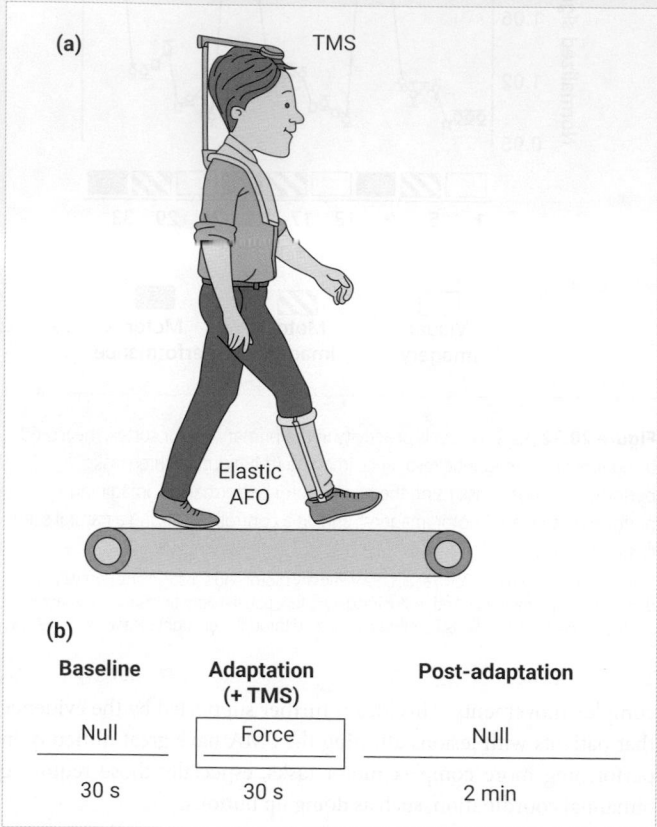

(a) TMS

Elastic AFO

(b)

Baseline	Adaptation (+ TMS)	Post-adaptation
Null	Force	Null
30 s	30 s	2 min

Figure 1 Walking set-up. (a) The subject is walking on the treadmill with an elastic band attached to the foot to resist raising the front of the foot around the ankle joint. The magnetic coil (TMS) is held in place over the stimulation site. (b) Each trial consisted of three phases: baseline, adaptation, and post-adaptation periods.

Source: J.T. Choi, L.J. Bouyer & J. Bo Nielsen. (2015) Disruption of locomotor adaptation with repetitive transcranial magnetic stimulation over the motor cortex. *Cerebral Cortex,* 25, 1981–6. Reproduced by permission of Oxford University Press. doi:10.1093/cercor/bhu015.

As controls, all subjects also undertook two further exercises: firstly, without the elastic band in place, to walk for 30 s without rTMS, 30 s with rTMS, and then 30 s without rTMS; and, secondly, as a repeat of the original test but with the rTMS moved away from M1 so it was stimulating either the somatosensory cortex (S1) or the premotor area (PMA).

What was the objective behind these two additional tests? See if you can work out the answer before reading on. The first exercise was to confirm that the rTMS had no effect on normal walking patterns, while the second one was to confirm that it was specifically M1 that was involved rather than other neighbouring areas of the brain.

Results and discussion

In the absence of rTMS, the subjects showed an initial decrease in ankle dorsiflexion when the band was in place (Figure 2a) (early adaptation) compared with the baseline but showed a recovery, towards normal levels, by the end of the 30 s (late adaptation). At the start of the post-adaptation phase the ankle showed increased dorsiflexion compared with the baseline, indicating that the subjects had learnt to compensate for the resistance and were initially still doing so in the first few trials following removal of the resistance. When the rTMS was applied over M1 during the adaptation phase, the subjects still showed a reduced dorsiflexion (Figure 2b). During the post-adaptation phase there was no overshoot in the extent of the ankle dorsiflexion, which recovered to the baseline level. This suggests that applying the rTMS to M1 had blocked the learned compensation process.

The authors tested the significance of the findings for the averaged data for all the subjects using a repeated measures analysis of variance (ANOVA) which indicated that the rTMS had resulted in a reduction in the after-effects of the removal of the band ($P<0.01$). Testing the control conditions showed that the rTMS to M1 had no significant effect on gait when no resistance was applied throughout the cycle (i.e. normal walking), nor did rTMS have any effect when applied to the other regions of the brain during the resistance trials.

The authors therefore concluded that the corticospinal system plays a role in learning patterns of adaptation of gait when walking conditions are varied.

Read the original work

Choi, J. T., Bouyer, L. J., & Nielsen, J. B. (2015) Disruption of locomotor adaptation with repetitive transcranial magnetic stimulation over the motor cortex. *Cereb. Cortex* **25**: 1981–6. doi:10.1093/cercor/bhu015

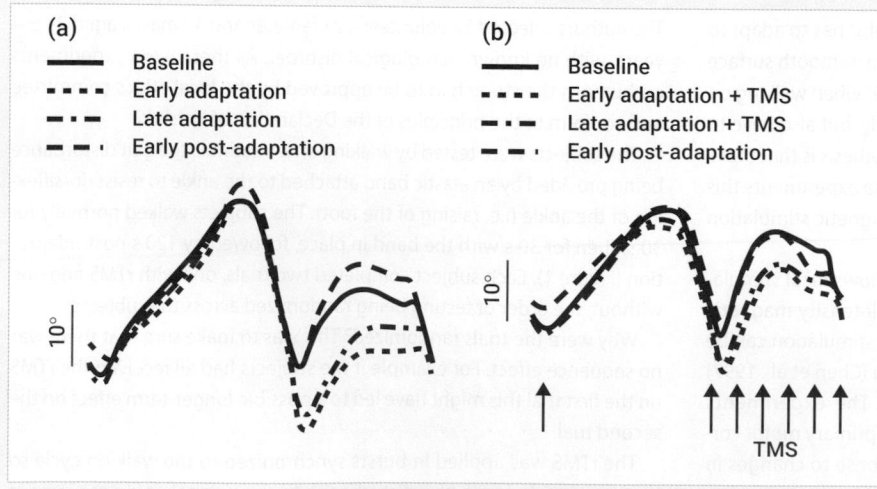

(a)
— Baseline
- - - Early adaptation
-·-·- Late adaptation
- - -· Early post-adaptation

(b)
— Baseline
- - - Early adaptation + TMS
-·-·- Late adaptation + TMS
- - -· Early post-adaptation

TMS

Figure 2 Adaptation of gait in response to application of ankle resistance in the absence (a) or presence (b) of rTMS applied to M1 during the adaptation phase. Note in (b) that there is no recovery of the dorsiflexion during the adaptation phase nor is there an overshoot in the amount of dorsiflexion following removal of the band.

Source: J.T. Choi, L.J. Bouyer & J. Bo Nielsen. (2015) Disruption of locomotor adaptation with repetitive transcranial magnetic stimulation over the motor cortex. *Cerebral Cortex,* 25, 1981–6. Reproduced by permission of Oxford University Press. doi:10.1093/cercor/bhu015.

Recordings from neurons within the lateral region of the PMA show that they are active prior to movements in a specific direction: if subjects are given a warning signal to make a movemen. but there is no indication of the direct,ion of the movement to be made, then the PMA neurons show little change in activity. However, if there is also an indication of the direction of the movement, then specific PMA neurons will show an increase in activity well before the onset of the movement. Which specific neurons are active will depend on the direction of the movement. Lesions to this region particularly affect the ability to perform targeted movements in response to a visual cue.

PAUSE AND THINK

Briefly compare the roles of the three main cortical motor areas: motor cortex, premotor cortex, and supplementary motor area.

Answer: The main descending output pathways are the corticospinal and corticobulbar tracts arising mainly from the primary motor cortex. Activation of each cortico-motor neuron generates a specific vector of movement, which represents the sum of the contractile activity of the group of muscles which it affects. The PMA appears to have an important role in movement timing and also in the directional planning of movements. Recordings from neurons within the lateral region of the PMA show that they are active prior to movements in a specific direction. The SMA has a particular role in the planning of complex movements, for example movements involving bimanual coordination.

Basal ganglia

Parkinson's disease is a relatively common disease of old age in which patients show a classic set of symptoms:

- slowness of movement and difficulty initiating movement;
- increased muscle tone, or rigidity;
- tremor at rest.

The underlying cause of Parkinson's disease has been linked to degeneration of neurons in the substantia nigra, one of the structures making up the group ref.rred to as the basal ganglia (Figure 20.24). Disorders of the basal ganglia give rise to deficits in movement control that are typically either *hypokinetic* (too little

movement) or *hyperkinetic* (too much movement): an example of the latter would be Huntington's disease.

The basal ganglia comprise a group of nuclei that are extensively interconnected and that are actively involved in the generation of movement and also in other processes, including memory and motivation. The main structures comprising the basal ganglia are the:

- striatum, comprising the:
  - caudate nucleus
  - putamen;
- globus pallidus;
- subthalamic nucleus;
- substantia nigra.

The inputs to the basal ganglia from the motor cortical areas are mainly routed through the striatum. There are several pathways taking slightly different routes through the basal ganglia, referred to as the hyperdirect, direct, and indirect pathways. The output is then routed through to the thalamus and back to the motor cortical areas.

1. At rest, the basal ganglia exert an inhibitory action on the thalamus, via the globus pallidus, which is maintained by the hyperdirect pathway. Therefore, there is no excitatory feedback onto the motor cortex and so movement does not take place.

2. Initiation of the movement is generated by excitatory output from the motor cortex to the striatum. This activates the direct pathway, which releases the thalamus from inhibition and so there is an excitatory action back onto the motor cortex, enabling movement.

3. This is followed by activation of the indirect pathway, which re-imposes the inhibitory output onto the thalamus, stopping movement.

As a generalization, therefore, we can say that the basal ganglia are involved in determining the speed and magnitude of voluntary movements.

If there is insufficient activation of the direct pathway, then movements will be difficult to initiate, be smaller, and be slower. This is the case in Parkinson's disease (see Clinical Box 20.2), where

---

CLINICAL BOX 20.2   **Parkinson's disease**

Parkinson's disease is relatively common in the elderly, affecting up to 2 per cent of those aged >75 years, although in a very small proportion of cases it can have a much earlier onset. The disease is characterized by slowness of movement (bradykinesia), muscle rigidity, and a pronounced tremor at rest that is often first seen in the hands.

 Go to the e-book to watch a video showing the signs of Parkinson's disease.

The disease is characterized by degeneration of a subset of neurons within one of the structures of the basal ganglia, the substantia nigra, that use dopamine as their neurotransmitter. These neurons project from the substantia nigra to the striatum (the nigro-striatal pathway).

For many years the most common therapeutic approach has been to treat patients with l-DOPA, which is a precursor of the neurotransmitter dopamine. Normally, this treatment will bring relief from the motor symptoms for several years, but the process of degeneration continues and over a period of 5 years the dosages have to be increased and about half of patients will start to develop complications with increasing fluctuations in their motor symptoms, including uncontrolled movements and alternation between being able to move and being immobile.

Some patients may be suitable for treatment with a surgical approach called deep-brain stimulation (DBS) (Figure 1). DBS involves the surgical implantation of electrodes into the basal ganglia, usually the subthalamic nucleus or the globus pallidus. During the surgery the exact location of the electrodes is determined in order to achieve the optimal effect. DBS affects the output patterns of the basal ganglia and, in many cases, has been shown to result in significant benefit to the patient's quality of life through improvements in their motor functions.

Read the original work

Schiefer, T. K., Matsumoto, J. Y., & Lee, K. H. (2011) Moving forward: advances in the treatment of movement disorders with deep brain stimulation. *Front. Integr. Neurosci.* **5**: 1–16.

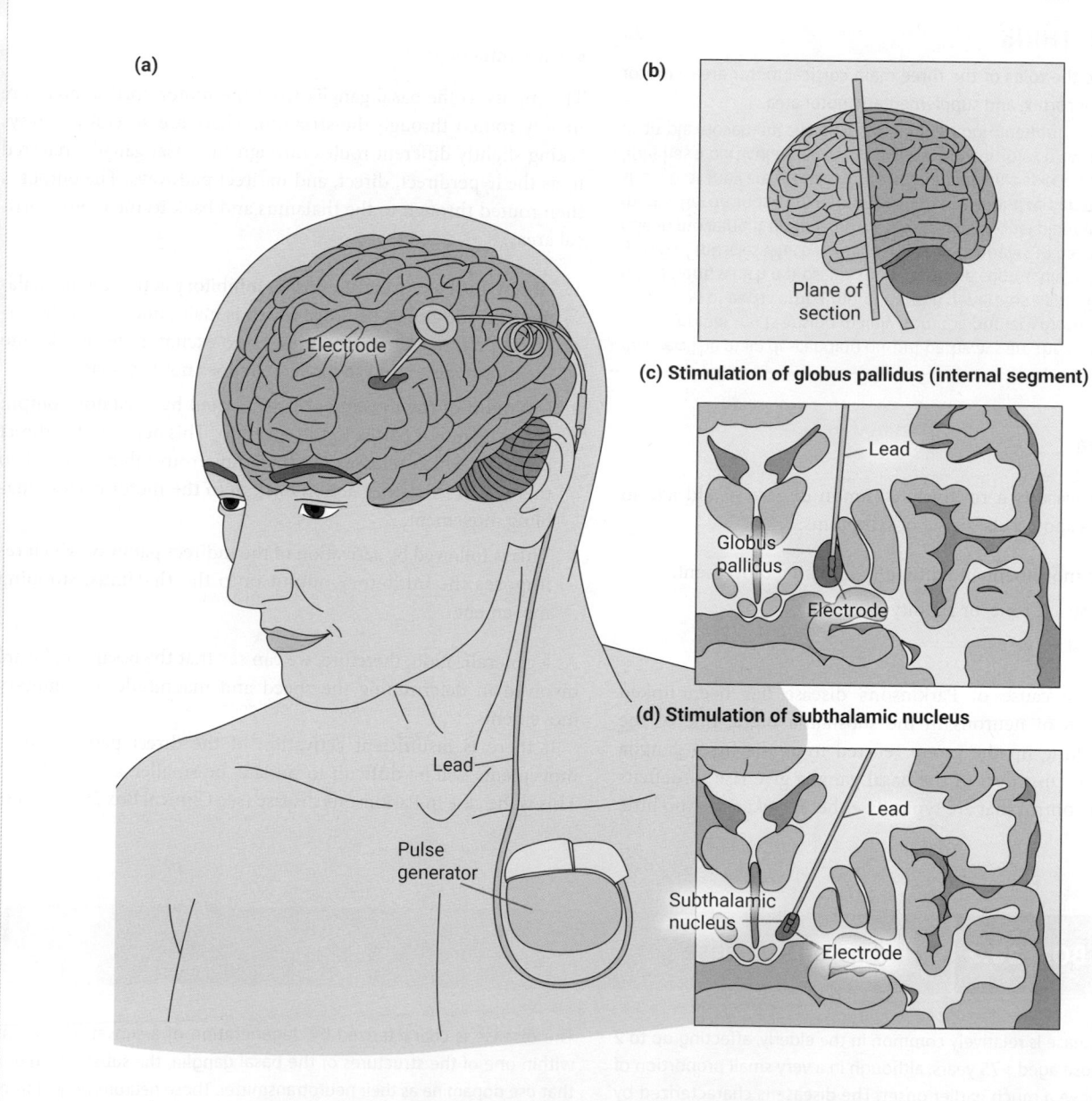

Figure 1 Deep-brain stimulation. Application of deep-brain stimulation showing the overall arrangement (a) and the placement of the electrodes in the globus pallidus (c) or subthalamic nucleus (d).

the degeneration of the neurons in the substantia nigra results in reduced activation of the direct pathway and so the thalamus is not released from inhibition.

The opposite is the case with Huntington's disease. In this case there is a mutation that leads to abnormal production of a protein called huntingtin, resulting in the protein being much longer than normal. This causes the protein molecules to clump together, which results in degeneration of some of the neurons in the striatum, leading to an excess of movement: patients are unable to keep still and make characteristic jerking movements and facial grimaces.

Cerebellum

The cerebellum is located at the base of the brain beneath the occipital lobe and posterior to the brainstem (see Figure 20.24). The surface of the cerebellum is highly folded, like the cerebral cortex, and the thin cerebellar cortex is densely packed with cells. The core of the cerebellum comprises the white matter of neuronal pathways within which are embedded the deep cerebellar nuclei that carry the outputs from the cerebellum to other regions of the CNS. The cerebellum is an important site of motor learning where plans for the sequencing of muscle activations are generated and can be output to the motor cortex via the thalamus.

The cerebellum is critically involved in the coordination of muscle activations: disorders of the cerebellum commonly result in **ataxia**, in which movements are uncoordinated and inaccurate (Figure 20.33). This is characterized by an inappropriate sequence of muscle activations during movement, often accompanied by co-contraction of antagonist muscles. Patients with ataxia may often show a stumbling gait because the sequence of muscle activations across the different joints of the legs is impaired.

PAUSE AND THINK

Compare the roles of the basal ganglia and the cerebellum in the voluntary control of movement.

Answer: The role of the basal ganglia can be generalized as determining the speed and magnitude of voluntary movements. The cerebellum is involved in the coordination of the sequence of muscle activations.

 Check your understanding of the concepts covered in this section by answering the questions in the e-book.

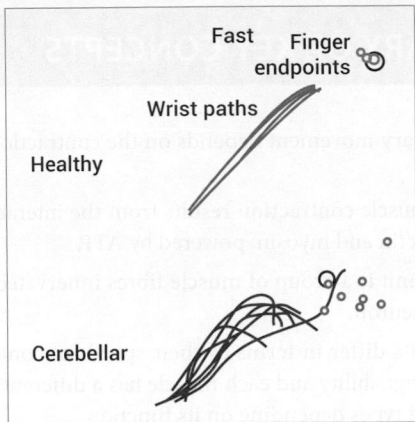

Figure 20.33 Fast pointing movements made by a healthy subject (top) and a cerebellar patient (bottom). The traces show overlays of several trials. The open circle shows the target point and the coloured dots the actual endpoint of the movement. The solid lines show the movements of the wrist in each trial. Note how the paths taken by the healthy subject are all very close together and direct and the target is reached accurately compared with the cerebellar patient. This type of movement disorder is termed dysmetria.

20.5 Conclusion

Voluntary movement is remarkably complicated and its successful completion requires the cooperation of a number of structures, both within the cerebral cortex and subcortical structures. To summarize the processes, the basal ganglia set the speed and amplitude of the movement while the cerebellum is responsible for the sequence of muscle activations. These structures both feed back to the motor cortex via the thalamus. Within the motor cortex, the SMA is particularly concerned with the organization of complex movements while the PMA has a key role in the planning of visually directed movements where there is a clear directional cue. The primary motor cortex represents the main output region with the selection of the neurons associated with the vectors of the required movement and their output via the corticospinal and corticobulbar tracts to activate the lower motor neurons that in turn innervate the motor units of the skeletal muscles. A remarkable feature of all of this is that the output is generated within less than a second of the initiation of the planning process. The movement control is highly dependent on sensory feedback from the eyes, from muscle receptors, and from joints.

The muscles themselves are complex in their structural and neural organization, so each skeletal muscle has a different architecture varying from a pennate arrangement, where the muscle fibres are short, through to the longitudinal muscles. Furthermore, each muscle has a different composition of motor unit subtypes. The architecture and the motor unit composition reflect the role the muscle plays in movement.

SUMMARY OF KEY CONCEPTS

- All voluntary movement depends on the contraction of skeletal muscle.

- Skeletal muscle contraction results from the interactions of the proteins actin and myosin, powered by ATP.

- A motor unit is a group of muscle fibres innervated by a single α-motor neuron.

- Motor units differ in terms of their speeds of contraction and relative fatiguability and each muscle has a different make-up of motor unit types depending on its function.

- Muscle–tendon–bone units act as lever systems.

- The simplest nerve circuits are the spinal reflexes, such as the stretch reflex, which plays an essential role in the regulation of posture and ongoing muscle contractions.

- The neural commands regulating voluntary movement are conveyed by the descending motor pathways from the motor cortical regions. In humans, the corticospinal and corticobulbar tracts control voluntary movements.

- The relative extent of the motor cortical representation of the different body regions directly relates to the precision of control.

- The basal ganglia and the cerebellum, along with the motor cortical areas, play key roles in movement planning.

Use the flashcards in the e-book to test your recall of key terms introduced in this chapter.

QUESTIONS

Looking for answers? Once you've answered these questions, follow the link in the e-book to the answer guidance and check your work.

Concepts and definitions

1. Draw a labelled diagram of the sarcomere.

2. Define the term motor unit.

3. Compare the structure and contractile mechanics of muscles that have a longitudinal arrangement of fibres versus those that are pennate.

4. Name the two descending motor pathways that originate from the motor cortex. Which areas of the body do they each control?

5. Which region of the motor control system is particularly involved with setting the size and speed of movements?

Apply the concepts

6. Describe the process of excitation–contraction coupling.

7. Describe the symptoms most commonly seen in Parkinson's disease. Which region of the basal ganglia is particularly affected?

8. Draw a diagram showing the neural circuitry of the stretch reflex. Give one example of a way in which the strength of the stretch reflex can be modified.

9. Compare the roles of the four main descending pathways of the extrapyramidal tracts.

10. Give an example of an eccentric and a concentric contraction of the quadriceps muscles of the thigh. Draw the type of lever the quadriceps forms across the knee joint.

Beyond the concepts

11. Describe how the stretch reflex and reciprocal inhibition work together in the regulation of posture.

12. Explain how muscle contracts. What is the significance of the different types of motor unit in regulating the force of contraction?

13. Compare and contrast the roles of the motor cortex, basal ganglia, and cerebellum in the control of voluntary movement.

14. Compare and contrast the anatomy and physiological roles of the different descending motor pathways.

15. Research two neurological diseases that affect the motor system and reflect on how they contribute to our understanding of the control of movement.

FURTHER READING

Bear, M. F., Connors, B., & Paradiso, M. (2015). *Neuroscience: Exploring the Brain*. 4th Edition. Wolters Kluwer.

Chapters 13 and 14 provide a more detailed overview of the central control of movement.

Frontera, W. R. & Ochala, J. (2015). Skeletal muscle: a brief review of structure and function. *Calcif. Tissue Int.* **96**: 183–95.

This is a very readable review of the structure and function of skeletal muscle.

Lemon, R. N. (2008). Descending pathways in motor control. *Annu. Rev. Neurosci.* **31**: 195–218.

Although a bit dated, this review article still provides an excellent overview of the organization and function of the descending motor pathways.

Pocock, G., Richards, C. D., & Richards, D. A. (2017). *Human Physiology*. 5th edition. Oxford University Press.

This text provides some more background detail to the understanding of muscle and movement control.

Purves, D., et al (2018). *Neuroscience*. 6th edition. Oxford University Press.

Chapters 16–19 provide a detailed overview of the motor system.

Yin, H. (2017). The basal ganglia in action. *Neuroscientist* **23**: 299–313.

This article provides a more detailed but very readable, account of the basal ganglia.

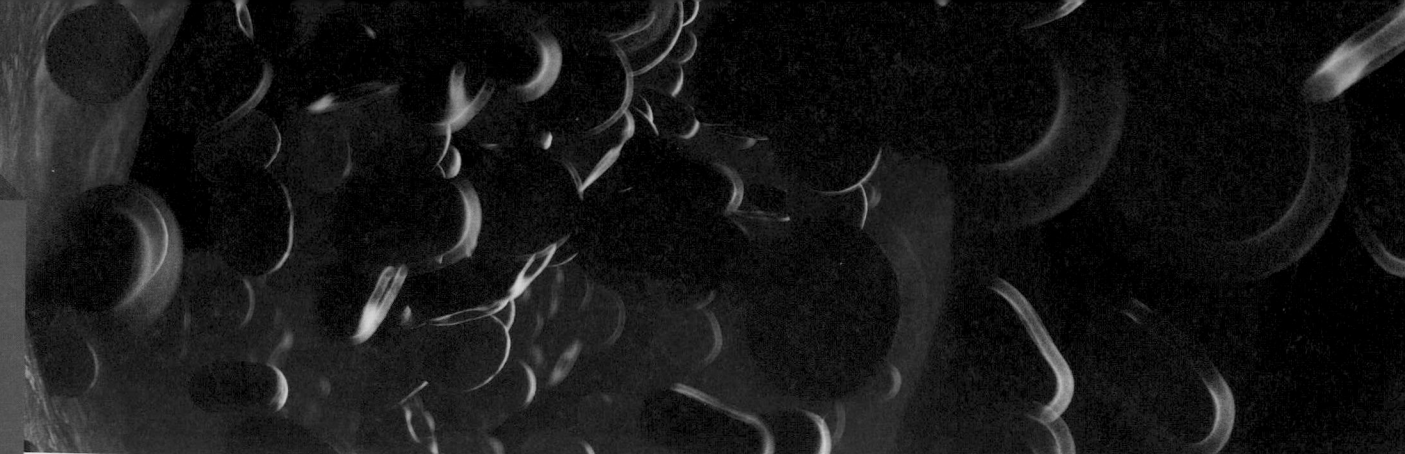

Cardiovascular System

Chapter contents

Introduction   718

21.1   Composition and function of blood   719

21.2   Vascular system   723

21.3   Heart   732

21.4   Integrated cardiovascular physiology—
        regulation of blood pressure   740

Watch the key concepts video in the e-book to prepare yourself for studying this chapter.

LEARNING OBJECTIVES

By the end of this chapter, you should be able to:

- Describe the composition of blood.
- Describe the production of blood cells.
- Describe the organization of the mammalian circulatory system.
- Compare the functional anatomy of a range of blood vessels.
- Explain how blood vessel function is controlled in regulating the distribution of blood.
- Outline the anatomy of the heart.
- Explain how the electrical activity of the heart is initiated and conducted.
- Describe the passage of blood through the heart.
- Explain the endocrine and neural factors that control cardiac function as physiological demand varies.
- Recall how blood pressure is regulated and explain some of the potential strategies to treat it when it is raised above normal.

Introduction

The human body is a complex organization of many billions of cells all of which are metabolically active. As such they utilize nutrients and generate waste products. The cardiovascular system acts to deliver nutrients to cells and remove waste products prior to excretion. It also plays a vital role in communication between cells and organ systems by transporting important signalling molecules such as hormones around the body.

At its simplest, the cardiovascular system consists of three elements: the heart, the vascular system, and the blood itself. While blood is a connective tissue (perhaps the ultimate connective tissue in that it links all the cells of the body), it will be considered in this chapter as part of the cardiovascular system.

- Heart: a pump that drives the blood around the body and can increase or decrease the flow of blood to meet the requirements of a range of needs.
- Vascular system: a system of tubing (blood vessels) through which flow can be regulated to provide the differential distribution of blood to regions of the body.
- Blood: a fluid in which nutrients, waste products, hormones, drugs, and cells (i.e. red blood cells, white blood cells, and blood platelets) are transported around the body.

In this chapter, we will consider the structure and function of each of these components and the role that the cardiovascular system has in helping to maintain homeostasis. Furthermore, we will consider how each of these components are controlled and think about what happens when control goes wrong and how that can be corrected.

21.1 Composition and function of blood

Blood contains a mixture of different cellular elements:

- **erythrocytes**—red blood cells;
- **leukocytes**—white blood cells;
- **platelets**—fragments of cells, also called thrombocytes.

These elements are suspended in a fluid called **plasma**. Typically, there are about 70 mL of blood per kg body weight in an adult. If a sample of blood is taken and centrifuged, then the cellular elements sink while the plasma remains at the top (Figure 21.1).

The proportion of blood volume occupied by the erythrocytes is called the **haematocrit**. This is about 45 per cent of the blood

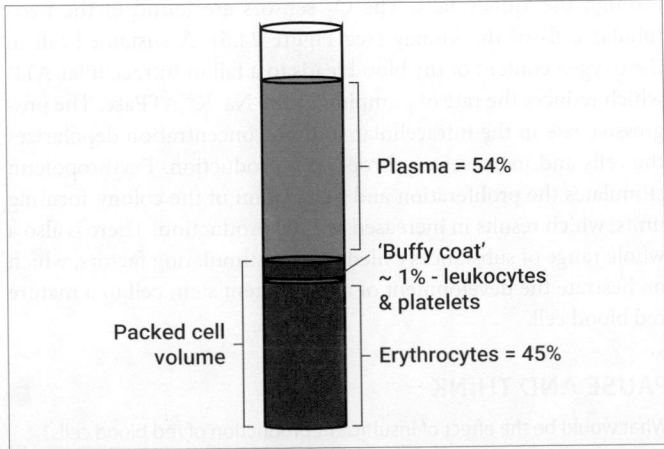

Plasma = 54%

'Buffy coat'
~ 1% - leukocytes
& platelets

Packed cell
volume

Erythrocytes = 45%

Figure 21.1 Composition of blood. Centrifugation of normal blood divides the blood into three layers: the erythrocytes and the plasma are the main layers, separated by a very thin layer, the 'buffy coat', which comprises leukocytes and platelets. The erythrocytes and the buffy coat make up the 'packed cell' volume.

volume in healthy adults. Plasma occupies about 54 per cent of the volume. At the interface between the erythrocytes and the plasma, is a very thin layer called the **buffy coat**, which is occupied by the leukocytes and the platelets.

Blood cells

Erythrocytes

Erythrocytes, also known as red blood cells, are the most abundant cell type in blood. In a healthy adult human, there are 4–6 $\times 10^{12}$ L^{-1} of blood. The principal role of these cells is to transport oxygen around the body by means of the oxygen transport pigment **haemoglobin**. This process is discussed more fully in Section 22.3 'Oxygen transport' and so we will only need a brief summary here.

When examined under the microscope red blood cells appear as biconcave discs because they lack a nucleus (termed anucleate). It is possible that the lack of the nucleus enables the cells to be a little more manoeuvrable as they pass through the smallest of blood vessels: the diameter of these vessels can be smaller than the diameter of the red cell, so they are deformed as they squeeze through. The biconcave shape also increases the surface area:volume ratio, maximizing oxygen exchange with the surrounding tissues.

The interior of red blood cells is densely packed with haemoglobin. Haemoglobin is a conjugate protein: this means it consists of a protein (globin) linked to a prosthetic group (haem). The structure of a typical haemoglobin molecule is shown in Figure 21.2, although there are a number of different variants of globin found in the body. The haemoglobin shown in Figure 21.2 is adult haemoglobin—it consists of two α and two β subunits. In the developing fetus, the β globin chains are replaced with γ globin chains. Myoglobin, which is found in skeletal muscle, acts as an oxygen store and has only a single haem group.

It is important to remember that oxygen 'binds' reversibly to the centre of the haem groups. Given the critical role that these cells have in oxygen transport, a reduction in the quantity of haemoglobin can compromise oxygen delivery to tissues—this is termed anaemia. Anaemia is defined as a haemoglobin concentration of less than 13.5 g L^{-1} in males and 11.5 g L^{-1} in females. There are various types of anaemia, which are considered in detail in Clinical Box 22.2.

A typical red blood cell has a life span of about 120 days before it is broken down and recycled. The signals that mark an individual red blood cell for destruction are not entirely clear, but they are thought to be linked to changes in their normal enzyme function.

What is the functional significance of the lack of a nucleus in the red blood cells? Take a moment to think about this question before you read on. As red blood cells lack a nucleus they contain no nuclear DNA and so cannot synthesize proteins. Therefore, there is no replacement of enzymes or structural proteins.

The breakdown of erythrocytes is effected by macrophages in the spleen and liver. Those in the spleen are particularly important in the breakdown of red blood cells. As the cells are broken down a protein found in plasma, called **transferrin**, transports the iron

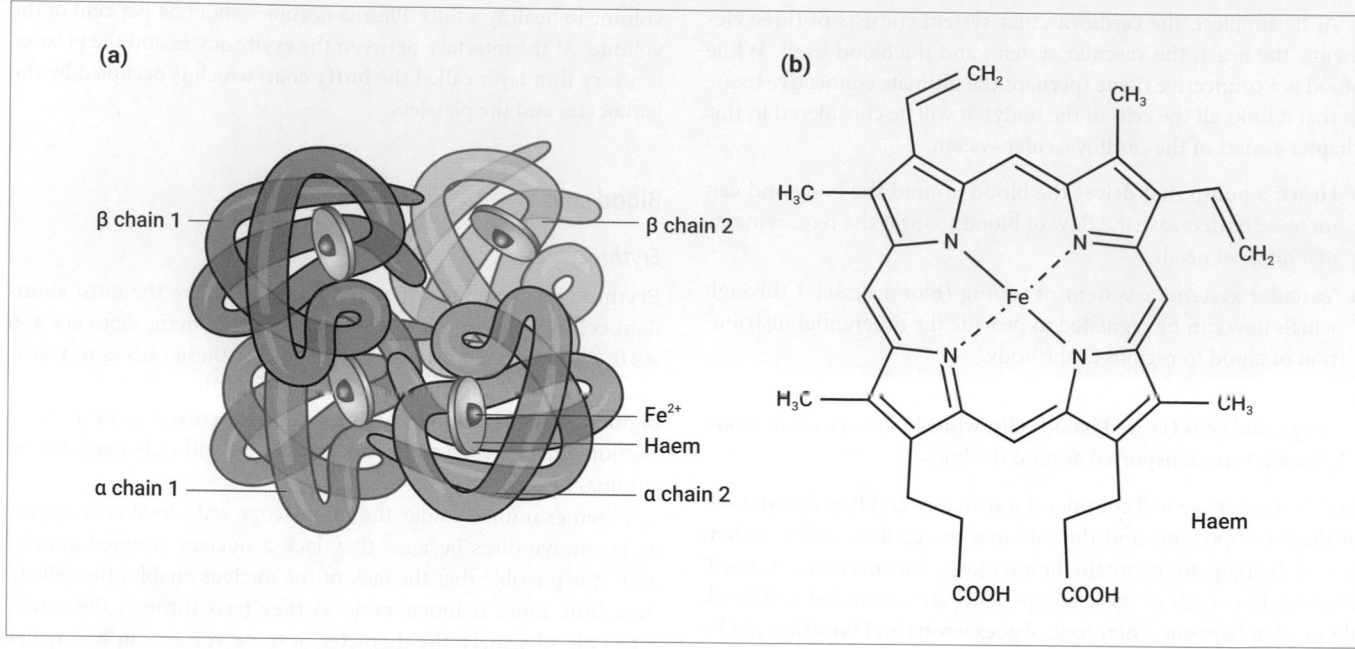

Figure 21.2 The structure of haemoglobin (Hb). Hb consists of four haem groups (the chemical structure shown in (b)) that are linked to four protein chains. Each component haem group and its associated globin chain is treated as a monomer. Therefore the completed structure is described as being a tetramer (four monomers combined together). The overall structure of Hb is shown in (a).

Source: Openstax CC BY 4.0 C J. Gordon Betts et al. *Anatomy and Physiology*, 2013. Download for free at: https://openstax.org/books/anatomy-and-physiology/pages/1-introduction

from the haemoglobin back to the bone marrow where it can be incorporated into new red blood cells. The remainder of the haem group is broken down to form a compound called **bilirubin** that is transported back to the liver. Following further metabolism, the bilirubin (now known as conjugated bilirubin) is released into the small intestine in bile. While the majority of this is absorbed in the small intestine, some enters the large intestine where the action of bacteria results in the formation of urobilinogen. This is lost from the body in faeces.

Given that erythrocytes have only a limited life span of about 120 days as described above, it is important that there is a continuous production of replacement red blood cells.

The formation of new red blood cells (and also white blood cells and platelets) is called **haemopoiesis**. Specifically, the production of new red blood cells is called **erythropoiesis**. This can occur in a range of locations in the fetus, in particular the liver. During early childhood, most of the production becomes confined to the bone marrow. By the time adulthood is reached, the number of bones that this occurs in is reduced, with the ribs, sternum, vertebrae, and pelvis being the main sites. The process of haemopoiesis is shown in overview in Figure 21.3. In summary, the bone marrow contains pluripotent stem cells, called haematopoietic stem cells: these are cells which have the potential to differentiate into any type of blood cell. In the case of red blood cells, these stem cells develop into a 'committed' stem cell called an erythroid cell.

Further maturation of these cells, including the synthesis of haemoglobin, results in the formation of **normoblasts**. Following this the nucleus is expelled from the cell, resulting in the formation of **reticulocytes**, which are now released into the circulation. Over

the next 2–5 days, these reticulocytes mature into fully functional erythrocytes (red blood cells). The whole development process from stem cell to mature erythrocyte takes about 2 weeks.

The regulation of red blood cell production is affected by a number of factors and can be limited by the availability of certain nutrients, in particular vitamin B_{12} and iron. The main endocrine stimulator of red cell production is the hormone erythropoietin (often referred to as EPO). This is a peptide hormone that is produced by the kidney when the oxygen content of the blood passing through the kidney falls. The O_2 sensors are found in the peritubular cells of the kidney (see Figure 24.6). A sustained fall in the oxygen content of the blood leads to a fall in intracellular ATP, which reduces the rate of pumping by the Na^+K^+ ATPase. The progressive rise in the intracellular sodium concentration depolarizes the cells and increases erythropoietin production. Erythropoietin stimulates the proliferation and maturation of the colony-forming units, which results in increased red cell production. There is also a whole range of substances called colony-stimulating factors, which orchestrate the development of a pluripotent stem cell to a mature red blood cell.

PAUSE AND THINK

What would be the effect of insufficient production of red blood cells?

Answer: If red cell production is compromised, then this will affect the capacity to transport oxygen leading to anaemia (Clinical Box 22.2). A major cause of anaemia worldwide is due to iron deficiency—genetic diseases such as sickle cell disease or thalassaemia result from disorders in haemoglobin production.

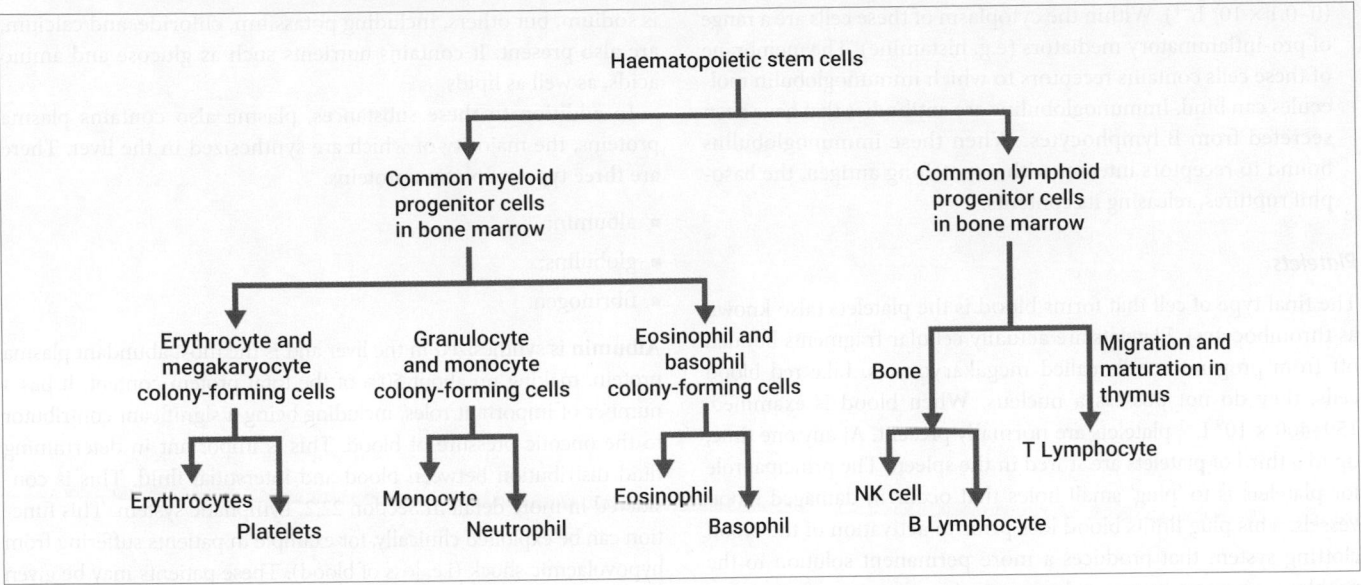

Figure 21.3 The formation of the cells found in blood. Note that there are additional steps between the colony-forming cells and the mature erythrocytes.

Source: Pocock et al. *Human Physiology*, 5th edn, 2017. Oxford University Press.

Leukocytes

Leukocytes, also known as white blood cells, constitute a group of cell types with an overall function of defence, forming part of the immune system (the immune system is considered in detail in Chapter 27). The appearance of these cells can be seen in Figure 21.4. So, what different types of white blood cell can we distinguish?

- **Neutrophils**—These form the largest proportion of white blood cells, accounting for 50–70% of all white blood cells. When measured, there are 2–7.5×10^9 L^{-1} of blood. Unlike red blood cells, neutrophils and other white blood cells retain a nucleus. The primary function of neutrophils is defence against invading micro-organisms (e.g. bacteria and fungi). They are attracted to sites of infection and inflammation due to their ability to sense chemoattractant molecules (e.g. some released from bacterial cell walls, as well as substances released by the host). Neutrophils contain a range of intracellular granules—the release of the contents of these results in bacterial killing. The mechanism of killing varies but generally involves the production of reactive oxygen species.

- **Lymphocytes**—constitute 20–40% of the total white blood cell count (1.5–3.5×10^9 L^{-1}). There are two types of lymphocytes: T lymphocytes and B lymphocytes. The development of T lymphocytes is dependent on the thymus and they participate in destroying cells that have been attacked by viruses. B lymphocytes (bone marrow-derived cells) are responsible for producing antibodies against pathogens.

- **Monocytes**—while significantly larger than neutrophils, monocytes constitute about 5% of the total white blood cell count (0.2–0.8×10^9 L^{-1}). Monocytes are phagocytic (i.e. they can engulf invading pathogens, particularly those that are intracellular). Interestingly, as part of the killing process, they may present components of the pathogen to T lymphocytes. Sometimes, they are referred to as antigen presenting cells.

- **Eosinophils**—form 1–5% of the total white blood cell count (0–0.4×10^9 L^{-1}). The primary role of this group of white blood cells is to offer defence against parasitic infections (e.g. worm infections).

- **Basophils**—constitute the smallest proportion of white blood cells, forming up to 0.2% of the total white blood cell count

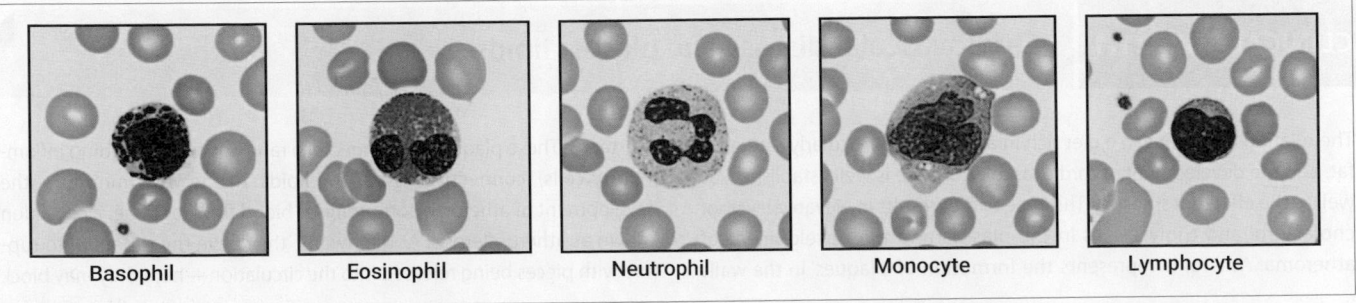

Figure 21.4 The appearance of the major classes of white blood cells (leukocytes). There are a number of different subtypes of each of these major classes.

Source: Moore et al. *Haematology*, 3rd edn. Oxford: Oxford University Press.

$(0{-}0.1 \times 10^9 \text{ L}^{-1})$. Within the cytoplasm of these cells are a range of pro-inflammatory mediators (e.g. histamine). The membrane of these cells contains receptors to which immunoglobulin molecules can bind. Immunoglobulins are antibodies that have been secreted from B lymphocytes. When these immunoglobulins bound to receptors interact with a matching antigen, the basophil ruptures, releasing its contents.

Platelets

The final type of cell that forms blood is the platelets (also known as thrombocytes). Platelets are actually cellular fragments budded off from progenitor cells called megakaryocytes. Like red blood cells, they do not possess a nucleus. When blood is examined, $150{-}400 \times 10^9 \text{ L}^{-1}$ platelets are normally present. At any one time, up to a third of platelets are stored in the spleen. The principal role for platelets is to 'plug' small holes that occur in damaged blood vessels. This plug limits blood loss prior to activation of the blood clotting system that produces a more permanent solution to the problem, a process known as **haemostasis**.

When blood vessels are damaged, the collagen-rich fibres that make up part of the blood vessel wall (see later in this chapter for further details) become exposed. This exposure results in platelets sticking to the collagen via an intermediary protein called von Willebrand factor. This in turn causes platelets to release the contents of cytoplasmic vesicles. The vesicles contain a variety of molecules, including ADP and 5-HT, that act locally to promote changes in the shape and metabolism of platelets. This is a process known as platelet activation. One outcome of platelet activation is that platelets become 'sticky', causing them to adhere to each other, a process known as platelet aggregation. This clumping of platelets results in the formation of a temporary platelet plug that provides an initial repair to the damaged blood vessel.

Plasma

Blood plasma constitutes the majority of blood volume (50–55%) and is essentially an aqueous solution that contains a range of substances dissolved in it. It is the medium by which substances (nutrients, waste substances, hormones, and drugs), as well as heat, are transported and distributed around the body.

While the precise composition of plasma may vary, over the long term it has a reasonably well-defined composition. Plasma contains a range of dissolved electrolytes of which the main one is sodium, but others, including potassium, chloride, and calcium, are also present. It contains nutrients such as glucose and amino acids, as well as lipids.

In addition to these substances, plasma also contains plasma proteins, the majority of which are synthesized in the liver. There are three types of plasma proteins:

- albumin;
- globulins;
- fibrinogen.

Albumin is synthesized in the liver and is the most abundant plasma protein, making up about 50% of the total protein content. It has a number of important roles, including being a significant contributor to the oncotic pressure of blood. This is important in determining fluid distribution between blood and interstitial fluid. This is considered in more detail in Section 22.2, Lymphatic system. This function can be exploited clinically, for example in patients suffering from hypovolaemic shock (i.e. loss of blood). These patients may be given infusions of albumin, or other solutions that contain large molecular weight compounds (e.g. dextran). Collectively, these infusions are known as colloidal solutions. The infused molecules (albumin, dextran) enter the blood and consequently increase oncotic pressure. This helps to keep water within the vascular system and goes some way to restoring a normal circulatory volume.

Another important role for albumin is that it can act as a carrier molecule for substances that do not dissolve in water. Examples of these include the thyroid hormone thyroxine, bilirubin, fatty acids, and many drugs such as penicillin.

Globulins may be divided into three groups—α, β, and γ—and have a diverse range of functions, including a transport role for substances (e.g. thyroid-binding globulin). The globulins, primarily γ-globulins, also known as immunoglobulins, have a role in providing an immune response directed against pathogens.

Fibrinogen is a protein involved in the blood clotting process.

It is possible to measure the levels of specific plasma proteins to give an insight into pathophysiology. For example, C-reactive protein may be used as a marker of inflammation, and measurement of specific lipoproteins, such as high- and low-density lipoproteins, gives us an insight into the likelihood of developing cardiovascular disease (see Clinical Box 21.1).

 Check your understanding of the concepts covered in this section by answering the questions in the e-book.

Cardiovascular disease and plasma lipids

The relationship between a diet rich in animal fat, particularly saturated fat, and the development of cardiovascular disease is well established, as well as the effects of smoking. This type of diet results in elevated levels of cholesterol and triglycerides in the plasma and the development of atheroma. Atheroma represents the formation of 'plaques' in the walls

of arteries. These plaques are formed of a range of cells (including inflammatory cells), connective tissue, and lipids. At the very minimum, the development of atheroma compromises blood flow to tissue, a condition known as atherosclerosis. At their worst, they have the potential to rupture with pieces being released into the circulation—here they may block other (smaller) blood vessels and result in the development of, for example, strokes or a cardiac infarction. Atherosclerosis is a major underlying cause of coronary artery disease and strokes.

Atherosclerosis affecting the coronary arteries can severely compromise heart function as the cardiac muscles receive an insufficient blood supply. This is often recognized as pain—angina pectoris—which may be triggered by exercise when the heart is required to beat faster.

At this point it is important to remember that lipids do not dissolve in the aqueous fluid that forms plasma; instead, they combine with proteins to form lipoproteins. There are several different ways to classify lipoproteins, one of which is their density. According to this, we get the following:

- Low-density lipoproteins (LDL)—essential, these transport cholesterol from the liver to tissues. When in excess, this promotes atheroma formation, so-called 'bad cholesterol'.

- High-density lipoproteins (HDL)—essentially, these transport cholesterol from the tissues to the liver where it is metabolized and excreted, so-called 'good cholesterol'.

- Very-low-density lipoproteins (VLDL)—mixture of cholesterol and triglycerides, the latter of which are released into the circulation as free fatty acids.

Therefore, it is essential that individuals maintain an appropriate (i.e. healthy) plasma profile of lipids—typically, total plasma cholesterol should be < 5 mM, while that of LDL should be < 3 mM and that of HDL > 1 mM. For the most part, eating a healthy diet and taking exercise should allow individuals to attain these targets.

However, for some individuals, this strategy fails to work so an alternative must be found. In these individuals, one approach to reducing plasma cholesterol levels (and therefore the risk of developing cardiovascular disease) is to take a drug from the statin family. The vast majority of our cholesterol is synthesized in the liver. It is a complex process, but the rate limiting step is controlled by the enzyme 3-hydroxy-3-methylglutaryl-coenzyme A reductase (HMG-CoA reductase). If this enzyme is blocked then liver synthesis of cholesterol is inhibited. Drugs which block this enzyme are known as statins, e.g. simvastatin and atorvastatin. As a consequence of inhibition, the number of LDL receptors on liver cells increases, therefore removing further cholesterol as LDL bind to them. Alongside these direct effects on plasma cholesterol levels, statins also have beneficial effects elsewhere, for example they stabilize plaques and reduce platelet aggregation. Equally, they have a range of unwanted adverse effects such as muscle pain and gastrointestinal upsets, but generally they are well tolerated and have been of significant benefit in reducing the risk of developing cardiovascular disease.

21.2 Vascular system

Circulatory systems

The vascular system represents the blood vessels that allow the delivery of nutrients, waste substances, electrolytes, hormones, and drugs around the body. In reality, we have two circulations (Figure 21.5). Although we will consider heart function in more detail later in this chapter, we need to think of the heart as a pump that can be divided into two.

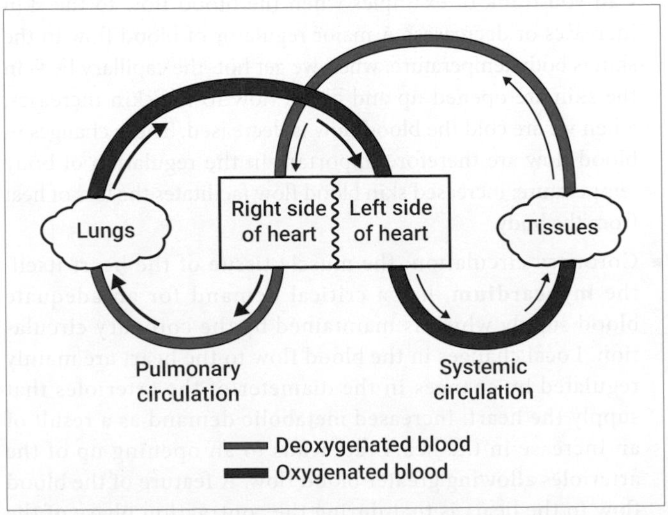

The right side of the heart pumps deoxygenated blood to the lungs via the **pulmonary circulation**, where it is oxygenated. This oxygenated blood is returned back to the left side of the heart from where it is pumped out via the **systemic circulation** delivering oxygen-rich blood to the tissues. The blood then returns back to the right side of the heart where the cycle begins again. Blood leaving the heart travels in **arteries**, while blood returning to the heart does so in **veins**. We will look at the structure and function of blood vessels later in this chapter. Before that, some general information regarding the pulmonary and systemic circulations will be useful to set the scene for this.

The pulmonary circulation

Deoxygenated, carbon dioxide-rich blood is carried back to the heart via the veins from the metabolizing tissues all over the body. These veins come together as the superior and inferior vena cavae (Figure 21.6), the so-called 'great veins' that bring the blood to the right atrium of the heart. The blood passes into the right ventricle of the heart, which then pumps it out via the pulmonary artery, to the lungs, where it is re-oxygenated before being delivered back to the systemic circulation. The mechanism by which gases are exchanged at the interface between the pulmonary capillaries and the alveoli is described in Section 22.2.

The distance between the heart and the lungs is very short and the driving blood pressures in the pulmonary circulation are significantly lower than in the systemic circulation, with peak pressures in the pulmonary circulation of 25 mmHg compared to 120 mmHg in the systemic circulation.

Figure 21.5 Schematic diagram showing the organization of the pulmonary and systemic circulations. Arrows indicate the direction of blood flow. Red shows oxygenated blood; blue, deoxygenated blood.

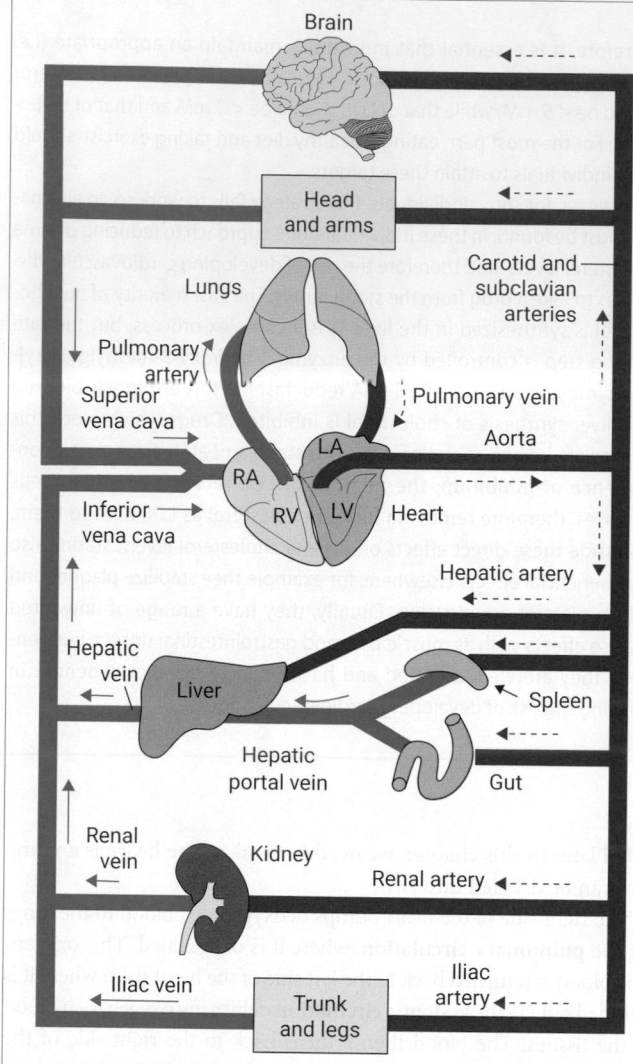

Figure 21.6 A simplified diagram showing the arrangement of the systemic and pulmonary circulations, illustrating some of the main arteries and veins.

Source: Pocock et al. *Human Physiology*, 5th edn, 2017. Oxford University Press.

PAUSE AND THINK

Looking at Figure 21.6, which of the great veins carries blood from the head back to the heart?

Answer: The superior vena cava brings together all the blood from the head and arms and carries it back to the right atrium.

The systemic circulation

The systemic circulation comprises the arteries, which deliver oxygen-rich blood to the tissues of the body; the **capillary** networks, which enable exchange of gases, nutrients, and waste products with the metabolizing tissues; and the veins, which return carbydioxide rich blood back from those tissues to the heart. Blood entering the systemic circulation, via the aorta, does so from the left ventricle under a significant pressure.

The principal arteries and veins of the human body are shown in Figure 21.6. As you might imagine, there are substantial differences in blood flow to different body regions which relate to the metabolic demand of those tissues. Figure 21.7 summarizes the relative distribution of blood flow to various organs in the body.

Within the systemic circulation in its entirety, we generally consider that there are also some special circulations—special in either their importance or the peculiarities of how they work. We can consider the following as examples of circulations that are special.

- Cerebral circulation: this is the blood supply to and from the brain. The brain has a high metabolic rate and therefore oxygen demand. Brain function can only be maintained for a few seconds in the absence of adequate blood supply, so it is essential to maintain a constant blood flow. The cerebral circulation is able to maintain a relatively constant blood flow in spite of wide changes in blood pressure around the rest of the body. This is called autoregulation. Blood flow to other structures (e.g. the kidneys) also displays autoregulation.

- Skeletal muscle circulation: this represents the blood flow to and from skeletal muscle. When we exercise the metabolic needs of skeletal muscles increase significantly. Therefore, there has to be a parallel increase in blood flow to meet this need. Indeed, skeletal muscle blood flow can increase up to 50× during strenuous exercise. Of particular importance is the fact that this increase is controlled primarily by local conditions of the muscle such as increased amounts of carbon dioxide and lactic acid, which lead to dilation of the arterioles.

▶ **We learn more about the cardiovascular response to exercise in Chapter 23.**

- Cutaneous circulation: the metabolic demands of the skin are relatively low, but the blood flow to the skin can be varied markedly.

Can you think of examples when the blood flow to the skin increases or decreases? A major regulator of blood flow to the skin is body temperature: when we get hot, the capillary beds in the skin are opened up and blood flow to the skin increases; when we are cold the blood flow is decreased. These changes in blood flow are therefore important in the regulation of body temperature: increased skin blood flow facilitates the loss of heat from the body.

- Coronary circulation: the muscle tissue of the heart itself, the **myocardium**, has a critical demand for an adequate blood supply which is maintained by the coronary circulation. Local changes in the blood flow to the heart are mainly regulated by changes in the diameter of the arterioles that supply the heart: increased metabolic demand as a result of an increase in the heart rate leads to an opening up of the arterioles allowing greater blood flow. A feature of the blood flow to the heart is that during the contraction phase of the cardiac cycle, the blood vessels are compressed leading to a temporary reduction in blood flow during each beat of the heart.

Figure 21.7 The regional distribution of blood flow to different body tissues and organs when at rest.

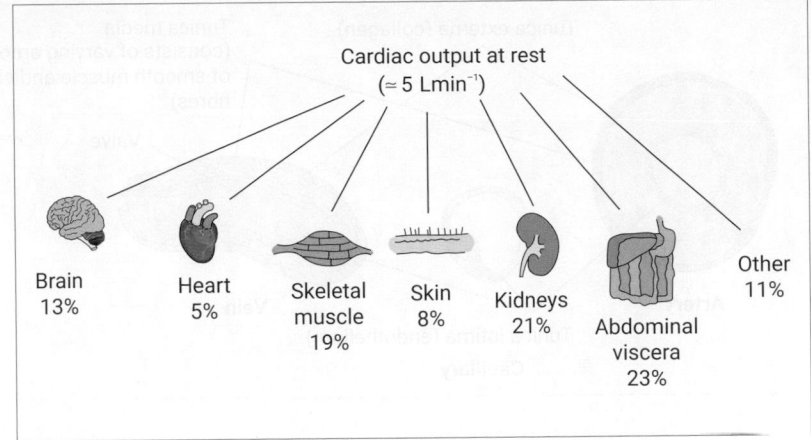

Cardiac output at rest
(≈ 5 Lmin^{-1})

Brain 13%

Heart 5%

Skeletal muscle 19%

Skin 8%

Kidneys 21%

Abdominal viscera 23%

Other 11%

- Fetal circulation: the circulation of blood to the fetus is special because the fetus exchanges gases, nutrients, and waste products with the maternal blood via the placenta. During fetal development, the lungs are not involved in gas exchange and there is a shunt (a vessel that cross-links the two sides of the heart) between the right and left sides of the heart so that the pulmonary circulation is mainly by-passed (see Section 26.9).

Lymphatic system

The lymphatic system is a network of capillaries and vessels linked to the venous system. When the blood flows through the capillaries some of the fluid in the blood passes out into the tissues to form the interstitial fluid. Most of this fluid returns to the capillaries and is returned to the heart via the venous system. Some of the interstitial fluid, however, remains and then drains into the lymphatic capillaries. As with the venous system, these lymphatic capillaries merge into larger vessels which eventually enter the venous system via the subclavian veins (see Figure 21.6).

The lymphatic system is also an important component of the immune system (Chapter 27).

Having considered some general ideas surrounding the circulation, it is now appropriate to consider looking at blood vessel structure and function in more detail.

Figure 21.8 and Table 21.1 illustrate the principal cross-sectional structural appearances of the three major types of blood vessel.

Arteries

When viewed in cross section, arteries can be seen to consist of three concentric layers of tissue (Figure 21.8).

- The outermost layer is called the **tunica adventitia** (also known as the tunica externa). This layer is formed entirely of a mixture of both dense and loose connective tissue.

- The middle layer of tissue is the **tunica media**. A structure called the external elastic lamina separates the tunica media from the tunica adventitia. Generally, the tunica media is the thickest layer. When examined microscopically, it can be seen to contain smooth muscle fibres that are arranged in both circular and spiral orientations around the lumen, supported by elastic tissue. This muscle, which is termed vascular smooth muscle, receives a nerve supply from the sympathetic nervous system. When this nervous supply is activated, the smooth muscle contracts. As it contracts, the diameter of the lumen through which blood flows is reduced, a process termed **vasoconstriction**. This means then

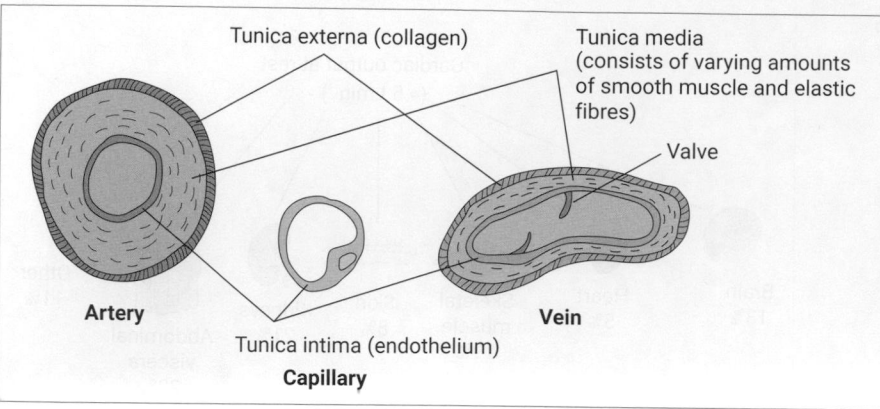

Tunica externa (collagen)
Tunica media (consists of varying amounts of smooth muscle and elastic fibres)
Valve
Tunica intima (endothelium)
Artery
Capillary
Vein

Figure 21.8 Comparison of the structure of the different types of blood vessel. Cross-sectional appearances of the three main types of blood vessel. Note the increased proportion of smooth muscle and elastic fibres in arteries compared to veins. By comparison, the capillaries consist of only a single layer of endothelial cells. Note these are not drawn to scale.

Table 21.1 The relative cross-sectional dimensions and composition of the blood vessels

	Artery	Arteriole	Capillary	Venule	Vein
Diameter	10–25 mm	30–300 µm	5–8 µm	5–200 µm	5–30 mm
Wall thickness	1–2 mm	20 µm	0.5–1.0 µm	2 µm	0.5–1.5 mm
Composition of wall (% of wall thickness)					
Endothelium	5	10	95	20	5
Smooth muscle	25	60	0	20	30
Elastic tissue	40	10	0	0	0
Collagenous tissue	30	20	5	60	65

that the pressure within the vessel increases and the blood flow reduces. Vasoconstriction, we shall see, is a major determinant of blood pressure within the cardiovascular system.

- The innermost layer of an artery is called the **tunica intima**. This is the thinnest layer, being only one cell thick. The cellular layer, called the endothelium, is in direct contact with blood in the lumen. The cells which form the endothelium are connected by tight junctions which prevent the loss of cells and large molecules from the blood. This endothelium 'sits' on a layer of connective tissue called the lamina propria. Surrounding this is a layer of elastic tissue called the internal elastic lamina, which separates the tunica intima from the tunica media.

The relative proportions of the tissues making up the walls of the arteries differ according to their location within the circulatory system. We can identify three main types of vessel in the arterial system:

- Elastic arteries are the largest type of artery and may be up to 25 mm in diameter (Figure 21.8). The largest of all elastic arteries in the human body is the aorta. This is the first artery in the systemic circulation that receives blood directly from the left ventricle when the heart contracts. In a similar manner, the first artery of the pulmonary circulation, the pulmonary artery, is an elastic artery. As the name suggests, the tunica media of these arteries has a high proportion of elastic tissue. Consequently, when the heart contracts and blood is ejected into the aorta, it responds by stretching due to blood entry and then recoiling due to the elastic fibres when the pressure drops as the heart relaxes. This means then that blood flow around the body is progressively converted from being pulsatile to a more continuous flow, therefore delivering a continuous supply of oxygen and nutrients to tissues and cells.

- Muscular arteries are small-to-medium-sized arteries, which may have a diameter up to 10 mm. They differ from elastic arteries in that their tunica media contains a higher proportion of vascular smooth muscle. These vessels are therefore more contractile than elastic arteries, and as such play an important role in both the regulation of blood pressure and the distribution of blood to different tissues. By varying the tone (i.e. the degree of contraction) of the smooth muscle, blood flow to tissues and organs can be increased or decreased.

- Arterioles are the smallest type of artery, with a diameter that ranges from 10 to 300 µm. The very smallest of arterioles are called metarterioles and it is from these that capillaries arise. Like muscular arteries, arterioles have a significant layer of smooth muscle and so have the potential to increase or decrease the diameter of the lumen and therefore control blood flow to tissues. The arterioles are the main resistance vessels of the circulatory system. We now know that there are a range of factors, such as increased CO_2 levels and increased temperature, that occur as a consequence of increased metabolic activity, which together with the influence of the sympathetic nervous system determine peripheral resistance and therefore blood flow to tissues. This means that the flow of blood to a particular tissue is responsive to the local metabolic needs of that tissue. This is in addition to the overall regulation of blood flow which is regulated by changes in activity of the endocrine and sympathetic nervous system.

Blood flow in arteries

We have seen that when the heart contracts, blood is forced out into the arterial system and flows around the body. Blood flow is therefore pressure driven: pressure increases when the heart contracts—**systole**—and then reduces as it relaxes—**diastole**. The greater the difference in pressure between the heart and capillaries (where pressure is very low), the greater the blood flow. Typically, the average pressure in the aorta is around 100 mm Hg, whereas in the capillaries it is about 20 mm Hg (Figure 21.9).

While the absolute pressures in the pulmonary circulation are much smaller than the systemic, at about 25 mm Hg, so is the resistance to blood flow. As a result, the volume of blood flow from the heart when it contracts is identical for both the pulmonary and systemic circulations.

Peripheral resistance, the resistance to blood flow in arterioles, is a major determinant of blood pressure and flow: if peripheral resistance is increased, then blood flow is reduced. We can express flow, then, very simply according to the following equation:

$$\text{Flow} = \frac{\text{pressure difference}}{\text{resistance}}$$

The French physicist Poiseuille investigated factors that influence resistance to flow. He summarized them in this equation:

$$R = \frac{8\eta L}{\pi r^4}$$

where:

- R = resistance to flow
- η = viscosity of liquid (in this case blood)
- L = length of tube (blood vessel)
- $\pi = 3.142$
- r = radius of tube.

Study the equation and think about what it tells us and why it is important for cardiovascular physiology. It tells us that even small changes in the diameter of the blood vessels may have profound effects on resistance, and therefore flow. This is because the

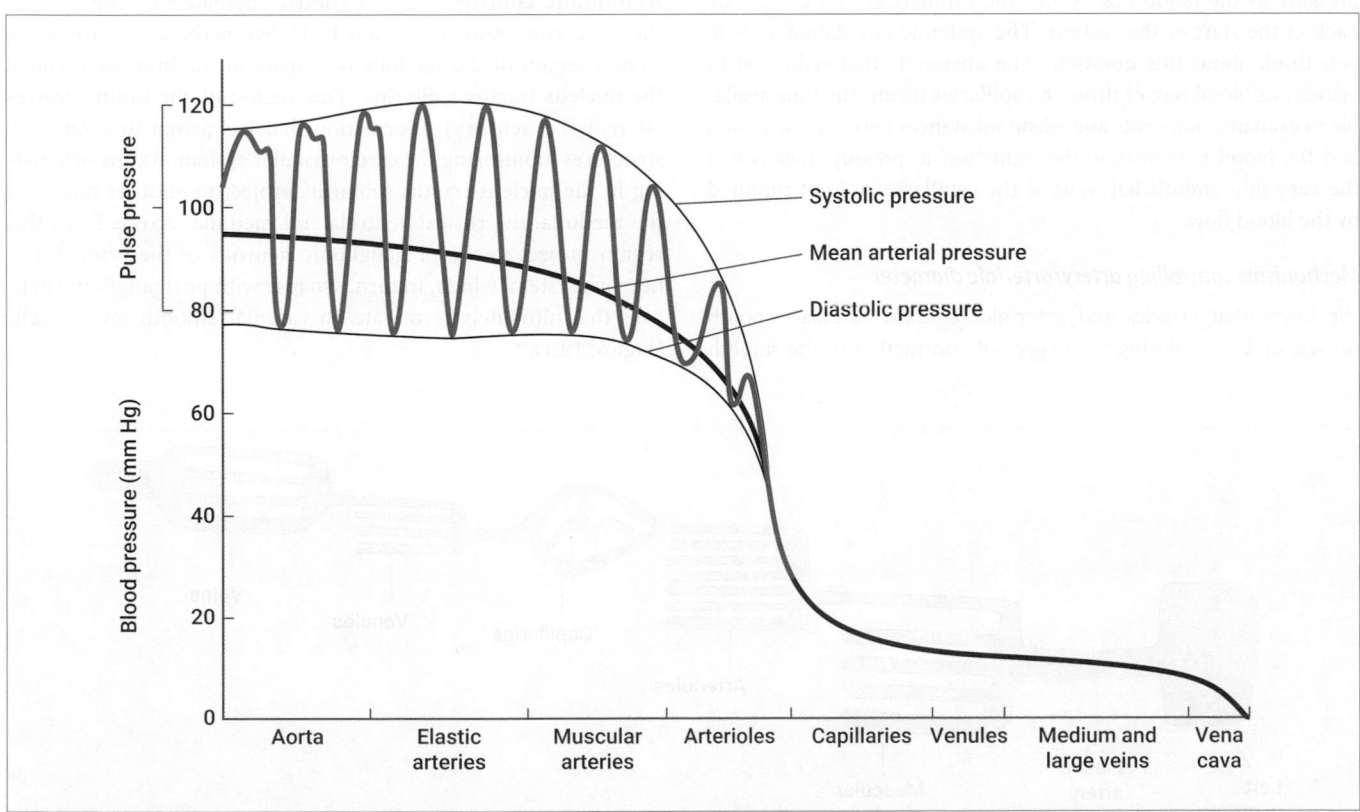

Figure 21.9 Variation in blood pressure around the circulatory system. Within the arterial side of the circulatory system, the blood pressure is pulsatile, reflecting the contraction (systole) and relaxation (diastole) of the heart.

resistance to flow is inversely proportional to the radius raised to the power of 4. This means then that if the diameter is doubled, then flow will increase 16 times ($2 \times 2 \times 2 \times 2 = 16$). Equally, halving the diameter will lead to a 16-fold decrease in blood flow.

Blood flow is the volume of blood that is moving per unit time, that is L s^{-1}. However, we may also be interested in how fast the blood is moving: displacement per unit time (i.e. m s^{-1}). There is a simple mathematical relationship that defines blood velocity:

$$\text{Velocity} = \frac{\text{flow}}{\text{cross-sectional area}}$$

In the body the flow of blood is determined by the volume that is ejected from the heart with each beat, this volume is called the **cardiac output**. Therefore, blood velocity can be rewritten as:

$$\text{Velocity} = \frac{\text{cardiac output}}{\text{cross-sectional area}}$$

Think for a minute about the cross-sectional area of the vascular system. We know that blood is ejected into the aorta, from there into smaller arteries, into arterioles, and then into capillaries—shown schematically in Figure 21.10. Because there are more and more blood vessels arising, even though the individual vessels are narrower, the total cross-sectional area increases dramatically. What this means then is that as blood moves from the aorta into the capillaries, its velocity decreases.

What is the significance of a reduction in blood velocity and pressure as the blood passes into the capillaries? You could look back at the start of the section 'The systemic circulation' to help you think about this question. The answer is that reduction in velocity as blood moves through capillaries means the time available to exchange nutrients and waste substances between the tissues and the blood is increased; the reduction in pressure means that the very thin endothelial walls of the capillaries are not ruptured by the blood flow.

Mechanisms controlling artery/arteriole diameter

We know that arteries and arterioles contain vascular smooth muscle and that altering the degree of contraction of the smooth muscle changes the diameter of the lumen of the vessel and therefore the volume of blood flowing through it. When the smooth muscle contracts, the lumen of the blood vessel reduces, a process called vasoconstriction. When the smooth muscle relaxes, the lumen increases and this is termed **vasodilation**.

Muscle contraction has been considered elsewhere in this book (see Section 20.1); however, it would be useful to quickly remind ourselves how this happens in vascular smooth muscle. Fundamentally, contraction occurs as actin and myosin filaments slide past each other. As in other types of muscle, the trigger for this process is a rise in intracellular calcium in the muscle cell. The influx of calcium may occur because of voltage-gated calcium channels opening, or it may be due to the opening of ligand-gated cation channels. Either way, the influx of calcium activates the calcium binding protein calmodulin, which in turn activates the enzyme myosin light-chain kinase. Once activated, this enzyme phosphorylates the heads of the myosin molecules, which, in turn, results in actin filaments being pulled alongside the myosin filaments, resulting in contraction.

We can think about control occurring in four ways:

- autonomic control;
- endocrine control;
- local control;
- myogenic control.

Autonomic control refers to control mediated by the sympathetic nervous system (Section 19.1). Sympathetic control arises from a region of the medulla oblongata in the brainstem called the nucleus tractus solitarius. This region of the brain receives afferent (i.e. sensory) information that has arisen in a range of structures monitoring the cardiovascular system. Axons originating in the nucleus tractus solitarius project to another region of the medulla, the rostral ventrolateral medulla. Axons from this region project to the preganglionic neurons of the sympathetic nervous system, which, in turn, synapse with postganglionic neurons that ultimately terminate on vascular smooth muscle cells (Figure 19.1).

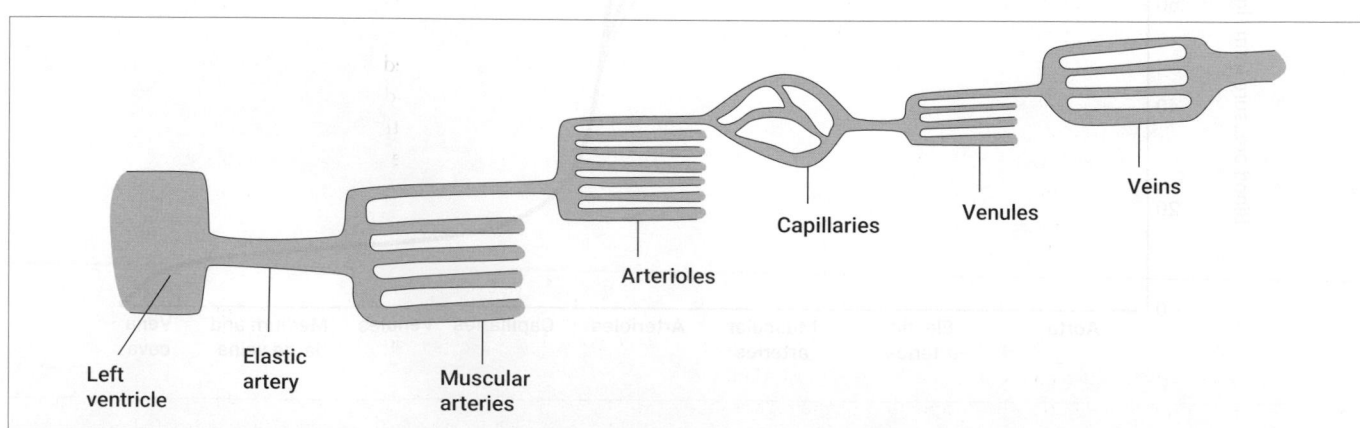

Figure 21.10 Schematic diagram showing the arrangement of the different types of blood vessels in the circulatory system. Despite a decrease in the diameter of the individual vessels, there is an overall increase in cross-sectional area because of the increase in the number of vessels in parallel.

The primary neurotransmitter released from sympathetic post-ganglionic neurons is noradrenaline. There are two main classes of adrenergic receptors: α and β. These receptors are part of the G-protein coupled receptor (GPCR) family of receptors (see Section 17.6). Both the α and β receptors have a number of subtypes, which show different distributions and have different actions.

The released noradrenaline interacts with α_1 adrenoreceptors on vascular smooth muscle to produce vasoconstriction. Vasodilation is achieved by removing this vasoconstrictor response. In contrast to sympathetic vasoconstrictor responses, there also exist some parasympathetic vasodilator fibres in specific tissues. These are seen in a range of structures such penile and clitoral erectile tissues.

There is a second type of adrenoreceptor, the β_2 receptor, that is found on some vascular smooth muscle, in particular in skeletal muscle, cardiac tissue, the lower airways (bronchioles), and the liver. When stimulated the β_2 receptors produce a sympathetic vasodilation and also dilation of the airways.

What would happen if the blood vessels of skeletal and cardiac muscle responded to adrenaline in the same way as other tissues? You could look back at Section 19.1 to answer this question. The release of adrenaline, from the adrenal glands, and noradrenaline by the sympathetic nervous system occur as part of the body's preparation system for the fight or flight response. The vasodilator response seen in cardiac and skeletal muscle increases blood flow to structures linked to an increase in demand for oxygen and nutrients. This is further enhanced by diverting the flow of blood from those vascular beds that show a vasoconstrictor response. The relaxation of the smooth muscles of the bronchioles also facilitates increased oxygen uptake.

Endocrine control refers to the control mediated by hormones released from a range of organs whose target structure is vascular smooth muscle. The principal hormones are:

- adrenaline;
- antidiuretic hormone (ADH);
- angiotensin II;
- atrial natriuretic peptide (ANP).

Adrenaline is released from the adrenal medulla when the sympathetic nervous system is activated. Adrenaline interacts with α_1 adrenoreceptors to produce vasoconstriction and with β_2 receptors to produce vasodilation in skeletal and cardiac muscle.

Antidiuretic hormone (ADH) is a hormone released from the posterior pituitary gland under the control of the hypothalamus. Its primary role is to increase water reabsorption in the collecting ducts of the kidney and in doing so it makes a significant contribution to regulating body fluid volume (Section 24.6). ADH is released when the body becomes dehydrated: increasing reabsorption of water leads to an increase in circulatory volume and therefore blood pressure. ADH may also be released when blood pressure drops significantly. In this situation, ADH interacts with V1a receptors to produce a vasoconstrictor response (ADH is also known as vasopressin—hence the designation of the receptors it interacts with as V1, V2, etc.).

Angiotensin II is produced from the activation of angiotensin I as it passes through the pulmonary circulation (Section 24.7). This,

in turn, is formed from the action of an enzyme called renin on the precursor angiotensinogen. In addition to its role in promoting the release of aldosterone to increase renal reabsorption of sodium chloride and water, angiotensin II is also a potent vasoconstrictor acting on G-protein coupled receptors on vascular smooth muscle. This pathway is exploited therapeutically through the use of ACE inhibitors (angiotensin converting enzyme inhibitors) to lower blood pressure in patients suffering from high blood pressure (hypertension)—see Clinical Box 21.3.

Atrial natriuretic peptide (ANP) is a hormone produced by the atria of the heart, primarily in response to stretching of the atria resulting from an increase in the circulating blood, which, in turn, increases blood pressure. Its action is to stimulate the renal excretion of sodium chloride and water, which reduces the extracellular fluid volume and therefore blood pressure. In addition to this role, it is also a vasodilator, albeit not particularly potent.

Local control of vascular smooth muscle refers to changes in tone produced by local factors, primarily locally produced substances, although factors such as temperature may also be considered here. This mechanism enables local regulation of the blood flow in response to local changes in metabolic demand.

Local control can produce both vasoconstrictor and vasodilator responses. For example, histamine is a local mediator that plays an important role in inflammation. In response to the release of histamine, arterioles dilate. Similar responses are produced by bradykinin and 5-hydroxytryptamine. In a similar manner, there are a group of compounds called prostaglandins, which are pro-inflammatory mediators that promote vasodilation. They are produced from arachidonic acid which itself is released from membrane phospholipids. Aspirin and other non-steroidal anti-inflammatory drugs inhibit the cyclooxygenase (COX) enzymes that catalyse the conversion of arachidonic acid to prostaglandins, thereby reducing the inflammatory response.

In addition to these locally produced mediators, we now also know that the endothelium (the innermost lining of the arterial system) is capable of producing and releasing biologically active molecules. Perhaps the best-known example of these is nitric oxide (NO). NO is produced in endothelial cells from the amino acid arginine by the action of the enzyme nitric oxide synthase (NOS). There are a range of factors such as the inflammatory mediators histamine and bradykinin which promote NO release, but having been released it diffuses to the tunica media. Once in this region it enters vascular smooth muscle cells and interferes with intracellular signalling systems, resulting in relaxation and therefore producing a vasodilator response. Equally, there are also compounds released from the endothelium which produce vasoconstriction, such as endothelin.

PAUSE AND THINK

Why does skin, where there is local inflammation, change colour, sometimes swell, and often feel warm to the touch?

Answer: The local mediators such as histamine and bradykinin released as part of the inflammatory response promote local vasodilation. As a result of the increased blood flow, the skin can appear to change colour (e.g. for people with white skin tones it appears red), as well as feeling warmer to the touch.

As we mentioned earlier, local factors related to increased metabolic activity may also lower the tone of vascular smooth muscle, causing vasodilation. These factors include an increase in temperature, low oxygen levels, high carbon dioxide levels (and therefore low pH levels), release of adenosine, and elevated potassium concentration. These mechanisms thereby enable localized regulation of blood flow in response to the local conditions. The ability of these factors to produce local vasodilation in response to increased metabolic activity is termed **active hyperaemia**.

Myogenic control is a property of the smooth muscle cells of the blood vessels whereby stretch of the smooth muscle cells triggers contraction of the muscle. This plays an important part in the autoregulation of blood flow through different tissues and organs, for example the kidney (see Section 24.2) such that the blood flow is maintained at a constant level, despite changes in the blood pressure in the body as a whole. An increase in systemic blood pressure will lead to increased blood flow and stretching of the smooth muscle opening stretch-activated ion channels. The resultant depolarization triggers the opening of voltage-gated calcium channels in the muscle membrane and the release of calcium from internal stores, resulting in muscle contraction which restores the blood flow to the previous levels. When the blood pressure falls, the ion channels close and the muscle relaxes.

Capillaries

Following delivery of oxygen and nutrient-rich blood to the tissues via the arterial systems, the final stage is the transfer of nutrients to the individual cells and, at the same time, the removal of waste products such as carbon dioxide. The blood vessels that allow this exchange to occur are the capillaries.

Capillaries originate from the very smallest of arterioles. The pre-capillary or terminal arterioles are no more than 10 μm in diameter and about 1 mm in length. In terms of their structure, capillaries consist only of a single layer of endothelial cells (the tunica intima) sitting on a basement membrane, as described previously. Despite this generalization in terms of structure, it is possible to distinguish three different types of capillary:

- Continuous capillaries are characterized by having endothelial cells that are connected together via so-called tight junctions. In many tissues, the tight junctions are incomplete, leaving intercellular clefts between some endothelial cells. Small molecules, including water, glucose, oxygen, and carbon dioxide, can move from blood to tissue and also in the opposite direction. This type of capillary is most common in the body. The only part of the body where the tight junctions are fully complete is the blood–brain barrier.

- Fenestrated capillaries are characterized by the presence of pores (fenestrae) within the endothelial cells. Fenestrated capillaries are common in the kidneys and small intestine and allow the movement of larger molecules.

- Discontinuous capillaries are also known as sinusoids. They share some similarities with fenestrated capillaries in that they are 'leaky', though their leakiness is greater in that blood cells and proteins may pass between the capillary and tissue. This type of capillary is found in the bone marrow, spleen, and liver.

Take a break here to think about these questions:

- How does the presence of discontinuous capillaries relate to the vascular function of these tissues and organs?

The bone marrow, spleen, and liver are all involved in the production or breakdown of the erythrocytes. These 'leaky' capillaries allow the movement of red blood cells, as well as large molecules, into and out of the blood stream.

- How are most small non-polar molecules exchanged between blood and tissues?

Most small molecules, such as oxygen and carbon dioxide, move from blood to tissue, or the opposite direction, by the process of diffusion. As you will remember, diffusion is the movement of substances from a region of higher concentration to a region of lower concentration (Section 16.2).

The process of diffusion can be summarized in Fick's equation:

$$J = -DA\frac{\Delta c}{x}$$

where:

- J = the rate of diffusion in molecules per unit time crossing the membrane, mol m^{-2} s^{-1}

- D = diffusion coefficient (a measure of the diffusing ability of a substance over time)

- A = the area of the membrane, m^2

- Δc = the concentration gradient across the membrane, the difference in the numbers of particles, m^{-4}

- x = the thickness of the membrane, m.

▶ **You can find a walkthrough of Fick's equation in Quantitative Toolkit 8, Case study 8.3.**

From this equation, then, it can be seen that one of the factors influencing how much of a substance diffuses is the area over which diffusion occurs. In terms of the vascular system, this is related to the capillary density: the greater the capillary density, the greater the amount of substance that can be moved. This is highest in structures such as the brain and heart, which have a high nutrient demand and are relatively intolerant of reduced supply compared with tissues such as skeletal muscle.

As indicated previously, the movement of substances between plasma and interstitial fluid occurs mainly by diffusion. Very small molecules and those that are highly lipophilic simply pass through the endothelial cells. This process is known as transcellular exchange. The movement of substances that are hydrophilic and/or larger substances is a little more problematical because they cannot pass through the lipid-rich plasma membrane. Therefore, these substances pass through the fenestrations of capillaries, or they may pass through the gaps between adjacent endothelial cells, so called paracellular exchange.

Against a background of the movement of substances between blood and tissues, what is involved in balancing the movement of water and solutes and their relative concentrations?

There is a range of physical factors that affect water movement. Of the two main variables the first is hydrostatic pressure: fluid moves from regions of high pressure to low pressure. The second

Figure 21.11 The origin and direction of forces involved in the movement of fluids into and out of capillaries (arrows indicate the direction of movement). Forces (a) and (d) operate to move fluid out of the capillary while forces (b) and (c) operate to move fluid into the capillary. The net force at any one time and location along the length of the capillary is ((a + d) − (b + c)). In some instances, the hydrostatic pressure in the interstitial fluid (c) becomes subatmospheric—this causes the direction of fluid flow to reverse from that shown above.

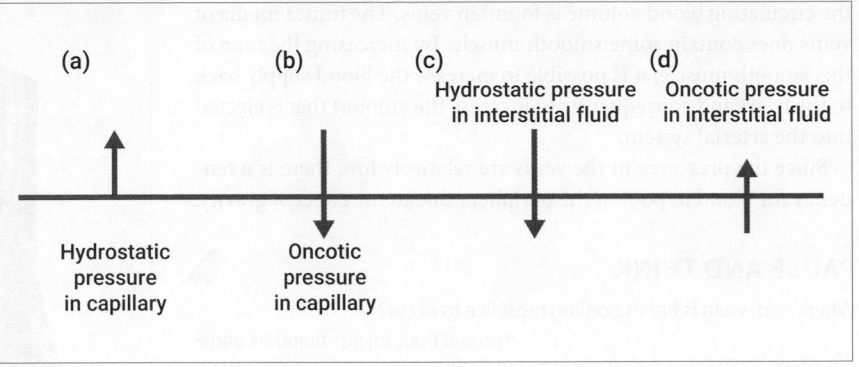

variable is oncotic pressure: water moves from regions of low solute concentration to regions of high solute concentration (essentially movement of water from a weak solution to a concentrated solution via osmosis). Osmotic movement of water is dependent, then, on the number of solute molecules dissolved in a solution—this is known as a colligative property. Because the major contributor to osmotic pressure in blood is proteins (remember that blood plasma contains proteins, such as albumins, which are unable to leave the circulation), the osmotic pressure is referred to as **oncotic pressure**. We can identify four different forces trying to move water as shown in Figure 21.11.

a. Blood pressure (hydrostatic pressure) is trying force fluid out of the capillary.

b. Plasma oncotic pressure is trying to pull fluid into the capillary.

c. Pressure in the interstitial fluid is trying to move fluid out of the capillary. Why should this happen? Logically, you might argue the opposite (i.e. it should be forcing fluid into the capillary). However, when the pressure is measured here it is found to be subatmospheric—this then effectively 'sucks' fluid out of the capillary. This subatmospheric pressure is a result of interstitial fluid entering lymphatic vessels.

d. Interstitial oncotic pressure is trying to pull fluid into the interstitial fluid. Remember that interstitial fluid is fluid which surrounds each and every cell in the body.

Therefore, the net direction of fluid movement is simply determined by the sum of these four forces. This balance is described as **Starling's hypothesis** of tissue fluid formation. Table 21.2 shows an example of the relative magnitudes of these forces at the arterial and venous ends of a capillary.

We can calculate what is happening at each end in terms of water movement: this can be done by calculating the difference between the outwardly and inwardly acting pressures at both the arterial

Table 21.2 The relative hydrostatic and oncotic pressures along the length of a capillary

Force (mmHg)	Arterial end	Venous end
Blood pressure (P_{Cap})	35	15
Interstitial pressure (I_P)	−3	−3
Plasma oncotic pressure (P_{Onc})	28	28
Interstitial oncotic pressure(I_{Onc})	9	9

and venous end of a capillary to establish the magnitude and direction of fluid movement. This can be calculated as follows:

$$\text{Filtration pressure at arterial end} = \left(P_{Cap}\right) + \left(I_{Onc}\right) - \left(I_P\right) - \left(P_{Onc}\right)$$
$$= 35 + 9 - (-3) - 28$$
$$= 19 \text{ mm Hg}$$
$$\text{Filtration pressure at venous end} = \left(P_{Cap}\right) + \left(I_{Onc}\right) - \left(I_P\right) - \left(P_{Onc}\right)$$
$$= 15 + 9 - (-3) - 28$$
$$= -1 \text{ mm Hg}$$

It can be seen then that at the arterial end there is a net pressure promoting fluid movement out of the capillary, while at the venous end the net pressure is promoting reabsorption of fluid. The filtration and reabsorption pressures are not identical: this suggests that more fluid passes out of the capillaries into the interstitial space than is reabsorbed. What then happens to the excess fluid? Any excess fluid in the interstitium is reabsorbed into the lymphatic system. Ultimately, the lymphatic vessels empty into the venous circulation, so overall the volume of fluid produced is identical to the volume reabsorbed.

In this type of idealized model, the fluid movement occurs out of the capillary at the arterial end and reabsorption occurs at the venous end. However, the relative balance of the forces is different in different tissues. For example, in the glomerulus of the kidney (Section 24.2) the hydrostatic pressure difference is such that filtration, outward movement of fluid, occurs along the length of the capillary.

Veins

Once blood has passed through the capillaries, having delivered oxygen and nutrients to the tissues and gained carbon dioxide, it then returns to the heart via the venous circulation. In a manner analogous to the arterial system, the smallest veins, those arising immediately from capillaries, are called **venules**. Venules converge and eventually form veins that return blood back to the heart.

Veins are composed of the same three concentric layers of tissues as arteries are. However, unlike arteries, the veins do not have to withstand high pressures: the pressures within the venous system are no higher than 20 mm Hg.

The middle layer, the tunica media, contains little or no elastic tissue and the walls of the veins themselves are relatively thin. This means that veins are easily distensible. Therefore, the veins are able to act as a reservoir for blood and at any one time, about 60% of the circulating blood volume is found in veins. The tunica media of veins does contain some smooth muscle. By increasing the tone of this smooth muscle, it is possible to increase the blood supply back to the heart and consequently to increase the amount that is ejected into the arterial system.

Since the pressures in the veins are relatively low, there is a tendency for blood to pool in the periphery due to the effect of gravity.

PAUSE AND THINK

Where and when is blood pooling most like to occur?

Answer: Pooling is most common in the lower legs and feet, especially when standing still for long periods.

Blood return to the heart is facilitated by three mechanisms:

- Veins have valves in them. The valves allow forward movement of blood moving towards the heart but are arranged such that they prevent any backflow of blood away from the heart.

- Blood flow back to the heart is also aided by the **muscle pump** (Figure 21.12). When the muscles of the leg contract, they squeeze on veins and force blood back towards the heart. Upon relaxation, the valves prevent the blood from flowing back.

- In a similar manner, blood flow in veins is also aided by the **respiratory pump**. During inspiration the pressure within the veins of the thoracic cavity is less than the pressure within the abdominal veins. This generates a pressure difference which helps to move blood back towards the heart. These two pumps are particularly significant during, for example, exercise where the increased metabolic needs of the body are met by increasing the rate and volume of blood ejected by the heart.

▶ **We discuss the operation of the cardiovascular system during exercise in more detail in Chapter 23.**

You can easily show the operation of the valves in your veins as follows: let your arm hang down so that the veins stand out on the back of your hand. Locate a branch in the vein. Pressing on the vein just above the branch, run another finger upwards, emptying the length of vein of blood. The vein will appear flat. When you release your finger from the bottom of the vein, it will refill with blood.

 Check your understanding of the concepts covered in this section by answering the questions in the e-book.

21.3 Heart

So far, we have described two components of the cardiovascular system: the blood and the vessels that blood flows in. The final component of the system is the heart, the pump that drives blood around the blood vessels.

The gross structure of the heart can be seen in Figure 21.13. The heart comprises four chambers: the upper chambers are called **atria**, which receive blood from the veins, and the bottom

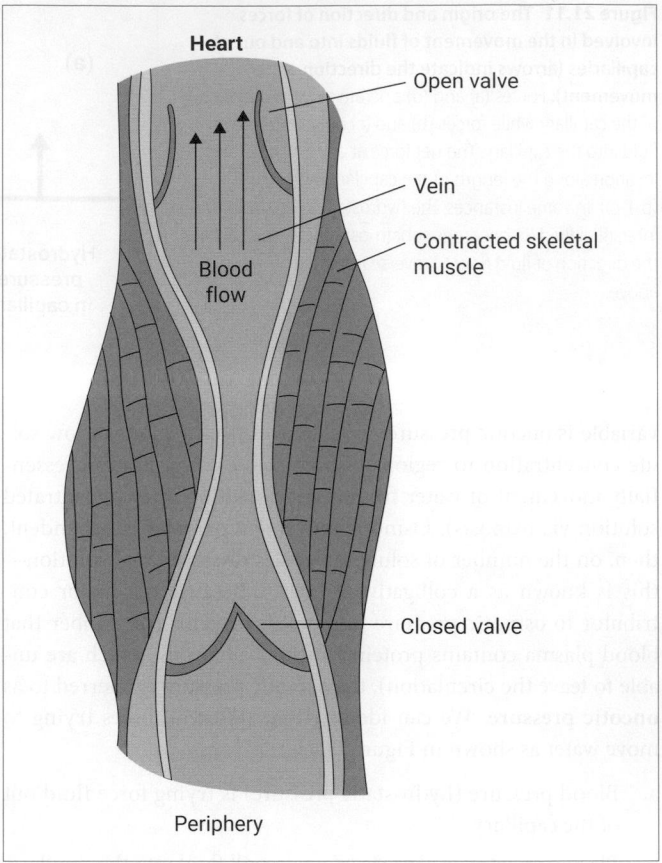

Figure 21.12 The muscle pump. When skeletal muscles surrounding veins contract the veins are compressed such that the pressure pushes the blood back towards the heart. When the muscle relaxes, backflow of the blood is prevented by the closure of the valves.

two chambers are called **ventricles**, which pump blood into the arteries. In reality, we can think of the heart as actually being two pumps operating in parallel. The right ventricle pumps blood into the pulmonary circulation, while the left ventricle pumps blood into the systemic circulation. Like the blood vessels, the heart is also formed from three layers of tissue:

- **endocardium**—the innermost layer, which is formed mainly of endothelial cells that are in contact with blood;

- **myocardium**—the middle, muscular layer;

- **epicardium**—the outer layer, which is formed of connective tissue and also makes up the inner layer of the **pericardium**, the tough connective tissue sac that encloses the heart and the origins of the main blood vessels.

The myocardium is formed mainly from cardiac myocytes, which are the muscle cells of the heart. Individual cardiac muscle cells are functionally linked to neighbouring cells by specialized cell–cell junctions called intercalated discs. We will see later in this chapter that this is of significance in that it allows the rapid transmission of action potentials across the entire myocardium and allows the heart to beat 'as one'. The microscopic appearance of cardiac muscle is shown in Figure 21.14.

Figure 21.13 The structure of the heart.

Source: Pocock et al. *Human Physiology*, 5th edn, 2017. Oxford University Press.

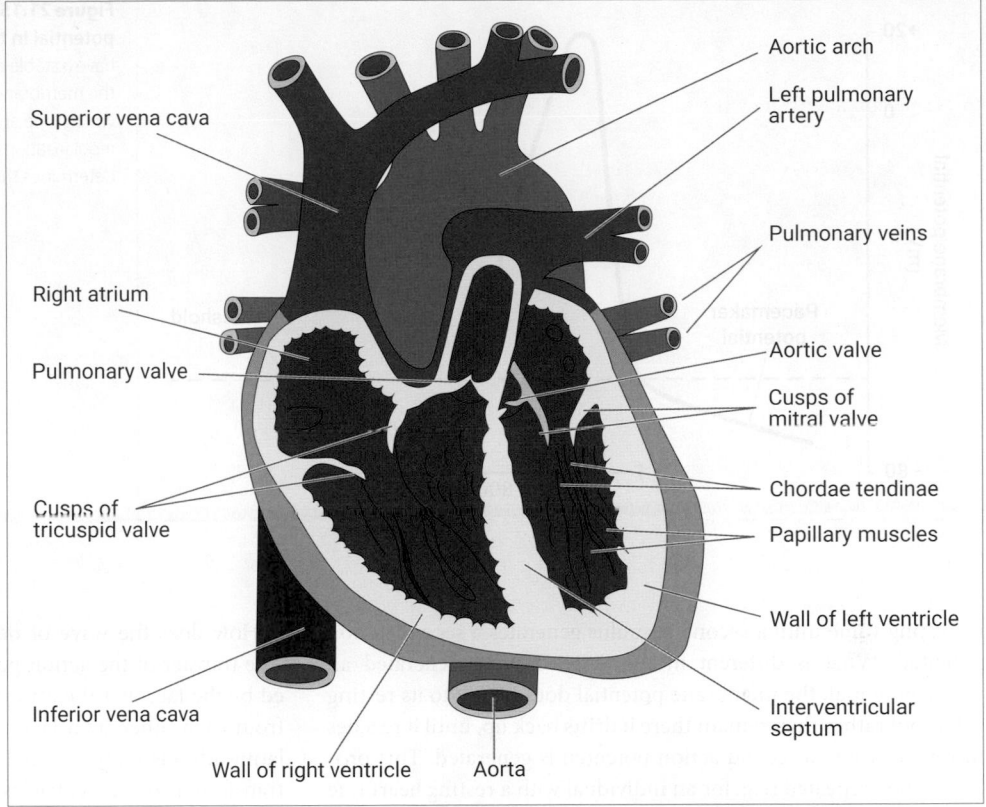

- Superior vena cava
- Right atrium
- Pulmonary valve
- Cusps of tricuspid valve
- Inferior vena cava
- Wall of right ventricle
- Aorta
- Aortic arch
- Left pulmonary artery
- Pulmonary veins
- Aortic valve
- Cusps of mitral valve
- Chordae tendinae
- Papillary muscles
- Wall of left ventricle
- Interventricular septum

Unidirectional blood flow from the atria to the ventricles is maintained by the **atrioventricular valves**. These are the **mitral valve** between the left atrium and the left ventricle and the **tricuspid valve** between the right atrium and the right ventricle. These valves prevent backflow from the ventricles into the atria during ventricular systole (contraction) and are stopped from being forced open by the chordae tendinae, which are anchored to the walls of the ventricles by the papillary muscles (Figure 21.13).

Likewise, the **pulmonary valve** and the **aortic valve** are located between the right and left ventricles and the pulmonary artery and aorta, respectively, and act to prevent backflow from the arteries into the ventricles when the ventricles relax.

Initiation of the heartbeat

In order for the heart to function as a pump, the myocytes need to contract. The general mechanisms describing muscle contraction are described elsewhere (Section 20.1), so will not be discussed in detail here. It is simply worth remembering that the process is driven by a rise in intracellular calcium and that contraction results from the cross-bridge activation between the actin and myosin filaments, with myosin passing over actin within the myocyte, resulting in the cell contracting. This process occurs with every beat of the heart and is triggered by the depolarization of the myocytes.

The process of depolarization begins in the atria and spreads across the myocytes of the atria and then the ventricles resulting in a coordinated contraction. The depolarization begins in a specialized region of the right atrium called the **sino-atrial node**

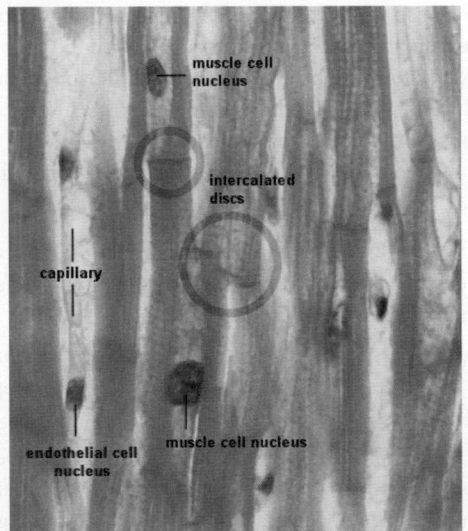

- muscle cell nucleus
- intercalated discs
- capillary
- endothelial cell nucleus
- muscle cell nucleus

Figure 21.14 Histological preparation showing the cardiac myocytes. The intercalated discs are the junctions between the myocytes that allow direct passage of the action potential from one cell to another.

(SAN). Cells in the SAN are characterized by having an unstable membrane potential. The membrane potential of these cells drifts upwards towards threshold as shown in Figure 21.15.

In a neuron, a stimulus (if it is of sufficient magnitude) will generate an action potential. Following the action potential the membrane will repolarize and the potential will return to, and stay at,

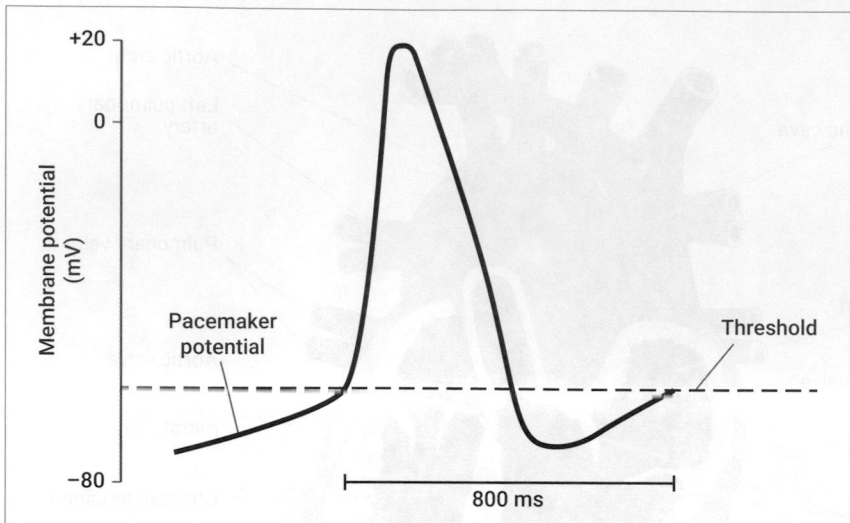

Figure 21.15 Intracellular recording of a single action potential in the sino-atrial node (SAN). The cell does not have a stable resting potential. Once the cell has repolarized, the membrane potential depolarizes towards threshold, at which time an action potential is generated. Following repolarization the cycle repeats. The rate of depolarization determines the pacemaker rate of the heart.

its resting value until a second stimulus generates a second action potential. What is different in the SAN? Having generated an action potential, the membrane potential does return to its resting value, but rather than remain there it drifts back up, until it reaches threshold, when a second action potential is generated. This process is then repeated (e.g. for an individual with a resting heart rate of 70 beats per minute, this process occurs 70 times each minute).

What causes the membrane potential to depolarize towards threshold? Pause while you consider this question. Given that the membrane potential of the cell must depolarize in order to generate an action potential, one of two things could happen: there could be an influx of positively charged ions, or the loss of positively charged ions from the cell could decrease.

The spontaneous activity in SAN cells actually occurs as result of both of the processes described above:

- When the membrane potential reaches about −60 mV, at the end of the repolarization, Na^+ channels open, allowing Na^+ ions to move in, initiating the depolarization phase.
- As the cells depolarize, T-type (transient) Ca^{2+} channels open, leading to further depolarization, which triggers the opening of L-type (long-lasting) Ca^{2+} channels.
- During the depolarization phase, the K^+ channels (opened during repolarization) continue to close. This therefore also contributes to the depolarization process by reducing the outflow of K^+ ions.
- When the depolarization reaches threshold more of the L-type Ca^{2+} channels open, triggering the rapid change in membrane potential of the action potential.
- Repolarization occurs as a result of the reopening of the K^+ channels and closure of the Ca^{2+} and Na^+ channels.

This intrinsic, spontaneous depolarization of cells in the SAN, which requires no external input to initiate it, initiates the contraction of the heart. The frequency with which cells in the SAN depolarize sets the rate of the contraction of the heart; hence, the SAN is also known as the pacemaker of the heart.

How does the wave of depolarization travel to the ventricles? The transfer of the action potential to the ventricles is complicated by the fact that the atria and ventricles are electrically isolated from each other by a layer of non-conducting, electrically insulating fibrous tissue, encircling the heart. The action potential is transferred to the ventricles by a specialized conducting system, the **atrio-ventricular node** (AVN) located at the base of the right atrium (Figure 21.16).

The AVN, which is essentially a collection of modified myocytes, passes through the insulation between the atria and ventricles and the action potential is then conducted through the atrio-ventricular bundle, called the **bundle of His**, which is also formed of modified myocytes. The passage of the action potential through the AVN is relatively slow—taking about 0.1 seconds.

What is the significance of the delay at the AVN? This delay allows time for blood in the atria to enter the ventricles before they contract.

The bundle of His lies between the left and right ventricles, in a region called the **interventricular septum**. The bundle splits into left and right bundle branches to supply the left and right ventricles, respectively. These branches terminate in a network of large fibres called the **Purkinje fibres**, which, in turn, terminate on myocytes in the ventricles. Depolarization of these cells results in a wave of ventricular contraction and ejection of blood from the heart. The contraction originates at the apex of the heart and travels towards the exit points of the aorta and pulmonary artery such that the maximum volume of blood is ejected from each ventricle with each heartbeat. The arrangement of the electrical conducting pathway in the heart is summarized in Figure 21.16.

As we have seen, the shape of the action potential at the SAN is different to the usual shape of an intracellular action potential recording from a neuron or skeletal muscle cell. In a similar manner, if we were to record intracellular action potentials from other regions of the heart, we would see recordings quite different to those seen in neurons. An intracellular recording from a ventricular myocyte is shown in Figure 21.17.

Figure 21.16 The flow of electrical activity through the heart. The heartbeat originates in the SAN and the wave of contraction passes through the atria. The action potential passes from the atria to the ventricles via the AVN. From the AVN, the wave of excitation passes down the bundle of His to terminate in the Purkinje fibres in the apex of the heart where the ventricular contraction is initiated, spreading upwards towards where the arteries exit.

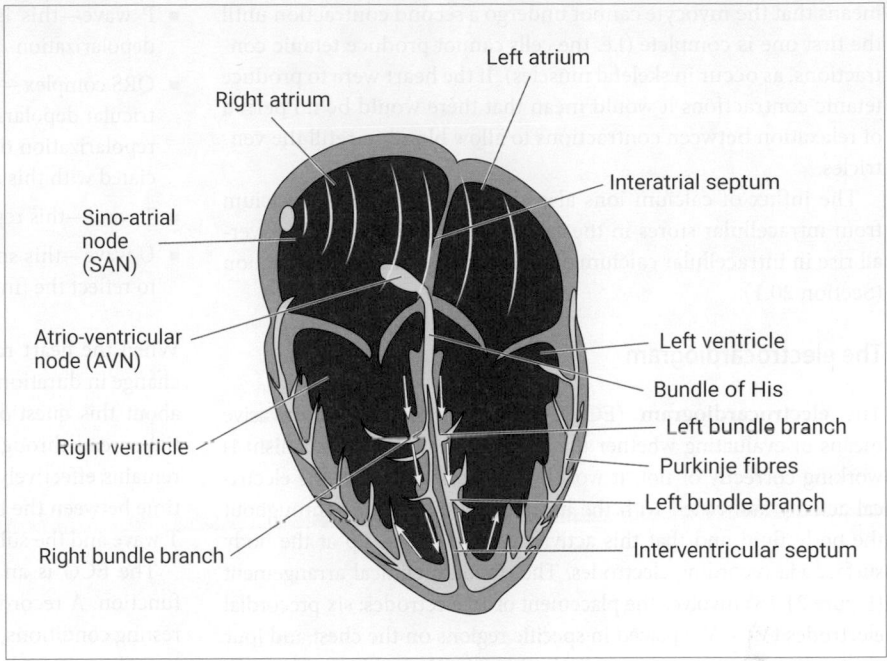

Figure 21.17 An intracellular recording of an action potential in a ventricular myocyte. See text for further details.

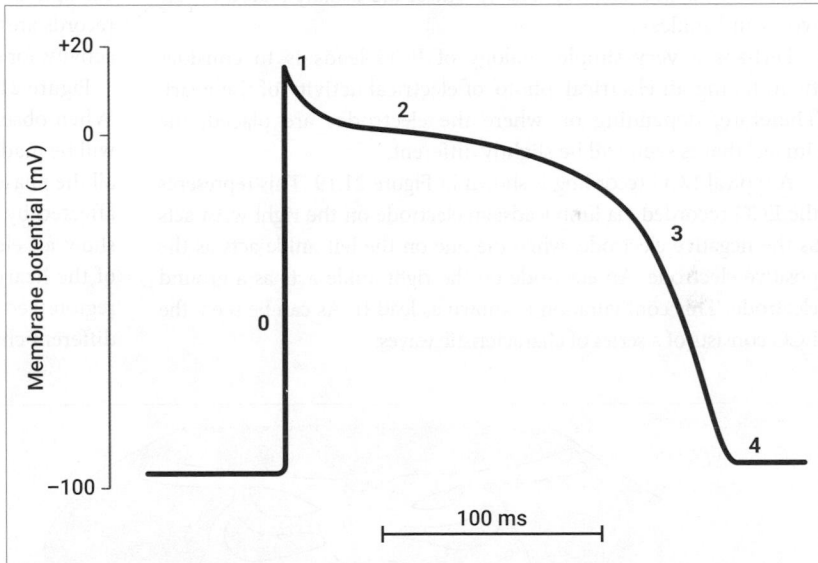

The most obvious difference between neuronal action potentials and cardiac action potentials is the duration of the response: an action potential in a ventricular myocyte lasts about 200–250 ms, about a 100-fold longer than an action potential in a neuron. Also, the shape of the ventricular action potential is quite different. Classically, the ventricular action potential is considered to consist of five phases, identified as 0–4:

■ Phase 0—this represents rapid depolarization of the cell, due to rapid influx of Na⁺ through voltage-gated Na⁺ channels. At the end of the upstroke, the Na⁺ channels close and inactivate but a set of calcium channels (L-type) are opened.

■ Phase 1—this represents the initial repolarization of the cell due to K⁺ efflux through a variety of channels. This is sometimes referred to as the 'notch'.

■ Phase 2—this is caused by a mixture of continuing influx of Ca²⁺ ions against a background of K⁺ efflux which occurs via a range of different types of K⁺ channel. This period of the action potential is sometimes referred to as the plateau and lasts about 100 ms.

■ Phase 3—ultimately, the Ca²⁺ channels become inactivated against a continued background of K⁺ efflux. Overall, there is a net outward current and therefore the cells repolarize.

■ Phase 4—sees the membrane potential return back to its resting value.

We know that the duration of the refractory period in ventricular myocytes, following the action potential, is almost the same as the time it takes them to contract. What is the physiological significance of this? Take a moment to think about this question. It means that the myocyte cannot undergo a second contraction until the first one is complete (i.e. the cells cannot produce tetanic contractions, as occur in skeletal muscles). If the heart were to produce tetanic contractions it would mean that there would be no period of relaxation between contractions to allow blood to refill the ventricles.

The influx of calcium ions also triggers the release of calcium from intracellular stores in the sarcoplasmic reticulum. The overall rise in intracellular calcium concentrations initiates contraction (Section 20.1).

The electrocardiogram

The **electrocardiogram** (ECG) provides a quick, non-invasive means of evaluating whether the cardiac conducting mechanism is working correctly or not. It works on the principle that the electrical activity associated with the heartbeat is transmitted throughout the body fluid and that this activity can be picked up at the body surface via recording electrodes. The standard clinical arrangement (Figure 21.18) involves the placement of 10 electrodes: six precordial electrodes ($V_1 - V_6$) placed in specific regions on the chest and four electrodes placed one on each limb (RA, LA, RL, and LL for the right and left arms and legs, respectively—these are usually placed on the wrists and ankles).

Perhaps a very simple analogy of ECG leads is to consider them taking an electrical 'photo' of electrical activity of the heart. Therefore, depending on where the electrodes are placed, the 'image' that is seen will be slightly different.

A typical ECG recording is shown in Figure 21.19. This represents the ECG recorded via limb leads: an electrode on the right wrist acts as the negative electrode, while the one on the left ankle acts as the positive electrode. An electrode on the right ankle acts as a ground electrode. This configuration is known as lead II. As can be seen, the ECG consists of a series of characteristic waves:

- P wave—this is the electrical activity associated with atrial depolarization.
- QRS complex—this is the electrical activity associated with ventricular depolarization. Also contained within this element is the repolarization of the atria, although the electrical changes associated with this are obscured by ventricular depolarization.
- T wave—this represents ventricular repolarization.
- U wave—this small wave is often not observed. It is considered to reflect the final stage of repolarization.

When the heart rate increases, which part of the ECG cycle will change in duration? Take a moment to look at the graph and think about this question. The conduction time for the action potential spread through the heart and its activation of the myocytes remains effectively unchanged; the period that will shorten is the time between the beats, which is the interval between the end of a T wave and the subsequent P wave.

The ECG is an invaluable tool in terms of evaluating cardiac function. A recording may be made for just a few minutes during resting conditions, during exercise testing, or so-called ambulatory recordings, which are long-term records of electrical activity, for example over a 24-hour period. Exercise testing and ambulatory records are used to reveal whether there are changes in the heart's activity, for example when the heart rate is elevated due to exercise.

Figure 21.19 shows a standard lead II record for a single cycle. When observing a trace over a given time period, observations will be made to see whether the heart rate is regular and whether all the phases of the cycle are present. For example, some patients affected by a heart attack, or myocardial infarction, commonly show an elevation of the ST segment due to the death of part of the heart muscle as a result of a loss of blood supply to that region (see Clinical Box 21.2 for examples of ECG recordings in different clinical conditions).

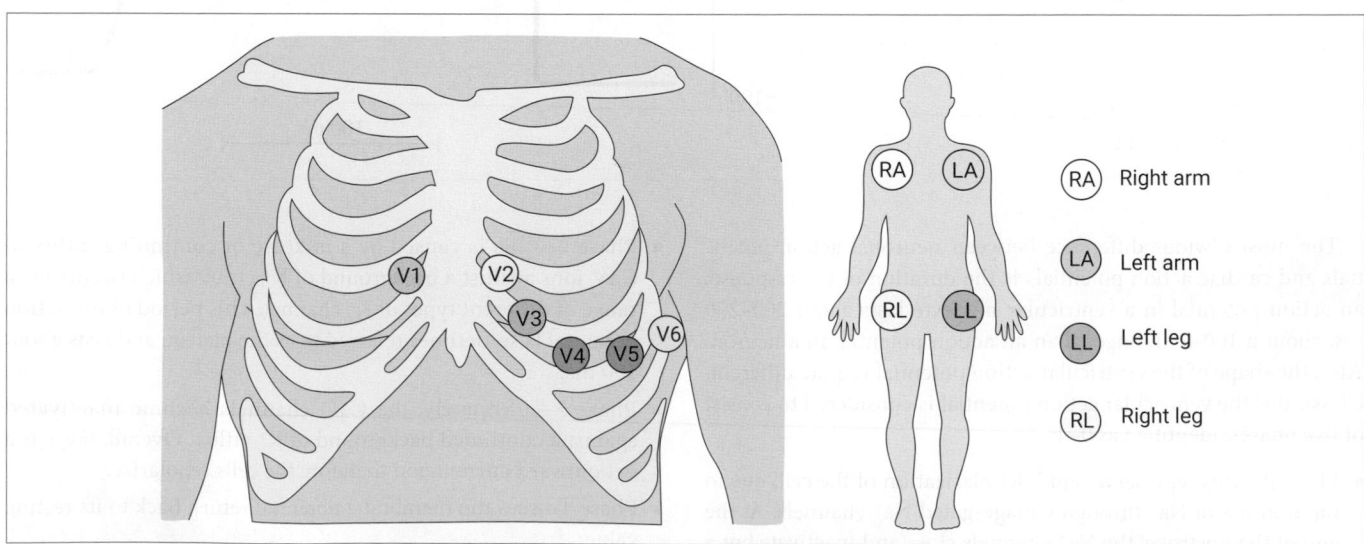

Figure 21.18 Electrode placement for the standard 12-lead electrocardiogram. There are six precordial electrodes ($V_1 - V_6$) and four limb electrodes. Combining the different electrodes together gives 12 'leads', which allow for different perspectives on the electrical activity of the heart.
Source: Cables and Sensors, LLC.

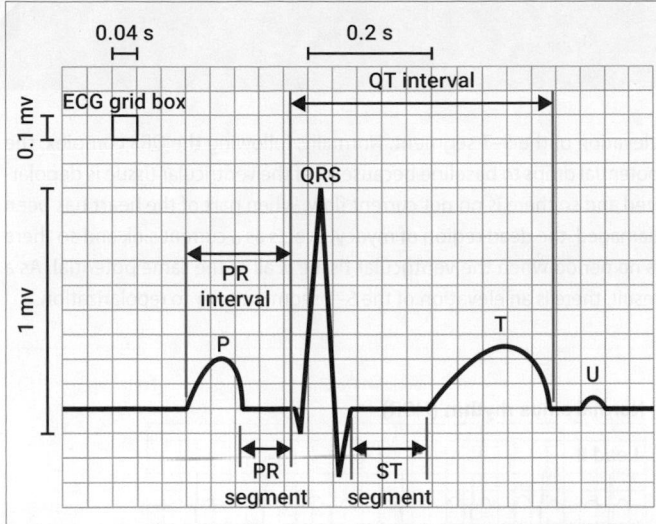

Figure 21.19 A typical ECG recording captured via a lead II configuration.

Source: The Children's Hospital of Philadelphia Research Institute. Image courtesy of Long QT syndrome Knowledge base. https://lqts.research.chop.edu/ecgfam.php

Combining electrical and mechanical activity—the cardiac cycle

So far we have described the electrical activity of the heart and the recognition that this activity gives rise to contraction of the cardiac muscle fibres. Linking these two aspects together gives rise to the cardiac cycle.

The cardiac cycle comprises two elements: **diastole** and **systole**.

Diastole refers to the period of time when the chambers of the heart are relaxed and filling with blood.

Systole refers to the contractile phases: atrial systole, when the atria contract and complete the filling of the ventricles, and then ventricular systole, when the ventricles are contracting and ejecting blood from the heart. A diagram illustrating the coordinated electrical and mechanical activities of the heart during a single cardiac cycle is shown in Figure 21.20. (Note that this diagram considers the cycle in the left side of the heart. The process is the same in the right side of the heart, although the systolic pressures are much lower.)

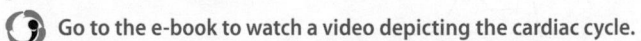 Go to the e-book to watch a video depicting the cardiac cycle.

The easiest way to consider the cardiac cycle is to consider each of the lines on Figure 21.20 in more detail.

- Ventricular pressure—this is a measure of the pressure in the ventricles. The pressure in the ventricles during diastole is very low. It rises a little as blood flows from the atria into the ventricle. As ventricular systole begins, so there is a large rise in pressure. This rise closes the atrio-ventricular valve, preventing backflow of blood to the atria. At this stage, both sets of valves are closed, so-called **iso-volumetric contraction**. The pressure in the left ventricle continues to rise until it exceeds that in the aorta, at which time the valve to the aorta opens, allowing blood to be ejected into the artery. The peak pressure within the left ventricle is about 120 mmHg.

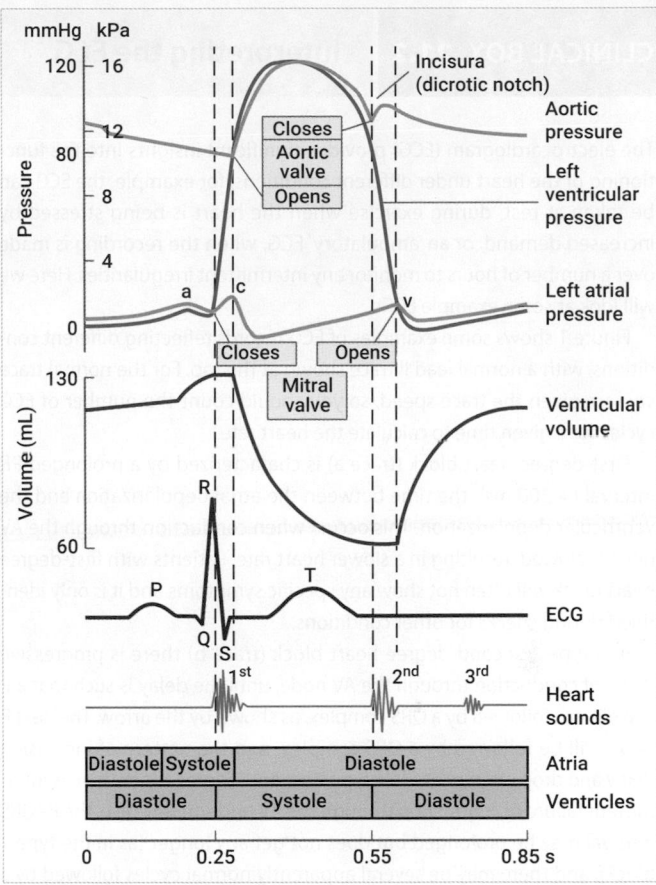

Figure 21.20 Illustration of a single cardiac cycle. This shows the relationship between electrical and mechanical (i.e. contractile) activity in the heart. Note that electrical activity precedes mechanical activity and that blood flow through the chambers of the heart and out into the arteries is determined by pressure gradients.

Source: Pocock et al. *Human Physiology*, 5th edn, 2017. Oxford University Press.

Once ventricular systole has completed, the ventricles re-enter diastole and relax. As the pressure drops, the pressure in the aorta exceeds that in the ventricle and so the aortic valve closes, preventing backflow into the ventricle. When the ventricular pressure falls below that in the atrium, the mitral valve opens and blood flows through from the atrium, restarting the cycle.

- Ventricular volume—this is the volume of blood in the ventricles. The volume of blood in the ventricles at the end of ventricular diastole (i.e. just before it contracts) is called the **end diastolic volume**. This volume is about 125 mL. As ventricular systole occurs, blood is ejected into the aorta or, in the case of the right ventricle, into the pulmonary artery. The volume of blood ejected is called the **stroke volume**. In healthy adults at rest, about 60 per cent of the end diastolic volume is ejected, giving a stroke volume at rest of about 75 mL. This means that at the end of systole there is still blood in the ventricle. This is called the **end systolic volume**.

- Aortic pressure—this is a measure of the pressure in the aorta. As the ventricles relax during diastole, pressure in the aorta declines,

CLINICAL BOX 21.2   Interpreting the ECG

The electrocardiogram (ECG) provides significant insights into the functioning of the heart under different conditions; for example, the ECG can be taken at rest, during exercise when the heart is being stressed by increased demand, or an 'ambulatory' ECG, when the recording is made over a number of hours to monitor any intermittent irregularities. Here we will look at some example ECGs.

Figure 1 shows some examples of ECG records reflecting different conditions, with a normal lead II trace shown at the top. For the normal trace you are given the trace speed, so you should count the number of ECG cycles for a given time to calculate the heart rate.

First-degree heart block (trace a) is characterized by a prolonged PR interval (> 200 ms), the time between the atrial depolarization and the ventricular depolarization. This occurs when conduction through the AV node is slowed, resulting in a slower heart rate. Patients with first-degree heart block will often not show any specific symptoms and it is only identified during checks for other conditions.

In a type 1 second-degree heart block (trace b) there is progressive delay of conduction through the AV node, until the delay is such that a P wave is not followed by a QRS complex, as shown by the arrow. The next P wave will be followed by a QRS complex, and the pattern of increasing delay and drop out repeats. In a type 2 second-degree block there is intermittent failure of conduction through the AV node. In this case, the P–QRS interval may be prolonged but does not get any longer (as in the type 1 block), and there may be several apparently normal cycles followed by a conduction failure. Such patients would be indicated for having a pacemaker implanted.

In third-degree heart block (trace c), there is no conduction through the AV node. In this case, the intrinsic pacemaker rhythm of the AV node will drive the ventricular contraction, but at a relatively slow rate. As a result, the P waves and the QRS complexes appear to be occurring independently of each other. Again, these patients would require a pacemaker.

The bottom three examples show how the ECG changes in other conditions. In the case of hyperkalaemia (abnormally high extracellular [K^+]), the ECG is characterized by a widening of the QRS complex and an extended T wave. Hyperkalaemia can be life-threatening as the elevated K^+ levels lead to atrial arrhythmias as a result of the changes in membrane potential.

Cardiac ischaemia is characterized by a delayed T wave (ventricular repolarization), or an inverted T wave depending on which of the coronary arteries is affected, whereas ischaemic injury, which results in death of part of the heart as a result of a 'heart attack', typically leads to an elevation of the S–T segment. Normally, following the QRS complex, the potential drops to baseline because all of the ventricular tissue is depolarized and so there is no net current flow. When part of the heart has been damaged, the dead region of myocytes acts as a current sink and so there is no period when the ventricular tissue is all at the same potential. As a result, there is an elevation of the S–T segment prior to repolarization.

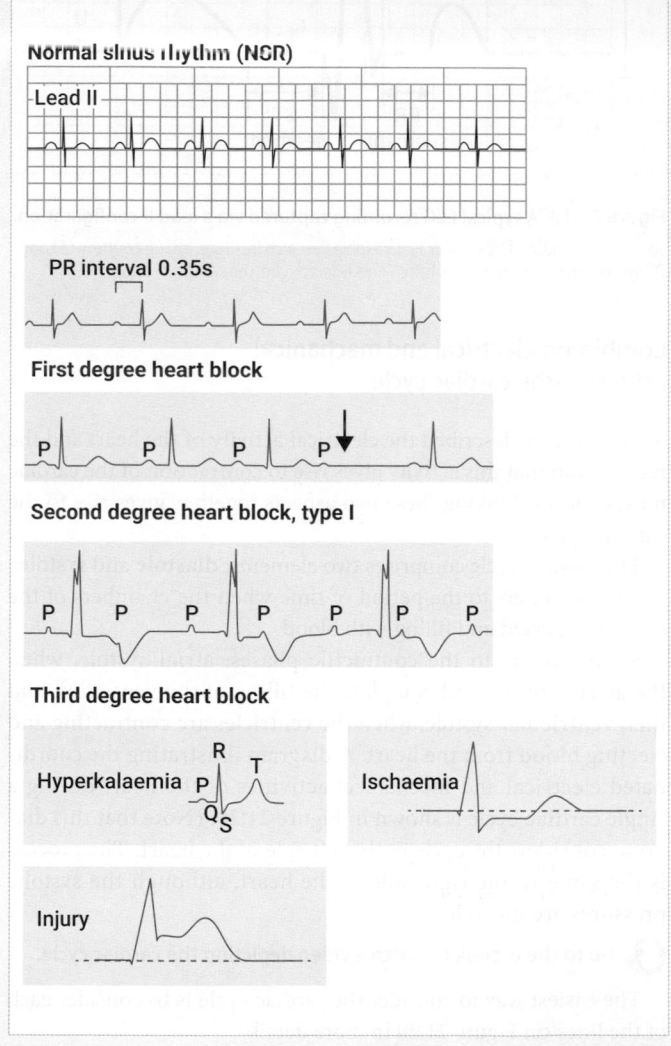

Figure 1   ECG records reflecting different conditions.

reaching a minimum value of about 80 mmHg. When ventricular systole occurs and the pressure inside the ventricle exceeds that in the aorta, the aortic valve opens and blood is ejected into the aorta. The pressure peaks in the aorta at about 120 mmHg. Subsequent diastole means that the pressure in the ventricle begins to decrease. At some point the pressure in the aorta exceeds that of the ventricle and the aortic valve closes, preventing backflow of blood into the ventricle. Remember from earlier in the chapter that the aorta is an elastic artery—as it recoils following systole, blood is moved onwards through the arterial system.

- Atrial pressure—this is the pressure in the atria. The pressure in the atria remains relatively low throughout the entire cardiac cycle. During diastole, when both the atria and the ventricles are relaxed, there is passive flow of the blood from the veins, through the atria, into the ventricles due to the very slight pressure gradient. Atrial systole precedes ventricular systole, so the contraction of the atria forces the remaining blood into the ventricles. When the ventricles contract during systole, this pressure gradient is reversed. This closes the atrio-ventricular valves. While the ventricles continue to contract, the atria relax (atrial diastole) and the pressure falls below that in the veins, so the atria start to fill again. When the pressure in the ventricles falls (ventricular diastole) it drops below that in the atria, so the atrio-ventricular valves open and blood enters the ventricle again.

- ECG—the ECG trace shows the timing of the electrical activity in relation to the mechanical changes.

- Heart sounds—these are the typical heart sounds that can be heard as the heart beats. The first sound is the closing of the atrio-ventricular valves at the start of iso-volumetric contraction. The second sound corresponds to the closing of the aortic and pulmonary valves at the start of diastole. These sounds are the 'lub-dub' sounds one can hear.

Cardiac output

So far, we have described the heart in terms of its ability to contract and eject blood into the systemic and pulmonary circulations. However, it is apparent that the heart has to meet a range of demands placed upon it.

Can you think of an example when your cardiovascular system has to respond to meet increased metabolic demands? A good example is the response seen when we exercise. Your heart beats both faster and more strongly, increasing the supply of oxygen and nutrients to the skeletal muscles (see Chapter 23).

The product of heart rate multiplied by the stroke volume gives us cardiac output:

$$\text{Cardiac output}\,(CO) = \text{heart rate}\,(HR) \times \text{stroke volume}\,(SV)$$

This is an important equation in cardiovascular physiology that should be remembered.

PAUSE AND THINK

If an individual's resting heart rate is 70 beats per minute, and their stroke volume is 75 mL, what is their cardiac output?

The equation is:

$$CO = HR \times SV$$

$$CO = 70 \times 75 = 5250\ \text{mL min}^{-1}\,\left(5.25\text{L min}^{-1}\right)$$

Answer: In this case:

Factors influencing heart rate

Heart rate can be influenced by both the autonomic nervous system and circulating hormones such as adrenaline.

The autonomic nervous system has two divisions, the sympathetic and parasympathetic (Section 19.1).

- The sympathetic nervous system releases noradrenaline into the heart. Noradrenaline mediates its effects on the heart via stimulation of β_1 adrenoceptors (see Section 21.2, 'Vascular System'), and increased levels of cAMP. This leads to increases in both the rate of contraction and also the force of contraction. The increase in rate is termed a **positive chronotropic effect** or **tachycardia**. This is achieved by increasing the rate at which pacemaker cells of the SAN reach threshold, and therefore generate action potentials. It also increases the rate at which they repolarize. The increase in force of contraction, which is termed a **positive inotropic effect**, is mediated via the β-adrenoceptors. Activation of these receptors leads to an increase in cAMP, which, in turn, increases the influx of calcium and so increases contractile force.

- The parasympathetic innervation is provided by the vagus nerve (10th cranial nerve). The vagus nerve releases acetylcholine, which acts to reduce the rate of contraction. The actions of acetylcholine are mediated by muscarinic M_2 receptors on the pacemaker cells in the SAN. This results in a **negative chronotropic effect**, or **bradycardia**, increasing the time it takes for these cells to reach threshold and therefore generate action potentials. Acetylcholine also slows down the transmission of cardiac action potentials as they pass through the atrioventricular node. At rest, the heart is maintained under tonic vagal inhibition, so an increase in heart rate can also be brought out by a reduction in vagal activity. In contrast to noradrenaline, acetylcholine has no effect on the contractile cells of the heart.

The endocrine system, in particular through the release of adrenaline, also acts on the heart. Adrenaline, released from the adrenal medulla, produces the same responses as sympathetic stimulation and achieves these effects through identical mechanisms. Other hormones, including the thyroid hormones, have positive chronotropic and inotropic actions.

Factors influencing stroke volume

Stroke volume is the volume of blood ejected from the ventricle with each beat of the heart. We have already identified a role for noradrenaline and adrenaline in increasing the stroke volume through increasing the force of ventricular contraction. What other mechanisms regulate stroke volume?

One factor that regulates stroke volume is the degree to which the ventricular myocytes are stretched prior to their contraction; this is termed intrinsic regulation. As a simple analogy, think about stretching an elastic band and then releasing it. The greater the stretch, the greater the force of contraction as it returns back to its pre-stretched size. This intrinsic regulation is sometimes called the **Frank–Starling relationship**, named after the two physiologists who identified it. In essence, it says that the more the ventricular myocytes are stretched, the greater the force of contraction. The degree of stretch is related to the end diastolic volume of blood: the greater the volume of blood in the ventricle at the beginning of systole, the greater the volume that will be ejected. The degree of stretch that the ventricle is put under is termed

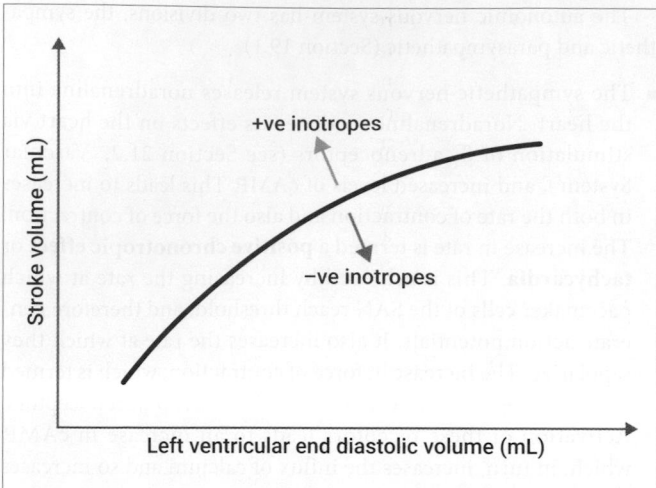

Figure 21.21 The relationship between left ventricular end diastolic volume (LVEV) and stroke volume (SV). As LVEV increases, so does SV, although the curve reaches a plateau and will decline if the preload stretching is excessive. The influences of positive inotropes, such as adrenaline, and negative inotropes, such as the β-blocker propranolol, are shown

its **preload**. This is demonstrated graphically in Figure 21.21. This intrinsic regulation balances the amount of blood returning to the heart (venous return) to the cardiac output: the greater the venous return, the greater will be the preload and so the force of contraction. In some disease states such as cardiac failure, this regulation breaks down and the cardiac output fails to meet the demands of the other organs of the body.

Stroke volume is also affected by the pressure in the aorta. In patients affected by clinically increased blood pressure—**hypertension**—the diastolic pressure in the aorta is increased. As a result, the pressure in the ventricles has to reach a higher level before the valve into the aorta opens and so less blood is ejected. The load imposed on the heart by the pressure in the aorta is termed the **afterload**.

 Check your understanding of the concepts covered in this section by answering the questions in the e-book.

21.4 Integrated cardiovascular physiology—regulation of blood pressure

We have now considered the elements of the cardiovascular system. One of the key concepts that have been considered on a number of occasions is the importance of pressure. This is of major importance because it is the difference in pressure around the system that is a major determinant of the movement of blood.

Blood pressure is the pressure within the cardiovascular system. Almost everyone has had their blood pressure measured at some stage, so what does it mean when we get the result back at 120/80 mmHg? The higher figure of 120 mmHg represents the pressure in the arterial system when the ventricles are contracting,

the **systolic pressure**. The lower figure of 80 mmHg represents the pressure when the ventricles are relaxing, the **diastolic pressure**.

How is blood pressure measured?

Blood pressure was traditionally measured in a non-invasive manner using a sphygmomanometer, which comprises an inflatable cuff connected to a column of mercury from which pressure measurements are taken (this is why the measurements are still quoted in mmHg rather than the SI units of pressure), and a stethoscope. The cuff is wrapped around the upper arm and inflated to a pressure that exceeds systolic pressure (e.g. 180 mmHg) and the stethoscope is placed over the brachial artery in the flat of the arm. This pressure prevents any blood flow in the artery and no sounds can be heard. As pressure in the cuff is gradually reduced, blood eventually begins to flow through the artery; this blood flow will be turbulent because of the pressure still on the artery reducing flow, and can be heard through the stethoscope. The pressure at which sounds are initially heard is the systolic pressure. As the cuff continues to deflate, the sounds get louder and then muffled before disappearing. The pressure at which the sounds disappear is the diastolic blood pressure. The sounds that are heard through the stethoscope are called the sounds of Korotkoff. Blood pressure is now more normally measured electronically using controlled inflation and deflation of the arterial cuff.

According to current guidelines, ideal blood pressure measured at rest is between 90/60 mmHg and 120/80 mmHg. An individual with a blood pressure within these ranges is said to be **normotensive**. If the resting pressures increase above 140/90 mmHg this is considered to be high blood pressure, termed hypertension, and it is associated with an increased risk of developing a stroke or heart attack or a range of other disorders. If the pressure is too low, this is termed **hypotension**, and is associated with an inability of blood to flow to tissues.

A variety of control mechanisms regulate and control blood pressure. Only one will be considered here, the baroreceptor reflex. We can quantify the regulation of blood pressure by the following equation:

$$\text{Blood pressure} = \text{cardiac output} \times \text{peripheral resistance}$$

Peripheral resistance is the sum of the resistance to blood flow at the level of the arteriole (i.e. it is a measure of the degree of vasoconstriction/vasodilation).

The baroreceptor reflex

Baroreceptors are specialized sensory receptors found in the aorta and the carotid sinus (the region where the common carotid artery bifurcates to give the internal and external branches). When blood pressure rises, the walls of these blood vessels are stretched and the baroreceptors respond by sending action potentials to the medulla oblongata in the brainstem. This leads to an increase in parasympathetic activity, which slows down the heart rate. As a result, there

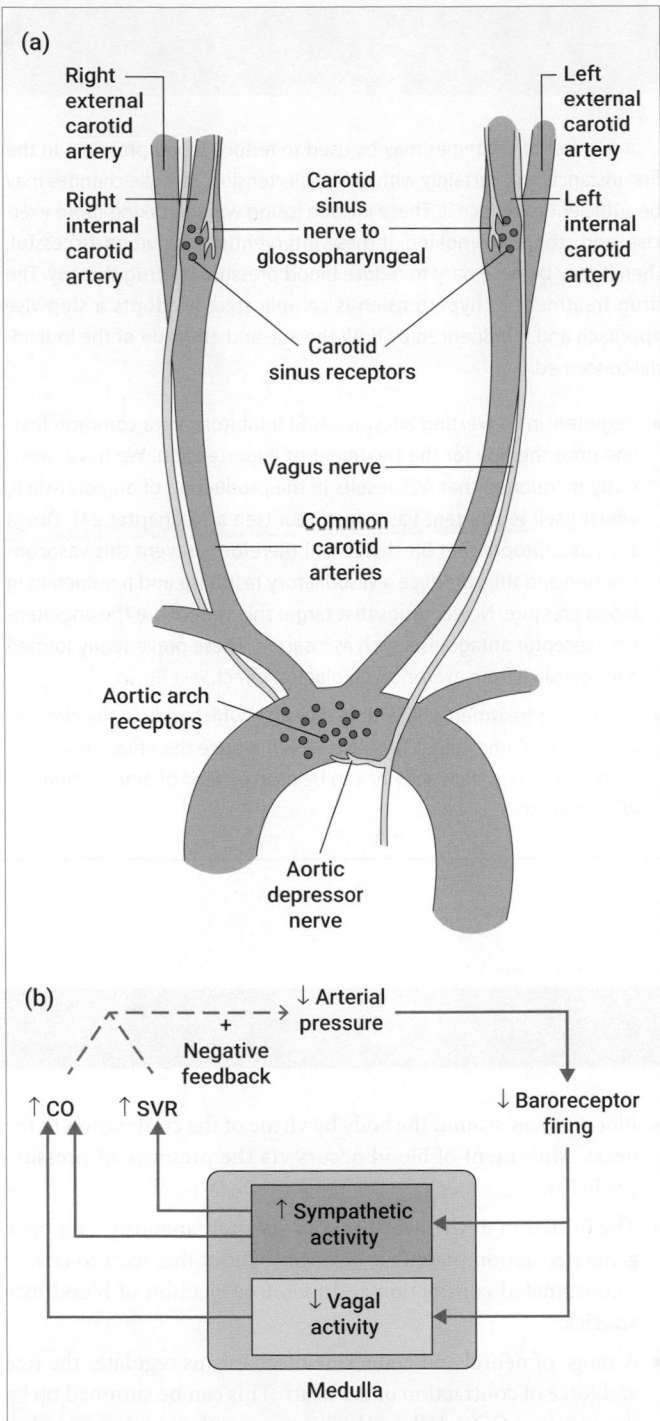

Figure 21.22 The baroreceptor reflex. (a) The location and innervation of baroreceptors and (b) their actions in response to a fall in blood pressure.

Source: Figure 32.4 from Barrett et al. *Ganong's Review of Medical Physiology*, 24th Edn. © McGraw Hill.

is a decrease in cardiac output, returning the blood pressure back to its desired value.

The opposite happens when a drop in blood pressure is detected: a fall in pressure results in both activation of the sympathetic

nervous system and a decrease in vagal parasympathetic activity. This increases both cardiac output, through an increase in heart rate and in contractility, and an increase in peripheral resistance, resulting in the blood pressure increasing back to its desired value. We also see an increase in venous return caused by the ability of the sympathetic nervous system to increase the contraction of smooth muscle in some veins, such as those in the gastro-intestinal tract. The increase in venous return increases the rate of filling of the heart and the preload, thereby increasing the force of contraction. A summary of the operation of the baroreflex is seen in Figure 21.22.

What do you think is the significance of the location of the baro-receptors? The baroreceptors are located in the aortic arch, meas-uring the blood pressure in the arterial system immediately leaving the left ventricle, and in the carotid sinus, measuring the blood pressure in the arteries supplying the brain.

The baroreceptor reflex is important in the short-term regula-tion of blood pressure.

In the longer term, the circulating volume of blood is a major determinant of blood pressure—in turn, a major determinant of this is the volume of urine produced. One of the regulators of circulating blood volume is antidiuretic hormone and the renin–angiotensin–aldosterone system (RAAS) (Section 24.6). The RAAS is activated in a number of different ways, for example when blood pressure drops, when activation of the sympathetic nervous system occurs, and when there is a reduction in tubular sodium levels in the kidney. A drop in blood pressure is detected by the juxtaglomerular appa-ratus of the kidneys. They respond by releasing a hormone called renin that acts on a plasma protein called angiotensinogen to pro-duce angiotensin I. As angiotensin I passes through the pulmo-nary circulation it is converted to angiotensin II by the activity of an enzyme called angiotensin converting enzyme. Angiotensin II is a potent vasoconstrictor—this helps restore blood pressure back to normal. In addition, it promotes the release of aldosterone from the adrenal cortex. Aldosterone increases the reabsorption of Na^+ and Cl^- from renal tubules. Because of this, there is also the re-absorption of water via osmosis. The consequence of this is that the circulating blood volume is increased, which further contributes to the restoration of a normal blood pressure.

However, in a significant number of individuals, the systems that control blood pressure fail and those individuals develop high blood pressure or hypertension. The cause and treatments for hypertension are considered in Clinical Box 21.3.

PAUSE AND THINK

Why do the baroreceptors not respond to hypertension and trigger a low-ering of the blood pressure?

Answer: The baroreceptors only respond to short-term changes in blood pressure. Hypertension develops slowly, typically over a number of years, and so the baroreceptors adapt to the progressively increasing pressure.

 Check your understanding of the concepts covered in this section by answering the questions in the e-book.

CLINICAL BOX 21.3 — Hypertension

Hypertension is a very common cardiovascular disorder, affecting one in four of the adult population. It is important that it is diagnosed and treated, as if it is left untreated there is an increased risk of having a stroke, heart attack, or developing renal failure. A normal healthy blood pressure is 120 mmHg systolic pressure and 80 mmHg diastolic pressure (often written as 120/80 mmHg). The thresholds for diagnosing hypertension have been lowered in recent years: stage 1 hypertension is defined as a systolic pressure of 130–139 mmHg and/or a diastolic pressure of 80–89 mmHg; stage 2 is a systolic pressure of >140 mmHg and/or a diastolic pressure of > 90 mmHg.

Knowing a little about cardiovascular physiology allows us to establish what the cause of hypertension may be. If either cardiac output or peripheral resistance increase for a prolonged period of time, then an individual will become hypertensive. What is known is that in the vast majority of cases it is an increase in peripheral resistance that causes an individual to become hypertensive.

Hypertension can be divided into two categories. The first of these is primary hypertension—here an individual is hypertensive, but has no other comorbidities that may account for it. Aside from increases in blood pressure associated with ageing, the major causes are related to lifestyle such as smoking and obesity. The second is secondary hypertension—in this case, there is an underlying pathology that results in hypertension, such as certain endocrine disorders.

A number of strategies may be used to reduce blood pressure. In the first instance, and certainly with mild hypertension, lifestyle changes may be sufficient to reduce it. These include losing weight, taking more exercise, and stopping smoking. If these interventions prove unsuccessful, then it may be necessary to reduce blood pressure by drug therapy. The drug treatment of hypertension is complicated: it adopts a stepwise approach and is influenced by both the age and ethnicity of the individual concerned.

- Angiotensin converting enzyme (ACE) inhibitors are a common first-line drug therapy for the treatment of hypertension. We have previously mentioned that ACE results in the production of angiotensin II, which itself is a potent vasoconstrictor (see also Chapter 24). Drugs such as captopril that block ACE will therefore prevent this vasoconstriction and thus produce a vasodilatory response and a reduction in blood pressure. Newer drugs that target this system are the angiotensin II receptor antagonists such as losartan. These prevent any formed angiotensin II from exerting its biological effect.

- Other drug treatments, including diuretics, which reduce the circulating blood volume, and β-blockers, which reduce the effect of sympathetic nervous system activity, can be used instead of or in addition to ACE inhibitors.

SUMMARY OF KEY CONCEPTS

- The overall function of the cardiovascular system is transport: nutrients, gases, and hormones are three examples of the materials transported.

- Blood is a mixture of different cell types suspended in a fluid matrix. The principal cells are erythrocytes, leukocytes, and platelets, and the fluid matrix is plasma. Each component has a specialized role.

- The vascular system comprises three different types of blood vessels: arteries, veins, and capillaries.

- Arteries transport blood away from the heart. Besides delivering nutrient-rich blood to tissues, they play a vital role in ensuring that blood is delivered to regions appropriate to metabolic demands.

- Veins return nutrient-poor blood to the heart. In addition to this role, the veins act as a reservoir for blood.

- Capillaries are the sites of exchange of nutrients and other substances between the blood and tissues.

- Blood moves around the body by virtue of the contraction of the heart. Movement of blood occurs via the presence of pressure gradients.

- The initiation of the heartbeat occurs spontaneously, and once generated action potentials pass throughout the heart to ensure a coordinated contraction and therefore ejection of blood into arteries.

- A range of neural and endocrine mechanisms regulates the rate and force of contraction of the heart. This can be summed up by the equation CO = HR × SV. This ensures that cardiac function meets metabolic requirements of the body.

- The baroreceptor reflex regulates blood pressure in the short term. The factors that influence it can be summarized in the following equation: BP = CO × TPR. In the longer term, the most important determinant of blood pressure is the circulating blood volume.

 Use the flashcards in the e-book to test your recall of key terms introduced in this chapter.

QUESTIONS

Looking for answers? Once you've answered these questions, follow the link in the e-book to the answer guidance and check your work.

Concepts and definitions

1. Describe the cellular and fluid components that together form blood.

2. Describe the differences, in terms of structure and pressures, of the systemic and pulmonary circulations.

3. Explain the role that the arterial system has in regulating blood flow to tissue.

4. Draw the typical appearance of an intracellular recording of an action potential in a ventricular myocyte. Label the diagram with the ion flow occurring during each phase.

5. Describe the role of baroreceptors in the regulation of blood pressure.

Apply the concepts

6. With reference to their structure, explain how arteries, veins, and capillaries are adapted for the role they play in the circulatory system.

7. Draw the typical ECG that is captured via lead II recording—how might this change when the sympathetic nervous system is activated?

8. Review the mechanisms that contribute to the control of arteriolar diameter.

9. Identify and describe the mechanisms that regulate cardiac output.

10. Describe the electrical and pressure changes that are observed in the heart during a single cardiac cycle.

Beyond the concepts

11. Compare and contrast the actions of acetylcholine and noradrenaline on cells in the sino-atrial node of the heart. What drug interventions could be made to treat an individual who is experiencing tachycardia?

12. Review the forces that result in the movement of fluid at arterial and venous ends of a capillary bed. How would these differ in an individual who is suffering from liver disease and an inability to produce plasma proteins?

13. What regulates blood pressure in the long term and how may this system be manipulated to treat hypertension?

14. Exercise increases the metabolic demands of muscular tissue—what changes are seen in the cardiovascular system to accommodate these increased demands?

15. Review the stages in the formation of red blood cells. What advice would you give to an athlete who is considering the illegal use of erythropoietin to enhance performance?

FURTHER READING

Aaronson, P. I., Ward, J. P. T., & Connolly, M. J. (2019) *Cardiovascular Medicine at a Glance*. 5th edition. Wiley Blackwell.
This text provides very useful, succinct coverage of cardiovascular medicine including key physiological principles.

Dampney, R. A. L. (2016) Central neural control of the cardiovascular system: current perspectives. *Adv. Physiol. Educ.* **40**: 283–96.
This brief review focuses on the central control of the cardiovascular system. It is easily readable and attempts to bring together the integrative nature of control systems.

Herring, N. & Paterson, D. J. (2018) *Levick's Introduction to Cardiovascular Physiology*. 6th edition. CRC Press.
This is an excellent, up-to-date overview of cardiovascular physiology suitable for undergraduate students.

Waller, D. G. & Sampson, A. P. (2018) *Medical Pharmacology and Therapeutics*. 5th edition. Elsevier.
The cardiovascular system section in this undergraduate textbook gives an excellent overview of the pathophysiology and treatment of a number of important cardiovascular diseases.

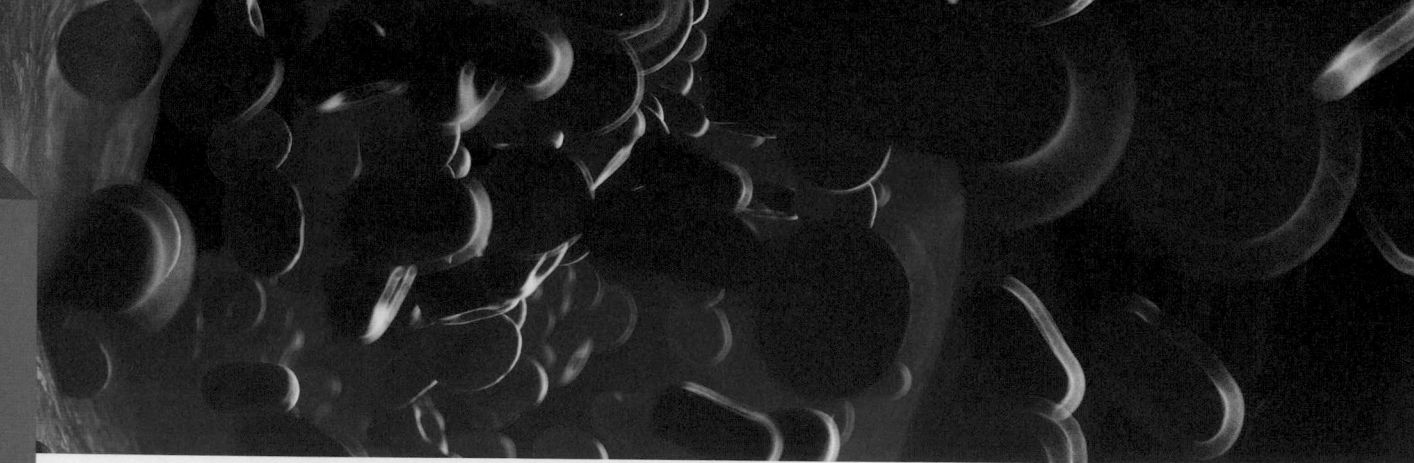

Respiratory System

Chapter contents

Introduction 745

22.1 Anatomy of the respiratory system 745

22.2 Ventilation of the lungs 749

22.3 Gas exchange 757

22.4 Control of ventilation 764

 Watch the key concepts video in the e-book to prepare yourself for studying this chapter.

LEARNING OBJECTIVES

By the end of this chapter you should be able to:

- Describe the structure and organization of the upper and lower airways.

- Explain how the diameter of the lower airways is regulated and its significance in lung function, including an explanation of the effects of asthma.

- Describe the operation of the muscles involved in inspiration and expiration.

- Explain the pressure changes in the thorax and their relationship to lung ventilation.

- Describe the work done to enable air movement, including the effects of airway resistance and lung compliance, and describe clinical examples of conditions affecting lung compliance.

- Explain the significance of surface tension within the alveoli and the role of surfactant.

- Describe the tests for lung function and the variables measured.

- Explain the differences in the partial pressures of oxygen and carbon dioxide in the different sections of the respiratory system and in the pulmonary circulation.

- Describe how oxygen is exchanged across the lung surface and transported in the blood.

- Explain the significance of the haemoglobin–oxygen dissociation curve and the associated Bohr shifts.

- Describe how carbon dioxide is transported in the blood, the role of the buffering systems, and their significance in regulating blood pH.

- Describe the neural control of ventilation and the role of chemoreceptors in measuring oxygen and carbon dioxide levels and pH of the blood.

- Explain the processes occurring during altitude acclimation.

Introduction

'It's as natural as breathing' and 'the breath of life' are very common expressions that sum up a process that forms so much of our everyday life that we take it completely for granted. We are very aware, though, that the prevention of breathing for more than a few minutes can be fatal for humans, as well as for many other mammals.

Breathing is an automatic process that normally proceeds in a rhythmic manner. Humans take about 12–15 breaths each minute, with each breath involving the movement of about 0.3–0.5 litres of air when at rest. This baseline process can be modified quite extensively: for example, during exercise both the rate and depth of breathing can be increased very significantly; on the other hand, we can also voluntarily control breathing in order to speak or play a wind instrument.

Respiration can be divided into two processes: cellular respiration and external respiration.

Cellular respiration is the biochemical process in cells that can be summarized as the production of adenosine triphosphate (ATP) from the breakdown of nutrients. In its simplest form, it can be written as the oxidation of glucose to release energy:

$$C_6H_{12}O_6 + 6O_2 \rightarrow 6CO_2 + 6H_2O \text{ plus energy}$$

The energy takes the form of the production of ATP from ADP + Pi.

▶ We learn more about energy metabolism in Chapter 7.

In order for cellular respiration to take place, the cells require a supply of oxygen and for carbon dioxide, the waste product, to be removed. This function is carried out by the circulatory system, which transports oxygen from the lungs to the tissues and CO_2 from the tissues to the lungs.

The process of **external respiration**, breathing, is the mechanism whereby gaseous exchange takes place with the air in the lungs to allow us to take in oxygen and breathe out carbon dioxide. As we will see, the concentrations of oxygen and carbon dioxide in the blood are tightly regulated, especially as the concentration of carbon dioxide is an important determinant of blood pH (see Section 22.3, 'Carbon dioxide transport').

External respiration, then, is the process of ventilating the lungs: that is **inspiration** (inhalation or breathing in) and **expiration** (exhalation, breathing out).

22.1 Anatomy of the respiratory system

The respiratory system comprises two main sets of structures (Figure 22.1):

- the conducting zone, which comprises the airways that carry the gases between the outside air; and

- the respiratory zone, which comprises the lungs, the organs where gas exchange takes place.

When we breathe in, air initially enters the respiratory system via the nose or the mouth. If you breathe in via the nose, the air passes into the nasal cavities, which are lined with epithelial cells that bear cilia. The cells of the nasal cavities also produce mucus which, in conjunction with the cilia, traps fine particles that may be inhaled. The cilia

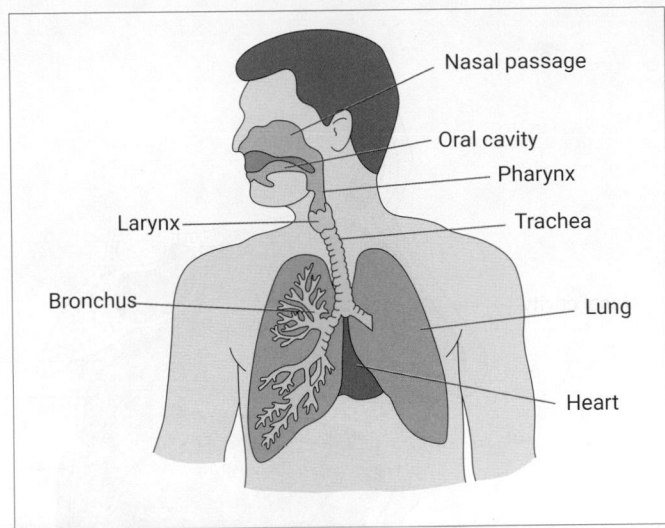

Figure 22.1 Structure of the respiratory system.
Source: Vecton/Shutterstock.com.

continually beat, sweeping the mucus plus the trapped particles to the back of the mouth where they are swallowed.

PAUSE AND THINK

Why might it be better to breathe through the nose rather than the mouth?

Answer: The cilia and mucus filter the air that enters the airways, removing small particles that would otherwise enter the lungs. The passage through the nose allows time for the air to be warmed towards body temperature and to be humidified. Also, the air passing through the nose engages with the olfactory system, providing the sense of smell (Chapter 18).

Are there times when you do find breathing through the mouth is more effective? Think about in what situations you might be more likely to breathe in through your mouth before reading on.

Answer: Other than when your nose is blocked as a result of having a cold, breathing through the mouth during exercise or when singing allows air to be taken in much more rapidly.

Upper airways

The large airway carrying inspired air down to the lungs is the **trachea** (Figure 22.1). To reach the trachea, the air first passes through the pharynx and larynx, which comprise the throat.

The pharynx is the upper part of the throat at the back of the mouth and is shared by both the digestive system (leading to the oesophagus) and the respiratory system (leading to the larynx). At the top of the larynx lies the hyoid bone (Figure 22.2). This is a curved bone that forms a site of attachment for a number of the muscles and ligaments of the throat, as well as of the tongue.

The epiglottis is a flap of fibrous tissue at the top of the larynx (Figure 22.2). During normal breathing, the flap stands open, but during swallowing, it closes to prevent entry of the food into the larynx. The reflex closure of the epiglottis is triggered by stimulation of touch receptors in the pharynx and uvula.

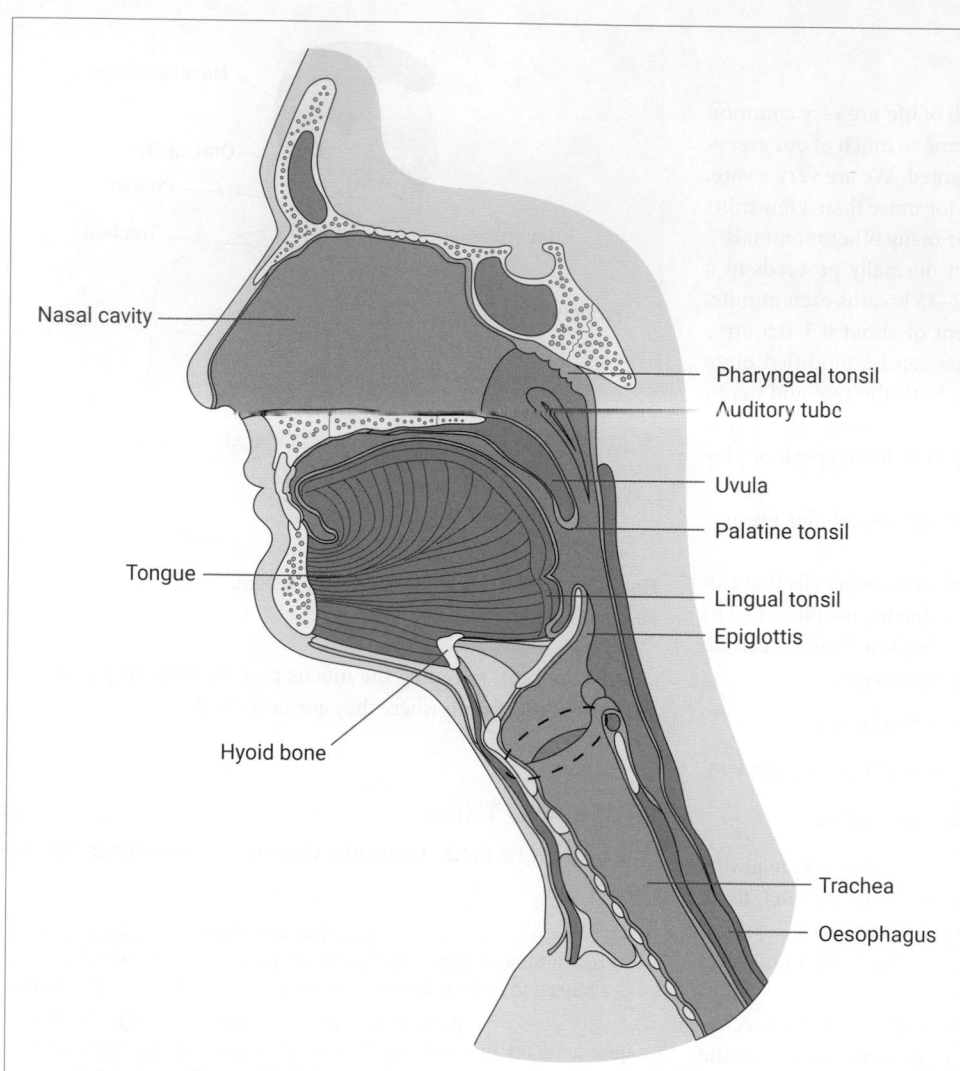

Figure 22.2 Anatomy of the mouth and upper airway.

Source: TheFreeDictionary © 2022 by Farlex, Inc.

Nasal cavity

Pharyngeal tonsil

Auditory tube

Uvula

Palatine tonsil

Tongue

Lingual tonsil

Epiglottis

Hyoid bone

Trachea

Oesophagus

If particles do enter the upper respiratory tract, then this will trigger the cough reflex (which you can read about in Section 22.2) to expel foreign bodies.

The larynx itself is made up of sections of cartilage that form the structure of the tube for the airway, with associated ligaments and muscles. Of these, the largest is the thyroid cartilage which is commonly known as 'Adam's apple', forming an identifiable structure on the anterior surface of the neck. The larynx is also referred to as the 'voice box' and this reflects one of its major functions, which is sound production.

PAUSE AND THINK

What prevents food from entering the respiratory tract?

Answer: This is role of the epiglottis, which closes the entry to the larynx during swallowing.

Sound production is dependent on paired mucous membranes, the vocal folds, that stretch horizontally across the larynx. The vocal folds and the space between them are referred to as the **glottis** (Figure 22.3). During sound production, the vocal folds are brought together partially closing off the air way. The generation of air pressure pushes the folds apart, causing them to vibrate. The pitch of the sound depends on the thickness of the folds and the tension in them, as well as the length of the larynx. As one of the secondary sexual characteristics, adult males have thicker vocal folds than adult females and children and hence a lower resonant frequency, giving a deeper tone. If you place your fingers (gently!) on your Adam's apple and sing high or low notes, you can feel the movement of the thyroid cartilage associated with the change in the resonance chamber of the larynx.

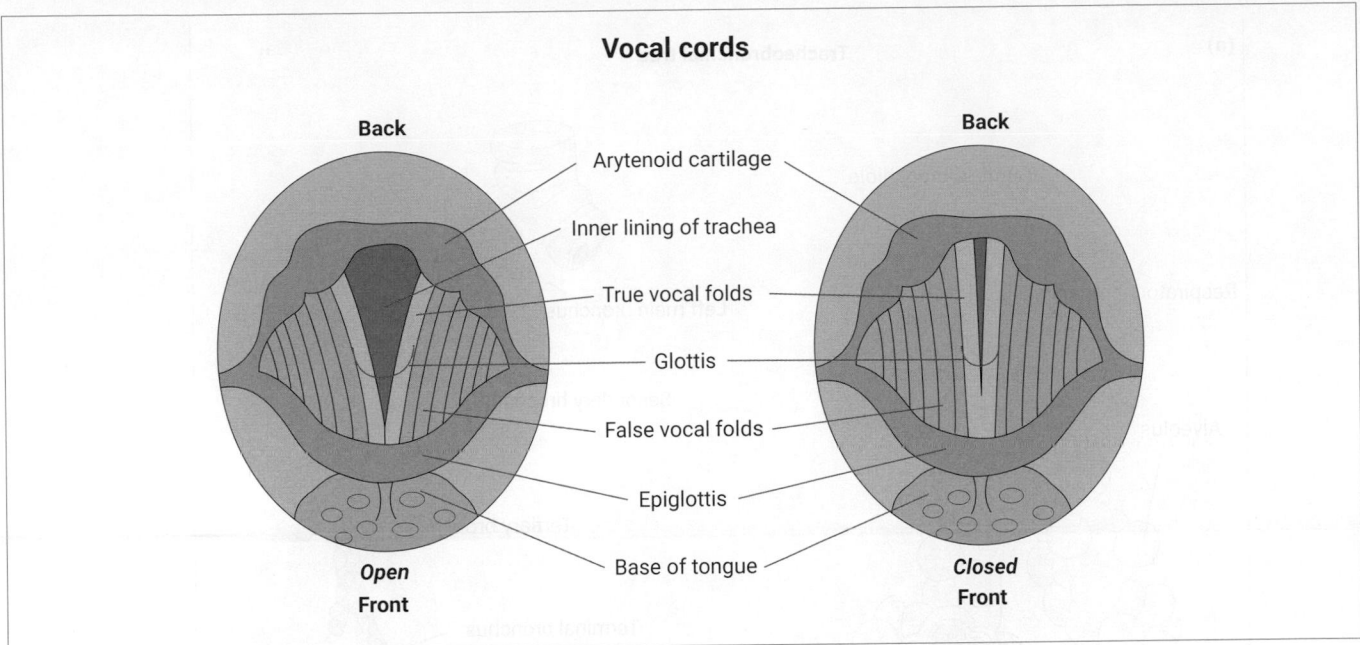

Figure 22.3 The glottis. Open and closed states of the vocal folds during sound production.
Source: Vectormine/Shutterstock.com.

PAUSE AND THINK

Why do you think people often 'lose their voice' during a respiratory tract infection?

Answer: The most common cause of voice loss is acute laryngitis, which results in inflammation of the vocal folds, so they no longer vibrate to produce sounds.

From the larynx, the inspired air passes into the main airway, the trachea (Figures 22.1 and 22.2). The trachea, or windpipe, is a flexible tube of connective tissue, which is held open by rings of cartilage. The trachea divides into the two bronchi, which feed into the right and left lungs.

Lower airways and alveoli

Each **bronchus** subdivides into a network of smaller bronchioles, ending in the terminal and respiratory bronchioles that feed into the **alveoli,** the tiny air sacs (Figure 22.4). As the bronchi decrease in diameter, there is less cartilage, which takes the form of small plates rather than rings.

During ventilation of the lung, there is resistance to the airflow. Airway resistance depends on the diameter of the airways and the velocity of the airflow. As the air passes down through the tracheo-bronchial tree, the resistance of the individual airways increases with their decreasing diameter. However, the profuse branching of the airways, which creates many thousands of bronchioles in

parallel, means that the total airway resistance in the lower part of the tree actually decreases.

The bronchioles, which are about 1 mm or less in diameter, have no cartilage and are lined with epithelial cells and smooth muscle so the diameter of these lower airways is under control of the autonomic nervous system. Parasympathetic stimulation via the vagus nerve (10th cranial nerve) leads to release of acetylcholine from the postganglionic fibres, which causes the airways to constrict by acting on the muscarinic cholinergic receptors (see Section 19.1). Constriction of the bronchioles will lead to an increase in airway resistance and so reduced air flow. The parasympathetic fibres also stimulate mucus secretion within the airways, which can further increase airway resistance due to narrowing.

In humans, unlike some other mammals, there is effectively no sympathetic adrenergic innervation of the smooth muscle to cause dilation of the airways, although beta-adrenergic receptors are present. Neurally stimulated bronchodilatation can also be caused by the non-adrenergic non-cholinergic (NANC) neurons of the parasympathetic nervous system, which have vasoactive intestinal peptide (VIP) and nitric oxide (NO) as the transmitters. Adrenaline can also act on the β_2 adrenergic receptors to cause relaxation of the smooth muscle and bronchodilation.

Can you think of a common condition in which there is excessive constriction of the airways? Take a moment to think before you read on. This is common in asthma where there can be hyperresponsiveness of the airways leading to obstruction of the airflow (see Clinical Box 22.1).

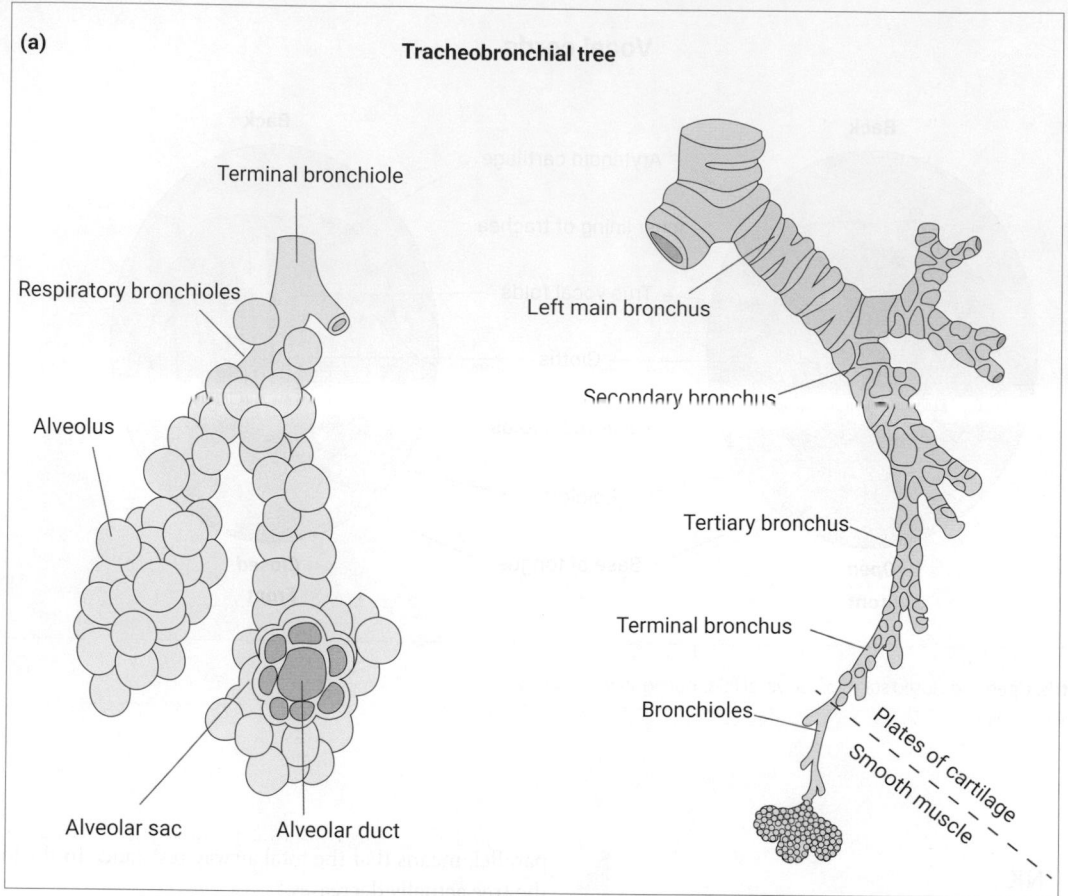

(a)

Tracheobronchial tree

Terminal bronchiole

Respiratory bronchioles

Alveolus

Alveolar sac

Alveolar duct

Left main bronchus

Secondary bronchus

Tertiary bronchus

Terminal bronchus

Bronchioles

Plates of cartilage

Smooth muscle

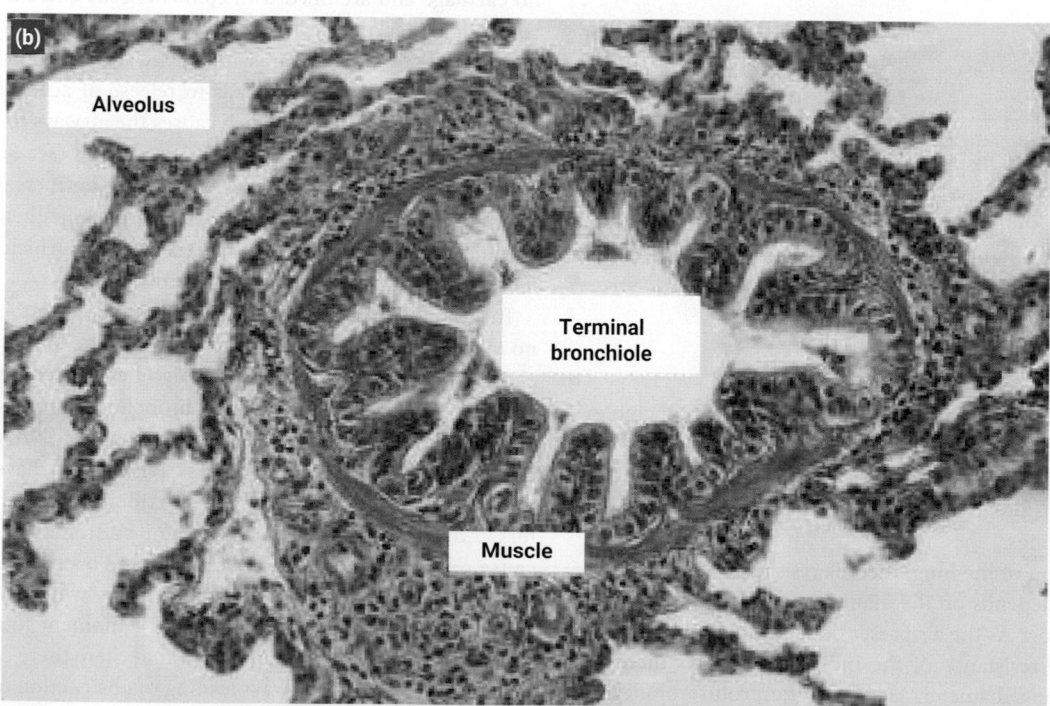

(b)

Alveolus

Terminal bronchiole

Muscle

Figure 22.4 Structure of the lower airways. (a) The progressive divisions of the airways; (b) shows the histology of a terminal bronchiole and associated alveoli.

Source: Courtesy Histology Guide © Faculty of Biological Sciences, University of Leeds. www.histology.leeds.ac.uk

Gas exchange, the take-up of oxygen by the blood and the removal of carbon dioxide, takes place in the respiratory bronchioles and the alveoli. The respiratory bronchioles and the alveoli are in very close contact with the network of capillaries arising from the pulmonary artery. The walls of the alveoli comprise connective tissue and elastic fibres, so the alveoli can stretch and relax back during breathing, and epithelial cells called **pneumocytes**.

The average alveolus is only about 200 μm in diameter, and there are of the order of 700 million alveoli in the lungs, giving a very large surface area for gas exchange: 70–80 m². The alveoli are lined with a thin film of fluid into which the oxygen dissolves before diffusing across the alveolar wall into the capillaries. Each alveolus is closely wrapped around by the pulmonary capillaries so that the diffusion barrier, from the alveolar space to the blood in the capillary, is only 0.2–0.5 μm.

There are two types of pneumocytes: the most numerous are the type I pneumocytes or squamous epithelial cells. These are thin epithelial cells that form the basic structure of the alveolus across which gas exchange occurs. The type II pneumocytes are larger epithelial cells which secrete **surfactant**, which is very important in reducing the surface tension of the fluid lining of the alveoli (see Section 22.2, Surface tension). The walls of the alveoli also have a population of macrophages that protect the lungs from antigenic agents such as bacteria.

 Check your understanding of the concepts covered in this section by answering the questions in the e-book.

22.2 Ventilation of the lungs

In the mammalian respiratory system, breathing is a tidal process: we breathe in and out through the same airways, so the air reverses direction of flow during each respiratory cycle. This contrasts with, for example, the gill system of fish where there is a constant flow of water across the gill surfaces. For mammals to breathe, the lungs have to expand to draw air in and then relax back to expel the air. The lungs themselves are elastic but have no contractile properties so ventilation of the lungs is affected by changes in the volume of the thorax, which are generated by the muscles within the chest: the **diaphragm** and the **intercostal** muscles in particular (Figure 22.5), though other thoracic and abdominal muscles can also contribute.

Ventilation of the lungs therefore involves work to stretch the lungs themselves and to overcome the resistance of movement of air through the airways.

Ventilatory movements

The lungs are located within the thoracic cavity, which is structurally maintained by the ribs, sternum, and vertebral column (Figure 22.5).

The right and left lungs are located within separate pleural cavities. The **pleural membranes** are two thin sheets of connective tissue—the outer, parietal pleural membrane and the inner, visceral pleural membrane. The outer membrane is attached to the chest wall and is separated from the inner membrane by a thin film of pleural fluid. The surface tension of the pleural fluid acts to hold the two membranes in close apposition and the fluid acts as a lubricant, enabling the two membranes to slide past each other during ventilatory movements.

Given what you have just read, what do you think happens if the pleural membranes are damaged? Puncturing the pleural membrane, as a result of a stab wound or a broken rib, for example can give rise to a pneumothorax, when air is able to pass into the thoracic cavity. This can lead to the collapse of a lung, which severely compromises its function for gas exchange. The two lungs are separately encased, so the lung on the other side will continue to function, but this is likely to be affected as a result of the overall changes in pressure within the chest cavity. If the wound site is sealed, then the bubble of air that is within the cavity will be reabsorbed over a few weeks, restoring lung function.

The main inspiratory muscles of ventilation are the:

- **Diaphragm**: a thin sheet of muscle that stretches across the floor of the rib cage and is innervated by the **phrenic nerve**. The diaphragm separates the thoracic cavity from the abdomen and has a number of holes in it allowing the passage of key structures including the oesophagus, aorta, and vena cava.

- **External intercostal muscles**: these muscles are located between the ribs and act on the rib cage.

A number of muscles in the upper thorax, such as the sternocleidomastoid and scalenus muscles (Figure 22.5), can act as accessory inspiratory muscles by acting to move the upper parts of the rib cage outwards and lift the sternum.

Inspiration

The action of breathing in is always an active process, with the number of muscles involved being related to the depth of inspiration. During 'quiet' breathing (for example, the breathing you are doing at the moment while sitting relaxed and reading) the predominant muscle action is that of the diaphragm (Figure 22.5). At rest, the diaphragm is dome-shaped; on contraction, the central region of the muscle flattens downwards (Figure 22.6). This increases the volume of the thoracic space in which the lungs sit. Because the lungs are located in the air-tight cavities formed by the pleural membranes, as these spaces expand, so the pressure around the lungs falls. The reduction in intrathoracic pressure causes the lungs to expand which, in turn, leads to a reduction in the pressure inside the lungs, drawing the air in (Figures 22.6 and 22.7). This will typically be occurring about 15 times per minute at rest.

 Go to the e-book to watch a video that depicts the mechanism of breathing.

When the ventilatory demand increases, the rate of ventilation can increase significantly and so also can the depth of ventilation, increasing the volume of air per breath (see 'Lung volumes' later in this section). The latter is achieved by larger contraction of the diaphragm and the contraction of more muscles to increase the expansion of the thoracic cavity and, therefore of the lungs. During quiet breathing the central region of the diaphragm will only move downwards about 1.5 cm, during deep inspiration this can increase to up to about 7 cm.

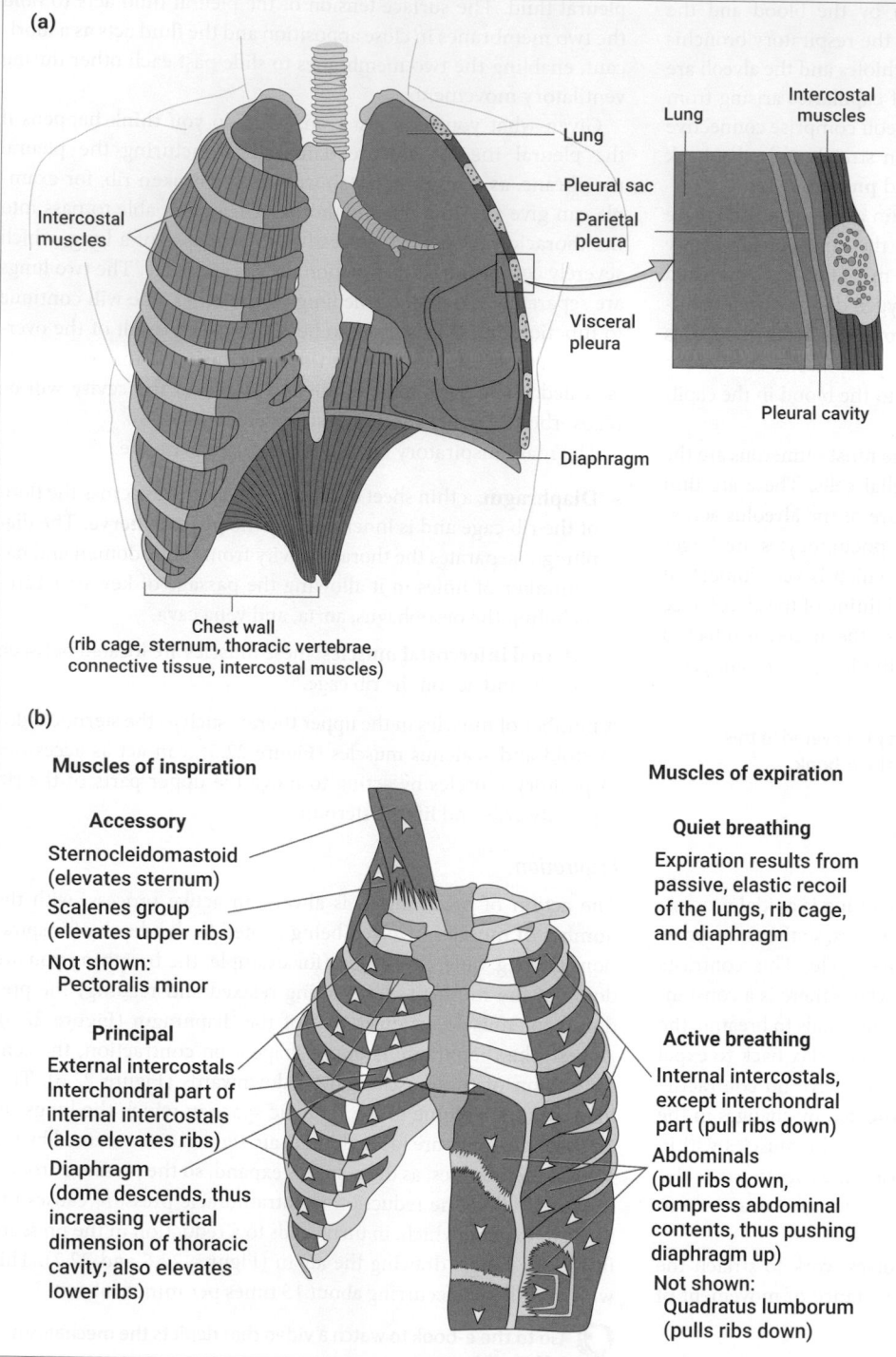

Figure 22.5 Thoracic anatomy. (a) Location of the lungs and associated muscles. (b) The roles of the different muscle groups.

Source: (a) Openstax CC BY 4.0 C J. Gordon Betts et al. *Anatomy and Physiology*, 2013. Access for free at: https://openstax.org/books/anatomy-and-physiology/pages/1-introduction Section URL: https://openstax.org/books/anatomy-and-physiology/pages/22-2-the-lungs; (b) Courtesy www.physio-pedia.com.

The second set of muscles that are involved are the external intercostal muscles. These muscles run obliquely from the lower surface of one rib to the upper surface of the rib below. When they contract, they draw the ribs upwards and outward and also move the sternum anteriorly (Figure 22.6). As a result, there is a greater expansion of the thoracic cavity leading to deeper inspiration.

Expiration

The process of breathing out, at rest, is a passive process: as the diaphragm relaxes it moves back upwards and relaxation of the external intercostals allows the rib cage to drop downwards and inwards (Figures 22.6 and 22.7). The rise in intrathoracic pressure along with the elastic recoil of the stretched lung tissue pushes the air out of the lungs.

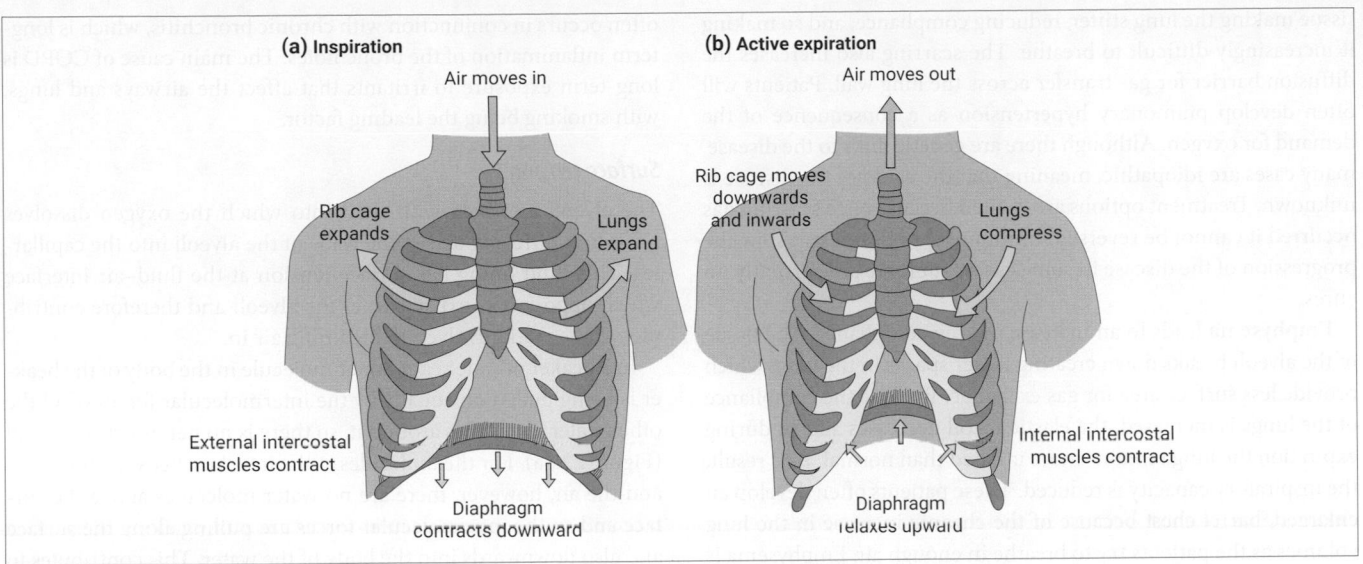

Figure 22.6 Mechanics of breathing. (a) Full inspiration is achieved by contraction of the diaphragm, which moves downwards, and of the external intercostal muscles, which move the ribs upwards and outwards and the sternum forwards. This can be assisted by accessory muscles acting on the upper chest wall. (b) Full expiration involves relaxation of the diaphragm and of the external intercostals and contraction of the internal intercostal muscles. This can be further assisted by contraction of the abdominal muscles pushing the diaphragm upwards.

Figure 22.7 Ventilatory cycle.

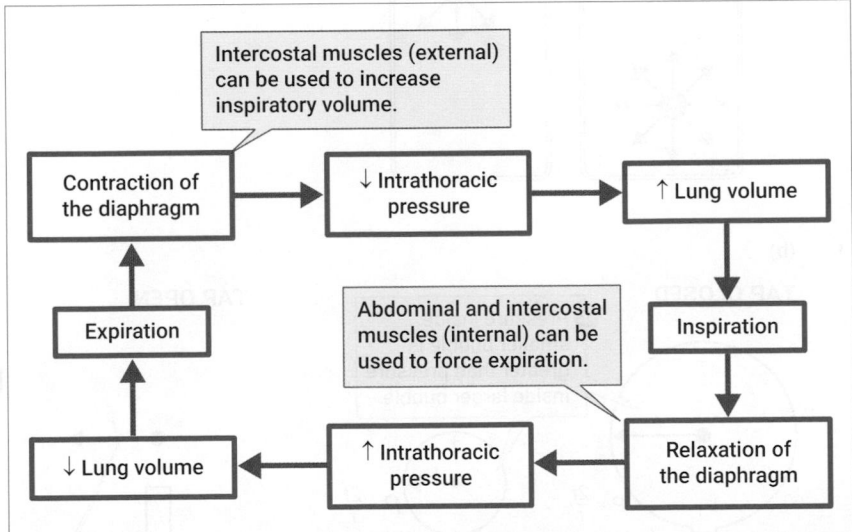

This passive expiration can be supplemented by active muscle contraction: the internal intercostal muscles contract and pull the rib cage inwards, while contraction of the abdominal muscles pushes the diaphragm upwards, so these two processes drive the air out of the lungs.

Lung compliance

During breathing movements the lung expands, drawing air in, and recoils back on breathing out. Therefore, work has to be done not only to move the air through the airways, but also to stretch the lung against the resistance of the tissue of the lung wall.

Compliance is a measure of how easy it is to stretch the lung. In other words, it is the change in lung volume as a function of the change in the pressure across the lung. There are two main factors that affect lung compliance:

- the tissue structure of the lung itself; and
- the **surface tension** of the fluid film lining the alveoli.

The compliance of the lung tissue is normally very high, but it can be affected in disease states that may reduce or increase lung compliance.

Pulmonary fibrosis is the most common form of a group of diseases called interstitial lung disease (ILD), which affect the interstitial tissue of the lung. Pulmonary fibrosis is characterized by a progressive shortness of breath, not just when exercising. It results from progressive scarring and thickening of the alveoli and lung tissue making the lung stiffer, reducing compliance, and so making it increasingly difficult to breathe. The scarring also increases the diffusion barrier for gas transfer across the lung wall. Patients will often develop pulmonary hypertension as a consequence of the demand for oxygen. Although there are genetic links to the disease, many cases are idiopathic, meaning that the cause of the disease is unknown. Treatment options are limited because once scarring has occurred it cannot be reversed. Immunosuppressants can slow the progression of the disease in some cases, but there are currently no cures.

Emphysema leads to an increase in lung compliance. The tissue of the alveoli breaks down creating larger spaces in the lung, which provide less surface area for gas exchange. Because the compliance of the lungs is increased, the elastic recoil decreases and so during expiration the lung remains more inflated than normal. As a result, the inspiratory capacity is reduced. These patients often develop an enlarged, barrel chest because of the chronic increase in the lung volumes as the patients try to breathe in enough air. Emphysema is classified as a chronic obstructive pulmonary disease (COPD) and

often occurs in conjunction with chronic bronchitis, which is long-term inflammation of the bronchioles. The main cause of COPD is long-term exposure to irritants that affect the airways and lungs, with smoking being the leading factor.

Surface tension

The alveoli are lined with fluid into which the oxygen dissolves prior to it diffusing across the walls of the alveoli into the capillaries. This fluid lining has surface tension at the fluid–air interface which opposes the expansion of the alveoli and therefore contributes to the work involved in breathing air in.

In a beaker of water each water molecule in the body of the beaker is being pulled on equally by the intermolecular forces of all the other water molecules around it, so there is no net direction of pull (Figure 22.8a). For the molecules at the interface between the water and the air, however, there are no water molecules above the surface and so the intermolecular forces are pulling along the surface and also downwards into the body of the water. This contributes to the meniscus seen across the surface of the water, which is not level

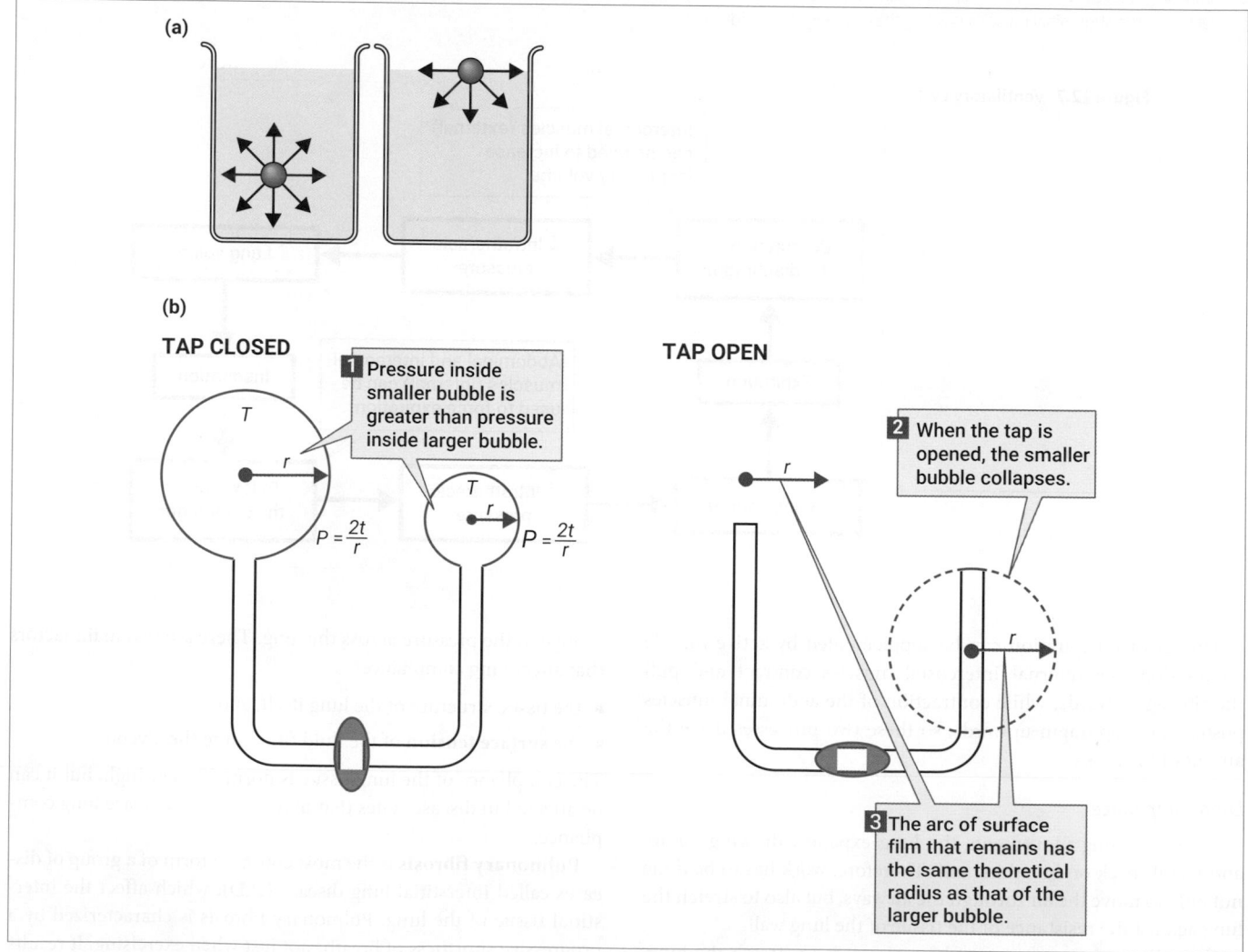

Figure 22.8 The forces involved in surface tension. (a) This shows the intermolecular forces acting on a water molecule in the centre of a beaker of water (left) and on the surface (right); (b) the effects of surface tension on two soap bubbles: see text for further details.

across the beaker but is highest at the edges and curves downwards towards the middle. This intermolecular attraction creates surface tension, which acts rather like an elastic membrane on the surface of the fluid: if you have ever seen an insect standing on the surface of a pond you can see the surface of the water indenting beneath the feet of the insect but it holds the overall weight.

If you have ever blown soap bubbles you will have observed the effect of surface tension which holds the bubbles' shape against the internal and external pressures acting on it. We can model each alveolus as a bubble (Figure 22.8b). From the Law of Laplace the pressure, P, acting on a stable bubble is given by the equation:

$$P = \frac{2T}{r}$$

where T is the surface tension and r is the radius of the bubble. The pressure P is the force of the gas inside the bubble, pushing the surface of the bubble outwards, while the surface tension is pulling the bubble inwards. In a stable bubble, these two forces are equal and opposite. If the bubble shrinks, then the gas becomes compressed and exerts greater outward force on the surface of the bubble, so the outward acting pressure is inversely proportional to the radius of the bubble.

What does this mean for the alveoli? Figure 22.8b shows an experiment in which two bubbles, of different sizes, have been created separately. If the tap connecting the two bubbles is opened (Figure 22.8c), then the gas from the smaller bubble will go into the larger one because of the higher pressure in the smaller bubble. As a consequence, the small bubble will collapse, leaving an arc of surface film which is of the same theoretical radius as that of the larger bubble. The lungs contain millions of alveoli of different sizes. If the surface tension acted as in the experiment we have just considered, then the small alveoli would collapse into the larger ones, greatly reducing the overall surface area for gas exchange.

The type II pneumocytes that form part of the alveolar walls secrete surfactant which comprises lipids, glycoproteins, and proteins. One of the main constituents is a phospholipid, dipalmitoyl phosphatidylcholine (DPPC), which acts like a detergent to reduce the surface tension of the fluid lining. This reduces the risk of small alveoli collapsing into the larger ones and also reduces the work required to inflate the alveoli on inspiration. Babies born very prematurely may suffer from neonatal respiratory distress syndrome (NRDS) because their lungs have not yet started producing surfactant and so they cannot breathe effectively. Surfactant production normally begins around weeks 24–28 of pregnancy and about half of all babies born before 28 weeks will develop NRDS and may require oxygen supplement and possibly ventilator support to enable them to breathe.

Cough reflex

The upper parts of the respiratory tract are richly supplied with receptors that respond, in particular, to touch, and also to triggers such as rapid changes in temperature (e.g. going outside on a cold day and taking a deep breath of cold air) or to certain chemical irritants. Coughing is also triggered by infections of the respiratory system.

Stimulation of the irritant receptors is transmitted by afferent neurons in the vagus nerve to the medulla of the brainstem. The action of the cough reflex is then triggered by a brief inspiration followed by closure of the glottis and simultaneous relaxation of the diaphragm with contraction of the internal intercostal and abdominal muscles leading to a rapid rise in intrathoracic pressure. Relaxation of the glottis then allows the expulsion of the air at high velocity, clearing the airways of the irritant particles.

Lung volumes

As we have seen, the volume of air moved during each breath can be varied quite significantly. This can be measured using a technique called **spirometry**, which can be used to measure both the lung volumes and the flow rate (Figure 22.9).

The subject initially breathes at rest through the spirometer for a few respiratory cycles to measure the resting tidal volume, and then is asked to make a maximal inspiration followed by a maximal expiration, the process being repeated for several cycles to obtain

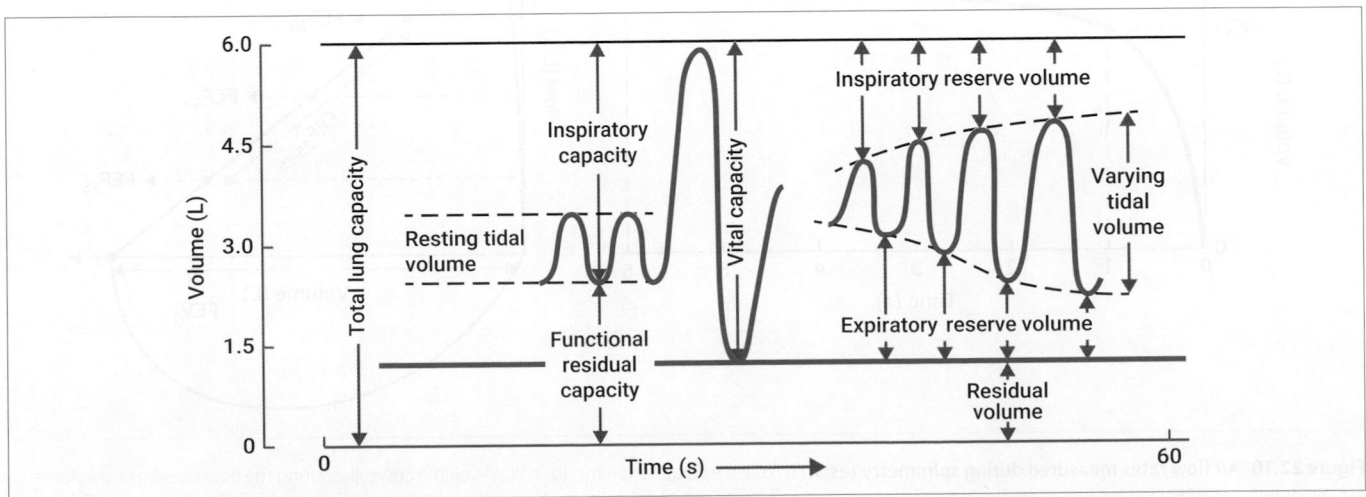

Figure 22.9 Lung volumes measure by spirometry.

Source: Pocock et al. *Human Physiology*, 5th edn, 2017. Oxford University Press.

average values. The following volume measures can be made through this procedure:

- **Tidal volume** (TV)—the volume of air moving in and out of the lungs during quiet breathing.
- **Inspiratory reserve volume** (IRV)—the maximum volume of air that can be inspired starting from the end of a normal inspiration.
- **Expiratory reserve volume** (ERV)—the maximum volume of air that can be expired starting from the end of a normal expiration.
- **Inspiratory capacity** (IC)—the maximum volume of air that can be inspired starting from the end of a normal expiration (TV + IRV).
- **Functional residual capacity** (EC)—the volume of air in the lungs following passive expiration (ERV + RV).
- **Vital capacity** (VC)—the maximum volume of air that can be breathed out starting from full inspiration.
- **Total lung capacity** (TC)—the total volume of the lung.
- **Residual volume** (RV)—the volume of air left in the lungs following maximal expiration, which cannot be measured by spirometry.

Why do you think there is some air left in the lungs at the end of a full expiration (RV)? You might find it useful to look back at Figure 22.6b when thinking about this question. It is because the rib cage retains a given volume at the end of expiration and so the lungs are always under a degree of tension exerted by the chamber created by the pleural membranes, so they cannot collapse fully, unless the pleura are punctured.

Table 22.1 Typical lung volumes

Volume of air (L)	Men	Women
Tidal volume	0.5	0.5
Vital capacity	5.0	3.7
Inspiratory reserve volume	3.1	1.9
Expiratory reserve volume	1.2	0.7

Spirometry values for clinical testing are measured against a standard data set, which takes account of variables that affect lung volumes, including sex, age, height, and ethnicity. Table 22.1 shows some averaged measures of lung volumes.

As well as measurement of the lung volumes, spirometry also enables measurement of the flow rate of the air during expiration, which enables assessment of airway function (Figure 22.10). The most common measures are:

- **Forced vital capacity** (FVC): the total volume of air that can be expired during a forced expiration following maximal inspiration. To give a more consistent standard, this is often expressed as FEV_6: the total volume of air expired after 6 seconds of a forced expiration following maximal inspiration.
- **Forced expiratory volume in 1 s** (FEV_1): the volume of air that can be expired during the first second of a forced expiration following maximal inspiration.
- **Peak expiratory flow** (PEF): the maximal flow rate, normally achieved at the start of expiration.
- **Forced expiratory flow** (FEF): the flow rate at specific time intervals, usually measured after 25% (FEF_{25}) and 75% (FEF_{75}) of the FVC has been expired.

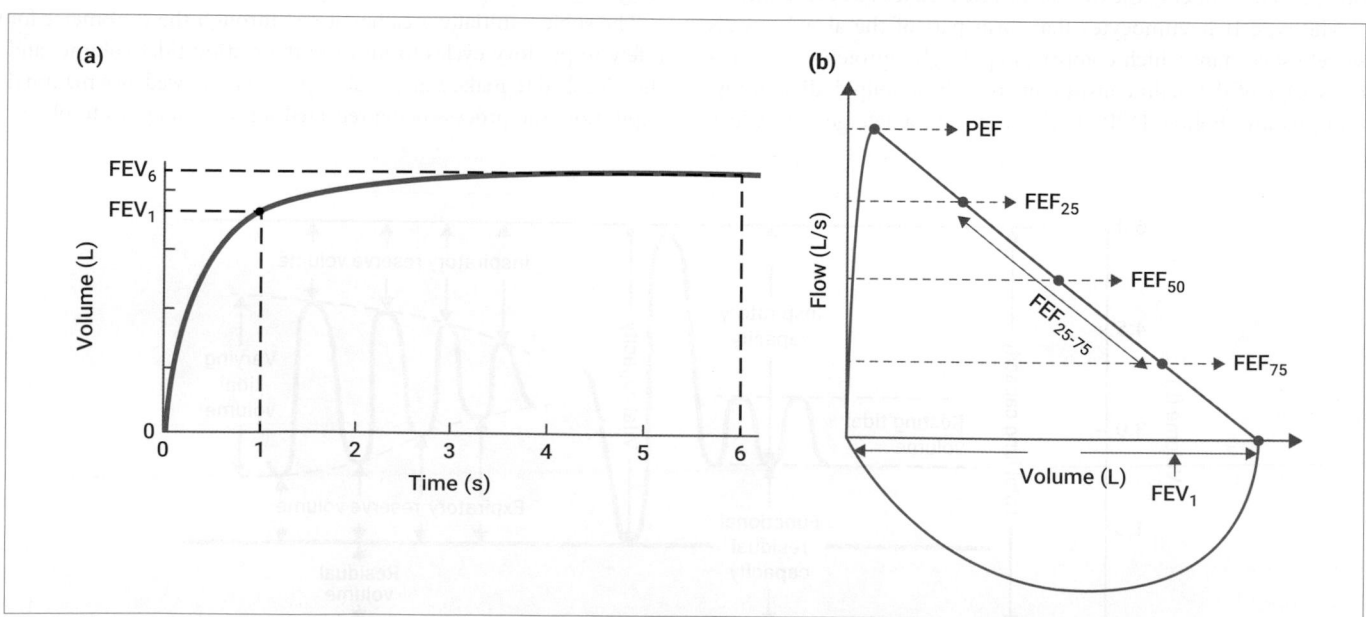

Figure 22.10 Air flow rates measured during spirometry tests. (a) Forced expiratory volume. (b) A flow–volume curve illustrating the flow rates against volume expired during a maximal forced expiration (the line above the *x*-axis) and inspiration (the line below the *x*-axis, which is recorded as negative expiratory flow).
Source: www.spirometry.guru

The ratio of FEV$_1$/FVC gives the proportion of air breathed out by the subject during the first second of a forced expiration: this is normally between 70 and 90%.

Airflow deficits

If the airways are narrowed, then the flow rate will be reduced as the flow of air is obstructed. The FEV$_1$/FVC ratio will be abnormally low: this is referred to as an **obstructive disorder** when the FEV$_1$ <80% of predicted. Examples of obstructive disorders would be asthma and COPD (see Clinical Box 22.1 for more details).

If the airflow is normal but the FVC is < 80% of the predicted values (i.e. the total volume of air expired is abnormally low), then there may be a **restrictive disorder**. An example of a condition giving rise to a restrictive disorder would be lung cancer, where the total volume of the lung available for air movements is reduced.

Air movements

As we have already discussed, the mammalian respiratory system involves tidal flows of air as the lungs expand and recoil back during inspiration and expiration, respectively. The only parts of the network of airways that take part in gas exchange are the alveoli and the terminal, respiratory bronchioles. The rest of the network,

---

CLINICAL BOX 22.1   Asthma

Asthma is a very common, long-term lung condition that affects over 5 million people in the UK. It often appears during childhood but can develop at any age. There are genetic traits that increase the probability of asthma, and asthma often occurs in association with allergic conditions such as hay fever or eczema, but it may also develop as a result of environmental exposure to irritants.

The characteristic symptoms are wheezing, persistent cough, breathlessness, and a feeling of tightness around the chest that interferes with breathing. Affected people may also experience asthma attacks during which the symptoms become significantly worse such that they may be unable to breathe due to airway constriction: such attacks result in about three deaths every day in the UK.

People affected by asthma show high sensitivity of their airways associated with inflammation such that the airways are narrowed (Figure 1) with thickening of the airway wall and an increase in mucus production, giving rise to an obstructive disorder. The airways may be sensitive to a number of 'triggers' that result in airway constriction, including pollen, animal fur, house dust mites, cigarette smoke, cold air, and respiratory tract infections.

Diagnosis will typically involve spirometry tests to evaluate the FEV$_1$/FVC ratio and also a peak expiratory test, which enable determination of the air flow through the lower airways. Further tests may be to measure the effect of a reliever inhaler on the peak flow (Figure 2) or to assess airway responsiveness in response to exposure specific triggers. Skin allergy tests may also be used to identify specific sensitivities.

Asthma is a chronic condition that needs to be managed. Where there are specific triggers that have been identified, then minimizing exposure is clearly important. Medication to relieve the symptoms takes a number of forms, the typical ones being:

- Reliever inhalers—these inhalers contain β$_2$-agonists that relax the smooth muscle of the lower airways to provide short-acting rapid relief

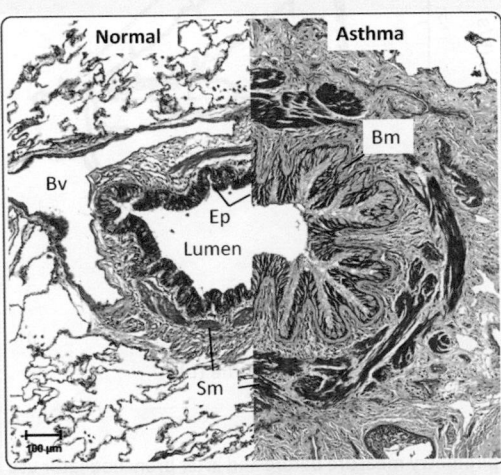

Figure 1 Histological preparations showing a normal airway (left panel) and one from a patient severely affected by asthma (right panel).

Source: Wadsworth, S. J., Yang, S. J., & Dorscheid, D. R. (2012). IL-13, asthma and glycosylation in airway epithelial repair. In: *Carbohydrates: Comprehensive Studies on Glycobiology and Glycotechnology*. IntechOpen. https://doi.org/10.5772/51970

from symptoms of breathlessness, resulting in an improvement in airway function (Figure 2 and Table 1).

- Preventer inhalers—these are usually prescribed when patients are using reliever inhalers on a regular basis. These contain inhaled corticosteroids that act to reduce the extent of inflammation and airway sensitivity.

Longer term treatment may also include long-acting reliever inhalers that contain long-acting dilators and/or β$_2$-agonists, or oral steroids in cases where the inhaled treatments are not sufficiently effective.

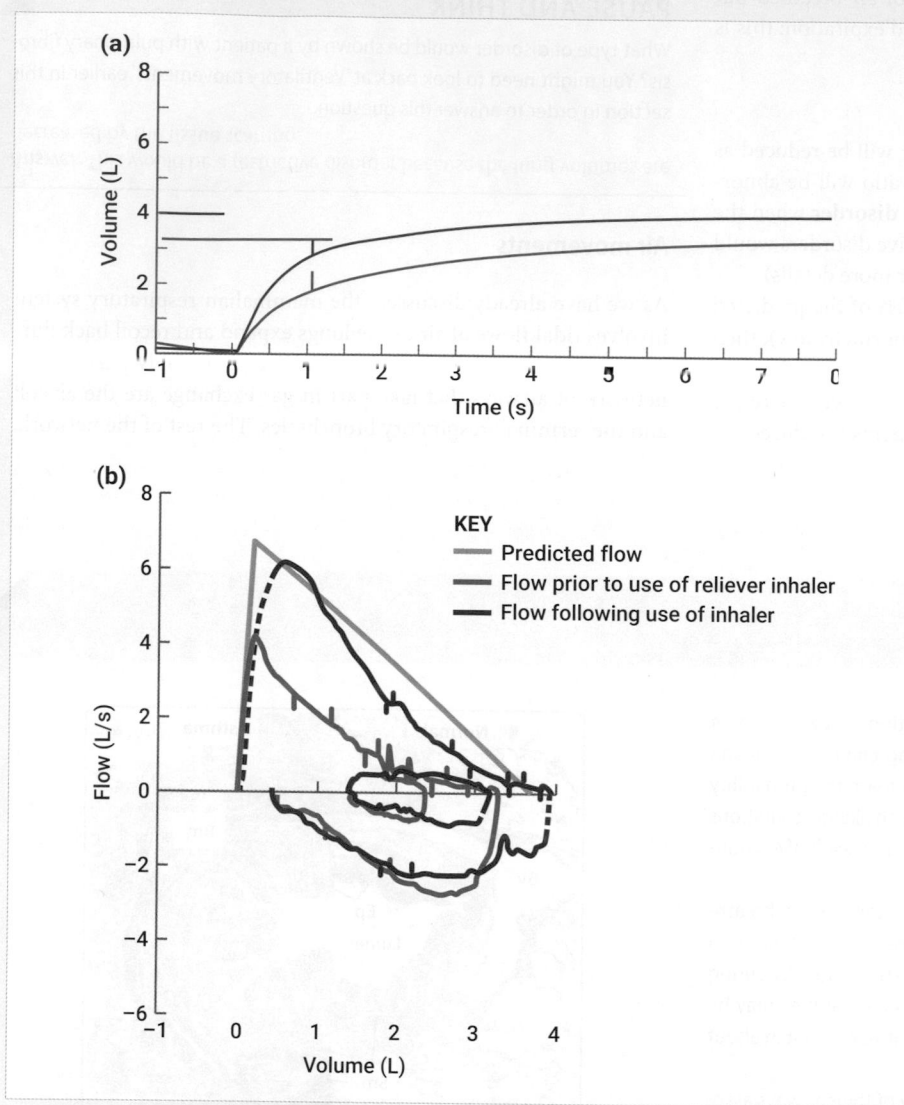

Figure 2 Spirometry test showing air flows for a patient with asthma. (a) The forced expiratory volume (FEV) of the person with asthma before (blue line) and after (red line) using a reliever inhaler. (b) The flow–volume curves before and after use of a reliever, demonstrating the marked increase in expiratory flow rate.

Source: Figure 1 from Kaplan A.G., et al. Diagnosis of asthma in adults. *CMAJ* 2009;181(10): E210–20.

KEY
— Predicted flow
— Flow prior to use of reliever inhaler
— Flow following use of inhaler

Table 1 Spirometry measure before and after bronchodilator use

Spirometry measure	Predicted	Before bronchodilator		After bronchodilator		% change
		Best	% of predicted	Best	% of predicted	
FVC (L)	3.70	3.30	89	3.95	107	20
FEV$_1$ (L)	2.94	1.80	61	2.76	94	53
Ratio (%) FEV$_1$/FVC	80	55	NA	70	NA	NA

FEV1, forced expiratory volume in 1 s; FVC, forced vital capacity; NA, not applicable.

the conducting airways which comprise the trachea, bronchi, and bronchioles all contain air which is moved to and fro during breathing but which does not take part in gas exchange. This volume of conducting airway is referred to as the **anatomical dead space** and typically represents a volume of about 150 mL, depending on body size. Therefore, of the resting tidal volume of 500 mL, 150 mL (33%) does not take part in gas exchange. The anatomical

dead space is a constant; therefore, during deeper breathing this relative proportion will decrease significantly.

There is also a **physiological dead space**, which is the anatomical dead space plus the volume of alveoli not involved in effective gas exchange. In healthy people, the physiological and anatomical dead spaces are very similar, but if the subject's lungs are poorly perfused with blood or the alveoli are not effectively involved in gas exchange, then the physiological dead space may be a significant volume.

The lungs never normally empty fully at the end of expiration: there is always the residual volume of air left, even following a forced expiration (Figure 22.9). The 'fresh' air that is breathed in therefore mixes with the residual air remaining in the lungs and lower airways, this residual air having already been involved in gas exchange. As a consequence, as we shall see in the next section, the proportion of oxygen in the air in the alveoli is lower than that in atmospheric air and the proportion of carbon dioxide is higher.

 Check your understanding of the concepts covered in this section by answering the questions in the e-book.

22.3 Gas exchange

So far we have considered the structures and processes involved in creating a flow of air in and out of the lungs. The next stage is the process whereby an exchange of gases takes place between the blood and the lungs: oxygen moves into the blood to be transported to the tissues to meet the metabolic demand, while carbon dioxide, a waste product of metabolic activity, is removed from the blood to be breathed out.

To be able to investigate these processes we must first consider the composition of the air we breathe.

Composition of air

The air we breathe is a mixture of gases. The typical composition of dry air is:

- 78.1% nitrogen;
- 20.9% oxygen;
- 0.04% carbon dioxide.

There are very small amounts of other gases such as argon. Air typically also contains about 1% of water vapour at sea level.

The SI unit for atmospheric pressure is the kiloPascal (kPa), where 1 Pa represents a pressure of 1 Nm^{-2}, although air pressure is often also quoted using the old barometric measure of mmHg (mm of mercury). In this chapter we will use the SI unit throughout, but some main values will also be given in mmHg for reference. The

atmospheric pressure decreases with height above sea level and also varies depending on the climatic conditions. The average atmospheric pressure at sea level is taken as 101.3 kPa (760 mmHg).

Each of the gases in the air contributes to the total atmospheric pressure as a function of the proportion of the gas in the air. This contribution is referred to as the **partial pressure**. Dalton's law states that the total pressure of a mixture of gases equals the sum of the partial pressures of those individual gases.

For example, the partial pressure of oxygen (Po_2) in the air at standard temperature and pressure is:

Proportion of oxygen in air is 20.9% or 0.209	Average atmospheric pressure at sea level

$$0.209 \times 101.3 = 21.17 \text{ kPa}$$

While the proportion of oxygen in atmospheric air is normally constant, the partial pressure will vary depending on the atmospheric pressure. Therefore, the air at high altitude will still contain 20.9% oxygen, but the partial pressure will be much lower due to the lower atmospheric pressure. For example, at the top of Mount Everest, the atmospheric pressure is only 33.7 kPa, so the partial pressure of oxygen is only:

$$0.209 \times 33.7 = 7.04 \text{ kPa}$$

Such low partial pressures make respiration very difficult: we will explore the effects of altitude in 'Altitude acclimatization'.

As discussed in Section 22.2, one effect of tidal breathing is that the 'fresh' air that we breathe in mixes with the air already in the bronchial tree and remaining in the lungs. So, the air in the alveoli that takes part in gas exchange has a different oxygen and carbon dioxide composition to the air in the atmosphere, as shown in Table 22.2.

PAUSE AND THINK

Why does the air we breathe out have more oxygen and less carbon dioxide in it than the air in the alveoli?

Answer: As we breathe out, the air from the alveoli mixes with the air in the upper airways that has not been involved in gas exchange and so the carbon dioxide is diluted and the oxygen enriched as part of that mixing process.

Table 22.2 Typical values of the partial pressures of oxygen and carbon dioxide at different levels in the respiratory tree

	Po_2 (kPa)	P_{CO_2} (kPa)	Po_2 (mmHg)	P_{CO_2} (mmHg)
Inspired air	21.2	0.04	158.8	0.3
Alveolar air	13.3	5.3	100	40
Arterial blood	13.3	5.3	100	40
Venous blood	5.3	6.6	40	50
Expired air	16.0	3.6	104	30

Gas exchange in the lung

Within the respiratory bronchioles and the alveoli gas exchange occurs, whereby oxygen diffuses down its pressure gradient into the capillaries that surround the alveoli and carbon dioxide passes from the capillaries into the alveoli (Figure 22.11).

Fick's Law (see Section 16.2) states that the rate of diffusion of a gas across a membrane is proportional to the difference in the partial pressure of the gas and is inversely proportional to the thickness of the membrane. The diffusion barrier of the lung–capillary interface is very thin, being only two cells thick (Figure 22.11). As shown in Table 22.2, even though the partial pressure of oxygen in the alveoli is well below that of the inspired air, it is still significantly greater than the partial pressure in the venous blood that will be entering the pulmonary capillaries.

For oxygen to diffuse into the blood, it first has to dissolve in the fluid lining of the alveoli. Oxygen is relatively poorly soluble in water: at core body temperature, 37 °C, there is approximately 3 mL of O_2 dissolved per litre (see Quantitative Toolkit 22.1 for the calculations).

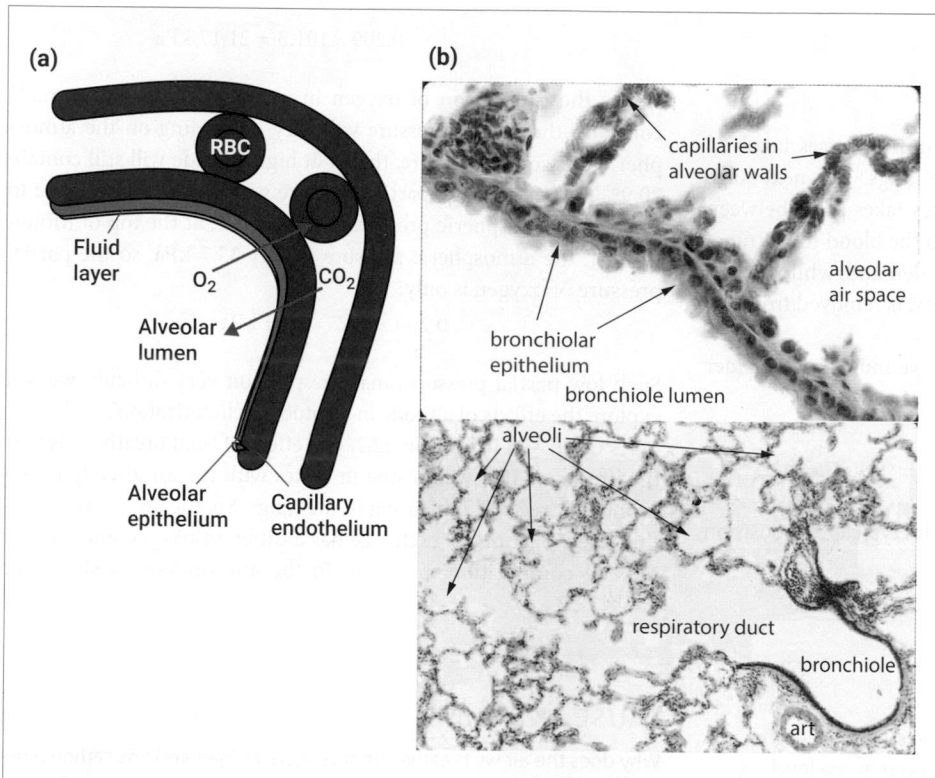

Figure 22.11 Gas exchange between the alveoli and the capillaries. (a) Diagram showing the relationship between an alveolus and a capillary. (b) Histological preparations showing the terminal bronchioles, alveoli, and capillaries. Note the single-cell thickness of the alveolar epithelium and the capillary endothelium, creating a minimal barrier to diffusion.
Source: (b) David G. King.

QUANTITATIVE TOOLKIT 22.1   **Calculation of dissolved oxygen per litre**

Henry's Law states that the amount of a gas dissolved in a fluid is proportional to the partial pressure of the gas. In the case of arterial blood, the partial pressure of oxygen is 13.3 kPa (Table 22.2). The amount of oxygen that dissolves is determined by its solubility coefficient, which is defined as the number of millilitres of oxygen that dissolve in 100 mL of plasma per kilopascal of partial pressure. In the case of oxygen, the coefficient is 0.0225 mL.

The amount of oxygen dissolved in 100 mL arterial blood is therefore:

$$0.0225 \times 13.3 = 0.299 \text{ mL}$$

which approximates to 3 mL L^{-1} of blood.

The volume of blood in an adult is about 5 L; therefore, the total volume of dissolved oxygen is less than 15 mL.

Why is the volume of dissolved oxygen actually less than 15 mL? Remember that the above calculation is based on arterial blood, with a partial pressure of 13.3 kPa. In venous blood, the partial pressure of oxygen can fall to 5.3 kPa, in which case the volume of dissolved oxygen in venous blood would only be 1.2 mL L^{-1}.

At rest, the body's demand for oxygen is approximately 250 mL min^{-1}. The total volume of oxygen dissolved in the blood is only 15 mL, which is less than one-twentieth of the actual demand.

What can we conclude from this? The volume of oxygen dissolved in the blood would be insufficient to sustain baseline metabolic activity and that the total surface area for respiratory exchange must be very large to enable sufficient oxygen to dissolve to meet the resting demand of 250 mL min^{-1}. Two key features of the mammalian respiratory system are a very large surface area of the lungs: estimates of total lung surface area range up to 140 m^2; and the presence of an oxygen transporting system, which is the haemoglobin found in the red blood cells (erythrocytes) (see 'Oxygen transport' later in this section).

Ventilation–perfusion matching

As we have seen from Table 22.2, significant partial pressure gradients favour the diffusion of oxygen from the alveoli into the capillaries and the diffusion of carbon dioxide from the capillaries into the alveoli. Maintaining these gradients depends on effective ventilation of the alveoli, thereby ensuring that there is a good supply of oxygen and removal of carbon dioxide. It also depends on effective perfusion of the capillaries, so that the blood flow around the alveoli is adequate to enable exchange to take place. At rest, an erythrocyte takes about 0.75 s to pass through the pulmonary capillary and this can fall to about 0.25 s during intense exercise. Even the most rapid transit time is normally sufficient to enable full equilibration of both oxygen and carbon dioxide between the alveoli and the blood.

PAUSE AND THINK

Can you think of a disease in which the equilibration could be compromised? You might to look back at Section 22.2 in order to answer this question.

Answer: In an interstitial lung disease (ILD), such as fibrosis, the wall of the lung becomes thickened which slows down the rate of diffusion and can lead to incomplete gas exchange.

The balance between ventilation, V′ (measured in L min^{-1}) and perfusion, Q′ (also L min^{-1}) is termed ventilation–perfusion matching and the V′/Q′ ratio is a measure of the effectiveness of the balance between these two variables. Ideally, the ratio would be 1, indicating a perfect match between ventilation and perfusion; in reality there is a degree of mismatch with the average value for the V′/Q′ ratio being 0.8 for the lung as a whole.

The pulmonary circulation, arising from the right ventricle of the heart, is at a much lower pressure than the systemic circulation, whereas the resting systolic/diastolic pressures in the aorta are about 120/70 mmHg, the pressures in the pulmonary artery are only about 25/8 mmHg. In such a low-pressure system the effects of gravity on blood flow are significant, so the lower parts of the lung tend to be better perfused with blood than the upper regions. This effect is increased because of the changes in the pressure in the lung: for example, during expiration, when the intrathoracic pressure rises the capillaries in the upper regions, where the diastolic pressure is low, will be compressed and blood flow can be occluded. The capillaries are then opened as the perfusion pressure rises during systole.

▶ We learn more about pulmonary circulation in Chapter 21.

The ventilation of the lung also varies in different regions: at end expiration, the alveoli at the base of the lung are at a smaller volume than those at the top of the lung. This is in part due to the pressure of the mass of the lung compressing the lower portions. During inspiration, therefore, the lower parts of the lung have greater capacity for expansion and therefore ventilation can lead to greater mixing of the gas in the alveoli.

Local regulatory mechanisms can operate to balance the V′/Q′ ratio within different areas of the lung. The most significant of these is smooth muscle constriction/dilation in both the blood vessels and the airways. Vasoconstriction of the blood vessels can be triggered by hypoxia, so blood flow is reduced to those areas of alveoli that have lower ventilation and therefore lower P_{O_2}. Likewise, where there are sections of the lung that may be underperfused but well ventilated, the P_{CO_2} will fall, leading to a rise in pH as the [H$^+$] falls. The fall in [H$^+$] stimulates airway constriction, so the ventilation will be directed towards those areas of the lung that are better perfused.

The ventilation–perfusion function can be measured using a V/Q scan; this may be required in a critical condition such as a suspected pulmonary embolism (a clot blocking off part of the vascular supply to the lung). The scan involves two tests run together: injection of a radioisotope into the bloodstream and inhalation of a radioactive gas. In a normal lung both radioactive tracers should be taken up by all areas of the lung; however, blockage of an artery or of an airway will reveal differences in the relative uptakes of the tracers in the affected areas of the lung.

Oxygen transport

The poor solubility of oxygen means that only about 12.5 mL of oxygen will dissolve in the 5 L of blood in the total circulation. Calculations of metabolic demand show that we need about 250 mL min^{-1} O$_2$ at rest; this can rise five-fold or more during exercise. Almost all the oxygen in the blood is carried bound to the respiratory pigment **haemoglobin** (Hb), which is located in the red blood cells, the erythrocytes. There are roughly 5 million erythrocytes per mm^3 of blood and the body contains about 750 g of haemoglobin.

The red blood cells make up about 45% of the volume of the blood and are produced from stem cells in the bone marrow (Section 21.1). As they develop they synthesize large amounts of haemoglobin, which occupies most of the cytoplasmic space. In the final stage of development, the cells lose their nuclei and, as a result, take on their characteristic doughnut appearance.

Haemoglobin is a complex protein molecule with a mass of 64,458 Da and is made up of four globin subunits: two α and two β subunits. Each subunit comprises a polypeptide chain linked to haem, which is made up of a protoporphyrin ring bound to a ferrous (Fe^{2+}) ion (Figure 22.12). It is the ferrous ion that reversibly binds to a molecule of oxygen. Therefore, each molecule of haemoglobin can bind four molecules of oxygen to form oxyhaemoglobin.

Cooperative binding

The binding of oxygen molecules to the Hb shows cooperativity. This means that as each molecule of oxygen binds to one of the Fe^{2+} in the Hb it has the effect of changing the quaternary structure of the Hb molecule, making it easier for the next molecule of oxygen

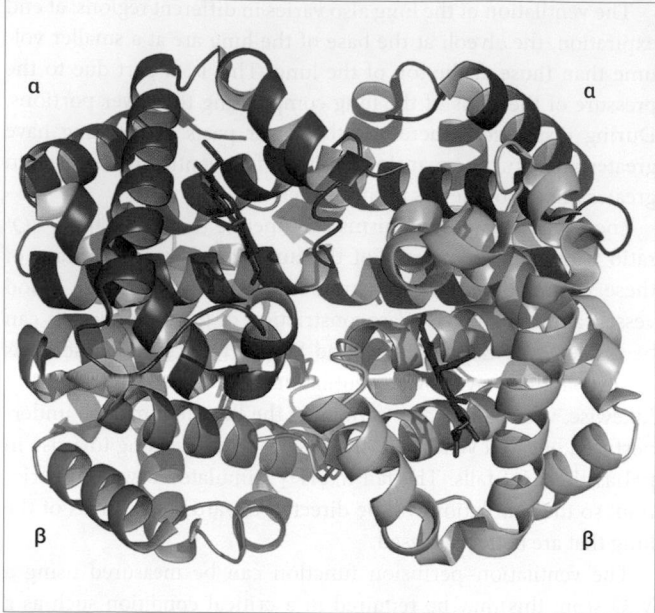

Figure 22.12 Diagram of a haemoglobin molecule. There are two α and two β subunits (labelled), each incorporating a haem complex with Fe²⁺ (shown here in pink). The structure of the haem group is illustrated in more detail in Figure 6.22.

Source: Snape, A. & Papachristodoulou, D. (2018). *Biochemistry and Molecular Biology*, 6th edn. Oxford University Press.

to bind. When Hb is deoxygenated, the quaternary structure limits the access of oxygen to the Fe²⁺ binding site; this is referred to as the *tense* form of the molecule. The change in the quaternary structure of the Hb as each oxygen binds in turn opens up access to the binding sites and the Hb is referred to as being in its *relaxed* state.

The cooperative binding of oxygen to Hb gives rise to a sigmoidal ('S' shaped) curve when the percentage saturation of Hb with oxygen is plotted against the partial pressure of oxygen (Figure 22.13).

The solid central line in Figure 22.13 shows the normal oxygen dissociation curve for Hb in adults. Cooperative binding means that the slope of the curve is relatively shallow at very low Po_2. As each molecule of oxygen binds, however, the curve steepens because the Hb relaxes further, facilitating the binding of subsequent molecules until all four binding sites are occupied and the Hb is saturated.

At 13.3 kPa, the partial pressure of oxygen in the lungs and arteries, the Hb is effectively 100% saturated. In venous blood under normal conditions, the Po_2 falls to about 5–6 kPa (Table 22.2) as the Hb has released oxygen to supply the metabolizing tissues. Note that this means the Hb is still >50% saturated with oxygen, so the binding process often only involves the binding and unbinding of one or two of the four oxygen molecules. As a result, the binding process is very rapid, taking place on the steep part of the dissociation curve.

As a note of caution, standard phrasing can be misleading: when we talk about 'deoxygenated blood' returning to the heart in the veins, it does not actually mean the blood has no oxygen at all. In fact, the blood returning to the right atrium to be pumped to the lungs will typically still be more than 50% saturated with oxygen.

PAUSE AND THINK

Figure 22.13 shows that the dissociation curve for fetal Hb (the purple line) lies to the left of the adult curve. What is the effect of this?

Answer: As the curve for fetal Hb lies to the left of that for adult Hb, it means the fetal Hb shows a higher affinity for oxygen: the P_{50} for adult Hb is normally 50% saturated (P_{50}) at 3.5 kPa, whereas the P_{50} for fetal Hb is 2.5 kPa. Therefore, the fetal Hb is effective at taking up oxygen from the maternal blood as the two circulations pass through the placenta.

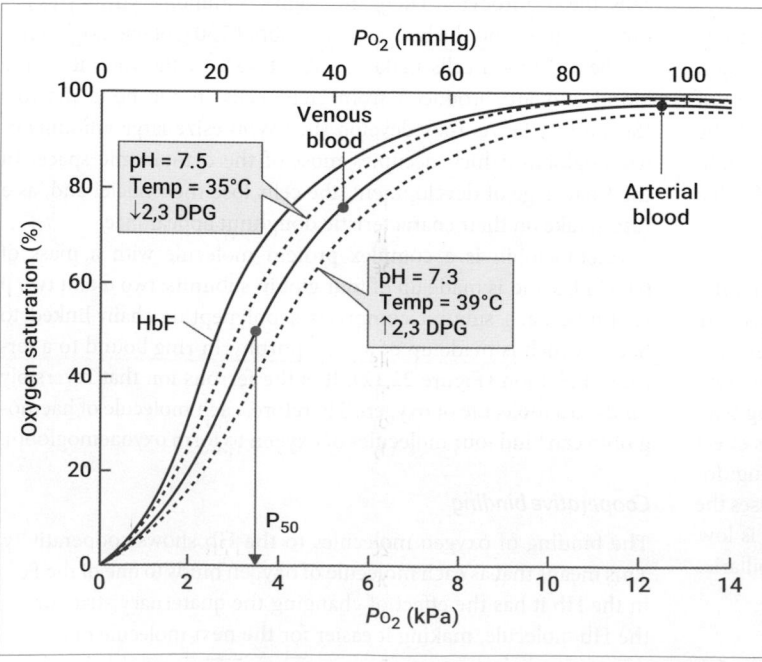

Figure 22.13 Oxygen dissociation curve. The graph shows the percentage saturation of haemoglobin with oxygen against the partial pressure of oxygen. The solid central line shows the normal curve for adult Hb. The solid line to the left shows the curve for fetal Hb (HbF).

The normal curve can shift to the left or right (the Bohr shift) depending on the pH and temperature.

The P_{50}, the partial pressure of oxygen at which the Hb is 50% saturated, is a standard measure of the affinity of oxygen binding to Hb.

Source: Thomas, C & Lumb, A (2012) Physiology of haemoglobin. Continuing Education in Anaesthesia, Critical Care & Pain, 12 (5). With permission from Elsevier.

Figure 22.13 also shows two dotted lines to the left and right of the normal curve. The affinity of Hb is affected by a range of factors, all of which cause very slight conformational changes to the quaternary structure of the Hb molecules, thereby affecting its binding.

Increases in temperature, acidity, and carbon dioxide concentration all cause the curve to move slightly to the right, the so-called **Bohr effect**.

Where are these conditions likely to be found? Think carefully about when there might be a local rise in temperature. Increases in temperature, acidity, and carbon dioxide concentration are all the results of metabolic activity; therefore, these conditions are most likely to be found in metabolizing tissues.

The Bohr effect gives rise to a reduced affinity of Hb for oxygen and so the Hb releases its oxygen more easily as the blood passes through areas of metabolic activity. Note that, in areas of high pH and lower [CO_2] the opposite is true and the affinity is higher, thereby facilitating binding: these are typical of the conditions found in the lungs. So, the Hb releases its oxygen more easily in the metabolizing tissues and binds it more rapidly in the lungs.

The dissociation curve is also affected by the concentration of 2, 3-diphosphoglycerate (2,3-DPG). This is a by-product of glycolysis (anaerobic metabolism), which also shifts the curve to the left (Figure 22.13); therefore, in conditions where the oxygen levels are very low, 2,3-DPG will be produced, leading to increased release of oxygen from the Hb.

We have seen that fetal Hb has different binding properties to adult Hb—are there any other oxygen-binding molecules? You could look back at Section 21.1 to answer this question. Myoglobin is found in skeletal muscle: this molecule has only one haem group, and so can bind only one oxygen molecule. Myoglobin has a very high affinity for oxygen and so acts as an effective oxygen store in muscles, releasing the oxygen when the muscles are actively contracting and the blood supply is insufficient to maintain an adequate oxygen supply.

As we have described, the continuous circulation of the red blood cells, which are packed with haemoglobin, is essential for supplying the oxygen requirements of the tissues. Insufficient levels of haemoglobin will lead to anaemia, which is a major health problem (see Clinical Box 22.2).

CLINICAL BOX 22.2   Anaemia

Anaemia is a condition in which there is insufficient functional haemoglobin in the blood to carry oxygen effectively. This may be caused by a number of factors, the most common being:

- blood loss;
- insufficient erythrocyte production;
- abnormalities of erythrocyte production;
- excessive breakdown of erythrocytes.

Anaemia represents a major health problem: in its 2015 report, the World Health Organization estimated that about 25% of the world's population suffers from anaemia, rising to over 40% of preschool-age children. In children, anaemia can have a significant impact on their development. About half of all the cases of anaemia are due to iron deficiency, which compromises erythrocyte production, and which is normally due either to insufficient absorption from the diet, most commonly due to poor nutrition, or excessive iron loss. In the UK about a quarter of teenage girls have low iron stores, with about 8% having iron-deficient anaemia, which is associated with insufficient dietary intake along with loss of iron as a result of menstrual bleeding.

The symptoms of anaemia vary but commonly include:

- fatigue;
- shortness of breath, especially during exercise;
- headaches;
- tachycardia (elevated heart rate);
- dizziness;
- pallor.

Globally, two other main causes of anaemia are:

- blood loss due to intestinal parasites (e.g. hookworm infection);
- genetic disorders of red cell production (e.g. sickle cell anaemia, β-thalassaemia).

β-thalassaemia is a genetic disorder in which the production of the β chain of haemoglobin is reduced such that the oxygen-carrying capacity of the blood is affected. The disease normally presents in early childhood and adversely affects development, leading to death in some cases if untreated. There is no cure, but the symptoms can be relieved by blood transfusions.

Sickle cell disease is so-called because the red blood cells are deformed and inflexible due to clumping of the haemoglobin, which also results in reduced oxygen carriage. Due to the inflexibility of the erythrocytes, there may be blockage of capillaries leading to ischaemia in the affected tissues. These so-called vaso-occlusive crises can be very painful and lead to organ damage.

Pernicious anaemia is an autoimmune disorder that results in reduced absorption of vitamin B12. Vitamin B12 is an essential component of the diet as it cannot be manufactured by the body and is necessary for the absorption of iron. For vitamin B12 to be absorbed in the ileum, it has to be bound to a protein called **intrinsic factor**, which is produced by the parietal cells of the stomach. In pernicious anaemia, antibodies are produced which damage the parietal cells leading to lack of intrinsic factor and so the body does not absorb iron from the diet. Patients can be treated by regular injections of vitamin B12.

Find out more

World Health Organization (2015) The Global Prevalence of Anaemia *in 2011* http://apps.who.int/iris/bitstream/10665/177094/1/9789241564960_eng.pdf

Carbon dioxide transport

Carbon dioxide is one of the main waste products of cellular respiration (Chapter 7). It is physiologically very important because the concentrations of CO_2 in the blood contribute significantly to the pH of the blood. The pH is tightly regulated because changes to pH have significant impact on physiological and metabolic processes, for example enzyme and neural functions (Chapter 16). As carbon dioxide is produced by the cells, it diffuses out into the capillaries and is transported via the veins to the lungs. Within the blood it is transported in three forms:

1. Dissolved in the plasma (~10%).
2. As carbonic acid/bicarbonate ions: $H_2CO_3/HCO_3^- + H^+$ (~ 60%).
3. Bound to haemoglobin, as carbaminohaemoglobin: $HbCO_2$ (~30%).

Bicarbonate and carbaminohaemoglobin systems therefore represent 90% of the CO_2 in the blood. These two systems operate as buffers: see Quantitative Toolkit 22.2.

In the case of the bicarbonate system, the CO_2 associates with water:

$$CO_2 + H_2O \rightleftharpoons H_2CO_3 \rightleftharpoons HCO_3^- + H^+$$

This reaction occurs relatively slowly in the plasma but is significantly faster in the erythrocytes due to the presence of the enzyme carbonic anhydrase (Figure 22.14). Within the erythrocyte, the H^+ concentration would build up because it is being produced rapidly by the action of carbonic anhydrase and it cannot diffuse out of the cell. However, H^+ can bind to the haemoglobin and this process is facilitated when Hb loses its oxygen. Therefore, as metabolic activity increases, the production of CO_2 increases and more O_2 is released from the Hb which makes it more effective at binding the H^+ ions reducing the fall in plasma pH.

QUANTITATIVE TOOLKIT 22.2   Acids, bases, buffers, and pH

pH literally means 'power of hydrogen' and is a measure of the hydrogen ion concentration.

$$pH = -\log\left[H^+\right]$$

The pH scale runs from 1 to 14, so pH 7 is neutral. A base has a pH >7, while an acid has a pH <7. As it is measured by a logarithmic scale, each pH unit represents a 10-fold change in H^+ concentration, so pH 5 is 10 times more acidic than pH 6. The greater the concentration of H^+ ions, $[H^+]$, the more acidic is the solution.

$$pH\,1 = 0.1\,M\,\text{hydrogen ion concentration}$$

$$pH\,7 = 0.0000001\,M\left(\text{more easily expressed as } 1 \times 10^{-7}\,M\right)$$

The pH of the blood is normally maintained at a slightly alkaline pH between 7.36 and 7.44. This is tightly regulated because changes in pH affect many metabolic processes.

Water dissociates into H^+ and OH^- ions:

$$H_2O \rightleftharpoons H^+ + OH^-$$

An acid is a substance that increases the $[H^+]$, whereas a base will increase $[OH^-]$. The stronger the acid or base, the more likely it is to dissociate to release H^+ or OH^- ions, respectively. Thus, hydrochloric acid (HCl) is a strong acid that dissociates easily in water to release H^+ ions, and sodium hydroxide (NaOH), 'caustic soda', is a strong base and releases OH^- ions. Substances that do not dissociate fully are referred to as weak acids or bases; for example, acetic acid (CH_3COOH—vinegar) is a weak acid.

A buffer is an aqueous solution that contains a weak acid and its conjugate base, or a weak base and its conjugate acid; it acts to resist changes to pH.

The carbonate–carbonic acid buffer is coupled to both the kidney and the respiratory system. Carbonic acid (H_2CO_3) is a weak acid and, in the blood, is in equilibrium with its conjugate base, the bicarbonate ion (HCO_3^-) and the hydrogen ion (H^+):

$$H_2CO_3 \rightleftharpoons HCO_3^- + H^+$$

When metabolic activity produces CO_2 it combines with water, leading to an increase in the $[H^+]$:

$$CO_2 + H_2O \rightleftharpoons H_2CO_3 \rightleftharpoons HCO_3^- + H^+$$

In the plasma $[H_2CO_3]$ is very low and $[HCO_3^-]$ is relatively high so when hydrogen ions are produced, the increase in acidity is buffered by the HCO_3^- forming carbonic acid; the change to pH is therefore resisted. The $[CO_2]$ and $[HCO_3^-]$ are tightly regulated by the respiratory system and the kidneys, respectively, thereby maintaining the pH homeostasis.

The pH of the blood with respect to the bicarbonate buffer system can be calculated using the **Henderson–Hasselbalch equation**.

If we consider the dissociation of carbonic acid, we can calculate the dissociation constant (K) for the reaction based on the relative concentrations of the acid and its conjugate base:

$$K = \frac{\left[H^+\right] \times \left[HCO_3^-\right]}{\left[H_2CO_3\right]}$$

This can be re-arranged as follows:

$$\left[H^+\right] = K \times \frac{\left[H_2CO_3\right]}{\left[HCO_3^-\right]}$$

We saw above that pH = −log[H$^+$]. Therefore we convert the equation into a log equation:

$$\log\left[H^+\right] = \log K + \log\left(\frac{\left[H_2CO_3\right]}{\left[HCO_3^-\right]}\right)$$

And then multiply both sides of the equation by −1; this gives us:

$$-\log\left[H^+\right] = -\log K - \log\left(\frac{\left[H_2CO_3\right]}{\left[HCO_3^-\right]}\right)$$

Substituting for pH:

$$pH = -\log K - \log\left(\frac{\left[H_2CO_3\right]}{\left[HCO_3^-\right]}\right)$$

Similarly, we can define the term −log K as pK_a

$$pH = pK_a - \log\left(\frac{\left[H_2CO_3\right]}{\left[HCO_3^-\right]}\right)$$

This is the formulation for the Henderson–Hasselbalch equation.

What is the significance of pK_a, the dissociation constant?

The pK_a is the pH at which the concentration of the acid equals that of the conjugate base, in other words:

$$[H_2CO_3] = [HCO_3^-]$$

A buffer is most effective when operating at pH values close to the pK_a. In the case of the bicarbonate system, the pK_a = 6.1, whereas blood pH is regulated between 7.36 and 7.44; therefore, the bicarbonate buffer is operating at pH values that are not very close to its pK_a.

How does this relate to the production of CO_2 during metabolic activity?

As we saw above, CO_2 combines with water to produce

$$CO_2 + H_2O \rightleftharpoons H_2CO_3 \rightleftharpoons HCO_3^- + H^+$$

Maintaining a pH of 7.4 and with a pK_a of 6.1, the ratio of [HCO$_3^-$] : [CO$_2$] is maintained at 20:1. If [CO$_2$] rises, and therefore pH falls, then lung ventilation increases to restore the balance as described in the next section.

What happens if the [HCO$_3^-$] falls?

[HCO$_3^-$] is also regulated by the kidneys which would respond by reabsorbing more HCO$_3^-$ (Chapter 24).

Figure 22.14 Carbon dioxide transport in the tissues and lungs.

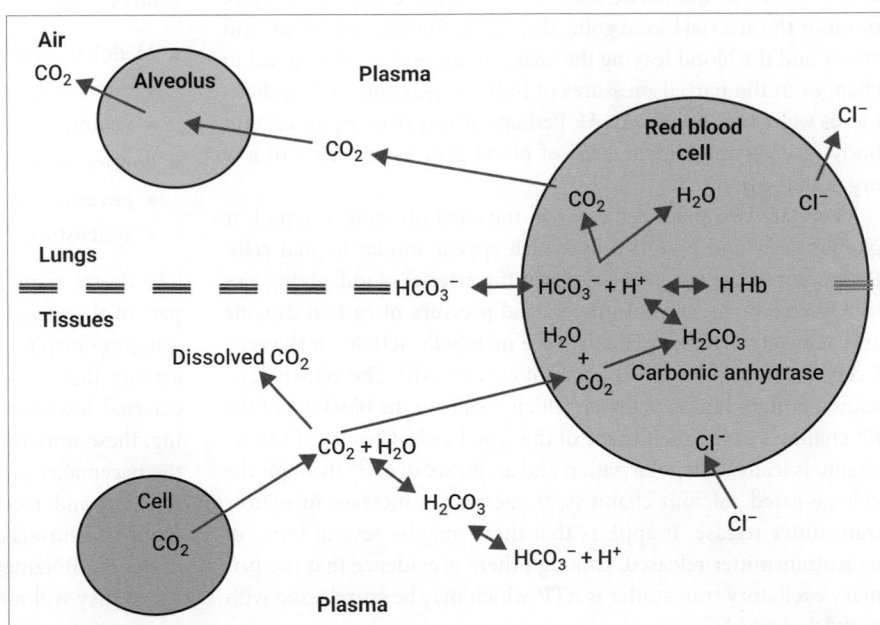

Note that as the negatively charged HCO$_3^-$ ions diffuse out of the erythrocyte down its concentration gradient (Figure 22.14). This would potentially create an electrical imbalance, with the inside of the erythrocyte losing negative charge. This is balanced by the 'chloride shift' whereby negatively charged chloride ions enter the erythrocyte to maintain the membrane potential.

When the blood reaches the lungs, the process operates in reverse: as CO_2 is lost from the plasma as a result of gas exchange,

so the buffer reactions operate in the opposite direction (Figure 22.14). Note also that as the Hb picks up oxygen, its affinity for the H$^+$ ions decreases, thereby facilitating the release of H$^+$ and the reformation of CO_2 which is blown off as we breathe out.

 Check your understanding of the concepts covered in this section by answering the questions in the e-book.

22.4 Control of ventilation

As we know, breathing is essential to life: prevention of lung ventilation for more than a few minutes leads rapidly to brain damage and death. The control of ventilation is tightly regulated by control centres in the brainstem. Notwithstanding this, we can exert some voluntary control over our breathing for short periods, for example in the way we use air flow to control speech, or voluntarily holding our breath. However, experience tells us that we can only hold our breath for a few tens of seconds before the urge to breathe becomes overwhelming.

Feedback systems, informed by chemoreceptors in the body, monitor the oxygen and carbon dioxide levels and control the rate and depth of ventilation on a minute-by-minute basis. There are also feedback mechanisms associated with the sensory feedback from muscles and joints related to exercise and also stretch receptors in the lungs themselves.

Chemoreceptors

There are two groups of chemoreceptors that monitor the oxygen and carbon dioxide levels in the body: the peripheral and central chemoreceptors.

The peripheral chemoreceptors are located in the carotid bodies of the carotid artery, which are the main chemoreceptors, and the aortic bodies of the aortic arch. These chemoreceptors therefore monitor the arterial blood going directly to the brain in the carotid artery and the blood leaving the heart via the aorta and respond to changes in the partial pressures of both oxygen and carbon dioxide, as well as to changes in pH. Perhaps unsurprisingly, the carotid body receives the highest level of blood flow for its mass of any organ in the body.

There are two main cell types in the chemoreceptors: type I or glomus cells, and type II cells, which appear similar to glial cells. Reduction in the levels of oxygen in the arterial blood, giving rise to hypoxia, or increases in the partial pressure of carbon dioxide and associated falls in pH, affect the metabolic activity in the type I cells leading to the production of cyclic GMP. The cGMP activates a protein kinase pathway, which results in the blocking of the K^+ channels in the membrane of the type I cells. Closure of the K^+ channels leads to depolarization and an influx of Ca^{2+} through the voltage-gated calcium channels, triggering an increase in neurotransmitter release. It appears that there may be several types of neurotransmitter released, although there is evidence that the primary excitatory transmitter is ATP, which may be co-released with acetylcholine (ACh).

The neurotransmitters released from the peripheral chemoreceptors lead to an increase in discharge in the afferent neurons of the autonomic pathways. The carotid bodies are connected via the glossopharyngeal nerve (9th cranial nerve) and the aortic via the vagus nerve (10th cranial nerve) to the medulla oblongata of the brainstem where they feed into the respiratory centres.

The central chemoreceptors are located close to the ventral surface of the medulla, lying very close to the brainstem respiratory centres. These receptors respond specifically to changes in the pH of the extracellular fluid, the **cerebrospinal fluid** (CSF). The **blood–brain barrier** is relatively impermeable to H^+ ions, so increases in blood [H^+] do not lead directly to a fall in the pH of the CSF. However, the blood–brain barrier is permeable to CO_2; therefore, rises in blood [CO_2] (hypercapnia), rapidly lead to an increase in CSF [CO_2] and a consequent fall in pH. The central chemoreceptors respond to the **hypercapnia** with an increase in ATP production, which, in turn, triggers the release of ACh.

The net action of activation of both the peripheral and the central chemoreceptors is to stimulate the respiratory centres, leading to an increase in rate and depth of lung ventilation.

Respiratory centres

During automatic respiration, lung ventilation is controlled by the rhythm of the respiratory centres in the brainstem. This is modulated by feedback from the chemoreceptors and also stretch receptors in the walls of the lungs. As we saw in Section 22.2, at rest inspiration is active, and mainly involves the contraction of the diaphragm, while expiration is passive. However, these processes can be extended to include the intercostal muscles and the abdominal wall during different activities when there is more expansive inspiration and also active expiration.

The ventilatory rhythms are controlled by two main respiratory centres (Figure 22.15), located in the medulla and pons:

- Medullary respiratory centre (primary respiratory centre):
  - dorsal respiratory group (DRG)
  - ventral respiratory group (VRG)
- Pontine respiratory centre:
  - pneumotaxic centre
  - apneustic centre.

The dorsal respiratory group of neurons is located in the dorsal part of the medulla (Figure 22.15) and is mainly concerned with initiating inspiration. These neurons show a rhythmic pattern of activity that activates the motor neurons of the diaphragm and external intercostal muscles. When the DRG neurons cease firing, these muscles relax and expiration takes place. The origin of the pacemaker activity of the DRG has been subject to significant research and recent models indicate that there is a specialized group of neurons called the **pre-Bötzinger complex**. The neurons of the pre-Bötzinger complex show spontaneous pacemaker activity, so they will maintain their pattern of discharge in the absence of any other inputs, and it is this firing pattern that appears to drive the DRG neurons, establishing the basic pattern of ventilation.

The ventral respiratory group of neurons is a mixed population, containing both inspiratory and expiratory neurons. This group of neurons is normally silent during quiet breathing, only becoming active when there is a demand for increased ventilation, involving deeper inspiration and forced expiration, for example during exercise.

The pontine groups of respiratory control neurons act to modulate the medullary neurons. These groups receive afferent inputs

Figure 22.15 Respiratory control centres in the brainstem.

Source: Openstax CC BY 4.0 C J. Gordon Betts et al. *Anatomy and Physiology*, 2013. Access for free at: https://openstax.org/books/anatomy-and-physiology/pages/1-introduction Section URL: https://openstax.org/books/anatomy-and-physiology/pages/22-2-the-lungs.

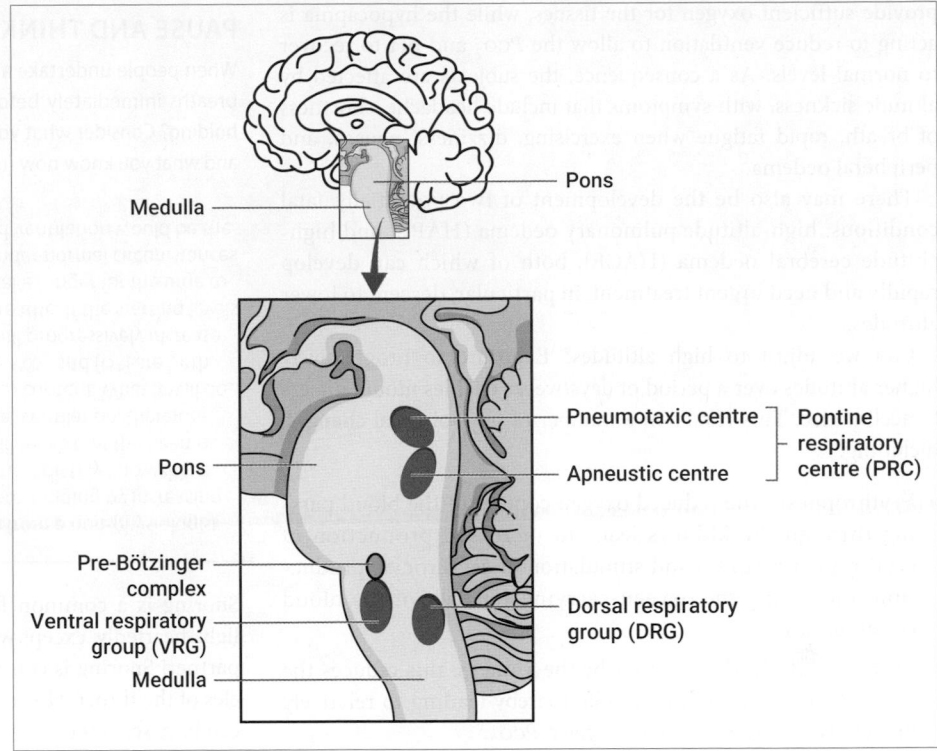

from both the chemoreceptors and the stretch receptors and are involved in modulating the ventilator pattern in response to afferent feedback. Activity in the apneustic centre prolongs inspiration, so that the breath is deeper, whereas activity in the pneumotaxic centre inhibits the apneustic centre and leads to termination of inspiration. One set of inputs to the pneumotaxic centre comes from the stretch receptors in the lung which trigger inhibition of inspiration when the lungs reach large volumes; this is termed the **Hering–Breuer reflex**.

Feedback modulation of respiration

Ventilation of the lungs is established as a rhythmic pattern, driven by the medullary respiratory centres. In the absence of any descending control or peripheral feedback this will maintain a basic respiratory rhythm. Under normal circumstances, though, respiration is tightly regulated and responds to changes in the physiological state, in particular the composition of the arterial blood and the CSF, as monitored by the chemoreceptors.

Increases in the partial pressure of CO_2 (hypercapnia) and in $[H^+]$ rapidly stimulate increases in the depth and rate of ventilation and also increase the cardiac output, as a result of chemoreceptor feedback particularly from the central chemoreceptors.

Likewise, decreases in the partial pressure of O_2 (**hypoxaemia**), detected by the peripheral chemoreceptors, will lead to increased ventilatory drive and increased cardiac output.

Both these responses, therefore, are negative feedback actions that will lead to an increase in the partial pressure of O_2 and a

decrease in the partial pressure of CO_2 in the blood, along with a decrease in $[H^+]$ to restore them to their set levels as a result of the increase in the rate and depth of ventilation and in the circulation of blood through the lungs and the tissues.

During most normal activities, these processes operate together; for example, during exercise there is typically an increase in oxygen demand by the tissues, an increase in CO_2 production and a decrease in pH, all of which drive increases in both ventilation and cardiac output. However, at high altitudes these control systems operate in opposition.

Altitude acclimatization

At high altitudes the proportion of oxygen in the air is still 20.9% (Section 22.3); however, the atmospheric pressure, and therefore the partial pressure of O_2, are reduced: at the top of Mount Everest the P_{O_2} is approximately one-third of that at sea level. On exposure to high altitude, therefore, we suffer from hypoxaemia, which results in an increase in ventilatory drive to increase the provision of O_2 to the blood. The effects of altitude typically become apparent at heights above 2500–3000 m.

If we increase the rate and depth of ventilation what will happen to the gas composition of the blood? Take a moment to think about this question. The oxygen levels will start to rise, but the increased ventilation will also result in a decrease in P_{CO_2} (hypocapnia) and a rise in pH, a condition referred to as **respiratory alkalosis**.

The two control systems therefore appear to be in opposition to each other: the hypoxaemia is driving an increase in respiration to provide sufficient oxygen for the tissues, while the hypocapnia is acting to reduce ventilation to allow the P_{CO_2} and pH to recover to normal levels. As a consequence, the subjects are affected by altitude sickness, with symptoms that include headache, shortness of breath, rapid fatigue when exercising, dizziness, nausea, and peripheral oedema.

There may also be the development of two potentially fatal conditions: high-altitude pulmonary oedema (HAPE) and high-altitude cerebral oedema (HACE), both of which can develop rapidly and need urgent treatment, in particular, descent to lower altitudes.

Can we adjust to high altitudes? Exposure to progressively higher altitudes over a period of days/weeks enables mountaineers to acclimatize. This results in a number of physiological changes, including:

- Erythropoesis: the reduced oxygen content of the blood passing through the kidneys leads to increased production of erythropoietin (EPO) and stimulation of erythrocyte production, increasing the oxygen-carrying capacity of the blood (Chapter 24).

- Excretion of bicarbonate ions by the kidneys: this reduces the buffering capacity of the plasma, thereby leading to relatively higher [H$^+$] concentration for a given P_{CO_2}.

- The peripheral CO_2 chemoreceptors will also reset around the lower P_{CO_2}.

Snoring is a common feature of sleeping and is usually viewed light-heartedly, except when your sleep is regularly disturbed by a partner! Snoring is commonly the result of relaxation of the muscles of the throat. However, in some cases such as sleep apnea, this can have serious clinical consequences (see Clinical Box 22.3).

 Check your understanding of the concepts covered in this section by answering the questions in the e-book.

CLINICAL BOX 22.3   Sleep apnea

Sleep apnea is a chronic condition that is characterized by repeated pauses in breathing during sleep. These pauses may occur up to 30 times an hour and are most commonly caused by collapse of the airways during sleep, as a result of the relaxation of the pharyngeal muscles—so-called obstructive sleep apnea (OSA). These pauses in breathing lead to partial or complete arousal from sleep, usually accompanied by a snorting sound. As a result, those people affected feel excessively tired and sleepy during the day, which can be dangerous, for example when driving.

About 40% of the population of the UK snore and it is estimated that some 2.5 million people suffer from OSA, with 80% of those being undiagnosed. OSA is associated with being overweight and untreated sufferers are also at risk of developing high blood pressure. The combination of snoring (typically reported by a partner who is also deprived of sleep) and excessive tiredness are indicative of OSA and sufferers are encouraged

to keep a sleep diary to record their sleep patterns at night, napping during the day, and how sleepy they feel at different times during the day.

Clinical diagnosis will often involve sleep studies to monitor sleeping patterns during the night. Detailed sleep study, **polysomnography**, can involve recording of the sleep patterns and breathing along with blood oxygen levels, heart rate, eye movements, and electroencephalogram (EEG). The EEG and eye movements allow determination of the depth and stage of sleep.

Depending on the severity of the OSA, treatments can take the form of lifestyle changes, such as losing weight, avoiding alcohol, and sleeping on one side rather on the back. Mouthpieces may also be used to help keep the airways open by adjusting the position of the lower jaw and tongue. Moderate to severe OSA can be treated using continuous positive airway pressure (CPAP) devices that fit over the mouth and nose and maintain a flow of air into the throat to help keep the airways open.

SUMMARY OF KEY CONCEPTS

- Ventilation of the lungs is an active process mainly driven by the contraction and relaxation of the diaphragm aided by the intercostal muscles.

- Lung ventilation can be compromised by obstructive deficits that affect airflow (e.g. asthma) and restrictive deficits that affect lung volumes (e.g. cancers).

- Gas exchange across the lung–capillary and the capillary–cell interfaces depends on diffusion.
- Oxygen is transported in the blood by the oxygen transporter haemoglobin which is found in the erythrocytes.
- Carbon dioxide concentrations in the blood play a key role in the regulation of blood pH.

- Peripheral and central chemoreceptors monitor the partial pressures of the respiratory gases and the pH of the blood.
- Respiratory centres in the medulla oblongata respond to feedback from the chemoreceptors and from stretch receptors in the lungs by modifying the rate and depth of ventilation.

 Use the flashcards in the e-book to test your recall of key terms introduced in this chapter.

QUESTIONS

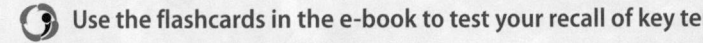 Looking for answers? Once you've answered these questions, follow the link in the e-book to the answer guidance and check your work.

Concepts and definitions

1. In which parts of the respiratory airways does gas exchange take place?

2. Which muscles are involved in ventilating the lungs?

3. What is meant by the term lung compliance and what are the two main factors that affect it?

4. What are the main forms in which carbon dioxide is transported in the blood?

5. Where are the chemoreceptors located and what do they monitor?

Apply the concepts

6. Draw the haemoglobin dissociation curve. What is the significance of the shifts of the curve in response to changes in pH and temperature?

7. Explain the mechanisms involved in increasing the ventilation of the lungs.

8. Why are very premature babies at risk of neonatal respiratory distress syndrome (NRDS)?

9. Explain the significance of measuring forced vital capacity (FVC) and forced expiratory volume (FEV).

10. What is a buffer and what are the main buffering systems in the blood?

Beyond the concepts

11. Compare the relative impacts of short- and long-term changes in the partial pressures in the blood of oxygen and carbon dioxide on lung ventilation.

12. Discuss the role of haemoglobin in gas transport. Describe a clinical condition in which the transport of oxygen is compromised.

13. How is lung ventilation controlled?

14. Discuss the physiological significance of carbon dioxide.

15. Describe the main physiological methods of measuring lung function and give examples of how the values may be affected by different lung diseases.

FURTHER READING

Crystal, G. J. & Pagel, P. S. (2020) The physiology of oxygen transport by the cardiovascular system: evolution of knowledge. *J. Cardiothorac Vasc. Anaesth.* **34**: 1142–51. https://doi.org/10.1053/j.jvca.2019.12.029
A review of the development of our understanding of blood gas transport.

Gaseous exchange. https://med.libretexts.org/Bookshelves/Veterinary_Medicine/Book%3A_Introductory_Animal_Physiology_(Hinic-Frlog)/2%3A_Gaseous_Exchange
This provides an online, readable resource covering all aspects of respiratory physiology. It does include consideration of other animals but is mainly mammalian and explains concepts very clearly.

Habibzadeh, F., Yadollahie, M., & Habibzadeh, P. (2021) *Pathophysiologic Basis of Acid–Base Disorders.* Springer.
The introductory chapters provide very clear accounts of the gas laws and acid–base balance. There are also good descriptions of acidosis and alkalosis.

Koeppen, B. & Stanton, B. (2017) *Berne & Levy Physiology.* 7th edition. Elsevier.
This is a detailed physiology reference book that provides very good coverage of the respiratory system in Section 5, Chapters 20–25.

Lumb, A. B. & Thomas, C. R. (2020) *Nunn and Lumb's Applied Respiratory Physiology.* 9th edition. Elsevier.
Provides a good overview of respiratory physiology in health and disease with extensive additional material in the eBook.

Ward, J. P. T., Ward, J., & Leach, R. M. (2015) *The Respiratory System at a Glance.* 4th edition. Wiley-Blackwell.
This is a very good, readable account that provides greater depth on respiratory physiology and clinical conditions.

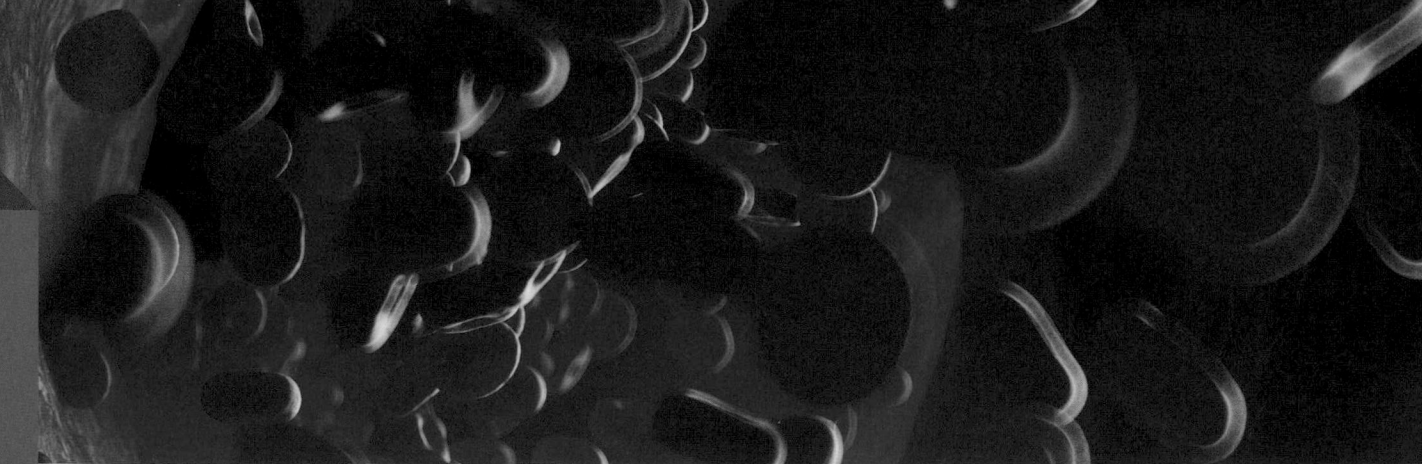

Exercise Physiology

Chapter contents

Introduction   768

23.1   Cardiovascular responses to exercise   769

23.2   Maximum oxygen uptake   775

23.3   Pulmonary responses to exercise   777

23.4   Exercise and acid–base balance   779

23.5   Exercise and environmental stressors   780

23.6   Physical activity, fitness, and health   782

Watch the key concepts video in the e-book to prepare yourself for studying this chapter.

LEARNING OBJECTIVES

By the end of this chapter you should be able to:

- Describe the effects of acute exercise on the cardiovascular and respiratory systems.

- Describe the effects of aerobic exercise training on cardiovascular and respiratory variables.

- Explain the mechanisms behind these observations.

- Explain the importance of maximum oxygen uptake (VO_{2max}).

- Describe and explain the effect of acute exercise on acid–base balance.

- Describe and explain the effects of exercise undertaken in different environmental conditions on physiological function.

Introduction

Physiological systems can adapt to numerous stimuli. Exercise, or physical activity in general, stresses almost every system in the human body and has received substantial attention in terms of scientific study in recent times. Exercise physiology involves studying the effect of exercise, with or without the addition of other stressors, on systems, organs, and tissues. As such, considering the effects of exercise on physiological function is of use for understanding regulation and limits of human function, as well as understanding and explaining the benefits of exercise or physical activity on health outcomes. In addition, understanding the effects of exercise on physiological function can lead to strategies to enhance adaptation to an exercise stimulus and lead to further improvements in function, in order to maximize exercise performance.

Undertaking a single bout of exercise results in significant changes in function of the cardiovascular, respiratory, muscular, gastrointestinal, renal, and nervous systems. Similarly, adaptation within most of these systems is possible following repeated bouts of exercise. The purpose of this chapter is to consider the effects of exercise on the cardiovascular and respiratory systems only to demonstrate how understanding of the function of these systems, covered in earlier chapters, is affected by exercise.

23.1 Cardiovascular responses to exercise

The fundamental role of the cardiovascular system is the provision of adequate oxygen and nutrient supply to individual cells and the maintenance of an appropriate environment in which to function. Exercise results in greater metabolic demand by muscular tissue and, as such, the cardiovascular system must respond to this increased demand. When considering the effects of exercise on cardiovascular function, it is important to consider the acute effects of both resistance (e.g. strength training) and aerobic (e.g. continuous) exercise, as well as the chronic adaptations that occur with repeated bouts of exercise.

Blood pressure

As discussed in Chapter 21, blood pressure within the cardiovascular system is important as this determines the direction and rate of blood flow throughout a closed circuit. **Systolic pressure** is the highest pressure observed within an artery during a **cardiac cycle** and occurs during ventricular contraction. **Diastolic pressure** is the lowest pressure observed within the artery and occurs during the ventricular relaxation phase of the cardiac cycle. Typical values in a healthy normotensive individual are 120 mmHg for systolic and 80 mmHg for diastolic pressure. **Mean arterial pressure** (MAP) is the average pressure across a cardiac cycle and is approximately 93 mmHg in a healthy normotensive individual as more time is spent in the relaxation phase than the contraction phase of the cycle. MAP is the product of **cardiac output** (CO) and **total peripheral resistance** (TPR), while CO is the product of **heart rate** (HR) and **stroke volume** (SV). Consequently, any intervention that influences HR, SV, or TPR will affect MAP. When considering the effects of exercise on blood pressure, it is important to consider the type of exercise that is being performed.

Blood pressure response to resistance exercise

Resistance exercise, such as weight training in the gym, usually consists of **concentric** (muscle shortening) and **eccentric** (muscle lengthening) activities (Section 20.2, Tendons and muscle contraction). Muscles that are undertaking these activities are supplied by vessels of the cardiovascular system. Particularly during concentric muscular contractions, those vessels lying within the muscles are physically compressed causing a large increase in TPR. This has two major effects:

1. It increases the MAP.
2. It reduces capillary blood flow and therefore **muscle perfusion**.

The second of these is particularly important as the reduction in muscle perfusion is related to the extent of force being produced by the muscle (i.e. the greater the extent of muscle contraction, the greater the reduction in muscle blood flow). This leads to a build-up of metabolites within the muscle tissue and resultant feedback from the muscle to the cardiovascular centre within the medulla of the brainstem, generating an increase in sympathetic nervous activity to restore muscle blood flow. This sympathetic activity leads to an increase in CO, further increasing MAP (Figure 23.1).

The **intensity** of muscle contraction is an important determinant of the extent of increase in MAP that occurs. Resistance exercise undertaken at relatively low intensities leads to relatively small increases in MAP, whereas exercise at high or maximal intensity can lead to very large increases in MAP. Indeed, it is not uncommon for systolic pressure to exceed 200 mmHg and diastolic pressure to exceed 150 mmHg during maximum voluntary contractions. Exercise that includes more of the body's muscle mass also leads to greater increases in MAP (i.e. contraction of a single small muscle will result in smaller increases in MAP than undertaking whole-body resistance exercise as this involves the activation of greater muscle volume).

Blood pressure responses to aerobic exercise

Aerobic exercise, such as running, involves regular cycles of muscular contraction and relaxation, and results in a substantial increase in metabolic demand. To meet this increase in metabolic demand, HR and SV are increased (the mechanisms explaining these observations are described later in this chapter), resulting in

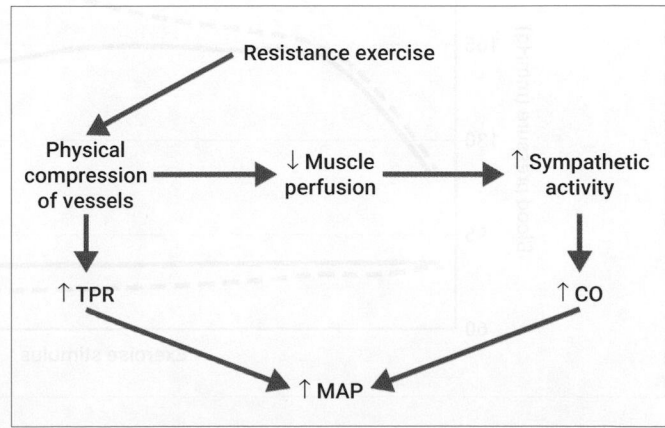

Figure 23.1 The mechanisms behind the effects of resistance exercise on mean arterial pressure.

an increase in CO. Based on the previously described relationship, this will lead to an increase in MAP. It is, however, of interest to consider how different types of aerobic exercise influence systolic and diastolic pressure, as well as MAP.

Continuous steady-state aerobic exercise, such as distance running or cycling, involves an individual working at a specific intensity of exercise for a given time. This is often expressed as a percentage of maximal heart rate or maximal oxygen uptake (VO_{2max}—see Scientific Process 23.1). During this type of exercise, CO increases rapidly at the start of the exercise stimulus, causing a rapid increase in systolic blood pressure, whereas diastolic blood pressure remains relatively unchanged. MAP, therefore, increases but only to approximately 120 mmHg in the early stages of exercise. As exercise continues, the continuous contraction of skeletal muscle leads to **vasodilation** of vessels, thereby increasing the blood flow to the active muscle tissues. As a result, TPR decreases, leading to a reduction in systolic blood pressure and MAP (Figure 23.2).

Continuous graded aerobic exercise is different in that it involves undertaking periods of aerobic exercise at a given intensity and then increasing that intensity at selected intervals, for example the subject may be working on a treadmill, starting off at a set speed on the flat for 3 minutes, then increasing the treadmill gradient (and therefore exercise intensity) by 1% every subsequent 3 minutes up until endurance limits are reached. Consequently, the effects of this on blood pressure are different to those of continuous steady-state aerobic exercise, as shown in Figure 23.2. In the early stages of this type of exercise, blood pressure response is similar to steady state exercise (i.e. a rapid increase in systolic blood pressure and MAP). As exercise intensity increases, systolic blood pressure also increases in a linear fashion to the extent that systolic blood pressure can reach 200 mmHg when undertaking maximal aerobic exercise. MAP therefore increases but not to the same extent, as diastolic pressure remains relatively stable throughout the increases in exercise intensity.

Following completion of both resistance and submaximal aerobic exercise, blood pressure is reduced compared to pre-exercise levels, which is observed for a number of hours after exercise. This observation is most likely the result of blood remaining within peripheral areas of the body and, therefore, reducing venous return and systemic blood pressure.

Cardiac output response to exercise

As described previously, CO is the product of HR and SV, so any intervention that affects either HR or SV will affect CO. CO is approximately 5 L min^{-1} at rest, but, due to the increase in metabolic demand that results from undertaking exercise, CO increases rapidly following the onset of the exercise stimulus. This is met by an increase in both HR and SV.

Intrinsic regulation of cardiac muscle is discussed in more detail in Chapter 21. It involves the conduction system of the heart and is instigated via pacemaker potentials being produced within the sino-atrial node (SAN). The rate at which pacemaker potentials occur within the SAN is influenced by the extent and type of central nervous system (CNS) activity that is present with parasympathetic activity reducing the frequency of pacemaker potentials and sympathetic activity, along with circulating levels of adrenaline, increasing the frequency of pacemaker potentials.

With no CNS input, the heart rate is approximately 100 beats per minute; however, in resting conditions, heart rate is usually between 60 and 80 beats per minute indicating that, at rest, the heart is under **vagal restraint**, through the parasympathetic nervous system. At the onset of exercise, parasympathetic activity is removed from the heart leading to an increase in pacemaker frequency and an increase in heart rate up to 100 beats per minute. If further increases are required, sympathetic activity increases following feedback to the cardiovascular centre in the brain from chemo- and mechanoreceptors in the periphery, which leads to an increased frequency of pacemaker potentials, therefore increasing heart rate. As exercise intensity increases, heart rate increases in a linear fashion (Figure 23.3a).

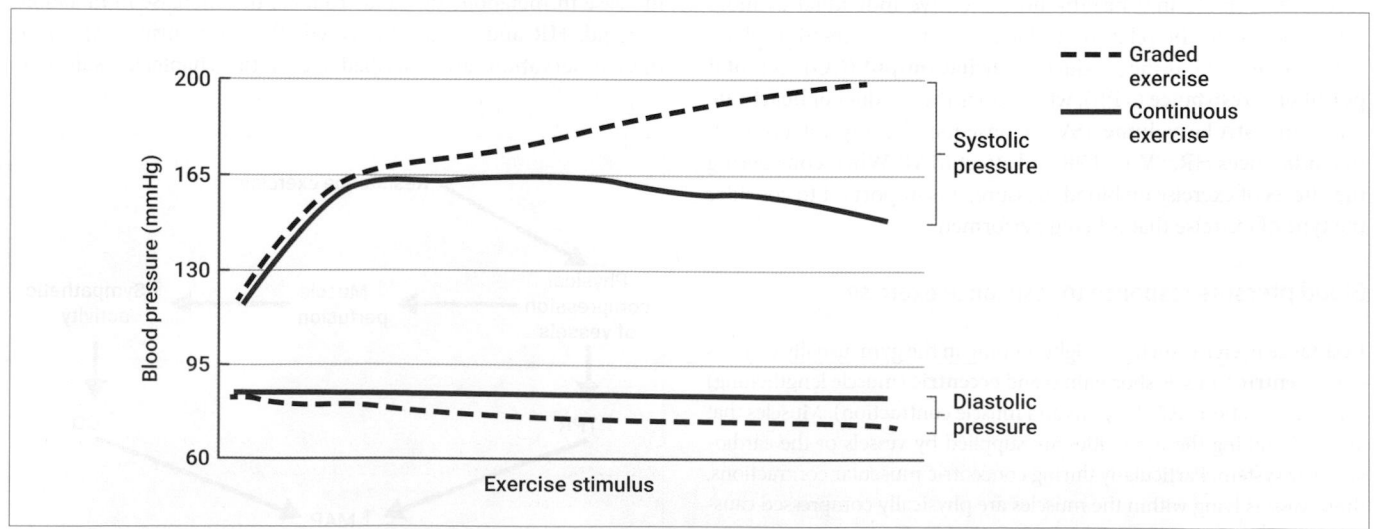

Figure 23.2 Systolic and diastolic blood pressure responses to graded and continuous aerobic exercise. The upper pair of lines shows changes in systolic pressure; the lower pair shows changes in diastolic pressure.

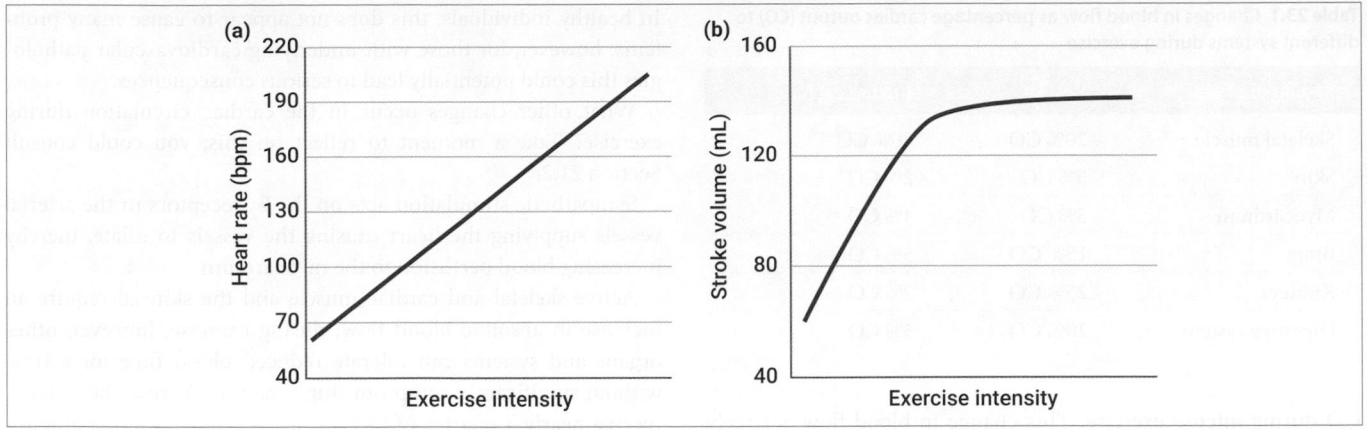

Figure 23.3 Exercise intensity. (a) Heart rate and (b) stroke volume response to increasing exercise intensity.

SV is the volume of blood ejected from the ventricles during one contraction and, as such, depends on the volume of blood within the ventricles before contraction and the volume of blood that can be emptied into the arteries during contraction. Frank–Starling's law (described in detail in Chapter 21), or **preload**, outlines the relationship between the force of contraction produced by the **myocardium** and the initial length of the muscle fibre in that the greater the amount of stretch on cardiac muscle fibres before contraction, the more force is produced. The main determinant of the extent of stretch on cardiac muscle fibres prior to myocardial contraction is the volume of blood within the ventricle, or **end diastolic volume**, which, in turn, is determined by the volume of blood that is returned to the heart in one cardiac cycle. During exercise, the increase in sympathetic nervous activity leads to constriction of veins which leads to increased venous return. This is supported by the rhythmic contraction of skeletal muscle, the 'muscle pump', which forces blood through the venous system and back to the heart. As a result, preload increases during aerobic exercise and this leads to an increase in SV (Figure 23.3b), which contributes to the increase in CO.

Afterload refers to the resistance of blood flow out of the left ventricle due to the pressure within the aorta: an increased systolic pressure reduces the pressure gradients between the left ventricle and aorta, and so should reduce SV. As outlined previously, exercise causes an increase in systolic blood pressure and, therefore, increases afterload, but SV still increases because the increase in preload exceeds the increase in afterload.

In addition to the effect of exercise on venous return and end diastolic volume, catecholamines (described in detail in Chapter 19) are released during exercise as a result of increased sympathetic activity acting directly on the heart and also causing increased secretion of adrenaline from the adrenal medulla. These act on the SAN, increasing heart rate, and on the myocardial tissue, in particular the β_1 adrenergic receptors, causing an increase in the speed and force of ventricular contraction leading to further increases in SV. SV increases until approximately 50% of maximal CO is reached and then plateaus with further increases in CO achieved through increases in HR (Figure 23.3b). A schematic

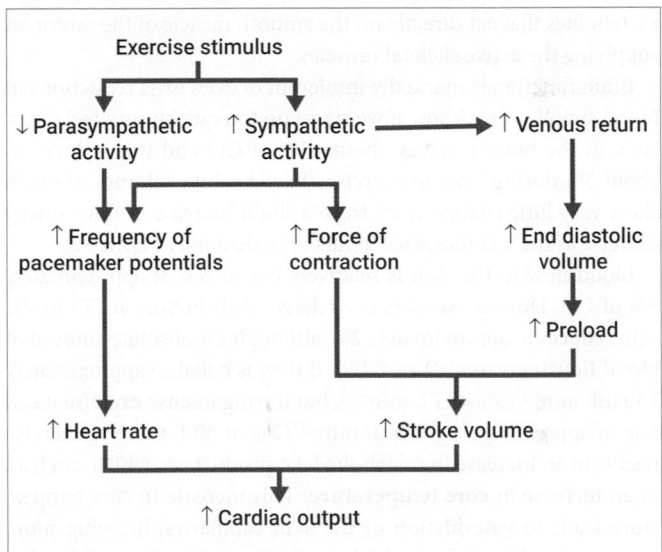

Figure 23.4 The main mechanisms behind increased cardiac output (CO) during endurance exercise.

outlining the mechanisms behind changes in CO during exercise is provided in Figure 23.4.

Effect of exercise on blood flow distribution

CO at rest is approximately 5 L min^{-1} but during maximal exercise this can increase to 30 L min^{-1} and higher in trained athletes. In addition to the large increase in CO that occurs during exercise, there are also changes in the proportion of CO that is distributed to different tissues (summarized in Table 23.1). The changes in distribution of blood flow occur due to **vasoconstriction** and vasodilation of vessels, as a result of central regulation of blood flow, as well as metabolic conditions surrounding active muscle, supplying different tissues.

At rest, approximately 20% of CO is distributed to muscular tissue but, due to the increased activity and metabolic demand produced by exercise, this can increase to approximately 80% of

Table 23.1 Changes in blood flow as percentage cardiac output (CO) to different systems during exercise

System or organ	Rest	Intense exercise
Skeletal muscle	20% CO	80% CO
Skin	5% CO	2% CO
Myocardium	5% CO	4% CO
Brain	15% CO	5% CO
Kidneys	25% CO	2% CO
Digestive system	20% CO	5% CO

CO during intense exercise. This change in blood flow is largely due to peripheral mechanisms leading to vasodilation of arterioles supplying skeletal muscle tissue. The main peripheral mechanism of blood flow during exercise is **active hyperaemia**, an increase in metabolic rate leading to an increase in the concentration of metabolites that act directly on the smooth muscle of the arterioles supplying the active skeletal muscles.

Brain function is markedly intolerant of even brief reductions in blood supply, as we know if we stand up too quickly and feel dizzy. At rest, the brain receives about 15% of CO and this reduces to about 5% during intense exercise. The absolute volumes, though, show very little change apart from a slight increase in flow during exercise as the CO increases to meet the demand of exercise.

Blood flow to the skin is relatively low at rest at approximately 5% of CO. During exercise, the relative distribution of CO to the skin reduces to approximately 2%, although the absolute amount of blood flow increases. At rest, blood flow is equal to approximately 250 mL min^{-1} (5% of 5 L min^{-1}), but during intense exercise it can rise to approximately 600 mL min^{-1} (2% of 30 L min^{-1}). Exercise results in an increase in metabolic heat production, which can lead to an increase in **core temperature**. This increase in core temperature leads to vasodilation of the skin capillaries, bringing more blood to the skin surface, which results in cooling. Heat is also dissipated by **sweating**, with the evaporation of water from the skin surface.

PAUSE AND THINK

Why do marathon runners drink during their races?

Answer: During sweating, water and electrolytes are also lost by evaporation from the surface of the skin, which can lead to dehydration and loss of key ions which, in turn, can affect cardiovascular function and exercise performance.

What about the heart itself? At rest, the myocardium receives approximately 5% of CO and this slightly reduces during intense exercise. The absolute quantities of blood received, though, are substantially increased from approximately 250 mL min^{-1} to 1 L min^{-1} as the metabolic need of cardiac muscle increases with increasing heart rate. An important consideration here is that with an increase in heart rate, the time for perfusion of myocardial tissue to occur is reduced, leading to an increased risk of ischaemia.

In healthy individuals, this does not appear to cause many problems; however, for those with underlying cardiovascular pathologies this could potentially lead to serious consequences.

What other changes occur in the cardiac circulation during exercise? Take a moment to reflect on this; you could consult Section 21.2.

Sympathetic stimulation acts on the β$_2$ receptors in the arterial vessels supplying the heart causing the vessels to dilate, thereby increasing blood perfusion to the myocardium.

Active skeletal and cardiac muscle and the skin all require an increase in absolute blood flow; during exercise, however, other organs and systems can tolerate reduced blood flow for a time without significantly compromising function. At rest, the kidneys receive nearly a quarter of CO but use a relatively small amount of the oxygen that is delivered. During exercise, vasoconstriction of arterioles supplying the kidneys results in significantly reduced blood flow (about 1–2% of CO).

Similarly, the digestive system receives a relatively large proportion of CO at rest and this is significantly reduced during exercise. Undertaking exercise at relatively high intensities seems to impair the ability of the stomach to empty ingested substances into the intestine, which may reduce the ability to absorb water and nutrients. It is thought that this can be explained primarily by a reduction in blood flow to the gastrointestinal system so this change in redistribution of blood flow is not without consequence. This is of particular interest in the area of exercise physiology as strategies to enhance exercise performance often involve ingestion of carbohydrates and water; however, their ingestion has limited impact if they are not absorbed from the intestinal tract.

Circulatory adaptations to aerobic exercise training

Undertaking exercise, or physical activity in general, is often advised as a component of weight loss strategies. This is despite evidence that undertaking only physical activity as a weight-loss strategy results in little reduction in body mass. There are three main reasons for this advice:

1. Evidence suggests that dietary caloric restriction together with an increased energy expenditure through physical activity results in the greatest extent of weight loss.

2. Incorporating physical activity into a weight-loss strategy has been shown to result in less weight regain following a reduction in body mass.

3. Undertaking regular physical activity results in many adaptations that are beneficial for health, regardless of weight loss.

These adaptations occur in most physiological systems and many of them involve the cardiovascular system.

Of the numerous cardiovascular adaptations that occur following a period of aerobic exercise training, one of the most important is the effect on the anatomy and contractility of the heart as this can explain various other adaptations that occur. Following aerobic exercise training, an increase in the volume of the left ventricle is observed compared to the pre-training state. Undertaking exercise sessions increases plasma volume (see below), which, in turn, increases end diastolic volume and stroke volume. Over time, this leads to an increase in left ventricular volume. In addition to the anatomical effects of exercise on the heart, aerobic exercise training also increases sensitivity to Ca^{2+}, a key component in **excitation-contraction coupling**, leading to an increase in contractile force.

Why does exercise lead to significant hypertrophy of the left ventricle but much less so of the right? To answer this question, think about the differences between the systemic and pulmonary circulation (Section 21.2). The changes in demand result in very marked changes in pressure within the systemic circulation, but although volume output through the pulmonary circulation increases in parallel there is very little change in pressure.

At this stage, it is important to consider why exercise-induced cardiac hypertrophy is beneficial for health, while disease-induced cardiac hypertrophy is problematic. First, the increase in heart size that accompanies exercise training is not so large as to be considered dysfunctional. Second, exercise is undertaken for relatively short durations so the stress on cardiac muscle is not as extensive as with chronic conditions such as **hypertension,** which carries additional risk of stroke and other cardiovascular complications. Third, other adaptations occur as a result of exercise, including an increase in cardiac blood flow and a decrease in resting heart rate, which do not occur in disease states. From a mechanistic perspective, exercise-induced hypertrophy is primarily the result of an increased demand for an increase in CO, whereas hypertension-induced hypertrophy is the result of an increase in afterload. The latter observation is important as the increase in contractile force in the hypertensive individual is only maintaining resting circulatory demands rather than enhancing them. Finally, it is important to note that exercise-induced increases in heart size are temporary and return to pre-exercise levels relatively quickly if regular aerobic exercise is not undertaken. This is not necessarily the case in disease states.

Another important consideration here is that these adaptations are specific to aerobic exercise training and not all exercise stimuli, for example resistance exercise training. As discussed, aerobic exercise training leads to an increase in the size of the left ventricular cavity; however, resistance exercise training has little effect on this variable. As discussed, the main effect of an acute bout of resistance exercise on the cardiovascular system is a large increase in systolic and diastolic blood pressure. Consequently, resistance training leads to an increase in myocardial thickness, but little effect on left ventricular cavity volume is observed. The reason for this is probably because aerobic exercise training leads to an increase in **plasma volume**, which is likely to affect left ventricular volume, as well as an increase in cardiac muscle size.

Plasma is the liquid component of the blood and a substantial proportion of the extracellular fluid (Section 21.1). It consists of many components, including proteins, electrolytes, glucose, and clotting factors, but is mainly formed of water. The main protein that is found within plasma is albumin, which has a major influence on **oncotic pressure**. Undertaking regular aerobic exercise leads to an increase in the synthesis of albumin which, in turn, leads to an increase in movement of water into the circulatory system. The end result is an increase in plasma volume which can be substantial and occurs relatively quickly. This has a direct influence

on heart structure as the increase in plasma volume assists with dilation of the left ventricular cavity, thereby contributing to the ventricular hypertrophy.

As discussed briefly earlier in this chapter, the SAN is constantly under vagal restraint leading to a resting heart rate of between 60 and 80 beats per minute. Aerobic exercise training leads to a resting increase in parasympathetic activity and a relatively small decrease in sympathetic activity of the CNS. At rest, this results in a reduced frequency of pacemaker potentials and a reduced resting heart rate. In trained individuals, this can be as low 40 to 50 beats per minute.

PAUSE AND THINK

While resting heart rate decreases following aerobic exercise training, resting CO remains similar or may slightly increase. What other variable contributes to the CO?

Answer: Cardiac output (CO) is the product of heart rate and stroke volume. The increase in resting CO resulting from exercise training is a result of an increase in stroke volume.

There are a number of potential explanations for the increase in resting stroke volume observed:

- the increase in left ventricular volume may be the main reason;
- the reduction in resting heart rate leads to an increased period of time within diastole allowing an increase in **ventricular filling** and, therefore, an increased end diastolic volume;
- improvements in contractility of cardiac muscle, as well as increases in Ca^{2+} sensitivity and reduced stiffness, may play a role in ensuring an increased stroke volume at rest.

The relationship between heart rate and exercise intensity essentially remains the same following aerobic exercise training (i.e. it is a linear relationship). The difference observed is the gradient of that relationship (Figure 23.5a). Maximum exercising heart rate seems to be unaffected by aerobic exercise training and may even be slightly reduced following aerobic training; however, the heart rate required to undertake a given task is lower in trained individuals compared to untrained individuals. Consequently, to meet the cardiac output demands of such an exercise task, stroke volume at a given intensity is higher in trained compared to untrained individuals (Figure 23.5b).

In both aerobically trained and untrained individuals, the onset of exercise leads to an increase in stroke volume. The mechanism for this observation is described in Section 23.1, Cardiac output response to exercise. As aerobic exercise training leads to previously described changes in heart anatomy, contractility, and heart rate, aerobic exercise training leads to different responses in trained versus untrained individuals. Stroke volume at all exercise intensities is greater in trained individuals compared to untrained individuals. In addition, trained individuals exhibit a greater increase in stroke volume in the early stages of exercise compared to untrained individuals. Maximal values for stroke volume occur at approximately 50% of maximal oxygen uptake (**VO_{2max}**) with further

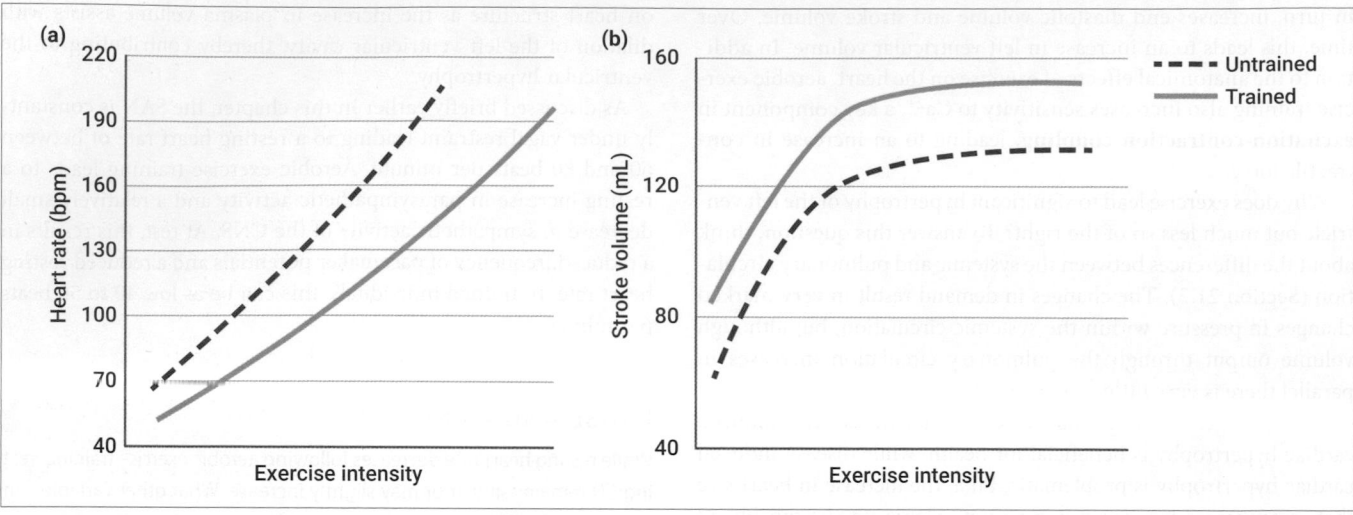

Figure 23.5 Exercise intensity in trained and untrained individuals. (a) Heart rate and (b) stroke volume responses to increasing exercise intensity in trained and untrained individuals.

increases in CO met primarily by an increase in heart rate. VO_{2max} is an important indicator of cardiorespiratory fitness (Section 23.2) and appears to be closely related to the maximal stroke volume that can be achieved during exercise.

Cardiac output is linearly related to oxygen delivery to the periphery and consumption by active tissue. Consequently, a greater CO enables a greater oxygen consumption and aerobic metabolism. Given that aerobic exercise training results in a substantial increase in stroke volume during exercise, maximum CO is also increased after repeated bouts of aerobic activity. Interestingly, CO at submaximal exercise intensities may be slightly lower in trained individuals compared to untrained individuals and this is likely to be due to a combination of alterations in blood flow and ability to utilize oxygen at the active tissue as a result of aerobic exercise training. One of the potential mechanisms for this is an increase in formation of new capillaries that supply skeletal muscle. This is termed **exercise-induced angiogenesis**. One of the key actors in this process is vascular endothelial growth factor (VEGF), which is seen to increase in muscle following exercise. It appears to be stimulated by a metabolic sensor, which is activated during exercise, called PGC1-α (peroxisome-proliferator-activated receptor-γ coactivator-1α).

Following a period of aerobic exercise training, distribution of blood to skeletal muscle during submaximal exercise remains similar to, or slightly lower than, pre-training levels; however, more blood flow is provided to skeletal muscle fibres that have high capacity for oxidative metabolism (type 1 fibres) than those that have relatively low capacity (type 2b fibres). During maximal exercise, skeletal muscle blood flow is increased following a period of aerobic training primarily due to greater distribution of blood flow away from non-active tissue and anatomical changes to the vascular system that favour increased blood flow.

Aerobic exercise training also increases the efficiency of oxygen extraction within active tissue. This is primarily due to changes within skeletal muscle that favour aerobic metabolism and an increased ability to utilize oxygen as a result of exercise-induced

angiogenesis (formation of new blood vessels). Important changes include an increase in mitochondrial size and density, increases in enzymes associated with fatty acid and carbohydrate metabolism, and increased ability to store glycogen within skeletal muscle.

Overall, aerobic exercise training triggers numerous adaptations within the cardiovascular system (summarized in Table 23.2) that result in the ability to increase oxygen delivery to active tissue and this, coupled with adaptations within skeletal muscle, leads to enhanced capacity to extract and use oxygen within the tissue.

 Check your understanding of the concepts covered in this section by answering the questions in the e-book.

Table 23.2 Summary of main cardiovascular adaptations to aerobic exercise training

Site	Adaptation
Heart	Increase in left ventricular mass
	Increase in resting and submaximal exercise stroke volume
	Decrease in resting and submaximal exercise heart rate
	Increase maximal stroke volume and cardiac output
Cardiac muscle	Increase calcium sensitivity
Blood	Increased albumin synthesis
	Increased plasma volume
Skeletal muscle	Increased capillary density
	Increased mitochondrial size and density
	Increased oxygen extraction
	Increased activity of aerobic enzymes

23.2 Maximum oxygen uptake

Maximum oxygen uptake (VO_{2max}) is a measure of the capacity of an individual to transport and utilize oxygen (see Scientific Process 23.1). It is considered to be a reliable indicator of a person's cardiorespiratory fitness and is also considered an **independent risk factor** for numerous disease states. VO_{2max} tends to be determined via graded exercise tests: there are many versions of these tests, but the most common involves an individual exercising for several minutes before intensity is increased. Measures, including heart rate and oxygen uptake (VO_2), are taken before intensity is increased again.

During the exercise test, when the individual is no longer able to continue exercising for the given time or when heart rate is near maximum predicted levels, this point is considered to be VO_{2max}. It is also possible to estimate this value from submaximal exercise heart rate and VO_2 responses. Ultimately, this value demonstrates the highest capacity for that individual to transport and use oxygen within the muscle and is, therefore, determined by the ability of the cardiovascular system to deliver oxygen to the muscle and the ability of the muscle to extract and utilize that oxygen. Consequently, a higher VO_{2max} indicates a greater capacity for transporting oxygen and greater cardiorespiratory fitness.

Endurance exercise training programmes and public health initiatives are often aimed at attempting to increase VO_{2max}.

SCIENTIFIC PROCESS 23.1   Assessing VO_{2max} in humans

Background

Measurement of the maximum oxygen uptake, the VO_{2max}, is widely used as a measure of the upper limit of a person's aerobic exercise levels. As such, it is used to assess the capacity of a broad range of individuals, from elite athletes to people affected by various diseases affecting the cardiovascular–respiratory systems. This measure was described by A.V. Hill and H. Lupton in 1923, based on measurements made on Hill himself when running, as:

> The rate of oxygen intake due to exercise increases as speed increases, reaching a maximum for the speeds beyond about 256 m/min. At this particular speed, for which no further increases in O_2 intake can occur, the heart, lungs, circulation, and the diffusion of oxygen to the active muscle-fibres have attained their maximum activity. At higher speeds the requirement of the body for oxygen is far higher but cannot be satisfied, and the oxygen debt continuously increases

The key criterion, therefore, has been based on the oxygen uptake, the VO_2, reaching a plateau level beyond which further increases in exercise demand do not result in any additional increase in VO_2. Because there are different ways of assessing VO_{2max}, it is not entirely clear whether the outcomes are equivalent. One of the most common approaches used is the continuously incrementing work rate. This approach is commonly employed using either a treadmill or cycle ergometer, with the work load being steadily increased over a relatively short period until VO_{2max} is reached. The advantage over previous methods is perceived to be that the tests could be carried out over a relatively short time period. The question posed by the researchers was whether the subjects would show a linear increase in VO_2, up to the point of reaching a plateau VO_2 level, i.e. the VO_{2max}, indicating maximal aerobic performance, during a trial of less than 20 min.

Methods

Seventy-one subjects aged 19–61 years participated after providing informed consent as required by the Local Research Ethics Committee in accordance with the Declaration of Helsinki.

PAUSE AND THINK

What is the role of the ethics approval?

Answer: Participants in a research study must be able to give informed consent prior to taking part. This means they must be informed regarding the nature of the study and any associated risks. It also must be made clear where there are patients involved that, if they decline to participate, there will be no impact on their clinical treatment.

Test 1: The subjects exercised on a cycle ergometer, which was set so that the power required increased at a rate of 15, 20, or 25 W min^{-1}. This was chosen to bring the subjects to fatigue within 10–15 min. Fatigue was defined as the point at which the subjects, despite encouragement from the experimenters, could no longer maintain a pedalling rate of 60 rpm.

Test 2: Thirty-eight of the subjects also completed a single constant-load test set at a level that represented 90% of the peak work rate as identified from the incremental trial.

Test 3: Six of the subjects also performed a series of constant load tests at different levels but all bringing them to the same fatigue limit.

The concentrations of oxygen and carbon dioxide in the respired air were measured breath by breath using a mass spectrometer. Heart rate and blood oxygen saturation, as measured using a pulse oximeter, were also measured.

Results and analysis

Figure 1 shows the typical linear increase in VO_2 with time for three subjects during the ramp increase in exercise intensity up to the point of fatigue. Only 12 subjects showed a plateau response at the end of the exercise (bottom trace; i.e. a decrease in slope indicating a point at which the VO_2 would no longer increase); 40 of the subjects showed a maintained linear response and 19 showed an actual increase in the slope at the end.

What do you think might be the interpretation of these results? The fact that only 12 of the 71 subjects showed a 'typical' plateau in the VO_2 at the point of fatigue could indicate that the incremental ramp test did not

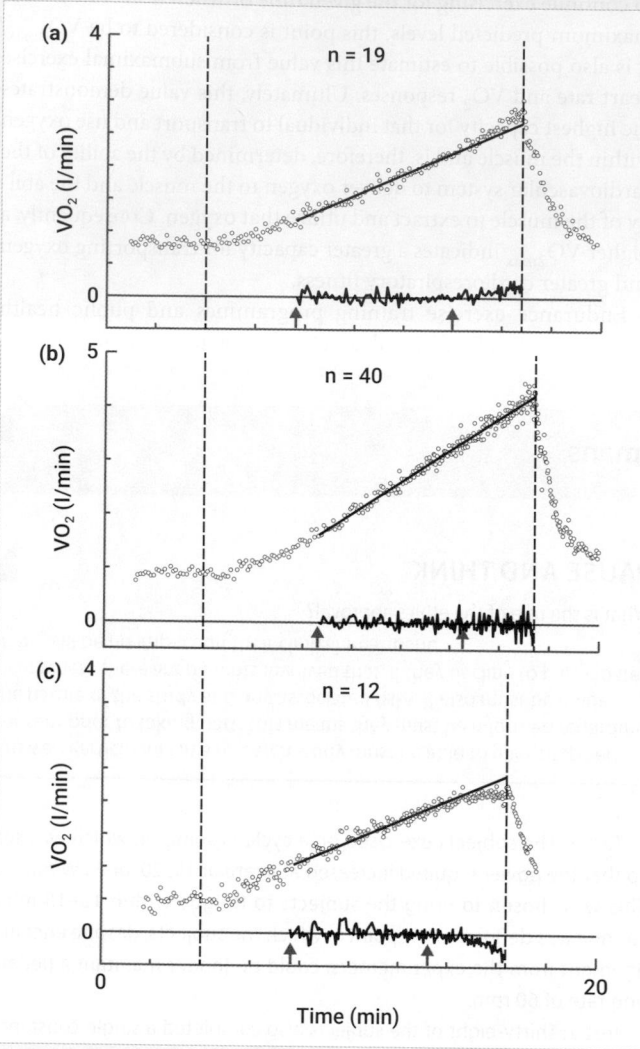

provide a reliable measure of VO_{2max}. This was tested by comparison of the results with those of the other two tests.

The plots of VO_2 against time for the two sets of constant load tests both showed the predicted plateaus corresponding to VO_{2max}. Importantly, though, the measures of VO_{2max} from these tests were the same as for the peak measures from the incremental ramp test. This is illustrated in Figure 2, which shows a linear correlation between the two sets of values for the subject population. The 95% confidence limits shown on the graph illustrate the correlation between the two tests. The peak VO_2 values from the two sets of tests were not significantly different ($P < 0.01$, ANOVA).

Conclusion

The comparison between the tests and the finding of comparable peak values for the VO_2 measurements indicate that the incremental ramp test is a valid measure for determining VO_{2max}.

Read the original work

Day, J. R., Rossiter, H. B., Coats, E. M., Skasick, A., & Whipp, B. J. (2003) The maximally attainable VO_2 during exercise in humans: the peak vs maximum issue. *J. Appl. Physiol.* **95**: 1901–7.

Figure 1 Examples of the VO_2 of three subjects during the incremental ramp test. The duration of the ramp is indicated by the vertical dotted lines. (a) Increase in the slope of the response at the end of the trial; (b) linear relationship continues; (c) plateau in VO_2 at the end of the ramp.

Source: Day, J.R., Rossiter, H.B., Coats, E.M., Skasick, A. & Whipp, B.J. (2003) The maximally attainable VO_2 during exercise in humans: the peak vs maximum issue. *Journal of Applied Physiology*, 95, 1901–7, Figure 1. © The American Physiological Society.

Figure 2 Linear correlation between the measured values of peak VO_2 (VO_{2max}) for the incremental ramp test and the constant load test.

Source: Day, J.R., Rossiter, H.B., Coats, E.M., Skasick, A. & Whipp, B.J. (2003) The maximally attainable VO_2 during exercise in humans: the peak vs maximum issue. *Journal of Applied Physiology*, 95, 1901–7, Figure 1. © The American Physiological Society.

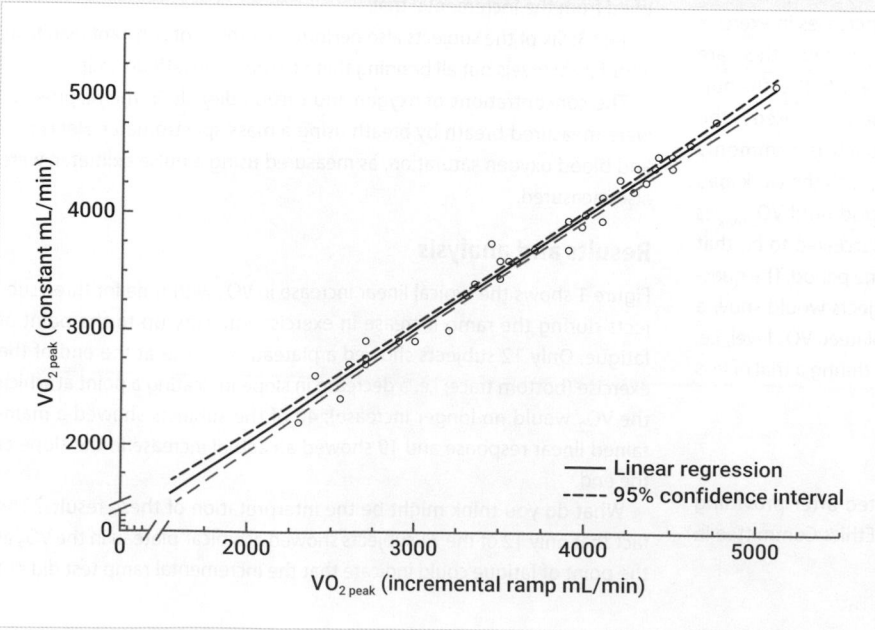

Training programmes rely on some simple principles: **specificity** and **overload**. Specificity refers to the exercise programme challenging the appropriate physiological system; for example, if the goal is to improve cardiorespiratory fitness, then aerobic exercise training rather than resistance training will be preferable. Specificity can also refer to the work done by individual muscle groups. Overload refers to the principle that a system has to be exercised to a greater level than normal in order to adapt. This is why exercise training programmes should be increased in intensity from time to time in order to ensure that adaptation still occurs. Ultimately, an aerobic exercise training programme at the correct intensity and of an appropriate duration leads to an increase in VO_{2max}.

PAUSE AND THINK

What are the primary drivers of an increase in VO_{2max} during an aerobic exercise training programme?

Answer: The increases in stroke volume and CO are the main drivers, with improved oxygen extraction and utilization also assisting.

As discussed in Section 23.1, aerobic exercise training leads to numerous adaptations within the cardiovascular system. In particular, there is an increase in stroke volume and CO, as well as improvements in oxygen extraction and ability to undertake aerobic metabolism. These adaptations are the main drivers for the increase in VO_{2max} that occurs following aerobic exercise training. Not all individuals will respond in the same way, however, as a large proportion of an individual's VO_{2max}, and their response to training, is genetically determined.

 Check your understanding of the concepts covered in this section by answering the questions in the e-book.

23.3 Pulmonary responses to exercise

As we have seen, acute exercise results in significant changes in cardiovascular function which maximize oxygen delivery to metabolically active tissue, and a period of aerobic exercise training enhances the ability of the cardiovascular system to deliver, extract, and utilize oxygen. For this to occur, oxygen has to be available for transport from the lungs. Consequently, acute exercise results in changes in the function of the respiratory system: the cardiovascular and respiratory systems are closely integrated and their responses to exercise should be considered with this fact in mind.

Control of ventilation

Detailed discussion of the control of **ventilation** is provided in Chapter 22; however, this section will provide a reminder of the key areas, to discuss how these result in an increased ventilatory response to exercise.

The processes of **inspiration** and **expiration** are the result of contraction and relaxation of the diaphragm, supplemented by the external and internal intercostals and the muscles of the abdominal wall. The diaphragm, as a skeletal muscle, is controlled via motor neurons of the phrenic nerve, which, in turn, are controlled via respiratory centres in the medulla oblongata and the pons within the brainstem. The function of the respiratory centres is affected by neural and humoral factors.

Neural factors that affect the respiratory centres include input from higher brain centres. With regard to exercise, perhaps the most important of these is the motor cortex, as direct innervation of skeletal muscle, and especially the diaphragm, leads to increased muscular contraction and increase in ventilation. In addition to this, feedback from skeletal muscle from the right ventricle and from joint receptors may also lead to an increase in ventilation. This will, again, be of particular importance during exercise so that active tissue provides direct feedback to the respiratory centre in order to increase ventilation and oxygen delivery in response to an increased demand.

Humoral factors that affect the function of the respiratory centres revolve around the function of chemoreceptors. Chemoreceptors (Chapter 22) respond to a change in the immediate environment causing a change in function. Central chemoreceptors are found in the medulla oblongata and specifically respond to changes in the composition of cerebrospinal fluid (CSF). Increases in **partial pressure** of carbon dioxide (Pco_2) and a fall in pH of the CSF lead to an increase in ventilation. Peripheral chemoreceptors are found in the aortic arch and the carotid bodies. Similar to central chemoreceptors, they respond to changes in Pco_2 and pH but of arterial blood rather than CSF. In addition to these stimuli, peripheral chemoreceptors also respond to changes in potassium concentration, core temperature, catecholamine levels, and arterial partial pressure of oxygen (Po_2). It is important to note that the driving force for an increase in ventilation in normal conditions is an increase in Pco_2 and not a reduction in Po_2.

The respiratory centres within the medulla oblongata and the pons respond to the neural and humoral feedback to change the rhythm produced by the three inspiratory centres: the pre-Bötzinger complex, the pneumataxic centre, and the caudal pons (Section 22.4 and Figure 22.15). The interaction between these three centres changes during exercise to ensure that ventilation matches oxygen demand.

Pulmonary response to acute exercise

Similarly to discussing the effects of exercise on blood pressure response, it is important to consider what happens to ventilation when undertaking either steady state or graded exercise. At rest, ventilation is approximately 5–6 L min^{-1}, but this can increase to >100 L min^{-1} in some cases in peak exercise.

At the beginning of steady-state aerobic exercise, a rapid increase in lung ventilation is observed, because of increases in both the rate of ventilation and the **tidal volume**. This plateaus relatively quickly and then remains stable as exercise intensity, and therefore metabolic demand, also remains stable. This initial increase in ventilation results mainly from an increase in sympathetic input to the respiratory centres. Interestingly, the initial increase in ventilation rate is not matched by control of arterial blood gases. In the early stages of steady-state exercise, arterial Po_2 is slightly reduced, while arterial Pco_2 is slightly increased. These return to normal levels

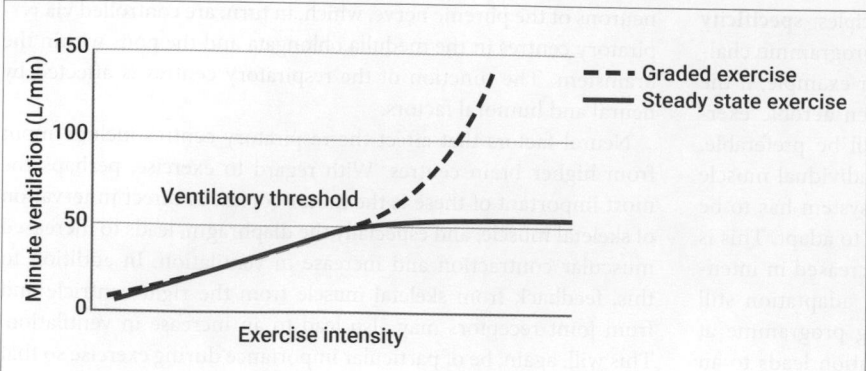

Figure 23.6 Minute ventilation (L min^{-1}) during steady-state and graded aerobic exercise.

and, after this, are tightly regulated within 1–2 min of beginning exercise. Following the initial rise in ventilation, feedback from the chemoreceptors regulates ventilation in relation to metabolic demand and control of arterial P_{O_2} and P_{CO_2}. When exercise is undertaken in a hot environment, ventilation rate is greater than when the same intensity of exercise is undertaken in ambient temperatures. P_{O_2} and P_{CO_2} are, however, the same. This is thought to be due to additional humoral feedback, including an increase in core temperature and circulating catecholamines, which results in increased activity of the respiratory centre.

During graded exercise, ventilation increases linearly with exercise intensity until between 50% and 70% of VO_{2max}, at which point ventilation begins to increase exponentially (Figure 23.6). This point is different between trained and untrained individuals and has been termed the '**ventilatory threshold**'. Unsurprisingly, as exercise intensity increases arterial P_{O_2} and pH reduce while arterial P_{CO_2} increases. While ventilation is increasing linearly, the activity of the respiratory centre is increased due to a combination of both neural and humoral factors. After the point of the ventilatory threshold, it has been postulated that the increase in ventilation leads to a decline in P_{CO_2} and the primary driver for ventilation becomes the increase in hydrogen ion concentration accumulation, and therefore a decrease in pH, following increased lactate production. This is the result of an increase in humoral feedback to the respiratory centre.

For many years, it has been suggested that training at a maximal steady-state speed is beneficial for exercise performance and this has traditionally been determined by measuring the **lactate threshold**: the exercise intensity at which blood lactate begins to rise exponentially rather than linearly. This is an invasive measurement, so it has been suggested that using the ventilatory threshold could be a proxy marker of the lactate threshold on the basis that the increase in lactate and, therefore, hydrogen ions leads to increased ventilation. The lactate threshold and the ventilatory threshold do not always occur at the same exercise intensity, suggesting that other factors in addition to pH influence ventilation during graded exercise.

Another potential explanation for the rapid rise in ventilation at high exercise intensities may be the increase in blood potassium concentration that occurs. Blood potassium levels significantly increase during high intensity exercise due to a net movement of potassium out of skeletal muscle during contraction. It is thought that this may stimulate peripheral chemoreceptors and increase respiratory rate. Exercising at high intensities also leads to an increase in body temperature and circulating catecholamines that may also contribute to this observation.

Pulmonary adaptations to aerobic exercise training

As outlined in Section 23.1, aerobic exercise training results in significant adaptations to the cardiovascular system that lead to improvement in delivery and extraction of oxygen. This, in turn, leads to increased physical fitness and potential health benefits. Although not covered in this chapter, aerobic training also leads to adaptations in many other physiological systems. By comparison, the respiratory system does not appear to significantly adapt to exercise training. A period of aerobic exercise training does not lead to changes in lung structure or anatomy, and it does not result in improvements in pulmonary gas diffusion capacity. It seems clear that the respiratory system is able to adapt to the demands of exercise without requiring additional capacity following exercise training; however, ventilation rate at submaximal exercise intensities is reduced following sustained exercise training. This is probably due to enhanced aerobic capacity within the muscle leading to less hydrogen ion production and, therefore, less feedback to the respiratory centres from peripheral chemoreceptors.

The respiratory system is also not the main factor that limits endurance exercise performance. In healthy individuals exercising at relatively high intensities, respiratory muscles do not appear to fatigue and, therefore, influence the rate of ventilation. When exercise intensity approaches maximal levels, though, it is possible that the respiratory system may be a limiting factor to performance due to respiratory muscle fatigue. Interestingly, in some elite athletes, arterial P_{O_2} decreases markedly at maximal exercise intensities, which is further evidence to suggest that the respiratory system can inhibit performance when exercising at very high intensities. Training the inspiratory respiratory muscles has a small, but significant, effect on exercise performance at high intensities. One potential mechanism for this observation is that respiratory muscle training may improve blood flow to these muscles during exercise and, therefore, prevent fatigue and improve performance.

Why do you think we sometimes get a stitch when we exercise? Take a moment to consider this before you read on. There are several theories as to why we get stitches when exercising. One of the main theories is that blood flow to the diaphragm is reduced causing it to cramp, but it is also thought this may be due to changes in potassium balance.

 Check your understanding of the concepts covered in this section by answering the questions in the e-book.

23.4 Exercise and acid–base balance

As discussed in Chapter 22 and Chapter 24, the regulation of blood pH is of critical importance and is maintained within very narrow limits to ensure normal physiological function. The regulation of blood pH is effected via the integrated response of buffering systems, the respiratory system, and the renal system. Reduction in blood pH, an **acidosis**, is the result of an increase in hydrogen ion concentration within the blood, whereas an increase in blood pH, an **alkalosis**, is due to a decrease in hydrogen ion concentration within the blood. These conditions are considered the result of changes in metabolism or respiratory function. Undertaking high-intensity exercise results in an increase in hydrogen ions due to the substantial increase in metabolic activity that occurs within active muscle. This, therefore, has an effect on blood pH and the regulation of acid–base balance.

Effect of exercise on acid–base balance

Undertaking high-intensity exercise results in a reduction in blood pH (i.e. it induces a metabolic acidosis). Hydrogen ions are released for several reasons. First, as a result of the increase in cellular respiration, CO_2 is produced within the active muscle and this, via the **carbonic acid–bicarbonate buffering system**, leads to an increase in hydrogen ion release. Secondly, the end product of anaerobic glycolysis, pyruvate, is converted to lactate during high-intensity exercise in order to regenerate NAD^+, which also leads to an increase in hydrogen ion release. Finally, the breakdown of ATP within the muscle directly increases hydrogen ion release within the cell. Consequently, during sustained exercise, P_{CO_2} in the blood may decrease as lung ventilation is driven by a change in pH.

Clearly, undertaking high-intensity exercise results in a reduction in blood pH through the mechanisms described above; however, it is unclear what effect this has on exercise performance. For many years, it was believed that the increase in hydrogen ion production directly resulted in muscular fatigue. It is not uncommon to hear sports commentators suggesting that athletes are getting tired because of 'the build-up of lactic acid'.

PAUSE AND THINK

Does lactic acid cause fatigue during exercise?

Answer: For many years, it was thought that hydrogen ions were the main cause of peripheral fatigue during exercise. In recent years, this belief has been challenged.

To an extent, this is correct as hydrogen ions can affect muscle contractile ability in a number of ways, but experiments on animal tissue indicate that the effect of hydrogen ions on muscular fatigue is high when the experimental temperature is relatively low. When experiments are repeated at normal body temperatures, the effect of hydrogen ions on muscle contractility is not as pronounced. Instead, it is thought that the build-up of inorganic phosphate from the breakdown of ATP may have more of an effect on muscle contractility and fatigue than the build-up of hydrogen ions.

The rise in hydrogen ion release during exercise does affect acid–base regulation, which is countered to a large extent by the muscle buffering systems. In particular, hydrogen ions are buffered by bicarbonate, phosphates, various cellular proteins, and carnosine so that the decline in muscle pH is attenuated. In addition, hydrogen ions are transported out of the muscle into the blood, where they are buffered by blood buffering systems. These include bicarbonate, haemoglobin, and some proteins (Section 22.3). As such, intra- and extracellular buffering systems are important in counteracting a rapid increase in hydrogen ions that occurs during high-intensity exercise. A period of high-intensity exercise training appears to enhance intracellular buffering mechanisms, providing a beneficial effect of training on acid–base balance regulation.

While buffering systems provide a rapid response to increases in hydrogen ion release, a reduction in blood pH still occurs during exercise. Consequently, the respiratory system also acts to regulate acid–base balance during high-intensity exercise. This is primarily driven by the carbonic acid–bicarbonate buffering system. This relationship, which is catalysed by the enzyme carbonic anhydrase, is described in the following equation:

$$CO_2 + H_2O \rightleftharpoons H_2CO_3 \rightleftharpoons H^+ + HCO_3^-$$

When P_{CO_2} is high there is an increase in carbonic acid. This readily dissociates into a hydrogen ion and a bicarbonate ion. Similarly, an increase in hydrogen ion accumulation that occurs as a result of cellular respiration during exercise leads to an increase in carbonic acid and a resultant increase in P_{CO_2}. CO_2 can then be expelled via the lungs, providing a means of regulating blood pH.

During exercise, the primary site of hydrogen ion release is within the muscle, so it is no surprise that the main factors that affect the extent of change in pH are the intensity of exercise, the duration of exercise, and the amount of muscle mass activated during exercise. In summary, muscle buffering systems counteract the rise in hydrogen ion production with excess hydrogen ions being shuttled out of the cell, at which point blood buffering systems also regulate pH and the respiratory system also ensures regulation by increasing ventilation. The renal system is not typically involved in the regulation of pH during exercise.

Nutritional supplements, exercise, and acid–base balance

Athletes and regular exercisers have long tried to enhance performance by supplementing their diets with nutritional supplements. The nutritional supplement industry in general, and in particular for exercise, is huge but not subject to the same rigorous product testing that is required in other industries. As such, many supplements that are available have not been shown to benefit exercise performance, particularly if an individual already has an adequate dietary intake, but they are still widely used.

Some interventions, though, *have* been shown to be beneficial to exercise performance. The ingestion of carbohydrate before exercise is beneficial for activities that rely on anaerobic ATP production because this maximizes muscle glycogen stores. Similarly, ingestion of carbohydrate during prolonged endurance exercise is beneficial as this provides an exogenous form of carbohydrate to be metabolized, sparing muscle glycogen until later in exercise. Supplementing the diet with creatine can be beneficial as this affects recovery from repeated bouts of high intensity exercise and, therefore, the efficiency of training. At the onset of exercise, phosphocreatine provides the majority of ATP required for muscular work; however, stores are depleted rapidly and cannot be resynthesized during exercise. Supplementation with creatine leads to an increase in intramuscular creatine concentration and therefore increases the rate of resynthesis of phosphocreatine during rest. Ingestion of protein after resistance exercise is beneficial for protein turnover and muscle mass.

Unfortunately, many nutritional supplements are not effective, or have not been shown to be effective, and the use of these supplements can be problematic due to documented contamination with banned substances in a number of cases.

One area that has received significant attention is whether there are nutritional supplements that can increase buffering capacity to attenuate the reduction in blood pH that occurs during high-intensity exercise and, therefore, improve performance.

Several variables should be considered when investigating the use of nutritional supplements in exercise. Perhaps the most significant of these, in the area of acid–base regulation during exercise, is whether an ingested substance is actually delivered to the target tissue and, if it is, whether meaningful increases in the availability of that substance occur. In many cases, these two key criteria are not met.

Given the importance of bicarbonate in both intra- and extracellular buffering capacity, it is no surprise that much attention has been given to whether the acute ingestion of sodium bicarbonate prior to exercise improves performance. Most studies in this area agree that ingestion of sodium bicarbonate does improve

high-intensity exercise performance, primarily through improvements in extracellular buffering capacity. Although this may be beneficial, ingestion of sodium bicarbonate leads to gastrointestinal distress in some people, therefore negating any potential benefits of its ingestion. Similarly, the ingestion of high doses of sodium citrate enhance extracellular buffering capacity, but may also cause gastrointestinal distress.

Interestingly, ingestion of beta-alanine, an amino acid precursor of carnosine, which enhances intracellular buffering capacity, has been shown to be beneficial for high-intensity exercise performance when ingested in large enough quantities to ensure delivery to the muscle. Unlike sodium bicarbonate and sodium citrate, beta-alanine supplementation does not appear to result in any major side effects. Any such supplements, though, should be only taken on expert advice.

 Check your understanding of the concepts covered in this section by answering the questions in the e-book.

23.5 Exercise and environmental stressors

It is clear from the content of this chapter that exercise is a major stressor to both the cardiovascular and respiratory systems. Exercise also results in changes in function to other physiological systems. Having considered the effects of exercise on the function of the cardiovascular and respiratory systems, it is now important to consider what effect the accumulation of stressors has on function. This section will consider how exercise undertaken in different environments, namely heat and altitude, affects physiological function and how this can affect exercise performance, as well as health.

Exercise in heat

As briefly discussed in Section 23.2, exercise results in an increase in metabolic heat production, which, in turn, increases core temperature. The increase in metabolic heat production is directly proportional to the intensity of exercise that is undertaken. To avoid large increases in core temperature (Figure 23.7a), heat must be lost

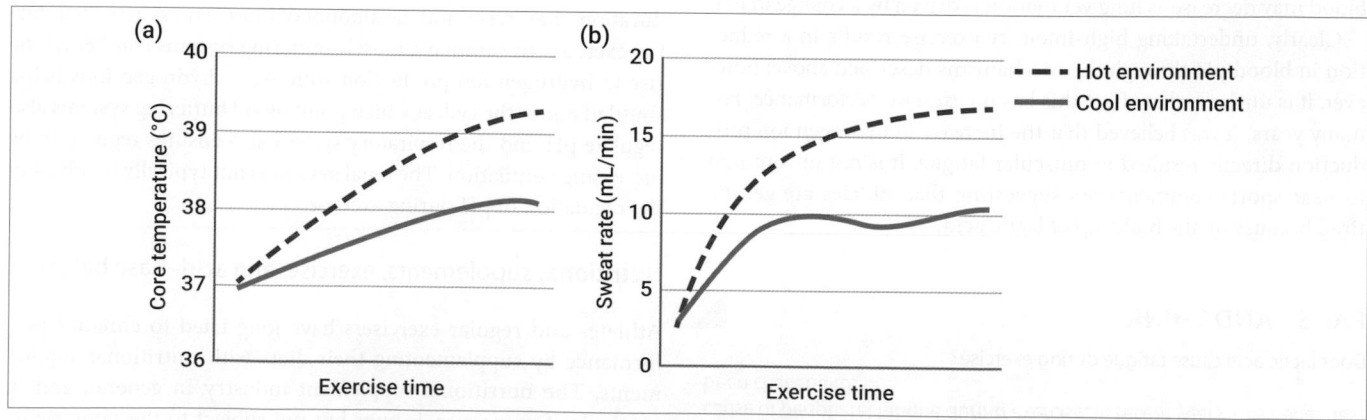

Figure 23.7 Exercise and core temperature. (a) Core temperature and (b) sweat rate during aerobic exercise in cool and hot environments.

from the body. The primary avenue for this increased heat loss during exercise is via sweating (Figure 23.7b), which, in the absence of fluid ingestion, reduces body water levels. Given that electrolytes, mainly sodium and potassium, are present in sweat, this also leads to changes in plasma electrolyte levels. Environmental temperature also influences heat loss mechanisms so it therefore follows that a combination of exercise and a high ambient temperature will result in a significant challenge to the body. The cardiovascular system is particularly affected by this combination.

The combination of heat stress and exercise significantly influences the cardiovascular system because it must meet the demands of active tissue, as well as increasing blood flow to the skin to dissipate metabolic heat. During submaximal exercise, CO is similar when activity is undertaken in either a hot or cold environment. As increasing sweat rate leads to a reduction in plasma volume, this leads to a reduction in SV when exercise at a given intensity is undertaken in the heat compared to ambient temperatures. In order to maintain the necessary CO at that exercise intensity, HR must increase. Consequently, there is less of a reserve for HR to increase during exercise in the heat and maximal CO is lower when compared to ambient temperatures.

As discussed in Section 23.1, exercise leads to the redistribution of blood away from non-active tissues. This effect is exacerbated when exercise is undertaken in the heat as blood flow to the active muscle and the skin is favoured over other tissues. In extreme situations, this can lead to significant complications within other systems.

PAUSE AND THINK

Based on what you have just read, what do you think is the effect of drinking fluid during exercise when it is hot?

Answer: Fluid ingestion during exercise replaces some of the water loss that occurs through sweating. This can potentially influence cardiovascular function by influencing the extent of reduction in plasma volume that occurs. It also affects the rating of perceived exertion of the person exercising as an individual's core temperature and skin temperature are two of the strongest physiological indicators that determine the effort that is required for a certain task.

It is possible to adapt to undertaking exercise in the heat. This is termed **heat acclimatization** and involves undertaking regular bouts of exercise in increased environmental temperature. As a result, compared to pre-acclimatization, skin blood flow is increased while sweating begins at an earlier time and to a greater extent, resulting in improved ability to attenuate the risk in core temperature that occurs while exercising in high ambient temperatures. This then ensures that submaximal exercise intensities are undertaken at a lower heart rate than when not acclimatized.

In recent years, there has been significant attention paid to the effects of hydration status on exercise performance. This is of particular interest when exercise is undertaken in the heat because of the increased sweat response and effect on circulatory function described above. Research undertaken in this area suggests that large acute deviations in water balance can significantly influence endurance exercise performance, although relatively small reductions in body water are unlikely to have much influence on performance or physiological function.

There are several potential explanations for these observations. The effect of exercise in the heat on circulatory function, described above, is one of the main explanations in that the reduction in maximal CO when exercising in a dehydrated state in the heat significantly influences exercise capacity. When dehydrated, blood flow to the skin also seems to be reduced, which may impact the ability to dissipate heat and further increase core temperature. In addition, there is evidence to suggest that the perceived exertion of undertaking submaximal exercise in the heat is greater when compared to the same intensity in ambient temperatures. This observation may, in part, be due to the increased HR observed at submaximal exercise intensities in the heat. Evidence also suggests that moderate levels of dehydration may lead to a reduction in cognitive ability and may also influence muscle function.

In addition to affecting exercise performance, exercise in the heat that results in significant reductions in body water can result in significant health problems. This usually requires large acute reductions in body water brought about by prolonged exercise in the heat without any fluid ingestion. Case studies have demonstrated that, in some situations, this can lead to collapse, significantly impaired cardiovascular and cognitive function, and **hypernatremia** (elevated plasma sodium concentrations). Conversely, other studies have demonstrated that over-ingestion of water during exercise can also result in impaired cognitive and cardiovascular function, as well as **hyponatremia** (reduced plasma sodium concentrations). This, in some cases, has proven to be fatal. This risk is well recognized and many endurance athletes drink fluids that contain a balanced ionic solution along with glucose during their events.

Exercise at altitude

When moving from sea level to altitude, the resultant reduction in atmospheric partial pressure of oxygen (Po_2) provides the main physiological challenge. This reduction in atmospheric Po_2 leads to a reduction in arterial oxygen availability and therefore a reduction in oxygen delivery to tissues. This presents a significant challenge at rest, as well as during exercise. Adaptations can be considered as acute or chronic, brought about as a result of acclimatization.

The initial response to a reduction in atmospheric and therefore arterial Po_2 is an increase in breathing rate. At sea level, the arterial partial pressure of carbon dioxide (Pco_2) is the main driver of respiratory response until arterial Po_2 is below certain limits. This occurs at altitude, at which point peripheral chemoreceptors within the aortic arch are stimulated, which results in an increase in breathing rate that is maintained for a substantial time after exposure. This hyperventilatory response leads to a reduction in Pco_2, resulting in an increase in blood pH and a respiratory induced alkalosis (Section 22.4, Altitude acclimatization).

When undertaking submaximal exercise following acute altitude exposure, CO is substantially increased to compensate for the reduction in arterial Po_2. This is the consequence of an increase in HR, stimulated by increased sympathetic stimulation at these exercise intensities, rather than an increase in SV. Altitude exposure therefore leads to a significant reduction in VO_{2max}.

CLINICAL BOX 23.1   The use of recombinant EPO by athletes

In recent years a great deal of attention has been given to the use of performance-enhancing drugs by athletes. In certain sporting situations, the use of recombinant EPO by athletes undertaking aerobic exercise has been observed. Essentially, the injection of recombinant EPO mimics the effects of altitude exposure and training. With altitude training, a reduction in oxygen delivery to the kidney leads to the release of endogenous EPO, resulting in an increase in oxygen carrying capacity. Injections of recombinant EPO directly act on bone marrow leading to an increase in red blood cell production. This is currently an anti-doping violation as it is considered an attempt to obtain an unfair advantage over competitors. In

terms of exercise performance, recombinant EPO injections have demonstrated an increase in haemoglobin concentration and an increase in maximal oxygen uptake, a key determinant of aerobic exercise performance, of approximately 10%. It is also associated with some potential health risks that are not necessarily associated with an increase in endogenous EPO secretion from altitude exposure. These include hypertension, and other cardiovascular disorders, which result from the increased viscosity of the blood due to the increase in haematocrit but no increase in plasma volume. These risks have been publicized in relation to the deaths at a young age of some high-profile athletes, for example, elite cyclists.

Over a period of altitude exposure, the respiratory-induced alkalosis is rectified via an increase in urinary bicarbonate excretion and conservation of H^+ ions, leading to a return to normal blood pH. Hyperventilation in response to reduced arterial P_{O_2} continues, but without the potentially negative effect on acid–base balance.

Perhaps the most significant longer-term adaptation to altitude acclimatization is an increase in the number of red blood cells and therefore the synthesis of haemoglobin. The formation of red blood cells is triggered by increased release of the hormone **erythropoietin** (EPO) from the kidney (Section 24.9): following altitude exposure, the reduction in arterial P_{O_2} leads to reduced oxygen delivery to the kidney, which, in turn, stimulates EPO release and an increased rate of formation of red blood cells from bone marrow. As a result, oxygen carrying capacity is increased. This, coupled with cellular adaptations that follow altitude acclimatization, leads to greater efficiency of oxygen delivery to ensure normal physiological function. In theory, this increase in oxygen carrying capacity should be beneficial for exercise performance at sea level and forms the basis for some elite athletes undertaking training at relatively high altitudes (but see Clinical Box 23.1).

Following acclimatization, CO at submaximal exercise remains greater than when at sea level, although the extent of this increase is less than in the early stages of altitude exposure. Plasma volume is reduced following acclimatization, perhaps because of an increased red blood cell mass, which leads to a reduction in SV during exercise, resulting in an increased HR at submaximal exercise intensities. Unsurprisingly, maximal CO is reduced after acclimatization and this is primarily due to the reduction in SV that occurs; however, increased parasympathetic activity following an altitude stay leads to a slight reduction in maximum exercising heart rate as well.

PAUSE AND THINK

Is exercise performance at sea level increased as a result of altitude acclimatization?

Answer: In theory, the increase in oxygen carrying capacity as a result of acclimatization should improve sea-level exercise performance, but this does not seem to be the case because of the negative effect that acclimatization to altitude has on heart rate, stroke volume, and CO.

Interestingly, a combination of non-training altitude exposure and training while not exposed to reduced atmospheric P_{O_2} seems to produce a combined effect that does enhance exercise performance at sea level. In this scenario, a period of altitude exposure results in the beneficial effects on EPO secretion, the increase in red blood cells, and an increased oxygen carrying capacity, and regularly returning to sea level to train ensures that the potentially negative impact on HR, SV, and CO are alleviated.

 Check your understanding of the concepts covered in this section by answering the questions in the e-book.

23.6 Physical activity, fitness, and health

In recent times, there has been significant attention paid to the role of physical inactivity in the development of numerous disease states and the role of physical activity and fitness in their prevention and treatment. This is particularly important given the large number of people classed as inactive. A detailed discussion of this topic is beyond the scope of this text; however, some key points are important to mention in light of information provided in this chapter.

Public health interventions tend to involve advising individuals to undertake regular exercise in order to reduce disease risk, though there is considerable debate over whether simply undertaking physical activity is sufficient to reduce mortality or whether that physical activity has to be of an intensity high enough to improve cardiorespiratory fitness, or VO_{2max}. Evidence suggests that increasing physical activity energy expenditure results in reduced relative risk of premature death; however, a low level of cardiorespiratory fitness is an independent risk factor for premature death and similar in risk to smoking, hypertension, and having high cholesterol. This would suggest that aerobic exercise should be of a level to improve cardiorespiratory fitness in order to improve health outcomes. This may be different when looking at the risk of different disease states.

CLINICAL BOX 23.2 Physical activity or physical fitness for health

A large amount of research has been conducted investigating whether improvements in physical fitness (as assessed by VO_{2max}) are necessary for improvements in health outcomes. Many of these are longitudinal, or observational, studies. Two well-known studies in this area are the Harvard Alumni Health Study (Paffenbarger et al., 1986) and the Aerobics Center Longitudinal Study (Blair et al., 1996).

The Harvard Alumni Health Study involved observation of 16,936 men with initial data collected in the 1960s. These data included general health characteristics, as well as indicators of levels of physical activity assessed via a questionnaire. By 1978, 1413 of the initial cohort of participants had died and when the researchers analysed the data there was a clear relationship between the amount of energy expended via physical activity and relative risk of death. This relationship was true when confounding variables such as smoking status, body composition, and age were considered. The results of this study suggested that increasing physical activity levels resulted in a reduced risk of all-cause mortality.

The Aerobics Center Longitudinal Study involved 10,224 men and 3120 women. Each person conducted a treadmill test to determine their level of physical fitness and they were grouped into quintiles, with individuals being classed as having low fitness if they were in the bottom 20% of all the people examined. Eight years later, the researchers determined the relationship between various factors and risk of all-cause mortality. When adjusted for age, low physical fitness was one of the most important risk factors alongside smoking status, hypertension, and high cholesterol. This indicates that low levels of physical fitness are an important consideration for all-cause mortality.

References

Blair, S. N., Kampert, J. B., Kohl, H. W., Barlow, C. E., Macera, C. A., Paffenberger, R. S., & Gibbons, L. A. (1996) Influences of cardiorespiratory fitness and other precursors on cardiovascular disease and all-cause mortality in men and women. *JAMA* **276**: 205–10. doi:10.1001/jama.1996.03540030039029

Paffenberger, R. S., Hyde, R., Wing, A. L., & Hsieh, C-C. (1986) Physical activity, all-cause mortality, and longevity of college alumni. *N. Engl. J. Med.* **314**: 605–13. doi: 10.1056/NEJM198603063141003

Considering obesity, the main goal of a weight loss programme is to create an energy imbalance that favours use of energy stores and fat loss. This can be achieved by caloric restriction and increasing physical activity energy expenditure so weight loss itself may not require high exercise intensities, although the improvement in cardiorespiratory function would be beneficial for co-morbidities associated with obesity. Risk of developing cardiovascular disease seems to be more closely associated with improvements in physical fitness rather than simply increasing energy expenditure. This would be consistent with the previous explanations on beneficial cardiovascular effects of aerobic exercise training. Improvements in both physical activity levels and physical fitness have been shown to be beneficial for risk of development of type 2 diabetes mellitus.

Ultimately, the relationship between physical activity and physical fitness is complicated and depends on numerous factors. As a result of the cardiovascular adaptations to exercise training, any aerobic exercise training that improves VO_{2max} is likely to be beneficial for numerous health outcomes, as well as exercise performance (Clinical Box 23.2).

 Check your understanding of the concepts covered in this section by answering the questions in the e-book.

SUMMARY OF KEY CONCEPTS

- Undertaking exercise results in significant stress on all physiological systems. This chapter has focused on the cardiovascular and respiratory systems, but it should not be inferred that these are the only systems affected by exercise.

- Exercise results in a significant increase in blood pressure; however, the extent of this increase depends on the type of exercise that is being undertaken.

- Undertaking aerobic exercise results in elevations in heart rate and stroke volume that lead to a substantial increase in CO to meet the metabolic demands of active tissue.

- During exercise, blood flow is redistributed to active tissue from other systems, which may affect their function in some cases.

- Undertaking a period of aerobic exercise training results in significant adaptations to the cardiovascular system. These include changes in heart anatomy and function, cardiac muscle contractile ability, blood distribution, and oxygen extraction.

- VO_{2max} is considered an independent risk factor for many disease states, as well as an important predictor of aerobic exercise performance.

- The onset of an exercise stimulus leads to acute increases in minute ventilation in order to provide oxygen delivery to the circulation.

- Adaptations to aerobic exercise training within the pulmonary system are limited.

- Exercise, particularly of a high intensity, leads to the release of hydrogen ions and a metabolic acidosis that is counteracted primarily by buffering systems and an increase in ventilation.

- There are nutritional supplements that may assist in the process of buffering hydrogen ions and improve exercise performance.

- Exercising in the heat adds an additional stress to the cardiovascular system as metabolic heat must be dissipated via sweating. If this does not occur, core temperature is increased, and this leads to fatigue.

- Exercising at altitude adds an additional stress due to the reduction in atmospheric partial pressure of oxygen. This leads to short-term effects on both the pulmonary and circulatory systems at rest and during exercise.

- Acclimatization to altitude leads to an increase in red blood cell production, which may enhance exercise performance in some situations.

Use the flashcards in the e-book to test your recall of key terms introduced in this chapter.

QUESTIONS

Looking for answers? Once you've answered these questions, follow the link in the e-book to the answer guidance and check your work.

Concepts and definitions

1. What is meant by the term mean arterial pressure (MAP)?

2. Why does cardiac preload increase during exercise?

3. What is meant by the term VO_{2max}?

4. Why does the heart, at rest, normally beat at a much lower rate than the intrinsic pacemaker rate?

5. What are the main changes in lung ventilation in response to exercise?

Apply the concepts

6. Compare the effects of preload and afterload on cardiac output.

7. Why does the left ventricle show much greater hypertrophy than the right ventricle in response to aerobic exercise training?

8. What are the main variables affecting the rate and depth of ventilation?

9. What factors contribute to the peripheral muscle fatigue developed during exercise?

10. Why have some endurance athletes been known to take EPO to enhance their performance?

Beyond the concepts

11. Summarize the acute effects of exercise on the cardiovascular and pulmonary systems.

12. Summarize the adaptations that occur within the cardiovascular and pulmonary systems to an aerobic exercise training programme.

13. Describe the effects of acute exercise on acid–base balance.

14. Describe the effects of acute heat and altitude exposure on cardiovascular and pulmonary function during exercise.

15. Discuss the significance of physical activity and associated fitness for maintaining health.

FURTHER READING

Fadel, P. J. (2015) Reflex control of the circulation during exercise. *Scand. J. Med. Sci. Sports* **25** (S4): 74–82.
This is a very readable review article looking at the sympathetic and parasympathetic responses to exercise by the cardiovascular system.

Lavie, C. J., Arena, R., Swift, D. L., et al. (2015) Exercise and the cardiovascular system. Clinical science and cardiovascular outcomes. *Circ. Res.* **117**: 207–19.
This is another very readable overview of the responses of the cardiovascular system to exercise.

Powers, S. K., Howley, E. T., & Quindry, J. (2020) *Exercise Physiology: Theory and Application to Fitness and Performance.* 11th edition. McGraw-Hill Education
A relatively broad text considering all aspects of exercise physiology, including a range of clinical examples and also the different forms of fitness testing.

Wilson, M. G., Ellison, G. M., & Cable, N. T. (2016) Basic science behind the cardiovascular benefits of exercise. *Br. J. Sports Med.* **50**: 93–9.
A review of the adaptations to training and consideration of the optimal exercise approaches.

Renal System

LEARNING OBJECTIVES

By the end of this chapter you should be able to:

- Describe the functional anatomy of the kidneys.
- Explain the mechanism of filtration of the blood and regulation of the filtration pressure.
- Compare the functions of the different sections of the nephron.
- Describe the mechanisms of reabsorption and secretion, with particular reference to the Na+ K+ ATPase pump.
- Explain the operation and significance of the loop of Henlé.
- Describe the regulation of water and ion balance.
- Describe the role of the kidney in the regulation of blood pressure.
- Describe the roles of the kidney as an endocrine organ.

Chapter contents

Introduction   785

24.1 Overview of kidney structure
      and function   786

24.2 Filtration in the renal corpuscle   787

24.3 The proximal convoluted tubule   789

24.4 The loop of Henlé   793

24.5 Distal convoluted tubule and collecting
      ducts   795

24.6 The kidney and blood pressure   797

24.7 The kidney and acid–base balance   798

24.8 Micturition   799

24.9 Endocrine functions of the kidney   800

 Watch the key concepts video in the e-book to prepare yourself for studying this chapter.

Introduction

The kidneys are usually thought of as the organs that 'clean' the blood and produce urine; however, the kidneys actually play a vital role in homeostasis, operating across a range of body systems. The two kidneys receive about 20% of the cardiac output: roughly one litre of blood per minute flows through the renal arteries and the flow is tightly controlled to maintain a regulated throughput.

The kidneys do 'clean' the blood by removing the waste products of metabolism by excretion in the urine, but they also play a critical role in the control of water and ion balance, in regulating the pH of the blood, controlling blood pressure, and in maintaining the production of red blood cells. As such, there is extensive interaction with the other body systems as a part of the overall control of the internal environment.

24.1 Overview of kidney structure and function

The two kidneys are located high in the abdominal space, either side of the midline of the body (Figure 24.1), and receive direct blood supply from the abdominal aorta, via the two renal arteries. The processed blood is returned to the heart via the renal veins that feed into the inferior vena cava. The output of the kidneys' function is **urine**, which is transported from the kidneys via the **ureter** to the **bladder**, where it is stored before being excreted via the **urethra**. Typically, the human kidneys process 180 L blood plasma a day and produce about 1.5 L of urine. Over 90% of the content of urine is water. The rest is mainly a mix of ions, and the waste products of metabolism, especially **urea**, which is produced as a result of protein metabolism. The actual volumes produced depend on a number of factors such as the level of hydration, the temperature, and whether or not you are exercising and thereby losing water through perspiration.

If the kidneys fail, what do you think is likely to happen? The answer is that urinary flow will decrease and so the body may retain too much water, which is often indicated by swelling of the feet and ankles; there will be a build-up of harmful waste in the blood, and the blood pressure may rise to high levels. Patients often have to undergo regular dialysis (see Clinical Box 24.1) or they may die.

The kidney is enclosed in a tough capsule of connective tissue that maintains the internal hydrostatic pressure (Figure 24.2).

Inside the capsule is an outer cortex, the renal cortex, which surrounds the renal medulla. At the core of the kidney is the renal pelvis where the urine collects before passing down to the ureter.

The functional unit of the kidney is the **nephron** (Figures 24.2 and 24.3). The number of nephrons varies significantly between species and, indeed, between individuals, but the average for the human kidney is about 900,000. Each nephron operates as a distinct unit processing the plasma by three mechanisms:

- tubular filtration;
- tubular re-absorption;
- tubular secretion.

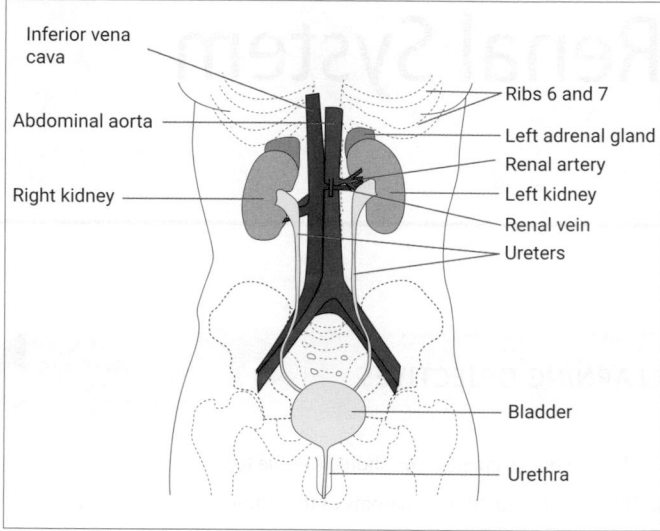

Figure 24.1 Location of the kidneys.
Source: Pocock et al. *Human Physiology*, 5th edn, 2017. Oxford University Press.

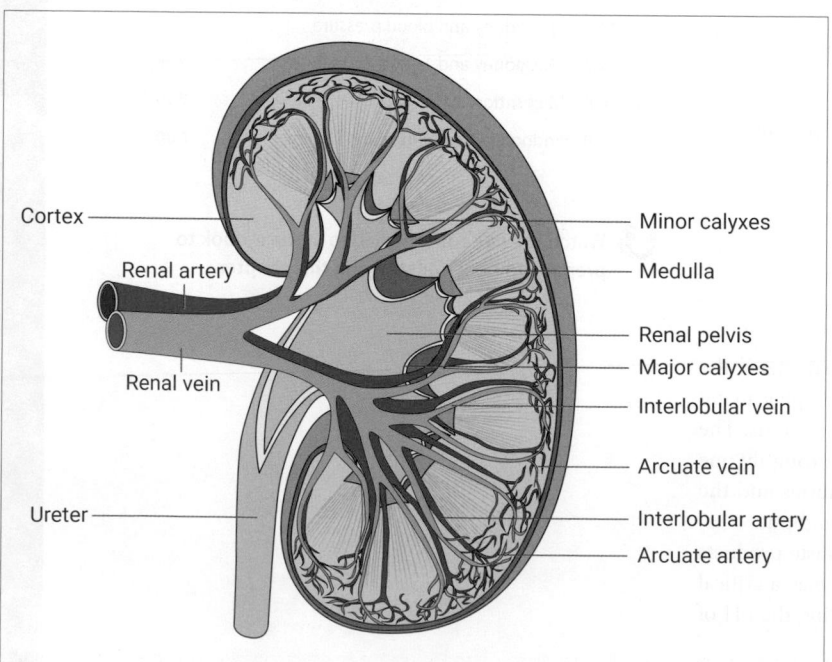

Figure 24.2 The kidney in cross-section.
Source: Sun Sun / Shutterstock.com.

Figure 24.3 The nephron.

Source: Pocock et al. *Human Physiology*, 5th edn, 2017. Oxford University Press.

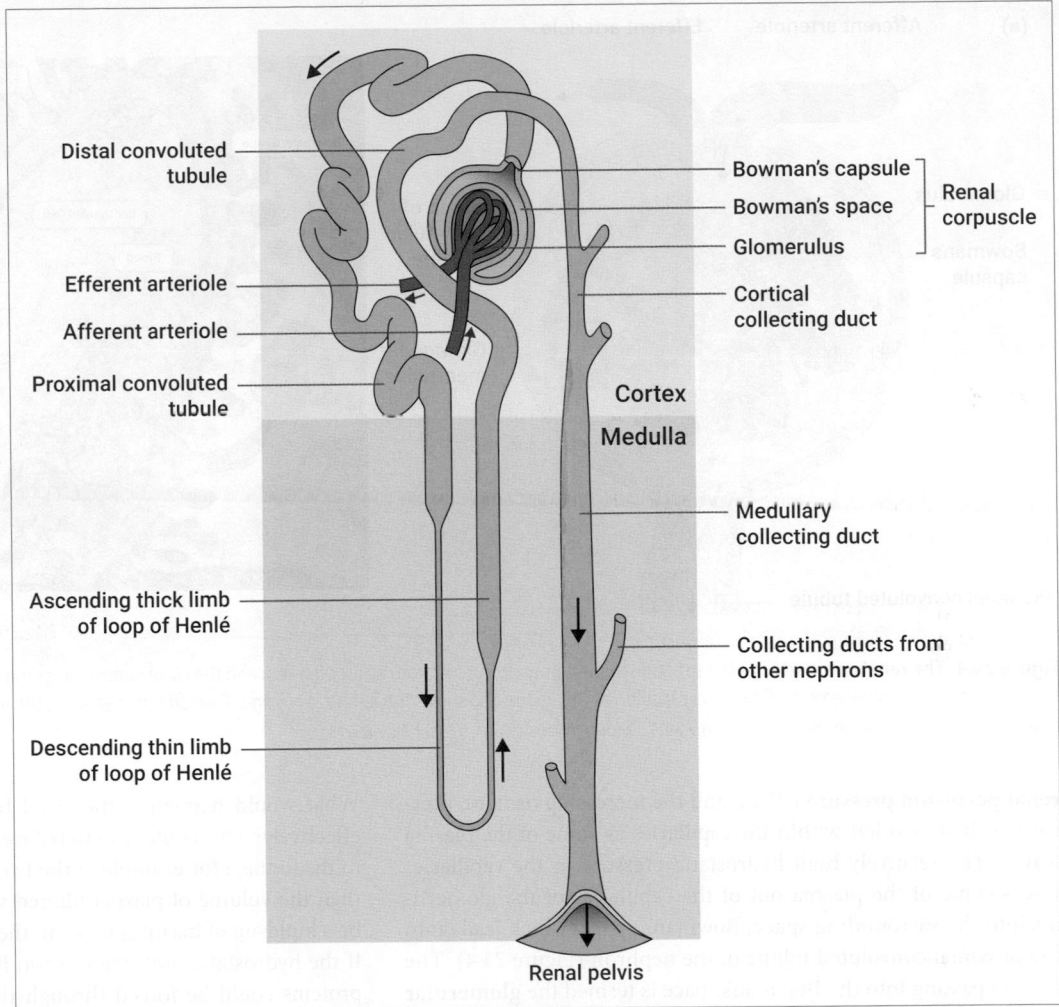

The nephron can be divided into five sections, each of which performs a different function:

1. Renal corpuscle.
2. Proximal convoluted tubule.
3. Loop of Henlé.
4. Distal convoluted tubule.
5. Collecting duct.

In the following sections we will be considering the processes involved in each part of the nephron as the composition of the fluid passed from the bloodstream is modified, eventually ending up as urine that is excreted.

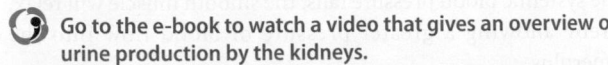 Go to the e-book to watch a video that gives an overview of urine production by the kidneys.

Check your understanding of the concepts covered in this section by answering the questions in the e-book.

24.2 Filtration in the renal corpuscle

The first stage of the kidney's operation is the filtration of the blood. This is a predominantly physical process whereby about 20% of the plasma from the renal arteries passes through the filters into the lumen of the nephron; the remainder of the plasma, along with all the red and white blood cells, platelets, and plasma proteins remains in the capillaries.

The filtration process takes place in the **renal corpuscle** of the nephron, located in the cortex of the kidney. The renal artery divides into numerous afferent arterioles, each of which enters the **Bowman's capsule** of a renal corpuscle (Figure 24.4). In turn, the afferent arteriole divides into many fine capillaries forming a network called the **glomerulus**. Bowman's capsule is rather like a cup that envelops the network of capillaries.

The endothelium of the capillaries comprises a single cell layer that is only about 0.5 µm thick and has numerous pores, called **fenestrations**. The glomerular basement membrane is, in turn, surrounded by the epithelial cells of the capsule, which form **podocytes**. The fenestrations of the endothelial cells and gaps between the podocytes create filtration slits which allow fluid to pass through, as through a sieve. Like a sieve, the size of the slits determines the size of the substances that can pass through, so large molecules such as proteins and blood cells and platelets are too large to pass through. Furthermore, the glycoproteins of the glomerular basement membrane surrounding the endothelial cells carry a negative charge, which tends to repel most proteins as these usually also carry a negative charge.

The key to filtration in the kidney is balance between the hydrostatic pressure of the blood entering via the afferent arteriole, the

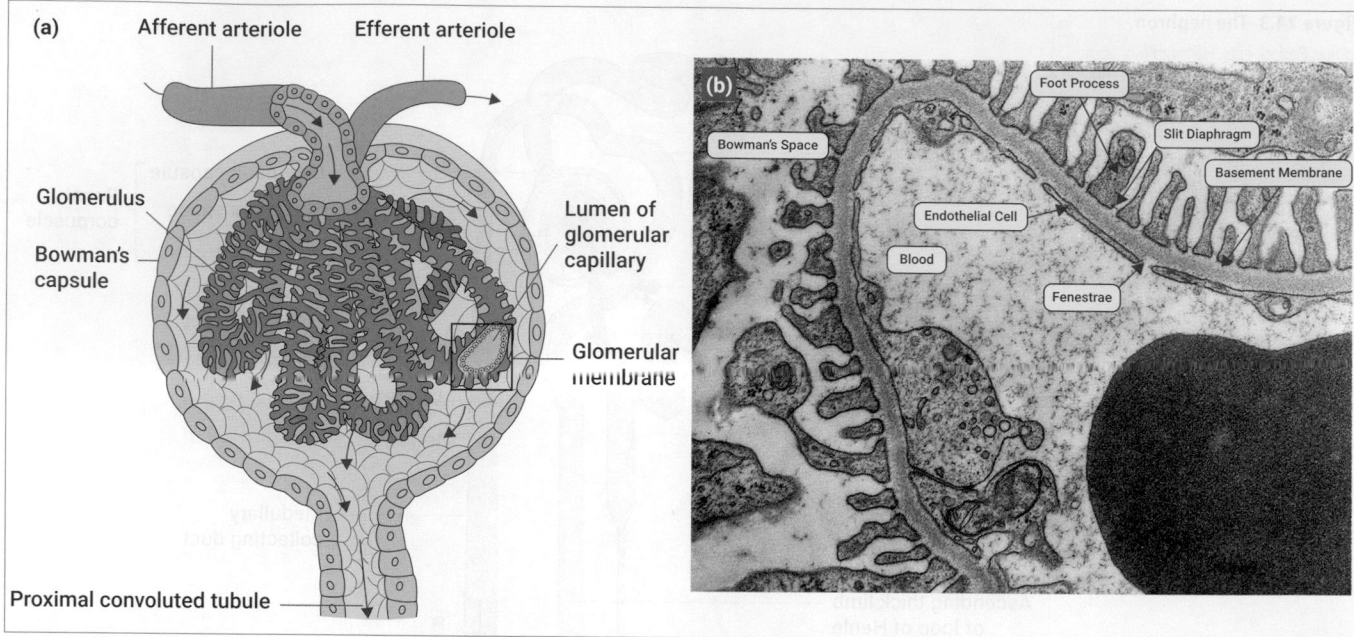

Figure 24.4 The renal corpuscle. (a) The structure of the corpuscle; (b) electron micrograph showing the close apposition of the endothelial cells of the capillary wall to the basement membrane of the Bowman's capsule. (b) This image also shows the fenestrations and filtration slits that act as the filtration sieve.

Source: (a) Pocock et al. *Human Physiology*, 5th edn, 2017. Oxford University Press; (b) Prof. Dr. Hany Marei.

renal perfusion pressure (RPP), and the increasing osmotic pressure of the blood left within the capillaries as some of the plasma leaves. The relatively high hydrostatic pressure in the capillaries forces some of the plasma out of the capillaries of the glomerulus into the surrounding space, Bowman's space, which leads into the proximal convoluted tubule of the nephron (Figure 24.4). The plasma passing into the Bowman's space is termed the **glomerular filtrate** and the rate at which it enters the space is the glomerular filtration rate (GFR).

What would happen if the capsule around the kidney was ruptured? Take a moment to think about this question before reading on.

If the capsule around the kidney is ruptured, the hydrostatic pressure within the kidney could be dissipated and so filtration would no longer take place. As the hydrostatic pressure drives the fluid out of the capillaries, however, the osmotic pressure of the plasma remaining in the capillaries rises due to the oncotic pressure exerted by the plasma proteins that are left behind. This then acts to limit the amount of plasma that enters the filtrate. (Refer back to our discussion of Starling's law of fluid formation in Section 21.2.)

Although both the hydrostatic and osmotic pressures in the arterial circulation can vary, the GFR is normally maintained within a narrow range by autoregulation.

The glomerular filtration rate

The GFR is regulated by two main processes that control the renal perfusion pressure at a local level:

- myogenic regulation of blood flow;
- tubuloglomerular feedback.

What would happen if the renal blood flow was not regulated effectively? This could lead to failure of kidney function or damage to the kidney; for example, if the hydrostatic pressure was too low, then the volume of plasma filtered would reduce and there could be a build-up of harmful waste in the blood and retention of water. If the hydrostatic pressure was too high, larger molecules such as proteins could be forced through the filtration pores, so protein would be lost from the body via the urine (proteinuria) and also there could be damage to the filtration membrane.

The hydrostatic pressure of the blood passing through the glomerulus depends on the difference between the pressures in the blood entering the glomerulus via the afferent arteriole and that leaving it via the efferent arteriole (Figure 24.4). The smooth muscle wall of the afferent arteriole can undergo **myogenic contraction**:

- If the blood pressure entering the afferent arteriole increases, the smooth muscle lining the arteriole is stretched, which triggers it to contract, restricting the flow of blood, so the increase in input pressure is not transferred to the capillaries of the glomerulus.

- If the systemic blood pressure falls, the smooth muscle will relax, thereby allowing a greater pressure of blood flow into the glomerulus.

Myogenic contraction of the afferent arterioles therefore directly acts to regulate the RPP to the glomerulus and therefore the hydrostatic pressure forcing the plasma into the Bowman's space.

Tubuloglomerular feedback is a feedback loop that is regulated by the flow of filtrate through the distal tubule of the nephron. The top of the ascending limb of the loop of Henlé of the nephron loops

Figure 24.5 The juxtaglomerular apparatus.
Source: © The Open University.

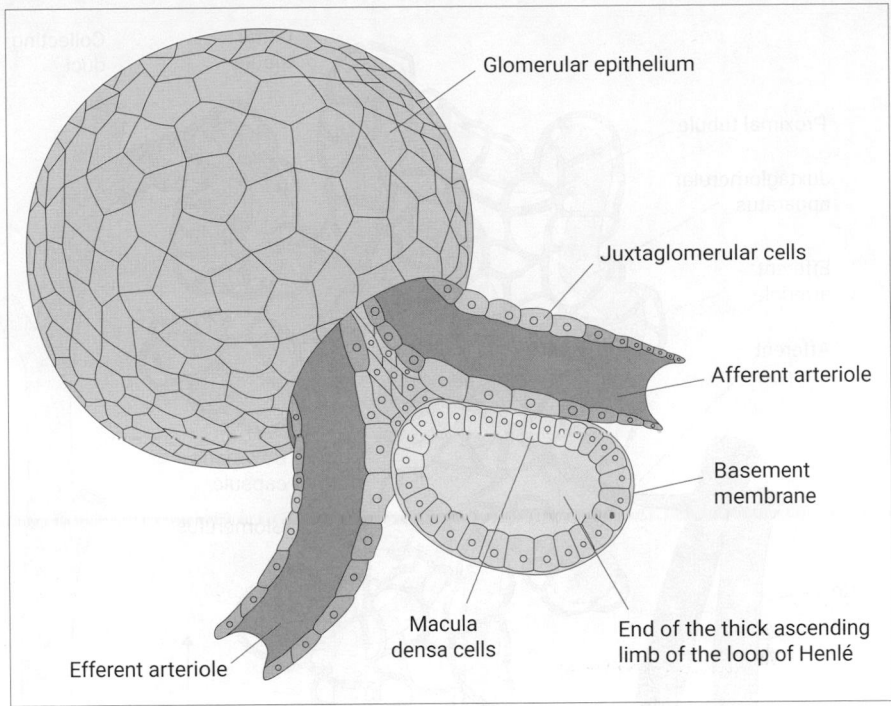

back so that it lies up against the afferent and efferent arterioles where they enter the renal corpuscle (Figures 24.2 and 24.5). At the top of the loop of Henlé, the tubule is lined by specialized cells, the **macula densa** cells.

An increase in GFR will lead to an increase in the delivery of NaCl through the loop of Henlé, as proportionately less is re-absorbed by the proximal tubule (Section 24.3). This leads, in turn, to more NaCl being reabsorbed by the macula densa cells. The macula densa cells, in turn, release ATP, which binds to puringeric re-ceptors on the adjacent afferent arteriole, causing it to vasoconstrict, and thereby reducing the renal blood flow and hence the GFR.

Note that the myogenic contraction and tubuloglomerular feed-back are localized mechanisms, as each one operates at the level of the individual nephron. The kidney also plays a major part in the systemic regulation of blood pressure (see Section 24.6), and its function is dependent on the effective regulation of the cardiovas-cular system; for example, patients suffering from heart failure may also experience renal failure because the blood pressure is insuf-ficient to maintain effective perfusion. As a result, these patients would show a build-up of toxic waste products in their blood.

PAUSE AND THINK

Why is it significant that the macula densa cells are located at the top of the loop of Henlé?

Answer: At this point they will be monitoring the fluid exiting the nephron and therefore can form an important part of the feedback sys-tem regulating the input flow—GFR.

 Check your understanding of the concepts covered in this section by answering the questions in the e-book.

24.3 The proximal convoluted tubule

All the nephron, including the **proximal convoluted tubule**, is surrounded by a network of capillaries, the peritubular capillaries (Figure 24.6). The capillaries arise from the efferent arterioles and therefore carry the blood that has remained in the blood vessels after passing through the renal corpuscle. Unusually, therefore, the blood flowing through the kidney vascular network passes through two sets of capillaries.

The filtrate that has passed into the Bowman's space enters the proximal convoluted tubule. This filtrate is similar in composition to plasma but lacks the blood cells and large molecules such as pro-teins. As the filtrate passes through the next stages of the nephron most of the substances are reabsorbed to re-enter the blood circu-lation via the renal vein (Table 24.1). Of the 180 L of plasma filtered by the kidneys only about 1.5 L is excreted as urine. Alongside the reabsorption processes, other substances are actively secreted into the filtrate, including hydrogen, ammonium and potassium ions, urea, creatinine, and some hormones and drugs (e.g. penicillin).

As can be seen from Table 24.1 the majority of the reabsorption takes place in the proximal convoluted tubule. Both reabsorption and secretion can occur via two routes (Figure 24.7):

- **Transcellular**—the substance is transported across the epithelial cell membrane adjacent to the lumen of the tubule (luminal membrane) to enter the cell and passes out into the interstitial space via the basolateral membrane. Transcellular transport is usually linked to active transport mechanisms, which use energy to transport the substance.

- **Paracellular**—the substance passes across the epithelial lining of the lumen by crossing the tight junctions and the intercellular spaces.

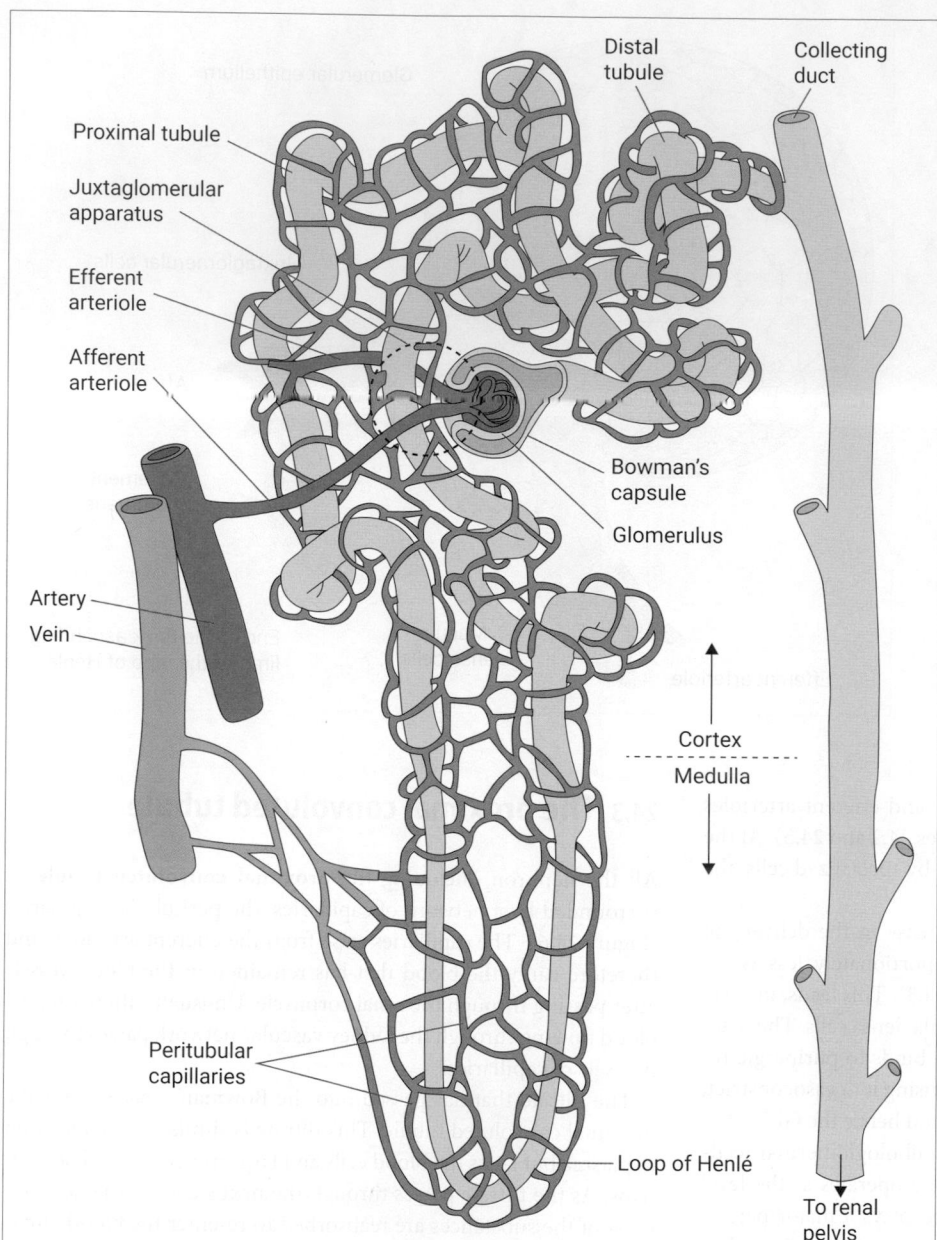

Figure 24.6 Peritubular capillary network.
Source: © The Open University.

Table 24.1 The proportions of major substances reabsorbed in the nephron

Substance	Total reabsorption (%)	Reabsorbed by the proximal convoluted tubule (%)	% Reabsorbed by the loop of Henlé (%)	% Reabsorbed by the distal convoluted tubule (%)	% Reabsorbed by the collecting duct (%)
Water	99	67	15	0	8–17
Sodium	99.5	67	25	7.5	0
Glucose	100	100	0	0	0
Amino acids	100	100	0	0	0
Urea	50	50	0	0	0

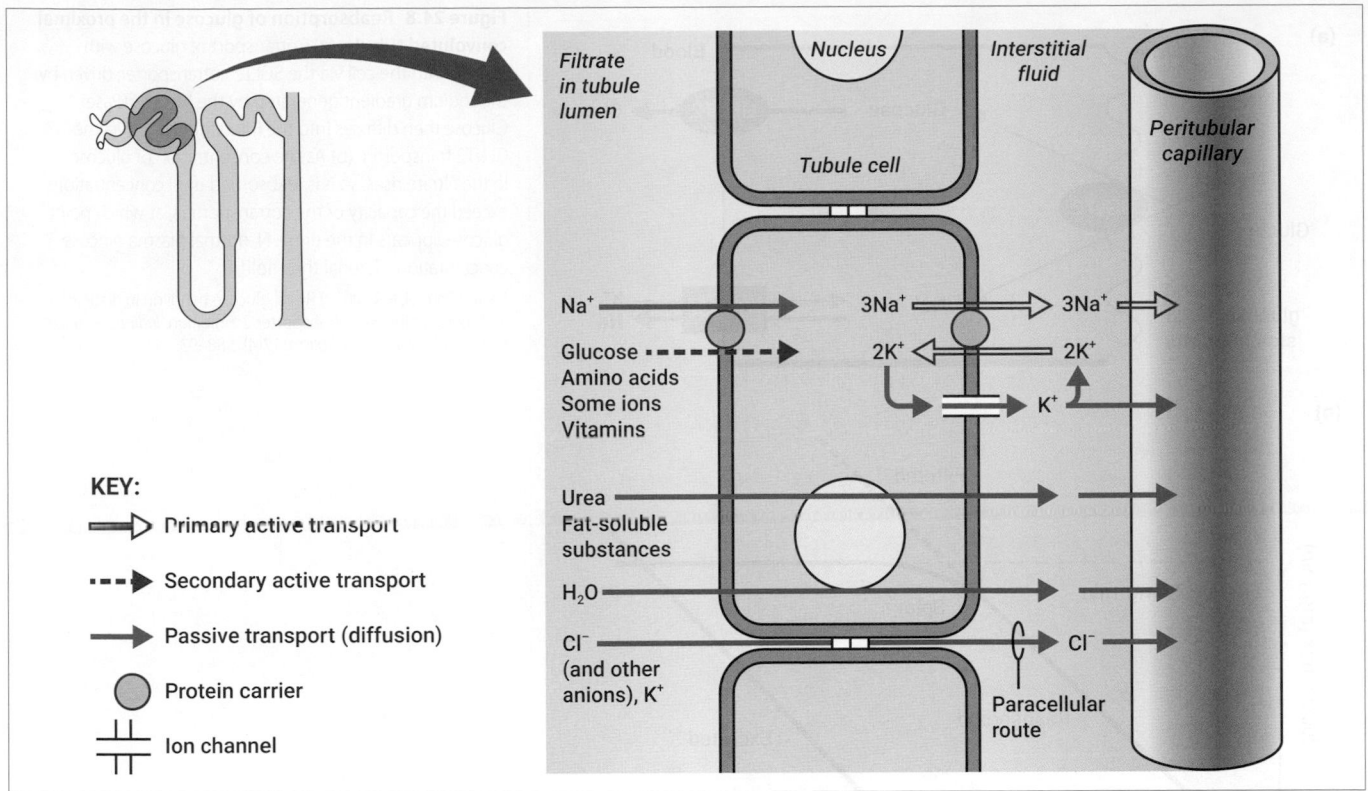

Figure 24.7 Processes of reabsorption in the proximal convoluted tubule.

Source: Oxford University Press.

Sodium reabsorption

The key to the reabsorption process is the action of the sodium–potassium ATPase ($Na^+K^+ATPase$) pump which is located on the basolateral membrane of the epithelial cells lining the tubules of the nephron.

The $Na^+K^+ATPase$ actively pumps Na^+ out of the luminal cells into the interstitial space, thereby creating a low Na^+ concentration within the cell. The process is similar to the mechanism in neurons with 3 Na^+ ions being pumped out in exchange for 2 K^+. The cells are permeable to potassium and so this ion diffuses back out of the cell, leaving no net increase in potassium concentration (Figure 24.7).

The low sodium concentration within the cell, along with the membrane potential of about −70 mV, creates an electrochemical gradient for sodium movement from the tubular lumen into the cell. This gradient is then the driver for much of the reabsorption of other substances.

Glucose reabsorption

As can be seen from Table 24.1, all the filtered glucose is normally reabsorbed and so no glucose appears in the urine. The driving force for the reabsorption of glucose comes from the sodium gradients. There are two transporter proteins on the luminal membrane: the sodium glucose cotransporters SGLT1 and SGLT2 (Figure 24.8;

see also Chapter 25). These cotransporters move Na^+ ions and glucose molecules into the tubule cells in a 1:1 ratio, with SGLT2 carrying about 90% of the glucose. Glucose then diffuses out of the cell, down its concentration gradient, via the glucose transporters GLUT1 and GLUT2, passing into the interstitial space and then into the peritubular capillaries.

Because the glucose reabsorption process depends on the presence of the cotransporters on the luminal membrane, there is a maximal capacity for reabsorption beyond which the transporters are saturated. When this occurs, glucose starts to appear in the urine (Figure 24.8, lower panel), a condition called **glycosuria**. The total capacity for reabsorption varies between nephrons, as a result of slight variations in the density of the SGLT cotransporters; as a consequence, there is variation between the theoretical maximal capacity for reabsorption and the actual capacity, a phenomenon referred to as **splay**.

Why might glucose levels exceed the normal capacity of the cotransporters? (You could also look at Section 25.2, where we consider the processes involved in the absorption of glucose, when thinking about this question.)

If the blood glucose levels are not tightly controlled they can rise and fall rapidly; for example, following a meal, relatively large amounts of glucose will be absorbed by the digestive system and enter the bloodstream. These are normally regulated by insulin, but if there is insufficient insulin secretion or the tissues are resistant to insulin then the glucose is not taken up by the liver and

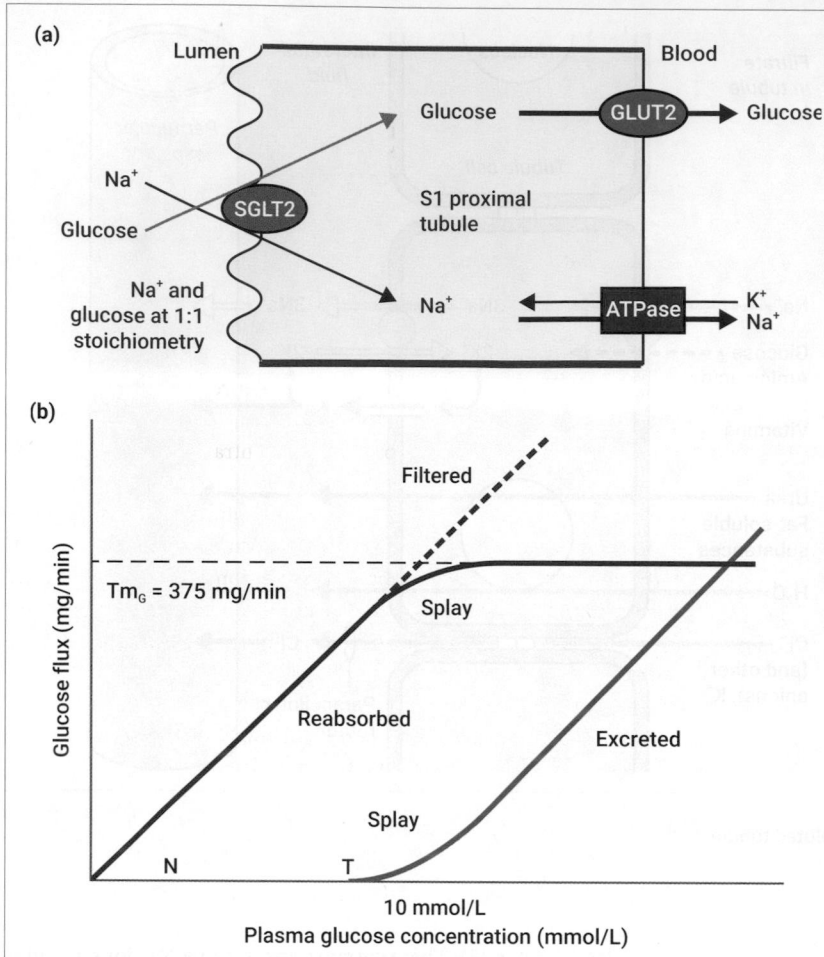

(a)

Lumen
Blood

Glucose — GLUT2 → Glucose

Na+
Glucose — SGLT2

S1 proximal tubule

Na+ and glucose at 1:1 stoichiometry

Na+ — ATPase → K+ / Na+

(b)

Tm_G = 375 mg/min

Filtered

Splay

Reabsorbed

Excreted

Splay

N   T

10 mmol/L
Plasma glucose concentration (mmol/L)

Glucose flux (mg/min)

Figure 24.8 Reabsorption of glucose in the proximal convoluted tubule. (a) Cotransport of glucose with sodium into the cell via the SGLT2 cotransporter, driven by the sodium gradient generated by the Na+K+ATPase. Glucose then diffuses into the interstitial space via the GLUT2 transporter. (b) As the concentration of glucose in the filtrate rises, so it is reabsorbed until concentrations exceed the capacity of the cotransporters, at which point glucose appears in the urine. N, normal plasma glucose concentrations; T, renal threshold.

Source: Poudel, R.R. (2013) Renal glucose handling in diabetes and sodium glucose cotransporter 2 inhibition. *Indian Journal of Endocrinology and Metabolism*, 17(4), 588–93.

muscles, and so the blood concentrations are elevated. These are features of diabetes.

Glycosuria is a classic sign of **diabetes mellitus**. This was first recognized by the sweet taste of the urine, hence the term mellitus, meaning sweet. In type 1 diabetes the pancreatic β cells fail to produce sufficient insulin, whereas in type 2 diabetes the liver and muscle, in particular, show insulin resistance and so do not take up the glucose from the blood (Section 19.3, Insulin). Diabetic patients show an increase in the expression of the glucose transporters, both SGLT and GLUT, thereby leading to an increase in the maximal capacity for reabsorption, but in the chronic condition this is not sufficient to prevent glucose loss in the urine.

Water reabsorption

About 67% of the water that is reabsorbed is through the proximal convoluted tubule (Table 24.1). Again, the driving force for reabsorption is the osmotic gradient generated by the active transport of Na+, associated with the movement of other solutes, includig glucose, HCO_3^-, an Cl^- from the filtrate into the interstitial spaces. The proximal tubule is highly permeable to water and so water flows into the interstitial spaces by both the transcellular and paracellular routes. The interstitial spaces are small in volume and so the flow of water creates a hydrostatic pressure that helps drive the movement of water and solutes into the capillaries.

Movement of other ions is also linked to the Na+ gradient: in the first half of the tubule the Na+ movements are linked with H+ extrusion from the luminal cells, by means of an antiporter exchange mechanism, whereby inward movement of a Na+ ion is coupled with outward movement of H+. There is also associated reabsorption of HCO_3^- ions. The balance of H+ extrusion and HCO_3^- reabsorption plays an important part in the acid–base balance of the blood. Amino acid reabsorption in the first half of the tubule is likewise linked with Na+ transport. In the second half of the tubule the Na+ reabsorption is particularly linked to chloride ion reabsorption.

The proximal convoluted tubule is also the site of secretion of a number of substances, including organic end-products of metabolism that are in the plasma. These include cyclic AMP, bile salts, urate, and creatinine. A wide range of drugs, including penicillin, salicylate (aspirin), and morphine, are also secreted. Many of the secreted substances are characterized by being transported in the plasma bound to plasma proteins and so they are not filtered. As a consequence, they can only be excreted in the urine if they are actively secreted from the peritubular capillaries into the filtrate.

What is the role of sodium ions in the transport of water and other substances across the luminal membrane?

Answer: The Na⁺K⁺ATPase pump maintains a concentration gradient of Na⁺ ions across the luminal cells by pumping Na⁺ ions out across the basal membrane. The concentration of Na⁺ ions inside the cells is therefore low with respect to the concentration in the lumen of the tubule. This gradient then facilitates the movement of other substances, either as cotransport (e.g. glucose and HCO_3^-) or antiport (e.g. H⁺ ions).

 Check your understanding of the concepts covered in this section by answering the questions in the e-book.

24.4 The loop of Henlé

As personal experience no doubt has shown you, the volume of urine produced each day varies significantly depending on your level of hydration. When you are dehydrated, urine production is low and the urine is very concentrated, typically a dark amber colour (mainly due to the presence of the breakdown products from bile salts). When you have drunk a lot of water and are well hydrated, the urine is dilute and may be almost colourless. The kidney, therefore, can produce urine that is hyperosmotic or hypoosmotic to the blood. The loop of Henlé plays a key role in the regulation of water balance by creating a steep osmotic gradient from the outer edge of the medulla through to the centre of the kidney.

The loop of Henlé is wrapped around by an extensive capillary network (Figure 24.6), the **vasa recta**, which is part of the peritubular capillary network arising from the efferent arteriole. The length of the loop of Henlé is related to the extent of the osmotic gradient; thus, cortical nephrons (with short loops) are associated with a lower osmotic potential than those that have long loops (juxtamedullary nephrons). The longer the loop of Henlé, the greater is the extent of the osmotic gradient.

Why might animals living in hot, dry climates tend to have long loops of Henlé?

Answer: These animals, such as the kangaroo rat or the camel, live in environments where water is scarce and tend to have long loops of Henlé, which enable the generation of a high osmotic gradient. As a result, they are able to produce very concentrated urine, thereby conserving water.

The key process in the creation of the osmotic gradient is again the active pumping of Na⁺ ions by the Na⁺K⁺ATPase in association with the relative permeability of the loop to water. Figure 24.9 shows the flows in and out of the loop.

The key features are:

- the descending limb is permeable to water but not to ions such as sodium;

- the ascending limb is impermeable to water but is permeable to NaCl;

- the upper part of the ascending limb has large numbers of Na⁺K⁺ATPase pumps that pump sodium ions out of the filtrate, with chloride ions following passively.

The filtrate that leaves the proximal convoluted tubule and enters the loop is isotonic, approximately 300 mosm L⁻¹. As the filtrate descends through the loop, water is drawn out by osmosis, creating an increasingly hypertonic filtrate: there can be a four-fold or more increase in concentration. This happens because the descending limb of the loop passes down through the medullary part of the kidney where there is an increasing osmotic concentration. Solutes do not follow the water into the interstitium because the descending limb is only permeable to water; therefore, the filtrate becomes increasing hypertonic.

What creates this high osmotic concentration in the medulla? This is where the Na⁺K⁺ATPase comes in: the pumps in the ascending limb pump sodium ions out of the limb, into the interstitium (Figure 24.9). This leads to the high osmotic concentration of the interstitium and the progressive dilution of the filtrate in the ascending limb because it is impermeable and so water cannot follow the sodium. So much sodium is pumped out that the filtrate leaving the ascending limb is actually hypotonic. The action of the pumps can be stimulated by the hormone **antidiuretic hormone** (ADH), which is released by the posterior pituitary in response to the detection of increased osmotic concentration of the blood by the hypothalamus (see Section 24.6). Increased osmotic concentration is indicative of dehydration and the action of ADH on the ascending limb is to increase Na⁺ transport, contributing to the concentration gradient in the interstitium (see Scientific Process 24.1).

As fresh filtrate enters from the proximal convoluted tubule it pushes the filtrate in the descending limb of the loop further down, while progressively losing water, so the fluid at the top is always isotonic while there is a progressive increase in concentration towards the bottom of the loop. Because the fluid at the bottom of the loop is highly concentrated, in the first stage of the ascending limb sodium and chloride ions can leave the loop passively (Figure 24.9) before the filtrate reaches the thick part of the ascending limb from which sodium ions are actively pumped.

The U-shaped loop with the filtrate moving in opposite directions and with increasing concentration on the descent and decreasing concentration on the ascent is referred to as the **countercurrent multiplier**.

 Go to the e-book to watch a video that explores how the loop of Henlé operates as a countercurrent multiplier.

Running parallel to the loop of Henlé is the vasa recta, the capillary network that forms a descending and an ascending loop in parallel to the loop of Henlé (Figure 24.6). The capillaries are freely permeable to both water and solutes and so are always at the same osmotic potential as the surrounding interstitial space. The flow of blood through the vasa recta is relatively slow, so there is sufficient time for the blood always to be equilibrated to the osmotic potential of the surrounding interstitium.

As a consequence, as blood flows down the descending limb of the capillary, deeper into the medulla, water flows out and solutes flow in, thereby increasing the concentration of the blood. In the

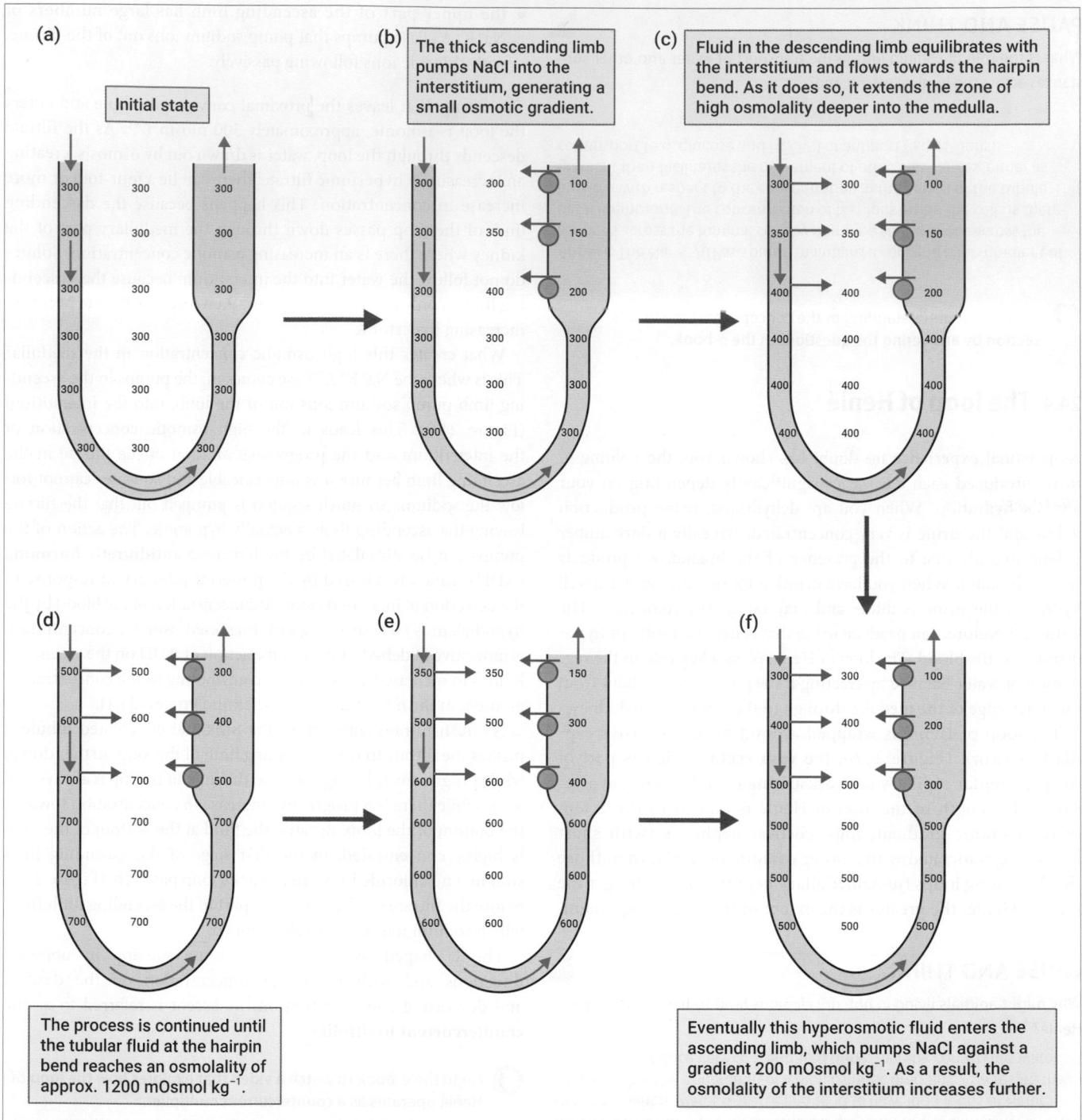

(a) Initial state

(b) The thick ascending limb pumps NaCl into the interstitium, generating a small osmotic gradient.

(c) Fluid in the descending limb equilibrates with the interstitium and flows towards the hairpin bend. As it does so, it extends the zone of high osmolality deeper into the medulla.

(d) The process is continued until the tubular fluid at the hairpin bend reaches an osmolality of approx. 1200 mOsmol kg^{-1}.

(e)

(f) Eventually this hyperosmotic fluid enters the ascending limb, which pumps NaCl against a gradient 200 mOsmol kg^{-1}. As a result, the osmolality of the interstitium increases further.

Figure 24.9 The loop of Henlé.

Source: Pocock et al. *Human Physiology*, 5th edn, 2017. Oxford University Press.

ascending limb, as it passes back up through the medulla towards the cortex, the solute concentration in the interstitium decreases and so solutes leave the ascending limb of the capillary and pass into the interstitium, and water travels in the opposite direction into the capillary.

You will notice, therefore, that at each point along the capillary, solutes are flowing out of the ascending limb and into the descending limb and water is flowing in the opposite direction: from the descending limb of the capillary into the ascending limb. As a result, the blood leaving the vasa recta is at almost the same

SCIENTIFIC PROCESS 24.1 **Antidiuretic hormone stimulates the reabsorption of sodium in the thick ascending limb of the loop of Henlé**

Background

The movement of sodium ions out of the ascending limb of the loop of Henlé is an essential part of the process of regulation of the concentration of urine (see Section 24.4). Sodium reabsorption can occur via the active pumping action of the $Na^+K^+ATPase$, but Na^+ can also pass out of the luminal fluid via the paracellular route, passing through the tight junctions between the cells via ion-selective proteins called claudins.

Hypothesis

It is known that ADH can stimulate the active transport of sodium from the thick part of the ascending limb via the transcellular route. The question asked by the researchers was whether ADH can also upregulate sodium transport via the paracellular route.

Methods

The experiments were performed on mice in accordance with German and Danish laws on animal welfare and with local ethical approval. Three groups of mice were studied:

1. Mice with normal access to standard food and water.
2. Mice fed with a water-restricted diet (WR).
3. Mice fed with a water-loaded diet (WL).

What was the purpose of including the WR and WL rats in the trial?

The researchers predicted that the WR rats would have a maintained high level of circulating ADH since they were dehydrated. This group therefore allowed them to test whether the response to ADH would be maintained over a long period of time and not just for the short periods when the tubes were perfused with ADH. The WL group would have low levels of ADH in circulation and therefore acted as a control.

After 5 days on the diets the mice were euthanized and the ascending limbs from the loops of Henlé were dissected out. These were then perfused with a standard physiological solution at 37 °C. The NaCl diffusion potential was measured while the thick section of the ascending limb was perfused with solutions with and without ADH. In a second set of experiments the drug furosemide was added to the perfusing solution. Furosemide is a diuretic drug that blocks the transcellular transport of sodium in the loop.

Results and conclusion

Transport of sodium from the lumen of the limb was increased in the presence of ADH. The addition of furosemide blocked the transcellular route, but the sodium transport was still increased above control levels in the presence of ADH. These findings indicate that the action of ADH increases sodium transport via both the transcellular and the paracellular routes.

In the case of the rats that had been subject to the water-restricted diet (WR), these showed an increase in baseline sodium transport compared with the WL rats, even in the absence of additional ADH in the perfusing solution. These findings show that chronic increase in ADH secretion, as in the WR rats, leads to a maintained increase in sodium transport from the lumen, across the thick ascending limb of the loop.

Read the original work

Himmerkus, N., Plain, A., Marques, R. D., Sonntag, S. R., Paliege, A., Leipziger, J., & Bleich, M. (2017) AVP dynamically increases paracellular Na^+ permeability and transcellular NaCl transport in the medullary thick ascending limb of Henle's loop. *Pflugers Arch.* **469**: 149–58. Doi: 10.1007/s00424-016-1915-5

---

osmotic potential as the blood entering it and there is no dissipation of the osmotic gradient in the medulla.

 Check your understanding of the concepts covered in this section by answering the questions in the e-book.

24.5 Distal convoluted tubule and collecting ducts

The filtrate leaving the loop of Henlé is hypotonic compared with the plasma and the filtrate that entered the loop. This dilute filtrate enters the distal convoluted tubule (Figure 24.3). The first part of the distal convoluted tubule is involved in further reabsorption of sodium by the same mechanisms as in the proximal convoluted tubule (Figure 24.7): there is a $Na^+K^+ATPase$ on the basal membrane which pumps Na^+ ions out of the luminal cells. This creates a low concentration of Na^+ ions in the cells which leads to

cotransport of Na^+ ions moving from the filtrate into the cells in combination with Cl^- ions via a symporter. This section of the distal tubule is impermeable to water and so the process of dilution of the filtrate continues.

The second section of the distal convoluted tubule again sees the reabsorption of Na^+, but the luminal cells also secrete K^+ into the filtrate and they can also secrete H^+ ions, thereby playing an important role in acid–base balance (Section 24.7).

The collecting duct and regulation of water reabsorption

The filtrate leaving the distal convoluted tubule enters the final stage of the nephron, the collecting duct (Figure 24.3). As its name suggests, the collecting duct collects the filtrate from a number of nephrons that feed into it and its output will be the urine that collects in the renal pelvis before being carried to the bladder via the ureter (Figures 24.1 and 24.2).

So far, as the filtrate has travelled along the nephron, the reabsorption of water has been unregulated as it passively follows sodium as it is reabsorbed. This accounts for over 80% of the water that is reabsorbed (Table 24.1). The collecting duct is the main region where water reabsorption is controlled: the filtrate entering the collecting duct is hypotonic, but the urine that is finally produced may be dilute or concentrated depending on the overall level of body hydration.

As can be seen from Figure 24.3, the collecting duct passes down from the renal cortex through the medulla. As it does so, it encounters the osmotic gradient created by the actions of the loop of Henlé. The permeability of the epithelium of the collecting duct to water determines how concentrated the urine is: if the collecting duct is permeable to water then water will be reabsorbed by osmosis, due to the higher interstitial osmotic concentration, thereby producing a concentrated urine. If the collecting duct is impermeable, then the water will not be reabsorbed and the urine will be dilute.

The permeability of the collecting duct is determined by the number of water channels, called **aquaporins**, that are present in the luminal membrane (Figure 24.10). There are 11 different types of aquaporin (AQP) found in different organs of the body. In the collecting duct AQP2 is found in the luminal membrane, while AQP3 and AQP4 are found in the basal membrane.

When AQP2 channels are present in the luminal membrane, water will be drawn by osmosis across into the luminal cell and then out into the interstitial space via AQP3 and AQP4. From there it will be drawn into the capillary network of the vasa recta. The density of AQP2 channels is regulated by ADH, also called vasopressin, (diuresis is the term for an increase in urine production). ADH acts on receptors on the basal membrane to stimulate the production of AQP2 proteins, which are inserted into

the luminal membrane forming channels for the water to pass through.

ADH is secreted by the posterior pituitary gland under control from the hypothalamus. When the body is well hydrated, ADH secretion is normally suppressed; however, when the body becomes dehydrated, ADH is secreted, leading to increased production of AQP2 channels, and more water is reabsorbed.

PAUSE AND THINK

How are the fluid levels, osmotic concentration of the blood, and blood pressure monitored centrally?

Answer: Two main ways are:

1. When the fluid levels decrease, the osmotic concentration of the blood will rise. This is detected by osmoreceptors located in the hypothalamus.

2. Reduced fluid levels will also lead to a reduction in blood pressure, which is monitored by the baroreceptors in the aorta and carotid arteries. Reduced stretch of the arterial wall reduces the sensory outflow from the baroreceptors.

Reduced baroreceptor activity and detection of an increased osmotic concentration both lead to increased ADH production and so water is conserved.

What other responses might be stimulated? Maybe think about times when your body needs to conserve water—how do you feel? The obvious behavioural response is a feeling of thirst, stimulating drinking, and thereby increasing water levels.

Thinking about the role of the kidney in regulating blood volumes and the relationship with blood pressure, why might patients with hypertension be given diuretics? One way of reducing blood pressure is to reduce the fluid volumes; therefore,

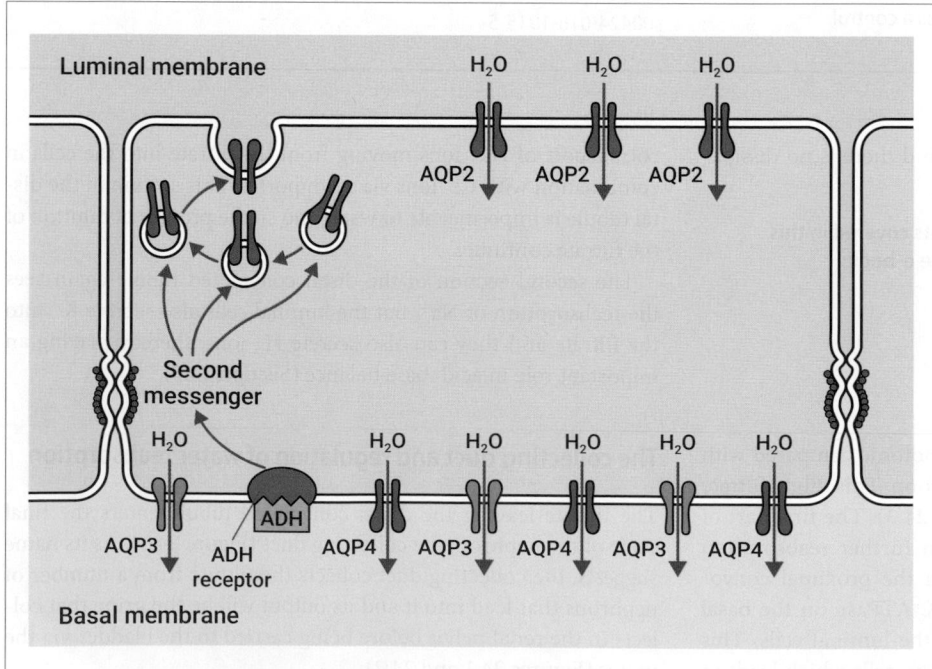

Figure 24.10 Aquaporins (AQP) in the collecting duct. The density of AQP2 channels in the luminal membrane is regulated by antidiuretic hormone, so controlling the amount of water that is reabsorbed.

Source: © The Open University.

giving diuretics reduces water reabsorption and so can lead to a reduced blood pressure. For example, furosemide reduces sodium reabsorption in the ascending limb of the loop of Henlé, so reducing the osmotic gradient around the collecting duct, or thiazide diuretics reduce sodium reabsorption in the distal convoluted tubule, thereby increasing the osmotic concentration of the fluid entering the collecting duct so that less water is drawn out by osmosis.

Summary of the journey along the nephron

Figure 24.11 shows the sections of the nephron from the initial filtration of the blood in the Bowman's capsule and the stages of reabsorption of key substances through to the production of urine via the collecting duct.

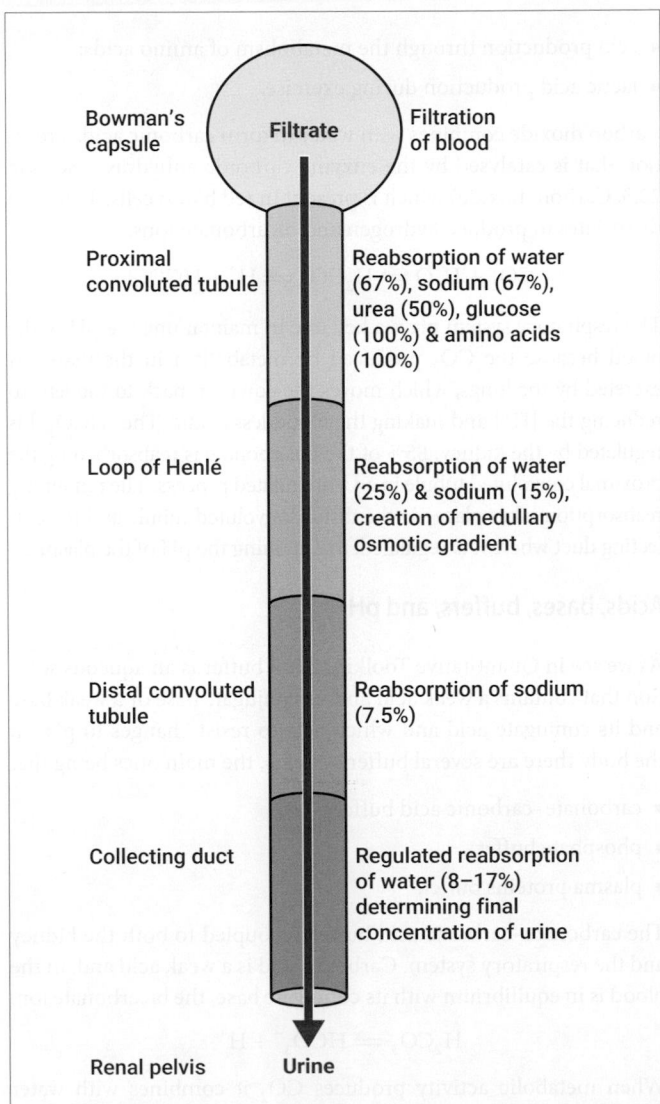

Figure 24.11 Stages of reabsorption along the length of the nephron. The substances being reabsorbed are returned to the vascular circulation via the peritubular capillaries.

 Go to the e-book to explore an interactive version of Figure 24.11.

 Check your understanding of the concepts covered in this section by answering the questions in the e-book.

24.6 The kidney and blood pressure

The kidney plays an important part in the regulation of blood pressure in a number of ways. We have already seen that urine production and blood pressure are related: reduced fluid volumes lead to a fall in blood pressure that results in ADH secretion, thereby conserving water and maintaining blood pressure. There are three main ways in which the kidneys are involved in the regulation of blood pressure:

1. Regulation of water reabsorption.
2. Regulation of sodium reabsorption.
3. Renin–angiotensin system.

The regulation of water and of sodium reabsorption have similar net outcomes by affecting the fluid volume of the body. Increases in sodium reabsorption increase the osmotic concentration of the blood and so lead to more water being reabsorbed. Table 24.2 shows the principal agents affecting reabsorption.

Renin–angiotensin system

The smooth muscle of the afferent arterioles is stretched when the blood pressure rises (Section 24.2), generating local responses that regulate GFR in the individual nephron. When the pressure of the blood in the arterioles falls the smooth muscle cells secrete **renin**. Renin is also secreted in response to sympathetic nerve activity and to a reduction in the flow of sodium past the macula densa, which is part of the tubulo-glomerular feedback system (Section 24.2). A fall in the sodium passing the macula densa indicates a reduction in GFR.

Renin is a proteolytic enzyme. As it circulates in the blood it cleaves **angiotensinogen**, which is synthesized by the liver, to produce **angiotensin I**. Angiotensin I, in turn, is cleaved by the enzyme **angiotensin converting enzyme** (ACE) to generate the active hormone **angiotensin II**. ACE is mainly found in the endothelial cells of the capillaries in the lungs. Angiotensin II is a potent effector in increasing blood pressure by:

1. Stimulating vasoconstriction of systemic arterioles.
2. Stimulating the adrenal cortex to release aldosterone.
3. Stimulating ADH secretion.
4. Increasing sodium reabsorption by the proximal convoluted tubule.

PAUSE AND THINK

Why might patients with hypertension be given ACE inhibitors such as ramipril to reduce their blood pressure?

Answer: ACE inhibitors block the action of angiotensin converting enzyme and so prevent angiotensin I being converted into angiotensin II.

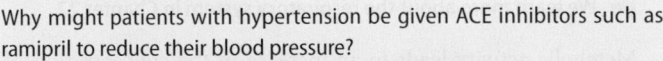

Table 24.2 Factors regulating the reabsorption of sodium and water

Agent	Stimulus	Main site(s) of action	Action
Angiotensin II	Renin secretion	Proximal convoluted tubule	Sodium and water reabsorption
Aldosterone	Aldosterone II	Ascending limb of the loop of Henlé	Sodium reabsorption
	Increased [K⁺]	Collecting duct	Sodium and water reabsorption
Atrial natriuretic peptide (ANP)	Increased blood pressure	Collecting duct	Decreased sodium and water reabsorption
	Increased fluid volumes		
Sympathetic nervous system (noradrenaline and adrenaline)	Decreased fluid volumes	Proximal convoluted tubule	Sodium and water reabsorption
		Ascending limb of the loop of Henlé	Sodium reabsorption
		Collecting duct	Sodium and water reabsorption
Antidiuretic hormone (ADH)	Increased blood osmolarity	Collecting duct	Water reabsorption
	Decreased fluid volumes		

Aldosterone is one of the hormones produced by the adrenal cortex. Aldosterone acts to increase the reabsorption of sodium by the cells of the ascending limb of the loop of Henlé and by the collecting duct. As a consequence the extracellular fluid volume is also increased due to water following the increased sodium reabsorption.

A further hormone that involves the kidney in its action to affect blood pressure is **atrial natriuretic peptide** (ANP). ANP is released by the muscle cells of the atria of the heart in response to stretching, associated with an increased fluid volume and therefore blood pressure. ANP acts in opposition to the renin–angiotensin system to:

1. Inhibit renin release.
2. Inhibit aldosterone secretion.
3. Inhibit sodium reabsorption by the collecting duct.
4. Inhibit ADH secretion.
5. Cause vasodilation of the renal afferent arterioles and constriction of the efferent arterioles, thereby increasing GFR.

ANP therefore increases the excretion of sodium, potassium, and water by the kidneys, inhibits the renin–angiotensin system, and so has a role in reducing blood pressure.

 Check your understanding of the concepts covered in this section by answering the questions in the e-book.

24.7 The kidney and acid–base balance

The regulation of blood pH is critical to the function of many enzymes. Normally, the pH is maintained between 7.36 and 7.44, and the kidney, along with other organ systems, in particular the respiratory system, plays a significant role in this function.

 We learn more about the respiratory system in Chapter 22.

Metabolic activity leads to an increase in the concentration of H⁺ ions through a number of routes, including:

- production of CO_2 through the metabolism of carbohydrates and fats;

- acid production through the metabolism of amino acids;
- lactic acid production during exercise.

Carbon dioxide combines with water to form carbonic acid, a reaction that is catalysed by the enzyme carbonic anhydrase (Section 22.3, Carbon dioxide) which is present in red blood cells. This then dissociates to produce hydrogen and bicarbonate ions:

$$CO_2 + H_2O \rightleftharpoons H_2CO_3 \rightleftharpoons H^+ + HCO_3^-$$

The respiratory system plays a key role in maintaining the pH of the blood because the CO_2 produced by metabolism in the tissues is excreted by the lungs, which moves the equation back to the left, so reducing the [H⁺] and making the blood less acidic. The [HCO₃⁻] is regulated by the kidney: 85% of the bicarbonate is reabsorbed by the proximal convoluted tubule in an unregulated process. The remaining reabsorption takes place via the distal convoluted tubule and the collecting duct where it is regulated, maintaining the pH of the plasma.

Acids, bases, buffers, and pH

As we see in Quantitative Toolkit 22.2, a buffer is an aqueous solution that contains a weak acid and its conjugate base or a weak base and its conjugate acid and which acts to resist changes to pH. In the body there are several buffer systems, the main ones being the:

- carbonate–carbonic acid buffer;
- phosphate buffer;
- plasma proteins buffer.

The carbonate–carbonic acid buffer is coupled to both the kidney and the respiratory system. Carbonic acid is a weak acid and, in the blood is in equilibrium with its conjugate base, the bicarbonate ion:

$$H_2CO_3 \rightleftharpoons HCO_3^- + H^+$$

When metabolic activity produces CO_2 it combines with water leading to an increase in the [H⁺]:

$$CO_2 + H_2O \rightleftharpoons H_2CO_3 \rightleftharpoons HCO_3^- + H^+$$

In the plasma [H₂CO₃] is low and [HCO₃⁻] is relatively high so when hydrogen ions are produced the increased acidity is

CLINICAL BOX 24.1 Kidney failure and renal dialysis

As we have seen, the kidneys play a key role in removing metabolic waste products, regulating the composition of the blood and controlling blood pressure. If their function is impaired, for example by an acute injury, or by chronic kidney disease (CKD), this can have serious consequences for the health of the patient. The most common causes of CKD include poorly controlled hypertension, diabetes mellitus, or polycystic kidney disease, which is a genetic disorder.

The symptoms of CKD include weight loss, oedema with swelling of the ankles or hands due to water retention, shortness of breath, and there may be blood in the urine. The typical tests for CKD include an estimation of GFR (eGFR), testing for the presence of blood or protein in the urine, and the ratio of albumin:creatine in the urine. The normal eGFR is above 90 mL min^{-1}: if the eGFR falls below this level, this is indicative of reduced function. Progression of the disease can be slowed by control of diet, weight loss, stopping smoking, and treatment to reduce hypertension.

If the eGFR falls too low and kidney function is severely impaired such that there is a build-up of toxic waste in the body, the patient may require dialysis. This involves artificially removing the waste products from the blood along with excess fluid that has accumulated. There are two main types of dialysis: haemodialysis and peritoneal dialysis.

Haemodialysis is the more common approach and involves circulating the blood through an external dialysis machine, which contains membranes that filter the blood, imitating the process that would occur in the glomerulus, and a dialysis fluid, which acts rather like the extracellular fluid, enabling exchange across a semipermeable membrane and which includes a counter-current mechanism (Figure 1). A permanent fistula may be used to enable access for collecting and returning the blood.

Patients undergoing haemodialysis will typically require treatment three times a week, for about 4 h each time. In between treatments, patients will normally gain weight due to fluid retention; therefore, the clinician can calculate the amount of fluid that needs to be removed during the dialysis process.

An alternative process, called peritoneal dialysis, involves flushing the abdomen with the dialysate and using the peritoneal membranes as the semipermeable filter membrane.

In the UK about 50,000 patients undergo dialysis or have a functioning kidney transplant. Among the biggest risks for these patients are infections associated with the treatment and with reduced immune system function.

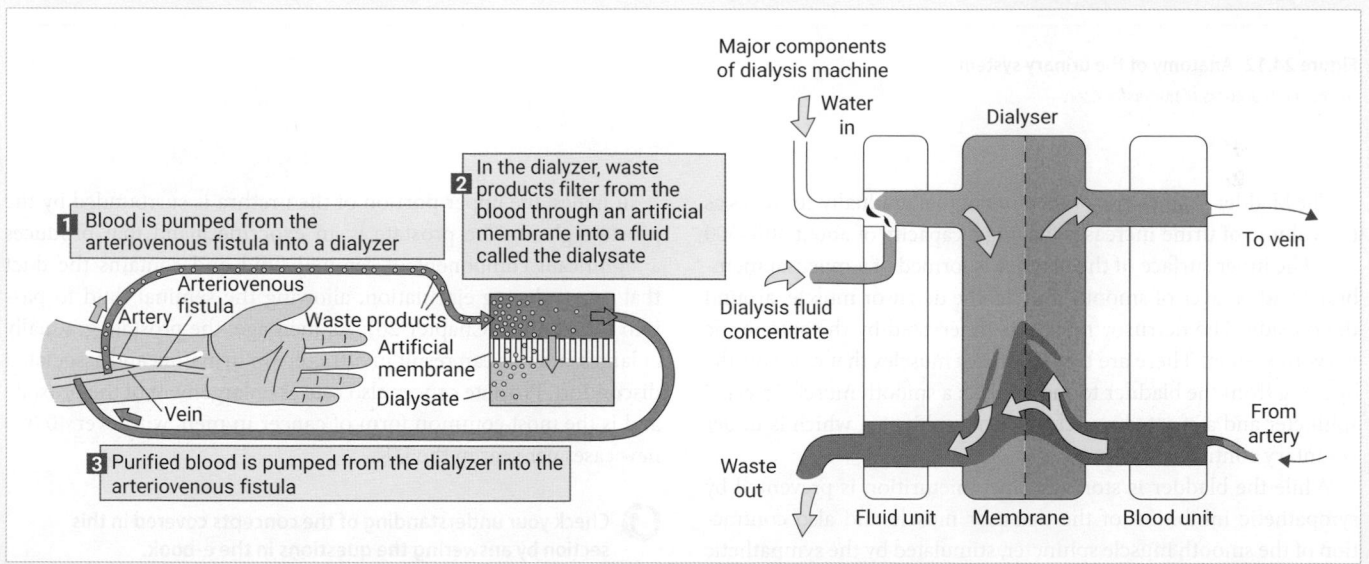

Figure 1 Haemodialysis. The right-hand panel shows the counter-current flow for the blood and the dialysate.
Source: © Legger/ Dreamstime.com.

buffered by the HCO_3^- forming carbonic acid; the change in pH is therefore resisted. The $[CO_2]$ and $[HCO_3^-]$ are tightly regulated by the respiratory system and the kidneys, respectively, thereby maintaining the pH homeostasis.

 Check your understanding of the concepts covered in this section by answering the questions in the e-book.

24.8 Micturition

The urine that leaves the collecting ducts passes into the renal pelvis (Figure 24.2) and then enters the ureter (Figures 24.1 and 24.12). The two ureters carry the urine to the **bladder**, where it is stored before being excreted via the urethra by the process of micturition.

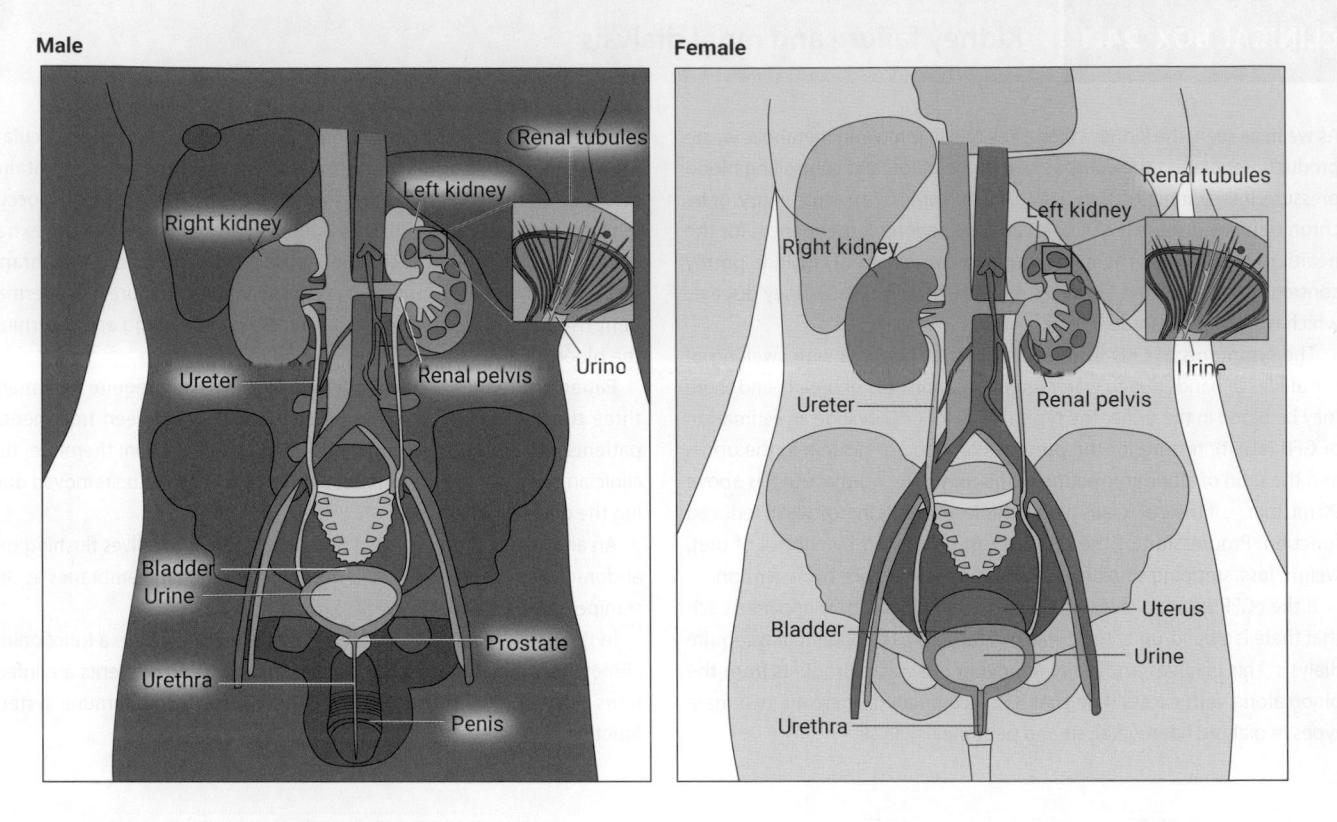

Figure 24.12 Anatomy of the urinary system.

Source: medicalstocks / Shutterstock.com.

The bladder is an elastic-walled organ that gradually stretches as the volume of urine increases, having a capacity of about 400–600 mL. The inner surface of the bladder is formed of a mucosal membrane with a layer of smooth muscle, the detrusor muscle, around the outside. The detrusor muscle is innervated by the autonomic nervous system. There are two sphincter muscles that close off the opening from the bladder to the urethra: a smooth muscle internal sphincter and a striated muscle external sphincter, which is under voluntary control.

While the bladder is storing urine, micturition is prevented by sympathetic inhibition of the detrusor muscle and also contraction of the smooth muscle sphincter, stimulated by the sympathetic nervous system, closing off the entrance to the urethra. The external sphincter is under voluntary control and can be stimulated to contract to prevent micturition.

As the bladder becomes full, stretch receptors in the wall of the bladder respond by signalling to the brainstem, which creates the urge to urinate. The parasympathetic nervous system triggers contraction of the detrusor muscle and contraction of the internal sphincter muscle is inhibited. When there is also voluntary relaxation of the external sphincter, the bladder can be emptied.

In males, the upper portion of the urethra is surrounded by the prostate gland. The prostate is an exocrine gland that produces a significant component of seminal fluid and contains the duct that opens during ejaculation, allowing the seminal fluid to pass into the urethra (Chapter 26). As men age, the prostate gradually enlarges and this can result in difficulty in urinating and associated discomfort. Prostate cancer also causes enlargement of the prostate and is the most common form of cancer in men, with over 40,000 new cases per year in the UK.

 Check your understanding of the concepts covered in this section by answering the questions in the e-book.

24.9 Endocrine functions of the kidney

The kidney has important endocrine functions and produces three hormones:

- renin;
- erythropoietin;
- calcitriol.

As we have already discussed in Section 24.6, renin, through its role in the renin–angiotensin system, has an important function in the regulation of blood pressure.

Erythropoietin

Erythropoietin is a glycoprotein cytokine that stimulates the production of the erythrocytes (red blood cells) by the bone marrow. The kidney is the major site of erythropoietin production, although some is also synthesized in the liver. There is normally a steady, low-level production of erythropoietin by the interstitial cells associated with the peritubular capillaries (Figure 24.6), which maintains the ongoing production of red blood cells. During periods of reduced oxygen transport, for example as a result of anaemia (see Clinical Box 22.2) or during altitude acclimation (Chapter 22), erythropoietin synthesis increases to stimulate more rapid red cell production.

PAUSE AND THINK

Why might endurance athletes train at high altitudes?

Answer: The hypoxic conditions lead to stimulation of erythropoietin synthesis and so increased red cell production. This will increase the total oxygen carrying capacity of the blood and so can result in improved endurance.

In recent years erythropoietin has been synthesized artificially and used as part of the treatment regime for patients with kidney failure to maintain red cell production. However, synthetic erythropoietin, often called EPO, has also had bad media coverage because it has been used as a performance enhancing drug by elite athletes, especially in high endurance events such as the Tour de France cycling competition. There is a risk associated with artificially increasing the red cell count because this also increases the viscosity of the blood, which can lead to cardiovascular disease (see Clinical Box 23.1).

Calcitriol

Calcitriol is produced by cells in the proximal convoluted tubule and, along with parathyroid hormone (PTH) and calcitonin, is one of the three hormones that regulate calcium homeostasis. The concentration of calcium in the extracellular fluid depends on the balance between absorption by the gastrointestinal tract, excretion via the kidneys, and resorption/uptake by bone (see Chapter 19).

Calcitriol acts to increase the levels of circulating calcium through three main routes:

- increasing the tubular reabsorption of calcium;
- increasing the uptake of calcium by the intestine by stimulation of production of calcium binding protein;
- increasing resorption of bone.

Calcitriol is a steroid hormone that is derived from **vitamin D**, with production being stimulated by PTH in response to lowered circulating concentrations of calcium. Vitamin D is derived from the diet and by cutaneous production in response to sunlight. A lack of vitamin D leads to hypocalcaemia due to reduced absorption of calcium. In children, this can affect bone growth, resulting in the bones being weak and bowed in shape with widened growth plates, a condition known as **rickets**, an X-ray of which you can see in Figure 24.13. In the elderly, hypocalcaemia can contribute to the development of osteoporosis, in which the bones become increasingly weak due to resorption of calcium, which makes them vulnerable to breakage.

The hormone calcitonin, which is produced by the thyroid gland, has opposing actions by inhibiting osteoclast activity in the bones, so preventing resorption of calcium and reducing reabsorption by the kidney, although the latter action appears to be limited in humans.

As well as its importance in skeletal structure, calcium homeostasis is also very important in maintaining the function of the nervous system: hypocalcaemia can lead to increased excitability of nerve and muscle, whereas hypercalcaemia decreases the excitability and can also result in cardiac arrhythmias.

 Check your understanding of the concepts covered in this section by answering the questions in the e-book.

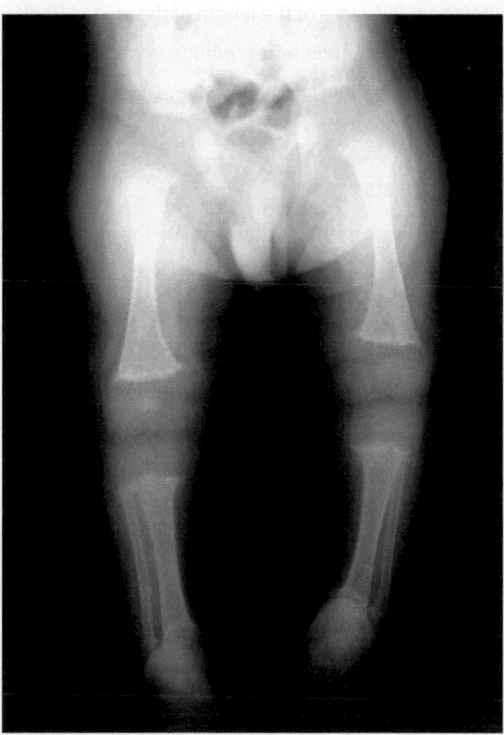

Figure 24.13 X-ray showing bowing of the legs in a young child with severe rickets.
Source: Science Photo Library / Alamy Stock Photo.

SUMMARY OF KEY CONCEPTS

- The kidney plays a key role in the regulation of water and ion balance in the body and the excretion of metabolic waste products.

- The process of filtration in the glomerulus involves regulation of the filtration pressure to maintain filtration flow through the nephron.

- Most of the reabsorption by the nephron takes place in the proximal convoluted tubule and is driven by the $Na^+K^+ATPase$ pump.

- The loop of Henlé is important in creating the osmotic gradient in the medulla of the kidney, through which the collecting ducts pass; longer loops are associated with the capacity to retain a greater proportion of the water through production of a more concentrated urine.

- The collecting duct is the region where regulation of water reabsorption takes place under the control of ADH.

- The kidney plays an important role in the control of blood pressure, both by regulating the amount of water that is reabsorbed and through the renin–angiotensin system.

Use the flashcards in the e-book to test your recall of key terms introduced in this chapter.

QUESTIONS

Looking for answers? Once you've answered these questions, follow the link in the e-book to the answer guidance and check your work.

Concepts and definitions

1. Where in the nephron does most of the reabsorption take place?

2. What is the significance of the length of the loop of Henlé?

3. What is the role of antidiuretic hormone in water balance?

4. Which hormone produced by the kidneys is involved in the stimulation of red cell production?

Apply the concepts

5. Describe the local processes that regulate the glomerular filtration rate.

6. Describe the process of glucose reabsorption.

7. Describe the role of the renin–angiotensin system in the regulation of blood pressure.

8. Explain why glucose may appear in the urine in type 1 diabetes.

9. Summarize the role of each section of the nephron in the production of urine.

Beyond the concepts

10. Compare the mechanisms involved in the local regulation of blood pressure to the kidney and the kidney's role in the systemic regulation of blood pressure.

11. The function of the nephron is entirely dependent on the operation of the Na^+K^+ ATPase pump. Discuss.

12. Discuss the mechanisms by which the kidney is involved in the regulation of water balance.

13. Discuss the interactions of the respiratory system and the kidney in regulating acid–base balance.

14. Explain how glomerular filtration rate is regulated. What is its role in affecting kidney function?

FURTHER READING

Koeppen, B. M. & Stanton, B. A. (2018) *Renal Physiology*. 6th edition. Elsevier.
This text provides a more in-depth consideration of the kidney and its function. There is a linked e-book and numerous links to associated clinical conditions.

Levey, A. S. & Inker, L. A. (2017) Assessment of glomerular filtration rate in health and disease: a state of the art review. *Clin. Pharmacol. Ther.* **102**: 405–19. https://doi.org/10.1002/cpt.729
This review article provides a more in-depth consideration of the role of GFR in kidney function and its control in health and disease.

O'Callaghan, C. (2016) *The Renal System at a Glance*. 4th edition. Wiley-Blackwell.
This provides an excellent, detailed summary of kidney function and its regulation.

Tubular reabsorption. Lumen Anatomy and Physiology II Module 9: The Urinary System. https://courses.lumenlearning.com/cuny-kbcc-ap2/chapter/tubular-reabsorption-no-content/
This website provides a very good summary of the processes of reabsorption within each section of the nephron.

Digestive System

CHAPTER 25

LEARNING OBJECTIVES

By the end of this chapter you should be able to:

- Describe the main anatomical sections of the digestive system.

- Relate structure to the main functions of each section of the digestive system.

- Describe the main histological features of the digestive system.

- Explain how ingested substances are moved from one region of the system to the next and how the motility of the different regions is controlled.

- Explain the processes of regulation and coordination of secretion of enzymes and other chemicals by the different regions of the digestive system.

- Describe the different mechanisms of absorption of the breakdown products of digestion.

- Explain the role of ancillary organs in the processes of digestion and absorption.

- Describe the mechanisms of glucose homeostasis and explain how it is controlled.

Chapter contents

	Introduction	803
25.1	Histology and regulation of the digestive tract	804
25.2	Anatomy of the digestive tract	805
25.3	Gut microbiome	822
25.4	The pancreas	823
25.5	The liver	826

🎬 Watch the key concepts video in the e-book to prepare yourself for studying this chapter.

Introduction

The digestive system is one of the most important systems in the human body, functioning to break down complex molecules into smaller ones that can be absorbed into the bloodstream, providing nutrients for the body as a whole. In its structure it is essentially a hollow tract running from the mouth to the anus and, consequently, whatever is in that tract is effectively outside of the body. The processes of digestion and absorption require the integration of complex motility, to ensure that ingested substances are moved from one part of the system to the next, along with regulation of secretion of a range of key chemicals and enzymes that break down the chemical structure of the complex molecules that are eaten.

Given that the contents of the digestive tract are essentially outside of the body, the system acts as a significant line of defence against foreign substances. As will be outlined during the chapter, a breakdown in this defence can lead to significant problems. The digestive system is also one of the largest endocrine organs in the body, and the hormones produced effect a plethora of functions from glucose homeostasis to appetite regulation. The system is also intrinsically involved in fluid and electrolyte balance as well as, of course, being responsible for elimination of waste products.

The main macronutrients within the human diet are carbohydrates, fats, and proteins. In general, carbohydrates, fats, and proteins cannot be absorbed in the form in which they are ingested and have to be broken down into smaller molecules before they can be absorbed. It is suggested that approximately 50–70% of total daily energy intake should be in the form of carbohydrate with no more than 10% from simple sugars, less than 30% of total daily energy intake should be in the form of fats, with 15–20% of total energy intake from proteins. In reality, a typical diet in somewhere like the UK does not tend to adhere to these guidelines, with a tendency towards highly processed foods that are often high in saturated fats and sugars.

The terms 'digestive system' and 'gastrointestinal system' are often used interchangeably. The digestive system refers to the whole tract from mouth to anus plus ancillary organs, whereas the gastrointestinal system refers only to the stomach and intestines. The purpose of this chapter is to consider the structure and function of the digestive system as a whole, including the digestive and absorptive processes that occur.

25.1 Histology and regulation of the digestive tract

Despite the length of the digestive tract, and the specialized functions of the different sections, the histology of the digestive system is relatively consistent. There are four main layers of digestive tract wall at all areas of the system, these are the:

- mucosa;
- submucosa;
- muscularis externa;
- serosa.

The thickness and structure of these layers vary depending on which section of the system is being considered. This links to regulatory function of the system as a whole. A schematic diagram showing the histological appearance of the gastrointestinal tract is shown in Figure 25.1 (see also Figure 25.7).

The mucosal layer of the digestive tract is the inner layer of the wall and consists of a layer of epithelium and a **lamina propria**.

PAUSE AND THINK

What is the significance of the lining of the gut being epithelium rather than endothelium?

Answer: Remember that the lumen of the gut is physiologically external to the body, and so the cells lining the lumen are epithelial as are the cells of the skin.

The type of epithelial tissue differs depending on the region. For example, stratified squamous epithelium, where the cells form a flat layer, is found in the oral cavity, whereas simply columnar epithelium (where the cells are elongated, forming a column-like

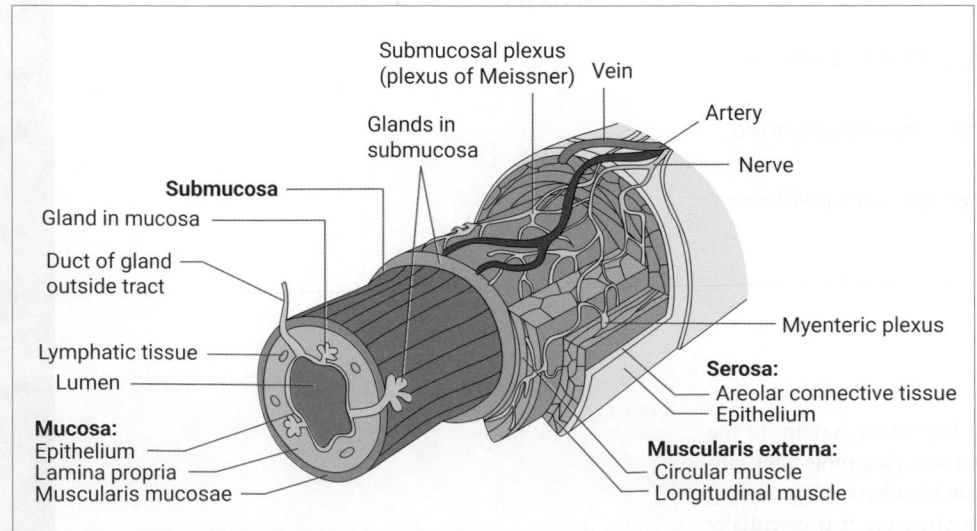

Figure 25.1 The principal regions and structures of a 'typical' region of the gastrointestinal tract.

Source: Openstax CC BY 4.0 C J. Gordon Betts et al. *Anatomy and Physiology*, 2013. Access for free at: https://openstax.org/books/anatomy-and-physiology/pages/1-introduction

Submucosal plexus (plexus of Meissner)   Vein
Glands in submucosa
Artery
Nerve
Submucosa
Gland in mucosa
Duct of gland outside tract
Myenteric plexus
Lymphatic tissue
Lumen
Serosa:
Areolar connective tissue
Epithelium
Mucosa:
Epithelium
Lamina propria
Muscularis mucosae
Muscularis externa:
Circular muscle
Longitudinal muscle

structure) is found in the large intestine. The lamina propria contains blood vessels and nerve endings, as well as glands, in some areas of the system. In addition, it also contains a layer of smooth muscle called the muscularis mucosae.

The **submucosa** is a layer of connective tissue, with some blood and lymphatic vessels, that effectively binds the mucosa to the muscularis externa. Perhaps the key component of the submucosal layer is the submucosal plexus. The submucosal plexus is found on the outer layer of the submucosa and consists of an intrinsic network of neurons. In particular, there are sensory neurons that can respond to changes in the environment within the intestinal lumen.

The **muscularis externa** is a layer of smooth muscle (which you can also read about in Chapter 20) that can vary substantially between areas of the digestive system. There are two layers of smooth muscle, an inner layer of circular muscle and an outer layer of longitudinal muscle. Contraction and relaxation of these smooth muscle layers within the muscularis externa are of central importance in terms of mechanical processing and transport of substances through the digestive tract. Also found within the muscularis externa, the myenteric plexus is a network of parasympathetic neurons and sympathetic postganglionic fibres (Section 19.1). The submucosal plexus and the myenteric plexus form the **enteric nervous system,** which controls many of the functions of the digestive system.

The **serosa** is a layer of membranous tissue adjacent to the muscularis externa and forms the boundary between the digestive tract wall and the peritoneal cavity. The serosa is only found within the stomach and the small and large intestines, whereas a layer of adventitia is found elsewhere.

The stomach and small intestine (see Section 25.2) have some specific histological characteristics. In particular, when empty, the mucosa of the stomach is folded into structures called **rugae** (Figure 25.8). The stomach can expand significantly when substances enter, with the rugae unfolding to accommodate the expansion. Within the small intestine, the mucosal layer of the digestive tract projects into a series of projections called **villi**. These villi are covered with microvilli, and this part of the small intestine is often referred to as the **brush border** membrane. The villi increase the surface area available for absorption of ingested material. Within each villus there are abundant capillary beds via which absorbed nutrients enter into the portal circulation for delivery to the liver.

The movement of ingested substances through the digestive tract is ultimately due to the contraction of smooth muscle within the muscularis externa and the mucosa. This is maintained in a rhythmic fashion by specialized cells within the digestive system called the interstitial cells of Cajal. These cells exhibit pacemaker potentials that result in contraction of smooth muscle without any external input, ensuring that a constant wave of smooth muscle contraction propels ingested substances through the digestive tract. The rate of smooth muscle contraction can be altered by various autonomic and hormonal factors that increase or decrease **motility** (ability to contract) through the system. These waves of smooth muscle contraction that propel substances through the digestive tract are termed **peristalsis** and are described in detail later in this chapter.

Regulation of the digestive system

A variety of factors can influence the way in which the digestive system operates; however, they can be broadly classed as neural activity, hormonal influence, and local factors.

Neural control of digestive system function is primarily related to small movements of substances through the tract. This usually involves activation of chemoreceptors and/or stretch receptors within the wall of the digestive tract that leads to a **myenteric reflex**, which regulates the peristaltic movements. These reflexes are classed as either short or long. Short reflexes involve activation of neurons within the myenteric plexus, which then leads to a change in smooth muscle activity, whereas long reflexes involve interneurons within the central nervous system resulting in changes in parasympathetic activity of neurons that have their synapses within the myenteric plexus. Consequently, long reflexes tend to result in relatively large changes in function, whereas smaller changes are controlled by short reflexes.

The digestive system secretes numerous hormones that can be both paracrine and endocrine in function. Some of these are discussed in more detail in Section 25.4. In terms of digestive system function, some of the hormones secreted from enteroendocrine glands within the epithelial layer of areas of the digestive system can either increase or decrease the sensitivity of smooth muscle to neural input. Consequently, they can increase or decrease the rate of smooth muscle contraction and movement of substances through the system.

Local factors provide the most immediate regulation of digestive system function. These factors tend to have relatively short-acting effects that are localized to the specific region in question. For instance, ingestion of large volumes of fluid expands the stomach, resulting in stretch of smooth muscle fibres, which can lead to contraction of those fibres and movement of fluid out of the stomach. Other factors such as the acidity, temperature, or composition of the contents of the intestinal lumen can induce similar effects. Similarly, the release of certain compounds from cells within the digestive system can lead to increased secretion of compounds from adjacent cells resulting in changes in function.

 Check your understanding of the concepts covered in this section by answering the questions in the e-book.

25.2 Anatomy of the digestive tract

The digestive tract is essentially a long muscular tube running from the mouth to the anus. A number of important, and distinct, anatomical structures within the digestive tract serve a variety of functions. The overall organization of the digestive tract is shown in Figure 25.2.

As shown in Figure 25.2, the main regions of the digestive system are:

- oral cavity;
- oesophagus;
- stomach;

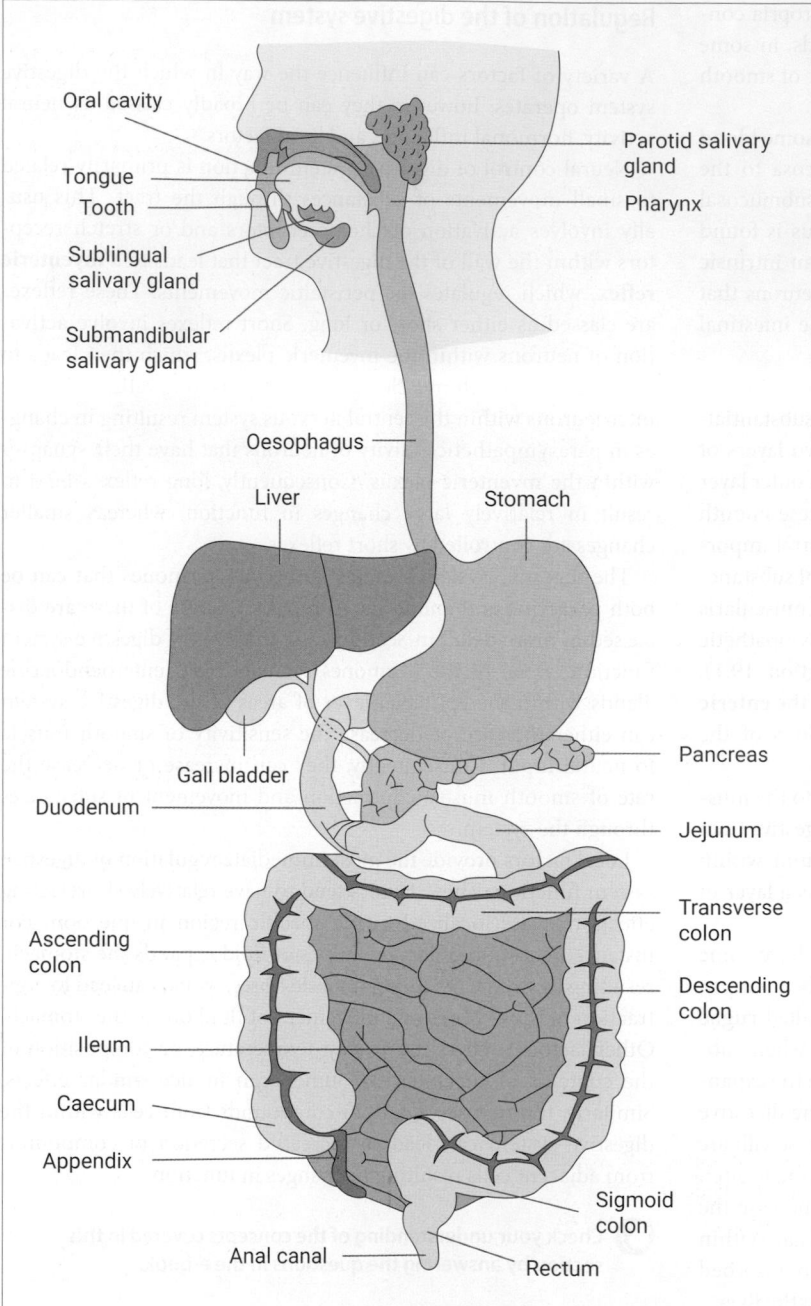

Figure 25.2 Organization of the digestive system showing principal structures and accessory organs.
Source: Pocock et al. *Human Physiology*, 5th edn, 2017. Oxford University Press.

- small intestine;
- large intestine;
- rectum;
- anus.

Some of these regions can be further subdivided, as we will see in the following sections.

Oral cavity

The digestive tract begins in the oral cavity which contains the tongue, the teeth, and the salivary glands (Figure 25.3). The oral cavity has several important functions. Ingested substances are processed mechanically via **mastication** (chewing) and through the action of the tongue. Lubrication is provided during this process by the addition of mucus and secretions from the **salivary glands**.

Some breakdown of large carbohydrates occurs in the oral cavity as a result of salivary secretions. The tongue contains touch, taste (Section 18.5, Taste and smell), and temperature receptors, which play an important sensory function as part of the role of the oral cavity in food processing.

The tongue and the salivary glands, or at least the secretions of the salivary glands, are located in the oral cavity. We will begin by considering the structure and function of the tongue.

Figure 25.3 Organization of the oral cavity.

Source: Purves et al., *Neuroscience*, 6th edn, 2017. Oxford University Press.

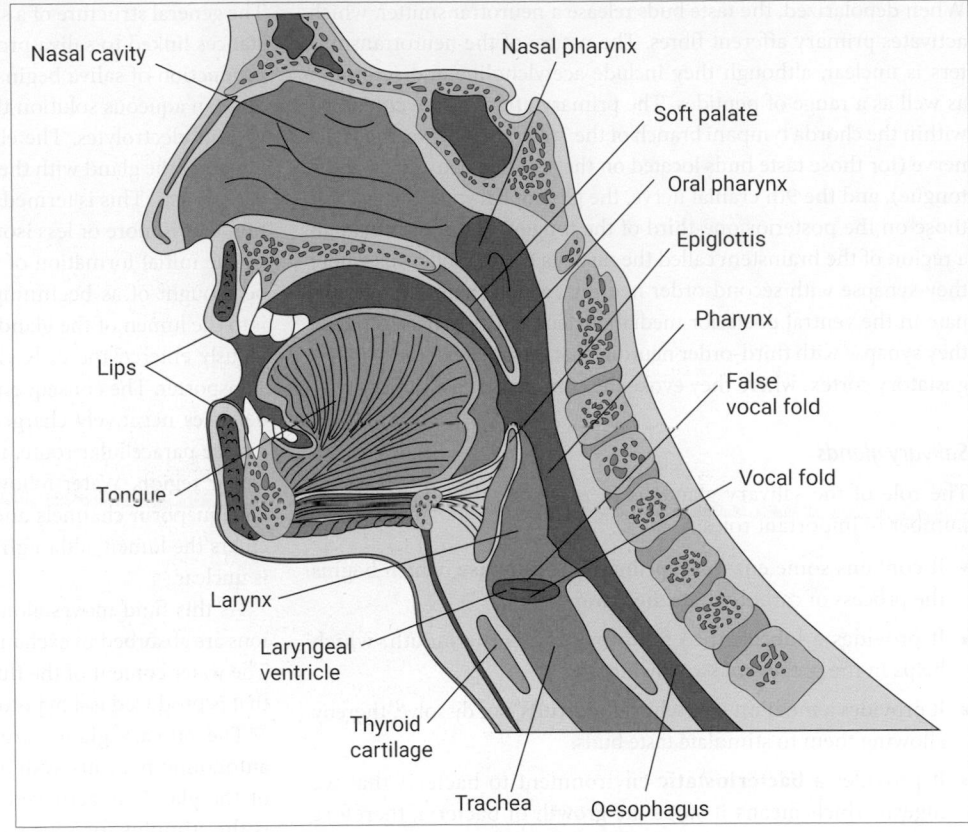

The tongue

The tongue is a skeletal (voluntary) muscle important for the sensation of taste as well as the mechanical processes of chewing and swallowing. The muscular activity of the tongue is essential for moving food around the mouth, mixing it, and bringing it between the teeth which physically break it down before it forms it into a **bolus** (a ball-like mixture of food and saliva) in preparation for swallowing. Taste is an important aspect of the food that we eat; for example, it can influence both what we eat and how much.

PAUSE AND THINK

Can you think of a further example of how the sense of taste can influence eating behaviour?

Answer: Think about the taste of food when it has 'gone off'. The unpleasant taste of food that has spoiled protects us against the ingestion of material that could potentially be harmful.

The sense of taste, **gustation**, is transduced by the presence of taste receptors located on the upper surface of the tongue. Taste receptors are located in taste buds on the tongue. As described in Section 18.5, Taste and smell, there are five basic types of taste:

- sour;
- bitter;
- salt;
- sweet;
- umami (the savoury taste associated with meat products, primarily associated with the amino acid L-glutamate).

While these individual tastes offer us valuable information (e.g. a bitter taste can be associated with spoiled food), the reality is that the taste of food is a product of the activation of all types of taste simultaneously. This is complemented by the additional stimulatory input from smell receptors, which also have a significant role in the overall sense of taste.

What effect does nasal congestion (e.g. associated with a cold), have on the sensation of taste? Take a moment to think about this question before you read on, and perhaps look at Figure 25.6. When an individual suffers with a cold, food tastes very bland. This is due to nasal congestion, which prevents aromas arising from foodstuffs stimulating olfactory receptors.

As you read in Chapter 18, different tastes are transduced by different physiological mechanisms, each of which is summarized here:

- salt—Na$^+$ channel opening produces a depolarization of the taste receptor;
- sour—proton entry into the taste bud results in inhibition of K$^+$ channels resulting in depolarization;
- bitter, sweet, and umami—activation of G-protein coupled receptors (GPCRs) results in a rise in intracellular Ca^{2+} concentration and depolarization.

When depolarized, the taste buds release a neurotransmitter, which activates primary afferent fibres. The nature of the neurotransmitters is unclear, although they include acetylcholine and serotonin as well as a range of peptides. The primary afferents are contained within the chorda tympani branch of the 7th cranial nerve, the facial nerve (for those taste buds located on the anterior two-thirds of the tongue), and the 9th cranial nerve, the glossopharyngeal nerve (for those on the posterior one-third of the tongue). They terminate in a region of the brainstem called the nucleus tractus solitarius where they synapse with second-order neurons, which themselves terminate in the ventral posterior medial nucleus of the thalamus. Here, they synapse with third-order neurons that terminate in the primary gustatory cortex, where they evoke the conscious sense of taste.

Salivary glands

The role of the salivary glands is to produce saliva. Saliva has a number of important roles:

- It contains some enzymes, primarily **α-amylase**, which begins the process of carbohydrate digestion.

- It provides a lubricant to food that enters the mouth, which helps in the process of swallowing.

- It provides a medium into which foodstuffs can dissolve, thereby allowing them to stimulate taste buds.

- It provides a **bacteriostatic** environment to bacteria that we ingest, which means it stops the growth of bacteria, therefore providing a defence against oral infections.

The enzyme α-amylase acts on complex carbohydrates, such as the polysaccharide starch, breaking them down into smaller carbohydrate molecules such as the disaccharide (two glucose molecules joined together) maltose and dextrins, which are a mix of polymers of glucose.

On average, a human produces about 1200–1500 mL of saliva every day. The secretions that eventually form saliva may be described as being either **serous**, which represents watery secretion, or **mucous**, which as the name suggests is a secretion rich in mucus. There are three principal pairs of salivary glands, as shown in Figure 25.4:

- **Parotid glands**. These glands are located in the cheeks, just in front and below the ear. These glands, like all the other salivary glands are ducted glands. The secretions of the parotid gland enter the mouth opposite the upper second molars. The parotid glands produce an entirely serous secretion, which contains a small amount of α-amylase. Despite being the largest salivary glands, these glands only account for about 25% of total saliva production.

- **Submandibular glands**. These glands, as the name suggests, lie beneath the mandible (jawbone). Their secretions, which are a mixture of both serous and mucous, enter the mouth via ducts that terminate near the lower incisors. These glands account for about 70% of total saliva production.

- **Sublingual glands**. These glands produce a mucus-rich secretion that enters the mouth below the tongue. Despite only producing 5% of total saliva production, it is the secretions of these glands that give saliva its characteristic sticky appearance.

The general structure of a salivary gland and the movement of substances linked to saliva production are shown in Figure 25.5. The production of saliva begins with the **acinar cells**. These cells produce an aqueous solution that contains the enzyme α-amylase and various electrolytes. The electrolytes are actively secreted into the lumen of the gland with the subsequent passive movement of water via osmosis. This is termed primary secretion and results in a solution that is more or less isotonic with plasma.

The initial formation of what will eventually become saliva can be thought of as beginning with the movement of chloride ions into the lumen of the gland through Cl^- channels, Cl^- having previously entered the cells via a transporter known as the NKCC1 transporter. The consequence of this is that the lumen of the gland becomes negatively charged, which in turn promotes Na^+ influx via the paracellular route, the gaps between individual cells in the acinar region. Water follows this movement, entering the lumen via aquaporin channels and a paracellular route. Equally, HCO_3^- enters the lumen, although the mechanism by which this happens is unclear.

As this fluid moves along the lumen of the gland Na^+ and Cl^- ions are absorbed in exchange for K^+ and HCO_3^- ions, respectively. The water content of the fluid remains the same, so the final saliva that is produced is a hypotonic solution.

The salivary glands are innervated by both divisions of the autonomic nervous system and it is through these that control of the glands is achieved. The presence of food in the mouth is the stimulus for saliva secretion, although other stimuli such as the thought or sight of food are also important: seeing and smelling food will initiate salivation even before anything has been eaten. Afferent fibres arising from the mouth terminate in salivatory centres located in the brainstem. Neurons here stimulate parasympathetic neurons, which form the efferent innervation of salivary glands. The effect of increased parasympathetic nervous stimulation is to increase saliva secretion through the release of acetylcholine. As part of this response, there is also an increase in blood flow to the salivary glands, which is mediated by the release of peptide hormones, vasointestinal peptide (VIP) and substance P. This also promotes the cells of the acinus to release bradykinin, which further increases blood flow. The result of parasympathetic stimulation, therefore, is a more serous, watery secretion.

Stimulation via the sympathetic nervous system leads to the release of noradrenaline, which results in an initial reduction in blood flow and the production of a more mucous, thicker secretion.

When we feel frightened, our mouths often 'dry up'. This happens due to the initiation of the 'fight or flight' response with activation of the sympathetic nervous system and release of the hormone adrenaline (Chapter 19), reducing the volume of saliva produced and increasing the viscosity of the saliva.

Pharynx, oesophagus, and swallowing

Food that is ingested is mechanically broken down in the mouth by the action of chewing, moistened, and, to a certain degree, the process of chemical digestion by enzymes is begun. The next stage of this process involves the swallowing of a bolus, allowing it to

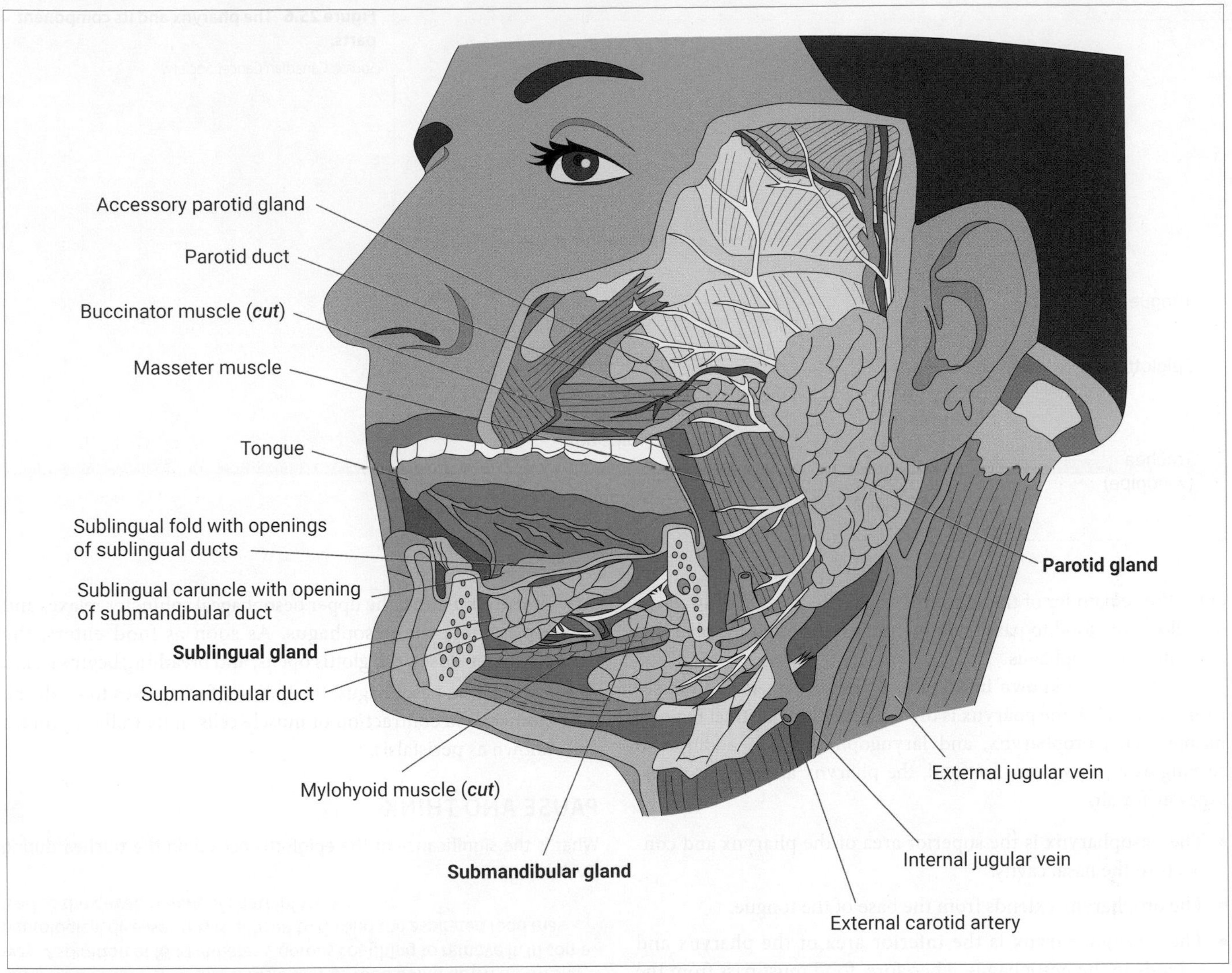

Accessory parotid gland

Parotid duct

Buccinator muscle (*cut*)

Masseter muscle

Tongue

Sublingual fold with openings
of sublingual ducts

Sublingual caruncle with opening
of submandibular duct

Sublingual gland

Submandibular duct

Parotid gland

External jugular vein

Internal jugular vein

Mylohyoid muscle (*cut*)

External carotid artery

Submandibular gland

Figure 25.4 Location of the three principal pairs of salivary glands (labelled here in bold).

Source: R.R. Hukkanen, S.M. Dintzis, P.M. Treuting. *Comparative Anatomy and Histology*, 2nd edn, 2018, 135–145, with permission from Elsevier.

Figure 25.5 Organization of a typical salivary gland indicating the principal ion movement.

Source: Pocock et al. *Human Physiology*, 5th edn, 2017. Oxford University Press.

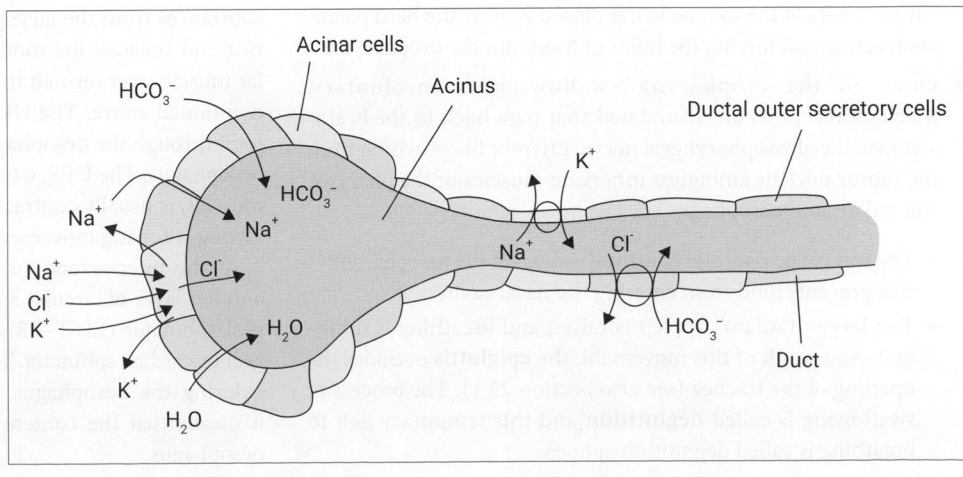

Acinar cells

Acinus

Ductal outer secretory cells

HCO_3^-

HCO_3^-

Na^+

K^+

Na^+

Cl^-

Na^+

Cl^-

Na^+

Cl^-

K^+

H_2O

HCO_3^-

K^+

Duct

H_2O

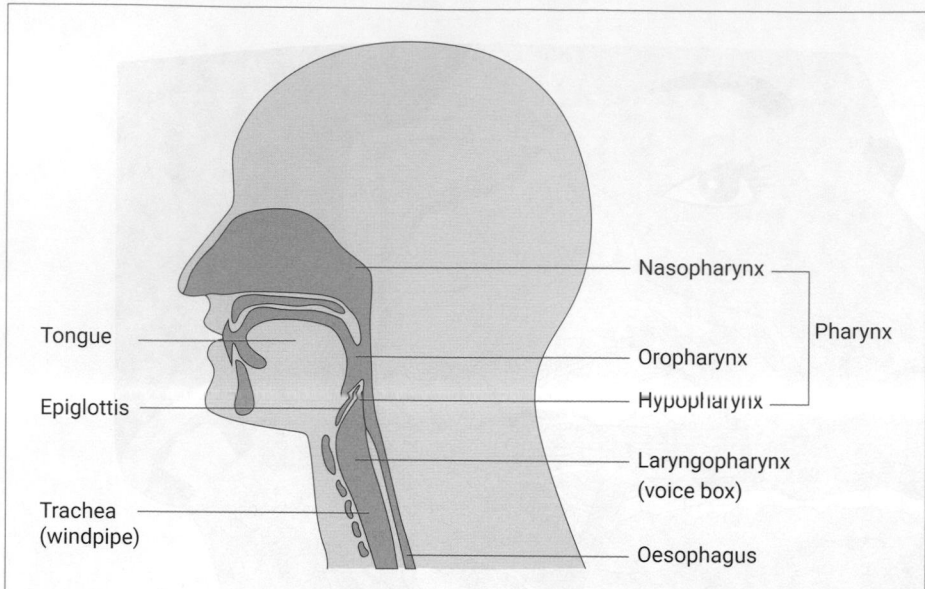

Figure 25.6 **The pharynx and its component parts.**

Source: Canadian Cancer Society.

enter the remainder of the digestive tract. The process of swallowing allows the food to pass from the mouth, through the pharynx, and into the oesophagus.

The pharynx is known informally as the throat. As you can see from Figure 25.6, the pharynx is divided into three distinct regions: nasopharynx, oropharynx, and laryngopharynx. In addition to serving as a passageway for food, the pharynx also acts as a passageway for air.

- The nasopharynx is the superior area of the pharynx and connects to the nasal cavity.

- The oropharynx extends from the base of the tongue.

- The laryngopharynx is the inferior area of the pharynx and extends to the oesophagus. Therefore, food must pass from the oral cavity and then through the oropharynx and laryngopharynx before entering the oesophagus.

The process of swallowing is complex and includes both voluntary and involuntary components:

- The initial phase of swallowing called the oral phase is voluntary. This consists of the tongue being placed against the hard palate, contracting and forcing the bolus of food into the oropharynx.

- Once in the oropharynx swallowing is involuntary. Mechanoreceptors are stimulated that pass back to the brainstem via the glossopharyngeal nerve. Efferent fibres arising from the motor nucleus ambiguus innervate muscles of the pharynx, soft palate, and oesophagus.

  - The soft palate rises and effectively closes off the nasopharynx— this prevents food from entering the nasal cavity.

  - The larynx ('Adam's apple') is raised and breathing is inhibited. As a result of this movement, the **epiglottis** occludes the opening of the trachea (see also Section 22.1). The process of swallowing is called **deglutition** and this temporary halt to breathing is called deglutition apnoea.

- At the same time, the upper oesophageal sphincter relaxes and food enters the oesophagus. As soon as food enters, the sphincter closes, the glottis opens, and breathing begins again.

- Once in the oesophagus, the bolus of food passes towards the stomach via contraction of muscle cells in its walls, a process known as peristalsis.

PAUSE AND THINK

What is the significance of the epiglottis occluding the trachea during swallowing?

Answer: Occluding the trachea means that food is unable to enter it and so possibly pass down into the lungs. This is referred to as aspiration of food; we all experience this occasionally as food going down the wrong way. Aspiration of food initiates vigorous coughing to remove it. In some neurological diseases this is difficult to do and the aspirated food may lead to the development of chest infections.

The process of swallowing transfers food into the oesophagus (Figure 25.7). The oesophagus is a hollow tube measuring approximately 25–30 cm in length. Its primary function is to transport substances from the laryngopharynx to the stomach. At the superior end (nearest the mouth) of the oesophagus, there is a circular muscle layer termed the upper oesophageal sphincter (UES) as mentioned above. The UES ensures one-directional movement of food through the oesophagus as well as preventing air entering the oesophagus. The UES, which is composed of a number of different muscles, is usually contracted but is triggered to relax during swallowing, allowing movement of food into the oesophagus.

At the inferior end (nearest the stomach) of the oesophagus, another layer of circular smooth muscle forms the lower oesophageal sphincter (LES)—this is sometimes called the gastroesophageal or cardiac sphincter. The LES prevents stomach contents from entering the oesophagus. If the LES fails to function correctly, it means that the contents of the stomach are able to enter the oesophagus.

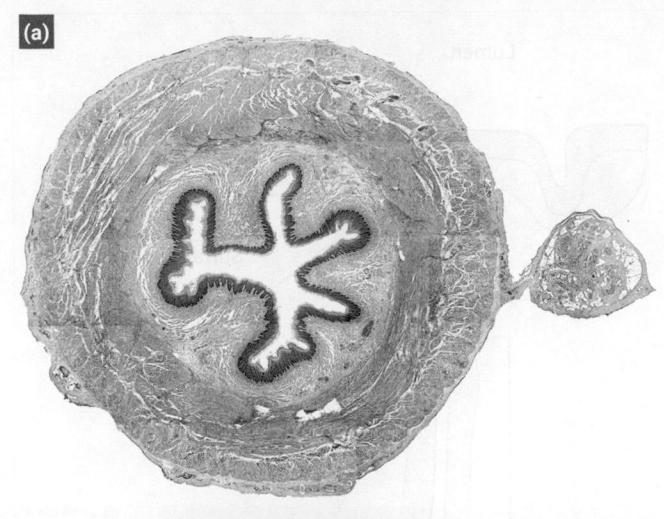

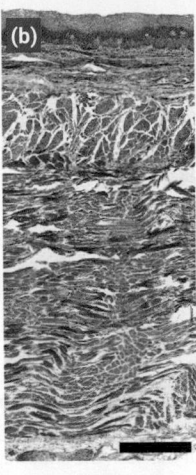

Figure 25.7 Structure of the oesophagus. (a) Low-power cross-section. Note how the oesophagus is folded: this enables a significant increase in the diameter of the lumen as the bolus of food passes through. (b) Higher-power view showing the layers of the wall. Note the layers are similar to those shown in Figure 25.1.

Source: (a) Jose Calvo/Science Photo Library; (b) Eye of Science/Science Photo Library.

The movement of stomach contents back into the oeosphagus is termed gastroesophageal reflux. The general term for disorders in which this happens is gastroesophageal reflux disorder (GORD). It is important because the acidic nature of the stomach contents may promote irritation and inflammation of the oesophagus which, if continued, may lead to ulceration. This is known more commonly as acid reflux and it leads to pain in the mid-sternal region of the chest as well as the throat. GORD is treated in a variety of ways, including the use of antacids, which neutralize the acidic contents of the stomach, or drugs, such as omeprazole, which block acid production.

Movement of ingested food through the oesophagus is achieved as a result of the waves of muscle contraction known as peristalsis.

Peristalsis

Peristalsis in the oesophagus is complex because the musculature of the oesophagus is divided into two regions: the upper part comprises striated muscle, whereas the lower part is smooth muscle. When the bolus is swallowed, the brainstem triggers contraction of the striated muscle through sequential activation of the lower motor neurons. The wave of contraction thus propels the bolus downwards.

In the lower part of the oesophagus, the smooth muscle is under both central and local control. The first stage of the process is simultaneous inhibition of the smooth muscle. This causes relaxation of the oesophagus, allowing the oesophagus to stretch as the bolus passes downwards. This is followed by a wave of excitation, causing the smooth muscle to constrict the oesophagus behind the bolus, which helps to propel the bolus downwards and prevents any backwards movement.

> **PAUSE AND THINK**
>
> What happens if you swallow food while standing on your head?
>
> *Answer:* The bolus of food still travels towards the stomach because the wave of contraction propels the bolus in one direction and prevents backward movement.

The wave of peristalsis triggered by swallowing is termed primary peristalsis. If any food is left in the oesophagus after the wave has passed, then the distention of the oesophageal wall will initiate a local reflex to trigger a second wave, secondary peristalsis.

Stomach

The stomach is a J-shaped organ that essentially acts as a reservoir for ingested substances in order to begin the main processes of digestion. The structure of the stomach is shown in Figure 25.8. The stomach itself can be split into four main regions.

- The **cardia** is a small section of the stomach found at the junction with the oesophagus. It contains numerous glands that secrete mucus and lubricate the area, minimizing the effect of any acidic contents re-entering the oesophagus.

- The **fundus** is the section of the stomach that lies above the cardia and it has an important function in increasing stomach capacity for ingested substances as well as containing numerous glands that secrete substances involved in the digestive process.

- The **body** of the stomach is the largest region of the stomach found inferiorly (below) to the fundus. It, like the fundus, contains numerous cells that secrete substances involved in digestion.

- The **pylorus** is the lower section of the stomach, the pyloric antrum, connecting the body to the small intestine. Between these two structures is the pyloric sphincter, which controls the one-directional movement from the stomach into the small intestine.

While in the stomach, ingested substances are broken down mechanically, by the waves of contractions of the stomach wall, as well as chemically via the secretion of hydrochloric acid and enzymes to form **chyme**, which is passed out of the stomach into the first section of the small intestine, the **duodenum**. Importantly, very little absorption occurs within the stomach and the movement

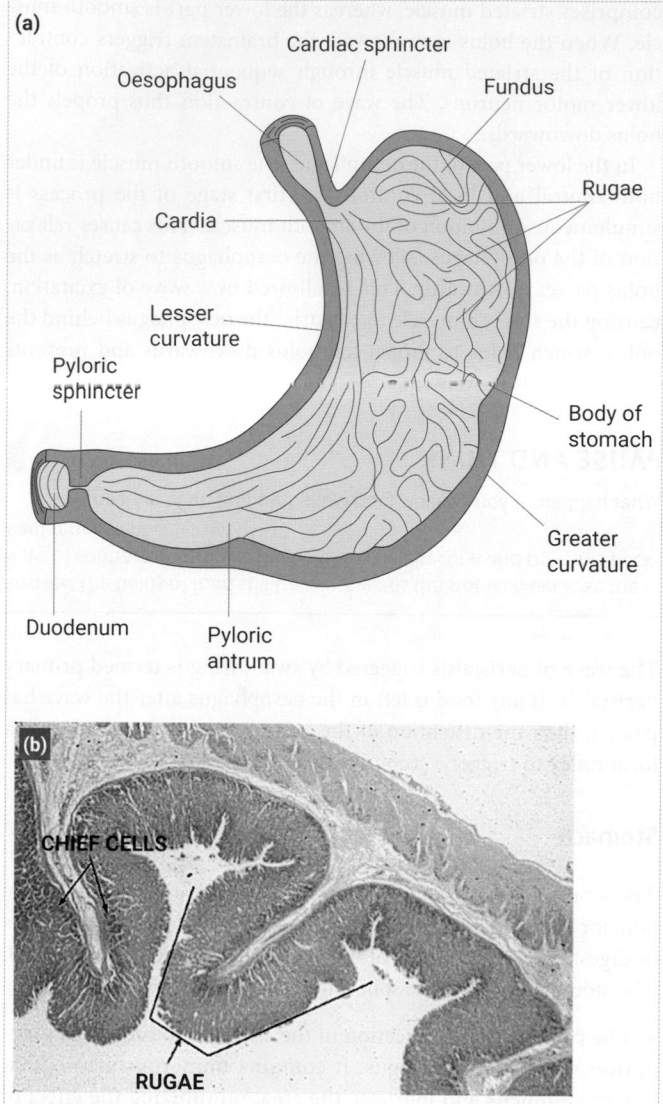

Figure 25.8 The stomach. (a) The principal regions of the stomach. (b) The folds of the rugae when the stomach is empty; these will unfold as the stomach fills, greatly increasing the overall volume.

Source: Pocock et al. *Human Physiology*, 5th edn, 2017. Oxford University Press.

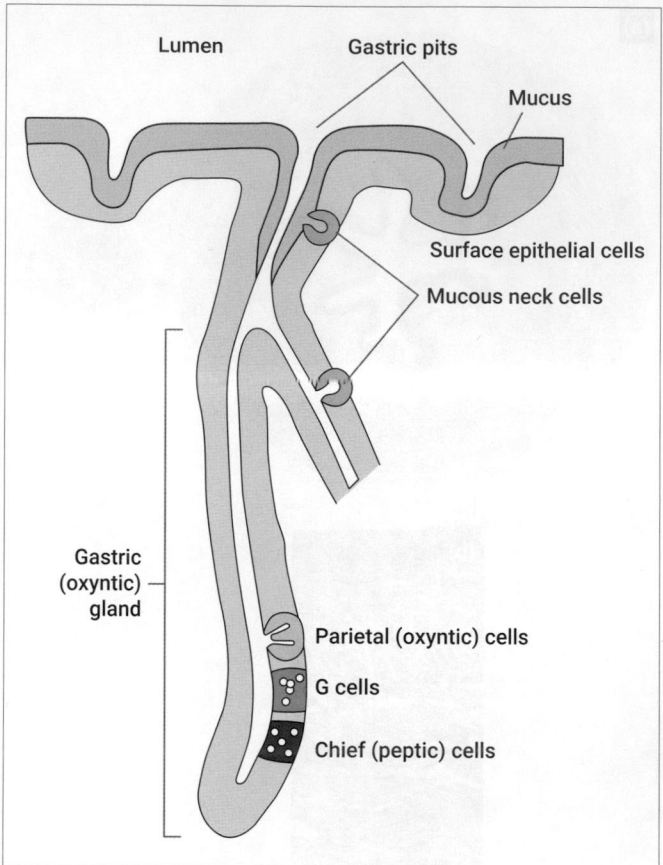

Figure 25.9 A 'typical' gastric gland with its constituent cell types. There is regional variation in the frequency of different cell types (e.g. parietal cells are common in the fundus and body but less so in the antral region). Gastric pits are the regions into which gastric gland secretions empty.

Source: Pocock et al. *Human Physiology*, 5th edn, 2017. Oxford University Press.

of ingested food into the intestine is an important consideration when determining how quickly a substance can be absorbed into the circulation.

Within the stomach, gastric glands are found interspersed between the epithelial cells that form the mucosal lining of the stomach. The structure of a typical gastric gland can be seen in Figure 25.9. Gastric glands consist of a number of different cell types.

- Parietal cells, sometimes called oxyntic cells. These cells secrete hydrochloric acid and intrinsic factor.
- Chief cells, which secrete pepsinogens. Pepsinogens are the inactive forms of proteolytic enzymes, which together are known as pepsin.

- G-cells, which secrete the hormone gastrin. This hormone is important in controlling secretions and motility in the digestive tract.
- Neck cells secrete mucus (as do the cells that form the epithelia of the stomach). This mucus is important in protecting the stomach from the acidic proteolytic secretions that it produces.

Let us now take a more detailed look at these cells and their secretions.

Parietal cells

As indicated previously, the major secretion from parietal cells is hydrochloric acid. A schematic diagram illustrating the production of HCl is shown in Figure 25.10. Under the influence of the enzyme carbonic anhydrase (see also Section 22.3), intracellular CO_2 combines with H_2O to form carbonic acid (H_2CO_3). In turn, H_2CO_3 combines with hydroxyl (OH^-) ions formed from the dissociation of water to form water and bicarbonate ions (HCO_3^-). In the basolateral membrane the HCO_3^- is removed from the cell in exchange for Cl^-, which in turn is exported from the cell into the lumen of the stomach via Cl^- channels and a K^+/Cl^- symporter.

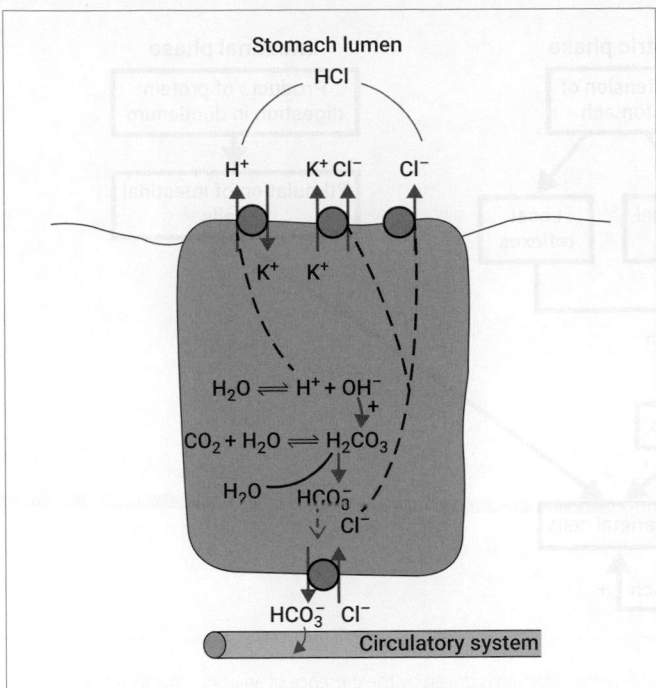

Stomach lumen

Figure 25.10 The production of hydrochloric acid by the parietal cells of the gastric glands.

Source: Pocock et al. *Human Physiology*, 5th edn, 2017. Oxford University Press.

The H$^+$ which has been formed from the dissociation of water in the cell is exported out of the cell into the lumen in exchange for the influx of K$^+$—this is termed the proton pump. The exported H$^+$ and Cl$^-$ combine in the lumen of the stomach to form HCl. The HCl formed has a pH of about 1–3.

As we will see shortly, the strongly acidic secretions of the stomach also contain proteolytic enzymes: what then prevents this fluid from destroying the tissue of the stomach?

A number of protective mechanisms protect the stomach.

- So-called tight junctions link epithelial cells of the stomach. This ensures that there can be no paracellular leakage of gastric secretions that may destroy underlying tissue. In essence, the gastric mucosa is watertight!

- The epithelial cells themselves, along with the neck cells, secrete an alkaline-rich mucus that forms a physical barrier against the actions of gastric secretions.

- Epithelial cells of the stomach also produce compounds called **prostaglandins**. Prostaglandins promote the production of the bicarbonate-rich mucosa, which lines the stomach wall. This also accounts for the reason why non-steroidal anti-inflammatory drugs (NSAIDs), such as aspirin, can have a deleterious effect on the stomach when taken over extended periods of time. NSAIDs work by inhibiting the production of prostaglandins and therefore compromise the protective effects of prostaglandins on the stomach.

Nonetheless, the hostile environment that is formed by the secretion of HCl in the stomach is beneficial to nutrition and digestion. The acidic environment:

- is essential for the activation of protease enzymes secreted by the chief cells of the gastric glands;

- helps in the digestion of some proteins such as the connective tissue that is found in meat;

- provides a defence mechanism against ingested micro-organisms; in destroying micro-organisms it prevents systemic infections from occurring.

In addition to their role in acid secretion, parietal cells also secrete a compound called **intrinsic factor**. Intrinsic factor is a glycoprotein that binds to vitamin B12. This combination protects vitamin B12 from being destroyed as it passes through the remainder of the digestive tract. Vitamin B12 is essential for the absorption of iron, required for the production of haemoglobin in red blood cells. A lack of intrinsic factor can result from an autoimmune condition in which parietal cells are destroyed, leading to reduced red cell production and a condition called pernicious anaemia (see Clinical Box 22.2).

Chief cells

The primary secretion of chief cells is the enzyme precursor **pepsinogen**. Pepsinogen, which is actually a number of proteolytic enzymes, is stored and released in an inactive form. The general term for this inactive form is a **zymogen**. In highly acidic environments, brought about by the secretion of hydrochloric acid from parietal cells, pepsinogen is converted to **pepsin**, the active form of the proteolytic enzyme. Pepsin is an endopeptidase—this means it attacks peptide bonds within a protein molecule. The result of this is that large protein molecules are broken down and converted into a number of smaller peptide molecules.

G-cells

The final cell type in gastric glands that will be considered is the G-cells. G-cells secrete the hormone **gastrin**, which directly influences the function of both parietal and chief cells. Although we refer to gastrin, it is important to note that the precursor molecule of gastrin (preprogastrin) is processed into differing sized fragments, all of which appear to have some physiological action. However, it appears that the primary effect is mediated by so-called G17. The effect of gastrin on the stomach is threefold:

- it promotes gastric acid secretion;

- it promotes pepsinogen secretion (and activation);

- it promotes epithelial growth.

The secretion of gastrin is increased by:

- the presence of the products of protein digestion in the stomach—in particular some amino acids such as phenylalanine are particularly potent via direct effect on G-cells;

- distension of the stomach, as would be associated with the intake of food;

- parasympathetic stimulation via the vagus nerve;

- a range of other endocrine and paracrine substances.

The factors controlling gastrin secretion described above are similar to the factors that control pepsinogen secretion. The control

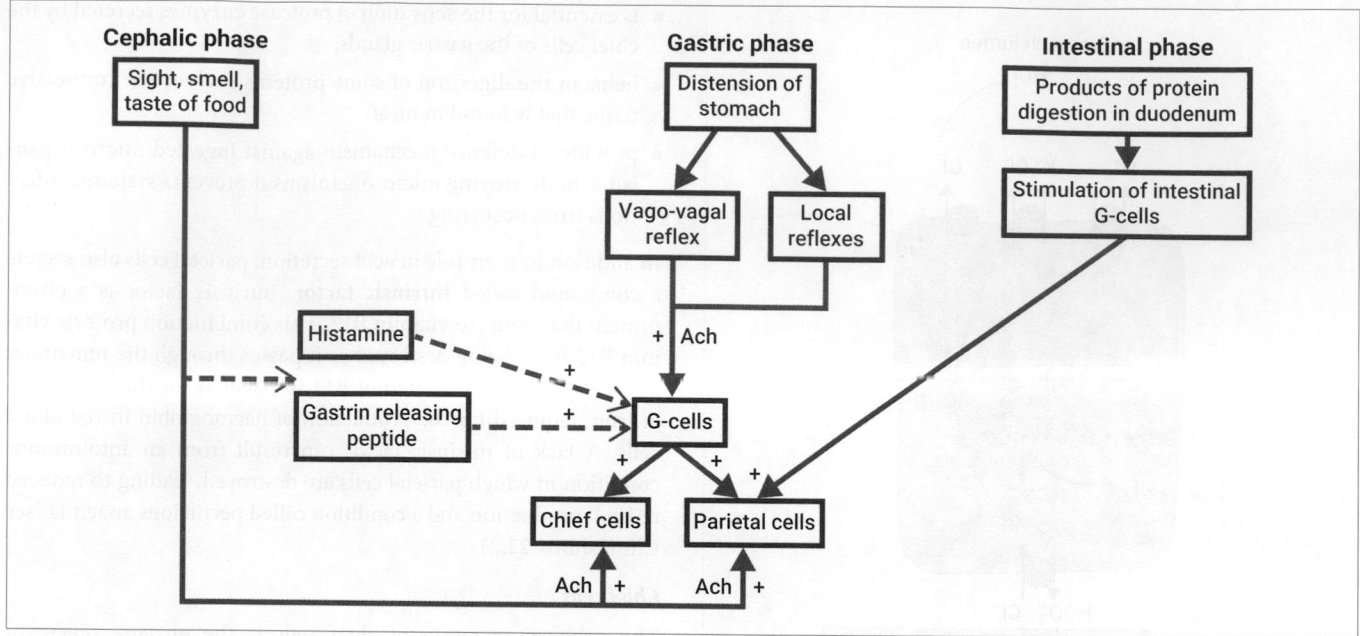

Figure 25.11 The major components that stimulate gastric secretions. Inhibition of gastric secretion is driven by the presence of an acidic chyme in the duodenum, which promotes the release of secretin from S-cells in the duodenum.

Source: Pocock et al. *Human Physiology*, 5th edn, 2017. Oxford University Press.

of secretions described so far is complex with a range of stimuli that have an influence. This is further complicated when different timings are considered; this gives rise to the idea of cephalic, gastric, and intestinal control. These different aspects are shown in Figure 25.11:

- the cephalic phase relates to the thought, sight, taste, or smell of food;
- the gastric phase relates to the presence of food in the stomach;
- the intestinal phase relates to the presence of chyme in the duodenum.

The cephalic and gastric phases lead to stimulation of secretions, while the intestinal phase has a two-fold action; initially, the chyme entering the duodenum stimulates the intestinal G-cells to release gastrin, which further stimulates gastric secretion. However, distention of the duodenum and lowering of the pH triggers the enterogastric reflex, which reduces gastric stimulation by the vagus nerve, reduces the gastric secretions, and closes the pyloric sphincter preventing further passage of chyme into the duodenum. The lowering of the pH also triggers the release of the hormone **secretin** that inhibits gastric secretion and stimulates bicarbonate secretion into the small intestine (see the next subsection).

Of recent interest is the observation that the hormone **ghrelin** is secreted from P/D1 cells within the fundus of the stomach. Ghrelin is currently the only hormone known to increase subjective feelings of hunger by activating pathways within the hypothalamus that stimulate neuropeptide Y secretion and activation of Y1 and Y5 receptors. Ghrelin therefore stimulates increased food intake and its levels in the blood are linked to the body clock, rising prior to

normal meal times and also when individuals are hungry. Ghrelin antagonizes the action of **leptin**. Leptin is a hormone produced by cells in adipose tissue that reduces food intake. This suggests a role in appetite regulation and energy balance for ghrelin. In addition to these well-studied observations, ghrelin also acts synergistically with growth hormone releasing hormone (see Section 19.3, Growth hormone) to increase growth hormone secretion, has been shown to inhibit insulin release (Section 25.4, Insulin), is thought to be involved in increasing gastric motility, and appears to have a role in modulating inflammatory cytokine release.

The stomach has an amazing capacity to expand in volume following food ingestion. This is achieved as a result of smooth muscle relaxation within the fundus and body of the stomach prior to food moving from the oesophagus into the stomach and the unfolding of the rugae. The process of receptive relaxation is initiated following swallowing and provides a reservoir in which ingested food is stored while the process of digestion begins. The motility of the stomach relies on peristaltic waves of smooth muscle contraction that begin within the body of the stomach and pushes stomach contents towards the antrum. Peristaltic contractions are the result of pacemaker potentials in smooth muscle fibres found within the stomach wall. The baseline number of pacemaker potentials ensures relatively low force production and slow rate of motility; however, the rate of potentials, and force of contractions, can be increased as a result of food intake, hormone secretion, and central nervous system input that lead to changes in gastric motility. In addition to this, feedback from the small intestine can reduce the rate of potentials and force of contractions, which will reduce gastric motility. The key factors are the presence of lipids within the duodenum, the acidity of chyme, and high osmolality of intestinal contents. These factors all reduce the secretion of HCl and pepsin secretion, which, in turn, alter the rate of pacemaker potentials and reduce gastric motility.

Small intestine

The small intestine is the site of the vast majority of absorption of ingested substances.

The movement of ingested substances from the stomach to the small intestine is a coordinated process involving contraction of stomach smooth muscle, opening of the pyloric sphincter, and relaxation of the smooth muscle in the first section of the small intestine. This process is called gastric emptying and is an important step in the process of nutrient absorption.

The small intestine is approximately 6 m in length and can be divided into three regions:

- duodenum;
- jejunum;
- ileum.

The **duodenum** is approximately 25–30 cm in length and its main function is to receive chyme from the stomach, as well as secretions from the pancreas and the liver. The **jejunum** is the main site of nutrient absorption with this part of the small intestine being approximately 2.5 m in length. The **ileum** is the longest segment of the small intestine at approximately 3.5 m in length and its primary function is to absorb vitamin B12 and bile salts. The ileocecal valve is a sphincter found at the end of the ileum that controls the movement of contents into the large intestine.

Within the small intestine, mucus is secreted from cells within the epithelium to lubricate the surfaces of the tract. Intestinal glands, known as **crypts of Lieberkühn**, are found at the base of intestinal villi and these produce brush border enzymes that are secreted into the intestinal lumen. The key action of these enzymes, as outlined below, is to assist with the digestive processes occurring within the small intestine that facilitate absorption of nutrients. In addition, the crypts of Lieberkühn also contain enteroendocrine cells that secrete numerous hormones that can influence the function of the digestive system. Hormones that have been identified include:

- 5-HT (also known as serotonin)—produced in enterochromaffin cells;
- secretin—produced in S-cells;
- somatostatin—produced in D-cells;
- cholecystokinin (CCK)—produced in I-cells.

The duodenum has its own specific secretory glands called **Brunner's glands** which secrete mucus when acidic contents are delivered from the stomach. The mucus secreted contains substantial amounts of bicarbonate ions, thus neutralizing the acid pH of the contents of the intestinal lumen and protecting the epithelial lining from erosion.

Motility of the small intestine

When there are relatively few contents within the intestinal lumen, a process called the migrating myoelectrical complex (MMC) occurs. The MMC is a period of peristaltic contractions that begins at the base of the stomach and travels along the small intestine. This process continues, with the start of the MMC moving down the digestive tract on each occasion, until food is ingested. The effect of the MMC is to move ingested substances down the intestinal tract. This maximizes absorption and removes any undigested material from the small intestine, preventing bacterial overgrowth. The hormone motilin is of paramount importance in this process

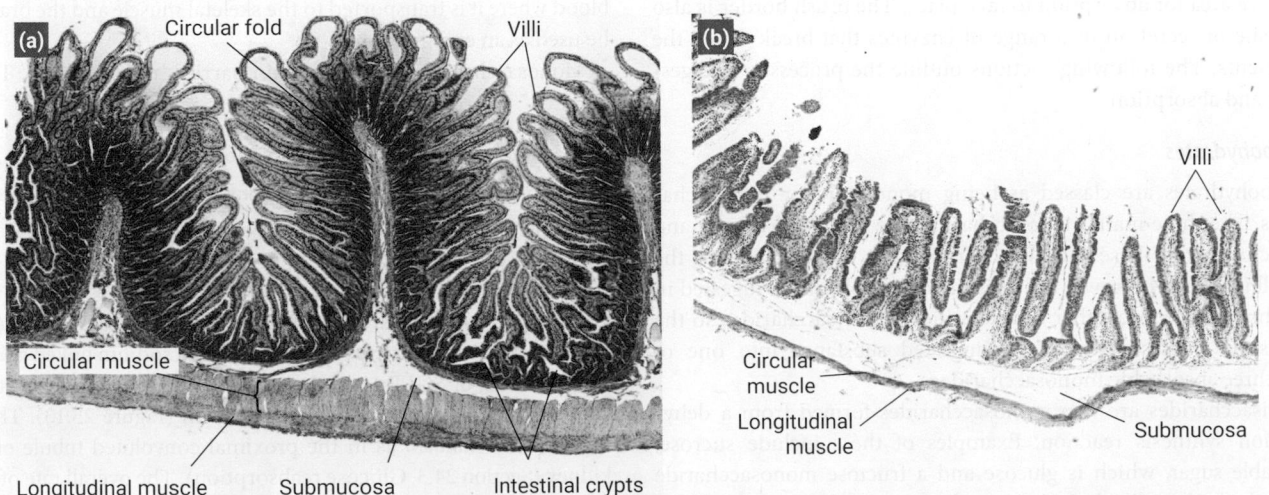

Figure 25.12 Histological cross-sections of the small intestine illustrating the expansion of the surface area by the presence of the villi. Although the structure is highly modified, it still shows the same four layers (Figure 25.1), the mucosa with columnar epithelium, the submucosa, the muscular layer, and the serosa.

Source: Pocock et al. *Human Physiology*, 5th edn, 2017. Oxford University Press.

as it appears to begin the process of the MMC when it binds to a motilin receptor. Interestingly, ghrelin (as discussed earlier) is structurally very similar to motilin and it is thought this is one reason why it may be involved in gastrointestinal motility.

Following ingestion of nutrients and the movement of chyme from the stomach into the small intestine, the MMC stops as a result of food ingestion stopping motilin release and a rhythmic process of contraction and relaxation of smooth muscle occurs. This process, known as **segmentation**, mixes the contents of the intestinal lumen and brings them into contact with the intestinal wall, facilitating absorption. Segmentation results from pacemaker potentials occurring within the circular smooth muscle found in the intestinal wall. These rhythmic contractions are relatively slow, meaning that the contents of the intestinal lumen are moved slowly through the small intestine, which maximizes the time for absorption of substances to occur. The rate of segmentation can increase or decrease in response to hormonal release as well as central nervous system activity. Increased parasympathetic activity increases the force produced during contractions, whereas sympathetic activity decreases it, resulting in changes in rate of small intestinal motility.

Intestinal digestion and absorption

With the exception of alcohol, no absorption of ingested substances occurs in the stomach, so absorption only commences after the chyme has entered the small intestine. The wall of the small intestine has a greatly expanded surface area: the luminal membrane of the **enterocytes** (the columnar epithelial cells) carries numerous villi, which are finger-like projections extending into the lumen of the small intestine (Figure 25.12). The surfaces of the villi themselves bear microvilli, which create a layer called the **brush border**. The microvilli, as their name suggests, are extremely small, typically 100 nm (remember 1 nm = 1×10^{-9} m) in diameter and a few hundred nm in length (to give an idea of scale, the microvilli are about 200 times smaller in diameter than a typical human hair). This structure provides a greatly enlarged surface area for absorption to take place. The brush border is also the site of secretion of a range of enzymes that break down the nutrients. The following sections outline the processes of digestion and absorption.

Carbohydrates

Carbohydrates are classed as being mono-, di-, or polysaccharides. The three main monosaccharides are glucose, fructose, and galactose. It is these monosaccharides that are absorbed in the small intestine; however, the majority of carbohydrates ingested in the human diet are either disaccharides or polysaccharides so the digestion process reduces the ingested substance into one of the three absorbable monosaccharides.

Disaccharides are two monosaccharides formed from a dehydration synthesis reaction. Examples of these include sucrose, or table sugar, which is glucose and a fructose monosaccharide joined together. Similarly, lactose (found in many dairy products) is formed from glucose and galactose monosaccharides. Maltose, found in beer, is two glucose monomers joined together. Following the ingestion of a disaccharide, enzymes are secreted from the brush border membrane of the small intestine that cleave the disaccharide into its constituent monosaccharides. Sucrose is cleaved by sucrase, while lactose and maltose are cleaved by lactase and maltase, respectively. The end result of this process is that glucose, galactose, and/or fructose are present within the small intestine and are then available for absorption. If, for some reason, it is not possible to cleave a disaccharide into its constituent monosaccharides the carbohydrate cannot be absorbed and is then passed into the large intestine.

PAUSE AND THINK

Given what you have just read, what do you think causes lactose intolerance?

Answer: Individuals suffering from lactose intolerance are unable to produce the enzyme lactase. As a result, ingestion of dairy products containing lactose leads to this sugar being moved to the large intestine where micro-organisms ferment the lactose, resulting in gas production and water movement into the intestine that ultimately causes diarrhoea.

Polysaccharides are long chains of monosaccharides. Examples of these found in the human diet include starch and non-starch polysaccharides such as fibre. Digestion of these substances begins in the mouth with secretion of salivary amylase and continues when in the stomach and the small intestine via the secretion of α-amylase from the pancreas. This results in the breakdown of some polysaccharides into disaccharides that are further hydrolysed to monosaccharides, as described above, which are then available for absorption. Some starches and most non-starch polysaccharides are resistant to digestion and are passed to the large intestine where they are fermented. This process is undertaken anaerobically by bacteria within the large intestine and produces by-products, the main ones being butyrate, acetate, and propionate. Butyrate is the main energy source for epithelial cells of the large intestine, whereas propionate is absorbed and metabolized by the liver. Acetate is absorbed from the intestine and ends up in the blood where it is transported to the skeletal muscle and the brain to be used as an energy source.

Monosaccharides have four main barriers to absorption. These are:

- a layer of mucus surrounding an enterocyte;
- the apical membrane of the enterocyte;
- use of the monosaccharide as an energy source;
- the basolateral membrane of the enterocyte.

Glucose and galactose are absorbed across the apical membrane of the enterocyte via an active transporter called sodium linked glucose transporter 1, or SGLT-1. As the name suggests, this transporter is closely linked to the cotransport of sodium (Figure 25.13). This is the same mechanism as in the proximal convoluted tubule of the kidney (Section 24.3, Glucose reabsorption). The overall rate of glucose absorption from the intestine is determined by the number of SGLT-1 transporters. The Na^+K^+ ATPase on the basolateral membrane generates a low intracellular concentration of Na^+ by pumping Na^+ ions out of the cell. This creates a concentration gradient

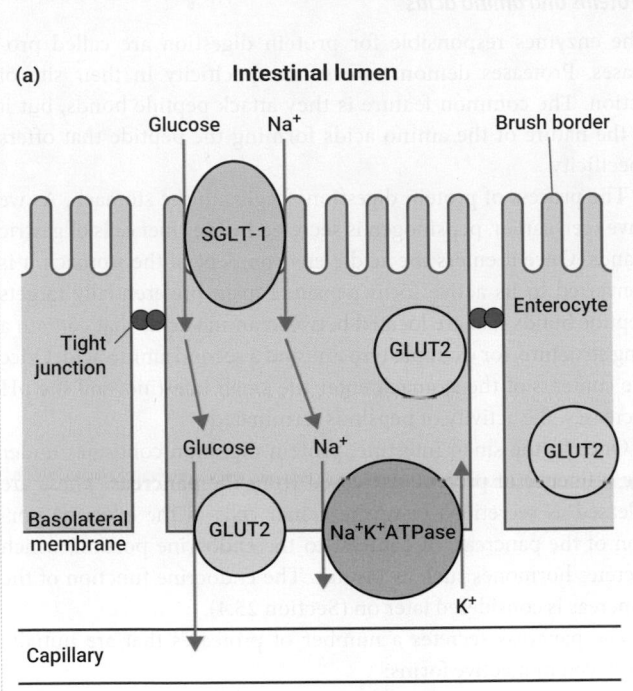

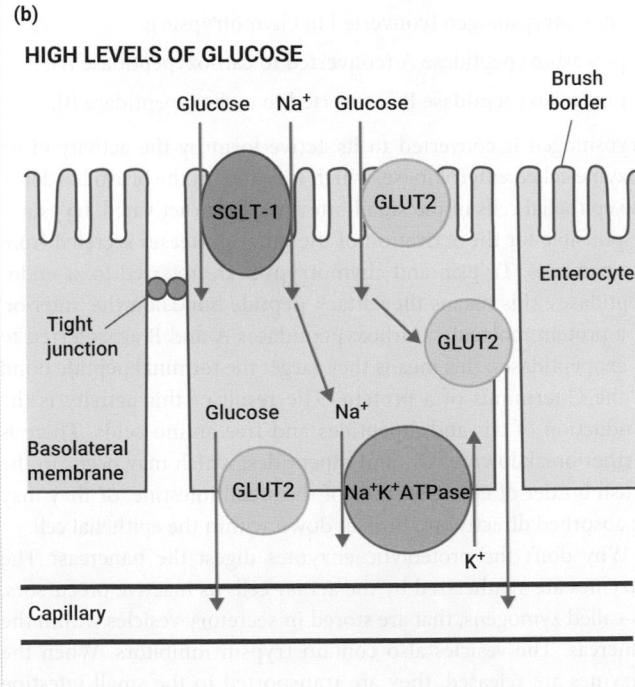

Figure 25.13 Glucose absorption in the small intestine. (a) The Na⁺K⁺ ATPase on the basolateral membrane generates a Na⁺ gradient by actively pumping Na⁺ ions out of the cell. This gradient then drives the cotransport of glucose with Na⁺ via the SGLT-1 transporter. The GLUT2 transporters then move the glucose out of the cell into the capillaries. (b) When luminal levels of glucose are high, the GLUT2 transporters are trafficked from the cytoplasm and inserted into the apical membrane, increasing the capacity for glucose absorption.

Source: Pocock et al. *Human Physiology*, 5th edn, 2017. Oxford University Press.

relative to the lumen of the intestine such that Na⁺ ions move into the enterocyte via the SGLT-1 transporter along with the glucose.

Fructose is absorbed across the apical membrane of the enterocyte by facilitated diffusion mediated by glucose transporter 5 (GLUT5). This is a passive transporter and, as such, has a maximum limit of absorption. In most people, that limit appears to be a rate of approximately 40 g of free fructose. Fructose is also used as an energy source by enterocytes to a greater extent than glucose. Glucose transporter 2 (GLUT2) also transports glucose, fructose, and galactose across the basal membrane of the enterocyte and into the circulation (Figure 25.13).

The amount of glucose in the lumen varies significantly: before eating, the levels will be low. During digestion of the meal, the levels will rise very rapidly. As the glucose levels rise, SGLT-1 transports the glucose across the apical membrane, but this activation also triggers the insertion of GLUT2 transporters into the apical membrane, thereby increasing the capacity for glucose absorption. As the glucose levels fall, the apical GLUT2 is inactivated and removed from the apical membrane again. This mechanism enables a rapid response to changing glucose levels and enables all the glucose to be absorbed, even though the amounts in the intestine vary significantly over time.

Lipids

The primary dietary source of lipids is triglycerides, with some phospholipids and cholesterol esters also being ingested. In a similar manner to carbohydrates, these substances cannot directly be absorbed by the intestine and need to be packaged in a certain way to be able to do this. In the case of triglycerides, these are broken down into glycerol and its fatty acids. Similarly, phospholipids and cholesterol esters are hydrolysed to lyso-phospholipids and free cholesterol, respectively, which liberates free fatty acids.

This process begins in the stomach where ingested lipids become emulsified. The lipids that are contained within foodstuffs are insoluble in the aqueous environment of the secretions of the digestive tract. This means they will have a tendency to coalesce (clump together), forming large lipid droplets. The enzymes responsible for lipid digestion are only able to attack the surface of a lipid droplet; therefore, if there are large droplets the lipid molecules on the interior of the droplet are not subject to digestion. Breaking the large droplets down into a number of smaller droplets increases the ability of lipases to digest lipids. Substances that break up large lipid droplets into smaller droplets are called **emulsifiers**. Bile salts act as an important emulsifier: these are produced in the liver and stored in the gallbladder before their release into the digestive tract, as will be discussed in more detail later in this chapter (Section 25.5). The physical mixing of the contents of the digestive tract further enhances the process of emulsification.

A variety of enzymes are responsible for the digestion of lipids:

- phospholipase A2 hydrolyses phospholipids;
- cholesterol ester hydrolase hydrolyses cholesterol esters;
- pancreatic lipase hydrolyses triglycerides resulting in the formation of free fatty acids and monoglycerides.

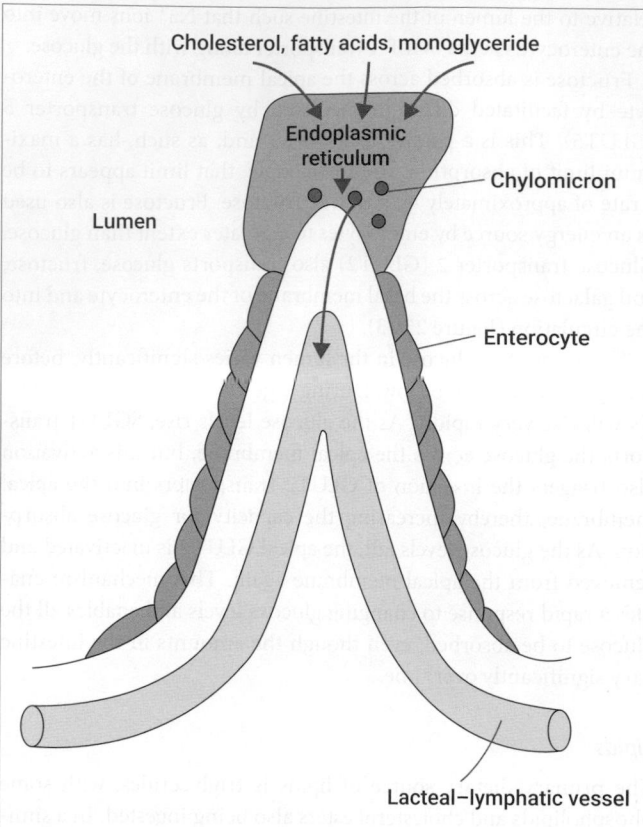

Figure 25.14 Lipid absorption in the small intestine. Under the action of the lipases, the products of lipid digestion form into chylomicrons in the endoplasmic reticulum of the enterocytes. These are transported into the lacteals, where they form chyle, which is then transported to the venous system.
Source: Pocock et al. *Human Physiology*, 5th edn, 2017. Oxford University Press.

The free fatty acids, monoglycerides, lyso-phospholipids, and cholesterol are packaged into a mixed micelle which is absorbed, without any bile acid, via diffusion into the enterocyte. The bile salts are absorbed in the terminal region of the ileum. Once into the enterocyte, short- and medium-chain fatty acids are directly absorbed via diffusion into the portal circulation; however, the majority of fatty acids reform triglycerides, cholesterol esters, and phospholipids, and, with the addition of apolipoproteins, are packaged into chylomicrons, which are then absorbed into the lymphatic circulation. This is shown schematically in Figure 25.14.

Interestingly, a number of pharmaceutical interventions for the treatment of obesity have focused on blocking intestinal absorption of ingested fats. One example of these known to result in significant weight loss blocks the activity of pancreatic lipase. This means that triglycerides are not hydrolysed to monoglycerides and fatty acids, resulting in less overall lipid absorption. Individuals taking this agent who also ingest large amounts of dietary fat suffer from side effects, such as diarrhoea, that lead to behavioural change that reduces dietary fat intake. Consequently, energy intake is reduced and body mass is reduced.

Proteins and amino acids

The enzymes responsible for protein digestion are called proteases. Proteases demonstrate some specificity in their site of action. The common feature is they attack peptide bonds, but it is the nature of the amino acids forming the peptide that offers specificity.

The process of protein digestions begins in the stomach. As we have seen earlier, pepsinogen is secreted by the chief cells of gastric glands. Once it enters the acidic environment of the stomach it is converted to its active form pepsin. Pepsin preferentially targets peptide bonds that are located between amino acids that contain a ring structure, for example tyrosine and a second amino acid. Once the contents of the stomach enter the small intestine, and the pH increases, the activity of pepsin is terminated.

Once in the small intestine, protein digestion continues under the influence of proteases released from the **pancreas**. These are released as secretions from the acinar cells in the exocrine portion of the pancreas, in contrast to the endocrine portion, which secretes hormones such as insulin. The endocrine function of the pancreas is considered later on (Section 25.4).

The pancreas secretes a number of proteases that are initially produced in inactive forms:

- trypsinogen (converted to trypsin);
- chymotrypsinogen (converted to chymotrypsin);
- procarboxypeptidase A (converted to carboxypeptidase A);
- procarboxypeptidase B (converted to carboxypeptidase B).

Trypsinogen is converted to its active form by the activity of an enzyme called enterokinase, which is found on the brush border of the epithelial cells of the small intestine. Once activated, trypsin is responsible for the activation of the other proteases secreted from the pancreas. Trypsin and chymotrypsin are referred to as endopeptidases: this means they attack peptide bonds on the 'interior' of a protein molecule. Carboxypeptidases A and B are referred to as exopeptidases: this means they target the terminal peptide bond at the C-terminus of a protein. The result of this activity is the production of tri- and dipeptides and free amino acids. There is further breakdown of tri- and dipeptides, which may occur in the brush border of epithelial cells of the small intestine, or they may be absorbed directly and broken down within the epithelial cell.

Why don't the proteolytic enzymes digest the pancreas? The enzymes are synthesized by the acinar cells as inactive precursors, so-called zymogens, that are stored in secretory vesicles within the pancreas. The vesicles also contain trypsin inhibitors. When the enzymes are released, they are transported to the small intestine via the pancreatic duct and the trypsinogen is converted to trypsin by the enterokinase. Pancreatitis is inflammation and damage to the pancreas as a result of activation of the enzymes within the pancreas. In its acute form this most commonly occurs either as a result of gallstones that increase the pressure in the duct or as a result of long-term alcohol abuse.

The dipeptides are absorbed by cells in the small intestine by means of the PepT1 transporter that is located in the brush border of the enterocytes. The PepT1 transporter transports these molecules into the cell via a secondary active transport mechanism.

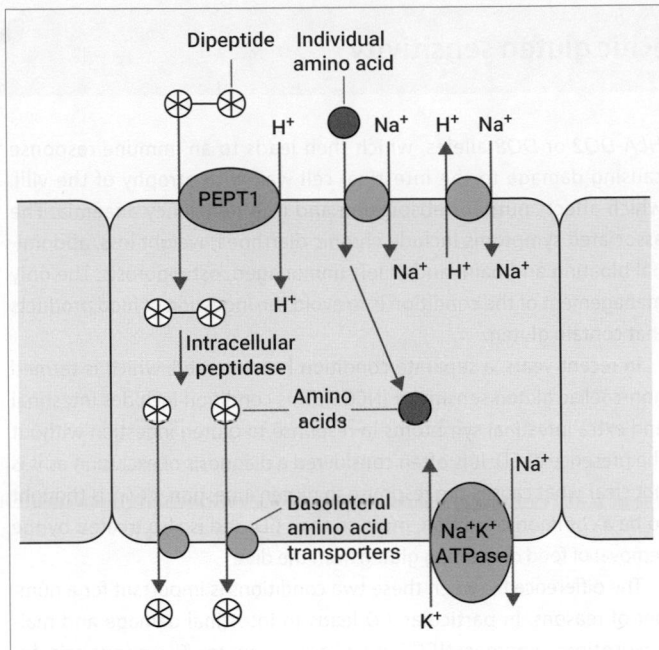

Figure 25.15 Absorption of dipeptides and individual amino acids from the lumen of the small intestine. The Na⁺K⁺ ATPase on the basolateral border actively pumps sodium ions out of the cell, generating the Na⁺ gradient that supports the cotransport processes across the apical membrane.

Source: Pocock et al. *Human Physiology*, 5th edn, 2017. Oxford University Press.

Initially, Na^+ is transported into the cell in exchange for a H^+ ion. The dipeptide is then transported into the cell, along with the H^+ ion. Once absorbed, the dipeptide is broken down into its constituent amino acids while the basolateral Na^+K^+ ATPase transporter removes the Na^+. Equally, the amino acids released are transported out of the cell via transporters on the basolateral side of the cell. We now know that there are at least five different types of amino acid transporter on the basolateral side of the cell, each of which displays some degree of specificity for the type of amino acid that it transports.

Individual amino acids are transported into the cell via transport molecules located in the brush border. These transport systems, of which there are at least six, depend upon the cotransport of Na^+. As with the transport of amino acids out of the cell, there is a degree of selectivity shown by each of these six transporters. A summary of the transport processes is shown in Figure 25.15.

Absorption of water

About 80% of the water in the digestive tract is absorbed through the wall of the small intestine. As noted above, sodium ions are actively pumped from the enterocytes into the extracellular space and from there into the blood stream. The accumulation of extracellular sodium generates an osmotic gradient so water is absorbed by osmosis via transcellular and paracellular routes.

Pancreatic juice

The exocrine secretions of the pancreas form what is called pancreatic juice. Pancreatic juice consists of an alkaline fluid within which are the enzymes responsible for the digestion of proteins, lipids, and carbohydrates. Each day about 1500 mL of pancreatic juice is secreted. Pancreatic juice is rich in bicarbonate ions and has a pH of ~8, so it acts to neutralize the acidic secretions that have entered the small intestine from the stomach and creates a pH environment in which the enzymes can function.

There are two main types of cell in the exocrine pancreas: the acinar cells and the duct cells.

- The acinar cells synthesize and secrete the enzymes present in the pancreatic juice.
- The duct cells are columnar epithelial cells that produce the bicarbonate-rich secretion.

In essence, bicarbonate ions (HCO_3^-) are actively cotransported from the blood across the basolateral membrane into the epithelial cells along with sodium ions. The Na^+ gradient for this cotransport is generated by the Na^+K^+ ATPase on the basolateral membrane. At the apical aspect of the cell bicarbonate ions are secreted into the lumen of the duct: this is mediated by two transporters, one that exchanges bicarbonate for chloride ions and one that cotransports bicarbonate with Cl^-. The cotransporter chloride ion channel involved is the cystic fibrosis transmembrane conductance regulator (CFTR) channel, mutations of which result in the development of cystic fibrosis.

The secretion of pancreatic juice is controlled by both neural and endocrine mechanisms, although the latter is probably most important. In a similar manner to secretions of the stomach, pancreatic secretions have cephalic, gastric, and intestinal phases. The cephalic phase is mediated via increased activity in the vagus nerve, while the gastric phase is mediated by the actions of gastrin. By far the most significant phase is the intestinal phase; in this case there is a major role for hormones secreted from the duodenal mucosa. The presence of acidic chyme in the small intestine promotes the release of secretin from the S-cells of the duodenal mucosa. In addition, the initial products of protein digestion stimulate the I-cells to release CCK. Both secretin and CCK stimulate the production of pancreatic juice.

Maintenance of the internal environment of the small intestine is essential to the process of absorption of nutrients. Factors that irritate the lining of the intestine, particularly those that damage the villi, have significant impact on the health of the individual. There are several such conditions, one common example being intolerance to gluten, as occurs in coeliac disease (see Clinical Box 25.1).

Endocrine functions of the small intestine

As we have seen, the small intestine contains a variety of enteroendocrine cells that secrete hormones that can affect digestive system function as well as influence the functioning of other systems. Some examples include:

- Gastrin is secreted from the duodenum in response to protein ingestion and appearance in the small intestine which provides feedback to the stomach and increases secretions from chief and parietal cells.
- Glucose-dependent insulinotropic peptide (GIP) is secreted from the duodenum in response to fat and glucose ingestion and

Coeliac disease and non-specific gluten sensitivity

Gluten is the main structural protein complex found in many commonly ingested food products that contain grains such as wheat, rye, and barley. Two main components of grains such as wheat are gliadins and glutenins, which contribute to the rising and elastic properties of dough in bread-making. In recent years, there has been a huge increase in the number of gluten-free food products that are available. Historically, these were needed for those suffering from coeliac disease (CD); however, a number of other conditions can be managed through the removal of products that contain gluten.

CD is a relatively common autoimmune condition affecting around 1 per cent of the population in Europe. As with all autoimmune conditions, CD is the result of a combination of genetic susceptibility, environmental factors, and the effect this has on immune function. Patients with CD present with specific variants of the human leukocyte antigen gene (*HLA-DQ2* and *HLA-DQ8*), which is responsible for regulating immune function on cell surface proteins. When carriers with these alleles ingest gluten it is digested to gliadin fragments which activate tissue transglutaminase (tTG), where it becomes deamidated. The deamidated gliadin fragment is presented to the antigen cell with

HLA-DQ2 or *DQ8* alleles, which then leads to an immune response causing damage to the intestinal cell wall with atrophy of the villi, which affects nutrient absorption and iron deficiency anaemia. The associated symptoms include chronic diarrhoea, weight loss, abdominal bloating and pain, and, if left unmanaged, osteoporosis. The only management of the condition is to avoid the ingestion of food products that contain gluten.

In recent years, a separate condition has emerged, which is termed non-coeliac gluten sensitivity (NCGS). This condition includes intestinal and extra-intestinal symptoms in response to gluten ingestion without the presence of CD. It is often considered a diagnosis of exclusion as it is not clear what causes the response to gluten ingestion. NCGS is thought to be a common condition, more so than CD, and is also treated by the removal of food containing gluten from the diet.

The difference between these two conditions is important for a number of reasons. In particular, CD leads to intestinal damage and malabsorption, whereas NCGS does not seem to. The prognosis for unmanaged CD is worse than for NCGS and, as CD is an autoimmune condition, it tends to aggregate in families, whereas NCGS does not.

appearance in the small intestine which reduces gastric motility and leads to an increase in insulin secretion from the pancreas.

- Cholecystokinin (CCK) is secreted from the I-cells of the duodenum in response to fat and protein entering the duodenum. It acts to increase the secretion of digestive enzymes from the acinar cells of the pancreas and potentially reduces the subjective feeling of hunger. CCK also stimulates the production and release of bile from the gallbladder (Section 25.5).

- Glucagon-like peptide 1 (GLP-1) is secreted from the distal ileum and the colon of the large intestine, primarily in response to carbohydrate intake that stimulates the secretion of insulin while also reducing subjective feelings of hunger.

- Glucagon-like peptide 2 (GLP-2) is also secreted from the distal ileum and colon in response to carbohydrate intake; however, this has trophic effects, including intestinal cell growth and repair.

- Secretin is released by the S-cells of the duodenum in response to lowering of the pH in the small intestine, as a result of acid chyme entering from the stomach. Its main actions include feedback inhibition of the parietal cells of the stomach, reducing the release of acid, and stimulation of the duct cells of the pancreas to increase the production of bicarbonate-rich pancreatic juice.

- Serotonin (5-HT) has widespread actions within the body. Within the digestive system it stimulates the production of mucus and also has an action to increase gut motility. In this context it is also associated with diarrhoea, in response to the presence or irritants in the digestive system—it is also linked to irritable bowel syndrome (IBS—see Clinical Box 25.2).

Large intestine

The large intestine is approximately 1.5 m in length and is separated into three sections, the **cecum**, the **colon**, and the **rectum**. Contents from the ileum enter the cecum through the **ileocecal valve** at which point compaction and the formation of **faeces** (sometimes also called stool) begins.

The colon is subdivided into four regions (Figure 25.16):

- ascending;
- transverse;
- descending;
- sigmoid colon.

While the majority of absorption occurs within the small intestine, some important absorptive functions occur within the colon. In particular, further absorption of water occurs within the colon as does absorption of important vitamins including vitamin K and vitamin B5. As in the small intestine, there is active absorption of sodium from the chyme with water following by osmosis.

The colon is also home to extremely large numbers of microorganisms, the gut microbiome, which recent evidence suggests can play an important role in a variety of disease states (see Section 25.3).

The large intestine shows characteristic patterns of motility:

- segmentation;
- antiperistalsis;
- mass movements.

Inflammatory bowel disease and irritable bowel syndrome

Inflammatory bowel disease (IBD) is an umbrella term for two similar conditions: ulcerative colitis (UC) and Crohn's disease (CD). Both UC and CD are autoimmune conditions that lead to relapsing periods of ulceration and inflammation of the wall of the digestive system causing pain, bloody diarrhoea, and mucus discharge. The difference between the two conditions is that UC is found only in the colon and the rectum, whereas CD can be anywhere in the digestive tract. In addition, ulcerations are deeper in CD compared to UC. Complications include perforation, fistula, abscess, obstruction, and increased risk of intestinal cancer, in addition to effects on other physiological systems. In general, the risk of complications is greater for CD than UC with the exception of risk of intestinal cancer.

The pathogenesis of IBD is complex and involves an interaction between genetics, environment, and immune function. It is thought that the role of genetics may be related to factors that influence tolerance of microbial load of the large intestine and there are a large number of environmental factors that have been suggested to increase risk of development of IBD.

These include diet and smoking status. Interestingly, studies have suggested that smoking protects against UC but increases risk of CD, which has led to the potential use of nicotine therapy to treat severe cases of UC.

Irritable bowel syndrome (IBS) is one of the most common functional gastrointestinal disorders observed in European and North American populations. It is often confused with IBD; however, the two disorders are very different. IBS is characterized by abdominal pain that is relieved upon defecation. Patients may be categorized as having diarrhoea-predominant or constipation-dominant IBS. In most cases, there is no anatomical or biochemical cause for the development of the condition and it is often diagnosed by exclusion of other conditions. Various factors, including changes in gastrointestinal transit, previous gastrointestinal infection, and visceral hypersensitivity, are thought to play a role in the pathogenesis of IBS. Alterations in serotonin production and release are common factors in many cases of IBS. Management of symptoms is the main treatment strategy for sufferers.

Figure 25.16 The colon of the large intestine. The diagram shows the bands of longitudinal smooth muscle (taeniae coli) and the haustra.

Source: Pocock et al. *Human Physiology*, 5th edn, 2017. Oxford University Press.

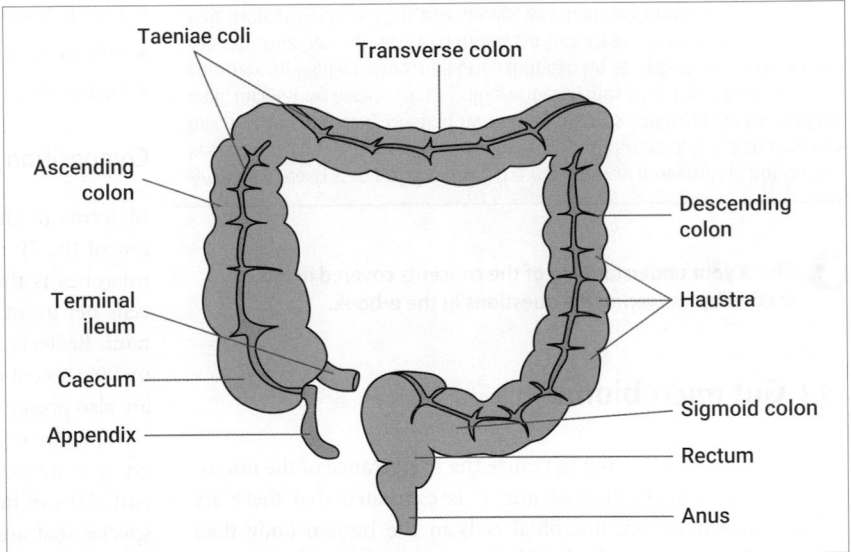

Segmentation contractions involve coordinated contraction of the circular and longitudinal muscles, which act to mix the chyme; combined with antiperistalsis these movements increase the time spent within the intestine and therefore the opportunity for absorption.

Mass movements are powerful peristaltic movements that act to move the intestinal contents into the rectum, forming the faeces. These occur several times a day, particularly following ingestion of a meal.

Rectum

The rectum is found at the end of the sigmoid colon and is the storage site of faeces. The last section of the rectum is called the

anal canal and the **anus** is the exit from this canal. The anal canal contains two sphincters. Both sphincters are maintained under a degree of tonic contraction ensuring the anus remains closed. The external anal sphincter consists of skeletal muscle and, as such, is under voluntary control. The internal anal sphincter is composed of smooth muscle and is not under voluntary control, being innervated by both the sympathetic and parasympathetic divisions of the autonomic nervous system. Sympathetic activation promotes contraction, while activation of the parasympathetic nervous system inhibits contraction (i.e. promotes its relaxation).

Unlike the small intestine, though, the propulsive movements of the large intestine are relatively slow. These movements move the contents of the large intestine towards its terminal region, the rectum. Periodically, mass movements occur, which empty large volumes of the large intestine. Because of these mass movements, the rectum becomes filled with faeces. As it fills with faeces, so the walls of the rectum become distended, initiating a defaecation reflex. Distension of the rectum activates the parasympathetic nervous system, which produces relaxation of the internal anal sphincter. Simultaneously, neural activity to the external anal sphincter is reduced, which results in its relaxation. Expulsion of faeces is aided by the contraction of the abdominal muscles and muscles of the thorax, including the diaphragm and glottis, a movement called the Valsalva manoeuvre. This causes an increase in intra-abdominal pressure that promotes the expulsion of faeces. At times when defaecation is inappropriate, the closure of the external anal sphincter is maintained under voluntary control. Consequent to this, a retro-peristaltic movement forces faeces back into the sigmoid colon. The ability to do this is a learned process that most people acquire during childhood. Ultimately, as a consequence of further mass movements, faeces are once again moved back into the rectum and the defaecation reflex begins once more.

PAUSE AND THINK

How do the structure and function of the digestive tract facilitate the absorption of ingested food?

Answer: In each stage of the tract, the mechanical movements not only keep the partially digested food moving along the tract, but also serve to mix and mechanically break down the foodstuffs and increase the surface area, facilitating access for the different enzymes that break down the complex molecules into small ones that can be absorbed. The cells lining the intestine have greatly expanded surface areas with extensive vascularization, providing a large area over which absorption can take place.

 Check your understanding of the concepts covered in this section by answering the questions in the e-book.

25.3 Gut microbiome

We are only just beginning to realize the importance of the microbial ecosystem in the human gut. It is estimated that there are at least 10 times more microbial cells in the human body than there are human cells. The healthy human gut has 10^{14} microbes comprising some 3000 different bacterial species alone. These include Gram-positive and Gram-negative bacteria, as well as archaea and eukaryotic microbes such as fungi and protozoa (see Module 2). There are also a vast number of bacteriophages (bacterial viruses). Together, this very complex community of microbes make up the '**microbiome**'. Altogether, the wet biomass weight of the microbiome is 1 kg and it is one of the most metabolically active areas of the human body; for example, the gut microbiome expresses 350 times more genes than the actual human genome.

We have a **mutualistic symbiotic** relationship with our gut microbes. The microbiome provides health benefits and produces vitamins (e.g. vitamin K, biotin, and folate) and short-chain fatty acids. The microbiome also stimulates the host immune system, aids in the inhibition of pathogenic microbes, transforms bioactive compounds (e.g. bile salts), and helps to maintain intestinal epithelium integrity. Enzymes produced by the microbiome also aid in digestion, especially for complex polysaccharides. We provide the microbiome with a consistent environment, with access to nutrients and a constant temperature. While there are benefits to the human host, certain circumstances, such as treatment with antibiotics, can cause an imbalance in the symbiosis leading to a dysbiosis, which can lead to effects such as:

- production of carcinogens;
- role in obesity and diabetes;
- involvement in inflammatory disease.

Composition of the microbiome

In terms of the bacterial content, the human gut has relatively few of the 70 recognized bacterial phyla present. The number of microbes is the biggest in the colon with 10^{11} to 10^{12} microbial cells per gram of faeces. Firmicutes and Bacteroidetes are dominant. Bacteria from the phyla Firmicutes and Bacteroidetes make up 90 per cent of the human gut bacteria. Species from other phyla are also present and are shown in Table 25.1.

While there are differences in the composition of an individual's gut, it is interesting to note that >50% of humans share 75 species and >90% of humans share 57 species. There are a core number of species that are common, but functionality can be carried out by different species.

Table 25.1 The main phyla and genus of species found in the human gut

Phylum	Genus
Firmicutes	*Lactobacillus, Clostridium, Staphylococcus, Enterococcus*
Bacteroidetes	*Bacteroides, Prevotella*
Actinobacteria	*Bifidobacterium*
Fusobacteria	*Fusobacterium*
Verrucomicrobia	*Akkermansia*
Proteobacteria	*Escherichia, Pseudomonas*

Source: Bliss E.S., Whiteside E. (2018). The gut-brain axis, the human gut microbiota and their integration in the development of obesity. *Front. Physiol.* 9, 900. CC BY 4.0

Development of the gut microbiome

The microbiota is established after we are born, as prior to this the fetus is surrounded by sterile amniotic fluid. How our gut is subsequently colonized is dependent on how we are born. Vaginally delivered babies have a microbiota that is dominated by the mother's vaginal and faecal bacteria. Babies born via a caesarean section (C-section) have a microbiota dominated by skin and bacteria from the hospital environment, and they have a reduced number of bifidobacteria. The biodiversity in babies born via C-section is decreased. These early colonizers of the human gut are called 'founder species' and the gut microbiome will not only develop from them, but will also be influenced by a number of internal and external factors, such as diet, host genetics, and gastrointestinal tract physiology. For example, there are differences in the microbiota of breast-fed and formula-fed babies. The microbiota changes again during weaning, including a reduction in the population of *Lactobacillus* and *Bifidobacterium*. The Bacteroidetes are prominent when we are young, but decline as we grow older, but the opposite of this trend can be seen with the Firmicutes phylum. The changes in the abundance of each phylum through our life stages can be seen in Figure 25.17.

Changing the gut microbiota

It is possible to change your gut microbiota. Prebiotics are dietary short-chain non-digestible carbohydrates such as inulin and fructo-oligosaccharides (FOS). When consumed, they stimulate growth of the existing gut microbiota such as *Bifidobacterium*. Probiotics, in contrast, are live micro-organisms that, when consumed in adequate numbers, will promote health benefits. There are many probiotic products available, from yoghurt-type drinks to food supplements. The microbiota can also be significantly altered during medical treatments, in particular with antibiotics (Figure 25.17).

Check your understanding of the concepts covered in this section by answering the questions in the e-book.

25.4 The pancreas

The role of the pancreas in controlling glucose

Glucose is the primary nutrient required for the production of ATP, which is the universal energy currency of cells. Therefore, it will come as no surprise that a considerable number of mechanisms exist that closely monitor and regulate glucose levels to

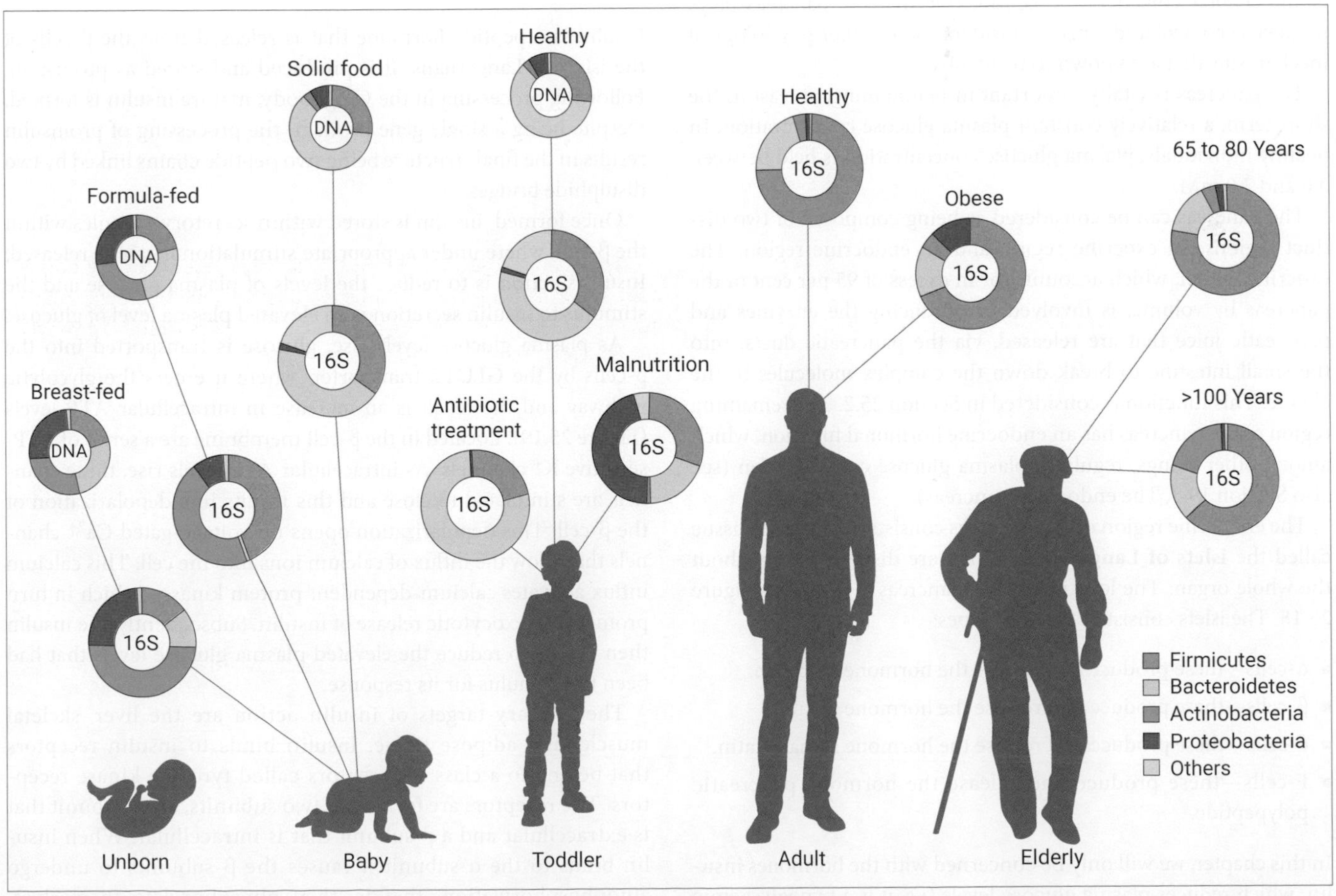

Figure 25.17 The human microbiota through life stages, as measured by 16S RNA or DNA.

Source: Ottman N., Smidt H., de Vos W.M., Belzer C. (2012). The function of our microbiota: who is out there and what do they do? *Front. Cell. Inf. Microbio.* 2,104.

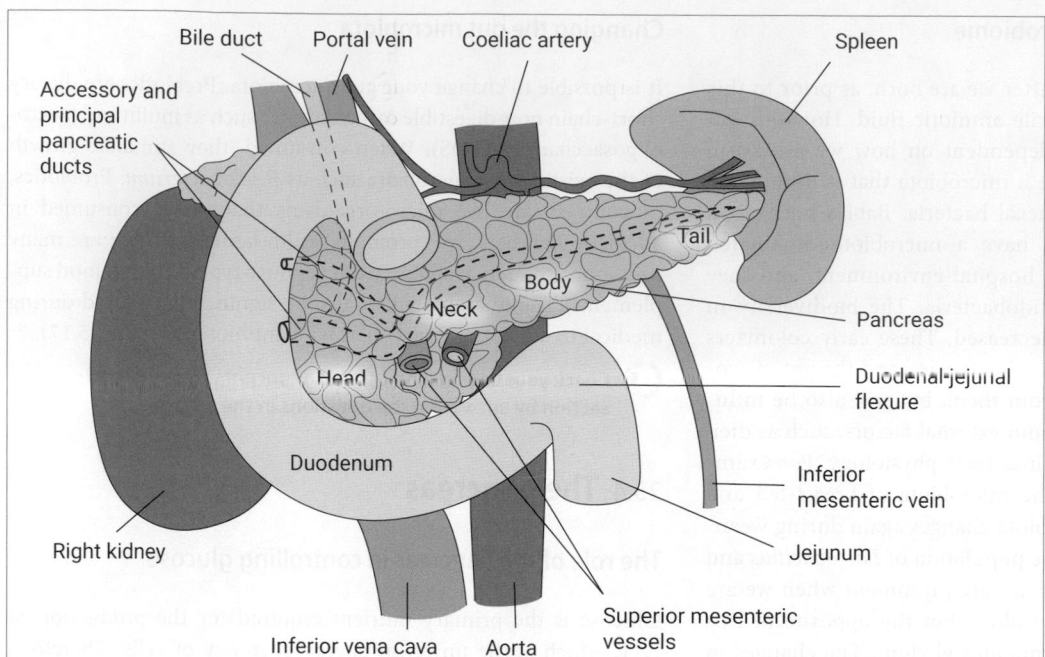

Figure 25.18 The location and organization of the pancreas.
Most pancreatic tissue is exocrine in nature—secretions from this part of the pancreas enter the digestive tract via the pancreatic ducts.
Source: Pocock et al, *Human Physiology*, 5th Edn, 2017. Oxford University Press.

maintain the required energy supply to the tissues. This section of the chapter considers the central role that the pancreas plays in glucose control and a brief consideration of other physiological mechanisms that are known to be involved.

The pancreas is vitally important in maintaining, at least in the short term, a relatively constant plasma glucose concentration. In healthy individuals, plasma glucose concentration is held between 3.0 and 7.0 mM.

The pancreas can be considered as being composed of two distinct regions, an exocrine region and an endocrine region. The exocrine region, which accounts for in excess of 95 per cent of the pancreas by volume, is involved in producing the enzymes and pancreatic juice that are released, via the pancreatic ducts, into the small intestine to break down the complex molecules in the chyme. This function is considered in Section 25.2. The remaining region of the pancreas has an endocrine hormonal function, which among other things, regulates plasma glucose concentration (see also Section 19.3, The endocrine pancreas).

The endocrine region of the pancreas consists of masses of tissue called the **islets of Langerhans**, which are dispersed throughout the whole organ. The location of the pancreas is shown in Figure 25.18. The islets consist of four cell types:

- α-cells—these produce and release the hormone glucagon.
- β-cells—these produce and release the hormone insulin.
- δ-cells—these produce and release the hormone somatostatin.
- F-cells—these produce and release the hormone pancreatic polypeptide.

In this chapter, we will only be concerned with the hormones insulin, which reduces plasma glucose levels (i.e. it is a hypoglycaemic hormone), and glucagon, which increases plasma glucose levels and is therefore a hyperglycaemic hormone.

Insulin

Insulin is a peptide hormone that is released from the β-cells of the islets of Langerhans. It is produced and stored as proinsulin. Following processing in the Golgi body, mature insulin is formed. Despite being a single gene product, the processing of proinsulin results in the final structure being two peptide chains linked by two disulphide bridges.

Once formed, insulin is stored within secretory granules within the β-cell, where under appropriate stimulation it can be released. Insulin's action is to reduce the levels of plasma glucose and the stimulus to insulin secretion is an elevated plasma level of glucose.

As plasma glucose levels rise, glucose is transported into the β-cells by the GLUT2 transporter, where it enters the glycolytic pathway and the result is an increase in intracellular ATP levels (Figure 25.19). Located in the β-cell membrane are a series of ATP-sensitive K^+ channels. As intracellular ATP levels rise, these channels are stimulated to close and this results in a depolarization of the β-cell. This depolarization opens up voltage-gated Ca^{2+} channels that allow the influx of calcium ions into the cell. This calcium influx activates calcium-dependent protein kinases, which in turn promote the exocytotic release of insulin. Subsequently, the insulin then begins to reduce the elevated plasma glucose levels that had been the stimulus for its response.

The primary targets of insulin action are the liver, skeletal muscle, and adipose tissue. Insulin binds to insulin receptors that belong to a class of receptors called tyrosine kinase receptors. The receptors are formed of two subunits, an α-subunit that is extracellular and a β-subunit that is intracellular. When insulin binds to the α-subunit it causes the β-subunits to undergo autophosphorylation, that is, they phosphorylate themselves. This initiates a series of intracellular signalling cascades that results in the insertion of preformed glucose transporter 4

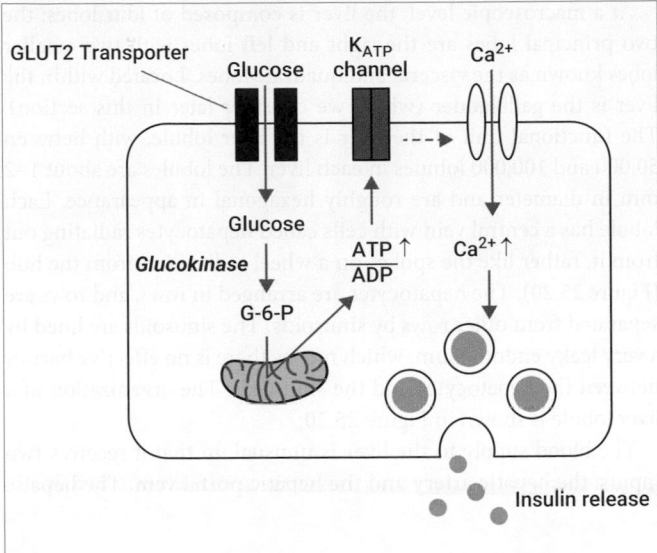

Figure 25.19 Stimulation of insulin release by glucose. Rising plasma glucose levels lead to glucose entering the β-cell via the GLUT2 transporter (GLUT1 and GLUT3 transporters may also be involved). The glucose is converted to glucose 6-phosphate (G-6-P) by glucokinase. In the mitochondria the G-6-P is converted to ATP, which binds to the ATP-sensitive potassium channel, leading to its closure. Closure of the K^+-leakage channel leads to depolarization and opening of L-type voltage-gated Ca^{2+} channels (see Section 19.3, The endocrine pancreas). The rise in intracellular Ca^{2+} results in release of insulin from the secretory vesicles.

Source: Mann E., Bellin M.D. (2016). Secretion of insulin in response to diet and hormones. Pancreapedia: Exocrine Pancreas Knowledge Base.

(GLUT4) molecules into the cell membranes of the cells that form these tissue.

What is the effect of inserting GLUT4 transporters into the cell membrane? GLUT4 increases the uptake of glucose into the cells, down its concentration gradient by facilitated diffusion, therefore reducing the plasma levels of glucose and returning them to a normal level.

Within the tissues, insulin has an anabolic action to stimulate storage processes; for example, in the liver insulin stimulates the action of the enzyme glycogen synthase leading to conversion of glucose to glycogen, which is stored in the liver. At the same time, the enzymes that act to convert glycogen back to glucose are inhibited. These actions also maintain the glucose concentration gradient from the plasma into the liver cell, facilitating the uptake process.

As the plasma glucose levels fall, the stimulus for insulin secretion also reduces. The final step in the conclusion of this process is the endocytotic removal of the previously inserted GLT4 transporters.

Insulin and elevated glucose levels also stimulate the release of the hormone somatostatin, which is produced in a number of organs, including the δ-cells of the pancreas. Somatostatin has a number of inhibitory actions: it is also known as growth hormone inhibiting hormone (GHIH). In the pancreas it feeds back to inhibit insulin release but also inhibits the release of the hormone glucagon.

What major clinical conditions are associated with failure to produce insulin or the development of non-responsiveness to insulin?

Diabetes mellitus, which literally translates to 'sweet urine', is a disorder of the pancreas that results in either a complete lack of insulin (type 1 DM) or a significant decrease in insulin secretions coupled with a resistance to the actions of insulin (type 2 DM). In the UK alone, there are over 3.5 million people with diabetes, a figure that is predicted to rise to 5 million by 2025. By far the vast majority (90%) of diabetic people are classified as having type 2 diabetes. In healthy, non-diabetic individuals, glucose is filtered at the kidney, but is subsequently reabsorbed back into the plasma. In diabetic people, the SGLT transporters in the proximal convoluted tubule can become saturated and so some of the filtered glucose appears in urine (see Figure 24.7), which is called glycosuria. Diabetes is discussed in more detail in Clinical Box 19.2.

Glucagon

The second pancreatic hormone with a key role in plasma glucose regulation is glucagon. Glucagon is secreted by the α-cells of the islets of Langerhans and it has a catabolic action that serves to increase the plasma glucose level when it is reduced. Like insulin, glucagon is a peptide hormone.

It is not fully clear how a reduction in plasma glucose stimulates glucagon secretion. However, it has been shown that insulin has an inhibitory action on glucagon secretion, which appears to be mediated by the inhibitory action of somatostatin.

Given that glucagon is a hyperglycaemic hormone it is reasonable to assume that its targets would be those aspects of metabolism that serve to increase plasma glucose levels. Glucagon binds to metabotropic GPCRs (G-protein coupled receptors—see Section 17.6, Metabotropic receptors); these are found in a wide variety of tissues, but there seems to be a concentration of them in the liver and kidneys. Binding of glucagon to the GPCR leads to an increased intracellular concentration of cAMP. This rise in cAMP activates the enzyme protein kinase A (PKA), which itself activates the enzyme phosphorylase kinase (PPK). PPK promotes the conversion of glycogen phosphorylase to its active form. The action of this last enzyme is to break down glycogen to release glucose 1-phosphate. This in turn is converted to glucose 6-phosphate by the enzyme phosphoglucomutase. Subsequently, glucose 6-phosphate can be dephosphorylated and the free glucose that is formed can be released into the blood. Therefore, this increases plasma glucose levels back towards their normal level.

Glucagon has other metabolic actions that complement its hyperglycaemic activity.

We highlighted earlier that neural tissue is fully dependent upon a continued supply of glucose to fuel ATP production. Periods of hypoglycaemia have the potential to compromise this requirement. Therefore, in times of hypoglycaemia, not only does glucagon promote release of glucose into the bloodstream, but it also promotes fatty acid metabolism. This means that those tissues that utilize fatty acids to produce ATP do so, therefore allowing the glucose that has been released into the blood to be preferentially utilized by the nervous system.

In addition to the change in pancreatic function elicited by hypoglycaemia and hyperglycaemia, there are a host of other mechanisms that also ensure that plasma glucose levels are held at an appropriate value. These include regulation by glucocorticoid, thyroid, and growth hormones as well as control by the autonomic nervous. Together, these mechanisms provide an integrated control of plasma glucose levels, ensuring that levels are appropriate depending upon body needs.

Between them, therefore insulin and glucagon are involved in the tight regulation of glucose levels, which in turn are critical to the energy regulation of the body.

 Check your understanding of the concepts covered in this section by answering the questions in the e-book.

25.5 The liver

The liver is the largest organ in the body and performs a range of tasks. Before we consider the roles of the liver, it would be useful to briefly consider its structure.

At a macroscopic level, the liver is composed of four lobes: the two principal lobes are the right and left lobes with two smaller lobes known as the visceral and quadrate lobes. Located within the liver is the gallbladder (which we consider later in this section). The functional unit of the liver is the liver lobule, with between 50,000 and 100,000 lobules in each liver. The lobules are about 1–2 mm in diameter and are roughly hexagonal in appearance. Each lobule has a central vein with cells called hepatocytes radiating out from it, rather like the spokes on a wheel radiate out from the hub (Figure 25.20). The hepatocytes are arranged in rows, and rows are separated from other rows by sinusoids. The sinusoids are lined by a very leaky endothelium, which means there is no effective barrier between the hepatocytes and the sinusoids. The organization of a liver lobule is shown in Figure 25.20.

The blood supply to the liver is unusual, in that it receives two inputs: the hepatic artery and the hepatic portal vein. The hepatic

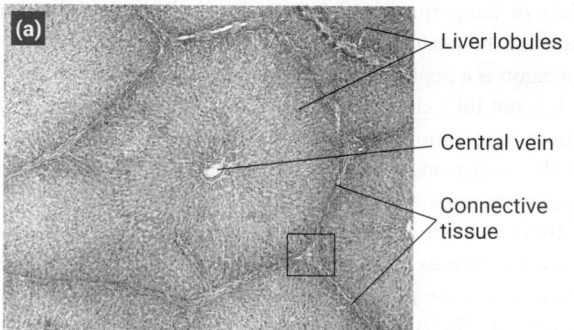

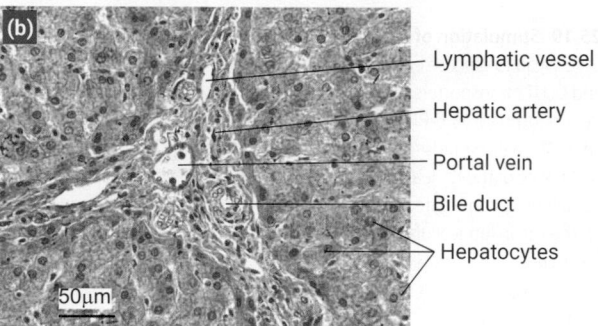

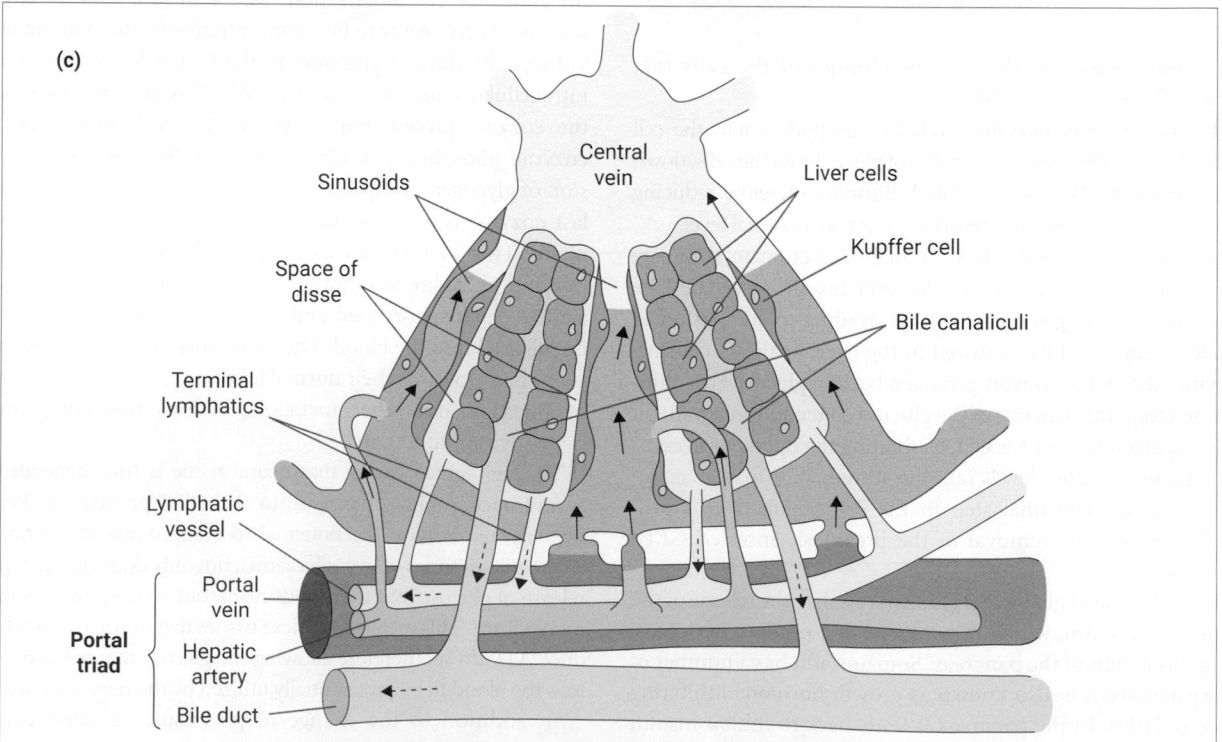

Figure 25.20 The organization of the liver. (a) A liver lobule at low magnification. (b) A higher-magnification image of the vessels and hepatocytes (the area of (a) indicated by the black box). (c) The organization of the lobule.

Source: Pocock et al. *Human Physiology*, 5th edn, 2017. Oxford University Press.

artery delivers oxygenated blood, while the hepatic portal vein delivers nutrient-rich but deoxygenated blood from the capillary networks of the intestines and stomach, as well as the spleen. Together, from these two sources, the liver receives about 25 per cent of the cardiac output. Ultimately, blood flows from these vessels into the sinusoids of the liver lobules. The sinusoids are lined with **Kupffer cells**, which are macrophages that are capable of removing material such as pathogens from blood.

What functions, then, does the liver perform? It performs a range of functions including:

- production of bile;
- regulation of metabolism;
- metabolism of waste/toxic products.

Each of these roles will now be considered.

Production of bile

Bile is important in the metabolism and absorption of lipids in the small intestine. It is formed in the hepatocytes and enters the bile canaliculi. Bile has an ionic composition similar to that of plasma. In addition, it contains:

- bile salts, which are produced from cholesterol;
- the bile pigments bilirubin and biliverdin, which are breakdown products from the haem complexes of red blood cells;
- cholesterol;
- lecithin;
- mucus.

As bile moves along the bile canaliculi towards the bile duct, its volume is increased by the addition of a bicarbonate-rich solution. The bile ducts converge and ultimately form the common bile duct that empties into the small intestine. As bile is formed, it may continuously enter the small intestine. Alternatively, it may enter the gallbladder (see the next subsection).

As indicated earlier (Section 25.2), bile is important in the metabolism of ingested lipids. An important component in this respect are the bile salts. The bile salts are formed from bile acids that in turn are produced from cholesterol. It is possible to distinguish two types of bile acids: primary bile acids (e.g. cholic acid) are produced in the hepatocytes; secondary bile acids (e.g. deoxycholic acid) are formed by the bacterial action of the primary acids in the small intestine. Bile salts are formed when primary bile acids are conjugated with amino acids (e.g. taurine and glycine).

The chemical nature of bile salts is that they contain both hydrophobic and hydrophilic components. The hydrophobic parts aggregate around the fat droplets forming structures called **micelles**, which have the hydrophilic components to the outside. These micellar structures are important in the emulsification of ingested lipids because the externally facing hydrophilic parts prevent re-aggregation of the fat droplets. Thus, the fat droplets expose a relatively large surface area for the action of the lipase enzymes.

Following release into the duodenum, the vast majority of bile salts are absorbed by an active transport process in the ileum and are returned back to the liver via the hepatic portal vein, referred to as the enterohepatic circulation. Once back in the liver, the bile salts are recycled and reused. A small proportion, however, are excreted in the faeces.

The gallbladder

Bile can be immediately released into the duodenum, where it aids lipid digestion and absorption. However, should the immediate release of bile be unnecessary (e.g. in between meals), it is stored in the gallbladder. The organization of the gallbladder is shown in Figure 25.21. The entry of material from the bile duct into the duodenum is controlled by variation in tone (the extent of contraction) of the **sphincter of Oddi**. In between meals the tone is high, essentially closing the sphincter and preventing release.

Once in the gallbladder, the bile is concentrated. This is achieved by absorbing electrolytes (e.g. Na^+, Cl^-, and HCO_3^-), and water from the lumen of the gallbladder, ultimately into the plasma. The ions are initially cotransported into the cells forming the mucosa of the gallbladder. At the basolateral surface of the cells, a number of transporters remove the ions from the cell

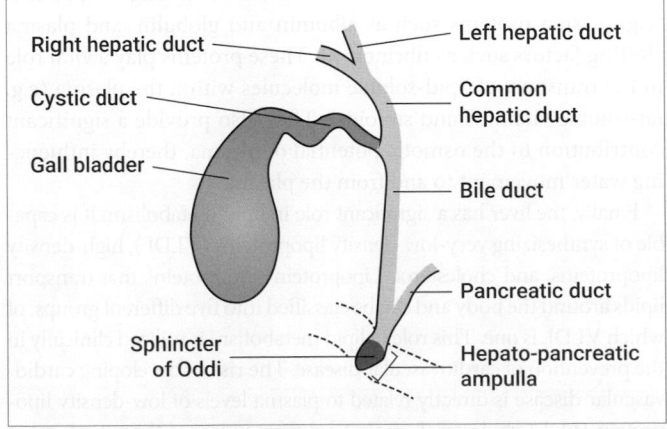

Figure 25.21 Location of gallbladder and liver and the routes of entry of their secretions into the digestive tract. Bile enters the duodenum via the ampulla of the bile duct. A sphincter (sphincter of Oddi) encircles the ampulla and therefore regulates the discharge of bile into the digestive tract.

Source: Pocock et al. *Human Physiology*, 5th edn, 2017. Oxford University Press.

into the surrounding interstitial fluid and from there into plasma. While the movement of ions is an active process, the movement of water is a purely passive process. As a consequence of this concentration, the concentration of bile salts in bile may increase 20-fold.

When food is next ingested the gallbladder begins to contract—this is particularly apparent when foods rich in lipids are eaten. The mechanism responsible for this contraction is the release of CCK from the enteroendocrine cells in the duodenum when a lipid-rich meal is eaten. In addition to causing the gallbladder to contract, CCK also relaxes the sphincter of Oddi, allowing bile to enter the duodenum.

So what are gallstones? Gallstones are most commonly slow-growing crystalline formations of cholesterol within the gallbladder. They can grow for many years and may remain asymptomatic. Sometimes, though, they will block the bile duct, giving rise to significant abdominal pain that can last for several hours.

Regulation of energy metabolism by the liver

The liver plays a vital role in contributing to the control of metabolism by virtue of the fact that it can synthesize, store, and break down nutrient sources.

Following a meal rich in carbohydrates, an elevated plasma glucose level results in the pancreas secreting insulin. Insulin, which exerts its effects via binding to tyrosine kinase receptors, stimulates the liver to remove glucose from the plasma. This glucose is then stored in the form of glycogen (see Section 25.4, Insulin). The liver is also capable of producing glucose from non-carbohydrate sources—this is known as **gluconeogenesis**. This process, which occurs during fasting, for example, results in the conversion of amino acids into glucose. The process begins with deamination of the amino acid and the remaining amino acid skeleton entering a biosynthetic pathway that results in the production of phosphoenol pyruvate, which in turn is synthesized into G-6-P and ultimately glucose.

In a similar manner to the production of glucose, the liver plays a pivotal role in the synthesis and secretion of a range of proteins (e.g. plasma proteins such as albumin and globulin, and plasma clotting factors such as fibrinogen). These proteins play a vital role in the transport of lipid-soluble molecules within the plasma (e.g. fat-soluble vitamins and steroids). They also provide a significant contribution to the osmotic potential of plasma, thereby influencing water movement to and from the plasma.

Finally, the liver has a significant role in lipid metabolism. It is capable of synthesizing very-low-density lipoproteins (VLDL), high-density lipoproteins, and cholesterol. Lipoproteins are proteins that transport lipids around the body and can be classified into five different groups, of which VLDL is one. This role in lipid metabolism is utilized clinically in the prevention of cardiovascular disease. The risk of developing cardiovascular disease is directly related to plasma levels of low-density lipoprotein (LDL) cholesterol: cholesterol forms 50% of these molecules. Any reduction in the levels of LDL cholesterol would reduce risk the risk of developing cardiovascular disease. Therefore, drugs called statins (e.g. atorvastatin) inhibit an enzyme needed for cholesterol synthesis—the enzyme is 3-hydroxy-3-methylglutaryl-coenzyme A (HMG-CoA). In doing this, plasma levels of cholesterol may be reduced by up to 50%, which results in a significant reduction in the risk of developing a stroke

or heart attack. It should be remembered, though, that cardiovascular disease is a multifactorial disease and lipid levels need to be reduced alongside other interventions (e.g. weight loss).

Metabolism of waste/toxic products

In addition to its varied roles in the biosynthesis of molecules, the liver has an important role in the breakdown, prior to elimination, of a range of molecules and, indeed, cells. In the case of the latter, the liver is responsible for the removal of aged red blood cells from the circulation, which it does in conjunction with the spleen. The iron at the centre of the haemoglobin molecule is recycled: it can be stored in the liver as ferritin or released and transported in the plasma with transferrin. The remainder of the haemoglobin molecule is metabolized to biliverdin and subsequently bilirubin. In the liver, bilirubin is conjugated with glucuronic acid. The conjugated bilirubin enters the canaliculi and then the bile duct, forming the bile pigments (see previous subsection on the gallbladder). Once in the small intestine, bacterial action converts it to urobilinogen. Some of this is subsequently reabsorbed; however, a proportion is excreted in the faeces. It is the presence of urobilinogen that gives faeces their characteristic colour.

Equally, the liver plays an important role in the breakdown of a variety of molecules. We mentioned earlier the process of gluconeogenesis and stated that the initial step in this process was the deamination of amino acids, but what is meant by this? Deamination is the removal of the $-NH_2$ group of an amino acid; this produces highly toxic ammonium ions, NH_4^+, which, if allowed to accumulate, have the potential to severely disrupt the acid–base balance. The liver converts ammonium ions into urea, which is non-toxic and water-soluble so it can be excreted in the urine. The conversion of ammonia to urea is described by the urea cycle, which is shown in Figure 25.22.

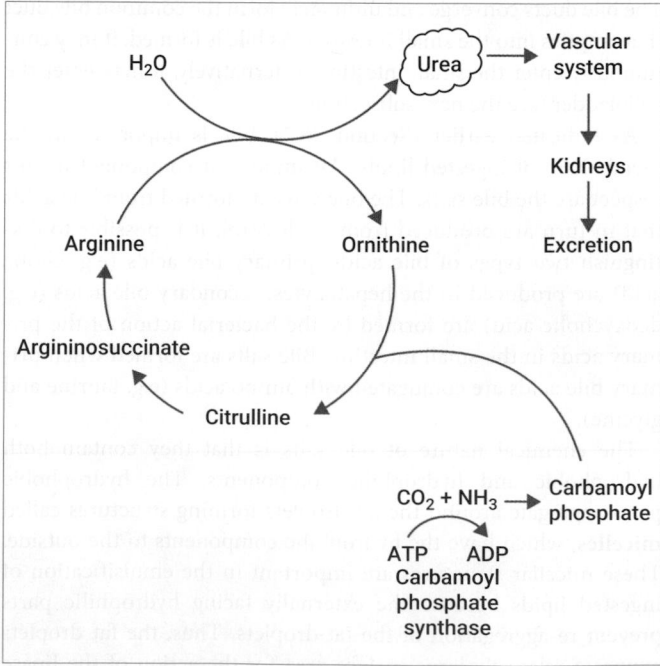

Figure 25.22 The formation of urea from ammonia. The urea is subsequently excreted in the urine.

The liver also plays a vital role in the metabolism of exogenous substances (i.e. substances that we ingest and which do not occur naturally within the body). A good example of this would be the involvement of the liver in drug metabolism. This process, which is known as biotransformation, takes active compounds, such as therapeutic drugs, and converts them to inactive compounds that are generally then much easier to excrete. Biotransformation occurs in two steps:

- step 1involves chemical modification of the drug (its oxidation or reduction for example), which requires the activity of one of the cytochrome P450 enzymes. Cytochrome P450 enzymes are a family of enzymes found in the endoplasmic reticulum of hepatocytes.

- step 2 involves the chemically modified drug being linked to another compound (e.g. glucuronic acid, a process known as conjugation) which renders it water-soluble with the potential to be excreted in the urine.

Check your understanding of the concepts covered in this section by answering the questions in the e-book.

SUMMARY OF KEY CONCEPTS

- The overall function of the digestive tract is to break down complex food molecules into simpler compounds that can be absorbed and utilized by the body.

- There are both macro- and micronutrients which are required in appropriate quantities to ensure health.

- Physical digestion of food occurs in the mouth via the action of chewing together with limited amounts of chemical digestion. The majority of chemical digestion occurs elsewhere in the digestive tract.

- Swallowing, a complex series of voluntary and involuntary movements, permits the passage of food into the digestive tract.

- The digestive tract produces a range of enzymes—proteases, carbohydrases, and lipases—that are responsible for the digestion of macronutrients. In some instances, the final aspects of digestion are coupled with absorption, so the distinction between digestion and absorption is not as clearcut as initially thought.

- The digestive tract produces a range of enzyme and other secretions that are capable of damaging the system; a range of mechanisms (e.g. the secretion of mucus and the storage and release of enzymes as inactive precursors), protects against this.

- Both the nervous and endocrine system contribute to the control of the digestive tract. In the case of the latter, there are a range of substances whose actions have yet to be fully characterized and that may prove to be of therapeutic value in treating some diseases of the digestive tract.

- Most of the nutrient absorption takes place through the brush border of the enterocytes of the small intestine.

Use the flashcards in the e-book to test your recall of key terms introduced in this chapter.

QUESTIONS

Looking for answers? Once you've answered these questions, follow the link in the e-book to the answer guidance and check your work.

Concepts and definitions

1. Describe the key macronutrients needed in our diet and identify the proportion that each should contribute to a healthy, balanced diet.

2. Draw a simple labelled diagram that illustrates the overall structure of the digestive tract.

3. Taste is a complex neurophysiological phenomenon. What are the primary sensations of taste that we perceive and what mechanisms allow us to perceive these tastes?

4. How does the pH change within the different regions of the gastro-intestinal tract?

5. What structural specializations of the cells lining the small intestine facilitate absorption of the products of digestion?

Apply the concepts

6. Review the role of sphincters in the digestive tract. For a named sphincter, identify the problems that may arise if the sphincter fails to function effectively.

7. Draw a labelled diagram of a typical gastric gland and identify a role for each of the cell types you identify.

8. What effect would the inability to produce gastric acid have on digestion?

9. Describe the digestion of a complex carbohydrate (e.g. starch) into its constituent components and identify how the products of digestion are absorbed.

10. Describe the role that the gallbladder and its secretions contribute to digestion.

Beyond the concepts

11. Distinguish between endo- and exopeptidases.

12. How is water absorbed from the digestive tract? What are the causes and consequences of inadequate water absorption?

13. Discuss the significance of the gut microbiome.

14. Compare and contrast the roles of insulin and glucagon in maintaining an appropriate plasma level of glucose.

15. Review the mechanisms that control the process of defaecation.

FURTHER READING

Barrett, K. E. (2013) *Gastrointestinal Physiology*. 2nd edition. McGraw Hill.
This is a general introductory text of gastrointestinal physiology suitable for undergraduate students. It covers all aspects considered in this chapter, together with a further consideration of gastrointestinal pathophysiology.

Fan, Y. & Pedersen, O. (2021) Gut microbiota in human metabolic health and disease. *Nat. Rev. Microbiol.* **19**: 55–71. https://www.nature.com/articles/s41579-020-0433-9
Although quite detailed, this article is readable and provides a very good, recent review of the topic.

Johnson, L. R. (2018) *Gastrointestinal Physiology*. Mosby Physiology Series, 9th edition. Elsevier.
This text provides a good understanding of gastrointestinal physiology in greater detail with a good focus on health and disease conditions.

Parikh, A. & Thevenin, C. (2021) Physiology, Gastrointestinal Hormonal Control. https://www.ncbi.nlm.nih.gov/books/NBK537284/
This article provides a very clear summary of the endocrine functions of the gastrointestinal tract.

Tobias, A. & Sadiq, N. M. (2020) Physiology, Gastrointestinal Nervous Control. https://www.ncbi.nlm.nih.gov/books/NBK545268/
This article provides a very clear summary of the structure and neural control of the gastrointestinal tract.

Waller, D. G & Sampson, A. P. (2018) *Medical Pharmacology and Therapeutics*. 5th edition. Elsevier.
The gastrointestinal section in this undergraduate textbook gives an excellent overview of the pathophysiology and treatment of various important gastrointestinal diseases.

Reproductive System

LEARNING OBJECTIVES

By the end of this chapter you should be able to:

- Describe the process of gamete production.

- Compare the processes of oogenesis and spermatogenesis and the relative timings of meiosis and mitosis.

- Describe the hormonal regulation of gamete production in males and females.

- Explain the roles of the female sex hormones in the regulation of the menstrual cycle by the hypothalamic–pituitary axis.

- Compare the reproductive anatomy of males and females.

- Describe the process of fertilization.

- Describe the establishment and maintenance of pregnancy.

- Explain the role of the placenta in supporting the development of the fetus.

- Explain the significance of the adaptations of the fetal circulation.

- Describe the process of birth.

Chapter contents

Introduction	831
26.1 Gamete production	832
26.2 Endocrine control	838
26.3 Hormonal changes during the menstrual cycle	844
26.4 Puberty	844
26.5 The menopause	846
26.6 Male reproductive anatomy	846
26.7 Female reproductive anatomy	848
26.8 Coitus and fertilization	850
26.9 Establishment and maintenance of pregnancy	850
26.10 Birth	857

 Watch the key concepts video in the e-book to prepare yourself for studying this chapter.

Introduction

Reproduction is a core physiological process, one that enables the diversity and continued existence of human life. As discussed in Topic 3, replication is one of the five functional qualities of life, and reproduction is a process of replication on a grand scale. You could argue that the drive to ensure the continuing survival of our genes

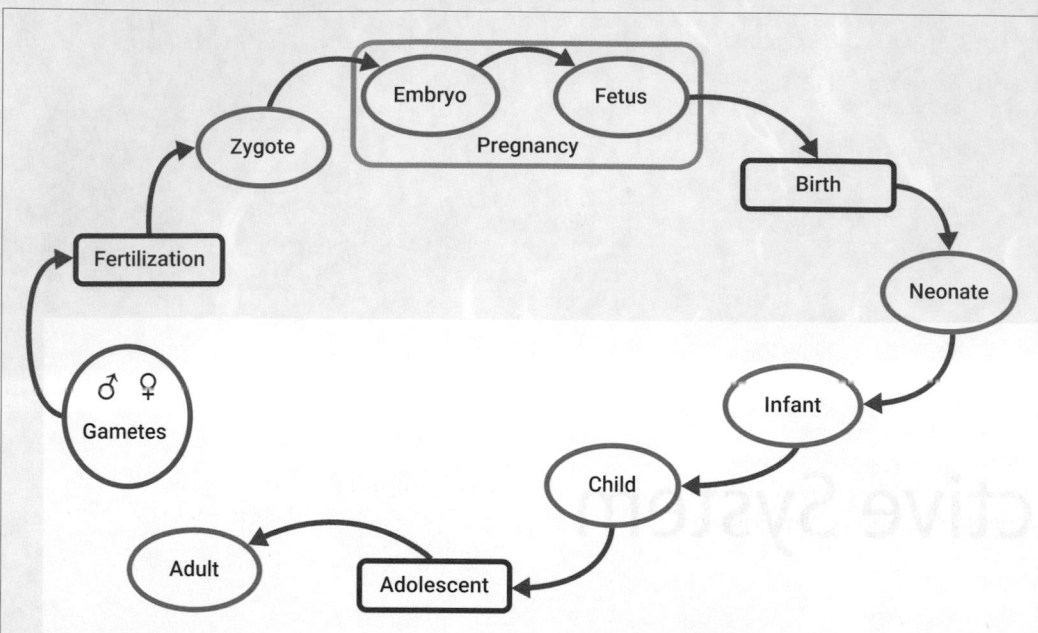

Figure 26.1 The reproductive cycle. Gametes produced by sexually mature male and female individuals meet and fertilization produces a zygote. The zygote develops in the mother's uterus through embryonic and fetal development during gestation. At birth, the neonate is expelled from the uterus and begins independent life. During the neonatal period and in infancy, the offspring is still dependent on its mother for nutritional support. Childhood, a period of sexual immaturity, is followed by adolescence, during which puberty occurs and sexual maturation is achieved

is the reason why we need to move, feed, grow, respire, excrete, and to process information from our environment.

Mammalian reproduction is a sexual process, meaning that there is mixing of genetic information from two individuals from different sexual groups; males and females. These sexual groups are classified according to their similar genetic, anatomical, and hormonal characteristics, some of which we will cover in this chapter. It is worth noting, however, that there may be significant variation in these characteristics within these groups, and there are individuals born with a mix of characteristics, sometimes called 'intersex' individuals, in human populations.

The sexual reproductive cycle (shown in Figure 26.1), involves a series of stages:

- Gamete production, which occurs in the gonads of both sexes in different ways; within the ovaries in females and the testes in males (see Section 26.1).

- Fertilization, which is achieved through sexual intercourse or coitus between a male and female (see Section 26.8).

- Pregnancy, during which females support the development of offspring inside their own bodies (see Section 26.9).

- Birth, where the fetus is expelled from the female body. Most mammals give birth to live young. One of the identifying features of mammal reproduction is the support of the young after birth with milk from specialized female **mammary glands** (see Section 26.10).

- Puberty: mammalian young are born in an immature state. What follows is a variable period of immaturity, which in humans is characterized by the neonatal, infant, and child stages. **Sexual maturity** is achieved through the process of **puberty**, at which time the individual becomes capable of reproduction (see Section 26.4).

We will discuss gamete production and the structure and physiology of each of the reproductive systems in detail in the following sections. In this chapter we will be focusing on human reproductive systems, and we use the terms 'male' (he) and 'female' (she) to refer to biological characteristics present at birth, rather than in any sense connected to gender. In other areas you may see these groups referred to as cis men or cis women: individuals who are not trans or intersex.

26.1 Gamete production

In most mammals, the mixing of genetic information is achieved by combining half of the genetic information of a male with half of the genetic information of a female. This genetic information is carried by **gametes (germ cells)**, which are stored in specialized organs called **gonads**.

Underpinning this process is a pair of complementary reproductive systems:

- In the male, the system comprises a pair of **testes** that are housed in the scrotum and produce the male gametes (**spermatozoa** or **sperm**). Accessory glands produce secretions to support the sperm, and there is a duct system connecting each testis to the urethra within the penis.

- In the female, the system comprises a pair of **ovaries** that produce the female gametes (**oocytes**). A pair of **Fallopian tubes** conduct the oocyte to the **uterus**, in which the fertilized oocyte can implant to undergo **gestation** during pregnancy. At the end of pregnancy a live offspring is delivered through the vagina. The mother will then support the continuing development of her offspring through the production of milk by the process of **lactation**.

The process through which gametes are produced in both females and males first involves the proliferation of germ cells. As we will see, though, the timings of gamete production in females (**oogenesis**) and males (**spermatogenesis**) are very different.

Within the gonad, the germ cells undergo a complex process of differentiation and development. In addition, development of the gametes involves a specialized form of cell division, **meiosis**, which creates gametes that have only a half complement of chromosomes (**haploid**) but allows for genetic reshuffling (Chapter 10). Meiosis occurs only during the production of the gametes and consists of one round of chromosome replication followed by two rounds of cell division, meiosis I and meiosis II, that apportion the chromosomes to four daughter cells (Figure 26.2).

Through meiosis, therefore, **diploid** germ cells containing two sets of the 23 chromosomes of the human genome are converted into gametes that each carry one set of 23 chromosomes, that is, they are haploid. Meiosis is key to sexual reproduction and to generating genetic diversity by the subsequent combination of the two sets of chromosomes from the two parents when fertilization takes place.

Do you know which parent determines the sex of the fetus? It is the male. Females normally have a pair of X chromosomes and therefore all the oogonia will have a single X chromosome following meiosis. Males normally have XY chromosomes and therefore each sperm will have either an X or a Y chromosome.

As we will see in Section 26.2, the process is coordinated by the **hypothalamic–pituitary–gonadal axis** (see also Section 19.3, The hypothalamic–pituitary axis), with cells within the gonads contributing to the regulation and management of the process.

Genetic variation

Two features of meiosis I contribute to the extent of genetic variation possible:

- First, independent assortment occurs as a result of random segregation of the chromosomes to the equator in the first meiotic division, simply because the chromosomes move to the nearest pole (anaphase I; Figure 26.3). Each daughter cell receives one chromosome from each homologous chromosome pair, but

Figure 26.2 Meiosis. The diagram illustrates the process for a single pair of chromosomes. The chromosomes first undergo replication, and in meiosis I the two replicated sets divide. Then in meiosis II, the replicated chromosomes separate so that each daughter cell has only half the number of chromosomes. This is the haploid form, compared with the diploid complement present in somatic cells (i.e. all cells other than the gametes).

Source: D. Papachristodoulou, A. Snape, W.H. Elliott, D.C. Elliott. *Biochemistry and Molecular Biology*, 6th edn, 2018. Oxford University Press.

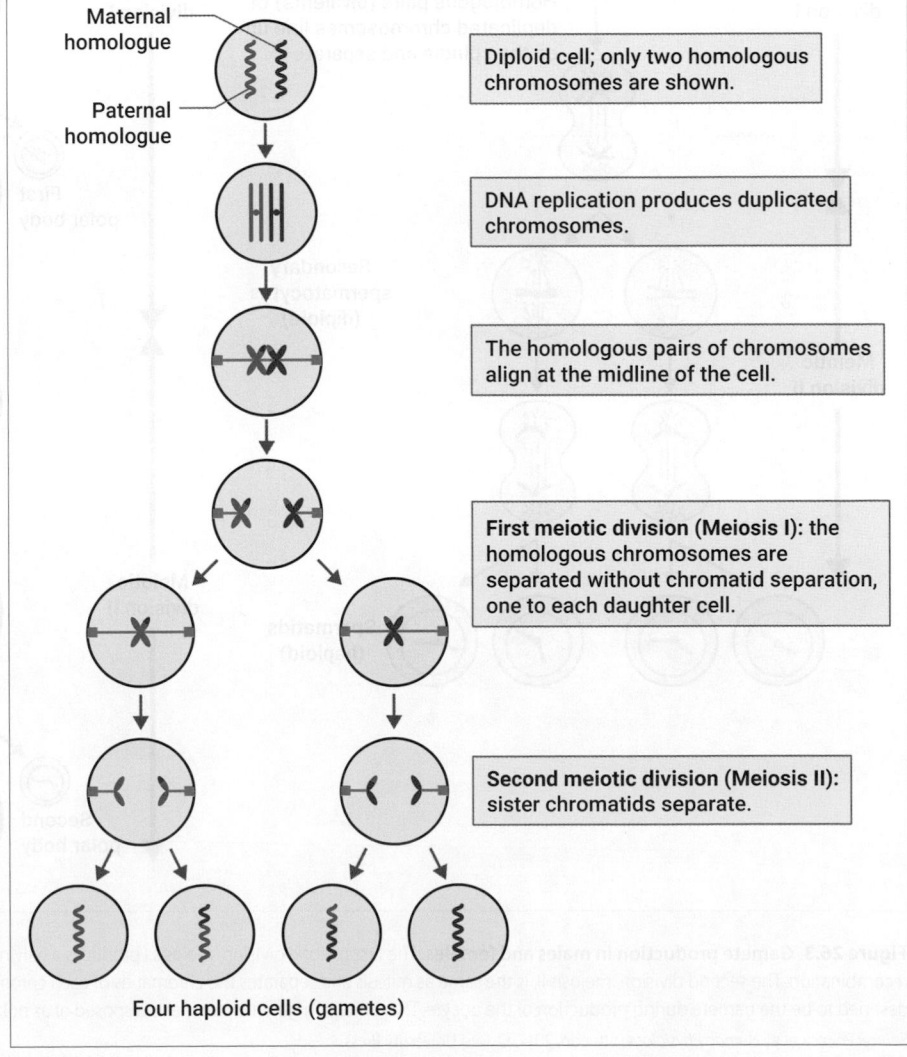

Maternal homologue
Paternal homologue

Diploid cell; only two homologous chromosomes are shown.

DNA replication produces duplicated chromosomes.

The homologous pairs of chromosomes align at the midline of the cell.

First meiotic division (Meiosis I): the homologous chromosomes are separated without chromatid separation, one to each daughter cell.

Second meiotic division (Meiosis II): sister chromatids separate.

Four haploid cells (gametes)

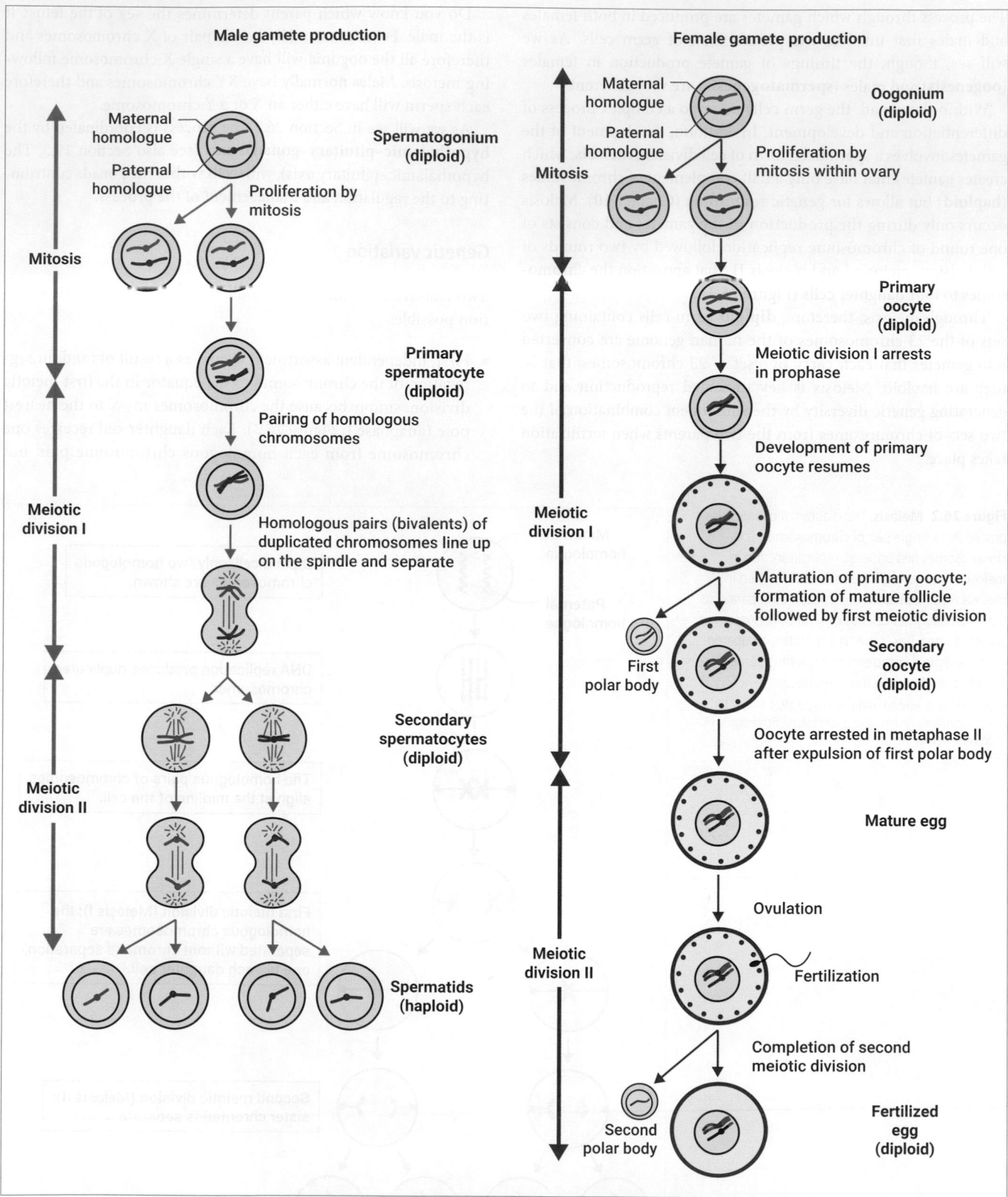

Figure 26.3 Gamete production in males and females. The first meiotic division, meiosis I produces a halving of the number of chromosomes and genetic recombination. The second division, meiosis II, is the same as mitosis and separates the chromatids of each chromosome. In the female, only one of the daughter cells is destined to be the gamete during production of the oocyte. The remaining genetic material is disposed of as polar bodies.

Source: Pocock et al. *Human Physiology*, 5th edn, 2017. Oxford University Press.

whether it is the one the parent organism inherited originally from its mother or its father is random. This means that, mathematically, the total possible combinations of the 23 chromosome pairs present in the human genome that can be inherited by each daughter cell at meiosis is 2^{23}, a number greater than 8 million.

■ Second, more variation is produced by the process of recombination, which is the exchange of genetic material that takes place between pairs of homologous chromosomes during the prophase I stage of meiosis I (Figures 26.2 and 26.3). The chromosomes break and join each other, at regions called the chiasmata, to exchange a portion of DNA.

Independent assortment and genetic recombination are covered in more detail in Chapter 10.

The molecular processes allowing for segregation of the correct chromosome complement to the daughter cells at each meiotic division are complex and, occasionally, **non-disjunction** occurs when the chromosomes fail to disjoin correctly, leading to an anomaly in the number of chromosomes in some of the daughter cells.

As shown in Figure 26.4, non-disjunction can occur during either meiosis I or II. Zygotes formed when fertilization involves a gamete with an anomalous number of chromosomes as a result of non-disjunction will inevitably have an anomalous number of chromosomes. If the gamete carries an extra copy of one

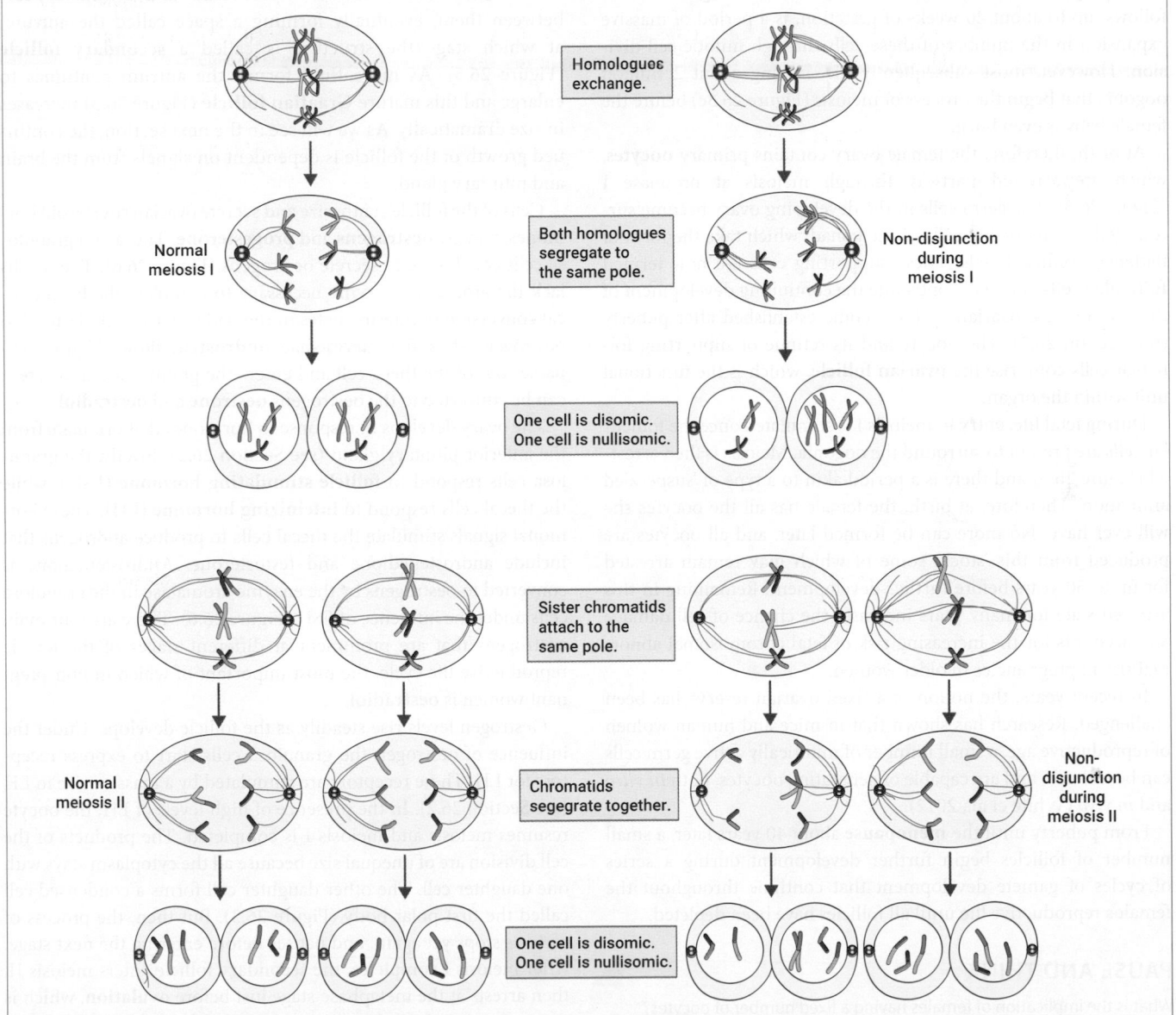

Figure 26.4 Non-disjunction resulting in the formation of gametes with anomalous numbers of chromosomes. The occurrence of non-disjunction in meiosis I or meiosis II results in different patterns of chromosome distribution in the final daughter cells. Disomy is the condition of having one or more chromosomes present twice in the gamete; nullisomy is the lack of a chromosome in the gamete.

Source: Meneely et al. *Genetics: genes, genomes, and evolution*, 2017. Oxford University Press.

chromosome, this is termed **trisomy**, while if a copy of a chromosome is missing, this is termed **monosomy**. Trisomy is occasionally compatible with life, while monosomy is nearly always lethal. Down syndrome, or trisomy 21, is the most common type of trisomy.

Oogenesis—the production of the female gamete, the oocyte

Shortly after her own conception, the female establishes the 'seed bank' for her gamete production much later in her life, once she has reached adulthood. During her own fetal development, germ cells colonize the cortex of the developing gonad, the ovary, of the developing female baby, and these cells become **oogonia**. What follows, up to about 20 weeks of gestation, is a period of massive expansion in the number of these cells through mitotic cell division. However, most subsequently die, leaving about 2 million oogonia that begin the process of meiosis (Figure 26.3c) before the female baby is even born.

At birth, therefore, the female ovary contains primary **oocytes**, which are arrested partway through meiosis at prophase I (Figure 26.3). The germ cells in the developing ovary become surrounded by supporting cells of the gonad, which take the form of flattened epithelial cells. These supporting cells are now termed **follicular cells**, and will contribute the continuing development of the oocyte once ovarian cycles become established after puberty (see Section 26.2). The oocyte and its retinue of supporting follicular cells comprise the **ovarian follicle**, which is the functional unit within the organ.

During fetal life, entry to meiosis I is stimulated once the follicular cells are present to surround the oogonia. Meiosis is then arrested (Figure 26.3) and there is a period akin to a type of 'suspended animation'. Therefore, at birth, the female has all the oocytes she will ever have. No more can be formed later, and all oocytes are produced from this 'stock', some of which may remain arrested for up to 50 years before further development. Remaining in this arrested state for many years increases the chance of cell damage and accounts for the increasing risk of fetal chromosomal abnormalities in pregnancies in older women.

In recent years, the notion of a fixed ovarian reserve has been challenged. Research has shown that in mice and human women of reproductive age, a small number of mitotically active germ cells can be isolated, and are capable of generating oocytes, both *in vitro* and *in vivo* (White et al., 2012).

From puberty until the **menopause** about 40 years later, a small number of follicles begin further development during a series of cycles of gamete development that continue throughout the female's reproductive life until all follicles have been depleted.

PAUSE AND THINK

What is the implication of females having a fixed number of oocytes?

Answer: Females have a limited time span in which to reproduce biologically. This time begins at puberty (sexual maturity) and ends at the menopause, when the female's reproductive capacity has been exhausted (Section 26.5).

Over the course of each cycle, a primary oocyte grows dramatically but does not restart meiosis. The follicular cells change shape to become cuboidal cells and then proliferate to form multiple-layered epithelium, giving rise to the **primary follicle** (Figure 26.5) that houses the primary oocyte. These cells are now termed the **granulosa cells** of the follicle. In addition, an outer coat of cells called the **theca** forms around the granulosa cells. The theca has two layers: an inner theca interna that is vascular; and an outer theca externa that forms a capsule for the follicle. The follicle becomes an endocrine 'power house' responsible for preparing the oocyte and producing the hormonal signals that support the function of the tissues of the female reproductive system, most notably, the uterus.

Granulosa cells continue to proliferate and a fluid appears between them, eventually forming a space called the antrum, at which stage the structure is called a **secondary follicle** (Figure 26.5). As more fluid forms, the antrum continues to enlarge and this mature **Graafian follicle** (Figure 26.5) increases in size dramatically. As we will see in the next section, the continued growth of the follicle is dependent on signals from the brain and pituitary gland.

Cells of the follicle synthesize and secrete ovarian sex steroid hormones, namely **oestrogens** and **progesterone**. Theca and granulosa cells collaborate to secrete oestrogens (Figure 26.6). Theca cells lack the aromatase enzyme necessary to complete the biochemical conversion of intermediates in the pathway from cholesterol to oestrogen. Thus, the intermediate **androstenedione** (Figure 26.6) passes out of the theca cell and enters the granulosa cell, where it can be converted to the oestrogens **oestrone** and **oestradiol**.

The ovary develops in response to hormones that originate from the anterior pituitary gland (see Section 26.2). Briefly, the granulosa cells respond to **follicle stimulating hormone** (FSH), while the thecal cells respond to **luteinizing hormone** (LH). These hormonal signals stimulate the thecal cells to produce androgens that include androstenedione and testosterone. Androstenedione is converted to oestrogens by the enzyme aromatase in the granulosa cells under the influence of FSH (Figure 26.6). There are four main oestrogens that are prominent at different stages of the female reproductive life cycle, the most important of which in non-pregnant women is **oestradiol**.

Oestrogen levels rise steadily as the follicle develops. Under the influence of oestrogen the granulosa cells start to express receptors for LH. These receptors are stimulated by a massive rise in LH (see Section 26.2). In the presence of high levels of LH, the oocyte resumes meiosis and meiosis I is completed. The products of the cell division are of unequal size because all the cytoplasm stays with one daughter cell. The other daughter cell forms a condensed cell called the first polar body (Figure 26.3). But then, the process of meiosis stops yet again, and pauses before entering the next stage. After meiosis I completes, the secondary follicle enters meiosis II, then arrests at the metaphase stage just before **ovulation**, which is the process by which the oocyte is expelled from the ovary.

At ovulation, the follicle size is approaching 25 mm. Collagenase activity in the follicle is stimulated by the high levels of LH, and the follicle ruptures. The oocyte is expelled from the ovary and is swept up by the fimbriae, finger-like projections that surround the

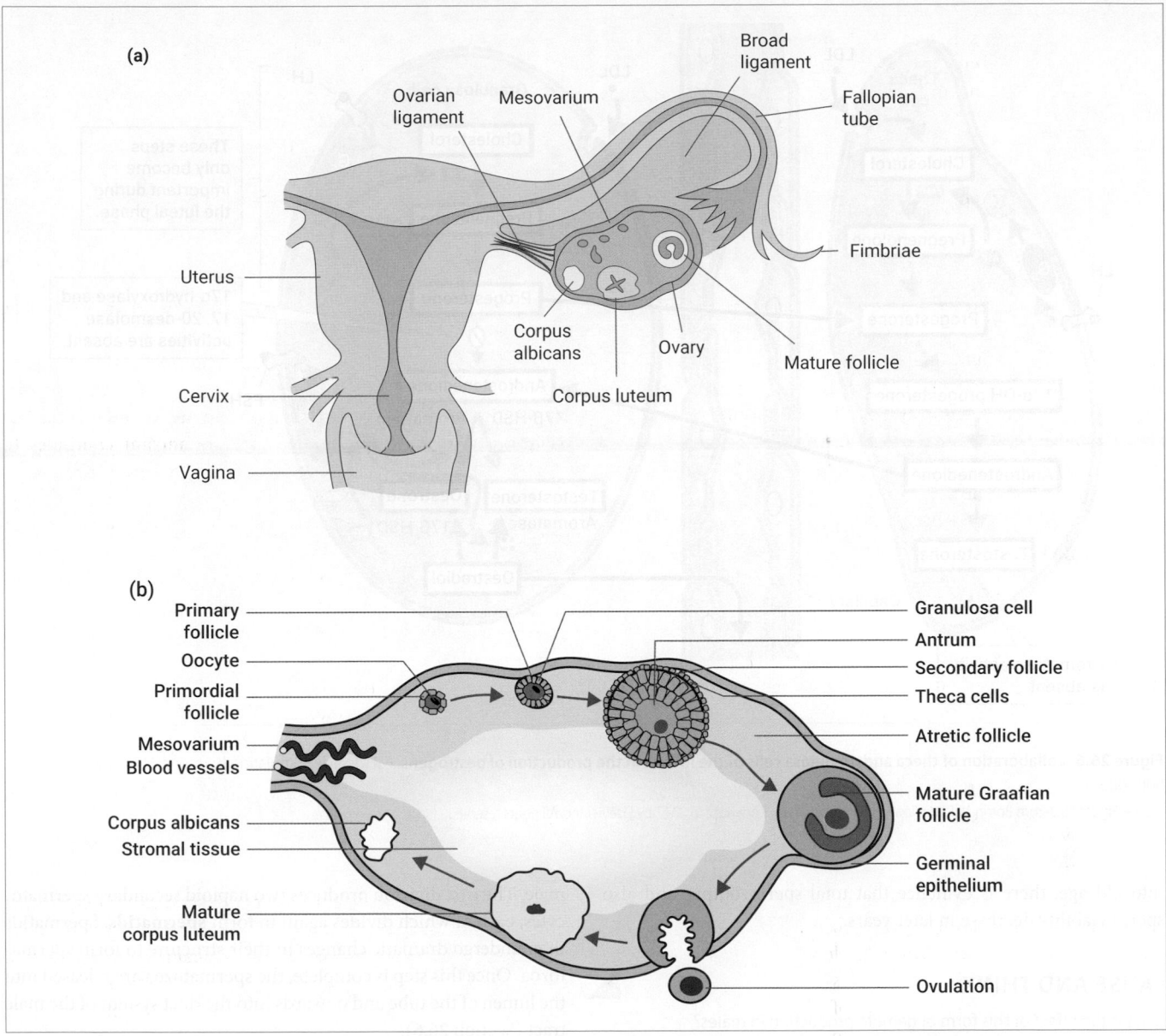

Figure 26.5 The cycle of ovarian follicle development. (b) Over the course of a female reproductive cycle, the follicle changes in appearance due to expansion and development of the follicular cells and growth of the oocyte. Once released from the ovary at ovulation, the oocyte is swept up by the fimbriae of the Fallopian tube (a).
Source: Pocock et al. *Human Physiology*, 5th edn, 2017. Oxford University Press.

'mouth' of the Fallopian tube (Figure 26.5). Meiosis II is completed in the oocyte only if fertilization occurs.

Spermatogenesis

The key difference between gametogenesis in females and males is that instead of one gamete per cycle, as happens in the female, in the male system hundreds of millions of gametes are produced in a day; this is a significant scale of production.

In the male, the process of gametogenesis does not begin in fetal life. Rather, in fetal life, the germ cells colonize structures known as **seminiferous cords**, which are converted at puberty (Section 26.4) into **seminiferous tubules** (Figure 26.7a). The seminiferous

tubules are composed of **Sertoli cells**, which surround the germ cells and support the developing male gametes. The 'germinal epithelium' (Figure 26.7c) comprises Sertoli cells and cells of the germ cell lineage at various stages in their development. As the germ cells develop, they rise from the basement membrane to the lumen (Figure 26.7c) within the seminiferous tubule. Interspersed between the tubules sit **Leydig cells** (Figure 26.7b, 'LC'), which secrete **testosterone**, the primary male sex steroid hormone.

In contrast to the female process, the germ cells in the male do not begin meiosis until puberty is reached. At puberty, spermatogenesis begins. The germ cells form spermatogonia stem cells (Figure 26.8), which undergo mitosis to maintain a population of self-regenerating stem cells. Although sperm production continues

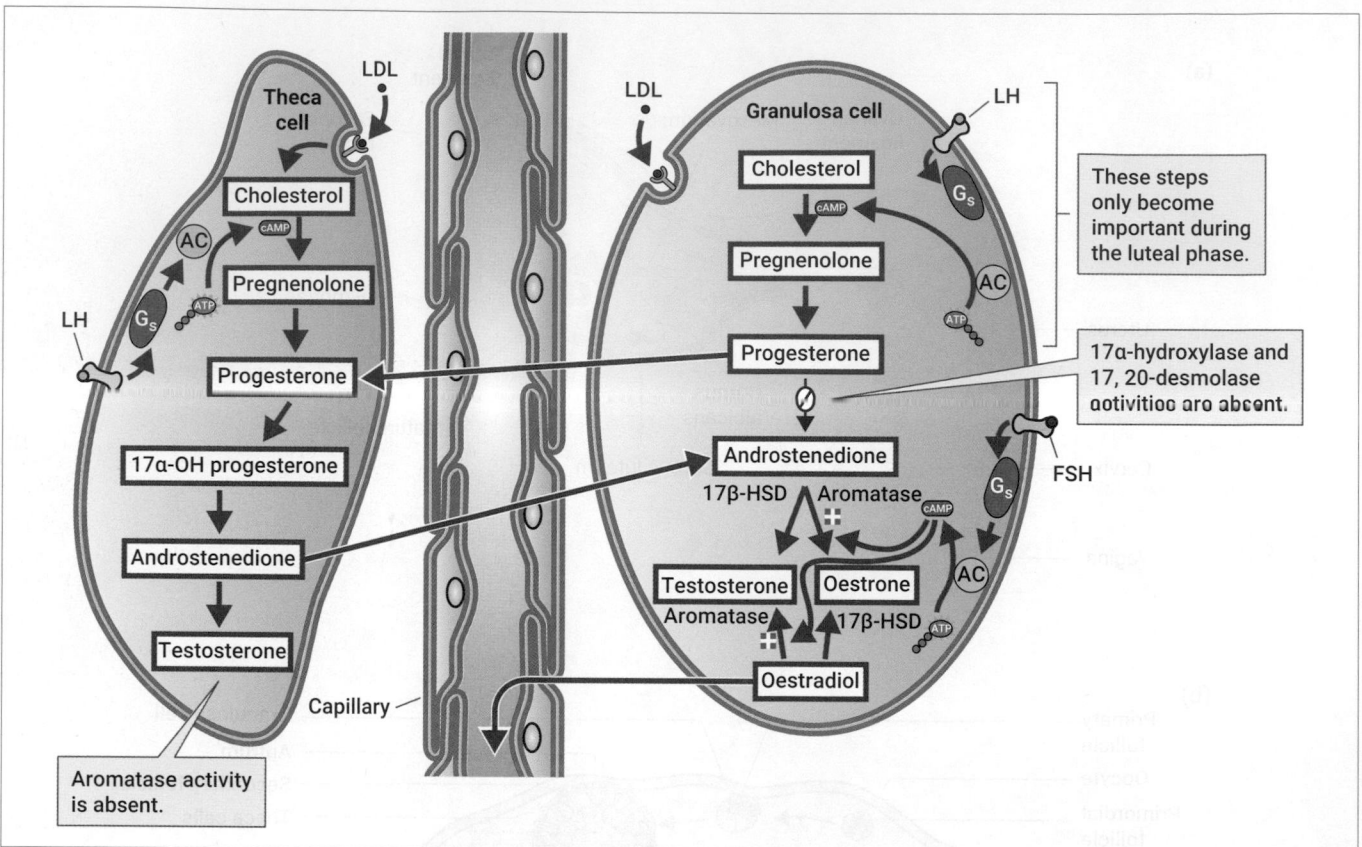

Figure 26.6 Collaboration of theca and granulosa cells of the follicle in the production of oestrogens. FSH, follicle stimulating hormone; LH, luteinizing hormone.

Source: Figure 55.9 from Boron, W.F. *Medical Physiology*, 3rd edn. Copyright © 2017 by Elsevier Inc. All rights reserved.

into old age, there is evidence that total sperm output and also sperm viability decrease in later years.

PAUSE AND THINK

What is the effect of this form of gamete production in males?

Answer: Not only is male fertility not time limited in the same way as it is in the female, but this 'production line' model of gamete production allows for production of large numbers of gametes on a continuous, rather than an intermittent, basis, although male fertility is likely to decline with age.

The process of spermatogenesis is outlined in Figure 26.8. From type A stem cells, groups of distinct cells—A1 spermatogonia—emerge at intervals, marking the beginning of spermatogenesis in that part of the tubule. These cells undergo differentiation to produce further type A cells (stem cells, thereby maintaining the pool of these cells) and also type B cells that are committed to differentiation to spermatozoa. Each type B spermatogonium then undergoes a fixed number of mitotic divisions to produce a clone (typically 64) of **primary spermatocytes**.

Primary spermatocytes move through the germinal epithelium towards the lumen. It is at this stage that meiosis begins in the male. The first division produces two haploid secondary spermatocytes, each of which divides again to form **spermatids**. Spermatids then undergo dramatic changes in their structure to form spermatozoa. Once this step is complete, the spermatozoa are released into the lumen of the tube and onwards into the duct system of the male tract (Section 26.6).

What do you think would happen if the process of spermatogenesis (or the spermatogenic cycle) were to happen at the same time in the whole of the testis? There would be peaks and gaps in production. This does not happen because the process begins at different times in different sections along the length of the tubule, so, at any given time, a particular section is producing mature spermatozoa: This phenomenon is known as the **spermatogenic wave**.

Check your understanding of the concepts covered in this section by answering the questions in the e-book.

26.2 Endocrine control

All the key processes required for successful reproduction are under endocrine control, which regulates the process through the controlled release of a variety of reproductive hormones.

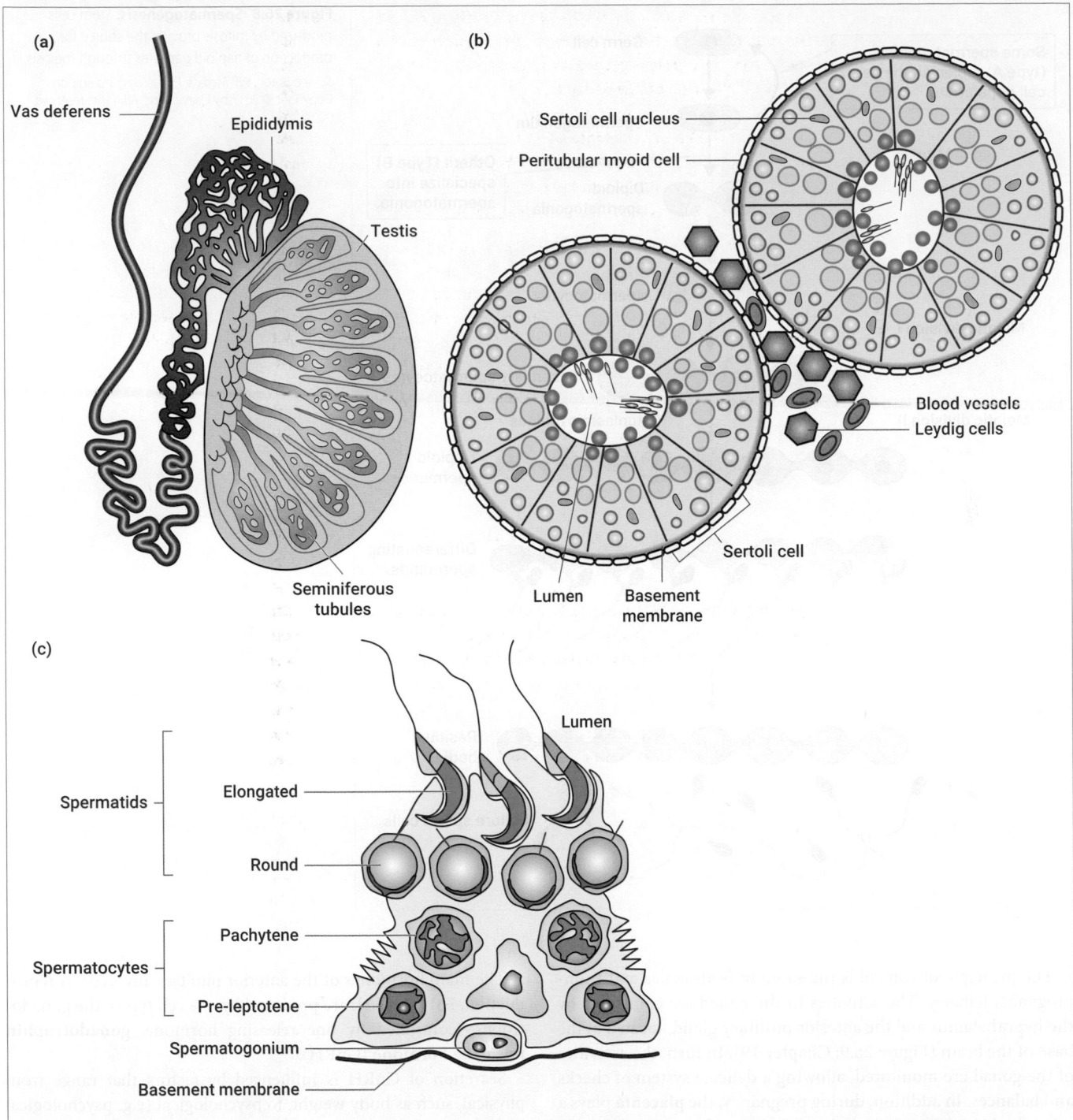

Figure 26.7 Seminiferous tubules. (a) Spermatogenesis occurs in seminiferous tubules within the testis. (b) The tubules are lined by Sertoli cells that surround the developing germ cells, (c) which move towards the lumen of the tubule as they approach completion of their development. Leydig cells occupy spaces between the tubules (b).

Source: H.J. Cooke et al. Mouse models of male infertility. *Nature Reviews Genetics*, 3, 790–801. Reprinted by permission from Springer Nature: Springer, © 2002.

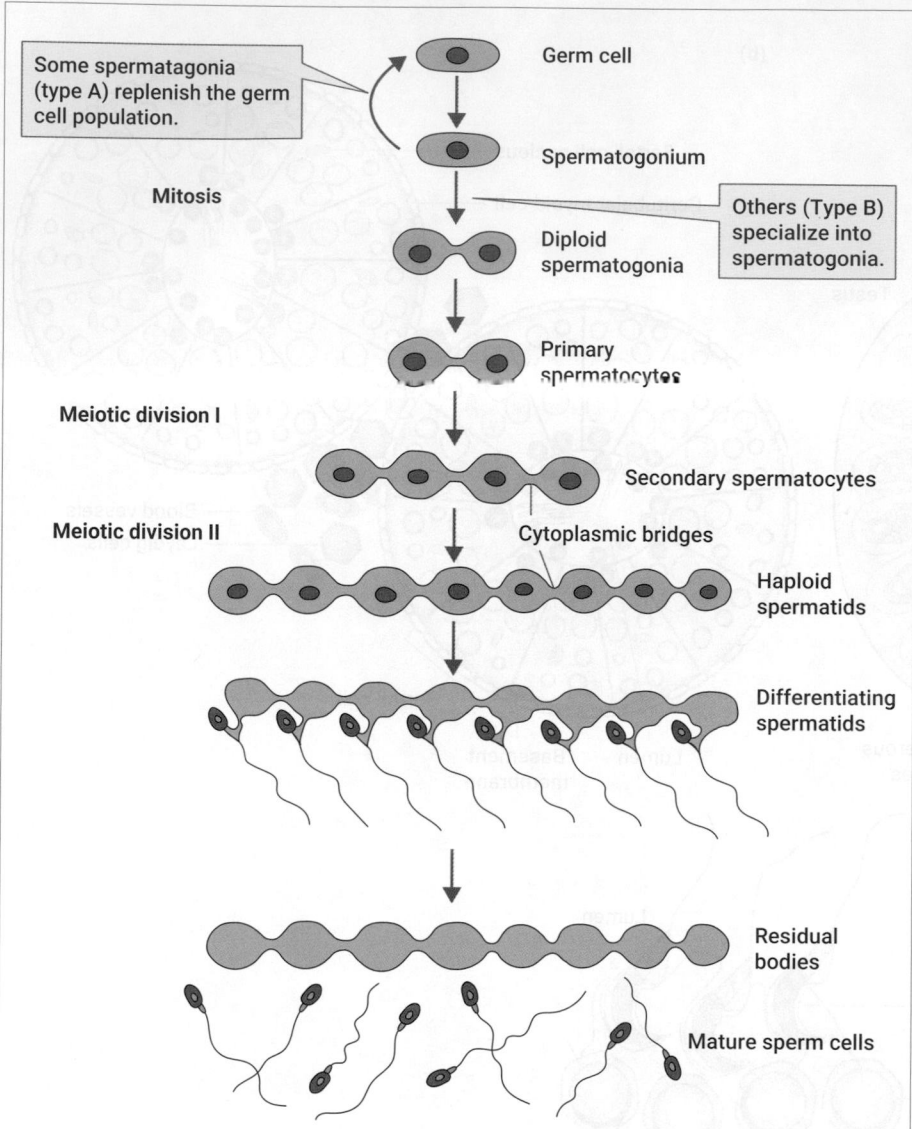

Some spermatagonia (type A) replenish the germ cell population.

Germ cell

Spermatogonium

Mitosis

Others (Type B) specialize into spermatogonia.

Diploid spermatogonia

Primary spermatocytes

Meiotic division I

Secondary spermatocytes

Meiotic division II

Cytoplasmic bridges

Haploid spermatids

Differentiating spermatids

Residual bodies

Mature sperm cells

Figure 26.8 Spermatogenesis. Stem cells produced by mitosis provide the source for production of haploid gametes through meiosis.
Source: Boron, WF. *Medical Physiology*, 3rd edition. Copyright © 2017 by Elsevier, Inc. All rights reserved.

The principle of control is the same in both males and (non-pregnant) females. The activities in the gonad are controlled by the hypothalamus and the anterior pituitary gland, located at the base of the brain (Figure 26.9; Chapter 19). In turn, the activities of the gonad are monitored, allowing a delicate system of checks and balances. In addition, during pregnancy, the **placenta** plays a critical role in the hormonal support of reproduction with its role in the maintenance and support of pregnancy (see Section 26.9).

Anterior pituitary gland

As set out in Chapter 19, the hypothalamic–pituitary axis has a wide-ranging influence on various tissues and processes of the body, including reproduction. The anterior pituitary gland has several cell types that secrete different hormones, most of which play a role in reproduction, either directly or indirectly. These hormones are summarized in Table 26.1.

The main hormones of the anterior pituitary involved in reproduction, FSH and LH, are produced by one cell type—the gonadotrophs—controlled by one releasing hormone, **gonadotrophin releasing hormone** (GnRH).

Secretion of GnRH is influenced by factors that range from physical, such as body weight, to psychological (e.g. psychological stress) and sensory inputs. Why might factors such as body weight affect sex hormone production? Take a moment to think about this question before reading on.

Mammalian reproduction is a biologically expensive process and the health of the animal, as well as environmental factors such as the availability of food, are important factors that impact on the regulation of the process. For example, in women who are severely undernourished, the normal ovulatory cycle may cease completely.

Hypothalamic neurons secrete GnRH in a pulsatile fashion, with a burst about once per hour. Gonadotrophs in the anterior pituitary release gonadotrophins only in response to these pulses of

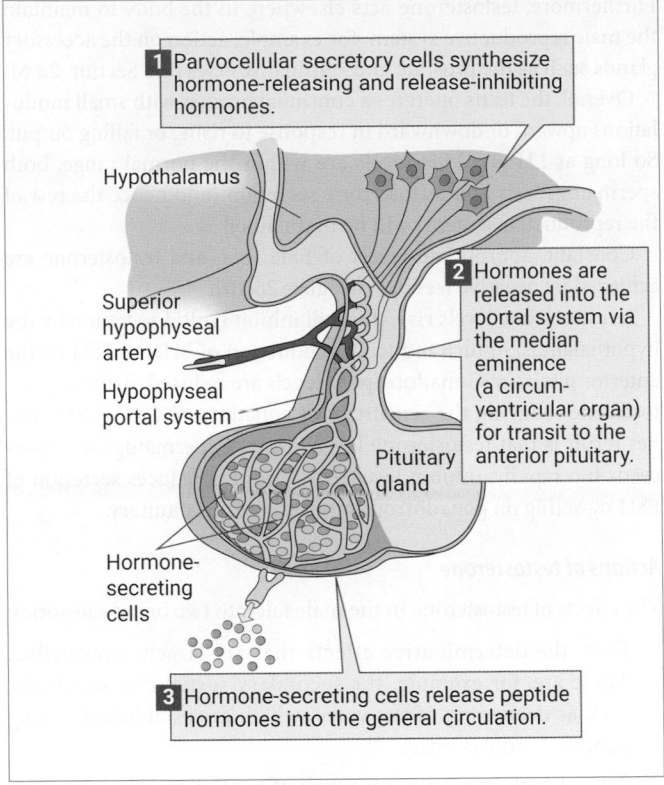

1 Parvocellular secretory cells synthesize hormone-releasing and release-inhibiting hormones.

Hypothalamus

Superior hypophyseal artery

Hypophyseal portal system

2 Hormones are released into the portal system via the median eminence (a circum-ventricular organ) for transit to the anterior pituitary.

Pituitary gland

Hormone-secreting cells

3 Hormone-secreting cells release peptide hormones into the general circulation.

Figure 26.9 Overview of the hypothalamus–pituitary–target organ axis.

Source: Figure 7.11 from Preston et al. *Lippincott's Illustrated Reviews: Physiology*, 2012. Copyright © Wolters Kluwer. Reproduced with permission of the Licensor through PLSclear.

GnRH. The role of FSH and LH, as shown in Table 26.1, is to direct the activities of the gonad. The gonad has two major functions:

- to produce gametes;
- to produce gonadal sex steroid hormones (primarily testosterone in the male, and oestrogen and progesterone in the female).

Without stimulation from FSH and LH, the gonad will not function.

The principle of control of this process is a **feedback loop** control mechanism (Figure 26.10). The products detected by the hypothalamus and anterior pituitary are the gonadal hormones inhibin and the sex steroid hormones. As the levels of these hormones in the vascular circulation rise, there is a reduction in the release of GnRH and also of FSH and LH.

Hypothalamus

In the male, testosterone reduces secretion of GnRH. In the female, oestrogen (principally oestradiol) at moderate concentrations reduces secretion of GnRH in a similar way to testosterone. Specifically, moderate levels of oestrogen reduce the amount of GnRH secreted per pulse. However, oestrogen alone at high concentration promotes the release of GnRH, producing a 'surge' of production. When progesterone is also present, it acts to reduce the frequency of pulses and thereby prevents high concentrations of oestrogen acting at the hypothalamus to produce a GnRH surge.

Anterior pituitary

Gonadotrophs in the anterior pituitary of males and females secrete both FSH and LH in response to pulsatile secretion of GnRH by

Table 26.1 Hormones of the anterior pituitary gland

Hormone	Secreted by cell type	Chemical nature	Main target tissue	Principle actions	Produced in response to
Follicle stimulating hormone	Gonadotrophs	Glycoprotein of two subunits, α and β	Gonad	Development of ovarian follicles, regulation of spermatogenesis	Gonadotrophin releasing hormone
Luteinizing hormone	Gonadotrophs	Glycoprotein of two subunits, α and β	Gonad	Triggers ovulation and formation of corpus luteum; and production of oestrogen and progesterone in the ovary, and of testosterone in the testis	Gonadotrophin releasing hormone
Thyroid stimulating hormone	Thyrotrophs	Glycoprotein of two subunits, α and β	Thyroid gland	Production of thyroid hormones	Thyrotrophin releasing hormone
Adrenocorticotropic hormone (ACTH)	Corticotrophs	Single peptide chain	Adrenal gland	Stimulates production of androgens and glucocorticoids from adrenal cortex	Corticotrophin releasing hormone
Growth hormone	Somatotrophs	Single peptide chain	Almost all tissues of the body	Stimulates body growth	Growth hormone releasing hormone
Prolactin	Lactotrophs	Single peptide chain	Mammary gland	Stimulates milk production and secretion	Absence of prolactin inhibitory hormone

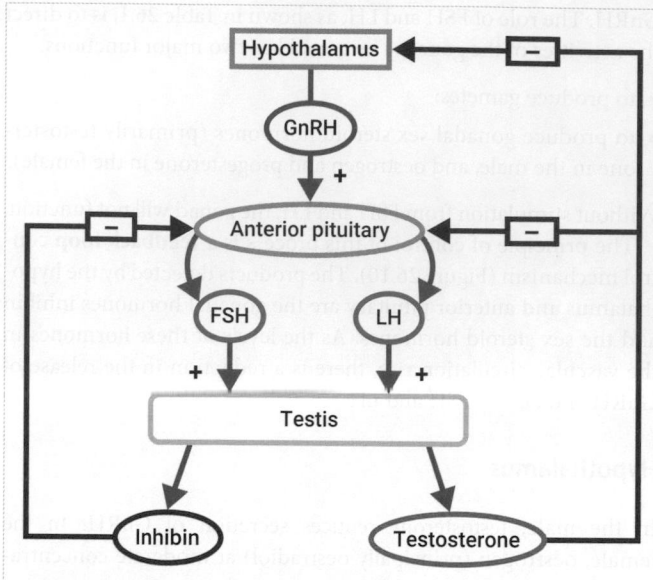

Figure 26.10 Flow loop diagram illustrating the male hypothalamic–pituitary–testis axis.

the hypothalamus. In the male, gonadotrophs are less sensitive to GnRH when testosterone levels are high, leading to a drop in the secretion of FSH and LH. In the female, oestrogen acts at the anterior pituitary in different ways, depending on its concentration, in the same way in which it acts at the hypothalamus. Therefore, the amount of FSH and LH secreted in response to each GnRH pulse is reduced when oestrogen is present in moderate concentration and promoted when oestrogen is high in concentration.

Inhibin, a protein hormone, is produced in the gonads by ovarian follicular granulosa cells in the female and by Sertoli cells in the male testis. These cells support development of the gamete; therefore, inhibin production will rise alongside development of the gamete, and can give an indication of how far that process has proceeded. Inhibin allows for an additional level of control, as it acts on the anterior pituitary gland to reduce production of FSH, the hormone that promotes growth of the granulosa and Sertoli cells.

The gonads I: male hypothalamic–pituitary–testis axis

The gonadotrophin hormones act on separate target cells in the gonad. In the male, FSH binds to receptors on Sertoli cells, while Leydig cells respond to LH. LH binding results in testosterone production by Leydig cells, and the amount of testosterone secreted is relatively constant. There is some evidence for modulation of LH responses within the testis through the action of inhibin and prolactin. FSH stimulates growth of the seminiferous epithelium (and consequently the testis) and leads to production of **androgen binding protein** by Sertoli cells; testosterone acts on Sertoli cells to promote spermatogenesis. Androgen binding protein ensures localized high testosterone concentration near to the developing sperm and this is essential for their development. FSH maintains Sertoli cells and makes them responsive to testosterone.

Furthermore, testosterone acts elsewhere in the body to maintain the male reproductive system; for example, action on the accessory glands such as the prostate and seminal vesicles (see Section 26.6).

Overall, the testis operates a continual process with small modulations upward or downward in response to rising or falling output. So long as LH and FSH levels are within the normal range, both spermatogenesis and testosterone secretion (and hence the rest of the reproductive system) will be maintained.

Constant, appropriate levels of FSH, LH, and testosterone are achieved by negative feedback (Figure 26.10).

If testosterone levels rise, this will inhibit GnRH secretion by the hypothalamus, in turn reducing production of LH and FSH by the anterior pituitary. Gonadotrophin levels are reduced further by testosterone reducing the sensitivity of gonadotrophs to GnRH. The net result is that testosterone levels drop. If spermatogenesis proceeds too rapidly, inhibin levels rise, and this reduces secretion of FSH by acting on gonadotrophs in the anterior pituitary.

Actions of testosterone

The effects of testosterone in the male fall into two broad categories:

1. First, the determinative effects that are largely irreversible. These are, for example, the secondary sexual characteristics, such as deepening of the voice, which are established during puberty at adolescence.

2. Second, testosterone has a range of regulatory effects that are highly reversible and therefore rely on continuous hormonal stimulation for their maintenance. Regulatory effects of testosterone include maintenance of the male internal genitalia (such as the accessory glands, prostate, and seminal vesicles necessary for the production of seminal fluid—see Section 26.6); metabolic effects, such as testosterone's anabolic action promoting increase in muscle mass and, consequently, power and stamina; and behavioural effects, such as aggression and sex drive.

The gonads II: female hypothalamic–pituitary–ovary axis

The model of gamete production in the female is very different to that in the male, and the female reproductive tract must also 'double-up' as a chamber to support the development of the fetus, so the control system operates in three main ways:

1. Controls gamete production, ensuring a single high-quality oocyte is produced one at a time.

2. Controls preparation of the chamber.

3. Builds in time for the embryo to signal its presence, if successful fertilization has taken place, before otherwise starting over again if no such signal is received.

Therefore, the events in the ovary take the form of a **cycle** rather than a continuous process, and follow the development of the stages of follicular development outlined in the subsection on oogenesis in Section 26.1.

The female reproductive cycle is a series of physiological events centred on ovulation. In the first phase, while the ovary is

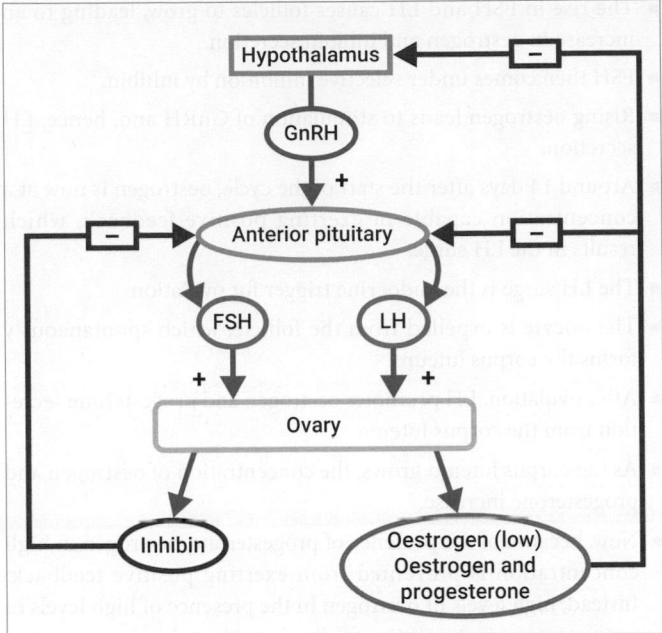

Figure 26.11 The hypothalamic–pituitary–ovary (HPO) axis in negative feedback. The HPO axis in negative feedback when oestrogen is present alone in low levels or in high levels in the presence of progesterone.

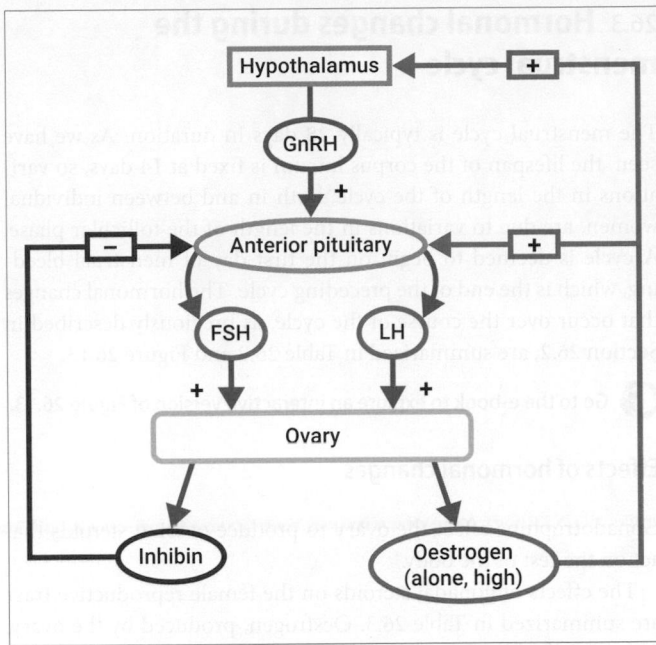

Figure 26.12 The hypothalamic–pituitary–ovary (HPO) axis in positive feedback. The HPO axis in positive feedback when oestrogen is present alone in high levels.

preparing the oocyte for ovulation, the uterus and Fallopian tubes are prepared in order to maximize the chance of fertilization.

During the antral phase of follicular development, cells of the theca interna bind LH, while granulosa cells bind FSH (Figure 26.5). LH stimulates thecal cells to produce androgens (principally androstenedione) and FSH stimulates granulosa cells to develop aromatase enzymes, which convert these androgens to oestrogens (Figure 26.6). Oestrogen at moderate concentrations has a negative feedback action on the hypothalamus and anterior pituitary (Figure 26.11). However, because the number of follicular cells is growing, more oestrogen is produced without the need for a corresponding increase in GnRH. Therefore, there is a gradual escalation of oestrogen production.

In the pre-ovulatory phase, LH receptors begin to be expressed in granulosa cells. The, by now very high, oestrogen concentration will lead to a 'surge' of LH through positive feedback, which will stimulate ovulation (Figure 26.12).

After ovulation, LH stimulates the remnant of the follicle, the **corpus luteum** (Figure 26.5), to secrete progesterone and oestrogen. As the corpus luteum grows, more steroids are produced without the need for a corresponding increase in LH. While oestrogen is still high, the presence of progesterone limits the capacity for high concentrations of oestrogen to positively feed back to the hypothalamus and anterior pituitary gland (Figure 26.11). Therefore, there is a switch back to negative feedback that ensures that no further follicular development is possible so long as the corpus luteum produces oestrogen and progesterone.

The corpus luteum remains functional for 14 days. What is the significance of this 14-day period? It allows time for a signal of sufficiently large amplitude to be received from cells of the developing placenta if fertilization has taken place. This period also allows for physiological changes in the tissues of the female reproductive tract to begin preparations for pregnancy.

If there is no signal, the cycle ends and a new one begins. If fertilization has occurred (see Section 26.8), **human chorionic gonadotrophin** (hCG), an analogue of LH, is secreted by the **syncytiotrophoblast** of the developing placenta, and supports the corpus luteum to continue to produce oestrogen and progesterone.

Therefore, two parallel cycles occur in the female:

- An **ovarian cycle,** which has an initial follicular phase when the follicle is developing, and which controls a cycle in the uterus, the **uterine cycle.** The ovarian cycle and the uterine cycle are collectively termed the **menstrual cycle.** The follicular phase in the ovary leads to proliferation of the endometrium—the proliferative phase of the uterine cycle.

- A **luteal phase,** governed by the corpus luteum. Following ovulation, the production of oestrogen and progesterone by the corpus luteum in the luteal phase of the ovarian cycle causes functional specialization of the endometrial glands in the secretory hase of the uterine cycle. The events of each phase are coordinated by gonadotrophins, effected by gonadal steroids.

Go to the e-book to watch a video that explores the menstrual cycle in more detail.

Check your understanding of the concepts covered in this section by answering the questions in the e-book.

26.3 Hormonal changes during the menstrual cycle

The menstrual cycle is typically 28 days in duration. As we have seen, the lifespan of the corpus luteum is fixed at 14 days, so variations in the length of the cycle, both in and between individual women, are due to variations in the length of the follicular phase. A cycle is deemed to begin on the first day of menstrual bleeding, which is the end of the preceding cycle. The hormonal changes that occur over the course of the cycle, as previously described in Section 26.2, are summarized in Table 26.2 and Figure 26.13.

 Go to the e-book to explore an interactive version of Figure 26.13.

Effects of hormonal changes

Gonadotrophins affect the ovary to produce ovarian steroids that act on the rest of the body.

The effects of gonadal steroids on the female reproductive tract are summarized in Table 26.3. Oestrogen, produced by the ovary, primes the cells of the **endometrium** (uterine lining) during its **proliferative** phase so that these cells are responsive to progesterone present during the **secretory** phase.

If fertilization does not occur, when the corpus luteum 'times out' there is a sudden fall in progesterone and oestrogen. This results in collapse of the elaborate secretory epithelium of the endometrium. The vascular supply to the endometrium, which developed under the influence of progesterone, undergoes spasm when progesterone levels drop. This spasm in the arterioles supplying the endometrium results in ischaemia to the tissue, which is then sloughed off as a menstrual bleed. Both the involution of the corpus luteum and the collapse of the endometrium occur through apoptotic cell death.

The hypothalamic–pituitary–ovarian axis summary

- At the beginning of the cycle, oestrogen, progesterone, and inhibin levels are low, which results in release of GnRH secretion from inhibition.
- LH and FSH will rise in response to the presence of GnRH pulses.
- The rise in FSH is greater, as low inhibin levels release FSH secretion from selective inhibition at the pituitary.

- The rise in FSH and LH causes follicles to grow, leading to an increase in oestrogen and inhibin secretion.
- FSH then comes under selective inhibition by inhibin.
- Rising oestrogen leads to stimulation of GnRH and, hence, LH secretion.
- Around 14 days after the start of the cycle, oestrogen is now at a concentration capable of exerting positive feedback, which results in the LH surge.
- The LH surge is the endocrine trigger for ovulation.
- The oocyte is expelled from the follicle, which spontaneously forms the corpus luteum.
- After ovulation, LH promotes oestrogen and progesterone secretion from the corpus luteum.
- As the corpus luteum grows, the concentration of oestrogen and progesterone increase.
- Now, because of the presence of progesterone, oestrogen in high concentration is prevented from exerting positive feedback. Instead, high levels of oestrogen in the presence of high levels of progesterone exert negative feedback on the axis.
- This phenomenon is the basis for the combined oral contraceptive pill, which contains oestrogen and progesterone in concentrations that act on the axis to put the body into a constant state of negative feedback on the hypothalamic–pituitary–ovarian axis.
- In the absence of pregnancy, the corpus luteum regresses spontaneously, which leads to a precipitous drop in concentration of progesterone and oestrogen levels.
- This event triggers a menstrual bleed and relieves inhibition on FSH and LH, so triggering the development of new follicles and the beginning of a new cycle.

 Check your understanding of the concepts covered in this section by answering the questions in the e-book.

26.4 Puberty

The genetic sex of an individual is determined during fetal life. However, the male and female reproductive systems remain inactive until puberty. On average, girls reach puberty before boys: 8–13 years for girls; 9–14 years for boys.

At the start of adolescence, there is a marked acceleration of growth in both boys and girls. Adult males generally end up taller

Table 26.2 Summary of hormonal changes during the menstrual cycle

Ovarian cycle stage	Gonadotrophins	Oestrogen	Progesterone
Early follicular	FSH and LH high	Low to rising	Low
Late follicular	FSH low LH rising	Rising	Low
Early luteal	Low	Rising	Rising
Late luteal	Low	High	High
End	Begin to rise	Fall	Fall

Figure 26.13 Hormonal fluctuations over the course of the menstrual cycle. The interplay between the different hormones is shown in relation to the development of the follicle. (a) Changes in circulating concentrations of FSH and LH. (b) The concentrations of oestrogen and progesterone. The development of the follicle within the ovary is correlated with the change in the uterus (endometrium). Menstrual bleeding occurs over days 0–4, followed by the proliferative phase up to ovulation at day 14 and then the secretory phase to day 28.

Source: Pocock et al. *Human Physiology*, 5th edn, 2017. Oxford University Press.

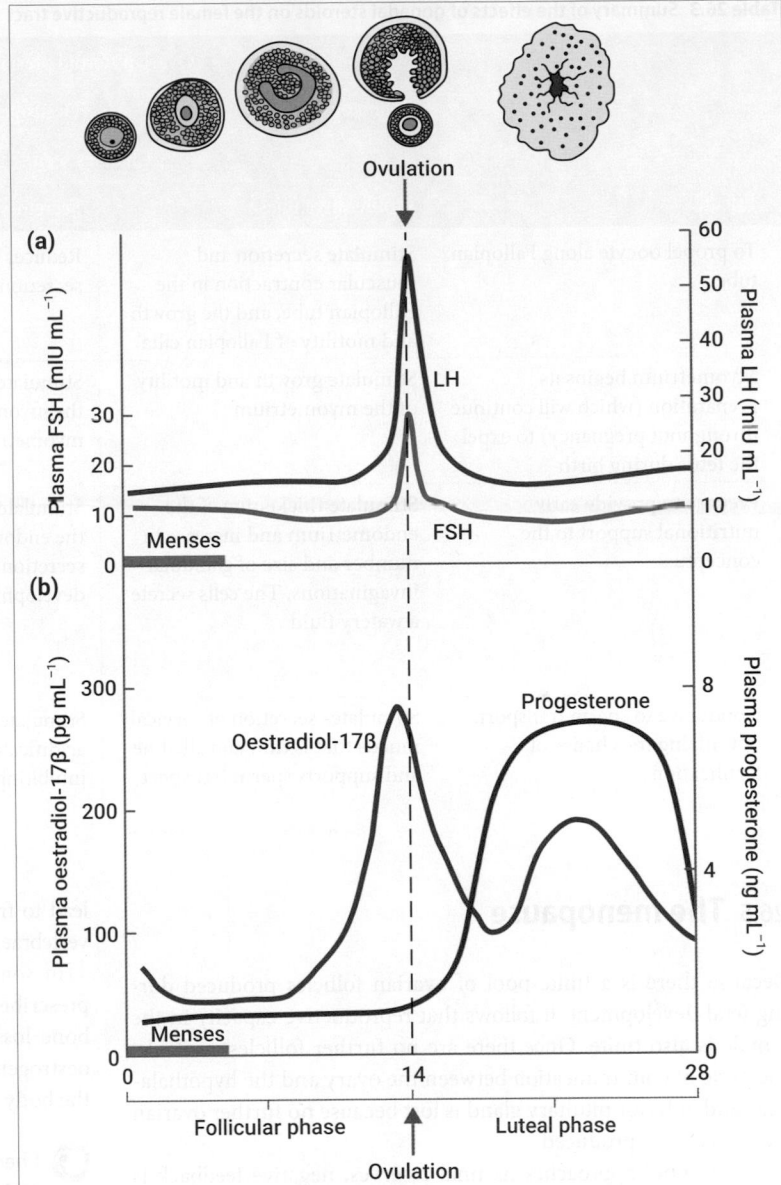

than adult females, because although the growth spurt starts later in boys it continues for longer. Once the growth spurt is complete, the **epiphyses** (cartilaginous growth plates) of the long bones fuse, thereby preventing any further growth in stature.

The secondary sexual characteristics develop under the influence of sex steroids. In the female, gonadal oestrogens influence breast development and female genital development. In the male, testicular androgens control development of genitalia and body hair and deepening of the voice. Pubic and axillary hair development is controlled by androgens from the adrenal glands.

Initiation of puberty

During childhood, GnRH secretion is low. The reasons why GnRH secretion is maintained low are subject to speculation. One explanation is that the developing hypothalamus is extremely sensitive to very low levels of sex steroid hormones. Puberty then arises

as this sensitivity decreases gradually on the approach to adolescence. Alternatively, the hypothalamic mechanisms for secretion of GnRH mature and steadily secrete more GnRH as childhood proceeds.

What is clear is that a wide range of factors influence the initiation of puberty, and the age of onset has been steadily decreasing over the last 150 years. Body weight is a critical factor because leptin, derived from body fat, can contribute to the maturation of hypothalamic secretion mechanisms. Children grow more quickly because of improved nutrition, achieving a critical weight sooner, and leading to onset of puberty earlier than in previous generations.

Precocious puberty, where puberty begins earlier than normal, can result from disruption, damage, or irritation of the central nervous system (CNS) as a result of meningitis or CNS tumours, indicating the critical role of the CNS in the initiation of puberty.

 Check your understanding of the concepts covered in this section by answering the questions in the e-book.

Table 26.3 Summary of the effects of gonadal steroids on the female reproductive tract

	Ovarian steroid present		
	Rising oestrogen	Oestrogen and progesterone	
	Changes seen		
Effect?	Proliferative phase	Secretory phase	Effect?
To propel oocyte along Fallopian tube	Stimulate secretion and muscular contraction in the Fallopian tube, and the growth and motility of Fallopian cilia	Reduces Fallopian tube motility, secretion, and cilia activity	Oocyte should now be in the lumen of the uterus
Myometrium begins its preparation (which will continue throughout pregnancy) to expel the fetus during birth	Stimulate growth and motility of the myometrium	Stimulates further thickening of the myometrium but reduces myometrial motility	Myometrial quiescence is essential to ensure the fetus is not expelled pre-term
Prepare to provide early nutritional support to the conceptus	Stimulate thickening of the endometrium and increase in number and size of glandular invaginations. The cells secrete a watery fluid	Stimulates further thickening of the endometrium, increased secretion, and in particular the development of spiral arteries	Elaboration of the spiral arterioles contributes to the development of a dedicated special circulation, the uteroplacental circulation, to support the fetus as its nutritional needs grow
Conducive to sperm transport, optimizing the chance of fertilization	Stimulates secretion of cervical mucus that is thin and alkaline and supports sperm transport	Stimulates thickening and acidification of cervical mucus, inhibiting sperm transport	Prevents access of further sperm or infectious agents if pregnancy occurs

26.5 The menopause

Because there is a finite pool of ovarian follicles produced during fetal development, it follows that reproductive capacity in the female is also finite. Once there are no further follicles available, the cycle of communication between the ovary and the hypothalamus and anterior pituitary gland is lost because no further ovarian steroids can be produced.

As the pool approaches its final reserves, negative feedback is lost, and the anterior pituitary produces ever-increasing amounts of FSH and LH to provoke a response in the ovary. Because the follicles are not producing inhibin either, FSH will increase to higher levels than LH. Therefore, menopause can be confirmed by measurement in serum of gonadotrophins, which will show elevated levels. Eventually, once all follicles are depleted, the hypothalamic–pituitary–ovary (HPO) axis will shut down completely, and FSH and LH levels will return to undetectable levels.

The average age for menopause in women is around 51 years, although the process may begin several months beforehand with increasing irregularity of ovarian and menstrual cycles. The impact of the menopause on an individual can be profound due to the significant changes in endocrine function; these can result in hot flushes, difficulty sleeping, low mood, and reduced interest in sexual intercourse.

Longer-term effects can include the development of **osteoporosis**, through which the bones become weakened as a result of resorption of calcium, linked to the low levels of circulating oestrogen. This can

lead to fractures, especially of the hip or wrist, and collapse of the vertebrae, resulting in an increasingly stooped posture.

In some cases, hormone replacement therapy (HRT) may be prescribed to reduce the severity of the symptoms and the rate of bone loss. This treatment typically comprises a combination of oestrogen and progesterone, thereby replacing the hormones that the body had been producing.

 Check your understanding of the concepts covered in this section by answering the questions in the e-book.

26.6 Male reproductive anatomy

The male reproductive system (Figure 26.14) comprises:

- the paired testes that produce spermatozoa (sperms) and contain cells that produce the sex hormones;
- a duct system that transports the spermatozoa to the urethra;
- accessory sex glands that supply the fluid components to the semen;
- the penis.

Testes and scrotum

The human testis is a compound tubular gland, organized into lobes in which are found the coils of seminiferous tubules, the site of spermatogenesis. Because the optimum temperature for

Figure 26.14 Male reproductive anatomy.
The male system comprises a pair of testes, which produce the sperm, housed in the scrotum, the prostate gland, seminal vesicles, which produce secretions to support the sperm, and the duct system connecting each testis to the urethra within the penis.

Source: Pocock et al. *Human Physiology*, 5th edn, 2017. Oxford University Press.

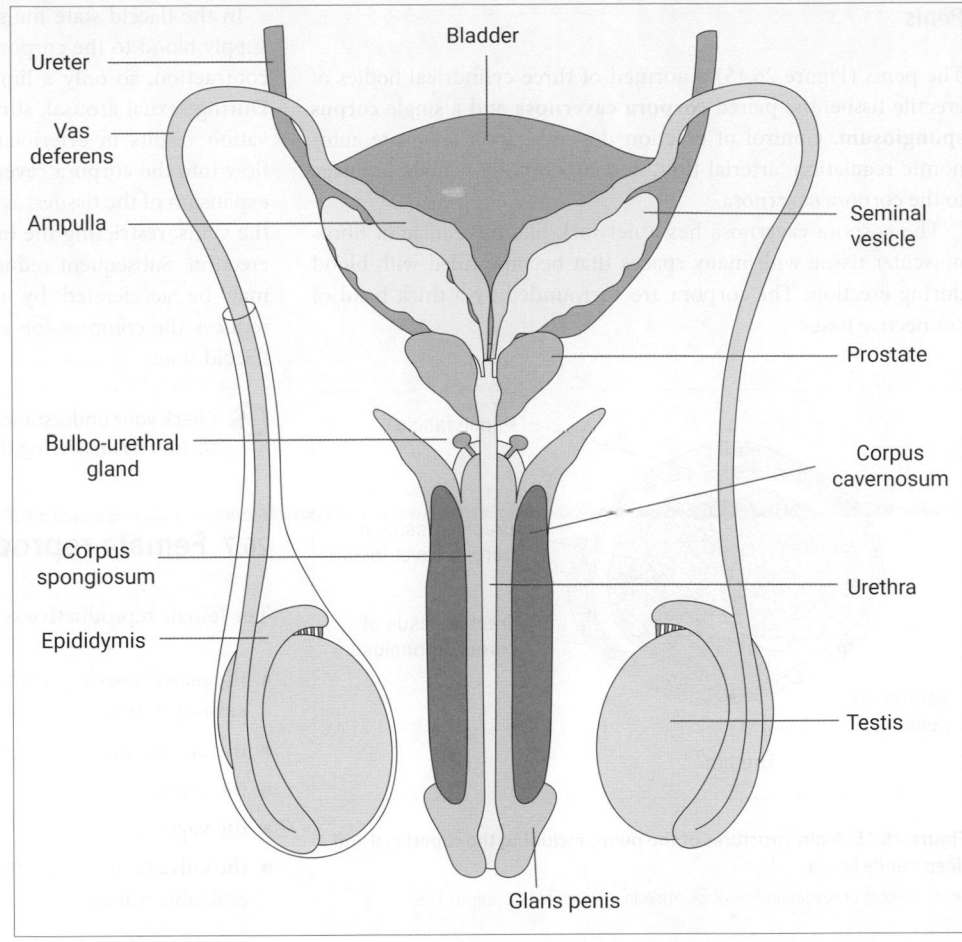

spermatogenesis is 2 to 3 °C lower than body temperature, the testes are effectively contained outside the body within the scrotum, which maintains the testes at this temperature. The testes descend from inside the body to the scrotum during fetal development.

Occasionally, the normal descent of the testis does not occur. In the resulting mal-descended testes, spermatogenesis is impaired because of the elevated temperature. Orchitis (inflammation of the testis) can occur following infection with the mumps virus after puberty. As a result, spermatogenesis can be impaired due to **atrophy** (wasting) of the seminiferous epithelium. Testosterone production is unaffected because the mumps virus affects the germinal epithelium rather than Leydig cells.

The duct system

Once mature, the sperm make their way from the testis out of the male body in order to encounter a female gamete. An extensive duct system with associated glands serves this purpose. A continuous tube connects the testis to the urethra. The sperm move from the seminiferous tubules through the efferent ducts, which conduct them into the **epididymis**, and onwards into the **vas deferens** (Figure 26.14).

During their passage through the epididymis, as they prepare to leave the scrotum, sperm undergo the final step in structural maturation to become motile. Sperm are then stored in the epididymis

until ejaculation. The terminal portion of the vas deferens on each side of the body gives rise to a seminal vesicle. The vas deferens on each side, with the duct of its associated seminal vesicle, merge with the urethra within the body of the **prostate gland**. The vas deferens has a smooth muscular coat that receives autonomic innervation. This permits rapid contractions that propel the tube's contents onwards into the urethra.

Accessory male sex glands

While sperm are themselves motile, they need a fluid medium to support their journey from the male tract into the female. This is provided by the **seminal fluid** (semen). The bulk of the seminal fluid, in which the sperm are suspended, is produced by the seminal vesicles, the prostate gland, and the bulbourethral glands (Figure 26.14). During sexual arousal and ejaculation, the secretions are emitted in a controlled sequence. The function of these glands is maintained only if adequate levels of testosterone continue to be produced. The seminal vesicles contribute more than half of the ejaculate volume. The prostate gland sits at the base of the urinary bladder, surrounding the proximal portion of the urethra. Prostatic secretions include prostaglandins, proteolytic enzymes, and citric acid. The bulbourethral glands, situated within the perineum, produce a clear watery secretion immediately before ejaculation that may lubricate the urethra.

Penis

The penis (Figure 26.15) is formed of three cylindrical bodies of erectile tissue: the paired **corpora cavernosa** and a single **corpus spongiosum**. Control of erection depends upon adequate autonomic regulation, arterial flow, and functioning venous drainage to the corpora cavernosa.

The corpora cavernosa has a network-like trabeculae of fibromuscular tissue with many spaces that become filled with blood during erection. The corpora are surrounded by a thick band of connective tissue.

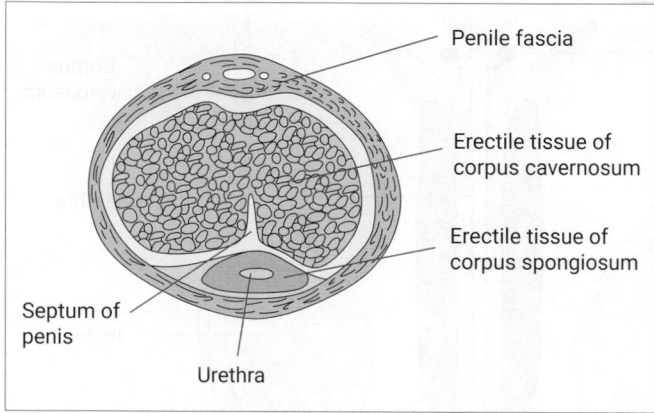

Figure 26.15 Main structures of the penis, including the superficial and deep penile fascia.

Source: Pocock et al. *Human Physiology*, 5th edn, 2017. Oxford University Press.

In the flaccid state the smooth muscle of the arterioles, which supply blood to the corpora cavernosa, is maintained under tonic contraction, so only a limited blood supply enters these tissues. During sexual arousal, stimulation of the parasympathetic innervation results in arteriolar dilation, leading to increased blood flow into the corpora cavernosa and the corpus spongiosum. The expansion of the tissues, as the blood enters them, also compresses the veins, restricting the outflow of blood and resulting in penile erection. Subsequent reduction in the arterial blood flow, which may be accelerated by sympathetic-induced vasoconstriction, reduces the compression of the veins, leading to a return to the flaccid state.

Check your understanding of the concepts covered in this section by answering the questions in the e-book.

26.7 Female reproductive anatomy

The female reproductive system comprises (Figure 26.16):

- the paired ovaries, which produce the female gametes (ova) and sex hormones;
- the uterine tubes (Fallopian tubes);
- the uterus;
- the vagina;
- the vulva (external genitalia), including the clitoris, labia majora, and labia minora.

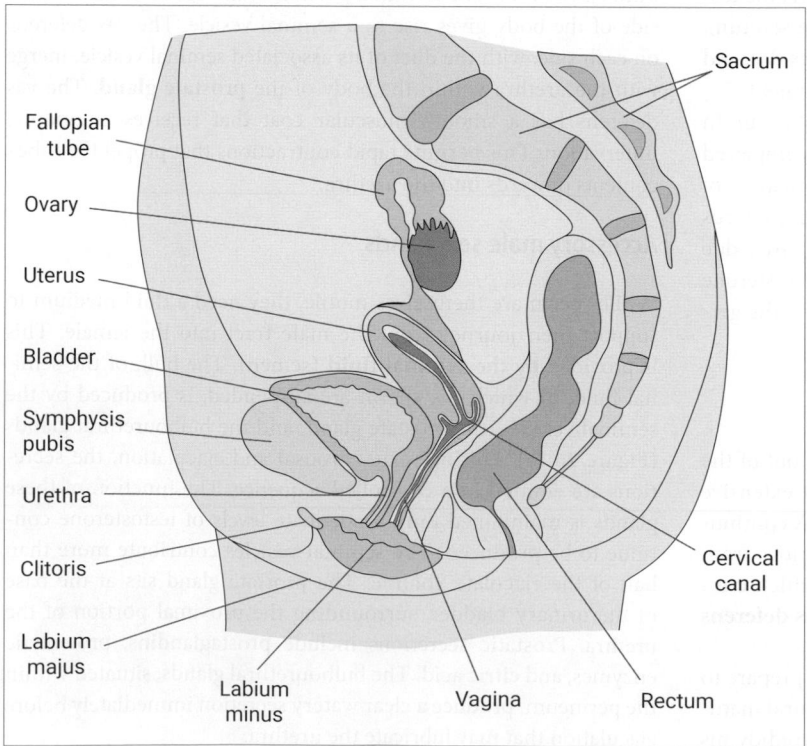

Figure 26.16 Female reproductive anatomy. The female system comprises the pair of ovaries that produce the oocyte, the uterus and Fallopian tubes, and the vagina, which opens into the vestibule of the vulva.

Source: Pocock et al. *Human Physiology*, 5th edn, 2017. Oxford University Press.

The paired human ovaries are small, ovoid organs located in the pelvis close to the lateral walls. Normal function of the ovary results in the development each month of a single ovulatory follicle (as described in Section 26.1).

Sometimes the ovary may develop ovarian cysts, which are usually derived from follicles that have not matured to an ovulatory follicle. Polycystic ovaries (with more than 10 cysts) are usually associated with infertility. Tumours of the ovary arise most commonly from epithelial components of the ovary or from germ cells. Ovarian cancer is often termed the 'silent killer' because such tumours are often very advanced, due to the ovaries' location deep in the pelvis, by the time the disease causes symptoms in the female.

Uterine (Fallopian) tubes

The uterine, or Fallopian, tube (which you can remind yourself about in relation to the rest of the female reproductive anatomy in Figures 26.5 and 26.16) is analogous to the duct system in the male. The Fallopian tube collects the oocyte, once expelled from the ovary, at ovulation and propels it along its length towards the body of the uterus. If fertilization is to occur, it is normally within the Fallopian tube that the oocyte encounters sperm.

Uterus and cervix

In the human, the uterus is a thick-walled muscular organ with a specialized lining. The muscular wall is termed the **myometrium** and the lining is termed the **endometrium** (Figure 26.17).

The endometrial and myometrial layers respond in different ways to steroids produced by the ovary. The muscular myometrium responds to the trophic stimulus by undergoing considerable **hypertrophy** (increase in cell size) and **hyperplasia** (increase in cell number). These processes occur in preparation for the force needed to expel the fetus at the end of pregnancy.

The endometrium is further divided into two layers. The deep basal layer represents a stem cell layer from which a new functional layer will develop at the start of each new cycle. Marked histological changes occur in the endometrium over the course of the menstrual cycle, reflective of the functional changes occurring (Figure 26.18).

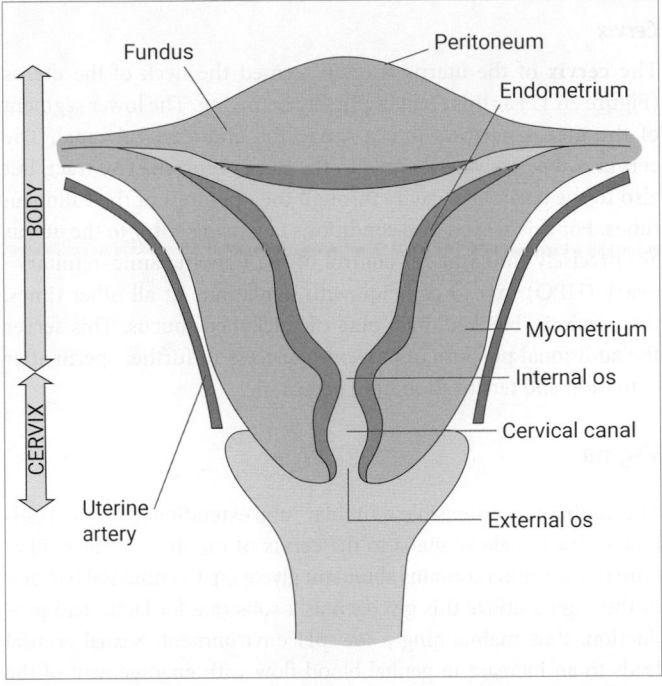

Figure 26.17 Anatomy of the uterus showing the muscular myometrium and the endometrial lining. The uterus narrows as it approaches the vagina at the cervix, which forms a canal, the endocervical canal.

Source: Reprinted from K. Abbas, S.D. Monaghan, I. Campbell. Uterine physiology. *Anaesthesia and Intensive Care Medicine*, 2016, 17, 346–8. With permission from Elsevier.

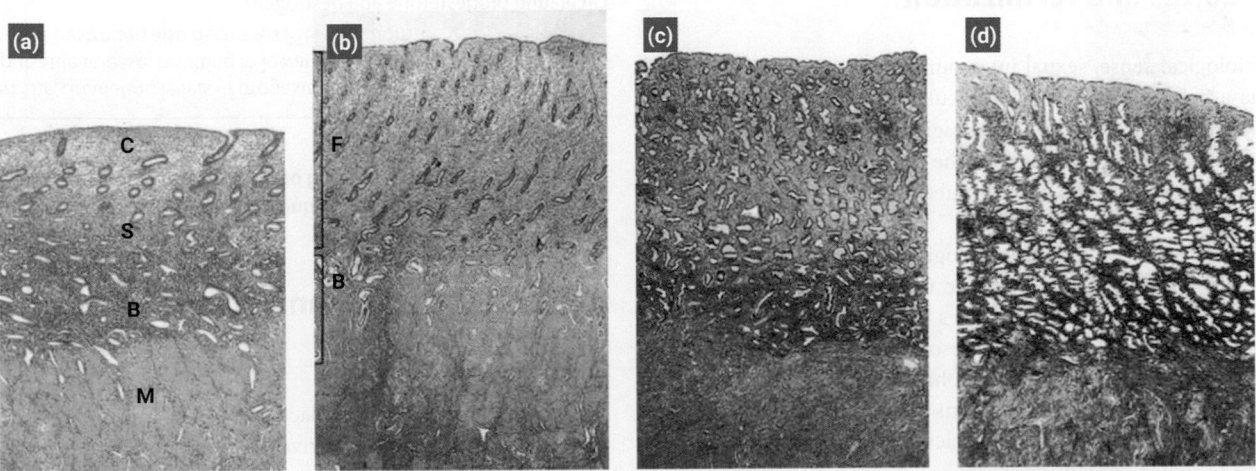

Figure 26.18 Endometrial changes in the uterine cycle. (a) Early proliferative: M, myometrium; B, basal layer; S, spongy layer of functional layer; C, compact layer of functional layer. (b) Late proliferative: B, basal layer; F, functional layer; (c) early secretory; (d) late secretory.

Source: Pocock et al. *Human Physiology*, 5th edn, 2017. Oxford University Press.

The endometrium consists of tubular glands that extend from the surface into the connective tissue (stroma). Most of the superficial part of the endometrium, called the stratum functionalis or functional layer, is subject to cyclical growth, degeneration, and shedding of the dead tissues. The deeper part of the endometrium (stratum basalis = basal layer), in the vicinity of the myometrium, does not exhibit these cyclical changes and is responsible for the regeneration of the upper endometrium.

Cervix

The **cervix** of the uterus is often termed the neck of the uterus (Figure 26.17) as this term implies a narrowing. The lower segment of the uterus narrows into a canal, the endocervical canal. The cervix is the gateway not only to the upper reproductive tract, but also to the peritoneal cavity through the openings of the Fallopian tubes. For this reason, the conditions favouring entry to the uterus are precisely timed under control of the hypothalamic–pituitary–ovary (HPO) axis to coincide with ovulation. At all other times, the cervix is 'blocked' by a plug of thickened mucus. This serves the additional purpose of preventing access to further sperm after ovulation and fertilization have occurred.

Vagina

The **vagina** is a distensible muscular tube extending from the vestibule of the female genitalia to the cervix of the uterus. The epithelium of the vagina contains abundant glycogen. Commensal bacteria in the vagina utilize this glycogen as a substrate for lactic acid production, thus maintaining a low pH environment. Sexual arousal leads to an increase in genital blood flow with engorgement of the labia and the clitoris. There is also an increase in the secretion of a watery mucus that lubricates the vagina, facilitating coitus, and partly buffering the acidic pH, which increases the survival of the sperm.

 Check your understanding of the concepts covered in this section by answering the questions in the e-book.

26.8 Coitus and fertilization

In a biological sense, sexual intercourse (**coitus**) occurs when the vagina accommodates a penis. Ejaculation from the penis results in the deposition of sperm in the vagina at the level of the cervix. From the cervix, a proportion of the sperm enter the uterus and reach the uterine tubes (Fallopian tubes), where they may encounter an oocyte.

The act of coitus involves physiological sexual excitement, which refers to both genital and systemic (e.g. rise in blood pressure) changes in both sexes. These changes involve the following phases: excitement, plateau, orgasm, and resolution. The climax of sexual excitement occurs in the orgasmic phase. In the female this can be accompanied by muscle contractions in the vagina and uterus. In the male this is accompanied by ejaculation and release from the urethra of semen containing sperm.

Immediately after ejaculation, the semen first coagulates due to the action of clotting factors present in semen—this is likely to help prevent sperm from being physically lost from the vagina. Later,

the semen liquefies by the action of enzymes derived from prostatic secretions. The small number of sperm that enter the uterus must travel a distance of some 15 to 20 centimetres to reach the Fallopian tube.

During their passage through the uterus to the Fallopian tube, the sperm undergo **capacitation** and the **acrosomal reaction**, the final stages in sperm maturation. At this stage, the sperm have full capacity to fertilize the ovum. Fertilization is not possible until capacitation has occurred. Capacitation is a process by which a glycoprotein coat is removed, allowing for changes in the sperm cell plasma membrane, and facilitating its fusion with the **zona pellucida**, the glycoprotein shell surrounding the oocyte (Figure 26.5). This process triggers the acrosomal reaction in the sperm, which results in release of proteolytic enzymes to allow the sperm to penetrate the thick zona pellucida.

By the time of ovulation, the oocyte has completed its first meiotic division (containing a haploid number of chromosomes and the bulk of the cytoplasm of the primary oocyte). After fusion of the sperm head with the oocyte, the oocyte finally undergoes the second meiotic division.

Only one sperm penetrates the cytoplasm of the ovum and its nucleus fuses with the nucleus of the oocyte. The product of this fusion is called the **zygote**. The entry of the sperm into the ovum sets off a series of events that prevent other sperm from entering the ovum. As a result of the fusion of the two haploid cells, the resultant zygote contains diploid chromosomes for normal development to proceed.

The zygote then begins the process of development by forming an **embryo**. The cellular interactions of the early embryo and the endometrium, and the establishment of the relationship between the embryonic and maternal tissues lead to the establishment of the early placenta. The placenta serves to support the growth and development of the fetus during pregnancy. To avoid a pregnancy, contraception may be used (see Clinical Box 26.1).

> ### PAUSE AND THINK
>
> Look back at Figure 26.11. What will be the effect of maintaining levels of circulating progesterone and oestrogen?
>
> *Answer:* The circulating levels of progesterone reduce the frequency of GnRH pulsatile release, resulting in lower secretions of FSH and LH. The presence of oestrogen also decreases FSH secretion.

 Check your understanding of the concepts covered in this section by answering the questions in the e-book.

26.9 Establishment and maintenance of pregnancy

Fertilization normally takes place in the Fallopian tube. The zygote travels down the tube and arrives between 4 and 5 days later as a **blastocyst**. The blastocyst begins the process of interacting with the endometrium around 6 days after fertilization to begin the process of **implantation**. The whole process of implantation, in all its complexity of cellular interactions involving cells of two individuals

---

CLINICAL BOX 26.1	**Contraception**

Contraception is action taken to prevent pregnancy occurring as a result of sexual intercourse; it can include fertility awareness and artificial contraception. There are various methods of artificial contraception available, the most common being:

- Barrier methods, such as the condom or diaphragm, which act as physical barriers to keep the sperm from encountering the oocyte; these may be reinforced by the application of spermicidal gel. Barriers such as condoms are also important in reducing the risk of contracting sexually transmitted infections (STIs)—see Clinical Box 26.2.
- The contraceptive pill, which prevents ovulation.
- Intrauterine devices (IUDs), which prevent implantation of the fertilized ovum.

Male and female sterilization represent permanent forms of artificial contraception:

- Male sterilization, which is most commonly through vasectomy, the cutting of the vas deferens, prevents sperm from being released.

- Female sterilization, through cutting or blocking the Fallopian tubes, prevents the oocyte from encountering the sperm.

The contraceptive pill (also referred to as 'the pill') takes various forms, one of the most common in current use being the combined oral contraceptive pill (COCP). The COCP is taken on a daily basis for 21 days with a 7-day pill-free interval (several common brands include a 'dummy' pill for the 7-day period to support continual practice). COCPs are taken routinely to control fertility by about 20 per cent of women in the UK aged between 16 and 49 years.

There are various formulations of COCP in terms of the proportions of the synthetic hormones that they include. The mode of action is similar in that they inhibit the development of follicles and prevent ovulation. As the name suggests, COCPs contain both oestrogen and progesterone.

As a result, follicular development is inhibited by the COCP. The prevention of the LH surge also prevents ovulation from occurring. A further action of progesterone in the COCP is to increase the viscosity of the mucous plug at the neck of the cervix, thereby blocking sperm from passing up towards the Fallopian tubes.

CLINICAL BOX 26.2	**Sexually transmitted infections**

Sexually transmitted infections (STIs) are, as the name suggests, a group of infections acquired through sexual contact. They may be bacterial or viral in origin, and can also include pubic lice and scabies (mites). STIs can be transmitted through vulvovaginal, penile, or anal contact. Infection may be passed from mother to child during birth.

STIs are highly prevalent worldwide. In recent years, there has been a marked increase in STIs with nearly 470,000 new cases diagnosed in England in 2019.

Bacterial STIs include chlamydia, gonorrhoea, and syphilis. Examples of viral infection include herpes simplex, human papilloma viruses (HPV), human immunodeficiency virus (HIV–AIDS), and viral hepatitis (A, B, and C).

Anyone who is sexually active is potentially at risk of contracting an STI, which underlines the importance of sexual health checks, especially with a new partner. This risk can be reduced by the effective use of barrier contraception, such as condoms.

The symptoms of STIs are dependent upon the pathogen. Many of the infections have both local and regional or systemic symptoms, as well as short- and long-term sequelae.

Chlamydia and gonorrhoea can both cause inflammation of the male and female uro-genital tracts. These infections may be asymptomatic, but can also cause dysuria—discomfort during urination, and vaginal or penile discharge. These infections can also spread locally to cause inflammation of the epidydimis and/or testes in males, or pelvic inflammatory disease (PID) in females, which can lead to damage to the Fallopian tubes and resulting in subfertility or ectopic pregnancy.

Syphilis, caused by the spirochaete bacterium *Treponema pallidum*, is an example of an STI that can have significant systemic symptoms. Primary syphilis initially presents as a solitary, painless genital ulcer. If untreated, this can progress to secondary syphilis weeks or months later, characterized by a widespread rash, hepatitis, glomerulonephritis, and neurological involvement, such as meningitis and cranial nerve palsies. Tertiary syphilis may present decades later with neurological and cardiovascular complications, which include dementia and psychosis, paraesthesia, aortitis, aortic regurgitation, and heart failure. Syphilis has been identified as the possible cause of the medical conditions affecting many well-known people, including Henry VIII and Lenin.

Herpes is characterized by a primary infection causing multiple, small, painful vesicles around the genitalia. The virus then lies dormant with periodic reactivation resulting in secondary infections, which are usually less symptomatic.

Ano-genital warts are caused by the human papilloma virus (HPV). Warts can be found around the external genitalia in both males and females, but they do not generally cause any other symptoms. However, warts may be particularly problematic in immunocompromised individuals, such as during pregnancy or people living with HIV. HPV types 16 and 18 are responsible for some cancers, such as cervical cancer. To protect against them, a vaccine is offered to boys and girls aged 12–13 years, and they are screened for as part of the National Cervical Cancer Screening Programme.

The diagnosis of STIs depends on the causative organism. The principles of diagnosis include taking swabs from potentially infected sites, urine samples, and blood tests. Swabs can be taken from ulcers, the vagina or cervix in females, or the penile urethra in males, and used for microscopy, polymerase chain reaction (PCR), or nucleic acid amplification testing (NAAT). Blood tests may be for antibodies or PCR.

If a positive test result is obtained, there are two key components to treatment: treating the primary individual, and tracing and treating their sexual contacts (contact tracing). The process of contact tracing can be performed by the individual with support from the sexual health clinic, or can be done directly by the clinic.

Chlamydia, gonorrhoea, and syphilis are all readily treatable with antibiotics, if detected early. Their later complications, however, may be irreversible. There is no cure for herpes, so treatment is aimed at symptomatic

relief in the form of topical local anaesthetic or antiviral tablets to reduce the frequency and duration of symptoms. Similarly, warts may be treated with topical therapies or removal. Viral hepatitis can be treated with antiviral drugs; however, some may go on to develop chronic infection. HIV is managed with antiretroviral medication; this can lead to an undetectable viral load, which means the individual will not develop many of the symptoms of HIV and will not pass on the infection.

Sexual and reproductive health is an important aspect of public health in terms of population education, screening, and vaccination, and it involves a wide range of healthcare settings. Finally, it is a key component of an individual's emotional, physical, social, and mental well-being.

For more information, visit: https://www.bashh.org/guidelines

---

(mother and fetus) and the genomes of three, is not complete until the end of the first trimester. Implantation allows for the development of a new circulation unique to pregnancy, involving the blood supply to the uterus and a newly forming circulation in the placenta that connects to the fetal circulation.

If fertilization is carried out *in vitro*, the embryo is maintained in culture until it has reached the blastocyst stage before it is transferred to the uterus.

The placenta has both maternal and fetal components and a good placenta determines a good pregnancy. It represents the interface between the mother and developing baby, and is the boundary across which nutrients are supplied from the mother and waste is removed from the fetus. It has its own developmental programme and its structure changes over time in ways that reflect the changes in function required of it.

The suppression of the HPO axis in the luteal phase (Table 26.2, Figure 26.13) prevents the menstrual cycle from restarting and beginning the process of gametogenesis anew. Gametogenesis should remain suppressed if a pregnancy is established, so that the reproductive system of the female can instead be dedicated to the support of the fetus. Once the placenta is sufficiently large, its production of oestrogen and progesterone takes over from that of the corpus luteum and maintains the suppression of the HPO axis.

Before that point is reached, the placenta, through its production of human chorionic gonadotrophin, an LH analogue, supports the continuing function of the corpus luteum of the ovary to produce the oestrogen and progesterone needed to ensure suppression of the HPO axis (see Section 26.2).

The transport function of the placenta is remarkable. The unit of exchange consists of **chorionic villi**, finger-like projections of the outer membrane enclosing the baby, the chorion (Figure 26.19). These villi consist of a core, containing fetal capillaries, which is covered by two epithelial layers. One of these epithelial layers is continuous and the other gradually takes the form of scattered cells as delivery approaches.

In humans, there are many chorionic villi, maximizing the increase in surface area available for exchange. These villi are

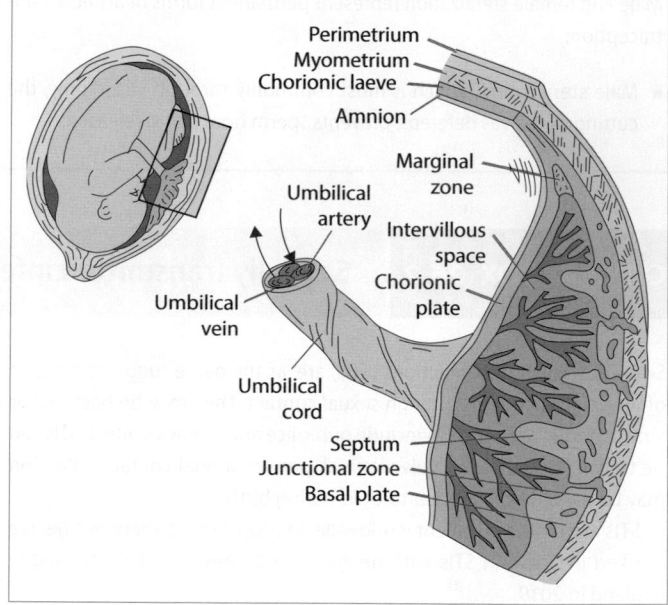

Figure 26.19 The structure of the placenta in humans.
Source: Figure 4.1 from Zakowski M. & Geller A. *Chestnut's Obstetric Anesthesia: Principles and Practice*, 2014. Elsevier.

bathed directly in maternal blood, supplied by the spiral arterioles of the endometrium. This arrangement minimizes the absolute number of cellular layers across which substances must cross. This is not without risk, however (see Scientific Process 26.1).

The placenta undertakes a number of key functions to support the growing fetus, including liver function, electrolyte transport, gas exchange, and transport of nutrients, such as glucose, amino acids, and iron. The processes involved are varied and include simple and facilitated diffusion, active transport, and receptor-mediated endocytosis.

In addition to nutrient supply and waste removal, the placenta is responsible for the provision of passive immunity that affords immune protection during the neonatal period.

SCIENTIFIC PROCESS 26.1   Research into the regulatory role of melatonin in the corpus luteum

Introduction

Following ovulation, the corpus luteum (CL) is stimulated by LH to secrete progesterone and oestrogen (Section 26.2) and the CL is maintained for about 14 days. If fertilization occurs, human chorionic gonadotropin (hCG) is secreted by the developing placenta, which maintains the CL, supporting the establishment and maintenance of early pregnancy.

Melatonin is synthesized by the pineal gland during darkness and plays an important role in the regulation of the circadian rhythm (the 'body clock'). Melatonin receptors are found in a wide range of tissues, and serum levels of melatonin have been shown to be elevated during the luteal phase of the menstrual cycle. The authors of this paper also previously showed that melatonin facilitated the restoration of a normal menstrual cycle in women affected by polycystic ovary syndrome (PCOS), which was linked to stimulation of progesterone production.

The authors therefore wanted to explore the possible effects of melatonin on the CL by measuring progesterone release, and also by measuring the release of prostaglandin E_2 (PgE$_2$), a local mediator that has a luteotrophic action—i.e. it stimulates the development and maintenance of the CL—and the release of prostaglandin $F_2\alpha$ (PgF$_2\alpha$), which has a luteolytic action (i.e. it stimulates the breakdown of the CL).

Methods

Ethical approval for the study was granted by the review board of the host university and informed consent was obtained from all participants.

Informed consent is essential so that all human subjects in research projects fully understand the nature of the research, any associated risks, and the use to be made of any tissues that are sampled. Participants sign a statement, confirming their understanding and consenting to the research.

A total of 25 CLs were obtained from normally menstruating women (25 to 38 years old) in the mid-luteal phase (days 5–6 from ovulation) at the time of surgery for non-endocrine gynaecological disease.

This approach to sampling is significant because the removal of the CLs is an invasive procedure and so could only be undertaken in individuals undergoing surgery for a medical reason. The CLs were all harvested at the same stage of development to ensure they would respond in the same way to any treatment. In this case, it was at a stage in the cycle when the levels of circulating melatonin are known to be relatively high.

The CLs were dissociated into separate cells, which were cultured and then divided into groups that were incubated with one of the following solutions:

1. Serum-free medium.
2. Human chorionic gonadotropin (hCG), 100 ng mL^{-1}.
3. Melatonin solutions: 100, 10, 1, 0.1, or 0.01 nM.
4. hCG plus melatonin solutions of 100, 10, 1, 0.1, or 0.01 nM.

Following incubation, the incubation media were assayed for progesterone and for PgE$_2$ and PgF$_2\alpha$.

What was the purpose of incubating with serum-free medium?

- This represents the control condition to monitor production of progesterone and for PgE$_2$ and PgF$_2\alpha$ in the absence of any stimulation.

What was the purpose of incubating with hCG?

- hCG is the factor, released by the placenta, which maintains the CL. Therefore, it acts as a comparator in this study to compare the action of melatonin against.

The CL cells were also examined using immunohistochemistry to look for the presence of melatonin receptors on the cell surface. In this technique, the cells are incubated with primary antibodies that have been created to react with the melatonin receptors. Following the incubation, the slides were stained with a reagent that reacts with the antibodies. Controls were also run using (a) staining but without the antibody; and (b) staining of control tissues that are known to carry the melatonin receptor.

Results

Figure 1a shows that incubation with hCG and with all concentrations of melatonin significantly increased the release of progesterone compared with the control CL cells, and hCG plus 0.1, 1, and 10 nM melatonin increased the release of progesterone above the levels seen for hCG alone.

Figure 2 shows that incubation with 1, 10, and 100 nM melatonin significantly increased the release of PgE$_2$ and all concentrations of melatonin reduced the release of PgF$_2\alpha$.

The immunohistochemistry revealed the presence of melatonin receptors on the membranes of the CL cells.

Conclusion

This study used combined chemical assay and immunohistochemistry approaches to explore the potential role of melatonin in supporting the maintenance of the CL. The results show that melatonin can act to support the maintenance of the CL through increasing the release of progesterone and of the luteotropic local factor PgE$_2$, while also decreasing the release of the luteolytic factor PgF$_2\alpha$. These findings correlate with demonstration of the presence of receptors for melatonin on the cell surface. These findings open up the possibility of using melatonin medication to support the establishment of early pregnancy.

Read the original work

Scarinci, E., Tropea, A., Notaristefano, G., et al. (2019) 'Hormone of darkness' and human reproductive process: direct regulatory role of melatonin in human corpus luteum. *J. Endocrinol. Invest.* **42**: 1191–7. https://doi.org/10.1007/s40618-019-01036-3

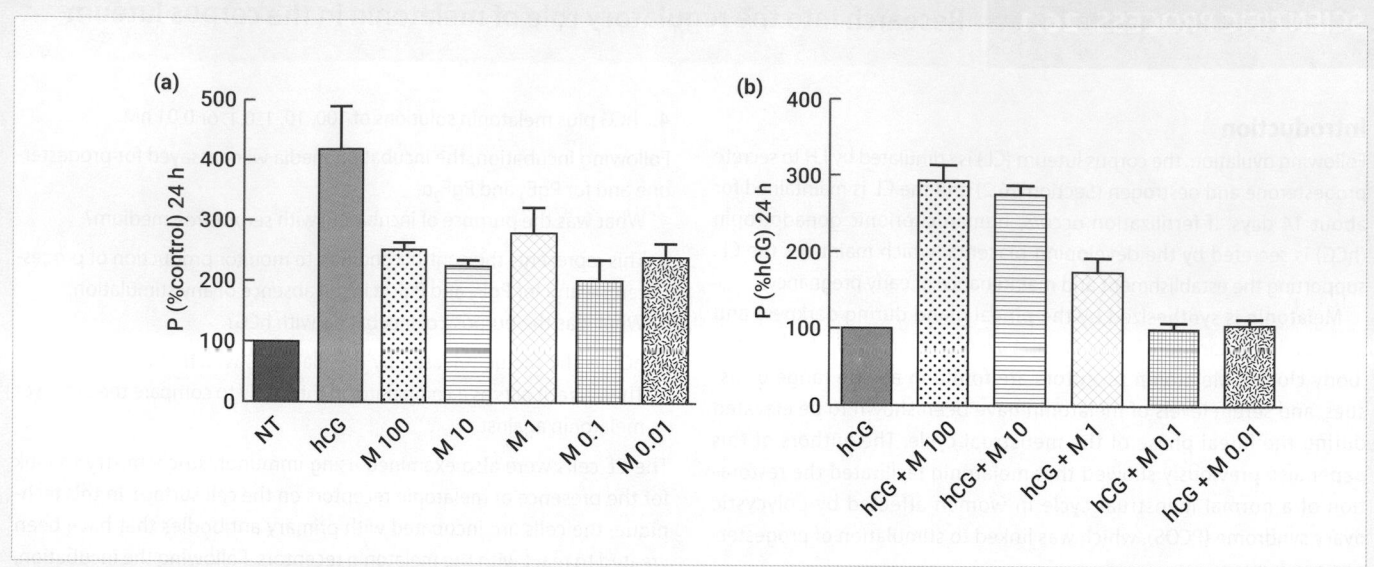

Figure 1 Progesterone secretion. (a) Changes in progesterone secretion by CL cells in response to incubation for 24 h with human chorionic gonadotropin (hCG), or different concentrations of melatonin, 100–0.01 nM. All responses are significantly greater than the control level (NT), $p < 0.001$. (b) Changes in progesterone secretion by CL cells in response to incubation for 24 h with human chorionic gonadotropin (hCG) alone or hCG *plus* melatonin. Melatonin at concentrations of 100, 10, and 1 nM significantly increased the release of progesterone compared with hCG alone.

Source: Scarinci E. et al. 'Hormone of darkness' and human reproductive process: direct regulatory role of melatonin in human corpus luteum. *Journal of Endocrinological Investigation*, 2019, 42, 1191–7. Reprinted by permission from Springer Nature: Springer.

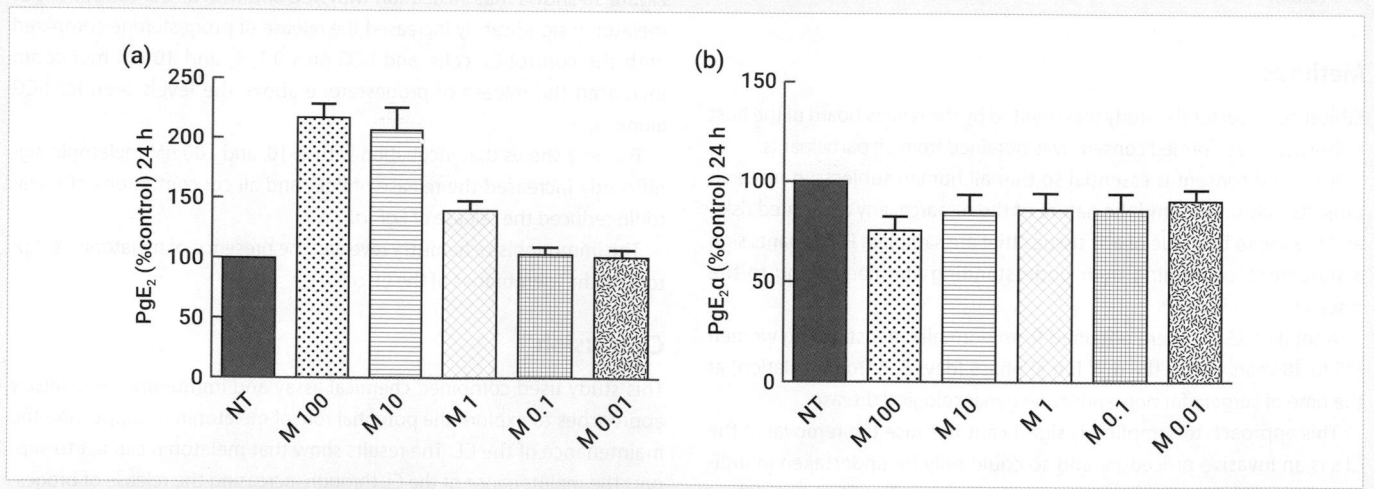

Figure 2 Effect of melatonin on the release of (a) PgE$_2$ and (b) PgF$_2$α. Melatonin at concentrations of 100, 10, and 1 nM significantly increased the release of PgE$_2$ and all concentrations of melatonin significantly decreased the release of PgF$_2$α.

Source: Scarinci E. et al. 'Hormone of darkness' and human reproductive process: direct regulatory role of melatonin in human corpus luteum. *Journal of Endocrinological Investigation*, 2019, 42, 1191–7. Reprinted by permission from Springer Nature: Springer.

Normally, the placenta is shed completely at **parturition**, and the massive blood supply to the implantation site is shut down through major myometrial contraction. If the placenta fragments after delivery, this can impair shut-down of the utero-placental circulation, and consequently cause significant post-partum haemorrhage.

The placenta therefore coordinates changes to maternal physiology to ensure that the developing baby is supported in all its needs.

Through its endocrine function, the placenta effects a wide range of physiological adaptions to the mother's body systems, as summarized in Table 26.4.

During pregnancy, there is an increase in lipolysis from the second trimester and an increase in maternal plasma free fatty acids on fasting. These changes serve to supply free fatty acids to provide a substrate for maternal metabolism, leaving glucose for the fetus.

Table 26.4 Summary of maternal physiological changes in pregnancy

Maternal system	Variable	Changes
Cardiovascular	Cardiac output	Increases, 40%
	Stroke volume	Increases, 35%
	Heart rate	Increases, 15%
	Systemic vascular resistance	Decreases, 25–30%
	Blood pressure	Decreases in first and second trimester, returns to normal in third trimester
Renal	Renal plasma flow	Increases, 60–80%
	Glomerular filtration rate	Increases, 55%
	Creatinine clearance	Increases, 40–50%
	Protein excretion	Increases, max. 300 mg/24 h
	Urea	Decreases, 50%
	Uric acid	Decreases, 33%
	Bicarbonate	Decreases, 18–22 mmol L^{-1}
	Creatinine	Decreases, 25–75 μmol L^{-1}
Respiratory	O$_2$ consumption	Increases, 20%
	Resting minute ventilation	Increases, 15%
	Tidal volume	Increases
	Respiratory rate	Unchanged
	Functional residual capacity	Decreases, in third trimester
	Vital capacity	Unchanged
	FEV$_1$	Unchanged
	Paco$_2$	Increases
	Pao$_2$	Decreases
Carbohydrate metabolism	Fasting blood glucose	Decreases
	Post-prandial blood glucose	Increases

Progesterone has wide-ranging effects on a number of tissues, notably smooth muscle, where it produces relaxation. The result is that during pregnancy, smooth muscle relaxation in the gastrointestinal tract by progesterone action can lead to delayed emptying, biliary tract stasis, and an increased risk of pancreatitis. Pregnancy is a state where the mother's blood is more likely to clot, with high levels of fibrin deposition at the implantation site (prothrombotic). In addition, there is increased fibrinogen and clotting factors, and reduced fibrinolysis. Added to this is stasis due to venodilation, which, in turn, leads to increased risk of thromboembolic disease during pregnancy.

These changes result from the combined actions of **human placental lactogen** (hPL; also known as human chorionic somatomammotrophin, hCS), oestrogen, progesterone, and prolactin.

Concurrently, preparations being made by the mother's body to ensure support of the baby after birth. Changes occur in the breasts, to ensure that lactation can start immediately after birth. Overall, there is a continuum that allows for an almost seamless transition from placental support, by means of the uteroplacental circulation and adaptations to maternal physiology, to independent life for the baby. At birth, although the neonate immediately takes responsibility for gas exchange and hepatobiliary function, there is a continuing need for nutritional and immune support, needs which are met by the mother's milk.

Pre-eclampsia is a common complication of pregnancy that can result in significant morbidity and mortality to both mother and baby. Pre-eclampsia affects up to 6% of pregnancies and usually develops after 24–26 weeks' gestation. The initial symptoms are hypertension and proteinuria (protein appearing in the urine—see Section 24.2). Once diagnosed, the woman will normally be monitored carefully because of the increased risk of serious complications, including convulsions or stroke in the mother and delayed growth of the fetus due to poor blood supply through the placenta. The pathophysiology of this condition is thought to lie, in part, in poor development of the placenta, leading to reduced blood supply. The result of this is signalling from the placenta that increases the maternal blood pressure to increase the uteroplacental blood supply, but which also results in development of maternal hypertension.

Fetal circulation

Oxygen- and nutrient-rich blood enters the fetal circulation via the umbilical vein, one of three vessels in the umbilical cord connecting the placenta to the fetus. In contrast to the situation in adults, this blood then enters the heart at the right side (Figure 26.20).

This arrangement requires a series of fetal circulatory shunts that can manage this supply and, critically, can be *reversed* at birth immediately to adopt the mature circulation.

The fetal lungs are non-functional and gas exchange occurs at the placenta. The first organ that newly oxygenated fetal blood would meet on arrival from the exchange membrane via the umbilical vein is the liver. Were shunts not available, this blood would pass through the liver and the lungs and mix with venous blood from the body and brain, losing most of its oxygen, before it reached the systematic arteries.

Thus, the fetal circulation takes the following path:

- umbilical venous blood is shunted around the liver by the **ductus venosus**;
- blood enters the inferior vena cava;
- blood enters the right atrium;
- shunts from right to left atrium via the **foramen ovale**, a hole in the interatrial septum;
- the output of the right heart flows through the **ductus arteriosus**, linking the pulmonary artery to the aorta;
- deoxygenated blood returns to the placenta in the umbilical arteries for reoxygenation.

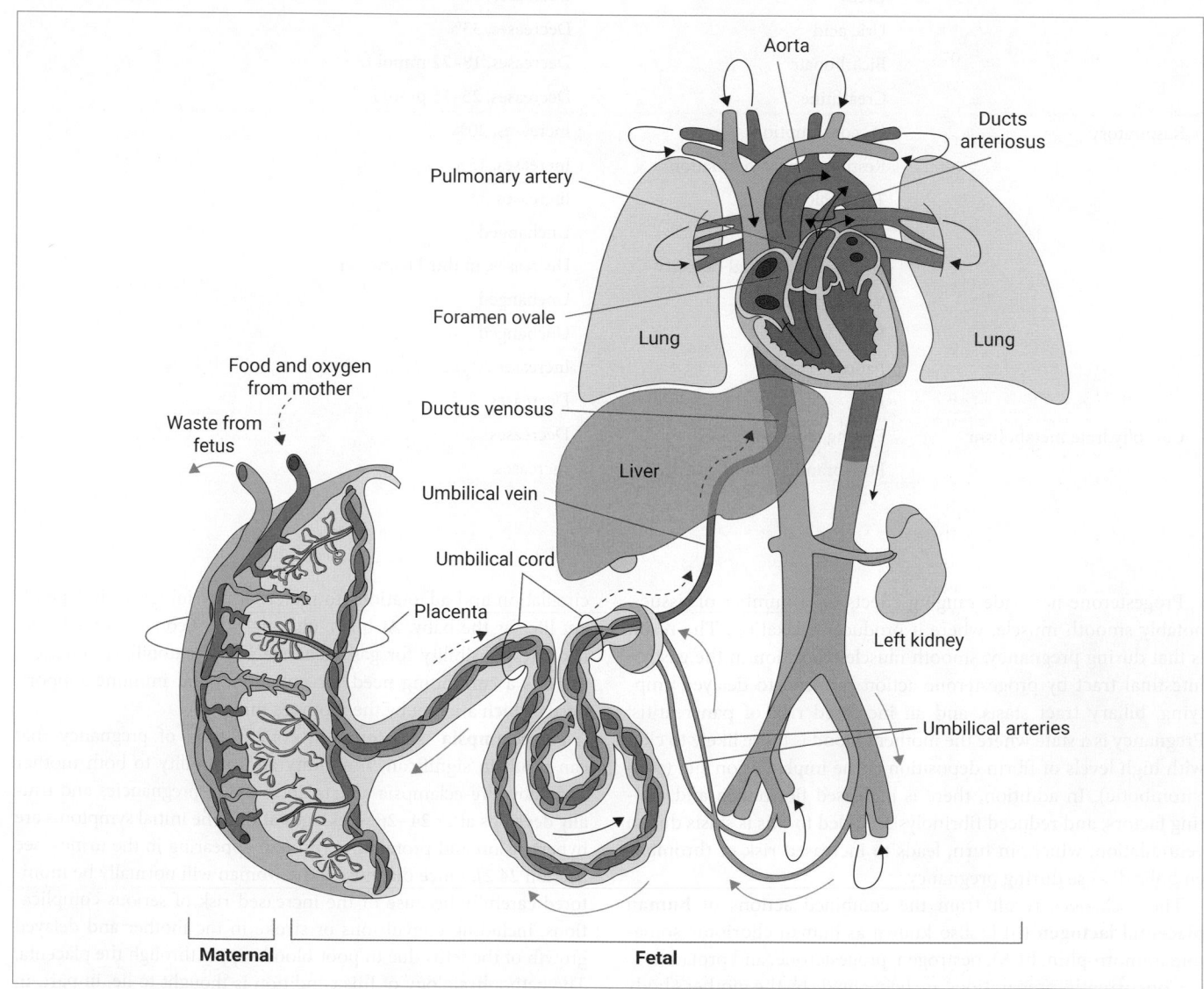

Figure 26.20 Fetal circulation. Circulation in the fetus requires shunts to manage the blood arriving from the placenta to the right atrium.

Source: R. Brusseau. Fetal Intervention and the EXIT procedure. In: C.J. Coté, J. Lerman, and B.J. Anderson (eds), *A Practice of Anesthesia for Infants and Children*, 6th edn, 2019. Elsevier.

This circulatory pattern is dependent on the pressure in the right atrium being greater than that in the left, and the pressure in the pulmonary artery being greater than that in the aorta. Both these conditions are achieved by the fact that fetal pulmonary resistance is high, offered by the collapsed fetal lungs.

Adaptation at birth

At birth, there are rapid, significant changes in the cardiovascular system associated with independent living.

With the first breath, there is a dramatic reduction in pulmonary vascular resistance and a dramatic rise in arterial oxygen saturation as the lungs inflate. The fall in pulmonary vascular resistance causes left atrial pressure to rise in respect to the right atrial pressure, so closing the foramen ovale—this then closes the shunt between the left and right atria.

Smooth muscle sensitive to high oxygen saturation in the wall of the ductus arteriosus contracts to close the ductus, so blood from the right ventricle flows into the pulmonary artery and is now routed through the lungs. So, both fetal shunts are rapidly closed just by taking the first breath. Initially, closure is physiological. Over time, the closure becomes anatomical through fusion of tissue layers.

The ductus venosus closes once the supply from the umbilical vein stops at delivery.

Fetal oxygen supply

Fetal blood has low Po_2, with an oxygen saturation of approximately 4 kPa compared to normal adult Po_2 of 11–13 kPa. However, there is a range of factors increasing fetal oxygen content. These include the fetal haemoglobin variant, whereby the fetal haemoglobin has a higher affinity for oxygen than does the adult form, and a raised fetal **haematocrit** (ratio of volume of red blood cells to volume of blood) compared to that in the adult. Fetal haemoglobin concentration is raised (fetal Hb = 166–175g L^{-1}, compared with 95–140 g L^{-1} at 2 years and 130–180 g L^{-1} in adult males). A double Bohr effect speeds up the process of oxygen transfer.

The double Bohr effect is covered in Chapter 22, but, as a reminder:

- As CO_2 passes into intervillous blood, the pH decreases.
  - Bohr effect.
- This decreases the affinity of Hb for O_2 on the maternal side of the circulation.
- At the same time as CO_2 is lost, the pH rises.
  - Bohr effect.
- This increases the affinity of Hb for O_2 on the fetal side of the circulation.

In addition, maternal CO_2 levels are lowered by hyperventilation stimulated by progesterone, which, overall, enables the fetus to have relatively normal Pco_2, as CO_2 will tend to travel down its concentration gradient across the placenta.

Fetal lungs

The respiratory system is one of the last systems of the body to develop. The cells capable of gas exchange and production of **surfactant** begin to appear only after 20 weeks of development. Surfactant acts to ensure that the alveoli of the lungs do not collapse, by lowering the surface tension in the alveoli (Section 22.2). The consequence is that neonates born very prematurely can have lungs that are insufficiently developed to function normally, which can result in neonatal respiratory distress syndrome.

Amniotic fluid

During development, the fetus is suspended in a watery environment, enclosed by the fetal membranes. This **amniotic fluid**, which is the product of fetal renal activity, reaches a maximum of 1 L around 38 weeks of gestation. Cells within the amniotic fluid are derived from the amnion and from the fetus. Therefore, analysis of the fluid can be made by **amniocentesis** (sampling of the fluid) and can be used, for example, to assess the presence of chromosomal abnormalities such as Down syndrome and neural tube defects.

 Check your understanding of the concepts covered in this section by answering the questions in the e-book.

26.10 Birth

The expulsion of the fetus requires a number of processes:

1. The creation of a birth canal, involving:
   a. Release of the structures that normally retain the fetus *in utero*
   b. Enlargement and realignment of the cervix and vagina.
2. Expulsion of the fetus.
3. Expulsion of the placenta and changes to minimize blood loss from the mother.

These processes are known as the first, second, and third stages of labour.

First stage of labour

The optimal fetal position in the pelvis (Figure 26.21) minimizes the diameter of the structure presented to the birth canal. The pelvic inlet is bounded posteriorly by the mother's pelvic girdle, a bony ring. The true diameter of this inlet is usually about 11 cm.

The fetus is normally retained in the uterus by a closed cervix and relative quiescence of the myometrium, a state maintained by the high progesterone concentrations seen during pregnancy. To create a birth canal, the cervix must dilate. Cervical dilatation is facilitated by biochemical changes in the extracellular matrix within the cervix, known as cervical ripening, but produced by forceful contractions of uterine smooth

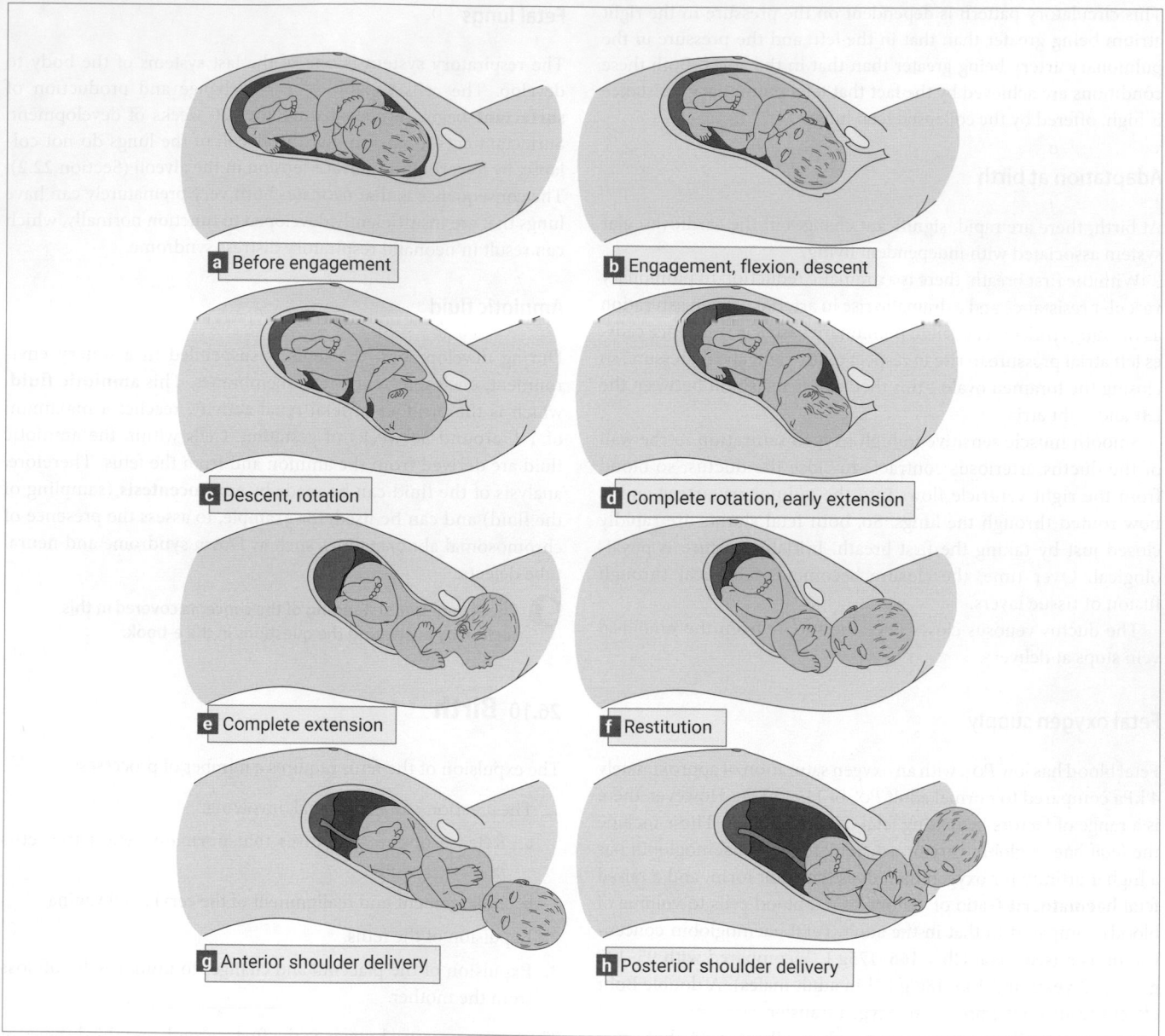

Figure 26.21 Cardinal stages of labour.

Source: Figure 11.11 from Landon et al. *Gabbe's Obstetrics: Normal and Problem Pregnancies*, 8th edn, 2021. Elsevier.

muscle. These contractions first thin the cervix ('effacement') and then dilate it.

Uterine smooth muscle

The myometrium is made up of bundles of smooth muscle cells. During pregnancy, the myometrium gets much thicker due primarily to hypertrophy. Force is generated by the usual actin and myosin intracellular machinery seen in smooth muscle cells (Chapter 20). Action potentials in the cell membrane lead to an increase in intracellular calcium ion concentration. Specialized gap junctions allow spread of action potentials from cell to cell, causing spread of coordinated contraction across the myometrium.

At the onset of labour there is a sudden increase in the frequency and force of contractions. Two hormones, **prostaglandins** and **oxytocin**, are implicated in this change. Prostaglandins are local mediators, while oxytocin is secreted from the posterior pituitary gland.

Hormonal mediators in parturition

Prostaglandins are biologically active lipids synthesized in most body tissues and act as local hormones. The endometrium is a major producer. Prostaglandin synthesis is controlled by changing the release of the enzyme phospholipase from liposomes. In many mammals, the major factor influencing this in the endometrium is an increase in the oestrogen:progesterone ratio.

If progesterone is high relative to oestrogen, prostaglandin synthesis is low. A fall in progesterone or rise in oestrogen increases prostaglandin synthesis. However, in humans, plasma progesterone does not generally fall at parturition. What is more likely to be the case in humans is local interconversion of steroids or expression of progesterone receptor subtypes to favour the activating subtype.

Prostaglandins have important roles in the mechanical events of parturition initiation.

Oxytocin acts by binding to receptors on uterine smooth muscle cells. At the end of pregnancy, myometrial cells become more sensitive to low circulating levels of the hormone.

The onset of labour is therefore associated with increased prostaglandin synthesis and release, stimulating more forceful contractions in conjunction with increased sensitivity to oxytocin. As labour progresses, contractions become more forceful and frequent.

At each contraction, muscle fibres shorten, but do not relax fully, a phenomenon known as **brachystasis**. As a consequence, the uterus shortens progressively as labour proceeds. This pushes the presenting part of the fetus into the birth canal and stretches the cervix over it.

Second stage of labour

The first stage of labour can often be of long duration, typically 8–12 h for a first pregnancy, and it ends when cervical dilatation reaches about 10 cm. The fetus is normally then expelled relatively quickly, typically in under 3 h. The cardinal stages of labour are shown in Figure 26.21.

The descended head flexes (Figure 26.21b) as it meets the pelvic floor, reducing the diameter of presentation. Internal rotation follows (Figure 26.21c). The head is then delivered, and as it emerges it rotates back to its original position and extends (Figure 26.21d and e). The shoulders then rotate, followed by the head (Figure 26.21f), and the shoulders are delivered (Figure 26.21g), followed by the rest of the fetus (Figure 26.21h). The second stage of labour ends with delivery of the fetus.

Third stage of labour

With the fetus removed, there is a powerful uterine contraction that separates the placenta from the uterus. The placenta and membranes are then expelled, normally within about 10 minutes. This completes the third stage of labour. This contraction of the uterus also compresses blood vessels and reduces bleeding through the formation of a so-called living ligature, formed as a result of the criss-cross arrangement of myometrial fibres.

Lactation

Mammary tissue (the tissue responsible for the production of milk) in humans is embedded in the breasts. Each lobe of the gland is made up of lobules of alveoli, blood vessels, and the lactiferous ducts that open at the nipple. The alveoli are the site of milk synthesis. Myoepithelial cells, which are smooth muscle cells responsible for milk let down, surround alveoli.

At puberty, breast development begins in the adolescent female, as oestrogens cause the ducts to sprout and branch and begin the first steps in the development of the alveoli.

With each menstrual cycle in the non-pregnant female, there are changes in the breast that occur under the influence of oestrogen and progesterone, produced by the ovary. Nevertheless, most development of mammary tissue in preparation for lactation occurs during pregnancy.

Rising titres of oestrogen and progesterone in early pregnancy stimulate a considerable hypertrophy of the ductular–lobular–alveolar system of the breast. From mid-pregnancy, cells are capable of milk secretion and contain substantial amounts of secretory material. However, milk is not secreted in significant quantities until after birth.

Milk is secreted in significant quantities from soon after birth. The composition varies with time and is summarized in Table 26.5.

In the first week, a protein-rich milk that is high in fat-soluble vitamins and immunoglobulins is produced. The composition serves to provide optimal nutrition to the baby and also to prime its system with a further complement of maternal antibodies to supplement those transferred across the placenta. This **colostrum** ensures that the vulnerable newborn gets tailor-made support in its first few days of life.

Over the next 2–3 weeks, immunoglobulin and total protein levels decline, while fat and sugar levels rise, to produce mature milk.

Control of milk secretion

During pregnancy a high progesterone:oestrogen ratio favours development of alveoli, but not secretion. After birth, changes in plasma ovarian steroid hormone concentrations lead to breast alveolar cells becoming responsive to prolactin, produced by the anterior pituitary (see Table 26.1). Stimulation of the nipple during suckling by the baby results in the relay of sensory information to the mother's hypothalamus. This results in a decrease in prolactin inhibiting hormone (dopamine) by the hypothalamus. This removes the tonic inhibition on the lactotrophs of the anterior pituitary, allowing release of prolactin. Therefore, suckling during one feed ensures production of milk for the next.

A second reflex promotes milk ejection from the breast, the 'let down' reflex. Stimulation of the nipple during suckling by the baby results in the relay of sensory information to the hypothalamus. This leads to a dramatic increase in the secretion of oxytocin from the

Table 26.5 Composition of human mature milk

Constituent	Percentage
Water	90
Lactose	7
Fat	2
Proteins	< 1
Minerals: Ca^{2+}, Fe, Mg, K, Na, P, S	< 1
Vitamins A, B, B_2, C, D, E, K	< 1
pH	7.0
Energy value	60–75 kcal/100 mL

posterior pituitary gland. This contracts myoepithelial cells, which eject the milk. Maintenance of lactation depends on regular suckling to promote prolactin secretion, and to remove accumulated milk.

 Check your understanding of the concepts covered in this section by answering the questions in the e-book.

SUMMARY OF KEY CONCEPTS

- Gamete production in males and females results from sequences of meiotic and mitotic divisions. The meiotic divisions result in the production of haploid gametes.

- In females, at birth the ovaries contain a population of primary oocytes, arrested part way through meiosis. From puberty through to the menopause, these will generate the follicles that develop each month to be released through ovulation.

- In males, the germ cells do not begin meiosis until puberty is reached, but from puberty onwards the testes produce millions of sperm every day.

- In both males and females, the activities in the gonads are regulated by the hormones released by the hypothalamus and anterior pituitary (the hypothalamic–pituitary axis).

- The key regulatory hormones in males and females are gonadotrophin releasing hormone (GnRH), from the hypothalamus, and follicle stimulating hormone (FSH) and luteinizing hormone (LH) from the anterior pituitary.

- The monthly cycle in the mature female involves both the maturation and release of the oocyte, and the preparation of the uterus and Fallopian tubes to facilitate fertilization and the implantation of the fertilized zygote.

- At fertilization, the fusion of the haploid sperm and oocyte result in the generation of a diploid zygote.

- The cells of the zygote start to divide and interact with the endometrium of the uterus to generate the placenta, which establishes the connection between the maternal and developing fetal blood circulations.

- Following birth, there are rapid changes in the fetal circulation, in particular in the heart and lungs, associated with independent living.

- The milk produced by the mother is not only nutrient rich, but also contains maternal antibodies that support the newborn's immune system.

 Use the flashcards in the e-book to test your recall of key terms introduced in this chapter.

QUESTIONS

 Looking for answers? Once you've answered these questions, follow the link in the e-book to the answer guidance and check your work.

Concepts and definitions

1. What is the process whereby the diploid germ cells divide to give rise to haploid gametes?

2. What are the main differences between the timings of gamete production in males and females?

3. What are the two parallel cycles occurring in the female? Briefly summarize the events of these two cycles.

4. Describe the endocrine changes occurring during the menopause.

5. What key changes take place in the sperm during their passage through the uterus to the Fallopian tube?

Apply the concepts

6. What features of meiosis contribute to genetic variation?

7. Describe and compare the endocrine operation of the hypothalamic–pituitary–testis axis and the hypothalamic–pituitary–ovary axis.

8. Review Figure 26.13 and Table 26.3. Describe in your own words the sequence of hormonal changes in relation to the ovulatory cycle.

9. Compare the processes of puberty, and their endocrine control, in the male and female.

10. How does the fetal blood circulation differ from that of the adult?

Beyond the concepts

11. Compare and contrast the endocrine regulation of gamete production in males and females.

12. Describe how the hypothalamic–pituitary–ovary axis acts to synchronize follicular development and the cyclical changes in the uterine endometrium.

13. Compare the circulatory system of the fetus with that of the adult. What changes occur at birth to facilitate independent living?

14. What endocrine processes enable maintenance of the fetus?

15. Compare and contrast the main methods of contraception. Why might there be a rationale for combining both condoms and the pill or IUDs?

FURTHER READING

Hayssen, V. (2020) Misconceptions about conception and other fallacies: historical bias in reproductive biology. *Integr. Comp. Biol.* **60**: 683–91.
This is a thought-provoking article that provides different perspectives on consideration of reproductive physiology.

Herbison, A. E. (2016) Control of puberty onset and fertility by gonadotropin-releasing hormone neurons. *Nat. Rev. Endocrinol.* **12**: 453–66.
A very readable account of the regulation of puberty, providing some more in-depth detail.

Johnson, M. (2018) *Essential Reproduction.* 8th edition. Wiley-Blackwell.
This is an undergraduate text that draws together all aspects of the biology of mammalian reproduction, including a wide range of clinical aspects.

Maybin, J. A. & Critchley, H. O. D. (2015) Menstrual physiology: implications for endometrial pathology and beyond. *Hum. Reprod. Update* **21**: 748–61.
This is a relatively short article that provides a good overview of menstrual physiology and pathophysiology. Although quite focused, it is very accessible. There is reflection on why menstruation occurs in humans as one of relatively few species.

Sanchez-Garrido, M. A & Tena-Sempere, M. (2013) Metabolic control of puberty: roles of leptin and kisspeptins. *Horm. Behav.* **64**: 187–94.

Schatten, H. (ed.) (2017) *Human Reproduction: Updates and New Horizons.* Wiley.
This is a more advanced book with chapters written by specialists. It is very useful for finding more in-depth details of reproductive physiology. There is also a focus on in vitro fertilization (IVF) and assistive reproductive technologies (ART).

White, Y. A. R., Dori, C., Woods, D. C., Takai, Y., Seki, O. I. H., & Tilly, J. L. (2012) Oocyte formation by mitotically active germ cells purified from ovaries of reproductive-age women. *Nat. Med.* **18**: 413–21.

White, B. A., Harrison, J. R., & Mehlmann, L. (2019) *Endocrine and Reproductive Physiology.* 5th edition, Mosby Physiology Series. Elsevier.
This text provides an advanced undergraduate view of endocrine physiology, although there is a specific focus on reproductive endocrinology.

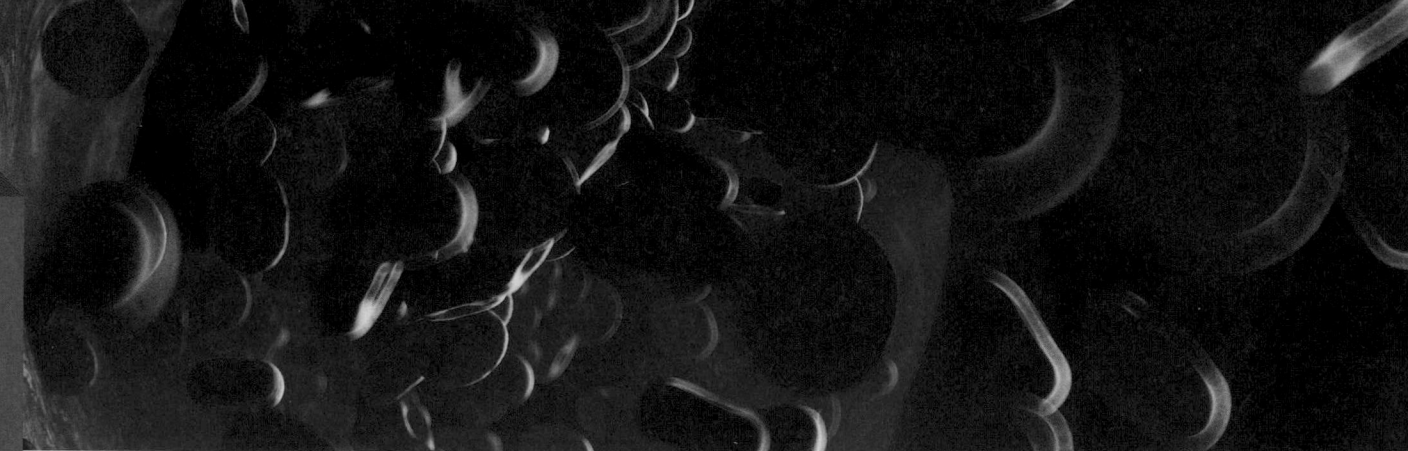

Immune System

Chapter contents

Introduction		862
27.1	Cells and organs of the immune system	863
27.2	The innate immune response	869
27.3	The adaptive immune response	876
27.4	Antigen recognition by lymphocyte receptors	885
27.5	Immunodeficiencies: diseases affecting how lymphocytes develop or function	895

🌀 Watch the key concepts video in the e-book to prepare yourself for studying this chapter.

LEARNING OBJECTIVES

By the end of this chapter you should be able to:

- Describe the cells and organs of the immune system and their main functions.
- Describe the features of an innate immune response to a pathogen.
- Describe the features of an adaptive immune response to a pathogen, and how this is followed by long-lasting immunity to the same pathogen.
- Explain the principles of vaccination and the stages of vaccine development.
- Describe how immune responses can cause allergy, transplant rejection, and autoimmune diseases.

Introduction

The immune system is designed to recognize and destroy anything that is dangerous to the body. This includes pathogens—those agents including viruses, bacteria, and parasites that can infect you and cause disease—and tumour cells, which lead to cancer. Your immune system extends throughout your whole body. It faces outward towards the barriers—skin and mucosal surfaces—that exist between you and your external environment and it is compartmentalized to allow the efficient production, maturation, and activation of key cells and mediators. These entities trigger pathways that ultimately destroy pathogens and cancer cells. In the absence of a functional immune system, individuals quickly become ill and, without the intervention of modern medicine, would die from an infection in a very short time.

▶ We learn more about pathogens in Chapter 14.

Given its complexity, it is not surprising that the immune system can go wrong from time to time, causing disorders such as allergies, in which the immune system reacts against harmless material in the environment, such as pollen or some foods. Other types of immune disorders are autoimmune diseases, such as rheumatoid arthritis, in which the immune system turns against a component of its own body. In the developed world, where many infectious diseases are controlled either by immunization or by antibiotics, allergies and autoimmune diseases are now a major health and economic burden, having an adverse impact on people's quality of life. Therefore, understanding how the immune system works is crucial to improving human health: whether to combat the infectious diseases that still kill millions of people in less-developed countries, or to find ways to prevent and treat the increasing prevalence of allergy and autoimmune disease in the developed world.

In this chapter, we explore the components of the immune system—its cells and organs—and how those components work together, both to recognize pathogens and to eliminate them from our bodies before they can cause serious harm. As we will see, the immune system comprises myriad cells and signalling molecules, which come together to form an intricate, sophisticated network. These myriad elements make it quite challenging to develop a big picture view of the immune system. (As you read through this chapter you will see individual components appearing several times in different contexts.) Do not let the detail overwhelm you, though: as you read through, just keep the big picture in mind.

27.1 Cells and organs of the immune system

The immune system is in a constant state of surveillance, keeping watch for threats that might compromise the body's health. The specific target of the immune system, the 'trigger' it is watching out for, and that it subsequently recognizes and responds to, is often referred to as an **antigen**. An antigen can be a pathogenic microorganism (bacterium, virus, or fungus), a toxin or other macromolecule, a vaccine, or even the body's own damaged or cancerous cells.

The immune system can be divided into two main functional parts: the **innate immune system**, and the **adaptive immune system**.

The innate immune system is present from birth. It includes the physical, protective barriers, such as the skin and the epithelial linings of the digestive tract, respiratory tract, and other internal surfaces that are exposed to microbes and other material from the environment. The rest of the innate immune system is composed of cells and molecules that can immediately, or very rapidly, detect an invading micro-organism, by recognizing features that are shared by various groups of pathogens but are absent in mammals, such as the components of bacterial cell walls. The innate immune system then responds by deploying pre-existing cells and molecules that attack the invaders.

By contrast, the adaptive immune system mounts highly individual responses to specific targets that it perceives as foreign, such as the particular type of pathogen causing an infection. For example, the innate immune system will recognize that a virus has infected the body and it will deploy a standard set of defences against the virus. It will not remember which virus it was, and if the same or another virus subsequently infects, the same set of standard defences come into play. However, the adaptive immune system recognizes the particular infecting virus, for example, measles, and makes a measles-specific response against it, which eventually clears the virus.

While we discover in Chapter 21 how the red blood cell is central to the functioning of our respiratory system, exploration of the immune system means entering the world of the white blood cell. The innate and adaptive immune systems each involve different members of the overall white blood cell community, which we will introduce shortly.

Beyond these cells, however, the innate immune system includes the physical, protective barriers, such as the skin and the epithelial linings of the digestive tract, respiratory tract, and other internal surfaces, that have tight junctions and are exposed to microbes and other material from the environment. The key feature of innate immune responses is the *speed* of that response: its cells and mediators are rapidly recruited and activated as the first line of defence. The responses are not specific to a particular organism, however, so a macrophage (a particular type of white blood cell) that is activated by a bacterial infection will be just as good at killing as a macrophage activated by yeast cells.

▶ We learn more about tight junctions in Chapter 28.

Adaptive immune responses depend on the activation of innate responses, which signal 'danger' and enable specific white blood cells (called lymphocytes) to be selected. The selected lymphocytes then proliferate and differentiate to target the pathogen for destruction. The most impressive characteristic following this primary activation is that some lymphocytes persist as memory cells so they can be activated much more quickly when a pathogen is detected again in the future. This faster future response is called the secondary response.

▶ We discuss this future response further in Section 27.3.

Immunological memory provides highly specific, rapidly responding, and often lifelong **immunity** to the specific pathogen, but not to any other pathogen. However, primary adaptive immune responses take at least a week to develop, during which time the pathogen could multiply and cause uncontrollable disease if not held in check by the innate immune system. In addition, the signals from innate responses will influence the quantity and quality of the specific immune response, and ensure that appropriate effectors are generated to eliminate specific pathogens. As such, the innate and adaptive immune systems operate in a tightly integrated way,

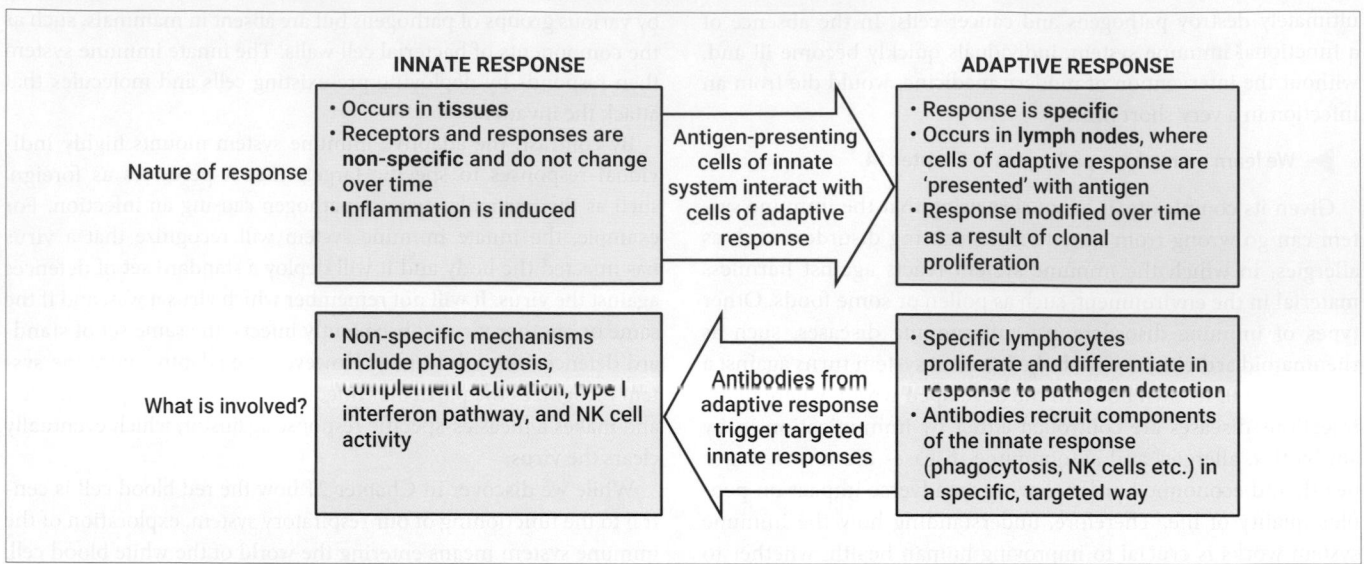

Figure 27.1 The innate and adaptive responses. These responses operate in concert, with the adaptive response recruiting elements of the innate response to mount a successful attack against pathogens. We explore all these points in more detail throughout this chapter.

as depicted in Figure 27.1. The proper operation of both systems is crucial to our physical wellbeing.

The innate immune system is impressive and successfully destroys most harmful micro-organisms we are exposed to. However, pathogens can mutate quickly, and develop virulence factors or immune evasion strategies that can rapidly overcome innate responses. Pathogenic micro-organisms, by definition, are those that cause damage and disease because they cannot be completely removed by the innate immune system. By contrast, complete removal is the strength of the adaptive immune system: it comprises lymphocytes expressing so many different, randomly generated receptors that at least some will be specific to the infecting pathogen—even a completely new virus, like SARS-CoV-2.

▶ **We learn more about virulence factors in Chapter 14.**

Humans cannot mutate as fast as viruses, but we can adapt our immune responses to specifically recognize any variant of a pathogen. This provides the powerful and focused response of adaptive immunity required to ultimately remove pathogens from the body. Babies who lack lymphocytes cannot make adaptive immune responses. They have severe combined immunodeficiency (SCID), will fail to thrive, and will be unable to survive beyond childhood unless treated with life-saving therapies, as we discuss further in Section 27.5.

PAUSE AND THINK

Where are immune responses started and why?

Answer: In tissues because this is where pathogens enter.

Cells of the immune system

The immune system comprises white blood cells or **leukocytes**, which originate in the bone marrow. There are different types of leukocyte, as shown in Figure 27.2, each of which has its own roles in immunity. These different leukocytes interact with each other, either directly or through the secretion of molecular signals called **cytokines**.

▶ **We learn more about cytokines in Section 27.2.**

The cells of the innate and adaptive immune systems originate from a common precursor in the bone marrow called haematopoietic stem cells, as illustrated in Figure 27.2. This cell type gives rise to red blood cells (see Section 21.1), as well as to the two main classes of leukocytes: the **myeloid** and the **lymphoid** lineages.

We see in Figure 27.2 how the innate immune system comprises all white blood cells *other than* lymphocytes. These cells express many **receptors** that can recognize general 'patterns' associated with pathogens, or 'danger' signals. These signals include components of the extracellular matrix (for example, heparan sulphate and fibrinogen), stress response proteins (for example, heat shock proteins), and proteins that modulate immunomodulatory proteins (for example, defensins and surfactant proteins A and D). These receptors have evolved over millions of years, and can be observed in many eukaryotic organisms, because they have provided a survival benefit to the host. You can imagine that these receptors have evolved to detect the 'enemy' in a broad sense.

By contrast, the adaptive immune system is only mediated by **lymphocytes** (T and B cells). The receptors on these cells are highly variable, making each receptor exquisitely selective for specific 'targets'. This specificity (and variability) is a consequence of the

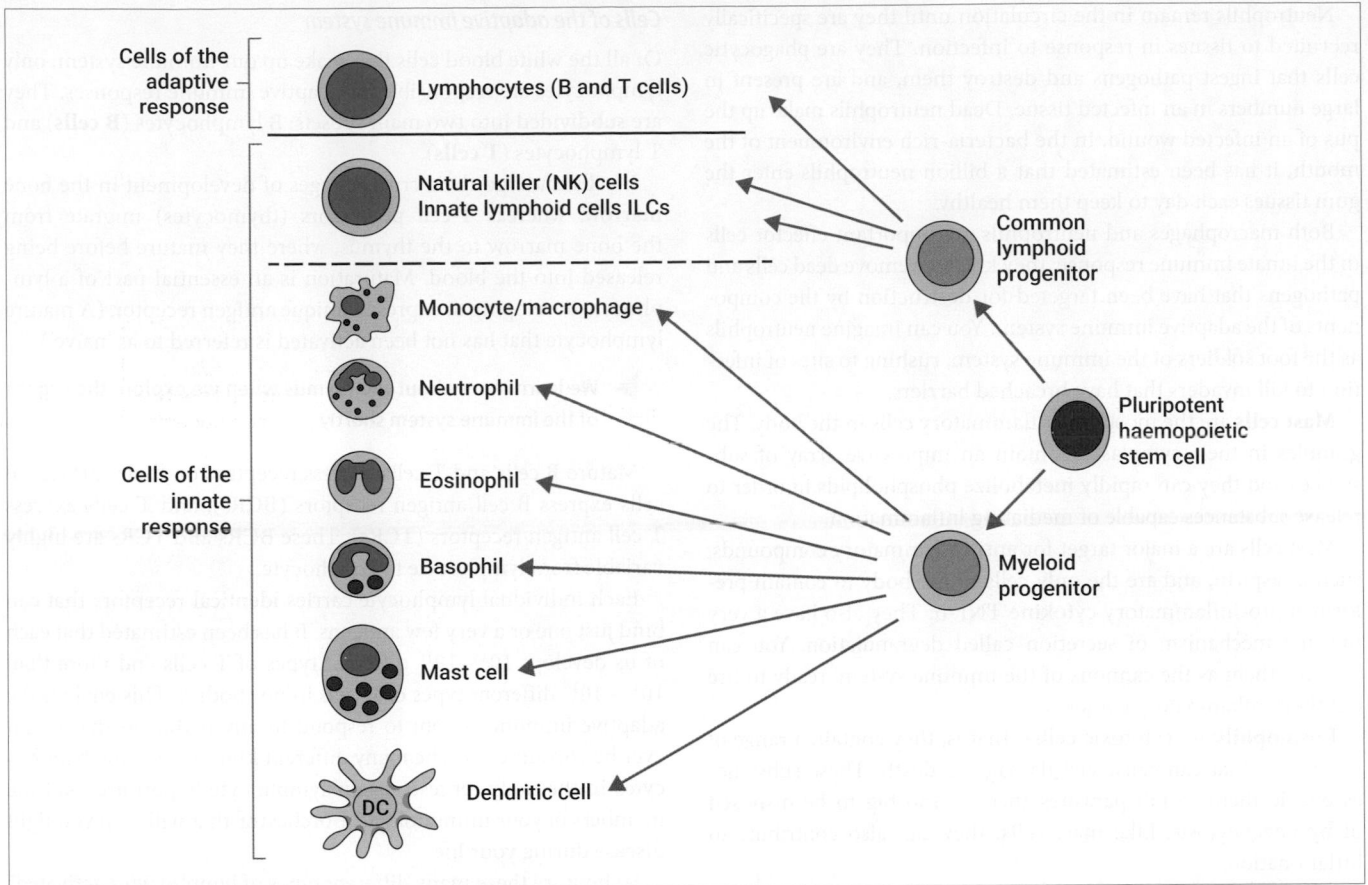

Figure 27.2 Cells of the immune system. White blood cells (leukocytes) are made in the bone marrow from myeloid or lymphoid progenitor cells (precursors). The majority are part of the innate system. Only lymphocytes (T and B cells) are responsible for adaptive immune responses.
Source: Cunningham A.C. (2016) *Thrive in Immunology*. Oxford University Press.

genes encoding these receptors being recombined during their development, as we discuss further in Section 27.3.

Let us now consider the cells of the innate and adaptive immune systems in a little more detail.

Cells of the innate immune system

The cells of the innate immune system arise from two cell lines: the myeloid and lymphoid cell lines. Look at Figure 27.2 and notice how the myeloid progenitor cell line gives rise to the following cell types:

- macrophages;
- neutrophils;
- dendritic cells;
- mast cells;
- basophils;
- eosinophils.

By contrast, the lymphoid lineage contributes the natural killer (NK) cells and innate lymphoid cells (ILCs).

So what role does each of these cell types play in the operation of the innate immune response?

Macrophages circulate in the blood as immature **monocytes** and undergo their final differentiation when they enter tissues, where they remain resident as long-lived cells. They are named according to their location—for example, macrophages can be called osteoclasts in the bone, microglial cells in the brain, and Kupffer cells in the liver. They are phagocytic cells that act as general scavengers in the body, ingesting and degrading dead and dying cells as well as micro-organisms.

▶ We learn more about phagocytosis in Section 27.2.

Macrophages also have a more specialized role in the immune system as they carry a battery of pattern recognition receptors on their surface that are able to detect the presence of pathogens and to initiate innate immune responses. You can imagine them as waste disposal units: they clear pathogens and dead cells, but they also stimulate inflammation by producing pro-inflammatory cytokines.

Neutrophils are short-lived cells and are the most abundant leukocyte in blood: one millilitre of blood contains between two and eight million neutrophils, each with a lifespan of less than six days. By contrast, there are around 150,000–600,000 monocytes per millilitre.

Neutrophils remain in the circulation until they are specifically recruited to tissues in response to infection. They are phagocytic cells that ingest pathogens and destroy them, and are present in large numbers in an infected tissue. Dead neutrophils make up the pus of an infected wound. In the bacteria-rich environment of the mouth, it has been estimated that a billion neutrophils enter the gum tissues each day to keep them healthy.

Both macrophages and neutrophils are important effector cells in the innate immune response: they kill and remove dead cells and pathogens that have been targeted for destruction by the components of the adaptive immune system. You can imagine neutrophils as the foot soldiers of the immune system, rushing to sites of infection to kill invaders that have breached barriers.

Mast cells are the most pro-inflammatory cells in the body. The granules in their cytoplasm contain an impressive array of substances and they can rapidly metabolize phospholipids in order to release substances capable of mediating inflammation.

Mast cells are a major target for anti-inflammatory compounds, such as aspirin, and are the only cells in the body to contain pre-formed pro-inflammatory cytokine TNF-α. They also have a very unusual mechanism of secretion called degranulation. You can imagine them as the cannons of the immune system, ready to fire out their inflammatory contents.

Eosinophils are cytotoxic cells—that is, they contain a range of substances that can cause cell damage or death. These substances enable them to kill parasites that are too big to be disposed of by phagocytosis. Like mast cells, they can also contribute to inflammation.

Natural killer (NK) cells are derived from the same lineage as the lymphocytes of the adaptive immune response but are distinct from them. NK cells are a type of innate 'killer cell' that can recognize virally infected or transformed body cells and kill them by inducing apoptosis (programmed cell death). They also produce a substance called interferon gamma (IFN-γ), which can be detected by macrophages and activate them to tidy up and eat the dead cells they have killed.

The **dendritic cell** (DC) is a myeloid cell that provides a bridge between the innate and adaptive immune responses. It is so called because of its long finger-like projections, which look like the dendrites of the neurons we encounter in Chapter 17. Dendritic cells reside in the tissues and, like macrophages, carry receptors on their surface that can detect pathogens. In particular, they are found in skin and mucosal surfaces where their dendritic projections can fit between epithelial cells and extend into the lumen of the lung and gastrointestinal tract, sampling those spaces to see if pathogens are present. In the skin, they are called Langerhans cells.

DCs are immature in tissue, but have an impressive ability to internalize material (via both pinocytosis—a type of endocytosis that sees molecules in the extracellular fluid being brought into the cell—and phagocytosis). They move to lymph nodes during the course of an immune response and mature there. Once mature, they are no longer capable of internalizing material, but can now activate T lymphocytes in lymph nodes. We learn more about the role of DCs in initiating an adaptive immune response—when they function as **antigen presenting cells**—in the next section.

▶ We learn more about endocytosis in Chapter 16.

Cells of the adaptive immune system

Of all the white blood cells that make up our immune system, only lymphocytes are responsible for adaptive immune responses. They are subdivided into two main subsets: B lymphocytes (**B cells**) and T lymphocytes (**T cells**).

B cells undergo their crucial stages of development in the bone marrow, whereas T-cell precursors (thymocytes) migrate from the bone marrow to the thymus, where they mature before being released into the blood. Maturation is an essential part of a lymphocyte developing to express a unique antigen receptor. (A mature lymphocyte that has not been activated is referred to as 'naïve'.)

▶ We learn more about the thymus when we explore the organs of the immune system shortly.

Mature B cells and T cells express receptors on their surfaces: B cells express B cell antigen receptors (BCRs), and T cells express T cell antigen receptors (TCRs). These BCRs and TCRs are highly variable from lymphocyte to lymphocyte.

Each individual lymphocyte carries identical receptors that can bind just one or a very few antigens. It has been estimated that each of us develops 10^{14}–10^{16} different types of T cells and more than 10^{12}–10^{13} different types of B cells in our bodies. This enables the adaptive immune system to respond to any pathogen that might ever be encountered. The many different kinds of specific lymphocytes in the body represent your lymphocyte 'repertoire'—all the members of your immunological 'orchestra' that will help you fight disease during your life.

So how are these many different types of lymphocytes activated? Lymphocyts are activated when the antigen receptor (TCR) that they express binds to a fragment of a pathogen (called an 'antigen' or 'immunogen') that has been presented to them. Following exposure to the right set of signals, they then differentiate into a variety of **effector T cells**, which all have specialized roles in the adaptive immune response.

We have just noted that lymphocytes are activated when a fragment of pathogen is *presented to them*—and this phrasing is important. T cells cannot respond *directly* to pathogens, but have to be activated by specialized antigen presenting cells (APCs). APCs essentially thrust pathogen fragments 'under the noses' of T cells: it is only when T cells encounter the pathogen they are sensitive to in this context that they can be activated. Dendritic cells are the most potent APCs and the only cell type that can activate a naïve T cell. Memory T cells are easier to activate and are responsible for our fast, secondary immune response. (Pause for a moment to note how we are already seeing an important interplay between the cells of innate and adaptive immune responses: DCs form part of the non-specific innate response, yet their activity as APCs is crucial for the proper function of T cells.)

▶ We discuss antigen recognition by lymphocytes in more detail in Section 27.4.

After they encounter their specific antigen, B cells differentiate into plasma cells, which secrete the **antibodies** of the adaptive immune response, as illustrated in Figure 27.3. Antibodies that are present in the intercellular fluid bind to pathogens or other antigens and target them for destruction by phagocytic cells and other effector mechanisms that are a feature of the innate immune

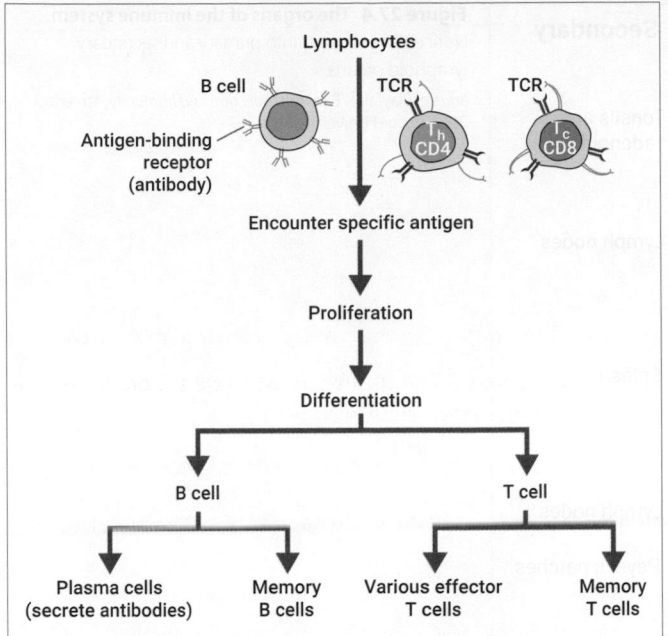

Figure 27.3 Activated lymphocytes differentiate into effector cells.
Lymphocytes express highly variable receptors—B cells express BCRs, while T cells express TCRs—which are selected by binding specific antigens (i.e. molecules derived from a pathogen). At the end of the immune response, memory B and T cells persist.

Source: Goldsby et al. *Kuby Immunology*, 4th edn.

response. (Notice, again, how the adaptive and immune responses act in concert: the adaptive response has specificity; it targets specific pathogens for destruction by 'labelling' them with antibodies. The innate response is non-specific: it will act on *any* pathogen that has been tagged for destruction.)

Antibodies are highly variable proteins that are a secreted version of the B cell's antigen receptor (BCR). In this way, the antibodies that are secreted by B cells are those that are a 'match' for the pathogens that originally activated those B cells.

More recently, **innate lymphoid cells** (ILCs) have been recognized as playing an important role in immune responses. They do not express lymphocyte receptors (TCRs) but they can produce a range of cytokines that resemble a subset of T cells, called helper T cells (whose activity we discuss further in Section 27.4). They are found in our barrier tissues (mucosal surfaces, skin) and lymphoid organs, and are classified according to the cytokines they produce. For example, ILC1 cells produce interferon gamma (IFN-γ)/tumour necrosis factor alpha (TNF-α); ILC2 cells produce interleukin (IL)-5 and IL-13; and ILC3 cells produce IL-17 and IL-22.

PAUSE AND THINK

1. Which cell types can *internalize* pathogens by phagocytosis?

   Answer: Macrophages, neutrophils, and dendritic cells.

2. What is the only white blood cell capable of adaptive immune responses?

   Answer: Lymphocyte.

Organs of the immune system

The immune system is compartmentalized. Leukocytes are made and mature in the primary lymphoid organs, which include the bone marrow and the thymus. Secondary lymphoid organs include lymph nodes, the spleen, and specialized structures found in the mucosa. These organs and their locations in the human body are illustrated in Figure 27.4.

The secondary lymphoid organs can be considered to be 'antigen capture units' and bring potential pathogens from the tissue into close contact with lymphocytes. This is important because there are so many different types of lymphocyte receptor (both TCR and BCR) that they must be concentrated in order to increase the likelihood of activating a useful lymphocyte with the right, specific receptor for the ability to fight any particular pathogen to be maximized.

The thymus is a unique primary lymphoid organ and is essential for the development of mature T cells. Patients who lack a thymus (e.g. as part of DiGeorge syndrome) are immunodeficient and unable to mount adaptive immune responses.

Secondary lymphoid tissues are important because they provide the locations in which adaptive immune responses are stimulated.

One secondary lymphoid organ is the **spleen**, a fist-sized organ located on the left side of the body below the stomach. Its function is to 'filter' the blood and mount adaptive immune responses to pathogens present in it.

The other major secondary immune tissues are the **lymph nodes** and the **mucosa-associated lymphoid tissues**. Lymph nodes are found throughout your body: you have over 500 linked via a lymphatic system. They show the generalized structure shown in Figure 27.5. Look at this figure and notice how a lymph node has a central medulla and outer cortex. Afferent lymphatics drain from tissue and efferent lymphatics enable traffic to the next lymph node. The lymph nodes represent the place in which an adaptive immune response starts: they capture pathogen from tissue and bring it into close contact with lymphocytes.

The mucosa-associated lymphoid tissues underlie the mucosal epithelia of the gut and the respiratory tract. The largest compartment of the mucosa-associated lymphoid tissues is the **gut-associated lymphoid tissue** (GALT), which contains small lymphoid organs called Peyer's patches, as well as more diffuse aggregations of lymphoid cells.

Lymph nodes and mucosa-associated lymphoid tissues are similar in internal organization, except for the way they receive antigens/immunogens derived from pathogens. Material enters the lymph nodes by way of the **lymphatic system**, which transports lymph, a straw-coloured fluid that drains from the tissues, and returns it to the blood (see Section 21.1). Lymphatic vessels, called **afferent lymphatics**, transport lymph from tissues (including the skin) to the lymph nodes. For example, if the skin is breached, microbial antigens will be received in the local lymph node via the afferent lymphatics.

In contrast, the GALT receives antigens directly from the gut through specialized epithelial cells called M cells, which overlay the Peyer's patches. M cells have the ability to absorb antigens derived from pathogens present in the gut lumen into the lymphoid tissue below.

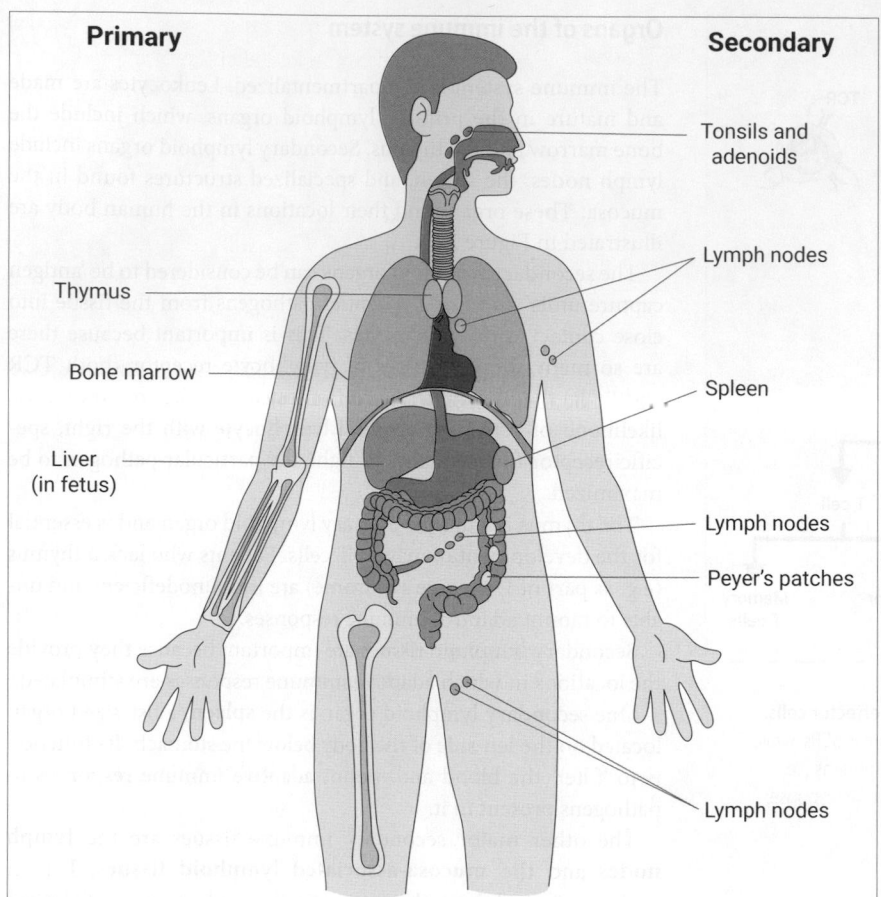

Primary

Secondary

- Tonsils and adenoids
- Lymph nodes
- Thymus
- Bone marrow
- Spleen
- Liver (in fetus)
- Lymph nodes
- Peyer's patches
- Lymph nodes

Figure 27.4 The organs of the immune system. Notice the grouping into primary and secondary lymphoid organs.

Source: Playfair & Bancroft. *Infection and Immunity*, 4th edn, 2013. Oxford University Press.

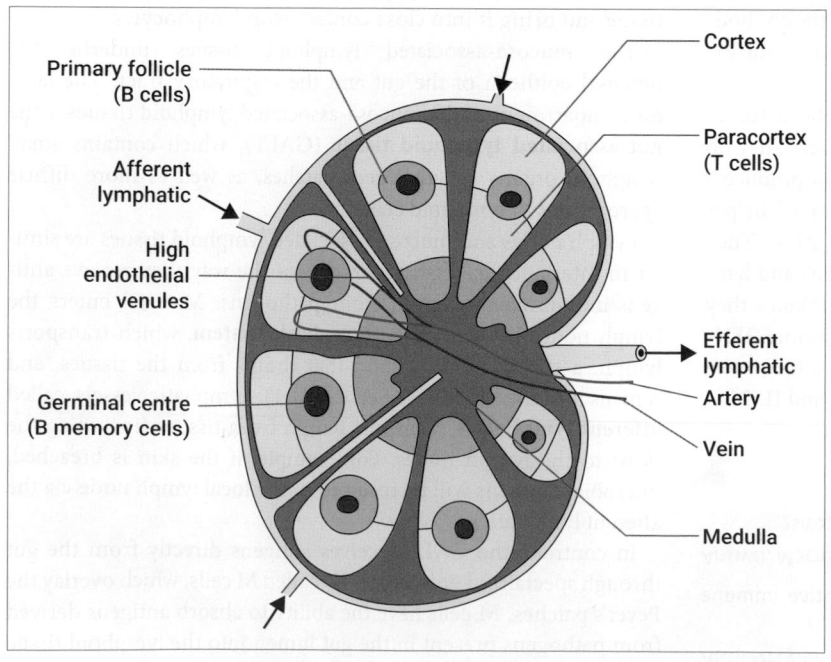

- Primary follicle (B cells)
- Afferent lymphatic
- High endothelial venules
- Germinal centre (B memory cells)
- Cortex
- Paracortex (T cells)
- Efferent lymphatic
- Artery
- Vein
- Medulla

Figure 27.5 Lymph nodes. The generalized structure of a lymph node, showing how it is connected to the rest of the lymphatic system via afferent and efferent lymphatic vessels.

Source: Playfair & Bancroft. *Infection and Immunity*, 4th edn, 2013. Oxford University Press.

A second category of lymphatics, the **efferent lymphatics**, receive cells and fluids from the lymphoid tissues, except for the spleen, which is connected to the blood system, and deliver them back to the blood system at the left side of the neck via the thoracic duct.

Naïve T cells continually circulate between the blood, secondary lymphoid tissues, the lymph, and back to the blood, looking to see if their receptors (TCRs) can bind to any pathogen-derived peptides presented by APCs (particularly dendritic cells) in the lymph node (see Figures 27.22 and 27.23 for more detail). By contrast, B cells tend to stay in follicles.

Lymphocytes enter the secondary lymphoid organs from the blood because there are specialized structures called high endothelial venules that lymphocytes will stick to. T cells will stay in the lymph node for about 18 hours, testing whether their specific receptor can bind to any APCs. If they are not activated (i.e. if they remain naïve), they will leave via the efferent lymphatics, which transports them to other lymph nodes, and eventually back to the blood to continue their circulation.

Activated and memory lymphocytes have different recirculation pathways, tend to stay close to the sites in the body where they were activated, and will therefore remain to do the important work of immunosurveillance.

PAUSE AND THINK

Why do lymph nodes swell?

Answer: When a pathogen or microbial antigen is delivered to the lymph node by the afferent lymphatic vessel, it will be recognized by the complementary B and T cells within the node. These cells will then rapidly proliferate, causing the node to swell.

 Check your understanding of the concepts covered in this section by answering the questions in the e-book.

27.2 The innate immune response

Having now explored the cells and organs in overview, it is time to consider how the immune system actually operates in practice. We begin by examining how it mounts an innate immune response.

Our first line of defence: protective barriers

Any potential pathogen has to breach our first line of defence—barrier sites that interact with the environment—before it encounters either the innate or adaptive immune system. These sites include our skin, digestive tract (mouth, throat, stomach, and intestines), and respiratory tract.

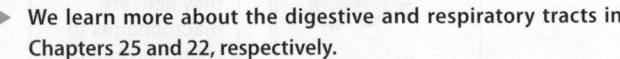 We learn more about the digestive and respiratory tracts in Chapters 25 and 22, respectively.

The skin

The skin is a flexible impermeable barrier. It is multi-layered, reinforced by the protein keratin, and is constantly being renewed. This renewal helps to prevent it being colonized by pathogens over time. Epithelial cells and infiltrating neutrophils can secrete small microbicidal peptides (for example, cathelicidins and β-defensins) in response to infection, in an attempt to suppress the entry of microbes across this barrier.

The respiratory tract

The lungs facilitate the exchange of oxygen and carbon dioxide between internal tissues and the surrounding environment (see Section 22.1). Goblet cells can secrete mucus, a mix of glycoproteins, which contain antibacterial molecules (e.g. lysozyme) that coat the surface to prevent pathogens from adhering. It is also lined with small hair-like structures called cilia, which constantly beat upwards in a wave-like motion, moving mucus towards the throat. The coughing reflex can then expel the mucus from the respiratory tract.

The gastrointestinal tract

The gastrointestinal tract digests food and absorbs nutrients and water (see Chapter 25). The low pH of the stomach prevents the growth of most micro-organisms, and goblet cells in the small intestine secret mucus that contains antimicrobial enzymes and peptides. The intestinal epithelium also self-renews to inhibit its colonization by pathogens. The surface of the small intestine normally contains finger-like protrusions called villi; their presence increases the surface area of the intestine, optimizing the absorption of food. However, during infection and inflammation, goblet cell numbers increase, and the villi wither away. This flat mucosa facilitates the flushing out of pathogens by peristalsis, which we discuss further later in this section.

All the linings we consider here are formed of an epithelial surface underlaid by connective tissues. Epithelial surfaces prevent pathogens from entering the body in different ways, depending on the structure and function of that surface.

How does the innate immune system recognize micro-organisms?

Once a micro-organism has breached one of our protective barriers, the innate immune system responds within hours. The response is identical each time the pathogen is encountered and there is no immunological memory.

The innate immune system depends on the recognition of molecular components of bacteria and viruses that are different from mammals, providing the immune system with the means of identifying sources of potential danger. (To pause for a moment, there has been long-standing debate about how to describe the way in which our immune system determines what to attack. Traditionally, we have talked in terms of the immune system distinguishing 'non-self' from 'self', such that it targets for attack anything that is not part of 'us'. But this is problematic when you consider that our bodies are home to myriad micro-organisms (in our guts, for example) that we rely on to remain healthy. The identification of 'danger' is another way of thinking about it.)

The molecular shapes unique to microbes that are recognized by the innate immune system are called **pathogen-associated molecular patterns** (PAMPs). PAMPs are recognized by receptors called **pattern recognition receptors** (PRRs), which are widely expressed by all innate immune cells. Our PRRs are 'biased' to the enemy and will recognize the differences in cell wall structure that exist between pathogens and host cells, and the organization of nucleic acids in pathogens. In addition, PRRs can be activated by damage-associated molecular patterns (DAMPs) or 'alarmins', which are triggered by noxious substances, or following certain types of cell death (including necrosis and pyroptosis, a form of cell death associated with inflammation).

▶ We learn more about the structure of the cell walls of microbes in Chapter 9.

The ancient physiological response to infection is local inflammation of the infected tissue, a process that is illustrated in Figure 27.6. Inflammation makes the endothelial lining of blood capillaries 'stickier' (by upregulating adhesion molecules) and more permeable than usual, enabling leukocytes and soluble mediators to infiltrate the infected tissue more easily. Cells in the inflamed tissue will also produce pro-inflammatory cytokines to activate both innate and adaptive immune responses. For example, dendritic cells will drain to lymph nodes more quickly, neutrophils will be rapidly recruited to tissue, and macrophages will be activated so they can kill more effectively.

In addition, some key pro-inflammatory cytokines will induce more widespread ('systemic') inflammation. These include TNF-α, IL-1β, and IL-6, which are also known as 'endogenous pyrogens', as they can increase your body temperature and induce fever. These cytokines can act locally, and also at other sites, including the bone marrow, where they induce the production of more leukocytes, and the liver, where they induce the acute phase response, as depicted

in Figure 27.7. The acute phase response amounts to the rapid production in the liver of a group of serum proteins. Ultimately, this response leads to the upregulation of cells and mediators that can recognize and destroy pathogens and clear infections from tissue. If the inflammatory response is unsuccessful, the tissue will become chronically inflamed; this may contribute to the pathological response.

Macrophages are the sentinels of the innate immune system and express numerous pattern recognition receptors on their surface that enable them to recognize and bind pathogens that have penetrated the protective barriers and entered tissues. Some of these receptors induce ingestion of the pathogen by **phagocytosis** (see later in this section); the ingested pathogen is then destroyed within the macrophage's lysosomes. Pathogen components bound to other types of PRR induce the macrophage to produce and secrete signalling proteins called cytokines, which affect the behaviour of nearby immune cells and non-immune cells, such as epithelial and endothelial cells, by acting at cell-surface cytokine receptors.

The role of cytokines

Cytokines are the intercellular messengers of the immune system, and are produced by many types of immune cells, including lymphocytes. They guide the progress of the immune response, ensuring that particular types of immune cells are activated as needed, and so stimulating or dampening down the response. The key cytokines are listed in Table 27.1.

Another group of small proteins produced by a wide variety of immune and non-immune cells are the subset of cytokines called the **chemokines**. Once secreted, chemokines diffuse to form a concentration gradient, along which cells that bear receptors for

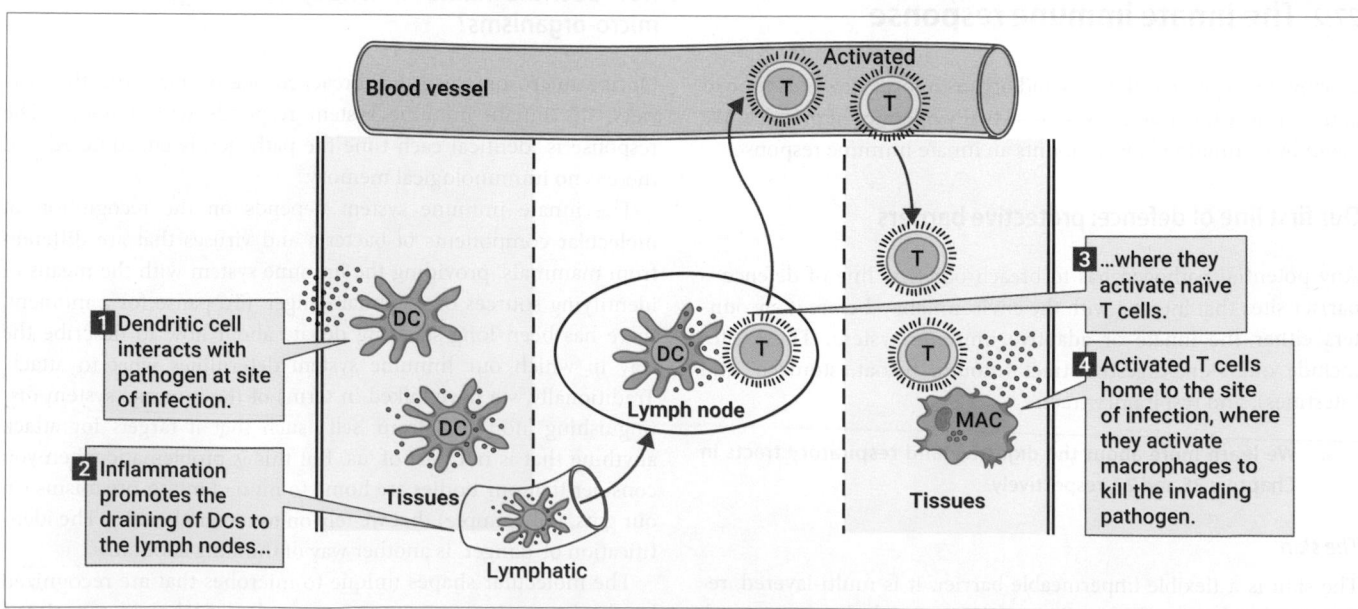

Figure 27.6 Inflammation. Inflammation promotes a range of cellular behaviours that eliminate infection more rapidly.

Source: Playfair & Bancroft. *Infection and Immunity*, 4th edn, 2013. Oxford University Press.

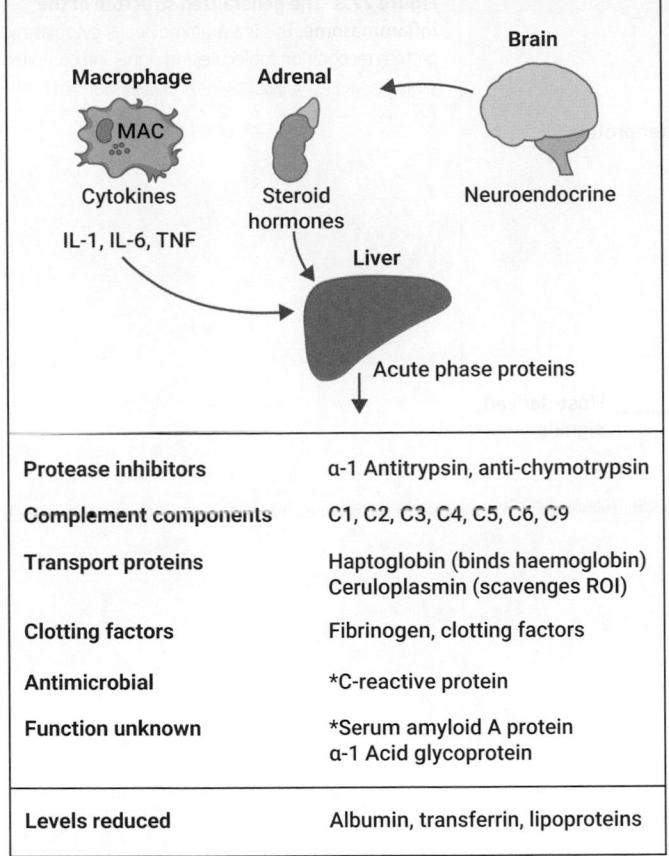

Protease inhibitors	α-1 Antitrypsin, anti-chymotrypsin
Complement components	C1, C2, C3, C4, C5, C6, C9
Transport proteins	Haptoglobin (binds haemoglobin) Ceruloplasmin (scavenges ROI)
Clotting factors	Fibrinogen, clotting factors
Antimicrobial	*C-reactive protein
Function unknown	*Serum amyloid A protein α-1 Acid glycoprotein
Levels reduced	Albumin, transferrin, lipoproteins

Figure 27.7 Systemic inflammation. Systemic inflammation leads to the acute phase response in the liver, which sees the expression of a range of proteins. *Source*: Playfair & Bancroft. *Infection and Immunity*, 4th edn, 2013. Oxford University Press.

that chemokine can migrate in a controlled direction. For example, locally produced chemokine CXCL8 or IL-8, secreted by cells within infected tissues, direct neutrophils to migrate from the blood into the tissue.

A key innate defence against viruses is the production of small proteins called **type I interferons** (IFN-α and IFN-β). These are produced by virally infected cells and help them and neighbouring cells to resist infection. In addition, a 'professional' interferon-producing cell called a plasmacytoid dendritic cell produces large amounts of type I interferons, which spread throughout the body and induce virus-resistance mechanisms in other cells.

How are danger signals passed on?

Some particularly important pattern recognition receptors can sense 'danger' and send signals to the nucleus. These include:

- **Toll-like receptors** (TLRs). These highly conserved transmembrane proteins are present on the cell surface or on intracellular organelles (for example, endosomes), where they encounter and recognize ingested pathogens and their breakdown products.

  TLRs are expressed by many cells of the innate immune system and also by epithelial cells. Ultimately, they lead to the activation of a highly conserved transcription factor, NF-κB. TLR stimulation can lead to the production of antimicrobial peptides, antiviral type I interferons, or cytokines and chemokines that induce inflammation, depending on the cell type and pathogen recognized.

- **Inflammasomes** are multimolecular complexes made of pattern recognition receptors, which assemble in the cytoplasm. They have the generalized structure depicted in Figure 27.8. Inflammasomes can be stimulated by PAMPs or DAMPs and lead to the activation of NF-κB, and the secretion of IL-1β and IL-18. Four key inflammasomes have been characterized to date: NLRP1, NLRP3, NLRC4, and AIM2.

Inflammasomes are expressed by a wide range of cells, including lymphocytes, macrophages, NK cells, and epithelial cells. An inflammasome activates similar signalling pathways as TLRs (for example, NF-κB), resulting in the activation of the cell and cytokine secretion.

Other groups of PRRs can recognize viral RNA and DNA in the cytoplasm. These include RIG-I-like receptors and the so-called 'stimulator of interferon genes' (STING), which detect cytoplasmic viral RNA and DNA, and mediate antiviral immunity by producing type I interferons.

Table 27.1 Key cytokines and their functions

Cytokine	Key functions
IL-1–IL-38	Immune effects
CXCL1–17 (alpha); CC1–28 (beta);	Mediate chemotactic responses (e.g. chemokines)
TNF-α, IL-1β, IL-6	Acute inflammation (local and systemic) Induction of fever and the acute phase response
IFN-γ (type II interferon)	Macrophage activation and local inflammation
IL-10, TGF-β	Anti-inflammatory
IFN-α, IFN-β (type I interferons) IFN-λ (type III interferons)	Anti-viral effects
GM-CSF	Haematological effects (e.g. colony stimulating factors)
TGF-β	Cell biological effects (e.g. growth factors)

Chemokines are named CCL or CXCL, depending on their structure; their receptors are named CCR and CXCR, respectively. IL, interleukin (an older term for some cytokines); IFN, interferon.

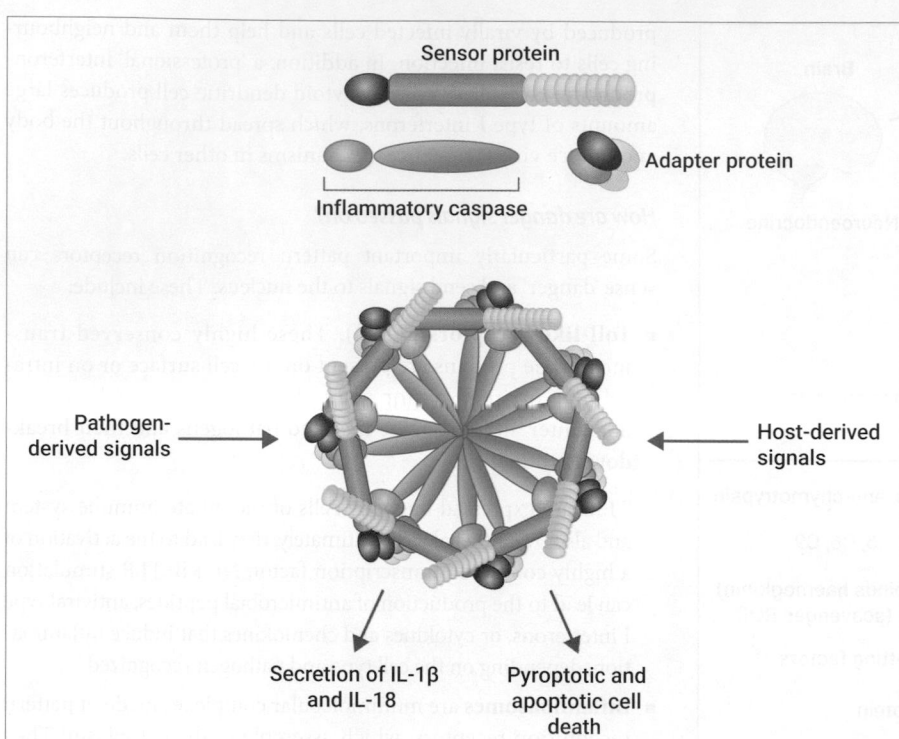

Sensor protein

Adapter protein

Inflammatory caspase

Pathogen-derived signals

Host-derived signals

Secretion of IL-1β and IL-18

Pyroptotic and apoptotic cell death

Figure 27.8 The generalized structure of the inflammasome. This is a multimolecular cytoplasmic pattern recognition molecule and signalling complex. *Source*: Davis et al. *Annual Review of Immunology*, 2011, 29, 707–35.

PAUSE AND THINK

What is the most important cytokine to stimulate an antiviral response?

Answer: Type 1 interferons (IFN), which are also known as IFN-α and IFN-β. A third type of interferon (type III) can also exert this function. Do not confuse with IFN-γ (type II interferon), which is produced by effector cells (e.g. NK cells and cytotoxic T lymphocytes). This interferon activates macrophages and promotes local inflammation.

The complement system

Complement is a group of more than 30 soluble proteins that are produced by the liver and circulate in the blood. These proteins form a cascade, which activate each other in sequence (much like a row of falling dominos). Complement is a major effector mechanism in destroying extracellular pathogens, and in promoting inflammation and phagocytosis. Complement can be activated in three ways; the **alternative pathway**, the **lectin pathway**, and the **classical pathway**. This activation is often referred to as the process of complement fixation. An overview of the complement system showing these three activation pathways is illustrated in Figure 27.9. Look at this figure, and notice the range of downstream effects of the complement system; these include phagocytosis, cell lysis, and inflammation.

The classical pathway was the first to be discovered but is the last to be activated, as it depends on antibodies generated, following adaptive immune responses (which we discuss in Section 27.3).

The alternative pathway and the lectin pathway are activated during innate immune responses.

There are two functional phases to the activation of the complement cascade. The first is the **amplification phase**, which varies in all three pathways, and converges with the formation of enzymes called C3 convertases on the pathogen surface. This enzyme leads to the deposition of large amounts of C3b on the pathogen cell wall and the release of pro-inflammatory C3a.

The second phase is the **lytic phase,** during which the C3 convertases generate a C5 convertase, which leads to the fixation of the terminal complement proteins (C5b, 6, 7, 8, 9n) into the cell membrane and the formation of a pore called a membrane attack complex. When this complex has formed, lysis of the pathogen may follow.

In addition, microbes coated with C3b are more easily phagocytosed by macrophages and other phagocytes that carry receptors for C3b. This is an advantage when dealing with bacteria with thick polysaccharide capsules, such as the pneumonia-causing *Streptococcus pneumoniae*, which phagocytes cannot ingest on their own.

Three outcomes of complement fixation contribute to the removal of the pathogen: inflammation; phagocytosis of the C3b-coated pathogen; and the formation of the membrane attack complex. Complement proteins C3a, C4a, and C5a will stimulate local inflammation, and will attract macrophages and neutrophils to the site of pathogen invasion. These phagocytes express receptors for C3b, and this process of C3b deposition and subsequent phagocytosis is known as **opsonization**.

The complement cascade must be tightly controlled to prevent excessive activation and damage to 'bystander' tissue. A number of

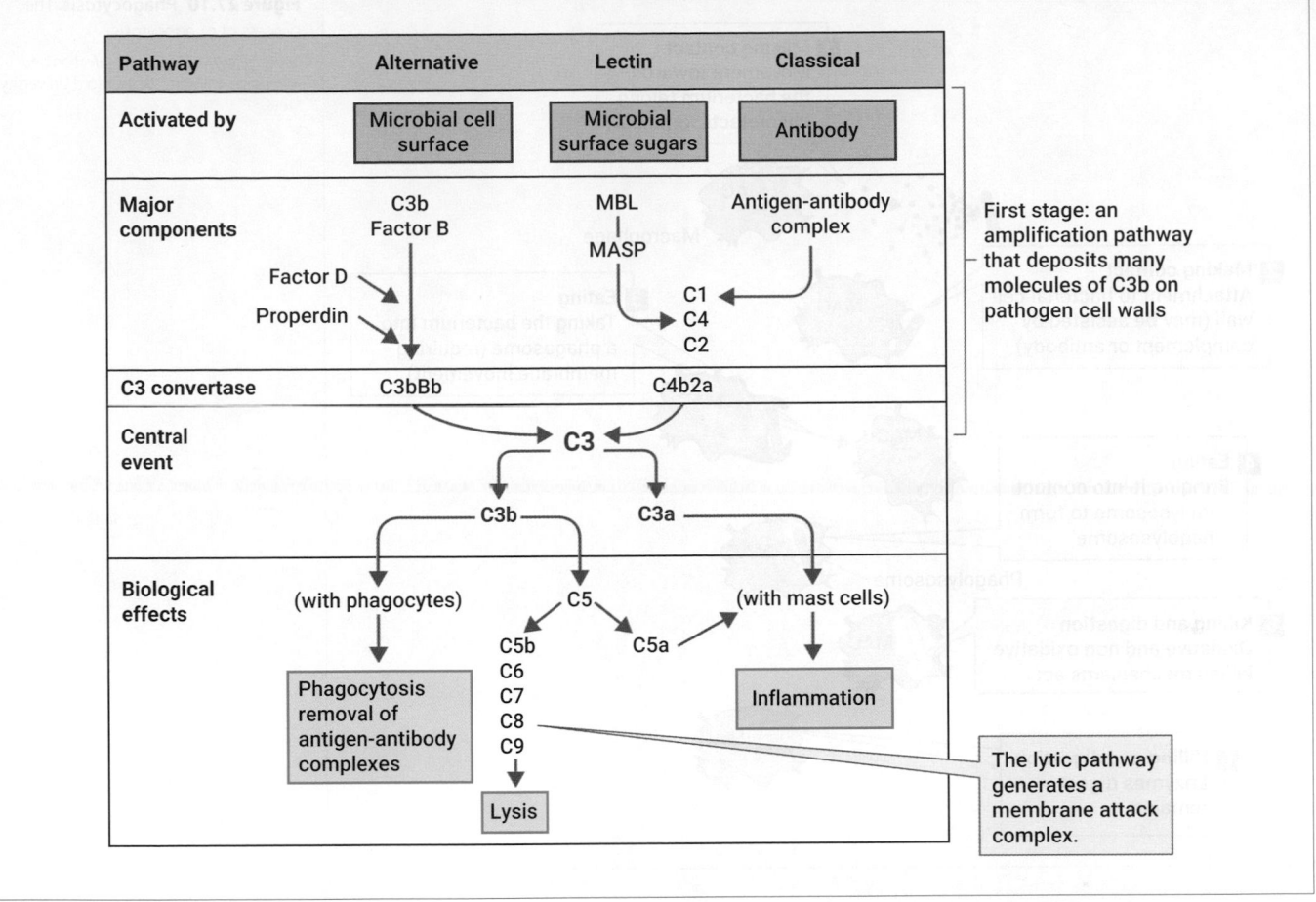

Figure 27.9 The complement system. The complement system can be activated by the innate immune response (alternative and lectin pathways) and the adaptive immune response (classical pathway mediated by antibodies). The names of individual complement components start with a C followed by a number (which reflects the order of their discovery rather than the order in which they act). In addition, the two parts of a complement protein that has been enzymatically cut and activated are given an additional 'a' or 'b' designation.

Source: Playfair & Bancroft. *Infection and Immunity*, 4th edn, 2013. Oxford University Press.

membrane-bound and soluble inhibitors interfere with the amplification and lytic phases of the pathway. For example, protectin (CD59) is a widely expressed membrane-bound protein that can bind to C9 and prevents the construction of the membrane attack complex.

PAUSE AND THINK

1. Which complement protein is central to the activation of the complement cascade?

   Answer: C3: C3b deposition amplifies the response.

2. What are the main effects of complement activation? (Hint: there are three.)

   Answer: Inflammation—stimulated by the release of complement components such as C3a, C4a, and C5a—enhances phagocytosis via opsonization of C3b-coated pathogens, and lysis by formation of a membrane attack complex.

Destruction of pathogens by macrophages and neutrophils

After C3b-coated pathogens have been ingested by phagocytes, they become enclosed in membrane vesicles called phagosomes. Phagosomes then fuse with lysosomes that contain digestive enzymes, forming a **phagolysosome**. The microbes are destroyed by digestion and any residual indigestible material is released into the intercellular spaces. This process of destruction is called phagocytosis; it is summarized in Figure 27.10. Dead or defective host cells are destroyed in the same way. Antimicrobial reactive oxygen species (ROS) and other free radicals are also generated in the phagolysosome, and can also help to kill the bacteria, as illustrated in Figure 27.11.

Neutrophils dominate the early phases of an immune response because they are present in large numbers in the blood and are rapidly attracted into infected tissue by chemokines, such as CXCL8, and complement (e.g. complement components C3a or C5a).

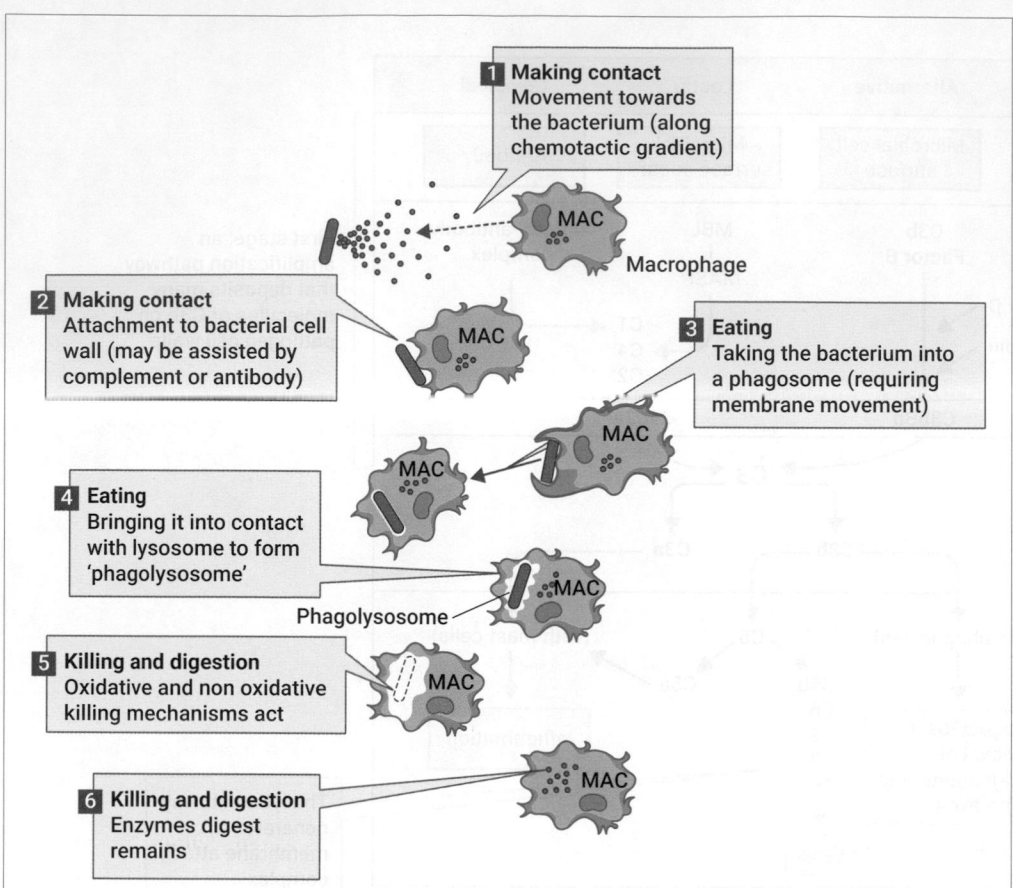

Figure 27.10 Phagocytosis. The process of phagocytosis.

Source: Playfair & Bancroft. *Infection and Immunity*, 4th edn, 2013. Oxford University Press.

1 Making contact
Movement towards the bacterium (along chemotactic gradient)

2 Making contact
Attachment to bacterial cell wall (may be assisted by complement or antibody)

3 Eating
Taking the bacterium into a phagosome (requiring membrane movement)

4 Eating
Bringing it into contact with lysosome to form 'phagolysosome'

Phagolysosome

5 Killing and digestion
Oxidative and non oxidative killing mechanisms act

6 Killing and digestion
Enzymes digest remains

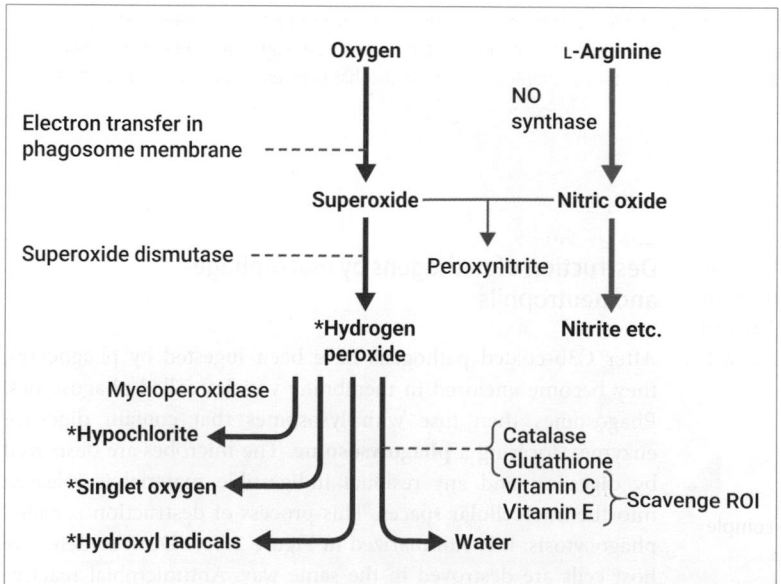

Figure 27.11 Reactive oxygen species. Phagocytosis can generate reactive oxygen species (denoted here with *) that are highly toxic to micro-organisms.

Source: Playfair & Bancroft. *Infection and Immunity*, 4th edn, 2013. Oxford University Press.

Oxygen

L-Arginine

NO synthase

Electron transfer in phagosome membrane

Superoxide ———— Nitric oxide

Superoxide dismutase

Peroxynitrite

*Hydrogen peroxide   Nitrite etc.

Myeloperoxidase

*Hypochlorite

*Singlet oxygen

Catalase
Glutathione
Vitamin C
Vitamin E } Scavenge ROI

*Hydroxyl radicals ——→ Water

Dying neutrophils can extrude their nuclear DNA to form meshworks called neutrophil extracellular traps (NETs), which also contain toxic components from the dying neutrophil. NETs trap and kill pathogens. Macrophages can phagocytose dead cells, including neutrophils.

PAUSE AND THINK

How many different types of phagocytic effectors are there?

Answer: Two main systems: the mononuclear phagocytic system (macrophages) and polymorphonuclear system (neutrophils).

Destruction of viruses by natural killer cells

Complement and phagocytosis can effectively destroy extracellular pathogens, including most bacteria and yeast. However, different strategies are needed to remove intracellular pathogens—that is, viruses. Pattern recognition receptors, particularly in the cytoplasm, bind to viral proteins or nucleic acid. This leads to the production of type I and III interferons: the type I interferons produced are IFN-α and IFN-β, while the type III interferon is IFN-λ. These interferons signal to neighbouring cells to upregulate antiviral defence mechanisms.

The interferons were named because they interfere with viral growth. Do not confuse them with IFN-γ (also called type II interferon), which is not directly antiviral, but is produced by effector cells that destroy virally infected cells.

The secreted interferons bind to interferon receptors, which are widely expressed on all body cells. This binding leads to cellular signalling and the subsequent transcription of genes called IFN-stimulated genes (ISGs) and IFN-regulated genes (IRGs). These genes increase our natural (or innate) resistance to viruses by upregulating the production of restriction factors, which restrict viral growth.

The most important effector to destroy infected (or transformed) body cells is the natural killer (NK) cell. Interferons (type I and type III) and pro-inflammatory cytokines (such as IL-2 and IL-12) stimulate NK cells to become **lymphokine activated killer** (LAK) cells. These cells can kill infected body cells via a tightly controlled recognition process.

Recognition is a two-stage process mediated by a combination of inhibitory receptors ('Don't kill me: I am normal!') and activating receptors ('Kill me: I am virally infected/transformed'), as illustrated in Figure 27.12.

All normal body cells express on their surface a type of molecule called a class I MHC (for major histocompatibility complex), about which we learn more in Section 27.3. Class I MHCs bind to the inhibitory receptors (IR) on NK cells to generate an inhibitory signal; when this signal is present, no killing takes place. As such, all body cells expressing class I MHC are protected from NK lysis.

The activating receptors only come into play if the inhibitory signal is *absent*—that is, if class I MHC expression is lost following viral infection, or in a transformed or stressed body cell. Stressed, infected, or transformed body cells express surface proteins called natural cytotoxicity receptors (NCRs), which mark them for destruction by NK cells. Only cells that *don't* bind the inhibitory receptor and *do* bind the activating receptor will be killed. The killing itself is the result of programmed cell death (apoptosis) being induced in the target cell.

NK effectors also secrete IFN-γ (type II interferon), which activates macrophages to remove the dead cells by phagocytosis.

Destruction of parasites by eosinophils

What happens if the pathogen is a multicellular parasite or a helminth (worm) that is too big to be removed by phagocytosis? These organisms stimulate mucosal surfaces to produce substances called 'alarmins' (such as IL-33), which signal tissue-resident leukocytes to release IL-5 and IL-13. This leads to the proliferation of goblet cells, which contributes to an increased production of antimicrobial peptides. These, in turn, trigger worm expulsion. It also leads to the recruitment of eosinophils, which are directly cytopathic to the parasites because they contain toxic proteins; these include eosinophil peroxidase, eosinophil cationic protein, and eosinophil-derived neurotoxin, among others.

Figure 27.12 Natural killer (NK) cells. NK cells mediate the killing of virally infected cells.

Source: Playfair & Bancroft. *Infection and Immunity*, 4th edn, 2013. Oxford University Press.

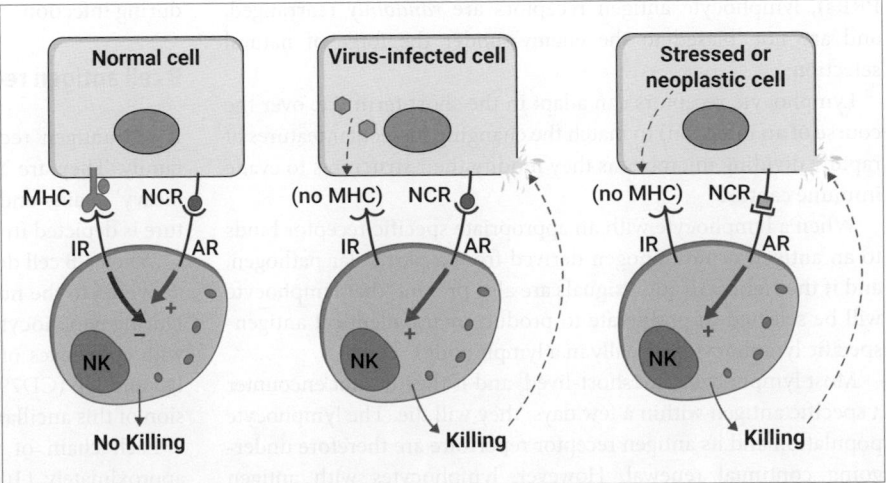

PAUSE AND THINK

How are large pathogens destroyed by the innate immune system?

Answer: Eosinophils contain toxic mediators. Inflammatory changes in the gut can lead to increased mucus production, flattening of villi, and worm expulsion.

 Check your understanding of the concepts covered in this section by answering the questions in the e-book.

27.3 The adaptive immune response

Innate immunity is found in all eukaryotic organisms. The cells that mediate this response carry pattern recognition receptors, which are biased to the enemy, and a common signalling pathway—the NF-κB pathway—leads to the activation of immune response genes and inflammation.

The subsequent evolution of the adaptive immune system, which is only present in vertebrates, permitted the highly specific recognition of pathogens on the basis that different pathogens have unique molecular shapes. But how does this specificity arise?

The answer lies in the receptors that cells of the adaptive immune system carry. The key cells of adaptive immunity are the lymphocytes, and this population contains a huge diversity of antigen receptors—both T cell receptors (TCR) and B cell receptors (BCR). The antigen recognized by a specific receptor is determined by the structure of that receptor which, in turn, is determined by the genes encoding it.

The huge diversity in receptor specificity arises because multiple gene segments can code for the same part of the lymphocyte antigen receptors (TCR and BCR). While all lymphocytes in a particular individual carry the same gene segments (because they share the same genome), a molecular mechanism, called somatic recombination, rearranges these gene segments in different ways in different lymphocytes. An enormous number of different combinations is possible across an entire lymphocyte population—a phenomenon called **combinatorial diversity**. These diverse combinations mean that different lymphocytes express their own unique receptors. Unlike the receptors of the innate system (the PRRs), lymphocyte antigen receptors are *randomly* rearranged, and are not 'biased to the enemy' under the force of natural selection.

Lymphocyte receptors can adapt in the short term (i.e. over the course of an infection) to match the changing molecular features of rapidly dividing microbes as they modify their structures to evade immune capture.

When a lymphocyte with an appropriate specific receptor binds to an antigen or immunogen derived from a particular pathogen, and if the right activation signals are also present, that lymphocyte will be selected to proliferate to produce many identical antigen-specific lymphocytes (usually in a lymph node).

Most lymphocytes are short-lived, and if they do not encounter a specific antigen within a few days, they will die. The lymphocyte population and its antigen receptor repertoire are therefore undergoing continual renewal. However, lymphocytes with antigen receptors that have been selected by pathogens, and have therefore been activated, are retained in the body to provide long-lived immunological memory.

The adaptive immune response that occurs on first encounter with a pathogen is called the **primary immune response**. This response is mediated by lymphocytes, which will control infection in 1–2 weeks. Once immunological memory has been generated, a subsequent encounter with the pathogen provokes a **secondary immune response**, which is much faster and more powerful than the primary response, and often will clear the pathogen without any symptoms becoming apparent. These two responses are illustrated in Figure 27.13. This two-phase response is the basis of **vaccination**, as we discuss further in Real World View 27.1.

Figure 27.13 illustrates the fundamental principles of the adaptive immune response.

- Specificity: A is different to B.
- Diversity: responses can be made to A *and* B.
- Escalating responses: the second exposure to A leads to a faster, more effective immune response.
- Memory: the primary immune response leads to memory cells specific to A.

Lymphocytes constantly circulate through the lymphoid tissues to increase the probability of an antigen receptor encountering its complementary antigen. Once this happens, lymphocytes proliferate to produce large numbers of lymphocytes with identical antigen specificity (a clone army), as illustrated in Figure 27.14. These lymphocytes then undergo final differentiation into effector cells, which cooperate with other leukocytes, such as macrophages, to destroy the pathogen.

B cells differentiate into antibody-producing (plasma) cells, while T cells differentiate into a variety of effector T cells with different functions. These include **cytotoxic T cells** (CTL) that kill virus-infected cells, and **T follicular helper cells** (Tfh) that interact with B cells to 'help' them produce antibodies.

The generation of adaptive immune response takes time, and this is why we suffer with the symptoms of infection for at least a week before recovering. It also explains why our lymph nodes swell during infection.

B cell antigen receptors and antibodies

B cell antigen receptors belong to the immunoglobulin protein family. They are Y-shaped molecules composed of two identical 'heavy' chains and two smaller identical 'light' chains; this structure is depicted in Figure 27.15.

When a B cell detects an antigen via its receptor, a signal must be delivered to the nucleus so that the cell is able to make a response. During lymphocyte development, the receptors become associated with complexes of proteins that can initiate signalling. These are Igα and Igβ (CD79a/b). Look at Figure 27.15 and notice the inclusion of this ancillary complex next to the BCR.

Each chain of the BCR is organized into Ig domains, each approximately 110 amino acids long. Each heavy chain and each

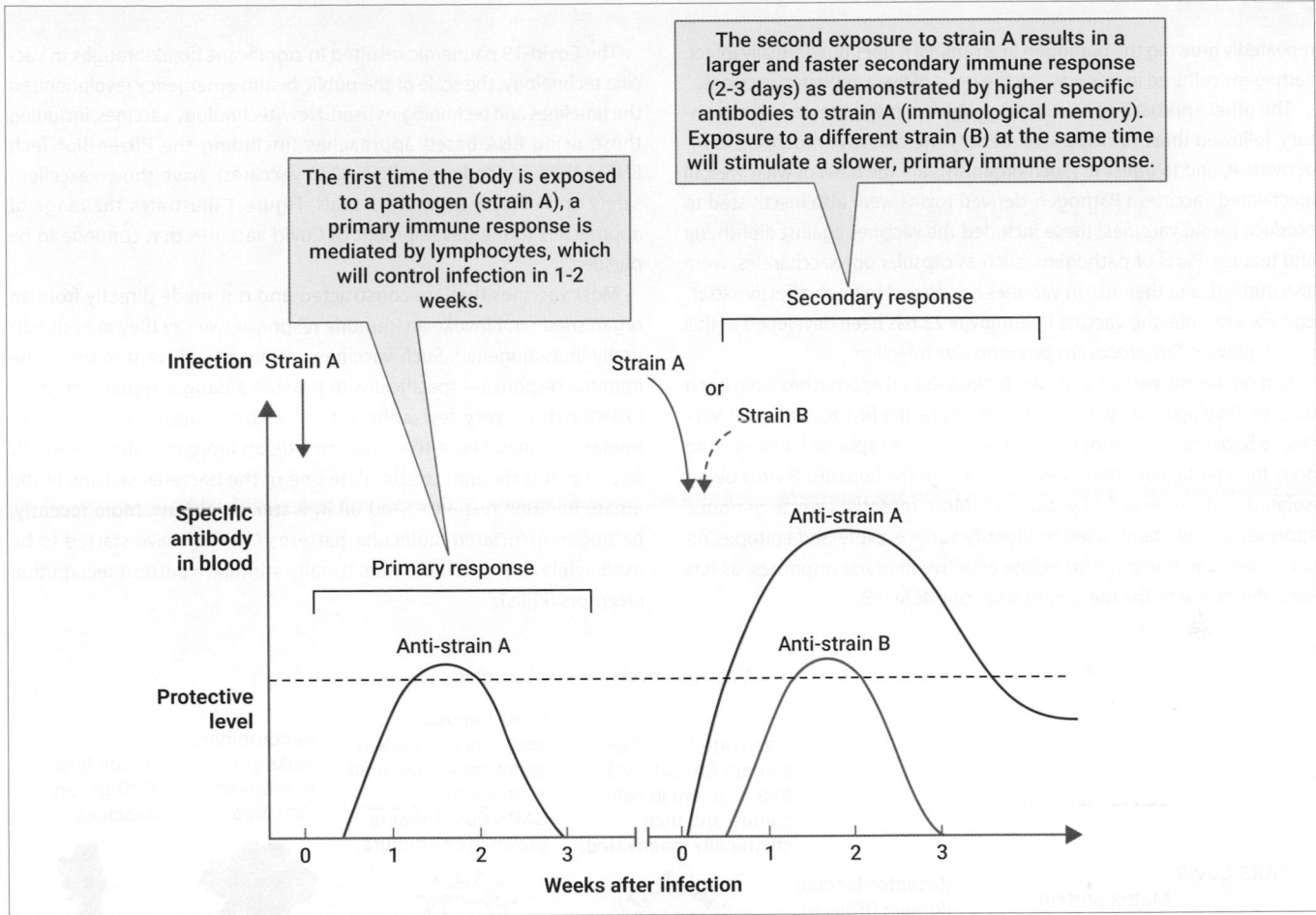

The second exposure to strain A results in a larger and faster secondary immune response (2-3 days) as demonstrated by higher specific antibodies to strain A (immunological memory). Exposure to a different strain (B) at the same time will stimulate a slower, primary immune response.

The first time the body is exposed to a pathogen (strain A), a primary immune response is mediated by lymphocytes, which will control infection in 1-2 weeks.

Figure 27.13 Adaptive immune response. The primary and secondary adaptive immune responses.

Source: Playfair & Bancroft. *Infection and Immunity*, 4th edn, 2013. Oxford University Press.

REAL WORLD VIEW 27.1   Vaccination

Vaccination to prevent disease has had a more positive impact on global health and wellbeing than any other medical advance. But how does vaccination work? Our immune systems confer on us natural immunity to a pathogen, following exposure to it. Vaccines aim to mimic this natural process by inducing a primary immune response. When an individual then gets exposed to the wild-type pathogen circulating in the environment, they make a rapid secondary immune response to it, which prevents the development of infection or serious illness.

Widespread immunization using vaccines has led to the reduction of communicable diseases, and the emergence of the concept of 'vaccine-preventable diseases', and this has revolutionized public health. Vaccines are given to healthy populations in order to prevent disease, and therefore there is great emphasis on vaccine safety to limit any potential harm.

The first vaccine was developed to prevent smallpox by Edward Jenner (1749–1823). It had been noticed that milkmaids who contracted the less severe bovine disease of cowpox rarely died of smallpox. Jenner tested

the hypothesis that exposure to cowpox protected against smallpox, by injecting pus from cowpox lesions from a milkmaid into a child, who recovered from the cowpox infection. Jenner then infected the child with a lethal dose of smallpox. Luckily, the child survived this experiment—which no ethics committee today would allow (sample size of 1!)—and the concept of vaccination was introduced.

The words 'vaccine' and 'vaccination' derive from Jenner's name for the cowpox virus, 'variolae vaccinae'. Vaccination against smallpox became widespread, and smallpox was finally declared eradicated in 1980, following a global immunization campaign led by the World Health Organization.

After Jenner, the next stages in vaccine development were initiated by the observations in the 19th century of Louis Pasteur and Robert Koch that diseases were caused by micro-organisms. Pasteur established methods of culturing pathogens such that the pathogen could no longer establish an infection, but was still able to induce a protective immune response. This procedure is called **attenuation** and is usually achieved by repeatedly growing the pathogen in an animal it does not normally infect. Pathogens cultured in this way are the basis of live attenuated vaccines.

The other approach to vaccination, which came in the early 20th century, followed three principles: to *identify* the causative organism, to *inactivate* it, and to *inject* it. (Such organisms are the basis of what we call inactivated vaccines.) Pathogen-derived toxins were also inactivated to produce toxoid vaccines; these included the vaccines against diphtheria and tetanus. Parts of pathogens, such as capsular polysaccharides, were also purified, and their use in vaccines has proved to be an effective strategy. For example, the vaccine Pneumovax 23 has been developed in this way to prevent *Streptococcus pneumoniae* infection.

In more recent years, molecular biology-based approaches have been used to develop novel vaccines. For example, the first recombinant vaccine to hepatitis B was produced in the 1980s—an approach that saw the gene for a particular antigen associated with the hepatitis B virus being isolated and expressed in yeast cells. More recently, reverse genomic approaches have been used to identify surface-expressed epitopes on pathogens and these can stimulate effective immune responses, as has been the case with the meningitis B vaccine 4CMenB.

The Covid-19 pandemic resulted in significant breakthroughs in vaccine technology: the scale of the public health emergency revolutionized the timelines and technologies used. New-technology vaccines, including those using RNA-based approaches (including the Pfizer-BioNTech BNT162b2 and Moderna mRNA-1273 vaccines), have shown excellent safety and efficacy in phase 3 trials. Figure 1 illustrates the range of approaches to the development of Covid vaccines that continue to be pursued.

Most vaccines that are constructed and not made directly from an organism do not invoke an immune response (we say they are not naturally immunogenic). Such vaccines require an adjuvant to boost the immune response—specifically, to provide a danger signal and cause inflammation. Very few adjuvants have been approved for use in human vaccines. Those that have include an inorganic aluminium salt (alum) that is thought to stimulate one of the bacterial sensors of the innate immune response, and oil in water emulsions. More recently, pathogen-associated molecular patterns (PAMPs) have started to be used, while RNA vaccines will naturally stimulate pattern recognition receptors (PRRs).

Vaccine development strategies:

Inactivated vaccines contain SARS-CoV-2 that is grown in cell culture and then chemically inactivated.

Live attenuated vaccines are made of genetically weakened versions of SARS-CoV-2 that is grown in cell culture.

Recombinant spike-protein-based vaccines

Recombinant RBD-based vaccines

SARS-CoV-2

Matrix protein

Receptor-binding domain (RBD) of spike protein

Spike protein

Envelope protein

Nucleoprotein and viral RNA

VLPs carry no genome but display the spike protein on their surface

Replication-incompetent vector vaccines cannot propagate in the cells of the vaccinated individual but express the spike protein within them.

Replication-competent vector vaccines can propagate to some extent in the cells of the vaccinated individual and express the spike protein within them.

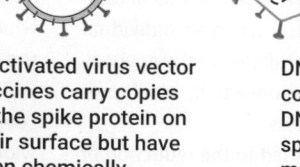

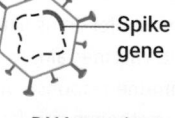

Spike gene

Inactivated virus vector vaccines carry copies of the spike protein on their surface but have been chemically inactivated.

DNA vaccines consist of plasmid DNA encoding the spike gene under a mammalian promoter.

RNA vaccines consist of RNA encoding the spike protein and are typically packaged in LNPs.

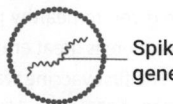

Spike gene

Figure 1 Covid-19 vaccines. Approaches used for SARS-CoV-2 vaccine development.

Source: Krammer, F. SARS-CoV-2 vaccines in development. *Nature*, 2020, 586, 516–27. Reprinted by permission from Springer Nature: Springer.

Figure 27.14 Lymphocyte receptor diversity leads to clonal selection. When an antigen is recognized by a specific receptor, the lymphocyte carrying that receptor undergoes clonal selection: it proliferates to generate an 'army' of clones, which all recognize the same antigen.

Source: Playfair & Bancroft. *Infection and Immunity*, 4th edn, 2013. Oxford University Press.

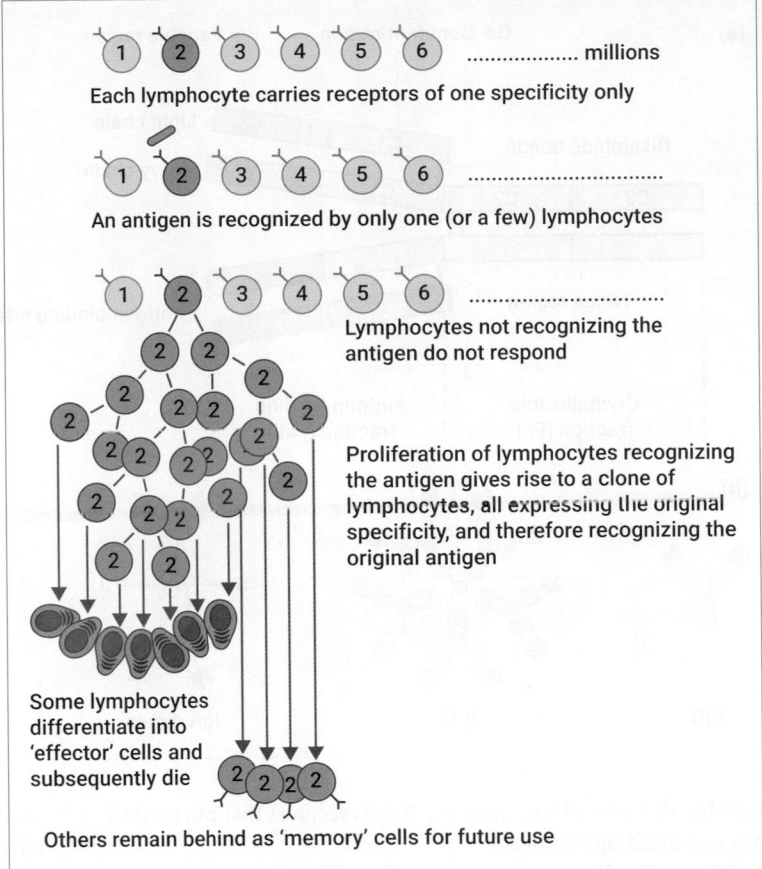

Each lymphocyte carries receptors of one specificity only

An antigen is recognized by only one (or a few) lymphocytes

Lymphocytes not recognizing the antigen do not respond

Proliferation of lymphocytes recognizing the antigen gives rise to a clone of lymphocytes, all expressing the original specificity, and therefore recognizing the original antigen

Some lymphocytes differentiate into 'effector' cells and subsequently die

Others remain behind as 'memory' cells for future use

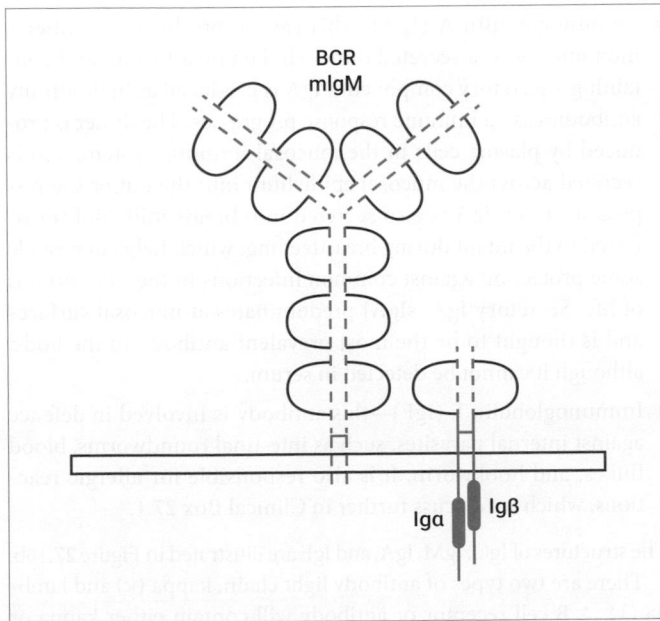

Figure 27.15 Protein structure of the B cell receptor (BCR). Illustration showing membrane-bound immunoglobulin and associated Igα and Igβ chains.

Source: From A. DeFranco, A. Weiss. Signal transduction by T- and B- lymphocyte antigen receptors. In: Ochs et al., *Primary Immunodeficiency Diseases: A Molecular and Genetic Approach*, 3rd edn (2013). By permission of Oxford University Press.

light chain comprises a **constant region** and a **variable region**, as illustrated in Figure 27.16a. The heavy chain contains 3–4 constant heavy (CH) domains and one variable heavy (VH) domain. The light chain contains a constant light (CL) and a variable light (VL) domain.

The variable regions form the tips of the Y, while the end of the constant region is held in the cell membrane. The tips of the Y are unique to each B cell receptor, and are the sites that recognize and bind antigen. B cell receptors therefore have two identical antigen-binding sites. Look at Figure 27.16a and notice how the antigen binding site is made up of a combination of the VH and VL domains. Also, notice how the protein as a whole has two defined 'fractions': the crystallizable fraction (Fc) and the antigen-binding fraction (Fab). The Fc fraction binds to Fc receptors, which tether the protein to a membrane surface and promote biological responses.

A large and complex antigen, such as a bacterium or virus, will contain many different molecular shapes, each of which is called an **epitope**. Epitopes are the molecular motifs recognized by the lymphocyte antigen receptors. Each different B cell receptor will recognize a different epitope, which can be a unique three-dimensional shape, formed by just a few amino acids on the surface of a protein, or a sugar, attached to a glycoprotein. In the region of 10^{11}–10^{12} different variants of the structure at the tips of the antigen receptor (i.e. the antigen-binding domain) probably exist, so a

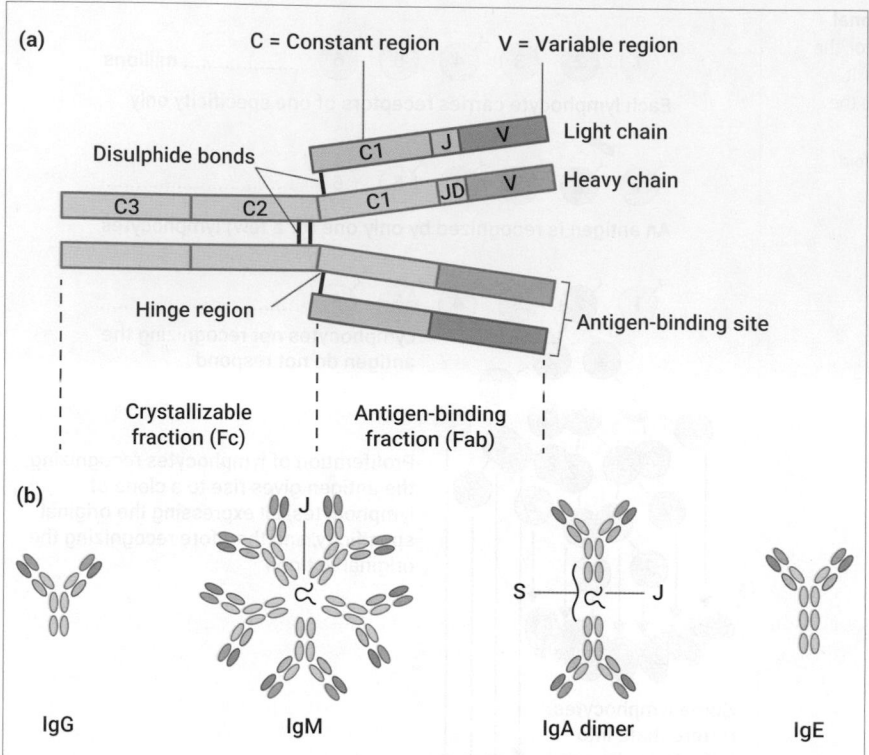

Figure 27.16 B cell antigen receptors. (a) The structure of a single antibody showing how it comprises two light and two heavy chains; each chain also comprises constant and variable regions. (b) The structures of the IgG, IgM, IgA, and IgE antibodies.

Source: Playfair & Bancroft. *Infection and Immunity*, 4th edn, 2013. Oxford University Press.

healthy adult should have multiple B cell receptors that can bind to any one invading pathogen.

When a B cell encounters its specific antigen, it undergoes terminal differentiation into a plasma cell, which is dedicated to producing antibodies. Antibodies are soluble secreted versions of the B cell receptor, with an identical antigen specificity.

What are the different types of antibody?

We note above how an immunoglobulin molecule—whether in the form of a membrane-bound BCR or a soluble antibody—comprises light and heavy chains that have constant and variable regions. As their name suggests, the constant regions of immunoglobulins are relatively invariant. There are five main types of heavy-chain constant region in immunoglobulins, which define the different classes of antibody that can be produced:

- Immunoglobulin G (IgG)—this is the most abundant antibody type circulating in the blood. IgG antibodies with high affinity for their specific antigen are produced as a primary immune response progresses. There are four subclasses of IgG, some of which are very effective at fixing complement (1–3). IgG antibodies can travel from the blood into tissues, including transfer across the placenta to a growing fetus. This provides newborn infants with some protection against common infections in the first months of life until their own immune system develops.

- Immunoglobulin M (IgM)—predominantly found in the blood, IgM is the first antibody produced in an adaptive immune response, and is also the antibody that recognizes repetitive epitopes on pathogens (like PAMPs). It is produced as a pentamer or hexamer of

antibody molecules (linked by a joining or J chain) and is very effective at fixing complement. It has a lower affinity for antigen.

- Immunoglobulin D (IgD)—this is produced in very small amounts. While its function is unknown, it is *thought* to be the main antigen receptor on naïve B cells.

- Immunoglobulin A (IgA)—this can be produced as either a monomer or as a secreted dimer (linked by a J chain and containing a secretory component). IgA is produced as high-affinity antibodies as an immune response progresses. The dimer is produced by plasma cells in the mucosal immune system, and is secreted across the mucosal epithelium into the gut or the respiratory tract. IgA is also secreted into breast milk and transferred to the infant during breastfeeding, which helps to provide some protection against common infections in the first months of life. Secretory IgA .sIgA) predominates at mucosal surfaces and is thought to be the most prevalent antibody in the body, although it cannot be detected in serum.

- Immunoglobulin E (IgE)—this antibody is involved in defence against internal parasites, such as intestinal roundworms, blood flukes, and hookworm. It is also responsible for allergic reactions, which we discuss further in Clinical Box 27.1.

The structures of IgG, IgM, IgA, and IgE are illustrated in Figure 27.16b.

There are two types of antibody light chain, kappa (κ) and lambda (λ). A B cell receptor or antibody will contain either kappa or lambda light chains, but not both. The light-chain constant region only contributes to the specificity of that antibody/receptor but does not affect the function of the antibody. The different functions of the antibody classes are instead conferred by the heavy-chain constant region.

CLINICAL BOX **27.1** | **IgE-mediated allergic responses**

Anyone who is allergic to pollen knows how miserable the summer months can be: continual sneezing, a runny nose, and weeping, swollen eyes make life difficult. Hay fever (allergic rhinitis) is caused by an allergic response. Allergy is a harmful response to an otherwise harmless environmental material (referred to as an **allergen**). It is caused by an adaptive immune response driven by a subset of helper T cells (Th2), which stimulate B cells to produce high-affinity IgE antibody specific to the allergen. This response evolved to expel parasites, such as helminths, which can be too big to phagocytose and so require an alternative effector pathway to remove them from the body.

It has been estimated that up to 40 per cent of the white population in the USA and UK have a genetic predisposition called **atopy**, which is a tendency to make unwanted IgE responses to common environmental allergens. These can be inhaled (for example, pollen, house dust mites, and animal dander—the material shed from an animal's skin and fur) or ingested (for example, peanuts, shellfish, and tree nuts). Individuals who are sensitive to allergens (i.e. those who are **atopic**) can show symptoms in the lung (as is the case with rhinitis and asthma) or the skin (as we see with atopic dermatitis and eczema). A sensitization to house dust mites can lead to asthma, which is the most common non-communicable disease in children.

Allergic symptoms, like any kind of inflammation, can be local or widespread (systemic). The most serious manifestation of a systemic allergic response is anaphylactic shock, which can be fatal. Unfortunately, an increasing number of children are developing allergies to food, particularly peanuts, leading to increasing hospitalizations, and even death.

The process by which allergens stimulate an allergic response is illustrated in Figure 1. Parasites and allergens activate mast cells, which express a high-affinity receptor called FcεR1 that bind the heavy chain of IgE. Most IgE in the body is bound to mast cells in connective tissue rather than circulating in the blood. When an atopic person is first exposed to an allergen, they can make a Th2 immune response that promotes the production of allergen-specific IgE antibodies (rather than the IgG subclass), which then bind to mast cell surfaces. This first exposure is known as **sensitization** and has no outward consequence. When the same allergen is re-encountered, however, it binds to the specific IgE on the mast cell surface. This triggers the mast cell to degranulate and release many pro-inflammatory mediators, including histamine and enzymes (serine esterases, and proteases such as chymase and tryptase), very rapidly. Allergic symptoms become evident within seconds. This is referred to as immediate (or type 1) hypersensitivity.

The inflammatory mediators cause local inflammation. For example, histamine causes the dilation of blood vessels, which enables leukocytes to enter the tissue more easily, and smooth muscle contraction. By contrast, the enzymes cause tissue damage. In addition, activated mast cells begin to synthesize and secrete prostaglandins and leukotrienes, which contribute to inflammation. Mast cells also secrete cytokines and chemokines that attract monocytes, macrophages, neutrophils, and related leukocytes (eosinophils) to the reaction site (Figure 1).

Figure 1 Illustration of the development of type 1 hypersensitivity. An allergen induces an immune response, which leads to the production of allergen-specific IgE (sensitization). These antibodies bind to mast cells in tissues. Re-exposure to the same allergen will cross-link specific IgE on the mast cell surface, stimulating immediate degranulation and release of many pro-inflammatory mediators, which cause local or systemic symptoms of inflammation.

Source: Playfair & Bancroft. *Infection and Immunity*, 4th edn, 2013. Oxford University Press.

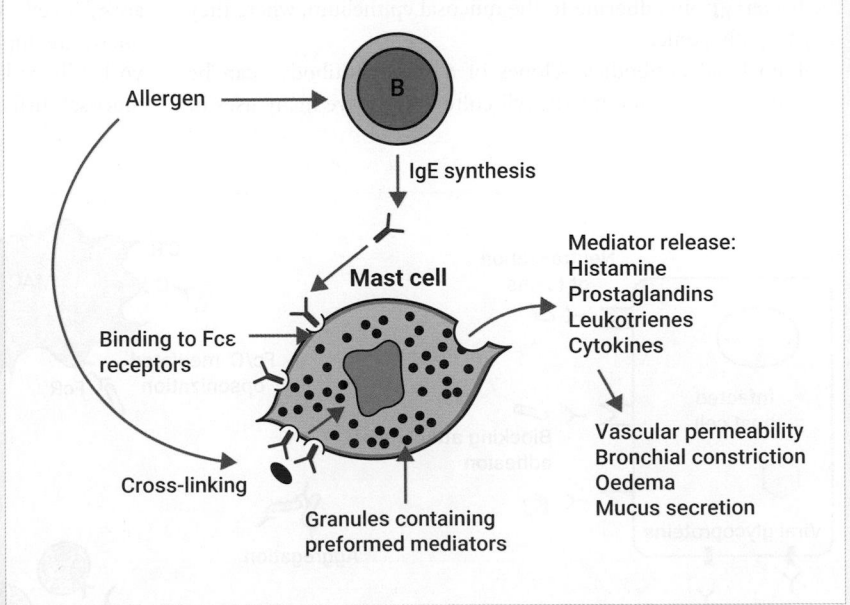

What functions do antibodies have?

Look at Figure 27.17, which illustrates the various ways in which antibodies help to ensure the successful operation of the immune system. These include:

- Neutralizing the effect of some threats to the body—for example, by binding to bacterial toxins and preventing them from binding to their receptors on host cells, or by binding to viruses and preventing them from infecting cells. Antibodies can block the entry of pathogens into cells and neutralize toxins and viruses.

- Activating a range of effector mechanisms, including NK cells (a phenomenon termed antibody-dependent cellular cytotoxicity (ADCC)).

- Making innate effector mechanisms specific—for example, targeting phagocytosis and complement-mediated lysis, and (via IgE) recruiting mast cells and eosinophils.

The multiple antigen-binding sites of antibodies also enable them to disarm pathogens by clumping them together (agglutination or aggregation). IgM is particularly effective at agglutination because the pentameric (five-unit) molecule has 10 antigen-binding sites. Phagocytic cells, such as macrophages, carry receptors that recognize the constant regions of antibodies, especially of IgG. These **Fc receptors** bind to the constant regions of antibodies that have bound a pathogen, and thus aid its phagocytosis and disposal in the same way that complement receptors aid the phagocytosis of a complement-coated pathogen (i.e. opsonization).

The aggregation of pathogens is also important to the way IgA functions. IgA is thought to neutralize toxins and coat bacteria, thereby causing them to clump together. This clumping prevents the bacteria from adhering to the mucosal epithelium, where they may be pathogenic.

Monoclonal antibodies—clones of a single antibody—can be produced in large amounts in cell culture and have many uses in biological research and therapeutics in medicine. Some of these applications are described in Clinical Box 27.2.

PAUSE AND THINK

What antibody …

a. Is the first to appear in a primary immune response?

b. Is responsible for protection at mucosal surfaces?

c. Associates with mast cells?

Answer: from the one

a. IgM.

b. Secreted IgA.

c. IgE.

T cell antigen receptors

T cells also express highly specific antigen receptors, which are members of the immunoglobulin superfamily. These receptors are not only smaller than the BCRs expressed on B cells, but also have variable and constant regions. T cell receptors are composed of two different chains, whose variable regions form a single antigen-binding site at the tip, as illustrated in Figure 27.18. There is no secreted form of the T cell receptor, and because its constant region has no function other than to support the variable region and anchor the chain in the T cell membrane, there are just a few different types of T cell receptor constant region.

T cell antigen receptors are composed of either α and β chains, in which case the T cell is called an αβ T cell, or γ and δ chains, in which case the T cell is called a γδ T cell. Most circulating T cells are αβ T cells. With this in mind, when we refer to T cells from now on, we are implicitly referring to αβ T cells, unless otherwise stated. γδ T cells are less diverse than αβ T cells and tend to be localized at mucosal surfaces.

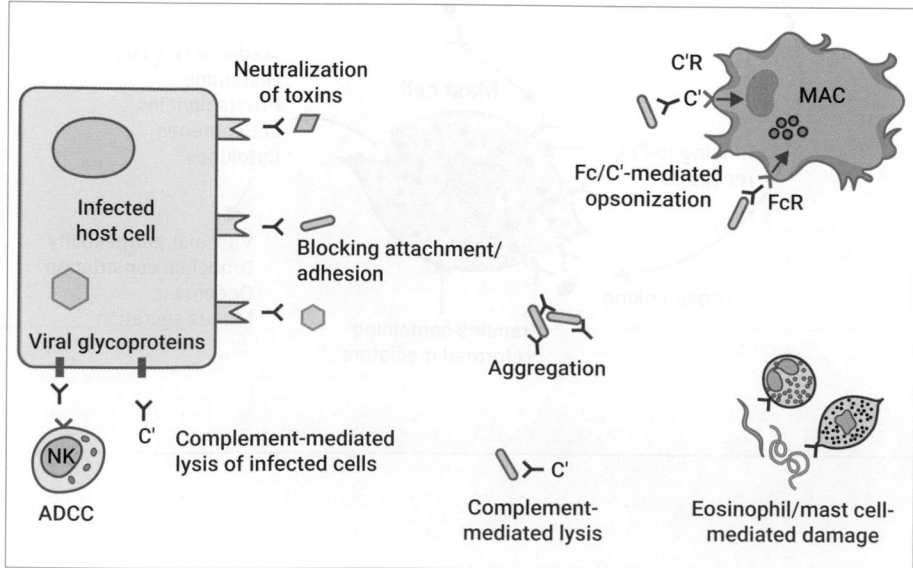

Figure 27.17 A summary of the effector functions of antibodies.

Source: Playfair & Bancroft. *Infection and Immunity*, 4th edn, 2013. Oxford University Press.

   Immunotherapy

Antibodies are highly specific molecules that can be used therapeutically to target immune cells and mediators in disease. Their use in this way has revolutionized the treatment of autoimmune disease (by turning off the immune response) and cancer (by turning *on* the immune response to destroy the tumour cells) and is an area of active research.

Therapeutic monoclonal antibodies

Unlike the natural antibodies found in blood or serum, which exhibit a huge range of specificities, monoclonal antibodies share the *same* specificity. They can be produced in large volumes in culture by growing clones of cells derived from just one specific B cell that has been fused to a tumour (myeloma) cell, making it immortal.

▶ **We learn more about cell culture in Chapter 12.**

Antibody engineering enables the characteristics of an antibody to be manipulated. Antibodies made in mice that have been engineered to display a certain specificity can be added to human immunoglobulin molecules.

You can recognize the type of antibody used therapeutically by its name. Chimeric antibodies that include the variable region from another animal (usually a mouse) and the constant region of human IgG have the suffix '-ximab'. For example, rituximab is used therapeutically against some autoimmune diseases to deplete B cells. By contrast, modern monoclonal antibodies are fully human and have the suffix '-umab'. For example, the anti-TNF-α monoclonal adalimumab is used to treat rheumatoid arthritis, while ipilimumab is used to treat melanoma and other solid tumours.

One of the first highly effective therapeutic monoclonal antibodies was the chimeric antibody infliximab, which targets TNF-α and is used to treat Crohn's disease, ulcerative colitis, rheumatoid arthritis, and other inflammatory conditions. The monoclonal antibody is given intravenously over a course of three injections. If this treatment appears successful, further infusions will be given approximately every eight weeks. (IgG has a half-life of less than 30 days, so these 'maintenance' infusions are required to keep the antibody circulating at therapeutically valuable levels.)

Immunotherapy is increasingly used to treat cancer by using antibodies to inhibit the negative immune regulation observed in successful tumours. Two immunologists, James P. Allison and Tasuku Honjo, were awarded the Nobel Prize for Physiology or Medicine in 2018 for their discovery of what the Nobel Assembly described as an 'entirely new principle for cancer therapy'. Remarkable remissions have been achieved for some individuals with cancers that were previously considered to be untreatable—for example, stage IV metastatic melanoma. To date, two immune regulators have been targeted clinically, although many more therapeutic agents that draw on the discoveries made by Allison and Honjo are in development.

Chimeric antigen receptor (CAR) T cell therapies

Another exciting way in which our immune cells have been harnessed to treat cancer is the development of CAR T cell therapy. This therapy sees genetic engineering used to generate T cell receptors that incorporate antibodies, conferring on them the ability to recognize other entities—specifically, tumour cells. These modified receptors are called chimeric antigen receptors (CARs), and the cells carrying them CAR T cells.

Anti-tumour CAR T cells have been engineered to be able to recognize an epitope on the tumour cell surface using a single-chain antibody variable region as the antigen receptor. Having recognized the epitope, they can then kill the target cell in response. A number of CAR T cells have now been approved to treat cancer, among them Yescarta (to treat B cell lymphoma) and Breyanzi (to treat certain types of non-Hodgkin lymphoma and diffuse large B cell lymphoma).

Like B cells, a signal must be delivered to the T cell nucleus when it recognizes an antigen. We see in Figure 27.16 how the BCR is associated with Igα and Igβ (CD79a/b); the equivalent for T cells is the formation of the TCR-CD3 complex, as shown in Figure 27.19. Look at this figure, and notice how the CD3 dimer can have three different forms, depending on the chains from which it is formed: CD3εγ, CD3εδ, and CD3ζζ. Look at the table at the end of this chapter to find out more about cluster of differentiation (CD) numbers.

How are diverse antigen receptors generated?

There are only around 20,000 genes in the human genome, so it would be impossible for a single gene to generate enough diversity to recognize every potential pathogen we might encounter. So how *is* the diversity that we see in reality made possible?

A highly diverse array of BCR and TCR with different variable regions is generated during lymphocyte development by a process of DNA rearrangement (somatic recombination). This occurs in each lymphocyte during its maturation in the primary lymphoid organs—the bone marrow (B cells) and thymus (T cells). Immature lymphocytes do not express receptors and contain the unrearranged 'germline' sequence that encodes their receptors. This sequence, found on chromosome 14, consists of three blocks of multiple gene segments that can each encode a part of the variable region. The gene segments are named V (for variable), D (for diversity), and J (for joining).

So, how does the DNA rearrangement unfold? In each cell, one gene segment from each block is randomly selected and joined together in the right order to produce a DNA sequence that encodes the variable region of the BCR heavy chain, as illustrated in Figure 27.19. This process of DNA rearrangement is called V(D)J recombination and relies on recombinase enzymes RAG-1 and RAG-2. These enzymes enable single gene segments from the multiple gene segments to connect to a constant domain to make a functional receptor gene. The remaining germline DNA is lost in a mature lymphocyte, and only the rearranged receptor is expressed.

A similar process generates the DNA sequences encoding the variable regions of the immunoglobulin light chains (which have just V and J segments) and the two chains of the T cell receptor. The α chain has V and J segments, while the β chain has V, D, and J segments. Note that V, D, and J are generic names for the gene segments. They will differ in DNA sequence in the genes for the different receptor chains.

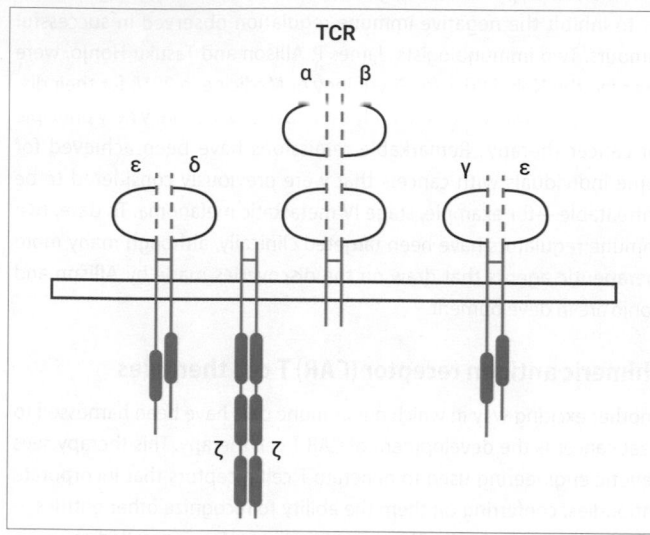

Figure 27.18 The T cell receptor (TCR). Illustration of the protein structure of the TCR showing the alpha and beta chain complexed with CD3 gamma, delta, epsilon, and zeta chains.

Source: From A. DeFranco, A. Weiss. Signal transduction by T- and B- lymphocyte antigen receptors. In: Ochs et al., *Primary Immunodeficiency Diseases: A Molecular and Genetic Approach*, 3rd edn (2013). By permission of Oxford University Press.

The diversity generated by the joining together of single gene segments from multiple alternatives in the germline into a single receptor gene is termed **combinatorial diversity**. Extra diversity is generated during the joining process by the insertion of random nucleotides at the junctions between gene segments. The additional diversity created by such nucleotide additions is referred to as **junctional diversity**. Further diversity is generated by the pairing of a heavy chain and a light chain. If successful rearrangement is not achieved, the lymphocyte will die.

Rearranged TCRs do not change following development and a naïve mature T cell will exit the thymus to recirculate between the blood and secondary lymphoid organs. By contrast, rearranged BCRs undergo a process called **affinity maturation** during the course of an immune response. Two enzymes can induce further point mutations in the BCR variable gene segments, which can cause sequence divergence and broaden the range of specific antibodies produced.

Why is it important to be able to generate so much diversity in both types of lymphocyte receptor (BCR/antibody and TCR)? Innate defences are biased to the enemy and focus on PAMPs, but pathogens can mutate far faster than us in order to avoid recognition and destruction. By contrast, BCRs and TCRs are extremely diverse receptors. Receptor rearrangement is done uniquely and randomly in every individual. This gives us great strength as a population, and enables us to respond to the entire world of pathogens: however they mutate to avoid recognition and destruction in one person, there will be no benefit in the next. Our bodies are exposed to so many possible pathogenic threats that it is essential for our immune systems to be able to recognize the epitopes of a wide range of pathogenic antigens for the appropriate adaptive immune response to be triggered, and for the threat to be eliminated.

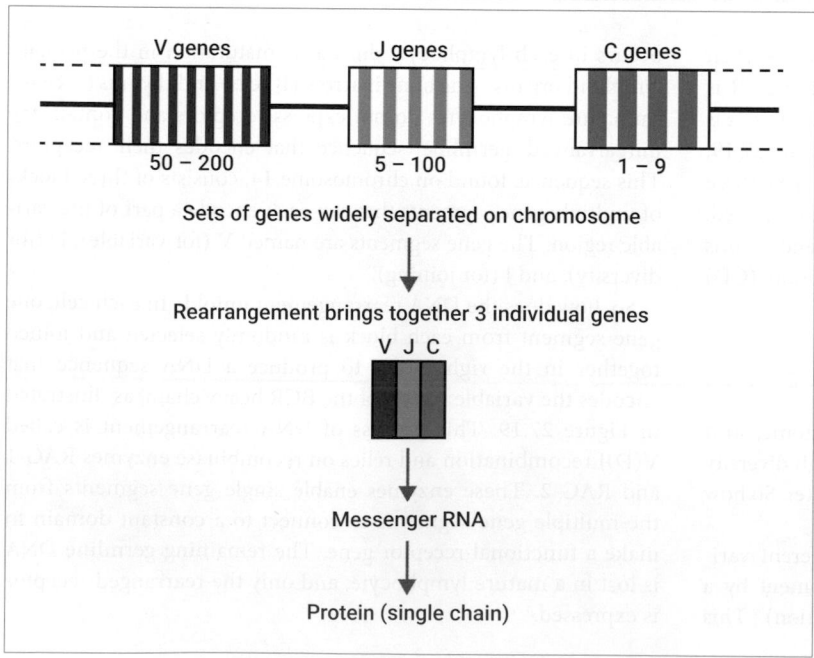

Figure 27.19 The V(D)J recombination process. The process is shown here for the immunoglobulin heavy chain. The precise numbers of segments can vary between different sources in the literature. This is largely due to differences that can occur in DNA sequences between individuals.

Source: Playfair & Bancroft. *Infection and Immunity*, 4th edn, 2013. Oxford University Press.

How are dysfunctional receptors eliminated?

The process of rearrangement is random, bringing the benefit of each of us having many different types of lymphocytes in our 'repertoire'. However, the risk is that some of these rearranged receptors could cross react with self-proteins, leading your immune system to recognize and destroy your own body tissue. To avoid this happening, once B cells have developed in the bone marrow, those carrying rearranged BCR that bind strongly to self-antigens are eliminated before they enter the circulation. This elimination process is called **clonal deletion**. Further elimination of potentially self-reactive B cells also happens outside the bone marrow if the naïve B cell encounters an antigen without the other signals necessary to activate the B cell being present.

Collectively, these processes aim to make the B cell repertoire **tolerant** of self-antigens. The process is incomplete, but an additional layer of control is added by B cells generally needing help from T cells (in the form of cytokines) to produce antibodies.

The rearrangement of TCR genes takes place in the thymus, and is followed by a selection process to ensure that mature naïve T cells are useful and not dangerous. Before exploring this selection process, we should pause to note that T cells do not bind to the three-dimensional structure of proteins in the way that B cells do, but rather to short linear peptides. T cells cannot recognize these peptides in isolation, however: they must be 'presented' to the TCR by a specialized peptide receptor called the **major histocompatibility complex** (MHC). As such, a T cell receptor recognizes the combination of MHC protein and its bound peptide, not just the peptide alone.

▶ We learn more about the major histocompatibility complex in the section 'T cell recognition of antigen'.

The T cell selection process has two strands: positive selection and negative selection. **Positive selection** selects for those rearranged TCRs that bind with *low affinity* to your self-MHC/peptide complexes.

Negative selection describes the deletion of those TCRs that can bind to self-MHC/peptide complexes with *high affinity*. These TCRs are dangerous and need to be deleted: they could cause your T cells to recognize and destroy your own tissue (autoimmunity). Those T cells with receptors that do not recognize self-MHC at all are removed during T cell development in the thymus.

This overall selection process is referred to as central tolerance. There are also other controls and regulatory mechanisms outside the thymus to prevent autoimmunity and these are called peripheral tolerance.

🌀 Check your understanding of the concepts covered in this section by answering the questions in the e-book.

27.4 Antigen recognition by lymphocyte receptors

We have now considered the structure of lymphocyte receptors, how receptor diversity is achieved, and how only 'competent' receptors are retained. But how do lymphocyte receptors actually recognize antigens?

B cell recognition of antigen

B cell receptors and their secreted antibodies directly recognize epitopes on the three-dimensional surface of intact 'native' antigens (i.e. their tertiary structures). Antigen-binding sites on the B cell receptor make non-covalent interactions with the epitope (including, for example, electrostatic or hydrophilic interactions). The interaction between receptor and antigen can be likened to that between a 'lock and key', with the antigen-binding site as the lock and the epitope as the key.

T cell recognition of antigen

We note above how T cells do not recognize 'native' antigen. Instead, they recognize peptides of around 10–24 amino acids long, which are derived from pathogen-derived proteins by an intracellular process called **antigen processing**. At the start of an adaptive immune response, antigen processing occurs within professional antigen-presenting cells, of which dendritic cells are the most important in the context of initiating an adaptive immune response. In the antigen-presenting cell, the peptides are bound to proteins of the MHC and the MHC–peptide complex travels to the cell surface, where it is accessible to T cells.

In the absence of an infection, MHC proteins bind self-peptides that are generated during the continual turnover of proteins within cells. This MHC/self-peptide combination will be ignored by the immune system in normal circumstances, because T cells have also undergone negative selection in the thymus that removes any T cells with receptors that bind too strongly to a combination of self-peptide and self-MHC.

It is hard to imagine that every single self-peptide from every developmental phase of your life could be in the thymus. Surely, many T cells have TCRs that are reactive for self-peptides on tissues *outside* the thymus? This conundrum is resolved by the expression in the thymus of a gene called autoimmune regulator (*AIRE*). This gene enables certain thymic epithelial cells to produce proteins usually restricted to other specific tissues. As a result, T cells in the thymus are exposed to an astonishingly wide range of self-peptides, and any T cells that bind strongly to these peptide–MHC complexes die by apoptosis (negative selection). This means that the repertoire of mature T cells should be self-tolerant and will not react to these self-peptides when encountered outside the thymus.

In humans, the MHC peptide receptors are called **human leukocyte antigens** (HLA).

There are two classes of MHC peptide receptors, class I and class II. The general structure of the class I and II receptors is illustrated in Figure 27.20. Class I MHC includes three proteins—HLA-A, HLA-B, and HLA-C—which are all expressed on the surface of every body cell that has a nucleus. Class I MHC typically present peptides that have been derived from intracellular pathogens present in the cell's cytosol. These are mainly viruses but may also include some intracellular bacteria, such as *Chlamydia*, *Rickettsia*, and *Mycobacteria*. They can also present tumour antigens, as the immune system functions to destroy cancer cells, not just pathogens.

The class II MHC peptide receptors have a more restricted distribution, and are mainly expressed on dendritic cells, macrophages,

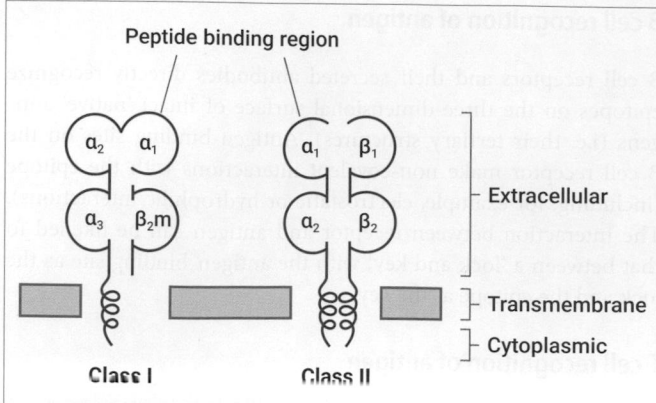

Figure 27.20 Major histocompatibility complex (MHC) peptide receptors. The general structure of class I and class II MHC peptide receptors.

Source: Cunningham A.C., *Thrive in Immunology*, 2016. Oxford University Press.

and B cells. These three cell types are collectively known as professional antigen-presenting cells because of their role in presenting pathogen-derived peptides to T cells. There are three human class II MHC peptide receptors: HLA-DP, HLA-DQ, and HLA-DR. They generally present extracellular peptides—for example, bacteria that live outside cells, and viruses that have not yet infected cells or have been shed from infected cells into the extracellular fluid.

Human HLA genes are the most polymorphic genes in the human population: there are thousands of gene variants (alleles) for each HLA. Class I alleles are more variable than class II alleles and the number increases as more studies are done. HLA alleles are also expressed co-dominantly, so you express two copies of each of your HLA A, B, and C proteins—one maternally coded and the other paternally coded. Consequently, there is great variation in HLA expression from one person to another, especially between genetically unrelated individuals, to the extent that the chance of you having the same HLA alleles as someone else is extremely low. The array of different HLA alleles a person expresses is known as their **tissue type**.

Exposure to other people's MHC peptide receptors provokes a powerful adaptive immune response, and so matching HLA types between donor and recipient before organ or tissue transplantation is extremely important to help to avoid rejection, as we discuss further in Clinical Box 27.3. The diversity among the MHC genes is not to frustrate transplant surgeons. Rather, it has evolved because the different HLA peptide receptors have different preferences for which peptides they bind to, thereby further diversifying the range of pathogen-derived peptides that can be presented to T cells.

Antigen recognition by T cells is aided by accessory molecules or **co-receptor** proteins expressed on the T cell surface, which associate closely with the T cell receptor, as illustrated in Figure 27.21. The co-receptors are called CD4 and CD8. Mature T cells express one or the other, but not both. As such, cells become either CD4-positive or CD8-positive during their development in the thymus.

CD4+ T cells are called **helper T cells** (Th) and play an essential role in helping to activate other immune effectors. CD4

co-receptors interact with class II MHC/extracellular peptide complexes, generally found on professional antigen-presenting cells (APC). This interaction helps Th cell TCRs to bind to the specific MHC/peptide complex on APCs.

CD8+ T cells can differentiate into cytotoxic T lymphocytes (CTL), which are highly specific killers. CD8 co-receptors interact with MHC class I/intracellular peptide complexes on all nucleated body cells and stabilize the interaction between CTL and their target cells. Look again at Figure 27.21 and notice how different T cells express different co-receptors, and recognize peptides presented on different MHC molecules.

PAUSE AND THINK

1. What do B cells and T cells recognize?

    Answer: B cells recognize the tertiary (three dimensional) structure of a pathogen. T cells recognize a short primary sequence of peptide derived from the pathogen presented in self MHC (the MHC–peptide complex).

2. What is the function of class I and class II MHC?

    Answer: Class I MHC is a peptide receptor, and presents intracellular peptides to CD8 T cells. Class II MHC is also a peptide receptor, but generally presents extracellular peptides to CD4 T cells.

Lymphocytes are activated in secondary lymphoid tissues

Mature naïve B cells and naïve CD4 and CD8 T cells leave the primary lymphoid organs, enter the circulation, and are delivered to the secondary lymphoid tissues, such as the lymph nodes, through the blood. Here, they may encounter their specific antigens and become activated/differentiate into effector lymphocytes.

The activation of CD4 T cells

The activation of naïve CD4 T cells in secondary lymphoid tissues can produce various functional types of effector CD4 T cell and is central to the activation of adaptive immune responses.

To see how an adaptive immune response is generated, imagine an extracellular bacterial pathogen that has entered the body through a wound in the skin. The bacteria will proliferate in the nutrient-rich tissue. However, multiple bacterial PAMPs will be recognized and will stimulate inflammation and innate immune responses. The wound will be sore, red, and warm because of the influx of cells and mediators. Rapid neutrophil infiltration will be seen as pus.

Dendritic cells (DCs) located in the soft tissues under the epidermis are constantly sampling their environment and will internalize the bacteria and its products—something they excel at. At the same time, inflammation will increase the rate at which DCs drain to the lymph node. DCs can also take up pathogens and pathogen products opsonized by complement; they also non-specifically ingest extracellular fluid and its contents in a continual process called **micropinocytosis**.

Once a pathogen has been ingested, proteins derived from them enter the endocytic pathway of DCs, where they are processed into

CLINICAL BOX 27.3 **Organ transplantation and graft-versus-host disease**

Transplantation is an effective treatment for end-stage organ failure. If your kidney, liver, or heart is so badly damaged that it can no longer function, you can receive a replacement from a donor. (While you can donate a kidney or a lobe of your liver, most transplant programmes rely on deceased donors.) However, a tissue or organ transplanted between two genetically unrelated people is invariably rejected. This is because unrelated individuals are unlikely to have identical tissue types because of the extensive polymorphism at the HLA locus, and so the recipient T cells mount an adaptive immune response against donor MHC on the graft (i.e. the HLA in humans, the tissue type).

Clinical transplantation is possible because of the development of immunosuppressive drugs, which specifically target T cell activation signalling pathways and suppress the T cell response against the donor tissue.

The transplantation of tissue between different members of the same species is called an **allograft**. The donor MHC is referred to as alloantigen, the response is 'alloreactive', and lymphocytes are allospecific. Other kinds of graft include an autograft (e.g. a skin graft from one part of the body to another), an isograft (a graft between genetically identical twins), and a xenograft (from one species to another).

About 1–10% of a recipient's T cells are alloreactive and directly recognize the donor MHC/peptide complexes. This is far higher than the number of T cells that would recognize a pathogen (estimated as 1 in a million or less) and explains the very strong immunological response. Patients who have had an organ transplant must take immunosuppressive drugs for life.

More difficult are bone marrow transplants or haematopoietic stem cell transplantation. The cells of a recipient's immune system will have been destroyed by chemotherapy or irradiation prior to transplant and the donor cells reconstitute the bone marrow. However, there is a high risk of graft-versus-host disease (GVHD). This condition sees lymphocytes from the graft attacking the healthy tissues of the transplant recipient. GVHD tends to target epithelial cells in the liver, skin, and the intestine, causing severe inflammation. Symptoms vary in severity, but include rash, cramps, and diarrhoea. It is very important to ensure there is a close match between the donor and recipient, and high-resolution genome sequencing is performed to find the best match.

Most solid organ transplants will survive rejection in the short term (i.e. acute rejection). The remaining challenge is to prevent chronic rejection—rejection over the longer term. The exact process of chronic rejection is not understood, but is thought to be a consequence of gradually accumulated vascular damage from multiple small, acute rejection events.

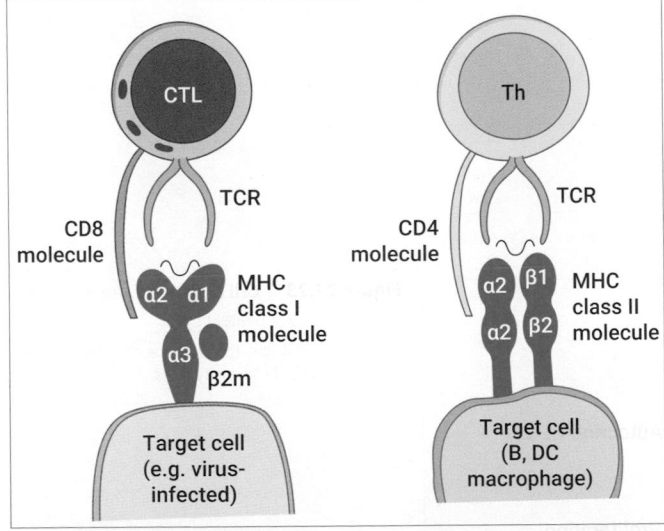

Figure 27.21 Accessory molecules aid T cell binding. T cell binding to MHC–peptide complexes is stabilized by accessory molecules CD8 and CD4.
Source: Playfair & Bancroft. *Infection and Immunity*, 4th edn, 2013. Oxford University Press.

peptides, loaded into MHC class II peptide receptors, and carried to the cell surface. This process is illustrated in Figure 27.22.

A danger signal will cause the DC to mature, making the cell less able to sample its environment, but stimulating it to focus on processing peptides from the bacteria in order to display them on their surface in lymph nodes for recirculating T cells to sample.

The activated DC—which now contains the pathogen, and is displaying the pathogen-derived peptides on its class II MHC—migrates to the nearest lymph node through an afferent lymphatic vessel that drains the tissue. On reaching a lymph node, it enters the T cell areas. In addition, intact pathogens and their breakdown products are also carried to the lymph node through the lymph.

As naïve CD4 T cells travel through the lymph node, they encounter DCs and 'scan' for peptides (presented on MHCs), all the while looking for a good fit with their TCR. If it recognizes a peptide that is being presented to it on a class II MHC peptide receptor, three signals are required for the T cell to become fully activated, as depicted in Figure 27.23:

- **Signal 1** is the specific recognition of MHC/peptide complex by the T cell receptor. Class II binding is stabilized by association of the CD4 co-receptor and class I by CD8 co-receptor (see Figure 27.21).

- **Signal 2** is the activation of a costimulatory pathway initiated by binding of the cell-surface proteins B7.1 and B7.2 on DCs to CD28 on the Th cell. Costimulation is activated by inflammation and innate immunity, so T cells should only be activated in the context of a danger signal and in response to an innate immune response.

- **Signal 3** is delivered to the T cell by cytokines present in the local environment as a result of the preceding innate immune response in tissue. Some of these cytokines are secreted by DCs, but others will have been secreted by macrophages and NK cells. This last step determines which differentiation pathway the activated CD4 T cell will take.

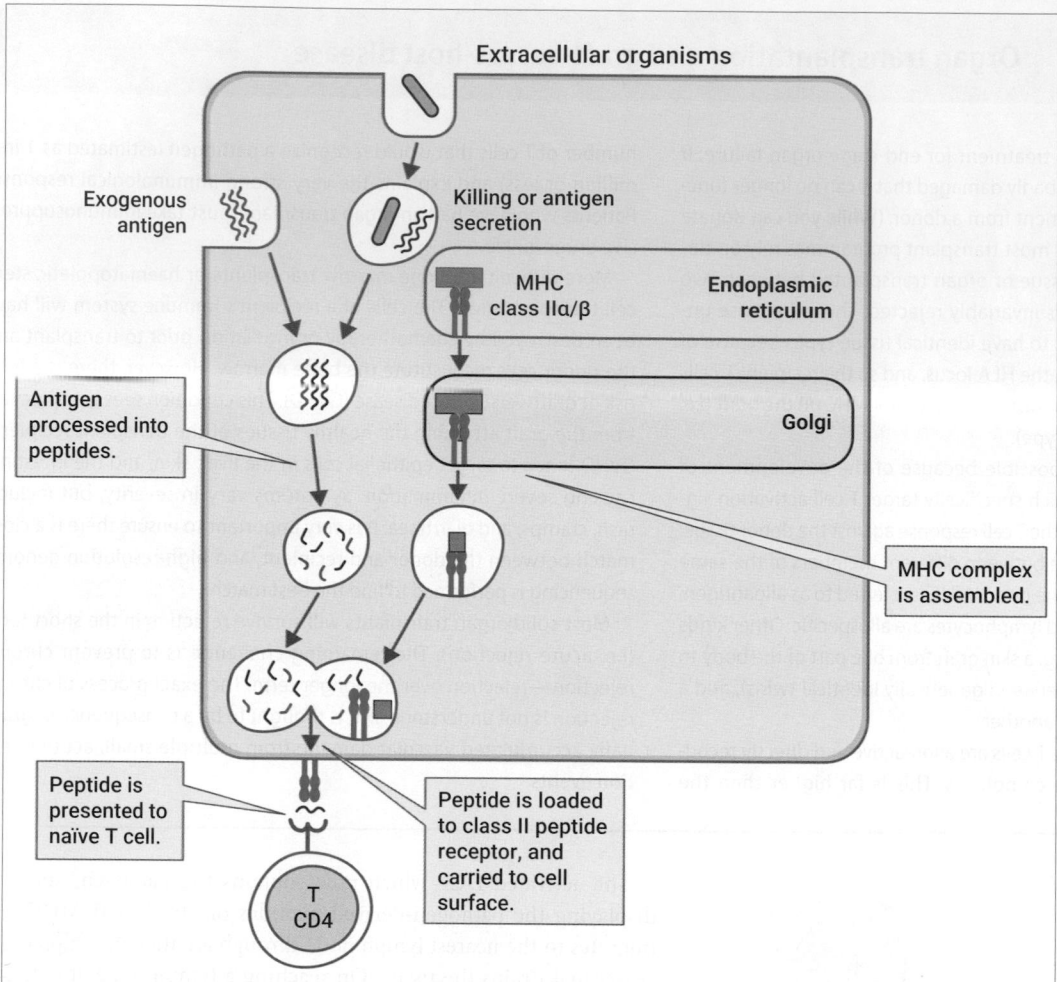

Figure 27.22 A dendritic cell ingests pathogens, and presents peptides derived from those pathogens to naïve T cells. Peptides must be loaded to a class II MHC peptide receptor on the surface of the dendritic cell to be recognized by the T cell.

Source: Playfair & Bancroft. *Infection and Immunity*, 4th edn, 2013. Oxford University Press.

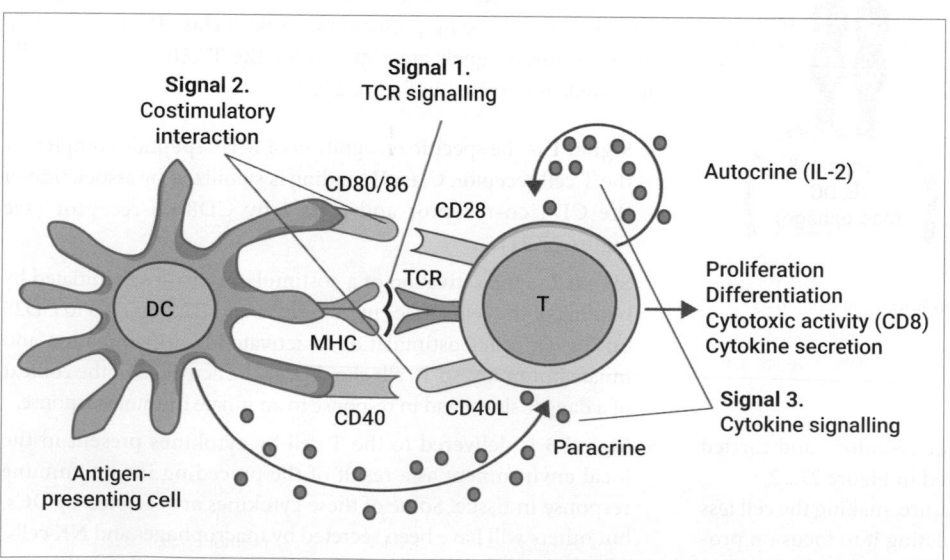

Figure 27.23 T cell activation. Three signals are needed to activate a T cell.

Source: Playfair & Bancroft. *Infection and Immunity*, 4th edn, 2013. Oxford University Press.

Directly following activation, the T cell secretes the cytokine IL-2 (essentially T cell growth factor) and upregulates its IL-2 receptor. This stimulates an autocrine pathway, which drives the T cell to proliferate. It now undergoes many rounds of cell division to form a clone of antigen-specific T cells with receptors specific to peptides derived from the infecting bacteria. These daughter cells then start to differentiate into CD4 effector T cells.

CD4 T cells can differentiate into five main effector types:

- T helper 1 cells (Th1);
- T helper 2 cells (Th2);
- T helper 17 cells (Th17);
- regulatory T cells (Treg);
- T follicular helper cells (Tfh).

The specific effector cell a CD4 T cell differentiates into depends on the cytokine signals it receives, as illustrated in Figure 27.24.

Some CD4 effector cells, such as Th1 and Th17, leave the lymph node and travel to the site of infection to carry out their functions. By contrast, Tfh cells and some Th2 cells stay in the lymph node, where they will help B cells produce high-affinity antibodies and memory cells. The differentiated T cells produce cytokines that help them to carry out their functions; they also secrete cytokines that enhance the differentiation of more T cells along the same pathway.

Let us now consider each of these T helper cells in a little more detail.

Th1 cells have a general role in promoting inflammation and can cooperate with macrophages to enable their destruction of pathogens by phagocytosis. So, what form does this 'cooperation' take? Infected macrophages display peptides derived from the bacterium on their class II MHC molecules. Th1 cells specific for these peptides bind to the antigen and produce cytokines and other molecules that activate the macrophages to a point at which they can destroy the bacteria. They also activate cytotoxic T lymphocytes (CTL) that can destroy virally infected body cells. Macrophages will also remove the dead cells by phagocytosis. A summary of Th1 immune responses is shown in Figure 27.25.

Th2 cells help B cells produce antibodies, especially of the IgE class, as illustrated in Figure 27.26. This immune response is an effective way to remove larger parasites, including multicellular helminths, which are too large to be phagocytosed. IgE specifically recruits and activates eosinophils, which are cytopathic, and mast cells, which promote inflammation and physiological changes that lead to worm expulsion.

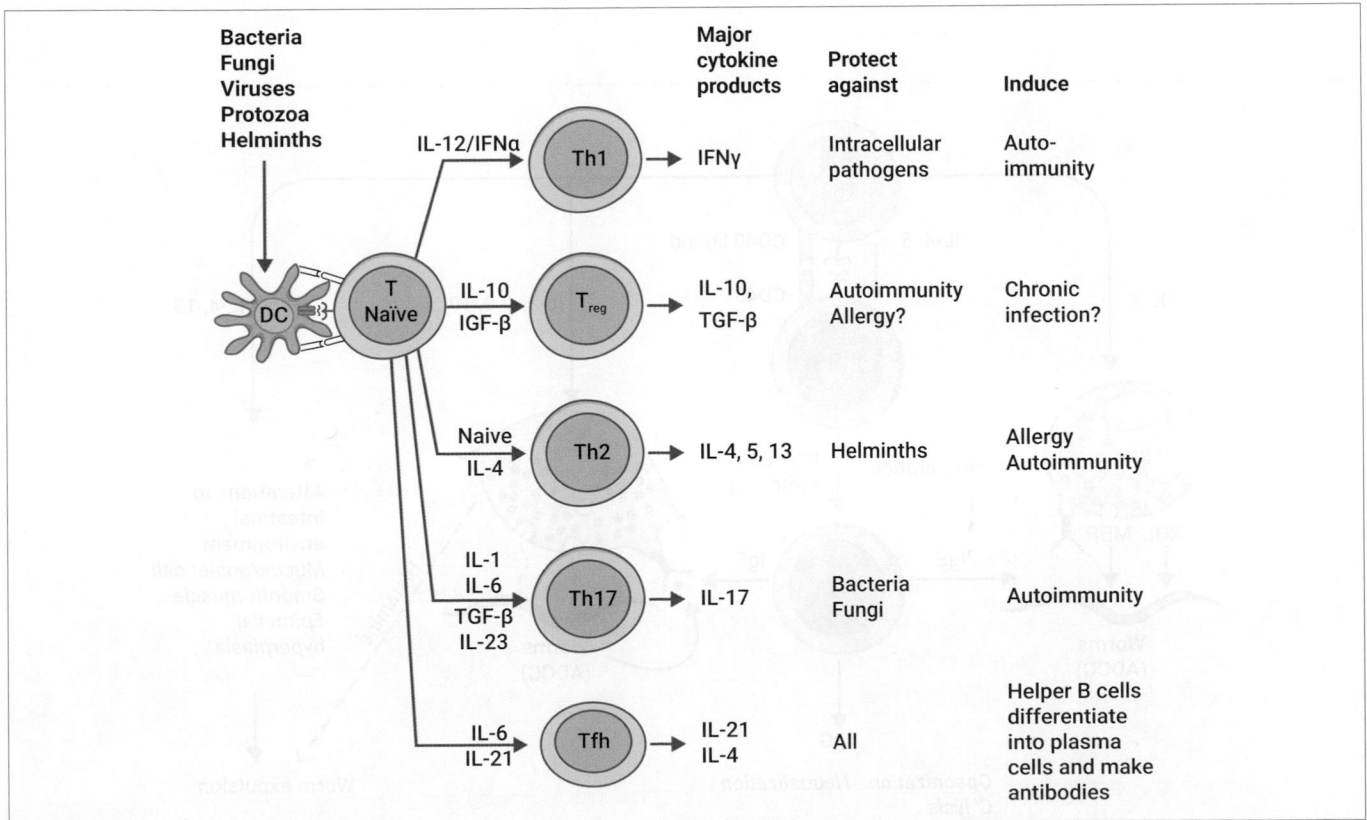

Figure 27.24 The main subsets of helper T cell and the signals that stimulate them. Different pathogens will stimulate different classes of helper T cells based on the cytokine signals received from tissue. For example, viruses will upregulate IL-12 and type I interferons in antigen-presenting cells and promote Th1 development. Th1 will activate cytotoxic T lymphocytes, which will destroy virally infected cells. By contrast, parasites will induce IL-4 secretion, which will promote Th2 development. This will promote IgE production and subsequent worm expulsion.

Source: Playfair & Bancroft. *Infection and Immunity*, 4th edn, 2013. Oxford University Press.

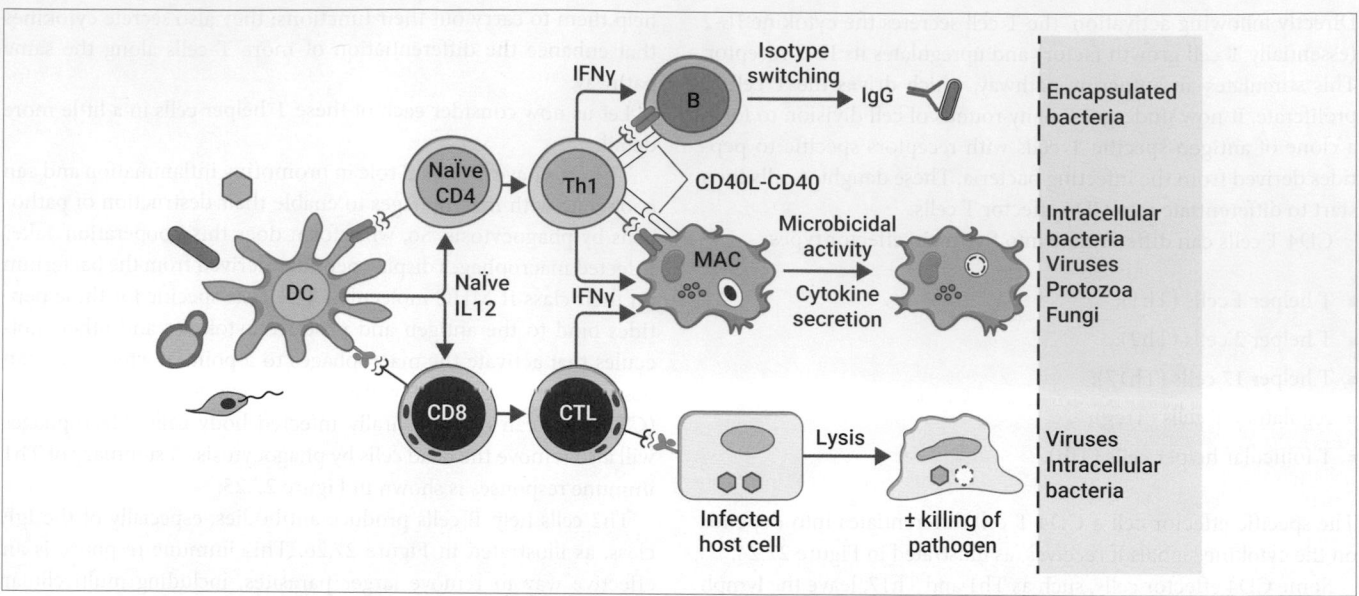

Figure 27.25 An illustration of the range of Th1 immune responses. Th1 immune responses are effective against a wide range of pathogens. They can stimulate B cells to produce IgG, activate macrophages and phagocytosis, and stimulate cytotoxic T cells (CTL) to destroy infected body cells.

Source: Playfair & Bancroft. *Infection and Immunity*, 4th edn, 2013. Oxford University Press.

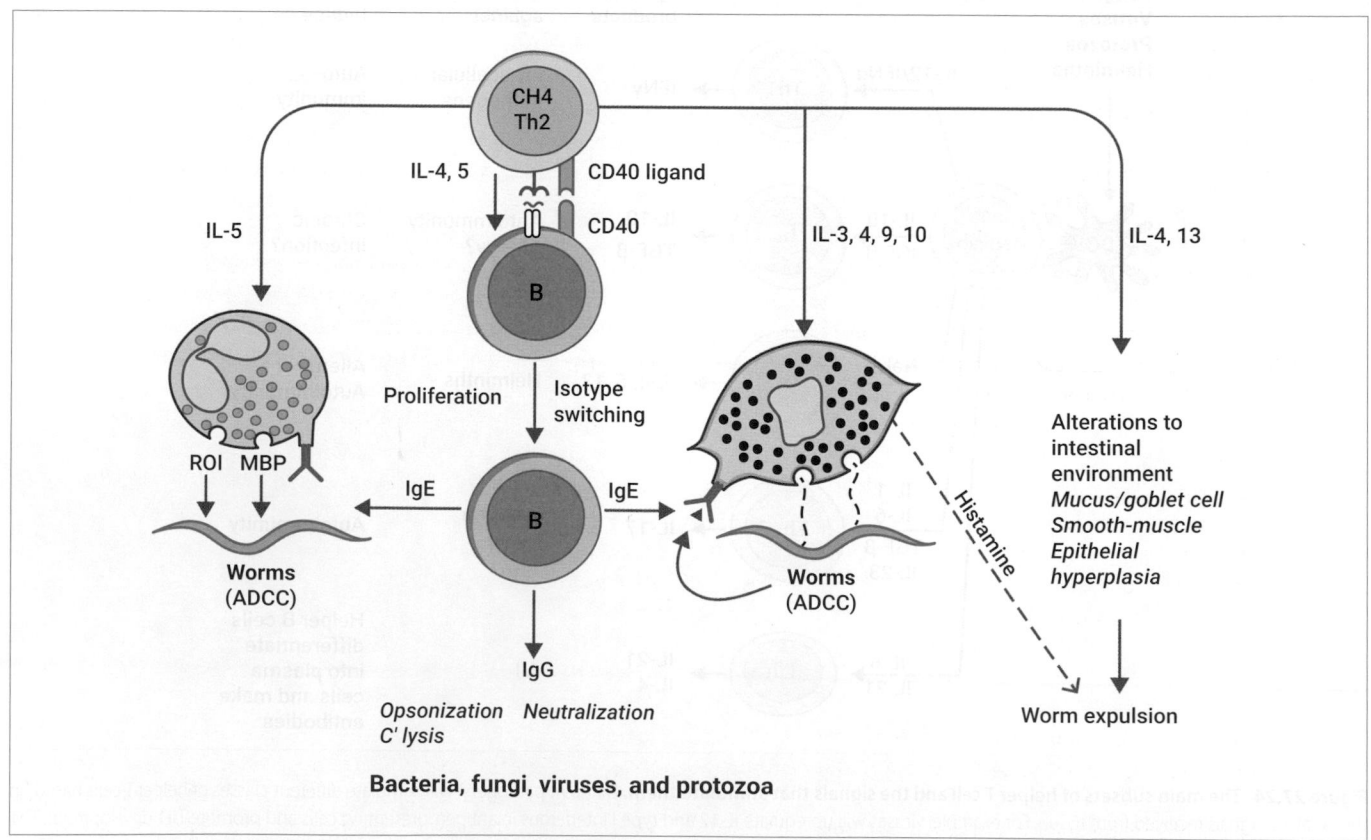

Figure 27.26 An illustration of Th2 immune responses.

Source: Playfair & Bancroft. *Infection and Immunity*, 4th edn, 2013. Oxford University Press.

Th17 cells produce the cytokine IL-17, which is important for health at barrier surfaces. IL-17 can enhance neutrophil responses and stimulate epithelial health through the induction of antimicrobial peptides.

Treg cells suppress immune responses. They also help to prevent autoimmunity, and help to bring an adaptive immune response to an end once it has run its course (see Clinical Box 27.4).

They are sometimes referred to as the immune 'police' for their ability to promote immune quiescence and to prevent immune-related damage. Some Treg cells differentiate in the thymus; others differentiate from naïve CD4 T cells outside the thymus in the presence of the anti-inflammatory cytokine TGFβ. Treg cells produce cytokines that are immunosuppressive and anti-inflammatory.

CLINICAL BOX 27.4 | Autoimmunity

Autoimmune diseases occur when immune self-tolerance breaks down and the immune system makes an adaptive immune response against the cells and tissues of its own body. This is equivalent to an army firing on itself—so-called 'friendly fire'—and causes chronic disease, which can be so damaging to tissues that the consequences can be life-changing, and even life-threatening. Autoimmunity is increasingly common and may affect up to 10% of the population.

Autoimmune diseases are categorized as either systemic or organ-specific. Systemic autoimmune diseases occur throughout the body. They include rheumatoid arthritis, which sees joints attacked in a non-specific way. As the name implies, organ-specific autoimmune diseases affect one main organ—for example, the thyroid (Graves' disease, Hashimoto's thyroiditis) or the pancreas (type 1 diabetes).

One of the first features of autoimmune disease to be discovered was the production of **autoantibodies**, that is, antibodies specific to self-proteins (**autoantigens**) that would normally be ignored by the immune system. Autoantibodies are produced in most autoimmune diseases. They are often diagnostic (i.e. they can act as a 'marker' that an autoimmune disease is occurring), and can also cause the symptoms of disease. For example, antibodies specific to the acetylcholine receptor interfere with receptor binding and lead to the symptoms of myasthenia gravis, which we discuss below.

Some autoimmune conditions can be treated by immunosuppressant drugs, such as methotrexate, azathioprine, mycophenolate mofetil, or ciclosporin. However, such treatments also lead to general immuno-suppression and a higher risk of infections. The pro-inflammatory cytokine tumour necrosis factor alpha (TNF-α) is involved in the inflammatory responses that are characteristic of many autoimmune diseases, and biologic drugs targeting TNF-α, such as the monoclonal antibody adalimumab, are now part of the therapeutic armoury. Monoclonal antibodies that remove B cells are also effective in some cases.

Why do some people develop autoimmune disease? It is thought to be a combination of genetic and environmental factors, and responses are usually initiated by some 'trigger' event such as an infection. (For example, it is thought that the immune response to a pathogen can cause an unfortunate, chance recognition of self-tissue in some people—a phenomenon called **molecular mimicry**.) This makes these diseases complicated to understand and treat. They are usually only diagnosed after the autoimmune response has caused some pathological effects, making it difficult to get close to the ultimate cause of these diseases and how they might be prevented.

Most autoimmune diseases are associated with particular HLA class I and class II peptide receptors, which act either as risk factors or protective factors. The association with HLA indicates that autoimmune diseases are T cell-dependent in origin, and that antigen presentation is involved.

Autoimmune conditions: some examples

More than 150 autoimmune conditions have been identified; we have selected just three examples that are representative of different presentations of autoimmunity.

Myasthenia gravis is a rare autoimmune disease that causes progressive muscle weakness following an autoimmune attack on the neuromuscular signalling system. In this disease, the autoantibodies produced directly cause the tissue damage. The most common cause of myasthenia gravis is IgG antibodies directed against the acetylcholine receptors at the neuromuscular junction, although other mediators of neuromuscular signals can also be a target. By binding to the receptors, the autoantibodies interfere directly with neuromuscular signalling.

People with myasthenia gravis can be treated with anticholinesterase drugs or immunosuppressants. Anticholinesterases block the enzymes that normally degrade the neurotransmitter acetylcholine, and its persistence at the neuromuscular junction allows some neuromuscular signalling to occur. B cell depletion therapy has proved effective for some patients who have not responded to other therapies.

Rheumatoid arthritis (RA) affects the synovial tissues that line the interior of joints. It most commonly presents as painful swelling of the joints in the hands and feet, although the knees and other joints may also be affected. The tissue damage results from inflammation of the synovial membrane inside the joint and subsequent erosion of bone and cartilage.

The inflamed joints contain CD4 and CD8 T cells, macrophages, monocytes, neutrophils, and plasma cells. The tissue damage is thought to be mediated by T cells and macrophages, which produce the pro-inflammatory cytokines IL-17 and TNF-α. Characteristic autoantibodies, called rheumatoid factor, are produced in rheumatoid arthritis, but their role in the disease is unknown. Indeed, rheumatoid factor is not a specific marker of RA: it can be seen in other systemic autoimmune rheumatoid diseases, including systemic lupus erythematosus (SLE) and connective tissue disease (CTD).

There is a strong genetic component to rheumatoid arthritis. For example, people expressing a particular allele of the DRB1 gene have a greater-than-usual risk of developing the disease.

Treatment for rheumatoid arthritis includes physiotherapy and drug treatments. The drug treatments range from non-steroidal anti-inflammatory drugs (NSAIDs), such as ibruprofen, which reduce pain and inflammation in less severe cases, through general immunosuppressant drugs, such as methotrexate in more severe cases, to the more recent introduction of biologic agents that target TNF-α and which have proven successful in suppressing symptoms and delaying progression in severe rheumatoid arthritis.

Multiple sclerosis is a neurodegenerative disease that results from immune attack on the myelin that sheathes the nerve cells in the white matter of the brain and spinal cord. Antigens from the brain do not normally reach the immune system, as they are confined behind the blood–brain barrier. In multiple sclerosis, however, it seems that some unknown trauma allows them to reach secondary lymphoid tissues, where they provoke the production of antigen-specific Th1 cells. These cells then return to the site of injury, pene-

trating the blood–brain barrier to set up an inflammation that also involves mast cells. The inflammation leads to demyelination and loss of nerve function by mechanisms that are still uncertain. Symptoms can include loss or blurring of vision, difficulties in walking, and paralysis of the limbs.

Drug treatments for multiple sclerosis, including monoclonal antibodies that deplete T cells and B cells, are available for the relapsing form, in which symptoms flare up and subside, followed by symptom-free periods as the myelin gets repaired. However, the most effective drugs carry the risk of serious side effects.

Tfh cells enter the follicles of secondary lymphoid tissues where they interact with B cells specific for the same antigen. This interaction helps the B cells to produce high-affinity antibodies, as we will discuss further shortly.

The activation of CD8 T cells

Naïve CD8 T cells produce only one main type of effector cell after they are activated by antigen encountered in secondary lymphoid tissues: they differentiate into cytotoxic T cells (CTL) with the aid of helper T (Th) cells. After differentiation, they leave the secondary lymphoid tissue and migrate to the site of infection, where they specifically kill infected body cells (for example, those infected by viruses).

Within the infected cell, pathogen-derived proteins in the cytosol are broken down into peptides by a catalytic structure called the proteasome. Peptides that have been generated by the proteasome, and subsequently released into the cytosol, are actively transported into the endoplasmic reticulum, where they associate with newly synthesized class I MHC proteins before being transported as a complex to the cell surface. Class I MHC proteins therefore present peptide antigens derived from intracellular pathogens and mark the cells for destruction, as illustrated in Figure 27.27.

The matching of a cytotoxic T cell to an infected target cell is achieved by the binding of CD8 co-receptors to class I MHC molecules, as depicted in Figure 27.28a. This means that when peptides from intracellular pathogens bound to class I MHC are recognized by a CD8 cell's T cell receptor, this interaction is stabilized by the CD8 co-receptor. Figure 27.28b illustrates how the effector CD8 T cell releases cytotoxic proteins onto the target cell, which neatly kill the cell by inducing apoptosis—a type of programmed cell death.

A single CD8 T cell can kill many infected cells in quick succession, leaving uninfected cells untouched. As almost all the cells of the body express class I MHC, any cell that is infected by a pathogen can be killed by CD8 cytotoxic T cells.

A proportion of activated CD8 T cells and CD4 T cells divide and enter the pool of 'memory' cells. If the same infection occurs again, the frequency with which the pathogen encounters antigen-specific T cells will be greater, and they will be easier to activate and stay close to where they were activated, making it possible to resolve the infection more rapidly. This is the secondary immune response (see Figure 27.13).

B cell activation and antibody production

The activation of B cells, when a specific antigen binds to the B cell receptor, can happen without T cells (it is 'T cell-independent') or can require T cell help (Figure 27.29). T cell-independent B cell responses are possible to repeating epitopes that are also pathogen-associated molecular patterns (PAMPs). These PAMPs can bind to B cell pattern recognition receptors and can also cross-link multiple antigen receptors on the B cell. For example, the repeating polysaccharides in bacterial cell walls can induce a strong enough signal in a B cell to enable differentiation into an antibody-producing plasma cell in the absence of T cell help. These T cell-independent B cell responses are therefore important contributors to the early clearance of bacterial infections.

A subset of B cells in the human spleen accumulates over the first few years of life and is particularly effective at making immune responses to T cell-independent antigens. Babies and young children, in whom this subset of B cells has yet to accumulate, are therefore more susceptible than adults are to infection with bacterial diseases, such as pneumococcal pneumonia.

T cell help to B cells is mainly given by Tfh cells, which stay in the lymphoid tissues. In a lymph node, for example, Tfh cells enter the follicles. At the follicle boundary they interact with naïve B cells, which have entered the lymph node from the blood and are travelling through the T cell areas. If the B cells recognize an epitope on the surface of a native antigen, they bind the antigen, internalize the receptor–antigen complex, degrade the antigen, and present antigen peptides bound to class II MHC on their surface. If these peptides are then recognized by Tfh cells, a B cell and a Tfh cell form a so-called 'cognate' pair and mutually signal to each other. This signalling results in the activation of the B cell, which then proliferates rapidly to form a **germinal centre** in the follicle. This overall process is illustrated in Figure 27.30.

Once activated, B cells temporarily stop expressing immunoglobulin on their surface, and undergo a process called **somatic hypermutation**. During this process, mutations are targeted to the parts of the rearranged variable-region gene that encode the antigen-binding site, and can result in a B cell receptor being expressed that binds the antigen more strongly, more weakly, or the same as the original receptor.

Figure 27.27 Class I MHC proteins. Class I MHC proteins present peptide antigens derived from intracellular pathogens to naïve CD8 T cells. *Source*: Playfair & Bancroft. *Infection and Immunity*, 4th edn, 2013. Oxford University Press.

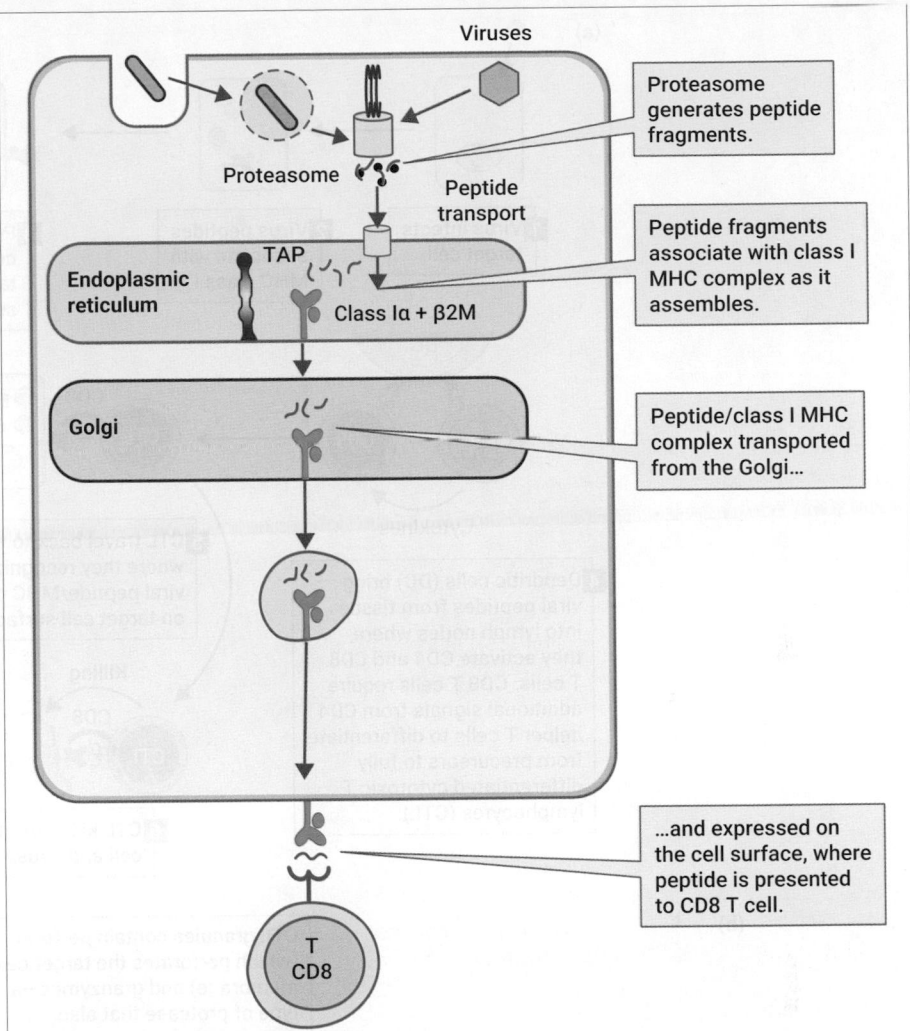

The B cells compete for antigen bound to specialized cells in the germinal centre: those with the highest affinity for antigen compete most effectively to bind it. They then internalize it, degrade it, and display antigen peptides on their surface, which enables them to interact further with Tfh cells and receive survival signals. B cells that cannot compete for antigen do not receive survival signals and die by apoptosis.

B cells making the best-fit antigen receptors then go on to either proliferate and differentiate into antibody-producing plasma cells, or go through additional rounds of mutation and selection. As the immune response progresses, the antibodies produced will become of higher and higher affinity for the specific pathogen, and so more effective at clearing it. This process is known as affinity maturation.

At the same time as undergoing affinity maturation, B cells are undergoing a process called **class switching** in the germinal centre. Class switching is a type of DNA recombination that sees the constant-region gene attached to the rearranged variable-region sequence being switched. This switching enables a B cell to change from the production of IgM (which is of relatively low affinity) to either IgG, IgA, or IgE (which are of higher affinity, and are more effective at clearing pathogens, albeit while requiring T cell help). Cytokines control which immunoglobulin constant-region heavy chain is switched to and is expressed. However, their pathogen specificity remains unchanged. This enables the different functional properties of the secreted antibodies to be exploited.

Germinal-centre B cells ultimately differentiate into either memory cells or plasma cells. Memory B cells produced at the end of a primary immune response are ready to undergo somatic hypermutation and class switching to produce high-affinity antibodies of a useful class if the pathogen is encountered again. Alternatively, B cells differentiate into plasma cells that secrete antibodies, which stay in the circulation for a considerable time.

B cells and T cells are usually specific for different epitopes derived from a given pathogen. B cells can recognize any component of the antigen that is accessible on the *surface* of the pathogen. By contrast, T cells can recognize a peptide (in combination with the peptide receptor, MHC) from any part of the structure, including the inside (which is not visible to the B cell). For a B cell to receive T cell help, therefore,

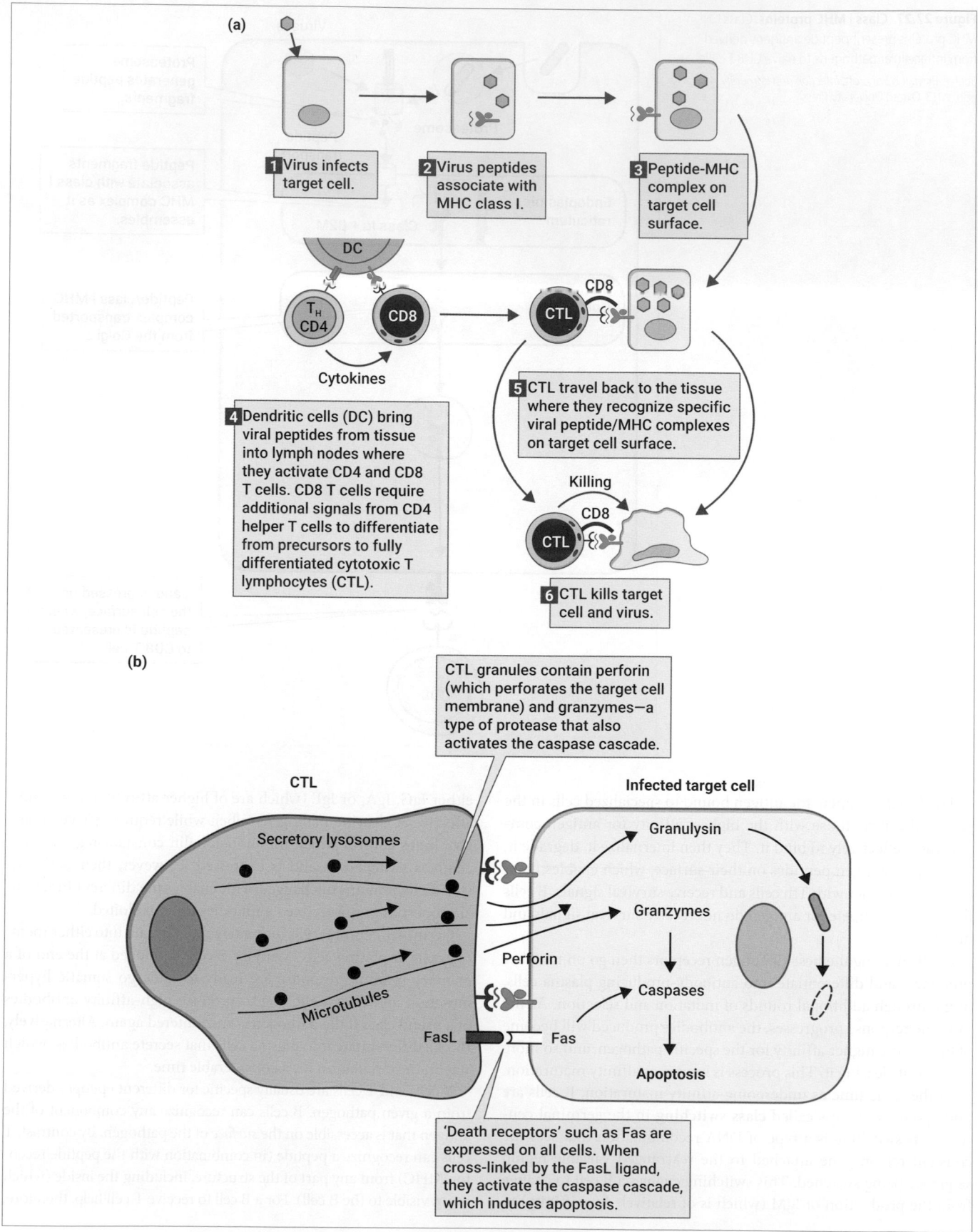

Figure 27.28 The mode of operation of cytotoxic T cells. (a) The destruction of virally infected cells by cytotoxic T lymphocytes (CTL). (b) CTL kill infected target cells by inducing programmed cell death.

Source: Playfair & Bancroft. *Infection and Immunity*, 4th edn, 2013. Oxford University Press.

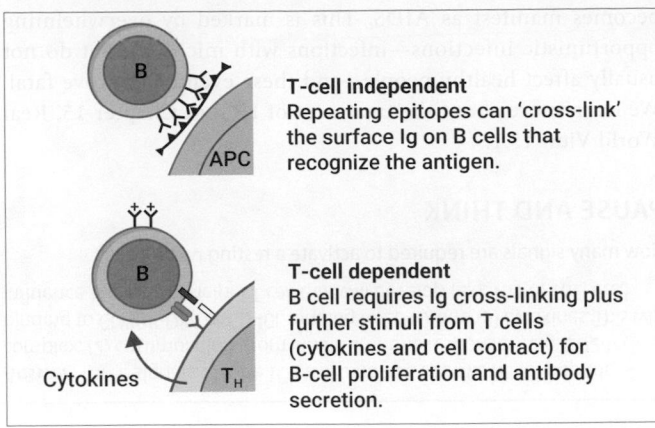

Figure 27.29 T cell-independent and T cell-dependent B cell activation.

Source: Playfair & Bancroft. *Infection and Immunity*, 4th edn, 2013. Oxford University Press.

T-cell independent
Repeating epitopes can 'cross-link' the surface Ig on B cells that recognize the antigen.

T-cell dependent
B cell requires Ig cross-linking plus further stimuli from T cells (cytokines and cell contact) for B-cell proliferation and antibody secretion.

the antigen it recognizes must have a protein component. This discovery has helped an effective vaccine to be produced against a common cause of pneumonia, the bacterium *Haemophilus influenzae*.

Check your understanding of the concepts covered in this section by answering the questions in the e-book.

27.5 Immunodeficiencies: diseases affecting how lymphocytes develop or function

Rarely, babies are born with genetic defects that affect the development or function of their lymphocytes. In the most severe cases, children rarely survive their first year, succumbing to repeated infections with any of the common micro-organisms they would normally encounter. Modern medical treatments, such as bone marrow transplants and gene therapy, now give such children a

Figure 27.30 The activation of B cells by helper T cells.

Source: Playfair & Bancroft. *Infection and Immunity*, 4th edn, 2013. Oxford University Press.

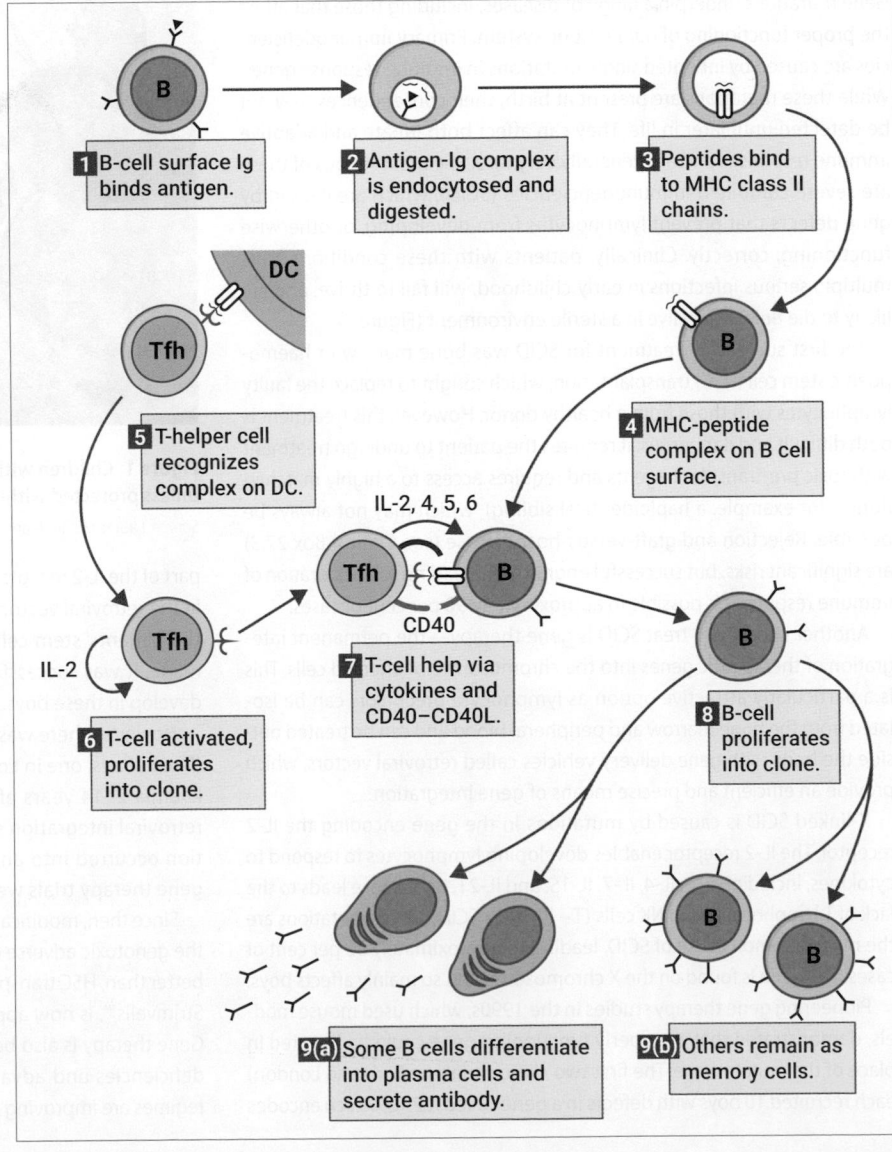

1 B-cell surface Ig binds antigen.

2 Antigen-Ig complex is endocytosed and digested.

3 Peptides bind to MHC class II chains.

4 MHC-peptide complex on B cell surface.

5 T-helper cell recognizes complex on DC.

IL-2, 4, 5, 6

CD40

6 T-cell activated, proliferates into clone.

7 T-cell help via cytokines and CD40–CD40L.

8 B-cell proliferates into clone.

IL-2

9(a) Some B cells differentiate into plasma cells and secrete antibody.

9(b) Others remain as memory cells.

chance of life, as we discuss further in Clinical Box 27.5. The rarity of the genetically determined severe immunodeficiencies attests to the vital function of the immune system in keeping us healthy. Immunodeficiency can also be caused by non-genetic factors, such as ageing or disease. The virus infection that has arguably had most impact on humanity in the past 40 years is the human immunodeficiency virus (HIV), which, if untreated, inevitably leads to a fatal immunodeficiency called AIDS (acquired immunodeficiency syndrome). HIV currently infects around 40 million people worldwide, and despite advances in treatment and prevention, over one million people become infected each year and around the same number die.

The immune deficiency that results from infection with HIV is caused by the efforts of CD8 T cells to eradicate virus-infected CD4 T cells. These efforts suppress, but cannot remove, the virus. If HIV infection is untreated, levels of CD4 T cells will gradually fall over a period of years until the immunodeficiency becomes manifest as AIDS. This is marked by overwhelming opportunistic infections—infections with microbes that do not usually affect healthy people—and these eventually prove fatal. We learn more about the treatment of HIV in Chapter 15, Real World View 15.1.

PAUSE AND THINK

How many signals are required to activate a resting naïve T cell?

Answer: Three signals: (1) The TCR will bind to a specific MHC–peptide complex; (2) costimulation (from innate immune responses), e.g. CD28 binding to CD80/CD86 or CD40L binding to CD40; and (3) cytokines. This will influence the type of helper T cell that will develop (e.g. TH1 or TH2, etc.).

 Check your understanding of the concepts covered in this section by answering the questions in the e-book.

CLINICAL BOX 27.5 — Gene therapy for the treatment of severe combined immunodeficiency

Gene mutations underpin a range of diseases, including those that affect the proper functioning of our immune system. Primary immunodeficiencies are caused by inherited single mutations in immune response genes. While these mutations are present at birth, their consequences may not be detected until later in life. They can affect both innate and adaptive immune responses and are generally very rare. The most serious of these are severe combined immunodeficiencies (SCID), which are caused by gene defects that prevent lymphocytes from developing, or otherwise functioning, correctly. Clinically, patients with these conditions have multiple serious infections in early childhood, will fail to thrive, and are likely to die unless they live in a sterile environment (Figure 1).

The first successful treatment for SCID was bone marrow or haemopoeitic stem cell (HSC) transplantation, which sought to replace the faulty lymphocytes with those from a healthy donor. However, this treatment is both difficult and dangerous: it requires the patient to undergo treatment with toxic pre-transplant agents and requires access to a highly matched donor (for example, a haploidentical sibling), which may not always be possible. Rejection and graft-versus-host disease (see Clinical Box 27.3) are significant risks, but successful engraftment leading to a restoration of immune responses is possible in approximately 90 per cent of cases.

Another strategy to treat SCID is gene therapy—the permanent integration of therapeutic genes into the chromosomes of affected cells. This is a particularly attractive option as lymphocyte precursors can be isolated from the bone marrow and peripheral blood and can be treated outside the body with gene delivery vehicles called retroviral vectors, which provide an efficient and precise means of gene integration.

X-linked SCID is caused by mutations in the gene encoding the IL-2 receptor. The IL-2 receptor enables developing lymphocytes to respond to cytokines, including IL-2, IL-4, IL-7, IL-15, and IL-21. Its absence leads to the lack of T lymphocytes and NK cells (T− B+ NK− SCID). These mutations are the most common cause of SCID, leading to approximately 30 per cent of cases. The gene is found on the X chromosome and so mainly affects boys.

Pioneering gene therapy studies in the 1990s, which used mouse models, demonstrated that a properly functioning gene could be inserted in place of the mutated one. The first two clinical trials (in Paris and London) each recruited 10 boys with defects in a gene called *IL2RG*, which encodes

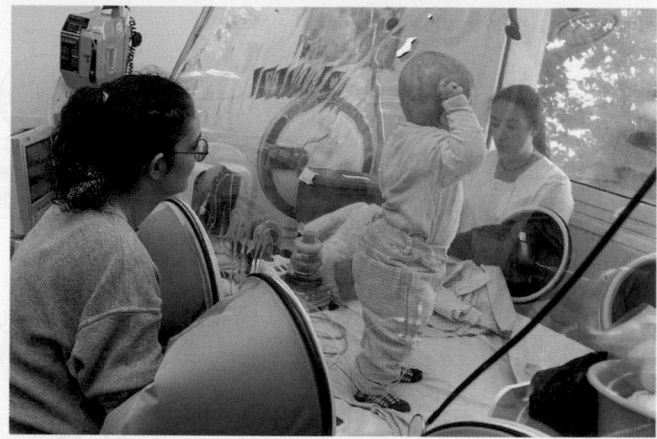

Figure 1 **Children with severe immunodeficiencies are unable to survive unless protected within a sterile environment.**
Source: Laurent/BSIP/Alamy.

part of the IL-2 receptor (the IL-2 gamma chain). IL2RG cDNA was included in the retroviral vector, with the hope that it would insert into the DNA of the patients' stem cells and enable the correct receptor proteins to be made. It was successful: gene therapy enabled mature lymphocytes to develop in these boys.

However, there was an unfortunate side effect that led to six patients (five in Paris, one in London) developing T cell acute lymphoblastic leukaemia 2–14 years after treatment. Further studies showed that the retroviral integration sites were not random (as expected), but integration occurred into an oncogene locus in five patients. Consequently, gene therapy trials were stopped until better vectors were developed.

Since then, modifications in process have led to good outcomes without the genotoxic adverse events, and outcomes are at least as good as, if not better than, HSC transplantation. The first gene therapy treatment for SCID, Strimvelis™, is now approved by regulatory bodies in the US and Europe. Gene therapy is also being applied to a wider range of primary immunodeficiencies and advances in vector technology and pre-conditioning regimes are improving patient outcomes for these once fatal diseases.

SUMMARY OF KEY CONCEPTS

- The immune system is designed to recognize and destroy agents that cause harm to the body.

- The immune system comprises primary lymphoid tissues, where the immune cells are generated, and secondary lymphoid tissues that contain mature immune cells and are where immune responses take place.

- Innate immune protection is the protection you are born with. It responds rapidly but identically on multiple challenges and has no memory.

- The adaptive immune system is dependent on lymphocytes that have a unique receptor for antigen.

- Lymphocytes respond to their specific antigens and generate immunological memory. This enables the immune system to

make faster and better responses on second and subsequent encounters with that antigen.

- The two main populations of lymphocytes are the B cells that produce antibodies and the T cells that have many different roles in helping immune responses.

- Vaccination is one of the major medical achievements of our time, and can provide lifelong immunity against a range of infectious diseases.

- A dysregulated immune response can cause disease—allergy, autoimmunity, inflammation—and contribute to pathology.

 Use the flashcards in the e-book to test your recall of key terms introduced in this chapter.

QUESTIONS

 Looking for answers? Once you've answered these questions, follow the link in the e-book to the answer guidance and check your work.

Concepts and definitions

1. What is the principle behind innate recognition of pathogens?

2. What is the critical first phase in defence against viruses?

3. How do you have so many different specific lymphocytes that can potentially recognize any pathogen you may be exposed to during the course of your life?

4. What signals are required to fully activate an effector T helper cell?

5. Describe the main antibody classes and their functions.

Apply the concepts

6. How does IFN-γ contribute to immune responses?

7. What would happen to a patient taking chemotherapeutic drugs that inhibit the proliferation of cancer cells?

8. What would happen if your thymus was removed at birth compared to adulthood?

9. How does vaccination exploit the adaptive immune response to protect us from disease?

10. What effect will antibodies that block costimulatory molecules have on immune responses?

Beyond the concepts

11. Why do you think helminth products are being developed to treat allergy? (Hint: what kind of immune response do we make to helminths?)

12. Fc receptors (FcRs) are responsible for many biological functions of antibodies. Why do you think there are no FcRs specific for IgM? (Hint: think about the properties of IgM and remember it is the first antibody produced during a primary immune response.)

13. Why do you think HLA-B/HLA-C bind to NK cell inhibitory receptors? (Hint: HLA-B/HLA-C are class I MHC molecules in humans.)

14. Discuss why immune-related adverse events occur following checkpoint blockade therapy (Clinical Box 27.2. Hint: successful tumours are anti-inflammatory environments.)

15. Discuss what Covid-19 has taught us about the immune responses to viruses in children and older adults. (Hint: the usual U-shaped curve for morbidity and mortality across the age spectrum does not fit this new virus. Death is directly proportional to older age and associated with excessive inflammation/reduced lymphocytes in lung tissue.)

FURTHER READING

Chaplin, D. D. (2010) Overview of the immune response. *J. Allergy Clin. Immunol.* **125**: S2–S23.
For more on the whole immune system.

Cunningham, A. C. (2016) *Thrive in Immunology.* Oxford University Press.
For simple revision on the subject, you also might like to consult this textbook.

Current Opinion in Immunology is an excellent resource to keep up to date with key new findings in the field. The editions are themed to cover the key areas (e.g. innate immunity, lymphocyte development and activation, vaccination, autoimmunity, allergy, and hypersensitivity), are written by experts in the field, and highlight the key publications (with some short explanations on why they are important).
https://www.sciencedirect.com/journal/current-opinion-in-immunology

Innate immunity

Johnston, R. B. (last updated 2020) An overview of the innate immune system.
https://www.uptodate.com/contents/an-overview-of-the-innate-immune-system

Adaptive immune system

Heimall, J. (last updated 2019) The adaptive cellular immune response: T cells and cytokines.
https://www.uptodate.com/contents/the-adaptive-cellular-immune-response-t-cells-and-cytokines

Immune diseases

The NIH in the UK and the CDC in the US have lists of diseases on their website, which include genetic immunological disease and which give accurate information in a publicly accessible way.
https://www.nihr.ac.uk/explore-nihr/specialties/ https://www.cdc.gov/

The WHO is good for up-to-date information on HIV, tropical diseases, and how immunization programmes are progressing.
https://www.who.int/health-topics/

New immunological therapeutics

This collection in the journal *Nature* covers new immunotherapy research into treating cancer. They are very accessible as the pieces are written by science writers mostly, rather than by researchers.
https://www.nature.com/collections/btwlvcpdls

Vaccines

The best source is The Green Book from the UK Government, which has the latest information on vaccines and vaccination procedures, for vaccine preventable infectious diseases in the UK.
https://www.gov.uk/government/collections/immunisation-against-infectious-disease-the-green-book

Some key CD numbers to know

CD number	Biochemical features	Cellular expression	Function
CD3	Made up of δ (20 kDa), ε (20 kDa), and γ (25–28 kDa) chains	T cells, thymocytes Used to identify T cells	Associated with the T cell antigen receptor (TCR) and contributes to T cell signalling and cellular activation
CD4	55 kDa transmembrane protein, member of the immunoglobulin superfamily domain containing four immunoglobulin-like extracellular domains	T helper cells, some thymocytes Some peripheral blood monocytes and tissue macrophages	Accessory molecule which aids helper T cells bind to class II MHC/peptide complexes and contributes to T cell signalling and activation Receptor for HIV gp120
CD8	Either a homodimer of two CD8α chains or a heterodimer of a CD8α and CD8β chain (32–34 kDa)	Cytotoxic T cells and their precursors, some thymocytes	Accessory molecule which aids cytotoxic T cells bind to class I MHC/peptide complexes
CD14	53–55 kDa membrane-bound glycosylated protein	All peripheral blood monocytes, most tissue macrophages, and some granulocytes Used to identify monocytes	Receptor for lipopolysaccharide (LPS) on Gram-negative bacteria
CD19	95 kDa transmembrane protein, member of the immunoglobulin superfamily domain containing two immunoglobulin-like extracellular domains	B cells Used to identify B cells	Part of a co-receptor (with CD21 and CD81) which forms a complex with the B cell antigen receptor (BCR) and contributes to the signalling process
CD25	45 kDa transmembrane protein, can exist as a homodimer, heterodimer (with IL-2R β CD122) or tripartite receptor (with IL-2Rβ and common γ chain CD132)	T cells, B cells, monocytes Used to identify activated cells Expressed by a subset of CD4+ regulatory T cells (Tregs)	IL-2 receptor alpha Dimeric CD25 forms the low-affinity IL-2 receptor (binds IL-2 but does not signal) The high affinity IL-2 receptor consists of IL-2α, IL-2β, and common γ chain (binds IL-2 and signals T cell proliferation)

CD28	44 kDa transmembrane protein expressed as a homodimer, member of the immunoglobulin superfamily domain containing a single immunoglobulin-like extracellular domain	T cells	Costimulatory molecule that binds to CD80 and CD86 on antigen presenting cells Provides signal 2 for T cell activation
CD34	105–120 kDa heavily glycosylated type I membrane protein (sialomucin)	Haematopoietic stem and progenitor cells Some endothelial cells Used to identify haematopoietic stem cells	Cell adhesion and signalling Binds to CD62L (l-selectin)
CD40	48 kDa type I transmembrane protein	B cells, macrophages, dendritic cells Some epithelial cells	Costimulatory molecule binds to CD40L (CD154) on activated helper T cells Leads to B cell activation and differentiation
CD45	Long single chain type I transmembrane molecule. Multiple isoforms exist which are differentially spliced (CD45RA 205–220 kDa, CD45RO 180 kDa, CD45RB 190–220 kDa)	All haematopoietic cells Used to identify leukocytes/bone marrow-derived cells CD45RA identifies naïve T cells CD45RO identifies activated or memory T cells	Intracellular tyrosine phosphatase, which plays a key role in B and T cell signalling and activation
CD56	Transmembrane protein (135–220 kDa). Multiple isoforms exist, member of the immunoglobulin superfamily domain containing five immunoglobulin-like extracellular domains	Natural killer (NK) cells Used to identify NK cells (CD3–CD56+ cells)	Adhesion molecule (neural cell adhesion molecule 1)
CD80 (also referred to as B7.1)	60 kDa transmembrane protein, member of the immunoglobulin superfamily domain containing two immunoglobulin-like extracellular domains	Activated B cells Some activated T cells and macrophages	Costimulatory molecule expressed by antigen presenting cells (APC), which provides signal 2 to T cells. Binds to CD28 (which leads to T cell activation) and CTLA4 (which can lead to activation induced cell death at the end of the immune response)
CD86 (also referred to as B7.2)	80 kDa transmembrane protein with a similar overall structure to CD80	Dendritic cells, monocytes, activated B cells	Costimulatory molecule expressed by antigen presenting cells (APC), which provides signal 2 to T cells. Binds to CD28 (which leads to T cell activation) and CTLA4 (which can lead to activation induced cell death at the end of the immune response)
CD127	52 kDa (isoform 1) transmembrane protein. Member of the immunoglobulin superfamily (four isoforms exist: others are 34 kDa, 30 kDa, and 29 kDa)	Lymphocyte precursors, T cells Not expressed in a subset of regulatory T cells (Tregs), which can be identified as CD4+ CD25+ CD127–	Cytokine receptor, also known as IL-7 receptor alpha Forms a heterodimer with the common gamma chain (CD132)

CD refers to 'cluster of differentiation' and simply identifies molecules expressed on immune (and other) cells that are identified by a group of monoclonal antibodies. CD numbers are assigned by international agreement at human leukocyte differentiation antigens' (HLDA) workshops. They are used to identify or phenotype cells and are generally (but not always) surface expressed.

MODULE FOUR

Organismal Diversity
Structure, Adaptation, and Survival

28 The Structure of Living Organisms

29 Body Plans

30 Interaction with the External Environment

31 Movement, Locomotion, and Migration

32 Defence Against Predation and Invasion

33 Reproduction and Development

MODULE FOUR

Biology is unique amongst the sciences because it studies things which are alive. So far as we know, our planet is the only place in the Universe where live objects exist. Life on Earth is magnificently abundant and astonishingly diverse, but it is also constantly changing. As biologists we strive to know where life came from, how it works, and why it has so many different forms.

Biology is the study of the diversity of life—but this presents a challenge: how shall we begin to understand life in all its varied forms, and how can we study it in an organized and productive way? It will help if we can identify some of the principles that underpin how all life operates, but we must also appreciate the uniqueness of individual organisms and understand how they survive and flourish.

Besides this, every organism must face the physical imperatives of our extraordinary planet. These include gravity, temperature, light, humidity, gases, and fluid movements, in addition to geological features and seasonal fluctuations. So, as well as studying what goes on inside cells and bodies, biologists investigate how each organism comes to terms with those external conditions and the many processes it cannot control.

In this module we will study biological organisms by looking at their structure, their life processes, and their interactions with the environments they inhabit. Our objective is to emphasize diversity while understanding how each individual organism meets the challenges it faces. To achieve this, we will organize material not by a system of classification but by the range of biological processes organisms use and the ways in which they engage with their surroundings.

Differences will be as informative as similarities, so you will find plants alongside animals, sponges alongside mammals, and fish alongside insects. Studying life this way will give us an appreciation of why it has been so remarkably successful in virtually every imaginable part of Earth's surface.

The information in this module is principally about eukaryotes—those forms of life based on cells with complex genetic material, membrane-enclosed compartments, and internal energy regulators. We discuss prokaryotes (bacteria) in more detail in Module 2.

Image: In adapting to survive in different environmental niches, many organisms have developed mutually beneficial associations. Here, a tree shrew licks the nectar from the lid of a giant pitcher plant on the rain-washed slopes of Mount Kinabalu, Borneo. In return, it uses the plant as a toilet, providing the plant with a vital source of nutrients. *Source:* Paul Williams/Science Photo Library.

The Structure of Living Organisms

LEARNING OBJECTIVES

By the end of this chapter you should be able to:

- Describe the basic structure of animal and plant cell membranes and the organelles they enclose.

- Explain how water moves through biological membranes and the physical forces which determine its rate of movement.

- Distinguish, using precise definitions, between osmosis, simple diffusion, facilitated diffusion, and active transport, noting the extent to which they are energy-dependent.

- List the functional characteristics of a cytoskeleton and describe the basic molecular structures which make up the cytoskeletons of plant and animal cells.

- Explain the role and importance of turgor in plant tissues and describe the basic mechanisms by which it is regulated.

- Distinguish between:

  a) unicellular, multicellular, and syncytial; and

  b) totipotency, pluripotency, and multipotency.

- Describe the structure of the following types of connections between animal cells and explain their characteristic functions: tight junctions, adherens junctions, desmosomes, gap junctions.

Chapter contents

Introduction   904

28.1   Cell membranes   904

28.2   Membrane transport: talking to the outside   907

28.3   The cytoskeleton   914

28.4   Volumetrics, pressures, and turgor in plants   917

28.5   Directionality and anchorage dependence   919

28.6   Colony and organism formation   922

28.7   Multicellular plants   927

28.8   Multicellular animals: Metazoa   930

28.9   Cell lineages and differentiation   933

28.10   Communication between cells   940

⟲ Watch the key concepts video in the e-book to prepare yourself for studying this chapter.

Introduction

Biological organisms are alive. Defining 'life' is a difficult thing to do (as we discover in Topic 3) but we can accept that it exists and we know what we mean when we speak of it. Life involves the uptake, storage, and transmission of information and energy, together with interactions between energy-rich molecules and the formation of complex structures. These events take place inside and between individual living units: cells, tissues, organs, bodies, and even whole communities. These units are what gives the biological world its identity.

The basic units of life, especially cells and the organelles inside them, are enclosed structures. They are tiny portions of the matter and energy of the Universe, bounded by the **membranes** which surround them. Each membrane, whether it is the wall of a bacterium, the cellulose wall of a plant cell, or the phospholipid membrane of an animal cell, or even if it surrounds a nucleus, a mitochondrion, or a chloroplast inside a cell, can be thought of as delineating and separating off part of the biological world.

Membranes isolate and restrict the space in which molecular interactions can take place. In an unrestricted, unenclosed space the physical process of **diffusion** (see Section 28.2) will eventually result in an evenly spread, low density, random distribution of molecules. Energy becomes dispersed and there is an increase in entropy (disorder). If the space is enclosed, molecular movements are spatially restricted, local concentrations of energy can be higher, and interactions become organized rather than chaotic.

Membranes enclose biological spaces but they are far from inert. They control the movement of molecules into and out of structures and they determine how each organism interacts with its external environment. Interactions include responses to chemicals and other external signals, attachments to other cells, anchorage to surfaces, movements, dispersal, and other dynamic processes. In complex organisms, membranes enable cells to link up and form tissues and organs.

In this chapter we consider how cells and multicellular structures work and how their boundary membranes regulate their external interactions. Our focus here is on eukaryotic life forms—plants and animals. (Module 2 examines prokaryotic life forms—bacteria and other microorganisms—and explains the distinction between eukaryotes and prokaryotes.) In later chapters we discuss how plant and animal bodies are structured, how they move, how they defend themselves, and how they reproduce.

Cells are self-contained units of life and what goes on inside them is revealed by investigating their biochemistry. We explore these topics in Modules 1 and 2. At the other end of the biological scale, Module 3 explains how the functions of cells and tissues underpin the physiology of the body, while Module 5 puts life in its environmental context.

28.1 Cell membranes

The membrane of a cell is often called the plasma membrane or **plasmalemma**, reflecting the fact that in complex multicellular animals it forms an interface with the surrounding *plasma*—the liquid part of blood. (*Lemma*, meaning husk or shell, comes from a Greek verb meaning 'to peel'.) Despite its origin, the term is applied to plant as well as animal cell membranes.

The plasma membrane is a **phospholipid bilayer**, as illustrated in Figure 9.6; it contains a complex collection of proteins and other molecules. These include enzymes, pores, pumps, transporters, and receptors. Some of these regulate how substances enter and leave the cell while others allow the cell to react to external signals, communicate with other cells or form multicellular structures.

Biological membranes are selective in the way they let molecules pass through. Molecular transport depends on electrical charge, solubility, and size and can be both passively and actively regulated. Regulation of the internal environment is crucial to cell function, especially in terms of water content, solute concentration, pH, the uptake of nutrients and excretion of waste products, and the response to hormones and other signals.

Chapter 9 explains the biochemistry of cell membranes in detail. An important point is that cell membranes represent a *lipid phase*, separating the *aqueous phase* of the cell cytoplasm from the *aqueous phase* of the extracellular environment. This feature of the membrane means that hydrophobic molecules can dissolve in and pass through the membrane relatively easily while hydrophilic molecules, including water itself, can only pass through with the aid of transporter molecules or channels.

In addition to the plasmalemma, in many organisms the cell membrane is surrounded by a cell wall. Such organisms include plants, fungi, most algae, and most prokaryotes. The cell membrane of many animal cells is surrounded by a cell coat, called the **glycocalyx**, made of **oligosaccharides**, or **glycans**, including glycoproteins and glycolipids. In cells where there is a lot of fluid movement in the external environment, such as in the gut, the respiratory system, the excretory system, and the reproductive system, there is also a layer of **mucus**.

Plant cells and their membranes

A distinctive characteristic of plant cells is the presence of a tough cell wall. Figure 28.1 shows how this cell wall lies on the outside of the plasma membrane, which is normally pressed tightly up against it. This gives plant cells a more robust external architecture than that found in animal cells. It increases the rigidity of soft tissues but it also allows complex three-dimensional morphologies, a feature which is important to cell differentiation and function of different types of plant cells.

The cell walls of ferns, mosses, and higher plants are built primarily from the polymer **cellulose**. It is slightly different in the lower plants such as algae. Cellulose is a polymer of glucose and closely related to **chitin**, the glucose polymer found in the cell walls of fungi and the exoskeletons of insects (see Section 9.5, Plant and fungal cell walls).

Cellulose is laid down in thin strands called **microfibrils**, by an enzyme complex called cellulose synthase. The cell wall is therefore composed of a network rather like a steel mesh. The arrangement of the microfibrils can vary. For example, Figure 28.2 illustrates how randomly orientated microfibrils allow the cell to expand in all directions, whereas parallel microfibrils allow expansion in a direction perpendicular to the plane of microfibril orientation, enabling cells to extend and take up complex three-dimensional shapes.

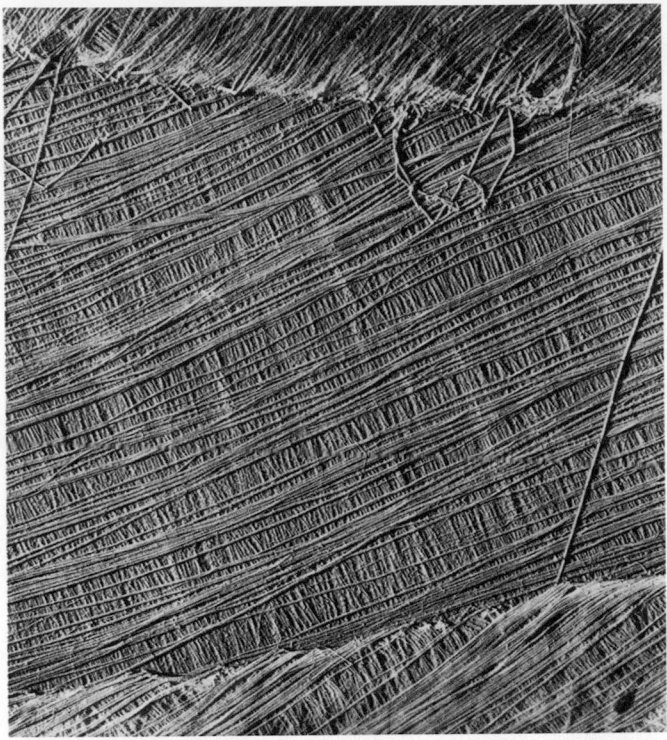

Figure 28.1 The plant cell wall. Scanning electron micrograph of an algal cell wall looking from the outside of the cell. The cell wall is made of a complex array of individual cellulose strands, called microfibrils, laid down in different orientations to produce a very tough structure.

Source: Biophoto Associates/Science Photo Library.

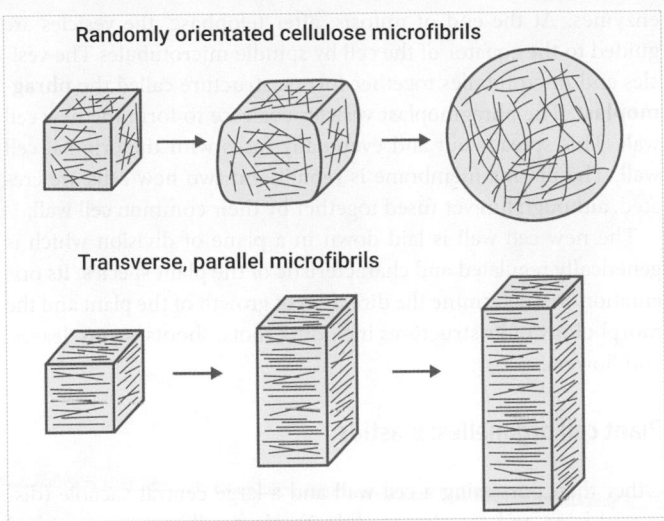

Figure 28.2 Orientation of cellulose microfibrils in plant cell wall. The nature of cell expansion is related to the orientation of cell wall microfibrils.

Source: Lincoln Taiz, Eduardo Zeiger, Ian Max Møller, and Angus Murphy, *Fundamentals of Plant Physiology*, 1st Edition, 2018, Oxford University Press.

In some algae, additional molecules are present in the cell wall: in species of red algae, **agarose** is a major component. This is extracted commercially from harvested algae for use as a gelling agent in the food industry and as a growth medium in microbiological research laboratories.

The presence of a complex cell wall means that cell division in plant cells is quite different from that in animal cells. In the final stages of *animal* cell division (which we discuss in more detail in Chapter 10), the plasma membrane develops a constriction point which eventually causes fission into two new separate cells. When a *plant* cell divides, the two daughter cells remain joined together and never move apart, as illustrated in Figure 28.3.

The interface between the dividing plant cells is constructed from collections of **vesicles**, made in the Golgi apparatus of the cell. These contain new cell wall components and their manufacturing

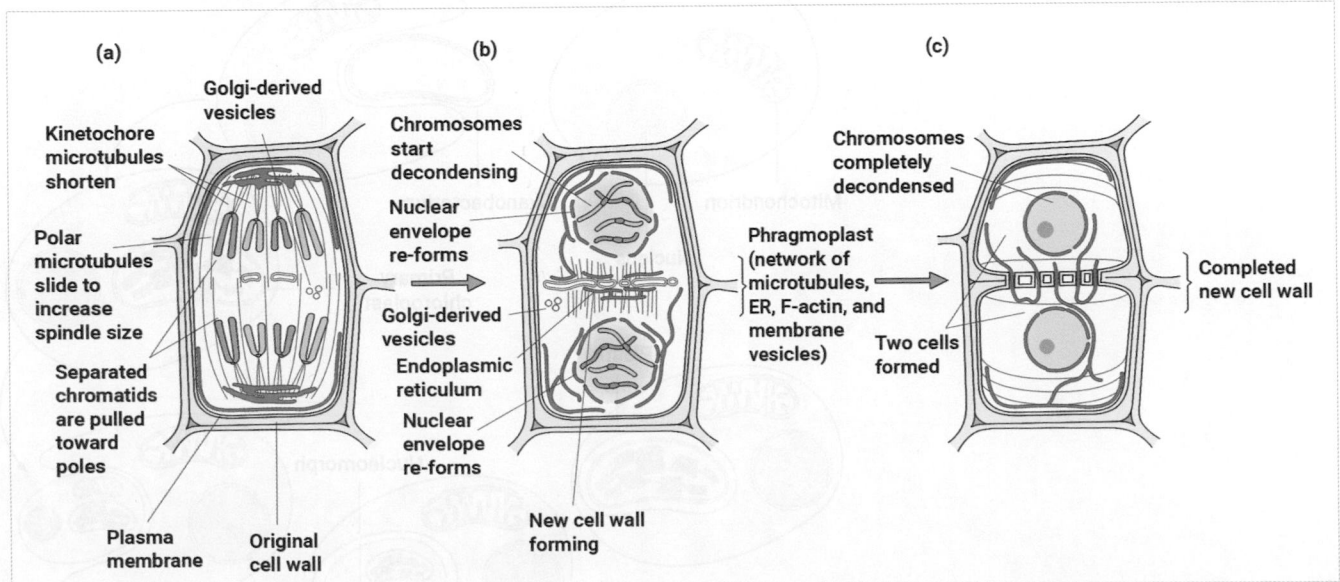

Figure 28.3 Plant cell division. After completion of mitosis (a), vesicles line up at the equator of the cell forming a phragmoplast (b). The vesicles coalesce and the phragmoplast spreads to fuse with the original cell wall (c).

Source: Lincoln Taiz, Eduardo Zeiger, Ian Max Møller, and Angus Murphy, *Fundamentals of Plant Physiology*, 1st Edition, 2018, Oxford University Press.

enzymes. At the end of mitosis, after telophase, the vesicles are guided to the equator of the cell by spindle microtubules. The vesicles and microtubules together form a structure called the **phragmoplast**. The phragmoplast vesicles coalesce to form the new cell wall. This spreads out and eventually fuses with the original cell wall. The plasma membrane is rebuilt and two new cells are created, although forever fused together by their common cell wall.

The new cell wall is laid down in a plane of division which is genetically regulated and characteristic of the plant species. Its orientation will determine the direction of growth of the plant and the morphology of its structures including roots, shoots, stems, leaves, and flowers.

Plant cell organelles: plastids

Other than possessing a cell wall and a large central vacuole (discussed later, and see also Module 2), plant cells are distinguished from animal cells by the presence of **plastids**. The best known and most studied plastids are **chloroplasts**. These occur in all green plant tissues, giving them colour by virtue of the pigment **chlorophyll**. Chloroplasts perform the fundamental process of photosynthesis in which light energy is transduced into chemical energy (as we discuss further in Chapter 7). This drives the capture of atmospheric carbon dioxide and enables plant growth by the metabolic production of new biomass.

Chloroplasts are around 5 µm in length and are easily viewed in leaf cells under the microscope, as shown in Figure 28.4. They divide and expand within the cytoplasm of leaf cells to produce populations of up to 200 per cell.

▶ **We learn more about the structure of chloroplasts in Chapter 9.**

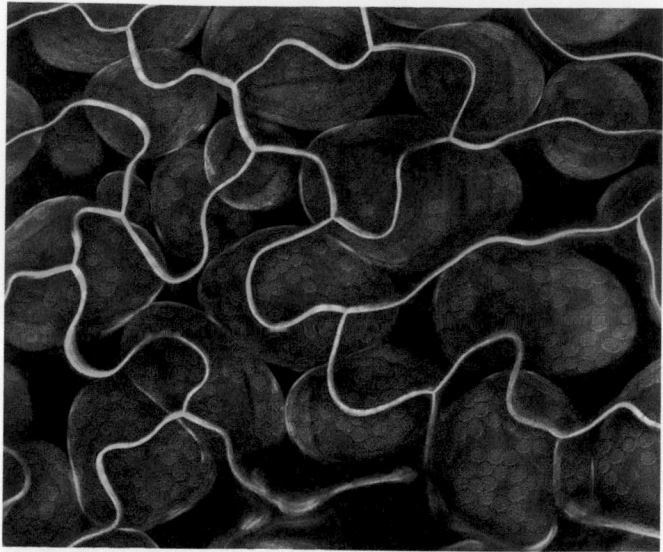

Figure 28.4 Chloroplasts. Mesophyll cells contain large populations of chloroplasts shown here with their green chlorophyll autofluorescing red as imaged by confocal laser scanning microscopy. The mesophyll cells are overlaid by the layer of epidermal cells which creates the skin of the leaf, the cell walls of which have been stained blue by a fluorescent marker.

Source: *Arabidopsis thaliana* plant cells containing chloroplasts, LM. Credit: Fernán Federici. Attribution 4.0 International (CC BY 4.0).

They are thought to have evolved within eukaryotic plant cells following the uptake of free-living photosynthetic prokaryotic organisms similar to the photosynthetic cyanobacteria which exist on the planet today. This evolutionary process is illustrated in Figure 28.5. You can see that it is a two-stage process in which

Figure 28.5 Endosymbiotic evolution of chloroplasts. Chloroplasts are thought to have evolved as a result of two endosymbiotic events, as depicted here.

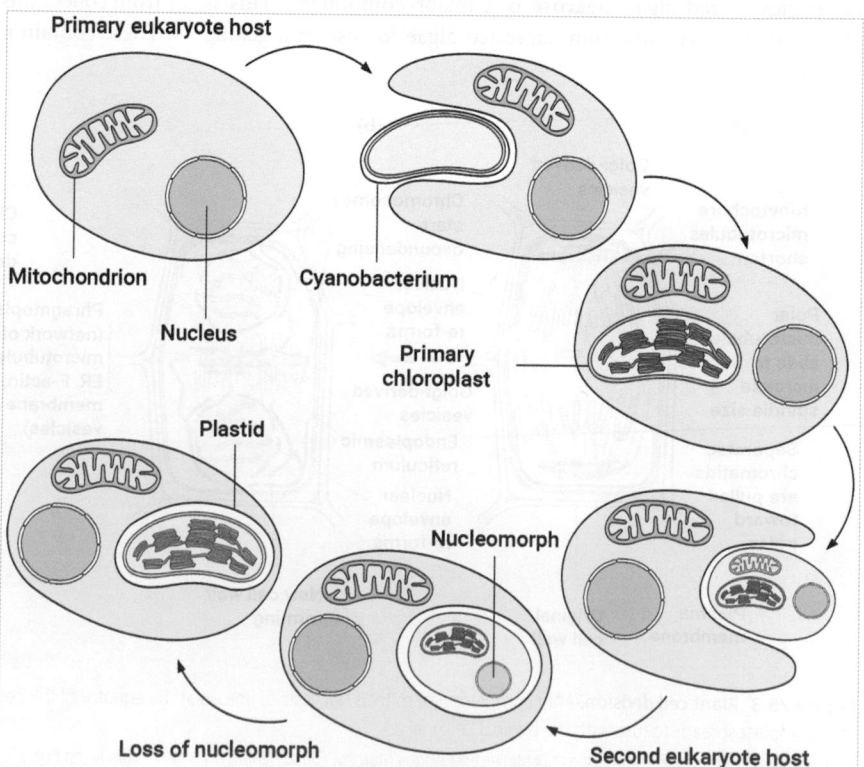

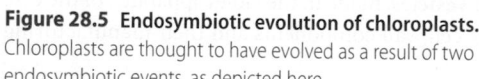

a cyanobacterium is taken up into a eukaryotic host, forming an alga-like structure containing a chloroplast, and this structure is taken up into a second eukaryotic host as a plastid.

Although this hypothesis, called **endosymbiosis**, was once highly contentious, modern evidence supports it. It is a mutual interaction which benefits both the chloroplast and the host cell which contains it.

Molecular studies of chloroplasts have shown many features which are prokaryotic in nature and consistent with the endosymbiotic hypothesis:

- Chloroplasts have their own DNA but many of the genes originally in the prokaryote's genome have moved into the cell's nuclear genome during evolution. These genes encode proteins which are translated in the cytoplasm and have to be imported back into the chloroplast.

- The chloroplast is surrounded by a double membrane, called the envelope; the inner membrane is similar to a bacterial membrane.

- Chloroplasts have their own ribosomes; these are prokaryotic in nature and different from those in eukaryotic cell cytoplasm.

- Chloroplasts divide within the cell's cytoplasm by means of binary fission, similar to the way in which bacteria divide. Indeed some of the proteins used in this process are the same as those used by modern bacteria to divide.

Chloroplasts and other types of plastid all originate from small undifferentiated plastids, found in small dividing plant cells, called **proplastids**. Proplastids differentiate according to the type of cell they are in. For example, in storage tubers like potato or cassava, they turn into **amyloplasts** in which carbohydrate is synthesized and stored in the form of starch grains. Similar amyloplasts are found in the storage tissues of seeds such as cereals. Indeed, all of the starch synthesized globally by plants grown for food accumulates in amyloplasts, laid down in roughly spherical grains, as depicted in Figure 28.6. The structure and composi-

Figure 28.7 Chloroplast to chromoplast transition during tomato fruit ripening. Green unripe fruit of tomato (*Solanum lycopersicum*) contain green chloroplasts in their cells but as they ripen, these chloroplasts undergo re-differentiation and change into chromoplasts. The green chlorophyll is degraded and red and orange carotenoid molecules accumulate in the chromoplasts which make the fruit red.

tion of starch grains are of major interest to the food industry because they influence how carbohydrate-rich raw materials are processed into foodstuffs.

Other types of plastids are pigmented and confer colour to plant structures such as petals or ripened fruit. **Chromoplasts** contain a variety of coloured molecules, often from the group called carotenoids, and tend to be yellow, orange, or red. An example of plastid differentiation is seen in ripening tomatoes where green chloroplasts in the unripe fruit turn into red chromoplasts in the ripe fruit; this process is illustrated in Figure 28.7. A similar process occurs in many other ripening fruit, including pumpkins, sweet corn, and peppers.

 Check your understanding of the concepts covered in this section by answering the questions in the e-book.

28.2 Membrane transport: talking to the outside

A cell's plasmalemma and cell wall enclose and define its internal biochemical space. However, all cells exchange material with their external environment. This constantly adjusts the biochemical conditions inside the cell and involves the exchange of metabolic resources and products. Materials which move across the membrane include water, gases, ions, nutrients, excretions, and secretions, as well as regulatory agents such as hormones and neurotransmitters.

In this section, we consider the general mechanisms of exchange across membranes and summarize some of their biochemical and regulatory characteristics.

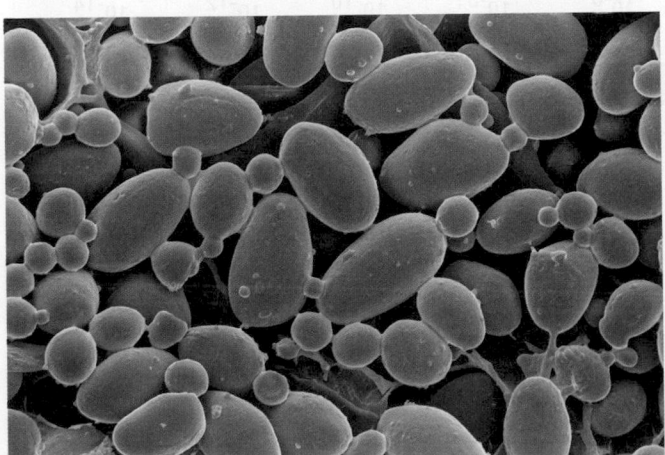

Figure 28.6 Scanning electron micrograph of starch grains in amyloplasts.

Source: Dr Jeremy Burgess/Science Photo Library.

Transport across membranes: general principles

It would be true to say that any biochemical molecule is capable of diffusing across any cell membrane, given sufficient time. However, that is not a useful statement because cells operate in the realities of biological time and exchange or restrict the passage of materials according to the demands of metabolic and physiological processes.

It is more useful to understand that molecules diffuse across membranes at different speeds as illustrated in Figure 28.8. Look at this figure, and notice that membranes are highly permeable to molecules such as gases and steroid hormones: such species diffuse rapidly. By contrast, membranes show low permeability to most ions: they diffuse slowly. Cells can ignore, tolerate, or live without molecules which diffuse extremely slowly. But diffusion is not the only mechanism: we also need to understand how cells can *actively* promote or reduce the movement of molecules which have particular metabolic importance.

The speed at which a molecule diffuses across a cell membrane depends on five variables:

1. the molecule's **permeability coefficient** (Figure 28.8),

2. its relative concentration on either side of the membrane (often called the **concentration gradient**),

3. the presence of any electrical charge, channels, or transporting mechanisms in the membrane,

4. the temperature of the local environment,

5. the thickness of the membrane.

The first of these variables reflects the chemical nature of the molecule, including its size and its ability to dissolve in the lipid at the centre of the cell membrane sandwich. Some features of this are shown in Figure 28.8. In general terms, fluid phases are either aqueous, in which hydrophilic (polar) molecules dissolve, or lipid, in which oils, fats, and other hydrophobic (non-polar) molecules dissolve. Although it is convenient to divide phases into two categories like this, it is better to envisage them as forming a continuum, from pure water to pure lipid, with all molecules showing a preference for solution at a particular point along it.

Generally speaking, most charged particles such as ions are virtually unable to diffuse through the lipid bilayer while small, uncharged particles, especially the respiratory gases, diffuse easily. In between these extremes, the diffusion of polar but uncharged molecules is greatly influenced by their size. Thus small molecules like urea diffuse moderately easily while larger molecules like glucose and amino acids move much more slowly and require biochemical assistance to be taken up in useful quantities by the cell.

Note from Figure 28.8 that the permeability coefficient is a velocity (distance divided by time, where distance is the thickness of the membrane) and that its scale of values is logarithmic over about 15 orders of magnitude. It is this enormous range which allows us to say that membranes are effectively *impermeable* to very slowly moving molecules.

The second variable, the concentration gradient, influences the *rate* of diffusion (Quantitative Tools 28.1). The actual flow rate of a molecule, expressed in moles/cm^2/s, is calculated by multiplying the difference in concentrations, in mol/cm^3 or mol/ml, by the permeability coefficient. This is discussed further in Quantitative Tools 28.1.

While variables 1 and 2 are defined by the chemistry of the molecule and the structure of the membrane, variable 3 reflects the phenotypic characteristics of the particular cell which the membrane encloses. In many cases, the presence of electrical charge, membrane channels, or transporters effectively overrides the physical effects of the first two variables.

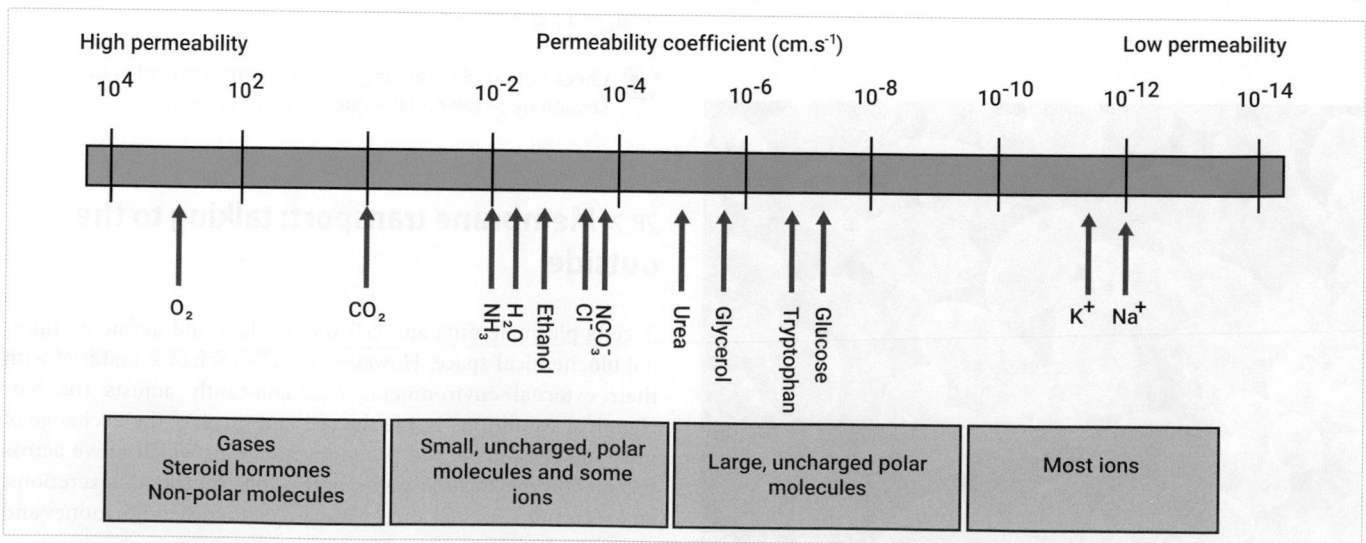

Figure 28.8 Membrane permeability. Approximate permeability coefficients for some important biological molecules, passing through a lipid bilayer membrane, and the general permeabilities of different classes of molecule. Note the logarithmic scale.

QUANTITATIVE TOOLS 28.1 Quantifying passive diffusion

(a) *Calculating the flow rate of glucose across an erythrocyte membrane*

The rate of flow of a molecule across a membrane by passive diffusion is the product of its concentration gradient and the membrane's permeability coefficient for that molecule.

This example is for glucose entering a red blood cell from plasma. It assumes that the glucose concentration inside an erythrocyte is effectively zero, so the gradient is the same as the plasma concentration:

Variable	Value	Unit
Gradient (concentration outside–concentration inside)	7×10^{-6}	mol cm^{-3}
Permeability coefficient (at 25°C)	2×10^{-4}	cm s^{-1}
Flow rate (gradient × p.c.)	14×10^{-10}	mol cm^{-2} s^{-1}

(b) *The Fick principle of passive membrane transport*

In physiological tissues, the actual rate of diffusion of a substance is influenced by the area of membrane available and its thickness. This is summarized in Fick's Law of Diffusion:

The rate of diffusion of a molecule through a membrane is directly proportional to the surface area and concentration gradient and inversely proportional to the resistance and thickness of the membrane.

In this definition, 'resistance' is the inverse of the permeability coefficient. Fick's Law can be expressed as an equation of proportionality, with permeability coefficient in the numerator rather than resistance in the denominator:

$$\text{Rate of diffusion} \propto \frac{\text{Surface area} \times \text{concentration gradient} \times \text{permeability coefficient}}{\text{Membrane thickness}}$$

Suitable units for the rate of diffusion would be mol cm^{-1} s^{-1}.

The fourth variable, temperature, becomes important when comparing organisms in different environments. Experimental estimations of the permeability coefficient are often made using synthetic membranes under highly controlled lab conditions, so before using a stated value we should be sure that it applies to real circumstances.

In mammals and birds, which maintain a relatively constant, above-ambient body temperature, membrane fluidity and molecular movement will be greater than in, say, a fish in Arctic waters. In general, molecules will move faster across membranes when conditions are warmer. The normal passage of glucose into a human erythrocyte will be slightly faster than calculated in Quantitative Tools 28.1 because the permeability coefficient was measured at 25 °C rather than 37 °C.

The final variable, membrane thickness, can be ignored in many situations (i.e. taken as unity in calculations) because the membranes of many different types of cell are essentially of similar thickness. However, it may become important where the passage of materials across multiple or complex membranes (say, where several cell layers are involved) is considered.

An example of this would be the passage of respiratory gasses across the membranes of lung alveoli: O_2 moves from the air drawn in to the lungs, across the flattened cells of the respiratory epithelium and then across the endothelial cells of the alveolar capillaries before entering the blood. CO_2 passes in the opposite direction. In cases such as these, the calculation of transport dynamics employs the Fick principle of diffusion (as discussed in Quantitative Tools 28.1) which does take membrane thickness into account.

Water movement across membranes

Most cell membranes are moderately permeable to water, as we see in Figure 28.8. **Osmosis** is the fundamental process by which water moves from one side of the membrane to the other and it is the *only* way of driving water between the compartments of tissues: there is no molecular pump for water. Because of this, cells regulate water movement by combining two other processes:

- changing the osmotic gradient by transferring osmotically active particles,
- adjusting the water permeability of the membrane by means of water channels.

Inorganic ions are often the osmotic particles which cells move to change the gradient, and some mechanisms for doing this are discussed later. Many cell membranes have water channel proteins which can be opened or closed to alter the flow rate of the water. These channels are called **aquaporins** (AQPs) and they form a large class of proteinaceous insertions in the membrane.

At least 12 types of AQP have been discovered, and molecules of this kind occur in all groups of prokaryotic and eukaryotic organisms. Each channel is made of six protein helices, twisted together in a bundle to form a narrow channel, as illustrated in Figure 28.9. This permits the passage of water molecules but not charged particles. Aquaporins are abundant in tissues which transport large amounts of water in a regulated fashion, such as the root hair cells of plants and the nephrons (tubules) of the mammalian kidney, but they are found in most tissues where water movement occurs.

A major pathway for water uptake from the soil by plant roots is through aquaporins in the plasma membrane of root hair cells. A difference in water potential (osmotic gradient) across the plasma membrane enables water to flow into the root hair cell via osmosis. It then flows through further internal root cells via **plasmodesmata** until it reaches the plant's plumbing system, the **xylem**, for

Figure 28.9 The structure and molecular function of aquaporins. Schematic representation of a cross section of the AQP1 channel. The pore has an 'hour glass' shape, with one vestibule open to the extracellular and the other the intracellular compartments. Osmotic difference stimulates the movement of water molecules in single file through the pore.

Source: Lincoln Taiz, Eduardo Zeiger, Ian Max Møller, and Angus Murphy, *Fundamentals of Plant Physiology*, 1st Edition, 2018, Oxford University Press.

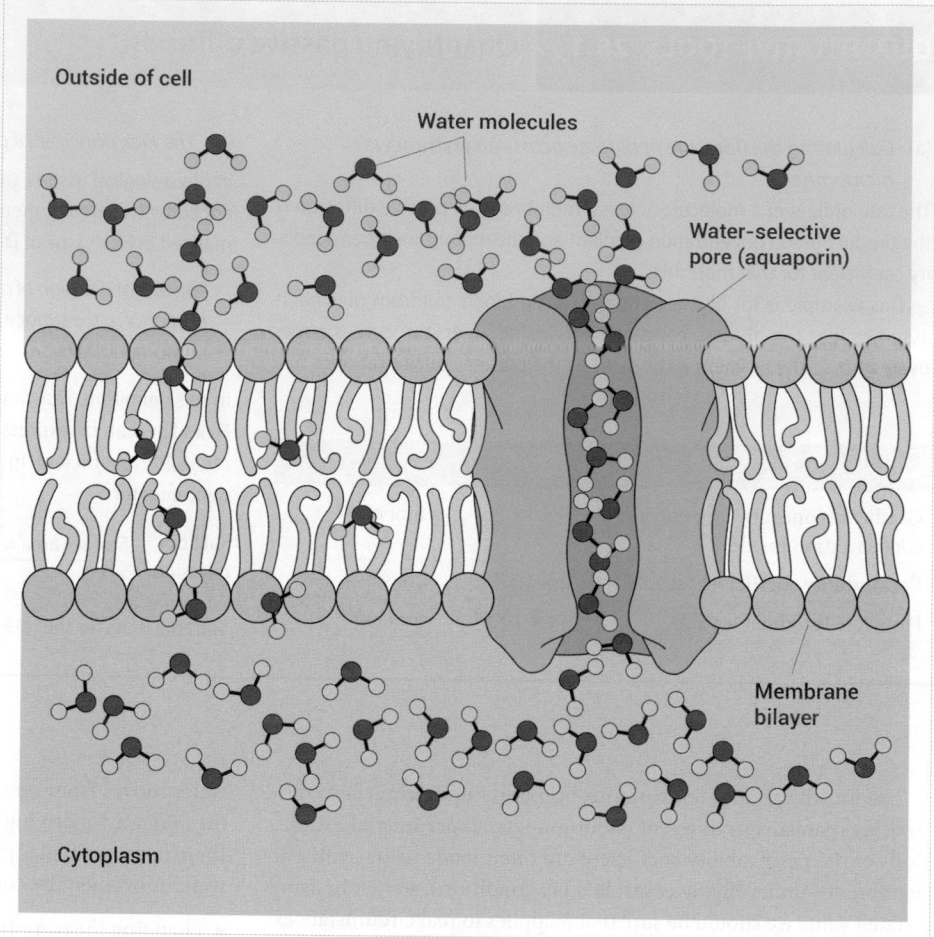

distribution around the organism. (Root structure and plasmodesmata are discussed later in this chapter.)

Plant AQPs can open or close in response to any environmental stress that the plant experiences. For example, in drought conditions when the soil is dry a specific serine residue in the aquaporin protein gets dephosphorylated. This closes the pore and water is conserved.

In complex animals, kidney **nephrons** reclaim much of the water which would otherwise be lost from the blood as the kidney makes urine. This reclamation process is illustrated in Figure 28.10. Cells lining the nephron, in regions called the distal convoluted tubule and collecting duct, actively push sodium ions into the blood in adjacent capillaries, stimulated by hormones such as **aldosterone** from the adrenal cortex. The resulting osmotic gradient encourages water to follow the ions into the blood. However, significant amounts of water can only move if the membrane is made permeable by the opening of AQP channels in the tubule cells.

 Go to the e-book to explore an interactive version of Figure 28.10.

Channel opening happens when AQP2, which is already present within the nephron cell, moves from the cytoplasm into the plasma membrane on the apical (capillary or blood) side of the cell. Other AQPs, especially AQP3 and AQP4, reside permanently in the cell membrane on the basal (urine) side. The AQP2 becomes inserted in the membrane when the cell is stimulated by the hormone

vasopressin (anti-diuretic hormone, ADH). This hormone is produced by the brain, for example during exercise, heat-induced dehydration, or shock, as well as in response to more subtle changes happening in the body.

The insertion of AQP2 into the membrane completes the connection, allowing transcellular water flow: water flows into the cell through AQPs 3 and 4, through the cytoplasm, and out through the AQP2. AQP1 in the walls of capillary cells in turn allows the water to enter the blood. (Look at Figure 28.10 and notice the involvement of AQPs in steps 5 and 6 of the process depicted.) The result is an increase in blood volume and a reduction in urine volume. It is one element of a complex of systems which regulate body water content and blood pressure.

Simple (passive) diffusion

When a membrane is freely permeable to something dissolved in the extracellular fluid or the cytoplasm, its speed of movement in either direction is dictated by the relative concentration of that solute on either side of the membrane. This type of movement requires no energy to drive it other than the intrinsic kinetic energy of the solute molecules themselves. It is therefore described as 'passive'.

The first graph in Figure 28.11 illustrates simple diffusion graphically: the rate of transport is linearly related to the difference in solute concentration (the '**gradient**'), provided there is no significant change on either side.

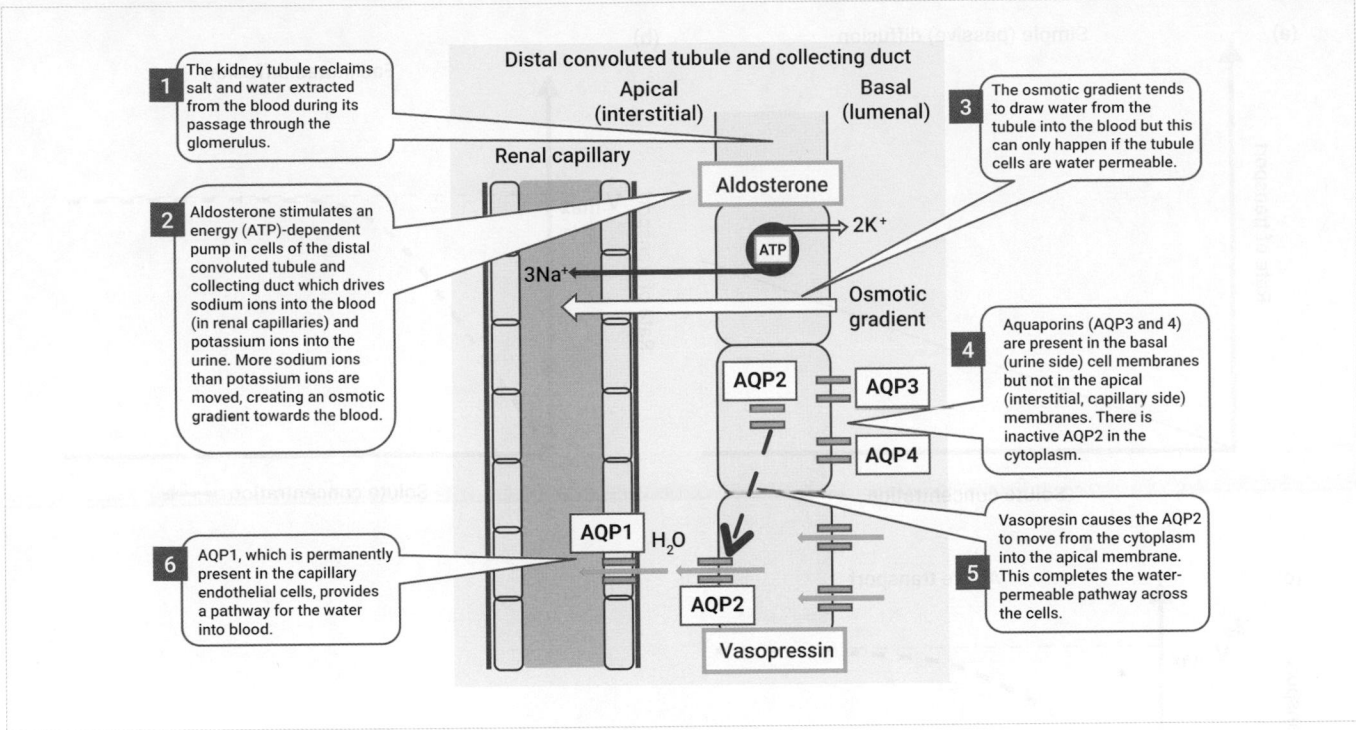

Figure 28.10 Hormonal regulation of salt and water reabsorption in the kidney tubule.

A good example of simple diffusion is the transfer of respiratory gases across the alveolar membranes of the lungs as we breathe. The rates of O_2 transfer from inspired air to blood and of CO_2 loss in the other direction depend entirely on the pressure gradients of the gases.

Respiratory membranes, such as those in mammalian lungs and the spiracles of insects, must be wet to permit the passage of gases. Biological membranes are always wet to a lesser or greater extent and solute movement is made possible by the random diffusion of molecules in the liquids on either side as well as through the membrane itself.

Facilitated diffusion

The natural tendency of a molecule to move down a concentration gradient can be enhanced by a pore in the membrane or by the presence of a transporting protein pathway. We see that water always moves passively, following gradients of osmotic pressure, and that AQP channels make for faster water flow with the potential for regulation. Water movement through AQPs is an example of **facilitated diffusion**.

Glucose is another molecule which is transported by facilitated diffusion. The **GLUT** family of transporter proteins enable the uptake of glucose into muscle, liver, fat, and other cells. One member of the family, GLUT 4, is notable for only being present in the membrane when the cell has been stimulated by insulin: it provides the principal mechanism by which that hormone reduces blood glucose concentrations.

Other examples of facilitated diffusion are those for the transport of glycerol (by a subclass of AQPs called aquaglyceroporins), and for the transport of urea in the kidney. There are very many others.

Molecular transporters like these are described as **uniport** because they allow the movement of a single solute, but more complex systems are available too. The intestinal absorption of glucose from digested food involves a glucose transporter which concurrently permits the movement of Na^+ ions in the same direction. This is called a **cotransporter**. The Na^+ moves passively down its concentration gradient, from gut lumen to capillary blood, and this usually provides a more powerful force than the glucose gradient. Because the Na^+ and glucose move together, glucose transport is largely independent of its own concentration and can even occur when its gradient is unfavourable.

The flow characteristics of facilitated diffusion are shown in Figure 28.11. It is in general faster than simple diffusion. The flow is initially linear but reaches a maximum velocity (V_{max}), determined by the capacity and availability of transporter proteins: the system can become saturated, producing the horizontal region at V_{max} in the graph.

Ion movement across membranes

As we have seen in Figure 28.8, the lipid nature of cell membranes makes them rather impermeable to charged particles, and especially to ions unless there are specific channels or carriers. The movement of ions is also influenced by any electrical potential difference,

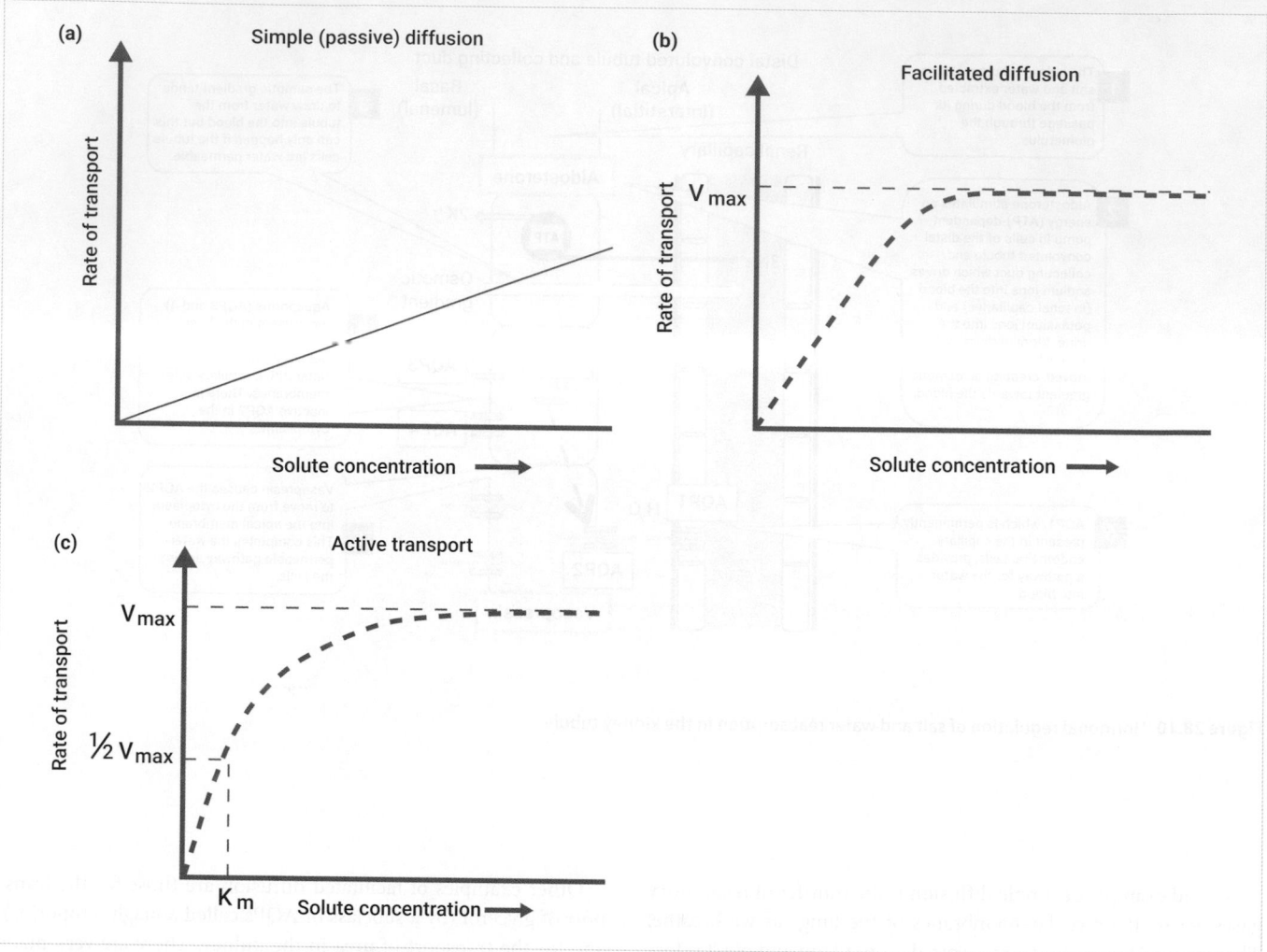

Figure 28.11 Rates of solute transport across membranes. (a) Simple (passive) diffusion. (b) Facilitated diffusion. (c) Active transport.

called the **membrane potential**, which may be present across the membrane. Most membranes have some sort of potential, with the insides of cells being negatively charged with respect to the outside.

In animals, membrane potentials are particularly high in excitable tissues such as nerve, muscle, and cardiac cells, and are the basis of their ability to propagate, transmit, and respond to electrical signals. All ion movements are affected by the concentration gradient too, of course. The combination of this with the membrane potential produces a net driving force for each ion called its **electrochemical gradient**.

▶ **We learn more about membrane potentials in Chapter 17.**

Some ions move down their electrochemical gradient by facilitated diffusion. An example of this with medical significance is the movement of chloride ions (Cl^-) through specific channels in cells of the lungs, skin, gut, pancreas, liver, and elsewhere. These channels may be absent or inactive in diseases such as cystic fibrosis,

resulting in membranes with disturbed ion and water balance and leading directly to the symptoms of the condition.

Ion pumping: an example of active transport

Frequently, cells move ions *against* the prevailing electrochemical gradient. This takes energy and is thus an example of **active transport**. Active transport occurs up a concentration gradient or where the membrane is otherwise impermeable to the solute. It is achieved by pumps located in the membrane. These are enzymic proteins which release the energy held in the phosphate bonds of ATP or some other high-energy source.

A common ATP-dependent pump is the Na^+/K^+ pump (often called Na^+/K^+ ATPase) found in many cell types, including excitable cells. This is an **antiporter** because the sodium is driven in one direction, usually out of the cell, at the same time as potassium is driven inward; this movement is depicted in Figure 28.12. The ratio of ion movements is three Na^+ out for two

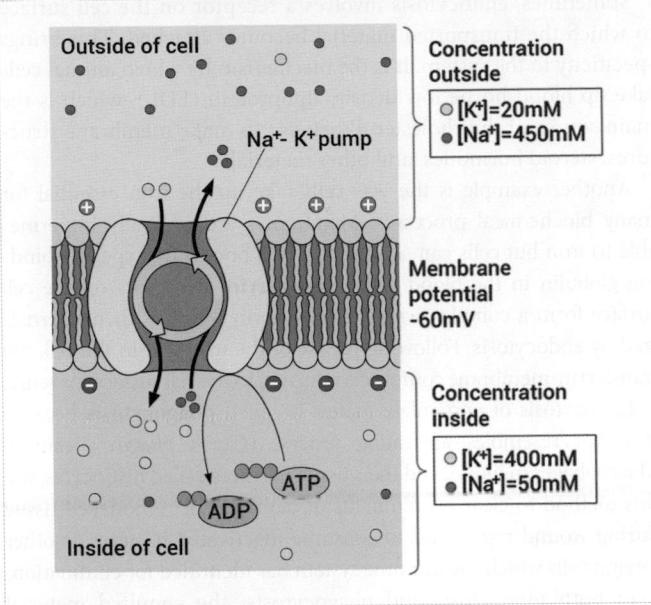

Figure 28.12 The steps involved in the operation of the Na⁺/K⁺ ATPase ion pump.

Source: Geoffrey Cooper, *The Cell, A Molecular Approach*, Eighth edition, 2018, Oxford University Press.

K⁺ in, resulting in a net outward movement of ions. This difference helps to maintain osmotic and electrical potentials across the cell membrane.

The kinetics of active transport reflect those of the pump enzyme, as shown in the third graph in Figure 28.11. As with facilitated diffusion, the rate of transport depends on the solute concentration and has a maximum velocity (V_{max}). However, the rate is never proportional to the concentration. The solute concentration at which the velocity is half its maximum is called K_m (analogous to the Michaelis constant in enzyme kinetics—see Chapter 6).

Pumps of this kind restore the electrical potential of heart cells following the membrane depolarization which is associated with their periodic contraction. They also maintain the internal volume of erythrocytes: the Na⁺ concentration of blood serum is much higher than that inside the cell and the tendency for inward ion leakage would increase the cells' osmotic pressure resulting in water uptake. Thus erythrocytes maintain their volume by actively exporting sodium ions. This needs to be taken into account when creating artificial solutions for physiological experiments. You can read more about this in Quantitative Tools 28.2.

▶ We learn more about ion pumps in Chapter 17.

QUANTITATIVE TOOLS 28.2   What is physiological saline?

For laboratory experiments involving living tissues we need a fluid environment—a medium—in which the process we are investigating will behave in as normal a way as possible. Cells and tissues have a range of demands which need to be met, including barometric pressure, temperature, pH, gas exchange, and osmotic pressure, besides needing nutrients and other essential chemicals.

The most complex media are those used in tissue culture. These may contain scores of highly purified ingredients in precisely weighed amounts, and variations in the recipe may be needed for particular cells, tissues, and organisms. Sterile incubators and bioreactors are used to control the physical environment.

A much simpler medium is 'physiological saline'. This regulates the osmotic environment and ensures that cells do not swell or shrink during experimentation. It is also valuable when preparing tissues for histological examination.

Saline is made from sodium chloride dissolved in water. How can we calculate the required concentration? We can take our cue from human blood: this bathes all the tissues of the body and provides an optimal osmotic environment. Red blood cells have a constant volume while immersed in it in the circulation.

A machine called an osmometer will show that blood plasma has an osmotic concentration of 290 milliosmoles/litre (abbreviated to 290

mOsm L⁻¹ or 290 mOsM.) All soluble particles, or solutes, contribute to the osmotic environment, irrespective of size and chemical identity. Thus the 'osmole' unit is simply a way of expressing the total molar concentration of soluble particles which are present. In our saline solution we will supply them all using sodium chloride.

The molecular weight of sodium chloride is 58.5, so one mole of it weighs 58.5 grams. The prefix 'milli' means 1/1000 (or 0.001 or 10⁻³), so we need 290 thousandths of an osmole in 1 litre of water:

$$290 \times 0.001 \times 58.5 = 16.97 \text{ grams of salt in 1 litre of distilled water}$$

However, when dissolved in water, NaCl splits into its two ions Na⁺ and Cl⁻. Both of these are osmotically active, so we would have 290 mOsm of Na⁺ and 290 mOsm of Cl⁻. Therefore, we actually need to make our solution half as strong:

$$8.48 \text{ g NaCl/litre}$$

This is physiological saline. In practice, it would be usual to include other chemicals, including something to regulate the pH. Phosphate salts can do this, but if we add these we need to make a corresponding reduction in the amount of NaCl to make up for their osmotic effect. Recipes for this, to make a fluid called phosphate-buffered saline or PBS, are readily available in laboratory manuals.

Insects often excrete excess potassium ions. Various types of ion pump are found in their excretory organs (**Malpighian tubules**) and gut. The caterpillar of the tobacco hornworm *Manduca sexta*, for example (see Figure 5.7), which is a devastating pest of tobacco crops, absorbs large amounts of potassium from its tobacco-leaf diet in the forward part of the gut. It disposes of this by returning it to later parts of the gut for elimination. Goblet-shaped cells in the epithelia of the mid-gut possess vesicles containing an ATPase which is a K^+/H^+ antiporter. This pushes K^+ ions, up a steep concentration gradient, from the **haemolymph** (the caterpillar's body fluid) out of the epithelial cells into the gut lumen. Protons move in the opposite direction. An adjacent H^+-transporting ATPase removes the protons from the vesicle, preventing them from accumulating and allowing the K^+/H^+ pump to continue working. This increases the acidity of the insect's haemolymph but mechanisms in its Malpighian tubules are available to restore pH balance. An interesting feature of this K^+/H^+ pump is that the enzyme adapts itself for efficient operation across a very wide range of K^+ gradients. In effect, it has a gearing system, allowing the animal to deal with leaves of different maturity.

Non-ionic active transport across membranes

The movement of ions against the electrochemical gradient is only one type of active transport. Energy-dependent systems are also necessary for the transfer of molecules and large particles to which the membrane is impenetrable or for which there is an adverse concentration gradient.

A principal means of achieving this is **endocytosis**, a process in which the cell membrane folds around the material to be transported and presents it for breakdown or further metabolism within the cell. Here, the energy is consumed by the cell's cytoskeleton (explained later) as it causes the formation of the projecting membrane pocket to engulf the material being taken up.

Endocytosis of liquid material is called **pinocytosis**. One example of this is the uptake of lipid droplets from digesta by epithelial cells of the small intestine. Another is the way in which epithelial cells of the thyroid gland take up thyroglobulin, a proteinaceous, iodine-rich colloid stored in the gland's follicles, before turning it in to the hormone thyroxine.

Sometimes, endocytosis involves a receptor on the cell surface to which the transported material becomes attached. This brings specificity to the system. It is the mechanism by which animal cells take up blood-borne low density lipoprotein (LDL), which is the main source of the cholesterol they use to make membrane structures, steroid hormones, and other materials.

Another example is the way cells take up the iron essential for many biochemical processes. Membranes are generally impermeable to iron but cells can acquire it if it is bound to a specific binding globulin in the blood called **transferrin**. Receptors on the cell surface form a complex with the transferrin which is then internalized by endocytosis. Following release of the iron within the cell, the transferrin-membrane complex is returned to the cell surface for reuse.

Endocytosis of particulate matter is called **phagocytosis** because it rather resembles an eating process (Greek *phago* = eating). Macrophages in blood, and their tissue partners called histiocytes, use this method to clear up remnants of damaged or remodelled tissue during wound repair and to consume inactivated bacteria or other foreign cells which the immune system has identified for elimination.

In both pinocytosis and phagocytosis, the engulfed material enters the cell cytoplasm enclosed in a ball of membrane, forming an intracellular vesicle called an **endosome**. A crucial step is usually the fusion of this vesicle with a lysosome. Enzymes in the lysosome break down the absorbed material and may then destroy the vesicle itself, releasing the contents for further biochemical processing by the cells. Alternatively, as in the LDL and iron uptake mechanisms described above, the membrane materials and receptors are recycled to the cell surface.

Some lysosomes, including those of macrophages, have an extremely low internal pH. This assists the denaturation and breakdown of endosome-enclosed proteins and can eliminate the toxicity of otherwise dangerous foreign materials.

Plant cell vacuoles have many characteristics similar to those of animal cell lysosomes, including the presence of a wide variety of enzymes and ion pumps which regulate the acidity and turgor of the cell. However, they have a diversity of other functions too, including the synthesis and breakdown of sugar and amino acid polymers. Some of these processes generate materials of value or interest to man including food starches and proteins, rubber latex, and opium.

🔄 **Go to the e-book to complete an activity associated with Figure 28.12.**

🔄 **Check your understanding of the concepts covered in this section by answering the questions in the e-book.**

28.3 The cytoskeleton

Cells possess a complex set of internal structural proteins which together comprise the cytoskeleton. The cytoskeleton has a large number of functions:

- It organizes the position of organelles within the cell.
- It provides pathways for the translocation of materials within the cell and for export.
- It provides the mechanical forces required for chromosome segregation and cell division.

- It allows the transmission of external mechanical signals into the cell.

- It forms the motile elements of flagellae and cilia and generates force.

In animal cells, the cytoskeleton also

- maintains cell shape and polarity but permits changes in response to external signals,

- carries the internal tensions necessary for cell movement and contraction,

- links up with the proteins which connect adjacent cells and transmits the forces which hold tissues together.

Broadly speaking, there are three classes of cytoskeletal element in eukaryotic cells, as summarized in Table 28.1. Plant cells contain **microtubules** and **microfilaments** while animal cells also possess **intermediate filaments**.

Table 28.1 The cytoskeletal structures of eukaryotic cells

Class	Occurrence	Morphology	Principal constituent proteins	Cellular location	Example functions
Microtubules	All eukaryotic cells	Hollow tubes made up of rotationally repeating subunits 25 nm dia.	Tubulin Kinesin Dynein	Structural proteins throughout the cytoplasm Centrioles, asters, and mitotic spindles in green algae and animal cells Cilia Flagellae	Pathways for organelle movement Defining cellular polarity, shape, and planes of division Positioning the endoplasmic reticulum and Golgi within the cell cytoplasm Division of nuclear material in mitosis and meiosis Fission and budding in yeast Cilial beating Flagellal waving Vertebrate sperm motility
Microfilaments	All eukaryotic cells	Single strands and polymers 6–8 nm dia.	Actin Myosin	Stress absorbing fibres throughout the cytoplasm Cortex of animal cells, adjacent to plasmalemma (actin) Contractile ring in cytoplasm of dividing animal cells Microvilli in brush borders Cochlear hair cells Thin (actin) and thick (myosin) filaments in muscle fibres	Linkage to cell–cell and cell–extracellular matrix adhesion proteins Cytokinesis Amoeboid movement in cell migration and phagocytosis Cytoplasmic streaming in plant and algal cells, protozoa, and slime moulds Movement of microvilli Acrosome reaction in invertebrate sperm
Intermediate filaments	Animal cells	Two double helices, twisted into a cable 8–10 nm dia.	Keratin (epithelial cells) Vimentin (mesenchymal cells, endothelial cells, erythrocytes) Desmin (striated and smooth muscle cells) Glial fibrillary acidic protein (GFAP; astrocytes) Neurofilament protein (neural cells) Nuclear lamins (all cells)	Cytoplasm Beneath the nuclear envelope A basketwork arrangement around the nucleus, extending into the cytoplasm Epithelial and skin surfaces, after other cellular material has died or eroded	Maintenance of cell shape and absorption of mechanical stress Nuclear positioning Organelle movement Linkage to cell–cell adhesion proteins in the plasmalemma Binds microfilaments together (desmin) Forms hard epithelial tissues such as hair, nails, feathers, and horn (keratin)

These elements are made of a range of different proteins, which have been highly conserved during evolution and are remarkably similar across all eukaryotic organisms. (Prokaryotic cells have a different sort of cytoskeleton, as we discover in Chapter 9.)

▶ **We explore the composition of microtubules, microfilaments, and intermediate filaments in Chapter 9.**

In animal cells, the microfilament protein actin is highly abundant in the cortex, immediately beneath the plasmalemma, and may account for as much as 5% of total cell protein. Actin and myosin, together with a complex of other proteins, make up the myofibrils of striated, cardiac, and smooth muscle cells.

▶ **We discuss the structure, function, and physiology of muscle further in Chapter 20.**

The intermediate filaments of animal cells show a great diversity of structures and functions, from the proteins which underpin the membrane surrounding the cell nucleus, to the constantly replenished banks of resilient protein which form the protective layers of the skin. All three classes of cytoskeletal protein take part in the complex process of cell division, often by becoming rapidly disassembled and reassembled or changing their alignment to facilitate the separation of chromosomes, organelles, cytoplasm, and plasmalemma.

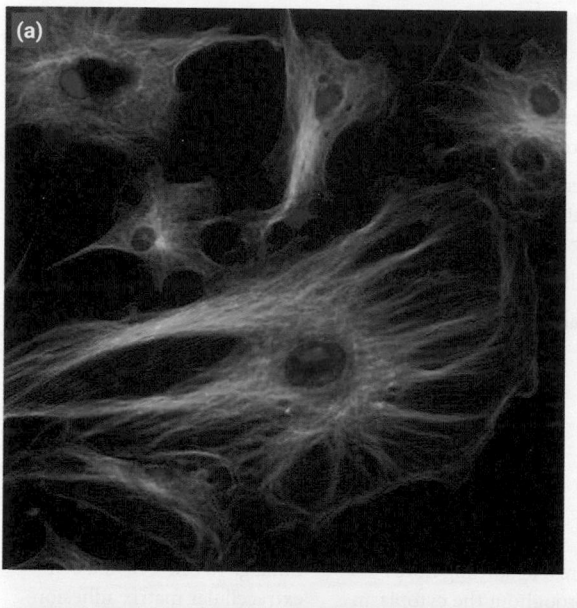

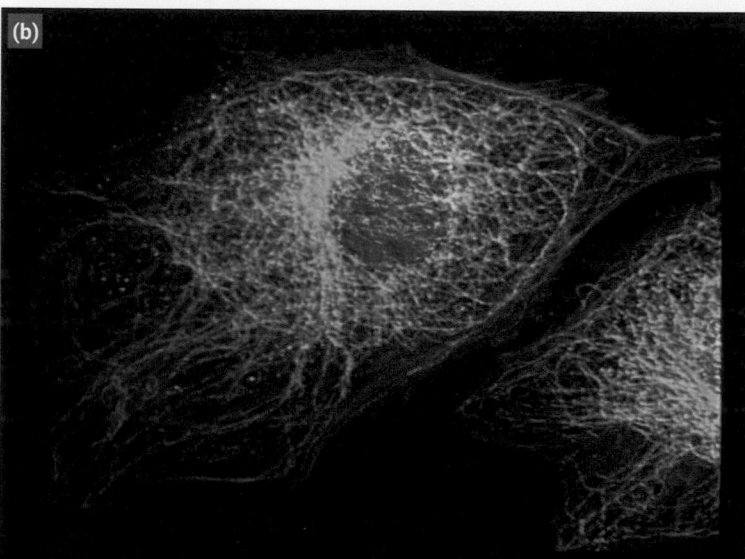

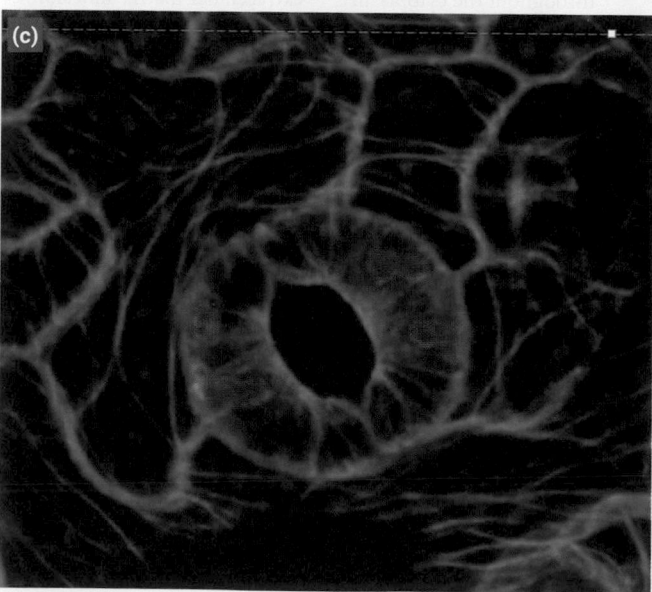

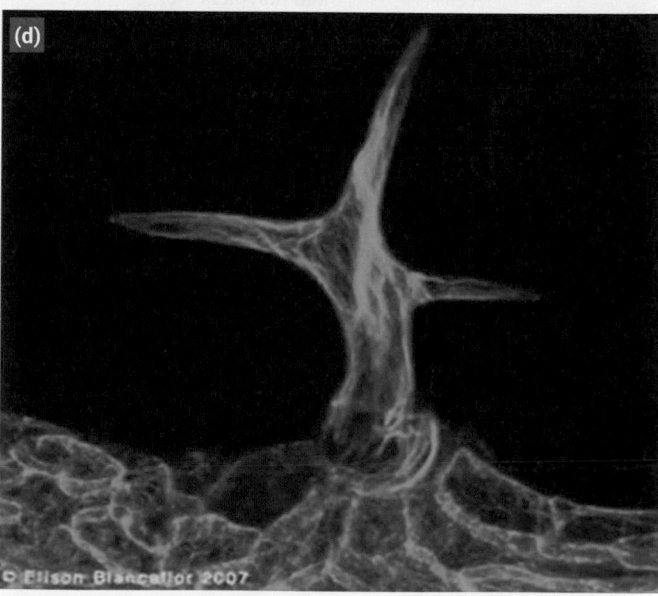

Figure 28.13   The cytoskeleton of (a, b) animal and (c, d) plant cells, highlighted by immunofluorescence. (a) Endothelial cells. (b) Skin cells. (c) Stomatal guard cells in leaf epidermis. (d) Leaf hair and surrounding epidermal cells.

Source: (a) Wikimedia Commons/image from the ImageJ-Programmpaket/Public Domain; (b) Dr Glyn Nelson, Newcastle University; (c) Gao XQ, Chen J., Wei PC et al. Array and distribution of actin filaments in guard cells contribute to the determination of stomatal aperture. Plant Cell Rep 27, 1655–1665 (2008). (d) Panel G of Figure 1 of Mathur et al. (1999). The actin cytoskeleton is required to elaborate and maintain spatial patterning during trichome cell morphogenesis in Arabidopsis thaliana. Development,126, 5559–5568.

Figure 28.13 shows elements of the cytoskeleton stained in different fluorescent colours. These beautiful pictures illustrate several important things besides the cellular events taking place:

- cytoskeletal proteins fill the cellular space,
- they suspend organelles such as the nucleus,
- they stretch right across the cells and link to anchorage points, and
- they align to planes of mechanical stress.

It is also evident from these pictures that eukaryotic cells are mechanically dynamic and far from being a disorganized cytoplasmic soup.

A conceptual model called tensegrity, depicted in Figure 28.14, helps us to appreciate the mechanical properties of the cytoskeleton. This comes from architecture and mechanical engineering and borrows two fundamental ideas about the transmission of forces: resistance to compression and the absorption of tension.

- Microtubules are like hollow, cylindrical poles, resisting deformation to produce a rigid framework.
- Microfilaments and intermediate filaments are more like flexible wires or ropes: they can bend to absorb distortion and they can take up tension by limited stretching.

Combining these elements creates a structure which is strong, flexible, and springy but holds its shape, rather like playground climbing equipment.

The tensegrity model emphasizes the cytoskeleton's role in managing cell shape, movement, and the transmission of forces.

In plant cells, most of the external structural integrity is provided by the cell wall: this provides three-dimensional structure and strength. Nevertheless, plant cells also have an internal cytoskeleton. This is made of proteins built into fibres or tubes in much the same way as in animal cells. Two classes of cytoskeletal elements are found in plant cells: microtubules, composed of heterodimer subunits of alpha and beta tubulin, and microfilaments, made of globular actin monomers. Plant cells lack the intermediate filaments found in animal cells, simply because the cellulose cell wall fulfils this role.

Although the microtubules and microfilaments within the plant cell contribute toward structural integrity, they are more important as transport highways, facilitating the movement of organelles, vesicles, and various cargoes around the cell as it goes about its day-to-day business.

 Check your understanding of the concepts covered in this section by answering the questions in the e-book.

28.4 Volumetrics, pressures, and turgor in plants

Young green plants do not have bones or other resilient structures and use a different strategy for resisting the force of gravity and remaining erect. Plant shoots grow upwards, to intercept the light and to exchange gas with the air in order to fix carbon dioxide. This aerial habit is facilitated by **turgid** tissues, a water-based support mechanism which has evolved in most land plants.

In longer-lived terrestrial plants, this structural support is aided by the laying down of extra thickenings and depositions within the cell wall of the water-carrying cells, the xylem in the stem, to generate wood. This greatly increases the tensile strength of the body of the plant, especially in the case of trees. Such adaptations are less crucial in aquatic plants since water is more buoyant than air and the weight of the plant is correspondingly reduced.

A major difference between plant and animal cell structure is the presence of a large central **vacuole**, filled with aqueous fluid and solutes. A plant cell vacuole is depicted in Figure 28.15; see also Module 2. This vacuole only forms after cell division has ceased: dividing cells in plant tissues tend to be small and lack vacuoles.

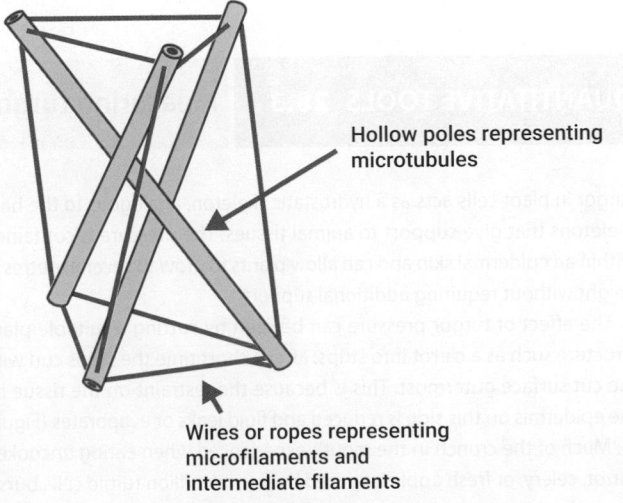

Hollow poles representing microtubules

Wires or ropes representing microfilaments and intermediate filaments

Figure 28.14 The cytoskeleton exhibits the principle of tensegrity in the same manner as a rope-based climbing frame.

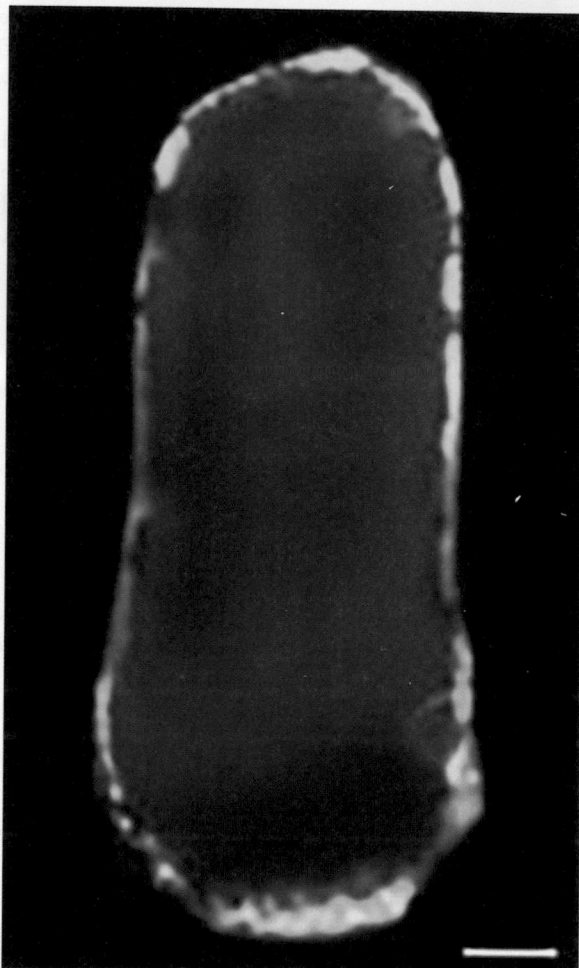

Figure 28.15 Plant cell vacuole. Plant cell sectioned through mid-plane and photographed under confocal microscopy. The large internal vacuole is surrounded by thin cytoplasm. Organelles are appressed against the cell wall. Most of the cell's volume is taken up by the vacuole.

Source: The cytoskeleton maintains organelle partitioning required for single-cell C-4 photosynthesis in Chenopodiaceae species. Chuong, Simon D. X., Franceschi, Vincent R., Edwards, Gerald E. *Plant Cell*, 18, 2207–2223, September 2006. American Society of Plant Physiologists.

The vacuole is much more significant to cell function than the smaller collections of vacuoles that occur in some animal cells. Its liquid content is under significant pressure, called **turgor pressure**. This develops in the cell because solutes are pumped across an inner membrane which surrounds a region of the cytoplasm. This is called the **tonoplast membrane** and it delimits the vacuole. The increased solute concentration raises the osmotic pressure of the cytoplasm, leading to an inward flow of water across the tonoplast membrane and into the vacuole.

The fluid pressure of the expanded vacuole generates an outward force on the cellulose cell wall. Because of its elastic rigidity, the cell wall absorbs the force and the cell becomes turgid. This is the basis for the rigidity of soft plant tissues such as leaves, stems, and petals; it is the main mechanism by which the aerial architecture of the plant is maintained, in the absence of wood. Turgidity can be directly measured and quantified, as discussed in Quantitative Tools 28.3.

When plants lose water under drought conditions, water is lost from the vacuole, the plasma membrane shrinks inwards from the cell wall and the cell becomes **plasmolysed**. The loss of turgidity leads to a reduction in tissue strength and wilting. If water is given to the plant in the short term, the system is reversible and the plant regains turgidity, but long-term drought eventually leads to plant death.

Damage to the tonoplast by ice crystals is one of the factors which limit the low-temperature storage of plant materials: the leaves of many edible plants become soft and mushy when taken from a freezer and thawed.

PAUSE AND THINK

Identify two major structural differences between plant cells and animal cells

Answer:
1. Plant cells walls contain cellulose.
2. Plant cells contain a fluid-filled vacuole called the tonoplast.

 Check your understanding of the concepts covered in this section by answering the questions in the e-book.

QUANTITATIVE TOOLS 28.3 Measuring turgor pressure in plant cells

Turgor in plant cells acts as a hydrostatic skeleton, analogous to the hard skeletons that give support to animal tissues. The pressure is contained within an epidermal skin and can allow plants to grow to several metres in height without requiring additional support.

The effect of turgor pressure can be seen by cutting a suitable plant structure such as a carrot into strips: after a short time the strips curl with the cut surface outermost. This is because the restraint on the tissue by the epidermis on this side is reduced and fluid leaks or evaporates (Figure 1). Much of the crunch in the mouth experienced when eating uncooked carrot, celery, or fresh apple is caused by several million turgid cells bursting as they are chewed.

The turgor pressure of individual plant cells can be measured directly by a probe (Figure 2). This is a thin glass capillary tube containing liquid, inserted into a single cell using a microscope and micromanipulators. As the probe is inserted, the pressure in the cell pushes the liquid back along the capillary. A manometer counter-balances and measures the force generated by the pressure.

Pressure is measured in pascals (Pa). Normal values for plant cells are in the range 0.5–1.5 MPa, which is several times the pressure in a car tyre. In cells such as algae, it can be up to 3.5 MPa, which is not far off that in a steam turbine in a power station. Note, however, that the volume of a plant cell is tiny so the pressure is distributed over a relatively large surface area.

Figure 1 The effect of turgor pressure on the shape of a carrot after cutting.

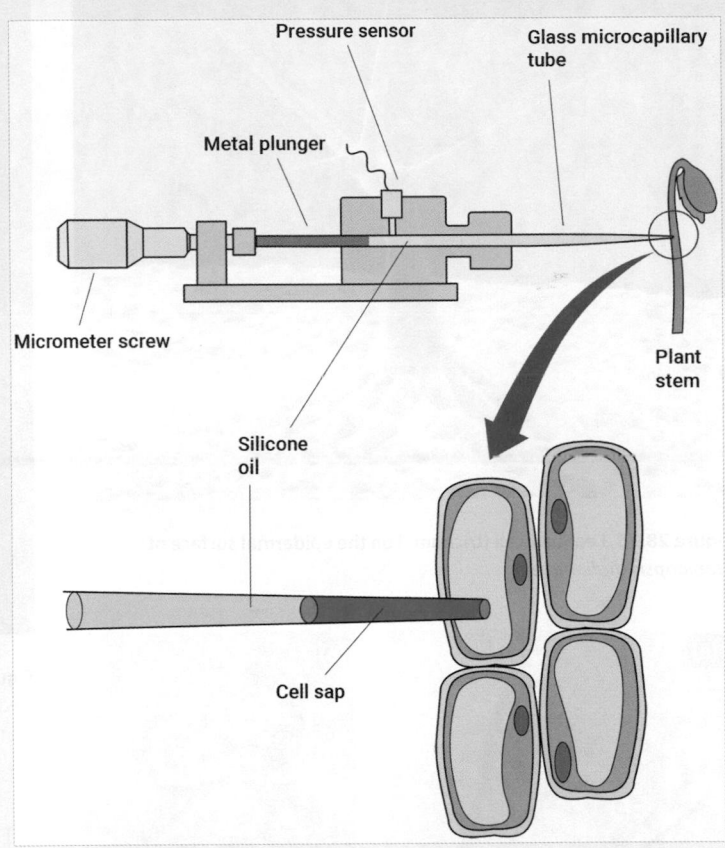

Figure 2 The measurement of turgor pressure.

Source: Plant Physiology and Development Sixth Edition, Taiz et al. 2014 Oxford University Press.

28.5 Directionality and anchorage dependence

The presence of a tough cell wall in plant cells presents a major problem as the plant grows and develops. Cell expansion and differentiation may involve growth in a particular direction or the development of a complex three-dimensional shape. For the cell wall to expand, the extent of cross-linking of cellulose microfibrils must be reduced so that the wall becomes plastic and can be deformed.

This process is largely controlled by the biochemistry of the cell wall. A reduction in pH caused by pumping in of protons increases the activity of a class of proteins called **expansins**. These enable the cellulose microfibrils to slip past each other, allowing the cell wall to expand. This is a complex process and if it does not take place evenly over the whole cell wall it creates tissues with irregular three-dimensional shapes.

Normally these shapes relate to some aspect of cell function. Good examples are leaf hairs, which in some cases develop as enormous single cells with several branched spikes sticking out dramatically from the leaf epidermis or the stem surface, as depicted in Figure 28.16.

Another example is the mesophyll cells in dicotyledonous leaves, which are located under the upper leaf surface. These elongate in a periclinal direction (perpendicular to the leaf surface) and generate a layer of palisade mesophyll cells with a long, thin morphology (Figure 28.17a). In contrast, mesophyll cells found in monocotyledonous leaves expand in localized areas of the cell surface to make highly lobed cells (Figure 28.17b).

The most extreme directional expansion in plant cells occurs when the cell only expands at one point, generating tip growth and the production of very long tubular cells. This type of growth is found in root hairs. They grow out from root epidermal cells as depiced in Figure 28.18, and maximize the surface area for water uptake.

It also happens in pollen tubes generated by germinating pollen on the stigmas of flowers. These tubes grow down between the cells of the style and facilitate fertilization of the ovule in the flower. In some cases, as in lily flowers, the pollen tube may be several centimetres in length, all generated by the tip growth of a single cell.

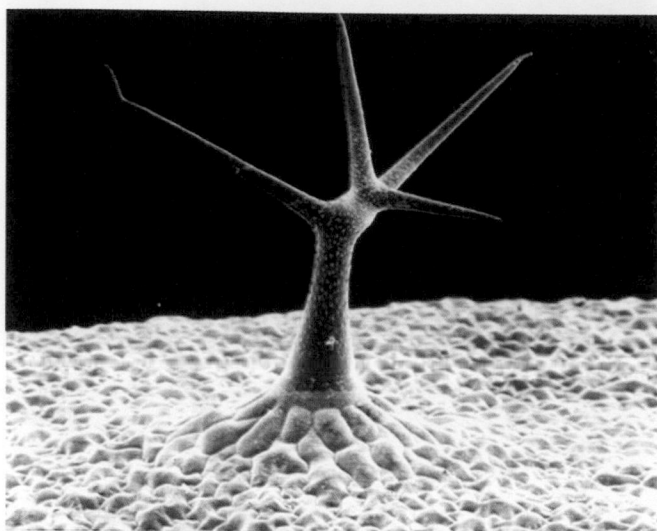

Figure 28.16 Leaf hair cell (trichome) on the epidermal surface of *Arabidopsis thaliana* leaf.

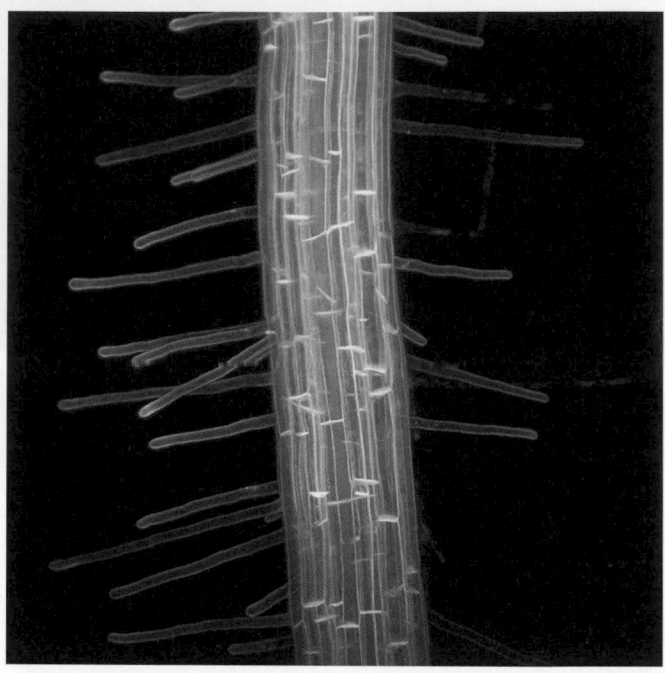

Figure 28.18 **Root hair cells.** Confocal microscope image of a live root of a seedling of *Arabidopsis thaliana*. The cell walls fluoresce; root hairs emanate from the main root.

Source: Anthony Bishopp, School of Bioscience, University of Nottingham.

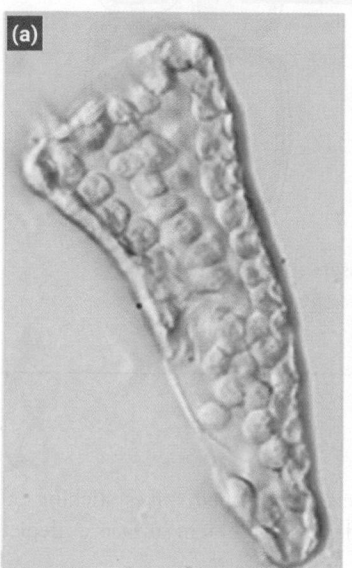

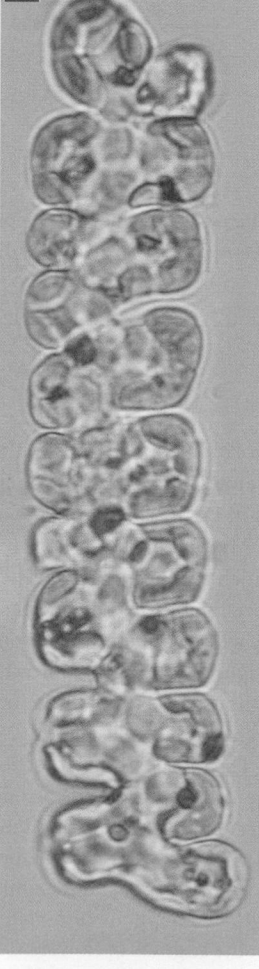

Figure 28.17 **Plant mesophyll cells.** (a) A palisade cell from a leaf of tomato (*Solanum lycopersicum*). [Scale bar = 10 μm.] (b) A lobed mesophyll cell from a leaf of the hexaploid bread wheat *Triticum aestivum*.

Source: (a) 'Mesophyll' Pyke, Kevin, Encyclopedia of Life Sciences Volume 15, June 2012, Figure 2.

Directionality in *animal* cells can be best understood by categorizing them in two groups: **anchorage-independent** (AI) and **anchorage-dependent** (AD).

Anchorage-independent cells include all those which circulate round the body, including erythrocytes, leukocytes, platelets, and many cells of the immune system such as natural killer cells, lymphocytes, and macrophages.

Anchorage-independent cells carry out their normal behaviours without needing to attach, either to each other or to a surface of some kind. They therefore lack an identifiable orientation (no 'up' or 'down', 'front' or 'back'). Nonetheless, they may have specific functions which do entail interactions, such as when erythrocytes collect in lymph nodes during an infection or when lymphocytes pass on information about invading organisms to macrophages.

The AI group also includes germ cells such as sperm, which exit the body and have a free living phase of existence until they fertilize an egg. Most invertebrate and all vertebrate sperm, however, do exhibit directional movement and thus have a clear orientation.

Most other animal cells fall into the AD category. In complex multicellular animals (metazoa), the highly specialized cells which make up each distinct tissue function according to their physical position. In other words, their biochemical and metabolic individuality depends on them being in an appropriate place and interacting with the cells and membranes which surround them.

In most cases, if AD cells are forced to live in an inappropriate environment (for example, during experimental cell culture) they will undergo changes of phenotype. If the conditions are especially inappropriate, for example if the cells are individually isolated or prevented from making surface attachments, they may dedifferentiate completely or even fail to survive.

These characteristics of AD cells demonstrate the *active* nature of the environmental interactions which they depend on. Towards the end of this chapter we discuss the nature of cell–cell and cell–membrane connections and point out some of the signalling and regulatory processes involved.

One result of these interactions is the adoption of polarity and orientation, or **directional habit**, which AI cells lack. Epithelial cells such as those of the gut demonstrate

this very well. The development of a brush border on the luminal surface of the cell, together with its associated absorptive and secretory biochemistry, comes about simply because it is anchored to a basement membrane and makes close connections with adjacent cells (by means of specialized junctions, described below), as illustrated in Figure 28.19. In other words, anchorage and position cause the cell to express its functional phenotype.

Figure 28.19 The orientation and polarity of epithelial cells. (a) The epithelial cells of the gut show a particular orientation and polarity because they are anchored to the basement membrane (b).

Source: Berne & Levy, Physiology, 7th Edition, Koeppen BM & Stanton BA, Figure 2.8, Copyright Elsevier 2018.

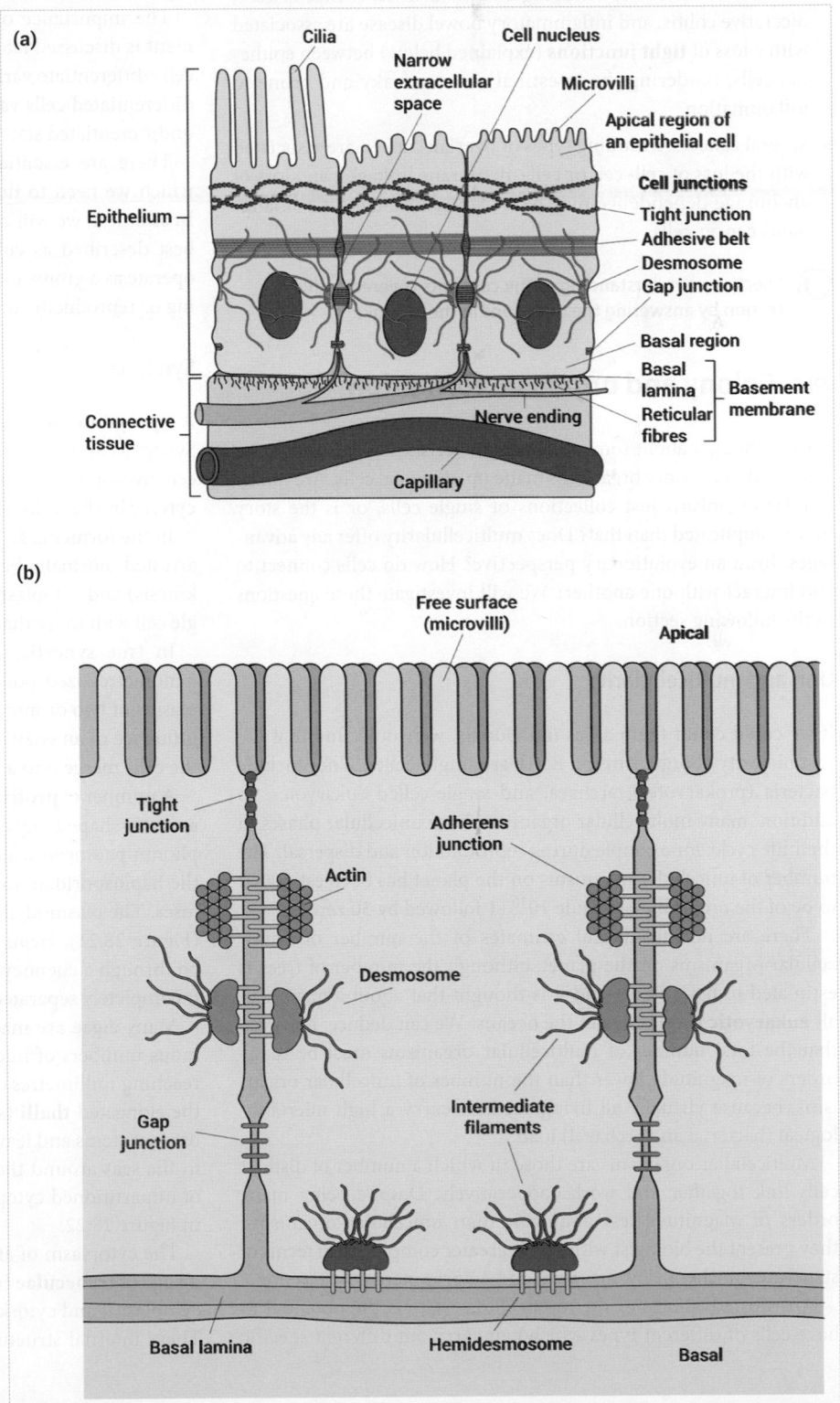

There are a number of human diseases which are characterized by cells losing their anchorage and/or polarity.

- *Epidermolysis bullosa* is a distressing inherited condition in which the skin is extremely fragile and subject to continual blistering. It arises because mutations in the genes for keratin or plectin mean that the dermis and epidermis lack the proteins which normally anchor the cells and resist abrasion.

- Diseases of the bowel including Crohn's disease, coeliac disease, ulcerative colitis, and inflammatory bowel disease are associated with a loss of **tight junctions** (explained below) between epithelial cells, rendering the intestinal mucosa leaky and prone to inflammation.

- Several cancers, including types of ovarian cancer, are associated with the loss of cell–cell or cell–membrane linkages, and loss of anchorage dependence may be a feature of metastatic (malignant) cancer cells.

 Check your understanding of the concepts covered in this section by answering the questions in the e-book.

28.6 Colony and organism formation

Having thought about some of the characteristics of living cells, we now need to consider organisms made up of several cells. Are multicellular organisms just collections of single cells, or is the story more complicated than that? Does multicellularity offer any advantages, from an evolutionary perspective? How do cells connect to and interact with one another? We will investigate these questions in the following section.

Uni- and multicellularity

If we could count them all as individuals, we would find that the vast majority of organisms on Earth are single-celled. They include bacteria (prokaryotes), archaea, and single-celled eukaryotes. In addition, many multicellular organisms have unicellular phases in their life cycle, for example during reproduction and dispersal. The number of unicellular organisms on the planet has been estimated to be of the order of magnitude 10^{30} (1 followed by 30 zeroes).

There are no meaningful estimates of the number of multicellular organisms on the planet, although the number of trees is estimated to be 3×10^{12} and it is thought that about a quarter of all **eukaryotic** species live in the oceans. We can deduce, however, that the total number of multicellular organisms must be many orders of magnitude lower than the number of unicellular organisms, because virtually all living surfaces carry a high microbiological (bacterial and archaeal) load.

Multicellular organisms are those in which a number of distinct cells link together and work cooperatively. Despite being many orders of magnitude less abundant than unicellular organisms, they present the biologist with much greater complexity in terms of structure, biochemistry, function, and environmental interaction.

A major advantage of the multicellular state is the potential to have cells of different types which can carry out different specific functions within the organism, and for subsets of cells to form multilayered and multifunctional tissues and organs.

These specialized structural and functional arrangements depend on cells differentiating themselves from one another along distinct developmental routes called **lineages**. For example, Figure 28.20 illustrates the lineages of the haematopoietic stem cell system. Look at this figure and notice how haematopoietic stem cells in bone marrow differentiate into several lineages of red, white, and immune cells.

The importance of cell lineages in plant and animal development is discussed later but it is true to say that the extent to which cells differentiate varies considerably between tissues. In addition, differentiated cells vary in their ability to revert to a more plastic, undifferentiated state.

There are essentially two types of multicellular organization which we need to understand: **syncytial** and **true multicellular**. In addition, we will consider the structure of organisms which are best described as **colonies**: collections of individual cells which operate as a group, even though each cell may be capable of surviving or reproducing on its own.

Syncytia

A **syncytium** is a cell which contains more than one nucleus. Syncytia can form in two different ways: by the interruption of the cell division process at a crucial point, forming a cell called a **coenocyte**, or by the fusion of two or more cells, forming a true syncytium.

In the former case, mitosis (described in detail in Chapter 10) is arrested, normally between the stages of nuclear division (karyokinesis) and cytoplasmic division (cytokinesis). The result is a single cell with more than one copy of its nuclear material.

In true syncytia, cells begin the fusion process by forming nanometre-sized pores. These establish a link between the cytoplasms of two or more adjacent cells. The pores expand, under the influence of an enzymic protein (a GTPase) called **dynamin**, until the cells merge into a single, multinucleated unit.

A number of **protists** are formed from syncytia. They include the amoeba-shaped algae called chlorarachniophytes, the plasmodiaphoran pathogen which causes the disease clubroot in plants, and the haplosporidian organisms which are parasites of marine molluscs. The plasmodial slime moulds, for example *Leocarpus fragilis* (Figure 28.21), *Hemitrichia serpula*, and *Physarum polycephalum*, go through a coenocytic feeding stage made up of vast numbers of incompletely separated cells.

Many algae are made up of giant coenocytes, often with enormous numbers of nuclei. Some of these cells are extremely large, reaching millimetres or even metres in diameter. It is possible that the elongated **thalli** (sing. thallus = a plant body not divided into distinct stems and leaves) of *Caulerpa cactoides*, a green alga found in the seas around the coast of Australia, have the largest volume of unpartitioned cytoplasm of any organism. This alga is depicted in Figure 28.22.

The cytoplasm of algal giant cells can be compartmentalized by means of **trabeculae** (rod-like parts of the cell wall, penetrating the cytoplasm) and cytoskeletal proteins, as illustrated in Figure 28.23. These internal structures control the distribution and movement

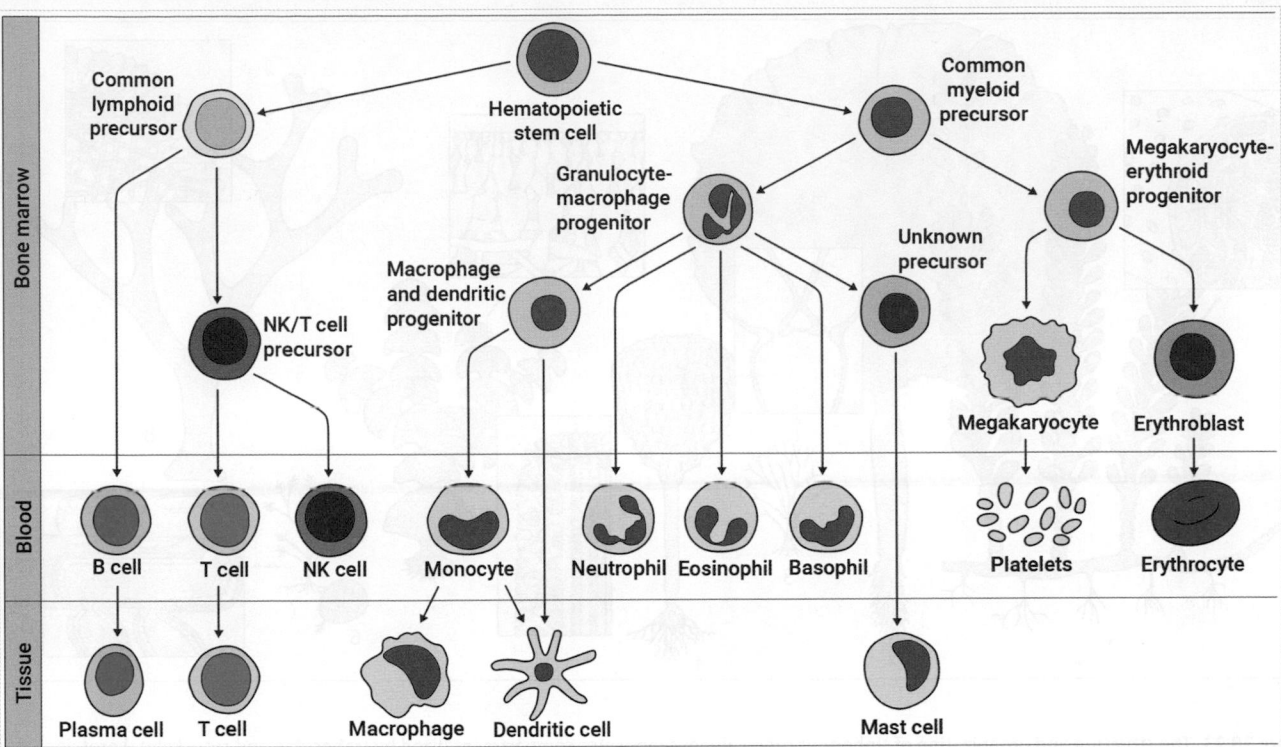

Figure 28.20 Cell lineages in the haematopoietic stem cell system. Haematopoietic stem cells in bone marrow are multipotent and differentiate into several lineages of red, white, and immune cells. They also divide in the undifferentiated state (not shown), so there is a constant supply of stem cells.

Figure 28.21 Plasmodial slime mold *Leocarpus fragilis* **on autumnal leaf and bark litter.**

Figure 28.22 Marine green alga *Caulerpa cactoides*.
Source: Image by: Julian Finn/Museum Victoria,/CC BY 3.0.

of organelles, including chloroplasts, and ensure that the nuclei remain distributed around the cell rather than clumped together.

Large syncytia are part of the normal state for many fungi. For example, filamentous ascomycete fungi such as *Neurospora crassa* (bread mould), the plant pathogen *Aspergillus nidulans*, and the vegetable decay fungus *Sclerotinia sclerotiorum* (shown in Figure 28.24) may form syncytial mycelia several centimetres in length

and containing millions of nuclei. Cytoplasmic material moves continuously through the mycelia at a rapid rate, with nuclei reaching speeds of up to 60 μm s^{-1}. They are good experimental models for investigating the mechanism of cytoplasmic flow within cells.

Mycelial syncytia may be formed by the fusion of cells from genetically distinct sources, combined with rapid nuclear division within the single mycelial cytoplasm. They are therefore genetic

Figure 28.23 The structure and organization of siphonous green algae: coenocytes compartmentalized by trabeculae and cytoskeletal proteins. Thalli of various green algae with insets (not to scale) showing their structure. 1. *Caulerpa actoides*; 2. *Avrainvillea gardineri*; 3. *Chlorodesmis* sp.; 4. *Penicillus captus*; 5. *Halimeda tuna*; 6. *Derbesia* sp.; 7. *Bryopsis plumosa*; 8. *Codium fragile*.

Source: Hurd CL, Harrison PJ, Bischof K, Lobban CS (2014) Seaweed Ecology and Physiology, Cambridge University Press.

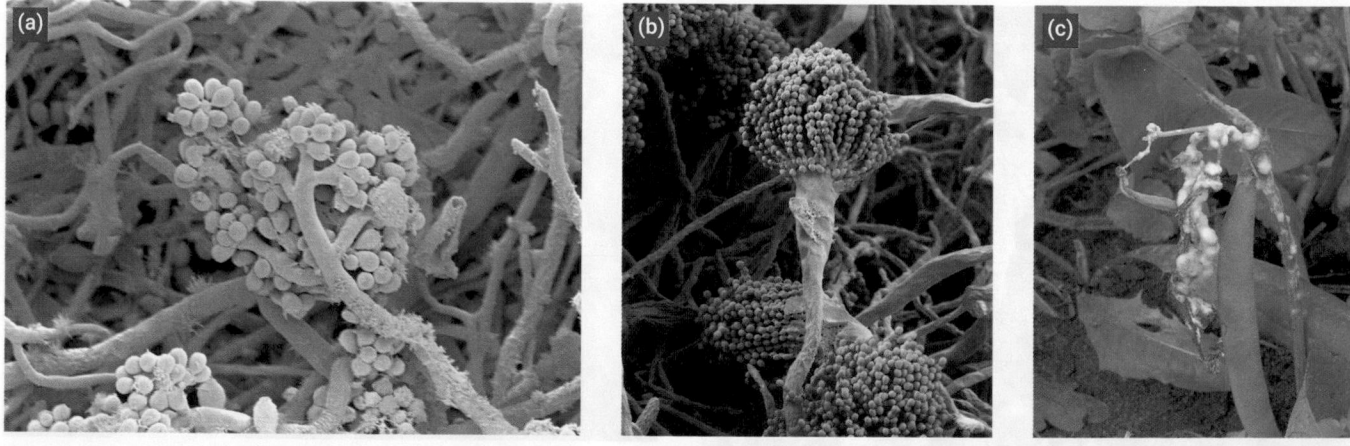

Figure 28.24 Three species of filamentous ascomycete fungi. (a) Electron microscope image of the fruiting body of the bread mould *Neurospora crassa*. (b) *Aspergillus nidulans*. (c) *Sclerotinia sclerotiorum* on *Phaseolus vulgaris* bushbean.

Source: (a) Wellcome Images. Fruiting body of bread mould. Credit: Liz Hirst, Medical Research Council. Attribution 4.0 International (CC BY 4.0). (b) Eye Of Science/Science Photo Library. (c) © Rasbak/Wikimedia Commons,/CC BY-SA 3.0 license.

chimeras, resulting from the transfer of genes between individuals, and they must depend on considerable tolerance of genetic diversity within the syncytial cytoplasm.

Understanding this process has practical importance, as it may determine the growth of fungi in variable environments, their ability to regenerate from single individuals, and the genetic basis of their resistance to fungicides. Studying fungal syncytia can also

be valuable in trying to understand syncytial structures in other life forms, for example in the embryos of fruit flies, zebra fish, and nematode worms, all of which are commonly used as experimental animals in biological research laboratories.

The formation of coenocytic syncytia is a normal part of development in many plants. Early endosperm development in some seeds, especially those of cereals such as wheat and rice, is coenocytic in that extensive nuclear division occurs, giving rise to a multinucleate tissue before cell walls are laid down to form a multicellular tissue (discussed in Chapter 33).

Other examples of coenocytic cells are found in the plants depicted in Figure 28.25a. These include the tapetum cells which nourish the pollen grains in the anthers of magnolia flowers, and articulated (jointed) laticifer cells of banana leaves which contain a type of latex. The sticky fluid of milkweed plants flows in disarticulated (continuous) cells, which are coenocytes.

During early stages of plant seed development, the storage tissue (endosperm) is coenocytic, as illustrated in Figure 28.25b: it results from many nuclear divisions without cell division. This stage is transient and the tissue becomes cellularized soon after.

Syncytia are also widespread in animals. Amongst the invertebrates, **rotifers** (a large group of microscopic, free living, aquatic and parasitic, **pseudocoelomate** animals; see Chapter 29) have syncytial ectodermal cells covering the main body region. Figure 28.26a shows a bdelloid rotifer and Figure 28.26b shows a scanning electron micrograph of its skin surface. Rotifers also have highly ciliated trinucleate excretory cells, called **protonephridia** and the early embryonic stages of some rotifers are complex syncytia. The ectodermal cells of nematode worms are formed from syncytia.

Syncytia occur in a variety of tissues in vertebrate animals. The process of cell fusion occurs at several crucial stages of vertebrate development including fertilization, when sperm fuses with egg. It also happens in early gestation: a **syncytiotrophoblast** forms from

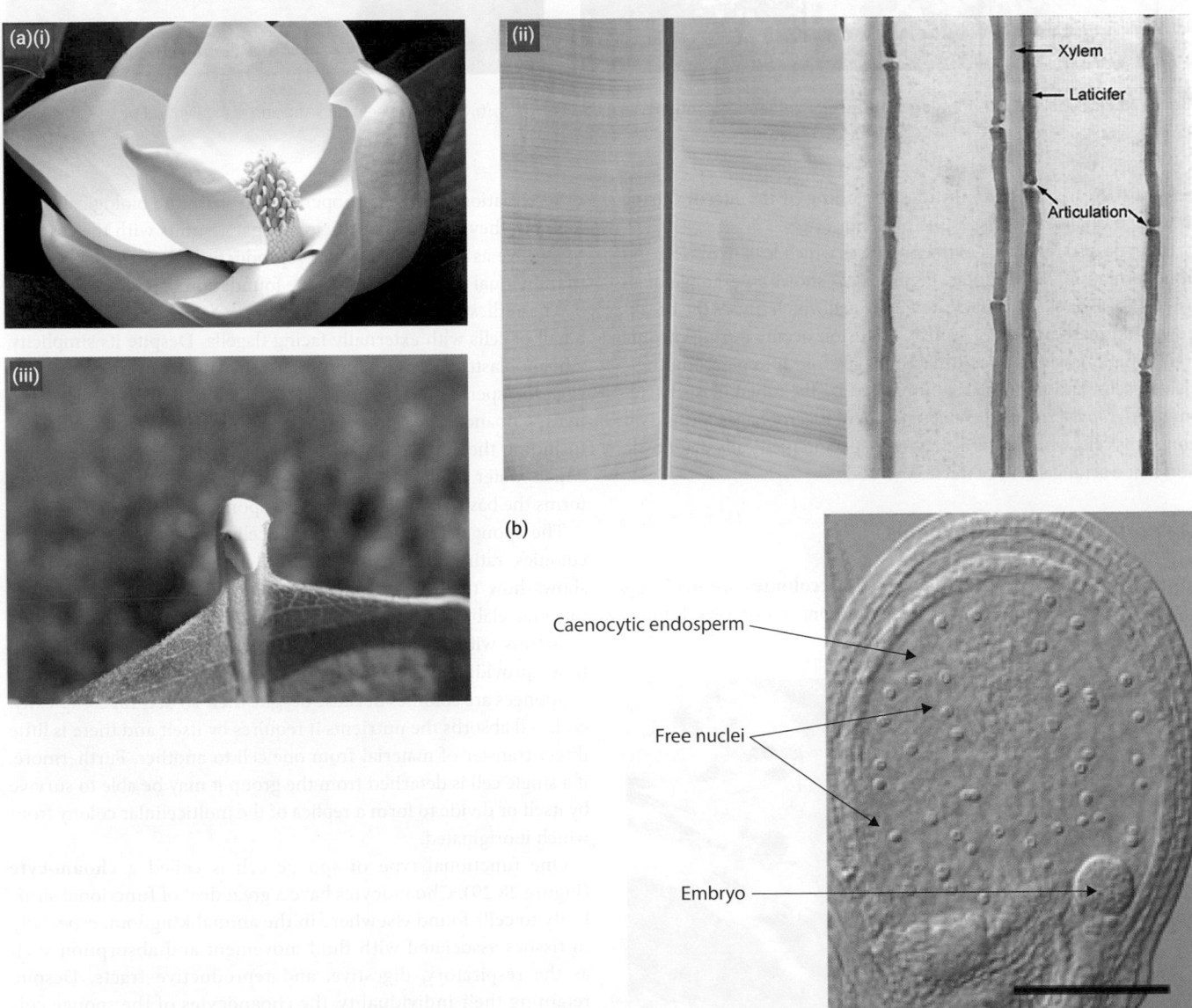

Figure 28.25 Three plants with coenocytic cells. (a)(i) *Magnolia grandiflora*. The anthers have coenocytic tapetum cells. (ii) Banana. The articulated laticifer cells in leaves are coenocytes. (iii) Milkweed (*Asclepias*). Sap emerging from the coenocytic cells of a cut stem. (b) Coenocytic endosperm during plant seed development.

Source: (ai) © Josep Renalias Lohen11/Wikimedia Commons,/CC BY-SA 3.0 license. (aii) © Hysocc/Wikimedia Commons,/CC BY-SA 3.0 license. (aiii) Corey Raimond. (b) The Arabidopsis CUL4–DDB1 complex interacts with MSI1 and is required to maintain MEDEA parental imprinting. Dumbliauskas E et al. (2011). EMBO Journal, 30, 731–743. Figure 2. John Wiley and Sons.

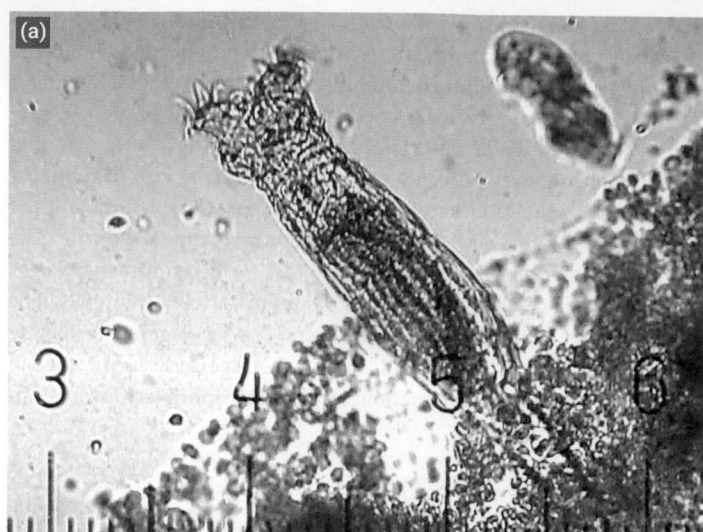
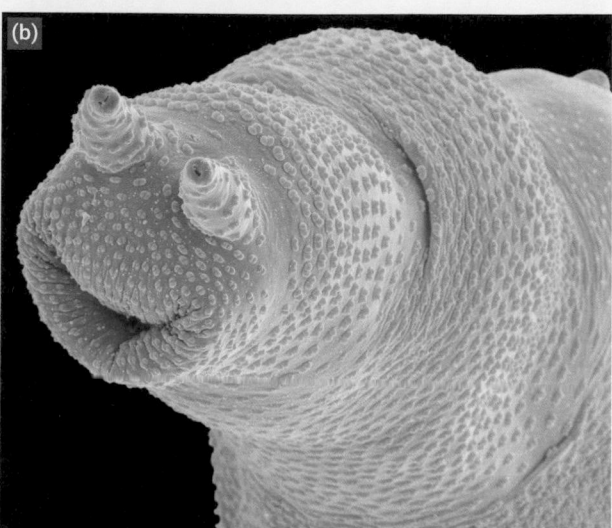

Figure 28.26 Syncytial cells forming the cuticular epidermis of rotifers. (a) A bdelloid rotifer. (b) Scanning electron micrograph of the skin surface.
Source: (a) Bob Blaylock at English Wikipedia, CC BY-SA 3.0. (b) Steve Gschmeissner/Science Photo Library.

cells of the early embryo, invades the lining of the uterus during implantation and establishes the placenta.

Syncytia also form part of the processes which lead to muscle and bone formation. For example, Figure 28.27 shows an example of an osteoclast, a giant, multinucleated bone cell which causes the breakdown and reabsorption of bone. Cell fusion occurs during chronic inflammation, when several macrophages fuse into giant multinucleated cells, and in lung alveoli infected by the tuberculosis bacterium (*Mycobacterium tuberculosis*), where macrophages form giant Langhans cells which have a characteristic and medically diagnostic horseshoe arrangement of nuclei.

Multicellular organisms: colonies

In general terms, organisms described as **colonies** are made up of collections of cells which have undergone no or very limited

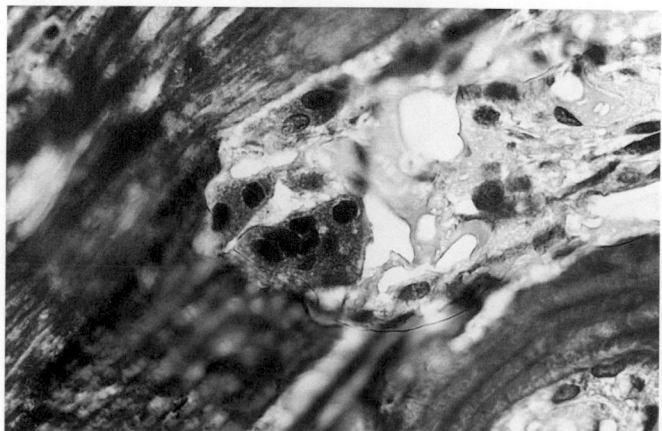

Figure 28.27 Syncytia and the formation of bone cells. Shown here are osteoclasts: giant, multinucleated bone cells which cause the breakdown and reabsorption of bone.
Source: Biophoto Associates/Science Photo Library.

differentiation. The cells cooperate and share the biological workload but they remain, in principle, separate units with the capacity to survive as individuals. Their reproduction also usually consists of individual dispersal prior to the foundation of a new colony.

The earliest type of colonial animal was probably Choanoblastea, a ball of cells with externally facing flagella. Despite its simplicity, Choanoblastea may have reproduced sexually by means of haploid eggs and sperm in the manner we are familiar with in modern animals. Choanoblastea probably evolved to have several cell layers, including the germ cells, and developed a sac or pocket through which water could be wafted by the flagella. This arrangement forms the basic structure of modern sponges.

The sponges (**Porifera**) are multicellular organisms which are colonies rather than truly multicellular animals. Figure 28.28 shows how they comprise cells arranged in clusters, sometimes forming elaborate and colourful structures. The resulting body is porous with channels through which the cilia-propelled fluid flows, providing currents of fluid for respiration and feeding.

Sponges are colonies because despite their structural complexity, each cell absorbs the nutrients it requires by itself and there is little direct transfer of material from one cell to another. Furthermore, if a single cell is detached from the group it may be able to survive by itself or divide to form a replica of the multicellular colony from which it originated.

One functional type of sponge cell is called a **choanocyte** (Figure 28.29). Choanocytes have a great deal of functional similarity to cells found elsewhere in the animal kingdom, especially in tissues associated with fluid movement and absorption such as the respiratory, digestive, and reproductive tracts. Despite retaining their individuality, the choanocytes of the sponge colony benefit from each other by forming channels for water to move through. This slows the passage of potential food particles, concentrating them in the vicinity of all the cells in the colony in a way which would not be possible for an individual cell. In other words, the cells benefit from cooperation but the interdependence of the group is largely physical rather than biochemical.

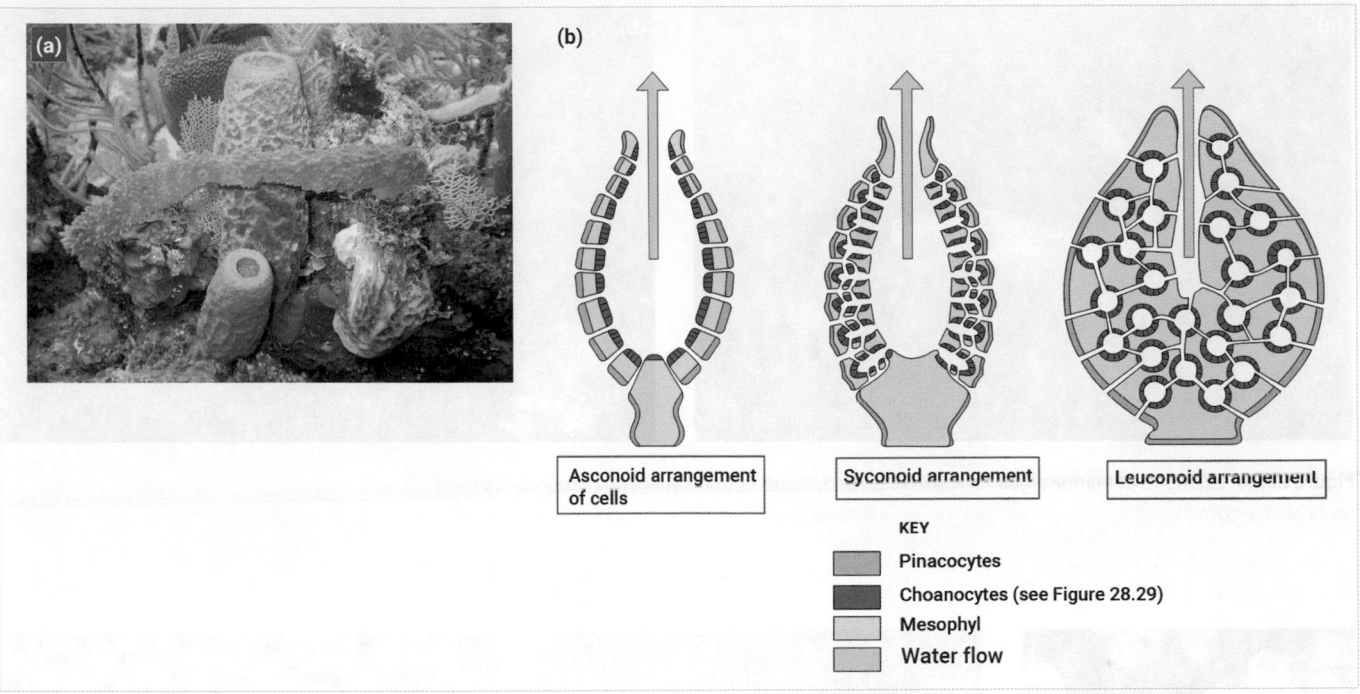

Figure 28.28 Sponges (Porifera) and their structure. (a) Sponge biodiversity and morphotypes at the lip of a wall site in 60 feet of water. Included are the yellow tube sponge, *Aplysina fistularis*, the purple vase sponge, *Niphates digitalis*, the red encrusting sponge, *Spiratrella coccinea*, and the gray rope sponge, *Callyspongia* sp. Caribbean Sea, Cayman Islands. (b) Sponge types: three different cell arrangements.
Source: (a) Twilight Zone Expedition Team 2007, NOAA-OE/CC BY 2.0. (b) Philcha/Wikimedia Commons/CC BY 3.0.

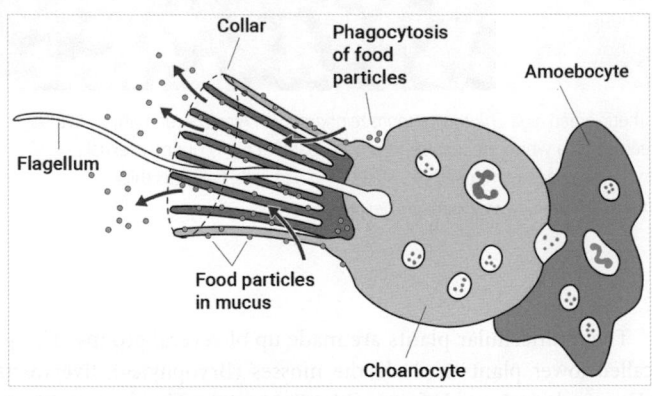

Figure 28.29 The structure of a choanocyte.
Source: © Steven M Carr.

PAUSE AND THINK

What is the difference, regarding membranes, between a syncytium and a multicellular colony

Answer: A syncytium has one membrane surrounding several nuclei and other intracellular organelles. A multicellular colony is made up of several cells, each with a membrane surrounding a single nucleus and associated organelles.

 Check your understanding of the concepts covered in this section by answering the questions in the e-book.

28.7 Multicellular plants

Plants—organisms which use photosynthesis to exploit energy from the sun—exist abundantly in single and multicellular forms. In this section we look at the structures of multicellular plants and consider how they develop.

Algae and the evolution of multicellular plants

A significant proportion of green photosynthetic organisms on the planet are single-celled organisms called **algae** or **chlorophytes**. They are largely aquatic, most living in fresh water but with some in sea water (green and brown marine algae, such as the kelp depicted in Figure 28.30). Algae have great diversity and levels of complexity and it is thought that the evolution of multicellular plant life on the planet began from a single celled form similar to modern green algae.

Green algae such as *Chlamydomonas* (Figure 28.31a) exist as single cells within huge populations. They have been the subject of extensive research as a model organism for studying the molecular biology of chloroplasts and the function of motile flagella.

Volvox (Figure 28.31b) is a green alga which forms spherical colonies of many thousands of individual cells, held in a gelatinous sphere composed of glycoprotein. The cells cooperate to the benefit of the whole *Volvox* colony. Each cell has a flagellum, and coordination of flagella movement within the colony allows directional movement of the whole Volvox sphere. Asexual **gonadia** are formed within the sphere: these are new *Volvox* colonies which will eventually leave the parent colony when it dies (Figure 28.31b).

Figure 28.30 Examples of marine algae. (a) Seal in kelp, Lundy Island, Bristol Channel. (b) Kelp forest, Shetland. *Source*: Simon Rogerson.

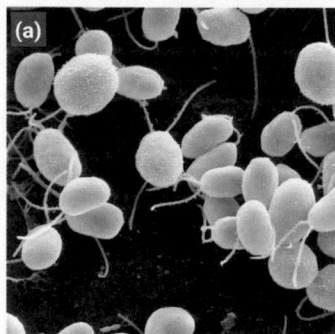
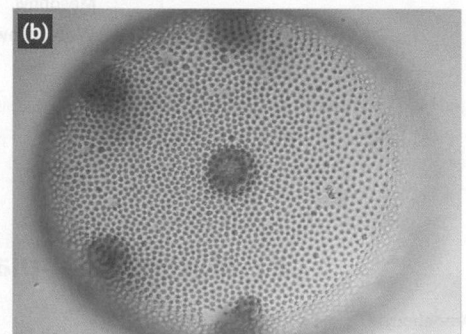

Figure 28.31 Colony and tissue formation in algae. (a) Single cells of the photosynthetic green alga *Chlamydomonas reinhardtii*. (b) A colony of many individual algal cells make up the spherical body of *Volvox aureus*, within which are several dark green gonidia which are released from the parent colony during asexual reproduction. (c) The multicellular green alga *Ulva lactuca*, known as sea lettuce, is a common seaweed composed of a sheet of joined cells, two cells thick.
Source: (a) Ami Images/Science Photo Library. (b) Photo by Y. Tsukii http://protist.i.hosei.ac.jp. (c) © Malcolm Storey, www.bioimages.org.uk.

Some algae are filamentous or sheet-like: the cells are physically joined and may represent the very early origins of the evolution of multicellular structures, as represented by the sea lettuce shown in Figure 28.31c.

It appears that multicellularity has emerged several times during the evolution of photosynthetic eukaryotic organisms. This has given rise to many types of multicellular algae, typified by the variety of seaweeds which are highly successful in coastal saline environments. Water provides much more buoyancy than air and supports the development of large colonies, especial in shallow, nutrient-rich waters where light is also plentiful. Kelp forests are highly efficient and amongst the most productive ecosystems on the planet (see Figure 28.30).

The evolution of multicellularity in photosynthetic organisms was probably a crucial point in the evolution of life forms on the planet, especially as they subsequently extended their range to include the terrestrial environment and started to grow upwards into the air. The generation of oxygen as a by-product of photosynthesis changed the composition of Earth's atmosphere and facilitated the evolution of animal life forms.

The multicellular plants are made up of several groups. The so called lower plants include the mosses (Bryophytes), liverworts (Hepatophytes), and ferns (Monilophytes). The higher plants (**Metaphytes**) comprise the flowering plants (**Angiosperms**) and the conifers (**Gymnosperms**). Plants in all of these groups have complex structures composed of many different types of cell. There are thought to be at least 350 thousand species of multicellular plant.

Meristems and plant development

Plants generate new cells by mitosis and cell division (cytokinesis). These processes are similar to those in animals except that plant cells are forever joined after cell division and thus always create a multicellular tissue of connected cells. Each specialized plant cell differentiates along a defined developmental pathway, having arisen initially from a progenitor cell produced in specialized areas within the plant called **meristems**. Meristem cells are a continual source of new cells, and are analogous to stem cells in animals.

The activity of meristems is illustrated in the flowering plants (Angiosperms). These plants fall into two groups according to whether they have one (mono) or two (di) cotyledons in their seeds:

- **monocotyledonous** (called monocots), including grasses, wheat, maize, sugar cane, orchids, and similar plants;

- **dicotyledonous** (dicots), including trees, shrubs, most wild and garden plants, and most vegetable and other non-cereal crop plants.

The arrangement of meristems is also different between the two groups. In monocots such as grasses, the meristem is at the base of the shoot, as illustrated in Figure 28.32. Newly generated cells move upwards into the developing leaf. If one cuts this kind of leaf, the basal meristem is unaffected and keeps making new cells. This is why mowing a lawn or allowing sheep to graze a field shortens the existing grass but does not prevent its continued growth: the basal meristems will regenerate more leaf.

In a simple dicot seedling, such as that also shown in Figure 28.32, there are two meristems. One is at the growing tip and is called the **shoot apical meristem** (SAM). The other is just behind the tip of the root and called the **root apical meristem** (RAM). The SAM gives rise to the entire above ground part of the plant while the RAM gives rise to the entire underground root system. Roots of monocot plants also have a RAM at each of their root tips which perform a similar role.

Other meristems develop as the plant grows, most noticeably in the **axils** where **leaf petioles** join the stem (see Figure 28.32). These axillary meristems remain quiescent while the main SAM is active. However, as the stem elongates the SAM moves further away; this removes a chemical suppression from the axillary meristems and they start to grow, producing new stems and leaves.

Suppression of other meristems by the SAM is the basis of the process called apical dominance. It is exploited by gardeners when pruning, to make plants bushier and inspire fresh growth. Cutting off the main shoot and removing the SAM induces new shoots and growth by the axillary meristems.

The cells generated by either the SAM or the RAM will develop into cells specific to the organ type that the meristem produces. However, in general plant cells can differentiate into any cell type. Even when fully differentiated they can still be made to dedifferentiate into basic stem cells and then eventually into a whole plant. This flexibility is called **totipotency** and is considered later under the general topic of stem cells.

PAUSE AND THINK

What survival advantage does a monocot plant gain from having a basal meristem?

Answer: If the above ground leaf material is eaten and cut off at ground level by a grazing animal the basal meristem survives and can produce more leaves.

This enables plants like grass to form huge monocultures, such as meadows, prairies, or savannah. They provide important, regenerating food sources for large numbers of grazing animals and are exploited by man in the production of cereals.

 Check your understanding of the concepts covered in this section by answering the questions in the e-book.

Figure 28.32 Plant meristems. The location of plant meristems in monocots and dicots.

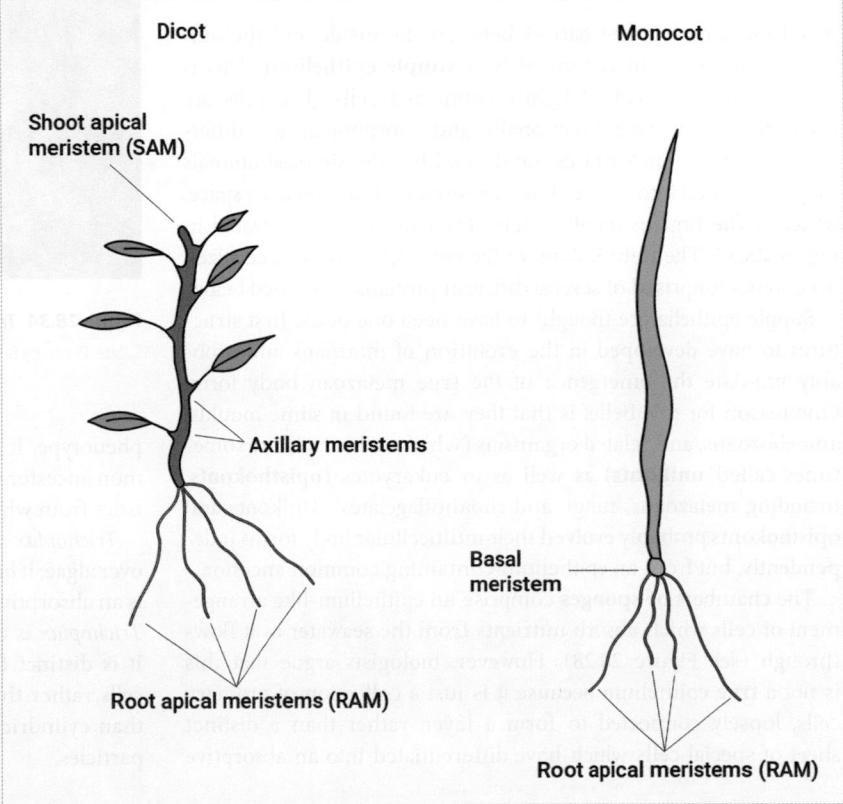

Dicot

Monocot

Shoot apical meristem (SAM)

Axillary meristems

Basal meristem

Root apical meristems (RAM)

Root apical meristems (RAM)

28.8 Multicellular animals: Metazoa

Metazoa, of which there are thought to be around eight million different species, are multicellular animals in which cells are organized into functionally distinct groups. The cells arrange themselves into distinct tissues and organs. They do this by forming complex junctions (explained below), by polarizing (adopting a directional orientation) and by becoming highly specialized for particular functions.

Thus metazoan cells share the biochemical workload and each becomes efficient at carrying out a limited range of processes (sensation, secretion, digestion, absorption, metabolism, contraction, defence, reproduction, etc.). Each cell only survives with the support and cooperation of others and they usually live or die as a group.

Communication systems (nerves, hormones, and other chemical signals) coordinate the development and activity of the differentiated cells, with the result that the whole organism exists as an integrated unit of life. However, this means that the majority of cells do not interact *directly* with the environment. Only those with externally exposed surfaces (skin, gut, respiratory system) participate directly in energy intake, excretion, gas exchange, or fluid exchange. The remaining cells are indirectly supplied, and in large animals this necessitates pulmonary (lung and gas distribution) and circulatory (fluid distribution) systems. Multicellular animals also have more complex reproductive and development systems than unicellular animals.

Metazoans have epithelia

In all the metazoa, the barrier between the inside and the outside of the organism is formed by a **simple epithelium**. This is a single-layered sheet of tightly connected cells. The cells are **polarized**—they have functionally and morphologically different top and bottom surfaces—and in all but the simplest animals they are formed into a tube. The tube surrounds a lumen, or space, which is the organism's alimentary tract or gut, as illustrated in Figure 28.33. The tight linking of the cells depends on specialized junctions, comprised of several different proteins, described below.

Simple epithelia are thought to have been one of the first structures to have developed in the evolution of mtazoans and probably pre-date the emergence of the true metazoan body form. One reason for this belief is that they are found in slime moulds, amoebozoans, and related organisms (which collectively are sometimes called **unikonts**) as well as in eukaryotes (**opisthokonts**, including metazoans, fungi, and choanoflagellates). Unikonts and opisthokonts probably evolved their multicellular body forms independently, but from an epithelium-containing common ancestor.

The chambers of **sponges** comprise an epithelium-like arrangement of cells which absorb nutrients from the seawater as it flows through (see Figure 28.28). However, biologists argue that this is not a true epithelium because it is just a collection of ordinary cells, loosely connected to form a layer, rather than a distinct sheet of special cells which have differentiated into an absorptive

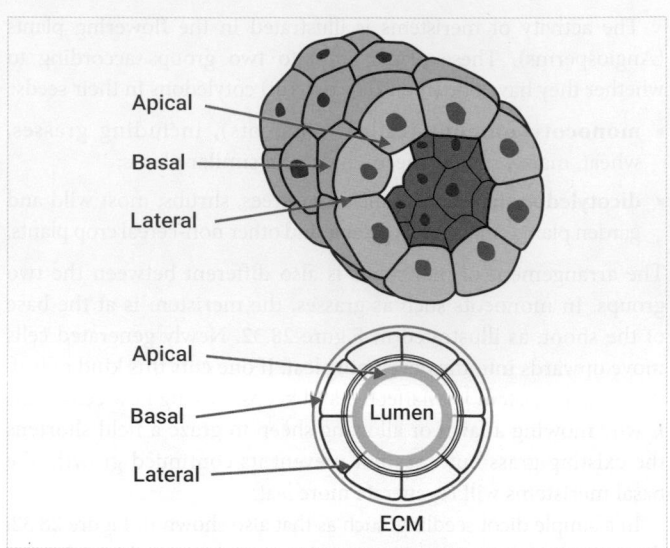

Figure 28.33 Cell linkage in a simple epithelium (gut).

Source: Current Topics in Membranes, 72, Miller et al., The Evolutionary Origin of Epithelial Cell–Cell Adhesion Mechanisms, 267–309, Copyright 2013, with permission from Elsevier.

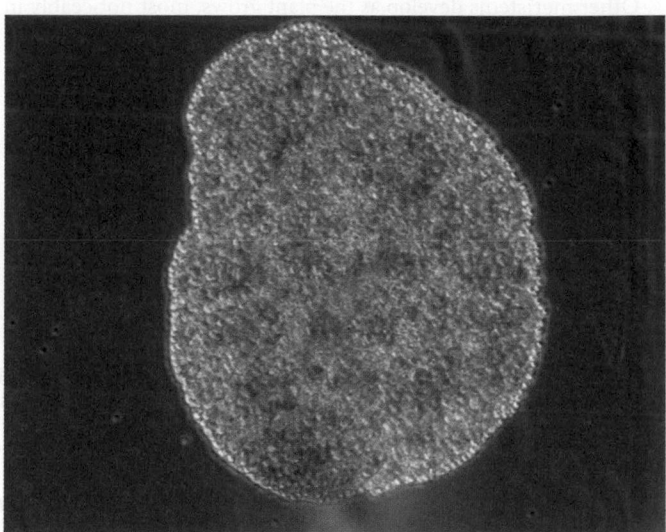

Figure 28.34 *Trichoplax adhaerens*.

Source: Photo by B. Schierwater.

phenotype. It is possible that sponges and metazoans had a common ancestor which possessed some of the fundamental characteristics from which a true epithelium could evolve.

Trichoplax adhaerens (Figure 28.34) is a flat organism that slides over algae; it has a ventral (lower) layer of epithelial-like cells, which is an absorptive surface, and one upper layer of non-epithelial cells. *Trichoplax* is one of only three members of the phylum Placozoa. It is distinct from the sponges in having two different layers of cells rather than one. The lower, absorptive surface is flat rather than cylindrical but may form a digestive cavity around nutrient particles.

Trichoplax may therefore be considered one of the simplest metazoans. All animals of greater complexity have the basic body plan of a cylindrical, absorptive internal epithelium as well as a true external epthelium. This underlying body plan, even when given greater complexity by the addition of layers of specialist cells, such as those of the circulatory, muscular, nervous, and other systems, can always be detected when analysing metazoan body structures (Chapter 29).

Epithelia have basement membranes: something to hold on to

The formation of epithelia was probably a crucial stage in the evolution of the metazoan body form. Evolutionary biologists recognize a number of other crucial events too, including the formation of **basement membranes** and the ability of cells to link themselves tightly together.

A basement membrane (sometimes called a basal lamina) is a non-cellular lining layer on an epithelium. It serves as a platform for the anchorage of the epithelial cells and enables them to achieve polarity (distinct top and bottom surfaces). Basement membranes separate cells into functional groups and tissue layers. They are therefore the basis of organ structures, compartmentalizing the body as well as providing structural support and physical resilience.

Basement membranes are frequent in all metazoa. Although they first emerged on the **serosal** (body) side of a nutrient-absorbing epithelium, facilitating the development of the absorptive and secretory functions associated with feeding and digestion, they are found throughout the bodies of modern complex animals. They surround major organs and tissues of all types as well as the endothelial cells which line the circulatory and lymphatic systems. They are also found in synapses, the junctions between nerve endings and muscle fibres.

A key component of all basement membranes is **collagen Type IV**. This is an immensely strong protein made from a three-ply, rope-like twist of strands, as depicted in Figure 28.35. There are around 20 different kinds of collagen but Type IV is unusual in having cross links between the strands. Together with end-linkages, this enables it to form sheet-like structures.

The other biochemical components of basement membranes are **laminin**, **nidogen**, and **perlecan**, formed into a complex protein mat, as illustrated in Figure 28.36. Laminin is a cross- or sword-shaped protein made of three long peptides, twisted around each other and held together by disulphide bonds. It self-assembles during tissue formation, helps to organize the other structural

Figure 28.35 The structure of collagen Type IV.

Source: Reprinted by permission from Springer Nature Customer Service Centre GmbH: Springer, Nature Reviews Cancer, Basement membranes: structure, assembly and role in tumour angiogenesis, Kalluri R © 2003.

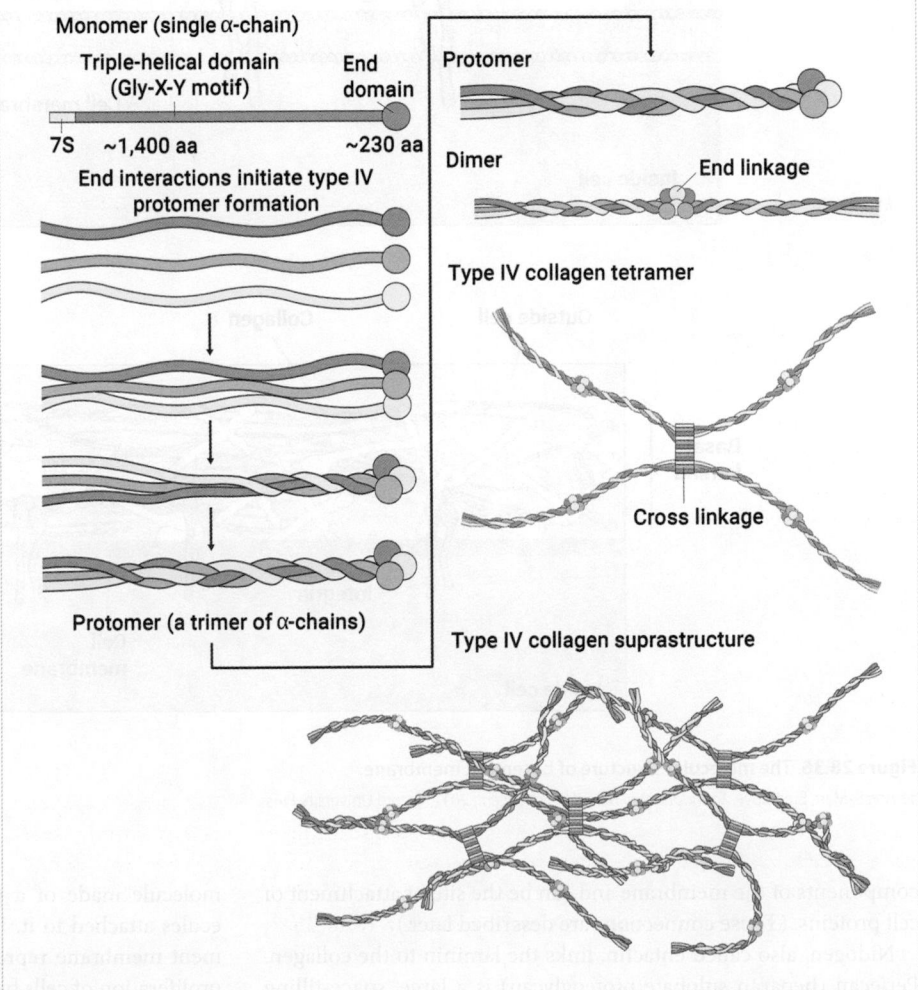

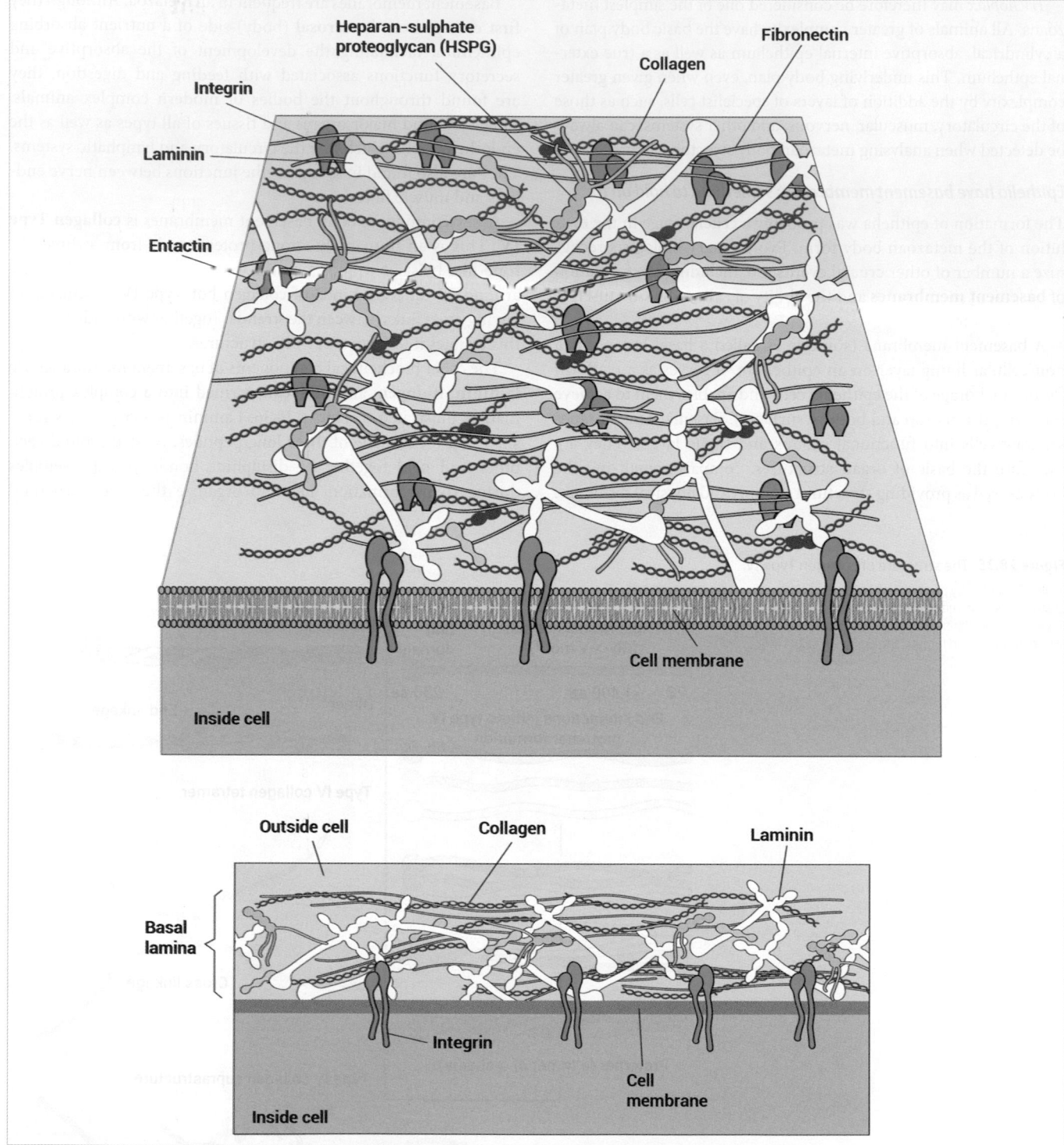

Figure 28.36 The molecular structure of basement membrane.

Source: S. Marc Breedlove, *Foundations of Neural Development*, 2017, Oxford University Press.

components of the membrane and can be the site of attachment of cell proteins. (These connections are described later.)

Nidogen, also called entactin, links the laminin to the collagen. Perlecan (heparan sulphate proteoglycan) is a large, space-filling molecule made of a protein core with glycosaminoglycan molecules attached to it. It helps to form the barrier which the basement membrane represents and can be crucial in preventing the proliferation of cells beyond the tissue boundary.

Figure 28.37 *Oscarella lobularis*, a thick, fleshy, marine homoscleromorph sponge with a jelly-like texture.

Source: © Bernard Picton.

The components of basement membranes have been detected in some sponges. One group of sponges called Homoscleromorpha (Figure 28.37) certainly possess collagen IV and have basement membrane-like structures, but other groups of Porifera (the Silicea and Calcarea) do not. The Homoscleromorpha therefore have some similarities with Eumetazoa which other Porifera do not possess.

As always in these cases, modern animals are not contemporary with ancestral animals and care must be taken not to think of existing Homoscleromorpha as a transition stage in the evolution of modern Eumetazoa. The modern sponges are considered to be descendants of a very old lineage of multicellular animals; in other words, they separated from the common ancestor they shared with other modern animals a very long time ago. The homoscleromorphs may have formed a distinct group of sponges shortly after that separation, before the splitting of the remaining ancestral sponges into the other groups of Porifera we recognize today. To that extent, the homoscleromorphs are probably closer cousins to the Eumetazoa (Gastraeozoa) than are the other sponges. The presence of epithelia and basement membranes may be evidence of this closer association, and it is also supported by analyses of molecular structures within these organisms.

PAUSE AND THINK

What is a 'serous' membrane and what is the 'serosal' side of an animal tissue

Answer: A membrane enclosing an organ of the body cavity (thorax or abdomen), usually made of a layer of cells attached to a basement membrane. The serosal side is thus the side of the tissue's cells which faces away from the other cells which make up the organ.

 Check your understanding of the concepts covered in this section by answering the questions in the e-book.

28.9 Cell lineages and differentiation

One of the most remarkable things in all biology is the way a single cell, itself the result of fertilization of an egg by a sperm or the product of some kind of budding or other propagation process, can be the source of all the different types of cell which make an organism. If all cells had the same functional characteristics, multicellular organisms could develop by cell division alone, although the degree of complexity and variety would be very limited. In all except the simplest multicellular organisms (sponges, for example), this is not the case: the cells in complex plants and animals take on particular functions in different body locations. A leaf cell has different characteristics from a root cell on the same plant. A skin cell carries out completely different biochemical functions from a liver cell in the same animal. How and why do these different functions emerge?

This process of cell specialization is called **differentiation**. It happens despite the fact that all cells in an organism (except the germ cells of reproduction) have the same genetic composition. Some cells, called stem cells, can divide over many generations and still permit specialization when circumstances are appropriate.

The question of what determines whether a cell will develop one set of functions rather than another, and what makes cells differentiate at particular stages of organism development, has intrigued biologists ever since cells were first recognized as the basic unit of life, following the invention of the microscope, in the 17th/18th century. It is thus one of the oldest questions in biological science.

It is equally fascinating to ask how differential cell development manages to happen, usually faultlessly, time after time, in the same way, in all the individuals of a species. Indeed, unless we carry out a molecular genetic analysis, we recognize and describe organisms by their **phenotype**: we define a species by its morphology and by the other characteristics which all its member individuals possess and which are the sum total of their reliably consistent but complex differentiation.

We might also legitimately ask whether cells have a *choice* about their route of specialization and whether, if the conditions allow, a differentiated cell can turn back and acquire a different set of characteristics.

To investigate these questions, we need to consider plants and animals separately. This is because they show some fundamental differences in how their cells divide and differentiate. One key difference, as we shall see, is that animal cells are capable of moving during organism development whereas plant cells are not. Position and direction have an important role to play in cell differentiation in both groups, but in rather different ways.

Cellular differentiation in plants

All plant cells are **totipotent**. This means that any cell can, in principle, differentiate into any other cell. In practice this doesn't normally happen because the **meristem cells** generate organs, such as roots, stems, and leaves, according to their location on the plant. Thus, although plant cells differentiate and replicate to form

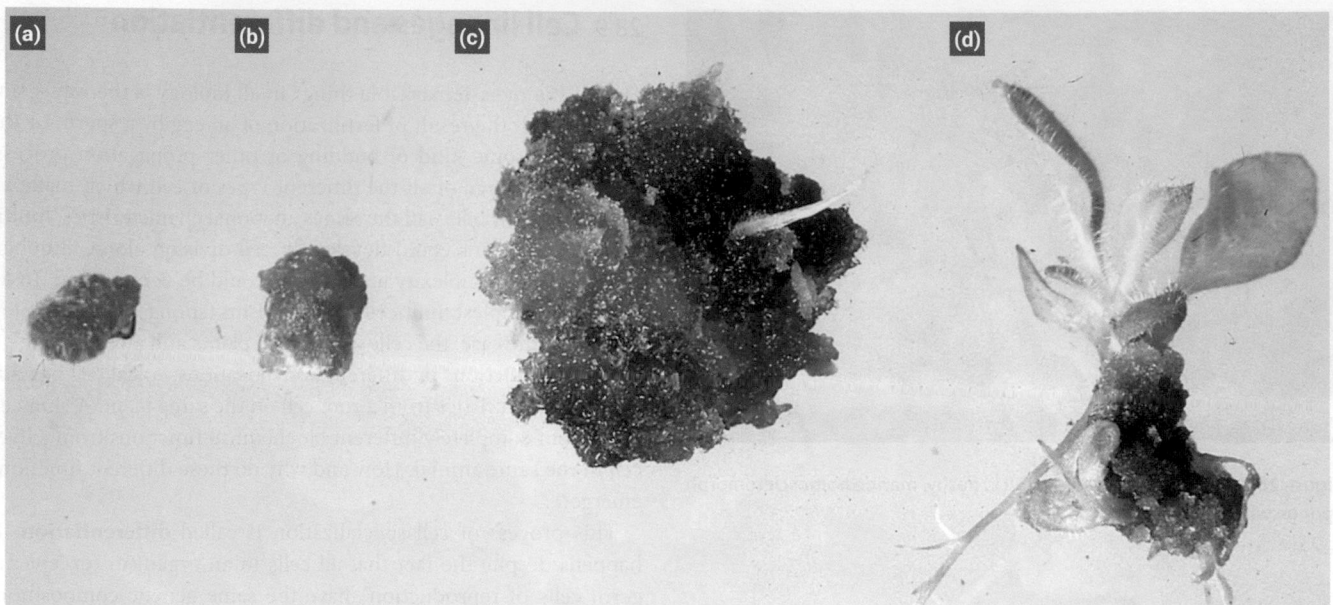

Figure 28.38 Regeneration of whole tobacco plant from undifferentiated callus. (a) Brown undifferentiated callus; (b) green callus containing photosynthetic cells; (c) initiation of roots; and (d) formation of complete tobacco plantlet with shoot and roots.

Source: Courtesy of Mike Davey, Plant and Crop Sciences, University of Nottingham.

specific tissues and organs, they do not become *irreversibly* committed to differentiation pathways (**lineages**) in quite the same way as animal cells do (see Cellular differentiation in animals).

The retention of totipotency by plant cells is the basis for the whole technology of plant micropropagation and laboratory-based tissue culture. It is relatively simple to take cells from plant tissues and regenerate large numbers of new whole plants from them. Indeed, Figure 28.38 shows how one cell can be induced to produce an entire mature plant, something that does not happen in normal animal systems.

In horticulture, plant propagation practices are based on the fact that tissues can redifferentiate and produce new sorts of organs. For instance taking cuttings and rooting them in compost involves the cells near the cut surface producing new root meristems and subsequently an array of new roots. Under tissue culture conditions, individual cells from almost anywhere in the plant can be coaxed to dedifferentiate back into a basic stem cell. These 'callus' cells can then develop in to new cell types and it is possible to produce entire plants, often in very large numbers for commercial or research purposes, as discussed in Real World View 28.1.

REAL WORLD VIEW 28.1 Micropropagation of plants

The micropropagation of plants is a major worldwide business, producing huge quantities each year for the horticulture and agriculture industries. Frederick Steward founded the science in the late 1950s and significant research since then has enabled the cloning of a wide variety of plant species.

The process is initiated with a small piece of plant tissue called an explant. This is cultured on agar using sterile techniques. The explant cells dedifferentiate and produce a masses of undifferentiated callus cells. The callus is then subcultured and induced to generate shoots and roots using various combinations of hormones, usually auxin and cytokinin. The new plantlets are transferred individually to new agar and grown on until large enough to be transferred to compost for conventional horticultural handling.

An efficient micropropagation system—can produce up to one million new plants from a single explant in one year. Valuable crops produced in this way include potato, banana, pineapple, orchids and chrysanthemums, strawberry, rhubarb, and pelargoniums. Some rare and ancient species, including valuable trees, have been saved from extinction by being propagated this way.

The use of shoot meristem tissue as a starting explant has the added advantage of being largely free of plant viruses, allowing the production of clean stock in species which are prone to infection.

Plant breeders value micropropagation techniques because they generate mostly identically plants, although there is some natural variation, termed somaclonal variation, which can be exploited to make new varieties.

Figure 1 Industrial micropropagation of orchids. (a) Micropropagation of orchid plantlets in sterile culture. (b) Orchid plantlets are transferred to sterile bark and grown on for 3 years before they flower. (c) Flowering orchid plants clonally propagated.

Source: (a) armmit/Shutterstock.com. (b) Piti Tan/Shutterstock.com. (c) nosonjai/Shutterstock.com.

Regions of newly divided undifferentiated cells on the flanks of the meristem dome are called **primordia**. The patterns of cell division and spatial development within primordia are highly ordered and under complex control, with the result that they form different tissues and organs of the mature plant.

We described earlier the arrangement of meristems in monocot and dicot plants. The SAM is made up of three layers, L1, L2, and L3, as indicated in Figure 28.39. These layers divide **anticlinally**—that is, the plane of the cell is perpendicular to the plane of the cell layer such that cells are pushed sideways within a layer rather than into the layers above or below. The rate of cell division increases, maintaining the meristem structure and moving cells to the flanks of the meristem dome from where primordia of new leaves are generated. The outer layer (L1) gives rise to all of the 'skin' of the plant, called the **epidermis**, and maintains itself as a separate **cell lineage**. Thus L1 cells differentiate into specific cell types such as leaf hairs (trichomes) and leaf pores (stomata).

The L2 and L3 layers give rise to cells internal to the leaf: mesophyll cells and vascular transport cells. They also account for the colour patterns in variegated leaves.

Cells in the L3 layer also generate the extending stem of the plant which, together with the pattern of leaf initiation on the meristem, will determine the overall architecture of the above ground parts of the plant. Later in development, the meristem may change from a vegetative state, making leaves, to an inflorescence meristem which generates flowers and all their parts.

Look at Figure 28.40, which depicts the structure of the root apical meristem. The cells of the root meristem give rise to the specific cells which make up the radial tissue layers of the root: the epidermis on the outside, the cortex, the endodermis, the pericycle, and

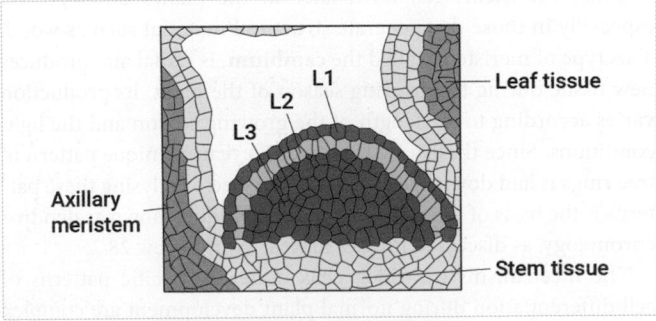

Figure 28.39 The structure of the shoot apical meristem.

Source: Reprinted from Trends in Plant Science Volume 5, Issue 3, Bowman and Eshed, Formation and maintenance of the shoot apical meristem, P110-115, Copyright 2000, with permission from Elsevier.

Figure 28.40 Root apical meristem.

Source: Images courtesy of Darren Wells, Plant and Crop Sciences, University of Nottingham.

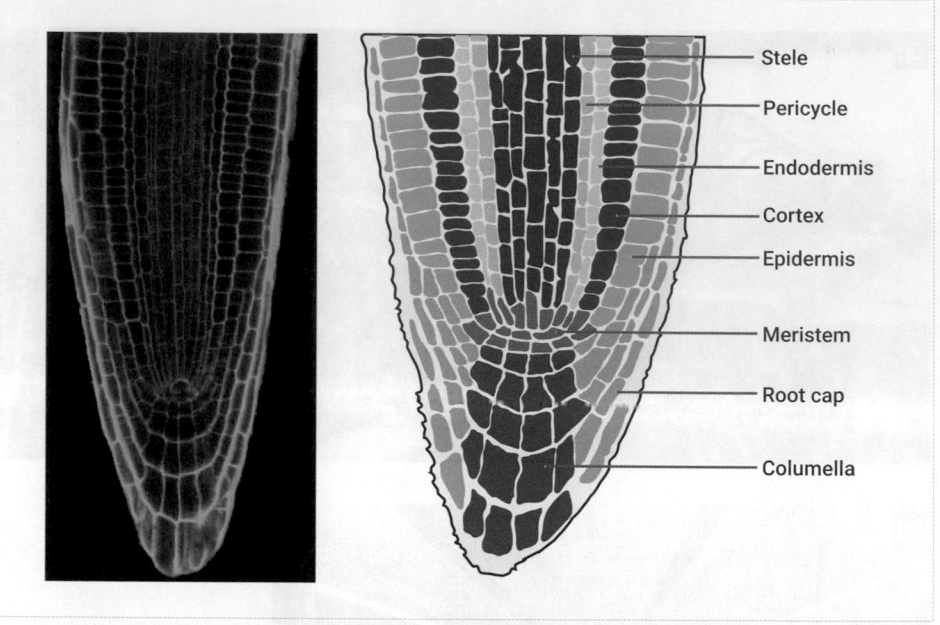

- Stele
- Pericycle
- Endodermis
- Cortex
- Epidermis
- Meristem
- Root cap
- Columella

the vascular tissue (stele) in the centre. In addition, at the furthest tip of the root are the columella cells, which have a gravity sensing function (see Chapter 30), and the root cap, which facilitates the passage of the root through the soil and is sloughed off as it pushes forward.

The root meristem does not generate specific organs like the SAM but simply makes more root. Figure 28.41 shows how a completely new root meristem can be initiated in existing root tissue, starting as a specific cell division within a pair of cells within the pericycle cell layer. Subsequent cell divisions then generate a new root meristem producing new root cells in the same way as the existing RAM. The new root meristem subsequently grows out between the cell layers of the main root cells and produces a new root, called a **lateral root**. More lateral roots can form on the first lateral roots.

Thus as roots grow, their development produces a hierarchical system. The complex architecture of underground plant root systems enables the plant to adapt its growth to local soil structures and to take up water and nutrients in an efficient manner.

Other meristems can form later in the plant's development, especially in those that generate structural material such as wood. This type of meristem, called the **cambium**, is radial and produces new tissue during the growing season of the plant. Its production varies according to the length of the growing season and the light conditions. Since these vary from year to year, a unique pattern of tree rings is laid down over the life of the tree. Analysing these patterns is the basis of the archaeological technique known as dendrochronology, as discussed further in Real World View 28.2.

The mechanisms by which cells undergo specific patterns of cell differentiation during normal plant development are complex and poorly understood. The pathway of differentiation of any cell appears to be determined by combination of its *lineage* and its *location*. Positional signals include the type and amount of chemical information coming from neighbouring cells, often called plant hormones (Chapter 30).

Cellular differentiation in animals

The story of cell differentiation in animals starts with a fertilized egg, or zygote. This is a single cell, usually with a diploid nucleus (containing two sets of chromosomes) set within cytoplasm alongside other organelles such as mitochondria. (For more detail on reproduction, see Chapter 33.)

To begin with, the zygote is covered with a membrane. In some animals (generally speaking, those which expel or lay eggs) the membrane remains in place for a long period of developmental time, until the young organism hatches. In other animals, including mammals, the membrane is short lived and 'hatching' happens early on, while the new individual is still a small group of cells and long before it is capable of independent survival. In many animals, including those of the *Xenopus* toad which is often used to investigate embryo development, the zygote cell has a distinct top, called the animal pole, and bottom, called the vegetal pole. In these animals the vegetal pole contains granules of yolk.

Irrespective of membrane duration and the presence of yolk, the zygote starts its development by rapid cell division. Each resulting cell itself divides (cleaves) but does not grow, and the process continues for about twelve cycles of division so that a solid cluster of small cells, called the **morula**, is formed. The polarity of the structure is retained during this initial phase of cell division such that cells at the 'top' of the morula relate to the animal pole of the zygote cell and those at the bottom relate to the vegetal pole, including the possession of the yolk.

This early distinction between different types of cell, which will persist in later stages of development, occurs because cell division is asymmetric: components of the cytoplasm called 'determinants' (some important proteins and strands of RNA), do not divide equally but become located in one daughter cell or the other.

The remaining stages of development, from the formation of an early embryo right through to the emergence of a fully formed animal, essentially rest on four processes, as summarized in Table 28.2. These processes—pattern formation, morphogenesis,

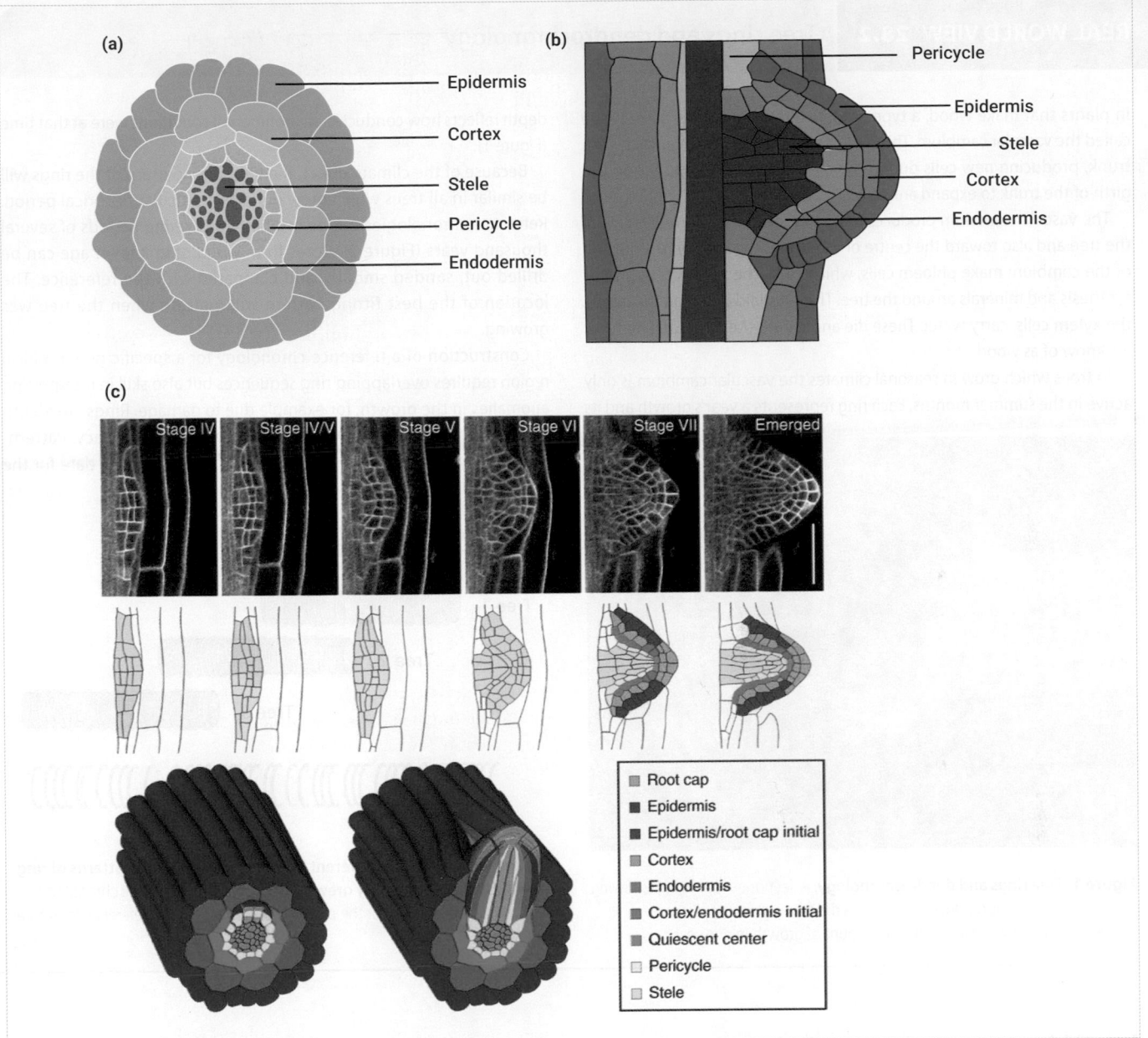

Figure 28.41 Lateral root meristem. (a) Radial structure of the root. (b) A new lateral root meristem forming from a specific cell division in a pericycle cell. (c) Confocal image of a lateral root emerging from the side of a primary root.

Source: (a–c) Reprinted from Morris, E. C., Griffiths, M., Golebiowska, A., et al. Shaping 3D Root System Architecture. Current Biology, 27(17), R919–R930, Copyright 2017, with permission from Elsevier.

cell differentiation, and growth—may occur concurrently and they often interact, with different combinations of activity at different developmental stages.

The development of body structure involves the development of a tube through the morula, which will form the animal's alimentary tract or gut, and the delineation of either two (ectoderm and endoderm) or three (ectoderm, mesoderm, endoderm) layers of cells from which all other tissues and organs will develop. Animals with two original layers are called **diploblasts**; those with three are **triploblasts** (Chapter 29).

The mechanism and exact arrangements of cells and tissues during this process vary a great deal between animal groups. In some cases, a number of ectodermal cells move inwards and undergo a transition into mesodermal cells, such as in the formation of the neural crest in chordates. Similarly, mesenchymal cells (connective tissue cells of mesodermal origin) may undergo transition into epithelial cells, for example in the formation of blood capillaries and other tubular structures.

Embryologists and developmental biologists are able to track individual embryonic cells and groups of cells to find out which

REAL WORLD VIEW 28.2   Tree rings and dendrochronology

In plants that make wood, a type of meristem forms in the stem tissue called the vascular cambium. This is a lateral meristem which encircles the trunk, producing new cells during the growing season and causing the girth of the trunk to expand annually.

The vascular cambium produces new cells toward the outer surface of the tree and also toward the centre of the trunk. The cells on the outside of the cambium make phloem cells, which carry the products of photosynthesis and minerals around the tree. The cells laid down on the inside, the xylem cells, carry water. These die and create the structural material we know of as wood.

In trees which grow in seasonal climates the vascular cambium is only active in the summer months. Each ring represents a year's growth and its depth reflects how conducive environmental conditions were at that time (Figure 1).

Because of the climate effect, the frequency pattern of the rings will be similar in all trees which grew during a particular historical period. Reference chronologies can be established covering periods of several thousand years (Figure 2). Cores from wood of unknown age can be drilled out, sanded smooth, and compared with the reference. The location of the best fitting pattern will indicate when the tree was growing.

Construction of a reference chronology for a specific geographical region requires overlapping ring sequences but also skill in interpreting anomalies in the growth, for example due to damage. Rings are identified under a microscope and measured to 0.1 mm accuracy. Pattern-matching algorithms are used to establish the most likely date for the sample.

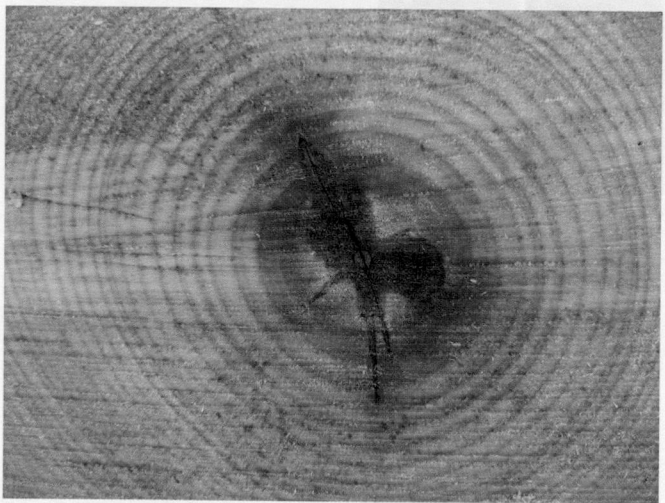

Figure 1   Tree rings and dendrochronology. A sectioned tree stump showing the sequence of rings produced annually as the tree grows. The distance between rings (arrows) represents the amount of growth in one year.

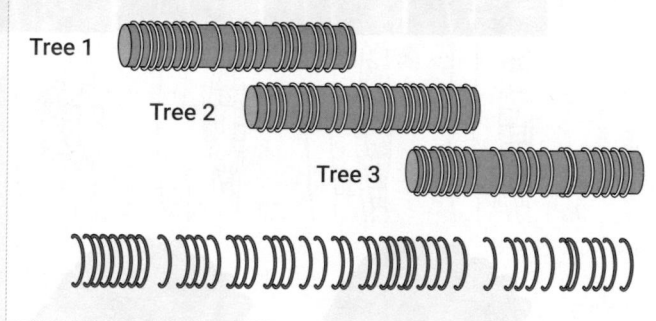

Figure 2   Samples from different trees have overlapping patterns of ring growth, showing that they grew contemporaneously. A full chronology (red) can be constructed for the period and used as a dating reference for future samples.

Table 28.2 Essential processes in organism development

Developmental process	Characteristics
Pattern formation	The establishment of a body plan. This includes cellular polarity and direction (top and bottom, front and back), the arrangement of tissue layers, and the emergence of axes of body construction (anterior to posterior, ventral to dorsal, radial symmetry, etc.)
Morphogenesis	The development of structural features of the body (gut, limbs, digits, brain, internal organs, etc.). This involves cell migration (the movement of cells from one part of the developing body to another), unequal rates of cell division in different localities, and apoptosis (programmed cell death)
Cell differentiation	The emergence of phenotypically distinct cells. Cells become biochemically, functionally, and structurally different from one another. They can often be grouped into distinct 'lineages': lines of common descent from an originating cell with particular characteristics
Growth	An increase in size. This happens mostly by an increase in cell number (cell multiplication or hyperplasia), but it can also occur by the accumulation of material inside cells (e.g. lipid storage in fat cells), by the formation of extracellular material (e.g. bone mineral, collagen), or by the accumulation of extracellular fluid (e.g. coelomic cavity fluid, eye humour, brain ventricular fluid)

structures they develop into in the mature animal body. In vertebrates, individual adult tissues can be traced back to germ layers, and something similar can be done in insects (Table 28.3). For echinoderms, snails, and worms, tracing leads back to individual cells (blastomeres) of the morula; differences depend on whether the cells originated in the animal or vegetal pole and by their position during subsequent rounds of cell division.

A range of processes are used in the laboratory to follow the origin, movement, and division of cells, as summarized in Table 28.3. Information gained in this way, which is called **lineage tracing** or **fate mapping**, is useful for understanding animal but also plant development. It can be an informative way of comparing organisms from different groups and it can provide clues to the evolutionary path which led to the existence of modern species. It can also be used by cancer specialists and other medical researchers to help in understanding the origins and progress of diseases, including some developmental disorders.

Stem cells

A **stem cell**, in an animal or plant, is an *undifferentiated* cell which has the capacity to divide and lead to the generation of cells with specific attributes or functions. The result of division of a stem cell may be symmetrical, producing daughter stem cells, or asymmetrical such that one daughter cell remains in the stem cell pool and the other differentiates.

In plants, stem cells are found in meristems. In animals, stem cells form the embryo (**embryonic stem cells**) but they are also present in adult tissues (**adult stem cells**) which are constantly regenerating, such as bone marrow, blood, epidermis, and gut, and in many other mature tissues as a basis for wound repair (see also Chapter 32).

The timing and nature of differentiation of stem cells is determined by a combination of external signals, including positional information (adjacency to other cells and **extracellular matrix**), hormones, growth factors, cytokines, and other types of biochemical signals.

If we recall that all the cells in an organism, with the exception of germ cells, have identical genes and DNA, 'differentiation' means the expression of a *subset* of those genes. The effect of this is to give each cell a set of functional characteristics, collectively called its phenotype. Thus we could say that stem cells have a minimal or poorly defined phenotype but have the flexibility to produce daughter cells with well-defined phenotypes.

Stem cells vary in their phenotypic potential: the possible types of cell into which they may differentiate. This set of cells defines a **lineage** because, looked at from the other direction, any set of differentiated cells have all emerged from a common ancestral cell.

In metazoa, the cells of the very early blastocyst, which give rise to all the later cells of the organism, have the utmost phenotypic potential and are referred to as **totipotent**. The haematopoietic stem cells in bone marrow, on the other hand, can differentiate into a various types of blood and immune cells (Figure 28.20) but not, for example, into nerve cells. Such cells are defined as **multipotent** and, because they are committed to differentiating in a particular direction, they can be described as **lineage-restricted**.

Table 28.4 explains the language used to describe the phenotypic potential of precursor cells (stem cells and progenitor cells), along with some examples from animals.

A general principle, particularly in complex animals, is that the process of differentiation is one way. In other words, cells reach a state of **terminal differentiation** and cannot, under normal circumstances, revert to the dedifferentiated state. However, this is not entirely true. Some animals (for example, cnidarians,

Table 28.3 Methods of cell lineage tracing

Method	Notes
Direct observation	The original method used by pioneer embryologists, after the invention of adequate light microscopes. Applicable to whole animals, if sufficiently small or transparent, and to cultured cells. Does not damage the cells but requires repeated, intensive observations with much opportunity for error
Cell labelling	Done with dyes and radioactive markers, to improve the microscopical tracking of cells. Depends on the label being non-damaging to the cells and capable of being transmitted during the cell division process. Labels can be misplaced and may spread; detection may require destructive sampling of the embryo
Genetic marking	Introducing gene-based markers, such as green fluorescent protein (GFP) into cells early in development, to follow their movement, replication and differentiation. Several different coloured markers can be used to create complex lineage traces. The method is difficult to do efficiently but markers are transmitted during mitosis and do not spread to neighbouring cells
Transplantation	Experimentally moving cells and tissues to a different position within the embryo to investigate the effect of location on cell differentiation and embryo development. Has long been used in developmental studies and remains a valuable experimental tool, especially when combined with marking/labelling techniques
Genetic mosaics	The experimental creation of genetic 'mosaics', cells with chromosomes from more than one animal with different characteristics. Has been especially useful for tracking cell and tissue movements in fruit flies and mice
Genetic manipulation	The incorporation of genetic switches and reporter genes (including GFP) into cells to track when genes are switched on or off and when key events in cell differentiation take place. It is used experimentally, switching genes on or off to examine the effect on cell fate and function

Table 28.4 Categories of precursor cell and their differentiation capacity

Category	Definition and capacity	Examples and products
Totipotent stem cells	Cells of the early blastocyst morula Able to differentiate into any other cell type	Give rise to all the cells of the body as well as those of the placenta and extra-embryonic membranes
Pluripotent stem cells	Cells of the early embryo and inner cell mass of mammalian blastocysts Able to differentiate into a wide range of cell types, but with the loss of some initial possibilities	Can give rise to all the cells of the body and to some extra-embryonic membranes but have lost the capacity to form the placenta
Multipotent stem cells	Tissue-limited stem cells, able to differentiate into a large but limited range of distinct cell types	Haematopoietic stem cells in bone marrow, generating several different lineages of blood and immune cells Mesenchymal stem cells of bone marrow, heart, adipose tissue, dental pulp, etc. Intestinal epithelial cells
Unipotent stem cells	Stem cells present in a single tissue and able to differentiate into a single type of cell	Spermatogonia in the testis, producing spermatocytes Dermal fibroblasts in epidermis, producing epidermal cells
Progenitor cells	Partially differentiated daughter cells of multipotent or unipotent stem cells, which have a limited capacity for division but whose daughter cells are the founders of subsequent lineages (and may themselves be progenitor cells)	Neuronal cells in the developing brain and spinal cord Erythroid, myeloid, and lymphoid cells of the haematopoietic system

planarians, starfish) can regenerate themselves completely from a single cell or small tissue fragment which has been excised from an existing body. The detached cell dedifferentiates and then goes through several rounds of division, after which the daughter cells **transdifferentiate** to reform the tissue layers and organ components of new organism.

Some animals (for example, urodele amphibians and some reptiles such as snakes, lizards, skinks, and geckos) can regenerate lost body parts such as limbs and tails; an example of such regeneration is depicted in Figure 28.42. This is most impressively demonstrated by salamanders which can regenerate whole limbs, complete with upper and lower parts and digits, from little more than a stump.

Figure 28.42 Green anole, *Anolis carolinensis,* in Hawaii. Note the regenerating tail stump.

Look at Figure 28.43, which shows how this regeneration unfolds over a series of weeks. This kind of regeneration seems to involve a transient form of cellular dedifferentiation although they may also bring into action a previously dormant stock of stem cells.

Mammalian tissues such as liver and heart have a limited ability to form replacement tissue after surgery or other damage, probably by also exploiting existing undifferentiated cells. Under laboratory conditions, some adult tissues including muscle may be persuaded to transdifferentiate to a different phenotype without an intervening dedifferentiation step.

Understanding these processes and determining the signals which initiate the dramatic changes in cell phenotype which must be necessary, has obvious therapeutic potential and has been the subject of a great deal of research. There is also a great deal of medical interest in artificially manipulating differentiated cells so that they might be used to treat illnesses.

 Check your understanding of the concepts covered in this section by answering the questions in the e-book.

28.10 Communication between cells

One of the advantages of multicellularity is that the cells in complex organisms can share the biological workload. Specialization is efficient, provided it is underpinned by cooperation. In fact, one definition of an organism could be a collection of cells whose functions are coordinated such that the whole is greater than the sum of the parts. Think of a football team: the forwards, the midfielders, and the goalkeeper all have distinct responsibilities. They are trained to perform specific tasks in specific places but they can only do this if other members of the team are also performing theirs. At the same

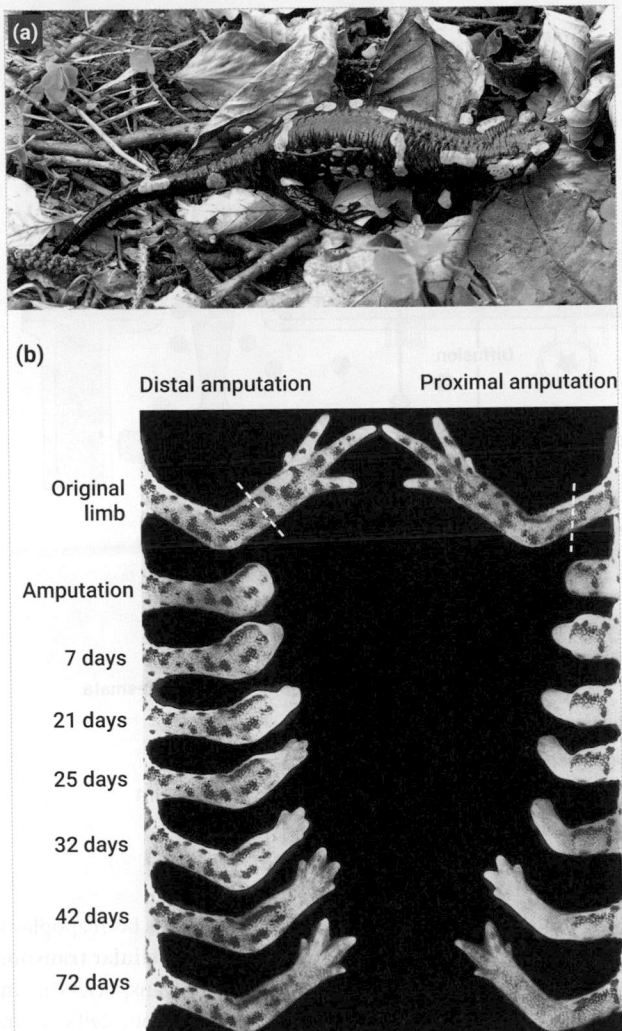

Figure 28.43 Limb regeneration in salamanders. (a) European fire salamander *Salamandra salamandra*. (b) Regeneration of a salamander forelimb. The amputation shown on the left was made below the elbow; the amputation on the right cut through the humerus. In both cases, the correct positional information is respecified.

Source: (b) Figure 16.23, Page 571 of Gilbert SF, Developmental Biology, 10th Edition, 2014. Reprinted from Principles of Regeneration, 1st edition, Richard J. Goss, Chapter 9, The Amphibian Limb, Figure 68, Page 142. Copyright 1969, with permission from Elsevier.

time, the team only functions as an effective unit if appropriate signals can pass amongst its members.

In multicellular plants and animals, communication between specialized cells happens at two levels: between adjacent cells and between cells separated by greater distances. Communication over distances is usually by the transmission of chemicals from one place to another, dissolved in some kind of moving fluid (blood, sap, gas, etc.). Plants and animals produce large numbers of small signalling molecules, including short peptides and other chemicals, and have hundreds of different receptor-like proteins in their membranes which can recognize the molecules secreted from nearby cells.

In complex animals, communication also takes place by the movement of electrical charge along nerve pathways. In the following sections we consider how *adjacent* cells connect to and

communicate with each other, partly in the formation of resilient tissue and organ structures but also in the coordination of their biochemical and physiological actions.

Making connections: plant cells

In multicellular plants, neighbouring cells have closely appressed and often fused cell walls, usually as a result of the cell division processes which led to their formation at a particular location. The plasma membranes of neighbouring cells, which are internal to the cell wall, are not in contact with each other and so an arrangement is required by which they can communicate with each other. They do this by three distinct routes, as depicted in Figure 28.44: apoplastic, symplastic, and transcellular transport.

The **apoplastic route** uses the area external to the plasma membrane. This is the **apoplast** and it includes the cell wall, the extracellular surfaces, and the intercellular spaces. In most tissues the extracellular surfaces are moist and allow the movement of molecules, and inside a leaf there are air spaces through which gases can diffuse. Thus when cells communicate this way, signalling molecules secreted by the plasma membrane move out through the cell wall and interact with receptors on nearby cell membranes or in neighbouring cell walls. The binding of a signalling molecule to a receptor initiates a **signal transduction pathway** within the cell.

Since all the cells in a tissue or organ are connected in this way, the apoplast acts as a corridor for communication over short or long distances. Cell–cell communication through the apoplast is significant not only in communication but also for how the cell interprets molecular information coming from its extracellular environment.

The second route, called the **symplastic route**, operates by direct cytoplasm-to-cytoplasm contact between neighbouring cells. It uses small pores, called **plasmodesmata**, in the cell wall separating adjacent cells. The structure of plasmodesmata is shown in Figure 28.45. Primary plasmodesmata form when new cell wall is laid down after mitosis and during cytokinesis.

During the construction of the cell plate, which forms the beginning of the cell wall between two newly divided cells, strands of endoplasmic reticulum are trapped and remain contiguous with that in the two neighbouring cells. The trapped endoplasmic reticulum forms a rod-like structure in the developing plasmodesmata called the **desmotubule**. It sits in the pore, which itself is lined by a contiguous plasma membrane linking the two neighbouring cells.

Secondary plasmodesmata also form between adjacent cells whose walls become appressed together during development but are not the result of a single cell division. The 'glue' that holds cell walls together is called the middle lamella (see Figure 28.45). Thus plasmodesmata create a continuity of cytoplasm between adjacent cells and facilitate the passage of molecular information. Indeed since all cells except stomatal guard cells possess large numbers of plasmodesmata in their cell walls, a plant could be considered as a single symplastic entity which moves information through intercommunicating cellular units.

The density of plasmodesmata varies between cells but can reach up to 10 per μm^2 of cell wall, meaning that a cell may have several

Figure 28.44 Three types of intercellular transport in plants: apoplastic transport, transcellular transport, and symplastic transport.

Source: Reproduced with permission of the Licensor through PLSclear. Auxin and other signals on the move in plants by Robert, Hélène S; Friml, Jirí. Nature Chemical Biology; Cambridge Vol. 5, Iss. 5, Figure 2 Page 327.

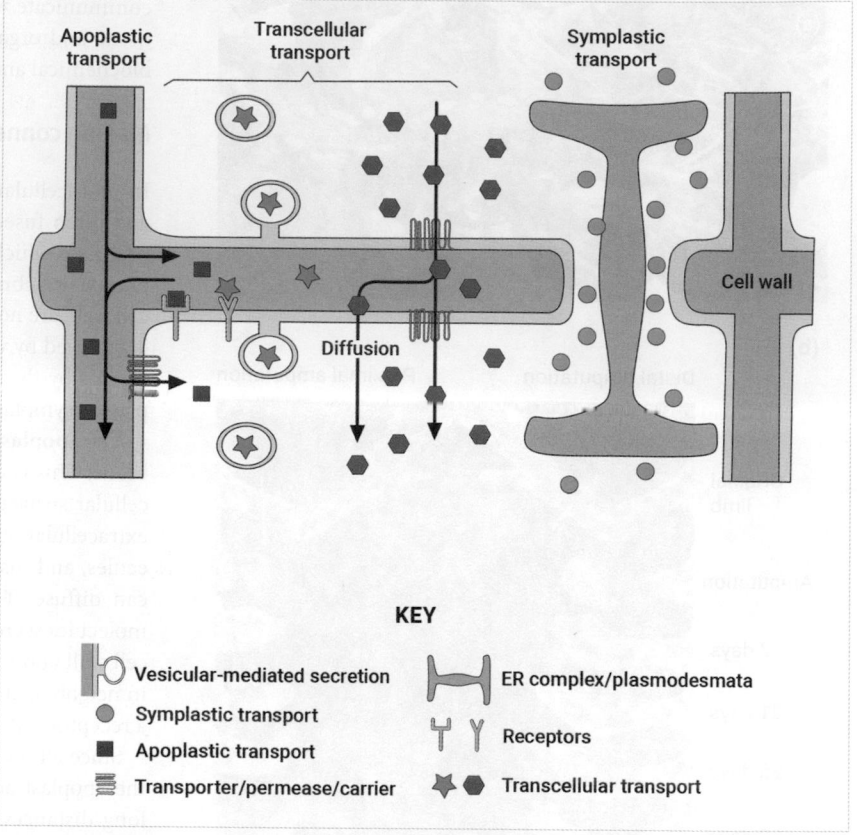

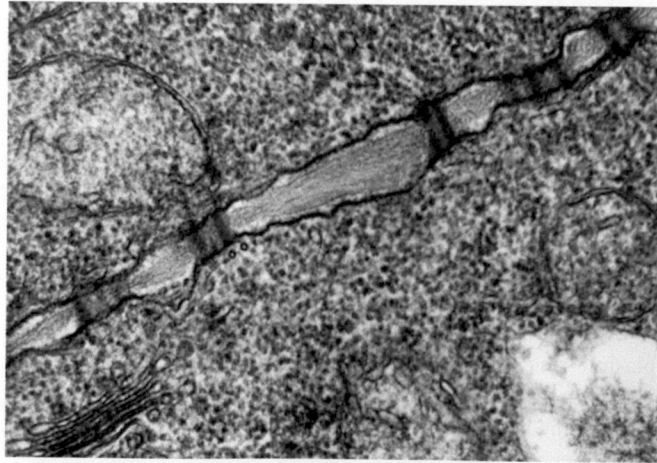

Figure 28.45 Plasmodesmata spanning the cell wall between two adjacent plant cells.

Source: Science Stock Photography/Science Photo Library.

hundred connections to its neighbours. The molecules transmitted between neighbouring cells influence cell function and determine the course of division and differentiation, especially in meristems and young tissues. They include **transcription factors** (proteins that control the expression of specific genes) and **microRNA** molecules (small non-coding RNAs which regulate gene expression at the post-transcriptional level), although the precise mechanism by which they move is unknown.

A third type of cellular communication involves both apoplastic and symplastic transmission and is termed **transcellular transport** (see Figure 28.44). This depends on specific transporters in the plasma membrane which export molecules from one cell's plasma membrane and import them across the plasma membrane of a neighbouring cell.

A classic example of transcellular transport is that of the plant hormone **auxin**, which moves between cells in a highly regulated manner, normally from the top of the plant toward the roots. The nature of this movement is summarized in Figure 28.46. In this kind of transport AUX/LAX proteins act as influx carriers, allowing auxin to enter a cell, while PGP proteins mediate export from the cell via ATP-dependent transport, and PIN proteins control export at one end of a cell, thus building a gradient of the hormone between cells. The flow of auxin and its localized concentration in different groups of cells is a crucial mechanism in many areas of plant development and physiology.

Making connections: animal cells

Besides the development of epithelia and basement membranes, another crucial development in the evolution of metazoa was the emergence of strong, sealing, potentially fluid-tight junctions between adjacent cells. Their evolutionary significance was that material secreted from cells could be contained in a restricted space, rather than flowing off into the surrounding water. This enabled the evolution of extracellular digestion (the breakdown of food materials outside the body), by enzymes secreted from the

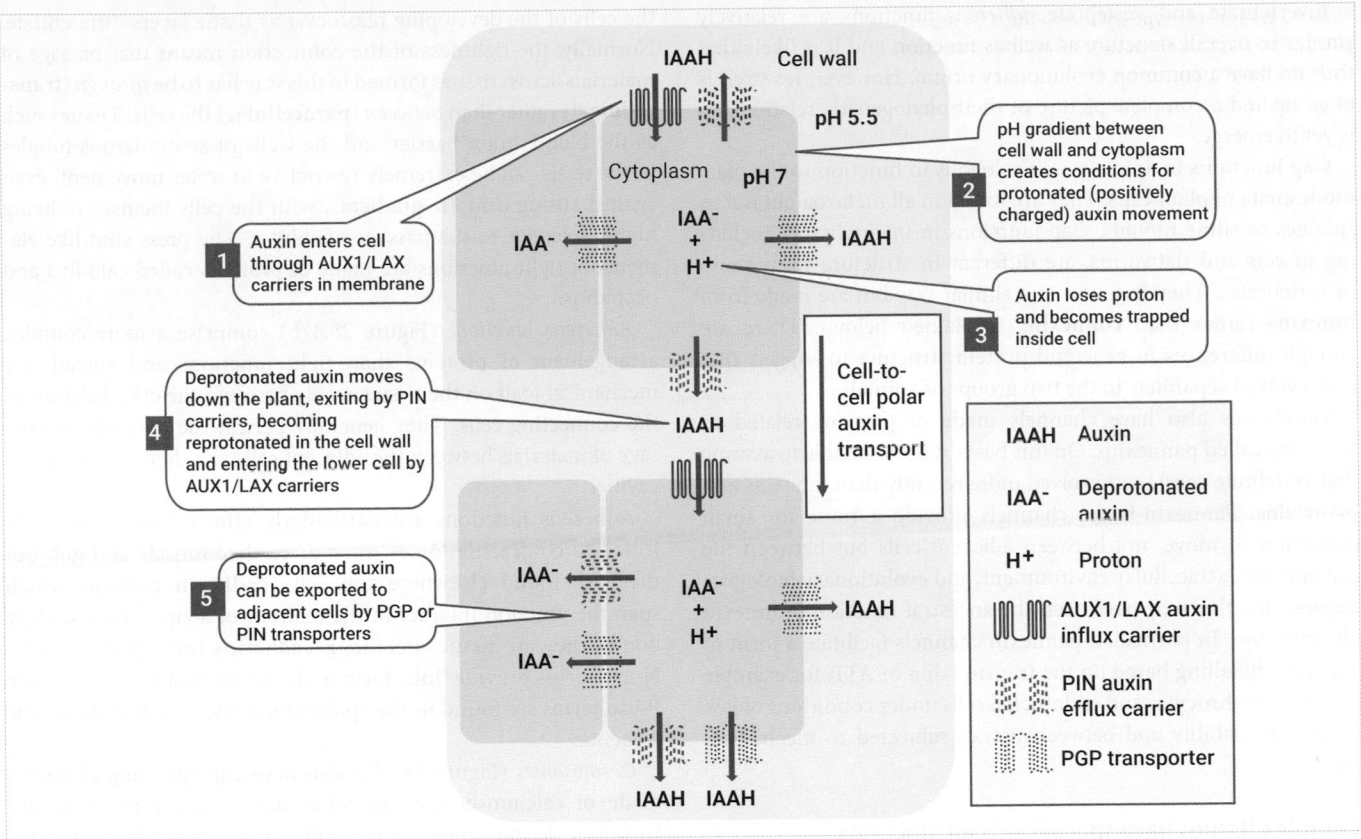

Figure 28.46 Intercellular polar transport of auxin hormone.

Source: Reproduced with permission of the Licensor through PLSclear. Auxin and other signals on the move in plants by Robert, Hélène S; Friml, Jirí. Nature Chemical Biology; Cambridge Vol. 5, Iss. 5, (May 2009) Figure 4 page 329: 325–32. DOI:10.1038/nchembio.170.

epithelial cells, prior to absorption across that same epithelium. This is a very efficient way of extracting nutrients and energy from food particles. It is the way the gastrointestinal tracts of all modern complex animals work.

Once again, we can look to the sponges for the first examples of junctions of this kind. Sponges of various types have both **occluding** and **adherens** junctions, as indicated in Table 28.5, although they are made of proteins with different molecular structures from their equivalents in metazoa. The exact nature of sponge occluding junctions remains unclear. However, they do form seals between adjacent cells, producing chambers where calcareous (Ca²⁺-containing) or siliceous (Si-containing) minerals can concentrate and precipitate into **spicules**.

The *occluding* junctions of invertebrates are analogous to the tight junctions of vertebrates (described below) in that they form fluid-impermeable barriers. They are present, for example, between the gut cells of insect larvae. They typically have a pleated, ladder-like structure when linking ectodermally derived epithelial cells but a smooth structure in between endodermally derived epithelial cells. These morphological features, as well as significant differences in protein and gene structures, suggest that the occluding junctions of vertebrates and invertebrates are only distantly, if at all, related.

Table 28.5 The intercellular junctions found in sponges, invertebrates, and vertebrates

Group and type of junction	Sponges	Other invertebrates	Vertebrates
Occluding junctions			
Septate	•	•	
Tight			•
Adherens junctions			
Adherens	•	•	
Spot (spot adherens)		•	
Zonal (belt desmosomes)		•	•
Macula adherens (spot desmosomes)			•
Gap junctions		•	•
Ig-CAMs	•	•	•

Invertebrate and vertebrate *adherens* junctions are relatively similar in overall structure as well as function and it is likely that they do have a common evolutionary origin. However, research is ongoing and a complete picture of their phylogenetic relationship is yet to emerge.

Gap junctions in animals are analogous in function to the plasmodesmata of plant cells. They are found in all metazoa but not in sponges or slime moulds. Gap junctions in invertebrates, including insects and flatworms, are different in structure from those in vertebrates. They function in a similar way but are made from **innexins** rather than **connexins** (explained below). There are enough differences in gene and protein structure to suggest that they evolved separately in the two groups of animals.

Vertebrates also have channels made of proteins related to innexins, called pannexins. On this basis, it is reasonable to assume that vertebrate connexins evolved more recently than innexins and pannexins. Pannexin-based channels provide a route for small molecules to move, not between adjacent cells but between the cell and the extracellular environment, and evolutionary biologists suggest that this may have been the ancestral function of innexin channels too. In particular, pannexin channels facilitate a form of cell–cell signalling based on the transmission of ATP, for example between erythrocytes and endothelial cells under conditions of low oxygen availability and between tissues subjected to mechanical stress.

Complex tissues have specialized cell junctions

The basolateral membranes of adjacent cells in epithelia are held very tightly together, besides being firmly attached to the basement membrane. This gives the tissue considerable strength and also allows for communication between the cells.

In complex animals, epithelia compartmentalize the organism's structure during development, but strong and functionally specialized cell–cell junctions are equally important to the structure and function of other types of tissue. They help to define the organs of the body, direct the conduction of fluids, and permit a complex range of interactions between cells.

Figures 28.47a–e illustrate the structure and function of five types of connections between cells: tight junctions, adherens junctions, desmosomes, gap junctions, and tunnelling nanotubes (TNTs). It is important to understand these different types of connections both structurally and functionally, so each illustration shows the junction from different points of view.

Each type of junction has a characteristic type of protein in its structure and many discoveries about junctions have been made on the basis of protein sequence analysis. These proteins have links to the cytoskeleton and other components inside the cell. Besides their structural roles, they transmit information and influence the phenotype, behaviour, and responsiveness of the connected cells. This signalling activity facilitates cell coordination within tissues and organs.

Tight junctions (Figure 28.47a) are found wherever complete isolation of body regions occurs, including the gut, between the endothelial cells which line the circulatory system, and between the cells of the developing blastocyst as tissue layers differentiate. Normally, the tightness of the connection means that passage of materials across tissues formed in this way has to be *through* (**transcellular**[1]) rather than *between* (**paracellular**) the cells. Tissues such as the blood–brain barrier and the walls of seminiferous tubules in the testis can be extremely restrictive to water movement, even against strong osmotic gradients, with the cells themselves being highly selective to the passage of solutes. The press-stud-like elements of tight junctions are made of proteins called caludins and occludins.

Adherens junctions (Figure 28.47b) comprise a more complex arrangement of proteins than tight junctions and spread any mechanical load on the tissue evenly between the cytoskeletons of the connecting cells. They generally permit the paracellular passage of materials between the adjacent cells and do not form sealed cavities.

Adherens junctions are particularly effective in tissues subjected to contractile forces, such as cardiac muscle and gut, but they are found elsewhere too. The **cadherin** proteins which span the junction interact in the manner of a zip to form a close bond. They are tissue-specific: E-cadherins link epithelial cells, N-cadherins provide links for muscle, nerve, and lens cells, while P-cadherins are found in the epidermis of the skin and also in the placenta.

Desmosomes (Figure 28.47c) also have gap-spanning elements, made of calcium-bonded glycol proteins similar to cadherins. However, desmosomes are generally arranged singly, rather like spot welds or rivets, and their intracellular connections are with the intermediate filaments of the cytoskeleton, such as the keratin of epithelial cells or desmin filaments in heart muscle cells.

Desmosomes form strong connections between cells and can create abrasion-resistant surfaces such as those of the skin and mucous membranes. A related type of connection, the **hemidesmosome**, occurs between epithelial cells and their basement membranes. Here, the adaptor protein plaque is inside the basal surface of the cell, linked to intermediate filaments, but the cadherin-like external proteins connect directly with laminin, anchoring the cell to its basement membrane.

Gap junctions (Figure 28.47d) make connections between adjacent cells but their role is principally functional rather than structural. They are channels made of clusters of six protein subunits arranged as a tube. The clusters are called **connexons** and each subunit is a **connexin**. To form the junction, two connexons, each embedded in a cell membrane, align themselves across the intercellular space. As a result, the cytoplasms of the two cells become directly connected, or contiguous, allowing small solutes such as ions and protons to pass directly from one cell to the next.

Gap junctions are not open permanently but can flip between open and closed states according to the presence of electrical, pH, or neurotransmitter signals. A good example of their function is

[1] This is a different use of the word from its plant cell communication context.

the transmission of the depolarized state between heart cells: this creates the coordinated wave of contraction characteristic of the cardiac cycle. Another example is in the islets of Langerhans of the pancreas: signals passing between insulin-secreting, glucagon-secreting, and other endocrine cells allow a coordinated and sensitive response to changes in blood glucose level.

Gap junctions are found between closely opposed cells in most tissues except skeletal muscle. They are absent from cells which exist in an unattached state such as erythrocytes and sperm.

Tunnelling nanotubes (*TNTs*; Figure 28.47e) are fine cytoplasmic connections between cells which principally have a communicative rather than a structural function. Although they have been observed by cell biologists for a long time, especially in colonies of cells cultured in the laboratory, their structure, composition, and likely roles have begun to be understood only recently.

In effect, TNTs make the cytoplasms of adjacent cells contiguous, so they could be thought of as creating a type of syncytium. However, the nuclear regions and major proportions of cytoplasm in each cell remain distinct and are not shared in the way that they are in true syncytia. A difference from the contiguity provided by gap junctions is that TNTs connect cells over much greater distances.

At the electron microscope level, TNT are found to be bundles of many fine fibres and strands, not all of which traverse the complete distance between the cells. They also contain organelles, including mitochondria, and appear to be a bridge for trafficking organelles between the cells without disrupting the cell membrane barrier.

The principal protein in TNTs is actin, the same material found in the microfilaments of the cytoskeleton. It appears that TNT actin is an extension of the cytoskeleton itself, thereby linking the scaffolding of the connected cells and forming pathways for molecular transmission and signalling.

TNTs are highly dynamic and often transient, constantly changing their length and thickness and adjusting the density of the connections between the cells.

A further type of intercellular connection depends on **immunoglobulin-like cell adhesion molecules**, or Ig-CAMs. This is a very large family of cell surface glycoproteins with structures resembling antibody molecules. They are typically embedded in the plasma membrane but form zipper-like links with

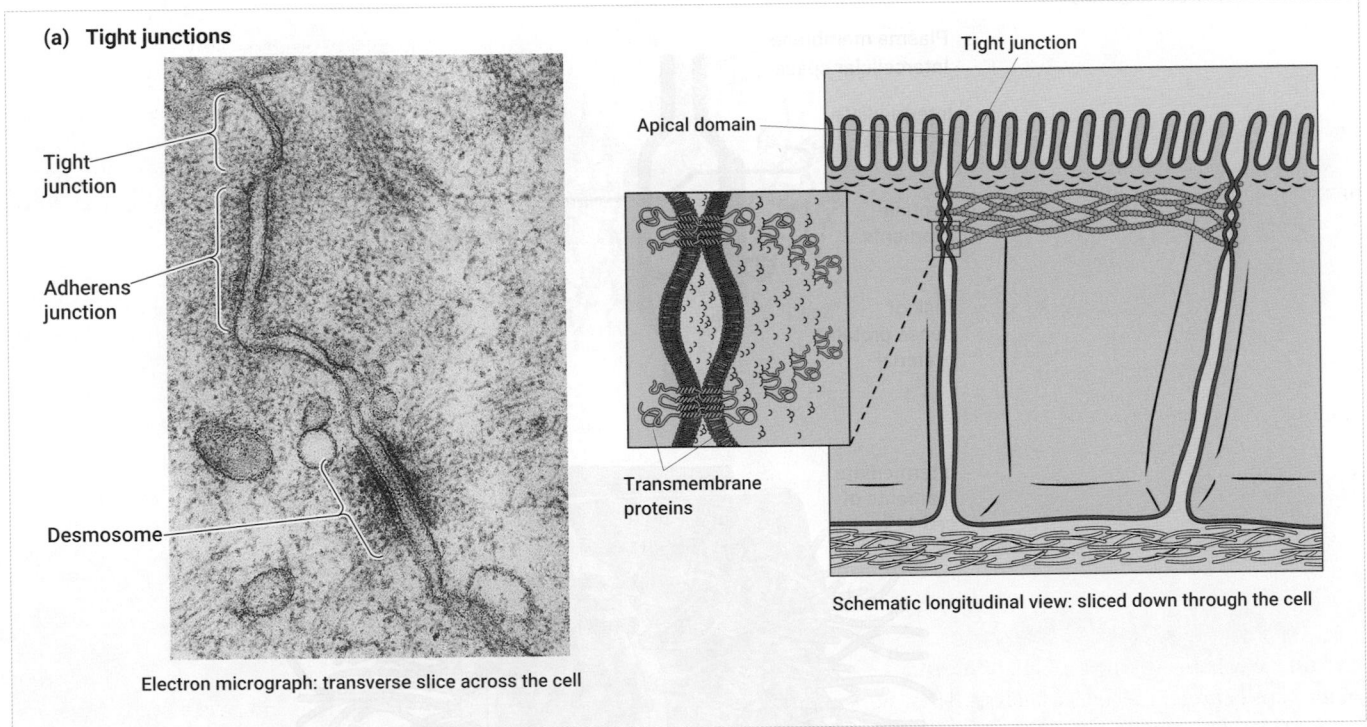

(a) Tight junctions

Tight junction

Adherens junction

Desmosome

Electron micrograph: transverse slice across the cell

Apical domain

Tight junction

Transmembrane proteins

Schematic longitudinal view: sliced down through the cell

Figure 28.47 The structure and function of five types of connections between cells. (a) Tight junctions. (b) Adherens junctions. (c) Desmosomes. (d) Gap junctions. (e) Tunnelling nanotubes (TNTs).

Source: (ai) Figure 16.20B (page 556) Chapter 16, Geoffrey Cooper, The Cell, A Molecular Approach, Eighth edition, 2018, Oxford University Press. (Bi) Mariana Ruiz LadyofHats/Wikimedia Commons/Public Domain. (biii) Reproduced with permission of the Licensor through PLSclear. Adherens junctions: from molecules to morphogenesis by Tony J. C. Harris & Ulrich Tepass Nature Reviews Molecular Cell Biology volume 11, pages 502–514 (2010) Figure 1. (ci) Boumphreyfr/Wikimedia Commons/CC BY-SA 3.0 https://creativecommons.org/licenses/by-sa/3.0/deed.en. (Ciii) CNX OpenStax/Wikimedia Commons/CC BY 4.0 https://creativecommons.org/licenses/by/4.0/deed.en. (di) Figure 16.21 from Geoffrey Cooper, The Cell, A Molecular Approach, Eighth edition, 2018, Oxford University Press. (diii) Figure 13.2 from Hill et al. Animal Physiology 4e, 2017, Oxford University Press. (ei) Cervantes DC & Zurzolo C (2019) Making Connections. The Biologist, 66(2), 10–13. Royal Society of Biology.

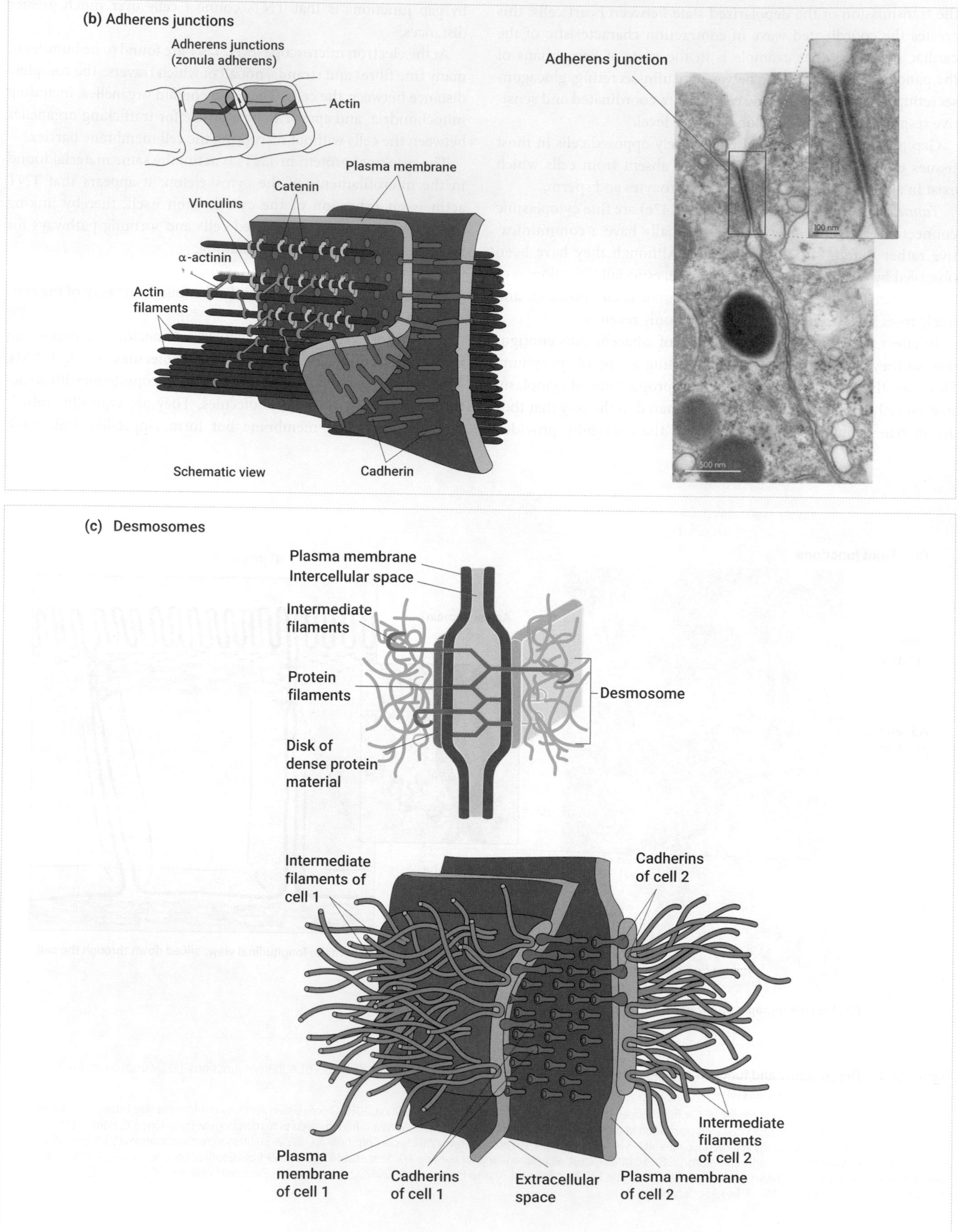

(b) Adherens junctions

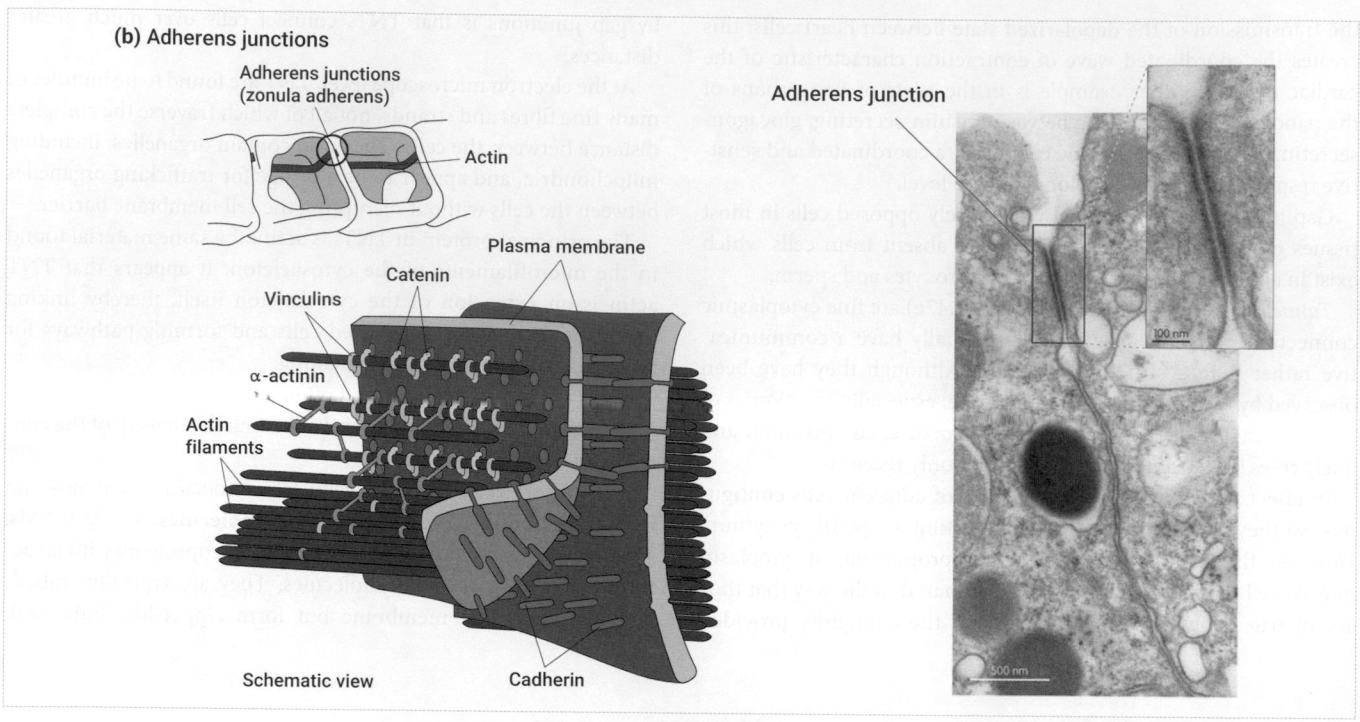

Adherens junctions
(zonula adherens)

Actin

Plasma membrane

Catenin

Vinculins

α-actinin

Actin
filaments

Schematic view

Cadherin

Adherens junction

100 nm

500 nm

(c) Desmosomes

Plasma membrane
Intercellular space

Intermediate
filaments

Protein
filaments

Disk of
dense protein
material

Desmosome

Intermediate
filaments of
cell 1

Cadherins
of cell 2

Plasma
membrane
of cell 1

Cadherins
of cell 1

Extracellular
space

Plasma membrane
of cell 2

Intermediate
filaments
of cell 2

Figure 28.47 (Continued)

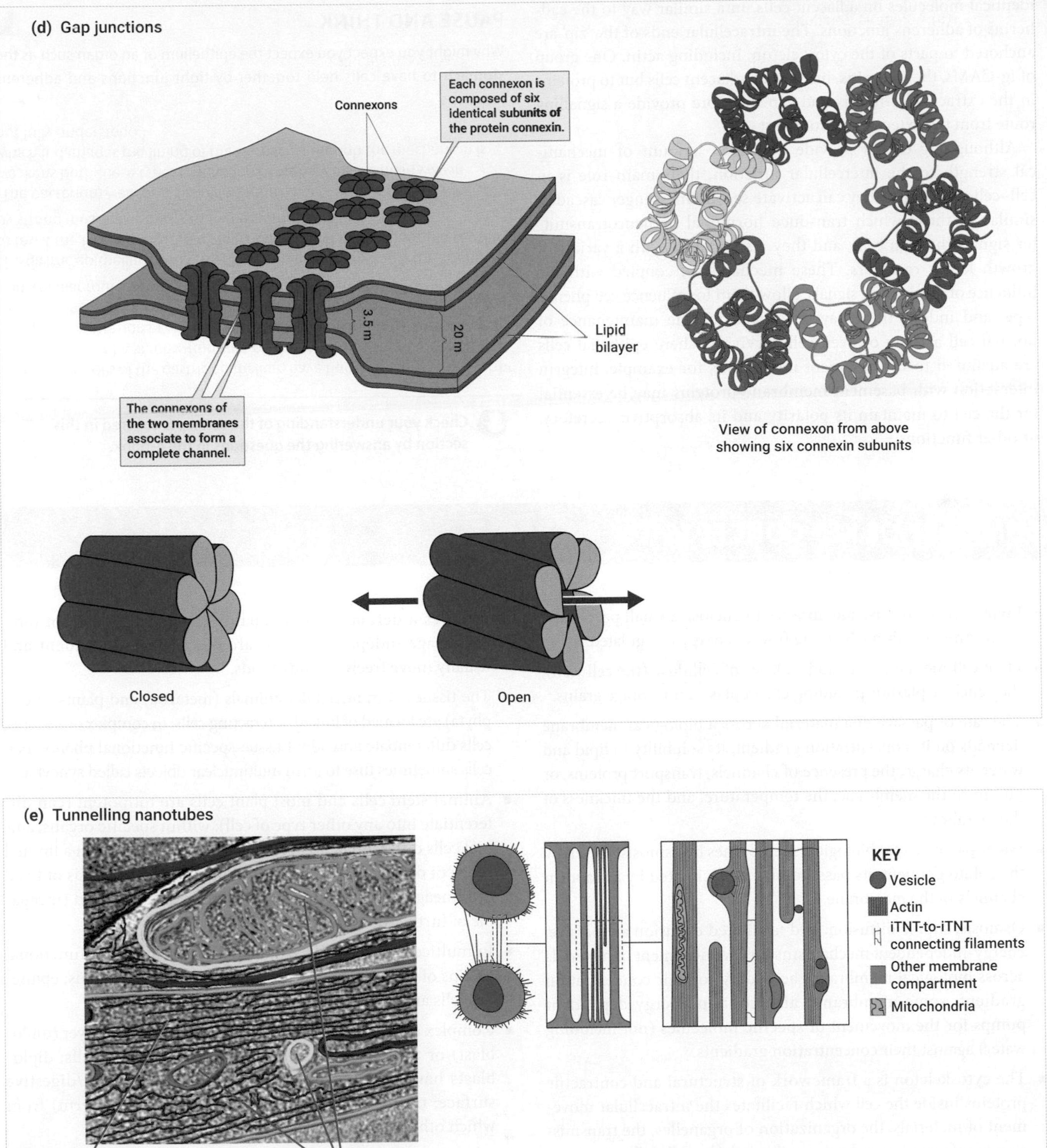

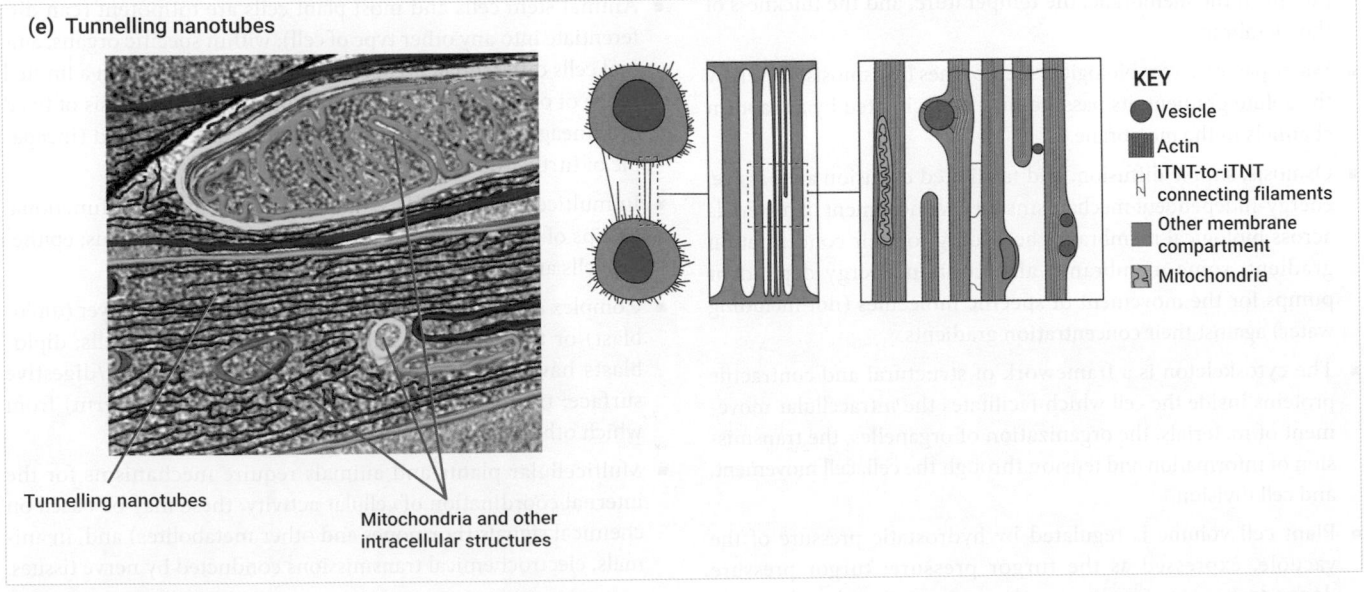

(d) Gap junctions

Connexons

Each connexon is composed of six identical subunits of the protein connexin.

The connexons of the two membranes associate to form a complete channel.

3.5 m

20 m

Lipid bilayer

View of connexon from above showing six connexin subunits

Closed

Open

(e) Tunnelling nanotubes

Tunnelling nanotubes

Mitochondria and other intracellular structures

KEY

Vesicle

Actin

iTNT-to-iTNT connecting filaments

Other membrane compartment

Mitochondria

Figure 28.47 (Continued)

identical molecules on adjacent cells, in a similar way to the cadherins of adherens junctions. The intracellular ends of the 'zip' are anchored to parts of the cytoskeleton, including actin. One group of Ig-CAMs, the **integrins**, link not to adjacent cells but to proteins in the extracellular matrix and can therefore provide a signalling route from the external environment.

Although Ig-CAMs provide a limited amount of mechanical strength to the intercellular junction, their main role is in cell–cell signalling. They can activate second messenger cascades, similar to those which transduce hormonal or neurotransmitter signals through cells, and they also interact with a variety of growth factor receptors. These mechanisms, coupled with the influence of mechanical signals, allow them to influence cell phenotype, and indeed they may be crucial for the maintenance of normal cell activity or even cell survival. Where epithelial cells are anchored to the basement membrane, for example, integrin interaction with basement membrane proteins may be essential for the cell to maintain its polarity and its absorptive, secretory, or other functions.

Check your understanding of the concepts covered in this section by answering the questions in the e-book.

SUMMARY OF KEY CONCEPTS

- Living organisms use membranes to enclose a small part of the environment within which the flow of energy is regulated.

- Plant cell membranes include a layer of cellulose (the cell wall); they enclose plastids including chloroplasts and storage grains.

- The rate of passage of a material across a biological membrane depends on its concentration gradient, its solubility in lipid and water, its charge, the presence of channels, transport proteins, or pumps in the membrane, the temperature, and the thickness of the membrane.

- Water passes across biological membranes by osmosis, driven by the solute gradient; its passage may be facilitated by aquaporin channels in the membrane.

- Osmosis, simple diffusion, and facilitated diffusion are passive, energy-independent mechanisms for the movement of materials across biological membranes according to their concentration gradient; some membranes also contain energy-dependent pumps for the movement of specific molecules (not including water) against their concentration gradients.

- The cytoskeleton is a framework of structural and contractile proteins inside the cell which facilitates the intracellular movement of materials, the organization of organelles, the transmission of information and tension through the cell, cell movement, and cell division.

- Plant cell volume is regulated by hydrostatic pressure of the vacuole, expressed as the turgor pressure; turgor pressure depends on osmosis; it provides support for the plant and facilitates the opening and closing of stomata.

- Anchorage-dependent animal cells require interaction with a surface to maintain their phenotype; that phenotype usually includes a definable directionality (front/back, top/bottom); anchorage-independent cells are not surface dependent and usually move freely in body fluids.

- The tissues of multicellular animals (metazoa) and plants (metaphyta) are formed of linked, interacting cells; in complex organisms, cells differentiate and adopt tissue-specific functional phenotypes; cells sometimes fuse to form multinuclear objects called syncytia.

- Animal stem cells and most plant cells are totipotent (can differentiate into any other type of cell); within specific organs, animal cells differentiate to become pluripotent (can form a limited range of other cell types) or multipotent (form the basis of fixed cell lineages), or they may be terminally differentiated (incapable of further differentiation).

- In multicellular animals, epithelia define and isolate functional groups of differentiated cells to form tissues and organs; epithelial cells are supported on a basement membrane.

- Complex animals may be constructed around a two-layer (diploblast) or three-layer (triploblast) arrangement of cells; diploblasts have an outer layer and an inner absorptive/digestive surface; triploblasts have an internal layer (mesoderm) from which other organs are derived.

- Multicellular plants and animals require mechanisms for the internal coordination of cellular activity; these may be based on chemical signals (hormones and other metabolites) and, in animals, electrochemical transmissions conducted by nerve tissues.

- Cell linkages take a number of forms (including tight junctions, adherens junctions, desmosomes, gap junctions) according to the need for bond strength, compartment sealing, paracellular movement of molecules, and direct intercellular communication.

 Use the flashcards in the e-book to test your recall of key terms introduced in this chapter.

QUESTIONS

Looking for answers? Once you've answered these questions, follow the link in the e-book to the answer guidance and check your work.

Concepts and definitions

1. Describe the role of membranes in defining living organisms.

2. What is osmosis?

3. List five types of membrane transport.

4. What is the difference between diploblasts and triploblasts?

5. What is turgor pressure in plants and what does it do?

6. How do cells link together in multicellular plants and animals?

Apply the concepts

7. Explain how ion and solute movements across a biological membrane can result in water redistribution.

8. What differences are there between sponges and true multicellular animals as regards cell differentiation and organization?

9. What is a cell lineage and how does the concept help in understanding the types of cells found in metazoan tissues?

10. Explain why complete plants can be grown in the lab from single cells or very small pieces of tissue.

11. What type of cell–cell junctions would you expect to find between the cells of the skin?

12. Explain why biologists consider the lumen of the gastro-intestinal tract to be *outside* the body.

Beyond the concepts

13. Which plant growth events are influenced by the hormone auxin? What other growth-influencing hormones are there in plants?

14. The cloning of Dolly the Sheep in the mid-1990s involved the dedifferentiation and reprogramming of cells from the mammary gland. How, in principle, was this achieved?

15. What structural characteristics is urbilateria (the last common ancestor of all bilateral animals) likely to have had and when may it have existed?

16. The presence of chloroplasts in plant cells may be the result of an endosymbiotic event early in plant evolution. What other cellular inclusions (plant and animal) may have resulted from this type of process? What is the evidence for and against it?

17. In the lungs, what determines the *rate* at which oxygen moves across the alveolar membranes, between air and blood, during normal respiration? What happens during intense exercise?

18. Most of the water in undigested food material is absorbed into the blood in the large intestine. What proportion of this water moves *through* the intestinal cells (transcellular transport) and what proportion passes *between* the cells (paracellular absorption)? Why do diseases such as cholera damage this process?

FURTHER READING

RBG Kew (2016). *The State of the World's Plants Report—2016*. Royal Botanic Gardens, Kew

A comprehensive status report on our knowledge of global vegetation: the number of plant species, recent discoveries, evolutionary relationships and plant genomes, the number of plants useful to man, the world's most important plant areas, global threats, and the effects of climate change, land use, and diseases.

Zhu, S., et al. (2016). Decimetre-scale multicellular eukaryotes from the 1.56-billion-year-old Gaoyuzhuang Formation in North China. *Nat. Commun.* 7: 11500.

Ancient origins of multicellular life. *Nature* 533: 441 (26 May 2016).

Life forms 'went large' a billion years ago https://www.bbc.co.uk/news/science-environment-36303051

The story of how multicellular, eukaryotic life emerged over 1.56 billion years ago. (The original research paper and two explanatory commentaries, from Nature and BBC News).

Kiecker, C., Bates, T. & Bell, E. (2016). Molecular specification of germ layers in vertebrate embryos. *Cell. Mol. Life Sci.* 73: 923–47.

A clear account of how cell layers form in vertebrate embryos and how individual tissues develop from them.

Pierre-Jerome, E., Drapek, C. & Benfey, P.N. (2018). Regulation of division and differentiation of plant stem cells. *Annu. Rev. Cell Dev. Biol.* 34: 289–310.

A comprehensive review of stem cells in plants, how they form in stems, roots, and shoots, and what controls their differentiation.

Anlas, A. & Nelson, C.M. (2018). Tissue mechanics regulates form, function, and dysfunction. *Curr. Opin. Cell Biol.* 54: 98–105.

How tissues are made from collections of cells, how differentiation is controlled, and the role of mechanical forces in tissue development.

Kretzschmar, K. & Watt, F.M. (2012). Lineage tracing. *Cell* 148: 33–45.

Tracing the lineages of animal cells, from the 19th century to the present day.

Body Plans

Chapter contents

Introduction   950
29.1 Form and function   951
29.2 Plant body plans   954
29.3 Animal body plans   964
29.4 Hard materials   991

Watch the key concepts video in the e-book to prepare yourself for studying this chapter.

LEARNING OBJECTIVES

By the end of this chapter you should be able to:

- Describe how the growth and forms of living organisms depend on physical and mathematical imperatives, shaped by the pressures of evolution.

- Explain the importance of cellular differentiation in multicellular organisms and plants.

- Describe the differences between monocotyledonous and dicotyledonous plants and explain the role of stomata in plant leaves.

- Account for the increasing levels of body complexity represented by diploblastic and triploblastic animals and appreciate the role of *HOX* genes.

- Know the difference between protostome and deuterostome animals and list the major forms of animal life that fall into each group.

- For chordates, explain the role of the notochord and pelvic girdles in limb development and the anatomical homology between limbs of species in different vertebrate groups.

- Describe the basic process of biomineralization and distinguish between its occurrence in shells, exoskeletons, test, bone, and horn.

Introduction

Much of the fascination of biology comes from the diversity of organisms around us, the variety of their habitats, and the wide range of sizes and shapes they represent. In this chapter we examine the structural features of living organisms and characteristics which allow them to survive and thrive. We consider how particular body structures may have emerged during evolution, the benefits and limitations associated with being large or small, and why complexity may give a survival advantage.

How big are biological organisms? If we work from single cells (prokaryotes or eukaryotes) up to the largest multicellular organisms (blue whale (Figure 29.1), giant redwood tree (Figure 29.2)) we travel more than 21 orders of magnitude in terms of volume and weight. We can extend the range even further if we include complex groups of coordinated individuals (termite nests, flocks of birds, human societies, or even complete ecosystems) which often behave like organisms.

In Chapter 28 we saw that biological structures are mini-environments, enclosed in membranes, and that cells are the basic units of organized life. Cells vary a great deal in size but there are limits to how large a cell can grow without encountering insurmountable physical problems. Large organisms therefore form as accumulations of cells. At the same time, complexity arises when subgroups of cells take on specific functions within an organism's body.

When individual cells grow larger they are said to undergo **hypertrophy**. When cells divide, causing a tissue or organ to get bigger, they are undergoing **hyperplasia**. Cells which change their structure or function by expressing different genes, as explained in Chapter 28, are said to **differentiate** (or to undergo **metaplasia**, although that term is used mostly for cancer and other disease states).

The achievement of large size coupled with increased complexity depends on all of these processes and they need to happen in a coordinated manner as the organism develops over its lifespan.

29.1 Form and function

A question which has fascinated biologists for a long time is whether the diversity of size, shape, and structure amongst organisms hides some general principles of biological form. Are there rules which have to be followed as organisms develop and as new forms of life evolve? Understanding these principles might be a way of understanding some of the rules of life itself, so it is easy to see why this question is of interest.

It seems most likely that such rules would come from physical and chemical imperatives—the universal and unavoidable 'laws' which govern the behaviour of all matter. Working out how these apply to life would help us to understand why organisms are the way they are. It might be possible to create a set of mathematical equations which apply right across the diversity of species and thus describe the dimensions of life.

There have been many attempts to work out what these principles could be and to assemble them into a guiding framework. Among the most famous are those of D'Arcy Thompson (*On Growth and Form*, 1917), Julian Huxley (*Problems of Relative Growth*, 1932), Samuel Brody (*Bioenergetics and Growth*, 1945), and Knut Schmidt-Nielsen (*Scaling, Why is Animal Size so Important?*, 1984), although fascination with the subject long precedes these important studies.

It is no coincidence that classic works like these often have 'growth' in the title, for it is in development, from a single cell to a mature individual, that the principles of magnitude and dimension can be explored. These authors assumed that the bodies of living beings have a natural perfection of shape and proportion, resulting from the pressures which determined their evolution. They believed that this could be expressed in the language of geometry and mathematics, often with close parallels to principles of architecture and engineering.

It has also been thought that the form of a body determines its functional and physiological characteristics. For example, palaeontologists may estimate the length, height, and weight of a dinosaur based on the discovery of a single, fossilized leg bone, and then go on to deduce facts about the animal's metabolism, nutrition, locomotion, and lifestyle. Similarly, if racehorse breeders understand how leg length, muscle structure, and heart capacity determine

Figure 29.1 A blue whale (*Balaenoptera musculus*).
Source: NOAA Photo Library/CC BY 2.0.

Figure 29.2 Giant redwood tree (*Sequoiadendron giganteum*).
Source: Lucky-photographer/Shutterstock.com.

sprint speed, they can design breeding and selection programmes to maximize the chance of producing winners.

Other investigators have extended the mathematical analysis of individual organisms into higher levels of biological organization, to populations of individuals and to the development and function of societies. For example, the ecological relationships between plants in a forest depend, ultimately, on the geometrical arrangement of leaves on stems and the lengths of roots, because these determine the efficiency with which each individual plant obtains light and water. This in turn may determine the invertebrate population in a local region and thus the ability of the whole forest to support colonies of mammals and other vertebrates.

Allometry

The science of body form and its scaling as organisms change in size is called allometry. Quantitative Tools 29.1 illustrates a traditional but still informative approach to this subject.

QUANTITATIVE TOOLS 29.1 — Size, volume, and surface area

The European mole (*Talpa europaea*) in Figure 1 could grow in three ways: get longer, get fatter, or get proportionately larger in all directions. What would happen to its **surface area** and **volume**?

We can approximate its shape using geometrical objects (ignoring its limbs and tail), as shown in Figure 2 by dividing its shape into the following:

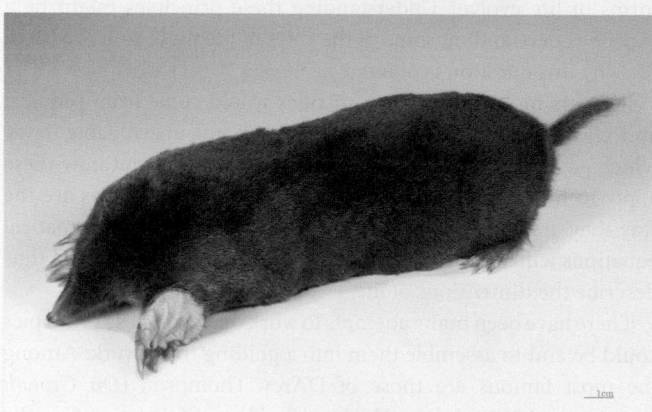

Figure 1 European mole (*Talpa europaea*).

Source: Didier Descouens/Wikimedia Commons/CC BY-SA 4.0.

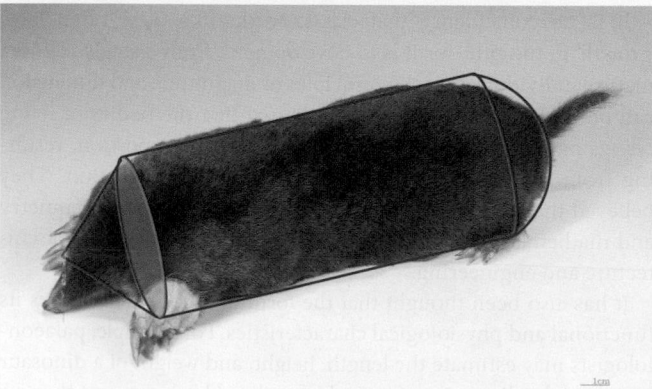

Figure 2 Approximating the shape of a mole using geometric objects.

- a cone (radius *r*, vertical height *r*)
- a cylinder (radius *r*, length *l*)
- a half sphere (radius *r*)

Table 1 shows some examples of how we could approximate surface area and volume.

When you look at Table 1, notice that:

- Doubling the mole's radius increases its surface area by about 2.5 times but increases its volume by nearly 50 times.
- Doubling the mole's body length increases its surface area by about 1.75 times and increases its volume by about the same (1.8 times).
- The ratio of surface area to volume *decreases* when the mole's radius increases but stays about the same when its body length increases.

The calculations show that an increase in radius has a much greater effect on body volume than it has on surface area, and that the ratio of surface area to volume gets smaller as the animal gets bigger. We can see the effect of radius more clearly if we plot surface area and volume over a range of body sizes. In these plots, the body length is kept constant at 5 cm and the radius is increased from 1 to 10 cm. (The mole would acquire a strange shape, but it demonstrates the effect!)

Look at the graphs (Figures 3 and 4). You can see in Figure 3 that the volume increases *exponentially* with the radius, and at a much faster rate than the surface area. Figure 4 demonstrates that as the animal gets larger the ratio of volume to surface area increases. (The lack of smoothness in the steep fall in surface area/body ratio, at the low radius end of the plot, is due to the rapidly diminishing effect of the nose cone and bottom half sphere on the overall body volume as the animal's radius gets bigger.)

The effects of increasing body length alone (not shown) are less interesting (rather like adding coins to a pile), especially for an animal with a relatively simple overall shape like a mole. In fact, animals generally grow in width *and* length. As we saw, a proportionate growth in width has a much greater effect on volume, and thus on body weight.

These results, of volume increasing more rapidly than surface area and of the ratio declining, illustrate some general principles. They apply to any organism, even if its shape does not lend itself to simple geometrical analysis.

Table 1 Ways to approximate surface area and volume

	Cone	Cylinder	Half sphere	Total If $r = 1$ cm and $l = 5$ cm	If $r = 2$ cm and $l = 5$ cm	If $r = 1$ cm and $l = 10$ cm	If $r = 2$ cm and $l = 10$ cm
Surface area (dimension L²)	$\pi r\sqrt{2r^2}$	$2\pi rl$	$2\pi r^2$	42.1 cm²	105.7 cm²	73.6 cm²	168.6 cm²
Volume (dimension L³)	$\pi\dfrac{r^3}{3}$	$\pi r^2 l$	$\dfrac{2}{3}\pi r^3$	19.2 cm³	94.0 cm³	34.6 cm³	150.8 cm³
Ratio	Surface area/volume:			2.2	1.1	2.2	1.1
	Volume/surface area:			0.46	0.89	0.47	0.94

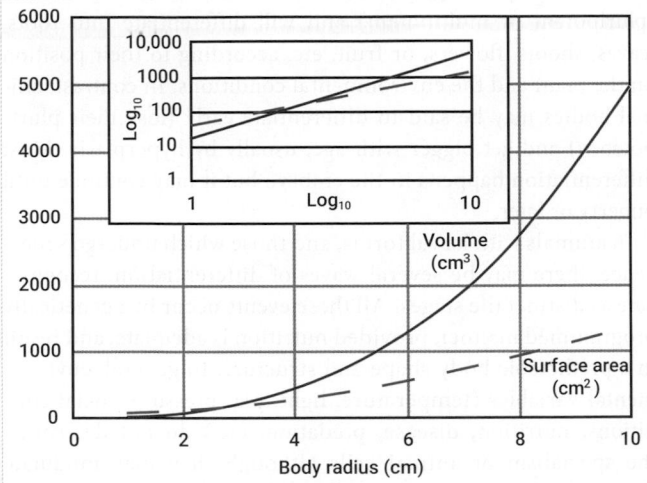

Figure 3 Trends in volume and surface area as body radius increases.

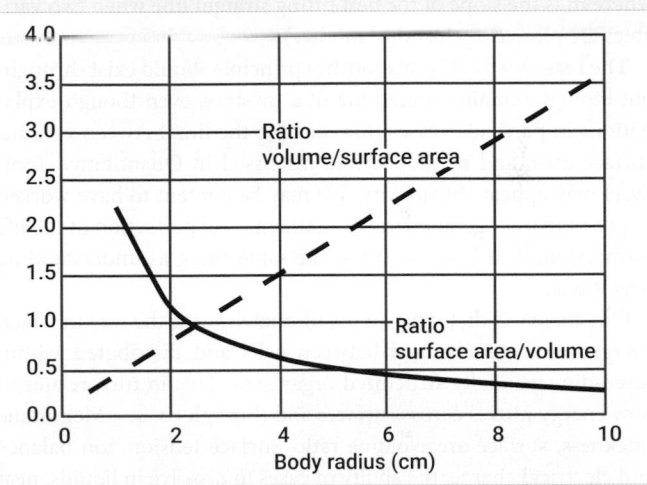

Figure 4 Trends in ratio of volume/surface area and surface area/volume as body radius increases.

Allometry takes things a step further. It helps not only to calculate overall body growth, but also to understand relationships between body components (the size of the brain, the length and thickness of limbs, for example), calculate speeds of movement, make comparisons between species, and understand how organisms may have changed over evolutionary time. The principles also have practical application, for example in estimating the appropriate dosages of drugs for individuals of different size and in the use of body mass index (weight/height²) to measure obesity.

Allometry recognizes two types of relationship, or *scaling*: isometric and allometric.

In **isometric scaling**, the growth of an organism (or differences between similar organisms) remains in proportion across the body. Isometric scaling follows a *square-cube law*. This means, for example, that if the length and other linear dimensions of the body are doubled, the surface area will increase four-fold and the volume (and therefore body weight) will increase eight-fold.

In our mole example, doubling *both* length and radius produced nearly a four-fold greater surface area (42.1 to 168.6 cm²) and nearly an eight-fold greater volume (19.2 to 150.8 cm³). 'Nearly' takes account of

the slightly disproportionate effects of the cone and half sphere in our model.

The formulae show that the law results from the effects of the square and cube terms: $2^2 = 4$ and $2^3 = 8$.

In perfect isometric scaling:

- volume scales proportionately with mass,
- surface area scales with mass to a power of 2/3 ($M^{0.75}$),
- length scales with mass to a power of 1/3 ($M^{0.25}$).

However, this is only approached when averaging large numbers of observations, and it always comes with a large amount of error.

In **allometric scaling**, which is the norm, the variables have a more complex relationship. Plots of allometric relationships can often be linearized by taking log values (graph inset), although the line of best fit always has an accompanying error.

A classic example is basal metabolic rate and body mass in mammals. Max Kleiber (1930s) established that metabolic rate (kcal/day) scales negatively with body mass (kg): BMR = 70M$^{0.75}$. This means that a large mammal (elephant) has a lower BMR than a small one (mouse) and a mature animal has a smaller BMR than a newborn one. Over a great many species, heart rate and oxygen consumption (proxies for metabolic rate) approximate a straight line when plotted against mass on logarithmic scales.

The author Jonathan Swift failed to appreciate the effects of allometric scaling. In his classic novel *Gulliver's Travels* (pub. 1726) the eponymous traveller finds himself in Lilliput whose inhabitants are 6 inches (15.2 cm) tall compared with his own 6 feet (183 cm). He consumed 12 times as much food as they did. Scaling by Kleiber's principle ($12^{0.75}$) suggests that he really needed less than 6.5 times as much.

It turns out that very many scaling relationships in plants and animals, whether concerned with overall body size, surface area, limb length, metabolic rates, gas exchange, locomotion, lifespan, or a host of other measurable characteristics, can be well approximated by powers of multiples of a quarter (so $m^{1/4}$, $m^{2/4}$, $m^{3/4}$, $m^{3/8}$ etc. where m is the slope of the best fitting straight line when two variables are plotted against one another).

The reason why this relationship principle should exist throughout biology remains something of a mystery, even though explanations in particular cases (for example the link between volume, surface area, and metabolic rate discussed in Quantitative Tools 29.1) may appear satisfactory. We may be content to have worked out the apparent geometrical or mathematical perfection of the life forms around us but that is not the same thing as understanding why it is so.

One theory is that it has to do, ultimately, with the way in which *energy* can be exchanged between cells and distributed within three-dimensionally structured organisms. This in turn relates to how energy moves across surfaces and through tissues. Membrane thickness, surface area:volume ratio, surface tension, ion balance and electrical charge, the ability of gases to dissolve in liquids, heat loss from surfaces, and many other variables all play a part; they are just some of the physical and chemical imperatives with which biological organisms have to comply.

Size and complexity in animals and plants

The complex bodies of *multicellular* plants and animals are achieved by a combination of growth and differentiation of cells in different regions. In all but the earliest stages of development, increases in overall body size occur mostly by hyperplasia (cell replication by **cytokinesis**, sometimes loosely called **mitosis**). Cells in certain locations (for example, nerve cells and fat cells in animals, vacuolated cells in plants) also increase in size (hypertrophy) or sometimes fuse (Chapter 28) but hyperplasia is the predominant mechanism of growth in most organisms.

The differentiation of cells in particular locations, called positional specialization, leads to the formation of particular structures and shapes (lungs and arms, say, or roots and leaves) as well as to functional differences within and between tissues and organs (for example plant meristems, skin epithelium, liver cells).

An organism's capacity for growth and positional specialization amongst its cells is determined genetically. The phenotype which emerges depends on the resources available and other environmental factors. One important and fundamental difference between plants and animals is the extent to which morphology is pre-determined in the germinating cell or embryo.

▶ The different 'potencies' of animal and plant cells on which this difference depends is discussed further in Chapter 28.

Generally speaking, plant cells are morphologically flexible (pluripotent or multipotent) and will differentiate into roots, leaves, shoots, flowers, or fruit, etc. according to their position on the plant and the environmental conditions. In contrast, animal bodies may be said to differentiate early (lose their pluripotency) and get bigger with age, usually by hyperplasia. Most differentiation happens in the embryo but it may continue until puberty or later.

In animals with larval forms, and those which undergo senescence, there may be several waves of differentiation, recognizable as distinct life stages. All these events occur in a genetically programmed manner, provided nutrition is adequate, and result in a predictable body shape and structure. In general, environmental variables (temperature, light, pH, pressure, social conditions, nutrition, disease, predation, etc.) do not determine the specialism of animal cells although they may modulate the eventual size, shape, and functional capacity of tissues and organs.

Plant bodies grow constantly and produce new structures throughout their life, and a mature plant can appear very different from its seedling. Final maturity may be reached in some long-lived plants decades after the basic shape has emerged, as in the case of many trees. In addition, plants have the ability to regenerate new parts of their structure regularly throughout their lifetime. Many do so in a seasonal manner; for example, deciduous shrubs and trees make a complete new set of leaves each year. The size and shape of a plant's leaf canopy will depend on the light conditions, whilst the direction and the extent of its root system may reflect water and nutrient availability. Flowering may be temperature, light, or water dependent.

 Check your understanding of the concepts covered in this section by answering the questions in the e-book.

29.2 Plant body plans

Modern plants are thought to have evolved from ancestral algae (Chapter 28). The body plan of multicellular algae such as seaweeds is relatively simple: just a flat **lamina** (tissue layer) with no specialized support or fluid-conducting tissues and no root. In large brown algae such as kelp (see Figure 28.30) the laminae may gather into thick, rope-like structures; these superficially resemble stems but have a solely supportive function, as do the extensions called holdfasts by which they are attached to rocks and other surfaces.

Greater complexity is found in the bryophytes, typified by mosses (Figure 29.3). Their photosynthetic tissues are **parenchymatous**, meaning they are soft and made of thin-walled, relatively undifferentiated cells. Mosses have simple root systems made of thin filaments, described as **rhizoid.** Bryophytes generally inhabit low lying, wet environments where all parts of the plant can gain fluid and nutrients without the need for conducting tissues or complex roots.

Vascular plants are those which do have structures for fluid conduction. Botanists call them 'higher' plants to distinguish them from the less complex algae and bryophytes. They include all the familiar grasses, shrubs, trees, and flowering plants that we see around us.

In vascular plants, water taken up from the soil by roots is moved up through the stems and lost from pores in the leaves. The water-conducting channel is called **xylem** and it forms an essentially contiguous system through the plant from roots to leaves. The ability to transport fluid from one region of the plant to another, including to great distances above the ground in some cases, has allowed vascular plants to colonize a much wider range of environments than the wetlands to which avascular plants are generally restricted.

The simplest vascular plants are the ferns (monilophytes; look at Figure 29.4 for an example). These have rhizoid underground root systems, linked to stems above ground called fronds. Fronds are vascular and provide the plant's photosynthetic and evaporative surfaces but are distinguished from the leaves of other plants by their fractally repeated arrangement of **pinnae** and their production of **sporangia** (asexual spores, involved in reproductive dispersal). Fronds grow directly from the rhizome rather than from other stems and emerge by unfurling from a characteristically scrolled 'fiddlehead' structure, as depicted in Figure 29.5.

The seed-producing plants (spermatophytes), include the gymnosperms (conifers and related plants with 'naked' seeds) and

Figure 29.4 Ferns (Tracheophyta).
Source: Image by jusuf111 from Pixabay

the angiosperms (flowering plants). They make up around 90% of all plant species and demonstrate great **plasticity** (capacity for variation) in their body plans.

A plant's basic *organization*, including species-specific or variety-specific features of shape, colour, **habit** (growth architecture), developmental cyclicity, seasonality, and environmental preference, is encoded in its genotype. Its *actual* shape, size, colour, and reproductive activity (production of offshoots, flowers, fruits, and seeds) are determined by the local climatic, nutritional, and attritional forces to which it is exposed. For domesticated and other harvested plants, this includes the activity of human agencies.

Much of this plasticity exploits the arrays of meristems (Chapter 28) which can be activated at various stages of life, for example to generate new stems or branches. In Chapter 28 we separated the flowering plants into monocotyledonous (which branch at ground level from basal meristems) and dicotyledonous (which form shoots, leaves, stems, and flowers from above ground meristems).

Monocots tend to be short lived but form extensive monocultures and have a great capacity for regeneration, for example after

Figure 29.3 Moss (Bryophyta).
Source: Image by PublicDomainPictures from Pixabay.

Figure 29.5 Fern fiddlehead (*Matteuccia struthiopteris*).
Source: Image by suju-foto from Pixabay.

grazing by animals. Dicots may develop more complex structures, maintained over repeated annual cycles, and can become extremely large. There are many examples of individual trees which are several thousand years old and which are probably amongst the oldest living individual organisms on Earth.

Many dicot plants produce secondary strengthening by depositing lignin in the form of wood (described below). In general, monocots do not do this. An exception is bamboo which grows tough new stems, called culms, from the base of the plant each year, as illustrated in Figure 29.6. Culms are composed of living vascular bundles containing sclerenchyma fibres embedded in a matrix of lignified parenchyma cells. They are hollow and straight and extend to great heights before branching occurs. Their rapid growth and the strength derived from their tubular form makes them ideal for scaffolding in the construction of buildings.

Seeds and the arrangement of leaves

The embryo plant of a dicot species such as a bean will develop inside the seed following flower pollination and fertilization (Chapter 33). A precise set of cell divisions arising from the zygote

Figure 29.6 Giant bamboo (*Dendrocalamus giganteus*).
Source: Mykola Ivashchenko/Shutterstock.com.

generate a new plant with a distinct apical-basal polarity: it has a shoot apical meristem at the top and a root meristem at the root tip. Look at Figure 29.7, which illustrates how a plant seedling passes through a number of distinct stages as it develops from a fertilized zygote. Notice the locations of the shoot apical meristem and root meristem on the seedling itself.

The early-stage embryo becomes desiccated as the seed dries out and may survive for very many years in that form. When conditions change to permit germination, the embryo resumes development and forms a seedling plant. This will have two meristems (SAM and RAM) and a stem (hypocotyl). It will also have two seed leaves, the cotyledons, which store molecules used by the developing seedling during the first stages of germination (Chapter 33).

Shortly after germination, new leaf primordia (groups of immature cells) develop on the flanks of the SAM (Chapter 28). Indeed some of the earliest leaf primordia may well be formed during the early development of the embryo inside the seed. They develop chloroplasts and begin photosynthesis as soon as they are exposed to light.

Leaf positioning

Two features of leaf primordia production on the SAM are fundamental to the eventual architecture of the mature plant. One is the rate of primordia production. This is called the **plastochron,** defined as the time between the initiations of two sequential primordia in the chronological sequence of leaves. The other is the exact positioning of the primordia on the flanks of the meristem. This is called **phyllotaxy** and refers to the position of each primordium relative to its predecessor in the sequence of leaf production. Leaf primordia form at a relatively constant rate in a fast growing plant.

Look at Figure 29.8, which shows the phyllotaxy exhibited by three different plants. Notice how the leaves on each plant have characteristic positioning. The phyllotactic pattern for a particular species is genetically determined. For example, leaf primordia can be initiated simultaneously on two sides of the meristem, 180° apart, producing leaves that are opposite each other in pairs. Alternatively, they can be initiated 180° apart but at different times, resulting in leaves on opposite sides of the stem but at different heights.

In some plants, primordia emerge at a constant angle of displacement which is other than 180°. This produces spiral phyllotaxy, as illustrated in Figure 29.9. The tightness of the spiral depends on the angle and the distance between primordia along the stem. Many plants with spiral phyllotaxy do not extend their stem significantly but develop a flat, rosette arrangement of leaves. The angle of divergence is relatively constant, normally 137.5° (see Figure 29.9b). Leaves in this arrangement show minimal shading of each other, maximizing the interception of light by the plant.

A plant will keep extending its stem and producing new leaves in a precise sequence and position while it continues to grow in a vegetative (non-reproductive) manner. Initially, the axillary meristems in each leaf axil (Chapter 28) are kept dormant by the dominant effect of the main shoot apical meristem. As the stem elongates, axillary meristems become more active and eventually initiate shoots of their own with the same phyllotactic arrangement of leaves as the main stem. The result is a bushy plant with several

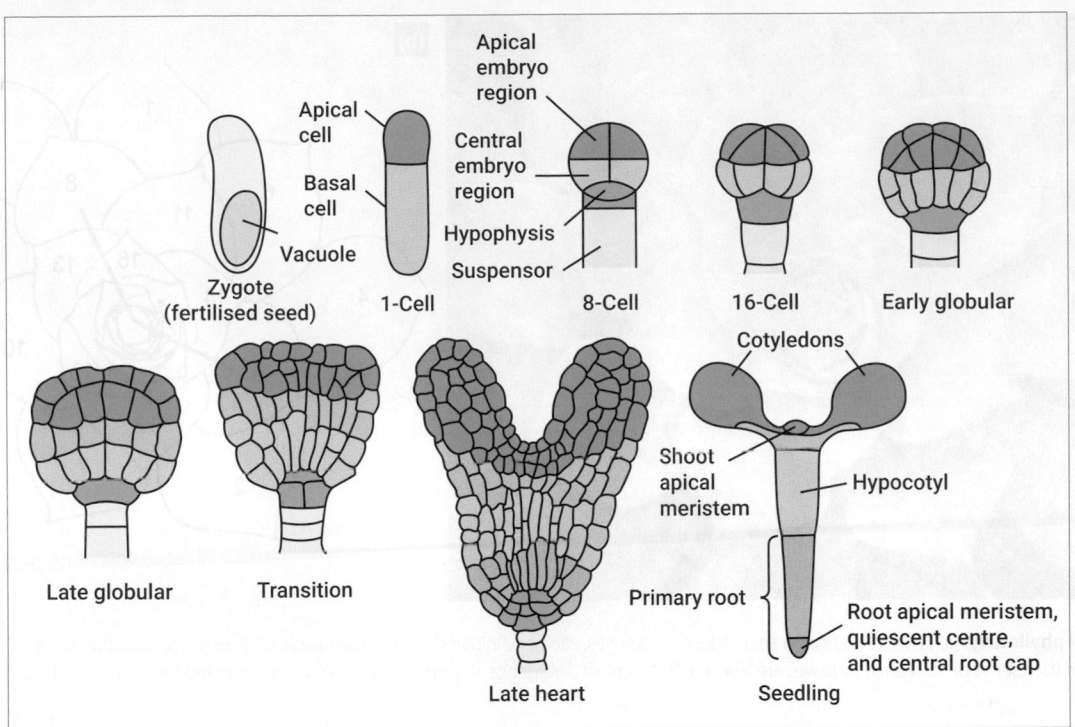

Figure 29.7 Stages of embryo development in a dicotyledonous plant.

Source: from Lincoln Taiz, Eduardo Zeiger, Ian Max Møller, and Angus Murphy, Fundamentals of Plant Physiology, 1st Edition, 2018, Oxford University Press.

Figure 29.8 Leaf phyllotaxy. (a) *Plectranthus* sp. have opposite leaves so that two leaves are initiated at the same time on opposite sides of the meristem, 180° apart. (b) Leaves of cleavers (*Galium arvense*) grow in a whorl of seven leaves which emerge at the same time around the meristem at about 53°apart. (c) Leaves of cherry laurel (*Prunus laurocerasus*) form on opposite sides of the meristem but at different times, such that they alternate on opposite sides of the stem.

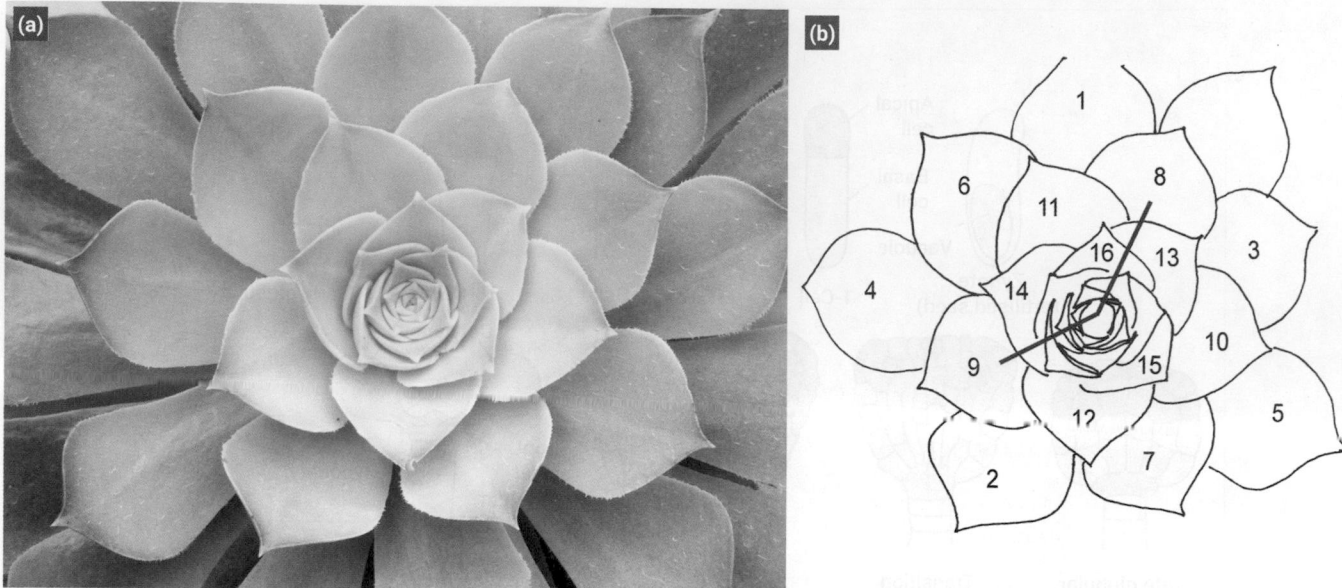

Figure 29.9 Spiral phyllotaxy. (a) Leaves of *Echeveria* are produced in a series, each one initiated at a constant angle of divergence from the previous one, creating a spiral. In the diagram (b) a series of 16 sequential leaves are labelled. The angle of divergence is approximately 137.5°, shown in red for leaves 8 and 9 in the series.

shoots, each headed by a shoot apical meristem making leaves and with a new axillary meristem in each leaf axil.

Shoots and branches

A key feature of the growth of axillary meristems, especially in trees and woody shrubs, is the angle at which new shoots branch out from the main stem. While main shoots usually grow directly upwards, side branches grow at a characteristic branch angle. Branching angles are genetically determined, producing the recognizable patterns of particular species, easily observed in deciduous trees in winter. For example, look at Figure 29.10 and notice the distinctive patterns shown by the three trees in the pictures, which range from near vertical to near horizontal.

The genetically prescribed angle is expressed as a reaction to the downward force of gravity. Branching angles are tightly controlled

Figure 29.10 Branching patterns of trees. (a) In poplar (*Populus* sp.), the branches are near to vertical. (b) In ash (*Fraxinus excelsior*) they are much less vertical. (c) In larch (*Larix* sp) they are nearer horizontal.

during early growth but are moderated in more mature trees by the weight and length of branches and the effects of light and climate. Much of the above ground architecture of dicot plants is determined by a combination of plastochron, phyllotaxy, and branching angle, as well as the hierarchy of stems and axillary meristems and their growth rates.

The complexity of tree branching, resulting from plant three-dimensional architecture, coupled with the communal growth of complex plant societies in forests, creates entire ecosystems of great importance to the planet. Many animals and insects make use of them, often over long periods of time, and they therefore support significant biodiversity.

PAUSE AND THINK

Look at Figures 1 and 2. Answer the questions for each figure. You will need a protractor to measure angles.

(a) Do you think the plant is a monocot or a dicot? Give your reasons.

(b) Measure the angles between successive leaves (centring your protractor on the red dot).

(c) Would you describe this as resulting in spiral phyllotaxy or is the leaf pattern more complex than that?

Note:

- Figure 1 has been photographed from one side. You can estimate the angle by taking the average between successive leaves.

- Figure 2 has been photographed from above. Concentrate on the four largest leaves only.

Figure 2 Plant viewed from above.

Figure 1 Plant viewed from the side.

Answers:

Figure 1

(a) Monocot, because all the leaves arise from the centre of the plant, presumably from a single, basal meristem.

(b) Average angle 128° (leaf 1–2 = 120°, 2–3 = 150°, leaf 3–4 = 115°).

(c) This appears to be an example of simple spiral phyllotaxy.

Figure 2

(a) Dicot, because the successive leaves arise as side shoots from meristems on the plant's main stem.

(b) Angle between leaves 1 and 2 = 155°; between leaves 3 and 4 = 143°; between leaves 1 and 3 = 58°.

(c) The leaves of this young plant clearly arise in pairs but the angle between them is not yet fixed. There is a separate, smaller angle between the successive pairs of leaves. This pattern of development is more complex than simple spiral phyllotaxy because the final arrangement of leaves in the mature plant will be the result of the interaction of the angles.

Trunks and bark

Plants which survive as individual organisms for long periods of time need to have correspondingly strong and resilient structures. Many perennial plants (plants which flower on an annual basis and live for more than two years) develop hard structural molecules,

in particular the class of organic polymers called **lignin**, which is synthesized in the walls of xylem cells in the vascular tissues.

The material we call wood is an accumulation of dead xylem cells strengthened by lignified cell walls and is generated from the cambial meristem (Chapter 28). It gives rise to twigs and stems in small shrubs and bushes, and in trees produces large trunks. Wood enables trees to grow to enormous size: without it, the tissues would be unable to be supported by turgor pressure beyond a couple of metres above ground level, due to the weight of the fluid column.

Larger shrubs and trees also develop bark as a protective outer coating. As a tree increases in diameter with yearly growth, the bark also grows but it may split. Bark phenotype is species-specific and can be used for identification, especially for tall forest trees where access to leaves may be difficult, or deciduous trees during winter.

Bark defends the trunk against many biotic and abiotic factors. Some trees, notably the cork oak (*Quercus suber*, Figure 29.11), also produce a thick layer of cork. This is made of **suberin** and is a hydrophobic foam deposit on the outermost layer of bark. It is water repellent and disease resistant but also provides thermal insulation. Cork is harvested commercially as a renewable resource over a cycle of years. Its properties are exploited in many products including wine bottle corks, flooring, insulation panels, and buoyancy devices.

Leaf morphology

Leaf shape and morphology are species characteristics which, like bark, are used taxonomically. The leaves of dicotyledonous plants are either simple or complex, with several leaflets making up a leaf.

Figure 29.11 Cork oak (*Quercus suber*).
Source: Image by Simon from Pixabay.

A range of different leaf morphologies is depicted in Figure 29.12; notice how these vary from a single leaf to multiple leaflets. Leaves act as solar panels enabling huge subcellular populations of chloroplasts to be exposed to as much sunlight as possible. They are also the site of gas exchange and water loss, through pores in the leaf epidermis called **stomata**. Thus the function of leaves entails a complex balance between maximizing light exposure, optimizing respiration, and minimizing transpirational water loss.

The shape and internal structure of a leaf are adapted to the environment in which the plant grows. The temperature of a leaf is affected by its shape: for example, the presence of lobes can assist in cooling because the greater length of edge relative to surface area increases the rate of water vapour loss and thus evaporative heat loss.

Leaf shape is also modified by a plant's age and by the light conditions in which it grows. On a large, mature, deciduous tree, leaves at the top of the canopy, called sun leaves, develop with a slightly different morphology (more palisade cells, a thicker cuticle and often with protective red pigments), compared with those produced deep down in the canopy (shade leaves). Leaves from monocot plants tend to be less varied in shape and normally have an elongated morphology, as depicted in Figure 29.13.

In the conifers (gymnosperms), many of which are adapted to cold, snowy, and potentially water-restricted environments, leaves are reduced to long thin structures called needles (Figure 29.14a). They have a small surface area, which reduces water loss, and a waxy cuticle which encourages snow shedding from branches.

The leaves of plants which grow in water (hydrophytes), such as water lilies, expand across the water surface and float by virtue of large internal airspaces within the leaf (Figure 29.14c). In contrast, the leaves of plants which live in dry environments (xerophytes), such as *Aloe vera*, are thick and succulent and often set on very short stems; such plants are capable of retaining water for an extended period. The leaves of plants adapted to the most extreme hot environments, such as cacti (Figure 29.14b), may be reduced to spines, also set on stems with a reduced surface area relative to their volume.

Carnivorous plants grow in wet, acidic conditions such as peat bogs where nitrogen is difficult to extract from the soil. They catch insects as a source of nutrition. Leaf morphology has evolved either into a fast moving trap, as in the Venus fly trap, or into a panel of sticky hairs, as in sundews, or has become folded into a trumpet-like tube containing digestive justices into which flies get trapped, as in pitcher plants (Figure 29.14d).

The leaves of climbing plants develop tendrils which allow the plant to climb up through the canopies of other plants (which is discussed further in the case of peas (*Pisum sativum*) in Real World View 29.1). Tendrils are sensitive to touch and undergo a thigmotropic response (bending the direction of growth) which enables them to curl around structures, cling on and climb upwards.

Leaf internal structure

The internal cellular architecture of a leaf facilitates photosynthesis while minimizing water loss from the plant. Gases diffuse into the leaf via stomatal pores. Stomata (which we discuss further in Chapter 30)

REAL WORLD VIEW 29.1   Exploitation of genetic variations

Peas (*Pisum sativum*) of two different types are widely grown as commercial crops. *Vining peas* are harvested young when sweet and are often frozen for storage before sale. *Dried peas*, which are harvested at maturity, are starchy and sold as 'mushy' peas or used in animal feed.

Non-cultivated wild peas climb up through scrub and have terminal tendrils which aid this growth habit. Their whole-plant phenotype makes them unsuitable for growing as a monoculture in fields, and they often collapse and rot on the ground in wet weather. A recessive mutation exists in peas called *afila* which causes the conversion of the leaflets into tendrils (see Figure 1). Plants homozygous for this mutation have no leaflets but only a mass of tendrils. Peas also have a leaf-like structure, called a stipule, which surrounds the petiole stem junction and is photosynthetically significant. Thus *afila* mutants can grow and function as well as wild-type peas. Tendril interaction between neighbouring plants in the field provides mutual support: the crop is less prone to rotting in wet weather and much more easily harvested.

Work at the John Innes Institute in Norwich UK in the 1970s and 1980s produced the first commercial varieties of the so called semi-leafless pea carrying the *afila* mutation (see Figure 2). Subsequently all dried pea varieties grown in the UK and many vining pea varieties have this phenotype today.

Figure 1 Pea leaf morphology. (a) Normal pea leaves are compound and consist of opposite pairs of leaflets with a terminal tendril. (b) Mutation in the *afila* gene causes leaflets to be converted into tendrils.

Figure 2 Field-grown pea crops. Normal wild-type peas (left) contrast with the semi-leafless phenotype (right).

---

are mostly on the lower surface, which reduces water loss by transpiration. Hydrophytes have stomata on the upper rather than the lower surface, so that gas exchange with the air can take place even though the leaf is floating.

Figure 29.15 shows how dicot leaves show a distinct dorsiventral axis in their structure whereas monocot leaves do not. The top surface of a dicot leaf such as lilac (*Syringa* sp.; Figure 29.15a) receives most sunlight while the lower surface is more shaded. The internal structure of the leaf reflects this: there are rows of **columnar palisade mesophyll** cells arrayed beneath the upper epidermis, with more spongy mesophyll cells beneath and internal to the epidermis of the lower leaf surface. Both types of mesophyll cells contain chloroplasts (Chapter 28). Stomata most commonly occur in the lower epidermis and have internal airspaces inside the leaf. The vascular tissues run through the centre of the leaf thickness.

In monocots such as maize (*Zea mays*; Figure 29.15b), the leaves are normally held erect and significant amounts of light fall on both surfaces. There is little difference in mesophyll structure through the leaf. The bundle sheath cells surrounding the vascular tissues are well developed; they contain chloroplasts and take part in C_4 photosynthesis.

▶ We discuss photosynthesis further in Chapter 7.

Root architecture

The underground part of the plant body is the root system. This plays two crucial roles in plant function: uptake of water and nutrients and physical anchorage of the plant in the substratum. The latter role is especially important in large plants such as trees, which would fall over without a substantial root system.

▶ We discuss the general structure of roots in Chapter 28.

The complexity of root systems varies enormously between different types of plants. For a particular plant species there is normally a balance between root biomass and the amount of above ground biomass. A combination of lateral root initiation and root

Figure 29.12 Variation in leaf morphology of dicotyledonous plants. (a) *Viburnum* sp.: simple and ovate. (b) *Geranium* sp.: extensively lobed with a toothed edge. (c) Holly (*Ilex* sp.): edge lobes evolved into spines. (d) Rose (*Rosa* sp.): compound, made of a series of opposite leaflets and a terminal leaflet.

elongation produces a species-specific root system, as depicted by the two examples in Figure 29.16. Complex root systems develop from the formation of lateral roots, and there is a hierarchy within the root network. At the smallest level are root hairs (see Figure 1.18) through which most of the water and nutrient uptake occurs.

The soil in which the plant grows has a profound effect on the extent of the root system; many biotic and abiotic factors play a role. Water content, compaction, pH, nutrient content, fungal ecosystems, and the presence of neighbouring root systems all affect the extent and complexity of a plant's root system.

The root systems of plants can be impressive. In one famous experiment, plants of rye (*Secale cereale*) grown in large containers for four months were dug up and the length of secondary, tertiary, and quaternary roots on each main root measured, along with counts of root hair density and length. There was more than 6000 km of root and over 10,000 km of root hair length. It was estimated that the total surface area interacting with the soil exceeded 600 m^2 (more than two tennis courts). Of course, larger plants have even more extensive root systems.

Tree roots are normally considered to spread a distance at least equal to the tree's height. However, they do not necessarily penetrate deeply: an analysis of trees blown over in storms in England showed that 95% were shallower than two metres in the soil. This is one reason why soil disturbance by trenching or building near to trees can adversely affect their health and stability.

🌀 **Check your understanding of the concepts covered in this section by answering the questions in the e-book.**

Figure 29.13 Variation in leaf morphology of monocotyledonous plants. (a) Onion (*Allium cepa*). (b) Maize (*Zea mays*). (c) Orchid (*Phalaenopsis* sp.).

Figure 29.14 Adaptions of leaf morphology to environment. (a) Pine. (b) Cactus (*Ferocactus* sp.). (c) Giant waterlily (*Victoria amazonica*). (d) Pitcher plants (*Nepenthes* sp.).

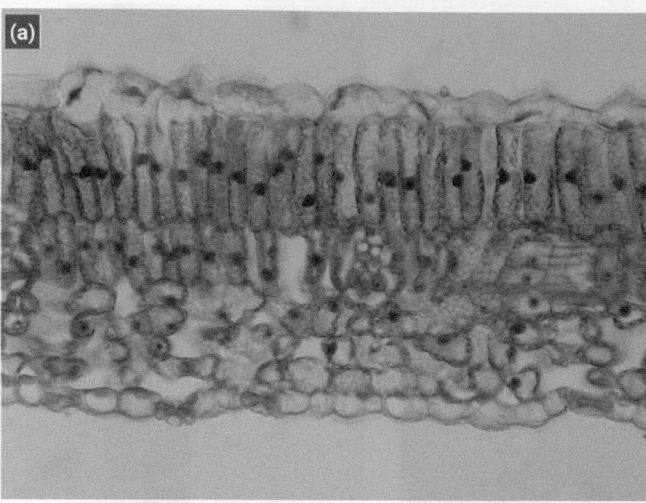
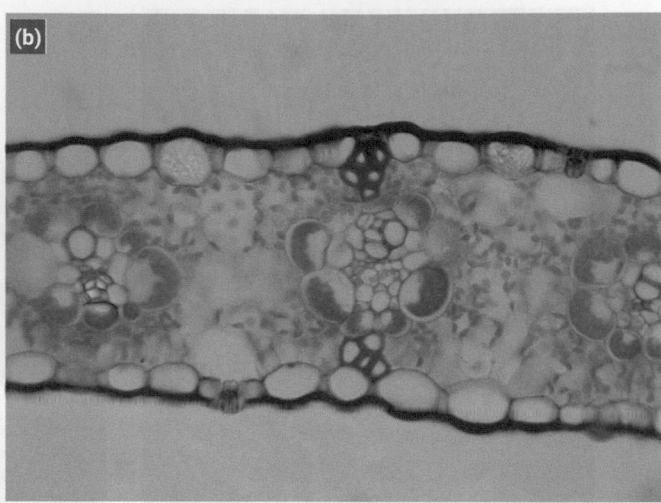

Figure 29.15 Leaf internal architecture (transverse sections). (a) Dicot (lilac). (b) Monocot (maize).
Source: Steven Kuensting.

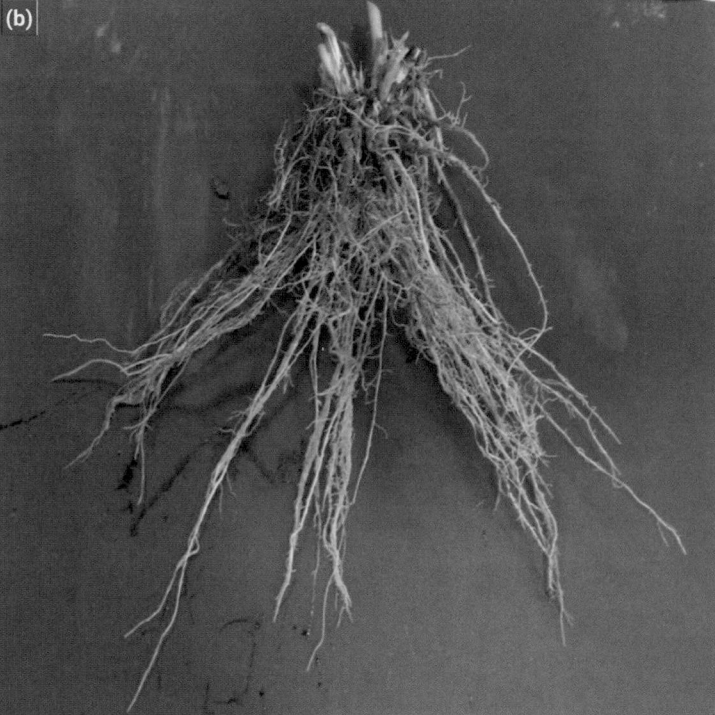

Figure 29.16 Variation in root system complexity. (a) Oil seed rape/canola (*Brassica napus*). (b) Wheat (*Triticum aestivum*).
Source: (a) Jeudy, C., Adrian, M., Baussard, C. et al. RhizoTubes as a new tool for high throughput imaging of plant root development and architecture: test, comparison with pot grown plants and validation. Figure 6a, Plant Methods, 12, 31 (2016). Creative Commons Attribution 4.0 International License. (b) Shaunagh Slack.

29.3 Animal body plans

As with plants, the bodies of all but the simplest animals are made of numerous cells working together in a coordinated manner. In Chapter 28 we saw that cells undergo differentiation in different regions of the body, giving them specialized functions.

In the case of sponges, the cells which comprise the colony become positionally specialized for functions such as chamber formation, fluid propulsion, nutrient absorption, and reproduction. However, they retain their underlying similarity so that any individual cell can, if isolated, found a new sponge colony.

We refer to all other multicellular animals as **metazoa**. The chief characteristic of metazoan cells is that they become *irreversibly* differentiated. They undergo a one-way specialization, a process which begins early in embryonic or larval life and may continue throughout the life of the animal.

The initial stages of cellular differentiation in metazoa lead to the formation of sheets or layers of cells, from which all subsequent functional specialization within the tissues of the body emerges. We can distinguish two main types of metazoan: those with bodies made of two layers of cells, called **diploblasts**, and those with three basic layers, called **triploblasts**. These cell layers are often called **germ layers** (but don't confuse them with *germ cells*; the eggs and sperm of reproduction).

Diploblasts

If we exclude the sponges, the diploblasts are the simplest multicellular animals. They are aquatic animals and comprise the phylum **Cnidaria**. They include corals (Anthozoa), sea anemones (Actinaria), jellyfish (Scyphozoa; Figure 29.17) and hydroids (or Hydrozoa; e.g. *Hydra*; Figure 29.18).

Cnidaria have hollow, essentially cylindrical bodies with a single mouth/anus opening. They also characteristically have stinging cells, called **cnidocytes** (*alt.* nematocysts, nematocytes) often arranged on tentacles around the opening. Cnidocytes develop from cnidoblast (nematoblast) cells embedded in the epithelium.

The hydroid body illustrates the typical diploblastic structure (see Figure 29.18). Its two layers of cells, are an external **ectoderm** or **epidermis** and an internal **endoderm**. These are both **epithelia** and each is usually just one cell thick. Between them is a thin, fluid-filled space, sometimes elaborated into a gelatinous matrix called the **mesogloea**. The mesogloea acts as a **basement membrane** on which the two cell layers rest.

The cells of the endoderm differentiate into complex glandular and digestive cells, turning the central chamber of the animal into

Figure 29.17 The lion's mane jellyfish (*Cyanea capillata*). This is the largest species of jellyfish (Cnidaria, Scyphozoa). Specimens have been recorded with bells 2 metres wide and tentacles 30 metres long.
Source: Derek Keats/CC BY 2.0.

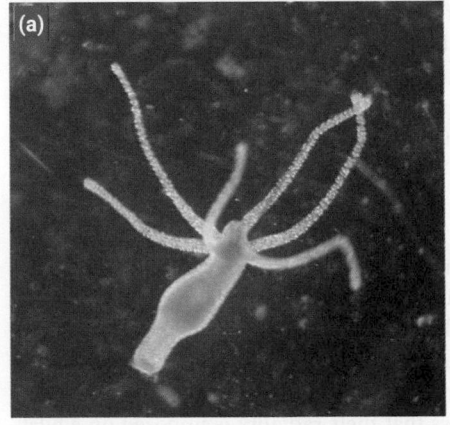

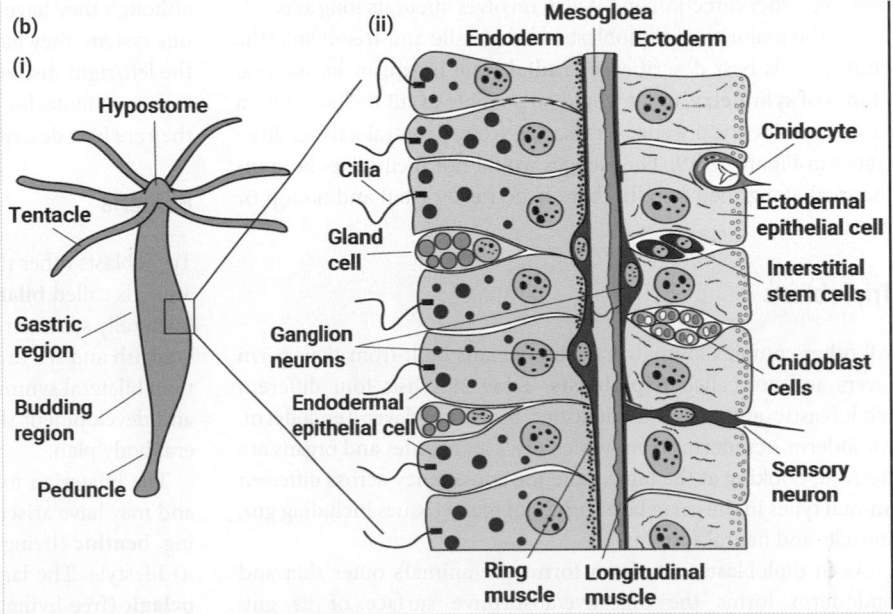

Figure 29.18 The structure and body plan of *Hydra*. (a) *Hydra vulgaris*. (b) Anatomy of a hydrozoan polyp such as *Hydra*: (i) body structure, with cylindrical body and tentacles surrounding a single opening (hypostome); (ii) The structure of the body wall with two layers of cells and mesogloea.
Source: (a) Corvana/Stephanie Guertin/Wikimedia Commons/CC BY 3.0. (b) Technau, Ulrich & Steele, Robert. (2011). Evolutionary crossroads in developmental biology: Cnidaria. Development (Cambridge, England). 138. 1447–58.

a digestive and absorptive structure. The endoderm is therefore sometimes called a **gastrodermis**.

The outer, ectodermal layer provides external protection and nervous communication. Some epidermal cells, especially around the opening, differentiate into cnidocytes—cells with coiled threads which are instantly discharged on physical contact and may sting their prey. Other epidermal cells have a sensory function. Epidermal cells have contractile elements at their base which allow the animal to wave them, when attached to the substratum, or bend in a swimming motion when dispersing.

Animals of this type have sessile and free-living (**vagile**) life stages, although one stage may predominate for most of the animal's life. The sessile form, called a **polyp**, is the main form for hydroids, anemones, and corals; they frequently form large, static colonies. The vagile, swimming form, called a medusa, is typical of jellyfish (see Figure 29.17). In most cases, sexual reproduction occurs during a polyp stage and involves the generation of medusae, although there is a great deal of variation and some stages may be transient. Asexual reproduction also occurs, from polyp or medusa forms depending on the type of animal.

The diploblast body begins its development as a ball of cells (a **morula**) and the central chamber forms by inward migration to form a blind tube, a process called **gastrulation**. The effect is rather like pushing your finger into a partially inflated balloon to create a two-layered structure. The point where the inward migration of cells begins is the point at which the animal's digestive opening or 'mouth' will form.

In sessile diploblasts the opening is at the top of the animal and the closed end forms a basal structure by which it is anchored to the substratum. With free-living diploblasts such as jellyfish, movement appears to be directional with the blind end of the central chamber orientated forward and the open end, surrounded by tentacles, following behind. However, the animal may actually move in either direction and it also revolves about its long axis.

For the majority of diploblasts, both sessile and free-living, the body plan is best described as **radial**. The organism has several **planes of symmetry** and it would be possible to cut the body into a number of identical wedges, radiating from a central axis, as illustrated in Figure 29.19. The wedges would not themselves be symmetrical along their length; there is no head or tail and no top or bottom.

Triploblasts

All other complex animals have body plans built from *three* germ layers and are called **triploblasts**. Table 29.1 lists four different triploblastic animals and identifies the tissue layer (endoderm, mesoderm, ectoderm) from which principal tissues and organs are derived. Looking at the table, note the consistency across different animal types in the germ layer origin of many tissues including gut, muscle, and neural systems.

As in diploblasts, ectoderm forms the animal's outer skin and endoderm forms the digestive/absorptive surface of its gut. However, these epithelia usually proliferate to become several cells thick. The third layer of cells, called **mesoderm**, lies between the ectoderm and endoderm. It emerges from cells which have

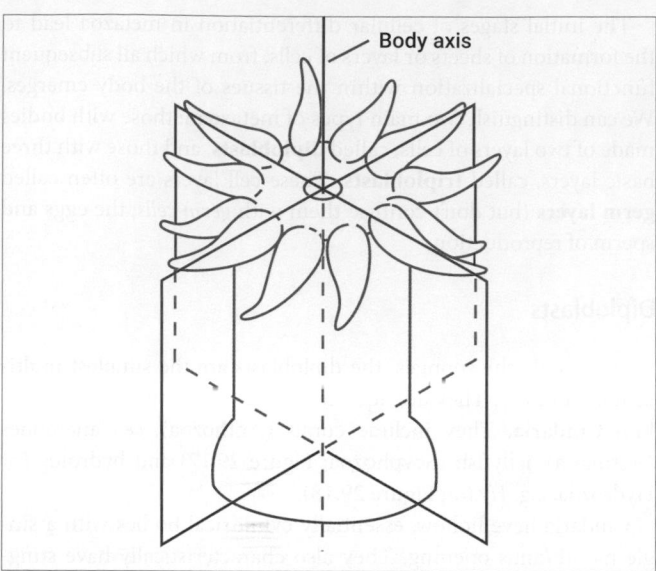

Figure 29.19 Radial symmetry in diploblasts.

Source: Richard C. Brusca, Wendy Moore, and Stephen M. Shuster. Invertebrates, Third Edition, 2016, Sinauer Associates, Oxford University Press.

proliferated deep inside the very early embryo, first as a single layer but almost immediately undergoing a great deal of proliferation and differentiation. The mesoderm develops into a highly diverse and bulky collection of tissues—muscle, bone, heart, circulatory, excretory, endocrine, reproductive organs, and others—eventually making up most of the internal structure of the body.

The simplest triploblasts are the ctenophores or comb jellies, examples of which are shown in Figure 29.20. These fine, translucent marine creatures are often found in plankton. There are about 120 species and they form a phylum of their own. This is because, although they have three layers of cells, a complex gut, and a nervous system, they have both the radial symmetry of diploblasts and the left/right distinction of more complex animals, represented by a pair of tentacles. This morphological combination means that they are best described as **biradial**.

Bilateria

Triploblasts other than ctenophores form a very large collection of animals called **bilateria**. They have a body plan which is basically bilaterally symmetrical. Some bilateria, especially the echinoderms (starfish and sea urchins) appear superficially to have radial rather than bilateral symmetry. However, close study of their embryology and development shows that their radiality is imposed on a bilateral body plan.

The bilaterian form is extremely ancient in terms of evolution and may have arisen when adult metazoa started to adopt a creeping, **benthic** (living on the sea bed but not necessarily attached to it) lifestyle. The larval stages of such animals probably remained **pelagic** (free-living, floating at mid-water levels).

They are thought to have evolved from a common ancestor known as **urbilateria** (*lit.* the original bilateral animal). There is no fossil evidence for this organism and to that extent it is a

Table 29.1 The tissue layer from which principal tissues and organs are derived in four different organisms

	Mouse (*Mus* spp.)	Sea urchin* (*Echinoidea* spp.)	Fruit fly (*Drosophila melanogaster*)	Earthworm (*Lumbricus terrestris*)
Animal type	Deuterostome, vertebrate	Deuterostome, echinoderm	Protostome, insect; open circulation	Protostome, annelid; closed circulation
Endodermal tissues	Gut Anterior pituitary gland Liver Lungs	Gut	Midgut	Gut
Mesodermal tissues	Skeleton Muscle Kidney Heart Urinary system Blood Endothelium Endocrine glands Gonads	Calcareous skeletal rods Muscle Gut	Muscle Heart Blood Circulatory system Endocrine glands Fat storage bodies Gonads	Muscle Vascular system Contracting blood vessels (pseudohearts) Blood Basement membrane (lack endothelial layer) Endocrine glands Gonads Septa
Ectodermal tissues	Teeth Skin epidermis Exocrine glands Mammary glands External genitalia and excretory structures Nervous system Posterior pituitary gland Eyes	Oral region External epithelium Neural structures	Foregut and hind gut Cuticle Exocrine glands Nervous system Wings and legs Eyes and antennae Spiracles and respiratory trachea External genitalia	Epidermis (cuticle) Exocrine glands Nervous system Chaetae Light sensitive tissues (eyes) Clitellum

*In sea urchin, the gut of the larval stage (where most studies have been done) is derived from 'endomesoderm', a combination of endodermal and mesodermal cells.

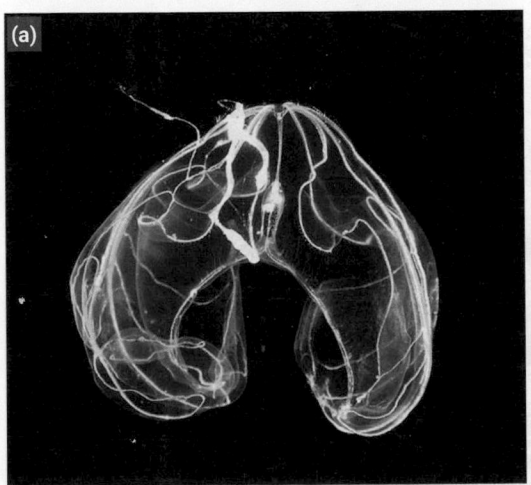

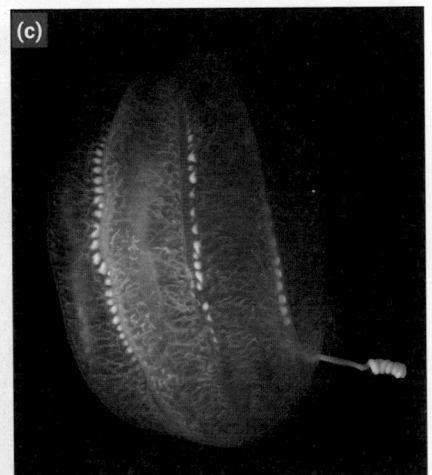

Figure 29.20 Examples of comb jellies (Ctenophora). (a) *Bathocyroe fosteri*. (b) *Mertensia ovum*. (c) *Aulacoctena* sp. (cydippid).

Source: (a) Photo courtesy of Marsh Youngbluth/NOAA/Public Domain. (b) Image courtesy of Arctic Exploration 2002, Kevin Raskoff, MBARI, NOAA/OER. (c) Kevin Raskoff. Credit: Hidden Ocean 2005 Expedition: NOAA Office of Ocean Exploration.

theoretical predecessor. However, there is also no evidence that bilateral symmetry evolved more than once, so it is reasonable to imagine it as a common ancestor.

The date when urbilateria lived is unknown but it may be as long as 670 million years ago. The earliest bilaterian in the fossil record, a slug-like creature called *Kimberella* (Figure 29.21), is dated to 555–558 million years ago, the late Precambrian period which preceded the great Cambrian explosion of life forms. *Kimberella* was once thought to be a slug-like animal but this is by no means certain as it lacks several of the characteristic features of living molluscs.

Bilaterians display front–back, top–bottom, and left–right in their structure, as illustrated in Figure 29.22. To be precise, such animals have an anterioposterior axis which forms a right angle with an apical–basal (or dorsal–ventral) axis. As a result, they have just one possible plane of symmetry: the left and right sides which are essentially mirror images of each other. To envisage this, take a straw and bend it to a right angle; now look at it end on and imagine splitting it into two halves down your line of sight. The halves will be mirror images, and the bend will mean that they can only be paired in one orientation.

Although the halves appear to be mirror images, this is only an approximation. Many of the internal organs of vertebrates (such as the heart and liver) develop asymmetrically and may also be positioned more to one side or surface than another. There will also be subtle variations in shape, size, colour, and muscular strength throughout the animal's body which detract from the mirror identity of the halves.

In some species, for example marine flat fish, the body and head twist during development such that virtually all the bilateral symmetry originally present in the early developmental stages is lost; this lack of bilateral symmetry is exhibited by the fish depicted in Figure 29.23a. The mechanism of this loss depends on chemicals related to skin pigmentation and vision which are released during metamorphosis from larva to adult. Bivalve molluscs such as oysters and scallops lose their symmetry by coming to rest on the sea bed on one side and developing much greater curvature in the upper shell, as illustrated in Figure 29.23b.

Bilaterian development—*which end is which?*

Triploblast development starts with a ball of cells called a blastocyst. The cells of the blastocyst surround an internal chamber (the **blastocoel**) and there is a distinct polarity (top and bottom, called the 'animal' and 'vegetal' poles) to the way they are arranged. Gastrulation starts with an **invagination** (an indentation) such that some cells from the outside, often at the junction between the animal and vegetal poles, move inwards.

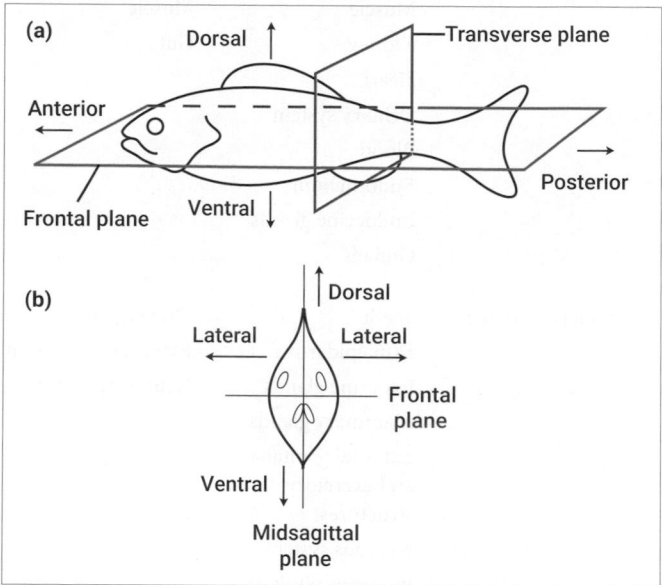

Figure 29.22 Planes of symmetry in bilaterian animals. (a) Lateral view. (b) Front view.

Source: Richard C. Brusca, Wendy Moore, and Stephen M. Shuster. *Invertebrates*, Third Edition, 2016. Sinauer Associates, Oxford University Press.

Figure 29.21 Kimberella: The earliest bilaterian fossil. (a) *Kimberella quadrata* fossil, from Pre-cambrian rocks, Flinders Ranges, Australia; (b) partial cast of Kimberella fossil.

Source: (a) Masahiro Miyasaka/Wikimedia Commons/CC BY-SA 4.0. (b) Verisimilus at English Wikipedia/Wikimedia Commons/CC BY-SA 3.0.

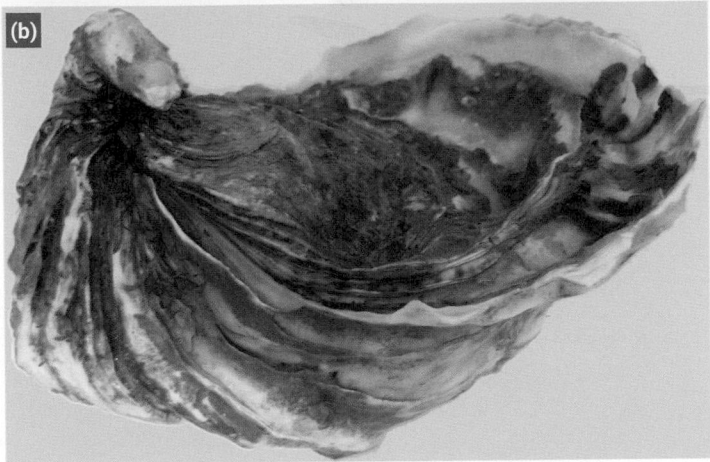

Figure 29.23 Loss of bilateral symmetry in different organisms. (a) Loss of bilateral symmetry in flat fish (Order *Pleuronectiformes*): Dover sole (*Solea solea*), left; and twospot flounder (*Bothus robinsi*), right. (b) Loss of bilateral symmetry in bivalve molluscs: Pacific oyster (*Crassostrea gigas*), left; and scallop (*Pecten albicans*), right. *Source*: (a, left) SEFSC Pascagoula Laboratory; Collection of Brandi Noble, NOAA/NMFS/SEFSC; (right) © Hans Hillewaert/Wikimedia Commons/CC BY-SA 4.0. (b, left) © David.Monniaux/ Wikimedia Commons/CC BY-SA 3.0; (right) © Hectonichus/Wikimedia Commons/CC BY-SA 3.0.

The details of gastrulation vary greatly between animal groups but in general terms the indentation proceeds right the way through the blastocyst, pushing the blastocoel to one side, and eventually connecting with a small indentation on the other side. The result is a tube with two openings, called the **archenteron**. It will eventually form the animal's **alimentary tract** (gut or gastrointestinal tract).

An important taxonomic distinction within the bilaterians depends on which end of the archenteron forms the animal's mouth and which forms the anus. In **protostomes** the initial end of the tube (the **blastopore**) develops into the mouth and the completion of the tube forms the anus. In **deuterostomes** it happens the other way round: the blastopore forms the anus. Figure 29.24 shows the animals which fall into each group.

The deuterostomes include the chordates (vertebrates and other animals with notochord), echinoderms, and tunicates. The protostome group includes all other bilaterian animals. As there is no evidence that one ancestral type subsequently changed into the other, we can say that, as chordates, humans are more closely related to starfish (echinoderm) than to squid (mollusc), ants (insect), or earthworms (annelid). Look at Figure 29.25, which represents the evolutionary relationship between different animal groups, and notice the relative positions of molluscs, annelids, echinoderms, and vertebrates.

The protostome/deuterostome distinction leads to further developmental and structural differences between the animals in the two groups:

- The development of a deuterostome blastocyst typically proceeds by **radial cleavage**: more or less continuous cell division at right angles to the plane of the surface of the egg. This happens right around the ball of cells but with asymmetries developing after the first few divisions.

- In protostomes there is much more developmental independence between the poles and across other regions of the blastocyst; cell division is by spiral cleavage (e.g. molluscs and annelids), in which daughter cells emerge at an oblique angle

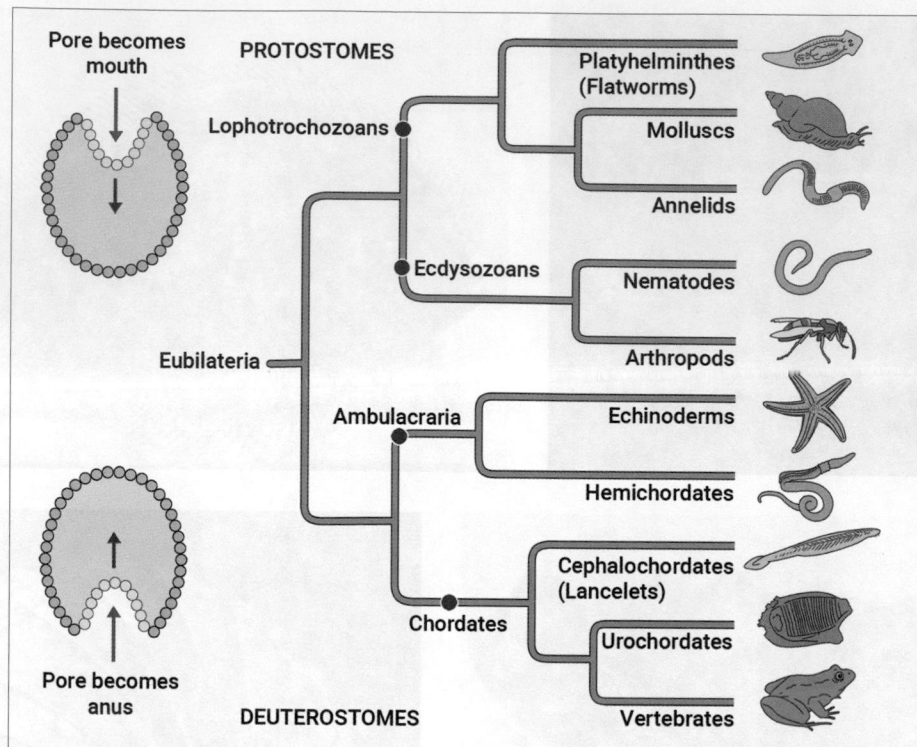

Figure 29.24 Protostomes and deuterostomes.

Source: Gilbert S.F. *Developmental Biology*, 10th Edition, (2014). Oxford University Press.

relative to the surface of the dividing cell, or by unequal cleavage (e.g. nematodes) in which the daughter cells have different sizes. These differing orientations of cell cleavage are illustrated in Figure 29.26.

■ Deuterostome embryos develop gill clefts. These are openings in the pharyngeal or neck region, behind the head. They develop into water inlets for filter-feeding in acorn worms, echinoderms, cephalochordates (amphioxus), and tunicates (sea squirts), into the respiratory gills of fishes and larval amphibians, and into hearing and balance organs in reptiles, birds, and mammals.

■ Many protostomes develop a hard surface on the *ectodermal* epithelium, such as the chitin coat of terrestrial arthropods, and it may become a mineralized **exoskeleton** (crustaceans) or shell (molluscs).

■ Mineralization in deuterostomes occurs internally: it leads either to the formation of skeletal bone (vertebrates) or to the production of an endoskeletal **test** (echinoderms), both of which are *mesodermal* structures.

▶ **We look at the processes of mineralization later in this chapter and at the mechanisms of body protection in Chapter 32.**

It is difficult to say when the division of the bilateria happened. It seems most likely that the first known eubilaterian, *Kimberella* (see Figure 29.21), was a protostome. If this is correct, the separation must have occurred before this animal appeared, so it probably took place rather early in bilaterian history. The earliest known deuterostome fossil is *Saccorhytus coronarius*, discovered in central China and believed to be 540 million years old. This fossil, together with an artist's impression of what the animal actually looked like, is shown in Figure 29.27.

Bilaterian structure

The emergence of the bilaterian form was probably associated with the presence of a particular set of genes called the *Hox* ('Homeobox') cluster. These genes code for **transcription factors** and thus regulate further gene expression. They are responsible for the patterning which is seen along the anterior–posterior axis

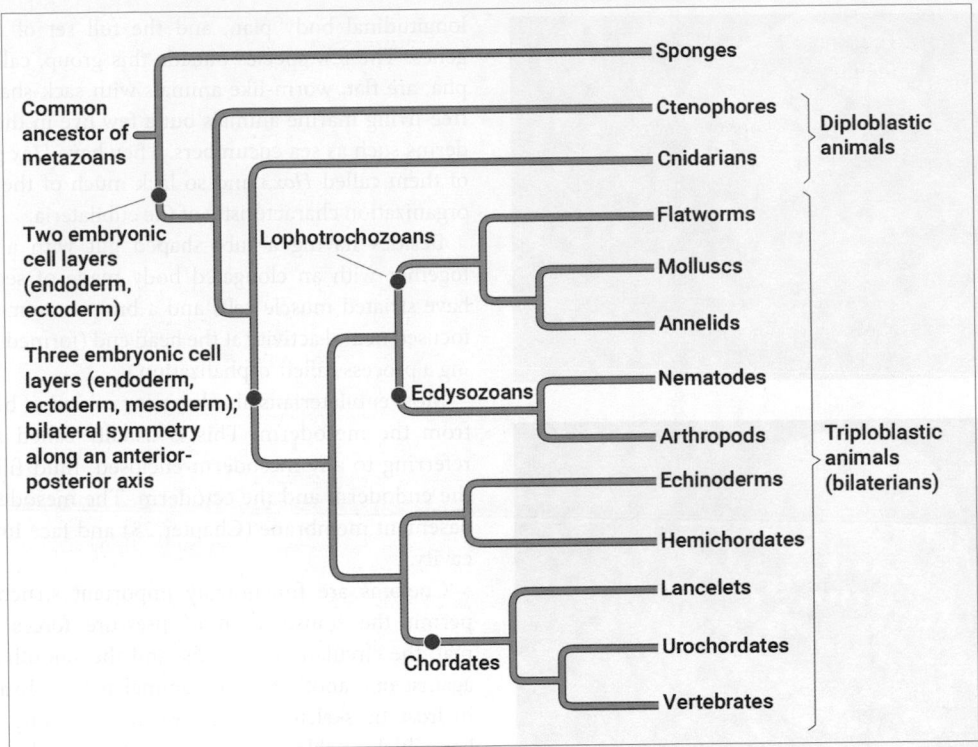

Figure 29.25 The phylogenetic relationships between animals.
Source: Gilbert S.F. *Developmental Biology*, 10th Edition, (2014). Oxford University Press.

Figure 29.26 Cell division during early development.
(Top) Cleavage orientations in three animal groups. (Bottom)
Changes in cell shape during the early development of a fertilized
egg into a blastula.
Source: Wolpert *et al.* (2015) *Principles of Development*,
5th ed. Oxford University Press.

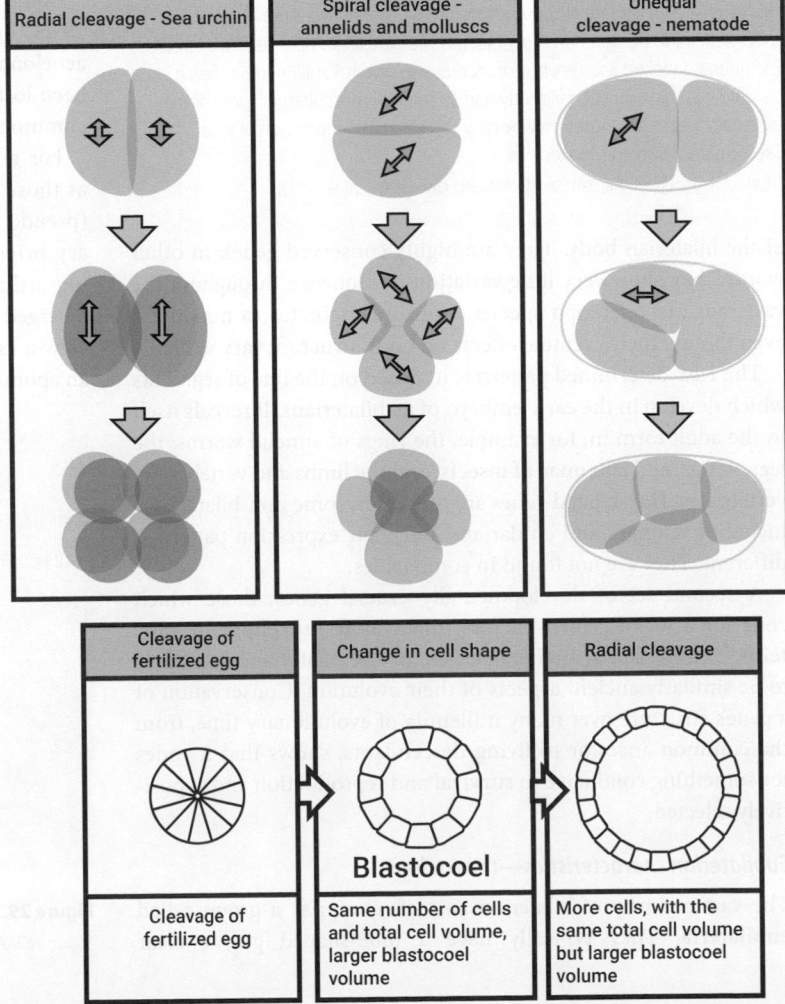

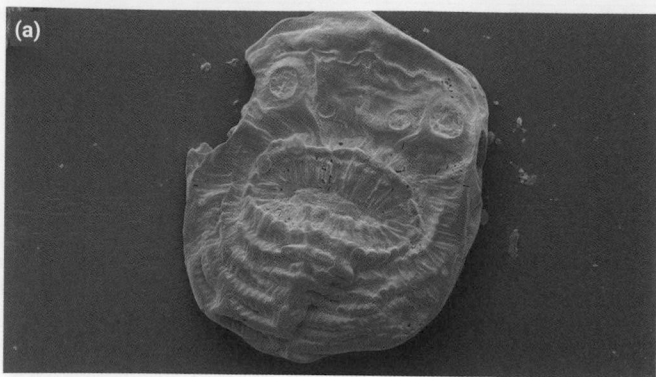

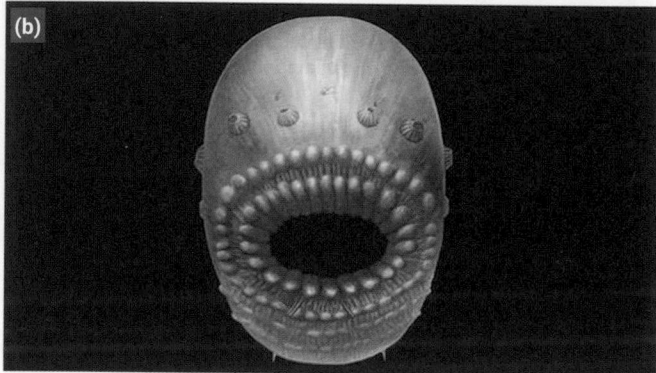

Figure 29.27 *Saccorhytus coronarius.* (a) This fossil of a sea animal, believed to be 540 million years old, was discovered in central China. It has the characteristics of a deuterostome, although the anal orifice is not visible in the specimen. If this designation is correct, it is the oldest example yet discovered. The animal, labelled *Saccorhytus coronarius*, was about 1 millimetre in size. It was covered in a thin skin and evidently had an internal musculature. Some of the externally visible structures may be gills. (b) An artist's impression of what Saccorhytus coronarius looked like.
Source: Images copyright Jian Han, Northwest University, China.

of the bilaterian body. They are highly conserved genes; in other words, they show very little variation in sequence throughout the vast range of bilaterian species, from invertebrates to mammals, even though their eventual effects on body structure vary widely.

The *Hox*-determined pattern is imposed on the line of segments which develop in the early embryo of all bilaterians. It reveals itself in the adult form in, for example, the rings of annelid worms, the legs, wings, and antennae of insects, and the limbs and vertebrae of vertebrates. *Hox*-related genes are present in some non-bilaterians, including sponges and cnidarians, but their expression pattern is different. They are not found in comb jellies.

A second set of developmentally crucial genes, those which code for a set of growth factors called **bone morphogenic proteins** (BMPs), are also characteristic of the bilaterians and likely to be similarly ancient aspects of their evolution. Conservation of a gene's structure over many millennia of evolutionary time, from the common ancestor to living descendants, shows that it codes for something conducive to survival and reproduction and is positively selected.

Eubilaterian characteristics—the coelom

The vast majority of bilaterian animals comprise a group called **eubilateria**. They typically have a tube-shaped gut, a clear

longitudinal body plan, and the full set of organizational *Hox* genes. The few species outside this group, called the acoelomorpha, are flat, worm-like animals with sack-shaped guts. Most are free-living marine animals but a few live in the tissues of echinoderms such as sea cucumbers. They have *Hox* genes but lack a set of them called *Hox3* and so lack much of the distinctive **somite** organization characteristic of the eubilateria.

Besides having a tube-shaped gut with a mouth and anus, together with an elongated body made of segments, eubilateria have striated muscle cells and a brain or some equivalent site of focused neural activity at the head end (formed in the embryo during a process called 'cephalization').

Most eubilaterians also have some kind of body cavity, derived from the mesoderm. This is usually called a **coelom**, a word referring to any mesoderm-enclosed, fluid-filled space between the endoderm and the ectoderm. The mesodermal cells sit on a basement membrane (Chapter 28) and face inwards towards the cavity.

Coeloms are functionally important structures because they permit the transmission of pressure forces through the animal, the circulation of fluids, and the smooth rubbing of organs against one another as the animal moves. Examples include the hydrostatic skeleton of earthworms (the long, fluid-filled chamber which enables it to push itself through soil), the spaces in which the **haemolymph** (blood) of insects flows, and the **thoracic** (chest), **pericardial** (heart), and **peritoneal** (abdominal) cavities of vertebrates.

Some bilaterians such as **platyhelminths** (flat worms, Figure 29.28) appear not to have a coelom (they are sometimes called acoelomate animals) but it is thought that the structure may have been lost in these animals as they evolved away from a coelomate common ancestor.

For a long time, the coelomic cavities of round worms, such as those of annelids and nematodes, were called **pseudocoeloms** (pseudo = false) to indicate that they had a distinct evolutionary origin from the coeloms of animals in other phyla including arthropods and vertebrates. It is now thought that coeloms emerged several times during animal evolution and it is best to view it as a valuable, functional, homoplastic structure rather than an apomorphy (a characteristic useful in taxonomy).

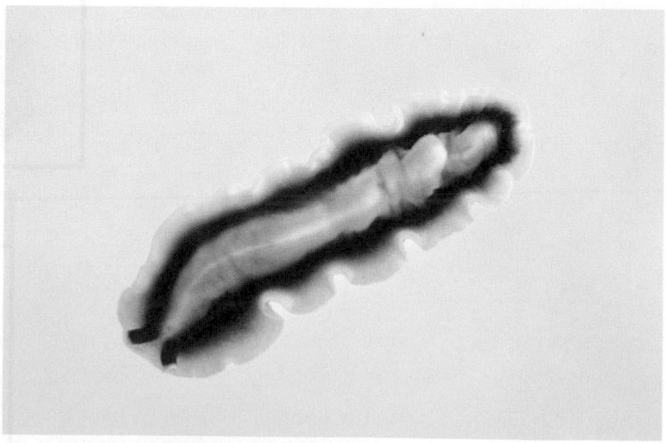

Figure 29.28 A flat worm (Platyhelminthes).
Source: NOAA/NMFS/PIFSC/ARMS.

PAUSE AND THINK

Identify the following animals as protostomes or deuterostomes. (If you are not sure what any of them are, find a picture and description online.)

- Sea eagle
- Edible crab
- Honey bee
- Wolf
- Starfish
- Hammerhead shark
- Kangaroo
- Oyster
- Octopus
- Sand dollar
- Leech
- Lamprey
- Centipede
- Garden snail
- Sea squirt

Answers:

- Sea eagle = deuterostome
- Edible crab = protostome
- Honey bee = protostome
- Wolf = deuterostome
- Starfish = deuterostome
- Hammerhead shark = deuteros-tome
- Kangaroo = deuterostome
- Octopus = deuterostome
- Oyster = protostome
- Octopus = protostome
- Sand dollar = deuterostome
- Leech = protostome
- Lamprey = deuterostome
- Centipede = protostome
- Garden snail = protostome
- Sea squirt = deuterostome.

Where do nervous systems come from?

The Cnidaria and Bilateria are sometimes grouped together as the Neuralia. Their common characteristic is the presence of an internal communication system in which information is transmitted along nerves in the form of electrical signals.

The emergence of nervous communication represents a major step in animal evolution. This is because it enables the coordination of the animal's cells to be coupled to purposeful interactions with the environment. We have no way of knowing whether the earliest neuralians were intelligent or even conscious in the sense that we use those words to describe ourselves. Nevertheless, we can imagine that the electrical interlinking of cells, the linking of sense organs to nerves, and the clustering of nerves in specific parts of the body quickly led to the development of a central processing region (a brain) capable of basic directional, decision-making, and memory functions.

The neuralian communication system consists of specialized cells called **neurones** (nerve cell bodies) with **axons** (long cytoplasmic extensions which make connections to other cells). In the simplest arrangement the neurones form a network. The propagation of signals happens by means of **action potentials** which are formed across the cell wall using Na^+/K^+ channels. Links between neurones and between neurones and other cells occur through electrical and chemical (neurotransmitter) **synapses**. Links to specialized muscle cells, which are also a neuralian characteristic, allow the animal to make complex, coordinated movements.

▶ **We learn more about action potentials in Chapters 17 and 28, and discuss synapses in Chapter 17.**

In addition to nerve and muscle cells, the neuralia have specialized sensory cells. These are often aggregated into complex sense organs which enable the animal to react rapidly to events in the external environment. The combination of these organs, a rapid transmission system, and a highly developed musculature would

have presented a strong adaptive advantage for ancestral neuralia and led to their conservation over evolutionary time.

Evolutionary biologists believe that cnidaria and bilateria came from a common ancestor and also that this ancestor was shared at a much earlier time during the Precambrian period with the ctenophores (and the placozoan *Trichoplax*). The ctenophores are much less well studied as a group (there is no whole genome available, for example), so this interpretation is by no means certain.

Based on these neuralian characteristics, we can see that the presence of a central nervous system and an accompanying set of sense organs is fundamental to the organization of all complex animals. One view, based on neuroanatomy and molecular analyses, is that nervous communication in animals first evolved alongside chemical detection. In other words, the olfactory sense may have been the first mechanism of environmental awareness to exploit this kind of transmission.

Nerves leading from olfactory cells may also have been the first to form a cluster, and thus the first to establish some kind of central nervous system. If this is correct, the complex central nervous systems and brains of living animals originated in the earliest neuralians. We don't have direct evidence about these ancestral creatures. However, even *Kimberella* had an anterior proboscis (potentially an olfactory organ) and a 'foot' and may well have interacted purposefully with its environment.

The origin of the nervous system in protostomes

There are two major groups of protostomes:

- Lophotrochozoa—worm-like animals, which have feeding tubes surrounded by sensory tentacles; they include the annelids such as earthworms and leeches.

- Ecdyzoa—animals with a great variety of shapes which grow by moulting; they include nematodes, crustaceans, and insects.

Figure 29.29 shows how the central nervous system in both groups consists of a ventral pair of nerves running the length of the body. At the mouth end of the animal, there is usually a cerebral ganglion (CG) consisting of a cluster of nerve cells, and a 'brain' formed from the nerve cord where it loops around that end of the gut.

In many animals, including insects (Ecdyzoa) and annelids (Lophotrochozoa), linkages between the pair of nerves occur at intervals along the length of the body, giving a rope ladder pattern. They are formed by neurones whose axons cross the midline of the animal.

The pattern of development and detailed arrangement is sometimes best seen in the larval forms of the animal, as depicted in Figure 29.30. The ventral nerve cord (VN) itself is formed early in embryo development, just after the mesoderm has emerged, and begins as a differentiation of the cells which make up the blastopore (called the circumblastoral ring). The ventral nerve cord is thus derived from the animal's ectoderm. The dorsal brain forms from the mouth end of the cord and the cerebral ganglia form from clusters of nerve cells adjacent to it. The nerve cord extends towards the anal end of the gut as the embryo elongates, guided by a set of 'pioneer cells' located at the posterior pole of the embryo. The anal loop may disappear and the paired cord may become a single tract.

In animals such as annelid worms the cord remains embedded in the rest of the ectodermal tissue with a basement membrane on its upper (inward-facing) side. In others, including molluscs and arthropods, it becomes completely detached from the rest of the ectoderm and is effectively suspended in the animal's coelomic or peritoneal cavity.

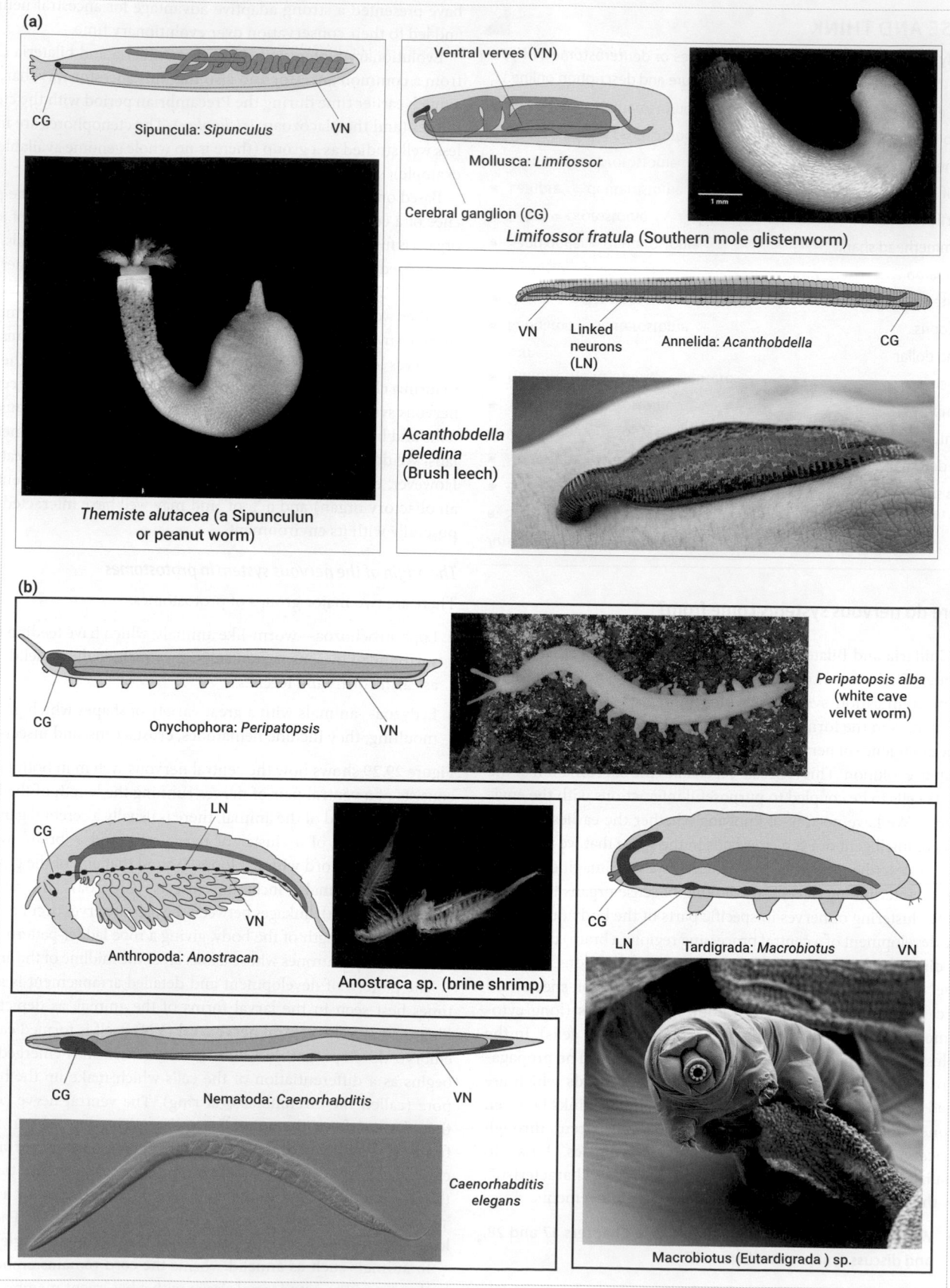

Figure 29.29 Structure of the central nervous system in protostomes. (a) Lophotrochozoans (worm-like protostomes with tentacled feeding tubes). (b) Ecdyzoans (protostomes that grow by periodic moulting).

Source: (a) Michael J. Boyle; K. Barwick/CSDMWWD; Martin Pelanek/Shutterstock.com (b) © Gonzalo Giribet; © Hans Hillewaert/Wikimedia Commons/CC BY-SA 4.0; Eye of Science/Science Photo Library; Zeynep F. Altun/Wikimedia Commons/CC BY-SA 2.5. Line drawings: Neilsen C. (2012) *Animal Evolution: Interrelationships of the Living Phyla*, 3rd ed. Oxford University Press.

PAUSE AND THINK

Using the information in Section 29.3, list some of the differences between protostomes and deuterostomes.

Feature	Protostomes	Deuterostomes
Embryonic blastopore	Forms the mouth	Forms the anus
Gill clefts	Not present	Develop into gills and other head and neck structures
Supporting/protective structures	External, formed of chitin or calcite, developed from ectoderm	Internal, formed of calcite (test) or bone (skeleton), derived from mesoderm
Nervous system	Pair of ventral nerve cords, derived from ectoderm	Central, dorsal nerve cord derived from ectoderm

Answers: Find out more about the protective role of hard materials in Chapter 32 and about the embryonic development of protostomes and deuterostomes in Chapter 33.

The origin of the nervous system in deuterostomes

The nervous systems of deuterostomes also form from ectodermal tissue. However, the embryology is rather different from that in protostomes. To describe the process any further it is necessary to subdivide this large and diverse collection of animals into several groups.

As shown in Figure 29.24, one branch of the deuterostomes, the Ambulacraria, includes the echinoderms (sea urchins, starfish, sea cucumbers) and the hemichordates (acorn worms). The other branch comprises the Chordata and includes the cephalochordates or lancelets (amphioxus), the urochordates or tunicates (sea squirts), and the vertebrates (fish, amphibians, reptiles, birds, and mammals).

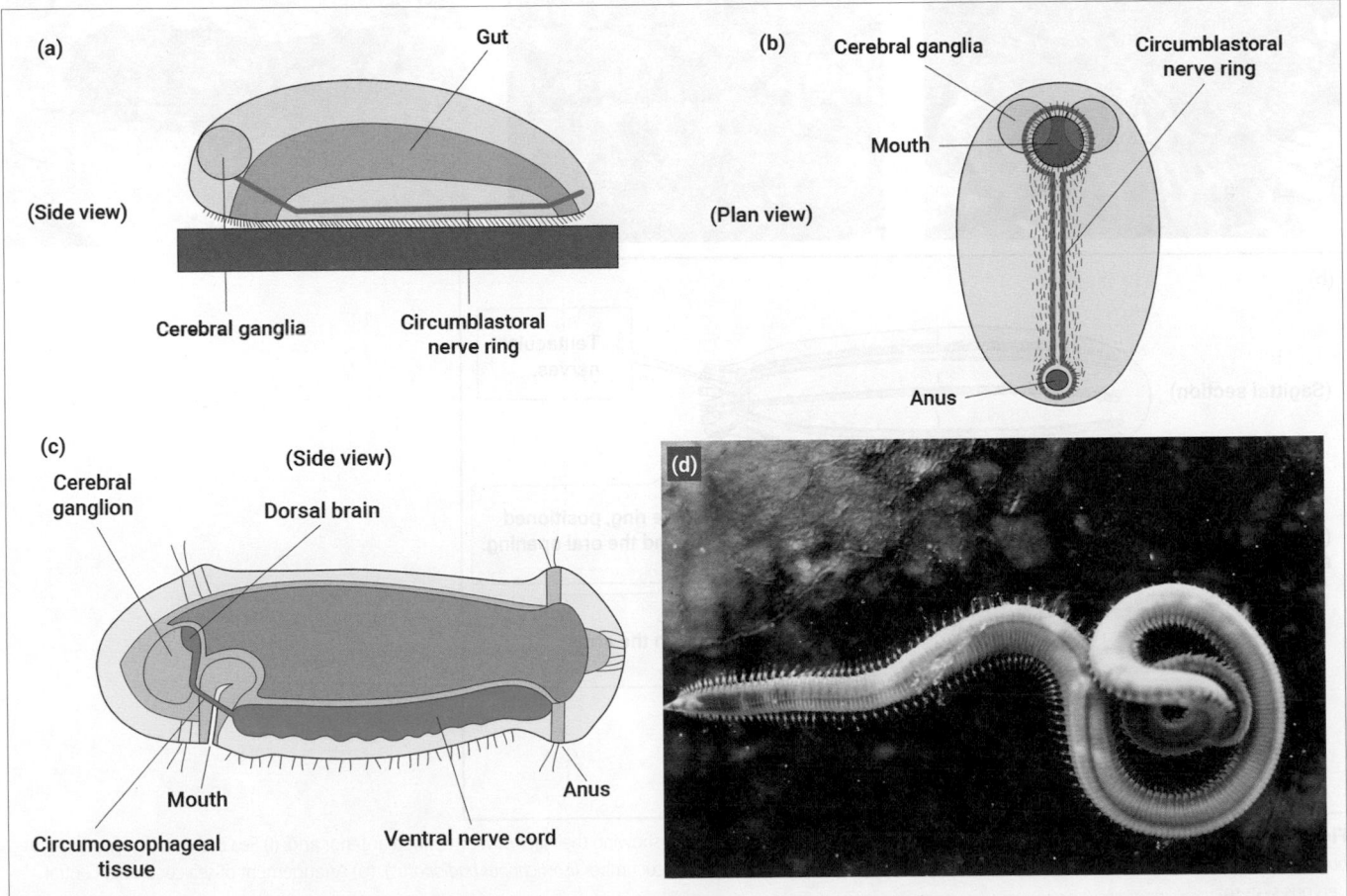

Figure 29.30 Generalized development of the protostome nervous system. (a) Side view and (b) plan view showing the general shape and location of cerebral ganglia and the circumblastoral nerve ring in relation to the gut of the early larval form. (c) Side view showing development in the later larval form prior to development into an adult. (d) An example protostome with this type of larval development: *Glycera* sp., a polychaete annelid.

Source: NOAA/Wikimedia Commons/Public Domain; Neilsen C. (2012) *Animal Evolution: Interrelationships of the Living Phyla,* 3rd ed. Oxford University Press.

The embryos of animals in the ambulacrarian branch develop a central nerve structure called a stomochord, derived from cells of the pharynx ('throat'). Remember that this end of the animal forms after the blastopore has pushed all the way through the embryo to form a mouth. Ambulacrarians pass through a free-swimming larval stage and the subsequent shape of the stomochord depends on the eventual shape of the adult animal. Thus it becomes elongated in the worm-like hemichordates but ring-like with radial branches in the echinoderms.

Ambulacraria do not develop a distinct brain, although in the hemichordates there is a ring structure (the 'collar cord') which has sporadic concentrations of neurones. They also have no distinct sense organs, although they have highly sensitive pressure-detecting epithelial cells all over the body surface. Echinoderms lack a well-defined anterior (head) end, as demonstrated by the examples in Figure 29.31a. Although they have complex nerve cords around the oesophagus and along the arms, they have no central coordinating centre. However, the main nerve structures do have distinct sensory and motor parts, separated by a basement membrane. These are linked across the membrane by synapses and short axons, providing the information transfer necessary for complex coordination. The nerve branches along the arms develop fine extensions and interact with the muscles and sense organs of the arms and pedicellaria (pincer-like structures on the surface which grab prey).

Animals in the chordate branch, which probably started to emerge more than 550 million years ago, all have a tadpole-like embryonic form at some stage of their life cycle. They are clearly distinguished from other animals by the presence of a **neural tube** and a **notochord**.

The neural tube is the line of nerve tissue along the length of the animal which forms the spinal cord and the brain. Like the ventral nerve cords in protostomes, this is an epidermal structure but instead of being formed at one end of the blastopore, it arises as an external fold along the length of the elongating embryo. It starts as a flat '**neural plate**' but then curls its edges up and over to produce a cylinder of cells surrounded by a strong basement membrane.

The notochord is a mesodermal structure, formed from the cells immediately below the neural plate. It is a long, stiff thread of cells

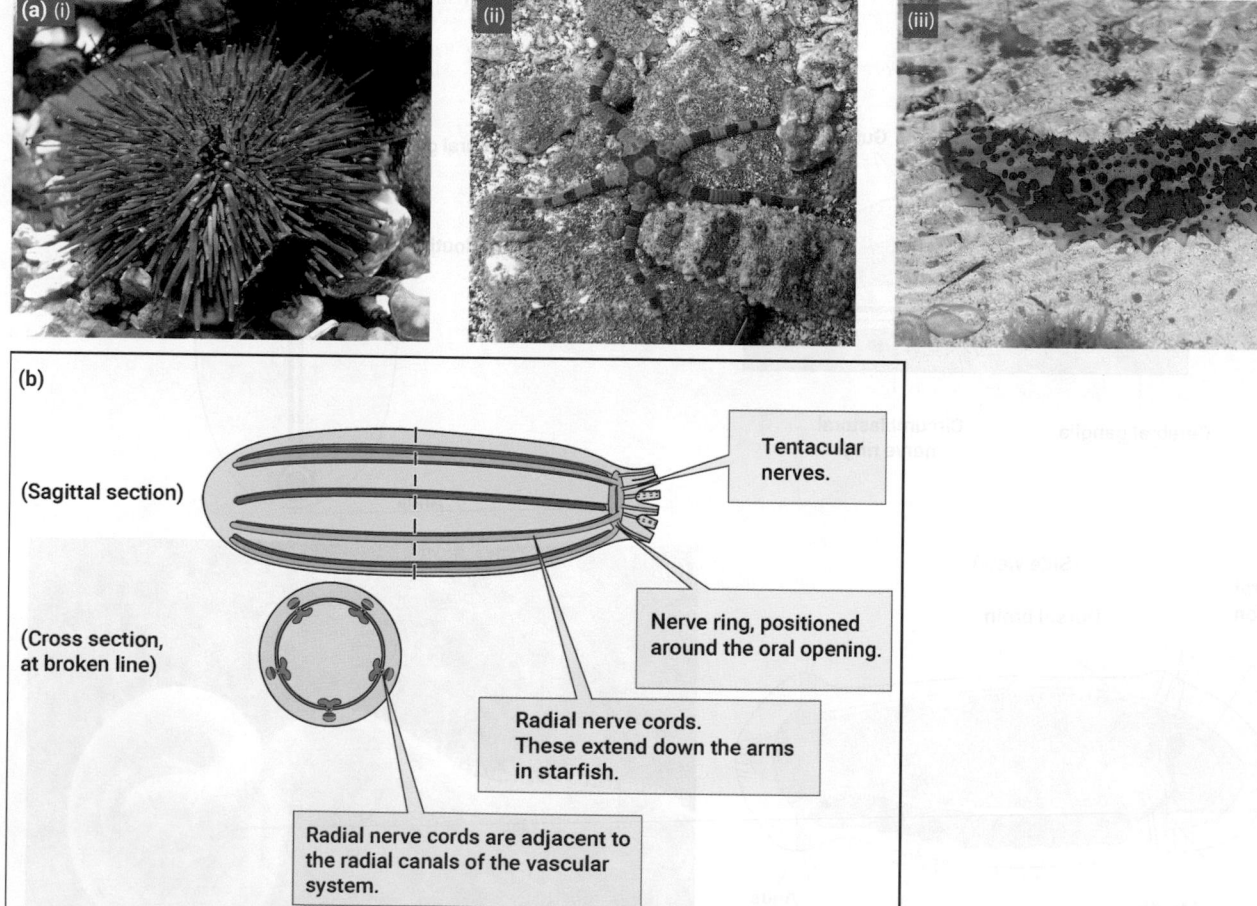

Figure 29.31 The nervous system of echinoderms. (a) Examples of echinoderms, showing their lack of well-defined anterior end. (i) Sea urchin (*Strongylocentrotus purpuratus*). (ii) Brittle star (*Ophiolepis superba*) and an adjacent sea cucumber. (iii) Sea cucumber (*Isostichopus badionotus*). (b) Arrangement of sea cucumber central nervous system.

Source: (ai) © Kirt L. Onthank/Wikimedia Commons/CC BY-SA 3.0. (aii) © Christophe Cadet/Wikimedia Commons/CC BY-SA 4.0. (aiii) © Hans Hillewaert/CC BY-SA 4.0. (b) Mashanov, V.S., Zueva, O.R., Heinzeller, T. et al. Front Zool 6, 11 (2009). CC BY SA 2/0.

and one of its roles is to direct the formation of the neural tube as the embryo elongates. The embryology of this process is explained in Figure 29.32 using amphibian development as an example. The notochord gives rigidity to the body and makes possible the development of a tail.

One hypothesis for the evolution of the notochord is that it allowed animals to burrow, much as amphioxus does today. Amphioxus (a cephalochordate; Figure 29.33a) is free living but spends much of its life partially buried in sand on the sea bed, as shown in Figure 29.33b. It has a relatively low metabolic rate compared with other chordates and uses its gills to strain the water for food particles rather than for breathing.

The amphioxus notochord is essentially a row of muscle cells along the entire length of the animal, as depicted in Figure 29.33c. It facilitates the undulatory or waving movements necessary for translocation and burrowing and for the movement of water during feeding. The neural tube has a small brain vesicle at the anterior end and also contains several types of photoreceptive cells, some of which are clustered into a pair of eye-like regions at the anterior end.

In the urochordates, the notochord is elongated in the larval stage, which is free-living. In adults, which are sessile and colonial, it becomes shortened and twisted into an elastic structure as the animal loses its tail and takes on a more barrel-like shape; this shape is illustrated in Figure 29.34. The brain is a cerebral ganglion which forms from the end of the neural tube just above the pharynx; it controls, amongst other things, the flow of water through the animal's syphons to allow the extraction of nutrient particles by the pharynx and subsequent digestion and excretion.

Urochordates are often called *tunicates* because of the cellulose-containing tunic which covers the body. Cellulose is a polysaccharide material normally considered a characteristic of plant, algal, and bacterial cells. It is thus unusual to find it in an animal,

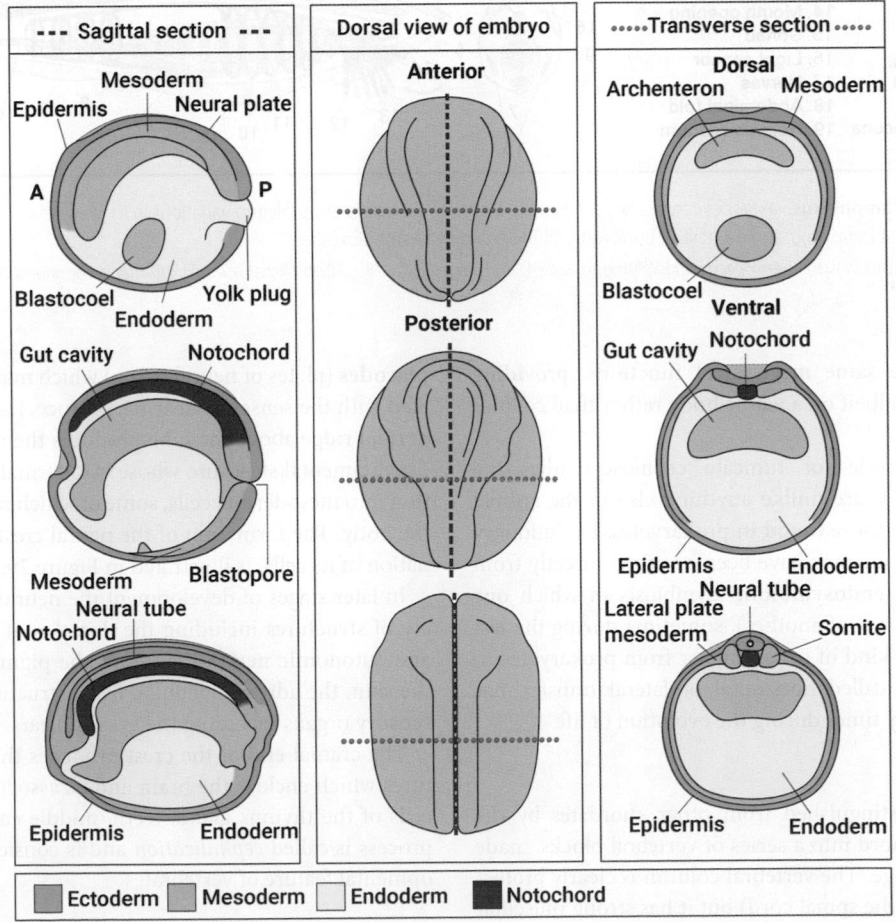

Figure 29.32 The process of neurulation (formation of the neural tube and notochord) in amphibians.

Source: Wolpert *et al.* (2015) *Principles of Development,* 5th ed. Oxford University Press.

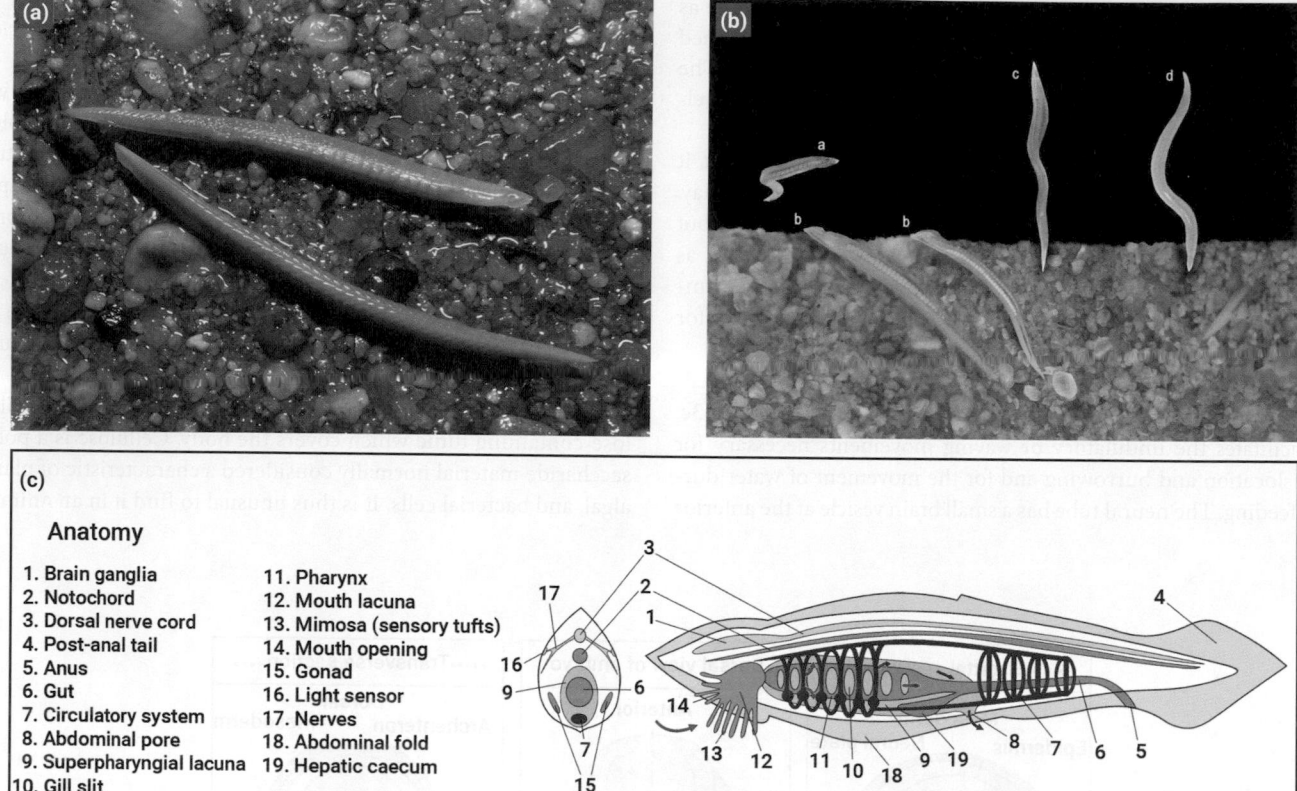

Figure 29.33 Anatomy of amphioxus. (a) *Branchiostoma lanceolatum*. (b) *Branchiostoma belcheri*, depicting variations in its lifestyle: a = swimming, b = head-up feeding position, c = headfirst burrowing, and d = tailfirst burrowing. (c) The anatomy of amphioxus.
Source: (a) © Vladimír Motyčka. (b) Jun-Yuan Chen. Palaeoworld, Volume 20, Issue 4, 2011, Pages 257–278. (c) Piotr Michał Jaworski/Wikimedia Commons/CC BY-SA 3.0.

although it serves the same mechanical functions, providing rigidity and resilience, albeit on a whole body rather than cellular scale.

The genes which code for tunicate cellulose-synthesizing enzymes, called *Ci-CesA*, are unlike anything else in the animal kingdom and more like those found in prokaryotes. Evolutionary biologists believe that they may have been acquired directly from a bacterial **parasite** or **endosymbiont** (symbiosis in which one partner lives inside the cell of another), sometime during the last 530 million years. This kind of gene transfer from prokaryotes to eukaryotes, sometimes called horizontal or lateral transfer, has probably occurred many times during the evolution of life.

Vertebrate chordates

The vertebrates are distinguished from other chordates by the transition of the notochord into a series of vertebral blocks, made either of bone or cartilage. The vertebral column is clearly protective of the neural tube (the spinal cord) but it has strong muscular connections and assists in distributing mechanical stresses through the body. In the case of bony skeletons, this development probably allowed vertebrates to leave the weight-supporting environment of the sea and colonize the land.

But vertebrates have other distinguishing features too. The neural tube develops an enlarged and highly complex anterior end, forming the brain. Closely related ectodermal structures form **placodes** (plates of neural tissue) which make the neurones associated with the senses of hearing, balance, taste, and smell. An ectodermal ridge above the tube produces the neural crest, a transient developmental structure whose ectodermal cells undergo a transition into mesodermal cells, some of which migrate to other parts of the body. The formation of the neural crest, and subsequent destination of its cells, is illustrated in Figure 29.35.

In later stages of development the neural crest generates a variety of structures including the dorsal root ganglia of the sensory and autonomic nervous systems, the pigment-containing cells of the skin, the adrenal medulla, facial structures, and parts of some sensory organs including the eyes and ears.

The cranial end of the crest produces the bone/cartilage structures which enclose the brain and its associated regions, as well as cells of the thymus gland, teeth, middle ear, and jaw. This whole process is called *cephalization* and is considered a defining developmental feature of vertebrates.

Brains and nervous system complexity

Although cephalization represents the extreme manifestation of brain development, we should not be misled into thinking that nervous system development has followed a single, goal-directed evolutionary pathway. One obvious piece of evidence against this comes from comparing the brains of vertebrates with those of protostomes such as cephalopod

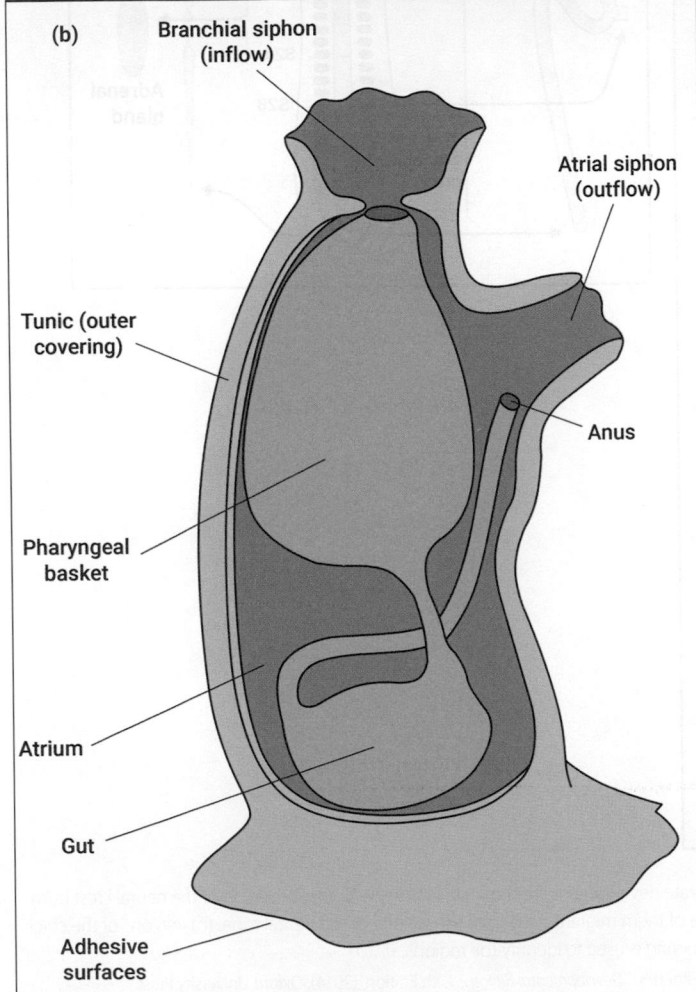

Figure 29.34 The nervous system of urochordates. (a) Examples of urochordates (tunicates). (b) General structure of urochordate nervous system.

Source: (a) Nick Hobgood/Wikimedia Commons/CC BY-SA 3.0; Steve Lonhart/NOAA MBNMS. (b) BF. Harvey Pough and Christine M. Janis. *Vertebrate Life*, 10th Edition, July 2018. Sinauer Associates, an imprint of Oxford University Press.

molluscs (octopus and squid) and insects (especially dipterans, such as fruit flies, and hymenopterans, such as bees). All of these animals have nervous systems with enlarged aggregations of networked neurones. All of them demonstrate what we can describe as intelligence, including prolonged memory,

a capacity to learn, reasoning skills, and complex individual and social behaviours.

Deuterostomes and protostomes shared a common bilaterian ancestor many hundreds of millions of years ago. It is virtually certain that this ancestral animal, whatever it was, lacked

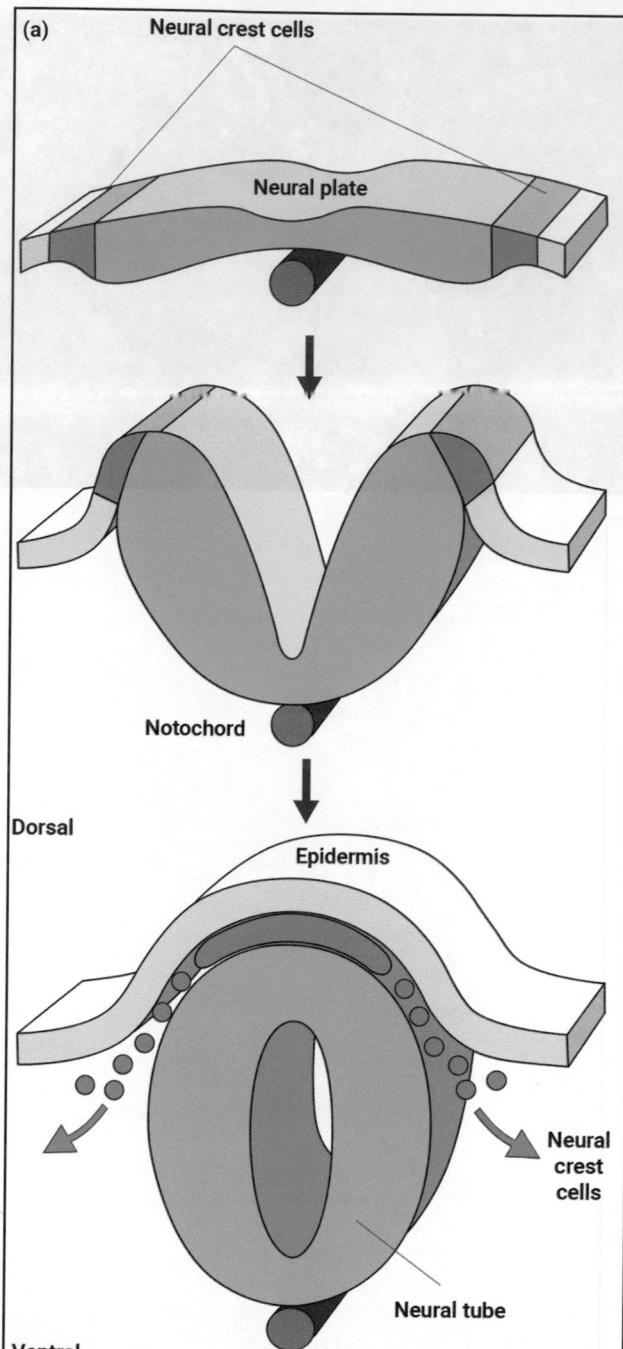

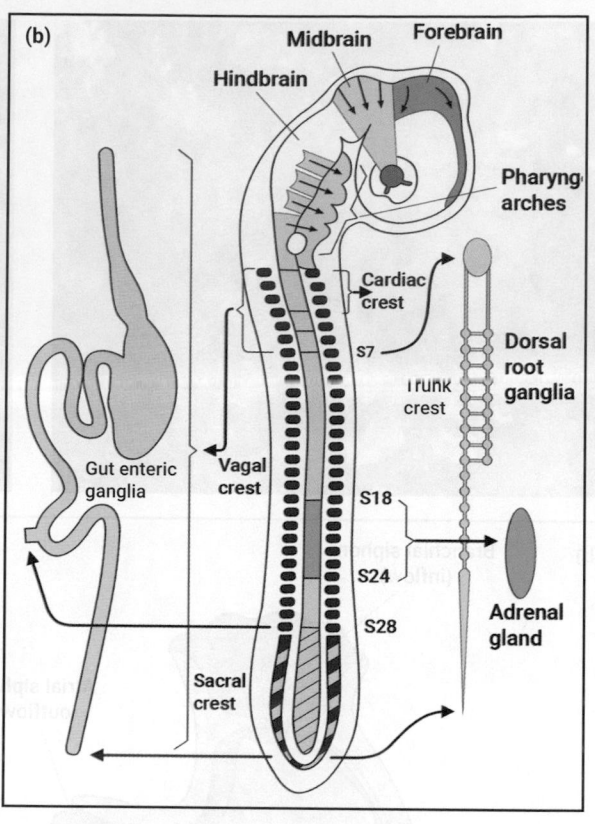

Figure 29.35 Formation of the vertebrate neural crest. Earlier stages of vertebrate development are shown in Figure 29.32. (a) Formation of the neural crest from the region above the closed neural tube. Neural crest cells are shown in green. Some of them migrate away from the vicinity of the neural tube. (b) Regions of the chick neural crest and the destinations of cells. The pink blocks are somites and their numbering is used to identify the regions.

Source: (a) Wolpert *et al*. (2015) *Principles of Development*, 5th ed. Oxford University Press. (b) Gilbert S.F. *Developmental Biology*, 10th Edition, (2014). Oxford University Press.

anything other than the simplest kind of neuralian internal communication system, perhaps with a cluster resembling a rudimentary brain. Molecular analyses have detected two brain- and memory-associated enzymes in animals of a broad range of living phyla representing both protostomes and deuterostomes. This suggests a common source of evolutionary development but there are several possible interpretations of the evidence and the exact ancestral relationship between the brains of living animals remains unclear.

Nevertheless the contemporary animal 'brain', even if it had its early form in a common ancestor, must have evolved to reach a high degree of complexity more than once. Even within the relatively closely related groups of vertebrates (fish, amphibians, reptiles, birds, and mammals), 'high intelligence', for example as demonstrated by crows, dolphins, and primates, also seems to have evolved several times.

Molecular analyses also support this interpretation of neuralian evolution if one considers sensory extensions of the brain such as the eye. The eyes of vertebrates and octopus, for example, are functionally and structurally similar, despite deuterostomes and protostomes being separate descendants of an ancestor which probably had a much simpler photoreceptive system.

▶ We learn more about eyes in Chapter 30.

Where do limbs come from?

Bilaterian animals from a wide range of different phyla possess limbs. Legs, arms, and wings facilitate movement (Chapter 32) and permit mobility in all of the major physical environments (on and in land, in water, and in the air). Limbs are put to other uses including sensation, nutrient acquisition, fluid propulsion, thermal regulation, defence, reproduction, communication, social interaction, and the construction of dwellings.

It is particularly interesting that limbs vary so widely in structure, anatomical origin, and number. Figure 29.36 arranges the bilaterians according to their broad taxonomic position and indicates the basic number of limbs in each case.

We see immediately that some groups have limbed species while others, even quite closely related, do not. This suggests that limbs

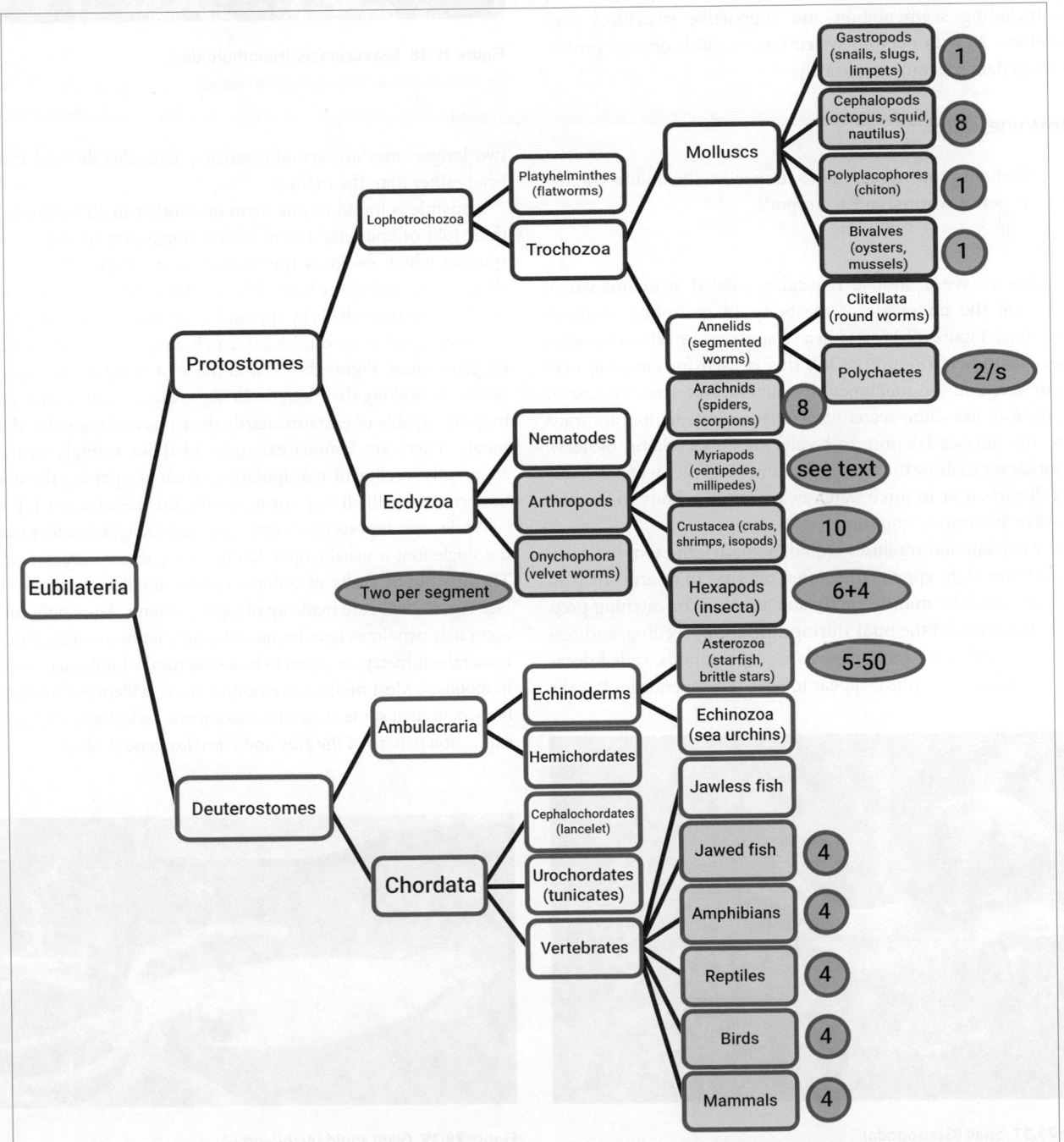

Figure 29.36 A taxonomic arrangement of bilaterian animals with limbs. Shaded groups are those in which all or virtually all species have limbs (see text for definition) or have a body plan in which limbs occur at some stage of development or adult life. Numbers indicate the number of limbs possessed by most representative species. 2/s = two per segment. Insects have six legs plus a basic plan of four wings.

have emerged several times during evolution and in some cases may have been lost. In other words, the presence and absence of limbs are homoplastic features of animal body plans.

Figure 29.36 also causes us to reflect on what we mean by a limb. We include appendages capable of flexural or rotational movement and which absorb tension or support weight at a distance from the main body. Some structures which are principally sensory, such as insect and crustacean antennae, originate as appendages to head segments and acquire limb-like functions.

To understand this great diversity, we need to consider the adaptive advantage of appendages in a range of different environments. This will also bring us an understanding of other aspects of body plans including segmentation and supportive structures such as skeletons. As with nervous systems, we should consider protostomes and deuterostomes separately.

Protostome limbs

Let us talk about protostomes first, and specifically molluscs, annelid worms, velvet worms, and arthropods.

Molluscs

All molluscs have a 'foot', a muscular, ciliated structure which develops on the underside of the body. In gastropod molluscs (snails, slugs; Figure 29.37) and in a separate group which includes the sea cucumbers (Figure 29.38), this is used for creeping over firm surfaces and for attachment by suction. In terrestrial gastropods, the foot has slime-secreting glands to create temporary trails and reduce surface friction. In bivalve molluscs (clams, oysters), the foot serves to draw the two shells tightly together when the animal is disturbed or to force water away from the body to create a propulsive 'swimming' motion.

In the cephalopod molluscs (squid, octopus), the foot has been extended into eight appendages, often referred to as arms or legs. These surround the **mantle cavity** and are used for catching prey, moving water round the body during swimming, feeding, and respiration, as well as for anchorage. Some cephalopods, called decapods (cuttlefish and squid), appear to have ten appendages but the

Figure 29.38 Sea cucumber (Holothuroidea).
Source: Image by Kevin McLoughlin from Pixabay.

two longer ones are actually sensory tentacles derived from the head rather than the mantle.

A mantle is found in one form or another in all molluscs. It is a thick fold of epithelial tissue with a tough cuticle and calcareous spicules which encloses the mantle cavity, between the external shell and the rest of the body. The mantle enables, for example, limpets to grow their shells by spreading calcified chitin along the rim.

Cephalopod arms can be extremely long (up to 8m in the largest giant squid; Figure 29.39) and those of octopus and squid possess suckers along their length. Being muscular rather than jointed, they are capable of extraordinarily dextrous and convoluted movements. There are famous examples of these animals performing apparently intelligent manipulations such as opening the doors of underwater shellfish traps or removing the lids of screw top jars.

While cephalopods have eight main appendages, in other molluscs the single foot is usually operated by four pairs of retractor muscles. The articulated shells of polyplacophore molluscs such as chitons (see Figure 29.69) are made up of eight elements. Thus eight (or more accurately two times four, because the early embryos of molluscs have bilateral symmetry), appears to be a structurally fundamental number in molluscs. Most molluscan embryos show evidence of an eight-way body pattern at some stage of development, reflecting a characteristic expression pattern of the *Hox* and *ParaHox* gene clusters.

Figure 29.37 Snail (Gastropoda).
Source: Image by azeret33 from Pixabay.

Figure 29.39 Giant squid (*Architeuthis*).
Source: Paulo Oliveira/Alamy Stock Photo.

Annelid worms

The annelid worms are a sister group to the molluscs whose bodies are constructed of ring-like segments. Marine polychaetes make up the majority of annelids, even though we may be most familiar with the terrestrial earthworm, *Lumbricus terrestris* (Figure 29.40). The annelid group also includes around 700 species of leeches.

The number of segments in annelids varies considerably, from 20–30 in Maldanidae (bamboo worms, which live in sand tubes), up to 60 in Opheliidae (short, maggot-like worms, sometimes found on sandy beaches), and as many as 350 in Cirratulidae (longer worms which live in mud or rock crevices). Leeches have 102 segments, while earthworms have 100–150 depending on species.

Amongst the annelids, limbs are found only in the polychaete worms and they are arranged on a segmental basis. The polychaete head usually bears antennae and/or **palps** (structures similar to tentacles). Suspension feeding polychaetes use these for trapping particles of food and feeding. Further down, the trunk of the worm normally has a pair of unjointed, fleshy appendages, called **parapodia**, on each segment.

Parapodia are protrusions of the body wall. They are usually **biramous** (see Arthropods), with each lobe containing a cluster of chitinous, bristle-like hairs called **chaetae**. Each parapodium has a set of diagonal muscles which start near the midline on the ventral side and extend out into the structure, as depicted in Figure 29.41. When the worm moves, the parapodia and their chaetae are extended so that the animal can push against the substratum and move or pivot; they are then retracted and removed from the substratum as the worm progresses in an undulatory motion.

Polychaete parapodia perform many functions besides surface locomotion. Some act as paddles enabling the worm to swim and others anchor the worm inside a tube of sand or mud. Those near the head may be adapted for breathing and be sites of ion transport as well as gas exchange. Other parapodia create water currents for feeding and are employed in defence. Touch receptors are concentrated around the head appendages and some of the parapodia.

Velvet worms

Another group of worms, the Onychophora (velvet worms, Figure 29.42) have appendages arranged on a segmental basis, although that is really where the similarity with polychaetes ends. The velvet worms are small group of terrestrial animals which live in tropical and temperate regions of the southern hemisphere. They are ecdyzoa (they shed their skin to grow) and thus a sister group to the arthropods (see Figure 29.36).

Velvet worms live amongst the leaf litter of forests and in other humid environments and are insectivorous, catching their prey in squirts of mucus. They have several curious characteristics, including unusual mating behaviours (in one species the male transfers sperm to the female by inserting his head in her vagina; in another, sperm are liberated from an open wound on the male's body caused by cells released by the female), giving birth to live young, and having, in some species, a gestational structure resembling a mammalian placenta.

Body segmentation in velvet worms is hidden insofar as it can be observed during development or by dissection but is not normally visible at the skin surface. The underlying segmentation dictates the presence of pairs of stubby 'feet', called **lobopods**, along the length of the body. There may be between 13 and 43 pairs, each may be single or duplicated, and many have claws. They are extremely flexible because their extension results from changes in local hydrostatic pressures as the animal bends and transmits forces through its body fluid. They can be retracted by small,

Figure 29.40 Terrestrial earthworm (*Lumbricus terrestris*).

Source: Image by Patricia Maine Degrave from Pixabay.

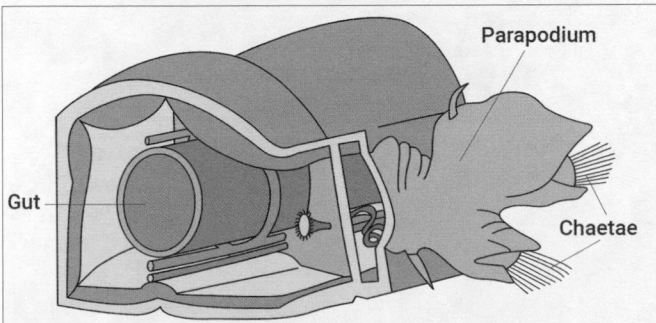

Figure 29.41 Cross sectional structure of a polychaete worm.

Source: © Hans Hillewaert/Wikimedia Commons/CC BY-SA 4.0.

Figure 29.42 Velvet worm (Udeonychophora).

Source: Geoff Gallice/Wikimedia Commons/CC BY-SA 2.0.

individually located muscles attached to tiny segmental structures called ventral organs.

Arthropods

The arthropods, of which there are four major subgroups (see Figure 29.36), are by far the largest group of limbed protostomes. In contrast to molluscs and worms, they have hard, jointed limbs. They first appear in the fossil record in the early-mid Cambrian period, up to 525 million years ago, and are characterized by a segmental body with an exoskeleton. This is a sheddable cuticle made of chitin, formed by the sclerotization or hardening of the external epithelium.

The characteristic limb structure of arthropods may have evolved as a way of increasing the efficiency of swimming or other movements in water. Evolution has allowed this diverse group of animals to colonize virtually all ecosystems, including the air through the development of wings.

Arthropods have paired appendages, associated with the segments of the body and controlled through connections to the animal's ventral nerve cords. During development, the body segments often have morphological and functional specialization (a process called *tagmosis*). The appendages at the anterior end are usually associated with a distinct head and mostly concerned with feeding and sensation. Those of the intermediate segments (the thorax in insects) are usually locomotory while those at the posterior end (abdomen) are excretory and reproductive. However, that basic body plan is extensively adapted, sometime to become scarcely recognizable.

Arthropod limbs are **uniramous** or **biramous**. Uniramous limbs, which are found in the myriapods, arachnids, and hexapods, are made of single lines of segments linked end-to-end rather like a string of beads. Biramous limbs, found in crustaceans and also seen in fossils of the extinct trilobites (also arthropods), comprise a branched pair of end-to-end linked segments.

The biramous condition is thought to be ancestral, with the different arthropod groups separately evolving the uniramous condition. Some species which are uniramous as adults, such as crabs (Figure 29.43), have biramous larval stages. The first antennae of crustaceans are uniramous.

The **arachnid** group of arthropods (spiders and scorpions) have just two main body segments: a posterior abdomen and an anterior cephalothorax. The cephalothorax is embryologically a fusion of the head and thoracic segments and is the location of the animal's appendages. The spider abdomen is a single segment, while that of scorpions is divided into seven subsegments.

Arachnids have four pairs of walking legs, each made of seven segments. They have no antennae but have a pair of prehensile pedipalps, each made of six segments, extending from the mouth region. In spiders, the final pedipalp segment is adapted for sensory or copulatory functions while in scorpions, it is clawed and used to grasp and immobilize prey and to carry out other feeding and defensive operations.

The **myriapods**, which include centipedes and millipedes (Figure 29.44), are all terrestrial animals. They have distinct anterior (head) and posterior segments but the segments in between, which may be numerous, remain distinct rather than fused into a thorax. The head segment is distinctive in these animals because of internal extensions of the hard external cuticle which are unconnected to similar structures in the later segments. This 'swinging tentorum' allows the head to move against the chewing movements of the mandibles, rather than being responsive to mandibular muscles as it is in other arthropods.

In centipedes, there is one pair of limbs per segment. Millipedes appear to have two pairs per segment, but that is really because the segments fuse together in pairs. Across the 13,000 or so myriapod species, the number of segments is highly variable, and leg numbers are estimated to vary from fewer than 10 to more than 750.

Crustaceans and **hexapods** are thought to be closely related arthropod groups (sometimes collectively called *pancrustaceans*). They probably diverged around 500 million years ago, in the Cambrian rather than Precambrian era, and it has been suggested that the hexapods represent an adaptation to terrestrial life rather than the aquatic life of most crustaceans. The branch of hexapods which includes the insects may have emerged even later than this—in the Silurian period around 450 million years ago—as a co-evolution with the appearance of vascular plants and the consequent rise in the oxygenation of the atmosphere.

Figure 29.43 Crab (Brachyura).
Source: Lieutenant Elizabeth Crapo, NOAA Corps.

Figure 29.44 Centipede (Chilopoda).
Source: Image by GLady from Pixabay.

Although some crabs spend periods of their life out of water, the only crustaceans which are fully land adapted are the Oniscidea (woodlice, Figure 29.45) group of isopods. Other isopods live in sea or fresh water and it is likely that the Oniscidea separated from an aquatic ancestor relatively recently—during the Carboniferous period, 300–350 million years ago.

Because crustaceans are such a large and structurally diverse group of animals, including *Daphnia*, barnacles, crayfish, shrimp, lobsters, crabs, woodlice, and many others, it is difficult to summarize their body plans and their limb arrangements. They show the general arthropod pattern of segments, usually 20 in total, grouped into three regions (head, thorax, and abdomen). However, these vary extensively across the species and can be challenging to interpret morphologically, especially in animals like barnacles (Figure 29.46) where the body is almost completely enclosed by exoskeletal plates.

Each segment of a crustacean may have appendages. The head segments produce one or two pairs of antennae as well as feeding appendages (mandibles and maxillae). The appendages from the thoracic segments develop into walking legs and feeding legs. In the largest group of crustaceans, the malacostraca, most animals have five pairs of legs, giving this group the name *decapods*. The anterior pair or pairs may have strong claws and become very large, as in lobsters and some crabs, taking on defensive as well as feeding functions. The appendages of the abdominal segments, called pleopods, may be adapted for rapid swimming (especially the final *uropods* and *telson* which form the fan-shaped tail of crayfish and lobsters), fertilization, and egg brooding, and some even have their own gills.

The hexapod (six-legged) group covers the largest number of arthropod species. They have colonized virtually every terrestrial and air environment, across almost the full range of the planet's temperatures and humidities, and many have developed parasitic and commensal relationships with other animals and even with plants.

As with the crustaceans, we cannot generalize about the diverse body plans represented amongst the hexapods but some broad morphological observations can be made. It is convenient to divide them into two broad groups:

- The *Entognatha* have internal mouthparts and are wingless; they include the Collembola (springtails, Figure 29.47), the Protura (the coneheads, a relatively recently discovered group of eyeless, soil-dwelling creatures), and the Diplura (bristletails; some taxonomists align these more with the insects).

- The *Ectognatha* (insects) have external mouthparts. The vast majority of the 10 million or so species have wings although a number have lost them during their recent evolution or may show them only in one sex or one life cycle stage.

The largest groups of insects are the Coleoptera (beetles), Lepidoptera (butterflies and moths), Diptera (two-winged insects including true flies, midges, gnats, and mosquitoes), Hymenoptera (insects with four, linked, membranous wings, including ants, termites, bees, wasps, and sawflies), Orthoptera (grasshoppers

Figure 29.45 Woodlice (Oniscidea).
Source: © Martin Cooper CC BY 2.0.

Figure 29.46 Barnacle (Cirripedia).
Source: © James St. John CC BY 2.0.

Figure 29.47 Springtails (Collembola).
Source: Donald Hobern Wikimedia Commons/CC BY-SA 2.0.

and locusts), and Hemiptera (true bugs, including aphids, cicadas, shield bugs, and leaf hoppers). This diversity is showcased in Figure 29.48.

Hexapods all show the head–thorax–abdomen body plan very clearly. The head region possesses sensory antennae, except in some of the ground dwelling entognatha. The six locomotory appendages are attached to the thoracic segments. The number is the same in entognatha and ectognatha but may have evolved separately in the two groups.

The wings of ectognatha are also thoracic structures. The basic plan, as seen in lepidoptera and hymenoptera, provides for two pairs of wings. In the diptera, the hind pair are replaced by *halteres*, short, knob-ended structures which vibrate rapidly during flight and provide the animal with stability, much like a gyroscope. In coleoptera, the fore wings are formed into hardened, protective wing cases; these are drawn back to liberate the hind wings as the insect prepares to fly.

Deuterostome limbs

Deuterostome animals divide into two broad groups, Ambulacraria and Chordata. We saw earlier that despite their similar embryological development in terms of the direction of gut formation, they have rather different final body plans. The ambulacraria, as represented by echinoderms, have an early embryonic stage which is bilaterally symmetrical but then approaches radial symmetry, usually around a pentameric arrangement. The chordates retain their bilateral symmetry throughout life. The difference extends to the main organs and structures of the body, including to the nature and formation of appendages.

Figure 29.48 The diversity of insects.

Source: Reprinted from *Current Biology*, 25, Issue 19. Michael S. Engel, Insect evolution, Pages R868–R872, Copyright 2015, with permission from Elsevier.

The ambulacraria

The radial arrangement of the echinoderm body (see Figure 29.31) produces canals with tube feet or podia. These are sucker-like structures which can be extended and retracted by changes in hydrostatic pressure originating in the animal's vascular system. They are used for manipulating objects (including food sources), sensation and defence, attachment to the substrate, and movement across surfaces.

Sea urchins (Echinoidea) have a roughly spherical body shape. The radial structure is apparent in the shell surface, although its underlying bilaterality is discernible. There are no projecting limbs. The podia are distributed over the entire body surface but with greater concentrations around the oral opening.

In brittle stars (Ophiuroidea; see Figure 29.31) and starfish (Asteroidea, Figure 29.49), the radiality is revealed in large projections called arms or rays. They become very long in some species and extremely fat and fleshy in others. There are usually but not always five of these: in species such as sun stars, they are paired and in individuals of all species there may be an irregular number as a result of regeneration following loss or damage.

The podia are arranged in lines along the undersides of the arms. The arms have a limited ability to flex and can bend around rocks and other surfaces, using podial suction to maintain attachment. This action is used to engulf and open the shells of food sources such as bivalve and gastropod molluscs and to orientate the mouth of the animal over other digestible material on the sea bed. Arm movement supported by podial suction also facilitates locomotion, so the arms have the function of limbs and are included in Figure 29.36.

The chordates

As we have seen, the chordate body plan is characterized by bilateral symmetry but also by the development of a notochord and a concentration of nerves at the head end. The vast majority of chordates are *vertebrates*; they have a column of hard tissue enclosing the notochord and other parts of the central nervous system. As well as physically protecting the neural tissues, this structure provides mechanical support for musculature and internal organs. In man and a few other primates, it supports upright posture and bipedalism.

We are used to vertebrates possessing jaws but not all of them do. The jawed vertebrates, called **gnathostomes** (the 'g' is silent), include amphibians, reptiles, birds, mammals, and most fish. A few types of fish (lampreys, hagfish, and some of their extinct relatives known from fossil evidence) do not have jaws and are called the **agnathans** (the 'ag' is not silent).

Gnathostomes, besides possessing jaws, have two pairs of appendages. They occur along the anterior–posterior axis of the animal in locations which have been highly conserved during evolution: the pectoral ('chest') and pelvic positions. Each pair has a supporting girdle of hard tissue with complex, rotationally flexible joints.

These appendages (arms, legs, wings) and their girdles are obvious features of tetrapod (sometimes called quadruped, especially for terrestrial species) vertebrates (amphibians, reptiles, birds, and mammals). Less obviously, the lateral fins of jawed fish are arranged in exactly the same way. Fish fins and tetrapod limbs are described as **homologous**: they resemble each other because of common evolutionary descent rather than convergence.

Agnathans are bony fish with gills but they lack pectoral and pelvic girdles and have no lateral fins. It is thought that the living agnathans, called cyclostomes, come from an early branching of the vertebrate line in the early Cambrian period, around 520 million years ago. In a subsequent evolutionary development, some agnathans such as hagfish (Figure 29.50) seem to have lost their vertebrae.

In gnathostomes, study of developmental molecular markers shows that the pectoral and pelvic appendages have **serial homology**. In other words, their development is genetically regulated in the same way at the two locations. Genes of the *T-box* family, expressed in embryo mesodermal cells, code for the production of

Figure 29.49 Starfish (Asteroidea).

Source: Image by Sophia Hilmar from Pixabay.

Figure 29.50 Hagfish (Myxini).

Source: Mark Conlin/Alamy Stock Photo.

a local hormone called fibroblast growth factor. This initiates cell division and the formation of limb buds.

Other *T-box* genes, together with genes from the *Hox* family (Chapter 28), control the subsequent patterning of limbs including the formation of bone segments and digits. These genes have become inactivated by mutations in limbless tetrapods, including snakes (Figure 29.51) and the group of amphibians represented by the slow worm (blind worm) *Anguis fragilis* (Figure 29.52).

Digits

We have seen that vertebrates have a basic tetrapod body plan, even if this may not be immediately obvious in some groups. Another common morphological feature of vertebrates is **pentadactyly**, the presence of five digits at the end of each limb. These too may be hidden or cryptic in some groups.

Evolutionary studies suggest that pentadactyly dates to the Devonian period, about 350 million years ago. Before that, tetrapods probably had six or seven digits and *their* ancestors may have had eight. We consider five to be the normal number of digits in living vertebrates but there are exceptions. For example, **perissodactyl** (odd toed) ungulate (hoofed) mammals usually do not have this number: modern horses have just one toe on each foot, rhinoceros have three, and tapirs have four on the front feet and three on the hind feet. Thus evolution has continued to play with the number.

The presence of more than five digits in a living animal usually implies a mutant or excessive inbreeding. However there is evidence that structures such as the claws of clawed frogs (*Xenopus tropicalis*, Figure 29.53) and moles (Figure 29.54) represent sixth digits. Figure 29.55 shows how elephants have six digital bones,

Figure 29.51 Snake (Serpentes).
Source: Image by Jarkko Mänty from Pixabay.

Figure 29.53 Clawed frog (*Xenopus tropicalis*).
Source: blickwinkel/Alamy Stock Photo.

Figure 29.52 Slow worm (or blind worm) (*Anguis fragilis*).
Source: Bernard Dupont/Flickr/CC BY-SA 2.0 https://creativecommons.org/licenses/by-sa/2.0/.

Figure 29.54 Mole (*Talpa*).
Source: Image by Dirk (Beeki®) Schumacher from Pixabay.

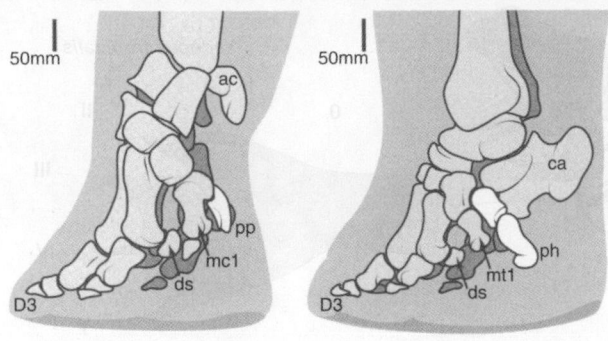

Figure 29.55 Anatomy of the elephant foot, showing a medial view of right feet. The manus is on the left, the pes is on the right. ac, accessorium (pisiform); ca, calcaneus; D3, third digit; ds, digital sesamoid(s); mc1, metacarpal I; mt, metatarsal I; ph, prehallux; pp, prepollex.

Source: Figure 1 from John R. Hutchinson, et al. (2011). From flat foot to fat foot: Structure, ontogeny, function, and evolution of elephant 'sixth toes'. *Science*, 334, 1699. Copyright © 2011, The American Association for the Advancement of Science.

Figure 29.56 Panda (*Ailuropoda melanoleuca*).
Source: Image by Jason Goh from Pixabay.

even though a smaller number (three to five according to species and foot) may be externally represented as toes. Pandas (Figure 29.56) grasp bamboo shoots using a sixth digit derived from a wrist bone called the radial sesamoid; this structure resembles an opposable thumb in its position and function.

These kinds of descriptions are interpretations. They depend on anatomical evidence and on the parallels we draw between the animals around us and the limited fossil evidence of their ancestors.

Developmental biologists have found that the digits of tetrapods mostly develop in a **postaxial** manner; that is to say, from the little finger towards the thumb. The conservation of this postaxial development across tetrapods is illustrated in Figure 29.57. There are exceptions to this, however, especially amongst certain amphibians (salamanders) and reptiles (lizards) whose digits develop **preaxially** (thumb towards little finger). One explanation is that ancestral tetrapods were anatomically ambiguous in this regard, although the picture is far from clear. Nevertheless, this may help us to understand animal variation and may underlie familiar irregularities such as the odd direction of the primate thumb in comparison to the fingers.

Wings

The wings of birds and flying mammals (bats) are evolutionary developments of vertebrate limbs and their digits which enable flight. They are *functionally* similar structures but they evolved separately, from a non-winged ancestor.

It is now widely accepted that birds are contemporary representatives of the animal group which included the terrestrial dinosaurs. The discovery of the fossil of a winged and feathered dinosaur, *Archaeopteryx*, in 1861 provided the first evidence of the dinosaur–bird ancestral link, and a great deal more fossil evidence has emerged subsequently.

Archaeopteryx lived about 150 million years ago. It lacked the deep sternum (keel) which supports the wing muscles in living birds and if it flew at all it probably did so inefficiently. Nevertheless, it suggests that flight-capable wings, feathers, and other anatomical features may have been present in its common ancestor with the birds, a representative of a subgroup of dinosaurs called the Aviale.

Interestingly, *Archaeopteryx* possessed teeth; no living birds have teeth and molecular genetic evidence suggests that the genes for dentition became switched off in the Aviale about 166 million years ago. Genetic analyses also show that a great diversification of the bird lineage occurred after the mass extinction event which killed off the terrestrial dinosaurs 66 million years ago. More than 10,000 avian species are known today.

Bird wings enable flight by creating a large surface over which air can flow to produce lift (see Chapter 31). In terms of general tetrapod anatomy, it is the bones of the forelimbs (humerus, radius, ulna) and some of the digits which have become extended, as illustrated in Figure 29.58. Feathers attached to the skin stretched over these further extend the aerofoil surface. There are also some adaptations of the final vertebrae to support tail feathers which have aerodynamic importance.

Bats (Figure 29.59) are the only mammals which fly by active propulsion over significant distances. A few other mammals, including sugar gliders and flying squirrels, glide for short distances between trees and food sources but are otherwise terrestrial or arboreal. Bats fall into two major groups: the megachiroptera or Old World fruit bats (186 species) and the microchiroptera or insectivorous bats (about 820 species). Bats evolved much more recently than birds, probably emerging in the Eocene, between 50 and 55 million years ago.

Bats create an aerofoil surface on each side of the body using a thin membrane of skin called the **patagium**. This is stretched out

Figure 29.57 Digital parallelism amongst vertebrates. The hindlimb digits of *Xenopus tropicalis* (a) and their likely relationships to the pectoral fin of lungfish, and limbs of *Ichthyostega* and mouse (b). The three anterior digits in *Ichthyostega* (in red) and the thumb of mouse probably derived from structures on the pre-axial side of ancestral fins. Digit 0 of the *Xenopus* hindlimb remains as a protruding claw.

Source: Woltering, J.M., Meyer, A. (2015) *Frontiers in Zoology*, 12, 23.

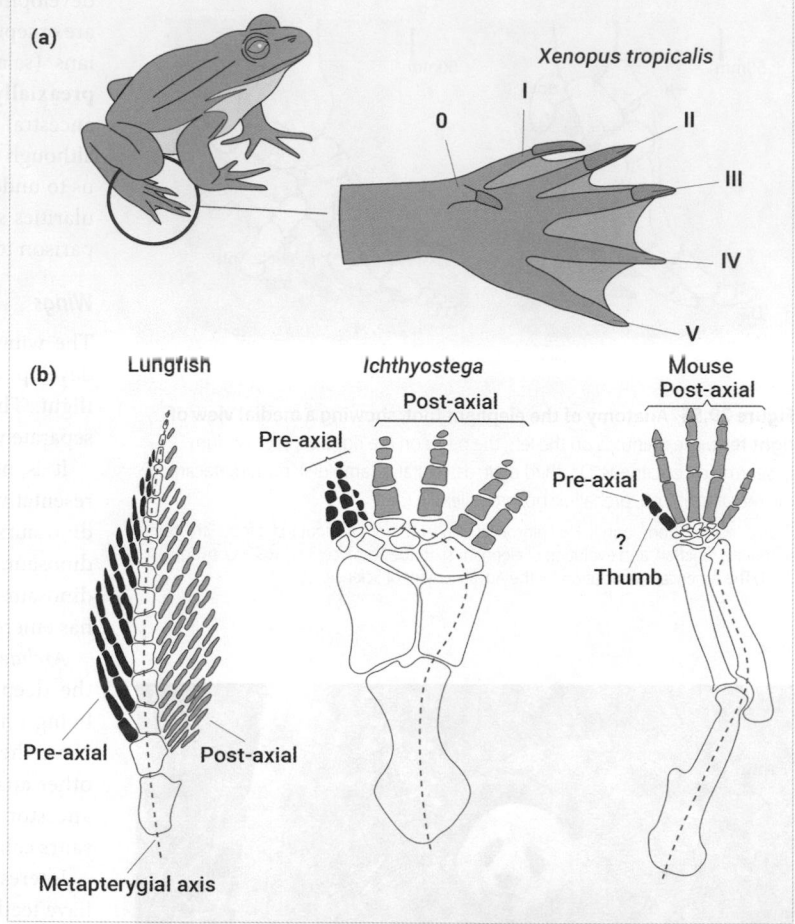

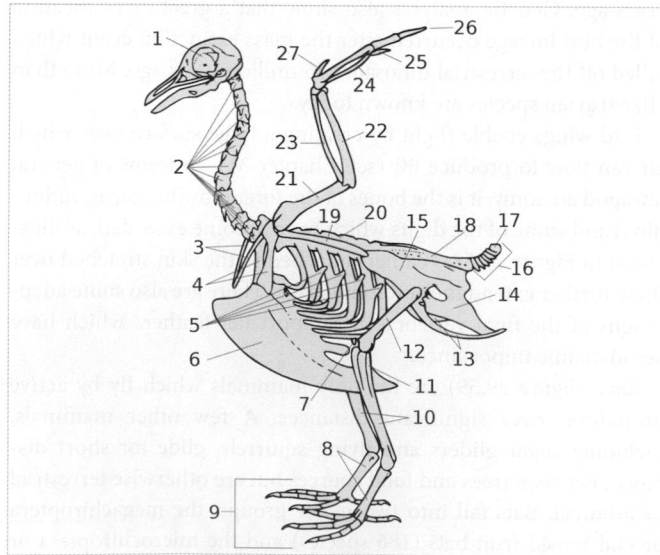

Figure 29.58 The bones of the avian skeleton. 1. Skull; 2. Cervical vertebrae; 3. Furcula; 4. Coracoid; 5. Uncinate process; 6. Keel; 7. Patella; 8. Tarsometatarsus; 9. Digits; 10, 11. Tibiotarsus; 12. Femur; 13. Ischium; 14. Pubis; 15. Illium; 16. Caudal vertebrae; 17. Pygostyle; 18. Synsacrum; 19. Scapula; 20. Lumbar vertebrae; 21. Humerus; 22. Ulna; 23. Radius; 24. Carpus; 25. Metacarpus; 26. Digits; 27. Alula.

Source: BIODIDAC/Mario modesto/Wiki Commons/CC BY 2.5.

Figure 29.59 Bat (Chiroptera).
Source: Shutterstock.

over four, highly elongated finger digits of the forelimb and then back to the hindlimb. In many species it also extends to the tail region. The humerus and radius of the forelimbs are also lengthened but the ulna and tibia of the hindlimb are reduced in length, presumably as a weight-saving adaptation for flight. The thumb digit is usually separate from the patagium. It has a claw, as do the digits of the hindlimb, enabling the animal to cling, hang, and crawl.

 Go to the e-book to complete an activity based on Figure 29.32.

 Check your understanding of the concepts covered in this section by answering the questions in the e-book.

29.4 Hard materials

Organisms representing virtually all forms of eukaryotic life develop hard body materials. With the exception of bone, these are essentially non-living tissues. They result either from the deposition of secretions, the layering of dead cells, or from the precipitation and accumulation of minerals.

Their obvious functions include support against gravity, resistance to air and water movements, protection from injury and predation, the facilitation of movement, and the acquisition of prey, food, and nutrients. They may also provide a reservoir of ions, where these are required for cellular and biochemical processes.

Hard materials in plants

Plants make their hardest materials from wood. The toughest woods are extremely resilient and can support immense weights. The Janka hardness test measures the force required to make an indentation in wood with a steel ball. It is useful for determining the best variety for particular purposes such as flooring. Ironwoods, such as Australian buloke, can take a force of over 22,000 Newtons (N). English oak takes 5000 N, while balsa can be dented by less than 450 N.

Other plants, including grasses, develop fibrous tissues with great tensile strength, the most extreme examples being found in the stems of bamboo plants. As well as being resistant to compression or extension, all these materials have a degree of flexibility which permits bending rather than breakage in the face of wind or mechanical distortion by animal movement.

Although mineralized structures are not generally found in plants, silicates accumulate in various plant parts and provide some strength. They are found in the leaves of monocotyledonous plants (grasses), some of which are extremely long. It explains the sharp edges of the leaves of some grasses and it may have a protective/defensive as well as supportive function.

The stems of some cereal plants such as wheat may become brittle under certain conditions of soil and nutrition, leaving them prone to 'lodging' in strong wind. The stem undergoes an irreversible bend and fails to support the weight of the ear. This prevents further maturation of the grain and makes harvesting difficult. Lodging is partly a genetic trait, so plant breeders have produced varieties which are less susceptible to it.

Hard materials in animals

Multicellular animals make hard materials from external cellular secretions, often coupled with minerals from the environment, and by depositing mineral onto internally secreted matrices (scaffolding or framework) of protein. The resulting structures are used for support, movement, physical protection, defence, and the acquisition of nutrients.

The two principal externally secreted materials are *chitin* and *keratin*:

- Chitin is a long chain polymer of polysaccharides, with a structure resembling that of cellulose. It is the main hard material in protostomes, forming a variety of carapaces, shells, exoskeletons, claws, mouthparts, sensory hairs, and wings. In crustaceans, it becomes mineralized. Chitin is the main component of fish scales and it is also produced in the cell walls of some algae and fungi, as depicted in Figure 29.60.

- Keratin is a fibrous protein, secreted by the epithelial cells of deuterostomes. It is the main constituent of hair, cornified skin (layers of dead skin cells), feathers, quills, nails, claws, beaks, hoofs, scales, and the body plates of chelonian reptiles (tortoises, turtles, terrapins). The horns of rhinoceros, which are illegally traded for inexplicably large sums of money, are made solely of keratin, as are the baleen plates with which mysticetian whales filter plankton from the sea.

Figure 29.60 The bracket fungus *Polyporus squamosus* (dryad's saddle) infesting the trunk of an ash tree (Carrog, Wales). The solid structures of the fungus which project from the tree contain chitin.

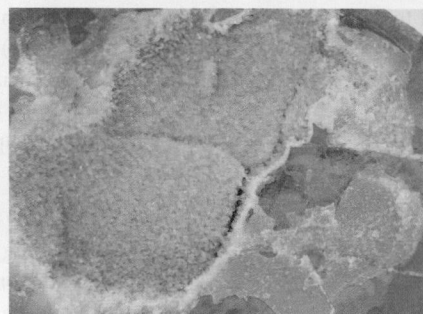

Figure 29.61 Calcite crystals in the centre of a septarian nodule, collected from cliffs of blue lias (Lyme Regis, UK). The object was probably formed 195–200 million years ago from the bodies of animals with calcareous shells, trapped in mud and subjected to heat and pressure. As the nodule subsequently solidified, the calcium carbonate recrystallized into calcite.

Biomineralization

The skeletons of coral, the shells of molluscs, the exoskeletons of crustaceans, the endoskeletons of echinoderms, and the bones, horns, teeth, and tusks of vertebrates all result from a process of **biomineralization** in which inorganic ions interact with organic matrices. The shells of birds' eggs are also biomineralized (see Chapter 33).

Calcium makes up 4.2% of the mass of Earth's crust and is the fifth most abundant element (after oxygen, silicon, aluminium, and iron), so it is a readily available resource. The principal calcium salts exploited by animals for biomineralization are carbonate and phosphates. Sulphates and silicates also occur.

Calcium's most common salt, carbonate, is soluble at low (acidic) pH but virtually insoluble in cold water at neutral and alkaline pH. This makes it ideal for making structures such as shells, provided animals have a means of absorbing, transporting, and precipitating it. The calcium phosphates of bone are more complex chemicals. They exist in a hydrated state and require complex cellular activity for deposition and reabsorption.

Calcium carbonate ($CaCO_3$) adopts two crystalline forms in biological systems: aragonite and calcite. In both cases individual crystals have eight corners and six faces, as would a square cube, but the symmetries of the shapes are different:

- In *aragonite* crystals, adjacent faces meet at right angles but the cube is lengthened in one or both of its linear directions (like a matchbox).

- In *calcite* crystals (described as trigonal-rhombohedral) the cube can be imagined as having been stretched along one of its diagonals and shortened along the other, so that the angles between the faces become 120° and 60° rather than 90°.

In both cases the crystals form columns and blocks. Aragonite mineral can form elaborate, flower-like structures, while the angular distortion of the calcite crystals means that they often form pointed structures. Calcite crystals can be found in the centres of septarian nodules, as depicted in Figure 29.61. These are rocks in which animal bodies have been fossilized under pressure within solidified lumps of clay, causing the body minerals to dissolve and slowly recrystallize.

Sea water is rich in inorganic ions, including calcium. Although most calcium salts have poor solubility, the anion and cation concentrations in most parts of the sea are below those necessary for spontaneous precipitation under normal conditions (as discussed in Real World View 29.2). Animals which use calcium carbonate encourage it to precipitate by creating an organic matrix, or scaffold, of mineral-binding protein. The local ion concentrations are increased and deposition occurs. During the process, the matrix proteins become permanently trapped within the mineralized structure.

Shells and exoskeletons

Calcium carbonate is found in the spicules of most sponges, the scales and otoliths (balance organs) of fish, and the coccolith plates which cover coccolithophore plankton cells (Figure 29.62). The exoskeletons of crustaceans, the testa (shells) of echinoderms, and the shells of molluscs, nearly all of which are sea creatures, principally use calcium carbonate as the hard material.

Figure 29.62 Coccolithophore plankton (Coccosphaerales).
Source: Dr David Furness/Wellcome Images CC BY-NC 4.0.

In crustaceans, most of the CaCO₃ precipitate is amorphous (non-crystalline). In molluscs it generally crystallizes as aragonite, but may form calcite in older layers. In echinoderms, sponges, and most other groups it forms calcite.

The amorphous form in crustaceans may reflect the fact that the exoskeleton needs to be replaced regularly as the animal grows: deposition has to be rapid if the animal is to avoid predation and there may be insufficient time for the relatively slow process of crystallization. Crayfish and lobsters retain balls of calcium carbonate in their stomachs, ready to supply calcium to the new exoskeleton as soon as it starts to form.

In molluscs (shell) and echinoderms (test) the hard material grows throughout life, either by repeated layering (easily seen in oyster shell, for example; see Figure 29.23b) or by continuous remodelling (cycles of deposition and removal) by enzymes from the surrounding cells (the mesodermal test of sea urchins, for

---

REAL WORLD VIEW 29.2 — Ocean acidification

Estimates suggest that about a third of the carbon dioxide that human activity has generated since the onset of industrialization (essentially by the generation of energy from fossil fuels) has been absorbed by the world's oceans. The result of this has been an increase in acidity in the upper layers of the ocean of about 0.11 pH unit.

The following equation displays the chemistry of the effect:

$$CO_{2(aq)} + H_2O \leftrightarrow H_2CO_3 \leftrightarrow HCO_3^- + H^+ \leftrightarrow CO_3^{2-} + 2H^+$$

It is predicted that there will be a further fall in the pH of the sea by the year 2100 of 0.3–0.5 units.

Sea-living animals with calcareous skeletons, bodies, and shells are at risk because carbonate salts ($CaCO_3$) have greater solubility under acid conditions. These organisms include molluscs, arthropods, and echinoderms, although the largest amounts of $CaCO_3$ in the ocean are in the bodies of small zooplankton (foraminifera and pteropoda), in marine algae (coccolithophores), in corals, and in the faecal pellets of fish (which drop to the sea floor).

All these organisms precipitate $CaCO_3$ through the following reaction:

$$Ca^{2+} + 2HCO_3^- \rightarrow CaCO_3\downarrow + H_2O + CO_2$$

It is driven predominantly towards the right by the precipitation of the $CaCO_3$. However, under acid conditions, the mineral can redissolve, driving the equation to the left. This is what happens to discarded animal bodies in the lower layers of the ocean, due to the acidity of decaying organic material. Acidification of the upper layers by increasing uptake of CO_2 also drives the equation towards the left, and that is the risk posed by human industrial activity.

One of the greatest effects of sea acidification may be on *cold* water coral reefs, especially those in the North Atlantic. These comprise species such as *Lophelia pertusa* or *Madrepora oculata* whose skeletons are made of aragonite, the softer form of $CaCO_3$. Figure 1 shows why the coral in this region is at particular risk.

In contrast, as seen in Figure 2, the waters of the Mediterranean are warm, especially salty, and relatively high in pH; they can sustain the structure of aragonite more effectively.

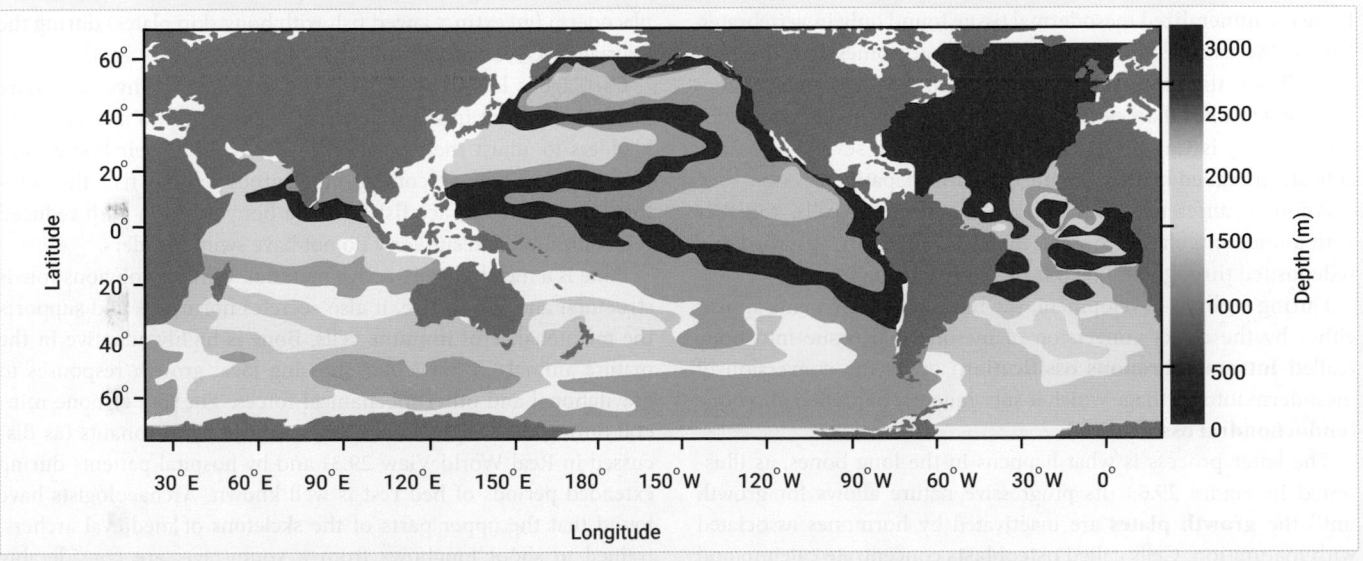

Figure 1 Aragonite saturation depth in the world's oceans. This is the theoretical depth at which, because of pressure, temperature, salinity, the buffering capacity of CO_3^{2-}, and the relative concentration of Ca^{2+}, solid aragonite is in thermodynamic equilibrium with its constituent ions. The greater the saturation depth the less stable aragonite in animal bodies will be at higher levels of the ocean.

Source: Reprinted with permission from Spectrophotometric measurements of the carbonate ion concentration: Aragonite saturation states in the Mediterranean Sea and Atlantic Ocean. Noelia M. Fajar, Maribel I. García-Ibáñez, Henar SanLeón-Bartolomé, Marta Álvarez, and Fiz F. Pérez. *Environmental Science & Technology*, 2015,49(19), 11679–87. Copyright 2015 American Chemical Society.

Figure 2 Studies of geochemical parameters in the Atlantic Ocean and Mediterranean Sea. The routes and dates of four oceanographic surveys are shown with their range of measurements. The inset table shows the measured values for salinity (g/kg), pH, aragonite saturation depth (AT; see Figure 1), and carbonate concentration (g/kg). SG: Strait of Gibraltar; SC: Strait of Sicily; DWBC: Deep Western Boundary Current.

Source: Reprinted with permission from Spectrophotometric measurements of the carbonate ion concentration: Aragonite saturation states in the Mediterranean Sea and Atlantic Ocean. Noelia M. Fajar, Maribel I. García-Ibáñez, Henar SanLeón-Bartolomé, Marta Álvarez, and Fiz F. Pérez. *Environmental Science & Technology*, 2015,49(19), 11679–87. Copyright 2015 American Chemical Society.

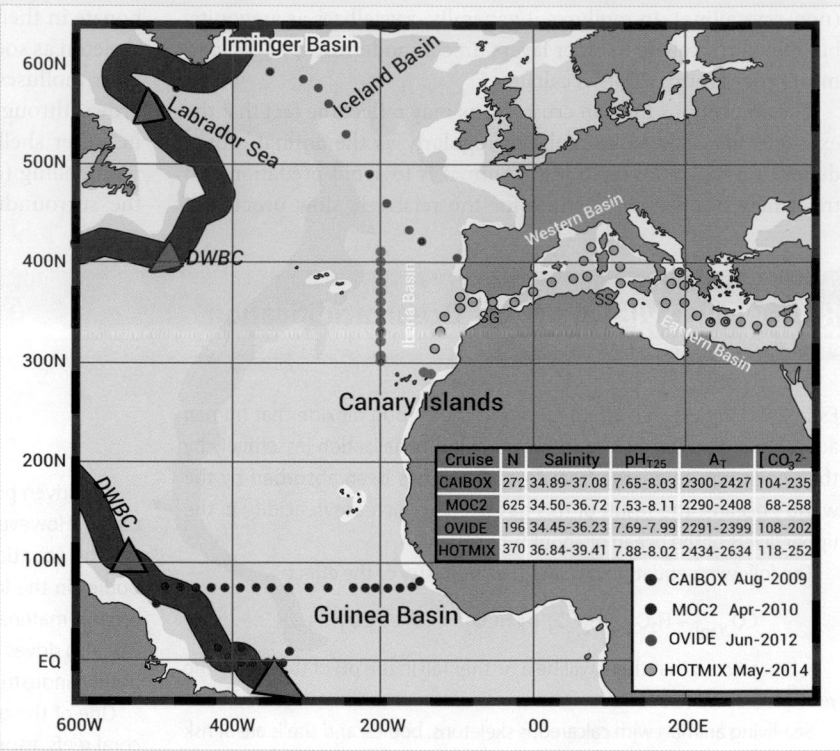

Cruise	N	Salinity	pH$_{T25}$	A$_T$	[CO$_3^{2-}$]
CAIBOX	272	34.89-37.08	7.65-8.03	2300-2427	104-235
MOC2	625	34.50-36.72	7.53-8.11	2290-2408	68-258
OVIDE	196	34.45-36.23	7.69-7.99	2291-2399	108-202
HOTMIX	370	36.84-39.41	7.88-8.02	2434-2634	118-252

● CAIBOX Aug-2009
● MOC2   Apr-2010
● OVIDE   Jun-2012
○ HOTMIX May-2014

example). These processes are slower and allow for crystallization. The mineral of sea urchin (echinoderm) spines, which are regenerable, defensive structures, starts in the amorphous form and gradually crystallizes to calcite.

Bones and cartilage

Bone is a mineralized mesodermal tissue found only in vertebrates. Unlike the exoskeletons and shells of protostomes and the tests of molluscs, the mineral component of bone is hydrated calcium phosphate rather than calcium carbonate. The general formula for this material is $Ca_{10}(PO_4)_6(OH)_2$ and there are several isoforms, collectively called hydroxyapatite or hydroxylapatite.

Another difference is that bone incorporates cells, together with a microcirculatory system. Bone is continually absorbed and redeposited throughout life; it is therefore a living tissue.

During embryo development, the first skeletal elements appear either by the direct conversion of mesodermal tissue into bone (called **intramembranous ossification**) or by the conversion of mesoderm into cartilage which is subsequently hardened into bone (**endochondral ossification**).

The latter process is what happens in the long bones, as illustrated in Figure 29.63. Its progressive nature allows for growth until the **growth plates** are inactivated by hormones associated with maturation. Cells called **osteoblasts** concentrate calcium and phosphate ions from the blood so that they precipitate on to the collagen of the cartilaginous matrix.

Cartilage is a hard, moderately flexible material made of cells called chondrocytes embedded in a matrix of proteoglycan (a macromolecule made of polysaccharide and protein) and elastin (a flexible, stretchable protein). In the joint-facing surfaces of

bones, parts of the ribs, the rings which keep the trachea open, the nasal septum, and ear pinnae, the cartilage remains non-ossified.

One group of vertebrates, the chondrichthyan fishes (sharks, rays, and skates) have skeletons made entirely of cartilage rather than bone. This group and their sister group of bony fish (the teleosts) are thought to have diverged from a common ancestral placoderm (an extinct jawed fish with bony skin plates) during the Silurian or Devonian period (400–440 million years ago).

Cartilage is less dense than bone, so chondrichthyes are more neutrally buoyant in sea water than teleosts and do not use swim bladders to adjust their position in the water. (Their bodies also tend to have a high fat content.) One group of bony fish, the notothenioids (Antarctic ice fish), have a bony skeleton with reduced mineralization and similarly do not have swim bladders.

Bone is a metabolically active material. While its obvious role is structural and supportive, it also secretes hormones and supports the maintenance of immune cells. Bone is highly adaptive in the mature animal, in particular showing local growth responses to gravitational and other mechanical forces. The loss of bone mineral during the weightlessness experienced by astronauts (as discussed in Real World View 29.3) and by hospital patients during extended periods of bed rest is well known. Archaeologists have found that the upper parts of the skeletons of medieval archers, trained to shoot longbows from a young age, are considerably thicker and heavier on the string drawing side.

Importantly, the skeleton also acts as a dynamic store of calcium. In addition to the osteoblasts which deposit mineral, bone contains osteoclasts and other types of cells which are able to release calcium and phosphate back into the circulation. All these cells are under close hormonal control, such that the blood calcium concentration

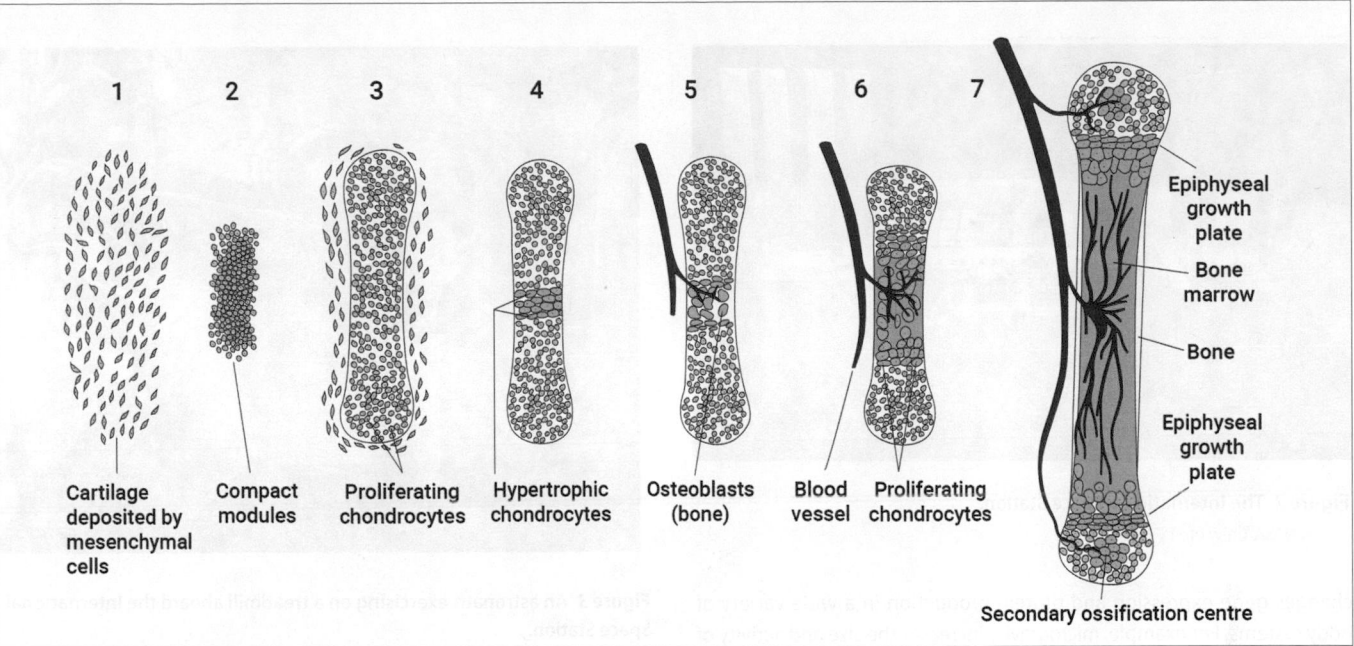

Figure 29.63 Stages in the mineralization and maturation of vertebrate bone.

Source: Gilbert SF. *Developmental Biology*, 10th Edition, (2014). Oxford University Press.

REAL WORLD VIEW 29.3   Bone density and gravity

It is well known that astronauts are prone to loss of bone mineral during their periods of weightlessness, because of the reduced gravitational stress on their skeletons. How big is the effect and how long does it last?

The chart in Figure 1 shows bone mineral density (BMD) in eight astronauts (seven male, one female; average age 45 years), who spent between 23 and 28 weeks on the International Space Station (see Figure 2).

Measurements were made on their spines and hips before they went into space, on their return to Earth, and 1 year and 2–4.5 years later.

Results (mean ± SEM) are expressed as a percentage of the pre-flight density. Trabecular bone is that part of the bone which is spongy and able to change rapidly in response to the mechanical forces transmitted by muscles.

Each astronaut undertook periods of exercise, including treadmill running and cycling, on 3–5 days per week while aboard the space station (Figure 3).

The graph suggests that the astronauts lost up to 15% of their skeletal mass despite initial health screening, exercise, and carefully controlled nutrition.

The data also show that the effects can be long lasting: pre-flight bone density may not have returned even 4.5 years after the flight. This has implications for the long-term health of astronauts, especially their susceptibility to injuries such as bone compression and fractures. It also suggests that particular care needs to be taken with astronauts who spend extended periods of time in space or who make repeated journeys.

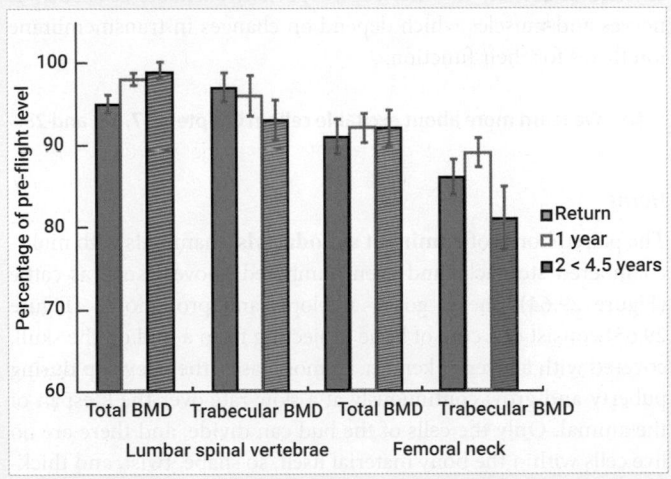

Figure 1   Bone mineral density (BMD) in eight astronauts after time spent on the International Space Station.

Source: Reprinted from Acta Astronautica 67 issues 1–2, Carpenter et al., Long-term changes in the density and structure of the human hip and spine after long-duration spaceflight 71–81. Copyright 2010, with permission from Elsevier.

Although the link between lack of physical stress on the bones and the loss of mineral density seems clear, its exact mechanism remains something of a mystery. Experiments on zebra fish, mice, and cultures of animal cells which have been flown in space suggest that microgravity

Figure 2 The International Space Station.

Source: NASA/Crew of STS-132.

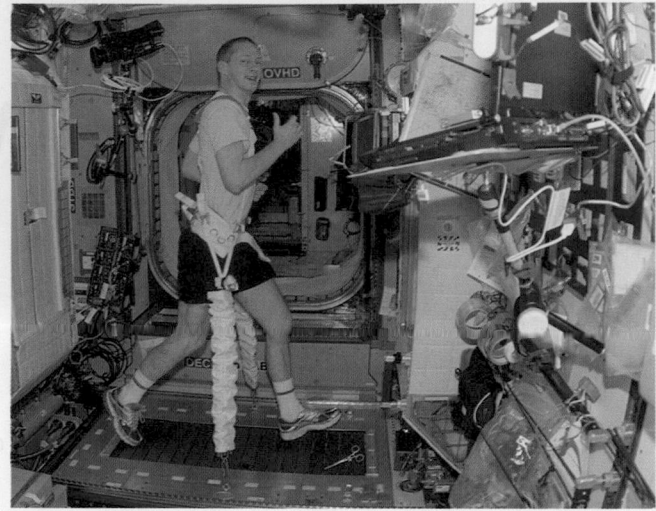

Figure 3 An astronaut exercising on a treadmill aboard the International Space Station.

Source: NASA.

changes gene expression and protein production in a wide variety of body systems. For example, microgravity increases the size and activity of osteoclasts, the hormonally regulated bone cells which during normal life on Earth allow calcium phosphate to be released from bone (for example during skeletal growth and in response to falls in blood calcium concentration). In the zebra fish, osteoclast mitochondria respond to changes in the expression of genes which control their function. Other studies show that microgravity reduces the differentiation of the mesenchymal stem cells from which bone cells are derived, acting through the cytoskeleton and through signalling molecules such as cytokines.

is held constant within very fine limits. This is crucial to the maintenance of activity in other cells, especially excitable cells such as nerves and muscles which depend on changes in transmembrane ion fluxes for their function.

▶ We learn more about excitable cells in Chapters 17, 18, and 28.

Horns

The paired horns of **ruminant artiodactyls** (mammals with multi-chambered stomachs and even-numbered hooves) such as cattle (Figure 29.64), sheep, goats, antelope, and pronghorns (Figure 29.65) consist of a core of bone projecting from a bud on the skull, covered with a layer of keratin. In most cases, they develop during puberty and grow continuously at a slow rate over the lifespan of the animal. Only the cells of the bud can divide, and there are no live cells within the bony material itself, so shape, twist, and thickness are entirely determined at the growth point near the skull. There is a good deal of sexual dimorphism and breed variation in the presence, size, and shape of these horns, especially where human domestication has been influential.

The antlers of deer (Cervidae, Figure 29.66) are not considered true 'horns' because they are deciduous (lost and regrown each year), often branched, and lack a keratin covering. The structure is very similar to bone. In the early stages of growth, the hard material is covered by the skin of the head, called 'velvet'. This supplies the antler with blood and nerves and enables growth

Figure 29.64 Cow (*Bos taurus*).

throughout the structure. The relatively soft, spongy bone deposited initially becomes harder and more compact as the mineralized matrix is remodelled.

Later in the annual cycle the velvet is lost and development of the antler ceases. This often coincides with the period of rut in which there is aggressive competition between males for status, territory, and breeding rights over females.

Figure 29.65 Pronghorn (*Antilocapra americana*).
Source: Image by David Mark from Pixabay.

Figure 29.66 Deer (Cervidae).
Source: Antranias Zimmer from Pixabay.

PAUSE AND THINK

(a) Calcium salts are often the basis of hard materials in animals. What are the main structural chemicals found in the following?

- Bone
- Antler
- Horn
- Nails and claws
- Cartilage
- Molluscan shell
- Echinoderm test
- Crustacean exoskeleton
- Insect exoskeleton

(b) What are the main hard materials in plants?

Answers:

(a)
- Bone: Hydrated calcium phosphate on a collagen matrix
- Antler: Hydrated calcium phosphate on a collagen matrix
- Horn: Keratin
- Nails and claws: Keratin
- Cartilage: Collagen and proteoglycan
- Molluscan shell: Calcium carbonate
- Echinoderm test: Calcium carbonate
- Crustacean exoskeleton: Calcium carbonate
- Insect exoskeleton: Chitin

(b) Lignin and suberin.

Very hard materials

The enamel of teeth and tusks (ivory, Figure 29.67) is, like bone, made of hydrated calcium phosphate (hydroxyapatite). This material is amongst the hardest known in the biological world, with a value of 5 on the Mohs scale of hardness. For comparison, pure calcite ($CaCO_3$) is Mohs 3, talc (magnesium silicate) is Mohs 1, and diamond is Mohs 10.

One important difference from bone is that tooth mineral is not laid down in collagen; other proteins provide the matrix for deposition but disappear once the mineralization has taken place, resulting in a much denser structure. Enamel also lacks cells and capillaries, so it can be considered a dead tissue once the final size and shape of the growing tooth has been determined. Tooth enamel can be eroded under acid conditions (produced by oral bacteria or acidic foodstuffs) but it is not naturally replaced from within the tooth itself. Artificial remineralization can be achieved by external application of calcium- or fluoride-rich materials to the surface, for example in toothpaste.

Figure 29.67 The tusks of an elephant are made of ivory, which is similar to bone, and is made of hydrated calcium phosphate (hydroxyapatite).
Source: Image by Les Bohlen from Pixabay.

Figure 29.68 Common limpet (*Patella vulgata*).

Source: Stefan Thiesen (wikimedia nick 'Buntrabe')/Wikimedia Commons/CC BY-SA 3.0.

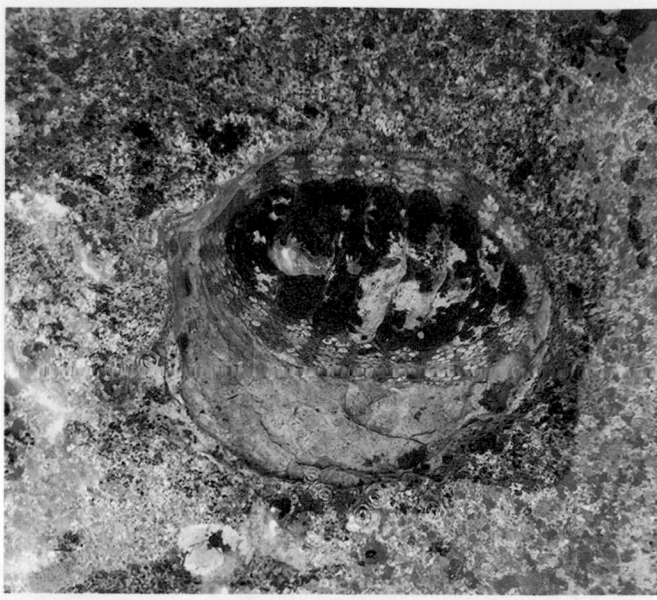

Figure 29.69 Rock erosion by a mollusc. This snakeskin chiton, *Sypharochiton pelliserpentis*, a polyplacophoran mollusc which lives on intertidal rocks, has eroded a dish-shaped well in the substrate while scraping up food materials (Auckland, New Zealand).

Some mammals, including rodents and lagomorphs (rabbits, hares), have teeth in which the enamel is on the outside rather than on the apical surface. Thus the dentine (the growing surface) is exposed rather than covered. This means that the animal's teeth grow continuously throughout life, with the edges being continually sharpened by abrasion against food materials.

The strongest biological material so far discovered is the tooth, or radula, of the common limpet, *Patella vulgata* (Figure 29.68). This gastropod mollusc attaches itself tightly to rocks by suction and uses its tooth to scrape food materials from the surface. The rock may be ground away in the process, accounting for the small hollows sometimes seen on limpet-rich rocks on the sea shore at low tide. Some marine polyplacophoran molluscs do the same, as depicted in Figure 29.69.

The radula material is goethite, an iron (ferric) oxyhydroxide, embedded in a chitin matrix. It is particularly interesting to structural engineers because its high tensile strength, which is considerably greater than that of spider silk, comes from the linear alignment of goethite nanofibres in the matrix.

 Check your understanding of the concepts covered in this section by answering the questions in the e-book.

SUMMARY OF KEY CONCEPTS

- The structures of living organisms are constrained by physical imperatives and are therefore potentially open to analysis according to mathematical principles.

- Multicellular organisms develop by the replication and differentiation of cells. Cells vary in their capacity for differentiation between organisms and at different stages of life.

- Plant structures (stems, leaves, roots, etc.) emerge at fixed points as the plant grows; they have evolved to be positioned for efficient function but are also influenced by environmental conditions.

- Vascular plant tissues use stomata to exchange gasses with the environment and to regulate the flow of water taken up through the roots.

- Complex animals originated from a bilaterian ancestor and evolved in two distinct branches according to whether the blastopore forms the mouth (protostomes) or anus (deuterostomes).

- Electrically active neural cells are present in all complex animals, usually connected by axons; they become clustered to form integrated systems and are linked to specialized clusters of sensory cells.

- In non-chordate deuterostomes the nervous system is organized as a ring around the body; in chordates the central nervous system is elongated along the body's long axis, usually supported and protected by hard tissue, and with a dominant brain at the anterior end.

- The limbs of protostomes emerge from the underlying segmental structure of the body. Those of deuterostomes are either radially arranged or emerge from structural girdles at two positions along the body's long axis.

- Living organisms exploit hard materials for support, protection, resource acquisition, and social interactions. These are formed from tissue proteins or other macromolecules or by mineralization, usually involving precipitated calcium salts.

 Use the flashcards in the e-book to test your recall of key terms introduced in this chapter.

QUESTIONS

 Looking for answers? Once you've answered these questions, follow the link in the e-book to the answer guidance and check your work.

Concepts and definitions

1. Define allometry and explain how it is used to interpret the body plans of living organisms.

2. What are stomata and where are they found?

3. What is the difference between monocotyledonous and dicotyledonous plants?

4. What distinguishes protostome from deuterostome animals?

5. What is meant by a plane of symmetry in an animal body?

6. What are the principal differences between bone and shell?

Apply the concepts

7. Explain why surface area:volume ratio is such an important concept in understanding the body forms of living organisms.

8. Explain the role of meristems in the development of plant structure and shape.

9. What are lignin and suberin and what role do they play in plant life?

10. What anatomical features define the chordate group of animals?

11. Explain how the brain arises as a prominent feature of the central nervous system of deuterostome animals and compare it with equivalent structures in protostomes.

12. Describe the segmental basis of limb development in different groups of protostome animals.

Beyond the concepts

13. Using internet resources, find a general map of the gill clefts of vertebrates; identify the clefts from which the ear structures of mammals originate and draw a rough sketch of the developmental process (this does not need to be anatomically precise!).

14. Using internet resources, summarize current ideas about the link between the limbs of aquatic vertebrates and those of terrestrial vertebrates; how many digits did the first land vertebrates have?

15. Find some examples of investigations into the land areas covered by tree roots and their leaf canopies. Is there thought to be a general allometric relationship?

16. Find a picture of *Archaeopteryx* and locate some of the anatomical features which suggest to that it was related to both birds and dinosaurs.

17. Flat fish (plaice, sole, flounder, etc.) have developmentally distorted heads with both eyes on the same side. At what stage of development does the distortion appear and what is currently understood about its genetic basis?

18. What do genes of the *Hox* family do? List as many different roles as you can find.

FURTHER READING

Dunn C. et al (2014). Animal phylogeny and its evolutionary implications. *Annu. Rev. Ecol. Evol. Syst.* **45**: 371–95.
A comprehensive survey of contemporary ideas about animal evolution. We are all more similar than we thought.

Liebeskind B.J. et al (2017). Evolution of animal neural systems. *Annu. Rev. Ecol. Evol. Syst.* **48**: 377–98.
How nervous systems came to be; signals, synapses, cells, and genes—a survey across the whole tree of animal life.

Roth G. (2015). Convergent evolution of complex brains and high intelligence. *Phil. Trans. R. Soc. B* **370**: 20150049. http://dx.doi.org/10.1098/rstb.2015.0049
Complex brains, how they evolved, and the link with intelligence (whatever that is).

Palovaara J. et al (2016). Tissue and organ initiation in the plant embryo: a first time for everything. *Annu. Rev. Cell Dev. Biol.* **32**: 47–75.
How big, complex plants develop from a few embryonic cells.

Lawson T. & Vialet-Chabrand S. (2019). Speedy stomata, photosynthesis and plant water use efficiency. *New Phytol.* **221**: 93–8.
How stomata control water and CO_2 movements in plants, fast and slow responses, and the implications for photosynthesis and crop yields.

Chen C.-H. & Poss K.D. (2017). Regeneration genetics. *Annu. Rev. Genet.* **51**: 63–82.
Limb regeneration, why it works or doesn't work in different animals, and the prospects for regenerative medicine.

Interaction with the External Environment

Chapter contents

Introduction   1001

30.1 Environmental pressures
      and evolutionary survival   1001

30.2 Light, geography, and the
      electromagnetic spectrum   1002

30.3 The sound environment   1018

30.4 The gravitational environment   1025

30.5 The magnetic and electrical
      environment   1029

30.6 The thermal environment   1030

30.7 The gaseous environment   1038

30.8 The aqueous environment   1044

Watch the key concepts video in the e-book to prepare yourself for studying this chapter.

LEARNING OBJECTIVES

By the end of this chapter, you should be able to:

- List the environmental variables which influence the survival and reproductive ability of organisms, and explain why they may be referred to as 'imperatives'.

- Describe the principal ways in which plants and animals are influenced by light.

- Describe the basic requirements for a light detection organ in animals, distinguish between photoreception and vision and describe the main stages of eye evolution.

- Describe some of the physiological structures which facilitate awareness of the sound environment.

- Describe the influence of gravity on plants and animals, including the differences between air and water environments.

- List the principal physical components of the thermal environment and be able to explain the connections between kinetic energy, heat, and temperature.

- Explain the concept of a zone of thermal neutrality and be able to draw a diagram relating environmental temperature, animal body temperature, and metabolic rate.

- Describe the basic laws that apply to gases and how they determine the physiology and life styles of animals and plants.

- Explain the concept of a zone of osmotic neutrality and describe, with suitable examples, some mechanisms of osmoregulation in marine and fresh-water organisms.

Introduction

In Chapter 28 we considered the functional units—the organisms—which make up life, including single-celled and multicellular plants and animals. We discovered that when collections of cells operate together they can form units of great complexity, with individual cells or groups of cells taking on specialist roles. In Chapter 29, we looked at the complex body structures which multicellular organisms make and considered some of the likely ancestral relationships between them. In this chapter we consider the fundamental components of the environment which influence life and we look at how organisms detect, react to, and exploit them.

30.1 Environmental pressures and evolutionary survival

Living organisms, their structures, and their life processes have evolved from ancestral organisms by a combination of variation and selection. Variation is located in genes and genetic identities are transmitted from one generation to the next by the mechanisms explained in Module 1. Selection is the process by which some individuals are able to survive and reproduce, while others are not.

The selection 'pressure' on an organism refers to the total effect of all the forces in the environment which threaten its survival and reproduction. These forces include food supply, competition, predation, disease, and physical factors. Animals and plants live or die according to whether their structure and function are compatible with the environment they are in.

It is essential to appreciate that forces of selection act on the *phenotype* of individuals. Genes are not under selection pressure directly. They are selected indirectly because they determine phenotype or permit various phenotypes to occur. This is why understanding the form and function of organisms is of such great importance to biologists. Modern organisms exist precisely because their ancestors had phenotypes which allowed them to tolerate their environmental conditions and survive long enough to reproduce.

Environmental variables

The components of the environment which influence life are called *environmental variables* because they are constantly changing and exert different influences under different circumstances. They are also *imperatives* because they are mostly impossible for an organisms to avoid or ignore. Table 30.1 shows the physical environmental variables which are of potential relevance to all living organisms.

The environmental variables which affect life are physical features of Earth and the cosmological environment in which Earth exists. The variations which organisms experience, and which they must tolerate if they are to survive, are essentially movements of energy in one form or another. In the ultimate analysis they are a legacy of our planet's history in the expanding universe. Light, heat, and sound are the most obvious variables, but subtler forces such as gravity, electricity, and magnetism are also influential.

In addition, organisms respond to biotic factors including nutrients, excreta, conspecifics, competitors, predators, territorial indicators, group density, migratory movements, parasites, and pathogens.

Plants and animals

Plants and animals cohabit Earth and are exposed to the same physical conditions. One important difference between them is their capacity for *active* and *passive* responses to environmental imperatives.

Plants are essentially sessile and are, literally, rooted to the spot for their lifetime—from a few days or weeks to hundreds of years. Unlike animals, they cannot use purposeful, reactive movements to avoid or select particular conditions. Plants may move during reproduction (seed dispersal, for example) or under the influence of water or air currents, but these are essentially passive, uncontrolled changes of location. As a result, the reactions of plants consist of alterations to growth, metabolism, and life cycle (flowering, leaf drop, etc.) rather than changes of location.

Most animals, especially in adult form, have the advantage of being able to make goal-directed movements towards or away from stimuli. They make choices about where they want to be and for how long, and it is usually possible to change the decision if a better alternative emerges. They are less constrained by their position than plants, especially when competing for space, food, or protection.

A second important difference between plants and animals is in the role of growth. Plants grow and develop in different directions, both above and below the soil, and will do so under the influence of a variety of stimuli. They may effectively translocate by this means, for example by spreading shoots or roots laterally across the soil or by 'climbing' a supportive structure towards the light.

But such opportunistic responses by plants are adaptive and limited: they are genetically determined, reflecting how their predecessors were selected according to the success or otherwise of *their* growth responses. Each morphological change is irreversible and leaves a permanent structural trace in the body of the organism.

Animals are no less sensitive to selection but have a much wider repertoire of possible responses to environmental pressures. They do not have to rely on a change of size or shape to make the best of their surroundings. Their main concern is the availability of sufficient metabolic energy in the form of food, together with water and air.

Beware of human interpretation

In considering the effects of environmental variables on animal and plant life we must avoid being limited by the extent of human experience. For example, many organisms detect wavelengths of light beyond our 'visible' spectrum and sound frequencies beyond our 'audible' range. Plants and animals often respond to physical cues to which humans have little direct sensitivity. Many organisms live in environments which humans cannot tolerate, flourishing in circumstances we consider to be extreme.

For these reasons, we need to be careful how we monitor, measure, and interpret the components of the physical environment.

Table 30.1 The physical environmental variables which are of potential relevance to all living organisms

Variable	Context
Light	Components of the electromagnetic spectrum
	Visible and non-visible wavelengths
	Variations in wavelength, intensity, angle of incidence, duration, polarization
Day length	As indicated by duration of light and/or dark periods, temperature, rainfall/humidity, tides, food availability
	Variations with time of year, latitude
Sound and mechanical features	Vibrations and particle movements in ambient fluid (air, water)
	Alterations caused by reflection, refraction, speed, and direction of fluid movement
	Changes of wavelength (frequency modulation) and/or volume (amplitude modulation)
	Direct contact with resistive features of the environment, perceived as pressure
	Wind, touch, tension, compression, sway, shake, rhythm
	Resistance of substrates (soil, sand, mud, rock)
Gravity	Earth's gravitational field
	Tidal movements due to sun and moon
Magnetic and electrical fields	Earth core magnetism
	Potential difference between organism and environment
	Action and membrane potentials
Temperature	Rate of transfer of heat between organism and environment by radiation, conduction, convection, and evaporation
	Variations with time of day, time of year, latitude, type of environment, location, orientation
	Modified by heat generation and evaporative water loss by organism
Atmospheric gas	Air movement, pressure, density
	Component gas partial pressures
Water and humidity	Flow rate, pressure, salinity, osmotic pressure, dissolved gas content, temperature, availability
	Physical state (fluid, ice, snow)
	Water vapour content of the atmosphere
	Variations with time of day, time of year, latitude, type of environment, location, air movement, temperature
	Modifications by heat generation and evaporative water loss by organism
Volatile and non-volatile chemicals and gases	Variations in identity and gradients in concentration
	Smell, taste, and texture
	pH, precipitated minerals, salts, anions, cations, individual elements, heavy metals
	Sugars, proteins, fats, complex organic molecules

We should avoid missing things which are not obvious to us or hard to detect. We also have to avoid limiting our studies to the human timescale: responses to the environment may be very fast or very slow. Human experience is no guide to what might be important for the survival, growth, and reproduction of other organisms.

In each of the following sections, we look first at some of the physics underpinning the most important environmental variables, and then consider how plants and animals interact with them.

 Check your understanding of the concepts covered in this section by answering the questions in the e-book.

30.2 Light, geography, and the electromagnetic spectrum

Light and its seasonal patterns influence most biological processes. Nearly all life that we know of depends on light from the sun, but its influence extends beyond this basic transfer of energy. We can think of there being a 'light climate', a complex and constantly varying set of conditions to which plants and animals are adapted.

It is easy to understand why light has such a large influence on living things. From one point of view, organisms comprise an immense food chain in which individuals compete for a share of the available energy. That energy has to be present in the habitat they occupy, transmitted in a form they can make use of. Inevitably, the success of individuals will be affected by seasonal and other variations in the energy supply.

From another point of view, the biosphere comprises a vast collection of organisms which are constantly growing, reproducing, and competing for limited space on the planet. The space they are fighting for includes access to light, as shown by the dense leaves in a forest canopy in Figure 30.1, a bloom of algae on the surface of a lake shown in Figure 30.2, or a cluster of reptiles basking on an exposed rock, as shown in Figure 30.3.

Figure 30.1 A forest canopy, pictured from below (Forest Research Institute, Kuala Lumpur, Malaysia).

Figure 30.2 An algal bloom, Lake Erie.

Source: NOAA CoastWatch.

Figure 30.3 Marine iguanas (*Amblyrhynchus cristatus*) basking on volcanic rocks, Galapagos Islands.

Source: Elizabeth Downes.

The light climate is the result of several components including **intensity**, **incident angle**, **wavelength**, and **polarization**. In addition, all parts of Earth experience the day–night cycle of illumination, and regions away from the equator experience seasonal distortions to the lengths of day and night over the course of the year. These factors interact to influence the behaviour, activity, reproduction, and survival of organisms.

The intensity of light reaching a particular location (also called **irradiance** or **insolation**) depends on the incident angle (the angle at which sunlight meets Earth's surface), its attenuation by the atmosphere, and the effects of seasonal variation at different latitudes. Figure 30.4 illustrates the global and seasonal variations in sunlight intensity, and explains how these effects interact.

Biologists and physicists use several different variables to represent light intensity. Light intensity can be thought of as the quantity of *light energy* (in Joules, J) received by an *area* of surface (for example, the skin of an animal or the leaf of a plant) in a given period of *time*. Light intensity may also be expressed for a particular part of the spectrum (a wavelength or set of wavelengths). Table 30.2 identifies the terminology used for these variables that represent light intensity, and their units of measurement.

As a practical application of this information, photosynthesis in plants is normally expressed as the number of photons hitting a unit area in unit time, called the 'photon flux density' (PFD). One mole of photons is called an Einstein (E), so the amount of light reaching a leaf is expressed as $\mu E\ m^{-2}\ s^{-1}$ (μE = microE or one millionth of an E). The maximum light intensity in England in mid-summer would be about 1500 $\mu E\ m^{-2}\ s^{-1}$ and at the equator about 2000 $\mu E\ m^{-2}\ s^{-1}$.

The intensity of light which organisms actually experience is greatly affected by environmental factors with absorptive properties. Particulates in the air absorb light to a considerable extent, even darkening Earth catastrophically in extreme cases. Periodically through history, mass extinctions of life have been caused by atmospheric pollution resulting from volcanic eruptions or asteroid strikes.

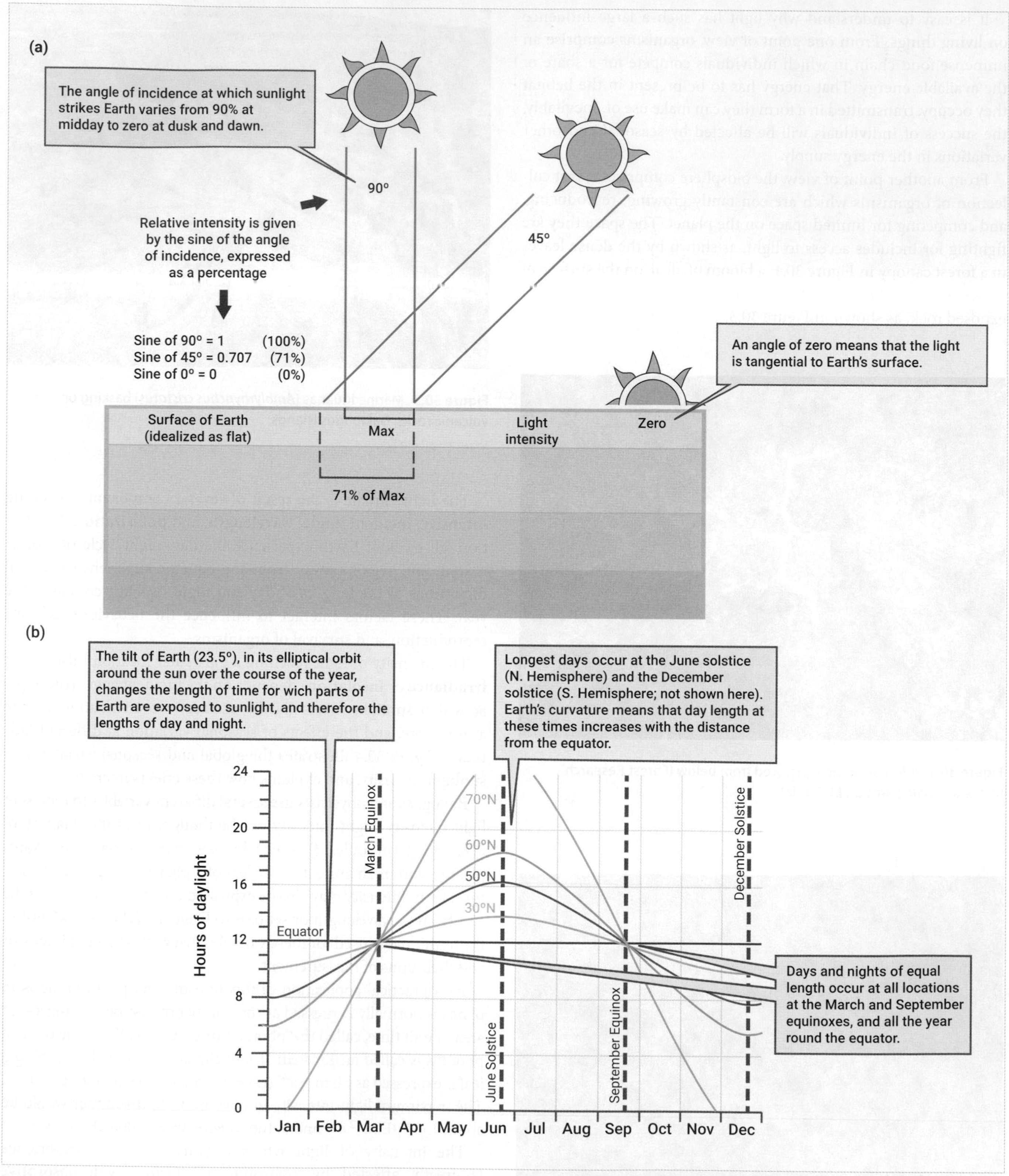

(a)

The angle of incidence at which sunlight strikes Earth varies from 90% at midday to zero at dusk and dawn.

90°

45°

Relative intensity is given by the sine of the angle of incidence, expressed as a percentage

Sine of 90° = 1   (100%)
Sine of 45° = 0.707 (71%)
Sine of 0° = 0   (0%)

An angle of zero means that the light is tangential to Earth's surface.

Surface of Earth (idealized as flat)

Max

Light intensity

Zero

71% of Max

(b)

The tilt of Earth (23.5°), in its elliptical orbit around the sun over the course of the year, changes the length of time for wich parts of Earth are exposed to sunlight, and therefore the lengths of day and night.

Longest days occur at the June solstice (N. Hemisphere) and the December solstice (S. Hemisphere; not shown here). Earth's curvature means that day length at these times increases with the distance from the equator.

Days and nights of equal length occur at all locations at the March and September equinoxes, and all the year round the equator.

March Equinox

June Solstice

September Equinox

December Solstice

70°N
60°N
50°N
30°N
Equator

Hours of daylight

24
20
16
12
8
4
0

Jan Feb Mar Apr May Jun Jul Aug Sep Oct Nov Dec

Figure 30.4 The origin of global and seasonal variations in sunlight intensity. (a) The impact of the incident angle. (b) The impact of the duration of daylight. (c) The impact of irradiance.

Source: (b) and (c) courtesy of Michael Pidwirny.

 Go to the e-book to explore an interactive version of Figure 30.4.

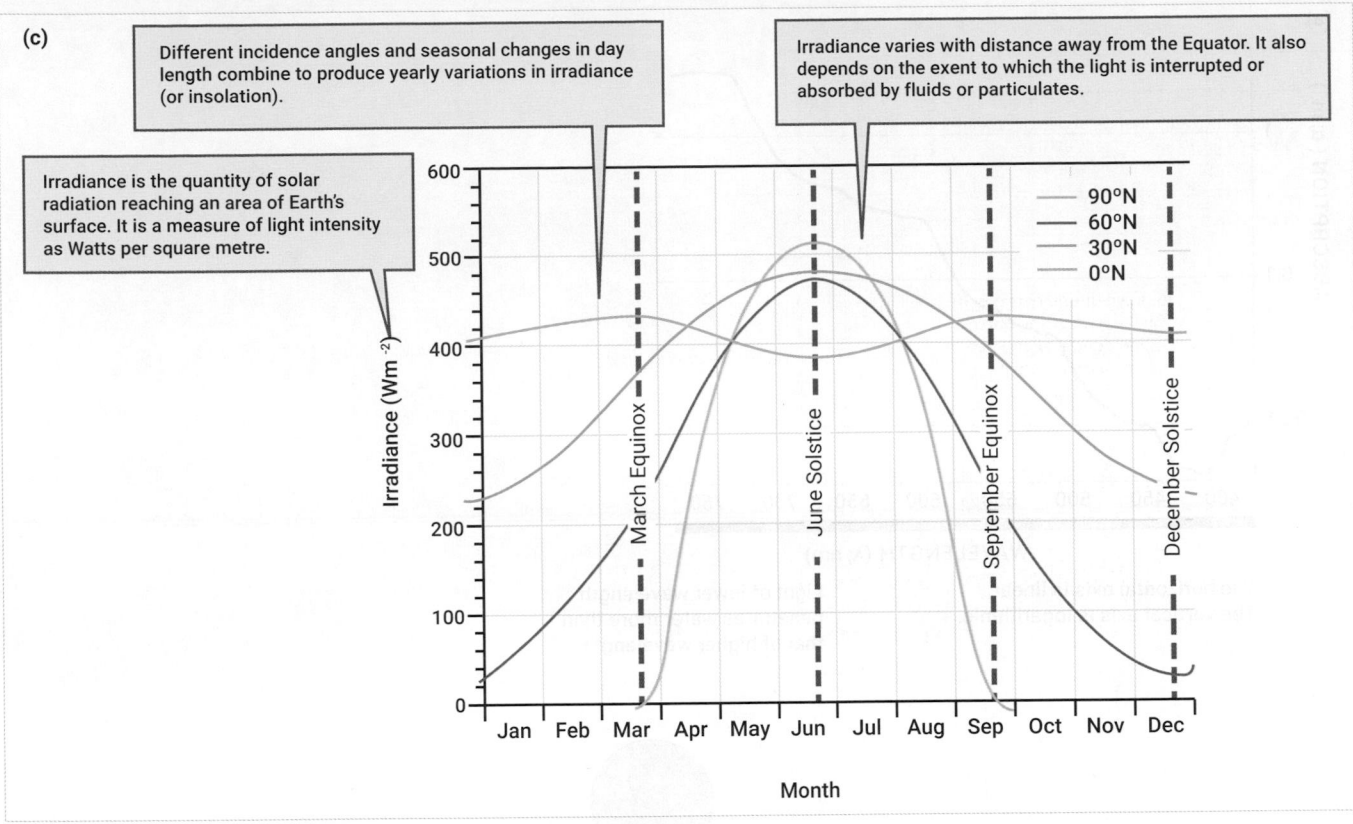

(c)

Different incidence angles and seasonal changes in day length combine to produce yearly variations in irradiance (or insolation).

Irradiance varies with distance away from the Equator. It also depends on the exent to which the light is interrupted or absorbed by fluids or particulates.

Irradiance is the quantity of solar radiation reaching an area of Earth's surface. It is a measure of light intensity as Watts per square metre.

Figure 30.4 Continued.

Table 30.2 Variables representing light intensity and their units

Variable	Unit
Radiant flux or radiant power	Watt (W) or Joule per second
Irradiance or insolation or light intensity	Watt per square metre (W/m²)
Spectral irradiance or spectral intensity	Watt per square metre (W/m³) at specified wavelength(s)

Even clean atmospheric air absorbs light, and the effect is increased by water vapour. Water vapour may reduce the amount of the sun's energy reaching the surface of Earth by as much as 70%. It is especially absorptive in the infrared region of the spectrum and it acts a greenhouse gas.

The absorption of light by free water (in lakes, rivers, and seas) is considerable. Figure 30.5a shows how all wavelengths of light are absorbed to a lesser or greater extent but those at the red end of the spectrum are absorbed particularly well. (Notice how absorption to the right of the graph—the red end of the spectrum—is higher than to the left, the blue end.) Different parts of the spectrum become absorbed at different depths: scuba divers and underwater photographers are familiar with the increasing magenta/blueness which this creates as they go deeper.

Figure 30.5b illustrates how the overall amount of light is reduced as the depth of the ocean increases. Most of the

visible light (400–700 nm) is absorbed in the first 200 m of depth. Photosynthetic plants are only found above this limit in what is called the photic or euphotic zone. *Abundant* plant life is usually only found the first few metres below the surface of lakes.

Below 200 m the light level resembles that at twilight, and by 1000 m there is essentially no sunlight penetration at all. Animals which spend their entire lives at these depths may have limited or no eyesight or may have eyes adapted to detect the bioluminescence produced by other organisms.

The penetration of light is also reduced by turbidity caused by plankton, algae, soil and sand particles, pollutants, and any other particulate matter. Sea ice both absorbs and scatters light. A combination of these effects means that a one metre layer of ice reduces the intensity of blue/green light (500 nm) by 10-fold and red light (700 nm) by 30-fold.

The structure of light

The white light from the sun which illuminates Earth is a mixture of colours and we are familiar with the way these become spread out in a rainbow, like the one in Figure 30.6. As shown in Figure 30.7, the colours of the rainbow are actually a subset of all the possible wavelengths of radiation, called the **electromagnetic spectrum**.

The range of wavelengths to which the human eye is directly sensitive and which make up the visible spectrum is from 400 to 700 nm. As you can see by the small section labelled as 'visible' in Figure 30.7, this range of wavelengths is rather limited. Other organisms may be

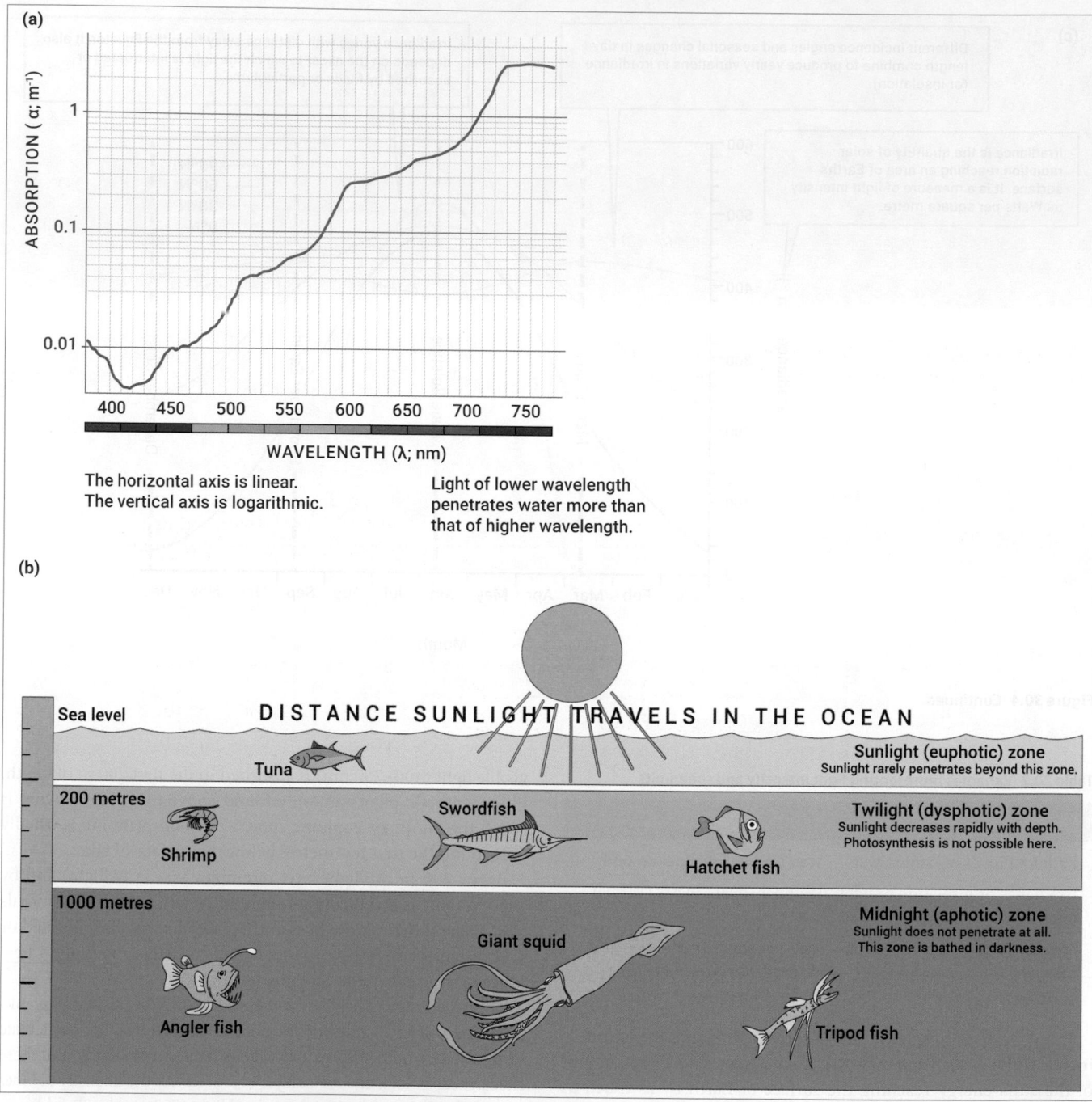

(a)

The horizontal axis is linear.
The vertical axis is logarithmic.

Light of lower wavelength penetrates water more than that of higher wavelength.

(b)

Figure 30.5 The absorption of light by free water. (a) The absorption by water of energy across the visible light spectrum. (b) The distance sunlight travels in the ocean.

Source: (a) Kebes/Wikimedia Commons/CC BY-SA 3.0. (b) NOAA.

able to detect wavelengths beyond that range, into the **ultraviolet** and **infrared** regions. As far as is known, regions of the spectrum beyond that are undetectable by any biological organisms. Humans can only detect them using technological instruments.

It is possible to express light intensity in terms of specific wavelengths—called the **spectral intensity** or **spectral irradiance**.

This parameter can be important where an organism is selective about the wavelengths to which it responds, for example during photosynthesis, or when the quality of light it receives is altered by other environmental factors.

The polarization of light refers to the angle at which it vibrates. Scattered sunlight normally vibrates equally in all directions at

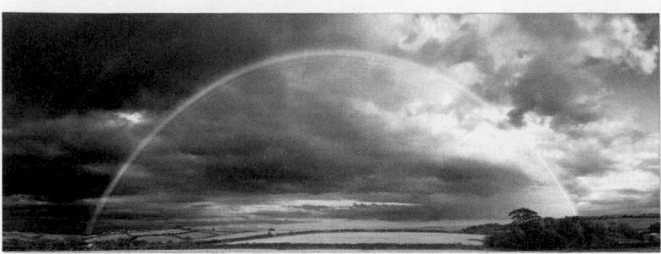

Figure 30.6 The white light from the sun, spread out in the form of a rainbow.
Source: Image by PublicDomainPictures from Pixabay.

right angles to its direction of travel. After being reflected from a surface or absorbed in a particular plane, the vibrations may occur predominantly in one direction.

PAUSE AND THINK

What factors determine how much of the sun's energy reaches a biological surface, such as a leaf or skin?

Answers:

Intensity
- time of day
- incident angle (affected by orientation, position, and season)
- wavelength
- polarization.

Attenuation
- absorption by air, water vapour, and ice
- scatter by particulates.

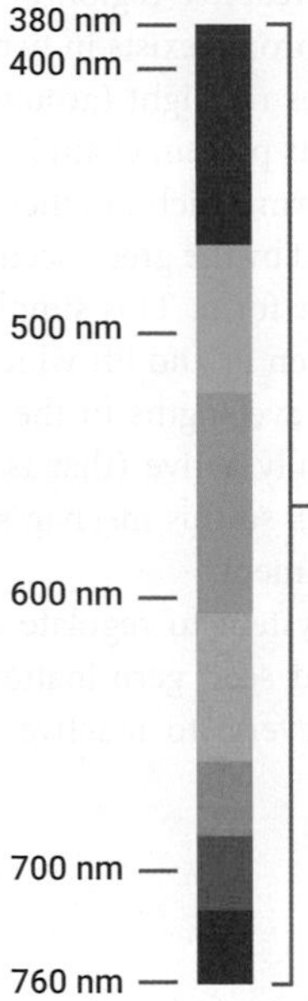

380 nm
400 nm
500 nm
600 nm
700 nm
760 nm
Visible

Class			Frequency	Wavelength	Energy
Ionizing radiation	Y	Gamma rays	300 EHz	1 pm	1.24 MeV
	HX	Hard X-rays	30 EHz	10 pm	124 keV
			3 EHz	100 pm	12.4 keV
	SX	Soft X-rays	300 PHz	1 nm	1.24 keV
	EUV	Extreme ultraviolet	30 PHz	10 nm	124 eV
Visible	NUV	Near ultraviolet	3 PHz	100 nm	12.4 eV
	NIR	Near infrared	300 THz	1 μm	1.24 eV
	MIR	Mid infrared	30 THz	10 μm	124 meV
	FIR	Far infrared	3 THz	100 μm	12.4 meV
Microwaves and radiowaves	EHF	Extremely high frequency	300 GHz	1 mm	1.24 meV
	SHF	Super high frequency	30 GHz	1 cm	124 μeV
	UHF	Ultra high frequency	3 GHz	1 dm	12.4 μeV
	VHF	Very high frequency	300 MHz	1 m	1.24 μeV
	HF	High frequency	30 MHz	10 m	124 neV
	MF	Medium frequency	3 MHz	100 m	12.4 neV
	LF	Low frequency	300 kHz	1 km	1.24 neV
	VLF	Very low frequency	30 kHz	10 km	124 peV
			3 kHz	100 km	12.4 peV
	ULF	Ultra low frequency	300 Hz	1 Mm	1.24 peV
	SLF	Super low frequency	30 Hz	10 Mm	124 feV
	ELF	Extremely low frequency	3 Hz	100 Mm	12.4 feV

Figure 30.7 The electromagnetic spectrum.

Navigation

Animals from a wide variety of groups use their ability to detect light polarity to navigate, locate food, avoid predators, and find shelter. Animals as diverse as mantis shrimps, fiddler crabs, dragonflies, dung beetles, zebra finches, curlews, and bats use light polarity to modify or calibrate other navigational systems such as those based on light direction and Earth's magnetic field.

The navigational skills of ants have been well studied and it is clear that they use a combination of environmental cues to find their way. When dragging food objects or leaves to their nests they maintain the intended direction whether travelling forwards or backwards. Thus, their navigation, based on external cues and reference points, works independently of body orientation.

Detection of light by plants

Plants use light in two quite distinct ways; to gain energy and to gain information. Plants absorb light energy and turn it into biochemical energy by photosynthesis (see Chapter 7). In terms of information, plants monitor the quality, intensity, duration, and angle of light which impinges on them, using this to direct patterns of growth, metabolism, and general behaviour.

The influence of light on *growth* is called **photomorphogenesis**. It includes all developmental changes in size and shape caused by alterations in the photoenvironment. The duration of the light period, the photoperiod, provides information on the time of year. This is because, at a particular latitude, day length is relatively consistent at any point in the calendar year. Although the absolute intensity of the light may be affected by atmospheric factors, the cyclic annual variation is essentially unaffected. Normally, plants monitor and respond to the duration of darkness rather than the duration of the light period.

The *quality* of the incident light in terms of its spectral composition (spectral intensity) can give information about the degree of shading, the placement and nature of neighbouring plants, and the time of day. The *direction* of the incident light influences the positioning of leaves and shoots, maximizing light interception and thus photosynthetic productivity.

Most living, non-woody plant tissues are light sensitive. This includes root tissues well beneath the surface because light can penetrate soil to a significant extent. Responses are facilitated by several photosensory (light-sensing) molecules within the cells. Figure 30.8 illustrates how photoreceptors are used by plants to monitor the light environment.

These various photosensory molecules detect light at particular ranges of wavelength. Sensitivity at the red end of the spectrum occurs through phytochrome (P) illustrated in Figure 30.8. Phytochrome is the major receptor for red and far red light, in the range 600–800 nm, as shown in Figure 30.9. (By convention, plant biologists refer to red light just outside the visible range as far red.) The full infrared range of the spectrum extends well beyond this, as we saw in Figure 30.7.

Phytochrome is a small protein complex containing a light-absorbing chromophore (a light-reactive region of the molecule) called phytochromobilin. Phytochrome exists in two interconvertible forms, Pr and Pfr. Pr absorbs red light (around 650 nm), as represented by the red region of its protein chain in the figure. On doing so, it changes into the Pfr form which can then absorb far red light (750 nm; sensitivity indicated by the green section of the molecule) and convert back to the Pr form. This simple interconversion creates an equilibrium between Pr and Pfr which corresponds to the ratio of red and far red wavelengths in the incident light. Only the Pfr form is biochemically active (that is to say, causes biochemical changes in the plant), so this mechanism enables the plant to monitor its photoenvironment.

Plants use the phytochrome system to regulate other processes including gene expression and seed germination. During the dark period, active Pfr slowly reverts to inactive Pr and awaits

Figure 30.8 Photoreceptors used by plants to monitor the light environment.

Source: Christian Fankhauser.

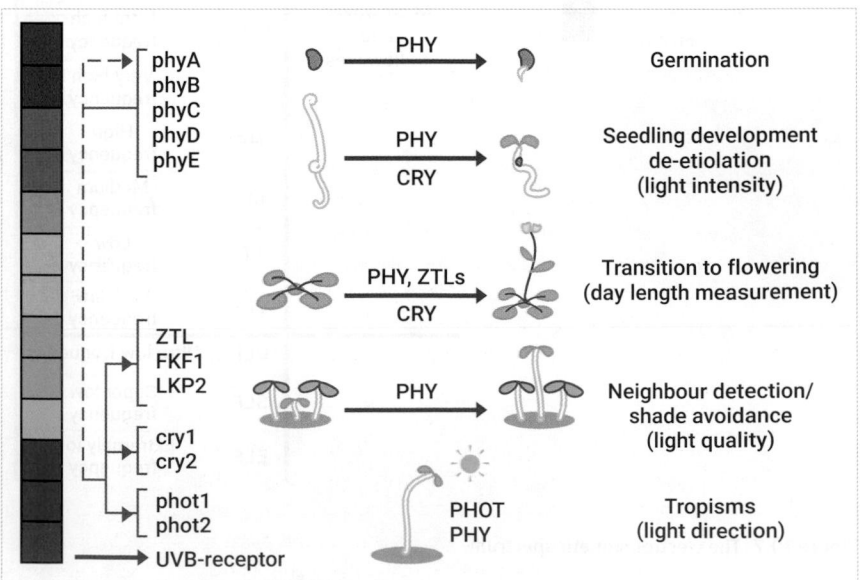

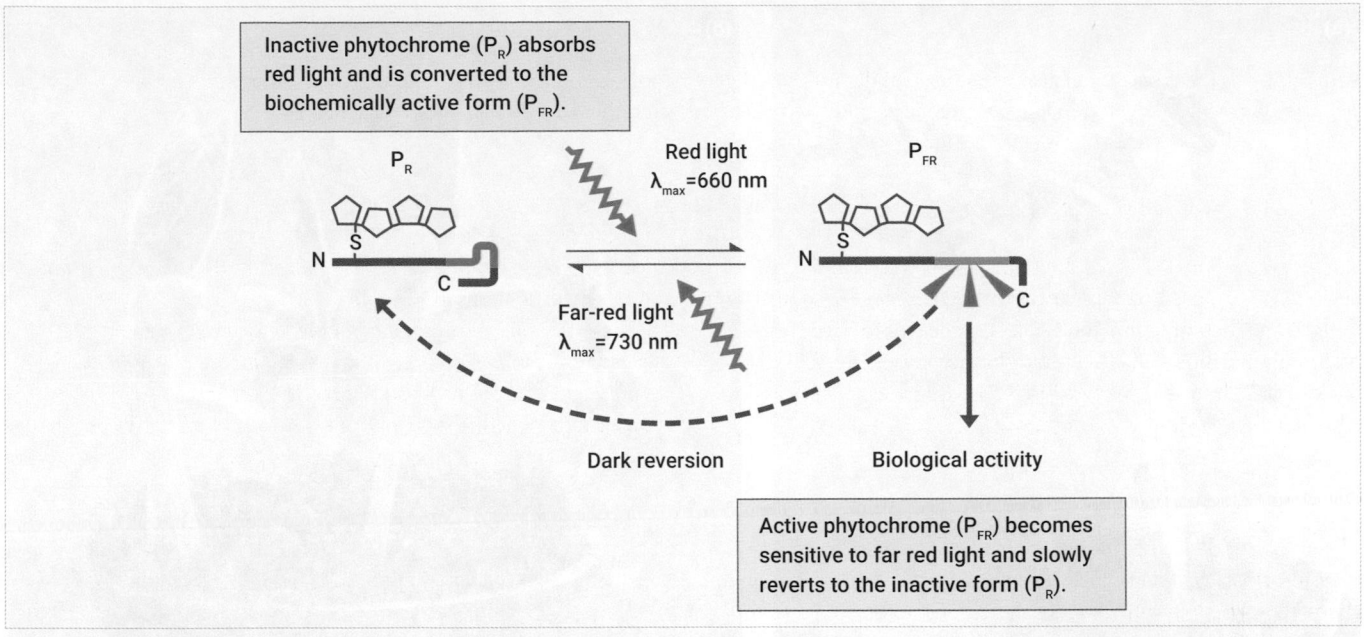

Inactive phytochrome (P_R) absorbs red light and is converted to the biochemically active form (P_{FR}).

P_R

Red light
λ_{max} =660 nm

P_{FR}

Far-red light
λ_{max} =730 nm

Dark reversion

Biological activity

Active phytochrome (P_{FR}) becomes sensitive to far red light and slowly reverts to the inactive form (P_R).

Figure 30.9 Phytochrome: the major receptor for red and far red light in plants.
Source: Pyke K. *Plastid Biology*. Chapter 6. Cambridge University Press 2009.

the reappearance of light for reactivation. The plant responds to the amount of Pfr remaining at dawn and thus to the length of the night, effectively making it aware of the time of year. In many plants, this is crucial in determining the switch from making new leaves to producing flowers or causing seeds to germinate, for example in spring as the nights get shorter (see Chapter 33).

Ratios of red and far red light vary greatly between different photoenvironments. This is especially the case under vegetation because the leaves above have already absorbed more red light than far red light. Thus phytochrome sensing allows the plant to react to the shade from neighbouring plants or objects. It responds by adapting its leaf growth, stem elongation, and overall architecture to maximize light interception. This light-directed growth process is called **shade avoidance**.

Plants also detect light at the blue end of the spectrum, between 300 and 500 nm. A major blue light receptor is cryptochrome. This molecule was first discovered in *Arabidopsis thaliana*, a commonly studied laboratory plant, and found to be similar to light-sensitive chemicals found in insects and mammals. It absorbs blue light between 390 and 500 nm and initiates alterations in gene expression. This drives several processes including the basic development of seedlings and preventing excessive stem elongation. By interacting with the phytochrome system, cryptochrome also regulates flowering time in relation to the day length.

The directed growth of young plants towards a better-lit environment is called **phototropism**. If you compare the two images of seedlings in Figure 30.10, you can see that the seedlings in Figure 30.10a are growing straight where light is mostly shining from above, whereas in Figure 30.10b, the seedlings are growing towards the light source shining from that direction. Cells expand to different extents on either side of the stem, causing the plant to curve towards the light. This process uses blue light as a cue, as first noted by Charles Darwin in his book *The Power of Movement in Plants* in 1880.

Studies on the molecular genetics of *Arabidopsis thaliana* show that two blue light photoreceptors, called phototropins, are involved in phototropism (Figure 30.8). Plant stems change their direction of growth by cells on either side increasing in size to different extents. This differential cell expansion is also known to be regulated by a concentration gradient of the hormone auxin. How phototropin-mediated sensing of blue light links to the growth response is unclear.

Roots are highly sensitive to the direction of gravity and hence grow downwards, but roots also respond to light. Root phototropism produces the opposite result from that in stems, causing them to grow downwards, away from the direction of light. Thus roots are said to be negatively phototropic and shoots positively phototropic.

Phototropin blue light receptors control several other important processes in the plant including stomatal opening in leaves, the movement of chloroplasts within leaf cells which optimizes their photoreception, and the positional growth of leaves which maximizes light interception.

The positional tracking of the sun across the sky, represented by changes in leaf and flower orientation, is called **heliotropism**. It involves changes to the angle at which leaves emerge and grow, and how flowers are positioned in relation to the sun through the day. In this way, leaves optimize their light interception and flowers maintain their warmth (an important factor in fertilization).

If a plant is in a complex three-dimensional canopy of mixed vegetation, leaf angle is adjusted in a way which optimizes light capture by avoiding shade. This response is caused by changes in the turgor pressure (water distribution: see Chapter 29) of motor

Figure 30.10 Phototropism in young seeding plants. Seedlings exposed to light from one direction (a) curve their stems and bend toward the light source whereas plants growing in light mostly from above grow straight upwards (b).

Figure 30.11 A natural herbaceous community.

cells, called **pulvini**, near the leaf base or at the base of the flower, rather than by growth.

Shade avoidance, phototropism, and heliotropism interact, with the result that the plant canopy becomes evenly distributed in 3D space and light capture is maximized, as depicted by the natural herbaceous community in Figure 30.11 and the forest canopy in Figure 30.1.

PAUSE AND THINK

Which wavelength ranges of the light spectrum do plants respond to and which processes are regulated by light?

Answers:

- 600–800 nm (red and far red)
- 390–500 nm (blue)
- Leaf and shoot positioning and angle
- Stem elongation
- Flower production
- Seed germination
- Stomatal opening
- Expression of genes controlling growth and photosynthesis.

Detection of light by animals

Unlike plants, animals do not use solar radiation directly for the generation of metabolic energy. Their interactions with light enable them to sense the environment, make behavioural decisions, and signal to other animals. Many animals also use solar radiation as a direct source of body heat, and light facilitates the production of vitamin D precursors in skin.

All classes of eukaryotic organisms have light detection mechanisms. Direct evidence for their presence in single-celled and simple multicellular animals often comes from observations of movement, either towards or away from a light source, in a process called **phototaxis**.

All animals react to light either with these simple movements or with more complex behaviours. In addition, many animals use light intensity as a reference for cyclic changes in the environment, such as day and night length over the seasons. This information may guide them to safe locations and food supplies, coordinate social interactions, and regulate internal biochemical and cellular rhythms.

What is an eye?

Eyes are found right across the metazoa. At its most basic, an eye consists of two cells—one containing a photosensitive chemical and acting as the light detector, and the other containing a shading pigment. The shading cell restricts the angle at which light can reach the detector cell, allowing the animal to determine the direction from which it comes.

In highly developed eyes, several such cells are clustered together, often with other structures which improve sensitivity and image resolution. The detection cells are linked to the animal's nervous system. Depending on its complexity, the animal may respond directly to changes in light quality (intensity, wavelength, polarization, movement) or may form a virtual (neurological) image of the objects which reflected the light.

There are three fundamental types of multicellular eyes, called **simple**, **compound**, and **complex**. Variants of each are found in different animal groups, and you can see some examples of these in Table 30.3. A remarkable feature is the variety to be found amongst

Table 30.3 Types of metazoan eye

Broad type	Specific type	Anatomy and mechanism	Animal group examples (Not all animals in these groups have the indicated type)
Simple	Ocelli	A unit comprising a photoreceptive cell coupled with a shading cell. Several units may be clustered together	Insect larvae
			Jellyfish
		Found on the abdominal segments, especially in crustaceans adapted to very dark conditions, but may be present along the whole length of the dorsal ganglia	Sea stars
			Planarians
			Larvae of annelids and molluscs
			Decapod crustaceans
Compound	Apposition	Compound eyes comprise an assembly of several thousand photoreceptive facets, called **ommatidia**, arranged over a curved (convex) surface. Each ommatidium has a tiny lens covering a retinal cell and secondary pigment cell	Arthropods
			Annelids
		Eyes of this type have a very wide range of view. They are highly sensitive to movement, especially to moving borders between light and shade, and they may detect the colour and polarization angle of reflected light	Bivalve molluscs
		Image resolution is low compared to that of non-compound eyes and it is unlikely that anything more than a basic shape can be discerned	
		Apposition eyes produce an inverted image	
	Superposition	As for apposition eyes but producing a right image by means of a reflective surface and a separate array of sensory cells. More sensitive in low light than apposition eyes	Decapod crustaceans
	Parabolic superposition	As for superposition eyes but with a parabolic reflector which concentrates the image to a single point	Crustaceans
Complex	Pit or stemma	100 or more cells (essentially ocelli), set in a small depression or pit, sometimes with reflective layer behind the cells	Arthropods, including insect larvae and imagos
		The pit usually has a small aperture which reduces the angles of light reaching the cells, much as in a pinhole camera	Gastropod and bivalve molluscs
		The pit may also contain a refractive gel which further concentrates the light	Sea stars
	Spherical lens	A pit eye, often with a complex retina, with a lens in the aperture made of a highly refractive material. There is usually a refractive humour between the lens and the retina. The lens reduces focal length and creates a sharper image	Annelids
			Gastropod, cephalopod, and chiton molluscs
			Insects

Table 30.3 Continued

Broad type	Specific type	Anatomy and mechanism	Animal group examples (Not all animals in these groups have the indicated type)
Complex	Multiple lens	More than one lens, either spread across the aperture or arranged in line (as in a telescope)	Copepods
		Lenses at the edge of the aperture produce a sharp image across the whole visual field. In-line lenses are poorly understood but may shorten the focal length and improve acuity	Spiders Raptors
	Refractive cornea	A non-spherical layer made of dense refractive material, located in front of the lens	All terrestrial vertebrates
		Improves focus by providing an initial correction to the light as it passes from the low refraction of air to the high refraction of the lens	Spiders Insect larvae
		A cornea is of no value in water because water already has a higher refraction than air, similar to that of the vitreous humour (this is why human divers have to wear masks which retain an air layer in front of the eye)	
	Reflector	A reflective surface inside the eye chamber which focuses the light to a single receptive point	Flat worms
		This is an alternative to a lens	Copepod and bivalve molluscs Fish

closely related animals and the coexistence of different eye types within groups or even in single species.

Simple eyes, or **ocelli** (singular **ocellus**), are clusters of photoreceptor cells, pigment cells, and support cells, sometimes arranged in a cup shape. Light can only enter through the cup opening, so even with these simple structures the animal gains a sense of light direction. The cup may contain a light-refracting gel, although the advantage of this is not fully understood in what is otherwise a rather basic light detector.

Compound eyes are made up of collections of distinct units called **ommatidia**. Figure 30.12b shows a compound eye of Antarctic krill.

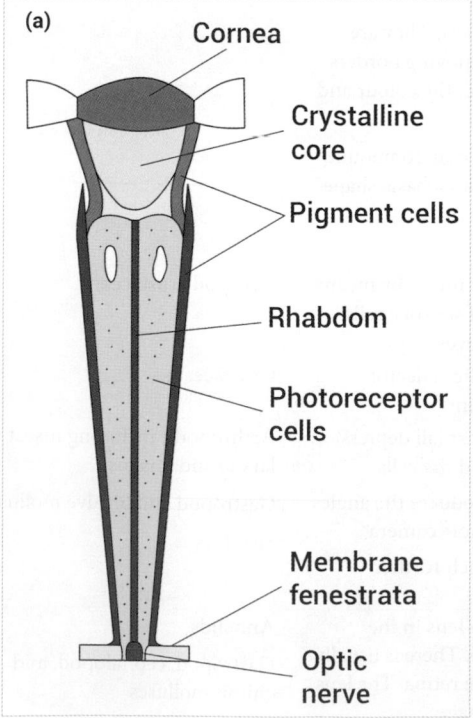

(a) Cornea
Crystalline core
Pigment cells
Rhabdom
Photoreceptor cells
Membrane fenestrata
Optic nerve

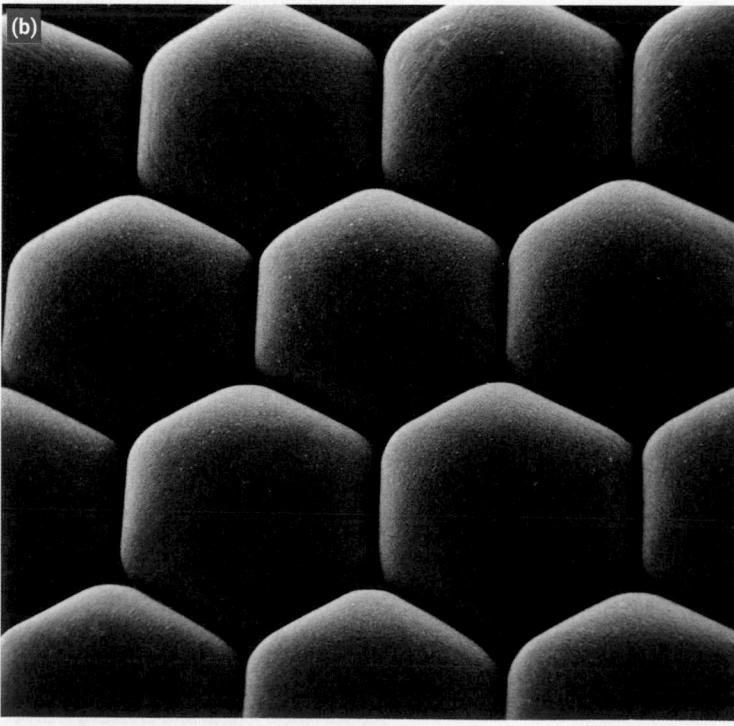

Figure 30.12 Ommatidia and the compound eye. (a) The structure of ommatidia. (b) The compound eye of Antarctic krill.
Source: (a) Nono64/Wikimedia Commons/CC BY-SA 3.0. (b) Uwe Kils/Wikimendia Commons/CC BY-SA 3.0.

The hexagons are cornea which lie at the top of the ommatidia, labelled in Figure 30.12a. These are found in polychaetes, bivalve molluscs, and arthropods, and the number of ommatidia is a species characteristic. Each ommatidium is really an elongated ocellus, so these eyes are sometimes described as rhabdomeric (Greek *rhabdos* = stick). Typically, each ommatidium has a hexagonal cross section, allowing them to be closely packed, with a cornea and a cone-shaped 'lens' above the photoreceptor cells.

Complex eyes are found in vertebrates and cephalopod molluscs; a range of examples of which you can see in Figure 30.13, including cattle (*Bos taurus*), octopus (*Octopus vulgaris*), and nautilus (*Nautilus pompilius*). The simplest form resembles a lens-less, pin-hole camera, but in most cases they have a cornea (a refractive covering), a lens, and an iris diaphragm. The iris diameter can be adjusted to change the amount of light which enters, like the aperture on a camera. The space between the cornea and the lens and that between the lens and the retina are filled with refractive gels (humours) which increase the quality of the light reaching the retina.

The photosensory cells in the complex eyes of vertebrates, rods and cones, are sometimes described as **ciliary** to distinguish them from the rhabdomeric cells of compound eyes. This refers to the fine, filamentous cilium between the outer and inner parts of the cell. Figure 30.14 shows the rods and cones in the mammalian eye; their characteristics are compared in Table 30.4. Ciliary cells are densely arranged such that the outer segments form a curved, light-receptive surface called the retina. The inner segments and their synaptic terminal attachments actually lie over the outer segments, so light has to pass through them to reach the sensitive region. The synaptic terminals communicate with the optic nerve which transmits electrical signals to the brain.

Rod-shaped cells detect variations in monochromatic light intensity (**scotopic** vision); they have high sensitivity and operate

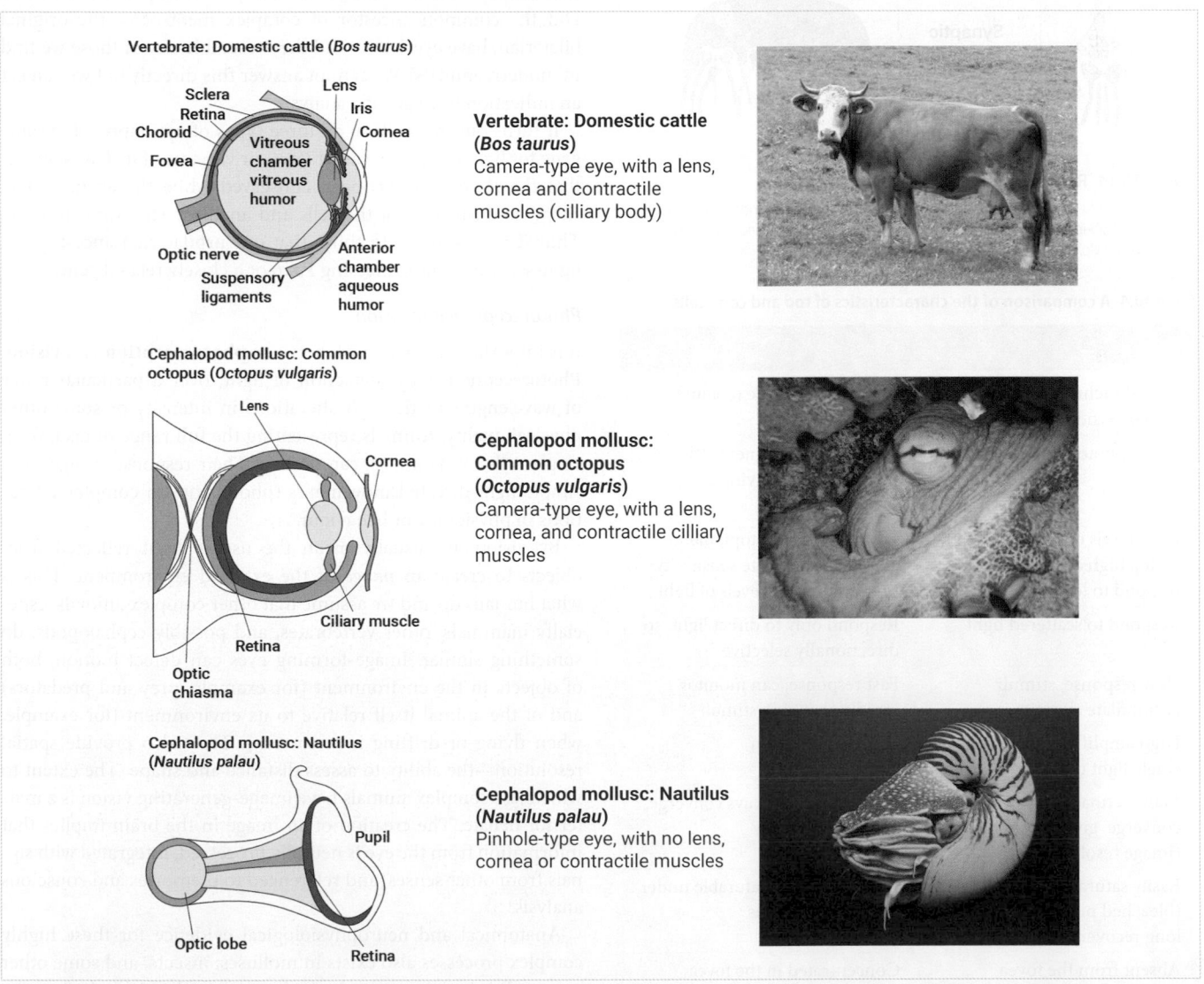

Vertebrate: Domestic cattle (*Bos taurus*)
Camera-type eye, with a lens, cornea and contractile muscles (cilliary body)

Cephalopod mollusc: Common octopus (*Octopus vulgaris*)
Camera-type eye, with a lens, cornea, and contractile cilliary muscles

Cephalopod mollusc: Nautilus (*Nautilus palau*)
Pinhole-type eye, with no lens, cornea or contractile muscles

Figure 30.13 Eye structure in vertebrates and cephalopod molluscs.

Source: (top left) Artwork by Holly Fischer/Wikimedia Commons/CC BY 3.0. (middle and bottom left) M. A. Yoshida, et al. Insights from Visual Systems, Integrative and Comparative Biology, Volume 55, Issue 6, December 2015, Pages 1070–1083, Oxford University Press. (middle right) Photographer: Becky A. Dayhuff, Environmental Educator/CC BY 2.0. (bottom right) Manuae/ Wikimedia Commons/CC BY-SA 3.0.

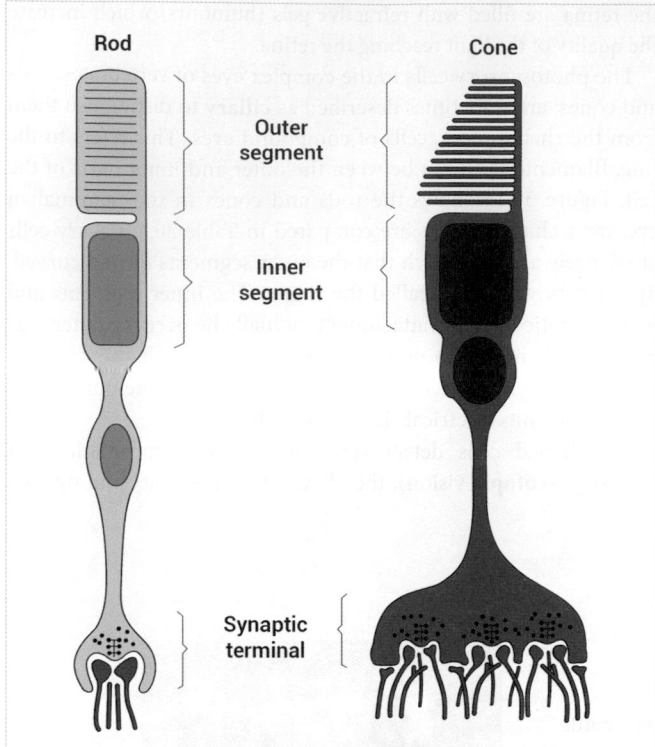

Figure 30.14 Rods and cones in the mammalian eye.

Source: Figure 2.11 from Jeremy Wolfe, Keith Kluender, Dennis Levi, Linda Bartoshuk, Rachel Herz, Roberta Klatzky, and Daniel Merfeld, 2020. *Sensation & Perception*, Sixth Edition, Oxford University Press.

Table 30.4 A comparison of the characteristics of rod and cone cells

Rods	Cones
Provide achromatic (non-colour) vision	Provide chromatic (colour) vision
Single photopigment	Several photopigments (three in human eye), giving a spectral range
High levels of photopigment, giving high absolute sensitivity: respond to low light	Low levels of photopigment, giving low absolute sensitivity: respond to high levels of light
Respond to scattered light	Respond only to direct light, so directionally selective
Slow response; stimuli accumulate over time	Fast response; can monitor rapidly changing stimuli
High amplification; detection of single light quanta is possible	Low amplification
Many retinal pathways converge, giving low acuity (image resolution)	Few retinal pathways converge, giving high acuity
Easily saturated response (bleached pigment), needing long recovery time	Response is not saturable under normal conditions
Absent from the fovea	Concentrated in the fovea
Loss leads to night-blindness	Loss leads to loss of vision

at low light levels. Cone-shaped cells detect light of different wavelengths (**photopic** vision) but are less sensitive and require higher light intensity. The central region of the retina, called the fovea, contains only cones and these are of a more slender shape.

▶ More information on the anatomy and function of the eye can be found in Chapter 18.

Eye evolution

Taxonomic, anatomical, and paleontological studies suggest that eyes have emerged on around 20 separate occasions during animal evolution. High resolution eyes may have emerged four times and eyes with lenses between seven and nine times. Many animals may possess more than one kind of light detection mechanism and there is no clear distinction of eye type between protostomes and deuterostomes.

This complex situation means that over the course of evolution there must have been widespread divergence and convergence in the development of mechanisms for photoreception and vision. Did the common ancestor of complex metazoans, the original bilaterian, have eyes? If so, did they resemble any of those we find in modern animals? We cannot answer this directly but we can get an indication from genetic analysis.

It turns out that cells in all three types of eye express the same gene for light reception, called *Pax6*, or variants of it. This suggests that the *Pax6* gene has been conserved while the arrangement, shape, and function of the cells and ancillary structures has not. Thus it is reasonable to think that our urbilaterian ancestor had light-sensitive cells expressing *Pax6* or a closely related gene.

Photoreception and vision

It is important to distinguish between **photoreception** and **vision**. Photoreception is the detection of light, over a particular range of wavelengths or through alterations in intensity or some other physical quality. Animals representing the full range of taxa, from single cells to mammals, can do this. Their responses range from simple light-directed movements (phototaxis) to complex alterations in physiology or behaviour.

By vision, we usually mean the use of light reflected from objects to create an *image* of the external environment. This is what humans do and we assume that other complex animals, especially mammals, other vertebrates, and possibly cephalopods, do something similar. Image-forming eyes can detect motion, both of objects in the environment (for example, prey and predators) and of the animal itself relative to its environment (for example, when flying or drifting in a current). They also provide spatial resolution—the ability to assess distance and shape. The extent to which less complex animals have image-generating vision is a matter for debate. The creation of an image in the brain implies that information from the eye is neurally processed, integrated with signals from other senses, and referenced to memories and conscious analysis.

Anatomical and neurophysiological evidence for these highly complex processes also exists in molluscs, insects, and some other groups. It is largely on this basis that we may legitimately describe these animals as being able to 'see'. Whether they create *mental* images of the kind that humans experience is probably an unanswerable question.

Detecting environmental changes

An eye requires at least two cells, one to detect light and one to indicate its direction, but not all light reception is even this complex. Animals including flat worms, crustaceans, and the aquatic larvae of annelids and molluscs have individual light receptive cells on their body surfaces enabling them to respond to light intensity. These connect to the central nervous system but are different from eyes because there are no accompanying shading cells. Thus they cannot detect light angle or contribute to the formation of visual images.

Such light receptive cells are found not only in simple animals, but in complex metazoa which also possess eyes and highly developed visual mechanisms. In birds, amphibians, reptiles, and fishes they are located just under the bones of the skull and also deep within the brain. Sufficient light penetrates the skull (through hair or feathers, as well as skin and bone) to allow variations in intensity over the year to cause changes in reproductive activity, moult, migration, and other features of individual and social behaviour.

Mammals including humans may have such cells within the eye structure, unassociated with the rods and cones which provide vision. People who suffer from seasonal affective disorder (SAD) apparently respond to the shortening days of autumn with uncontrollable changes in mood. There are also records of neurologically blind or visually impaired individuals reacting to changes in day length. The mechanisms of these effects are not fully understood, but they demonstrate light awareness which is distinct from vision.

The pineal gland

The pineal gland is a part of the brain which is well known to be light sensitive. In some animals (fishes, amphibians, reptiles) it is *directly* sensitive. In others, especially mammals and birds, it receives light-induced signals *indirectly* from the eye.

The pineal gland is located between the cerebral hemispheres, either just under the skull or deeper down in the brain. It secretes a hormone, **melatonin**, which influences the cyclic activity of cells throughout the body. Melatonin also influences other body-regulating hormonal systems such as the thyroid gland and its secretions.

The synthesis of melatonin is inhibited by light. Thus the pineal secretes it during the dark period, the duration of secretion reflecting the changing length of night over the course of the year. In birds, melatonin is also secreted by the retina and by some extra-retinal photoreceptive cells in the brain.

In mammals, the pineal is linked to the brain's 'clock', located in the **suprachiasmatic nucleus** of the **hypothalamus**. Because of this connection, the cycles of the body are governed by an underlying rhythm which is modified by the actual amount of daylight. This constant adjustment of the clock is called **entrainment**.

Experiments on human volunteers, typically held in caves or underground bunkers, have shown that the daily switch between night and day is crucial to entrainment. If subjects are held for extended periods away from natural light and without other environmental cues, entrainment is lost. Cycles of behaviour such as those of sleep, appetite, and emotion, as well as rhythms in body temperature, excretion, and hormone secretion, start to 'free-run' and become extended. On return to normal conditions, the subjects think that much less time has passed than is actually the case.

Melatonin influences reproductive, social, and behavioural activities in most vertebrates. Sheep and other **artiodactyls** become reproductively active as days lengthen in autumn. Birds may begin to migrate as day lengths change and this is often linked to cycles in reproduction and feather moult. **Hibernation** and **aestivation** in animals living in extreme environments are similarly regulated.

How does light detection work?

In each of the three types of eye that we listed in Table 30.3, the light-sensitive cells contain **visual pigments**. A visual pigment is a combination of a protein, called an **opsin**, and a vitamin-derived, light-sensitive chemical called a **chromophore**. One very common chromophore is retinal, which is chemically related to vitamin A.

When light interacts with the photosensitive chemical, there is a change in bond energy and molecular structure, often involving a protein in the cell membrane called a G-protein. In photoreceptive cells linked to the nervous system, the cell transduces this reaction into nerve impulses. The nervous system integrates and transmits the impulses, initiating a physiological response by the animal.

The spectral sensitivity of a visual pigment (the range of wavelengths it reacts to) depends on which amino acids in the opsin protein interact with the chromophore. Opsins show an enormous variety in their amino acid structure, as prescribed by their DNA code. This means that light sensitivity and vision are under genetic control and open to selection pressure. Spectral sensitivity can thus be specific to animal groups or even individual species.

Table 30.5 lists the main types of chromophore chemicals found in animals as well as other light-sensitive chemicals found in animals and plants. The presence of similar chemicals in diverse organisms is unsurprising given that the basic process of converting radiant energy into chemical energy can be carried out efficiently by only a limited range of biological molecules. Several phytochromes and rhodopsins are also found in bacteria.

In many cases, especially in complex metazoa, colour detection depends on a palette of pigments with different spectral sensitivities. The light sensitivity for individual animals can be plotted along the scale of wavelengths to identify their maximal responses. It often happens that they have separate peaks in the blue, green, and red regions. These can be related to the identification of specific, coloured features in the environment (food supplies, predators, potential mates, landmarks) or for their eyes to operate optimally under different light conditions (as we discuss further in Scientific Process 30.1).

In addition to eyes, there is an even simpler form of light detection, found for example in the pigmented 'eye spots' of some sponge larvae. Although these animals have no nervous system, they use phototaxis to locate suitable locations in which to settle. The pigmented cells, arranged in a ring around the larva's posterior end, possess cilia which operate as light-responsive rudders.

Table 30.5 Common photoreceptive molecules in eukaryotic organisms

Chemical	Organisms and location	*Sensitivity* and function
Photopsins (cone opsins)	Vertebrate retina, cone cells	*Range of colour wavelengths*
		Colour vision
		Formation of visual images
Rhodopsins (rod opsins)	Vertebrate retina, rod cells	*Green–blue light*
	Invertebrates	Low light sensitivity and night vision
		Formation of visual images
Melanopsin (OPN4)	Vertebrate pineal, retinal ganglion cells and iris cells; some cone cells	*Blue light in the visible range*
		Regulation of circadian rhythms
		Mediation of pupillary reflex
Neuropsin (OPN5)	Vertebrate eye and neural tissues	*UV light*
	Invertebrates including echinoderms and tardigrades	
Pinopsin	Pineal gland in birds and reptiles	*Blue light*
		Control of melatonin synthesis
Cryptochromes (CRY1, CRY2)	Sponges	*Blue light*
	Mammals	Magnetic field reception
	Insects	Moderation of clock gene activation
	Plants	Phototaxis
		Phototropism
Channelrhodopsin	Green algae	*Blue/green light*
Chlamyopsin		Mediate phototaxis
Volvoxopsin		
UVR8	Plants	*UV-B light*
		Mediates UV-induced stress responses
Phototropin	Plants	*Blue light*
		Mediates phototropism and stomatal opening
Phytochrome	Plants	*Red and far red*
		Seasonal coordination of flowering and seed germination

SCIENTIFIC PROCESS 30.1	**Photoreceptors in Australian ants**

Background and research question

Ants have compound eyes and rely heavily on vision for navigation of their environment. For many years, they were thought to be unusual amongst Hymenoptera in having dichromatic (two-colour) rather than trichromatic (three-colour) vision. Is this really the case?

Hypothesis

Yuri Ogawa and colleagues set out to test the hypothesis that ants have more complex colour vision than previously believed. They investigated

the vision systems of two very similar species of Australian ant of the *Myrmecia* genus, one of which, *M. vindex*, is nocturnal while the other, *M. croslandi*, is diurnal.

Methods

Using ants immobilized in wax, the team used platinum electrodes to record electrical activity coming from the surface of the eye when exposed to light across a range of wavelengths. They determined *relative spectral sensitivity* by comparing how many photons were needed to produce the same response as an equivalent amount of white light.

Results

Figure 1 shows the spectral sensitivity of the nocturnal *M. vindex* (blue line) and the diurnal *M. croslandi* (red line). The species have roughly similar sensitivity curves, with peaks in the UV (<400 nm) and green (530 nm) regions, but *M. vindex* has greater sensitivity below 500 nm and its UV peak is at a slightly longer wavelength.

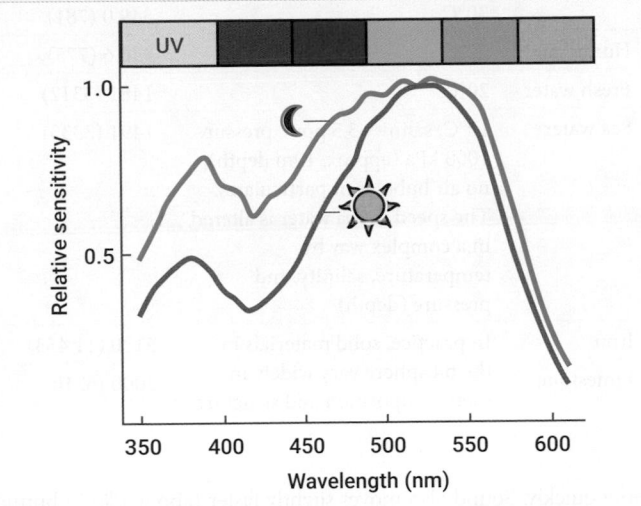

Figure 1 The spectral sensitivity of animal eyes: Australian ants.

Source: Republished with permission of Proceedings of the Royal Society Biological Sciences, from Ogawa Y, Falkowski M, Narendra A, Zeil J, Hemmi JM. (2015) Three spectrally distinct photoreceptors in diurnal and nocturnal Australian ants. Proc. R. Soc. B 282, Figure 1(a) Page 3: 20150673. http://dx.doi.org/10.1098/rspb.2015.0673; permission conveyed through Copyright Clearance Center, Inc.

This result is consistent with nocturnal ants having only ultraviolet light available for navigation, whereas diurnal ants can use a greater range of wavelengths.

Methods

Ogawa's team also investigated the ants' photoreceptors to determine which photopigments were responsible for the spectral sensitivity of the eyes. To do this they needed to test how the eyes responded to narrow and broad ranges of wavelengths.

They did this by:

a) adapting the eyes to different coloured light before measuring the spectral sensitivity with the surface electrodes;

b) recording directly from ommatidia using electrodes placed inside the cells.

Results

Figure 2 shows the results from the nocturnal *M. vindex*. The eye's response depends on three types of photopigment, sensitive in the green, blue, and UV regions. The large amount of overlap between the responses shows that electrical signals coming from individual photoreceptors are coupled to produce integrated signals.

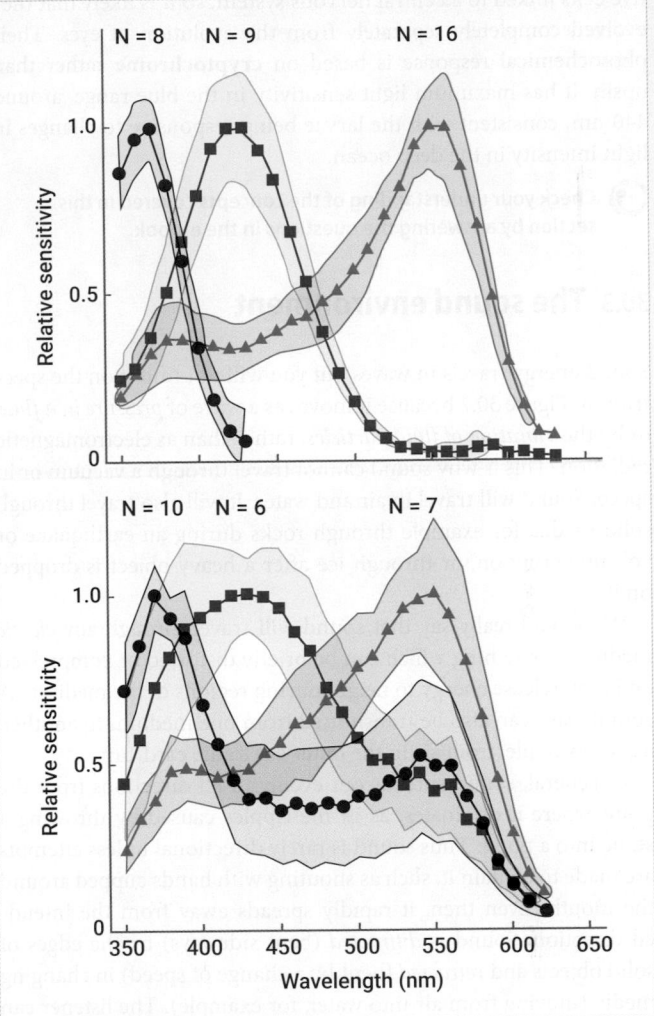

Figure 2 Results from the nocturnal *M. vindex*.

Source: Republished with permission of Proceedings of the Royal Society Biological Sciences, from Ogawa Y, Falkowski M, Narendra A, Zeil J, Hemmi JM. (2015) Three spectrally distinct photoreceptors in diurnal and nocturnal Australian ants. Proc. R. Soc. B 282, Figure 2 Page 3: 20150673. http://dx.doi.org/10.1098/rspb.2015.0673; permission conveyed through Copyright Clearance Center, Inc.

Conclusions

The eyes of these Australian ants have three types of photoreceptor, rather than the two as previously thought, and overlap provides broad spectral sensitivity.

The paper ends with this comment:

'The evolution of trichromacy in arthropods, including insects, predates the evolution of flowers and we propose here that the most fundamental common need for colour vision may well have been in the context of landmark guidance.'

Read the original paper

Ogawa Y, Falkowski M, Narendra A, Zeil J, Hemmi JM (2015). Three spectrally distinct photoreceptors in diurnal and nocturnal Australian ants. *Proc. R. Soc. B 282*: 20150673. *http://dx.doi.org/10.1098/rspb.2015.0673*

These cells do not express the *Pax6* gene found in photoreceptive cells linked to a central nervous system, so it is likely that they evolved completely separately from the evolution of eyes. Their photochemical response is based on **cryptochrome** rather than opsin. It has maximum light sensitivity in the blue range, around 440 nm, consistent with the larvae being responsive to changes in light intensity in the deep ocean.

 Check your understanding of the concepts covered in this section by answering the questions in the e-book.

30.3 The sound environment

Sound energy travels in waves but you will not find it on the spectrum in Figure 30.7 because it moves as a wave of *pressure in a fluid* or by the *vibration of fluid particles*, rather than as electromagnetic radiation. This is why sound cannot travel through a vacuum or in space. Sound will travel in air and water. It will also travel through solid media, for example through rocks during an earthquake or volcanic eruption, or through ice after a heavy object is dropped on it.

We should really say that sound will travel through any *elastic* medium: something which can be briefly distorted or compressed and then release energy to neighbouring regions of the medium. A sound wave can also be transmitted from one medium to another, as, for example, from air in the outer ear to the eardrum.

In general, sound spreads out evenly in all directions from the point where it originates, as in the ripples caused by throwing a stone into a pond. Thus sound is rarely directional unless attempts are made to contain it, such as shouting with hands cupped around the mouth. Even then, it rapidly spreads away from the intended direction. Sound is *diffracted* (bent sideways) by the edges of solid objects and *refracted* (bent by a change of speed) in changing media (moving from air into water, for example). The listener can often tell which direction a sound comes from but may be misled by diffraction and refraction, as well as by reflection and echoes. Any *concurrent movement* of the medium itself, for example, wind blowing towards or away from the listener, will also influence how it travels.

Let's consider how fast sound travels in different media. Table 30.6 shows its velocity (speed in a given direction) in air and water and in solid materials for comparison. We see that sound travels more than four times more quickly in fresh or sea water than in air, and two to three times more quickly in solid materials such as iron or stone than in water.

In general terms, stiffer media transmit the sound more rapidly, so sound travels faster in solids than in liquids and faster in liquids than in gases. The *stiffness* (technically the *elastic modulus*) of a medium can be thought of as its resistance to distortion.

The velocity of sound in a particular medium (gas, liquid, or solid) depends on several physical attributes including its *molecular weight*. Heavier molecules have greater inertia and take more time to get moving, so the higher the molecular weight, the lower the speed. For example, taking a breath of helium raises the pitch of the voice because helium atoms are lighter than air (nitrogen and oxygen) atoms and conduct the sound wave to the listener

Table 30.6 The velocity of sound in different media

Medium	Conditions (sea level, unless indicated)	Velocity of sound, m/s (miles/h)
Dry air	−10°C	325.2 (727)
	0°C	331.3 (741)
	10°C	337.3 (755)
	20°C	343.2 (768)
	30°C	349.0 (781)
Humid air	20°C	346.6 (775)
Fresh water	20°C	1481 (3312)
Sea water	10°C, salinity 3.5 ppm, pressure 1000 kPa (approx. 10m depth), no air bubbles or particulates. The speed in sea water is altered in a complex way by temperature, salinity, and pressure (depth)	1491 (3335)
Iron	In practice, solid materials in the biosphere vary widely in their composition and structure	5120 (11,453)
Limestone		3000 (6710)

more quickly. Sound also moves slightly faster (about 1%) in humid air than dry air (as shown by the data for air at 20°C in Table 30.6) because some nitrogen and oxygen atoms are replaced by lighter water molecules.

Table 30.6 also shows that warmer air conducts sound more rapidly than cooler air. If temperature remains constant, *atmospheric pressure* has no effect on the speed of sound: it is essentially the same at the top of a mountain as at the bottom. This is because pressure and density are inversely related: increasing the pressure (which would make the sound faster) compresses the air and increases its density (which would make the sound slower); the two effects cancel each other out.

The pitch of a sound depends on its *frequency* (the number of waves passing a point in a given period of time). Frequency is the inverse of *wavelength* because shorter waves will pass more often than longer ones. The wavelength of sound is usually given in metres (m). The frequency is given in cycles per second and written as $c.s^{-1}$ or Hz. The Hz symbol stands for Hertz and is named after a German physicist famous for his fundamental research.

The *volume* of a sound depends on its *amplitude*, essentially how much displacement of the medium there is as the wave passes through. Loud sounds contain more energy than soft sounds, and the *intensity* of a sound (the power it represents, in $W.cm^{-2}$) increases with the square of its amplitude. An animal which increases the volume of the sound it makes does not thereby alter its pitch, although *producing* sounds of different volume may entail changes of pitch or changes in the complexity of the wave structure.

Acoustic engineers measure sound intensity in decibels (db; one tenth of a bel). The decibel scale is logarithmic because of the large range it covers. It compares sound intensities to an arbitrary reference value of 10^{-16} $W.cm^{-2}$ which is roughly the faintest sound detectable by the human ear. The maximum tolerable sound intensity is around 120 db (10^{-4} W.cm^{-2}).

We talk of sounds being loud or soft, but these are subjective terms. They describe the *sensation* of sound and are affected by pitch and quality, by the level of background noise, and by the position of the listener. Loudness is roughly proportional to the logarithm of intensity but the scale is far from linear.

PAUSE AND THINK

The following equations relate to the physics of sound. Express them in simple sentences.

[d is fluid density in g/L; °K is absolute temperature; A is amplitude in m; λ is wavelength in m; I is intensity in W.cm^{-2}]

$$\text{Speed} \propto \frac{1}{d^2}$$

$$\text{Speed} \propto \frac{1}{\sqrt{°K}}$$

$$\text{Pitch} = \frac{1}{\lambda}$$

$$\text{Intensity} \propto A^2$$

$$\text{Volume} \propto \text{Log}_{10}I(approx.)$$

Answers:

$\text{Speed} \propto \frac{1}{d^2}$ Speed is inversely proportional to the density of the medium.

$\text{Speed} \propto \frac{1}{\sqrt{°K}}$ Speed is inversely proportional to the square root of the absolute temperature.

$\text{Pitch} = \frac{1}{\lambda}$ Pitch (frequency) is the inverse of wavelength.

$\text{Intensity} \propto A^2$ Intensity is proportional to the square of the amplitude.

$\text{Volume} \propto \text{Log}_{10}I(approx.)$ Volume is approximately proportional to the log of the intensity.

Plants and the sound environment

It is not at all obvious that plants 'hear', in the sense that we use that term with animals, but there is evidence that they can respond to environmental sounds. The niches which plants occupy are full of sounds which provide potentially useful information, from the buzzing of insects, the singing of birds, and the footfall of herbivores, to the rush of the wind and the noise of an approaching storm. It is reasonable to suppose that plants have ways of responding which are beneficial to their survival.

As we have seen, sound moves by the oscillation of particles in a fluid or solid medium. The vibrational energy can be transmitted to any object, not just to an ear or other complex listening device. Plants respond to touch through mechanosensors in their cell membranes (see next section), and the membranes vibrate when exposed to sound waves, so there is no reason in principle why they should not be acoustically sensitive.

Anecdotal reports suggest that plants respond to music or to 'vibrations' emitted by attentive and affectionate carers. Such pseudo-scientific information cannot be tested in a meaningful way but it is possible to look for direct effects of sounds on aspects of plant performance.

Here are three examples of plants responding to sound in ways which confer clear survival, reproductive, or productivity advantages:

- 'Buzz pollination' is a widespread phenomenon among flowering plants in which their anthers release pollen only in response to the characteristic wing vibrations of specific insects. This may have evolved from the competition between plants and insects, reducing pollen removal by non-pollinating 'pollen thieves' and maximizing the value of true pollinators.

- *Arabidopsis thaliana* responds to sounds made by nearby leaf-chewing caterpillars by producing higher than normal quantities of caterpillar-repellent chemicals.

- Seed germination rates, oxygen uptake, and growth rates in several commercially valuable vegetables, including okra, zucchini, Chinese cabbage, and field beans, can be increased by a variety of naturally occurring sounds. Experimenters have isolated the most effective frequencies and suggest that these have ecological significance in terms of their interactions with animals.

There is no evidence for dedicated sound sensors in plants but a possible mechanism for these effects is a change in the ion permeability of the plasma membrane. In particular, calcium ion channels may become more porous when vibrations change the tension of the membrane.

Changes in calcium ion concentration within the cell's cytoplasm regulate a variety of cellular processes, mediated through signal transduction pathways. Changes caused by sound may well operate in this manner, although this is yet to be clearly demonstrated.

A more indirect but no less adaptive type of 'acoustic' response is shown by tropical plants which rely on bats for pollination. Their petals have evolved to be highly reflective of the ultrasound calls of their bat pollinators. The sound can be reflected from a wide range of angles, enabling the bat to detect the plant from some distance away. Other bat-dependent plants have hairy leaves: these reflect high frequency calls but attenuate low frequency background noises, thereby 'sparkling' within the soundscape and drawing the bat's attention.

Plants and mechanosensing

Plants are highly sensitive to touch and pressure and change their growth patterns in response, using mechanosensitive ion channels in the cell membrane to initiate the reaction.

Soil is a heterogenous medium and a root growing through it will encounter a variety of textures and barriers. Look at Figure 30.15. In Figure 30.15a, the root is growing downwards on an agar plate. As the root grows, it might reach a solid obstacle in the way, such as the white mass in Figure 30.15b. After touch-sensing, differential expansion of cells further up the root (rather than at the tip) cause the growth to curve and move away from the obstacle, as you

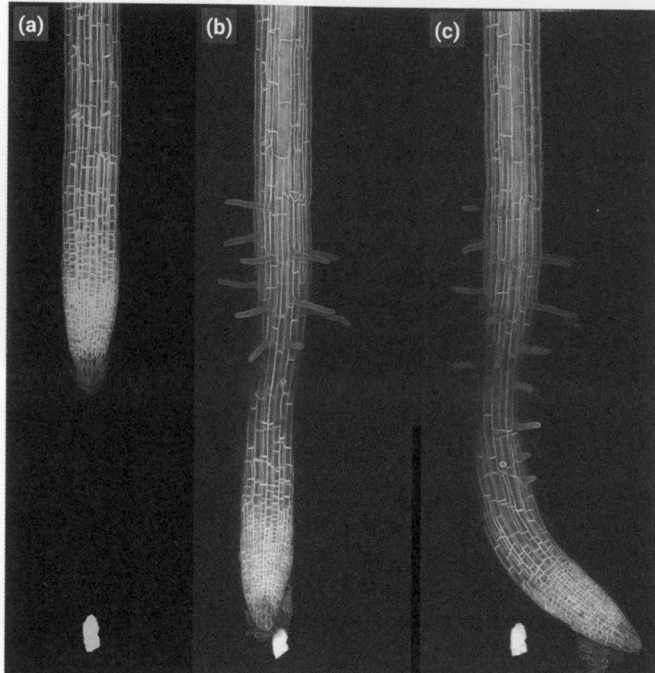

Figure 30.15 Root thigmotropism. Real time imaging of *Arabidopsis thaliana* root growth using a confocal laser scanning microscope. (a) Root grows downwards on an agar plate. (b) It touches a solid obstacle in its way. (c) After touch sensing, differential expansion of cells further up the root cause the growth to curve and move away from the obstacle.

Source: Photographs courtesy of Darren Wells, Plant and Crop Sciences, University of Nottingham.

Figure 30.16 Thigmotropic responses of climbing plants. (a) Hop (*Humulus lupulus*). (b) Cucumber.

Figure 30.17 Thigmonastic response of a carnivorous plant—the Venus flytrap (*Dionaea muscipula*).

can see by the bend in the root in Figure 30.15c. The root will curve round the obstacle and continue to grow along it until the root can recommence its downward growth. This is called a **thigmotropic** response.

Climbing plants also use thigmotropic responses when their stems touch a support. Figure 30.16a shows a hop plant (*Humulus lupulus*), and Figure 30.16b shows a cucumber plant (*Cucumis sativus*), which are two examples of climbing plants. Those with thin stems, such as the hop, can wrap around a supporting structure, forming a spiral as they grow upwards. Thick-stemmed climbing plants, such as the cucumber, but typically also beans and lianas, produce tendrils to gain support.

In both cases contact between the plant and the support initiates a thigmotropic response. Cells in the tissue nearest to and touching the support sense the contact and reduce their rate of expansion. The relatively greater growth on the other side of the stem causes it to curve around the support and gain a hold, enabling the plant to climb. Tendrils like the one in Figure 30.16b are highly sensitive to touch, supposedly more so than human skin, and quickly coil or spiral around the supporting structure.

A different kind of mechano-sensitivity is shown in plants such as mimosa (*Mimosa pudica*) and the Venus fly trap (*Dionaea muscipula*). Figure 30.17 shows that the response is mechanical flexing rather than redirectional tissue growth. It is rapid and reversible and results from a change in fluid flow within the leaf

material. Such responses are described as **thigmonastic** (Greek *thigmo* = touch, and *nastos* = squeezed together; the term **photonastic** is used for the night-time, low light-induced closure of the flowers and leaves of some other plants.)

In the case of *Mimosa*, leaflets detect vibration as well as touch and respond by folding downwards. This reduces their surface area and protects the plant from grazing insects. The leaves of the Venus fly trap have evolved to close suddenly when surface hairs detect the presence of insects. The insect is digested and the plant gains nutrients, especially nitrogen, from the catch. Fly trap plants typically live on acidic bogs and this insectivorous strategy improves their nutrient balance.

In both these cases, movement is achieved by a **pulvinus** structure, a swelling at the base of the leaflet where it meets the stem. The mechanical stimulation of the leaflet surface leads to an action potential (a flow of ions across the cell membrane). This is received by the pulvinus and causes a rapid change of turgor pressure in motor cells, quickly altering leaf angle as a result. These are probably the fastest movements found anywhere in the plant kingdom.

Animals and the sound environment

Animals are highly sensitive to the sound environment. Those which 'hear' have a specific organ or part of the body which detects the oscillatory movement of particles of gas, liquid, or solid media as they transmit sound waves. Animals without such specific organs can also detect vibrations propagated as pressure waves through the medium. As with plants, the ability to sense sounds is an aspect of a broader category of sensitivity to movement, generally called **mechanoreception**.

Mechanoreceptors are cells or groups of cells that translate mechanical stimuli into electrical impulses, which can then be interpreted by the nervous system. Receptors may be specialized for the detection of touch and pressure, to perceive position or equilibrium, or to detect cues such as water movements.

Touch and pressure

The mechanoreceptors which detect internal pressure changes are known as baroreceptors. Tactile receptors detect touch, pressure, and vibration at the body surface.

▶ We learn more about baroreceptors in Chapters 21 and 24.

Vertebrate tactile receptors are sensory cells located in the epidermis and other tissues. **Pacinian corpuscles** and **Merkel's discs** have large and small receptive fields, respectively. Pacinian corpuscles lie deep within skin, joints, muscles, and organs and are especially sensitive to vibrations. Merkel's discs are much nearer the surface of the skin and sense light touch and pressure.

Invertebrate touch receptors are structurally different from those of vertebrates. **Trichoid sensilla** are hair-like projections from the cuticle which sense touch through displacement.

Vertebrates and invertebrates also have **proprioreceptors (proprioceptors)** which detect body position through changes in pressure and movement in muscles, tendons, and joints. They are involved in movement coordination and in bodily self-awareness.

Equilibrium and hearing

Mechanoreceptors are also involved in the sensing of equilibrium (balance) and in hearing. Equilibrium sensing is the ability to detect the position of the body in relation to gravity, i.e. which way is up or down.

Invertebrates have separate systems for equilibrium sensing and hearing. Figure 30.18 shows a **statocyst**, which is a sensitive structure found in many invertebrates that senses equilibrium. These are hollow, fluid-filled cavities containing a layer of mechanosensory neurones with sensory hairs called setae. Dense particles of calcium carbonate called statoliths migrate around the statocyst as the animal moves. The neurones respond to statolith movement and initiate an electrical signal in the nervous system.

Vertebrates use mechanoreceptors called **hair cells** for both equilibrium and hearing. Figure 30.19 shows the structure of a typical hair cell found in the majority of vertebrates. These hair cells generally consist of a single, long **kinocilium** surrounded by numerous

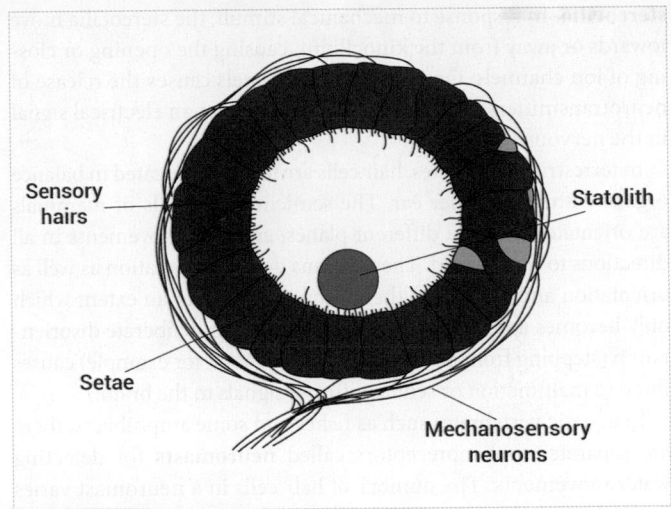

Figure 30.18 Diagram of a statocyst, the balance-sensitive structure found in many invertebrates.
Source: Davis, W. J. (1968)/Wikimedia Commons/Public Domain.

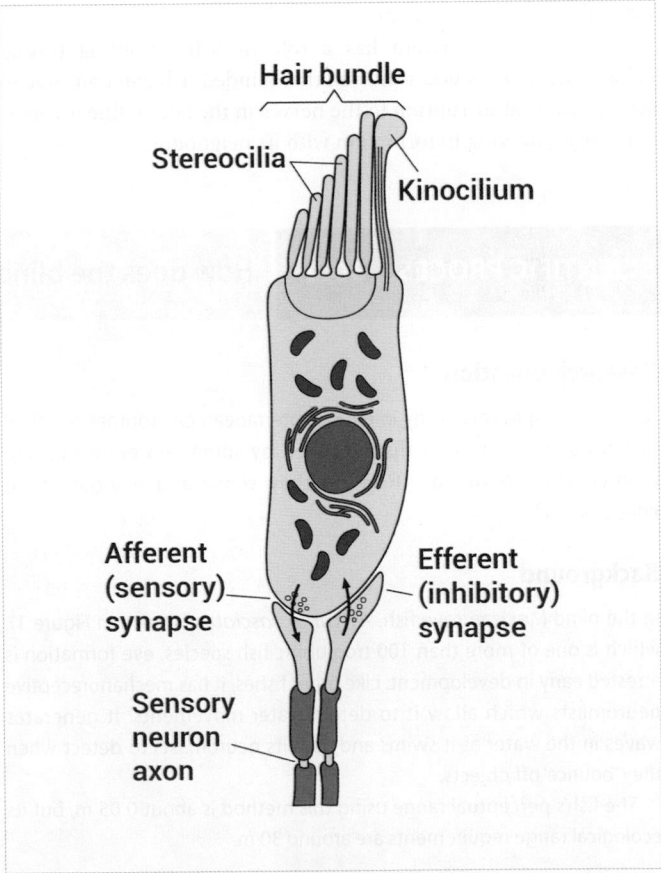

Figure 30.19 The structure of a typical hair cell found in the majority of vertebrates. The hair cell is activated if the stereocilia are bent towards the kinocilium.
Source: Richard W. Hill, Gordon A. Wyse, and Margaret Anderson. *Animal Physiology*, Fourth Edition, 2016. Oxford University Press.

stereocilia. In response to mechanical stimuli, the stereocilia move towards or away from the kinocilium, causing the opening or closing of ion channels. Opening of ion channels causes the release of neurotransmitters, detection of which generates an electrical signal in the nervous system.

In terrestrial vertebrates, hair cells are primarily located in balance organs within the inner ear. The semicircular canals of mammals are orientated in three different planes, enabling movements in all directions to be detected. These organs detect acceleration as well as orientation and we rely on them subconsciously to an extent which only becomes apparent when disease, injury, or deliberate disorientation (stepping from a playground roundabout, for example) causes them to malfunction or send confusing signals to the brain.

In aquatic vertebrates such as fishes and some amphibians, there are separate mechanoreceptors called **neuromasts** for detecting water movements. The number of hair cells in a neuromast varies depending on the species, and they are usually covered by a gelatinous cap.

In fishes, neuromasts can be scattered over the skin or arranged in a line, called the lateral line, running down the length of the body. In fast-swimming fishes, like the pollack (*Pollachius pollachius*) in Figure 30.20, the lateral line may rise above the pectoral fins, potentially avoiding any confusion from disruption of water flow by fin movement.

The neuromast system has a role in fish schooling (group behaviour). Early studies showed that blinded fish can continue to school, but that disruption to the nerves in the lateral line prevents a fish from moving in formation with its neighbours.

Figure 30.20 A Pollack (*Pollachius pollachius*) with a clearly visible lateral line extending from the tail, over the pectoral fin to the top of the gill cover.
Source: Citron/CC-BY-SA-3.0.

Blind cave fish use deflections of waves created by their own movement through the water off objects in their environment to create a spatial map of their environment. The detection of water movement through neuromasts is also important for orientation in a water current and detecting food and potential predators. You can read more about blind cave fish in Scientific Process 30.2.

| **SCIENTIFIC PROCESS 30.2** | **How does the blind cave fish 'see' its environment?** |

Research question

Animals living permanently in dark, subterranean environments, called troglobionts, are usually distinguished by some degree of eye and pigmentation reduction. How do they sense and navigate their environment?

Background

In the blind Mexican cave fish, *Astyanax fasciatus* (shown in Figure 1), which is one of more than 100 troglobitic fish species, eye formation is arrested early in development. Like other fishes, it has mechanoreceptive neuromasts which allow it to detect water movements. It generates waves in the water as it swims and uses its neuromasts to detect when they 'bounce' off objects.

The fish's perceptual range using this method is about 0.05 m, but its ecological range requirements are around 30 m.

Hypothesis

In 2004, Theresa Burt de Perera and colleagues from the University of Oxford wondered whether a blind cave fish can learn where objects are in

Figure 1 Mexican cave fish, *Astyanax fasciatus*.
Source: Daniel Castranova, NICHD/NIH.

its environment and obtain a bigger 'picture' of its surroundings? She tested the hypothesis that the fish can encode ('remember') the order of landmark sequences in their internal representation of space.

Methods

Burt de Perera measured the swimming speed of fish placed individually in a circular channel. The channel contained symmetrical but geometrically different landmarks (plastic bricks) which could be moved to create novel environments for consecutive tests. The landmarks were positioned at distances beyond the fish's close perceptual range so that complex navigation would be necessary. The circular channel meant that the fish had to swim along the landmark sequence (clockwise or anticlockwise) rather than across the centre.

Results

Burt de Perera observed that mean pek swimming velocity increased when fish were placed into a novel environment and that it decreased as the fish became familiar with their new environment, as depicted by the graph in Figure 2.

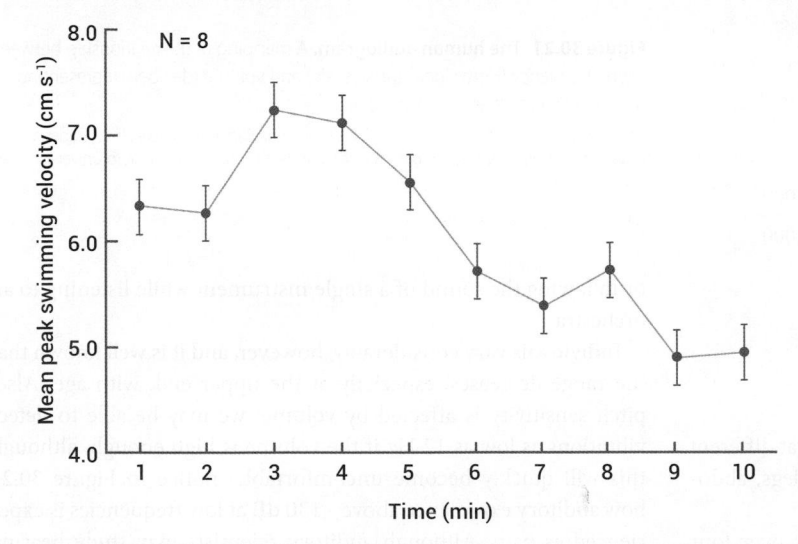

Conclusion

The increased swimming speed in a new environment followed by a reduction as the environment becomes more familiar is taken as evidence that the fish can memorize its surroundings and can notice when landmarks change.

In addition to being able to detect objects in their near vicinity, blind cave fish can learn the order of landmarks, store this information in the form of a spatial map, and detect any changes in the environment. The fish cannot see but can sense its environment and work out how landmarks are spatially joined together.

Read the original paper

Burt de Perera, T. (2004). Fish can encode order in their spatial map. *Proceedings of the Royal Society of London B* **271**: 2131–2134.

Figure 2 Mean swimming velocity of 10 fish over a 10 minute period in pilot tests (± S.E.). Peak swimming speed is reached between 2 and 5 minutes. As fish become familiar with their environment their swimming speed reduces to a low and relatively stable level.

Source: Republished with permission of Proceedings of the Royal Society Biological Sciences, from Figure 1, Page 2132 from Burt de Perera, T. (2004). Fish can encode order in their spatial map. Proceedings of the Royal Society of London B 271: 2131–2134. See https://royalsocietypublishing.org/doi/10.1098/rspb.2004.2867; permission conveyed through Copyright Clearance Center, Inc.

Hearing

Crustaceans and molluscs are amongst the animals most sensitive to low frequency ranges but hearing ranges do not neatly reflect the taxonomic distribution of animal forms (Table 30.7). Some of the lowest frequency sounds are detected by mammals, but so are some of the highest—well up into the **ultrasonic** range (thousands of kHz) in bats and cetaceans.

The hearing ranges of animals and their regions of greatest sensitivity have evolved under environmental pressures associated with lifestyle and survival. They especially reflect the exploitation of the sound environment to detect prey, avoid predation, and communicate between individuals.

The ability of crustaceans (crabs and lobsters) and molluscs to detect low frequency sounds, such as those produced by the

movements of fishes and other sea creatures, may be relevant to understanding the effects of sound pollution caused by marine engines, drilling rigs, and other human devices.

In animals with complex social structures, there is an extraordinary subtlety in the types of sounds used and a corresponding ability to detect minute differences, with human speech probably being the most complex sound world of all. Besides variations in volume and pitch, some animals exploit mechanisms similar to those used in radio transmission and sonar technology, including frequency modulation (FM), echo delay, and Doppler shift, to visualize their environment through sound.

Hearing in invertebrates

The ability of invertebrates to hear is very variable and some can detect both vibrations and sound. The most sensitive organ for

Table 30.7 The approximate auditory ranges for animals of different groups

Group	Animal	Sensitivity range (Hz)
Mammals	Human	20–20,000
	Rat	400–75,000
	Cat	55–79,000
	Dog	64–44,000
	Elephant	17–10,000
	Little brown bat	10,300–115,000
	Bottlenose dolphin	150–150,000
Amphibians	Bullfrog	100–3000
Reptiles	Gecko	150–5600
	Pygopod lizard	1000–20,000
Birds	Chicken	125–2000
	Canary	250–8000
	Barn owl	100–12,000
Fish	Catfish (bony)	50–6000
	Sharks (cartilaginous)	100–1400
Molluscs	Octopus	0.1–1000
	Cuttlefish	20–600
Arthropods	Greater wax moth	20,000–300,000
	Gypsy moth	25,000–150,000
	Mosquito	100–2000
	Norway lobster	20–180
	Hermit crab	5–400

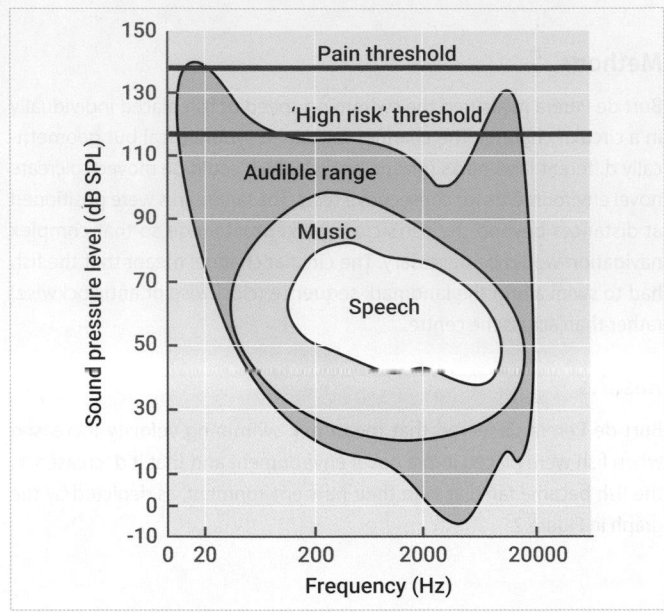

Figure 30.21 The human audiogram. A mapping of the relationship between sound frequency (Hertz; logarithmic scale) and volume (decibels, representing the sound pressure level; linear scale).

Source: Jeremy Wolfe, Keith Kluender, Dennis Levi, Linda Bartoshuk, Rachel Herz, Roberta Klatzky, and Daniel Merfeld. *Sensation & Perception*, Sixth Edition, 2020, Oxford University Press.

detecting sound in insects is the tympanal organ, found at different locations all over an insect's body including on their legs, abdomen, and thorax.

Charles Darwin, who famously studied earthworms over four decades, observed that worm activity was unaffected by the loudest sounds he could make on his piano *unless* their container was placed directly on the instrument. Subsequent studies have confirmed that earthworms have no auditory apparatus, so cannot 'hear', but that they are highly responsive to changes in pressure. This can be demonstrated by stamping persistently on a wet lawn in spring or autumn: after a while worms may begin to emerge through the topsoil onto the surface.

Hearing in vertebrates

The human auditory system is the best studied of all systems of hearing. Its range of sound detection lies approximately between 20 Hz and 20 kHz (20,000 Hz), as illustrated in Figure 30.21. In general, lower frequencies require greater amplification if they are to be detected, but the relationship is far from simple. In practice, the ability to hear a sound is also affected by its quality (the purity of the wave form) and background noise. The human brain is particularly good at filtering the information coming from the ears: for example, concentrating on a single voice in a crowded room

or following the sound of a single instrument while listening to an orchestra.

Individuals vary considerably, however, and it is well known that the range decreases, especially at the upper end, with age. Also, pitch sensitivity is affected by volume: we may be able to detect vibrations as low as 12 Hz if the volume is high enough, although this will quickly become uncomfortable; notice in Figure 30.21 how auditory experience above ~130 dB at low frequencies is experienced as pain. Although auditory scientists may study hearing under ideal conditions, in practice what we hear depends on the nature and purity of the sound, the resonance of the environment, and the level of background noise. Chapter 18 describes the structure of the human ear and the mechanisms of hearing.

The human range is quite limited compared to those of many other animals. The overall range of values across animal groups covers at least seven orders of magnitude (0.1 Hz in octopus to 150 kHz in dolphin) and when plotting them graphically it is conventional to use a logarithmic scale, as in Figure 30.21. Note that animals vary in the extent of their ranges, as well as in their minimum and maximum sensitivities. It is also the case that hearing sensitivity is not uniform over the entire range—there is typically a narrow region in which hearing is most acute.

Ear structures

The middle ears of mammals possess three auditory **ossicles**. These tiny bones transmit sound-induced vibrations from the eardrum to the sensory system and auditory nerve in the inner ear. The bones increase auditory sensitivity and allow most mammalian species to detect higher frequencies than other vertebrates (amphibians, reptiles, birds) which have only a single ossicle.

The ears of bony fishes comprise a complex of small bones called the Weberian apparatus which are derived from parts of the skeleton. The apparatus is connected directly to the fish's swim bladder, enabling detection of pressure waves in the water in addition to the oscillatory vibrations of water molecules. This increases the range of their hearing and sounds to which they can respond. Cartilaginous fishes have similar ear structures but have no swim bladders, so they can only detect particle-transmitted sounds.

Insects have two main types of hearing systems. Most insects, like many other arthropods, have **tympanal** ears in which a cuticular membrane is stretched over part of the body containing an air sac. An associated sensory nerve connects to the central nervous system. Tympanal ears are sensitive to both particle vibrations and pressure waves and can detect frequencies up to 300 kHz or more. They pick up communication sounds over large distances but may also intercept the echolocation signals of predatory insectivorous bats.

In contrast, insects including mosquitoes, fruit flies, and bees have auditory organs located in their antennae. These operate over short distances and are sensitive to particle velocities at frequencies below 1 kHz. They can detect the wing beats of other individuals and are important for social organization and reproductive behaviour.

Echolocation

Echolocation, sometimes called **sonar**, refers to the ability of some mammals, birds, and insects to sense the environment by emitting tones and neurally analysing the reflected sound. Information obtained this way is used for navigation, foraging, predator avoidance, and social communication. Much of the effectiveness of animal sonar relies on the brain being able to analyse and compare the auditory responses of the two ears.

Bats famously locate insect prey using sonar, exploiting frequencies far above the range which the human ear can detect. Each species of bat has its own characteristic frequency or set of frequencies and people who listen for bats with electronic detectors (which scale the sound down to human auditory frequencies) often identify them by that criterion alone. Bat calls may be composed of several elements:

- FM (frequency modulation) calls which sweep over a range of wavelengths. These are best for locating the distance of a target and for distinguishing targets which are close together.

- CF (constant frequency) calls comprising single tones, often with multiple harmonics. These are used for tracking moving targets, such as flying insects. The bat can adjust the sound to the speed of its prey, and its brain (the auditory cortex) can compensate for any Doppler effects as the insect flaps its wings or as the relative speed of the two animals changes.

- Click patterns. These are best for finding the range of solid surfaces and for foraging: the click frequency typically increases as the landing surface gets closer or as the bat homes in on its prey.

Among other mammals, the toothed whales (dolphins, killer whales, sperm whales) produce high frequency, forward directed clicks from the head. These sounds are focused by the 'melon', the large ball of fatty tissue which is prominent on the front of the head. Reflected sounds are picked up by fatty regions around the lower jaw and the ears. Baleen whales communicate socially by low frequency sounds, which can travel very long distances through the water, but they probably do not use sonar as it would be unhelpful in finding the plankton on which they feed.

Shrews, tenrecs, and rats use a form of echolocation to navigate their environments, and some birds may do the same. These are less sophisticated sonar systems than those of bats or whales. They are probably of limited use for hunting or defence, although cave-dwelling swifts may use sonar to catch flying insects. Humans with sight impairment can be trained to use echolocation; some individuals claim that it can nearly replace vision for safely navigating complex urban settings.

 Check your understanding of the concepts covered in this section by answering the questions in the e-book.

30.4 The gravitational environment

Gravity is the fundamental force in the space–time continuum which gives matter weight. It operates between all objects in the universe and becomes especially noticeable when there is a large size difference, as between Earth and our bodies.

The gravitational effect of Earth causes objects to accelerate towards its centre at a rate of 9.8 m.s^{-2}. This is sufficient to hold all living and non-living things to the surface of the planet, against the tangential outward momentum caused by its rotation. Fluids (water) and atmospheric gases are similarly retained, leading to the ambient pressures which we can measure in the environment.

The gravitational effects of the sun and the moon, as orbital movements change their positions relative to Earth, cause bulk movements of the sea and result in tides. Tidal movements ('high' and 'low' tides) occur nearly twice in each 24-h cycle, unless geographical features cause over-riding water flows in particular localities. Variations in tide height occur over months and years depending on the proximities of the sun and moon to Earth. Their degree of alignment determines whether their individual influences on the sea are additive, resulting in high 'spring' tides (sun and moon on the same side of Earth; the name has nothing to do with the season of spring) and low 'neap' tides (when they are on opposite sides).

Plants and gravity

Gravity has a profound effect on plant growth and development. Without **graviperception** (sensitivity to gravitational forces), plants would be unable to utilize essential environmental resources. Roots are positively **gravitropic**: they grow in th direction of gravity, toward the centre of the planet. Shoots are negatively gravitropic and grow against the direction of gravity.

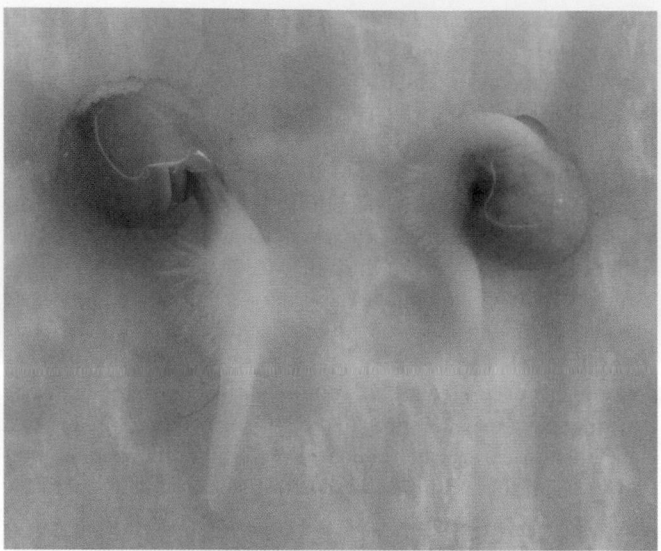

Figure 30.22 Root gravitropism in germinating seedlings. Germinating seeds of mustard (*Sinapsis alba*) show how different orientations of the seed require the emerging root to compensate its growth and change direction so that the root grows eventually grows downwards.

Figure 30.23 Stem gravitropism. Shoots of Russian sage (*Perovskia atriplicifolia*) that were originally upright but have been knocked down by wind reorientate their growth direction in a negatively gravitropic manner.

Graviperception is carried out by **statoliths**, specialized plastids (see Chapter 28) which accumulate starch grains and thus become dense and heavy within the cytoplasm. Graviperceptive cells in roots, are located at the root tip and termed **columella** cells. In the stem, the statoliths are present in the parenchyma cells surrounding the vascular tissues. In both cases, the statoliths falling to the bottom of the cell under the influence of gravity initiate a signal transduction pathway which determines the direction of organ growth. Figure 30.22 shows root gravitropism in two germinating seedlings.

The hormone **auxin** mediates the growth pattern in response to gravity. Its concentration across the stem or root changes in response to statolith orientation. Cells respond to auxin by expanding at different rates, causing curvature in the tissue and growth in the appropriate direction.

This effect is best observed in herbaceous plants in early summer when they show extensive growth; when blown down flat by wind or rain, they will curve back up to maintain an erect habit, as in Figure 30.23. Gravity sensing in trees ensures that their main trunk grows upwards, a habit best seen in large conifer trees where the main trunk is usually perfectly vertical.

Seedlings are especially sensitive to gravity, resulting in the opposite growth directions for the new shoots and roots. Seeds scattered on the soil in random orientation always achieve upward shoot growth and downward root growth as they germinate.

The effects of gravity on plant development and growth have been demonstrated by experiments in the weightless conditions of space (as discussed further in Real World View 30.1). They would be important considerations were mankind ever to try to colonize other planets and produce food plants.

PAUSE AND THINK

Consider these two observations:

1. The stem of a hyacinth bulb which comes into flower on my window ledge in the spring bends towards the light; I have to turn it regularly to avoid this happening.

2. Outside in my garden, a small apple tree which stands against a high brick wall has a main trunk which grows vertically upwards.

Suggest appropriate technical phrases to describe the behaviour of these plants.

Answers: The hyacinth stem is positively phototropic and the effect of this overrides the negative geotropism which otherwise dictates the way it grows.

The apple tree trunk is negatively geotropic and the effect of this overrides any positive phototropism the tree may possess.

Aquatic animals and gravity

A fundamental difference between animals and plants, as discussed at the start of this section, lies in their capacity for independent movement. This means that the responses of most animals to gravity do not need to involve growth. Nevertheless, animals are highly sensitive to gravity, usually showing responses in both body orientation and the direction of movement.

Free-swimming animals with radially symmetrical bodies, such as jellyfish and the planktonic larvae of many invertebrates, appear not to show a preference for their orientation *about* their longitudinal axis. However, they do show gravitational awareness in their choice of axis position, either horizontal to Earth's surface or with a distinctive 'head up' position in the water.

Animals which live in water maintain buoyancy as a resistance to the downward accelerative force of gravity. In essence, *flotation* at the surface requires the displacement of a weight of water at least

REAL WORLD VIEW 30.1   Growing plants in space

The evolution of plants over the last 400 million years has occurred under the influence of Earth's gravitational field. But how would plants behave if gravity was not present? And would it be possible to grow plants, especially food plants, on space stations, or even on other planets . . .?

Growing plants in conditions of low gravity may lead to discoveries about gravisensing and plant development. Experiments can be done on Earth's surface but they are difficult to design, carry out, and control. However, the development of manned space flight and orbiting space stations in the past few decades has enabled researchers to answer these questions, and explore how plants respond to microgravity conditions.

Plant growth in the absence of gravity

Let's think about the first question: how would plants behave if gravity was not present? Researchers used *Arabidopsis thaliana*, or rice seeds, to try and answer this question. When the rice seeds are grown in the dark in microgravity, the researchers found that shoot and root orientation becomes random, as depicted in Figure 1. When they are grown in microgravity but with directional illumination, roots grow away from the light and shoots grow towards it, just as on Earth.

These experiments reveal that the strong negatively phototropic response of roots is hidden on Earth by their positive gravitropic response. This means that the environmental control of root and shoot directionality is more complex than was first thought.

Growing food plants in space

What about growing food plants in space? Culture systems on orbiting space stations have enabled the growth of leafy vegetables such as mizuna and lettuce, as you can see in Figure 2. Researchers found that they develop perfectly well with automated water and nutrient supply systems, illuminated by banks of LEDs. These vegetables have been eaten by astronauts!

On the ground

Rice   In microgravity   Arabidopsis

Figure 1 Rice (a,b) and *Arabidopsis* (c,d) seedlings grown in the dark in microgravity (b,d) on a Space Shuttle flight compared to those grown on the ground (a,c) under 1 g gravity.

Source: Courtesy of Takayuki Hoson.

Figure 2 Four week old Chinese cabbage plants growing on the International Space Station.

Source: NASA.

Plants have also been grown from seed through to seed, demonstrating the potential for continual cultivation. In fact, there may be several advantages to growing plants on long space trips:

- Fixation and reuse of the carbon dioxide breathed out by the astronauts.
- Reductions in weight of cargo required to take pre-made food from Earth.
- Psychological benefits to the crew in having something to look after on a long trip.

The next steps for future space travel would be to develop systems for growing plants with high biomass, which could feed astronauts over many months, say on a trip to Mars. In the meantime, researchers will need to consider a more significant logistical problem: they will need to consider the design of plant culture systems on the moon, or other planets, where conditions are different from those on Earth. Besides zero gravity and the lack of atmosphere, there are higher levels of harmful radiation, low temperatures, different light patterns, and unpredictable supplies of water and nutrients.

equivalent to the body's weight (Archimedes' principle). The displacement counteracts the pressure of water on the body. A simple way to understand this is to take a ball of modelling clay and drop it into a bowl of water: it will sink. If the clay is flattened and formed into a boat shape, it can be made to float. The weight of the clay is unchanged but the amount of water displaced is greater.

Animals which are neutrally buoyant in mid-water, such as fishes, displace an *exactly equivalent* amount of water. Neutral buoyancy is the minimal energy state because no effort is required to adjust the depth, and it is the condition in which swimming is most efficient. However, moving to other depths creates a challenge: gravity causes the water pressure to increase with depth, compressing any air spaces in the body and making the animal less buoyant.

Bony fishes deal with this problem by altering their swim bladder volume (see Chapter 31). The swim bladder is a gas-filled bag in the dorsal part of the body cavity, made of strong collagen. The fish adjusts the volume of the bladder to achieve neutral buoyancy at whatever depth it swims, accumulating gas as it goes deeper and releasing the expanding air as it moves to shallower water.

Cartilaginous fishes have no swim bladders. They generally have less dense bodies than bony fishes, due to a higher tissue lipid content. This enables them to achieve neutral buoyancy by swimming movements—in much the same way as a bird stays airborne by flying—within a limited range of depths. Large pelagic sharks have to keep swimming, otherwise they will sink.

▶ For more on buoyancy, see Chapter 31.

Air-breathing vertebrates which dive, including seals, turtles, sea snakes, penguins, and other marine birds, breathe out immediately before submerging. This reduces their buoyancy and thus the effort required to swim down. Some of these animals can adjust the amount of air expelled according to the anticipated depth of the dive. As they dive, air in the lungs and any trapped in

feathers or hair becomes compressed, reducing its buoyant effects. The animal may therefore achieve neutral buoyancy at depth, once the effort of submerging has been overcome.

Airborne animals and gravity

Animals which fly resist the effect of gravity and, like aquatic animals, achieve neutral buoyancy to stay airborne. However, the density of air (1.23 kg.m^{-3} at sea level) is 800 times less than that of water (1000 kg.m^{-3}) and only the tiniest animals, such as small flies, can displace a sufficient weight of air to exploit this effect. They still need to beat their wings to hover.

Larger fliers, including insects, birds, bats, and gliding marsupials, exploit additional upward forces—those produced by thermal or other air currents or by aerodynamic forces across the surfaces of wings, whether static or beating (see Chapter 31). All these approaches require the animal to be constantly moving, so entail a significant energy penalty.

Very large or heavy birds have a particular problem in getting airborne. They may need a lengthy 'run' to take off into the wind or may have to make use of inclines. Their wing beats are typically faster and more forceful as they establish an aerodynamic position. Large water birds such as swans and geese use their webbed feet to achieve some additional propulsion over the water surface as they attempt to get airborne.

Terrestrial animals and gravity

For animals on the ground, whose weight is supported neither by water nor air, gravity imposes mechanical stresses which must be resisted entirely by structural adaptations within the body. Small, soft-bodied terrestrial animals such as worms and slugs rely on the strength of the external skin (**integument**) and the cohesion of internal tissues to maintain body shape. In these cases, hydrostatic support is provided by maintaining the body fluids at a pressure slightly above ambient, rather like keeping a sausage skin tight by filling it with water.

The extent to which this is adaptively successful is essentially a surface area:volume ratio problem (see Chapter 29). Animals which are basically round in cross section, even if elongated, can rely solely on hydrostatic support provided the ratio between the tubal surface area and body volume is high. If the animal gets larger by increasing its diameter, the ratio falls and the pressure (force per unit area) put on the integument by the body fluid increases. The body starts to flatten and the limiting point will be reached either when the tissues of the integument can no longer bear the strain generated by the fluid or when sideways roll of the body becomes unmanageable.

This purely mechanical effect puts a size limit on soft-bodied terrestrial animals. To reach a greater size, the body requires additional support in the form of a hard skeleton, either external (as in arthropods) or internal (as in vertebrates). But hard materials are dense and add weight, so the skeleton itself contributes to the gravitational burden which must be supported.

The legs of elephants, the heaviest terrestrial animals currently living, are limited in length by a compromise between bone strength (largely determined by width and density) and the additional weight. Longer bones are more inclined to bend and break under stress. It is likely that the largest size reached by dinosaurs was determined by these mechanical imperatives. Many such animals grew disproportionately in length rather than height, presumably as an offset against the limited ability of longer bones to resist distortion.

 Check your understanding of the concepts covered in this section by answering the questions in the e-book.

30.5 **The magnetic and electrical environment**

The magnetic forces on Earth are much weaker than those of gravity but are nevertheless important in environmental sensing and the responses of organisms. They result from the presence of a very large quantity of molten iron in Earth's core.

The magnetic poles align closely but not exactly to the axial poles of the planet. We arbitrarily designate the Arctic pole as 'North' and the Antarctic pole as 'South' and, by a strange convention, label the ends of compass needles which point to them in that way. (Like poles repel and opposite poles attract, so they should really be labelled the other way round!)

Earth's magnetic field is far from uniformly distributed over the planet's surface and it is moving gradually westwards due to the flow of the molten iron mass. The overall polarity has reversed, or flipped, several times in the planet's history, most recently some 781,000 years ago, in the middle of the period occupied by our hominid relative *Homo erectus*; the next flip may be imminent.

The strength of a magnetic field (actually, 'magnetic flux density') is expressed in Tesla (T). The magnetic field at Earth's surface is relatively weak—between 33 µT (at the Equator, acting horizontally) and 67 µT (at the poles, acting vertically) but it is enough to protect us from the charged particles of solar winds. For comparison,

an iron bar can be magnetized up to 1.6 T and the field in a medical MRI scanner may be in the 0.5–3.0 T range.

All living organisms are exposed to Earth's magnetic field but designing experiments to find out if they respond to it can be very difficult. It is possible to shield the organism using iron plates and then use electric coils to create an artificial field in the appropriate range (10–100 µT), but there are many technical difficulties to overcome, especially to ensure that other environmental influences such as heat, light, and sound are not changed. Many experiments have been poorly controlled and given unreliable results, but a few have been well designed and produced interesting effects.

Plants and magnetism

Magnetic fields lower in strength than that of Earth cause pea, bean, and other plant seedlings to grow with slightly longer shoots than normal. This is due to cell elongation and may result from increased water uptake or changes in membrane function and metabolism.

Slightly higher field strengths increase the rate of seed germination in broad beans, soy beans, and barley, and can also increase the dry matter content (i.e. the non-water material of the tissues) of seedlings. As with geotropism, the hormone auxin may mediate these responses. Magnetic fields affect the transition of *Arabidopsis* plants to the flowering state, with molecular analyses suggesting that genes associated with cryptochrome signalling (see Section 30.2) are involved.

Much higher magnetic fields will produce larger effects on photosynthesis, flower structure, and the biochemical content of cells. These do not realistically represent conditions on Earth so it is hard to say if the results are relevant, although some may have practical applications in plant husbandry.

The best-controlled experiments show that the effects of magnetic fields interact with those of gravity. This makes experiments on seedlings taken into space, where both gravitational and magnetic strengths are virtually zero, hard to interpret.

There is no consensus about the adaptive advantage to plants of being responsive to magnetic fields. Plants are sessile and do not navigate as animals do, but it could be useful for them to have a reference against which to compare daily and yearly changes in light direction.

Animals and magnetism

There is a great deal of evidence, experimental and anecdotal, that animals perceive Earth's magnetic field and respond to it. As with plant experiments, laboratory investigations involve artificial magnetic fields. Field investigations may involve attaching a bar magnet to an unrestrained animal and observing its behaviour. The bar magnet creates a stronger local field than that of Earth but its range is restricted to the individual animal being tested.

Magnetoreception has been demonstrated in animals which migrate long distances, including several birds, or change their geographical location at particular times in their lives (sea turtles, bats), but also in non-migratory animals (flies, Lepidoptera, chickens, mole rats, lobsters, newts, frog tadpoles). Homing pigeons rely on magnetic information as well as on topographical features and the direction of sunlight to navigate their way home.

Thus it is very likely that magnetoreception is a widespread animal phenomenon and many further examples remain to be discovered.

Sensing electricity

Electric fields can provide animals with information about their environment, particularly the location of other prey animals. All animals give off weak electrical signals, caused by rhythmic muscle contractions. These can allow electroreceptive predators to seek out prey.

Electroreceptive animals include fish, amphibians, dolphins, monotremes, and several other vertebrates. They depend on an electrically conductive medium—water or wet soil.

Sharks and rays use electroreception to hunt other animals. As you can see in Figure 30.24, their receptors are located in specialized organs, the **Ampullae of Lorenzini**, which are located on their heads and pectoral fins. They vary in number, with as many as 2800 around the head of the hammerhead shark.

Until very recently, electroreception had only been demonstrated in vertebrates. It is now known that bumblebees use sensory hairs to detect electric fields. They can do this in dry air, so the mechanism does not require direct electrical conduction. It is likely that there are many more animals using this sense that we are yet to discover.

Go to the e-book to explore more about electroreception in bumblebees.

Check your understanding of the concepts covered in this section by answering the questions in the e-book.

30.6 The thermal environment

'Heat' is the energy in molecules which makes them vibrate. This is a form of 'kinetic' energy because it can be transformed immediately into 'work'. (It is equivalent to 'free' energy in thermodynamics). Kinetic energy should be distinguished from 'potential' energy: this is the energy in chemical bonds, released in response to combustion or metabolism, or that which emerges as an object falls due to gravity.

All molecules possess kinetic energy all the time and the hotter they are the more they have. The molecules of a substance at the absolute zero temperature (−273.15 °C or 0 K) would be completely still, with no kinetic energy at all. But this zero point is logically and physically impossible to reach.

Temperature is our way of indicating the amount of heat energy held in a substance, a body, or an environmental system. The First Law of Thermodynamics states that energy cannot be created or destroyed. The Second Law, the law relating to **entropy**, tells us that heat energy can only be transferred from a warm body to a cooler one. In other words, as the temperature of one body decreases, the temperature of another must increase, by an amount equivalent to the heat which has been swapped. The exceptions to this are when energy is stored or liberated in a different form (a chemical change) or when there is a change of state (solid to liquid or liquid to gas).

It is important for biologists to understand energy conservation and the difference between heat and temperature. Imagine sawing a wooden block into two equal parts. If the original block had a uniform temperature of 10 °C and the parts are also at 10 °C, then each part has half the heat but the same temperature as the original block. The total heat is the same.

But imagine too that the sawing process heats up the saw blade and produces a pile of warm sawdust. Where does this extra heat come from? It comes from the muscular energy of the arm moving the saw (which came from energy in food, which came from

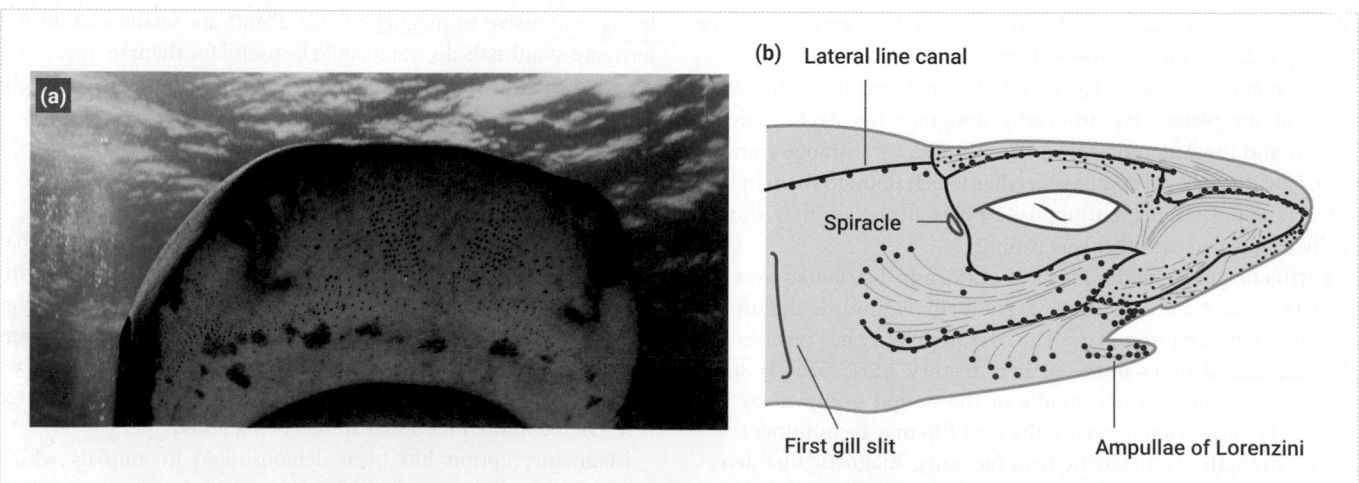

(b) Lateral line canal

Spiracle

First gill slit

Ampullae of Lorenzini

(a)

Figure 30.24 The electroreceptive system of a shark: Ampullae of Lorenzini. (a) Lorenzini pores on the snout of a tiger shark. (b) The locations of Ampullae of Lorenzini along the side of a shark's head.
Source: (a) Albert kok/Wikimedia Commons/CC BY-SA 3.0. (b) Chris_huh/Wikimedia Commons/Public Domain.

the sun) and from the breaking of molecular bonds in the wood fibres as the saw cuts through them. Thus some kinetic energy has come from outside the block and some potential energy stored chemically within the wood has been converted into kinetic, heat energy.

Most of the energy stored in our block could then be released into the environment by burning it. (This is essentially what nutritionists do to determine the energy content of food.) The warmth of the resulting ash would represent the remaining kinetic energy. This will decrease as the ash cools to the temperature of the environment, and the environment will warm up by an equivalent amount.

Biological organisms participate in transferring energy from one object to another and in its conversion between kinetic and potential states. Energy must be conserved, so it should always be (theoretically) possible to account for all of it. Energy conversion in organisms is mostly accounted for within metabolism (see Modules 1 and 3). Transfer and conversion are never 100% efficient, so metabolic processes always release some heat energy to the environment.

Energy transfer, especially between organisms and their environment, is crucial to biological function and survival, so we need to understand its basic features.

Transfer of heat energy

Free heat energy can move in *four* ways: **conduction**, **convection**, **radiation**, and **evaporation**. All of these are used by biological organisms. Figure 30.25a illustrates the basic principles of each method of heat transfer.

- Conduction: Transmission of heat by transfer of kinetic energy, from particles (atoms/molecules) with more (warmer, shown red in the figure) to those with less (cooler, blue). Kinetic energy is represented by the purple arrow.
- Convection: Movement of heat from one location to another by means of a moving fluid.
- Radiation: Transmission of energy as long wavelength electromagnetic waves (infrared, 10^{-6}–10^{-3} m). Requires no physical medium.
- Evaporation: Transmission of energy by change of state (normally liquid to gas). The liquid and gas are at the same temperature.

Let's look at each of these in a bit more detail. We will keep referring to the illustrations in Figure 30.25, so keep this to hand.

Conduction

Conduction is energy transfer by the kinetic movement of adjacent molecules, essentially as they bump into one another. Molecules with greater energy (warmer, moving more) transfer some of what they have to those with less (cooler, moving less).

Consider the example of touching a cold rock on a beach. When you touch that cold rock, some of the energy in your fingers is conducted to the rock. Your fingers lose heat and the molecules in your skin vibrate more slowly, while the rock warms up and its

molecules vibrate slightly faster. The rock 'feels' cold because sensors in your skin detect the energy movement.

The *rate* at which heat is transferred by conduction, however, depends on the difference in temperature: the bigger the difference, the faster it moves. This is Newton's Law of Cooling, which is illustrated in Figure 30.26. What this means is that the rate of transfer slows down as the objects approach similar temperatures. We subjectively feel how cold an object is by the speed at which we lose heat to it.

Convection

Convection is the transfer of heat by the physical repositioning of molecules—their movement from one location to another. This happens to a significant extent only in fluids (liquids and gases); solids do not permit biologically significant molecular translocation. The molecules in fluids are constantly moving and warmer ones travel further and faster than cooler ones. Moving molecules can mix, a process called **diffusion**, so the temperature of the fluid eventually becomes uniform throughout. How long this actually takes depends on:

- the type of fluid (liquid or gas)
- its physical properties (viscosity, density)
- its chemistry (especially its ability to mix with and dissolve in other fluids)
- the temperature difference between the warm and cold regions
- the presence of any additional stirring forces (winds, currents, animals).

Radiation

Radiation is transfer of heat in electromagnetic form. Energy from the sun reaches Earth in this manner. Radiation moves in straight lines and can pass through a vacuum because, unlike convection and conduction, it does not involve the movement of molecules or other particles. In the electromagnetic spectrum (which is illustrated in Figure 30.7), thermal radiation occurs mostly in the infrared region, just beyond the long wavelength end of the visible spectrum. All bodies emit radiant heat—called thermal radiation—to a lesser or greater extent, so transfer of heat by radiation between two objects occurs in both directions. Figure 30.25b therefore illustrates the *net* direction of heat energy transfer. The amount of energy transmitted by thermal radiation depends on:

- the relative temperatures of the emitting and absorbing bodies
- their colours
- their surface qualities (reflectivity).

Black, dull surfaces are better emitters than coloured, shiny, polished, or white ones, as demonstrated by thermal imaging (which you can see in Figure 30.27). Black, dull surfaces are also better at absorbing thermal radiation, so a good emitter is also a good absorber, as shown in Figure 30.25b. These properties are of great importance in the emission and absorption of heat by, for example, animal pelts or human clothing. An 'ideal' black body absorbs and radiates all wavelengths perfectly, although in practice no such objects exist in the biological world.

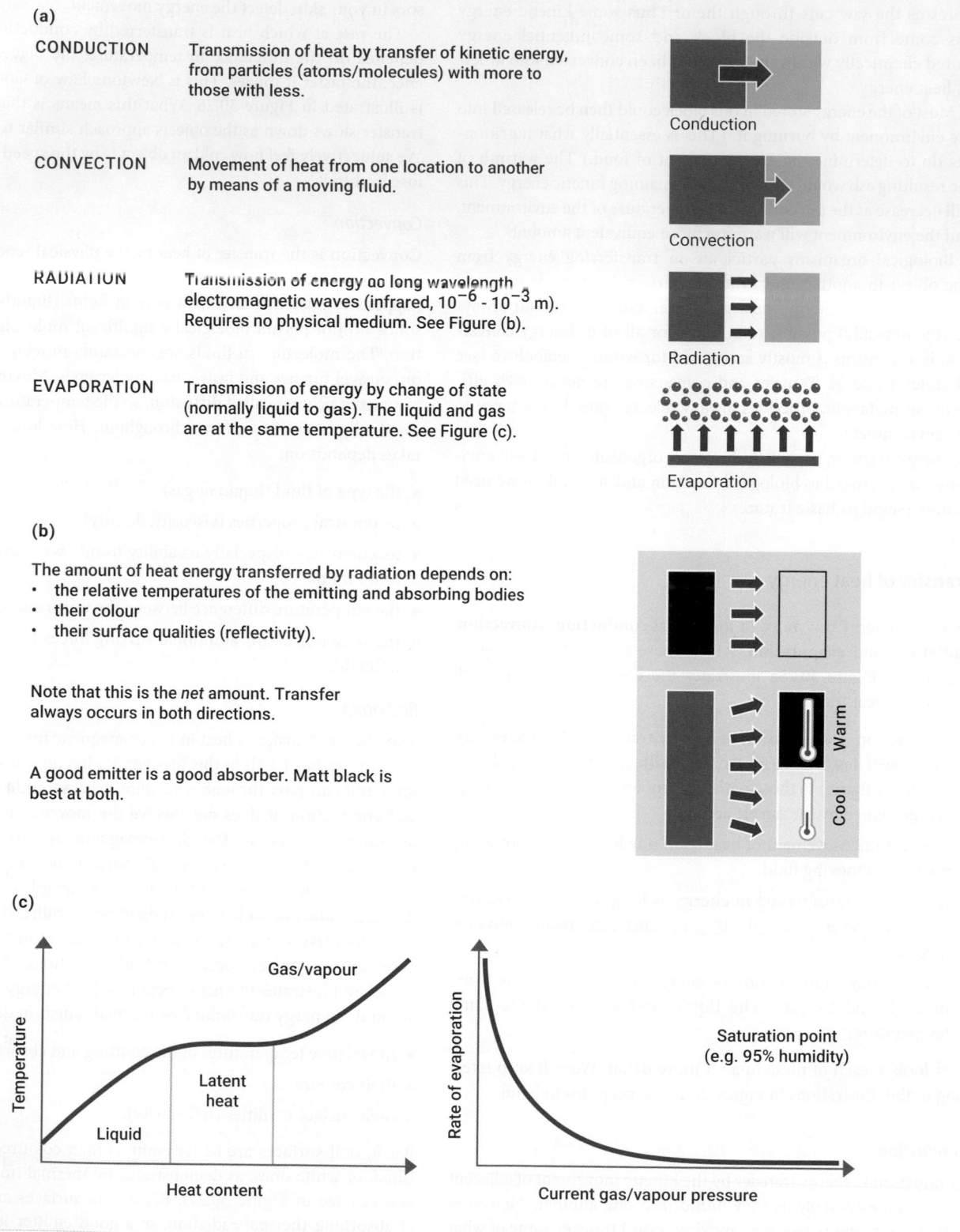

(a)

CONDUCTION Transmission of heat by transfer of kinetic energy, from particles (atoms/molecules) with more to those with less.

Conduction

CONVECTION Movement of heat from one location to another by means of a moving fluid.

Convection

RADIATION Transmission of energy as long wavelength electromagnetic waves (infrared, 10^{-6} - 10^{-3} m). Requires no physical medium. See Figure (b).

Radiation

EVAPORATION Transmission of energy by change of state (normally liquid to gas). The liquid and gas are at the same temperature. See Figure (c).

Evaporation

(b)

The amount of heat energy transferred by radiation depends on:
• the relative temperatures of the emitting and absorbing bodies
• their colour
• their surface qualities (reflectivity).

Note that this is the *net* amount. Transfer always occurs in both directions.

A good emitter is a good absorber. Matt black is best at both.

Warm

Cool

(c)

Temperature

Gas/vapour

Latent heat

Liquid

Heat content

Rate of evaporation

Saturation point (e.g. 95% humidity)

Current gas/vapour pressure

Figure 30.25 The movement of free heat energy. (a) Methods of heat transfer. (b) Heat transfer by radiation. (c) Evaporation is a change of state (liquid to gas) at constant temperature. The energy required, which changes the way the molecules move, is called 'latent heat'.

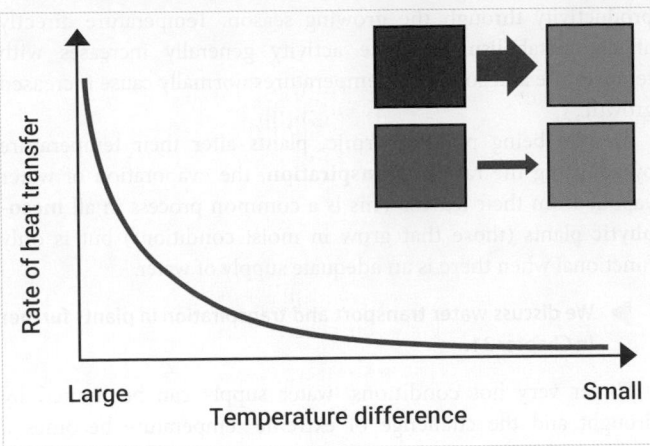

Figure 30.26 Newton's Law of cooling. The rate of transfer of heat from one body to another is proportional to the difference in their temperatures.

Evaporation

Heat transfer by evaporation occurs when a fluid changes state from a liquid into a gas or vapour. Energy is released by the *change of state itself*, even though the gas or vapour has the same temperature as the liquid from which it formed, as illustrated in the left graph of Figure 30.25c. This is why animals can lose heat energy by sweating, even in environments where conductive and convective heat loss are impossible because the air temperature is higher than body temperature.

In physical terms, the rate of evaporation of a liquid depends its volatility and the current pressure of the resulting gas. As more liquid evaporates, the gas pressure rises and the rate of evaporation

declines. The maximum gas pressure is the **saturation point**, where the gas starts to return to liquid, as shown in the right-hand graph in Figure 30.25c.

In biological systems, **water** turns into **water vapour** in air. We generally speak of **humidity** instead of water vapour pressure. When the humidity reaches about 95% the air is saturated and the vapour starts to turn into steam, fog, and eventually rain ('precipitation'). The saturation point depends on air temperature and atmospheric pressure. It is lower in cold temperatures and at lower atmospheric pressures. Under the right conditions, water vapour can turn directly into ice (frost) and vice versa.

Living organisms and the thermal environment

Across the planet, the range of environmental temperatures in which organisms may live extends from about −60 °C in the coldest Arctic and Antarctic regions to as much as 400 °C around areas of thermal activity such as hot springs and gas vents. Most life occurs well within these extremes.

At very low temperatures, an organism can survive if it can retain sufficient heat for metabolic and other process to occur and also avoid tissue damage caused by freezing. At very high temperatures, organisms must be able to avoid desiccation, metabolic disturbances, and burning. At less extreme temperatures, organisms still require protection from thermal damage, from sudden changes, and from adverse effects of cyclic environmental events.

For life to exist, the crucial factor is not the environmental temperature but *the rate of heat transfer*—the rate of heat loss or heat gain. As we have seen, this depends on the *difference* between the organism's temperature and that of its surroundings.

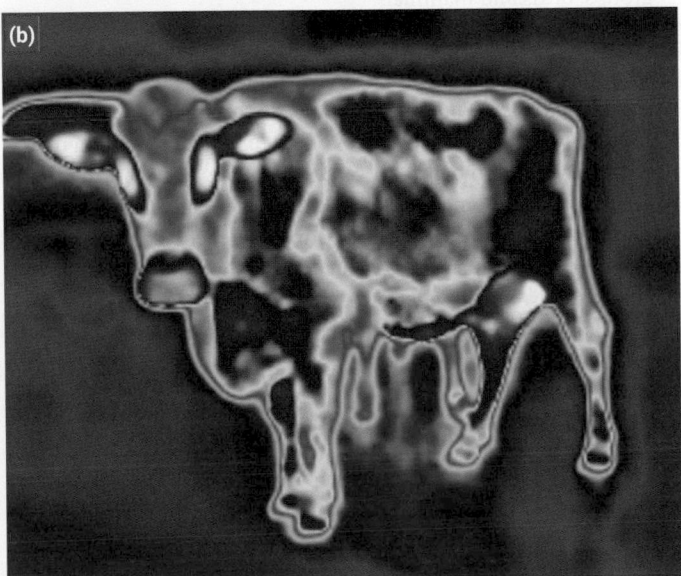

Figure 30.27 Thermal imaging. (a) A normal colour photo. (b) The same image taken with a thermal detection camera.
Source: Courtesy of Dr Matthew Bell, University of Nottingham.

Organisms have two options for cltering the rate of heat transfer between themselves and the environment: adjusting their insulation and changing their body temperature.

- Insulation against conduction is achieved by avoiding direct contact, reducing contact surface areas, or choosing surfaces with lower conduction rates (**conductivity**).

- Insulation against convection depends on reducing the flow of surrounding fluids and avoiding fast moving fluids.

- Insulation against radiation is achieved by finding cover and by altering surface colours and textures.

- Insulation against evaporative heat loss involves physiological reduction of body fluid loss and avoidance of exposure to environmental water.

Plants and animals use all of these mechanisms. Changing the body temperature is more complex because it depends on the organism's metabolism and its environmental circumstances:

- **Poikilothermic** (Greek *poikilos* = varied) organisms allow their temperature to fluctuate with that of their surroundings. In general terms, all plants fall into this category. Protists, parasites, many small aquatic invertebrates, and some soil-living terrestrial animals are poikilotherms, especially those with small bodies and high surface area/volume ratios where actively adjusting body heat would be energetically too costly.

- **Ectothermic** animals use environmental energy to *raise* their body temperature. They include insects, molluscs, fishes, amphibians, and reptiles.

- **Endothermic** animals (homeotherms or homoiotherms) use metabolic energy to *raise* their body temperature (often in addition to environmental heat). They comprise two distinctive groups: birds and mammals.

Body temperature in true poikilotherm organisms changes *passively*. Ectotherms and endotherms adjust their body temperature *actively* and *constantly*. Ectotherms do this through behaviour, changing location to locate sources of heat or cold or adapting their periods of activity to daily and yearly temperature cycles. In other words, they retain some control over their thermal balance.

The effort required to make adjustments depends on the extent to which the organism can reduce the temperature difference. Animals have a much wider repertoire of adaptations and adjustments than plants and have evolved diverse strategies to change the rate of heat transfer.

Plants and the thermal environment

Plants are poikilothermic and may experience huge variations in temperature. Being sessile, they cannot move away from temperature extremes as animals do; they have to remain where they are and withstand the temperature, be it hot or cold. Plant life has evolved to survive in habitats representing more than a 100 °C difference in thermal exposure (−60 °C to +60 °C). Adaptations involve changes in metabolism, physiology, and morphology.

For plants that live in temperate conditions, day-to-day variations in temperature have a profound effect on growth and productivity through the growing season. Temperature directly affects metabolism: enzyme activity generally increases with temperature and so higher temperatures normally cause increased growth.

Despite being poikilothermic, plants alter their temperature by adjusting the rate of **transpiration**, the evaporation of water vapour from their leaves. This is a common process in all **mesophytic** plants (those that grow in moist conditions) but is only functional when there is an adequate supply of water.

▶ We discuss water transport and transpiration in plants further in Chapter 31.

Under very hot conditions, water supply can be limited by drought and the challenge of extreme temperature becomes a problem of water availability. For plants that live in very cold climates, the major problem is not the low temperature itself but the freezing of water. Ice crystals form in the cells and cause structural damage, which normally results in cell death.

A mechanical response by some plants to extreme radiation is the reorientation of leaves. **Paraheliotropism** describes a change in the orientation of leaves so that they are parallel to the direction of sunlight. This minimizes the interception of light and reduces temperature. The converse of this is **diaheliotropism**: leaves are orientated at 90° to the incident radiation, maximizing light interception and heat uptake. Some plants move between these two strategies depending on the availability of water. Both are driven by changes in turgor pressure of motor cells at the leaf base in the pulvinus (see Section 30.3).

Temperature sensing by plants

Plants continuously monitor their thermal environment and use the information to regulate developmental processes:

- At latitudes with distinct seasons, a long spell of cold weather during the winter may be needed to induce flowering. This process is called **vernalization**.

- In species that grow in cold climates, seeds may require a period of cold exposure before they will germinate. This is called **stratification**.

- Pollination and fertilization of flowers are highly sensitive to temperature.

▶ We discuss these processes and their adaptive significance in Chapter 33.

The mechanisms of temperature sensing involves a combination of thermally induced changes in protein conformation, membrane fluidity, enzyme activity, and cytoskeletal dynamics. These processes occur in every cell, so every cell acts as a temperature sensor. Changes are transduced by cellular signalling pathways and influence gene expression and protein accumulation.

At the molecular level, exposure of plants to above-optimal temperatures for a short period of time will lead to the rapid synthesis and accumulation of **heat shock proteins**. These so-called chaperone molecules protect newly synthesized proteins from heat damage or denaturation (loss of shape and function). Heat shock proteins are expressed in response to a range of environment stresses that the plant might be subjected to, including drought, cold shock, and exposure to UV light.

Morphological adaptations by plants

Plants that grow in hot climates often have a morphology which minimizes temperature gain. The absorption of radiation can be reduced by having leaves and/or stems which are light in colour. Alternatively, leaves may be covered in short, soft, grey-coloured hairs which produce a shiny, reflective effect, called a **pubescent** surface. Examples of plants with reflective, pubescent leaves, are species of *Senecio*, which you can see in Figure 30.28.

Leaves may evolve an irregular or dissected outline, giving a greater ratio of edge to surface area; this increases the rates of transpiration from the leaf surface and thus the efficiency of cooling. A lobed leaf will have a temperature closer to air temperature, as compared with a smooth-shaped leaf of the same area which will be hotter.

At the other extreme are the evergreen conifers (gymnosperms). They thrive in cold environments and may form the climax vegetation in most of the world's cold snow bound latitudes. Their needle-shaped leaves have a very small tissue volume and a thick waxy cuticle which prevent water loss in strong winds. Internally, the central vein of each needle is protected by a layer of cells which can move water to the outer surfaces where it freezes, rather than freezing within the cells. The thin shape of the needles and the overall conical shape of the tree itself prevents long-term cover by snow and reduces the likelihood of structural breakage by heavy snow.

Deciduous tree species which inhabit cold or temperate climates lose their leaves in autumn, leaving dormant buds at the end of branches which break out into new leaves the following spring.

The survival of these buds without freezing, in air temperatures which may approach −40 °C, is achieved by supercooling. Ice forms in the extracellular spaces in the bud, termed the **apoplast**, producing an ice sink which draws water from the cells and prevents them from freezing. These buds remain in a supercooled state throughout the winter but, ironically, are much more at risk from being damaged by low temperature in early spring: after the supercooled state is lost, a sudden cold spell is likely to lead to bud death.

Animals and the thermal environment

A good way to envisage an organism's relationship with its thermal environment is to consider its zone of thermal neutrality (ZTN), and the ways in which it can raise or lower body temperature for normal function and survival.

A ZTN diagram, such as Figure 30.29a, shows the general relationship between environmental temperature, body temperature, and metabolic effort in biological organisms. Exact temperatures depend on the type of organism and its environment.

In the comfortable region, the organism 'feels' neither hot nor cold. Towards either end of the zone, it may feel warm or cool but tolerates this without making significant adjustments to metabolic rate. Beyond the upper and lower critical temperatures, the organism must use significant quantities of energy to either warm or cool itself—to counteract heat loss or gain from the environment—and there is a consequent increase in metabolic rate.

These principles apply to *all* living organisms although their ability to regulate temperature varies greatly. Poikilothermic and ectothermic animals typically have a narrow ZTN (remember, the zone is a range of environmental temperatures, not body temperatures). This can restrict their geographical movement or impose other behavioural limitations, especially as seasonal changes are taking place.

Homeotherms (mammals and birds) have a wide ZTN and can maintain a constant temperature outside this range too. In other words, they are especially resistant to temperature change. Look at

Figure 30.28 Plants with reflective leaves. These two species of *Senecio* (members of the daisy family) have leaves covered in soft, grey hairs which reflect sunlight, reducing the absorption of radiant heat. (a) *Senecio cineraria* 'Silver Dust'. (b) *Senecio haworthii* Woolly Senecio.

Source: (a) Image by Eugen Visan from Pixabay. (b) Winfried Bruenken (Amrum)/Wikimedia Commons/CC BY-SA 2.5.

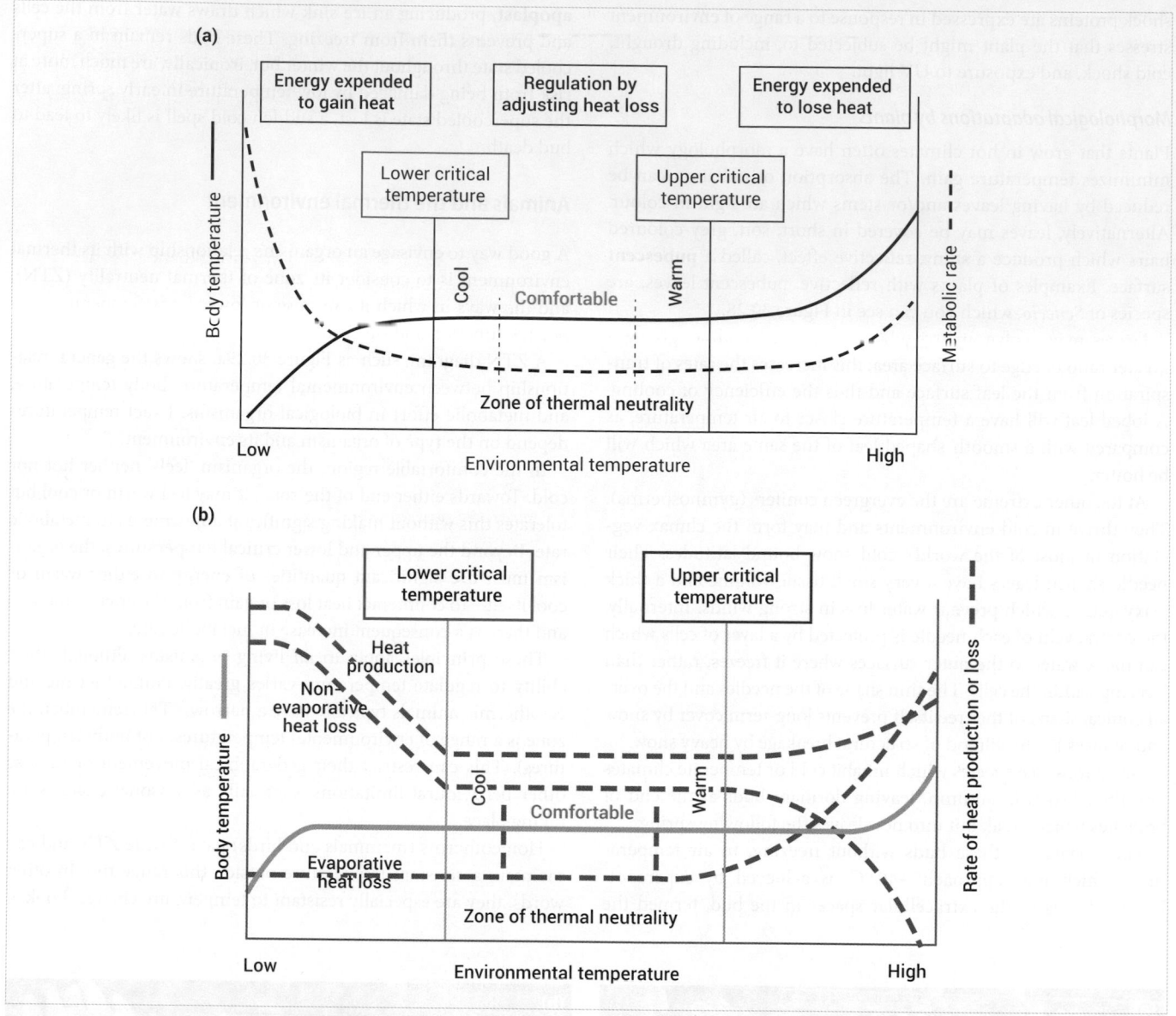

Figure 30.29 The zone of thermal neutrality. (a) The zone of thermal neutrality in relation to an organism's temperature. (b) The ZTN for *homeothermic* animals. Note how body temperature stays constant well beyond the ZTN.

Figure 30.29b, which depicts the ZTN for homeothermic animals, and notice how the body temperature remains constant either side of the ZTN. Deep body temperature, which for mammals is typically in the range of 36–40 °C and for birds 40–42 °C, normally varies by only ±1 °C or less. This is an important component of overall body homeostasis. It has helped to make homeotherms extraordinarily successful, colonizing a wide range of environments right across the planet.

Mechanisms of thermoregulation in animals

The vast range of methods which animals have for increasing and decreasing the temperature difference between their bodies and

the environment are summarized in Table 30.8, grouped by the heat transfer mechanisms discussed above.

The mechanisms available depend on the animal's environment and the temperature gradient. In many cases, for example in moving fluids, conduction and convection are both effective. In others, there may have to be a compromise between transfers by radiation and those through convection and conduction. Such balances can be achieved by changing the characteristics of pelt (feathers, fur), altering blood flow to different regions of the body, adapting the amount of insulation by subcutaneous fat, and by varying behaviour.

Evaporation allows heat loss when there is no difference in temperature between the body and the air, and even when the air

Table 30.8 Thermoregulation in animals, by adjustments to heat loss and gain

| | To raise body temperature: | | To lower body temperature: | |
	reduce heat loss by	increase heat gain by	increase heat loss by	reduce heat gain by
Conduction	Reducing area of body surface in contact with cold objects; Moving to warmer water; Huddling in groups; Increasing thermal barriers; Employing counter-current flows	Increasing area of body surface in contact with warm objects; Moving to warmer water; Reducing thermal barriers	Increasing area of body surface in contact with cold objects; Moving to cooler water; Reducing thermal barriers; Employing counter-current flows	Reducing area of body surface in contact with warm objects; Moving to cooler water; Spreading the group; Increasing thermal barrier
Convection	Avoiding moving air/water; Increasing depth and density of thermal insulation; Reducing area of exposed body surface; Reducing blood flow to body surface; Employing counter-current flows; Huddling in groups	Decreasing depth and density of thermal insulation; Increasing area of exposed body surface	Seeking moving air/water or causing these fluids to move; Increasing area of exposed body surface; Increasing blood flow to body surface; Employing counter-current flows	Spreading the group; Reducing area of exposed body surface
Radiation	Reducing area of exposed body surface; Adopting darker coloration; Reducing blood flow to body surface; Huddling in groups; Reducing metabolic rate	Moving to region of high radiant flux (e.g. direct sunlight); Increasing area of exposed body surface; Adopting darker coloration; Reducing reflective barriers; Spreading the group	Moving away from region of high radiant flux (by finding shade); Increasing area of exposed body surface; Increasing blood flow to body surface; Adopting darker coloration; Spreading the group; Raising metabolic rate	Moving away from region of high radiant flux (by finding shade); Reducing area of exposed body surface; Adopting lighter coloration; Increasing reflective barriers; Spreading the group
Evaporation	Reducing sweating and panting; Reducing blood flow to body surface; Reducing exposure to water and wet environments		Increasing sweating and panting; Increasing blood flow to body surface; Increasing exposure to water and wet environments	

is warmer. This appears to go against the principle that heat can only move from warm (high kinetic energy) to cold (low kinetic energy), but in fact it does not because the vapour increases the number of molecules in the air and also disperses the heat energy at a lower density than when it was in liquid water.

The effectiveness of evaporative heat loss depends on how saturated with water vapour the air already is: it works better in low humidity than high humidity. The air immediately around the evaporative surface (say, the skin) rapidly increases in saturation. Causing it to move away, using a fan or by exposure to wind, replaces it with less saturated air and allows further evaporation.

In addition to the mechanisms shown in Table 30.8, some animals have ways to make heat within the body. This is **endogenous** production and is the biochemical conversion of molecular bond energy into heat. It happens mostly in homeotherms although a few ectotherms can do it to a limited extent. There are essentially three mechanisms: shivering, raising the rate of metabolism, and increasing biochemical activity in fat.

Shivering consists of rapid, low amplitude muscular twitching. It is controlled autonomically and has a small effect in the short term. Raising the metabolic rate by general motor activity (for example, by increasing activity and exercise) is rather more effective.

The processes of feeding and digestion also produce small amounts of heat, by a combination of raised metabolic rate and the energy released as the food is digested and absorbed.

Exercise raises the metabolic rate and also increases respiration rate and sweating. This leads to greater evaporative heat loss, so the heat gain to the body is always compromised. In general, any increase in body temperature also increases the *rate* of heat loss to the environment as demanded by Newton's Law (which is depicted in Figure 30.26). This extra loss can be reduced by redirecting blood flow away from the surface of the body and by increasing thermal insulation, but it is never completely avoided.

In cold water, conductive and convective heat losses are very high (water conducts heat about 25 time faster than air, even when still). They cannot be compensated for by muscular exercise, so if you fall into the sea from a boat it is best to stay huddled up and as still as possible until you are rescued. In small mammals and in the young of some larger mammals, there is a special kind of subcutaneous fat which can directly generate heat energy. This is called brown fat or brown adipose tissue (BAT) and the process is called **non-shivering thermogenesis**. The fat is brown because it is especially well supplied with mitochondria, but these mitochondria perform a modified process of electron transfer which leads to the production of heat rather than ATP.

Non-shivering thermogenesis is especially important in small mammals. Their high surface area:volume ratio gives them a greater tendency to lose body heat, especially by convection and evaporation, when compared with larger mammals. They also have higher rates of overall metabolism (heat production per unit body weight). BAT turns some of this extra metabolism into heat, and it can be especially important in cold seasons and when animals emerge from hibernation. Birds and other non-mammalian vertebrates do not have BAT.

PAUSE AND THINK

Look at Figure 30.29, which illustrates the concept of the zone of thermal neutrality. Using the information in the text if you need to remind yourself, note down the key reasons why the zone is relatively narrow for poikilothermic animals and wide for homeothermic animals.

To test yourself further, consider the thermal environments of some example animals, such as:

■ Antarctic penguins

■ desert lizards

■ Arctic cod

■ earthworms.

You might want to research the lifestyles and behaviours of these species online to help you here.

Answers: The ZNT is relatively narrow for poikilothermic animals because their body temperature is directly influenced by the environmental temperature and there are limits to the range of temperatures at which body physiology can function efficiently and adequately.

The ZNT for homeotherms is wider because they can, within broad limits, defend their body temperature against fluctuations in environmental temperature. Over the range of the ZNT, homeotherm body temperature is held more or less constant and optimal.

Examples:

■ Antarctic penguin (homeotherm): Maintains body warmth metabolically despite living in an environment which is close to zero degrees centigrade and subjected to strong winds, and includes periods spent in the sea as well as on land.

■ Desert lizard (poikilotherm): Exposed to wide variations in temperature (low at night, hot during the day) which force the animal to change its behaviour and limit its activity on a diurnal basis.

■ Arctic cod (poikilotherm): Spends entire life in a wet environment which is cold but subject to relatively limited temperature change. Many enzymatic processes in fish operate optimally at lower temperatures than equivalents in homeothermic vertebrates.

■ Earthworm (poikilotherm): Spends most of its life in a soil environment which is subject to relatively limited and slow variations in temperature. May change its activity and behaviour over the year to accommodate slowly changing temperatures.

 Check your understanding of the concepts covered in this section by answering the questions in the e-book.

30.7 **The gaseous environment**

In terms of general metabolism, animals use oxygen to burn food and excrete carbon dioxide, while plants take up carbon dioxide to generate carbohydrate by photosynthesis and excrete oxygen. These familiar facts highlight the significance of the gaseous environment to all forms of life. Given that about four fifths of the air is inert nitrogen, they also remind us that only a small proportion of the air is biochemically relevant to life.

Besides their importance in biochemistry, the gases around us and those dissolved in water in the aqueous environment in which much of life exists, have physical effects on life and its processes. Therefore, as with the other parts of the physical environment, we need to know something about how gases behave if we are to understand how life works.

Gas pressures and the gas laws

The air above us, which is essentially about 16 km thick, has weight (that is to say, its mass is gravitationally attracted to Earth) and therefore exerts a pressure. The weight of the air column above a square cm of Earth's surface is about 1 kg (or, more correctly, 9.8 N). We define the pressure this represents as *1 Atmosphere*

(1 ATM or 1 Bar). Other ways of expressing it are *1 Torr, 101.3 kilo-Pascals* (*kP*; where a *Pascal* is a *Newton* per square metre), and *760 mmHg* (representing the weight of a 760 mm column of mercury).

The actual pressure we experience depends on local weather conditions—essentially the result of air movements around the planet coupled with the effects of temperature—and how high we are above Earth's surface. Earth is neither smooth nor round, so it is convenient to express the height in relation to the level of the sea. The pressure at the top of Ben Nevis, Scotland, which is 1372 m above sea level, is about 0.85 ATM (644 mmHg), while at the top of Mount McKinley, Alaska, 6096 m above sea level, it is about 0.46 ATM (349 mmHg).

If we go the other way, into a deep cave or mine, the air pressure will be higher than at sea level. (A few bodies of water are also a little below sea level.) However, the effect of this is relatively small and has little practical effect on life.

If we go under the sea, the pressure increases due to the weight of the water added to that of the air—and this does have a significant effect. As a rough guide, the pressure of water is 1 ATM for every 10 m of descent, so the total pressure at 20 m is about 3 ATM.

Air is a mixture of nitrogen, oxygen, and carbon dioxide, plus tiny amounts of other gases which biologists can ignore, and usually some water vapour (Table 30.9). A law relating to gas mixtures (Dalton's Law) states that the pressure of each component gas depends on the proportion it contributes to the mixture. These are called partial pressures; the total pressure is the sum of the partial pressures.

Humid air at sea level with a total pressure of 1 ATM (760 mmHg) might have the following components and partial pressures: Figure 30.30 shows how these partial pressures are affected by altitude and subsea depth.

Temperature has an effect on gas volume, and thus on gas pressure. Charles' Law tell us that at a constant pressure, the volume (*V*) of a gas is proportional to its absolute temperature (*T*, in K):

$$V \alpha T$$

so doubling the absolute temperature would double the volume.

Boyle's Law says that at a constant temperature, the pressure (*P*) of a gas is inversely proportional to its volume:

$$P \alpha 1/V$$

so halving the volume would double the pressure.

These statements can be combined and, with the inclusion of a constant, made into an equation:

$$PV = nRT \quad \text{or} \quad nR = PV/T$$

Table 30.9 Composition of air and partial pressures

		Partial pressure	
Gas	Proportion	Bar	mmHg
Air	100%	1	760
Nitrogen	78%	0.78	593
Oxygen	21%	0.21	160
Carbon dioxide	0.04%	0	0.3
Water vapour	1%	0.01	8

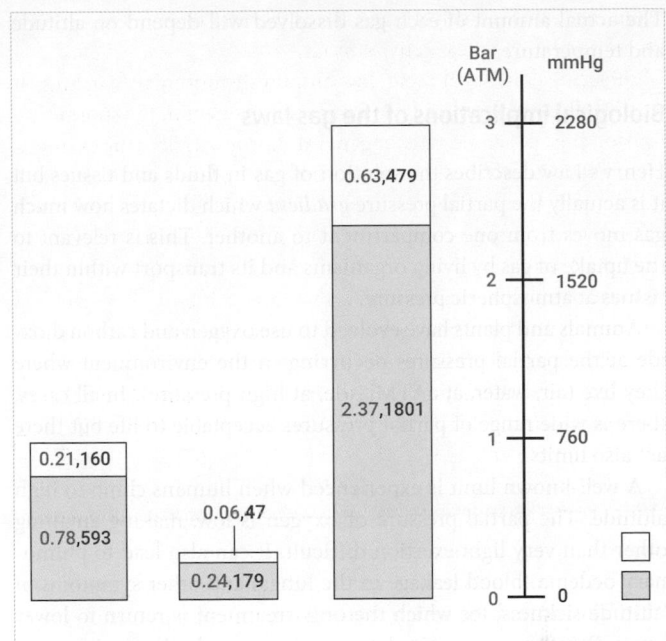

Figure 30.30 Gas pressures and partial pressures. Partial pressures (orange: *bar*; blue: *mmHg*) of gases in a sample of air which contains (by volume) 78% nitrogen, 21% oxygen, 0.04% carbon dioxide, and 1% water vapour, when placed at sea level, at the top of Everest, or 20 m under water. The total pressure of the air is the sum of the partial pressures. The partial pressures of carbon dioxide and water vapour are too low to illustrate on this scale.

The constant *nR* is an expression of the physical nature of the gas (the number of molecules and their energy) and need not detain us. The essential point is that it must remain *constant*, so any change to *P*, to *V*, or to *T* must be associated with compensatory changes in the other variables. This is just what the two Laws are saying. Note that *T* is the *absolute* temperature in degrees Kelvin (K; the temperature in °C plus 273.15).

Another biologically important feature of gases is the way they dissolve in liquids. The law which applies here is that of Henry: it says that the amount of a gas which is dissolved in a liquid is proportional to the partial pressure of the gas. We see the effect of this when unscrewing the cap of a bottle of fizzy pop: as the pressure is reduced, the gas (usually carbon dioxide) comes out of solution in the form of bubbles.

Besides being proportional to partial pressure, the actual amount of gas which dissolves depends on the nature of the liquid (in biological systems this is usually water or a water-based fluid but it can also be a tissue such as fat), the solubility of the gas in that tissue (this is a constant), and the temperature. In general, gases dissolve better at lower temperatures than at high ones.

One result of Henry's Law is that when a liquid is exposed to a gas, a pressure equilibrium is rapidly established between the free gas and the dissolved gas. It is usually the partial pressures we are interested in and there will be a separate equilibrium for each gas in the mixture. For example, there will always be nitrogen, oxygen, and a small amount of carbon dioxide dissolved in lake, river, and sea water as a result of its exposure to the atmosphere.

The actual amount of each gas dissolved will depend on altitude and temperature.

Biological implications of the gas laws

Henry's Law describes the solution of gas in fluids and tissues but it is actually the partial pressure *gradient* which dictates how much gas moves from one compartment to another. This is relevant to the uptake of gas by living organisms and its transport within their tissues at atmospheric pressure.

Animals and plants have evolved to use oxygen and carbon dioxide at the partial pressures occurring in the environment where they live (air, water, at aATMitude, at high pressure). In all cases, there is wide range of partial pressures acceptable to life but there are also limits.

A well-known limit is experienced when humans climb to high altitude. The partial pressure of oxygen is low, making anything other than very light exertion difficult. It can also lead to pulmonary oedema (blood leakage in the lungs) and other symptoms of 'altitude sickness', for which the only treatment is return to lower levels. Breathing pressurized oxygen from a cylinder may prevent the problem. This also explains why oxygen is supplied in an aircraft cabin if the pressure drops in an emergency.

A few human societies have adapted to life at high altitude, for example in the Andes. Individuals may develop very large lung capacities and high blood haemoglobin levels, facilitating additional oxygen carriage in the blood. This kind of physiological adaptation is exploited by athletes who train at high altitude.

Even at sea level, one of the limits on exercise intensity is the ability of the lungs and circulation to supply oxygen to the muscles: theoretically, if the oxygen partial pressure was higher at atmospheric pressure, the maximum possible intensity of exercise would be greater.

Animals which live permanently at high pressures, for example at great depth in the ocean, have high levels of gas dissolved in their tissues. Nitrogen and oxygen dissolve particularly well in fat and nerve tissue. This is not a problem unless the animal is brought rapidly to the surface—the gases come out of solution, producing bubbles in tissues.

Mammals which dive to great depths, including seals and whales, allow their lungs to become compressed. This reduces the chance of bubble formation but the animals also have circulatory shunts which prevent the capillary fluid from damaging the alveoli. Some can stay down for lengthy periods of time—30–40 minutes—but seem to be resistant to the bubble formation and tissue damage which might be expected when they surface.

Some diving animals manage gas release behaviourally by staging their ascents, in much the same way that scuba divers do to avoid decompression illness (the 'bends').

Gas exchange in plants

Most living plant tissues are relatively thin and gas exchange with the atmosphere occurs by diffusion. This is a passive process, driven by the partial pressure gradient of the gas.

For example, carbon dioxide from the air enters the leaves through pores in the epidermis called **stomata**, which you can see

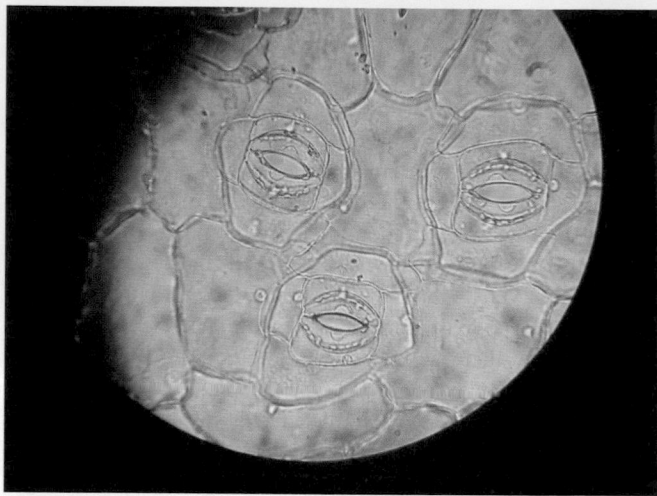

Figure 30.31 Stomatal pores in the surface of a *Commelina* sp. leaf, viewed through a simple light microscope. In this picture, the guard cells are curved so that the stomata are open.

in Figure 30.31. Once the air enters through the stomatal pores, it dissolves in cellular fluids and becomes available to chloroplasts for the production of carbohydrates. The biochemical capture and fixation of the carbon dioxide reduces its partial pressure gradient and more gas enters.

Leaf tissues contain air spaces which gases can diffuse in and out of easily. Oxygen, which is the waste product of photosynthesis produced in the chloroplast by the light-induced breakdown of water molecules (see Chapter 7), also diffuses out of the leaf down a concentration gradient in the opposite direction to CO_2.

Even in slightly thicker parts of the plant such as roots, in which oxygen uptake is required for respiration, the gas simply diffuses into the tissues. The rate at which this happens is sufficient for normal tissue function provided the soil is sufficiently aerated. If the soil becomes waterlogged, oxygen diffusion is greatly reduced. Roots quickly become anaerobic and may die if the situation pertains for a significant period of time. This is a major problem when fields of agricultural crops become flooded.

In plants which normally live in wet soils (e.g., paddy rice, pond and bog plants) the roots contain air spaces termed **aerenchyma**. These penetrate through the internal tissues of the root, facilitating better diffusion of oxygen into the tissues.

Water vapour is another gas whose exchange between leaves and the external atmosphere is of great importance to plants. Earlier we described how transpiration of water vapour can reduce leaf temperature by evaporative cooling, but transpiration through stomata is also the end point of the pathway by which water is translocated through the plant from the soil. Some 50% of the water falling as rain on the planet is transpired into the atmosphere through leaf stomata.

Gas exchange is a significant problem for plants which live permanently in water. This is partly because of the slow diffusion rates of gases in water but also because carbon dioxide equilibrates with bicarbonate ions in water, as summarized in the following equation, and becomes less available to the plant.

$$CO_2 + H_2O \leftrightarrow H_2CO_3 \leftrightarrow H^+HCO_3^-$$

Many water plants use the enzyme **carbonic anhydrase** to drive the equilibrium to the left, back towards the free CO_2. Many single-celled photosynthetic organisms such as photosynthetic bacteria and algae use a range of biochemical mechanisms to concentrate CO_2 or bicarbonate at the site of RuBisCO, the primary carbon-fixing enzyme in photosynthesis (see Chapter 7).

Plants interact with the gaseous environment in two other significant ways. Plants emit huge quantities of volatile organic compounds into the atmosphere, the most abundant being monoterpene and isoprenes. Isoprenes may confer greater thermal tolerance to the leaf and also increase tolerance to reactive oxygen species, although the underlying mechanism is unclear. On a global scale, isoprene emission by plants dwarfs the generation of pollutant hydrocarbons into the atmosphere by human activity. Some plants have developed a symbiotic relationship in their roots with nitrogen-fixing bacteria called **rhizobia**. These plants are called legumes and are typified by beans, peas, alfalfa, lupins, and clover. They use the nitrogen as a major part of their nutrition. The process happens in nodules on the roots, as you can see in Figure 30.32, where rhizobia use the enzyme nitrogenase to capture atmospheric nitrogen and produce ammonia.

The nitrogenase enzyme is easily poisoned by atmospheric oxygen, so the plant also synthesizes leghaemoglobin in these nodules, a molecule closely related to the haemoglobin in animal blood but with a ten times greater affinity for oxygen. The leghaemoglobin buffers oxygen levels within the nodule, enabling sufficient oxygen for bacterial cells to respire but preventing an excess which would interfere with nitrogenase function.

Gas exchange in animals

In very small animals, including protists and small invertebrates, which have a large surface area:volume ratio, gas can move in (oxygen) and out (carbon dioxide) of cells directly from the surrounding medium (usually water), but always dictated by the gas partial pressure gradients and the relative solubility of gases in the external medium and cell cytoplasm.

Figure 30.32 Nodules on a root system, formed by rhizobia bacteria in association with the plant.

In slightly larger multicellular animals, such as round worms and insects, the surface area is insufficient to supply air to all parts of the body. The air reaches cells inside the body through tubes called spiracles, and there is often a fluid (haemolymph) in which the gases can dissolve for distribution around the body tissues.

In vertebrates, gas exchange takes place in gills or lungs and the movement of air in and out of the body is assisted by water movements or ventilation. The uptake of oxygen relies entirely on the gradient between the partial pressure in inspired (alveoli) or water-borne (gill surface) gas and that in the deoxygenated blood returning from the body tissues. The reverse transfer of carbon dioxide is similarly gradient-driven. There is no such thing as a pump for moving gases across membranes.

Many animals have evolved strategies to survive periods of low oxygen partial pressure (hypoxia) or even to survive in the absence of oxygen (anoxia). Some environments regularly become hypoxic or anoxic, for example, underground burrows, stagnant ponds, and, as we discuss earlier in this section, high altitudes. Nematode worms, such as those in Figure 30.33, and other forms of life, have even been found living in deep mines where the water has extremely low levels of oxygen.

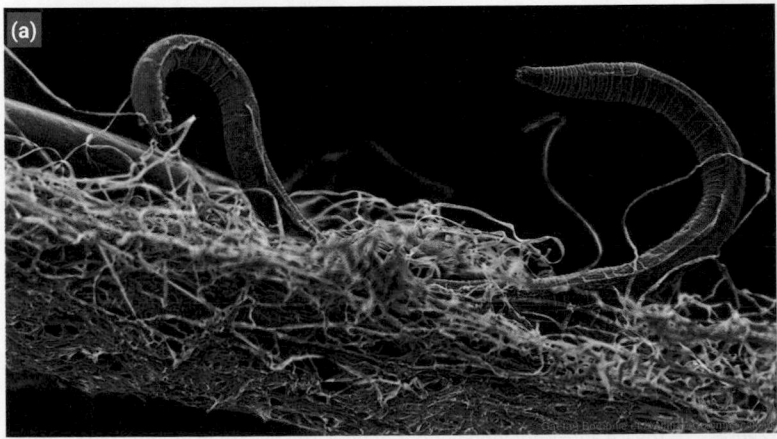
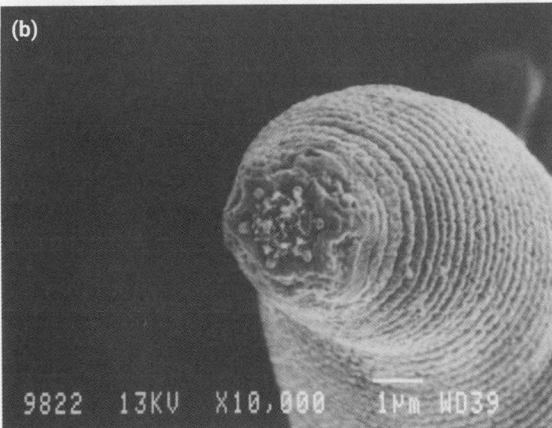

Figure 30.33 Deep mine nematodes. (a) A nematode (*Poikilolaimus* sp.) found 0.9 miles (1.4 km down) in Kopanang gold mine, South Africa. (b) *Halicephalobus mephisto*, found 0.9 miles (1.3 km) down in Beatrix gold mine. At such depths, temperatures reach 37 °C, hotter than most terrestrial nematodes can survive.
Source: (a) Courtesy Gaetan Borgonie/Extreme Life Isyensya. (b) Image courtesy Gaetan Borgonie, University Ghent.

One of the most extremeranoxic environments occurs in temperate lakes that freeze over in the winter: oxygen is prevented from dissolving in the water by an ice layer that can reach several metres thick. The problem is compounded by the way the ice also blocks sunlight (see Section 30.2) and prevents aquatic plants from photosynthesizing. The concentration of oxygen in the water falls quickly as it is used up by organisms, and the partial pressure may remain very low for many months.

Some animals survive these conditions by using anaerobic pathways to meet their energy demands. They can also reduce their energy requirement by avoiding excessive activity or reducing their metabolic rate, although some ATP must be produced to maintain vital functions, particularly brain activity.

These animals use alternative, anaerobic pathways to generate ATP, usually through glycolysis. The main end point of glycolysis is lactic acid (lactate) but its acidity can cause damage to tissues if it accumulates, so it is not a simple solution to the problem of low oxygen.

The crucian carp and related goldfish (*Carassius* spp.) can produce alternative end points to glycolysis which are less toxic than lactate (which we discuss further in Scientific Process 30.3). Alternatively, the lactic acid can be stored in a way which limits toxicity and tissue damage until oxygen becomes available. For example, the anoxia-tolerant painted turtle (*Chrysemys picta*) can sequester lactate in its shell and use shell calcium carbonate to buffer its acidic effects.

SCIENTIFIC PROCESS 30.3   How do goldfish survive without oxygen?

Research question

During periods of low temperature, goldfish can stay alive for several days without access to oxygen. How do they do this?

Background

Goldfish (*Carassius auratus*), like the one in Figure 1, are commonly kept as pets, and have an impressive ability to survive anoxia, especially in very cold and freezing water.

Figure 1 The goldfish *Carassius auratus*.
Source: Wikimedia Commons/CC BY-SA 3.0

In the absence of oxygen, vertebrates normally generate ATP by an anaerobic metabolic pathway called glycolysis. This creates a problem because the major end-product, lactate, is not easily excreted and causes tissue damage if it accumulates.

In the 1950s it was found that goldfish can survive anaerobically without producing lactate. The mechanism of this was not understood.

Hypothesis

In 1980, Shoubridge and Hochachka from the University of British Columbia investigated the hypothesis that goldfish use anaerobic pathways to survive anoxia but break down (catabolize) lactate into a different end-product which could then be excreted into the water.

Methods

They exposed four goldfish to carbon monoxide to remove any residual oxygen, then held them in anoxic water at 4 °C. They also held four goldfish as controls in water with normal levels of oxygen. After 12 hours, they measured levels of lactate and ethanol in the tissues of the fish and also analysed the content of the water.

Results

The graph in Figure 2 shows that the goldfish that experienced anoxic conditions had higher levels of lactate in their tissues than the controls. Anoxic goldfish also had ethanol in their tissues whereas control goldfish did not, and anoxic goldfish had excreted ethanol into the water.

Conclusion

The goldfish were surviving in the absence of oxygen using glycolysis. Although lactate was being produced, the goldfish were converting it to ethanol and excreting this into the water. This avoided the damaging effects of lactate accumulation.

Subsequent work by other researchers has shown that another species, the crucian carp (*Carassius carassius*) like the one in Figure 3, can also produce ethanol as an end-product of glycolysis. In winter, the lakes that carp inhabit are covered in a thick layer of ice, making the water beneath the ice anoxic for long periods of time. The release of ethanol to the water comes at a significant energetic cost. The carp have evolved to build up

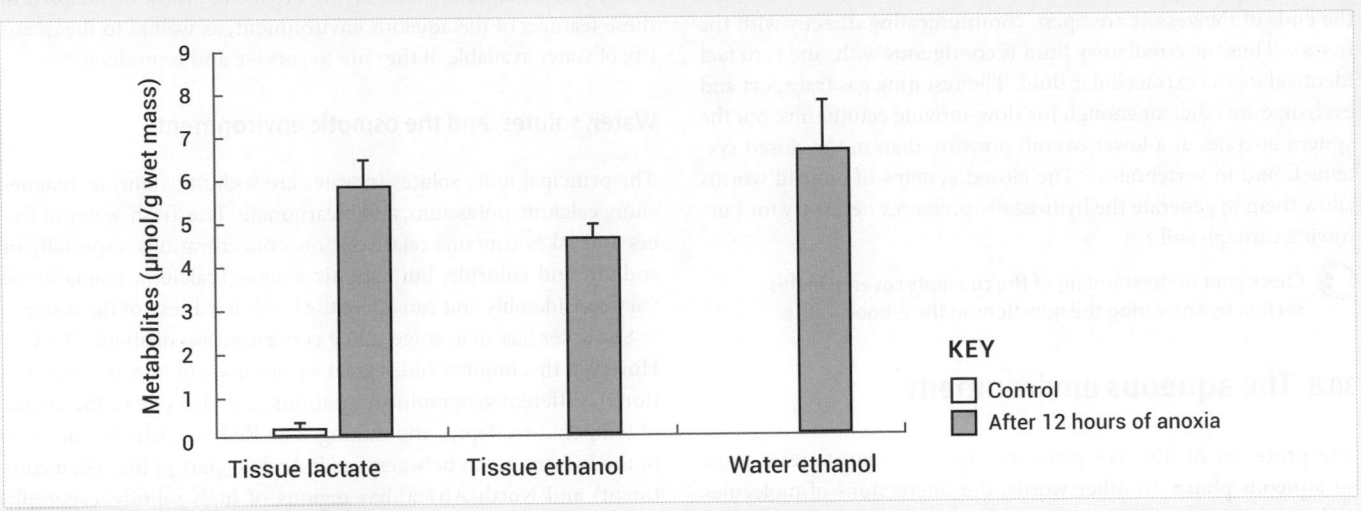

Figure 2 **Metabolite changes in whole goldfish and surrounding water in control (white) and after 12 hours of anoxia (yellow) at 4 °C.** Data are means ± S.E. N=4. Ethanol was undetectable in controls.

Source: Data redrawn from Table 1 of Shoubridge, E.A. & Hochachka, P.W. (1980). Ethanol: Novel end product of vertebrate anaerobic metabolism. *Science*, 209, 308–309.

Figure 3 **The crucian carp,** *Carassius carassius.*
Source: Viridiflavus/Wikimedia Commons/CC BY-SA 3.0.

large energy stores in the form of glycogen during the preceding autumn, to be turned into ethanol during periods of anoxia.

Does the crucian carp get drunk?

Blood ethanol levels of ethanol in the carp remain below 10 mM which is equal to 46 mg/100 ml. To put this in context, the current drink-drive limit in England is 80 mg/100 ml.

Read the original work

Shoubridge, E.A. & Hochachka, P.W. (1980) Ethanol: Novel end product of vertebrate anaerobic metabolism. *Science*, **209**, 308–309.

Johnston, I.A. & Bernard, L.M. (1983) Utilization of the ethanol pathway in carp following exposure to anoxia. *J. Exp. Biol.*, **104**, 73–78.

Nilsson, G. E. & Lutz, P. L. (2004) Anoxia tolerant brains. *J. Cerebr. Blood Flow Metab.*, **24**, 475–486.

Prosser, C.L., Barr, L.M., Pinc, R.D., & Lauer, C.Y. (1957) Acclimation of goldfish to low concentrations of oxygen. *Physiol. Zool.*, **30**, 137–141.

Gas transport in animals

The amounts of oxygen and carbon dioxide dissolved in blood are determined by partial pressures and solubilities. The solubility of oxygen in water is relatively low and the amount of gas actually *carried* by the blood is increased by the presence of pigments, such as haemoglobin located in the erythrocytes (red blood cells) of vertebrates. The pigment binds the gas molecules, taking them out of free solution in the blood plasma and carrying them as an intracellular molecular complex.

Haemoglobin and similar pigments are characterized by their ability to carry very large amounts of oxygen, and the binding of oxygen increases as the amount being carried increases. Haemoglobin also carries carbon dioxide although most of this

gas is carried in the aqueous fraction of the blood, buffered by the bicarbonate system. The complex biochemistry and physiology underlying these processes are described in Module 3.

All vertebrates and a few other animals (annelid worms, for example) have **closed** circulations. Their blood flows within entirely closed systems of arteries, veins, and capillaries, pumped by the heart. Gas exchange at the lungs and within tissues occurs across cells, called endothelial cells, which line all these vessels. The action of the heart, supported by membranes in the vessels, maintains a high pressure in the system. This increases the partial pressures of the gases, and thus the overall gas-carrying capacity of the blood, which increases physiological efficiency.

Other complex animals (molluscs, echinoderms, arthropods) have **open** circulations. The circulating fluid is carried in vessels, propelled by pumps or by the movements of the animal's body, but the ends of the vessels are open, communicating directly with the tissues. Thus the circulatory fluid is contiguous with, and is in fact identical to, the extracellular fluid. The resulting gas transport and exchange are efficient enough for slow-moving ectotherms, but the system operates at a lower overall pressure than in the closed systems found in vertebrates. (The closed systems of annelid worms allow them to generate the hydrostatic pressures necessary for burrowing through soil.)

 Check your understanding of the concepts covered in this section by answering the questions in the e-book.

30.8 The aqueous environment

The processes of life take place in what biochemists refer to as an **aqueous phase**. In other words, the interactions of molecules and the exchange of energy between them happen in water-based environments. Water-based life is the only kind we know of, which is why the presence of water is considered a pre-requisite for the existence of life on other planets. As explained in Chapter 28, cells are essentially enclosed, self-contained regions of the biosphere, and one of their features is the aqueous intracellular fluid in which biochemical processes occur.

The movement of water across membranes and between cells is always a *passive* process: it does not consume metabolic energy and there are no 'water pumps'. It happens by osmosis, across membranes, or by molecular diffusion between contiguous compartments (Chapter 28). The only way that organisms can regulate it is by controlling the osmotic gradient, by setting up water-resistant barriers or by altering the way contiguous fluids mix.

Rates of osmosis and diffusion depend on solute concentrations. In addition, water dissolves gases and chemicals, is subject to variations in pH and temperature, flows between compartments, and

undergoes changes of state. Living organisms must be adapted to these features of the aqueous environment, as well as to the quantity of water available, if they are to survive and reproduce.

Water, solutes, and the osmotic environment

The principal ionic solutes in water are sodium, chloride, magnesium, calcium, potassium, and bicarbonate. The 'fresh' water of rivers and lakes contains relatively low concentrations, especially of sodium and chloride, but cationic solutes (calcium, magnesium) vary considerably and can affect the local 'hardness' of the water.

Sea water has an average solute concentration of about 35 g.L^{-1}. However, this number hides great variations, not only of composition in different geographical locations, but also due to the effects of temperature, depth, and mixing. The Red Sea, which is an inlet of the Indian Ocean between Saudi Arabia (part of the Asian continent) and North Africa, has regions of high salinity, especially at the northern end where it may be 41 g.L^{-1}. This is caused by the high rate of evaporation near the equator and relatively slow replenishment from the Indian Ocean at the Gulf of Aden in the south.

Salt lakes, some of which support specialized forms of life, have very high salinities. The Great Salt Lake in the United States has a salinity between 50 and 270 g.L^{-1} depending on its water level, and the Dead Sea, which is a saturated salt solution, has been measured at 340 g.L^{-1}.

Water close to the outfall from rivers, such as in estuaries, deltas, and mangrove swamps, is called brackish water and is essentially a mixture of sea and fresh water. The proportions of each in the mixture depend on the location and the state of the tide, and the composition is also affected by the flow rate of the river and the quantity of solutes brought from upstream.

Table 30.10 shows some representative values for solute concentrations and osmolarities of different types of water. (For

Table 30.10 Water solute composition and osmolarity

Solute (mMol.L^{-1})	Soft lake water	River water	Hard river water	Sea water	Salt lake water
Sodium	0.18	0.40	6.05	633	831
Magnesium	0.15	0.31	0.65	6	2277
Calcium	0.22	0.51	4.95	32	577
Potassium	–	0.04	0.11	16	150
Chloride	0.03	0.23	13.29	623	6589
Sulphate	0.09	0.21	1.38	53	8
Bicarbonate	0.43	1.10	1.37	3	–
Total solutes	**1.10**	**2.80**	**27.81**	**1366**	**10,432**
NaCl equivalent (g.L^{-1})	0.32	0.82	0.81	40	305
Molarity (mMol.L^{-1})	0.55	1.40	13.90	683	5216
Osmolarity (mOsM.L^{-1})	**1.10**	**2.80**	**27.80**	**1366**	**10,432**

Data adapted from: Schmidt-Nielsen K., 1997, *Animal Physiology: Adaptation and Environment*, 5th ed., Cambridge University Press. Table 8.1, page 302 and Table 8.2 Page 303.

comparison, human plasma has an osmolarity of 290 mMol.L^{-1}.) Living organisms are found in all these types of water and exhibit a range of osmoregulatory adaptations to these diverse environments.

When you look at Table 30.10, be aware that the molecular weight of NaCl is 58.5, so a 1.0 mol solution contains 58.5g of NaCl in 1.0 L of water. For an explanation of why the osmolarity is twice the molarity, see Chapter 28.

Osmoregulation in plants

Plants do not have osmoregulatory organs in the sense that we mean it for animals but they are osmotically responsive and can regulate the flow of water through their structures. The principal challenge for plants is to maintain turgor pressure, to keep cells and tissues rigid (see Chapter 29). This enables the plant to retain an erect structure, especially young plants where hard supporting tissues are yet to develop.

Water is taken up from the soil by cells in the epidermis of the root, primarily the root hair cells. It passes down an osmotic gradient until it reaches the transport system, the xylem, and is moved through the plant body via the vascular network (Chapter 31).

From an individual cell's point of view, water is readily available on the outer surface of the cell wall, which in most tissues is saturated with water. Within a leaf, the unloading of water from the xylem (the leaf's vascular network) enables a film of water to cover the surfaces of the mesophyll cells, making water readily available for uptake by the cell.

Water can pass into the cell, down an osmotic gradient, by means of distinct water channels called aquaporins (Chapter 28). The osmotic gradient is determined by the solute concentration within the cytoplasm. This in turn is controlled by an array of transport proteins in the plasma membrane, regulating the traffic of a wide variety of metabolites and ions into and out of the cell.

The turgor pressure resulting from the movement of water creates hydrostatic pressure in the cell's vacuole, pushing the entire cell contents against the internal face of the cell wall. Turgor pressure is also controlled by mechanoreceptors in the various cell membranes. These monitor the physical stretching of the membrane and adjust the water uptake. Without such a regulatory system, even with a tough cell wall the cell could burst.

Loss of water by plant cells can be catastrophic. In droughted tissues, the turgor pressure is reduced and the plasma membrane shrinks away from the cell wall. This leads to a state of **plasmolysis** in which the tissues wilt. Some plants can tolerate this state for varying lengths of time but if it is prolonged, cell and tissue death is the likely outcome.

Marine plants

Many plants, such as marine plants and those in coastal areas, live in environments where they experience high concentrations of salt, primarily sodium chloride. Increasingly, agricultural crop plants have to be grown in soils in which salt concentrations have risen as a result of ground water extraction. The increasing salinity of agricultural soils is a major problem worldwide.

Plants which tolerate high salt concentrations and are adapted to such conditions are termed **halophytes**. They have mechanisms to cope with extreme salt stress. Some plants pump the salt back out of the cell using transporter protein pumps in the plasma membrane, but this strategy is energetically costly to the cell. Other plants secrete salt into specialized tissues or onto the cell surface. Plants which are less able to get rid of salt may synthesize molecules called **osmolytes**. These equilibrate the osmotic difference between the cell cytoplasm and the external environment. The resulting balance enables water to be taken up into the cell.

Osmoregulation in animals

Across the animal kingdom, organisms take a variety of approaches to dealing with the osmotic environment, much as they do with the thermal environment. The extent of osmotic homeostasis ranges from allowing body fluid composition to change passively in response to external variations, so that the body and external fluids are always **isosmotic**, to rigidly maintaining a constant internal osmotic concentration whatever the environmental situation.

Animals in the first group are sometimes called **osmoconformers** and those in the second group **osmoregulators**. In fact, these terms are not especially helpful because all animals can be placed somewhere on a spectrum between the extremes. Terrestrial animals, which are not constantly exposed to free water, generally fall in the latter group although their tolerance of internal variation can still be wide. Clearly, animals demonstrate different degrees of biochemical sensitivity, and have correspondingly different levels of physiological complexity in their responses.

The effect of environmental osmotic concentration on body volume depends on the osmotic *gradient* (which we also discuss in Chapter 28). If there is an osmotic gradient between the environment and the body, water will move in or out of the animal and there will be an increase or decrease in body volume. This process will be influenced by any barriers to water movement in and out of the body, and the action of energy-dependent, physiological systems by which the animal can alter the water or salt content of the body. ('Salt' in this context means any osmotically active solute.)

As with thermal regulation, it is helpful to think of each animal having a 'zone of osmotic neutrality', the range of environmental conditions in which its body volume does not change beyond the limits which its biochemistry can tolerate. This concept is depicted in Figure 30.34. The zone may be wide or narrow, depending on the complexity of the animal and the type of environment in which it lives. Outside this zone, large amounts of energy are required to restore the osmotic balance; at the extremes, life for the animal becomes impossible.

Protist osmoregulation

Animals with very large surface area:volume ratios are especially susceptible to changes in environmental solute concentration.

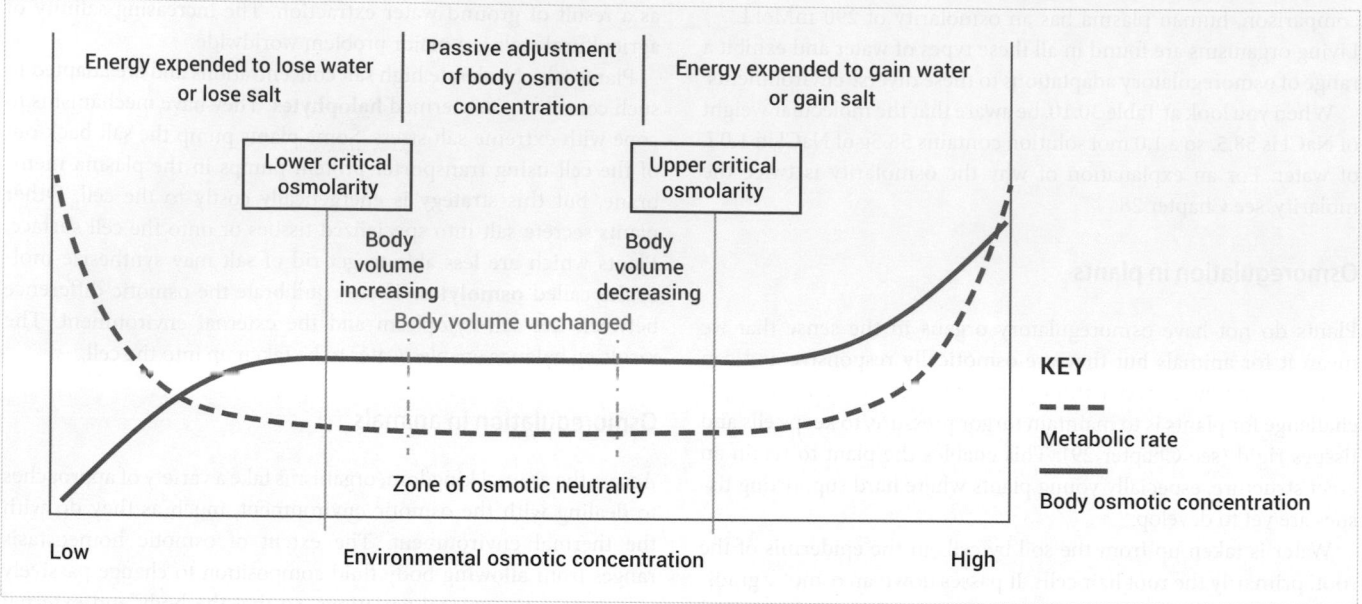

Figure 30.34 The zone of osmotic neutrality. The general relationship between environmental osmotic concentration, body osmotic concentration, and metabolic effort in animals. The effect of environmental osmotic concentration on body volume depends on the osmotic *gradient* (see Chapter 28), any barriers to water movement in and out of the body, and the action of energy-dependent, physiological systems by which the animal can alter the water or salt content of the body. 'Salt' in this context means any osmotically active solute.

Single-celled organisms fall into this group. A few have water-resistant cell walls but many have external membranes with only limited resistances 24 and to water movement. Protists have a limited zone ofosmotic neutrality because a small loss or uptake of water makes a big difference to volume. They may make rapid responses to sudden changes in volume and can adapt to longer term changes by altering the expression of regulatory genes.

Protists occupy a very wide range of aqueous environments. Free-living protists, such as amoebae and paramecium, mostly occupy fresh water and are exposed to *hypo-osmotic* environments. Marine protists and gut-dwelling parasitic protists (e.g. trypanosomes) have to survive in largely *hyperosmotic* conditions (Figure 30.35). Parasites may also experience a variety of environments, especially if they have complex life cycles involving a number of different vector and host species, or if they survive for a period of dormancy between hosts, for example as a soil cyst.

Most protists adjust their internal water content by means of contractile vacuoles. Vacuoles manage the excretion of nitrogenous and other wastes. A chamber containing the waste material is formed within the cytoplasm and moves to the external membrane where it discharges its contents to the outside. Water is lost during this process, the amount being adjusted according to osmotic requirements. In addition, ions (Na^+, K^+) and organic solutes (amino acids, sorbitol, and others) can be expelled from the cytoplasm through osmotically sensitive membrane channels, thereby adjusting the osmolyte content of the cell.

Protists living in variable hyperosmotic conditions have an additional mechanism involving acidic vesicles, which is illustrated in Figure 30.36. Water can be stored in contractile vacuoles (CV).

Acidic vesicles contain phosphorus compounds, cations, and basic amino acids (AA), and are similar to the lysosomes found in metazoan cells. They store osmotically active solutes (in a complex called PolyP) and release them rapidly into the cytoplasm when the cell is under hyperosmotic stress (rapidly losing water) to redress the osmotic gradient. These systems use energy from ATP. Water movement, across the external membrane and across membranes within the animal, takes place through aquaporin channels (see Chapter 29 for an explanation of how these work). There are also specific membrane channels for ions and other solutes exchanged with the external environment.

Metazoan osmoregulation

In multicellular animals, excretory and osmoregulatory processes are usually located in clusters of cells or, in more complex animals, in discrete structures such as nephridia in invertebrates and nephrons in vertebrates. These are excretory tubules which communicate between the body fluids and the outside. The kidneys of vertebrates are essentially collections of nephrons.

The tissues in these organs, whatever their complexity, always carry out at least two specific types of process:

1. They transport ions and other solutes from one side of a membrane to the other, so that local concentrations on either side can be adjusted and osmotic gradients created.

2. They control the rate at which water moves across the membrane in response to the osmotic gradient caused by the transported solutes.

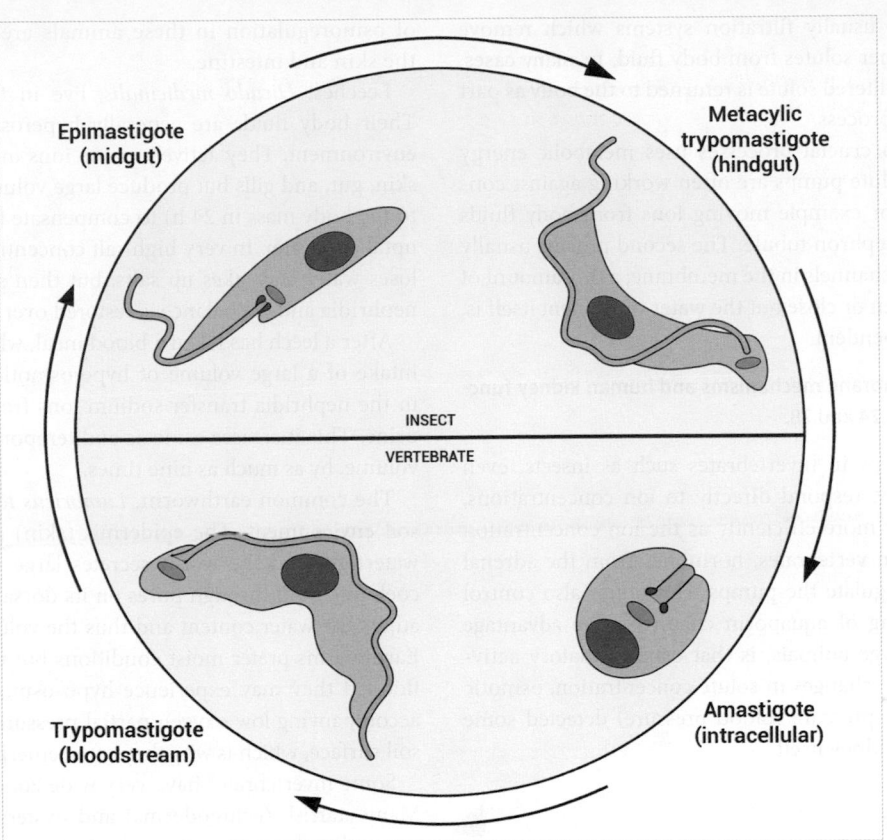

Figure 30.35 The life cycle of *Trypanosoma cruzi.*

Source: Reprinted from International Review of Cell and Molecular Biology, 305, Docampo et al., New insights into roles of acidocalcisomes and contractile vacuole complex in osmoregulation in protists, 69–113, Copyright 2013, with permission from Elsevier.

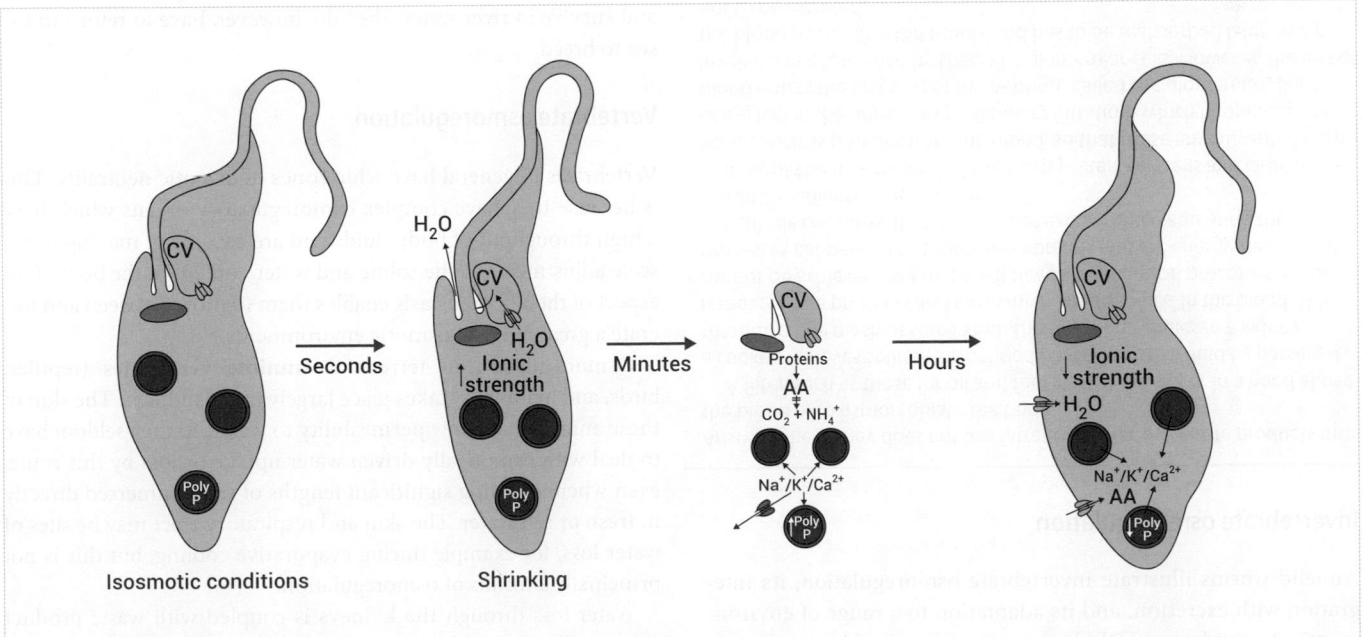

Figure 30.36 The response of protists to hyperosmotic stress. When exposed to hyperosmotic conditions cells rapidly shrink before regaining their original size some time later.

Source: Reprinted from International Review of Cell and Molecular Biology, 305, Docampo et al., New insights into roles of acidocalcisomes and contractile vacuole complex in osmoregulation in protists, 69–113, Copyright 2013, with permission from Elsevier.

In addition, there are usually filtration systems which remove waste materials and other solutes from body fluid. In many cases, much of the originally filtered solute is returned to the body as part of the osmoregulatory process.

The first of the two crucial processes uses metabolic energy because the ion and solute pumps are often working against concentration gradients, for example moving ions from body fluids into the lumen of the nephron tubule. The second process usually depends on aquaporin channels in the membrane; a tiny amount of energy makes these open or close but the water movement itself is, as always, energy-independent.

▶ For more on membrane mechanisms and human kidney function, see Chapters 24 and 28.

The membrane pumps in invertebrates such as insects, even though they use energy, respond directly to ion concentrations, often working faster or more efficiently as the ion concentration goes up on one side. In vertebrates, hormones from the adrenal gland and elsewhere regulate the pumps. Hormones also control the opening and closing of aquaporin channels. The advantage of this, especially in large animals, is that osmoregulatory activity can be influenced by changes in solute concentration, osmotic pressure, or hydrostatic pressure (blood pressure) detected some distance way from the kidney itself.

PAUSE AND THINK

The kidney is sometimes colloquially referred to as a 'filtration' organ. Based on what you have learned from the text, do you think this is correct?

Answer: The kidney does not use filtration to remove waste products from the blood or to osmoregulate the body.

A filter, such as those for oil and fuel in a car engine or in a hood above a cooker, removes solid or suspended particles from a fluid by passing it through a fine mesh of some kind. The nephron tubules in a kidney remove waste products and ions which are *dissolved* in the blood. This cannot be achieved by a mesh—it needs biochemical processes. These processes happen in the tubule membranes and are energy-dependent. The tubule concurrently adjusts water excretion by regulating the osmotic gradients which are formed.

It *would* be true to say that the kidney filters out cells and large molecules such as proteins from the blood, so that these are prevented from being lost in the urine. This preliminary filtration, which happens as the blood enters the first part of the nephron called the glomerulus, does involve a mesh, formed by specialist glomerular cells. However, the rest of the blood passes into the tubule and has to be reabsorbed later on if it is not to be excreted.

You will find more about the mechanism of kidney function in Chapter 24.

Invertebrate osmoregulation

Annelid worms illustrate invertebrate osmoregulation, its integration with excretion, and its adaptation to a range of environments. Annelids generally have a water-permeable integument but the junctions between cells are resistant to ion movements, so losses of salt by this route are prevented. The principal organs of osmoregulation in these animals are **nephridia**, along with the skin and intestine.

Leeches, *Hirudo medicinalis*, live in fresh to brackish water. Their body fluids are generally hyperosmotic compared to the environment. They actively pump ions into the body through the skin, gut, and gills but produce large volumes of urine (equivalent to the body mass in 24 h) to compensate for the resulting osmotic uptake of water. In very high salt concentrations the leech initially loses water and takes up salts, but then salts are excreted by the nephridia and the balance is restored over a period of a few days.

After a leech has taken a blood meal, which represents the rapid intake of a large volume of hyperosmotic fluid, canalicular cells in the nephridia transfer sodium ions from the body fluid to the urine. This increases osmosis and temporarily increases the urine volume, by as much as nine times.

The common earthworm, *Lumbricus terrestris*, lives in a moist soil environment. The epidermis (skin) of an earthworm is not watertight and the worm secretes large amounts of mucus and coelomic fluid through pores on its dorsal surface. The nephridia adjust the water content and thus the volume of the worm's body. Earthworms prefer moist conditions but if their burrows become flooded they may experience hypo-osmotic stress. This and the accompanying low oxygen partial pressure may drive them to the soil surface, which is why they often emerge after heavy rain.

Some invertebrates have very wide zones of osmotic neutrality. Many starfish (echinoderms) and oysters (bivalve mollusc) survive well in brackish water because they can, within limits, change the osmotic concentration of their body fluids to that of the surrounding water. Bivalves can also shut their shells completely if the external osmolarity becomes unacceptable. Chinese mitten crabs (*Eriocheir* spp.) normally live in the sea but can penetrate estuaries and survive in river water; they do, however, have to return to the sea to breed.

Vertebrate osmoregulation

Vertebrates in general have wide zones of osmotic neutrality. This is because they have complex osmoregulatory organs which have a high throughput of body fluids and are capable of making large-scale adjustments to the solute and water content of the body. This aspect of their homeostasis enables them to move between and tolerate a great range of osmotic environments.

Osmoregulation in terrestrial **amniote** vertebrates (reptiles, birds, and mammals) takes place largely in the kidneys. The skin of these animals has a low permeability to water, so they seldom have to deal with osmotically driven water uptake or loss by this route, even when spending significant lengths of time immersed directly in fresh or salt water. The skin and respiratory tract may be sites of water loss, for example during evaporative cooling, but this is not principally a means of osmoregulation.

Water loss through the kidneys is coupled with waste product excretion but is highly tuned to the overall hydration of the body. Water intake by the oral route is similarly regulated according to need, through the complex behavioural mechanism of thirst.

Overall, osmoregulation is coordinated by the brain in conjunction with several endocrine glands (pituitary gland, adrenal gland, heart) and neural reflexes emanating from pressure and osmoreceptors located in the vasculature.

In birds, osmoregulation takes place in the kidneys but they have additional ways of excreting salt and conserving water. The ducts of the bird's urinary system, which lead waste material away from the kidneys, exit not to the outside as they do in mammals but to the hind part of the gastrointestinal tract. Here, much of the water is reabsorbed and the urea is turned into uric acid. This is combined with faecal matter from digestion and all the waste material is voided through a single vent below the tail feather region, called the cloaca.

Marine birds, whose fish diet is rich in salt, excrete salt, sometimes in a virtually crystalline form, through nasal glands. The ions are concentrated by means of energy-dependent Na–K–Cl co-transporters in the epithelial membranes of the glands, under the control of the parasympathetic nervous system.

Marine vertebrates

Fishes fall into two broad categories. Those which are restricted to one particular osmotic environment, be it salt or fresh water, are called **stenohaline**. They have a narrow zone of osmotic neutrality compared to animals in the other group, called **euryhaline**. The latter can live across a broad range of salinities due to highly developed osmoregulatory systems.

Most *freshwater* teleosts (bony fishes) are stenohaline. Their internal osmotic concentration is higher than that of their surroundings so they tend to gain water. The gills actively take up salt from the environment by means of mitochondria-rich cells. Water will diffuse into the fish by osmosis, so it excretes a hypotonic (dilute) urine. This process of osmoregulation in fresh-water fish is illustrated in Figure 30.37b.

Marine teleosts are stenohaline and have an internal osmotic concentration *lower* than that of the surrounding seawater, so tend to lose water and gain salt. They actively excrete salt from the gills. The process of osmoregulation in salt water fish is illustrated in Figure 30.37a.

Stenohaline marine cartilaginous fishes conserve water by a different mechanism. They retain urea in their blood at a relatively high concentration, raising its osmolarity to that of sea water or greater. For example, sharks have a blood solute concentration greater than 1000 mOsm.L^{-1}, slightly above that of sea

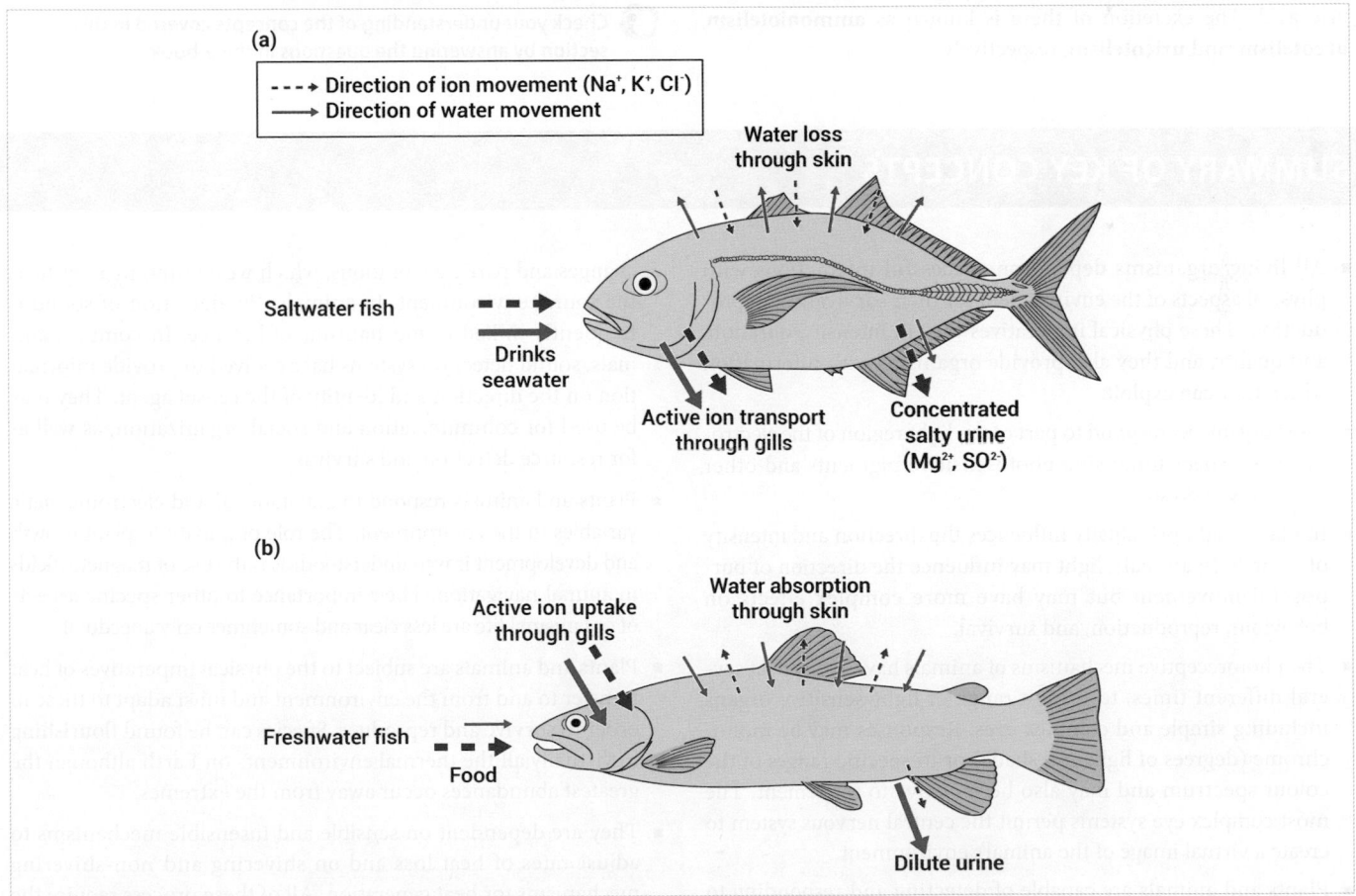

Figure 30.37 Osmoregulatory mechanisms in stenohaline teleosts. (a) Saltwater fish. (b) Freshwater fish.

Source: (a) Raver, Duane/NOAA; modified by Biezl translation improved by User:smartse/Wikimedia Commons/Public Domain. (b) Kare Kare modified by Biezl translation improved by smartse/Wikimedia Commons/CC BY-SA 3.0.

water, and do not drink water as other fishes do. Urea is damaging to living tissues so the fishes retain trimethylamine oxide as a protectant. This product of metabolism is normally excreted in other vertebrates but is a useful protein stabilizer when urea levels are high.

Examples of euryhaline fishes include flounder, lampreys, salmon, and eels. Many such species move from one type of water to another as part of their annual breeding cycles and must accommodate to hyperosmotic and hypo-osmotic environments. They can adjust both the amount of water they drink and the extent to which their urine is diluted. Their gills can pump Na^+ and Cl^- ions in either direction (in or out of the body), according to the type of water they are in. The necessary physiological adjustments may take days or weeks, so occur gradually just before or just as the fishes migrate.

Nitrogenous waste excretion

The regulation of salt and water balance in animals is often associated with their processes for excreting nitrogenous wastes. These materials result from the breakdown of proteins. In vertebrates there are three main forms of waste nitrogen: ammonia, urea, and uric acid. The excretion of these is known as **ammoniotelism**, **ureotelism**, and **uricotelism**, respectively.

Ammoniotelism is the least energetically expensive way for an animal to excrete nitrogen, but this is also the most potentially toxic form. For animals that are surrounded by water, ammonia is quickly diluted and transported away.

This is not an option for terrestrial animals. Mammals and birds use energy to convert ammonia into less toxic materials—urea and uric acid, respectively—before excretion. Urea requires water for elimination whereas uric acid can be excreted in a solid, essentially dry form.

The majority of fishes excrete ammonia across their gills. For fishes held in a confined volume of water such as a home aquarium, there is a risk of toxicity. Fishes which inhabit naturally restricted waters have evolved to produce the less toxic urea as an end product of protein breakdown, instead of ammonia.

 Go to the e-book to explore more about urea pulses in toadfish.

The cartilaginous fishes produce urea rather than ammonia. However, the majority of it is retained and used in osmoregulation. Animals that have both aquatic and terrestrial life stages, such as frogs, may be ammoniotelic during the aquatic phase and then switch to ureotely following metamorphosis.

 Check your understanding of the concepts covered in this section by answering the questions in the e-book.

SUMMARY OF KEY CONCEPTS

- All living organisms depend on successful interactions with physical aspects of the environment for their survival and reproduction. These physical imperatives vary in intensity, duration, and quality, and they also provide organisms with information which they can exploit.

- Most organisms respond to part of the light region of the electromagnetic spectrum, using photosensitive pigments and other chemicals to do so.

- In plants, light principally influences the direction and intensity of growth. In animals, light may influence the direction of purposeful movement but may have more complex effects on behaviour, reproduction, and survival.

- The photoreceptive mechanisms of animals have evolved, at several different times, to form a range of light-sensitive organs including simple and complex eyes. Responses may be monochrome (degrees of light and shade) or to specific ranges of the colour spectrum and may also be sensitive to movement. The most complex eye systems permit the central nervous system to create a virtual image of the animal's environment.

- Plants and animals are capable of detecting and responding to mechanical changes in the environment, in the form of pressure changes and particle vibrations, which we commonly refer to as the sound environment. In animals, the detection of sound is frequently linked to mechanisms of balance. In complex animals, sound detection systems have evolved to provide information on the direction and identity of the causal agent. They may be used for communication and social organization, as well as for resource detection and survival.

- Plants and animals respond to gravitational and electromagnetic variables in the environment. The role of gravity in plant growth and development is well understood, as is the use of magnetic fields in animal navigation. Their importance to other specific aspects of organismal life are less clear and sometimes only anecdotal.

- Plants and animals are subject to the physical imperatives of heat transfer to and from the environment and must adapt to these in order to survive and reproduce. Species can be found flourishing in virtually all the thermal environments on Earth although the greatest abundances occur away from the extremes.

- They are dependent on sensible and insensible mechanisms to adjust rates of heat loss and on shivering and non-shivering mechanisms for heat generation. All of these process require the expenditure of metabolic energy.

- Every species of organism has a ZTN, a range of environmental temperatures in which it survives with minimal expenditure of metabolic energy. Such zones vary greatly in their temperature range and are influenced by the extent to which the organism is capable of adjusting its own body temperature.

- Mammals and birds are the only animals capable of maintaining body temperatures within fine limits in the face of large differences in environmental temperature; they therefore have very wide zones of thermal neutrality and have been successful in an equivalently large range of habitats. However, this adaptability comes at a very high metabolic cost.

- Every organism can similarly be said to have a zone of osmotic neutrality, a range of environmental osmotic pressures in which it maintains the volume of its body and cells with minimal expenditure of metabolic energy. Complex animals maintain their water and electrolyte balance using specialist organs of osmoregulation, usually associated with mechanisms of waste excretion.

- Plants and animals are also subject to the imperatives of the gaseous environment, including physical determinants related to volume/temperature/pressure, partial pressure, solubility in fluids, and transfer across membranes. Adaptation to these imperatives is required for respiration and thermoregulation as well as for survival in air or in water at different depths.

 Use the flashcards in the e-book to test your recall of key terms introduced in this chapter.

QUESTIONS

 Looking for answers? Once you've answered these questions, follow the link in the e-book to the answer guidance and check your work.

Concepts and definitions

1. List at least five environmental imperatives to which organisms must be adapted if they are to survive and reproduce successfully.

2. What is a tropism?

3. Which parts of the electromagnetic spectrum do living organisms respond to?

4. What is the difference between sensible and insensible heat loss?

5. What is a zone of thermal neutrality?

6. What variables determine the rate at which oxygen passes over a biological membrane?

Apply the concepts

7. Explain why seals have a thick layer of blubber beneath the skin but fish in the same environment do not.

8. How do plants adjust their rate of water loss under different climatic conditions?

9. How can a) reptiles and b) mammals avoid an increase in body temperature in very hot desert environments?

10. What happens to a) the fraction of oxygen and b) the partial pressure of oxygen in the air at i) high altitude, ii) down a deep mine, iii) in deep water?

11. Explain how animals might use electromagnetic signals for navigation.

12. Explain why metabolic energy is necessary for osmoregulation in complex animals.

Beyond the concepts

13. Find two physiological reasons why the length of a snorkel must be limited to 50 cm.

14. Sketch a design for an experiment to determine which wavelengths of light determine the direction of plant seedling growth.

15. Using internet resources, list at least four ways in which bats use echolocation to find food resources.

16. Sound travels further in water than air. Over what distances are marine animals thought to be able to communicate using sound? Why might human activity on the oceans be disturbing to marine animals in this context?

17. What are the spectral sensitivities of the photopigments in the human eye? What is the situation in individuals described as 'colour blind'?

18. Bar-headed geese have been observed to fly over the Himalayas, including Everest. What physiological challenges do they face in doing this and how do they adjust to them?

FURTHER READING

McKinley, M., et al. (2018) Integrating competing demands of osmoregulatory and thermoregulatory homeostasis. *Physiology* **33**: 170–81.

A review of how mammals adapt to different temperatures and deal with the problem of maintaining appropriate water balance while sweating or panting.

Reitman, M.L. (2018) Of mice and men—environmental temperature, body temperature, and treatment of obesity. *FEBS Lett.* **592**: 1905–2196.

Thermoregulation in small and large animals, and the role of brown fat in non-shivering thermogenesis.

Lopez, D., et al. (2014) Gravity sensing, a largely misunderstood trigger of plant orientated growth. *Front. Plant Sci.*, **5**: Article 610.

How plants detect gravity and how their roots, shoots, and stems respond to it.

Maier, J.A.M., et al. (2015) The impact of microgravity and hypergravity on endothelial cells. *Biomed. Res. Int.* **2015**: Article ID 434803, http://dx.doi.org/10.1155/2015/434803

The effects of very high and very low gravity on animal cells, using endothelial cells as the focus. Describes some intriguing experiments and the possible effects of space travel.

Nilsson, D.-E. (2009) The evolution of eyes and visually guided behaviour *Phil. Trans. R. Soc. B* **364**: 2833–47.

A review of the key steps in the evolution of light detection and vision: sensitive cells, pigments, and complex eyes.

Porter, M. (2016) Beyond the eye: molecular evolution of extraocular photoreception. *Integr. Comp. Biol.* **56**: 842–52.

Light detection mechanisms not involving eyes, including a survey of the range of photosensors used by biological organisms from prokaryotes to eukaryotes, plants to animals.

Movement, Locomotion, and Migration

LEARNING OBJECTIVES

By the end of this chapter, you should be able to:

- Give a general account of the age of Earth and the approximate time when life began, and describe how the drift of land masses has shaped, and continues to shape, the surfaces available for colonization by life.

- Explain the difference between native and alien species (as commonly described) and the role of human activity in species distribution.

- List and describe, with suitable examples, at least three types of animal migration.

- Describe the principal mechanisms of plant dispersal.

- Explain the pendulum-like basis of limbed movement and identify the physical parameters which determine the speed and efficiency of locomotion.

- Explain the aerofoil basis of flying and describe the effects of thrust, lift, and drag on wings.

- Explain the physical demands of movement through water, including the role of buoyancy, and describe, with suitable examples, the adaptations of fish and other animals.

Chapter contents

Introduction		1054
31.1	Distribution of organisms on Earth	1054
31.2	Native or alien?	1060
31.3	Migration	1068
31.4	Plant dispersal	1072
31.5	Mechanisms of active locomotion	1077
31.6	Movement without limbs: worms	1077
31.7	Movement without limbs: snakes	1080
31.8	Walking and running	1080
31.9	Flying and gliding	1088
31.10	Swimming	1093

Watch the key concepts video in the e-book to prepare yourself for studying this chapter.

Introduction

In the first three chapters of this module we considered the nature and structure of biological organisms. We saw that living things enclose regions of the biosphere and are defined by how they exploit the resources of Earth.

There we also examined the basic functional units of life: individual cells, assemblies of similar cells, or complex collections of differentiated cells. We looked at some general principles of body structure and organization, and then considered the multitude of ways in which organisms interact with and adapt to environmental variables and physical imperatives.

Life forms are distributed right across Earth, occupying every possible type of niche on the ground, in water, and in the air. What we need to think about now is how that ubiquitous distribution came about. What were the forces which led to it, and what are the mechanisms by which organisms move around their environments?

To investigate this, we will find it necessary to make constant adjustments to our scale of focus: we will need to see why and how individual organisms move from one location to another while, at the same time, understanding how this is representative of much larger scale migrations of life forms around the planet. The movement of organisms happens spatially—from place to place—but it also happens in time and has done so throughout the history of life.

Many free-living individual organisms have no choice about when and where they move: their position, orientation, and location are entirely determined by environmental forces such as wind, tidal flows, and gravity. In other words, their movements and distribution take place *passively*.

For other organisms, body structure and lifestyle have evolved to provide strong, physical resistance to some of the environmental forces which might change their location: think of coral reefs, trees, and limpets.

More complex free-living organisms may respond to changes in their immediate surroundings by *actively* determining where to position themselves and for how long. They make reactive, purposeful, or even conscious decisions about where they should be, influenced by resource availability, access to mates, or avoidance of predators and disadvantageous conditions.

Sometimes, as with insect swarms, annual bird migrations, or the seasonal movements of herds of ungulates, the relocation of individuals happens on a massive and dramatic scale.

At an even larger scale, geological and environmental forces, often operating over timespans of millennia, have determined how life is dispersed. They can help to explain why life is where it is in different parts of the planet. Why, for example, does New Zealand have no **indigenous** terrestrial mammals, whereas marsupials occur in both Australia and America?

We must also recognize that human kind, with our high abundance and widespread distribution, our ability to undertake rapid long-distance travel, and our enormous technological skill, represent a major force in the dispersal, concentration, and redistribution of other forms of life.

31.1 Distribution of organisms on Earth

Before we can understand the forces which have led to the distribution of different life forms across Earth, we need some appreciation of the history of our planet and how its land masses and seas came to be distributed in the way we observe them today.

Palaeontological movements

Palaeontologists currently believe that life began on Earth around 4300 million years ago (note that from here we will abbreviate 'million years ago' to 'Ma', which means 'mega-annum'). This was about 100 million years after liquid water appeared on Earth. All life that we know of requires water and so its earliest forms, whatever they were, must have developed in the early oceans or in other aqueous environments.

▶ We explore the emergence of life on Earth in more detail in Topic 2.

At that time, as now, Earth's **crust** was a combination of **continental** and **oceanic** regions. The continental regions include areas of land and those parts of the land which slope into shallow shelves and reefs beneath the water. The oceanic regions are those which form the floors of deep oceans, including parts of the crust where material is emanating from Earth's **mantle**. The mantle and crust are collectively known as the **lithosphere**. Figure 31.1 depicts the structure of Earth, from the lithosphere down to the mantle and core of our planet.

A key feature of this structural arrangement is the presence of **tectonic plates**, which you can see outlined in Figure 31.2. These are large, coherent masses of crust which, over long periods of geological time, drift apart and collide. Their movement is driven by the upwelling of the mantle material, on which they can be thought of as floating. This leads to the constant recycling of the crust: new oceanic crust is created at the mid-ocean ridges (divergent plate boundaries) and destroyed where existing material sinks below the continental crust (convergent plate boundaries). As a result, the oldest ocean floor has a maximum age of around 200 million years.

Boundaries between the plates are often sites of volcanic activity and earthquakes. They are characterized by massive, unstable irregularities such as faults, rift valleys, trenches, subduction zones, and mountain ranges.

The gradual movement of tectonic plates is often referred to as continental drift. It has taken place over the lifetime of Earth and is still continuing. Large aggregations of land masses once formed supercontinents—extended areas of contiguous land on which we can map the origins of the continents, islands, and other land masses which exist today.

The existence of these land masses is of interest to biologists because it tells us much about the fundamental distribution of life forms. They are of particular significance in understanding terrestrial life because they potentially allowed it to spread unhindered by large bodies of water.

Figure 31.3 summarizes some key events in the likely timeline of life on Earth. We see that although multicellular organisms first

Figure 31.1 A cutaway diagram showing the structure of Earth, from core to lithosphere. The lithosphere includes the crust and the upper layers of the mantle. It is the region where geological activity takes place, including the drifting of continental land masses.

Source: Srimadhav/Wikimedia Commons/Public Domain.

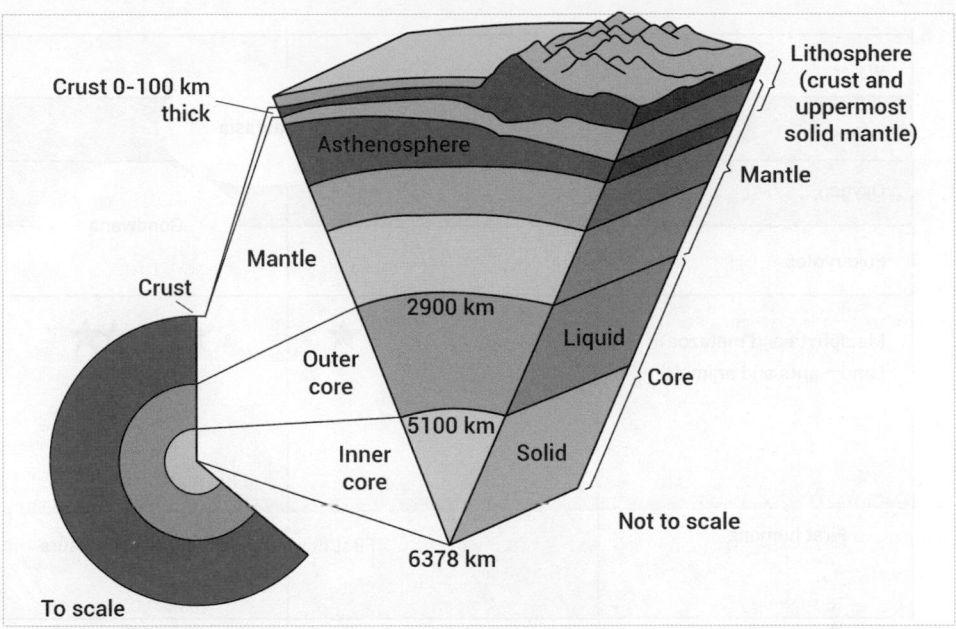

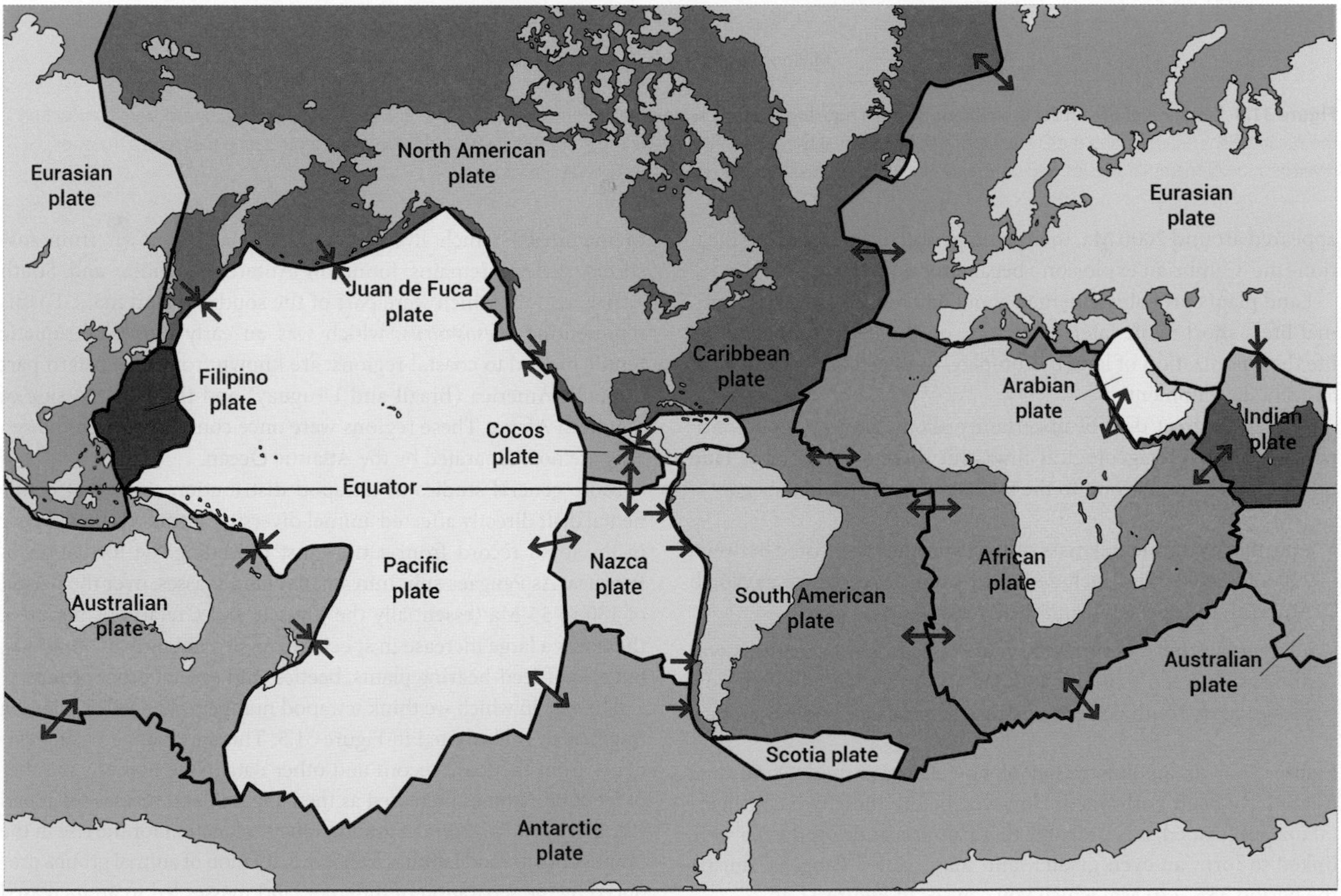

Figure 31.2 Earth's uppermost mantle and crust (collectively called the lithosphere) breaks up to form a series of tectonic plates which move around the surface. This figure shows the plates as they appear today. 'Divergent' plate boundaries are illustrated by arrows pulling in opposite directions (←→). These appear in the middle of the oceans, the two most prominent being the Mid-Atlantic Ridge and the East Pacific Rise. Note that these divergent plate boundaries tend to run north to south, suggesting an intimate relationship with Earth's rotation. Along these boundaries the seafloor spreads, renewing the crust with fresh material. At a 'convergent' boundary (→←) the crust of one plate is sucked down into the upper mantle, thrusting the crust of the other plate upwards to form a mountain range.

Source: Baggott J. (2015) *Origins: The Scientific Study of Creation*. Oxford University Press.

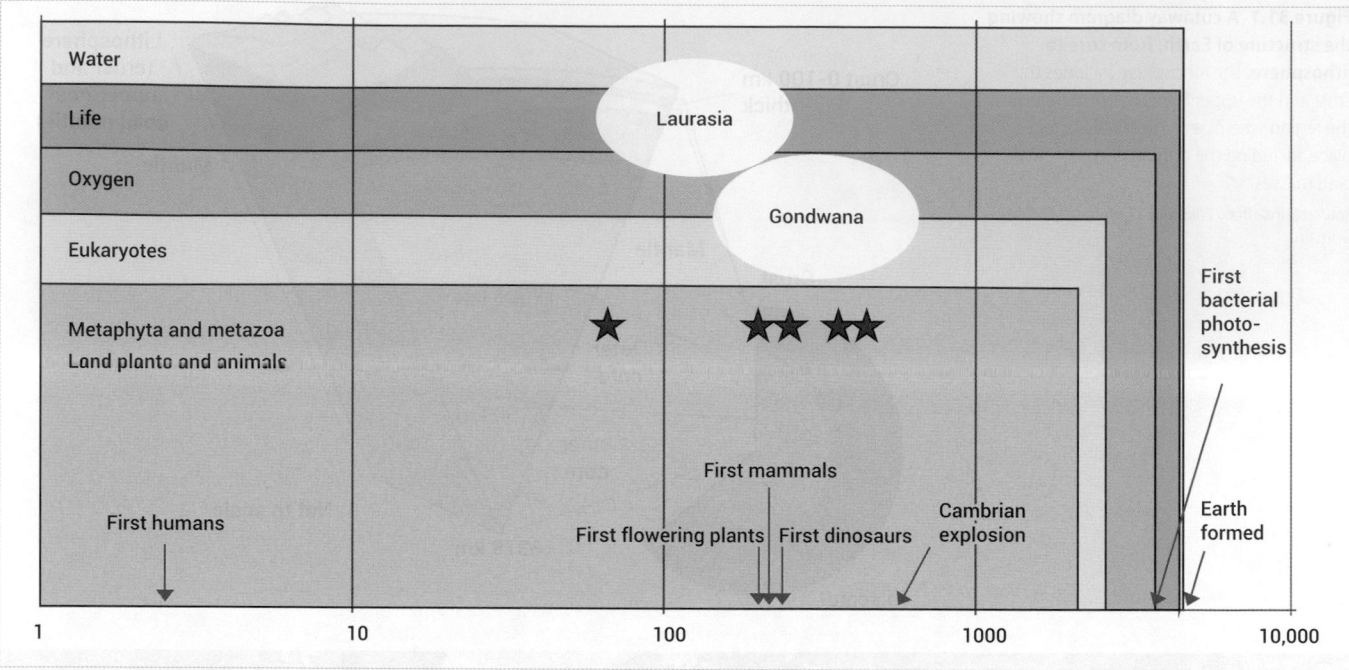

Millions of years ago (logarithmic scale)

Figure 31.3 Timeline of life on Earth (millions of years ago; logarithmic scale). The most recent extinction event, 66 Ma, was initiated by an asteroid impact on the Yucatan Peninsula, forming the Chicxulub Crater. Dinosaurs subsequently declined, leaving only those ancestral to birds and reptiles, and there was a massive diversification of mammal species. Red stars represent major extinction event (more than 75% species lost).

appeared around 2000 Ma, their main period of species diversification (the 'Cambrian explosion') began around 541 Ma.

Land plants probably emerged about 450 Ma and terrestrial animal life a short while later. Therefore, on the overall timescale of life the colonization of land by complex, multicellular organisms is a recent development.

There is a great deal of uncertainty about the way supercontinents evolved over geological time, but there are two large land masses which are relevant to the time since terrestrial life began:

- a northern continental mass called Laurasia; this existed between 250 and 56 Ma and included what we now recognize as North America, Europe, Asia, and the Arctic;
- a larger southern continental mass called Gondwana; this was present between 620 and 132 Ma and included what we now recognize as South America, Africa, Australasia, and Antarctica.

Figure 31.4a is an illustration of how Laurasia and Gondwana might have been positioned. However, before these two continental masses formed, it is probable that Laurasia and Gondwana were linked to form an even greater land mass, called Pangaea, during the period 335–170 Ma, which you can see in Figure 31.4b.

Distribution, diversity, and the fossil record

The fossil record supports the view that these large land masses facilitated the terrestrial distribution of life. *Lystrosaurus*, a herbivorous therapsid (reptilian vertebrates possibly ancestral to mammals) which lived about 250 Ma, is known from fossilized skeletal remains found in Antarctica, India, and South Africa, regions which were part of the southern land mass. Fossils representing *Mesosaurus*, which was an early (300 Ma) aquatic reptile limited to coastal regions, are known from the Eastern part of South America (Brazil and Uruguay) and the Western side of Southern Africa. These regions were once connected, even though they are now separated by the Atlantic Ocean.

More general studies of tetrapod distribution show that continental drift directly affected animal diversity. Tetrapods first appear in the fossil record from a time just preceding the formation of Pangaea. As Pangaea split into smaller land masses, over the period of 180 to 55 Ma (essentially the Jurassic and Cretaceous periods), there was a large increase in species diversity, not only of tetrapods, but also of seed-bearing plants, beetles, and several other groups.

The way in which we think tetrapod numbers changed as Pangaea fragmented is illustrated in Figure 31.5. This was mapped using evidence from the fossil record and other data. Note how the number of tetrapod families increased as the Pangaeal land masses separated (the black solid line on the graph). One explanation for the rise in the number of tetrapod families is that the isolation of animal groups produced by the separation of different land masses led to an increased rate of formation of new species. This process is called **speciation**.

🌀 Go to the e-book to explore an interactive version of Figure 31.5.

However, careful analysis of the evidence and the use of theoretical models suggest that this is not a simple story. The fragmentation of Pangaea affected water and air movements around

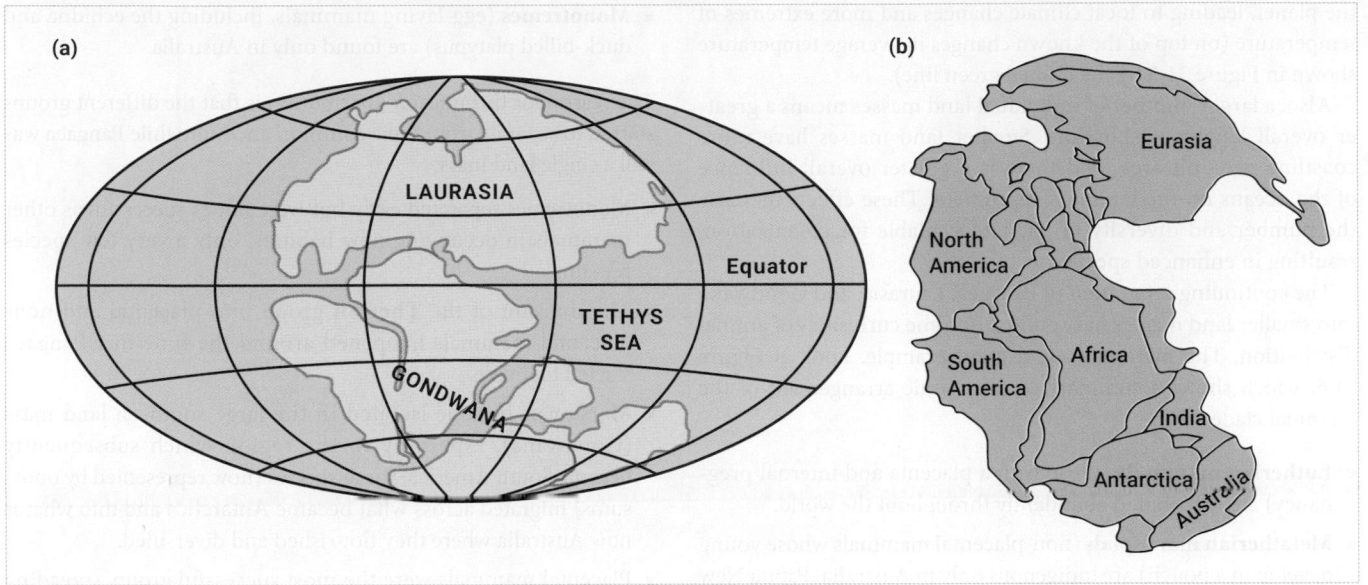

Figure 31.4 The evolution of the supercontinents. (a) The likely arrangement of the plates forming the supercontinent of Pangaea during the Triassic period, some 200 Ma. The major land masses of Laurasia (north) and Gondwana (south) were transiently linked between 335 and 170 Ma. (b) A mapping of the probable locations of current land masses within the supercontinent of Pangaea.

Source: (a) Lennart Kudling/Wikimedia Commons/CC BY 3.0; (b) Fama Clamosa/Wikimedia Commons/CC BY-SA 4.0.

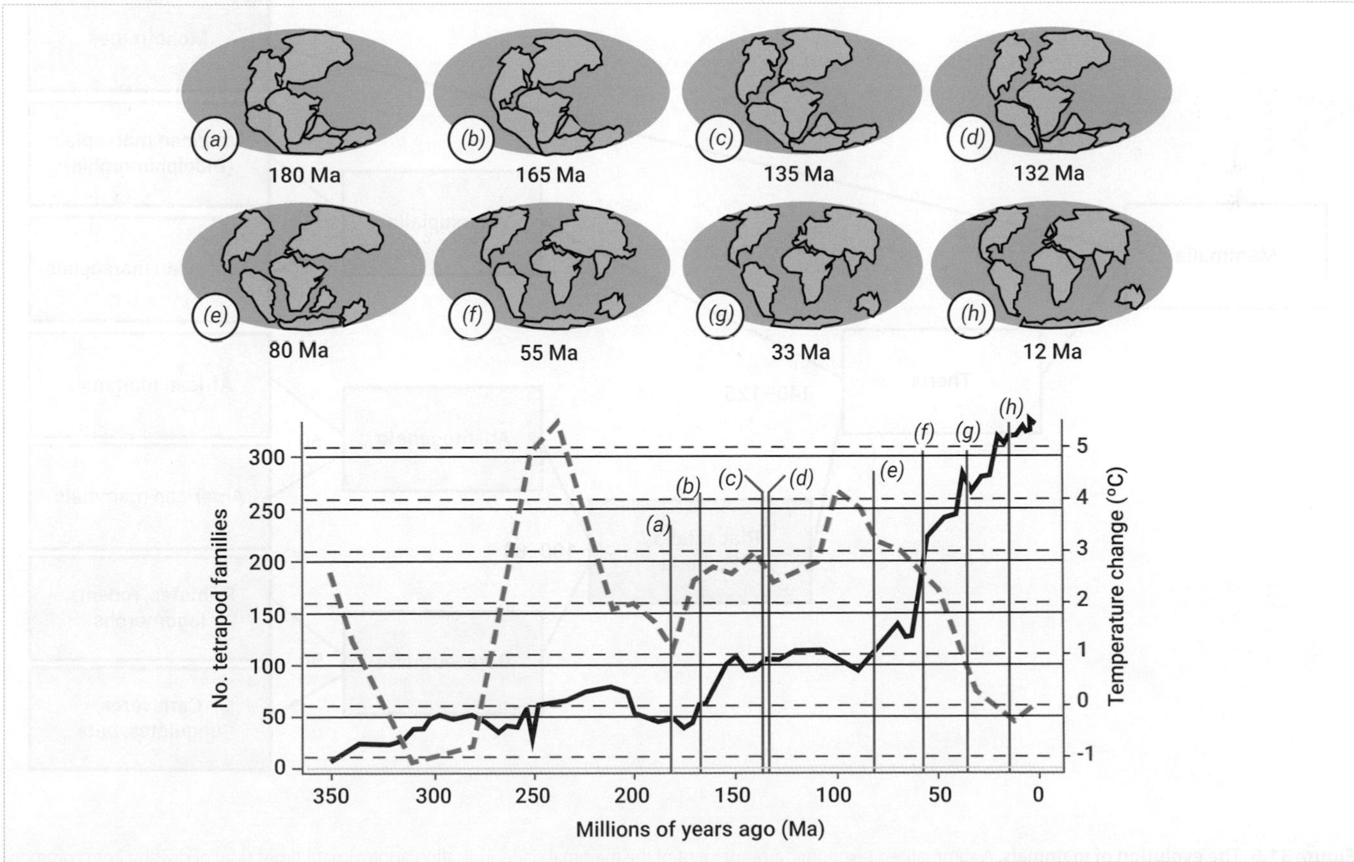

Figure 31.5 Pangaea diversity. Tetrapod diversity (number of tetrapod families; black line) and the change in global average temperature (dashed green line) during the fragmentation of Pangaea.

Source: Adapted from: Jordan MR. *et al.* (2016) Quantifying the effects of the break up of Pangaea on global terrestrial diversification with neutral theory. *Phil Trans R Soc Biol*, 371 (1691).

the planet, leading to local climate changes and more extremes of temperature (on top of the known changes in average temperature shown in Figure 31.5 by the dashed green line).

Also, a larger number of individual land masses means a greater overall length of shoreline. Smaller land masses have more coastline per unit area, and there is a greater overall influence of the oceans on the land and its climate. These effects increase the number and diversity of habitats available for colonization, resulting in enhanced speciation.

The continuing separation of Pangaea, Laurasia, and Gondwana into smaller land masses has resulted in some curiosities of animal distribution. The mammals are a good example. Look at Figure 31.6, which shows a summarized taxonomic arrangement of the mammal clade:

- **Eutherian mammals** (those with a placenta and internal pregnancy) are distributed abundantly throughout the world.
- **Metatherian marsupials** (non-placental mammals whose young develop in a pouch) are indigenous only in Australia, Papua New Guinea, and South America.

- **Monotremes** (egg-laying mammals, including the echidna and duck-billed platypus) are found only in Australia.

The reason for this uneven distribution is that the different groups started to separate from their common ancestor while Pangaea was still a single land mass:

- Monotremes separated early but were not as successful as other mammals in occupying new habitats; only a very few species remain.
- The division of the **Therian** group into placental and non-placental mammals happened around the time that Pangaea started to divide.
- Marsupials became isolated in the large southern land mass (Gondwana), especially in the region which subsequently became South America; these species (now represented by opossums) migrated across what became Antarctica and into what is now Australia where they flourished and diversified.
- Placental mammals were the most successful group, spreading through most parts of the Northern land mass called Laurasia.

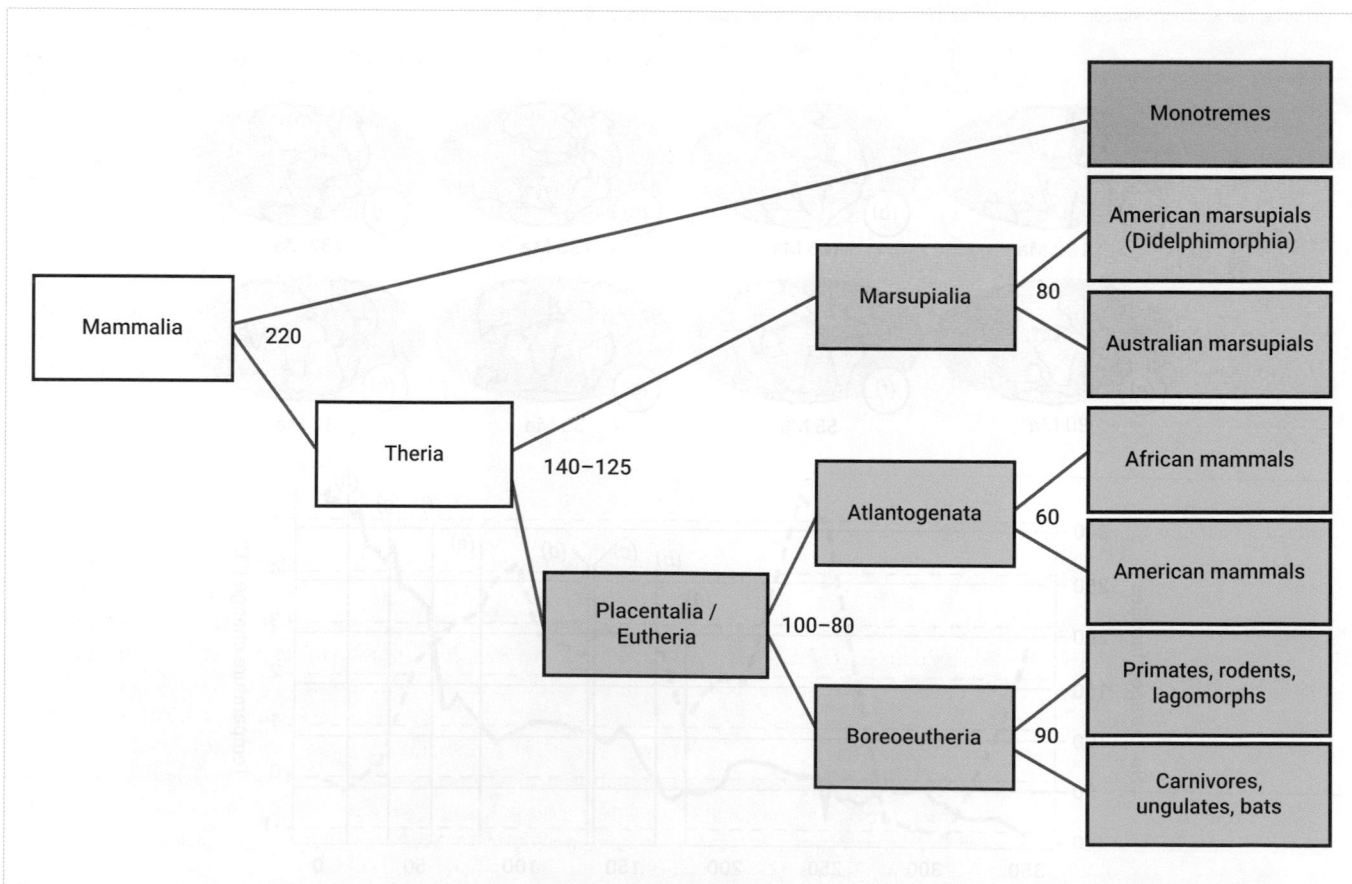

Figure 31.6 The evolution of mammals. A summarized taxonomic arrangement of the mammal clade, indicating approximate times (Ma) of division from common ancestors.

PAUSE AND THINK

Identify your current geographical location and, using Figures 31.2, 31.3, and 31.4, find the land mass it used to be part of, and its probable location, at the following times:

- When life on Earth began.
- When eukaryotes first evolved.
- At the time of the Cambrian explosion.
- When the first mammals appeared.
- When the first flowering plants appeared.

Example answers:

Nottingham, England

- When life on Earth began: Eurasian plate
- When Eukaryotes first evolved: Laurasia
- At the time of the Cambrian explosion: Laurasia
- When the first mammals appeared: Laurasia/Gondwana
- When the first flowering plants appeared: Laurasia/Gondwana

Hamburg, Germany

- When life on Earth began: Eurasian plate
- When Eukaryotes first evolved: Laurasia
- At the time of the Cambrian explosion: Laurasia
- When the first mammals appeared: Laurasia/Gondwana
- When the first flowering plants appeared: Laurasia/Gondwana

Melbourne, Australia

- When life on Earth began: Australian plate
- When eukaryotes first evolved: Gondwana
- At the time of the Cambrian explosion: Gondwana
- When the first mammals appeared: Laurasia/Gondwana
- When the first flowering plants appeared: Laurasia/Gondwana

Islands and diversity

So, why are there no indigenous land mammals in New Zealand? New Zealand and the island of New Caledonia started to separate from Australia around 83 Ma, as part of a small continental plate called Zealandia. As Figure 31.3 shows, this was long after the major groups of mammals had evolved.

Evidence from St Bathans on New Zealand's South Island shows that a mammal representing a group ancestral to both placental and marsupial mammals once lived there. The same ancestral group existed in South America. As there is no fossil evidence of placental or marsupial mammals in New Zealand, these groups evidently failed to survive after the Zealandia separation. The reason is unclear, but the most likely explanation is that extremes of climate on the small island, compared with those on larger land masses, did not allow them to establish themselves.

New Zealand has other indigenous mammals (bats, whales, dolphins, seals, sea lions), but these would not have been prevented from colonizing the islands by *land* separation. There are also indigenous vertebrates, including birds and reptiles. More recently, man has introduced, deliberately or accidentally, rats, mice, rabbits, deer, agricultural species, and domestic pets. The brushtail possum (*Trichosurus vulpecula*), a marsupial brought by man from Australia in the 19th century, is now considered a significant pest.

Other islands that illustrate the effect of land isolation on wildlife are Madagascar and Hawaii.

Madagascar is situated in the Indian Ocean, 400 km off the East coast of Africa. About 90 per cent of Madagascar's animal and plant species are **endemic** and found nowhere else on Earth. Its unique fauna includes primates (lemurs), insectivores (tenrecs, solenodons, and golden moles), carnivores (mongooses), rodents, and bats. Fossil evidence shows that many other vertebrates, including the flightless elephant bird, which is illustrated in Figure 31.7, once lived there. Invertebrates include oligochaete worms, beetles, moths, butterflies, and spiders.

Madagascar is a remnant of Gondwana, representing the junction between what are now Africa and India. Its west coast began to split from East Africa about 182 Ma and its east coast from India about 90 Ma. It has been relatively isolated as a land mass for a sufficiently long period of time that species have evolved in the absence of the predators and competitors which might have influenced them on the larger continents of which it was previously part. In addition, some species have evolved to fill empty niches in ways that their cousins elsewhere have not.

Madagascar's unique species can be aligned to ancestral groups found in Africa and India to trace their likely time of arrival on the island. Ancestors of the larger vertebrates may have survived on the island as it split away. Ancestors of others, such as the insectivores, may have arrived on rafts of vegetation from Africa when the geographical distance was still short. The bats, which can fly but often only for a limited distance, may be descendants of ancestors whose range became progressively constrained by the continental drift. More recent arrivals, including rodents and domestic animals, probably came with man, beginning 4000 years ago but especially with migrations from Indonesia some 3000–2000 years ago and later interbreeding with settlers from Africa.

The Hawaiian islands are located in the Pacific Ocean, 3200 km from the west coast of the United States. These islands arose not by splitting from a land mass, but as remnants of volcanoes (the Hawaiian–Emperor seamount chain) during the westward movement of the Pacific continental plate. The islands are still expanding at the south-eastern end as more lava emerges. The north-westerly islands in the chain are thus the oldest, having first emerged around 28 Ma. The youngest and largest island, Hawaii itself, began to emerge about 1 Ma.

The only mammals on the Hawaiian islands are those introduced by man, but there is a great diversity of birds, small terrestrial reptiles, snails, insects, and spiders. The reptiles,

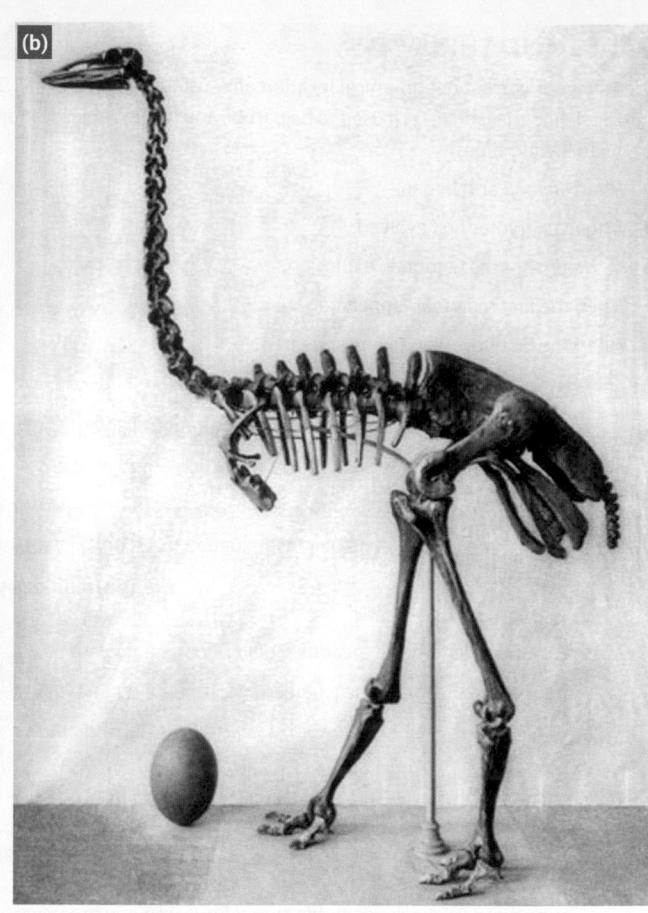

Figure 31.7 The Madagascan elephant bird. (a) An artist's reconstruction of the Madagascan elephant bird (*Aepyornis maximus*) which became extinct in the late 17th or very early 18th century. It is known from bone fragments left on the island and a small number of reconstructed skeletons (b). The bird was flightless and stood over 3 m tall. Its eggs, the largest known, weighed about 10 kg.

Source: (a) Jaime Chirinos; (b) The History Collection/Alamy Stock Photo.

including chameleons, geckos, and lizards, are also introduced species. Invertebrates were probably carried there on vegetation drifting on sea currents, and also as passengers on migrating birds. Hawaii's isolation has produced remarkable diversification in a relatively short time. For example, over 10,000 species of insects are now considered as 'native' to the islands.

Similar isolation-induced speciation effects have occurred on other islands, famously including the birds and reptiles of the Galapagos off the west coast of South America and the massive 'dragon' monitor lizards of the Indonesian Komodo island.

The effects of isolation on the diversification and spread of species fascinated the 19th century naturalist and explorer Alfred Russel Wallace and contributed to his development, with Charles Darwin, of the theory of evolution by natural selection.

🌐 Follow the link in the e-book to explore more about islands and diversity, and the work of Alfred Russel Wallace.

🌐 Go to the e-book to complete an activity that will help you engage with the information presented in Figure 31.3.

🌐 Check your understanding of the concepts covered in this section by answering the questions in the e-book.

31.2 **Native or alien?**

We are used to thinking of organisms as occurring more naturally in some locations rather than others. Indeed, biologists and those interested in wildlife management and conservation often distinguish between 'indigenous' or 'native' species and those which are 'alien' or 'introduced'. The latter are sometimes called 'invading' species, especially if they are unwanted or if they displace or compete with species already present. Such designations may become the basis for protection, conservation, and eradication policies.

So, what do these descriptors really mean? Consider these examples:

■ Many Australians were disappointed to learn, from studies published in 2016, that their beloved marsupials are not true 'natives' and probably originated in what is now South America when land masses were contiguous.

■ In the United Kingdom, the grey squirrel, *Sciurus carolinensis*, introduced from North America, is sometimes seen as a pest or even classed as vermin. Figure 31.8 shows the changes of squirrel distribution between 1945 and 2010, and as you can see, the

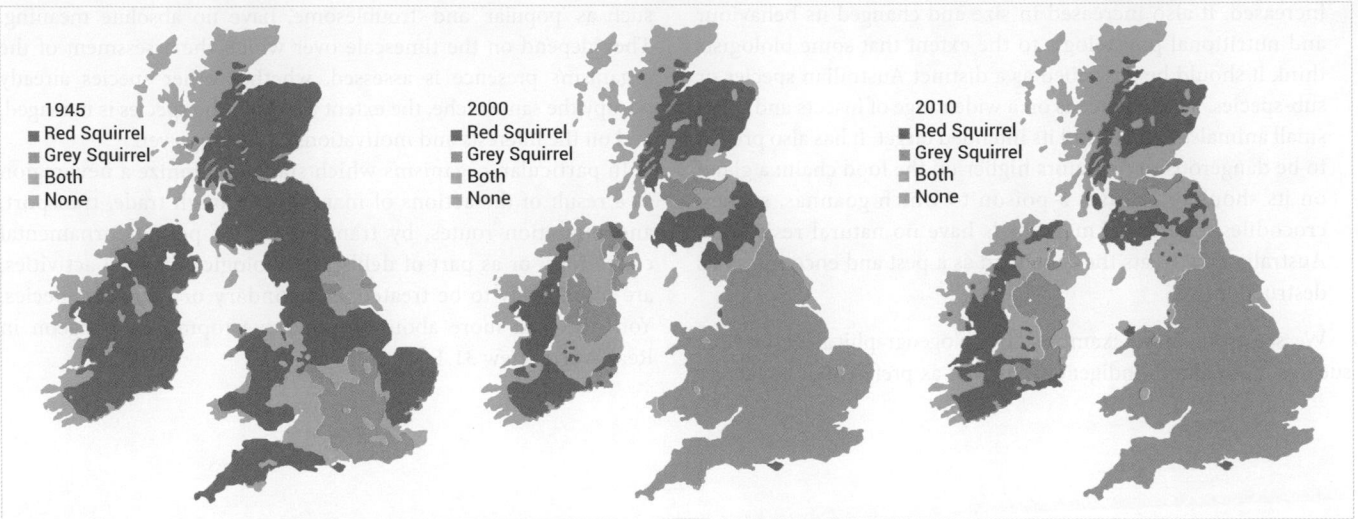

Figure 31.8 Changes in UK squirrel distribution since 1945.
Source: © Craig Shuttleworth/RSST.

grey squirrel has significantly outcompeted the red squirrel, *Sciurus vulgaris*, during this time period. This is because the grey squirrel can pass on a lethal viral disease to the red squirrel but is itself immune.

- The ornamental shrub, *Rhododendron ponticum* (like the one in Figure 31.9), which is a popular ornamental shrub in many large gardens in Britain, was introduced from mainland Europe in the mid-18th century. It has spread very successfully and is now considered a troublesome invasive species in many parts of the UK.

- The cane toad (*Rhinella marina*), which you can see in Figure 31.10, is an introduced species which got out of hand and turned into a major problem. In the 1930s, the toad was exported around the world from Puerto Rico where it was a biological regulator of cane beetle grubs. When introduced to the northern parts of Australia, it bred prolifically and its numbers rapidly

Figure 31.9 Rhododendron (*Rhododendron pictum*).
Source: Ryan Somma/CC BY-SA 2.0.

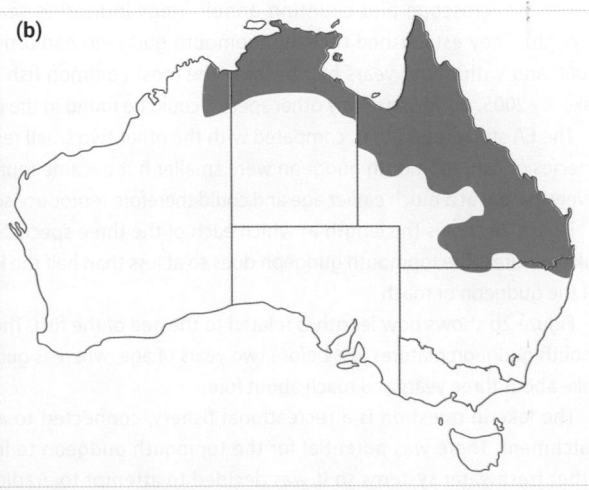

Figure 31.10 The cane toad. (a) The cane toad (*Rhinella marina*). (b) Map of Australia showing the extent (shaded area) of cane toad distribution in 2016. The original introduction was in a limited region of the north-east (Queensland) coast.

Source: (a) Ondrej Prosicky/Shutterstock.com; (b) Figure 3 from Reid Tingley, et al. (2017) New weapons in the toad toolkit: a review of methods to control and mitigate the biodiversity impacts of invasive cane toads (*Rhinella marina*). *The Quarterly Review of Biology*, 92:123–49. © 2017 The University of Chicago.

increased. It also increased in size and changed its behaviour and nutritional physiology, to the extent that some biologists think it should be described as a distinct Australian species or sub-species. The toad feeds on a wide range of insects and other small animals, well beyond its intended target. It has also proved to be dangerous to predators higher up the food chain: a gland on its shoulder releases a poison to which goannas, snakes, crocodiles, and small marsupials have no natural resistance. Australia now treats the cane toad as a pest and encourages its destruction.

We see from these examples that biogeographical descriptors such as 'native' and 'indigenous', as well as preferential adjectives such as 'popular' and 'troublesome', have no absolute meaning. They depend on the timescale over which the assessment of the organism's presence is assessed, whether other species already occupy the same niche, the extent to which the species is managed, and on the interests and motivations of the observer.

In particular, organisms which start to colonize a new region as a result of the actions of mankind, through trade, transport, and migration routes, by transmission of pets or ornamental collections, or as part of deliberate biological control activities, are more likely to be treated as secondary or invading species. You can read more about the invasive topmouth gudgeon in Real World View 31.1.

REAL WORLD VIEW 31.1   Invasion of the topmouth gudgeon

In 2002, anglers fishing in a freshwater lake in the UK reported a new species that they had not seen before. The lake was previously home to two small fish species (roach, *Rutilus rutilus*, and gudgeon, *Gobio gobio*) and was also stocked with larger bream (*Abramis brama*), tench (*Tinca tinca*), and carp (*Cyprinus carpio*) for recreational fishing. The fish was identified as a topmouth gudgeon (*Pseudorasbora parva*).

The topmouth gudgeon, pictured in Figure 1, is an Asian cyprinid fish. It was accidentally released in Europe in 1960 during the movement of Chinese carp for fish farming. It is now considered the most invasive fish species in Europe. Its success has been due to its early maturity, batch spawning, and nest guarding. The topmouth gudgeon causes ecological problems by out-competing native species and carrying diseases which can spread to other fish populations.

Scientists from the Environment Agency (EA) sampled fish in the lake using nets and electrofishing. They assessed the fish they caught for length, sex, and gonad mass. They also aged the fish by viewing scales under a microscope and counting annuli (rings indicating seasonal growth). They established that the topmouth gudgeon had arrived in 2000 and within four years had become the most common fish in the lake. By 2005, no young of any other species could be found in the lake.

The EA study found that, compared with the other two small resident species of fish, topmouth gudgeon were smaller but became reproductively mature at a much earlier age and could therefore reproduce sooner.

Figure 2a shows the length at which each of the three species in the lake matures. The topmouth gudgeon does so at less than half the length of the gudgeon or roach.

Figure 2b shows how length is related to the age of the fish. The topmouth gudgeon matures just before two years of age, whereas gudgeon take about three years and roach about four.

The lake in question is a recreational fishery, connected to a river catchment. There was potential for the topmouth gudgeon to invade other freshwater systems so it was decided to attempt to eradicate it from the lake.

Figure 1 The topmouth gudgeon (*Pseudorasbora parva*).
Source: Seotaro/Wikimedia Commons/CC BY-SA 3.0.

Resident fish species were removed and held off-site, and the lake was then treated with rotenone (a piscicide or fish poison). Dead fish were removed and once the rotenone had broken down, resident species were released back into the lake. The eradication was successful, allowing the native fish populations to recover.

Removing invasive aquatic species like this can be very costly and is difficult to do unless they exist in a contained area such as a lake.

Read the original work

Britton, J.R. & Brazier, M. (2006). Eradicating the invasive topmouth gudgeon, *Pseudorasbora parva*, from a recreational fishery in northern England. *Fisheries Management and Ecology*, **13**, 329–335.

Britton, J.R., Davies, G.D., Brazier, M. & Pinder, A.C. (2006). A case study on the population ecology of a topmouth gudgeon (*Pseudorasbora parva*) population in the UK and the implications for native fish communities. *Aquatic Conservation: Marine and Freshwater Ecosystems*, **16**, 1–11.

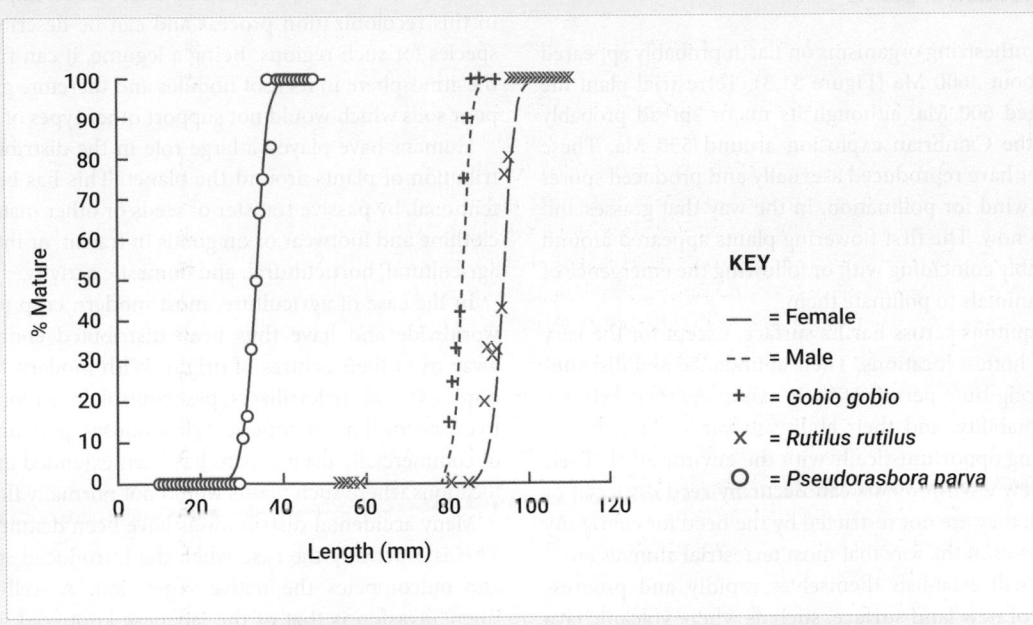

Figure 2 (a) Relationship between length and percentage of mature individuals for topmouth gudgeon (O), roach (x), and gudgeon (+).

Source: Figure 2 of Britton JR, Davies GD, Brazier M & Pinder AC (2007). A case study on the population ecology of a topmouth gudgeon (*Pseudorasbora parva*) population in the UK and the implications for native fish communities. *Aquatic Conservation: Marine and Freshwater Ecosystems* 16,1–11. John Wiley and Sons.

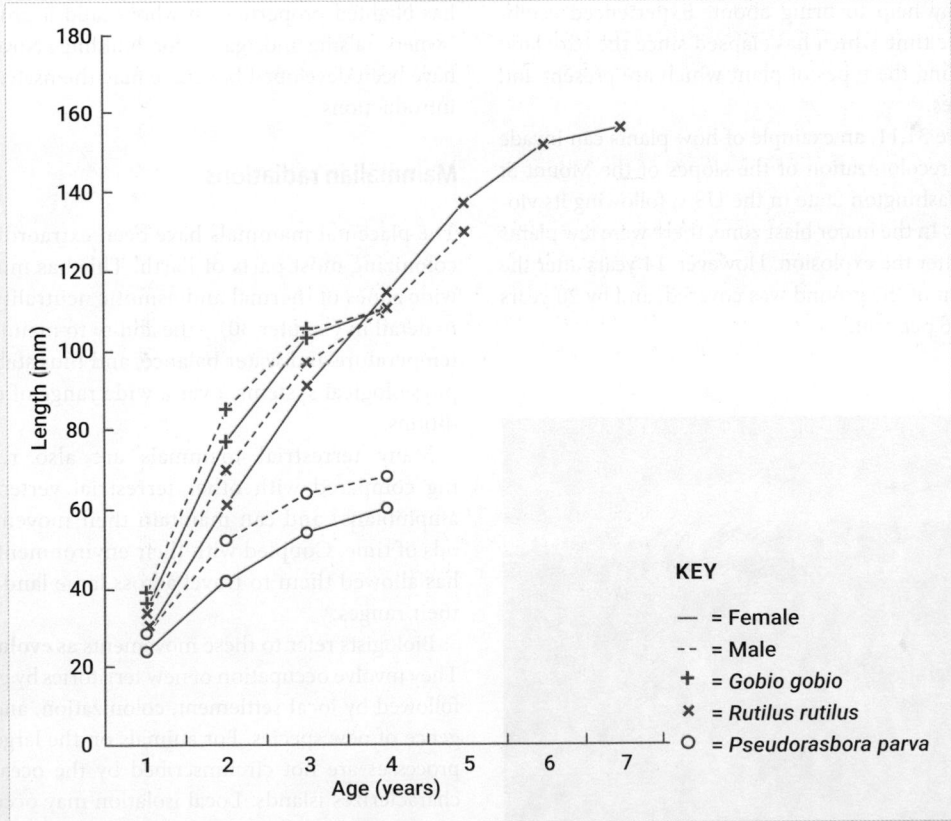

Figure 2 (b) Mean length and age of female (solid lines) and male (dashed lines) for topmouth gudgeon (O), roach (x), and gudgeon (+) in March 2005.

Source: Figure 2 of Britton JR, Davies GD, Brazier M & Pinder AC (2007). A case study on the population ecology of a topmouth gudgeon (*Pseudorasbora parva*) population in the UK and the implications for native fish communities. *Aquatic Conservation: Marine and Freshwater Ecosystems* 16,1–11. John Wiley and Sons.

Planetary distribution of plants

The first photosynthesizing organisms on Earth probably appeared in the oceans, about 3600 Ma (Figure 31.3). Terrestrial plant life may have emerged 600 Ma, although its major spread probably coincided with the Cambrian explosion around 550 Ma. These plants must either have reproduced asexually and produced spores or depended on wind for pollination, in the way that grasses and related plants do now. The first flowering plants appeared around 200 Ma, presumably coinciding with or following the emergence of insects or other animals to pollinate them.

Plants are ubiquitous across Earth's surface, except for the very coldest and very hottest locations. Their abundance and distribution reflect the long time period of their existence, their extreme variety and adaptability, and their ability to survive and reproduce by interacting opportunistically with the environment. Their migration into new environments can occur by seed dispersal or vegetative spread: they are not restricted by the need for contiguity between land masses in the way that most terrestrial animals are.

Plant species will establish themselves rapidly and progressively in regions of new land surface, such as where volcanic lava and ash have covered existing land, or where bare areas emerge from landslips, cliff falls, dune shifting, or erosion. Each wave of vegetation benefits from the accumulation of remains from the previous one, its nutrient potential, and the stabilization of the surface which it may help to bring about. Experienced ecologists can estimate the time which has elapsed since the bare land appeared by examining the types of plant which are present and their relative densities.

As shown in Figure 31.11, an example of how plants can invade new territory is the recolonization of the slopes of the Mount St Helens volcano in Washington State in the USA, following its violent eruption in 1981. In the major blast zone, there were few plants present three years after the explosion. However, 14 years after the explosion, 38 per cent of the ground was covered, and by 20 years coverage was up to 66 per cent.

Figure 31.11 Prairie lupine on Mount St Helens.
Source: oksana.perkins/Shutterstock.com.

The prairie lupine (*Lupinus lepidus*) was an important first plant in this recolonization process and can be described as a pioneer species for such regions. Being a legume, it can fix nitrogen from the atmosphere in its root nodules and therefore grow in nutrient-poor soils which would not support other types of plant.

Humans have played a large role in the distribution and redistribution of plants around the planet. This has been either unintentional, by passive transfer of seeds or other material attached to clothing and footwear or on goods in transit, or intentional during agricultural, horticultural, and domestic activity.

In the case of agriculture, most modern crop plants are grown worldwide and have thus been distributed enormous distances away from their centres of origin. With modern husbandry techniques, including fertilizers, pest control, irrigation, and the ability to create local microclimates (glass houses, polytunnels), the range of commercially useful crops has been extended into geographical locations where such plants would not normally flourish.

Many accidental distributions have been distinctly detrimental. This is especially the case when the introduced species outgrows and outcompetes the native vegetation. A well-known case of 'alien' invasion is that of the Japanese knotweed which arrived at Kew Gardens in London, UK, via Holland in 1850. Over the next 150 years, the plant spread widely through the UK and proved very difficult to control. It is now classed as an illegal plant and has to be reported to the authorities if found. Its invasive potential has blighted properties on whose land it grows, even preventing owners raising mortgages for building. Novel biocontrol agents have been developed but these may themselves be classed as alien introductions.

Mammalian radiations

The placental mammals have been extraordinarily successful in colonizing most parts of Earth. This has much to do with their wide zones of thermal and osmotic neutrality (which we discuss in detail in Chapter 30)—the ability to maintain a constant body temperature and water balance, and thus stable biochemical and physiological systems, over a wide range of environmental conditions.

Many terrestrial mammals are also relatively fast moving compared with other terrestrial vertebrates (reptiles and amphibians) and can maintain their movements for long periods of time. Coupled with their environmental adaptability, this has allowed them to travel across large land masses and extend their ranges.

Biologists refer to these movements as **evolutionary radiations**. They involve occupation of new territories by groups of individuals followed by local settlement, colonization, and the gradual emergence of new species. For animals on the large land masses, these processes are not circumscribed by the oceanic isolation which characterizes islands. Local isolation may occur for other reasons (terrain, climate, food resources, predators), but in general the distribution continues over large distances.

Adaptive radiation is a general term often used to incorporate all species diversification associated with range expansion. Some biologists confine the use of the term to situations where a single lineage of organisms (such as the lizards on Hawaii, the finches famously studied by Charles Darwin on the Galapagos, or the Australian cane toad described above) diversifies rapidly due to changing local conditions within a limited region such as an island. This kind of speciation happens over a shorter period of time than it does with evolutionary radiation.

Regarding mammals, if we step back through the 220 Ma since they first appeared on the planet (take another look at the timeline in Figure 31.3 if you need a reminder), we can see how they have distributed themselves over the large continental land masses derived from Gondwana and Laurasia (which are illustrated in Figures 31.4 and 31.5).

The major expansion in mammalian diversity began around 66 Ma, after the Chicxulub extinction event and the destruction of the non-flying dinosaurs (Figure 31.3). Two major lines of placental mammals established themselves: the **Atlantogenata** in the African/North American region and the **Boreoeutheria** in the European/Asian region. These two lines form the basis for the large groupings of placental mammals present today, such as those listed in Figure 31.6, although there has been a great deal of overlap between them since their ancestors separated.

The broad groupings within the two lines, sometimes called super-orders, summarize their likely ancestral history: the Atlantogenata, probably originating in America, formed separate groups in America (**Xenarthra**) and Africa (**Afrotheria**), while the Boreoeutheria, distributed over Europe and Asia, split into the primates, rodents, and lagomorphs (**Euarchontoglires**) and the carnivores, ungulates, and bats (**Laurasiatheria**).

The essential point about these divisions is that they are based, as we have seen, on the availability of contiguous land masses at the time when ancestral species existed. Prior to the modern understanding of continental drift which began in the late 1960s, biologists tried in vain to explain the entire distribution of terrestrial mammals around the world on the basis of migration across land bridges between *unchanging* continents. The early ancestral mammalian radiations which followed the decline of the dinosaurs can only be explained if land masses which are now separate were once joined.

Land bridges

Land bridges have played an important role in mammalian distribution in the relatively recent past. For example, the Bering Strait in the North Pacific between the far Eastern tip of Asia (now the Russian Federation) and the extreme western tip of North America (now Alaska) was accessible with lower sea levels, as part of a broad region some 1000 km wide called Beringia. Horses, camels, bears, and many other terrestrial creatures including man may have crossed this area during the most recent ice ages (135,000–30,000 years ago) when the sea levels were lower and much of this area was grassland.

This route probably explains why some large, hoofed, cold-adapted mammals such as reindeer (caribou; *Rangifer tarandus*) and elk (moose; *Alces alces*) are found right across the most northerly regions of North America, Northern Europe, and Siberia. Herding activity by man may have played a large part in directing their distribution, but they also migrate very large distances on an annual basis as food availability changes with the seasonal advance and retreat of the snow line. The extent of migration changed as the climate changed, leading to wider distribution.

The remains of woolly mammoths, which flourished until around 10,000 years ago and were predated by man, have been found well preserved in the permafrosts (non-fluctuating ice sheets) of Siberia and there is evidence of their existence in other parts of Asia and in Central America. Remains of hard tissues, including tusks, have been found in Alaska and in parts of eastern England, indicating the relatively recent connections between land masses which are now separated.

Sea routes

Aquatic life or that which floats is not dependent on land connections for distribution, although the distance between land masses may have a great effect on the speed and eventual success of colonization. Human activity which changes the direction of water courses or links otherwise separate bodies of water can inadvertently lead to species spread, which is discussed further in Real World View 31.2.

REAL WORLD VIEW 31.2 | **Invasive fish species in the Mediterranean Sea**

As you can see in Figure 1, when the Suez Canal opened in 1869, the Red Sea became connected to the Mediterranean Sea, making it possible for marine organisms to migrate from one to the other. Of the 800 non-indigenous species found in the Mediterranean, about half probably arrived from the warmer waters of the Red Sea, conducted through the canal on the prevailing northward current of water. They are called **Lessepsian** or **Erythrean migrants**.

Some of these migrants represent a health hazard: a poisonous species of jellyfish (*Rhopilema nomadica*), and the silverstripe blaasop (*Lagocephalus sceleratus*), a species of pufferfish which can be lethal if eaten.

Over 96 fish species have moved to the Mediterranean in this way. Figures 2 and 3 show two examples of such species: the dusky spinefoot (*Siganus rivulatus*), shown in Figure 2, and the marbled spinefoot (*Siganus luridus*), shown in Figure 3, are highly effective invaders. In fact, they account for up to 95 per cent of herbivore abundance in the Mediterranean. Their migration from the tropical, coral-dominated environment of the Red Sea to the cooler, rocky-reef-dominated

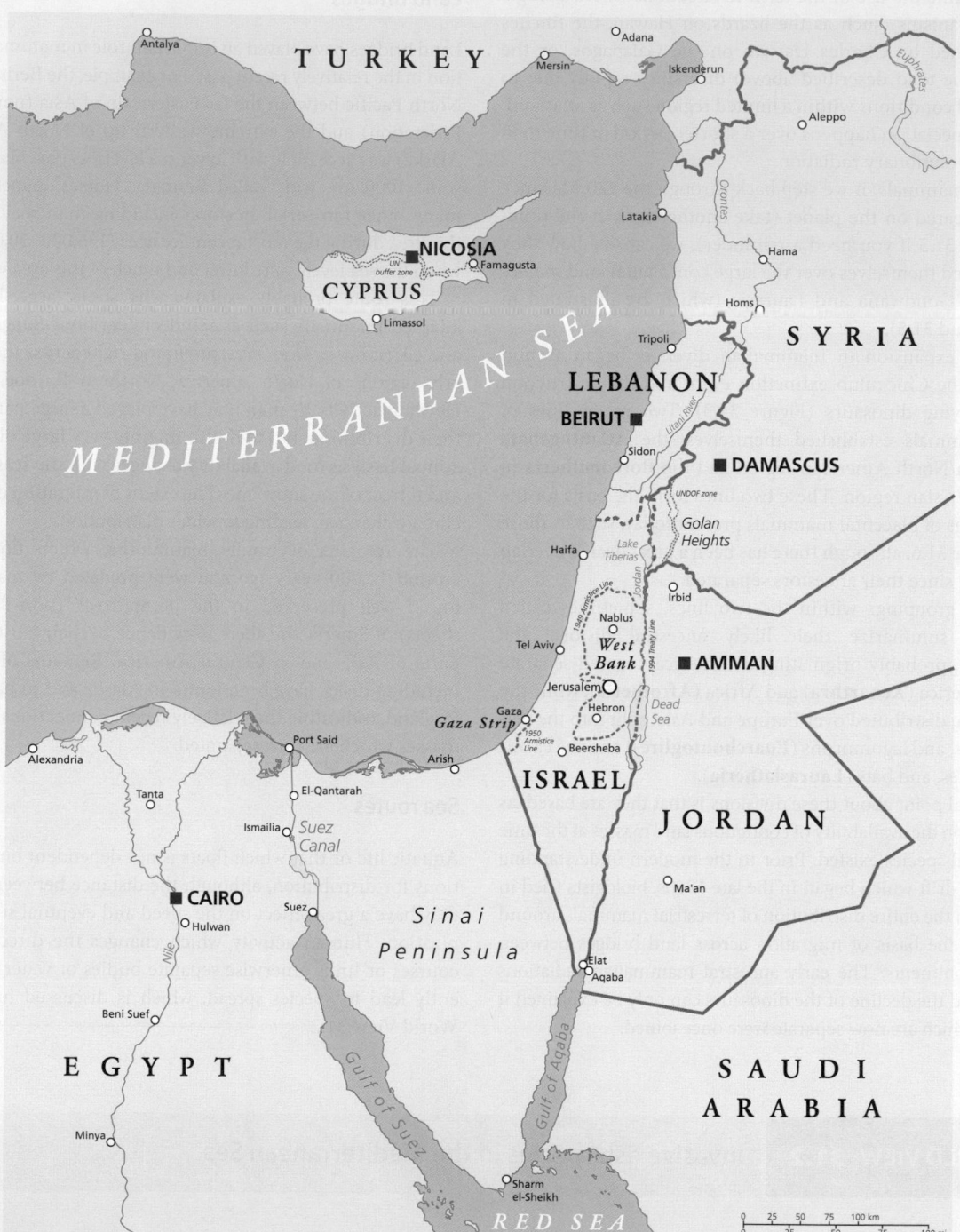

Figure 1 The Suez Canal connects the Red Sea with the Mediterranean Sea.
Source: Peter Hermes Furian/Shutterstock.com.

Mediterranean was facilitated by niche availability and diet adaptability, plus their effective defences against predation.

The Mediterranean has few other herbivores but an abundance of algae compared with the Red Sea. *Siganus* will forage on almost any type of algae, especially in the cold season. Although predation may be high during their larval stages, they are defended as adults by numerous sharp and venomous spines lining the dorsal region of the body.

The migration of *Siganus* has had ecological, evolutionary, and socio-economic impacts. Their voracious, non-selective consumption of algae has changed the pattern of light penetration through the water and led to

Figure 2 Dusky spinefoot (*Siganus luridus*).
Source: LABETAA Andre/Shutterstock.com.

Figure 3 Marbled spinefoot (*Siganus rivulatus*).
Source: Alexey Masliy/Shutterstock.com.

barren, debris-filled areas. This has reduced ecological complexity and depleted numbers of native fish and other animals.

Nevertheless, the fishing industry has benefited from *Siganus* species. Despite their toxicity, they have become a valuable commercial resource

across the Mediterranean from Lebanon to Greece. Careful regulation of fishing may be the key to managing *Siganus*, allowing natural algal beds and diverse animal species to recover.

This is an example of the unexpected ecological impact of human activity. However, it also shows that scientific understanding may suggest ways of tackling and ameliorating the problem.

Acknowledgement: Thanks to Steph Heyworth for this information.

Although the redistribution of animals which live in the oceans is, in principle, unconstrained by land masses, there are still many reasons why species or groups of species may be found in limited regions. These include the locations of ocean currents and the effects of tidal movements, temperature, light, salinity, food supplies, and predators.

Following the March 2011 Japanese earthquake and subsequent tsunami, ocean currents in the Pacific distributed nearly 300 animal and plant species from Asia to North America, a distance of some 7000 km. (The ocean currents that triggered the tsunami are illustrated in Figure 31.12.) They hitchhiked on floating items of man-made debris, including boats, plastic bottles, ropes, and even

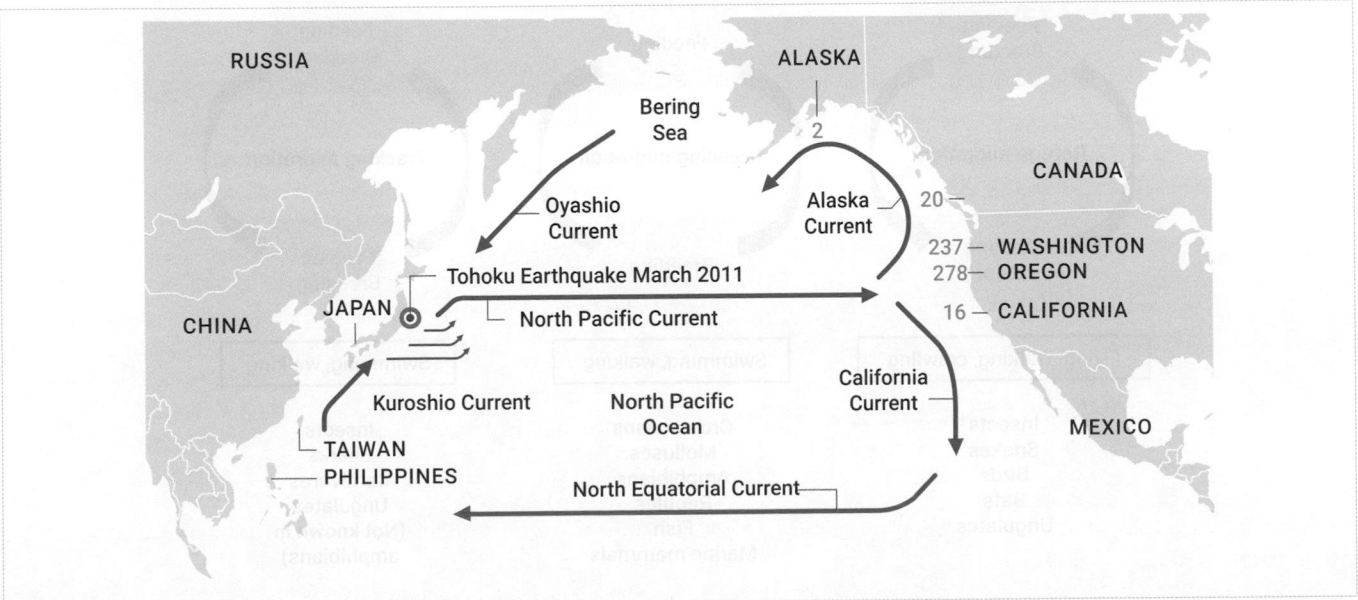

Figure 31.12 Hitchhikers on post-tsunami debris. After the 2011 Tōhoku earthquake and tsunami, ocean currents transported around 300 species to new locations around the Pacific on human-made debris. Numbers in green indicate the approximate number of species recorded on debris in Canada and states in the US.
Source: Steven L. Chown (2017). Human-made objects swept out to sea after the 2011 Tōhoku earthquake carried over 300 species to new locations. *Science*, 357 (6358), 1356.

a large section of floating dock. Ecologists are concerned that these species, introduced to new locations, may lack natural predators and become troublesome invaders.

Airborne life forms, including insects, birds, and bats, can also move freely and their distributions do not show the effects of continental history in the same way, yet distance and environmental factors may still limit their spread.

 Check your understanding of the concepts covered in this section by answering the questions in the e-book.

31.3 **Migration**

Some organisms have preferred locations for breeding and feeding, and many undertake movements from one location to another on a seasonal basis. This is called **migration**, defined as *a seasonally repeated, round trip movement among two or more locations*. Sometimes, movements over periods of 24 hours or repeated movements occupying a number of days are included in the definition. Whatever the timescale, migration is adaptive: it is a behavioural response to particular conditions by which a species increases its fitness for survival, growth, and reproductive success.

Migratory species are found in all the major groups of vertebrates and in many invertebrate groups too. They occur all around the globe and in all types of environment: aerial, terrestrial, and aquatic. Sometimes, the presence or absence of a migrant at particular times of year will have a major effect on the local environment and its ecosystem, especially if large numbers of individuals are involved.

To understand why animals migrate, it is helpful to establish three different categories, which are illustrated in Figure 31.13, and explained here:

- **Refuge** migrations are those in which the animal has a preferred home location but moves away at certain times of the year to find temporarily more favourable conditions, perhaps of food or climate or to avoid unfavourable conditions or predators.

- Animals with **breeding** migrations have two home locations, one for foraging and one for breeding; they will often show a much lower degree of male and female interaction in the foraging location.

- Animals showing **tracking** migrations have no home location but move continuously to follow food resources, such as gradually changing vegetation or prey density, or to avoid migrating predators; they eventually return to earlier locations, so this type of migration still has the characteristic of repetition.

These categories are not mutually exclusive, for a species may migrate for a complex of reasons, but they help to identify the different pressures which drive relocation. Among the environmental signals which initiate seasonal migrations, sometimes called **proximal drivers**, photoperiod and temperature often play the greatest role. For refuge and breeding migrators, the level of nutrition and body condition can be the main cue. Among insects and several other groups of invertebrates, sudden changes in population number and the consequent changes in food availability seem to trigger mass movement.

The following are three examples of animals whose migratory activity has been particularly well studied and which seem to be mainly driven by one type of pressure.

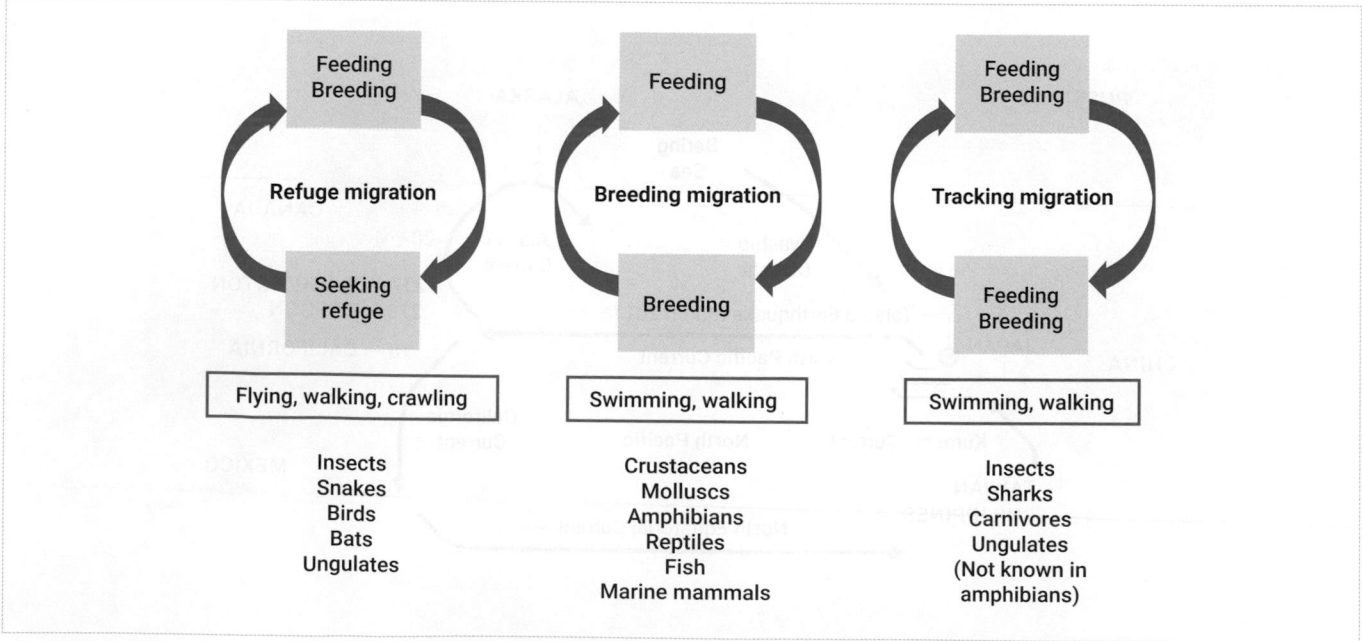

Figure 31.13 Types of migration. Three categories of animal migration, with examples of animal groups whose species frequently show these behaviours and their typical means of locomotion.

Source: Reprinted by permission from Springer Nature: Evolutionary Ecology 'Drivers of animal migration and implications for changing environments.' Shaw AK © 2016.

Refuge migration: zooplankton

Zooplankton is a collective term for the abundant small animal life suspended in the oceans. It includes not only microscopic organisms and larvae, but can also include larger animals such as small jellyfish, molluscs such as pteropods and crustaceans (copepods, isopods, and amphipods), and even juvenile fish. The crustaceans are sometimes referred to separately as krill, but that term is more generally used when the zooplankton are being viewed as a food source for larger animals. You can see some examples of zooplankton in Figure 31.14.

Zooplankton drift freely in ocean currents and in large bodies of water such as fjords and lagoons. Large accumulations move up and down the water column each day in what is thought to be the largest daily migration of biomass on Earth. In addition to water currents, this is caused by the need to avoid predators during the light of day, as well as to feed on phytoplankton (plant plankton) in the surface waters.

This daily movement, down into the deep water in the morning and then rising to the surface when the sun sets, is called **diel vertical migration (DVM)**. Individuals may move tens to hundreds of metres up or down the water column in the space of only a few hours. Figure 31.15 shows a recording of the diurnal vertical migration of krill in Lurefjorden, Norway, which fits this pattern. You can see that the densities of krill in different parts of the water column, detected by ultrasound and represented by the colour scale, change over the 24-hour period.

Breeding migration: fur seals

The Northern fur seal (*Callorhinus ursinus*) is a marine mammal which travels large distances, but whose location varies with the time of year and the annual reproductive cycle. Figure 31.16 shows the satellite-tracked locations of male and female Northern fur seals in the Northern Pacific Ocean during the winter (non-breeding) season. For most of the year, males and females live in separate parts of the North Pacific Ocean. As you can see by the blue lines on Figure 31.16, the males are found mostly in the Bering Sea between Alaska and Russia and the North Pacific, while the females live more to the south and east, in the warmer waters of the Gulf of Alaska and the California coast, as shown by the orange lines in Figure 31.16.

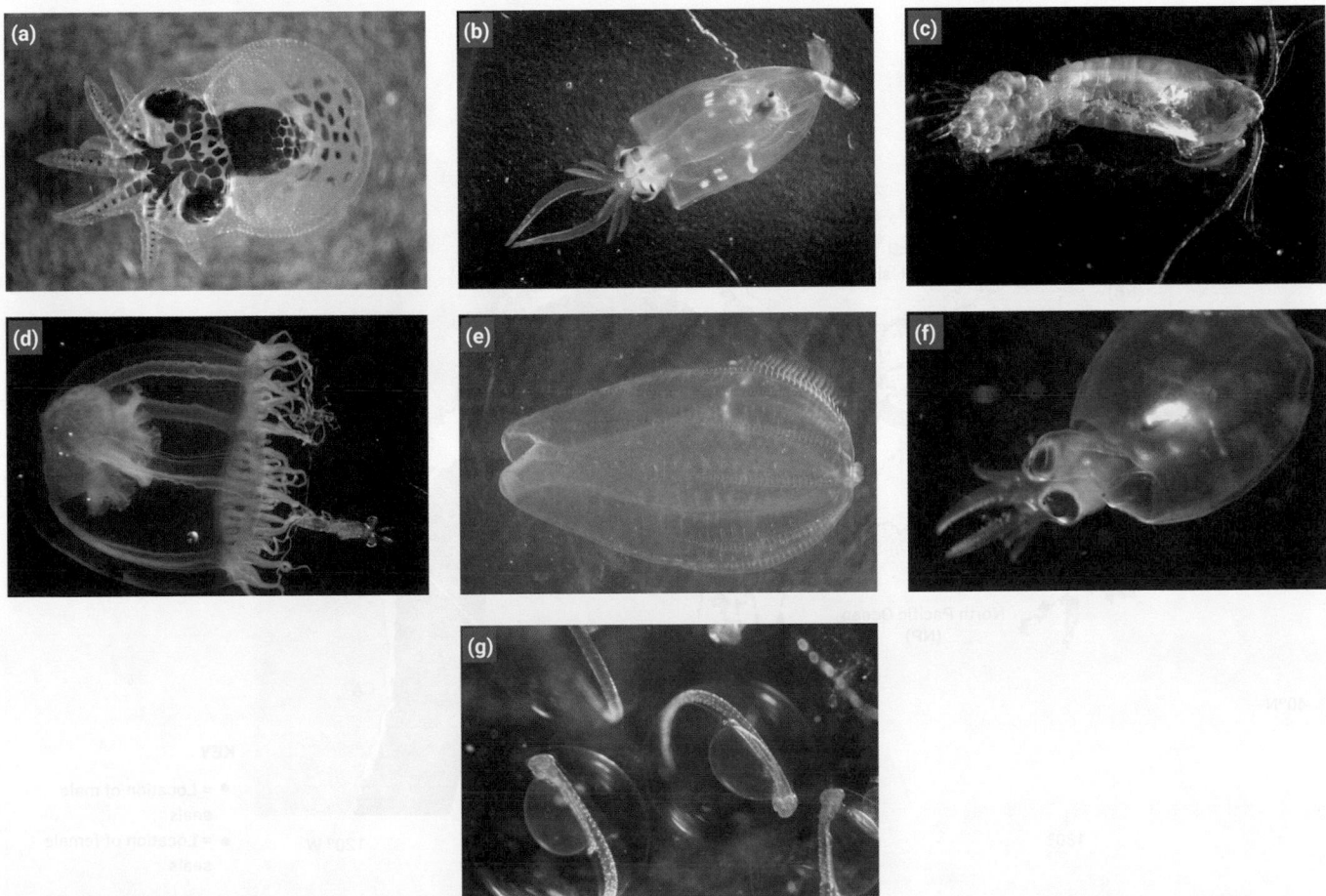

Figure 31.14 Examples of zooplankton. (a) Octopus or squid larva; (b) squid larva; (c) copepod with eggs; (d) teensy jellyfish; (e) comb jellyfish; (f) squid larva; (g) fish eggs with developing embryonic fish seen under the microscope.

Source: Matt Wilson/Jay Clark, NOAA NMFS AFSC.

Figure 31.15 Diurnal vertical migration of krill in Lurefjorden, Norway. The scale on the right is a false colour representation of biomass density as recorded by echosounding (decibels, dB). The graph shows that the dense biomass, which is mostly due to krill, moves to greater depths during the period of daylight and returns to shallower waters at dusk. The numbered regions (1–4) show the depths at which helmet jellyfish (*Periphylla periphylla*) gather to feed.

Source: Beyond the average: diverse individual migration patterns in a population of mesopelagic jellyfish. Thor A. Klevjer, Anders Røstad, Josefin Titelman, et al. *Limnology and Oceanography*, 2011; Vol 56:6. © 2011 by the Association for the Sciences of Limnology and Oceanography Inc., John Wiley and Sons.

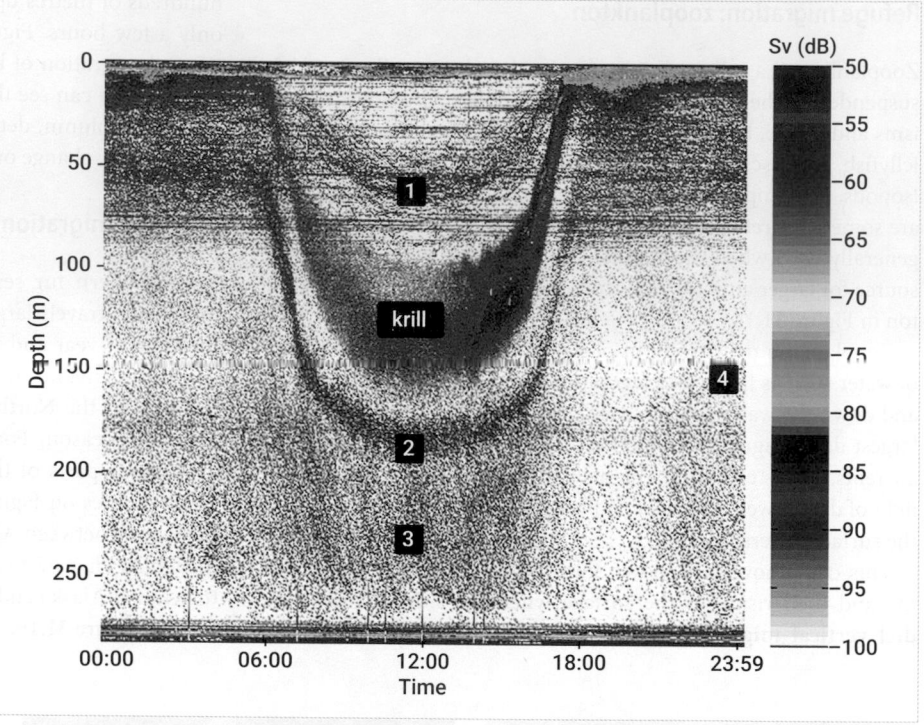

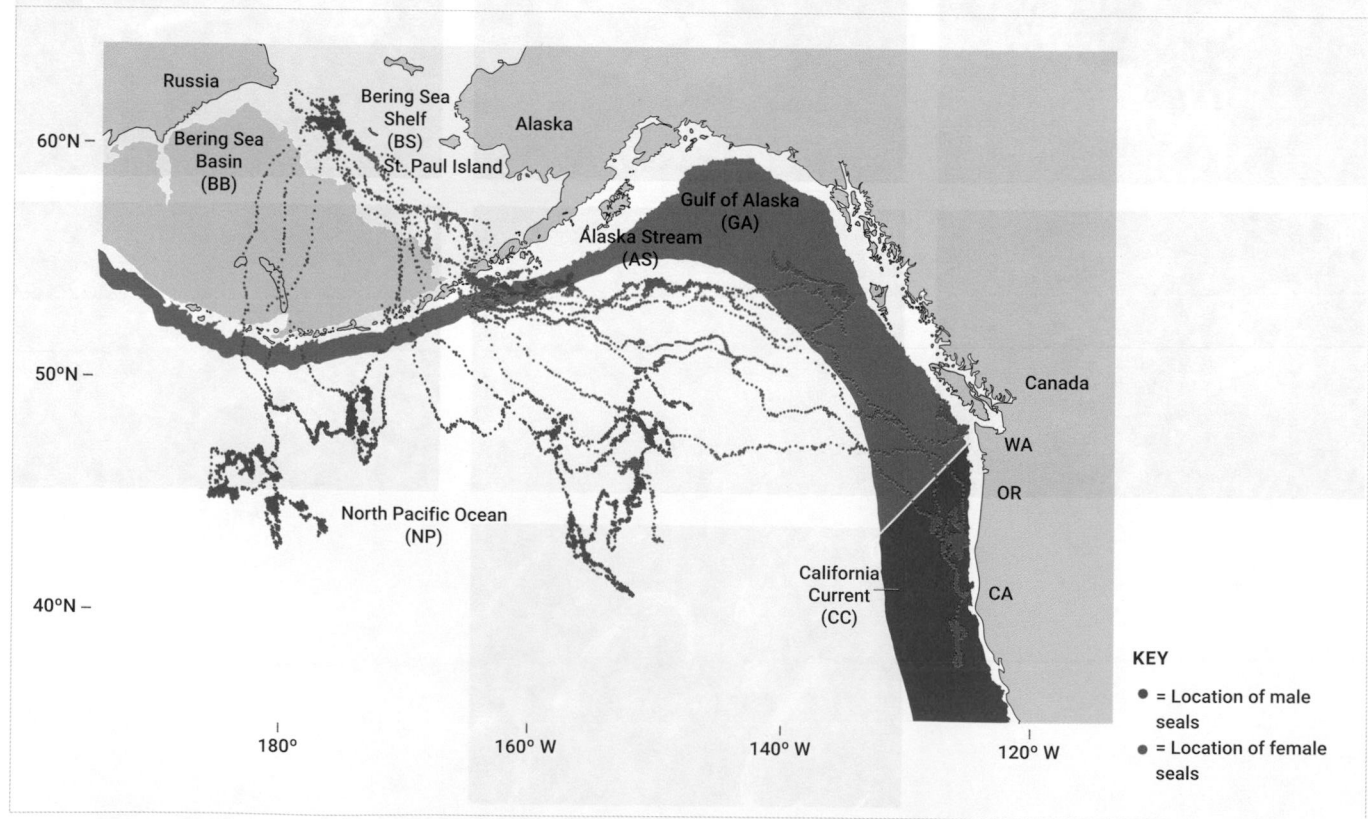

Figure 31.16 Where do male and female fur seals forage for food? Satellite-tracked locations of male (blue) and female (orange) northern fur seals in the northern Pacific Ocean during the winter (non-breeding) season. Each dot represents a position, estimated at 6-hour intervals.

Source: The Sun, Moon, Wind, and Biological Imperative–Shaping Contrasting Wintertime Migration and Foraging Strategies of Adult Male and Female Northern Fur Seals (*Callorhinus ursinus*). Sterling et al. *PLoS ONE* 9(4) April 10, 2014.

In the summer, from mid-May to August, males come ashore to occupy breeding sites called rookeries, on the coasts of Alaska, Russia, and California. From mid-June, females come ashore in the same locations and immediately give birth to single pups. The males, which have vigorously fought over and defended their territories, mate with the females within a week of the mass birth event and then depart to the ocean.

The young are supported for 4–5 months, with each mother making repeated trips of 1–2 weeks out sea to feed, returning to nurse its pup on land for 1–2 days. After weaning, the young and their mothers return independently to the ocean. If the post-partum mating was successful, the adult females will already be 4–5 months pregnant. This might suggest that the total gestation period is a year; however, the implantation of the embryo is delayed and its development does not begin until around the time of return to the sea, so the embryo only takes about eight months to develop. We discuss this more in Chapter 33.

Male seals are large and well insulated, and can survive in the cold waters of the far northern Pacific. Females are smaller and survive best in more southerly seas fed by the warm California current. Satellite-linked tracking collars show that the sexes avoid competition by foraging in separate areas of the ocean. Both sexes dive to where their main fish prey reside but this is deeper for the males (average 80 m) than the females (30 m) because of the different sea conditions. In general, larger animals dive deeper and for longer than smaller animals.

Tracking migration: reindeer

Reindeer (*Rangifer tarandus*), shown in Figure 31.17, migrate as a herd naturally in the spring and autumn, making use of different grazing ranges at the different times of year. The reindeer (caribou) of Arctic Alaska may move 5000 km or more over the course of a year, making them the furthest travelled of any terrestrial mammal.

In winter, reindeer will occupy habitats with abundant lichen and limited snow cover. The quality of nutrition here is low but adequate. Their summer pastures have a rich variety of high-quality vegetation which can support reproduction. Particular locations may be chosen for calving, with the animals returning to these locations every year. This means that the reindeer migratory cycle is closely aligned to the seasons and is associated with precise and predictable patterns of breeding behaviour.

The reindeer populations of Scandinavia provide an interesting example of human interaction with a naturally migratory animal. Animals are exploited for milk, meat, and hides, with evidence of herding and management going back at least to the late Iron Age. They are central to the culture and economy of the Sámi people, and may have facilitated the transition from a hunter-gatherer to a pastoral way of life in the early Middle Ages.

The Sámi traditionally allow their reindeer to migrate freely between their winter and summer sites but follow them closely to enable milking and culling. Gentle herding techniques facilitate ownership and controlled breeding. The population can be funnelled over the easiest terrain and towards the most productive grazing.

Multifactorial migration

For some animals, migration may have causes and consequences which do not lend themselves to simple categorization. Birds and fish, with their enormous varieties of size, speed, habitat, diet, and lifestyle, show migratory behaviours which can often only be understood at the species level. In addition, the pronounced effects on birds and fish of the physical characteristics of the fluids in which they live, also influences their collective behaviour.

Among fish, there are species which migrate both *between* and *within* fresh- and seawater habitats. They do so for a variety of reproductive and non-reproductive reasons. Examples that represent all three categories of animal migration are listed in Table 31.1.

To accomplish this, diadromous fish need to adjust to dramatic changes in osmotic pressure, using physiological mechanisms (which we discuss in Chapter 30). Migratory fish of all types, but

Figure 31.17 Caribou. (a) Male caribou (reindeer; *Rangifer tarandus*). (b) Caribou herd in Alaska.
Source: (a) Dean Biggins (U.S. Fish and Wildlife Service); (b) United States Fish and Wildlife Service.

Table 31.1 Types of fish migration and example species

Broad type	Migration range	Nature of migration	Examples
Diadromous	Fresh water ↔ Sea water	Anadromy (spawn in fresh water; adults live and forage in sea water)	Salmonids (salmon, trout, char), sturgeon, lamprey, and some other temperate species
		Catadromy (spawn in sea water; adults live and forage in fresh water)	Eels, mullets, and some other tropical/subtropical species
		Amphidromy (movements between fresh and sea water, not directly linked with breeding cycles)	Galaxiids (includes species which are principally found in fresh and sea water; abundant in the southern hemisphere)
Potomadromous	Fresh water ↔ Fresh water	Distinct wintering, feeding, and spawning regions, or movement for other reasons	Roach (*Rutilus rutilus*): avoids predators by occupying lakes in summer and rivers in winter
			Grayling (*Thymallus arcticus*): spawns in streams in summer and moves to deeper, non-freezing waters in winter
Oceanomadromous	Sea water ↔ Sea water	Distinct wintering, feeding, and spawning regions, or movement for other reasons	Mackerel (*Scomber scombrus*) and tuna (*Thunnus thynnus*); spawn in low latitudes and feed in high latitudes for complex reasons of sea temperature, salinity, and currents
			Capelin (*Mallotus villosus*); spawn in shallow coastal waters and feed in deeper off-shore waters
			Atlantic cod (*Gadus morhua*); vary between coastal and offshore locations according to sea temperature and wind strength
			Whale sharks (*Rhincodon typus*) and basking sharks (*Cetorhinus maximus*) follow prey densities

especially those which feed and spawn in different environments, also typically undergo large changes in body composition (weight and fat content) over the annual cycle.

Migration is a common feature of animal lifestyles across all groups and in all types of habitat. There may be several different stimuli for movement, and animals may gain more than one type of adaptive advantage. In interpreting what is going on, it can be particularly difficult to understand how animals consistently find their way, often over very large distances, and how the mass movements of many individuals are coordinated.

PAUSE AND THINK

 Consider some further examples online of species where there seems to be more than one proximal driver for migration or where the reasons for migration are not fully understood:

- jellyfish
- green turtles
- monarch butterflies.

Study these and, referring to Figure 31.13, decide what you think the most important driver or drivers might be.

In thinking about the descriptions in these panels, you may feel that the categorization of migratory behaviours in Figure 31.13 is simplistic or open to debate. It is wise to remember that they are all human interpretations of natural phenomena. Our understanding may change as more information becomes available from further research.

 Check your understanding of the concepts covered in this section by answering the questions in the e-book.

31.4 Plant dispersal

Plants are sessile organisms and do not move independently like animals. They can neither migrate nor disperse in a self-directed or purposeful way using their own energy and senses. Their movement depends either on growth, on the surrounding physical environment, or on the actions of animals.

The plant body may simply grow through the environment, invading new soil, and over a period of time establish new territory. This is a common dispersal strategy for many invasive weeds and other plants which have vigorous underground root systems. This strategy normally restricts plants to moving locally, say across a field over several years, so its effectiveness is limited.

An example of this is bracken. Bracken is a fern, a non-flowering vascular plant. It demonstrates how plants can move through the environment and cause serious problems with land usage, as discussed in Real World View 31.3.

Plants most often disperse themselves as part of the reproductive process. Movement by asexual propagation involves the shedding of parts of the plant body, to be moved through the environment by some external force called a **vector**. The liberated plant parts establish themselves elsewhere, develop into new plants, and eventually mature.

REAL WORLD VIEW 31.3   Bracken—a problem plant on the move

Bracken (*Pteridium aquilinum*) is a fern. Its growth habit and success demonstrate how plants can move through the environment and cause serious problems with land usage.

Bracken grows from thick underground stem systems called rhizomes which extend horizontally in the top 50 cm of soil. Each rhizome has an abundance of new potential growth points called buds which can produce new shoots and fronds.

Many buds are dormant but the bracken will rapidly send new shoots above the ground if existing shoots are cut or damaged. As it grows, the edge of the bracken stand can move forward and encroach new land at a rate of over 2 m per year. Once it becomes established, normally in acidic upland soils, it creates a huge biomass below ground with over 8 kg of rhizome per square metre. It is very difficult to eradicate and, left unchecked, will invade land used for stock grazing.

Few animals eat bracken but it is toxic and livestock that eat too many young fronds can become ill. They need to be managed away from potentially dangerous sources, for example by using walls to define safe grazing areas, as you can see in Figure 1.

Bracken control is a major problem in upland areas. Regular cutting and spraying, as demonstrated in Figure 2, are only moderately effective in keeping it under control.

Figure 1 Bracken and land usage. The use of walls to separate grazing livestock from bracken.
Source: Louise A Heusinkveld/Alamy Stock Photo.

Figure 2 The control of bracken by cutting.
Source: Photimageon/Alamy Stock Photo.

An example is Canadian pondweed (*Elodea canadensis*), a water plant native to North and South America. A small piece of it was introduced into Ireland in 1836 as a fragment on an imported log from Canada; it rapidly spread across Europe, first occurring in England in 1841, and soon began to clog up waterways. It is now common in freshwater throughout Europe. Its spread is facilitated by the passage of boats, especially canal boats, through river and canal systems. Vessels carry broken pieces of pondweed with them and distribute it widely.

The main reproductive method for plants to disperse is by the distribution of seeds (see Chapter 33). The sole function of a seed is to carry genetic information to a new location, germinate, and form the basis of a new, seed-producing plant. Ideally, the new location will be some distance from the parent plant and most seeds show some morphological adaptations which facilitate dispersal by abiotic or biotic vectors. Seeds take a myriad of different forms and morphologies according to the different species of plant.

Dispersal by wind

Many plants produce seeds which are very light in weight and have structures which enable them to be picked up by moving air and blown around in the aerial environment. Such seeds are so efficient in their dispersal that they can reach high altitudes and spread long distances, across oceans and between continents. Plants well known for efficient wind-dispersed seeds include cotton (*Gossypium* spp.), poplar trees (*Populus* spp.), and dandelion (*Taraxacum officinale*) like the ones in Figure 31.18.

In these species, the seeds dry out during the process of dehiscence (explosion from the pod) and become very light in weight, in the region of 0.5 mg each. The plume of fibres attached to them acts as a parachute and extends the time of suspension in air currents. The fibres associated with the cotton seed are the basis for the spun cotton thread used for human clothing.

Figure 31.18 Wind dispersal. (a) Mature flower head of dandelion (*Taraxacum officinale*) consisting of numerous seeds; each has a white feathery parachute which enables it to be picked up by the wind from the flower head and dispersed widely. (b) The success of this seed dispersal mechanism shown by the abundance of dandelions in this field in Derbyshire, UK.

Such seeds are blown about randomly in the environment, so the chances of one landing in a suitable place to germinate are low. The **fecundity** of such plants (seeds produced per plant) is enormous, increasing the chance that some will eventually germinate in favourable conditions and successfully disperse the plant. This is an extreme reproductive strategy (r-strategy), which we will discuss in more detail in Chapter 33.

Other species use the wind in a subtler way. Some plants, especially large trees, have winged attachments on their seeds. When they are shed at a great height, the wing prolongs the descent of the seed to the earth allowing the seed to glide or helicopter some distance from the parental tree. Such winged seeds are called **samaras**, and you can see examples of these in Figure 31.19.

Figure 31.19 Winged seeds. (a) Sycamore (*Acer pseudoplatanus*) and (b) lime (*Tilia* spp.).

Other plants use the wind to disperse their seeds by a physical blowing action akin to that of a pepper pot. The poppy (*Papaver* spp.) produces loose seeds within a head at the top of a tall stalk; these are shaken out through holes when the dry seed head is agitated in the wind, as shown on the poppy seed heads in Figure 31.20.

An extreme example of wind-aided distribution is where the wind moves the entire plant body, which itself contains viable seeds. This is used by species generically called tumbleweed which inhabit dry, arid environments. They are blown across the land and seeds are shed from the dead plant body as they pass. Some become tenacious weeds.

Dispersal by water

Some seeds can be dispersed by being shed into moving water and transported downstream in rivers or across seas and oceans. Plants using this method of dispersal are coconut (*Cocos nucifera*), like the ones in Figure 31.21, and mangrove (*Rhizophora*). Both have large hollow seeds containing air which are extremely buoyant.

Such seeds are structurally resilient and may float for a long time before being washed up on a distant beach and germinating. Seeds of the sea bean (*Entada gigas*) can be carried by ocean currents from the Southern hemisphere continents to Europe by this method.

Plants which live partially or totally submerged in water often distribute themselves by breaking off fragments of the plant body, as with the Canadian pondweed mentioned earlier. *Hydrilla verticillata*, a highly successful waterweed in the USA, originally spread from South East Asia with human help via the aquarium plant trade. It grows vigorously and is now widely distributed, blocking waterways and clogging boat propellers.

Distribution by mechanical propulsion

In many plant species, the fruit structure bearing the seeds dries out and seeds are released explosively in a process called dehiscence. The fruit has different tissue layers and shapes of cells, such that significant potential energy is stored as it dries. When the fruit ruptures the seeds can be flung considerable distances and at speeds of up to 12 m.s^{-1}.

Examples of this type of seed pod occur in peas (like the sweet pea plants (*Lathyrus latifolius*) in Figure 31.22), and beans and the closely related silique in mustard and cabbage plants and other members of the Brassicaceae. These structures probably evolved from seeds forming on the mid-vein of a leaf which then folded over to form a protective pod. When dry, the pod splits and may flick the seeds some distance. During the domestication of wild plants, such dispersal mechanisms were disadvantageous because the projecting seeds represented a loss of crop on the ground prior to harvest. They have therefore been bred out or diminished in many commercial varieties.

Another example of seed dispersal under pressure is the squirting cucumber (*Ecballium elaterium*). When the fruit is ripe it ejects a mucus containing the seeds from its detached end.

 Go to the e-book to watch a video showing how seed dispersal occurs in the squirting cucumber (*Ecballium elaterium*).

Dispersal by animals

Dispersal by animals plays a quantitatively important role in the distribution of plants through the environment. Many plant seeds have evolved to exploit this mechanism, either by developing a means of becoming passively attached to the external coat or feathers of animals, or by forming a fleshy coat (fruit) which is attractive to animals as food.

Figure 31.20 Poppy seed heads. Loose seeds are positioned at the head at the top of the tall stalk of the poppy plant.

Figure 31.21 Seeds of the common coconut (*Cocos nucifera*). After falling into the sea or river from the palm they can float on sea currents until washed ashore where they can germinate.

Source: Kongfha C/Shutterstock.com.

Figure 31.22 Seed pods. Seed pods of sweet pea (a; *Lathyrus latifolius*) are constructed of two halves which twist as they separate (b) throwing the seeds some distance from the parent plant.

Some fruits become passively attached to distributing animals by having a 'sticky' surface, normally consisting of small hooked hairs which catch on animals or birds as they move through vegetation. The attachment is highly tenacious, as any countryside walker who has found seeds on socks and boots after an autumn walk will confirm. This is illustrated in Figure 31.23.

The attachment of seeds to the feathers of migratory birds is a remarkably effective strategy for long distance dispersal. The seeds of sessile plants may be distributed between hemispheres and around the planet by this means and it helps to explain the colonization by flora of remote oceanic islands.

The island of Surtsey off the south-west coast of Iceland, which was formed by volcanic eruption between 1963 and 1967, has an abundance of plant species which probably arrived this way. A single migratory bird was found to carry seeds of 30 different plant species among its feathers.

Seeds packed inside fleshy fruits use the attractiveness of the structure as a food source to achieve dispersal by animals. As the newly produced seeds mature, the lure of the fruit may be increased by changes in colour and increasing sweetness. The ingested seeds may pass through the animal's gut untouched by digestive processes and be excreted in the faeces some distance away.

Figure 31.23 Examples of burs. (a) Burdock (*Arctium lappa*). (b) Common weed known as cleavers or sticky willy (*Galium aparine*).

Figure 31.24 Examples of fleshy fruits. (a) Tomato (*Solanum lycopersicum*) fruit turning from green to red and becoming softer and sweeter as it ripens. (b) Blackberry (*Rubus fruticosus*) (c) Blackbird eating berries of a cotoneaster shrub in winter.

The domestication and agricultural exploitation of such fruits as the tomato and blackberry in Figure 31.24 has made them a major part of the human diet, as well as that of animals such as birds. Indeed, humans and a few other animals (including some other primates, guinea pigs, and fruit bats) have evolved an absolute metabolic dependence on the ascorbic acid (vitamin C) which these foods supply. The seeds of tomatoes can survive the human digestive system and sewage treatment processes, and germinate where solid waste is deposited.

Other plants produce seeds with significant stores of high-energy molecules, necessary for their future germination, which are of high nutritional value for animals. Nuts, which are high in fat and protein, are a major food source for a wide variety of mammals and birds.

Some animals harvest these and store them hidden within the environment for later consumption. Fortunately for the plant, many fail to be recovered and subsequently germinate in places distant from the mother plant. Grey squirrels routinely hide a wide variety of nuts and tree seeds, especially acorns of the oak tree (*Quercus* spp.) and seeds of horse chestnut (*Aesculus* spp.), and forget where they put them. This represents a major dispersal mechanism for these plants.

 Check your understanding of the concepts covered in this section by answering the questions in the e-book.

31.5 Mechanisms of active locomotion

Terrestrial animals which crawl, walk, run, or swing through vegetation *actively* propel their body mass against the much larger mass of Earth. Animals which fly, swim in water, or push their way through soils or other fluid materials (mud, sand, leaf litter) do this by relying on the viscous resistance of the medium in which they move. They may also generate an upward thrust against the pull of gravity or against the prevailing movements of the fluid (winds and tides).

These are all mechanical processes in which the muscular forces generating the locomotion of the animal through its limbs are matched by equal and opposite forces absorbed by the planet. The muscular force comes from biochemical energy obtained from food, which is derived ultimately from the sun.

Isaac Newton's Laws of Motion apply to all kinds of active locomotion. When you walk or run, your movement continues for as long as your muscles continue to contract and relax. As you stride forward, Newton's Third Law requires that Earth makes a small movement in the other direction (just as you push the track away from you when running on a treadmill).

The basic equation for Force (*F*, in Newtons, N) shows that it depends on the mass (*m*, in kg) of the body and its acceleration (*a*, in metres per second per second).

$$F = ma$$

The equation uses acceleration rather than speed because, by Newton's First Law, force is required to make motion *change* rather than *continue*. If this is confusing, remember that motion on Earth is constantly resisted by friction, air resistance, and gravity; continuous locomotion can be viewed as comprising perpetual changes from stationary to moving.

The exact nature of the force, the musculature involved, and the way in which locomotion is achieved depends on the structure of the animal and its type of movement. Animals may use more than one type of locomotion and adapt their movements to maximize efficiency when circumstances change.

 Check your understanding of the concepts covered in this section by answering the questions in the e-book.

31.6 Movement without limbs: worms

In this section, we turn our attention to movement without limbs. There are three main methods of movement to discuss, including peristalsis, segmental progression, and thrashing. We will start with peristalsis.

Peristalsis

For some organisms, legs would only get in the way. Not having limbs allows annelid worms, such as earthworms, to move unhindered through small spaces and burrow in sediment. Annelids move using a method called **peristalsis**: their segments expand and contract in waves along the length of the body, transferring the body mass forward in progressive phases. Body segmentation allows portions of muscles to contract independently of those in neighbouring segments.

The common earthworm has a hydrostatic skeleton held in place by circular and longitudinal muscles. Look at Figure 31.25, which

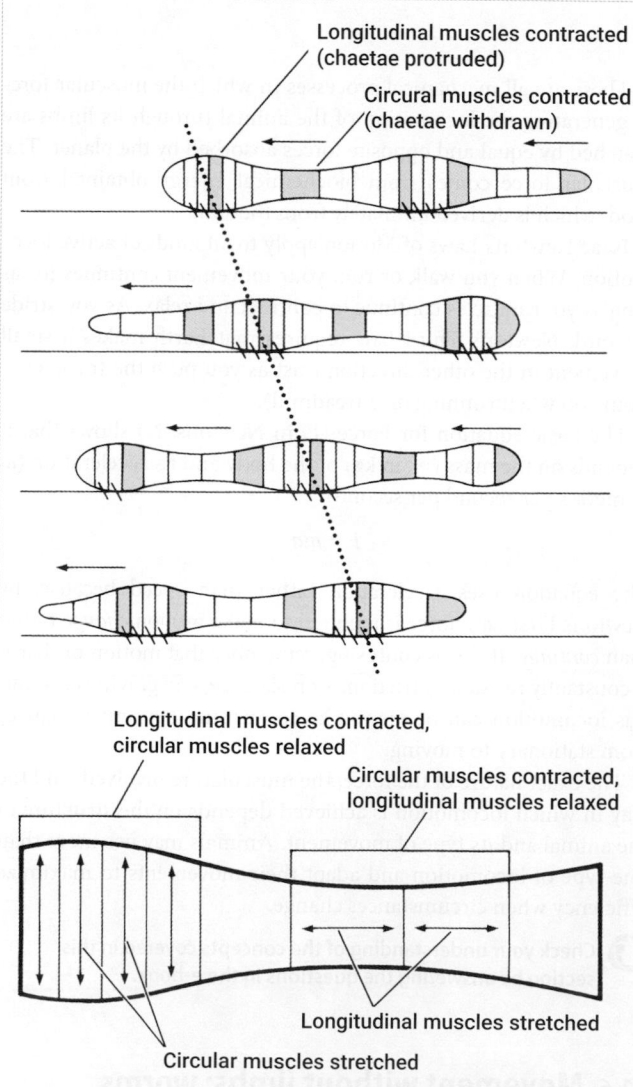

Figure 31.25 Peristaltic locomotion of the earthworm. Locomotion is achieved by sequential contraction and relaxation of circular and longitudinal muscles, passing in waves along the body.

Source: Brusca, Moore and Shuster (2016) *Invertebrates*, 3rd edition. Sunderland, Massachusetts U.S.A. Sinauer Associates/Oxford University Press.

illustrates the way in which the earthworm moves. Notice how the worm begins its forward movement by contracting the circular muscles at the anterior end; this lifts up the head and stretches it forward. The wave of circular contraction passes along the body towards the posterior end, making the worm longer and thinner. The longitudinal muscles then contract in a wave making the worm shorter and fatter.

During the longitudinal muscle contraction, the worm extends small hairs called **setae** from its cuticle and pushes them into the soil. This increases the body's frictional resistance and prevents it from sliding backwards. The setae are withdrawn as the circular muscles contract.

Segmental progression

Polychaetes (bristle worms) are also annelids but differ from earthworms in that their bodies often have pairs of fleshy appendages called parapodia attached to each segment. These parapodia possess bristles called **chaetae**. Polychaete worms can be divided into three groups according to their degree of locomotion: sessile, motile, and discretely motile:

- Sessile polychaetes, including fanworms, remain anchored, partially buried in mud or sand along shorelines. They do not burrow horizontally but may form an encasing, vertical tube for protection. They forage by sending out retractable, feather-like tentacles to collect plankton or may use paddle-like parapodia to waft water through the burrow.

- Motile polychaetes, such as those in the genus *Nereis*, burrow, crawl, and swim. They have complete septae between segments, enabling each segment to be hydraulically isolated from the next. The circular muscles maintain hydrostatic pressure in the segments, as with earthworms, but the longitudinal muscles on each side of the segment contract and relax alternately. Look at Figure 31.26, which illustrates how polychaetes move. Notice how as each lateral force is cancelled out a horizontal wave of contraction, an undulation, passes from anterior to posterior and the worm moves forward by pressing back on the substrate. The parapodia move forward on the contracted side and backward on the other side. The parapodia on the crest of the wave are extended fully and their chaetae are pushed into the substrate. As the wave crest passes, the parapodia are lifted off and the chaetae retracted. Variations in the length and amplitude of the waves of contraction produce different rates of crawling or swimming.

- Discretely motile polychaetes, such as the lugworm *Arenicola*, construct burrows or dwell in tubes. They have well developed circular and longitudinal muscles, and their movement is also by peristalsis. Those that live in tubes usually have limited parapodia and complete septa. The parapodia have uncini, specialized hooks used to anchor the worm onto the sides of tubes. By contracting their longitudinal muscles these worms can withdraw rapidly into their tubes when disturbed or threatened.

 Go to the e-book to watch a video showing how polychaetes move.

Figure 31.26 Polychaete movement.

Source: Brusca, Moore and Shuster (2016) *Invertebrates*, 3rd edition. Sinauer Associates, an imprint of Oxford University Press.

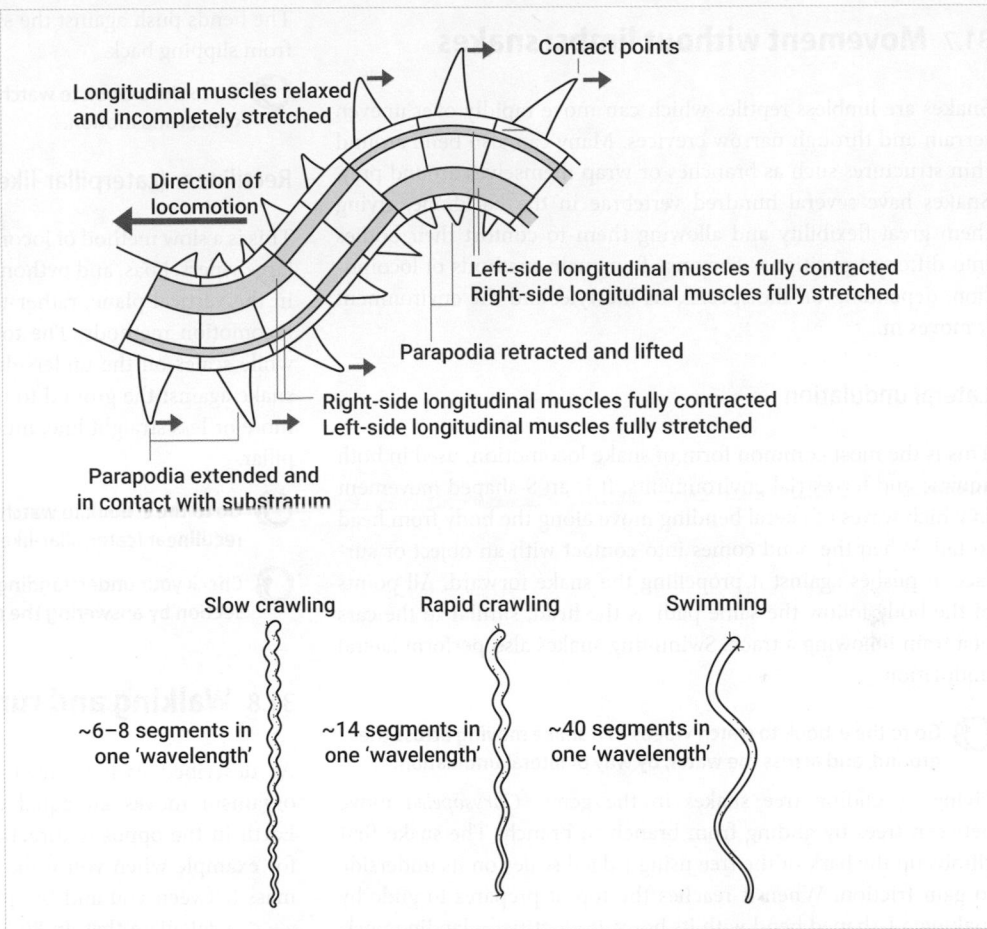

Thrashing

Unlike the annelids, nematode worms only possess longitudinal muscles. These muscles are unusual as they have both contractile and non-contractile regions, and the worm moves by contracting them in waves. Figure 31.27 illustrates how each half section of the animal contracts and relaxes alternately, producing lateral waves which travel from the anterior to the posterior end of the body. The resulting bending and thrashing propels the worm through sediment or water.

Some nematodes have spines, groves, ridges, or glands on their cuticle to allow them to grip onto the substratum. Nematodes with long, thin **flagelliform** (whip-like) tails anchor the tip of the tail to sediment particles with mucus secreted from a caudal gland. They move forward from the anchor by extending the tail or quickly pull the body into the sediment by coiling the tail.

- Go to the e-book to explore an interactive version of Figure 31.27.
- Go to the e-book to watch a video of a nematode swimming to help you understand how they move.
- Check your understanding of the concepts covered in this section by answering the questions in the e-book.

Figure 31.27 Nematode movement.

Source: Brusca, Moore and Shuster (2016) *Invertebrates*, 3rd edition. Sunderland, Massachusetts U.S.A. Sinauer Associates/Oxford University Press.

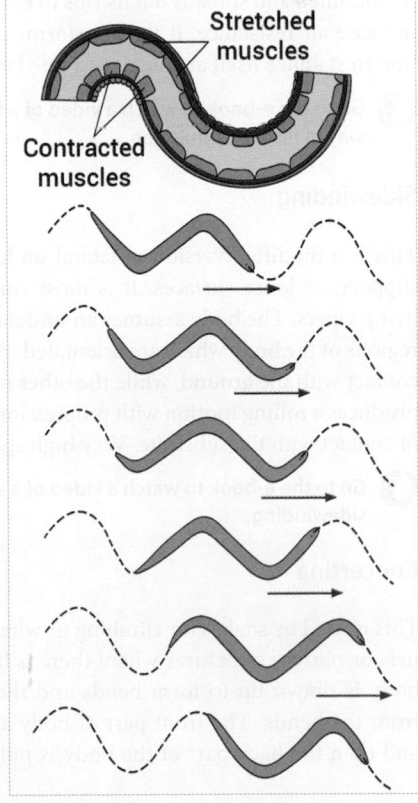

31.7 Movement without limbs: snakes

Snakes are limbless reptiles which can move rapidly over uneven terrain and through narrow crevices. Many can also bend around thin structures such as branches or wrap themselves around prey. Snakes have several hundred vertebrae in the backbone, giving them great flexibility and allowing them to contort their bodies into different positions. They use four main methods of locomotion, depending on the species, its lifestyle, and the environment it moves in.

Lateral undulation

This is the most common form of snake locomotion, used in both aquatic and terrestrial environments. It is an S-shaped movement in which waves of lateral bending move along the body from head to tail. When the bend comes into contact with an object or surface, it pushes against it propelling the snake forward. All points of the body follow the same path as the head, similar to the cars of a train following a track. Swimming snakes also perform lateral undulation.

 Go to the e-book to watch videos of a snake moving across the ground, and across the water, by way of lateral undulation.

Flying or gliding tree snakes in the genus *Chrysopelea* move between trees by gliding from branch to branch. The snake first climbs up the bark of the tree using ridged scales on its underside to gain friction. When it reaches the top, it prepares to glide by making a J-shaped bend with its body. It identifies a landing position and pushes its body away from the tree. As it falls, it sucks in its abdomen and spreads out its ribs to enlarge the surface area and increase air resistance. It then performs continual lateral undulation to stabilize itself and achieve a safe landing.

 Go to the e-book to watch a video of a flying snake moving by way of lateral undulation.

Sidewinding

This is a modified version of lateral undulation used on smooth, slippery, or loose surfaces. It is most commonly seen in desert-living vipers. The body assumes an undulating position and all the regions of the body which are orientated in one direction remain in contact with the ground, while the other regions are lifted up. This produces a rolling motion with progression based on several points of contact with the substrate. Very high speeds can be achieved.

 Go to the e-book to watch a video of a snake moving by sidewinding.

Concertina

This is used by snakes for climbing or when crawling through tunnels or narrow structures where there is little room to move. The body is drawn up to form bends and then straightened forward from the bends. The front part of body anchors onto the surface and then the back part of the body is pulled back up into bends.

The bends push against the sides of the orifice and stop the snake from slipping back.

 Go to the e-book to watch a video of a snake moving using a concertina motion.

Rectilinear (caterpillar-like)

This is a slow method of locomotion, used by larger species such as large vipers, boas, and pythons. The body is contracted into curves in the vertical plane, rather than the lateral curves used in other locomotion methods. The top of each part of the curve is lifted while scales on the underside grip onto the ground, pushing the snake against the ground to move forward. The snake moves in a more or less straight line, much in the manner of a looping caterpillar.

 Go to the e-book to watch a video of a snake moving by way of a rectilinear (caterpillar-like) movement.

 Check your understanding of the concepts covered in this section by answering the questions in the e-book.

31.8 Walking and running

As described earlier, Newton's Third Law dictates that as an organism moves an equal amount of force is transferred to Earth in the opposite direction. The movement made by Earth, for example when you walk, is tiny because of the difference in mass between you and the planet. Using the equation for force, we can calculate that an 80 kg person walking westwards round the Equator at 4.5 km.h^{-1} (1.25 m.s^{-1}) produces 100 N of force. That force applied to Earth weighing around 5×10^{24} kg, gives it an eastwards acceleration of about 1.7×10^{-23} m.s^{-2}, which of course is negligible.

The ability of the body to transfer force to Earth depends on frictional resistance at Earth's surface. If friction is low or absent, as, for example, on wet ice, soft mud, a steep sand dune, or loose scree on a mountainside, no progression is possible. Many animals create additional resistance by gripping with fingers, toes, claws, and tails, or, as we have seen with snakes, scales.

Animals with four feet rather than two gain extra traction by distributing their body weight over a larger surface area and by making greater adjustments to variations in the quality and angles of surfaces.

Locomotion: pendulums and efficiency

The mechanical principles of bipedal and quadrupedal locomotion have been studied in great depth. Particular interest in bipedal motion arises from the need to treat medical conditions such as spinal injuries and strokes, to design prostheses for amputees, to improve geriatric support, to understand paediatric development, and from the desire to improve athletic performance.

Animal conservationists, farmers, and those who breed and train animals for sport need to be able to model quadrupedal locomotion, and there is increasing interest in all types of animal

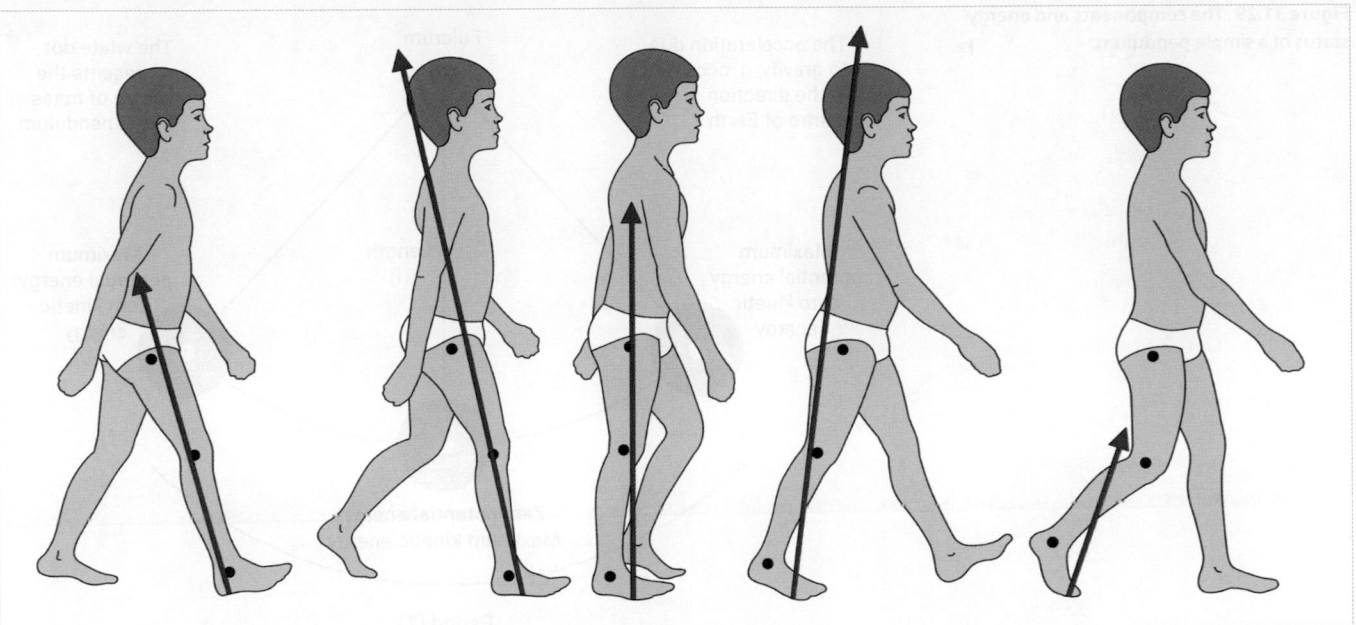

Figure 31.28 Human walking. Notice how the net force vector (the sum of forces through the main joints—the black dots) changes over a complete walking cycle. The net vector force is represented by the arrow on each image.

Source: Reprinted from Vaughan CL. (2003) Theories of bipedal walking: an odyssey. *Journal of Biomechanics*, 36(4), 513–23. Copyright 2003, with permission from Elsevier.

movement, both bipedal and quadrupedal, from technologists designing humanoid or locomotory robots.

In all these cases the challenge is to understand how the body mass is transmitted from stride to stride and how this is achieved with the greatest efficiency, balance, agility, speed, and duration. The efficiency of locomotion can be defined as getting the maximum speed or distance for the energy consumed.

It has been known for a long time that individual humans continually adjust their step rate and stride length to minimize the energy required for the distance travelled. This is achieved in such a way that the reaction with the ground is minimized. As we walk, the net force vector, which is the sum of forces passing through the joints and a combination of magnitude and direction, fluctuates over the point of contact with the ground, as illustrated by the arrows in Figure 31.28. The pressures which lead to evolutionary change have resulted in energetically efficient movements.

When an animal walks on land or swings between the branches of a tree its body mass is briefly but repeatedly lifted against gravity and then repositioned a little further away. Overall efficiency comes from minimizing the energy needed for the lift and the transfer, and by relying on momentum as much as possible. The mathematical description of this turns out to be immensely complex, but one way to begin analysing it is to think of the movements of limbs and bodies as those of a series of pendulums.

A pendulum, as illustrated in Figure 31.29, is an object which swings from a fixed point (**fulcrum**) in such a way that it continually transfers its energy from *potential* to *kinetic*. The energy of the swing movement is transmitted through the object's centre of mass—the point at which the effects of gravity are focused

(colloquially called 'the centre of gravity')—and the object's motion describes an arc (part of a circle) around the fulcrum.

 Go to the e-book to explore an interactive version of Figure 31.29.

Galileo and others in the 16th and 17th centuries showed that the period (*T*, sec) of swing of a pendulum (the time it takes to move from one extreme position to the other) depends not on the mass of the object, but on the distance (length, *l*, m) between the object and its fulcrum. This is shown in the following equation (where *g* is the acceleration due to gravity, m.s^{-2}):

$$T = 2\pi \sqrt{\frac{l}{g}}$$

PAUSE AND THINK

Demonstrate to yourself, by simple experiment,

a) that the period of a pendulum depends on its length and not on the mass of the bob;

b) that the kinetic energy of a pendulum bob is zero at the limit of its swing.

What happens if you try to jump off the swing at its point of maximum displacement?

Answer: One way to do this would be to find a pair of adjacent swings of equal length in a playground and observe how they swing when used by people of different weights. Do you move together or at different rates?

Figure 31.29 The components and energy status of a simple pendulum.

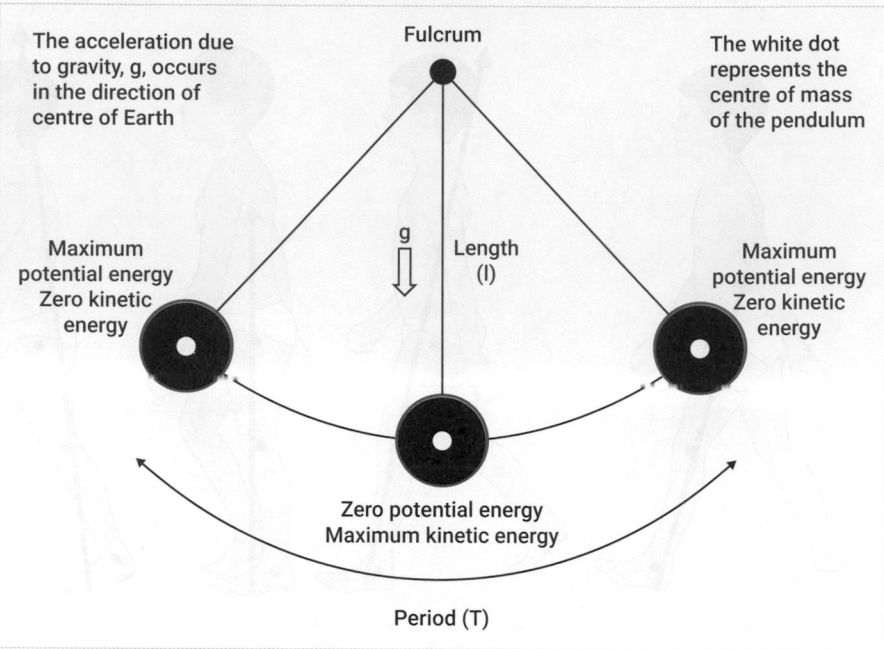

The acceleration due to gravity, g, occurs in the direction of centre of Earth

Fulcrum

The white dot represents the centre of mass of the pendulum

Maximum potential energy Zero kinetic energy

g   Length (l)

Maximum potential energy Zero kinetic energy

Zero potential energy Maximum kinetic energy

Period (T)

The importance of this in terrestrial animal locomotion is that movement depends on the lengths of limbs, the positions of joints, and the location of the body's centre of mass (its height above the ground or distance from a brachial swinging point). The mass of the body does not by itself affect the *mechanics* of movement, but it will influence the energy required to start and stop the locomotion (see the equation for force, above), the surface resistance of feet, hoofs, and hands, and the overall muscular construction of the body.

With bipedal walking, one foot stays on the ground during each stride while the other transfers the body weight to its new position. The stationary foot is a pivot over which the weight passes as the other foot moves from behind to in front. The stride itself consists of a swinging motion in which the moving leg is suspended from the pelvis.

The whole system resembles a pair of unequal pendulums. The first is an inverted pendulum (rather like the rocking arm of a metronome) in which the centre of mass is some where near the umbilicus (belly button) and the fulcrum is the ankle pivot. The second consists of the leg swinging from its fulcrum at the hip. The exact centre of mass of the leg varies with musculature, bone structure, and other anatomical factors.

The pendulums interact in each complete stride with mass being continually transferred between them. One result of this is that the centre of mass undergoes a roughly sinusoidal (wave-like) oscillation, with the highest point at the middle of the leg swing and the lowest point at the maximum extent of the stride when both feet are on the ground.

Part of the complexity of the mathematical analysis comes from the fact that joints allow limbs to bend at certain positions. They are therefore far from being simple pendulums. Further complications come from the lateral twisting of the pelvis (or shoulders during brachiation) and the forward rocking motion of the foot; these adjustments move the positions of the fulcrums themselves, both linearly and laterally. Each of the interacting pendulums is therefore made up of several interacting components.

Animals exploit these additional movements to reposition the centre of mass during each stride or swing. They flatten the arc of movement, reducing the vertical oscillation, and also adjust stride length and speed, and provide lateral balance as motion proceeds. This leads to optimal efficiency and step-by-step adaptation to surface irregularities. Humans exploit it when carrying heavy loads by keeping the additional mass as close to the central vertical axis of the body as possible, perhaps by placing it above the head and walking as erect as possible, such as the woman in Figure 31.30.

To illustrate these effects geometrically or explain them mathematically would take us far outside our scope. We can appreciate, however, the importance of momentum (kinetic energy operating in a particular direction). During a walking stride, the mass of the body starts to fall forward. The muscles and foot of the stationary leg control the first part of the movement, but the body's centre of

Figure 31.30 Women in Mysore, India, balancing baskets on their heads. The weight of the load is transferred vertically downward through the body, minimizing the carrying effort.
Source: JeremyRichards/Shutterstock.com.

mass then moves beyond the area which the foot can support. The repositioning of the other (moving) foot and leg arrest the body's fall, turning it into horizontal motion.

The body mass acquires momentum from this forward movement. Only a small fraction of momentum is transferred down the leg, through the foot, and into the ground. The majority of the momentum generates propulsion for the next stride, and there is

also some recoil energy stored in the tendons and muscles of the limb (acting like a spring).

Something similar happens when primates swing by their arms between branches, which is called brachiating, and illustrated in Figure 31.31. Notice how the energy of the swing moves between the vertical and horizontal directions, whether the animal is swinging between adjacent supports or from a single support.

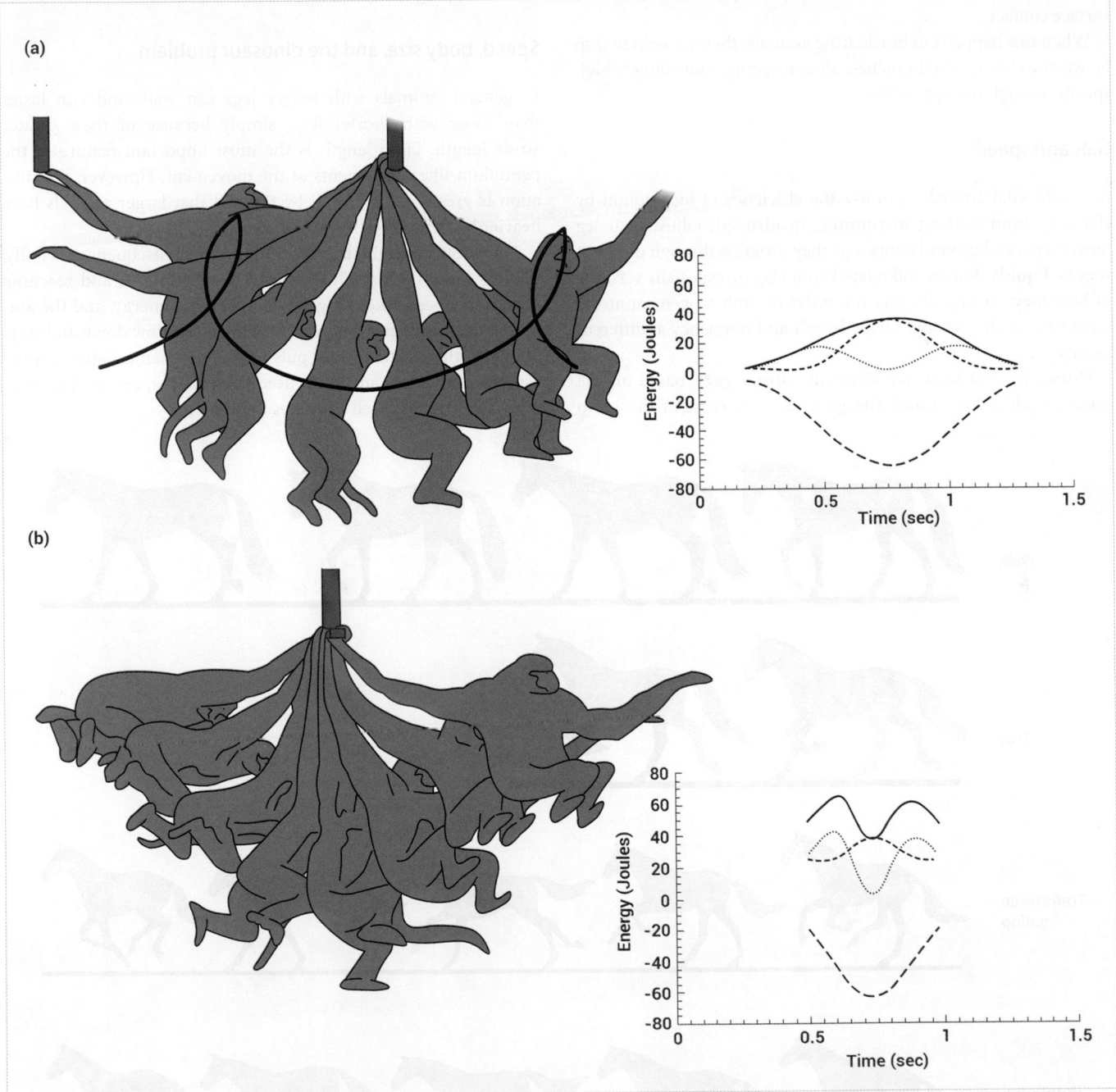

Figure 31.31 The mechanics of brachiation. Tracings from film of a lar gibbon (*Hylobates lar*) swinging between supports representing tree branches. Graphs show the energy in the animal's body over time. Thinner curves show vertical (solid) and horizontal (dashed) components of the animal's movement. (a) Gibbon transfers its weight from one support to the next, all within reaching distance. Its centre of mass is indicated by the looping line on the trace. (b) Gibbon uses a single support when a leap is required.

Source: Bertram, J.E. and Chang, Y.-H. (2001). Mechanical energy oscillations of two brachiation gaits: Measurement and simulation. *Am J Phys Anthropol.*, 115, 319-26. John Wiley and Sons.

Provided the pendulums in the system are balanced, this becomes a highly efficient means of locomotion.

Whether walking or brachiating, an animal can change its forward speed by increasing the propulsive force provided by the muscles of the stationary limb in each stride. At a certain speed, the momentum extends the distance between the steps (or swings) to the point where it becomes more efficient to insert a small forward jump. This is now a run (or leap) rather than a walk (or swing), and the forward movement of the body has periods without any surface contact.

When this happens in brachiating animals, they are seen to leap between widely spaced branches, often reaching astonishingly high speeds through the vegetation.

Gait and speed

Just as bipedal animals optimize the efficiency of locomotion by changing from walking to running, quadrupeds adjust their leg movements and ground contact as they progress through different speeds. **Equids** (horses and related animals) are especially versatile in how they do this, altering the order of limb movement at the same time as they change stride length and frequency at different speeds.

Horses have at least five different natural gaits, based on the order in which feet touch the ground, the extent of the jump (period with no ground contact), and the degree of asymmetry between the limbs on either side of the body. Look at Figure 31.32 and notice how horses walk using a four-beat rhythm, trot using a two-beat rhythm, as well as canter and gallop in two different ways with three-beat rhythms (or something very close to it). The reasons for changing between the natural gaits during free movement have to do with efficiency, but they may also minimize the load on the musculoskeletal system and reduce the risk of slipping.

Speed, body size, and the dinosaur problem

In general, animals with longer legs can walk and run faster than those with shorter legs, simply because of their greater stride length. Limb length is the most important feature of the pendulum-like components of the movement. However, the situation is greatly complicated by the fact that larger animals have heavier bodies.

The greater weight has two important consequences. Firstly, there is more inertia. This means that starting off and reaching maximum speed takes longer and uses more energy, and the animal finds it correspondingly more difficult to slow down and stop. Secondly, the greater weight puts greater mechanical stress on the legs and hips; bones and limbs need to be thicker, denser, and more muscular, and this itself increases the weight.

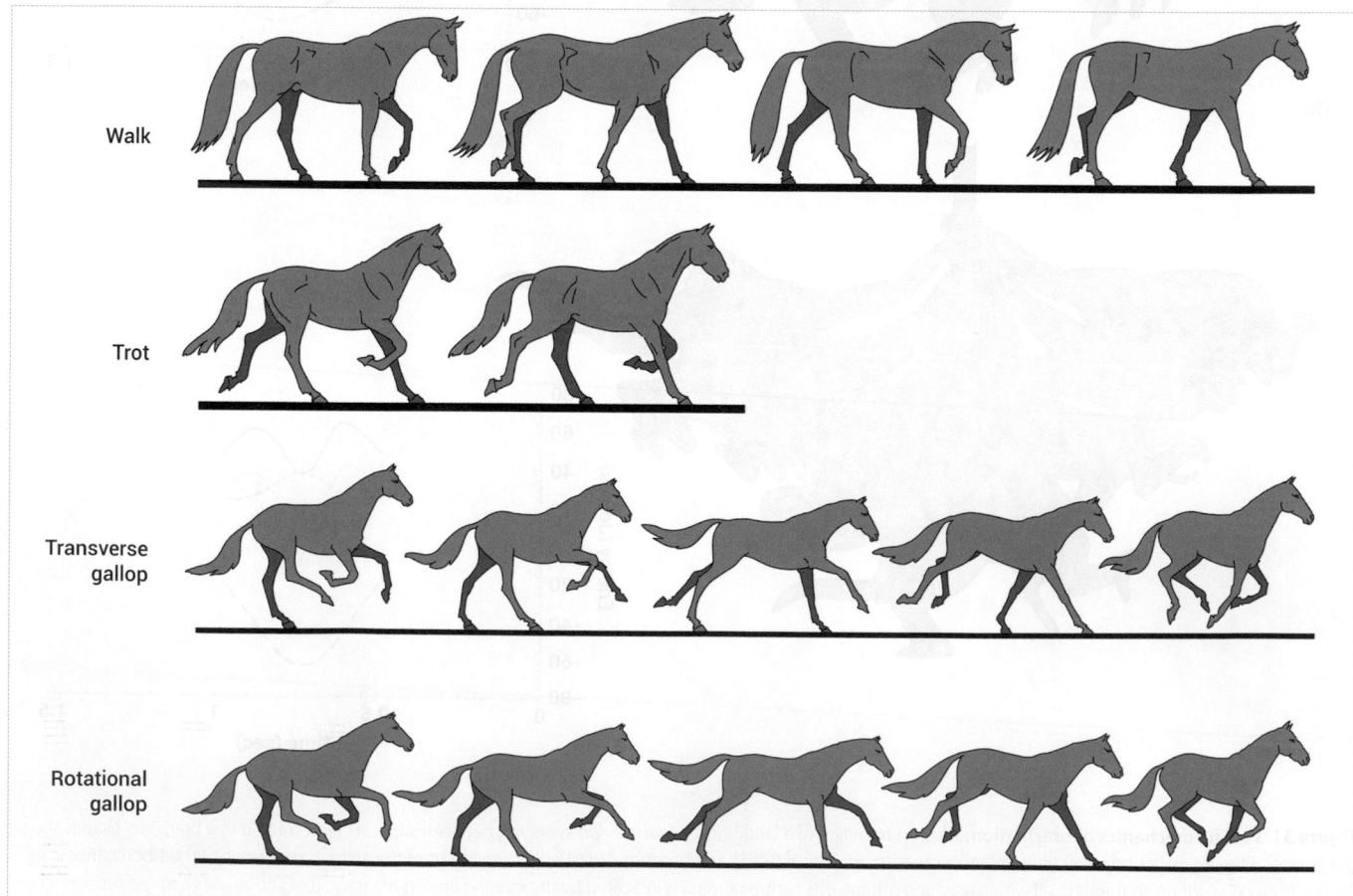

Figure 31.32 Four natural gaits of the horse.

Source: Figures 2 and 3 from Clayton HM (2016). Biomechanics of the exercising horse. *Journal of Animal Science*, 94, 4076–86. Oxford University Press.

Figure 31.33 The cheetah (*Acinonyx jubatus*).

Source: Malene Thyssen (http://commons.wikimedia.org/wiki/User:Malene)/CC BY-SA 3.0.

Terrestrial animals which are especially noted for their high sprint speeds, such as cheetahs which can reach 120 km.h^{-1}, have evolved to optimize the complex relationship between stride length, body size, and muscle energy production. As demonstrated in Figure 31.33, cheetahs extend their stride by jumping and have muscles which can generate large amounts of energy very quickly. They also flex the back to increase the stride length and have elastic mechanisms in the body structure which makes energy use more efficient. Their overall body size is limited by the need to accelerate rapidly, yet they require a minimum leg length to generate a long stride and to allow for a muscular arrangement which can produce extended jumps in their gait.

The cheetah's muscles rely on anaerobic metabolism when sprinting and become exhausted very quickly, so the animal is unable to maintain high speeds for long and requires frequent rests. Thermoregulatory limitations also come into play, especially the need to avoid overheating.

▶ **You can read more about anaerobic respiration in Chapter 37.**

Overall, animals which depend predominantly on speed for catching prey or escaping predators, including the fastest flyers (falcons) and the fastest swimmers (marlin), tend to be of medium size when compared with others in their broad taxonomic groupings.

We are familiar with similar differences in body conformation and performance in human athletes specialized for different disciplines.

- Sprinters may be taller than average, and have strong skeletons and bulky musculature with a high proportion of fast twitch fibres. (See Chapter 20 for more on muscle fibre types, metabolism, and function.) They achieve rapid acceleration through short bursts of intense power, but their muscles generate lactic acid which needs to be dispersed by subsequent rest and hyperventilation.

- Long-distance runners tend also to be tall with longer-than-average stride lengths but have a lighter musculature with abundant slow twitch, fatigue-resistant fibres. Their rate of acceleration is much lower than that of sprinters, but their muscles depend mostly on aerobic metabolism.

- Athletes specializing in strength disciplines (throwing, wrestling, lifting) are not usually concerned with speed over the ground, even if rapid reaction times are advantageous. They develop muscular bulk and strength, sometimes with additional non-muscular mass, and turn the greater weight to advantage by deploying the inertia it generates.

For the very largest terrestrial animals, the problem of skeletal strength becomes overwhelming in limiting their speed of movement. Complex mathematical modelling, taking account of leg and stride length, body weight, bone strength, and the non-linear relationships between size, strength, and contraction rate in muscles, shows that *Tyrannosaurus rex*, with a total body weight of 7 tonnes or more, probably had a top speed of up to 20 km.h^{-1}—a slow, plodding pace (Figure 31.34). Had it moved any faster, its bones would have been shattered by the inertial forces of its immense weight.

Figure 31.34 *Tyrannosaurus rex*. The therapsid dinosaur *Tyrannosaurus rex* (a) as a skeleton (in the American Museum of Natural History, New York) and (b) reconstructed as a model (Natural History Museum, London).

Source: (a) J.M. Luijt/Wikimedia Commons/CC BY-SA 2.5 NL. (b) Marcin Floryan/Wikimedia Commons/CC BY-SA 2.5.

Reptilian locomotion

Quadruped reptiles have evolved a gait which is different from that of most other vertebrates. The limbs of a typical lizard project outwards from the body, as shown in Figure 31.35, in contrast to the downward projection seen in mammals. The femurs of the hindlimbs make an angle with the pelvis, and the 'knee' junction between the femur and the lower leg bones is angled rather than straight.

This arrangement supports the weight of the body in the form of an arch, with the limbs acting as transverse struts. The animal's feet are spread wide apart and the distance between them can be several times the width of the body. The tendency for the feet to splay outwards is counteracted by their traction against the surface, represented as an inwardly directed frictional force.

The arched form works well for relatively light-bodied animals but would not support a heavy body. The angled alignment of the leg struts must meet *above* the centre of gravity, otherwise the arch will not be strong enough. This limits the body size to which these types of animal can evolve. Many large ancestral reptiles and reptilian dinosaurs had a more 'mammalian' arrangement with the feet positioned directly under the body.

One advantage of the lizard's limb arrangement is in climbing. The distance between the body's centre of gravity and the substratum (a wall, rock face, or tree trunk, for example) is small: the animal's weight is also spread out and it can cling on without leaning outwards.

Small lizards and geckos can easily negotiate vertical and overhanging surfaces. Take a look at Figure 31.36. Notice in Figure 31.36b

that the tokay gecko climbs with a paired-limb gait and twists its body to maximize the effective stride length. They also have adaptations to their feet which facilitate grip on smooth surfaces, which you can see in Figure 31.36c.

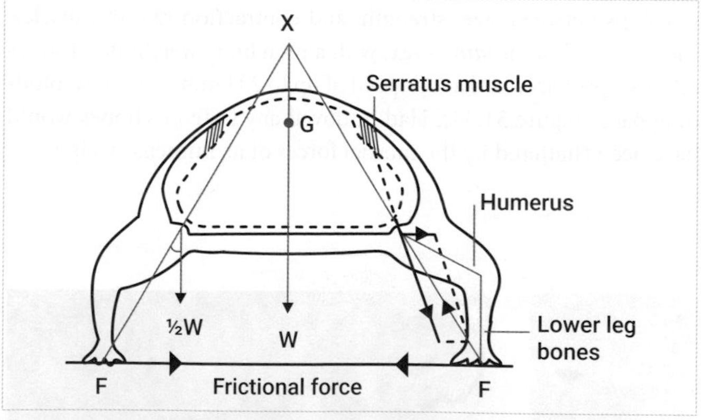

In larger climbing reptiles, such as chameleons, the legs have acquired a further bend, such that the feet are back under the body, placed almost together. The rounded legs wrap themselves around tree branches or leaf stalks, enabling the animal to balance on structures which are much narrower than its body. The toes have pronounced curves to improve grip on rounded structures.

Very large terrestrial reptiles which do not climb (alligators, crocodiles, komodo dragons) have wide bodies and widely arched, almost horizontally projecting limbs. They rest for long periods with the belly flat on the ground and may also support their body weight by spending time in the water. The active and resting postures of these animals are completely different from those of mammals.

The reptilian limb arrangement provides balance and stability, but what is the effect on locomotion? The gait of lizards and other reptiles with laterally projecting limbs is described as a **sprawl**. The feet cannot move directly forward under the body; instead, the legs are lifted and rotated forwards in something resembling a rowing action. The joint between the lower part of the leg and the foot is particularly flexible and allows for a forward flick at the end of the rotation. This extends the stride length a little, but the stride still depends mostly on the radius and length of the arc described by the leg.

 Go to the e-book to watch a video of a gecko moving along and notice its projecting limbs, which we describe as a sprawl.

All this means that the animal can walk or run but cannot gallop in the way horses and other quadruped mammals do. Variations in speed are mostly produced by varying the frequency of leg movements. Even at speed, the run involves very little time spent out of contact with the ground—there is no extended jump within the stride. The run is really a four-beat trot in which a single foot is always in contact with the ground.

A few lizards, especially among the **varanids** (including monitors and komodo dragons) overcome the stride length restriction by becoming bipedal. The gould's monitor, or sand monitor (*Varanus gouldii*) in Figure 31.37, is an example of a bipedal varanid. They lift the front of the body, balancing kangaroo-like on the tail, and then walk or run on the hindlegs with a greatly extended stride, much as humans and some other apes do. They may even do this on water.

 Check your understanding of the concepts covered in this section by answering the questions in the e-book.

Figure 31.35 Architecture of quadruped reptile body. The body's centre of gravity (G) is within the body cavity, its exact position depending on the structure and shape of the individual species. The total weight (W) is distributed along strut lines, equally through the legs and on to the feet (F). The strut lines from an arch, intersecting above the centre of gravity (X).

Source: Reprinted from: A.P Russell, V Bels (2001) Biomechanics and kinematics of limb-based locomotion in lizards: review, synthesis and prospectus. *Comparative Biochemistry and Physiology, Part A: Molecular & Integrative Physiology*, 131(1), 89-112. Copyright 2001, with permission from Elsevier. https://www.sciencedirect.com/science/article/pii/S109564330100469X?via%3Dihub

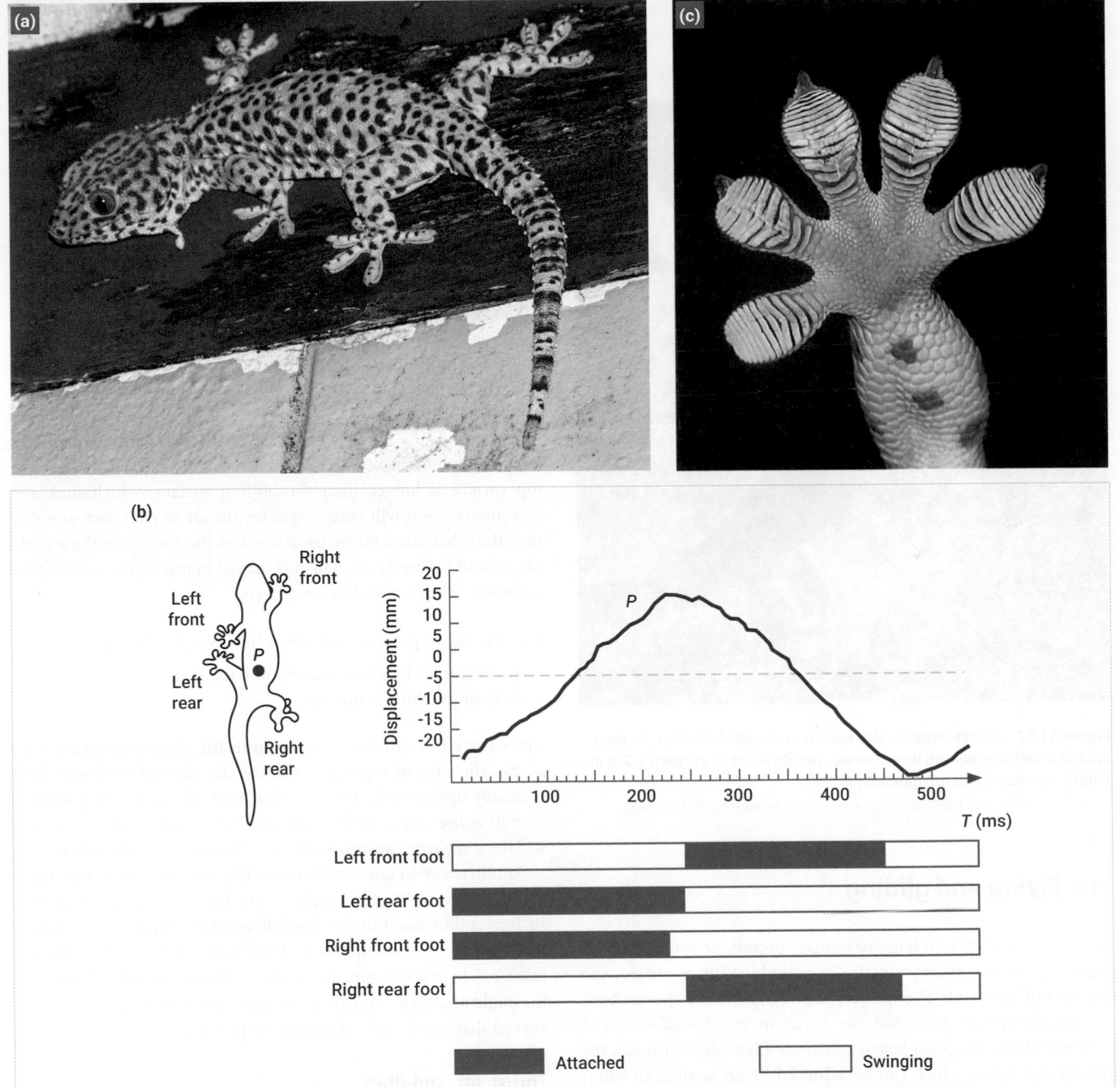

Figure 31.36 How does a gecko climb a wall without falling off? Geckos and other small lizards are noted for their agility and their ability to cling to vertical or even overhanging surfaces. A tokay gecko (*Gekko gecko*) (a) moves up a vertical surface using its legs in contralateral pairs, bending its body to extend its stride length (b). The plot line shows how far point *P* is displaced to the left and the right as the gecko moves. The tokay gecko can attach itself, even to smooth surfaces such as glass, by means of adhesive pads on its toes (c).

Source: (a) PumpkinSky/Wikimedia Commons/CC BY-SA 4.0. (b) Reprinted by permission from Springer: Wang, W., Li, X., Wu, S. *et al.* (2017) Effects of pendular waist on gecko's climbing: Dynamic gait, analytical model and bio-inspired robot. *Journal of Bionic Engineering*, 14, 191–201. Copyright 2017. (c) David Clements/Wikimedia Commons/Public Domain.

Figure 31.37 Gould's monitor, also known as the sand monitor, *Varanus gouldii*, a carnivorous and insectivorous reptile which may reach 1.5 m in length, can stand and run bipedally.

Source: Donna/CC BY 2.0 http://www.flickr.com/photos/geowombats/136601260/

31.9 Flying and gliding

When you step on to a large passenger aircraft, or watch one flying overhead, you place your trust in a combination of physics and the skill of aeronautical engineers. The largest such aircraft built to date, the Airbus A380-800, has a maximum take-off weight of 573 metric tonnes. Such heavier-than-air flight demonstrates the remarkable forces which can be achieved by an aerofoil (a wing) passing at speed through the air.

Animals which fly—insects, birds, bats, and the extinct pterosaurs of the dinosaur age—make use of the same fundamental physical forces as an aircraft. Aeroplanes are larger and faster than flying animals, as well as heavier, but those differences are essentially matters of scale. Animal wings often beat or flap, sometimes very fast, in a way that aircraft wings do not, but that too can be a deceptive difference when the underlying physical principles are analysed.

Organisms and parts of organisms which move *passively* through the air (wind-borne plant seeds, falling leaves) or glide (sugar gliders, flying squirrels, flying fish, flying snakes, and flying frogs), and those which move through water (fish, turtles, aquatic mammals, penguins, **pelagic** seabirds, and most other forms of **non-sessile** aquatic life) exploit the same basic principles.

To understand how animals fly, we need to understand some of the basic principles of aerofoils. Like terrestrial locomotion, flight can quickly become a complex mathematical topic, but the fundamentals can be appreciated with basic equations and simple, descriptive analysis.

A basic aerofoil

A wing is an aerofoil: a sheet of material which passes through the air in such a way that the air flows over it (parallel to its surface rather than acting as a ram). Air passes over the top and bottom surfaces, so the wing can be thought of as splitting the air into two regions. In a symmetrical aerofoil, where the top and bottom surfaces have equal length, as seen in the first diagram in Figure 31.38, there is no lift.

However, if the wing surfaces are of unequal length from front to back—for example, if the wing is slightly curved so that the top surface is longer than the bottom surface, which makes it asymmetrical—it will take longer for the air to pass over one side than the other. An asymmetrical aerofoil, like the one in the second diagram in Figure 31.38, which is curved in this way is said to have a **camber**. The camber has two effects:

- it produces a pressure difference between the two surfaces, and
- the longer surface will face more frictional resistance from the air than the shorter surface.

These two effects produce **aerodynamic lift**: the pressure difference pushes the wing at right angles to the direction of air flow, which is usually upwards; the frictional force acts along the wing surface, over its entire area, and the difference between the top and bottom will draw the wing more towards one direction than the other.

Another way to get aerodynamic lift, even from a symmetrical aerofoil, is to tilt it at an angle to the direction of air movement, such as is illustrated in the third diagram in Figure 31.38. The air will again be split unequally and will flow unevenly over the surfaces, producing pressure and friction. The angle of the tilt is called the **angle of attack**. Animal wings use both of these mechanisms—curved shape and angle of attack—to produce lift.

Thrust, lift, and drag

When a wing lifts a body, the aerodynamic lifting force acts in the opposite direction to the gravitational force on the body (its weight). The movement of air over the wing is caused either by forced propulsion (the aircraft's engines) or by the body's existing momentum (running, jumping, or flapping for take-off, followed by gliding and further flapping). This force is called **thrust** and it acts in the direction in which the body is moving. The force acting in the opposite direction to the thrust is called **drag**, as depicted in Figure 31.39. In principle:

- flight is possible when lift is equal to or exceeds weight;
- forward movement occurs when thrust exceeds drag.

Figure 31.38 Aerofoil shape and the generation of lift.

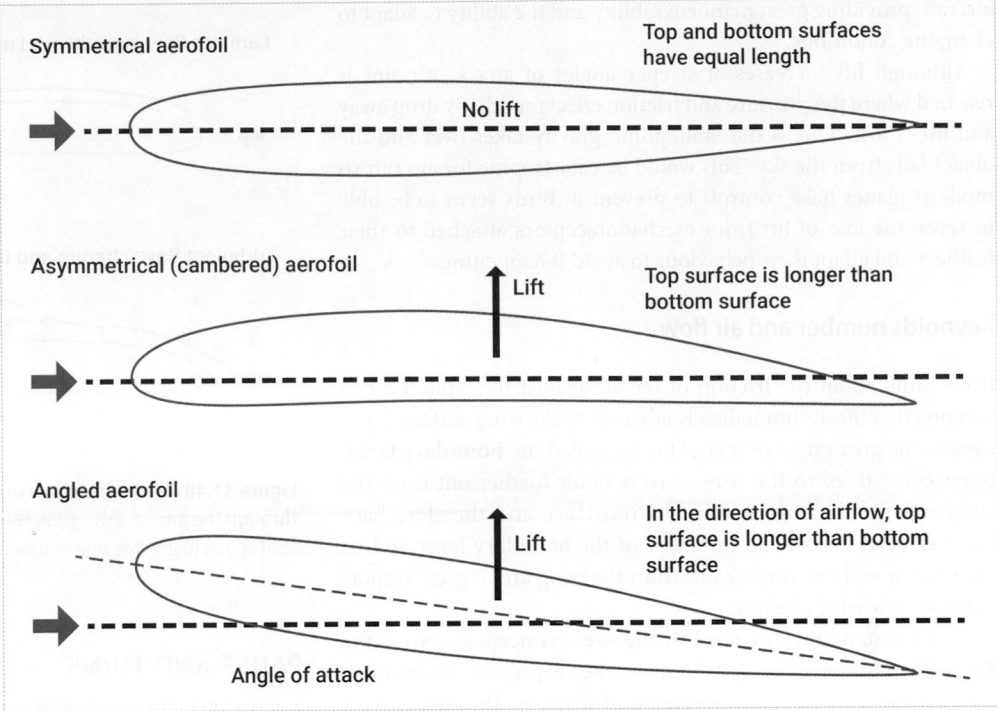

Symmetrical aerofoil

Top and bottom surfaces have equal length

No lift

Asymmetrical (cambered) aerofoil

Lift

Top surface is longer than bottom surface

Angled aerofoil

Lift

In the direction of airflow, top surface is longer than bottom surface

Angle of attack

Figure 31.39 The opposing forces of thrust and drag.

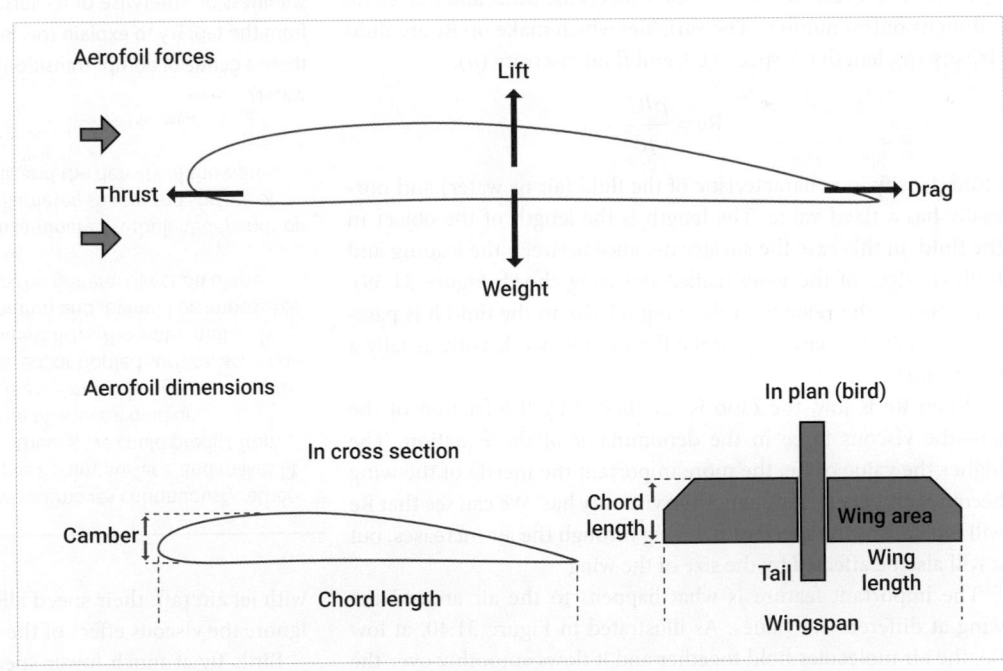

Aerofoil forces

Lift

Thrust

Drag

Weight

Aerofoil dimensions

In plan (bird)

In cross section

Camber

Chord length

Chord length

Wing area

Tail

Wing length

Wingspan

Weight depends on the body's mass, the force of gravity, and the buoyancy produced by the air. Mass is essentially a set property of the animal, and gravity and buoyancy can also be thought of as fixed, so weight cannot be altered during flight. Thrust depends on the body's speed relative to the air. Lift and drag depend on the wing's angle of attack and its overall size and shape.

Thrust, lift, and drag can all be adjusted during flight within the limits of the animal's physiology and the structure and morphology of its wings. This is exactly like an aeroplane changing its air speed (thrust), its angle of flight (angle of attack), and the degree to which its various wing panels are extended. However, the wings of birds and other animals are much more flexible than those of aircraft, providing great manoeuvrability and the ability to adapt to changing conditions.

Although lift increases at steeper angles of attack, a point is reached where the pressure and friction effects suddenly drop away and lift is lost. This is the 'stall' point: gravity takes over and the object falls from the sky. This would be catastrophic for aircraft so modern planes have controls to prevent it. Birds seem to be able to sense the loss of lift from mechanoreceptors attached to their feathers and adapt their behaviour to avoid it happening.

Reynolds number and air flow

If we think about the friction of the air against the wing, we can imagine that the air immediately adjacent to the wing surface experiences the greatest resistance. This air, called the **boundary layer**, essentially 'sticks' to the wing. Layers of air further out from the wing experience progressively less resistance and therefore have less frictional effect. The thickness of the boundary layer and its interaction with air further out from the wing are of great importance in how wings behave.

To understand this, we need an engineers' concept known as the **Reynolds number**. Reynolds number (Re) expresses the relationship between the force of movement of the wing through the air, called the inertial force, and the frictional force produced by the air's viscosity. It divides one by the other to produce a ratio. Both elements are forces, so the division cancels the units and makes Re a dimensionless number. The variables which make up Re are fluid density (ρ), length (l), speed (U), and fluid viscosity (μ):

$$\text{Re} = \frac{\rho l U}{\mu}$$

Fluid density is a characteristic of the fluid (air or water) and normally has a fixed value. The length is the length of the object in the fluid, in this case the surface distance between the leading and trailing edges of the wing (called the wing chord; Figure 31.39). The speed is the velocity of the wing relative to the fluid it is passing through. The viscosity of the fluid is, like its density, usually a fixed value.

When Re is low, the ratio is dominated by the friction of the air—the viscous force in the denominator of the equation. The higher the value of Re, the more important the inertia of the wing becomes and the less influence the viscosity has. We can see that Re will increase as the speed of the wing through the air increases, but it will also be affected by the size of the wing.

The important feature is what happens to the air around the wing at different Re values. As illustrated in Figure 31.40, at low Re, the air molecules hold together and it flows smoothly over the wing from front to back. This is called **laminar flow**. In laminar flow it is theoretically possible to predict accurately where any molecule of air will be from one moment to the next.

As the Re increases, a point is quickly reached (at about Re = 10^2 for air) at which the arrangement of the molecules becomes unstable and chaotic. Vortices appear in the fluid and its flow becomes **turbulent**: the moment-to-moment position of molecules cannot be predicted.

At very high Re (above about 10^6), the turbulent air near the wing becomes detached from the surface. This is what happens

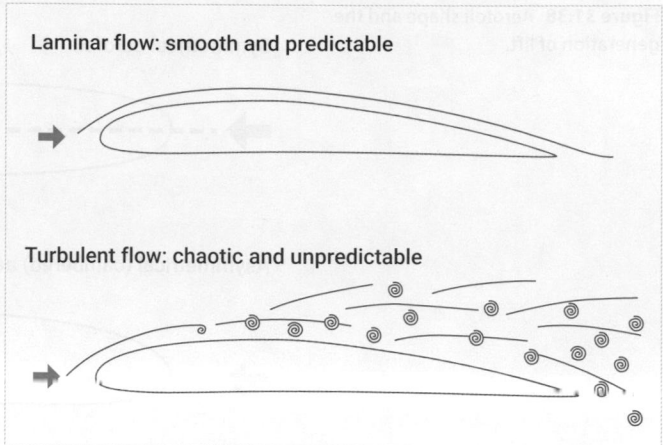

Figure 31.40 Types of air flow over an aerofoil, in the direction of travel through the air. For a wing, turbulent flow also occurs as vortices at the tip, at right angles to the direction of travel (not shown).

PAUSE AND THINK

Using a kitchen tap, adjust the water flow in finely controlled stages to move between continuous but very slow flow to fast flow. Describe what you see at each stage regarding the transparency of the water column, the shininess or otherwise of its surface, and the noise it makes as it issues from the tap. Try to explain this in terms of laminar and turbulent flow. Is there a gentle or abrupt transition between the states of movement of the water?

Answer: At very low flow, provided the water moves continuously rather than as drops, the water is transparent, has a shiny surface, and makes little noise. This is laminar flow in which, in theory, we could predict from moment to moment where each water molecule of water would be.

As the rate of flow is increased, there is an abrupt change to a translucent appearance with a non-shiny, confused, or rippled surface, accompanied by an increase in noise. This is turbulent flow. The water molecules are now moving randomly within the column and it would be impossible to predict their movement (other than in the general direction of the water column).

The point at which flow changes from laminar to turbulent depends on a large number of factor, including the diameter, shape, and surface of the tap orifice, the temperature of the air, and the density of the water.

with jet aircraft: their speed allows aircraft engineers to essentially ignore the viscous effect of the air.

Birds fly at much lower speeds than aircraft. They also change their behaviour, between take-off and landing, and between flapping, gliding, swooping, and other forms of flight. Most birds operate in a 'moderate' range of Re (1×10^4 to 1.5×10^5) where the boundary layer can have a significant effect. The air flow is turbulent rather than laminar but the Re is not so high that the boundary layer becomes permanently separated from the wing surface.

Very small changes in wing shape and other parameters can make the air detach and reattach or sometimes ripple over the surface. Birds have highly flexible wings of different shapes and sizes.

Figure 31.41 The general shape and relative sizes of representative bird species. The number in parentheses is the ratio of the wingspan to the mean chord length of the wings.

Source: Adapted from Taylor, G.K. & Thomas, A.L.R. (2014) *Evolutionary Biomechanics*. Oxford: Oxford University Press.

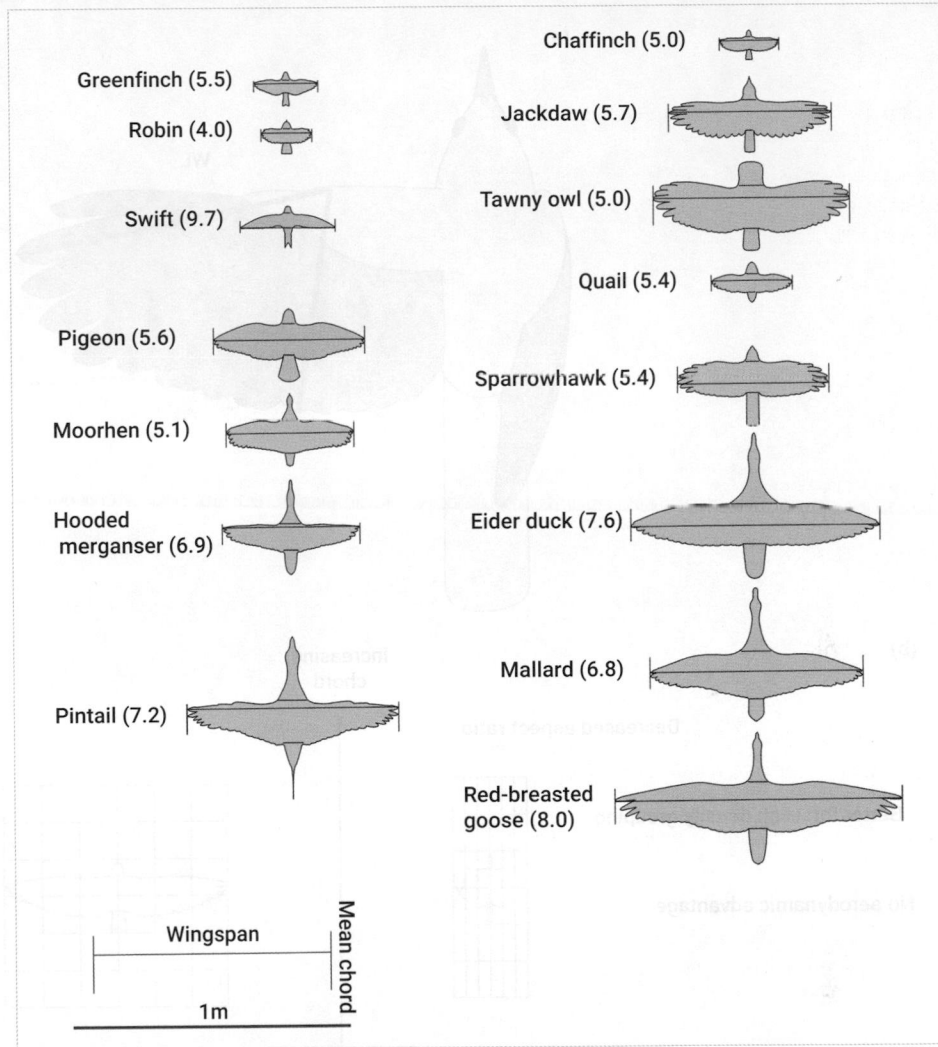

The feathers bend, twist, and vibrate, and they also alter the frictional influence of the air. Viscous effects are especially important towards the trailing edge of the wing; many birds have free feathers in this location and at the wing tips which disperse the air and reduce the drag. The bird's body and legs are also affected by air flow and some birds have 'loose' feathers in these regions too.

All of this means that birds are influenced by a balance of inertial and viscous forces, acting in extremely complex and variable ways, with unpredictable effects on aerodynamics. Biologists have struggled to describe bird flight in mathematical models, although some important principles are known.

Principles of efficient bird flight

The efficiency of flight increases as the ratio of lift to drag increases. Lift depends partly on wing area and on its ratio to body weight (called wing loading). However, drag is proportional to the surface area of the wing and other parts of the body.

Drag has been found experimentally to be proportional to body mass. This means that smaller birds with larger wings are more efficient flyers than larger birds with smaller wings. However, the difference is more marked at low speeds than at high speeds. Small birds can take to the air more easily than large ones, but once a large bird is airborne its greater power and speed make it much more aerodynamically efficient.

The importance of these general principles is greatly affected by the behaviour of the bird and the type of flying it engages in. Figure 31.41 compares the outlines of different types of bird, illustrating some relationships between wingspan, wing area, and overall body form.

For example, sustaining long distance flight is known to be improved by having wings with a high aspect ratio. This is measured as a hand-wing index (HWI), as illustrated in Figure 31.42a.

Figure 31.42a shows how the HWI is calculated from the wing length (WL, the distance from the carpal joint to the tip of the longest primary feather) and the secondary length (SL, the distance from the carpal joint to the tip of the first secondary feather) using the formula:

$$HWI = 100 \times \left(\frac{(WL - SL)}{WL} \right)$$

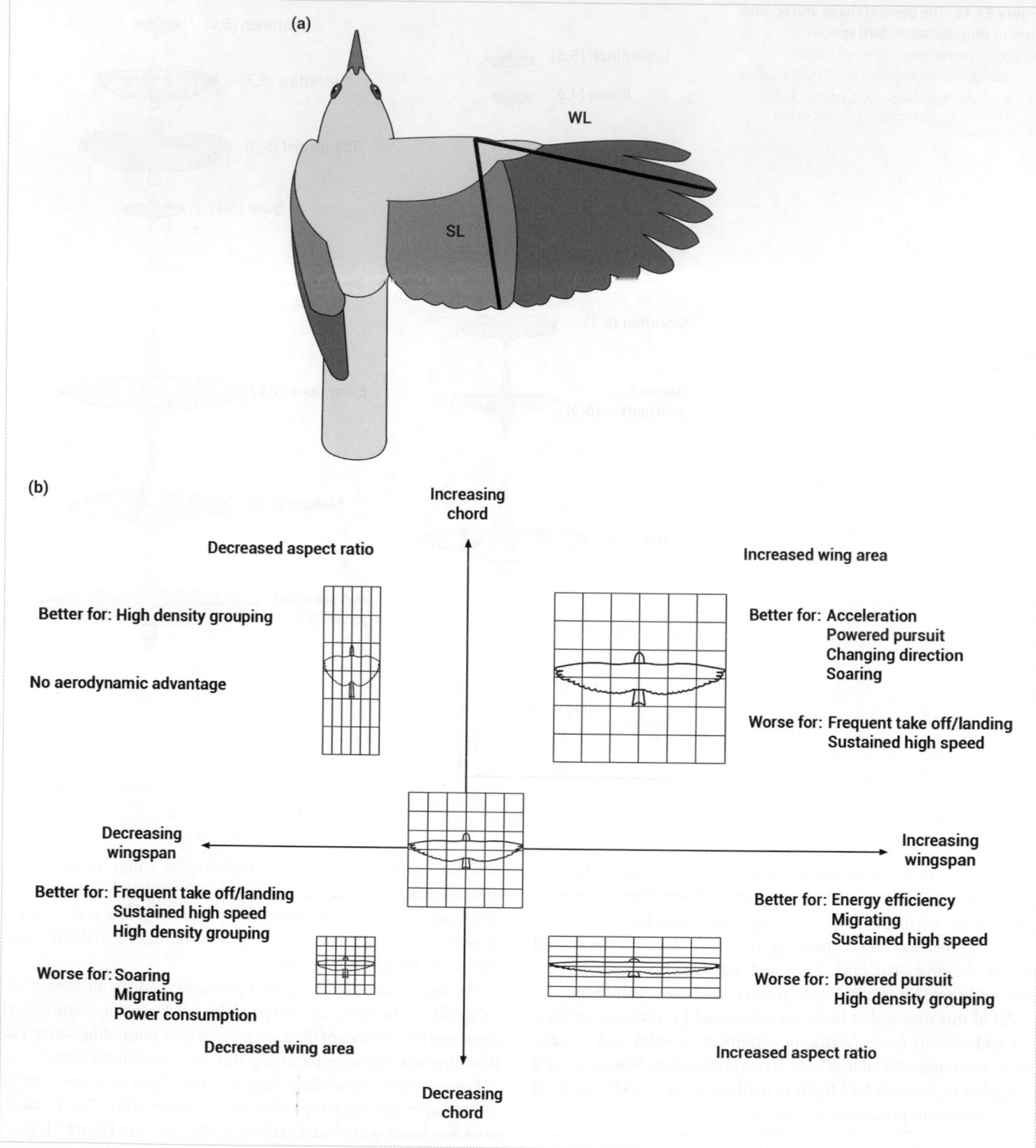

Figure 31.42 Bird aspect ratios. (a) Measuring wing aspect ratio: the Hand-Wing Index (HWI). WL = wing length; SL = secondary length. (b) Wing aspect ratio and the aerodynamics of bird flight.

Source: (a) Republished with permission of Proceedings of the Royal Society Biological Sciences. Adapted from Figure 2 of Claramunt, S. *et al*. (2011) High dispersal ability inhibits speciation in a continental radiation of passerine birds. *Proc Roy Soc B*, 283, DOI: 10.1098/rspb.2016.1922. https://royalsocietypublishing.org/doi/full/10.1098/rspb.2011.1922; permission conveyed through Copyright Clearance Center, Inc. (b) Adapted from Taylor, G.K. & Thomas, A.L.R. (2014) *Evolutionary Biomechanics*. Oxford: Oxford University Press.

Migratory birds tend to have higher HWI values. This can have implications for the rate at which new species evolve in new locations.

The HWI shows us that bird wings can be considered according to wingspan and chord length, which influence overall wing area and aspect ratio, as illustrated in Figure 31.42b. Each of these can determine adaptive advantage under particular lifestyles when combined with body size and wing loading.

Some general principles are:

- Decreased wing area gives increased speed.
- Increased wing area gives increased force (and so better acceleration), reduced sink rate (and so better soaring), and reduced power requirements.
- Increased aspect ratio gives increased energy efficiency and reduced power requirements.
- Decreased aspect ratio (short, stubby wings) gives no aerodynamic advantage.

All of these are greatly influenced by the bird's weight and thus by wing loading, and there are advantages and disadvantages to each strategy. As in all biological circumstances, pressure comes from predators, from the search for food and water, and from the need to survive and reproduce in different habitats, so the trade-offs manifest themselves in the great diversity of bird types which we see around us.

Flapping flight

The majority of insects, birds, and bats flap their wings to generate thrust in flight. Flapping is also used when animals hover. In all flying animals, aerodynamic forces depend on the phase (take-off, landing, cruising, hovering, changing direction, or height) and speed of flight, as well as the size and shape of the animal. Because flapping is a rapid movement (extremely so in some insects), much higher Re values apply to the movements of air over the wing surfaces than with static aerofoil wings. The aerodynamics of flapping wings are poorly understood, but some basic principles are known.

Studies of insect wings show that each flapping movement is made up of four distinct phases. The down beat (phase 1) and up beat (phase 3) are separated by very brief phases in which the lower surface of the wing turns to face upwards (phase 2) and downwards (phase 4). This rotation of the wing during each stroke has a number of effects:

1. It changes the angle of attack and ensures that the front edge always leads.
2. The air closest to the wing surface oscillates between acceleration and deceleration. The resulting changes in pressure act as if mass was being added to the wing, increasing its inertia and adding to the amount of aerodynamic force generated.
3. The boundary layer of air at the leading edge becomes separated from the wing and becomes a vortex (a spinning curl of air). This essentially rolls back down the wing as it moves forward and prevents it from losing lift and stalling. In many insects, the wings from each side come into contact ('clap') at each end of the stroke; this can cancel out the vortices,

preventing further turbulence, and it also adds inertia (called 'fling') to the movement.

4. The air around the wing circulates, with the upper and lower regions eventually reuniting smoothly at the trailing edge. This has complex effects on lift but has the advantage of reducing the drag.

Many flying insects have two pairs of wings. Studies on the most adept flyers, such as dragonflies, show that the wing pairs are coordinated, acting together to direct the air flow in the most efficient manner. This makes them especially agile, allowing them to change instantly from hovering to fast forward flight, to make extremely tight turns, and even to fly backwards and sideways.

The true flies, or diptera, have one pair of wings. The hind pair are replaced by balance organs called **halteres**—short, club-like structures which vibrate and generate rotational stability, somewhat like a gyroscope. This facilitates hovering and the manoeuvrability needed, for example, to penetrate flowers during feeding.

Birds and bats engaging in flapping flight make use of some or all of the effects described for insects. For example, listen out for the wing claps when pigeons and rock doves take to the air. With birds, the flow of air around the wing is especially complex due to the flexibility of the feathers and their ability to disturb the boundary layer (the scales on insect wings may do this too).

🔊 Go to the e-book to complete an activity that encourages you to engage further with Figure 31.42.

🔊 Check your understanding of the concepts covered in this section by answering the questions in the e-book.

31.10 Swimming

There are many similarities between flying and swimming. Both involve movement through a fluid and many of the mathematical principles associated with aerofoils also apply to swimming. The forces of thrust, lift, and drag occur as an animal moves through the water, although the balance between them is altered.

Water is about 800 times denser than air. This provides buoyancy and reduces the need for aquatic animals to generate lift. Many aquatic animals are neutrally buoyant or can easily control their buoyancy without a movement-derived lifting force (we discuss buoyancy in more detail later in this section). Because lift is not required, more energy is available to generate thrust. This is important because the higher density of water produces more drag.

All the swimming techniques used by animals involve the transfer of force resulting from muscle contractions to the water. The simplest example of how muscle contraction generates thrust can be seen in jellyfish. The adult medusa has an umbrella-like membranous but muscular bell. Rhythmic contractions of circular muscles in the bell force water out along the line of the tentacles, generating thrust in a forwards or upwards direction. In some jellyfish, elastic fibres provide an antagonistic force to help to restore the bell shape as the circular muscles relax. The bell muscles are not strong, but they can give the animal enough power to resist moderate currents and purposefully change position in the water.

Cephalopods (octopus, squid, cuttlefish) swim by jet propulsion, expelling water from their mantle. Contraction of circular muscles forces water out through a tube called a siphon. The direction of the resulting thrust depends on which way the siphon is pointed. Relaxation of these muscles and contraction of radial muscles then refills the mantle cavity with water.

Many aquatic vertebrates, including the majority of fishes, swim using undulations of their body. These generate waves of movement from the head towards the tail, called sinusoidal waves. They enable the body to push against the water and generate thrust. Similar swimming movements are seen in aquatic reptiles such as sea snakes and crocodiles. In fishes, the most extreme form of this type of swimming is seen in eels, and called **anguilliform** swimming (after the eel genus name, *Anguilla*).

In fishes, the muscles used for swimming are formed of distinct bands called **myomeres**. As an eel swims, waves of muscle contraction move backwards from the head, through alternate contractions of myomeres on either side of the body. This swimming style is efficient at low speeds, but drag increases considerably as the animal goes faster.

In contrast, some of the fastest swimming fish, such as tuna, keep their body relatively rigid and move their tail from side to side in an oscillating movement. In this type of swimming the caudal (tail) fin oscillates in the vertical plain and generates thrust as it comes to the end of each stroke.

In aquatic mammals such as whales and dolphins, the tail fluke generates propulsive force by moving in a horizontal plain. Seals achieve a similar movement by holding their hindlimbs together when they swim, to form a tail-like appendage.

Principles of efficient swimming in fish

Just as the aspect ratio of a bird's wing indicates the style and efficiency of its flight, the aspect ratio of a fish's fins provides information on the type of locomotion that it uses. Figure 31.43 shows an example of the shape of a caudal fin, and explains how to calculate fin aspect ratio, by calculating the span²/area.

The aspect ratio is often of greatest importance in the caudal fin. Table 31.2 lists how different aspect ratios of caudal fins might lend themselves to different adaptations in fish.

Using similar terminology to that used for flight, the general principles are:

- Sustained fast swimming is facilitated by a high aspect ratio in the caudal fin.

- For a given aspect ratio, an elliptical distribution of area produces the least amount of drag.

- A decreased aspect ratio (short, wide caudal fin) is less likely to result in a stall at a high angle of attack.

- Lower aspect ratios are better for acceleration and manoeuvring.

In most fast-swimming bony fish, the caudal fin is homocercal, meaning that the upper and lower lobes are symmetrical, as in the example in Figure 31.44b. In contrast, the majority of sharks have heterocercal caudal fins in which the upper lobe is larger than the lower lobe and the vertebral column extends into the upper lobe. You can see an example of this in Figure 31.44a. This shape provides both lift and thrust when moved side to side, although the mechanics of shark locomotion are poorly understood. In some other fish, such as the flying fish shown in Figure 31.44d, the lower lobe is longer than the upper lobe.

Undulations of the body in the absence of a significant caudal fin, or the oscillation of a caudal fin with a rigid body, can be viewed as two extremes of a continuum of swimming styles among fishes. Most use a combination of body undulations and tail oscillations to generate thrust. The mode of swimming may also change as an individual changes its speed of locomotion.

The diversity of fish body shapes is far greater than in birds, and fishes have a variety of other fins that can play a greater or lesser role in swimming. Pectoral fins can generate thrust. Wrasses (family Labridae) swim primarily using their pectoral fins in a rowing motion, known as labriform swimming. Skates and rays swim by undulations of their pectoral fins.

Figure 31.43 Measuring the fin aspect ratio.

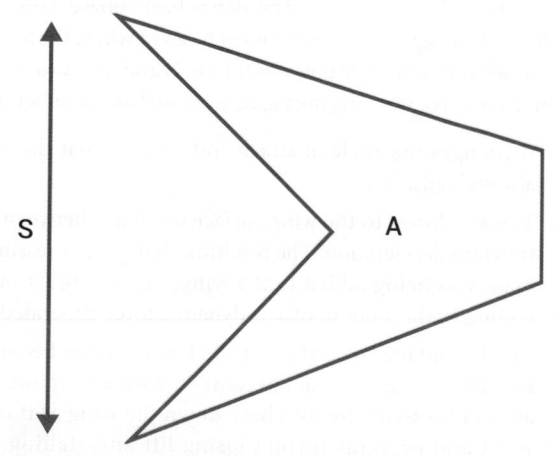

The AR is calculated from the span of the fin (S; distance between each tip of the fin out-stretched) and the fin area (A) using the formula:

$AR = S^2/A$

S

A

Table 31.2 Different fin aspect ratios and their influence on swimming behaviour. The key dimensions of fish fins are span, shape, overall area, and aspect ratio (AR). This table illustrates caudal fins, but similar principles apply to pectoral fins.

Aspect Ratio	Good for.......	Example
High	Fast-cruising speeds Reduced surface area reduces frictional drag and pointed tips reduce pressure drag	Sailfish
	Sustained swimming	Yellow and blue back fusilier
Low	Manoeuvrability Fast acceleration	Pinecone Fish

Source: Av. Rajat Kreation/Shutterstock.com; Nick Hobgood/Wikimedia Commons/CC BY-SA 3.0; Tomarin/Wikimedia Commons/CC BY 2.1 JP.

Figure 31.44 Fish caudal fins. (a) Heterocercal caudal fin in a grey reef shark: the upper lobe is bigger than the lower lobe. (b) Homocercal caudal fin: the upper and lower lobes are the same size. (c) Hypocercal caudal fin in a flying fish: the lower lobe is bigger than the upper lobe.
Source: (a) Jean-Lou Justine/Wikimedia Commons/CC BY-SA 3.0; (b) W.A. Djatmiko (Wie146)/Wikimedia Commons/CC BY-SA 3.0; (c) NOAA.

Overcoming drag

Animals moving through water experience greater drag than those moving through air. A streamlined shape will greatly reduce drag and it is therefore not surprising that the majority of fish have evolved to be **fusiform**. Often referred to as 'fish-shaped', a fusiform object can be thought of as a cylinder which is tapered at both ends, and has a maximal width equal to 25 per cent of its length. The head of a fish often comes to a point at the front, and the body narrows to a point at the back, known as the caudal peduncle, where the caudal fin meets the body.

Fish that have the ability to swim fast are generally fusiform. A similarly beneficial shape is also seen in whales and dolphins. Any parts of the body that protrude will disrupt water flow and increase drag, which is why many fast swimming fishes tuck their pectoral fins tightly against their bodies.

There are many strangely shaped fishes, which do not depend on speed for survival. Figure 31.45 shows how they may:

- be well camouflaged—an example being the leafy seadragon, a fish of the seahorse family which lives among seaweed;

- have poisonous spines—an example being the red lionfish;

- use sheer bulk to avoid predation by all but the largest sea creatures—an example being the sunfish, a large fish which feeds on jellyfish and is prey only to large whales and sharks. Its lateral flattening reduces drag but it is relatively slow moving.

As with birds, we can use Re to predict the type of water flow occurring at the boundary layer where water meets the body of the fish or its fins. When Re is low, the ratio is dominated by the friction of the water—the viscous force in the denominator of the equation. The higher the value of Re, the more important the inertia of the fish becomes and the less influence the viscosity has.

Looking again at the Re equation in Section 31.9 we can see that the number increases as speed through the water increases, but it will also be affected by the size of the fish. For most fishes, Re does not exceed 10^5, so laminar flow is maintained. (Note that although this Re is higher than that experienced by birds, the Re at which transition from laminar to turbulent flow occurs is different due to the higher density and viscosity of water.) However, for larger, faster-swimming species, boundary layers can transition from laminar to turbulent flows.

Delaying the transition of the boundary layer from laminar to turbulent will reduce drag. Shark skin is covered in **dermal denticles**, similar to fish scales, and it is believed that the shape of these reduces drag by reducing turbulence in the boundary layer. Indeed, the shape of dermal denticles has inspired the development of a new drag-reducing coating for use on the hulls of boats.

It should not be forgotten that some animals, such as seabirds, can both fly through the air *and* swim through the water. In Figure 31.46, you can see a flying fish. Flying fish have evolved an unusual **hypocercal** shape of caudal fin which provides lift at the water surface. Although there are many similarities between flying and swimming, any animal that does both faces a trade-off in optimizing locomotion in two different media. This means that many birds that swim have secondarily lost the ability to fly; these include penguins such as the one shown in Figure 31.46.

PAUSE AND THINK

An unusually large number of swimming world records were broken in 2008–2009, following the introduction of full body swimsuits made from materials such as polyurethane. The suits were close-fitting and also trapped tiny bubbles of air against the body surface. What is likely to have been the advantage of these suits?

Answer:

The suits probably:

a) reduced the swimmer's drag in the water by smoothing out their shape;

b) created a slippery boundary layer of air between the body surface and the water;

c) increased the swimmer's buoyancy, due to the trapped air, thereby increasing the amount of the body which passed through air rather than water.

The effects on speed were small but significant given the margins of swimming records. The use of such suits is now banned in international competition. Some previous suit designs had microhairs on the surface which disturbed the boundary layer and reduced drag.

Figure 31.45 Examples of bony fish with non-streamlined bodies which do not depend on speed through the water for feeding, evasion of predators, or other aspects of lifestyle. (a) Leafy seadragon (*Phycodurus eques*). (b) Red lionfish (*Pterois volitans*). (c) Common sunfish or mola (*Mola mola*).

Source: (a) Robb (Katzili at de.wikipedia)/Wikimedia Commons/CC BY-SA 3.0; (b) Bernard Dupont/Wikimedia Commons/CC BY-SA 2.0; (c) Per-Ola Norman/Wikimedia Commons/Public Domain.

Figure 31.46 Flying fish and swimming birds. (a) A flying fish (*Exocoetus* sp.) taking off from the sea surface. (b) A gentoo penguin (*Pygoscelis papua*) swimming underwater at Nagasaki Penguin Aquarium, Nagasaki, Japan. It has identification bands on its wings.
Source: (a) NOAA/Wikimedia Commons/Public Domain; (b) Ken FUNAKOSHI/Wikimedia Commons/CC BY-SA 2.0.

Buoyancy

As we have seen, aquatic animals can generate lift via muscle movements. This is known as dynamic lift, where the animal uses energy to maintain its position in the water column. The animal must swim constantly or it will sink. Alternatively, an animal can employ static lift, altering its density to match that of the surrounding medium and achieving neutral buoyancy.

The most obvious way to increase buoyancy is to reduce the density of the body. Specific gravity (SG) is the ratio of the density of a substance to the density of fresh water at the same temperature. Table 31.3 provides a list of different substances and their specific gravities. Notice that the SG of fresh water is taken as 1.0 (SG has no units), so substances which are denser than fresh water have a SG > 1.0. They will sink when placed in fresh water.

Sea water has a specific gravity of 1.026, so is slightly denser than fresh water due to its salt content. An organism which is neutrally buoyant in sea water would sink in fresh water; conversely, an organism which is neutrally buoyant in fresh water will find it impossible to dive below the surface in sea water.

Buoyancy increases if there is a reduction in structures made of the relatively dense calcium carbonate. In fact, the majority of aquatic organisms that have a calcium carbonate exoskeleton (crabs and other arthropods) or shell (molluscs and echinoderms) are **benthic**. They do not spend a lot of time swimming because it is energetically costly to do so. (The nautilus is an exception to this, which we will discuss shortly.)

An alternative to reducing the amount of dense substances within the body is to store buoyant compounds. The skeletons of sharks and rays are made of unmineralized **cartilage** (a type of **collagen**) rather than bone. Some sharks improve their buoyancy by storing lipids such as squalene and diacylglyceryl ethers in their livers.

Table 31.3 Table of specific gravities

Substance	Specific gravity
Fresh water	1
Sea water	1.026
Muscle	1.06
Cartilage	1.1
Fat	0.9
Squalene	0.86
Diacylglyceryl ethers	0.89
Calcium carbonate (aragonite)	2.93
Calcium phosphate (apatite)	3.2

Note

- The value for sea water varies with location (see Chapter 30).
- Values for muscle, cartilage, and fat are indicative and vary with composition.
- No value is given for bone because it varies greatly between species, according to its location in the skeleton and whether it is spongy (light and porous) or compact (heavy). The value for calcium phosphate represents the specific gravity of its mineral component.

Squalene has a lower specific gravity than other fats and oils and is therefore ideal as a stored buoyancy aid.

The most common method that aquatic organisms use to control their buoyancy is storage of gas. The mechanism varies, but the principle is the same.

- Some organisms use buoyancy in a passive manner: they have a simple gas float that keeps them at or near the water surface. *Sargassum* is a brown alga which forms large floating mats. Small, round structures within the algae, known as pneumatocysts, are filled with air and allow them to float. Another example is the siphonophore genus *Physalia*, which includes the Portuguese man of war. You can see examples of both brown algae and the Portuguese man of war in Figure 31.47. Notice the small round gas bladders in the algae, and the gas floats in the Portuguese man of war.

- Other aquatic organisms actively adjust the amount of gas that they store and regulate their buoyancy, allowing them to move up and down in the water column. A few animals do this by passing gas into and out of a rigid storage chamber. An example

is the nautilus in Figure 31.48, which is a cephalopod with a spiral, snail-like shell made of calcium carbonate. Despite its high mineral density, the shell is constructed of chambers, called siphuncles, which the animal can fill with air or nitrogen to alter its buoyancy. The animal adds siphuncles to the shell as it grows.

- An alternative to storing gas in a rigid compartment is to use a flexible chamber made of strong collagen, called a swim bladder. This has the benefit of taking up less body space but creates an additional problem when dealing with changes in pressure as the animal moves up and down the water column: as the pressure

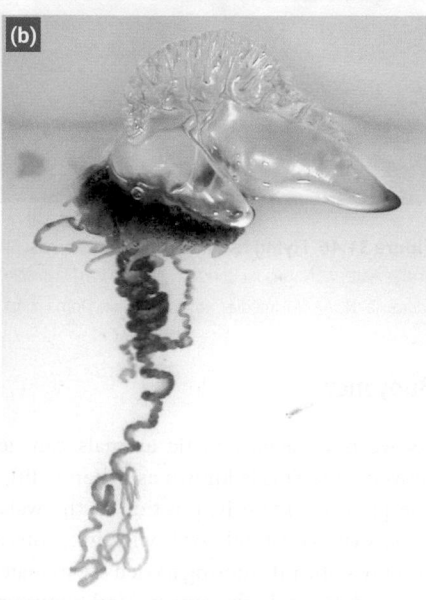

Figure 31.47 Simple gas flotation. (a) *Sargassum* sp., showing gas bladders that allow the algae to float at or near the water surface. (b) *Physalia physalis*, Portuguese man of war. A gas float supports colonies of siphonophores, which hang below the water surface.
Source: (a) Bogdan Giuşcă/Wikimedia Commons/Public Domain; (b) Image courtesy of Islands in the Sea 2002, NOAA/OER.

Figure 31.48 Nautilus. (a) In marine environment. (b) Shell in section.
Source: (a) Manuae/Wimikedia Commons/CC BY-SA 3.0; (b) Chris 73/Wikimedia Commons/CC BY-SA 3.0.

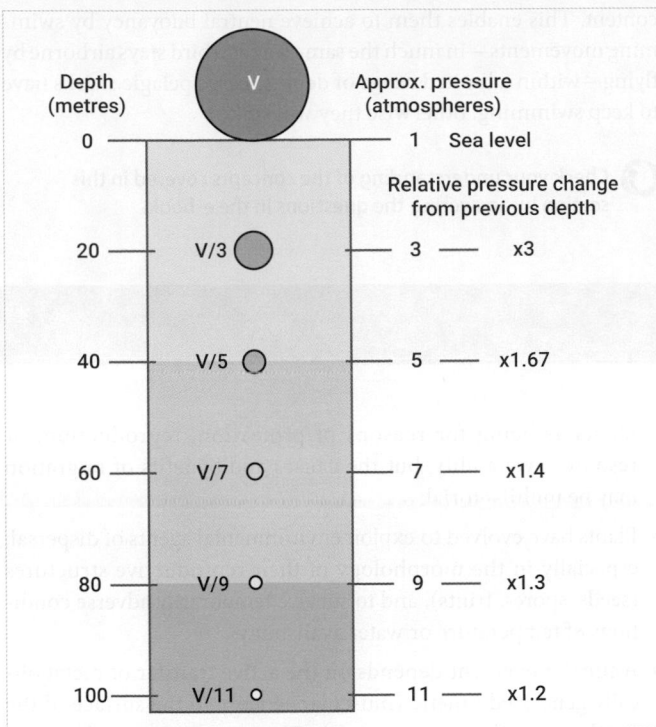

Figure 31.49 Water pressure at depth. Note that the pressure at sea level increases by about one atmosphere for each 10 m of depth; volume is inversely proportional to pressure (as described by a physical law called Boyle's Law); greater *relative* changes in pressure, thus of volume, take place nearer the surface.

decreases as an organism ascends the water column, so the swim bladder must be able to increase in size. Look at Figure 31.49, which illustrates this concept using a balloon as an analogy; notice how the size of the balloon reduces as the depth increases. To obtain neutral buoyancy, a fish must inflate its swim bladder to about 5 per cent of its body volume in sea water and about 7 per cent in fresh water.

Most bony fishes possess a swim bladder, located in the body cavity, usually just below the vertebral column. There are two types: **physoclistous** and **physostomous**.

- Physoclistous fishes (mainly marine teleosts) have no connection between the swim bladder and the gut but fill the bladder by secreting gas into it from the blood stream. Look at Figure 31.50, which shows a physoclistous swim bladder and associated vasculature. Blood is supplied to a gas gland by a *rete mirabile*, a network of capillaries which acts as a counter-current exchange mechanism for gas transfer between arterial and venous blood. The gas is trapped, preventing loss from the swim bladder. When lactic acid is released by the gas gland, the resulting decrease in pH causes haemoglobin to release oxygen. This increases the pressure in the gas gland and gas diffuses into the swim bladder. Although pressure within the swim bladder is higher than in surrounding tissues, gas cannot easily diffuse out due to a layer of guanine platelets in the swim bladder wall. Gas is released through a muscular valve, known as the **ovale**, and returned to the blood circulation.

- Physostomous fishes fill their swim bladders by swallowing air at the water surface and passing it from the gut directly into the bladder through a pneumatic duct. Air is released by passing it back out to the external environment via the gut, often by burping. This method of filling the swim bladder is restrictive in that the fish has to be at the water surface to gulp air. As with the balloon analogy in Figure 31.49, during any foray to depth, the volume of the swim bladder will decrease, neutral buoyancy will be lost, and the fish will start to sink. Therefore, some physostomous fishes also employ the mechanisms of inflating their swim bladder used by physoclistous fishes.

If a physoclistous fish that has filled its swim bladder at depth is caught in a fishing net and hauled quickly to the surface, it may not have time to release gas in response to the very rapid change in pressure. This is why deep sea fishes brought to the surface may have an inverted swim bladder, projecting from the mouth, or the bladder and internal organs may be destroyed by bursting.

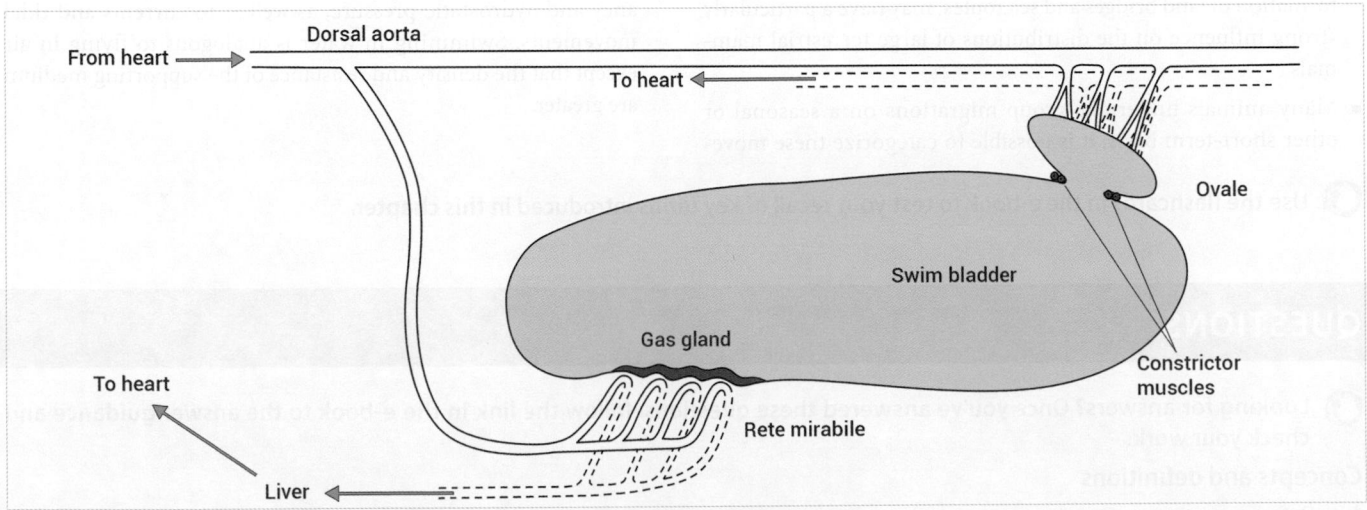

Figure 31.50 A physoclistous swim bladder.

Most fishes which spend their whole lives at great depths (below 1000 m) do not have swim bladders. The explanation for this is shown in Figure 31.49: the greater the depth, the smaller the *relative* change in pressure when moving a given distance up or down in the water column.

Cartilaginous fishes have no swim bladders. They generally have less dense bodies than bony fishes, due to a higher tissue lipid content. This enables them to achieve neutral buoyancy by swimming movements—in much the same way as a bird stays airborne by flying—within a limited range of depths. Large pelagic sharks have to keep swimming, otherwise they will sink.

 Check your understanding of the concepts covered in this section by answering the questions in the e-book.

SUMMARY OF KEY CONCEPTS

- Organisms move across the surface of Earth both spatially and through time; the planetary distributions of organisms which we currently observe are the result of historical movements, habitat change, and adaptation.

- Life probably began on Earth about 4.3 billion years ago. The movement of land masses as the planet developed has shaped and positioned the surfaces available for colonization.

- Islands and other isolated land masses often create environments with limited resources and exposure to a reduced range of predators but may present extremes of climate and unique local habitats.

- Plant distribution is essentially the result of passive movements (caused by wind, water, and the activities of animals), coupled with local growth and colonization.

- Animal distributions are usually the result of active, purposeful, or consciously driven movements, in addition to those caused by environmental agents.

- Evolutionary radiation refers to the colonization of new territory by an organism followed by the emergence of new species as a result of adaptation to different environmental conditions. Adaptive radiation has a more general meaning but often refers to short-term movements of groups of organisms which do not necessarily involve species change.

- The separation and coalescence of land masses, including the formation of land bridges and sea routes, may have a particularly strong influence on the distributions of large terrestrial mammals.

- Many animals undertake group migrations on a seasonal or other short-term basis; it is possible to categorize these move-

ments as being for reasons of protection, reproduction, or resource availability, but the causes and benefits of migration may be multifactorial.

- Plants have evolved to exploit environmental agents of dispersal, especially in the morphology of their reproductive structures (seeds, spores, fruits), and to survive temporarily adverse conditions of temperature or water availability.

- Animal movement depends on the active transfer of metabolically generated kinetic (muscular) energy to the surface of the Earth or its fluids (air, water). The transfer may take place through the whole body (limbless) or through focal structures (limbs) in contact with the substratum.

- Limbed movement takes a variety of forms across the animal kingdom but can be analysed according to the principles of a pendulum or combination of pendulums. The speed and efficiency of movement depend on pendulum length, fulcrum position, and gait.

- Animals that fly (insects, birds, a few mammals) have wings that exploit the basic principles of an aerofoil; their movements can be interpreted according to the effects of thrust, lift, and drag.

- Wing shape determines the characteristics of flight and has evolved in relation to each species' lifestyle and habitat.

- Animals which live in water are exposed to the forces of buoyancy and hydrostatic pressure, as well as to currents and tidal movements. Swimming in water is analogous to flying in air except that the density and resistance of the supporting medium are greater.

 Use the flashcards in the e-book to test your recall of key terms introduced in this chapter.

QUESTIONS

Looking for answers? Once you've answered these questions, follow the link in the e-book to the answer guidance and check your work.

Concepts and definitions

1. Define speciation.

2. What is a tectonic plate?

3. Explain two meanings of **species radiation**.

4. Define migration.

5. What are peristalsis and **undulation** in the context of animal movement?

6. Define aerofoil lift and drag and distinguish between laminar and turbulent flow in a fluid.

Apply the concepts

7. Explain how island isolation might influence speciation.

8. Why is it advantageous for plants to exploit moving features of their environment (fluids, animals)?

9. What limits the speed at which an elephant can run?

10. Explain why some birds and insects flap their wings.

11. Explain how bird wings have evolved to avoid the effects of the boundary layer of air.

12. Explain how fish body surfaces have evolved to avoid the effects of the boundary layer of water.

Beyond the concepts

13. Using a ruler or tape, measure the likely pendulum components in one of your legs. Also measure your stride length when a) walking and b) running at different speeds. Explain how the same leg dimensions allow you to move at different speeds.

14. Look out of the window or stand outside. For the next bird you see, roughly sketch its wing shape and decide the most likely nature of its flight pattern(s). If the bird is flying, compare your conclusion with what you observe.

15. Eucalyptus trees are endemic to Australasia and other regions of the Southern hemisphere but not to the Northern hemisphere. Why might this be?

16. Using online resources, obtain some information on the history and distributions of the white-clawed crayfish (*Austropotamobius pallipes*) and the signal crayfish *(Pacifastacus leniusculus)* in the UK. Would you describe each of these species as native or alien? What reasons could be used to argue for or against your conclusions?

17. Would it be correct to describe the European eel (*Anguilla anguilla*) as a migratory fish? What is its life story and what determines its global location at different times of year?

18. Charles Darwin was interested in the movement of plant seeds around the world in ocean currents and the importance of this for plant distribution. How did he attempt to find out about this?

FURTHER READING

Aldheeb, M.A., et al. (2016) A review on aerodynamics of non-flapping bird wings. *J. Aerosp. Technol. Manag., São José dos Campos* 8: 7–17.

Chin, D.D. & Lentink, D. (2016) Flapping wing aerodynamics: from insects to vertebrates. *J. Exp. Biol.* **219**: 920–32.

Clayton, H.M. (2016) Biomechanics of the exercising horse. *J. Anim. Sci.* **94**: 4076–86.
The biomechanics of horse movements and how to analyse it.

Croteau, E. K. (2010) Causes and consequences of dispersal in plants and animals. *Nat. Educ. Knowl.* **3**: 12.
The causes and effects of dispersal at the level of individuals, populations, and species.

Jordan, M.R.S., Barraclough, T.G. & Rosindell, J. (2015) Quantifying the effects of the break up of Pangaea on global terrestrial diversification with neutral theory. *Phil. Trans. R. Soc. B*, **371**: 20150221. http://dx.doi.org/10.1098/rstb.2015.0221
Continental movements and their influence on biodiversity: a detailed commentary and two original articles, one descriptive and the other quantitative.

Lees, J.J., Dimitriadis, G. & Nudds, R.L. (2016) The influence of flight style on the aerodynamic properties of avian wings as fixed lifting surfaces. *PeerJ*, **4**: e2495. doi: 10.7717/peerj.2495
How non-flapping and flapping wings work and how wing aspect ratios are analysed.

Lindborg, R. (2007) Evaluating the distribution of plant life-history traits in relation to current and historical landscape configurations. *J Ecol.* **95**: 555–64.
A readable review of the complex interactions between plant characteristics (especially seed types and structures), types of landscape, environmental changes, and plant distribution.

Meyns, P., Bruijn, S.M. & Duysens, J. (2013) The how and why of arm swing during human walking. *Gait Posture* **38**: 555–62.
Why do we swing our arms when we walk?

Reppert, S.M. & de Roode, J.C. (2018) Demystifying monarch butterfly migration. *Curr. Biol.* **28**: R1009–22.
The 'stunning' story of monarch butterflies and their long distance, multi-generational migrations; how do they find their way and what environmental cues do they use?

Sellers, W.I., et al. (2017) Investigating the running abilities of *Tyrannosaurus rex* using stress-constrained multibody dynamic analysis. *PeerJ*, **5**: e3420. doi: 10.7717/peerj.3420
How fast could T. rex run? Could it run at all?

Shaw, A.K. (2016) Drivers of animal migration and implications in changing environments. *Evol. Ecol.* **30**: 991–1007.
A review paper on the basic biology of animal migration, what it is, why it happens, and its occurrence in different animal groups.

Shepard, E.L., Ross, A.N. & Portugal, S.J. (2016) Moving in a moving medium: new perspectives on flight. *Phil. Trans. R. Soc. B*, **371**: Issue 1704 (September).
A collection of reviews on flight in birds, insects, and bats, covering evolution, dynamics, physiology, and migration.

Tenenbaum, D. (2017, May 18) As continents continue moving, study suggests effects on biodiversity. https://phys.org/news/2017-05-continents-effects-biodiversity.html

Zaffos, A., Finnegan, S. & Peters, S.E. (2017) Plate tectonic regulation of global marine animal diversity. *PNAS* **114**: 5653–8.

Defence against Predation and Invasion

LEARNING OBJECTIVES

By the end of this chapter, you should be able to:

- Explain why defending biological space is a defining characteristic of life and facilitates the maintenance of an organism's biological identity.

- Describe how knowledge of a plant's external and internal defence mechanisms can be exploited in the breeding of crops and other useful plants.

- Evaluate the integumental structure of an animal as a defensive barrier while allowing for its other physiological properties.

- List, with basic understanding: (a) four phases of the animal response to integumental injury, (b) four stages of the inflammatory response, (c) the three components of the triple response, and (d) three enzyme cascades which these processes initiate.

- Describe the role of surface coloration in camouflage, deception, and mimicry, and be able to apply the terminology commonly used to describe these morphological features.

- Know the distinction between toxins, poisons, and venoms, and be able to evaluate, for any specified organism, which of these terms best describes its external secretions.

Chapter contents

Introduction		1103
32.1	Protection from invasion: plants	1103
32.2	Protection from invasion: animals	1109
32.3	The response to injury	1121
32.4	External defences	1128
32.5	Toxins, poisons, and venoms	1134
32.6	Chemoreception and predator avoidance	1136

Watch the key concepts video in the e-book to prepare yourself for studying this chapter.

Introduction

In the first chapter of this module we saw that all living organisms enclose a region of the biosphere. They adopt a small part of Earth's environment within which they carry out biochemical reactions and regulate the flow of energy. This allows them to grow, develop, and reproduce.

For single-celled organisms, the enclosed region is defined by the cell membrane. The membrane restricts the free movement of molecules but possesses transport mechanisms which regulate the flow of materials in and out of the cell. This is how the organism controls its interactions with the external environment, as well as its internal life processes.

The cells which make up complex, multicellular organisms do the same thing, making use of complex membranes and cell walls. They greatly increase the size of the enclosed region by joining together to form tissues, organs, and bodies. The cells cooperate by differentiating, taking on specialized biochemical and physiological tasks on behalf of the whole animal or plant.

At a larger scale, organisms group together to form communities. Now the enclosed region is a complex ecosystem, influenced by the activities of many individuals.

In all these cases, the enclosed region of the biosphere needs to be defended. Without some form of protection, biological material will be absorbed back into the environment and the organism will lose its identity. This is also what happens when an organism dies.

A key characteristic of life, therefore, is the ability to defend an enclosed region of the biosphere. A dead cell, animal, or plant does not do this. It is quickly invaded by micro-organisms and then consumed by opportunistic, scavenging organisms. Its decay releases chemicals and stored energy, making them available for other organisms to use.

An alive but inadequately defended organism is similarly susceptible to take-over by other forms of life: it will be predated for food or colonized by parasites. At all levels of biology, there is a constant arms race between defence and invasion, between protection and predation. This creates an evolutionarily important environmental pressure, and it also sets up food chains and results in the constant recycling of biochemical resources and energy.

Defence mechanisms often overlap with those employed for other purposes such as resource protection, shelter, feeding, reproduction, and community cohesion. In all cases, they are phenotypic adaptations which enhance survival and increase the chance that genes will be passed on.

In this chapter we focus on *external* defence mechanisms, including the physical resilience offered by integuments and their ability to repair themselves, as well as the colours, chemicals, and behaviours which give organisms camouflage and protection.

Multicellular organisms also have *internal* defence mechanisms. These are initiated when the integument is damaged or breached. They confer resistance to colonization by pathogens and typically comprise an innate, reactive component coupled with an adaptive component whose responses are acquired and improved by experience. They are illustrated by the human immune system, described in Chapter 27.

32.1 Protection from invasion: plants

Plants are essentially sessile in their habit. This means that they are passive in their interactions with other organisms and have defensive rather than active or offensive ways of protecting themselves. Some micro-organisms can colonize parts of plants for mutual benefit, and a few plants are parasitic or hemiparasitic on other plants, but most instances of exploitation by other species occur when the plant becomes a food source or when micro-organisms cause infection and disease.

Plants trap energy from the sun and turn it into carbohydrate through photosynthesis. Animals cannot do this, so plants are the primary source of energy in most food chains. Plants are constantly under attack from organisms which consume their leaves, stems, flowers, seeds, or roots. This process is termed **herbivory**.

The extent of herbivory on any given plant varies considerably but it can be highly damaging. Damage may be caused by large numbers of small animals grazing, such as caterpillars on a cabbage plant, or a single large animal such as an elephant eating branches of a tree. Plants are remarkably resilient and can easily tolerate a moderate degree of damage to their structures by grazing.

Most plants respond to low levels of herbivory by simply growing more plant body in the form of new shoots, stems, and leaves. Excessive damage will simply kill the plant, a situation observed in dramatic fashion when plagues of locusts decimate green foliage as they move across the land in vast numbers, as shown in Figure 32.1. The local ecological effect of intense gazing can be dramatic because the widespread removal of one type of plant creates clear, competition-free spaces for others to flourish, such as in the case of Dartmoor ponies on coastal grassland in Cornwall, as shown in Figure 32.2.

The structural morphology of many types of plants enables them to grow despite being grazed: axillary meristems can generate new growing shoots from ground level (see Chapter 29). A good example is in the evolution of grasslands: they are a **climax community** (a community of organisms with a relatively stable structure and composition) dominated by grasses and constantly grazed by animals as their primary source of food. The basal meristem of grasses (see Chapters 28 and 29) allows the constant regeneration of new leaf tissue and the plant survives.

Figure 32.1 A swarm of migrating locusts in the Malagasy Republic.
Source: Pav-Pro Photography Ltd/Shutterstock.com.

Figure 32.2 Grazing herbivores: Dartmoor ponies on coastal grassland in Cornwall.

Source: Phillip Capper/Wikimedia Commons/CC BY 2.0.

Despite this in-built resilience, plants have evolved a variety of systems to minimize the likelihood of being eaten by herbivores. These strategies fall into two groupings; physical adaptations and biochemical adaptations. Investment in defence mechanisms of any sort comes at an energetic cost to the plant; it is advantageous to deter as much herbivory as possible in order to survive.

Physical defences against herbivory

Many plant species produce a variety of sharp pointed external structures which deter grazing animals. Spines, thorns, and prickles vary enormously in shape, size, and abundance, as you can see from the examples in Figure 32.3. These structures are highly effective in preventing large-scale herbivory by grazing mammals but they do little to deter insect grazing.

Members of the Rosaceae such as brambles (*Rubus fruticosus*) and roses (*Rosa* spp.), which you can see in Figure 32.3c, are well

Figure 32.3 Strategies to deter grazing. (a) Thorns on gooseberry (*Ribes uva-crispa*); (b) spines on cactus (*Cactaceae* spp.); (c) prickles on stems of roses (*Rosa* spp.); and (d) prickles formed on the edge of thistle leaves (*Cynareae* spp.).

known for their vicious prickles which can easily draw blood on human skin. Two shrubs native to Europe, hawthorn (*Crataegus monogyna*) and blackthorn (*Prunus spinosa*), both produce very sharp thorns, as suggested in their names. They have been used for centuries in hedging, to delineate field boundaries and for making living barriers which livestock cannot cross.

Another well-known groups of plants with sharp spines are the cacti (*Cactaceae* spp.); as shown in Figure 32.3b. Their leaf structures have evolved into spines, a strategy which not only deters hebivores, but also reduces surface area and shades the plant from the high light levels in their natural habitat.

Cacti live in **xerophytic** conditions where a grazing mammal would find little else to eat, so having an extensive array of deterrent spines is crucial to survival. Cacti such as the *Opuntia* spp. shown in Figure 32.4, produce short stiff hairs called **glochids**, which are very thin and barbed. They break off in large numbers when touched and can cause severe irritation in a grazing animal's mouth or on its skin.

In some tree species, such as horse chestnut (*Aesculus hippocastanum*) pictured in Figure 32.5, sharp spines form on the surface of fruiting structures, defending seeds while they are developing.

Figure 32.4 The cactus *Opuntia microdasys*.

Chemical defences against herbivory

The complex biochemistry of plants allows them synthesize a vast array of molecules. This synthesis is called **plant secondary metabolism**. Each species' spectrum of secondary metabolites is unique and any potential herbivore needs to be able to deal with them when grazing.

Among the chemicals produced are alkaloids, phenolics, flavonoids, cyanogenic glucosides, and proteinase inhibitors. Many of these molecules have been exploited during human history in the form of drugs and medicines, and they underpin much of modern pharmacology.

The effect of plant metabolites on herbivores varies considerably. Many are harmless, while others are highly toxic and potentially

Figure 32.5 Spiny fruit cases which deter herbivory of the developing seed. (a) Developing fruit of the horse chestnut tree (*Aesculus hippocastanum*); (b) developing fruit of the thorn apple (*Datura stramonium*).

lethal. Most of the world's crop plants have been carefully selected and bred to minimize or completely remove any dangers.

In wild and undomesticated ecological settings, the precise spectrum of molecules produced can be of great significance, particularly in interactions between plants and insects. Plant secondary metabolism has evolved in parallel with the evolution of insects, leading to species specialization:

- As depicted in Figure 32.6, caterpillars of the peacock butterfly feed on the leaves of stinging nettle and are unaffected by the acid contained in leaf hairs which is an irritant for human skin. Caterpillars of the cinnabar moth (*Tyria jacobaeae*) feed quite happily on ragwort (*Jacobaea vulgaris*), a common grassland plant which is highly toxic to mammals, especially horses. Indeed they are so effective at eating ragwort that the moth has been introduced into some countries as a biocontrol agent for the plant.

- As depicted in Figure 32.7, the caterpillar of the tobacco hornworm moth, *Manduca sexta*, is a commercially significant pest of tobacco crops for the way it feeds on the leaves of young plants. It has evolved a digestive mechanism which absorbs but immediately excretes the extremely high quantities of potassium ions which tobacco plants contain. The potassium in the leaves is toxic to other insect predators which cannot adequately excrete it.

In these examples and many others, the adult butterfly or moth lays its eggs on a specific plant species, which the hatched caterpillars then feed on voraciously. The caterpillar's metabolism has evolved to cope with the specific cocktail of secondary metabolites that its host plant produces.

Figure 32.6 Caterpillar herbivory. (a) Caterpillars of the peacock butterfly (*Aglais io*) feed mainly on stinging nettle. (b) Caterpillars of the cinnabar moth (*Tyria jacobaeae*) feed mainly on ragwort (*Jacobaea vulgaris*), a plant which is toxic to mammals.

Some plant molecules reduce the palatability and digestibility of tissue. Such biochemical defences against herbivores are not static. The deterrent molecules may be synthesized constitutively, so they are present at all times in the plant tissue, or may be synthesized as an induced response when a herbivore starts nibbling. The presence of a nibbling insect on a leaf may initiate the release

Figure 32.7 The tobacco hornworm, *Manduca sexta*. (a) Larva (caterpillar). (b) Adult (moth).
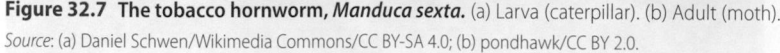
Source: (a) Daniel Schwen/Wikimedia Commons/CC BY-SA 4.0; (b) pondhawk/CC BY 2.0.

of signals to adjacent tissues, upregulating metabolism prophylactically and thereby minimizing the overall destruction of the plant.

In induced defence responses such as this, the signal which results from a plant being grazed can be transmitted over a long distance, effectively warning neighbouring plants of imminent attack. Figure 32.8 illustrates how such systems allow for signals to be transmitted to other parts of the same plant (intraspecific signalling) or to other plants nearby, including those of other species (interspecific signalling). The signals produced include hormones, extra-floral nectar (EFN; secretions from parts of the plant other than the flower), and gaseous molecules called volatile organic compounds (VOCs). VOCs travel through the air, either to leaves on the same plant or to those on others.

Signalling can also occur underground when roots are nibbled by soil-borne herbivores. Similarly, the fact that aphids are feeding on the leaves of one plant can be communicated to other nearby plants through the network of underground hyphae that interlink the roots of plant communities which live with mycorrhizal fungi.

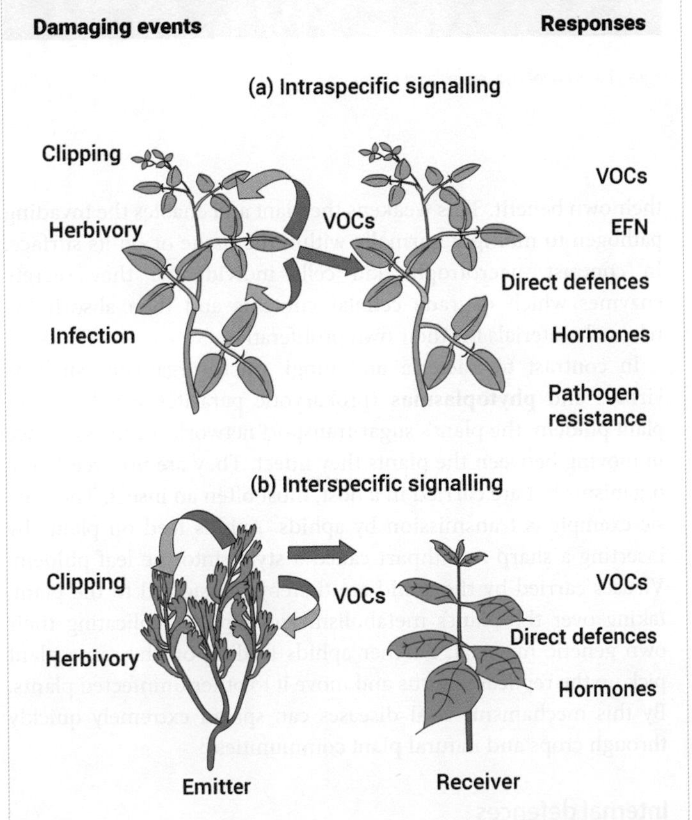

Some plants take the defence response a stage further by attracting insects which are the natural predators of the herbivorous insects they are being attacked by. The attracted insect may lay its own eggs within the bodies of the herbivorous insect.

An alternative type of chemical defence occurs by means of stinging hairs on stems and leaves, such as the stinging nettle (*Urtica dioica*) in Figure 32.9. These structures are a development of trichomes or leaf hairs. They are small, often multicellular structures, with a brittle, sharp, pointed cell at the end; a close-up of which you can see in Figure 32.9b. When this is broken on contact with skin, it exudes a cocktail of molecules which cause irritation and an allergic reaction. This mechanism is highly effective at stopping grazing by large animals.

Stinging nettle trichomes produce oxalic and tartaric acid. The effect of these is relatively transient and benign. More serious stings are inflicted by other members of the nettle family found in Australia, most notably the giant stinging tree (*Dendrocnide excelsa*) and the gympie-gympie (*Dendrocnide moroides*). Leaves of these inflict very serious stings, the pain from which is agonizing and can last days or months. The active molecule, called moroidin, is an octapeptide (a chain of eight amino acids) with an unusual structure containing tryptophan and histidine.

Although trichomes are very common on leaves and stems of the majority of plant species, not many of them cause stinging. In addition to the nettle family (Urticaceae), only plants of the Hydrophyllaceae and Euphorbiaceae families of angiosperms have stinging hairs. This suggests that as a defence strategy its adaptive advantage is limited.

Another mode of defence is for plant tissues to accumulate sharp crystalline structures internally which can cause damage to the mouthparts of potential herbivores. These are called **raphides** and are made of needle-shaped crystals of calcium oxalate, synthesized in cells called **idioblasts**.

Other examples of defensive chemicals include the oxalic acid found in rhubarb leaves and the odorous chemicals produced by herbs and spices. In the case of herbs and spices, man has selected those which are of medicinal or culinary value and developed strains with intense or concentrated characteristics. The plants remain poisonous or obnoxious to other organisms.

Some plants pressurize the idioblast so that the raphides shoot out at the attacker when touched or when eating commences. The tropical plants *Dieffenbachia* spp., commonly grown as houseplants, are also known as 'dumb canes' because eating them expels raphides which cause severe reactions in the mouth and throat and a temporary loss of speech.

Defence against invasion

Plants face a considerable threat from disease-causing microorganisms in the environment. The exterior surfaces of plants

Damaging events   **Responses**

(a) Intraspecific signalling

Clipping

Herbivory   VOCs

Infection

VOCs
EFN
Direct defences
Hormones
Pathogen resistance

(b) Interspecific signalling

Clipping

Herbivory   VOCs

VOCs
Direct defences
Hormones

Emitter   Receiver

Figure 32.8 Plant-to-plant communication in defence against herbivory and other damage.

Source: Reprinted from Heil, M; Karban, R (2010). Explaining evolution of plant communication by airborne signals. *Trends in Ecology & Evolution*, 25, 3137–44. Copyright 2010, with permission from Elsevier.

Figure 32.9 The stinging nettle (*Urtica dioica*). (a) Ungrazed plant in flower. (b) Stinging hairs cover the leaves and stems.

and the surrounding environment, including the soil, contain huge numbers of bacteria and fungi. All of the world's crop plants are constantly challenged by micro-organismal diseases, so plant pathology is a major topic of academic and applied research.

▶ **We learn more about plant pathology in Chapter 14.**

There are two general types of pathological micro-organisms:

- **Biotrophs**: parasitic micro-organisms which live within the plant tissues, exploiting them as a food source but not killing the plant.

- **Necrotrophs**: micro-organisms which kill the plant and feed off of its decaying tissues.

For invasion to occur, the first stage is for the bacteria or fungal cell to gain entry to the internal plant tissues. The surfaces of most plant tissues are relatively well protected, often by a layer of wax, called the cuticle, or by tougher tissues such as bark. Nevertheless, infectious micro-organisms can enter through the stomatal pores in the surface epidermal cell layer of leaves and stems or through wounds caused by breakage of the plant tissues. Figure 32.10 illustrates three fungal infections, on decaying wood (a necrotroph), on a living chrysanthemum leaf (biotroph), and on the skin of an orange (necrotroph). Note that the necrotroph infections may occur after the plant has died and lead to the decay and recycling of material.

The result of an insect grazing on a leaf provides the perfect entry point for such diseases. Once inside the plant the strategies of biotrophs and necrotrophs are different. Biotrophs steal plant nutrients by parasitizing the host's metabolism and use them to

their own benefit. This weakens the plant and enables the invading pathogen to multiply normally within the tissue or on its surface. In contrast, necrotrophs kill cells individually: they secrete enzymes which degrade cellular contents and then absorb the released materials for their own proliferation.

In contrast to bacteria and fungi, micro-organisms such as viruses and **phytoplasmas** (prokaryotic parasites which live in plant phloem, the plant's sugar transport network) need assistance in moving between the plants they infect. They are not free-living organisms but are carried in a host, most often an insect. The classic example is transmission by aphids. Aphids feed on plants by inserting a sharp mouthpart called a **stylet** into the leaf phloem. Viruses carried by the aphid are thereby transferred to the plant, taking over the plant's metabolism and rapidly replicating their own genetic material. Further aphids feeding on the same plant pick up the replicated virus and move it to other uninfected plants. By this mechanism, viral diseases can spread extremely quickly through crops and natural plant communities.

Internal defences

Most plants possess some degree of innate immunity to invading pathogens. The plant recognizes molecules on the surface of bacterial and fungal pathogens, called **pathogen-associated molecular patterns**. Recognition induces the expression of genes in the plant which encode molecules capable of inhibiting pathogen spread.

During the course of evolution, an arms race has developed between host plants and invading pathogens. Successful pathogens

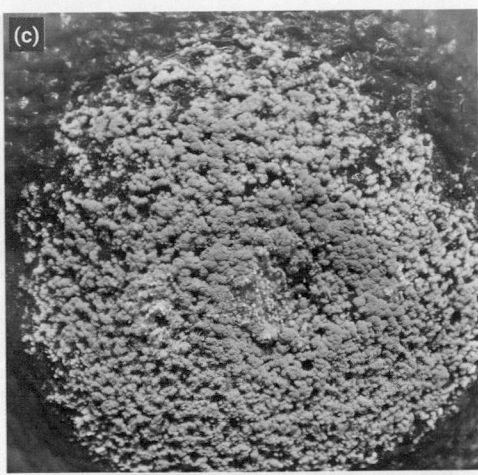

Figure 32.10 Plant diseases. (a) Hairy curtain crust fungus (*Stereum hirsutum*). (b) Chrysanthemum brown rust (*Puccinia chrysanthemi*). (c) Mould on an orange.
Source: Photographs by Kevin Pyke.

have evolved new molecules, termed **elicitors** or **effectors**, which overcome the plant's innate resistance mechanisms. In turn, some plants achieve resistance through genetic mechanisms that recognize the elicitor molecules and disarm them.

This reciprocal adaptation is sometimes referred to as a **gene for gene** process: the pathogen continually evolves to evade the host's recognition system, while the host continually evolves to outmanoeuvre the pathogen. Knowledge of these systems can be of great importance in agriculture where the loss of crops during both growth and post-harvest is economically significant. Crop breeding programmes often exploit and enhance the resistance side of this fine genetic balance.

Wound repair

Plants have a limited capacity to repair themselves after damage. Figure 32.11 shows evidence of the ability of trees to repair themselves. To do this, trees and shrubs can compartmentalize parts of their structure which have been invaded by pathogens. Dead and damaged branches or stems are allowed to break off from the main structure. This is a regular process and there are examples of trees which have survived this way for many hundreds of years. The shedding of pathogen-damaged branches, normally in high winds in winter storms, cleans the tree of dead tissue. It then regenerates new tissues at the growing points on its structure.

After wounding, trees and many other plants produce a barrier tissue which may prevent or reduce further pathogen invasion. This tissue is an outgrowth around the wounded or cut surface and is called a **callus**, which you can see as the bulging regenerated tissue in Figures 32.11a and 32.11c. Callus tissue is easy to observe on old trees where branches have been lost or on fruit trees which have been pruned. It forms a raised ring around the end of the severed stem and may even enclose it completely. In the past, these structures were commonly referred to as burrs. The woody tissue so formed has an oblique or angled grain, as in the sectioning through a burr in Figure 32.11b, and is much sought after by wood workers and craftspeople as it can be highly attractive when cut and polished.

PAUSE AND THINK

When a plant repairs itself after mechanical or biological damage, the cells at the repair site may form stem, leaves, flowers, roots, or other structures depending on the type of plant and the location of the damage. What does this tell us about the ability of plant cells to differentiate? You might like to look at Section 28.9 to help you answer this question

Answer: All plant cells are totipotent: they are, in principle, capable of differentiating from their current phenotype into any other. (Whether they do so depends on their position and, in particular, the formation of meristems.) Most animal cells are unable to do this.

Check your understanding of the concepts covered in this section by answering the questions in the e-book.

32.2 Protection from invasion: animals

As we discuss in Chapter 28, organisms survive by enclosing and protecting a region of biosphere. All animals, whether diploblasts, triploblasts, or complex metazoa, have some form of defensive external membrane or covering. This is called the integument. In this section we consider the structure and function of the integuments of complex animals.

The integument

The integument defines the boundary of an animal's body and is its first line of defence against the outside environment, predators, and invading organisms. It is a dynamic tissue, for it needs to grow and must allow for re-shaping as the animal develops. It must also be capable of self-repair following injury. The integument is usually some kind of flexible but resilient skin, but in some animals it incorporates a stiff protein, while in others it accumulates precipitated mineral from the environment to form a hard structure (shells, exoskeletons, carapaces, and testa, as described in Chapter 29).

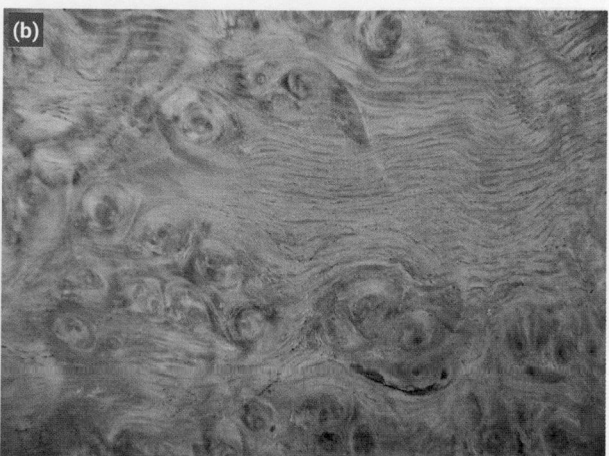

Figure 32.11 Tree burrs. (a) Burrs on the trunk of a mature tree following damage. (b) Sectioning through a burr produces highly distinctive patterning which is much used in veneers and other wood-based decoration. (c) Callus formed by birch stem around fencing wire.

The integument has numerous functions in addition to mechanical defence, so some complex adaptations have evolved. These other functions include:

- sensation;
- resistance to or regulation of the passage of water;
- resistance to or regulation of the passage of gases;
- resistance to the entry of micro-organisms and parasites;
- physical resilience and withstanding abrasion;
- insulation and maintenance of body temperature;
- coloration;
- protection against ultraviolet light;
- signalling and communication;
- reception of light energy;
- anchorage; and
- flight and other forms of movement.

Taking all this into account, we see that the integument is an essential organ with a highly complex physiology. An interesting feature of all animal integuments is the presence of a layer of multipotent stem cells. These are cells which are continually dividing but which can differentiate into a range of functional types. The result is a structure composed of a dynamic mixture of cells, each with specific role, but with the capacity to repair itself after injury.

In order to understand the structure of animal integuments, it is helpful to organize the metazoa (multicellular animals) in the way we have done in other chapters (Chapters 28, 29, and 33):

- Sponges (Porifera) have the least complex body organization, with very limited cellular differentiation and virtually no organ structure.
- Animals more complex than sponges have fully differentiated cells formed into layers:
  - Coelenterates (hydra, cnidaria) have *two* cell layers, one of which forms the environment-facing boundary of the body.
  - All other metazoa have *three* cell layers (germ layers), called endoderm, mesoderm, and ectoderm.
- The three-layered animals are subdivided according to how the gut forms in relation to the invagination of the blastocyst:
  - Protostomes (invagination forms the mouth): worms, molluscs, and arthropods.
  - Deuterostomes (invagination forms the anus): echinoderms, tunicates, and vertebrates (fish, amphibians, reptiles, birds, mammals).

Porifera

In a typical sponge, cells take on different tasks, but they all retain the ability to form a new individual sponge if separated from the whole. This developmental plasticity has implications for the way in which a sponge recovers from mechanical injury.

Figure 32.12 shows the general organization of sponge tissue, including how the functionally distinct cells are arranged in relation to the external environment and the flow of water.

The pinacocytes are identified as exo- or endo- according to whether they face outwards or inwards. Porocytes delineate the rim of the pores by which water enters the structure, drawn in by the wave movements of flagellae on choanocytes.

Some sponges have structural proteins (collagen) and hard materials (spicules) embedded in the structure. However, there is no basal lamina beneath the cells and cell–cell adhesions have limited functionality. Overall, the organization of a sponge's environment-facing tissue is rather simple compared with that of the integuments of multilayered animals.

Coelenterates

Figure 32.13 shows how the cnidaria and other ctenophora have bodies composed of two monolayers of cells, entoderm (or endoderm) and ectoderm, with a non-cellular mesogleal layer between them. The mesoglea contains contractile elements (myofibrils) and fibres of connective tissue which provide the animal with movement and strength.

The ectoderm contains sensory cells, which detect mechanical movements and water currents, nettle cells (nematocysts) and cnidoblasts, which react to stimulation by ejecting barbed or venomous coiled threads, and interstitial cells which provide support. Each of these cell types differentiates from pluripotent primary germ cells within the ectoderm layer.

Worms (helminthes)

The platyhelminthes (flatworms), annelids (round worms and polychaetes), and nematodes are among the simplest of the animals, with three germ cell layers. Each family has its own characteristic integumentary features, but they all have epithelia which are particularly interactive with the environment.

In round worms and nematodes, the ectodermal cells typically form an epidermis, with cytoplasmic extensions or microvilli reaching outwards through a protective structural layer, as illustrated in Figure 32.14. On the external surface there is a secreted layer of slimy carbohydrate called the glycocalyx.

In parasitic flatworms the extended cells are derived from the mesoderm, rather than ectoderm, and there is a basement membrane between the two regions. The external layer is biochemically active and absorptive, and may react immunologically with the host tissues. It frequently contains spines which allow the animal to anchor itself in the host tissue. Free-living worms have a layered cuticle. This is physically resilient and also biochemically active.

In both groups, the external surface is the site of gas exchange and has an important osmoregulatory role (Chapter 30). Embedded in the epidermis are layers of muscular and collagen fibres which facilitate movement and support the body's hydrostatic pressure. Polychaetes and other free-living worms have locomotory appendages protruding from the surface (Chapters 29 and 31).

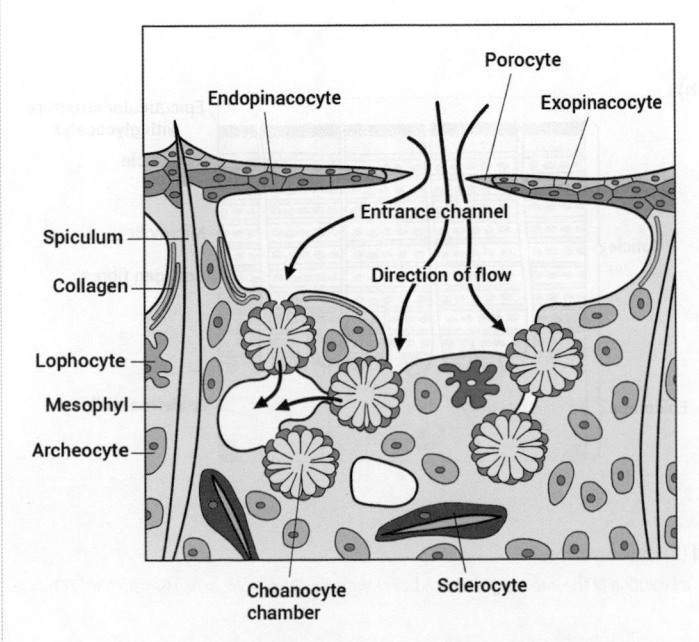

Figure 32.12 Organization of the environment-facing cells of a sponge.

Source: Adapted from Schempp C. et al. (2009) Dermatology in the Darwin anniversary. Part 1: Evolution of the integument. *Journal der Deutschen Dermatologischen Gesellschaft*, 7(9), 750–757. John Wiley and Sons.

Figure 32.13 Organization of the ectoderm and entoderm (endoderm) of *Hydra*, a cnidarian.

Source: Adapted from Schempp C. et al. (2009) Dermatology in the Darwin anniversary. Part 1: Evolution of the integument. *Journal der Deutschen Dermatologischen Gesellschaft*, 7(9), 750–757. John Wiley and Sons.

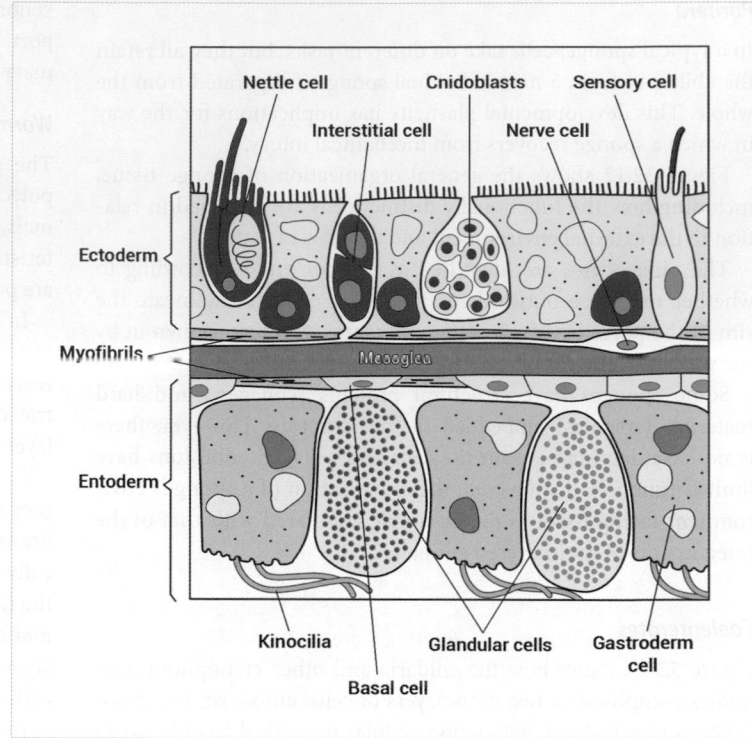

Arthropods

The integument of insects and other arthropods is made of an epidermal cell layer beneath a layered cuticle, as shown in Figure 32.15. The cuticle forms a hard exoskeleton out of a polysaccharide called chitin. Chitin is a polymer of *N*-acetylglucosamine, derived from glucose. (Besides forming the arthropod exoskeleton, chitin is a key structural component of fungi.)

The chitin layers are secreted by the epidermis and remain attached to it by hemidesmosomes (Chapter 28). The exoskeleton cannot expand along the plane of the body wall and will not stretch outwards, so the animal can only grow by moulting: splitting and sloughing off the exoskeleton and laying down a new one. The pupal case of metamorphic insects is made of chitin and the imago emerges by splitting it open.

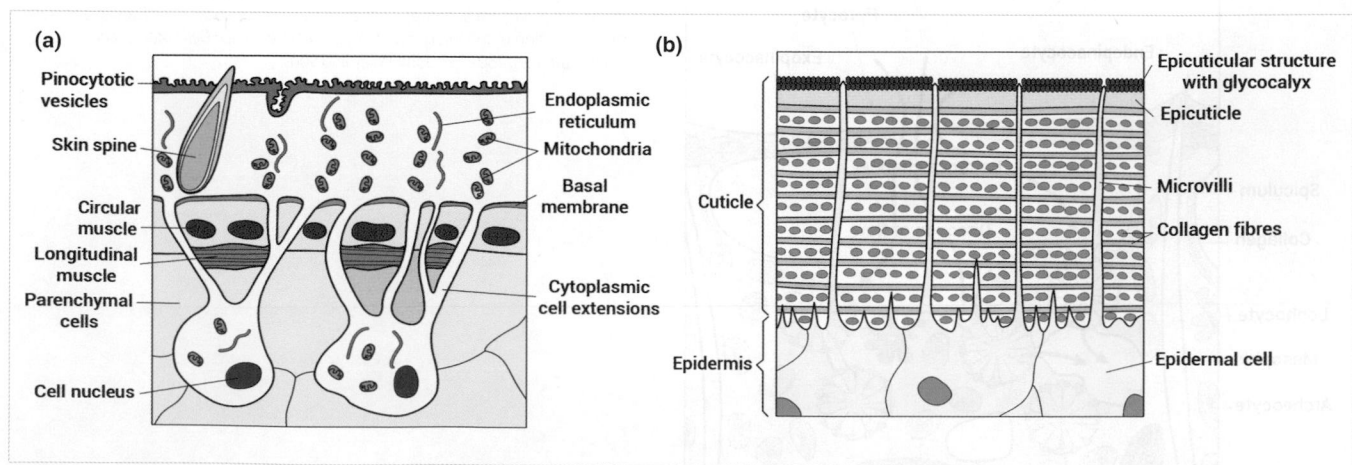

Figure 32.14 Organization of the epithelium of (a) parasitic flat worms and (b) roundworms.

Source: Adapted from Schempp C. et al. (2009) Dermatology in the Darwin anniversary. Part 1: Evolution of the integument. *Journal der Deutschen Dermatologischen Gesellschaft*, 7(9), 750–757. John Wiley and Sons.

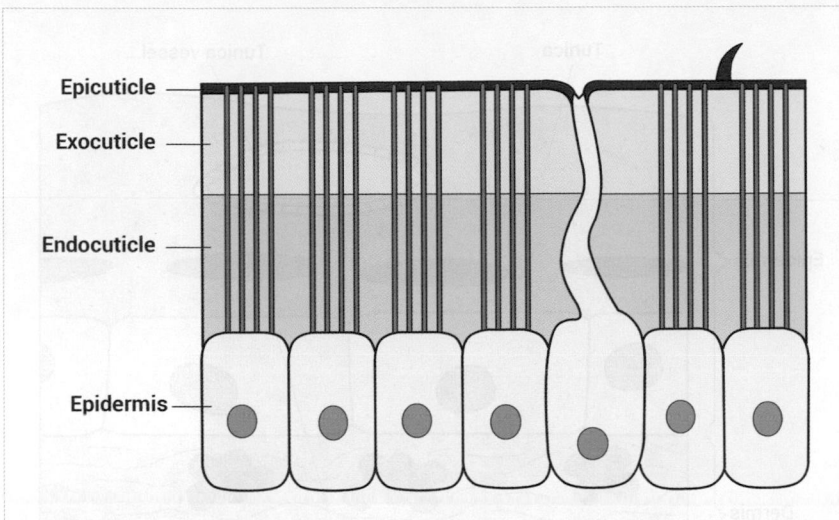

Figure 32.15 Organization of the epithelium of insects and other arthropods.

In insects such as beetles, the chitinous layer is particularly tough and resembles a shell, especially over the abdomen and wing cases. The integument is impenetrable to gases. Terrestrial arthropods have breathing tubes called spiracles along the abdomen which allow air to penetrate the body.

The outermost layer, called the epicuticle, contains little chitin but is rich in lipids, waxes, polyphenols, and denatured proteins, forming an essentially waterproof barrier. Some of these chemicals give the animal colour, while others may be distasteful to predators. Some insects have fine molecular gradations within the layer that make the surface iridescent.

The integument of crustacean arthropods (crabs and related animals) has a similar basic structure but there is a minimal amount of chitin and an accumulation of calcium carbonate (Chapter 29). The structural defence created by this hard material is often enhanced by spines and sharp ridges.

Echinoderms

The echinoderm integument is made of a single layer of epidermal cells resting on a basal matrix which includes a basement membrane, as illustrated in Figure 32.16.

The integument grows as the cells divide, increasing the area but not the thickness of the epidermis. The cells also differentiate into a range of supporting, sensory, secretory, phagocytic (ingesting), and pigment types. In many species there are also sensory and protective spine cells, usually aligned in radially symmetrical ridges over the body and limbs.

In starfish, sea urchins, and related animals, further mechanical protection comes from a hard shell made of calcium carbonate. This is strictly a **test** rather than a shell because it forms in the mesoderm, below the basement membrane, rather than the ectoderm.

Tunicates

Tunicates are non-vertebrate marine chordates. They have a unique form of integument which incorporates a cellulose-like material called tunicin. Cellulose is the material that forms the walls of plant cells and it is not otherwise found in animals. It may have emerged in tunicates as a result of gene transfer from a symbiotic plant cell such an alga.

The tunicate epithelium is a single layer of epidermal cells supported on a dermis, as shown in Figure 32.17. The epidermal cells

Figure 32.16 Organization of the epidermal monolayer of echinoderms.

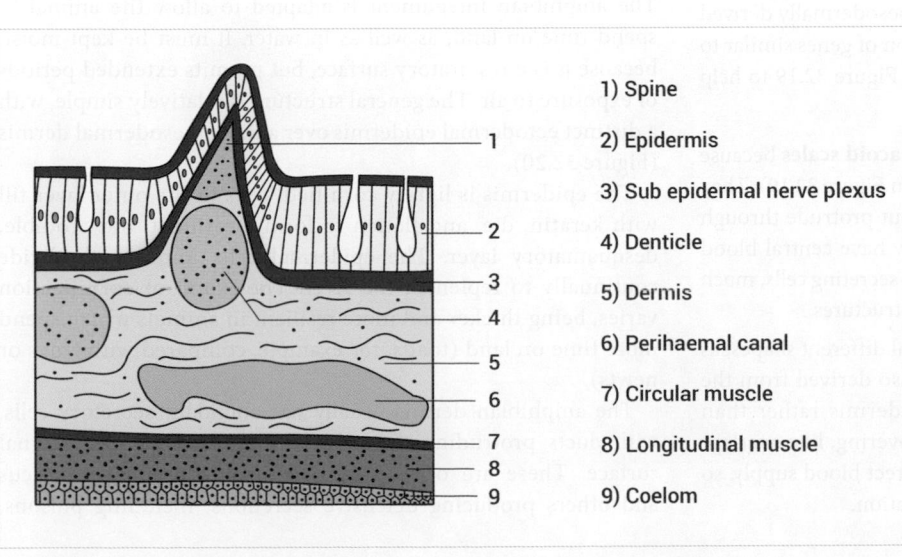

1) Spine

2) Epidermis

3) Sub epidermal nerve plexus

4) Denticle

5) Dermis

6) Perihaemal canal

7) Circular muscle

8) Longitudinal muscle

9) Coelom

Figure 32.17 Organization of the tunicate epithelium, including the cellulose-rich tunica.

Source: Adapted from Schempp C. et al. (2009) Dermatology in the Darwin anniversary. Part 1: Evolution of the integument. *Journal der Deutschen Dermatologischen Gesellschaft*, 7(9), 750–757. John Wiley and Sons.

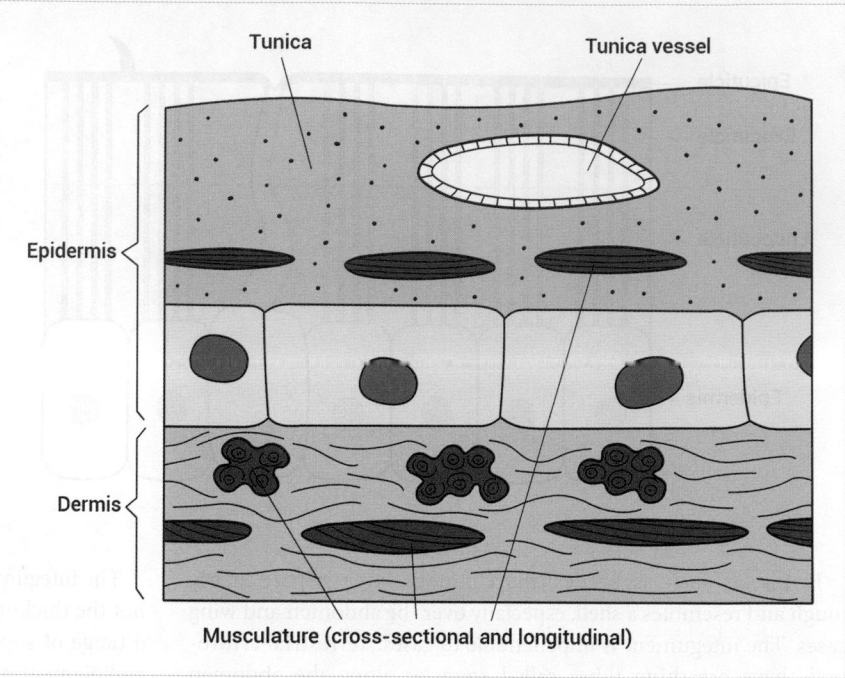

secrete tunicin, forming a resilient, but highly flexible, translucent outer coat with the texture of a thin plastic sheet. The epidermis and the dermis contain contractile muscular fibres.

Apart from the cellulose, the tunicate integument resembles that of other chordates in having a distinct dermis and epidermis. However, the epidermis is non-desquamatory: it does not continually fall away and is not continually replenished by further layers from underneath as it is in vertebrates.

Fish

Fish integuments, like those of all other vertebrates, have a mesodermally derived dermis and a multi-layered, epidermally derived external layer, separated by a basement membrane; this structure is shown in Figure 32.18. The epidermal cells divide slowly but continually, replacing those lost from the external surface.

The important protective features of fish integuments are the scales which cover the epidermis. These are mesodermally derived and their production results from the activation of genes similar to those of teeth and hair in mammals. Look at Figure 32.19 to help you here.

- In cartilaginous fish, the scales are called **placoid scales** because of their overall plate-like shape (the top line in Figure 32.19). They originate within the dermal corium layer but protrude through the epidermis to the external surface. They have central blood capillaries and a layer of dentin- and enamel-secreting cells, much like teeth, and are thus metabolically active structures.

- The scales of bony fish (teleosts) take several different shapes, as shown in Figure 32.19. These scales are also derived from the dermis but push up from under the epidermis rather than through it. They form a stacked, layered covering, like overlapping tiles on a house roof. They have no direct blood supply, so they are not metabolically active after formation.

The principal structural material in scales of both groups of fish is keratin, the protein that also makes the hair, nails, and claws of mammals and birds.

Within the epithelial layers are mucus-secreting cells and pigment cells (melanophores). If you have ever picked up a fish, you will know that the scales are covered in a layer of mucus making the animal feel 'slimy'. The mucus provides additional defence against pathogens and mechanical injury.

Fish can increase mucus production in response to stress and disease. Some, such as the hagfish, excrete copious amounts of mucus which aids in defence from predators (which is discussed further in Scientific Process 32.1). Parrotfishes secrete mucus cocoons around themselves at night, protecting them from predators and parasites.

Amphibians

The amphibian integument is adapted to allow the animal to spend time on land, as well as in water. It must be kept moist, because it is a respiratory surface, but permits extended periods of exposure to air. The general structure is relatively simple, with a distinct ectodermal epidermis over a thick mesodermal dermis (Figure 32.20).

The epidermis is lightly cornified: cells in the outer layer fill with keratin, die, and flatten to form a resilient, but erodible, desquamatory layer. The epidermal cells underneath divide continually to replenish the layer. The extent of cornification varies, being thicker and more resilient in animals which spend more time on land (toads, for example, compared with frogs or newts).

The amphibian dermis usually has abundant secretory cells, with ducts protruding through the epidermis to the external surface. These are of various types, some producing mucus and others producing defensive secretions, including poisons,

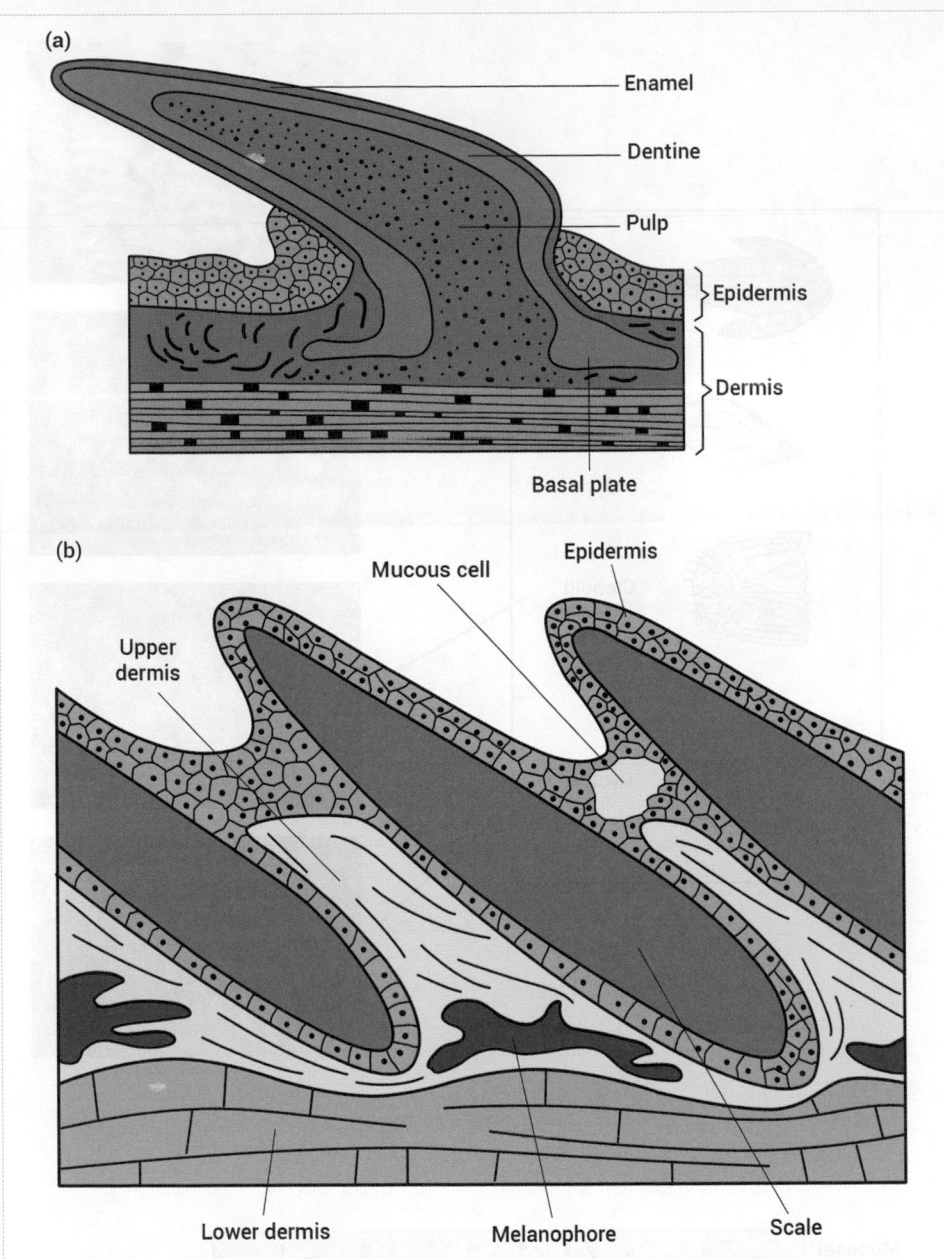

Figure 32.18 Organization of the integument. Organization of the integument in (a) cartilaginous and (b) bony fish, including dermal and epidermal layers and scales.

analgesics, and other pharmacologically active substances. Dispersed between the secretory cells are structural fibres (collagen, muscle) and blood capillaries, and also immune cells, nerve cells, and pigment cells.

Reptiles

Reptiles have a highly cornified, horny outer epidermis, with several layers of dead, flattened cells which can be completely impermeable to water. This has allowed many reptile groups to adopt a completely terrestrial lifestyle, with some species occupying the hottest and driest of habitats. The dermis is correspondingly free of secretory cells (Figure 32.21).

The cornified epidermis typically forms defensive, flat, backward-pointing scales called squamous scales, as seen, for example, in snakes and other non-limbed reptiles. Some snakes and quadruped reptiles, including crocodiles and alligators, develop an additional layer of bony scales, called osteoderms, within the dermal corium. These increase the resistance and impenetrability of the integument but still permit limited flexibility. However, the animal can only grow by periodically shedding its skin.

Turtles and tortoises have no epidermal scales but their dermal scales develop into thick, wide plates called scutes, which become cornified on the outer surface. These merge to form a robust, keratin shell which almost completely encloses the body and makes the animal virtually impregnable to carnivorous predators. The shell is inflexible, but it can grow continuously by division of the innermost layer of dermal cells and the gradual expansion of the merged plates.

Figure 32.19 Four types of fish scale. Placoid scales are found in the Chondrichthyes (cartilaginous fish). The other three types are found in the Osteichthyes (bony fish).

Source: (placoid) Pascal Deynat/Odontobase/Wikimedia Commons/CC BY-SA 3.0; (ganoid) Alessandro Mancini/Alamy Stock Photo; (ctenoid) Steve Lowry/Science Photo Library; (cycloid) Nosyrevy/Shutterstock.com.

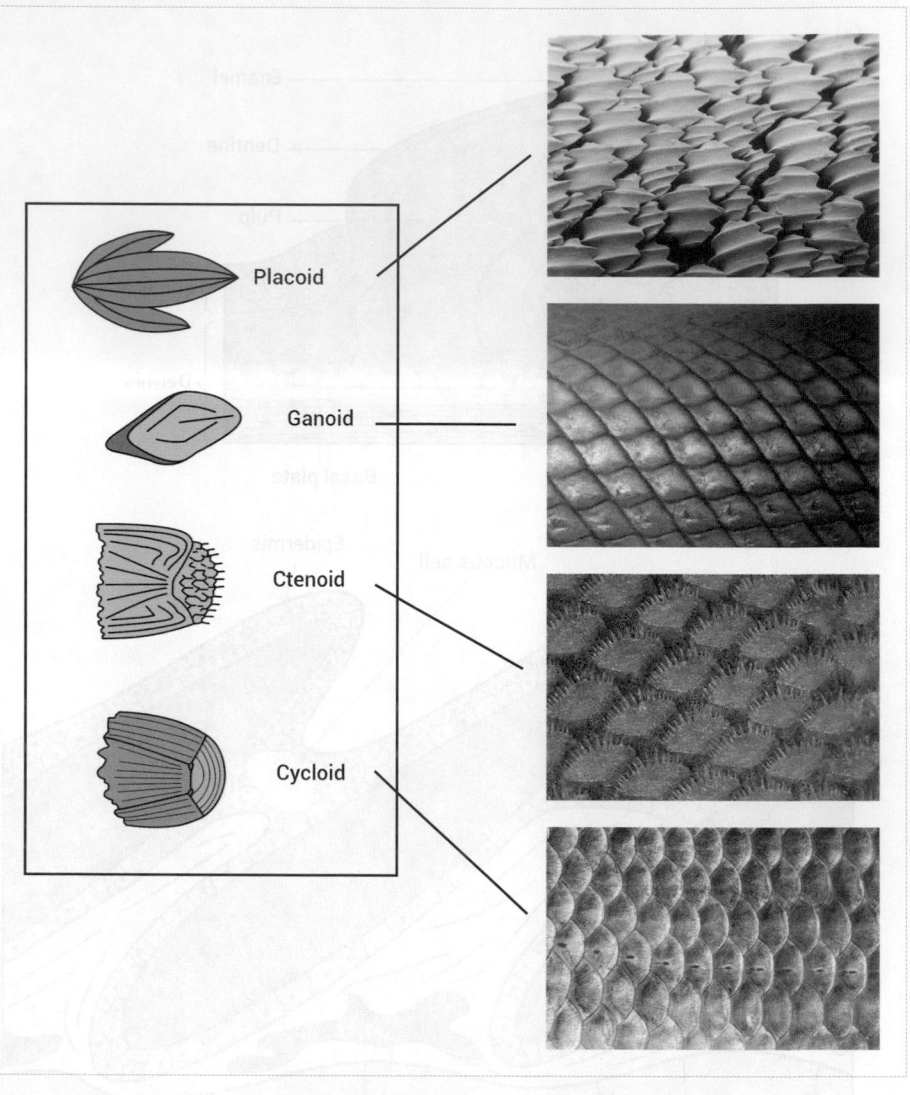

Figure 32.20 Organization of the amphibian (toad) integument including cornified epidermal and secretory dermal layers.

Source: Varga JFA, Bui-Marinos MP, Katzenback BA (2019) Frog skin innate immune defences: sensing and surviving pathogens. *Front Immunol*, 9, 3128. doi: 10.3389/fimmu.2018.03128/CC BY 4.0/https://creativecommons.org/licenses/by/4.0/.

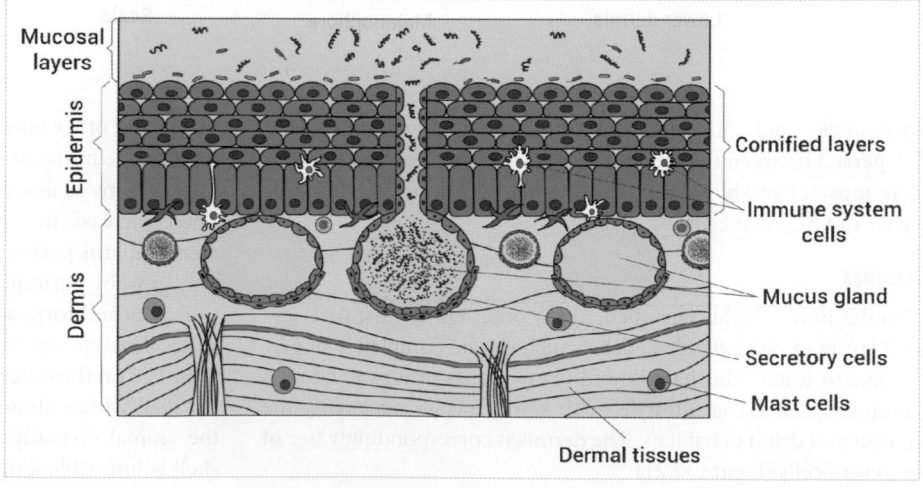

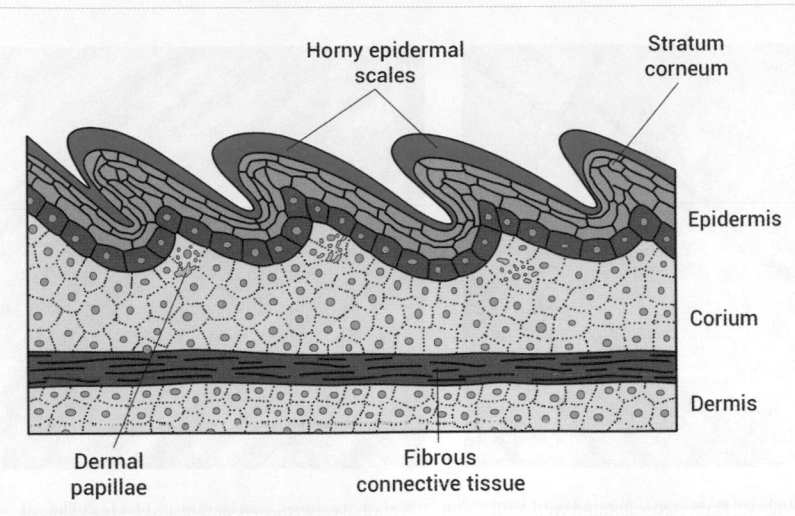

Figure 32.21 Two types of reptilian integument, showing epidermal squamous scales and dermal bony scales.

Labels: Horny epidermal scales · Stratum corneum · Epidermis · Corium · Dermis · Dermal papillae · Fibrous connective tissue

---

SCIENTIFIC PROCESS 32.1 — Hagfish slime

Research question

Why do hagfish produce slime?

Background

Hagfishes (family Myxinidae) are jawless fishes that feed on dead or dying animals. You can see a photograph of the Pacific hagfish in Figure 1. The way hagfish feed is that they use their set of rasping teeth which allow them to tear off pieces of flesh. Not much is known about their biology as they tend to live in deeper parts of the ocean.

Hagfish possess 90 to 200 slime glands, arranged in lines running down either side of their body. Hagfish slime is different from the mucus secreted by other fish species: it dilutes in sea water three times more than other fish mucus and has elastic properties. They produce a large quantity of slime, as demonstrated in Figure 2.

Hypothesis

Hagfish produce slime as a defence mechanism.

Methods

In 2011, Zintern and colleagues used baited remote underwater stereo-video (BRUVS) to film hagfish behaviour in the deep sea.

Results

Their video footage of a variety of predators attempting to bite a hagfish shows that hagfish can secrete slime as a defence mechanism against predators. It is also clear that when the hagfish is bitten, its mucus irritates the mouth and gills of its predator, which then releases the hagfish.

Read the original work

Zintzen, V., Roberts, C.D., Anderson, M.J., Stewart, A.L., Struthers, C.D., Harvey, E.S. (2011). Hagfish predatory behaviour and slime defence mechanism. *Scientific Reports*, **1**, 131.

Figure 1 Pacific hagfish.
Source: Brandon Cole Marine Photography/Alamy Stock Photo.

Figure 2 Pacific hagfish showing the amount of slime they reproduce.
Source: Photograph by Andra Zommers.

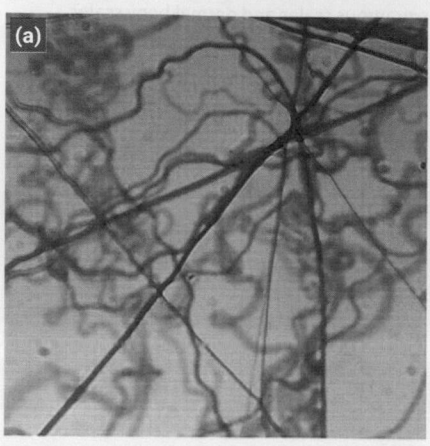 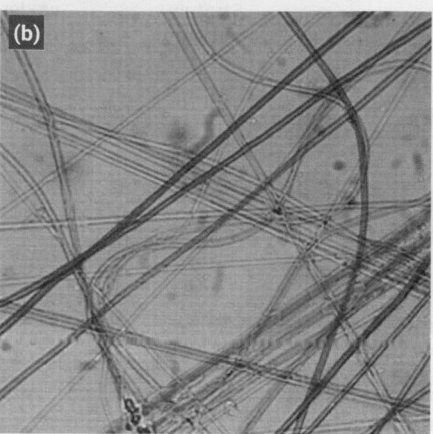 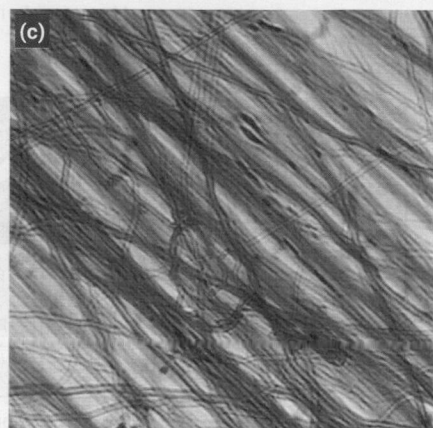

Figure 3 Images of slime threads. Threads (a) from unperturbed slime; (b) from slime agitated with a rule after collection (i.e. perturbed slime); and (c) bundling of slime threads in perturbed slime. Scale bar = 10 μm.

Source: Fudge, D.S., Levy, N., Chiu, S. and Gosline, J.M. (2005). Composition, morphology and mechanics of hagfish slime. *Journal of Experimental Biology*, 208, 4613–4625. https://jeb. biologists.org/content/208/24/4613.

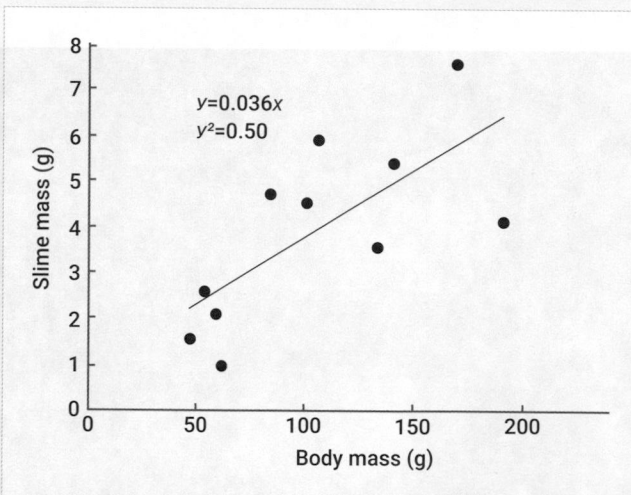

$y = 0.036x$

$y^2 = 0.50$

Figure 4 Relationship between body mass and slime mass of hagfish.

Source: Fudge, D.S., Levy, N., Chiu, S. and Gosline, J.M. (2005). Composition, morphology and mechanics of hagfish slime. *Journal of Experimental Biology*, 208, 4613–4625. https:// jeb.biologists.org/content/208/24/4613.

Research question

Which components of the slime give it its physical characteristics?

Hypothesis

The properties of the slime depend on its mucin and protein constituents.

Methods

In 2017, Fudge and colleagues from the University of British Columbia carried out a variety of chemical and physical analyses on hagfish slime to determine its properties.

Results

Slime was 99.996% sea water, 0.0015% mucin, and 0.002% threads. This concentration of mucin is very *low* compared with the amounts in other types of mucus. Individual slime threads were soft and elastic when subjected to low strains but became strong and extensible under high strains. Overall, protein threads (as shown in Figure 3) dominate the mechanics of hagfish slime and provide elasticity. The mucins give additional viscosity.

Further observations

By stimulating the slime glands of anaesthetized hagfish, Fudge and colleagues found a linear relationship between the size of the hagfish and the amount that it could produce, as demonstrated in Figure 4. They estimated that the amount of slime a hagfish stores is equivalent to about 3–4% of its body mass.

Conclusions

- Hagfish produce slime as a defence against predation.
- They produce more as they grow.
- The extreme elastic properties of the slime are due mostly to low concentrations of mucins rather than proteins.

Read the original work

Fudge, D.S., Levy, N., Chiu, S. and Gosline, J.M. (2005). Composition, morphology and mechanics of hagfish slime. *Journal of Experimental Biology*, **208**, 4613–4625.

Birds

Birds have a relatively simple epidermis and dermis but develop feathers as a unique type of integumentary appendage. The genes that allow birds to form feathers have been identified and it is likely that these first appeared in the common ancestor of birds and the alligators, which are their nearest reptile cousins. Presumably the ancestor did not have feathers, so the gene may have had another function which became adapted for feather production.

The bird epidermis is made of a few layers of cells, with the outer ones becoming cornified and flaky. It has muscle and structural fibres but, over most of the body, no secretory or sweat glands. A feather forms as a junctional structure between the dermis and epidermis (Figure 32.22). The dermis develops a pulpy, tubular papilla which pushes the epidermis up to form a sheathed barb. The barb keratinizes and pushes out from the sheath, leaving a follicle as its point of insertion within the skin.

The feather is made entirely of keratin but may incorporate small quantities of pigments. Epidermal cells at the base of the follicle remain metabolically active. They allow the feathers to grow but they also release them periodically through the year at times of moult.

Although avian skin is essentially non-secretory, birds have oil-secreting glands. These are typically paired structures positioned beneath the tail feathers. Birds groom themselves continually while at rest, using the beak to spread oil over the feathers. This is essential for the removal of dirt particles and small parasites. The oils also have a waterproofing effect on the fine elements of the feather.

Mammals

Like birds, mammals are endothermic and have an integumentary covering which assists in thermoregulation, among other important functions. The mammalian covering is hair rather than feathers, and the presence of this material over at least some parts of the body surface is one of the diagnostic features of the group. In addition, the integument includes a subcutaneous layer of fat with insulatory properties, as shown in Figure 32.23.

Figure 32.23 also illustrated how mammalian skin has a dermis and a continually replicating, desquamatory epidermis (in keeping with most other vertebrates). The thicknesses of each of these layers varies considerably between mammalian groups and among species, as does the thickness of the fat layer.

In small, heavily fur-covered mammals (rodents, insectivores, small primates), the outer layers of the epidermis are lightly keratinized. In some of the very large terrestrial mammals (elephant, rhinoceros), the hair covering is minimal over most of the body surface, but there is heavy keratinization and the skin has a leathery texture.

There is no essential biochemical difference between hair, fur, and wool, even though these terms tend to be used in an

Figure 32.22 Stages in the development of a bird's feather, showing the origin of the structure at the junction of the epidermis and dermis (corium).

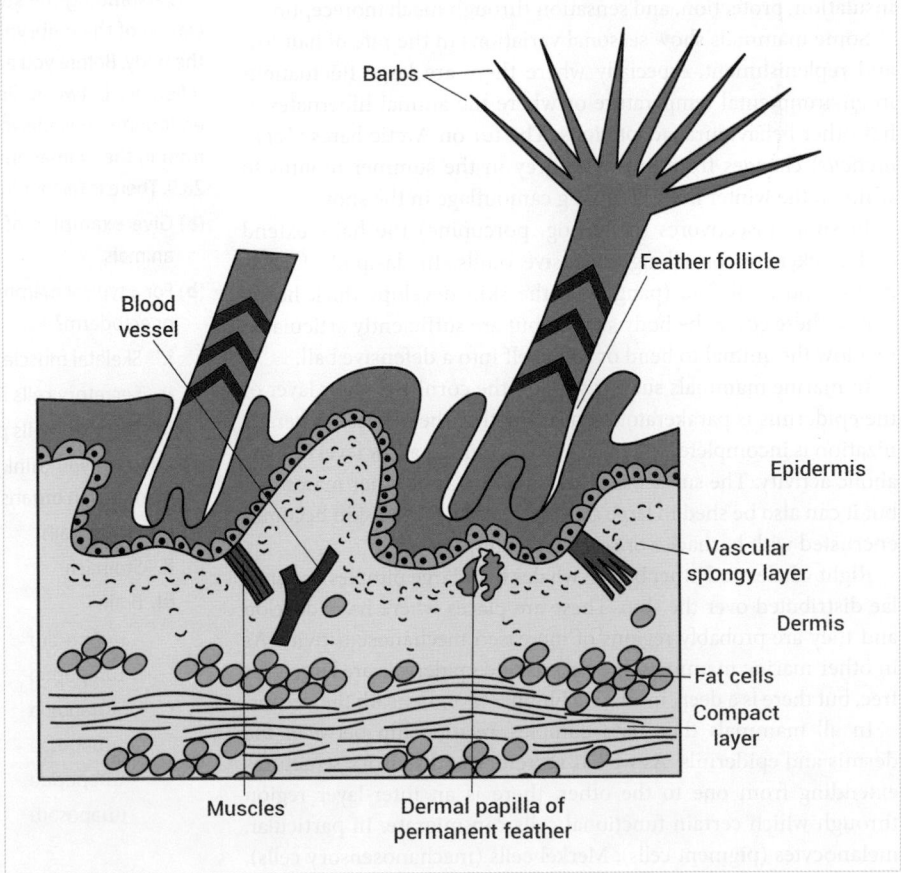

Figure 32.23 Mammalian skin structure and hair.

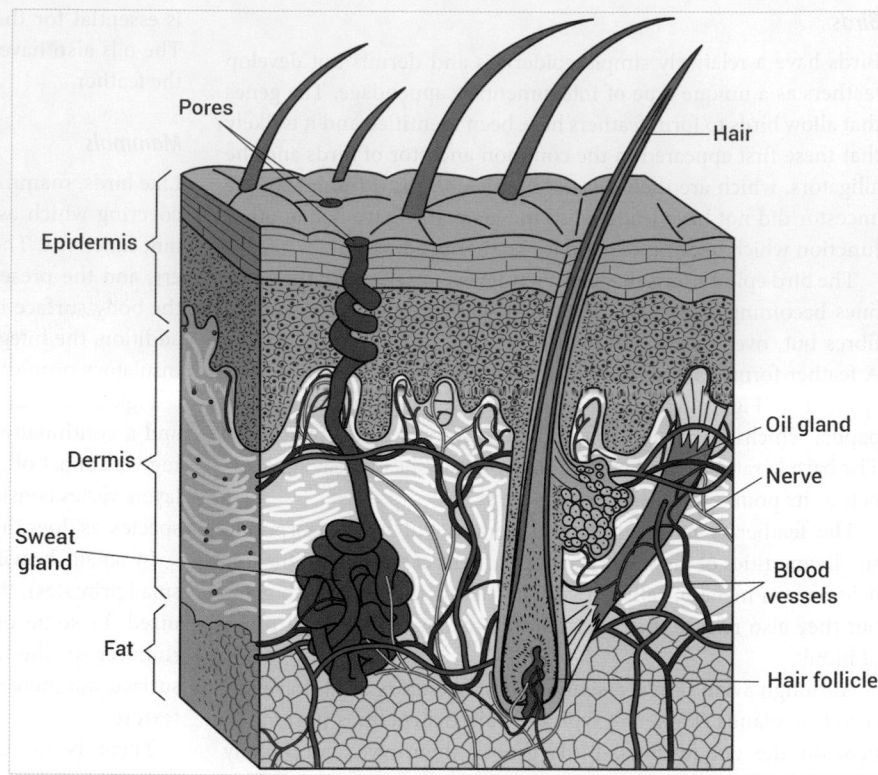

Pores

Hair

Epidermis

Oil gland

Nerve

Dermis

Sweat gland

Blood vessels

Fat

Hair follicle

animal-specific context. However, the hairs in an animal pelt vary in length and thickness, and in the extent to which they provide insulation, protection, and sensation through mechanoreception.

Some mammals show seasonal variations in the rate of hair loss and replenishment, especially where there are large fluctuations in environmental temperature or where the animal hibernates or has other behavioural adaptations. The fur on Arctic hares (*Lepus arcticus*) changes from brown or grey in the summer months to white in the winter months, giving camouflage in the snow.

In some insectivores (hedgehog, porcupine) the hairs extend and thicken to form long, defensive quills. In dasipoda (armadillos) and pholidota (pangolins) the skin develops thick horny plates; these cover the body surface but are sufficiently articulated to allow the animal to bend or roll itself into a defensive ball.

In marine mammals such as whales, the cornified outer layer of the epidermis is parakeratotic, meaning that the process of keratinization is incomplete, with some cells retaining a low level of metabolic activity. The surface is desquamatory, as in other mammals, but it can also be shed in large sheets, especially if the skin becomes encrusted with barnacles or small parasites.

Right whales and other baleen whales have large pimples or papillae distributed over the skin. These are places where hairs develop and they are probably regions of increased mechanosensitivity. As in other marine mammals, the dermis and epidermis are largely fat free, but there is a deep, insulating blubber layer beneath the dermis.

In all mammals there is a complex relationship between the dermis and epidermis. As well as secretory and sensory structures extending from one to the other, there is an inter-layer region through which certain functional cells can migrate. In particular, melanocytes (pigment cells), Merkel cells (mechanosensory cells), and Langerhans cells (immune cells) can pass from the dermis into the lowest layers of the epidermis.

PAUSE AND THINK

Understanding the structure of animal integuments requires an appreciation of the embryological origins of the layers of cells which make up the body. Before you answer the questions, make sure you understand the difference between diploblasts and triploblasts, as well as the meaning of endoderm, mesoderm, and ectoderm, and where these cell layers come from in the animal embryo. (You can find information on this in Section 28.9. There is further information on embryo development in Chapter 33.)

(a) Give examples of two diploblastic animals and two triploblastic animals.

(b) For a typical mammal, are the following tissues endoderm, mesoderm, or ectoderm?

  i. Skeletal muscle

  ii. Secretory cells lining the small intestine

  iii. Secretory cells producing sweat

(c) What do you think is the embryological origin of each the following mammalian organs?

  i. Finger nails

  ii. Stomach

  iii. Brain

Answers:

(a) Diploblasts: hydra, jellyfish
Triploblasts: honey bee, pangolin

(b) i. Skeletal muscle — mesoderm
ii. Secretory cells lining the small intestine — endoderm
iii. Secretory cells producing sweat — ectoderm

(c) i. Finger nails — ectoderm
ii. Stomach — endoderm
iii. Brain — mesoderm

The extended integument

Across the animal kingdom, there are examples of organisms that enhance their defences by exploiting resources available from the environment. Humans are especially adept at this: we use minerals, fossil products, plant fibres, and animal tissues to construct elaborate and complex buildings and to create equally innovative and resilient clothing. In fact, most humans spend virtually their entire lives covered in and enclosed by protective materials.

This familiar feature of human behaviour represents the creation of functional extensions to our integument. We find it difficult or impossible to rely on skin and hair alone for defence against high and low temperatures, wind, water, humidity, pressure changes, hard surfaces, sunlight, noise, predatory animals, abrasive plants, toxic chemicals, and pathogenic micro-organisms. We supplement the protection provided by skin in almost every way imaginable, and we depend absolutely on invented integuments when venturing away from Earth and into space.

It is not difficult to find examples of animals which do something similar:

- Hermit crabs live permanently in the discarded shells of gastropod molluscs, vacating them only briefly to jump into larger ones as they grow.
- Several different crab species, known as decorator crabs, add items from their environment to their carapace to provide camouflage. Attached objects include pieces of algae, bits of shell or sediment, and even other organisms. They may also enhance their defence against predation by adding toxic sessile animals to their integument. An example of a decorator crab which has covered its carapace with a stinging hydrozoan is shown in Figure 32.24.
- Several species of soft-bodied polychaete worms construct tubes to protect themselves from predators and harsh environmental conditions, and to provide support for their feeding appendages. For example, see the trumpet worm inside a tube constructed of sand grains and shell fragments in Figure 32.25a, the alvinella emerging from a tube in Figure 32.25b, and the Christmas tree worm both out of its tube, and hidden in coral tubes in Figure 32.25c and d.
- The anemonefish (*Amphiprion ocellaris*) found in warm water coral reefs lives among the protective tentacles of the purple anemone (*Heteractis magnifica*).
- Ants, termites, wasps, and other hymenopteran insects build complex, extended dwellings from soil, leaves, and their own secretions in which to create complex social structures, breed, and rear their young.
- Blind mole rats (*Spalax ehrenbergi*) spend their entire lives in underground tunnels in similarly complex communities.
- Most birds build nests, some of them of remarkable architectural form and strength, not to inhabit themselves, but in which to incubate eggs and nurture newly hatched young.
- Many reptiles take a similar approach, building nest-like mounds for their offspring in which they can regulate temperature and humidity.

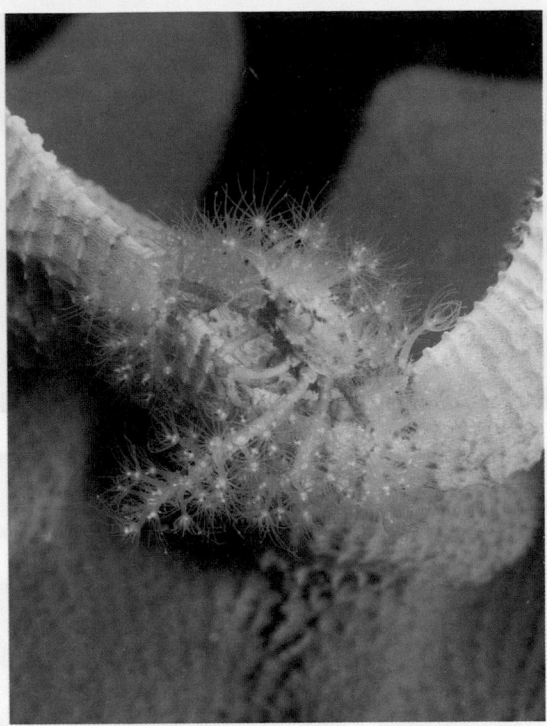

Figure 32.24 A decorator crab which has covered its carapace with a stinging hydrozoan.

Source: Nhobgood Nick Hobgood/Wikimedia Commons/CC BY-SA 3.0.

- Chimpanzees and some other primates use leaves and branches to create sleeping places among the limbs of trees.

In all these cases, the resulting protection exceeds that which the integument alone can provide.

Check your understanding of the concepts covered in this section by answering the questions in the e-book.

32.3 The response to injury

The capacity for self-repair after injury is a common feature of animal integuments. This comes partly from the continual division of epidermal cells and the replacement of desquamated or abraded outer layers. However, that process is relatively slow and only permits repairs to minor, superficial injuries.

Significant mechanical damage can only be repaired by structurally reshaping the affected area and replacing cells with specialized phenotypes. This is called **tissue remodelling**. It involves clearing away damaged cells and proteins, and then replacing tissue to regain shape and functionality.

Damage is also likely to compromise the integument's role as a protective barrier against infection, so the repair process has evolved to include automatic stimulation of the animal's immune system.

We can divide the response to injury into four phases:

- Inflammation.
- Haemostasis.

Figure 32.25 Polychaete worm tubes. (a) *Lagis koreni* (trumpet worm) inside a tube composed of sand grains and shell fragments, below *L. koreni* removed from tube. (b) *Alvinella* emerging from tube. (c) *Spirobranchus giganteus* (Christmas tree worm). (d) *Spirobranchus giganteus* on coral.
Source: (a) © Hans Hillewaert/Wikimedia Commons/CC BY-SA 4.0; (b) Frank Starmer/CC BY-SA 2.0; (c) Ifremer/Phare; (d) Photo by L. Gunton.

- Wound repair.
- Tissue remodelling.

Unsurprisingly, we understand most about these processes in humans, but we can assume that they are broadly similar in other vertebrates, even if exact mechanisms vary.

Inflammation and haemostasis are rapid, automatic responses to injury which precede, but also initiate, the repair process itself. Inflammation also initiates and boosts the **immune response** (see Chapter 27). This has two principal components:

- The **innate** component comes into play whenever the animal is invaded by a foreign organism or is exposed to an **antigen** associated with a cell, protein, or other macromolecule which is not part of itself ('non-self'). The response is rapid but not specific to the identity of the antigen.

- The **adaptive** component responds to foreign antigens in a slower but specific manner and the response can be graded according to the magnitude of the invasion. In addition, the adaptive system usually confers future protection on the animal by leaving a memory of the antigen, called **acquired immunity**; this means that the response to a subsequent infection is more rapid and effective.

Inflammation and subsequent events provoke the innate immune system into action, and they also create conditions in blood and tissues that are conducive to an effective adaptive response.

Figure 32.26 places each of the phases of the response to injury on a timeline. The response begins almost immediately after damage has occurred but continues over an extended period. Its actual duration depends on the nature and extent of the injury, but it can last for years, with a great deal of variation between individuals.

Go to the e-book to explore an interactive version of Figure 32.26.

Inflammation

Inflammation is a familiar process, but it has some distinct and well-defined characteristics. It is defined as a local, reflex response to damage or infection. It is *local* because it is normally confined to the site of injury and the area immediately surrounding it, rather than the whole body. It is a *reflex* because it happens automatically and uncontrollably.

Inflammation is normally beneficial because it initiates wound repair and prepares the immune response. Sometimes, inflammatory reactions occur inappropriately and lead to discomfort and disease. The reflex nature of inflammation also means that its intensity is not always proportionate to the severity of the initiating cause.

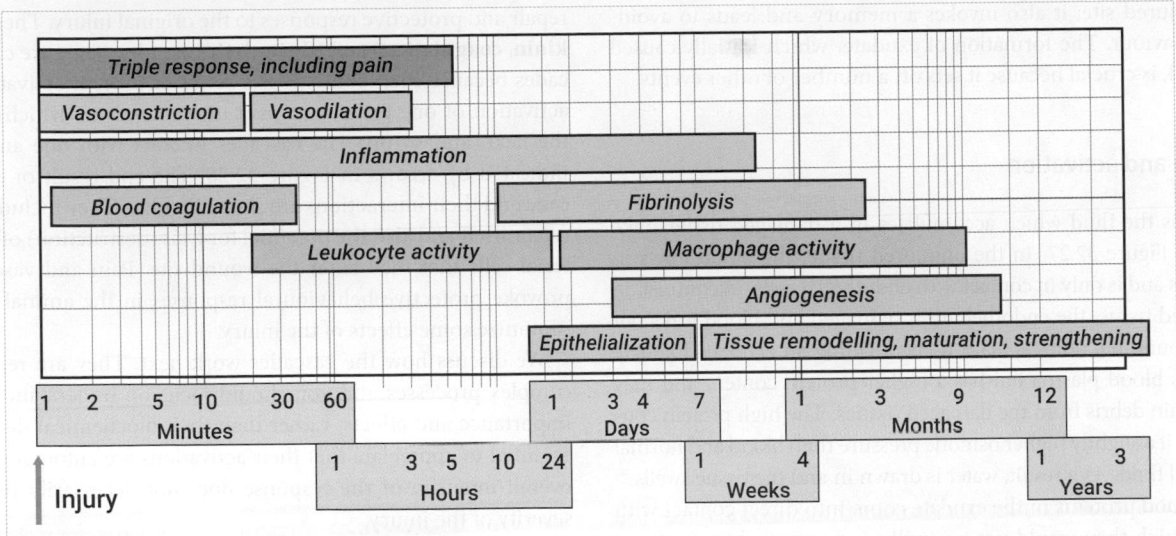

Figure 32.26 Timescale of events in the repair of a skin wound. The phases are shown on logarithmic scales because the rate of change is initially rapid and then slows.

Examples of undesirable conditions involving inflammation include rashes, reactions to stings and bites, eczema, asthma, allergies, and arthritis. In extreme situations, some of these can become dangerous or life threatening. Several types of anti-inflammatory medicines are available, some for treating minor irritations and others for more serious reactions.

If we take a small cut to the skin as an example injury, the immediate response of the damaged area involves the so-called **rubor** (redness), **calor** (heat), **dolor** (pain), and **tumour** (swelling) events. These familiar signs represent the first stages of the inflammatory response. They result from a complex group of processes involving endothelial cells (cells which line the blood vessels and all other parts of the circulatory system), cells in the blood and surrounding tissues, nerves, muscles, and cell-signalling chemicals.

To make sense of this, we can divide the initial inflammatory response into two stages.

1. The **white reaction**. This happens within a few seconds and is seen as a sudden, transient reduction in capillary blood flow in the immediate environment of the cut. It is due to the closure of tiny sphincter muscles in venous and arteriolar capillaries, as a reflex response to the mechanical stimulation of the injury.

2. About 10–15 seconds after the injury comes the so-called 'triple response', first described in detail by Thomas Lewis in 1927. It has three components: red reaction, wheal, and flare, which are detailed in Table 32.1. These develop sequentially over a period of minutes but may remain detectable for several hours or even days.

Elements of the triple response have immediate adaptive value. The swelling may partially occlude the injured tissue and reduce blood loss, although this will only be significant for small cuts. The pain produces awareness of the injury and encourages protection

Table 32.1 The components of the triple response, during the initial stages of the inflammatory response to an injury

Component	Characteristics	Cause
Red reaction (erythema, skin reddening)	Reddening of the local skin up to 1 mm distant from the site of injury	Capillary dilation; a direct response of muscle fibres in the capillary walls to the physical pressure of the injury
Wheal (oedema, the accumulation of fluid)	Raised area incorporating and extending beyond the region of erythema	Increased capillary permeability, due to a) the direct effects of the injury and b) to the contraction of endothelial cells stimulated by histamine and other chemicals secreted by mast cells (connective tissue cells which release inflammatory chemicals); results in extravasation (leakage) of fluid (exudate) into tissues, causing them to swell
Flare	Spreading of the reddened, swollen area; local pain	Arteriolar dilation due to an axon reflex, initiated by sensory receptors in the skin and transmitted via nerves to and back from the spinal cord

The pain is due partly to the reflex and partly to chemicals released by mast cells and other cells in the local area |

of the injured site; it also invokes a memory and leads to avoidance behaviour. The formation of exudate, which initially caused the wheal, is crucial because it sets off a number of other events.

Exudate and activation

Exudate is the fluid which accumulates in a damaged tissue. Take a look at Figure 32.27. In the uninjured tissue, the blood flows in capillaries and is only in contact with endothelial cells. In contrast, in the injured tissue, the endothelium is damaged and blood leaks out into surrounding, non-capillary tissue forming an exudate. Exudate resembles blood plasma but has a higher protein content and may also contain debris from the damaged tissues. The high protein content gives it a slightly higher osmotic pressure than blood and normal interstitial fluid. As a result, water is drawn in and the tissue swells.

The blood proteins in the exudate come into direct contact with tissues which they would not normally encounter when confined within the circulation. This activates a range of other tissues and cells, including:

- *endothelial cells* of damaged capillaries, which participate in wound repair;
- *macrophages* (*histiocytes*) in tissues, which can consume and destroy invading micro-organisms and also get rid of fragments of damaged tissue; and
- *fibroblasts*, which make collagen, fibronectin, and other structural proteins, and which also secrete enzymes necessary for wound repair and tissue remodelling.

You can see from Figure 32.28 that the formation of exudate also initiates some essential biological processes associated with wound repair and protective responses to the original injury. These are the **kinin**, **coagulation**, and **fibrinolytic** cascades. They are called cascades because each consists of a series of protein activations: the activation of one protein turns it into an enzyme which activates the next, and so on. The cascades interact with one another, as indicated by arrows in Figure 32.28. The end result of each cascade and their interactions are shown in blue. They include wound repair itself and also the potential for lysis (destruction) of any bacterial cells that may enter the wound site. Pain and vasodilation provoke protective behavioural responses in the animal and can minimize some effects of the injury.

We discuss how the cascades work next. They are remarkably complex processes and you should focus on understanding their importance and effects, rather than their biochemical details. It is essential to appreciate that their activations are automatic and the overall intensity of the response does not necessarily reflect the severity of the injury.

The kinin cascade

Kinins are a group of proteins formed from blood globulin called **kininogens** (precursors in cascades often have the suffix *–ogen*; others have the prefix *pro-*). An important kinin is **bradykinin**, a small peptide with direct and indirect effects. It causes the relaxation of smooth muscle fibres in blood capillaries, thereby increasing their permeability and enhancing the 'tumour' component of the inflammatory response, and stimulates pain receptors in the skin. One of its indirect actions is to stimulate the production of **prostaglandins** (members of the eicosanoid group of chemicals) and related chemicals at the site of injury. These enhance the pain experience (see Chapter 19) and cause some of the other unwanted symptoms associated with inflammation such as bronchospasm

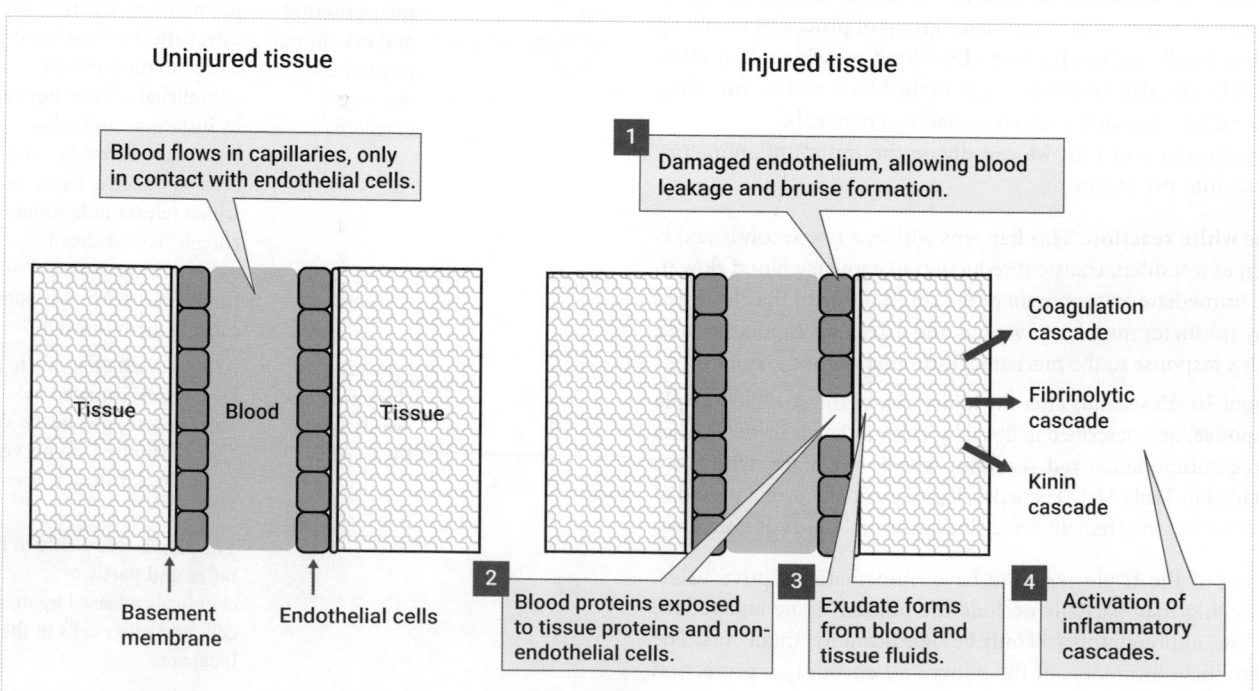

Figure 32.27 Formation of exudate.

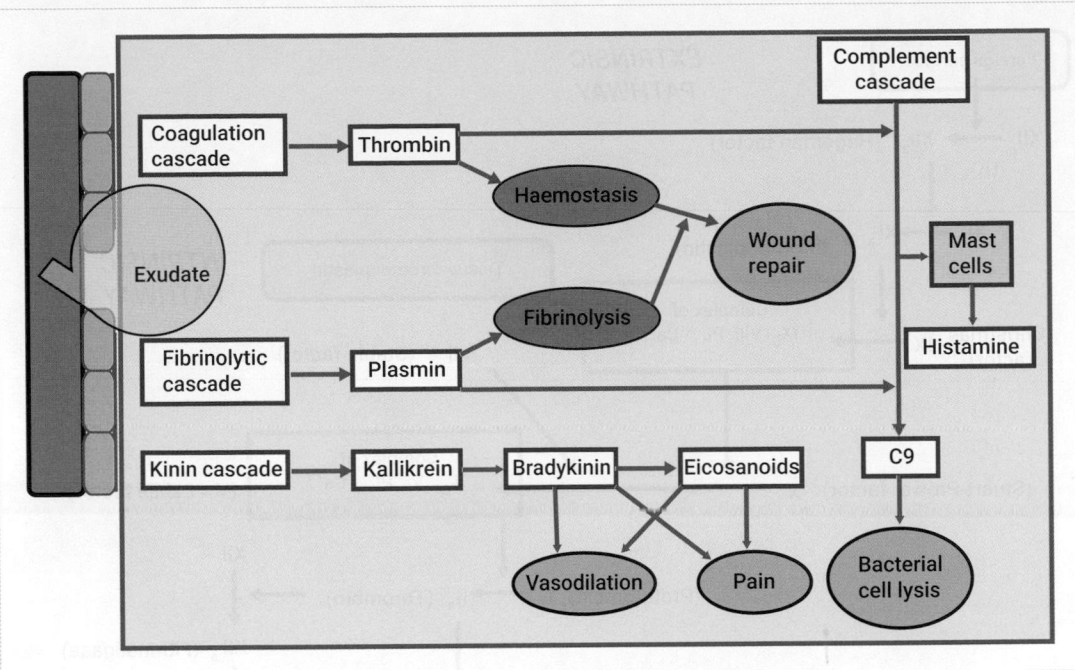

Figure 32.28 Enzyme cascades initiated by the formation of exudate following a tissue injury. Each black arrow represents several steps of sequential enzyme activation. These events take place at the site of exudate formation, on the body surface if bleeding has occurred and within the circulatory system.

(restriction of the airway) and hyperaemia (increased blood flow due to expanding blood vessels).

There is medical interest in understanding these mechanisms for they offer pharmaceutical routes to treating conditions such as rheumatoid arthritis, asthma, and eczema, as well as controlling generalized pain. Aspirin, paracetamol, and ibuprofen produce pain relief and reduce inflammation partly by preventing the synthesis of prostaglandins.

The coagulation cascade

This is the process by which the blood clots and stops flowing. Its adaptive value is clear: if the flow of blood from an open wound did not stop, fluid, proteins, and cells would be rapidly lost from the body. The loss of fluid volume from a severe injury can result in medical 'shock' (a sudden fall in blood pressure), which quickly becomes life threatening; it is one reason why first aid procedures include its mechanical prevention.

But the coagulation process also needs to deal with internal injuries: those in which blood leaks from the circulation but does not escape the body, such as after a forceful but non-cutting impact. At the same time, blood needs to keep circulating in the rest of the body, and of course it needs to flow normally when there is no injury at all.

How is it that the blood remains fluid under normal circumstances, yet stops flowing almost immediately when there is an injury? The answer rests in the complexity of the coagulation cascade and the highly specific circumstances under which its reactions can be initiated. To understand this, we need to start at the end of the process and work back to the beginning.

Activation of the coagulation cascade leads to the formation of a clot or **thrombus**. This consists of a large, polymeric protein called **fibrin** within which small, non-nucleated blood cells called **platelets** are trapped. Fibrin results from the activation of its inactive precursor called **fibrinogen**. Fibrinogen is one of the largest proteins in the blood, so large in fact that under normal circumstances it only just stays in solution in the plasma.

As soon as it is activated, molecules of fibrin start to stick together and precipitate, trapping platelets as they do so. Fibrin sticks to tissue proteins, especially **fibronectin** and **collagen**, so the thrombus forms against the damaged tissue at the point where the blood has leaked from its capillaries. With external injuries, this becomes a scab; for internal ones, it forms a bruise.

Stepping back, what activates fibronectin? This is the action of an enzyme called **thrombin**, the activated form of the blood protein **prothrombin**. (Prothrombin, like most inactive precursors in the cascades, is made by the liver and is always present in the blood.) A second enzyme, **fibrinoligase**, strengthens the bonds between the reacting fibrin molecules, solidifying the thrombus and making it highly resistant to solution or mechanical destruction.

These are the final steps in the coagulation cascade. Figure 32.29 shows the main preceding steps and hints at some of its immense complexity. Its elements have traditionally been given both names and numbers (not in their order of activation, unfortunately, but in their order of discovery).

Without focusing on each individual step, we see that the cascade has two, alternative initiation points. These are the **extrinsic** and **intrinsic pathways** and they explain why the system reacts to both external and internal injuries.

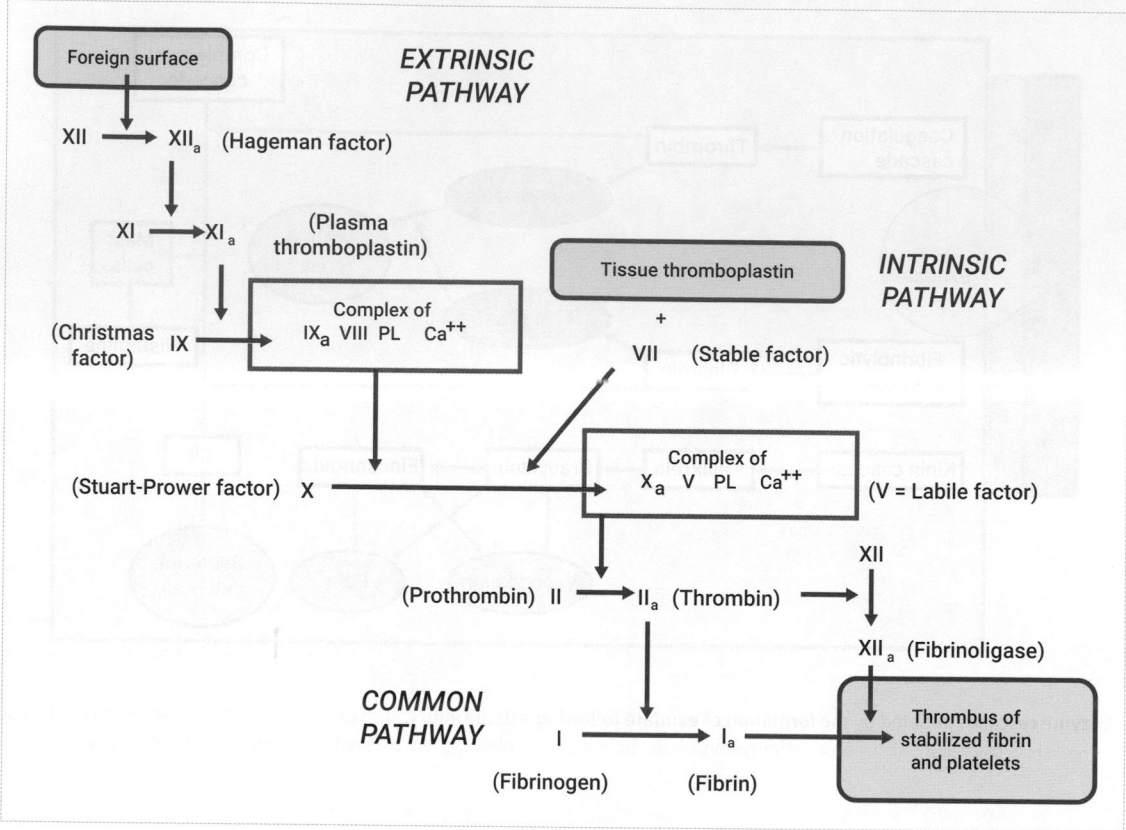

Figure 32.29 **The blood coagulation cascade.** In this diagram, each element of the cascade is identified by its number along with its most frequently used name (most elements have more than one name). The 'a' subscript indicates that the element has changed from its inactive to its activated state.

- The intrinsic pathway was so called because all its elements are present in the blood. It is activated when the blood comes into contact with an external or foreign surface, such the collagen of broken skin, the glass or plastic of a blood sample tube, or even air.

- The extrinsic pathway is activated in damaged tissues and is initiated by a factor (tissue thromboplastin) released by cells in the walls of damaged blood vessels and leukocytes (white blood cells). Depending on the nature of the injury, both pathways are likely to be activated and many physiologists consider the distinction to be unnecessary.

Unless one or other of the pathway conditions is met, the cascade remains inactive and the blood flows normally. The only exceptions are certain diseases, medical procedures, or states of unavoidable inactivity (prolonged bed rest, long airline flights) which cause a **thrombosis**: a clot *within* the circulation itself.

Factor VIII in the complex of the extrinsic pathway is antihaemophilic factor. Its absence leads to haemophilia, an inherited disease in which blood fails to clot or clots very slowly. The condition is carried in females but expressed in males.

The platelets entrapped in the thrombus assist in its adhesion to the damaged tissue. During entrapment and adhesion, platelets become activated. They secrete several molecules, some of which enhance the coagulation process and promote inflammation and tissue repair.

PAUSE AND THINK

Blood that is taken from donors and used for transfusion needs to be collected and stored in a fluid condition. Chemicals called *anticoagulants* are used to prevent it from clotting. Many of these chemicals work by binding to calcium ions and inactivating them in a process called chelation. Why do these chemicals stop blood from clotting? Use Figure 32.29 to help you

Answer:

The last step of the intrinsic pathway and the first step of the common pathway require the formation of a complex including calcium ions (Ca²⁺) and phospholipid (PL, from blood platelet membranes). If the calcium ions are inactivated, the complex cannot form and the rest of the coagulation cascade cannot be activated. Examples of anticoagulants which work this way are EDTA, citrate, and heparin.

The fibrinolytic cascade

Fibrinolysis is the process of thrombus breakdown which occurs as a wound is being repaired. It is stimulated by exudate (which we see in Figure 32.28) but works on a longer time scale than coagulation.

Fibrin is resistant to most protease enzymes, but it is broken down by a specific enzyme called **plasmin**. The fibrinolytic cascade results in the activation of plasminogen, an inactive precursor made in the liver, to plasmin. Plasmin destroys the linkages between the fibrin molecules, allowing them to dissolve and for the thrombus to fall away.

Plasmin also activates another set of enzymes called **metallo-proteinases**. They include some which can break down collagen. Collagen breakdown, followed by secretion of new collagen by fibroblasts and other tissue cells, is central to the tissue remodelling by which the damaged tissue is repaired.

All these enzyme systems have their own regulators and inhibitors. Some clot preventing and 'clot busting' pharmaceuticals aim to interfere with these processes.

Links to immune defences

The main products of the coagulation and fibrinolytic cascades, thrombin and plasmin, have a further interaction with the **complement** cascade, which is indicated in Figure 32.28. The complement cascade is part of the innate immune system, activated whenever there is a bacterial or other infection. One of its effects is to cause the lysis (break up) of bacterial and other invading cells by destroying the integrity of their cell walls.

There is a clear adaptive advantage in promoting complement activity at a time of injury because of the likelihood of microbiological ingress through the wound. Importantly, the complement system is activated *irrespective* of the nature of the invading organism: it does not depend on the activation of the rest of the immune system.

Wound repair

Coagulation and fibrinolysis eventually lead to wound repair. The relative timescales of events are indicated in Figure 32.26, but the sequence itself will be familiar to you from the formation and shedding of scabs after a cut or scratch to the skin.

Returning damaged tissue to its original state can take a long time and, unless the injury is superficial, there may always be a scar at the healed site. The actual shaping (remodelling) of the repair is regulated by growth factors and hormones secreted by cells and tissues in the vicinity.

If the injury is severe, or if it has caused a restriction in oxygen supply to the tissues, the endothelial cells of damaged capillaries may be stimulated to proliferate, so that new vessels begin to sprout and grow. This process is called **angiogenesis**.

In addition, epithelial stem cells can become activated in such a way that they differentiate into fibroblasts, hair cells, sweat gland cells, pigment cells, and others. There is good evidence that skin and other epithelial tissues retain a 'memory' of their original condition, so that the site of repair comes eventually to resemble its pre-damage state.

The invertebrate immune system

The 1.3 million or so *invertebrate* species represent a vast diversity of organisms and there is an equivalent diversity in how they defend themselves against microbiological invasion and disease. Invertebrates have a wide range of lifestyles and requirements for protection. For example, the immune system of the clam *Arctica islandica*, which lives in the sea and has been reported to live for 500 years, must be very different from that of an adult butterfly, which is terrestrial and lives for only a few days.

We focus here on the immune system of arthropods—animals with an exoskeleton, segmented bodies, and six or more jointed legs. They include insects, crustaceans, spiders, and their relatives. As with vertebrates, we need to distinguish between innate immunity—that which comes into play automatically, irrespective of the nature of the invading organism—and specific immunity in which the response is tailored to specific antigens and pathogens. Similarly, there are parts of the system which involve cells and parts which depend on macromolecules carried by the body fluid (haemolymph), called humoral factors.

The innate immunity of invertebrates is broad and non-specific. Pattern recognition receptors (PRRs) detect pathogen-associated molecular patterns (PAMPs) found on invading pathogens. The PRRs are encoded in DNA and therefore part of every species' genetic identity. They are proteins, either bound to the surfaces of cells or dissolved in body fluid. They initiate both cellular and humoral defences, as indicated in Figure 32.30.

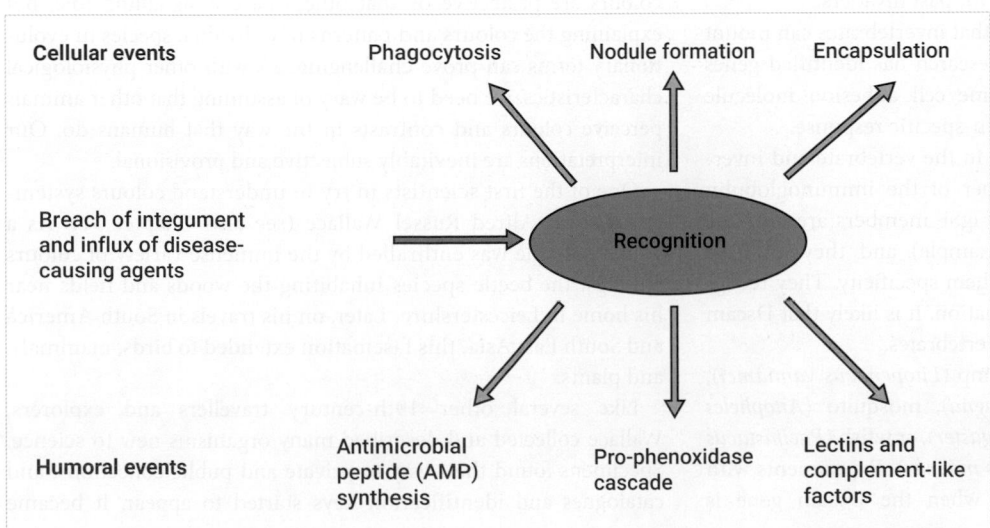

Figure 32.30 Cellular and humoral immune defence strategies in invertebrates.

Source: Adapted from Rowley A. F., Powell A. (2007) Invertebrate immune systems–specific, quasi-specific, or nonspecific? *Journal of Immunology*, 179(11), 7209–7214, Figure 1. Copyright 2007. The American Association of Immunologists, Inc.

Cellular events — Phagocytosis — Nodule formation — Encapsulation

Breach of integument and influx of disease-causing agents → Recognition

Humoral events — Antimicrobial peptide (AMP) synthesis — Pro-phenoxidase cascade — Lectins and complement-like factors

Cellular events depend on the activation of specific cells and include:

- phagocytosis, which is the engulfing and destruction of foreign cells and particles;
- nodule formation, in which a protective growth develops at the site of injury; and
- encapsulation, in which a potentially dangerous particle or foreign material is inactivated by being surrounded and isolated.

Humoral events depend on chemical reactions rather than cells. They include:

- the synthesis of antimicrobial peptides;
- activation of the prophenoloxidase cascade, which creates chemicals capable of inactivating pathogenic cells; and
- the synthesis and secretion of lectins and complement factors which can enhance the effectiveness of cellular events.

The main method of immune defence in arthropods involves **haemocytes**, cells that are carried in the haemolymph. Haemocytes engulf and digest invading organisms such as protozoa, bacteria, fungi, and viruses. They also engulf multicellular parasites and other organisms, forming a thick layer around them which isolates them from the rest of the host's body. Haemocytes also aid in the coagulation of haemolymph if the animal is injured.

Arthropod haemolymph also contains antimicrobial peptides (AMPs). These are small molecules which respond rapidly to invading pathogens, causing holes to form in the membranes of microbial cells so that they lyse. Antimicrobial enzymes contained in lysosomes are also involved, and there are carbohydrate-binding proteins called lectins, either free in the blood or associated with the haemocytes, which act as receptors and effectors. Finally, an enzyme cascade, the prophenoloxidase cascade, produces cytotoxic (cell-killing) and opsonic factors. Opsonic factors make the foreign material attractive to haemocytes and enhance phagocytosis (engulfing, followed by lysosomal destruction).

The above are all non-specific defence mechanisms. Invertebrates do not have the types of lymphocytes found in vertebrates and so it has been suggested that they lack a highly specific, adaptive immune system, including a memory for past invaders.

However, there is genetic evidence that invertebrates can mount a more specific immune response. Research has identified genes such as that for the Down syndrome cell adhesion molecule (Dscam), which may enable a pathogen-specific response.

Dscam, known for its involvement in the vertebrate and invertebrate nervous systems, is a member of the immunoglobulin superfamily (IgSF) of proteins. Many IgSF members are immune molecules (immunoglobulins, for example) and they all have multiple possible forms which gives them specificity. They recognize PAMPs and promote their elimination. It is likely that Dscam has a similar or related function in invertebrates.

Dscam has been detected in a shrimp (*Litopenaeus vannamei*), planktonic crustacean (*Daphnia magna*), mosquito (*Anopheles gambiae*), fruit fly (*Drosophila melanogaster*), crayfish (*Pacifastacus leniusculus*), and shore crab (*Carcinus maenas*). Experiments with *Drosophila melanogaster* show that when the Dscam gene is

silenced, the ability to phagocytose foreign pathogens is reduced. The mechanism behind the specific recognition of PAMPs remains unclear, but these findings point to an immune system that does not have to rely on innate, non-specific responses.

 Check your understanding of the concepts covered in this section by answering the questions in the e-book.

32.4 External defences

We have seen how the integument creates a protective barrier to infection and invasion and how tissues respond to injury if that barrier is breached. If we look at organisms from a wider, external viewpoint, we see that they have many other structural and behavioural ways of defending themselves from injury, invasion, and predation.

Camouflage, deception, and mimicry

Why are organisms coloured? This is a simple question with complicated answers. As humans, we take pleasure in the colours of the natural world, expressing wonder when they are especially bright or varied and often describing them in terms of beauty or attractiveness. We sometimes impose our subjective, aesthetic values on the colours of animals and plants when we breed or select them for productive, ornamental, or other purposes. Figure 32.31 shows an example of human intervention in the selective breeding of *Alstroemeria* (Peruvian lily) to create a range of colours for ornamental purposes.

Although we may enjoy the capacity of organisms to express surface coloration, colours are not there principally for our delight. The natural ability of an organism to reflect distinct wavelengths of light, to express surface patterning, or to show different levels of contrast must offer some underlying adaptive advantages. Like any other phenotypic characteristic, colour is the result of selection pressure—an adaptation that gives protection, increases survival, and improves the likelihood of successful reproduction.

This question has been debated by biologists for many decades and many uncertainties remain. It may seem obvious that some colours are protective or that others have a signalling role, but explaining the colours and patterns of individual species in evolutionary terms can prove challenging. As with other physiological characteristics, we need to be wary of assuming that other animals perceive colours and contrasts in the way that humans do. Our interpretations are inevitably subjective and provisional.

One of the first scientists to try to understand colours systematically was Alfred Russel Wallace (see also Chapter 31). As a young man, he was enthralled by the immense variety of colours amongst the beetle species inhabiting the woods and fields near his home in Leicestershire. Later, on his travels in South America and South East Asia, this fascination extended to birds, mammals, and plants.

Like several other 19th-century travellers and explorers, Wallace collected and described many organisms new to science. Specimens found their way to private and public collections, and catalogues and identification keys started to appear. It became

Figure 32.31 Varieties of *Alstroemeria* (Peruvian lily), a popular ornamental flowering plant.

Source: (a) Image by Jacques GAIMARD from Pixabay. (b) Image by Manfred Richter from Pixabay. (c) Image by Pablo Valerio from Pixabay. (d) Image by Kevin Cannings from Pixabay. (e) Image by Chesna from Pixabay. (f) Image by tarsan099 from Pixabay.

possible to arrange organisms in a logical, taxonomic way based on the growing awareness of the principles of natural selection set out by Wallace and Charles Darwin. It rapidly became clear that colour was a fundamental and important biological characteristic that needed to be understood and explained. Table 32.2 lists Wallace's categorization of coloration in living organisms; we explore some of these categorizations in more detail below.

Are signals honest or dishonest?

We can consider the signals an organism sends out to be either *honest* or *dishonest*: do they represent the organism's status accurately (in which case they are honest) or do they deceive a potential recipient of the signal (in which case they are dishonest)?

There are two broad types of signalling, one of which is dishonest and one that is honest.

Crypsis refers to the adoption of camouflage or behaving in a way that discourages detection. It is clearly a dishonest strategy. The potential predator may be deceived by an organism's similarity to other features of the environment causing it to miss the prey completely, as, for example, when a stick insect is indistinguishable from the vegetation on which it sits. Alternatively, the predator may be confused in some other way, as when a zebra's stripes obscure the visual systems of insect pests.

Aposematism refers to the pairing of a warning coloration with a repellent or lethal defence such as a venom or poison. This is widespread in nature, both within the animal kingdom (signalling between animal species) and between kingdoms in the way plants and fungi signal to animals. Such signals are honest, because the brightly coloured prey really would be dangerous to the predator.

Signals beyond coloration

We now know a great deal more than Wallace did about the diversity of life forms which inhabit Earth, and especially about animal groups such as fish and other marine organisms about which he and his contemporaries knew very little. In addition, we have a greater appreciation of the ecological interactions between species and between organisms and their environments.

We also know a great deal more than Wallace could possibly have known about the cellular and biochemical mechanisms of colour pigment generation (in skin, hair, feathers, etc.) and the genetic processes by which colours and patterns are inherited.

Despite this, Wallace's categories are valuable as guides to interpretation and have largely stood the test of time. A contemporary analysis uses a more comprehensive, formal terminology and it usually extends to the whole array of signals which an organism may employ, for attraction, communication, and reproduction, as well as for defence. Look at Table 32.3, which summarizes this terminology, and Figure 32.32, which depicts some of the signalling strategies denoted by this terminology. (We will explore the terminology itself in more detail in a moment.) Overall, this analysis recognizes that coloration may be just one component of an organism's signal repertoire: it may be combined with morphological, behavioural, vocal, or olfactory signals.

Evolutionary biologists are particularly interested in the effectiveness of these strategies and try to understand how stable and heritable characteristics such as colour might be. The production of signals of all kinds, and sometimes the suppression of them, takes energy. The benefits, in terms of predator avoidance, resource acquisition, or successful reproduction, must outweigh the costs if the signal is to persist within the population.

Table 32.2 Alfred Russel Wallace's categorization of coloration in living organisms

Wallace's coloration category and its alternative name	Explanation	Wallace's examples
1. Protective colours **Crypsis** (see Figure 32.32a)	Camouflage is achieved by resemblance to the prevailing environment or a specific background element Pigmentation may be uniform or patterned, may be different when viewed from above or below, or may be appropriate for nocturnal or diurnal lifestyles	White animals in the Arctic Sandy-coloured animals in deserts Birds green in tropics but brown in temperate regions Nocturnal animals being dusky grey Birds' eggs matching their background Transparent invertebrates in oceans Animals with dark dorsal and light ventral surfaces

2. Warning colours
 	
   a. of specially protected species
 	
Aposematism (see Figure 32.32bi)
 	
   b. of otherwise defenceless species which mimic those in category 2a (see Figure 32.32bii)
 	
Batesian mimicry (see Figure 32.32cii)
	Conspicuous colour gives organism protection by indicating toxicity or other danger. Potential predators learn to avoid them
	
	An organism that would not itself be dangerous to a predator protects itself by adopting the conspicuous coloration of one which does pose a threat. The similarity of coloration does not need to be exact
	
	Depends on i) the two organisms inhabiting the same area, ii) there being a limited range of mimicked ('model') species, and iii) on there being many more individuals of the model organism than of the mimicking organism
		Brightly coloured insect larvae, especially those of butterflies and moths
		
		Many examples among beetles, hymenoptera (ants, bees, and wasps), snakes, and birds

3. Sexual colours **Sexual dichromatism**	A difference in coloration between the sexes. One sex may be brightly coloured or ostentatiously ornamented, while the other is not May have to do with sexual competition—the competition for mates or breeding opportunities—or with one sex (often but not always the female) being less conspicuous while nesting or rearing young Wallace himself was uncertain about the sexual competition aspect of this but accepted that coloration and ornamentation may indicate status or health	Many examples among birds
4. Typical colours **Species polymorphism**	Conspicuous coloration in both sexes Enables recognition of one species by another, avoiding predation, conflict, mating errors, and hybridization Allows individuals within a species to recognize each other and form a coherent social group. The coloration of individuals may change with age, reproductive status, or social position	Many examples among birds
5. Attractive colours	Occurs in angiosperms (flowering plants) only Associated with attracting pollinators and with the distribution of seeds by fruit-eating animals Coloration may be generally attractive to animals (especially insects, birds, and mammals) or may have evolved to attract a specific or limited range of species The crypsis commonly found in nut and seed colourings, in contrast to the conspicuousness of the fruits that contain them, may assist in their safe dispersal for subsequent germination	Flowering and fruiting plants and trees

Table 32.3 The terminology of organismal coloration and signalling strategies

Type of signal	Descriptive term	Characteristics	Function	Example
Camouflaging	Cryptic	Dishonest	Evasion of predators	Figure 32.32a
Revealing (sematic)	Aposematic	Honest	Defence; warning to predators of the organism's inherent danger	Figure 32.32b(i)
		Dishonest (Müllerian mimicry)	Defence; organism presents an inherent danger but predators are deterred by mimicking an honest aposematic organism	Figure 32.32b(ii)
	Parasematic	Dishonest	Defence; organism appears dangerous to a potential predator but is not	Figure 32.32c(i)
		Dishonest (Batesian mimicry)	Defence; organism presents no inherent danger but it mimics an honest aposematic organism to deter predators	Figure32.32c(ii)
		Honest (Zahavian signalling)	Signalling of fitness; organism uses coloration as a reliable indication of its 'handicap'—its likely capability for defence or reproduction	Figure 32.32c(iii)
	Episematic	Honest	Recognition and communication; for social or reproductive purposes	Figure 32.32c, d

The episemmatic categorization is applicable to angiosperms, which use flowers to attract pollinators or use fruit to attract distributors of seeds.

Interpreting signals: fitness and adaptation

It is reasonable to ask why *all* organisms have not evolved to look dangerous. Surely that would be in their best defensive interests? Some non-dangerous organisms do indeed exhibit false signals (called dishonest **parasematism**), either by appearing to be dangerous when they are not or by adopting the features of a dangerous species living in the same environment (see Figure 32.32c). The latter approach is called **Batesian mimicry**, which is mentioned in Table 32.2 as a type of warning colour approach to defence. Batesian mimicry is named after H. W. Bates, who introduced Alfred Russel Wallace to the beetles of Leicestershire and who travelled with him to South America.

For other organisms, there may be significant disadvantages in parasematic signalling. It is energetically costly to be colourful and it may also interfere with the ability to hide, obtain food, thermoregulate effectively, or communicate with conspecifics for social or reproductive purposes.

There is also a more general arms race argument: the more widespread falsely deterrent signals occur, the less notice is taken of them and the less individually effective each signal becomes. Looked at this way we can see that each species evolves with its own balances between honesty and dishonesty and between crypsis and revelation, driven by the existential pressures of survival and reproduction.

It was realized by Fritz Müller at the end of the 19th century that some animals exaggerate the risk they present to predators by adopting the coloration or other signals of more dangerous or aggressive organisms living in the same environment. **Müllerian mimicry** differs from Batesian mimicry in that there really is some potential danger to the predator. This type of deception could be described as acquired exaggeration. It is thought to have evolved from the statistical advantage of resembling and mixing in with other dangerous species: it reduces the chance of being taken.

Some biologists view Batesian and Müllerian mimicry as forms of crypsis. In both cases, the organism is protected by the way its appearance has evolved to converge with that of surrounding species. It reduces the chance of being predated by blending into the background of real danger created by the organisms which surround it. If you look at Wallace's conditions for Batesian mimicry (Table 32.2), you can see that he fully appreciated this dilution effect.

The mechanisms of convergence seem to be similar in Batesian and Müllerian mimicry, and genetically similar to crypsis, so the distinction may be just in their names. As with all other attempts to understand evolution, this is a matter for interpretation and we should not become excessively concerned with labels.

Follow the link from the e-book to learn more about the genetics of mimicry.

Some honest signals do not relate directly to defence or the avoidance of danger from predators. **Zahavian signalling** is indirect: rather than sending specific information about protective or other mechanisms, it is an external indicator of the organism's health or fitness.

Zahavi, who first described this category of signalling in the 1970s, described it as indicating a 'handicap', much as a golfer might represent his or her level of skill in the game. Only fit individuals will have sufficient metabolic resources to generate a worthwhile Zahavian signal. This makes it an adaptive feature and maintains the honesty of the signal through the generations. A florid signal that is a significant drain on resources or which interferes with

Figure 32.32 Examples of the coloration and signalling strategies identified in Tables 32.2 and 32.3. (a) Crypsis. A leaf insect of the *Phyllium* genus, camouflaged by resemblance to the leaf on which it rests. (b) Aposematism. (i) Fire salamander (*Salamandra salamandra*), on Austrian hillside, showing honest aposematism. This amphibian exudes toxic chemicals from skin glands, often concentrated in the yellow-coloured regions. Its yellow/black coloration is highly visible to potential predators. (ii) Similarity between Viceroy butterfly (*Limenitis archippus*, left) and monarch butterfly (*Danaus plexippus*, right), both of which taste noxious to predators. An example of Müllerian mimicry. (c) Parasematism. (i) Tiger beetle (*Cicindela chinensis*) exhibiting dishonest parasematism. Many have large eyes and are highly coloured; some have false eyespots. (ii) The harmless hoverfly (*Eupeodes corollae*, left) has shape and markings which resemble those of stinging wasps (Vespidae, inset). This is an example of Batesian mimicry. (iii) Male peacock (Indian peafowl, *Pavo cristatus*), displaying its tail. (d) Multifactorial. Red deer stag (*Cervus elaphus*) in the breeding season: large antlers, loud vocalization, and aggressive behaviour. These signals are aposematic (defending territory), Zahavian (indicative of strength and fitness), and episematic (establishing position in social and reproductive hierarchies).

Source: (a) Pavel Kirillov/CC BY-SA 2.0. (bii) Viceroy butterfly: D. Gordon E. Robertson/Wikimedia Commons/CC BY-SA 3.0, and Monarch butterfly: Captain-tucker/Wikimedia Commons/CC BY-SA 3.0; (ci) Didier Descouens/Wikimedia Commons/CC BY-SA 3; (cii) Thomas Bresson/CC BY 2.0; (cii) Kurayba/Flickr/CC BY-SA 2.0/https://creativecommons.org/licenses/by-sa/2.0/; (ciii) image by Allan Lau from Pixabay; (d) Image by Ana Gic from Pixabay.

normal function, becomes a representation of health, strength, or prowess. It shows that the individual possesses more resources than it needs for its own marginal survival.

The male peacock's tail is a good example. The abundantly coloured tail feathers grow excessively, preventing flight, hindering rapid movements on the ground, and making the bird especially conspicuous. The tail can be raised in a magnificent but physiologically valueless display. Female peacocks may choose between males according to the exuberance of their tails, using it as a proxy indicator of health and the likelihood that mating will lead to successful reproduction. This is an example of **sexual selection**.

PAUSE AND THINK

If the male peacock's tail is an example of Zahavian signalling, used by females in the selection of male partners, do you think it is necessary for the female bird to *consciously* associate the quality of the tail with the health and fitness of the male bird?

Answer: It is not possible for us to know how a female peacock thinks when selecting a mate. However, it is not *necessary* for her to be making a conscious or reasoned decision in terms of male fitness and the quality of chicks she may subsequently produce. In evolutionary terms, provided the tail is a *reliable* indicator of fitness, those females which are, for whatever reason, attracted to well-developed tails will tend to reproduce more successfully and pass on their genes. Equally, the more frequently chosen males will tend to pass on *their* genes, including for fitness and large tails. If you think about it for a moment, it is clear that this is a self-reinforcing process of mate selection, irrespective of how the male and female birds are motivated. This is reassuring because, as objective biologists, we should try to avoid imputing purpose or reason where it is impossible for us to know about it.

The example of the red deer stag in Figure 32.32d shows that signals may fall into more than one category. It reminds us that the categorization is an *interpretation* only: it helps us understand why such signals are adaptive, but we can never be sure that our explanation is correct or complete.

The *evolution* of some signals can be particularly difficult to explain. The following is a list of some questions that have continued to trouble biologists since the foundational work of Wallace, Darwin, Bates, Müller, and other 19th-century pioneers:

- How do predators come to associate warning signals with adverse prey characteristics (toxins and poisons, physical danger, etc.), especially when the prey is lethal? Can they use taste or some other clue for information?

- How is this information transmitted to offspring or conspecifics? Can it be inherited or learnt by example?

- Do predators learn about differences in the balance between the toxin and nutrient content of individual prey organisms within a population? Do they bargain reward against risk?

- How are cryptic and sematic signals maintained across many generations? What balance is necessary between successful and unsuccessful predation or parasitism for this to happen?

Camouflage and mimicry in plants

The strategies of camouflage and mimicry, as forms of defence, are found much less frequently in plants than in animals. However, there are examples of camouflage in plants where the plant body grows to resemble its near environment.

Plants of the genus *Lithops* grow in arid rough terrain in South Africa; the plant body takes on the form and coloration of stones present on the ground, as you can see in Figure 32.33a. They merge well with their surrounding environment and are very difficult to see, being easily spotted only when they flower. They are often called 'living stones'.

True mimicry is rare in plants but one example is *Boquila trifoliolata*. This tropical vine grows up a wide variety of different trees and shrubs and its leaf morphology is highly variable, as shown in Figure 32.33b. Its leaf shape appears to mimic that of its host tree. Such a strategy presumably reduces the chances of its leaves being eaten when it is surrounded by lots of others of a similar shape.

 Check your understanding of the concepts covered in this section by answering the questions in the e-book.

Figure 32.33 Plant camouflage and mimicry, deterring herbivory. (a) *Lithops* blending with the stony background. (b) Tropical vine leaves mimic leaves of the host plant.
Source: (b) Reprinted from Gianoli E, Carrasco-Urra F. (2014) Leaf mimicry in a climbing plant protects against herbivory. *Current Biology*, 24:9, 984-7. Copyright 2014, with permission from Elsevier.

Research question

Why do fireworms sting?

Background

Fireworms, like the one in Figure 1, are polychaetes in the family Amphinomidae. On contact with human skin they cause a burning sensation, leaving a painful red rash and inflammation.

Figure 2 Light microscope image of chaetae from *Eurythoe complanata*.

Source: Tilic, E., Pauli, B., Bartolomaeus, T. (2017) Getting to the root of fireworms' stinging chaetae—chaetal arrangement and ultrastructure of *Eurythoe complanata* (Pallas, 1766; Amphinomida). *Journal of Morphology*, 278, 865–76. https://doi.org/10.1002/jmor.20680

Figure 1 The fireworm, *Eurythoe complanata*.

Source: Tilic, E., Pauli, B., Bartolomaeus, T. (2017) Getting to the root of fireworms' stinging chaetae—chaetal arrangement and ultrastructure of *Eurythoe complanata* (Pallas, 1766; Amphinomida). *Journal of Morphology*, 278, 865–76. https://doi.org/10.1002/jmor.20680

Methods

Tilic and colleagues used transmission electron microscopy to study the internal structure of *Eurythoe complanata* chaetae.

Results

Fireworm chaetae are *not* hollow. They are solid and contain calcareous material. The painful reaction is caused by chaetae piercing the skin and breaking off, causing direct damage and inflammation.

Conclusions

The fireworm sting is a skin reaction to the penetration of a physical object, not to a poison.

Read the original work

Tilic, E. et al (2017) Getting to the root of fireworms' stinging chaetae—chaetal arrangement and ultrastructure of *Eurythoe complanata* (Pallas, 1766) (Amphinomida). *Journal of Morphology*, 278, 865–876.

Hypothesis

Abundant chaetae, or bristles (like the ones in Figure 2), protrude from the worm's body. The chaetae are hollow and contain poison. On contact with epidermis, they inject poison, like hundreds of tiny hypodermic needles.

32.5 Toxins, poisons, and venoms

We have seen that many organisms defend themselves by producing chemicals that are dangerous or distasteful to potential predators. Such chemicals are employed in a range of different defensive strategies, and they are also used in catching prey and for other purposes. Some of these products become dangerous when ingested, while others are actively deployed using specialized cells or structures. We need to use precise terminology to describe them.

The word **toxin** is a general one. It has a broad chemical definition, but within biology it can cover almost any chemical produced by one organism which is troubling or dangerous to another. It can be a deterrent by smell or taste or it can inflict suffering ranging from mild discomfort or irritation (for example, a nettle sting), through moderate or severe pain (some jellyfish stings) to immobilization and death (some snake bites). Toxins can be effective at low concentrations, but the intensity or speed of effect is usually proportional to the quantity of chemical transferred.

Poisons are toxins acquired *passively*, for example by ingestion (the epithelial secretions of some caterpillars which are dangerous to birds, or a death cap fungus whose toxins cause human liver damage) or by touch (the bufotoxins in the skin of amphibians such as the cane toad).

Venoms, in contrast, are toxins transferred *actively* during mechanical injury, through a wound, or by means of a specialist physiological device (the fangs of a snake, the tail of a scorpion, or the abdominal stings of insects).

Poisons mostly have a defensive role, while venoms are usually used in deliberate attack, often for predation, but sometimes also for active defence. Venoms are often stored in exocrine glands until release. They include the salivary secretions used by vampire bats, and the neurotoxic secretions used by spiders to immobilize web-caught insects. They can also be taken to include the stinging nematocyst cells of jellyfish and other cnidaria.

Not all stinging structures deliver poisons. In some cases the structure itself is the source of irritation. This is one way in which annelids protect themselves, for, unlike invertebrates such as arthropods, they do not possess a hard, protective exoskeleton. You can read more about non-poisonous stinging structures in Scientific Process 32.2.

Unlike the method of the fireworms in Scientific Process 32.2, other animals secrete venom. However, there are only a few types of mammal are known to secrete venoms. These are:

- Three species of vampire bat (*Diphylla ecaudata*, *Desmodus rotundus*, and *Diaemus youngi*), which are Microchiroptera.

- The platypus (*Ornithorhynchus anatinus*), which is a monotreme.

- The Haitian and Cuban solenodons (*Solenodon paradoxus* and *Atopogale cubana*), which comprise a family of their own.

- The European water shrew (*Neomys fodiens*) and American short-tailed shrew (*Blarina brevicauda*) which are Insectivora.

- The slow lorises (*Nycticebus coucang* and *N. pygmaeus*) which are in the Lorisidae family of primates.

The wide variety of families which these animals represent shows that the venomous trait has evolved independently several times. It is thus an example of convergent evolution.

 Follow the link from the e-book to explore more about venomous animals, and the example of the slow loris.

Whilst many organisms manufacture their toxins themselves, within glands or skins cells, there are also some which acquire chemicals from their food or environment and exploit them for toxic effect.

- Poison dart or poison arrow frogs of the Dendrobatidae family (such as those shown in Figure 32.34) are highly coloured rainforest animals of central and south America which accumulate lipophilic alkaloids called batrachotoxins from arthropods in their diet. They are themselves immune to the poison but secrete it through the skin as a defence against predators.

- Some crab species exploit the nematocysts of sea anemones by attaching entire anemones to their limbs or shells.

- The common European hedgehog (*Erinaceus europaeus*) can lick poisons from the skins of the toads it consumes on to its spines, rendering themselves toxic to large mammal predators such as badgers and foxes.

- The African crested rat (*Lophiomys imhausi*) protects itself by smearing toxic substances from the roots and branches of the *Acokanthera* tree on to its crest fur, where the hairs are especially spongy and absorbent. The animal's coloration also resembles that of a porcupine, in another example of Müllerian mimicry.

Besides poisons and venoms, some biologists identify a third category of biological toxins called **toxungens**, which includes chemicals actively transferred by a means not involving physical injury. Transfer can be by spraying (for example, some ants and earwigs eject toxins from abdominal glands when disturbed), spitting (cobras project venom into the faces of predators), or smearing (some termites spread abdominal glues that immobilize and kill ants and other nest invaders). The toxungen becomes applied to the surface of the victim's body and its chemical structure may enable it to penetrate skin or mucous membranes such as the respiratory tract or eyes.

We can see from this that toxins can be classified according to their strategy of use (passive or active) and their biological effect, and also by their source of secretion and their means of transfer.

Some chemical defences rely on distaste or sensory rejection, rather than on toxicity:

- Skunks, which are members of the Mephitidae family of carnivores, spray a penetrating, malodorous deterrent from a pair of anal scent glands when attacked. The smell is detectable for several kilometres and notoriously difficult to remove from skin or

Figure 32.34 Poison dart frogs of Central and South America. (a) *Dendrobates tinctorius 'azureus'*. (b) *Dendrobates leucomelas*.

Source: (a) H. Zell/Wikimedia Commons/CC BY-SA 3.0; (b) Dmitrij Rodionov (DR)/Wikimedia Commons/CC BY-SA 4.0.

clothing. The animal has a limited supply, however, and will exhibit deterrent behaviours (foot stamping and threat posturing) before resorting to spraying.

- Sea birds such as the fulmar will vomit an obnoxious and highly water-resistant fluid on predators, including humans, which raid their nests for eggs. The oily material originates from glands in the bird's proventriculus (equivalent to a stomach).

- Among the insects, earwigs attacked by anoles (lizards) will emit a deterrent odour which resembles the smell of rotting flesh. The attacker is repelled and learns to avoid the insect, at least for several days.

PAUSE AND THINK

The use of defensive poisons, venoms, and deterrent chemicals by animals depends on the exploitation of several different physiological processes. List what these might be.

Answer:
- Sensory awareness of potential predators.
- Biochemical synthesis of noxious or signalling chemicals.
- Secretion of fluids by integumentary structures, parts of the gastro-intestinal tract, the respiratory tract, or other organs.
- Processes for deploying the deterrent (may exploit the respiratory, excretory, and reproductive systems).
- Individual and social behaviours for the appropriate positioning of chemicals.

 Check your understanding of the concepts covered in this section by answering the questions in the e-book.

32.6 Chemoreception and predator avoidance

Many animals rely on chemical signalling to find food and mates, and they also use it to avoid and deter predators. The deterrence of predators by foul smells released by prey, or the detection of the smell of a predator by its potential prey, relies on the sense of chemoreception.

Chemical signals released into the environment, in contrast to light and sound signals (Chapter 30), can be long-lasting. They also have the distinctive quality of forming a gradient from the source of the signal. This means that they have a degree of directionality. However, because chemical signals travel in air or water, they are easily affected by wind or currents in the external medium.

Chemoreception includes the senses of smell (olfaction) and taste (gustation). In terrestrial animals, smell and taste are easy to distinguish; we think of smells as chemicals in the air and tastes as chemicals detected by the mouth. The olfactory system detects a broad range of signals, including pheromones, while the gustatory system detects cues associated particularly with feeding.

For many animals this distinction is less meaningful, with many aquatic animals having chemoreceptors all over their body. It is helpful to view defence by chemoreception as one of several, overlapping ways in which chemical signalling is exploited by animals.

Olfaction

The main odour-detecting cells in vertebrates are the **olfactory receptor cells** (or olfactory receptor neurones). These are bipolar sensory neurones with a ciliated apical end and an axon running to the **olfactory bulb**. Vertebrates normally possess two olfactory bulbs, which are multi-layered, rounded, neural structures with different types of neurones in each layer. The olfactory bulbs pass information about odours to other parts of the brain (see Chapter 22).

In mammals, the apical ends of sensory neurones are found in the roof of the nasal cavity, but their exact location varies across the vertebrates. Fish usually possess two olfactory pits on the snout known as olfactory rosettes. Most tetrapods have external nares that lead to nasal passages and the nasal cavity.

Some terrestrial tetrapods possess an additional olfactory organ known as the **vomeronasal** organ. This detects **pheromones**, environmental chemicals released by conspecifics (other individuals in the group). Pheromones facilitate between-individual communication, especially during mating and in the establishment of social hierarchies. In reptiles the vomeronasal organ is known as Jacobson's organ. It exists as a separate pit into which the tongue can deliver chemicals.

Invertebrates use olfactory receptor neurones, similar in structure to those of vertebrates. Olfactory sensilla often have several pores to the external environment which allow odorants to come into contact with the olfactory receptor neurones. In arthropods they are packaged in hair-like sensilla found all over the body. In aquatic crustaceans, sensilla occur together in 'tufts' known as **aesthetascs** located on the antennae.

There are different types of sensilla, described according to their shape (straight, curved, asymmetric) or role as guard, companion, or sensory hairs, as illustrated in Figure 32.35 for two representative arthropods, the Caribbean spiny lobster and alfalfa plant bug.

Gustation

Unlike the olfactory system, which can detect thousands of different odorants, the gustatory system has a much more limited ability to discriminate between chemicals. In humans we talk about five main 'tastes': salty, bitter, sweet, sour, and *umami* ('savoury' or 'meaty'). Gustatory sensitivity declines with age (as does olfactory sensitivity), which may explain why our appreciation of flavours changes over time and why children tend not to like spicy or salty food.

Tastes are detected by taste buds. These comprise a number of taste receptor cells, clustered together in an onion-shaped structure. Figure 32.36a is a light micrograph of a taste bud, and Figure 32.36b is a schematic representation of a taste bud, both from the palate of the rainbow trout. You can see in Figure 32.36 the onion-shaped structure. Unlike olfactory receptor cells, however, taste receptor cells are epithelial cells rather than neurones. The receptors that detect sweet, bitter, and umami belong to the T1R and T2R G-protein coupled receptor family (GPCR). Binding of a ligand (i.e. a **tastant**) causes activation of these receptors and, via second messenger pathways, leads to the depolarization of the receptor cell membrane. The cell releases neurotransmitters onto gustatory neurones, which continue the signalling pathway.

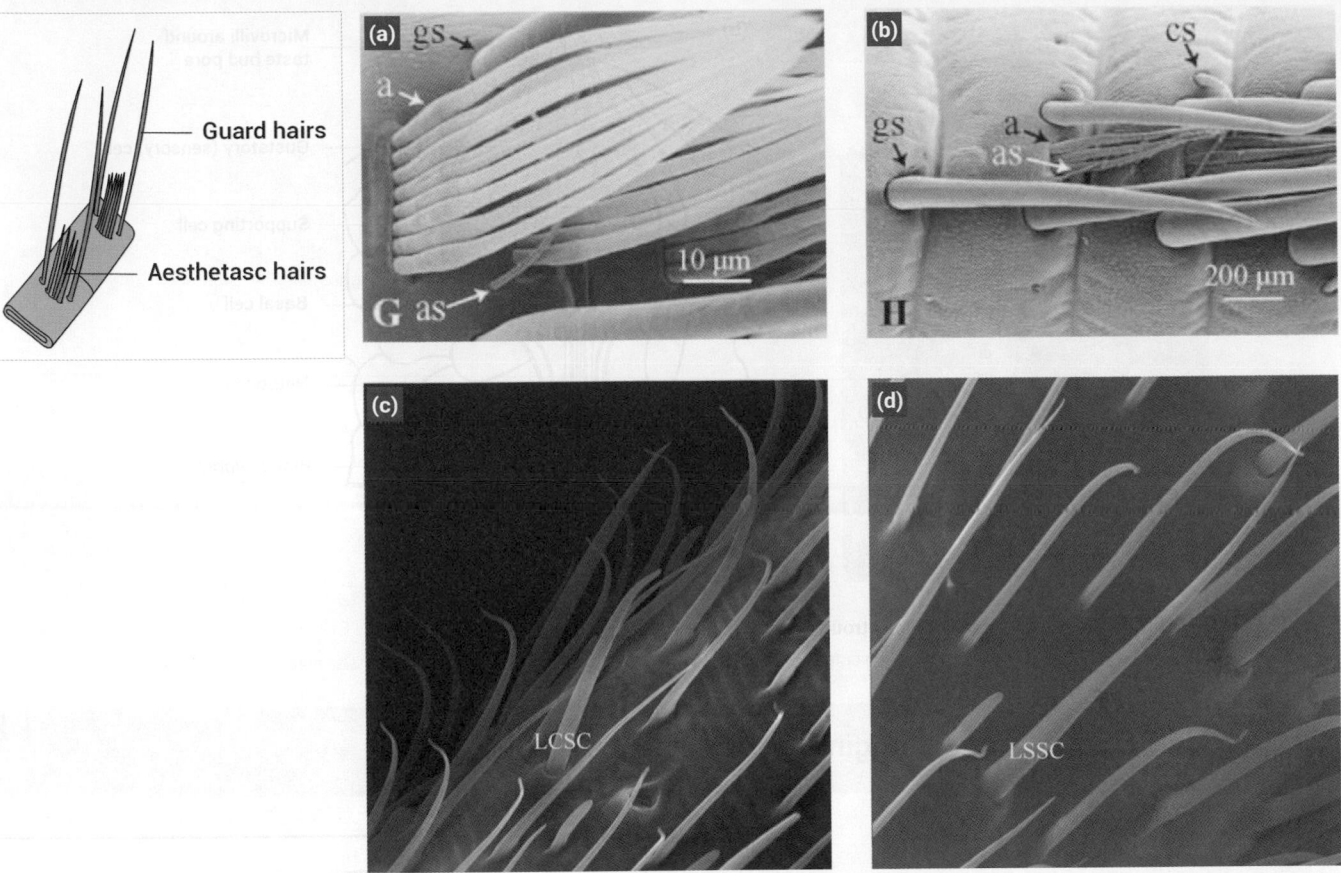

Figure 32.35 Scanning electron micrographs of invertebrate sensillae. (a) Aesthetasc (a), guard sensilla (gs), and asymmetric sensilla (as); (b) aesthetasc (a), guard sensilla (gs), asymmetric sensilla (as), and companion sensilla (cs) on flagella from a Caribbean spiny lobster, *Panulirus argus*. The sketch shows how the hairs are arranged on a crustacean antenna. Long curved (c) and straight (d) sensilla trichodea on flagella from an alfalfa plant bug, *Adelphocoris lineolatus*.

Source: (a,b) Cate, H.S., Derby, C.D. (2001) Morphology and distribution of setae on the antennules of the Caribbean spiny lobster Panulirus argus reveal new types of bimodal chemo-mechanosensilla. *Cell Tissue Res*, 304, 439–54. Reprinted by permission from Springer © 2001. (c,d) Gu *et al.* (2012) Functional characterizations of chemosensory proteins of the alfalfa plant bug *Adelphocoris lineolatus* indicate their involvement in host recognition. *PLoS One*, 7. https://doi.org/10.1371/journal.pone.0042871.

More than one taste receptor cell may synapse with a single taste neurone. This may partially explain the vast range of tastes that can be distinguished, although, as with olfaction, the situation is complex and far from fully understood.

Invertebrates detect tastes in the same way that they detect odorants, using sensilla. Gustatory sensilla are very similar to olfactory sensilla except they have one external opening (pore). Whatever the animal is tasting has to pass through this single pore to the sensory neurones, so direct contact with the chemical source is usually needed. Taste sensilla are found on mouthparts, and can also be located on a variety of other appendages, including legs, tentacles, and wings.

Using chemoreception to avoid predation

If an animal is to recognize the presence of a predator by olfaction, the chemical smell of the predator must be consistent and reliable. Furthermore, if the chemical signal persists in the environment, either for a long period of time or across a large space, then it loses its information value and is not very useful to the animal.

Many mammals avoid the smell of their predators by altering their behaviour. Predator odours may be released from faeces, urine, skin, or fur, and can cause a variety of behavioural responses. If an animal smells a predator, it may respond by minimizing its activity, keeping still to avoid attracting attention. Additionally, reducing activities such as foraging, feeding, and grooming allows the animal to focus its attention on watching for predators.

In fishes, the most researched predator warning signal is alarm substance (**Schreckstoff**), first discovered by von Frisch in 1938. He found that if the skin of a minnow (*Phoxinus phoxinus*) was damaged, the release of alarm substance into the surrounding water caused a behavioural fright reaction in other minnows. Alarm substance has now been demonstrated in a range of fish species and tentatively identified as the chemical hypoxanthine-3(*N*)-oxide.

The release of alarm substance by a prey fish when it is injured by a predator is unlikely to be of benefit to the prey fish itself but acts as a warning to nearby conspecifics. Predators may also release alarm substance if they have consumed a particular prey species. You can read more about the way in which some prey fish can even change their body shape to avoid predation in Scientific Process 32.3.

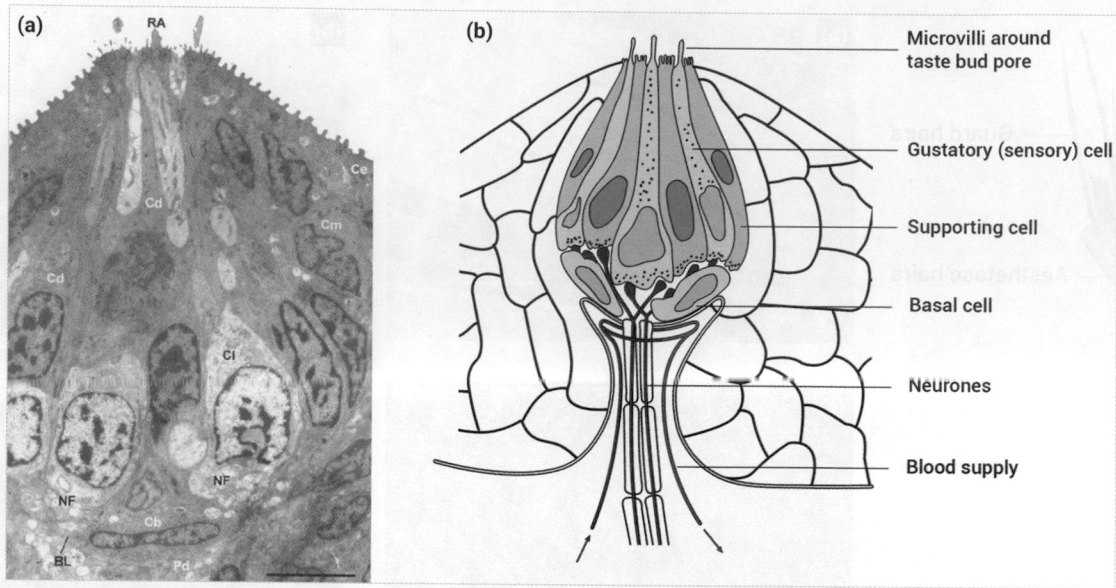

Figure 32.36 Taste bud from the palate of a rainbow trout. Light micrograph (a) and schematic representation (b).

Source: (a) Reutter & Witt (2004) Are there efferent synapses in fish taste buds? *Journal of Neurocytology*, 33, 647–56. Reprinted by permission from Springer © 2004.

SCIENTIFIC PROCESS 32.3 — Changing shape to avoid predation

Research question

What are the costs and benefits of changing body shape?

Background

The crucian carp (*Carassius carassius*), which you can see in Figure 1, changes its body shape in the presence of predators; a response is called phenotypic plasticity. A deeper body shape makes the carp more difficult to eat: in laboratory studies pike prefer feeding on shallow-bodied carp. While a deeper body may be less appetizing, greater body depth increases drag and may compromise swimming ability.

Hypothesis

A deeper-bodied morphology influences the ability of crucian carp to swim away from a predator.

Methods

Paolo Domenici and colleagues took carp from ponds with and without pike, placed them individually in a circular arena and measured characteristics of their swimming behaviour. Once the fish had acclimated to the arena, they were startled by the release of a novel object dropped into the tank.

Results

Compared with shallow-bodied carp, deep-bodied carp from the pond containing predators had a higher drag coefficient, making sustained swimming more energetically expensive. However, they also had a higher speed and acceleration in response to the disturbance.

Conclusion

The deeper body shape may be less appetizing but it also gives the fish a better performance in fast-start responses; this may be advantageous in avoiding predation.

Figure 1 Crucian carp taken from ponds (a) with and (b) without a major predator (pike, *Esox lucius*). Scale bars = 10mm.

Source: Domenici, P., Turesson, H., Brodersen, J. & Brönmark, C. (2008). Predator-induced morphology enhances escape locomotion in crucian carp. *Proc. Royal. Soc. B.* 275, 195-201. https://www.ncbi.nlm.nih.gov/pmc/articles/PMC2596180.

Read the original work

Domenici, P., Turesson, H., Brodersen, J. & Brönmark, C. (2008). Predator-induced morphology enhances escape locomotion in crucian carp. *Proc. Royal. Soc. B.*, 275, 195–201.

Chemoreception in territorial defence and conflict avoidance

Many mammals use smells to mark their territory. Scent marking enables solitary individuals living within the same home range to communicate with each other, helping to avoid conflict over territory and resource ownership. Scents include urine, faeces, and glandular secretions released through facial rubbing. Scent marking is most common near territory boundaries and increases during the reproductive season when animals are looking for mates.

Other animals use scent marking:

■ Bumblebees leave short-lived repellent smells on flowers that they have visited, preventing themselves or other bumblebees from visiting that flower again until the food source has been replenished.

■ During social fighting, some fishes and crustaceans release urine, with greater amounts produced as aggression increases.

Chemical signals in urine convey information about an animal's opponent, including its dominance status, and therefore could help an individual decide whether to continue fighting or to concede defeat before physical injury occurs.

 Follow the link in the e-book to explore more about the dominance hierarchies in fishes.

 Check your understanding of the concepts covered in this section by answering the questions in the e-book.

SUMMARY OF KEY CONCEPTS

■ All living organisms have mechanisms by which they defend the parts of the biosphere that they occupy. If these mechanisms are inadequate or fail, the organism is invaded by parasites or pathogens, is consumed as a food source or dies and decays.

■ Existential and reproductive imperatives mean that organisms and their predators are constantly evolving in an arms race between defence and invasion, protection and predation.

■ Plants cannot escape their predators and are under threat of consumption by animals, called herbivory. They depend for survival on physical deterrents (spikes, thorns, impenetrable structure), products of secondary metabolism (aversive chemicals, poisons, stings), and the regenerative properties of their tissues.

■ Plants defend against micro-organisms and invertebrates by producing chemicals that specifically resist pathogenic elicitors or effectors, often through gene for gene adaptations. Selective breeding and genetic manipulation can exploit these processes to increase the productivity of food crops.

■ An animal's integument forms a protective barrier but must allow for the controlled exchange of fluids, gases, and nutrients. Integuments are also involved in insulation, coloration, communication, interception of light energy, locomotion, and other functions.

■ Integuments typically have a layered structure, based on ectoderm and mesoderm. They permit growth (except in arthropods, where periodic moulting and replacement occur) and surface regeneration. The nature of the integumental layers and the variety of cells they contain are defined by the animal group. The outer layers of most vertebrates are decidual and elaborated into scales, hairs, or feathers.

■ The response to integumental injury in vertebrates comprises inflammation, haemostasis, wound repair, and tissue remodelling. Inflammation, which is a local reflex at the site of injury, is typified by redness, heat, pain, and swelling. An initial, transient drop in blood flow is followed by a triple response of redness, swelling, and spreading hyperaemia.

■ The pain, swelling, and other features of inflammation result from the formation of exudate coupled with the secretion of chemicals by cells in local tissues and blood.

■ Haemostasis, wound repair, and tissue remodelling result from the initiation of enzyme cascades in local tissues and blood. These cascades also promote responses in the immune system which are adaptive against the potential invasion of micro-organisms.

■ The colours of organisms can be assumed to have evolved for their adaptive value. Coloration may provide camouflage, indicate danger to predators (aposematism), or act as a signal to others in the group. Group-directed signals provide information on fitness, status, or reproductive condition, and permit sexual selection.

■ Camouflaging coloration enables an organism to hide in its environment, but the concept of crypsis can be extended to a broader understanding of deception, including Batesian mimicry (lacking intrinsic danger but resembling other, dangerous species in the vicinity) and Müllerian mimicry (being dangerous but resembling *more* dangerous species).

■ Toxins are chemicals produced by organisms which deter, injure, or kill predators. Poisons are toxins transferred passively, by ingestion or touch, while venoms are toxins that are secreted actively and usually transferred by mechanical injury such as a bite. Poisons and venoms are common among many groups of invertebrates and vertebrates, but only a very few mammals are known to be venomous.

■ Potential predators detect toxins and warning chemicals using sensory systems related to those of small and taste. It remains unclear how predators learn to avoid lethal toxins and how such information is passed between conspecifics and between generations.

 Use the flashcards in the e-book to test your recall of key terms introduced in this chapter.

QUESTIONS

 Looking for answers? Once you've answered these questions, follow the link in the e-book to the answer guidance and check your work.

Concepts and definitions

1. Define herbivory and give several examples using animals representing different animal groups and environments.

2. Define and distinguish between biotrophic and necrotrophic micro-organisms.

3. Define integument and list some of its functions. How might its protective functions interact with or be limited by its other physiological functions?

4. Define and describe the differences between dermis and epidermis.

5. Define inflammation, haemostasis, wound repair, and tissue remodelling as components of the response to injury.

6. Define crypsis and aposematism.

Apply the concepts

7. Explain what distinguishes the integuments of mammals and birds from those of other vertebrates.

8. Explain how secondary metabolism in plants forms the basis of protection against predators and pathogens.

9. Explain the concept of gene for gene adaptations between organisms and their pathogens and predators.

10. Explain how the response to injury initiates elements of the immune response and discuss the likely adaptive advantages of this link.

11. What is the difference between a poison and a venom? Find examples of each, other than those presented in the main text.

Beyond the concepts

12. What are the key differences and similarities between vertebrate and invertebrate immune systems? (You may wish to refer to Chapter 27 for more information about the vertebrate immune system.)

13. Explain the difference between Batesian and Müllerian mimicry and find some examples of each, other than those presented in the main text.

14. What might the adaptive advantages be to an animal of honest and dishonest signalling? Find examples of each from as wide a range of organisms as you can, avoiding those presented in the main text.

15. In what possible ways might a species learn to avoid a toxic or lethally dangerous organism? In what ways might this information be transmitted horizontally (between individuals in the population) or vertically (from individuals to their offspring)? In considering the possibilities, think how you might test them experimentally.

16. Discuss the possible value or otherwise of the term toxunguent.

FURTHER READING

To find out more about coloration, the history of its interpretation and modern insights, good places to begin would be:

Caro, T. (2017) Wallace on coloration: contemporary perspective and unresolved insights. *Trends Ecol. Evol.* **32**: 23–30.

Caro, T. (2009) Contrasting coloration in terrestrial mammals. *Philos. Trans. R. Soc. B* **364**: 537–48.

Caro, T., Stoddard, M.C. & Stuart-Fox, D. [Eds.] (2017) Theme issue: Animal coloration: production, perception, function and application. *Philos. Trans. R. Soc. B* **372**, Issue 1724 https://royalsocietypublishing.org/toc/rstb/372/1724
A series of reviews and research articles on the biochemistry, physiology, and genetics of coloration, how colours are perceived, their function, and evolution.

Caro, T. & Allen, W.L. (2017) Interspecific visual signalling in animals and plants: a functional classification *Philos. Trans. R. Soc. B* **372**: 20160344. http://dx.doi.org/10.1098/rstb.2016.0344
A broad review of visual signals, camouflage, and mimicry; a functional assessment of different methods of visual communication.

Skelhorn, J., Halpin, C.G. & Rowe, C. (2016) Learning about aposematic prey. *Behav. Ecol.* **27**: 955–64. doi:10.1093/beheco/arw009
An accessible guide to aposematism, warning signals, and deception, including a brief historical account of the topic back to Darwin and Wallace.

Roeder, A. & Harrison, J. (2019) To regenerate or not to regenerate: factors that drive plant regeneration. *Curr. Opin. Plant Biol.* **47**: 138–50. https://www.sciencedirect.com/science/article/pii/S1369526618300517
A comprehensive guide to plant regeneration which explains why a complete plant can grow from single cell taken from a leaf, stem, or root.

Keener, A.B. (2016) Holding their ground. *Scientist*, Feb. 1, 2016. https://www.the-scientist.com/features/holding-their-ground-34128 and https://www.the-scientist.com/infographics/plant-immunity-34130
A summary of plant immune systems and disease resistance in crops.

Gurtner, G.C., et al. (2008) Wound repair and regeneration. *Nature* **453**: 314–21.
A broad review of wound repair processes, tissue regeneration, and the underlying cellular activity, including how it happens in other mammals and invertebrates.

Cerenius, L. & Söderhäll, K. (2013) Variable immune molecules in invertebrates. *J. Exp. Biol.* **216**: 4313–19.
A summary of invertebrate immune mechanisms which shows that they are more complex than previously believed and may include specific or adaptive components.

Nelsen, D.R. et al (2014) Poisons, toxungens, and venoms: redefining and classifying toxic biological secretions and the organisms that employ them. *Biol. Rev.* **89**: 450–65.
A comprehensive review of the biology and taxonomy of poisons and related chemicals, and how animals and plants make use of them.

Reproduction and Development

LEARNING OBJECTIVES

By the end of this chapter, you should be able to:

- Explain the concept of a reproductive strategy and describe some of its elements using example plants and animals.

- Know the meaning of sexual differentiation and be able to account for sex determination by genetic or environmental factors in different groups of animals.

- Define the terms used to describe reproduction, including fertility, fecundity, infertility, and reproductive efficiency.

- Apply the concept of r- and K-strategies to reproduction in any plant or animal species, appreciating the role of parental investment in the success of offspring.

- Describe the essential differences in reproduction between angiosperms and gymnosperms.

- Outline the fundamental structure of an animal spermatozoon and describe the basic processes of spermatogenesis and capacitation.

- Describe some characteristics of invertebrate reproduction and outline the developmental sequence in holometabolous and hemimetabolous insects.

- List some of the key differences between amniotes and anamniotes, and between placental and non-placental mammals, and know the principal animal groups in each category.

Chapter contents

Introduction		1142
33.1	Reproduction and sex	1142
33.2	Measuring reproduction: fecundity and fertility	1156
33.3	Reproduction in plants	1158
33.4	Reproduction in animals	1167
33.5	Development in amniotes	1174
33.6	Sperm	1176
33.7	Annelid reproduction	1179
33.8	Insect reproduction	1181
33.9	Fish reproduction	1184
33.10	Amphibian reproduction	1190
33.11	Mammalian reproduction strategies	1191

Watch the key concepts video in the e-book to prepare yourself for studying this chapter.

Introduction

In this module we have been studying the characteristic features of living organisms. At the start we saw that all organisms, whether simple or complex, enclose a region of biological space in which the flow of energy is regulated. In Chapter 32 we looked at how that space is defended, noting that if defence fails the organism dies and the energy stored in its structure is taken up and used elsewhere in the biosphere. In between, we investigated the variety of structural forms which organisms adopt, how they interact with their environment, and how they move around it.

It is difficult to give a precise, comprehensive, and uncontroversial definition of 'life' even though we usually know what we mean when we use the word. Besides the enclosure and defence of a region of biological space, most biologists would agree that all living organisms are characterized by an ability to replicate themselves and by being in a state of continual development throughout their lifespan. This is the subject of this chapter.

Reproduction, growth, and differentiation (the specialization of functions within an organism) all happen at the cellular level, even if the results are seen as changes in the whole organism. Genes are fundamental to all these processes because they define an organism's *phenotypic potential*. They are also the units of biological information which are replicated when an organism reproduces itself.

We need to consider:

- the different forms of replication which happen in single-celled and multicellular organisms

- where and in what form genetic information is held so that it can be transmitted from one generation to the next

- the range of strategies which organisms employ to achieve successful reproduction

- how organisms are in a state of continual change and developmental during their lifespan.

More detail on genes and how they work can be found in Chapter 4 and Chapter 35. We discuss cell biology in Module 2. We consider reproduction in humans in Chapter 26, where you will also find further information about the production of gametes, fertilization, pregnancy, and early development. This chapter considers reproductive processes in eukaryotic organisms only; we explore reproduction in prokaryotes in Chapter 10.

33.1 **Reproduction and sex**

Every organism alive today is a descendent of a previous generation, with ancestors spanning back across the ages. But where did this chain of life begetting life begin?

Ancestry and its meaning

Biologists currently assume that life on Earth began with the emergence of a single living organism, probably at least 3.5 billion years ago. The nature of this first example of life is unknown and probably impossible to imagine, not least because the location of its emergence is still uncertain. Indeed, biologists argue at great length about what they would accept as defining the original living thing—the first biological organism—and how it came into existence as a functional unit.

▶ We discuss the emergence of life futher in Topic 2.

Be that as it may, biologists also assume that everything now living is descended from that single, foundational organism. For this to be correct, at least three things must have occurred:

1. The first organism replicated itself.

2. There has been an uninterrupted sequence of subsequent replication events up to the present day.

3. Organisms resulting from replication have been the subject of variation.

The only feasible alternative would be for life to have originated more than once. But if that were the case we might expect to find forms of life with unique chemical structures and systems of metabolism based on something other than the water–carbon–nitrogen–phosphorus system with which we are familiar.

No such separate forms have ever been reliably detected. If more than one original form had occurred and they had somehow combined to generate life as we currently know it, we would be forced to speculate whether the merged forms had arisen independently, so the result would be essentially the same.

Given these assumptions, logic dictates that everything alive today, in all its myriad shapes and varieties, is ancestrally related. We may find it easy to imagine, for example, that all living mammals are descended from some ancestral mammal and, tracing things further back, that all vertebrates, all deuterostomes, all triploblasts, and all metazoans share a single, individual, multicellular ancestor.

But we should go further and accept that multicellular plants and multicellular animals originated from some earlier kinds of organism and that these can claim their own ancestry somewhere among single-celled forms of eukaryotic life. Presumably some even earlier connection links eukaryotes to prokaryotes, or to an ancestor of both which was neither.

This logic provides the basis for creating endlessly diverging family trees of life (for example, the fractally arranged *One Zoom*: http://www.onezoom.org/) and permits us to assign distant cousin status to all currently living animals, plants, protists, and microbes.

Note that trees of biological genealogy are intended to illustrate ancestral relationships. It is a further step to assume that organisms located within a particular region of the tree (a set of branches or twigs) share common genetic, structural, morphological, behavioural, or other characteristics.

▶ For further discussion of ancestry and its relationship to the way biologists describe, name, and group living organisms, see Topic 5 and also Chapter 11.

Replication, reproduction, and reproductive strategies

Replication and the resulting ancestral connection imply the continual inheritance of genetic material down the generations. **Reproduction** is the term we use to describe the processes by which replication and inheritance take place.

The complete set of reproductive characteristics of a plant or animal constitute its **reproductive strategy**. These characteristics include:

- its use of sexual or asexual procedures;
- the presence or otherwise of distinct sexes;
- the relationship between the timing of reproduction and stages of body development;
- the dependence of reproductive activity on environmental signals;
- cycles of gamete production and the numbers of gametes produced;
- methods of mating and fertilization;
- the extent to which a fertilized zygote or embryo is protected by its parents;
- parental involvement in the early development of offspring and their upbringing; and
- the numbers of offspring produced per organism, per litter, per reproductive event, per lifetime, or per year.

(The meanings of words such as **gamete**, **zygote**, and **embryo** will become clear as we proceed.)

The complexity and variety of reproductive strategies which we see among different types of organism evolved under three pressures:

1. to optimize the chance that successful mating and fertilization will take place;

2. to maximize the chance that offspring will be produced and develop under environmental conditions conducive to their own survival and eventual reproduction; and

3. to enable effective parental and social support.

The evolutionary force that these pressures represent is the replication of DNA and the transmission of the information it carries. An organism's reproductive strategy is part of its phenotype. Successful phenotypes are those which allow their underlying genotypes to survive and proliferate.

Gene transmission and ploidy

In eukaryotic organisms, DNA is located in chromosomes. The chromosomal status of reproducing cells is a convenient way of understanding their reproductive condition and the amount of DNA they contain. **Ploidy** refers to the number of complete sets of chromosomes in a cell and is represented as [2n] for **diploid** and [n] for **haploid**. Diploid means that there is a full complement of *paired* chromosomes, and therefore paired alleles for each genetic characteristic. Haploid means that there is just one of each chromosome pair, so half the full amount of DNA.

▶ We learn more about ploidy in Chapter 4.

There are essentially two ways in which an organism can reproduce itself: by **asexual** and **sexual** processes. Asexual processes, also called **clonal** reproduction, include fission and budding in single-celled and multicellular organisms and parthenogenesis in multicellular organisms.

In *asexual* systems, the offspring are genetically identical to their parents. In the case of single cells, either a diploid [2n] cell replicates its genetic material and divides to produce two diploid 'daughter' cells, or a haploid [n] cell divides to produce two haploid daughter cells.

▶ We discuss cell division in more detail in Chapter 10.

In multicellular organisms, asexual reproduction generates eggs which can turn into genetically identical replicas of the parent, either diploid or haploid. Sometimes, a pair of haploid cells combine to produce a diploid offspring.

$$[n] + [n] \rightarrow [2n]$$

In *sexual* reproduction, the parent organism generates haploid [n] germ cells in the form of gametes (typically, eggs and sperm). Two such germ cells combine to form a diploid offspring. In contrast to the asexual state, the genetic compositions of the two germ cells differ because they usually come from different individuals (which we call male and female), and further variation occurs when the germ cells combine.

In most sexually reproducing complex organisms, germ cell production occurs in specialized tissues or in organs of reproduction called gonads.

Sexual and asexual phases

Many organisms, both unicellular and multicellular, have asexual and sexual phases in their lifecycles. Figure 33.1 illustrates the asexual and sexual phases exhibited by the fern, while Figure 33.2 depicts the alternating phases exhibited by the aphid. In fact, organisms which we think of as reproducing sexually actually pass on their genes through a sequence of haploid and diploid states, as we shall see.

The lengths of sexual and asexual phases and of haploid and diploid states vary considerably between different types of organisms. This makes it something of a challenge for biologists to generalize about how reproduction occurs, even within related groups of organisms, let alone across the great diversity of life.

A further complexity is that there are both advantages and disadvantages to the haploid and diploid states with regard to the expression of genes, protection against harmful mutations, adaptation, and evolutionary success.

Before considering the mechanisms of reproduction in more detail, we can make the following broad general statements about sexually and asexually generated offspring.

- Asexual fission results in the formation of two identical cells at the expense of the single cell from which they originated. In contrast, when gametes are produced from a multicellular parent, the original, parental collection of cells—the reproducing organism—remains.

- As a result of this, daughter cells produced by fission are never contemporary with their parent cell. When reproduction occurs through germ cells, different generations can and usually do co-exist.

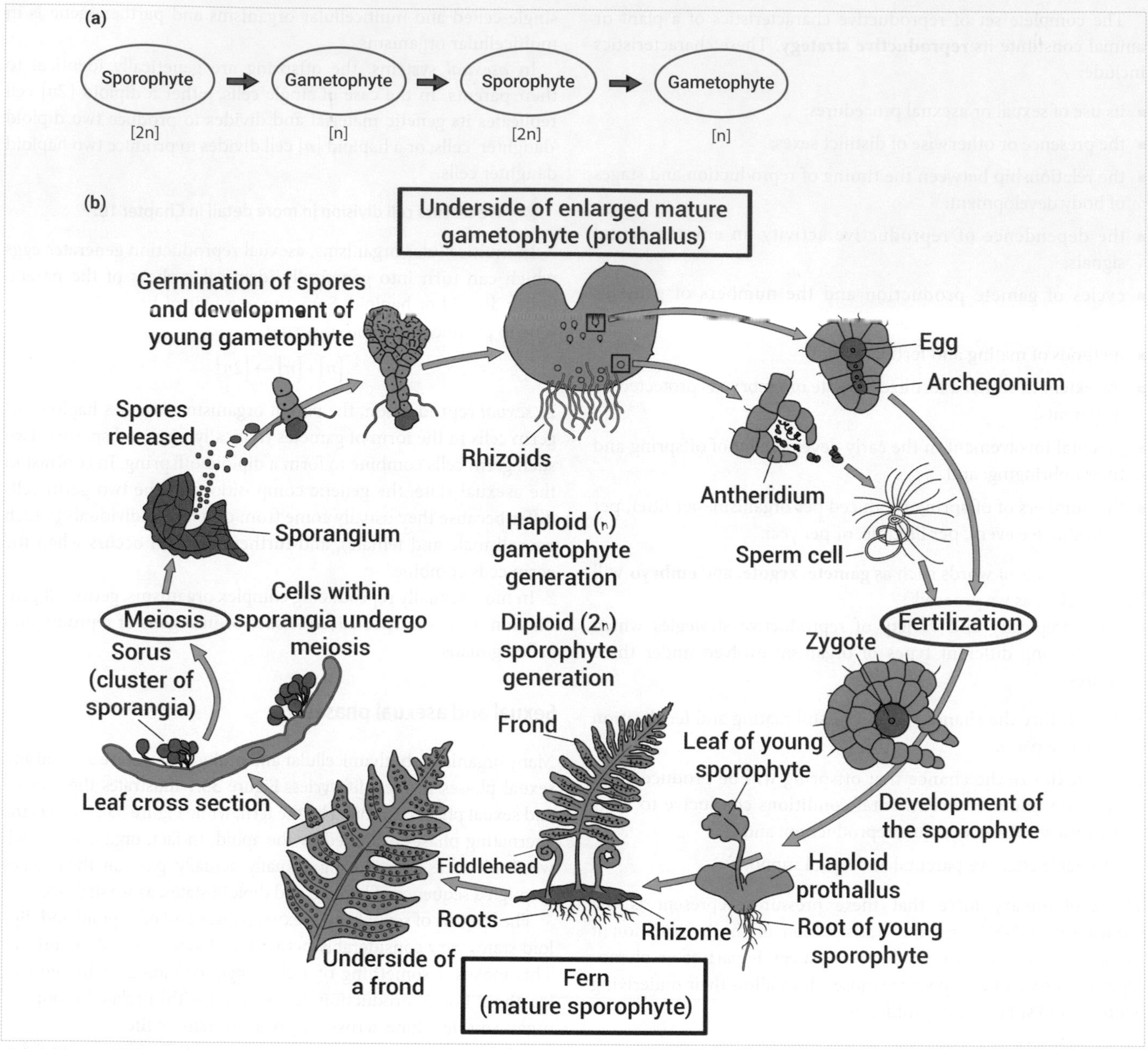

(a)

Sporophyte → Gametophyte → Sporophyte → Gametophyte
[2n]   [n]   [2n]   [n]

(b)

Underside of enlarged mature gametophyte (prothallus)

Germination of spores and development of young gametophyte

Spores released

Sporangium

Meiosis

Cells within sporangia undergo meiosis

Sorus (cluster of sporangia)

Leaf cross section

Rhizoids

Egg

Archegonium

Antheridium

Sperm cell

Haploid (n) gametophyte generation

Diploid (2n) sporophyte generation

Frond

Fertilization

Zygote

Leaf of young sporophyte

Development of the sporophyte

Fiddlehead

Roots

Rhizome

Haploid prothallus

Root of young sporophyte

Underside of a frond

Fern (mature sporophyte)

Figure 33.1 Reproduction in ferns. (a) Ferns reproduce in a sequence of diploid (asexual) and haploid (sexual) stages. (b) For conceptualization, it is usually illustrated as a cycle.

Go to the e-book to explore interactive versions of Figures 33.1 and 33.2.

- If we consider a cell as a vehicle for the transmission of its genetic material, those which reproduce by fission never die (unless by external causes), while multicellular organisms that reproduce by germ cells, whether sexual or asexual, eventually always do.

- The cells which result from asexual reproduction are always genetically identical. Those which result from sexual reproduction are genetically different.

- A diploid organism which reproduces by sexual means discards half of its DNA when each haploid gamete is formed; to form a new individual it relies on the remaining DNA combining with that from another haploid gamete.

- The specialization of germ cell-producing tissues in multicellular organisms means that the production of genetic material for transmission occurs in a limited region of the organism's body (ovary, testis, anther, or equivalent). This is one reason why neither parental phenotypic characteristics acquired from the environment nor genetic mutations occurring in non-germ cells can be inherited through the 'germ line' during normal reproductive processes: such changes are not passed on to the next generation.

Because two different individuals are normally involved in sexual reproduction, there is the potential to increase the amount of genetic diversity within the population.

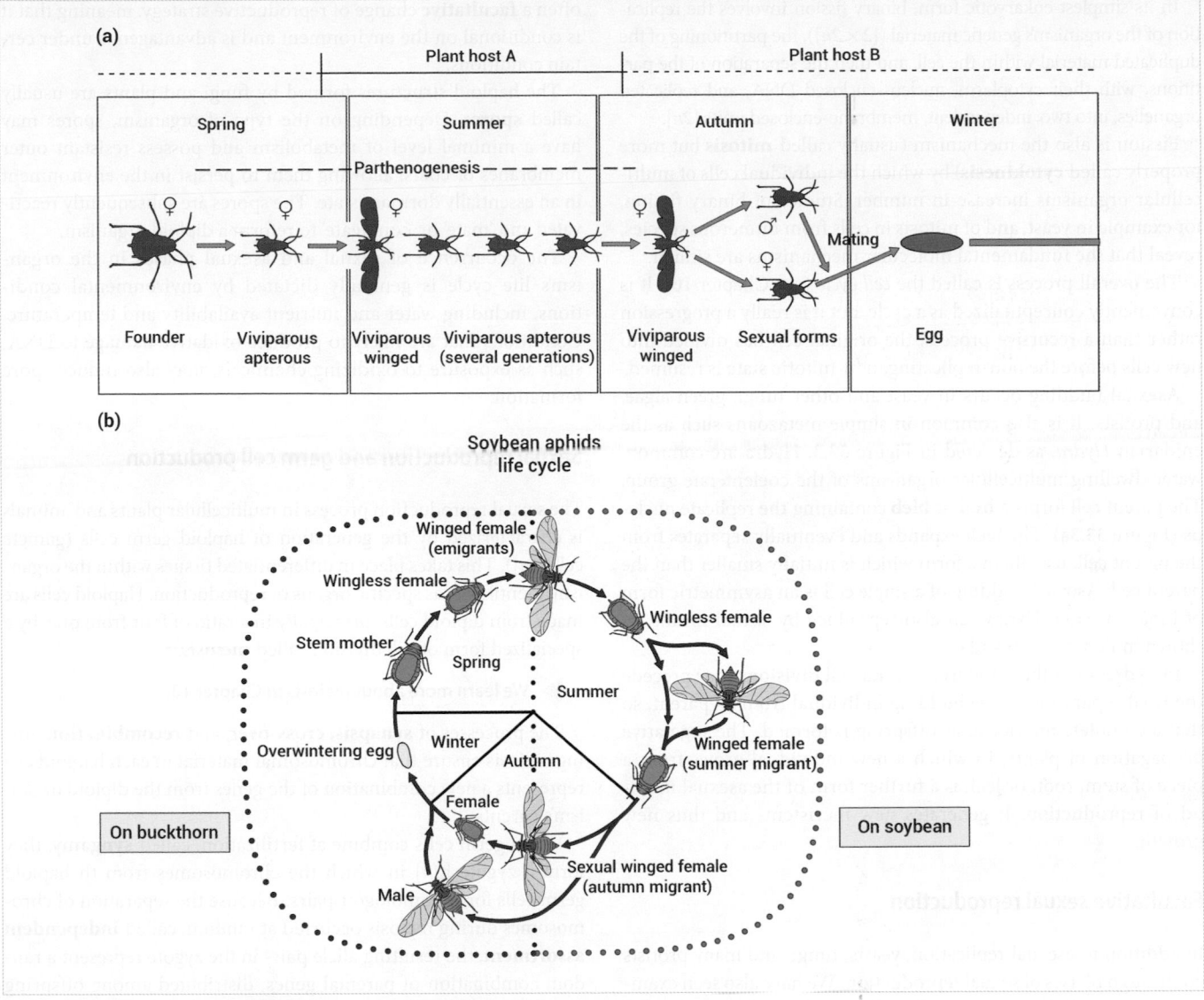

Figure 33.2 The aphid life cycle. (a) Aphids are noted for their complex life cycles. Their reproduction alternates between asexual and sexual phases, and they also appear in winged and flightless forms. (b) The soybean aphid alternates between two plant hosts.
Source: (b) David Voegtlin, University of Illinois at Urbana-Champaign.

Biologists describe this advantage, and the additional resilience it produces against disease and other environmental challenges, as 'hybrid vigour'. Conversely, if a population contains too few individuals for sufficient variation to be available, making organisms less able to survive or breed, they speak of 'inbreeding depression'.

Animal and plant breeders deliberately undertake selective 'outcrossing' to diversify the parental stock from which their selected varieties are derived.

Some biologists believe that sexual reproduction evolved because of the advantage that increased variability offers to the population. At the level of the individual organism, small variations in the identities of inherited genes may be particularly valuable in maintaining resistance to parasites. Of course, parasites themselves are constantly evolving to overcome host resistance. The resulting arms race represents a significant evolutionary pressure.

Asexual cellular fission and budding

Cellular fission and budding are processes in which each resulting cell has the same amount of DNA as the cell from which it originated. Binary fission occurs in many protists, some yeasts and other fungi, as well as in prokaryotes and archaea. It also occurs in plants, although the new cells remain attached to one another. The 'binary' descriptor implies that the resulting cells are equal in size and completely replace the original, parent individual.

In its simplest eukaryotic form, binary fission involves the replication of the organism's genetic material ([2 × *2n*]), the partitioning of the duplicated material within the cell, and then the separation of the partitions, with their cytoplasm, nucleus-enclosed DNA, and replicated organelles, into two, independent, membrane-enclosed units [*2n*].

Fission is also the mechanism (usually called **mitosis** but more properly called **cytokinesis**) by which the individual cells of multicellular organisms increase in number. Studies of binary fission, for example in yeast, and of mitosis in cells from numerous species, reveal that the fundamental molecular mechanisms are similar.

The overall process is called the *cell cycle* (see Chapter 10). It is conveniently conceptualized as a cycle, but it is really a progression rather than a recursive process: the original cell has divided into new cells before the non-replicating, non-mitotic state is resumed.

Asexual budding occurs in yeast and other fungi, green algae, and protists. It is also common in simple metazoans such as the cnidarian *Hydra*, as depicted in Figure 33.3. Hydra are common, water-dwelling multicellular organisms of the coelenterate group. The parent cell forms a **bud** or **bleb** containing the replicate nucleus (Figure 33.3a). The bleb expands and eventually separates from the parent cell, usually in a form which is initially smaller than the parent cell. Asexual budding of a single cell is an asymmetric form of binary fission. Hydra can also reproduce by sexual means, as shown in Figure 33.3b and c.

In hydra and other cnidaria, several cell divisions may precede the final separation of the budding individual from its parent, so that a complete multicellular offspring is formed. The vegetative propagation of plants, in which a new individual grows from a piece of stem, root, or leaf, is a further form of the asexual method of reproduction. It generates new meristems and thus new growth.

Facultative sexual reproduction

In addition to asexual replication, yeasts, fungi, and many protists go through phases of sexual reproduction. We have also seen examples of multicellular organisms that progress through a sequence of asexual and sexual states (see Figure 33.1 and Figure 33.2). This is often a **facultative** change of reproductive strategy, meaning that it is conditional on the environment and is advantageous under certain conditions.

The haploid structures formed by fungi and plants are usually called **spores**. Depending on the type of organism, spores may have a minimal level of metabolism and possess resistant outer membranes or coats, allowing them to persist in the environment in an essentially dormant state. The spores are subsequently reactivated and 'mate' or 'conjugate' to reform a diploid organism.

The occurrence of sexual and asexual phases in the organism's life cycle is generally dictated by environmental conditions, including water and nutrient availability and temperature. Conditions that are likely to produce oxidative damage to DNA, such as exposure to oxidizing chemicals, may also induce spore formation.

Sexual reproduction and germ cell production

The sexual reproduction process in multicellular plants and animals is characterized by the generation of haploid germ cells (gamete cells; [*n*]). This takes place in differentiated tissues within the organism, identifiable as specific organs of reproduction. Haploid cells are made from diploid cells, principally in a ratio of four from one, by a specialized form of cell division called **meiosis**.

 We learn more about meiosis in Chapter 10.

The processes of **synapsis, cross-over**, and **recombination** during meiosis ensure that chromosomal material in each haploid cell represents a new combination of the genes from the diploid organism's parents.

When germ cells combine at fertilization, called **syngamy**, they form a zygote [*2n*] in which the chromosomes from th haploid germ cells form homologouspairs. Because the separation of chromosomes during meiosis occurred at random, called **independent assortment**, the resulting allele pairs in the zygote represent a random combination of parental genes, distributed among offspring on a fixed odds basis according to the principles of Mendelian genetics (see Chapter 4).

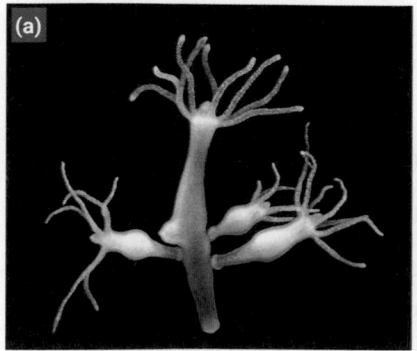 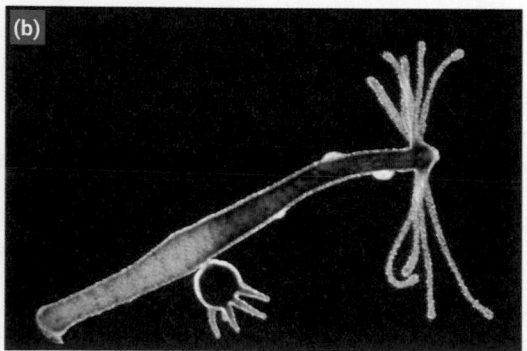 

Figure 33.3 Hydra reproduction. (a) Brown hydra with mature and immature buds. (b) Male hydra with multiple testes on the body column. (c) Female hydra with a fertilized egg on the left and an asexual bud on the right.

Source: (a) Lebendkulturen.de/Shutterstock; (b) Alamy; (c) Copyright Carolina Biological Supply Company. Used by permission only.

A typical flowering plant which reproduces sexually has a complex life cycle, passing alternately through haploid and diploid generations. The diploid state, which may be represented by a large structure such as a tree or shrub, is called a **sporophyte** because it makes spores, called pollen, for dispersal. Spores are single cells that can germinate if exposed to the right environment, developing by themselves into complete new plants.

We focus more on plant reproduction later in this chapter, but the sexual reproductive cycle of a typical flowering plant is shown in Figure 33.4. This is called the **gametophytic** phase because the diploid seeds it generates for dispersal result from the combination of haploid gametes. The example in the figure represents just one species and there are many variations among different groups of plants. They are all complex processes, both physiologically and genetically.

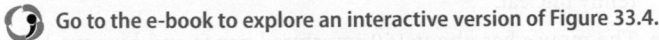 **Go to the e-book to explore an interactive version of Figure 33.4.**

In animals in which the sexes are separate, male gametes (sperm, [*n*]) are produced in the testes and female gametes (ova or eggs, [*n*]) are produced in the ovaries. These organs are normally the only places in the animal where meiotic division occurs. The haploid gametes are stored in the gonadal tissues or associated tubules until being released for fertilization.

- In male animals, it is usual for sperm to be manufactured continuously within the testis throughout the male's reproductive lifespan, although often with seasonal variations in quantity or with environmentally and socially determined periods of interruption.

- In females, the ovary produces eggs from a cluster of primordial germ cells which is established when the ovarian tissue is first

formed. Thus, the female is provided with a finite, although usually excessive, store of germ cells. The maturation of these cells as eggs may be associated with the mitotic proliferation of adjacent **somatic** ovarian cells, with the result that complex, multicellular structures are formed within the gonad and around the germ cell. The release of female germ cells for fertilization is often cyclical and seasonal, and may be strongly affected by developmental, nutritional, and social factors.

We see from this that in both sexes, but especially in females, haploid germ cells can exist in a dormant condition for a long time. In animals with internal fertilization, there may be a further period of storage inside the recipient organism's reproductive tract.

In both sexes, the generation of mature, biochemically active germ cells and their release for fertilization is usually accompanied by characteristic behaviours and by related physiological changes in the rest of the animal's body.

Parthenogenesis

Parthenogenesis, also called **apomixis** in plants, is usually defined as reproduction by a female organism which does not involve fertilization by a male. In genetic terms, it is a form of asexual reproduction in which a female embryo or seed is produced without combination.

The absence of fertilization in parthenogenesis means that the syngamy which characterizes sexual reproduction does not occur. As a result, parthenogenetic offspring are usually genetically identical to their mothers, with either a full ([*2n*], called 'full clones') or half ([*n*], called 'half-clones') complement of maternal alleles. Offspring are normally female (XX if diploid, or XO if haploid), although there can be exceptions to this in organisms where sex determination is different from the usual XY system (see later in this section).

Plants which reproduce by apomixis (examples are dandelion (*Taraxacum* spp.), garlic (*Allium sativum*), and blackberry (*Rubus fruticosus*)), produce seeds without requiring the fusion of gametes. The resulting embryos are genetically identical to the parent plant.

Besides apomixis in plants, parthenogenesis occurs naturally in a wide range of animals. Protostome examples include many insects and several other arthropods, velvet worms, rotifers, flatworms, and gastropod molluscs. In some of these, especially among the insects, the offspring result from **automixis**, a form of meiosis in which the haploid cells subsequently fuse to make a diploid but genetically uniform organism.

Examples of parthenogenesis also occur amongst the deuterostomes. They include several amphibians (especially salamanders, but also some hybrid frog species), some lizards and snakes, and a number of shark species such as the hammerhead. A few domesticated birds (chicken, turkey, pigeon) have been observed to reproduce parthenogenetically, but this may be an artefact of intensive breeding rather than a natural occurrence.

Parthenogenesis has not been reliably reported in mammals, except under experimental conditions. Dolly the sheep was produced by cloning and nuclear transfer, a process in which the nucleus from a non-gamete cell, in this case from the mammary gland, was de-differentiated, reprogrammed, and transferred to an egg whose original nucleus had been removed. This is rather different from using gametes alone to produce offspring.

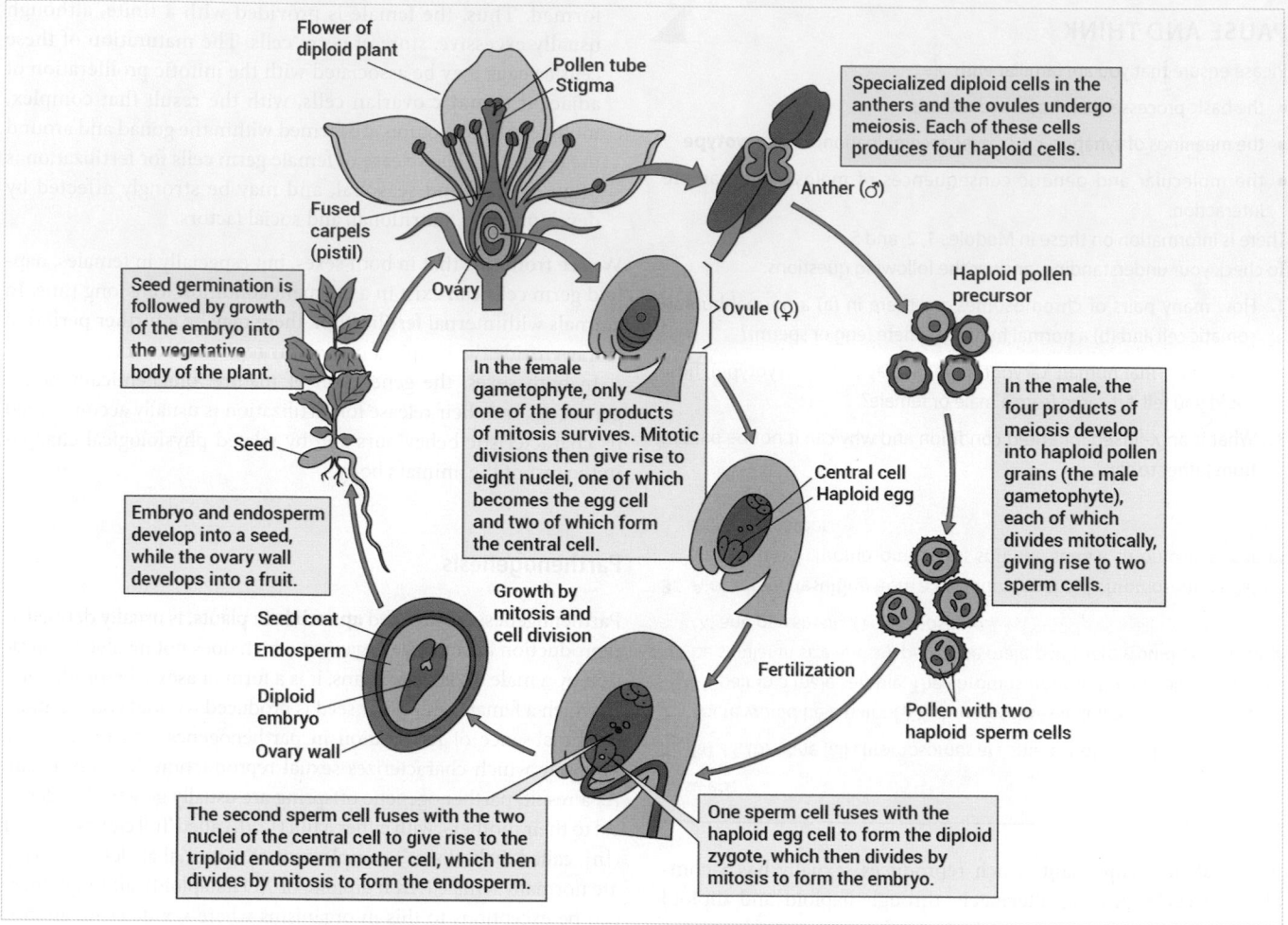

Flower of diploid plant

Pollen tube
Stigma

Specialized diploid cells in the anthers and the ovules undergo meiosis. Each of these cells produces four haploid cells.

Fused carpels (pistil)

Anther (♂)

Ovary

Ovule (♀)

Haploid pollen precursor

Seed germination is followed by growth of the embryo into the vegetative body of the plant.

In the female gametophyte, only one of the four products of mitosis survives. Mitotic divisions then give rise to eight nuclei, one of which becomes the egg cell and two of which form the central cell.

In the male, the four products of meiosis develop into haploid pollen grains (the male gametophyte), each of which divides mitotically, giving rise to two sperm cells.

Seed

Central cell
Haploid egg

Embryo and endosperm develop into a seed, while the ovary wall develops into a fruit.

Growth by mitosis and cell division

Fertilization

Seed coat
Endosperm
Diploid embryo
Ovary wall

Pollen with two haploid sperm cells

The second sperm cell fuses with the two nuclei of the central cell to give rise to the triploid endosperm mother cell, which then divides by mitosis to form the endosperm.

One sperm cell fuses with the haploid egg cell to form the diploid zygote, which then divides by mitosis to form the embryo.

Figure 33.4 The life cycle of the cherry (*Prunus* spp.).

Source: Maarten J. Chrispeels and Paul Gepts [Eds.]. Plants, Genes, and Agriculture, Sustainability through Biotechnology, 2017. Oxford University Press.

PAUSE AND THINK

What is the essential genetic difference between reproduction which involves gamete fertilization and that which does not (parthenogenesis)?

Answer: In gamete fertilization, the genetic material of the offspring comes from two parents rather than one. In parthenogenesis, offspring contain the genetic material from one parental individual only.

Same-sex reproduction

If same-sex gametes could be combined, the offspring would have two mothers or two fathers, or perhaps come from a single father or a single mother. Generating diploid young in this unisexual way would require the artificial combination of two eggs or two sperm. In mammals, this creates a problem to do with a particular set of genes called **imprinted genes**:

- During normal mammalian reproduction, the expression of some genes is regulated before mating and fertilization, during the time the gametes spend in the respective parental

reproductive tract (eggs in the ovary and oviduct, sperm in the testis and seminiferous tubules).

- This regulation is an **epigenetic** process, called **imprinting**, and it results in those particular genes being turned off or turned on. Imprinting is necessary for the generation of successful mammalian offspring, for reasons which are currently unclear.

- The mouse has about 150 imprinted genes and humans have about half that number. With one exception, they exist in either the female or the male gamete, not both. Some imprinted genes are important for early embryo development and maturation; others are associated with aspects of adult life including metabolism, neurological function, and behaviour. They can be the basis of severe genetic diseases, including some cancers, if expressed inappropriately.

Imprinted genes complicate the problem of unisexual reproduction and could be one reason why it has not evolved naturally in mammals. Complex laboratory manipulation is needed to control them if parthenogenetic offspring are to be produced artificially (see Scientific Process 33.1).

SCIENTIFIC PROCESS 33.1 | **Experimental mammalian parthenogenesis**

Research question

Can mammalian gametes be manipulated in the lab to achieve parthenogenetic offspring?

Background information

To produce a successful parthenogenetic mammalian embryo in the lab, imprinted genes would need to be switched off in the gametes which are being combined.

Hypothesis

Zhi-Kun Li and colleagues at the Chinese Academy of Sciences investigated whether 'bi-maternal' and 'bi-paternal' mice could be produced from embryonic stem cells if imprinted genes were manipulated.

They *tested* the following hypothesis:

Suppressing imprinted genes in male or female mouse embryonic stem cells would allow them to combine parthenogenetically and develop into healthy offspring.

They *predicted* that some imprinted genes would be essential for embryo development but others could be suppressed without adverse effect; the genes would be specific for bi-maternal or bi-paternal embryos.

Methods

Figure 1 shows the scheme of Li et al.'s experiments.

- They took haploid embryonic stem cells from mouse sperm and eggs and cultured them in the lab rather than allowing them to mature in the animal (pink panel). This resulted in the stem cell genomes being 'demethylated' and therefore lacking the pattern of imprinted gene expression which they would normally possess.

- To produce *bi-maternal* mice (blue panel), they injected cultured female stem cells into normal oocytes from which the nuclei had been removed. These oocytes were implanted, at the MII stage of meiosis, into adult mothers. Before this, the team deleted genes which would normally be imprinted, to determine which ones controlled developmentally crucial traits.

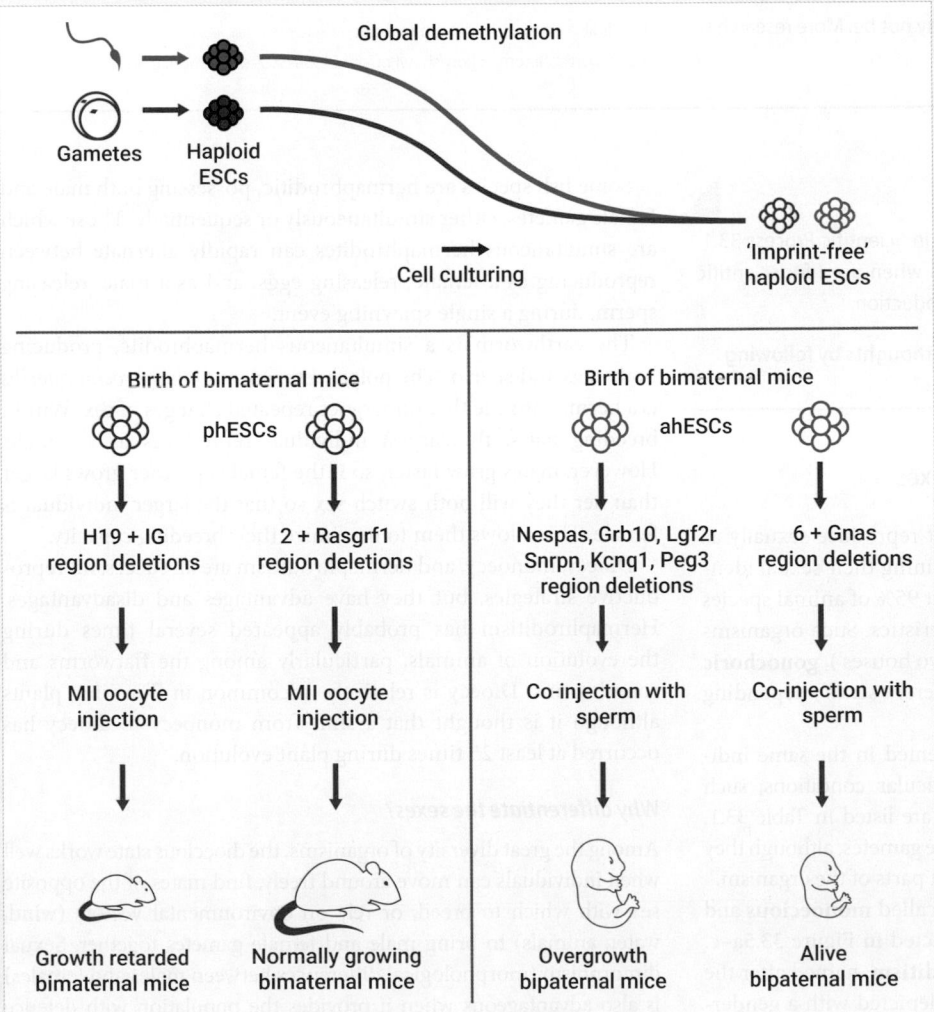

Figure 1 The experimental scheme through which bimaternal and bipaternal mice were produced.

Source: Reprinted from Li *et al*. Generation of bimaternal and bipaternal mice from hypomethylated haploid ESCs with imprinting region deletions. *Cell Stem Cell*, 23(5), 666–76. Copyright 2018, with permission from Elsevier.

- To produce *bi-paternal* mice (yellow panel), they injected the enucleated oocytes with the male stem cells along with normal sperm and implanted them into host mothers, again having previously deleted some key genes.

How was the gene manipulation achieved?

Gene deletions in the cultured stem cells were performed by CRISPR-Cas9 gene editing technology.

Results

- Despite many losses, the host mothers gave birth successfully to bi-maternal and bi-paternal offspring

- As Figure 1 indicates, some gene deletions adversely affected the growth of the pups and some developed other abnormalities, but some pups appeared normal.

- The bi-paternal pups all died soon after birth, but at least one of the bi-maternal offspring matured to adulthood and went on to produce her own offspring (Figure 2).

Conclusions

1. Unisexual reproduction is possible in mammals, provided imprinted genes are regulated.

2. Specific imprinted genes, in male and female gametes, are crucial for normal embryo development but others may not be. More research is needed to understand this.

Figure 2 A bi-maternal mouse with her own offspring.

Source: Photograph by Leyun Wang, Institute of Zoology, Chinese Academy of Sciences.

Read the original research paper at

https://www.nature.com/articles/nrg3766.pdf

For a review of imprinted genes, see:

https://www.nationalgeographic.com/science/2018/10/news-gene-editing-crispr-mice-stem-cells

There is also a non-technical summary at:

https://www.cell.com/action/showPdf?pii=S1934-5909%2818%2930441-7

PAUSE AND THINK

Think a bit further about the kind of research in Scientific Process 33.1. Consider whether there are any ethical issues when used for scientific interest or to manipulate human or animal reproduction.

 You can find some ideas to prompt your thoughts by following the link in the e-book.

Differentiated and undifferentiated sexes

We are used to thinking of organisms that reproduce sexually as either male or female, with individuals retaining their sexual identity throughout their lifespan. Indeed, about 95% of animal species and a few plant species have these characteristics. Such organisms are called **dioecious** (from the Greek for 'two houses'), **gonochoric** ('separate offspring'), or **heterothallic** ('different sexes') depending on the type of organisms being discussed.

Other organisms have both sexes represented in the same individual or cn change their sex under particular conditions; such strategies—and examples exhibiting them—are listed in Table 33.1. Individuals can produce both male and female gametes, although they may do so at different times or from different parts of the organism.

Plants which have this characteristic are called **monoecious** and there are many common examples, as depicted in Figure 33.5a–c. The general term in animals is **hermaphroditism**, named after the Greek god Aphrodite who was classically depicted with a gender-ambiguous body.

Some fish species are hermaphroditic, possessing both male and female gametes, either simultaneously or sequentially. Those which are simultaneous hermaphrodites can rapidly alternate between reproducing as a female, releasing eggs, and as a male, releasing sperm, during a single spawning event.

The earthworm is a simultaneous hermaphrodite, producing both eggs and sperm. The polychaete worm *Ophryotrocha puerilis* is a hermaphrodite that undergoes repeated changes of sex. Within breeding pairs, the largest individual reproduces as a female. However, males grow faster, so if the female's partner grows larger than her they will both switch sex so that the larger individual is female. This allows them to maximize their breeding capacity.

Dioecy, monoecy, and hermaphroditism are all successful reproductive strategies, but they have advantages and disadvantages. Hermaphroditism has probably appeared several times during the evolution of animals, particularly among the flatworms and roundworms. Dioecy is relatively uncommon in flowering plants although it is thought that a shift from monoecy to dioecy has occurred at least 25 times during plant evolution.

Why differentiate the sexes?

Among the great diversity of organisms, the dioecious state works well when individuals can move around freely, find mates of the opposite sex with which to breed, or rely on environmental vectors (wind, water, animals) to bring male and female gametes together. Sexual dimorphism (morphological differences between males and females) is also advantageous when it provides the population with defence and protection or helps in obtaining food and other resources.

Table 33.1 Non-dioecious reproductive strategies

Type of strategy	Description	Examples
Sequential hermaphroditism	Individuals start life as one sex but change to the other, or express their different sexes sequentially, often due to position in social hierarchy, predator pressure, avoidance of self-fertilization, or an absence of other sexes in the population	
Protandry	Animals: individual starts as male and changes to female	Coral reef clownfish, *Amphiprion* spp.
	Animals: males emerge before females	Emerald ash borer, *Agrilus planipennis*
	Plants: plant produces male pollen before female stigma is receptive to pollination	Foxtail lily, *Eremurus altaicus*
Protogyny	Animals: individual starts as female and changes to male	Ballan wrasse, *Labrus bergylta*
	Plants: flower pistils mature before anthers produce pollen	Arum lily, *Zantedeschia aethiopica*
	Plants: production of colours and scents to attract pollinators varies with development of male and female flower parts	Kettle trap flower, *Aristolochia gigantea*
Bi-directional	Individual can become male or female, irrespective of initial state	Angelfish, *Centropyge ferrugata*
Suppressed	Individual is male or female, but expression of its sex, and thus of its breeding ability, is suppressed and it is effectively neutered. Such species are often described as cooperative breeders	Naked mole rat, *Heterocephalus glaber*
		Paper wasp, *Polistes dominula*
Simultaneous hermaphroditism	Individuals may be defined as male or female but are capable of making both male and female gametes, at the same time or sequentially	
Self-incompatible	Gametes from a single individual are incapable or rarely capable of self-fertilization, for reasons of timing, morphology, or genetic incompatibility	Garden snail, *Cornu aspersum*
		European yew, *Taxus baccata*
Self-compatible (autogamy)	Gametes from the same individual are capable of self- as well as non-self-fertilization	Nematode worm, *Caenorhabditis elegans*
		Rock cress, *Arabidopsis thaliana*

Figure 33.5 Monoecious and dioecious plants. Hazel (*Corylus avellana*) is monoecious: catkins (a) are male flowers, producing pollen that is wind distributed; it pollinates female flowers (b) on the same or other hazel trees; nuts (c) form on each hazel tree after fertilization. Holly (*Ilex aquifolium*) is dioecious: male flowers (d) only occur on male holly plants; their pollen is distributed onto holly flowers on female holly plants (e). Fertilization of the female flowers results in red berries borne only on female plants (f).

Source: (b) AJ Cann/Flickr/CC BY-SA 2.0; (c) image by Emilian Robert Vicol from Pixabay; (d) sunsets_for_you/Flickr/CC BY-SA 2.0; (e) H. Zell/Wikimedia Commons/CC BY-SA 3.0.

Separate sexes can work less well for organisms in widely distributed, very low-density populations where the energetic costs of finding a suitable mate are high. It can also be problematic for sessile organisms, unless they have an efficient germ cell dispersal system.

If individuals have both male and female characteristics, or have male and female parts, any member of the population can, in principle, mate with any other individual, rather than requiring a member of the opposite sex. This reduces the challenge of finding mates but increases the genetic problem of inbreeding. Overall, increased reproductive opportunity and inbreeding depression represent conflicting evolutionary pressures.

Several solutions to the problem of inbreeding have evolved, right across the plant and animal kingdoms. They include the generation of partner-attracting signals, such as colours, odours, sounds, and behaviours, the use of signals to attract pollinators, and the development of distinct patterns of social behaviour conducive to male-female interaction.

Does self-fertilization occur?

Although hermaphroditic animals and plants apparently possesses the ability to self-fertilize, this is not usually the preferred option as it severely reduces genetic variation among offspring. Simultaneous hermaphrodites will often come together in pairs and release their eggs and sperm in a way that maximizes cross-fertilization. Several species of flowering plants have anatomical features that make self-fertilization impossible or unlikely, a matter which famously intrigued Charles Darwin.

In his book *The Different Forms of Flowers on Plants of the Same Species* (1877), Charles Darwin noted that the flowers of primula (*Primula veris*) existed in two different forms; his illustration of these forms is shown in Figure 33.6a. In the *long-styled form*, the stigma occurs at the top of the flower and the anthers near the middle. In the *short-styled form*, the anthers are at the top and the stigma near the middle. He wondered if the different forms prevented self-fertilization. He artificially cross-fertilized the two forms to see whether the seeds produced would be fertile using a method he depicted in Figure 33.6b. Crosses between plants where the anthers and stigma were at the different levels produced large numbers of fertile seeds. Crosses between plants with similar structure produced small numbers of infertile seeds.

Some monoecious plants and hermaphrodite animals do mate with themselves: in these cases, all individuals within a population are reproductively compatible with themselves, as well as with each other.

Self-fertilization is much more common in plants than in animals, occurring in about half of all species, and may have evolved when species or variety hybrids became geographically isolated. Among animals, it is frequently found in flatworms and nematodes. It is also common in fungi and algae.

Pseudohermaphrodites

The word hermaphrodite, or pseudohermaphrodite, has been used with a distinct meaning in the human context. It refers to **intersex** individuals with incomplete, ambiguous, or non-gender-distinct genitalia, or those who, for genetic or developmental reasons, have external genitalia that do not reflect their gonadal gender. In most cases, individuals are genetically male but appear female due to undescended or inactive testes and/or insensitivity to male sex hormones. They may have an incomplete or undifferentiated reproductive tract and are infertile.

Pseudohermaphroditism occurs as an abnormality in some animals, especially canines, usually where a female has male-like genitalia due to over-exposure to male sex hormones. This may occur during development or result from endocrine disease.

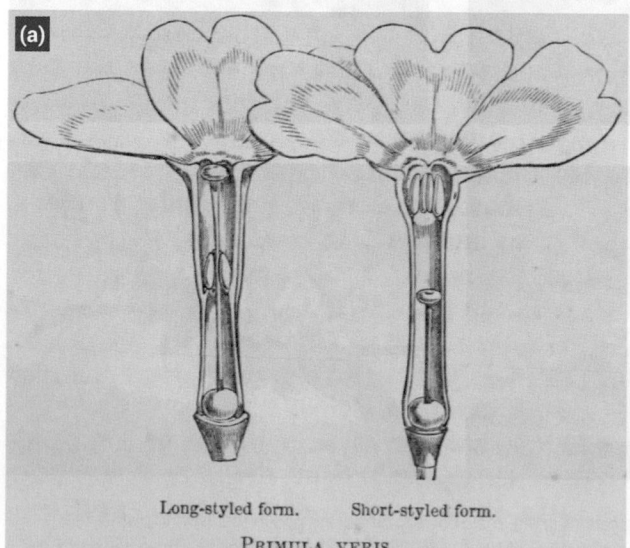

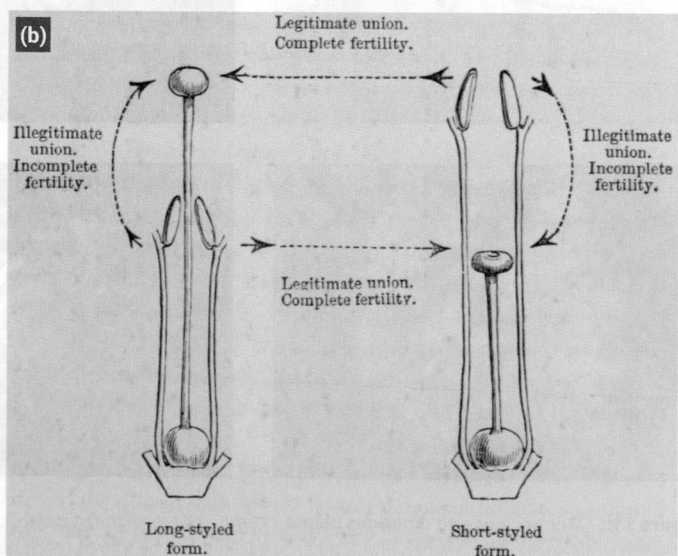

Figure 33.6 The two different forms of primula flowers. (a) The two different forms as observed by Charles Darwin. (b) The method Darwin adopted to explore cross-fertilization between the two forms.

The spotted hyena (*Crocuta crocuta*) is unusual in that normal females have a clitoris shaped like a penis and expanded labia which resemble a scrotum. Hyenas show very little sexual dimorphism, at least to human view, and are sometimes described as psuedohermaphrodite. Nevertheless, the sexes are functionally and genetically distinct.

Sex determination

Sexual reproduction is widespread among eukaryotic organisms, but there are many different ways in which sex differences are determined. The *Tree of Sex Database* (http://treeofsex.org/) holds information on the diversity of sexual types and karyotypes (chromosomal identity and arrangement) for over 25,000 species of plants, 12,000 species of invertebrate animals, and 2000 species of vertebrates.

Biologists can interrogate this data to study important questions such as the rates at which sexual and non-sexual species diversify, the likely effects of environmental change on sex ratios and species extinction rates, and the possible ways in which species may fuse or hybridize.

This raises the question of how sexual differentiation occurs. It turns out that the distinction between male and female may occur at the gene level—called **genetic sex determination (GSD)** or depend on external influences—called **environmental sex determination (ESD)**.

Genetic determination of sex

For the vast majority of organisms, the sex of offspring is determined by the genes they acquired during zygote formation. There is a variety of sex-determining karyotypes (pairings of sex chromosomes) and there is also variation within groups as to which sex results from the **heterogametic** (XY or ZW) and which from the **homogametic** (XX or ZZ) arrangement.

There are three functional stages in the sex determination process:

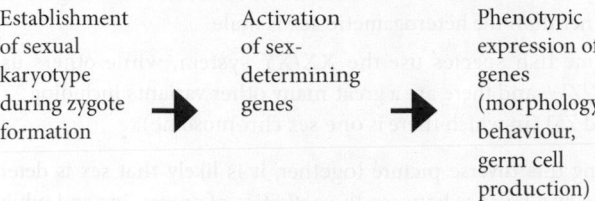

Establishment of sexual karyotype during zygote formation → Activation of sex-determining genes → Phenotypic expression of genes (morphology, behaviour, germ cell production)

The phenotypic result—the characteristic male-ness or female-ness of the individual—depends on the expression of sex-determining genes and their interaction with other genes and cells. As the organism grows and develops, the interaction becomes more complex, sometimes producing extreme sexual dimorphism.

In vertebrate animals, the activation of sex-determining genes early in the life of the embryo leads to the differentiation of the gonads. The early testis and early ovary produce different hormones. These dictate the development of the rest of the reproductive system and also shape the rest of the individual's body, both as an embryo and in later life.

The main sex-determining genes have been identified and there are a number of distinct systems, as summarized in Figure 33.7:

- Mammals have the XX/XY system in which the 'male' Y chromosome possesses a gene called *SRY* (for 'Sex Region of the Y'). *SRY* codes for a transcription factor (SOX9 protein), the presence of which leads to the formation of a testis in the early embryo. (We learn more about the *SRY* gene in Chapter 4.) This early testis produces hormones (testosterone and AMH) which cause the formation of a male reproductive tract. In the absence of *SRY* (XX) and these hormones the embryo develops as female.

- Birds and some reptiles use the ZZ/ZW system in which the Z chromosome has a gene called *Dmrt1*. *Dmrt1*, which is quite

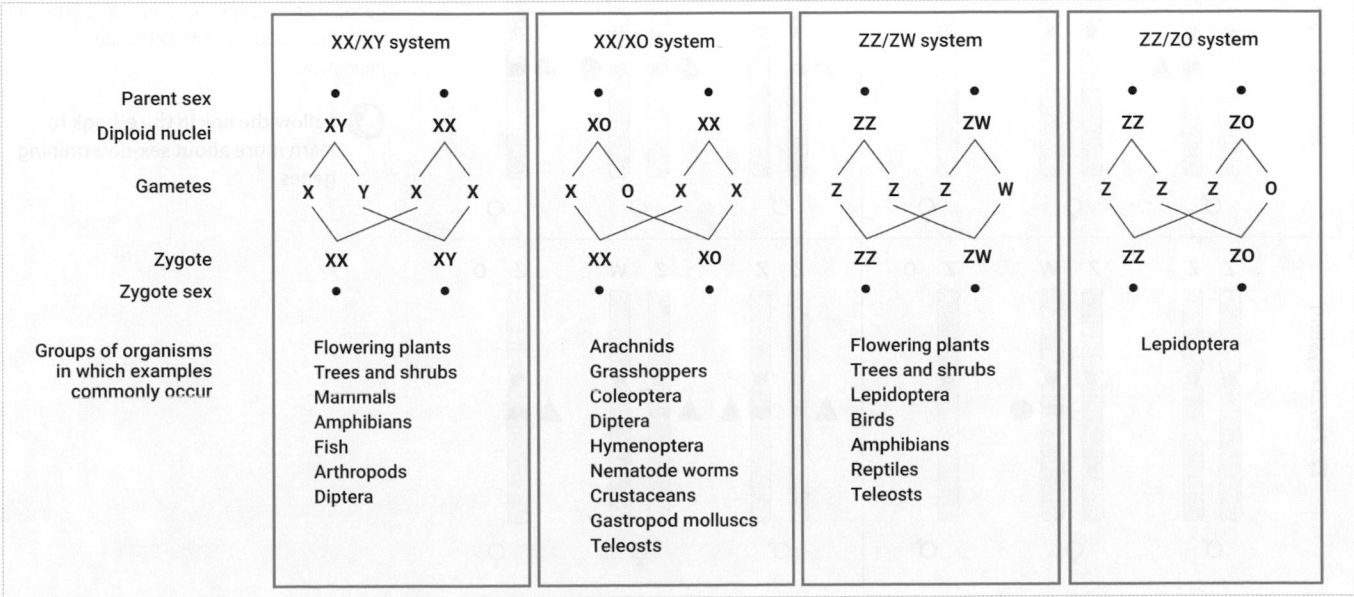

Figure 33.7 Sex chromosomes and karyotypes exhibited by different organisms.

different from *SRY*, produces a male animal when present in high amount (ZZ rather than ZW), so that the homogametic rather than the heterogametic sex is male.

- Some fish species use the XX/XY system, while others use ZZ/ZW, and there are a great many other variants including XO and ZO (in which there is one sex chromosome).

Putting this diverse picture together, it is likely that sex is determined by a balance between the activities of promoting and inhibiting genes. There are two interpretations of how this might work. These are called the 'Active Y/W' and 'Dosage' modes, and are depicted in Figure 33.8.

- In the Active Y or Active W modes, one of the sex chromosomes possesses a gene whose presence or absence on the Y (male) or W (female) chromosome determines the outcome.

- In the Dosage mode, genes on the sex chromosomes result in male or female depending on the whether it is expressed at a low (from one chromosome) or high (from two chromosomes) level. A low level produces one sex while a high level produces the other sex.

Note that the two modes produce different outcomes when only one sex chromosome is present (XO or ZO).

Environmental determination of sex (ESD)

In some animals, especially reptiles and fish, the sex of offspring is greatly influenced by the environment. The most important environmental factors are temperature, pH, population density, and visual cues from other individuals. Such animals have sex-determining genes, but their effects are over-ridden by the environmental signal, early in development or later.

Among reptiles, nest temperature influences the sex of offspring in all crocodilian and tuatara species, in three quarters of turtles, and in about a quarter of lizard species. The temperature at which eggs are first incubated determines the proportions of males and females which eventually hatch, as illustrated for three different reptiles in Figure 33.9. Depending on the species, warmer temperatures produce more females or more males, or one sex may predominate at temperatures towards the extremes of the range. Critical temperatures for each species can be determined experimentally but may vary over the reproductive season.

Experiments show that temperature acts epigenetically: it alters the transcription of the sex-determining genes before the embryo's gonads have formed. The result is that different amounts of sex hormones, especially oestrogens, are present in the egg as the embryo develops.

Adult alligators and other reptiles appear to control the temperature of their nests deliberately by adjusting the insulating plant material from which they are constructed. This suggests that the parent animals are aware of the environmental temperature and respond purposefully to it.

The adaptive advantage of reacting in this way, which changes the sexual make-up of the local population, is unclear. It may have to do with the availability of resources for the young and the avoidance of predators. Alternatively, it may reflect the reproductive advantage of one sex over the other when the dispersal of males and females varies according to season and climate.

The sex of some fish is determined by the size of the shoal and interactions between individuals. The sexual identity of adults can change after a period of a few days of exposure to the appropriate social conditions, as illustrated in Figure 33.10. The fish changes its morphological and behavioural phenotype, as well as

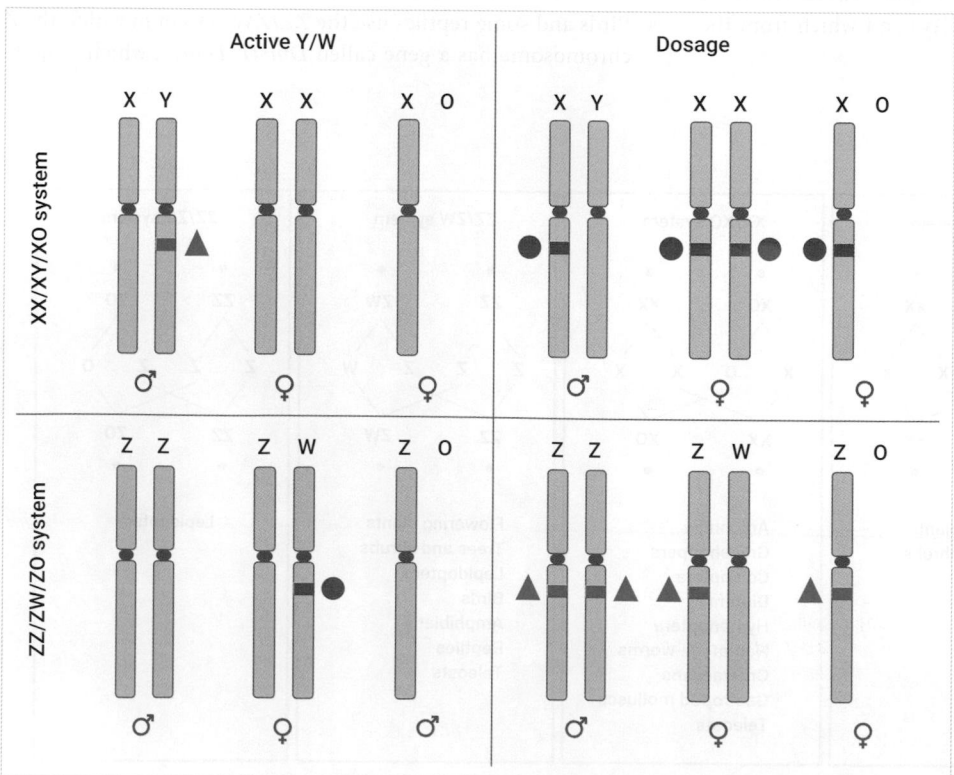

Figure 33.8 Two ways in which sex chromosomes may determine sex: the Active Y/W and the Dosage systems. Grey rods are sex chromosomes, with black dots representing centromeres. Red and blue bars represent female- and male-determining genes. Triangles and circles represent activated promoting factors which determine phenotype.

Follow the link in the e-book to learn more about sex-determining genes.

Figure 33.9 Three modes of temperature sex determination in reptiles. (a) American alligator (*Alligator mississippiensis*). (b) Red-eared slider turtle (*Trachemys scripta*). (c) Leopard gecko (*Eublepharis macularius*). The diamond indicates the *critical temperature*—the approximate temperature at which there is an equal number of male and female offspring in the litter.

Source: (a) Postdlf/Wikimedia Commons/CC BY-SA 3.0; (b) Photo by Greg Hume/Wikimedia Commons/CC BY-SA 3.0; (c) George Chernilevsky/Wikimedia Commons/Public Domain.

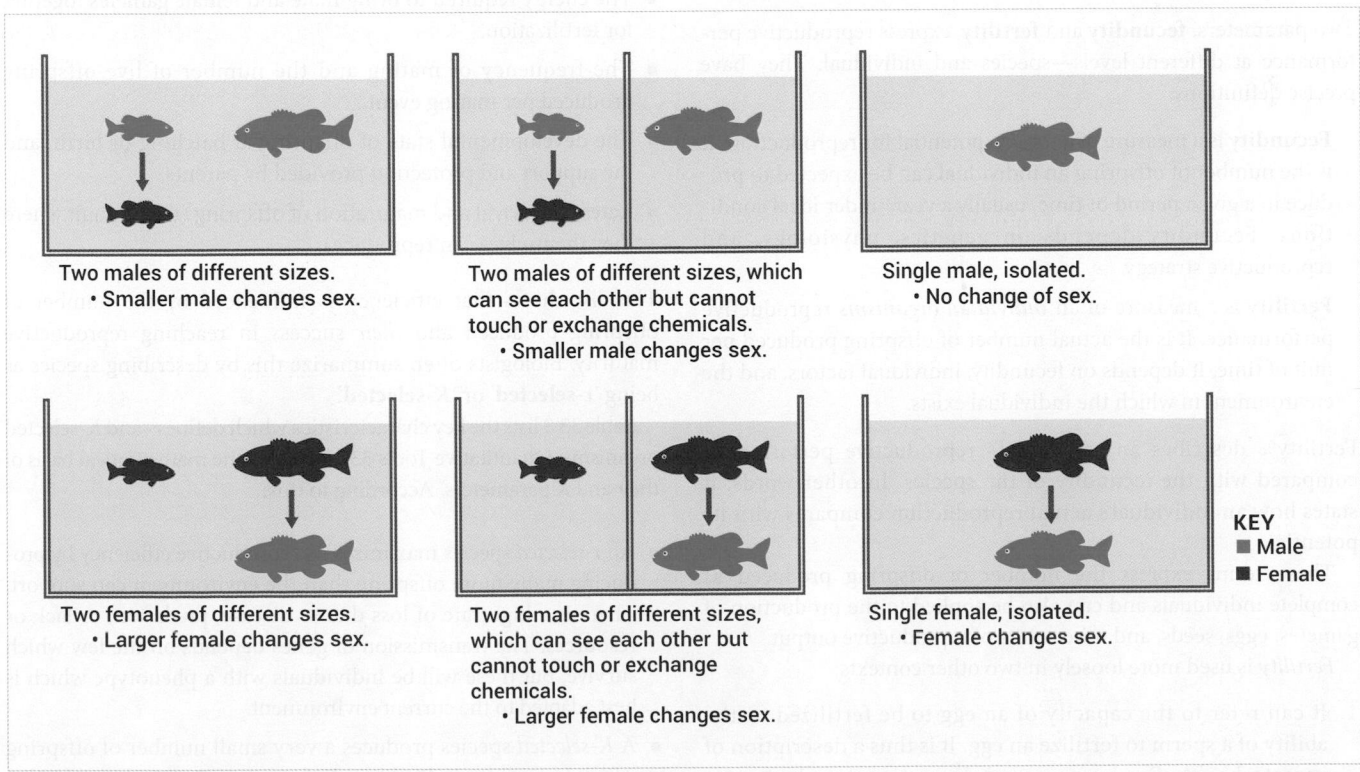

Figure 33.10 Sex change in fish. Six types of sex determination that may occur in some species of fish as a result of social cues.

Source: Capel, B. (2017) Vertebrate sex determination: evolutionary plasticity of a fundamental switch. *Nature Reviews: Genetics*, 18, 675–89. Reprinted by permission from Springer Nature ©2017.

the type of gametes (eggs or sperm) it produces. Visual cues are responsible and the mechanism is at the level of the brain and the endocrine system.

Individuals which mature first as males and later become females are called **protandrous**. Examples are the anemone fishes (clownfish). Individuals which mature first as females and later become males are called **protogynous**. Examples include the parrotfishes. A 'size-advantage' hypothesis has been put forward to explain adaptive value of these sex changes.

Follow the link in the e-book to learn more about clownfish, and specifically how they can change sex.

Check your understanding of the concepts covered in this section by answering the questions in the e-book.

33.2 Measuring reproduction: fecundity and fertility

Biologists often find it useful to be able to measure the rate at which organisms reproduce. It can be helpful to know how fast a population of organisms might grow, to model the capacity of an environment to support them, or to know how best to manage numbers and species diversity. It can also be helpful to know how the reproduction of an individual or group compares with the potential of its species or population.

Definitions and calculations

Two parameters, **fecundity** and **fertility**, express reproductive performance at different levels—species and individual. They have precise definitions:

Fecundity is a measure of a *species'* potential for reproduction. It is the number of offspring an individual can be expected to produce in a given period of time, usually a year, under ideal conditions. Fecundity depends on genetics, physiology, and reproductive strategy.

Fertility is a measure of an *individual organism's* reproductive performance. It is the actual number of offspring produced per unit of time. It depends on fecundity, individual factors, and the environment in which the individual exists.

Fertility% describes an individual's reproductive performance compared with the fecundity of the species. In other words, it states how an individual's actual reproduction compares with its potential.

These terms express the number of offspring produced as complete individuals and can also be applied to the production of gametes, eggs, seeds, and other units of reproductive output.

Fertility is used more loosely in two other contexts:

1. It can refer to the capacity of an egg to be fertilized or the ability of a sperm to fertilize an egg. It is thus a description of gamete *health*. For humans and domestic animals, it may describe the quantity and quality of sperm present in a male's ejaculate. An individual organism that is reproductively immature, suffers a reduction or temporary loss of fertility, or has reached the end of its reproductive lifespan may be described as **sub-fertile** or **infertile**. An organism that is completely unable to reproduce would be described as **sterile**.

2. In agricultural, horticultural, and ecological contexts, fertility can refer to the ability of soil to support plant life. 'Fertilizers' are materials that supply minerals and other nutrients and increase soil productivity. They do not influence plant reproductive processes directly, but may increase the number and size of individual plants, flowers, seeds, grains, and fruits.

Reproductive efficiency and parental investment

The **reproductive efficiency** of an organism is a relative measure of the energy and time required to produce offspring which are themselves capable of reproduction. In other words, it describes the total effort required to transmit genetic information from one generation to the next. For context, reproductive characteristics for a range of mammals are summarized in Table 33.2.

An individual's reproductive efficiency depends on its reproductive strategy, its environment, and its reproductive success rate. It is thus a combination of fecundity and fertility.

Factors affecting reproductive efficiency include:

- The energy required to produce gametes, and the frequency of gamete production.
- The energy required to bring male and female gametes together for fertilization.
- The frequency of mating and the number of live offspring produced per mating event.
- The developmental state of offspring at hatching or birth, and the support and protection provided by parents.
- Rates of survival and maturation of offspring (to the point where they themselves can reproduce).

This list shows that efficiency is influenced by the number of offspring produced and *their* success in reaching reproductive maturity. Biologists often summarize this by describing species as being '*r*-selected' or '*K*-selected'.

Table 33.3 lists the key characteristics which define *r*- and *K*-selected organisms. Quantitative Tools 33.1 explains the mathematical basis of the *r* and *K* parameters. According to this:

- An *r-selected* species maximizes its reproductive efficiency by producing many more offspring than the environment can support. There is a high rate of loss due to disease, predation, or lack or resources. The transmission of genes depends on the few which survive, but these will be individuals with a phenotype which is best adapted to the current environment.
- A *K-selected* species produces a very small number of offspring which have a strong likelihood of survival. Survival may be due to advanced development at the time of birth or high parental investment in the form of protection and nutritional support.

K-selected species exploit predictable aspects of the environment, such as seasonal changes in resources, or have complex behaviours which avoid predation. However, they may be at great risk if the environment changes significantly.

Put simply, *r*-selected and *K*-selected organisms depend, respectively, on high fecundity with a poor rate of survival or low fecundity with a high rate of survival. A useful analogy might be the difference between a shotgun (many tiny pellets scattering the target) and a rifle (a single large bullet requiring exact aim).

The *r*- and *K*- descriptors are applicable to plants and animals. They are also used to describe parts of the reproductive process, including gamete production and distribution. It is important to understand that they are *relative* descriptors which form a continuum. They enable species comparisons to be made, not absolute measurements.

Among plants, the annual and perennial weed species which rapidly colonize disturbed or open ground, such as grasses, vetches, docks, and nettles, are *r*-selected. Examples of *K*-selected species include perennial shrubs and deciduous trees. The former produce lots of small seeds and spread rapidly but transiently, often being supplanted by longer-lived species. The latter produce fewer seeds but once established may grow to a very large size and dominate the landscape.

Table 33.2 Reproductive characteristics of some mammals (representative values)

Mammal	Order	Gestation length	Neonatal mass	Litter size	Length of lactation	Inter-birth interval
Human	Primate	259 d	3000–3300 g	1	0.5–3 y	15 m (no lactation) 27 m (with lactation)
African elephant	Proboscoidea	20–22 m	70–100 kg	1	2 y	2–9 y
Giraffe	Artiodactyla	450 d	40–60 kg	1	6 m	14–22 m
White rhinoceros	Perissodactyla	15–18 m	40–50 kg	1	1 y	2–5 y
Domestic cattle	Artiodactyla	279–290 d	30–40 kg	1	10 m	2 y
Domestic sheep	Artiodactyla	145–152 d	3.5–5 kg	1–3	90–150 d	1 y
Wild boar	Artiodactyla	4 m	500–800 g	1–10	4 m	230 d
Reindeer	Artiodactyla	200–210 d	3–7 kg	1	2–5 m	1 y
Ferret	Carnivora	40–43 d	7–10 g	5–11	6–8 w	20 w
Horse	Perissodactyla	340 d	80 kg	1	1 y	2 y
European rabbit	Lagomorpha	28–30 d	40–45 g	4–7	4 w	29 d
Norwegian rat	Rodentia	20–22 d	5–6 g	5–12	4 w	30 d
House mouse	Rodentia	18–20 d	1 g	5–10	3 w	22–58 d
Domestic dog	Carnivora	58–68 d	200–500 g	3–8	5–8 w	1 y
Bottlenose dolphin	Cetacea	12 m	11–16 kg	1	1.1.5 y	2 y

d, days; m, months; w, weeks; y, years

Table 33.3 Lifestyle characteristics of organisms with r- and K-strategies for reproduction

Characteristic	r-selected	K-selected
Population size	High, variable, and prone to rapid fluctuation	Low and relatively constant
Habitat	Unstable and frequently disturbed	Relatively stable and predictable
Competition from other species for space and resources	Low or variable	High
Resistance to diseases and predators	Low	High
Impact on the environment	Low	High
Lifespan	Short	Long
Growth and development	Rapid growth, early maturity	Slow growth, late maturity
Body size	Small	Large
Parental involvement (animals)	Brief	Extended/intense

Biologists describe the reproductive strategies of organisms as being 'r-selected' or 'K-selected'. In general terms, an r-selected species produces large numbers of offspring (or gametes) of which only a few survive. A K-selected species produces very few offspring at each reproductive event, but the survival rate is high. Every species can be positioned somewhere on a continuum between the extremes of the two strategies. The descriptors *r* and *K* come from parameters in an idealized population growth curve (see Figure 1).

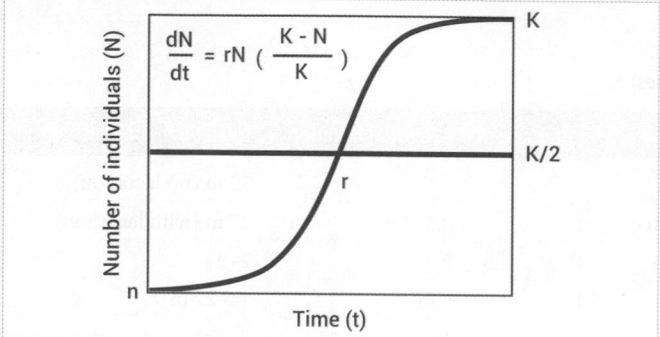

Figure 1 An idealized population growth curve.

The figure shows that, over time, the number of individuals in a freely reproducing population will grow from n, the number of founding individuals, to *K*, the maximum number which the resources of the environment can sustain.

A graph of the rate of growth of the population has an S- or sigmoid shape. This is because at any point in time, the birth rate (r; the number of new individuals) depends on the current size of the population (N) and how close it is to the maximum size (K). The birth rate is at its maximum when N = K/2. Mathematically, a curve of this shape is described as logistic. Its equation is shown in the inset to the figure.

Although this is a mathematical representation, the concept is a theoretical one and it would be unusual for an organism to be *quantified* in this way. The value of the concept is that it allows the reproductive strategies of different organisms to be to be compared *qualitatively*, while understanding where the energy of reproductive activity is directed.

The word 'selected' is used as a reminder that the characteristics of living organisms, including their reproductive strategies, have evolved as a result of selection under the pressures of the environment. It is equally acceptable to describe an organism as having an *r*- or *K*-strategy.

Among animals, *r*-selected species are typified by insects and other invertebrates which reproduce abundantly if environmental conditions allow. They may form large colonies, with their environmental impact reflecting their population density rather than the resource requirements of individuals. They reach maturity early and there may be several generations within a year. Vertebrate animals tend towards *K*-selection: they produce fewer young which take longer to mature, often depending on parental support, and have longer lifespans.

The contrast between *r*- and *K*- can be easily seen by comparing, for example, small rodents and large pachyderms (see Table 33.2).

- Rats and mice have repeated, short cycles of reproduction and short gestation periods; they produce large litters of immature but rapidly developing young which may themselves be able to reproduce within a few weeks, although the survival rate of pups may be low.

- Elephant and rhinoceros reproduce sporadically at intervals of several years, have highly extended gestation periods, and deliver young which, although relatively well developed, require many years of parental and societal support before being capable of independent survival and reproduction.

Describing reproductive strategies in this way reveals that there is more than one way for an organism to achieve successful reproduction, more than one way for genetic information to be transmitted. It also reminds us that the pressures of natural selection act directly on the reproductive process.

 Check your understanding of the concepts covered in this section by answering the questions in the e-book.

33.3 Reproduction in plants

In this section we will consider the mechanisms of reproduction in Spermatophytes—plants which produce seeds. (Figure 33.11 shows the broad taxonomic organization of the plant kingdom.) This group includes flowering plants (**angiosperms**, sometimes called **anthophyta**) and non-flowering plants (**gymnosperms**).

Many plants have the ability to reproduce asexually: fragments of tissue (leaves, branches, stems, roots) may break off and grow into new plants. For some plants this is the main way of achieving distribution and colonization of new territory. It is a form of clonal reproduction (or propagation), leading to new individuals which are genetically identical to the original plant.

Spermatophytes overcome the genetic restrictions of clonal replication by using sexual reproduction. This encourages variation and gives them a considerable adaptive advantage when the environment changes.

Reproduction in angiosperms

The flowers that characterize angiosperms have evolved to attract the attention of animal pollinators. In addition, many species produce fruits which attract animal vectors and facilitate distribution. Humans depend, directly or indirectly, on many hundreds of different types for food, fibre, and ornamentation, so they have huge economic importance.

The diversity and abundance of angiosperm species demonstrate that flower production is an immensely successful reproductive

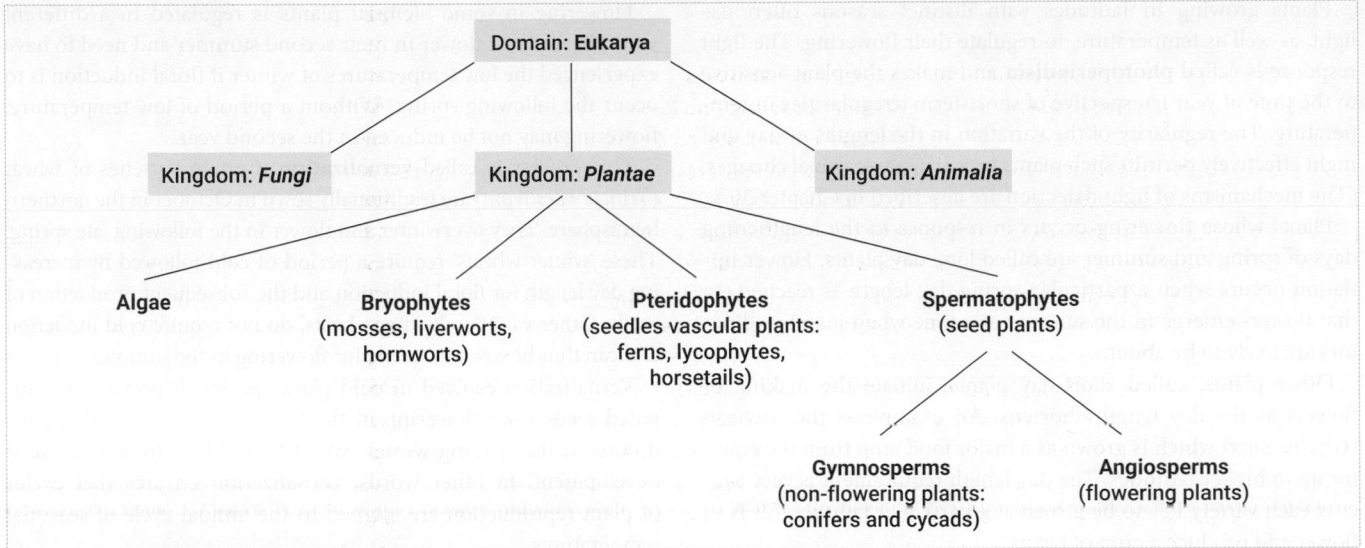

Figure 33.11 The principal plant groups.

strategy. Angiosperms can reproduce prolifically, but they can also adapt rapidly to changing environmental conditions. Human activity, including selection for productive traits, husbandry, agriculture, and the effects of industrial processes, exerts enormous pressure on angiosperm survival and adaptation.

Although all angiosperms produce flowers, there is great diversity and complexity in the mechanisms they use. The genetics of gamete production remains relatively poorly understood for many types of plant, and in most species there is a complex interplay between asexual and sexual phases of reproduction. Superimposed on this is the sensitivity of flowering plants to seasonal and other environmental influences.

Flowers and flowering

All angiosperms have periods of non-reproductive growth, called **vegetative growth**, during which they produce stems and leaves generated by shoot apical meristems (see Chapter 29). At some point in the life cycle, a change occurs in some or all of the shoot apical meristems: they stop making leaves and stems, change into floral meristems, and start making the various organs that make up flowers.

The relationship between these two growth phases varies enormously between different plant species.

- Plants called **annuals** grow vegetatively for a short period after seed germination. This normally occurs in spring if climatic conditions are suitable. They then flower during the summer, produce seeds, and die. Thus, their entire life cycle lasts less than a year.
- **Biennial** plants grow vegetatively for one season, pass the winter ('overwinter') in a dormant state, then flower the following year and subsequently produce seeds and die.
- The longer-lived **perennial** plants flower each year and produce seeds annually. When the plant is young there may be several years of non-flowering growth until it reaches a mature state. Examples of this are long-lived trees, which may take several years to attain maturity and commence flowering.

When to flower?

Plants continuously monitor their environment and time their flowering in response. Temperature and light are the most important variables, but some plants also respond to water availability and the activity of herbivorous animals.

Most angiosperms respond to seasonal variations in temperature. Other developmental processes, including growth and leaf production, are also thermally linked such that they happen faster at higher temperatures.

The timing of flowering is related to the *thermal history* of the plant's immediate environment. This is expressed as **thermal time** and is the product of temperature and time, expressed in units of degree days:

$$\text{Thermal time}(°\text{Cd}) = \text{Average daily temperature}(°\text{C}) \times \text{Number of days}(\text{d})$$

Plants track the accumulation of thermal time and will flower accordingly. Thermal time is also used as a practical tool for understanding and managing the development of crop plants (Real World View 33.1).

Flowering may be delayed under cool conditions. This can have a noticeable effect on the time when fruits appear, for example in the availability of apples for harvesting or the presence of hedgerow fruits such as sloes and blackberries. Records of flowering times of spring plants maintained over recent centuries show clearly that increased spring temperatures associated with climate change have resulted in earlier flowering times.

Short-lived annual plants flower a short time after germination, in response to rising spring temperatures. Some perennial plants which flower very early in spring, such as primroses (*Primula vulgaris*), respond to the accumulation of thermal time but the flowers which appear were actually made the preceding autumn, simply expanding into mature flowers when conditions allow.

Plants growing in latitudes with distinct seasons often use light, as well as temperature, to regulate their flowering. The light response is called **photoperiodism** and makes the plant sensitive to the time of year irrespective of short-term irregularities in temperature. The regularity of the variation in the lengths of day and night effectively permits such plants to anticipate seasonal changes. (The mechanisms of light detection are described in Chapter 30.)

Plants whose flowering occurs in response to the lengthening days of spring and summer are called long day plants. Flower initiation occurs when a particular spring day length is reached, so that flowers emerge in the summer at a time when insect pollinators are likely to be about.

Other plants, called short-day plants, initiate the making of flowers as the day length shortens. An example is the soybean (*Glycine max*) which is grown as a major food crop from the equator up to higher latitudes. The day length requirement is very precise: each variety has to be grown at a particular latitude if it is to flower and produce a crop of beans.

Flowering in some biennial plants is regulated in a different way. These plants flower in their second summer and need to have experienced the low temperatures of winter if floral induction is to occur the following spring. Without a period of low temperature, flowering may not be induced in the second year.

This process is called **vernalization**. Certain varieties of wheat (*Triticum aestivum*) are traditionally sown in October in the northern hemisphere. They overwinter and flower in the following late spring. These 'winter wheats' require a period of cold followed by increasing day length for floral induction and the subsequent production of seeds. Other varieties, 'spring wheats', do not require cold induction and can thus be sown in spring for flowering in the summer.

Vernalization evolved in wild plant species. It prevents germinated seeds from flowering in the later autumn, given that conditions in the ensuing winter would be unlikely to support seed development. In other words, vernalization ensures that cycles of plant reproduction are aligned to the annual cycle of seasonal temperatures.

REAL WORLD VIEW 33.1   Thermal time and growing degree days in plant development

Plants are essentially poikilothermic and so their tissue temperature largely follows that of the surrounding environment. Their growth and development is greatly affected by changes in ambient temperature. The effects of temperature are cumulative (represented by thermal time, as described in the main text) but they depend on the plant experiencing particular temperatures for certain minimum periods. Figures 1 and 2 illustrate this for a typical cereal crop.

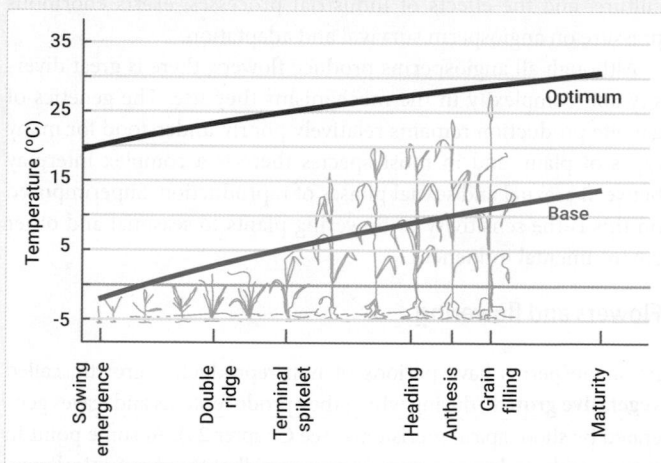

Figure 2 The minimal and optimal temperatures required to initiate specific developmental events and support growth during the life of a typical cereal crop.

Source: Food and Agriculture Organization of the United Nations. Reproduced with permission.

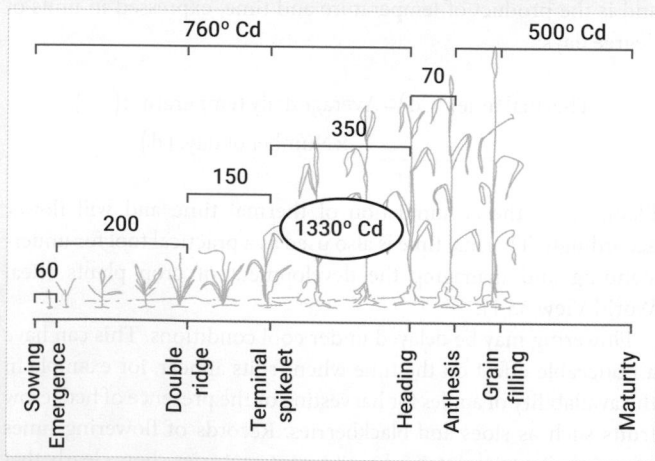

Figure 1 The minimum thermal times (°Cd) required to initiate specific developmental events during the life of a typical cereal crop. The temperatures used to calculate each value are daily averages (maximum + minimum, divided by 2). The value in red is the total for the complete development of the plant.

Source: Food and Agriculture Organization of the United Nations. Reproduced with permission.

A more accurate assessment of thermal time, used by plant scientist and horticulturalists, uses the calculation of growing degree days (GDD) as an integral:

$$GDD = \int (T - Tbase)dt$$

Here, *T* is the measured temperature and *Tbase* is a basal temperature such as 10 °C or 5 °C. The basal temperature is set as the minimum temperature likely to be experienced under local climatic conditions. For some tropical plants, flowering or fruiting may require a high basal temperature such as 30 °C.

For practical purposes, GDD can be calculated to a close approximation as:

$$GDD = \max\left(\frac{T\max + T\min}{2} - Tbase, 0\right)$$

Table 1 shows GDDs for some trees and crops.

Measurements of GDD and the long-term collection of data allow the creation of predictive models which are of great value to plant growers and crop producers.

They can also be used to determine when plants may be most susceptible to adverse condition, including drought and insect pests, and the data can be built into decisions about when best to apply treatments such as pesticides and fertilizers.

Table 1 Growing degree days (GDD) for some trees and crops

Plant	Species	Growing degree days, *Tbase = 10 °C*
Witch-hazel	*Hamamelis* spp.	Begins flowering at <1 GDD
Red maple	*Acer rubrum*	Begins flowering at 1–27 GDD
Norway maple	*Acer platanoides*	Begins flowering at 30–50 GDD
White ash	*Fraxinus americana*	Begins flowering at 30–50 GDD
Crabapple	*Malus* spp.	Begins flowering at 50–80 GDD
Horsechestnut	*Aesculus hippocastanum*	Begins flowering at 80–110 GDD
Privet	*Ligustrum* spp.	Begins flowering at 330–400 GDD
Elderberry	*Sambucus canadensis*	Begins flowering at 330–400 GDD
Maize	*Zea mays*	800–1400 GDD to crop maturity
Field beans	*Phaseolus vulgaris*	1100–1300 GDD to maturity depending on cultivar and soil conditions
Sugar beet	*Beta vulgaris*	130 GDD to emergence and 1400–1500 GDD to maturity
Barley	*Hordeum vulgare*	125–16 (GDD)2 GDD to emergence and 1290–1540 GDD to maturity
Wheat (Hard Red—a spring-sown variety)	*Triticum aestivum*	143–178 GDD to emergence and 1550–1680 GDD to maturity
Oats	*Avena sativa*	1500–1750 GDD to maturity

Flower morphology

Despite their huge diversity of colours and shapes, from the very simple to the very complex, all flowers have the same basic functional structure. They possess organs producing male and female gametes, together with a mechanism enabling the two to come together and fuse in the process of fertilization.

The structure of a typical, simple flower is shown in Figure 33.12. Simple flowers are formed as four concentric whorls of organs. The outermost whorl contains the sepals, the next contains the petals, the next contains the stamens, and the innermost whorl contains the carpel. Examples of simple flowers are shown in Figure 33.13. Some more extreme morphologies are shown in Figure 33.14.

Horticulturalists exploit this capacity for variation to produce plants with elaborate shapes and dramatic colours. Some natural mutations also alter morphology, the most common being 'double' flowers in which a mutation in a gene associated with flower development causes stamens to develop as petals, as illustrated in Figure 33.15.

Some plants have several flowers grouped together on a single structure, as depicted in Figure 33.16. This structure is termed an

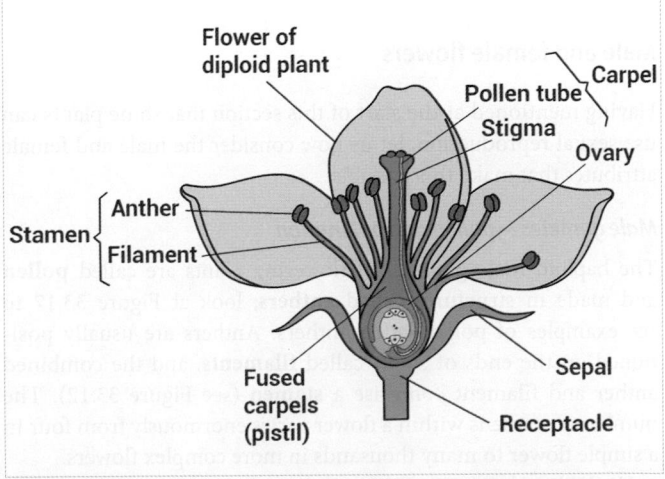

Figure 33.12 The basic structure of a simple angiosperm flower.

Source: From Maarten J. Chrispeels and Paul Gepts [Eds.] Plants, Genes, and Agriculture, Sustainability through Biotechnology, 2017. Oxford University Press.

Figure 33.13 Examples of simple flowers. Each has the same basic morphology: a central carpel with a stigma surrounded by a collection of stamens with varying numbers of petals on the outside. (a) Marsh marigold (*Caltha palustris*). (b) Cactus (*Echinopsis* sp.). (c) Blue poppy (*Meconopsis grandis*).

Figure 33.14 Extreme variation in floral morphology. (a) Orchid species. (b) Bird of paradise flower (*Strelitzia reginae*). (c) Passion flower (*Passiflora* sp.).

inflorescence. Such structures can be large and contain thousands of individual flowers, making a visually dramatic display. Insects attracted to the inflorescence can move from flower to flower with great efficiency.

Male and female flowers

Having mentioned at the start of this section that some plants can use sexual reproduction, let us now consider the male and female attributes that make this possible.

Male gametes—pollen and pollination

The haploid male gametes of flowering plants are called **pollen** and made in structures called **anthers**; look at Figure 33.17 to see examples of pollen-laden anthers. Anthers are usually positioned on the ends of stalks called **filaments**, and the combined anther and filament comprise a **stamen** (see Figure 33.12). The number of stamens within a flower varies enormously from four in a simple flower to many thousands in more complex flowers.

Haploid male gametes are generated by meiosis in the anther. Subsequent divisions occur by mitosis and lead to the production of mature pollen grains with haploid nuclei. Each anther contains two pollen grains (see Figure 33.4). Pollination of a flower and subsequent fertilization require grains to be released from the anther and transferred to the surface of the female part of the flower, the stigma.

In contrast to the process in animals, sexual reproduction in plants does not involve mate selection or any other purposeful behavioural activity. Instead, flowering plants have evolved ways of distributing pollen far and wide across the environment. This is an r-strategy that increases the chance of successful pollination, either on the same plant or on a different plant. Thus, individual flowers may produce millions of pollen grains (see Figure 33.17).

The simplest strategy is to allow wind to disperse the pollen. This happens in grasses and many tree species. Flowers which are wind pollinated are typically insignificant and lack coloured petals. The presence of vast amounts of this type of pollen in the atmosphere at certain times of year is responsible for hay fever in individuals who are allergic to it.

Plants which depend on insect vectors to disperse pollen have evolved elaborately shaped petals and brightly coloured flowers (see Figure 33.13). Assuming that the insects gain some benefit from visiting the flowers, logic suggests that flowering plants and insects must have co-evolved. In some cases the relationship is species-specific. Nevertheless, as with wind pollination, dispersal by insects is a passive process over which the individual plant has no direct control.

Figure 33.15 Mutations in homeotic genes controlling flower development. (a) Simple flower of the dog rose (*Rosa canina*), with five petals and many stamens surrounding the central pistil. (b) Highly bred modern rose, selected for showier flowers with more petals. Numerous stamens have been replaced by petals.

Figure 33.16 Floral inflorescences containing many individual flowers. (a) Horse chestnut (*Aesculus hippocastanum*). (b) *Buddleia* sp. (c) *Allium* sp. (d) *Hoya carnosa*.

In addition to coloration, many flowers encourage insects to collect pollen by producing scents from the petals. Others have glands called nectaries, which secrete an enticing sugary liquid (see Figure 33.16).

Scents comprise a complex cocktail of volatile organic compounds, related to the defensive secretions which deter herbivorous animals (Chapter 32). In plants such as honeysuckle (*Lonicera* spp.) scent production is closely aligned to the circadian cycle so that more is emitted, for example, in the evening when moths are most likely to be about.

Scents that are attractive to humans are heavily selected by plant breeders, but not all plant scents are considered pleasant. Figure 33.18 shows the titan arum, *Amorphophallus titanum*, from the rainforests of western Sumatra and Java, which has an elongated flower that emits a smell of rotting flesh. This attracts beetles and flies, which help with pollination.

The pollen grains of many species are remarkably resilient and, if not involved in fertilization, can remain in the environment for very long periods of time. Figure 33.19 illustrates how they have characteristic shapes and textures which can be used to identify the plant species from which they came. Archaeologists and palaeontologists make use of them to understand and date the sites they investigate (Real World View 33.2).

Female gametes: ovules and fertilization

The female part of the flower, the carpel, consists of ovules within an ovary which is normally situated toward the base of the flower (see Figure 33.12). Within each ovule, meiosis leads to the production of eight haploid nuclei. One of these nuclei is destined to produce the zygote. The others play a role in controlling fertilization.

The stigma and style comprise the **pistil**. Pistils have a wide variety of morphologies but all perform the same function of bringing the male and female gametes together. In some large flowers such as cacti (Figure 33.13) or lilies (Figure 33.17), the style may be several centimetres long.

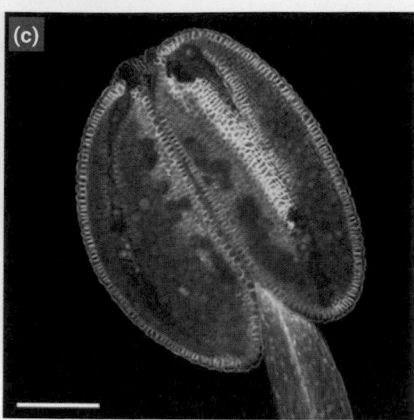

Figure 33.17 Pollen development and the production of pollen in anthers. (a) Flowers such as this lily produce huge amounts of orange-brown pollen, shed from their elongated anthers. It drops onto the white flower petals and is an attractive food for hoverflies. (b) The model lab plant *Arabidopsis thaliana* has four stamens surrounding the central stigma. (c) When viewed in cross section on a confocal microscope the individual pollen grains within the anther can be seen (scale bar = 100 microns).

Source: (b, c) Yang et al. (2017) Transcription factor MYB26 is key to spatial specificity in anther secondary thickening formation. *Plant Physiology*, 175, 333–350. Oxford University Press.

Figure 33.18 The titan arum, *Amorphophallus titanum*. The inflorescence can be more than 3 metres in height.

Source: Sailing moose/Wikimedia Commons/CC BY-SA 4.0.

When pollen lands on the stigma surface at the top of the style it grows a pollen tube, as illustrated in Figure 33.20. This grows down between the cells of the style and fuses with the ovule at the flower's base. The two haploid pollen nuclei descend through the pollen tube and enter the ovule.

On entering the ovule, one pollen nucleus fuses with one of the haploid maternal nuclei to form a diploid zygote, as depicted in Figure 33.21. A second fusion of nuclei occurs in the ovule between the second haploid nucleus from the pollen grain and two haploid nuclei from the ovule. This forms a triploid [3*n*] nucleus which subsequently divides by mitosis and generates storage tissues within the developing seed.

Single-sexed flowers

The flowers described above are bisexual in that they possess both male and female organs. Some plants, including marrows (*Cucurbita pepo*) and cucumbers (*Cucumis sativus*), the common hazel (*Corylus avellana*), and the smooth alder (*Alnus serrulata*; Figure 33.22), have separate male and female flowers on the *same* plant. Such plants are described as monoecious (see Figure 33.5).

Where the male and female flowers are borne on *separate* plants, the species is described as dioecious. Examples are willow (*Salix alba*), cannabis (*Cannabis* spp.), and holly (*Ilex aquifolium*; Figure 33.5). Male and female plants must grow sufficiently close to one another to permit wind or insect pollination. A good understanding of monoecious and dioecious plants is important in horticulture when seeking to obtain fertilized seeds, vegetables, and fruit.

The developing embryo and endosperm

Following fertilization, the zygote develops by a precise series of cell divisions, resulting in an embryo plant within the developing seed. An important phase of development is the deposition of storage material, called **endosperm**, which will be utilized when the seed subsequently germinates.

Figure 33.19 The diversity of pollen grains. The pollen grains of individual plant species have characteristic shapes and their exines (outer walls) are sculpted into characteristic patterns and textures. This is evident in electron micrographs of pollen grain coats extracted from bedrock 66 million years old, around the time when flowering plants first appeared.

Source: Antoine Bercovici.

REAL WORLD VIEW 33.2   Pollen and archaeology

The outer wall of a pollen grain, called the **exine**, is made of a biopolymer called **sporopollenin**. This is an inert material which is extremely resistant to enzymic degradation and therefore to destruction in the environment. As a result, pollen grains may survive in the environment for extremely long periods of time under the right conditions.

The stability of pollen grains and the diagnostic reliability of their structures make them valuable as contextual evidence in forensic, archaeological, and palaeontological investigations. With knowledge of the original plant's biology (its preferred climate or soil, or its time of flowering), it is possible to work out the conditions under which the pollen was deposited and establish useful details about the site from which it came. This science is called **palynology**.

Information from pollen grains, coupled with other data such as that from tree rings and mineral deposits, is revealing details of past climatic events and regional ecologies, sometimes going back many thousands of years.

Research has been carried out on the environmental history of the lowland peat moors of Somerset, England, known as the Somerset Levels (see Figure 1). Microscopic analysis of pollen in core samples from deep in the peat reveals the types of land that were present at different times in different areas of the Levels:

- Pollen from lime, oak, and elm trees indicates mixed deciduous woodland.
- Floodplain woodland, which is wetter, has more hazel pollen.
- Mixed carr woodland, which is much wetter, has pollen from alder, birch, hazel, and Scots pine.

Figure 1 The Somerset Levels from Glastonbury Tor.
Source: Arpingstone/Wikimedia Commons/Public Domain.

- Samples from marsh land contain mostly grass pollen.
- Samples consisting almost entirely of cereal pollen are evidence of land clearance and arable farming.

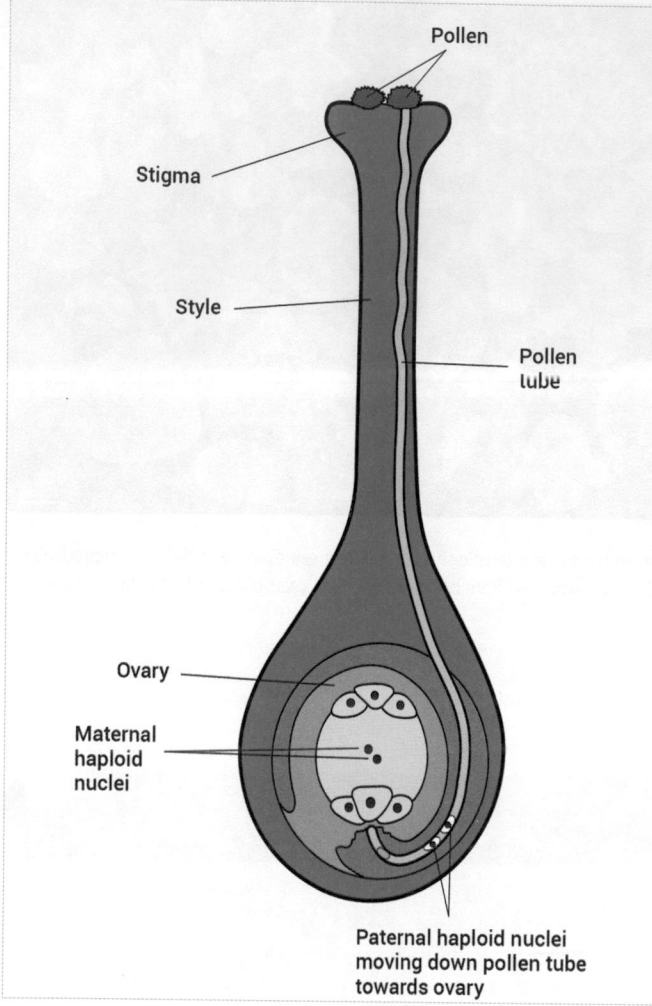

Figure 33.20 Pollen tube growth. A pollen tube grows when pollen lands on the stigma surface.

Source: Lincoln Taiz, Eduardo Zeiger, Ian M. Møller, and Angus Murphy. *Plant Physiology and Development*, Sixth Edition, 2014. Oxford University Press.

Endosperm is produced in different ways in monocotyledonous and dicotyledonous plants.

- In monocots, such as the major crop plants of wheat, rice, and maize, endosperm is formed by the triploid nucleus dividing and passing through a coenocytic phase before making cell walls and becoming cellular (see Figure 28.25). The synthesized molecules vary in composition between different plant species but, in general, are rich in either carbohydrate (starch), lipids, or storage proteins. Wheat storage proteins have elastic properties, exploited in dough during bread making. Maize endosperm contains a high proportion of lipid; this is extracted as oil and used in cooking.

- In dicot plants, the coenocytic phase is very short-lived and there is correspondingly little endosperm. Storage molecules accumulate in the cotyledons inside the developing seed. In soybeans, the stored material contains large amounts of oil and protein, making the seeds valuable as food for humans and animals.

In all plant species, seed storage molecules are degraded during germination, providing the seedling plant with nutrients and energy before it can photosynthesize independently.

Seeds, fecundity, and plant flowering

We defined fecundity as a measure of reproductive potential in terms of the number of offspring produced. With flowering plants we can interpret this as the number of seeds produced per reproductive event. Each seed develops from a single fertilized ovule within the base of the flower and each requires an individual pollen grain to grow down to it and carry out the fertilization event. Across the plant kingdom, r- and K-strategies are evident:

- In the peach (*Prunus persica*) and related drupe fruit trees, there is a single fertilization event resulting in a fruit with a single seed.

- In a pea flower (*Pisum sativum*), there are usually about eight ovules in the ovary and eight pea seeds in the resulting pea pod.

- The flowers of poppy (*Papaver* spp.) have a large number of individual ovules, producing several thousand seeds from an individual flower head.

Large, longer-lived perennial plants show a great deal of variability in when to flower, how much to flower, how many flowers become fertilized, and how many fruit develop. These outcomes are greatly influenced by climate, nutrition, and predation with the result that the plant's fertility% may be unpredictable from season to season.

Some trees appear to monitor and adjust the number of seeds and fruits that develop during the reproductive season. This is frequently seen in apple trees where, after large numbers of fruits start to develop, the tree aborts a significant proportion; they fall from the tree in what is called the June drop. The tree goes on to support a reduced number of larger, healthy fruit in something that amounts to a quality control system.

Some perennial plants do not flower every year. Bamboo only flowers on rare occasions, often decades apart. Other plants only produce significant numbers of fertilized seeds at intervals over their lifespan. In forest trees such as oak and beech, these are the so-called *mast years* when there is a huge crop of seeds; they typically occur every 5–10 years and are difficult to anticipate without a detailed understanding of the plant variety and local conditions.

The ability of plant seeds to survive in their dormant state, sometimes for very many years, has been exploited by conservationists who are anxious to protect the world's biodiversity. The Millennium Seed Bank at Kew in London is a large depository of seeds from plants native to Britain. It also contains seeds from around the world, provided by collectors over many decades. By 2020 it aimed to house 25% of the world's plant species.

Reproduction in gymnosperms

The gymnosperms are a highly successful group of plants, mostly trees, noted for their prolific production of seeds, their ability to flourish in extreme weather conditions, and their tolerance of poor soils. They form the extensive communities of coniferous forests typically found at the higher and lower latitudes of our planet.

Figure 33.21 Double fertilization and the production of zygote and endosperm.

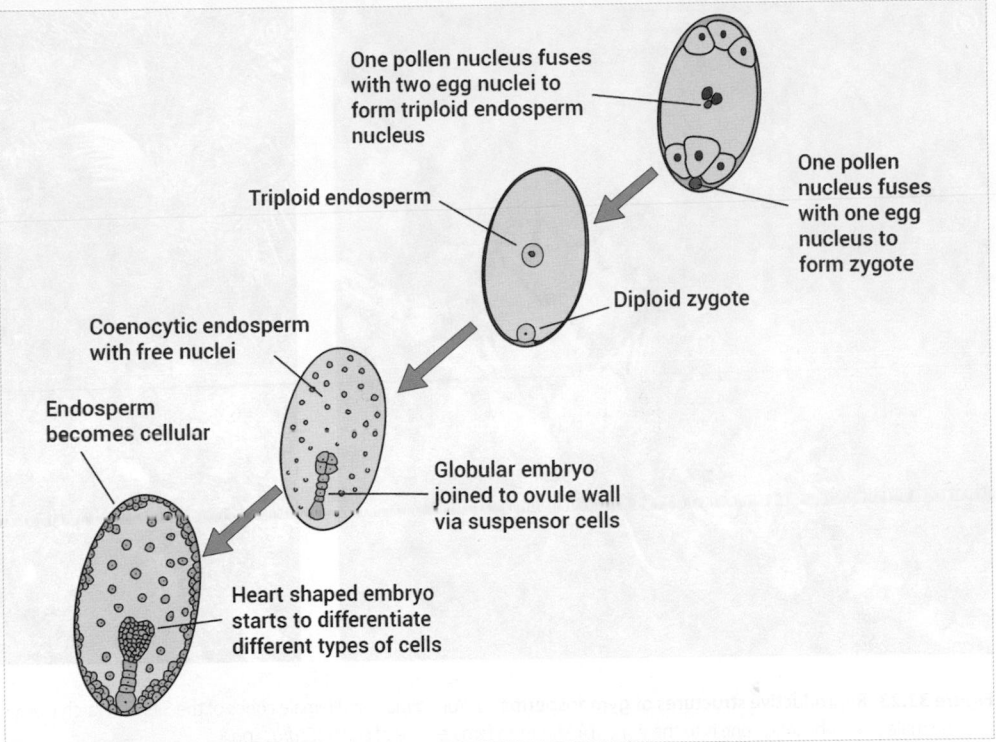

One pollen nucleus fuses with two egg nuclei to form triploid endosperm nucleus

One pollen nucleus fuses with one egg nucleus to form zygote

Triploid endosperm

Diploid zygote

Coenocytic endosperm with free nuclei

Globular embryo joined to ovule wall via suspensor cells

Endosperm becomes cellular

Heart shaped embryo starts to differentiate different types of cells

Figure 33.22 Catkins of the smooth alder (*Alnus serrulata*).

Source: Axel Kristinsson/Flickr/CC BY 2.0.

The common taxonomic characteristic of gymnosperms is that their reproductive structures, called ovules, are not enclosed in a fruiting structure (Greek *gymno* = 'naked'). They do not produce elaborate flowers and are essentially wind-pollinated. A gymnosperm's equivalent of a flower could be said to be the woody structures called cones, such as those illustrated in Figure 33.23. Male and female cones reside on a single tree, so the plants are monoecious, with wind moving the pollen between the cones on the same tree or others.

The ovules on the female cones are not protected or enclosed as they are in the base of angiosperm flowers but are exposed within the scales of the cone. Gamete production, pollination, and pollen tube growth occur in much the same way as they do in angiosperms (see Figure 33.4), although there is only a single fertilization to form the zygote and no triploid endosperm.

There is much variation in the speed of reproduction in gymnosperms. In most conifers, the production of a seed occurs within one year, but in pine trees (*Pinus* spp.) seed production takes two years to complete. Dispersal of the seeds from within the female cones is sometimes facilitated by birds and animals. Some birds move cones and hoard them. In ecosystems dominated by conifers, fire may be a required physical element: fire-scorched cones open up their scales and allow seeds to be shed and dispersed.

 Check your understanding of the concepts covered in this section by answering the questions in the e-book.

33.4 Reproduction in animals

Most metazoan animals reproduce by combining haploid gametes from male (sperm, produced in the testis) and female (egg, produced in the ovary) individuals. This process is called **fertilization** and it results in the production of a diploid cell called a zygote, which then develops into an embryo.

What is an egg?

Biologists use the word 'egg' to mean two different but related things:

1. The entity which contains the female gamete, carries it away from the ovary, and makes it available for fertilization by sperm. This process occurs in virtually all animals that reproduce by

Figure 33.23 Reproductive structures of gymnosperms. (a) Young male and female cones of the Siberian larch (*Larix sibirica*). The pink female cone is to the left and the male pollen-bearing cone is to the right. (b) Maturing female cone of cedar (*Cedrus* spp.).

Source: (a) Konstantin39/Shutterstock.com; (b) K Pyke and K Ballinger.

sexual mechanisms, and in many parthenogenetic animals as well. The egg comprises a single cell containing a haploid nucleus, called the **germ cell**. This is usually enclosed in **somatic cells** (non-germ, diploid, 'body' cells) during its development within the ovary. It is released from the ovary in a process called **ovulation**. Fertilization takes place inside or outside the body depending on the type of animal.

2. An enclosed structure, ejected from the mother's reproductive tract, in which an *already fertilized zygote* starts its development outside the parental body. Such eggs are describes as cleidoic because they have a have a substantial external membrane (Greek *kleistos* = closed). At the end of the early development period, the membrane ruptures and a young embryo or larva hatches as a free-living organism.

In birds, mammals, and some invertebrates, gamete-containing eggs do not leave the female body but are fertilized internally by sperm following copulation. Embryo development then either proceeds internally, as in mammals, or a cleidoic egg is formed and released for development outside the body. Mammals and other animals which deliver live young are described as **viviparous**. Those which produce cleidoic eggs are **oviparous**.

In contrast, many invertebrates, especially arthropods and echinoderms, as well as teleost (bony) fish and amphibians release gamete-containing eggs into the environment, to be fertilized outside the body. This represents an r-strategy because the eggs are usually produced in large numbers with a low probability of fertilization and subsequent survival.

Oviparous eggs, with their resilient coverings, provide nutrition and physical protection to the zygote and developing embryo. They

may also prevent desiccation and microbiological infection, while permitting sufficient gas exchange for the developing embryo to respire. They include the familiar eggs of birds, but also those of reptiles and elasmobranch (cartilaginous) fishes. This represents a K-strategy because limited numbers are produced and offspring emerge from the egg at an advanced stage of development, sometimes capable of independent or near independent existence.

Types of egg

Oviparous eggs, because they support the early development of offspring within an enclosed unit, contain nutrient in the form of yolk. Egg types vary in the amount of yolk they contain and can be categorized according to the role it plays during early embryonic development. The categorization extends to non-cleidoic eggs because there are many structural similarities.

The following are the main types of egg. The 'lecithal' stem comes from *lekithos*, the Greek word for yolk.

Microlecithal (alecithal, homolecithal, oligolecithal, isolecithal)

Microlecithal eggs contain very little or no yolk, although some nutritional material may be present in the cell's cytoplasm. This kind of egg is found in all eutherian mammals and also in the marine non-vertebrate chordates (amphioxus and the tunicates) and in echinoderms.

In the mammalian ovary, each germ cell is enclosed in a ball of somatic cells called a follicle, as illustrated in Figure 33.24. Follicles form early in the life of the ovary and remain for some time in a dormant state with one layer of cells surrounding the germ cell inside a membrane. At an appropriate point in the life of the female, and

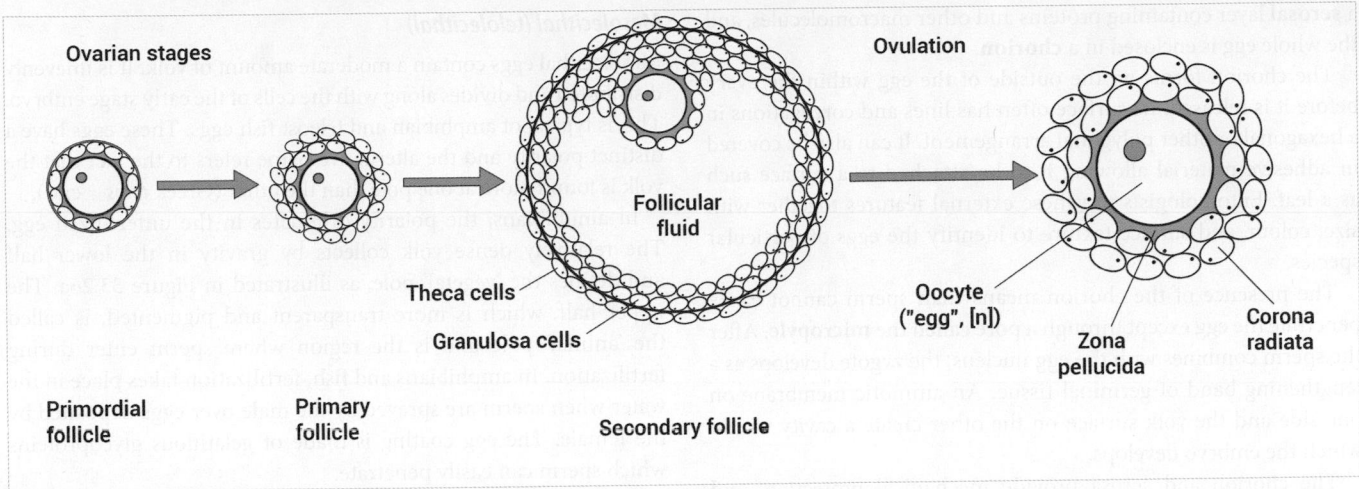

Figure 33.24 Origin and formation of the mammalian oocyte (microlecithal).

often on a cyclical basis, a limited number of follicles will increase in size and become available for ovulation (see also Chapter 26).

Follicle growth is initially due to an increase in the number of somatic cells on both sides of the membrane. As the follicle gets larger, fluid is secreted amongst the cells and it expands as a fluid-filled chamber. The oocyte (germ cell) is usually positioned to one side, embedded in a layer of cells called the **cumulus oophorus**. Between the germ cell nucleus and the cumulus cells is a layer of protein called the **zona pellucida**.

At ovulation, the wall of the follicle is eroded by enzymes and the fluid content is released into the female reproductive tract. The oocyte, with the zona pellucida and some cumulus cells, is washed with the fluid into one of the Fallopian tubes which leads to the uterus. If the female has mated, spermatozoa may have progressed up the tube and can fertilize the egg.

Fertilization usually requires the zona layer to have expanded by fluid accretion, and this process detaches most or all of the remaining cumulus cells from the structure. The first stages of embryo development, forming a blastula, take place inside the zona pellucida membrane. The embryo 'hatches' when this membrane breaks down, allowing it to implant in the uterine endometrium.

Follow the link in the e-book to learn more about microlecithal development.

Centrolecithal

In centrolecithal eggs, both the yolk and the haploid nucleus are located centrally within the structure. Most insects and other arthropods produce eggs with this structure, as do several other invertebrates including coelenterates (Cnidaria and Ctenophora).

Figure 33.25 shows the structure of an insect egg. Prior to fertilization the yolk occupies most of the volume, with just a small amount of other cytoplasmic material. The yolk, cytoplasm, and nucleus are surrounded by a **vitelline membrane**. Around this is

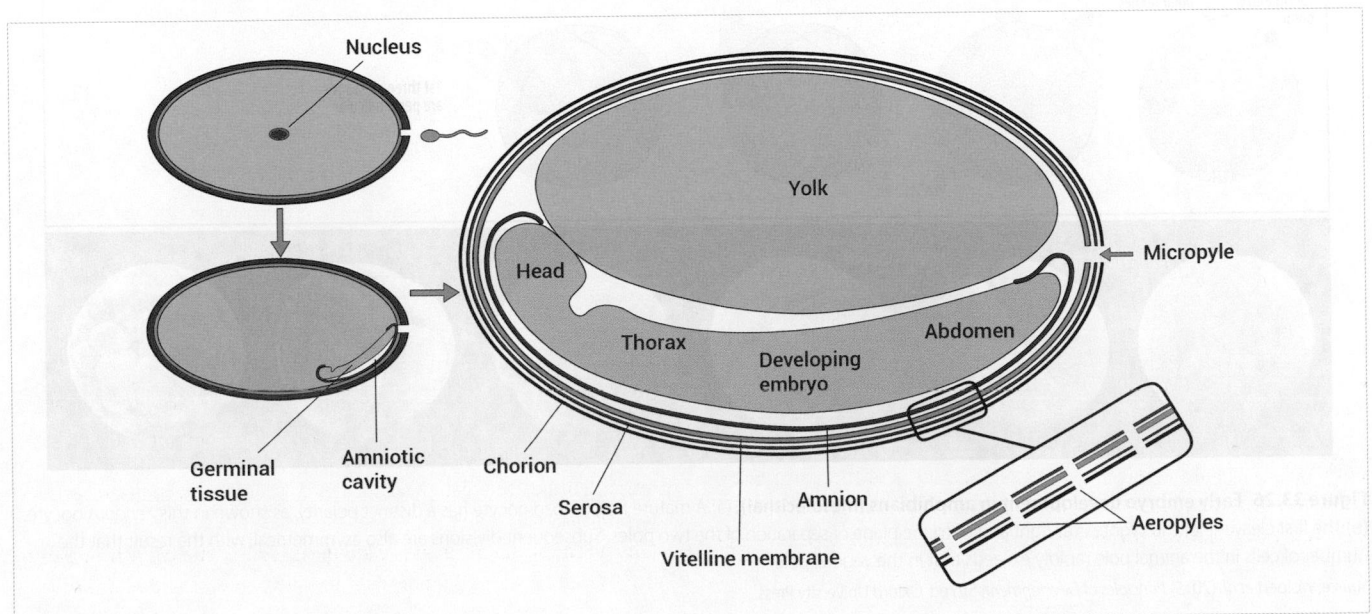

Figure 33.25 The general structure of an insect egg—an example of a centrolecithal egg.

a **serosal** layer containing proteins and other macromolecules, and the whole egg is enclosed in a **chorion**.

The chorion forms on the outside of the egg within the ovary before it is released. Its surface often has lines and corrugations in a hexagonal or other polygonal arrangement. It can also be covered in adhesive material allowing it to be attached to a surface such as a leaf. Entomologists use these external features together with size, colour, and surface texture to identify the eggs of particular species.

The presence of the chorion means than sperm cannot easily penetrate the egg except through a pore called the **micropyle**. After the sperm combines with the egg nucleus, the zygote develops as a lengthening band of germinal tissue. An amniotic membrane on one side and the yolk surface on the other create a cavity within which the embryo develops.

The chorion and serosa provide mechanical protection, and also prevent dehydration and microbiological infiltration. Gas exchange by the developing embryo occurs through pores in the membrane complex called **aeropyles**.

Mesolecithal (telolecithal)

Mesolecithal eggs contain a moderate amount of yolk. It is unevenly distributed and divides along with the cells of the early stage embryo. This is typical of amphibian and teleost fish eggs. These eggs have a distinct polarity and the alternative name refers to the fact that the yolk is found more at one pole than the other (Greek *telos* = end).

In amphibians, the polarity originates in the unfertilized egg. The relatively dense yolk collects by gravity in the lower half and defines the 'vegetal' pole, as illustrated in Figure 33.26a. The upper half, which is more transparent and pigmented, is called the 'animal' pole and is the region where sperm enter during fertilization. In amphibians and fish, fertilization takes place in the water when sperm are sprayed by the male over eggs deposited by the female. The egg coating is made of gelatinous glycoproteins which sperm can easily penetrate.

The distinction between the animal and vegetal poles persists as embryo development proceeds. The newly fertilized egg divides in the vertical plane at the first mitosis, but in subsequent

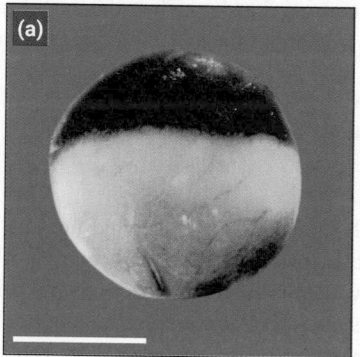

(a)

(b)

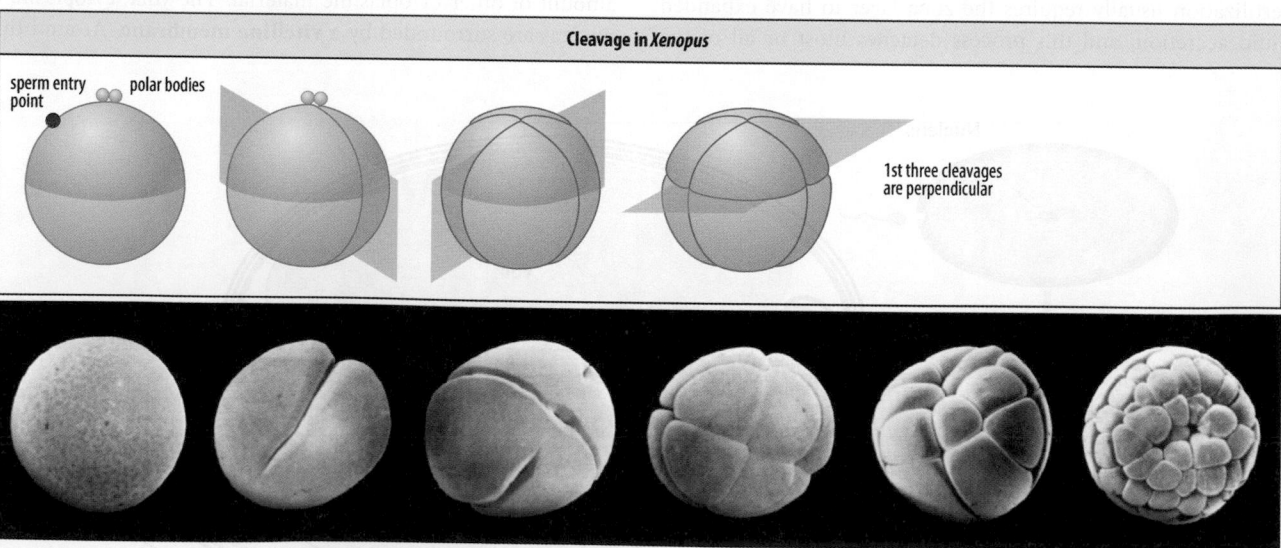

Figure 33.26 Early embryo development in amphibians (mesolecithal). (a) A mature frog or toad oocyte has a distinct polarity, as shown in this *Xenopus* oocyte. (b) The first cleavage of the zygote is at right angles to the plane of separation of the two poles. Subsequent divisions are also asymmetrical, with the result that the number of cells in the animal pole rapidly exceeds that in the vegetal pole.

Source: Wolpert *et al.* (2015) *Principles of Development*, 5th ed. Oxford University Press.

divisions the cells in the animal hemisphere divide more rapidly than those in the vegetal hemisphere. This is partly because yolk takes longer than cytoplasm to divide, and also because the genes that control mitosis and cellular organization are differentially expressed in the two parts.

The asymmetry persists as subsequent divisions occur, resulting in a blastocyst with fewer, larger blastomeres in the vegetal pole and more, smaller blastomeres in the animal pole; look at Figure 33.26b and notice the asymmetry that is apparent, particularly in the fifth image in the sequence shown. The vegetal cells contain most of the yolk and provide nutrition for the animal cells as the blastocysts forms. They are also the site where the blastopore begins to form, at the initiation of gastrulation. Thus, the original polarity of the egg is fundamental to the development of the embryo.

Macrolecithal (megalecithal, polylecithal)

Macrolecithal eggs have a very large amount of yolk and are typical of birds, reptiles, and monotreme mammals; an example is illustrated in Figure 33.27. The egg supports an extended period of embryo development during which the yolk is completely consumed.

Figure 33.27 The structure and formation of a chicken's egg—an example of a macrolecithal egg.

Source: Michael J.F. Barresi and Scott F. Gilbert. *Developmental Biology*, 12th edition, 2019. Oxford University Press

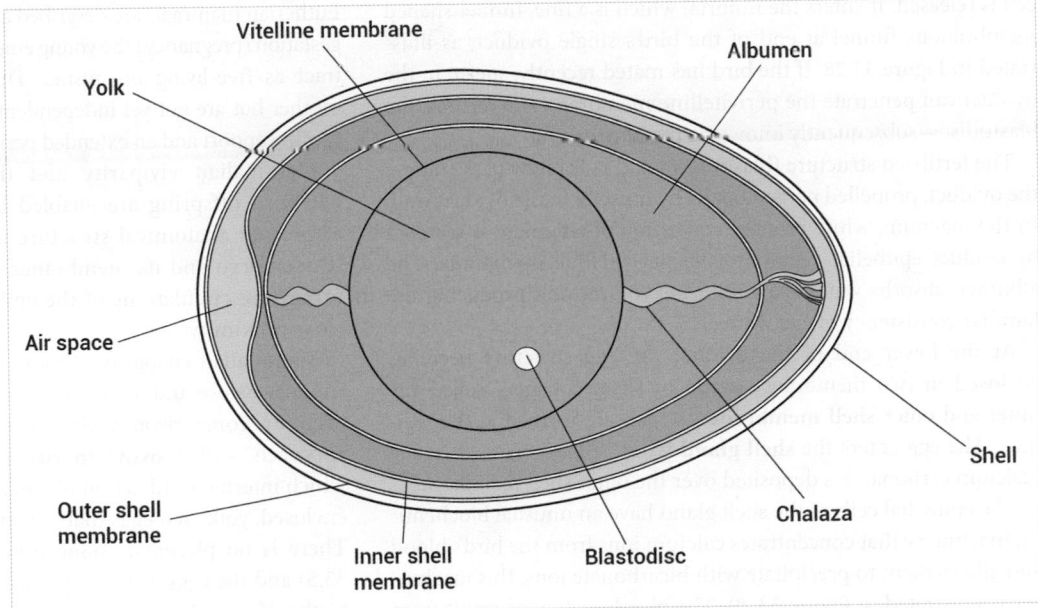

Figure 33.28 The avian oviduct, where egg components are formed. Follicles grow, mature, and ovulate sequentially. The follicle remnant (somatic cells and membranes) left after ovulation does not form a functional structure.

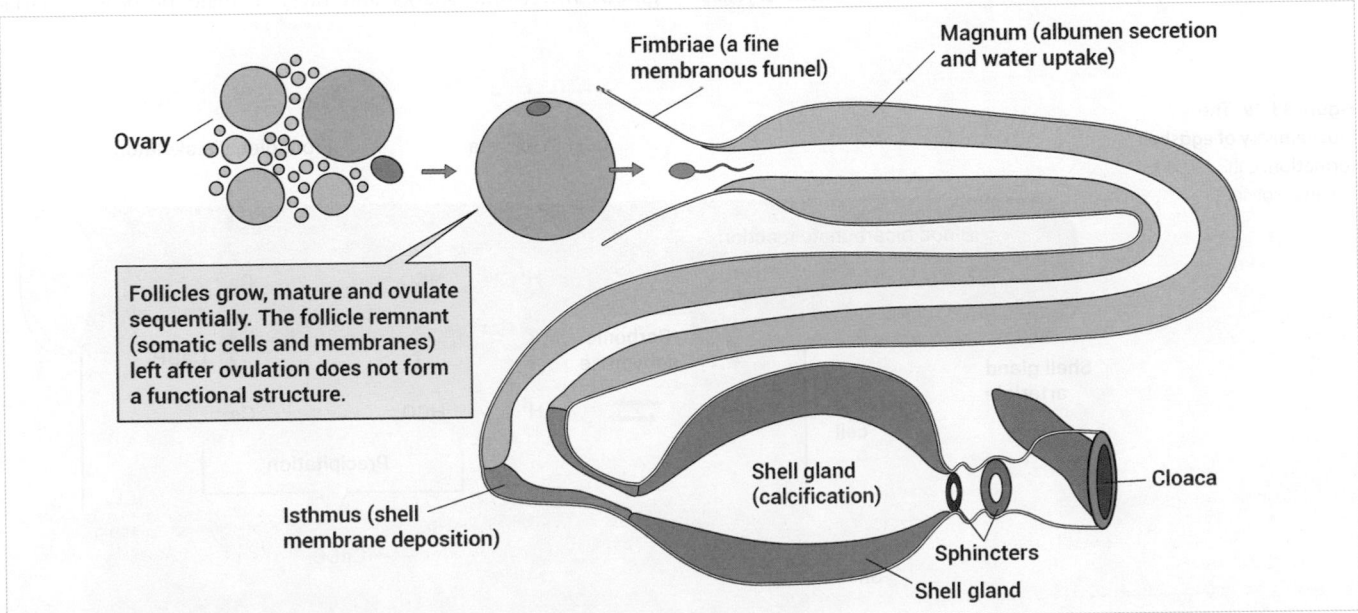

In birds, germ cells develop sequentially as follicles within the ovary and grow by the deposition of yolk material. Yolk is a complex mixture of proteins and lipoproteins, together with minerals and pigments. It is made in the liver and transported to the ovarian follicles in the bloodstream. In a domestic chicken, the quantity of yolk deposited in a single follicle may double in 24 hours.

Each germ cell comprises a single sphere of yolk, with a haploid nucleus called the **blastodisc** aligned to one side, enclosed in an extremely fine perivitelline membrane (equivalent to the zona pellucida of microlecithal eggs).

At ovulation, enzyme activity in the follicle membrane creates an aperture called a **stigma** through which the yolk-laden germ cell is released. It enters the fimbria, which is a fine, funnel-shaped membranous funnel at end of the bird's single oviduct, as illustrated in Figure 33.28. If the bird has mated recently, sperm in the oviduct can penetrate the perivitelline membrane and fertilize the blastodisc—subsequently known as the **blastoderm**.

The fertilized structure (blastoderm and yolk) now passes along the oviduct, propelled peristaltically by muscles in the oviduct wall. In the magnum, white protein, consisting of albumen, is secreted by oviduct epithelial cells onto the perivitelline membrane. The albumen absorbs water, expanding its volume and producing the familiar consistency of egg white.

At the lower end of the oviduct the egg structure becomes enclosed in two membranes made of sheet collagen, called the inner and outer **shell membranes**, which are formed at the isthmus. The egg enters the **shell gland** where a hard crust of calcite (calcium carbonate) is deposited over the outer shell membrane.

The epithelial cells of the shell gland have an unusual biochemical machinery that concentrates calcium ions from the bird's blood and allows them to precipitate with bicarbonate ions; this machinery is illustrated in Figure 33.29. The bicarbonate ions result from the normal solution of carbon dioxide in the water fraction of the blood. The shell material forms as insoluble calcite crystals

growing outwards from seeding points on the surface of the outer shell membrane.

 Follow the link in the e-book to learn more about the structure and function of eggshells.

In contrast to a mesolecithal egg, the yolk in a fertile macrolecithal egg remains undivided as the embryo develops from the zygote. The yolk is the principal nutritional resource for the embryo but the water and albumen from the white are also consumed.

Viviparity

Eutherian mammals are described as viviparous because at the end of gestation (pregnancy) the young emerge from the female reproductive tract as free-living organisms. They become separated from the mother but are not yet independent because they require lactational (milk) support and an extended period of parental protection.

Mammalian viviparity and the advanced development of newborn offspring are enabled by the presence of a placenta. This is an anatomical structure formed between the conceptus (the embryo and its membranes) and the uterine lining, which brings the circulations of the embryo and the mother into very close proximity.

Mammalian viviparity, with its intense dependence on parental support before and after birth, is an extreme form of K-strategy, even by comparison with oviparity. There is an intermediate approach, called **ovoviviparity** (ovovivipary, or ovivipary), in which internal fertilization of female gametes results in membrane enclosed, yolk-rich eggs that are retained inside the mother's body. There is no placental connection with the mother (see Section 33.5) and the eggs eventually hatch internally, so the mother gives birth to free-living young.

This occurs in some species of reptile (slow worms) and cartilaginous fish (some sharks and rays). Female porbeagle sharks

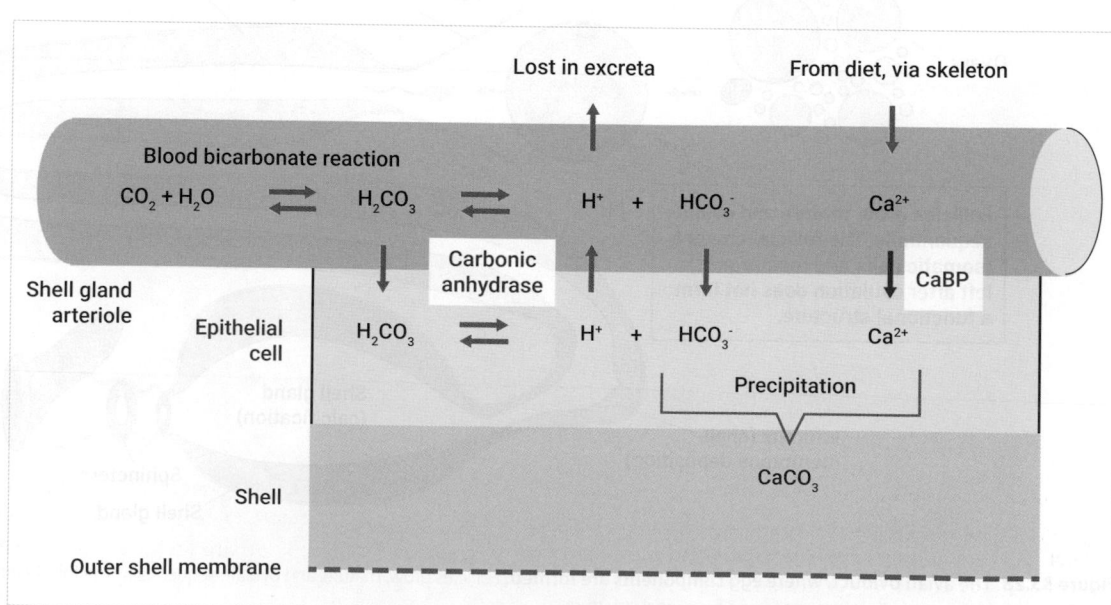

Figure 33.29 The biochemistry of eggshell formation. CaBP, calcium-binding protein.

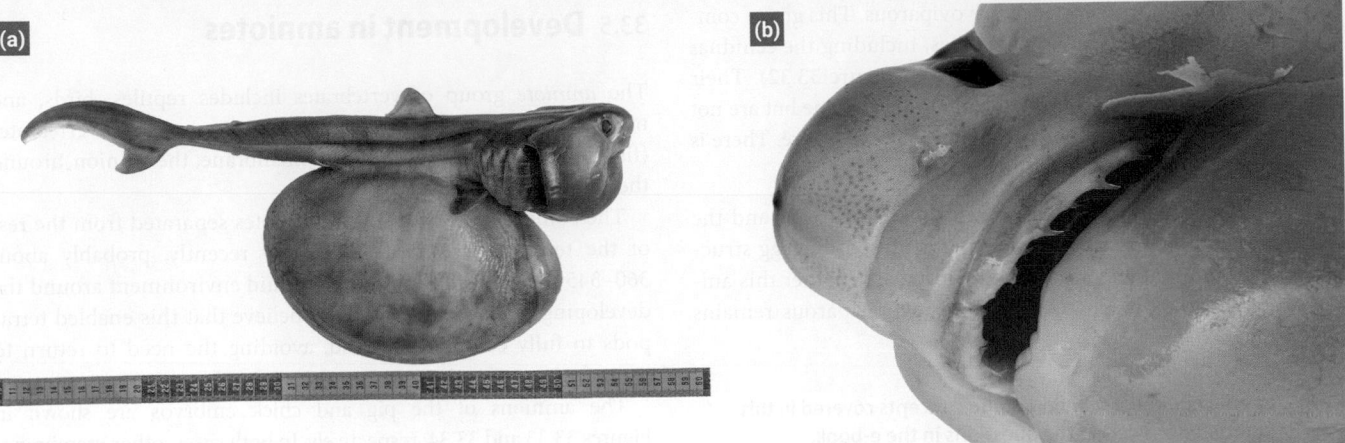

Figure 33.30 The porbeagle embryo. The pictures are anterior views of the heads of (a) a 9.6 cm embryo and (b) a 26.4 cm female embryo showing the functional fangs.
Source: © RLLord Sealordphotography.

Figure 33.31 Seahorse reproduction. Male sea horses incubate their partner's eggs in an abdominal pouch. Male (left) and female *Hippocampus reidi*.
Source: Andreas März/Flickr/CC BY 2.0.

(*Lamna nasus*; Figure 33.30) have two uteri in which their embryos develop. At the stage when the shark embryo develops internal gills, it also grows massive recurved fangs with which it tears open and consumes any remaining eggs in the uterus, a process called **oophagy**. The additional yolk supports further growth and the young shark is born in an advanced, free-living state.

Male sea horses (*Hippocampus* spp.) and the related pipefishes (Syngnathinae), incubate their partner's eggs in an abdominal pouch, as depicted in Figure 33.31. As a result, they effectively give birth to live young. During breeding, females pass their eggs to a male who fertilizes them and places them, along with some sea water, in a specialized pouch in their abdomen. The male seals the pouch which protects the eggs from the external environment. The pouch contains antibacterial and antifungal molecules which help to protect the young.

A few amphibians, including the fork-tongued frogs of Sulawesi (Dicroglossidae), Darwin's frog of South America, and the recently extinct *Rheobatrachus* frogs of Queensland, also *appear* to be viviparous because the parents provide brood areas for their tadpoles inside the alimentary tract.

Eggs and the evolution of viviparity

Biologists have wondered about the evolutionary relationship between oviparity and viviparity. The question is either 'which came first?' or 'has either strategy evolved more than once?' As with all biological conundrums like this, the problem lies in knowing what strategy was used by the common ancestors of living animals.

Evidence from fossil sites in China suggests that an early long-necked dinosaur of the group that includes reptiles and birds gave birth to live young. This may have been an adaptation to an aquatic lifestyle, in contrast to the land-based lifestyle of egg-laying reptiles and birds. However, reptiles and birds are closely related and probably a poor guide to the ancestors of vertebrates as a whole.

The mammals (milk-producing animals) are also a confusing group as far as reproduction is concerned. Most mammals are eutherian: their eggs are fertilized internally and the early embryo development is supported by a placenta. But two other groups of mammals have different characteristics:

1. Marsupial (pouched) mammals produce cleidoic eggs with a relatively thick coating, called the shell coat. This prevents the embryo from making the close contact with the mother's uterus seen in eutherian mammals. The membranes between the embryo and the shell coat have blood capillaries, allowing for some exchange with the mother's uterus, but the embryo's initial development occurs within this enclosed structure. There may be a brief period of placenta-type connection in some marsupials (wombats, Tasmanian devils, opossums, and dunnarts), but direct internal support from the mother is very limited and it is absent in the macropods (wallabies and kangaroos). The embryo emerges (hatches) from the shell coat just a few hours or days before it is born and is ejected from the uterus at a very early stage of development compared with eutherian young. Subsequent maturation, supported by suckling, takes place external to the mother's body in a pouch.

2. Prototherians, or monotremes, are oviparous. This group comprises a very small number of species, including the echidnas and the duck-billed platypus (shown in Figure 33.32). Their eggs are enclosed in a complex, leathery membrane but are not hardened with mineral in the way that birds' eggs are. There is no internal connection with the mother's body.

One hypothesis is that the shell coat of marsupial eggs and the complex membrane of monotreme eggs represent the egg structure of the common ancestor of all mammals. Whether this animal was oviparous, viviparous, or even ovoviviparous remains uncertain.

 Check your understanding of the concepts covered in this section by answering the questions in the e-book.

33.5 Development in amniotes

The *amniote* group of vertebrates includes reptiles, birds, and mammals. They are distinguished from *anamniote* vertebrates (fish and amphibians) by having a membrane, the amnion, around the conceptus as it develops.

The common ancestor of the amniotes separated from the rest of the tetrapod vertebrates relatively recently, probably about 360–345 Ma. The amnion creates a fluid environment around the developing embryo and biologists believe that this enabled tetrapods to fully colonize the land, avoiding the need to return to water for reproduction.

The amnions of the pig and chick embryos are shown in Figures 33.33 and 33.34, respectively. In both cases, other membranes

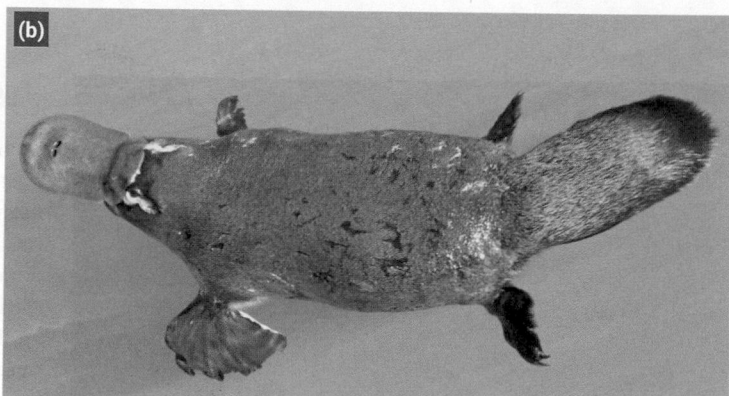

Figure 33.32 Oviparous monotreme mammals. (a) Echidna (*Tachyglossus aculeatus*). (b) Duck-billed platypus (*Ornithorhynchus anatinus*).
Source: (a) Image by pen_ash from Pixabay; (b) Brisbane City Council/Wikimedia Commons/CC BY 2.0.

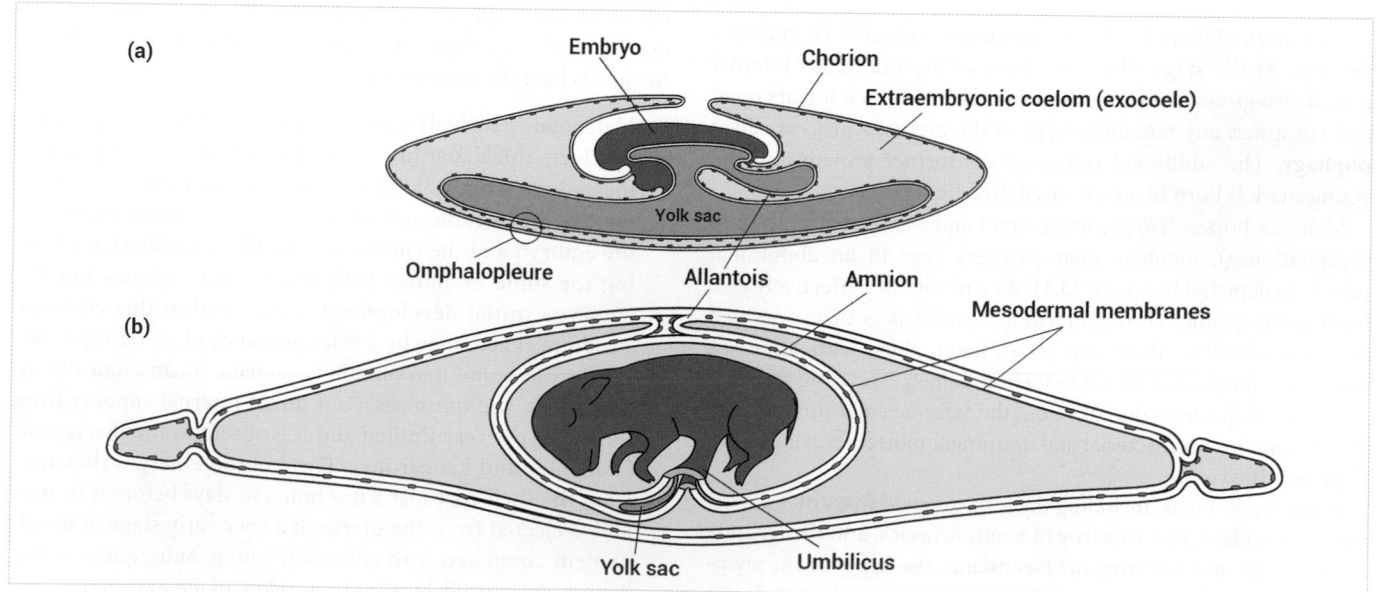

Figure 33.33 Stages of pig embryo development. (a) At one week, and (b) at two weeks of development.
Source: (b) Chordate morphology by Malcolm Jollie 1962; Image courtesy of Biodiversoty Heritage Library.

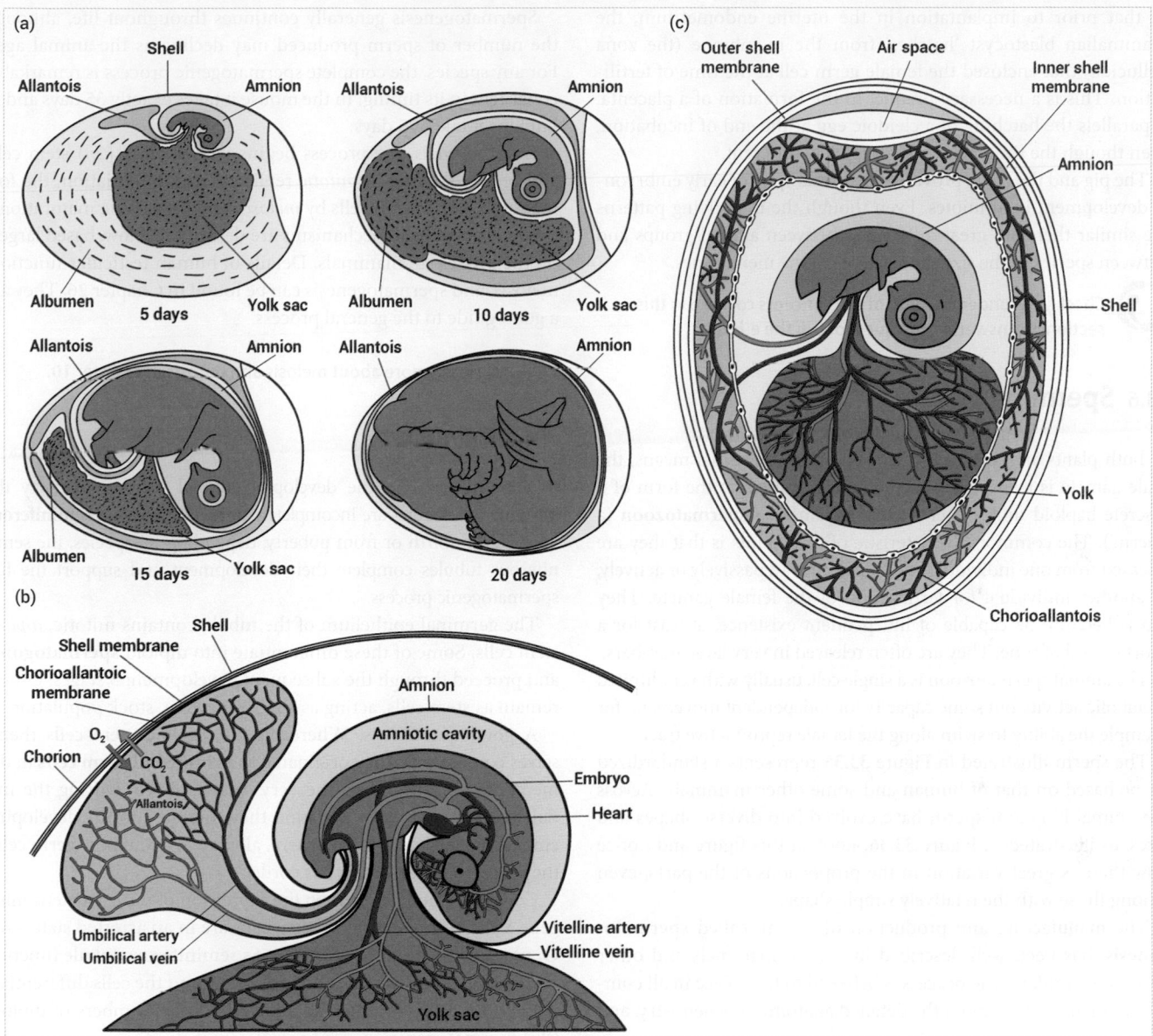

Figure 33.34 Stages of chick embryo and membrane development. (a) Development within the egg at 5–20 days of incubation. (b) Chorioallantois formation and yolk sac vessels at about 4 days. (c) Arrangement of embryo, membranes, and blood vessels during the last week of development (hatching at 20–22 days). *Source*: (b) Wolpert *et al.* (2015) *Principles of Development*, 5th ed. Oxford University Press.

called the allantois and chorion lead away from the amnion-enclosed embryo and fuse to facilitate nutrient supply and gas exchange:

- In the pig, the amnion forms as a fold around the early embryo and completely encloses it, while mesodermal membranes form a fluid chamber called the exocoele, as shown in Figure 33.33a. The placenta—which is the interface with the mother's circulatory system—forms initially from the yolk sac, a two-layered structure called the omphalopleure (Figure 33.33b). This is later displaced by the allantois (Figure 33.33b). Thus, the placenta changes from chorio-vitelline (chorion + yolk sac membrane) to allanto-chorionic (allantois + chorion).

- In the chick, as shown in Figure 33.34, the amnion also surrounds the early embryo, remaining in place for the whole of

development, but there is no placenta and the yolk sac provides nutrition for the embryo for the whole of its development. A chorioallantoic membrane (chorion + allantois), which has a good blood supply, forms around the inside face of the inner shell membrane. It exchanges gases (carbon dioxide and oxygen) through the shell and also takes up calcium to form the shell and to supply mineral for the chick's bone development. You can also see in Figure 33.34 that the blunt end of the egg has an air space which assists gas exchange; the shell itself is porous to air and water

By comparing Figures 33.33 and 33.34 we see that, despite the oviparous/viviparous distinction between birds and mammals, the embryonic membranes are similar. There is a further similarity in that prior to implantation in the uterine endometrium, the mammalian blastocyst 'hatches' from the membrane (the zona pellucida) that enclosed the female germ cell at the time of fertilization. This is a necessary prequel to the formation of a placenta. It parallels the hatching of a cleidoic egg at the end of incubation, even though the timescales are completely different.

The pig and the chick provide typical examples of early embryonic development in amniotes. Even though the underlying patterns are similar there are great differences between animal groups and between species in the sizes and extents of the membranes.

 Check your understanding of the concepts covered in this section by answering the questions in the e-book.

33.6 Sperm

In both plants and animals which reproduce by sexual means, the male gamete is transported to the female gamete in the form of a discrete haploid unit, either a pollen grain or a **spermatozoon** (a 'sperm'). The common characteristic of these units is that they are released from one individual and find their way, passively or actively, to another individual for fertilization of the female gamete. They must therefore be capable of independent existence, at least for a short period of time. They are often released in very large numbers.

The animal spermatozoon is a single cell, usually with very limited metabolic activity but some capacity for independent movement, for example the ability to swim along the female reproductive tract.

The sperm illustrated in Figure 33.35 represents a standardized shape based on that of human and some other mammals. Across the animal kingdom sperm have evolved into diverse shapes and sizes, as illustrated in Figure 33.36. Look at this figure and notice how there is great variation in the proportions of the parts, even among those with the relatively simple shape.

The manufacture and production of sperm, called **spermatogenesis**, has been well described in several mammals and other species. A similar basic process is believed to take place in all complex animals, even though the detailed anatomy, biochemistry, and timing vary extensively.

Spermatogenesis generally continues throughout life, although the number of sperm produced may decline as the animal ages. For any species, the complete spermatogenic process is remarkably consistent in its timing. In the mouse it takes exactly 35 days and in humans it takes 75 days.

In all animals, the process begins with diploid (*2n*) stem cells and proceeds through *mitotic* replication, differentiation, the formation of haploid (*n*) cells by *meiosis*, and then some maturational stages. The general mechanisms are described below, based largely on the process in mammals. Details of human testicular function, meiosis, and spermatogenesis can be found in Chapter 26. They are a good guide to the general process.

▶ We learn more about meiosis and mitosis in Chapter 10.

The origin of sperm

In the fetus and in the developing animal prior to puberty the seminiferous tubules are incompletely formed and called **seminiferous cords**. From birth or from puberty, depending on species, the seminiferous tubules complete their development and support the full spermatogenic process.

The germinal epithelium of the tubule contains mitotic, *diploid* stem cells. Some of these differentiate into diploid **spermatogonia** and proceed through the subsequent developmental stages. Others remain as stem cells, acting as a self-renewing, stock population.

A good question is: Where did the diploid stem cells themselves come from? They originated as **primordial germ cells** in the membranes surrounding the very early embryo. During the initial stages of embryo maturation they migrated to the developing embryonic testicular tissue where, along with local non-germ cells, they formed the seminiferous cords.

Cord formation depends on the expression of the sex-determining gene *SRY*. At this stage, the germ cells are in an arrested state—the G_1 phase of the mitotic cell cycle. As seminiferous tubule function is established, mitotic division is resumed and the cells differentiate so that the testis becomes populated by large numbers of diploid (*2n*) spermatogonia.

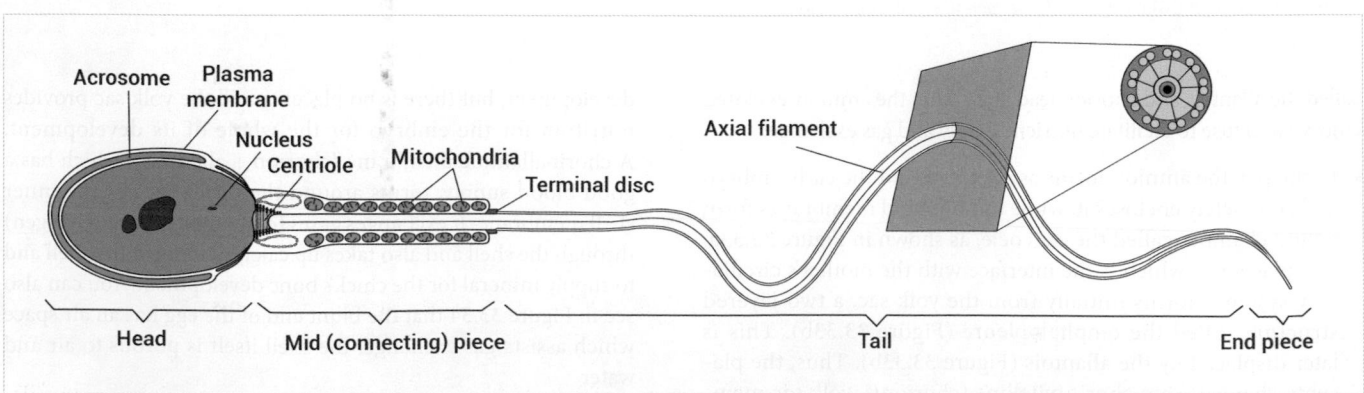

Figure 33.35 The structure of a human sperm.
Source: Mariana Ruiz Lady of Hats/Wimimedia Commons/Public Domain.

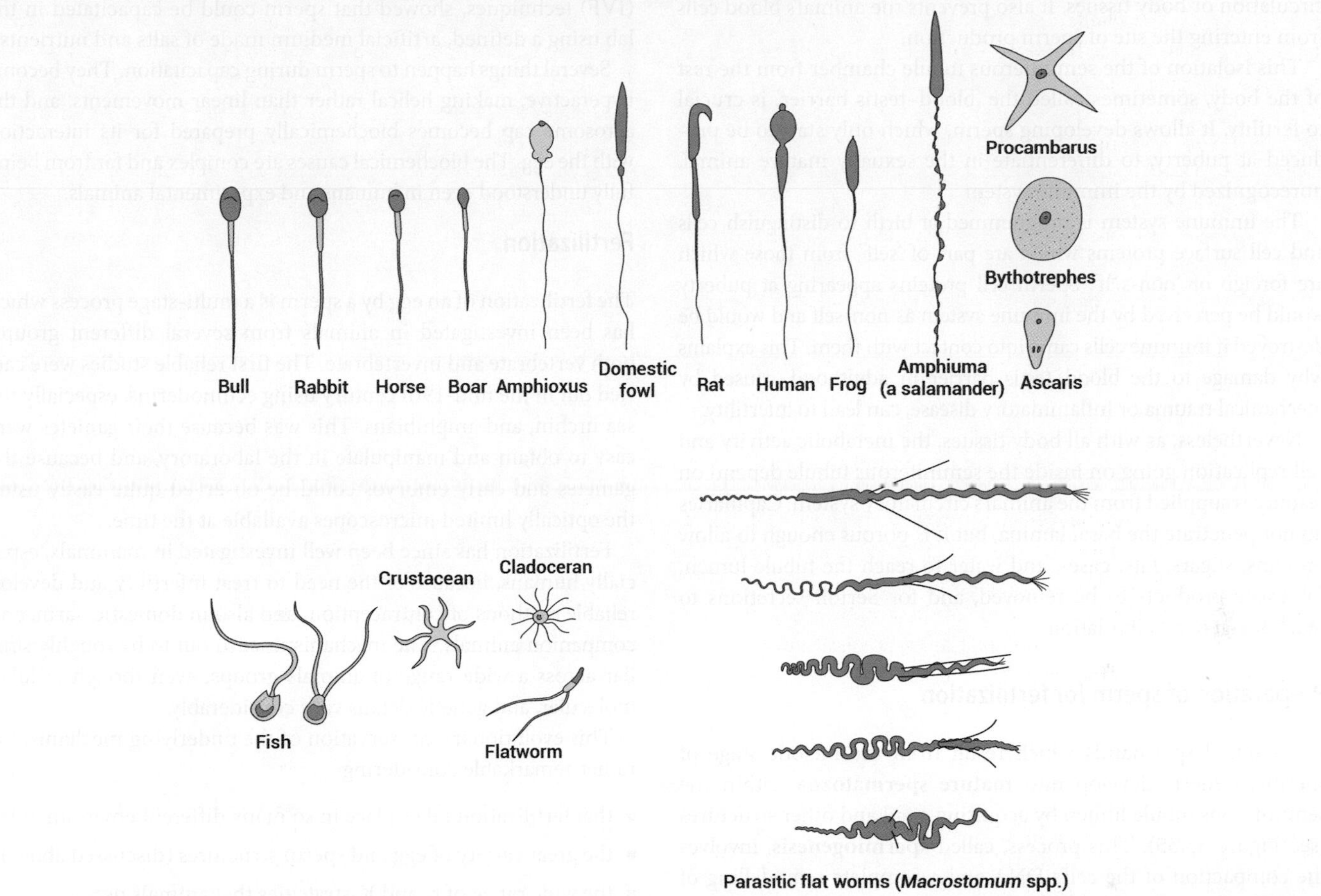

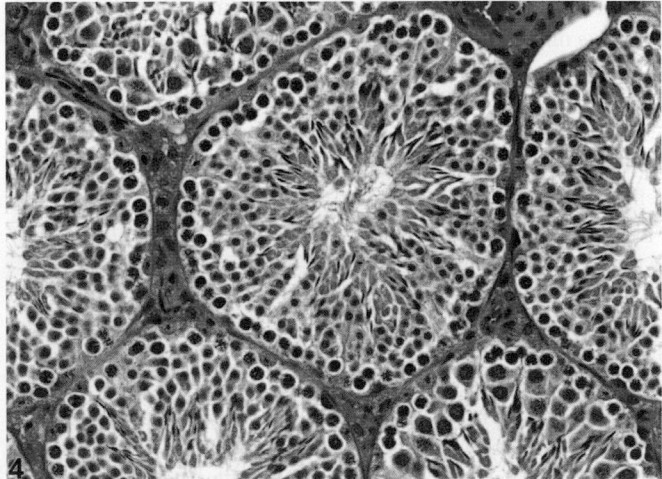

Figure 33.36 Animal sperm shapes.

Source: The many styles of sperm. *Nature*, 469, 269. Reprinted by permission from Springer ©2011.

In mammals, spermatogonia typically undergo six mitotic divisions and thus increase their numbers considerably. They now move inwards and by increasing their distance from the tubule membrane become committed to division by *meiosis*.

The meiotic divisions produce (*meiosis I*) **primary spermatocytes** (*4n*), then secondary spermatocytes (*2n*), and finally (*meiosis II*) haploid **spermatids** (*n*). This meiotic process differs in two important ways from that by which female germ cells are formed:

- there are no polar bodies, so *one* primary spermatocyte (from one spermatogonium) produces *four* spermatids (in females, one primary oocyte produces one fertilized oocyte);

- there is no meiotic arrest, so spermatid production is continuous provided that the testis and reproductive axis are functional (in females, *meiosis I* is only completed at ovulation and *meiosis II* occurs at fertilization).

Seminiferous tubules and the blood–testis barrier

A cross-section of a seminiferous tubule is shown in Figure 33.37. A notable feature of the seminiferous tubule is that its outermost layer is an especially resistant, **complex basal lamina**. There are

Figure 33.37 Mouse seminiferous tubules in cross section.

Source: Image courtesy of the National Toxicology Program.

also tight junctions (see Chapter 28) between the Sertoli cells close to where they rest on the basal lamina. This prevents developing or mature sperm cells from moving out of the tubule into the animal's circulation or body tissues. It also prevents the animal's blood cells from entering the site of sperm production.

This isolation of the seminiferous tubule chamber from the rest of the body, sometimes called the 'blood–testis barrier', is crucial to fertility. It allows developing sperm, which only start to be produced at puberty, to differentiate in the sexually mature animal, unrecognized by the immune system.

The immune system is programmed at birth to distinguish cells and cell surface proteins which are part of 'self' from those which are foreign or 'non-self'. Sperm cell proteins appearing at puberty would be perceived by the immune system as non-self and would be destroyed if immune cells came into contact with them. This explains why damage to the blood–testis barrier in adulthood, caused by mechanical trauma or inflammatory disease, can lead to infertility.

Nevertheless, as with all body tissues, the metabolic activity and cell replication going on inside the seminiferous tubule depend on resources supplied from the animal's circulatory system. Capillaries do not penetrate the basal lamina, but it is porous enough to allow proteins, sugars, fats, gases, and water to reach the tubule lumen, for waste products to be removed, and for Sertoli secretions to reach the general circulation.

Preparation of sperm for fertilization

The haploid spermatids which result from the meiotic stage of spermatogenesis develop into **mature spermatozoa** within the seminiferous tubule lumen by acquiring a tail and other structures (see Figure 33.35). This process, called **spermiogenesis**, involves the compaction of the cell's DNA and a complete remodelling of the cell cytoplasm.

A great many new genes are expressed during spermiogenesis. They include genes for the formation of the acrosome (a cap over the head region, derived from the cell's Golgi apparatus), genes which give the sperm its correct shape (abnormally shaped sperm often occur, especially in the tail regions), and other genes which organize mitochondria to give the energy for movement.

Sperm leaving the seminiferous tubules are still non-motile and biochemically unable to fertilize an egg. The next step in their maturation occurs in a structure adjacent to the testis called the **epididymis**. This is a set of highly coiled storage tubes of great length (1 m in mouse, 3 m in rat, and 6 m in human) in which the sperm become concentrated by about 10-fold due to the osmotic removal of fluid from the tubal lumen.

The epididymis provides a low pH, low Ca^{2+} environment and also contains a unique mixture of proteins. The sperm remain largely quiescent (non-mobile) during storage, which may last for many days or weeks. However, the extreme conditions in the epididymis promote cAMP production and other biochemical changes which ensure that the movements of the sperm after ejaculation will be strong and properly coordinated for forward movement.

The final step in sperm maturation is **capacitation**. This takes place not in the male, but in the female reproductive tract. Experiments on rodents and rabbits in the 1950s showed that sperm only had the capacity to fertilize an egg if they had spent several hours within the female's Fallopian tubes. Subsequent experiments, which had a direct bearing on modern in vitro fertilization (IVF) techniques, showed that sperm could be capacitated in the lab using a defined, artificial medium made of salts and nutrients.

Several things happen to sperm during capacitation. They become hyperactive, making helical rather than linear movements, and the acrosome cap becomes biochemically prepared for its interaction with the egg. The biochemical causes are complex and far from being fully understood even in humans and experimental animals.

Fertilization

The fertilization of an egg by a sperm is a multi-stage process which has been investigated in animals from several different groups, both vertebrate and invertebrate. The first reliable studies were carried out in the mid-19th century using echinoderms, especially the sea urchin, and amphibians. This was because their gametes were easy to obtain and manipulate in the laboratory, and because the gametes and early embryos could be observed quite easily using the optically limited microscopes available at the time.

Fertilization has since been well investigated in mammals, especially humans, because of the need to treat infertility and develop reliable methods of contraception, and also in domestic, farm, and companion animals. The mechanisms turn out to be roughly similar across a wide range of animals groups, even though cellular, molecular, and genetic details vary considerably.

This evolutionary conservation of the underlying mechanism is rather remarkable considering:

- that fertilization takes place in so many different environments
- the great variety of egg and sperm structures (discussed above)
- the wide range of r- and K-strategies that animals use.

Whatever the context, animal fertilization always takes place in a liquid medium, whether that is the open sea or fresh water or internally within fluids secreted by the female's reproductive tract.

A further common characteristic is that the ovum is normally fertilized by a single sperm. There are usually a great many sperm at the site of fertilization and a biochemical mechanism has evolved within the egg that prevents multiple fertilizations.

In many species, fertilization is an opportunity for mate selection (of males by females). A common reproductive strategy is for females to mate with more than one male. This is called **polyandry**. The result is that sperm from different males will compete for the opportunity to fertilize a single egg.

Mate selection is a common behavioural activity right across the animal kingdom. It starts with attraction and courtship, and often involves elaborate behavioural interactions between individuals, such as those described below for certain species of fish. Where females mate sequentially with more than one male, the strength and motility of the sperm may decide the outcome.

In animals with internal fertilization such as birds, sperm may be stored within the female tract for an extended period of time, giving a single male the chance to fertilize several eggs. The sperms of rival males which mate subsequently may be able to displace them, making physical competition between sperm part of the overall competition for reproductive success between males.

Competition is therefore a strong evolutionary mechanism for maintaining the quality of sperm. The animal with the *longest*

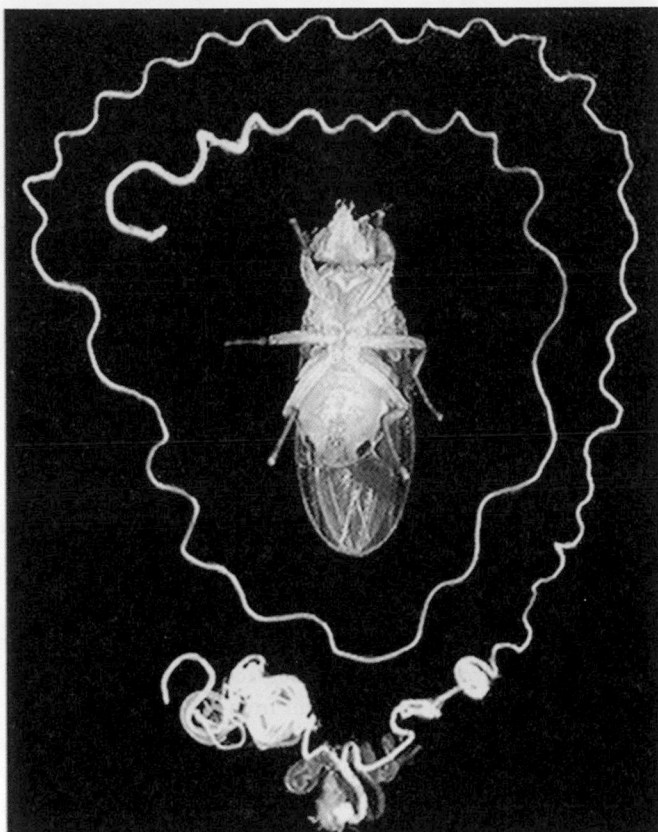

Figure 33.38 Fruit fly sperm. The long sperm of the fruit fly (*Drosophila melanogaster*), shown in comparison with the male fly. The structure around the fly is actually the dissected testis but the sperm typically occupies its whole length.

Source: Photo by Scott Pitnick.

sperm is believed to be the fruit fly (*Drosophila melanogaster*). The fly itself is only a couple of millimetres in length but its sperm has a very long tail and may approach six centimetres, as shown in Figure 33.38. Longer sperm presumably make for greater speed or strength or in some other way maximize the chance of fertilization.

In other vertebrates, including mammals, success may depend on the sheer volume of the ejaculate and the overwhelming number of sperm it contains. Fertilization generally has the characteristics of an r-strategy—many sperm competing for a single egg—even when the subsequent events of reproduction may have K- characteristics.

Stages of fertilization

In all types of animals, fertilization has the following stages:

1. Initial contact between the sperm and the surface of the egg.
2. Sperm binding and the acrosome reaction.
3. Penetration of the egg coat.
4. **Syngamy**: fusion of the egg and the sperm.
5. Intracellular sperm–egg interactions, including genetic combination.

The initial contact between sperm and egg, even when it takes place internally within the female tract, is essentially a chance event. In

mammals and birds, sperm find their way along the tract by following chemical and temperature signals. This is against the flow of female tubal fluid, so sperm must be efficient and persistent swimmers. Interaction with the egg normally occurs in the upper part of the tract; in large species this means that the sperm travel a considerable distance.

In mammals, the egg is surrounded by a jelly-like protein coat, the zona pellucida, and in some species there is also a layer of cells derived from the ovarian follicle, called the **cumulus cells**. There are equivalent coverings in other species. Thus, the sperm has to penetrate a significant physical barrier.

The zona pellucida, or its equivalent, contains several unique proteins and glycoproteins with which the head of the sperm can interact. A crucial part of the second stage of fertilization is the **acrosome reaction** in which the head of the sperm secretes enzymes which can penetrate the zona.

The acrosome reaction itself is initiated by the first contact between sperm and zona and is accompanied by an increase in the motile activity of the sperm. This assists physical penetration of the zona at the point where the enzymes have weakened the zonal proteins. During this event, the head of the sperm may separate from the tail.

Penetration by a single sperm is immediately followed by a biochemical change in the egg which has the effect of making the egg impenetrable to other sperm and preventing **polyspermy**. The biochemistry of this is complex, but it involves the release of Ca^{2+} ions within the egg.

Syngamy occurs when the membrane of the sperm head fuses with the egg plasma membrane. Receptors on the sperm and egg membranes make this fusion possible and what follows is a complex of intracellular sperm–egg interactions involving changes in pH and Ca^{2+} ion concentrations and the activation of signal transduction pathways. As a result, the centrioles from the sperm are incorporated into the egg and chromosomes from the two gametes come together. The final stage of meiosis in the egg, which had been suspended up to this point, is resumed, including the extrusion of a second polar body so that the resulting zygote is diploid. Subsequent events in the zygote occur through mitosis.

 Check your understanding of the concepts covered in this section by answering the questions in the e-book.

33.7 Annelid reproduction

The annelid worms are a large group of terrestrial and aquatic species. As we saw in Chapter 30, even terrestrial species such as the common earthworm need a wet environment in which to live, and they can adapt to environments with a range of osmotic conditions.

Annelids use asexual and sexual methods of reproduction. In some cases, segments break off from the parent worm and divide clonally to form new individuals. In others, some of the water-dwelling annelids engage in mass spawning events where thousands of gametes are released into the water column.

The following describes the reproduction of polychaete worms (the bristle worms). They illustrate many of the genetic and physiological principles discussed earlier in this chapter.

Polychaetes: asexual reproduction

Polychaetes can produce genetically identical forms of themselves by asexual reproduction. This is **schizotomy** and it occurs when the worm's body divides and new parts are regenerated. A new individual may detach from the parent body (**paratomy**) or the body may break apart and regenerate missing sections (**architomy**). Architomy is more widespread, the most extreme example being worms of the Cirratulidae family, which fragment repeatedly until single segments remain. Segments originating from the middle of the body go on and develop into adult worms.

This reproductive process is distinct from the ability of some worms to regenerate themselves from injured sections of body.

 Go to the e-book to watch a video of a flatworm regenerating to help you understand this process.

Polychaetes: sexual reproduction

Polychaete gametes are formed by specialized germ cells in the lining of the coelom (body cavity). The germ cells are grouped together in the germinal epithelium attached to septa (divisions between segments) in anterior segments. Gametes are released into the coelom where they drift freely and mature in the coelomic fluid. When the individual is sexually mature the coelom becomes packed with eggs and sperm.

Gametes exit the body in a variety of ways. The worm may develop **coelomoducts** or **gonoducts** in the segments: these are ciliated funnels through which eggs and sperm travel to the external environment. The coelomoducts can also join up with the nephridia (excretory ducts; see Chapter 30), so that the gametes exit via the nephridiopores, or the sperm can exit through anal apertures. Alternatively the worm's body wall ruptures and releases the gametes.

Most segments of the worm can produce gametes, although different segments of the same worm can produce different gametes. For example, fan worms (family Sabellidae; Figure 33.39a) are hermaphrodites, producing eggs from anterior segments and sperm from posterior segments.

In other species, individuals can be male at one time in their life and female at another. The hermaphroditic bristle worm *Ophryotrocha diadema* (Figure 33.39b) mates in pairs after a period of courtship. Each worm takes its turn to be the male (the one that fertilizes the eggs) and the other to be the female (the one that produces the eggs). Producing eggs takes much more energy than producing sperm so this strategy enhances the reproductive success of both individuals while reducing the overall cost of egg production.

Direct transfer of sperm from one individual to another is common amongst polychaetes. Sperm are packaged in parcels called **spermatophores** which protect them from the external environment. They enter the female either by direct insemination into a receptacle or they can be hypodermically inserted through the body wall. Hypodermic insertion occurs in some members of the

Figure 33.39 Polychaete worms. (a) Feather duster worm (Family: Sabellidae). (b) Bristle worm, *Ophryotrocha diadema*. (c) Australian terebellid worm (*Loimia* sp.). (d) *Myrianida prolifera* (Family: Syllidae).
Source: (a) Nhobgood Nick Hobgood/Wikimedia Commons/CC BY-SA 3.0; (b, d) Arne Nygren/Sjøfartsmuseet Akvariet Gøteborg/Wikimedia Commons/CC BY-SA 4.0; (c) © Queensland Museum, Gary Cranitch.

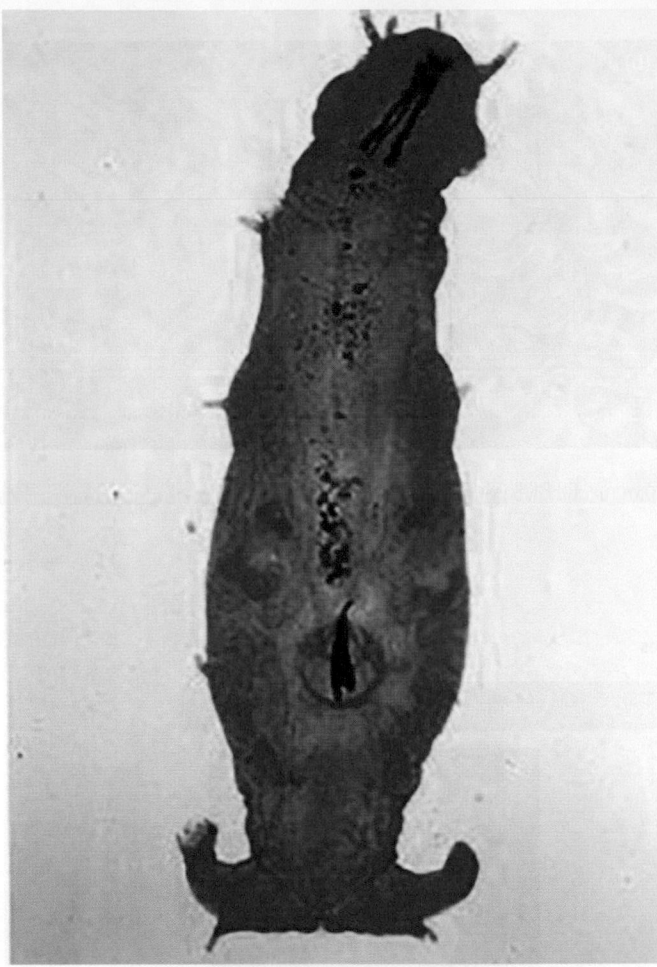

Figure 33.40 *Stratiodrilus robustus*, a polychaete worm of the family Histriobdellidae.

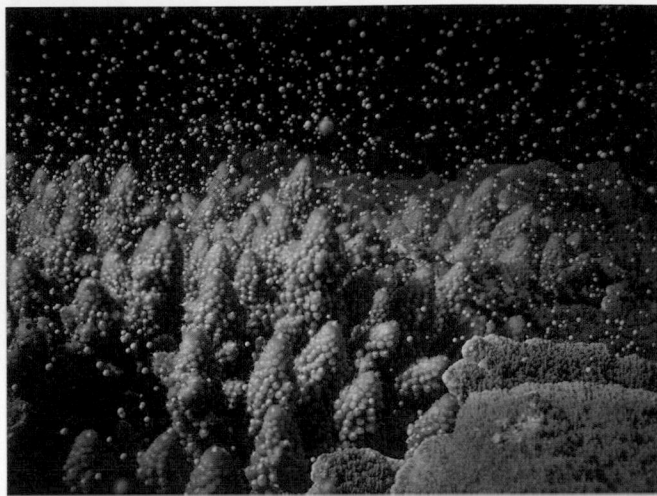

Figure 33.41 An upside-down 'snowstorm': coral mass spawning. Corals on the Great Barrier Reef produce eggs and sperm and release them together in a spectacular mass synchronized event. This annual event occurs after a full moon in October, at night during a neap tide when currents are reduced, enabling the eggs and sperm to meet.

Source: Coral Brunner/Shutterstock.com.

family Histriobdellidae, such as that shown in Figure 33.40. The worm's penis is reinforced with chitin to facilitate penetration of the female's body wall.

Sometimes, spermatophores are released directly into the water and the female either collects them or the sperm swim to the female. Females may gather the sperm using their feeding tentacles. This occurs in some of the terebellid worms. In the syllid species *Myrianida prolifera* individuals perform a nuptial dance after which the male releases sperm covered in mucus which he sticks to the female's back.

Worms for dinner? Mass spawning events

Polychaetes can also free-spawn so that their gametes are fertilized externally. However, as with other free-spawning animals, this leaves things to chance and adults are unable to take care of their larvae. Individuals may reduce losses of their own gametes and increase the probability of fertilization by synchronizing their spawning; a dramatic illustration of such synchronized spawning is shown in Figure 33.41.

Benthic (bottom-dwelling) species may swim up to the surface of the water to reproduce. One method of achieving this is **epitoky** in which a sexually immature worm becomes mature and pelagic (free-swimming).

Palolo worms (family Eunicidae) live in corals and, like corals, take part in mass spawning events triggered by the phases of the moon. During spawning the rear of the worm breaks off, leaving the head in the coral; the back end swims to the surface of the water where it releases thousands of gametes.

People in Samoa take advantage of this: they set off in canoes or wade through the water with nets to scoop the gametes and worms out of the water, as depicted in Figure 33.42. The eggs and sperm are then dried to make a flour which is used in biscuits. The worms can be eaten raw, fried in oil, baked into a loaf with coconut milk and onions, or served on toast according to preference.

 Check your understanding of the concepts covered in this section by answering the questions in the e-book.

33.8 **Insect reproduction**

We described earlier the formation and structure of the insect egg. After fertilization, the zygote develops into a larval form. Insect larvae are of five broad types, as summarized in Table 33.4.

A remarkable feature of insect reproduction is **metamorphosis**, the ability of larvae to change their body shape and size. Metamorphosis has been critical in insect evolution because it has allowed different life stages to exploit different food sources and habitats. One result of this is that adults and juveniles are not in direct competition with each other.

There are three types of metamorphosis in insects:

- incomplete metamorphosis (hemimetaboly);
- complete metamorphosis (holometaboly); and
- no metamorphosis (larvae look exactly the same as adults, e.g. silverfish).

Figure 33.42 A Palolo worm spawning event. (a) Samoan islanders collecting Palolo worms. (b) Palolo worms in a basket.
Source: (a) nik wheeler/Alamy Stock Photo; (b) dpa picture alliance/Alamy Stock Photo.

Table 33.4 Insect larvae can be grouped into five broad types

Type	Insect groups	Example
Eruciform	Lepidoptera	(a)
	Mecoptera	
	Coleoptera	
	Hymenoptera (Symphyta)	
Scarabaeiform	Coleoptera (Scarabaeoidea)	(b)
Campodeiform	Coleoptera	(c)
	Trichoptera	
	Neuroptera	
Elateriform	Coleoptera (Elateridae)	(d)

Type	Insect groups	Example
Vermiform	Diptera	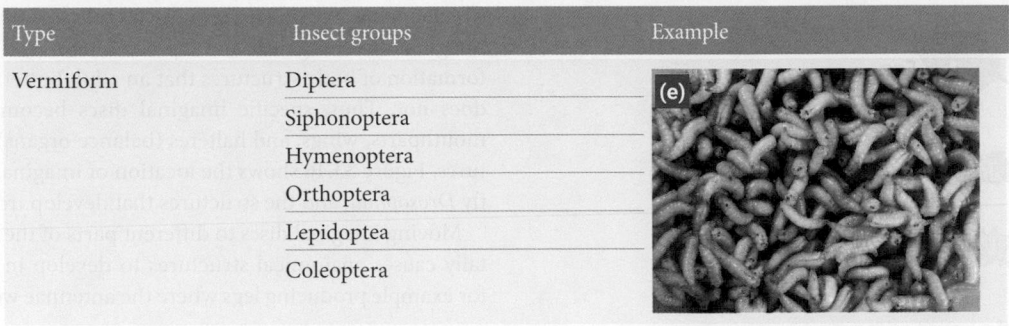
	Siphonoptera	
	Hymenoptera	
	Orthoptera	
	Lepidoptea	
	Coleoptera	

Source: (a) animantis/Flickr/CC BY 2.0; (b) Gilles San Martin/Flickr/CC BY-SA 2.0/https://creativecommons.org/licenses/by-sa/2.0/; (c) Katja Schulz from Washington, D. C., USA/Wikimedia Commons/CC BY 2.0 https://creativecommons.org/licenses/by/2.0/; (d) Katja Schulz/Flickr/CC BY 2.0 https://creativecommons.org/licenses/by/2.0/; (e) Dalius Baranauskas/Wikimedia Commons/CC BY-SA 3.0.

Incomplete metamorphosis occurs in grasshoppers, cockroaches, dragonflies, and several other groups, as depicted in Figure 33.43. The immature stages, called nymphs, closely resemble the adult but may lack wings and fully developed reproductive organs. The nymphs grow in a series of stages, called instars. After each stage they moult their exoskeleton and grow. The number of instars varies between species.

Complete metamorphosis, such as that depicted in Figure 33.44, is seen in bees, wasps, flies, butterflies, and related insects. The immature stages, the larvae, are morphologically very different to the adults. The larvae eat voraciously and grow, before forming an inactive pupa or chrysalis. After a fixed period of time during which the tissues are reorganized, the insect emerges from the pupa as an adult. In the monarch butterfly, for example, it takes 10 to 14 days for the adult to emerge.

Go to the e-book to explore an interactive version of Figure 33.44.

Figure 33.43 Incomplete metamorphosis. The life cycle of a grasshopper.
Source: Open Door Website www.saburchill.com.

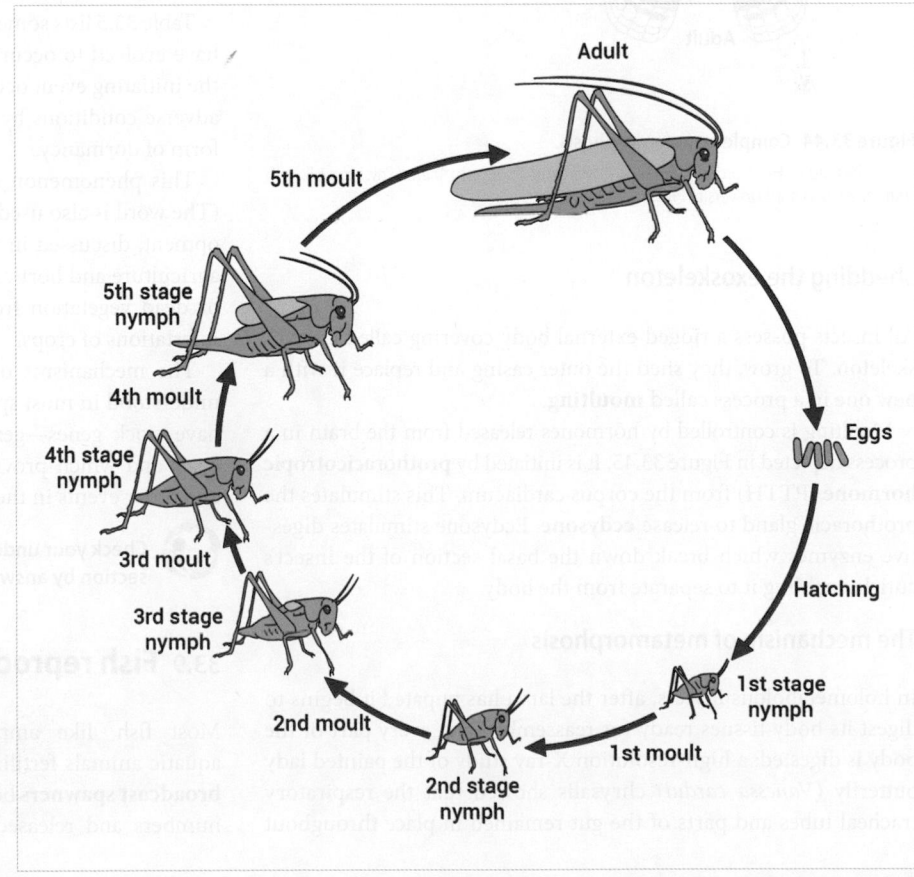

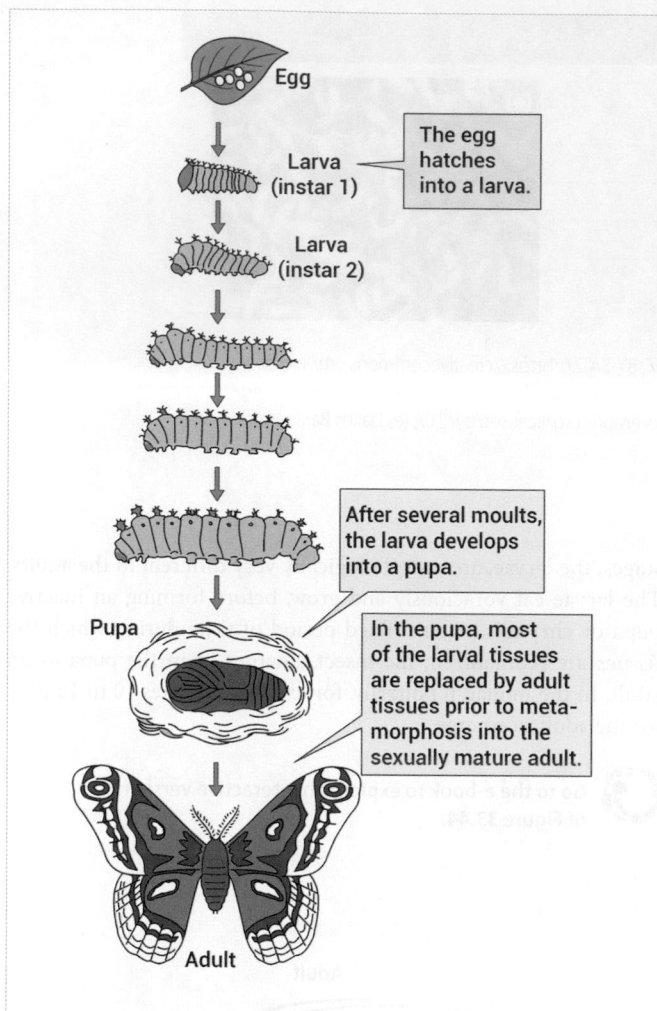

Figure 33.44 Complete metamorphosis.

Source: Richard W. Hill, Gordon A. Wyse, and Margaret Anderson. *Animal Physiology*, Fourth Edition, 2016. Oxford University Press.

Shedding the exoskeleton

All insects possess a ridged external body covering called an exoskeleton. To grow, they shed the outer casing and replace it with a new one in a process called **moulting**.

Moulting is controlled by hormones released from the brain in a process depicted in Figure 33.45. It is initiated by **prothoracicotropic hormone** (PTTH) from the corpus cardiacum. This stimulates the prothoracic gland to release **ecdysone**. Ecdysone stimulates digestive enzymes which break down the basal section of the insect's cuticle enabling it to separate from the body.

The mechanism of metamorphosis

In holometabolous insects, after the larva has pupated it begins to digest its body tissues ready for reassembly. Not every part of the body is digested: a high-resolution X-ray study of the painted lady butterfly (*Vanessa cardui*) chrysalis showed that the respiratory tracheal tubes and parts of the gut remained in place throughout

development. Specialized groups of cells inside the larva, called imaginal discs, organize the rapid cell divisions necessary for the formation of body structures that an adult butterfly has but a larva does not. Thus, specific imaginal discs become legs, antennae, mouthparts, wings, and halteres (balance organs) and other structures. Figure 33.46 shows the location of imaginal discs in the fruit fly *Drosophila*, and the structures that develop from each.

Moving imaginal discs to different parts of the pupa experimentally causes anatomical structures to develop in the wrong place, for example producing legs where the antennae would normally be.

What controls metamorphosis?

Juvenile hormone (JH) is released from the corpus allatum. This hormone allows the insect to grow but keeps it at the same life stage. When it begins to pupate or become an adult, the amount of JH released decreases substantially or it may not be produced at all.

Juvenile hormone is used in some pesticides to prevent insects from pupating into an adult: when sprayed with it, the larvae continue to grow and become very large. They never reach sexual maturity and thus are prevented from breeding and producing offspring.

Insect diapause

In many insects, the *timing* and *length* of the stages of metamorphosis are strongly influenced by environmental conditions. Certain stages of the life cycle, especially the egg and larval stages, are sensitive to predictable, or at least probable, environmental events such as extremes of temperature and limited availability of food and water.

Table 33.5 lists some examples. It shows that the actual delay may have evolved to occur at a stage subsequent to that during which the initiating event occurred. This enables the insect to survive the adverse conditions by extending part of its normal life cycle as a form of dormancy.

This phenomenon in insects is often referred to as **diapause**. (The word is also used for delays in mammalian embryonic development, discussed in Section 33.11.) It is of particular interest in agriculture and horticulture where insect pests may survive in soil or dead vegetation from one year to the next, causing repeated infestations of crops.

The mechanisms of insect diapause are complex and poorly understood in most species. Like most animals and plants, insects have clock genes—genes whose expression occurs on a cyclical basis and which provide the underlying timing of metamorphic and other events in the animal's life cycle.

 Check your understanding of the concepts covered in this section by answering the questions in the e-book.

33.9 Fish reproduction

Most fish, like amphibians, echinoderms, and many other aquatic animals fertilize their eggs externally. They are known as **broadcast spawners** because eggs and sperm are produced in great numbers and released into the water. Fertilization depends on

Figure 33.45 Hormonal control of insect development.

Source: Richard W. Hill, Gordon A. Wyse, and Margaret Anderson. *Animal Physiology*, Fourth Edition, 2016. Oxford University Press.

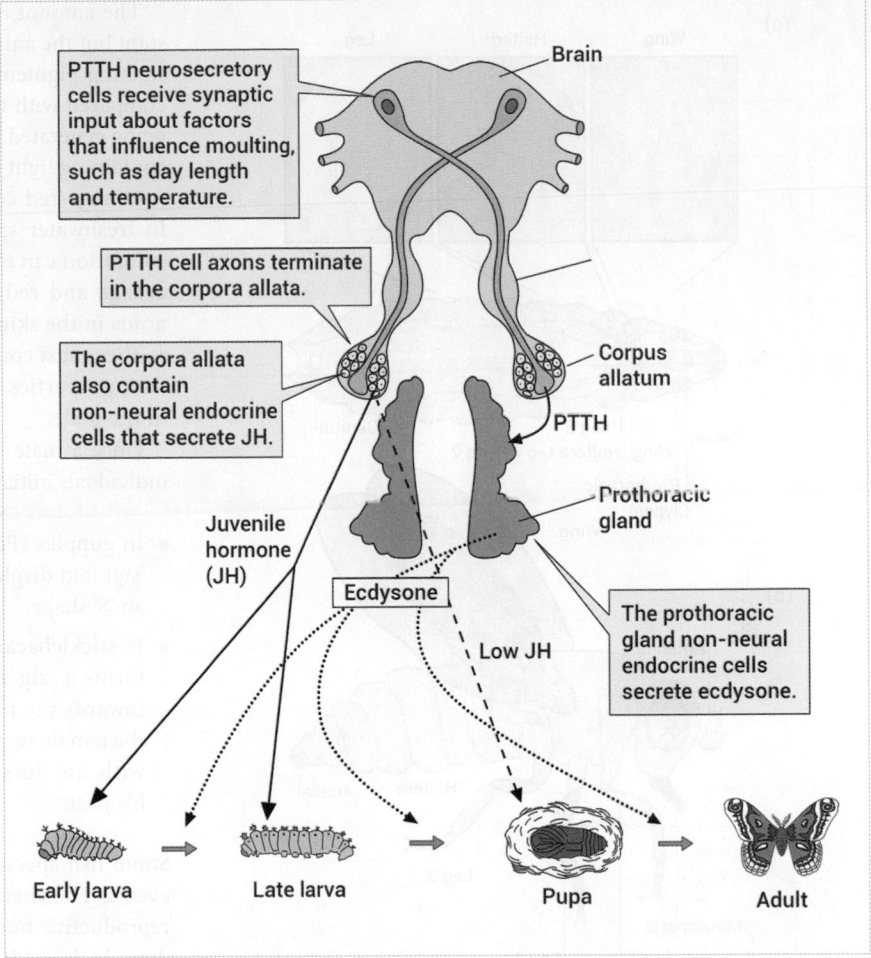

PTTH neurosecretory cells receive synaptic input about factors that influence moulting, such as day length and temperature.

PTTH cell axons terminate in the corpora allata.

The corpora allata also contain non-neural endocrine cells that secrete JH.

Brain

Corpus allatum

PTTH

Prothoracic gland

Juvenile hormone (JH)

Ecdysone

Low JH

The prothoracic gland non-neural endocrine cells secrete ecdysone.

Early larva   Late larva   Pupa   Adult

chance interaction and may be affected by water currents. Zygotes are then at the mercy of predators and other environmental threats. Broadcast spawning is an extreme example of r-strategy.

Although fertilization is left to chance, social interactions can play an important part in the determination of sexual identity and function. Fish often live in large, highly ordered, complex societies. About a quarter of fish species show some sort of parental care for their young. The rest do not, leaving their eggs, sperm, and zygotes unprotected.

Although in most species parent fish have little direct involvement with their own young after mating has taken place, their influence over the success of their offspring can be significant. They determine the quality of the eggs produced, the identity of mates, and the genetic identity of their offspring. They also decide the optimal time for mating and select the best breeding site.

Attracting and courting a mate

Before courtship can begin, an individual must attract a mate using chemical, visual, acoustic, or electrical signals. In those species which show sexual dimorphism there may be differences in size, colour, and behavioural signals. Such differences evolve as a result of **sexual selection**.

- The South American pulse fish *Brachypopomus pinnicaudatus* (Figure 33.47a) communicates using discharges from an electric organ. Discharges are twice as long in males as in females, and females are attracted by longer-lasting ones.

- Male toadfishes (Figure 33.47b) sit on the bottom and produce courtship calls which females assess on the basis of sound quality.

- Swordtails (Figure 33.47c) use olfactory cues to attract mates.

Water can transmit chemical and sound signals over much greater distances than visual signals (see Chapter 32).

Body surface coloration is important in improving the likelihood of successful reproduction. Within the vertebrate integument, colour can be determined by structural compounds which selectively scatter light and pigments which absorb part of the light spectrum. In some animals, including fish, pigments are contained in chromatophore cells.

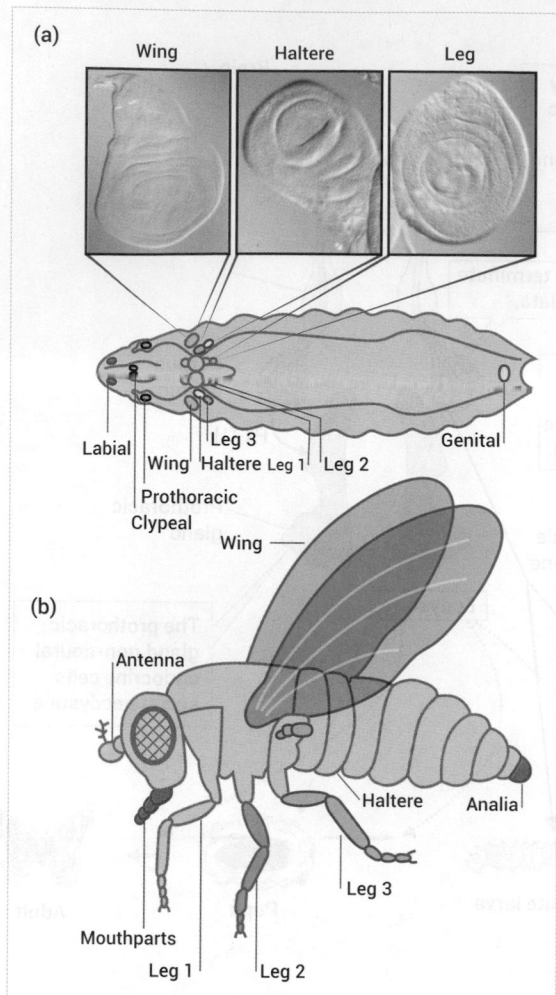

Figure 33.46 Imaginal discs in the fruit fly *Drosophila*. (a) The location of imaginal discs. (b) The structures that develop from each disc.

Source: Ginés Morata (2001) How drosophila appendages develop. *Nature Reviews Molecular Cell Biology*, 2, 89–97. Reprinted by permission from Springer Nature, © 2001.

The amount of pigment in a chromatophore is generally constant but the animal can change its colour by expanding and contracting pigment distribution. The range of pigments is limited compared with the diversity of colours on display, many colours being generated by combinations of pigment cells overlaying cells containing light-reflecting structures.

Orange/red colours are often used by fish to attract females. In freshwater species (Figure 33.47d, e) the brightness of the coloration can indicate their ability to find food. This is because orange and red pigments are produced by depositing carotenoids in the skin. Animals cannot synthesize these compounds, so they must come from the diet. Carotenoids also have antioxidant properties, so brighter coloration may signal good health in the male.

Once a mate has been attracted, communication between the individuals initiates courtship behaviour.

- In guppies (Poeciliidae; Figure 33.47d), the male performs a 'sigmoid display' in front of the female, bending his body into an 'S' shape.

- In sticklebacks (Gasterosteidae; Figure 33.47e), the male performs a 'zig-zag' dance in which he jumps away from and towards the female with his mouth open and spines erect. If the female responds, he will start biting and prick her stomach with his dorsal spines before subsequently leading her to his nest.

Some fish species use **pheromones** to coordinate reproductive events. The chemicals released often have a similar structure to reproductive hormones, especially sex steroids and prostaglandins. Male goldfish (Cyprinidae; Figure 33.47f) respond to the changing odour of reproductively mature females with a suite of behavioural and physiological responses. These coordinate reproductive activity, including increases in milt volume and sexual behaviours. In salmon, a combination of vibrational and visual cues is necessary to induce spawning.

Table 33.5 Insects with environmentally sensitive stages in their life cycles

Scientific name	Common name	Sensitive stage	Diapause
Diatraea grandiosella	Southwestern corn borer	Early larval	Late larval
Sarcophaga crassipalpis	Flesh fly	Early larval	Pupa
Sarcophaga argyrostoma	Flesh fly	Mid to late larval	Pupa
Manduca sexta	Tobacco hornworm	Late embryonic (egg) to late larval	Pupa
Leptinotarsa decemlineata	Colorado potato beetle	Early adult	Late adult
Bombyx mori	Silkworm	Late embryonic (egg) to early larval	Embryonic
Lymantria dispar	Gypsy moth	Late embryonic	Late embryonic
Danaus plexippus	Monarch butterfly	Early adulthood	Adulthood
Acronicta rumicis	Knot grass moth	Mid larval	Mid larval
Cydia pomonella	Codling moth	Early to mid larval	Mid larval
Gynaephora groenlandica	Arctic woolly bear moth	Mid larval	Mid larval

Figure 33.47 Courtship cues in different species of fish. (a) South American pulse fish, *Brachyhypopomus pinnicaudatus*. (b) Oyster toadfish, *Opsanus tau*. (c) Swordtail, *Xiphophorus helleri*. (d) Guppy, *Poecilia reticulata*. (e) Goldfish, *Carassius auratus*. (f) Stickleback, *Gasterosteus aculeatus*.

Source: (a) Photograph by Marcelo Loureiro; (b) Noelweathers/Flickr/CC BY 2.0 https://creativecommons.org/licenses/by/2.0/; (c) Wojciech J. Płuciennik/Wikimedia Commons/CC BY-SA 4.0; (d) image by Zucky123 from Pixabay; (e) image by Hans Braxmeier from Pixabay; (f) Jack Wolf/Flickr/CC BY-ND 2.0 https://creativecommons.org/licenses/by-nd/2.0/.

Migration and spawning

Many fishes migrate to reproduce, in some cases moving from fresh water to sea water (**catadromous migration**) or vice versa (**anadromous migration**; Chapter 31). Salmon are well known to migrate from their sea water feeding grounds back to the freshwater streams where they were born in order to reproduce.

In moving between sea and fresh water, individuals must adapt their osmoregulation (see Chapter 30). In addition to this considerable physiological challenge, returning to spawn in a fresh water stream may entail a long, up-river journey. Pacific salmon make

Figure 33.48 Fish nests. (a) Female stickleback over a nest with a male in attendance outside the nest. (b) Gourami bubble nest. (c) Bluehead chub nest.
Source. (a) Nature Picture Library/Alamy Stock Photo; (b) Dunkelfalke/Wikimedia Commons/CC BY 2.0 https://creativecommons.org/licenses/by/2.0/.

this exhausting trip only once: when they arrive back at their birth place they spawn and then die.

Animals which spawn once are known as **semelparous**. Other species, such as the Atlantic salmon, will make the migration from sea water to fresh water more than once in their life time; this is **iteroparity**.

Salmon locate their home stream using their sense of smell. As they leave and migrate to sea as juveniles, they imprint (memorize) olfactory cues associated with their home and then orientate themselves to find their way back home to spawn. Olfactory cues are used by other migratory fish such as the sea lamprey (*Petromyzon marinus*). Like salmon, they migrate from sea to fresh water but locate suitable spawning sites by following pheromones released by stream-dwelling larvae.

Parental care

Distributive spawning is the common strategy for fish, but there are also examples of oviparity, ovoviviparity, and viviparity. In ovoviviparous and viviparous species, offspring are protected during development by being inside the parent, prior to their release as free-living individuals.

For oviparous fishes, there is no internal protection, but this does not always mean that eggs are abandoned. Parents perform a variety of behaviours that improve the survival chances of their offspring. This is mostly paternal (male-only) but can be maternal (female-only) or biparental (both parents).

Creating a safe place for eggs to develop once they are laid is probably the simplest form of parental care seen in fishes. Some species create nests to protect their young, as depicted in Figure 33.48:

- Salmon and trout simply dig holes in the substrate in which to bury their eggs.
- Hornyhead chub (*Nocomis biguttatus*) use the substrate to construct large mounds.
- Sticklebacks construct nests from plant material.
- Siamese fighting fish form bubble nests at the water surface.

A parent fish remaining to defend its eggs will usually tend them, removing dead ones and debris to keep the developing embryos clean. They may also fan the eggs with their tail to maintain a healthy flow of oxygenated water.

Some fishes use a more mobile form of protection through mouth-brooding. After fertilization the parent picks up the eggs and incubates them in the mouth until they hatch. In the Nile tilapia (*Oreochromis niloticus*), parents provide offspring with nutrition after hatching by secreting mucus within their mouths for the fry to feed on. Discus fish (*Symphysodon discus*) parents feed newly hatched fry on mucus secretions from their bodies (Scientific Process 33.2).

SCIENTIFIC PROCESS 33.2 | **Discus fish mucus and the risks of parental care**

Research question

Does the transfer of nutrients from parents to offspring carry any risks?

Background information

Discus fish (*Symphysodon* spp., Figure 1) are cichlid fish which live in the Amazon. They have an unusual parental care mechanism in which both parents produce a mucus layer over their bodies from which the young feed.

The mucus is high in protein and has elevated levels of IgM antibody, providing the offspring with passive immunity during early development. The mucus of wild discus fish has elevated concentrations of ions; this is valuable for the offspring as the waters of the Amazon are particularly ion-poor.

There is a substantial energetic cost in providing this food supply. In aquarium-bred fish, parents which feed their offspring for longer than a week have fewer subsequent broods. Parents share the cost: when one parent has had enough of its offspring biting at its mucus, it will flick its

Figure 1 Male and female discus with their fry.

Source: Photo courtesy of Dr Richard Maunder.

body and transfer them to the other parent. The mucus contains elevated levels of the hormone cortisol, which may reflect the parental stress of mucus provision.

Hypothesis

Richard Maunder, from the University of Plymouth, with colleagues from Brazil and Scotland, studied the parental behaviour of discus fish. Excretion of toxins in mucus is a part of a fish's natural defence, so they wondered whether pollutants were passed, via the mucus, to the fry during feeding. This could be a significant problem in a contaminated environment.

They *tested* the following hypothesis:

In a polluted environment, parents who feed their offspring with bodily secretions risk passing on concentrated toxins.

They *predicted* that pollutants such as heavy metals would be concentrated in the parental mucus.

Methods

Discus fish were allowed to breed in an aquarium tank. Over the 3-week period after hatching during which the fry fed on parental mucus, parents were removed from the tank in turn for periods of 90 minutes and placed in cadmium (Cd)-free or Cd-contaminated water, before being returned to the main tank to rejoin their fry. The researchers took care to avoid exposing the fry to the treatment water and to avoid any rejection of the fry by the parents.

Samples of parental mucus were obtained for Cd analysis by mass spectrometry, at intervals over the duration of the test. Fry were also taken to measure their total body Cd content. The Cd level in the main tank water remained below the limits of detection.

Results

Figure 2 shows the Cd content of parental mucus at four time points after exposure to control or Cd-containing water. Figure 3 shows the

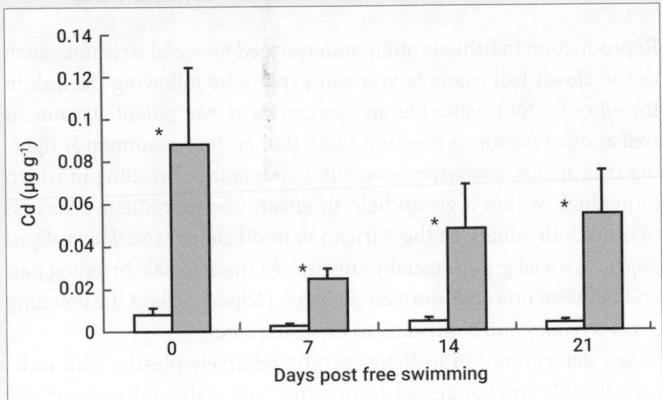

Figure 2 The Cd content of parental mucus at four time points after exposure to control (Cd-free, white bars) or Cd-containing water (shaded bars).

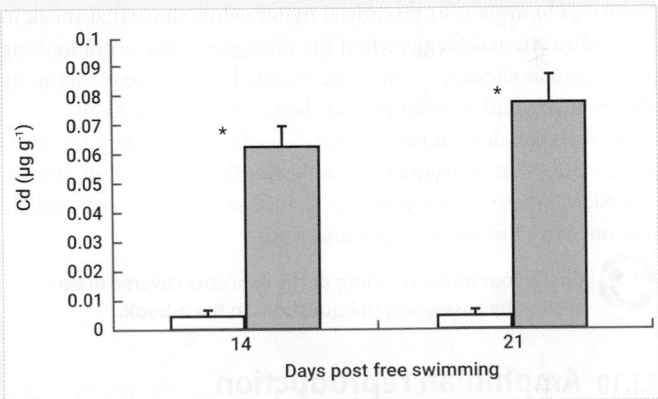

Figure 3 The whole-body Cd content of fry, 2 weeks and 3 weeks after their parents had been exposed to control (white bars) or Cd-containing water (shaded bars).

whole body Cd content of fry, 2 weeks and 3 weeks after their parents had been exposed to control or Cd-containing water. Each value is the mean ± standard error of mean for eight fish (Figure 2) or four fry (Figure 3). Asterisks indicate significant differences between control and treatment.

Conclusion

Cd in water can be transferred in concentrated form from discus fish parents to their offspring through the mucus upon which the fry feed.

Contamination of the discus fish's natural environment turns their unusual reproductive strategy into a source of toxicity for offspring.

Read the original work

Maunder, R.J., Buckley, J., Val, A.L., and Sloman, K.A. (2013). A toxic diet: transfer of contaminants to offspring through a parental care mechanism. *Journal of Experimental Biology*, **216**, 3587–3590.

Social structures and hierarchies in fish communities

Reproduction in fishes is often underpinned by social structures such as the clown fish example you can explore by following the link in the e-book. Dominance hierarchies can form over potential mates, as well as other resources (Section 32.6). Rather than continuously fighting over mates, some species exhibit cooperative breeding in which individuals within a group help to ensure the reproductive success of other individuals. In the African daffodil cichlid (*Neolamprologus pulcher*), social groups usually consist of a male/female breeding pair and between one and fourteen 'helpers'. Helpers defend the breeding pair's territory and contribute to offspring care.

Sex determination in fishes can be relatively plastic, with individuals able to change sex both simultaneously and sequentially (see Section 33.1). Further plasticity is seen in species which display **alternative reproductive tactics (ART)**. This is where variation in behaviour and phenotype occurs within a sex, usually males, resulting in two different reproductive strategies. Males who engage in dominance contests and compete for females are known as bourgeois or territorial males, while those that sneak in to fertilize a female's eggs when the bourgeois male is not looking are known as sneaker or parasitic males. The two are of distinctly different size and morphology as depicted in Figure 33.49.

As with sex determination, ART can be *fixed*, where the individual uses that reproductive tactic for the whole of its life, or *reversible*, where individuals can change tactic either sequentially (i.e. once in a lifetime) or back and forth.

 Check your understanding of the concepts covered in this section by answering the questions in the e-book.

33.10 Amphibian reproduction

Amphibians are tetrapod vertebrates which can live on land as adults but need water to reproduce. Amphibian eggs are mesolecithal and, like those of fish, are fertilized externally. Male and female behaviours change at certain times of year to increase the chance that sperm and eggs will be released in close proximity.

Most amphibians reproduce in large bodies of fresh water such as ponds, lakes, or slow-moving rivers. Each fertilized egg is initially encased in a large gel-like mass. On hatching, it develops first into a larval form, such as the well-known tadpoles of frogs. The larva has an extended tail but no legs, and has an obligatory aqueous lifestyle. It breathes through external gills, and gas exchange only occurs at the gill surface if it is wet.

As the larva grows, it metamorphoses by losing its tail and developing legs. The external gills are also lost and it moves to internal lung-based respiration. The rates of larval development and metamorphosis are influenced by environmental temperature and mediated internally by iodine-containing hormones similar to those produced by the thyroid gland in mammals.

A small number of frog species have adapted to rain forest conditions by reproducing in the limited amounts of water which collect on plant surfaces. Others have taken this adaptation further by completing the larval stage entirely within the egg; they hatch directly into the adult form and can thus be said to have circumvented the water obligation.

One group of amphibians, the salamanders, have an extended larval stage. The larvae become partially terrestrial and may be carnivorous, feeding on small insects and other invertebrates, but eventually mature into an adult form.

The axolotls (family Amblystomatidae), which are a sub-group of amphibians related to salamanders, remain in the larval stage for the whole of their lifespan. They have limbs but retain a tail and are recognizable by their large external gills. The reliance on gill breathing means that they are completely aquatic. They reach sexual maturity and reproduce in the larval form.

The continuation of the larval form, exemplified by salamanders and axolotls, is referred to as **neoteny**. It is associated with a reduced secretion of thyroid hormones and can be halted experimentally by treating the animals with iodine. Environmental iodine

Figure 33.49 Goby males. The parental or bourgeois male is the large fish and the sneaker male is the small fish. (a) Side-on view; (b) front-on view.
Source: Hossein Mehdi.

levels are usually low in the regions where these animals live and the neotenic condition may have evolved as an adaptation. They have also retained an unusual ability to regenerate limbs and tails following injury by predators.

 Check your understanding of the concepts covered in this section by answering the questions in the e-book.

33.11 Mammalian reproduction strategies

We have already discussed and illustrated several aspects of reproduction in mammals. Reproduction in the other amniote vertebrates (reptiles and birds) employs many of the same basic mechanisms, although there are a great many variations even between closely related taxonomic groups. All aspects of reproduction show specialization in individual species and it is unwise to treat familiar or well-studied examples as being in some way normal or representative.

Mammals in general are noted for the extended support given by parents to their young (many non-mammals do this too). This intense parental investment of time and energy in the young represents an extreme K-strategy when compared with most other animals. Part of this is the nutritional support provided for all mammalian young in the form of lactation.

Details of the biochemistry and physiology of lactation in humans can be found in Chapter 26. That information is a good guide to the basic mechanisms which occur in all mammals. The principal variations in lactation among the mammalia consist of:

1. the number of young which can be nursed at the same time; this is largely determined by the structure of the mammary glands and number of teats;

2. the duration of the lactation period from birth to weaning;

3. the composition of the milk, especially its energy density which is mostly represented by its fat content; and

4. the extent to which milk provides passive immunity to the newborn offspring by passing on some of the mother's immune proteins.

In many mammals, including to a limited extent in humans, lactation is associated with a temporary loss of female fertility. This is because some of the hormones released during suckling, especially prolactin, have an inhibitory effect on the brain and other parts of the reproductive axis. This interruption helps to define the minimum time interval between birth and subsequent conception. It influences the frequency of reproduction and is therefore a key component of the animal's fecundity.

We are so familiar with human reproduction, and with reproductive processes in domestic and agriculturally important mammals, that we may think of them as representative of mammals as a whole. But this can be misleading. As with all other groups of animals, the 4300 or so species of mammal show enormous variation, even between closely related groups.

Cycles, oestrus, and seasonality

The human menstrual cycle, in which the endometrial lining is regularly shed if there is no implantation of a fertilized egg (see Chapter 26) is an extremely unusual reproductive strategy. The only other animals in which uterine bleeding occurs if there is no conception are the other apes (chimpanzees, gorilla, orang-utan), some Old World monkeys, elephant-shrews (Macroscelidae, small mammals with long snouts, related to the insectivores), the Cairo spiny mouse (*Acomys cahirinus*, a small African rodent), and the black mastiff bat (*Molossus rufus*).

Its rare occurrence in such distantly related mammals shows that it has arisen by convergent evolution. It is best thought of as just one variant in the overall diversity of female reproductive physiologies and of mammalian reproductive strategies in general.

Here are some other variants. (See Chapter 26 for technical explanations in the human context.)

- Reproductive cycles can occur repeatedly throughout the year or be confined to certain seasons (determined by day length, water or food availability, opportunities for social interaction between males and females, and other factors).

- Females may have periods of behavioural and physiological receptivity—known as 'heat' or oestrus—or may be continuously receptive.

- Ovulation can occur spontaneously in each cycle or may be induced by physical stimulation of the female reproductive tract during copulation.

- Each spontaneous cycle or stimulated ovulation may result in the release of a single egg or several.

- The follicular (growth of follicles up to ovulation) and luteal (development and function of the progesterone-secreting corpus luteum) phases of the ovarian cycle can be unequal or roughly equal in length, and they can be sequential or overlapping.

- If fertilization and implantation occur, the function of the ovary changes: instead of developing oestrogen-secreting follicles with mature eggs, the corpus luteum is active for an extended period, secreting the progesterone which supports the uterus during pregnancy. The initiation of this change of function is called the **maternal recognition of pregnancy** and is a very complex process. In some animals it is caused by a positive hormonal signal coming from the conceptus to the ovary. In others the conceptus inhibits a negative (i.e. corpus luteum inhibiting) signal coming from the uterus.

- In some mammals, the corpus luteum is required for most of pregnancy. In others it functions for a short period after which its hormonal role is taken over by the placenta. In yet others, several corpora lutea are needed over the course of pregnancy.

- Unlike humans, few if any other mammals have an extended, menopause-like period of female reproductive senescence in the later part of life.

When comparing the human cycle to those in other animals, be aware that the days of the human cycle are conventionally numbered from the first day of menstrual bleeding while those of other animals are numbered from the day of ovulation.

To help you understand some of the variation in reproduction between some common mammals, do some research to answer the following questions:

(a) Is ovulation in each of the following animals spontaneous or induced by mating?

- Domestic cat
- Rabbit
- Horse
- Sheep.

(b) Are the follicular and luteal phases of the ovarian cycle of the cow sequential (as in humans) or overlapping? How long is the cycle and on which day does ovulation usually occur?

(c) Is reproduction in the domestic pig highly seasonal (as in sheep) or non-seasonal (as in cows)?

(d) What is the average litter size of the following animals:

- Rat
- Dog
- Elephant
- Seven-banded armadillo.

Answers:

(a) Domestic cat: induced; rabbit: induced; horse: spontaneous; sheep: spontaneous.

(b) The phases are overlapping. The cycle averages 21 days and ovulation is on day 0.

(c) Non-seasonal (but sometimes will vary over the year).

(d) Rat: 8–18; dog: 3–5 (varies greatly with breed); elephant: 1; seven-banded armadillo: 8–15 (all identical).

Pregnancy and diapause

The course of normal human pregnancy is extremely predictable: birth can be anticipated to occur roughly 40 weeks after the last menstrual period. Statistics from western countries show a median pregnancy length, from ovulation to birth, of 268 ± 18 days. Medical professionals can make use of this information in monitoring the progress of pregnancy and deciding whether to intervene.

This reliability exists because there are generally no delays between the fertilization of the egg by the sperm, the passage of the fertilized egg from the Fallopian tube to the uterus, its implantation in the uterine endometrium, the establishment of a placenta, the development and growth of the fetus, and the initiation of parturition.

Many other mammals do not show this consistency. There are numerous examples of where interruptions occur, quite normally, at all of these stages. The general term used for this is **embryonic diapause**, meaning a period of delayed development of the offspring prior to birth.

Diapause in insects (Section 33.8) is an environmentally induced delay or extension to a stage in metamorphosis. The usage of the term in mammals is similar in so far as it describes an interruption to the reproductive process, even though mechanisms of reproduction in the two animal groups are completely different.

Embryonic diapause, like other aspects of reproductive strategy, has evolved more than once and examples have been found in over 130 species representing many different mammalian orders. In some it always happens at the same stage of gestation and for a fixed period of time. This is called **obligate diapause**. In others, its occurrence, timing, and length are variable, regulated by photoperiod, ambient temperature, or food supplies, or by the concurrent suckling of an earlier offspring. This is **facultative diapause**.

In practice, the distinction between obligate and facultative diapause is probably an oversimplification: diapause is just more variable in some species than in others. In all cases, it must be interpreted in the context of the animal's lifestyle and for its value in promoting successful reproduction. One major advantage can be that the timings of gestation and birth are separated from the time of mating. Some biologists liken it to plant seeds remaining dormant until conditions for germination are optimal, or the ability of birds to adjust the rate at which the embryos inside their eggs develop by altering their incubation behaviour.

Look at Figure 33.50 for some examples of mammalian diapause; we will consider each of these examples in turn.

Northern fur seal (Callorhinus ursinus)

The breeding season runs from the end of June to the beginning of August, but each female is receptive to the male for just one day. After mating, the fertilized egg moves from the mother's oviduct to her uterus but remains there in a dormant state for about 3.5 months. In early November, the zona pellucida is shed ('hatching'), the embryo implants, and fetal development begins. The complete gestation period from mating to birth takes about 360 days and the resulting pup is suckled for 3–4 months. This strategy means that mothers mate during the week after they have given birth and while they are starting to lactate. For most of the year, fur seals travel widely over the seas of the Northern hemisphere. The population comes together, for birth and mating, at limited locations on the Alaskan coast for a very brief period of time. The embryo's diapause, which is thought to be controlled by photoperiodic cues, allows the mother to carry its next offspring right through the year, even though the time required for fetal development is only 8–9 months. It also separates the physiological demands of lactation from those of pregnancy.

European roe deer (Capreolus capreolus)

The roe deer is the only artiodactyl to have a diapause. It is a seasonal adaptation which allows fawns to be born in spring when they have the best chance of survival. The rut (competitive sexual behaviour and mating) occurs in mid-summer when males and females are in peak physiological condition. The fertilized egg moves immediately to the uterus but, unlike the fur seal, hatches from its zona pellucida within a few days. A 4–5-month period of diapause then ensues, ending in late December or early

Figure 33.50 Mammals which have reproductive diapause. (a) Northern fur seal (*Callorhinus ursinus*). Male with harem. (b) Roe deer (*Capreolus capreolus*). Male (buck) and female (doe). (c) Tammar wallaby (*Macropus eugenii*). Mother with pouch young. (d) Little brown bat (*Myotis lucifugus*). (e) Rufous horseshoe bat (*Rhinolophus rouxii*).
Source: (a) M. Boylan/Wikimedia Commons/Public Domain; (b) Jojo/Wikimedia Commons/CC BY-SA 3.0; (c) LCAT Productions/Shutterstock.com; (d) Andy Reago & Chrissy McClarren/Flickr/CC BY 2.0 https://creativecommons.org/licenses/by/2.0/; (e) Aditya Joshi/Wikimedia Commons/CC BY-SA 3.0.

January when the embryo implants and establishes a placenta. Fetal development takes about 5 months and the mother gives birth in May or June. A characteristic of roe deer diapause is that the hatched blastocyst is not completely inactive: it has a low but significant rate of metabolism and some cell division occurs. The rate of development increases rapidly as soon as implantation takes place. Roe deer diapause is therefore a delayed implantation.

Tammar wallaby (*Macropus eugenii*)

The macropod marsupials (kangaroos and wallabies) are well known for giving birth to altricial offspring which need a long period of development in the mother's pouch before they can survive independently. Tammar wallabies mate in January (Southern hemisphere summer), immediately after the females have given birth and acquired new pouch young. The presence of the continuously suckling joey in the pouch inhibits the development of the newly fertilized embryo through the autumn. The embryo's quiescence after the winter solstice is also maintained by lengthening days, resulting in a diapause of up to 330 days. For most of that time the embryo floats free within the uterus. The joey will start to emerge from the pouch in October but may return to suckle at intervals until it becomes completely independent in late December. The resulting loss of the lactation stimulus at the teat initiates hormonal changes in the mother which reactivate the embryo in the uterus; its development there lasts less than four weeks. If the joey 'at foot' is lost, say to disease or predation, reactivation of the quiescent embryo may be brought forward.

Hibernating bats

Two groups of insectivorous bats, the Vespertilionidae (350 species, found throughout the world) and the Rhinolophidae (horseshoe bats, about 65 species in the tropics and Australasia), include species which use diapause to facilitate reproduction despite long periods of winter hibernation.

- In the little brown bat (*Myotis lucifugus*), a North American vespertilionid, females come into heat (oestrus) in late summer, mate and immediately enter their winter hibernation. The sperm deposited by the male are stored in the female tract until her arousal in the spring. She then ovulates, the egg is fertilized, and gestation begins. Young are born between May and July, and suckled for 2–3 weeks. This strategy therefore uses sperm storage and delayed ovulation.

- In *Rhinolophus rouxii*, an Asian horseshoe bat, mating also takes place before hibernation but the female ovulates at that time. The fertilized egg remains free in the reproductive tract until the female is aroused in the spring: implantation then occurs, followed by pregnancy, birth, and lactation. This is a true embryonic diapause, based on delayed implantation.

These contrasting strategies solve the hibernation problem in different ways but maximize reproductive opportunities. In several species, newborn young develop rapidly and may themselves mate during their first year. Many other reproductive adaptations are found in bats, sometimes related to migratory patterns and the availability of food resources.

🔗 Follow the link in the e-book to learn about an especially curious case of extreme diapause: that of the armadillo.

🔗 Check your understanding of the concepts covered in this section by answering the questions in the e-book.

SUMMARY OF KEY CONCEPTS

- The continuity of life on Earth depends on organisms being able to reproduce successfully.

- Every species of organism has a distinctive reproductive strategy which has evolved under the pressure to replicate its genes successfully.

- Genes are replicated by cell division and through the production of gametes; replicating structures possess a fixed proportion of the parent organism's genes, defined by ploidy.

- Asexual and sexual strategies for reproduction can both be successful but have advantages and disadvantages; organisms may alternate between them at different stages of their life cycles.

- In sexual reproduction, gametes originate from specific anatomical structures; their isolation contributes to the inability of organisms to inherit characteristics which their parents acquired from the environment.

- An organism's reproductive success may be defined by the extent to which its actual fertility approaches its fecundity.

- Reproductive strategies, and the components which comprise them, may be interpreted according to their r- (high rate of production with low success) and K- (low rate of production with high success) characteristics; r and K are both relative terms.

- Non-flowering plants (gymnosperms) depend principally on wind and other environmental components for the transmission of gametes; flowering plants (angiosperms) have evolved complex structures to attract insects and other animals for the transmission of gametes.

- Plant fertilization and the development of seeds and fruit depend on complex structural changes taking place within reproductive structures; most plants are also capable of vegetative (non-sexual) propagation.

- The general structure of an animal egg depends on the reproductive characteristics of the species and can be broadly characterized according to the structure of the yolk and membranes and the mode of development of the embryo.

- Animal spermatozoa, which are usually produced in very large numbers, transmit the male gamete in a form which has limited metabolic capacity and requires specific biochemical activation.

- In amniote vertebrates (reptiles, birds, mammals) the early embryo exchanges nutrients and gases through a membrane called the amnion; the exchange may occur at the egg surface or, in mammals, with the maternal metabolism, usually through a placenta.

- Animal reproduction is characterized by its enormous variability between species; the well-known reproductive systems of humans and domestic species are a poor guide to detailed reproductive events in other mammals.

- Many insects and mammals use diapause or other interruptions to adapt their reproductive events to specific environmental and social conditions.

🔗 Use the flashcards in the e-book to test your recall of key terms introduced in this chapter.

QUESTIONS

Looking for answers? Once you've answered these questions, follow the link in the e-book to the answer guidance and check your work.

Concept and definitions

1. Explain the concept of reproductive strategy.

2. Explain what is meant by ploidy.

3. Explain the concept of r- and K-strategies.

4. Distinguish between amniote and anamniote vertebrates.

5. What is a placenta and which groups of animals possess them?

Apply the concepts

6. Distinguish between sexual and asexual reproduction in terms of gametes and ploidy, explain how each method of reproduction offers the potential for variation, and give some advantages and disadvantages.

7. Describe some of the main reproductive differences between angiosperms and gymnosperms; sketch the reproductive life cycle of a typical plant in each case.

8. Sketch the general structure of four different types of egg, find two examples of animals which produce them, and explain how their structure and development relate to the animals' reproductive strategies.

9. Describe the basic life cycles of hemimetabolous and holometabolous insects, find two species examples of each from different insect groups, and identify the role of each developmental stage in the reproduction of the animal.

10. Find some examples, other than those listed in the text, of animals which show embryonic diapause and explain how this may assist in their reproductive success.

Beyond the concepts

11. Explain why sexual reproduction may have evolved in multicellular eukaryotic organisms; in developing your ideas, remember that reproductive systems involve the transmission of genes but that environmental pressures act on phenotype.

12. Find three examples of plants which depend on insects or other animals for fertilization and/or seed dispersal; in each case consider the extent to which the reproductive interaction between the plant and the animal may have evolved to become close or even obligatory.

13. Explain why the shells of bird eggs need to be porous; find the approximate dimensions of (a) a blackbird's egg and (b) an ostrich's egg, and using basic arithmetic calculate the surface area:volume ratio of each; based on your calculations, what difference would you expect there to be in the porosity of the shells of the two species?

14. Using examples from this text or available online, compare insect diapause with embryonic diapause in mammals. What adaptive advantages does diapause provide in each case and are there any similarities in the physiological mechanisms by which they occur?

15. Starting from the production of gametes, compare the reproduction of fish, amphibians, and reptiles in as many ways as you can.

FURTHER READING

Aanen, D., Beekman, M., & Kokko, H. [Eds.] (2016) Theme Issue: Weird sex: the underappreciated diversity of sexual reproduction. *Phil. Trans. R. Soc. B* **371**: Issue 1706 https://royalsocietypublishing.org/toc/rstb/371/1706
Reviews on everything to do with the diversity of reproduction and reproductive strategies, from parthenogenesis and IVF to sex in fungi and diapause.

Tree of Sex Consortium (2014) Tree of sex: A database of sexual systems. *Nature Scientific Data* **1**: 140015. doi: 10.1038/sdata.2014.152
A survey of reproductive systems and a source for finding out more about species.

Sankaranarayanan, S. & Higashiyama, T. (2018) Capacitation in plant and animal fertilization. *Trends Plant Sci* **23**: 129–39.

De Jonge, C. (2017) Biological basis for human capacitation—revisited. *Hum. Reprod. Update* **23**: 289–99.
Sperm capacitation: plants and animals compared . . . and humans.

Kersten, B. et al. (2017) Genomics of sex determination in dioecious tress and woody plants. *Trees* **31**: 1113–25.
Sex determination in plants, including an explanation of sex determining genes and their dosage effects.

Liu, H. et al. (2017) Sexual plasticity: A fishy tale. *Mol. Reprod. Dev.* **84**: 171–94.
The strange story of how fish change sex according to the company they're in.

Stoddard, M.C. et al. (2017) Avian egg shape: Form, function, and evolution. *Science* **356**: 1249–54.
Why are bird eggs asymmetrical and why do they vary so much in shape? Answers from evolution to do with flight, as well as physical imperatives.

Renfree, M.B. & Fenelon, J.C. (2017) The enigma of embryonic diapause. *Development* **144**: 3199–210. doi:10.1242/dev.148213
Embryonic diapause in mammals: which animals have it, how it happens, and the advantages it brings. Includes examples of reproductive timelines.

Skulachev, V.P. et al. (2017) Neoteny, prolongation of youth: from naked mole rats to 'naked apes' (humans). *Physiol. Rev.* **97**: 699–720.
A review of neoteny in mammals, why it has evolved and its relationship with ageing. Naked mole rats are the strangest mammals (except for humans!)

Weller, S.G. (2009) Invited Review: The different forms of flowers—what have we learned since Darwin? *Bot. J. Linnean Soc.* **160**: 249–61.
The form and function of flowers: includes an explanation of Charles Darwin's drawings and what we've learned in the century and half since then.

MODULE FIVE

Organisms in their Environments

34 Fundamental Concepts: Ecology, Evolution, Species, and Speciation

35 Genes: Evolutionary Change in Alleles, Genotypes, and Phenotypes

36 Populations: Quantifying Demographics and Modelling Change

37 Communities: Species Interactions and Biodiversity Metrics

38 Ecosystems: Abiotic Interactions and Environmental Processes

39 Challenges: Key Threats to Ecosystems

40 Solutions: Managing, Conserving, and Restoring Ecosystems

Image: A polar bear sitting on an ice floe in the Svalbard Archipelago, Norway. Human activity is fundamentally shaping the way the planet behaves as a global ecosystem, threatening the survival of species such as this. *Source:* Peter J. Raymond/Science Photo Library.

MODULE FIVE

The Nuvvuagittuq Greenstone rocks in Quebec, Canada have a remarkable claim to fame: they contain the first evidence of life on Earth. This evidence takes the form of fossilized remains that date to around 3.8 billion years ago. If we fast-forward through those 3.8 billion years we move through multiple geological time periods, each of which is characterized by the distinctive makeup of the rocks formed during these periods, and fossils within those rocks. This geological record allows us to trace major changes in the development and functioning of our planet, from the formation of chemical elements to the passing of ice ages. They also give us evidence of life's origins and the subsequent formation of new species, of evolutionary change, and of extinction events—most notably of dinosaurs 65 million years ago.

But we are now entering a new geological age: the Anthropocene or 'human period'. The recognition of this new age reflects the fact that the existence and activity of humans is now so profound that it will be indelibly imprinted in Earth's geological and fossil record. The Anthropocene is perhaps more commonly known as an 'Ecological Emergency'. Many environmental organizations and governmental bodies have now formally declared Ecological Emergency status, recognizing the profound pressures being placed on ecological systems by a range of factors including human-accelerated climate change, habitat loss and degradation, intensive agriculture, human population growth, urbanization, and pollution.

The net result of this Ecological Emergency is that biodiversity is declining and the processes that regulate ecosystem function—upon which we rely for a healthy environment—are becoming degraded. Almost nothing we do is without ecological, environmental, and evolutionary consequence: the addition of nitrogen and phosphorous fertilizers to fields is negatively affecting aquatic wildlife; greenhouse gas emissions are leading to warmer temperatures that are changing a whole host of animal behaviours, including patterns of bird migration; and demand for palm oil is driving the deforestation of tropical rainforest and the loss of iconic mammals such as orangutans.

So, how do we address these problems, which are both complex and urgent? First, we need to understand the evolutionary and ecological processes that we explore in Chapters 34–38. Second, we need to research the challenges that species, habitats, and ecosystems face as a consequence of human activity and other stressors (Chapter 39). Finally, we need to create solutions to address these problems—and have the willingness to implement them (Chapter 40). Put simply, the study of ecology and evolution has never been more important.

Fundamental Concepts

Ecology, Evolution, Species, and Speciation

Chapter contents

Introduction	1199
34.1 Ecology: a scientific approach to complexity	1200
34.2 Evolution: the lynchpin of ecology	1203
34.3 The species: a key concept in evolutionary ecology	1214
34.4 Bringing together ecology and evolution	1218

LEARNING OBJECTIVES

By the end of this chapter you should be able to:

- Define 'ecology' and 'evolution', and explain the historical development of the two biological disciplines.

- Describe, and be able to apply, the scientific method within the context of ecological research.

- Outline the theory of evolution by natural selection, and explain how and why selection drives evolutionary change.

- Appreciate the many different ways in which 'species' can be defined, and evaluate how speciation events lead to the evolution of new species.

- Explore the science of ecology, the science of evolution, and the bidirectional links between ecological and evolutionary processes (explaining how changes in ecology can drive evolution, and how and why the evolution of individuals has impacts at population, species, community, and ecosystem levels).

 Watch the key concepts video in the e-book to prepare yourself for studying this chapter.

Introduction

Planet Earth is a complex place. Millions of species—encompassing bacteria, protists, fungi, plants, and animals—are involved in complicated interwoven relationships with one another and with the physical environment that surrounds them. Plants use atmospheric carbon dioxide and sunlight to photosynthesize, chemical molecules are exchanged and transformed through biogeochemical cycles, organisms feed on other organisms, habitats change, and species evolve. These processes occur in an astonishing variety of habitats, from deserts to tropical rainforests, polar icecaps to savanna grasslands, mountain tops to deep-sea trenches, saltwater oceans to freshwater lakes. Humans add to this natural complexity: we change the physical environment, create new habitats such as fields and plantation woodlands, move organisms to new locations, and manage species by harvesting, culling, and conserving them.

Understanding this complexity is at the very centre of ecology. In this chapter, we will explore how ecologists examine current relationships between species and the environment in which they live, and also how we need to appreciate evolutionary processes to understand both how this complexity arose and how it is maintained. This will give us strong foundations upon which to build in the following chapters within this module.

34.1 Ecology: a scientific approach to complexity

The first manned mission to the moon, Apollo 8, established a stable orbit on 24 December 1968. In a live broadcast, command module pilot Jim Lovell said 'The vast loneliness [of space] is awe-inspiring . . . the Earth from here is a grand oasis in the big vastness of space'. That evening, crew member Bill Anders captured a photograph that encapsulated that sentiment: Earth positioned above the lunar horizon (see Figure 34.1). This photograph became known as 'Earthrise' and was the first time any person had seen Earth as a single planet against a barren lunar landscape. The Earthrise image led to the notion of Spaceship Earth—Earth being a single complex entity that is largely self-contained—and served to highlight its inherent vulnerability.

Based on our current knowledge, planet Earth supports the only life in the solar system. Life on planet Earth includes an astounding diversity of species ranging from unicellular prokaryotes such as bacteria to protozoa, diatoms, algae, fungi, plants, nematodes, molluscs, crustaceans, insects, arachnids, echinoderms, amphibians, reptiles, fish, mammals, and birds, which are shown in Figure 34.2. The actual number of species is difficult to quantify. If we ignore the single-celled prokaryotic eubacteria and archaea (and also ignore viruses, which lack metabolism and cannot function or replicate outside the cell of a host organism) and focus only on eukaryotic species, one of the most robust estimates (from 2011) is that there are some 8.7 million species currently extant, plus or minus 1.3 million. This might seem somewhat vague—and it is—but it is a significant improvement on previous calculations, which ranged from 3 million to 100 million. It also highlights the extent of our ignorance: the estimate suggests that 86% of predicted land-based species and 91% of predicted aquatic species are yet to be discovered.

As we saw in Chapter 4, trying to define, group, and classify species is extremely complex and trying to do so raises a lot of

Figure 34.1 Earthrise. Although the first photographs of Earth from space were actually taken by unmanned rockets as early as 1946, the famous Earthrise image gained popularity in 1968 as the first photograph of Earth from space, probably because of the widespread interest in the first manned mission to the moon. In this image, Earth is positioned with south to the left and north to the right. Antarctica is top left and West Africa is just visible to the right.

Source: NASA.

interesting questions. Why are there so many different species? Why are some species common and some species rare? Why are different species found in different places? What is it about the relationships between organisms and their environment that produces the complex patterns of life on Earth? **Ecology** is the part of the biosciences that tries to answer these questions to explain the patterns and processes of life on Earth. Fundamentally, ecology focuses on **interactions**:

1. the interactions between organisms
2. the interactions between organisms and their physical surroundings.

The word ecology itself comes from the Greek οἶκος meaning 'home environment' and λογία meaning 'study of'. The term was first used by Ernst Haeckel, who is pictured in Figure 34.3, in 1866 as Ökologie in his native German. However, as we shall see in Section 1.2, Haeckel was not the first scientist, nor the only one of his contemporaries, to be thinking ecologically.

Describing ecological complexity

Ecology is an interesting science—and an important one—but it is not an easy one. The fundamental challenge for ecologists is complexity. The biosphere contains billions of genetically distinct individuals, divided into millions of species (known and unknown), all interacting with each other and with a wide range of different and changeable physical environments. As scientists, we need to not only make sense of that ecological complexity, but

Figure 34.2 The diversity of species on Earth. Earth supports an estimated 8.7 million eukaryotic species from a wide range of taxonomic groups.

Source: Shih-Hao Liao/Alamy Stock Photo.

Figure 34.3 Dr Ernst Haeckel.
He first coined the term 'ecology' and defined it in 1866.

Source: U.S. National Library of Medicine.

including climate change. One way of approaching this is to describe ecological complexity through a hierarchy of ecological levels:

- The gene: the genetic basis of life.
- The individual: the unique product of a specific combination of genes.
- The species: a group of genetically similar individuals.
- The population: a collection of organisms of the same species that inhabit a defined area over a specific period of time.
- The community: a collection of individuals belonging to multiple species populations that co-occur in a specific geographical area.
- The ecosystem: a community of species that interact with one another and their environment to produce a complex system.
- The biosphere: the sum of Earth's ecosystems, communities, populations, species, individuals, and genes.

also to be able to apply that understanding to the real world to manage and conserve species, habitats, and ecological processes in the face of a number of significant and pressing challenges,

Researching and modelling ecological complexity

In 1904, the British plant ecologist George Tansley—who, along with Ernst Haeckel, is regarded as one of the founders of ecology—voiced the opinion that ecology, at that time, was too descriptive and lacking in systematic methodology. His concern, shared by the American ecologist Frederic Clements, was that ecology lacked an **experimental approach** and was not based on

scientific analysis. Things have changed substantially during the intervening 120 years. Today, ecology is grounded in the **scientific method**—a way of researching ecological patterns and understanding how these have arisen rather than simply describing the end result. We covered the scientific method in general in Life and Its Exploration, but now we need to consider this approach specifically within ecological contexts. This is explained in Experimental Toolkit 34.1.

EXPERIMENTAL TOOLKIT 34.1 | **An ecological approach to the scientific method**

The task for an ecologist—like any scientist—is to ask and answer specific questions that are appropriate to the level in the hierarchy of ecological complexity being studied. Those questions might focus upon a gene's role in nutrient processing in bacteria, the consequences of the loss of a patch of grassland for butterfly populations, or the effects of deforestation in tropical rainforest ecosystems on the atmosphere. Most ecological research involves some or all of the linked steps outlined here.

1: Observation

Most ecological research starts with detailed and meticulous observation: numbers, distributions, interactions, behaviours, and so on. A preliminary analysis to summarize our observations graphically or numerically may help to identify patterns that are worthy of further study.

2: Question

Research questions arise out of observations and preliminary analysis of those observations. A research question might be specific (e.g. Why do zebra have black and white stripes?) or can be open ended (e.g. How is it best to manage a reserve in Africa to optimize the environment for a zebra population?)

3: Hypothesis or prediction

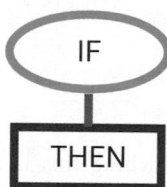

A research hypothesis is a plausible explanation for patterns observed—an educated guess—that often takes the form of a statement (e.g. zebra have black and white stripes because this reduces risk of predation by confusing predators through disruptive coloration). Hypotheses are usually specific and directional, for example 'variable x has a positive effect of variable y' rather than 'variable x and y are linked'. In some cases, we might have insufficient knowledge about a particular species or ecosystem to make a meaningful research hypothesis. In this case, we might make multiple predictions that we can then test (e.g. zebra stripes might have arisen for camouflage, for thermoregulation because micro-scale air currents are generated between black and white areas, or to reduce predation).

4a: Experiment

Testing a hypothesis or prediction usually means collecting data. This can mean designing and performing an experiment. An experiment involves us **manipulating** an aspect of the study system and recording the results of that manipulation. Ecology experiments may take place in the same place as the observation (a field experiment) or in a simplified version of the ecological system (a laboratory experiment). Field experiments are closer to the ecological realities involved, but they can be expensive, and it can be difficult to separate the things we are interested in—as defined by our research question—from other interactions in that ecological system (these are called **confounding factors** because they confound the relationship between the factor we are manipulating and the result of that manipulation).

4b: Non-manipulative testing

While most research projects in cell, molecular, physiological, and organismal biology can be conducted using manipulative experiments, in some ecological cases it is not possible, logistically feasible, or ethically acceptable to manipulate the factor of interest. In these cases, we collect data observationally in the field and analyse the results, factoring in potential confounding factors if necessary. For example, it would not be ethically acceptable to alter patterning of multiple zebra and release individuals next to lion to test whether this affects predation. We could, however, examine predation rates of zebra at a number of different field locations and link that to stripe patterns in the zebra species/sub-species at that location (factoring in the relative density of predators and then the abundance of zebra in relation to other prey species of a similar size) to determine whether predation probability was linked to stripe intensity and pattern.

5: Analysis

Analysis of data from a manipulative study is usually fairly straightforward because—if well done—an experiment should remove all variability from the system except for the single factor under consideration. This means that we can use statistical tests such as t-tests (for differences) or correlation (for relationships). Analysis of data from a non-manipulative study can be harder because there are more factors involved. This might mean we have to manipulate data before analysing statistically (e.g. to calculate proportional predation rate per lion per year) or using more complex statistical approaches that allow us to analyse multiple factors at the same time to replicate reality (multivariate analysis).

▶ Refer to Quantitative Toolkit 1 (Understanding data) and Quantitative Toolkit 4 (Ratio and proportion) to help your understanding.

6: Modelling

$$N = \frac{(M + 1)(C + 1)}{(R + 1)} - 1$$

Finally, we may be able to use our results to develop a model that reflects our understanding of a particular interaction or process. In many cases, these are mathematical models that enable us to generate additional hypotheses, research questions, and predictions. Alternatively, the models we produce can summarize procedures or decision-making processes and thus be used to guide our interventions into ecological systems to protect, manage, conserve, or restore them (see Chapter 40).

Although the steps here show the scientific method for ecology as a linear process, in reality it can be iterative and cyclical. In some cases, well-planned observational studies or manipulative experiments don't work and we have to start step 4 again. In other cases, the data collection goes well but the findings that arise from the data analysis in step 5 do not support the hypothesis/predictions we made in step 3. In this case, we might need to return to steps 1–3 to generate alternative hypotheses and the whole scientific process starts again.

If you need a reminder about statistics or data handling, take a look at the Quantitative Toolkits on Understanding data (Quantitative Toolkit 1), Describing data (Quantitative Toolkit 3), Understanding samples (Quantitative Toolkit 5), and Assessing patterns (Quantitative Toolkit 7).

PAUSE AND THINK

You have *observed* that the leaves of plants seem to emerge earlier in spring in years when spring temperature is higher. You *predict* that higher February temperatures control leaf emergence. How might you study this in the laboratory via a manipulative experiment for a small species such as creeping buttercup (*Ranunculus repens*) and observationally in the field for a large species such as an English oak tree (*Quercus robur*)? What are the pros and cons of each approach?

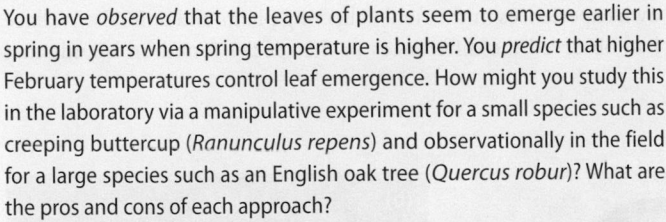

Answer: In the lab, you could easily artificially increase the temperature of plants housed in plastic or glass treatment areas and you could control other variables such as moisture. However, you would need to ensure that the plants still received the same amount of daylight as they would outside, and ideally have a day:night cycle that changed over time with days lengthening naturally. Humidity might be completely different to what would be experienced naturally and might be hard to replicate. In the field, you cannot separate temperature from wind, rain, presence of other species, etc. You therefore have a lot more going on that you need to analyse and that might make any link between temperature and leaf emergence harder to detect. In terms of timescale, you could have multiple experimental set-ups at different temperatures in the lab to gather a lot of data in one year. In the field, you would only have one set-up per site (natural weather, natural leaf burst) and so would have to do the work over many years—or have many study sites across which temperatures were variable in one year—to get sufficient data to analyse.

🌀 **Check your understanding** of the concepts covered in this section by answering the questions in the e-book.

34.2 Evolution: the lynchpin of ecology

In his 1973 essay, the Ukrainian-American geneticist Theodosius Dobzhansky famously noted that '*nothing in biology makes sense, except in the light of evolution*'. The essential role of evolution in underpinning many biological concepts is the reason that we cover the topic as a broad concept in Topic 3. It is certainly true that our understanding of ecology—at all levels of the ecological hierarchy—is closely linked to our understanding of evolutionary processes. Nothing exemplifies this better than another famous quote, this time in one of the most important scientific books ever written:

> *It is interesting to contemplate an entangled bank, clothed with many plants of many kinds, with birds singing on the bushes, with various insects flitting about, and with worms crawling through the damp earth, and to reflect that these elaborately constructed forms, so different from each other, and dependent on each other in so complex a manner, have all been produced by laws acting around us.*

Charles Darwin, *On the Origin of Species*, published in 1859.

Encapsulated within Darwin's metaphor of the 'tangled bank' (as depicted in Figure 34.4) is the notion that the natural world is an elaborate and complex network of interactions that combine both **biotic** (the living component of the natural world) and **abiotic** (non-living) components. As we saw in Section 34.1, this

Figure 34.4 Darwin's Tangled Bank is a metaphor for the complexity and interrelatedness of the natural world.

Source: Goodenough and Hart (2017) *Applied Ecology: Monitoring, Managing, and Conserving*. Oxford, UK: Oxford University Press. Reproduced with permission of the Licensor through PLSClear.

interrelatedness is fundamental to the science of ecology. The fact that this is recognized in the first book ever written on evolution, seven years before the concept of 'ecology' had actually been defined by the German biologist Ernst Haeckel in 1866, indicates just how central evolution is to ecology.

The aim of this section is to consider evolution from the standpoint of how it links to ecological patterns and processes. We focus on basic concepts of evolution and the effects of this in the species, populations, communities, and ecosystems we see in the world today—in other words the ecological *results* of evolution. We cover the *processes* by which evolution acts to create these results in Chapter 35.

The development of evolutionary theory

Until the end of the eighteenth century, biologists studied collections of living specimens in zoos and botanical gardens or dead specimens in museums and herbariums. These collections allowed them to name organisms (an activity called **taxonomy**) and classify them in various ways (an activity called **systematics**). There was, however, little in the way of experimental ecology that could account for the patterns observed. Some individuals speculated that species might have 'evolved' over time through natural processes, but there was no generally accepted theory to explain how that might have happened.

▶ We learn more about taxonomy and systematics in Topics 4 and 5.

The first coherent theory of evolutionary change was developed by two biologists who, working independently, based their models on an ecological understanding of the interactions between organisms, and between organisms and their surroundings, gained from extensive observations in the field. These biologists were Charles Darwin and Alfred Russel Wallace (see Figure 34.5).

In 1831, at the age of 22, Charles Darwin was given the opportunity of a lifetime: he was invited to join the ship HMS *Beagle* as an unpaid naturalist for a five-year journey around the world. The trip allowed Darwin to observe the interactions between organisms and their surroundings in a range of different environments. After the trip, he spent the next 20 years thinking through the implications of his observations to come to the conclusion that the patterns in nature were the consequence of an evolutionary process he termed natural selection.

In 1858, Alfred Russel Wallace wrote to Darwin, from an island in Indonesia, setting out a similar set of conclusions based on his own experiences in the field collecting plants and animals. Wallace had observed the differences in the species found on islands that

Figure 34.5 The founders of evolution, and excellent early ecologists. (a) Charles Darwin in 1840 in a portrait by George Richmond; (b) Alfred Russel Wallace in a photograph taken in Singapore in 1862.

Source: (a) Science History Images/Alamy Stock Photo; (b) Alfred Russel Wallace and James Marchant (1916), *Alfred Russel Wallace; letters and reminiscences*. University of Toronto - Gerstein Science Information Centre.

were very close together geographically. He was particularly struck by the difference between the islands of Bali and Lombok, which were separated by a sea channel just 37 km wide, but which supported completely different assemblages of species. This observation lead Wallace to hypothesize that species had developed differently over time in different places.

The letter prompted Darwin to act, and the two men presented a paper setting out their theory of evolution by natural selection at a meeting of the Linnaean Society of London in July 1858. Darwin set the theory out in more detail the following year in a book entitled *On the Origin of Species by Means of Natural Selection* (1859).

Both Darwin and Wallace had adopted the scientific method approach that we saw in Experimental Toolkit 1.1: they firstly observed, they secondarily questioned, and they thirdly hypothesized. Indeed, the theory of natural selection is based on three key observations followed by two hypotheses.

- **Observation 1: Most natural populations have the potential to grow at a rate that the environment cannot sustain:** Darwin was influenced by *An Essay on the Principle of Population*, which had been published by Thomas Robert Malthus in 1798. Malthus noted that human populations have the potential to grow geometrically, doubling in size every 25 years or so (1, 2, 4, 8, 16, etc.). In contrast, the ability of the environment to sustain the population through food production would—at best—grow arithmetically (1, 2, 3, 4, 5, etc.). Malthus predicted a future of starvation unless steps were taken to restrict the rate of growth and/or increase our ability to exploit the environment for food. Darwin was aware that humans were just one example of a wider ecological phenomenon. For example, in spring and summer a female cabbage aphid (small organisms that you can see in Figure 34.6) can produce 5–10 genetically identical daughters every day without the need for sexual reproduction: a process known as **parthenogenesis**. Each daughter is born containing the embryos of the next generation, can give birth within a week, and will go on to produce somewhere between 5 and 10 offspring a day for up to 30 days. The resulting potential for population growth led the entomologist Richard Harrington to note that, in a year, aphids have the potential to form a layer 149 km deep over the surface of Earth.

Figure 34.6 Cabbage aphids (*Brevicoryne brassicae*).

Source: Holger Kirk/Shutterstock.

- **Observation 2: The potential for population growth results in competition to survive and reproduce:** Earth is not covered in cabbage aphids. Aphid populations vary throughout a year but are relatively stable in the long term. This stability is a combination of factors such as competition for food and predation. These factors mean that many aphids do not survive to breed, while those that do differ in reproductive output and contribute different numbers of daughters to the next generation. Darwin recognized the role of competition in populations and termed this the 'struggle for existence'. In reality, though, as some individuals will not survive long enough to reproduce, and those that do will not be equally successful in the reproductive process, it might have been better to say the 'struggle to survive and reproduce' and thereby combine the concepts of predation and competition.

- **Observation 3: There is variation in every population and some of this variation can be passed from parents to offspring:** Not all cabbage aphids are the same—they differ in characteristics. Because of the difference in specific characteristics (for example, feeding efficiency or predator avoidance), some aphids are more likely to survive than others.

These three observations led Darwin to two crucial hypotheses:

- **Hypothesis 1:** Individuals that survive and reproduce will tend to pass on to the next generation those key characteristics that increase the chances of survival and reproduction.

- **Hypothesis 2:** Over time, the characteristics that increase the chances of survival and reproduction will become more common in the population (and those characteristics that reduce the likelihood of survival and reproduction will become less common). Darwin used the term **natural selection** for this process.

Key evolution terminology: traits, genotypes, and phenotypes

We call the characteristics of an organism that we can observe and measure 'traits'. Traits can be developmental (e.g. speed of growth), anatomical (e.g. tongue length), physiological (e.g. breathing capacity), or behavioural (e.g. hiding behaviour for predator avoidance). The expression of a trait in an individual organism is its **phenotype**. Because we cannot list all the characteristics of a given organism, we tend to focus on those that are of interest to us: for example, the enzymes involved in a particular biochemical pathway, different fur colour or coat patterns, or the distance an individual is able to travel in search of a mate. The genetic make-up of an individual is called its **genotype**. As with the phenotype, we don't tend to characterize entire genomes to assess genotype, but often focus instead on the role of specific genes.

Individuals in a population differ from each other—Observation 3—and we call the differences associated with a particular trait in a given population the **phenotypic variation** for that trait. Sometimes a specific genotype may produce a different phenotype under different environment conditions: a phenomenon known as **phenotypic plasticity**. To make matters more complex, different genotypes interact with environmental conditions in different ways. These genotype/environment interactions can make it difficult to predict the phenotypes that will be produced by a particular genotype without knowing something about the environmental variation involved. In much the same way, it is difficult to predict the phenotypes associated with particular environmental conditions without knowing something about the genotypic variation involved.

Gregor Mendel's pioneering work in genetics, as discussed in Chapter 4, was based on **simple traits**: traits in which a small number of discrete phenotypes are determined by single (or very few) changes in specific individual genes with little influence from the environment. Single-gene traits tend to show **discontinuous variation**. In other words they are 'long' or 'short'; 'dark' or 'light'; 'round' or 'square', rather than varying in length from long to short, varying in colour from dark to light, or varying in shape from round to square: there is no spectrum, just several (often two) distinct states. For example, some of Mendel's experiments involved pea plants (*Pisum sativum*) that produced either smooth peas or wrinkled peas. In Figure 34.7a, you can see the smooth peas on the left, and the wrinkled peas on the right:

Simple trait = pea appearance
Phenotypes (discontinuous variation) = smooth OR wrinkled

Although simple traits are easier to understand conceptually, most of the traits that occur in living organisms are **complex traits**: a range of phenotypes are determined by multiple changes in multiple genes. For these traits, the phenotypes we observe are the result of complex interactions between genotype and environment and thus complex traits show a continuous variation. In other words, colour and size are variable along a spectrum. Shell colour in the coquina clam (*Donax variabilis*) is an example of a complex trait. You can see the variations in shell colour in the coquina clam in Figure 34.7b.

As we will discover in Chapter 35, evolution is fundamentally a change in the frequencies of phenotypes in a particular population over time, which arise as a result of a change in the frequencies of particular genotypes (and genes) that underpin that. Such change is usually in response to selection as a driver of evolution.

Key evolution drivers: selection

Evolution is fundamentally a genetic response to the abiotic and/or biotic environment. We will consider the genetic *processes* involved in Chapter 35 but let us now consider the ways in which environmental stimuli can *result* in evolution.

Natural selection

Natural selection is the differential survival and reproduction of different phenotypes in the face of the challenges posed by a particular environment. It can take different forms in different populations, in different places, and at different times. We usually sub-divide natural selection into:

1. Stabilizing selection.
2. Directional selection.
3. Disruptive selection.

Stabilizing selection occurs if individuals that are close to the phenotypic mean (i.e. those which have a fairly typical level of a particular trait such as colour or size for that particular species) are more likely to survive and reproduce. When this happens, it acts to reduce phenotypic and genotypic variance in a population. Stabilizing selection is usually observed in populations that are well matched to stable environments. Figure 34.8 shows an example of stabilizing selection in a yellow jawfish (*Opistognathus aurifrons*): this mouth-brooding fish species carries its young in its

Figure 34.7 Examples of phenotypic variation. Phenotypic variation arising from: (a) simple traits (smooth or wrinkled peas: *Pisum sativum*); and (b) complex traits (shell colour and pattern in the coquina clam: *Donax variabilis*). The second half of the scientific name of the clam (*variabilis*) reflects the phenotypic variation observed in the shells of this species.

Source: (a) MShieldsPhotos/Alamy Stock Photo; (b) Debivort/CC BY-SA 3.0/Wikimedia Commons.

Figure 34.8 Stabilizing selection. Brood size in a yellowhead jawfish (*Opistognathus aurifrons*).

Source: Fotograferen.net/Alamy Stock Photo.

mouth and so the optimal number of offspring is the maximum number that an adult can carry in its mouth. If the brood size is too small, there is a high likelihood that no young will survive to maturity because normal predation levels are greater than the chance that enough individuals will survive to sustain the population. Conversely, if the brood size is too large, the parents are less likely to be able to protect the young from predators and their *individual* predation risk increases to the point at which predation is greater than survival.

Directional selection occurs if individuals with a marginal phenotype are more likely to survive and reproduce than individuals that are close to the phenotypic mean. For example, take a look at Figure 34.9. In stage one, where the antibiotic-resistant phenotype first appears, individuals with this phenotype are outnumbered by individuals with the antibiotic-sensitive phenotype. In stage two, when bacteria are exposed to an antibiotic, most of the antibiotic-sensitive bacteria die before they can reproduce. Finally, in stage three, individuals with the antibiotic-resistant phenotype survive and reproduce, so that the resistant phenotype—and the genes associated with resistance—become more frequent in subsequent generations. This is known as antibiotic resistance and is seen, for example, in MRSA whereby methicillin-resistant *Staphylococcus aureus* (MRSA) have a competitive advantage over methicillin-sensitive *Staphylococcus aureus* (MSSA). Directional selection is usually observed in populations that are poorly matched to their environment, often because the environment has changed in some way (in the case of bacteria, because of the increasingly widespread use of antibiotics, which is a selective disadvantage to non-resistant bacteria).

🌀 Go to the e-book to explore an interactive version of Figure 34.9.

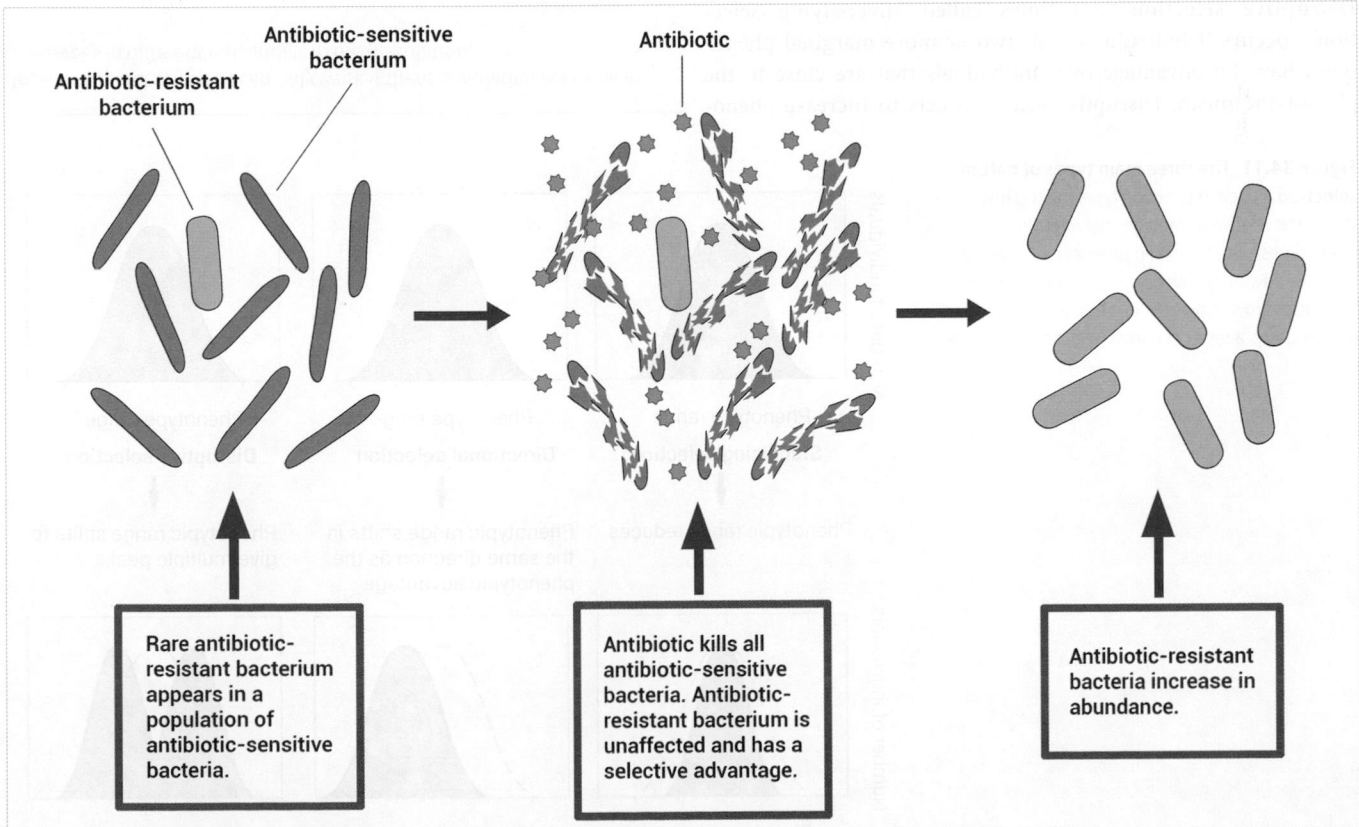

Antibiotic-resistant bacterium

Antibiotic-sensitive bacterium

Antibiotic

Rare antibiotic-resistant bacterium appears in a population of antibiotic-sensitive bacteria.

Antibiotic kills all antibiotic-sensitive bacteria. Antibiotic-resistant bacterium is unaffected and has a selective advantage.

Antibiotic-resistant bacteria increase in abundance.

Figure 34.9 Directional selection. The development of antibiotic resistance in bacteria.

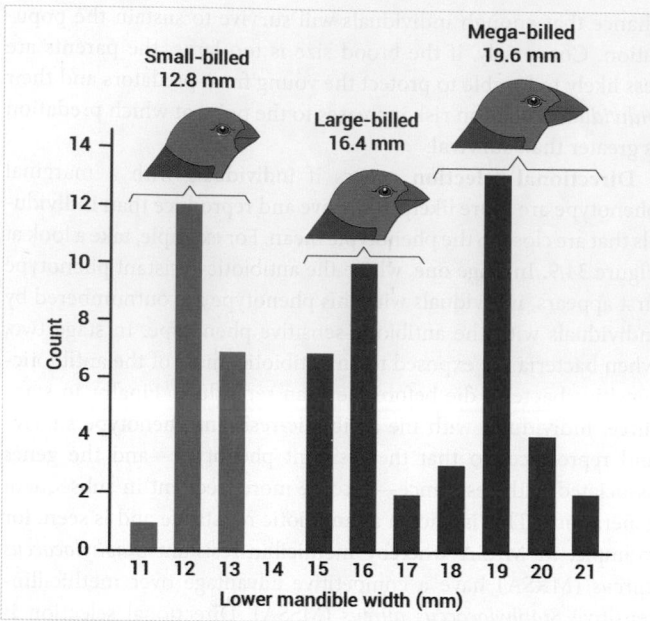

Figure 34.10 Disruptive succession in black-bellied seedcrackers (*Pyrenestes ostrinus*). Frequency histogram showing the three different bill morphologies based on lower mandible width (LMW) measurements.

Source: Reproduced from von Holdt, B.M., Kartzinel, R.Y., Huber, C.D. et al. Growth factor gene IGF1 is associated with bill size in the black-bellied seedcracker Pyrenestes ostrinus. Nat Commun 9, 4855 (2018). https://doi.org/10.1038/s41467-018-07374-9. Distributed under the terms of the Creative Commons Attribution 4.0 International License (CC BY 4.0). https://creativecommons.org/licenses/by/4.0/.

Disruptive selection—sometimes called 'diversifying selection'—occurs if individuals with two or more marginal phenotypes have an advantage over individuals that are close to the phenotypic mean. Disruptive selection acts to increase pheno-

typic and genotypic variance in a population as the genes for the extreme phenotypes increase in frequency. Disruptive selection can occur in dense populations where competition between individuals forces different individuals to exploit different parts of the environment. One example is the evolution of distinct bill types in a population of black-bellied seedcrackers (*Pyrenestes ostrinus*) in Cameroon, where there are three different morphs (distinctive 'types' of the same species). Figure 34.10 shows a frequency histogram showing the three different bill morphologies based on lower mandible width (LMW) measurements: birds with small bills feed on soft-seeded sedges (*Scleria goossensii*), birds with large bills feed on hard-seeded sedges (*S. verrucosa*), and birds with mega bills specialize on very hard-seeded sedges (*S. racemosa*). Birds that have intermediate bills between two of these three categories (LMW = 14 mm or 17–18 mm in Figure 34.10) are less efficient at feeding on any type of seeds and so are less successful, and are thus at a selective disadvantage and therefore uncommon.

The three different types of natural succession—stabilizing, directional, and disruptive—are summarized diagrammatically in Figure 34.11.

Go to the e-book to explore an interactive version of Figure 34.11.

PAUSE AND THINK

What one thing does there need to be in a population for natural selection to occur?

Answer: Natural selection can only work if there is variation available in genotypes in different individuals in the population.

Figure 34.11 The three main types of natural selection. For each selection type, the top row shows the relative advantages between phenotypes, with the green phenotypes having a selective advantage over the red phenotypes, and the bottom row shows the result of those differences as a result of natural selection over time.

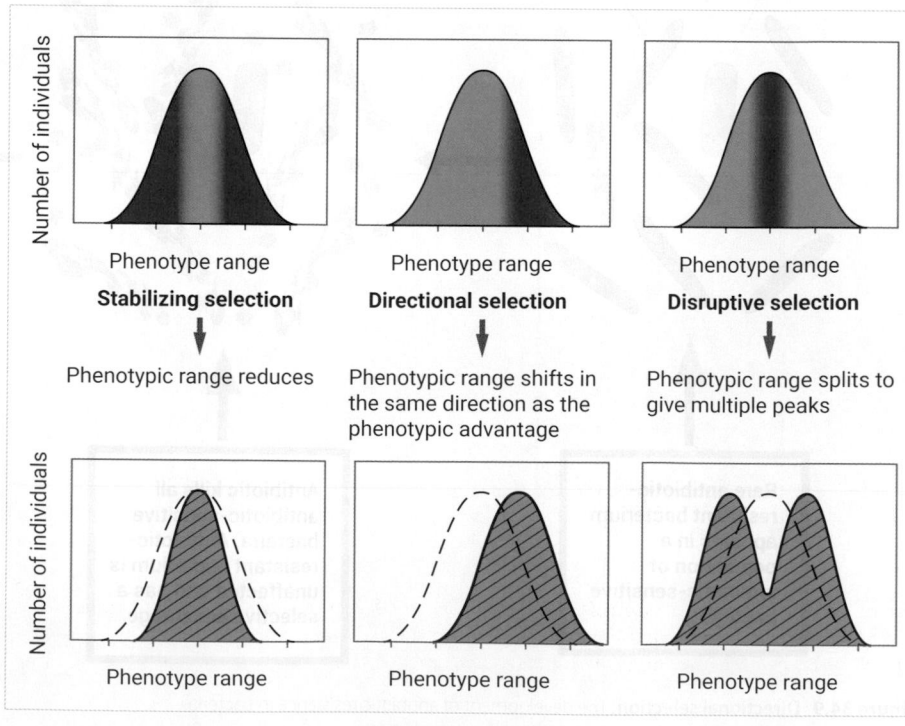

Sexual selection

Natural selection acts on the ability of an organism to survive and reproduce. One form of natural selection, **sexual selection**, however, acts directly on traits that affect the ability of an individual to reproduce by:

- Changing traits associated with the ability to attract, select, and/or retain mates (e.g. sexual ornamentation).
- Changing traits associated with successful fertilization of an egg by sperm (e.g. sperm swimming speed).

We often think of natural selection as a process that improves the chances of survival and reproduction, but sexual selection can involve trade-offs between reproduction and survival. This is exemplified by the male Wilson's bird of paradise (*Cicinnurus respublica*), as shown in Figure 34.12, where plumage not only plays an important role in attracting females, but also makes the bird more visible to predators.

In many cases, sexual selection and natural selection interact with one another, with traits becoming larger or more extravagant through sexual selection to improve reproduction and natural selection then acting to 'put the brakes on' when the magnitude of sexual traits reduces survival more than it increases reproductive success.

Artificial selection

Humans have a significant effect on the environment and the species within it. This impact has included the domestication of plants and animals through the process of **artificial selection**. Like natural selection, artificial selection involves differences in the survival and reproduction of individuals with different genotypes, but the individuals that survive and reproduce are selected by human intervention rather than a natural process such as competition.

Artificial selection is a form of directional selection, but the very high level of control exercised over which individuals are allowed to reproduce means that it can be very efficient at producing

Figure 34.12 A male Wilson's bird of paradise. The plumage of the male Wilson's bird of paradise (*Cicinnurus respublica*) plays an important role in attracting females—but it also makes the bird more visible to predators.

Source: Gabbro/Alamy Srs.Sotock Photo.

significant changes in the phenotypes observed in a population over a relatively short period of time. We discuss artificial selection processes and the phenotypic diversity that can be created more in Scientific Process 34.1.

Results of selection: genotypic adaptations

As we will see in more detail in Chapter 38, all organisms have a set of conditions in which they can survive and reproduce at an optimum level. In other conditions, they may be unable to survive and reproduce at all, or they may survive and reproduce less well. An **adaptation** is a change in an organism's biology that results in an improvement in its evolutionary fitness—its reproductive success—in a particular environment.

Most organisms are able to react to the pressures associated with a particular environment, in the short term, through the **phenotypic response**. This involves rapid, often reversible, changes in the phenotype of the individuals that can be anatomical, developmental, physiological, or behavioural. This is necessary because environments can change quickly, and individuals need to be able to respond just as quickly. A simple example of a behavioural phenotypic response is when you look at the weather forecast before dressing in the morning to select suitable clothes for the temperature that you are likely to experience that day, while an example of physiological response is improved cardio-respiratory function after regular exercise training.

In the longer term, if selection is consistent in type and direction, individuals will respond to environmental change through a process of **genotypic adaptation**: changes in phenotypes underpinned by changes in genotypes in the population driven by natural selection.

It is important to recognize that the ability of individuals to respond to environmental change through phenotypic responses, and the ability of populations to respond through genotypic adaptation, are *both* the result of evolution. In environments that are changeable, organisms often evolve considerable ability to respond to that environment phenotypically. In this way, the *ability* to be able to respond phenotypically often has a genetic basis but the *actual response* itself is non-genetic. In contrast, environments that are relatively consistent might favour individuals that exhibit less phenotypic flexibility but that are genetically well-adapted to the environment concerned.

> ## PAUSE AND THINK
>
> If an environment that has been stable for many years suddenly changes substantially (e.g. an area that has been bog for hundreds years becomes suddenly drier due to local hydrological changes), would it be better for individuals of a species to be genetically well-adapted to bog conditions or genetically predisposed to be able to respond to change phenotypically?
>
> *Answer:* In cases of rapid and dramatic change in the environment, species where individuals are phenotypically flexible are more likely to be able to successfully respond to change in the short term than are species where individuals are genetically well-adapted to the original environment.

SCIENTIFIC PROCESS 34.1   Artificial selection in dogs

Background

All breeds of dogs are the same species: *Canis lupus*. In fact, they are all the same sub-species: *Canis lupus familiaris*. The grey wolf is also the same species but has a range of different sub-subspecies, of which *Canis lupus lupus* is the most widespread. Although the domestication of dogs began over 14,000 years ago, the extreme phenotypic diversity exhibited among breeds has originated much more

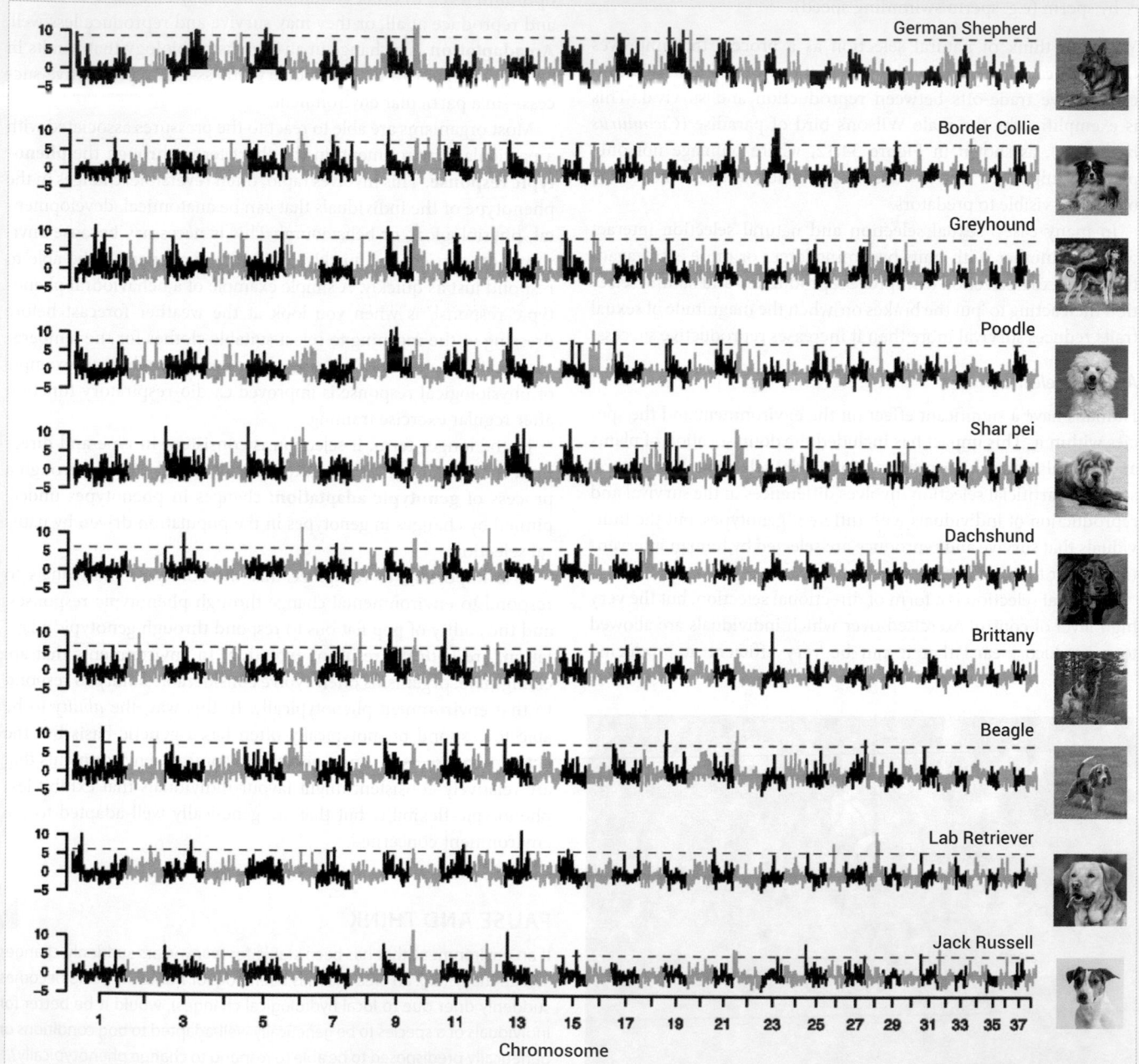

Figure 1 Genome-wide differences in dog breeds across all 39 chromosomes (each chromosome shown in alternating grey/black coloration). The d_i metric on the y-axis is a measure of genetic similarity, which measured the difference between each part of the genome for each breed. The dashed red line denotes the 99th percentile (upper limit that only 1% of measures were outside) for each breed; any loci breaching this are extremely different from the average.

Source: Reproduced with permission from J.M Akey et al. (2010) Tracking footprints of artificial selection in the dog genome. *PNAS*, 107(3): 1160–1165. https://doi.org/10.1073/pnas.0909918107.

recently. This has largely been the result of intense artificial selection through highly selective breeding practices.

Aim

The canine genome, shaped by centuries of strong (artificial) selection, is a good system through which to explore the genetic changes leading to phenotypic variation and the mechanistic basis of rapid short-term evolution. Most studies focus on specific individual genes, but Joshua Akey and colleagues used **genome-wide scans for selection** (GWSS) to examine genetic differences across the entire genome between breeds.

Methods

Akey collected DNA from blood and cheek swab samples from individual 275 dogs representing 10 phenotypically diverse breeds. Each sample was then genotyped. The researchers devised a metric of genetic similarity called d_i, which measured the difference between each part of the genome for each breed (breed i) relative to the genome-wide average excluding that breed. Thus, if d_i for a particular genetic region (termed locus; plural loci) is 0 for a given breed, that breed is genetically identical to the average for that locus. The further away from 0 d_i is (either positively or negatively), the larger the genetic dissimilarity. This metric is well suited to detecting selection for a particular breed, or subset of breeds, and establishing the direction of change relative to the mean (positive or negative).

Results

Genomic distribution across all 39 canine chromosomes for each of the 10 dog breeds was shown graphically using the d_i metric (Figure 1). This allowed the authors to identify 155 genomic regions (loci) that had strong signatures of recent artificial selection, and which were thus candidate genes controlling those traits that vary most among breeds, including size, coat colour and texture, skeletal morphology, and physiology. This included a set of three genes—*RSPO2*, *FGF5*, and *KRT71*—that appear to influence coat phenotypes.

Discussion

The extensive phenotypic diversity between dog breeds has long been recognized as a unique opportunity for studying evolutionary change, especially change caused by artificial selection. However, much of this phenotypic variation has been hard to study through single-gene approaches (the 'segregation problem'). Here, whole-genome mapping has allowed geneticists to identify 155 regions of the canine genome that exhibit signatures of artificial selection. This includes all genes previously thought to be important in canine phenotypes plus many previously unconsidered candidate genes that contribute to phenotypic variation among breeds.

Extension question

In what other species or species groups might genome-wide scans for selection be useful for establishing the genetic basis of different phenotypes caused by artificial selection?

Read the original paper

Akey, J.M., Ruhe, A.L., Akey, D.T., Wong, A.K., Connelly, C.F., Madeoy, J., Nicholas, T.J. and Neff, M.W. (2010). Tracking footprints of artificial selection in the dog genome. *PNAS*, 107(3), 1160–1165.

Results of genotypic adaptations: evolutionary fitness

Evolutionary changes do not occur 'for the good of the species', they occur because they confer benefit to the individual involved because the genotypic adaptations are useful either to them or to their potential offspring. For example, if we accept that the neck length of giraffes (*Giraffa* species) has evolved in relation to accessing food resources (there are other theories), giraffes have not evolved long necks because the *species* would be able to access food resources out of reach of other African herbivores, but rather because *individuals* with long necks had a competitive advantage over those with short necks and were more likely to successfully survive and reproduce.

In our consideration of natural selection so far, we have stressed the importance of survival and reproduction, but these are actually two aspects of the same thing. What really matters are the consequences of survival *and* reproduction: **reproductive success**—the number of offspring a given individual contributes to the next generation relative to the number contributed by other members of the population. This is important as it determines the proportional extent to which that individual's genotype is represented in subsequent generations. Biologists use the term fitness—often, and more precisely, **evolutionary fitness**—as a measure of the genetic contribution of one individual to the next generation (and future generations). An individual can increase its evolutionary fitness by increasing its own reproductive success (**direct fitness**) or by adopting behaviours that will increase the reproductive success of close relatives that share much of its genotype (**indirect fitness**).

Indirect fitness can lead to a special type of natural selection—**kin selection**—that acts on traits that improve the overall reproductive success of related members of a group. Kin selection can result in behaviours that *seem* to be altruistic because individuals appear to sacrifice their own direct fitness in favour of indirect fitness of kin. These concepts are explained in Figure 34.13, which shows that direct fitness is conferred by the simple process of having more offspring and grand offspring (a), whereas indirect fitness benefits (b) can be gained when one individual helps another individual's breeding success *as long as* those individuals are closely related (i.e. have very simply genotypes).

Indirect fitness strategies are often seen in social insects, whereby workers sacrifice their own reproductive success to promote the reproductive success of the queen. For example, the exploding ant (*Colobopsis explodens*) was first discovered in 2014 living in the tops of trees in Borneo (Figure 34.14). The workers explode when the colony is attacked, covering that attacker in a toxic, sticky fluid

Figure 34.13 Direct and indirect fitness. Direct fitness (a) involves an individual passing on their genotype to subsequent generations directly through their own offspring and grand offspring. Indirect fitness (b) involves an individual helping a close relative passing on their (very similar) genotype to subsequent generations. In both cases, more offspring and grand offspring = higher fitness.

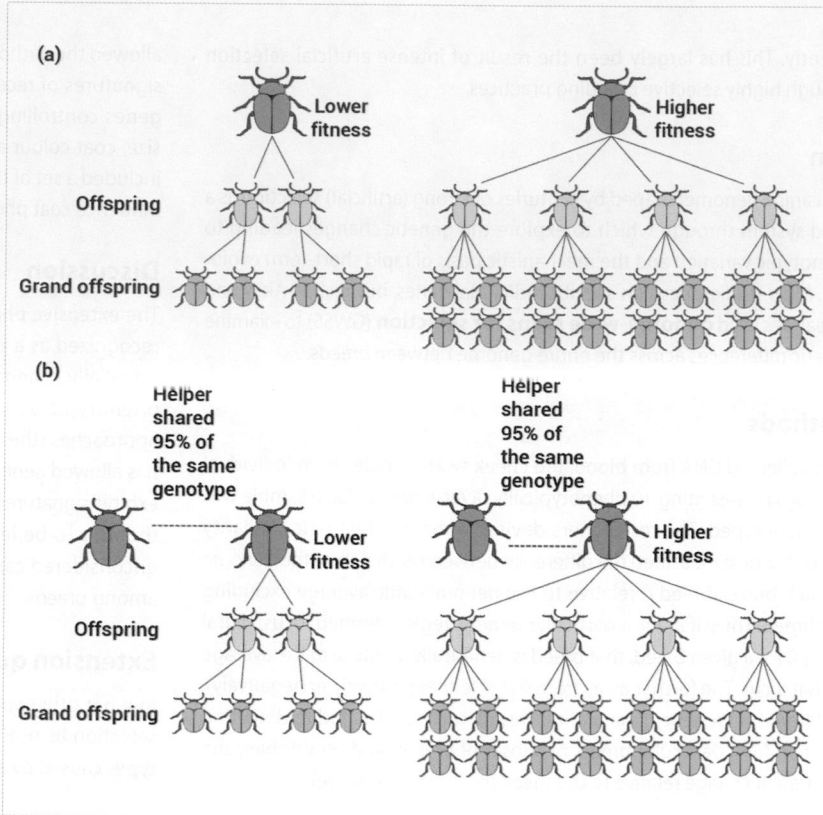

Figure 34.14 The exploding ant (*Colobopsis explodens*).

Source: Reproduced from Laciny A, et al. (2018) *Colobopsis explodens* sp. n., model species for studies on "exploding ants" (Hymenoptera, Formicidae), with biological notes and first illustrations of males of the *Colobopsis cylindrica* group. *ZooKeys*, 751: 1–40. https://doi.org/10.3897/zookeys.751.22661. Distributed under the terms of the Creative Commons Attribution License (CC BY 4.0). https://creativecommons.org/licenses/by/4.0/.

in a process that kills the workers concerned: this is called **autothysis**. The worker ants do not reproduce (no direct fitness) but because all of the ants in the colony are close relatives, the workers' genotype will be passed to the next generation indirectly through the queen, which produces huge numbers of offspring. Protecting the queen—even at the expense of the worker's own life—thus provides a high level of indirect fitness for the worker ant. In this way, autothysis is not altruistic: there is no 'for the good of the species' argument, it is simply an individual action to improve individual fitness (but in this case indirect fitness).

We can define—and measure—an individual's **inclusive fitness** as the combined results of its direct fitness and its indirect fitness. In practice, this would involve counting an individual's offspring, if any (the direct fitness component), and the number of offspring produced by relatives who have had their reproductive success improved after taking into account a measure of the similarity of their genotypes (the indirect fitness component).

It should be noted that a genotypic adaptation to one aspect of the environment may reduce an individual's ability to adapt genotypically to another: there can be a complicated set of trade-offs. For example, as we will see in Chapter 38, the crab plover (*Dromas ardeola*) has to balance the competing selection pressures for long legs (to access food resources from lakes) with selection pressures for short legs that are better suited to digging depressions in the sand that they need to create refuges from the heat of the sun. We can infer these different selection pressures, and their relative strengths, using **optimality theory**: this assumes that the phenotypes expressed represent the best compromise between competing pressures to ensure optimal long-term inclusive fitness.

Observing the outcomes of evolution

As we have seen, evolutionary processes act on individual organisms and, as such, the individual is the **unit of selection**. It is possible to talk about evolutionary fitness at levels *lower* than the individual—the extent to which a particular gene is passed from one generation to the next—but we should not talk about evolution at a level *higher* than the individual. Technically, therefore, we should not refer to 'a population evolving' or 'a species evolving'. What evolves is successive generations of *individuals* that comprise a population or a species. However, we often observe the outcomes of individual-level evolutionary change within populations or species when, for example, we note that the birds that inhabit one island have larger bills than the birds (or the same species or a similar species) that inhabit another island. Indeed, we often use the level at which evolutionary change is evident to classify the type of evolution that has occurred:

- **Microevolution:** evolution resulting in changes being evident *between different populations*.
- **Macroevolution:** evolution resulting in changes *between different species*, including the production of new species through **speciation events**.

We will discuss the concept of speciation in Section 34.3. Before this, though, let's think about evolution in humans by examining recent research on microevolution in humans (Scientific Process 34.2) and the considering some common misconceptions about evolution by natural selection that still persist today (Real World View 34.1).

 Check your understanding of the concepts covered in this section by answering the questions in the e-book.

SCIENTIFIC PROCESS 34.2   Microevolution in human dive reflex

Introduction

The Bajau people—the Sea Nomads of Indonesia—have lived a sea-based life for over 1000 years. They obtain food by free diving for up to 5 minutes to depths of over 70 m and some individuals spend up to 60% of their waking time under water. Figure 1 shows the hunting technique used by the Bajau people. Like all humans (*Homo sapiens*), the Bajau have a **mammalian diving reflex**. This means that if breath is held and the face is submerged underwater, the individual's heart rate slows, arterial blood pressure increases, blood is diverted away from the limbs, and the spleen contracts to supply more oxygen-rich red blood cells.

Hypothesis

Free diving is inherently dangerous—even experienced free divers can lose consciousness and drown. There is thus a strong selection pressure for the dive reflex to become more pronounced (and thus more beneficial) in the Bajau population compared to other human populations.

Figure 1 A Bajau diver hunting.

Source: Timothy Allen/The Image Bank Unreleased/Getty Images.

Methods

Melissa Ilardo and colleagues at the University of Copenhagen initially compared the spleen sizes of Bajau people to another local population, the land-based Saluan people, using ultrasound scans. Confounding factors (Experimental Toolkit 34.1), such as sex, age, and body mass, were factored into this analysis. The task of identifying the results of natural selection was undertaken using **genome-wide scans for selection** (GWSS), in which unusual patterns of genotypic diversity in a population are linked to selection pressures and adaptive phenotypes (see also Scientific Process 34.1 where the same process was used to examine genetic differences between breeds of dog). In the Bajau study, GWSS was used to compare the genomes of the Bajau and Saluan participants, using DNA extracted from saliva samples, with genome sequences from a reference population—the Han Chinese. This allowed atypical patterns of genetic variation in the Bajau, which might be linked to their diving abilities, to be identified. The final step was to scan the DNA results to try to identify the specific areas (genetic loci) that might be responsible for any difference in spleen size.

Results

The spleens of the Bajau were found to be around 50% bigger than those of the Saluan (as shown graphically in Figure 2). This difference occurred throughout the population, including in Bajuan individuals who don't dive. This suggests that it is not a straightforward phenotypic response (repeated diving by an individual), but is instead the result of a genotypic adaptation (a genetic change). The results of the DNA analysis suggested that the Bajau are genetically closer to the Saluan than to most other Asian populations, and that both diverged from an ancestral population about 16,000 years ago. The genome scans identified a number of genes that might be linked to diving in the Bajau. In particular gene *PDE10A* was found to be linked to both thyroid function and spleen size. Thyroid hormones regulate the production of red blood cells, so selection for this particular genotype could provide an advantage for the Bajau in two ways: more red blood cells to carry oxygen around the body and a larger spleen in which to store them.

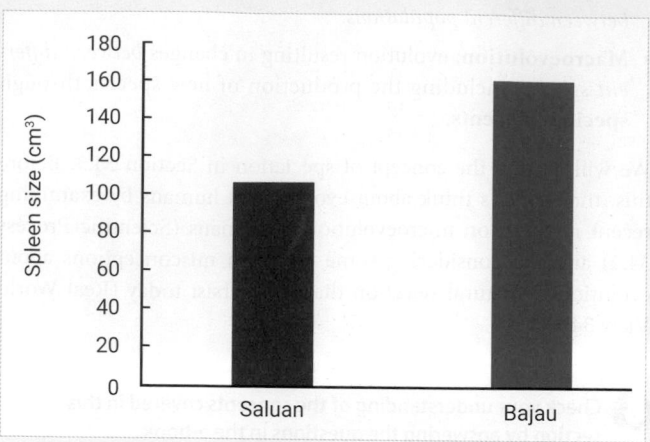

Figure 2 Mean spleen size for Saluan and Bajaun people.

Source: Reproduced with permission from M. A. Ilardo et al. (2018) Physiological and Genetic Adaptations to Diving in Sea Nomads. *Cell*, 173(3): 569–580.e15. Copyright © 2018, Elsevier Inc.

Discussion

The results suggest that the Bajau have several specific adaptations linked to diving, showing that their biology and their lifestyle have been co-evolving for the last 1000 years. In particular, natural selection on genetic variants in the *PDE10A* gene has increased spleen size in the Bajau, providing them with more red blood cells and a bigger reservoir in which to store them until they're needed when, and if, individuals free dive.

Extension question

What other dive-associated traits could be compared between the Bajau and Saluan?

REAL WORLD VIEW 34.1   **Misconceptions about evolution by natural selection**

Ask someone to find a single image to exemplify 'evolution' and they will probably return with something similar to Figure 1. However, although most people know about evolution, the topic can be contentious and divisive to some of those with strong religious beliefs. There is still substantial resistance to the teaching of evolutionary science in some contexts, and there can be a reluctance to using public funds for research and public engagement projects that have an explicitly evolutionary focus. Some of the challenges to evolutionary theory rest on the following common misconceptions.

Evolution is just a theory: All scientific understandings are theories: the problem here is the word 'just'. A theory—like a model—is a detailed set of hypotheses that has been supported by extensive evidence. Would scientists be prepared to abandon or change a theory that was challenged by new evidence? Yes. Does any such evidence exist to challenge the core of evolutionary theory? No.

Evolution by natural selection is a random process: Evolution by natural selection involves random and non-random processes. Mutation, which we will explore in Chapter 35, is important in generating genetic variation, and is random. Natural selection, however, is anything but a random process.

Evolution by natural selection should result in perfection: Natural selection has to work with the phenotypes that are present in a population, and is more likely to alter existing traits than produce new ones. An adaptation that is useful in one environment at one time may be useless—even harmful—in a different environment at a different time. An individual's fitness does not depend on being perfect, it depends on being as well adapted (or slightly better adapted) as the individuals with which it competes.

Everything must be an adaptation: It is a common mistake to think that all of the characteristics of an organism are adaptive in some way. Some characteristics are the results of chance events or the result of our evolutionary history. For example, we have five digits—five fingers and five toes—because the first jawed vertebrates had five digits and there has not been a strong selection pressure for this to change.

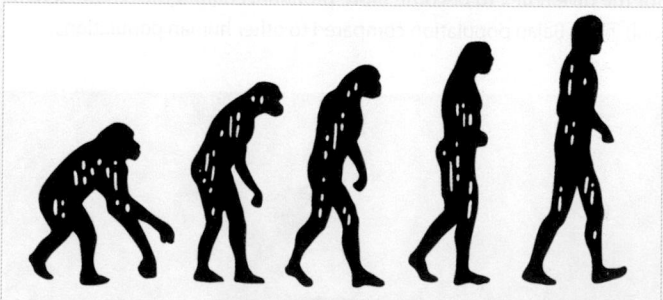

Figure 1 **Hominoid evolution.** The classic depiction of the evolutionary process.

Source: MoreVector/Shutterstock.

34.3 The species: a key concept in evolutionary ecology

The species is the fundamental unit of evolutionary ecology. In this section, we will first examine how we define a species and then how species arise as a result of macroevolution (see Section 34.2).

What is a species?

We can define a species in terms of:

- **Reproduction:** 'species' = a group of individual organisms that can reproduce with each other to produce viable, fertile, and fit offspring.

- **Morphology:** 'species' = a group of individual organisms that share a distinctive set of attributes (i.e. they have a similar *phenotype*).
- **Genetics:** 'species' = a group of individual organisms that share an evolutionary history and are thus more genetically similar to one another than they are to individuals of another species (i.e. they have a similar *genotype*).

The biological species concept

The **biological species concept** defines a species in terms of reproductive isolation: as a group of interbreeding individuals that is reproductively isolated from other individuals. This is based on the fact that all the members of the species are—potentially—able to reproduce with each other to produce offspring that are:

- Viable: able to survive and reproduce.
- Fertile: able to reproduce with other members of the same species to produce viable offspring themselves.
- Fit: have a similar ultimate inclusive fitness as other individuals in the same population.

In order for viable, fertile, and fit offspring to be produced, the parent individuals must be genetically similar to one another and share a common ancestry (be descended from a single ancestral population)—we will think more about these two points in Chapter 35. There are some cases in which members of different (but closely related) species can reproduce to produce **hybrids**. According to the strictest interpretation of the biological species concept, if individuals from two 'species' can interbreed to produce hybrids, they should properly be regarded as individuals from the same species. However, in many—but not all—cases, hybrids are not fertile or have lower reproductive success/inclusive fitness than other individuals, which is where the 'fertile and fit' aspects of the biological species concept definition become relevant.

The prevalence of hybrids is often constrained in nature because two species that hybridize only overlap in a small geographical area. For example, the 'grolar bear' (Figure 34.15) is a cross between a grizzly bear (the North American brown bear: *Ursus arctos*) and a polar bear (*Ursus maritimus*). Grolar bears have been found previously in captivity as a result of bears being kept in mixed-species

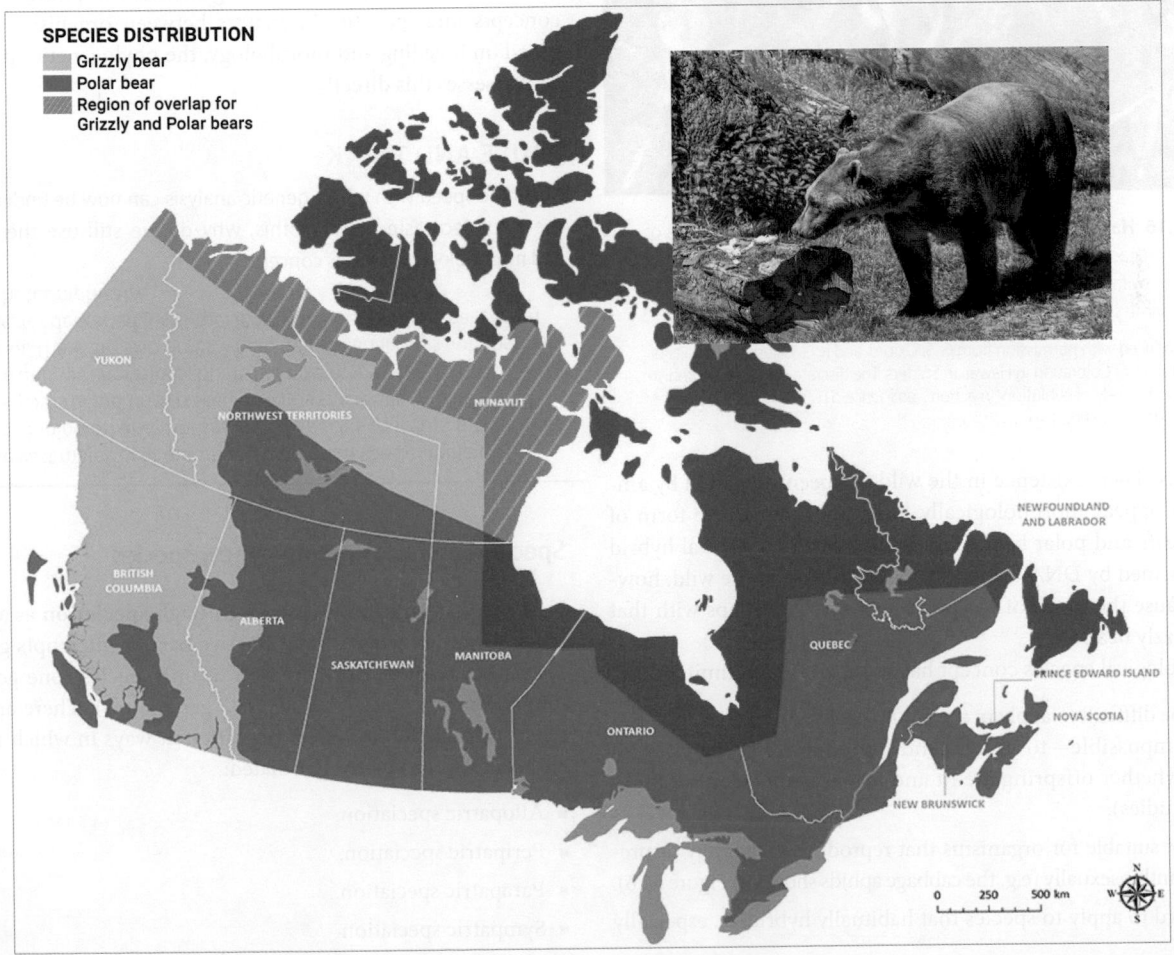

SPECIES DISTRIBUTION
- Grizzly bear
- Polar bear
- Region of overlap for Grizzly and Polar bears

Figure 34.15 Prevalence of grolar bears. Grolar bears (*Ursus arctos* × *Ursus maritimus*) are very rare in the wild, partly because the ranges of grizzly bears (*U. arctos*) and polar bears (*U. maritimus*) have limited overlap.

Source: (bear image) Arterra Picture Library/Alamy Stock Photo.

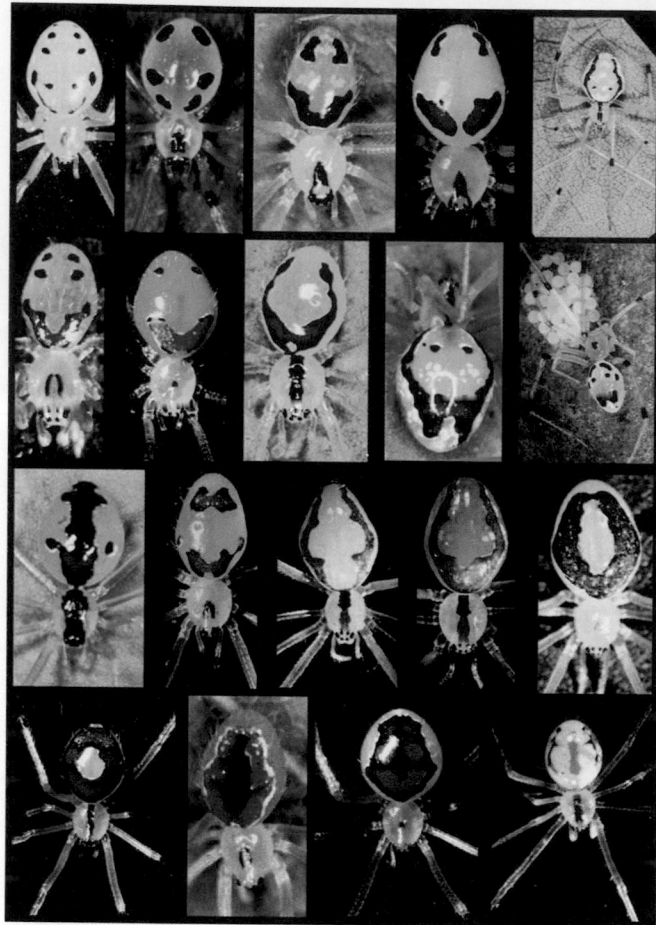

Figure 34.16 Hawaiian happy face spiders. Identification on the basis of phenotypic characteristics can be difficult for polymorphic species such as Hawaiian happy face spiders (*Theridion grallator*), where individuals look different but are the same species.

Source: Reproduced with permission from G. S. Oxford, and R. G. Gillespie. Portraits of Evolution: Studies of Coloration in Hawaiian Spiders: The discrete color polymorphisms in spiders allow the study of evolution "in action". *BioScience*, 51(7): 521–528. Copyright © 2001, Oxford University Press.

enclosures. Their existence in the wild has been suggested by animals that appear morphologically to be an intermediate form of grizzly bears and polar bears, and in 2006 the first natural hybrid was confirmed by DNA testing. Hybrids are rare in the wild, however, because the range of the polar bear rarely overlaps with that of the grizzly bear.

The biological species concept has some important limitations:

- It can be difficult to apply in the field where it can be challenging—often impossible—to understand reproductive biology (especially whether offspring are fit and fertile as this requires long-term studies).

- It is not suitable for organisms that reproduce exclusively or predominantly asexually (e.g. the cabbage aphids shown in Figure 34.6).

- It is hard to apply to species that habitually hybridize, especially plants.

- It can be difficult to apply to organisms that are now extinct, unless there are excellent biological records for that species.

The morphological species concepts

The **morphological species concept** defines a species on the basis of phenotypic characteristics: in most cases, physical appearance. It can be misleading, especially when members of the same species are **polymorphic** and vary in appearance (e.g. the Hawaiian happy face spider (*Theridion grallator*): Figure 34.16) or where individuals of two genetically different species look very similar. Its advantage is that it doesn't depend on any particular knowledge about the genetics of the organisms or their ability to reproduce with each other. As such, it is comparatively easy to use both in the field and for asexual and extinct species.

In some cases, biologists distinguish between the **biospecies** (species defined by reproductive isolation by the biological species concept) and the **morphospecies** (species defined by phenotypic characteristics by the morphological species concept).

The phylogenetic species concept

The **phylogenetic (or evolutionary) species concept** explicitly recognizes that the distinction between separate species is the consequence of a process of evolutionary change. It defines a species as the smallest set of organisms that share a common ancestral population and can be distinguished in some way from other sets. In other words, whereas the biological and morphological species concepts infer genetic differences between organisms indirectly based on breeding and morphology, the phylogenetic species concept assesses this directly.

PAUSE AND THINK

Given the speed with which genetic analysis can now be undertaken and the ever-decreasing cost of this, why do we still use the biological and morphological species concepts?

Answer: the vast majority of our understanding of species dynamics is based on the biological and morphological species concepts. Increasingly, genetic analysis is being undertaken on 'known' species, and this often results in taxonomic alterations, but it will take many years to revisit and revise all taxa. There are also times where genetic analysis is not possible, for example, for degraded fossil specimens or where species have been recorded photographically.

Speciation: the formation of new species

New species come into existence through **speciation** as a result of macroevolution when a reproductive barrier interrupts gene flow. Speciation results in the division of one species (one gene pool) into two new species (two separate gene pools). There are several types of speciation to reflect the different ways in which individuals become reproductively isolated:

- Allopatric speciation.
- Peripatric speciation.
- Parapatric speciation.
- Sympatric speciation.

These are detailed next and then summarized in Figure 34.17.

Go to the e-book to explore an interactive version of Figure 34.17.

Figure 34.17 Types of speciation process.

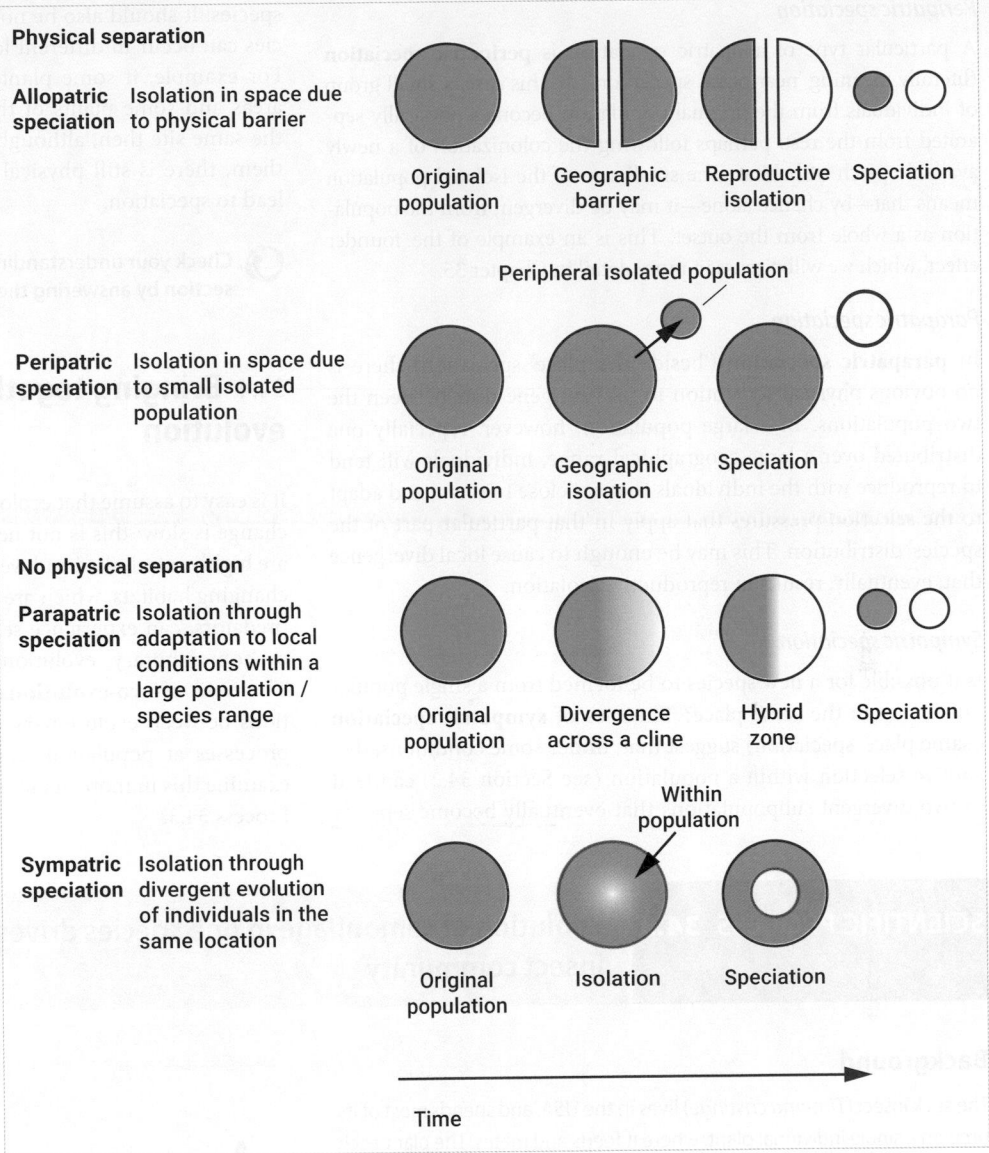

Physical separation

Allopatric speciation Isolation in space due to physical barrier

Original population — Geographic barrier — Reproductive isolation — Speciation

Peripatric speciation Isolation in space due to small isolated population

Peripheral isolated population

Original population — Geographic isolation — Speciation

No physical separation

Parapatric speciation Isolation through adaptation to local conditions within a large population / species range

Original population — Divergence across a cline — Hybrid zone — Speciation

Within population

Sympatric speciation Isolation through divergent evolution of individuals in the same location

Original population — Isolation — Speciation

Time

Allopatric speciation

Allopatric speciation is probably the easiest form of speciation for us to understand conceptually. It occurs when a single population is separated into two (or more) separate populations by a physical barrier such as a mountain range, valley, sea, or river. These barriers can also be created by humans rather than being natural; for example, roads, fences, or large areas of land that have been developed or changed from the original habitat type. Allopatric speciation always involves a clear physical separation—allopatric literally means 'different place'—but it is important to remember that physical separation does not always lead to allopatric speciation. The three key points are:

1. Whether or not some individuals move between populations (i.e. whether or not there is gene flow). Gene flow might occur when the separated populations are still geographically close, especially for species that are highly mobile (e.g. birds that fly) or that disperse over considerable distances (e.g. wind-

dispersed seeds). The chance of speciation decreases as gene flow increases; you could actually argue that if there is gene flow there is not full reproductive isolation even if the populations themselves are physically isolated.

2. Whether the selection pressures in the two new populations are very similar to one another. When selection pressures are similar, individuals in both populations might respond in similar ways and so evolutionary change is observable at the species level but is consistent between the two populations so speciation does not occur.

3. Whether the physical separation of the two populations is recent/temporary or long-established. If separation is recent and/or temporary, speciation might not occur. In this case, the timescale should be considered relative to the generation time of the species involved: a separation of 100 years is just four generations for an African elephant (*Loxodonta africana*) but 3600 generations for a fruit fly (*Drosophila melanogaster*).

Peripatric speciation

A particular type of allopatric speciation is **peripatric speciation** (literally meaning 'near place' speciation). In this case, a small group of individuals from the original population becomes physically separated from the rest: perhaps following the colonization of a newly available patch of habitat. The small size of the isolated population means that—by chance alone—it may be divergent from the population as a whole from the outset. This is an example of the 'founder effect', which we will discuss in more detail in Chapter 35.

Parapatric speciation

In **parapatric speciation** ('beside the place' speciation) there is no obvious physical separation to prevent gene flow between the two populations. In a large population, however, especially one distributed over a large geographical range, individuals will tend to reproduce with the individuals that are close to them, and adapt to the selection pressures that apply in that particular part of the species' distribution. This may be enough to cause local divergence that, eventually, results in reproductive isolation.

Sympatric speciation

Is it possible for a new species to be formed from a single population living in the same place? Theories of **sympatric speciation** ('same place' speciation) suggest that, under some conditions, disruptive selection within a population (see Section 34.2) can lead to two divergent subpopulations that eventually become separate

species. It should also be noted that different individuals of a species can occur in different locations within the same overall area. For example, if some plants of the same species grow in damp areas and some plants of the same species grow in dry areas of the same site then, although there is no physical barrier between them, there is still physical separation on a microscale that can lead to speciation.

 Check your understanding of the concepts covered in this section by answering the questions in the e-book.

34.4 Bringing together ecology and evolution

It is easy to assume that ecological change is rapid and evolutionary change is slow: this is not necessarily true. Ecology and evolution are highly interlinked and we now know that populations in new or changing habitats, which are facing new resources, competitors, or predators, can experience selection pressures that produce 'rapid' or 'contemporary' evolution over months rather than centuries. The field of **eco-evolutionary dynamics** examines the interactions between evolutionary change of individuals and ecological processes at population, community, and ecosystem levels. We examine this in more detail for an invertebrate species in Scientific Process 34.3.

SCIENTIFIC PROCESS 34.3   Evolution of camouflage in one species drives change in an entire insect community

Background

The stick insect (*Timema cristinae*) lives in the USA, and spends most of its time on a single individual plant, where it feeds and mates. The plant each individual stick insect uses is either a green-bark ceanothus (*Ceanothus spinosus*), which has short rounded leaves, or a chamise (*Adenostoma fasciculatum*), which has long thin leaves. There are two stick insect morphs: a green morph that is well-camouflaged against the green bark of the ceanothus and a striped morph that is well-camouflaged on the chamise, where the central stripe breaks up the outline of the body when the insect is oriented along the woody part of the plant (Figure 1).

Method

Researchers from the University of Sheffield manipulated the level of adaptation by moving insects between plants to create populations when adaptation was strong (mostly well-camouflaged insects), weak (mostly poorly camouflaged insects) or somewhere in the middle (a mix of well and poorly camouflaged insects).

Results and discussion

After a month, as might be expected, there were larger numbers of stick insects on plants where the adaptive match between insects and plants was good. This demonstrates that natural selection can drive change in the *population* in a matter of months. Interestingly, the researchers also found

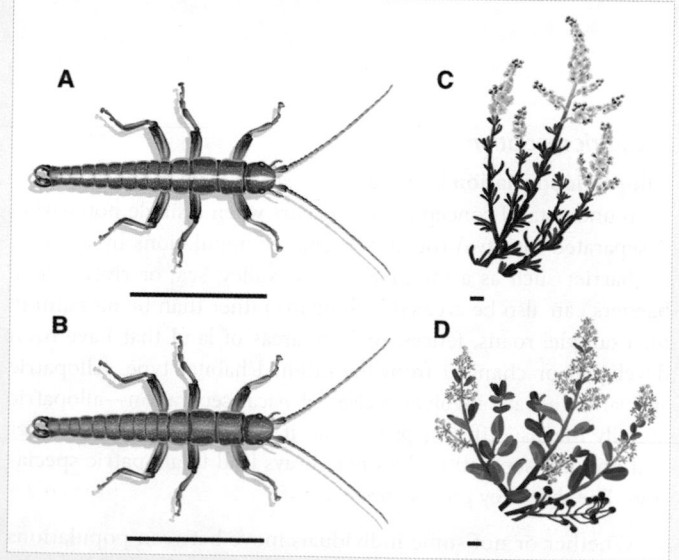

Figure 1 ***Timema cristinae* morphs and their respective host plants.**
A = striped morph, B = green morph, C = chamise (*Adenostoma fasciculatum*), and D = ceanothus (*Ceanothus spinosus*). Black bars depict approximate 1 cm scale.

Source: Reproduced with permission from T.E. Farkas, et al. (2013) Evolution of Camouflage Drives Rapid Ecological Change in an Insect Community. *Current Biology*, 23(19): 1835–1843. Copyright © 2013, Elsevier Ltd. Published by Elsevier Inc.

that the plants containing poorly camouflaged stick insects had smaller and less diverse communities of other arthropods (insects and spiders) and less herbivory damage resulting from feeding insects. This was initially surprising, given that the stick insects compete with many of these other species for food (and so we might expect there to be a greater number of other species when stick insect populations were lower). The answer to this lies in predation pressure. The poorly camouflaged stick insects attracted bird predators that preyed upon all arthropod species, not just the stick insects (if cages were placed around the plants, to keep birds out, there was no decrease in the numbers of other arthropods and no change in herbivory

damage for the plant). This shows that evolution by natural selection of individuals in one species can, when the adaptation becomes maladaptive (i.e. becomes counterproductive—in this case where camouflage is no longer effective), modify the whole *community* and fundamentally change *ecosystem* processes of competition, herbivory, and predation.

Extension question

In this case, the change in the match between the stick insect and the plant species was created artificially by the researchers. Under what circumstances might this occur naturally?

In Section 34.1, we discussed the ecological hierarchy from gene to biosphere. We have used this to structure the rest of the chapters in this module.

■ The gene: concept and outcomes covered in this chapter; processes covered in Chapter 35.

■ The individual: concept and outcomes covered in this chapter; processes covered in Chapter 35.

■ The species: this chapter.

■ The population: a collection of organisms of the same species that inhabit a defined area over a specific period of time (Chapter 36).

■ The community: a collection of individuals belonging to multiple species populations that co-occur in a specific geographical area (Chapter 37).

■ The ecosystem: a community of species that interact with one another and their environment to produce a complex system (Chapter 38).

We end the module by considering challenges to ecosystems in (Chapter 39) and possible solutions to these challenges through protection, management, conservation, and restoration in (Chapter 40).

 Check your understanding of the concepts covered in this section by answering the questions in the e-book.

SUMMARY OF KEY CONCEPTS

■ The discipline of **ecology** is a scientific approach to the interactions between organisms (within and between species), and the interactions between organisms and their physical environment.

■ Living systems can be divided into levels of **ecological complexity** from the gene to the biosphere.

■ The two central principles of evolution by natural selection are: (1) that individuals that survive and reproduce will pass on to the next generation the key characteristics that increase the chances of survival and reproduction; and (2) that over time these characteristics will become more common in the population.

■ There are various forms of selection: **natural selection** can be stabilizing, directional, or disruptive; **sexual selection** that acts on traits that affect reproduction directly; and **artificial selection** that involves human intervention.

■ If selection is consistent in type and direction, individuals will respond to environmental change through a process of **genotypic adaptation**: changes in phenotypes underpinned by changes in genotypes.

■ Most organisms are also able to react to the pressures associated with a particular environment through a **phenotypic response** that does not cause genetic change.

■ Evolutionary changes do not occur 'for the good of the species'; they occur because they confer benefit to the individual involved.

■ **Evolutionary fitness** is the genetic contribution of one individual to the next generation relative to the other members of the population and thus the proportional extent to which that individual's genotype is represented in subsequent generations.

■ An individual can increase its evolutionary fitness by increasing its own reproductive success (**direct fitness**) or by increasing the reproductive success of relatives (**indirect fitness**).

■ Indirect fitness can lead to a special type of natural selection—**kin selection**—that acts on traits that improve the overall reproductive success of related members of a group.

■ Species can be defined reproductively (which individuals breed), morphologically (how individuals look—the phenotype), or phylogenetically (the genotype).

■ New species arise through **speciation**: this is split into allopatric and peripatric (where populations are separated by a physical barrier) and parapatric and sympatric (where there is no physical barrier).

■ Evolutionary changes that affect individuals can have impacts not only at the level of that species, or a particular population of that species, but also for the wider ecological community and ecosystem, for example by altering patterns of competition, herbivory, and predation.

 Use the flashcards in the e-book to test your recall of key terms introduced in this chapter.

QUESTIONS

 Looking for answers? Once you've answered these questions, follow the link in the e-book to the answer guidance and check your work.

Concepts and definitions

1. What is ecology?

2. List the seven levels of ecological complexity in order from smallest to largest.

3. What is the difference between genotype and phenotype?

4. Explain the difference between stabilizing, directional, and disruptive selection.

5. How does a phenotypic response differ from a genotypic adaptation?

Apply the concepts

6. Why does artificial selection typically create evolutionary change more quickly than natural selection?

7. Why would there be a stronger selection for kin selection when individuals share 95% of their genotype rather than, say, 85% of their genotype?

8. Explain why simple traits, controlled by one gene, tend to result in discontinuous variation (e.g. trait type 'a' or trait type 'z'), whereas complex traits, controlled by multiple genes, tend to result in continuous variation (a range of trait types from 'a' to 'z').

9. Why is it more important to consider the number of grand offspring an individual has as a measure of its evolutionary fitness rather than simply the number of offspring?

10. You are interested in why individuals of a species in an island population and individuals of the same species in a mainland population appear to be the same—in other words, why allopatric speciation has not taken place. You later find out that the island has repeatedly been temporarily joined to the mainland by a land bridge (an emergent strip of land) when the sea level has dropped. Why might this explain the lack of allopatric speciation in this case?

Beyond the concepts

11. Find three examples of species where sexual selection has acted upon the ability to attract/select/retain mates (e.g. sexual ornamentation) and three examples of species where sexual selection has acted upon traits associated with successful fertilization of an egg by sperm (e.g. sperm swimming speed).

12. Kin selection is usually considered with reference to social insects such as ants (which we covered in the chapter), bees, wasps, and termites. Find an example of a species outside of these taxa where there is kin selection—individuals helping close family members to maximize the reproductive success of the family member rather than maximizing their own individual reproductive success—and explain why it occurs in that species.

13. You are in a debate on whether or not altruism exists. Come up with three key arguments from a biological perspective as to why altruism does not exist, using examples to support your arguments.

14. When a group of species that have been classified using the biological or morphological species concepts are analysed genetically using the phylogenetic species concept, we often need to make changes to species classifications and scientific names. In some cases, we even need to create new genera or families because individuals that we thought were closely related turn out to be much less similar genetically than expected. Would you expect differences in taxonomic classification based on whether the biological/morphological or phylogenetic species concept has been used? Which version of the taxonomy do you think is more accurate?

15. Find a real-world example or case study for each of the following: allopatric speciation, peripatric speciation, parapatric speciation, and sympatric speciation.

FURTHER READING

Courchamp, F., & Bradshaw, C. J. (2018). 100 articles every ecologist should read. *Nat. Ecol. Evol.* **2**: 395.
Key papers in ecology present a great reading list for all ecology students.

Darwin, C.R. (1859) *On the Origin of Species by Means of Natural Selection, or the Preservation of Favoured Races in the Struggle for Life.* London: John Murray.
Natural selection as outlined by Darwin.

Hosken, D. J. & House, C. M. (2011) Sexual selection. *Curr. Biol.* **21**: 62–65.
Very clear overview of sexual selection.

Quammen, D. (1996) *The Song of the Dodo: Island Biogeography in an Age of Extinctions.* New York: Scribner.
A very readable account of the development of evolutionary theory.

Dobzhansky, T. (1973) Nothing in biology makes sense, except in the light of evolution. *Am. Biol. Teach.* **35**: 125–129.
The centrality of evolution to ecology biology.

Mayhew, P. (2006) *Discovering Evolutionary Ecology.* Oxford: Oxford University Press.
How to bring together ecology and evolution.

Genes

Evolutionary Change in Alleles, Genotypes, and Phenotypes

Chapter contents

Introduction   1222

35.1 Describing variation: phenotypes, genotypes, and alleles   1222

35.2 Analysing variation: phenotypes, genotypes, and alleles   1223

35.3 Modelling variation: Hardy–Weinberg modelling   1226

35.4 Genetic change and evolution   1231

 Watch the key concepts video in the e-book to prepare yourself for studying this chapter.

Introduction

Charles Darwin died before the structure and function of DNA was understood. Darwin saw the results of evolution, and devised the theory of natural selection to explain those results, but knew of no plausible mechanism for the origin or transmission of heritable variation (i.e. traits that could be inherited) between parents, offspring, and grand offspring on which his theory depended.

We now know that natural selection acts on genotype (the genetic make-up of an individual) and that evolution is, in essence, a change in genetic make-up of a population over time resulting from specific selection pressures. This process of evolutionary genetic change always acts on individuals. Sometimes, the results of evolution can be evident at species level, including the emergence of new species through speciation, in a process we call **macroevolution**. However, evolution can also involve the accumulation of changes in the frequencies of genes and alleles *within* a population or species in a process we call **microevolution**. As such, evolution is not just the development of new species from existing ones; it is also the smaller changes that occur within the individuals that make up a population or a species from generation to generation.

▶ **We learn more about speciation in Chapter 34.**

Examples of macroevolution and microevolution are shown in Figure 35.1: the variation in shape of the bill in different species of Darwin's finch (Thraupidae) due to specialization of different food sources is an example of macroevolution, while the variation in colour of the peppered moth (*Biston betularia*), which enables the darker moth to be better camouflaged against predators, is an example of microevolution.

35.1 Describing variation: phenotypes, genotypes, and alleles

As we saw in Chapter 34, individuals in a population differ from each other. To understand this variation, we need to be able to observe and quantify it. We call the traits (i.e. the characteristics) of an organism that we can see and measure its **phenotype**, and the differences in a particular trait between individuals are referred to as the **phenotypic variation** for that trait. Individuals in a population also differ in their **genotype**—their genetic make-up—and we call differences between individuals **genotypic variation**.

A population in the genetic sense is not just a group of individuals, but a set of genes, and each of those genes may be available in one or more variants: we call these **alleles**. The act of reproduction passes genes from generation to generation, while **sexual reproduction** also allows alleles to be brought together in new combinations. Before we go on to look at ways of analysing the genotypic and phenotypic variation of a population, let's briefly consider our key terminology and concepts in Table 35.1.

When a diploid organism has a copy of the *same* allele on the chromosome inherited from its mother and the chromosome inherited from its father, it is homozygous for that allele regardless of whether these are alleles are both dominant or both recessive. If an organism has *different* alleles on the chromosome inherited from its mother and the chromosome inherited from its father, it is heterozygous for that allele. Look at Figure 35.2 to help you understand this concept: in all diploid organisms, phenotype will typically be controlled by a **dominant allele** (represented by the white square in Figure 35.2) because a phenotype Mcontrolled by a **recessive allele** (represented by the white circle in Figure 35.2) will develop only if the recessive allele is present on both copies of

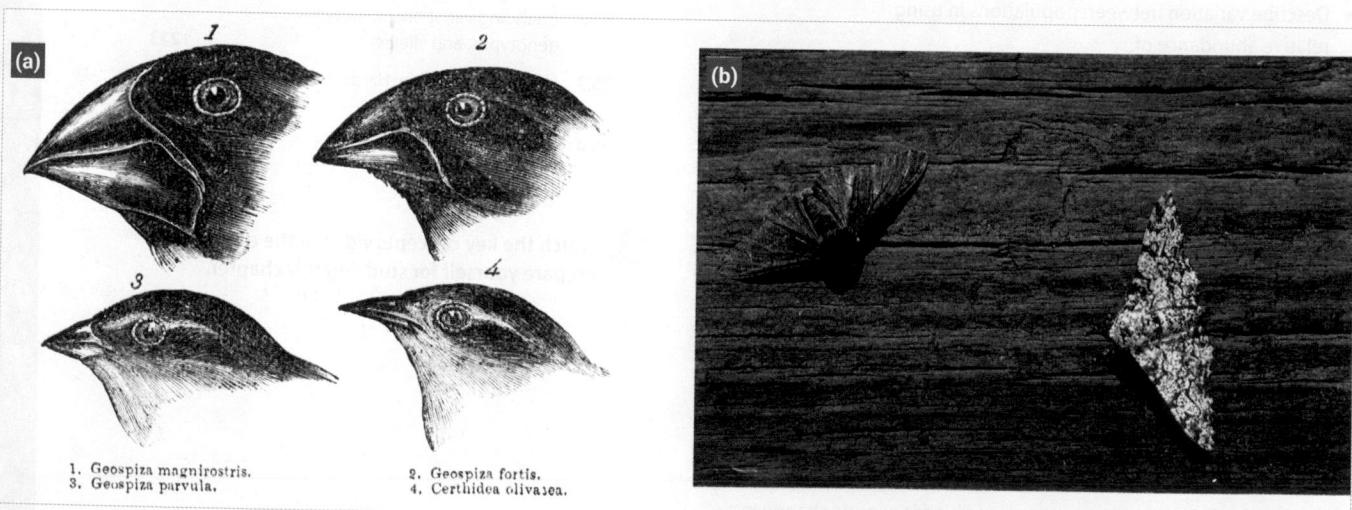

(a)

1. Geospiza magnirostris.
2. Geospiza fortis.
3. Geospiza parvula.
4. Certhidea olivacea.

(b)

Figure 35.1 Classic examples of evolution. (a) Macroevolution is exemplified by Darwin's finches (Thraupidae), a group of birds inhabiting the Galapagos islands that differ in bill morphology following speciation from one common ancestor; and (b) microevolution is exemplified by the peppered moth (*Biston betularia*) when the frequency of naturally occurring dark morphs of the species increased during the Industrial Revolution because they had superior camouflage against tree trunks that were darkened by soot and thus a lower predation risk.

Source: (a) C. Darwin (1845) Journal of researches into the geology and natural history of the various countries visited by H.M.S. *Beagle* round the world; (b) Bill Coster IN/Alamy Stock Photo.

Table 35.1 Key terminology and concepts in evolution

Term	Definition
Genome	The total genetic material of an organism
Chromosome	A DNA molecule that carries all (prokaryotes) or part (eukaryotes) of the genome
Gene	A specific part of DNA found at a specific position (locus—plural loci) on a chromosome that codes for an RNA molecule or protein
Allele	Distinctive variants of a specific gene
Diploid	Organisms that contain cells with two complete sets of chromosomes—and two sets of genes—one set from each parent, that were bought together by the fusion of gametes (eggs and sperm) at fertilization during sexual reproduction
Haploid	Organisms that contain cells with one set of chromosomes—and one set of genes—from one parent as a result of asexual reproduction
Dominant allele	An allele that produces the same distinctive phenotype regardless of whether it is paired with the same allele in a diploid organism (**homozygous**) or a different allele (**heterozygous**)
Recessive allele	An allele that produces its distinctive phenotype only when it is paired with the same allele in a diploid organism

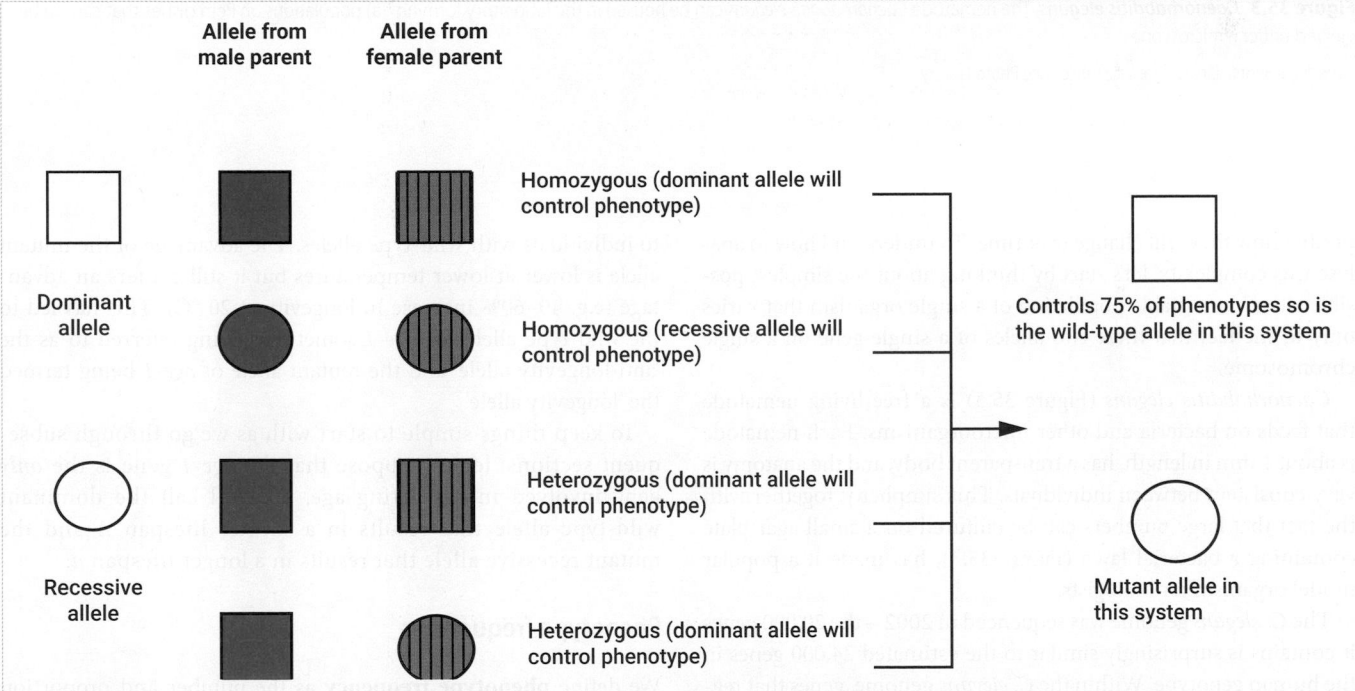

Figure 35.2 Allele inheritance in diploid organisms when there are two alleles, one dominant and one recessive, of a single gene.

the relevant chromosome (the single row in Figure 35.2 with the circle)—dominant homozygous and heterozygous individuals will all have the dominant allele phenotype (the three rows in Figure 35.2 with squares). The allele of a gene that encodes the phenotype that is most common in a particular natural population is known as the **wild-type allele**, while any alternative allele is referred to as a **mutant allele**.

 Check your understanding of the concepts covered in this section by answering the questions in the e-book.

35.2 Analysing variation: phenotypes, genotypes, and alleles

The genetic information in a population (represented by all the genes and alleles that the population contains) is known as the **gene pool**. Another way of thinking about a gene pool is as the collection of all the alleles of all the genes in the population. The science of **population genetics** focuses on analysing allele frequencies within a given population and trying to understand and

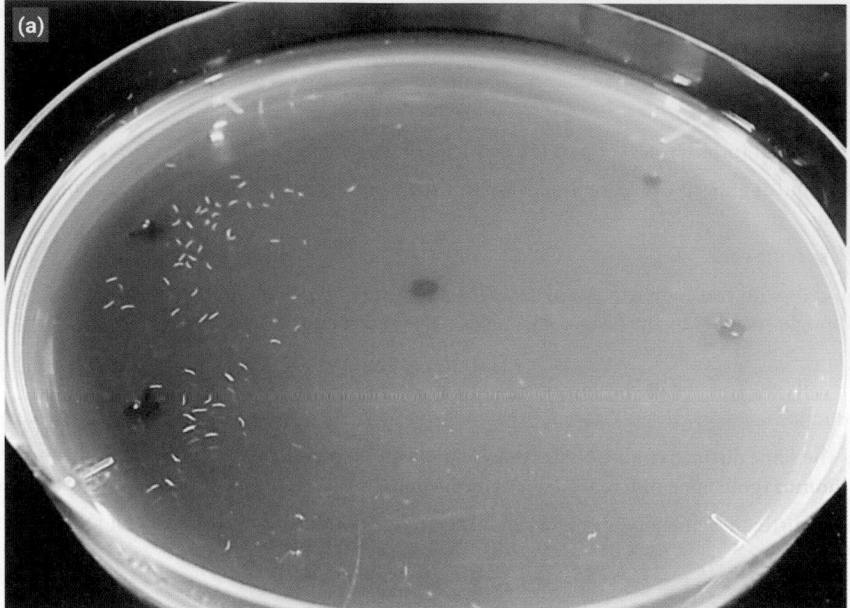

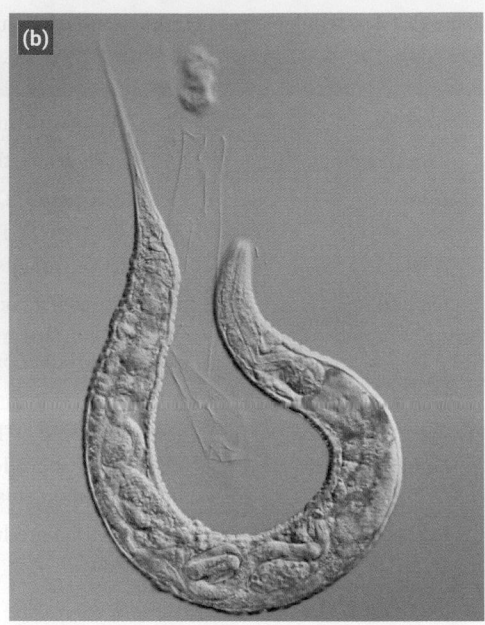

Figure 35.3 *Caenorhabditis elegans.* The nematode *Caenorhabditis elegans* can be housed in the laboratory forming: (a) populations on Petri dishes that can (b) be viewed under a microscope.

Source: (a) Hirotsu Bio Science Inc.; (b) Science Photo Library.

predict how they will change over time. To understand how to ana-lyse this complexity, let's start by thinking about the simplest pos-sible situation: a stable population of a single organism that varies only in the fact that there two alleles of a single gene on a single chromosome.

Caenorhabditis elegans (Figure 35.3) is a free-living nematode that feeds on bacteria and other microorganisms. Each nematode is about 1 mm in length, has a transparent body, and the anatomy is very consistent between individuals. This simplicity, together with the fact that large numbers can be cultured on a small agar plate containing a bacterial lawn (Figure 35.3), has made it a popular model organism for biologists.

The *C. elegans* genome was sequenced in 2002—the 20,000 genes it contains is surprisingly similar to the estimated 24,000 genes in the human genotype. Within the *C. elegans* genome, genes that reg-ulate ageing, **gerontogenes**, were discovered. The first gerontogene to be identified for nematodes was named *age-1* and is involved in DNA repair. Nematode lifespans are controlled by gerontogenes (including *age-1*), environmental temperature, and the interactions between gerontogenes and temperature. Individuals with the nor-mal—wild-type—allele for the gene *age-1* have a mean lifespan of 15 days at 25 °C with the oldest individuals living until 22 days. However, a mutation in *age-1* has resulted in a recessive allele—*age-1(hx546)*. Unusually, this mutant allele is associated with *greater* longevity: individuals that are homozygous for the allele (i.e. have two copies of *age-1(hx546)*) have a mean lifespan of 23 days and a maximum lifespan of 46 days at 25 °C. This represents an increase of 53% in mean longevity and 109% in maximum longevity relative

to individuals with wild-type alleles. The advantage of the mutant allele is lower at lower temperatures but it still confers an advan-tage (e.g. 40–60% increase in longevity at 20 °C). This has led to the wild-type allele for *age-1* sometimes being referred to as the 'anti-longevity allele' and the mutant allele of *age-1* being termed the 'longevity allele'.

To keep things simple to start with as we go through subse-quent sections, let us suppose that the *age-1* gene is the *only* gene involved in regulating age. We will call the dominant wild-type allele that results in a shorter lifespan *A* and the mutant recessive allele that results in a longer lifespan *a*.

Phenotype frequencies

We define **phenotype frequency** as the number and proportion of individuals in a population that have a particular phenotype. Consider a laboratory population of 15,000 nematodes housed on 30 agar plates, with each plate containing about 500 individu-als. There are two phenotypes: nematodes with a longer lifespan (homozygous *aa* individuals affected by the recessive mutation) and nematodes with a normal lifespan (all of the others, who will be either homozygous *AA* for the wild-type allele or heterozygous *Aa*). Because *A* is the dominant allele, there is no observable dif-ference in the lifespans of the homozygous *AA* individuals or the heterozygous *Aa* individuals. It wouldn't be practical to measure the lifespans of all of individuals in our *C. elegans* population, so we select 1% of individuals to study at random: we assume that this sample of 150 is representative of the overall population.

Table 35.2 The number and relative frequency of nematodes with each phenotype in a sample of 150 individuals

Phenotype	Number	Relative frequency	Calculation
Caenorhabditis elegans with a normal lifespan	132	0.88	132/150 = 0.88
Caenorhabditis elegans with a longer lifespan	18	0.12	18/150 = 0.12
Total sample	150	1.00	

Table 35.3 The number and relative frequency of nematodes with each genotype in a sample of 150 individuals

Phenotypes				Genotypes			
Name	Number	Relative frequency	Calculation	Name	Number	Relative frequency	Calculation
Normal	132	0.88	132/150 = 0.88	*AA*	84	0.56	84/150 = 0.56
				Aa	48	0.32	48/150 = 0.32
Long-lived	18	0.12	18/150 = 0.12	*aa*	18	0.12	18/150 = 0.12
Total	150	1.00		Total	150	1.00	

To measure the phenotypic variance in our population, we need to calculate the relative frequency of each phenotype, as shown in Table 35.2: the otal of the relative frequencies will always be 1 because all individuals must have one of the phenotypes.

▶ If you need some additional support, read Quantitative Toolkit 4.

PAUSE AND THINK

Let's suppose that in another sample of the same laboratory population of nematodes 30 out of 150 individuals are long-lived, while the remaining 120 are short-lived. What are the relative phenotypic frequencies in this sample and would you be concerned that this sample differs in numbers (and thus frequencies) of the phenotypes compared to the sample summarized n Table 35.2?

Answer: The relative frequency of the long-lived phenotype in this sample would be 0.20 (30/150); the relative frequency of the short-lived phenotype would be 0.80 (120/150). There is a reasonable difference in the relative phenotype frequencies in this sample when compared to the original sample shown in Table 35.2. This might indicate that sampling of just 1% of the population is not sufficient for the estimate to be accurate. This illustrates the importance of sample size in scientific work.

Genotype frequencies

Now that we have quantified the phenotypes of the nematode population, we can start to investigate the genotypes that result in these phenotypes. It is not possible for us to distinguish between the wild-type *AA* and heterogeneous *Aa* genotypes by simply looking at the phenotypes of the nematodes (in this case, by quantifying lifespan) because *AA* and *Aa* individuals have similar longevity. It is, however, possible for us to use a range of genetic techniques, including DNA sequencing, to distinguish these genotypes.

As with phenotype frequencies, we can calculate the **genotype frequency** by dividing the number of each genotype by the total number of individuals in the sample. We can now add that information on this nematode population. Table 35.3 shows the same phenotype information as in Table 35.2 but with the additional genotype information. As with the phenotype frequencies, the genotype frequencies add up to 1 as all the nematodes in our sample must have one of the three possible genotypes: *AA*, *Aa*, or *aa*.

▶ If you need some additional support, read Quantitative Toolkit 4.

Allele frequencies

Our genotype data tell us the relative frequencies of the three possible genotypes *AA*, *Aa*, and *aa* in our small population of nematodes. This allows us to calculate the number of the two alleles—*A* and *a*—in our population on the basis that:

- Each *AA* genotype contains two *A* alleles.
- Each *Aa* genotype contains one *A* allele and one *a* allele.
- Each *aa* genotype contains two *a* alleles.

In our sample, therefore, the total number of *A* alleles is double the number of *AA* genotypes plus the number of *Aa* genotypes [*A* alleles = (2 × 84) + 48 = 216], while the total number of *a* alleles is two times the number of *aa* genotypes plus the number of *Aa* genotypes [*a* alleles = (2 × 18) + 48 = 84].

The total number of alleles is 2× the number of individuals in the population, as each individual has two copies of this gene and the gene exists in only two alleles: *A* and *a*. As with phenotype and genotype, however, a breakdown of relative **allele frequency** in the population is a more useful measure than the total number of each allele represented in that population. To calculate relative allele frequency, whereby the frequency of each allele is expressed as a proportion of all the alleles in the population, we divide the number we have for each allele by the number of alleles in the population as a whole, as shown in Table 35.4.

Table 35.4 The number and relative frequency of alleles *A* and *a* in a sample of 150 nematodes

Alleles	Number	Relative frequency	Calculation
A	216	0.72	216/300
a	84	0.28	84/300
Total	300	1.00	

PAUSE AND THINK

Let's go back to our second nematode sample, where we had 30 individuals with the short-lived phenotype and 120 individuals with the long-lived phenotype. We quantify genotype and find that there are 30 individuals with *aa* genotype (as we would expect because there were 30 individuals with the long-lived phenotype). Of the remaining 120 individuals, 75 were *AA* and 45 were *Aa*. Use these values to calculate genotype frequencies and allele frequencies in this second nematode sample.

Answer:

Genotypes:

Genotypes	Number	Relative frequency	Calculation
AA	75	0.5	75/150
Aa	45	0.3	45/150
aa	30	0.2	30/150
Total:	150	1.0	

Alleles: Remember the number of alleles (300) is double the number of individuals (150). The number of *A* alleles is two times the number of *AA* genotypes plus the number of *Aa* genotypes: $(2 \times 75) + 45 = 195$ alleles. The number of *a* alleles is two times the number of *aa* genotypes plus the number of *Aa* genotypes: $(2 \times 30) + 45 = 105$ alleles. The relative frequency of the *A* allele is 0.65 (195/300; the relative frequency of the *a* allele is 0.35 (105/300).

Check your understanding of the concepts covered in this section by answering the questions in the e-book.

35.3 Modelling variation: Hardy–Weinberg modelling

In 1908 a British mathematician, G. H. Hardy, and a German doctor, Wilhelm Weinberg, independently described a mathematical way of modelling allele, genotype, and phenotype frequencies. This model allows us to link the genotype frequencies and allele frequencies for a single gene in a population that is genetically stable (i.e. a population that is not evolving or, more specifically, not evolving with respect to the specific trait being studied).

Remember that in a diploid organism, the genotype of an individual is determined by the fusion of an egg containing one allele with a sperm cell containing one allele, such that the individual has two sets of each gene (two alleles), which might be the same

(homozygous) or different (heterozygous). As this is a model, we'll replace the population-level frequencies of allele *A* and allele *a* with the letters *p* and *q*, respectively (note that *p* in this context has nothing to do with statistical probability).

If the proportion of *A* alleles in the population is *p*, then the probability that any egg, selected at random, will contain allele *A* is *p* as well. The same argument goes for sperm. If the proportion of *A* alleles in the population is *p*, then the probability that any sperm, selected at random, contains allele *A* is also *p*. If the probability of an egg containing *A* is *p* and the probability of a sperm containing *A* is *p*, then the probability of an egg containing allele *A* fusing with a sperm containing allele *A* to form a homozygous *AA* zygote is $p \times p = p^2$. Similarly, for a homozygous *aa* zygote, the probability of an egg containing the *a* allele is *q* and the probability of a sperm containing an *a* allele is *q*, so the probability of a homozygous *aa* zygote is $q \times q = q^2$.

For heterozygotes (*Aa*), the same rules apply here but the situation is more complex because a heterozygous zygote can occur from the fusing of an *A* egg with an *a* sperm or an *a* egg with an *A* sperm. The probability of an *A* egg fusing with an *a* sperm is $p \times q$. The probability of an *a* egg fusing with an *A* sperm is $q \times p$. The probability of producing an *Aa* heterozygote is $(p \times q) + (q \times p) = 2pq$ because $p \times q$ and $q \times p$ are identical in a mathematical context (just like 3×2 and 2×3 both equal 6).

Punnett squares

The different genotypes that can result in the next generation as a result of two individuals reproducing sexually are often shown diagrammatically in a Punnett square (named after Reginald Punnett, 1875–1967). The basic Punnett square system is shown in Figure 35.4. When there are only three genotypes (as in our nematode population, for example)—*AA*, *Aa*, and *aa*—and they occur in the proportions p^2, $2pq$, and q^2, respectively. All of the individuals in

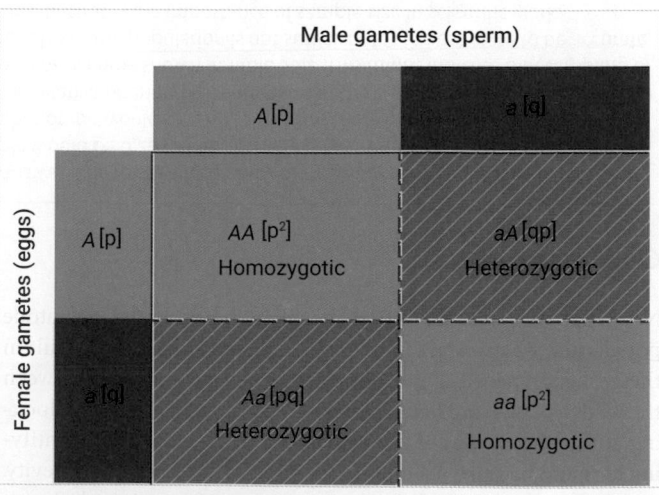

Figure 35.4 A Punnett square showing the possible combinations of alleles *A* (= *p*) and *a* (= *q*). The associated probabilities are in square brackets. The homozygotic individuals deriving from *A* alleles are shown in green; the homozygotic individuals deriving from *a* alleles are shown in red; the heterozygotic individuals are shown in orange.

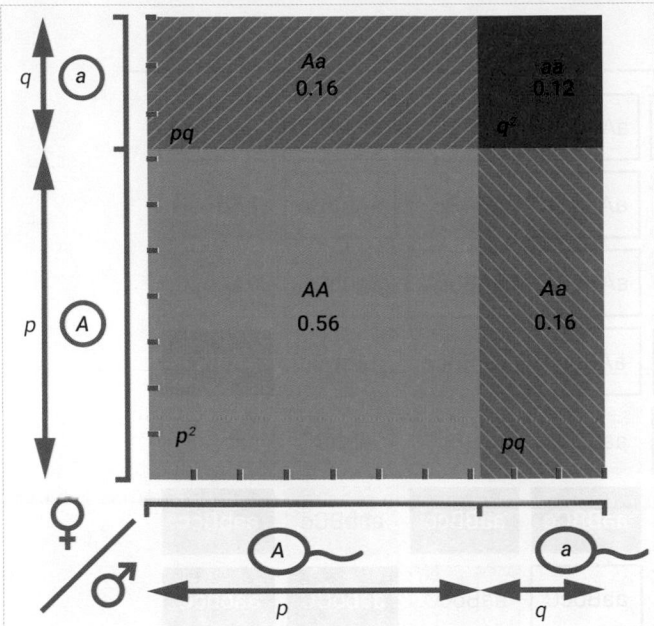

Figure 35.5 Allele frequencies. Diagrammatic representation of allele frequencies of the nematode data from Table 35.4 shown using the *p* and *q* modelling notation from Figure 35.4 where *A* = *p* and *a* = *q*. Note that for both male (*x*-axis) and female (*y*-axis) *p* is 0.72 and *q* is 0.28 as calculated in Table 35.4. The vertical line projected from the meeting point of *p* and *q* on the *x*-axis and the horizontal line projected from the meeting point of *p* and *q* on the *y*-axis give us four squares for the genotype frequencies, which are exactly what we calculated previously in Table 35.3. (Note that for *Aa* we have two squares here, one for *pq* and one for *qp*, which together sum to 0.32, the same value as for *Aa* in Table 35.3.)

any population must have one of those three genotypes, in the relative proportions indicated by the allele frequencies above, so:

$$p^2 + 2pq + q^2 = 1$$

When examining a Punnett square, note that we have added another assumption here: we have assumed that alleles are passed from parents into eggs and sperm in a random manner, and that the fusion of eggs and sperm to form zygotes is also random. This assumption is necessary for the model, and our laboratory-based *C. elegans* population is likely to meet this assumption, but, as we'll see in Section 35.4, this assumption does not always hold true.

If we return to the nematode example, we can use the data from Tables 35.3 and 35.4 and the Punnett square concept (Figure 35.4) to draw up a diagrammatic representation of both the allele frequencies and genotype frequencies in our laboratory population, whereby the length of the sides relate to the allele relative frequency in males (*x*-axis) and in females (*y*-axis) so that the size of each sub-square relates to the relative frequency of the genotypes in the population (Figure 35.5).

Punnett squares can be much more complex than shown in Figures 35.4 and 35.5 because often there are multiple genes involved in a creating a specific phenotype. For example, in nematodes there are actually several genes involved in longevity besides *age-1*, including *daf-2*, *daf-7*, *daf-16*, *rle-1*, and *math-33*. Let's assume the first three of these genes are the most important, that each only has two alleles:

- *age-1* = *AA* (short-lived), *Aa* (short-lived), *aa* (long-lived)—as per previous examples.

- *daf-2* = *BB* (short-lived), *Bb* (short-lived), *bb* (long-lived).
- *daf-7* = *CC* (short-lived), *Cc* (short-lived), *cc* (long-lived).

There are eight allele combinations that are possible from each of the two parents, giving 64 possible genotypes for the resultant offspring. A nematode will be long-lived to some extent if it is homozygous for the recessive allele of *any* of these three genes, but there are three possibilities:

- Recessive for all three genes = ultra long-lived.
- Recessive for two of the three genes = very long-lived.
- Recessive for one of the three genes = long-lived.

The Punnett square for this scenario is shown in Figure 35.6, which shows there is one possible combination resulting in offspring with an ultra long-lived genotype (the black box in the bottom right corner of Figure 35.6), 10 combinations resulting in offspring with a very long-lived genotype (the dark-pink boxes in Figure 35.6), 25 combinations resulting in offspring with long-lived genotype (the light-pink boxes in Figure 35.6), and 28 combinations resulting in offspring with a short-lived genotype (the white boxes in Figure 35.6).

Hardy–Weinberg assumptions

The Hardy–Weinberg model can only be applied to organisms that are diploid (not haploid) and that reproduce sexually. Moreover, it can only be used in organisms where heterozygotes have a phenotype that reflects the dominant allele. Heterozygotes expressing a phenotype that reflects both alleles (**co-dominance**) or a blending of alleles (**incomplete dominance**) (Figure 35.7) cannot be modelled in this way.

Even for organisms that are suitable for Hardy–Weinberg modelling because they are haploid, reproduce sexually, and have phenotypes reflective of normal allele dominance, there are five main assumptions that need to be met:

1. The alleles being modelled must not be changing due to mutation (**no mutation**).

2. The alleles being modelled must not be changing due to evolution through natural selection (**no evolution**).

3. The population must be large enough to minimize the possibility of allele frequencies varying from generation to generation as the result of chance fluctuations (**no genetic drift**). Moreover, allele frequencies must be equal for male and female and, in practical terms, the population must also be large enough to prevent sampling errors when we analyse the allele frequencies.

4. Mating must be random throughout the population with regard to the specific trait being modelled. This means that any individual with a specific trait must be equally likely to mate with any other individual in the population regardless of that trait (**random mating**).

5. The population must be closed with no movement of individuals into the population (immigration) and no movement of individuals out of the population (emigration). This means that genes neither enter nor leave the population (**no gene flow**).

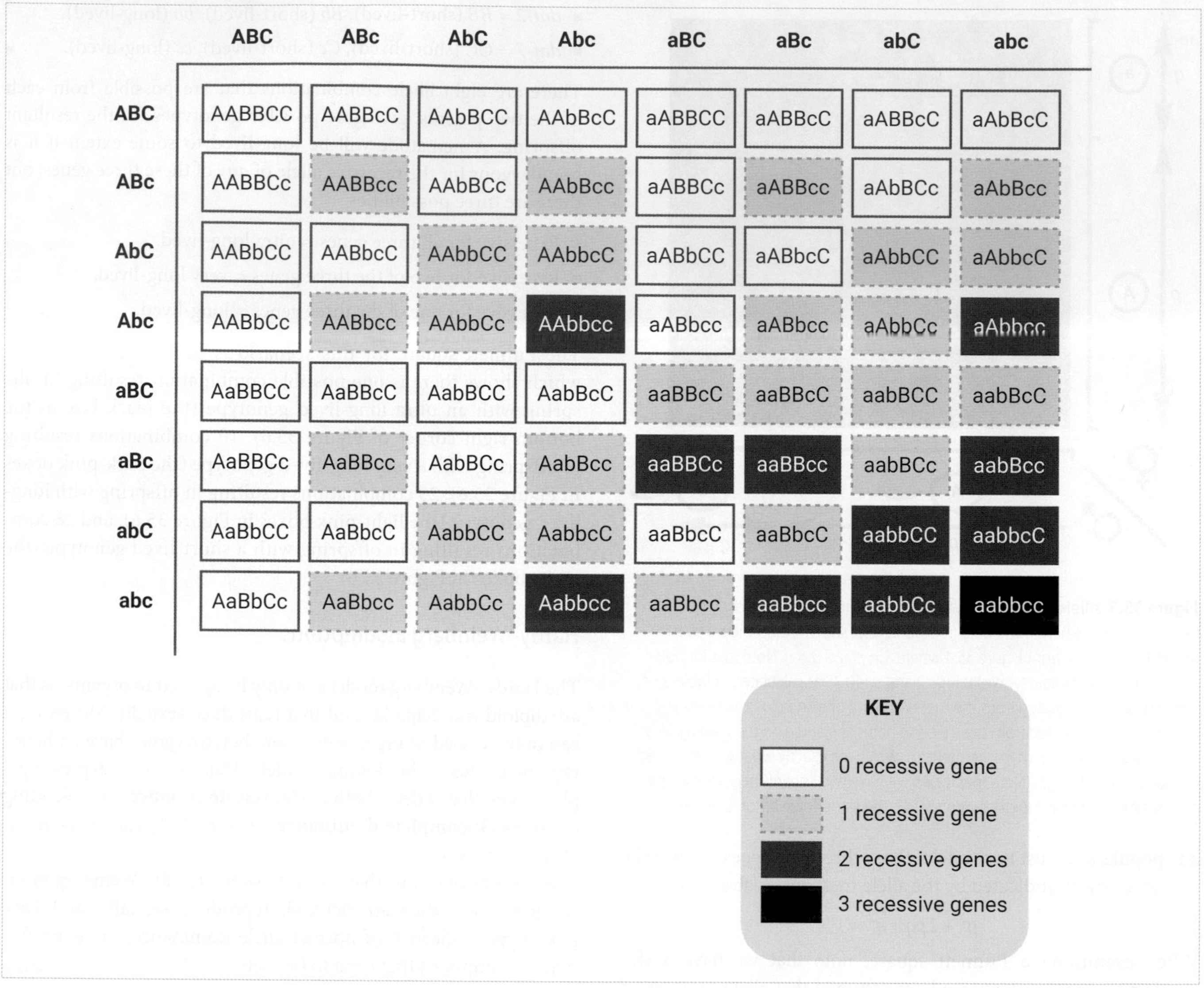

	ABC	ABc	AbC	Abc	aBC	aBc	abC	abc
ABC	AABBCC	AABBCc	AAbBCC	AAbBcC	aABBCC	aABBcC	aABBcC	aAbBcC
ABc	AABBCc	AABBcc	AAbBCc	AAbBcc	aABBCc	aABBcc	aAbBCc	aAbBcc
AbC	AABbCC	AABbcC	AAbbCC	AAbbcC	aABbCC	aABbcC	aAbbCC	aAbbcC
Abc	AABbCc	AABbcc	AAbbCc	AAbbcc	aABbcc	aABbcc	aAbbCc	aAbbcc
aBC	AaBBCC	AaBBcC	AabBCC	AabBcC	aaBBcC	aaBBcC	aabBCC	aabBcC
aBc	AaBBCc	AaBBcc	AabBCc	AabBcc	aaBBCc	aaBBcc	aabBCc	aabBcc
abC	AaBbCC	AaBbcC	AabbCC	AabbcC	aaBbcC	aaBbcC	aabbCC	aabbcC
abc	AaBbCc	AaBbcc	AabbCc	Aabbcc	aaBbcc	aaBbcc	aabbCc	aabbcc

KEY

- ☐ 0 recessive gene
- ☐ 1 recessive gene
- ☐ 2 recessive genes
- ■ 3 recessive genes

Figure 35.6 A Punnett square showing the possible combinations of alleles *A* and *a*, *B* and *b* and *C* and *c*. The shading reflects genotypes homozygous for recessive (longevity) alleles; see Key for how many genes are recessive within the overall genotype.

Under the strictest definition of Hardy–Weinberg, the genotypes being modelled should also not be responsible for any differences in survival or reproductive success. However, in reality, this assumption is very restrictive as most genotypes influence survival and/or reproductive success, often in ways that are unclear. Indeed, the nematode example we have been using throughout this chapter immediately violates this assumption as the genes we have been considering are specifically linked to longevity.

Populations that meet all of the Hardy–Weinberg assumptions are said to be in a **Hardy–Weinberg equilibrium**. For populations in Hardy–Weinberg equilibrium, we can link allele frequencies to genotype frequencies in a population because it is genetically stable so that allele frequencies, and genotype frequencies, do not change from generation to generation. However, the assumptions of Hardy–Weinberg are so restrictive that very few natural

populations meet them, as we discover in Real World View 35.1 for Tay–Sachs disease.

PAUSE AND THINK

Scientists have discovered a new genetic disease. The disease is caused by a recessive allele (g). The incidence of the syndrome in a particular population is 4% or four people in every 100. Predict the frequencies of the wild-type allele G and the mutant allele g in this population at this time.

Answer: The frequency of affected individuals (q^2 – genotype gg) = 4/100 = 0.04 so the frequency of the recessive allele (q) = 0.2. As p + q = 1, the frequency of the dominant allele (G) = 1 – 0.2 = 0.8.

 Check your understanding of the concepts covered in this section by answering the questions in the e-book.

Figure 35.7 Dominance, co-dominance, and incomplete dominance. Normally phenotype is determined by the dominant allele wherever that appears (e.g. (a) red petals for plants homozygous for dominant allele and plants that are heterozygous) with recessive allele only affecting genotype when the individual is homozygous for recessive allele (e.g. (b) white petals). Co-dominance and incomplete dominance are forms of intermediate inheritance in which one allele for a specific trait is not completely expressed over its paired allele. This results in a phenotype in which the expressed physical trait is a combination of the phenotypes of both alleles. In co-dominance the two traits are visible separately (e.g. (c) red petals and white petals in this example) while incomplete dominance blends the traits (e.g. (d) pink petals in this example).

Source: Shutterstock.

REAL WORLD VIEW 35.1   **Tay–Sachs disease**

Tay–Sachs disease is a recessive genetic disorder of humans. The most common form, infantile Tay–Sachs disease, causes progressive nerve damage and usually results in death before 4 years of age. Tay–Sachs is caused by a genetic mutation in the *HEXA* gene on chromosome 15. As a recessive disorder, only individuals with two mutant Tay–Sachs alleles (i.e. recessive alleles inherited from both mother and father) will have the disease: there is a 1 in 4 (25%) chance of this occurring if both parents have one copy of the mutant allele. Individuals that are heterozygous, with a single Tay–Sachs allele, do not exhibit symptoms but are **carriers** of the disease and would pass the disease to any children they subsequently have if they mated with another carrier (otherwise, their children would have a 50% chance of themselves being carriers).

If we code the dominant wild-type allele as *T* and the recessive mutant Tay–Sachs allele linked to the disease phenotype as *t*, we can show this **autosomal recessive** system diagrammatically, where asymptomatic carriers of the disease with one mutant allele and one normal wild-type allele (heterozygotic individuals) are coded *Tt*, unaffected homozygous individuals are coded *TT*, and affected homozygotic individuals are coded *tt*, as shown in Figure 1.

Tay–Sachs disease does not occur uniformly. Although it can occur in any country and any ethnic group, it is particularly associated with four genetic groups of people that are geographically focused in specific locations (as shown in Figure 2):

1. Ashkenazi Jews: a genetic group originally from Eastern Europe in which the disease incidence is about 1 in 3500 live births.

2. Cajans: an ethnic group mainly living in the US state of Louisiana. In this group, the *HEXA* mutation has been traced back to a single founder couple who initially lived in France in the 18th century.

3. Amish populations in the US state of Pennsylvania.

4. French Canadians: an ethic group around Québec whose ancestry is 16th-century French settlers.

The most common *HEXA* mutation that causes Tay–Sachs disease is the insertion of an additional strand of DNA, the nucleotide sequence TATC, in the gene after 1278 normal base pairs (called *1278insTATC*), However, there are at least 100 other *HEXA* mutations that can cause Tay–Sachs. This means the disease is an example of **compound heterozygosity**, whereby disease can potentially result from the inheritance of two *unrelated* mutations in the HEXA gene, one from each parent. This means that an individual can have two recessive alleles for the same gene, but with those two alleles being different from each other (e.g. both alleles might be mutated but at different locations).

The concept of compound heterozygosity is important, firstly because it explains why there is an adult-onset form of the disease that is progressive but seldom fatal, and secondly, because it has important implications when we consider Hardy–Weinberg. Populations in which Tay–Sachs is

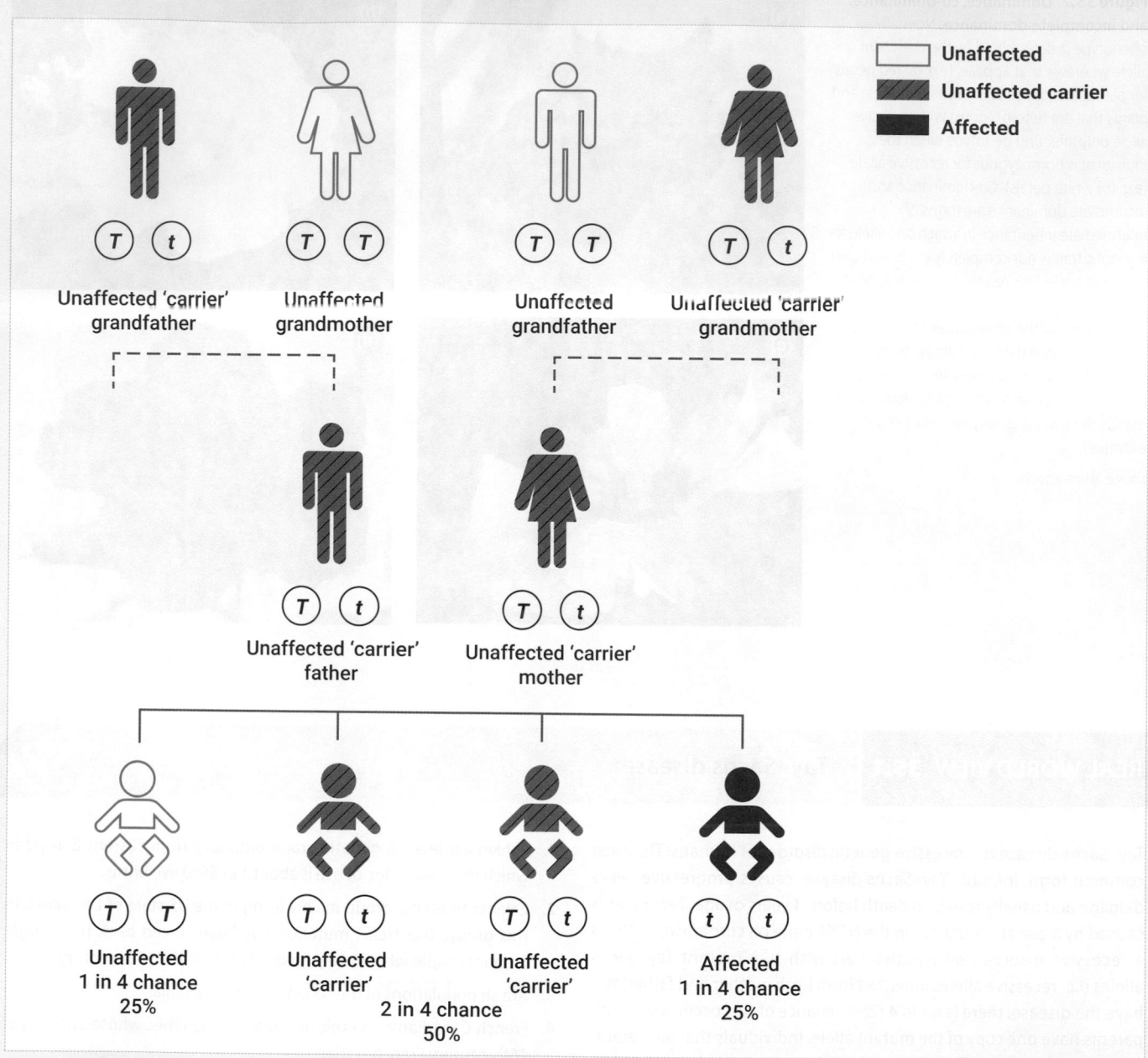

Figure 1 **Autosomal recessive inheritance of Tay–Sachs disease.** Colour coding matches our Punnett square (Figure 35.4) but note that we have replaced *A* (generic notation for a gene) with *T* (notation for genes linked to Tay–Sachs disease).

particularly prevalent are not likely to be in Hardy–Weinberg equilibrium in relation to the disease as several key assumptions are violated:

- **No change due to mutation:** The compound heterozygosity of Tay–Sachs means that the *HEXA* gene is not subject to a singl35 mutation, but rather to a range of mutations that are likely to still be occurring.

- **No gene flow:** We can tell there is gene flow because, although Tay–Sachs is particularly prevalent in some ethnic populations and geographical locations, it also occurs outside these.

- **Random mating:** Mating is highly unlikely to be random throughout the population. Individuals with the homozygous disease phenotype

(genotype *tt*) are likely to die before reproductive age. Individuals with the heterozygous carrier phenotype (genotype *Tt*) might either selectively choose mates that are not carriers (the organization Dor Yeshorim carries out an anonymous screening programme) or, if carrier status is unknown or two known carriers choose to reproduce, they might undergo prenatal genetic testing to determine whether the fetus is genotype *tt* and undergo selective abortion if it is.

Moreover, the genotypes involved *are* linked to differences in survival and reproductive success. Tay–Sachs disease (genotype *tt*) is usually fatal and most people with the condition die before they reach reproductive age.

There is also evidence that carriers (genotype *Tt*) are partially protected against some diseases, in particular tuberculosis (which might explain why the Tay–Sachs allele hasn't been removed from the population by natural selection).

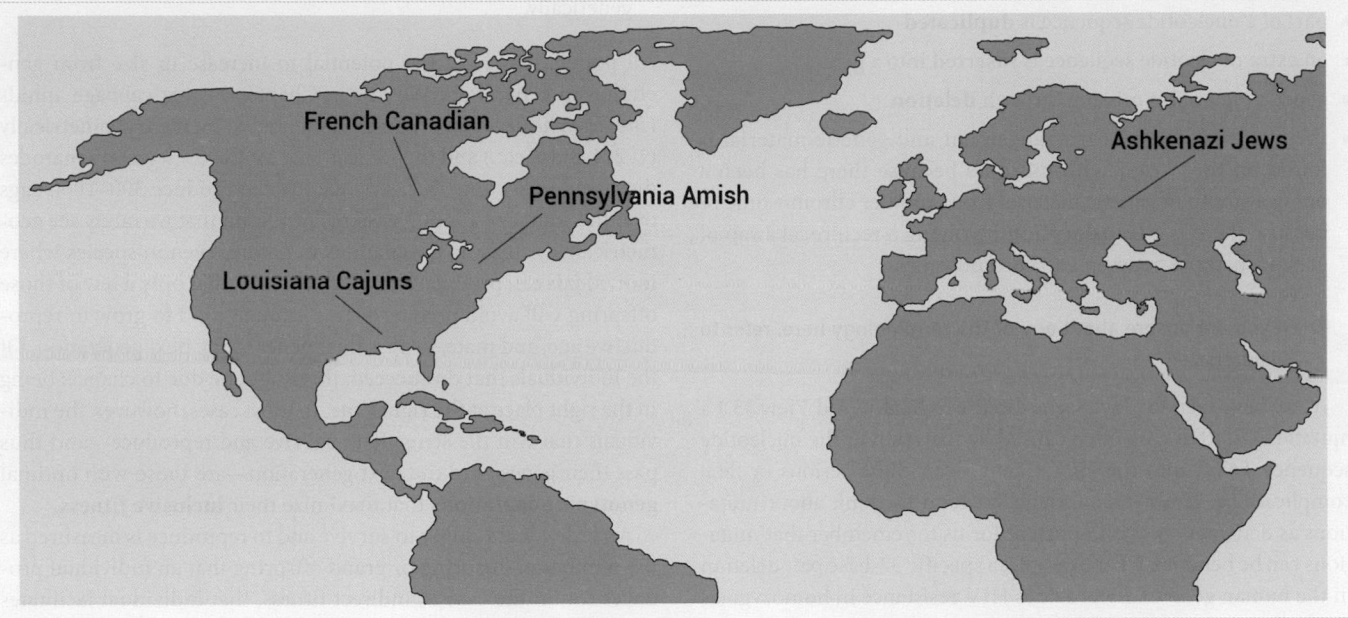

Figure 2 Geographical distribution of the main ethic groups where Tay–Sachs is most prevalent, but incidence of the disease is global.
Source: Shutterstock/Roman Vinokhodov.

35.4 Genetic change and evolution

We can use the concept of the Hardy–Weinberg equilibrium to compare and contrast populations of a species where there is no genetic change with populations of the same species in which there is genetic change. As we have already seen, allele frequencies will remain stable from generation to generation only if the assumptions on which Hardy–Weinberg is based are met. Evolution is a change in allele frequencies from generation to generation—in other words, the processes that drive evolution are processes that violate the assumptions of the Hardy–Weinberg equilibrium. This means that we can translate the criteria that underpin the Hardy–Weinberg equilibrium into processes that drive evolutionary change.

1. **Mutation:** violates the assumption that alleles must not be under change due to mutation.

2. **Selection:** violates the assumption that alleles must not be under change due to evolution.

3. **Genetic drift:** violates the assumption that alleles must not fluctuate randomly between generations.

4. **Non-random mating:** violates the assumption of random mating.

5. **Gene flow:** violates the assumption that the population being modelled is a closed population.

In the sections that follow we'll look at each of these processes of evolutionary change in turn.

Mutation

A **mutation** is a permanent change in an organism's DNA: this might be a change in the sequence of a gene or in the physical arrangement of the genetic material on a chromosome. Mutations result from errors during DNA replication, damage to existing DNA caused by exposure to a **mutagen** (a mutation-generating factor such as the chemical ethidium bromide, ultraviolet light, or radiation), or a problem with DNA repair. Mutations can involve loss of genetic material, but it is far more common to encounter the rearrangement of linear gene sequences or the rearrangement of chromosomes on which specific genetical material is located. Mutations occur when:

■ single nucleotides become **substituted** one for another either because of a:

  ● *transition* that exchanges purine for purine (A with G) or pyrimidine for pyrimidine (C with T); or

- *transversion* that exchanges purine for pyrimidine or pyrimidine for purine (C or T with A or G)

■ part of a nucleotide sequence is **inverted** so it becomes reversed end-to-end

■ part of a nucleotide sequence is **duplicated**

■ an extra nucleotide sequence is **inserted** into a gene

■ genetic material is missing through **deletion**

■ there is chromosomal rearrangement and genetic material is found on the 'wrong' chromosome because there has been a one-way jump of genetic material from another chromosome or because there is a **translocation** involving a reciprocal swap of genetic material between two chromosomes.

▶ **If you are unsure about any of the terminology here, refer to Chapters 2–4.**

As we have seen for Tay–Sachs disease in Real World View 35.1 a mutation—in that case often caused by insertion of the nucleotide sequence TATC into the *HEXA* gene—can cause serious or fatal complications. However, although we tend to think about mutations as detrimental, it is important for us to remember that mutations can be beneficial. For example, a specific 32-base pair deletion in the human gene *CCR5* provides HIV resistance in homozygotes and delays the onset of AIDS in heterozygotes. Most important in the context of this chapter, though, is that mutations are the ultimate source of genetic variation and the only way—apart from gene flow when the incoming individuals have different genotypes to those individuals already in the population—in which new alleles can occur within a population. Mutations are thus the raw material for natural selection. Without mutation, there would be no genetic variation and no prospect of evolution.

Natural selection

As discussed in Chapter 34, natural selection is the differential survival and reproduction of different phenotypes due to differences in how well adapted different phenotypes are to a specific set of environmental conditions. Some individuals will have a phenotype that is well adapted to a specific environment and this increases their likelihood of surviving and reproducing, and that are thus more likely to pass on their genes. In contrast, other individuals will have a phenotype that decreases their likelihood of surviving and reproducing, and that are thus less likely to pass on their genes. Because these phenotypes are determined—at least in part—by genotypes and thus by alleles, natural selection has the potential to alter genotype frequencies and allele frequencies in a population over time.

For selection to take place in a population and change allele frequencies:

■ There must be genetic variation in the individuals within the population due to mutation and/or movement of individuals (and thus genetic material) into the population.

■ Some of that genetic variation must affect the survival and reproductive success of individuals.

■ The environment typically needs to be fairly stable so that the advantages that some phenotypes (and thus genotypes and alleles) have over others remain fairly consistent to allow the need for selection (selection pressures) to be reflected genetically.

All populations have the potential to increase in size from generation to generation. We saw in Chapter 34 that cabbage aphids (*Brevicoryne brassicae*) have the potential to increase geometrically (1, 2, 4, 8, 16, etc.) and the same is true for the *C. elegans* nematodes discussed in Section 35.2: a nematode can produce 300–1000 eggs over the course of about 3 weeks. The reason that we rarely see geometrically increasing populations in nature, even in species where individuals can produce lots of offspring, is that only a few of those offspring will avoid predators, find enough food to grow to reproductive age, and mate, to pass their genes to the next generation. Of the individuals that do succeed, this might be due to chance: being in the right place at the right time. In most cases, however, the individuals that win the struggle to survive and reproduce—and thus pass their genes onto the next generation—are those with optimal **genotypic adaptations** that maximize their **inclusive fitness**.

An individual's ability to survive and to reproduce is measured as the number of offspring or grand offspring that an individual produces (or, in the case of indirect fitness, that individual facilitates indirectly). This is the concept of fitness at the level of the individual and, as we saw in Chapter 34, we term this **absolute fitness**. We can also look at the fitness of a genotype by examining the fitness of all individuals with a specific genotype compared with the mean fitness of all the individuals in the population with a different genotype: we term this **relative fitness**.

▶ **We learn more about these concepts in Chapter 34.**

Although the concept of absolute fitness is arguably more straightforward conceptually, it can be difficult in practice to link fitness to genotype on an individual basis. This is because although a specific genotype may confer an advantage on the individuals that possess it, this might be offset by the effects of pure chance (e.g. accidents or random predation), which complicates the fitness–genotype link. If we look at all the individuals in the population that possess the genotype, however, things become clearer and the random fluctuations caused by chance events for specific individuals become less important. For these reasons, in practice, we tend to calculate relative fitness.

Let's look at the concept of relative fitness in a specific example: nematodes that can swim at three different speeds (three different phenotypes), with each phenotype resulting from a different genotype (dominant homozygotic *FF* for fast swimmers, heterozygotic *Ff* (also equivalent to *fF*) for slow swimmers, and recessive homozygotic *ff* for very slow swimmers). For the moment, we'll assume that swimming speed has no effect on reproductive success, and focus on its impact on survival rate. The very slow swimmers are the most likely to be caught by predators, while the fast swimmers are less likely to be caught by predators. In other words, there is phenotypic variation in the population, which is linked to the genotype of each individual, and that variation affects the survival of individuals. If we have data on the percentage of nematodes with each

Table 35.5 Nematode survival in relation to swimming speed phenotype and underlying genotype

Phenotype (swim speed)	Genotype	Survival	Relative fitness	Rationale/calculation
Fast	*FF*	30%	1.0	Relative fitness of 1.0 by default (highest survival)
Slow	*Ff*	17%	0.57	Survival of individuals of this genotype (17 out of 100) divided by survival of individuals with highest relative fitness (30) = 0.57
Very slow	*ff*	5%	0.17	Survival of individuals with this genotype (5 out of 100) divided by survival of individuals with highest relative fitness (30) = 0.17

genotype that survives to reproductive age, we can calculate the relative fitness of each genotype, as shown in Table 35.5. (If you need a reminder about this, take a look at Quantitative Toolkit 4: Ratio and proportion.)

As we learnt in Chapter 34, there are three common patterns of natural selection based on the links between phenotypes and fitness: directional selection, stabilizing selection, and disruptive selection. We now need to link these concepts to relative fitness, genotype frequencies, and allele frequencies.

- In **directional selection,** extreme phenotypes have the highest relative fitness: this means that, over time, allele frequencies will shift towards the homozygous genotype responsible for that phenotype (that is what is occurring in this example—directional selection towards *FF*). The wild-type alleles thus increase (*F* here).

- In **stabilizing selection**, it is an intermediate phenotype that has the highest relative fitness so allele frequencies will tend to move towards the heterozygous genotype (which would be *Ff* in this example).

- In **disruptive selection,** the phenotypes furthest from the mean have the highest relative fitness and the genotypes that are the most extreme in separate directions (which would be *FF* and *ff* in this example) increase.

PAUSE AND THINK

In the nematode example, presented in Table 35.5, is the selection likely to be directional, stabilizing, or disruptive?

Answer: The data suggest we have a case of directional selection towards fast swimmers FF.

Modelling natural selection: heritability

In our consideration of phenotype, genotype, and alleles so far in this chapter, we have largely assumed that phenotype is entirely dependent on genotype. However, as we saw in Chapter 34, phenotypic variation actually results from a *combination* of genotypic variation *and* non-genetic **phenotypic responses** to environment. The effect of environment on phenotype, in addition to its effect on genotype, can complicate things when we are trying to understand the origin of the differences that we observe. For example, a

higher percentage of the people born in the UK in the year 2000 will live to be 100 compared to those born in the UK in the year 1900. Some of this change might be related to genes linked to survival and reproductive success, but the bulk of the increase will be the result of environment, especially diet and medical advances.

We touched on this briefly in Section 35.3 when we noted that nematode lifespans are controlled by not only genes such as *age-1, daf-2, daf-7, daf-16, rle-1,* and *math-33*, but also by environmental temperature. Moreover, the interaction between genetics and the environment is also important with the effect of the mutant *age-1* genes being greater at higher temperatures. In other words, the phenotypic variance we observe in the population (σ^2_P) is the result of overall genotypic variance (σ^2_G) *and* response to environmental variance (σ^2_V). We can express this in an equation:

$$\sigma^2_P = \sigma^2_G + \sigma^2_V$$

where:

- σ^2_P = phenotypic variance
- σ^2_G = genetic variance
- σ^2_V = (response to) environmental variance.

The ability of natural selection to bring about long-term evolutionary change depends on the extent to which phenotypic differences are linked to genetic differences. Because of this, we often need to quantify the relative contribution of genes and environment to the phenotypic variation observed in a population. The proportion of phenotypic variation for a particular trait that results from genetic differences rather than environmental factors (or random variation) is called **heritability**:

$$h^2 = \sigma^2_G / \sigma^2_P$$

where:

- h^2 = heritability
- σ^2_G = genetic variance
- σ^2_P = phenotypic variance.

Heritability is quantified on a scale from 0 (genetic differences play no role in phenotypic variation) to 1 (genetic differences are responsible for all phenotypic variation)—the process of quantifying this is explained in Quantitative Tools 35.1.

QUANTITATIVE TOOLS 35.1   Calculating heritability

To visualize heritability, we can create a scatter plot where all the data points on the graph are individuals, the *y*-axis is the value of that trait for that individual and the *x*-axis is the value of that same trait for the parents of that individual. Because diploid individuals have two parents, we take the mid-point value (the median) for that trait for both parents.

Let's imagine we have data on the number of days 150 large blue butterflies (*Phengaris arion*) spend in their adult life stage (adult lifespan, measured in days) and the same data for both the parents of each butterfly. If we draw a scatterplot based on these data, we would have 150 data points (i.e. one point per butterfly). Individual adult lifespan (number of days survived as an adult by each of our focal butterflies) would be shown by the vertical placement of each data point (i.e. how far up the *y*-axis a point is), while the median of parental adult lifespan (the mid-point between the number of days survived by the mother as an adult and the number of days survived by the father as an adult) would be shown by the horizontal placement of each data point (i.e. how far along the *x*-axis a point is), as shown in Figure 1.

We can take the scatterplot approach further by standardizing the data so that the mean of each variable is 0 the standard deviation (a measure of the variability around the mean) is 1. This means we can directly compare the heritability of the survival of nematodes measured in days with, for example, the heritability of leaf length in oak trees measured in cm or the heritability of robin weight measured in grams despite the fact that the species differ, the variable of interest differs, and the units (days, cm, g) differ.

To standardize data, we take each data point and subtract the mean (which we will have previously calculated on all data points) and then divide that answer by the standard deviation (which we will have

previously calculated on all data points). Once we have standardized the data, we plot this on a second scatterplot (Figure 2) and although the actual numbers on the axes will be different compared to the original unstandardized data, the pattern will be exactly the same (compare the standardized axes in Figure 2, which are both relative scales centred on 0, with the original measurement in Figure 1).

The next step is to calculate the line of best fit through the data points. This is often achieved by selecting an option when graphing the data, but this is actually displaying the result of a statistical analysis called **regression**. Regression involves calculating a line of best fit mathematically as being the straight line that minimizes the sums of the distances between each data point and the line. Assuming the line of best fit is positive (i.e. goes upwards diagonally) the gradient of the line can be expressed as a number between 0 and 1 and it is this number that is our measure of heritability. In our butterfly data, therefore, our heritability of adult lifespan is 0.8157. This means that for every unit of 1 on the *x*-axis we go up 0.8157 units on the *y*-axis (Figure 2). Note that if our gradient was exactly 1, we would have a completely symmetrical data set whereby for every unit of 1 on the *x*-axis we go up 1 unit on the *y*-axis and the line of best fit would be a perfect 45° diagonal.

When a line of best fit is calculated through regression analysis, it is usual to report an **intercept** in addition to the gradient. The intercept is the number on the *y*-axis the line of best fit passes through when *x* is equal to 0. In this case, the equation for the line of best fit is actually $y = 0.8157x$.

We can convert the initial heritability value, calculated using the slope of the line of best fit from a regression analysis, into a heritability

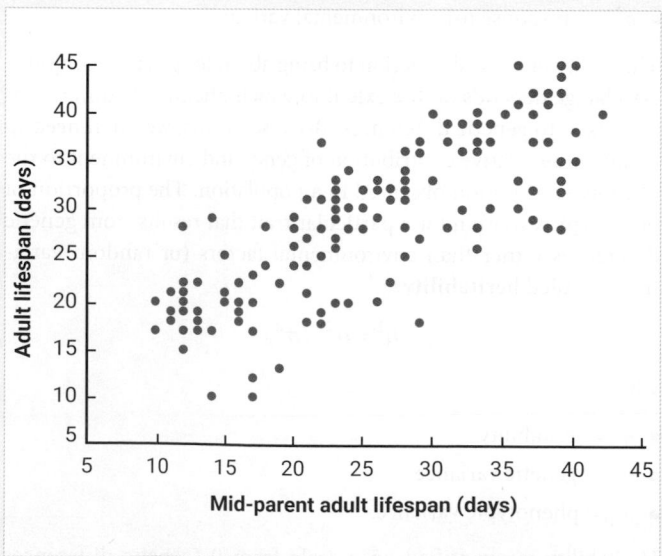

Figure 1 Scatterplot of the lifespan of each individual (*y*-axis) against the average (median) lifespan of its parents (*x*-axis).

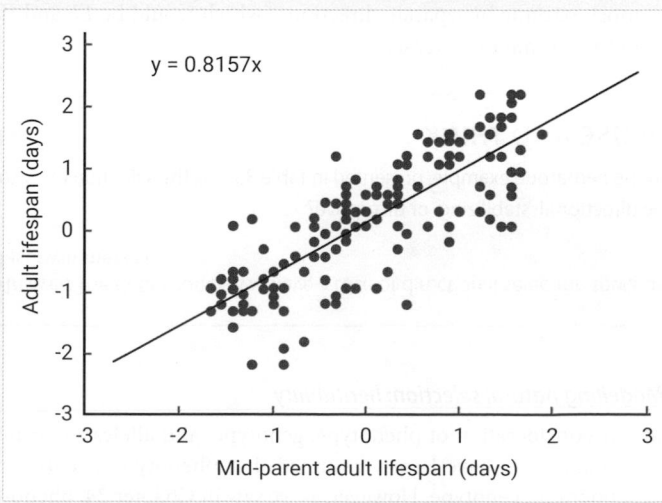

Figure 2 Scatterplot of the lifespan of each individual (*y*-axis) against the average (median) lifespan of its parents (*x*-axis) after both variables had been standardized so that the mean of each variable = 0 and the standard deviation of each variable = 1.

percentage by multiplying it by 100. The simple rule for assessing how much phenotypic variation is due to genetic variation and how much is due to environmental variation (or chance) is, therefore:

- Genetic component = heritability score converted to a percentage. Our heritability value of 0.8157 calculated here would become 81.57% when 0.8157 is multiplied by 100.

- Environment component = the reciprocal of the genetic component (i.e. 100% minus genetic component % calculated above). In the case of our butterfly data, this would be 18.43%.

You can read more about this standardization in Qualitative Toolkit 3, regression in Quantitative Toolkit 7, and the equation of a line in Quantitative Toolkit 8.

 Go to the e-book to get a copy of the data featured in this panel to explore for yourself.

PAUSE AND THINK

You have data on the weights of 10 buffalo calves at birth. You also know the weights of the mother of each of the calves when she was born. You want assess the heritability of birth weight in this species. The paternity of each buffalo is unclear so you can only use maternal weight here rather than mid-parent weight. Using the data in the table and the method details in Quantitative Tools 35.1, standardize both offspring birth weight and maternal birth weight variables (remember, the calculation for this is to take each data point, subtract the mean, and then divide that answer by the standard deviation). Once you have done this, create a scatterplot of the standardized data and fit a trend line to calculate heritability.

 Go to the e-book to watch a walk-through of the calculations involved in answering this Pause and Think.

	Maternal weight at birth (kg)	Offspring weight at birth (kg)
	40	39
	38	39
	37	38
	39	39
	40	40
	41	40
	42	43
	41	42
	39	38
	40	41
Mean	39.70	39.90
Standard deviation	1.49	1.66

All complex traits are the result of genetic and environmental influences. It is important to remember that heritability is a measure of the extent to which the phenotypic *variance* of a trait in a population can be explained by the genotypic *variance* and thus:

- Heritability cannot describe how much of a particular individual's phenotype has been determined by their genetics. Heritability is a feature of *a specific trait* in a *group of individuals* that make up a population, not the individuals themselves (remember that the scatterplots in Figures 1 and 2 of Quantitative Tools 35.1 plotted all individuals on one graph and heritability was calculated at the population level using all these data points rather than for each individual separately)

- A heritability score of 1 does not mean that the environment doesn't affect the trait, just as a heritability of 0 doesn't mean that genetics doesn't affect the trait. We are quantifying *phenotypic variability* between individuals in a population and how this relates to *genetic variability* at a population level not how genotype links to phenotype directly for specific individuals.

- Heritability is not a fixed property of a particular trait: it is a property of that trait in a particular population in a particular environment at a particular time. Thus, the heritability of leaf length in a population of oak trees in a UK woodland in the 1900s might be very different to the heritability of leaf length in a population of oak trees in a French woodland in the 2000s.

- Different traits will have different heritabilities, even in the same population.

Modelling natural selection: selection differentials

A selection differential is a measure of direction and magnitude of the selection pressure on a particular trait of one individual in the population relative to the mean of other individuals in that population, taking into account a measure of direct fitness. This is easiest to understand by means of an example, so let's explore the quantification and interpretation of selection differentials of the timing of bird breeding in response to climate change in Scientific Process 35.1.

SCIENTIFIC PROCESS 35.1 — Selection for early laying in birds in response to climate change

Background

Phenology refers to the timing of a seasonal biological event: the timing of tree leaf burst, for example, or the timing of amphibian spawning. Phenology is not fixed because the timing of seasonal events differs in response to local conditions that change over space (e.g. in relation to habitat quality) or over time (e.g. in relation to changing temperatures). The date on which the first egg in a clutch is laid is a commonly used measure of bird breeding phenology. Previous studies have established that lay dates are getting earlier for many bird species nesting in temperate areas, such as the UK. This has been attributed to warming spring temperatures, related to climate change, leading to caterpillars emerging earlier, and birds responding by laying earlier as they need to synchronize chick demand for caterpillars with peak supply of caterpillars.

Aim

In 2011, a research team from the University of Gloucestershire set out to quantify long-term changes in breeding phenology for six UK songbirds: four resident species—blue tit (*Cyanistes caeruleus*), great tit (*Parus major*), coal tit (*Periparus ater*), and nuthatch (*Sitta europaea*)—as well as the migratory pied flycatcher (*Ficedula hypoleuca*) and redstart (*Phoenicurus phoenicurus*). The team then related observed changes in breeding phenology with the direction/intensity of selection pressure for the timing of laying and, more importantly, change in the intensity of that selection pressure.

Methods

The researchers used data from almost 7000 bird breeding attempts over 31 years (1974–2004). Selection pressure was quantified based on the differences in breeding success of birds that breed at different times of the

breeding season. To do this, standardized annual selection differentials were calculated for each of the six species in each of the 31 years. This involved subtracting the annual mean lay date for all individuals in the population from the mean lay date of all individuals in the population after the lay date of each nest had been divided by the number of offspring to fledge per nest. This result was then divided by the population standard deviation of lay date in that year. This was done on the basis that:

- If there is no impact on the timing of laying on success, the selection differential would be 0.

- If birds laying earlier had higher success relative to late-laying individuals (i.e. there was directional selection pressure to lay early), the selection differential would be negative, with the strength of that negative selection being shown by the magnitude of that value relative to zero (lower = stronger).

- If birds laying later had higher success relative to early-laying individuals (i.e. there was directional selection pressure to lay late), the selection differential would be positive, with the strength of that positive selection being shown by the magnitude of that value relative to zero (higher = stronger).

Results and discussion

Three of the six bird species studied, blue tit, great tit, and nuthatch (pictured in Figure 1), were found to be laying earlier as spring temperatures increased. For all of these species, selection differentials were negative in all years (i.e. there was always a directional selection pressure to lay early), but the selection intensity for early laying was consistent over time rather than getting stronger. This suggests that the individuals in these populations are responding appropriately and fully to climate change: there is consistent selection to lay early, birds are laying earlier, the selection pressure is not getting stronger, and therefore the rate of observed change is keeping pace with the need to change.

In two of the six species—the migratory pied flycatcher and redstart (Figure 1)—there was no observed change in laying date, but the selection for early laying was getting stronger over time. This strengthening selection for early laying suggests that individuals of these species *should* be laying earlier but the lack of observed change shows that they are failing to track climate change.

Finally, for the sixth species, the resident coal tit (pictured in Figure 1), there was no observed change in laying date and selection differentials were very varied. This suggests that there was no directional selection for this species.

Extension question

Why might migratory pied flycatchers and redstarts be laying at the same time of year each year—and thus not apparently responding to climatic change—when there is directional selection for early laying that is consistent and getting stronger each year?

Read the original paper

Goodenough, A.E., Hart, A.G., and Stafford, R. (2010). Is adjustment of breeding phenology keeping pace with the need for change? Linking observed response in woodland birds to changes in temperature and selection pressure. *Climatic Change* 102: 687–697.

Tracking climate change effectively: changes in observed laying date, no change in directional selection to lay early.

Blue tit

Great tit

Nuthatch

Not tracking climate change effectively: no change in observed laying date despite strenghthening directional selection to lay early.

Pied Flycatcher

Redstart

Stabilising selection: no change in observed laying date and no change in selection over time

Coal tit

Figure 1 Linking observed changes in the timing of egg-laying in six woodland songbirds to changes in selection pressure for the timing of breeding.

Source: all from Pixabay.com, except pied flycatcher from Jesus Giraldo Gutierrez/Shutterstock.com.

Genetic drift

Natural selection changes the genotype and allele frequencies in a population by increasing the likelihood that beneficial alleles will be passed to the next generation and decreasing the likelihood that detrimental alleles will be passed to the next generation. This is not a random process: some individuals will have alleles that increase survival and reproductive success—like our long-lived and fast-swimmer nematodes, for example. Over time, those alleles will become more frequent in the population. Other individuals will have alleles that, relative to others in the population, decrease the likelihood of survival and reproductive success—like our short-lived and slow-swimmer nematodes. These alleles will thus become less frequent in the population.

However, some individuals might pass their detrimental alleles to the next generation simply because they were 'lucky', while others with alleles that should be beneficial might die before reproduction because they were 'unlucky'. In the case of our nematodes, for example, some individuals with the long-lived phenotype might, just by chance, happen to live in a patch of water that dries up. Others may have the slow swimmer phenotype but, just by chance, avoid predators and survive to mate and pass on their genes to the next generation. Here, the genes that are passed on will be the genes of the 'lucky' individuals, regardless of how well adapted those individuals, and ths those alleles, are to the environment. The change in the allele frequencies of the population from generation to generation that result from this random process is called **genetic drift**. Another way to phrase this is that drift is the random selection of reproductive individuals.

Like natural selection, genetic drift leads to changes in allele frequencies and therefore to evolution, but unlike natural selection it doesn't result from, or drive, genetic adaptation of organisms to their environments. In other words, the changes in allele frequencies are not linked to individuals' ability to survive or reproduce:

they are the result of chance events, which can radically alter allele frequency in different populations even if they were identical initially, as shown in Figure 35.8, with multiple simulations of change in allele frequency over time showing the profound effects of different (random) variations due to chance events. It might help to think of natural selection as being like swimming—purposeful movement—and genetic drift as being like drifting with the current, which might move in different directions in different areas (populations).

The effect of genetic drift in large populations is usually negligible, unless considered over a very long period of time. This is because the effects of chance events ('lucky' individuals surviving) are usually evened out across the population as a whole. The effects of genetic drift in a small population, however, can be considerable because random change in allele frequencies will have a larger effect on the overall gene pool (because each individual's alleles represent a larger proportion of the alleles in a small population than in a large population). In other words, there are fewer individuals to buffer the effects of non-adaptive changes in allele frequency.

Pronounced effects due to genetic drift often result from:

- **Population bottleneck:** a large population declines and the resultant smaller gene pool is not fully representative of the gene pool in the original population.

- **Founder effects:** a few individuals from one population colonize a new location, such as an isolated island. The new population that results has a gene pool that is not fully representative of the gene pool in the original population.

Population bottlenecks

If a large population declines so only a few individuals remain, it is likely (by chance alone) that those individuals will not be fully representative of the allele frequencies in the original population.

Figure 35.8 Genetic drift. A computer simulation of genetic drift for an allele *A* in 6 different populations (each shown by a different colour) with no selective pressure. The initial frequency of allele *A* was set at 50% (or 0.5) in each population.

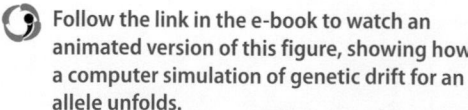 Follow the link in the e-book to watch an animated version of this figure, showing how a computer simulation of genetic drift for an allele unfolds.

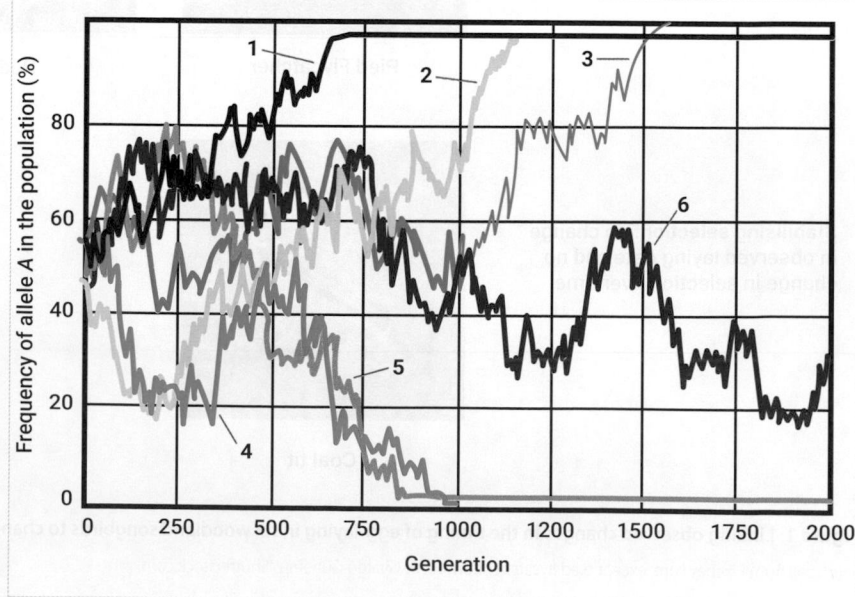

Figure 35.9 Phenotypes of peas. The phenotypes ((a) smooth or (b) wrinkled) and the underlying genotype (*RR, Rr,* or *rr*) upon which Mendel based a lot of his pioneering genetics work in what we now term the field of Mendelian genetics.

Source: Goodenough and Hart (2017), *Applied Ecology: Monitoring, managing, and conserving.* Oxford: Oxford University Press. Reproduced with permission of the Licensor through PLSClear.

 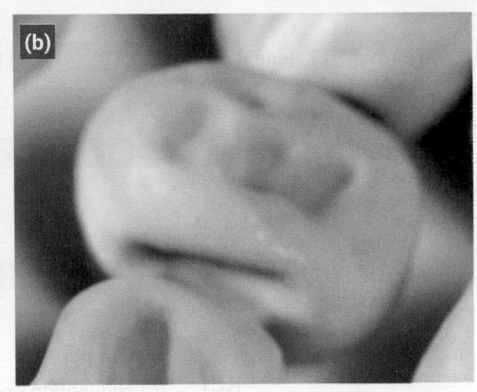

Let's think back to Figure 34.7 and the example pea plants (*Pisum sativum*) that produced either smooth peas or wrinkled peas. This is an example of a simple trait (i.e. caused by one gene and resulting in a small number of possible phenotypes). In this case, the genetic difference is an inserted fragment of genetic material, about 800 base pairs long, on the *SBEI* gene. The insertion causes the wrinkled phenotype caused by being homozygous for the recessive mutant allele (*rr*) as distinct from the rounded phenotype caused by being homozygous for the dominant wild-type allele (*RR*) or heterozygous (*Rr*) as shown in Figure 35.9.

Suppose that we have a large population of peas where the recessive *r* allele that is linked to the wrinkled phenotype is rare with an occurrence of 0.001 (1 in 1000) but the pea population then crashes because of disease (e.g. wilt caused by the *Fusarium* soilborne fungus). If only two plants survive, one of which, by chance, is heterozygous (*Rr*) and the other is homozygous for the mutant allele (*rr*), the allele frequency of *r* in the new, smaller population is now 0.75. The allele frequency has changed dramatically—from 0.001 in the original large population to 0.75 in the remnant smaller population—as a result of pure chance.

PAUSE AND THINK

Calculate the allele frequency for r in the smaller pea population if one of individuals that survived by chance was heterozygous (Rr) and the other was homozygous for the wild-type allele (RR).

Answer: r in the new population would have a frequency of 0.25 (there are 1× r and 3 × R alleles in total between the surviving individuals; 1/4 = 0.25).

Founder effect

The **founder effect** is a specific form of population bottleneck that occurs when a few individuals *colonize a new location* and are isolated from the original population through that process. The resultant small population is caused not by the original population declining to give a remnant population, but rather by a new (small) population becoming established in a separate location. Again, the allele frequency in the new small population is unlikely to be fully representative of the original large population.

An example of a founder effect is seen in the Afrikaner population in South Africa. The Afrikaner population is descended primarily from 344 settlers from the Netherlands in 1658. One of those initial Dutch settlers had Huntington's disease, which is caused by excessive duplication (too many repeats) of a trinucleotide sequence (CAG, which codes for the amino acid glutamine) on the *HTT* gene, as shown in Figure 35.10. The highest prevalence of Huntington's disease in Europe, including in the Netherlands, is 7 per 100,000 (allele frequency of 0.00007). Today, the Afrikaner population has an unusually high frequency of Huntington's disease because the allele frequency of the mutated *HTT* gene in the original Dutch colonist population was 1 in 344 (allele frequency of 0.003)—all known cases of the disease in the current population have been traced back to this one common ancestor.

Impacts of population bottlenecks and founder populations

Genetic drift contributes to evolution in a number of ways, especially when linked to situations when the population size is small and genetic variation of the population is reduced. At its most extreme, genetic drift can:

- Change disease prevalence or resistance in a population and thus affect survival to reproductive age (e.g. Huntington's in Afrikaners).

- Decrease the ability of a population to respond, through natural selection, to a change in the environment as the raw material of natural selection—genetic variability—is reduced. This could ultimately put a population (or even species) at risk of extinction.

- Cause peripatric speciation because it (randomly) increases the genetic differentiation between two or more populations. This is especially true if founder populations not only have a different *gene pool* relative to the original population, but are also subject to different *selection pressures* relative to the original population.

▶ You can read more about peripatric speciation in Chapter 34.

Non-random mating

The three evolutionary drivers we have considered so far in this section (mutation, natural selection, and genetic drift) all involve

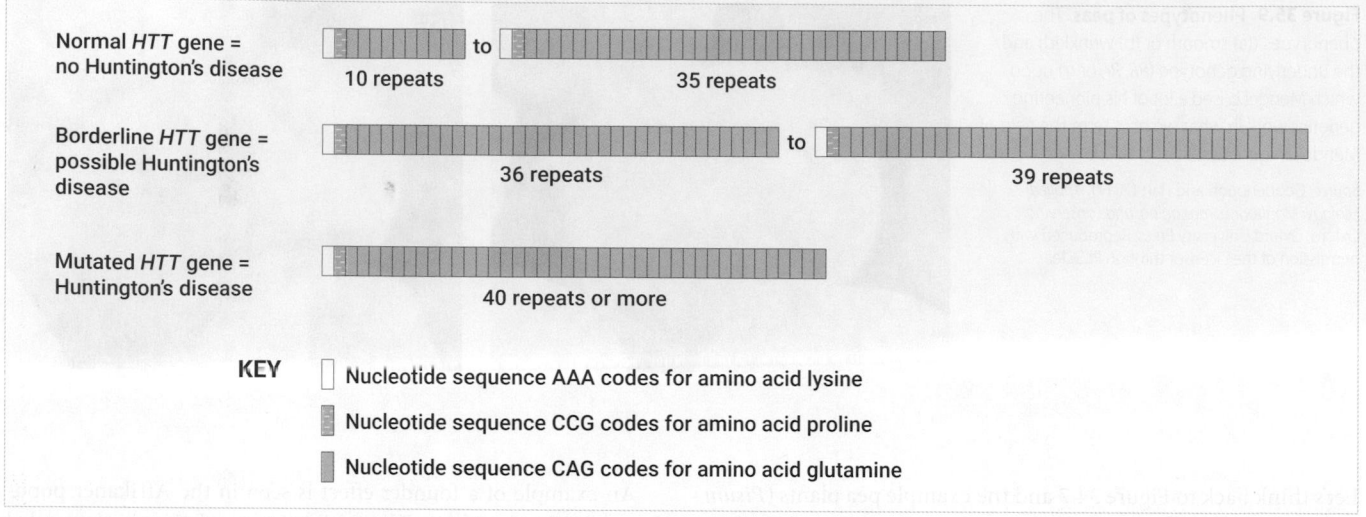

Figure 35.10 Huntington's disease. Whether the *HTT* gene leads to Huntington's disease or not depends on the number of CAG repeats and thus the length of the glutamine amino acid chain.

changes in allele frequency because of genetic processes. Each of these drivers contravenes a Hardy–Weinberg equilibrium assumption that we considered in Section 35.3:

1. The alleles being modelled must not be changing due to mutation (**no mutation**).
2. The alleles being modelled must not be changing due to evolution through natural selection (**no evolution**).
3. The population must be large enough to minimize the possibility of allele frequencies varying from generation to generation as the result of chance fluctuations (**no genetic drift**).

One of the other assumptions of the Hardy–Weinberg equilibrium is that all the individuals in the population mate with each other on a random basis so that—in theory at least—the chance of any individual mating with any other individual of the opposite sex is not linked to genotypes and/or phenotypes. In reality, however, individuals tend to be very careful about mate choice.

We use the term **assortative mating** for systems in which males and females with particular genotypes and/or phenotypes mate with each other more or less frequently than would be expected under a random system.

- If organisms select mates with *similar* genotypes and/or phenotypes more often than expected by chance, this is **positive assortative mating**. The effect of positive assortative mating will be a progressive increase in the number of homozygous genotypes in the population and a decrease in the number of heterozygous genotypes. A good example here is that tall females often tend to partner with tall males and produce taller-than-average children. In this case, there is a genetic component (there are >700 genes associated with height, including *ACAN*, *GH1*, and *FGFR3* so the inheritance system here is **polygenic inheritance**), but there is also an assertive mating component.

- If organisms select mates with *different* genotypes and/or phenotypes we call it **negative assortative mating** (sometimes **disassortative mating**). The effect of negative assertive mating will be maintenance of, or an increase in, the number of heterozygous genotypes. A good example is seen in feral pigeons (*Columba livia*) where individuals choose their mate partly based on them having different plumage relative to their own. (Note too that the ability to do this also means that individual pigeons must know what they themselves look like (the concept of self-recognition), which is a higher-order cognitive skill.)

Assortative mating includes non-random mate choice arising as a result of sexual selection, as discussed in Chapter 34.

Gene flow

Closed populations are populations in which there is no movement of individuals (and thus genes) in or out of the population. Figure 35.11 shows examples of closed populations, including those restricted by human influence, such as the fruit flies in a laboratory (Figure 35.11a) and captive animals like this Arabian leopard (Figure 35.11b), and also the naturally occurring isolation of a population of duck-billed platypus (*Ornithorhynchus anatinus*) that is reproductively isolated because the separation distance to the next closest population is greater than the distance an individual can travel, and the Devil's Hole pupfish (*Cyprinodon diabolis*), which only occur in one (highly restricted) location. However, most populations are open and thus subject to the processes of immigration (new individuals entering the population from another population) and emigration (individuals leaving the population). This movement of individuals, and alleles, from population to population results in changes in allele frequencies: a process known as **gene flow** (sometimes called allele flow).

Figure 35.11 Closed populations. (a) Laboratory populations of fruit flies (*Drosophila melanogaster*); (b) captive populations of Arabian leopards (*Panthera pardus nimr*); (c) a population of duck-billed platypus (*Ornithorhynchus anatinus*) on King Island off the Australian coast that is isolated as separation distance to the next closest population is greater than the distance an individual can travel; and (d) endemic or micro-endemic species that only occur in one (highly restricted) location such as the Devil's Hole pupfish (*Cyprinodon diabolis*) living in one thermal spring in Nevada, USA.

Source: (a) Sundry Photography/Shutterstock; (b) Yosyhiro/Shutterstock.com; (c) Stone Nature Photography/Alamy Stock Photo; (d) worldswildlifewonders/Shutterstock.com.

▶ **You can read more about the Arabian leopard and Devil's Hole pupfish in Chapter 36.**

Gene flow can have significant evolutionary consequences because the process can act to reduce the genetic differences between populations. This means that if the population of a species at location C comprises individuals that have immigrated from populations at location A and location B, the population at location C will be influenced by the genotypes of individuals from location A (and thus the selection pressures operating at location A that influence that genotype) and the genotypes of individuals from location B (and thus the selection pressures operating at location B). This can act to constrain evolution through natural selection among individuals in location C because the genotypic variation inserted into the population via gene flow weakens the influence of directional succession in that location.

Analysis of allele frequencies can do more than capture genetic variation within a single population; it can be used to compare populations and quantify evolutionary linkages. For example, genetic variation has been used to reconstruct the migratory history of the current population of India based on 132 genetic samples from 25 groups that differed in language family, and geographical location (Figure 35.12). The results suggest that there were two genetically distinct ancient populations in India rather than one: the Ancestral North Indian (ANI) population (genetically close to Middle-Easterners, Central Asians, and Europeans) and the Ancestral South Indian (ASI) population. Most modern Indians have 39–71% ANI ancestry. In contrast, the current populations of the offshore Andaman Islands (the Onge and Great Andamanese) had strong ASI ancestry and no apparent ANI ancestry, suggesting that these two populations have had little gene flow historically and have thus been closed populations, probably due to their geographical isolation.

Figure 35.12 Indian groups genotyped. The 23 mainland groups predominantly have Ancestral North Indian ancestry, the Onge and Great Andamanese groups inhabiting the offshore Andaman Islands have Ancestral South Indian ancestry. The colour of the marker indicates the language family of each group.

Source: Copyright © 2009, Springer Nature.

 Check your understanding of the concepts covered in this section by answering the questions in the e-book.

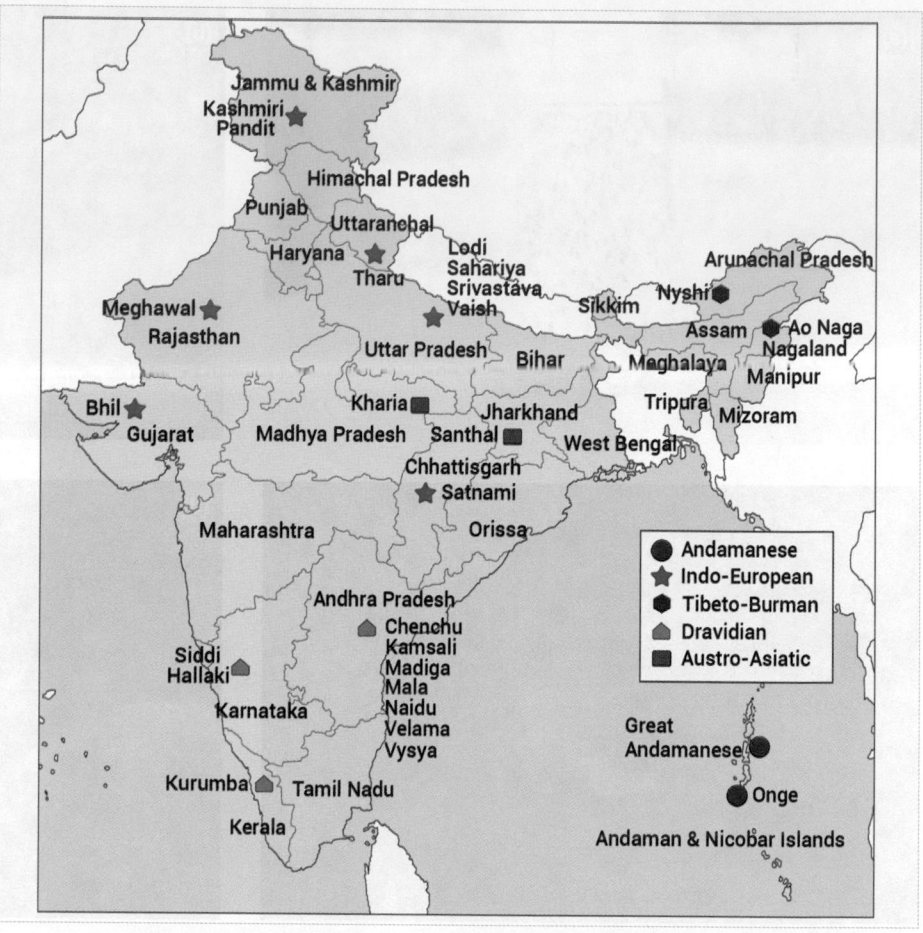

SUMMARY OF KEY CONCEPTS

- Populations contain **variation**, and that variation can be analysed in terms of what we can observe about organisms (their **phenotypes**) and the genetic make-up of organisms (their **genotypes**).

- The genotype in a specific individual is determined by which variants (**alleles**) of particular genes the organism possesses. We can quantify all this variation by determining **phenotype frequencies**, **genotype frequencies**, and **allele frequencies**.

- The **Hardy–Weinberg equation** allows us to link genotype frequencies and allele frequencies in a stable population: that is, a population in which genotype and allele frequencies are not changing due to mutation, evolution by natural selection, genetic drift, non-random mating, and gene flow. It also allows us to test whether evolution is occurring in a given population.

- Evolution is a change in the frequencies of genotypes and alleles over time.

- The main drivers of **evolutionary change**, the main processes responsible for changes in allele frequencies in a given population, are mutation, natural selection, genetic drift, non-random mating, and gene flow.

- Mutations provide the raw material for evolution—they create genetic variability in a population—but they can also play a role in changes in allele frequencies in a population over time.

- Natural selection is the process by which alleles that increase an individual's survival rate or reproductive rate (its fitness) become more common in the population, while alleles that decrease an individual's survival rate or reproductive rate become less common.

- Genetic drift is a change in the frequency of an allele due to random events: the chance occurrences that determine whether a given individual survives and reproduces. It is more of an issue in smaller populations, where the effect of chance events can be more marked.

- Non-random mating can involve assortative mating, in which males and females with particular genotypes and/or phenotypes mate with each other more or less frequently than would be expected by chance.

- Gene flow involves individuals (and thus genes) moving between populations.

 Use the flashcards in the e-book to test your recall of key terms introduced in this chapter.

QUESTIONS

Looking for answers? Once you've answered these questions, follow the link in the e-book to the answer guidance and check your work.

Concepts and definitions

1. What is meant by phenotype and how does this differ from genotype?

2. What is meant by a wild-type allele and how does this differ from a mutant allele?

3. What is the difference between incomplete dominance and co-dominance?

4. List the five assumptions of Hardy–Weinberg equilibrium.

5. Define heritability.

Apply the concepts

6. Human blood groups are A (allele I^A), B (allele I^B), or O (allele i). A and B are co-dominant, while O is recessive so only people with genotype ii have O type blood. From this information, list the two genotypes that could give type A blood, the two genotypes that could give type B blood, and the single genotype that gives type AB blood.

7. Draw up a Punnett square to model what would happen when an individual mouse that is heterozygous for two genes linked to tail length ($AaBb$) mates with an individual that is homozygous for the recessive allele of both genes ($aabb$).

8. You are asked whether a particular population is in Hardy–Weinberg equilibrium. You know that the population in question is subject to arrival of new individuals through the process of immigration. Why does this mean that the population cannot be in Hardy–Weinberg equilibrium?

9. Why does calculating heritability of a phenotypic trait using regression analysis usually involve establishing the relationship between the phenotype trait of each individual and the median (mid-point) of both parents for the same trait, rather than just considering one parent?

10. Genetic drift is a change in allele frequencies between generations due to chance. How can these random changes, which by definition must be non-adaptive, cause evolutionary change?

Beyond the concepts

11. Autoimmune lymphoproliferative syndrome (ALPS) is a human autoimmune disease that causes lymphocytes—a type of white blood cell—to build up in the body. ALPS is normally caused by an abnormality on the *FAS* gene. A child presents with symptoms of ALPS and subsequent testing confirms the diagnosis. Neither of the child's biological parents suffers from the disease and genetic testing shows that neither of them is a carrier. What has happened and what is the correct term for this situation?

12. The Gouldian finch (*Erythrura gouldiae*) is a small songbird that lives in Australia. There are three main colour morphs (variants): birds with black heads, birds with red heads, and birds with yellow heads. Gouldian finches display positive assortative mating, whereby they tend to mate with an individual of the same colour morph. What is the likely explanation for this and why might this ultimately lead to speciation?

13. The white-throated sparrow (*Zonotrichia albicollis*) is an American sparrow that comes in two colour morphs (variants): tan and white. The species is an excellent example of negative assortative (disassortative) mating whereby white morphs tend to mate with tan morphs and vice versa. Tan birds (male or female) tend to invest more in parental care than white birds, but white females are more dominant. How do these observations explain the mating system observed in this species?

14. Find an example of where inbreeding has increased the incidence of an autosomal recessive disease.

15. 'Natural selection is more important than genetic drift in terms of driving evolution'—to what extent do you agree with this statement?

FURTHER READING

Dawkins, R. (2006) *Climbing Mount Improbable*. London: Penguin Books.
   A readable account of how natural selection adapts organisms to their environments.

Prothero, D. (2017) *Evolution: What the Fossils Say and Why It Matters* (second edition). New York: Columbia University Press.
   The importance of the fossil record in understanding evolution.

Orr, H.A. (2009) Fitness and its role in evolutionary genetics. *Nat. Rev. Genet.* **10**: 531–539.
   Discussion of the concept of fitness, and the links between natural selection and fitness.

Populations
Quantifying Demographics and Modelling Change

Chapter contents

Introduction   1244

36.1   Population demographics   1245

36.2   Types of population   1251

36.3   Methods of quantifying population demographics   1253

36.4   Modelling population change   1257

36.5   Modelling life-history strategies   1265

 Watch the key concepts video in the e-book to prepare yourself for studying this chapter.

LEARNING OBJECTIVES

By the end of this chapter you should be able to:

- Describe populations in terms of:
  - abundance (number and density of individuals);
  - spatial arrangement (geographical distribution of different populations; dispersion of organisms within an individual population); and
  - population structure, including age and sex ratio.
- Differentiate between open and closed populations.
- Define 'metapopulation' and explain the importance of gene flow between populations.
- Use life tables effectively to visualize data on survival and fecundity, and be able to use these data to model population change over time.
- Describe key methods for surveying populations in the field, including quadrats, point counts, transect surveys, and camera trapping.
- Calculate a population estimate based on capture–mark–recapture data.

Introduction

As we saw in Chapter 35, a population is a collection of organisms of the same species that inhabit a defined area over a specific period of time. There are many different reasons that biologists study populations, including:

- Establishing the likely impact of ecosystem threats on the viability of a specific population.

- Reducing or buffering impacts of non-native species on native species.

- Controlling populations of a pest species.

- Quantifying how many individuals (and of what age or sex) can be harvested from a population without reducing the long-term viability of that population, for example, through managed offtake such as fishing quotas or sustainable hunting.

- Assessing the need for, or effectiveness of, conservation action.

- Quantifying how many individuals would be needed to successfully establish a new population of a species following a species reintroduction scheme.

We will come back to these applied questions later in this module when we consider threats to populations, communities, and ecosystems in Chapter 39, and management and conservation solutions in Chapter 40. Before we can successfully manage a population in an applied setting, though, we need a thorough understanding of the processes that underpin and regulate population dynamics.

36.1 Population demographics

Studying populations can be undertaken at a range of spatial scales, from the very local (e.g. the common frog (*Rana temporaria*) population of a garden pond) to the global (e.g. global population of blue tits (*Cyanistes caeruleus*)). In order to understand populations, and the processes that drive and regulate them, there are several aspects of a population that we might want to quantify. The quantitative description of a population is called **demography**, and the science involved is called demographics. Key demographic factors include:

1. **Population size:** how many individuals there are.

2. **Population density:** how many individuals there are per unit area (e.g. per km^2).

3. **Population distribution:** the geographical distribution of the population (possibly relative to other populations of the same species).

4. **Population dispersion:** the spatial arrangement of individuals in the population.

5. **Population structure and life-history parameters:** sex ratio, age ratio, fertility rates, survivorship.

We explore these issues further in the next few subsections using a range of different examples.

Population size

The most fundamental piece of information about a population is the number of individuals it contains: the population size (N). Conceptually, this is very straightforward. In practice, however, biologists encounter a number of important challenges when trying to determine population size in natural environments, which in turn have consequences for our ability to understand and manage those populations.

Challenge one: spatial context

Our working definition of a population, as defined in the introduction to this chapter, is 'a collection of organisms of the same species that inhabit a defined area over a specific period of time'. This definition requires us to define the spatial boundaries within which a group of individuals live in order to classify this group as a population.

In some cases, defining the spatial limits of a population can be straightforward. For example, studying the wild population of the Devil's Hole pupfish (*Cyprinodon diabolis*), which we first encountered in Figure 35.11 is straightforward spatially because this species only occurs in one thermal hot spring in Nevada, USA: this is an example of a naturally delimited population. In other cases, humans have erected barriers to encompass a population (e.g. the lion population (*Panthera leo*) of Kruger National Park in South Africa is anthropogenically delimited by the park boundary fence). However, defining spatial boundaries is not always easy because most populations do not occur in discrete areas. In such cases, demographers often focus on genetic or ecological discontinuities, often associated with physical aspects of the environment, to identify a more-or-less distinct population occupying a clearly defined area: for example, Moorish idol fish (*Zanclus cornutus*) associated with one tropical reef or the bluebells (*Hyacinthoides non-scripta*) in one patch of ancient woodland. This process can be complicated by the fact that individual organisms may move from place to place (including moving into and out of the study area). This might happen because they are **nomadic** and move across vast areas, or because they are **migratory** and have separate winter and summer locations.

▶ We learn more about migration in Chapter 31.

Challenge two: temporal contexts

Just as the definition of a population requires us to consider the spatial boundary for a group of organisms, we also need to decide the time period under consideration. This temporal boundary will usually be determined by the nature of the organism and its lifespan: so, for example, we might study bacteria populations over hours or days, plankton populations over weeks and months, and big cat populations over years or decades.

Challenge three: defining the individual

For some populations, it can be difficult to recognize or even define the individual organisms that comprise the population. When we think about a population we normally think about a collection of **unitary organisms**: that is, organisms that exist as single, discrete individuals with a predictable and consistent pattern of development (for example, insects, reptiles, birds, and mammals) and that can therefore be counted numerically. It is important to remember, however, that some species take the form of **modular organisms**, which grow through the continued production of new parts or modules (fungi or marine corals are good examples). In these cases, it is not possible to directly count the individuals visually. The most appropriate way of measuring population size will depend on our research questions but often involves using other metrics of population rather than size.

Figure 36.1 shows populations for four different species that not only differ dramatically in typical population size, but that also exemplify the challenges inherent in quantifying population size.

Population density

Population density is usually the number of individuals per unit area, although in some cases it is quantified as the number of individ-

uals per unit volume (e.g. number of diatoms per litre of river water; number of *Plasmodium* parasites per millilitre of blood) or number of individuals per unit weight (e.g. microbes per gram of soil).

Population density can sometimes be a more useful metric than population size. For example, when surveying modular organisms such as coral, establishing the population size is challenging (as explained in Figure 36.1), but it is possible to quantify the percentage density fairly easily (e.g. percentage of a survey plot covered by coral), and this can act as a proxy for population size. Population density is also a useful method of standardizing population estimates of populations that inhabit areas of different sizes so they can

Figure 36.1 Examples of species where defining population size is challenging. (a) The diatom (*Cymbella janischii*) is found in rivers in the Pacific Northwest of the USA. It usually occurs in low numbers but, in response to environmental conditions, the population can grow dramatically over hours and days to form thick mats that cover sections of the river bottom: any measurement of population, therefore, is a temporary 'snapshot'. (b) Red-billed quelea (*Quelea quelea*) live in huge populations that are nomadic—their distribution follows food availability. (c) Many fungi that have fruiting bodies that grow above ground (mushrooms and toadstools) actually have much of their biomass below ground. When fruiting bodies grow in close proximity, including in the 'fairy rings' produced by the Scotch bonnet (*Marasmius oreades*) shown here, there might appear to be many individuals but actually we are seeing multiple short-lived appendages of a single long-lived modular organism. (d) Coral is made up of thousands of tiny polyps; each polyp secretes a hard outer skeleton of calcium carbonate that attaches either to rock or the dead skeletons of other polyps to create a coral colony. These polyp conglomerates slowly lay the limestone foundation for coral reefs so what we see as 'a coral' is actually multiple individuals.

Source: (a) Courtesy of Sarah Spaulding, US Geological Survey, diatoms.org; (b) Thomas Clode/Shutterstock; (c) Stan Rohrer/Alamy Stock Photo; (d) Timothy Baxter/Shutterstock.

be directly compared. For example, it would be hard to compare the number of grey wolves (*Canis lupus*) in Yellowstone National Park (9000 km²) with the number of wolves in Denali National Park in Alaska (25,000 km²) because population size would be skewed by area, but we could calculate number of wolves per 1000 km² by dividing the Yellowstone estimate by 9 and the Denali estimate by 25 so that the figures are directly comparable.

In other cases, population size and density might both be quantified because they give different information. For example, when conservation biologists are examining the viability of a population such as the tiger (*Panthera tigris*) in the Sundarban mangrove forest on the Bangladeshi–Indian border (which we consider later in Scientific Process 36.3):

- **Population size** is used to assess whether there are sufficient individuals and a large enough gene pool for the population to be viable.

- **Population density** provides extra information such as how likely animals are to encounter one another to mate, as well as how easily disease and parasites might be transmitted between individuals.

Population distribution

At its simplest, the range of a species is the geographical area in which it occurs. Species such as the blue whale (*Balaenoptera musculus*) that have a very large range are called **cosmopolitan species**. At the opposite end of the spectrum are species with a

restricted range: these are called **endemic species**. For example, the ring-tailed lemur (*Lemur catta*), like all lemurs, is only found on the island of Madagascar, while the English whitebeam (*Sorbus anglica*) is a species of tree that is only found in the UK. Species that are found at a single site are called **micro-endemic species**. One of the best-known examples is the Devil's Hole pupfish, which we considered earlier as being one of the few examples of a species where the population is clearly spatially defined—in this case defined by the sides of the only thermal spring in which it naturally occurs.

The range of a species will depend on its evolutionary history (as we explore in Chapters 34 and 35) and its ecological interactions with other species and the environment (which we will consider in Chapters 37 and 38, respectively). Most species have a **fundamental range** (the spatial area with suitable conditions to support a species and which it could theoretically inhabit; sometimes called a potential range) and a **realized range** (the area it actually does inhabit; sometimes called an actual range). Most species have a realized range that is a subset of their fundamental range. Within its range, a species can have a **continuous distribution** (no substantial gaps), or a **discontinuous distribution** (notable areas of absence). The difference between these distribution types is displayed in Figure 36.2, which shows the current (small) discontinuous range of the *Pan* genus—chimpanzees and bonobos—and the historic (large) continuous range. Large gaps in distribution are called **disjunctions**.

When distributions are examined at a very fine scale, very few species will have an absolutely continuous distribution as there will be gaps between different populations. The usual demarcation as

Figure 36.2 Discontinuous and continuous distribution of chimpanzees. Current discontinuous distribution of chimpanzees (*Pan troglodytes*) and bonobos (*Pan paniscus*) shown in brown relative to the historic continuous range (shown in yellow).

Source: Goodenough and Hart (2017) *Applied Ecology: Monitoring, Managing, and Conserving*. Oxford: Oxford University Press. Reproduced with permission of the Licensor through PLSClear. Bonobo: GUDKOV ANDREY/Shutterstock.com. Chimpanzee: Abeselom Zerit/Shutterstock.com.

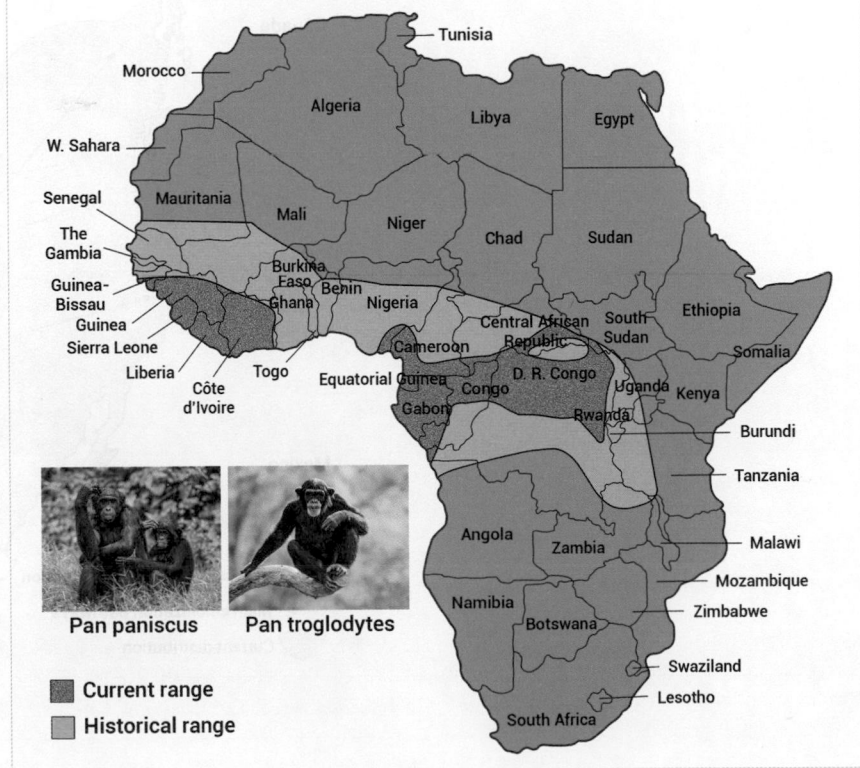

Pan paniscus   Pan troglodytes

■ Current range
■ Historical range

to whether these disjunctions are big enough to classify a distribution as discontinuous (and the populations as discrete) is whether they are greater than the normal dispersal capabilities of that species. If gap size exceeds dispersal capacities, distributions are usually considered discontinuous, especially if there are many such gaps. Species that are migratory can have a discontinuous range if summer and winter areas do not overlap.

The range of a species or a particular population can also change over time. This change might be natural and short-term due to individuals migrating between a breeding range and a non-breeding range (e.g. grey whale (*Eschrichtius robustus*) in the Pacific) or natural and long-term (e.g. the adonis blue butterfly (*Polyommatus bellargus*) has spread naturally northwards in the UK over recent years).

Humans also change the range of species by moving them—intentionally or accidentally—from one part of the world to another. The European starling (*Sturnus vulgaris*) was introduced

to America in 1890 when about 60 individuals were released in Central Park in New York by Eugene Schieffelin, who wanted to introduce all the species mentioned in Shakespeare's plays to America (starlings being mentioned in Henry IV Part 1). The population became established in New York and subsequently expanded to cover the USA, Canada, and Mexico, as shown by the arrows in Figure 36.3.

▶ We explore other examples of non-native species in Chapter 31 and will discuss the issues surrounding the problems and management of non-native species in more detail in Chapters 39 and 40.

In some cases, the range of a species can contract because of threats such as habitat loss. Quantifying realized range size, and change in realized range size, form an essential part of the International Union for Conservation of Nature (IUCN) species-at-risk lists, which are discussed in Chapter 40.

Figure 36.3 Change in range over time. Range change for an introduced species, the European starling (*Sturnus vulgaris*), from an initial population of 60 in 1890 in New York to >200 million across USA, Canada, and Mexico.

Source: Goodenough and Hart (2017) *Applied Ecology: Monitoring, managing, and conserving.* Oxford: Oxford University Press. Reproduced with permission of the Licensor through PLSClear. Photo: arjma/Shutterstock.com.

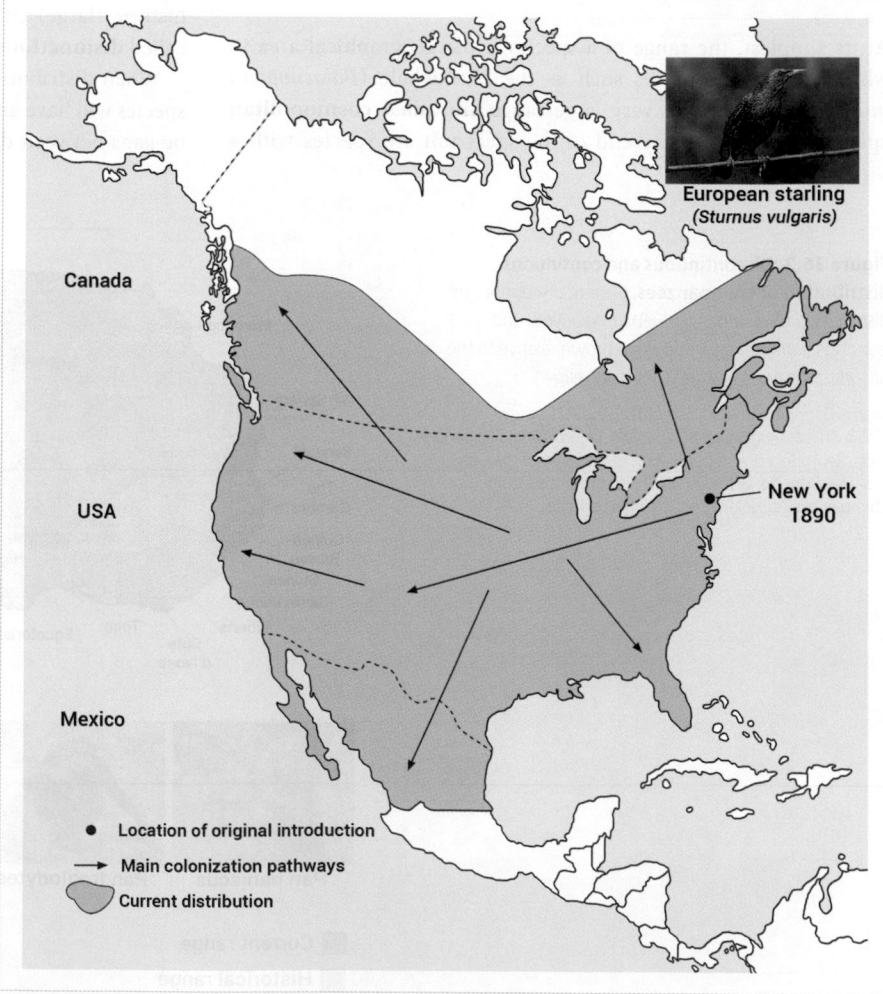

European starling
(*Sturnus vulgaris*)

Canada

USA

Mexico

New York
1890

● Location of original introduction

→ Main colonization pathways

⬤ Current distribution

PAUSE AND THINK

Why are endemic species, and especially micro-endemic species, more prone to extinction than cosmopolitan species?

Answer: If there is a population crash for an endemic or micro-endemic species in the only geographical area they inhabit, that would have consequences for the entire population. For a micro-endemic species, extinction at one site would mean extinction from the biosphere as a whole. Also, although there are many counter-examples, range size can be indicative of population size (larger range size = bigger population size).

Population dispersion

Dispersion is the spatial arrangement of the individuals within a given population (note that this should not be confused with 'dispersal', which refers to the movement of individuals). In nature, biologists recognize three main dispersion patterns: clustered, uniform, and random, which are illustrated in Figure 36.4. (It is worth noting that we discuss intraspecific competition, which we mention briefly in Figure 36.4, in more detail in Chapter 37.)

Go to the e-book to explore an interactive version of Figure 36.4.

Many organisms cluster together because it benefits individual survival or reproductive success. Spatial analysis can be used to investigate these non-random patterns and to test hypotheses about the costs and benefits involved. For example, the larvae of the acorn barnacle (*Semibalanus balanoides*) swim in the plankton layer before finding a place to settle and metamorphosing into fixed, filter-feeding adults. The arrangement of the adults is determined by larval abundance, larval choice in where to settle, and post-settlement mortality. Adult barnacles usually cluster together. This suggests that the benefits of a clustered dispersion (e.g. availability of mates, protection from abiotic challenges, such as desiccation or wave damage, reduced predation pressure) outweigh the costs (e.g. competition between individuals for space, food, and other resources).

Population structure and life-history parameters

In addition to knowing something about the size, density, distribution, and dispersion of a population, it can be useful to know something about the **population structure**. Two fundamental factors to consider when assessing population structure are the **sex ratio** and **age profile**. When suitable data are available, these two parameters can be conveniently plotted on a graph called a population pyramid. Figure 36.5 depicts a population pyramid for the closed population example used in Chapter 35, Figure 35.11: the

Figure 36.4 Spatial arrangement of individuals withing a population.
(a) Clustered; (b) uniform; and
(c) random.

Source: Paula Cobleigh/Shutterstock.com; JeremyRichards/Shutterstock.com; olpo/Shutterstock.com.

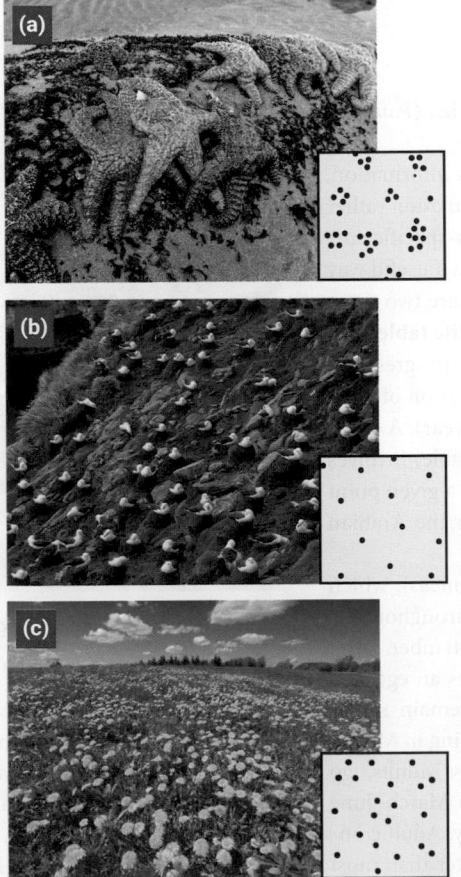

Clustered: Individuals are grouped together into small groups that usually reflect the uneven distribution of biotic and abiotic factors. Organisms may also organize themselves into clumped populations for other reasons: for example, social interactions or protection from predators. However, there is often substantial competition between individuals within the clusters (intraspecific competition).

Uniform: This pattern of population dispersion is observed in fairly homogenousenvironments when resources are evenly distributed but when the individuals in the population influence each other in a negative way: for example, by competing to defend a territory, such as a nesting site.

Random: Random distribution tends to be seen when the individuals are not strongly influenced by abiotic factors (or when those abiotic factors themselves are not clustered) andwhere there is little competition between individuals.

Figure 36.5 Age:sex pyramid for the Arabian leopard (*Panthera pardus nimr*) captive population on 31 December 2009. The *y*-axis represents age class (in years in this example) and the *x*-axis shows the number of animals represented in each age group, with males on the left and females on the right giving us 26 age:sex classes in total (13 age classes, each spanning two years, for each sex). The line represents the shape that this age:sex profile should be for the population to be considered stable.

Source: Reproduced with permission from J. Budd & K. Leus. The Arabian Leopard *Panthera pardus nimr* conservation breeding programme. *Zoology in the Middle East*, 54(supp.3): 141–150. Copyright (c) 2011, Rights managed by Taylor & Francis. https://doi.org/10.1080/09397140.2011.10648905. Photo: Anmar T/Shutterstock.

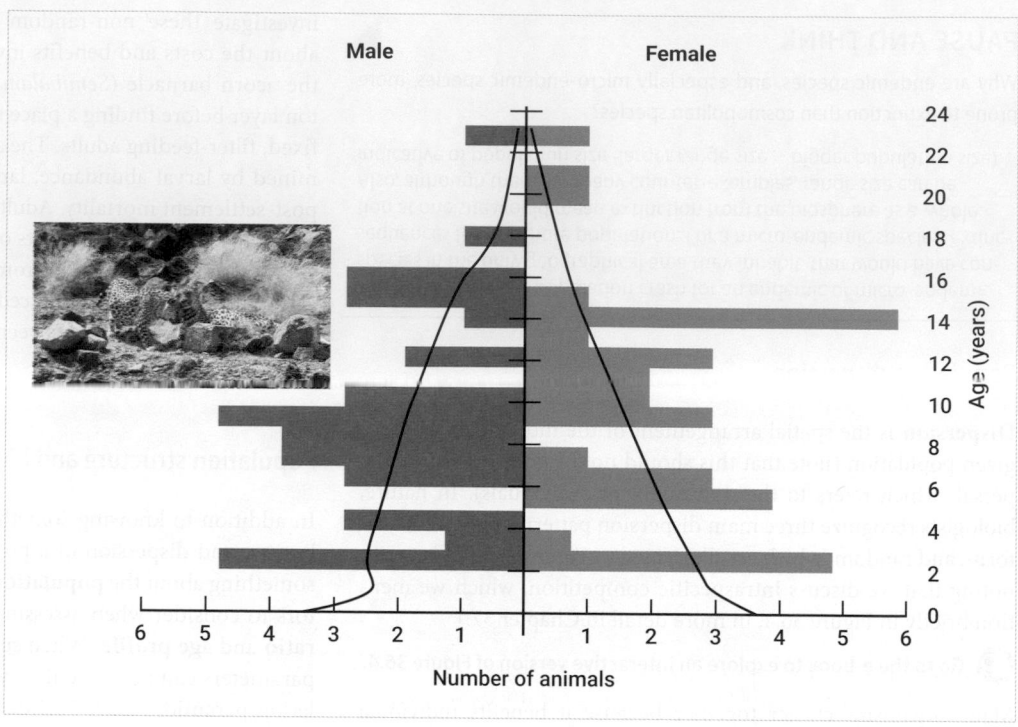

captive population of the Arabian leopard subspecies (*Panthera pardus nimr*).

While a population pyramid gives the basic age/sex information, it focuses purely on an overall description of the population rather than on the *processes* driving these patterns (e.g. sex-specific survival, age-specific reproductive output). A **life table** is a useful way of setting out this more detailed information. There are two types of life table: cohort tables and static tables. A **cohort life table** (also called a generation or horizontal life table) follows the progress of a specific group—a cohort—made up of a single generation of individuals born at a specified time (e.g. within the same year). A **static life table** (also called a time-specific or vertical life table) involves a description—a snapshot—of a whole population at a given point in time (similar to the snapshot age:sex pyramid for the Arabian leopard shown in Figure 36.5).

The common field grasshopper (*Chorthippus brunneus*), which is pictured in Figure 36.6, is found in dry grassland throughout the UK. Towards the end of the summer, usually in September, adult males and adult females mate and each female buries an egg pod just beneath the surface of the ground. The eggs remain in the underground pod throughout the winter before hatching in March. The young grasshoppers, which look like wingless adults, go through several instar stages, separated by moults, in March–June before emerging from the final moult as adults in July. Adult common field grasshoppers can survive cold weather better than most grasshoppers, but few individuals born in any summer will survive beyond December. This means that the common field grasshopper is an example of a species where the generations do not overlap

Figure 36.6 The common field grasshopper (*Chorthippus brunneus*).

Source: Jörg Hempel/CC BY-SA 3.0 DE/Wikimedia Commons.

(i.e. all adults seen in summer and autumn will have hatched in the spring of the same year and all will die that same winter); the species only exists in egg form in January of each calendar year (in the UK).

Table 36.1 shows a cohort life table for a study population of the common field grasshopper. The data contained in a life table will depend on the organism concerned and the purpose of the study. This organism has a life history that involves a number of discrete

Table 36.1 A cohort life table for a population of the common field grasshopper (*Chorthippus brunneus*) that combines a survival schedule with a fecundity schedule

1: Stage	2: Survival schedule	3: Survivorship	4: Fecundity schedule	5: Individual fecundity	6: Fecundity relative to survival
	Number of individuals at start of stage	Proportion of original cohort surviving	Number of eggs produced at stage	Number of eggs per surviving individual	Number of eggs per original individual
X	n_x	l_x	f_x	$m_x = f_x/n_x$	$l_x \times m_x$
0 (Jan; egg)	44,000	1.000	–	–	–
1 (March; instar I)	3513	0.080	–	–	–
2 (April; instar II)	2529	0.058	–	–	–
3 (May; instar III)	1922	0.044	–	–	–
4 (June; instar IV)	1461	0.033	–	–	–
5 (July–Sept; adult)	1300	0.030	22,617	17	0.51

Source: Based on original data from Richards, O.W. and Waloff, N. (1954) Studies on the biology and population dynamics of British grasshoppers, *Anti-Locust Bulletin*, 17, 1–182.

life stages (egg, four successive instars, adult) that occur in different months, such that it is more useful to break down the information for each life stage than for arbitrary age classes.

Take a few moments to familiarize yourself with the column headings of Table 36.1 and the data shown.

- Column 1 sets out the names of the stages of the grasshopper lifecycle and the calendar month(s) in which each instar occurs—note that there is no overlapping here and no adults survive the winter.

- Column 2 is the number of individuals at the start of each stage: the **survival schedule (n_x)**.

- Column 3 is calculated from the survival schedule by dividing the data in the survival schedule (i.e. the number of individuals surviving at each stage) by the initial population size at the start of the cohort (the number of eggs in this case). This gives proportion of individuals surviving to each stage or **survivorship (l_x)**. In this example, 0.080 for instar I is calculated by dividing the 3513 individuals recorded at that life stage by the 44,000 individuals there were at the start of the population.

- Column 4 is the number of offspring produced during each stage: the **fecundity schedule (f_x)**.

- Column 5 is calculated by dividing the fecundity schedule by the survival schedule to give the average reproductive output per individual: **individual fecundity (m_x)**. In this example, the 22,617 eggs produced by the adult population in life stage 5 are divided by the 1300 grasshoppers still alive at this stage.

- Column 6 is the product of both the fecundity of the individuals at that stage in the population (m_x) and the probability of surviving to get to that stage (l_x) and is calculated by multiplying these numbers together. In this example, the calculation is 17 eggs per surviving adult (m_x) multiplied by the probability of becoming a surviving adult (l_x) of 0.030 so $l_x m_x = 0.51$. We will discuss this metric more in the context of net reproductive rates in Section 36.4, both for these grasshopper data and more generally.

Note that we have introduced concepts relating to fecundity here. To find out more about this concept, see Chapter 33.

PAUSE AND THINK

Calculate individual fecundity (m_x) for adult grasshoppers using the survival schedule (n_x) data in Table 36.1 but changing the number of eggs (the fertility schedule: f_x) from 22,617 eggs to 75,565 eggs.

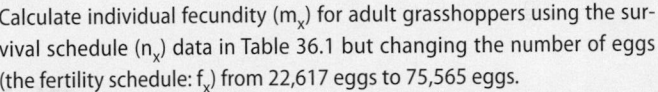

Answer: The calculation is 75,565 eggs produced by the adult population in life stage 5 divided by the 1300 grasshoppers still alive at this stage, which is 58 eggs per grasshopper instead of the value of 17 in Table 36.1.

 Check your understanding of the concepts covered in this section by answering the questions in the e-book.

36.2 Types of population

We tend to consider two main types of populations based on movement of individuals. As we discussed briefly in Chapter 35, we refer to these as closed populations and open populations:

- **Closed populations:** In closed populations, population size, population density, and age/sex demographics are regulated solely by birth rate and death rate—there is no movement of individuals between populations. Truly closed populations are relatively rare in nature but they can occur. Examples include micro-endemic species that only occur at one site (such as the Devil's Hole pupfish that we considered in Section 36.1 and in Figure 35.11), the population of red squirrels (*Sciurus vulgaris*) on Brownsea Island in Dorset, UK, or the fenced population of black rhinoceros (*Diceros bicornis*) in Etosha National Park, Namibia (see Figure 35.11 in Chapter 35 for other examples).

- **Open populations:** Open populations are regulated not only by birth and death rates but also by the movement of individuals between populations. We refer to movement of individuals into a population as **immigration** and movement of individuals out of a population as **emigration** (a useful way to remember these terms is that *i*mmigration is movement *i*nto the population whereas *e*migration is *e*xit from the population). Note that these concepts are different from an individual organism moving between winter and summer geographical areas, which is repeated bi-directional movement via migration as covered in Chapter 31. We are talking here about one-off, one-direction movement of an individual from one population (from which it is an emigrant) to another (to which it is an immigrant).

The difference between open and closed populations is summarized in Figure 36.7.

In addition to categorizing populations as open or closed, we sometimes consider the populations at a regional scale. A regional group of connected populations—a population of populations—is known as a **metapopulation**. The populations within a metapopulation, sometimes called subpopulations, usually inhabit distinct

patches of appropriate habitat (akin to virtual islands) separated by a matrix of less suitable habitat (akin to the sea). For a group of populations to be considered a metapopulation there needs to be movement of individuals (and thus genes) between the populations through the process of individual movement. This means that the subpopulations are, by their very definition, open populations.

▶ We discuss the movement of individual organisms in Chapter 31.

The study of metapopulations is complex because researchers often need to bring together knowledge of population dynamics (considered throughout this chapter) and knowledge of gene flow (which is discussed in Chapter 35) to analyse population structures based on a molecular profile, as explained in Scientific Process 36.1.

Figure 36.7 Closed and open populations.
(a) Closed populations are regulated by birth and death, whereas (b) open populations are regulated by birth, death, and movement of individuals in and out of the population.

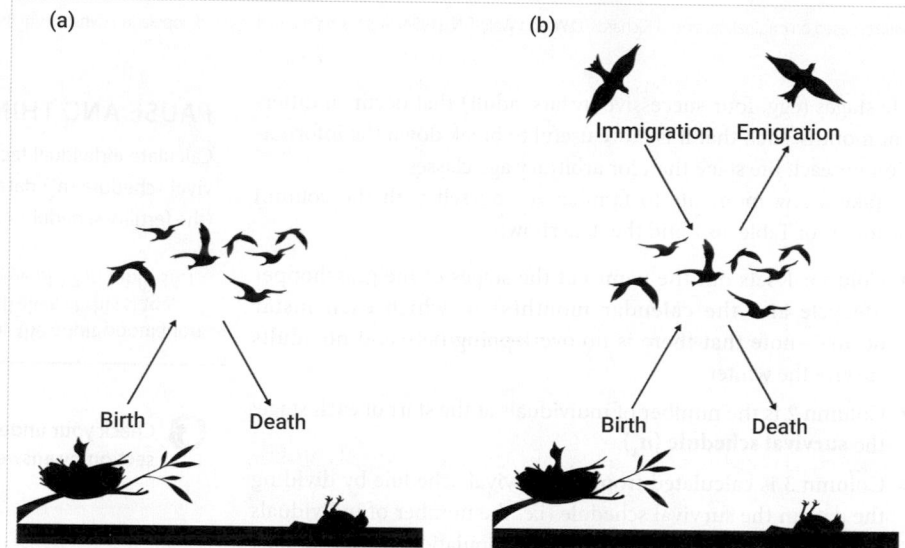

SCIENTIFIC PROCESS 36.1   Metapopulation dynamics and gene flow in northern goshawks

Background and hypothesis

The northern goshawk (*Accipiter gentilis*) is a large bird of prey (Figure 1) that breeds in coastal Canada and Alaska—both mainland forest patches and near-shore islands. Given that goshawks can fly, it might be hypothesized that these different populations actually constitute a metapopulation. Against this possibility is the fact that, in other geographical areas, goshawks exhibit high site fidelity with up to 97% of birds coming back to the same area to nest each year. Sarah Sonsthagen and colleagues decided to test these competing hypotheses by quantifying between-population movements.

Methods

Blood samples were taken from 332 nesting birds in nine different populations. In the lab, total genomic DNA was extracted from the blood and a specific region of the mitochondrial DNA (mtDNA) was amplified using polymerase chain reaction (PCR). This created numerous copies of the

Figure 1 Northern goshawk (*Accipiter gentilis*).
Source: Jesus Giraldo Gutierrez/Shutterstock.

target control region, which could be separated from the rest of the DNA using gel electrophoresis. This uses an electrical current to separate DNA through a gel that acts rather like a sieve that work horizontally using an electrical current rather than vertically using gravity. The relevant section of DNA (the amplified region) was then extracted from the gel and sequenced.

The reason that mtDNA was used in this study (and indeed is most commonly used for this type of research) is that mtDNA is passed down the maternal line as an exact copy between generations. In this way, each individual within a population should have exactly the same mtDNA as its mother, grandmother, great-grandmother, etc. However, over time, mutations occur. This means that the match in the mtDNA between two individuals becomes a genetic clock that indicates how related those two individuals are (more differences = less closely related). Analysis of mtDNA of multiple individuals within a population allows the examination of patterns of evolution. By extension, therefore, analysis of mtDNA from multiple individuals in several different populations allows the examination of spatial movement between those populations with genetic dissimilarity being a proxy for movement of individuals over time.

In this case, the researchers examined the differences in mtDNA between all 332 birds as one big group, and then between the mtDNA of the birds in each of the nine populations separately, to establish whether or not individuals had historically moved between populations, and, where movement had occurred, to establish the direction of movements to identify the main immigration and emigration routes for each subpopulation.

Results

As depicted in Figure 2 using arrows, four populations (coded CBC, KUP, KIS, and POW) were identified as having high historical emigration rates among populations (shown by many arrows pointing away from these populations in Figure 2), while VAN and REV had high immigration rates (shown by many arrows pointing towards these populations in Figure 2). Two other populations, coded ADM and LFD, both experienced moderate rates of immigration and emigration (in Figure 2, there are a few arrows pointing away from and towards these populations). Population HG appeared to be relatively isolated from other populations as no gene flow was observed between this population and any

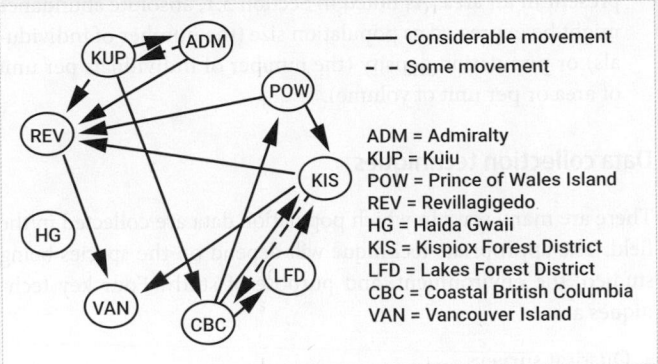

Figure 2 Metapopulation dynamics for nine populations of northern goshawks. The diagram shows the direction of movement of individuals (arrow) and relative importance of immigration or emigration.

of the others (there are no arrows pointing away from or towards this population in Figure 2).

Discussion

This analysis demonstrates that there is a well-functioning metapopulation framework with long-term gene flow between eight of the nine populations studied. It also revealed that gene flow is often asymmetric with some populations having high emigration (**source populations**) and some having high immigration (**sink populations**).

Extension question

Using mtDNA is an effective way of tracking population linkages over evolutionary time. What research methods might be suited to examining movement between populations of this large bird over a much shorter timescale (e.g. 5 years)?

Read the original paper

Sonsthagen, S.A., McClaren, E.L., Doyle, F.I., Titus, K., Sage, G.K., Wilson, R.E., Gust, J.R., and Talbot, S.L. (2012) Identification of metapopulation dynamics among Northern Goshawks of the Alexander Archipelago, Alaska, and Coastal British Columbia. *Conserv. Genet.* **13**: 1045–1057.

 Check your understanding of the concepts covered in this section by answering the questions in the e-book.

36.3 Methods of quantifying population demographics

Quantifying the demographics of a population, and analysing change in the population over space or time, requires detailed, accurate, and robust field data. Surveying populations in the field is one of those aspects of biology that is straightforward in theory but challenging in practice.

Data types

There are several types of population data that might be collected, including, in order of complexity:

1. **Presence/absence:** This is the easiest parameter to measure—simply whether a species occurs in a given area or is likely (on the best available evidence) not to occur in that area. This is useful for assessing the geographical range of a species, or the spatial confines of a particular population, but not for any advanced modelling of population dynamics.

2. **Relative abundance:** This is the abundance of species relative to one another based on count data, percentage coverage, or abundance ranks. It is useful for assessing community structure, but is not normally used for detailed understanding of a specific population within that community. The exception to this is when quadrats are used to generate percentage cover data as a proxy for population density.

3. **Absolute abundance:** This is the actual number of individuals present in an area. As noted in Section 3.1, absolute abundance might be expressed as population size (the number of individuals) or population density (the number of individuals per unit of area or per unit of volume).

Data collection techniques

There are many ways in which population data are collected in the field. The appropriate technique will depend on the species being studied, the environment, and purpose of study. Four key techniques are:

- Quadrat surveys.
- Point count surveys.
- Transect surveys.
- Capture–mark–recapture analysis.

Quadrat surveys

Quadrat sampling involves marking out a small spatial area as a sampling unit. The sampling unit is typically subdivided into smaller sub-units to facilitate data collection as shown in Figure 36.8. Data are usually collected as percentage coverage or prevalence (number of sub-units in which a species is present) to give information on relative abundance or density based on coverage

rather than number. In some cases, the dispersion of individuals within the quadrat is mapped to give information about spatial arrangement of individuals within the population.

Quadrats are a useful way of collecting data on ground layer vegetation (often using a 1 m × 1 m quadrat) but can also be used to study other species in other situations: for example, lichen coverage on trees or rocks (typically 10 cm × 10 cm), seaweed in intertidal zones (typically 2 m × 2 m), rocky shore molluscs and crustaceans (typically using a 20 cm × 20 cm quadrat), or even for corals, where an underwater quadrat is used (Figure 36.8).

Point count surveys

Whereas quadrat surveys are generally on a species' coverage at specific points, point counts tend to involve more detailed fieldwork to count individuals at, or from, specific sample points. These points might be laid out across the study area either systematically, strategically (e.g. in relation to the different habitats at that site or to target particular areas), or randomly. The exact methods that are used at each point might vary considerably, from trapping animals (e.g. using pitfall traps for ground-dwelling invertebrates, light traps for moths, or Longworth traps for small mammals), to checking pre-set refugia (e.g. metal or carpet sheets for reptiles), to simple counts of the number of individuals of a species can be seen or heard from that point (e.g. birds, large mammals, and bats when acoustic detectors to transform ultrasonic echolocation calls

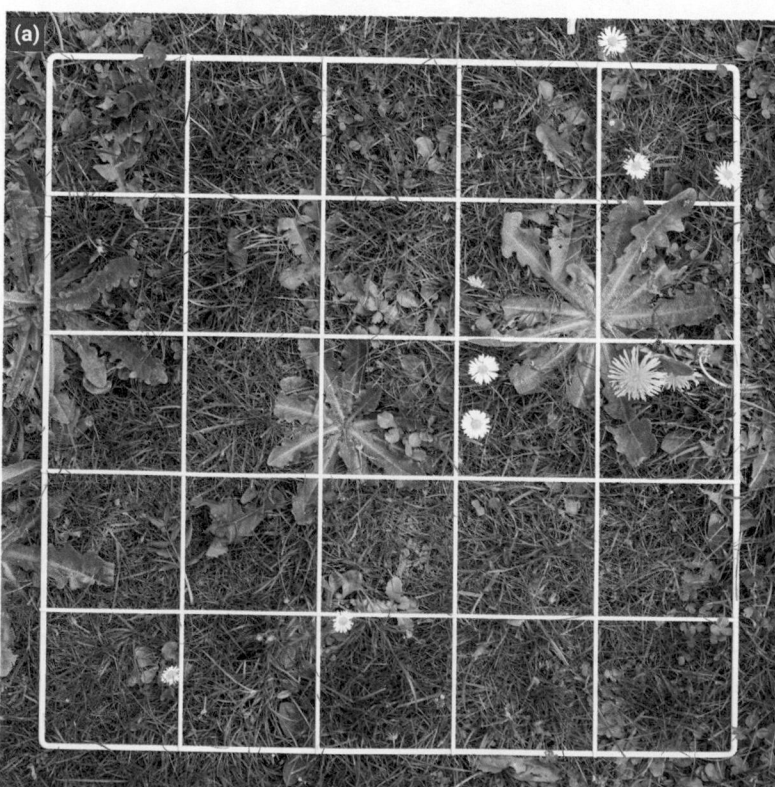

Figure 36.8 Quadrats are usually square sampling grids and subdivided into sections. Quadrats can be used to sample species in: (a) terrestrial environments for species such as plants or lichens; and (b) underwater environments, for example to quantify species within coral reefs.

Source: (a) Science Photo Library; (b) Alamy.

Figure 36.9 Camera traps can be set up in a range of environments and configurations to take images, including nocturnal images. Here, (a) shows two cameras mounted together with the top unit set to record photographs and the bottom unit to capture video footage; both units are mounted on a metal support within an acacia bush in South Africa overlooking an animal burrow; (b) shows the back of the camera trap units showing the setup options; and (c) shows an aardvark emerging at night and captured with infrared setting on the top unit.

to within the range of the human ear). See Further Reading recommendations to find out more about specific survey types.

One particular form of point counts involves using passive camera traps as a way of collecting data on populations without the biologist physically being present. This can be a real advantage for populations that occur at very low density (so the likelihood of an individual being encountered using traditional point counts or transects is low) or where the species concerned is hard to survey (e.g. nocturnal). The advantage of **camera trapping** is that animals are not captured or disturbed in any way, making this a perfect example of **non-destructive** and **non-invasive** sampling. Camera traps operate continually and silently, and many have a night mode using an infrared light source. They can usually be set so that a photograph or a short video sequence is recorded at the approach of an animal using motion sensitivity technology. Camera traps should be deployed at points where they are more likely to capture species activity so spatial distribution is likely to be strategic and targeted (rather than systematic or random). Depending on the species, this might be:

- Near a known den or feeding area or at a water hole.
- Next to a pathway or wildlife corridor.
- Next to a boundary such as a hedge or ditch.
- At a site where there are unconfirmed sightings or indirect evidence of species' presence.

Camera traps are excellent for confirming species presence (see Figure 36.9). When used for populations where animals are uniquely identifiable from fur/plumage patterns (e.g. jaguar (*Panthera onca*) or brown hyena (*Hyaena brunnea*)), abundance can also be estimated; we explore this for tigers (*Panthera tigris*) in Scientific Process 36.2.

Transect surveys

Transects are another commonly used method for surveying populations. The simplest form of transect is a **strip transect**, which is effectively a quadrat stretched into a long strip. Like quadrats,

strip transects sample a proportion of the study area. An observer travels along a predetermined strip of land at a set speed and all observations of a given species within a pre-set distance from the transect line are recorded to provide an estimate of population density. Both visual and acoustic records can be used for this sort of survey, depending on the species involved. Strip transects can be very effective as they are quick and simple to do and need no special equipment. However, they make one major assumption: that all individuals in the strip are recorded. This causes a dilemma. The thinner the strip, the more likely it is that all individuals are detected and recorded, but numerous trips will be needed to cover a reasonable area. The wider the strip, the better the sample size but the more chance there is that individuals will be missed and population density underestimated.

Line transects solve this dilemma by recording the perpendicular (right angle) distance from the transect to every animal observed. The logic of this is that humans are not perfect—they will miss sightings. At zero distance (on the line of travel) all objects will be recorded. The further away an object is, the more likely it will be missed. We assume that we see all the animals that are very close to the transect, and the probability of detection declines as distance increases: this is a **detection function**. Figure 36.10 explains the process by which this is calculated for European hares (*Lepus europaeus*), a species that is declining but where there are still important populations in some agricultural landscapes in the UK.

Once the detection function is known, we can work out a suitable strip transect width. In the example shown in Figure 36.10, 11 out of the 15 hares that were present within 150 m either side of the surveyor were seen—this is a 73.3% detection rate. If multiple strip transects of 300 m wide (= half-width 150 m) are undertaken as necessary to cover the whole area, and the overall number of animals detected is summed, we can re-scale the estimate to allow for the missed individuals:

Estimated population size = Number of animals observed/detection probability as a proportion.

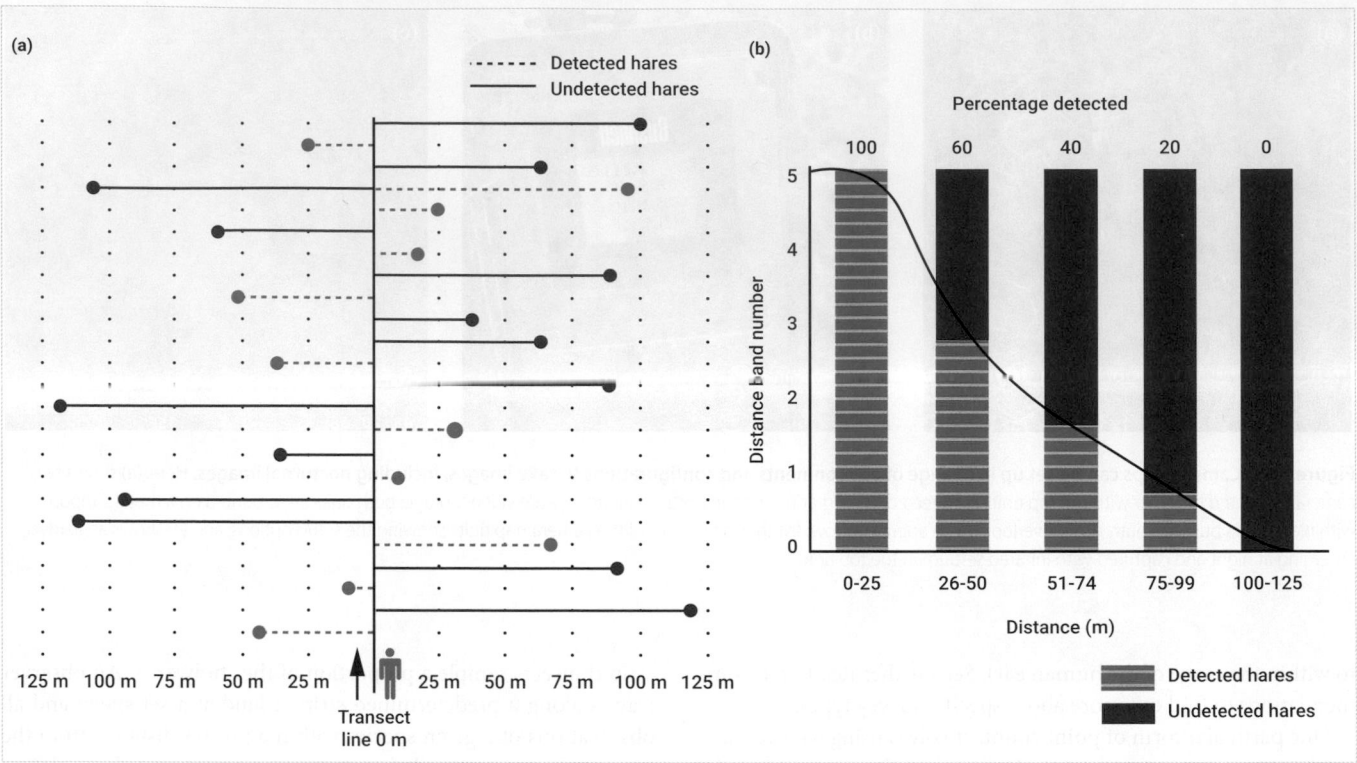

Figure 36.10 The detection function in European hares (*Lepus europaeus*). (a) A walked transect through a population of hares, with all animals close to the transect line being seen (100% detection) but detection declining with distance: individuals that were detected from the transect line (11 out of 25) are shown in green, together with the perpendicular distance of each individual to the transect line, while individuals that were not detected (14 out of 25) are shown in red. (b) Five distance bands have been created so that we can group the number seen into successive distance categories and plot this on a histogram to visualize detection decay—you can cross-reference this information by counting the number of hares that were detected in each distance band (remember to sum both the left and right side of the transect). In this case, as is typical, the detection curve—the black line superimposed on the histogram—takes the form of a logistic decay curve.

Source: WildlifeWorld/Shutterstock.com.

If our overall area was 1500 m wide and we did five transects, each seeing 20 hares, the estimated population size would be: 136 animals (100 animals observed/0.733 detection probability = 136.43 rounded down to the nearest whole animal). If our overall area was 3000 m wide but, because of time constraints, we only did five transects, we would re-scale the overall estimate for the surveyed area (doubling 136 animals to 272 animals in this case).

The above workings assume that the animals are dispersed uniformly, or, if they are random or clustered, that our transect locations are **representative** of the entire population. This is always important in **surveying** (examining part of a population and upscaling it to cover the whole population) rather than **censusing** (examining the whole population). Censusing is the more accurate method for assessing population size and density as every individual is specifically included, but this is not usually logistically possible given the size of the area, the number of individuals, and the fact that animals move around as well as the surveyor.

Capture–mark–recapture

Capture–mark–recapture (CMR) is used to estimate population size in numerous taxa, including mammals, fish, reptiles, and insects. CMR is often used for temporal monitoring or to estimate

initial population size to inform offtake quotas, which is discussed further in Chapter 40. The basic premise is to quantify population size using the ratio of marked to unmarked individuals. To do this, multiple individuals of the same species are captured and marked before being released back into the original, wild population. The marks do not necessarily need to identify one individual from another (**unique/individual marking**), but they must indicate individuals that have been caught previously (**batch marking**). An example of marking is shown in Figure 36.11 for European hedgehogs (*Erinaceus europaeus*).

After release of all the marked individuals, the population is re-sampled by re-trapping and the ratio of previously caught (marked) individuals to new (unmarked) individuals is recorded. Because this ratio in the sample should be the same as in the whole population, assuming our CMR protocol has been designed to capture a representative selection of individuals in a representative area, it can be used to estimate overall population size. The simplest CMR method is the **Lincoln–Peterson index**, which only requires one capture stage, one marking stage, and one recapture stage. The equation is:

$$N = \frac{M \times C}{R}$$

Figure 36.11 Spine-marking for a European hedgehog (*Erinaceus europaeus*). The same position and colour can be used to mark animals in relation to site and year (batch marking) or, for small-scale studies, to identify animals individually. Here, the yellow spine tags will be trimmed before release.

Source: Lucy Bearman-Brown, used with kind permission.

where:

- N = the number of individuals in the population
- M = the number of individuals captured initially and *M*arked
- C = the total number of individuals *C*aptured in the second sample
- R = the number of individuals *R*ecaptured (i.e. the number of marked individuals in the second sample).

So, if 20 individuals were captured and marked initially, and, in the second capture 18 were caught of which 10 were marked, the estimated population size (N) would be:

$$N = \frac{20 \times 18}{10} = 36$$

The major assumption here is that the population is closed (no immigration or emigration: see Section 36.2) and that N is stable over time either because there is no reproduction or mortality between surveys (or that birth rate and death rate are equal and so cancel one another out). There are specialist methods when these assumptions do not hold, including the Jolly–Seber method (see Further Reading).

Another key assumption is that the initial trapping and marking process does not **bias** the second sample. This means that marked individuals should not be more likely to be trapped than unmarked individuals (trap-happy animals can occur when traps are baited if individuals associate traps with food), and should not be less likely to be trapped than unmarked animals (trap-shy animals can occur if individuals associate traps with stress). If different groups of individuals have different 'trapability' (e.g. if juveniles have a different likelihood of entering traps relative to adults), methods with multiple recapture periods, such as the Burnham and Overton method (see Further Reading), might be necessary.

The type of tagging mark depends on species and length of study, as well as ethical and legal considerations. Marks are either permanent/temporary and unique/batch. Examples of marking include fur clipping for small mammals (usually batch, temporary), dye or paint for invertebrates (usually batch, usually temporary), ear

notching for large mammals (unique or batch, permanent), or adding a leg or wing tag for birds (usually unique, permanent). In some cases, marks are big enough to be viable when the individual is encountered without it having to be trapped so that capture–mark–resight can be used instead of capture–mark–recapture.

 Check your understanding of the concepts covered in this section by answering the questions in the e-book.

36.4 Modelling population change

Over time, populations may increase in size, decrease in size, or remain unchanged. The ability to understand, quantify, and model changes in the size of a population—as well as change in other population parameters—is central to the study of population ecology and practical management of endangered populations.

Quantifying population change between two survey periods

Once we have data on a population from survey work, we can start to assess change over time. In some cases, we might know the size or density of a specific population at two particular points in time and we simply want to quantify the direction and magnitude of change between those census points. We can do this using a metric called the **population multiplication rate** usually represented by λ (sometimes called finite rate of change), which is simply the size of the population (N) the *second* time we measured it (N_2) divided by the size of the population the *first* time we measured it (N_1):

$$\lambda = \frac{N_2}{N_1}$$

- If the population has remained unchanged then $N_2 = N_1$ so $\lambda = 1$.
- If the population has increased between measurements ($N_2 > N_1$) then λ will be > 1 (a growing population).
- If the population has decreased between measurements ($N_2 < N_1$) then λ will be < 1 (a shrinking population).

If we consider the tiger population size estimates for 2010 and 2012 in Table 1 of Scientific Process 36.2 we can assess the population

SCIENTIFIC PROCESS 36.2   Establishing the population size of Bengal tigers in the Sundarban mangrove forest

Background

The range of the tiger (*Panthera tigris*) (Figure 1) has been severely reduced by habitat loss. The Sundarban mangrove forest covers over 10,000 km² of southern Bangladesh and the Indian state of West Bengal. It is the largest single block of tidal mangrove forest in the world (an aerial view of which you can see in Figure 2), it is a UNESCO World Heritage site, and it is thought to support one of the largest tiger populations in India and Bangladesh.

Research question

Previous estimates of tiger density in the Sundarban had ranged very considerably from as low as 0.7 animals/100 km² to as high as 23.5 animals/100 km². An accurate estimate of the Sundarban tiger population is important in order to conserve the species in this area and limit conflicts between tigers and the ever-growing local human population. This is not an easy task, though, as each tiger can have a home range of over 10 km² and the habitat consists of dense mangrove forest criss-crossed by tidal channels that are hard to navigate by boat.

Methods

Manjari Roy and her colleagues at the Wildlife Institute of India needed to adapt the normal methods of measuring population size for this particular species and area. Surveying from the water was considered too risky given that small boats would be needed to navigate channels and tigers have been known to attack these. Instead, Roy and her colleagues started by fitting four tigers with radio-tracking collars to get a better idea of the tigers' distribution and home ranges. They found that individual tigers rarely crossed channels more than 1 km in width and used this information to determine two sample areas known as effective trapping areas (ETAs).

Once the ETAs were established, the researchers conducted two surveys—in 2010 for ETA one and in 2012 for ETA two—by setting up camera traps to estimate tiger numbers in specific locations, indicated by the dots and triangles on the map in Figure 3. This was done on the basis that each tiger could be individually identified by its unique stripe pattern. Thus, for each 'detection event' (camera trap image of a tiger), it would be possible to establish whether that tiger had been detected before, either on that camera trap or another one in the study area, or whether it was a new individual. The ratio between unique new and unique previously seen individuals could be calculated in a similar way to the ratio between marked and unmarked individuals in a capture–mark–recapture (CMR) framework to estimate population size. The number of camera traps was proportional to the ETA size (Table 1).

Two different CMR frameworks were used. The first was based solely on the ratio of unique new and unique previously seen individuals. The second used a computer modelling approach called spatially explicit capture–recapture (SECR) because it used the locations of each detection event to estimate the number of tigers within the sample area. The SECR approach is particularly appropriate for free-ranging animals that occupy a large area. The model is based on the probability of an animal being detected and the distance between the detection event and an estimation of the centre of that specific animal's home range (the spatial area an individual habitually uses).

Results

The population estimate for the ETA surveyed in 2012 is a lot higher than that of the ETA surveyed in 2010 (Table 1), but the area of this second ETA is also much higher. The population densities calculated for the two ETAs, which, by definition, factor in the size of the area, are much less variable than the population size estimates.

Figure 1 Bengal tiger in the Sundarban mangrove forest.

Source: Thomas and Pat Leeson/Science Photo Library.

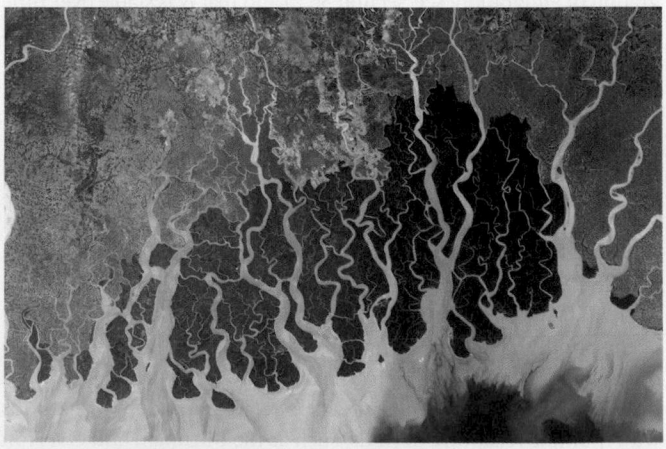

Figure 2 The Sundarban mangrove forest.

Source: NASA image created by Jesse Allen, Earth Observatory, using data obtained from the University of Maryland's Global Land Cover Facility. https://earthobservatory.nasa.gov/images/7028/sundarbans-bangladesh

Figure 3 Map of the 2010 and 2012 effective trapping areas and camera trap locations.

Source: Reproduced from M. Roy, et al. (2016) Demystifying the Sundarban tiger: novel application of conventional population estimation methods in a unique ecosystem. *Popul Ecol*, 58, 81–89. https://doi.org/10.1007/s10144-015-0527-9. Distributed under the terms of the Creative Commons Attribution 4.0 International (CC BY 4.0).

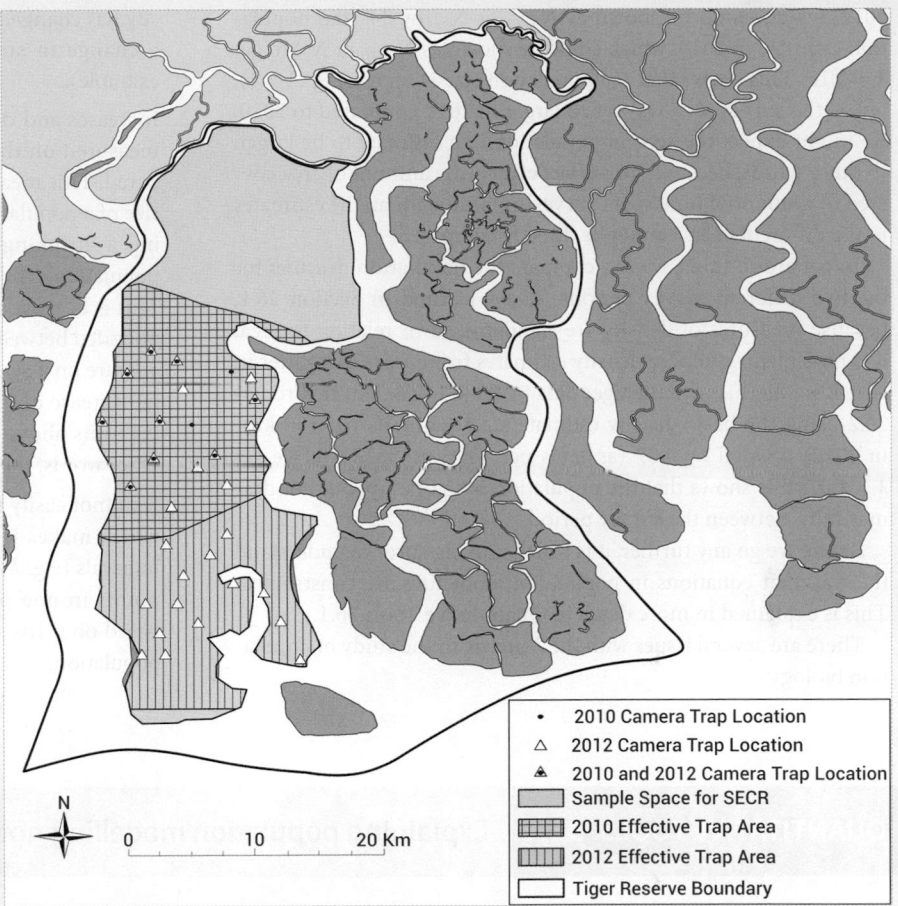

- 2010 Camera Trap Location
- 2012 Camera Trap Location
- 2010 and 2012 Camera Trap Location
- Sample Space for SECR
- 2010 Effective Trap Area
- 2012 Effective Trap Area
- Tiger Reserve Boundary

N

0   10   20 Km

Table 1 Tiger population metrics based on the effective trapping area (ETA) surveys in 2010 and 2012

	ETA1: 2010	ETA2: 2012
Total numbers of photographic traps	11	29
Number of unique tigers photographed	10	22
Number of unique tigers re-photographed	6	18
Population estimate (N)	11	24
Effective trapping area (A)	270 km^2	518 km^2
Population density (N/A)	4.07/100 km^2	4.63/100 km^2
Population density: spatially explicit	4.08/100 km^2	5.81/100 km^2

Discussion

This study gave a much more accurate estimate of population size and population density than had been possible previously. Moreover, the density estimates for the two different ETAs were fairly similar, which increases our confidence in the result. It was a careful study with ETAs defined based on the ecology and behaviour of the animals being studied. The spatially explicit modelling approach also allowed the researchers to compensate for a patchy habitat and the complex probabilities of capturing and then recapturing individual animals. Unfortunately, the estimates of tiger population size and density from this study are at the lower end of previously estimated values, but these can now inform conservation initiatives and provide a baseline for future monitoring.

Extension question

How could the methods and sample sizes of this study have been further improved to give an even more reliable and robust population estimate?

Read the original paper

Roy, M. et al. (2010) Demystifying the Sundarban tiger: novel application of conventional population estimation methods in a unique ecosystem. *Popul. Ecol.* **58**: 81–89.

multiplication rate by dividing 24 (the population estimate in 2012, which could be shown as N_{2012} or N_2) by 11 (the population estimate in 2010, which could be shown as N_{2010} or N_1) to get $\lambda = 2.18$. This shows that the population size is increasing ($\lambda > 1$), but as the survey area was much larger in 2012 compared to 2010, we might expect the second population size estimate to be larger. In other words, because our **survey effort** (the amount of area covered or amount of time spent surveying) is not equal, the estimates in the different years cannot be directly compared.

To get around this, we can compare the population *densities* for the two different survey periods. As we learned in Section 36.1, because measures of density are standardized in relation to area, it is possible to compare density estimates from areas that differ in size. If we do this with the tiger data in Table 1 of Scientific Process 36.2 using the basic density estimate: 4.63 (animals per 100 km^2 in 2012) divided by 4.07 (animals per 100 km^2 in 2010) we get $\lambda = 1.07$; this shows that the population has gone up, but not dramatically, between the survey periods.

Before we go any further, it is worth making sure we understand the way that equations in population modelling are constructed. This is explained in more detail in Quantitative Tools 36.1.

There are several issues with the λ metric in the study of population biology:

1. It is purely descriptive; it tells us *how* the population size or density has changed but not *why* (we could not determine whether a change in survival or fecundity was driving change, for example).

2. Increases and decreases in the population rate of increase are measured on different scales and so cannot be directly compared. This means that although the direction of change in the size of a population is clear from whether λ is below 1 (decreasing; a shrinking population) or above 1 (increasing; a growing population), the magnitude of the change is harder to interpret. This is because a shrinking population will have a value that is bounded between 0 and 1 while a growing population will have a figure on a scale from 1 to infinity. This means that a λ of 0.6 (a decrease of 0.4 units below 1) and a λ of 1.4 (an increase of 0.4 units above 1) are not the same magnitude of change. In other words, the metric is asymmetric.

3. It cannot easily be converted from one unit of time to another, which makes it hard to compare estimates on different time intervals (e.g. λ based on a 5-year interval between two census points in one population is not directly comparable with λ based on a 10-year interval between census points in another population).

QUANTITATIVE TOOLS 36.1   Explaining population modelling notation

We have seen that the population multiplication rate (λ) is calculated using the equation:

$$\lambda = \frac{N_2}{N_1}$$

Here the subscript numbers after N refer to the first (N_1) and second (N_2) time we measured the size of the population. We could as easily have used different subscripts for the measurements of population sizes at different times. For instance, we could measure the difference between the eighth and ninth time we measured the population size:

$$\lambda = \frac{N_9}{N_8}$$

Or if we measured a population in different years the equation might take a format similar to this:

$$\lambda = \frac{N_{2020}}{N_{2019}}$$

Alternatively, if we measured the population size of a species such as the free-living nematode *Caenorhabditis elegans*, which we saw in Chapter 35 has a very short generation time, measuring λ over days, weeks, or months might be most appropriate, for example:

$$\lambda = \frac{N_{June}}{N_{May}}$$

The common theme here is that we are always looking at the difference between one time period and a second time period, with the actual size of the time period being optimized for the organism being studied.

The convention is to use the letter t to denote a time period because this means that we can standardize our equations. Thus, if t is a time period when we first measured the population, our second census point becomes $t+1$ (we could also show earlier census periods as $t-1$). Written this way, t can represent years, months, days, hours, or minutes, or an interval relevant to the population under consideration (e.g. generation time for that species) and the $+1$ simply shows that we are moving forward in time by one unit.

We can write a general equation for population multiplication rate (λ) once we realize that it is calculated from the population size (N) in a particular time period (t) divided by the population size in the previous time period. Therefore:

$$\lambda = \frac{N_{t+1}}{N_t} \text{ or } \lambda = \frac{N_t}{N_{t-1}}$$

A special case of N_t is the starting population before any increase or decrease happens. This can be represented as the population in time period zero (N_0).

In addition, λ only works if we have at least two estimates of population at different times. In reality, we often only have *one* population estimate and we want to use that to *predict* how the population will change. This requires the use of a predictive modelling approach rather than simply documenting change between two census points.

Basic mathematical prediction of population change

One of the easiest ways of modelling population change for a closed population (i.e. a population with no immigration or emigration) is to consider the **net reproductive rate per generation** (R_0). R_0 values can be interpreted in the same way as the population multiplication rate (λ): a value of $R_0 = 1$ indicates an unchanging population, $R_0 > 1$ indicates a growing population and $R_0 < 1$ indicates a shrinking population. The key difference between R_0 and λ is that R_0 values can be calculated *predictively* from just one population census point.

We can calculate R_0 from the data in a life table by multiplying survivorship data for each life stage (l_x) by individual fecundity at each life stage (m_x) to quantify fecundity relative to survival to give $l_x m_x$ at each life stage, and summing these values to give R_0, thus:

$$R_0 = \Sigma l_x m_x$$

where:

- R_0 = net reproductive rate
- Σ = 'the sum of . . .'
- l_x = survivorship: the proportion of the individual cohort surviving to a given stage or age x
- m_x = individual fecundity: the number of offspring (or eggs) per surviving individual at each stage or age x.

In the grasshopper life table (Table 36.1), we calculated $l_x m_x$ just once—multiplying 17 eggs per surviving adult (l_x) by multiplying the 0.030 probability of each grasshopper in the initial population becoming a surviving adult (m_x). This gave us $l_x m_x = 0.51$. Because there is only a *single* life stage (time period) when reproduction occurs, R_0 in this case is exactly equal to the single $l_x m_x$ value (i.e. $R_0 = l_x m_x$ rather than $R_0 = \Sigma l_x m_x$). As the $R_0 < 1$ in this case, we would predict that the population is shrinking (in fact, the R_0 of 0.51 means that there is only about one replacement individual for every two individuals in the original population).

Grasshoppers are an example of a **semelparous species**, which means that they have a single reproductive episode over their lifetime. Because this is synchronized for all individuals in the population, and there is no adult overwinter survival, generations do not overlap. This is often the case for insects that undergo metamorphosis. In such cases, R_0 will *always* be equivalent to the $l_x m_x$ in the single life stage during which reproduction occurs. In addition, for a semelparous species with non-overlapping generations, the population multiplication rate is equivalent to the net reproductive rate per generation ($\lambda = R_0$). We can show this with a quick calculation based on the grasshopper data shown in Table 36.1.

Our starting population of grasshoppers was 44,000 (eggs). This value can be referred to as N_t (the size of the population the first

time we measured it); the population the following year (t_{+1}) was 22,617 (eggs)—refer back to Quantitative Tools 36.1 if you are unsure about this notation. That gives us:

$$N_t = 44{,}000$$

$$N_{t+1} = 22{,}617$$

$$\lambda = \frac{N_{t+1}}{N_t} = \frac{22{,}617}{44{,}000} = 0.51$$

You will see that λ being 0.51 is exactly the same as our previous figure for R_0, thus confirming that $\lambda = R_0$ in this case.

However, most species (including humans) are not semelparous, but instead have multiple reproductive episodes and, because these are not synchronized, generations overlap. The term for this is **iteroparous**. In these cases, we need to calculate the $l_x m_x$ for each life stage or age class separately in a life table, and then sum these together (i.e. $R_0 = \Sigma l_x m_x$ rather than $R_0 = l_x m_x$).

As an example, look at Table 36.2. This is a life table of the Galapagos cactus finch (*Geospiza scandens*), which shows female survivorship (l_x, expressed as a proportion of original females that survive), individual female fecundity (m_x, average number of daughters per surviving female), and fecundity relative to survival ($l_x m_x$). In this case, as is quite common, we are looking at the number of *female* offspring per *female* rather than using total offspring (male and female) per individual.

Table 36.2 Life table of the Galapagos cactus finch (*Geospiza scandens*)

Stage	Female survivorship	Individual female fecundity	Fecundity relative to survival
Age class in years	Proportion of original females surviving	Average number of daughters per surviving female	
X	l_x	m_x	$l_x m_x$
0	1.000	0	0
1	0.512	0.364	0.186
2	0.279	0.187	0.052
3	0.279	1.438	0.401
4	0.209	0.833	0.174
5	0.209	0.500	0.104
6	0.209	0.833	0.174
7	0.209	0.250	0.052
8	0.209	3.333	0.696
9	0.139	0.125	0.017
10	0.070	0	0
11	0.070	0	0
12	0.070	3.500	0.245
13	0	—	—

Source: Adapted from Grant, P.R and Grant, B.R. (1992) Demography and the Genetically Effective Sizes of Two Populations of Darwin's Finches, *Ecology*, **73**, 766–784.

Ideally, a life table should always be compiled for the *specific* population that we are modelling as survival and fecundity will vary for different populations of the same species. However, in reality, we often use standardized species-level estimates for parameters such as survivorship and fecundity schedules from other populations. Although this is less robust, it means that it is possible to create a fairly accurate population model for our study population using a single number—the initial population size—to predict likely change, with the other necessary information coming from knowledge of the species from previous studies of other populations. This is often much more practical and means that we can model populations quickly without collecting data over many years in the field. This is the big advantage of R_0 relative to λ.

Population growth

Populations of all species have the potential to grow rapidly when resources are freely available. The change in the size of a population over time will depend on births, deaths, immigration, and emigration. For the sake of simplicity, let's ignore the effects of immigration and emigration, and focus instead on a closed population where birth and death rates are the only controlling factors. We will further simplify things by assuming that survival is consistent (i.e. death rate is a constant) so that reproductive rate is the only factor controlling population size.

Discrete breeding events and geometric growth

Consider the grasshopper example that we have been using throughout this chapter. We are going to make a simple change to the fecundity schedule presented in Table 36.1 so that, at stage 5

(adult), there are now 88,000 eggs produced rather than 22,617: this alters individual fecundity and fecundity relative to survival. These changes are shown in Table 36.3. The population now has a net reproductive rate per generation (R_0) of 2 and is thus growing. We can check that value for the population multiplication rate (λ) is also 2 (remember that λ and R_0 are the same in semelparous species with non-overlapping generations) thus: 88,000/44,000 = 2.

If this population is in an environment where there is absolutely no limit on the resources available, and we assume that there is no immigration or emigration of grasshoppers, we can model what might happen to this population over several years. (Remember also that because there is only one generation per year, the number of years in the model in this case is also equal to the number of generations.) Table 36.4 shows what a hypothetical population of grasshoppers might look like over time if the survival and fecundity are as per Table 36.3.

The population size is growing very quickly: it has increased 10-fold in 11 generations as shown by the growth curve in Figure 36.12. We call this type of population growth **geometric growth** that occurs over discrete time periods. This is a pattern whereby the population in each generation is equal to the population of the generation before multiplied by the R_0 of 2. For example, looking at the first and third column of Table 36.4, the adult population of generation 1 ($N_1 = 2600$) can be calculated by multiplying the adult population in generation 0 ($N_0 = 1300$) by 2. In the same way, the adult population in generation 2 ($N_2 = 5200$) can be calculated by multiplying the adult population in generation 1 ($N_1 = 2600$) by 2, and so on.

We can generalize our current equation, which is generation specific, to apply to all generations. We do this by noting that the population size of the next generation (N_{t+1}) is equal to the population of the current generation (N_t) multiplied by 2 (2 being the net reproductive rate per generation, R_0, for this population) as follows:

$$N_{t+1} = N_t \times 2 \quad \text{or} \quad N_{t+1} = N_t \times R_0$$

We can now predict the size of the population in any generation if we know the size of the previous generation and the net reproductive rate (R_0). This is a simple population growth model for geometric growth over discrete time periods. Models are very useful as they allow us to make predictions. Currently, our model only works by using the information in one generation to predict

Table 36.3 An adjusted cohort life table for the common field grasshopper from Table 36.1, with numerical modifications highlighted in bold

1: Stage	2: Survival schedule	3: Survivorship	4: Fecundity schedule	5: Individual fecundity	6: Fecundity relative to survival
	Number of individuals at start of stage	Proportion of original cohort surviving	Number of eggs produced at stage	Number of eggs per surviving individual	Number of eggs per original individual
X	n_x	l_x	f_x	$m_x = f_x / n_x$	$l_x \times m_x$
0 (egg)	44,000	1.000	–	–	–
1 (instar I)	3513	0.080	–	–	–
2 (instar II)	2529	0.058	–	–	–
3 (instar III)	1922	0.044	–	–	–
4 (instar IV)	1461	0.033	–	–	–
5 (adult)	1300	0.030	**88,000**	67.7	**2.0**

Table 36.4 Growth of a hypothetical population of grasshoppers using parameters from Table 36.3 in an environment where there is no resource limit, immigration, or emigration. Remember that all adults die after producing eggs (so no generation will contain adults of the previous generation) and that, as there is only one generation per annum, year and generation are synonymous. See Quantitative Tools 36.1 for details of notation for multiplication rate

Time period or generation (t)	Number of eggs per time period (N_t)	Surviving adults per time period (N_i)	New eggs produced (N_{t+1})	Population multiplication rate (λ), equivalent to net reproductive rate per generation (R_0), calculated by N_{t+1}/N_t
0	44,000	1300	88,000	2
1	88,000	2600	176,000	2
2	176,000	5200	352,000	2
3	352,000	10,400	704,000	2
4	704,000	20,800	1,408,000	2
5	1,408,000	41,600	2,816,000	2
6	2,816,000	83,200	5,632,000	2
7	5,632,000	166,400	11,264,000	2
8	11,264,000	332,800	22,528,000	2
9	22,528,000	665,600	45,056,000	2
10	45,056,000	1,331,200	90,112,000	2

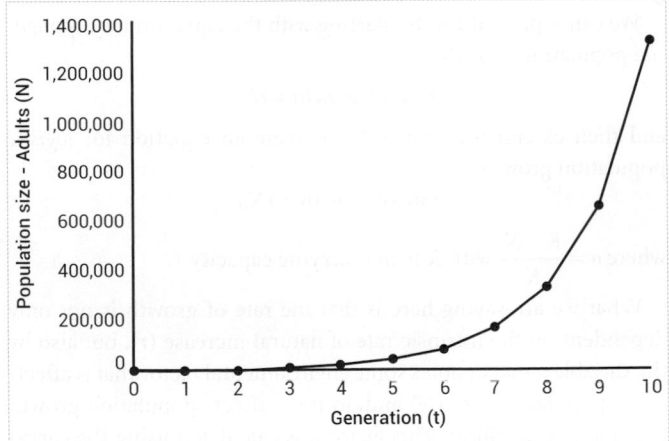

Figure 36.12 Geometric growth. Geometric growth over discrete time periods in a population of grasshoppers described in Table 36.4.

Source: Graph by Toby Carter, used with kind permission.

forward a single generation more. This is somewhat restrictive so we often opt to extend the model to become more generalized (and thus more useful) by using the just initial starting population size (N_0), the generation number (t) and the net reproductive rate per generation (R_0), as shown in Table 36.5.

Continuous breeding and exponential growth

We now need to consider the situation where, instead of there being a single breeding event in each year for our grasshoppers, they could breed at any time of year (i.e. reproduction occurs continuously rather in discrete annual jumps). The number of eggs produced, and the number of adults to hatch from these eggs, will remain the same each generation but the graph for the population will have a smooth curve (as shown in Figure 36.13).

The rate of growth for a continuously breeding population is exponential if no resource is in short supply (i.e. no limits to

population growth) and the rate of growth is proportional to the population size:

$$rate\ of\ growth = rN$$

where:

- r is the **intrinsic rate of natural increase**
- N is the population size.

The intrinsic rate of natural increase (r) is mathematically superior to λ and R_0. Whereas λ and R_0 vary around a threshold of 1 (i.e. values above 1 indicate a growing population and values below 1 a decreasing population), the intrinsic rate of natural increase (r) varies around zero. Therefore, for r, positive growth rates indicate a growing population and negative growth rates a shrinking population. These values are symmetrical about zero so that a rate of increase of 0.5 (i.e. +0.5) is exactly equal to a rate of decrease of −0.5. In addition, r can convert easily from one unit of time to another allowing for easy comparisons.

Limited resources and logistic growth

We have now considered the situations where a population will grow indefinitely if breeding occurs in discrete events (geometric growth), or if breeding is continuous (exponential growth), when a population is entirely unlimited by natural resources. However, in almost all circumstances, this scenario is unlikely. Normally, we would expect the resources available to be limited and therefore the number of individuals in the population has an impact on population growth because individuals are in competition with one another. This type of competition is called **intraspecific competition** because it occurs between individuals of the same species (we will cover this in more detail in Chapter 37). Intraspecific competition normally has minimal impact when the population is small in comparison to the available resources and greater impact when the population is large in comparison to the available resources.

Table 36.5 Growth of the same hypothetical population of grasshoppers from Table 36.4 but now generalized to show the relationship to R_0 in each generation. Remember that $R_0 = 2$ for this population

Generation (t)	Number of adults per generation (N_t)	Calculation of number of adults for each generation (N_t) (number version)	Calculation of number of adults for each generation (N_t) (equation version)*
0	1300	N_0 (starting population)	
1	2600	$= 1300 \times 2$	$= N_0 \times R_0{}^1$
2	5200	$= 1300 \times 2 \times 2$	$= N_0 \times R_0{}^2$
3	10,400	$= 1300 \times 2 \times 2 \times 2$	$= N_0 \times R_0{}^3$
4	20,800	$= 1300 \times 2 \times 2 \times 2 \times 2$	$= N_0 \times R_0{}^4$
5	41,600	$= 1300 \times 2 \times 2 \times 2 \times 2 \times 2$	$= N_0 \times R_0{}^5$

* The calculations for the number of adults in each generation (N_t) using the equation specific to the generation (the final column of the table) can be generalized for all generations as $N_t = N_0 \times R_0{}^t$ (refer back to Quantitative Tools 36.1 if you need an explanation of the notation used here).

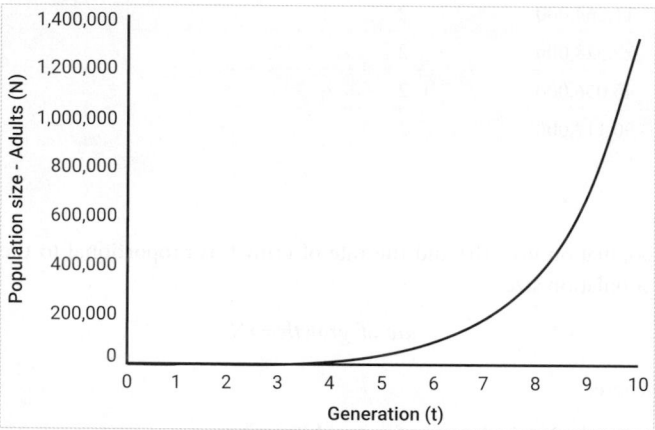

Figure 36.13 Exponential growth. Growth in a population of grasshoppers described in Table 36.4 and shown in Figure 36.12 but now represented as continuous population growth with a smoothed exponential curve rather than growth over distinct time periods.

Source: Graph by Toby Carter, used with kind permission.

To understand what might happen when resource limitation is factored into models of population growth, we will return to our grasshopper data. Let's assume that our grasshopper population is now living in an environment that can only support 500,000 adults in total. The maximum population size the resources in a specific environment can support is called the **carrying capacity** and is usually represented by K. In this example, therefore, $K = 500,000$ grasshoppers, but our geometrically predicted grasshopper population size after 11 generations, based on the starting population of 44,000 grasshoppers, is more than 500,000 (note that the number

of surviving adults in generation 9 of Table 36.4 is 665,600, which is higher than 500,000).

In any situation where there is a fixed carrying capacity, we expect intraspecific competition to have an impact on the *rate* at which the population grows. This would mean that population growth will slow as the population size gets close to the carrying capacity in a pattern of population growth called **logistic population growth**.

We can represent this by starting with the equation for *exponential* population growth:

$$rate\ of\ growth = rN$$

and then extend this principle to create an equation for *logistic* population growth:

$$rate\ of\ growth = rNa$$

where $a = \dfrac{K - N}{K}$ with K being carrying capacity.

What we are saying here is that the rate of growth is not only dependent on the intrinsic rate of natural increase (r), but also by the variable a that denotes some environmental factor that is affected by population size (N) and, in turn, affects population growth (e.g. food availability). This factor a is calculated using the carrying capacity (K) and the population size (N). We would expect a to have little effect on growth when the population is small (i.e. growth would be close to 100% of what we would expect and so the effect of variable a is negligible). Conversely, a will have maximum effect, essentially halting growth, when the population gets close to the carrying capacity (i.e. growth would be close to 0% and so the variable a will be considerable). To understand how we calculate a, see Table 36.6.

Table 36.6 Illustration of how a (the multiplier for r) is calculated, with a carrying capacity (K) equal to 100 and N standing for population size

N	Resource used ($= N/K$)	Impact on r	Magnitude of multiplier for r (a)	Multiplier for r $a = [K-N]/K$
1	$1/100 = 0.01\ (= 1/K)$	Small	Large	$(100-0)/100 = 1$
50	$50/100 = 0.5\ (= 50/K)$	Intermediate	Intermediate	$(100-50)/100 = 0.5$
100	$100/100 = 1\ (= 100/K)$	High	Small	$(100-100)/100 = 0$

Table 36.7 Net reproductive rates (R_0) and population sizes for population growth of grasshoppers in successive generations using the logistic equation for R_0 given above, with a carrying capacity (K) of 500,000 based on a starting population of 1300 adults and an initial R_0 of 2, both as per Table 36.4. In this case, we are modelling over 21 generations rather than 11

Generation (t)	Adjusted R_0	Population size (N_t) for adults
0	2.00	1300.0
1	1.99	2593.3
2	1.99	5159.8
3	1.98	10,214.1
4	1.96	20,019.2
5	1.92	38,497.1
6	1.86	71,489.9
7	1.75	125,094.0
8	1.60	200,120.3
9	1.43	285,837.0
10	1.27	363,735.7
11	1.16	421,119.2
12	1.09	457,182.1
13	1.04	477,633.4
14	1.02	488,560.8
15	1.01	494,214.2
16	1.01	497,090.3
17	1.00	498,540.9
18	1.00	499,269.4
19	1.00	499,634.4
20	1.00	499,817.1

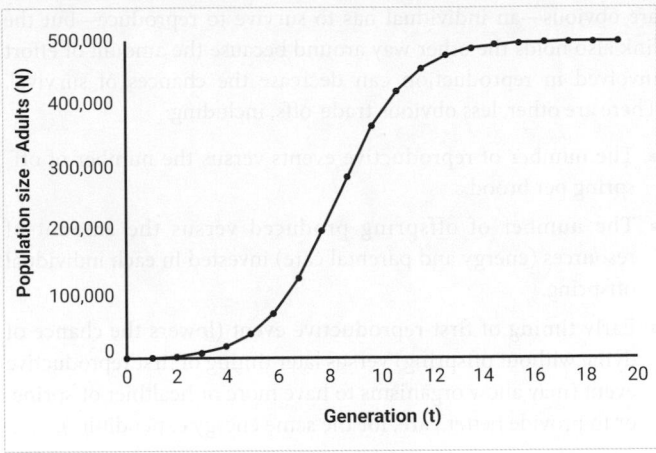

Figure 36.14 Logistic growth. Logistic growth in a population of grasshoppers described in Table 36.4 and shown in Figure 36.13 but now under a constraint of a maximum carrying capacity of 500,000 and showing logistic growth.

Source: Graph by Toby Carter, used with kind permission.

 Check your understanding of the concepts covered in this section by answering the questions in the e-book.

36.5 Modelling life-history strategies

An organism's life history is the sequence of events related to survival probability and reproductive effort that occur between birth and death. Life-history patterns evolve by natural selection, which is discussed in Chapter 34. They represent an 'optimization' of growth, survival, and reproduction for that particular species. There are many characteristics that summarize investments in growth, reproduction, and survivorship and these are called life-history traits. The seven traits generally considered to be the most important are:

1. Size at birth.

2. Growth rate (and whether there is one life form or whether the species metamorphoses).

3. Age and size at maturity.

4. Number, size, and sex ratio of offspring.

5. Age- and size-specific reproductive investments.

6. Age- and size-specific mortality schedules.

7. Length of life.

With the exception of the parameters relating to size and growth, all life-history traits fundamentally link to population biology and can be summarized in a life table (refer back to Table 36.1 and Table 36.2).

Trade-offs

There are always trade-offs in life-history strategies because survival and fecundity are inherently interlinked. Some of these links

If our grasshoppers are all equal in resource needs, each individual would use 1/500,000th of the resources (remember that our carrying capacity (K) is 500,000). If we assume that the limiting reduction (a) in resources affects reproduction, reproductive rate will decrease as the population increases. When the population approaches the carrying capacity, therefore, the growth rate will essentially be zero, such that the population is only breeding to maintain itself. We can then recalculate successive population sizes for grasshoppers operating under the above constraint of $K = 500,000$, with R_0 being adjusted relative to the population size for each successive population as shown in Table 36.7, and then graph this as shown in Figure 36.14.

 Go to the e-book to explore an interactive figure, which compares the graphs in Figures 36.12 and 36.14 to show how growth varies under unconstrained and constrained conditions.

are obvious—an individual has to survive to reproduce—but the link also holds the other way around because the amount of effort involved in reproduction can decrease the chances of survival. There are other, less obvious trade-offs, including:

- The number of reproductive events versus the number of offspring per brood.

- The number of offspring produced versus the amount of resources (energy and parental care) invested in each individual offspring.

- Early timing of first reproductive event (lowers the chance of dying without offspring) versus later timing of first reproductive event (may allow organisms to have more or healthier offspring, or to provide better care, for the same energy expenditure).

Survivorship curves

One way of investigating the life history of an organism is to plot the information about age-specific survival in a life table as a **survivorship curve**. This is a graphical representative of the numbers (or proportion/percentage) of individuals that survive to each stage or age group, usually plotted using a logarithmic scale. This is shown in Figure 36.15 for grasshopper data in Table 36.1 and Galapagos cactus finch data in Table 36.2.

There is a lot of variation in the survivorship curves observed under natural conditions, but they are conventionally categorized into three main types that reflect the different life-history strategies of the species concerned. Take a look at Figure 36.16 for an example. Type I curves represent a low level of mortality (high survival) initially and over most of the life history, with a steep drop in survival rates towards old age: humans are a good example of this type of survivorship curve. Type III curves are the opposite:

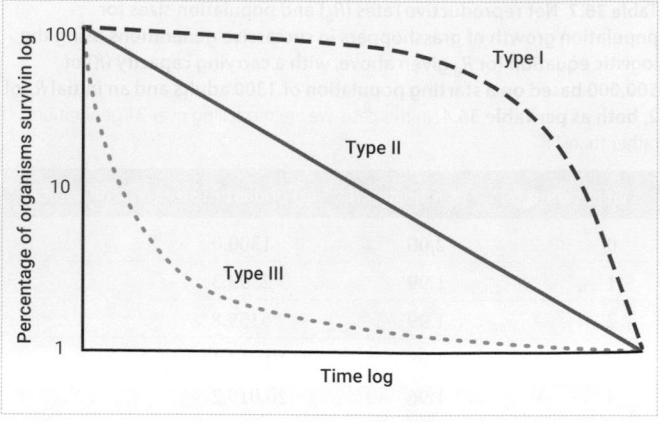

Figure 36.16 The three main types of survivorship curve.

high mortality (low survival) in the early stages of the life history with gradual mortality followed by a fairly good life expectancy for those that survive. Looking back to Figure 36.15, there the green line shows the survivorship curve for the grasshopper data in Table 36.1, we can see that this species most closely resembles a Type III curve in Figure 36.16, which is typical for many inspect species. Type II curves represent a position between these two extremes, with a constant survival rate over the life history as a whole. This is the curve type that the blue line in Figure 36.15, which shows the survivorship curve for the finch data in Table 36.2, most closely resembles.

K-strategists and r-strategists

Organisms with life histories that produce Type I curves are known as *K*-strategists: these organisms produce few offspring (or at least few offspring per reproductive event) but invest substantial resources in their survival. Populations of many large mammals, including human beings, adopt this strategy. It tends to be most common in predictable environments, where sudden density-independent events are rare, and the population size can approach the carrying capacity (*K*) of the habitat (hence the labels *K*-strategist or *K*-selection).

Type III curves are associated with *r*-strategists, like our grasshoppers, which produce lots of offspring, often very quickly in single or rapid-sequence reproductive events, but invest few resources in their survival. This strategy tends to be common in unpredictable environments: why put a lot of investment into a small number of offspring when you could lose them as the result of unpredictable—density-independent—environmental events? An organism that lives in an unstable environment is likely to have population size that is relatively small in comparison with the potential carrying capacity (*K*), but the population growth rate (*r*) can be substantial (hence the labels *r*-strategist or *r*-selection).

▶ **For more information on *r*- and *K*-strategists, see Chapter 33.**

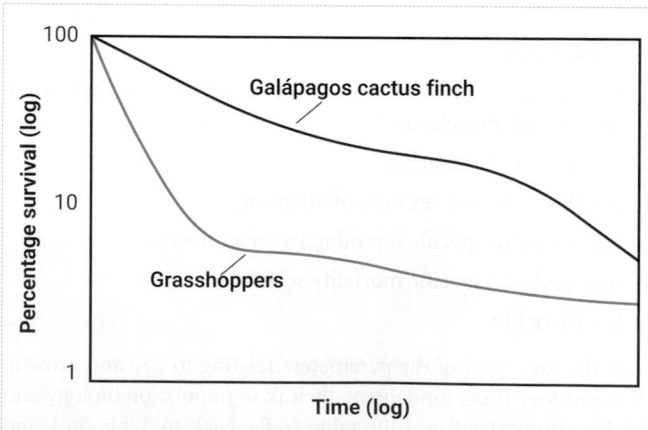

Figure 36.15 Survivorship curves for grasshoppers and Galapagos cactus finch. Data have been taken from Table 36.1 and Table 36.2 but with proportions converted to percentages and with time rescaled so that six life stages (grasshoppers) and 13 years (finches) can be shown on the same *x*-axis.

PAUSE AND THINK

Using the options in parentheses, complete the statements below, using a single word in each case, to indicate the main life-history traits you would expect for *r*-strategist species and *K*-strategist species:

Trait	*r*-strategist	*K*-strategist
Life span (short or long?)		
Age of first reproductive event (early or late?)		
Number of offspring produced per reproductive event (few or many?)		
Gestation period/seed germination period (short or long?)		
Size of offspring (small or large?)		
Amount of parental care provided (little or lots?)		

Trait	*r*-strategist	*K*-strategist
Life span (short or long?)	Short	Long
Age of first reproductive event (early or late?)	Early	Late
Number of offspring produced per reproductive event (few or many?)	Many	Few
Gestation period/seed germination period (short or long?)	Short	Long
Size of offspring (small or large?)	Small	Large
Amount of parental care provided (little or lots?)	Little	Lots

Answer:

 Check your understanding of the concepts covered in this section by answering the questions in the e-book.

SUMMARY OF KEY CONCEPTS

- We typically summarize information about a population in terms of **size** (number of individuals), **density** (number of individuals per unit of area or per unit of volume), **spatial arrangement**, and **structure** (age:sex ratio).

- A **life table** is used to summarize population structure. A **cohort life table** is used for a specific group—a cohort—made up of a single generation of individuals born at a specified time (e.g. within the same year). A **static life table** is used to give a snapshot view of a whole population at a particular point in time.

- Populations that are subject only to the effects of birth and death are termed **closed populations**, whereas those where individuals also move in and out of the population through immigration and emigration are termed **open populations**.

- A **metapopulation** is a regional group of subpopulations where there is known movement between the subpopulations so the whole unit is interconnected. Metapopulations are, by their very definition, open.

- When surveying populations, we might be aiming to confirm **presence**, suggest likely **absence**, quantify abundance in comparison to abundance of other species (**relative abundance**), or quantify the actual number of individuals (**absolute abundance**) to calculate population size/density.

- Key field methods for surveying species' populations include **point count surveys** either directly or using a remote method such as camera trapping, **transect surveys**, and **quadrats**.

- If distance bands are built into transect surveys, and we know the **detection function** for a specific species, the abundance of individuals from transect surveying can be used to calculate population density.

- Calculation of population size can often be undertaking using **capture–mark–recapture**, whereby animals are caught (stage 1), marked and released (stage 2), and the ratio of marked to unmarked individuals in another trapping event (stage 3) is used to quantify population size, for example, using **Lincoln's Index.**

- **Population multiplication rate** is usually represented by λ and is the size of the population the *second* time we measured it divided by the size of the population the *first* time we measured it ($\lambda = N_2/N_1$).

- **Net reproductive rate per generation (R_0)** is calculated by multiplying survival each life stage (l_x) by individual fecundity at each life stage (m_x) to quantify fecundity relative to survival at each life stage ($l_x m_x$), and summing these values to give R_0.

- The maximum population size the resources in a specific environment can support is called the **carrying capacity** and is usually represented by K.

- **Geometric growth** over discrete time periods is a pattern of population growth when reproduction occurs in a series of discrete events, generations to not overlap, and the population *is not* limited by resources (so the ratio of change in a population is consistent between generations).

- **Exponential growth** is a pattern of population growth when reproduction can occur continually, generations overlap, and the population *is not* limited by resources.

- **Logistic growth** is a pattern of population growth when reproduction can occur continually, generations overlap, but the population *is* limited by resources.

- **Life-history traits** are characteristics that summarize investments in growth, reproduction, and survivorship in a species (e.g. number of offspring per reproductive event, number of reproductive events, longevity).

- **r-strategist** species have many offspring, high offspring mortality, and short longevity.

- **K-strategist** species have few offspring, low offspring mortality, and long longevity.

 Use the flashcards in the e-book to test your recall of key terms introduced in this chapter.

QUESTIONS

 Looking for answers? Once you've answered these questions, follow the link in the e-book to the answer guidance and check your work.

Concepts and definitions

1. What is the difference between population size and population density?

2. What is the difference between a cohort life table and a static life table?

3. How do open populations and closed populations differ?

4. What is a metapopulation?

5. What does carrying capacity (K) mean?

Apply the concepts

6. If you studied a population carefully over a period of 10 years and found that the birth rate and death rate were both static over time, but the population size (and density) had increased over the same time period, what population process must be occurring?

7. When undertaking a capture–mark–recapture study, why is it important, both scientifically and ethically, to ensure that the mark does not increase predation risk or affect mate choice?

8. Given that the intrinsic rate of natural increase (r) is mathematically superior to the population multiplication rate (λ) and the net reproductive rate per generation (R_0), and all three metrics are used to quantify/predict change in population size over time, why do we still use λ and R_0?

9. List the factors that could determine carrying capacity for a species (in other words, what finite resources might individuals of the same species compete for)?

10. A female mosquito (*Anopheles quadrimaculatus*) can lay 100–300 eggs per night. Based on this fact alone, is the species an *r*-strategist or a *K*-strategist?

Beyond the concepts

11. Find a map of global human population density. You will see that *Homo sapiens* has become a cosmopolitan species since dispersing beyond East Africa, but the population density is clustered rather than uniform. What abiotic and biotic factors explain why the human population is simultaneously so cosmopolitan and so clustered?

12. You need to survey absolute abundance of water beetles in a large pond in order to estimate population size. What techniques might be best to do this?

13. A new, very secretive, species of lemur has been discovered in a remote part of Madagascar. The only thing known about its behaviour is that it seems to like the fruit of one particular species of tree. What methods might you use to estimate population size and density for this species?

14. Compare and contrast exponential and logistic population growth and explain why we rarely see exponential growth of a population within a natural ecosystem.

15. What is the carrying capacity for planet Earth for humans? Explain your answer and discuss why this is a hard question to answer.

FURTHER READING

Sutherland, W.J. (2006) *Ecological Census Techniques: A Handbook.*
  Cambridge: Cambridge University Press
  *A guide to census methods for species surveying and
  capture–mark–recapture.*
Rowcliffe , J.M. et al. (2011) Quantifying the sensitivity of camera traps: an
  adapted distance sampling approach. *Meth. Ecol. Evol.* **2**: 464–476.
  Camera trap methods for population quantification.

de Boer, R.J. (2018) *Modeling Population Dynamics.* Utrecht University
  https://tbb.bio.uu.nl/rdb/books/mpd.pdf
  Next steps in population modelling.

Communities

Species Interactions and Biodiversity Metrics

Chapter contents

Introduction		1270
37.1	Ecological interactions	1271
37.2	Classifying communities	1280
37.3	Species community metrics	1280
37.4	Communities over space	1284
37.5	Communities over time	1286

Watch the key concepts video in the e-book to prepare yourself for studying this chapter.

LEARNING OBJECTIVES

By the end of this chapter you should be able to:

- Identify and be able to discuss a range of interactions that occur:
  - between individuals of the same species (intraspecific); and
  - between individuals of different species (interspecific).
- Explore how species populations come together to form integrated ecological communities and explain how communities are defined and classified.
- Understand how to calculate some of the key metrics used to describe communities and assess change in those communities in relation to:
  - species richness;
  - species diversity; and
  - species similarity.
- Describe how species communities change:
  - over space through species–area relationships and global diversity gradients; and
  - over time through ecological succession.

Introduction

Ecological communities are collections of individuals belonging to multiple different species that co-occur in a specific geographical location. In essence, therefore, an ecological community is a collection of populations that interact through a complex series of relationships called **ecological interactions**. A community thus becomes a level of organization in the ecological hierarchy that we have been following in this module: it is a step above the level of the **population** (which we explore in Chapter 36) because it involves multiple species rather than a single species, but it is a step below the level of the **ecosystem** because it does not include interactions between species and their physical environment (which we cover in Chapter 38).

37.1 Ecological interactions

The study of ecology is fundamentally the study of interactions. There are three types of interaction: (1) between individuals of the same species (**intraspecific interactions**); (2) between individuals of different species (**interspecific interactions**); and (3) between individuals and their environment. We consider the first two of these in this section; the third is covered in Chapter 38 when we discuss ecosystem processes.

Ecological interactions can typically be divided into four main groups:

1. Feeding interactions of **herbivory** (animal eating plant) and **predation** (animal eating animal).

2. **Competition** between two or more individuals for finite resources needed to survive, grow, and reproduce.

3. Interactions that occur as a result of a specific mode of living, mainly forms of **parasitism**.

4. Interactions that occur to the benefit of either both parties involved in the interaction (**mutualism**) or to the benefit of one without negatively affecting the other (**commensalism**).

We will consider each of these interaction types in more detail in the following subsections. As we do that, it is important to remember that interactions can occur either as the result of the basic biology of the species involved (e.g. diet; reproductive strategy) or as a result of specific behaviour undertaken by the species involved. We will draw out examples of these interaction pathways in the following text, before summarizing interaction types at the end of the section.

Herbivory and predation

Some species can generate their own food. These are termed **autotrophs** (literally meaning self-feeder) and comprise algae, most plants, and some bacteria, including cyanobacteria. Autotrophs all contain chlorophyll, a series of related green pigments that allows them, in the presence of light, to fix carbon through the process of **photosynthesis**. This essentially means that these species can generate their own food, and thus their own energy.

▶ The biological processes involved in photosynthesis and energy fixation are covered in Chapter 7 and we consider the importance of this process from an ecosystems perspective in Chapter 38.

All species that are not autotrophic are **heterotrophs** (literally meaning other-feeder). Heterotrophs rely on other species for nutrition and usually this reliance takes the form of a direct feeding relationship. If a direct feeding relationship involves an animal eating a plant, we term the animal the **herbivore** and the consumption process is known as herbivory. Herbivory is one of the most common ecological interactions and there are examples of herbivore species from most taxonomic groups, including mammals, birds, fish, reptiles, molluscs, and insects (Figure 37.1). Some herbivores only eat plants, and are therefore **obligate herbivores,** while others incorporate plant material as part of a wider diet, which also includes eating other organisms such as animals and fungi. These species can be termed **facultative herbivores**, although the term **omnivore** is more common. In some cases, an animal will consume all parts of a plant in a single event, but usually they 'nibble' at select parts of a plant and then move on. If this nibbling involves ground-based species such as grasses (i.e. head-down feeding), this is referred to as **grazing**. If trees and shrubs are consumed (i.e. head-up feeding), the term **browsing** is used.

When an animal consumes another animal, this process is often referred to as predation. More correctly, though, predation involves not only consumption of meat, but also the process of an animal (the **predator**) killing an animal (the **prey**). Thus, while the process of animal-eating-animal is related to basic biology—dietary needs—the study of predation also involves consideration of the way in which prey is caught so predator behaviour is also important. For example, some species harness the element of surprise to effectively catch prey, and are often referred to as **ambush predators**. Ambush predators can adopt a sit-and-wait strategy, possibly using camouflage or concealment so that potential prey does not see them. A good example of a species group using concealment behaviour is trapdoor spiders (Figure 37.2a). Other ambush predators create structures in which they ensnare prey. Perhaps the most obvious example of a trap-building ambush predator is a spider that has constructed a web, but trap-building

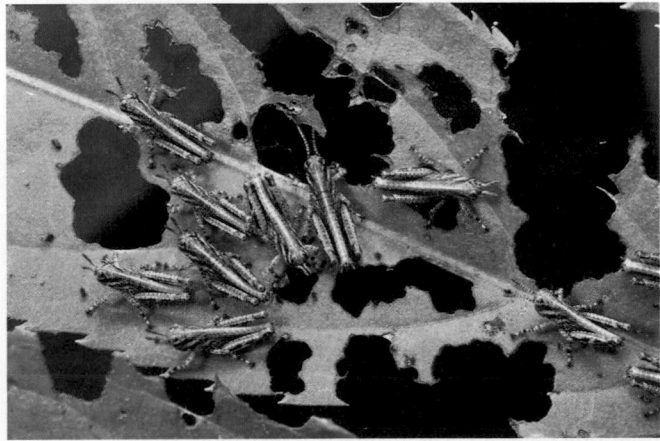

Figure 37.1 Grazing insects. We often think of herbivores as large mammals but insects are actually one of the most ecologically important grazers, especially species such as locusts (Orthoptera), shown here, and the larval forms of butterflies and moths (Lepidoptera).

Source: Awei/Shutterstock.

Figure 37.2 Predation strategies can be immensely varied and include concealment, trap-building, aggressive mimicry and cooperative hunting.
(a) Trapdoor spiders create a burrow covered by a hinged trapdoor from where they can launch a surprise attack. (b) Some species of antlion (Myrmeleontidae) build prey traps—circular steep-sided craters of loose soil and the antlion buries itself in the soil at the bottom waiting for prey to be guided to it in miniature soil 'avalanches'—one of the simplest and most efficient traps in the animal kingdom. (c) Female fireflies (*Photuris* spp.) copy the light signals of other species, thereby attracting male fireflies that they capture and eat. (d) Many species hunt in groups; here meat-eater ants (*Iridomyrmex purpureus*) have worked cooperatively to capture and kill a cicada.

Source: (a) Pong Wira/Shutterstock; (b) photograph by Adam Hart, used with kind permission; (c) James E. Lloyd/Animals Animals/agefotostock; (d) jjron, GFDL 1.2 <http://www.gnu.org/licenses/old-licenses/fdl-1.2.html>, via Wikimedia Commons.

behaviours are also used by other species as shown in Figure 37.2b. Other ambush predators use stealth to stalk their prey and then wait until a good opportunity to pounce; cats such as cougars (*Puma concolor*) are experts in using this behaviour. Other predators use attract potential prey by mimicking individuals of that species searching for mates (**aggressive mimicry**) or hunt in groups in **cooperative hunting** to either improve success rates or tackle prey that is too large to hunt alone: examples of these strategies are shown in Figure 37.2c–d.

In Chapters 34 and 35, we tended to think about evolution as something that occurred to individuals of one species in isolation from other species. The reality is rather more complex because it is often the results of evolutionary change through natural selection in one species that becomes the stimulus for evolutionary change through natural selection in another. We term this co-evolution. Predators and prey are often engaged in a co-evolutionary battle, especially in situations when predators/prey

interactions are highly focused (i.e. a single predator primarily preying upon a single prey species). When this occurs, it leads to a tug-of-war scenario because whenever natural selection acts on the prey species to evolve a mechanism to better evade predation (**genotypic adaptations** such as better camouflage or anatomical/physiological changes to allow the species to run faster), the natural selection also acts on the predator species to evolve better mechanisms for catching the prey (counter-adaptations such as better attuned eyesight or faster acceleration). This is known as an **evolutionary arms race**, and is the evolutionary ecological equivalent of two nations increasing their military power in response to one another's actions.

▶ Refer to Chapter 34 if you need a reminder about natural selection processes.

Occasionally, the adaptions of a prey species can seem out of step with current predator pressure. In these instances, it is possible that

Figure 37.3 Evolutionary arms race. The probable evolutionary arms race between the American cheetah and the pronghorn is still evident in the extant species (pronghorn), even though the putative cause of the adaptation (American cheetah) is extinct.

Source: (top) © Roman Yevseyev; (bottom) Breck P. Kent/Shutterstock; Anton_Ivanov/Shutterstock.

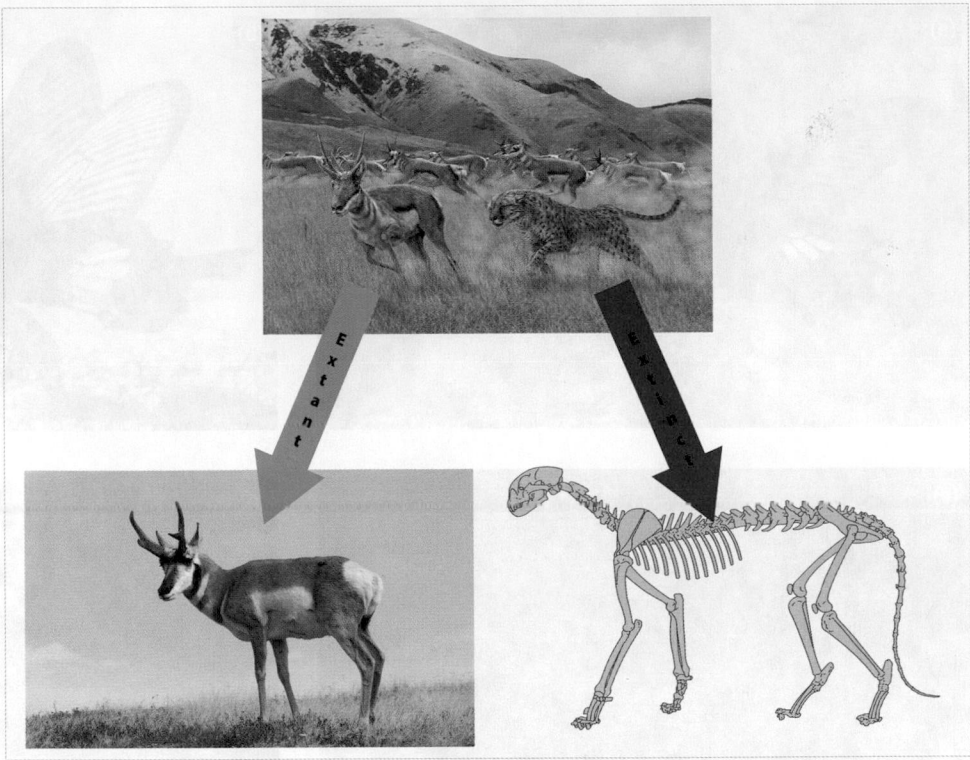

the current situation is a reflection of past predation pressures. One example is the pronghorn antelope (*Antilocapra americana*) in the western USA, which has a top speed of almost 100 kilometres per hour, far in excess of the speed necessary to outpace its current predators such as wolves and coyotes (*Canis lupus* and *C. latrans*). It is likely that this ability evolved in response to the now-extinct American cheetah (*Miracinonyx trumani*), fossils that have been found in several places, including Natural Trap Cave in Wyoming. If this is indeed the case, pronghorn speed is effectively an **evolutionary hangover** of a historical evolutionary arms race (original interaction between cheetah and pronghorn shown at the top of Figure 37.3; American cheetah have now become extinct, but we can infer the historical interaction from contemporary analysis of pronghorn speed).

PAUSE AND THINK

What are the problems of concluding that the pronghorn has adapted to allow high-speed locomotion because of an extinct big cat predator? If the pronghorn's speed is an evolutionary hangover, wouldn't we expect it to lose this ability over time if the cause no longer exists?

Answer: We have an incomplete picture of the cheetah's ecology because it is now extinct. The fossils that have been found show it had long legs and enlarged nasal openings, which both suggest speed, but we cannot be sure of this. We know that the cheetah and the ancestral pronghorn overlapped in time, but we are unsure if they overlapped in habitat and we have no information on behaviour or diet. We would not necessarily expect the adaptation to disappear—it might have become a proxy for mate selection or it might simply be that if there is no cost to being able to run fast, there is no counter selection pressure to 'un-evolve' this ability.

In addition to prey species having specific anti-predator strategies, many prey species also have generic predator defences. Some species secrete distasteful substances or are poisonous (e.g. cane toads: *Rhinella marina*), others are brightly coloured (e.g. cinnabar moth caterpillars: *Tyria jacobaeae*), have patterning that deters predators (several adult moths and butterflies have patterns on their wings that look like eyes, to suggest high vigilance), or have coloration or patterning mimicking another species that is poisonous (this is called **Batesian mimicry**, which is discussed in more detail in Chapter 32, but some examples of this are provided in Figure 37.4).

Predation and herbivory are both typically regarded as examples of interactions where one organism gains (the consumer gains nutrition and thus energy) and the other organism is negatively affected (a loss of biomass in the case of a grazed/browsed plant; loss of life for true prey species). As always in ecology, though, there are counter-examples and two specific subtypes of herbivory nicely illustrate this:

1. **Frugivory:** a frugivore is a species that feeds largely or exclusively on fruit. There are some examples of this occurring at temperate latitudes (i.e. the UK, mainland Europe, USA), but most examples come from the tropics where there is a source of fruit all year round. Although from a plant perspective loss of fruit is a loss of biomass, the fruit is effectively sacrificial. The plant offers this as an enticement to, and reward for, frugivorous species acting as seed dispersers for the seeds within the fruit. In this case, far from being negatively affected, the plant gains because its seeds are deposited away from the parent plant so that offspring do

Figure 37.4 Batesian mimicry often occurs between relatively closely related species. For example: (a) the common mormon butterfly (*Papilio polytes*) is a visual mimic for another butterfly species, (b) the common rose (*Pachliopta aristolochiae*), which is toxic. Sometimes mimicry even involves different taxonomic groups, for example: (c) chicks of the cinereous mourner bird (*Laniocera hypopyrra*) visually mimic (d) toxic co-occurring caterpillars.

Source: (a) Mrs Nuch Sribuanoy Shutterstock; (b) Robert Ang/Shutterstock.com; (c) and (d) https://www.journals.uchicago.edu/doi/10.1086/679106.

not directly compete with the parent plant for space, light, food, and water.

2. **Nectivory:** a nectivore is a species that feeds largely or exclusively on nectar. As with frugivores, most nectivores occur in tropical environments where there are some plants in flower all year. Generally, nectar is produced by plants to attract species who then act as pollinators, with the nectar being the 'reward'. This is typically very effective. However, some nectivorous species have developed adaptations that allow them to access the nectar without entering the flower (as shown by the specialized bills of the scintillant hummingbird (*Selasphorus scintilla*) shown in Figure 37.5a, and the slaty flowerpiercer (*Diglossa plumbea*) shown in Figure 37.5b). In these cases, the plant loses energetically costly nectar without reaping any pollination benefit.

All instances of herbivory are, by their very nature, interspecific—they involve both a plant and an animal. Most examples of predation are also interspecific because predator and prey are usually from different taxa. This is not always the case, however, as some species exhibit **cannibalistic** behaviour and the predation relationship then becomes intraspecific. We have already seen one example of this, when female fireflies eat males (Figure 37.2c); other examples include cockroaches (Blattodea) and Mormon crickets (*Anabrus simplex*). One specific form of cannibalism is **sexual cannibalism**. This occurs when one sex, usually the female, consumes the male during or immediately after copulation and is perhaps most famously exemplified by widow spiders, including redback spiders (*Latrodectus hasseltii*) in Australia, and tropical and sub-tropical predatory praying mantids (Mantidae).

Competition

All living organisms need space and other resources to survive, grow, and reproduce. Because resources are not finite, individuals are in competition either with other individuals of the same species

Figure 37.5 Many species of bird feed on nectar but feeding strategies differ. (a) A scintillant hummingbird (*Selasphorus scintilla*) hovers in front of each flower and probes with its long bill, thus acting as a pollinator, while (b) slaty flowerpiercers (*Diglossa plumbea*) pierce flowers from behind with their needle-like bills to 'steal' nectar without fulfilling a pollinator role.

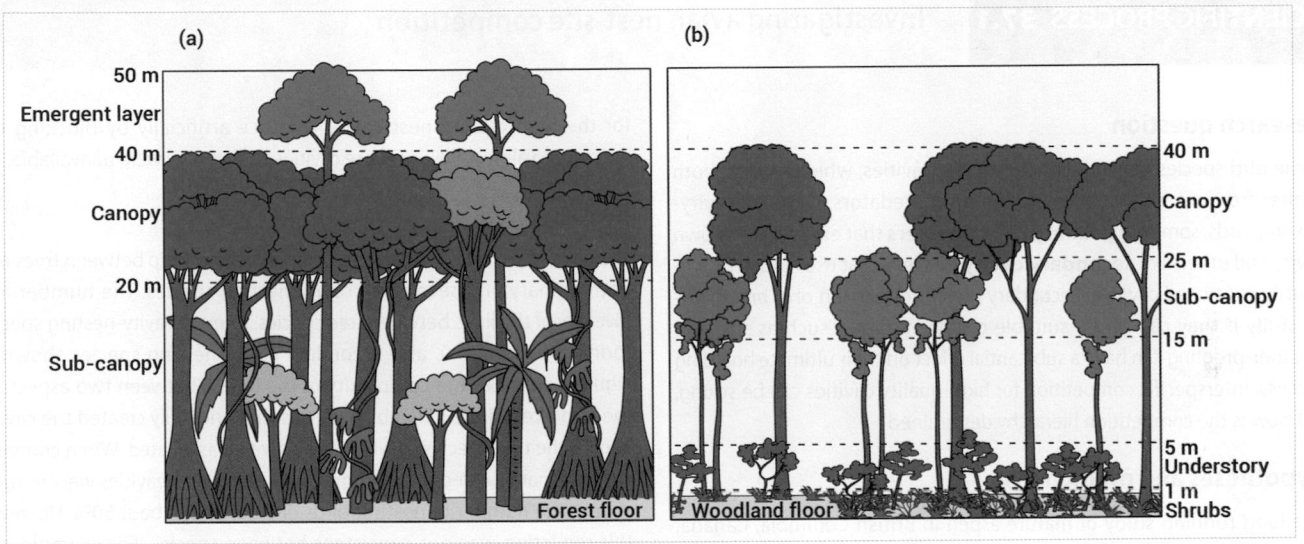

Figure 37.6 The vertical structure of woodlands can differ based on environmental conditions. (a) Tropical rainforest structure is determined by competition for light being very high, which is not the case for (b) temperate woodlands.

(**intraspecific competition**) and/or individuals of different species (**interspecific competition**).

Intraspecific competition is important in determining some of the patterns within populations that we explored in Chapter 36. In particular, it is a fundamental driver of the logarithmic growth curve that is typical of populations where growth is fundamentally limited by **carrying capacity** (K). Intraspecific competition normally has minimal impact when the population is small in comparison to the available resources, and greater impact when the population is large in comparison to the available resources. Population growth is therefore largely unchecked by intraspecific competition initially but will gradually begin to slow as the population size increases; eventually plateauing as the carrying capacity

is reached, as shown in Figure 36.14 by the growth initially being rapid but then flattening and becoming static.

While intraspecific competition is thus crucial for understanding *populations*, interspecific competition is key to understanding *communities*. When organisms of two or more species require similar resources, the competitive power of those different organisms will be one of the regulators of the relative abundance of each, and thus the **species structure** of the community. This is especially true in the case of predator–prey cycles, which we explore further in Section 37.5.

Interspecific competition can also affect **habitat structure**. Figure 37.6 shows two different examples of habitat structure involving woodland habitats: a tropical rainforest (Figure 37.6a) and a temperate woodland (Figure 37.6b). One of the most valuable resources for plants in a tropical rainforest environment is sunlight because the dense vegetation means that very little light penetrates the canopy. This means that the vast majority of plants are either quick-growing trees that 'beat the rush' to fill any gaps in the canopy or emerge beyond it or are climber species that use trees as support to reach the canopy layer (note dense canopy, emergent trees, and climbers in Figure 37.6a, together with the relatively sparse understory). In this way, the overall vegetation structure is fundamentally related to intense competition for light. This contrasts with more temperate deciduous woodlands, where trees often grow more slowly and where there is often a rich flora below the canopy making use of the relatively abundant light in this vegetation stratum (note the abundant vegetation below the canopy layer in Figure 37.6b).

There are two main types of competition:

1. **Interference competition:** one individual dominates a particular resource by actively preventing another individual accessing it (e.g. through aggressive defence of nest site or food resource). Interference competition is thus fundamentally driven by behaviour and is sometimes termed 'contest competition'.

2. **Exploitative competition:** one individual denies another individual access to a resource as a result of its own use of that resource. For example, if a tree is already growing in an area, it is monopolizing that space resource by virtue of its presence alone. Similarly, if a hawk consumes all the small mammals in a given plot of land, there is no prey resource for an owl in the same area. This type of competition occurs because of basic biological processes (growing, feeding, nesting, etc.) and is sometimes termed 'scramble competition'.

SCIENTIFIC PROCESS 37.1   Investigating avian nest-site competition

Research question

Some bird species make their nests in tree cavities, which provide both shelter from weather and protection from predators. Of these cavity-nesting birds, some are **primary cavity nesters** that excavate their own cavity and others are **secondary cavity nesters** that use natural or previously excavated cavities. Secondary cavity nesters can only breed successfully if they can find a suitable cavity: variables such as size and weather-proofing can have a substantial effect on their ultimate breeding success. Interspecific competition for high-quality cavities can be strong, but how is the competition hierarchy determined?

Hypotheses and predictions

In a long-running study of mature aspen in British Columbia, Canada, Kathryn Aitken and Kathy Martin conducted a two-stage research project to firstly identify competitive interactions in secondary cavity-nesting birds and then to establish the competitive hierarchy by increasing competition artificially. They were specifically examining the predictions that: (1) higher competition would decrease in the overall avian community, which would be indicative that the overall community was nest-site limited; and (2) that this change occurred equally for all species in the community rather than at different levels for different species.

Methods

In the first project (Martin et al., 2004), the authors studied what species used which cavities over a 2-year period. Through this process, they identified which of the cavities were preferred (i.e. used in both years of the baseline study), which they termed 'high-quality' cavities. In the second stage of the study (Aitken and Martin, 2008), the researchers increased competition

for the high-quality nest-cavity resource artificially by blocking the entrance to almost 50% of these cavities so that they were unavailable.

Results

The initial 2-year study showed a complex relationship between trees and both primary and secondary cavity-nesting species. The number and diversity of the links between tree species, primary cavity-nesting species (primary excavators), and secondary cavity-nesting species shown in Figure 1 indicates the complexity of the links between two aspects of the resource: (1) the initial bird species that originally created the cavity; and (2) the tree species in which that cavity was located. When competition was manipulated, the authors found that if nest cavities were in short supply, the number of nesting birds decreased by about 50%. However, this reduction was not consistent between species. For example, the breeding population of European starling (*Sturnus vulgaris*) decreased by 89%, while the abundance of nesting tree swallows (*Tachycineta bicolor*) did not change, and numbers of nesting mountain bluebird (*Sialia currucoides*) actually increased.

Conclusion

The findings of the second stage of the study were unexpected. It had been assumed that all species would be affected negatively (possibly to differing extents), but the fact that some species *increased* was initially surprising. What seems to have happened is that the bluebird, which is a **subordinate species** (i.e. one that is low down the competition hierarchy) actually benefited from the very substantial decrease in the starling breeding population, which usually dominates the nest cavity resource. The authors concluded that adaptability in nest site preferences of

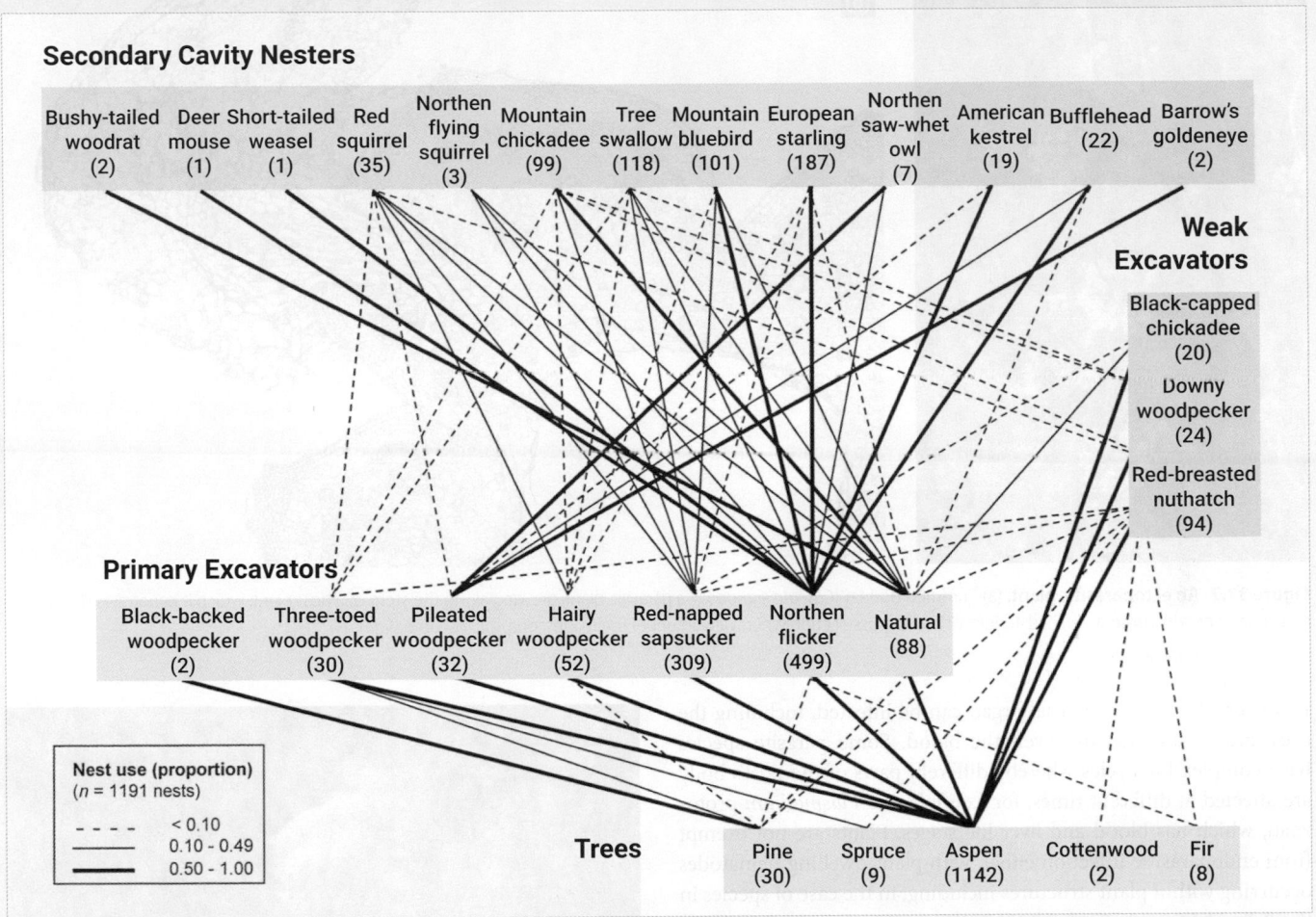

Figure 1 Interactions between secondary cavity-nesting bird species in Canada and both the primary cavity-nester (excavator) and the tree species in which the cavity occurs.

Source: Reproduced with permission from K. Martin, et al. (2004) Nest Sites and Nest Webs for Cavity-Nesting Communities in Interior British Columbia, Canada: Nest Characteristics and Niche Partitioning. *The Condor*, 106(1): 5–19. Copyright © 2004, Oxford University Press. https://doi.org/10.1093/condor/106.1.5.

subordinate cavity nesters might enable them to better contend with natural variation in availability of resources, such as nest cavities, and thus to cope better with interspecific competition.

Extension question

What do you think would happen if the number of cavities was increased (e.g. by putting up nestboxes) so that the availability of suitable nest cavities was no longer limiting?

Read the original papers

Martin, K., Aitken, K. E., & Wiebe, K. L. (2004) Nest sites and nest webs for cavity-nesting communities in interior British Columbia, Canada: nest characteristics and niche partitioning. *Condor* **106**: 5–19.

Aitken, K. E. & Martin, K. (2008) Resource selection plasticity and community responses to experimental reduction of a critical resource. *Ecology* **89**: 971–980.

Parasitism

We have already noted that most heterotrophs obtain nutrition from herbivory or predation. Another way of obtaining food is through parasitism. This is where one species—the **host**—supports other species that either live inside them (**endoparasites**)

or on them (**ectoparasites**). This type of association, though, goes beyond being just a feeding relationship because parasitism is an entire mode of life for parasitic species, either throughout their whole life or for a particular life stage.

In animals, endoparasite examples include tapeworms and roundworms in the gut, lungworms in the lung, and liver fluke in the liver.

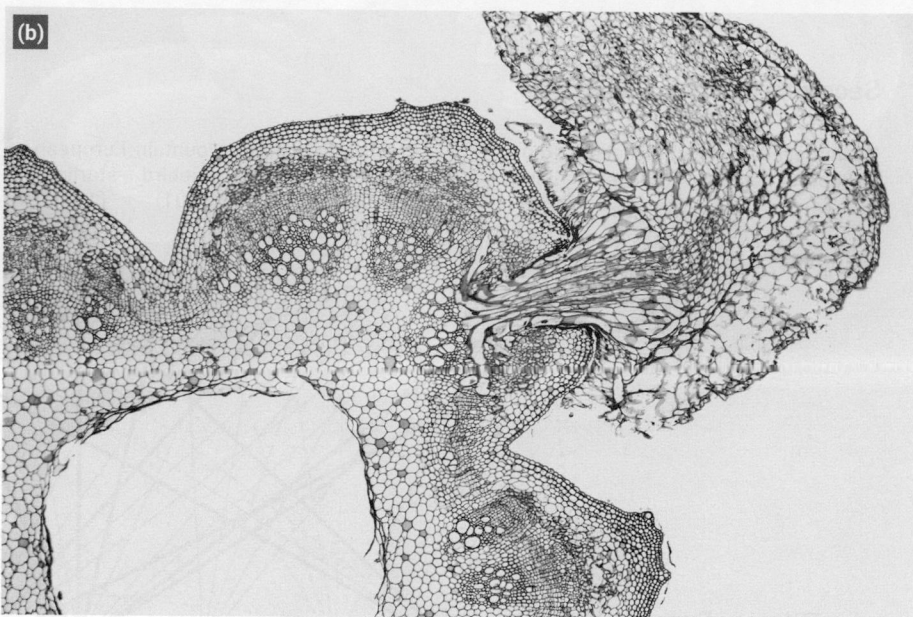

Figure 37.7 An ectoparasitic plant. (a) Common dodder (*Cuscuta europaea*), a parasitic plant, growing along the stem of a host plant, which it penetrates using haustoria to obtain nutrients (b) haustorium visible in cross-section when viewed under the microscope. microscopic slide).

Source: blickwinkel/Alamy Stock Photo.

However, almost any internal organ can be infected, including the eyes, brain, kidneys, and even the blood. Some parasite species have complex life cycles whereby different parts of the host's body are affected at different times; for example, the *Plasmodium* protozoan, which has blood and liver life stages. Plants are not exempt from endoparasitic infection either, with plant-dwelling nematodes occurring within plant structures including, in the case of species in the *Belonolaimus* genus, deep within the roots.

Ectoparasites are equally varied. For animals, they include lice, ticks, mites, fleas, and fly larvae that typically live within hair, fur, or feathers. Most such species feed on blood, although some also feed on skin cells. Ectoparasitic plants in the UK include common dodder (*Cuscuta europaea*) (Figure 37.7a), which parasitizes a range of other plant species including stinging nettles (*Urtica dioica*), and the interestingly named broomrape (*Orobanche minor*), which parasitizes species in the pea and daisy families. Parasitic plants often lack chlorophyll and true roots, instead having **haustoria**, which are root-like structures that penetrate into the tissue of the host plant (Figure 37.7b). The unusual biology of these plants means that they are incapable of photosynthesizing, and so are uncommon examples of heterotrophic plants.

While most parasitic plants parasitize other plants (within-taxon parasitism), some parasitic plants parasitize fungi. Good examples are the bird's nest orchid (*Neottia nidus-avis*), and beech-drop (*Monotropa hypopitys*), which is shown in Figure 37.8. Both beech-drop and bird's nest orchid are excellent examples of non-photosynthetic **myco-heterotrophs** (i.e. species that obtain their nutrients and energy from fungi). There are also examples of plant species that are an ecological 'halfway house' between a parasite and a free-living species. For example, mistletoe (*Viscum album*) obtains part of its food from the tree in which it grows and part from carbon, which it fixes during photosynthesis. In this way, it is both an autotroph and a heterotroph, and is referred to as a **hemiparasite**.

Figure 37.8 Beech-drop. Beech-drop (*Monotropa hypopitys*) does not contain any chlorophyll and thus looks almost white or brown in colour; this is typical of parasitic plants. In this case, the plant depends on soil fungi, which are often themselves associated with beech trees (*Fagus sylvatica*) for food.

Source: Photo by Ian Boyd, used with kind permission.

One particularly extreme form of parasitism is exhibited by **parasitoids**. Whereas parasites usually have a negative impact on their host, effects are rarely lethal for the simple reason that a parasite depends on its host being alive if that host is to continue supporting it. **Parasitoidism** is distinguished from other forms of parasitism by the fact that the interaction is ultimately fatal for the host. The classic example here is most of the wasps in the Ichneumonidae family. Adult females lay eggs on or in the body of another invertebrate host, often butterfly or moth caterpillars or beetle larvae, and after hatching, the ichneumonid larva consumes its host, ultimately killing it. For this reason, parasitoidism is sometimes regarded as being on the boundary between parasitism and predation.

Ecology is a complex topic and sometimes classifying ecological traits and phenomena can be challenging. Defining parasitism and parasites is a good case in point. For example:

- Some plant species have both parasitic and free living characteristics (hemiparasites) and are thus both autotrophic and heterotrophic; classifications that would normally be regarded as mutually exclusive.

- There is debate about whether insect species such as mosquitoes, which are generally free-living but take blood meals, should be regarded as parasites. E.O. Wilson, one of the most eminent biodiversity scientists, has defined parasites as 'predators that eat prey in units of less than one'. If we use this definition, mosquitoes would be classed as parasites, especially because the blood meals that a female mosquito takes are vital to the development of her eggs (and thus her life cycle cannot be completed without this). However, we could also make the argument that mosquitoes are **micropredators**—or even **micrograzers** if we extend the definition of a grazer to include blood being 'nibbled'.

- Just because one species occurs on another does not necessarily mean that it is a parasite. Sometimes a species lives on, or is transported by, another species without taking any resources from it. These are examples of **commensalism** and we will consider these in the next subsection.

Mutualism and commensalism

The interactions of predation, herbivory, competition, and parasitism are all generally one-sided: one individual benefits while another is disadvantaged. In some cases, though, there are interactions where both individuals benefit (**mutualism**) or where one benefits and the other is unaffected (**commensalism**). Sometimes these relationships are grouped under the term of **symbiosis**, and this term often becomes confounded with the concept of 'beneficial' ecological interactions. However, symbiosis literally means 'living together' so while both mutualism and commensalism are symbiotic relationships, so too are other (negative) interactions, including parasitism.

We tend to refer to two different types of mutualism. Mutualisms where the species involved can survive without one another are termed **facultative mutualisms** (sometimes termed **protocooperation**). A good example is afforded by insect-pollinated plants that are pollinated by more than one insect species, which, in turn, visit multiple plant species to obtain nectar—an individual plant species will benefit from an individual pollinator species, but neither is solely reliant on the other.

Figure 37.9 Obligate mutualism. An example of obligate mutualism is provided by the interaction between Pacific clownfish (*Amphiprion pacificus*) and the magnificent sea anemone (*Heteractis magnifica*). The anemone's stinging tentacles reduce predation risk for the clownfish, which itself is protected from stings by a layer of mucus, while the clownfish cleans and fertilizes the anemone.
Source: cbpix/Shutterstock.

In contrast, **obligate mutualism** describes situations where neither species can survive without the other. For example, nitrogen-fixing bacteria in the *Rhizobium* genus often occur within plant root nodules of legume species such as bird's foot trefoil (*Lotus corniculatus*), which itself could not grow in nitrogen-poor environments such as chalk grasslands without being able to utilize the nitrogen made available through the bacteria. Another classic example of obligate mutualism is the interaction between Pacific clownfish (*Amphiprion pacificus*) and the magnificent sea anemone (*Heteractis magnifica*) as depicted in Figure 37.9.

In the case of commensalism, one species benefits at no cost to the other. Common types of commensalism include:

- **Phoresy** is one species transporting another at no energetic cost; a good example of this 'hitchhiking' is pseudoscorpions being transported by insects including flies and wasps.

- **Inquilinism** is the term used when one species uses another for shelter or support at no cost to the host species. This would be the case for ivy (*Hedera helix*) growing on a tree trunk without tapping into that host tree for nutrients (i.e. not a plant ectoparasite) not vying with it for light (i.e. not a competitor).

Summarizing interaction types

Throughout Section 37.1, we have considered a wide range of ecological interactions. These interactions are summarized in Figure 37.10, where you are reminded about the key types of ecological interaction and how these differ in terms of the outcomes (positive or negative or neutral) for the species involved in each type of interaction. It is also important to note that sometimes species that co-occur spatially do so without obvious interaction and without obvious positive or negative implications for either

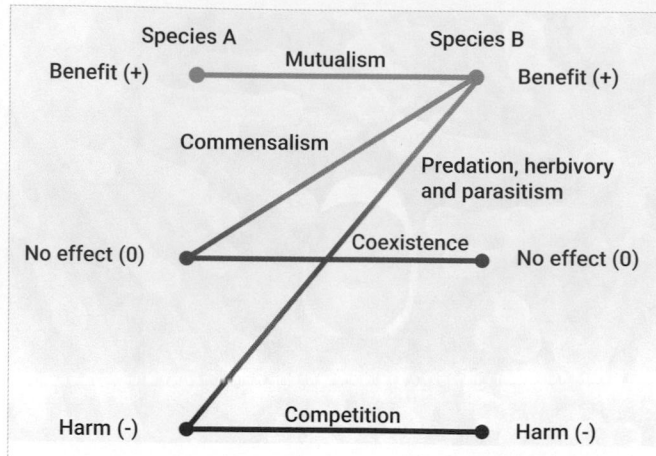

Figure 37.10 The potential interactions between two species that co-occur in a community. Coexistence and competition (and, rarely, predation in the case of cannabalism) can also occur between two organisms of the same species.

species: we term this **coexistence**. This occurs as a result of niche partitioning, as we see in Chapter 38.

PAUSE AND THINK

Examine Figure 37.10 and research at least one example of each of interaction type from the terrestrial environment, and at least one example from the marine environment, that have not already been covered in this chapter. In all cases, identify whether the interaction has arisen because of the species' basic biology (life-history traits, etc.) or as a result of specific behaviours.

 Go to the e-book to complete an activity related to Figure 37.10.

 Check your understanding of the concepts covered in this section by answering the questions in the e-book.

37.2 Classifying communities

Ecological communities can be classified in several different ways:

- **Habitat based:** The species community found in a specific habitat (e.g. a specific lake or pond) or, at a wider spatial scale, the assemblage of species found in a generic habitat (e.g. ancient woodland across mainland Europe).

- **Taxon based:** All species within a particular taxonomic group in a specific location (the plant community of a specific field, for example, or the aquatic invertebrate community of a specific stream). Sometimes we highlight a few dominant or characteristic species in the community to describe it, as in the case of *Fraxinus excelsior—Acer campestre—Mercurialis perennis* woodland (i.e. woodland characterized by ash, field maple, and dog's mercury), for example. This is the premise of the National Vegetation Classification (NVC) process in the UK, which is outlined further in Real World View 4.1.

- **Time and taxon based:** Sometimes we refine the taxon-based approach to reference a particular time period—the wintering bird community, for example, or the spring butterfly community.

- **Interaction based:** Species linked by a particular interaction type, for example, a predator–prey community.

 Check your understanding of the concepts covered in this section by answering the questions in the e-book.

37.3 Species community metrics

Although we can use highly detailed methods to classify the community type, as exemplified by NVC detailed in Real World View 37.1, these approaches necessitate a substantial amount of field effort, both to draw up the initial classification system and key, and then at each site that needs to be classified. Moreover, they are also

---

REAL WORLD VIEW 37.1 | **National Vegetation Classification concept and method**

The National Vegetation Classification (NVC) method was developed in the UK during the 1990s to classify vegetation communities. The same idea is used in many other countries, including the USA (NVC), Canada (CNVC), and the Czech Republic (National Phytosociological Relevés). Although the terminology and the specifics of the NVC methodology differ between countries, the basic approach is very similar.

The ultimate aim of using NVC at a site is to determine which of the pre-defined vegetation categories best describes the actual vegetation community of that site. The pre-defined vegetation communities are based on field data from a wide range of sites that have been previously analysed to group similar vegetation communities together. In this way, the pre-defined vegetation communities can be thought of as a series of umbrellas under which the vegetation communities of specific sites can sit. As an example, woodland and scrub (W) communities comprise 25

different broad types (W1 to W25). The first 18 of these are full woodland, which is defined as a community with a full canopy, while the last seven are scrub communities. Within most of these communities, there are sub-communities to refine the broad classification and give additional detail (e.g. W2a woodland or W5c woodland). In total, there are 73 sub-communities within the 25 main W communities. Creation of the broad communities and the more specific sub-communities was based on vegetation surveys of 2648 woodlands across the country.

In total, there are 286 recognized NVC communities that sit with 12 broad NVC types:

- Woodland (W1–W25)

- Mires (M1–M38)

- Heaths (H1–H22)

- Mesotrophic (reasonably fertile) grasslands (MG1–MG13)
- Calcicolous (alkaline) grasslands (CG1–CG14)
- Calcifugous and montane (acidic/upland) communities (U1–U21)
- Open habitats (OV1–OV42)
- Aquatic habitats (A1–A24)
- Swamps and fens (S1–S28)
- Shingle, strandline, and sand dune communities (SD1–SD19)
- Salt-marsh communities (SM1–SM28)
- Maritime cliff communities (MC1–MC12).

The process of determining the correct NVC category for a specific vegetation community starts with an ecologist visiting the site to create a **species inventory** (literally a list of what species are present). This is done by surveying a suitably sized plot or series of sub-plots, each of which has to be completely within one main habitat and as homogenous as possible. The most complex habitat to survey is woodland because vegetation typically occurs in multiple vertical layers or **strata**. In this case, we use nested plots: a large plot of 50 m × 50 m is used to sample the trees and smaller plots within this are used to sample shrubs (woody species that do not form a canopy, such as hawthorn (*Crataegus monogyna*) and bramble (*Rubus fruticosus*)), field layer (long vegetation that can be walked through, such as long grass and ferns), and the ground layer (short vegetation that people walk on rather than through, including short grasses and mosses). A diagram of the nested plot arrangement is shown in Figure 1.

 Go to the e-book to explore an interactive version of Figure 1.

When the vegetation inventory has been made, the abundance of each species is recorded, usually using a 10-point ranking scale called DOMIN (1 = 1–2 individuals with no measurable cover; 2 = several individuals but no measurable cover; 3 = 1–5% cover; 4 = 5–10% cover; 5 = 11–25% cover; 6 = 26–33% cover; 7 = 34–50% cover; 8 = 51–75% cover; 9 = 76–90% cover; 10 = 91–100% cover).

Once the field data have been collected, the vegetation community is keyed out using habitat-specific keys (Rodwell, 1991–2000 in the UK). These work in the same way as a taxonomic key for keying out an unknown species, with the recorder working through a series of questions and selecting the most appropriate answer at each point until a final classification is reached. In the case of UK woodlands, for example, the first question is used to determine if the vegetation community is true woodland or scrub, the second is used to determine if the habitat is wet or dry, and the subsequent questions are based on the exact species that are present and in what quantities to decide upon the best-fit species community and (usually) the best-fit sub-community within that.

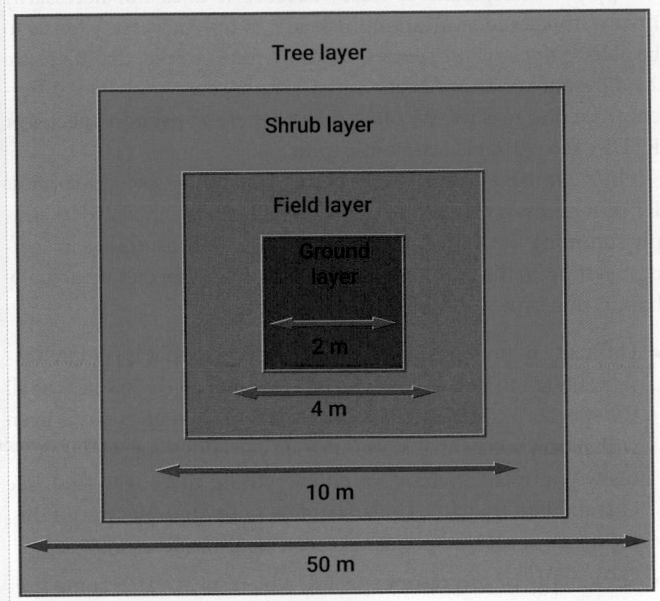

Figure 1 Nested vegetation survey plots for NVC. In woodland, usually all plots are needed in order to survey the different vegetation strata at an appropriate spatial scale. If NVC is being used in more open habitats, the smaller quadrats are used as appropriate. For instance, if NVC is used in heathland it might be appropriate to have 10 × 10, 4 × 4 and 2 × 2 metre nested plots, whereas in short grassland communities only the 2 × 2 metre plot would be needed.

The final community is given an alphanumeric code, for example:

Overall code = W8b made up thus:

Habitat type (alpha) = W (Woodland)

Community (numeric) = Number 8 (*Fraxinus excelsior—Acer campestre—Mercurialis perennis* (ash, field maple, dog's mercury))

Sub-community (alpha) = b (characterized by *Anemone nemorosa* (wood anemone)

Extension question

Why is a nested plot sometimes needed for multi-strata habitats rather than us just picking an intermediately sized plot (e.g. 10 × 10m) and surveying all strata in that area?

Read the original work

Rodwell, J.S. (1991–2000) *British Plant Communities. Vols 1–5.* Cambridge: Cambridge University Press.

highly specialized to particular types of community in particular geographical locations (plant communities in the UK in the case of NVC. Accordingly, it is useful for us to be able to understand and apply more general species community metrics to describe and compare aspects of community structure.

Species richness

One of the most simple, but arguably most powerful, community metrics is a simple count of the number of species in a community (or number of genera, families, etc.) This is referred to as

species richness. The species richness metric is extremely simple to apply—conceptually at least—because it does not necessarily involve species identification. Instead it is only necessary for us to be able to *differentiate* species; to know, for example, that there are 23 different species of beetle in an area because they look different from one another. We often refer to these as **pseudo-species** or RTUs—recognizable taxonomic units.

Once we have assessed all species (including pseudo-species/ RTUs if necessary) in an area, species richness is calculated by simply summing these together. However, as with most things in ecology, getting to the point of knowing how many species we actually have is deceptively difficult. Issues include:

- Difficulty in undertaking ecological surveying that is comprehensive enough to find at least one individual of all the species (or all the species in the target taxonomic group) in a given area. Even with a very substantial amount of survey effort, we can never be really confident we have found everything there is to find (see Quantitative Toolkit 5: Understanding sampling). Because of this, most records of species richness are actually underestimates.

- Seasonality of **migratory** species, which means that some species in a community are only present at particular times of the year (which might be outside of when we have undertaken the ecological surveys).

- Difficulty in distinguishing between **cryptic species** (i.e. individuals that are morphologically almost identical to each other but belong to different species; as shown in Figure 37.11). When there are a lot of cryptic or closely related species in an area, species richness is typically underestimated.

- Ensuring that there is no double-counting of species that are **polymorphic** (i.e. different individuals look different from one another but are the same species) or that undergo **metamorphosis** and thus have life stages that look totally different (see examples in Figure 37.11).

Species diversity

Species richness can also be called **alpha diversity** (α) and is a measure of diversity *within* sites—the *local* species pool. The two other scales at which species diversity can be considered are:

- **Gamma diversity** (ϒ): the species diversity of an entire landscape (i.e. *regional* species pool across all sites).

- **Beta diversity** (β): beta diversity is diversity *between* sites. There are several different ways that beta diversity can be calculated, but **Whittaker's beta diversity** is the most straightforward: this is the ratio between the alpha diversity of a specific site and the gamma diversity of the landscape in which that site is located (β = ϒ/α).

Simple examples of alpha, beta, and gamma metrics are given in Figure 37.12.

For ease of interpretation, the beta and gamma diversity values provided in Figure 37.12 focused on number of species. However,

Figure 37.11 Cryptic, polymorphic, and metamorphosing species. Although assessing species richness is conceptually easy—a simple count of the different species present—it can be hard to assess what actually constitutes a species by simple observation (the biological species concept is discussed in Chapter 35). It can be especially difficult to do this for cryptic species that look the same but differ, for example, (a) the wood white (*Leptidea sinapis*) versus (b) cryptic wood white (*Leptidea juvernica*). Conversely, there are occasions where organisms can look different from one another but are actually either different morphs of the same species (for example, (c) and (d) are different colour morphs of the harlequin ladybird (*Harmonia axyridis*)) or different life stages of the same species (for example, (e) and (f) show the difference between common frog (*Rana temporaria*) tadpoles and adults).

Source: (a) imageBROKER/Alamy Stock Photo; (b) Oliver Smart/Alamy Stock Photo; (c) PHOTO FUN/Shutterstock.com; (d) Thijs de Graaf/ Shutterstock.com; (e) Eric Isselee/Shutterstock; (f) Rudmer Zwerver/Shutterstock.

Site 1

Alpha diversity = 3

Whittaker's beta diversity = 1.67 (5/3)

Site 2

Alpha diversity = 2

Whittaker's beta diversity = 2.5 (5/2)

Site 3

Alpha diversity = 4

Whittaker's beta diversity = 1.25 (5/4)

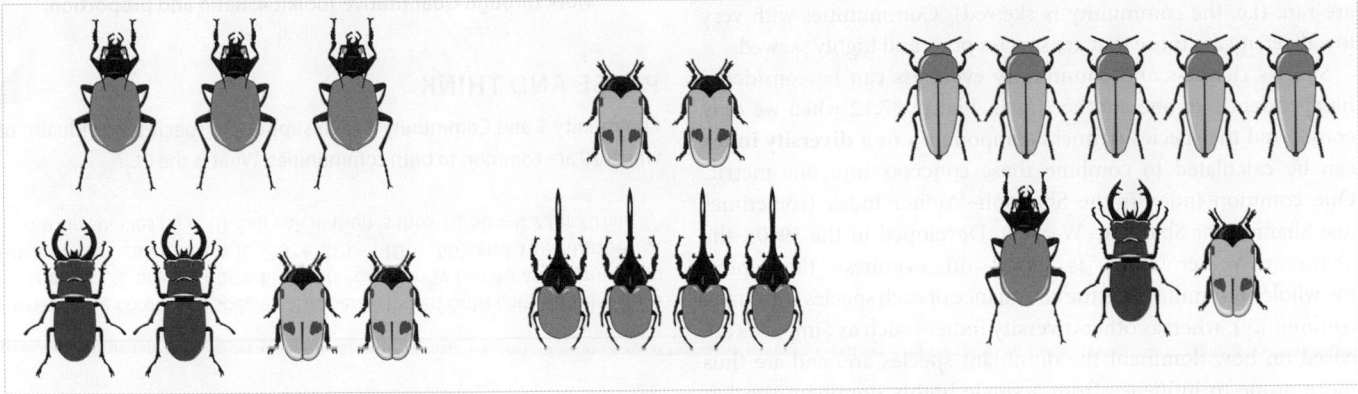

Figure 37.12 Alpha (α), beta (Υ), and gamma (β) diversity values for beetles for three sites in a single landscape. Alpha values give the species richness at each site, the single gamma values give total species found across all sites (five in this case because there are five different beetle species overall), and the beta value shows the alpha diversity of each site relative to the gamma diversity of the landscape (i.e. Whittaker's beta diversity calculated as β = Υ/α).

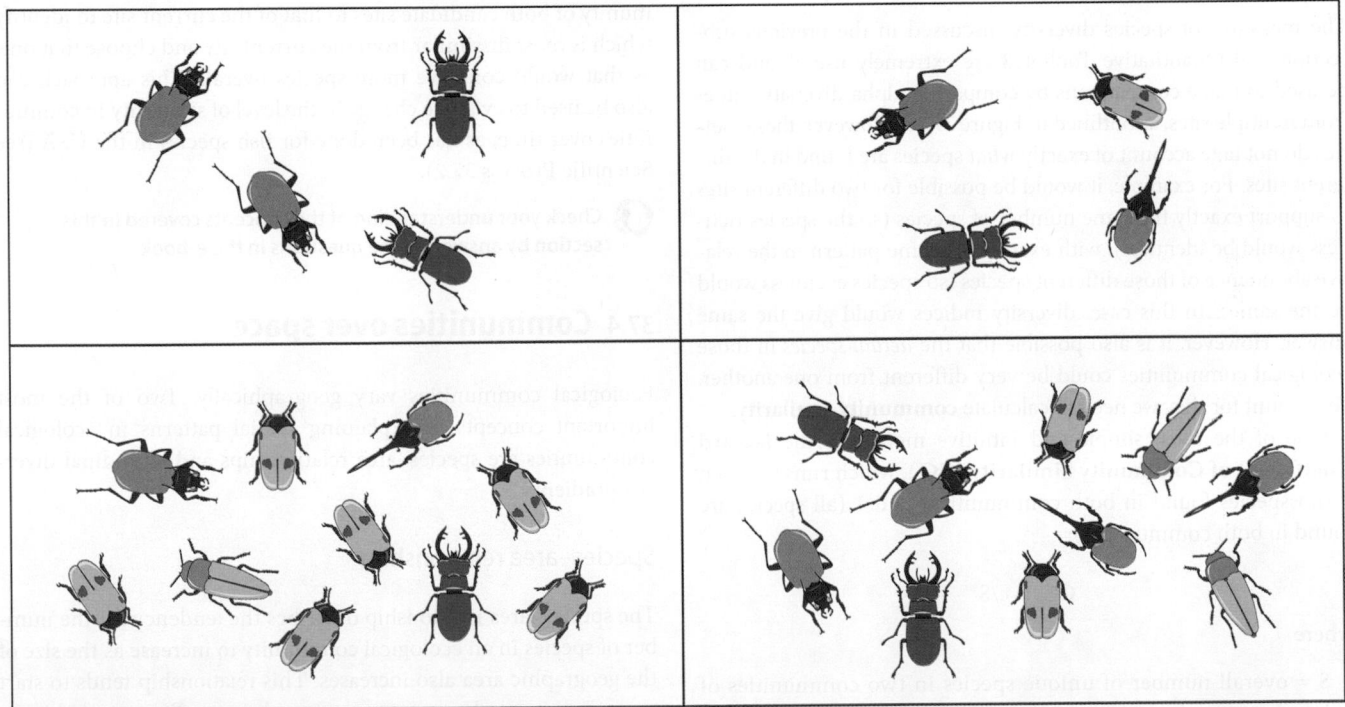

Figure 37.13 Species diversity at beta and gamma scales combines information on species richness *and* how the abundance of each species is relative to one another and thus the evenness of the community. In both the top and bottom pairs of images, the community depicted on the right would lead to a higher diversity assessment than the community depicted on the left. The top pair show the differences driven by species richness—both communities have four individuals but the community on the right has a greater number of species (n = 4) than the one on the left (n = 2). The bottom pair shows the differences between species evenness—both communities have 10 individuals and five species represented, but the one on the left has an unequal number of each (skewed distribution), whereas the one on the right has an equal number of each (uniform or even distribution).

beta and gamma diversity measures usually involve going a step beyond considering species richness. Diversity at these spatial scales is a combination of two parameters: the number of different species (species richness) *and* the relative abundance of each

of those species and how uniform this is (species evenness). Beta and gamma species diversity thus combines two different concepts—richness and evenness—as illustrated in Figure 37.13. High diversity is created when communities have numerous species (i.e. species rich) with the majority of species having approximately equal abundance (i.e. the community is even). Species diversity is lower when either the community has relative few species (i.e. species poor) or where the community is highly skewed such that a small number of species dominate the community, while the rest are rare (i.e. the community is skewed). Communities with very low diversity are those that are species poor and highly skewed.

Species richness and community evenness can be considered independently of one another (as in Figure 37.12 when we only considered the species richness component), or a **diversity index** can be calculated to combine these concepts into one metric. One common index is the Shannon–Weiner index (sometimes just Shannon or Shannon–Weaver). Developed in the 1940s, the Shannon–Weiner index is based on evenness throughout the whole community (i.e. the abundance of each species within the community), whereas other diversity indices such as Simpson's are based on how dominant the dominant species are, and are thus more prone to influence from a single highly dominant species. The Shannon–Weiner index is detailed in Quantitative Toolkit 8: Formulae and equations.

Species similarity and overlap

The measures of species diversity discussed in the previous subsection and Quantitative Toolkit 8 are extremely useful, and can be used to make comparisons by comparing alpha diversity values from multiple sites, as outlined in Figure 37.13. However, these metrics do not take account of exactly *what* species are found in the different sites. For example, it would be possible for two different sites to support exactly the same number of species (so the species richness would be identical), with exactly the same pattern in the relative abundance of those different species (so species evenness would be the same). In this case, diversity indices would give the same answer. However, it is also possible that the *actual species* in those ecological communities could be very different from one another. To account for this, we need to calculate **community similarity**.

One of the most simple and intuitive metrics is the **Jaccard Coefficient of Community Similarity (CC$_j$)**, which runs between 0 (no species found in both communities) and 1 (all species are found in both communities):

$$CC_j = c/S$$

where

- S = overall number of unique species in two communities of interest
- c = number of species common to both communities of interest.

Suppose there are two sites that we want to compare: site 1 supports 75 species and site 2 supports 15 species. Our total number of species is 90, but we find that 10 species are found at both sites and this takes the number of *unique* species down to 80. Our equation is then 10/80 = 0.125. This number is often easier to interpret if we multiply it by 100 to give a percentage. In this case, we are saying

that only a small minority of species (12.5%) are found in both communities. The index is pairwise so to compare three communities, for example, three tests are needed (1 and 2, 1 and 3, 2 and 3).

▶ If you need a reminder about proportions and percentages, work through Quantitative Toolkit 4: Ratio and proportion.

The CC$_j$ index can be useful when determining conservation actions. Imagine that a species-rich grassland site is already protected by a conservation organization and that an opportunity has arisen to protect more grassland in the same area, by purchasing *one* of two possible sites. It might make sense to compare the community of both candidate sites to that of the current site to identify which is most dissimilar from the current site and choose that one, as that would conserve more species overall. This approach can also be used to evaluate change in the level of similarity in communities over time, as has been done for fish species in the USA (see Scientific Process 37.2).

 Check your understanding of the concepts covered in this section by answering the questions in the e-book.

37.4 Communities over space

Ecological communities vary geographically. Two of the most important concepts in explaining spatial patterns in ecological communities are species–area relationships and latitudinal diversity gradients.

Species–area relationships

The species–area relationship describes the tendency for the number of species in an ecological community to increase as the size of the geographic area also increases. This relationship tends to start strongly but gets less pronounced as the overall area gets bigger, which you can see by the way the logarithmic curve starts to plateau in Figure 37.14. This makes intuitive sense. If you were to survey any patch of grass and looked at a 50 cm by 50 cm quadrat you would find a few different grass and/or grassland plant species—say four in total. If you increased the size of the survey plot to 1 m by 1 m, (a four-fold increase in area) the total number of species that recorded would typically rise quite considerably, maybe to 10 or so. If you increase the size of the plot to 2 m by 2 m (also a four-fold increase

SCIENTIFIC PROCESS 37.2 Change in similarity of fish communities in different states in the USA

Research question

One concern about introducing non-native species to a new environment is that this process might cause **biotic homogenization**, whereby ecological communities that have been different historically become more similar over time. This makes logical sense but there is little numerical evidence for this.

Hypotheses and predictions

In the first national-scale study of biotic homogenization, Frank Rahel (2000) examined fish species communities of the 48 contiguous states in continental USA (i.e. excluding Alaska and Hawaii). He hypothesized that, over time, state fish communities would have become more similar to one another, because non-native species became more common and more widespread (thus becoming a greater proportion of all fish species and covering multiple states).

Methods

Rahel used the Jaccard Coefficient of Community Similarity (CC_j) to compare the historical similarity in fish species between each and every US state relative to each and every other US state. This involved 1128 pairwise combinations (i.e. Alabama–Arizona, Alabama–Arkansas, Alabama–California, etc.). Historical similarity was defined using *native* fish using CC_j for every combination of states. The same exercise was done to compare current fish communities (including non-native species) again for every combination of states. Finally, the change in similarity was calculated by subtracting the current similarity of fish communities (say 0.7) from the historical similarity of fish communities (say 0.8) give the CC_j change for each pair of states (0.1 in this case).

Results

There was strong evidence for biotic homogenization because almost 90% of the changes in fish community similarity were positive (i.e. states had become more similar in their fish communities over time), as shown in Figure 1. The mean increase in similarity was 7.2%. The 89 pairs of states that historically had zero similarity (no species in common) now have an average similarity of 12.2%. For example, Arizona and Montana went from 0% historical similarity to 26.8% current similarity. Rahel determined that

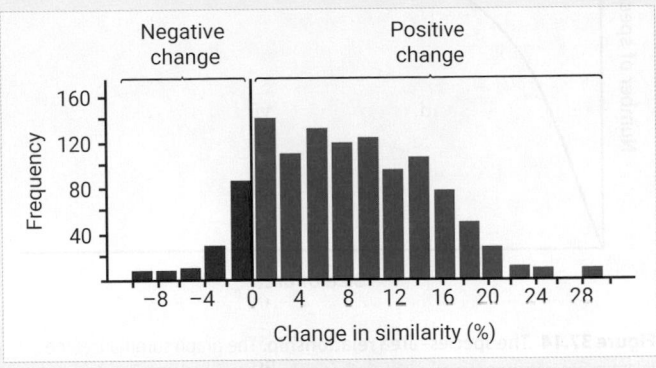

Figure 1 Change in the level of similarity between fish communities.
Change in similarity in different US states between European settlement of North America and now using CC_j (based on 1128 comparisons through which every state was compared to every other state). Positive changes indicate increasing similarity.

Source: Reproduced with permission from F. J. Rahel. (2000) Homogenization of Fish Faunas Across the United States. *Science*, 288(5467): 854–856. copyright © 2000, The American Association for the Advancement of Science.

increases in similarity were usually because of new non-native species coming in rather than state-specific loss of (different) native species.

Conclusion

Biotic homogenization is occurring in US fish communities due to introduction of non-native species.

Extension question

CC_j is a useful tool to get a numeric estimation of biotic homogenization, but it does not tell us what is driving that change. Other than species introductions and local extinctions, what other community processes could cause temporal change in species similarity between different locations over time?

Read the original paper

Rahel, F. J. (2000). Homogenization of fish faunas across the United States. *Scienc* **288**: 854–856.

in area), you would get more species again, but maybe not quite as many new ones as the first size jump. If you increased the size of the survey plot again to 4 m by 4 m (another four-fold increase), you might add just one or two species to the **cumulative total**. At some point, you would likely find that even if you keep on increasing the area, no more new species are located as the species number reaches the **asymptote**.

 We learn more about quadrat surveys in Chapter 36.

The species–area relationship is fairly robust: it occurs in most environments and for most taxonomic groups. However, the gradient of the relationship, how long it takes to reach the asymptote, and the actual number of species involved, varies substantially.

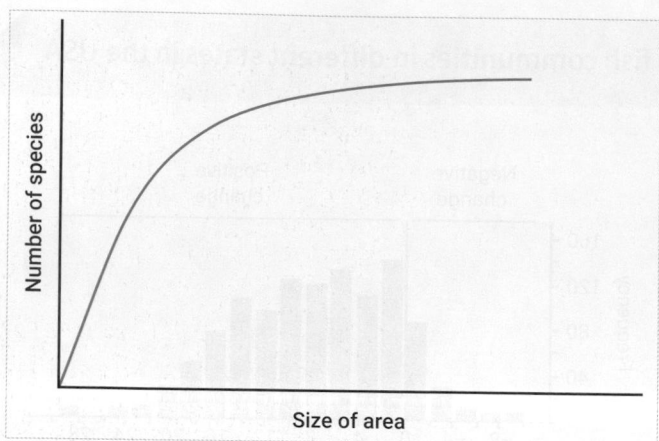

Figure 37.14 The species–area relationship. The graph summarizes the relationship between the area size (independent variable shown on the *x*-axis) and the number of species found in those areas (dependent variable shown on the *y*-axis): it usually takes the form of a logarithmic curve.

PAUSE AND THINK

Describe the likely differences in tree species–area curves for: (1) a temperate woodland in the UK, compared with (2) a tropical rainforest in Borneo. How would the shape of the curves be likely to differ? How would the actual numbers of the *x*-axis (area size) and *y*-axis (number of species) differ?

Answer: In the UK temperate woodland, it would be possible to find all the species fairly quickly, so the curve would be steep and the asymptote would be reached quickly, whereas in a Bornean tropical rainforest, the overall number of species is much higher and many are very scattered spatially so the curve would be shallower and more elongated. The total number of species found would be higher in the tropical rainforest (so the *y*-axis would be bigger), but a considerable area would need to be surveyed before the asymptote is finally reached so the *x*-axis would also be much longer.

Global patterns in community complexity

In addition to small-scale patterns in species richness over space (e.g. species–area relationships), there are also important global patterns in species richness. The most well-known of these is the **latitudinal diversity gradient**, which describes the pattern in species richness with latitude that results from more species occurring in the tropics than the poles. The species richness differences between equatorial regions and higher latitudes can be very substantial. For example, in the 'Pause and think' at the end of the previous section, you were asked to consider the examples of a UK woodland and a Bornean tropical rainforest. The whole of the UK, which is located about 55°N of the equator, only has 85 recognized tree species despite having 3.19 million hectares of woodland. This contrasts with a typical woodland plot of just 4 hectares at Danum Valley, a research area in Malaysian Borneo, which is located almost on the equator, which supports around 390 tree species out of 5446 tree species supported by Malaysia as a whole (Malaysia being about one-third bigger than the UK).

Latitudinal diversity gradients occur in many taxonomic groups, as shown for mammals in Figure 37.15. The general pattern often tends to be steeper in the southern hemisphere and shallower in the northern hemisphere. In the northern hemisphere, there is

---

also a slight additional peak around 35–40 °N, which coincides with the species-rich Mediterranean ecosystem in Europe and the chaparral ecosystem in California (also discussed in Chapter 38).

 Check your understanding of the concepts covered in this section by answering the questions in the e-book.

37.5 Communities over time

The ecological community at a given spatial location is not a fixed entity; rather it is fluid and prone to change over time. Two key processes that lead to temporal change in communities are **predator–prey cycles** and **ecological succession**.

Predator–prey cycles

In Chapter 36, we learned about many of the processes that control species' populations are effectively internal to that particular species. However, as we can now appreciate, species populations (and thus species population processes) do not occur in isolation from one another. Thus, what happens in the population of one species can impact the population of another species. As we saw in Section 37.1 and Scientific Process 37.1, changes in competitive interactions over time can lead to population changes for either or both species involved. Temporal change in predator–prey dynamics can also drive population changes, especially when this happens cyclically.

The simplest predator–prey population change model is the **Lotka–Volterra model**, originally proposed independently by Alfred Lotka and Vito Volterra in the 1920s. The model is cyclical because the population of a predator increases in response to an increase in prey numbers. This increase in predator population size holds until the point at which the increased predation pressure on the prey causes the prey population to decline. This decrease in prey means that the predator's demand for food outstrips supply and the predator population then crashes. With this reduction in predation—a concept termed **predator release**—the prey population increases once again and the whole cycle restarts. The 'boom-and-bust' cycle of the predator is always slightly behind the boom-and-bust cycle of the prey, such that the predator species population will peak after the prey species population has peaked, as shown by the 'predator' peaks happening just after the 'prey' peaks in Figure 37.16. The **lag time** between predator and prey peaks depends on the biology and life-history traits of the species involved.

Ecological succession

Land that has never previously been vegetated can occur as a result of processes such as a volcanic eruption (lava fields), new land emerging from the sea as a result of isostatic rebound (upward movement of land masses after ice sheet melting removing the weight that had previously caused isostatic depression), glacier retreat exposing bare rock, and areas where windblown sand or silt (loess) is deposited.

Assuming the new land is fairly stable, it gradually starts to become vegetated, initially by **pioneer** plant species. Pioneer species are usually *r*-strategists (Chapter 36) that are adapted to exploit new opportunities where competition is low and they quickly become established. If the new environment remains disturbed, the community might remain dominated by pioneers. However,

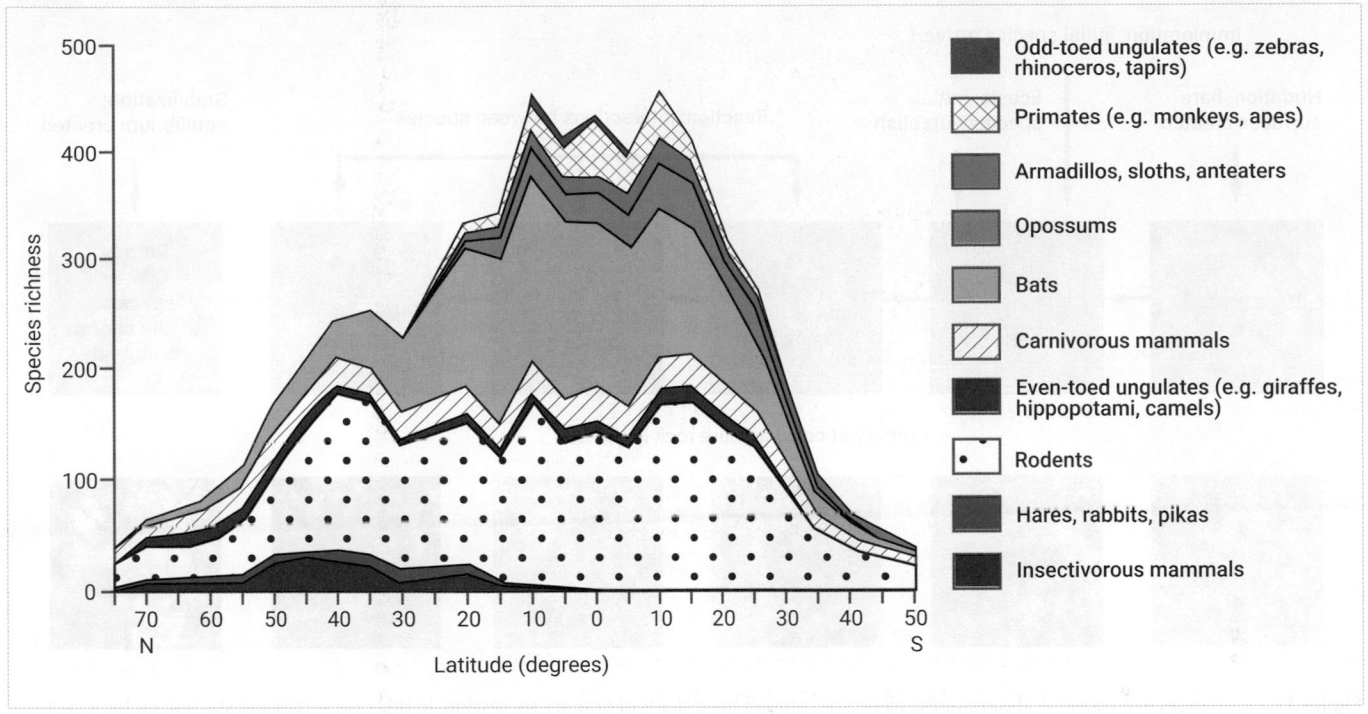

Figure 37.15 The latitudinal diversity gradient for orders within the class Mammalia. Note the different gradient steepness between northern and southern hemispheres.

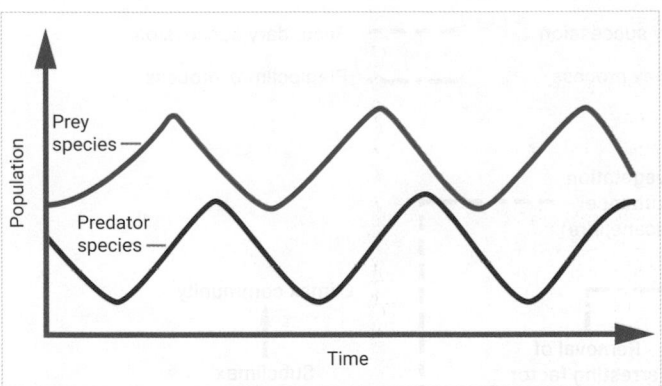

Figure 37.16 Lotka–Volterra model. This is showing the linked population changes in a predator species and a prey species at the same location, which is controlled through temporal changes in predation pressure.

given some stability, the community composition generally changes as early colonizers are eventually out-competed by K-strategist species, often including long-lived shrubs and trees. A change in community structure over time on land that has never previously been vegetated is called **primary succession**. Where succession follows a disturbance in an established community (e.g. where the land has been vegetated before but has been denuded by a storm or clear-felled by humans, but crucially the soil remains *in situ*) we use the term **secondary succession**.

Ecological succession was first conceptualized by Frederic Clements in 1916 and so is sometimes referred to as **Clementsian succession**. Clements noticed that new land in any part of the world tended to go through a series of vegetative stages, which he

termed seres, culminating in a fairly predictable vegetation community. He suggested that primary successional processes were more-or-less predictable, starting with bare ground or water (first stage of Figure 37.17), going through a succession of communities (seres 1–4 in Figure 37.17) and culminating in a **climatic climax community** determined by geographical location (sere 5 in Figure 37.17). Broadleaved woodland is the usual climax community over most of the UK, for example, except in parts of Scotland where coniferous woodland becomes typical. A real-world example of this successional process is illustrated using photographs in Figure 37.17 for succession on bare rock in the UK. The dynamic and interconnected nature of ecosystems means that as vegetation changes, so too will faunal species.

🌀 Go to the e-book to explore an interactive version of Figure 37.17.

In the 1930s, Arthur Tansley revised and extended Clements' original ideas by changing the concept of succession from being a **monoclimax** process (i.e. one that would always end in the same climax community in the same climatic zone) to being a **polyclimax** process (i.e. one where there are several possible climax endpoints). This reflects the fact that local non-climatic factors can affect the final community, including those related to soil or hydrology. It should also be noted that sometimes an event occurs that prevents the theoretical final end community from ever being reached, or that the climax community changes after it has been reached. These processes are shown graphically in Figure 37.18.

🌀 Go to the e-book to explore an interactive version of Figure 37.18.

🌀 Check your understanding of the concepts covered in this section by answering the questions in the e-book.

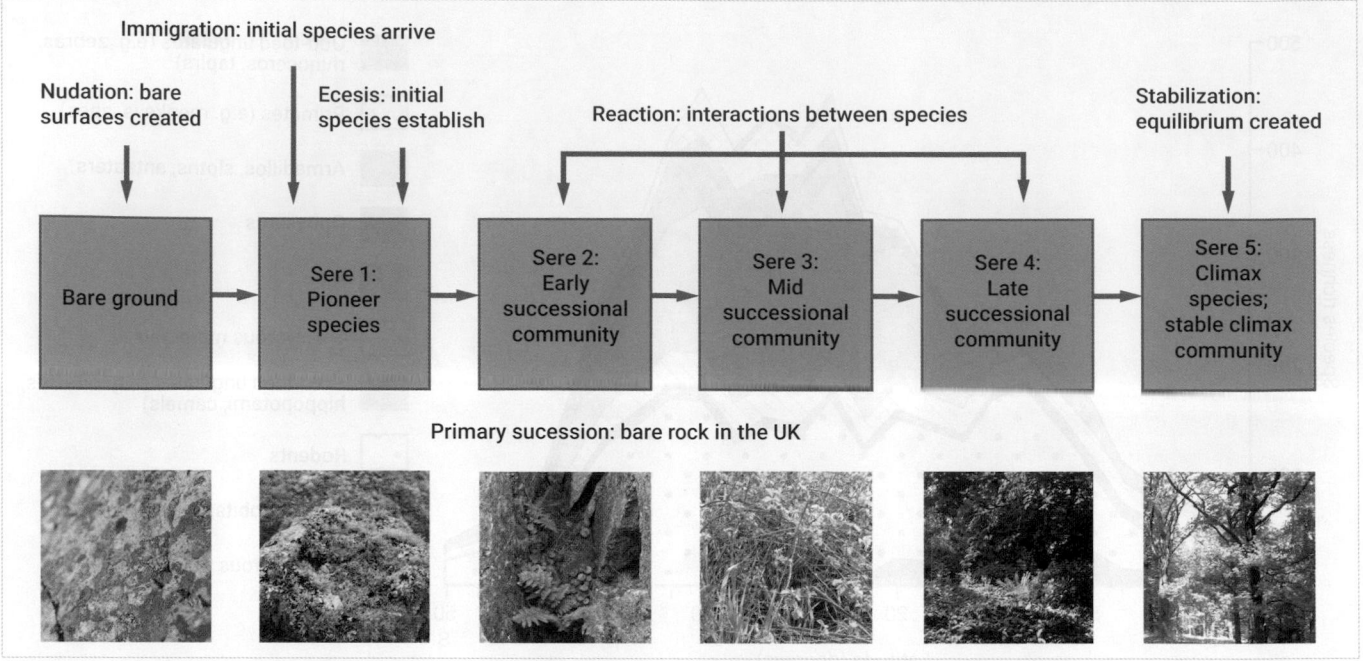

Figure 37.17 Changes in an ecological community over time through the process of primary succession, in this case succession starting on bare rock (a lithosere). Over a considerable amount of time, the community becomes more complex, both in terms of its structure and the species that it supports.

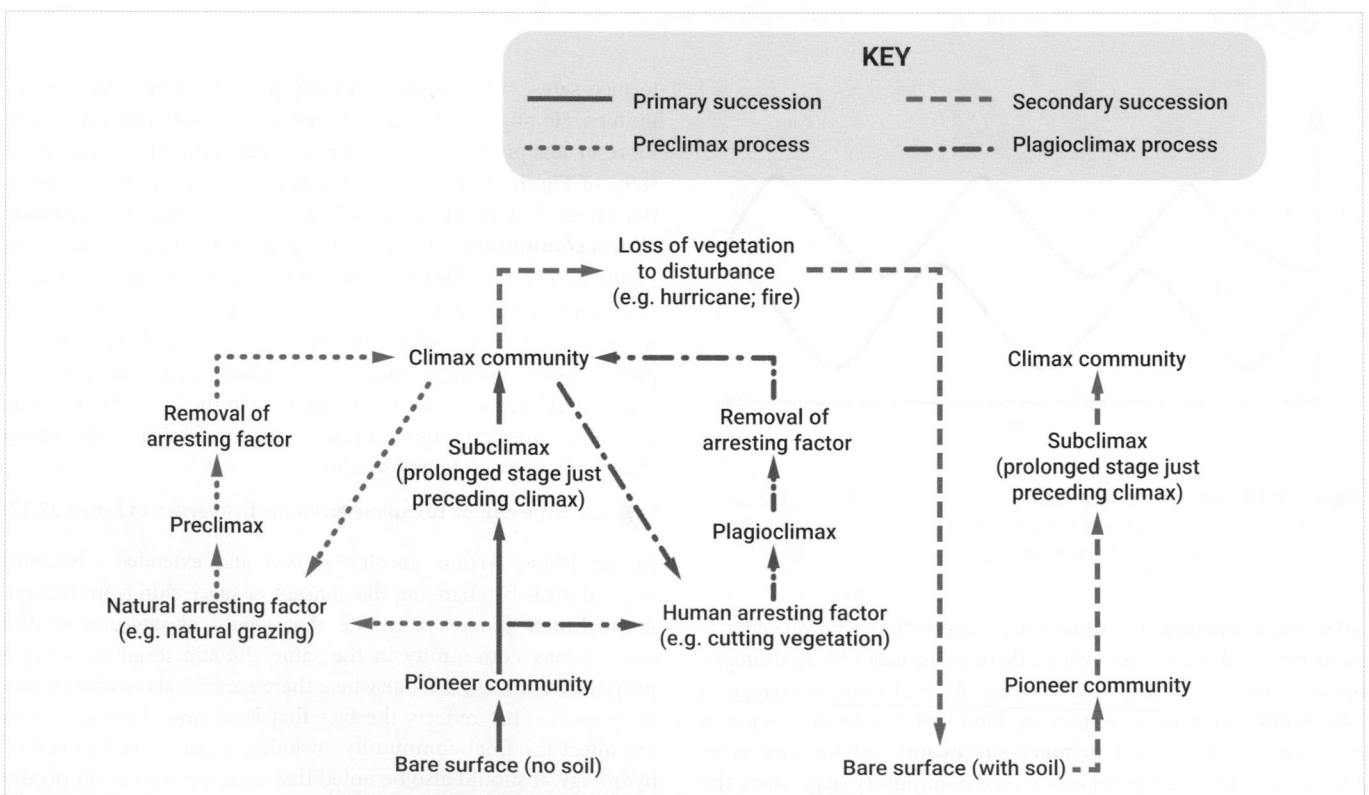

Figure 37.18 The different processes in ecological succession. Primary succession, which should always happen when new land is exposed, is shown by the solid black line. Secondary succession, which occurs after disturbance after the original climax has been reached (or when primary succession is well advanced) is shown by the dashed orange line: this may or may not occur. At any point in the original primary succession, an arresting factor can halt the successional process. This arresting factor could be natural (dashed green line) or caused by human activities (dashed blue line) and again may or may not occur.

SUMMARY OF KEY CONCEPTS

- Ecological communities are collections of co-occurring species that interact with one another either directly or indirectly.

- Interactions can involve feeding relationships (predation and herbivory), competing for finite resources, or can be relationships that are mutually beneficial.

- Some interactions are so fundamental that they interlink with the mode of life for at least one of the species involved (e.g. parasitism) or affect population sizes (e.g. predator–prey cycles).

- We often summarize ecological interactions by thinking about the impact of a given interaction upon each species involved and whether that is positive, negative, or neutral.

- Interactions can be the result of the basic biology of the species involved (e.g. diet) or can involve species' behaviour (e.g. hunting strategies).

- There are many different ways that ecological communities can be classified, but most classifications are based on habitat, taxon, or interaction, possibly with time factored in.

- Species richness—the number of species in a particular habitat or area—is one of the most useful metrics by which to summarize and compare ecological communities.

- Species diversity metrics, which combine information on species richness and the relative number of individuals in each of those species, are also very useful. There are many diversity indices; Shannon–Weiner is one of the most common.

- Comparing two (or more) communities as regards the actual species involved can be undertaken using measures of community similarity, such as Jaccard's Coefficient of Community Similarity.

- Species communities are not fixed entities; they change over time and space.

- Species richness tends to increase with area (species–area relationship) and is also associated with latitude, with more species occurring in the tropics than in polar regions.

- Over time ecological communities can change through community-level processes such as predator–prey cycles or because of natural changes such as ecological succession.

 Use the flashcards in the e-book to test your recall of key terms introduced in this chapter.

QUESTIONS

 Looking for answers? Once you've answered these questions, follow the link in the e-book to the answer guidance and check your work.

Concepts and definitions

1. What two interactions can be grouped together into 'feeding interactions'?

2. What is the difference between mutualism and commensalism?

3. What is the difference between 'species richness' and 'species diversity'?

4. What is the species–area relationship?

5. What is the difference between primary and secondary succession?

Apply the concepts

6. Draw up a list of the advantages and disadvantages of being a heterotrophic (parasitic or hemiparasitic) plant.

7. The pronghorn (*Antilocapra americana*) (Figure 37.3) was presented as an example of a situation when we can only understand an aspect of a species' current biology or behaviour by examining historical species interactions. Find another example

where a species' current biology or behaviour is the legacy of a historical species interaction, such as a predator–prey relationship, which no longer occurs today.

8. Batesian mimicry typically involves one species evolving to look like another species (typically one that is toxic in order to reduce predation risk), as shown in Figure 37.4, but it could involve one species evolving to sound or smell like another. Find an example of acoustic (sound) or olfactory (smell) Batesian mimicry.

9. Explain why many ecological communities are becoming less distinctive relative to others (i.e. why is biotic homogenization occurring?).

10. Find a real-world example of the Lotka–Volterra predator–prey cycle shown in Figure 37.16.

Beyond the concepts

11. Species interactions often involve a positive or a negative outcome for the species involved but this is not always the case. Explain why

and how species within an ecological community might simply coexist without directly interacting with one another.

12. Why are species that are strongly (or entirely) reliant on the presence of another species because of a mutualistic interaction potentially more vulnerable to environmental change or disturbance?

13. If you were managing a collection of small individual nature reserves in the same geographical area, why might it be more important to optimize gamma diversity (i.e. the species richness of all sites

together at a landscape scale) than the alpha diversity (species richness) of each individual site?

14. As presented in Figure 37.14, the species–area relationship is a logarithmic curve. Sometimes we use log_{10} to transform both the *x*-axis and the *y*-axis. How would this affect the way that the relationship appears graphically and why might we want to sometimes do this?

15. The latitudinal diversity gradient shown in Figure 37.15 shows that tropical regions typically have more species than polar regions. Why might tropical rainforests, in particular, have such high species richness?

FURTHER READING

Dhondt, A.A. (2012) *Interspecific Competition in Birds*. Oxford: Oxford University Press.
Focus on competition in one well-studied taxonomic group.

Těšitel, J. (2016) Functional biology of parasitic plants: a review. *Plant Ecol. Evol.* 149: 5–20.
Review of parasitic plant biology.

Bronstein, J.L. (2015) *Mutualism*. Oxford: Oxford University Press.
Mutualisms overview.

Mokkonen, M. & Lindstedt, C. (2016) The evolutionary ecology of deception. *Biol. Rev.* 91: 1020–1035.
Overview of mimicry.

Whittaker, R.H. (1972) Evolution and measurement of species diversity. *Taxon* 21: 213–251.
Introduction to alpha, beta, and gamma diversity.

McGuinness, K.A. (1984) Species–area curves. *Biol. Rev.* 59: 423–440.
Review of species–area relationships.

Lamanna, C., et al. (2014) Functional trait space and the latitudinal diversity gradient. *Proc. Natl. Acad. Sci. U. S. A.* 111: 13745–13750.
Critical analysis of processes involved in latitudinal diversity gradients.

Cutler, N. (2010) Long-term primary succession: a comparison of non-spatial and spatially explicit inferential techniques. *Plant Ecol.* 208: 123–136.
Overview of primary succession.

Ecosystems

Abiotic Interactions and Environmental Processes

LEARNING OBJECTIVES

By the end of this chapter you should be able to:

- Define 'ecosystem' and be able to outline the main abiotic and biotic factors that differ between ecosystems (and that are thus useful in ecosystem classification).

- Explore and appreciate the range of ways that organisms interact with their environment, adapt to that environment, or modify the environment.

- Appreciate the key role of temperature and water availability in underpinning ecosystem dynamics.

- Explain how species' population dynamics and distributions are controlled by environmental factors.

- Define the concept of 'ecosystem engineers' to explain how some species actively change their environment.

- Describe key ecosystem processes, including:
  - primary and secondary productivity;
  - energy flows and trophic structures, including food chains and food webs; and
  - biochemical nutrient cycling.

- Appreciate and critique the concepts of ecological tipping points and ecosystem breakdown.

- Define 'Anthropocene' and contextualize this in relation to previous mass extinction events.

Chapter contents

Introduction | 1292
38.1 Defining ecosystems | 1292
38.2 Abiotic factors that define and differentiate ecosystems | 1293
38.3 Species–environment interactions: tolerance ranges, niches, and adaptations | 1297
38.4 Ecosystem processes | 1301
38.5 Keystone species and ecological engineers | 1311
38.6 Ecological tipping points, ecosystem collapse, and mass extinctions | 1313

Watch the key concepts video in the e-book to prepare yourself for studying this chapter.

Introduction

In Chapter 36, we explored how different individuals of the same species co-occurring in the same geographical area form species *populations*. In Chapter 37, we broadened this to consider how co-occurring populations interact with one another to form ecological *communities*. However, we have not yet considered how ecological communities (and the species within them) interact with their environment in *ecosystems* through processes such as photosynthesis, energy exchange, and nutrient cycling; how species can adapt to better fit their environment; or how, in some cases, species can adapt an environment to better fit them through the process of ecosystem engineering. We will consider all these processes in this chapter but, before we go any further, it is worth reminding ourselves of the ecological hierarchy of population → community → ecosystem that we first considered in Chapter 34. This is shown in Figure 38.1 using coral reef as an example.

38.1 Defining ecosystems

The term ecosystem was first used by Arthur Tansley in 1935, who defined it as 'the interactions between organisms and their abiotic and biotic environments'. To unpack this, **abiotic** refers to anything that is not living (i.e. the physical and chemical aspects of the environment), while **biotic** means anything relating to biology (i.e. the biotic interactions covered in Chapter 37). An ecosystem is thus the sum of the living organisms in an area, the physical and chemical environment of that area, and the interactions, relationships, and processes that link these.

'Biomes' and 'ecosystems' are related concepts that overlap but the terms are often (incorrectly) used interchangeably. It is important that we define these terms and use them with precision:

- **Biomes:** biomes are very large-scale *habitat types* that have broad similarity because of the **climatic zone** in which they are situated. Examples include tropical rainforests, arid deserts, and tundra. Most biomes are **multi-continental** (i.e. occur on more than one continent): tropical rainforest, for instance, is found in central and south America, Africa, Asia, and Australia. Although the exact species that inhabit the same biome can differ substantially across geographical areas, there is similarity in terms of vegetation structure and overall climate. We can map biomes to see their distribution globally. Maps can be drawn at different levels of complexity and use slightly different terminology, but Figure 38.2 provides a typical example.

- **Ecosystems:** ecosystems can range in spatial scale from national (sub-continental) to highly spatially restricted; they also range from generic to highly specific. For example, UK oak woodland is an ecosystem (a broad-scale, generic ecosystem), but so too are the ditches of a specific wetland site, dry stone walls, old trees with deadwood used by wood-boring beetles, and even individual bird nests, which often support a miniature ecosystem of parasitic and free-living invertebrates.

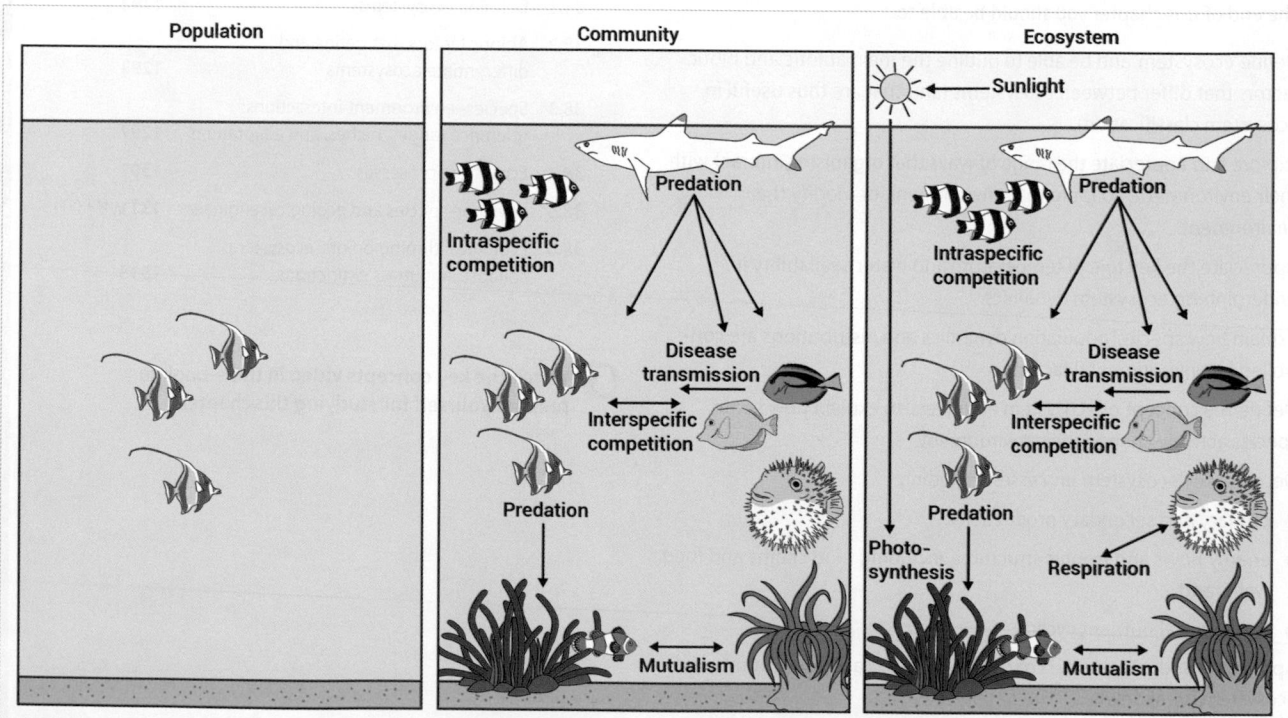

Figure 38.1 Populations, communities, and ecosystems. The differences between the concepts of a species population (multiple individuals of the same species), an ecological community (multiple species that interact with one another), and an ecosystem (a complex network of individuals, species, and links—both species–species links such as mutualism and species–environment links such as photosynthesis and respiration).

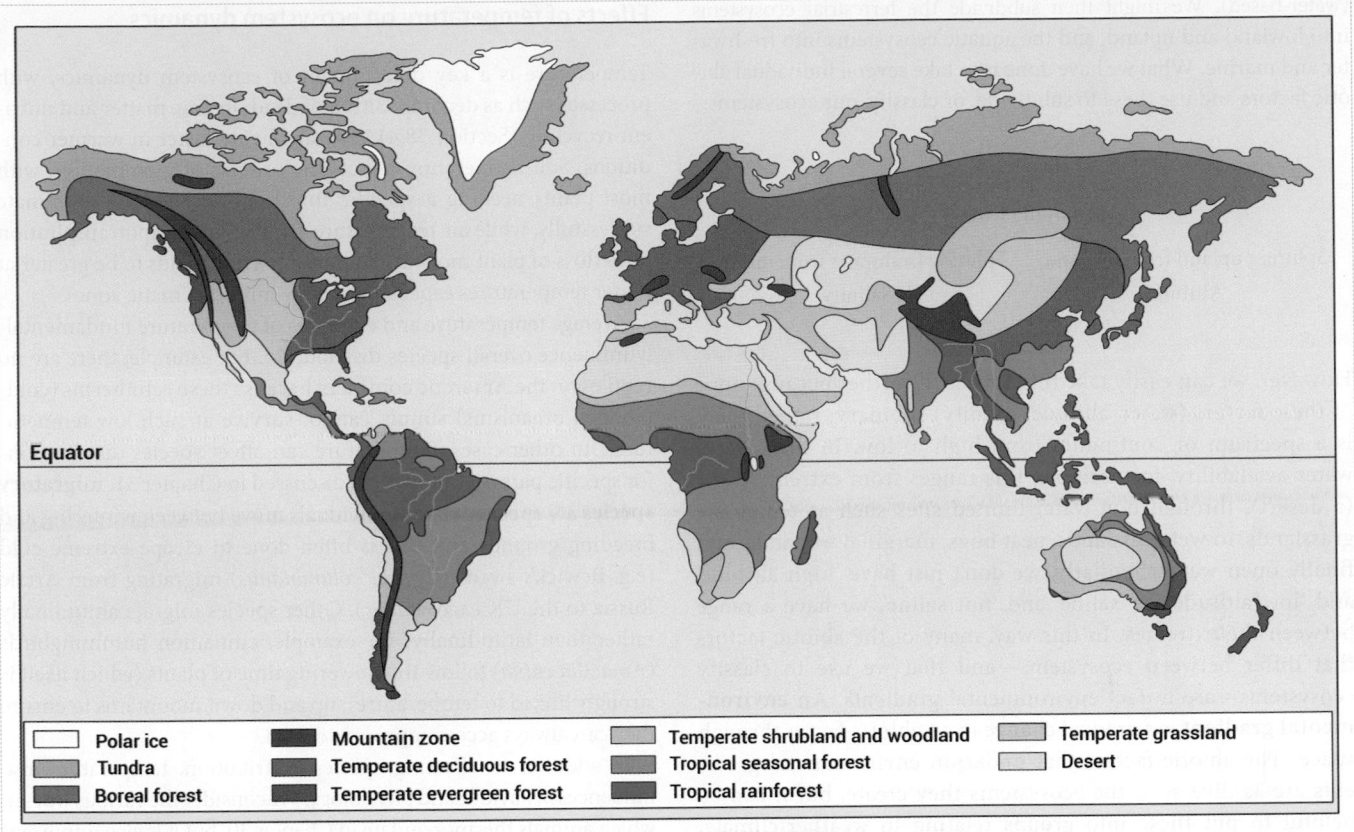

Figure 38.2 Major biomes of the world.

Legend:
- Polar ice
- Tundra
- Boreal forest
- Mountain zone
- Temperate deciduous forest
- Temperate evergreen forest
- Temperate shrubland and woodland
- Tropical seasonal forest
- Tropical rainforest
- Temperate grassland
- Desert

Source: © 2012. The Trustees of Columbia University in the City of New York. Center for International Earth Science Information Network (CIESIN)/Columbia University. 2012. National Aggregates of Geospatial Data Collection: Population, Landscape, And Climate Estimates, Version 3 (PLACE III). Palisades, NY: NASA Socioeconomic Data and Applications Center (SEDAC). http://sedac.ciesin.columbia.edu/data/set/nagdc-population-landscape-climate-estimates-v3. Distributed under the Creative Commons Attribution 3.0 Unported (CC BY 3.0). https://creativecommons.org/licenses/by/3.0/.

Although biomes differ in their environmental conditions, and indeed are fundamentally controlled by climate, the main way that they differ visually is often in their vegetation. In many ways, biomes can best be viewed as large-scale habitats or vegetation zones. Our focus, therefore, is on the *end result* of a particular set of circumstances (climate dynamics, etc.). This contrasts with ecosystems, when we tend to focus upon the outcome of abiotic and biotic processes *and* upon those processes themselves, and at a smaller spatial scale. We still tend to refer to ecosystems in terms of the habitat involved (a 'spruce woodland ecosystem', for example), but this is underpinned by a focus on the interlinked processes that create, control, and regulate that habitat.

As highlighted in Chapter 34, we also have the concept of the biosphere. This is a global concept and relates to all parts of Earth that support life, from the deepest ocean trench to the highest mountain peak. It is, in effect, the sum of all life on Earth and thus effectively the sum of the world's ecosystems and biomes.

PAUSE AND THINK

Complete this statement by entering the words 'ecosystems' and 'biomes' as appropriate: 'Consideration of very large-scale habitat types, based on climate and the effect this has on vegetation, is the study of ...., whereas the study of the abiotic and biotic processes and their outcomes at a sub-continental (sometimes very localized) spatial scale is the study of ....'

Answer: 'Biomes', and then 'ecosystems'.

 Check your understanding of the concepts covered in this section by answering the questions in the e-book.

38.2 Abiotic factors that define and differentiate ecosystems

One simple way that we can classify ecosystems is to divide those that are terrestrial (land-based) from those that are aquatic (water-based). We might then subdivide the terrestrial ecosystems into lowland and upland, and the aquatic ecosystems into freshwater and marine. What we have done is to take several individual abiotic factors and use these to subdivide, or classify, our ecosystems:

Splitting terrestrial from aquatic
Amount of water

Splitting upland from lowland   Splitting freshwater from marine
Altitude   Salinity

However, we can easily take this approach further because none of these factors (water, altitude, salinity) is binary; rather, there is a spectrum or continuum from high to low. In the case of water availability, for example, this ranges from extremely low (a desert), through non-water-limited sites such as temperate grasslands, to wet grasslands, peat bogs, marginal wetlands, and finally open water. Similarly, we don't just have 'high altitude' and 'low altitude', or 'saline' and 'not saline', we have a range between the extremes. In this way, many of the abiotic factors that differ between ecosystems—and that we use to classify ecosystems—are in fact environmental gradients. An **environmental gradient** is a gradual change in an abiotic factor through space. The abiotic factors that underpin environmental gradients are as diverse at the ecosystems they create, but it can be helpful to put these into groups relating to weather/climate, geomorphology (the physical features of Earth's surface), and soil/water parameters (Table 38.1).

Table 38.1 Key abiotic factors that define and differentiate ecosystems

Weather and climate	Geomorphology
Precipitation: amount, seasonality, intensity, availability*	Latitude
	Altitude
	Geology
Temperature: minimum, maximum, average*	Topography: slope gradient, direction of slope
Humidity	
Wind: sustained strength, gust strength	

Soil/water parameters	Miscellaneous
Soil: depth, texture, structure, moisture	Sunlight: hours, intensity, seasonality
Water: depth, clarity	Fire
Chemistry: pH and salinity, nutrient availability, heavy metals	Exposure (combines altitude, topography, wind)
Oxygen levels	Growing season (combines precipitation, temperature and light)

Many of these factors link together or interact with one another, for example, altitude and latitude both affect temperature. Asterisks denote abiotic factors that are discussed in more detail in subsequent sections.

Effects of temperature on ecosystem dynamics

Temperature is a key determinant of ecosystem dynamics, with processes such as decomposition of dead organic matter and nutrient recycling (Section 38.4) generally being faster in warmer conditions. Soil temperature affects the process of germination, with most plants needing a specific threshold for seeds to germinate successfully, while air temperature affects plant evapotranspiration rates (loss of plant moisture to the air), which tends to be greater at hotter temperatures especially in non-humid climatic zones.

Average temperature and extremes of temperature fundamentally influence overall species distribution. For example, there are no reptiles on the Antarctic continent because these ectotherms (cold-blooded organisms) simply cannot survive at such low temperatures. In other cases, temperature can affect species distributions for specific parts of the year. As discussed in Chapter 31, **migratory species** are species where individuals move between wintering and breeding grounds and this is often done to escape extreme cold (e.g. Bewick's swans (*Cygnus columbianus*) migrating from Arctic Russia to the UK each winter). Other species migrate altitudinally rather than latitudinally; for example, cinnamon hummingbirds (*Amazilia rutila*) follow the flowering time of plants (which itself is strongly linked to temperature), up and down mountains to ensure they can always access appropriate food.

In addition to affecting species' distribution, temperature also influences behaviour and physiology. We consider the various ways in which animals thermoregulate in Chapter 30, but it is also important to consider how temperature can affect ecology more generally. For example, some mammals **hibernate** during cold weather and their physiological processes go into 'slow-mo' mode: breathing and heart rate drop and metabolism slows to enable survival in the absence of food (this is also touched upon in Chapter 30). Hibernation is generally seen in small mammals such as dormice (*Muscardinus avellanarius*) and European hedgehogs (*Erinaceus europaeus*), but it can occur in larger species such as black bears (*Ursus americanus*) when females actually give birth during hibernation.

▶ **We learn more about hibernation in Chapter 30.**

A few bird species go into a hibernation-like state called **torpor** to survive when there is little food. One such species is the common poorwill (*Phalaenoptilus nuttallii*) an American nightjar species, which can slow its metabolism when insect prey is less abundant during winter. True hibernation is restricted to endotherms (warm-blooded organisms), but some endothermic reptiles exhibit **brumation**: this is similar to hibernation but the 'sleep' is lighter and physiological changes are less pronounced. Some individuals that metamorphose, such as some butterfly species that are shown in Figure 38.3, overwinter in non-active life stages (e.g. as eggs or pupae). This differs from hibernation because behaviour and physiology remain unchanged by temperature directly; rather, the timing of the non-active life stage is simply synchronous with colder temperatures.

In some cases, dormancy is partial, with individuals waking periodically. The Madagascan fat-tailed dwarf lemur (*Cheirogaleus medius*) has a particularly unusual form of hibernation where some individuals essentially become ectothermic and 'sleep through', while others wake periodically. We examine this in more detail in Scientific Process 38.1.

Figure 38.3 UK butterflies often overwinter in non-active life stages that are not strongly linked to temperature or temperature-dependent food sources such as flowering plants. For example, (a) silver-spotted skippers (*Hesperia comma*) overwinter as eggs, while (b) Duke of Burgundy (*Hamearis lucina*) overwinter as chrysalises.

SCIENTIFIC PROCESS 38.1 — Becoming ectothermic: an unusual hibernation strategy in lemurs

Research question

The Madagascan fat-tailed dwarf lemur (*Cheirogaleus medius*) (such as the one shown in Figure 1) had been observed using tree holes, seemingly for hibernation. In 2004, Kathrin Dausmann and colleagues set out to find out more about the unusual case of a tropical species hibernating and how temperature was regulated in different individuals.

Methods

The study took place a dry deciduous forest in west Madagascar. Each of the 53 study lemurs were fitted with a temperature sensor attached to a collar, which sent details of body temperature (T_b) and heart rate to a remote receiver. Some animals were also fitted with a radio transmitter so their exact location could be determined. Ambient temperature (T_a) and tree hole temperature (T_h) were measured using additional temperature loggers.

Results

The study found that lemurs hibernate in tree holes for up to seven months of the year but that many undergo regular spells of awakening. For lemurs using poorly insulated holes, body, hole, and ambient temperature all closely matched one another (shown in the top graph in Figure 2). This means that the animal's body temperature tracked the ambient/hole temperature similar to what we would expect for an ectothermic species such as a lizard. As long as temperatures periodically reached over 30 °C, the animal would stay in a dormant state for weeks at a time, with its metabolic processes using just 2% of the energy than would be the case when the same animal was awake and active. However, for lemurs using well-insulated holes temperature traces for ambient conditions varied more than the temperature trace for the hole, while the body temperatures differed substantially from both ambient and hole (shown in the bottom graph in Figure 2) because it was being regulated internally by the animal (endothermicly) rather than being controlled by the environment (ectothermicly). The animal's body temperature in such circumstances typically remained consistent at about 25 °C, but the animal would wake for periods of 6 hours or so once a week to warm itself up.

Figure 1 Madagascan fat-tailed dwarf lemur.

Source: Frans Lanting Studio/Alamy Stock Photo.

Conclusion

This study was the first physiological confirmation of prolonged hibernation by a tropical mammal, as well as the first proof of hibernation in a primate. This demonstrates that animals can hibernate at comparatively high temperatures but still be hypometabolic (i.e. have very slow

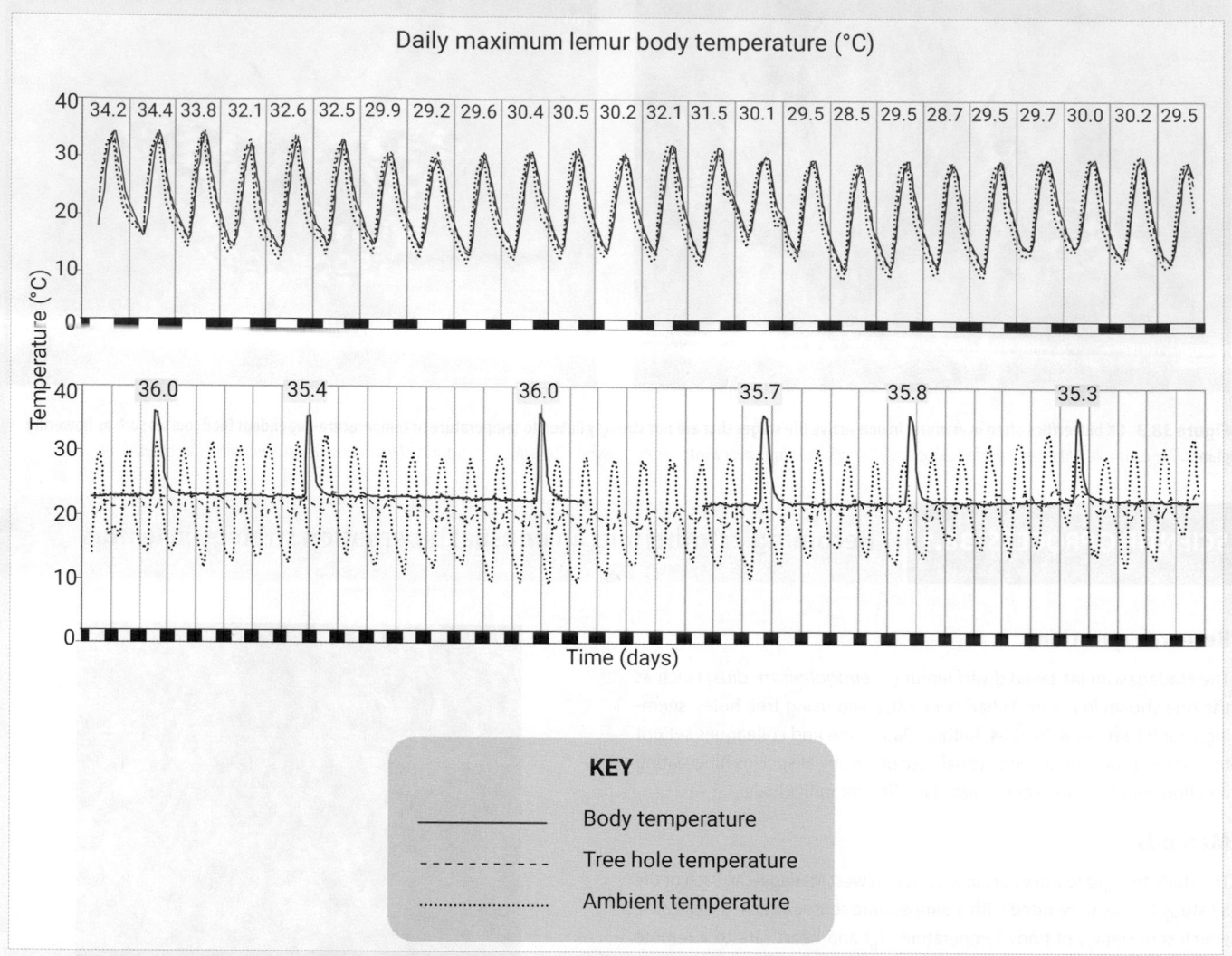

Figure 2 Body temperature of Madagascan fat-tailed dwarf lemurs during hibernation. The animal's body temperature (red traces) was monitored continuously in tree holes that were poorly insulated (top) or well insulated (bottom); tree hole temperature (black traces) and ambient temperature (grey traces) were also measured. Data are shown in 24 hr blocks. Numbers (top) give the daily maximum lemur body temperature.

Source: Reproduced with permission from Dausmann, K., et al. (2004) Hibernation in a tropical primate. *Nature,* 429: 825–826. Copyright © 2004, Springer Nature. https://doi. org/10.1038/429825a.

metabolic processes) and can essentially become ectothermic during hibernation.

Extension question

We have considered dormancy from the perspective of avoiding an extreme of temperature. In preparation for the section on the effects of

water availability on ecosystem dynamics, find an example of a species that becomes dormant to avoid drought.

Read the original paper

Dausmann, K. H., Glos, J., Ganzhorn, J. U., & Heldmaier, G. (2004). Hibernation in a tropical primate. *Nature* 429: 825–826.

Some species can also enter a form of 'summer hibernation' to escape extremely hot conditions. This is called **aestivation** and has been found, for example, in bogong moths (*Agrotis infusa*), an Australian species of moth that aestivates in cold caves, which act as refugia during hot summer months, as shown in Figure 38.4 (note the large aggregation of moths in the main image; a single individual is also shown).

Knowledge of hibernation, aestivation, and migration strategies is vital in practical ecology. When we study population biology (which

you can read about in Chapter 36), it is vital that we remember that surveying of presence and abundance need to be undertaken at a time of year when the target species is both present and active. Similarly, any studies of interspecific relationships at a community level (which are discussed in Chapter 37), is also linked to time of year since in relation to what species are present and active within the community. There can be practical implications in relation to conservation, too, which are covered in Chapter 40. For example,

Figure 38.4 Gregarious aestivation. A rare example of gregarious aestivation (group summer 'hibernation') is undertaken by Bogong moths (*Agrotis infusa*) in Australian caves where densities can exceed 17,000 moths per square metre.

Source: Reproduced from Warrant E, et al. (2016) The Australian Bogong Moth *Agrotis infusa*: A Long-Distance Nocturnal Navigator. *Front. Behav. Neurosci.*, 10: 77. doi: 10.3389/fnbeh.2016.00077. Copyright © 2016 The Authors. Distributed under the terms of the Creative Commons Attribution License (CC BY 4.0). https://creativecommons.org/licenses/by/4.0/. Photo by Eric Warrant. Main photo: Auscape International Pty Ltd/Alamy Stock Photo.

locating these sites used for gregarious aestivation or hibernation activity can be important as large numbers of individuals can be lost if a single refugia is destroyed.

Effects of water availability on ecosystem dynamics

In plants, water is a vital component of photosynthesis and thus energy creation (the other two being carbon dioxide and sunlight). Water is also vital for the structure of the plants by keeping cells **turgid** not **flaccid**: this is the reason one of the first signs of drought in many non-woody species is that the plant starts to wilt.

▶ **We learn more about the role of water in plant physiology in Chapter 30.**

In addition, all **vascular plants** (i.e. plants with an internal circulatory system, which covers most plant species other than mosses, liverworts, and algae) also need water to transport nutrients around the plant structure. For animals, water is vital for many physiological processes. This includes temperature regulation, regulation of pH, and removal of waste products from the body through urination and digestion.

▶ **We discuss the renal system further in Chapter 24.**

As discussed in Chapter 2, water also plays a vital role in energy production through hydrolysis of adenosine triphosphate (ATP), where its presence causes the atomic bond to break, releasing one phosphate ion and producing adenosine diphosphate (ADP), which can then be further hydrolysed to produce energy.

Given how important water is, it is not surprising that the precipitation in an area is a key determinant of ecosystem form and function.

Ecosystems that are water-limited are usually referred to as **xeric** ecosystems and tend to be inhabited by species that are drought-adapted (Section 38.3). It is important to note that xeric ecosystems can include polar deserts or high-altitude alpine ecosystems where the moisture present is locked up in snow or ice. **Mesic** ecosystems, by contrast, are ecosystems that are terrestrial but not water-limited, while **hydric** ecosystems are permanently or seasonally saturated by water.

 Check your understanding of the concepts covered in this section by answering the questions in the e-book.

38.3 Species–environment interactions: tolerance ranges, niches, and adaptations

The spatial distribution of species is fundamentally controlled by two factors:

1. The environmental conditions (temperature, water availability, etc.) that species can tolerate.

2. Where those environmental conditions are found.

We can consider this on a small scale (e.g. the distribution of flowering plants at a single woodland site), a regional or national scale, or even an international scale. There are, of course, other factors influencing a species' distribution, including its dispersal power (how far individuals of a species can move either under their own power via **active dispersal** or by harnessing environmental features such as air or water currents in **passive dispersal**), but, for now, let's keep things simple by focusing on abiotic factors.

Tolerance range

The reason that species' distributions are so tightly linked to abiotic conditions and environmental gradients is because every species has a specific **tolerance range** for a whole range of environmental parameters, such as those outlined in Table 38.1. For example, a plant species might only be able to survive in environments in which the temperature is between −5 and 30 °C, where the soil is reasonably fertile, there is water available throughout the year, and where light levels are sufficient to allow photosynthesis to occur for at least 6 hours per day during the growing season. The plant thus has a tolerance range for each individual abiotic factor—temperature, soil fertility, moisture, light, and so on—within which it can survive and reproduce.

As shown in Figure 38.5, a species' tolerance range for each individual abiotic factor is complex. The central part of the species' overall tolerance range is the **optimal range** where the species can maximize survival and reproduction. Near to the extremes of the tolerance range (i.e. close to the minimum and maximum levels of how much water a species needs in this example), there will be physiological stress. This means that normal physiological processes such as photosynthesis, respiration, and circulation are occurring under stress and either demand more energy, or are less effective, than under optimum conditions. A species can survive, but it does so at the edge of its physiological capabilities. Population size is likely to be lower in this physiological stress zone than under optimal conditions because it is likely to be outcompeted by species better suited to such conditions or because reproduction and population growth rate potential is compromised.

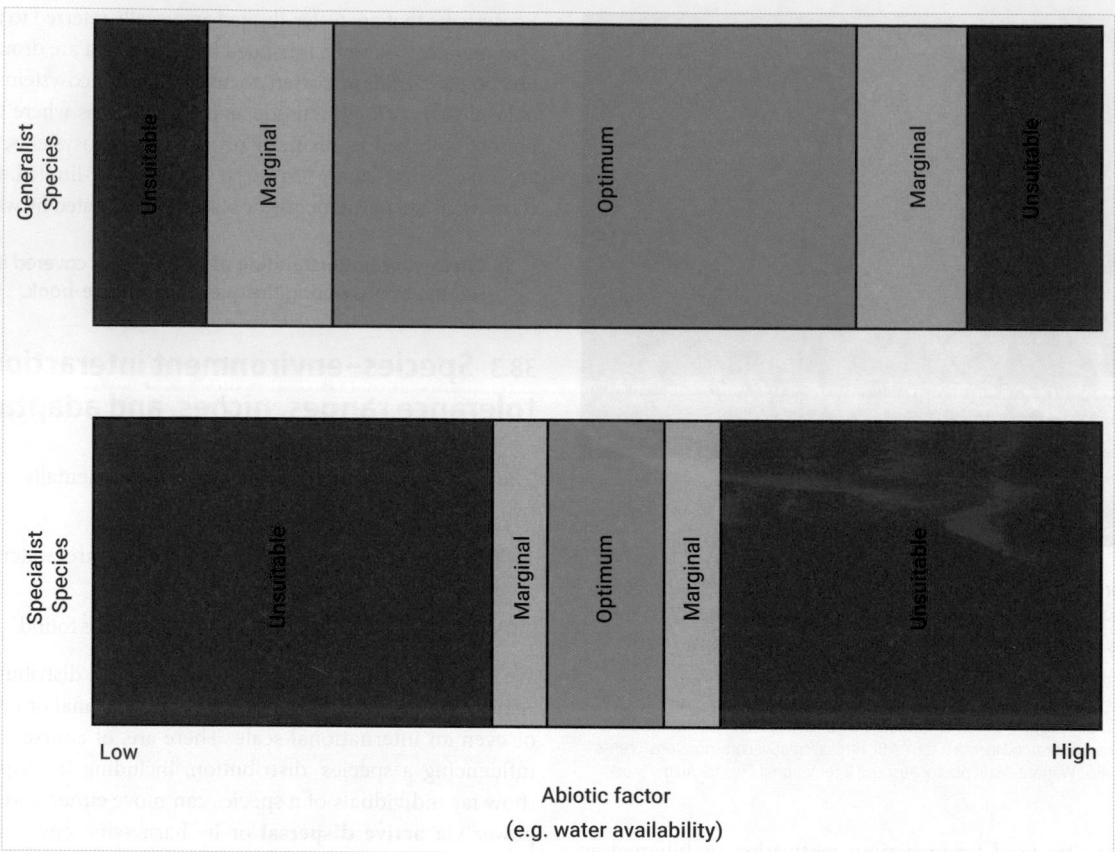

Figure 38.5 Tolerance range. Tolerance range for a single abiotic factor (in this case water availability) showing the optimum range (green), the marginal range where the species would be under physiological stress and would occur only in low numbers (amber), and areas outside the species' survival range (red).

▶ **Read more about species competition in Chapter 37, and about reproduction and population growth rate potential in Chapter 36.**

Generalist species have wider tolerance ranges for abiotic factors (shown at the top of Figure 38.5 by the larger optimal (green) range), while specialist species have a much narrower tolerance range (shown at the bottom of Figure 38.5 by the smaller optimal (green) range). This also helps to explain why generalist species tend to have more cosmopolitan distributions than specialist species.

Niches

While tolerance range is a useful concept, it does have its limitations. The real world is complex and considering each abiotic factor in isolation from one another is not very realistic. Because of this, we tend to talk more about **niches** than tolerance ranges. The concept of the niche was formally developed by Hutchinson in 1957, taking account of work done in the early 1900s by Charles Elton, among others, with a niche being the sum of the environmental and ecological conditions that a species needs to survive. Niches are, by their very nature, multifaceted and encompass tolerance ranges for numerous different variables.

Let's unpack the concept of a niche a little further. If a species' niche is controlled by just two factors, say temperature and water availability, it is conceptually straightforward to define that niche. This can be visualized on a graph with two axes, where the x-axis is one factor

(temperature) and the y-axis the other (water). It is possible, as shown in Figure 38.6, to draw a shape on the graph that encompasses the theoretical space with respect to temperature and water availability in which our hypothetical species can live. In other words, the shape would depict a two-dimensional niche. It would then be possible to add in a third factor/dimension—for example, the amount of sunlight needed. Returning to our graph, this third dimension would be represented by the z-axis, and we could draw a three-dimensional volume that would represent the niche with respect to these three factors. This is shown in Figure 38.6. Although it is not easy to plot a fourth or higher dimension, it would be possible to keep adding axes to incorporate more and more abiotic factors with the resultant niche 'shape' (what mathematicians call a delimited hyper-volume) becoming ever more complex.

The overall niche is referred to as the **fundamental niche**, which is depicted by the large outer circle in Figure 38.7. However, most species only occur in a subset of their fundamental niche and this is referred to as the **realized niche**, which is depicted as the small inner circle in Figure 38.6. The reason that the realized niche is typically smaller than the fundamental niche is that species interact with other species in ways that can constrain niche breadth. For example, a species could be in competition for a particular resource which might mean its full fundamental niche cannot be exploited but instead has to be **partitioned** (divided between) between two or more species. Thus, while the fundamental niche is based on the physiological capabilities of the species concerned and species

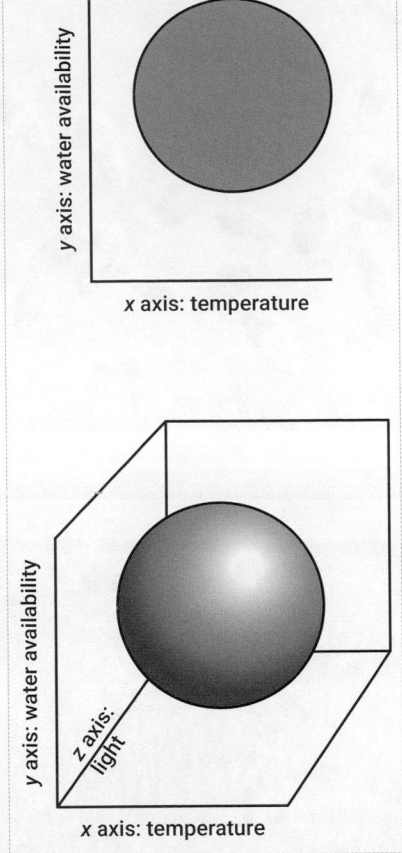

Figure 38.6 Two- and three-dimensional niches. Hypothetical diagram of a two-dimensional niche (top) and three-dimensional niche (bottom).

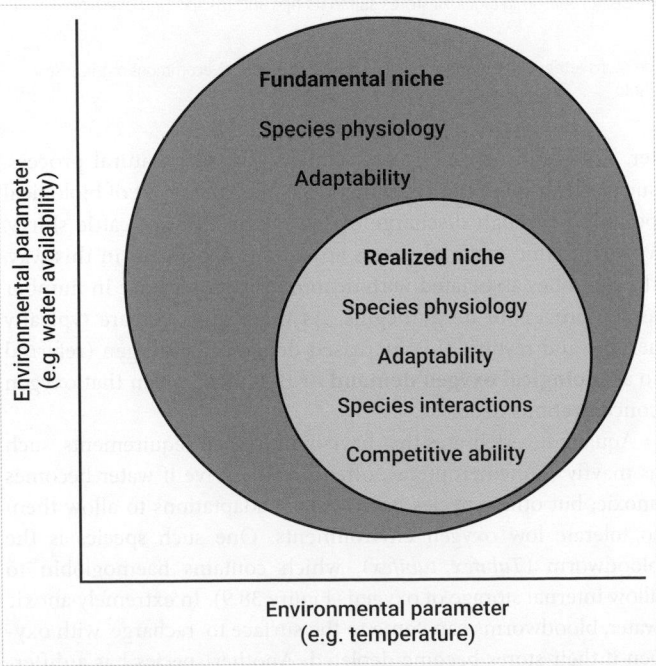

Figure 38.7 Realized and fundamental niches. A species' realized niche (where it actually occurs) is almost always a subset of its fundamental niche (where it could occur physiologically).

adaptability, the realized niche is based on physiology, adaptability, and interactions with other species (and thus what other species are present and their relative competitive ability).

Limiting factors

Although niche theory involves considering many different variables, it is important to remember that not all factors are equally important in determining if a species can occur in an area. For example, we might have a species of flowering plant that needs a particular thermal range, a particular moisture range, and a particular level of nitrogen in order to grow in an area. If we have a geographical location that is within the plant's tolerance for moisture and temperature, but which is extremely nutrient poor, it is the nutrient level that would stop the plant from growing there. Nutrient level would thus be the **limiting factor** for that particular species at that location—it is the abiotic factor that literally limits that species' occurrence. The same plant could be limited by other factors in other locations, for example, it might be that a second site has suitable conditions in terms of temperature and nutrients but is too wet to allow the species to grow there.

Species adaptations

Although abiotic factors have an important role in controlling and regulating where specific species can occur, this is not solely a one-way process. Most—if not all—species have evolved specific adaptations that allow them to extend their tolerance range for specific abiotic factors and thus broaden their overall realized niche. We consider how selection pressures can cause evolution through natural selection and result in **genotypic adaptations** in Chapter 34, as well as ways that an organism can adjust its phenotype in a non-genetic way via **phenotypic plasticity**. We now need to consider the important of these in an ecosystem context.

Adaptations are often most dramatic, and thus most obvious, for species inhabiting what are typically regarded as 'extreme' or 'harsh' environments (such as deserts, polar regions, acidic lakes, or sulphur-rich hot springs). Species inhabiting these environments (**extremophiles**) have often evolved to become highly specialized.

We tend to refer to species adaptations as being in one of the following categories:

- **Physical:** a physical change to the anatomy of an organism (usually genetic).

- **Physiological:** a change to the internal systems and biological processes of an organism (usually genetic).

- **Behavioural:** a change in the way an organism responds to, and interacts with, its external environment (phenotypic plasticity).

Let's consider some examples of extreme ecosystems and the different types of adaptations that particular species have evolved to succeed in them.

Highly saline ecosystems: physical and physiological adaptations of plants

Most plant species are **glycophytes** that can only tolerate a very small amount of salt and thus cannot grow in saline ecosystems. As

Figure 38.8 Halophytic adaptations to saline ecosystems. Halophytic adaptations to saline ecosystems include (a) the use of mannitol by *Fucus* seaweeds to alter osmosis and (b) the use of glycerol by phytoplankton in the *Dunaliella* genus to alter osmosis. (c) *Salicornia bigelovii* stores salt in its tips and (d) *Spartina alterniflora* excretes salt..

Source: (a) Mark Heighes/Shutterstock; (b) © CSIRO. Distributed under the terms of the Creative Commons Attribution 3.0 Unported (CC BY 3.0). https://creativecommons.org/licenses/by/3.0/; (c) Marco Schmidt/CC BY-SA 2.5/Wikimedia Commons; (d) Jennifer Wright/Alamy Stock Photo.

an example, rice plants can tolerate 1–3 grams per litre (g/l) of dissolved salt (sea water, for comparison, contains about 40 g/l). Some plants, though, are **halophytes**, which means they are highly salt tolerant and can grow at salinities of up to 70 mg/l.

There are three basic strategies a halophytic plant can utilize: reducing salt uptake, increasing salt output, or storing salt internally somewhere it does not interfere with the normal plant physiology. To decrease uptake of salt, plants can utilize organic compounds such as mannitol and glycerol that raise internal osmotic pressure to reduce saline intake: this is a physiological adaptation. Other halophytic plants take up salt but excrete this via specially developed cells in an adaptation that involves a physical change to the plant *and* altered physiology. Finally, other halophytic plants use selective storage, whereby salt is diverted internally to the tips of the plant to become concentrated which, in some cases, are then shed. Examples of all these mechanisms are given in Figure 38.8.

Anoxic aquatic ecosystems: physical adaptations of invertebrates

Aquatic ecosystems can become **anoxic** (i.e. very low in dissolved oxygen levels) because of an influx of dead organic mat-

ter. This dead organic matter can be part of a natural process, such as leaf-fall in the autumn, or it can be the result of biological pollution through discharge of sewage or input of cattle slurry. When organic material enters an aquatic ecosystem in this way, the microbes associated with decomposition increase in number as the process of decay begins. As these microbes are typically aerobic and respire, this increased demand for oxygen (referred to as **biological oxygen demand** or BOD) can mean that oxygen concentrations decrease.

Aquatic invertebrates that have high oxygen requirements, such as mayfly (Ephemeroptera), often don't survive if water becomes anoxic, but other species have specific adaptations to allow them to tolerate low oxygen environments. One such species is the bloodworm (*Tubifex tubifex*), which contains haemoglobin to allow internal storage of oxygen (Figure 38.9). In extremely anoxic water, bloodworms can come to the surface to 'recharge' with oxygen if their stores become depleted. Another species has a different adaptation: drone fly larvae (*Eristalis tenax*), which are often called rat-tailed maggots in their larval form, have a siphon that acts as a snorkel, allowing the aquatic larvae to remain submerged by breathing air as shown in Figure 38.9.

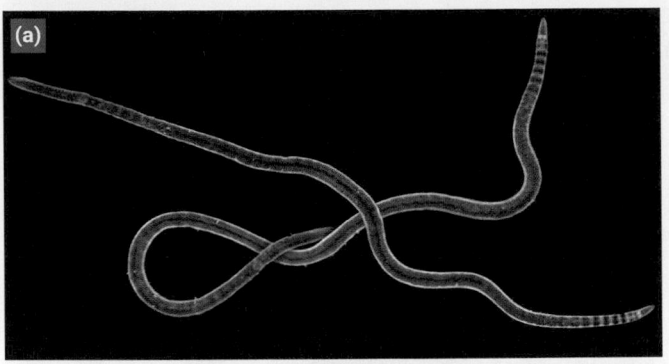

Figure 38.9 Different species have different adaptations to anoxic aquatic ecosystems. (a) Bloodworms (*Tubifex tubifex*) have haemoglobin to help store oxygen, while (b) rat-tailed maggots (*Eristalis tenax*) breathe air through a 'snorkel' while remaining submerged in water.

Source: Nature Photographers Ltd/Alamy Stock Photo.

Desert ecosystems: physical and behavioural adaptations of birds

Birds living in desert environments have two key challenges: lack of water and intense heat. The Namaqua sandgrouse (*Pterocles namaqua*) (Figure 38.10a) overcomes the lack of water surrounding its desert nests by flying to water holes and soaking specially adapted feathers on its stomach and breast in the water. These feathers are modified to be super-absorbent so that the adult birds can fly back to their nest, even if this is some distance away, and their chicks can drink the water directly from their feathers. This is a physical adaptation as it involves a structural change to feather anatomy. Crab plovers (*Dromas ardeola*) (Figure 38.10b) have a behavioural adaptation to escape intense heat. They dig down into the sand to access cooler (and often slightly damp) sand below the surface to create sun refugia. Their legs are not well-adapted for this activity, being rather too long for easy sand excavation, but they are instead well adapted for wading in shallow water to access food. In this way, any selection pressure for the birds to evolve short legs that are better adapted for digging would face a counter selection pressure for long legs better adapted to foraging in water.

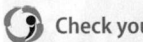 **Check your understanding of the concepts covered in this section by answering the questions in the e-book.**

38.4 Ecosystem processes

Ecosystem processes involve anything that relates to energy generation and transfer (encompassing photosynthesis, energy flows, food chains, and food webs) or movement and storage of chemical compounds (encompassing nutrient recycling and bioaccumulation).

Productivity

Productivity is the amount of energy *generated* or *transferred* within an ecosystem. It is usually expressed as mass per area over time, for instance grams per metre squared per day ($g\ m^{-2}\ d^{-1}$) or per year ($g\ m^{-2}\ yr^{-1}$). **Primary productivity** is the energy *generated* by plants and algae containing chlorophyll (**autotrophs**) via the process of photosynthesis whereby energy harvested from the sun is used to produce glucose from carbon dioxide and water. This process is discussed in Chapter 2. There is one exception to this, which is worth noting: **chemosynthesis**, which is discussed in Scientific Process 38.2.

Figure 38.10 Physical and behavioural adaptations of birds. (a) Namaqua sandgrouse (*Pterocles namaqua*) have specially adapted water-absorbent feathers. (b) Crab plovers (*Dromas ardeola*) use their unsuitably long legs to dig down into baking hot sand to reach cooler substrate.

Source: (b) Martin Mecnarowski/Shutterstock.com.

Background

As we see in Experimental Toolkit 34.1, the majority of (good) ecological research that is eventually published and highly cited starts with a clear research question designed to answer a known gap in knowledge. The researchers are also usually able to make hypotheses or predictions about what they might find because there is already enough known about the scientific topic to make an 'educated guess'. Things don't always happen this way, though, and some of our greatest discoveries are little more than scientific accidents.

If we had been studying ecosystems before 1977, we would have said that the only way for energy to enter an ecosystem was via photosynthesis and, ergo, ecosystems could only occur where there was sunlight. However, in May 1976, a research expedition led by Peter Lonsdale and Ray Weiss off the Galapagos Islands found a spike in sea temperature. A remotely operated camera was deployed and photographed a seafloor fissure at the location of the temperature spike. Intriguingly, there appeared to be life in the area, including what appeared to be crabs and bivalves. A return expedition, in February 1977, recorded the first definitive images of life over 8000 feet below the surface and in complete darkness. Within 48 hours, these initial images had been verified by a research team in a submersible—called Alvin—led by Jack Corliss (pictured in Figure 1).

Research question

What the Lonsdale–Weiss and Corliss teams had accidentally discovered was a hydrothermal vent ecosystem, powered not by carbon and sunlight, but by inorganic compounds. This single discovery opened a whole new area of science. Key questions included what other species were involved, how the species interacted to form a community, and, perhaps most intriguingly of all, what chemical process took the place of photosynthesis to provide the energy to power that ecosystem (and allow crabs, bivalves, and other species to survive).

Methods

The methods used to answer these research questions over the last four decades have involved *in situ* surveying by remote cameras and submersibles, as well as extensive work in labs. Particularly important has been the ability to use submarines to retrieve specimens from the ocean floor for detailed *ex situ* analysis in laboratories. One of the first samples retrieved was of the large red-plumed vestimentiferan tube worm (*Riftia pachyptila*), which was studied by then-student Colleen Cavanaugh in the 1980s.

Results

In 21 of 31 specimens of *Riftia* examined by Cavanaugh, crystals of elemental sulphur were found. This observation suggested that the species 'had capacity for chemoautotrophic oxidation of sulphide as an internal source of nutrition'—in other words that the species was an autotroph and could manufacture its own food but, crucially, using inorganic compounds. What has been discovered was **chemosynthesis**—the process by which an inorganic compound, often hydrogen sulphide, is converted into energy without the need for sunlight:

$$CO_2 + O_2 + 4H_2S \rightarrow C_6H_{12}O_6 + 4S + 3H_2O$$

Carbon dioxide + water + hydrogen sulphide → glucose + sulphur + water

Figure 1 Jack Dymond, Jack Corliss, and John Edmond with the submersible 'Alvin' on the 1977 discovery expedition.

Source: Courtesy Bob Collier, used with kind permission.

Figure 2 Large concentration of *Riftia* with some of the other species associated with hydrothermal vent ecosystems—anemones and mussels.

Source: © NOAA. Distributed under the terms of the Creative Commons Attribution-ShareAlike 2.0 Generic (CC BY-SA 2.0). https://creativecommons.org/licenses/by-sa/2.0/.

Conclusion

The scientific process starts with observation. In this case, the observation of life on the ocean floor around hydrothermal vents was incidental to the discovery of the vents themselves. Following many different studies, at different locations (and subsequently in different laboratories), we now know that hydrothermal vents support whole communities of different species that occur in close proximity around the vents and interact with one another, as pictured in Figure 2. At the centre of this is the mutualistic relationship between chemoautotrophic bacteria and organisms such as tube worms first identified by Cavanaugh from the very first Galápagos samples.

Extension question

Find an example of an ecosystem based on chemosynthesis that uses inorganic compounds other than hydrogen sulphide.

Read the original paper

Cavanaugh, C.M., Gardiner, S.L., Jones, M.L., Jannasch, H.W., & Waterbury, J.B. (1981). Prokaryotic cells in the hydrothermal vent tube worm *Riftia pachyptila* Jones: possible chemoautotrophic symbionts. *Science* 213: 340–342.

- **Gross primary production** (GPP) is the total amount of solar energy converted to chemical energy during photosynthesis. GPP potential depends on several factors, but the amount and intensity of sunlight are vital. This means that just as there is a **latitudinal gradient** in species richness (see Chapter 37), there can also be a latitudinal gradient in GPP potential, with greater *annual* potential in the tropics where light intensity is higher (although GPP potential in polar regions can be considerable at specific times of the year when there is 24 hr daylight). In aquatic ecosystems, the depth and clarity of water is also important because this affects light penetration and thus photosynthetic potential. In the terrestrial environment, water availability is also crucial as water is a vital component of photosynthesis.

- **Net primary production** (NPP) is the fraction of energy originally fixed through photosynthesis that is actually *available* to autotrophs to convert into new biomass after routine processes (respiration, transport of water and nutrients, and cell repair) have been carried out. Plants sometimes allocate NPP to producing new growth, thereby making the parent plant bigger, taller, or stronger: this is direct use of NPP. Alternatively, plants can use NPP indirectly through investment in reproduction (i.e. energy used to produce flowers, nectar, pollen, fruits, and seeds).

PAUSE AND THINK

Summarize how we would calculate net primary productivity from gross primary productivity.

Answer: NPP = GPP minus energy used for plant maintenance.

Energy transfer within ecosystems

Except in the very rare cases of chemosynthesis, all energy within an ecosystem comes from the sun as converted by autotrophs—chlorophyll-containing plants, algae, and bacteria—into carbon-containing compounds. Autotrophs really are the 'powerhouses' of ecosystems: without them there would be no energy generated, and no energy to transfer, and the whole system would break down.

We have already seen come across the idea that energy transfer in an ecosystem is not an efficient process, when we defined GPP and NPP. In this case, energy transfer between autotrophs and heterotrophs is reduced because plants need to use some of the energy that they generate to sustain themselves. The same is true for **secondary productivity**—the energy *transferred* between species that do not generate their own food. In fact, the ten percent law of energy transfer in ecosystems, introduced by Raymond Lindeman (1942),

states that only 10% of the energy in one organism is available for a second organism that consumes it (i.e. 90% of energy is lost). This loss of energy occurs because all heterotrophs use energy to maintain their physiological functions (respiration, circulation, digestion, reproduction, etc.), for movement, and, for **endotherms** (warm-blooded animals), to maintain their own body temperature. Energy is also lost because when one organism consumes another not all body parts are usually consumed and digestion is not fully efficient at absorbing all possible energy and nutrients.

Trophic levels: food chains

Understanding feeding relationships is key to understanding energy transfer in ecosystems. Some energy transfer between autotrophs and heterotrophs is direct, through herbivory (e.g. rabbits grazing on grass, butterflies drinking nectar, birds eating fruit), plant parasitism, or plant decomposition (e.g. wood-boring beetles and fungi consuming deadwood after the plant has died). Refer to Chapter 37 if you need a reminder about these species interactions. In other cases, heterotrophs obtain autotrophic energy *indirectly* by consuming heterotrophs that had themselves previously consumed autotrophs. This might involve the processes of predation, animal parasitism, or animal decomposition. At their most simple, feeding relationships can be summarized by a food chain. There are two main types of food chain:

1. **Grazing food chain:** Occurs where energy comes directly from autotrophs being grazed by herbivores: the food chain is herbivore-based.

2. **Detritus food chain:** Occurs where most energy comes from dead organic material that is broken down by microbial action and invertebrates: the food chain is detritivore-based.

As shown in Figure 38.11, a grazing food chain starts with an autotroph as a primary (energy) producer, which in our example is the oak (*Quercus robur*). The primary producer is consumed by a herbivore—the primary consumer, which in our example is the oak moth (*Tortrix* spp.). The primary producer is, in turn, consumed by a carnivore—the secondary consumer—which in our example is the great tit (*Parus major*). In most food chains, the secondary consumer then becomes prey for another (top) predator, which is the tertiary consumer. In our example, the tertiary consumer is the sparrowhawk (*Accipiter nisus*). Each of these stages in the food chain is a **trophic level** (trophē from the Greek meaning feeding).

As shown in Figure 38.12, a detritus food chain is similar to the grazing food chain shown in Figure 38.11 except that rather than

Figure 38.11 Grazing food chain process. An example from an English woodland.

Source: Images from Shutterstock.com

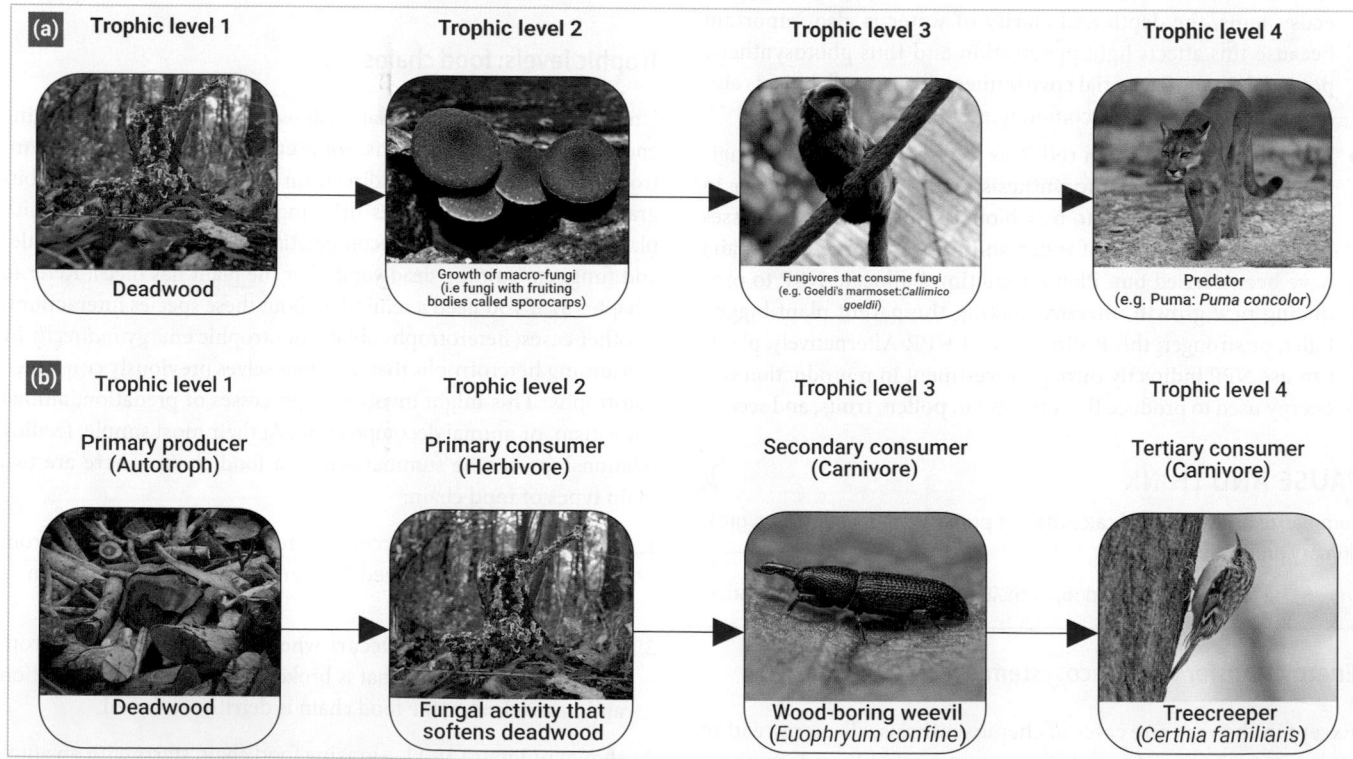

Figure 38.12 Detritus food chains. Detritus food chains created either (a) when fungi are a direct food source for subsequent trophic levels; or (b) when they indirectly facilitate invertebrate detritivores to access dead organic matter.

Source: (a): (i) wisawa222/Shutterstock.com; (ii) akslocum/Shutterstock.com; (iii) DejaVuDesigns/Shutterstock.com; (iv) Perla Sofia/Shutterstock.com; (b): (i) wisawa222/Shutterstock.com; (ii) Giuma/Shutterstock.com; (iii) thatmacroguy/Shutterstock.com; (iv) Ihor Hvozdetskyi/Shutterstock.com.

starting with a living autotroph, which is still continuing to fix solar energy through photosynthesis, a detritus food chain starts with dead organic matter (DOM): in our example in Figure 38.12, the DOM is deadwood, but it could equally be fallen leaves, rotting woody debris, or dead animals. This DOM is utilized as an energy source by a detritivore. Detritivores include fungi, as in our example, as well as bacteria and protozoa. In some cases, detritivores provide a direct source of food for organisms in the next trophic level because they are directly consumed, as shown in Figure 38.12a where a primate such as Goeldi's marmoset (*Callimico goeldii*) eats detritivore fungi and is then consumed by a larger predator, in our example a puma (*Puma concolor*). In other cases, such as the example in Figure 38.12b, microbes power a detritus food chain by breaking down DOM, which then allows this resource to be used by invertebrate detritivores such as beetle larvae, which become prey for birds.

Detritus food chains are indispensable in ecosystems to break down DOM and recycle 'second-hand' energy (and nutrients). That concept of recycling is key to understanding how detritus food chains work: the organisms in lower trophic levels act to *release* the energy stored in DOM through previous photosynthesis when the plant was alive, rather than *creating* it directly.

Because energy transfer within food chains is inefficient, some people argue is it is more environmentally friendly to be vegetarian or vegan rather than including meat and/or fish in human diet. This is discussed in Real World View 38.1.

PAUSE AND THINK

Why are most food chains constrained to around four trophic levels rather than there being many more 'links in the chain' between autotrophs (or dead organic matter) and the top carnivore?

Answer: The poor efficiency of energy transfer within food chains means that if there are more than a few links in the chain, very little of the energy originally fixed by the autotrophs would be transferred to the top and the food chain would eventually break down.

Trophic levels: food webs

Food chains, such as those shown in Figures 38.11 and 38.12, over-simplify reality as they assume that each species only eats—and is only eaten by—one other species. In other words, the concept of a food chain assumes that feeding relationships are a linear process of highly specific predator–prey interactions, each depicted by a

---

REAL WORLD VIEW 38.1   **Vegetarianism from an ecosystem and ecological footprint perspective**

People become vegan (abstaining from any animal product) or vegetarian (avoiding eating meat—and sometimes fish—but continuing to eat animal milk and eggs) for many different reasons. For some people, dietary choices stem from religious conviction; for others dietary changes are motivated by health considerations to reduce fat, cholesterol, or sugar intake. Animal rights, ethical considerations, emotional attachment to animals, or concern over husbandry or slaughter conditions can also be key drivers.

For some people, there is an ecosystem perspective here, too. From a purely ecological perspective, if energy is lost within and between each trophic level (Section 38.3), it makes sense to obtain as much energy as possible from as low down the food chain as possible. With an increasing human population resulting in an ever-increasing number of people to feed, the issue of how sustainable it is to 'waste' energy by eating animal products (which are, by definition, higher up the food chain) rather than plants is becoming increasingly pressing.

Indeed, our diet is one of the most important factors in our **ecological footprint** (Figure 1). Our ecological footprint is the sum of the natural resources we use to sustain our way of life. Ecological footprints are often based on the geographical area of biologically productive land needed to provide everything we use: fruit, vegetables, grains, fish, meat, wood, fibre for clothing, fossil fuels, generation or alternative energy, raw materials for the products that we buy, absorption of carbon dioxide from our activities, and space for buildings and roads. Obviously, there are many ways that we can reduce our ecological footprint, but decreasing the dietary intake of animals and animal products is a powerful one. Not only does this reduce energy being wasted in food chains, but it also means that food needs can be met from a smaller land area and using much less water.

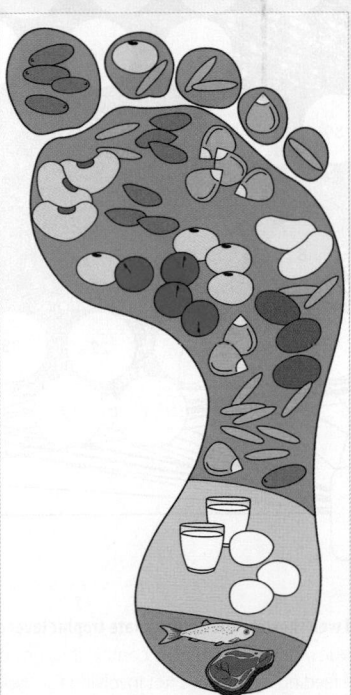

Figure 1 Ecological footprint. Our diets constitute a lot of our ecological footprint and the size of this can be decreased by following Linderman's 10% rule and deriving foods from as low down the food chain as possible (i.e. plant-based products).

Extension question

What is your opinion on use of ecosystem perspectives when discussing vegetarianism or veganism? Is it food for thought or are there counterarguments?

single line in food chain diagrams, such as those in Figures 38.11 and 38.12. In reality, however:

1. Ecosystems typically have multiple species at each trophic level.

2. There are multiple predator–prey relationships connecting those species.

3. Grazing and detritus food chains interweave.

4. Some species occupy different trophic levels in different contexts.

5. Some species are consumed by species from more than one trophic level, so that trophic levels are sometimes bypassed.

To summarize these highly complex situations, we use a food web, such as that shown in Figure 38.13, whereby each species becomes a 'node' in a complex network of feeding interactions, each shown by a line.

Although food webs are conceptually straightforward, the reality that they represent can be incredibly intricate, as highlighted by the many connections in Figure 38.13. This means that they can be difficult to draw, especially when feeding interactions between species in an ecosystem are complex or unclear. This often occurs for ecosystems that:

■ Contain species that are **biphasic** (i.e. have two or more distinct life phases); for example, newts are aquatic during the breeding

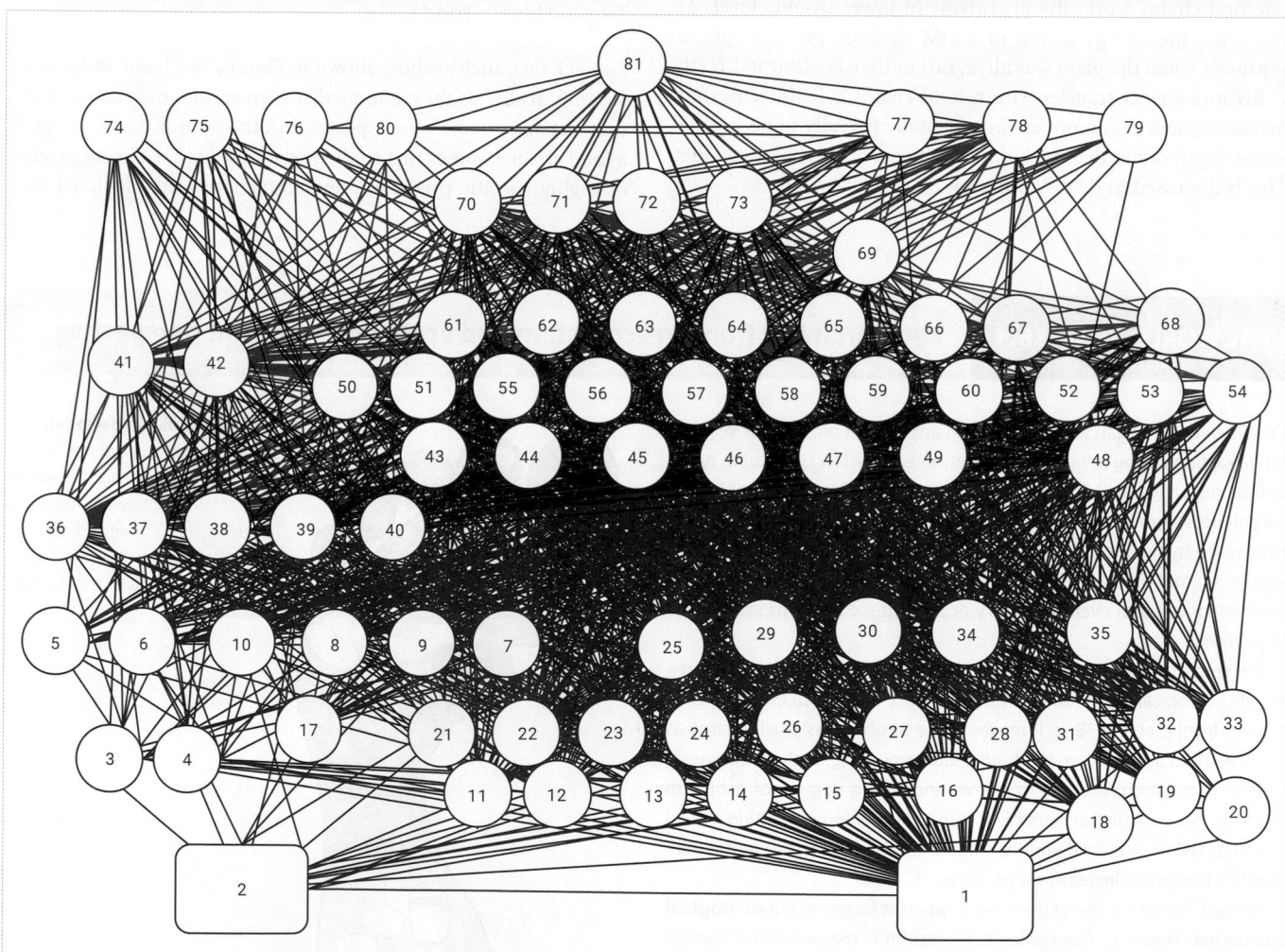

Figure 38.13 Northwest Atlantic food web showing the approximate trophic level of each species, with the highest trophic level at the top of the web. Pelagic (open-ocean) organisms generally occur on the left side of the web, whereas benthic (bottom-dwelling) organisms generally occur on the right. Red lines indicate feeding interactions involving fish while black lines indicate feeding interactions not involving fish. Note that the two main start points in the web, numbers one and two, are detritus (particulate dead organic matter) and phytoplankton (photosynthetic microscopic marine algae); in other words, the entire web involves the interweaving of grazing and detritus food chains.

The species represented are: 1 = detritus, 2 = phytoplankton, 3 = *Calanus* sp., 4 = other copepods, 5 = ctenophores, 6 = chaetognatha (i.e. arrow worms), 7 = jellyfish, 8 = euphasiids, 9 = *Crangon* sp., 10 = mysids, 11 = pandalids, 12 = other decapods, 13 = gammarids, 14 = hyperiids, 15 = caprellids, 16 = isopods, 17 = pteropods, 18 = cumaceans, 19 = mantis shrimps, 20 = tunicates, 21 = porifera, 22 = *Cancer* crabs, 23 = other crabs, 24 = lobster, 25 = hydroids, 26 = corals and anemones, 27 = polychaetes, 28 = other worms, 29 = starfish, 30 = brittle stars, 31 = sea cucumbers, 32 = scallops, 33 = clams and mussels, 34 = snails, 35 = urchins, 36 = sand lance, 37 = Atlantic herring, 38 = alewife, 39 = Atlantic mackerel, 40 = butterfish, 41 = Loligo, 42 = Illex, 43 = pollock, 44 = silver hake, 45 = spotted hake, 46 = white hake, 47 = red hake, 48 = Atlantic cod, 49 = haddock, 50 = sea raven, 51 = longhorn sculpin, 52 = little skate, 53 = winter skate, 54 = thorny skate, 55 = ocean pout, 56 = cusk, 57 = wolfish, 58 = cunner, 59 = sea robins, 60 = redfish, 61 = yellowtail flounder, 62 = windowpane flounder, 63 = summer flounder, 64 = witch flounder, 65 = four-spot flounder, 66 = winter flounder, 67 = American plaice, 68 = American halibut, 69 = smooth dogfish, 70 = spiny dogfish, 71 = goosefish, 72 = weakfish, 73 = bluefish, 74 = baleen whales, 75 = toothed whales and porpoises, 76 = seals, 77 = migratory scombrids, 78 = migratory sharks, 79 = migratory billfish, 80 = birds, 81 = humans.

Source: Reproduced with permission from Link, J. (2002) Does food web theory work for marine ecosystems? *Mar Ecol Prog Ser*, 230: 1–9. doi:10.3354/meps230001.

season but often surprisingly terrestrial during the non-breeding season.

- Contain species that undergo metamorphosis and feed on different species during those different life stages; for example, the larval food plant of a caterpillar is often completely different to the nectar food plant of adult butterflies of the same species.

- Contain species that are migratory and thus only occur in that specific ecosystem for part of the year.

- Are marginal or transitional in nature (e.g. wetlands) where species often rely both on terrestrial and aquatic environments for food.

- Are undergoing a state of change or transition, either because of the introduction of new species or the local loss of a native species, which means that predator–prey relationships change.

One method that can be useful to research and understand food chains is stable isotope analysis where we use chemical signatures to disentangle feeding relationships and food webs. This is explored in more detail in Scientific Process 38.3.

SCIENTIFIC PROCESS 38.3 — Using stable isotope analysis to understand food webs

Research question

The Arctic fox (*Vulpes lagopus*) (Figure 1) is a widespread top predator in Arctic tundra ecosystems. This species could potentially act as an indicator for monitoring the vertebrate prey of these environments, but only if feeding interactions are well understood. Ehrich et al. (2015) aimed to improve understanding of Arctic fox food webs through chemical analysis of the fur of this top predator.

Hypotheses and predictions

Stable isotopes are different 'types' of the same chemical element. While the number of protons defines the element itself (e.g. nitrogen, carbon, etc.) and the sum of the protons and neutrons gives the atomic mass, the number of neutrons defines the isotope of that element. Most carbon is carbon-12 (^{12}C), but some carbon is carbon-13 (^{13}C) because it has an extra neutron; the ratio of these can be expressed as δ^{13}C. We can do the same thing for 'typical' nitrogen ^{14}N and 'rare' nitrogen ^{15}N to give δ^{15}N. All these isotopes are stable—they do not change over time—and their ratio in predator tissues reflect the resources assimilated over a certain period. If prey species consumed have different isotopic signatures, we can then examine the chemical 'fingerprint' of a predator to get insights into dietary composition. The Arctic fox is the only terrestrial mammalian predator endemic to the Arctic. The authors hypothesized that the diet of the Arctic fox, as inferred from stable isotope analysis, could be used to monitor changes in vertebrate prey communities in tundra food webs.

Methods

Fur is metabolically inert once it is grown (i.e. the stable isotope signature is 'locked in' to that specific section of hair, effectively like a time capsule). Samples of Arctic fox fur were collected at fox dens at six different sites (highlighted by the stars in Figure 2). This fur would have been growing in

autumn/winter of the previous year and so its isotopic signature would reflect that period. In order to understand how the isotopic signature of fox fur samples was reflective of diet, the team needed to understand the isotopic fingerprint of potential prey. They thus collected tissue samples from small rodents, medium-sized terrestrial mammals such as hare (*Lepus*), large herbivores such as reindeer (*Rangifer tarandus*) and muskox (*Ovibos moschatus*), marine species (including fish), and birds.

Results

There was considerable variability in the fox isotopic signatures for both nitrogen (δ^{15}N) and carbon (δ^{13}C) between the six study sites and over different years. Fox stable isotopic signatures were plotted on scatter plots for each location as polygons (i.e. shapes that encapsulated the profiles of all individual foxes sampled), while main prey items were added as individual data points. This is shown in Figure 2.

Conclusion

The variability in fox isotopic signatures in space and time reflects the potential for using stable isotope analysis to track changes in diet and thus changes in prey resource. A major limitation, however, is that changes in diet can only be detected if different prey species have different signatures. Here isotopic signatures of hares and collared lemmings overlapped at the Yamal site, as did the chemical signature of *Microtus* voles with many bird species. This means that while the fox isotopic signatures for Yamal look very consistent between years in Figure 2, it might simply be that diet and prey availability changes are undetectable using this method. At the other extreme, where fox isotopic signatures are wide and there are numerous potential prey species, it can be hard to disentangle what prey 'ingredients' have gone into creating fox isotopic signatures (e.g. at the Wrangel site; Figure 2).

Extension question

To what extent do you think stable isotope analysis is a useful tool in understanding food webs? What are the advantages and what are the limitations?

Read the original paper

Ehrich, D., Ims, R.A., Yoccoz, N.G., Lecomte, N., Killengreen, S.T., Fuglei, E., et al. (2015). What can stable isotope analysis of top predator tissues contribute to monitoring of tundra ecosystems? *Ecosystems* 18: 404–416.

Figure 1 Arctic fox (*Vulpes lagopus*). Arctic fox have a wide dietary niche, but often hunt lemmings when they are available.

Source: Nature Picture Library/Alamy Stock Photo.

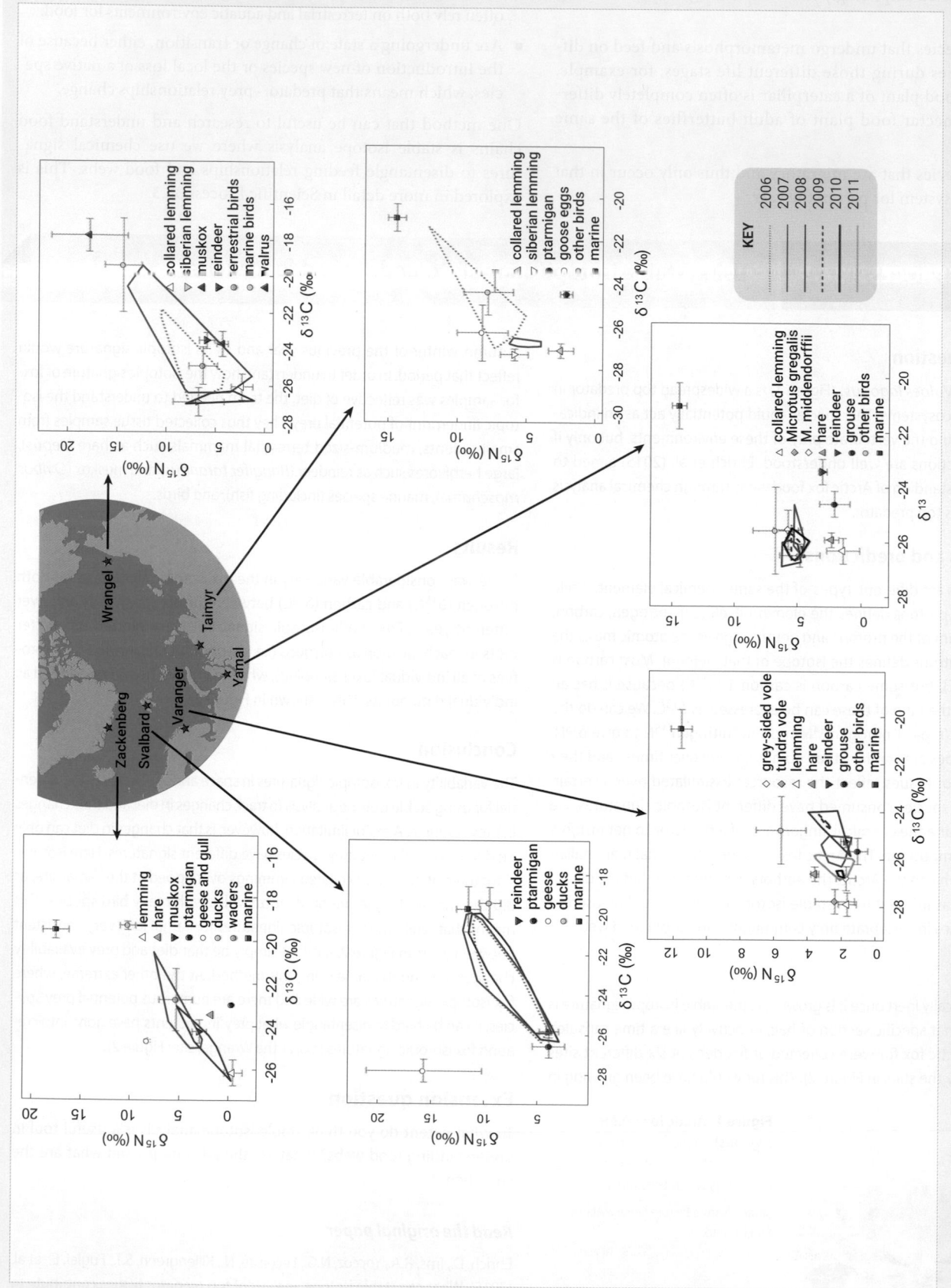

Figure 2 Six study sites in the Arctic tundra where isotopic signatures of arctic foxes and their main prey items were sampled. The graphs show carbon stable isotopic signature (relative abundance of ^{13}C in relation to ^{12}C on the x-axis and relative abundance of ^{15}N to ^{14}N on the y-axis) for foxes as polygon shapes in different colours for different years (each polygon encompassing the isotopic signatures of all sampled foxes that year) and key prey items shown as individual data points with x- and y-error bars to show variability.

Source: Reproduced with permission from Ehrich, D., et al. (2015) What can stable isotope analysis of top predator tissues contribute to monitoring of tundra ecosystems? *Ecosystems* 18:404–416. Copyright © 2015, Springer Nature. https://doi.org/10.1007/s10021-014-9834-9.

Ecological pyramids

Ecologists often use diagrams to understand ecosystems. In addition to food chains and food webs, another way of summarizing the complexity of ecosystems is using ecological pyramids. There are three main types:

- Pyramid of numbers (number of organisms at each trophic level).
- Pyramid of biomass (amount of biomass at each trophic level).
- Pyramid of energy (amount of energy at each trophic level).

A **pyramid of numbers** is conceptually the most straightforward—a simple count of the number of individuals within populations of all species within each trophic level of a particular ecosystem. Most pyramids of numbers are traditional pyramid-shaped (i.e. broad base tapering towards the top). However, because no account is taken of biomass, it is possible to have kite-shaped pyramids where the first trophic level is dominated by a few large autotrophs, or an inverted pyramid when many parasites depend on a single host.

A **pyramid of biomass** is harder to construct but more ecologically realistic because the relative mass of organisms at each trophic level is taken into account, rather than focusing solely on number. Most pyramids of biomass are traditional upright pyramids, but occasionally they can be inverted. This occurs, for example, in some aquatic ecosystems when producers have low biomass but a very rapid turnover (short lifespan, high reproduction) and thus the biomass of higher trophic levels is often greater *at a specific point in time* than the biomass of the producers.

A **pyramid of energy** covers the amount of energy within each trophic level rather than the number or biomass of the species involves. These different pyramid types are shown, with examples, in Figure 38.14.

Nutrient cycling

Nutrient cycling is the storage and transfer of a nutrient within an ecosystem and it is an essential part of ecosystem function. As these processes involve biological, geological, and chemical processes, they are often known as **biogeochemical cycles**. The main nutrients we focus on when describing biogeochemical cycles are carbon and nitrogen. We discussed these from a microbial perspective in Chapter 13, where we outlined the processes involved. We complement this here by considering the importance of the cycles from an ecosystem perspective.

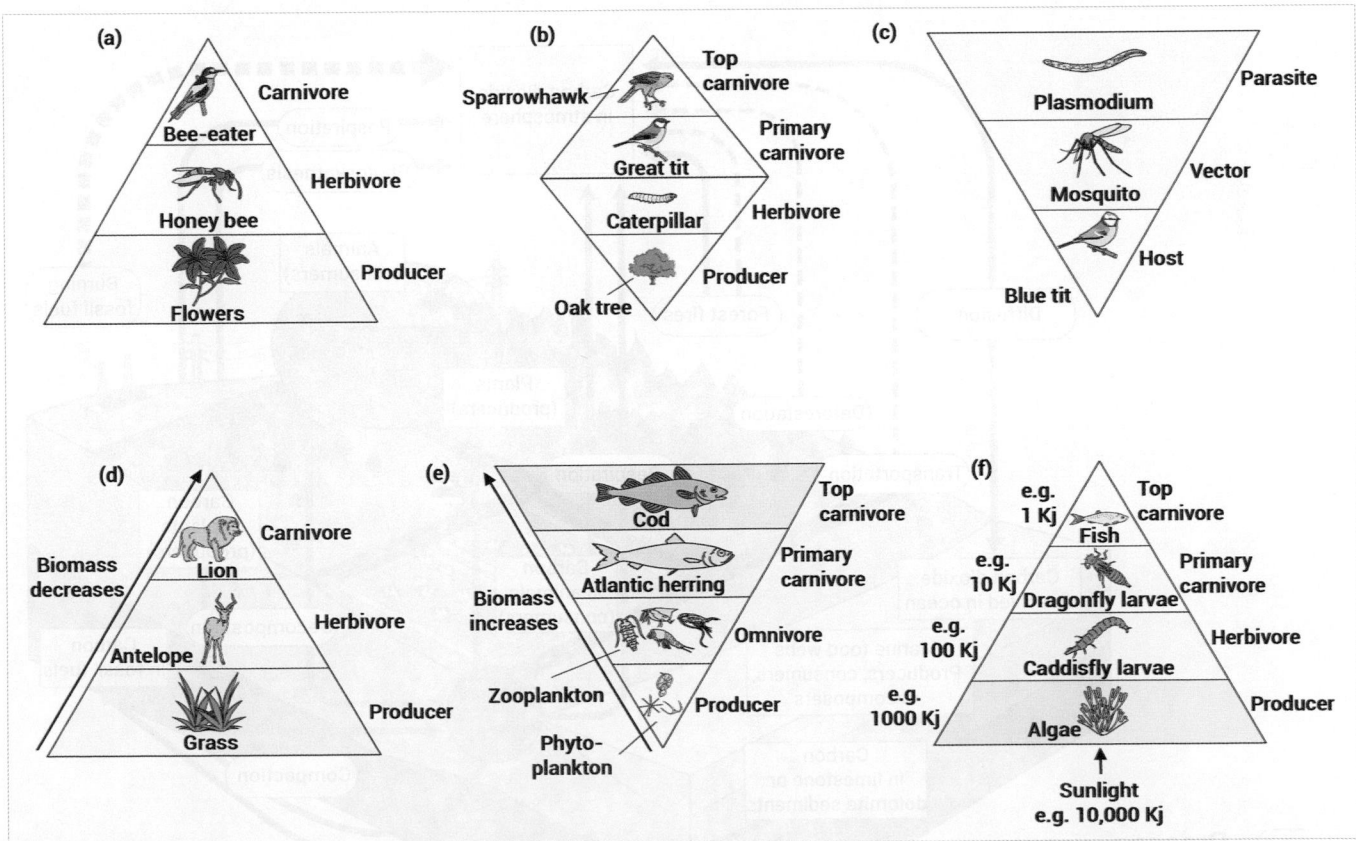

Figure 38.14 Types of ecological pyramids drawn to represent patterns in trophic levels. Pyramids of numbers (a–c) can be typical, upright pyramids, kite-shaped pyramids where a few large autotrophs dominate, or inverted pyramids in parasitic food chains. Pyramids of biomass (d, e) are usually upright pyramids as producers are generally the largest trophic level by biomass, but occasionally they can be inverted when producers have low biomass but rapid turnover. Pyramids of energy (f) are always upright because energy only enters the ecosystem at the first trophic level.

Carbon cycle

Carbon is the element of life that is contained in all living species. The carbon cycle summarizes a series of interactions between biotic and abiotic ecosystem components. Carbon within biota is the result of photosynthesis by autotrophs, which converts carbon dioxide from the air into glucose in the presence of sunlight and water. Once plant-based metabolic processes have occurred, residual glucose is converted into new plant biomass (this is net primary productivity, which we covered at the start of Section 38.4). Carbon is transferred as energy through feeding interactions summarized by food chains and food webs (recall Figures 38.11–38.13). The carbon stored within living organisms is, in most cases, temporary as it is released back into the atmosphere through the processes of respiration and decomposition. Other big carbon stores include soils, sedimentary rocks such as limestone, peat, and fossil fuels; atmospheric return pathways include erosion, weathering, soil cultivation, burning of organic matter, and, for peat bogs, desiccation (drying out). Other key carbon pathways include bidirectional carbon exchange with oceans, which are big carbon stores themselves, creation of rock through sedimentation, and volcanic eruptions. The carbon cycle is shown diagrammatically in Figure 38.15.

It can be helpful for us to think about carbon cycles as being in two parts: slow carbon exchange and fast carbon exchange.

Photosynthesis, respiration, decay, and ocean surface to atmosphere bidirectional exchange occur over reasonably short timescales, whereas sedimentation, weathering, and deep ocean exchange occurs over much longer timescales. This distinction is becoming blurred, however, because the carbon stored in medium- to long-term stores (woodland timber and fossil fuels, respectively) is being released over short timescales; it is this that is leading to human-accelerated climate change, the impacts of which we explore in Chapter 39.

Nitrogen cycle

Nitrogen gases (N_2) form the majority (about 78%) of the air in the atmosphere. This nitrogen is not directly **bioavailable**—in other words, it cannot be accessed and used directly by most species. For nitrogen to become accessible, it needs to be converted into ions that can be taken up by plants and, through them, be used by animals. These ions are nitrites (NO_2^-), nitrates (NO_3^-), and ammonium (NH_4^+) and the processes through which they are converted from gaseous compounds include nitrogen fixation by nitrogen-fixing bacteria, which often occur in close symbiotic association with plant roots (Section 38.5). Plants can then assimilate (take up) bioavailable nitrogen, which in turn is passed up the food chain. When plants and animals die, nitrogen is returned to the soil through ammonification (conversion into ammonia) by microbial action.

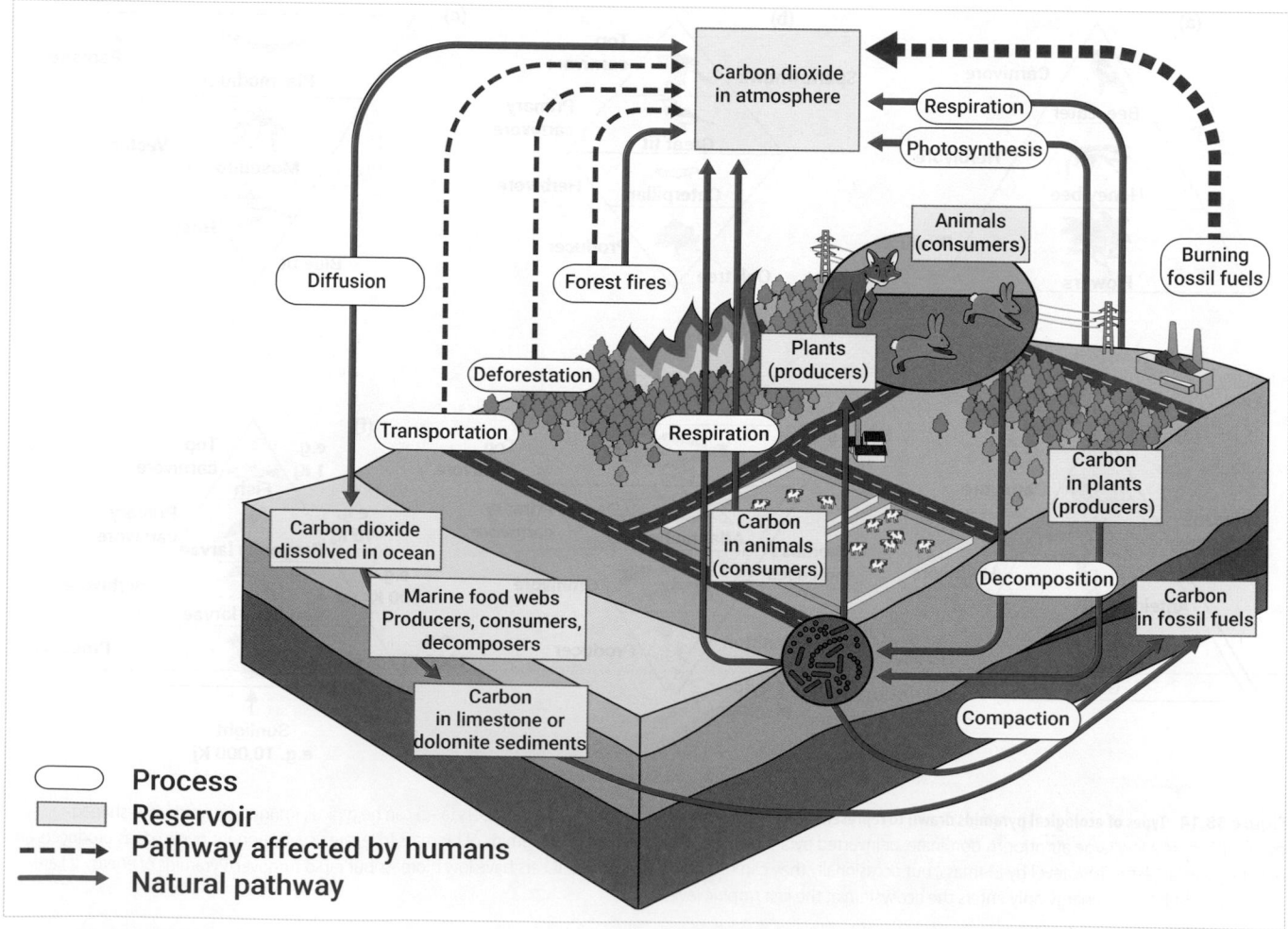

Figure 38.15 The carbon cycle.

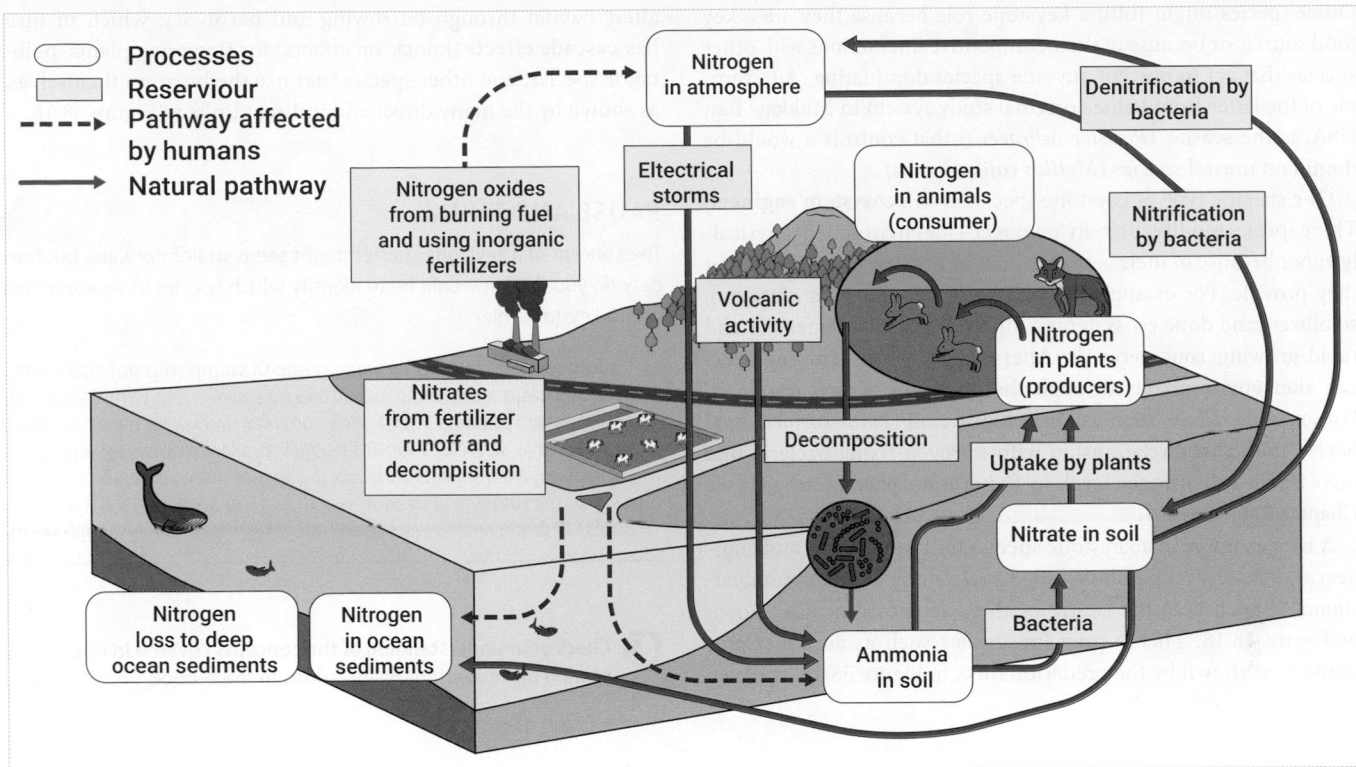

Figure 38.16 The nitrogen cycle.

After this, the nitrogen is returned to the atmosphere as N_2, through action by denitrifying (nitrate-reducing) bacteria such as *Paracoccus denitrificans*, which essentially 'undo' the process of nitrifying bacteria. The nitrogen cycle is shown diagrammatically in Figure 38.16.

Go to the e-book to complete an activity associated with Figure 38.11.

Check your understanding of the concepts covered in this section by answering the questions in the e-book.

38.5 Keystone species and ecological engineers

Throughout this chapter, the vital role of autotrophs such as algae and plants as the 'powerhouses' of ecosystems from an energy perspective has been highlighted. Other than that, though, there has been an implicit assumption that all other species in an ecosystem are equal. This is not the case. Some species have an especially important role within an ecosystem and we call these species **keystone species** (similar to a keystone in architecture as explained in Figure 38.17). By their very definition, therefore, variation in abundance of a keystone species has a greater impact than variation in abundance of non-keystone species in the same ecosystem.

Some species have a keystone role because of their position in a food chain/web (Section 38.4) and thus their trophic interactions. Top or apex predators that prey upon a predator in lower trophic levels (mesopredators) reduce the predation risk for the species the mesopredators would normally prey upon: this process is called **mesopredator release**. This can be essential to keep an ecosystem balanced ecologically and is one of the reasons why many

of the conservation schemes to reintroduce species focus on apex predators.

▶ We discuss a conservation scheme to reintroduce sea otters to restore their keystone role in Chapter 39.

Figure 38.17 The keystone species concept. The keystone species concept was first outlined by Paine in 1966, who used the term 'keystone' to link to the concept of an architectural keystone—the stone at the apex of an arch that is essential for its structural integrity. The parallel with ecosystem structure is that although an ecosystem can involve many species, some are ecologically more important to that ecosystem than others, just as some stones are more important for the arch than others.

Other species might fulfil a keystone role because they are a key food source, or because of their competitive interactions with other species that act to prevent any one species dominating. An example of the latter from Paine's original study system in Mukkaw Bay, USA, is the seastar (*Pisaster ochraceus*) that controls a would-be dominant mussel species (*Mytilus californianus*).

One specific type of keystone species is an **ecosystem engineer**. These species modify (literally 'engineer') the environment physically either because of their very existence or because of resources that they provide. For example, marram grass (*Ammophila arenaria*), stabilizes sand dune ecosystems simply because of its extensive and rapid-growing root network. Alternatively, ecosystem engineers can alter environments through the provision of new resources. For example, plants such as bird's foot trefoil (*Lotus corniculatus*) have a mutualistic relationship with nitrogen-fixing bacteria that increase the soil nitrogen levels by fixing atmospheric nitrogen (see Chapter 37).

A nice example of a keystone species for both trophic and engineering reasons is the black-tailed prairie dog (*Cynomys ludovicianus*), which is at the centre of the ecological linkages shown in Figure 38.18. This is prey for species such as coyote (*Canis latrans*), as shown by the predation links in Figure 38.18, and also alters habitat through burrowing and herbivory, which in turn has **cascade effects** (knock-on effects) for flowering plants, pollinator species, and other species that use the burrows themselves, as shown by the many direct and indirect links in Figure 38.18.

PAUSE AND THINK

The concept of a keystone species might seem straightforward, but how easy do you think it would be to identify which species in an ecosystem have a keystone role?

Answer: Identification of a keystone species as actually being keystone is often only possible after decline or loss from an ecosystem. Attempts to document general characteristics of keystone species have foundered due to the diversity of ways in which a species can act as a keystone. Moreover, even for known keystone species, the effect of presence (and thus absence) of that species on ecosystem dynamics is often context dependent and thus differs in different locations or at different times.

Check your understanding of the concepts covered in this section by answering the questions in the e-book.

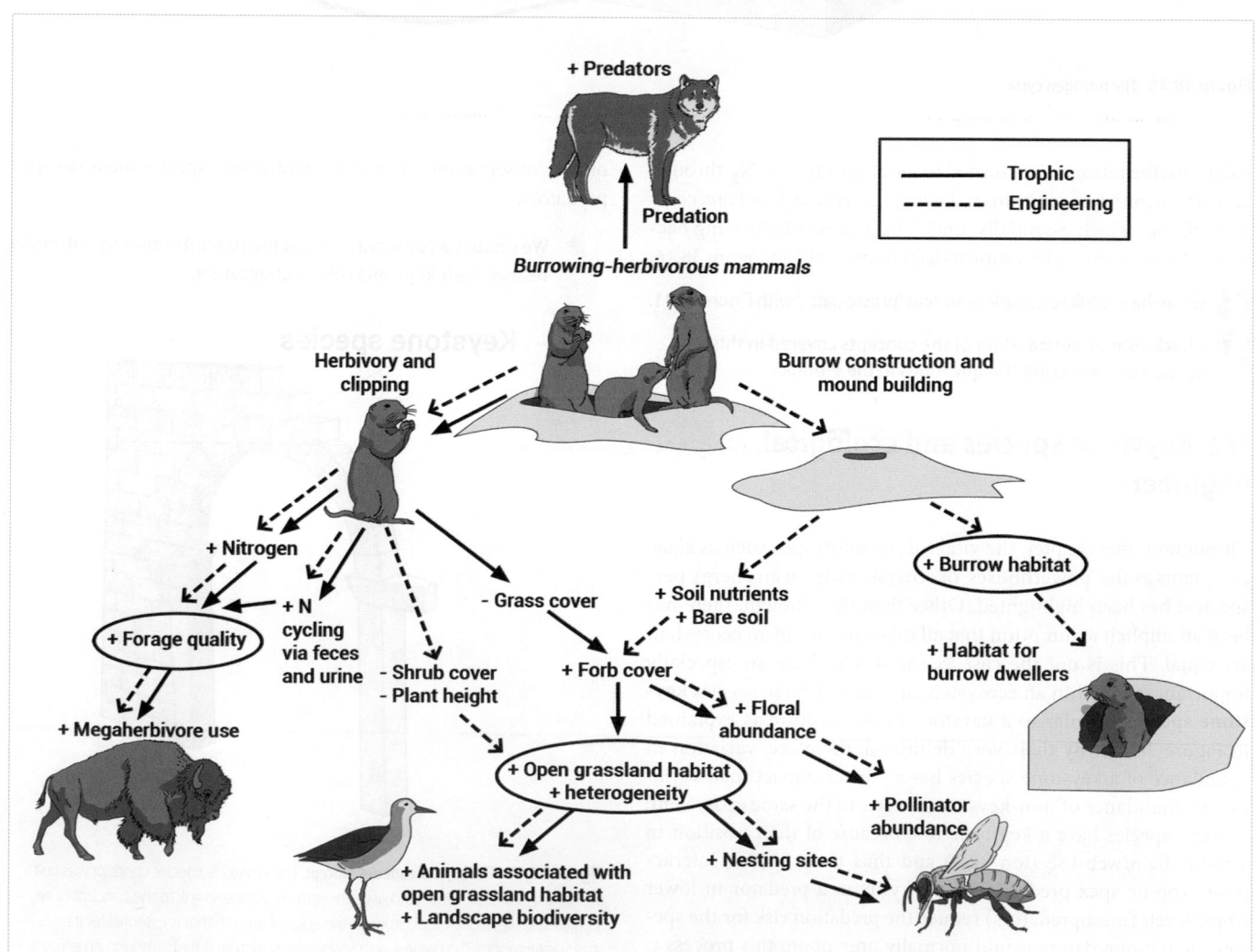

Figure 38.18 The black-tailed prairie dog (*Cynomys ludovicianus*) is a keystone species both through trophic and engineering interactions.

38.6 Ecological tipping points, ecosystem collapse, and mass extinctions

As we discuss in Chapter 34, both the evolution of new species and the extinction of current species are part of biology. All species will, at some point, become extinct: species extinction is a natural part of ecology. As such, well-functioning ecosystems exhibit species turnover and the loss of a species from an ecosystem is not something that, in isolation, is necessarily an ecological issue. However, we need to distinguish here between natural and human-accelerated extinction, as well as remembering that the loss of a *keystone species* could become an ecological tipping point and potentially ultimately lead to ecosystem collapse.

An **ecological tipping point** is the point at which an ecosystem experiences a long-lasting change that has tangible consequences—it can be thought of as the ecological point of no return towards ecosystem collapse. When this happens, it tends to have significant impacts on biodiversity and ecosystem regulation. Ecological tipping points can be triggered by extinction if that extinction involves keystone species (or multiple species that together have a significant role in ecosystem functioning) because of the magnitude of cascade effects (Section 38.5) on both biotic and abiotic environments. Alternatively, tipping points might be due to cascade effects from an environmental change, and we discuss examples of this in Chapter 39.

Tipping points tend to involve a change that is self-perpetuating through **positive feedback loops**. For example, deforestation might reduce regional rainfall, which increases fire risk, which causes forest dieback and further reductions in rainfall because of reduced evapotranspiration; or the loss of an ecosystem engineer

might cause habitat degradation and consequential loss of other species. There is often a specific environmental or ecological threshold beyond which the ecosystem abruptly shifts state (e.g. a specific level of deforestation or loss of a specific species), although predicting this is extremely challenging.

Different ecosystems have different levels of **ecological resilience** to ecological tipping points. Quite often, complex ecosystems have higher resilience and are thus less vulnerable. We might think this is counter-intuitive, but looking back at the food web of a marine ecosystem depicted in Figure 38.13 can help us understand it. Although some of the species in that are absolutely vital (the producer species one and two in particular), in many cases loss of one species would not have too much of an effect because of the number of species and links involved. Compare this with a simple ecosystem where there are just a few species (such as the one shown in Figure 38.11), where the loss of any one species would likely have a much greater impact.

While ecological tipping points and subsequent ecosystem collapse tend to occur at a local or regional scale, sometimes catastrophic events occur at a global (or at least trans-continental) scale. Such events can cause multiple ecosystems to collapse simultaneously over what is, in evolutionary terms, a very short timescale and result in the extinction of a very large number of species. These are referred to as **mass extinction events** (MEEs). Analysis of the fossil record suggests that we have had five previous mass extinction events in Earth's history (*x*-axis of Figure 38.19), when a high percentage of both flora and fauna species were lost (*y*-axis of Figure 38.19). The last MEE was the extinction of the dinosaurs at the Cretaceous–Paleogene (K–Pg) boundary (also called the K–T or Cretaceous–Tertiary boundary) about 66 million years ago, likely due to a meteorite strike. The biggest MEE was the end-Permian

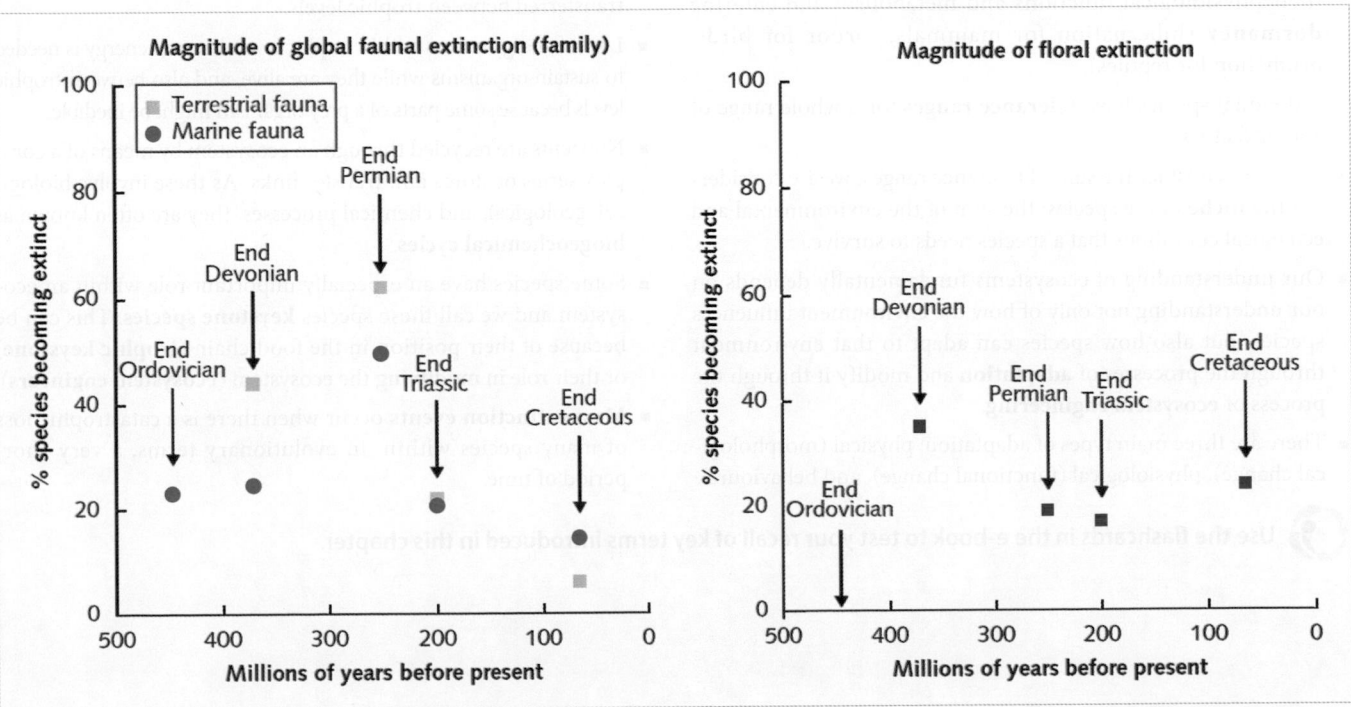

Figure 38.19 Past mass extinction events.

Source: Goodenough and Hart (2017), *Applied Ecology: Monitoring, managing, and conserving.* Oxford: Oxford University Press. Reproduced with permission of the Licensor through PLSClear.

event (251 million years ago), possibly due to volcanic eruptions, when up to 95% of species became extinct. The deficit in global biodiversity from the end-Permian event took an estimated 100 million years to be redressed through new species arising from speciation processes.

Despite gaps in the fossil record that make calculation of 'natural background' extinction rates complex, current rates are, on average, 1000× higher than the background norm. This has led many scientists to conclude that we are entering, or indeed have already entered, the sixth MEE. What makes the sixth MEE unique is that it is the first to be brought about by the dominance of a single species: mankind. Indeed, some palaeontologists go so far as to say that we are in a new geological age, the **Anthropocene,** because our effect has been so great that it will be recorded in the fossil record. This means that we are living in an age where there are considerable challenges to ecosystems, and considerable need to find and implement solutions to these challenges. We will consider some of these challenges and potential solutions in Chapters 39 and 40, respectively.

 Check your understanding of the concepts covered in this section by answering the questions in the e-book.

SUMMARY OF KEY CONCEPTS

- Ecosystems are made up of the organisms in a particular location and the **interactions** between one another and their environment.

- Ecosystems thus comprise living organisms in an area, the physical and chemical environment of that area, and the interactions, relationships, and processes that link these.

- Abiotic factors (e.g. temperature, water availability, sunlight) vary over space to create environmental gradients, which affect species distribution and ecology.

- Where an abiotic factor changes seasonally in a given area beyond the tolerance range of a given species, individuals of that species might avoid extreme conditions by moving away from that area temporarily (**migration**) or by slowing down their physiological functions and metabolism and entering **dormancy** (hibernation for mammals, torpor for birds, brumation for reptiles).

- Individual species have **tolerance ranges** for a whole range of abiotic factors.

- When we consider the sum of tolerance ranges, we are considering the **niche** of the species: the sum of the environmental and ecological conditions that a species needs to survive.

- Our understanding of ecosystems fundamentally depends on our understanding not only of how the environment influences species, but also how species can adapt to that environment through the processes of **adaptation** and modify it through the process of **ecosystem engineering**.

- There are three main types of adaptation: physical (morphological change), physiological (functional change), and behavioural.

- Autotrophs are the main powerhouse behind ecosystems since they convert solar energy to chemical energy through the process of photosynthesis in the presence of light: this is **primary productivity**.

- Heterotrophs depend on the energy fixed by autotrophs through feeding interactions such as herbivory and predation, decomposition, or parasitism: these are called **trophic interactions** because they involve an interaction between two organisms at different **trophic levels**.

- We can summarize trophic levels and interactions in an ecosystem through **food chain** and **food web** diagrams.

- Energy flows through food chains/webs from the primary producers (autotrophs) up to the top predator species, but this process is not energy efficient: only about **10% of energy** is transferred between trophic levels.

- Loss of energy occurs within trophic levels because energy is needed to sustain organisms while they are alive, and also between trophic levels because some parts of a prey organism might be inedible.

- Nutrients are recycled through an ecosystem by means of a complex series of stores and transfer links. As these involve biological, geological, and chemical processes, they are often known as **biogeochemical cycles**.

- Some species have an especially important role within an ecosystem and we call these species **keystone species**. This can be because of their position in the food chain (**trophic keystone**) or their role in modifying the ecosystem (**ecosystem engineers**).

- **Mass extinction events** occur when there is a catastrophic loss of many species within, in evolutionary terms, a very short period of time.

 Use the flashcards in the e-book to test your recall of key terms introduced in this chapter.

QUESTIONS

Looking for answers? Once you've answered these questions, follow the link in the e-book to the answer guidance and check your work.

Concepts and definitions

1. What is the difference between a fundamental niche and a realized niche?

2. What is the difference between primary and secondary productivity?

3. How do photosynthesis and chemosynthesis differ?

4. On average, what percentage of energy in one trophic level is transferred to the next trophic level?

5. How do pyramids of numbers, pyramids of biomass, and pyramids of energy differ?

Apply the concepts

6. Why are food webs usually much more representative of real-world ecosystems than food chains? What types of ecosystems would be most likely to have simple food chains rather than complex food webs?

7. Why can a pyramid of energy never be inverted (it is always broader at the base than the top)?

8. Human activity is altering the carbon cycle (Section 38.4). How is human activity altering the nitrogen cycle?

9. Why can the loss of a top or apex predator from an ecosystem potentially have profound effects on species in lower tropic levels and thus the overall ecosystem?

10. What factors can lead to mass extinction events? Are there any common causal factors?

Beyond the concepts

11. List the defining features of a tropical rainforest biome and a boreal forest biome then compare and contrast the similarities and differences of these.

12. Explain the different strategies that species use to avoid, or reduce the effects of, high temperatures.

13. Marine mammals living in cold water such as beluga (*Delphinapterus leucas*) and narwhal (*Monodon monoceros*) need specific adaptations. What are these and are they predominantly physical, physiological, or behavioural?

14. Find an example of a keystone species that has been lost from an ecosystem and the implications that this has had on other species within that ecosystem.

15. It is often difficult to get a reliable measure of extinction rates from the fossil record to quantify the 'natural background' extinction rate to understand mass extinction events. Why is this and does it matter?

FURTHER READING

Hobbs, R.J., et al. (2014) Managing the whole landscape: historical, hybrid, and novel ecosystems. *Front. Ecol. Environ.* **12**: 557–564.
Overview of non-natural and human-modified ecosystems.

Pocheville, A. (2015) The ecological niche: history and recent controversies. In: *Handbook of evolutionary thinking in the sciences* (pp. 547–586). Springer, Dordrecht.
Overview of niches.

Link, J. (2002) Does food web theory work for marine ecosystems? *Marine Ecol. Prog. Series* **230**: 1–9.
Review of food web concepts and application in marine environments.

Mills, L.S., Soule, M.E., & Doak, D.F. (1993) The keystone species concept in ecology and conservation. *Bioscience* **43**: 219–224.
Comprehensive overview of keystone species concepts.

Barnosky, A., et al. (2011) Has the Earth's sixth mass extinction already arrived? *Nature* **471**: 51.

Payne, J.L., & Clapham, M.E. (2012) End-Permian mass extinction in the oceans: an ancient analog for the twenty-first century? *Annu. Rev. Earth Plan. Sci.* **40**: 89–111.
Mass extinction events review and consideration of the next one.

Challenges

Key Threats to Ecosystems

Chapter contents

Introduction		1317
39.1	Identifying and classifying threats	1317
39.2	Human population growth	1319
39.3	Resource use and emissions	1320
39.4	Habitat loss, degradation, and fragmentation	1326
39.5	Introduction of non-native species	1331
39.6	Implications of ecosystem threats	1331

 Watch the key concepts video in the e-book to prepare yourself for studying this chapter.

LEARNING OBJECTIVES

By the end of this chapter you should be able to:

- Identify abiotic, biotic, and human processes that affect—and sometimes threaten—ecosystems.

- Assess and classify threats in terms of origin, frequency, severity, magnitude, timescales, and type of impact.

- Describe why growth in human population and demand are fundamental challenges that underpin many other threats to ecosystems.

- Explain the impacts of ecosystem threats on ecology, specifically:
  - unsustainable resource use, including the impacts of releasing pollutants to air, land, and water;
  - the causes of, and impacts resulting from, habitat loss, degradation, and fragmentation of habitats; and
  - the introduction of non-native species and their impacts on native species.

- Appreciate the processes that lead to species decline or extinction.

- Explain what factors can predict extinction risk and outline how these are used in species at risk decision-making strategies.

- Identify examples of ecosystem services and appreciate the impact of degradation of such services to overall ecosystem functioning.

Introduction

Environments are complex and dynamic entities that are constantly in a state of flux. Natural abiotic processes such as storms, volcanic eruptions, tsunamis, mudslides, and forest fires can radically change ecosystem function. Changes in biotic conditions or processes, for example caused by disease outbreaks or the arrival of non-native species, can also have profound effects. These reasonably rapid changes are superimposed upon long-term global processes: climate change, plate tectonic movement, and the evolution and extinction of species.

As we see at the end of Chapter 38, **mass extinction events** (MEEs) describe the process whereby a substantial proportion of the world's species become extinct over what, at a geological time-scale, is a very short period. The most well-known MEE is the loss of the dinosaurs between the Cretaceous and Tertiary geological ages, but many scientists consider that we are now entering a sixth MEE: the **Anthropocene**. This name reflects the fact that the impact of human activity is now so profound it will be seen within the fossil record—the first time that a single species has had such an effect.

Understanding the natural and anthropogenic threats to ecosystems, and their effects, is complicated because all ecosystem processes are interconnected. Moreover, the same threat can manifest very differently, and have very different impacts, in different ecosystems. However, it is this very complexity that makes practical ecology and ecosystem management such an exciting, engaging, and unique topic within biology—and one which is highly topical.

39.1 Identifying and classifying threats

In Chapter 38 we explore some of the numerous factors that can affect ecosystems. Changes in these factors can become challenges that ultimately threaten ecosystems and affect species (causing population decline and sometimes extinction), habitats (causing habitat loss or degradation), or that can compromise ecosystem functioning.

Threats can be loosely grouped into three categories based on the origin of the threat:

1. **Abiotic:** factors related to the natural physical environment.
2. **Biotic:** factors related to natural interactions between species.
3. **Anthropogenic:** human-caused.

In reality, things are not quite as simple this three-point classification system because threats can be the result of a combination of factors. For example, humans could introduce a non-native predator into an ecosystem (Section 39.5), or human-accelerated climate chang (Section 39.3) could increase the growth rate of a competitively superior plant to the detriment of other species. However, this classification system does give us some clarity when considering the many threats to ecosystems and their relative importance.

Exploring key concepts through two natural disaster examples

Abiotic events suchaareplace with closed-up em dashstorms, volcanic eruptions, tsunamis, mudslides, and forest fires tend to occur quickly and have obvious, immediate visual effects on the environment and often receive prominence in the media. However, biotic threats can be equally important and the potential for non-native species to invade a new area provides a useful example of this. Once in their new environment, these species can affect native species through processes such as competition, predation, herbivory, and parasitism. These impacts can result in population decline of specific species and even loss of species from specific sites.

▶ **We discuss competition, predation, herbivory, and parasitism in Chapter 37.**

Let's consider two very different natural disasters that, between them, encompass both abiotic and biotic threats to ecosystems:

- **Example 1:** On 18 May 1980, Mount St Helens, a volcano in the North West United Sates, erupted. This event had a dramatic impact on the landscape, habitats, and species in the surrounding area. Land was quickly engulfed by lava and pyroclastic rock flows (which are still evident in the landscape today in the solidified lava that once flowed from the volcano, as shown in Figure 39.1), while falling ash covered land up to 300 km from the volcano itself. A vast area of habitat, including tens of thousands of acres of mature forest, was lost. It was estimated by Washington's Department of Game that nearly 7000 large animals (primarily deer, elk, and bear) perished in the area most affected by the eruption, as well as many smaller animals.

- **Example 2:** The Japanese Tōhoku tsunami is another natural event that had substantial ecosystem consequences. On 11 March 2011, the fourth largest earthquake ever recorded (9.1 on the Richter scale) under the ocean floor caused a 4 m high wave that engulfed over 56,000 hectares of land. From an ecological perspective, salt-water ingress affected terrestrial and freshwater habitats across the inundation zone. However, there were also ecological effects thousands of kilometres away because pier pilings, buoys, and even small boats, travelled over 8000 km across the Pacific and were washed ashore on the western seaboard of America, especially in Oregon. This might sound more like an

Figure 39.1 The ecosystem effects of the Mount St Helens volcanic explosion in 1980.

Source: Photo by Ray Schauweker/CC BY 3.0/Wikimedia Commons.

environmental threat than an ecological one, but numerous marine species, especially the Japanese oyster (*Crassostrea gigas*), were transported on the debris and were thus introduced into a new ecosystem outside their native range (Figure 39.2).

Using the Mount St Helens volcanic eruption and Japanese tsunami case studies, and drawing in a few other examples, let's explore key concepts and terminology relating to ecosystem threats.

Frequency of threat

The 1980 volcanic eruption and 2011 tsunami were both specific single events, but if we examine a longer time span we find that these events are episodic: Mount St Helens has had at least four

Figure 39.2 Species can be transported vast distances on debris. Species can 'raft' or 'hitch-hike' across oceans on debris from storms and tsunamis, such as species moving from Japan to the USA on debris resulting from the Japanese Tōhoku tsunami in 2011.

Source: Copyright Hatfield Marine Science Center Oregon State University.

major eruptions in the last 500 years, while the first recorded tsunami in Japan was on 29 November 684 AD. **Episodic events** are those that are repeated but not necessarily at regular intervals; other threats might be **periodic events** (repeated at regular intervals), or, by contrast, can be **isolated events** or **continuous events**.

Impact severity versus impact magnitude

Despite the obvious ecosystem impact on the area immediately surrounding Mount St Helens, global impacts were negligible. As far as can be ascertained, no species became extinct and no habitats were lost at a biosphere level. This leads us to a couple of important concepts:

1. **Impact severity:** how substantial is the effect?

2. **Impact magnitude:** how large is the spatial scale over which the effect occurs?

In many cases, a threat results in severe impacts but these have a relatively low magnitude—at least when considered at a global scale. The Japanese tsunami is a counter-example here because although local impacts were severe, the magnitude was also higher than might have been predicted at the time because of marine debris—and the species thus transported—washing up on coastlines halfway around the world 5 years after the event.

It should be noted that abiotic and biotic threats can also become **cumulative** over space. This could occur if, for example, many individual forest fires occur: the magnitude of each one might be modest, but if the area affected by all fires is added together, the magnitude increases substantially. Cumulative effects can also occur over time; for example, one unusually cold and wet breeding season that affects the breeding success of seabirds might have a low impact severity but a succession of such breeding seasons in consecutive years would cumulatively have a more severe impact.

Timescales, lag time, and duration of impact

Some impacts occur immediately after an event and are temporary, while others are persistent or even permanent. Natural disasters such as volcanic eruptions and tsunamis tend, by their very nature, to be sudden events, but can have long-term effects. Even 40 years on, the effects of the Mount St Helens eruption could be seen in the landscapes, habitats, and species communities of the surrounding area, but regeneration is occurring. Impacts are thus long-lasting but are unlikely to be permanent. The Japanese tsunami occurred more recently, but it is likely that the effect of the salination of coastal areas will still be evident at least a hundred years into the future. The translocation of non-native species to America after the tsunami is an interesting example of an impact that was not immediate (there was a **lag time** of several years before debris was washed up and a second lag while non-native species were becoming established) but where effects are likely to be permanent.

In contrast, ecosystems can be threatened by abiotic and biotic factors that change gradually over time. This is neatly exemplified by the difference between weather and climate. Weather events such as intense rainfall, drought, or hurricanes happen relatively quickly over a period of hours, days, or weeks; changes in climate happen very gradually over a period of many years. Both can challenge ecosystems, and both can have long-term effects, but the difference from a biological perspective is that when a change is gradual, there is more chance for organisms to respond, either through **genotypic adaptation** as a result of evolution by natural selection, or by **phenotypic plasticity** (see Chapter 34).

Direct versus indirect impacts

In some cases, an event will have a **direct ecosystem effect**. Such effects, for example vegetation being covered by a lava flow, can be difficult to predict and manage but they are, conceptually at least, easy to understand. Spatially, they are also easy to study because they occur in the same geographical place as the original event. However, there can also be instances where effects are spatially removed from the location of the original natural disaster. These **indirect ecosystem effects** can be almost impossible to predict. For example, the non-native oyster (*Crassostrea gigas*) introduced to the USA as a result of the Japanese tsunami is having substantial impacts on native species through competition interactions, which is presenting an urgent and complex management conundrum.

Cascade effects

If indirect effects are hard to predict, cascade effects are even harder. Cascade effects can be thought of as a row of dominoes where each tile has been placed carefully on its edge adjacent to one another. Under natural conditions, the row of dominoes is stable. However, if something affects any one domino and it falls over, that phenomenon is repeated and repeated as domino after domino falls over. The more dominoes there are in the system and the closer together they are, the more the pattern spreads.

Ash trees (especially European ash *Fraxinus excelsior*) are prone to a disease called ash dieback, which is caused by the *Hymenoscyphus fraxineus* fungus. First reported in Poland in the 1990s, the disease has spread widely throughout Europe. In ash-dominated woodland, the first effects of ash dieback are felt—not surprisingly—by ash trees, initially losing leaves and suffering from bark lesions like those shown in Figure 39.3, and then with whole trees dying.

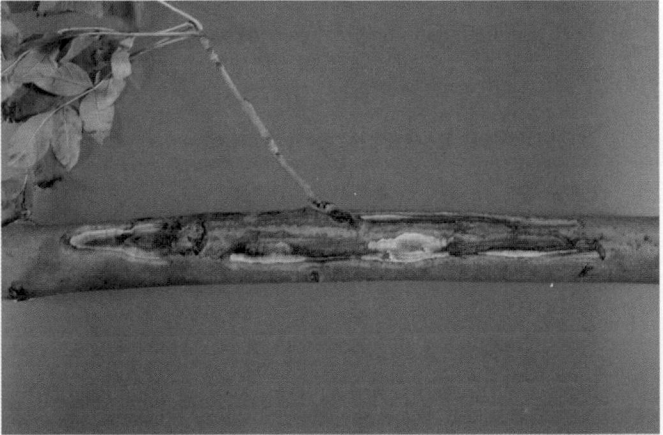

Figure 39.3 Bark lesions are a symptom of ash dieback.

Source: Courtesy of The Food and Environment Research Agency (FERA), Crown Copyright, used under Open Government Licence.

However, there are many species that are in a **mutualistic** or **commensal** relationship with ash, including other vascular plants, lichens, fungi, invertebrates, birds, and mammals.

 We discuss mutualistic and commensal relationships in Chapter 37.

These species might suffer cascade ('knock-on') effects. For example, if a fungus has adapted only to grow on ash trees and ash trees decline or disappear, the fungus will also decline or disappear. Moreover, a reduction in ash trees will allow other species of tree to move into the ecosystem, which might previously have been excluded by **competition**, and they might facilitate the arrival of other species with which they have a symbiotic link. Over time, therefore, a relatively simple biotic change—a natural outbreak of a disease that directly affects one species—could change an entire ecosystem.

Anthropogenic threats

Superimposed on all the natural threats to ecosystems are threats caused by, or increased by, humans and human activities. These effects take a variety of forms and can operate at a range of spatial scales from global to national, regional, and local. It is these factors that we will consider in detail throughout the rest of this chapter. As with natural threats, anthropogenic threats can:

- Differ in frequency: isolated, episodic, periodic, or continuous.
- Have a severe impact or a minor one (i.e. vary in severity).
- Occur over a wide area or a small one (i.e. vary in magnitude).
- Occur suddenly, gradually, or after a lag time.
- Have short-term or long-term effects.
- Be direct, indirect, or cumulative.
- Have cascade effects.

Check your understanding of the concepts covered in this section by answering the questions in the e-book.

39.2 Human population growth

We discuss population growth in Chapter 36 from the standpoint of animals and plants, but we also need to consider human population growth. There are now more people alive on the planet than ever before and the projection is that the population will continue to rise steeply into the next century, despite annual growth levels decreasing (note the cumulative population still increasing in Figure 39.4, while the annual growth rate is decreasing). Not only is human **population size** increasing, so too is **population distribution** and, in many places, **population density**. In other words, human population growth means that people are living in more areas, and at much greater concentrations, than would have historically been the case. The human population is currently almost 8 billion, up from around 2 billion 100 years ago. If we take a longer-term view, today's population is almost 2000 times larger than it was 12,000 years ago, at the end of the last ice age, when the world's population was around 4 million (<50% of the current population of London).

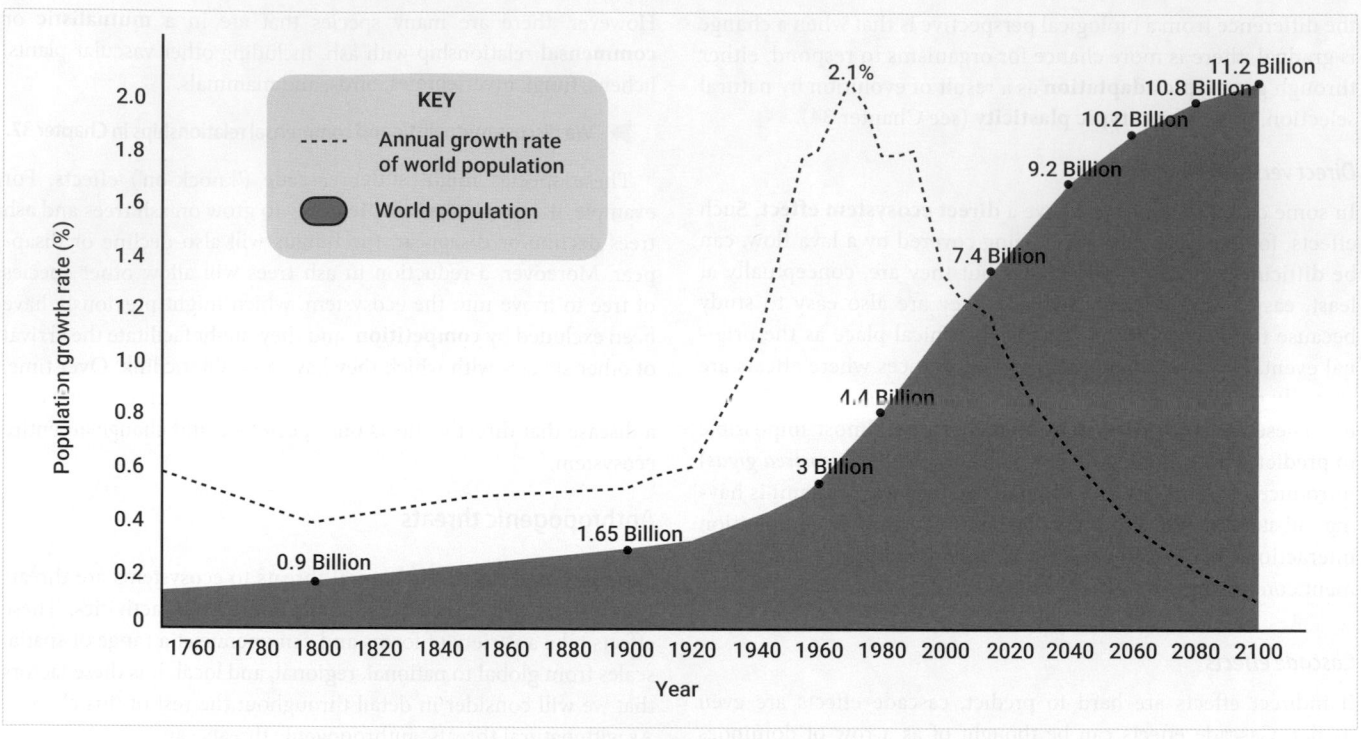

Figure 39.4 Human population size and growth rate since the mid-1700s. Population size can increase even after the growth rate decreases because growth rate (at a global scale) is effectively number of births per year minus number of deaths per year. Suppose the birth rate drops from three children to two children per couple of child-bearing age but that the death rate remains consistent. This would decrease growth rate per capita but, because we have already had a population boom, and there are more couples of child-bearing age around, the overall number of children born around the world—and thus the overall population size—still increases.

Source: Max Roser/OurWorldInData/CC BY 4.0.

How human population leads to ecosystem threats

Although human population growth is important it is not (usually) the simple presence of people that threatens ecosystems, but rather what those people do in order to meet their needs, wants, and aspirations. Because people need shelter, urbanization is increasing and that can result in loss of natural habitat. Because people need feeding, agricultural production is increasing spatially and marginal lands that are often unsuitable for food production are increasingly being farmed, which can result in **desertification** or **salinization**. The intensity of agriculture is also increasing rapidly, and in many cases fertilizers are overused to increase the yield causing **eutrophication** (nutrient enrichment of ecosystems); pesticide use can also reduce invertebrate species richness and abundance. Because people also need water, water abstraction from rivers, lakes, and groundwater is increasing and management of water, for example by building dams or diverting rivers, is becoming more common. Because people also need goods and services, industrialization—and associated pollution—is rapidly increasing as is infrastructure to allow people to move, and materials and goods to be transported, around the globe.

We can divide anthropogenic threats into three main groups, which we will consider in subsequent sections:

1. Resource use and emissions:
   a. Non-renewable, semi-renewable, and renewable resources.
   b. Impacts of extraction, transport, and use (including emissions of greenhouse gases and pollutants).
2. Habitat loss, fragmentation, and degradation.
3. Introduction of non-native species.

 Check your understanding of the concepts covered in this section by answering the questions in the e-book.

39.3 Resource use and emissions

Humans use a wide range of resources. This can cause ecosystem threats through extraction, transportation, and use, especially through emission of by-products connected with use of non-renewable resources.

Resource types

Natural resources tend to be put into one of three groups. At one end, there are non-renewable resources that will not regrow or regenerate after they have been used (or at least not over human timescales). Prime examples of these **non-renewable**

Figure 39.5 Examples of non-renewable, semi-renewable, and renewable resources.

Source: (a) Imeleca (Leonid Meleca)/Shutterstock.com; (b) andriano.cz/Shutterstock.com; (c) majeczka/Shutterstock.com; (d) oleandra/Shutterstock.com; (e) Gunnar Pippel/Shutterstock.com; (f) Anton Starikov/Shutterstock.com; (g) SantiPhotoSS/Shutterstock.com; (h) Kletr/Shutterstock.com; (i) zhao jiankang/Shutterstock.com; (j) Kletr/Shutterstock.com; (k) ChiccoDodiFC/Shutterstock.com.

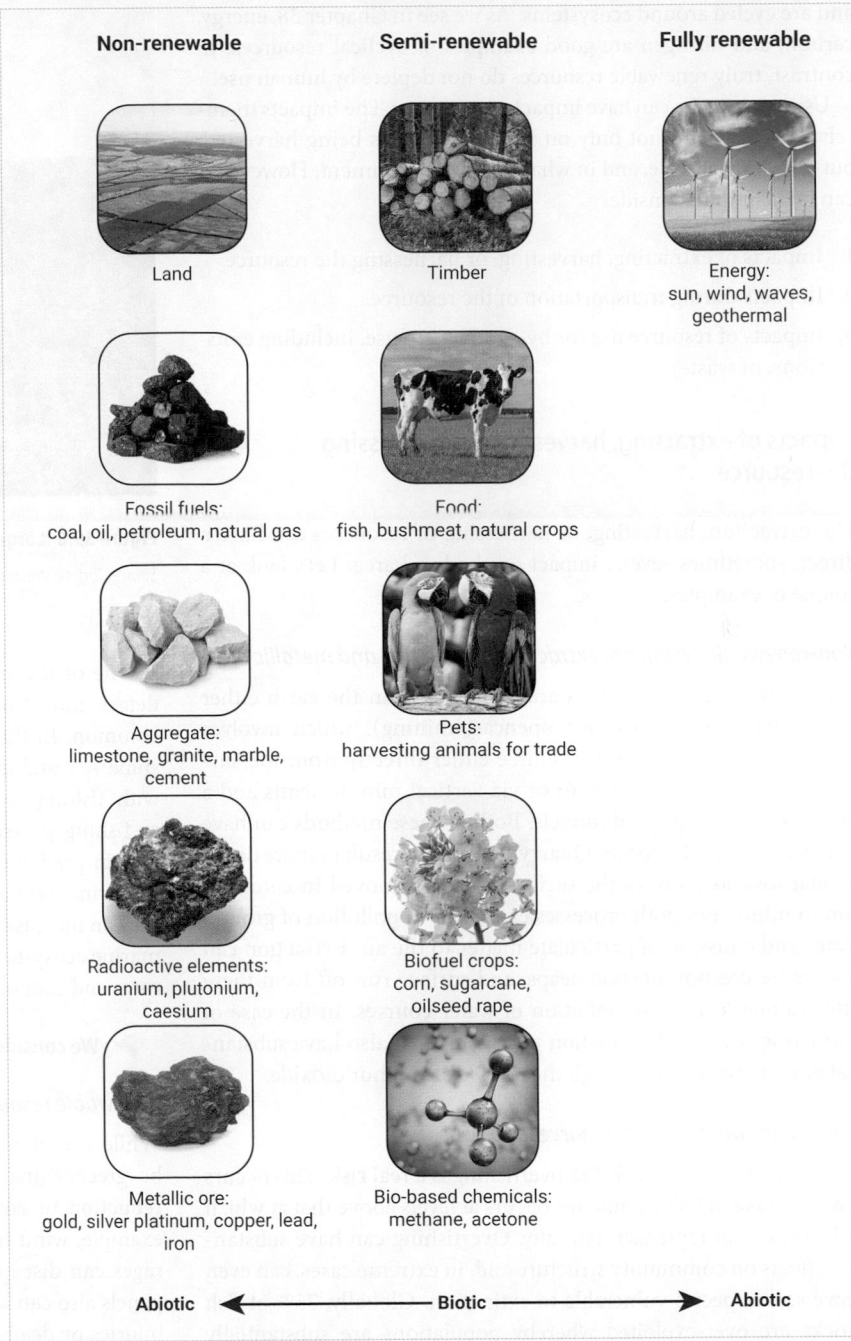

resources are fossil fuels such as coal and oil: when they are gone, they are gone (see left of Figure 39.5 for other examples; note non-renewable resources are usually abiotic). Technically, *any* use of non-renewable resources is unsustainable because these resources are finite, but use can be carefully managed. At the other end of the spectrum are truly **renewable resources** that are almost inexhaustible in supply, such as wind, waves, and sunlight (see right of Figure 39.5; note renewable resources are usually abiotic). There are also **semi-renewable resources**, which are resources that can be sustainably harvested as long as they are actively managed. Semi-renewable resources tend to be biotic: two of the best

examples are timber and fish (see the middle of Figure 39.5 for other examples). When semi-renewable resources are harvested appropriately it is possible for use to be sustainable, but it also possible to over-exploit these resources and have a detrimental impact on species, habitats, and even entire ecosystems.

Most non-renewable or semi-renewable resources deplete with human use. Exceptions are land, which is not destroyed but simply modified (and, in some cases, degraded), and radioactive elements that decay naturally into non-radioactive elements such as heavy metals. There are also resources that are non-renewable in the sense that they cannot be created or destroyed, but that can change form and are cycled around ecosystems. As we see in Chapter 38, energy, carbon, and nitrogen are good examples of cyclical resources. In contrast, truly renewable resources do not deplete by human use.

Using resources can have impacts on ecology. The impacts themselves will depend not only on what resource is being harvested, but also at what scale, and in what sort of environment. However, it can be helpful to consider:

1. Impacts of extracting, harvesting, or harnessing the resource.
2. Impacts during transportation of the resource.
3. Impacts of resource use (or by-products of use, including emissions, or waste).

Impacts of extracting, harvesting, or harnessing the resource

The extraction, harvesting, or harnessing of resources can have a direct, sometimes severe, impact on the local area. Let's look at a couple of examples.

Non-renewable resources: extracting aggregates and metallic ores

Aggregates and metallic ores are extracted from the earth either using quarrying (also called opencast mining), which involves accessing the subterranean resource either directly from the surface (as shown in Figure 39.6) or via vertical mining shafts and a network of underground tunnels. Both of these methods can have notable ecological impacts. Quarrying tends to result in more direct habitat loss, as more of the surface layer is removed in comparison to minin, but both processes can result in pollution of ground water and emission of particulate matter to the air. Extraction can also cause creation of spoil heaps, and surface run-off from these after rainfall can cause pollution of water courses. In the case of extracting ores, post-extraction processing can also have substantial effects, especially through the release of sulphur dioxide.

Semi-renewable resources: harvesting fish

Fishing can be sustainable but overfishing is a real risk. This occurs when offtake of fish by humans occurs at levels above that at which fish stocks can replenish naturally. Overfishing can have substantial effects on community structure and, in extreme cases, can even leave some species vulnerable to extinction. Globally, 75% of fish stocks are over-exploited whereby populations are substantially below replacement level (see Chapter 36).

Figure 39.6 Large-scale quarrying can result in substantial loss of habitat.
Source: Jpgola/Wikimedia Commons/CC BY-SA 3.0.

One of the main problems is that overfishing can be hard to detect until it becomes critical. Complex cascade effects are also common. In the Persian Gulf, for example, populations of short-spine sea urchin (*Echinometra mathaei*) are positively correlated with fishing levels, such that the number of urchins increases as fishing pressure increases. This is probably because a reduction in predator fish such as wrasse (Labridae) and emperor fish (Lethrinidae) has resulted in predator release for the urchins, and thus an increase in population. Fishing pressure not only threatens marine ecosystems from an ecological perspective, but also the cultures and economies that depend on them.

▶ We consider possible solutions to overfishing in Chapter 40.

Renewable resources: harnessing 'green' energy

While use of renewable energy resources is often considered to be 'greener' than using non-renewable fossil fuels because of the reduction in emissions, there can be ecological side effects. For example, wind turbines can cause mortality of bats and tidal barrages can disrupt fish migration. As shown in Figure 39.7, solar panels also can look surprisingly similar to water bodies and cause injuries or death to waterbirds that crash land on them after mistaking them for areas of open water.

Figure 39.7 Solar panels in solar farms can be mistaken for lakes by birds.
Source: Photo by Steve Bittinger. Distributed under the terms of the Creative Commons Attribution 2.0 Generic (CC BY 2.0). https://creativecommons.org/licenses/by/2.0/.

PAUSE AND THINK

Biofuel crops such as maize and oilseed rape can reduce use of fossil fuels for energy. Although burning biofuels still releases greenhouse gases, emissions are usually substantially lower. However, there can be ecosystem impacts arising from actually growing biofuels to harness biofuel potential. What might these be?

Answer: Uses up land—can result in habitat loss, change, or degradation; can result in creation of a monoculture (lots of the same crop grown in one area rather than a patchwork of different crops); can require substantial amounts of water abstraction with potential impacts for water courses.

Impacts during transportation of the resource

Once resources have been obtained, they usually need transporting to the consumer (often via industrial processing plants and retail sites). Extraction will usually involve mechanized transportation, at least for part of the journey, which itself utilizes fossil fuels. Transporting energy harnessed from wind or sun will usually be done via power lines; this has less impact, but installing the pylons can result in some habitat loss. Usually, transportation is fairly predictable in terms of impact on ecosystems, but sometimes there can be unexpected events that can cause catastrophic damage to ecosystems. These concepts are exemplified for a pollution incident that occurred during oil transportation in a vulnerable environment, Alaska, in Real World View 39.1.

REAL WORLD VIEW 39.1   Oil pollution in Alaska

When the Alaska Purchase Treaty was signed by the United States Senate and the Russian Empire on 30 March 1867, most Americans thought Alaska was a frozen wasteland. Actually, Alaska is a natural resources hotspot and the most economically valuable resource is the Prudhoe Bay oil fields off the north coast: the largest oil resource in North America.

Tapping the oil in Prudhoe Bay, however, was far from easy because of the remote location. The only feasible way to transport any oil extracted was via an overland pipeline so the 800-mile-long trans-Alaska pipeline (shown in Figure 1), was built to move the oil from Prudhoe Bay in the north to Valdez in the south. The pipeline's construction was beset by ecological concerns. Among these were the possible disruption to vulnerable vegetation and potential disruption to caribou (*Rangifer tarandus*) migration, which led to crossing points being installed at key locations. The biggest concern, however, was oil spills. This concern has been largely unfounded along the main pipeline. The potential for things to go wrong was demonstrated in dramatic fashion on 24 March 1989, when the *Exxon*

Valdez oil tanker ran aground in Prince William Sound, where the pipeline terminates and oil is transferred to ships for its onward journey.

Nearly 38,000 m^3 of oil was released into the previously pristine waters, eventually covering over 2000 km of coastline and a vast area of ocean (Figure 2). The ecological and environmental impacts were colossal. Immediate effects included the deaths of 250,000 seabirds, at least 2800 sea otters (*Enhydra lutris*), 300 harbour seals (*Phoca vitulina*), and 250 bald eagles (*Haliaeetus leucocephalus*), as well as countless fish, shellfish, and other species. Not only were the effects dramatic, they were also persistent: a 2003 study found oil in the area almost 15 years after the original spill. The oil's persistence was partly due to the large quantity of oil lost, partly to the lack of an immediate response to contain the area affected and treat the initial oil slick, and partly due to the very cold waters in Alaska that slow the degradation process by limiting microbial action.

By 2019—the 30th anniversary of the disaster—the area had largely recovered. The area once again supports mammals such as Steller sealion

Figure 1 The trans-Alaska pipeline.

Source: Luca Galuzzi - www.galuzzi.it/Wikimedia Commons/Attribution-ShareAlike 2.5 Generic (CC BY-SA 2.5).

Figure 2 The *Exxon Valdez* oil spill covered a huge area of sea and coastline.

Source: RGB Ventures/SuperStock/Alamy Stock Photo.

Figure 3 Steller sealion (*Eumetopias jubatus*) in Prince William Sound in 2018.

(*Eumetopias jubatus*) (Figure 3), which is indicative of a healthy ecosystem. This recovery, however, should not distract us from remembering that oil spills can be one of the most dramatic and detrimental anthropogenic environmental impacts in severity and magnitude, especially where vulnerable locations are concerned.

Extension question

Research how preventing and treating marine oil spills has advanced since the *Exxon Valdez* disaster. Does this adequately reduce the threat of something like this happening again or should more be done?

Reference: general overview of the Exxon Valdez oil spill

Klasner, F., Sholly, K., & Pfeiffenberge, J. (2009) 20 years later . . . Exxon Valdez oil spill. US National Park Service. https://www.nps.gov/kefj/learn/nature/upload/kefj_evos_1989-2009_qa.pdf

Impacts of resource use

Use of resources often causes **pollution**. This can take many forms depending on the resource and how it is used but can include heavy metals such as mercury being released into soils, volatile organic compounds (VOCs) being released into the air, and microplastics being released into the aquatic environment. In some cases, pollution is direct. For example, pesticides can be a highly effective way to manage pests, as we discuss in Chapter 40, but their application can have negative effects on non-target species. In other cases, pollution effects arise indirectly through disruption to natural biogeochemical cycles. The two most obvious impacts are emission of greenhouse gases as a by-product of fossil fuels combustion causing climate change (carbon cycle) and addition of nutrients into an ecosystem causing the process of eutrophication (nitrogen cycle).

Greenhouse gases and climate change: processes and effects

At a global scale, probably the most serious emissions involve **greenhouse gases,** especially carbon dioxide. Carbon dioxide is released as a by-product of burning non-renewable fossil fuels, which releases carbon previously sequestered (stored) in coal, petroleum, and natural gas, as well as that stored in semi-renewable resources such as peat and timber. This impacts the carbon cycle (which we discuss in Chapter 38), because although carbon is neither being created nor destroyed, it is changing form and its location within the biosphere. An increasing level of carbon dioxide (as well as other gases such as methane) in the atmosphere acts to trap heat from sunlight in a layer around Earth, thus raising its temperature and contributing to **human-accelerated climate change**, which is a change in climate cause by humans rather than cyclical natural processes. We sometimes use the term 'greenhouse effect' for human-accelerated climate change because carbon dioxide forms a layer around Earth, as shown in Figure 39.8, which traps heat below it in much the same way that the glass in a greenhouse traps heat within it.

Human-accelerated climate change is having a profound effect on ecology, especially through its influences on how species

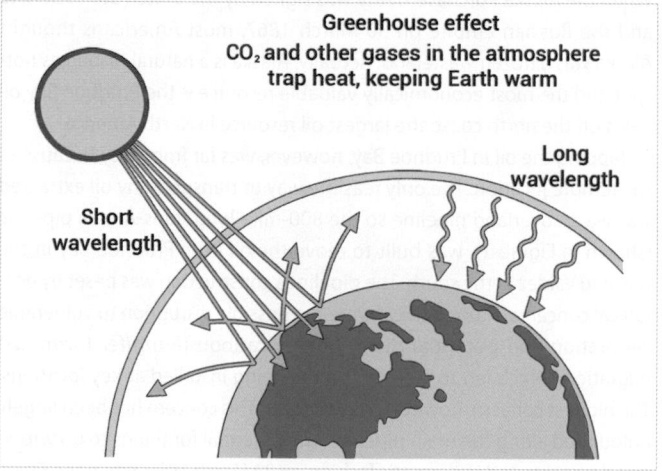

Figure 39.8 The greenhouse effect.

interact with one another, which is central to species communities and underpins well-functioning ecosystems. Climate change can affect species' interaction by affecting geographical distribution and/or by altering the timing of seasonal activities.

▶ We learn more about species communities in Chapter 37 and ecosystem function in Chapter 38.

1. **Climate-induced changes in species' distributions:** As we see in Chapter 37, species have tolerance ranges for abiotic parameters such as temperature and precipitation. This means that the range of a given species is partly determined by climatic factors both directly and through climate influences on other species upon which they depend. Consequently, change in climate often results in change in species' range (see Chapter 36). This is not a new phenomenon but has previously been a gradual process because climate change itself has been slow. In contrast, human-accelerated climate change is occurring much faster than the historical norm and the ranges of many species are

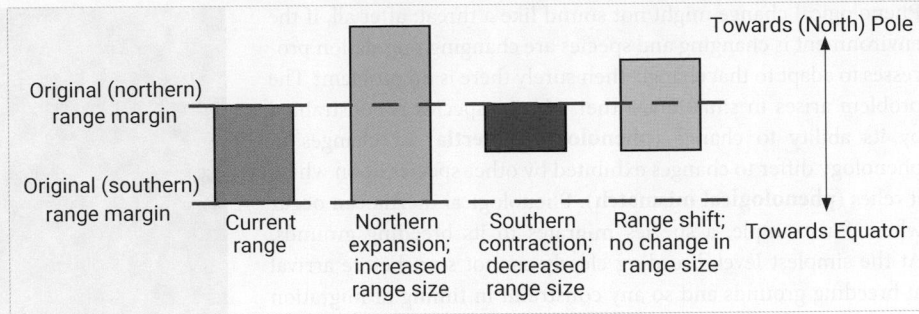

Figure 39.9 Climate-induced change. Possible range change scenarios due to human-accelerated climate change in the northern hemisphere relative to the current situation (grey). In some cases, ranges will expand (green), in others they will contract (red), and in the remainder they will remain unchanged (amber).

already altering substantially. Ranges can extend (northwards in the northern hemisphere) and thus get bigger, as shown by the first change scenario (Figure 39.9, in green), or they can contract at the (southern) edge and so get smaller, as shown by the second change scenario (Figure 39.9, in red). Alternatively, they can shift in their entirety while remaining the same size, as shown in the final scenario (Figure 39.9, in amber). Some species are also moving altitudinally, with mountain tops being the altitudinal equivalent of polar regions.

Go to the e-book to explore an interactive version of Figure 39.9.

If a species is able to respond fully and completely to the need to shift distribution (i.e. if it is physically and behaviourally able to move and there is suitable habitat in the new climatic zone), effects on that species are likely to be minimal. However, some organisms are constrained in their ability because physical movement of individuals or successive generations is constrained (non-mobile species or species with low dispersal power) or because the distance is too great (or the intervening terrain too inhospitable) to make the journey possible. Moreover, a species' range is determined by a lot more than just climate. If the new 'climatic window' does not coincide with appropriate habitat or food sources, for example, this is likely to mean the impacts of climate-induced distribution shifts are substantial, possibly even leading to extinction.

PAUSE AND THINK

The Scottish crossbill (*Loxia scotica*) is a species of finch that only feeds on pine cones, which it pulls apart with its highly specialized bill. Under climate change scenarios, the climatic window for this species is likely to move from Scotland to Iceland. Why is this likely to be a problem?

Answer: There are almost no trees in Iceland (and those that do occur are usually very stunted birch or rowan) so crossbills would not have a suitable food resource.

2. **Climate-induced changes in species' phenology:** Phenology refers to the time at which species undertake annual events. This includes, for example, the timing of when trees burst into leaf, amphibians spawn, birds breed, and mammals hibernate. In animals, phenological processes are largely controlled by an organism's **circannual rhythm**, which is the annual version of the **circadian rhythm** or 'body clock'. The circannual rhythm is

largely controlled by the **endocrine** (hormone) system, which itself is influenced by environmental stimuli. The main environmental stimulus, at least for species that live outside the tropics, is change in day length, which, of course, is a proxy for changes in seasons. For example, in birds, increasing day length decreases the amount of melatonin produced by the pineal gland. This in turn stimulates the production of testosterone (males) or prolactin (females), which encourages breeding (see Chapter 30). Because of this, sunlight is often referred to as the ultimate **zeitgeber** or time-giver. However, in some instances there are other zeitgebers, mainly to fine-tune phenological processes. In many cases, species use temperature to adjust their phenology. This makes logical sense as many annual processes relate to temperature either directly (e.g. hibernation, which is discussed in Chapter 38) or indirectly (e.g. through influences on seasonal food resources needed to sustain migration or breeding activity). The phenology of many species is changing as a result of climatic change. For spring-based events such as bird breeding, timing is tending to get earlier (for example, the timing of egg-laying in nuthatches (*Sitta europaea*) happening in late April, as opposed to early May, as shown in Figure 39.10), while timing is tending to get later for autumn/winter events such as fungi fruiting and the start of mammal hibernation. If you need a reminder about reading the graph in Figure 39.10, work through Quantitative Toolkit 7: Assessing patterns.

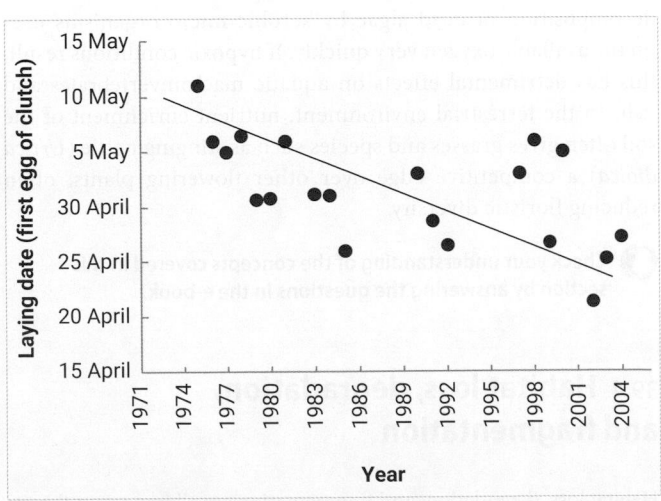

Figure 39.10 Phenological change. Change in the annual mean timing of egg-laying by nuthatch (*Sitta europaea*) in a UK woodland.

Phenological change might not sound like a threat; after all, if the environment is changing and species are changing population processes to adapt to that change, then surely there is no problem? The problem arises in situations either when a species is constrained by its ability to change (**phenological inertia**) or changes in phenology differ to changes exhibited by other species upon which it relies (**phenological mismatch**). Phenological inertia can occur when, for example, a species migrates to its breeding grounds. At the simplest level, breeding clearly cannot start before arrival at breeding grounds and so any constraint in timing of migration will also constrain ability to adapt timing of breeding. Phenological mismatches occur when an organism is out of step with its environment. To use another avian example, birds should synchronize breeding so peak demand for food from their young coincides with peak supply. For many species this peak supply is caterpillars, which have been hatching earlier in more recent years. However, if caterpillars emerge 2 weeks earlier now relative to 25 years ago but the birds are only breeding one week earlier, they are no longer fully synchronized. Such mismatches can have profound effects on breeding success and, ultimately, population size. For example, some Dutch populations of pied flycatchers (*Ficedula hypoleuca*), a small insectivorous bird, have declined by up to 90% primarily because of mismatches with food affecting breeding and subsequent population-level effects.

 For more on this topic, see Chapter 36.

Nutrient enrichment: processes and effects

The enrichment of nutrients in an ecosystem is called eutrophication. This can occur through the application of fertilizer, which is usually rich in nitrogen, phosphorus, and potassium; release of water from residential areas containing detergent (soap and washing powder that are high in phosphorus); or input of animal or human effluent into an ecosystem. In aquatic habitats, eutrophication can cause algal blooms, which is when algae populations increase exponentially (see Chapter 36) and become visible with the naked eye, as shown by the green 'scum' on the surface of the lake caused by the very high abundance of algae in Figure 39.11. This can rapidly de-oxygenate the water because decomposition of dead algae by aerobic micro-organisms uses up the available oxygen very quickly. If hypoxic conditions result, this has detrimental effects on aquatic macroinvertebrates and fish. In the terrestrial environment, nutrient enrichment of the soil often gives grasses and species such as stinging nettle (*Urtica dioica*) a competitive edge over other flowering plants, often reducing floristic diversity.

 Check your understanding of the concepts covered in this section by answering the questions in the e-book.

39.4 Habitat loss, degradation, and fragmentation

Habitat loss, degradation, and fragmentation are all important threats to global biodiversity, as we shall see in the next few subsections.

Figure 39.11 An algal bloom resulting from eutrophication. Blooms are often worse in still waters because nutrients do not move downstream with the current, and in summer, when water temperature is higher.

Source: Chris Craggs/Alamy Stock Photo.

Habitat loss and habitat change

Habitat loss refers to the process of converting land from habitats of high ecological value to human-constructed habitats. This generally involves hard infrastructure such as residential and industrial areas, but can also involve plantations that replace natural ecosystems. Such land uses generally have lower ecological value—sometimes substantially so—than the one they replaced but it should be noted that although we might use the term 'habitat loss', it is not strictly accurate as most land uses are habitats for something. Urban areas, for example, usually contain gardens, industrial areas often have trees that support a whole range of invertebrates, and agricultural land generally has hedges, ditches, streams, or ponds. Even walls, fences, and gravestones are habitats for mosses and lichens. As such, what we are really talking about here is *habitat change* rather than *habitat loss*. Habitat change often arises as a result of conversion of land for urban development or agriculture.

Of course, any change that presents an ecosystem challenge for one species can present an ecological opportunity for another. This is especially the case with generalist species, which can sometimes flourish in urban environments. For example, red foxes (*Vulpes vulpes*) and black rats (*Rattus rattus*) often find that the waste associated with humans provides excellent resources. Even species that are more specialized can adapt well. For example, in natural environments, peregrine falcons (*Falco peregrinus*) nest on cliff ledges but have adapted to using window ledges and air conditioning units on tall buildings, church towers, and even on bridges in the urban environment, as shown in Figure 39.12.

Figure 39.12 Some species, such as this peregrine falcon (*Falco peregrinus*), can make urban areas home.

Source: Metropolitan Transportation Authority/Flickr/CC BY 2.0.

Deforestation

Deforestation is the removal of trees from an area through clear-felling (rather than selective thinning). Sometimes deforestation occurs in plantation woodlands specifically managed for timber production, in which case it usually involves plantation areas that are reseeded after clear-felling. Often, however, deforestation involves harvesting trees outside of managed areas or clear-felling to make way for urbanization or agriculture. This last can include demand for land for cash crops such as soya beans or palm oil. Even if an area of woodland is replanted or allowed to naturally recolonize, it can take a long time for all the species that were originally present to move back. In some instances, such as clear-felling a primary (original) rainforest, the secondary (replacement) forest that grows back can be much poorer ecologically than what was lost. Even where a new plant species community is similar to the original, the new trees cannot always support the variety of species that the old ones did, simply because their age and size does not permit a full range of deadwood-associated species to colonize them.

Loss of vegetative cover through deforestation can lead to **soil erosion**. This process occurs when the upper layers of soil are displaced by water or wind. This is a common occurrence when land is stripped of vegetation because the soil is exposed to the elements and the lack of a root system means that the soil is not bound together, especially in areas with high-intensity rainfall or on slopes (compare the left and right sides of Figure 39.13 to see the difference that vegetation makes to soil erosion rates). The sediment that is removed through soil erosion can also be problematic, especially when it is deposited in local water courses through the process of **sedimentation**, as shown for the river on the left of Figure 39.13.

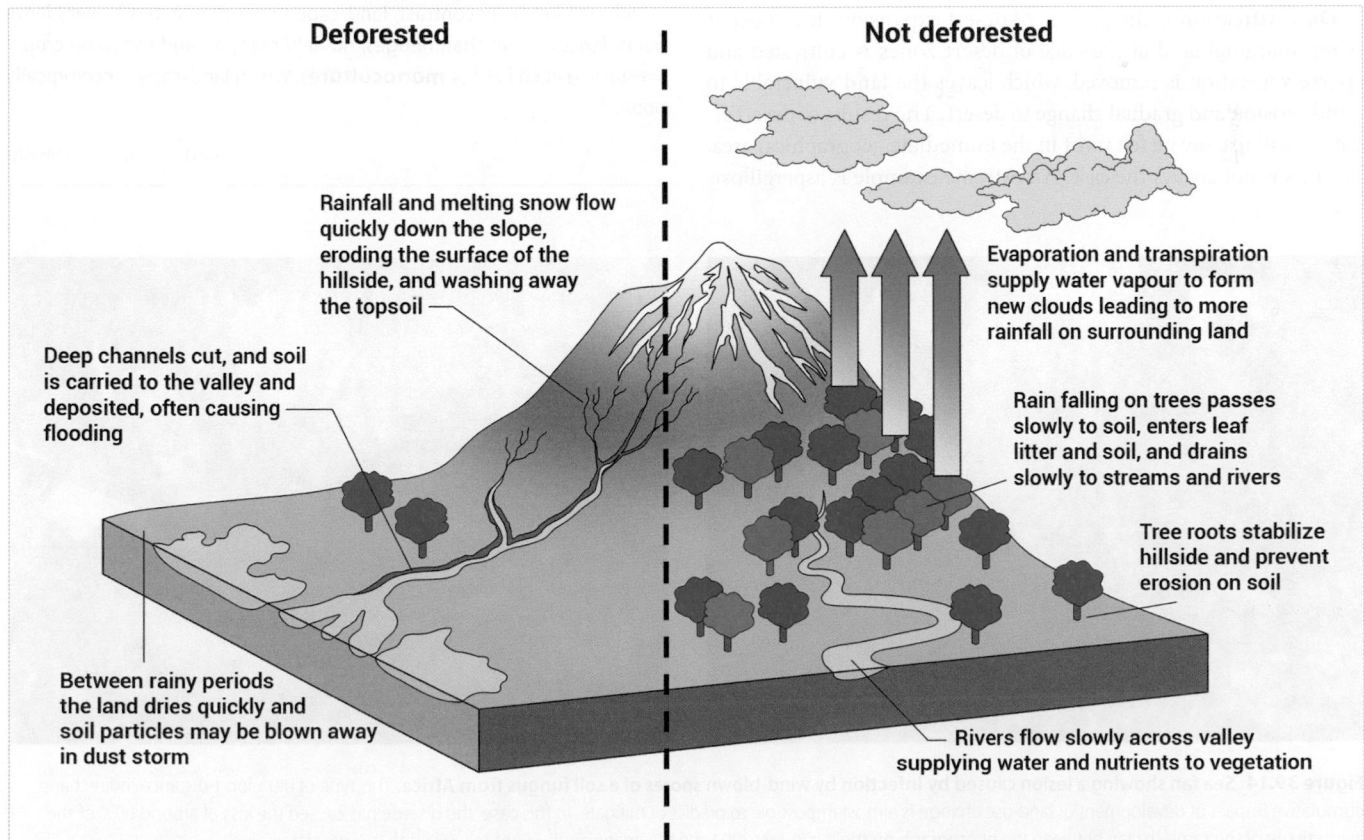

Deforested | **Not deforested**

Rainfall and melting snow flow quickly down the slope, eroding the surface of the hillside, and washing away the topsoil

Deep channels cut, and soil is carried to the valley and deposited, often causing flooding

Between rainy periods the land dries quickly and soil particles may be blown away in dust storm

Evaporation and transpiration supply water vapour to form new clouds leading to more rainfall on surrounding land

Rain falling on trees passes slowly to soil, enters leaf litter and soil, and drains slowly to streams and rivers

Tree roots stabilize hillside and prevent erosion on soil

Rivers flow slowly across valley supplying water and nutrients to vegetation

Figure 39.13 The linked processes of deforestation, soil erosion, and sedimentation.

Habitat degradation

The term habitat degradation is used to describe a change in the ecological condition of a habitat that results in loss of biodiversity or a decline in habitat function. It is not usually a substantial change in habitat type; rather it is the reduction in the condition of land and the habitat type it supports. Numerous processes can result in habitat degradation at a local scale. For example, failure to maintain open areas (glades and rides) of woodland by routine vegetation cutting might cause flowering plants to decline, with cascade effects on insects. Similarly, pond sedimentation can reduce water depth and impact aquatic biodiversity. However, on a large scale, one of the main reasons land degrades is because of changes in agricultural management.

Changes in agriculture leading to habitat degradation

Some areas are extremely sensitive to grazing pressure: timing, animal species, and stocking levels can all result in land becoming either overgrazed or undergrazed. Undergrazing can result in rank grasses (strongly competitive species with little nutritional value) becoming dominant over other grass species and grassland plants because their population is not being sufficiently kept in check by grazers. Overgrazing, by contrast, can remove plant growth too quickly and mean that disturbance-prone species do not establish. In severe cases, overgrazing can create bare patches of land that are then susceptible to soil erosion (refer to Figure 39.13) or even desertification.

Desertification is the process of desert expansion. It can occur when marginal land at the edge of desert zones is cultivated and sparse vegetation is removed, which leaves the land vulnerable to wind erosion and gradual change to desert. The results of desertification will usually be felt most in the immediate geographical area, but this is not always the case. An extreme example is aspergillosis

disease in coral sea fans in reefs in the Caribbean and Florida Keys caused by the soil fungus *Aspergillus sydowii* (compare the original healthy image of the sea fan on the left of Figure 39.14 with the image of the same sea fan after *Aspergillus* infection on the right of Figure 39.14). The fungus is transported via trade winds across the Atlantic in dust from the Sahel region of Africa as a result of desertification. This is an example not only of ecosystem challenges being indirect and geographically removed from their point of origin, but also a completely different ecosystem being affected.

Farming and irrigating marginal land in hot, dry, climates can lead to **salinization**. This occurs when water evaporation results in land having elevated salt levels, especially at the surface. This can affect growth of natural vegetation, as well as crops. **Agricultural intensification** can also degrade habitats. This is the process by which farmers aim to increase yield, often by increasing the use of fertilizers and pesticides. Use of these chemicals have their own impacts (e.g. death of non-target species; eutrophication), but their existence can also mean that areas of habitat on farmland that would not otherwise have been profitable to farm because of low soil fertility are cultivated. Intensification can also mean removing 'wasted' areas such as hedgerows.

PAUSE AND THINK

Consider two hypothetical agricultural landscapes. In landscape one, there are lots of small fields bounded by species-rich hedgerows, wide field margins, and low-intensity farming with different crops or different livestock in different fields. By contrast, landscape two is dominated by very large fields, fences rather than hedges, no field margins, and the same crops/livestock in each field (a **monoculture**). Which landscape is ecologically poorer?

Answer: Landscape two.

Figure 39.14 Sea fan showing a lesion caused by infection by wind-blown spores of a soil fungus from Africa. This type of ultra-long-distance indirect and cumulative impact of development or land-use change is almost impossible to predict or mitigate. In this case, the disease has caused the loss of around 60% of the living tissue of the same sea fan between the photograph on the left in July 2002 and the image on the right taken less than a year later.

Source: Reproduced with permission from C.D. Harvell (2011) Impacts of aspergillosis on sea fan coral demography: modeling a moving target. *Ecological Monographs*, 81(1): 123–139. Copyright © 2011 by the Ecological Society of America. https://doi.org/10.1890/09-1178.1.

Habitat fragmentation

Habitat fragmentation is the process by which a landscape becomes increasingly patchy. Let's imagine two landscapes: the first is entirely forested, while the second was *originally* entirely forested but now has many 'interruptions', perhaps being criss-crossed with roads and punctuated by urban settlements and agricultural fields. Whereas the first landscape is almost entirely one habitat, the second landscape is fragmented into lots of smaller patches. Fragmentation thus encompasses two concepts, the amount of habitat lost, and the extent of the isolation of the remaining habitat patches, which are explained below and shown graphically in Figure 39.15.

1. **The amount of habitat lost:** Fragmentation of a landscape invariably leads to a loss of the specific habitat which originally comprised that landscape. Loss can be minimal—for example a narrow road bisecting a habitat might not use much space—but

where fragmentation involves a change in land use that takes up a large area, habitat loss can be substantial.

2. **The isolation of remaining habitat patches:** Where fragmentation is substantial, the remaining patches of the original habitat can become very isolated from one another because there is a considerable distance between each patch.

The impacts of habitat fragmentation depend on its extent—both in terms of percentage of original habitat loss and the distance between the remaining patches (which itself affects patch isolation). The problem of patch isolation is reduced if patches are connected, a factor known as **connectivity**. For example, if patches of woodland are linked by hedgerows, species within the woodland patches can use the hedgerows as corridors of similar habitat along which to travel.

▶ We consider ways we can improve connectivity to reduce issues of fragmentation in more detail in Chapter 40.

Some species are at particularly high risk of negative effects from fragmentation. This includes **specialist species** that have very specific habitat requirements and species that have low mobility or dispersal power and so cannot easily move from an isolated patch of land. On the contrary, **generalist species** and species with high mobility and dispersal power will generally be less badly affected.

▶ See Chapter 38 for a reminder of specialist versus generalist species.

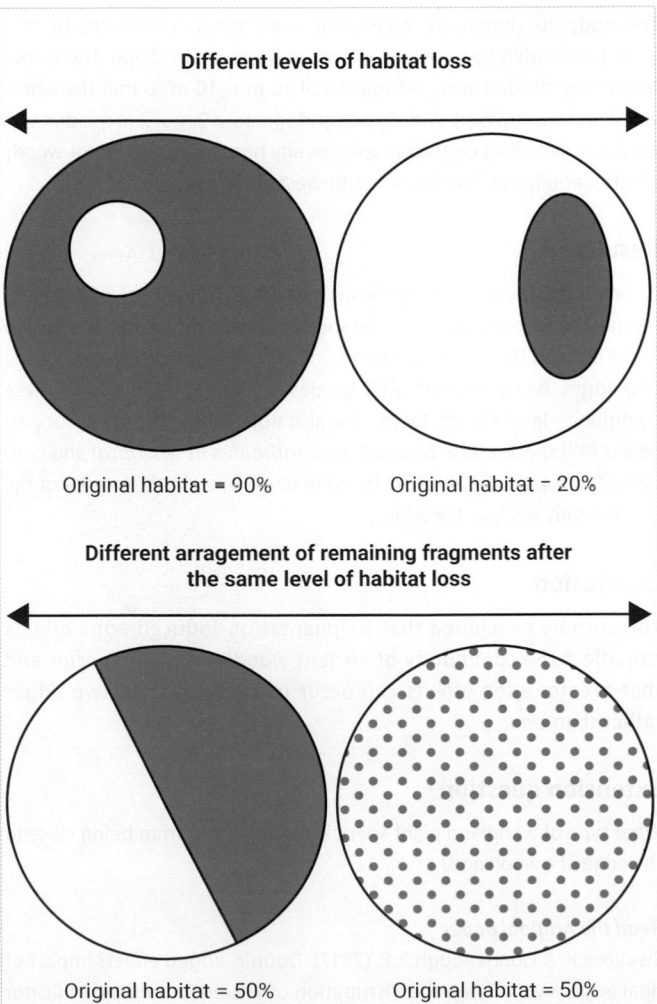

Different levels of habitat loss

Original habitat = 90%   Original habitat = 20%

Different arrangement of remaining fragments after the same level of habitat loss

Original habitat = 50%   Original habitat = 50%

Figure 39.15 Fragmentation. The process of fragmentation affects both the amount of the original habitat lost (top) and the spatial arrangement of remaining habitat (bottom).

PAUSE AND THINK

Why would habitat fragmentation usually have more of an impact on specialist species than generalist ones? And more of an impact on species with restricted movement and dispersal power compared to highly mobile species?

Answer: More specialist species are less likely to be able to survive in the 'gaps' between the original habitat patches and will likely become concentrated in the remaining patches (if these are big enough to support them), whereas generalist species might be able to survive in the new habitat, as well as in the remaining fragments of the old habitat. Species with high mobility—especially those that can fly—are likely to be less affected as they can move quickly between the remaining fragments.

Having patches of fragmented habitat can increase **edge effects**. Edge effects occur when abiotic and biotic conditions at the edge of a patch or fragment of a particular habitat differ from those in the interior. For example, in a woodland patch, the edge will probably be lighter and wetter because of increased exposure. Disturbance is likely to be higher and competitio, predation, and disease risk could be greater too. By contrast, the interior of the woodland will be darker and much less prone to environmental disturbance. As fragmentation increases, the ratio between edge and interior becomes increasing swayed towards there being more 'edge' and less 'interior', and this is likely to affect species communities. This is explored in more detail in Scientific Process 39.1 for a particular group of plant species.

 Check your understanding of the concepts covered in this section by answering the questions in the e-book.

| SCIENTIFIC PROCESS 39.1 | Investigating edge effects on ancient woodland indicator plants |

Background

Some species of plant are regarded as ancient woodland indicator (AWI) species. These are plants that are particularly associated with ancient woodland (i.e. woodland that is over 400 years old) and include species such as lesser celandine (*Ranunculus ficaria*) and wood anemone (*Anemone nemorosa*), as well as bluebell (*Hyacinthoides non-scripta*) and red campion (*Silene dioica*) (Figure 1). These species have low colonization potential due to poor seed production, low dispersal capability, and short-term persistence in the seed bank. They are also vulnerable to disturbance, either abiotically (weather, fire, etc.) or biotically (prone to being outcompeted, lower tolerance for herbivory), and tend to be shade-loving.

Figure 1 Ancient woodland indicator species include red campion (*Silene dioica*).

Hypothesis

Based on the life-history traits of AWIs, Swallow and Goodenough (2017) predicted that proximity to woodland edge would affect the AWI community. The specific predictions were: (1) that AWI species richness would be lowest near the edge and increase towards the middle of the plot; and (2) that species communities close to two edges rather than one (i.e. at the corner of a plot) would be subject to greater edge effects.

Methods

The study site comprised two ancient woodland patches in the UK that were surrounded by farmland and roughly square in shape. The woodlands were divided into grid squares of 10 m × 10 m so that the whole woodland was covered. Within each grid square, a 2 m × 2 m quadrat was used to collect data on the presence of any herbaceous and semi-woody plants identified as AWIs in the south-west of the UK.

Results

As predicted, there was a significant difference in the species communities nearer the edge with ≤ 2 AWI species found at the edges versus ≥ 6 in the interior. The corners of the woodland patches, which were close to two edges, had the lowest AWI species richness. In addition to these community-level effects, there were also noticeable differences for particular AWI species, with bluebell (*Hyacinthoides non-scripta*) and herb paris (*Paris quadrifolia*) being found in abundance in the interior of the plot but only rarely at the edge.

Conclusion

The authors concluded that fragmentation-induced edge effects can affect the community of ancient woodland plant species and that greater edge effects can occur in areas close to two edges rather than one.

Extension question

What type of woodland plant species might benefit from being close to the edge of a woodland?

Read the original paper
Swallow, K. & Goodenough A.E. (2017). Double-edged effect? Impact of dual edge proximity on the distribution of ancient woodland indicator plant species in a fragmented habitat. *Community Ecol.* 18: 31–36.

39.5 Introduction of non-native species

There are many ways that humans can change species communities, but one direct way is via the introduction of non-native species. A **non-native species** is any species that has been translocated to a new area, outside its normal native range. This is usually as a result of human activity, although this is not always the case (remember the example of non-native marine species being introduced to America from Japan on tsunami debris that we considered in Section 39.1). Where introductions are the result of human-aided translocations, this can be deliberate, for example a plant being introduced because of its ornamental value, or it can be accidental as in the case of non-native spiders hitching a ride as stowaways in shipments of bananas and turning up in the local supermarket.

Although the vast majority of non-native species do not establish in their new location (either because they are never released into the wild or because ecological and environmental conditions are not conducive to establishment), non-native species are still a common and widespread part of the flora and fauna of most countries. Indeed, in some areas such as New Zealand, almost 50% of plant species are non-native. Of the species that not only arrive in an area but go on to thrive in their new environment, the majority will have little, if any, effect on native species. However, a minority will become invasive.

Invasive species are non-native species that have spread widely and rapidly within their introduced range, increased substantially in abundance, and are now regarded as a pest because of negative interactions with native species and/or damage to the wider environment (because many non-native species are not invasive, these terms should not be confounded). One of the reasons that non-native species can become invasive is that they have often left behind those species that, in their original range, were keeping their population in check through competition, parasitism, or predation. Having escaped these '**natural enemies**' a non-native species has the potential to expand rapidly in both range and population size.

Invasive species can have a range of impacts on native species. The four main impacts are:

- Interspecific competition (e.g. non-native marine algae *Caulerpa taxifolia* outcompeting native marine plants in the Mediterranean)—see Chapter 37 for more information on types of interspecific competition.

- Increase herbivory or predation pressure (e.g. predation of native birds and mammals by non-native rats (*Rattus rattus/norvegicus*) on Pacific islands)—see Chapter 37 for more information on herbivory and predation.

- Vectoring disease (e.g. non-native birds being a reservoir for avian malaria in Hawaii, which is then spread to native species by mosquitoes).

- Genetic swamping of a native species via hybridization that can ultimately mean that there are no 'pure' examples, genetically speaking, of a native species (e.g. non-native ruddy

Figure 39.16 Hybridization. The ruddy duck is a stunning bird and presents no problem in its native range but it hybridized with the closely related white-headed duck in its non-native range.

Source: Allan Hack/Flickr/CC BY-ND 2.0.

ducks (*Oxyura jamaicensis*), shown in Figure 39.16, hybridizing with the rare and declining white-headed duck (*Oxyura leucocephala*) in Spain). You can read more about hybridization in Chapter 34.

PAUSE AND THINK

Why are non-native predators predisposed to affect native prey species?

Answer: Because native species have co-evolved, predator–prey interactions are often highly developed with native prey species having anti-predator defence strategies that are effective against native predators as we see in Chapter 37. When non-native predators are introduced, native species might not have anti-predator strategies that are effective and so are more likely to be negatively affected at a population level.

However, we should remember that non-native species can provide ecological benefits to native species. For example, many non-native species can act as valuable nectar and pollen resources; native bumblebees across Europe benefit from the abundant nectar of buddleia (*Buddleja davidii*), which we will be considering in Section 39.6. The reverse can also hold true; for example, pollination of native prickly parrot-peas (*Dillwynia juniperina*) in Australia is undertaken primarily by the non-native honeybee (*Apis mellifera*).

 Check your understanding of the concepts covered in this section by answering the questions in the e-book.

39.6 Implications of ecosystem threats

As might be expected given the range of ways that anthropogenic activity can create ecosystem challenges, the effects of these challenges are similarly varied. We have already explored degradation, fragmentation, and loss of habitats through processes such as urbanization, deforestation, desertification, soil erosion, and intensification of agriculture (Section 39.4). We have also considered impacts of climate change on species' distribution and phenology (Section 39.3) and shifts in species communities as a result of non-native species being introduced (Section 39.5). However, although we have touched on the effects of threats on the populations of specific species, we have not yet considered this in detail nor explored the processes by which this can happen. Moreover, we haven't considered the impact that changes to ecosystem functions might have. These topics will be considered in this section, before we move onto Chapter 40 in which we will consider possible solutions to ecosystem threats.

Species effects: decline and extinction

Some ecosystem threats have implications on specific species. Threats can be direct, indirect, or cumulative, or occur as a result of cascade effects (Section 39.1). Sometimes species' populations decline as a result, possibly as a result of impacts on adult survival, reproductive success, or offspring fitness. Negative effects can occur because of a specific threat (e.g. mortality increasing if a non-native predator is introduced), or because of a reduction in geographic range of a species or the carrying capacity of sites within that range.

As we discuss in Chapter 36, the **carrying capacity** (K) of an area is the number of individuals of a particular species that can be supported by the resources in that area, and this is not fixed in time or space. In other words, the carrying capacity of a particular site can go down if the site becomes degraded or fragmented (it can also be increased through effective management and conservation strategies—see Chapter 40). Similarly, the carrying capacity of different areas will differ—not all woodlands can support the same number of fallow deer (*Dama dama*), for example, as they might differ in size, food availability, predation risk, disease risk, level of competition, etc. Generally, ecosystem challenges that reduce land area, split habitat up via fragmentation, or erode the natural resources of an area will ultimately reduce carrying capacity. If there is one specific variable that is a **limiting factor** (see Chapter 38), which acts to limit carrying capacity, specific and targeted management could potentially be highly effective.

In some cases, a threat can result in the loss of a species either for a specific site or globally. Technically, **extinction** is defined as the complete loss of a species from the entire biosphere, such as, for example, occurred for the dodo (*Raphus cucullatus*) and the woolly mammoth (*Mammuthus primigenius*). Sometimes we can be in a situation where a species is lost from a particular area—say a specific country or even a specific site—but still occurs in other areas and so is not actually extinct. The correct term for this form of 'local extinction' is **extirpation**.

Extinction (or extirpation) comes about because of a series of linked processes called an **extinction vortex**. The term 'vortex' is used to signify a process that gathers speed in one inevitable direction, like water going down a plughole. The root cause can be environmental and thus driven by factors such as environmental disturbance or habitat loss (top two vortex diagrams in Figure 39.17), or genetic and thus driven by processes such as inbreeding, which is more likely to occur for a small, isolated population (bottom two vortex diagrams in Figure 39.17).

🌐 Go to the e-book to explore an interactive version of Figure 39.17.

Very often, species that are at risk of extinction as a result of ecosystem threats are in that position because a number of different factors combine to produce a cumulative effect. For example, the Mauritius kestrel (*Falco punctatus*), a falcon endemic to the island of Mauritius in the Indian Ocean, was once a common species. Decline was due to numerous interacting factors, including introduction of non-native species such as small Indian mongoose (*Herpestes auropunctatus*) and use of pesticides to control mosquitoes. The most important factor, however, was habitat loss and extreme fragmentation caused by deforestation. Although kestrels can and do fly, individuals are very philopatric (i.e. they stay in the area in which they were born) and so do not disperse far. As a result, even where populations did survive, they were very small and genetically isolated.

In order to predict likely extinctions, there are several species-at-risk schemes, including the International Union for the Conservation of Nature IUCN red list, as discussed in Real World View 39.2.

Ecosystem service effects: degradation and compromise

An **ecosystem service** is an environmental function or process undertaken within an ecosystem that is of benefit to humans. Common examples include water filtration, flood regulation, food supply, and decomposition of waste. Two of the most important ecosystem services are carbon sequestration and pollination.

Carbon sequestration is the process by which carbon from the atmosphere is captured and becomes 'locked up' in long-term stores, such as trees, fossil fuels, or peat bogs. This process is particularly important given global warming as part of the reason for climate change is the release of carbon from such stores (primarily by burning of fossil fuels, but also by other processes such as allowing peat bogs to dry up). Conserving bogs, and planting trees, can also be a mitigation method to buffer the impacts of rising carbon dioxide levels in the atmosphere.

Pollination, the transference of pollen from one plant to another, is vital for the majority of plants to reproduce. Pollination can be undertaken by a range of different taxa including insects, especially bees (Figure 39.18) and butterflies, as well as some birds and mammals. As we learn in Chapter 37, pollination is not really a

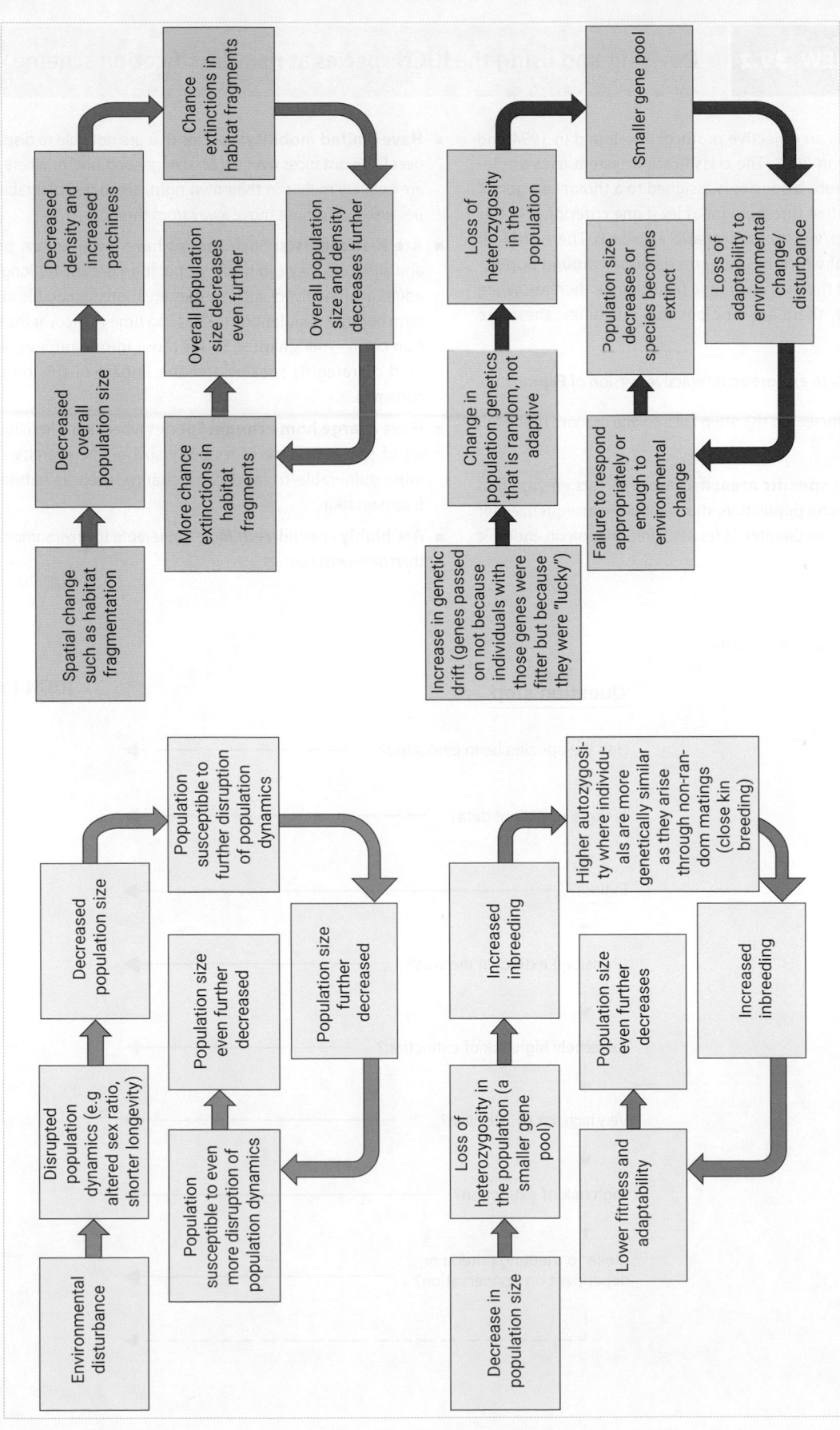

Figure 39.17 Extinction vortices—the process by which extinction can happen. The top two vortex scenarios are environmental in origin (i.e. occur because of a specific environmental disturbance or change), whereas the bottom two are triggered by a change within a species population itself.

The IUCN red list uses an objective protocol developed in 1994 and revised substantially in 2001. The classification system uses a rule-based approach whereby a species is assigned to a threat category if it meets the quantitative threshold for at least one criterion—in this way it often takes the 'worst case scenario' approach. There are four broad criteria, each of which has sub-criteria, based around population size (and change therein) and range (and change therein). When a species is assessed, there are nine possible outcomes; these are shown in Figure 1.

 Go to the e-book to explore an interactive version of Figure 1.

Species that are at an increased risk often fall into one or more of the following categories:

- **Are endemic to a specific area:** If localized threats emerge that are detrimental to one population, this will have consequences for the entire species—see Chapter 36 for more information on endemic species.

- **Have limited mobility:** Species that are not able to disperse rapidly over large distances over successive generations, or where individuals are not very mobile in their own right, are more vulnerable to threats because they cannot move away from them.

- **Are _K_-strategists:** Such species have lower annual productivity and although they can buffer population decline for longer because adults are long-lived, such species are more vulnerable in the longer term because populations take a long time to recover from a population crash—see Chapter 36 for more information on _K_-strategist (and _r_-strategist) species and the impact of this on population dynamics.

- **Have a large home range:** Species where each individual needs a lot of space because of its body size or the scarcity of prey are more vulnerable to landscape change such as habitat loss and fragmentation.

- **Are highly specialized:** More vulnerable to environmental changes than generalist species.

Figure 1 IUCN classification rankings for extinction risk. Green arrows signify a 'yes' to the question step; red arrows signify a 'no'.

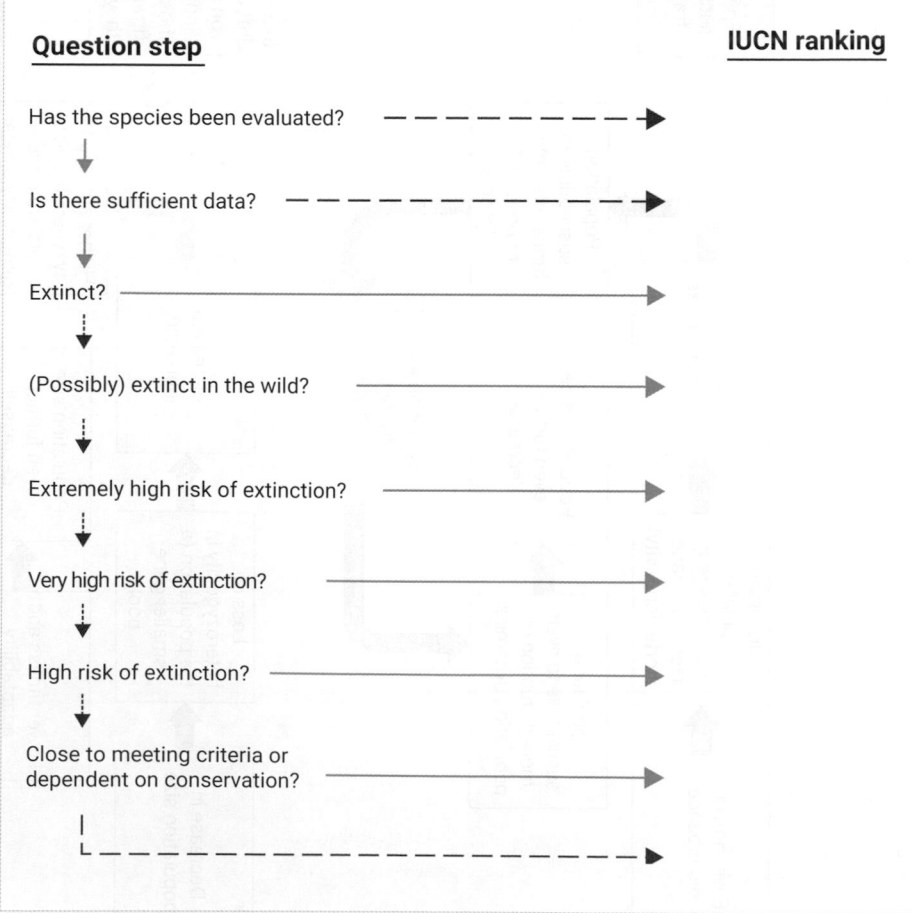

■ **Are not particularly adaptable:** There are two main ways that species can adapt to environmental change, either through evolution over successive generations or a process called **phenotypic plasticity**. Plasticity is the degree to which an individual can change in response to the environment in a non-genetic way: plants flowering early in a warm year can be a plastic response to local conditions, for example. In the same way, individuals can modify the timing of reproduction according to local conditions, or adjust diet in response to prey availability. See Chapter 34 for more information on evolution through natural selection versus phenotypic plasticity.

■ The cheetah (*Acinonyx jubatus*), shown in Figure 2, has been classified as VU (Vulnerable) for the last nine IUCN assessments. At the time of the last assessment in 2014, there were approximately 6500 individuals in 29 sub-populations, many of which were highly fragmented, small, and isolated. The overall population trend relative to the previous assessment was 'decline'. The assessment also recorded that the current geographical range was smaller than the historical range, with the northern African distribution having contracted particularly notably. Neither of these two factors was considered substantial enough to move the species to EN (Endangered). Threats recorded were: human–wildlife conflict (especially in agricultural areas and areas where high-speed roads cross cheetah home ranges); hunting (often as a result of being caught in snares intended for antelope); and war and civil unrest. Inbreeding resulting in low genetic diversity is also a problem (see Chapter 35 for more discussion on this topic).

Extension question

Read the detailed information on the IUCN global cheetah assessment at https://www.iucnredlist.org/species/219/50649567#population and

Figure 2 The cheetah (*Acinonyx jubatus*) has been ranked as Vulnerable on the IUCN red list.

compare this to the assessment for the Asiatic cheetah subspecies (*Acinonyx jubatus venaticus*) https://www.iucnredlist.org/species/220/13035342, which is classified as CR—Critically Endangered. Why is there such a difference in the classification rankings? What factors drive this?

Reference

Mace, G.M., Collar, N.J., Gaston, K.J., Hilton-Taylor, C., Akcakaya, H.R., Leader-Williams, N., et al. (2008) Quantification of extinction risk: IUCN's system for classifying threatened species. *Conserv. Biol.* **22**: 1424–1442.

Figure 39.18 The common carder bee (*Bombus pascuorum*) often relies on non-native flowers such as buddleia for nectar, especially late in the season.

'service' as this would imply an altruistic action whereas actually it is a side-effect of pollinators visiting plants to obtain nectar. However, that subtlety aside, it is certainly true to say that without pollination as an ecosystem service there would be catastrophic effects on plant life, not only in natural environments, but in agriculture too.

 Check your understanding of the concepts covered in this section by answering the questions in the e-book.

SUMMARY OF KEY CONCEPTS

- Challenges to ecosystems can be natural (**abiotic** or **biotic**) or **anthropogenic** in origin.

- Natural abiotic challenges include sudden **natural disasters** such as storms and tsunamis, as well as more gradual change such as natural **climate change**.

- Natural biotic challenges include ecological interactions, such as **competition** and **predation**.

- Anthropogenic challenges often involve **changes to land use** through urbanization or infrastructure development, resource use, pollution, and introduction of non-native species.

- Some impacts of ecosystem threats can be felt a long way geographically from their point of origin and in completely different ecosystems.

- The majority of anthropogenic challenges stem from the very large human population size, which is substantially bigger than at any other time in history.

- Although human population growth is important it is not (usually) the simple presence of people that threatens ecosystems, but rather what those people do to meet their needs, wants, and aspirations.

- We can classify resources as **non-renewable**, **semi-renewable**, or **renewable**.

- Impacts can occur during resource extraction/harvesting, transportation, or usage.

- Pollution is often an important consequence of resource use.

- Globally, climate change—mainly due to burning of fossil fuels—is a key threat and this can have implications for species' distributions and phenology.

- **Habitat loss** and **fragmentation** can cause issues for any species by reduction in the land area of a particular habitat and breaking up what is left into smaller more isolated patches. Fragmentation particularly affects **specialist species** and/or species with limited dispersal.

- Changes to land often results in **habitat degradation** through the processes of deforestation, soil erosion, desertification, salinization, sedimentation, and desiccation.

- Not all **non-native species** that are introduced to an area (naturally or by human **translocation**) establish themselves in the wild, but, if they do, they can have substantial effects on native species.

- Extinction processes are often linked together in a downwards spiral called an **extinction vortex**; vortices can be driven by environmental or genetic degradation.

- Factors that increase a species' **vulnerability to extinction** include how geographically widespread it is, how specialist it is, and how adaptable it is.

- In addition to challenges affecting species and habitats, vital **ecosystem services** such as carbon sequestration and pollination can also be affected by natural or anthropogenic change to ecosystems.

 Use the flashcards in the e-book to test your recall of key terms introduced in this chapter.

QUESTIONS

 Looking for answers? Once you've answered these questions, follow the link in the e-book to the answer guidance and check your work.

Concepts and definitions

1. What is the difference between non-renewable, semi-renewable, and renewable resources?

2. What is the difference between a 'non-native species' and an 'invasive species', and why should these terms not be used interchangeably?

3. What two concepts are encapsulated by the notion of habitat fragmentation?

4. What does endemic mean and why are endemic species often at greater risk of extinction?

5. What is the definition of an ecosystem service?

Apply the concepts

6. Find a recent example of a widespread forest fire and make a list of its direct, indirect, and cascade impacts.

7. What are the likely *long-term* impacts of ash dieback disease on species, habitats, and ecosystem services?

8. Find an example of an area where desertification is threatening species, habitat, and ecosystem services.

9. Summarize why some species are altering their phenology in response to climate change but others are not.

10. Search the IUCN red list (www.iucnredlist.org) to find a plant species that is Critically Endangered or Endangered (the top rankings) and draw up a brief profile to explain the main threats to that species.

Beyond the concepts

11. We examined the impact of edge effects on woodland plant species, especially ancient woodland species, in Scientific Process 39.1. Explain how edge effects will manifest in pond or lake ecosystems,

both in terms of abiotic parameters such as temperature and the effect of this on species community and ecosystem functioning.

12. Some species adapt well to urban environments. What biological or ecological traits do such species often have?

13. We touched on the long-range effects of desertification in Africa on coral reef health in Florida (Figure 39.14). Find another example of a long-range impact of a specific ecosystem threat.

14. Draw up a list of five tropical rainforest species that only tend to be associated with primary (virgin) rainforest rather than secondary

rainforest and thus tend to be lost as soon as primary rainforest has been cleared through deforestation, even if the area is replanted or becomes naturally recolonized. What biological and ecological trait(s) does each species have that explain its lack of ability to occur in secondary forests?

15. You have been asked to prepare a 60-second media piece on the importance of ecosystem services to ecosystem function and human livelihoods. What key facts and examples do you want to get across?

FURTHER READING

Middleton, N. (2013) *The Global Casino*, fifth edition. Abingdon: Routledge.
    Overview of threats to ecosystems
Goodenough, A.E. (2010) Are the ecological impacts of alien species misrepresented? *Community Ecol.* **11**: 13–21.
    General overview of ecological impacts of non-native species
Ingegnoli, V. (2002) *Landscape Ecology: A Widening Foundation*. Dordrecht: Springer, Netherlands.
    Excellent introduction to landscape elements and landscape ecology

Keith, D.A., McCarthy, M.A., Regan, H., Regan, T., Bowles, C., Drill, C., et al. (2004) Protocols for listing threatened species can forecast extinction. *Ecol. Lett.* **7**: 1101–1108.
    Predicting extinction risk
Everard, M (2017) *Ecosystem Services: Key Issues*. Abingdon: Routledge.
    Overview of ecosystem services

Solutions

Managing, Conserving, and Restoring Ecosystems

Chapter contents

Introduction   1339

40.1   Types of solution   1339

40.2   Protection   1339

40.3   Management of ecological problems   1343

40.4   Conservation   1345

40.5   Restoration   1354

Watch the key concepts video in the e-book to prepare yourself for studying this chapter.

LEARNING OBJECTIVES

By the end of this chapter you should be able to:

- Evaluate a range of options for protecting, managing, conserving, and restoring ecosystems.

- Appreciate that complex ecological problems often need complex management solutions.

- Compare and contrast passive management (custodial initiatives) with active management (on-ground action), as well as explain the difference between legislation and policy in an ecological context.

- Apply understanding of population dynamics and ecological interactions to real-world situations to devise appropriate strategies to solve specific problems, including those posed by:

  - pests and disease; and

  - non-native species introductions.

- Compare and contrast *in situ* and *ex situ* approaches to conservation.

- Appreciate the role of Minimum Viable Conservation and Population Viability Analysis in conservation.

- Explain the range of reasons why species might be reintroduced into their former range and the process by which this is undertaken.

Introduction

As we discussed in Chapter 39, threats to ecosystems are numerous and widespread: virtually nowhere on the planet is devoid of human impact. The severity of threats, and their ecosystem implications, is also increasing. This is largely because, in addition to the many natural threats to ecosystems, which have existed for millennia, we have added the largest threat of all: humans. However, although the need to take action to prevent ecosystems being compromised further has arguably never been greater, our ability to take effective action is, as we shall see in this chapter, also at an all-time high.

40.1 Types of solution

Many different actions can be taken to address ecosystem threats but we can largely group them under four main headings:

1. **Protection:** Protection of an area is possible either administratively by designating it as a legally protected area or via on-ground initiatives such as putting up fences to prevent grazing animals accessing certain areas. Management of when and how people can extract or harvest natural resources, for example via mining concessions and fishing quotas, would also fall under this category, as would legally protecting species and tackling wildlife crime.

2. **Management:** Ecosystem management can take place in a conservation context (see next point) but it is important to recognize that management of ecosystems is much broader than 'just' conservation. For example, management of forestry plantations, agricultural land, and urban green spaces often demands some knowledge of biodiversity and ecosystem functioning, but ecological and environmental considerations are very much secondary relative to timber harvesting, food production, or leisure activities, which become the main management motivation. In other instances, the motivation might be ecological but still remain outside conservation (e.g. management of pests or non-native species).

3. **Conservation:** This involves preventing species from declining or becoming extinct and/or ensuring that habitats do not become lost, degraded, or fragmented. *In situ* conservation (conservation in the wild) might focus on specific sites, often areas that are legally protected, or take a wider countryside approach to conserve biodiversity, habitats, and ecosystem services at a landscape scale. *Ex situ* conservation (maintenance of species in captivity) can also be important for ensuring long-term survival of animals or plants.

4. **Restoration:** At a habitat level, this can involve recreating habitats previously found at a particular site. This might be needed for transitional habitats that have not been maintained due to lack of appropriate management (e.g. undergrazed grassland that has become scrub encroached) or to reverse long-term degradation. Restoration would also cover remediation; the cleaning up of environmental contamination from pollutants. At a species level, restoration ecology covers reintroduction of species into their former native range. In theory, restoration provides huge opportunities—it is the ecological equivalent of being able to click the 'undo button'—but it is often extremely expensive, time-consuming, and difficult to achieve. It is always better to prevent habitats from becoming degraded and prevent species being lost from an area in the first place rather than to embark upon ecological restoration. A further challenge of restoration is that it depends on there being extensive knowledge of previous conditions as it is those previous conditions that are the target of restorative action.

Each of these four concepts will be considered separately in this chapter (Sections 40.2–40.5), but it is important to remember that, in reality, they are often used in combination to tackle complex ecosystem threats.

 Check your understanding of the concepts covered in this section by answering the questions in the e-book.

40.2 Protection

Unlike ecological management, conservation, and restoration, which all tend to have a strong active element, protection is often achieved through **custodial initiatives**. It includes protection given to specific species and sites through legislation, as well as specific directives and policy initiatives such as Biodiversity Action Plans (BAPs). Such approaches are sometimes called **passive management** because there is very little practical work involved (other than potentially a site-level assessment to assess how a site meets protection criteria, or obtaining baseline data on a species). However, the term passive management is rather misleading as it can imply it is easier than, or inferior to, active management. In fact, obtaining suitable legislative protection for a site or species is both complex and time-consuming.

Legislation

Legislation that pertains to *in situ* protection and management can be divided into **species-focused legislation** and **site-focused legislation**. In some countries, these are covered in separate Acts, in others there is a single Act with multiple parts. Failure to comply with legislation is typically punished either using fines or custodial sentences, depending on the country and nature of the offence.

Species-focused legislation

Most countries have national legislation that sets out the 'standard' protection for all species (or all species within specified taxonomic groups). There might also be annexes to the main provision, which detail species with additional 'top up' protection (e.g. because they meet specific rarity, importance, or sensitivity criteria) or species that are exempt from standard protection (e.g. because they are pests). Some countries also make specific provision for management of non-native species, either as an annex to the main national Act that covers species management (as is done in the UK) or in separate legislation (as occurs in the USA).

The main legislation in the UK is the **Wildlife and Countryside Act (1981)**. Species-focused protection is dealt with in Part 1 of the Act, with additional protection or exemption dealt with in 17 Schedules. The most important of these are:

- Schedule 1: Birds with additional protection.

- Schedule 2: Birds that may be caught, removed, or killed.

- Schedule 5: Animals with additional protection (including all reptiles, most amphibians, all bats, some land mammals).

- Schedule 8: Plants that are protected.

- Schedule 9: Non-native species (Part 1 = animals; Part 2 = plants).

Such provision paves the way for legally enforceable penalties for people who kill protected species, destroy habitat of protected species, or allow non-native species to spread either through deliberate action or as a result of not taking specified precautions. For example, it is an offence under Schedule 9 to plant Himalayan balsam (*Impatiens glandulifera*; Figure 40.1) or to carelessly facilitate spread of the plant by not cleaning machinery. Any plant material that is cut back and the soil in which the plant grows also needs to be treated as hazardous waste.

All legislation needs carefully controlled mechanisms to allow exemptions to be granted. This is particularly necessary with regard to development when situations often arise that require action to be taken that would normally be illegal. For example, a recent housing development in South Wales was only viable if a large pond on site was filled in. Normally, this would not have been an issue, but this particular pond was a breeding site for great crested newts (*Triturus cristatus*; Figure 40.2). As this species has additional specific protection under Schedule 5, deliberately filling in the pond, or ordering that action, would have been an illegal offence. The developer successfully applied for a licence to undertake this work on the basis that suitable **mitigation** is put in place so that there was no net loss for biodiversity as a result of the action. In this case, suitable mitigation involved all individual newts being trapped by a trained and licensed ecologist and **translocated** (moved carefully following a species-specific protocol) to a suitable pond elsewhere.

In addition to national-level species legislation, we need to be aware of the international **Convention on International Trade in Endangered Species of Wild Fauna and Flora (CITES)** regulations, which control global movement of animals and plants. The Convention was drafted in 1963 and adopted in 1973, and aims to ensure that international trade does not threaten species' survival. Currently, 180 countries have signed up and trade in more than 35,000 species and sub-species is now regulated. Species covered by CITES might be listed because:

1. they are threatened with extinction (e.g. manatees *Trichechus* spp.)

2. trade must be controlled in order to avoid harvesting (e.g. king cobra *Ophiophagus hannah*)

3. they are protected in at least one country that has signed up to CITES and so partial protection is needed.

As with national legislation, permission is sometimes required to 'break' CITES laws without punishment. This most often happens when animals or plants are being moved between countries for the purposes of reintroduction or between zoos in different countries as part of a captive breeding programme. Such actions are normally permitted under the exceptional circumstances rule, but still require an export licence (and often an import licence) to be granted. It should be noted that CITES regulations, although well-intentioned,

Figure 40.1 It is illegal to allow some non-native species to spread. It is illegal in the UK to 'wantonly or carelessly' allow Himalayan balsam (*Impatiens glandulifera*) to spread because of its non-native, invasive status.

Source: Maja Dumat/Flickr/CC BY 2.0.

Figure 40.2 Some species have legal protection. The great crested newt (*Triturus cristatus*) is legally protected in the UK under the Wildlife and Countryside Act, 1981.

Source: Martin Pelanek/Shutterstock.com.

are arguably not always necessarily in a species' best interest. For example, banning trade in white rhino (*Ceratotherium simum*) covers their horn, too (Figure 40.3). As it is illegal to trade the very considerable quantities of horn already stored from legal de-horning activities, the only source of horn is from living animals, which means that poaching rates and black-market trading are at an all-time high. Extensive attempts to reduce demand have so far been ineffective. Thus, although contentious and in some ways paradoxical, it could be argued that allowing trade in horn that has been legally and safely removed would discourage poaching.

Site-focused legislation

Site-focused legislation covers the designation and management of legally **protected areas**. Most countries have several different types of **statutory designation** and thus several different types of protected areas. For example, many countries have National Parks, which tend to be one of the highest levels of statutory designation and tend to cover a large area, as well as sites that are much smaller or important in a local context rather than a national one. There

are also designations that occur at an international level rather than a national one. The legislative structures behind designations can be both complex and convoluted, as demonstrated for the UK in Table 40.1.

PAUSE AND THINK

Can it ever be right to allow trade (and effectively support de-horning) of rhino? Is it an imperfect solution but a pragmatic one given the extremely high rates of poaching which, left unchecked, will drive the species to extinction in the wild within the next decade? Does the fact that horn is a renewable resource, growing like human fingernails, affect your decision as long as harvesting is done humanely, ethically, and sustainably? Is it better to see horn removed from a live animal better than seeing animals killed for their horn, or is that not acceptable in your opinion?

*Answer: You might want to read Biggs et al. (2013) Legal trade of Africa's rhino horns. Science **33**: 1038–1039.*

Table 40.1 Protected area designations in the UK at local, national, and international levels (numbers relate to 2019 data)

Level	Name	Details	Legislative framework
Local	Local Nature Reserves (LNR; number depends on county)	Sites of local importance for biodiversity and to engage local people with nature	Designated and managed by local authorities
National	National Park (n = 15)	Generally notified for aesthetic value but give protection to ecosystems through planning restrictions	Notified under National Parks and Access to Countryside Act 1949 (and amendments thereto in the Environment Act 1995 and Countryside and Rights of Way Act 2000) [Also National Parks (Scotland) Act 2000]
National	Area of Outstanding Natural Beauty (AONB; n = 46)	As above	As above except National Parks (Scotland) Act 2000
National	National Nature Reserves (NNR; n = 224 in England)	Most important ecological sites in UK	National Parks and Access to the Countryside Act 1949; Wildlife and Countryside Act 1981 (as amended by Natural Environment and Rural Communities Act 2006)
National	Site of Special Scientific Interest (SSSI; ASSI Area of Special Scientific Interest in N. Ireland; number fluctuates but ~6500)	Listed for biological and/or geological features and value	Originally notified under National Parks and Access to the Countryside Act 1949. All SSSIs existing in 1981 were re-notified under Wildlife and Countryside Act 1981; sites designated since have been through the latter Act (as amended by Natural Environment and Rural Communities Act 2006). Improved provisions for management of SSSIs introduced by Countryside and Rights of Way Act 2000 (in England and Wales), Nature Conservation (Scotland) Act 2004 and Wildlife and Natural Environment (Scotland) Act 2010.
International	Ramsar sites (n = 175)	Wetlands of international importance	Designated under International Convention on Wetlands of International Importance
International	Biosphere reserves (n = 6)	Sites displaying a balanced relationship between people and environment; educating and inspiring the community	UNESCO
International	World Heritage Sites (n = 31)	Usually applied to cultural heritage but can be applied to unique ecosystems	UNESCO (World Heritage Convention)

Figure 40.3 It is illegal to trade in rhino horn. White rhinoceros (*Ceratotherium simum*) are frequently poached for their horn, which is made of keratin (the same material as human fingernails) but is believed by some to have medicinal qualities and is thus much in demand on the black market. Recorded poaching levels have escalated rapidly since the mid-2000s.

Figure 40.4 Sites (Areas) of Special Scientific Interest. Some very rare species, such as adder's tongue spearwort (*Ranunculus ophioglossifolius*), have entire sites designated to protect them.
Source: Photo by Margerita I. M. Wilson, used with kind permission.

Sites (Areas) of Special Scientific Interest are designated for particular features at that site. From a biological perspective, this can be for the presence or high abundance of a particular species (or suite of species), an especially good example of a particular species community, or an especially good example of a specific habitat. SSSIs do not have to be large to be effective. For example, at just under 3 hectares, one of the smallest SSSIs in the UK is Badgeworth in Gloucestershire designated for an aquatic buttercup, the extremely rare adder's tongue spearwort (*Ranunculus ophioglossifolius*) shown in Figure 40.4. Although globally this species is not threatened (it is of Least Concern, the lowest level of extinction risk globally, according to the IUCN, as discussed in Real World View 39.2), it is at very high risk of extirpation (local extinction) in the UK.

Policy

Legislation is supported and underpinned by policy. We can think of **policy** as a set of guiding principles or protocols on a particular topic, which can be aspirational (and thus characterized by

targets) or regulatory (and thus characterized by thresholds). One policy of particular relevance to biodiversity and ecosystems is the **Convention on Biological Diversity**. This international policy framework arose from the Earth Summit in Brazil in 1992 and has become known as the Rio Protocol. Many countries also have national-level policy frameworks that specifically affect the environment or ones that hav40 impacts on the environment (e.g. mineral extraction, waste, forestry, and agricultural policies).

Policy and legislation can sometimes interweave when aspects of policy become legally binding. One example of where this has happened is in fisheries policy, where fishing quotas have become enshrined in legislation. This is explored further in Real World View 40.1.

Active protection

Although we have noted that protection is usually a custodial activity, it can sometimes involve action on the ground. For example, to combat poaching, it might be necessary to use active

---

REAL WORLD VIEW 40.1   Fishing quotas and other methods to manage sustainable harvesting

As we see in Chapter 39, semi-renewable resources can be used sustainably but only if harvesting is carefully managed. This is especially true for animal resources, where harvesting must not exceed **net reproductive rate per generation** (R_0) (see Chapter 36). For fish populations, there are several controls that might be put in place to prevent overfishing, in theory at least:

- Limiting the total number or total weight harvested of a species.
- Limiting or restricting harvesting of specific age/sex/size individuals within a harvested species (e.g. minimum mesh sizes in fishing nets to protect juveniles from exploitation).

- Limiting the number and/or equipment (e.g. limiting the number of fishing boats).
- Limiting when individuals can be harvested (e.g. away from fish spawning periods).

Fishing quotas are often based on 'total allowable catch' (TAC) for each species. However, if these are just issued with an annual (or seasonal) upper limit, each fisherman can be pressured to bring in the fish before the TAC is reached and the fishery is closed for the season. Fierce competition between fishermen at the start of the season can also result in investment in additional boats, making the industry highly inefficient and burdening local fishing economies. There are also safety concerns if boats put to sea in bad weather to get as much of the catch as possible. Regulations such as per-trip catch limits, days-at-sea limits, and shortened seasons can be used to slow the pace of fishing, but, in fact, they often exacerbate the race and subsequent ecological damage. Individual fishing quotas (IFQ) are often fairer and give each fisher a stake in the fish stocks, and thus an interest in the long-term sustainability of fish populations. They also mean that fishers no longer have to race to maximize their catch. Research in New Zealand has suggested that IFQ fishers are also more likely to fund ecological research as they recognize that better ecological data can guide decisions to improve the future stability and value of the fishery.

It is also possible to limit locations from where fish can be harvested. For example, there has been considerable debate about whether fishing should be allowed anywhere in Marine Protected Areas (MPAs). As of 2015, fishing is allowed in at least 94% of MPAs and is only completely banned in full Marine Reserves, which have to be no-take zones. Fishing in MPAs might be allowed because overfishing is not recognized as a threat in that particular area or, in some cases, allowing some fishing may be a compromise to get support for other biodiversity action or protection.

Extension question

Imagine that you have to prepare a very short briefing note for a government minister about to chair a debate on the wisdom or otherwise of allowing fishing in Marine Protected Areas. What key points might you expect the minister to hear from both sides of the debate?

measures such as **anti-poaching patrols**, often with armed personnel and trained anti-poaching canines, such as the one shown in Figure 40.5. Other methods include using GPS tags on animals that alert reserve wardens to animals that move—or are taken—beyond a pre-defined geographical area. These approaches are sadly becoming commonplace for species that are poached for specific commodities as part of illegal worldwide trade (rather than at subsistence level for meat). Key examples are rhino poached for their horn, as discussed above, pangolin (*Manidae* spp.) taken for their scales, and elephants (*Elephantidae* spp.) targeted for ivory.

Figure 40.5 An anti-poaching patrol. Two members of the Anti-Poaching Unit (APU) at Mankwe Wildlife Reserve, South Africa, with a trained APU canine.
Source: Photo by Lynne MacTavish, used with kind permission.

 Check your understanding of the concepts covered in this section by answering the questions in the e-book.

40.3 Management of ecological problems

Ecological management is not only about conservation. It can often be about managing ecological problems, including the need to manage a pest or disease damaging other species or human livelihoods, or problems arising from the introduction of a non-native species.

Pests and disease

An ecological pest is a species that has a detrimental impact on something that is valued by humans. This might be humans themselves (e.g. through effects on health or economy), or an aspect of the ecosystem that humans hold in high regard, such as a charismatic species. Organisms can be regarded as pests because they:

- Cause a **disease** in humans or farm/companion animals (e.g. **micro-organisms**, including bacteria, fungi, and viruses that can cause disease directly; **parasites** such as *Plasmodium*, which causes malaria; or species that contain toxins such as giant hogweed (*Heracleum mantegazzianum*), which causes skin irritation).

- Are a **vector** (transport mechanism) for a disease-causing micro-organism or parasite (e.g. fleas that vector plague and mosquitoes that vector the malaria-causing parasite).

- Are a **reservoir host** (the ultimate source) of a disease-causing micro-organism or parasite (e.g. many bats).

- Destroy agricultural crops, ornamental/garden plants through **herbivory** (e.g. crickets, grasshoppers, and locusts (Orthoptera) can wipe out fields of crops very quickly).

- Consume domestic or companion animals through **predation** (e.g. wolves (*Canis lupus*), which can kill human-raised animals and, occasionally, humans themselves).

- Cause damage by **competing** with agricultural crops or ornamental/garden plants or cause damage to human infrastructure (e.g. Japanese knotweed (*Fallopia japonica*) has very extensive root systems that can undermine roads).

▶ **Refer back to Chapter 37 if you need a reminder about any of these types of interactions.**

In many cases, pests are not managed directly. This might be because direct management is not possible or not practicable, because the costs of doing so outweigh potential gains, or because doing so would be counter-productive (i.e. when resistance to treatment is likely, or when removal of the pest individuals will only be temporary before new individuals arrive to fill the void). In such cases, pest management focuses on treating (or compensating for) the problem, rather than tackling the root cause of the issue. For example, in the case of livestock threatened by a native predator, management is likely to focus on creating predator-proof enclosures and compensating farmers for any losses, rather than culling the predator. Similarly, for some diseases, the focus is on treating symptoms rather than removing the causal agents or vectors. Occasionally, some diseases can be managed by treating the likely target of a disease prophylactically (i.e. to prevent infection), as in the case of giving humans yellow fever vaccines to reduce the chances of someone contracting the disease even if they are bitten by an infected mosquito.

Where the decision is taken to control pests themselves, there are several possible courses of action:

1. **Physical removal:** Includes cutting, burning, or uprooting (plants), as well as culling and trapping to translocate individuals (animals).

2. **Containment:** Spatially containing the pest species, for example by using barriers.

3. **Chemical management:** Use of pesticides and herbicides. Should ideally have a mode of action specific to the pest species rather than a general one.

4. **Biological pest management:** Releasing a predator, competitor, parasite, or disease to reduce the abundance of the pest species. Must be highly specific to the pest species so that non-target species are not 'attacked'.

Managing pests is not without issue. Despite careful research, there can be effects of treatments on **non-target species**. One common example of non-target effects occurs where anthelminthic drugs are given to farm animals suffering from intestinal worms. A large proportion of this drug, or active metabolites of it, is excreted and has a detrimental impact on soil biota. This is an example of a direct effect (as discussed in Chapter 39).

One of the most well-known examples of an indirect non-target effect is the insecticide dichlorodiphenyltrichloroethane (DDT), which was used to kill mosquitoes to reduce the incidence of diseases such as malaria and yellow fever. DDT was extremely effective, but, unfortunately, it also **bioaccumulated** in the food chain. As we see in Chapter 38, top predators consume numerous prey organisms from lower trophic levels in order to meet their energy requirements. This partly reflects the loss of energy between each trophic level and partly reflects the fact that top predators tend to be fairly long-lived *K*-strategist species.

▶ **We learn more about this topic in Chapters 33 and 36.**

However, while the majority of energy is lost within and between trophic levels, heavy metals such as mercury do not biodegrade, but instead build up in food chains over time. The lack of degradation means that the chemical lod within each individual is effectively the sum of that compound ever ingested by that individual *and* its prey and toxicity can result. In this case, bioaccumulation resulted in top predators having very high levels of DDT because every time they consumed prey containing DDT, the non-biodegradable chemical was stored in their body. The increase is demonstrated in the food chain illustrated in in Figure 40.6, where you can see DDT increasing from 0.000003 parts per million (ppm) in the water in which zooplankton live, to 25 ppm once it reaches the fish-eating birds of prey. This caused neurological issues and, for birds, resulted in very thin eggshells that severely compromised reproductive success.

Even when the target species is the only species directly affected, there can be ecological challenges to overcome. The primary issue is that of **resistance** to chemical control measures. This is similar to antibiotic resistance that changes, for example, *Staphylococcus aureus* into methicillin-resistant *Staphylococcus aureus* (MRSA), which is much harder to treat because traditional antibiotics are ineffective.

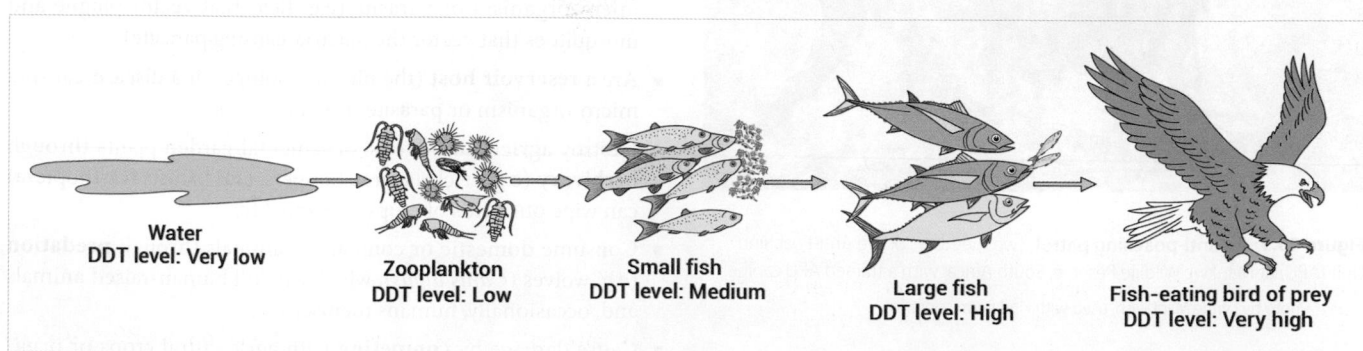

Water
DDT level: Very low

Zooplankton
DDT level: Low

Small fish
DDT level: Medium

Large fish
DDT level: High

Fish-eating bird of prey
DDT level: Very high

Figure 40.6 Bioaccumulation. Bioaccumulation in a food chain occurs when top predations accumulate non-biodegradable chemicals from their prey.

See Chapter 34 for more detail on this subject from an evolutionary perspective.

The same issue can occur with pests evolving resistance to a pesticide. Many pesticides rely on targeting specific chemical pathways with the pest's body. If some individuals have slightly different pathways, they might be able to tolerate that pesticide. If they can then breed and pass on that ability as a **heritable trait** (Chapter 35) multiple individuals in the populations could quickly become impossible to manage using that particular chemical approach. This is explained further in Figure 40.7, which shows that because individuals that can tolerate pesticides best (i.e. are resistant to their effects) are the individuals that will survive and breed, gradually the population will become dominated by resistant individuals that are, because of that pesticide resistance, very hard to control.

Go to the e-book to explore an interactive version of Figure 40.7.

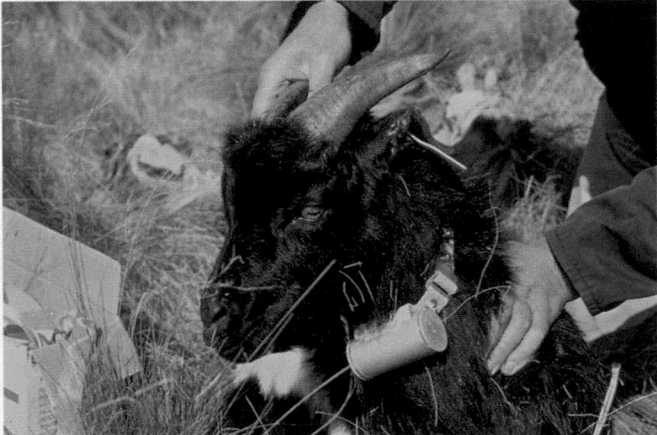

Figure 40.7   Pesticide resistance. How the process of pesticide resistance develops.

Source: Goodenough and Hart (2017), *Applied Ecology: Monitoring, managing, and conserving*. Oxford: Oxford University Press. Reproduced with permission of the Licensor through PLSClear.

PAUSE AND THINK

Why does chemical resistance only become a problem if the factor responsible for resistance is a heritable trait?

Answer: If a few individuals are unaffected by the application of a chemical and breed, but cannot specifically pass the tolerance of the pesticide onto their offspring, the population size is likely to remain quite low even if some of the offspring randomly have the same ability to tolerate the chemical as their parents. However, if the tolerance is heritable, the breeding adults will likely all have the trait (as these are the only individuals surviving after treatment) and thus most offspring will also have this trait, as will their offspring, etc. In other words, the problem of resistance becomes a problem at the population level rather than the individual level.

Non-native species

As we discuss in Chapter 39, invasive non-native species can have significant impacts on native species that mean that they need to be eradicated or controlled. The best possible action is, of course, to prevent translocation in the first place, using appropriate biosecurity protocols. Where species do break through such defences, there is a good chance of successful eradication if management is undertaken promptly soon after establishment when the population size is low and spatially restricted. Once a population has become established, eradication is unlikely (except in rare cases on islands) but spatial containment might be possible.

After a non-native species has become established, measures to control or reduce populations become complex, costly, and less likely to succeed. Most control measures are similar to those used for general pest species, but occasionally more innovative methods are employed. For example, in New Zealand, sterile male goats are fitted with a tracking collar and then released. These individuals then join local herds of feral goats (*Capra hircus*), which can then be located by hunters through the radio collar, as shown in Figure 40.8.

Check your understanding of the concepts covered in this section by answering the questions in the e-book.

Figure 40.8   Animals can be tracked with radio collars. Judas goats with individual tracking collars are used to alert authorities to the location of feral goat populations.

40.4 **Conservation**

Fundamentally, conservation is about working to safeguard species, habitats, and ecosystems from destruction or degradation. Unlike **preservation**, which aims to preserve a landscape and its biodiversity in its 'natural' state without any interference from humans, the founding principle of **conservation** is that of active management of species and habitats. In many ways, conservation is much more complex than preservation because it requires us to make decisions about management priorities, to set targets, and to decide on the best methods to achieve those targets.

In situ versus *ex situ* conservation

It can be useful to divide conservation into two general approaches: *in situ* and *ex situ*. *In situ* approaches involve conservation in the wild and encompass all on-site techniques for conserving species, as well as all habitat-based conservation. Conversely, *ex situ* approaches involve conservation in captivity and encompass all off-site techniques (including captive collections such as zoos and botanical gardens). *Ex situ* conservation has the advantage of moving species away from any threats they are facing in the wild to an artificial environment, but requires that artificial environment to be suitable for the species' needs to ensure adequate husbandry, appropriate welfare, and to encourage successful reproduction. However, conserving a species *in situ* often means that conservation benefits a whole range of co-occurring species at the same time, as well as acting to protect habitat; ecological functions of species are also maintained; these are referred to as **cascade benefits**. In terms of costs, *in situ* conservation is generally cheaper than *ex situ* conservation, but can be harder to fund.

In some cases, *in situ* and *ex situ* approaches are used at different times, in different places, or to meet different aims. For example, for the Grand Cayman blue iguana (*Cyclura lewisi*), shown in Figure 40.9, insurance populations were created *ex situ*, while *in situ* habitat management and invasive species control was undertaken. Even a single conservation initiativ can involve both approaches. For example, **species reintroduction** always involves *in situ* work—the actual reintroduction plus any pre-reintroduction management and monitoring—but can also have an *ex situ* element if the reintroduced individuals come from captive breeding

Figure 40.9 *In situ* and *ex situ* conservation. The Grand Cayman blue iguana (*Cyclura lewisi*) is a species that has benefited from both *in situ* and *ex situ* conservation.

Source: Photo by C. Phifer (CC BY 2.0).

programmes. The majority of this section focuses upon *in situ* conservation, but *ex situ* approaches are covered briefly.

Conservation decision making

In an ideal world, all species and habitats in need of conservation action would receive conservation action. In reality, however, conservation resources (money, time, space, and expertise) are not infinite and demand is always likely to outstrip supply. Accordingly, we need to make decisions regarding where conservation resources should be deployed. Prioritizing conservation resources is effectively a system of triage: assessing the need for, and the likely benefits of, action in a given situation. Some of the candidates for prioritization are given in Table 40.2 before we then move on to consider species-focused conservation and habitat-focused conservation.

PAUSE AND THINK

Conservation triage is based on humans making decisions that can be subjective and, at times, even controversial. Consider the suggestions for prioritization of conservation resources for species given in Table 40.2 and think of at least one counter-argument for each suggestion.

Answer: Counter-arguments for species priorities could include (1) extinction risk—species that are critically endangered will usually need considerable resources to save, so it might be better focusing on species before they reach this point; (2) keystone—it can be very difficult to identify keystone species before they are lost from an ecosystem and if they cannot be identified they cannot be prioritized; (3) flagship—often the species at greatest risk or the most useful to ecosystems are not charismatic and so focusing on the charismatic few could divert conservation resources from where they would be most useful; (4) umbrella—if the umbrella species that 'fronts' a conservation initiative becomes extinct this can compromise an entire conservation initiative.

Species-focused conservation

Species-focused conservation is conservation that specifically aims to protect a particular species or group of species. This requires knowledge of the threats to that species, but usually starts with a detailed assessment of the species' biology, its current population size, and recent population trends. This is important because we often need to know whether a species (globally or for a specific population) is viable if appropriate conservation is put in place or whether it is too small to be recoverable. Early attempts to answer this question adopted a one-size-fits-all approach. The most common generalization was the **50–500 rule**, which stipulated that a population needed a minimum of 50 individuals to survive in the short term and a minimum of 500 individuals for long-term maintenance of genetic variability and adaptability.

While the 50–500 rule approach was successful in terms of focusing biologists' minds on the issue of population viability and the implications of that for conservation management, it was far too simplistic. This is because, as we see in Chapter 36, the population-level processes operating for an **r-strategist** such as a water flea (*Daphnia magna*) would be very different to those for

Table 40.2 Possible candidates for prioritization of conservation resources

High priorities for resources: species	High priorities for resources: habitats
Species at imminent risk of extinction: One way of prioritizing resource allocation is to give it to those species most in need—those that are listed as Critically Endangered on the IUCN red list that we cover in Chapter 39, for example. This would cover species such as the black rhino (*Diceros bicornis*). If not prioritized, such species could quickly become extinct.	**Areas with high threat level:** This is the habitat equivalent of prioritizing species that are critically endangered. If such areas are not prioritized, biodiversity and ecosystem function might be irretrievably lost or damaged.
Keystone species: As we see in Chapter 38 keystone species are especially important within an ecosystem. Prioritizing such species increases the chances of maintaining a well-functioning ecosystem.	**Areas of high species richness:** The rationale here is a simple one—if conservation efforts are invested in areas that support a large number of different species, more species will benefit than if those same resources were diverted to areas with lower species richness (see Chapter 37).
Flagship species: These are a charismatic species to which people specifically relate—such as the bald eagle (*Haliaeetus leucocephalus*) in the USA. They can be thought of as the 'poster boys' of conservation and can be useful both for fundraising and environmental education.	**Areas of high endemism:** Endemic species are those that only occur in one specific area (see Chapters 36 and 37). Conservation of endemic species is, or at least should be, a conservation priority simply because the loss of such a species from the area where it is endemic is not extirpation, but extinction: local loss equals global loss (see Chapter 39). At a global level, this often involves conserving **oceanic islands** (i.e. islands that have arisen because of volcanic or coralline activity and never been part of the mainland).
Umbrella species: These are species that live in habitats also inhabited by multiple other species, have large home ranges, or live in particularly rare or vulnerable habitat. The idea is that, in directing resources to *in situ* conservation of such species, the habitat in which that species occurs will also be protected, as will naturally co-occurring species. In this way, the focal species becomes a protective 'umbrella' for wider ecology. A good example is the Florida panther (*Puma concolor coryi*), which inhabits mixed-species swamp forest, including the Everglades.	**Areas with distinctive communities:** As we covered in Chapter 37, some areas have species communities that, although probably not unique, are very different from communities in other sites. If we can ensure that conservation resources are allocated to as many sites with distinctive communities as possible, we maximize the chances of covering as many different species globally as possible.

a *K*-strategist such as a blue whale (*Balaenoptera musculus*). In recognition of this, two new concepts emerged:

1. **Minimum Viable Population (MVP):** The number of individuals necessary for a population of a given species to have a 90–95% chance of surviving long term (usually 100–1000 years into the future). This number is *specific to an individual species* rather than being a general rule-of-thumb figure like the 50–500 rule. The MVP is important for conservation action in three main ways as it determines: (1) whether there is a need for conservation action and how quickly; (2) the chances of conservation being successful; and (3) the best approach (*in situ* if possible; *ex situ* if the population is below or close to the viability threshold).

2. **Population Viability Analysis (PVA):** The computer modelling process that calculates the MVP. Undertaking a PVA requires a considerable amount of species-specific information on the following parameters:

   - **Population structure:** Optimal sex ratio, age structure, breeding strategy (monogamy, polygyny, etc.).
   - **Productivity:** Number of broods per year, number of young per brood, proportion of males and females typically in the breeding pool, age at first reproduction, and age at last reproduction.
   - **Mortality:** Longevity, survival rates per year for both sexes and each age class
   - **Genetic diversity:** History of population bottlenecks.

   - **Spatial relationships between populations:** Dispersal rate, dispersal frequency, and direction of movement.
   - **Carrying capacity:** Habitat quality and suitability, size of habitat patch.

PVA modelling develops the concepts we encounter in Chapter 36 for predicting population size over time based on survival and fecundity schedules detailed in life tables that are used to calculate net reproductive rate per generation (R_0) in populations where there is a maximum upper limit to population as defined by carrying capacity (K). Here, however, the approach has been extended to also allow for genetics (see Chapter 35), movement between populations within a metapopulation framework (see Chapter 36), and more detailed information about population structure. It also permits modelling of harvesting (Real World View 40.1). Undertaking the PVA process might also be useful in indicating what conservation should focus upon. For example, if mortality is particularly high then measures to improve survival might be indicated, while if fecundity is low then measures aimed at improving reproductive success might be useful.

Improving survival

There are several ways of improving survival of species *in situ*. The first step is always to try to understand any threats that are increasing mortality. These could include, for example, habitat degradation, lack of food, non-natural fire regimes, non-native species, high predation or disease risks, or poaching. Only once threats have been identified can suitable conservation strategies be devised and implemented. Figure 40.10 outlines a hierarchy of species-focused conservation action, starting with the best option of eliminating the threat, and working through to the least-preferable option of moving the population under threat away from the threat, which is expensive and can lead to unforeseen ecological consequences.

Improving reproductive success

As highlighted in Figure 40.10, a key conservation strategy is to improve species' reproductive success. This might be because population decline is caused directly by poor or declining productivity, or because improving productivity is a way to buffer the effect of increased mortality. The second of these stems from the interplay between the four key population processes identified in Chapter 36: birth rate, death rate, emigration, and immigration. In other words, if mortality or emigration are higher than the species norm and cannot be decreased, increasing reproductive success (or facilitating immigration) can be an important way to stabilize a population.

Methods of improving reproduction are highly species-specific and location-specific. In some instances, the key is optimal habitat management of vital sites in order to create and maintain the necessary conditions for breeding. For example, many butterfly species, such as the brimstone butterfly (*Gonepteryx rhamni*) in Figure 40.11, have a single larval food plant and will only lay their eggs on that one species. For the brimstone butterfly, that plant is buckthorn (*Rhamnus cathartica*), which you can also see in Figure 40.11. In this instance, habitat management needs to promote a high abundance of the food plant species, with each plant being in good condition and available at the optimal time of year for the female to lay her eggs successfully, the caterpillars to hatch successfully, and ultimately feed, grow, and pupate successfully.

In other cases, we can facilitate successful reproduction by adding infrastructure such as bird nestboxes, bat roost sites, dormouse boxes, and reptile refugia. These might be necessary because of a lack of natural nest sites or high competition for them (see Scientific Process 37.1), or to reduce parasite levels or predation risk (see Chapter 37). It is vital that when nesting microhabitats are added, they are not only optimal in their own right, but are

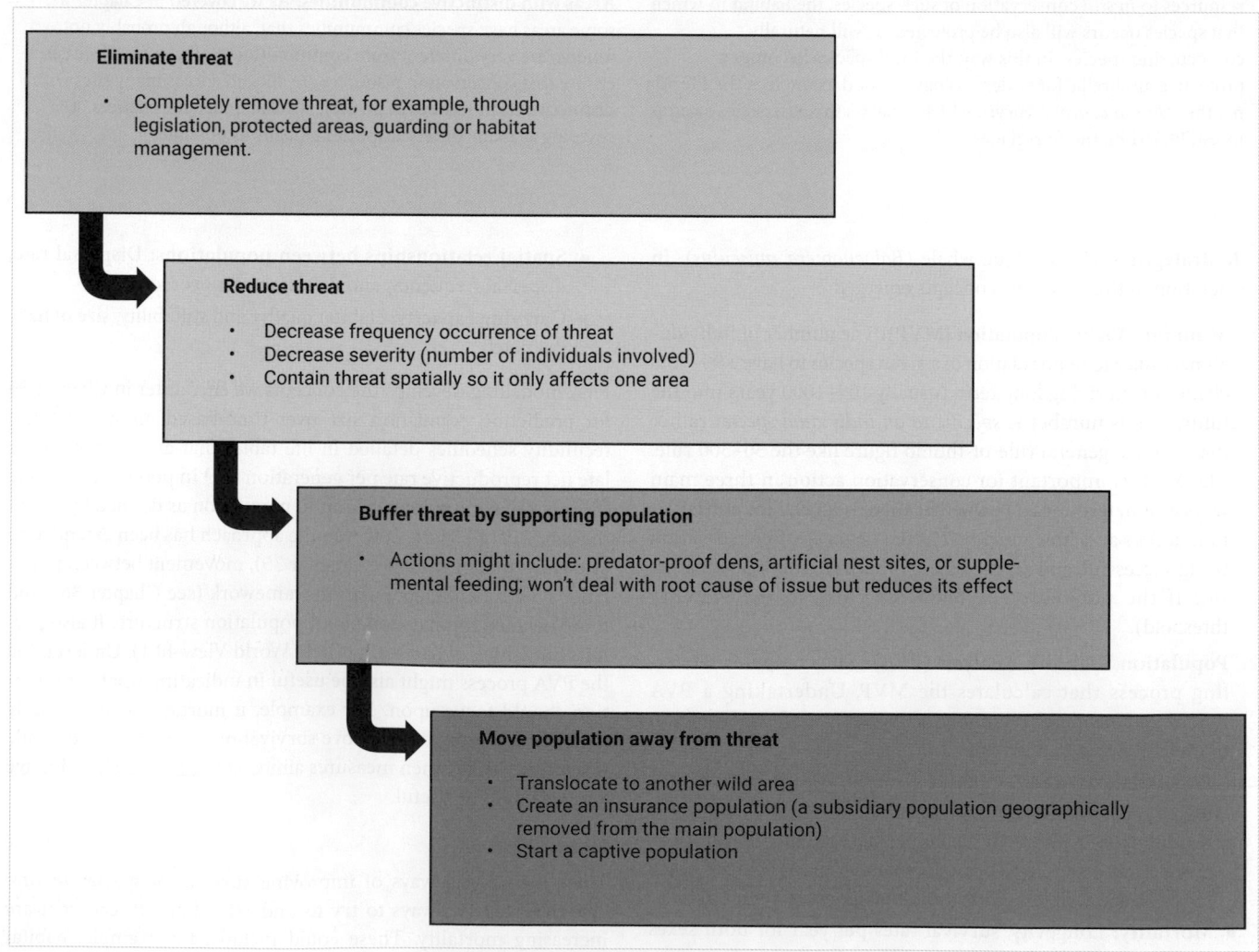

Figure 40.10 Species-focused conservation. Hierarchy of species-focused conservation action.

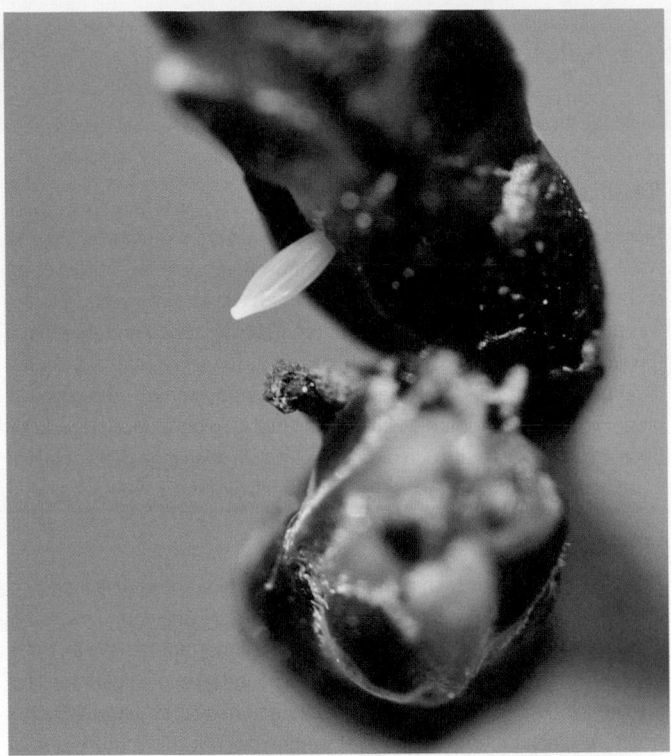

Figure 40.11 A single egg of a brimstone butterfly (*Gonepteryx rhamni*) laid on the larval host plant buckthorn (*Rhamnus cathartica*).

also located optimally within a site in relation to habitat, disturbance, food, and water sources. Such issues are discussed for placement of nestboxes for a migratory songbird, the pied flycatcher (*Ficedula hypoleuca*), in Scientific Process 40.1.

Culling

It might initially seem counterintuitive to consider culling in a chapter on conservation, but it can be an important tool in the conservation toolbox. Culling can be undertaken for a variety of reasons. Conceptually, the most straightforward reason is culling a species that is having a negative impact on a conservation target species. This is most likely to occur in cases where an invasive species has been introduced to an area outside its native range and is then preying on, or outcompeting, a rare native species. A common example is the introduction to the UK of grey squirrels (*Sciurus carolinensis*) from America, which has had a devastating effect on native red squirrels (*Sciurus vulgaris*) because of out-competition and vectoring parapox virus, or non-native ruddy ducks (*Oxyura jamaicensis*) hybridizing with the rare and declining white-headed duck (*Oxyura leucocephala*) in Spain.

▶ You can read more about this topic in Chapter 39.

More rarely, there can be instances of culling one native species to reduce its effects on another (more endangered) native species. This approach has been taken in eastern Canada where native gulls (*Larus* spp.) have been culled to increase fledging success of common terns (*Sterna hirundo*) (Figure 40.12).

On occasion, it can be necessary to cull individuals of a species that is actually the target of conservation action. This is most likely

Figure 40.12 The culling of one native species can be done to protect another, more endangered one. Tern chicks are very vulnerable to predation by gulls and so, at some key tern colonies, gulls are being controlled to reduce effects on tern populations.

to occur to deal with overpopulation in a particular area when failure to reduce population density would cause widespread morbidity or mortality.

▶ You can read more about population density in Chapter 36.

Such action would usually only be undertaken if other methods of reducing population—primarily translocation of excess animals to another location—are not viable. This could occur for large animals that are expensive or impractical to translocate (or suffer high mortality or morbidity from translocation), or where it would be difficult to move the 'right' animals to ensure a negligible effect on dominance hierarchies and family group structures.

Moving from single-species to multi-species approaches

Historically, many conservation schemes have been based around a single focal species. **Single-species conservation** is, conceptually at least, fairly straightforward as there is one clear goal and often a reasonably straightforward set of actions to reach that goal. However, over the past couple of decades, species-oriented conservation has increasingly moved towards a **multi-species conservation** approach, especially within *in situ* contexts.

Multi-species conservation is often more realistic ecologically as it encapsulates concepts of species interactions and also reflects the fact that many sites will support several different species needing conservation action. It does, however, pose challenges when different species have different management requirements, especially when these are opposing. Other difficulties can arise when species' interactions have negative consequences for one of the species involved. This second issue occurs for conservation biologists who manage heathland nature reserves in Dorset for the UK's two rarest reptile species: smooth snakes (*Coronella austriaca*) (Figure 40.13a) and sand lizards (*Lacerta agilis*) (Figure 40.13b). These species are both conservation priorities, but there is a cyclical predator–prey relationship (see Chapter 37) whereby smooth snakes prey upon sand lizards. This means that although multi-species approaches are essential, they are also problematic.

SCIENTIFIC PROCESS 40.1 | Quantifying pied flycatcher habitat to optimize conservation interventions

Research question

Habitat is a key determinant of breeding success in many species. Identifying the factors that influence success is important to inform conservation strategies, especially for species that are declining rapidly such as the pied flycatcher (*Ficedula hypoleuca*), which is a small migratory songbird shown in Figure 1. A UK study in 2014 was the first to provide a detailed and quantitative analysis of the habitat–productivity relationships for this species.

Hypothesis and predictions

Two main hypotheses were made based on previous, non-scientific observations of the species' needs and ad hoc data. Firstly, it was predicted that vegetation structure and species assemblage would influence bird breeding success, with areas dominated by oak (*Quercus*) being predicted to be optimal. Secondly, it was predicted that the location of nestboxes relative to landscape features would be a key determinant of breeding success, with an anticipated negative relationship between success and distance to footpaths due to disturbance from people.

Methods

The study was undertaken at a nationally important breeding site with the UK's longest-running nestbox monitoring scheme with 137 nests studied over a 5-year period. Each nest was mapped in relation to the nearest road, footpath, and water source, and 22 habitat variables were measured to describe microhabitat (tree supporting the nestbox, nestbox height, nestbox orientation) and mesohabitat (vegetation data in a circular plot centred on the nestbox). Breeding data were collected by a weekly inspection of all nestboxes between April and July each year. This allowed recording of the number of eggs laid, the number of young to hatch, and the number of young to fledge.

Results

More successful nests tended to be located away from footpaths, close to water, and in areas with a few large trees (rather than numerous smaller ones) and abundant saplings. Sapling species richness was also important. Success was positively related to abundance of oak (*Quercus robur*) and silver birch (*Betula pendula*), and negatively related to beech (*Fagus sylvatica*), sycamore (*Acer pseudoplatanus*), and bracken (*Pteridium aquilinum*). Success was lower in boxes facing south-southwest and higher in boxes located on taller (more mature) trees. Although pied flycatchers are often regarded as birds of grazed (open) woodland, success was not related directly to grazing regime.

Conclusion

The finding of this study challenged, to some extent, the widely held (but not previously tested) view that pied flycatchers are only found in grazed woodland. Instead, the work showed that it is important to have an open habitat, with a lot of 'free air' between the ground layer of vegetation and the canopy (Figure 2 shows a typical favoured habitat structure). Such a structure can sometimes be created via grazing, but this is not essential and actually, in some cases, might be detrimental as sapling species richness is important, probably because of the role of saplings in supporting a rich caterpillar food resource for the adult birds to feed their chicks.

Extension question

Find a scientific peer-reviewed paper of your own choice that has researched species–habitat interactions for a focal species and summarize how conservation for that species should be informed by the research findings

Read the original paper

Goodenough, A. E. (2014). Effects of habitat on breeding success in a declining migrant songbird: the case of Pied Flycatcher *Ficedula hypoleuca. Acta Ornithol.* **49**: 157–173.

Figure 1 A pied flycatcher (*Ficedula hypoleuca*).

Source: Anton MirMar/Shutterstock.

Figure 2 Nagshead RSPB reserve in the Forest of Dean provides excellent habitat for pied flycatchers to breed successfully.

PAUSE AND THINK

What approaches could you use to conserve smooth snakes and sand lizards? Why would it be important to understand predator–prey cycles (refer back to Chapter 37 if you need a reminder of this) when monitoring temporal trends in species populations to assess effectiveness of management?

Answer: Ideally, a large enough area would be protected to ensure that the population sizes of both species are large enough for natural predator–prey cycles to occur without threatening either species. It would be important to understand the predator–prey cycle because this usually results in peaks and troughs in populations and it would be important that these natural changes are not mistaken for long-term changes that demand conservation action. Zonation approaches can be helpful to allow some lizard populations to be managed in snake-free environments.

Figure 40.13 Conservation priorities: predator and prey. Predator and prey—and both conservation priorities: (a) smooth snake (*Coronella austriaca*); and (b) sand lizard (*Lacerta agilis*).

Source: (a) Juanan Alonso/Shutterstock.com; (b) Dmitry Fch/Shutterstock.com.

Habitat-focused conservation

Species-focused conservation often necessitates management of habitats so that they are suitable for specific target species. Habitat conservation for its own sake rather than for the species it supports is usually reserved for small habitats that are critically endangered, or that perform a useful **ecosystem service** (Chapter 38). Examples could include flower-rich meadows that support numerous pollinators or sand dune ecosystems which, in addition to their unique ecological characteristics, play a key part in coastal defences.

Site-based conservation

At a site level, there is much that can be done to manage habitats, either to conserve those habitats in their own right, or to conserve the species that depend on them. Depending on the site and the aims of the conservation work, site managers can work to create new habitat, improve habitat, or maintain habitat.

Creating new habitat

Habitats can be created at a range of spatial scales, ranging from microscale (such as a deadwood pile for specialist invertebrates, bryophytes, and fungi) to mesoscale (such as ponds) to macroscale (such as plantation woodlands). Often a site-specific management plan calls for a particular type of habitat to be created for a specific species or a group of ecologically similar species. For example, if a marginal wetland site is being managed so that it attracts more

wading bird species, installing **scrapes** (shallow depressions in the ground where water can accumulate up to the depth of a few centimetres to provide foraging habitats) might be an appropriate habitat creation strategy. Figure 40.14 shows a cross-sectional view of a scrape that would be used for wading birds. The scrape itself is on the left of the figure, where you can see the deeper depression in the ground for the water to accumulate.

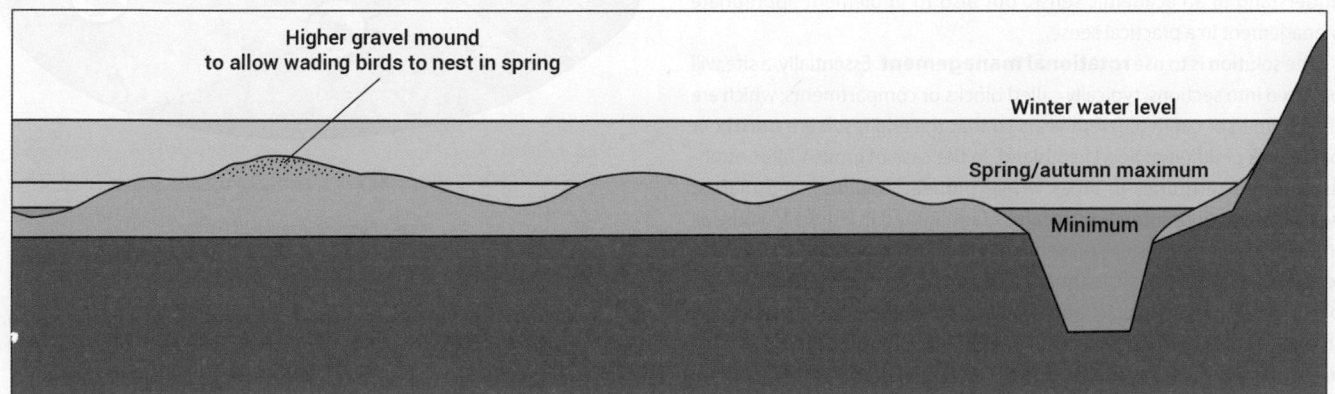

Figure 40.14 An example of habitat creation. A cross-sectional view of a scrape for wading birds.

In other instances, our goal might simply be to increase the richness and diversity of habitats. Generally, the more habitats there are in an area, the more **ecological niches** are created, and the greater the diversity of species the site can support

 We learn more about ecological niche concepts in Chapter 37.

Thus, having matrix of different habitats is often beneficial. In the real world, however, there is often a trade-off as each habitat patch must be capable of supporting the Minimum Viable Population (MVP) or higher of focal species. In other words, while having more habitat patches generally means more species, each habitat patch must be large enough so that its carrying capacity (the number of individuals of a given species that it can support) is not below the MVP. Although we often think of habitat matrices as involving very different habitats (say woodland, grassland, and heathland), in reality, differences might be more subtle, for instance involving different ages of the same habitat. An example for heathland is given in Real World View 40.2.

Improving habitat

Improvement to habitat generally tends to be specific to the needs of a particular species as 'improvement' is often highly species specific. For example, coppicing trees (cutting back tree growth to ground level to encourage regrowth in multiple stems, as shown in Figure 40.15, rather than a single trunk) is an excellent way to improve habitat for hazel dormouse (*Muscardinus avellanarius*) because this species needs structural complexity so it can move arboreally. This same intervention will have negative consequences for other species that need, for example, a more open area between the canopy layer and the ground layer (e.g. pied flycatcher *Ficedula hypoleuca*, covered in Scientific Process 40.1). It is vital, therefore, that the aims of a site, and what species are prioritized, are considered before management begins, as these decisions will drive management in different directions.

REAL WORLD VIEW 40.2 | **Rotational heathland management**

Heathland is often maintained through periodic management of vegetation by burning or cutting. If this is not undertaken, heathland plants such as heather (Ericaceae spp.) are replaced by gorse (*Ulex* spp.) and small trees and, eventually, the heathland goes through the process of ecological succession to become woodland or dense scrub.

The decision of whether to cut or to burn generally depends on logistics such as site access and manpower. However, whichever method is chosen, it is not possible to manage the whole site at once, as heathland-dependent organisms will have no habitat left. In addition, some species, such as red grouse (*Lagopus lagopus scotica*), need heathland in a range of different successional states within close proximity. Young heather (pioneer phase, <8 years) is a key food source. Berry-producing plants (such as bilberry *Vaccinium myrtillus*) and insects in mid-successional heather (building phase, 8–20 years) are vital during breeding. Grouse nests occur in mature stages (>20 years) where cover is greatest. As many heathland sites are still managed for grouse, either for commercial shoots or conservation reasons, such habitat requirements are vital to not only understand in an academic sense, but also to implement appropriate management in a practical sense.

The solution is to use **rotational management**. Essentially, a site will be dived into sections, typically called blocks or compartments, which are then burned or cut in different years so that the end result is a **matrix** or patchwork of different aged heathland. In the case of grouse, birds establish nesting territories in areas where pioneer, building, and mature heathland sections intersect (the areas highlighted in Figure 1, where all three heathland types are present together) to fulfil both nesting and feeding requirements. It is thus important to ensure that the habitat blocks or compartments are appropriately sized and sited. If patches are smaller, there will be more three-way intersections (and potentially a higher carrying capacity for the site and more birds), but equally there is a limit to how small it is possible to go in terms of the logistics of cutting/burning and to ensure that habitat patches are large enough to contain the necessary food resources and provide sufficient shelter for the birds. In terms of placement, siting intersections away from footpaths is useful to maximize the number of potentially usable habitat intersections.

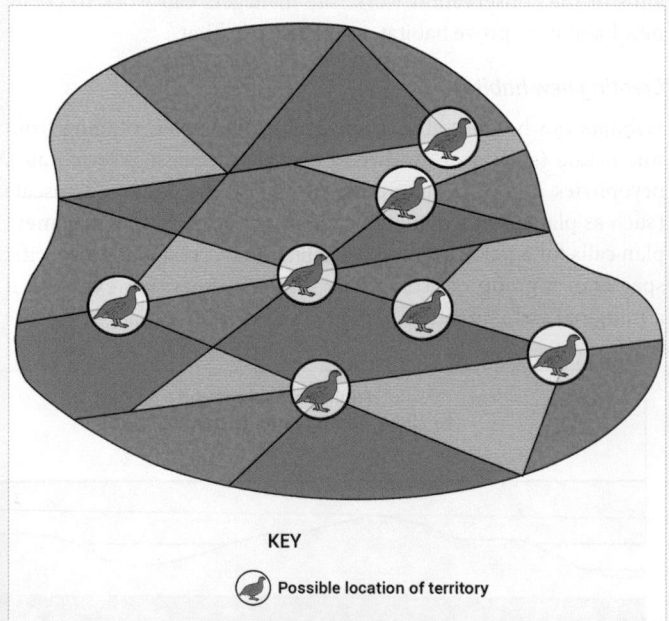

KEY

🐦 **Possible location of territory**

Figure 1 A matrix of heathland created by rotational cutting or burning. The intersections between the patches are potential territory locations, some of which are indicated.

Figure 40.15 Coppice management used to be undertaken to supply wood for fencing. Management is now undertaken to conserve coppice-adapted species.

Source: T.P. Holland/Flickr/CC BY 2.0.

Maintaining existing habitat

The mainstay of site-based conservation is the maintenance of existing habitat. This is often considered 'business as usual' within site-specific management plans, but this should not be taken to suggest that it is unimportant—far from it. Once habitats have been created or improved as necessary, they need regular management to maintain their effectiveness. This is especially important for transient habitats such as ponds or grassland which, left unmanaged, will undergo the natural process of **ecological succession** and become scrub and often, ultimately, woodland (see Chapter 37 for more on this topic). Methods for **arresting succession** depend on the type of habitat involved, as well as logistical and financial constraints, but always involve creating disturbance. In the terrestrial environment, such disturbance is usually grazing, cutting, or burning vegetation (as discussed in Real World View 40.1). In the aquatic environment, arresting succession might involve clearing vegetation at the edges of a wetland to prevent marginal creep.

Landscape-based conservation

Although protected areas and nature reserves are often viewed as key sites, and a lot of conservation management is focused on them, they do have limitations. Firstly, even if all such sites are fully protected and have optimal management (which is rarely the case), the total area protected will still only be a very small percentage of overall landmass. Indeed, other than New Caledonia, a French territory comprising dozens of islands in the South Pacific (Figure 40.16), and Slovenia in Europe, no country had >50% of its landmass listed as protected in the 2012 census by the World Bank. Secondly, and

Figure 40.16 Landscape-based conservation. New Caledonia in the Pacific is the most highly protected area for wildlife by land area.

Source: Thomas Cuelho/Flickr/CC BY 2.0.

even more importantly, because of the way that protected areas are distributed, the land that is protected is always split into **virtual islands**. This reflects the fact that reserves and protected areas are basically isolated entities in a 'sea' of other land uses. This can, for some species, prevent individuals from moving between reserves and protected areas, which can have several impacts:

- It reduces gene flow, which, in severe cases, can result in **inbreeding.**

- It increases the risk of **chance extinctions** in the different areas because the smaller populations are more at risk of stochastic processes. **Stochastic processes** are basically chance events that affect species' populations. These processes can stem from natural fluctuations in productivity and mortality (demographic stochasticity) or fluctuations in the abiotic and biotic environment (environmental stochasticity).

- It decreases natural dispersal and thus the chances of natural recolonization if a species is lost from a site.

There are other key limitations that affect protected areas and nature reserves, all of which are linked to **off-site pressures**. As discussed in Chapter 39, climate change is shifting species' distributions, often pushing whole ranges towards higher latitudes (i.e. north in the northern hemisphere). This could have a notable effect on protected areas if species' ranges move so that that target species no longer occur in the sites that are managed specifically for them. Other off-site pressures include fire, flooding and drought, pollution, and the spread of disease.

The impact that such off-site pressures can have is exemplified in one of the most iconic American National Parks, Yellowstone, being placed on the World Heritage Site in Danger List in 1995. This was partly due to contamination upstream having diffuse effects within the Park and partly due to Yellowstone bison being infected with brucellosis, which could spread this to neighbouring farmland. This second is a rare example of a problem emanating *from* a protected area causing management to be invoked *within* that protected area to the detriment of species protected within it, in this case through culling. Happily, in the case of Yellowstone, both issues were resolved and the site was then removed from the at-risk register, but this case study highlights the vulnerability of all individual sites, regardless of whether they are protected.

Connectivity

One way that we can decrease the vulnerability of individual sites is to increase **connectivity** between them. We thought about the vital role of connectivity in Chapter 39 in relation to habitat fragmentation. Improving connectivity can be done in three main ways as listed below in desirability order and shown graphically in Figure 40.17.

1. **Linear corridors:** Constructing specific linear habitats such as hedgerows to assist species movement.

2. **Landscape corridors:** Ensuring that similar areas of land are managed so there are 'pathways' across the landscape—such as a contiguous line of fields between two grassland nature reserves.

3. **Stepping stone islands:** Forming small patches of land that species can use in a series of 'jumps' between two larger nature reserves.

Improving connectivity basically allows species to 'join the dots' more easily. This can be essential for movement of individuals, and thus dispersal and gene flow because a metapopulation forms rather than separate populations being entirely isolated (also discussed in Chapter 36). Landscapes with high connectivity have more resilient ecological networks, especially given the need for species movement in response to climate change (see Chapter 39). It is also possible to improve the **permeability** of landscape features such as roads, which can otherwise become barriers for species, by providing green land bridges and underpasses to help animals cross from one side to the other.

In addition to improving connectivity between core areas such as nature reserves, other wider-countryside conservation initiatives include ecoregion (whole-landscape) management and agri-environment schemes.

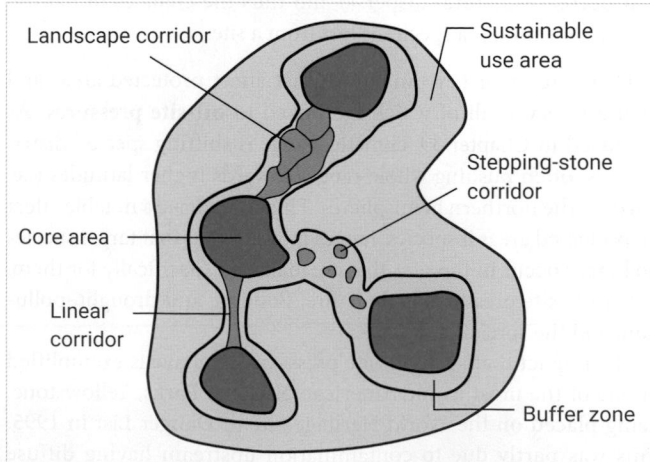

Figure 40.17 Methods of improving conductivity in the landscape.

Source: Goodenough and Hart (2017), *Applied Ecology: Monitoring, managing, and conserving*. Oxford: Oxford University Press. Reproduced with permission of the Licensor through PLSClear.

Whole-landscape management

Although human activity tends to occur in administrative units such as countries and counties, ecosystems, habitats, and species do not follow the same spatial arrangement. Accordingly, there is an increasing drive to manage areas as **ecoregions**, which are geographically distinct areas with similar environmental conditions and common species communities. Although some ecoregions can be extremely large (for example the 2000-mile-long Yellowstone to Yukon ecoregion in the American and Canadian Rockies), the same basic principle can be applied on a much smaller scale for a group of woodlands, for example, or a single floodplain region.

Agri-environment schemes

Agri-environments schemes generally involve farmers undertaking initiatives such as planting hedgerows, creating margins at the edge of fields that are not ploughed or harvested or adding ponds. Other actions can involve altering the timing of activities, for example, harvesting crops later than would be ideal to allow wild flowers to set seed or ground-nesting birds to complete their breeding cycle. Farmers are paid a subsidy to fund work and/or compensate for loss of earnings.

Ex situ conservation

In many ways, *ex situ* conservation is very different to *in situ* approaches. However, although the environment is very different, many of the actual processes and underlying principles involved are very similar. This is exemplified in Real World View 40.3, which examines *ex situ* conservation in ring-tailed lemurs (*Lemur catta*).

 Check your understanding of the concepts covered in this section by answering the questions in the e-book.

40.5 Restoration

Restoration is the process by which a site is managed to recreate a previous, better condition. This can involve removing contamination to resolve pollution issues (**remediation**), restoring previous habitats and landscapes, or reintroducing species.

Remediation

There are many different ways that we can remove pollutants from the environment. Some of these are environmental techniques, such as physical removal or **chemical stabilization**, which is when polluting compounds are converted to a form that is less soluble and therefore less likely to harm organisms. However, many approaches make use of species' biochemical pathways or ecological processes in specific processes that all come under the heading of **bioremediation**.

Ring-tailed lemurs (*Lemur catta*), like all lemur species, are native to Madagascar, off the east coast of southern Africa. The species had a large range, but is now restricted to fragmented and isolated patches where population density is very low. The species is currently listed as Endangered, the second most threatened category, on the IUCN red list (see Chapter 39 for more on this topic). This assessment of extinction risk reflects a decrease in population size of more than 50% in the last 36 years (three generations). The main threats in the wild are hunting and habitat loss, both of which are difficult to control or buffer. Because of this, and in addition to *in situ* conservation management (for example in six national parks: Andohahela, Andringitra, Isalo, Tsimanampetsotsa, Zombltse, and Vohibasia), there is now a sizable captive population of this species spread through numerous collections in several countries. In 2009, this captive population was estimated to be 2500 individuals. This is sizable in comparison to many other species represented in zoos, and stems partly because the species is relatively small and easy to keep in captivity, and partly because of its public appeal.

Within zoological collections, lemurs are normally housed in troops (social groups) often in mixed-species enclosures with other lemur species. They generally need access to both an indoor area (vital at night in most parts of the world) and an outdoor area for use in the day. Lemurs are used to hot conditions and are susceptible to hypothermia in colder environments; heat lamps are often used to mitigate this problem. As an intelligent primate, lemurs need stimulation, and this is often provided through enrichment, for example, by hiding food or providing swings, rope walks, and aerial platforms. They also need 3D enclosures that enable and facilitate climbing to reflect their natural environment. Increasingly, lemurs are being housed in walk-through enclosures to bring them ever closer to the public. This can provide excellent visitor experiences, but care needs to be taken to ensure that visitors do not become a cause of stress or induce stereotypical or non-desirable behaviours such as overgrooming or reduce resting and feeding.

Lemurs breed successfully in captivity (Figure 1), but it is vital to ensure that genetic diversity is maintained and inbreeding is avoided. In many ways, having multiple captive populations of lemurs is equivalent to having multiple isolated wild populations—virtual island effects are likely

to arise without effective management. In the case of captive breeding, this is managed through a **studbook**—the life history and thus relatedness of each captive lemur as far as is known (captive-bred animals) or as far as can be ascertained (wild-caught animals). This means that, at a global level, animals can be moved between collections to ensure that no one genetic lineage becomes over-represented. Because lemurs are relatively small and easy to transport, movement of 'breeding stock' usually involves moving individuals, but in some larger animals, moving sperm and using artificial insemination can be used as an alternative.

Extension question

Mixed-species enclosures are common for lemurs. What additional factors will need to be considered for enclosure design and husbandry for a mixed-species enclosure rather than a single-species enclosure?

Further reading

IUCN Lemur factsheet: http://www.iucnredlist.org/details/11496/0

Figure 1 Ring-tailed lemurs breed readily in captivity providing that enclosure design and diet are suitable.

Bioremediation: metabolic breakdown

Probably the most common form of bioremediation uses micro-organisms (usually bacteria but occasionally fungi) to metabolize (biochemically breakdown) pollutants. In some instances, this process is undertaken by micro-organisms already in a specific area, but sometimes micro-organism growth is **biostimulated** to increase numbers. Such biostimulation can involve adding oxygen by adding a fountain to still water or a weir to moving water to increase mechanical mixing, raising temperatures, increasing nutrient levels, or optimizing pH to create the perfect conditions for the target micro-organism species. In other situations, **bioaugmentation** is used, whereby the necessary micro-organisms are cultured in a laboratory and added to the environment artificially.

PAUSE AND THINK

Under what circumstances would micro-organism bioremediation be particularly challenging? If you need a hint, refer back to the *Exxon Valdez* oil spill in Real World View 39.1

Answer: Micro-organism bioremediation is challenging when environmental conditions are harsh, as that reduces the number of micro-organism species that are capable of metabolizing the relevant pollutant and that are adapted to survive in that environment. Remediating areas affected by marine oil spills is particularly difficult in cold waters, such as those around Alaska where the *Exxon Valdez* oil spill occurred, for example.

Bioremediation: bioaccumulation

Bioaccumulation involves using species, particularly plants, to accumulate pollutant molecules and store large quantities in their tissues. When plants are the bioaccumulating agent, the process is referred to as **phytoaccumulation** or **phytoextraction**. Plants rooted in contaminated soil take up pollutants that are then dispersed through the roots, leaves, and stems. Where root mass is large, plants can become hyperaccumulators and pollution uptake can be considerable. Pollution problems that can be resolved in this way include soil pollution by metals (including zinc, cadmium, cobalt, nickel, lead, arsenic, and copper), explosives (including TNT), and radioactive chemical elements such as uranium.

Bioremediation: rhizofiltration

Plants can also be useful species for remediation of aquatic systems, especially via the process of **rhizofiltration**, which is illustrated in Figure 40.18. Rhizofiltration can only occur in flowing water. It involves the rhizomes (root masses, often in the rhizosphere immediately above the sediment) filtering out pollutants, often through the action of micro-organisms associated with these structures. This can be useful both for heavy metals and organic pollutants, including sewage. Humans can harness plant power by creating specific phytoremediation systems, for example reedbed systems typically using *Phragmites* spp. to clean up polluted or waste water, including output from houses.

Habitat restoration

Habitat restoration can be undertaken at a small scale to recreate previous (better) ecological conditions. This might be necessary when the habitat is a transitional one that has not been maintained due to lack of appropriate management (e.g. undergrazed grassland that has become scrub encroached) or to reverse long-term degradation. Habitat restoration can also be undertaken at a landscape scale, for example to recreate wilderness areas through rewilding.

Rewilding

Rewilding is a landscape-level approach. It seeks to establish core areas of habitat that can be linked by corridors wherever possible to form a 'wilderness network' and thus differs from many conventional *in situ* conservation projects, which often deal with specific sites of small scale and in local contexts. Central to the rewilding approach is the restoration of the 'wild' coupled with an absence or extreme reduction of future management. Just as with reintroduction (see below), the 're' is crucial: the approach is not just about

letting nature run its course but rather about restoring 'the wilderness' that existed previously at that location.

Rewilding is straightforward in theory, but defining what the initial condition actually *was* can be problematic. This means that, even conceptually, rewilding is often more of an ideal than a reality that can be recreated. There are also issues around the amount of land needed for the approach to work, especially as the land in question should have little, if any, habitation, agriculture, or forestry activity. Not all rewilding is actually feasible because the original ecosystem processes that were important previously in creating and maintaining the wild landscape are difficult or impossible to recreate. An example is that grazing of coastal floodplains would historically have been undertaken by aurochs (*Bos primigenius*), a large wild ox. This species is now extinct so any rewilding attempts that depend on aurochs-like grazing would need to use a surrogate species that fulfils a similar ecological role—heck cattle would be a possibility in this case.

Species reintroduction

As we discuss in Chapter 39, extirpation is the process by which a species becomes extinct from a particular site, county, or country but still exists globally and so is not actually extinct. A good example is the Eurasian lynx (*Lynx lynx*), which is extirpated in the UK but still occurs in mainlad Europe (and, indeed, in captivity as part of *ex situ* conservation efforts).

Reintroduction involves humans releasing a species into a site in its *former native range* from which it has become extirpated. Any species that is the subject of a *reintroduction* must thus have previously existed in the area into which it is being released, even if this was a long time ago. The aims of all reintroduction programmes should be to create a new population that:

- exceeds the Minimum Viable Population
- is free-ranging
- is self-sustaining with little or no ongoing management.

There are several reasons why we might decide to reintroduce a species. Some reintroductions—the majority in fact—are motivated by species considerations. These might involve aiding the conservation of the species by creating a wild population to avoid the species occurring only in captivity (e.g. the reintroduction of scimitar-horned oryx (*Oryx dammah*)) or to increase the overall global population of a species, especially when space has become a limiting factor in *ex situ* environments (e.g. Przewalski's horse (*Equus przewalskii*) reintroduced to Mongolia).

Figure 40.18 Rhizofiltration. Typical reedbed setup to filter wastewater using plants such as *Phragmites*.

Source: A. Goodenough, *Applied Ecology: Monitoring, managing, and conserving*. Oxford: Oxford University Press. Reproduced with permission of the Licensor through PLSClear.

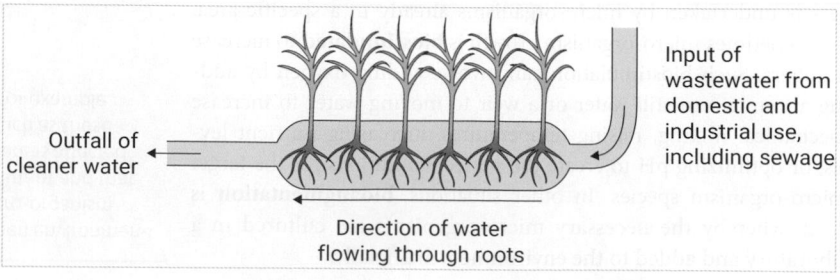

Outfall of cleaner water

Input of wastewater from domestic and industrial use, including sewage

Direction of water flowing through roots

Alternatively, reintroductions can be motivated by the role that a species has in an ecosystem. For example, the Eurasian beaver (*Castor fiber*), is an **ecosystem engineer** (see Chapter 38), which alters riparian habitat to create shallow pools used by a range of aquatic species, and has been reintroduced in France, Germany, Spain, Belgium, and the UK. **Keystone species** (species that are vital to ecosystem functioning) are also common reintroduction targets.

▶ **For more on keystone species see Chapter 38.**

For example, red-rumped agoutis (*Dasyprocta leporina*) have a pivotal role in an ecosystem by virtue of their seed dispersal function and were reintroduced to Tijuca National Park, Brazil, for this reason.

Reintroducing keystone species or ecosystem engineers can act to rebalance the ecosystem as the ecological functions originally undertaken by these species recommence. A classic example is the sea otter (*Enhydra lutris*) (Figure 40.19), which became extirpated in British Columbia, Canada. Their absence led to a population explosion in their main prey, sea urchins (Echinoidea), which decreased the abundance of kelp and had cascade effects on fish biodiversity. Once sea otters were reintroduced to British Columbia this rebalanced the ecosystem.

Figure 40.19 The sea otter is a keystone species. The sea otter (*Enhydra lutris*) is a keystone species as a result of its predator–prey interactions. This meant the species was a prime target for reintroduction following extirpation in Canada. As with any reintroduction scheme, release will only happen once the original causes of decline—hunting for the fur trade in this case—have been resolved.
Source: rbrown10/Shutterstock.

Feasibility study
- Draw up a species profile – is enough information known to attempt reintroduction?
- Assess original threats and drivers of extirpation – what is known?
- Complete a generic risk assessment – what existing species/habitats could be harmed?
- Is there general political, financial and public support for the reintroduction?

Locate release stock
- Decide if wild-caught individuals will be translocated or if captive animals will be used.
- Identify actual wild/captive population.
- Ensure individuals are genetically close to the extirpated population as possible, this will be based on the best available evidence including analysis of museum specimens.

Decide on release site
- Identify site(s) for release and assess suitability.
- Perform any pre-release management work or elimination of threats.
- Educate and engage with local stakeholders and general public.

Actual and post reintroduction
- Capture individuals (if wild-caught stock being used) – permits might be necessary.
- Translocate individuals – permits and/or quarantine might be necessary.
- Release individuals.
- Add in any habituation management (e.g. supplemental food during establishment).
- Monitor new population and any impacts on the other species or the environment.

Figure 40.20 Reintroduction scheme. The main stages in species reintroduction schemes.

 Check your understanding of the concepts covered in this section by answering the questions in the e-book.

The process of reintroduction

There is a well-defined set of codes of practice for reintroduction. The International Union for the Conservation of Nature (IUCN) Species Survival Commission (SSC) set up a Reintroduction Specialist Group (RSG) in 1988. This group created a set of comprehensive guidelines on reintroductions covering every step of the process from initial idea, through harvesting and release, to post-release monitoring and management. The Association of Zoos and Aquaria has also devised guidelines focusing specifically on reintroductions that use captive-born individuals or those that have been in captivity between capture and release. The basic process for species reintroduction is shown in Figure 40.20.

SUMMARY OF KEY CONCEPTS

- We can typically divide the solutions that we use to address the threats to ecosystems discussed in Chapter 39 into one of four groups: protection, management of problems, conservation, and restoration.

- **Custodial measures** to protect species and specific sites through legislation and policy structures is vital.

- Breaking wildlife law or ordering it to be broken is a **punishable offence**. There is a **licence system** in place to allow people to undertake activities that would otherwise be illegal, but this is tightly controlled and regulated.

- **Active protection** can be important in some cases, for example to combat poaching.

- Management of ecosystems is not solely about conservation—sometimes we need to **manage problems**.

- There are three main ways to manage pests—**physical removal**, **chemical treatment**, or **biological control**. Containment can also be important.

- Just as energy flows through a food chain, so too do some chemical compounds that do not degrade. This is the process of **bioaccumulation,** which can have toxic effects of organisms in higher trophic levels.

- Conservation can occur in the wild (**in situ**) or in captivity (**ex situ**); both have advantages and disadvantages.

- Conservation can be **habitat-focused** but is more usually **species-focused** (although that often involves some form of habitat management for a species' needs).

- We are gradually moving from **single-species conservation** to **multi-species conservation** approaches.

- We are only likely to be able to successfully conserve a species population when we understand its biology and threats; a **population viability analysis** can be extremely helpful.

- Although **protected areas** and **nature reserves** are important, connecting these at a landscape scale is vital.

- Other **landscape-scale conservation** approaches, such as working with farmers, are important.

- Species can assist in the **remediation** of polluted land and water, especially micro-organisms and plants.

- In addition to remediation, **recreating habitats, rewilding landscapes,** and **reintroducing species** are key aspects of restoration ecology.

 Use the flashcards in the e-book to test your recall of key terms introduced in this chapter.

QUESTIONS

 Looking for answers? Once you've answered these questions, follow the link in the e-book to the answer guidance and check your work.

Concepts and definitions

1. Explain the differences between legislation and policy in an ecological management/conservation context.

2. How do 'total allowable catch' and 'individual fishing quotas' differ? Which do you think is preferable?

3. How would you define an ecological pest?

4. What is meant by a 'landscape-scale' conservation project?

5. Explain the difference between species introduction and species *re*introduction.

Apply the concepts

6. Choose one species on Schedule 5 (protected animal) or Schedule 8 (protected plant) of the United Kingdom's Wildlife and Countryside Act 1981 and research why that species has been listed and what activities the listing would preclude without a licence.

7. Find an example of where chemical pest control has gone wrong, either because non-target species have been affected or because the target species has become resistant to the chemical. Could these impacts have been predicted beforehand? What lessons can we learn to apply in the future?

8. For non-radioactive compounds that bioaccumulate, species in higher trophic levels are more at risk from toxic chemicals than species in lower trophic levels. However, for radioactive elements position in the food chain is arguably less important, at least for radioactive elements that decay rapidly. Why?

9. We noted that preventing the translocation of non-native species to new areas is usually a much more effective form of management than trying to eradicate, remove, or buffer issues after species establishment. For a country of your choice, research their biosecurity policies and practices and evaluate their effectiveness.

10. Why are keystone species and ecosystem engineers more often reintroduced into ecosystems than other species? Find an example of a reintroduction, anywhere in the world, involving either a keystone species or ecosystem engineer not covered in this chapter and explain the specific reasons for its reintroduction.

Beyond the concepts

11. One of the key tasks of wildlife legislation is to prevent negative effects of development on ecology. Why is it important that the necessary surveys and mitigation are not undertaken by developers themselves?

12. We touched briefly in this chapter on preservation rather than conservation. Given that preservation is about maintaining an area in its current state and devoid of human influence, do you think that it is possible for this concept to be applied in the 21st century?

13. Supposing that you had the power to 'save' one species from extinction just by naming it, what species would you choose and why?

14. Given that reintroductions tend to cost substantially more money than conserving species before they become extirpated, some people might argue that they should not be undertaken and the money used to help multiple other species instead. What do you think and why?

15. One of the often-cited examples of successful rewilding is Oostvaardersplassen in the Netherlands. Research what has been done at this site and decide whether, in your opinion, it meets the criteria for true 'rewilding'.

FURTHER READING

Goodenough, A.E. & Hart, A.G. (2017) *Applied Ecology: Monitoring, Managing and Conserving.* Oxford: Oxford University Press.
Comprehensive overview of managing and conserving ecology

Zavaleta, E.S., Hobbs, R.J., & Mooney, H.A. (2001) Viewing invasive species removal in a whole-ecosystem context. *Trends Ecol. Evol.* **16**: 454–459.
Non-native species management

Wallington, T.J., Hobbs, R.J., & Moore, S.A. (2005) Implications of current ecological thinking for biodiversity conservation: A review of the salient issues. *Ecol. Soc.* **10**: 16.
Review of key principles and challenges in conservation

Mace, G., Possingham, H., & Leader-Williams, N. (2007) Prioritising choices in conservation. In: *Key Topics in Conservation Biology*, ed. D.W. MacDonald and K. Service, Oxford: Blackwell, pp. 17–34.
Overview of conservation priorities and decision making

Seddon, P.J., Griffiths, C.J., Soorae, P.S., & Armstrong, D.P. (2014) Reversing defaunation: restoring species in a changing world. *Science* **345**: 406–412.
Review of reintroductions

Index

Note: tables and figures are indicated by an italic *t* and *f* following the page number. Greek characters are filed under their English spellings, e.g. 'β' is filed under 'beta'.

Numbers

5-hydroxytryptamine (5-HT, serotonin) 729, 815, 820
50–500 rule 1346
95 per cent confidence intervals 97

A

A-bands 688*f*, 689*f*
abacavir 570
abiotic factors 1292, 1294*t*
  limiting factors 1299
  tolerance ranges 1297–8
ABO blood group system 239
absolute abundance 1254
absolute fitness 1232
absolute refractory period 602
absolute zero 186
abundance 1253–4
accommodation (eye) 633, 664
ACE (angiotensin-converting enzyme) inhibitors 310–11, 311*f*
acetic acid fermentation 510
acetylcholine 657, 659*f*
  effect on airways 747
  effect on cardiac output 739
  effect on heart rate 611
  at neuromuscular junctions 692, 693
acetylcholine receptors 660–1
acetyl-coenzyme A (acetyl CoA) 326, 326*f*
  production from pyruvate 332*f*
  reaction with oxaloacetate 333*f*
Achilles tendon 699*f*
aciclovir 570–1
acid–base balance 762, 798–9
  effect of exercise 779–80
  and nutritional supplements 780
  role of nephron 792, 795
acidic vesicles 1046
acidophiles 439, 441, 480
acidosis 779
acids 169
  strong and weak 170–1
acinar cells 808, 809*f*
aconitase 333*f*
acquired immunity 1122
acromegaly 672
acrosome reaction 850, 1179
actin 296, 298, 299*f*, 394, 688, 688*f*, 689, 915*t*, 916
  crossbridge cycle 689–91, 690*f*
  in TNTs 945, 947*f*
Actinobacteria 439
action potentials 595–6, 600–4, 620, 973
  changes in ionic conductance 601*f*

conduction 604–5, 604*f*
conduction velocity 605
ion movements 602–4
refractory periods 602
sino-atrial node 734*f*
ventricular myocytes 735*f*
voltage-gated sodium channel 602, 603*f*
activation energy 167, 187
active hyperaemia 730, 772
active sites 306*f*
active transport 12*f*, 585, 912–14, 912*f*
  non-ionic 914
  protein-mediated 588–90
acute phase response 870, 871*f*
adaptations 1209, 1299
  to anoxia 1300, 1301*f*
  to desert environments 1301
  to gravity 1028–9
  leaf morphology 1300*f*
  to salinity 1300*f*
  to thermal environment 1035
  see also thermophiles
adaptive immune system 863–4, 864*f*, 876–85, 1122
  antibodies 880, 882, 893
  antigen recognition 885–95
  B cell activation 892–3
  B cell antigen receptors 876, 878–9
  cells of 864–5, 865*f*, 866
  elimination of dysfunctional receptors 885
  fundamental principles 876
  generation of diverse antigen receptors 883–4
  IgE-mediated allergic responses 881
  primary and secondary responses 876, 877*f*
  T cell activation 886–92
  T cell antigen receptors 882–3
  vaccination 877–8
adaptive radiation 1064–5
adenine (A) 27, 176
  structure 177*f*, 200*f*
adenoids 868*f*
adenosine 177*f*
adenylyl cyclase 611
adherens junctions 401*f*, 943*t*, 944, 945*f*, 946*f*
adhesins 539
adhesion factors 526, 536*t*
adrenal glands
  catecholamines 677*t*
  chromaffin cells 655, 657, 659
  cortex 676, 677*f*
  corticosteroids 676–7, 677*t*
  location and structure 677*f*

medulla 677–8, 677*f*
adrenaline (epinephrine) 657, 659, 677–8
  effect on airways 747
  effect on cardiac output 739, 771
  effect on cardiac pacemaker 662*f*
  effect on circulatory system 729
  effect on glycogen metabolism 315*f*
  synthesis from tyrosine 559, 660*f*
adrenergic neurons 657
adrenergic receptors 661
  vascular smooth muscle 728–9
adrenergic transmission 659–60, 661*f*
  drug effects 664
adrenocorticotrophic hormone (ACTH) 669*f*, 676, 841*t*
adult stem cells 939
aerenchyma 1040
aerobic exercise
  blood pressure responses 769–70
  cardiovascular adaptations to training 772–4, 774*t*
  pulmonary adaptations to training 778–9
Aerobics Center Longitudinal Study 783
aerofoils 1088, 1089*f*
aeropyles 1169*f*, 1170
aesthetascs 1136, 1137*f*
aestivation 1015, 1296
afferent arterioles 787*f*, 788*f*
  myogenic contraction 788
afferent neurons 595, 619, 621–2, 623*f*
affinity maturation 884
aflatoxins 538–9
Africa, emergence of humans from 395*f*
Afrikaner population 1239
Afrotheria 38*f*, 1065
afterload 740, 771
agar 480
agarose 905
age-1 gene 1224
age:sex pyramids 1250*f*
aggrecan 402
aggregating data 105
agnathans 987
agonists 657
agouti gene 239, 241*f*
agri-environment schemes 1354
Agrobacterium tumefaciens 412*f*, 435
  horizontal gene transfer 411–12
AIDS (acquired immunodeficiency syndrome) 896
air, composition of 757, 1039
air pressure 1038–9
airflow defects 755

airways
  anatomy 745–9, 745*f*, 748*f*
  asthma 755*f*
  autonomic control 663
  see also respiratory system
alanine (Ala, A) 288*f*
  enantiomers 291*f*
  structure 179*f*, 286*f*
alarm substance 1138
alarmins (DAMPs) 870, 875
albinism 243*f*
albumen 1171*f*
albumin 722
alcoholic beverages 511, 512*t*
  brewing beer 499, 512–13
alcohols 158*t*
aldehydes 158*t*
aldolase 330, 331*f*
aldosterone 676–7, 798*t*, 910, 911*f*
algae 466, 927–8, 928*f*
  body plans 954–5
  brown 462
  buoyancy 1097, 1098*f*
  cell wall 905
  coenocytes 922–3, 924*f*
  golden 463–4
  green 467
  red 466–7
  toxic 464*f*
algal blooms 461*f*, 464*f*, 1003*f*, 1326*f*
alien species *see* invasive species
alimentary tract development 969
aliphatic compounds 157
alkaliphiles 480
alkalosis 779
alkenes 158*t*
all-or-nothing events 600
allele frequency 1225–6
alleles 225–6, 1222–3, 1223*t*
  dominant and recessive 241, 243, 1222, 1223*f*, 1229*f*
  multiple 239
allergens 881
allergic reactions 863, 881
  to fungi 538
allografts 887
allometric scaling 953–4
allometry 953
allopatric speciation 47–8, 1217*f*
allosteric enzymes 312–14, 313*f*, 329
allosteric regulators 314*f*
allostery, haemoglobin 301–2, 302*f*
α actinin 689
α amylase 808, 816
alpha diversity 1282, 1283*f*
α helixes 293, 294*f*
  in glycophorins 304
α-ketoglutarate dehydrogenase 333*f*
α-motor neurons 695*f*

α proteobacteria 435
Alport syndrome 239
alternative reproductive tactics
    (ART) 1190
alternative splicing 262*f*
altitude acclimatization 765, 782
Alveolates 459, 461
alveoli 748*f*, 749
    gas exchange 758*f*
    surface tension 752–3
Alzheimer disease 297*f*
amatoxins 538
Ambulacraria 975
    limbs 987
    nervous systems 976*f*
ambush predators 1271–2
amides 158*t*
amines 158*t*
amino acid derivatives 665
amino acids 179
    absorption in small intestine 819*f*
    deamination 828
    dehydration condensation
        reaction 28*f*
    enantiomers 290–1, 291*f*
    industrial production 505
    linking to tRNA (activation) 266*f*
    oxidation 329
    prebiotic synthesis 26*f*
    reabsorption in the nephron 790*t*,
        792, 797*f*
    residues 286–7
    sequence comparisons 318, 320*f*
    side chain interactions 289–90,
        290*f*
    structure 286–91, 286*f*, 288*f*
aminoacyl-tRNA synthetases 266
ammonification 520
ammoniotelism 1050
amniocentesis 857
amnion 1174*f*, 1175*f*
amniotes, embryo
    development 1174–6, 1174*f*,
        1175*f*
amniotic fluid 857
amoebic dysentery 542*t*
amoebic meningoencephalitis 542*t*
Amoebozoa 465–6
Amphibians
    Batrachochytrium dendrobatidis
        infection 450*f*
    cane toad 1061–2, 1061*f*
    clawed frog (Xenopus
        tropicalis) 988*f*, 990*f*
    eggs 1170–1, 1170*f*
    great crested newt 1340*f*
    integument 1114–15, 1116*f*
    poison dart frog 1135*f*
    reproduction 1190–1
    salamanders 940, 941*f*, 1190–1
amphioxus 977, 978*f*
amphipathic molecules 173, 370
amphitrichous arrangement of
    flagella 381*f*
amphotericin B 537*f*
ampulla 646, 647*f*
ampullae of Lorenzini 1030*f*
amylase 808, 816
amyloids 297
amyloplasts 400, 907

Anabaena spp. 434, 435*f*, 518
anabolic reactions 185
anadromous migration 1187
anaemia 719, 761
anaerobic pathways 1042–3
anaerobic respiration 337–8
    Thermotoga subterranea 433
anaesthetics 627
anal canal 806*f*, 821–2, 821*f*
analgesics 627–8
analogous characters 59
anammox reaction 520
anaphase 231, 233*f*, 416*f*
anaphase I 422
anaphase II 423
ancestors 57
ancestry 1142
anchorage-dependent (AD)
    cells 920–1
anchorage-independent (AI)
    cells 920
ancient woodland indicator
    (AWI) 1330
androgen binding protein 842
androstenedione 836
aneuploidies 227, 228*t*, 418, 423–4
Anfinsen, Christian 295
Anfinsen cage 296
angiogenesis 1127
    exercise-induced 774
angiosperms (flowering plants) 955
    fecundity 1166
    life cycle 1148*f*
    reproduction 1158–66
    see also dicotyledons;
        monocotyledons
angiotensin converting enzyme
    (ACE) 797
angiotensin converting enzyme
    (ACE) inhibitors 742
angiotensin I 797
angiotensin II 729, 797, 798*t*
angiotensinogen 797
anguilliform swimming 1094
animal body plans 964–5
    bilaterian 966, 968, 970, 972
    diploblasts 965–6
    limbs 981–91
    nervous systems 973–81
    triploblasts 966
animal cells
    differentiation 936–7, 939–40
    directionality and
        anchorage 920–2
animal defences
    chemical deterrents 1135–6
    external defences 1128–33
    integument 1109–21
    predator avoidance 1137–8
    response to injury 1121–7
    toxins, poisons, and
        venoms 1134–5
    see also immune system
animals
    effects of gravity 1026–9
    electroreception 1030
    light detection 1010–18
    magnetoreception 1029–30
    responses to environmental
        pressures 1001

responses to sound 1021–5
    thermal environments 1035–8
annelid worms (round worms) 983
    integument 1111, 1112*f*
    osmoregulation 1048
    reproduction 1179–81
annual plants 1159
anoxic environments 1041–2, 1300
antagonists 657
anterior corticospinal tract 709
anterior pituitary gland 667*f*
    endocrine cell types 669*t*
anterior pituitary hormones 668,
    840–1, 841*t*
    negative feedback loops 660, 670*f*
    prolactin 859
anthers 1161*f*, 1162, 1164*f*
anthrax 439
Anthropocene 1314, 1317
anthropogenic threats
    habitat loss, degradation, and
        fragmentation 1326–31
    human population growth
        1319–20, 1320*f*
    implications of 1331–5
    resource use and
        emissions 1320–6
antibacterial drugs see antibiotics
antibiotic resistance 38, 206, 428,
    500, 501, 531, 533, 1207*f*
    β-lactamase 529–30
antibiotics 500, 501*t*, 531–2
    antifungal 537
    development of new drugs 501–2
    mechanisms of action 532–3,
        532*f*
antibodies 866–7, 880*f*
    functions 882*f*
    immunotherapy 883
    production of 893
anticlinal cell division 935
anticoagulants 1126
anticodons 264, 265*f*
antidiuretic hormone (ADH,
    vasopressin) 666, 667*f*, 670,
    671*f*, 729, 798*t*, 910, 911*f*
    action on aquaporins 796
    action on loop of Henlé 793, 795
    regulation of secretion 796
antigen presenting cells (APCs) 866
antigen processing 885
antigen receptors
    B cell receptors (BCRs) 876, 878–9
    diversity 883–4
    elimination of dysfunctional
        receptors 885
    T cell receptors (TCRs) 882–3
antigen recognition 885
antigens 863, 866
anti-logs 75
antimicrobial peptides (AMPs) 1128
antimycotics 537
anti-poaching patrols 1343*f*
antiporters 912–13
antiretroviral therapy (ART) 571
antiviral therapy 569–71
antlers 996, 997*f*
ants
    leaf cutter 501–2, 501*f*
    photoreceptors 1016–17

aorta 726, 733*f*
aortic pressure 737–8, 737*f*
aortic valve 733*f*
aphids 1205*f*
    life cycle 1145*f*
    reproduction 1205
    virus transmission 1108
API strips 46*f*
apical dominance 929
apical growth, fungal hyphae 446–7
apicomplexans 461, 462*f*, 540*f*
apomixis 1147
apoplast 1035
apoplastic transport 941, 942*f*
apoptosis 396
aposematism 1129, 1130*t*, 1132*f*
appendix 806*f*, 821*f*
appetite regulation 814
apple scab 552*f*
aquaporins (AQPs) 587, 909–10,
    910*f*, 1048
    action in kidney 796*f*, 910, 911*f*
    action of ADH 670
aquatic animals
    and gravity 1026, 1028
    see also fishes
aqueous environment 1044
aqueous humour 632*f*
Aquificales 433
Arabidopsis thaliana 1009, 1019
    growth in microgravity 1027*f*
arachnids 984
aragonite 992, 993
Archaea 50*f*, 54*f*, 383, 439–40
    cell morphology and
        structure 383–4
    Crenarchaeota 441–2
    Euryarchaeota 442, 444
    extreme environments 440*f*,
        443–4
    halophiles 444–5
    Korarchaeota 440
    methanogens 443–4
    Nanoarchaeota 440
    phylogenetic tree 441*f*
    similarities with Eukarya 384
archaeal phospholipids 373*f*
archaeobacteria 16, 17*f*
Archaeopteryx 989
archamoebae 465–6, 466*f*
archentron 969
architomy 1180
Areas of Outstanding Natural Beauty
    (AONB) 1341*t*
arginine (Arg, R) 288*f*
Aristotle 43, 48
arms races 1145, 1272, 1273*f*
    plants and pathogens 1108–9
aromatic compounds 157
arteries 723, 725–6, 725*f*
    blood flow 727–8
    control of diameter 728–30
    structure 725–6, 726*f*
    types of 726–7
arterioles 726–7
arthropods
    immune system 1127–8
    integument 1112–13, 1113*f*
    limbs 984–6
    see also insects

Arthrospira spp. 435
arthrospores 537*f*
artificial selection 39, 1209
   dogs 1210–11
Artiodactyls 57, 58*f*
Ascomycota 451–3
ascorbic acid (vitamin C) 303
ascospores 453*f*
ascus 453
aseptic technique 470
asexual phases 1143, 1144*f*, 1145*f*
asexual (clonal) reproduction 1143
   cellular fission and
      budding 1145–6
   parthenogenesis 1147, 1149
   polychaete worms 1180
ash dieback disease 449, 550*f*, 1319*f*
asparagine (Asn, N) 288*f*
aspartic acid (Asp, D) 288*f*, 505
aspect ratios
   birds 1092*f*
   fins 1094*f*, 1095*f*
aspergillosis 535–6, 1328
Aspergillus spp.
   A. fumigatus 535*f*
   A. nidulans 923, 924*f*
   A. oryzae 511
   use in biotechnology 504
aspirin 311–12, 729
assortative mating 1240
asthma 755–6, 881
astigmatism 633, 634*f*
astronauts, bone mineral
     density 995–6
ataxia 716
atherosclerosis 722–3
athletes, body conformation 1085
athlete's foot (tinea pedis) 534, 535*f*
Atlantogenata 1065
atmosphere
   formation of 21
   as source of organic molecules 23
atmospheric pressure 757
atomic structure 150–1, 151*f*
atopy 881
ATP (adenosine triphosphate)
     189–90, 328
   coupled reactions 191
   energy release from 191*f*
   generation in electron transport
     chain 335*f*, 336–7
   generation in glycolysis 330, 331*f*,
     332
   generation in photosynthesis 339
   generation in TCA cycle 333*f*
   inhibition of phosphofructase 314
   in muscle contraction 690–1
   regeneration of 191–2
   role in energy economy of a
     cell 190*f*
   structure 190*f*
   yield from oxidation of
     glucose 337
ATP synthase 335, 336*f*
atria 732, 733*f*
atrial natriuretic peptide (ANP) 729,
     798
atrial pressure 737*f*, 739
atrio-ventricular node (AVN) 734,
     735*f*

atrophy 697
atropine 664
attachment, viruses 560, 561*f*
   inhibitors of 570
attenuation of pathogens 877–8
attenuation of transcription 281
audiometry 646
auditory nerve 641*f*, 644
autoantibodies 891
autoclaves 470, 471*f*
autocrine signalling 614, 665, 666*f*
autogenic inhibition 704*f*
autoimmune diseases 863
autoimmunity 891–2
autoinducers 382
automixis 1147
autonomic nervous system
     (ANS) 595, 654–5
   control of airways 663, 747
   control of bladder 662–3
   control of cardiac output 662, 739,
     770–1
   control of eye 663–4
   control of gland secretion 664
   control of liver 664
   control of vascular smooth
     muscle 728–9
   drug effects 664
   effect on insulin secretion 681
   effect on saliva secretion 808
   neurotransmission 657–60
   organization of 655–7
   receptors 660–1
autophagy 392*f*
autosomal dominant conditions 243,
     245*f*
autosomal recessive conditions 243,
     244*f*
   cystic fibrosis 358, 591–2
   hereditary deafness 247*f*
   Tay–Sachs disease 1229–30
autosomes 225, 226*f*
autothysis 1211–12
autotrophs 1271, 1301
   and carbon cycle 1309
   chemoautotrophs 1302–3
   in ecological pyramids 1308–9
   in food chains 1303–4
   hemiparasites 1278, 1279
auxin 942, 1026
   intercellular polar transport
     943*f*
availability bias 9*t*
Avery, Oswald 203–4
Avogadro's number (N_A) 121*t*
axil filaments (flagella) 380–1, 381*f*
axolotls 1190–1
axon hillock 610*f*
axons 595, 596*f*, 973
   diameter 605
axoplasm 596
   ion concentrations 597*f*

B
B cell receptors (BCRs) 876, 878–9
   clonal deletion 885
   generation of diversity 883–4
   protein structure 878*f*
B lymphocytes (B cells) 721, 866–7,
     867*f*, 869

activation 892–3, 895*f*
   antigen recognition 885
   class switching 893
   differentiation 880
bacilli 374*f*
Bacillus spp. 438
   B. anthracis 526, 528*f*
   B. thuringiensis 514
bacteria 17*f*, 50*f*, 54*f*, 373
   antibiotic resistance 38, 206
   biofilms 383
   classification and relatedness 46
   DNA replication 213–16
   endospores 382*f*
   Gram stain 376–7
   Gram-negative 377–8, 379*t*
   Gram-positive 377, 379*t*
   gut microbiome 822–3
   horizontal gene transfer 409–12
   lacking a cell wall 379
   L-forms 379, 380*f*
   quorum sensing 382–3
   secretion systems 551*t*
   terminology 433*t*
bacterial cell wall 375–6
   Gram-negative bacteria 379*f*,
     380*f*
   Gram-positive bacteria 378*f*,
     380*f*
   growth of 376*f*
   S-layers 376, 377*f*
bacterial cells 374*f*
   chemical composition 172
   cytoskeleton 374*t*
   flagella 380–1, 381*f*
   gas vesicles 381–2
   glycocalyx 380
   magnetosomes 382*f*
   morphologies 374*f*
   nucleoid 373
   pili 381
   plasmids 374
   relative dry weight
     composition 16*f*
   ribosomes 375*f*
bacterial diversity
   Aquificales and
     Thermotogales 433
   Chlamydiae 437
   Chloroflexi 433–4
   cyanobacteria 434–5
   Firmicutes and
     actinobacteria 437–9
   main phyla 440*t*
   phylogenetic tree 432*f*
   proteobacteria 435–6
   Spirochaetes 437
   Thermus and *Deinococcus* 433
bacterial pathogens 526, 530*f*
   Helicobacter pylori 531
   Neisseria gonorrhoeae 531
   phage therapy 572–3
   phage typing 572*f*
   plant diseases 551
   Staphylococcus aureus 529–30,
     530*f*
   Streptococcus pneumoniae 530–1,
     530*f*
   transmission of 530
   virulence factors 526–8

bacterial reproduction
   binary fission 406–8, 406*f*
   budding 408, 409*f*
   fragmentation 408
bacterial toxins 439, 693
bacteriophages (phages) 377, 557*f*,
     567–9
   lytic and lysogenic cycles 569*f*
   phage therapy 572–3
   phage typing 572*f*
   plaques 570*f*
   transduction 410*f*
bacteriorhodopsin 445
bactoprenol 376
Bajau people 1213–14
balance 645–8, 1021
   vestibulospinal tract 709
balance organs 1022
balantidiasis 542*t*
Balantidium coli 540*f*
Baltimore, David 560
bamboo 956*f*
bar (column) graphs 109*f*
bar model, ratios 88
bark 960
barnacle 985*f*
baroreceptor reflex 740–1, 741*f*
baroreceptors 1021
basal (core) promoter, DNA 275
basal ganglia 706*f*, 713–15
basal metabolic rate (BMR) 953–4
basal promotor 258
base 2 73*f*
base excision repair (BER) 220
base pairing 176, 178*f*, 181, 201–2
   proofreading 216, 217*f*
   wobble pairing 265
basement membrane (basement
     lamina) 931–3
   mesogloea 965*f*
   molecular structure 932*f*
bases 169–70
   conjugate 171
Basidiomycota 454, 455*f*
basidiospores 454, 455*f*
basilar membrane 642, 643*f*
basophils 721–2, 721*f*, 865*f*
batch culture 484, 485*f*, 504
   vinegar production 510
Batesian mimicry 1130*t*, 1131, 1132*f*,
     1273, 1274*f*
Batrachochytrium dendrobatidis
     450*f*
bats 989–91, 990*f*, 1193*f*
   echolocation 1025
   embryonic diapause 1194
   pollination by 1019
Bayesian statistics 117
bears, grolar 1215–16, 1215*f*
beech-drop 1278*f*
Beer–Lambert Law 126
beer production 499, 512–13
behavioural ecology 41
benthic zone 430, 431*f*
benzene 157
Bernard, Claude 582
beta-alanine supplementation 780
β barrel structure 304, 305*f*
beta-blockers (beta-adrenoceptor
     antagonists) 664

beta diversity (β) 1282, 1283*f*
β-haemolysis 437*f*
β-lactamase 529–30, 533
β-lactams 532, 533*f*
β proteobacteria 436
β sheets 293, 294*f*
   in porins 304
β-thalassaemia 262–3, 761
between-subject design 103
bias 9*t*, 98, 102
bicarbonate ions
   regulation by kidneys 798
   secretion by pancreas 819
bicarbonate buffering system 762–3,
   779, 798
biennial plants 1159
Bifidobacterium spp. 504
bilateria 966, 968
   development 968–70
   loss of bilateral symmetry 969*f*
   planes of symmetry 968*f*
   protostomes and
      deuterostomes 970*f*
   structure 970, 972
bile 720, 827
bile duct 824*f*
bile pigments 827, 828
bile salts 817, 827
bilirubin 720, 828
binary data 64
binary fission 406–8, 406*f*, 1145–6
   study of 407
   yeasts 453, 454*f*
binomial distribution 82, 84*t*
binomial nomenclature 43–4
   examples 45*t*
bioaccumulation 1344
bioaugmentation 1355
biodiversity *see* diversity
biofilms 381, 383, 429, 514, 526, 528*f*
biofuels 505–6, 1323
biogeochemical cycles 517, 1309
   carbon cycle 520–1, 1310, 1310*f*
   nitrogen cycle 517–20, 1310, 1311*f*
bioinformatics 316, 318, 359–61
   applications 321
biological control 514–15
   use of viruses 573
biological hazard levels 472
biological nomenclature 427
biological oxygen demand
   (BOD) 516, 1300
biological species concept 1215–16
bioluminescence 461*f*
biomes 1292, 1293*f*
biomineralization 992
bioreactor systems 484, 485*f*
bioremediation 515, 1354–6
biospecies 1216
biosphere 1201
Biosphere reserves 1341*t*
biostimulation 1355
biotechnology 499
   biological control 514–15
   bioremediation 515
   chemicals and fuels 504–6
   composting 516–17
   food and drink production 507–13
   microbial polysaccharides 513–14
   pharmaceutical industry 499–504

wastewater treatment 515–16
biotic factors 1292
biotic homogenization 1285
biotransformation 829
biotrophs 447, 1108
bipedal locomotion 1080–4
   human walking 1081*f*
1, 3 biphosphoglycerate 330, 331*f*
biramous limbs 983, 984
birds 989
   adaptations 1301*f*
   aspect ratios 1092*f*
   breeding phenology 1236–7
   eggs 1171–2, 1171*f*
   embryo development 1175*f*
   flight *see* flying and gliding
   integument 1119
   nest-site competition 1276–7
   osmoregulation 1049
   oviduct 1171*f*
   shapes and relative sizes 1091*f*
   skeleton 990*f*
   wings 989
birth 857–9, 858*f*
bivariate analyses 105
black-bellied seedcrackers 1208*f*
black wart of potato 449, 450*f*
bladder 786*f*, 800*f*
   autonomic control 662–3
blast diseases 552
blastocoel 968, 969, 971*f*
blastocyst 850, 968–70
blastodisc 1171*f*, 1172
blights 552
blind cave fish 1022–3, 1022*f*
blinding 104
blood 719
   anaemia 761
   carbon dioxide transport 762–3,
      763*f*
   coagulation cascade 1125–6, 1126*f*
   composition of 719*f*
   oxygen transport 759–61
   plasma 722
blood–brain barrier 764
blood cells
   erythrocytes 719–20
   haemopoiesis 720, 721*f*
   leukocytes 721–2
   platelets 613, 722
   see also lymphocytes
blood flow distribution, effect of
   exercise 771–2
blood groups 226, 239
blood pressure 727*f*, 740
   hypertension 742
   measurement 740
   regulation of 740–1, 789, 797–8
   responses to exercise 769–70
   blood–testis barrier 1178
blood velocity 728
blood vessels
   arrangement of 728*f*
   arteries 725–30
   capillaries 730–1
   dimensions and composition 726*t*
   structure 726*f*
   veins 731–2
body of the stomach 811, 812*f*
body plans

animals *see* animal body plans
   plants *see* plant body plans
body rhythms 1015
body shape maintenance 1028–9
Bohr effect 760*f*, 761
Boltzmann constant (k_b) 121*t*
bond energies 161–2, 163*t*
   and chemical reactions 165–7
bond rotation 156*f*
bond strengths 163*t*
bonds
   covalent 151–2, 152*f*
   hydrogen 163, 164*f*
   ionic 162*f*
   non-covalent 161, 162–3
   van der Waals interactions 162–3
bone marrow 720, 868*f*
bone marrow transplantation 887
bone mineral density (BMD) 995–6
bone morphogenic proteins
   (BMPs) 972
bones 994–6
   effect of hypocalcaemia 801
   mineralization and
      maturation 995*f*
bonobos 1247*f*
bony fishes (teleosts)
   ears 1025
   osmoregulation 1049
   reproduction 1168
   scales 1114, 1115*f*, 1116*f*
   swim bladder 1028, 1099–1100
Boroeutheria 1065
Borrelia burgdorferi 437
Botrytis cinerea (noble rot) 513*f*
botulinum toxin (Botox) 439,
   528, 693
bovine spongiform encephalopathy
   (BSE) 297
Bowman's capsule 787*f*, 788*f*
box plots 112*f*
boxes 257, 258–9
Boyle's Law 1039
brachiation 1083–4, 1083*f*
brachystasis 859
bracken 1072, 1073*f*
brackish water 1044
bradykinin 628*f*, 729, 1124
brain
   auditory pathway 644, 645*f*
   blood flow during exercise 772
   cerebellum 706*f*, 716
   cerebral circulation 724
   evolution 978–81
   model of 12*f*
   motor areas 706*f*
   respiratory centres 765*f*
   sensory areas 623–4, 623*f*
   visual cortex 637–9, 638*f*
branching angles 958–9
Brazilian pit viper 311*f*
bread making 499, 510
breasts 859
breathing 745, 749
   control of 764–5
   expiration 750–1
   inspiration 749–50
   lung compliance 751–2
   mechanics of 751*f*
   muscles involved 750*f*

responses to exercise 777–8
   ventilatory cycle 751*f*
breeding migration 1068*f*
   fur seals 1069–71, 1070*f*
brewing beer 499, 512–13
brightfield microscopy 489, 494*f*
broadcast spawning 1184–90
bronchi (singular bronchus) 747, 748*f*
bronchioles 747, 748*f*, 749
broomrape 1278
brown algae (Phaenophytes) 462
brown fat (brown adipose tissue,
   BAT) 1038
Brownian motion 585
browsing 1271
brumation 1294
Brunner's glands 815
brush border 805, 816
bryophytes 955*f*
budding 408, 409*f*, 1146
   viruses 558, 564
   yeasts 453–4, 454*f*
buffering 762–3, 779, 798
buffers 171–2
bulbourethral glands 847*f*
Buller's drop 454, 455*f*
bundle of His 734, 735*f*
buoyancy 1026, 1028, 1097–100
burs 1076*f*
butanal 159*f*
butanone 159*f*
butt rot 515*f*
butterflies 1294, 1295*f*
   food plants 1348, 1349*f*
   life cycle 1184*f*
buzz pollination 1019

C
CAAT box 258, 275
cacti 960, 1105*f*
cadherins 944, 946*f*
caecum 806*f*, 821*f*
   see also large intestine
Caenorhabditis elegans 1224*f*
calcite 992*f*
calcitonin 674, 682, 801
calcitriol 682, 801
calcium carbonate ($CaCO_3$) 992
   shells and exoskeletons 992–4
calcium homeostasis 681–2, 682*f*, 801
calcium ions (Ca^{2+})
   investigations in live cells 493–4,
      495*f*
   and skeletal muscle function
      691–2, 693
calcium storage 994, 996
calculations 127
callus (tree burrs) 1109, 1110*f*
Calvin cycle 340
cambium 936, 938
camera traps 1255*f*, 1258, 1259*f*
camouflage 1130*t*, 1131*t*
   animals 1129, 1132*f*, 1222*f*
   plants 1133*f*
cAMP (cyclic adenosine
   monophosphate) 611
Campylobacter jejuni 436
Canadian pondweed 1073
cancers 922
   cervical 851

Warburg effect 338
Candida albicans 447*f*, 451, 536
    virulence factors 536*t*
candidiasis 536*f*
cane toad 1061–2, 1061*f*
cannibalism 1274
Cannon, Walter 582
capacitation of sperm 850, 1178
capillaries 728, 730–1
    fluid movement into and out
        of 730–1, 731*f*
    hydrostatic and oncotic
        pressures 731*t*
    structure 726*f*
capillary density 730
capillary networks 724
capping, mRNA 260, 261*f*
capsids 556*f*, 557*f*
    assembly 563, 564*f*
    T number 558*f*
capsomeres 557
capsule, bacterial 380, 526, 528*f*
captopril 311*f*
capture–mark–recapture
        (CMR) 1256–7
carbohydrates 173
    digestion and absorption 816–17
    and exercise 780
    photosynthesis 340*f*
    structure 174*f*
carbon
    bonding 152*f*
    properties of 22
    valence 153*t*
carbon cycle 520, 521*f*, 1310, 1310*f*
carbon dioxide (CO$_2$)
    effect on ocean pH 172
    gas exchange 758–9
    greenhouse effect 1324
carbon dioxide transport 762–3,
        763*f*
carbon frameworks 153, 154*f*
carbon isotope studies 18
carbon sequestration 1332
carbonate–carbonic acid buffer
        762–3, 798–9
carbonic acid–bicarbonate buffering
        system 779
carbonic anhydrase 762, 779, 798,
        1041
carbonyl functional group 157, 158*t*
carboxylic acids 157, 158*t*
carboxypeptidases 818
carboxysomes 434
cardiac cycle 737–9, 737*f*
cardiac muscle 687, 732, 733*f*
    action potentials 735*f*
cardiac output 728, 739–40
    autonomic control 662
    effect of aerobic exercise
        training 774
    response to exercise 770–1, 771*f*
cardiac sphincter (LES) 811, 812*f*
cardiorespiratory fitness 782–3
cardiovascular disease 722–3, 828
cardiovascular system 718–19
    adaptations to aerobic exercise
        training 772–4, 774*t*
    arteries 725–30, 725*f*

blood 719–23
blood pressure 740–2
capillaries 730–1
cardiac cycle 737–9, 737*f*
cardiac output 739–40
    changes in pregnancy 855*t*
circulatory systems 723–5, 723*f*,
        724*f*
electrocardiogram (ECG) 736,
        737*f*, 738*f*
heart 732–6
responses to exercise 769–72
veins 731–2
carnivorous plants 960
carotid bodies 764
carotid sinus 740, 741*f*
carpel 1161*f*, 1163
carrier proteins 588*f*
carriers of genetic conditions 243
carriers of infectious diseases 529
carrying capacity (*K*) 143, 1264,
        1275
    effect of ecosystem
        challenges 1332
cartilage 994
cartilaginous fishes (elasmobranchs/
        chondrichthyes)
    buoyancy 1028, 1100
    ears 1025
    osmoregulation 1049–50
    scales 1114, 1115*f*, 1116*f*
Cas9 endonuclease 357, 358*f*
cascade effects 1311, 1319
catabolic reactions 185
catabolite gene activator protein
        (CAP) 274
catadromous migration 1187
catalysts 167
    see also enzymes
catecholamines 659, 677–8, 677*t*
    synthesis from tyrosine 559, 660*f*
    see also adrenaline; noradrenaline
categorical data 64
caterpillar herbivory 1106*f*
caudal fins 1094, 1095*f*
caudate nucleus 706*f*, 713
Caulerpa cactoides 922, 923*f*
causation bias 9*t*
cause and effect 105–6
CCR5 antagonists 570
CD numbers 898–9
CD4+ T cells 886
    activation 886–8
    differentiation 889*f*
    functions 889–92, 890*f*
CD8+ T cells 886
    activation 892, 893*f*
cell–cell contacts 401*f*
cell counts 481, 483*f*
    dry weight 484
    haemocytometers 482*f*
    optical density measurement 483–4,
        484*f*
    total 481, 482*f*
cell culture flasks 474*f*
cell cultures 474–5
    ethical issues 476
    growth and storage 478
    how they grow 475

microbial *see* microbial cell
        cultures
    plant cells 479*f*
    requirements 475, 477*t*, 478
    uses of 475
    viewing and measurement 478–9
cell cycle 412, 413*f*, 1146
    key checkpoints 417*f*
cell differentiation 277, 933, 938*t*
    in animals 936–7, 939–40
    haematopoietic stem cell
        system 923*f*
    in plants 933–6, 937*f*
cell division 231
    DNA packaging 210, 212*f*
    plant cells 905–6, 905*f*
    regulation 277
    see also cytokinesis; meiosis;
        mitosis
cell growth 73*f*
cell junctions 943–8
cell lineage tracing 939*t*
cell–matrix contacts 401*f*
cell membranes 904
    see also plasma membrane
cell sizes 368, 369*f*, 385–6
cell theory 368
cell wall 904
    archaea 384
    bacterial 375–6
    diatoms (frustule) 462, 463*f*
    fungal 398–9
    plants 397–8, 904–6, 905*f*
cell wall proteins (CWPs) 397
cells
    Archaea 383–4
    bacterial 373–83
    diversity 368
    eukaryotes 384–401
    first identification of 368*f*
    phospholipid membrane 369–73
    prokaryotic and eukaryotic 369*t*
    specialized 401
    surface area to volume ratio 368
cellular communication 940–1
    animal cells 942–8
    plant cells 941–2
cellular fission 1145–6
cellular respiration 745
    see also anaerobic respiration;
        glucose oxidation
cellular signalling 614–16
    role in gene control 279, 280*f*
cellulose 173, 397*f*, 904
    structure 174*f*, 398*f*
    tunicates 978, 1113
censuses 1256
centipedes 984
central dogma 253
central nervous system (CNS) 595
    see also brain; spinal cord
central tendency 80*t*
central vacuole 400*f*
centrolecithal eggs 1169–70, 1169*f*
centromeres 209, 231, 414*f*
centrosomes (microtubule organizing
        centre) 231, 233*f*, 393–4, 415
cephalization 978
cephalopods 982

eyes 1013*f*
locomotion 1094
cephalosporins 533*f*
Cercozoa 464*f*
cerebellum 706*f*, 716
cerebral circulation 724
cervical cancer 851
cervix 837*f*, 848*f*, 849*f*, 850
    dilatation during labour 857–8
Cetaceans 57, 58*f*
CFTR gene mutations 591–2
chaetae 983*f*, 1078
Chagas' disease 459, 547
Chain, Ernst 500*f*
chain-termination method (Sanger
        sequencing) 349–50, 351*f*
change of free energy (ΔG) 186–7
channelrhodopsin 1016*t*
channels 587–8, 599
chaperone proteins 295–6
Chargaff, Erwin 201
Charles' Law 1039
charophytes 467
charts *see* graphs
cheese production 511
cheetah 1085*f*, 1273*f*, 1335*f*
chemical defences, plants 1105–7
chemical energy 185
chemical equilibrium point 187
chemical reactions 164, 185
    activation energy 187, 188*f*
    catalysts 167
    energy changes 165–7, 186–7
    equilibrium 187
    oxidation and reduction 188–9
    reversibility and
        equilibrium 164–5
chemical synapses 606, 607*f*
    ligand-gated channels 608–9
    metabotropic receptors 610–14
    neurotransmitters 606–7, 608*f*
    synaptic integration 609–10,
        609*f*
chemiosmosis 335–6, 339
chemokines 870–1
chemoreception 1136
    predator avoidance 1137–8
    territorial defence 1139
chemoreceptors 764, 777
chemosynthesis 1302–3
chemotaxis 380
chiasmata 420, 421*f*, 835
chief (peptic) cells 812*f*, 813
chimeric antigen receptor (CAR)
        T cell therapies 883
chimeric mice 278*f*
chimpanzees 1247*f*
chiral molecules 158
chirality 26
chitin 173, 398, 991, 1112
    structure 174*f*
    structure of NAG 399*f*
Chlamydia 437*f*, 851, 852
Chlamydiae 437
Chlamydomonas 927, 928*f*
chlamyopsin 1016*t*
Chloroflexi 433–4, 434*f*
chlorophylls 339, 399
chlorophytes 467, 927

chloroplasts 339*f*, 399–400, 906–7, 906*f*
  DNA 206
  origins of 385, 408, 434, 906–7, 906*f*
  structure 399*f*
chlorosis 553
Choanoblastea 926
choanocytes 926, 927*f*
cholecystokinin (CCK) 815, 819, 820, 828
cholera 11, 569
cholesterol 173, 502, 828
  role in membranes 371, 372*f*
  structure 176*f*, 371*f*
cholinergic neurons 657
cholinergic transmission 657, 659*f*
  drug effects 664
chondrichthyan fishes 994
chordae tendineae 733*f*
Chordata 975
  limbs 987–91
  nervous systems 976
  see also vertebrates
chorion 1169*f*, 1170, 1174*f*
chorionic villi 852
choroid 632*f*
chromaffin cells 655, 657, 659
chromatids 231, 233*f*, 234*f*
chromatin 210, 211*f*, 212*f*, 387*f*
chromatin modification 276
chromatin remodelling 276
chromophores 1015, 1016*t*
chromoplasts 907*f*
chromosomes 206, 207*f*, 225, 387, 1223
  crossing over 234, 235–8, 235*f*, 236*f*
  DNA packaging 210, 211*f*
  human chromosome 1 240*f*
  numerical abnormalities 226–7, 228*t*, 418, 423–4, 835–6
  replication 412–14, 414*f*
  translocation 424*f*, 1232
  X and Y 833
chronic kidney disease 799
chronic obstructive pulmonary disease (COPD) 752
chronotropic effects 739
Chrysopelea spp. 1080
chylomicrons 818
chyme 811
chymotrypsin 818
Chytridiomycota (chytrids) 449–50
  Batrachochytrium dendrobatidis 450*f*
ciclosporin A 502
cilia 869
ciliary body 632*f*
ciliary cells 1013
ciliary muscles 632*f*, 633
ciliates 461, 462*f*, 540
circadian rhythms 582, 1325
circannual rhythms 1325
circulatory systems 723–5, 723*f*, 1044–5
  fetal circulation 856–7, 856*f*
  lymphatic system 725
  pulmonary circulation 723–4
cis–trans isomers 156*f*, 157, 160*f*

peptide bonds 293*f*
CITES regulations 1340–1
citrate synthase 333*f*
citric acid 504*f*
clades 53, 56
cladistics 53, 428
cladograms 53, 55*f*, 56*f*
Clarin gene (*CLRN1* and *CLRN2*) mutations 246–7
class switching 893
classes/class intervals 82
classical (numerical) taxonomy 44, 428
classical genetics *see* Mendelian genetics
classification of organisms *see* taxonomy
clathrin 560, 562*f*, 590–1
Claviceps purpurea 538, 539*f*
clawed frog (*Xenopus tropicalis*) 988*f*, 990*f*
cleidoic eggs 1168
Clements, Frederic 1202
Clementsian succession 1287
climate change 1324, 1353
  effects on species' distributions 1324–5, 1324*f*
  historical changes in temperature 1057*f*
  phenological change 1325*f*
  see also global warming
climax communities 1103, 1287
clitoris 848*f*
clonal (asexual) reproduction 1143
clonal deletion 885
cloned DNA 347
  adapter sequences 349, 350*f*
clonidine 664
cloning 1147
closed circulations 1044
closed populations 1240, 1241*f*, 1251, 1252*f*
Clostridium spp. 438, 518
  C. botulinum 528
clustered dispersion 1249*f*
clustering illusion bias 9*t*
Cnidaria 965–6
  integument 1111, 1112*f*
cnidocytes (nematocysts) 965
coagulation, changes in pregnancy 855
coagulation cascade 1125–6, 1126*f*
coat colour, environmental influences 239, 241*f*
cocci 374*f*
Coccidiodes immitis 536, 537*f*
coccidioidomycosis (San Joaquin Valley fever) 536–7
coccidiosis 461
coccolithophore plankton 992*f*
cochlea 640–2, 641*f*, 642*f*, 643*f*
cochlear nerve 642*f*
coding strand, DNA 255
co-dominance 1227, 1229*f*
co-dominant alleles 239
codons 254–5
coelacanth 57, 59*f*
coeliac disease 820
coelomoducts 1180

coeloms 972
coenocytes 922–3, 924*f*
  in plants 924–5, 925*f*
coenzymes 308
  coenzyme A (CoA, CoA-SH) 326*f*
  coenzyme Q (ubiquinone) 334, 335
  NAD⁺ 193
coexistence 1279–80, 1280*f*
cohesins 412
cohort life tables 1250, 1251*t*
coitus (sexual intercourse) 850
collagen 296, 298, 302–3, 402
  structure 304*f*, 402*f*, 931*f*
collecting ducts 787*f*, 795–7
  reabsorption of major substances 790*t*
colon 806*f*, 820, 821*f*
  see also large intestine
colonies 926
coloration 1128–9, 1135*f*
  fish 1185–6
  Wallace's categorization 1130*t*
colostrum 859
colour blindness 244
colour detection 1015
colour opponency 636
colour vision 634–5, 636
columella cells 1026
columnar palisade mesophyll 961
combinatorial diversity 876, 884
comets 22, 23*f*
commensalism 526, 1279, 1280*f*
communities 1201, 1270–1, 1292*f*
  biotic homogenization 1285
  classification of 1280
  temporal changes 1286–8
community similarity 1284, 1285
competence, bacterial cells 410
competition 1274–6, 1280*f*, 1343
  avian nest-site competition 1276–7
  interference and exploitative 1276
  interspecific 1275–6
  intraspecific 1263, 1275
competitive and non-competitive inhibitors 311, 312*f*
  effect on enzyme kinetics 313*f*
competitive antagonists 657
complement cascade 872–3, 873*f*, 1125*f*, 1127
complementary DNA (cDNA) 353
  preparation from mRNA 353, 355*f*
  protein production 356
  quantitative (real-time) PCR 354, 355*f*
complex eyes 1011*t*, 1013–14, 1013*f*
complexity of species 58–9
compliance, lungs 751–2
composting 516–17
compound eyes 1011*t*, 1012–13, 1012*f*
compound heterozygosity 1229
compounds 150
concatemers 565
concentration gradients 908
concentrations 87, 88–9, 90–1
  effect on reversible reactions 165, 166*f*

concentric muscle contractions 700*f*
concentricolide 452
concertina movements, snakes 1080
condensing data 105
conductance 601
conducting system of the heart 733–6, 735*f*
conduction of heat 1031, 1032*f*
  thermoregulation 1037*t*
cones 634–5, 636*f*, 1014*f*
confidence intervals 97, 113*f*
confirmation bias 9*t*
conflict avoidance 1139
confocal microscopy 492–3, 493*f*
conformations of molecules 156
confounding variables 67, 106
congenital malformations
  causes of hearing loss 239
  chromosomal numerical abnormalities 226–7, 228*t*
  and NAD deficiency 194
  see also genetic disease
conidia 453, 535*f*
conidiophores 452–3, 453*f*
conifers 960
conjugate bases 171
conjugation 381, 410–11, 411*f*
conjunctiva 632*f*
connective tissue 687
connectivity 1329, 1354*f*
connexin mutations 246
connexins 606, 944
connexons 606, 944, 947*f*
conservation 1339, 1345–6
  decision making 1346
  ex situ 1354
  and genetic variation 40
  habitat-focused 1351–4
  landscape-based 1353–4
  predator–prey relationships 1349, 1351*f*
  prioritization 1347*t*
  site-based 1351–3
  in situ versus *ex situ* 1346
  species-focused 1346–50
constants 121*t*, 126
continental drift 1054
continuity 32
continuous capillaries 730
continuous culture 484, 485*f*
continuous data 64
continuous growth model 141
contraception 851
contractile vacuoles 1046
control groups 103
controlled experiments 102
convection 1031, 1032*f*
  thermoregulation 1037*t*
Convention on Biological Diversity (Rio Protocol) 1342
convergent evolution 59
cooperative hunting 1272*f*
coral reefs 7*f*
  impact of ocean acidification 993
co-receptors 886
cork 960
cornea 632*f*, 633
corona radiata 1169*f*
coronary circulation 724

changes during exercise 772
coronaviruses 558
corpus albicans 837f
corpus callosum 706f
corpus cavernosum 847f, 848f
corpus luteum 837f, 843, 844, 1191
  effects of melatonin 853–4
corpus spongiosum 847f, 848f
correlated variables 67
correlational (observational)
    studies 67, 68t
corroboration 13
corticobulbar tract 705, 709
corticospinal tract 705, 707f, 708f,
    709
  neuron numbers, relationship to
      dexterity 709f
corticosteroids 676
cortisol 676
cortisone, industrial production 500,
    502
Corynebacterium diphtheriae 528
cosmopolitan species 1247
cotransporters 589, 911
cotyledons 956
cough reflex 753
countercurrent multiplier 793–5, 794f
coupled reactions 191
courtship, fish 1185–6, 1187f
covalent bonds 151–2, 152f
  bond rotation 156f
  polar 153, 154f
  single, double, and triple 155f
  valence 152–3
covalent modification of
      enzymes 314, 315f, 329–30
Covid-19 (SARS-Co-2 virus) 343–4
  spike proteins 559
  vaccines 559, 878f
crab plover 1212
cramp balls 451, 452
C-reactive protein 722
creatine supplementation 780
Crenarchaeota 441–2, 442f
Creutzfeldt–Jakob disease (CJD) 297
cribriform plate 648, 649f
Crick, Francis 204, 253
CRISPR-Cas9 357, 358f
cristae 396f
critical evaluation 13
critical significance levels 115
crop plants
  artificial selection 39f
  genetic engineering 39–40
crossbridge cycle, striated
      muscle 689–91, 690f
crossing over 234, 235–8, 235f, 236f,
    419, 420, 421f, 836
crucian carp 1043f
  anoxia survival 1042–3
crustaceans
  body plans 984–5
  exoskeletons 992–3
  extended integument 1121f
  integument 1113
crustose lichens 457f
crypsis 1129, 1130t, 1132f
cryptic species 1282f
cryptochromes 1009, 1016t, 1018

cryptosporidiosis 542t
crypts of Lieberkühn 815
crystal violet 490
ctenophores (comb jellies) 37f, 966,
    967f
  integument 1111
culling 1349
culms 956
culture media
  eukaryotic cells 475, 477t
  microbial cells 479–80, 480t
cumulative ecosystem effects 1318
curiosity 6
cyanobacteria 434–5, 517–18
  as origin of chloroplasts 385,
      906–7
  Spirulina (Dihé) 435, 508f
cyclic adenosine monophosphate
      (cAMP) 612f
cyclin-dependent kinase inhibitors
      (CDKIs, CKIs) 418
cyclin-dependent kinases (CDKs,
      CDCs) 417–18
cyclins 417–18
  concentrations during the cell
      cycle 418f
cycliophora 37f
cyclosporidiosis 542t
cysteine (Cys, C) 287–8, 288f
  disulphide bonds 289f
cystic fibrosis 358, 591–2, 912
  main organs affected 591f
cysts, protozoan 540, 541f
cytochrome oxidase 335
cytochrome P450 enzymes 829
cytochromes 334
  cytochrome c 334–5
cytokines 864, 870, 871t
  role in T cell activation 887
cytokinesis (cytosolic division) 232,
    233f, 416f, 954
cytoplasm 384
cytosine (C) 27, 176, 203
  structure 200f
cytosine methylation 280f
cytoskeleton 916f
  bacteria 374t
  eukaryotic cells 393–4
  functions 914–15
  structures of 915–16, 915t
  tensegrity model 917f
cytotoxic T lymphocytes (CTL) 886,
    892
  mode of operation 894f
cytotoxins 528

D

D-isomers 158
DAG (diacyl glycerol) 612
dalton (Da) 172, 285
damage-associated molecular patterns
      (DAMPs, alarmins) 870, 875
dandelion 1073, 1074f
darifenacin 664
dark current 634
Darwin, Charles 34, 35, 48, 224,
    1024, 1060, 1152, 1204f, 1222
  'tangled bank' metaphor 1203–4,
    1204f

theory of natural selection 1204–5
Darwin's finches 1222f
data
  independence of 104–5
  levels of measurement 64
  units 64–6
data analysis 1203
data collection techniques 1254–7
data description 79–80
  central tendency 80
  frequency distributions 82, 83t
  measures of variability 80–1, 81t
data types 1253–4
data (dependent) variables 66
daughter chromosomes 231
Dawkins, Richard 35
DDT (dichlorodiphenyltrichloroethane)
    1344
dead space 756–7
deafness
  genetic causes 246–7
  pleiotropy 239
death cap 446f, 454, 538
death phase, cell cultures 485f
decibel scale 1018–19
deciduous trees 1035
decompression sickness (the
      'bends') 1040
decorator crabs 1121f
decussation
  corticospinal tract 707
  optic chiasm 636–7
deductive reasoning 8, 9f, 10
deep-brain stimulation (DBS) 714f
deep-branching bacteria 433
deer 997f
defecation 822
defence 32
defence mechanisms
  physical barriers 863
  plants 553
  see also immune system
deforestation 1327f
degenerate code 255
dehydratases 332
dehydration, effect on exercise
      performance 781
dehydration condensation
      reaction 28f
Deinococcus spp. 433, 434f
δ proteobacteria 436
delta symbol (Δ) 138
demographic factors 1245
  population density 1246–7
  population dispersion 1249
  population distribution 1247–8
  population size 1245–6
  denaturation of proteins 295, 317
dendrites 595, 596f, 609
dendritic cells (DCs) 865f, 866, 870f
  role in T cell activation 886–7,
    888f, 894f
dendrochronology 938f
dendrograms 44, 46f
dengue fever 13f
denitrification 520
deoxyribose 27f, 199f
dependent variables 66–7
depolarization 600f

dermatophytes 534
descriptive statistics 79–80, 96
  central tendency 80
  frequency distributions 82, 83t
  measures of variability 80–1, 81t
desert ecosystems 1301
desertification 1320, 1328
designed (experimental) studies 67,
    68t
desmin 915t
desmosomes 401f, 944, 945f,
    946f
desmotubules 941
Desulfovibrio spp. 436
detection function 1255, 1256f
detritus food chains 1303–5, 1304f
detrusor muscle 800
deuteranopia (red–green colour
      blindness) 244
deuteromycetes (fungi
      imperfecti) 448
deuterostomes 969–70, 970f
  limbs 986–91
  nervous systems 975–81
  subdivisions 975
development
  amniotes 1174–6, 1174f, 1175f
  bilateria 968–70
  cell differentiation 936–7, 939
  cleavage orientations 971f
  essential processes 938t
  nervous systems 973–81, 977f,
      980f
diabetes mellitus 680–1, 792, 825
diadromous fish 1071, 1072t
diaheliotropism 1034
dialysis 799f
diapause 1184, 1186t
  embryonic 1192–4
diaphragm 749, 750f, 751f
diastole 727, 737f
diastolic pressure 740, 769
  responses to exercise 769–70,
      770f
diatoms 461–2, 463f
diazotrophs 517–19, 518t
dicotyledons (dicots) 929f, 956
  leaf internal structure 964f
  leaf morphology 962f
Dictyostelium discoideum 465, 466f
dieback 449, 550f
diel vertical migration (DVM) 1069,
    1070f
diet 804
  and exercise 779–80
differential interference contrast
      (DIC) microscopy 493
differentiation 933, 951, 954
  animal cells 936–7, 939–40
  in plants 933–6, 937f
diffusion 31, 585–7, 586f, 730, 1031,
    1044
  in alveoli 758
  facilitated 588f, 911, 912f
  Fick's law 126, 127, 909
  influencing factors 908–9
  simple (passive) 910–11, 912f
diffusion coefficient (D) 586
diffusion flux 586–7

digestive system 803–4
  anatomy 805, 806f
  gut microbiome 822–3
  histology 804–5
  large intestine 820–2
  liver 826–9
  oral cavity 806–8
  pancreas 823–6
  peristalsis 805
  pharynx 810f
  regulation of 805
  small intestine 815–20
  stomach 811–15
  swallowing 810–11
  typical regions and structures 804f
digital parallelism 989, 990f
digits 988–9
Dihé 508f
dihydropyridine receptors
  (DHPRs) 691, 692f
dihydroxyacetone phosphate 330,
  331f
dilutions 91t
dimorphism 535, 536t
  yeasts 453
dinitrophenol 336, 337
dinoflagellates 459, 460f, 461
dioecious organisms 1150, 1151f,
  1164
diphtheria 528
diploblasts 937, 965–6, 971f
diplococci 374f
diploid cells 1143
diploid chromosome number
  (2n) 225, 418–19, 1223t
dipoles 162–3
direct fitness 1211, 1212f
direct proportion 122
  Beer–Lambert Law 126
  Fick's Law 126
directional selection 1207, 1208f,
  1233
disaccharides 816
discontinuous capillaries
  (sinusoids) 730, 826f
discontinuous distributions 1247–8,
  1247f
discrete data 64
discus fish, parental care 1188–9
disease triangle 547f
disease vectors 1343
disinfection 471
disjunctions 1247
dispersion 1249
disruptive (diversifying)
  selection 1208f, 1233
dissemination 14
dissociation constant (K_a) 171
distal convoluted tubule 787f, 795
  peritubular capillary network 790f
  reabsorption of major
    substances 790t
distribution of animals
  land bridges 1065
  mammalian radiations 1064–5
  migration 1068–72
  relationship to continental
    drift 1056–8
  sea routes 1065, 1067–8

distribution of plants 1064
  dispersal 1072–7
distributions 1247–8
disulphide bonds 288, 289f, 290f
diuretics 796–7
diversity 37–8, 40, 42, 1200, 1201f,
  1282–4
  of antigen receptors 876, 883–4
  Archaea 439–45
  bacterial 432–9
  of cells 368
  fungi 445–58
  insects 986f
  island species 1059–60
  protists 458–67
diversity index 125, 1284
diving, decompression
  management 1040
diving reflex 1213
division, symbols for 121
divisome 407
Dmrt1 gene 1153–4
DNA (deoxyribonucleic acid) 26, 27f,
  173, 176, 197–8, 387
  5' and 3' ends 201t, 256–7
  adaptations of thermophiles 443–4
  base pairing 201–2
  complementary (cDNA) 353–6
  double helix 204–5, 205f
  evidence of role as genetic
    material 203–4
  hydrogen bonding 181
  mitochondrial (mtDNA) 206,
    207f, 395, 396, 1253
  phosphodiester bonds 201t
  spontaneous base pairing in
    solution 206
  structural differences from
    RNA 202–3
  structure 12f, 178f, 198f, 202f
  supercoiling 213f
DNA damage 219–20, 219f
DNA ligase 214
DNA methylation 280, 281f
DNA packaging 210, 211f, 212f
DNA polymerases 213–14, 215,
  216f
  eukaryotic 217–18
  in PCR 346–7
  proofreading action 216, 217f
DNA profiling 209–10
DNA recognition sequences 345f
DNA repair 220
DNA replication 178f
  comparison in prokaryotes and
    eukaryotes 217–18
  direction of 214
  elongation reaction 214f
  'end replication' problem 218f
  mismatch repair 217f, 218
  polarity problem 214f
  prokaryotic cells 213–16
  proofreading 216, 217f, 218
  replication forks 215f
  replication origins 211–12, 213f
  semi-conservative replication 210,
    212f
  topoisomerases 213
DNA sequencing 349, 351f

chain-termination method (Sanger
  sequencing) 349–50, 351f
next-generation technologies 351–3,
  352f
whole-genome sequences 353,
  354f
DNA technology 344
  applications 353–9
  cloned DNA storage 347, 349f
  electrophoresis 344–5, 346f
  PCR 345–7, 347f, 349
  recombinant DNA 347, 348f
  restriction enzymes 344
  use of viruses 573
DNA transposons 208f
DNA viruses 564–6
Dobzhansky, Theodosius 1203
dodder 1278f
dogs
  genome-wide differences 1210–11,
    1210f
  selective breeding 44
dolutegravir 571
domains 49
dominant alleles 243, 1222, 1223f,
  1229f
dominant phenotypes 228
dopamine 859
dormancy 1294
dorsal column-medial lemniscal
  pathway 621–4, 623f
dorsal root ganglions neurons 619
double-blind studies 104
double covalent bonds 155f
  in fatty acids 371
double helix structure, DNA 204–5,
  205f
double-strand breaks, DNA 220
doubling time (generation time) 140,
  488
Down syndrome (trisomy 21) 228t,
  424, 836
drug metabolism 829
dry rot 454
dry weight determination 484
Dscam (Down syndrome cell
  adhesion molecule) 1128
duck-billed platypus 57, 59f
duckweeds, taxonomy 57, 58f
ductus arteriosus 856f, 857
ductus venosus 856f, 857
duodenum 806f, 815, 824f
  see also small intestine
Dutch elm disease 452, 453f, 552f
dynamic equilibrium 598
dynein 298, 915t

E

e (Euler's number) 75–6, 121t,
  122, 488
ear
  anatomy 641f
  balance 645–8
  hearing 639–43
ear structures 1024–5
Earth
  asteroid impacts 20
  atmosphere 21
  differentiation of 19f

formation of 18–19
  internal structure 19
  oceans 21
  plate tectonics 19–21, 20–21f
  structure of 1054, 1055f
Earthrise image 1200f
earthworm 983f
  Darwin's studies 1024
  osmoregulation 1048
  peristaltic locomotion 1078f
eccentric muscle contractions 700
ecdysone 1184, 1185f
echinoderms
  integument 1113f
  limbs 987
  nervous system 976f
echolocation (sonar) 1025
eco-evolutionary dynamics 1218
ecological communities *see*
  communities
ecological footprints 1305f
ecological interactions 1271,
  1279–80, 1280f
  competition 1274–6
  herbivory and predation 1271–4
  mutualism and
    commensalism 1279
  parasitism 1277–9
ecological management 1339
  non-native species 1345
  pests and disease 1343–5
ecological pyramids 1309, 1309f
ecological relationships 952
ecological resilience 1313
ecological succession 1286–7, 1288f
ecological tipping points 1313
ecology 1200
  complexity 1200–1
  founders of 1201f, 1202
  hierarchy of levels 1201
  scientific method 1202–3
ecoregions 1354
ecosystem challenges
  anthropogenic 1319
  classification of 1317
  direct and indirect effects 1319
  habitat loss, degradation, and
    fragmentation 1326–31
  human population
    growth 1319–20
  impact severity and impact
    magnitude 1318
  implications of 1331–5
  natural disasters 1317–18, 1318f
  non-native species 1331
  resource use and
    emissions 1320–6
  timescales 1318–19
  types of solution 1339
ecosystem collapse 1313
ecosystem engineers 1311, 1357
ecosystem processes 1301
ecosystem services 1332, 1335
ecosystems 1201, 1292f
  abiotic factors 1294–7
  classification of 1293–4
  energy transfer 1303
  keystone species 1311–12, 1312f,
    1357

nutrient cycling 1309–10
productivity 1301, 1303
species–environment
   interactions 1297–301
trophic levels 1303–8, 1304f
ectoderm 965f, 966
ectodermal tissues 967f
Ectognatha
   body plans 985–6
   see also insects
ectoparasites 1277, 1278
ectoplasm 465
ectothermic organisms 1034
edge effects 1329–30
Edwards syndrome (trisomy 18) 228t
effect sizes 116
efferent neurons 595
eggs 1167–8
   types of 1168–73
   see also oocytes
eggshell formation 1172f
Eimeria spp. 461
elastic arteries 726
elastin 296
electrical synapses 605–6, 607f
electrocardiogram (ECG) 736, 737f
   electrode placement 736f
   interpretation 738f
electrochemical gradient 912
electromagnetic spectrum 1005–6,
   1007f
electron delocalization 156–7
electron microscopy 494–6
electron shells 151
electron transport 192–4
electron transport system
   (chain) 325, 326–8, 327f,
   328f, 334
   chemiosmosis 335–6
   electron carriers 334–5, 334f
   oxidative phosphorylation 336–7
   proton ejection 336
electronegativity 153, 155t
electrons 150, 151
electrophilic substances 192
electrophoresis 344–5
   DNA sequencing 350, 351f
electroreception 1030f
elementary bodies 437
elements 150
elephant, foot anatomy 989f
elephant bird 1060f
elongation factors 269–70
embryo development,
   amniotes 1174–6, 1174f, 1175f
embryology
   cell differentiation 936–7, 939
   essential processes in
      development 938t
   nervous systems 973–81, 977f,
      980f
   see also development
embryonic diapause 1192–4
embryonic stem (ES) cells 356, 939
   emergence of life 1054, 1142
   complex biomolecules 25–9
   fossil studies 16, 18f
   life forms 28
   origins of organic molecules 22–5

phylogenetic studies 16, 17f
emigration 1251–2
emphysema 752
emulsifiers 817
enalapril 311f
enamel 997
enantiomers (optical isomers) 157–8,
   160f, 161
   amino acids 290–1, 291f
   biological significance 161
ENCODE project 359
end diastolic volume 771
'end replication' problem, DNA 218f
endangered species
   IUCN red list 1334–5
   legislation 1340–1
endemic species 1247
endergonic reactions 166, 186f
endocardium 732
endocrine signalling 614, 665, 666f
endocrine system 665f, 797–8, 800–1
   adrenal glands 676–8
   calcium and phosphate
      homeostasis 681–2
   control of reproduction 838,
      840–3
   control of vascular smooth
      muscle 729–30
   hormone immunoassays 683
   hormone types 665–6
   hypothalamic–pituitary axis 666–8
   kidneys 800–1
   small intestine 819–20
   thyroid gland 672–6
endocytosis 392f, 590–1, 590f, 914
   receptor-mediated (RME) 560,
      561f, 562f
endodermal tissues 965f, 967f
endogenous heat production 1037–8
endomembrane system 388–9
   Golgi apparatus 389–90, 390f
   lysosomes 391–2, 392f
   protein trafficking 390–1, 391f
endometrium 844
   uterine cycle 849–50, 849f
endomysium 687f, 688
endoparasites 1277–8
endoplasm 465
endoplasmic reticulum (ER) 389f
endorphins 628
endosomes 561f, 562f, 914
endosperm 1164, 1166, 1167f
   coenocytic 925f
endospores 382f, 439, 470
endosymbiosis 384–5, 978
   origin of chloroplasts 906–7, 906f
endothelium 726
endothermic organisms 1034
endotoxins 378, 527–8, 528t
energy
   definition of 185
   forms of 185
   Gibbs' free energy 186–7
   laws of thermodynamics 31, 185–6
energy changes 165–7
energy conservation 1030–1
energy cycle in life 189f
energy levels, electrons 151
energy sources 185

energy transfer 31, 1030–1
   within ecosystems 1303
   heat transfer 1031–3, 1032f
energy transformations 185
enfuvirtide 570
enhancers 275
enolase 332
Entamoeba spp.
   E. coli 466f
   E. histolytica 541f
enteric nervous system (ENS) 595,
   657, 805
enterotoxins 528
enthalpy changes (ΔH) 166, 185
Entognatha 985, 986
entrainment 1015
entropy 31, 1030
entropy changes (ΔS) 185–6
envelopes, viral 558, 564
environmental gradients 1294
environmental sex determination
   (ESD) 1154–6, 1155f
environmental variables 1001, 1002t
   human interpretation of 1001–2
   plant and animal responses 1001
   see also aqueous environment;
      electroreception; gaseous
      environment; gravity; heat; light;
      magnetic fields; sound
enzyme-catalysed reactions, effect of
   substrate concentration 313f
enzyme inhibitors 310–12
   competitive and non-
      competitive 312f, 313f
enzyme kinetics 309–10, 311f
enzyme regulation 314, 315f, 329–30
enzymes 167, 180, 187, 188f, 298, 306
   active site 306f
   allosteric 312–14
   cofactors 308
   conformational changes 181
   covalent modification 314, 315f
   in DNA replication 211, 213–14,
      215, 216f, 217–18
   'induced fit' model 306–7, 306f
   industrial applications 310
   industrial production 505t
   isoenzymes 308
   kinases 191
   'lock-and-key' model 306f
   lysosomal 391
   mechanisms of action 307–8, 307f
   pH and temperature effects 308f,
      309
eosinophils 721f, 865f, 866, 875
epicardium 732
epidermolysis bullosa 922
epididymis 839f, 847f, 1178
epigenetic modification 32
epigenetic processes
   imprinting 1148–9
   temperature sex
      determination 1154, 1155f
epiglottis 745, 746f, 807f, 810
epimysium 687f
epithelia 930–1, 930f
   basement membranes 931–3
   Cnidaria 965–6, 965f
   triploblasts 966

epithelial cells, orientation and
   polarity 920–2, 921f
epitoky 1181
epitopes 878
ε (epsilon) proteobacteria 436
equations 121–2
   rearranging 127–31
equilibrium, chemical 164–5, 187
equilibrium potential 597–8
   Nernst equation 131, 193, 599–600
equilibrium sensing 1021–2
equivalent ratios 87–9
ergotism (St Anthony's fire) 538
error bars 97f, 111f
error plots 112f
errors, Type I and Type II 115–16
erythrocytes (red blood cells) 719–
   20, 759
   aged, removal from
      circulation 828
   anaerobic respiration 338
   development, GATA-1 278–9
erythropoiesis 720
   altitude acclimatization 765, 782
erythropoietin (EPO) 720, 801
   recombinant, use by athletes 782
Escherichia coli (E. coli) 436f
   DNA replication 213–16
   mutant studies 407, 408
esters 158t
estradiol 176f
ethanoic acid 171
ethanol
   industrial production 504, 506
   production in fermentation 338,
      339f
ether linkages 373f
ethers 158t
ethics 14
   cell lines 476
   gene therapy 357, 358, 359
ethylene oxide 471
Euarchontoglires 1065
euchromatin 210
euglenids 459
Euglenozoa 458–9
Eukarya 50f, 54f
   phylogenetic tree 460f
eukaryotes 17f
   membranes 31
eukaryotic cell division
   interphase 412–14
   meiosis 418–23
   mitosis 414–18
   see also meiosis; mitosis
eukaryotic cells 384
   central vacuole 400f
   chloroplasts 399–400
   comparison with prokaryotic
      cells 369t
   cytoskeleton 393–4
   endomembrane system 388–91
   fungal cell wall 398–9
   lysosomes 391–2, 392f
   mitochondria 394–6
   nucleus 387–8, 387f
   organelles 384–5
   peroxisomes 400–1

eukaryotic cells (*Cont.*)
  plant cell wall 397–8
  ribosomes 388*f*
  sizes 369*f*, 385–6
Euler's number (*e*) 75–6, 121*t*, 122
European mole (*Talpa europea*) 952*f*
Euryarchaeota 442, 444
euryhaline fishes 1049, 1050
Eustachian tube 640, 641*f*, 642*f*
eutherian mammals,
    distribution 1058
eutrophication 520, 1320, 1325–6,
    1326*f*
evaporation 1032*f*, 1033
  thermoregulation 1036–7, 1037*t*
evolution 33, 1203, 1222
  antibiotic resistance 38
  arms races 1108–9, 1145, 1272,
    1273*f*
  beginnings of 34–5
  brain 978–81
  causes of change 35
  convergent 59
  definition of 34
  drivers of 1206–9
  eco-evolutionary
    dynamics 1218–19
  eyes 1014
  founders of the theory 1204–5,
    1204*f*
  genotypic adaptations 1209
  human 39
  human dive reflex 1213–14
  on islands 1059–60
  mammals 1058*f*
  misconceptions 1214
  of multicellularity 928
  non-directed nature of 58–9
  observing the outcomes of 1213
  origins of organelles 384–5
  proteomic and genomic
    studies 318, 320–1
  study of 36–40
  terminology 1205–6, 1223*t*
  theories and models 11
  of viviparity 1173–4
evolutionary change, processes
    of 1231–41
evolutionary chronometer (molecular
    clock) 49, 50, 51*f*, 428
evolutionary developmental biology
    ('evo-devo') 40–1
evolutionary distance 51
evolutionary fitness 1211–12
evolutionary pressures 1143, 1145
evolutionary radiations 1064
evolutionary relationships 53
exchange (antiport) 589
excision repair mechanisms,
    DNA 220
excitation–contraction coupling 688,
    691
excitatory post-synaptic potentials
    (EPSPs) 608
  synaptic integration 609–10
excystation 540, 541*f*
exercise-induced angiogenesis 774
exercise physiology 768–9
  acid–base balance 779–80

at altitude 781–2
  cardiovascular responses 769–74
  fitness and health 782–3
  in heat 780–1
  hydration status 781
  maximum oxygen uptake 775–7
  nutritional supplements and
    diet 779–80
  pulmonary responses 777–9
exercise training
  cardiovascular adaptations 772–4,
    774*t*
  pulmonary adaptations 778–9
  specificity and overload 777
exergonic reactions 166, 186*f*
exocytosis 591
exons 207
exoskeletons 992–4
  arthropods 1112–13
  impact of ocean acidification 993
  insect moulting 1183
exotoxins 377, 527, 528, 528*t*
expansins 919
experimental design 100–1
  adding variables 105
  bias and inaccuracy 102
  control 102
  implications for interpretation 105
  measurement error 101–2, 102*f*,
    103*f*, 104
  power analyses 104–5
  randomization 103
  variability and noise 101–2
experimental studies 67, 68*t*, 1202
expiration 750–1, 751*f*, 777
expiratory reserve volume
    (ERV) 753*f*, 754
explanatory (independent)
    variables 66
explants 475
exploding ant 1211–12, 1212*f*
exploitative competition 1276
exponential decay 125*t*, 142–3
exponential equations 122, 124*t*
  rearranging 129–30
exponential growth 139–40, 486–8,
    1263, 1264*f*
  continuous growth model 141
  reproduction number (R) 140–1
exponential phase, cell cultures 485*f*
expression cloning 613–14
extended phenotypes 35, 36*f*
external (part-to-whole) ratios 87
  concentrations 90–1
  relating to fractions and
    percentages 89*t*
external respiration 745
  see also breathing
extinction 1332
  chance extinctions 1353
  mass extinction events 1313–14
extinction risk, IUCN red list 1334–5
extinction vortices 1332, 1333*f*
extirpation 1332, 1356
extracellular matrix (ECM) 401–2
  key components 402*f*
extracellular polymeric substance
    (EPS) 383
extrapyramidal tracts 709

extremophiles 1299
  Archaea 443–5
  methanogens 443–4
  see also acidophiles; halophiles;
    thermophiles
exudates 1124*f*
eye, human
  autonomic control 663–4
  cornea and lens 633–4
  focusing the image 633*f*
  refractive errors 633, 634*f*
  retina 634–6, 635*f*
  rods and cones 636*f*
  structure 632*f*
eye spots 1015, 1018
eyes 1011–14
  evolution 1014
  light detection 1015
  photoreception and vision 632,
    1014–15
  types of 1011*t*

F

F_1 generation 228
facilitated diffusion 588*f*, 911, 912*f*
F-actin
  microscopic studies 539*f*
  see also actin
factor VIII 244, 1126
factors (independent variables) 66
facultative diapause 1192
facultative herbivores 1271
facultative mutualism
    (protocooperation) 1279
facultative parasites 534, 548
facultative sexual reproduction
    1146
FAD (flavin adenine
    dinucleotide) 193*f*
$FADH_2$ 335*f*, 336
fairy rings 454, 456*f*
Fallopian tubes 837*f*, 848*f*, 849
Faraday constant (*F*) 121*t*
fascial compartments 687*f*
fascicles 687*f*, 688
fats *see* lipids
fatty acids 26, 168*f*, 173
  in lipid bilayers 371
  micelles and liposomes 28
  oxidation 188, 329
Fc receptors 882
feathers 1119*f*
fecundity 1156, 1251
  angiosperms 1166
fed-batch culture 484, 485*f*
feedback
  control of ventilation 765
  hypothalamic–pituitary axis 668,
    670*f*
  hypothalamic–pituitary–gonad
    axes 841, 842*f*, 843*f*
  negative 329, 582–3
  positive 583, 1313
  proprioception 629–31
  tubuloglomerular 788–9
female reproductive system 848–50,
    848*f*
  anatomy 848–50, 848*f*
  birth 857–9

effects of gonadal steroids 846*t*
  hypothalamic–pituitary–ovarian
    axis 842–3, 843*f*
  lactation 859
  menopause 846
  menstrual cycle 844, 845*f*
  oogenesis 836–7
  ovarian follicle development 837*f*
  pregnancy 850, 852–7
  puberty 844–5
fenestrated capillaries 730
  glomerulus 787
fermentation 338, 339*f*, 499
  acetic acid production 510
  bread making 510
  brewing beer 499, 512–13
  wine production 513
ferns 955*f*
  reproduction 1144*f*
fertility 1156
fertilization 1167
  broadcast spawning 1184–90
  mammalian 850, 1169, 1178–9
fertilizers 1156
fetal circulation 725, 856–7, 856*f*
  adaptation at birth 857
fetal haemoglobin 761, 857
fetal lungs 857
fever 584
fibrin 1125
fibrinogen 722, 1125
fibrinogliase 1125
fibrinolytic cascade 1125*f*, 1126–7
fibronectin 1125
fibrous proteins 298, 299*f*
  collagen 302–3
Fick's law of diffusion 126, 127,
    586–7, 730, 909
'fight or flight' response 582, 654,
    655, 664
figures *see* graphs
filamentous bacteria 374*f*
filaments 1161*f*, 1162
filtration
in kidneys 787–9
  as sterilization method 471*f*
filtration pressures, capillaries 731
fimbriae 381
Fimicutes 437–9
fins, aspect ratios, 1094*f*, 1095*f*
fireworms 1134*f*
first-order afferents
  nociception 626, 627*f*
  touch 621, 623*f*
first-order reactions 142*f*
fishes
  body shapes 1096*f*
  buoyancy 1028
  courtship 1185–6, 1187*f*
  ear structures 1025
  eggs 1170–1
  integument 1114, 1115*f*
  invasive 1065–7
  lateral line 1022*f*
  migration 1071–2, 1072*t*
  nests 1188*f*
  osmoregulation 1049–50, 1049*f*
  overfishing 1322
  parental care 1188–9

predator avoidance 1137, 1138*f*
reproduction 1184–90
scales 1114, 1115*f*, 1116*f*
sex determination 1154–6, 1155*f*
social structures 1190
swimming *see* swimming
see also bony fishes (teleosts);
cartilaginous fishes
(elasmobranchs/
chondrichthyes)
fishing quotas 1342–3
fission *see* binary fission
fitness 1211–12, 1232–3
fixed factors 105
flagella (singular flagellum) 380–1,
381*f*
flatworms *see* platyhelminths
Fleming, Alexander 38, 500*f*
flight 1028
Florey, Howard 500*f*
flowering, timing of 1159–60
flowers
female gametes 1163–4
fertilization 1164
male gametes 1162–3
morphological variation 1161–2,
1162*f*, 1163*f*, 1164*f*
scent production 1163
single-sexed 1164
structure 1161*f*
fluid mosaic model 372*f*
fluorescence-activated cell sorting
(FACS) machine 479*f*
fluorescence microscopy 493, 494*f*
fluorescent probes 494, 495*f*
fluorescent staining 491–2, 492*f*
fly agaric 454, 455*f*, 538
flying and gliding
aerofoils 1088, 1089*f*
efficiency 1091, 1093
flapping flight 1093
Reynolds number and air
flow 1090
thrust, lift, and drag 1088–90,
1089*f*
flying fish 1096, 1097*f*
FMN (flavin mononucleotide) 193,
334
structure 194*f*
foliate papillae 648, 650*f*
foliose lichens 457*f*
follicle stimulating hormone
(FSH) 669*f*, 836, 841*t*
actions in female 843
actions in male 842
control of secretion 841–2
menopausal levels 846
menstrual cycle 844, 845*f*
follicles, ovarian 836, 837*f*
fomites 529
food chains 1303–5, 1304*f*
food oxidation 189
food webs 1305–7, 1306*f*
stable isotope analysis 1307–8
foramen ovale 856*f*, 857
forced expiratory flow (FEF) 754*f*
forced expiratory volume in 1 s
(FEV$_1$) 754*f*
forced vital capacity (FVC) 754

forest canopies 1003*f*
form
aspect ratios 1092*f*, 1094*f*, 1095*f*
general principles of 951
relationship to function 951–2
scaling relationships 953–4
size and complexity 954
of societies 952
surface area to volume ratio 952–
3, 953*f*
formose reaction 27*f*
formulas 122
Beer–Lambert Law 126
Fick's laws of diffusion 126, 127
rearranging 127–31
Shannon–Weiner diversity
index 125
and shapes of graphs 124*t*
fossil record 16, 18*f*, 37, 1056
founder effects 1238, 1239
fovea 632*f*, 634
Fox, George 49
fractions, relating to ratios 89*t*
fragmentation
bacteria 408
of habitats 1329*f*
Frankia spp. 518
Franklin, Rosalind 204
Frank–Starling relationship 739–40
free energy (G, Gibbs' free
energy) 186–7
frequency codes 596
frequency distribution graphs 108
clustering and stacking 110*f*
frequency distribution tables 82*t*
frequency distributions 80, 82
construction of 83*t*
examples 84*t*
freshwater environments 430
see also lakes; rivers
fructose 159*f*
absorption in small intestine 817
fructose 1,6-bisphosphate 330, 331*f*
fructose-6-phosphate (F6P) 330, 331*f*
interconversion to glucose
6-phosphate 164–5*f*, 166*f*
frugivory 1273–4
fruit fly
imaginal discs 1186*f*
sperm length 1179*f*
fruits, seed dispersal 1077*f*
frustule 462, 463*f*
fruticose lichens 457*f*, 458
Fts (filamentous temperature
sensitive) proteins 406*f*, 407,
408
fucoxanthin 463
fumarase 333*f*
functional genomics 360
functional groups 157, 158*t*
functional magnetic resonance
imaging (fMRI) 624–5
functional residual capacity 753*f*
fundamental niche 1298, 1299*f*
fundamental ranges 1247
fundus of the stomach 811, 812*f*
fungal allergens 538
fungal cell wall 398–9
structure 399*f*

fungal diversity
Ascomycota 451–3
Basidiomycota 454, 455*f*
Chytridiomycota 449–50
five major phyla 449*f*
Glomeromycota 451
lichens 457–8, 459*f*
yeasts 453–4
Zygomycota 450–1
fungal pathogens 534
plant diseases 547–53
subcutaneous infections 534–5
superficial infections 534, 535*f*
systemic infections 535–7
fungal poisoning 538–9
fungal spores 448
arthrospores 537*f*
ascospores 453*f*
basidiospores 454, 455*f*
conidia 453, 535
zoospores 449, 450*f*, 456
zygospores 450–1, 451*f*
fungi 445, 446*f*
in Asian food production 511
bioremediation 515
in cheese production 511
culture collections 474*f*
deuteromycetes 448
dual nomenclature 449
life cycle 448*f*
mushroom production 508–9
mycoparasites 514–15
mycorrhizal relationships 447
nutritional types 447
single-cell protein 507–8
structure 445–7, 447*f*
syncytial mycelia 923–4, 924*f*
fungiform papillae 648, 650*f*
fur seals (*Callorhinus ursinus*) 1193*f*
embryonic diapause 1192
migration 1069–71, 1070*f*
Fusarium spp. 508, 552
fusion of viruses 560, 561*f*
inhibitors of 570

G

G$_1$ phase of interphase 413*t*
G$_2$ phase of interphase 413*t*
g-actin 394
G-cells 812*f*, 813
G-protein coupled receptors
(GPCRs) 610, 615
activation 611–12, 611*f*
intracellular actions 612*f*
P2Y12 receptor 613–14
G-proteins (GTP-binding
proteins) 298, 610–11
effects of 612, 614
gaits 1084
Galapagos cactus finch 1260*f*
gallbladder 806*f*, 827–8, 827*f*
gallstones 828
gamete production 832–3, 834*f*
genetic variation 833, 835–6
oogenesis 836–7
spermatogenesis 837–8, 839*f*,
840*f*
gametes 419
gamma diversity (Υ) 1282, 1283*f*

γ globulins (immunoglobulins) 722,
876, 879*f*, 880*f*, 893
γ motor neurons 630*f*
γ proteobacteria 436
ganciclovir 570–1
ganglia (singular ganglion) 655*f*
gap junctions 401*f*, 606, 607*f*, 944–5,
947*f*
gas constant (*R*) 121*t*
gas exchange 757–8, 909
in animals 1041–4
in the lung 758–9
partial pressures of oxygen and
carbon dioxide 757*t*
in plants 1040–1
simple (passive) diffusion 911
ventilation–perfusion
matching 759
gas laws 1039
biological implications 1040
gas pressures 1038–9
gas transport 1044–5
gas vesicles 381–2
gaseous environment 1038
gastric glands 812*f*
gastric pits 812*f*
gastrin 812, 813, 819
gastrodermis 966
gastroesophageal reflux disorder
(GORD) 811
gastroesophageal sphincter (cardiac
sphincter, LES) 810, 812*f*
gastrointestinal system 804
as a protective barrier 869
typical regions and structures 804*f*
see also digestive system
gastrulation 966, 968–70
GATA-1 275, 278
gated channels 588, 599
GC box 258, 275
geckos 1086, 1087*f*
gel electrophoresis 316–17, 344–5,
346*f*
DNA sequencing 350, 351*f*
PAGE apparatus 316*f*
Gelsinger, Jesse 358–9
gene dosage 227
gene duplication 318
gene expression 252–3
order of events 253*f*
in prokaryotes and eukaryotes 254
transcription 255–63
translation 264–72
gene flow 1240–1
goshawks 1252–3
gene for gene process 1109
gene guns 359
gene pool 1223–4
gene regulation
for cell differentiation 277
DNA elements involved 275*f*
DNA methylation 280, 281*f*
in eukaryotes 275–9
lac operon 272–5
post-transcriptional 280–1
transcription factors 272*f*
gene targeting 356–7
gene therapy 357, 358–9, 573, 896
general transcription factors 258

generation rates (doubling times) 140, 408, 488
generic part of a name 43, 44
genes 32, 207, 1222–3
   alleles 225–6
genetic code 176, 254–5, 254t
genetic disease 241, 243
   chromosomal numerical abnormalities 226–7, 228t
   hereditary deafness 246–7
   pedigree analysis 244, 246, 248f
   single-gene disorders 243–4
genetic drift 35, 1238–9, 1238f
genetic engineering 39–40
   horizontal gene transfer 411–12
genetic heterogeneity 239
genetic maps 239
   human chromosome 1 240f
genetic sex determination (GSD) 1153–4
genetic variation 32–3, 833, 835–6
   role of meiosis 234–5, 420–2, 1146
genetically modified (GM) crops 359
genome editing 357, 358f
genome sizes 207, 208t
genomes 197
   organization in cells 206, 207f
   repetitive sequences 207, 209
   transposons 208–9
   viral 558
genome-wide scans for selection (GWSS)
   Bajau people 1213
   dog breeds 1210–11
genomic libraries 349f
genomic studies 359–61
   and study of evolution 318, 320–1
genotype 32, 227, 1205
genotype–environment interactions 1205–6
genotype frequency 1225
genotypic adaptations 1209
genotypic variation 1222
genus 427
geometric growth 1262–3, 1263f, 1264t
germ cell production 1146–7
germ theory of disease 11, 524–5
germinal centres 892–3
germination 956
   influence of sound 1019
gerontogenes 1224
gerontoplast 400
ghrelin 678, 814, 816
giant redwood tree 951f
giant squid 982f
Giardia lamblia 546f
giardiasis 541, 542t, 545–6
Gibbs' free energy (G) 186–7
   Nernst equation 193
Gibson assembly 349, 350f
gigantism 672
gland secretion, autonomic control 664
glial cells 595, 605
glial fibrillary acidic protein (GFAP) 915t
global warming 5
   Foote's prediction 10

historical changes 1057f
post-industrial temperature changes 6f
see also climate change
globular proteins 298
   oxygen-binding 299–302
globulins 722
globus pallidus 706f, 713
glochids 1105f
Gloeotrichia spp. 517–18
Glomeromycota 451
glomerular filtration rate (GFR) 788–9
glomerulus 787, 788f
glottis 746, 747f
glucagon 678, 679f, 681, 823–6
glucagon-like peptides (GLPs) 820
glucans 398
glucocorticoid receptor 616
glucokinase 308
gluconeogenesis 828
gluconic acid 504
glucose 159f, 173
   absorption in small intestine 816–17, 817f
   control of blood levels 679f, 791, 823–6
   oxidation 188
   reabsorption in the nephron 790t, 791–2, 792f, 797f
   structure 174f
   transport across membranes 588, 589, 911
glucose 6-phosphate (G6P) 330, 331f
   interconversion to fructose-6-phosphate 164–5f, 166f
glucose-dependent insulinotropic peptide (GIP) 819–20
glucose oxidation
   ATP yield 337
   main stages 325–8, 328f
   net result 325f
   see also electron transport system; glycolysis; tricarboxylic acid (TCA) cycle
glucose transporters (GLUT proteins) 817f, 911
   GLUT1, GLUT2 791, 792f
   GLUT4 825
glutamate, structure 179f
glutamic acid (Glu, E) 288f, 505
glutamine (Gln, Q) 288f
gluten sensitivity 820
glyceraldehyde-3-phosphate 330, 331f
glycerol 173, 369
   industrial production 504
glycerophospholipids 173
   structure 175f
glycine (Gly, G) 288f, 289
glycocalyx 380, 904
glycogen 173, 329, 825
   regulation of metabolism 314, 315f
   structure 174f
glycolipids 370
   structure 371f
glycolysis 325, 328f, 330–3, 331f
   anaerobic 337–8, 338f
   coupled reactions 191

phosphofructokinase 312
   reversible reactions 166f
   Warburg effect 338
glycophorins 303–4
   structure 305f
glycophytes 1299
glycoproteins
   gp 120 560, 570
   hormones 665
glycosaminoglycans (GAGs) 402
glycosuria 791–2
gnathostomes 987
goethite 998
golden algae 463–4
goldfish 1042f
   anoxia survival 1042–3
Golgi apparatus 389–90, 390f
Golgi tendon organ 631f
gonadotrophin releasing hormone (GnRH) 840–2
   role in puberty 845
Gondwana 1056, 1057f
gonochoric organisms 1150
gonorrhoea 531, 851, 852
goshawks 1252–3, 1252f
Gould's monitor 1086, 1088f
Graafian follicle 836, 837f
gradient (slope) 136–7
   estimation of 138f
graft-versus-host disease (GVHD) 887
Graham's law of diffusion 586
Gram, Hans Christian 377
Gram stain 375, 376–7, 378f, 490, 492f
Gram-negative bacteria 377–8, 379f, 379t, 380f
   phylogenetic tree 432f
Gram-positive bacteria 377, 378f, 379t, 380f
   phylogenetic tree 432f
grana (singular granum) 399f
granulosa cells 836, 837f, 838f, 842
graphs 108
   box plots 112f
   connecting lines 112f
   error plots 112f
   estimating slope of 138f
   of frequency distributions 109, 110f
   gradient (slope) 136–7, 136f
   main components 108t, 109f
   model lines 111, 112f
   relating to formulas 122, 124t
   of two variables 109, 110f, 111f, 112f
grasshoppers 1250f
   cohort life table 1250f, 1251t
   life cycle 1183f
   population modelling 1260
Grave's disease 676
graviperception 1026
gravitational potential energy 185
gravitropism 1025–6, 1026f
gravity 1025
   and airborne animals 1028
   and aquatic animals 1026, 1028
   plant growth in absence of 1027–8
   plant responses 1025

and terrestrial animals 1028–9
grazing 1271f
grazing food chains 1303, 1304f
grazing pressure 1328
great crested newts 1340f
green algae 467
green fluorescent protein (GFP) 493, 495f
greenhouse gases 1324
Grey, George Otto 475, 476
GroEL/GroES system 296f
gross primary production (GPP) 1303
grouping (independent) variables 66
growth 930t, 951
growth hormone 669f, 670–1, 841t
   deficiency of 672
   excess secretion 672
growth hormone releasing hormone (GHRH) 668, 669f, 670–1
growth kinetics, microbial cell cultures 486–8, 487f
growth plates 994, 995f
guanine (G) 27, 176
   structure 200f
gustation (taste) 648, 650f, 807–8, 1136–7
   taste buds 1139f
gut-associated lymphoid tissue (GALT) 867
gut development 969
gut epithelium 921f, 930f
gut microbiome 822–3
   changes through life stages 823f
   main phyla and genus of species 822t
gymnosperms 955, 960
   reproduction 1166–7, 1168f

H

H-zones 689
habitat change 1326
habitat creation 1351–2, 1351f
habitat degradation 1327–8
habitat-focused conservation
   landscape-based 1353–4
   site-based 1351–3
habitat fragmentation 1329f
habitat improvement 1352
habitat loss 1248, 1326–7
habitat maintenance 1353
habitat restoration 1356
habitat structure 1275–6
habituation 104
Haeckel, Ernst 1201f
haem 334, 396
   structure 301f
haematocrit 719
haematopoietic stem cell system 923f
haematoxylin–eosin (H&E) stain 490, 492f
haemocytes 1128
haemocytometers 481, 482f
haemodialysis 799f
haemoglobin 295, 299–300, 313, 719, 759, 1044
   allostery 301–2, 302f

amino acid sequence 318, 320*f*
anaemia 761
carbon dioxide transport 762
cooperative binding of
    oxygen 759–60
fetal 302, 857
mutations 302
oxygen dissociation curves 760–1,
    760*f*
oxygen saturation curve 300–1,
    301*f*
sickle cell disease 302–3
structure 300*f*, 301*f*, 720*f*, 760*f*
haemolymph 914, 1128
haemophilia 244, 1126
haemopoiesis 720, 721*f*
haemostasis 722
hagfish 987*f*, 1117*f*
    slime production 1117–18
hair 1119–20
hair cells 1021–2, 1021*f*
Halobacterium salinarum 442–3,
    444*f*, 445*f*
halophiles 442–3, 444–5
    adaptations 443
halophytes 1045, 1300*f*
Haloquadratum walsbyi 383*f*, 384,
    445*f*
halteres 1093
hand-wing index (HWI) 1091–3,
    1092*f*
haploid cells 1143
haploid chromosome number
    (n) 225, 419, 1223*t*
happy face spider 1216*f*
hard materials
    in animals 991–8
    in plants 991
Hardy–Weinberg equilibrium 1228,
    1231
Hardy–Weinberg model 1226
    assumptions 1227–8
Hartig net 454
Harvard Alumni Health Study 783
Hashimoto's thyroiditis 676
haustoria 458, 550*f*, 1278*f*
haustra 821*f*
Hawaiian islands 1059–60
health, physical activity and
    fitness 782–3
hearing 639, 639–40, 1023
    auditory ranges 1024*t*
    central pathways 644, 645*f*
    ear structures 1024–5
    human audiogram 1024*f*
    inner ear 641*f*, 642, 642–3
    in invertebrates 1023–4
    middle ear 640, 641*f*, 642*f*
    sound waves 639, 640*f*
    in vertebrates 1024
hearing loss 646
    genetic causes 239
heart 723
    cardiac cycle 737–9, 737*f*
    cardiac output 739–40
    conducting system 733–6, 735*f*
    coronary circulation 724
    electrocardiogram (ECG) 736,
        737*f*, 738*f*

structure 732–3, 733*f*
    see also cardiac muscle
heart attacks 738
heart block 738
heart rate
    autonomic control 662
    effect of aerobic exercise
        training 773
    response to exercise 770, 771*f*
heart sounds 737*f*, 739
heat 1030
    and exercise performance 780–1
    thermal environment 1033–8
heat acclimatization 781
    see also thermophiles
heat insulation 1034
heat shock proteins 444, 1034–5
heat transfer 1031–3, 1032*f*
heathland management 1352
height variation 241, 242*f*
HeLa cells 475*f*, 476
helical bacteria 374*f*
helicases 211, 215*f*
Helicobacter spp.
    H. hepaticus 436
    H. pylori 527*f*, 530*f*, 531
heliotropism 1009
Helminthosporium victoriae 550–1
helper T cells (Th cells) 886
    activation 886–8
    B cell activation 892, 895*f*
    differentiation 889*f*
    functions 889–92, 890*f*
hemicellulose 397*f*, 398
hemidesmosomes 401*f*
hemiparasites 1278
Henderson–Hasselbalch
    equation 171, 762–3
Henry's Law 1039
hepadnaviruses 566, 568*f*
hepatitis viruses 563*t*
hepatocytes 826*f*
herbivory 1103–4, 1271, 1280*f*
    chemical defences 1105–7
    frugivory 1273–4
    nectivory 1274, 1275*f*
Hering–Breuer reflex 765
heritability 1233–6
    calculation of 1234–5
hermaphroditism 1150, 1151*t*, 1152,
    1180
herpesviruses 563*t*, 851, 852
    capsid assembly 563, 564*f*
heterochromatin 210
heterocysts 435*f*, 518, 519*f*
heterothallic organisms 1150
heterotrimeric proteins 610
heterotrophs 1271
heterozygosity 231, 1222
heuristics 8
hexapods 984
    body plans 985–6
hexokinase 306*f*, 308, 330, 331*f*
hibernation 1015, 1294
    lemurs 1295*f*
high altitudes 1040
    acclimatization to 765
    exercise physiology 781–2
high-density lipoproteins (HDL) 723

high efficiency particulate air (HEPA)
    filters 471
high-energy phosphate
    compounds 189
    see also ATP
highly active antiretroviral therapy
    (HAART) 571
Himalayan balsam 1340*f*
hippocampus 706*f*
histamine 628*f*, 729, 881
histidine (His, H) 288*f*
histograms 82, 97*f*, 109*f*
histone acetyl transferases
    (HATs) 276–7
histone deacetylases (HDACs) 277
histones 210, 211*f*
Histoplasma capsulatum 536, 537*f*
histoplasmosis (Darling's
    disease) 536
HIV (human immunodeficiency
    virus) 560, 563*t*, 571, 896
    replication 566
HLA (human leukocyte
    antigens) 885–6, 886*f*, 891
holotypes 44
homeostasis 32, 582–3, 654
    blood pressure regulation 740–1
    calcium and phosphorus 681–2,
        801
    control of ventilation 764–5
    osmolality regulation 670, 671*f*
    plasma glucose control 679*f*,
        823–6
    thermoregulation 582, 583*f*, 584,
        724, 772, 780–1
    water conservation 796
homologous appendages 987
homologous chromosome pairs 225
homologous genes 360*f*
homologous proteins 318, 320*f*
homologous recombination,
    DNA 220
homoplasy 59
Homoscleomorpha 933*f*
homozygosity 231, 1222
honey agaric 454
Hooke, Robert 368, 489
horizontal gene transfer 409
    conjugation 410–11, 411*f*
    transduction 410
    transformation 409–10
    transposition 409, 409*f*
    use in genetic engineering
        411–12
hormone replacement therapy
    (HRT) 846
hormones 298
    adrenal 676–8
    calcium and phosphorus
        homeostasis 681–2
    hypothalamus and
        pituitary 666–72
    immunoassays 683
    pancreatic 678–81
    solubility 665–6
    thyroid 672–6
    types of 665–6
    types of hormonal
        communication 666*f*

horns 996, 997*f*
horses, gaits 1084*f*
host tropism 558
Hox ('Homeobox') genes 970, 972
human chorionic gonadotrophin
    (HCG) 843, 852
human genome 207–9, 208*f*, 239,
    349, 353, 354*f*
Human Genome Project 351
human immunodeficiency virus *see*
    HIV
human papilloma virus (HPV) 851
human population growth 1319–20,
    1320*f*
humans 39
humans, 'Out of Africa'
    hypothesis 395*f*
humic material 429, 430*f*
humidity 1033
Huntington disease (HD) 243, 297,
    715, 1239, 1240*f*
hyaluronidase 529
hybrid vigour 1145
hybridization 40, 45, 47*f*, 1215–16
Hydra 965*f*
    integument 1112*f*
    reproduction 1146
hydric ecosystems 1297
hydrochloric acid production,
    stomach 812–13, 813*f*
hydrogen, valence 153*t*
hydrogen bonds 163, 164*f*
    in DNA and RNA 181, 202*f*
    in proteins 290*f*
    in water 167, 168*f*
hydrogen ion concentrations 170
hydrogen molecules 152*f*
hydronium ion 170
hydrophilic interactions 167
hydrophobic interactions 168*f*, 169*f*
hydrophytes 960
hydrothermal vents 24, 24*f*,
    1302–3
hydroxyl functional group 157, 158*t*
hyenas 1153
Hymenoscyphus fraxineus 552*f*
hyoid bone 745, 746*f*
hyperalgesia 627
hyperkalaemia 738
hypernatraemia 781
hyperopia (far sightedness) 633,
    634*f*
hyperplasia 951, 954
hyperpolarization 600*f*, 601
hypersensitivity reactions 881*f*
    see also allergic reactions
hypertension 740, 742, 773
    diuretics 796–7
hyperthermophiles 433, 480, 481*f*
    adaptations 443
    Archaea 440–1
hypertrophy 951, 954
    left ventricle 773
    skeletal muscle 688, 697
hyphae 445–6, 446*f*
hypocholesterolaemic drugs 502
hypocotyl 956, 957*f*
hyponatraemia 781
hypotension 740

hypothalamic–pituitary axis 666–8, 669f, 841f
control of pituitary hormones 668–70
negative feedback 668, 670f
hypothalamic–pituitary–adrenal axis 678f
hypothalamic–pituitary–ovarian axis 842–3, 843f, 844
suppression in pregnancy 852
hypothalamic–pituitary–testis axis 842f
hypothalamus 584, 841, 1015
location of 667f
hypotheses 10, 1202
hypothesis testing 10–11

I

I-bands 688f, 689f
ibotenic acid 538f
ice sheet melting 6f
idioblasts 1107
IgA 880f, 882
IgA protease 526
IgD 880f
IgE 880f
production, role of Th2 cells 889, 890f
IgE-mediated allergic responses 881
IgG 880f
IgM 880f, 882
ileocaecal valve 815
ileum 806f, 815, 821f
see also small intestine
imaginal discs 1184, 1186f
immigration 1251–2
immunassays 683
immune response 1122
immune system 862–4
adaptive immune response 864f, 876–85
autoimmunity 891–2
cells of 864–7, 865f
innate immune response 864f, 869–75
invertebrates 1127–8
organs of 867–9, 868f
immunity 863
immunodeficiencies 895–6
immunoglobulin-like cell adhesion molecules (Ig-CAMs) 945, 948
immunoglobulins 722, 876, 879f, 880f
class switching 893
immunohistochemistry 492f
immunological memory 863, 876, 892
immunosuppressants 502, 887, 891
immunosurveillance 869
immunotherapy 883
implantation 850, 852
imprinted genes 32, 1148–9
inbreeding depression 40, 1145, 1152
inclusive fitness 1212
incomplete dominance 1227, 1229f
independence of data 104–5
independent assortment of chromosomes 422f
independent variables 66–7

Indian population, migratory history 1241, 1242f
indices (singular index) 71, 73f
multiplying and dividing 123t
indigenous species 1060–2
indirect fitness 35, 36f, 1211–12, 1212f
induced dipoles 162–3
induced pluripotent stem cells (iPSCs) 359
inductive reasoning 8, 9f
infectious diseases
as cause of death 524t
sexually transmitted infections 851–2
transmission of 529
see also pathogens
inflammasomes 871, 872f
inflammation 870f, 1122–5
acute phase response 871f
allergic responses 881f
role of Th 1 cells 889, 890f
inflammatory bowel disease (IBD) 821
influenza viruses 560, 562f
infundibulum (pituitary stalk) 666, 667f
inheritance 224
inherited disorders
cystic fibrosis 358, 591–2
haemophilia 244, 1126
hereditary deafness 247f
Huntington disease 243, 297, 715, 1239, 1240f
sickle cell disease 302–3, 761
Tay–Sachs disease 1229–30
thalassaemias 262–3
inhibin 842
inhibitory post-synaptic potentials (IPSPs) 608, 610
initial segment, axon 610f
initiation factors 269–70
initiator (Inr) 258
injury, response to
animals 1121–7, 1123f
plants 1109, 1110f
innate immune system 863–4, 864f
cells of 864, 865–6, 865f
complement system 872–3, 873f
eosinophils 875
macrophages and neutrophils 873–4
natural killer cells 875
protective barriers 869
recognition of micro-organisms 869–71
innate immunity 1122
invertebrates 1127–8
innate lymphoid cells (ILCs) 867
inner ear 640, 641f, 642, 642f
innervation ratio, motor units 695
innexins 944
inosine (I) 266
inotropic effects 739
inquilinism 1279
insects
body plans 985–6
diapause 1184, 1186t
diversity 986f

eggs 1169–70, 1169f
flight 1093
hearing systems 1025
hormonal control of development 1185f
integument 1112–13, 1113f
larvae 1182t
life cycles 1145f, 1183f, 1184f
metamorphosis 1181–4, 1183f, 1184f
moulting 1183–4
see also ants; fruit fly; grasshopper
inspiration 749–50, 751f, 777
see also breathing
inspiratory capacity 753f
inspiratory reserve volume (IRV) 753f, 754
instantaneous rate of change 138–9
insulation 1034
insulators 275f
insulin 298, 824–5, 828
control of secretion 679, 681
diabetes mellitus 680–1, 792, 825
functions 679f
recombinant 502, 503f
stimulation of release 825f
structure 289f, 503f, 678f
insulin-like growth factors (IGFs, somatomedins) 670, 671
insulin receptor 679, 824
integrase inhibitors 571
integrins 948
integument 1109
amphibians 1114–15, 1116f
arthropods 1112–13, 1113f
birds 1119
coelenterates 1111, 1112f
echinoderms 1113f
extension of 1121f, 1122f
fish 1114, 1115f
functions 1110
mammals 1119–20
reptiles 1115, 1117f
response to injury 1121–7, 1123f
sponges (Porifera) 1111f
tunicates 1113–14, 1114f
worms 1111
see also skin
intercostal muscles 749, 750f, 751f
inter-dependent variables 67
interference competition 1276
interferons 871, 875
interferon gamma 866
intergenic sequences 207
interleukins 871t
IL-17 891
intermediate filaments 393f, 394, 915t
internal capsule 706f, 707
interphase 233f, 412–14
stages of 413t
interquartile range 81t
intersex 1152–3
interspecific competition 1275–6
interspecific interactions 1271
interstitial cells of Cajal 805
interventricular septum 733f, 734, 735f
intrafusal muscle fibres 630

intraspecific competition 1263, 1275
intraspecific interactions 1271
intrinsic factor 761, 812, 813
intrinsic rate of natural increase (r) 141
introduced species 1248
see also non-native species
introns 207, 261
invasive species 1060–2, 1331, 1349
in the Mediterranean Sea 1065–7
plants 1064
topmouth gudgeon 1062–3
inverse agonists 657
inverse processes 127
inverse proportion 124t
Fick's Law 126
invertebrates
immune system 1127–8
osmoregulation 1048
ion channels 587–8, 599
ion pumps 912–14, 913f
ionic bonds 162f
in proteins 290f
ions
transport across membranes 911–14
see also bicarbonate ions; calcium ions; potassium ions; sodium ions
IP$_3$ (inositol 1,4,5-triphosphate) 612
iris 632f, 634
radial and sphincter muscles 663f
iron, membrane transport 914
iron deficiency 761
iron–sulphur complexes 334
irreversible reactions 188
irritable bowel syndrome (IBS) 821
isidia 458
island species 1059–60
islets of Langerhans 678, 824
isocitrate dehydrogenase 333f
isoelectric focusing 317–18, 317f
isoenzymes (isozymes) 308
isolation, virtual islands 1353
isoleucine (Ile, I) 288f
isomers
biological significance 161
and bond rotation 156
stereoisomers 157–8, 160f
structural 157, 159f
isometric muscle contractions 700f
isometric scaling 953
isoprenes 1041
isoprenoid groups 443f
isotonic muscle contractions 700
itaconic acid 504
iteroparity 1188, 1260
IUCN red list classification scheme 1334–5
ivory 997f

J

Jaccard Coefficient of Community Similarity (CC$_j$) 1284, 1285
Japan, Tohoku tsunami 1317–18
Japanese knotweed 1064
jejunum 806f, 815
see also small intestine

jellyfish 965f
  fluorescence 493, 495f
  locomotion 1093
Jenner, Edward 877
junctional diversity 884
juvenile hormone (JH) 1184, 1185f
juxtacrine signalling 614
juxtaglomerular apparatus 789f

K

K+/H+ pump 914
K-selection 1156–8, 1168, 1266
  mammals 1191
  viviparity 1172
K562 cells 475f
karyogamy 448
karyograms 226f
karyotype 225
kelp 462–3, 463f, 928f
Kelvin temperature scale 186
keratin 296, 298, 394, 915t, 991
ketones 158t
keystone species 1311–12, 1311f, 1357
kidney failure 799
kidneys 785–7
  acid–base balance 798–9
  anatomical location 786f
  blood pressure regulation 797–8
  changes in pregnancy 855t
  collecting ducts 795–7
  distal convoluted tubule 795
  endocrine functions 800–1
  filtration process 787–9
  loop of Henlé 793–5, 794f
  osmoregulation 1048–9
  proximal convoluted tubule 789–92
  renin–angiotensin–aldosterone system (RAAS) 741
  structure 786f
  summary of nephron function 797f
  water reclamation 910, 911f
Kimberella 968f, 970, 973
kin selection 35, 1211–12
kinase-linked receptors 615
kinases 191
kinesin 298, 915t
kinetic energy 185, 1030
kinetochores 415f
kinetoplastids 458–9
kinetoplasts 459, 460f
King Alfred's cakes 451, 452
kinin cascade 1124–5, 1125f
kininogens 1124
kinocilium 1021f, 1022
Klebsiella spp. 518
Klinefelter syndrome (47, XXY) 228t
knee-jerk reflex 702f, 703
knockout mice 356–7, 357f
Koch, Robert 877
Koch's postulates 524–6, 525f
  and Helicobacter pylori 527
Korarchaeota 440, 442f
Krebs cycle see tricarboxylic acid (TCA) cycle
Kupffer cells 826f, 827
kuru 297

L

L-form bacteria 379
  reproductive strategies 380f
L-glutamic acid 505
L-isomers 158
labelled lines 595, 620
laboratory categories 472, 473t
labour 857–9, 858f
lac operon 272–5
  expression 274f
  structure 273f
lac repressor protein 273f
Lacks, Henrietta 476f
lacrimal glands 664
lactase 39
lactate
  anaerobic respiration 337–8, 337f
  production during exercise 697
lactate threshold 778
lactation 859–60, 1191
Lactobacillus spp. 437f, 504
  L. sanfrancisco 510
lactose 816
lag phase, cell cultures 485f
lagging strand, DNA replication 214, 215f
lakes
  microbial populations 430
  solute composition and osmolarity 1044t
  zones of 430, 431f
laminar flow 1090f
laminar flow cabinets 471, 472f
laminin 931–2
land bridges 1065
landscape-based conservation 1353–4
Langerhans cells 866
large intestine 806f, 820, 821f
  motility 820–1
  rectum 821–2
larynx 745, 746, 807f
last universal common ancestor (LUCA) 16
latent heat 1032f
lateral corticospinal tract 707, 708f, 709
lateral geniculate nucleus (LGN) 636, 637f
lateral line 1022f
lateral root meristem 937f
lateral roots 936
lateral undulation, snakes 1080
latitudinal diversity gradients 1286, 1287f
Laurasia 1056, 1057f
Laurasiatheria 1065
leading strand, DNA replication 214, 215f
leaf cutter ants 501–2, 501f
leaf morphology 960
  dicotyledons 962f
  monocotyledons 963f
  peas 961f
leaf positioning 956
leaf primordia 956
leaves
  adaptations 963f, 1035f
  internal structure 960, 964f

leeches, osmoregulation 1048
left ventricular hypertrophy 773
leghaemoglobin 519–20, 1041f
legislation
  site-focused 1341–2
  species-focused 1339–41
leguminous plants 435f, 518–20, 519f, 520f
Leishmania spp. 458–9
leishmaniasis 541, 543–5
  global distribution 542f, 544f
  parasite life cycle 545f
lemurs 1295f, 1355
  hibernation 1294, 1295–6
lens 632f, 633
Leocarpus fragilis 922, 923f
leptin 360f, 361, 814
leucine (Leu, L) 288f
leukocytes (white blood cells) 719, 721–2, 721f, 863, 864, 865f
  see also lymphocytes
levers 700, 701f
Leydig cells 837, 839f
libraries
  cDNA 353
  genomic 347, 349f
lichens 457–8
  uses of 459f
life
  definition of 30, 904
  emergence of see emergence of life
  qualities of 30–3
  timeline of 1056f
life cycles
  aphids 1145f
  butterfly 1184f
  Entamoeba histolytica 541f
  ferns 1144f
  flowering plants 1148f
  fungal 448f
  Giardia lamblia 546f
  grasshopper 1183f
  Leishmania spp. 545f
  sexual and asexual phases 1143, 1144f, 1145f
  Trypanosoma cruzi 1047f
life-history strategies
  K-strategists and r-strategists 1158, 1266
  survivorship curves 1266f
  trade-offs 1265–6
life-history traits 1265
life tables 1250, 1260–1, 1260f
ligand-gated channels 588, 608–9
ligands 181, 298, 657
light
  detection by animals 1010–18
    see also eye, human; eye spots; eyes; vision
  detection by plants 1008–10
  electromagnetic spectrum 1005–6, 1007f
  polarization 1006–7
  speed of (c) 121t
  use in navigation 1008
light absorption 1005, 1006f
light climate 1002–3
light intensity (irradiance, insolation) 1003, 1006

global and seasonal variations 1004f
  variables and units 1005t
lignin 397, 959
limbs 981–2
  deuterostomes 986–91
  digits 988–9, 990f
  as levers 700, 701f
  protostomes 982–6
  taxonomy of 981f
  wings 989–91
limiting factors 1299
limnetic zone 430, 431f
limpet 998f
Lincoln–Peterson index 1256–7
line graphs 111f, 112f
line transects 1255
lineage-restricted cells 939
linear scales 72, 73f, 140f
LINEs (long interspersed elements) 209
lines of best fit 1234f
Lineweaver–Burk plot 310
linked genes 235, 236f, 237–8
Linnaeus, Carl 43–4, 48, 49, 52
lion's mane jellyfish 965f
lipases 817
  industrial applications 310
lipid metabolism 828
lipid rafts 372
lipids 26, 173
  digestion and absorption 817–18, 818f, 827
  oxidation of 328–9
  structure 175f
lipoproteins 723, 828, 914
liposomes 28
Listeria monocytogenes 437
lithosphere 1054, 1055f
littoral zone 430, 431f
liver 806f, 868f
  autonomic control 664
  bile production 827
  blood supply 826–7
  metabolism of waste/toxic products 828–9
  regulation of energy metabolism 828
  structure 826f
lizards, locomotion 1086
lobopods 983
locomotion 1080
  flying and gliding 1088–93
  muscle contraction 699–700
  reptiles 1086, 1087f, 1088f
  resonant frequencies 699
  role of tendons 699
  snakes 1080
  swimming 1093–100
  traction 1080
  walking and running 699, 1080–6
  worms 1077–9
  see also motor system
lodging, cereal plants 991
logarithmic expressions
  multiplying and dividing 123t
  rearranging 130–1
logarithmic scales 72, 73f, 140f
  comparing sizes 74f
  pH scale 74f

logarithms 75, 122
  natural 76
logical fallacies 9t
logical reasoning 8, 10
  inductive versus deductive 9f
logistic growth 143f, 1264–5, 1265f
long-latency stretch reflex 704
longitudinal muscles 698f
loop of Henlé 787f, 793–5, 794f
  peritubular capillary network 790f
  reabsorption of major
    substances 790t
lophotrichous arrangement of
    flagella 381f
Lotka–Volterra model 1286, 1287f
low-density lipoprotein (LDL) 723,
    828, 914
lower motor neurons 705, 707f
lower oesophageal sphincter
    (LES) 810, 812f
lugworm 1078
lung compliance 751–2
lung volumes 753–5, 753f
lungs
  airways and alveoli 747–9, 748f
  anatomy 749, 750f
  asthma 755–6
  emphysema 752
  fetal 857
  gas exchange 758–9
  pulmonary fibrosis 751–2
  surface tension 752–3
  ventilation of 749–53
Luria–Bertani (LB) broth 480t
luteinizing hormone (LH) 669f, 836,
    841t
  actions in female 843
  actions in male 842
  control of secretion 841–2
  menopausal levels 846
  menstrual cycle 844, 845f
Lyme disease 437
lymph nodes 867, 868f, 892
  T cell activation 887
lymphatic system 725, 867, 869
  lacteals 818f
lymphocyte activation 866
  B cells 892–3
  CD4+ T cells 886–7, 888f
  CD8+ T cells 892, 893f
lymphocytes 721f, 863, 864, 865f,
    866–7, 869
  antigen receptor diversity 883–4
  antigen receptors see B cell
    receptors; T cell receptors
  CD numbers 898–9
  clonal selection 876, 879f
  cytotoxic T cells 894f
  differentiation into effector
    cells 867f, 889f
  see also B lymphocytes; T
    lymphocytes
lymphokine activated killer (LAK)
    cells 875
lysine (Lys, K) 288t
  structure 179f
lysis 375
lysogenic cycle, bacteriophages
    567–8

lysosomes 391–2, 392f, 914
Lystrosaurus 1056
lytic cycle, bacteriophages 568

M
M cells 867
M-lines 689f
M protein 564
macroevolution 1213, 1222f
macrolecithal eggs 1171–2, 1171f
macromolecular interactions 180–1
macromolecules 172–3
  see also carbohydrates; lipids;
    nucleic acids; proteins
macrophages 863, 865f, 870, 874
macula 634, 635f
macula densa 789
Madagascar 1059
  elephant bird (Aepyomis
    maximus) 1060f
Magnaporthe oryzae 550f, 552
magnetic fields 19, 1029
  animal responses 1029–30
  plant responses 1029
magnetoreception 1029–30
magnetosomes 382f
magnification 90f
Magnolia grandiflora 925f
maize 961
major histocompatibility complex
    (MHC) 875, 885, 892, 893f
  MHC peptide receptors
    (HLA) 885–6, 886f, 891
malaria 541, 542t
  vector control 547
malate dehydrogenase 333f
MALDI (matrix-assisted laser-
    desorption ionization) 318, 319f
male reproductive system
  anatomy 846–8
  hypothalamic–pituitary–testis
    axis 842f
  puberty 844–5
  spermatogenesis 837–40
Malthus, Thomas Robert 1205
maltose 816
mammalian radiations 1064–5
mammals
  embryo development 1174f, 1175
  evolution 1058f
  integument 1119–20
  reproductive characteristics 1157t
  reproductive strategies 1191–4
manipulated (experimental)
    studies 67, 68t
mantle, molluscs 982
maraviroc 570
Margulis, Lynn 385
marine environment
  microbial populations 430–1
  see also oceans
marine iguana 1003f
marine plants, osmoregulation 1045
marking 1256, 1257f
Marshall, Barry 527
marsupials 1173
mass extinction events 1313–14,
    1313f, 1317
mass movements, large intestine 821

mass spawning events 1181f, 1182f
mass spectrometry 318, 319f
massively parallel sequencing 352–3
  Illumina technique 352f
mast cells 865f, 866, 881f
Mastigophora 540
matched measures 98
maternal recognition of
    pregnancy 1191
mathematical models 11
maximum oxygen uptake
    (VO$_{2max}$) 775–7
  assessment in humans 775–6
mean 80t
mean arterial pressure (MAP) 769
  response to aerobic exercise 770
  response to resistance
    exercise 769f
measles virus 563t
measurement error 101–2, 102f, 103f
  reduction 104
mechanoreceptors 1021
  equilibrium sensing 1021–2
  neuromasts 1022
  tactile receptors 1021
mechanosensing, plants 1019–20
medial geniculate nucleus 644
medial lemniscus 622, 623f
median 80t
Mediterranean Sea, invasive
    species 1065–7
medulla oblongata 621, 622, 623f,
    706f
medullary pyramids 707
medusa forms 965f, 966
meiosis 231, 232, 234, 235f, 236f, 419,
    833f, 834f
  comparison with mitosis 420f
  control of 423
  errors 423
  gamete production 832–3, 834f
  genetic variation 32–3, 35, 234–5,
    420–2, 833, 835–6, 1146
  meiosis I 419–22
  meiosis II 422–3
  non-disjunction 835f
  oogenesis 836–7
  stages of 419f
Meissner's corpuscle 621
melanopsin (OPN4) 1016t
melatonin 1015
  effects on corpus luteum 853–4
membrane lipids 173
  structure 175f
membrane permeability 908f
membrane potentials 596, 597f,
    911–12
membrane proteins 298–9, 305f
  glycophorins and porins 303–4
membrane transport 585–91
  active transport 912–14, 912f
  facilitated diffusion 911, 912f
  influencing factors 908–9
  ions 911–14
  non-ionic active transport 914
  simple (passive) diffusion 910–11,
    912f
  water movement 909–10, 1044
membranes 31–2, 904

respiratory 909, 911
  see also plasma membrane
memory cells 892, 893
Mendel, Gregor 225f, 1206
  later experiments 231, 232f
  Law of Independent
    Assortment 231
  modern explanation of
    results 230–1
  monohybrid cross
    experiments 227–30, 229f
Mendelian genetics 225
  extensions and
    refinements 239–41
menopause 846
menstrual cycle 843, 844, 845f, 1191
  breast changes 859
  effects of gonadal steroids 846f
meristems 928–9, 933, 935–6, 939,
    955
  lateral root meristem 937f
  seedlings 956, 957f
Merkel's discs 1021
mesenchymal cells 937
mesic ecosystems 1297
mesoderm 966
mesodermal cells 937
mesodermal tissues 967f
mesogloea 965f
mesolecithal eggs 1170–1, 1170f
mesophiles 480, 481f
mesopredator release 1310
Mesosaurus 1056
messenger RNA (mRNA) 176, 178,
    253t, 264
  capping 260, 261f
  cDNA production 353, 355f
  poly(A) tail 263f
  polycistronic 271f
  post-transcriptional
    modification 258, 259f
  Shine–Dalgarno sequence 268f,
    269
  short life of 256
  splicing 261–2, 261f
  stem-loop structure 258f
  synthesis of (transcription)
    255–63, 255f
meta-analyses 116
metabolism 31, 185
  changes in pregnancy 855t
  role of the liver 828
metabotropic receptors 610–14
metagenomics 428–9
metalloproteinases 1127
metamorphosis
  amphibians 1190
  insects 1181, 1183, 1184f
metaphase 422
  meiosis 234, 422
  mitosis 231, 233f, 415–16, 415f
metaphase checkpoint 416
metaphase plate 415
metaplasia 951
metapopulation dynamics 1252–3
metapopulations 1252
Metarhizium anisopliae 515
metatherian animals, distribution
    1058

metazoa 930
  body plans 964–5
  epithelia 930–3, 930f
  osmoregulation 1046, 1048–50
*Methanobrevibacter
  ruminantium* 444f
methanogens 443–4
methanol, industrial production 506
methicillin-resistant *Staphylococcus
  aureus* (MRSA) 500, 530
methionine (Met, M) 288f
methionyl-tRNAs 269
methyltransferases 280, 281f
miasma theory 11
micelles 28, 818, 827
Michaelis–Menten equation 129,
  309–10
  rearranging 128
microbes
  biological hazard levels 472
  gut microbiome 822–3
  oil content 506t
  plant defences 1107–9
  recognition by innate immune
    system 869–71, 875f
  terminology 433t
  see also bacteria; fungi; viruses
microbial biotechnology 499
  biological control 514–15
  bioremediation 515
  chemicals and fuels 504–6
  composting 516–17
  food and drink production 507–13
  pharmaceutical industry 499–504
  wastewater treatment 515–16
microbial cell cultures
  bioreactor systems 484, 485f
  growth kinetics 486–8, 487f
  growth phases 484–6, 485f
  media 479–80, 480t
  temperature and pH 480, 481f
microbial culture collections 473–4
  fungi 474f
microbial habitats 429
  freshwater 430
  marine 430–1
  soil 429–30
microbial pathogens *see* pathogens
microbial polysaccharides 513–14
  commercial uses 514t
  Xanthomonas campestris 514f
microbial safety cabinets 471
microbial taxonomy 428
microbiomes 429
micro-endemic species 1247
microevolution 1213, 1222f
  human dive reflex 1213–14
microfibrils 904, 905f
microfilaments 393f, 394, 915t
microfossils 18f
microlecithal eggs 1168–9
micro-organisms *see* microbes
micropropagation of plants 934f, 935f
micropyle 1169f, 1170
microsatellite sequences 209
microscopy 488–9
  cell counts 481, 482f
  compound light microscope 490f
  confocal imaging 492–3, 493f

early microscopes 489f
electron microscopy 494–6
how to use a microscope 491
phalloidin 539f
resolution 490, 491f
stains 490–2, 492f
super-resolution methods 496
viewing live samples 493–4
microtubule organizing centre
  (MTOC) *see* centrosomes
microtubules 393–4, 393f, 415f, 915t
  dynamic nature 394f
microvilli 816
micturition 800
middle ear 640, 641f, 642f
migrating myoelectrical complex
  (MMC) 815–16
migration 1068, 1294
  breeding 1069–71, 1070f
  fish 1187–8
  human populations 1241, 1242f
  multifactorial 1071–2
  refuge 1069, 1070f
  tracking 1071
  types of 1068f
milk composition 859t
milk secretion 859–60
milkweed, coenocytic cells 925f
Miller, Stanley 23, 24f
millipedes 984f
mimicry 1131, 1132f, 1273, 1274f
  aggressive 1272f
  Batesian 1130f, 1131, 1132f, 1273,
    1274f
  Müllerian 1131, 1132f, 1135
  plants 1133f
mimosa 1020f
Min proteins 407, 408
mini-cells 408
Minimum Viable Population
  (MVP) 1347
miscarriage 194
mismatch repair 217f, 218
mistletoe 1278
Mitchell, Peter 335, 336
mitochondria 325, 326f, 385, 394
  origins of 385
  structure 396f
mitochondrial disease 337
mitochondrial DNA (mtDNA) 206,
  207f, 395, 396, 1253
mitosis 231–2, 233f
  anaphase 416f
  comparison with meiosis 420f
  control of 417–18
  cytokinesis 416f
  errors in 418
  metaphase 415–16, 415f
  overview 414f
  prometaphase 415
  prophase 415
  telophase 416
mitotic spindle 231, 233f, 415
mitral valve 733f
mixed statistical models 105
MN blood group system 226, 239
mode 80t
model lines (trend lines) 111–12, 112f
modelling 1203

models 11–12, 12f
  predictive use 13f
modular proteins 304, 320f
mole (mammal) 988f
molecular clock 49, 50, 51f, 428
molecular mimicry 891
molecular sieving 344
molecular (cladistic) taxonomy 428
molecules 151
  organic 153
  three-dimensional shapes 153
moles (amount of a substance) 90–1
molluscs
  limbs 982
  shells 992, 993–4
monoclonal antibodies 882, 883
monocotyledons (monocots) 929f,
  955–6
  leaf internal structure 964f
  leaf morphology 963f
monocytes 721f, 865f
monoecious plants 1150, 1151f, 1152,
  1164
monophyletic groupings 53
monosaccharides 173, 816
  see also glucose; ribose; sucrose
monosomies 227, 228t, 418, 423f,
  424, 836
monotremes 1174f
  distribution 1058
  eggs 1171–2
monotrichous arrangement of
  flagella 381f
mono-unsaturated fatty acids 173
moon
  effect on Earth 19
  formation of 18–19
morphogenesis 938t
morphological species concept 1216
morula 936–7
mosquitos 1279
mosses 955f
motilin 815–16
motion
  bacterial 380
  Newton's Laws 1077
  see also locomotion; motor system
motor cortex 706f, 710–13
  somatotopic representation 708f
  transcranial magnetic
    stimulation 710–12
motor end plates 692
motor neurons 692, 694–5
motor proteins 298
motor system 705, 709
  basal ganglia 713–15
  cerebellum 716
  conscious movement 707f
  descending pathways 705, 707–9,
    707f, 708f, 710f
  upper and lower motor neuron
    lesions 705
motor units 694–6
moulds 446f
  see also fungi
moulting, insects 1183–4
Mount St Helens eruption 1317,
  1318f
  recolonization 1064

movement
  bacterial 380
  see also locomotion; motor system
mucosa, digestive tract 804–5, 804f,
  815f
mucosa-associated lymphoid
  tissues 867
mucus 904
  discus fish 1188–9
  hagfish slime 1117f, 1118f
  secretion by fish 1114
mucus-secreting cells
  Brunner's glands 815
  respiratory system 869
  stomach 812f
Müllerian mimicry 1131, 1132f, 1135
multicellular animals (metazoa) 930
multicellular organisms 922
  colonies 926
multicellular plants
  algae 927–8, 928f
  evolution 928
  meristems 928–9
multi-drug resistance 428, 533
multifactorial inheritance 241
multiple sclerosis 892
multipotent stem cells 939, 940f
multi-species conservation 1349
multivariate analyses 105, 117
Mus (mouse) clade 56f
muscarinic cholinergic receptors 660,
  661
muscimol 538f
muscle contractions 181, 700f
muscle fibres 687f
  slow and fast 695–6, 696t, 697
  structure 688–9
muscle pump 732f
muscle spindles 629–31, 630f
  stretch reflexes 703
muscle types 687
muscles 686–7
  blood flow during exercise 771–2,
    774
  resistance exercise, cardiovascular
    responses 769
  see also cardiac muscle; skeletal
    muscle; smooth muscle
muscles of ventilation 749, 750f, 751f
muscular arteries 726
muscularis externa 804f, 805
mushroom poisoning 538
mushroom production 508–9
mutagens 1231
mutant alleles 1223, 1224
mutation 32, 35, 1231–2
mutualism 526, 1279f, 1280f
  lichens 457–8
myasthenia gravis 891
mycelium 445
mycetism (mushroom poisoning) 538
mycetomas 535
mycobionts 457
mycofiltration 515
myco-heterotrophs 1278f
mycology 445
  see also fungi
mycoparasites 514–15
Mycoplasma 16f, 379

mycorrhizal fungi 447, 454
mycorrhizosphere 447
mycotoxicosis 538
myelin sheath 605, 606f
myenteric plexus 804f, 805
myenteric reflex 805
myocardium 732, 733f
  see also cardiac muscle
myofibrils 688f
myoglobin 299–300, 761
  amino acid sequence 318, 320f
  oxygen saturation curve 300–1, 301f
  structure 300f
myomeres 1094
myomesin 689
myometrium 849f, 858
myopia (near sightedness) 633, 634f
myosatellite cells 688
myosin 298, 688–9, 688f, 915t, 916
  crossbridge cycle 689–91, 690f
  thick chain 689f
myriapods 984
myxomatosis 573

N

Na$^+$ K$^+$ ATPase pump 588–9, 589f, 600
  distal convoluted tubule 795
  loop of Henlé 793
  proximal convoluted tubule 791
Na$^+$/K$^+$ pump 912–13, 913f
N-acetylglucosamine (NAG) 375–6, 375f, 399f
N-acetylmuramic acid (NAM) 375–6, 375f
NAD$^+$ (nicotinamide adenine dinucleotide) 192–3, 308
  deficiency of 194
  reduction in glycolysis 331f
  reduction in TCA cycle 333f
  structure 192f
NADH 326, 327, 328
  oxidation in electron transport chain 335f, 336
  production in glycolysis 332
NADP+ 339
Naegleria fowleri 540f
Nägeli, Carl von 49
naked virions 558
naming conventions 43–4, 427
  examples 45t
  fungi 448–9
  misleading names 57
  viruses 556t
Nanoarchaeota 440, 442f
National Parks 1341t
National Vegetation Classification (NVC) 1280–1
native species 1060–2
natural disasters 1317–18, 1318f
natural killer (NK) cells 865f, 866, 875, 899
natural logarithms (log$_e$, ln) 76, 122
natural selection 35, 1205, 1206–8, 1232–3
  heritability 1233–6
  misconceptions 1214
  misunderstanding and rejection of 36–7

selection differentials 1236–7
supporting evidence 37
  see also evolution
natural (observational) studies 67, 68t
nature reserves 1341t
nautilus 1098f
navigation 1008
  echolocation 1025
  fish 1023
Neanderthal hominids 44
necrotrophs 447, 1108
nectivory 1274, 1275f
negative assortative mating 1240
negative feedback 582–3
  control of ventilation 765
  hypothalamic–pituitary axis 668, 670f
  hypothalamic–pituitary–adrenal axis 678f
  hypothalamic–pituitary–ovary axis 843f
  hypothalamic–pituitary–testis axis 842f
  stretch reflexes 703
neglected tropical diseases (NTDs) 541, 543
Neisseria gonorrhoeae 436f, 530f, 531
nematodes 1041f
  Caenorhabditis elegans 1224f
  integument 1111
  locomotion 1079f
Neocallimastix spp. 449
neonatal respiratory distress syndrome (NRDS) 753
neostigmine 664
neoteny 1190–1
nephridia 1046, 1048
nephrons 786–7, 787f, 1046, 1048
  collecting ducts 795–7
  distal convoluted tubule 795
  filtration process 787–9
  loop of Henlé 793–5, 794f
  peritubular capillary network 790f
  proximal convoluted tubule 789–92
  reabsorption of major substances 790t
  summary of function 797f
  water reclamation 910, 911f
Nereis spp., locomotion 1078
Nernst equation 193, 599–600
  rearranging 131
nervous system, mammalian 595
  action potentials 600–4
  myelin sheath 605, 606f
  resting potential 596–600
  signal conduction 595–6, 604–5, 604f
  transmission between neurons 605–14
  see also autonomic nervous system; brain; enteric nervous system; parasympathetic nervous system; sensory system; spinal cord; sympathetic nervous system
nervous systems 973
  deuterostomes 975–81
  protostomes 973, 974f, 975f
  vertebrates 978–81

nest-site competition 1276–7
net primary production (NPP) 1303
net reproductive rate per generation (R_0) 1260–1
neural crest 978, 980f
neural plate 976, 976–7, 980f
neural tube 976, 977f, 980f
Neuralia 973
neurocrine signalling 665, 666, 666f
neurofilament proteins 394, 915t
neuromasts 1022
neuromuscular junctions 692–3
neurones 595, 596f, 973
neuropsin (OPN5) 1016t
Neurospora crassa 923, 924f
neurotoxins 528
neurotransmission 605–14
  autonomic nervous system 657–60, 661f
neurotransmitters 595
  autonomic nervous system 657
  at neuromuscular junctions 692
  release of 606, 608f
  removal from synaptic cleft 608–9
  types of 607
neurulation 977f
neutrons 150
neutrophil extracellular traps (NETs) 874
neutrophils 721f, 865–6, 865f, 873–4
nevirapine 571
New Zealand 1059
Newton, Isaac 1077
niacin 194
niches 1298–9, 1299f
nicotine 664
nicotinic cholinergic receptors 660–1, 693
nidogen (entactin) 932
nitric oxide (NO) 729
nitrification 520
Nitrobacter spp. 520
nitrogen, valence 153t
nitrogen cycle 517–20, 517f, 1310, 1311f
nitrogen fixation 434–5, 517, 1041
  global sources 518f
  Rhizobium–legume symbiosis 518–20, 519f
nitrogen-fixing bacteria 1279
nitrogenase 1041f
nitrogenous bases 199
  base pairing 201
  structure 200f
nitrogenous waste excretion 1050
Nitrosomonas spp. 436, 520
Nitrosopumilus maritimus 441, 442f
noble rot (*Botrytis cinerea*) 513f
nociception 626
nociceptors 626t
nodes of Ranvier 605, 606f
noise 101
nomenclature *see* naming conventions
nominal data 64
non-coding strand, DNA 255
non-competitive antagonists 657
non-covalent bonds 161, 162–3
non-disjunction 418, 423f, 835f
non-haem iron proteins 334

non-homologous end-joining mechanism, DNA 220
non-native species 1248, 1331, 1335f, 1349
  management of 1345
  see also invasive species
non-random mating 1240
non-shivering thermogenesis 1038
non-steroidal anti-inflammatory drugs (NSAIDs) 584, 627, 666, 729, 813
noradrenaline (norepinephrine) 657, 659, 677
  effect on cardiac output 739
  effect on vascular smooth muscle 728–9
  synthesis from tyrosine 559, 660f
nori 467f
normal distribution 82, 84t, 109f
normoblasts 720
normotension 740
nosocomial infections 529
Nostoc spp. 518, 519f
notochord 976–7, 977f
nuclear lamins 915t
nuclear pores 387f
nucleic acids 173, 176–9
  see also DNA; RNA
nucleobases 26–7
nucleocapsid 557
nucleoid 373
nucleolus 387f, 388
nucleosides 199
  nomenclature 201t
nucleosomes 210, 211f
  and DNA replication 218
nucleotide excision repair (NER) 220
nucleotides 176, 199
  inosine 266
  nomenclature 201t
  structure 177f, 199f, 200f
nucleus of a cell 384, 387–8, 387f
  origins of 385
nucleus of an atom 150
null hypothesis significance tests (NHSTs) 104, 114
  alternatives to 117
  choosing a test 114f
  effect sizes 116
  interpretation guidelines 116t
  parametric versus non-parametric options 115f
  P-values 115
  reporting 116–17
  statistical power 116
  Type I and Type II errors 115–16
null statistical hypothesis 10–11
Nuremberg Code 14
nutrient cycles 517, 1309
  carbon cycle 520–1, 1310, 1310f
  nitrogen cycle 517–20, 1310, 1311f
nutritional supplements 779–80
nystatin 537f

O

oat blight 550–1
obesity 783
objective grouping 43–4
obligate diapause 1192

obligate herbivores 1271
obligate mutualism 1279
obligate parasites 547–8, 548*f*
observational studies 67, 68*t*
observer effects 104
obstructive respiratory disorders 755
obstructive sleep apnea (OSA) 765
occluding junctions 943*t*
ocean currents 1067*f*
oceanomadromous fish 1072*t*
oceans
    acidification 6*f*, 172, 993–4
    light penetration 1005, 1006
    microbial populations 430–1
    origins and development 21
    sea level rise 6*f*
    solute composition and
        osmolarity 1044*t*
    warming 6*f*
ocelli (singular ocellus) 1011*t*, 1012
octamer complexes, histone 210
octet rule 152
ocular dominance columns 637
oculocutaneous albinism type 1
    (OCA1) 243*f*
oesophagus 746*f*, 806*f*, 810
    peristalsis 811
    structure 811*f*
oestradiol 836
oestrogens 836
    effect on GnRH secretion 841
    effects on female reproductive
        tract 846*f*
    menstrual cycle 844, 845*f*
    production in ovaries 838*f*
oestrone 836
oestrus 1191
oil pollution 1323
Okazaki fragments 214, 215*f*, 216*f*,
    564
oleic acid 168*f*
olfaction (smell) 1136
olfactory system 648, 649*f*
oligodendrocytes 605
ommatidia 1011*t*, 1012–13, 1012*f*
omnivores 1271
oncotic pressure 731
oocytes 834*f*, 836
    meiosis 423, 424
    origin and formation 1168–9,
        1169*f*
oogenesis 834*f*, 836–7
oogonia 834*f*, 836
oomycetes 456
open circulations 1044–5
open populations 1251–2, 1252*f*
operons 271
    lac 272–5
opioids 627, 628
opisthokonts 930
opportunistic pathogens 526
opsins 1015
opsonization 872
optic chiasm 636–7, 637*f*
optic disc (blind spot) 632*f*, 635*f*
optic nerve 635*f*, 636
optic radiation 637*f*
optical density measurement 483–4,
    484*f*

optical isomers *see* enantiomers
optimal ranges 1297, 1298*f*
optimality theory 1212
oral cavity 806–8, 807*f*, 809*f*
orchitis 847
orders of magnitude 71*f*, 72*f*
ordinal data 64
organ of Corti 639, 642
organ systems 653–4, 654*f*
    cardiovascular system 718–42
    digestive system 803–29
    endocrine system 665–84
    immune system 862–99
    nervous system 594–614
    renal system 784–802
    reproductive system 831–60
    respiratory system 744–67
    sensory systems 618–51
organ transplantation 887
organelles 384
    benefits of 385–6
    chloroplasts 399–400
    mitochondria 394–6
    origins of 384–5
    see also chloroplasts; mitochondria
organic molecules 153, 154*f*
    electron delocalization 156–7
    functional groups 157, 158*t*
    naming conventions 155–6
    origins of 22–5
    see also covalent bonds
organoids 475
orientation of cells 921*f*
orthogonal variables 67
Oscarella lobularis 933*f*
Oscillatoria 435
osmoconformers 1045
osmolality regulation 670, 671*f*
osmolytes 1045
osmoreceptors 670
osmoregulation
    in animals 1045–50
    in plants 1045
osmoregulators 1045
osmosis 587, 909–10, 1044
osmotic gradient, loop of Henlé 793–
    5, 794*f*
ossicles 639, 640, 641*f*, 642*f*, 1024–5
ossification 994, 995*f*
osteoblasts 994
osteoclasts 926*f*
osteoporosis 801, 846
otolith organs 646, 647*f*
ova 1147
    see also eggs; oocytes
oval window 640, 641*f*, 642
ovarian cycle 843
ovaries
    avian 1171*f*, 1172
    of flowers 1161*f*, 1163
    mammalian 837*f*, 848*f*, 849
    oogenesis 834*f*, 836–7
overload 777
oviduct
    avian 1171*f*, 1172
    see also Fallopian tube
oviparous animals 1168
ovoviviparity 1172–3
ovulation 836–7, 843

oxaloacetate 333*f*
oxidation 188–9
    of food 189
    see also glucose oxidation
oxidative decarboxylation of
    pyruvate 325–6, 327*f*
oxidative phosphorylation 192, 194,
    328, 334, 336
    location in mitochondria 396
    relationship to
        photosynthesis 339*f*
oxygen
    as an electron acceptor 192
    gas exchange 758
    partial pressure of 757
    production in photosynthesis 340
    valence 153*t*
    VO$_{2max}$ 775–7
oxygen dissociation curves 760–1,
    760*f*
oxygen transport 759–61
oxytocin 666, 667*f*, 670, 859
oyster mushroom 509

P

P-values 115
    reporting 117
P wave 736, 737*f*
P2Y12 receptor 613–14
pacemaker (sino-atrial node) 733–4
    actions of adrenaline 662*f*
Pacinian corpuscles 1021
pain 626
    control of 627–8
    sensation of 627
    withdrawal reflex 704–5, 704*f*
pain gate hypothesis 628
pain pathways 626, 627*f*
pairing of samples 98
palmitic acid 168*f*
Palolo worms, mass spawning 1181,
    1182*f*
palps 983
palynology 1165
pancreas 806*f*, 823–4
    anatomy 824*f*
    endocrine functions 678–81
    glucagon 825–6
    insulin 824–5
    protease production 818
pancreatic juice 819
pancreatic polypeptide 678
pancreatitis 818
panda 989*f*
Pangaea 1056, 1057*f*
    fragmentation of 1057*f*
pannexins 944
papillary muscles 733*f*
paracellular reabsorption 789
paracrine signalling 614, 665, 666*f*
paradigm shifts 11
paraheliotropism 1034
Paramecium spp. 457*f*, 462*f*
parameters 96
parametric tests 114
parapatric speciation 1217*f*, 1218
parapodia 983*f*, 1078, 1079*f*
pararetroviruses 566–7
parasematism 1131, 1132*f*

parasites 1277–9, 1280*f*, 1343
    destruction by eosinophils 875
parasitoids 1278–9
parasympathetic nervous system 595,
    655
    connections 658*f*
    effect on airways 663, 747
    effect on bladder 663
    effect on cardiac output 662, 739,
        770
    effect on eye 663–4
    effect on insulin secretion 681
    effect on saliva secretion 808
    neurotransmission 657, 659
    organization of 657
parathyroid glands 681
parathyroid hormone (PTH) 681–2
paratomy 1180
parenchymatous tissues 955
parental care, fish 1188–9
parietal cells 812–13, 812*f*
Parkinson's disease 713–14
parotid gland 806*f*, 808, 809*f*
parthenogenesis 1147, 1205
    experimental 1149
partial pressures 1039*f*
    of oxygen and carbon dioxide 757
parts per million (ppm) 87, 88–9
part-to-part (internal) ratios 86–7
    representations for 87*t*
part-to-whole (external) ratios 87
    concentrations 90–1
    relating to fractions and
        percentages 89*t*
    representations for 87*t*
parturition (birth) 857–9, 858*f*
parvoviruses 563*t*
passive protein-mediated
    transport 587–8
passive transport 585
Pasteur, Louis 499, 877
pasteurization 470
patagium 989
Patau syndrome (trisomy 13) 228*t*
patch isolation 1329
pathogen-associated molecular
    patterns (PAMPs) 869, 892,
    1108
pathogens 523, 1343
    bacterial 526–33
    definition 524
    disease triangle 547*f*
    fungal 534–7
    Koch's postulates 524–6
    plant diseases 547–53
    protozoan 539–47
    sites of infection 524*f*
    transmission of 529
    viral *see* viruses
    virulence factors 526–8
pattern formation 938*t*
pattern recognition receptors
    (PRRs) 870, 871, 1127
Pax6 gene 1014
PCR (polymerase chain
    reaction) 344, 345–7, 347*f*
    adapting DNA for cloning 349, 350*f*
    quantitative (real-time) 354, 355*f*
    RT-PCR 354

peak expiratory flow (PEF) 754*f*
peas
 leaf morphology 961*f*
 Mendel's experiments 225*f*,
  227–31, 232*f*, 1206, 1239*f*
pectin 397*f*, 398
pedigree analysis 246, 246–7, 248*f*
peer review 13, 14
pelagic zone 430
pelvic splanchnic nerves 657
pendulums 1082*f*
 model of locomotion 1081–3
penicillins 38, 500
 structure 533*f*
Penicillium spp.
 P. camemberti 446*f*, 511*f*
 P. chrysogenum 451
 P. roqueforti 511
penis 847*f*
 function 848
 structure 848*f*
pennate muscles 698*f*
pentadactyly 988
pentose sugars 199*f*
peplomers 558
peppered moth 1222*f*
pepsin 309, 813, 818
pepsinogens 812, 813
PepT1 transporter 818
peptide bonds 28*f*, 179*f*, 264*f*, 286*f*
 cis and *trans* configurations 293*f*
 electron delocalization 157
 partial double-bond
  character 291–2
peptides 179, 285
 directionality 286
peptidoglycan 375–6, 375*f*, 532
 and Gram stain 377, 378*f*, 379*f*
peptidyl transferase reaction 269
percentages 89*t*
perception 619
 see also sensory system
perennial plants 1159
perimysium 687*f*, 688, 688*f*
perinuclear space 387
periodic acid–Schiff's (PAS)
 stain 491, 492*f*
periodic table 150*f*
peripatric speciation 1217*f*, 1218
peripheral nervous system (PNS) 595
peripheral resistance 727
periplasm 377, 378
peristalsis 805, 1078*f*
 in oesophagus 811
 in small intestine 815–16
 in stomach 814
peritrichous arrangement of
 flagella 381*f*
perlecan (heparan sulphate
 proteoglycan) 932
permeability coefficient 908*f*
permeability of landscape
 features 1354
pernicious anaemia 761
peroxisomes 400–1
 structure 400*f*
pesticide resistance 1345*f*
pesticides 1344
pests 1343

management of 1344–5
Petri dishes 480*f*
Peyer's patches 867, 868*f*
pH 170*f*, 762
 of blood, effect of exercise 779
 and dissociation of a weak acid 171
 effect on enzyme activity 308*f*, 309
 optima for microbe growth 480,
  481*f*
pH calculations 75, 77
pH scale 74*f*
Phaenophytes (brown algae) 462
phage therapy 572–3
phage typing 572*f*
phagocytosis 392*f*, 590*f*, 873, 874*f*, 914
phagolysosomes 873, 874*f*
phalloidin 539*f*
phallotoxins 538
phantom limb pain 626
pharmaceutical industry 499
 antibiotics 500, 501*t*
 hypocholesterolaemic drugs 502
 immunosuppressants 502
 probiotic products 504
 steroids 500, 502
 use of recombinant DNA
  technology 502, 503*f*
pharynx 745, 746*f*, 807*f*, 810*f*
phase-contrast microscopy 493, 494*f*
phase locking 644
phenological inertia 1325
phenological mismatches 1325
phenology
 bird egg laying 1236–7
 climate-induced changes 1325*f*
phenotype 32, 227, 1205, 1222
 environmental influences 239, 241*f*
 extended 35, 36*f*
phenotypic frequency 1224–5
phenotypic plasticity 35, 36*f*, 1205,
  1299
phenotypic responses 1209
phenotypic switching 536*t*
phenotypic variation 1205, 1206*f*, 1222
phenylalanine (Phe, F) 288*f*
 structure 179*f*
phenylephrine 664
pheromones 1136, 1186
 fungal 451, 454
pheroplasts 379
phoresy 1279
phosphatidylcholine 370*f*
phosphodiester bonds 201*t*
phosphoenolpyruvate 330, 331*f*
phosphofructokinase 312, 314, 330,
  331*f*
phosphoglucose isomerase 330, 331*f*
2-phosphoglycerate 330, 331*f*
3-phosphoglycerate 330, 331*f*
phosphoglycerate kinase 330, 331*f*
phosphoglycerate mutase 330, 331*f*
phospholipid bilayers 173, 175*f*, 370–1,
  372*f*, 584, 585*f*, 597, 598*f*, 904
phospholipid membrane *see* plasma
  membrane
phospholipids 26*f*, 369, 584
 adaptations of extremophiles 443
 phosphatidylcholine 370*f*
phosphoric acid anhydrides 158*t*

phosphoric acid esters 158*t*
phosphorus, valence 153
phosphorus homeostasis 681–2
photic zone 430, 431*f*
photomorphogenesis 1008
photon flux density (PFD) 1003
photonastic responses 1020
photoperiodism 1160
photopic vision 1014
photopsins 1016*t*
photoreception 1014–15
 see also vision
photoreceptive molecules 1015,
  1016*t*
photoreceptors 634–5, 636*f*
 animals 1015
 in Australian ants 1016–17
 plants 1008–9, 1008*f*
 see also eye, human; eye spots; eyes
photosynthesis 189, 338–9, 1297,
  1301
 Calvin cycle ('dark reactions') 340
 and carbon cycle 1309, 1310*f*
 gross primary production 1306
 light reactions 339, 340*f*
 net primary production 1306
 relationship to oxidative
  phosphorylation 339*f*
 reversible reactions 166*f*
 role of 340
Photosystem II 434
phototaxis 380
phototropins 1009, 1016*t*
phototropism 1009, 1010*f*
phragmatoplast 905*f*, 906
phycobionts 457
phycocyanin 466
phycoerythrin 466
Phycomyces blakesleeanus 451*f*, 452*f*
phyllotaxy 956, 957*f*, 958*f*
phylogenetic (ancestral) studies 16
phylogenetic relationships 52–6
phylogenetic species concept 48,
  1216
phylogenetic trees 17*f*, 50*f*, 51, 52*f*,
  54*f*, 56, 427*f*, 1142
 Archaea 441*f*
 Bacteria 432*f*
 Eukarya 460*f*
 relationships between
  animals 971*f*
physical activity 782–3
physical defences
 animals 1109–21
 plants 1104–5, 1104*f*, 1105*f*
physiological saline 913
physiology 581–2
 autonomic nervous system 654–64
 cardiovascular system 718–42
 cell membranes 584–92
 cellular signalling 614–16
 digestive system 803–29
 endocrine system 665–84
 exercise physiology 768–84
 homeostasis 582–4
 immune system 862–99
 muscle and movement 686–716
 nervous system 594–614
 renal system 784–802

reproductive system 831–60
respiratory system 744–67
sensory systems 618–51
physoclistous fishes 1099
physostomous fishes 1099
phytoalexins 553
phytoanticipins 553
phytochrome 1008–9, 1009*f*, 1016*t*
Phytophthora infestans 456*f*, 548*t*
phytoplasmas 548*t*, 1108, *1108*
pi (π) 121*t*
pied flycatcher 1350*f*
pili (singular pilus) 381
 sex pilus 410, 411*f*
Pilobolus 431
pilocarpine 664
pineal gland 1015
pinna 639, 641*f*
pinnae, ferns 955
pinocytosis 590*f*, 914
pinopsin 1016*t*
pioneer species 1286
pirenzepine 664
pistil 1161*f*, 1163
pituitary disorders 671–2
pituitary gland 666–72, 841*f*
 location of 667*f*
pituitary hormones
 anterior pituitary hormones
  840–1, 841*t*, 859
 hypothalamic control 668–70,
  670*f*
 posterior pituitary hormones 670
pK_a 171, 763
placebo effect 103, 106, 628
placenta 843, 856*f*, 1175
 expulsion of 859
 functions 852, 854
 structure 852*f*
placodes 978
planned (experimental) studies 67,
  68*f*
plant adaptations
 halophytes 1300*f*
 to hot climates 1035
 leaf morphology 963*f*
plant body plans 954–6
 embryo and seedling
  development 956, 957*f*
 leaf internal structure 960, 964*f*
 leaf morphology 960, 962*f*, 963*f*
 leaf positioning 956, 957*f*, 958*f*
 root architecture 961–2, 964*f*
 shoots and branches 958–9
 trunks and bark 959–60
plant cell cultures 479*f*
plant cell differentiation 933–6, 937*f*
plant cell wall 397–8
 structure 397*f*
plant cells
 aquaporins 909–10
 cell division 905–6, 905*f*
 cell wall 904–6, 905*f*
 directional expansion 919, 920*f*
 plastids 906–7
 vacuoles 914, 917–18, 918*f*
plant defences
 camouflage and mimicry 1133*f*
 against herbivory 1104–7

against invasion 1107–9
toxins and poisons 1134–5
plant diseases 547, 548t, 1109f
    blast diseases 552
    dieback 552
    establishment of infection 550–1
    hyperplastic effects 553
    hypoplastic effects 552–3
    penetration of the host 548–50
    plant responses 553
    rot and soft rot diseases 551–2
    spots and blights 552
    types of parasitism 547–8
    wilting 552
plant dispersal 1072–7
plant propagation 934f, 935f
plant responses
    to environmental pressures 1001
    gravitropism 1025–6
    to magnetic fields 1029
    to sound 1019
plant secondary metabolism
        1105–7
plant signalling 1107f
plant turgor 917–18, 919f
plants
    first appearance on Earth 1064
    light detection 1008–10
    mechanosensing 1019–20
    osmoregulation 1045
    planetary distribution 1064
    principal groups 1159f
    and thermal environment 1034–5,
        1160–1
    wound repair 1109
plasma 719, 722
plasma cells 866, 880, 893
plasma glucose control 679f,
        791, 823–6
plasma lipids 722–3
plasma membrane (phospholipid
        membrane) 584–5, 597, 598f
    archaeal 373, 384
    associated proteins 372–3
    components 369–70
    fluid mosaic model 372f
    structure 370–3, 372f, 585f
    transport across 585–91
    see also membrane transport
plasma proteins 722
plasma volume, effect of exercise
        training 773
plasmalemma 904
plasmalogens 400
plasmids 206, 207f, 374
    conjugation 410–11, 411f
    recombinant DNA
        technology 347, 348f, 359,
        411–12
plasmin 1126–7
plasmodesmata 397f, 398, 909, 941,
        942f
plasmodium 465, 540f
plasmogamy 448, 454
plasmolysis 918, 1045
plastids 385, 400–1, 906–7
    see also chloroplasts
plastochron 956
plate tectonics 19–21, 20–21f

platelets (thrombocytes) 613, 719,
        722, 1125, 1126
platyhelminths (flat worms) 38, 972f
    integument 1111, 1112f
pleiotropy 239
pleural membranes 749, 750f
ploidy 1143
pluripotency 954
pluripotent stem cells 940f
pneumocytes 749, 753
pneumolysin 531
pneumothorax 749
podocytes 787
poikilothermic organisms 1034
point counts 1254–5
poison dart frogs 1135f
poisons 1134–5
    fungal 538–9
Poisson distribution 82, 84t
polar bodies 834f
polar bonds 153, 154f
polar molecules 153, 154f, 162
    water 167, 168f
polarity of cells 920–2, 921f, 930
    zygote and embryo 936
polarity problem, DNA
        replication 214f
polarization of light 1006–7
pollen 1162, 1164f
    and archaeology 1165
    diversity of 1165f
pollen tubes 919, 1164, 1166f
pollination 1161–2, 1274, 1332, 1335
    influence of sound 1019
pollution 1324
    bioaccumulation 1356
    oil 1323
poly(A) tail, mRNA 263f
polyacrylamide gel electrophoresis
        (PAGE) 316f, 317f
polyandry 1178
polychaete worms 983f, 1180f, 1181f
    reproduction 1179–81
    segmental progression 1078, 1079f
    tube construction 1121, 1122f
polycistronic mRNA 271f, 563
polygenic inheritance 241
polymerase chain reaction (PCR)
        50–1, 574
polymerization of biomolecules 27
polymorphism 1216f, 1282f
polynucleotides
    structure 177f
    see also DNA; RNA
polyomaviruses 560
polypeptides 179f, 285
    directionality 286
polyphyletic groupings 53
polyps 966
polysaccharides 173
    digestion and absorption 816
    see also cellulose; chitin; glycogen;
        starch
polysomes 271
polyunsaturated fatty acids 173
poppy, seed dispersal 1075f
population bottlenecks 1238–9
population change 1257
    quantification 1257, 1260–1

population demographics 1245–51,
        1258–9
    quantification 1253–7
population density 1246–7
    quantification of changes 1260
population dispersion 1249
population distribution 1247–8
population genetics 1224
population growth 73f, 1205, 1262
    continuous breeding 1263
    discrete breeding events 1262–3,
        1264t
    human 1319–20, 1320f
    with limited resources 1263–5
population modelling
    notation 1260
    R_0 1260–1
population multiplication rate 1257,
        1260
population pyramids 1249, 1250f
population size 1245–6, 1247
    challenges 1246f
population structure 1249–51
population types 1251–2
Population Viability Analysis
        (PVA) 1347
populations 96, 1201, 1244, 1292f
    reasons for study of 1245
    source and sink 1253
porbeagle sharks 1172–3, 1173f
Porifera see sponges
porins 299, 303–4, 378
    structure 305f
portal vein 824f, 826f, 827
Portuguese man of war 1098f
positive assortative mating 1240
positive feedback 583
    hypothalamic–pituitary–ovary
        axis 843f, 844
    tipping points 1313
postaxial development of digits 989,
        990f
posterior pituitary hormones 670
postganglionic neurons 655f
    parasympathetic 657
    sympathetic 655
post-transcriptional gene
        regulation 280–1
potassium (K^+) ions
    blood levels during exercise 778
    and equilibrium potential 598f
    hyperkalaemia 738
potassium leakage channels 587f,
        599f
potato blight 456, 548t
potomadrous fish 1072t
powdery mildew 548f
power analyses 104, 116
power functions 122, 125t
power stroke, crossbridge cycle 690f
prairie lupine 1064f
prazosin 664
preaxial development of digits 989
prebiotics 823
pre-Bötzinger complex 764, 765f
pre-central gyrus 706f
precocious puberty 845
predation 1271–3, 1280f
predation strategies 1272f

predator avoidance 1137
    body shape change 1138f
predator defences 1273
predator–prey cycles 1286, 1287f
predator release 1286
predictor (independent) variables 66
pre-eclampsia 855
prefixes 72t
prefrontal cortex 706f
preganglionic neurons 655f
    parasympathetic 657
    sympathetic 655
pregnancy
    amniotic fluid 857
    birth 857–9, 858f
    embryonic diapause 1192–4
    fetal circulation 725, 856–7, 856f
    fetal oxygen supply 857
    implantation 850, 852
    maternal recognition of 1191
    NAD deficiency 194
    physiological changes 854–5, 855t
    placenta 852f, 854
    pre-eclampsia 855
preload 739–40, 771
premotor cortex 706f, 710, 711, 713
pre-mRNA (primary transcript) 258,
        259f
presbyopia 633
preservation 1345–6
prestin 643
primary follicle 836, 837f
primary immune response 876, 877f
primary motor cortex (M1) 706f
primary protein structure 291, 292f
    determination of folding 295,
        296f
primary succession 1287, 1288f
primase 214, 215f, 218
primers, DNA viruses 564
primitivity 57–8, 59f
primordia 935
prion diseases 297
probability 89
probiotics 504, 823
Prochlorococcus 431f
productivity 1301, 1303
profundal zone 430, 431f
progenitor cells 940f
progesterone 836
    effect on GnRH secretion 841
    effects during pregnancy 855
    effects on female reproductive
        tract 846f
    menstrual cycle 844, 845f
    production in ovaries 838f
    structure 176f
prokaryotic cells 16f
    binary fission 406–8, 406f
    comparison with eukaryotic
        cells 369t
    differences from eukaryotic
        cells 384
    horizontal gene transfer 409–12
    membranes 31
    sizes 369f
    see also archaea; bacteria
prolactin 669f, 841t, 859–60
prolate structure 567

proline (Pro, P) 288*f*, 289, 293
  in collagen 303
  structure 290*f*
prometaphase 231, 415
prometaphase I 421
prometaphase II 422
promoter region, DNA 256*f*, 259*f*, 275*f*
pronghorn antelope 997*f*, 1273*f*
proofreading, DNA replication 216, 217*f*, 218
prophase 231
  meiosis 234, 419–20, 422
  mitosis 415
proplastids 907
proportion 90
proportional to symbol (∝) 122
proprioception 629, 701–2
proprioreceptors
    (proprioceptors) 1021
  Golgi tendon organ 631*f*
  muscle spindles 629–31, 630*f*
prostaglandins 584, 628*f*, 666, 729, 813, 1124–5
  role in parturition 858–9
prostate gland 800*f*, 847*f*
protandrous organisms 1156
protease inhibitors 571
proteases 818
proteasome 892, 893*f*
protected area designations 1341*t*
protection 1339
  active 1342–3
  fishing quotas 1342–3
  policy 1342
  site-focused legislation 1341–2
  species-focused
    legislation 1339–41
protein domains 304, 305*f*
  domain shuffling 321
  and modular construction of
    proteins 320*f*
protein kinases 314, 315*f*, 318
protein phosphatases 314
protein spikes 556*f*, 558
  SARS-CoV-2 virus 559
protein structure 285, 287*f*
  amino acid side chain
    interactions 289–90, 290*f*
  determination of folding 295, 296*f*
  influence of peptide bond 291–2
  levels of 291, 292*f*
  modular 320*f*
  molecular chaperones 295–6
  primary 291, 292*f*
  quaternary 292*f*, 295
  relationship to function 296, 298
  secondary 292*f*, 293, 294*f*
  stability of folded proteins 293, 295
  tertiary 292*f*
protein synthesis 198
  chemistry of 264*f*
  elongation phase 269, 270*f*
  initiation and elongation
    factors 269
  initiation phase 268–9
  polysomes 271
  role of the liver 828

termination 270
  see also translation
protein trafficking 390–1, 391*f*
proteins 179–80
  adaptations of extremophiles 444
  collagen 302–3
  conformational changes 181
  cytoskeleton 374*t*
  digestion and absorption 813, 818–19
  diversity of 285*f*
  functions 181, 298*f*
  homologous 318, 320*f*
  membrane-embedded 303–4, 305*f*, 372–3, 584–5, 585*f*
  oxygen-binding 299–302
  in plasma 722
  size 285
  structural classes 298
  structure 180*f*
proteoglycans 402
proteome 315
proteomic studies 315–16
  applications 321
  gel electrophoresis 316–17
  isoelectric focusing 317–18, 317*f*
  mass spectrometry 318, 319*f*
  and study of evolution 318, 320–1
prothoracicotropic hormone
    (PTTH) 1184, 1185*f*
prothrombin 1125
protist diversity 458
  algae 466–7
  Alveolates 459, 461
  Amoebozoa 465–6
  Cercozoa 464*f*
  Euglenozoa 458–9
  Stramenopiles 461–4
protists
  osmoregulation 1045–6, 1047*f*
  syncytia 922
protocells 26, 28
Protoctista 456
protons 150, 151
protoplasts 379
protostomes 969–70, 970*f*
  nervous systems 973, 974*f*, 975*f*
protozoa 539
  life cycles 541*f*, 545*f*, 546*f*
protozoan diseases 539–41, 542*t*
  giardiasis 545–6
  leishmaniasis 541, 543–5
  treatment 546–7
  vector control 547
proximal convoluted tubule 787*f*
  calcitriol production 801
  glucose reabsorption 791–2, 792*f*
  peritubular capillary network 790*f*
  reabsorption of major
    substances 790*t*, 791*f*
  secreted substances 792
  sodium reabsorption 791
  transcellular and paracellular
    reabsorption 789
  water reabsorption 792
proximal drivers 1068
pseudocoeloms 972
pseudohermaphroditism 1152–3
pseudopeptidoglycan 384

pseudoplasmodium 465
pseudopodia 465
pseudoreplication 104
pseudo-species 1282
psychrophiles 439, 480, 481*f*
puberty 844–5
  breast development 859
pulmonary circulation 723, 724*f*
pulmonary fibrosis 751–2
pulmonary valve 733*f*
pulvini 1010
pulvinus structures 1020
Punnett square 230*f*, 231, 232*f*, 237*f*, 1226–7, 1228*f*
pupil, autonomic control 663–4
pupillary reflex 634
purines 26–7, 199
  structure 200*f*
Purkinje fibres 734, 735*f*
putamen 706*f*, 713
pylorus 811, 812*f*
pyramidal tracts *see* corticospinal
    tract
pyramids, ecological 1309, 1309*f*
pyrimidines 26, 27, 199
  structure 200*f*
pyrogens 584
Pyrolobus spp. 441
pyruvate
  conversion into acetyl-CoA 332*f*
  conversion to lactate 337–8, 337*f*
  fermentation 338, 339*f*
  oxidative decarboxylation 325–6, 327*f*
  production in glycolysis 325*f*, 331*f*, 332
pyruvate decarboxylase 338
pyruvate dehydrogenase 326, 332*f*
pyruvate kinase 332
  structural domains 304, 305*f*

Q

QRS complex 736, 737*f*
quadrats 1254*f*
quadrupedal locomotion 1080–4
quantitative (real-time) PCR 354, 355*f*
quarrying 1322*f*
quasi-experimental studies 67, 68*t*
quaternary protein structure 291, 292*f*, 295
quills 1120
quin2 494
Quorn™ 508
quorum sensing 382–3, 528

R

R_0 1260–1
radial cleavage 969, 971*f*
radial symmetry 966*f*
radiation 1031, 1032*f*
  thermoregulation 1037*t*
  use in sterilization 470–1
Ramsar sites 1341*t*
random dispersion 1249*f*
random factors 105
random sampling 98
randomization 103

and number of variables 105
range changes 1248*f*
  climate-induced 1324–5, 1325*f*
range of data 81*t*
ranges 1247–8
raphides 1107
rates 136–7
rates of change 137–9
  exponential decay 142–3
  exponential growth 139–42
  logistic growth 143*f*
  over distance 138
ratio squares 66, 88
ratios
  comparing 87–9
  concentrations and dilutions 90–1
  equivalent 87
  part-to-part 86–7
  part-to-whole 87
  and proportion 90
  relating to fractions and
    percentages 89*t*
  representations for 87*t*
Rayleigh criterion 490
reaction rates 137, 142–3
reactive oxygen species 873, 874*f*
reading frames, ribosomes 268
realized niche 1298, 1299*f*
realized ranges 1247
rearranging formulas 127–31
receptive fields 620
receptor potentials 620
receptor-mediated endocytosis 560, 561*f*, 590*f*
receptors 298
  of immune system 864–5, 866, 876
  see also B cell receptors; T cell
    receptors
  sensory systems 618–51
recessive alleles 243, 1222, 1223*f*
  Tay–Sachs disease 1229–30
recessive phenotypes 228
reciprocal inhibition 703
recognizable taxonomic units
    (RTUs) 1282
recombinant chromosomes 235, 236*f*, 237
recombinant DNA 344, 347, 348*f*
  adapter sequences 350*f*
recombinant DNA technology
  applications 353–9, 411–12, 502, 503*f*
  in genomics and
    bioinformatics 359–61
  rEPO 782
recombinants 420
recombination frequencies 239
rectangular hyperbola 124*t*
rectilinear locomotion, snakes 1080
rectum 806*f*, 821–2, 821*f*
  see also large intestine
red algae (rhodophytes) 466–7
  nori 467*f*
red blood cells (erythrocytes) 719–20, 759
  aged, removal from
    circulation 828
  anaerobic respiration 338
  development, GATA-1 278–9

red–green colour blindness 244
red list (endangered) species 1334–5
red nucleus 706f
red tides 461f
redox potential value (E'₀) 192
    Nernst equation 193
reduction 188
reductive evolution 385
reflex responses
    baroreceptor reflex 740–1, 741f
    cough reflex 753
    myenteric reflex 805
    pupillary reflex 634
    stretch reflexes 702–4
    vestibulo-ocular reflex 648
    withdrawal reflex 626, 704–5, 704f
refractory periods 602
refuge migration 1068f
    zooplankton 1069, 1070f
regeneration 940f, 941f, 954, 1191
regression analysis 111, 1234
regulatory genes 40–1, 41f
reindeer (caribou) 1071f
    migration 1071
reintroduction of species 1346, 1356–8
related samples 98
relative abundance 1253
relative fitness 1232–3
relative refractory period 602
release factors 270
remediation 1354–6
renal corpuscle 787f, 788f
    filtration process 787–9
renal perfusion pressure (RPP) 787–8
renal system
    acid–base balance 798–9
    blood pressure regulation 797–8
    distal convoluted tubule and
        collecting ducts 795–7
    endocrine functions 800–1
    filtration 787–9
    kidney structure and
        function 786–7
    loop of Henlé 793–5
    micturition 799–800
    proximal convoluted
        tubule 789–92
renewable energy 1322–3
renin 797
renin–angiotensin–aldosterone
    system (RAAS) 741, 797–8
rennin 511
repeated measures 98
replication 32
replication forks 212, 213f, 215f
    DNA viruses 565
replication origins 211–12, 213f
repolarization 600f
reproducibility 13–14
reproduction 1167f
    amphibians 1190–1
    angiosperms 1158–66
    annelid worms 1179–81
    asexual cellular fission and
        budding 1145–6
    definition 1142
    development in amniotes 1174–6,
        1175f
    eggs 1167–72

facultative sexual
    reproduction 1146
fecundity and fertility 1156–8
ferns 1144f
fish 1184–90
    germ cell production 1146–7
    gymnosperms 1166–7, 1168f
    insects 1181–4
    parthenogenesis 1147, 1149
    same-sex 1148–50
    sperm 1176–9
    viviparity 1172–4
reproduction number (R) 140–1
reproductive characteristics of
    mammals 1157t
reproductive cycles 832f, 1191
reproductive efficiency 1156
reproductive strategies 1143
    angiosperms 1166
    fish 1190
    mammals 1191–4
    mate selection 1178
    non-dioecious 1151t
    r- and K- selection 1156–8
    and succession 1286–7
reproductive success 1211
    facilitation of 1348–9
reproductive system, human
    birth 857–9, 858f
    coitus and fertilization 850
    endocrine control 838, 840–3
    female anatomy 837f, 848–50
    gamete production 832–8
    lactation 859–60
    male anatomy 846–8
    male and female systems 832
    menopause 846
    menstrual cycle 844, 845f
    pregnancy 852–7
    puberty 844–5
reptiles
    body architecture 1086f
    eggs 1171–2
    integument 1115, 1117f
    locomotion 1086, 1087f, 1088f
    sex determination 1154, 1155f
research hypothesis testing 10
reservoir hosts 529, 1343
residual volume (RV) 753f, 754
resistance exercise 773
    blood pressure responses 769f
resolution 490, 491f
resonance stabilization 156–7
resource limitation 1263–5, 1275
resources
    impacts of extraction, harvesting,
        or harnessing 1322–3
    impacts of transportation 1323
    impacts of use 1324–6
    renewable and non-
        renewable 1320–2, 1321f
respiration 325, 745
    anaerobic 337–8
    see also glucose oxidation
respiratory alkalosis 765
respiratory bronchioles 749
respiratory centres 764–5, 765, 777
respiratory pump 732
respiratory system, human

adaptations to exercise
    training 778–9
air movements 755–7
airflow defects 755
altitude acclimatization 765
anatomy 745–9, 745f
asthma 755–6
changes in pregnancy 855t
control of ventilation 764–5
fetal lungs 857
gas exchange 758
lung volumes 753–5
as a protective barrier 869
responses to exercise 777–8
sleep apnea 765
ventilation of the lungs 749–53
response (dependent) variables 66
'rest and digest' responses 655
resting potential 596–600
restoration 1339
    of habitats 1356
    remediation 1354–6
    species reintroduction 1356–8
restriction enzymes (restriction
    endonucleases) 344, 345f, 573
restrictive respiratory disorders 755
reticulate bodies 437
reticulocytes 720
reticulospinal tract 705, 708f, 709,
    710f
retina 632f, 634–6
    blood vessels and macula 635f
    cellular structure 635f
retinal 156, 1015
retinotopic map 637
retrotransposons 208f, 209
retroviruses 253
    replication 566, 567f
reverse causation 105–6
reverse T3 675f
reverse transcriptase 208, 353
reverse transcriptase inhibitors 570
reverse transcription 253, 566
reverse transcription PCR
    (RT-PCR) 574
reversible reactions 164–5, 187, 188
    direction of 165, 166f
    energy considerations 166–7
rewinding 1356
Reynolds number (Re) 1090, 1096
rheumatoid arthritis (RA) 891–2
rhizobia 1041
Rhizobium spp. 435, 518–20, 1279
rhizofiltration 1356f
rhizoids 955
Rhizopus spp. 500, 502f, 511, 551
rhizosphere 430
rhododendron 1061f
rhodopsins 634, 635, 1016t
ribose 27f, 199f
ribosomal RNA (rRNA) 49, 50–1,
    253t, 267f
ribosomes 176, 264, 266–8, 267f
    bacterial 375f
    eukaryotic 388f
    molecular clock 428
    polysomes 271
    reading frame 268
    as target of antibiotics 532

ribozymes 269
rice blast 550f, 552
rickets 801f
Rickettsia 435
rigor mortis 691
ring-tailed lemurs 1355
ringworm (tinea corporis) 534f
rivers
    microbial populations 430
    solute composition and
        osmolarity 1044t
RNA (ribonucleic acid) 26, 27f, 173,
    176, 178, 198
    base pairing 201–2
    classes of 253t
    hydrogen bonding 181
    phosphodiester bonds 201t
    ribosomal 49, 50–1
    self-propagating molecules 28
    structural differences from
        DNA 202–3
    see also messenger RNA; ribosomal
        RNA; transfer RNA
RNA interference (RNAi) 281f
RNA polymerase 255–6, 258
    positioning of 256
    reaction catalysed by 256f
    Rpol II 259, 260f
RNA viruses 198
    replication 566
rods 634, 635, 636f, 1013, 1014f
roe deer, embryonic diapause 1192–3,
    1193f
root apical meristem (RAM) 929f,
    935–6
    structure 936f
root architecture 961–2, 964f
root hair cells 919, 920f
    water uptake 909–10
rotational site management 1352f
rotaviruses 563t
rotifers, syncytia 925, 926f
rots 551–2
rough endoplasmic reticulum 389f
round window 641f
r-selection 1156–8, 1168, 1266
RT-PCR 354
rubisco (ribulose-1,5-biphosphate
    carboxylase/oxygenase) 340
rubrospinal tract 705, 708f, 709,
    710f
Ruffini endings 621
rugae 805, 812f
ruminant artiodactyls 996
ryanodine receptors 691, 692f

S

S phase of interphase 412, 413t
Saccharomyces cerevisiae 447, 451,
    454f, 499, 510
Saccorhytus coronarius 970, 972f
saccule 646, 647f
salamanders 1190–1
    regeneration 940, 941f
salbutamol 663f
saline, physiological 913
salinization 1320, 1328
saliva 808
salivary glands 806f, 808, 809f

salmon 1187–8
Salmonella spp. 436
　S. typhi 529
salt lakes 1044*t*
saltatory conduction 605
same-sex reproduction 1148–50
sample error 96
　implications of 113
sample quality 96–7, 98*f*
sample size 96, 97*f*, 104*t*
samples 96
　bias 98
　non-representative 105
　related and unrelated 98
sandflies 542, 544*f*
Sanger sequencing (chain
　termination method) 349–50,
　351*f*
Saprolegnia spp. 456*f*, 457*f*
saprophytes 447
Sarcodina 540
sarcolemma 688*f*
sarcomere 688*f*, 689*f*
　structure 688–9
sarcoplasmic reticulum Ca²⁺ ATPase
　(SERCA) 692
sarcoplasmic reticulum 688
sarin 609
saturated fatty acids 173, 371
saturation point 1033
scaffolding proteins 563
scale data 64
scales of study 7*f*
scaling 953–4
scanning electron microscopy
　(SEM) 495–6, 496*f*
Scarpa's ganglion 647*f*
scatterplots 111, 111*f*, 1234*f*
scent marking 1139
schizotomy 1180
Schleiden, Matthias 368
Schwann, Theodore 368
Schwann cells 605
scientific approach 6, 8
　logical reasoning 8, 10
　mathematical models 11
　research hypotheses 10
　team science 13–14
　theories and models 11–12
　ways of thinking 8
scientific method 10, 1202–3
scientific notation 71
scientific premise 10
sclera 632*f*
sclerotia 538
Sclerotinia sclerotiorum 923, 924*f*
scotopic vision 1013–14
SDS-PAGE 317*f*
sea bean 1075
sea cucumber 982*f*
sea level rise 6*f*
sea otter 1357*f*
sea water, composition and
　osmolarity 1044*t*
seahorses 1173*f*
second messenger systems 611–12
second-order afferents
　nociception 626, 627*f*
　touch 622, 623*f*

secondary follicle 836, 837*f*
secondary immune response 876,
　877*f*
secondary lymphoid organs 867
secondary protein structure 291,
　292*f*, 293, 294*f*
secondary sexual characteristics 845
secondary succession 1287, 1288*f*
secretin 814, 815, 819, 820
sedimentation 1327
seed banks 1166
seed dispersal 1073, 1273–4
　by animals 1075–7
　by mechanical propulsion 1075,
　　1076*f*
　by water 1075
　by wind 1073–5, 1074*f*
seed pods 1075, 1076*f*
seedling development 956, 957*f*
segmental progression 1078, 1079*f*
segmental reflexes 704
segmentation
　large intestine 821
　small intestine 816
segregation of alleles 230
segregation of daughter
　chromosomes 231, 233*f*
selection 32–3, 1001
　artificial 39*f*
　natural 35, 36–7, 36*f*
selection differentials 1236–7
selective breeding, dogs 44, 47*f*
self-fertilization 1152
sella turcica 666
semelparity 1188, 1260
semicircular canals 641*f*, 642*f*, 646–8,
　647*f*
semi-conservative replication of
　DNA 210, 212*f*
seminal fluid (semen) 847, 850
seminal vesicles 847*f*
seminiferous cords 1176
seminiferous tubules 837, 839*f*,
　1176–8, 1177*f*
Senecio 1035*f*
sense strand, DNA 255
sensillae 1136, 1137*f*
sensitization to allergens 881
sensory stimuli 619–20
sensory system 618–19
　adaptation to stimuli 619, 628
　balance 645–8
　general sensation 619–20
　hearing 639–45
　organization of 619*f*
　pain 626–8
　proprioception 629–31, 701–2
　smell 648, 649*f*
　taste 648, 650*f*, 807–8
　temperature 628–9
　touch 620–5
　vision 632–9
septa, fungal hyphae 446*f*, 447*f*
septarian nodules 992*f*
sequencing 428
　metagenomics 428–9
　see also DNA sequencing
serial dilution 481, 483*f*
serial homology 987

serine (Ser, S) 288*f*
　structure 179*f*
serosa 804*f*, 805
serosal membranes 931
serotonin (5-HT) 729, 815, 820
Sertoli cells 837, 839*f*, 842
severe combined immunodeficiency
　disease (SCID) 358, 359, 864,
　896
sex chromosomes 225, 226*f*, 1153*f*
sex determination 833
　Active Y/W and Dosage
　　systems 1154*f*
　environmental (ESD) 1154–6,
　　1155*f*
　genetic (GSD) 1153–4
　Sry gene 226, 227*f*
sex-linked inheritance 243–4, 245*f*
sexual cannibalism 1274
sexual dichromatism 1130*t*
sexual differentiation 1150–3
sexual intercourse (coitus) 850
sexual phases 1143, 1144*f*, 1145*f*
sexual reproduction 1143, 1144
　advantages 1144–5
　facultative 1146
　germ cell production 1146–7
　see also flowers; reproductive
　　system
sexual selection 35, 36*f*, 1133, 1185,
　1209*f*
sexually transmitted infections
　(STIs) 851–2
shade avoidance 1009, 1010
Shannon–Weiner diversity
　index 125, 1284
sharks
　buoyancy 1097, 1100
　dermal denticles 1096
shell gland 1171*f*, 1172
shells 992–4
　impact of ocean acidification 993
　tortoises and turtles 1115
Shine–Dalgarno sequence 268*f*, 269
shivering 1038
shoot apical meristem (SAM) 929*f*
　structure 935*f*
'shotgun' sequencing 353
SI (International System of)
　units 64*t*
sickle cell disease 302–3, 761
siderophores 527
sidewinding 1080
Siganus spp. 1065–7, 1067*f*
sigma notation (Σ) 121
sigma replication, DNA viruses 565,
　566*f*
sigmoid colon 806*f*, 821*f*
signal sequences, proteins 390–1,
　391*f*
signal transduction pathways 941
signalling 665
　cellular 614–16
　receptor types 615–16
　see also endocrine system; nervous
　　system
signalling molecules 298, 614
signals
　honesty of 1129, 1131*t*

interpretation of 1131, 1133
silencers (S) 275*f*, 276
silicon 22
similarity matrices 46*f*
simple (passive) diffusion 910–11,
　912*f*
simple epithelia 930*f*
simple eyes (ocelli) 1011*t*, 1012
simple sequence repeats 209
simulations 12
SINEs (short interspersed
　elements) 209
single-cell protein (SCP) 507–8
single covalent bonds 155*f*
single-gene disorders 241
sink populations 1253
sino-atrial node (SAN,
　pacemaker) 733–6, 735*f*
sinusoids (discontinuous
　capillaries) 730
　liver 826*f*
site-based conservation 1351–3
　habitat creation 1351–2
　habitat improvement 1352
　habitat maintenance 1353
Sites of Special Scientific Interest
　(SSSIs) 1341*t*, 1342
skeletal muscle 687–8
　architecture 698*f*
　blood flow 724
　contractile force 696–7
　contraction 699–700
　crossbridge cycle 689–91, 690*f*
　effect of training 697
　excitation–contraction
　　coupling 691
　human muscle groups 694*f*
　motor units 694–6
　muscle fibre structure 688–9
　neuromuscular junctions 692–3
　regulation of sarcomere
　　shortening 691
　relaxation 691–2
　sliding filament mechanism 689
　slow and fast fibres 695–6, 696*t*,
　　697
　structure 687*f*
skeleton
　birds 990*f*
　bone 994–6
skin 1119–20
　blood flow 724
　blood flow during exercise 772
　as a protective barrier 869
　sensory receptors 621, 622*f*
　structure 1120*f*
　see also integument
skunks 1135–6
S-layers, bacterial cells 376, 377*f*
sleep apnea 765
sleeping sickness 459, 542*t*
sliding filament mechanism of muscle
　contraction 689
slime layer 380
slime moulds 465*f*, 922, 923*f*
　sorocarp 466*f*
slow worm 988*f*
small intestine 806*f*, 815
　digestion and absorption 816–19

endocrine functions 819–20
  histology 815f
  motility 815–16
  pancreatic juice 819
  villi 805, 816
smell (olfaction) 648, 649f, 1136
smooth endoplasmic reticulum 389f
smooth muscle 687
  airways 747, 748f
  digestive tract 804f, 805, 815f
  vascular 725, 726f, 728–30
snakes 988f
  locomotion 1080
snakeskin chiton 998f
SNARE proteins 391, 392f, 606, 608f
snoring 765
snRNPs 262
social structures, fish 1190
sodium chloride (NaCl) 162f
sodium glucose cotransporters (SGLT1, SGLT2) 791, 792f, 816, 817f
sodium ions (Na⁺)
  factors regulating reabsorption 798t
  plasma concentrations during exercise 781
  reabsorption in the nephron 790t, 791, 797f
soft rot diseases 551t
soil composition 429
soil erosion 1327
soil microbial population 429–30
soil types 430t
solar farms 1322f
Solar System, origins 15, 18
solid-state fermentation 509f
solutions, concentrations and dilutions 87, 88–9, 90–1
somatic hypermutation 892
somatic nervous system 595
somatic sensation 619
somatosensory cortex (S1) 622–4, 623f
somatostatin (GHIH) 668, 669f, 670–1, 678, 815, 825
somatotopic representation
  motor cortex 708f
  somatosensory cortex 623f, 624
soredia 458
sorocarp 465, 466f
sound
  animal responses 1021–5
  intensity 1018–19
  pitch and volume 1018
  plant responses 1019
  transmission of 1018
  velocity in different media 1018t
  see also hearing
sound production 746
sound waves 639, 640f
source populations 1253
soy sauce (shoyu) 511
speciation 1056, 1059–60, 1213, 1216–18, 1217f
species
  biological species concept 1215–16
  definitions of 44–5, 47–8, 1214–16

morphological species concept 1216
  new discoveries 37f
  number of 37, 1200, 1201f
  phylogenetic species concept 1216
  phylogenetic definition 48
species–area relationships 1284–5, 1286f
species diversity 1282–4
species–environment interactions 1297–301
  adaptions 1299–301
  limiting factors 1299
  niches 1298–9
  tolerance ranges 1297–8
species-focused conservation 1346–7
  culling 1349
  hierarchy of action 1348f
  improving reproductive success 1348–9
  improving survival 1347–8
  multi-species approaches 1349
species names 427
species polymorphism 1130t
species richness 1281–2, 1284
  latitudinal diversity gradient 1286, 1287f
specific gravity (SG) 1097t
specific growth rate (μ) 141
specific heat of water 167
specific part of a name 43, 44
specificity, exercise training 777
spectral intensity (spectral irradiance) 1006
spectral sensitivity 1015
  Australian ants 1017f
speed 1084–5
  graphs of 136–7, 136f
speed of light in a vacuum (c) 121t
sperm (spermatozoa) 1147, 1176
  fertilization 1178–9
  maturation 850
  origin of 1176–7
  preparation for fertilization 1178
  shapes of 1177f, 1179f
spermatids 838, 839f, 840f
spermatocytes 838, 839f, 840f
spermatogenesis 834f, 837–8, 839f, 840f, 1176–7
spermatogenic wave 838
spermatophores 1180–1
sphincters
  anal 821–2
  gastroesophageal (cardia) 810, 812f
  ileocaecal valve 815
  of Oddi 827f, 828
  pyloric 811, 812f
  upper oesophageal 810
sphingosine 369
sphygmomanometer 740
spike proteins 556f, 558
  SARS-CoV-2 virus 559
spinal cord
  motor pathways 705, 707–9, 707f, 708f
  sensory pathways 621, 623f, 626, 627f
spinal cord injuries 626

spinal reflexes 702–5
spinothalamic tract 626, 627f
spiracles 1113
spiral cleavage 969–70, 971f
spirochetes 374f, 437f
spirometry 753–5
  asthma 756f
Spirulina 435, 508
spleen 719, 824f, 867, 868f, 892
spliceosomes 262
splicing 207, 261–2, 261f
split genes 207
sponges (Porifera) 926, 927f, 930, 933, 964
  integument 1111f
  occluding and adherens junctions 943
spongy mesophyll 961
spontaneous reactions 166, 186
sporangia 450, 451f, 452f, 955
spores 1146, 1147
  endospores 382f, 439, 470
  fungal 448, 449, 450–1, 453, 454, 456, 535, 537f
  sporangia 955
sporophytes 1147
sporopollenin 467
Sporozoa 540
spots (plant diseases) 552
springtail 985f
squirrels
  distribution in UK 1061f
  seed dispersal 1077
SRY (Sex Region of the Y) gene 1153
Sry gene 226, 227f
stabilizing selection 1206–7, 1208f, 1233
stable isotope analysis 1307–8
stains see Gram stain 490–2, 492f
stamens 1161f, 1162
standard deviation (s) 81t
standard errors 96
standard form 71
standard free energy change of a reaction (ΔG⁰) 187
stapedius muscle 640, 641f
staphylococcal enterotoxin B (SEB) 528
Staphylococcus spp. 437
  S. aureus 528f, 529–30, 530f
  see also methicillin-resistant Staphylococcus aureus
staphylokinase 529
starch 173
  digestion and absorption 816
  storage in amyloplasts 907f
  structure 174f
starfish (Asteroidea) 987f
Starling's hypothesis 731
static life tables 1250
statins 502, 723, 828
stationary phase, cell cultures 485f
statistical control 103
statistical interaction 105
statistical models 11, 12f
statistical power 116
  power analyses 104
statistical significance 113
  null hypothesis significance

tests 114–17
statistical software 115
statistics 96
statocysts 1021f
statoliths 1026
stem cells 939
  categories 940f
  meristems 928–9
stem-loop structure
  mRNA 258f
  tRNA 265f
stenohaline fishes 1049–50
Stentor spp. 461
stereocilia 642–3, 644f, 647f, 1021f, 1022
stereoisomers 157–8
  cis-trans isomers 157, 160f
  enantiomers 157–8, 160f, 161f
steric hindrance 292, 293f
sterigma 454
sterility 1156
sterilization 470–1
  male and female 851
steroid receptors 615–16, 615f
steroids (steroid hormones) 665
  corticosteroids 676
  industrial production 500, 502
  structure 176f
  see also aldosterone; oestrogens; progesterone; testosterone
sterols 370
stick insects 1218–19
stigma 1161f, 1163
stinging plants 1106f, 1107, 1108f
stitch, exercise-induced 779
stochastic processes 1353
stomach 806f, 811–12
  chief cells 813
  control of secretions 813–15, 814f
  emulsification of lipids 817
  gastric glands 812f
  G-cells 813
  motility 814
  parietal cells 812–13
  pepsin 818
  regions of 812f
  rugae 805
stomata 549f, 960–1, 1040f
  as entry points for pathogens 548–9
Stramenopiles 461–4, 462f
strand displacement, DNA virus replication 566
stratification 1034
Strecker, Adolph 26
streptococci 374f
Streptococcus spp. 437f
  S. pneumoniae 203–4, 530–1, 530f
Streptomyces spp. 439, 500, 502
stretch reflexes 703–4
  autogenic inhibition 704f
  knee-jerk response 702f
  long-latency 704
  reciprocal inhibition 703f
striated muscle 687, 688f
  see also cardiac muscle; skeletal muscle
strip transects 1255–6

stroke volume 739–40, 771f
  effect of aerobic exercise
    training 773
stromatolites 383f
strong acids 170
structural isomers 157, 159f
structural proteins 296
study types 67, 68t
style 1163
suberin 960
subjective grouping 43
sublingual gland 806f, 808, 809f
submandibular gland 806f, 808, 809f
submucosa, digestive tract 804f, 805, 815f
submucosal plexus (of Meissner) 804f, 805
substantia nigra 706f, 713
subthalamic nucleus 706f, 713
succession 1286–7, 1288f
succinate dehydrogenase 333f
succinyl-CoA synthetase 333f
sucrose 816
sugars 26, 173, 816
  deoxyribose 27f
  formose reaction 27f
  ribose 27f
  see also glucose; monosaccharides
Sulfolobus solfataricus 441, 442f, 480
sulphur, valence 153
summation (Σ) 121
  muscle contraction 696
supercoiling of DNA 213f
supercontinents 1054, 1056, 1057f
superior olivary nuclei 644
supplementary motor cortex 706f, 710, 711
suprachiasmatic nucleus 1015
surface area estimation 952, 953f
surface area to volume ratio 385–6, 386f, 952–3, 953f, 1029
  cells 368
surface tension 752–3, 752f
surfactant 749, 753, 857
surveys see data collection techniques
survivorship curves 1266f
swallowing (deglutition) 810–11
sweat glands 1120f
sweating 781
  autonomic control 664
swim bladder 1028, 1098–100, 1099f
swimming 1093–4
  birds 1097f
  buoyancy 1097–100
  efficiency 1094
  overcoming drag 1096
symbiosis 1279
  lichens 457–8
sympathetic nervous system 595, 655
  connections 656f
  control of cardiac output 739
  effect on airways 663
  effect on bladder 662–3
  effect on cardiac output 662, 771
  effect on eye 663–4
  effect on gland secretion 664
  effect on insulin secretion 681
  effect on liver 664
  effect on saliva secretion 808

effect on sodium and water
    reabsorption 798t
  effect on vascular smooth
    muscle 728–9
  neurotransmission 659–60
  organization of 655–7
sympathomimetic drugs 664
sympatric speciation 1217f, 1218
sympatry 47
symplastic transport 941–2, 942f
synapomorphys 53
synapses 595, 605, 607f, 973
  adrenergic 660, 661f
  chemical 606–14
  cholinergic 657, 659
  electrical 605–6
  neuromuscular junctions 692–3
synapsis 234, 235f, 420
synaptic cleft 606, 607f
synaptic integration 609–10, 609f
synaptotagmin 606, 608f
Synchytrium endobioticum 449, 450f
syncytia 922–3
  in animals 925–6, 926f
  fungal 923–4
  in plants 924–5, 925f
  skeletal muscle 688
syncytiotrophoblast 925–6
Synechococcus 431f
syngamy 1179
syphilis 437, 851, 852
systematics see taxonomy
systemic circulation 723, 724–5, 724f
  regional distribution of blood
    flow 725f
systole 727, 737f
systolic pressure 740, 769
  responses to exercise 769–70, 770f
System 1 and System 2 thinking 8t

T

T-box genes 987–8
T cell receptors (TCRs) 882–3, 884f
  generation of diversity 883–4
  selection process 885
T lymphocytes (T cells) 721, 866–7, 867f, 869
  activation 886–7, 888f, 892, 893f
  antigen recognition 885
  binding to MHC 887f
  CAR T cell therapy 883
  co-receptors 886
  cytotoxic, mode of operation 894f
  differentiation 889f
  Th cell functions 889–92, 890f
T (triangulation) number, viral
    capsids 557, 558f
t-tests 104t
T-(transverse) tubules 688, 691, 692f
T wave 736, 737f
tachycardia 739
tacrolimus 502
tactile receptors 1021
tangents 138
Tansley, Arthur 1292
Tansley, George 1202
taste 648, 807–8
taste buds 648, 650f, 808, 1136, 1139f

TATA box 258, 259f
TATA box-binding protein 259
taxa (singular taxon) 43, 48–9
taxis 380
taxonomy 37–8, 42–3, 427
  bacteria 46
  controversies 56
  domains 49
  fungi 448–9
  higher taxa 48–9, 49f
  interpretation of organismal
    relationships 52–6
  microbial 428
  naming conventions 44
  objective grouping 43–4
  pitfalls 57–9
  ribosomal RNA analysis 50–1
  species 44–8
  subjective grouping 43
  viruses 556
Tay–Sachs disease 1229–31, 1232
team science 13–14
technological advances 6
tectonic plates 1054, 1055f
tectospinal tract 705, 708f, 709, 710f
teeth 997–8
telomerase 218
telomere shortening 218–19
telomeres 209
telophase 232, 416
telophase I 422
telophase II 423
tempeh 511
temperature 1030
  adaptations to heat 1301
    see also thermophiles
  effect on ecosystems 1294–7
  effect on enzyme activity 308f, 309
  effect on membrane transport 909
  effects on plant growth 1160–1
  thermoregulation 582, 583f, 584, 724, 772, 781
temperature receptors 628–9
temperature sensing, plants 1034–5
temperature sex determination 1154, 1155f
template strand, DNA 255
temporal changes
  ecological succession 1286–7, 1288f
  predator–prey cycles 1286, 1287f
tendon organ 631f
  stretch reflexes 704f
tendons 698–9
  Achilles tendon 699f
tendrils 960, 961f
tensegrity model of cytoskeleton 917f
tensor tympani muscle 640, 641f
terminal boutons 692
terminal cisternae 688, 691, 692f
terminal differentiation 939
termination sequences, DNA 257f, 258
terrestrial earthworm 983f
  Darwin's studies 1024
  osmoregulation 1048
  peristaltic locomotion 1078f
territorial defence 1139
tertiary protein structure 291, 292f, 293, 295, 295f

test (dependent) variables 66
testis
  anatomy 846–7, 847f
  spermatogenesis 834f, 837–8, 839f, 840f
testosterone 837
  actions of 842
  effect on GnRH secretion 841
  structure 176f
tests 1113
tetanic contraction (tetany) 696f, 697
tetanus 438
Tfh cells 892
TFIID complex 259, 260f
Th cells see helper T cells
thalamus 622, 623f, 706f
  lateral geniculate nucleus 637f
thalassaemias 262–3
thalidomide 161f
thallus, lichens
  morphology 457–8, 457f
  structure 458f
theca cells 836, 837f, 838f
theoretical frequency distributions 82
  examples 84t
theories 11
thermal energy 185
thermal environment 1033f
  animal responses 1035–8
  plant adaptations 1035f
  plant responses 1034–5
thermal imaging 1033f
thermal radiation 1031
thermodynamics, laws of 31, 185, 1030
thermophiles 439, 480, 481f
  adaptations 443
  Archaea 440–1
  bacteria 428, 433–4
thermoregulation
  animals 582, 583f, 584, 1036–8, 1037t
  blood flow 724
  during exercise 772
  exercise in heat 780–1
  plants 1034, 1035
Thermotoga maritima 428, 433f
Thermotogales 433
Thermus spp. 433
theta replication, DNA viruses 565, 566f
thick filaments, skeletal
  muscle 688–9
thigmonastic responses 1020f
thigmotropic responses 1020f
thin filaments, skeletal muscle 689
thinking, System 1 and System 2 8t
thiols 158f
third (confounding) variables 67, 106
third-order afferents
  nociception 626, 627f
  touch 622, 623f
thirst 670
thrashing locomotion,
    nematodes 1079f
threonine (Thr, T) 288f
threshold level, action
    potentials 600f, 601
thrombin 1125

thrombosis 613, 1126
thrombus 1125
thylakoids 339f, 399f, 400
thymine (T) 176, 203
  structure 200f
thymus 867, 868f
thyroid cartilage (Adam's apple) 746
thyroid disease 676
thyroid gland 674
  calcitonin 682
thyroid hormone receptors 616, 672f, 673–4
thyroid hormones 672–4
  control of secretion 676
  physiological actions 674
  relationships between 675f
  synthesis 674–6, 675f
thyroid stimulating hormone (TSH) 669f, 676, 841t
thyrotropin releasing hormone (TRH) 668, 669f, 676
thyroxine (T4) 672, 675f
  see also thyroid hormones
tidal volume (TV) 753f, 754
tides 1025
tiger, population size measurement 1258–9
tight junctions 401f, 813, 944, 945f
time series 117
tinea 534
tipping points 1313
tissue culture 475
  plant cells 934f
  see also cell cultures
tissue culture hoods 478f
tissue damage, response to 628f
tissue fluid formation 731
tissue remodelling 1121
tissue types 886
tissues 401
titin 285, 689
TNF-α 871t
tobacco hornworm 914
tobacco mosaic virus 553f
tolerance ranges 1297–8, 1298f
toll-like receptors (TLRs) 871
tongue 807–8, 807f
tonoplast membrane 918
tonotropic representation 644
tonsils 746f, 868f
topmouth gudgeon 1062–3
topoisomerases 213
torpor 1294
total cell counts 481, 482f
total lung capacity (TC) 753f, 754
totipotency, plant cells 929, 933–4, 934f
totipotent stem cells 939, 940f
touch 620
  acuity 620–1, 621f
  functional imaging study 624–5
  sensory pathways 621–4, 623f
toxins 527–8, 529, 1134–5
  bacterial 439, 693
  comparison of exotoxins and endotoxins 528t
  fungal 538–9, 550–1
toxoplasmosis 461, 542t
toxungens 1135

trachea 745, 746f
trachoma 437
tracking migration 1068f
  reindeer 1071
trade-offs 1265–6
  sexual selection 35
traits 1205
  inheritance of 224–5
    see also Mendelian genetics
  simple and complex 1206
transcellular reabsorption 789
transcellular transport 942f
  auxin 943f
transcranial magnetic stimulation (TMS) 710
  study of control of walking 711–12
transcription 198, 253, 255f
  capping 260, 261f
  elongation phase 259, 260f
  initiation phase 257–9
  poly(A) tail 263f
  pre-mRNA modification 258, 259f
  RNA polymerase 255–6
  RNA polymerase positioning 256
  short life of mRNA 256
  splicing 261–2, 261f
  TFIID complex 259, 260f
  viral 560–1, 563t
transcription factors 258–9, 272f, 942
  activation of 279, 280f
  in eukaryotes 275–9
  families 273f
  GATA-1 278–9
  lac operon 272–5
transcription initiation complex 276f
transdifferentiation 940
transduction 410f, 620
  bacteriophage lysogeny 569f
transects 1255
transfer RNA (tRNA) 202f, 253t, 264–6
  base pairing 265f
  linking to amino acids 266f
  for methionine 269
  structure 265f
transferrin 719–20, 914
transformation 409–10, 410f
transgenic mice 356–7, 357f
transient dipoles 162, 163f
translation 198, 253, 264
  amino acid activation 266f
  elongation phase 269, 270f
  initiation and elongation factors 269
  initiation phase 268–9, 268f
  overview 264
  polysomes 271
  in prokaryotes and eukaryotes 271–2
  ribosomes 266–9, 267f
  termination 270
  transfer RNAs 264–6, 265f
  viral 563
translocation 424f
transmission electron microscopy (TEM) 495, 496f
transmission genetics see Mendelian genetics

transpiration 1034, 1040
transport across plasma membranes 585–91
  see also membrane transport
transport proteins 298
transposition 409f
transposons ('jumping genes') 208–9, 208f, 409f
tree ferns 57, 59f
tree rings 938f
trees of life see phylogenetic trees
Treg cells 889, 891
trend lines (model lines) 111–12, 112f
Treponema pallidum 437
triacylglycerol (TAG) 328–9
triads, skeletal muscle 688, 691
tricarboxylic acid (TCA) cycle (Krebs cycle) 325, 326, 327f, 328f, 332–3, 333f
  inputs and outputs 333f
trichoid sensilla 1021
trichomes 1107
trichomoniasis 542t
Trichoplax adhaerens 930f
tricuspid valve 733f
trigeminal pathways 622
triglycerides (triacylglycerols) 173
  structure 175f
triiodothyronine (T3) 672, 675f
  see also thyroid hormones
triose phosphate isomerase 330, 331f
triple covalent bonds 155f
triple response 1123t
triple superhelix, collagen 303
triploblasts 937, 966, 971f
trisomies 227, 228t, 418, 423f, 424, 836
  Down syndrome 424
trophic levels 1303–8, 1304f
  ecological pyramids 1309, 1309f
  food chains 1303–5
  food webs 1305–8
trophozoites 540f, 541f, 546f
tropic hormones 668
tropisms see heliotropism; phototropism 1034
tropomyosin 689
troponin complex 689, 691
truffles 451, 452f
trypanosomes 459, 460f, 540
  Trypanosoma cruzi 547, 1047f
trypsin 818
tryptophan (Trp, W) 288f
Tsien, Roger 494
tsunamis 1317–18
tuberculosis, multi-drug resistance 428
tubulin 296, 298, 299f, 393, 394f, 915t
tubulinids 465, 466f
tubuloglomerular feedback 788–9, 797
tunica adventitia (tunica externa) 725, 726f
tunica intima 726f
tunica media 725, 726f
tunicates, integument 1113–14, 1114f

tunicin 1113
tunnelling nanotubes (TNTs) 945, 947f
turbidity, cell growth measurement 483–4
turbulent flow 1090f
turgor 917–18, 1297
turgor pressure 400, 918, 1045
  effect on shape 919f
  measurement 919f
Turner syndrome (45, X) 228t
tusks 997f
two-point discrimination 620–1
tympanal ears 1025
tympanum (ear drum) 639, 641f
Type I and Type II errors 115–16
typhoid fever 529
Tyrannosaurus rex 1085f
tyrosine (Tyr, Y) 287, 288f
  catecholamine synthesis 559, 660f

U

U wave 736, 737f
ubiquinone (coenzyme Q, Q) 334, 335
ubiquitin 591
Ulva lactuca 928f
umbilical cord 856f
uncoating, viruses 560
unicellular organisms 922
uniform dispersion 1249f
unikonts 930
uniport transporters 911
unipotent stem cells 940f
uniramous limbs 984
units 64–5, 64t
  in calculations 127
  converting between prefix-unit combinations 65–6, 66f
univariate analyses 105
unrelated samples 98
upper motor neurons 705, 707f
upper oesophageal sphincter (UES) 810
upstream control elements 258, 259f, 275
uracil (U) 27, 176, 203
  structure 200f
urbilateria 966, 968
urea 786
  reabsorption in the nephron 790t
urea cycle 828f
urease 531
ureotelism 1050
ureters 786f, 799, 800f
urethra 786f, 800f
  female 848f
  male 847f, 848f
uricotelism 1050
urinary system 799–800, 800f
urobilinogen 828
urochordates (tunicates) 977–8, 979f
Usher syndrome 239, 246
uterine cycle 843, 849f
uterus 837f, 848f, 849–50, 849f
utricle 646, 647f
UVR8 1016t

V

vacuoles 400
  plant cells 914, 917–18, 918f
  protists 1046
vagal restraint 770
vagina 848f, 850
vagus nerve 657
valence (valency) 152–3
  of elements found in biological
    molecules 153t
valine (Val, V) 288f
vallate papillae 648, 650f
valves
  in heart 733f
  in veins 732f
van der Waals interactions
  (bonds) 162
  in proteins 290f
van Leeuwenhoek, Anton 368, 489
variability, measures of 80–1, 81t
variables 64, 66
  confounding 67
  dependent and independent 66–7
  inclusion in experimental
    design 105
  regulated and unregulated 583
variance (s^2) 81t
variation, genetic 1001, 1205–6
  analysis of 1223–6
  description of 1222
  Hardy–Weinberg model 1226,
    1227–8
  loss of 40
  Punnett square 1226–7, 1228f
  role of meiosis 32–3, 35, 234–5,
    420–2, 833, 835–6, 1146
  sources of 67, 101
vas deferens 847f
vasa recta 793
vascular plants 955
vascular smooth muscle 725, 726f
  control of 728–30
vasoconstriction 725–6, 728–30
vasodilation 728–30
vasopressin see antidiuretic hormone
V(D)J recombination 883, 884f
vectors 1072, 1343
vegetarianism/veganism, ecological
  perspective 1305f
vegetative growth 1159
vegetative propagation 1146
veins 723, 725f, 731–2
  muscle pump 732f
  structure 726f
velvet worms 983–4, 983f
vena cava 733f
venoms 1134–5
ventilation–perfusion matching 759
ventilatory cycle 751f
ventilatory movements 749–52
  see also breathing
ventilatory threshold 778f
ventricles 732, 733f
ventricular action potential 735f
ventricular pressure 737f
ventricular volume 737f
venules 731
Venus flytrap 1020f
vernalization 1034, 1160

vertebrates
  nervous system 978–81
  osmoregulation 1048–50
vertical gene transfer 406
very-low-density lipoproteins
  (VLDL) 723, 828
vesicles, spontaneous formation 26
vestibular nerve 647f
vestibular system 646–8, 647f
vestibulospinal tract 705, 708f, 709,
  710f
viable cell counts 481, 483f
vibrios 374f
  Vibrio cholerae 569
victorin 550–1
villi 803, 815f, 816
vimentin 915t
vinegar (acetic acid) production 510
Virchow, Rudolf 368
virions 555, 556
  naked 558
virtual islands 1353
virulence factors 524, 526–8, 530, 531
  Candida albicans 536t
  protozoan 539
  transmission of 569
virus replication 559–60, 561f
  assembly 563–4
  attachment and fusion 560
  DNA viruses 564–6
  entry and uncoating 560, 562f
  envelope formation 564
  genome replication 563
  retroviruses and
    pararetroviruses 566–7, 567f,
    568f
  RNA viruses 566
  transcription 560–1
  translation 563
viruses 555
  antiviral therapy 569–71
  bacteriophages 567
  culture methods 560
  destruction by natural killer
    cells 875
  genomes 558
  myxomatosis 573
  naming conventions 556t
  pest control application 573
  phage therapy 572–3
  phage typing 572f
  plant diseases 551
  release from host cell 564
  RNA genomes 198
  sizes 556
  structure 556–8, 556f
  use in molecular biology 573
visceral sensation 619
vision 632
  cornea and lens 633–4
  dark adaptation 635
  distinction from
    photoreception 1014–15
  eye structure 632f
  focusing the image 633f
  refractive errors 633–4
  retina 634–6
  transduction process 636f
  see also eye spots; eyes

visual cortex 637–9, 638f
visual illusions 619f, 632f
visual pathway 636–7, 637f
visual pigments 1015
vital capacity (VC) 753f, 754
vitamins
  industrial production 505, 506t
  niacin 194
  vitamin B12 761
  vitamin C (ascorbic acid) 303
  vitamin D 682, 801
vitelline membrane 1169f, 1171f
vitreous humour 632f
viviparity 1172–4
viviparous animals 1168
vocal folds (vocal cords) 746, 747f,
  807f
volcanic eruptions 1317, 1318f
volley coding 644
voltage-gated channels 588
voltage-gated sodium
  channels 601–2
  structure 603f
volume estimation 952f, 953f
Volvox 467f, 927, 928f
vomeronasal organ 1136
Vorticella spp. 457f
V'/Q' ratio 759

W

walking and running 1080–4
  gait and speed 1084
  human walking 1081f
  role of tendons 699
  speed 1084–5
  TMS studies of motor
    cortex 711–12
wallaby, embryonic diapause 1193–4,
  1193f
Wallace, Alfred Russell 35, 1060,
  1128–9, 1204–5, 1204f
  categorization of coloration 1130t
Warburg effect 338
Warren, Robin 527
waste products
  metabolism by the liver 828–9
  see also urea
wastewater treatment 515–16, 516f
water 22, 1297
  absorption in small intestine 819
  factors regulating
    reabsorption 798t
  hydrophilic and hydrophobic
    interactions 167–9, 168f, 169f
  lack of, adaptation to 1301
  light absorption 1005, 1006f
  movement across
    membranes 909–10, 1044
  polarity 154f, 167, 168f
  pressure at depth 1099f
  reabsorption in the nephron 790t,
    792, 793–7, 797f
  seed dispersal 1075
  solute composition and
    osmolarity 1044t
  solvent properties 167, 168f
  specific heat 167
waterborne infections 515–16
Watson, James 204

Watson–Crick base pairs 201, 202f
weak acids 170–1
Weberian apparatus 1025
weight reduction 772, 783
  pharmaceutical interventions 818
whales
  echolocation 1025
  skin 1120
white blood cells (leukocytes) 719,
  721–2, 721f, 863, 864, 865f
  see also B lymphocytes;
    lymphocytes; T lymphocytes
white reaction 1123
white rhinoceros 1341, 1342f
Whittaker, Robert H. 49
Whittaker's beta diversity 1282,
  1283f
whole-genome sequences 353, 354f
wild-type alleles 1223
Wildlife and Countryside Act
  (1981) 1339–40
wilt diseases 552
wind-dispersal of seeds 1073–5,
  1074f
wine production 513
winged seeds (samaras) 1074f
wings 989–91
  as an aerofoil 1088, 1089f
  air flow over 1090–1, 1090f
  aspect ratios 1092f
  flapping flight 1093
  hand-wing index (HWI) 1091–3,
    1092f
  insects 1093
  thrust, lift, and drag 1088–90,
    1089f
withdrawal reflex 626, 704–5, 704f
within-subject design 103
wobble base pairing 265–6
Woese, Carl 49
wood 959–60, 991
wood lice (Oniscsidea) 985f
woodland habitat structures 1275–6,
  1275f
woolly mammoths 1065
World Heritage Sites 1341t
worms
  integument 1111, 1112f
  methods of movement 1077–9
  velvet worms 983–4, 983f
  see also earthworms; nematodes;
    platyheminths; polychaete
    worms
Woronin bodies 446
wound repair 1127
wound response
  animals 1121–7, 1123f
  plants 1109, 1110f

X

X chromosome 225, 226f, 833
X-linked inheritance 243–4, 245f,
  359
Xanthomonas spp. 513, 514f, 552
Xenarthra 1065
xeric ecosystems 1297
xeroderma pigmentosum (XP) 220f
xerophytes 960, 1105f
XX/XY system 1153f

xylem 909
xyloglucan 398

Y

Y chromosome 225, 226f, 833
   Sry gene 226, 227f
yeasts 447f, 453–4
   bread making 510
   brewing beer 499, 512–13
   budding 408
   dimorphism 535
   pathogens 536

Saccharomyces cerevisiae 447, 451,
   454f, 499, 510
   single-cell protein 507–8
   wine production 513
yellow jawfish 1206–7, 1207f
Yellowstone National Park
   1353–4
Yersinia pestis 436

Z

Z-discs 688f, 689
Z ring 406f, 407

Zahavian signalling 1131, 1132f, 1133
zeitgebers 1325
Zernike, Frits 493
zinc finger proteins 273f
zona pellucida 850
zonal pellucida 1169f, 1179
zone of osmotic neutrality 1045,
   1046f, 1048
zone of thermal neutrality (ZTN)
   1035–6, 1036f
zoonotic diseases 529
zooplankton 1069f

migration 1069
   refuge migration 1070f
zoosporangium 456f
zoospores 449, 450f, 456
Zygomycota 450–1
zygospores 450–1, 451f
zygote 850, 936, 1146
zygote development,
   angiosperms 1164, 1166,
   1167f
zymogens 813
ZZ/ZW system 1153f